Handbook of Cyanobacteria

Handbook of Cyanobacteria

T. A. SARMA
Formerly Professor of Botany
Department of Botany, Punjabi University
Patiala (Punjab), India

CRC Press
Taylor & Francis Group
Boca Raton London New York

CRC Press is an imprint of the
Taylor & Francis Group, an **informa** business
A SCIENCE PUBLISHERS BOOK

CRC Press
Taylor & Francis Group
6000 Broken Sound Parkway NW, Suite 300
Boca Raton, FL 33487-2742

Cover illustration courtesy of Prof. G.L. Tiwari

Printed in the United States of America on acid-free paper

International Standard Book Number: 978-1-57808-800-3 (Hardback)

Library of Congress Cataloging-in-Publication Data

Sarma, T. A., 1945-
Handbook of cyanobacteria / author, T.A. Sarma.
p. cm.
Includes bibliographical references and index.
ISBN 978-1-57808-800-3 (hardcover : alk. paper) 1. Cyanobacteria.
I. Title.

QR99.63.S27 2013
579.3'9--dc23

2012036515

Visit the Taylor & Francis Web site at
http://www.taylorandfrancis.com

CRC Press Web site at
http://www.crcpress.com

Science Publishers Web site at
http://www.scipub.net

To my wife

Jaya

for

the inspiration, encouragement
and patience

Preface

I have been deeply fascinated by the cyanobacteria ever since my student days. As a teacher and researcher, I have always looked for a resource book that contains comprehensive information about cyanobacteria. Many universities have included cyanobacteria as a specialization at the post-graduate level. The travails faced by the research students in accessing literature are too diverse. Many of my colleagues are of the view that there is a necessity of a single reference book for research and teaching. 'Handbook of Cyanobacteria' is meant to fill such a role. A lot of new information has been generated consequent upon the publication of the first genome sequence of an autotroph *Synechocystis* sp. PCC 6803 and the understanding of every aspect of the biology of these organisms has undergone a sea change. This Handbook begins with a summary of the current knowledge on the taxonomy, phylogeny and evolution of cyanobacteria followed by the sequenced genomes, differentiation of akinetes and heterocyst. The mechanisms of cellular movements (gliding, swimming and twitching movements) exhibited by various cyanobacteria in order to adjust to their environmental niches and the operation of the circadian rhythms have been considered. The cyanobacterial symbiosis, cyanophages and cyanobacterial toxins are the other important areas covered followed by a discussion on stress responses (salinity, temperature, desiccation and oxidative). These ten important areas truly reflect the essential features of the biology of these organisms and also aptly justify this Handbook.

It has been quite an exhilarating work for me to sort out and read the literature (up to December, 2011) and present it in a simple language. The availability and the choice of illustrations had been quite demanding. I am grateful to the editors/publishers of the Journals of Open Access category and to the individual scientists who permitted me to use the figures that appeared in the research papers published by them in these Journals. Many of the illustrations have been provided by individual scientists and Government Agencies from their websites. All the illustrations are duly acknowledged in the captions provided with the respective figures.

At this juncture, I express my deep sense of gratitude to a host of my teachers, colleagues and students who richly contributed to my professional growth. Institutional support from Punjabi University is gratefully acknowledged.

I earnestly hope that this Handbook adequately serves the purpose for which it is intended and will be useful to students as well as teachers. I would greatly appreciate receiving suggestions for general improvement of this Handbook.

T. A. SARMA
(ta.sarma@gmail.com, tangiralas.com)
(May, 2012)

Contents

CHAPTER 1

TAXONOMY, PHYLOGENY AND EVOLUTION

I. TAXONOMY

A classification provides a platform to identify a particular organism to the level of species. The art of classification is known as taxonomy and phylogeny helps to draw evolutionary relationships between the organisms classified. So a phylogenetic system of classification serves a better purpose for classification as well as inferring evolutionary relationships. The earliest taxonomic treatises considered cyanobacteria as an algal group under the general name blue-green algae (Thuret, 1875; Bornet and Flahault, 1888; Gomont, 1892). Traditionally thus, the blue-green algae have been classified along with eukaryotic algae under the International Code of Botanical Nomenclature (ICBN). As per rules of ICBN, a type species has to be deposited in a recognized Herbarium either in a dried state or fixed algal material. Due to the limitation in the availability of pure culture strains and the paucity of information on their biochemical and genetic features, the earlier workers depended mostly on morphological characteristics. Due to the absence of membrane-bound cell organelles, the blue-green algae were considered closer to bacteria (Cohn, 1872, 1875). Subsequent to the publication of taxonomic keys on heterocystous forms (Bornet and Flahault, 1888) and the Oscillatoriaceae (Gomont, 1892), a comprehensive account on the taxonomy of these organisms has been presented by Geitler (1932) wherein 145 genera encompassing 1300 species have been classified into 20 Families under three Orders. The "Geitlerian" approach provided new impetus for a number of taxonomists to classify them (Fritsch, 1945; Drouet and Daily, 1956; Golubic, 1967; Bourelly, 1970 and many others) from time to time.

Drouet and Daily (1956) revised Geitler's classification and brought down the number of genera to 24 with 62 species based on their observations on the blue-green algae collected from all over the world. The taxonomic criteria developed for identification have been based on morphological features and they further attributed the diversity in morphology due to the existence of ecophenes or ecotypes. According to them, the blue-green algal species is highly polymorphic and the phenotypic changes lack a genetic basis. That is polymorphism is assumed to be purely phenotypic. The ecophenes thus represent form-variations caused by environmental factors rather than due to genetic differences. Thus the species concept envisaged by Drouet predicts that a particular species should be morphologically identical when grown under the same environmental conditions and shows a similar change in morphology with changing environmental conditions while at the same time the genotype remains unchanged. On the basis of taxonomic keys available from time to time (Geitler, 1932; Drouet and Daily, 1956; Desikachary, 1959; Drouet, 1981; Komárek and Anagnostidis, 1989), a number of workers enumerated the cyanobacterial flora from wide variety of locations and habitats from all over the world.

The classification proposed by Fritsch (1945) has been mainly based on morphological characters. Of the five orders recognized, three have been placed together on the basis of absence of hormogones (Chroococcales, Chamaesiphonales and Pleurocapsales) and the other two Orders *Nostocales* and *Stigonematales* that form the hormogones have been placed together. Chroococcales are characterized by unicellular or colonial (commonly palmelloid) forms that multiply by cell division and by endospore formation. The members included in the Order Chamaesiphonales are unicellular or colonial epiphytes or lithophytes exhibiting marked polarity and reproduce by the formation of endospores or exospores. The Pleurocapsales comprise of heterotrichous filamentous types without heterocysts and reproduce by the formation of endospores. The members of Nostocales are either non-heterocystous or heterocystous filamentous forms. Some of these show false-branching. Multiplication is by means of hormogones, hormocysts and akinetes. Order Stigonematales comprises of heterotrichous filamentous types that exhibit true branching and possess heterocysts. The presence of pit-connections between the cells is another characteristic feature of these forms. The multiplication is by hormogonia, hormocysts and more rarely by akinetes. A few representative types of cyanobacteria identified on the basis of traditional taxonomic keys have been presented in Figs. 1 to 8.

A classification of cyanobacteria under ICBN has been proposed (Anagnostidis and Komárek, 1985) and revised (Komárek and Anagnostidis, 1989) that took into account the Bacteriological code as well. In this classification they recognized four orders- Chroococcales, Oscillatoriales, Nostocales and Stigonematales, which are further subdivided into families, subfamilies, genera and species. They emphasized that the taxonomic keys developed by them would be greatly helpful for ecologists to study the diversity of cyanobacteria from their natural habitats. Many objective features and characteristics for distinguishing various genera and species have been defined by them. Due to the application of both botanical and bacteriological criteria, an attempt has been made to reconcile the differences between the botanical and bacteriological codes. The taxonomic criteria based on botanical code included cell morphology, cell division, polarity, methods of false-branching, tapering

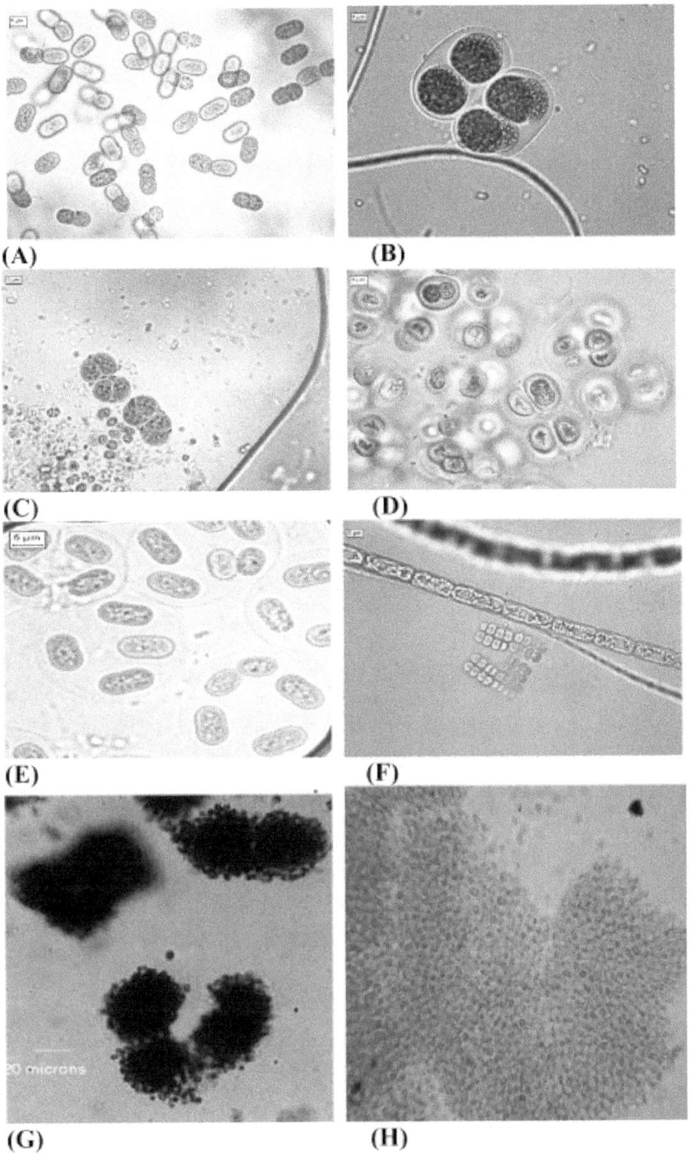

Figure 1: Members of Chroococcales. *Aphanothece* sp. (A), *Chroococcus* sp. (B and C), *Gloeocapsa* sp. (D), *Gloeothece* sp. (E), *Merismopedia* sp. (F), colonies of *Microcystis* sp. (G), colony of *Microcystis aeruginosa* (H). Magnification bar in A to F represents 5 μm, G 20 μm and H x100. Pictures A to F courtesy G. L. Tiwari, Department of Botany, University of Allahabad, Allahabad-211002, India. (G) Courtesy Roger Burks (University of California at Riverside), Mark Schneegurt (Wichita State University) and Cyanosite (www-cyanosite.bio.purdue.edu). (H) Courtesy Jens Dahlmann (Friedrich Schiller University, Jena), Mark Schneegurt (Wichita State University) and Cyanosite (www-cyanosite.bio.purdue.edu).

and hormogonia formation. Furthermore, they also used information obtained from bacteriological methods including molecular markers when required. The concomitant application of both ICBN and bacteriological code has resulted in the proper classification of this group (Garcia-Pichel *et al.*, 1998). However, Anagnostidis and Komárek (1988) cautioned that the re-classification of cyanobacteria

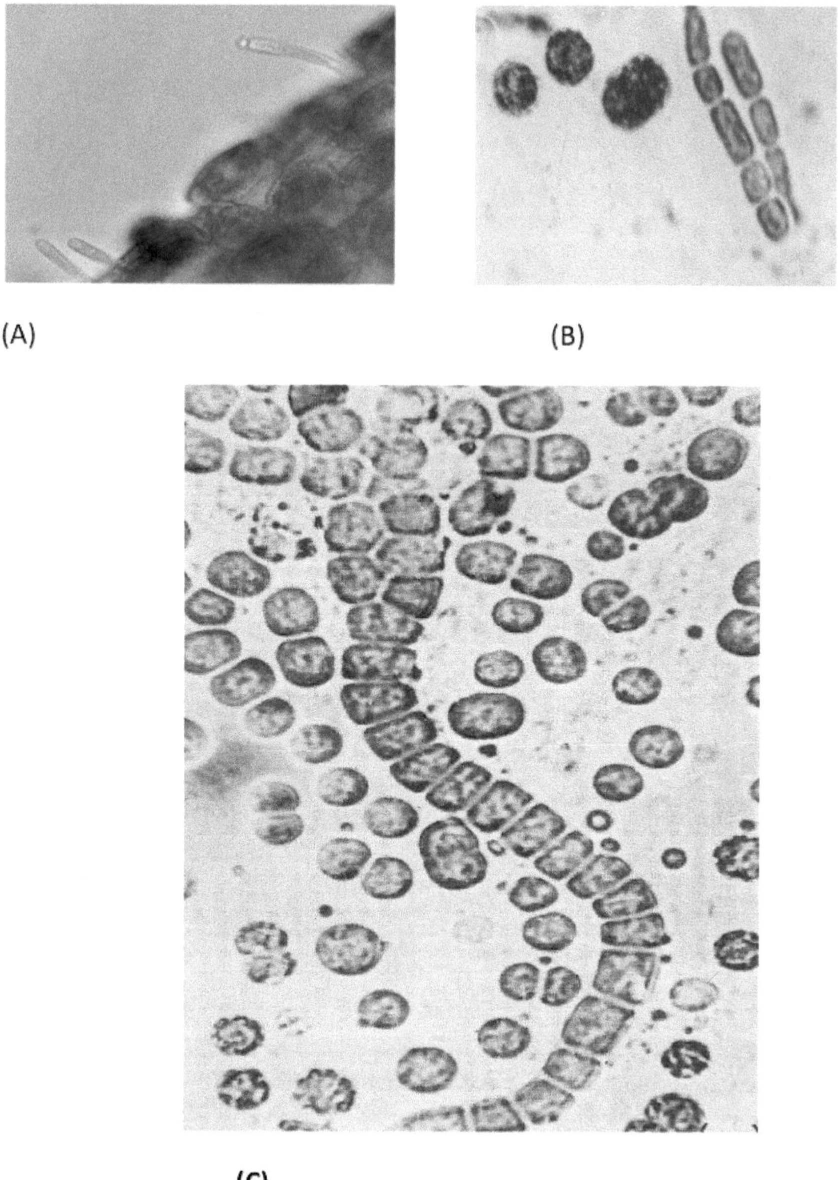

Figure 2: Members of Chamaesiphonales. *Chamaesiphon* sp. (A), *Stichosiphon indica* in cultures, two sporangia juxtaposed on agar plates (B), a sporangium releasing baeocytes in two rows from liquid cultures (C). Magnification of pictures (B) and (C) at x 1125 and x 675, respectively. Picture A courtesy Mark T. Aubel, GreenWater Laboratories/CyanoLab, 205, Zeagler Drive, Suite 302, Palatka, Florida 32177, USA. (http://www.greenwaterlab.com/photo_algal.htm). Pictures B and C are from Sarma (1981).

is difficult for a number of reasons because for example (i) traditional taxonomic criteria were developed over a century ago and were predicated on characters such as false-branching and sheath characteristics; (ii) the morphology of cyanobacteria is largely influenced by environmental factors;

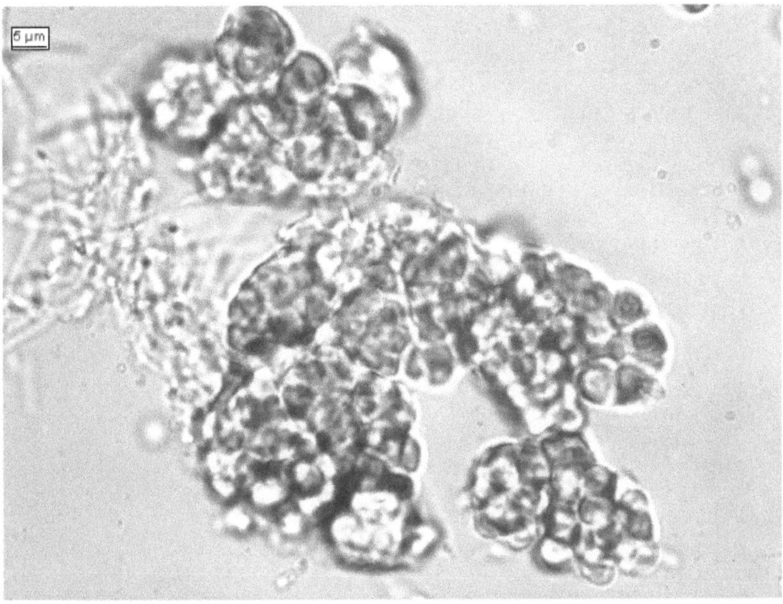

Figure 3: *Pleurocapsa* sp. from nature. Magnification bar represents 5 µm. Picture courtesy G. L. Tiwari, Department of Botany, University of Allahabad, Allahabad-211002, India.

(iii) the genetic diversity of collected strains within a culture may be decreased by selective culturing conditions; (iv) the species concept depended on the taxonomist and (v) the conservative approach of a number of taxonomists to adopt the new molecular methods. A number of proposals have been made for unification of nomenclature in case of botanical and bacteriological codes (Oren, 2004; Oren and Tindall, 2005; Hoffmann *et al.*, 2005).

Hoffmann *et al.* (2005) presented a classification scheme under ICBN on a polyphasic approach by taking into account morphological, ultrastructural, ecological and genetic relationships (mainly 16S rRNA gene sequences). This is an attempt to merge the Botanical as well as Bacteriological Codes. The Cyanobacteria are given the rank of a Class and designated them as Cyanoprokaryota with three sub-classes Synechococcineae, Oscillatorineae and Nostocineae. The Synechococcineae with two orders Synechococcales and Pseudoanabaenales consists of both coccoid and filamentous forms, respectively. The second sub-class Oscillatorineae comprises coccoid (Chroococcales) and filamentous forms (Orders Phormidiales and Oscillatoriales). The unification of heterocystous cyanobacteria into one sub-class Nostocineae with a single order Nostocales is a unique feature. The inclusion of Prochlorophyta in cyanobacterial classification system and the absence of distinction into coccoid and filamentous forms at the highest sub-class level are the highlights of this classification. The distinction into sub-classes is based on arrangement of thylakoids and the presence of differentiated cells. Chroococcales with unicellular or colonial members are characterized by radial arrangement of thylakoids. The non-heterocystous filamentous forms have been divided into three orders based on the nature of filaments and the arrangement of thylakoids. The Oscillatoriales possess large filamentous forms with radial arrangement of thylakoids while Pseudoanabaenales consist of thin filamentous forms in which the thylakoids are arranged parallel to the cell surface. The order Nostocales possesses filamentous, heterocystous, unbranched and branched members.

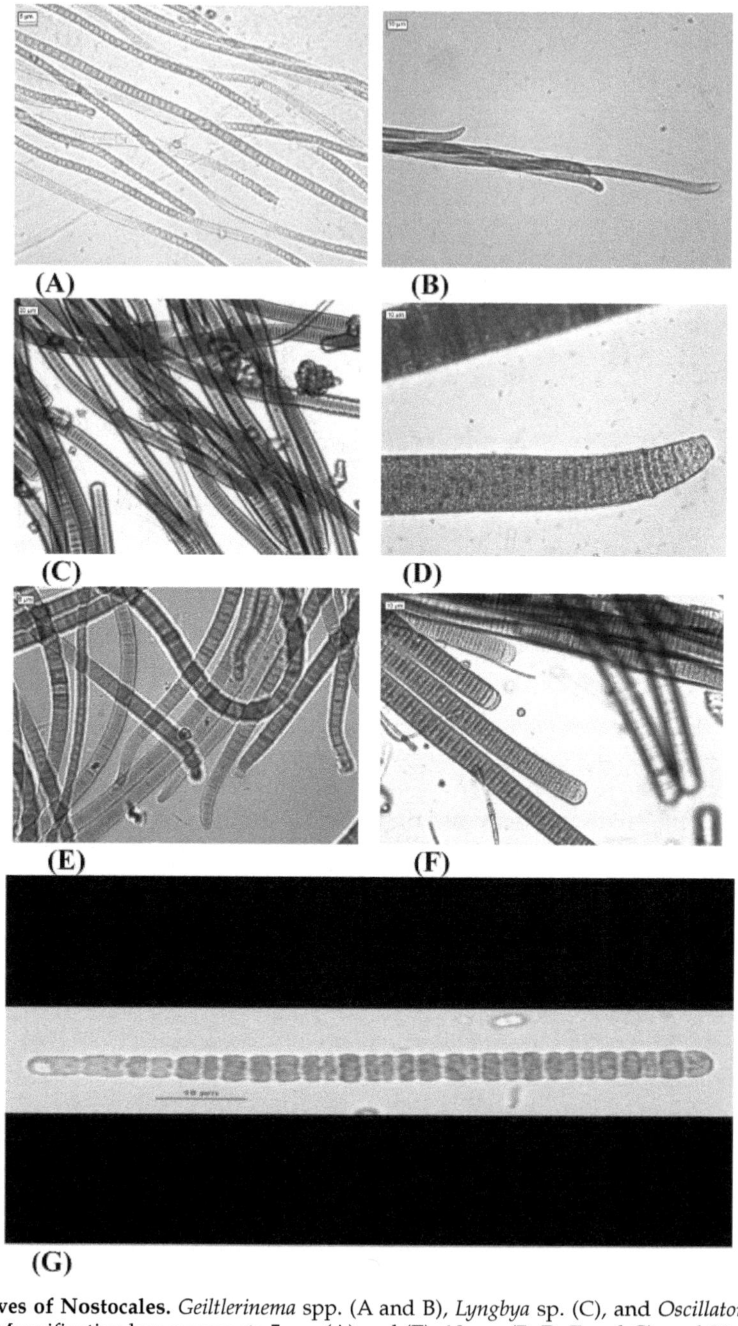

Figure 4: Representatives of Nostocales. *Geitlerinema* spp. (A and B), *Lyngbya* sp. (C), and *Oscillatoria* spp. (D, E and F), *Trichodesmium* sp. (G). Magnification bar represents 5 µm (A) and (E), 10 µm (B, D, F and G), and 20 µm (C). Pictures A to F courtesy G. L. Tiwari, Department of Botany, University of Allahabad, Allahabad-211002, India. Picture (G) courtesy P. Znachor, Laboratory of Phytoplankton Ecology, Institute of Hydrobiology, Biology Centre ASCR, Na Sádkách , 37005 České Budějovice, Czech Republic. www.FytoPlankton.cz.

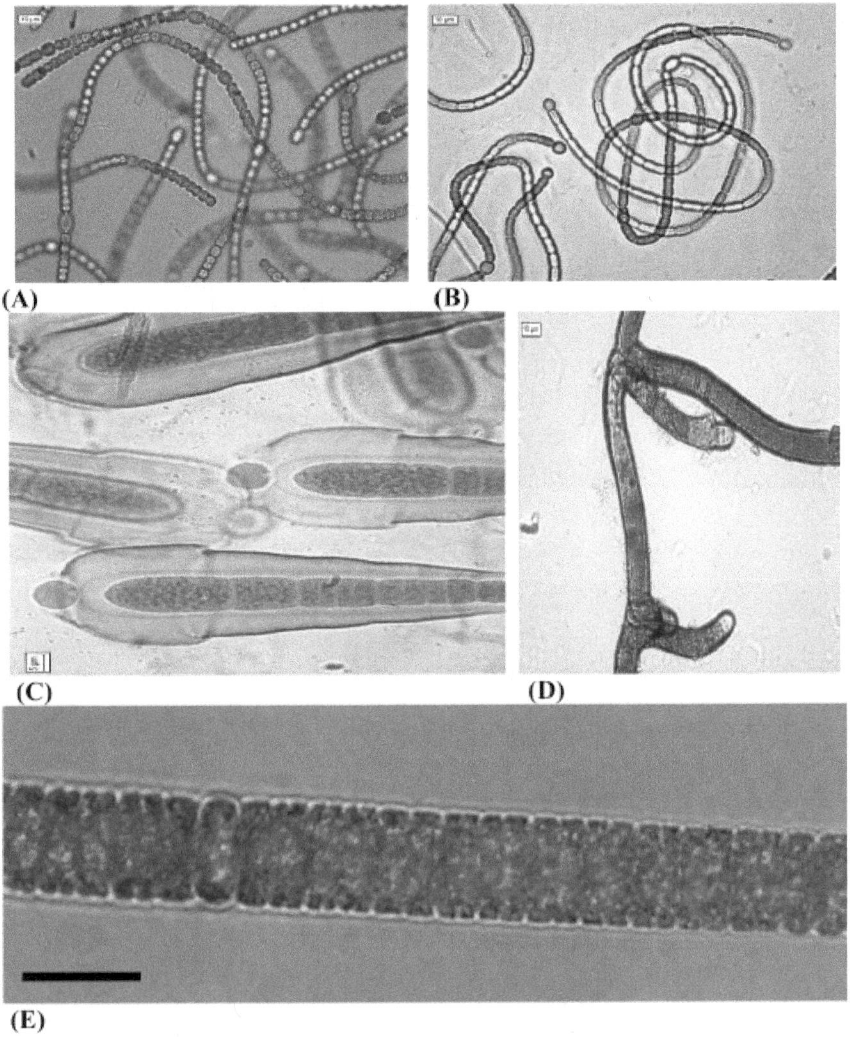

Figure 5: Members of Nostocales. *Anabaena* sp. (A), *Nostoc* sp. (B), *Gloeotrichia* sp. (C), *Scytonema* sp. (D) and *Nodularia spumigena* (E). Magnification bar represents 5 μm (C), and 10 μm (A, B, D and E). Pictures A to D courtesy G. L. Tiwari, Department of Botany, University of Allahabad, Allahabad-211002, India. Picture E courtesy Christina Esplund, Linnaeus University, School of Natural Sciences, Department of Marine Ecology, SE-391 82 Kalmar, Sweden.

Some distinctive features of the classification of Hoffmann *et al.* (2005) are: (i) Coccoid species of *Merismopedia* have been separated on the basis of cell size (small-celled Merismopedias included in Synechococcales of the first sub-class Synechococcineae and large-celled Merismopedias included in Chroococcales of the second sub-class Oscillatorineae). (ii) The members of heterocystous, true-branched members have been separated into three families of Nostocales on the basis of type of branching. Members with Y-shaped branching have been placed in Symphonemataceae while those with T-shaped branching pattern are included in Hapalosiphonaceae. *Stigonema* which exhibits multiseriate branching was placed in Stigonemataceae. However, Komárek (2005) pointed out that (i) molecular (phylogenetic) data should be accepted as the basic criterion for the classification; (ii)

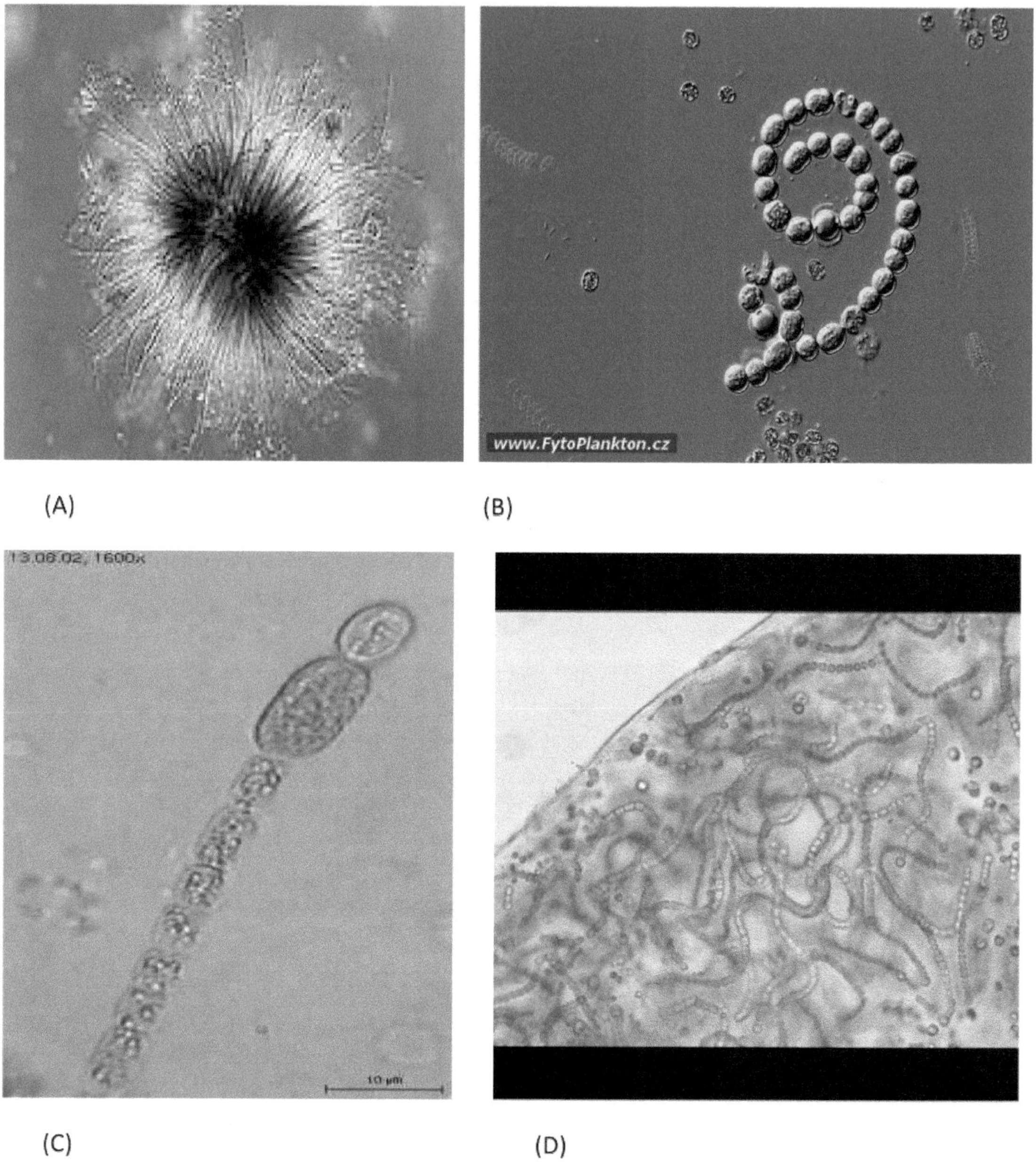

Figure 6: Members of Nostocales. A colony of *Gloeotrichia echinulata* (A), *Anabaena spiroides* (B), *Cylindrospermum* sp. with terminal heterocyst and sub-terminal akinete (C) and a portion of colony of *Nostoc commune* magnified (D). Magnification of pictures (A), (B) x 200; (C) 10 µm and pictures (A) and (B) in Nomarski contrast. Pictures (A) and(B) courtesy P. Znachor, Laboratory of Phytoplankton Ecology, Institute of Hydrobiology, Biology Centre ASCR, Na Sádkách , 37005 České Budějovice, Czech Republic. www.FytoPlankton.cz. Picture (C) courtesy Ralf Wagner (Dusseldorf), Mark Schneegurt (Wichita State University) and Cyanosite (www-cyanosite.bio.purdue.edu). Picture (D) courtesy Mark T. Aubel, GreenWater Laboratories/ CyanoLab, 205, Zeagler Drive, Suite 302, Palatka, Florida 32177, USA. (http://www.greenwaterlab.com/photo_algal.htm).

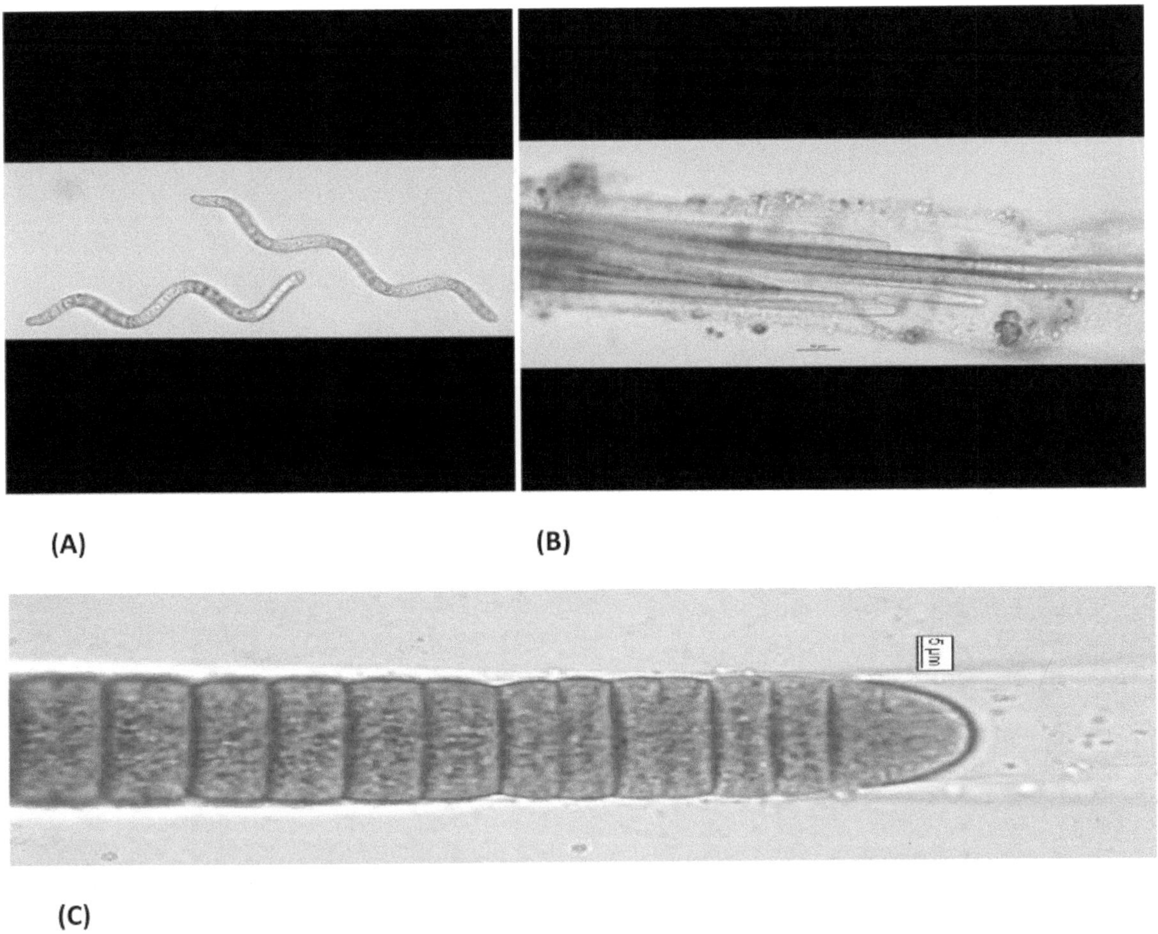

Figure 7: Members of Nostocales. Trichomes of *Arthrospira maxima* (A), *Microcoleus lacustris* (B) and *Porphyrosiphon* sp. (C). Magnification bar in picture (B) 10 μm and picture (C) 5 μm. Pictures (A) and (B) courtesy Mark T. Aubel, GreenWater Laboratories/CyanoLab, 205, Zeagler Drive, Suite 302, Palatka, Florida 32177, USA. (http://www.greenwaterlab.com/photo_algal.htm). Picture (C) courtesy G. L. Tiwari, Department of Botany, University of Allahabad, Allahabad-211002, India.

numerous clusters identified on the basis of 16S rRNA sequencing are more or less congruent with traditional taxonomic assignments. In this connection mention may be made of Oscillatorian genera *Planktothrix*, *Tychonema*, *Microcoleus*, *Trichodesmium*, *Arthrospira* and *Spirulina*. These are distinguishable on the basis of morphological features as well as their 16S rRNA sequence divergence is near 95% as suggested for distinguishing bacterial genera (Wayne *et al.*, 1987; Stackebrandt and Goebel, 1994). Similarly, in heterocystous genera *Cylindrospermum*, *Cylindrospermopsis* and *Anabaenopsis* are distinct morphologically and on molecular phylogeny.

Based on the recommendations of the subcommittee on Phototrophic Bacteria of the International Committee on Systematic Bacteriology (ICSB) of the International Association of Microbiological Societies (IAMS), the cyanobacteria have been included under the Ist Division followed by the bacteria in the IInd division in the eigth edition of Bergey's Manual of Determinative Bacteriology (Buchanan and Gibbons, 1974). However, looking into the advances made in the physiology, biochemistry

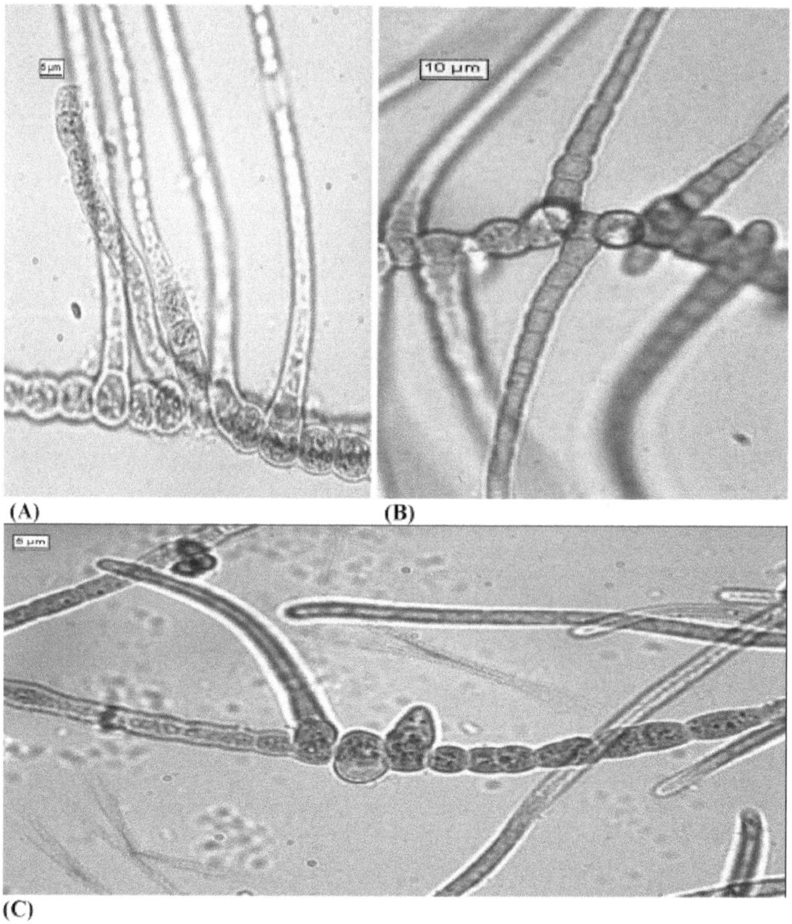

Figure 8: Members of Stigonematales. *Fischerella* sp. (A), *Hapalosiphon* sp. (B), *Westiellopsis* sp. (C), Magnification bar represents 5 µm (A and C) and 10 µm (C). Pictures courtesy G. L. Tiwari, Department of Botany, University of Allahabad, Allahabad-211002, India.

and genetic characterization of cyanobacteria based on pure culture techniques, Stanier *et al.* (1978) proposed the inclusion of cyanobacteria under the provisions of International Code of Nomenclature of Bacteria. In doing so, they emphasized that properties common to and distinctive to all bacteria may be extended to cyanobacteria. Accordingly, in bacterial taxonomy (Rule 18a) the type species and (through type species) genus is represented by living cultures of a strain so designated. They also cited Bacteriological Code of a Rule (18h) that permits the replacement of cultures by a description, a preserved specimen or an illustration of strains giving them the status of holotypes as has been permitted in case of myxobacteria. However, this provision seems to have been withdrawn with effect from November, 2000 (De Vos and Trüper, 2000). Rippka *et al.* (1979) suggested a classification based on bacteriological code and divided the cyanobacteria into five sections. In Section I and Section II members belonging to the orders Chroococcales; Chamaesiphonales and Pleurocapsales as suggested in the Botanical Code have been redistributed. Section III entirely consists of non-heterocystous filamentous cyanobacteria and Sections IV and V comprise heterocystous, unbranched and branched representatives, respectively. The following is the classification scheme presented by Rippka *et al.* (1979) for cyanobacteria.

Section I (Unicellular): This section consists of unicellular cyanobacteria that reproduce by binary fission or by budding and separated into three groups on the basis of baeocyte (a new term coined to represent the erstwhile endospores or exospores) formation. The first group comprises of *Synechococcus* Nageli 1849 (thylakoids present; sheath absent); *Gloeothece* Nageli 1849 (thylakoids present; sheath present) and *Gloeobacter* Rippka, Waterbury & Cohen-Bazire 1974 (thylakoids absent; sheath absent) that form baeocytes by binary transverse division. The second group comprises unicellular members that produce baeocytes in two or three successive planes and represented by *Synechocystis* Sauvageau 1892 (thylakoids present; sheath absent) and *Gloeocapsa* Kützing 1843 (thylakoids present; sheath present). The third group is represented by *Chamaesiphon* Braun & Grunow 1895 emend Geitler 1925 that reproduces by budding from one pole of the ovoidal cell.

Section II (Unicellular): This section is divided into three groups. The first group is represented by members that are unicellular and reproduce by the formation of baeocytes by multiple fission and the baeocytes may be motile or immotile. Examples are *Dermocarpa* Crouan & Crouan 1858 (motile baeocytes) and *Xenococcus* Thuret 1880 (immotile baeocytes). The members of the second group reproduce both by binary fission and multiple fission and represented by *Dermocarpella* Lemmermann 1907; *Myxosarcina* Printz 1921; *Chroococcidiopsis* Geitler 1933. *Pleurocapsa*-group with binary fission yields irregular cellular aggregates and forms pseudofilamentous structures with the formation of motile baeocytes by multiple fission.

Section III (Filamentous): Filamentous non-heterocystous cyanobacteria that divide in only one plane with helical (*Spirulina* Turpin 1827) or straight trichomes (*Oscillatoria* Vaucher 1803) are included here. The other non-heterocystous cyanobacteria (*Lyngbya*, *Phormidium* and *Plectonema*) have been merged into LPP group A: *Pseudoanabaena* Lauterborn 1915 and LPP group B.

Section IV (Filamentous): This section consists of filamentous cyanobacteria that form heterocysts and divide in only one plane. On the basis of hormogone formation, cell size and shape two groups are recognized. The members of the first group reproduce by random trichome breakage to form hormogones that are indistinguishable in cell size and shape from main filaments and some form akinetes. *Anabaena* Bory de St. Vincent 1822, *Nodularia* Mertens 1822 and *Cylindrospermum* Kützing 1843 are the representatives. The members of the second group form hormogones with rapid gliding motility and possess smaller cells with gas vacuoles. *Nostoc* Vaucher 1803, *Scytonema* Agardh 1824 and *Calothrix* Agardh 1824 are the genera recognized.

Section V (Filamentous): These are represented by heterocystous, branched forms that divide in more than one plane, form hormogones and some differentiate akinetes. *Chlorogloeopsis* Mitra & Pandey 1966 and *Fischerella* Gomont 1895 are the representatives.

The scheme of Rippka *et al.* (1979) was adopted and modified in Bergey's Manual of Systematic Bacteriology (Boone and Castenholz, 2001). The cyanobacteria constitute a monophyletic group within Eubacteria and are closely related to purple bacteria and gram-positive bacteria. At present, the phylum cyanobacteria includes two groups of oxygenic photosynthetic bacteria i.e. chlorophyll-*b/a* containing Prochlorales (Prochlorophyta), and cyanobacteria (Castenholz, 2001). In this Manual five Subsections ("Orders") have been recognized. Subsection I consists of unicellular or non-filamentous aggregates of cells (or in other words colonial forms) that reproduce by binary fission in one, two or three planes, symmetric or asymmetric or by budding (corresponding to the Order Chroococcales under ICBN and Section I of Rippka *et al.*, 1979). Subsection I though includes *Chroococcus, Cyanobacterium, Cyanobium, Dactylococcopsis, Microcystis* but differs from traditional Chroococcales by the inclusion of the two Prochlorophytes, *Prochloron* and *Prochlorococcus* and *Chamaesiphon* that reproduces by budding. Subsection II comprises unicellular or colonial forms that reproduce by internal multiple fissions with production of daughter cells smaller than the parent or by multiple fission plus binary fission (corresponds to the orders Chamaesiphonales and Pleurocapsales under ICBN and Section II of Rippka *et al.*, 1979). Subsection III is composed of filamentous forms where the trichomes do not possess differentiated cells like heterocysts or akinetes (corresponds to the non-heterocystous families under the order Nostocales under ICBN and Section III of Rippka *et al.*, 1979). Subsection IV characteristically possesses forms with trichomes differentiating into heterocysts when the concentrations of nitrate or ammonium in the medium is low and some differentiate akinetes as well (corresponds to heterocystous Nostocales under ICBN and Section IV of Rippka *et al.*, 1979). Subsection V consists of true branching representatives that develop multiseriate trichomes and are heterocystous and some produce akinetes (corresponds to the order Stigonematales under ICBN and Section V of Rippka *et al.*, 1979). Gibbons and Murray (1978) suggested that the cyanobacteria be treated under the order 'Cyanobacteriales' and integrated with photobacteria. The name of the order was not based on a designated name of genus or species. Although a special request was made by these workers to the Judicial Commission of International Committee on Systematic Bacteriology, their proposal was not accepted (Holt, 1979). In order to pave the way for the creation of this order under Bacteriological Code, Rippka and Cohen-Bazire (1983) created two genera *Cyanobium* and *Cyanobacterium*. The type species of the genus *Cyanobacterium* as *Cyanobacterium stanieri* (strain PCC 7202; ATCC 29140; in honour of late R. Y. Stanier) should fulfill the condition for the creation of the order Cyanobacteriales under the Bacteriological Code. One of the intriguing aspects of this classification is the recognition of 'form genera' a term that has no standing either under ICBN or Bacteriological Code (Oren, 2004). The concept of 'form species' has been questioned by Whitton and Potts (2000). The number of genera validly published under Bacteriological Code is only six and 13 species names have been proposed in the validation lists published by International Journal of Systematic and Evolutionary Microbiology or International Journal of Systematic Bacteriology.

Cavalier-Smith (2002) classified all bacteria into a single Kingdom, i.e. Kingdom Bacteria. The ancestors for the development of cyanobacteria have been identified as Negibacteria (Sub-Kingdom I) in which two Infrakingdoms Eobacteria (Infrakingdom I) and Glycobacteria (Infrakingdom II) have been recognized. Cyanobacteria are included in Division I that has been divided into two Sub-divisions, Gloeobacteria (Sub-division I; with the only known genus *Gloeobacter* having phycobilisomes but with out thylakoids) and Phycobacteria (Sub-division II; consisting of all the traditional blue-green algae and Prochlorophytes). He formally validated all the five traditional cyanobacterial orders (valid under ICBN) under the Bacteriological Code under two Classes.

Class I: Chroobacteria: Based on the genus *Chroococcus*; unicellular, palmelloid, colonial or with filaments lacking heterocysts.

Order 1: Chroococcales

***Order 2: Pleurocapsales*:** Colonial or filamentous, reproducing by intramural multiple fission to yield small unicellular dispersal stages.

***Order 3: Oscillatoriales*:** Unbranched linear filaments without heterocysts, cells typically shorter than broad.

Class II: Hormogoneae (ex. Thuret 1875)

Order 1: Nostocales: *Nostoc*, filaments that multiply vegetatively by hormogonia usually with heterocysts, unbranched.

Order 2: Stigonematales: *Stigonema*, branched, heterocystous forms.

II. PHYLOGENY

Phylogenetic relationships tell us the manner in which the different groups of organisms have evolved during the course of time. In case of eukaryotes (both animals and plants), macromorphological features (organismal biology) form the main basis for drawing phylogenetic relationships. In case of prokaryotes, the range of morphological features being very limited, a number other corollary features have been taken into consideration. Along with morphological features, immunological, physiological, biochemical, ecological as well as genetic features have been taken as the supporting taxonomic features to identify bacteria and draw their phylogeny. This has given rise to a polyphasic approach to taxonomy. In this connection, a single gene whose sequence is highly conserved and at the same time shows variable sequences that could form the basis for identification can be selected. The first choice fell on 16S rRNA gene. The projection of this gene as molecular marker for drawing phylogenetic relationships has revolutionized thinking in biological world with the culmination of the classification of living organisms into bacterial, archaebacterial and eukaryotic domains (Woese *et al.*, 1990). This triggered a large number of studies all world over and generated a wealth of knowledge. These studies have clearly pointed out that the sequencing of 16S rRNA gene falls short of differentiating bacteria up to or below the level of species. It turned out that the rRNA operon that consists of internally transcribed sequences (ITS) between 16S and 23S rRNA could well serve as a good parameter. This has no doubt helped in the identification of bacteria up to strain level and the resolving power of this ITS is found to be greater than simply sequencing of 16S rRNA. Alternative searches for other ideal genes suited for this purpose resulted in the identification of some housekeeping genes such as RNA polymerase gene, cell division genes and genes involved in DNA repair. Taking 16S rRNA phylogeny as the basis, comparisons have been made between other gene markers selected and the resulting phylogenies. This gave rise to the concept of multi-locus sequence typing (MLST). Another most promising, reliable but very expensive approach is whole genome sequencing of bacteria and comparisons between the bacterial species in question. The studies on genomics have converged with phylogeny resulting in an interesting area known as Phylogenomics (Eisen, 1998; Eisen and Fraser, 2003). All these developments are detailed below.

1) 16S rRNA gene

The 16S rRNA gene has all the characteristics for a phylogenetic marker gene because of its universal distribution in prokaryotes, functional consistency, the presence of variable and conserved regions and high information content (Woese, 1987; Ludwig and Klenk, 2001; Figs. 9 and 10). Other attributes of 16S rRNA gene are (i) its suitable length of about 1500 bp, (ii) the presence of long highly conserved regions useful for measuring distant phylogenetic relationships, (iii) the presence of sufficient variable regions to measure close relationships, (iv) not prone for rapid sequence change and (v) not liable for getting transferred from one organism to another through lateral gene transfer (LGT). Woese (1987) considered it an excellent molecular chronometer for finding out evolutionary relationships among all living organisms. On the basis of 16S rRNA gene sequences, Woese *et al.* (1990) proposed a three

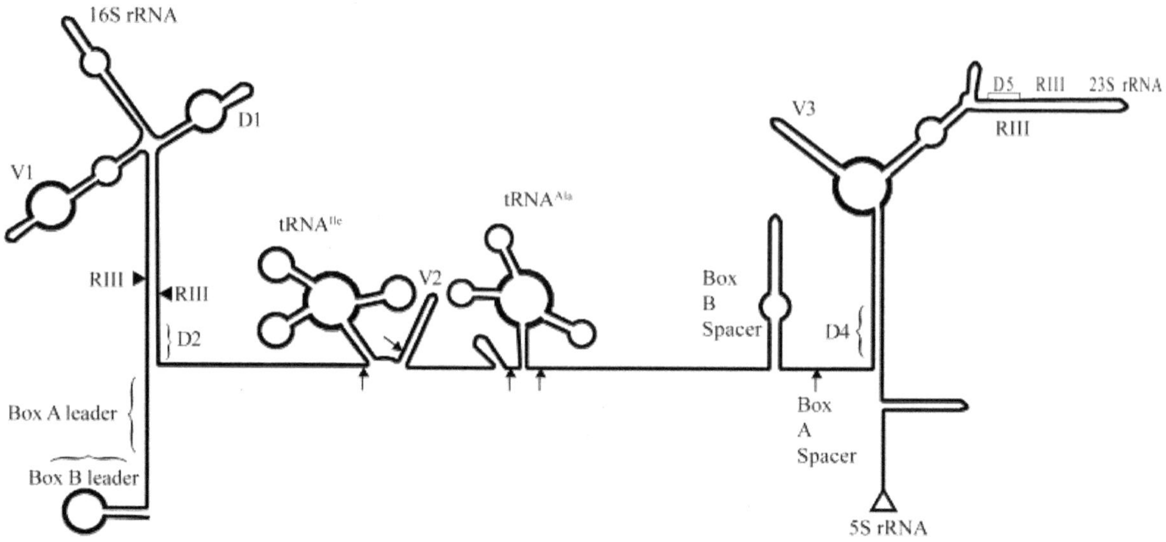

Figure 9: Diagramatic representation (not drawn to scale) of rRNA operon of *E. coli*. D1, D2, D4 and D5 represent conserved domains and V1, V2 and V3 represent variable regions. Based on the molecular structure as in Iteman *et al.* (2000).

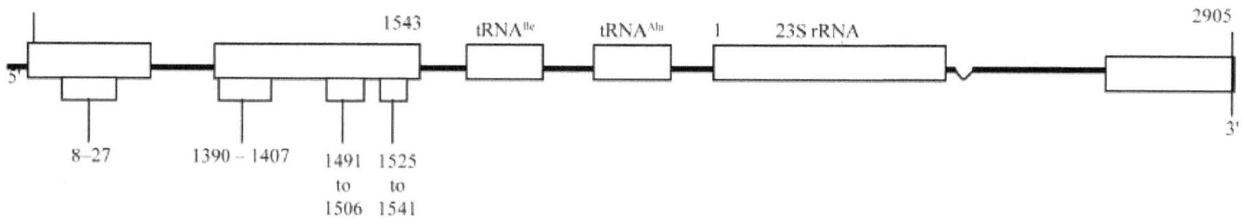

Figure 10: Line drawing (not drawn to scale) of rRNA operon of *E. coli* showing specific regions of 16S rRNA that are generally amplified in the identification of prokaryotes. There are six other regions in 23S rRNA also (not shown) utilized for this purpose. Modified from Gürtler and Stanisch (1996).

domain classification consisting of Bacteria, Archaea and Eukarya (Fig. 11). Studies on this molecule have generated a huge public data base (RDP II: http:// rdp.cme.msu.edu/ containing 481,650 16S rRNAs as on 13th February, 2008). It is generally agreed that a 16S rDNA sequence similarity of 96–97% and DNA:DNA hybridization value of 70% relative binding (RB) with ΔT_m 5°C represent lower boundaries for the distinguishing bacterial species (Wayne *et al.*, 1987). Fox *et al.* (1992) reported that three psychrophilic strains of *Bacillus* (*B. globisporus* W25^T Type strain; *B. psychrophilus* W16A and W5) that are phenotypically very similar exhibited 99.5% 16S rRNA gene sequence similarity but differed from each other on the basis of DNA:DNA hybridization studies. They concluded that identity of 16S rRNA sequences does not constitute a sufficient criterion to guarantee species identity. A number of proposals have been made for altering the lower limits for the delineation of a species in bacteria. An RB value of 50% at a ΔT_m 7°C has been suggested by Rosselló-Mora and Amann (2001) for distinguishing species because even if 16S rDNA sequence similarity was as high as 98–99% the total DNA homology was as low as 10–40% RB. Similarly, Keswani and Whitman (2001) observed that the relationship between RB and 16S rDNA sequence similarity greatly varied within the same subphylum. In view of these, it appears that there is a need to alter the upper and lower limits of RB values for species delimitation in prokaryotes.

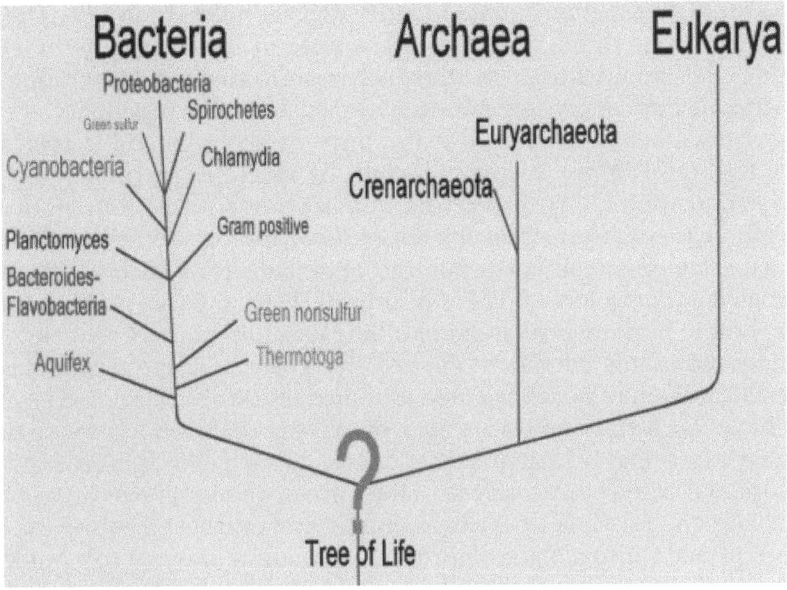

Figure 11: Three domain classification of living world based on 16S rRNA sequencing.

The 16S rRNA gene sequences are easy to align that allow comparison between strains. This is a crucial step in phylogenetic analysis (Swofford *et al.*, 1996). In alignment, the sequences are arranged in such a manner by inserting gaps so that the homologous positions of sequences are placed in the same columns of the data matrix. A number of computer programmes have been used for aligning the sequences, e.g. GenBank or EBI (European Bioinformatics Institute), Ribosomal Data Base Project (RDP) and Antwerpen Projects. Further, a variety of individual software tools for sequence editing, alignment and phylogenetic analyses are available from different software packages such as phylogeny inference package (PHYLIP; Felsenstein, 1989), Maximum likelihood (ML method; FastDNAml; Olsen *et al.*, 1994); molecular phylogenetics based on Maximum-likelihood (Molphy-Version 2.3), multiple sequence alignment (ClustalW; Thompson *et al.*, 1994; Chenna *et al.*, 2003) and ARB software (Ludwig *et al.*, 2004; accessible at http://www.arb-home.de). Although ARB software initially was designed for rRNA data, it can now be used for any nucleic acid or amino acid sequence as well. This software is helpful in aligning the sequences through primary structure data that contains processed aligned sequences for comparison. It can be linked via local or worldwide networks. Tools for data import and export, primary and secondary structure editing, specific hybridization probe design are the additional features of the ARB software. ClustalW and ClustalX version 2.0 multiple sequence alignment programmes have been completely rewritten in C++ and made available at http://www.ebi.ac.uk/tools/clustalw2 by Larkin *et al.* (2007).

At this stage, it is required to know the definitions of gene terminology used while drawing molecular phylogenetic relationships: (i) homologues are those genes that have descended from a common ancestor (e.g. all globin genes); (ii) orthologues are those homologous genes that have diverged from each other after speciation events (e.g. human β- and chimp β-globin); (iii) paralogues are those homologous genes that have diverged from each other after gene duplication events (e.g. β- and gama-globin) and (iv) xenologs are those homologous genes that have diverged from each other after LGT events (e.g. antibiotic resistance genes in bacteria). Positional homology is a specific amino acid or nucleotide positions in different genes. The methods of predicting gene function when homologues have multiple functions are to identify highest hit and top hits. The uncharacterized gene is assigned the function of the gene that is identified as the highest hit by a similarity search programme. Top 10+ hits are identified for the uncharacterized gene. Depending on the degree of consensus of the top hits, the query sequence is assigned a specific function.

Evolutionary trends are represented in the form of a phylogenetic tree. It is a graphical representation of the evolutionary history of genes or species. The branching pattern of a tree (topology) displays the evolutionary relationships of the strains. A tree consists of nodes and edges. The nodes correspond to organisms and the edges show their relationships. The terminal nodes possess only one linking edge and correspond to organisms for which we have data. These are commonly designated as operational taxonomic units (OTUs). Internal nodes possess three linking edges and correspond to hypothetical ancestor of one set of organisms. One unique internal node with only two edges represents the root of the tree, the common ancestor to all taxa.

The phylogenetic trees may be represented as rooted or unrooted. Rooted or unrooted trees are identical except that in the former nodes are present while in the latter nodes are absent but possess only two connecting edges (Fig. 12). All proposed tree-building methods fall into two general classes: algorithmic and optimal criteria approaches. In the former, a unique tree can be built by a series of steps while in the latter all possible trees are examined and one best meeting the certain criteria is chosen. In the algorithmic approach, the algorithm plays a fundamental role but in optimal criteria approach the algorithm simply is a tool to evaluate the criteria. Algorithmic approaches are known as distance methods because they compute a tree via a distance matrix composed of the distances between each pair of sequences. A distance based algorithm is basically a procedure for building a tree based on the matrix. If the outputted tree is to be additive it must obey four basic rules, i.e. (i) all distances are positive; (ii) a distance between two points can be zero only when the two points are actually the same; (iii) the distances are symmetrical and (iv) there are no shortcuts in the tree, i.e. the distance between a-c cannot be longer than the sum distances of a-b and b-c. In addition, a tree point condition is applied that is distance a-b cannot be larger than their distance to a third point (the maximum distances a-c and b-c). Unweighted Pair Group Method (UPGMA) is an ultrametric tree-building algorithm that follows all the rules mentioned earlier. UPGMA proceeds by inferring one ancestral sequence per step. In the first round UPGMA selects the least distant pair of sequences (or one of them), summarizes their distance as the first branches of a new tree and recalculates the entire matrix with the pair as one entity (taking the mean distances). After N-1 steps (where N is the number of sequences) the matrix is reduced to just one element. The last inferred ancestor is taken as the root of the tree. Distance methods such as neighbor joining (NJ) use pair-wise distances (i.e. the number of base differences between two sequences), calculated from aligned sequences and usually

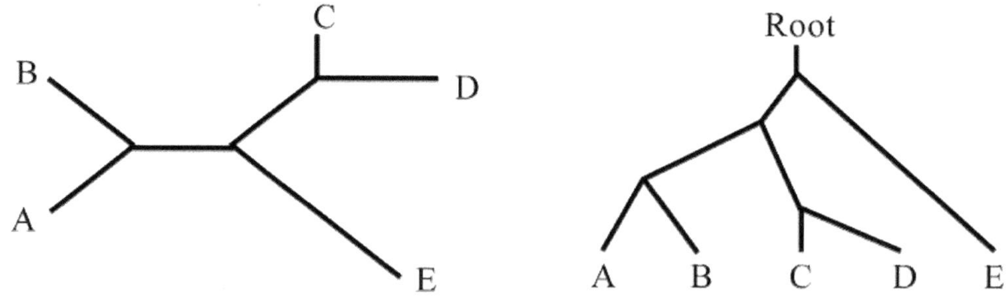

Figure 12: Representation of unrooted (left) and rooted (right) phylogenetic trees.

corrected to evolutionary distances within a substitution mode (Saitou and Nei, 1987). The sequences with the shortest distances are clustered together in a tree, where the tree length is optimized to correspond to the distance matrix. The root is not specified by a NJ tree. It is very common to identify the root by adding a sequence to the set that is known to be more different from all the others that is known as an out group. The position of the out group on the tree indicates the root. For example, in the case of the human mitochondria, a chimpanzee outgroup was used to identify the root.

There are two methods under optimal criteria approach. These are maximum parsimony (MP) and maximum likelihood (ML) methods (Nei and Kumar, 2000; Ludwig and Klenk, 2001). The MP method uses the actual sequence data instead of distances and searches for the tree(s) with minimum length, i.e. topology of the tree can be explained with a minimum number of transformations from one character state to another. A number of possible trees are compared and each is given a score that is a reflection of minimum number of character changes (e.g. amino acid substitutions) that would be required over evolutionary time to fit the sequences in that tree. Comparing the number of inferred mutations for each possible tree, we define an informative site as one that favours one of the possible trees in terms of fewer mutations. The optimal tree is considered to be the one requiring the fewest changes, the most parsimonious tree.

ML method is similar to MP method in that possible trees are compared and given a score. The score is based on how likely the given sequences would have evolved in a particular tree given a model of amino acid or nucleotide substitution probabilities. It estimates the likelihood for tree topology that could have resulted in the sequence alignment under the given model of evolution and searches for the tree with maximum likelihood (Swofford *et al.*, 1996; Nei and Kumar, 2000). Bootstrapping is a statistical method in which resamplings are made within the original multiple sequence aligned and new data sets are made. Starting with the multiple sequence alignment it builds a random multiple sequence alignment by sampling with replacement the different sites of the original multiple alignment. For a large number, say 10,000, of such pseudo multiple sequence alignments a tree is built by the same method used to build the original trees. Each node of the tree can be given a bootstrap percentage indicating how frequently those species joined by that node group together in different trees. It is particularly convenient because it does not require anything more than the actual data used to build the tree. It can also be used in conjunction with any tree-building method. Based upon the 10,000 trees, bootstrapping can be used to annotate the internal branches of the original tree. For each internal branch, we count how many times the same particular "split" is found among the 10,000 trees. The higher this number the more one is confident in the robustness of the split.

2) Molecular techniques based on 16S rRNA gene

A number of analytical methods are now available for measuring the variations in the sequences of 16S rDNA. Pure cultures of microorganisms whose identification has been made on a polyphasic approach (that is based on morphological, biochemical, physiological, phenotypic and chemotaxonomic criteria) can be subjected to 16S rRNA gene sequence analysis and phylogenetic relationships can be drawn by any one of the methods described above. During recent years, ecological criteria are given equal importance with molecular markers in the identification and characterization of strains of bacteria into ecotypes (Whitaker *et al.*, 2003; Whitaker, 2006; Martiny *et al.*, 2006; Staley, 1999, 2006; Cohan, 2006; Fraser *et al.*, 2008). To investigate the prokaryotic diversity of a given ecosystem, culture-independent approach known as metagenomic analysis has been developed. According to Whitman *et al.* (1998) the prokaryotic diversity assumes greater significance as their cellular production rate is estimated to be 1.7×10^{30} cells per year and their carbon content equals to nearly 60–100% of that present in plants. Furthermore, their contribution is highest in the open oceans. Thus the newly emerging metagenomic analysis has to its credit more than 100 projects in the GOLD web site, Genomes OnLine Database [February, 2008; http://www.Genomesonline.org/gold (gi)]. Of these, 31 projects have already been completed. By now, nearly 120,000 sequences are deposited in databases such as RDP. Mention may be made of four large environmental sequencing projects that have been completed from acid mine drainage (Tyson *et al.*, 2004), Sargasso Sea (Venter *et al.*, 2004), farm soil (Tringe *et al.*, 2005) and whale falls (Tringe *et al.*, 2005). In terms of minimum number of species Sargasso Sea is at first position (with 1000 species) wheras in terms of estimated total number of species farm soil occupied the first place with >3000 species. Likewise, in terms of total number of ORFs annotated Sargasso Sea showed maximum (>1000,000) followed by farm soil (>180,000) (Foerstner *et al.*, 2006). There are seven techniques identified for analysis of 16S rRNA gene. These are as follows.

(i) Oligonucleotide cataloguing: This is one of the earliest techniques used to identify bacteria. The 16S rRNAs from individual bacteria is digested with ribonuclease T1 (which cleaves specifically at G residues) and produces short oligonucleotides of lengths up to 20 or so. A collection of these from a given rRNA consists of oligonucleotide catalogue (Fox *et al.*, 1977). A specific pattern characteristic of a given species is obtained that enables one to arrange bacteria into phylogenetic groupings. The data are represented in terms of binary association coefficients (called as S_{AB} values) defined as the ratio of twice the sum of bases in oligonucleotides in two catalogues. Because of its failure to identify bacteria beyond their phyla and branching orders or their subdivisions, sequencing approach received more attention subsequently (Woese, 1987).

ii) Sequencing of 16S rRNA gene: Total community DNA or DNA extracted from identified cultures is used as a template for polymerase chain reaction (PCR) amplification of 16S rRNA genes with universal or domain specific primers. A clone library is constructed for genes encoding rRNA (rDNAs) and is screened for finding out the sequence differences and phylogenetic relationships are drawn. By applying this method one can assess the richness and evenness of a community qualitatively based on the number of unique clones and the relative frequencies of each "ribotype". Schmidt *et al.* (1991) described a technique for analysis of marine picoplankton community by 16S rRNA gene cloning and sequencing bypassing culture of the organisms. In this method, natural picoplankton populations are collected and filtered by tangential flow filtration. The collected biomass is subjected to DNA extraction and DNA is purified by CsCl gradient centrifugation. The DNA obtained is fragmented, size-fractionated and cloned into bacteriophage lambda. Out of 3.2 x 10^4 recombinant phages, 38 clones of lambda phage containing the 16S rDNA were screened by hybridization with "mixed kingdom" 16S rRNA probes. Ribosomal DNA clones are then sorted, PCR amplified with r-DNA specific primers. Single strand rDNA clones were subjected to sequencing to obtain nucleotide pattern. Then phylogenetic characterization of the constituents of the population revealed the identity of 4 cyanobacterial, 11 proteobacterial and 1 eukaryote 16S rDNA sequences. An alternative method for 16S rDNA sequencing that is quite faster and cheaper is Random Sequence Reads (RSRs). RSRs are quite suitable for comparative study of the biodiversity of a metagenomic library. In this method, the sequence analysis of oligonucleotide frequencies represented in the genomes permits the identification of different species (Karlin and Burge, 1995; Abe *et al.*, 2003). Manichanh *et al.* (2008) compared RSR method against 16S rDNA sequencing for estimating the biodiversity of a metagenomic library. The sequences of the oligonucleotides are compared with a database of prokaryotic sequences (GenBank-prok) using BLASTN. They applied this local alignment programme to find out the closest relative for each of the RSRs out of a total of 516,770 entries. By taking into account the top Blast hit, they identified the species to which the sequence belongs and performed the taxonomic assignment. Pyrosequencing is another technique that is dependent on the characterization of PCR products of 16S rDNA. This is a cheap, labour saving method that does not involve cloning of the gene. The original instrument known as GS20 Instrument that has been fabricated by Margulies *et al.* (2005) sequences approximately 100 bases. The current GS FLX platform can sequence 250 bases and the next generation GS XLR instrument is expected to sequence nearly 400 bases. Liu *et al.* (2008) performed accurate taxonomic assignments from the 16S rRNA gene sequences produced by highly parallel pyrosequencers.

iii) 16S rRNA-23S rRNA operon copy number and the internally transcribed spacer (ITS) sequences: Bacterial rRNA genes are commonly present in a single operon in the order of 16S rRNA-23S rRNA-5S rRNA each of which is separated by one ITS region (Srivastava and Schlessinger, 1990). One or two tRNA genes (either $tRNA^{Glu}/tRNA^{Ala}$ or both $tRNA^{Ala}$ and $tRNA^{Ile}$) are often present in the 16S rRNA-23S rRNA ITS region. However, in certain bacterial species like *Thermus thermophilus* (Srivastava and Schlessinger, 1990), *Leptospira interrogans* (Fukunaga and Mifuchi, 1989), *Borrelia* spp. (Ojaimi *et al.*, 1994) and *Wobachia pipientis* (Bensaadi-Merchermek *et al.*, 1995) the rRNA genes are separated into two distinct 16S rRNA and 23S rRNA operons. In case of bacteria, great variation in between the number of ITS regions and alleles of the rRNA operon has been found. For example, in *Escherichia coli* there are 7 copies of operons coding for the three rRNAs (Morgan *et al.*, 1977) whereas in *Bacillus subtilis* (Loughney *et al.*, 1982) and *Clostridium perfringens* (Garnier *et al.*, 1991) 10 copies each of the operons coding for the three rRNAs are present. On the other hand, one copy of the operon is present in *Mycobacterium* sp. (Bercovier *et al.*, 1986) whereas one to two copies of the operon may be present in *Mycoplasma* (Amikam *et al.*, 1984). Moreover, greater heterogeneity in these multiple copies has also been observed. In *E. coli* where these copies are named as rrnA, rrnB, rrnC, rrnD, rrnE, rrnG rrnH, the 16S rRNA-23S rRNA ITS region in the operons rrnB, rrnC, rrnE and rrnG contains a gene coding for $tRNA^{glu\text{-}2}$ where as operons rrnA, rrnD and rrnH have genes for $tRNA^{Ile\text{-}1}$ and $tRNA^{Ala\text{-}IB}$ (Antón *et al.*, 1998). However, majority of gram-positive bacterial species studied (15 out of 19) have no tRNA genes at all in the ITS region (Gürtler and Stanisich, 1996). This shows that there is a great variation in the number, length and composition of the 16S-23S rDNA spacer regions of diverse range of eubacteria and archaebacteria. As a part of Human Microbiome Project (accessible at http://nihroadmap.nih.gov/hmp), the diversity of 16S rRNA genes in 883 prokaryotic genomes (including 13 from cyanobacteria) revealed 568 unique sequences. Of these, 425 species showed 2 to 15 copies of rRNA genes per genome. There is every possibility for taxonomic mis-classification, based exclusively on 16S rRNA, of a number of species into more than one species as it had happened with seven bacterial species associated with human disease development (Pei *et al.*, 2010). The diversity of 23S rRNA genes in 184 prokaryotic species has been determined by Pei *et al.* (2009) who found multiple 23S rRNA genes in 113 genomes. Significantly, eight of the species exhibited intragenomic variation in the 23S rRNA gene sequences. Intervening sequences ranging in length from 9 to 1471 nucleotides were found in seven species. In case of *Anabaena* sp. strain PCC 7120 there are four 23S rRNA genes of which rrnA 23S is 4299-nucleotides long whereas the rest of the three each are 2828-nucleotides long, the difference in the lengths is due to the presence of 1471-nucleotides long intervening sequence that encodes a transposase. The second organism that has a transposase gene as an intervening sequence, within the 23S rRNA gene, is *Deinococcus radiodurans* (Pei *et al.*, 2009). Rastogi *et al.* (2009) constructed phylogenetic trees based on rRNA copy numbers and genome sizes in which colour codes given with increasing rRNA copy numbers and genome sizes can be correlated with the colour codes given on the phylogenetic tree constructed with the 31 housekeeping genes from 578 bacterial genomes. This will be useful in determining microbial diversity and reduces chances for mis-identification due to the existence of variable rRNA copy numbers.

A number of cyanobacteria have been studied for the sequences of the 16S rRNA-23S rRNA ITS domain. Three types of ITS regions have been identified. The first type contains both $tRNA^{Ile}$ and $tRNA^{Ala}$ sequences. These are found in *Anabaena* sp. (Lu, 1999), *Arthrospira* PCC 7345 (Nelissen *et al.*, 1994), *Nostoc* sp. (Lu, 1999), *Synechococcus* PCC 6301 ("*Anacystis*", Tomioka and Sugiura, 1984), and *Trichodesmium* NIBB 1067 (Wilmotte *et al.*, 1994). The second type of ITS regions contain only $tRNA^{Ile}$ found in 47 strains of *Microcystis* (Otsuka *et al.*, 1999), *Spirulina* PCC 6313 (Nelissen *et al.*, 1994) and the unicellular *Synechocystis* sp. strain PCC 6803 (Kaneko *et al.*, 1996). Kaneko *et al.* (1996) identified two rRNA operons both consisting of $tRNA^{Ile}$ present in inverted repeats in *Synechocystis* sp. strain PCC 6803. The third type of ITS regions do not contain any tRNA as noted in *Nodularia* BCNO D9427 (Hayes and Barker, 1997). The 16S rRNA-23S rRNA ITS sequences of two other cyanobacteria *Mastigocladus* HTF strain PCC 7518 of uncertain identity and that of the cyanelle *Cyanophora paradoxa* have also been determined (Iteman *et al.*, 2000). There is a great size variation in ITS regions of cyanobacterial origin. They vary from 354 to 545 nucleotides with the exception of 287 nucleotides in the cyanelle *C. paradoxa.* The existence of multiple rRNA operons was for the first time reported in three species of *Anabaena* and one species of *Nostoc* (Nichols *et al.*, 1982). The existence of four rrn operons in *Anabaena* sp. strain PCC 7120 was reported by Ligon *et al.* (1991). Three PCR products have been generated from the 16S rRNA-23S rRNA ITS region of *Anabaena* sp. strain PCC 7120. Of these, two of them represented the true ITS regions while the third fragment was longer (with 512 nucleotides) and possessed $tRNA^{Ile}$ and $tRNA^{Ala}$ genes separated by a large stem-loop structure (Iteman *et al.*, 2000). Further, these workers identified highly conserved motifs important for folding and maturation of rRNA transcripts (homologous to bacterial antiterminators boxB-boxA) that can serve as potential targets of PCR primers and oligonucleotide probes for detection and identification of cyanobacteria. The 16S rRNA-23S rRNA ITS regions of *Anabaena* sp. strain PCC 7120, *Mastigocladus* HTF strain PCC 7518 and *C. paradoxa* contain $tRNA^{Ile}$ and $tRNA^{Ala}$. Boyer *et al.* (2001) examined variability in the ITS regions among multiple rRNA operons in five species of cyanobacteria (*Scytonema hyalinum, Toplypothrix distorta, Calothrix parietina, Coelodesmium wrangelii* and a new genus designated by isolates SRS 6 and SRS 70). The ITS regions from these organisms transcribe two tRNA molecules $tRNA^{Ile}$ and $tRNA^{Ala}$. So in this respect, these species belong to the first type of ITS region described above. However, *C. parietina* and *S. hyalinum* also contained ITS regions without tRNA. In addition, *S. hyalinum* contained two non-coding ITS regions that are identical in length but differed in their sequences. The species included in their study has wider representation taxonomically because three of the species belonged to three of the four families of Nostocales. Boyer *et al.* (2001) cautioned that although the potential of ITS region as a tool for studying molecular systematic and population genetics is great but the presence of multiple non-identical rRNA operons poses a problem, especially in sequence comparison and secondly when restriction digests of PCR products are analyzed. So as to enable investigators to align and compare the highly variable ITS region, D'Auria *et al.* (2006) created a software known as Word Count based (IWoCS) system with 32,061,819 entries of which 482 sequences pertain to cyanobacterial ITS sequences. Of these, 204 sequences pertain to Chroococcales, 56 of Nostocales, 60 of Oscillatoriales, 67 of Prochlorales, 2 of Stigonematales and 93 unidentified/unknown sequences. This provides a novel tool for the use of ITS sequences in typing and identification of bacteria and cyanobacteria.

iv) Amplified rDNA restriction analysis (ARDRA): Ever since this method was discovered by Grimont and Grimont (1986) more than 2000 research papers have been published on 16S rRNA gene restriction patterns leading to the establishment of 'ribotyping' as potential taxonomic tool for the enumeration of bacterial diversity. The molecular genetic basis of ribotyping has been elucidated (Bouchet *et al.*, 2008). This is considered to be a rapid method for comparisons of rDNAs (Laguerre *et al.*, 1994; Moyer *et al.*, 1994). It involves digestion of rDNAs (obtained by PCR amplification by using universal primers) with restriction enzymes that have 4–bp recognition sites. The restriction digests are analysed by agarose gels. For community analysis, the large number of fragments can be resolved by using polyacrylamide gels to produce community specific pattern (Martínez-Murcia *et al.*, 1995; Massol-Deya *et al.*, 1995). The disadvantage with this method is that it is of limited use for detecting specific phylogenetic groups or for estimating species richness or evenness.

A variation of the above method is to subject the whole genomic DNA for restriction analysis instead of a single gene as in ARDRA. This is termed as amplified fragment length polymorphisms (AFLPs). Naturally occurring DNA polymorphisms are basically due to point mutations or rearrangements (i.e. insertions and deletions) in the DNA. AFLP analysis is based on selective amplification of DNA restriction fragments (Vos *et al.*, 1995). Restriction enzyme digestion of genomic DNA is follwed by PCR amplification and the products are subjected to separation by electrophoresis. This enables us to detect the presence or absence of such polymorphisms by the presence or absence of banding patterns. The variation in banding patterns may directly reflect the genetic relationships between the bacterial strains examined. So these banding patterns in otherwords represent genomic fingerprints allowing numerical analysis for characterization and identification. Genome-wide variations in the strains can be scanned by AFLP. Thus it will greatly help in the resolution of taxonomic problems which are otherwise problematic and answers for these are difficult to obtain through morphological features or by other molecular methods. The broad taxonomic applicability of this technique has been tested in case of bacteria (Huys *et al.*, 1996) and fungi (Majer *et al.*, 1996). The advantage of AFLP over other techniques is that multiple bands are derived from all over the genome. This prevents over interpretation or misinterpretation due to point mutations or single-locus recombination, which may affect other genotypic characteristics. The main disadvantage of AFLP markers is that alleles are not easily recognized (Majer *et al.*, 1998). Due to the inherent advantages of its utility, reproducibility and efficiency, the AFLP technique has been used to investigate cyanobacterial populations as well (Janssen *et al.*, 1996).

RAPD fingerprinting in conjunction with PCR has been used to identify microorganisms to the strain level (Welsh and McClelland, 1990). The sequences frequently used are short tandemly repeated repetitive sequences (STRR). These are of the nature of heptanucleotide sequences that are present nearly at the rate of 100 copies per genome of *Calothrix* sp. strain PCC 7601. These can be used for special identification in the taxonomy of heterocystous cyanobacteria. Three types of the STRR sequences identified are STRR1 [CCCCA(A/G)T], STRR2 [TT(G/T)GTCA] and STRR3 [CAACAGT]. Of these, the former two are present 100 copies each per *Calothrix* genome (Mazel *et al.*, 1990). These can be used as oligonucleotide probes (Rouhiainen *et al.*, 1995) or as primers for the generation of PCR-amplified DNA profiles (Rasmussen and Svenning, 1998) for the identification of toxic cyanobacteria in cultures as well as from field samples. Besides the STRR sequences, the repetitive extragenic palindrome sequences (REP) and Enterobacterial Repetitive Intergenic consensus sequences (ERIC) (Sharples and Lloyd, 1990; Hulton *et al.*, 1991; Wilson and Sharp, 2006) have been used for the identification of bacteria. DNA polymorphisms in phycocyanin locus were used to generate RAPD fingerprinting profiles to differentiate members of the genera *Anabaena* and *Microcystis* (Neilan *et al.*, 1995). Symbiotic and free-living cyanobacterial cultures were identified by PCR fingerprinting (Weiwen *et al.*, 2002) and genotypes of *Microcystis* were discriminated by RAPD fingerprinting technique (Nishihara *et al.*, 1997).

v) Terminal restriction fragment length polymorphism (T-RFLP): The fifth method is just an extension of the second method described by Liu *et al.* (1997) where the initial steps of DNA isolation, PCR amplification and digestion with restriction enzymes are similar to those used for ARDRA. However, in this method one of the primers used is labelled at 5′-terminus with a fluorescent dye and the terminal restriction fragments (T-RFs) are separated by electrophoresis and their fluorescence measured. T-RFLP analysis provides quantitative data about each T-RF detected, including size in base pairs and intensity of fluorescence (represented by peak height). Due to terminal RFLPs (T-RFLPs) found in 16S rDNAs, this method can be used with DNA from complex microbial communities. It is important to select proper PCR primers as well as restriction enzymes (Fig. 13). Computer simulated analysis of T-RFLP for 1,002 eubacterial sequences, 686 sequences could be PCR-amplified and

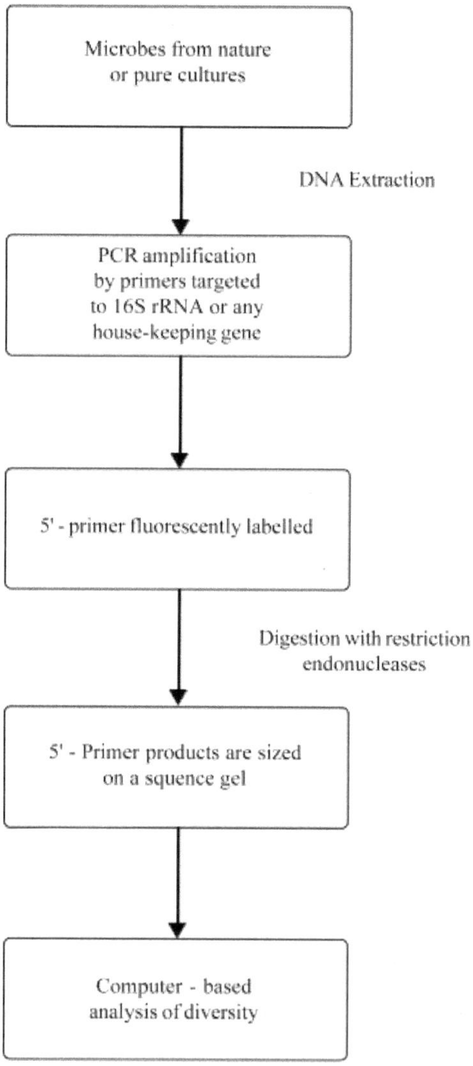

Figure 13: Flow-chart depicting different steps in T-RFLP technique.

classified into 233 unique T-RFLPs or ribotypes. Using this method, Liu *et al.* (1997) were able to distinguish all bacterial strains in a model bacterial community. It has been shown to be an effective technique for discriminating microbial communities in a range of environments (Tiedje *et al.*, 1999) and also found to be relatively stable to variability in PCR conditions (Osborn *et al.*, 2000; Ramakrishnan *et al.*, 2000). A web-based research tool for microbial community analysis was developed by Marsh *et al.* (2000) for T-RFLP analysis and was located at the Ribosomal Database Project website (http://www.cme.msu:edu/RDP/html/analyses.html). According to these workers, it is important to know (i) the type of restriction enzyme(s) that will provide the most discriminating activity for estimates of population diversity, (ii) the enzymes that will provide the best resolution for the best phylogenetic group and (iii) the particular primer-enzyme combination which will be optimal for the community under investigation. Each unique T-RF is considered equivalent to one OTU and so can be equated to a single species/strain within a given community (Moesender *et al.*, 2001). Thus T-RFLP technique can be used to assess spatial and temporal changes in 16S rDNAs from microbial communities (Kitts, 2001; Osborn *et al.*, 2000; Dunbar *et al.*, 2001; Sessitsch *et al.*, 2001). T-RFLP technique is highly sensitive and helpful in acquiring data very rapidly compared to other methods such as denaturing gradient gel electrophoresis (DGGE) or ARDRA (Ferris and Ward, 1997; Moesender *et al.*, 1999; Muyzer, 1999; Horz *et al.*, 2001). Blackwood *et al.* (2003) are of the opinion that T-RFLP data provide inaccurate estimates of true diversity in microbial communities. Wherever significant differences in T-RFLP diversity indices have been found, all such work should be reinterpreted by the application of a correction factor TRF-E_{var} as a reflection of differences in community composition rather than a true difference in community. In addition, molecular profiling methods such as these normally represent only "dominant" organisms in the community (that constitute less than 1% of the community) and rare species are not represented. Detecting the diversity of rare species is also important because in some microbial communities the vast majority of the diversity is constituted by the rare species (Gans *et al.*, 2005; Pedrós-Alió, 2006). There is every possibility for the generation of T-RFs of the same size from multiple taxa that are distantly related. In that case, the diversity of the dominant taxa will be underestimated (Dunbar *et al.*, 2001; Engebretson and Moyer, 2003; Blackwood and Buyer, 2007). Blackwood *et al.* (2007) advocated the physical capture of T-RFLP using a biotynylated primer and streptavidin-coated beads. They were able to show that the physical capture method described by them created similar profiles such as those generated by fluorescent T-RFLP. When sequencing of such biotynylated captured T-RFs was done, most of the sequences did not match with those already present in database suggesting that these belonged to rare species, although the T-RFs were of the same size. So these workers emphasized that the T-RFs should best be identified on the basis of sequencing rather than by comparing their sizes to T-RFs of computer digests (of database sequences). They selected T-RFLP of bacterial ribosomal gene because of its popularity, availability of database of sequences and bioinformatic tools. Engebretson *et al.* (2003) explored the possibility of resolving single populations in model communities by using selected restriction endonucleases. They also measured the success of restriction endonucleases in detecting sequence variants from model communities that showed variations in species abundance. From the database of gene sequences, the restriction endonucleases have been classified on the basis of their ability to resolve T-RFs. Of the 18 restriction endonucleases tested, *BstUI*, *DdeI*, *Sau 96I* and *MspI* showed highest resolving potential of identifying single populations in model communities. All restriction endonucleases used by these workers could generate T-RFs of more than 70% OTUs at richness values greater than 50 OTUs per model community.

A controversy seems to have arisen on the suitability of fingerprinting methods for the community studies. While community fingerprinting methods such as automated ribosomal intergenic spacer analysis (ARISA) and T-RFLP are useful for comparative analyses, they are not useful to assess the phylotype richness or community evenness because the values for these have been found to be variable and appear to be dependent on restriction enzyme used for generating the profile. Although T-RFLP provides a rapid means for comparing the relationships between bacterial communities, it may not be suitable for highly complex communities (Dunbar *et al.*, 2000). Subsequently, these workers assessed phylogenetic relationship of every group of 16S rDNA sequences that yielded the TRFs of the same size. When replicate samples (aliquots of a single restriction digest) were compared the profiles substantially differed from one another suggesting that the results are not reproducible. However, phylogenetic inferences will be most effective if TRF profiles represent only a single bacterial division or a smaller group. The analytical precision of this method is doubtful as about 85% of the 169 TRF profiles were not reproducible when nine replicate samples from a single soil sample were analysed. Such methods are inherently limited by their detection threshold or, more precisely, by their dynamic range (Dunbar *et al.*, 2001). Danovaro *et al.* (2006) compared microbial community diversity and richness estimates obtained by using ARISA with T-RFLP of 16S rRNA genes. They showed that ARISA estimates of bacterial species richness are always higher than those obtained using T-RFLP analysis. Bent *et al.* (2007) argued that the number of peaks detected in either T-RFLP or ARISA assays grossly underestimate the actual richness of any community. According to them, the two methods cannot be treated at par, as the resolving power of each of these is different and thus can partition the community differently. Moreover, the observed categories do not correspond to the named taxonomic identities. They concluded that the "current microbial community fingerprinting methods cannot provide reliable diversity indices" and also added that it is not understandable why the investigators do not realize the limitations of such methods and hastened to add that "the emperor has no clothes". Danavaro *et al.* (2007), however, observed that T-RFLP analysis and ARISA in fact lead us to clearly distinguishable taxonomic categories. It is now well realized that the 16S rRNA gene alone is not sufficient to define phylogenetic relationships among closely related species (Normand *et al.*, 1996; Guasp *et al.*, 2000; García-Martinez *et al.*, 2001; Song *et al.*, 2004; Brown *et al.*, 2005). In view of this, 16S rRNA-23S rRNA ITS region emerged as an important molecular marker for discrimination to the species level and even within species (Guasp *et al.*, 2000; Goncalves *et al.*, 2002; Xu and Cote, 2003; Song *et al.*, 2004; Kwon *et al.*, 2005). The criticism of Bent *et al.* (2007) to ARISA and T-RFLP can also be extended to any fingerprinting technique including DGGE/thermal gradient gel electrophoresis (TGGE). They concluded that all fingerprinting techniques are always useful in comparing biodiversity of different habitats and ecosystems. Fingerprinting techniques are akin to prĕt-a-porter clothes. Although such a dress is not as glamorous as the dress of a top stylist but the fact remains that at least "it does not leave you naked".

vi) PCR and DGGE: The sixth method is dependent on DNA melting behavior. In DGGE, chemicals such as urea or formamide gradients are used to separate the PCR products. This has been developed by Muyzer *et al.* (1993, 1995) in which cellular/environmental DNA sample is subjected to PCR amplification by universal or specific domain primers to identify 16S rRNA gene or any other target gene. The amplified gene products are then subjected to DGGE, stained and visualized for profile and data analysis. The total number of bands represents the total community structure and each band is represented by the individual species/strain component and the intensity of the band in turn reflects the abundance of particular species/strain. This technique involves sequence-dependent separation of PCR products which can be used to assess the genotypic diversity in environmental samples. Isolated cultures can be characterized by DGGE and can be assigned to field populations based on their DGGE profiles (Fig. 14). There are two advantages of this method. This method helps to (i) assess community structure and (ii) provides relatively crude estimates of species diversity. The disadvantages are that it involves comparative insensitive staining techniques and does not provide information concerning specific phylogenetic groups that comprise a microbial community. GC fractionation is shown to enhance bacterial community diversity assessment and the detection of minority populations of bacteria is quite feasible by DGGE. Applying this technique, Holben *et al.* (2004) subjected bacterial community DNA (total bacterial community of cecum of broiler chicken worldwide) to GC fractionation by differential density imposed by AT-dependent DNA binding dye bis-benzimidazole. Such GC-fractionated DNA was PCR amplified and subjected to DGGE analysis. By this approach, it was possible to identify a number of phylotypes that were not recovered using traditional random cloning and sequencing approach. Also directed cloning and sequencing of individual bands from DGGE lanes corresponding to G+C fractions allowed detection of numerous phylotypes that were not recovered by other traditional methods. Using PCR-DGGE, total community structure of cyanobacteria has been determined in the mats inhabiting the intertidal

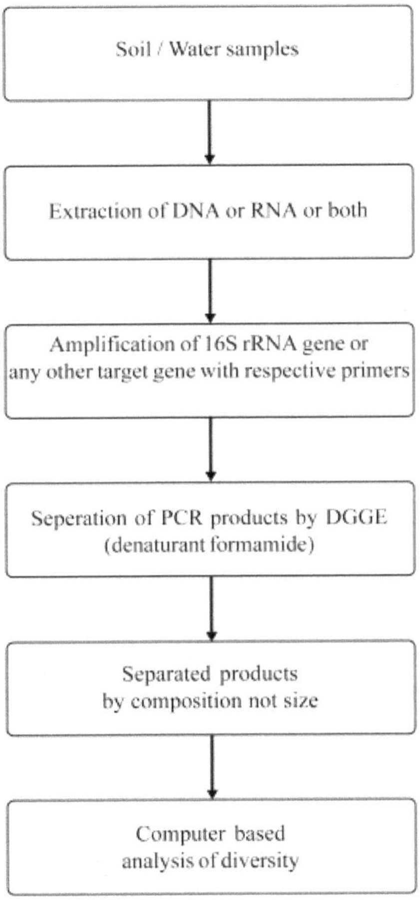

Figure 14: Flow-chart showing different steps in PCR-DGGE.

zone of oil-polluted coast of Saudi Arabia (Al-Thukair *et al.*, 2007). This technique can provide a wealth of information on the spatio-temporal distribution of soil microorganisms and solve ecological puzzles relating to their nutritional interactions (Nakatsu, 2007). DGGE profiles of rRNA ITS region have helped in understanding genotypic variation in *Microcystis* species in relation to microcystin concentration in freshwater lakes (Kardinaal *et al.*, 2007).

vii) DNA microarray technology: It is potentially a valuable tool for quantitative, systematic and detailed studies of microbial communities. It can measure thousands of distinct DNA sequences simultaneously and requires a low sample size. Identification of microbes can be based on PCR amplification of rDNA genes with the help of oligonucletide primers or native rRNAs hybridized to oligonucleotide microarrays. Initial studies were mainly based on small-scale microarrays for profiling specific bacterial species of interest from diverse environments (Guschin *et al.*, 1997; Small *et al.*, 2001; Koizumi *et al.*, 2002; Liu *et al.*, 2001; Loy *et al.*, 2002; Wang *et al.*, 2002, 2004; Castiglioni *et al.*, 2004). El Fantroussi *et al.* (2003) assessed environmental microbial populations by extracting directly rRNA and hybridizing with oligonucleotide microarrays with out PCR amplification. The hybridization patterns were reproducible and exhibited variation in eustuarine sediments differing in salinities. A thermal denaturation curve (i.e. melting profile) for each probe-target duplex indicated the specificity of hybridization. Palmer *et al.* (2006) designed 10,462 16S rDNA probes (of which 7167 are unique sequences) for recognition of diverse microbial populations through this technique. Bacterial species occurring in less than 0.1% in complex mixtures could be easily identified. The microbial diversity assessed with this technique is comparable to the community profile revealed with 16S rDNA sequences of the same species. So this technique is quite well suited to characterize complex microbial communities. The advantages of this technique are that it is quite rapid, reproducible and can even detect bacterial species missed by sequencing of >600 clones. However, the limitations of this technique are that it can only detect species and taxonomic groups for which probes have been successfully designed and it cannot directly discover novel species which can only be possible by the sequencing technique.

By the application of the above molecular techniques, the taxonomy and phylogeny of various cyanobacteria has been investigated *in vitro* (based on cultures) and *in vivo* (field samples). In many cases a correlation has been done between the cultures and field samples.

3) 16S rRNA gene sequencing in cyanobacteria

Cyanobacteria form a monophyletic cluster among eubacteria (Woese, 1987; Garrity and Holt, 2001). The cyanobacterial cluster also contains the plastids of eukaryotes (Giovannoni *et al.*, 1988; Wilmotte and Golubic, 1991; Turner, 1997). Based on phylogenetic analysis of the 16S rRNA gene sequences it has been concluded that the diversification of cyanobacteria occurred within a short time (Giovannoni *et al.*, 1988; Wilmotte and Herdman, 2001). Another important evolutionary event is the recognition of the polyphyletic nature of the Prochlorophyta (Urbach *et al.*, 1992) and their clustering with cyanobacteria (Wilmotte, 1994; Palenik and Swift, 1996). This suggests that the Prochlorophyta and cyanobacteria shared a common ancestor and the recognition of Prochlorophyta as a separate group. Dideoxynucleotide-terminated sequencing, using reverse transcriptase and oligodeoxynucleotide primers (complementary to *E. coli* 16S rRNA sequence positions 519 to 536, 907 to 926, 1392 to 1406 bp) revealed that unicellular cyanobacteria of Section I and the non-heterocystous, filamentous organisms of Section III are dispersed through out the phylogenetic tree suggesting that these morphotypes have multiple evolutionary origins. This is not consistent with their traditional classification. However, heterocystous cyanobacteria form a monophyletic cluster or group (Giovannoni *et al.*, 1988; Turner, 1997; Wilmotte and Herdman, 2001). This group includes members of both Sections IV and V defined by Rippka *et al.* (1979) corresponding to the orders Nostocales and Stigonematales, respectively under ICBN.

A number of workers have put forward proposals for the recognition of phylogenetic lineages or clusters that are not consistent with the classifications under ICBN or that of Rippka *et al.* (1979) or the Bergey's Manual. Wilmotte and Herdman (2001) recognised 14 phylogenetic clusters among cyanobacteria mainly based on 16S rRNA gene sequencing. This formed a part of Bergey's Manual. Other workers recognized seven (Honda *et al.*, 1999), ten (Turner *et al.*, 1999) and five (Seo and Yokota, 2003; Tomitani *et al.*, 2006) phylogenetic clusters among cyanobacteria. They took into account sequence comparisons of a number of housekeeping genes besides 16S rDNA sequences to draw phylogenetic relationships. These are presented below.

Honda *et al.* (1999) detected seven major evolutionary lineages in cyanobacteria (including Prochlorophycean algae) based on 16S rRNA gene sequence analysis which has further been supported by phylogenetic relationships based on other genes such as *psbA*, *rbcL*, *rnpB*, *rpoC* and *tufA*. Five new *Synechococcus* 16S rRNA gene sequences (belonging to *Synechococcus* PCC 7001, PCC 7003, PCC 7109, PCC 7117 and PCC 7335) have been assigned to Marine Cluster (MC)-B and MC-C but were separated into three lineages. They suggested that organisms classified in the genus *Synechococcus* have evolved diversely and should be reclassified into several independent taxonomic units. Further, the *Synechococcus* strains and filamentous cyanobacteria make a monophyletic group reflecting a convergent evolution of multicellular organization.

Turner *et al.* (1999) recognized 10 monophyletic clusters in cyanobacteria based on 16S rRNA gene sequence analysis. *Nostoc* (NOST), *Pseudoanabaena*, *Oscillatoria* (OSC) and *Synechococcus* clusters were common in the studies of Turner *et al.* (1999) and those studied by Tomitani *et al.* (2006). The *Synechocystis*/*Pleurocapsa*/*Microcystis* sequence group identified by Turner *et al.* (1999) does not derive much support from the work of Tomitani *et al.* (2006), although *Prochloron* and *Pleurocapsalean* cyanobacteria formed a cluster. Turner *et al.* (2001) compared phylogeny, based on 16S rRNA gene sequencing by using ML method, of strains of nitrogen-fixing unicellular cyanobacteria. Three independent lines of descent have been identified without any correlation between aerobic versus anaerobic nitrogen-fixing activity. *Cyanothece* PCC 7418 has a sequence similarity of 100% with the strains assigned to *Dactylococcopsis* and *Euhalothece*. Strains assigned to *Cyanothece* are polyphyletic with four strains (including *Synechococcus* RF-1 that has been redesignated as *Cyanothece* PCC 8801) falling into three distinct sequence clusters containing strains assigned to other genera.

Tomitani *et al.* (2006) conducted a detailed phylogenetic analyses of 20 cyanobacterial strains belonging to 15 diverse representative filamentous taxa on the basis of *16S rRNA*, *rbcL* and *hetR* gene sequences in which comparison between ML, NJ and MP methods has been made. Filamentous cyanobacteria belonging to subsection III appear to be mixed with unicellular species of the subsections I and II indicating their polyphyletic origin. Further certain of the strains e.g. *Prochloron* and those belonging to subsection II form clusters in all the trees constructed with ML, NJ and MP methods. Heterocystous cyanobacteria of subsections IV and V form a monophyletic clade in phylogenetic trees constructed by the ML and NJ methods (with a bootstrap value of 97%) and to a lesser extent by MP method where the bootstrap values of 66% were found. Further the monophyly of heterocystous cyanobacteria is supported by *rbcL* phylogeny in all the three phylogenetic trees constructed by NJ, MP and ML methods. Likewise, the phylogenetic analysis based on *hetR* gene supports the monophyly of subsection V by NJ and MP methods with bootstrap values of 99% and

96% respectively. The monophyletic nature of subsection V in *hetR* trees is consistent with 16S rRNA phylogeny but not with analyses of *rbcL, nifH* and *nifD* phylogenies.

Wilmotte and Herdman (2001) considered that it is quite premature to treat taxonomy of cyanobacteria on a phylogenetic basis. They further cautioned that due to two reasons the studies on taxonomy and molecular phylogeny of cyanobacteria suffered a set back. The first is that certain strains of cyanobacteria investigated from culture collections have not been properly identified and secondly, the data on 16S rRNA sequences on some strains originated from morphologically unidentified members. That is why the identification of a cyanobacterium on polyphasic approach should be done first before initiating studies on molecular phylogeny. Another important aspect is to have as much ecological information as possible on cyanobacteria at the time of collection and identification on a polyphasic approach. For this would greatly help in the diversity assays of cyanobacteria from different environments based on molecular markers and even if some sequences have been identified on the basis of culture independent approach, they can then be assigned to a probable strain of close proximity based on the habitat. Wilmotte and Herdman (2001) have identified 14 lineages or clusters in cyanobacteria by comparing 16S rRNA gene sequences and phylogenetic relationships drawn on the basis of ML and NJ methods and bootstrap analysis involved 500 resamplings. These are as follows: (i) heterocystous cluster, (ii) *Prochlorococcus marinus, Synechococcus, Cyanobium* and the sequences from the Sargasso Sea; (iii) *Prochlorothrix hollandica* lineage, (iv) the marine *Leptolyngbya* lineage, (v) freshwater "*Leptolyngbya*" lineage, (vi) the *Pseudoanabaena* cluster, (vii) the halotolerant unicellular strains, (viii) the lineage of baeocyte-forming strains, (ix) the *Spirulina* cluster, (x) *Synechococcus* PCC 7002 and *Leptolyngbya fragile* PCC 7376 lineage, (xi) the *Synechocystis* PCC 6906-*Microcystis* lineage, (xii) the *Trichodesmium-Oscillatoria* PCC 7515 lineage, (xiii) the *Arthrospira* lineage and (xiv) the *Geitlerinema* lineage. Of these 14 clusters, cluster (vii) comprising the halotolerant strains has been suggested to be quite unique that is based on a polyphasic approach. Garcia-Pichel *et al.* (1998) compared the morphological, physiological features and 16S rRNA gene sequences of unicellular, extremely halotolerant and moderately thermophilic cyanobacteria. All these share a number of chemotaxonomic markers including phycobilins, carotenoids and possess mycosporine-like amino acids as well. These are variable in their cell size and form colonies that exhibit two planes of division and reproduce by baeocyte formation (strains MPI 96N303 and MPI 96N304). Some are colonial but do not reproduce by baeocytes and cells divide in a single plane (MPI 96P402, MPI 96P605, MPI 96AL03 and MPI 96AH13). Besides these, *Cyanothece* PCC 7418, *Dactylococcopsis* PCC 8305, *Aphanothece halophytica* ATCC 43922, *Cyanothece* MPI 95AH10 are also included in this analysis. Taxonomically, all strains possibly belong to *Cyanothece* according to Castenholz and Waterbury (1989) but *Aphanothece* or *Cyanothece* according to Anagnostidis and Komárek (1986). On the basis of NJ phylogenetic tree analysis all these strains are placed in a monophyletic cluster. This is the first group of cyanobacteria that are closely related with each other on the basis of 16S rRNA gene sequencing and a set of physiological characteristics. According to Wilmotte and Herdman (2001) this study 'is a nice example of a polyphasic approach to the taxonomy of cyanobacteria'.

Seo and Yokota (2003) have drawn comprehensive phylogenetic relationships amongst cyanobacteria (consisting of 3 strains of Chroococcales, 5 strains of Pleurocapsales, 7 strains of Oscillatoriales, 7 strains of Nostocales and 2 strains of Stigonematales) based on 16S rRNA, *gyrB* (DNA gyrase subunit B), *rpoC1*(DNA-dependent RNA polymerase γ-subunit) and *rpoD1*(principal σ factor of *E. coli* σ^{70} type DNA-dependent RNA polymerase) gene sequences that revealed 5 subgroups of cyanobacteria. They designed PCR primers for *gyrB, rpoC1* and *rpoD1* genes by using consensus amino acid sequences registered in GenBank. Phylogenetic relationships drawn based on *gyrB, rpoC1* and *rpoD1* sequences supported 16S rRNA gene sequence based phylogeny. Phylogenetic trees based on sequences of these genes and the combined amino acid sequences deduced from *rpoC1* and *rpoD1* suggested that LGT of rRNA might be suspected in case of *Synechocystis* sp. strain PCC 6803. The 24 cyanobacterial sequences have been resolved into five clusters (A-E). Cluster A comprises sequences from *Anabaena cylindrica, Anabaena variabilis, Anabaenopsis circularis, Chlorogloeopsis* sp. PCC 6718, *Chroococcidiopsis* sp., *Fischerella muscicola* and *Nodularia spumigena.* Cluster B consists of *Oscillatoria agardhii, Oscillatoria nigro-viridis, Phormidium* sp. and *Spirulina platensis* while Cluster C is composed of *Microcystis wesenbergii* NIES-104, *Pleurocapsa* sp. PCC 7327, *Pleurocapsa* sp., *Stanieria cyanosphaera* PCC 7437, *Synechocystis* sp. strain PCC 6803 and *Xenococcus* sp. Cluster D consists of *Pseudoanabaena* sp. PCC 7367, *Pseudoanabaena* sp. PCC 7403 and *Synechococcus* sp. PCC 7001. Cluster E is represented by *Leptolyngbya* PCC 7375 and *Synechococcus leopoliensis* IAMM-6.

A number of studies have been devoted to resolve taxonomic disparities and also draw phylogenetic relationships amongst different groups of cyanobacteria. Of the 14 clusters recognized on the basis of 16S rRNA gene sequences in cyanobacteria (Wilmotte and Herdman, 2001), such attempts were made with cyanobacteria belonging to six clusters. The taxonomic resolution of strains of cluster I comprising of heterocystous members (*Anabaena, Aphanizomenon, Cylindrospermopsis,* planktic *Anabaena, Planktothrix*), cluster II of *Synechococcus*, cluster V of freshwater *Leptolyngbya* lineage, cluster VI of *Pseudoanabaena*, cluster XII of *Trichodesmium-Oscillatoria* PCC 7515 lineage and cluster XIV of *Geitlerinema* lineage has been undertaken during recent years. These are summarized below.

i) Heterocystous cyanobacterial cluster: (Cluster I as per Wilmotte and Herdman, 2001): Studies on the phylogeny of the heterocystous cyanobacteria based on 16S rRNA gene sequencing revealed that these form a monophyletic cluster or group (Giovannoni *et al.*, 1988; Turner, 1997; Wilmotte and Herdman, 2001). Within this cluster, phylogenetic relationships of planktonic heterocystous cyanobacteria based on the 16S rRNA gene sequence comparisons (Lyra *et al.*, 1997; 2001; Barker *et al.*, 1999; Beltran and Neilan, 2000; Lehtimäki *et al.*, 2000; Li *et al.*, 2000; Moffit *et al.*, 2001; Saker and Neilan, 2001), DNA-dependent RNA polymerase gene (*rpoC1*; Fergusson and Saint, 2000; Wilson *et al.*, 2000), the intergenic spacer region of the phycocyanin operon (PC-IGS; Neilan *et al.*, 1995; Baker *et al.*, 1999; 2000; Bolch *et al.*, 1999) and the IGS between two copies of the *gvpA* encoding major structural gas vesicle protein (Barker *et al.*, 1999) have been drawn.The extent of morphological diversity among the planktonic heterocystous members has been depicted (Figs. 15, 16 and 17). Iteman *et al.* (2002) reported the existence of five monophyletic clusters among heterocystous cyanobacteria. Cluster 1 comprises forms assigned to *Anabaena* and *Aphanizomenon*. Cluster II comprises strains of *Anabaena*. Cluster III is composed of *Anabaena circinalis* and *A. affinis* (Beltran and Neilan, 2000). Cluster IV and V possess planktonic and non-planktonic strains of *Nodularia* and *Cylindrospermopsis raciborskii*, respectively. Although on the basis of morphological characateristics or heterocyst differentiation patterns the members of the last two clusters can be identified but 16S rRNA gene sequence similarity is so high that all of them can be assigned to a single species. Alternatively, it can also be stated that the resolving power of the 16S rRNA is not sufficient enough to make a distinction of these members. Keeping this in view, Iteman *et al.* (2002) compared the 16S rRNA gene sequences and the 16S rRNA-23S rRNA ITS sequences of 11 planktonic members (*Anabaenopsis flos-aquae* PCC 9302, PCC 9332, PCC 9349, PCC 7905; *Anabaenopsis* sp. strains PCC 9215, PCC 9216, PCC 9608, *Anabaenopsis elenkinii* PCC 9420, *Cyanospira capsulata* PCC 9502, *Cyanospira rippkae* PCC 9501 and *Nodularia* sp. PCC 9350) that formed a distinct subclade within the monophyletic clade of heterocystous cyanobacteria with the exception of strains of *Cylindrospermopsis*. Five distinct phylogenetic clusters could be identified with in this subclade. Strains of *Anabaena* and *Aphanizomenon* are dispersed through three clusters

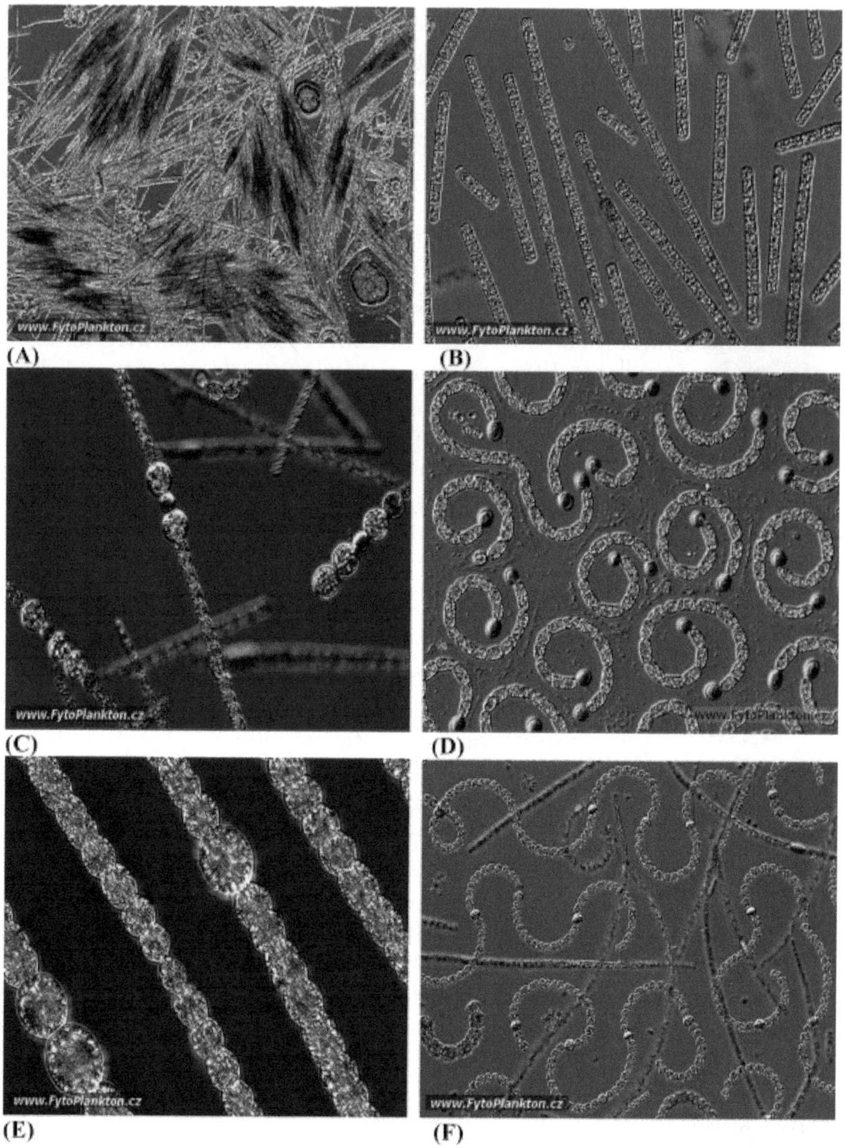

Figure 15: Representatives of planktonic cyanobacteria from different water bodies of Czech Republic. Colonies (A, 300x) and individual filaments (B, 600x) of *Aphanizomenon klebahnii*; *Aphanizomenon aphanizomenoides* (C, 600x), *Anabaenopsis* cf.*elenkinii* (D, 400x), *Anabaena planctonica* (E, 300x), and *Anabaena flos-aquae* (F, 300x). All are Nomarski contrast images downloaded from www.FytoPlankton.cz with the kind permission of Dr. P. Znachor, Laboratory of Phytoplankton Ecology, Institute of Hydrobiology, Biology Centre, ASCR, Na Sádkách, 37005 České Budějovice, Czech Republic.

that are totally not consistent with their generic and species identities. However, all the heterocystous strains contained up to five rrn operons with variable ITS regions in all the taxa investigated. As observed in case of bacteria, in cyanobacteria too, the RB values and 16S rRNA sequence similarity are at variance (Lachance, 1981). Iteman *et al.* (2002) showed that two strains of *Nodularia*, i.e. *Nodularia* PCC 7804 and PCC 7314 exhibited 98.3% 16S rDNA sequence similarity but only 65% RB (ΔT_m 6

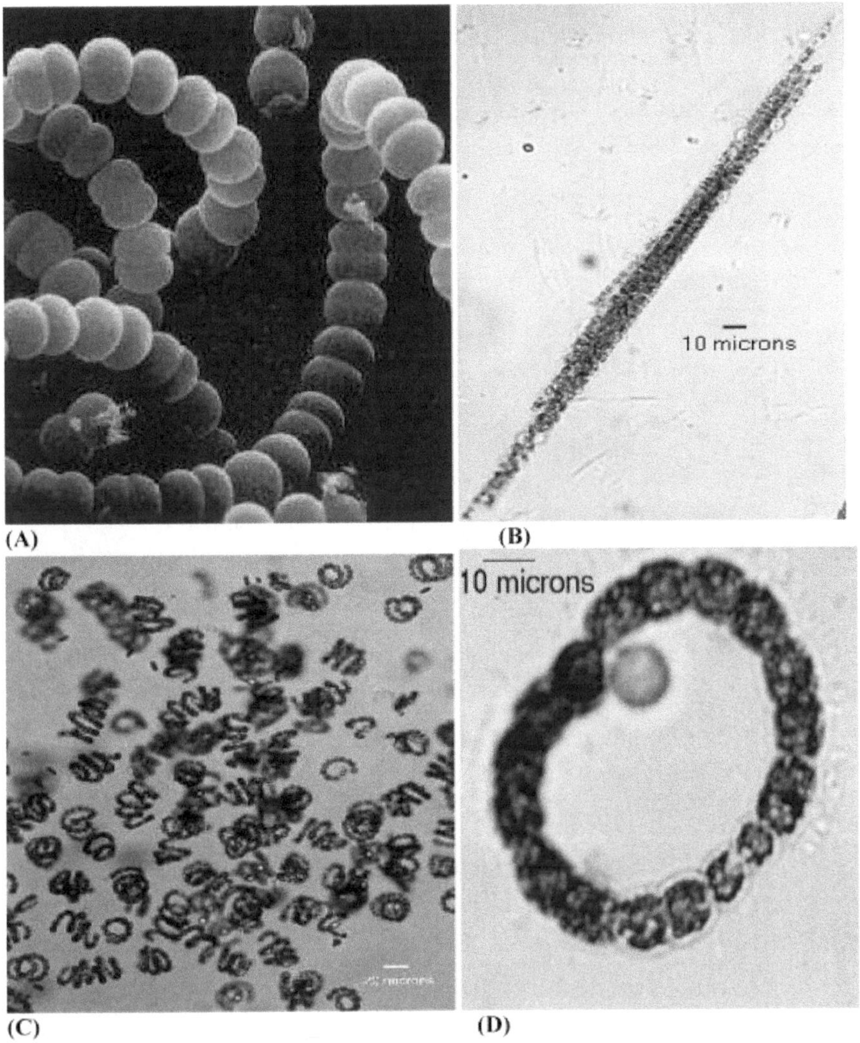

Figure 16: *Anabaena flos-aquae* (A), *Aphanizomenon flos-aquae* (B), *Anabaena spiroides* (C) and *Anabaenopsis circularis* (D). Magnification of A is x2500 in the rest the magnification bar represents 10 µm (B and D) and 20 µm (C). Picture A courtesy Wayne Carmichael (Wright State University), Mark Schneegurt (Wichita State University) and Cyanosite (www-cyanosite.bio.purdue.edu) and. Pictures B, C and D courtesy of Roger Burks (University of California at Riverside), Mark Schneegurt (Wichita State University) and Cyanosite (www.cyanosite.bio.purdue.edu).

°C). In contrast, *Calothrix* sp. strains PCC 7102 and PCC 7709 exhibited a lower 16S rDNA sequence similarity (96.4%) but higher RB (74%) at ΔT_m 4°C. *Anabaena* PCC 7804 exhibited 95.1% and 95.6% 16S rDNA similarity with *Cylindrospermum* PCC 7417 and *Nodularia* PCC 7804 but the total DNA homology is 23% and 19%, respectively. Due to the complexity involved in the clustering of the strains of *Anabaena* and *Aphanizomenon*, a phylogenetic comparison of these two genera (26 strains of the former and 14 strains of the latter) based on sequencing of 16S rRNA gene, 16S rRNA-23S rRNA ITS1 region and *rbcXL* gene has been done by Gugger *et al.* (2002). Planktic *Anabaena* strains were indistinguishable from *Aphanizomenon*. These two genera have been found to be polyphyletic

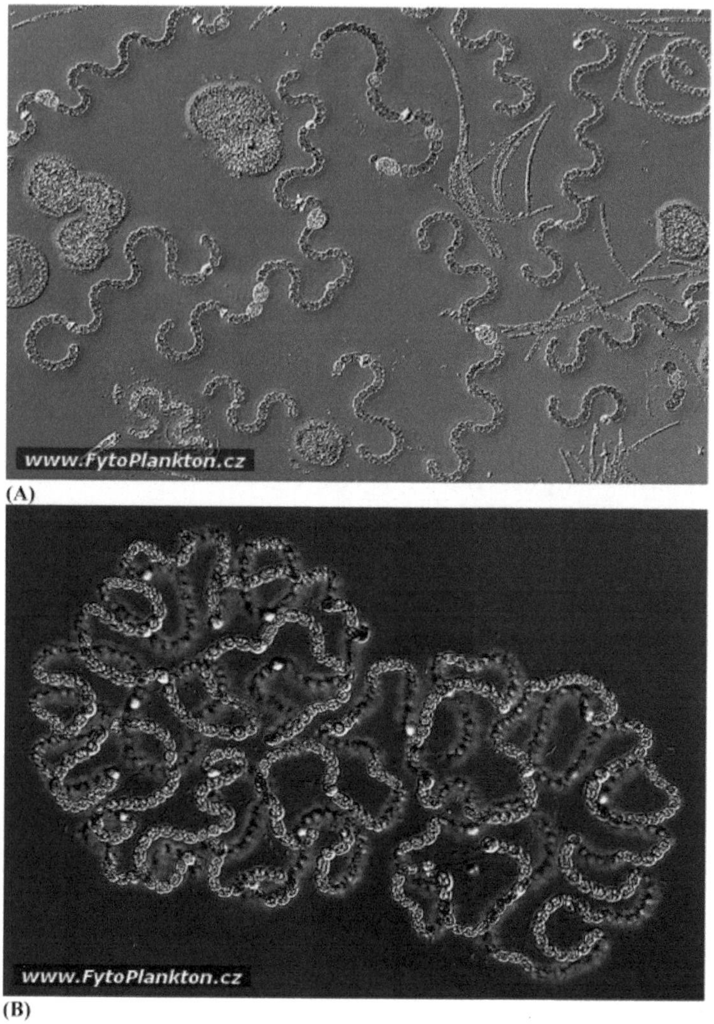

(B)

Figure 17: Representatives of planktonic cyanobacteria from different water bodies of Czech Republic. Nomarski contrast pictures of *Anabaena crassa* (A) and *Anabaena lammermannii* (B) with a magnification of both at x200. Pictures courtesy P. Znachor, Laboratory of Phytoplankton Ecology, Institute of Hydrobiology, Biology Centre ASCR, Na Sádkách, 37005 České Budějovice, Czech Republic, www.FytoPlankton.cz.

on the basis of comparison of 16S rRNA gene sequences. The sequencing of *ITS1* and *rbcXL* regions showed that planktic *Anabaena* is heterogeneous as the toxic *Anabaena* strains were clustered together in all methods of analyses but were grouped with non-toxic *Anabaena*/*Aphanizomenon* strains. Due to their molecular proximity, strains of planktic *Anabaena* and *Aphanizomenon* isolates appear to belong to the same genus, despite their morphological differences. Accordingly, this necessitates a taxonomic revision of these two genera.

Gugger and Hoffmann (2004) presented evidences for polyphyletic nature of true branching cyanobacteria (belonging to subsection V of cyanobacteria of Bergey's Manual and order Stigonematales). Strains belonging to the genera *Fischerella* (Born. et Flah.) Gom.1895 (*F. muscicola*

PCC 7414, *Fischerella* sp. 1711 and *Fischerella* sp. SAG 2027) *Hapalosiphon* Näg. in Kütz. ex Born.et Flah.1886 (*Hapalosiphon* sp. 804-1), *Mastigocladopsis* Iyeng. et Desik. 1946 (*Mastigocladopsis repens* MORA), *Nostochopsis* Wood. ex Born. et Flah. 1886 (*Nostochopsis* sp. 89-45), *Stigonema* Ag. ex Born. et Flah. 1886 (*Stigonema ocellatum* SAG 48.90), *Symphyonema* Jao 1944 (*Symphyonema* sp. 1269.1 and *Symphyonema* sp. 1517), *Symphyonemopsis* Tiwari et Mitra 1969 (*Symphyonemopsis* sp. VAPOR1) and *Westiellopsis* Janet 1941 (*Westiellopsis prolifica* SAG 16.93, *Westiellopsis* SAG 23.96, *Westiellopsis* sp. 1590-1, *Westiellopsis* sp. 1590-2, *Westiellopsis* sp. 89-785/4 and *Westiellopsis* sp. 985-1) have been selected that differed from one another in the type of branching (Y vs T-branching) and the position of heterocysts (intercalary, lateral-sessile and lateral-pedicellate). A phylogenetic study of these strains based on 16S rRNA gene sequencing following NJ method revealed four clusters (Clusters 1, 2, 3 and 5). Clusters 2 and 3 are always basal to Cluster 1 and all these exhibit T-branching whereas Cluster 5 is composed of Y-branching types such as *Symphyonema* and *Symphyonemopsis*. These results do not correlate with the traditional as well as bacteriological classifications. Other heterocystous unbranched members (*Nodularia, Anabaena, Aphanizomenon, Nostoc, Cylindrospermopsis*) are present in Cluster 4 whereas *Scytonema hofmanni* PCC 7110 and *Scytonema* sp. are present in Cluster 6.

Rajaniemi *et al.* (2005) investigated the molecular phylogeny of heterocystous cyanobacteria and included *Anabaena, Aphanizomenon, Trichormus* and *Nostoc* in their study. Most of the 51 strains selected were representatives of the planktic *Anabaena* and *Aphanizomenon*. The former was represented by 10 species (*A. crassa* (Lemm.) Kom.-Legn. et CronB., *A. circinalis* Rabenh. ex Born. et Flah., *A. planctonica* Brunnth., *A. mucosa* Kom.-Legn. et Eloranta, *A. spiroides* Kleb., *A. smithii* (Kom.) M. Watan., *A. sigmoidea* Nyg., *A. flos-aquae* [Lyngb.] Bre´b. ex Born. et Flah., *A. cf. circinalis* var. *macrospora* and *A. lemmermannii* Richt) while the latter by three species [*Aph. flos-aquae* Ralfs ex Born. et Flah., *Aph. gracile* (Lemm.) Lemm. and *Aph. issatschenkoi* (Usač) Prošk.-Lavr.,] as identified according to traditional morphological criteria (Geitler, 1932; Desikachary, 1959; Komárek and Anagnostidis, 1989). It is pertinent to mention here that the genus *Trichormus* was created by Komárek and Anagnostidis (1989) by transferring three species of *Anabaena* (*A. variabilis* Born. et Flah, *A. azollae* Strasb. and *A. doliolum* Bharadw.) on the basis of akinete development. In addition, nine benthic species of *Anabaena* and three of *Trichormus* (*T. variabilis* (Born. et Flah.) Kom. et Anag., *T. azollae* (Strasb.) Kom. et Anag., *T. doliolum* (Bharadw.) Kom. et Anag) and *Nostoc* sp. have been included based on traditional morphological criteria. On the basis of 16S rRNA, *rpoB* and *rbcLX* gene sequence phylogenetic analyses, Rajaniemi *et al.* (2005) identified six clusters. Cluster I is characteristic in consisting of *Anabaena* and *Aphanizomenon* strains as well as five benthic *Anabaena* strains (BECID22, BECID32, XP6B, Itu34S7 and 277) that lack gas vesicles got intermixed with planktic *Anabaena* and *Aphanizomenon* in Cluster I with high bootstrap support. These results thus do not support the separation of planktic and benthic *Anabaena/Aphanizomenon* strains as has been done earlier by Iteman *et al.* (2002) who found a subcluster of planktic heterocystous cyanobacteria with the exception of *Cylindrospermopsis*. Cluster I is further subdivided into nine subclusters (i.e. A-I). On the basis of cutoff points suggested for bacterial species and genus, 97.5% and 95% 16S rRNA sequence identity, respectively, these can be divided into more than three (or up to nine) species. NJ method based on 16S rRNA gene sequences showed clustering of *Anabaena, Aphanizomenon, Trichormus* and *Nostoc* with bootstrap support of over 65% for NJ, MP and ML methods of analyses. Studies on principal component analysis of morphological characters revealed that shape and size of akinetes afforded an important parameter for differentiating *Anabaena, Aphanizomenon, Trichormus* and *Nostoc* strains. The species of *Trichormus* reognised by Komárek and Anagnostidis (1989) did not form a monophyletic cluster. Accordingly, the three species of *Anabaena* transferred by Komárek and Anagnostidis (1989) to designate *Trichormus* probably are distinct species. It is interesting to note that benthic *Anabaena/*

Trichormus strains were placed outside Cluster I which contained all planktic and five benthic *Anabaena* as well as *Aphanizomenon* strains.

Restriction analysis of 16S-23S ITS region has been put to use for phylogenetic analysis of strains of non-photosynthetic eubacterial genera (Navarro *et al.*, 1992; Vinuesa *et al.*, 1998), cyanobacteria belonging to various orders (Lu *et al.*, 1997) and among the genera of heterocystous cyanobacteria (West and Adams, 1997). Iteman *et al.* (2002) determined the 16S rDNA sequences (*A. flos-aquae* PCC 9302, PCC 9332, *Aph. flos-aquae* PCC 7905, *Nodularia* PCC 9350, *Anabaenopsis* sp. PCC 9215 and *Cyanospira rippkae* PCC 9501) and the phylogenetic trees based on both ML and distance-based methods yielded similar results. With the exception of *C. raciborskii* all the strains could be assigned to five clusters. Although 16S rRNA sequence comparisons justify generic assignments but this is not suitable for resolving differences below the level of genus (Turner, 1997). But on the basis of 16S rRNA sequence comparisons a close relationship existed between *Aphanizomenon* and toxic *Anabaena* strains from various geographical locations. The RFLP patterns of PCC strains studied by these workers do not justify their generic separation into *Anabaena flos-aquae* and *Aph. flos-aquae* as has been done on the basis of morphological grounds or 16S rDNA sequences. Some other workers also found closer relationships between strains of *Anabaena* and *Aphanizomenon* based on 16S rDNA RFLP studies (Lyra *et al.*, 1997, 2001; Lehtimäki *et al.*, 2000). Strains of *Nodularia* clustered with the strains of *Anabaena* and *Aphanizomenon* based on sequences of 16S rDNA as well as RFLP studies presented by Iteman *et al.* (2002) though the *Nodularia* strains found to group with strains of genus *Nostoc* (Lyra *et al.*, 2000; Lehtimäki *et al.*, 2000). Komárek and Komárkova (2006) suggested that on the basis of close resemblances of planktonic *Anabaena* strains to those of *Aphanizomenon* based on molecular phylogenetic analyses (Iteman *et al.*, 2002; Gugger *et al.*, 2002; Rajaniemi *et al.*, 2005a,b; Willame *et al.*, 2006) the generic name *Aphanizomenon* should be retained as the valid name for this large polymorphic group consisting of all planktonic *Anabaena* and *Aphanizomenon* strains. But according to combined morphological and molecular criteria, the group is so morphologically diversified that can be divisible into different generic units (*Aph. flos-aquae, Aph. gracile, Aph. issatschenkoi* and *Aph. volzii*-sub-clusters) with 16S rDNA sequence similarity of about or less than 95%. These conclusions derive support from the work of Wu *et al.* (2010) who conducted morphological identification of 53 strains of *Aphanizomenon* from China and assigned them to three morphotypes, i.e. *Aph. flos-aquae, Aph. gracile* and *Aph. issatschenkoi*. Phylogenetic analysis of the strains based on concatenated alignment of 16S rDNA, *rbcLX* and *cpcBA*-IGS sequences led the recognition of three clades corresponding to the three morphotypes.

ii) *Synechococcus* cluster: (Cluster II as per Wilmotte and Herdman, 2001): Six strain clusters of *Synechococcus* have been defined by Waterbury and Rippka (1989) based primarily on differences in G+C content (mol%) and habitat (freshwater and marine). These strain clusters represent provisional genera and are designated as follows: *Cyanobacterium* cluster; *Synechococcus* cluster, *Cyanobium* cluster, MC-A, MC-B and MC-C. *Synechococcus* isolates maintained at the Pasteur Collection Centre (PCC) have been divided into five clusters that do not correspond with the above six clusters. (Rippka and Herdman, 1992). Cultured *Synechococcus* strains from Japanese lakes and those from PCC were subjected to PCR amplification and DNA sequencing of 16S rDNA and the *cpcBA-IGS* and the flanking regions by Robertson *et al.* (2001). A phylogenetic analysis of *Synechococcus* strains revealed seven clusters that reflected phycobilin content. One strain of *Synechococcus* (PCC 7902) that stands apart is only loosely affiliated to cyanobacterial lineage where no other cyanobacterial sequences were found. An additional cyanobacterial lineage is represented by two hotspring *Synechococcus* isolates (PCC 6716 and PCC 6717), with these *Synechococcus elongatus* (Toray) PCC 6301 and other strains are grouped together. These results signify that the genus *Synechococcus* is not monophyletic as was stated earlier by Urbach *et al.* (1998) and Honda *et al.* (1999). According to Robertson *et al.* (2001), *Synechococcus* sp. strain PCC 6716 and *Synechococcus* sp. strain PCC 6717 should be classified as members of a different genus from that of *Synechococcus* sp. strain PCC 6301. These may represent early morphological types of cyanobacteria and forerunners for the evolution of other cyanobacterial morphological types. In this connection, it is important to know that *Synechococcus* sp. strains PCC 6716 and PCC 6717 and *S. elongatus* (Toray) have been isolated from hotsprings. However, as per Bergey's Manual of Systematic Bacteriology, *Synechococcus* sp. strains PCC 6716 and PCC 6717 represent members of the same species as *Synechococcus* PCC 6301. They retained the seven lineages described by Honda *et al.* (1999) and added an eigth one for the *S. elongatus* (Toray), *Synechococcus* sp. strains PCC 6716 and PCC 6717. Of the other groups, Group 6 is most complex and consists of three subgroups 6a, 6b and 6c. Group 6a comprises of *Synechococcus* isolates from saline lakes of Antarctica and strains of *Synechococcus* from MC-A and MC-B (PCC 7805, WH 8103, and WH 8101; Vincent *et al.*, 1999). Group 6b includes strains from MC-B and *Cyanobium* cluster. Along with these, strains of MC-A (*Synechococcus* PCC 7805 and WH 8103) and *Prochlorococcus marinus* also fall into this subgroup (Honda *et al.*, 1999; Vincent *et al.*, 1999). Subgroup 6c includes three strains of *Synechococcus* sp. (PCC 6301, PCC 7942 and PCC7943) and species of *Microcystis* (*M. elebans* NIES 42 and *M. holsatica* NIES 43, now treated as members of *Synechococcus*) that lack gas vacuoles. They further preferred the creation of new genera for the subgroups 6a, 6b and 6c that contain most of the *Synechococcus* isolates and also for *Synechococcus* members forming group 8. They concluded that some of the members of *Synechococcus* are misclassified members (such as PCC 7902 found in group 4, PCC 7309, PCC 7117, PCC 7002 found in group 5 and PCC 7335 present in group 7).

Fuller *et al.* (2003) identified three new novel clades of *Synechococcus* on the basis of 16S rDNA sequencing. One of these clades contains halotolerant isolates lacking PE and strains that are not capable of utilizing nitrate as sole source of nitrogen. This clade is clustered with MC-A. Cloning and sequencing of the two copies of 16S rRNA of genes present in *Synechococcus* sp. strain WH 7803 and a comparison of the same with *Synechococcus* sp. strain WH 8102 revealed them to be identical. *In situ* community structure of marine *Synechococcus* populations in the Red Sea, at the time of their abundance, predominantly was represented by genotypes of a single clade and these are common representatives isolated into culture. These accounted for a very minor component of *Synechococcus* population diversity. *Prochlorococcus* contains divinylchlorophyll *a* (Chl a_2) and both monovinyl and divinylchlorophyll *b* as the major photosynthetic pigment in contrast to chlorophyll *a* and phycobiliproteins that are typical of cyanobacteria (Chisholm *et al.*, 1988, 1992; Goericke and Repeta, 1992). Two genetically and physiologically distinct ecotypes have been recognized that have implications on their ecological distribution. High-B/A (previously designated as low-light adapted) isolates have larger ratios of Chl b/a_2 and are able to grow in extremely low radiances where low-B/A isolates (previously designated as high-light adapted strains) are incapable of growth. Low-B/A isolates have lower Chl b/a_2 ratios and are able to grow maximally at higher light intensities where high-B/A isolates are inhibited (Moore and Chisholm, 1999). On the basis of 16S rDNA sequences strains of low-B/A ecotype are very closely related to each other (with 99% identity supported by bootstrap values; Urbach *et al.*, 1998; Moore and Chisholm, 1998; Rocap *et al.*, 1999). At the same time strains of high-B/A ecotypes have a lower degree of identity in their 16S rDNA sequence (97 to 98%) and are not monophyletic but at least form three independent branches (Rocap *et al.*, 1999). These also exhibit a high degree of sequence similarity to the *Synechococcus* strains of MC-A.

The resolution of Prochlorococcus-Synechococcus by 16S rRNA-23S rRNA ITS regions: Rocap *et al.* (2002) resolved *Prochlorococcus* and *Synechococcus* ecotypes on the basis of 16S rRNA-23S rRNA ITS sequence homology. They have examined 32 *Prochlorococcus* strains and 25 *Synechococcus* isolates from all over the world. These exhibited a variation in length and G+C content of the spacer. All of them uniformly possessed $tRNA^{Leu}$ and $tRNA^{Ala}$ genes. The results further justify the recognition of high-B/A and low-B/A strains of *Prochlorococcus* and six clades of MC-A of *Synechococcus*. Interestingly, three of them are associated with a particular phenotype (motility, chromatic adaptation and lack of phycourobilin).

The diversity of marine *Synechococcus* strains occurring in Chesapeake Bay has been determined by 16S rRNA-23S rRNA ITS sequences (Chen *et al.*, 2006). A phylogenetic analysis of 82 picocyanobacterial ITS sequences revealed that the majority of the sequences were affiliated with either MC-A or MC-B. Among the 22 environmental clones isolated from Chesapeake Bay (CB) waters, 13 of them clustered with MC-A while the rest 9 clustered with MC-B. They further identified 16 sub-clusters (having more than 95% sequence identity) with a bootstrap value of 100 in MC-B. Eleven of these sub-clusters overlapped with those already reported in literature (Rocap *et al.*, 2002; Ernst *et al.*, 2003) whereas five subclusters (CB1 to CB5) have been considered to be novel and unique to Chesapeake Bay. Of these, at least two subclusters (CB4 and CB5) could be defined. The subcluster CB-4 consisted of five PE-rich *Synechococcus* strains (CB0206, CB0207, CB0208, CB 0209 and CB0210) that were closely related to PC-rich *Synechococcus* strains (CB0101, CB0102, CB0202 and CB 0204). It is interesting to note that a close relationship between *Synechococcus* PE-rich and PC-rich strains has also been derived from *rbcL* phylogeny. Further the separation of MC-A and MC-B is also supported by *rbcL* phylogeny. The interoperon variation in *Synechococcus* is not problematic because of the existence of two identical rRNA operons. However, the length of heterogeneity in ITS region is sufficient to distinguish the different *Synechococcus* strains. Based on ITS phylogeny and GC content, they suggested that *Synechococcus* sp. strain WH 8007 should serve as the reference strain for MC-B but not *Synechococcus* sp. strain WH5701 as suggested earlier (Urbach *et al.*, 1998; Fuller *et al.*, 2003).

The phylogeny of *Synechococcus* strains based on 16S rRNA gene sequences revealed that this botanical form-genus is distributed into five of the eight cyanobacterial lineages (Honda *et al.*, 1999; Robertson *et al.*, 2001). Two or more strains of *Synechococcus* distributed into each of these five clusters exhibited 99.4-100% 16S rRNA sequence identity. One of these five clades is represented by picophytoplanktons in which along with PE-rich and PC-rich strains of *Synechococcus*, strains of *Prochlorococcus* have been clustered as revealed by the phylogenetic relationships drawn on the basis of 16S rRNA gene sequences as well as DNA-dependent RNA polymerase gene sequences (Palenik and Swift, 1996; Urbach *et al.*, 1998). Strains of *Synechococcus*, i.e. WH7803, PCC 7001, PCC 6307 that are representatives of oceanic, marine and freshwaters, respectively differ from one another in the length of ITS of the rrn operon and also differ from *P. marinus* subsp. *pastoris* PCC 9511. However, these four cyanobacteria are closely related on the basis of 16S rDNA phylogeny. A comparison of the sequences of the ITS region of these four strains with those available in the database for *Prochlorococcus* MED4 (CCMP 1378) and MIT9313 revealed that HL-adapted *Prochlorococcus* clade differed from those of LL-adapted clade by the length of ITS (Laloui *et al.*, 2002). Ernst *et al.* (2003) studied the RFLP patterns of *psbA* genes of 19 picoplanktonic cyanobacteria and the phylogenetic analysis is not consistent with the original distinction of isolates based on pigmentation. In view of this, it was concluded that selection pressure plays a great role in shaping the populations and the diversity in *Synechococcus* strains reflects ecosystem-specific adaptations. On the basis of *cpcBA*-IGS spacer region, Crosbie *et al.* (2003) showed that some closely related forms of picocyanobacteria are widely dispersed and that the conclusions of Ernst *et al.* (2003) are quite premature and not based on appropriate population size required for sampling. They further showed the existence of at least six to seven clusters of non-marine picocyanobacteria with in picophytoplankton clade which further supports global dispersal of some closely related picocyanobacterial genotypes.

Cells of *Microcystis* are coccoid, tend to aggregate into colonies, possess gas vacuoles and are present in amorphous mucilage or circumscribed sheath. Watanabe (1996) recognized five different morpho-species (*M. aeruginosa, M. novacekii, M. ichthyoblabe, M. viridis* and *M. wesenbergii*) from within the bloom-forming species of *Microcystis* on the basis colony, sheath characteristics and the mode of arrangement of cells. The molecular and chemotaxonomic data point out that each of these five morphospecies is non-monophyletic and is quite similar genetically and biochemically. Otsuka *et al.* (2001) proposed that these five morphospecies should be merged into one species i.e. *M. aeruginosa*. A number of attempts have been made to distinguish toxic strains of *M. aeruginosa* from the non-toxic ones. These are based on 16S rRNA gene sequencing (Neilan *et al.*, 1997; Lyra *et al.*, 2001; Tillett *et al.*, 2001), 16S rRNA-23S rRNA ITS sequencing (Otsuka *et al.*, 1999; Janse *et al.*, 2003), *cpcBA-IGS* sequences (Neilan *et al.*, 1995; Tillett *et al.*, 2001) and the microcystin (*mcy*) gene, (Nishizawa *et al.*, 1999; Tillett *et al.*, 2001; Kurmayer *et al.*, 2002). These observations point towards the absence of any relationship between genetic similarity and toxicity of a strain. This is more evident when *mcy* was chosen as a marker for distinguishing between the toxic and non-toxic strains where it was revealed that a single *mcy* genotype consisted of both toxic and non-toxic strains. To overcome these inconsistencies, Tanabe *et al.* (2007) adopted MLST approach for phylogenetic analysis of *Microcystis* based on seven housekeeping genes such as *ftsZ* (governing cell division protein FtsZ), *glnA* (glutamine synthetase), *gltx* (glutamyl-tRNA synthetase), *gyrB* (DNA gyrase subunit B), *pgi* (glucose-6-phosphate isomerase), *recA* (encodes recombination protein RecA) and *tpi* (triose phosphate isomerae). Phylogenetic trees constructed on the basis of NJ method revealed five clusters of *Microcystis*. Cluster A consisted of toxic strains, Cluster B had both toxic as well as non-toxic strains, Cluster C , Cluster D and Cluster E possessed non-toxic strains but with the exception of Cluster D where the strains possessed *mcyG* gene.

Based on the morphological criteria, ten species recognized in Europe [*M. aeruginosa* (Kützing) Kützing, *M. viridis* (A. Braun in Rabenhorst) Lammermann, *M. wesenbergii* Komárek in Kontratieva, *M. novacekii* (Komárek) Compère, *M. icththyoblabe* (Kützing), *M. flos-aquae* (Wittrock) Kirchner, *M. natans* (Lemmermann) ex Skuja, *M. firma* (Kützing) Schmidle, *M. smithii* (Kützing et Anagnostidis) and *M. botrys* (Teiling)] are resolved by 16S rRNA gene sequences with less than 1% divergence (Otsuka *et al.*, 1998; Boyer *et al.*, 2001). Phylogenetic trees of the strains in the genus *Microcystis* based on direct sequencing of 16S rRNA-23S rRNA ITS region did not match with the morphological characteristics defined for delineating the various species (Otsuka *et al.*, 1999). In view of this, El Herry *et al.* (2008) coined the term 'morphospecies' instead of species. The RFLP patterns of the 16S-23S rDNA ITS regions of morphospecies of *Microcystis* (MCYS-LB01, MCYS-LB02 and LB03) isolated from Lebna Dam in Tunisia exhibited a similar pattern suggesting that they represent intraspecific varaiations or phenotypic variations but did not agree with previous assignments made in respect of MCYS-LB01 and LB02 to *M. aeruginosa* or MCYS-LB03 to *M. wesenbergii*.

iii) Freshwater *Leptolyngbya* lineage (V lineage according to Wilmotte and Herdman, 2001): The genus *Leptolyngbya* was created by Anagnostidis and Komárek (1988) to include a large number of *Oscillatoria* spp. with trichomes up to 3 μm. Some of these strains have been included in LPP-group B by Rippka *et al.* (1979). *Leptolyngbya foveolarum* Komárek 1964/112 and *L. boryanum* PCC 73110 that cluster here share morphological and 16S rRNA gene sequence identity. In addition to *L. foveolarum*

Komárek 1964/112 and *L. boryanum* PCC 73110, Nelissen *et al.* (1996) also described the presence of *Oscillatoria amphigranulata* CCCNZ-Concert-Oa-1, *Leptolyngbya minutum* D5 and the marine strains *L. ectocarpi* PCC 7375, *L. ectocarpi* N 182 and *L. ectocarpi* CCAP 1462/5 in this cluster but poorly supported by bootstrap values. However, *L. foveolarum* Komárek 1964/112 and *L. boryanum* PCC 73110 showed 99.5% 16S rDNA sequence similarity with 100% bootstrap support. On the basis of phylogenetic analysis, Nelissen *et al.* (1996) are of the opinion that *O. amphigranulata* CCCNZ-Concert-Oa-1 should be assigned to the genus *Leptolyngbya* despite the fact that *O. amphigranulata* Van Goor was assigned to the genus *Limnothrix* by Anagnostidis and Komárek (1988). Castenholz (1992) suggested that the morphological descriptions of *O. amphigranulata* CCCNZ-Concert-Oa-1 and *O. amphigranulata* Van Goor are different and that the latter should be assigned to a new genus and species altogether. The 16S rRNA gene sequences of *Phormidium* M-99 and *Oscillatoria* M-117 are shown to be identical to those of this lineage but Ishida *et al.* (1997) have not documented the morphological identification for these strains.

A polyphasic taxonomic approach has been undertaken to resolve 75 strains of Oscillatorioid cyanobacteria by taking into consideration morphological characters, growth temperature, salinity tolerance, complementary chromatic adaptation (CCA), DNA base composition and 16S rDNA analysis. Six groups have been identified on the basis of 16S rDNA sequencing, all of which are clearly separated from the type species *Oscillatoria princeps* Gomont NIVA CYA 150. That is why Suda *et al.* (2002) recommended that these strains should be classified into separate genera other than *Oscillatoria*. Groups I-III are closely related to each other and Groups IV-VI are distinct from one another and from Groups I-III. Group I is subdivided into strains producing PC (Group I-pc) and PE in addition to PC (Group I-pe). The mol% G+C content of Groups I and II is 39.5 ± 0.4 and 39.5 ± 0.2 respectively whereas Group III showed a mol% G+C of 40.0 ± 0.2. A slightly higher mol% G+C of 44.0 ± 0.4 and 44.0 ± 0.6 is found in Groups IV and V, respectively. The similarity of 1361 bp region of 16S rDNA is in the range of 95.4–96.8% for Groups I-III and fatty acid composition of these strains is similar. Strains of Group IV exhibited a similarity of 89.7% and the fatty acid composition of this group is different. These results are supported by DNA:DNA hybridization studies.. Group I consisted of 58 strains with a RB value of 99.2%. Groups II and III were represented by 5 and 4 strains with RB values at 98.4% and 99.8%, respectively. Group IV was represented by 6 strains and these within each other exhibited an RB value of 99.1%. Groups V and VI are represented by one strain each with RB values of 91.9% and 86.4%, respectively with other groups. On the basis of overall assessment, strains of group I-pc have been treated as *Planktothrix agardhii* and those of group I-pe as *Planktothrix rubescens*. They also considered it appropriate to treat *O. mougeotii* Kützing ex Forti as a synonym of *P. rubescens* (*O. rubescens*). Amended or new taxonomic descriptions (for the subgroups and groups) for *P. agardhii* (type strain NIES 204^T), *P. rubescens* (type strain CCAP $1459/22^T$), *Planktothrix pseudoagardhii* sp. nov. (type strain T 1-8-4^T), *Planktothrix mougeotii* (type strain TR1-5^T), *Planktothricoides raciborskii* gen nov.comb. nov. (type strain NIES 207^T), *Tychonema bourrellyi* (type strain CCAP $1459/11B^T$) and *Limnothrix redekei* (type strain NIVA CYA $277/1^T$) have been presented.

The studies of Thacker and Paul (2004) on morphological, chemical and genetic diversity of *Lyngbya* spp. and *Symploca* spp. revealed that 16S rDNA sequence analysis though strongly supported monophyly of *Lyngbya* and *Symploca* as well as monophyly of *L. bouillonii* and *L. majuscula* but it does not explain the chemical variability among the *Lyngbya* species.

(iv) *Pseudoanabaena* cluster (VI lineage according to Wilmotte and Herdman, 2001): In most of the strains of *Pseudoanabaena* the trichomes are less than 4 µm width with cells longer than broad and possess constricted cross walls (Castenholz, 2001; Komárek, 2003). Some of the strains exhibit

CCA by regulating the levels of PC and PE according to the available spectrum of light. The presence of gliding motion and the ability to fix atmospheric nitrogen are the other traits exhibited by some other strains of this genus. Due to the closer resemblances of *Pseudoanabaena* spp. with those of *Limnothrix* as well as their occurrence together makes their identification slightly difficult. Acinas *et al.* (2009) conducted a phylogenetic study of 28 *Pseudoanabaena* strains from Baltic Sea (BS) and the Albufera de Valencia (AV) on the basis of sequencing of 16S rRNA and 23S rRNA genes, their *ITS1* region, the *cpcBA* operon, the *IGS* between *cpcA* and *cpcB* and the *nifH*. These studies revealed the existence of specific clusters for strains from BS and AV and a third cluster contained a mixture of strains from both the geographical regions that exhibited CCA. Although the analyses from 16S rRNA and 23S rRNA genes are consistent but the analyses from other loci indicated extensive genetic recombination between strains. 16S rDNA analysis revealed clustering of certain *Pseudoanabaena* with strains of *Limnothrix* including *L. redekei*. These observations are consistent with the earlier findings of Wilmotte and Herdman (2001). This cluster is commonly called as *Pseudoanabaena / Limnothrix* group. Several environmental sequences as well as other isolates including the type strain *Pseudoanabaena* PCC 7408 and several strains of *Limnothrix* from Lake Loosdrecht along with *Limnothrix redekei* CCAP 1443/1 and *Limnothrix redekei* CCAP 227/1 are members of this lineage.

v) *Trichodesmium-Oscillatoria* PCC 7515 lineage (XII lineage according to Wilmotte and Herdman, 2001): So far five species of *Trichodesmium* have been described (Wille, 1904; Sournia, 1968) and these have been further confirmed by both morphological as well as ultrastructural features. These five species of *Trichodesmium* are, *T. thiebautii, T. erythaeum, T. tenue, T. contortum,* and *T. hildebrandtii* (Janson, 1995). There is very low genetic diversity amongst these strains as revealed by molecular approaches of sequence analysis of *nifH, hetR,* and 16S rRNA (Carpenter, 1983; Zehr *et al.*, 1990; Ben-Porath *et al.*, 1993). Wilmotte *et al.* (1994) showed that *Trichodesmium* sp. strain NIBB 1067 is closely related to *Oscillatoria sancta* PCC 7515 showing 94.9% 16S rRNA gene sequence similarity. Genes encoding $tRNA^{Ile}$ and $tRNA^{Ala}$ are present in the 16S rRNA-23S rRNA ITS region of *Trichodesmium*. On the basis of 16S rDNA and *hetR* sequences, Janson *et al.* (1999) were able to resolve three clades in *Trichdesmium* containing (i) *T. thiebautii* and *T. hildebrandtii*, (ii) *T. contortum* and *T. tenue*, and (iii) *T. erythraeum*. However, the *hetR* sequenceing provided a better resolution between the species *T. erythraeum* and *T. thiebautii* than analysis of the 16S rDNA region gave. Orcutt *et al.* (2002) investigated the genetic diversity of *Trichodesmium* spp. by using three independent techniques that provide high resolution. A PCR-based DNA fingerprinting method using base pair extended short oligonucleotide primers for HIP1 was used to distinguish *Trichodesmium* spp. This technique was used to distinguish a number of strains of cyanobacteria (Robinson *et al.*, 1995; Smith *et al.*, 1998). Further, DGGE analysis of a fragment of *hetR* gene and the sequencing of 16S rRNA-23S rRNA ITS region were also conducted to reveal the genetic diversity in this genus. Natural populations of *Trichodesmium* spp. (from Bermuda in the Sargasso Sea and North Australia in the Arafura and Coral Seas) and some culture isolates (from Sargasso Sea and the Indaian Ocean) were subjected to diversity studies by employing the above three methods. Although a remarkable similarity existed in between cultures of *T. thiebautii, T. hildebrandtii, T. tenue* and *Katagnymene spiralis*, low genetic diversity was revealed in the strains from the two hemispheres. Two major clades are recognized and largest genetic variation was found in between the strains of *T. erythraeum* (Orcutt *et al.*, 2002).

vi) *Geitlerinema* lineage (XIV cluster according to Wilmotte and Herdman, 2001): Bittencourt-Oliveira *et al.* (2009) conducted a phylogenetic study of ten strains of *Geitlerinema* [six of *Geitlerinema amphibium* (C. Agardh ex Gomont) Anag. and four of *Geltlerinema unigranulatum* (Rama N. Singh) Komárek et M.T.P. Azevedo] on the basis of PC-IGS sequencing. Two strains (*G. unigranulatum*

BBCUSP 352 and *G. unigranulatum* BBCUSP 94) have been found to be distinct from all others despite their morphological similarity. These two strains could be assigned to a different species of *Geitlerinema*. However, on the basis of morphological features, cell ultrastructure and *PC-IGS* sequences, *G. amphibium* and *G. unigranulatum* are indistinguishable and so can be merged into a single species (Figs. 18, 19 and 20). Phylogenetic trees constructed by MP, NJ, ML and Bayesian analyses revealed similar pattern in the clades as well as the clades remained the same by the rooted and unrooted trees.

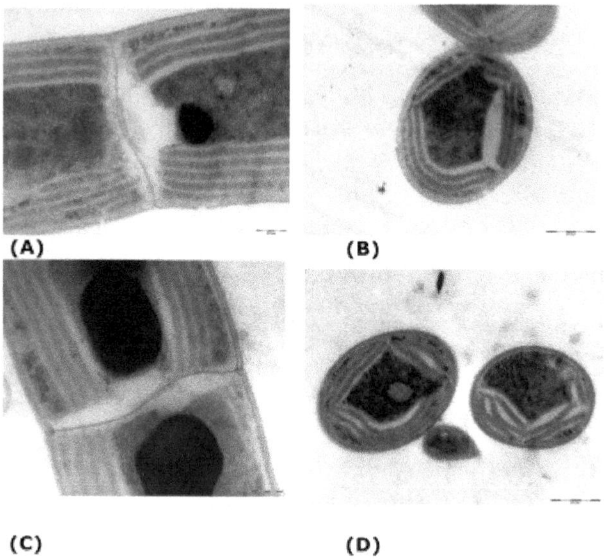

Figure 18: Electron micrographs of *Geitlerinema amphibium* longitudinal and cross sections of strains BCCUSP91 (A and B) and BCCUSP85 (C and D), respectively. Scale bar 500 nm. Photographs courtesy M. D. C. Bittencourt-Olievera, Department of Biological Sciences, Luiz de Queiroz College of Agriculture, University of São Paulo, Piracicaba 13418-900, SP, Brazil.

PCR and DGGE analysis in cyanobacteria

Any gene sequence that shows considerable sequence heterogeneity can be chosen as a molecular marker for assessing the microbial diversity through this method. DGGE analysis of 16S rRNA gene segment of cyanobacterial populations has been done (Nübel *et al.*, 1997; Garcia-Pichel, 2000; Abed and Garcia-Pichel, 2001). Due to the fact that 16S rRNA gene falls short of resolving differences between closely related organisms, a number of workers consider 16S rRNA-23S rRNA ITS region to be very suitable for high resolution analysis because of greater degree of sequence heterogeneity and the availability of published rRNA-ITS sequences. DGGE has been used for analysis of *Synechococcus* (Becker *et al.*, 2002) and *Aphanizomenon* (Laamanen *et al.*, 2002) strains on the basis of 16S rRNA-23S rRNA ITS sequences. Amplification of *ITSa* region gave rise to one sharp band occupying an unique position in the gel in all the cyanobacteria studied. However, only *Aph. gracile* yielded two bands. The amplification products of *ITSb* and *ITSc* regions that were longer when subjected to DGGE gave rise to one (in case of *Synechococcus*, *Synechocystis*, *Lyngbya*, *Pseudoanabaena*, *Trichodesmium* and *Prochlorothrix*), two (in case of *Leptolyngbya*) and two to four bands (in case of most of the strains

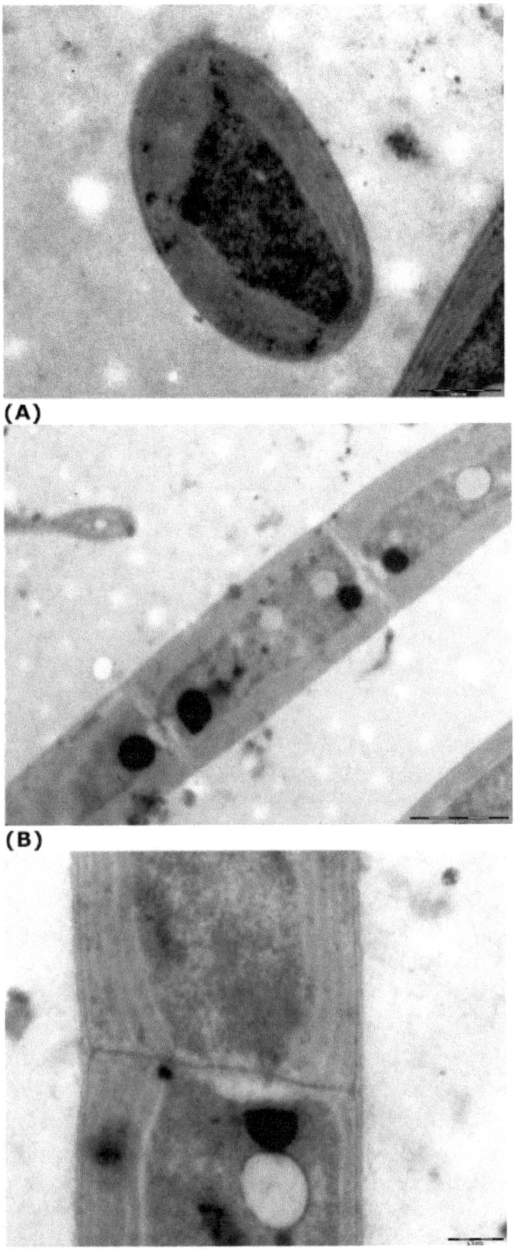

(A) (B) (C)

Figure 19: Electron micrographs of *G. unigranulatum* BCCUSP352 strain cross (A) and longitudinal sections (B and C). Scale bar 500 nm. Photographs courtesy M. D. C. Bittencourt-Olievera, Department of Biological Sciences, Luiz de Queiroz College of Agriculture, University of São Paulo, Piracicaba 13418-900, SP, Brazil.

from genera *Anabaena*, *Aphanizomenon*, *Planktothrix*, *Cylindrospermopsis* and *Gloeotrichia*) (Janse *et al.*, 2003). DGGE of *hetR* gene for resolving the differences in between strains of *Trichodesmium* and *Nostoc* (Palinska *et al.*, 1996; Orcutt *et al.*, 2002) and *nifH* for determining the diversity of nitrogen-fixing

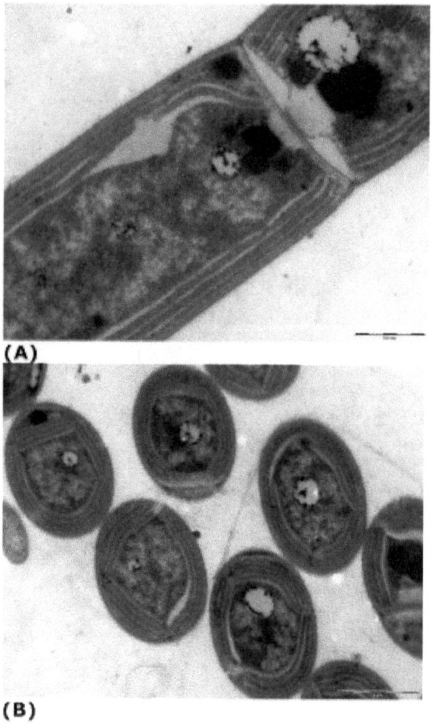

Figure 20: Electron micrographs of *G. unigranulatum* BCCUSP47 strain longitudinal (A) and cross sections (B). Scale bar 500 nm. Photographs courtesy M. D. C. Bittencourt-Olievera, Department of Biological Sciences, Luiz de Queiroz College of Agriculture, University of São Paulo, Piracicaba 13418-900, SP, Brazil.

organisms (Lovell *et al.*, 2001) have been employed. Berrendero *et al.* (2008) compared phylogenetic relartionships based on 16S rRNA gene sequences with the variability in PC operon and the intervening spacer sequence (*cpcBA-IGS*) between strains of *Rivularia* and *Calothrix* by temperature-dependent DGGE. The high variability of the *PC-IGS* among the genotypes was not consistent with the morphologies observed. However, the phylogenetic relationship drawn in case of *cpcBA-IGS* is largely consistent with 16S rRNA gene sequences.

Phylogenetic analysis based on the sequences of the nitrogen-fixing gene *nifD* from a number of representatives of subsection IV and subsection V reflected the monophyly of the heterocystous cyanobacteria. Neither of these subsections is monophyletic because of intermixing of the strains with two sister clades. Clade I is composed of two smaller branches Clade 1a and Clade 1b. Clade 1a consisted of representatives of *Calothrix* (IV.II; strain PCC 7102), *Scytonema* (IV.I; PCC 7814 and PCC 7110), *Fischerella* (V; strains PCC 7603, PCC 7414, PCC 1903) and *Chlorogloeopsis* (V; strain PCC 6718) and Clade 1b was represented by *Nostoc* (IV.I; strains PCC 7120, PCC 6720, PCC 7423), *Cylindrospermum* (IV.I; PCC 7604) and *Chlorogloeopsis* (V; PCC 6912). Clade 2 is composed of strains of *Calothrix, Nostoc, Anabaena* (IV.I), *Nodularia* (IV.I), *Cylindrospermum* and *Tolypothrix* (IV.II). Intermingling of strains of *Calothrix* (PCC 7507) with those of *Nostoc commune* UTEX 584 and sister *Anabaena* strains (PCC 7122, PCC 7108 and ATCC 33047) has been noted in Clade 2. The MP and NJ trees were virtually identical to ML tree (Henson *et al.*, 2004).

4) Diversity of cyanobacteria determined on the basis of molecular markers: During recent years, diversity studies of cyanobacteria based on sequencing of 16S rRNA, 16S rRNA-23S rRNA ITS region, ARDRA, T-RFLP and PCR with DGGE techniques have been made circumventing the collection-and culture-based studies (i.e. metagenomic analysis). In some of the cases, the molecular taxonomic data have been correlated with the cultures of cyanobacteria raised from the same habitats. Some of these are summarized here.

The molecular techniques described above play a meaningful role in assessing the diversity of cyanobacteria from a wide variety of ecosystems (Kumari *et al.*, 2009). These require the extraction of nucleic acids and do not allow the identification of individual cells from the environmental samples. However, the identification of whole fixed cells by *in situ* hybridization with rRNA targeted oligonucleotide probes has been advocated (Amann *et al.*, 1995). This technique is so powerful that cyanobacteria in deeper layers of microbial mats can be identified. One disadvantage is that the autofluorescence exhibited by cyanobacterial cells due to the presence of chlorophyll disrupts the fluorescence label associated with oligonucleotide probes upon hybridization within the target cells. To overcome this problem, Schönhuber *et al.* (1999) developed a non-fluorescent assay method, based on horseradish peroxidase (HRP)-labelled oligonucleotides, for identification of cyanobacterial cultures as well as cyanobacteria existing in microbial mats. With the help of cyanobacteria-specific primers designed by them (CYA 361, CYA 664, CYA 762, CIV/V 1342 together with EUB 338 that targets most bacteria) and dot blot hybridization, the cells can be detected by the formation of a coloured precipitate by oxidative polymerization of the non-fluorescent substrate diaminobenzidine. Additionally, they also evaluated the automatic fluorescence detection system based on enzymatic signal amplification (Schönhuber *et al.*, 1997; Van Gijlswijk *et al.*, 1996). This is known as tyramide signal amplification (TSA) system that involves covalent binding of fluorochrome tyramide substrate molecules to electron rich moieties such as tyrosine and tryptophan in the HRP-containing cells. It was possible to identify *in situ* cyanobacterial diversity from microbial mats from alpine lake in Switzerland. According to these workers, this method complements the traditional methods based on phenotype and the identification of cyanobacteria will be more reliable. Another approach, suggested to overcome the problem of autofluorescence of the cells, is to hybridize with biotin-labelled probes and allow reaction with a streptavidin-HRP conjugate, prior to detection employing TSA. This has successfully been done in case of bacteria (Lebaron *et al.*, 1997). With the help of probe CYA664 (5′-GGAATTCCCTCTGCCCC-3′) that targets all known marine species of *Synechococcus* and *Prochlorococcus* and in combination with TSA, closely related *Prochlorococcus* genotypes belonging to HL-adapted and LL- adapted were shown to be distributed at different depths of North Atlantic Ocean and the Red Sea. While LL-adapted strains are distributed in deeper layers in both these waters, the HL-adapted *Prochlorococcus* populations not only exhibited distinct distribution patterns for each region but also are distributed at different depths (West *et al.*, 2001). However, the cell fixation step and permeabilization of cells using alcohol and/or detergents used in these studies prior to hybridization with labelled 16S rRNA-targeted probe sometimes damages the cells. This is especially true for *Prochlorococcus* or *Synechococcus* cells from oceanic samples. In light of this, another method promising in the analysis of individual cells is whole-cell hybridization with fluorescently labelled rRNA-targeted oligonucleotide probes combined with flow cytometry (Amann *et al.*, 1990). This has been successfully achieved to quantify individual cells from cultures of *Synechococcus* (Chisholm, 1992). A relationship between the rRNA content of *Synechococcus* cells and growth rate has been demonstrated by Binder and Liu (1998). At slow growth rates, the rRNA content almost remained constant but at intermediate growth rates the rRNA content per cell increased proportionally with growth rate and a corresponding decrease in its content was noted at saturating light intensities

(Binder and Liu, 1998). So this method would be ideally suited for estimating the growth rates of such organisms *in situ* as well. Another approach for *in situ* identification is the use of peptide-nucleic acid (PNA) probes. PNA possesses a homomorphous pseudopeptide backbone instead of sugarphosphate backbone of DNA (Egholm *et al.*, 1993; Sen and Nilsson, 1998). The hybridization product of PNA and RNA exhibits higher thermal stability than the DNA-RNA duplex. Due to this inherent advantage, Worden *et al.* (2000) used PNA probes to detect rRNA in intact cells of *Prochlorococcus* and *Synechococcus* in combination with flow cytometry. It would thus be easier to detect these cells in mixed microbial communities during field studies.

Rudi *et al.* (2000) employed oligonucleotide labelled probes targeting 16S rDNA followed by hybridization of these probes to their respective complements spotted on to a solid support (DNA array). With the help of 10 such species-specific probes developed, it was possible to identify relative abundance of cyanobacteria occurring in mesotrophic and eutrophic lakes. There was a good correlation between the diversity of cyanobacterial species determined as per traditional taxonomy and by molecular methods of analysis. Likewise, Wilson *et al.* (2000) designed species-specific primers for the amplification of a 609-bp region of *rpoC1* gene by PCR from 19 cultured isolates of *C. raciborskii*. Sequence analysis revealed that all isolates belonged to the same species. Further, RAPD and STRR sequence profiles revealed a greater heterogeneity among *C. raciborskii* strains. The identification of *C. raciborskii* alone from purified DNA samples as well as environmental samples was quite possible by the amplification of specific region of *rpoC1* gene. This would be greatly helpful in the identification of this toxin-producing cyanobacterium from natural waters. Abed *et al.* (2002a) studied the diversity of benthic cyanobacterial mats inhabiting polluted waters of Wadi Gaza (Gaza Strip, Palestine) and monitored community changes of these mats in the laboratory consequent upon degradation of petroleum compounds. PCR amplified products of 16S rRNA gene, with cyanobacteria-specific primers CYA 359F (with 40 nucleotide clamp at 5′-end) and CYA 781R, were subjected to DGGE at 60 °C. DNA from these bands was sequenced, aligned and phylogenetic tree construction made by different methods in the ARB software. *Phormidium* and *Oscillatoria*-like cyanobacterial morphotypes dominated the polluted habitats and the *in vitro* samples were dominated by a *Phormidium*-like new cyanobacterium named as *Halomicronema*. Samples collected from microbial mats, maintained in artificial ponds in Eilat (Israel), when subjected to DGGE also revealed the existence of *Halomicronema* in large numbers. By adopting a polyphasic taxonomic approach (morphology, ultrastucture, 16S rRNA-based phylogeny, temperature and salinity physiology and chemotaxonomic markers like the presence of special carotenoids and mycosporine-like amino acids) *Halomicronema excentricum gen. et sp. nov.* has further been characterized (Abed *et al.*, 2002b).

Taton *et al.* (2003) presented cyanobacterial diversity patterns in natural and artificial microbial mats of Lake Fryxell (McMurdo Dry Valleys, Antarctica) on the basis of morphological and molecular methods. DNA isolated from environmental samples was subjected to ARDRA analysis after PCR amplification of 16S rDNA and fragments of 422 bp long generated by seminested PCR have been subjected to DGGE, followed by sequencing. They have identified 16S rRNA-23S rRNA ITS sequences from environmental samples for the first time. Molecular diversity appeared to be more than microscopic diversity. Phylogenetic analysis revealed 11 lineages of which 3 have been found to be exclusive to Antarctica and two of these were found to be novel. Microscopy revealed 8 morphotypes (*Nostoc*, *Nodularia* cf. *harveyana*, *Oscillatoria* cf. *subproboscidea*, *Phormidium* cf. *autumnale*, *Schizothrix* sp., *Leptolyngbya* 1 and *Leptolyngbya* 2 with rare presence of *Hydrocoryne* cf. *spongiosa* and *Phormidium* cf. *autumnale*). *Phormidium* cf. *autumnale* has been observed as a minor component of microbial mats *in vivo* but dominated *in vitro* with the absence of other heterocystous members like

Nodularia cf. *harveyana* and *Hydrocoryne* cf. *spongiosa*. Jungblut *et al.* (2005) studied the diversity of cyanobacterial mat communities from meltwater ponds with variable salinities located at McMurdo Ice Shelf, Antarctica. Callejas *et al.* (2010) presented ARDRA analysis of benthic cyanobacterial mat community from King George Island, Antarctica leading to the identification 16S rRNA genes similar to those of *T. bourrellyi* and *T. bornetii* for the first time. These accounted for 9.4% of the total clones whereas 7.6% clones of the PCR amplified 16S rRNA gene library resembled the phylotypes described earlier from Lake Fryxell (Taton *et al.*, 2003) and McMurdo Ice Shelf, Antarctica (Jungblut *et al.*, 2005).

Cyanobacterial ecotypic diversity of microbial mats, growing in a hotspring at 68°C, has been determined by 16S rRNA-23S rRNA gene sequence analysis. The existence of physiologically distinct populations of *Synechococcus*, adapted to differing light conditions, is a characteristic feature of these microbial mats (Ferris *et al.*, 2003). Blackband disease (BBD) is a disease of corals that occurs worldwide and is caused by sulfide-rich microbial mat dominated by filamentous cyanobacteria. The identity of these organisms was traced by cloning and sequencing of 16S rRNA gene using universal primers. Myers *et al.* (2007) identified four isolates of *Geitlerinema* and three related to the genus *Leptolyngbya*. However, DGGE of 26 BBD samples using cyanobacteria-specific primers revealed that the most common representative is a species of *Oscillatoria*. The cultures showed the property of sulfide-resistant oxygenic photosynthesis and the production of the toxin microcystin as the root cause of the disease.

The distribution of unicellular diazotrophic cyanobacteria along a transect of Arabian Sea has been studied by PCR amplification of 16S rRNA gene by employing an oligonucleotide primer, NITRO 821R specific for unicellular diazotrophs in combination with cyanobacteria-plastid specific forward primer CYA 359F. The distribution patterns suggested that the organisms were preferentially located in warm oligotrophic subsurface waters and temperature appeared to be the key factor governing their distribution (Mazard *et al.*, 2004).

Song *et al.* (2005) studied the biodiversity and seasonal variation of cyanobacterial assemblages in paddy fields in Fujian, China by a semi-nested PCR, followed by DGGE analysis. The 24 cyanobacterial phylotypes observed varied with the season and growth phase of rice crop. Sompong *et al.* (2008) determined cyanobacterial diversity from six hot springs of Thailand by extracting DNA, and the 16S rRNA genes were identified by PCR amplication by employing CYA 359F and CYA 781R primers and separated the DNA by subjecting to DGGE with 7M urea and 10% V/V formamide as denaturants. Thirty eight distinct bands were reamplified and sequenced. They were able to identify *Oscillatoria* lineage (01), *Phormidium* lineage (P1), *Synechococcus* lineage (S3), *Oscillatoria* lineage (02), *Synechococcus* lineage (S2), *Synechococcus* lineage (S1) and *Phormidium* lineage (P2). A comparative survey of cyanobacterial flora has been conducted by light microscopy and PCR-DGGE of 16S rRNA gene of the cyanobacterial flora from rice fields in eastern Uttar Pradesh by Srivastava *et al.* (2009). The diversity of cyanobacteria as revealed by light microscopic identifications based on traditional taxonomy (Fig. 21) corresponded with the observations on PCR-DGGE. From a total of 51 bands selected for sequencing only 31 DGGE bands proved to be significant for phylogenetic analysis. NJ method of analysis conducted revealed the existence of 6 clades. The distribution of cyanobacteria appeared to depend on the salinity of soils. Of the six clades, Clade 1 is represented by filamentous, heterocystous cyanobacteria such as *Anabaena, Nostoc, Gloeotrichia* and *Cylindrospermum*. Clade 2 is composed of branched members like *Hapalosiphon* and *Fischerella*. Clade 3 consisted of cyanobacteria from normal soils. Clade 4 is composed of *Aulosira, Phormidium, Lyngbya* with occasional presence of *Anabaena oryzae* and *A. anomala*. Clade 5 is represented by *Oscillatoria* and *Gloeothece* from saline soils while the distribution of *Aulosira, Nostoc, Anabaena* and *Hapalosiphon* depended over a wide

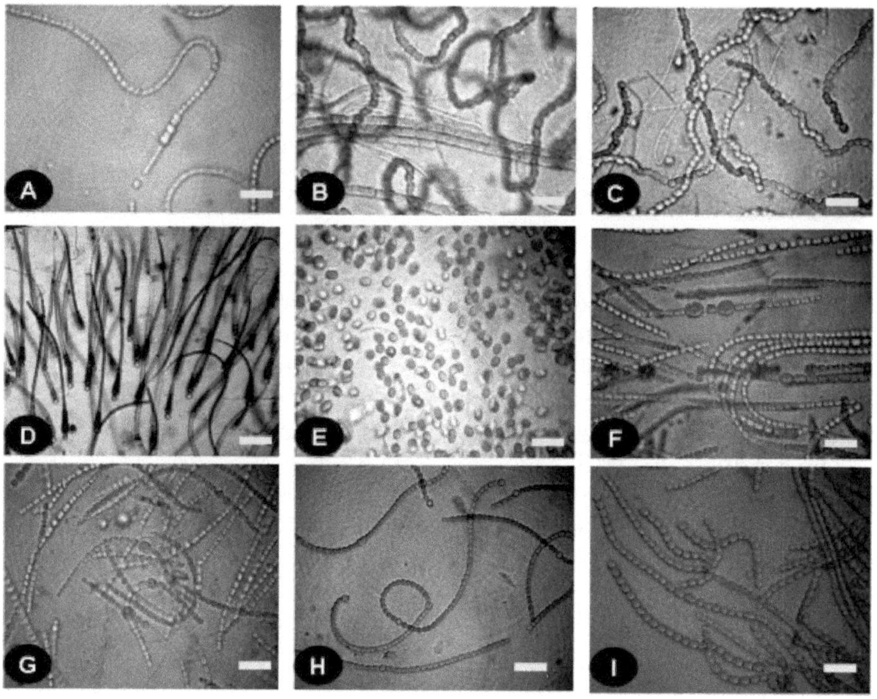

Figure 21: Community of cyanobacteria collected from different rice fields as seen in microscope (resolution 40×). Some of the cyanobacterial genera that constituted the community were *Anabaena* (A, C and F), *Aulosira* (B), *Gloeotrichia* (D), *Aphanothece* (E), *Nostoc* (G and H) and *Hapalosiphon* (I). Bars, 10 μm. With the kind permission of L. C. Rai, Center of Advanced Study in Botany, Banaras Hindu University, Varanasi-221005, India and B. A. Neilan, School of Biotechnology and Biomolecular Science, The University of New South Wales, Sydney, NSW 2052, Australia. [Srivastava *et al.* (2009) *Saline systems* **5:** 4; doi :10.1186/1746-1448.5-4].

range of salinities. The diversity of marine picocyanobacteria (*Prochlorococcus* and *Synechococcus*) as assessed by DGGE and T-RFLP of 16S-23S rRNA ITS region revealed the presence of more than one band or T-RF per strain or cloned sequence suggesting the existence of anomalies due to the formation of secondary structures. Modelling of ITS structures at different temperatures of PCR revealed the presence of hairpin loops. So accordingly, Lavin *et al.* (2008) modified T-RFLP method by the addition of labelled ITS primer in the last two cycles of the PCR that resulted in most cases only one T-RF per ribotype. Natural picophytoplanktonic communities in marine waters of Chile showed that it is possible to identify several phylogenetic lineages within the genera *Prochlorococcus* and *Synechococcus*. This study shows that T-RFLP with the proposed modification is suitable for studying diversity of natural populations of cyanobacteria.

The assessment of genetic diversity of geographically unrelated *M. aeruginosa* strains using AFLP has been made by Oberholster *et al.* (2005). Phylogenetic analysis of AFLP polymorphisms of 23 strains of *Microcystis* revealed the existence of four clusters. Cluster 1 is composed of *Microcystis* strains (NIES) from Japan whereas Cluster 2 is represented by European strains. South African strains (from northern parts) formed a group in Cluster 3 and Cluster 4 consisted of strains from central and southern parts of South Africa along with strains from USA. The superior discriminative power of this technique in the differentiation of different genotypes of *Microcystis* has been stressed. Floyd *et al.* (2005) analysed the holdings of prokaryotes at the American Type Culture Collection (ATCC) with regard to the geographical and environmental habitats from which they were isolated. A comparison of the selected habitats in the recently published literature on culture-independent assays on microbial diversity revealed as to how the cultured members existing in ATCC are distributed in nature. Of the 37 environmental habitats, isolates from soil were maximum (27.3%), followed by those from host-associated, terrestrial, plant (16.0%) and lastly from food and food-processing associated (7.3%) habitats. The origin of nearly 41.4% of the cultures (2,209 entries) is not known. Isolates from N. America were 24.1% (1289 entries) followed by those from Europe (14.7% and 784 entries) and lastly from Asia (11.5% and 612 entries). A comparison of the number of known microbial species in biodiversity hotspots from various continents revealed maximum number of species (314) from Europe and Central Asia, followed by 138 microbial species from Mainland Asia (Indo-Burma region of hotspot). Microcystin synthetase (*mcy A* gene) was chosen as the molecular marker for the quantification of toxic cyanobacteria from polluted waters with the help of a novel primer MSR-2R by conducting a real-time PCR. A direct correlation existed between the number of cells in the culture and those determined by PCR. The detection limit has been found to be 8.8 cells per reaction. The gene could be successfully amplified and quantified from environmental samples (Furukawa *et al.*, 2006).

Kim *et al.* (2006) determined the cyanobacterial diversity during algal blooms in Daechung reservoir, Korea on the basis of *cpcBA-IGS* sequences from environmental samples. Sequences of *Microcystis*, *Aphanizomenon* (*Anabaena*), *Pseudoanabaena* and *Planktothrix* (*Oscillatoria*)-like groups and three other previously reported *cpcBA-IGS* sequences have been detected during the peak of bloom growth. The succession of the cyanobacteria based on *cpcBA-IGS* sequences is further supported by PCR-DGGE of 16S rRNA genes. Boutte *et al.* (2006) tested the utility of specific 16S rRNA cyanobacterial primers CYA 359F, CYA 781R(a) and CYA 781R(b) to assess molecular diversity of cyanobacteria from their natural habitats. CYA 781R(a) preferentially targeted filamentous cyanobacteria whereas CYA 781R(b) targeted unicellular cyanobacteria. Environmental *nifH* (encoding reductase subunit of nitrogenase) gene sequences from the biological soil crusts of Colorado Plateau, USA revealed the existence of three distinct clades of heterocystous cyanobacteria. These sequences matched up to nearly 89% with the *nifH* gene sequences of cultured heterocystous cyanobacteria *Nostoc commune*, *Scytonema hyalinum*, *Tolypothrix* and *Spirirestis*. These constituted the dominant members of this habitat (Yeager *et al.*, 2007).

Valério *et al.* (2009) identified 118 cyanobacterial isolates from Portuguese freshwater reservoirs belonging to Chroococcales (54), Oscillatoriales (15) and Nostocales (49). Cluster analysis of STRR and LTRR sequences, PCR fingerprinting profiles were obtained and these were subjected to molecular identification at the species level. Based on16S rRNA and *rpoC1* phylogenies a diagnostic key for the identification of cyanobacteria based on 16S rRNA-ARDRA, ITS amplification and ITS-ARDRA has been proposed. *Planktothrix agardhii* showed lowest diversity and *Aph. flos-aquae* has the highest diversity. Prabina *et al.* (2005) has used RAPD fingerprinting as a tool for checking genetic purity of cyanobacterial strains. In most of these studies, more than ten primers were used to generate RAPD patterns. Phylogenetic analysis of 12 cyanobacterial strains belonging to Oscillatoriaceae (8 of *Oscillatoria* and 4 of *Lyngbya*) by random amplification of polymorphic DNA revealed the presence of two major clusters. One of these clusters is composed of *O. obscura* and in the second the rest of the cyanobacterial strains have clustered together (Perumal *et al.*, 2009).

5) The species concept in bacteria and cyanobacteria

As per Bergey's Manual of Systematic Bacteriology the definitions given for a bacterial genus and species are reproduced below.

Genus: "The bacterial genus is usually a well-defined group that is clearly separated from other genera, and then through descriptions of genera in the 1984 edition of Bergey's Manual exemplify the depth to which this taxonomic group is usually known. However, there is so far no general agreement on the definition of a genus in bacterial taxonomy, and considerable subjectivity is involved at the genus level. Indeed, what is perceived to be a genus by one person may be perceived as being merely a species by another systematist."

Species: "A bacterial species may be regarded as a collection of strains that share many features in common and differ considerably from other strains. One strain of a species is designated as the type strain; this strain serves as the name-bearer strain of the species and is the permanent example of the species, i.e. the reference specimen for the name. The type strain has great importance for classification at the species level, because a species consists of the type strain and all other strains that are considered to be sufficiently similar to it as to warrant inclusion with it in the species...."

A species represents a group of organisms (strains) in a taxonomic rank (Stackebrandt and Goebel 1994) and forms the basic unit of a classification system. It is a tool for describing diversity. The species name should convey about the phenotypic features of an organism and about the relationships to other organisms. We will first consider the concepts that have been developed during recent years in defining species in bacteria. Later on the species concept for cyanobacteria will be discussed. As cyanobacteria form part of phylum eubacteria, the species concept of bacteria may as well be extended to cyanobacteria. However, cyanobacteria are morphologically highly divergent in comparison to most other bacteria, and consequently, it has been suggested that morphological features be given more weight in the species definition (Castenholz and Norris 2005). It would be in the fitness of things if we state here species limit set for percentage of RB derived from DNA-DNA hybridization studies. Strains that show ~70% RB are considered to belong to the same species (Wayne *et al.*, 1987). As this condition cannot be applied with ease in all cases and since it would be difficult for cultivating certain strains, alternatives to this 'golden standard' have been looked for. Konstantinidis and Tiedje (2005) observed that those strains which exhibit 70% RB [or in other words those that exhibit 95% average nucleotide identity (ANI) over all shared genes] have been shown to be extremely similar in their 16S rRNA gene sequences. On the contrary, strains that have almost identical 16S rRNA gene sequences may not be closely related either on the basis of their RB (Rosselló-Móra and Amann, 2001) or ANI (Konstantinidis and Tiedje, 2005). If two strains have less than 97% sequence identity of 16S rRNA they are assigned to different species but added to this the percentage of their RB is still required in order to identify them as separate species. Since ANI represents average nucleotide identity of the total genome sequence shared between two strains, it is exceptionally a robust and sensitive method for measuring evolutionary relatedness. Strains that show more than 60% ANI exhibit greater than 97% 16S rRNA gene sequence identity. Due to a strong correlation between 95% ANI and 70% RB, Konstantinidis and Tiedje (2005) seem to favour retention of 70% RB criterion as it brings together relatively homogeneous set of strains which share at least 80% genetic homology. As ANI is a genome-derived parameter, all conserved genes between any two genomes are likely to show high degree of genetic relatedness. Moreover, there have been no reports on the organisms that show less than 98.5% identity in their 16S rRNA gene sequences and show at the same time 70% RB (Stackebrandt and Ebers, 2006) or 95% ANI. Instead of basing the phylogenetic studies on a single gene sequence like 16S rRNA or alternatively comparing the entire genome sequences, a reasonable compromise is to select at least seven or eight core gene sequences (housekeeping genes) that can provide robust phylogeny of organisms (MLST approach; Maiden *et al.*, 1998; Gevers *et al.*, 2005). Zeigler *et al.* (2003) tested the usefulness of 32 protein-coding gene sequences, distributed widely in bacteria, in assigning bacterial strains to species. A single housekeeping gene such as *recN* (that encodes recombination/repair protein) is found to be sufficient for this purpose as strains that exhibited 96% similarity in their *recN* sequences also showed >70% similarity in their genome sequences and could be assigned with 95% confidence level to the same species. If two strains or isolates exhibit <84% similarity in the *recN* sequences, the overall similarity in the genome sequences has been found to be <70% and with the same confidence level they could be assigned to different species. The resolving power could be enhanced by the inclusion of two more genes such as *dnaX* (that encodes DNA polymerase III subunits γ, τ) and *thdF* (that encodes GTP-binding protein) in the analysis. The other housekeeping genes such as *uvrC* (excision nuclease subunit C), *uvrB* (excision nuclease subunit B), *ruvB* (Holliday Junction helicase subunit A), *dnaB* (replicative DNA helicase), *dnaJ* (chaperone with DnaK), *ftsZ* (tubulin-like cell division protein), *atpA* (ATP synthase F1,α-subunit) and *atpD* (ATP synthase F1,β-subunit) are among the other 32 genes selected by them for comparing bacterial genomes by pair-wise comparisons to arrive upon their conclusions. According to Konstantinidis and Tiedje (2007) such genes show robust phylogenetic signal and give 70–95% values of ANI. The working hypothesis for the bacterial species concept is that bacteria form coherent genome clusters with distinctive phenotypic features and the clusters seem to have been created by two major forces i.e. selection and recombination. So during selection of genes for MLST approach one should take into account LGT and recombination between the genes when one wants to identify strains with in a given species. Konstantinidis *et al.* (2006) showed that selection of three such best performing genes could provide robust phylogeny and accurate prediction of ANI values. The most favourable species concepts for bacteria are biological (Mayr, 1997) and the phylogenetic concepts (Hull, 1997). These two are followed by the ecotype concept introduced by Cohan (2002) which is a variation of the phylogenetic concept. Cohan (2006) stressed the need to develop a theory-based approach to define bacterial species rather than simply recognizing the genotypic diversity based on a single molecular marker like 16S rRNA gene sequencing and demarcate bacterial species into many clusters. In doing so, the sequence clusters or OTUs should be clearly be correlated to the ecotypes. Any sequence based phylogeny is likely to contain a number of subclusters and it is not clear which level of cluster corresponds to ecotypes. According to him, a conceptual framework based on the evolutionary dynamics of bacterial populations for estimating the number of ecotypes within a community and then identifying them has been proposed. Studies made on community phylogeny on three groups of organisms and their habitats revealed interesting results. The ecotypes identified in case of *Bacillus subtilis-B. licheniformis* occurring in Evolutionary Canyon III of Negev Desert closely corresponded to the existing species and subspecies that formed a clade in sequence-based identification. In contrast, world-wide clinical and environmental isolates of *Legionella pneumophila* could be distinguishable into 14 ecotypes that had less 16S rDNA sequence divergence. Likewise, 8 ecotypes of *Synechococcus* in A-A′ Clade also exhibited less 16S rDNA sequence divergence. However, in *Synechococcus* A-A′ Clade microhabitat distribution reveals one ecotype harbouring in the upper photic zone at 68°C while another ecotype is specialized to the lower photic zone suggesting that it is temperature that makes the two ecotypes distinct from one another. So in identifying a species under polyphasic taxonomy, the diagnostic ecological features would prove more meaningful. To the extent the stable ecotype model is correct, the different clusters can be equated to different ecotypes. If some sequence clusters are not falling in the same ecotype as the most closely related strains under the species such clusters can be considered as different species.

The study of bacterial species in relation to geography is gaining interest during the past few years and is known as bacterial biogeography (Staley, 1999; Staley and Gosink, 1999; Martiny *et al.*, 2006). Evidences are in favour of the occurrence geovars of bacteria particularly in extreme environments. Staley (2006) extended the species concept based on ecological information by introducing a combined Genomic Phylogenetic Species Concept (GPSC) that is based on genomic and MLST approaches. The basic principle is that every bacterial species grows in a particular niche of its own and no two species occupy the same niche. In molecular terms, the functional proteins and enzymes produced by that particular species are unique and a true reflection of the niche. The core housekeeping genes of the species will also reflect the evolutionary pattern. In case there is a change in the niche, accordingly a genetic drift and adaptation to the new niche is a consequential event. Because GPSC takes into account sequence-based information (e.g. protein, RNA, DNA), it is more practical, rapid, readily accessible from anywhere and slowly gaining universality. To elaborate the concept of GPSC, Staley (2006) cited *Prochlorococcus* as one of the examples where a comparison between HL-adapted (MED4) and LL-adapted (MIT 9313) strains revealed interesting features. On the basis of 16S rRNA phylogeny both strains seem to have descended from a common LL-adapted ancestor. There is a great difference in the size of the genomes of the two strains. MED4 has a genome of 1.65 Mb (with a mol % G+C of 30.8) whereas MIT 9313 possesses a genome size of 2.41 Mb (with a mol% G+C of 50.7). These differences are in fact major in nature that warrants their separation into not only separate species but also into two different genera. However, their commonality is the existence of 1350 genes with additional of 364 genes in case of MED4 and 923 genes in case of MIT 9313. It is suggested that these additional genes either may have been acquired from their common ancestor (retained and duplicated) or via LGT. Although these two strains differ from each other by 0.8% of 16S rDNA sequence, the median sequence identity of the shared genes is only 78%. These two strains are considered as distinct ecotypes and GPSC concept can be readily applied to these strains.

Whitaker (2003) investigated the relationship between genetic and geographical distance in combined analysis of nine genomic loci of archaebacteria collected from globally distant geothermal regions. There appeared to be a significant distance effect that showed a distinct relationship between genetic and geographical distance among populations. Whitaker (2006) discussed allopatry as a speciation mechanism in microorganisms in which both environmental selection and geographical isolation can lead to differentiation among populations.

III. EVOLUTION OF BACTERIA AND CYANOBACTERIA

Charles Darwin (1859) presented in his classical work on "The Origin of Species" a theory of natural selection and envisaged that all organisms have evolved from a common ancestor and diverged from one another during the course of evolution. Accordingly, a universal "Tree of Life" (TOL) can be constructed representing gradual evolution of organisms. Although the concept of the existence of a common ancestor was accepted by many of his contemporaries and subsequent followers the theory of natural selection was not readily accepted. But due to the objections raised by many workers, he later modified his theory to the extent that the driving force for natural selection is spontaneous variations. One of his staunch oppents of the theory of natural selection was Thomas Morgan who strongly advocated a theory of "mutationism" that explains the evolutionary process as chiefly due to beneficial mutations. Simply because a number of mutations are of deleterious nature, this theory of Morgan was not accepted by a number of leading personalities (Fisher, 1930;

Wright, 1931; Haldane, 1932; Dobzhansky, 1937, 1951). These scientists had been ardent supporters of Darwin's theory and later this came to be known as neo-Darwinism. Zuckerkandl and Pauling (1965) were the first who put forward the idea that DNA and protein sequences can be used to draw evolutionary relationships and it is possible to construct a TOL. The studies on molecular evolution gave importance to mutations since it is due to the mutations the change in the sequence of DNA or proteins is affected. But this idea was also not acceptable to neo-Darwinians as according to them mutations have nothing to do with the phenotypic evolution. Although it was recognized that the basic underlying force for evolution is mutation and selection but subtle distinction has been made between phenotypic evolution and molecular evolution. It was noted that selection plays greater role in phenotypic evolution rather than in molecular evolution. The neutral theory of molecular evolution put forward by Kimura (1980) emphasizes 'phyletic gradualism' as the basis of evolution where the evolution of molecular characters is gradual and clock-like and that of the morphological characters can also evolve in a gradual fashion. Another aspect that lends support to this theory is mutations that occur in neutrally-evolving genes are either indifferent or negative (Kimura, 1983). Proteins encoded by neutrally-evolving genes may even show more than 70% sequence divergence but still retain the conserved function and the three-dimensional structure. It means due to a strong selection pressure the structure and function of such proteins is maintained. Thus in case of macroevolutionary characters as well there is a possibility for the retention of the same protein structure and function. However, Kimura's neutral theory suffered due to two drawbacks. The first one is that he took into account nucleotide substitutions in the whole genome instead of taking into account gene as a unit because nucleotide substitutions in the non-coding regions have no relevance in either mutation or selection. Secondly, the definition of selective neutrality took into account the probability of fixation or the time until fixation of mutation. Many molecular biologists consider a neutral or nearly neutral mutation that does not have appreciable effect on the gene function. The views expressed by King and Jukes (1969) on neutral mutations seem to be more convincing. They argued that though mutation is the primary force for evolution, the role of natural selection is to eliminate harmful mutations. According to them proteins that have rigid functional and structural constraint show lower amino acid substitutions than the proteins requiring weak functional constraints. Additional molecular data on the sequences of functionally important proteins lends support to this theory than to Kimura's computation of natural selection. Thus the concept of neutral mutations conceived by Kimura and convincingly put forward by King and Jukes is contrary to the Darwin's positive selection of adaptations but by stabilizing or purifying selection eliminates deleterious mutations allowing the fixation of neutral mutations by genetic drift. It is through the evolution of multigene families involving gene duplications and gene loss that an effect at phenotypic level would be reflected (Nei and Rooney, 2005). Taking into account these facts and other supportive molecular data Nei (1983, 1984) proposed the term neomutationism which was later called by him as neoclassical theory of evolution (Nei, 1987). Nei (2005) discussed at length the evolution of multigene families and their likely impact in shaping the evolution of new genetic systems.

The term 'phylogeny' was for the first time coined by E. Haeckel who followed Darwin's concepts and constructed phylogenetic trees for various groups of organisms. For many scientists who supported Charles Darwin the TOL is a biological fact while someothers believe TOL to be a hypothesis that is yet to be proved. In view of the progress achieved in the molecular biology of prokaryotes the Darwinian concept needed a re-examination. Some of the relevant issues that require attention are as follows: (i) would it be possible to test the Darwin's theory in the context of molecular phylogenetic data obtained during recent years on prokaryotes? (ii) the phylogenetic trees built on the basis of molecular sequences would be congruent with the phylogenetic trees built on the basis of phenotypic (morphological) features or organismal biology, (iii) can the sequence-based phylogenies provide answers and proofs of the reality of macroevolution? and (iv) how far LGT (also designated as horizontal gene transfer by some workers) affects the tree-building?

LGT is found more frequently in prokaryotes than in eukaryotes. LGT occurs in prokaryotes by means of transformation, conjugation and transduction. Genomic and phylogenetic analyses have established beyond doubt the ubiquitous occurrence of LGT among prokaryotes. Among very close relatives, homologous recombination is a great force to cause sequence divergence in many bacteria. Homologous recombination is so widespread among the strains of a designated species that even the housekeeping genes are not even spared. LGT events, on the other hand, occur across domains as for example, from Archaea to Bacteria and from Bacteria to nuclear genomes of eukaryotes. Sequencing of complete genomes and their analysis has contributed to a better understanding of LGT. On the basis of genome comparisons of bacteria and archaea, it is possible to demonstrate that major fractions of genes have been acquired through LGT events in many prokaryotes. It is significant to note that in a number of pathogenic bacteria (pathogenicity islands) and symbiotic bacteria (symbiosis islands) over 30% of the genomes are acquired through events of LGT (Hacker and Kaper, 2000; Ochman and Moran, 2001; Perna *et al.*, 2001). There are very few orthologous gene sets whose history is free of LGT (Gogarten *et al.*, 2002; Gogarten and Townsend, 2005). Even genes such as those of ribosomal proteins and RNA polymerases have not been spared (Brochier *et al.*, 2000; Makarova *et al.*, 2001; Iyer *et al.*, 2004). Based on the sequencing of *Arabidopsis*, chloroplast and cyanobacterial genomes, it was possible to identify thousands of genes of cyanobacterial endosymbiont (the chloroplast) into the host (plant) genome through LGT (Martin *et al.*, 2002). Another 'big bang' that occurred due to LGT is the evolution of mitochondria in eukaryotes. Alpha-proteobacterial ancestor of mitochondria probably invaded an archaebacterial host triggering the genesis of a eukaryotic cell that signalled the formation of endomembrane system, cytoskeleton and the nucleus (Martin and Muller, 1998; Martin and Koonin, 2006).

There are at least three approaches to determine LGT events. The first one is to determine the atypical sequence characteristics such as codon usage bias and G+C content. Because codon usage and G+C are species-specific, it was possible to identify genes up to 18% in *E. coli* as laterally transferred genes (Lawrence and Ochman, 1998). Recent studies of Sorek *et al.* (2007) are indicative of the fact that genuine barriers for LGT in this model organism are quite low as it virtually accepted all prokaryotic genes offered to it *in vitro*. Likewise, approximately 24% of genes in *Thermotoga maritima* have come through LGT events (Garcia-Vallvé *et al.*, 2000). The second method to detect LGT is to reconstruct a phylogeny of genes. By this approach, it is possible to compare phylogenetic trees among individual genes within closely related species and recognize unusual origins of genes. The third alternative is to detect the presence/absence of genes in closely related bacteria. Though it is quite not possible to detect the number of insertion and deletion events due to gene rearrangements, it is possible to know the total number of genes due to insertions/deletions. Hao and Golding (2004) analysed 50 bacterial complete genomes from nine groups, mapped the number of gene insertions, deletions and gene duplications onto the phylogeny of the species based on 16S rRNA phylogeny. The results revealed that the number of gene insertions was much larger in the external branches of the trees and they greatly outnumbered the occurrence of deletions. At the same time, the genome sizes of these bacteria remained roughly constant suggesting that these insertions were specific to an organism and were lost before related species could evolve. Beiko and Charlebois (2007) developed a new software program known as EvolSimulator that can perform gene-, protein- and genome-scale evolutionary processes that occur in prokaryotes. It can simultaneously create lineages, gene families and sequence substitutions that can measure the levels of LGT events. Dagan *et al.* (2008)

assorted 539,723 proteins encoded among 181 prokaryotic sequenced genomes to obtain matrices of shared genes. The results revealed the existence of a network of shared genes and the genomes in the network shared at least one gene family and thus are interconnected with each other forming a complete network. They suggested that on the average 81±15% of the genes in each genome were involved in LGT. Cyanobacteria possess 27±20% of genes acquired in their genomes through LGT. The above account substantiates that the genomes of all organisms have evolved through LGT. If that be so, how far is it feasible to trace the evolution of all organisms from a common ancestor? Is it possible to construct a universal TOL? In addition to LGT, certain methodological problems that lead to misleading phylogenies have been highlighted by Moreira and Philippe (2000). These are (i) loss of phylogenetic signal by the accumulation of overlapping mutations, (ii) incongruity between the real evolutionary process and the assumed models of sequence evolution and (iii) differences of evolutionary rates among species or among positions within a sequence. According to Soria-Carrasco and Castresana (2008) the inconsistencies or discrepancies that are generally encountered in phylogenetic trees may also be due to phylogenetic artifacts or orthology assignment problems instead of LGT events. They traced phylogenetic inconsistencies in the three domains of life and the inconsistencies in phylogenetic trees of Proteobacteria and eukaryotes were more similar when compared with the reference tree constructed on the basis of MLST approach.

All possible explanations have been put forward for building a universal TOL. Wolf *et al.* (2002) suggested that LGT and differential gene loss are the major evolutionary phenomena in prokaryotes. On the basis of complete gene sets it would be possible to construct a TOL. Phylogenetic analysis of 220,240 proteins from 144 sequenced genomes resulted in the construction of a super tree of prokaryotes that is consistent with 16S rDNA and other molecular sequence trees. Evidences have been presented for the existence of LGT particularly among strains of *Prochlorococcus* and *Synechococcus* (Beiko *et al.*, 2005). Clemente *et al.* (2006) created a data base (at address http://www.jaist.ac.jp/~clemente/cgi-bin/phylo.pl.) for free public use meant for phylogenetic analysis based on non-genomic data. This site consists of information on 13 organisms, 25 metabolic pathways and >1000 enzymes, compounds and reactions. A sort of pseudo-alignment of metabolic pathways is conducted that differs from a full alignment in the mapping being not necessarily one to one. The pseudo-alignment from different organisms was done in order to assess their structural similarity. The pseudo-alignment involved both graphic representation of each metabolic pathway as well as a similarity measure between individual reactions, enzymes and compounds present in the pathway. Using 61 of their 68 pathways, a phylogenetic analysis separated photosynthetic cyanobacteria from eukaryotic photosynthetic organisms. However, within cyanobacteria, it was not possible to distinguish the Chroococcales (*Synechocystis*, *Synechococcus* and *Thermosynechococcus elongatus*) from the Prochlorales (*P. marinus*, *P. marinus* sub. sp. *pastoris*), the Nostocales (*Anabaena*) and the Gloeobacterales (*Gloeobacter violaceus*). In case of other bacteria the phylogeny drawn from such analysis was found to be consistent and reasonably close to the correct taxonomy based on sequence data available with NCBI. Cottret *et al.* (2010) created a webserver known as Met Explore (freely accessible for public use at http://metexplore.toulouse.inra.fr) having genome-scale reconstructed metabolic networks to link metabolites identified in untargeted metabolic experiments.

Beiko *et al.* (2005) are of the view that there might infact be no universal TOL and attempts to construct such a universal TOL are futile in view of the rampant LGT events. In support of this they presented a phylogenetic analysis of 220,240 proteins from 144 sequenced genomes from which it was possible to construct a super tree that is remarkably congruent with 16S rDNA and other molecular sequence trees. But the protein trees that are strongly incongruent with a super tree node are discordant and provide evidence of LGT. This is more significantly noted in thermophiles, pathogens and cyanobacteria. Doolittle and Bapteste (2007) questioned the fundamental belief over the existence of a universal TOL and proposed an alternative path for tree-building known as "pattern pluralism". This theory recognizes the existence of different evolutionary models for different taxa or at different levels or for different purposes. According to them, it provides a good alternative instead of searching for an impractical universal TOL. But Koonin (2009) is of the view that the evolutionary tree introduced by Darwin need not be abandoned altogether because (i) it is possible to construct phylogenetic trees representing accurately the evolution in individual gene families, and (ii) congruent trees can be obtained for large sets of orthologous genes. There are two alternatives. One is to represent a highly consensus tree of highly conserved genes showing the LGT events. The second is to present a complex network with cross-connections representing the phases of LGT that cannot be represented as a tree. Eisen (1998) and Eisen and Fraser (2003) emphasized the need to combine the studies on whole genome sequences (genomics) and the evolutionary studies (phylogeny) together into a new composite area known as 'phylogenomics'. According to them, LGT being a rare event has not distorted the structure of the tree. The diversity of prokaryotic genomes seems to be very enormous as revealed by the presence of good number of new genes in every sequenced genome (Tettelin *et al.*, 2005). The total number of genes present in a given species is termed as 'pangenome'.

The root of the TOL: The classification proposed by Woese *et al.* (1990) envisages that the rooting of the TOL is between the eubacteria and archaebacteria and that the eukaryotes were identified as specific relatives of archaebacteria. It means the archaebacteria constitute a more recent group than that of eubacteria. Then which are the ancestors of eubacteria and archaebacteria that might have constituted the root from which evolution had taken place? The eubacterial group consists of gram-positive, gram-negative bacteria and cyanobacteria. So it means the prokaryotes consist of two-distinct and non-overlapping (i.e. monophyletic) groups that are not specifically related to each other than either was to the eukaryotes. Based on the presence of specific sequences of amino acids in proteins, known as 'signature sequences' (also termed as 'indels'), Gupta (1998) emphasized that: (a) Archaebacteria are monophyletic and are distinct from other prokaryotes. This conclusion is further supported by phylogenies based on gene sequences of 16s rRNA, DNA replication, transcription, translation, t-RNA splicing, histones and theTcp-1 chaperonins. (b) Based on the signature sequences present in proteins such as HSP70, glutamine synthetase 1 (GSI) and glutamate-1-semialdehyde 2,1-aminomutase the archaebacteria are closely related to gram-positive bacteria. (c) Signature sequences also indicate that *Deinococcus* and *Thermus* group of bacteria are intermediates in the transition from gram-positive to gram-negative bacteria. (d) Among the gram-negative bacteria, cyanobacteria constitute one of the deepest branching divisions specifically related to *Deinococcus-Thermus* group. This close relationship between these two groups is also revealed by the signature sequences unique to proteins such as DnaJ-Hsp40 and elongation factor TE-Ts sequences. These results are suggestive of a common ancestor for these two groups of prokaryotes. However, due to the existence of common signature sequences in other proteins such as Hsp60, acetolactate synthase and asparginyl-tRNA synthetase cyanobacteria are shown to be closer to gram-negative bacteria as well. So exclusive of *Deinococcus-Thermus* group, cyanobacteria might have had a common ancestor with gram-negative bacteria. These evidences further point out to the suggestion that: (i) *Deinococcus-Thermus* group of bacteria are more ancestral than cyanobacteria; (ii) *Deinococcus-Thermus* group and cyanobacteria are not the ancestors of other gram-negative bacteria but branched off from the early ancestors. It is also probable that extensive LGT events might have taken place in between *Deinococcus-Thermus* group and cyanobacteria. The extensive sequence data further point out that the primary division within prokaryotes is not between archaebacteria and eubacteria (as suggested by Woese and coworkers) but between organisms that have either a monoderm cell structure (i.e. prokaryotic cells surrounded by a single membrane which includes all archaebacteria and gram-positive bacteria) or a diderm cell structure (i.e. prokaryotic cell structure surrounded by an outer membrane which includes all the gram-negative bacteria). Gupta (1998) further suggested that the root of the prokaryotic tree or TOL lies in between (or within) archaebacteria and gram-positive bacteria. In this connection, mention may be made of a eubacterial species *T. maritima* which had been treated as a gram-negative bacterium but later was shown to be a gram-positive one. So along with archaebacteria the gram-positive bacteria are infact represent ancestral lineages based on Earth's geological history and the hot and anaerobic conditions prevailed at that time were conducive for the growth of these two groups of organisms. Further it has been concluded that the ancestral eukaryotic cell is not a direct descendent of the archaebacterial lineage but is in fact a chimera that resulted from a unique fusion event between two different groups of prokaryotes such as a thermoacidophilic archaebacterium (monoderm) and a gram-negative eubacterium (diderm) followed by the integration of their genomes. Thus all eukaryotes (both amitochondriate and aplastidic) might have acquired the gene contributions from both these sources and retained them.

On the other hand, Cavalier-Smith (2002) presented a rooted TOL depicting the evolution of different bacterial groups from a root consisting of gram-negative bacteria. So in this respect the concepts evolved by Cavalier-Smith (2001) mostly agree with the conclusions of Woese *et al.* (1990) in recognizing the archaebacteria as a more recent group whose ancestors have given rise to archaebacteria on the one hand and the earliest eukaryotic organisms on the other. However, the difference lies in recognizing the root of the TOL. Accordingly, the roots identified by the three groups of workers are (i) ancestors between eubacteria and archaebacteria (Woese *et al.*, 1990), (ii) ancestors between archaebacteria and gram-positive bacteria (Gupta, 1998) and (iii) ancestors between archaebacteria and gram-negative bacteria (Cavalier-Smith, 2001).

According to Cavalier-Smith (2002) life on planet Earth most likely has evolved ~3850 million years (My) ago. Negibacteria (in Latin 'Negi' means negative, i.e. gram-negative bacteria) with a double peptidoglycan layer of cell wall, porins, acyl ester phospholipids, photosynthesis and chlorosomes appeared at the base ~3400 My ago that formed the root for the TOL. Evolution diverged into cyanobacteria and Eobacteria (in Latin 'Eo' means dawn or early). At about ~2500 My ago cyanobacterial branch gave rise to Eubacterial group comprising Posibacteria ('Posi' for gram-positive bacteria with endospores-Endobacteria) and Actinobacteria (comprising Arthrobacteria, Streptomycetes and Arbobacteria). Thermophilic ancestors of Eubacteria called as Neomura gave rise (~850 My ago) to Archaebacteria and Eukaryota (Fig. 22). The line from Eobacteria diverged separately into Spirochaetes (~2500 My ago after cyanobacteria have evolved), Planctobacteria and Proteobacteria. It is the Proteobacteria that probably formed the ancestors for mitochondrial development during the time of appearance of primitive eukaryotes which had mesophily and phagotrophy as their properties. In this bacterial megaclassification he recognized one Kingdom (as per Bacteriological Code) instead of a Domain and Enterobacteriales as the Type order. He considered the term as 'Eubacteria' as a useful grade name but not treated in his classification as a taxon. The Bacteria are divided into two new subkingdoms. Subkingdom I consists of Negibacteria (with a cell envelope of two distinct genetic membranes) and Unibacteria (consisting of new phyla Archaebacteria and Posibacteria). A number of morphological, palaentological and molecular data have been integrated to present a unified picture of bacterial cell evolution despite occasional LGT events. Archaebacteria and eukaryotes comprise the clade neomura. There are some characters unique (8 of them) to Archaebacteria of which mention may be made of the absence of Hsp90 chaperone, the splitting of RNA polymerase into two proteins and

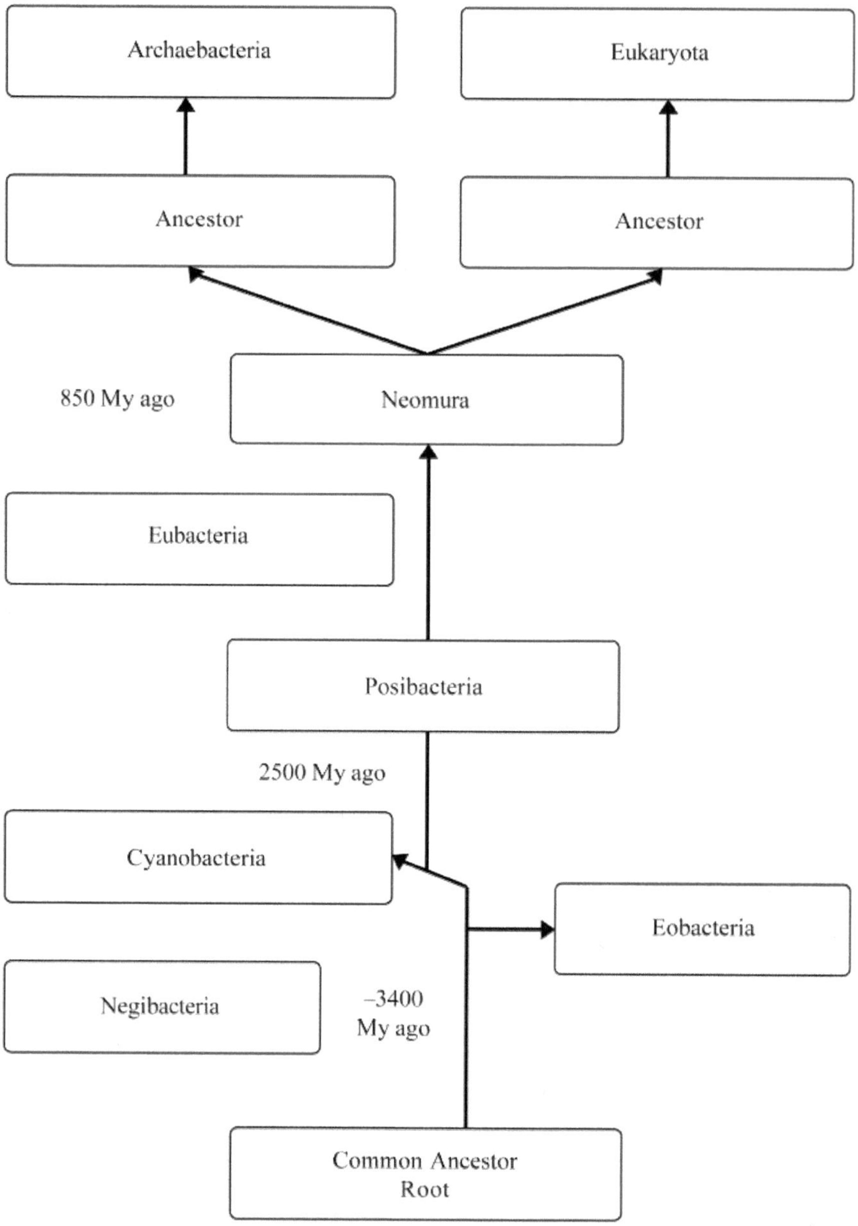

Figure 22: Phylogenetic tree with a common ancestor for the evolution of living organisms. Based on Cavalier-Smith (2002).

the splitting of glutamine synthetase into three separate proteins. Archaebacteria share 19 characters with eukaryotes that are not found in eubacteria.

Fossil evidences suggest that life on the planet Earth might have had its origins in the Precambrian eon and during the subsequent span of 542 My the evolution of the present day flora and fauna could be traced through the Palaeozoic, Mesozoic and Coenozoic ages, collectively known as

Phanerozoic age. According to the norms set by the International Commission on Stratigraphy, seven chronostratigraphic units have been recognized in the Precambrian period. These are Palaeoproterozoic (2000–1650 My), Lower Mesoproterozoic (1650–1200 My), Upper Mesoproterozoic (1200–1030 My), Lower Neoproterozoic (1030–850 My), Upper Neoproterozoic (850–630 My), Lower Edicaran (630-550 My) and Upper Edicaran (550–542 My) units (Sergeev, 2009). The evolution and distribution of life in the Precambrian era has been reviewed (Sharma and Shukla, 2009). Fossil microflora, dating to Late Precambrian, from Bitter Spring Formation of Central Australia has been reported (Schopf, 1968). Schopf and Walter (1983) reported simple trichomes of Oscillatorian cyanobacteria from ~2800 My Fortescne Group, Western Australia. Further evidences on the existence of microfossils of the early Archaean Apex Chert (Schopf, 1993) and Laser-Raman imagery of the Earth's earliest fossils (Schopf *et al.*, 2002) have been presented. Some of the fossils of cyanobacteria have been traced to ~2,100 My from Franceville Group of Gabon formations. These are represented mostly by coccoid, star-like, filament, tubular forms and nostocacean akinetes i.e. *Archaeoellipsoides* (Amard and Bertrand-Sarfati, 1997). *Archaeoellipsoides* is a well preserved representative occurring abundantly in the ~1500 My Cherts from Billyakh group of Siberia (Golubic *et al.*, 1995). This genus has also been found in the silicified carbamates of the ~1650 My Amelia Dolomite of northern Australia (Page *et al.*, 2000) as well as 1631±5-My Cherts from the Kheinjua Formation in India (Srivastava, 2005).

Rudi *et al.* (1998) studied the evolutionary rates in a number of strains of cyanobacteria (16 strains of *Microcystis*, 6 strains of *Tychonema*, 10 strains of *Planktothrix* and 12 strains of *Nostoc*) by comparing sequence divergence in the neutrally-evolving genes such as *rbcL* and *rbcX* and compared them with 16S rRNA phylogeny. While there was low sequence divergence within *Microcystis*, *Tychonema* and *Planktothrix* for *rbcLX* and 16S rDNA, strains of *Nostoc* revealed three genetically clustered lineages for these genes. The 16S rDNA and *rbcLX* phylogenies were not congruent with those in the clustered groups. The evolutionary model presented by them explains as to how the modern species has evolved by sequence homogenization involving extensive exchange of genetic material for neutrally-evolving genes. Further, they emphasized that the macroevolutionary characters or morphological characters have been stable for long periods if a comparison is made between the fossils and present day representatives of cyanobacteria. This is likely to be achieved by exchange of genetic material between phyletically closely related strains of cyanobacteria for neutrally-evolving genes. This is supported by the existence of greater sequence homogeneity for the closely related cyanobacteria and the exchange of genetic material probably stabilizes the function and structure of proteins. According to Rudi *et al.* (1998) their model of gene transfer explains the morphological similarities between the fossil and the recent species.

Due to selection pressures operative in the populations of pure cultures, the genotype that dominates is other than the dominant genotype in the natural population at the time of collection. So a culture thus no longer represents the dominant genotype supposed to represent the type specimen. Due to this departure, how far is it justified to extrapolate the results from the laboratory to the field samples? Additionally, if phylogenetic analysis of field samples is done directly (circumventing the process of culture techniques) by metagenomics, a situation arises where we have a sequence identified without any proper identification of the organism based on classical morphological or a polyphasic approach (Ward *et al.*, 1994; Amann *et al.*, 1995). It means at the time the type species is proposed, it has to be genetically characterized. Now the situation is that whatever type species for the representative taxa of cyanobacteria we have now, they lack genetic characterization. Palinska *et al.* (2006) presented evidences to show that historic identifications made on the basis of classical taxonomy could be confirmed by subjecting herbarium specimens to microscopic (light and electron microscopy) and molecular methods of analysis (16S rRNA gene amplification and its sequence comparisons). They obtained materials deposited at the Herbariums of University of Vienna (HUW) and Swedish Museum of Natural History, Stockholm (SH) by workers other than those who discovered them in the corresponding years mentioned after the name of the species. They included representatives of unicellular [*Microcystis* (*Polycystis*) *aeruginosa* Kützing 1885 and *Chroococcus turgidus* Kützing 1850 of Section I], filamentous, non-heterocystous [*Microcoleus chthonoplastes* (Mertens) Zanardini ex Gomont 1895 and *Trichodesmium erythraeum* Ehrenberg ex Bornet et Flahault 1884 of Section III] and filamentous, heterocystous forms [*Nostoc muscorum* Agardh ex Bornet et Flahault 1886 and *Nodularia spumigena* Bornet et Flahault 1906 of Section IV of Rippka *et al.* (1979)]. *T. erythraeum* was obtained from SH and the rest were from HUW. They are of the opinion that though these specimens were not characterized genetically at the time collection, the preserved type materials in all these cases had the advantage of permanently retaining the characteristics of the specimens noted at the time of collection. In this connection, the suggestions of Castenholz and Norris (2005) are worth recalling who emphasized that the morphological features be given more weightage in the species definition of cyanobacteria.

Zhaxybayeva *et al.* (2006) developed a tool that helps in comparing phylogenetic trees constructed for each of the orthologous genes detected in four taxa at a time. This method is known as "embedded quartet decomposition analysis" or "quartet decomposition". They included 11 completely sequenced genomes (*Anabaena* sp. strain PCC 7120, *T. erythraeum* IMS101, *Synechocystis* sp. strain PCC 6803, *P. marinus* CCMP1375, *P. marinus* MED4, *P. marinus* MIT9313, marine *Synechococcus* WH 8102, *T. elongatus* BP-1, *Gloeobacter violaceus* PCC 7421; *Nostoc punctiforme* ATCC 29133 and *Crocosphaera watsonii* WH8501) in their study that represent freshwater, marine and hotspring cyanobacteria. A comparison of 1128 protein-coding gene families from these genomes revealed the existence of 879 homologues in selected prokaryotic genomes and the remaining 249 data sets have been found to be "cyanobacteria-specific". According to them, these "cyanobacteria-specific" homologue gene sequences are not suitable for interphylum transfers but 700 of the 879 gene data sets are highly resolved and 80% of these are distributed in cyanobacteria. Of the 700 genes, 540 support cyanobacteria as a coherent group (accounting to ≈77%) while the remaining 160 gene sequences (≈23%) either have sequences interspersed among cyanobacteria or some cyanobacterial gene sequences being found elsewhere. This suggests for a possibility of transfer events to or from cyanobacteria. Zhaxybayeva *et al.* (2009) extended the embedded quartet scatter plots to understand gene flow in the intertwined evolutionary histories of 18 *P. marinus* and *Synechococcus* genomes. The conclusions are that: (i) the HL-adapted *P. marinus* strains constitute a monphyletic clade; (ii) the LL-adapted *P. marinus* strains form a paraphyletic clade; (iii) frequent transfer of at least 16 gene families took place between HL- and LL-adapted strains and (iv) high way of gene sharing occurred between *Synechococcus* spp. and LL-adapted *P. marinus* strains to the extent of erosion of genus boundaries, supporting the obserevations of Beiko *et al.* (2005). According to them, however, these events have not disturbed the *Prochlorococcus*-specific ecological adaptations. Other estimates of LGT point out that cyanobacteria might have acquired between 9.5% and 16.6% genes through LGT (Ochman *et al.*, 2000; Nakamura *et al.*, 2004). These estimates are on a lower side when compared to 27 ± 20% of genes acquired through LGT by cyanobacteria as mentioned earlier (Dagan *et al.*, 2008).

IV. THE OTHER SIDE OF THE COIN

The impact of 16S rDNA sequencing in prokaryotic world is so enormous for the simple reason it has generated a great wealth of information on microbial diversity, community structure and phylogenetic relationships from diverse environments and geographic regions. Moreso, the impact seems to have left an impression in the minds of many investigators that it is the key for unraveling the taxonomic disputes of cultivable and uncultivable organisms. A lot of confusion seems to have been generated on the existence of a number of "computer bacteroids" based on the existence of only sequences from metagenomic analyses. There are no nomenclatural guidelines existing in bacterial taxonomy for recognizing only sequences as equivalent to organisms. Murray and Schleifer (1994) stated that international nomenclatural code "is not able to provide sensible regulation of nomenclature for new taxa defined by very limited data". It is pertinent to point out that the original bacteriological code written in 1957 and the considerations of the Judicial Commission (both in 1976 and 1990) have not forseen a situation for consideration of bacterial taxa based on only molecular sequences. There are a number of pitfalls in this area. Some of the fears expressed are that (i) the molecular sequences are based on samples of relatively small size, (ii) a sampling error can not be ruled out, (iii) likewise, errors in sequencing can not be ruled out, (iv) specially with reference to 16S rRNA gene sequencing studies, contaminating 16S rDNA in the sample can not be overlooked, (v) a number of workers considered only a single strain of a taxon such as a genus yet the picture presented was for the entire taxon and even the validity of the adjacent taxa was questioned. Gest (1999) presented "a primer for the new millennium" to lift bacterial classification and taxonomy from its existing "chaotic status". He quoted extensively the opinions of many scientists on the role of molecular markers and the validity of phylogeny based taxonomy. According to him there is a complete disregard to bacterial nutrition, physiology and biochemistry based on which many basic principles in cell biology and biochemistry have been enunciated.

The current situation in cyanobacterial taxonomy is more disconcerting for both the traditional as well as the molecular taxonomist. The first major requirement is a classification that resolves the guidelines of both ICBN and International Code of Nomenclature of Prokaryotes (ICNP). Moreover, the guidelines for nomenclature and a polyphasic approach to the taxonomy of cyanobacteria have not been clearly defined. Pinevich (2008) summarized and highlighted the paradoxes of biodiversity, phylogeny and taxonomy of cyanobacteria. Though it is needless to repeat here the inconsistencies in cyanobacterial taxonomy, one taxonomic riddle could not be resolved despite the progress that has been achieved by the application of molecular markers. This relates to the identification of *Anabaena* sp. strain PCC 7120. More than half a dozen research papers have been published in support of it being designated as *Anabaena* sp. strain PCC 7120 while as many number of research papers have been published in support of it being redesignated as *Nostoc* sp. strain PCC 7120. The present trend is that many workers started representing it as *Anabaena/Nostoc* sp. strain PCC 7120. What is the signal that is sent to the traditional taxonomist who finds it difficult to enter into the woods of phylogenetic trees? Oren (2011) highlighted the inadequacies in the resolution of guidelines between ICBN and ICNP and suggested that 'the guidelines for the characterization of prokaryotic strains for taxonomic purposes advocated by Tindall *et al.* (2010) be adopted for nomenclatural guidelines of cyanobacteria'. Unless a major international effort is undertaken, the cyanobacterial taxonomy continues to lie in the same state for another century. It is because it took more than 60 years to identify and classify very few cyanobacterial strains on a polyphasic approach (Oren, 2011).

LITERATURE CITED

Abe, T., Kanaya, S., Kinouchi, M., Ichiba, Y., Kozuki, T., and Ikemura, T. (2003) Informatics for unveiling hidden genome signatures. *Genome Res* **13:** 693–702.

Abed, R. M. M., and Garcia-Pichel, F. (2001) Long-term compositional changes after transplant in a microbial mat cyanobacterial community revealed using a polyphasic approach. *Environ Microbiol* **3:** 53–62.

Abed, R. M. M., Garcia-Pichel, F., and Hernández-Mariné, M. (2002a) Polyphasic characterization of benthic, moderately halophilic, moderately thermophilic cyanobacteria with very thin trichomes and the proposal of *Halomicronema excentricum* gen. nov., sp. nov. *Arch Microbiol* **177:** 361–370.

Abed, R. M. M., Safi, N. M. D., Köster, J., De Beer, D., El-Nahal, Y., Rullkötter, J., and Garcia-Pichel, F. (2002b) Microbial diversity of a heavily polluted microbial mat and its community changes following degradation of petroleum compounds. *Appl Environ Microbiol* **68:** 1674–1683.

Acinas, S. G., Haverkamp, T. H. A., Huisman, J., and Stal, L. J. (2009) Phenotypic and genotypic diversification of *Pseudoanabaena* spp. (Cyanobacteria). *ISME J* **3:** 31–46.

Al-Thukair, A. A., Abed, R. M. M., and Mohamed, L. (2007) Microbial community of cyanobacteria mats in the intertidal zone of oil-polluted coast of Saudi Arabia. *Mar Pollut Bull* **54:** 173–179.

Amann, R. I., Binder, B. J., Olson, R. J., Chisholm, S. W., Devereux, R., and Stahl, D. A. (1990) Combination of 16S rRNA-targeted oligonucleotide probes with flow cytometry for analyzing mixed microbial populations. *Appl Environ Microbiol* **56:** 1919–1925.

Amann, R. I., Ludwig, W., and Schleifer, K.-H. (1995) Phylogenetic identification and *in situ* detection and identification of individual microbial cells with out cultivation. *Microbiol Rev* **59:** 143–169.

Amard, B., and Bertrand-Sarfati, J. (1997) Microfossils in 2000 Ma old cherty stromatolites of the Franceville Group, Gabon. *Precambrian Res* **81:** 197–221.

Amikam, D., Glaser, G., and Razin, S. (1984) Mycoplasmas *(Mollicutes)* have a low number of rRNA genes. *J Bacteriol* **158:** 376–378.

Anagnostidis, K., and Komárek, J. (1985) Modern approach to the classification system of the Cyanophytes 1: Introduction. *Arch Hydrobiol* **71:** 291–302.

Anagnostidis, K., and Komárek, J. (1988) Modern approach to the classification system of cyanophytes. 3. Oscillatoriales. *Arch Hydrobiol* suppl **80:** 327–472.

Antón, A. I., Martínez-murcia, A. J., and Rodríguezvalera, F. (1998) Sequence diversity in the 16S-23S intergenic spacer region (ISR) of the rRNA operons in representatives of the *Escherichia coli* ECOR collection. *J Mol Evol* **47:** 62–72.

Barker, G.L., Hayes, P.K., O'Mahony, S.L., Vacharapiyasophon, P. and Walsby, A. E. (1999) A molecular and phenotypic analysis of *Nodularia* (cyanobacteria) from the Baltic Sea. *J Phycol* **35:** 931–937.

Barker, G. L., Handley, B. A., Vacharapiyasophon, P., Stevens, J. R., and Hayes, P. K. (2000) Allele-specific PCR shows that genetic exchange occurs among genetically diverse *Nodularia* (cyanobacteria) filaments in the Baltic Sea. *Microbiology* **146:** 2865–2875.

Beiko, R. G., and Charlebois, R. L. (2007) A simulation test bed for hypotheses of genome evolution. *Bioinformatics* **23:** 825–831.

Beiko, R. G., Harlow, T. J., and Ragan, M. A. (2005) Highways of gene sharing in prokaryotes. *Proc Natl Acad Sci USA* **102:** 14332–14337.

Beltran, E. C., and Neilan, B. A. (2000). Geographical segregation of the neurotoxin-producing cyanobacterium *Anabaena circinalis*. *Appl Environ Microbiol* **66:** 4468–4474.

Ben-Porath, J., Carpenter, E. J., and Zehr, J. P. (1993) Genotypic relationships on *Trichodesmium* (Cyanophyceae) based on *nifH* sequence comparisons. *J Phycol* **29:** 806–810.

Bensaadi-Merchermek, N., Salvado, J.-C., Cagnon, C., Karama, S., and Mouchés, C. (1995). Characterization of the unlinked 16S rDNA and 23S-5S rRNA operon of *Wolbachia pipientis*, a prokaryotic parasite of insect gonads. *Gene* **165:** 81–86.

Benson, D.A., Karsch-Mizrachi, I., Lipman, D. J., Ostell, J., Rapp, B. A., and Wheeler, D. L. (2002) GenBank. *Nucleic Acids Res* **30:** 17–20.

Bent, B. J., Pierson, J. D., and Forney, L. J. (2007) Measuring richness based on microbial community fingerprints: the Emperor has no clothes. *Appl Environ Microbiol* **73:** 2399–2401.

Bercovier, H., Kafri, O., and Sela, S. (1986) Mycobacteria possess a surprisingly small number of ribosomal RNA genes in relation to the size of their genome. *Biochem Biophys Res Commun* **136:** 1136–1141.

Berrendero, E., Perona, E., and Mateo, P. (2008) Genetic and morphological characterization of *Rivularia* and *Calothrix* (Nostocales, Cyanobacteria) from running water. *Int J Syst Evol Microbiol* **58:** 447–460.

Binder, B. J., and Liu, Y. C. (1998) Growth rate regulation of rRNA content of a marine *Synechococcus* (cyanobacterium) strain. *Appl Environ Microbiol* **64:** 3346–3351.

Bittencourt-Oliveira, M. do. C., Moura, A. do. N., Oliveira, M. C. de., and Massola Jr., N. S. (2009) *Geitlerinema* species (Oscillatoriales, Cyanobacteria) revealed cellular morphology, ultrastructure and DNA sequencing. *J Phycol* **45:** 716–725.

Blackwood, C. B., and Buyer, J. S. (2007) Evaluating the physical capture method of terminal restriction fragment length polymorphism for comparison of soil microbial communities. *Soil Biol Biochem* **39:** 590–599.

Blackwood, C. B., Marsh, T., Kim, S-H., and Paul, E. A. (2003) Terminal restriction fragment length polymorphism data analysis for quantitative comparison of microbial communities. *Appl Environ MIcrobiol* **69:** 926–932.

Blackwood, C. B., Hudleston, D., Zak, D. R., and Buyer, J. S. (2007) Interpreting ecological diversity indices applied to terminal restriction length polymorphism data: Insights simulated microbial communities. *Appl Environ Microbiol* **73:** 5276–5283.

Bolch, C. J. S., Orr, P., Jones, G. J., and Blackburn, S. J. (1999) Genetic, morphological and toxicological variation among globally distributed strains of *Nodularia* (Cyanobacteria). *J Phycol* **35:** 339–355.

Boone, D. R., and Castenholz, R. W. (Eds.) (2001). *Bergey's Manual of Systematic Bacteriology*, 2nd edn., Vol. 1. Springer-Verlag, New York, NY, USA.

Bornet, E., and Flahault, C. (1888) Revision des Nostocacées. *Ann Sci Nat Bot* **3:** 323–381, **4:** 343–373, **5:** 51–129, **7:** 177–262.

Boucher, Y., Douady, C. J., Sharma, A. K., Kamekura, M., and Doolittle, W. F. (2004) Intragenomic heterogeneity and intergenomic recombination among haloarchaeal rRNA genes. *J Bacteriol* **186:** 3980–3990.

Bouchet, V., Huot, H., and Goldstein, R. (2008) Molecular genetic basis of ribotyping. *Clinical Microbiol Rev* **21:** 262–273.

Bourelly, P. (1970) *Les algues d'eau douce III*. N Boubeé et cie, Paris, pp.512.

Boutte, C., Grubisic, S., Balthasart, P., and Wilmotte, A. (2006) Testing of primers for the study of cyanobacterial molecular diversity by DGGE. *J Microbiol Methods* **65:** 542–550.

Boyer, S. L., Flechtner, V. R., and Johansen, J. R. (2001) Is the 16S–23S rRNA Internal transcribed spacer region a good tool for use in molecular systematics and population genetics? A case study in cyanobacteria. *Mol Biol Evol* **18:** 1057–1069.

Brochier, C., Philippe, H. and Moreira, D. (2000) The evolutionary history of ribosomal protein RpS14: horizontal gene transfer at the heart of the ribosome. *Trends Genet* **16:** 529–533.

Brown, M. V., Schwalbach, M. S., Hewson, I., and Fuhrman, J. A. (2005) Coupling 16S-ITS rDNA clone libraries and automated ribosomal intergenic spacer analysis to show marine microbial diversity: development and application to a time series. *Environ Microbiol* **7:** 1466–1479.

Buchanan, R. E., and Gibbons, N. E. (Eds.) (1974) *Bergey's Manual of Deterinative Bacteriology*. Williams and Wilkins Co., Baltimore, MD, USA.

Callejas, C., Gill, P. R., Catalán, A. I., Aziz, G., Castro-Sowinski, S., and Batista, S. (2010) Phylotype diversity in a benthic cyanobacterial mat community on King George Island, maritime Antarctica. *World J Microbiol Biotechnol* doi: 10.10007/s11274-010-0578-1

Carpenter, E. J. (1983) Nitrogen fixation by marine *Oscillatoria* (*Trichodesmium*) in the world's oceans. In: Carpenter, E. J., and Capone, D.G. (Eds.) *Nitrogen in the Marine Environment*. Academic Press, New York, NY, USA. pp. 65–103.

Castenholz, R. W. (1992) Species usage, concept, and evolution in the cyanobacteria (blue–green algae). *J Phycol* **28:** 737–745.

Castenholz, R. W. (2001) Phylum BX. Cyanobacteria. In: Boone, D. R., and Castenholz, R. W. (Eds.) *Bergey's Manual of Systematic Bacteriology*, 2nd edn., Vol. 1,: Springer-Verlag, New York, NY, USA. pp. 473–487.

Castenholz, R. W., and Norris, T. B. (2005) Revisionary concepts of species in the cyanobacteria and their applications. *Arch Hydrobiol Algol Stud* **117:** 53–69.

Castenholz, R. W., and Waterbury, J. B. (1989) Group I. Cyanobacteria. In: Krieg, N. R., and Holt ,J. G. (Eds.) *Bergey's Manual of Systematic Bacteriology*, vol. 3: Williams & Wilkins, Baltimore, MD, USA. pp. 1710–1728.

Castiglioni, B., Rizzi, E., Frosini, A., Sivonen, K., Rajaniemi, P., Rantala, A., Mugnai, M.A., Ventura, S., Wilmotte, A., Boutte, C., Grubisic, S., Balthasart, P., Consolandi, O., Bordoni, R., Mezzelani, A., Battaglia, C., and De Bellis, G. (2004) Development of a universal microarray based on the ligation detection reaction and 16SrRNA gene polymorphism to target diversity of cyanobacteria. *Appl Environ Microbiol* **70:** 7161–7172.

Cavalier-Smith, T. (2001) Obcells as proto-organisms: membrane heredity, lithophosphorylation, and the origins of the genetic code, the first cells, and photosynthesis. *J Mol Evol* **53:** 555–595.

Cavalier-Smith, T. (2002) The neomuran origin of archaebacteria, the negibacterial root of the universal tree and bacterial megaclassification. *Int J Syst and Evol Microbiol* **52:** 7–76.

Chen, F., Wang, K., Kan, J., Suzuki, M. T., and Wommack, K. E. (2006) Diverse and unique picocyanobacteria in Chesapeake Bay, revealed by 16S–23S rRNA internal transcribed spacer sequences. *Appl Environ Microbiol* **72:** 2239–2243.

Chen, X., and Widger, W. R. (1993) Physical genome map of the unicellular cyanobacterium *Synechococcus* sp. strain PCC 7002. *J Bacteriol* **175:** 5106–5116.

Chenna, R., Sugawara, H., Koike, T., Lopez, R., Gibson, T. J., Higgins, D. G., and Thompson, J. D. (2003) Multiple sequence alignment with the Clustal series of programs. *Nucleic Acids Res* **31:** 3497–3500.

Chisholm, S. W. (1992) Phytoplankton size. In: Falkowski, P. G. and Woodhead, A. D. (Eds.) *Primary Productivity and Biogeochemical Cycles in the Sea*. Plenum Press, New York, NY, USA. pp. 213–237.

Chisholm, S. W., Frankel, S. L., Goericke, R., Olson, R. J., Palenik, B., Waterbury, J. B., West-Johnsrud, L., and Zettler, E. R. (1992) *Prochlorococcus marinus* nov. gen. nov. sp .: an oxyphototrophic marine prokaryote containing divinyl chlorophyll *a* and *b*. *Arch Microbiol* **157:** 297–300.

Chisholm, S. W., Olson, R. J., Zettler, E. R., Goericke, R., Waterbury, J., and Welschmeyer, N. (1988) A novel free-living prochlorophyte abundant in the oceanic euphotic zone. *Nature* (London) **334:** 340–343.

Clemente, J. C., Satou, K., and Valiente, G. (2006) Phylogenetic reconstruction from non-genomic data. *Bioinformatics* **23:** e110-e115. doi: 10/bioinformatics/bt1307

Cohan, F. M. (2002) What are bacterial species? *Annu Rev Microbiol* **56:** 457–487. doi: 10.1146/annurev.micro.56.012302.160634

Cohan, F. M. (2006) Towards a conceptual and operational union of bacterial systematic, ecology and evolution. *Phil Trans R Soc* B **361:** 1985–1996.

Cohn, F. (1872) Grundzüge einer nuen natürlichen Anordnung der Kryptogamischen Pflanzen. *Jarhb Schles Ges Vaterl KuH* **49:** 83–89.

Cohn, F. (1875) Untersuchungen über Bakterien. II. *Beitr Biol Pflanz* **1:** 141–207.

Cottret, L., Wildridge, D., Vinson, F., Barett, M. P., Charles, H., Sagot, M. F., and Jourdan, F. (2010) Met Explore: a webserver to link metabolic experiments and genome-scale metabolic net works. *Nucleic Acids Res* **38:** W132–W137. doi: 10.1093/nar/gicq312

Crosbie, N. D., Pöckl, M., and Weisse, T. (2003) Dispersal and phylogenetic diversity of nonmarine picocyanobacteria, inferred from 16S rRNA gene and *cpc*-intergenic spacer sequence analysis. *Appl Environ Microbiol* **69:** 5716–5721.

D'Auria, G., Pushker, R., and Rodríguez-Valera, F. (2006) IWoCS: analyzing ribosomal intergenic transcribed spacers configuration and taxonomic relationships. *Bioinformatics* **22:** 527–531.

Dagan, T., Artzy-Randrup, Y., and Martin, W. (2008) Modular networks and cumulative impact of lateral transfer in prokaryote genome evolution. *Proc Natl Acad Sci USA* **105:** 10039–10044.

Danovaro, R., Luna, G. M., Dell'Anno, A., and Pietrangeli, B. (2006) Comparison of two fingerprinting techniques, terminal restriction fragment length polymorphism and automated ribosomal intergenic spacer analysis, for determination of bacterial diversity in aquatic environments. *Appl Environ Microbiol* **72:** 5982–5989.

Darwin, C. (1859) *On the Origin of Species*. Murray, London.

Darwin, C. (1872) *The Origin of Species*. 6th edn. Murray, London.

Desikachary, T. V. (1959) *Cyanophyta*. Indian Council Agricultural Research, New Delhi.

De Vos, P., and Trüper, H. G. (2000) Judicial Commission of the International Committee on Systematic Bacteriology. IXth International (IUMS) Congress of Bacteriology and Applied Microbiology. Minutes of the meetings, 14, 15 and 18 August 1999, Sydney, Australia. *Int J Syst Evol Microbiol* **50:** 2239–2244.

Dobzhansky, T. (1937) *Genetics and the Origin of Species*. Columbia University Press, New York, NY, USA.

Dobzhansky, T. (1951) *Genetics and the Origin of Species*. 2nd edn. Columbia University Press, New York, NY, USA.

Doolittle, R. F., Feng, D. F., Tsang, S., Cho, G., and Little, E. (1996) Determining divergence times of the major kingdoms of living organisms with a protein clock. *Science* **271:** 470–477.

Doolittle, W. F., and Bapteste, E. (2007) Pattern pluralism and the tree of life hypothesis. *Proc Natl Acad Sci USA* **104:** 2043–2049.

Drouet, F. (1981) Revision of the Stigonemataceae with a summary of the classification of the blue-green algae. *Nova Hedwigia* **66**: 1–221.

Drouet, F., and Daily, W. A. (1956) Revision of the coccoid Myxophyceae. *Butler University Botanical Studies* **12:** 1–218.

Dunbar, J., Ticknor, L. O., and Kuske, C. R. (2000) Assessment of microbial diversity in four southwestern United States soils by 16S rRNA gene terminal restriction fragment analysis. *Appl Environ Microbiol* **66:** 2943–2950.

Dunbar, J., Ticknor, L. O., and Kuske, C. R. (2001) Phylogenetic specificity and reproducibility and new method for analysis of terminal restriction fragment profiles of 16S rRNA genes from bacterial communities. *Appl Environ Microbiol* **67:** 190–197.

Egholm, M., Buchardt, O., Christensen, L., Behrens, C., Freler, S. M., Driver, D. A., Berg, R. H., Kim, S. K., Norden, B., and Nielsen, P. E. (1993) PNA hybridizes to complementary oligonucleotides obeying the Watson-Crick hydrogen bonding rules. *Nature* (London) **365:** 266–268.

Eisen, J. A. (1998) Phylogenomics: Improving functional predictions for uncharacterized genes by evolutionary analysis. *Genome Res* **8:** 163–167.

Eisen, J. A., and Fraser, C. M. (2003) Phylogenomics: Intersection of evolution and genomics. *Science* **300:** 1706–1707.

El Fantroussi, S., Urakawa, H., Bernhard, A. E., Kelly, J. J., Noble, P. A., Smidt, H., Yershov, G. M. and Stahl, D. A. (2003) Direct profiling of environmental microbial populations by thermal dissociation analysis of native rRNAs hybridized to oligonucleotide microarrays. *Appl Environ Microbiol* **69:** 2377–2382.

El Herry, S., Nasri, H., and Bouaïcha, N. (2008) Morphological and phylogenetic analysis of colonies of *Microcystis* morphospecies isolated from the Lebna Dam, Tunisia. *Afr J Microbiol Res* **2:** 340–348.
Engebretson, J. J., and Moyer, C. L. (2003) Fidelity of select restriction endonucleases in determining microbial diversity by terminal-restriction fragment length polymorphism. *Appl Environ Microbiol* **69:** 4823–4829.
Ernst, A., Marschall, P., and Postius, C. (1995) Genetic diversity among *Synechococcus* spp. (cyanobacteria) isolated from the pelagial of Lake Constance. *FEMS Microbiol Ecol* **17:** 197–204.
Ernst, A., Becker, S., Wollenzien, U. I. A., and Postius, C. (2003) Ecosystem dependent adaptive radiations of picocyanobacteria inferred from 16S rRNA and ITS-1 sequence analysis. *Microbiology* **149**: 217–228.
Fantroussi, S. E., Urakawa, H., Bernhard, A. E., Kelly, J. J., Noble, P. A., Smidt, H., Yershov, G. M., and Stahl, D. A. (2003) Direct profiling of environmental microbial populations by thermal dissociation analysis of native rRNAs hybridized to oligonucleotide microarrays. *Appl Environ Microbiol* **69:** 2377–2382.
Felsenstein, J. (1989) PHYLIP-Phylogeny Inference Package (version 3.2). *Cladistics* **5:** 164–166.
Fergusson, K. M., and Saint, C. P. (2000) Molecular phylogeny of *Anabaena circinalis* and its identification in environmental samples by PCR. *Appl Environ Microbiol* **66:** 4145–4148.
Ferris, M. J., and Ward, D. M. (1997) Seasonal distributions of dominant 16S rRNA defined populations in a hot spring microbial mat examined by denaturing gradient gel electrophoresis. *Appl Environ Microbiol* **63:** 1375–1381.
Ferris, M. J., Kühl, M., Wieland, A., and Ward, D. M. (2003) Cyanobacterial ecotypes in different optical microenvironments of a 68 °C hotspring mat community revealed by 16S-23S rRNA internal transcribed spacer region variation. *Appl Environ Microbiol* **69:** 2893–2898.
Fisher, R. A. (1930) *Genetical Theory of Natural Selection*. Oxford University Press, Oxford., England, UK.
Floyd, M. M., Tang, J., Kane, M., and Emerson, D. (2005) Captured diversity in a culture collection. Case study of the geographic and habitat distributions of environmental isolates held at the American Type Culture Collection. *Appl Environ Microbiol* **71:** 2813–2823.
Foerstner, K. U., von Mering, C., and Bork, P. (2006) Comparative analysis of environmental sequences: potential and challenges. *Phil Trans R Soc* B **361:** 519–523.
Fraser, C., Alm, E. J., Polz, M. F., Spratt, B. G., and Hanage, W. P. (2008) The bacterial species challenge: making sense of genetic and ecological diversity. *Science* **323:** 741–746.
Fritsch, F. E. (1945) *The Structure and Reproduction of Algae, vol II.* Cambridge University Press, Cambridge, England, UK.
Fox, G. E., Peckman, K. J., and Woese, C. R. (1977) Comparative cataloging of 16S ribosomal ribonucleic acid: molecular approach to prokaryotic systematics. *Int J Syst Bacteriol* **27:** 44–57.
Fox, G. E., Wisotzkey, J. D., and Jurtshuk Jr., P. (1992) How close is close: 16S rRNA sequence identity may not be sufficient to guarantee species identity. *Int J Syst Bacteriol* **42:** 166–170.
Fukunaga, M., and Mifuchi, I. (1989). Unique organization of *Leptospira interrogans* rRNA genes. *J Bacteriol* **171:** 5763–5767.
Fuller, N. J., Marie, M. D., Partensky, F., Vaulot, D., Post, A. F., and Scanlan, D. J. (2003) Clade-specific 16S ribosomal DNA oligonucleotides reveal the predominance of a single marine *Synechococcus* clade throughout a stratified water column in the Red Sea. *Appl Environ Microbiol* **69:** 2430–2443.
Furukawa, K., Noda, N., Tsureda, S., Saito, T., Itayama, T., and Inamori, Y. (2006) Highly sensitive real-time PCR assay for quantification of toxic cyanobacteria based on microcystin synthase A gene. *J Biosci and Bioeng* **102:** 90–96.
Gans, J., Wolinsky, M., and Dunbar, J. (2005) Computational improvements reveal great bacterial diversity and high metal toxicity in soil. *Science* **309:** 1387–1390.
García-Martínez, J., Bescós, I., Rodríguez-Sala, J. J., and Rodríguez- Valera, F. (2001) RISSC: a novel database for ribosomal 16S–23S RNA genes spacer regions. *Nucleic Acids Res* **29:** 178–180.
Garcia-Pichel, F. (2000) Cyanobacteria. In: Lederberg, J. (Ed.) *Encyclopedia of Microbiology*. 2nd edn. Academic Press, San Diego, CA, USA. pp. 907–927.
Garcia-Pichel, F., Nübel, U., and Muyzer, G. (1998) The phylogeny of unicellular, extremely halotolerant cyanobacteria. *Arch Microbiol* **169:** 469–482.
Garcia-Vallvé, S., Romeu, A., and Palau, P. (2000) Horizantal gene transfer in bacterial and archael complete genomes. *Genome Res* **10:** 1719–1725.
Garnier, T., Canard, B., and Cole, S. T. (1991). Cloning, mapping, and molecular characterization of the rRNA operons of *Clostridium perfringens*. *J Bacteriol* **173:** 5431–5438.
Garrity, G. M., and Holt, J. G. (2001) The road map to the manual. In: Boone, D. R., and Castenholz, R. W. (Eds.) *Bergey's Manual of Systematic Bacteriology.* 2nd edn. Springer-Verlag, New York, NY, USA. pp. 119–155.
Geitler, L. (1932) Cyanophyceae. In: Kolkwitz, R. (Ed.) *Rabenhorst's Kryptogamenflora von Deutschland,* O¨ sterreich und der Schweiz, Vol 14. Akademische Verlagsgesellschaft, Leipzig, Germany. pp 1–1196.
Gest, H. (1999) Bacterial classification and taxonomy: "a primer" for the new millennium. *Microbiology Today* **26:** 70–72.
Gevers, D., Cohan, F. M., Lawrence, J. G., Spratt, B. G., Coenye, T., Feil, E. J., Stackebrandt, E., Van de Peer, Y., Vandamme, P., Thompson, F. L., and Swings, J. (2005) Opinion: re-evaluating prokaryotic species. *Nat Rev Microbiol* **3:** 733–739.

Gibbons, N. E., and Murray, R. G. E. (1978) Validation of Cyanobacteriales Stanier in Gibbons and Murray 1978 as a new order of the kingdom Procaryotae Murray 1968, and of the use of neuter plural endings for Photobacteria and Scotobacteria classes nov. Gibbons and Murray 1978. Request for an opinion. *Int J Syst Bacteriol* **28:** 332–333.

Giovannoni, S. J., Turner, S., Olsen, G. J., Barns, S., Lane, D. J., and Pace, N. R. (1988) Evolutionary relationships among cyanobacteria and green chloroplasts. *J Bacteriol* **170:** 3584–3592.

Goericke, R., and Repeta, D. J. (1992) The pigments of *Prochlorococcus marinus*: the presence of divinyl-chlorophyll *a* and *b* in a marine prokaryote. *Limnol Oceanogr* **37:** 425–433.

Gogarten, J. P., and Townsend, J. P. (2005) Horizontal gene transfer, genome innovation and evolution. *Nat Rev Microbiol* 3: 679–687.

Gogarten, J. P., Doolittle, W. F. and Lawrence, J.G. (2002) Prokaryotic evolution in light of gene transfer. *Mol Biol Evol* **19:** 2226–2238.

Gogarten, J. P., Olendzenski, L., Hilario, E., Simon, C., and Holsinger, K. E. (1996) Dating the cenancester of organisms. *Science* **274:** 1750–1751.

Golubic, S. (1967) *Algenvegetation der Felsen, eine oekologische Algenstudie im dinarischen Karstgebiet* (*Algal Vegetation on Cliffs, an Ecological Study of Algae in the Dinaric karstic Region*). "Binnengewasser" **23:** 1-183. Schweizerbart'sche Verlagsbuchhabdlung, Stuttgart, Germany.

Golubic, S., and Hofmann, H. J. (1976) Comparison of Holocene and mid- Precambrian Entophysalidaceae (Cyanophyta) in stromatolitic algal mats: cell division and degradation. *J Paleontol* **50:** 1074–1082.

Golubic, S., Sergeev, V. N., and Knoll, A. H. (1995) Mesoproterozoic *Archaeoellipsoides*: akinetes of heterocystous cyanobacteria. *Lethaia* **28:** 285–298.

Gomont, M. (1892) Monographie des Oscillariées. *Ann Sci Nat Ser Bot* **15:** 263–368. **16:** 91–264.

Goncalves, E. R., and Rosato. Y. B. (2002) Phylogenetic analysis of *Xanthomonas* species based upon 16S–23S rDNA intergenic spacer sequences. *Int J Syst Evol Microbiol* **52:** 355–361.

Gotelli, N. J., and Colwell, R. K. (2001) Quantifying biodiversity: procedures and pitfalls in the measurement and comparison of species richness. *Ecol Lett* **4:** 379–391.

Gould, S. J., and Eldredge, N. (1993) Punctuated equilibrium comes of age. *Nature* (London) **366:** 223–227.

Grimont, F., and Grimont, P. A. (1986) Ribosomal ribonucleic acid gene restriction patterns as potential taxonomic tools. *Ann Inst Pasteur Microbiol* **137B:** 165–175.

Guasp, C., Moore, E. R. B., Lalucat, J., and Bennasar, A. (2000) Utility of internally transcribed 16S-23S rDNA spacer regions for the definition of *Pseudomonas stutzeri* genomovars and other *Pseudomonas* species. *Int J Syst Evol Microbiol* **50:** 1629–1639.

Gugger, M. F., and Hoffmann, L. (2004) Polyphyly of the true branching cyanobacteria (Stigonematales). *Int J Syst Evol Microbiol* **54:** 349–357.

Gugger, M., Lyra, C., Henriksen, P., Couté, A., Humbert, J.-F., and Sivonen, K. (2002) Phylogenetic comparison of the cyanobacterial genera *Anabaena* and *Aphanizomenon*. *Int J Syst Evol Microbiol* **52:** 1–14.

Gupta, R. S. (1998) Protein phylogenies and signature sequences: A reappraisal of evolutionary relationships among Archaebacteria, Eubacteria and Eukaryotes. *Microbiol Mol Biol Rev* **62:** 1435–1401

Gürtler, V., and Stanisich, V. (1996) New approaches to typing and identification of bacteria using the 16S-23S rDNA spacer region. *Microbiology* **142:** 3–16.

Guschin, D., Mobarry, B., Proudnikov, D., Stahl, D., Rittmann, B., and Mirzabekov, A. (1997) Oligonucleotide microchips as genosensors for determinative and environmental studies in microbiology. *Appl Environ Microbiol* **63:** 2397–2402.

Hacker, J., and Kaper, J.B. (2000) Pathogenicity islands and the evolution of microbes. *Annu Rev Microbiol* **54:** 641–679.

Haldane, J. B. S. (1932) *The Causes of Evolution*. Princeton University Press, Princeton, N.J. USA.

Hanage, W. P., Fraser, C., and Spratt, B. G. (2006) Sequences, sequence clusters and bacterial species. *Phil Trans R Soc* B **361:** 1917–1927. doi: 10.1098/rstb.2006.1917

Hao, W., and Golding, G. B. (2004) Patterns of bacterial gene movement. *Mol Biol Evol* **21:** 1294–1307.

Hayes, P. K., and Barker, G. L. (1997) Genetic diversity within Baltic Sea populations of *Nodularia* (Cyanobacteria). *J Phycol* **33:** 919–923.

Henson, B. J., Hesselbrock, S. M., Watson, L. E., and Barnum, S. R. (2004) Molecular phylogeny of the heterocystous cyanobacteria (Subsections IV and V) based on *nifD*. *Int J Syst Evol Microbiol* **54:** 493–497.

Herdman, M., Janvier, J. B., Waterbury, J. B., Rippka, R., Stanier, R. Y., and Mandel, M. (1979). Deoxyribonucleic acid base composition of cyanobacteria. *J Gen Microbiol* **111:** 63–71.

Hoffmann, L., and Castenholz, R. W. (2001) Subsection V. (Formerly Stigonematales Geitler 1925). In: Boone, D. R., and Castenholz, R. W. (Eds.) *Bergey's Manual of Systematic Bacteriology*, 2nd edn., vol. 1, Springer-Verlag, New York, NY, USA. pp. 589–599.

Hoffmann L., Kaštovský J., and Komárek, J. (2005) Proposal of cyanobacterial system, 2004. In: Komárek J., and Anagnostidis K. (2005): Cyanoprokaryota, 2. Teil/2nd Part: Oscillatoriales. In: Büdel B., Krienitz L., Gärtner G., and Schagerl, M. (Eds.) Süsswasserflora von Mitteleuropa 19/2, Elsevier/Spektrum, Heidelberg, Germany. pp. 657–660.

Holben, W. E., Feris, K. P., Kettunen, A., and Apajalahti, J. H. A. (2004) GC Fractionation enhances microbial community diversity assessment and detection of minority populations of bacteria by denaturing gradient gel electrophoresis. *Appl Environ Microbiol* **70:** 2263–2270.

Holt, J. G. (1979). International Committee on Systematic Bacteriology. Judicial Commission. Minutes of the Meeting, 3 September 1978, Munich, West Germany. *Int J Syst Bacteriol* **29**, 267–269.

Holt, J. G., Krieg, N. R., Sneath, P. H. A., Staley, J. T., and Williams, S. T. (1994) Group 11. Oxygenic phototrophic bacteria. In: Holt, J. G. (Ed.) *Bergey's Manual of Determinative Bacteriology*, 9th edn., Williams & Wilkins, Baltimore, MD, USA. pp. 377–425.

Honda, D., Yokota, A., and Sugiyama, J. (1999) Detection of seven major evolutionary lineages in cyanobacteria based on the 16S rRNA gene sequence analysis with new sequences of five marine *Synechococcus* strains. *J Mol Evol* **48:** 723–739.

Horz, H., Yimga, M. T., and Liesack, W. (2001) Detection of methanotroph diversity on roots of submerged rice plants by molecular retrieval of *pmoA*, *mmoX*, *mxaF*, and 16S rRNA and ribosomal DNA, including *pmoA*-based terminal restriction fragment length polymorphism profiling. *Appl Environ Microbiol* **67:** 4177–4185.

Hull, D. L. (1997) *Species: the units of biodiversity. The ideal species concept—and why we can't get it*. Chapman and Hall, London, UK.

Hulton, C. S. J., Higgins, C. F., and Sharp, M. P. (1991). ERIC sequences: a novel family of repetitive elements in the genomes of *Escherichia coli*, *Salmonella typhimurium* and other enterobacteria. *Mol Microbiol* **5:** 825–834.

Huys, G., Coopman, R., Janssen, P., and Kersters, K. (1996) High resolution genotypic analysis of the genus *Aeromonas* by AFLP fingerprinting. *Intl J Syst Bacteriol* **46:** 572–580.

Ishida, T., Yokota, A., and Sugiyama, J. (1997) Phylogenetic relationships of filamentous cyanobacterial taxa inferred from 16S rRNA sequence divergence. *J Gen Appl Microbiolol* **43:** 237–241.

Iteman, I., Rippka, R., Tandeau de Marsac, N., and Herdman, M. (2000) Comparison of conserved structural and regulatory domains within divergent 16S rRNA–23S rRNA spacer sequences of cyanobacteria. *Microbiology* **146:** 1275–1286.

Iteman, I., Rippka, R., Tandeau de Marsac, N., and Herdman, M. (2002) rDNA analyses of planktonic heterocystous cyanobacteria including members of the genera *Anabaenopsis* and *Cyanospira*. *Microbiology* **148:** 481–496.

Iyer, L.M., Koonin, E.V. and Aravind, L. (2004) Evolution of bacterial RNA polymerase: implications for large-scale bacterial phylogeny, domain accretion, and horizontal gene transfer. *Gene* **335:** 73–88.

Jain, R., Rivera, M.C., and Lake, J.A. (1999) Horizontal gene transfer among genomes: The complexity hypothesis. *Proc Natl Acad Sci USA* **96:** 3801–3806.

Janse, I., Meima, M., Kardinaal, W. E. A., and Zwart, G. (2003) High-resolution differentiation of cyanobacteria by using rRNA-internal transcribed spacer denaturing gradient gel electrophoresis. *Appl Environ Microbiol* **69:** 6634–6643.

Janson, S., Bergman, B., Carpenter, E. J., Giovannoni, S., and Vergin, K. 1999 Genetic analysis of natural populations of the marine diazotrophic cyanobacterium *Trichodesmium*. *FEMS Microbiol Ecol* **30:** 57–65.

Janson, S., Siddiqui, P. J. A., Walsby, A. E., Romans, K. M., Carpenter, E. J., and Bergman, B. (1995) Cytomorphological characterization of the planktonic diazotrophic cyanobacteria *Trichodesmium* spp. from the Indian Ocean and Caribbean and Sargasso Seas. *J Phycol* **31:** 463–477.

Janssen, P., Coopman, R., Huys, G., Swings, J., Bleeker, H., Vos, P., Zabeau, M., and Kersters, K. (1996) Evaluation of the DNA fingerprinting methods: AFLP as a new tool in bacterial taxonomy. *Microbiology* **142:** 1881–1893.

Jaspers, E., and Overmann, K. (2004) Ecological significance of microdiversity: identical 16S rRNA gene sequences can be found in bacteria with highly divergent genomes and ecophysiologies. *Appl Environ Microbiol* **70:** 4831–4839.

Jungblut, A-D., Hawes, I., Mountfort, D., Hitzfeld, B., Dietrich, D. R., Burns, B. P., and Neilan, B. A. (2005) Diversity with in cyanobacterial mat communities in variable salinity meltwater ponds of McMurdo Ice Shelf, Antarctica. *Environ Microbiol* **7:** 519–529. doi: 10.1111/j.1462-2920.2005.00717.x

Kaneko, T., Sato, S., Kotani, H., Tanaka, A., Asamizu, E., Nakamura, Y., Miyajima, N., Hirosawa, M., Sugiura, M., Sasamoto, S., Kimura, T., Hosouchi, T., Matsuno, A., Muraki, A., Nakazaki, N., Naruo, K., Okumura, S., Shimpo, S., Takeuchi, C., Wada, T., Watanabe, A., Yamada, M. Yasuda, M., and Tabata, S. (1996) Sequence analysis of the genome of the unicellular cyanobacterium *Synechocystis* sp. strain PCC 6803. II. Sequence determination of the entire genome and assignment of potential protein-coding regions. *DNA Res* **3:** 109–136.

Kardinaal, W. E. A., Janse, I., Kamst-van Agterveld, M., Meima, M., Snoek, J., Mur, L. R., Huisman, J., Zwart, G., and Visser, P. M. (2007) *Microcystis* genotype succession in relation to microcystin concentrations in freshwater lakes. *Aquat Microb Ecol* **48:** 1–12.

Karlin, S., and Burge, C. (1995) Dinucleotide relative abundance extremes: a genomic signature. *Trends Genet* **11:** 283–290.

Keswani, J., and Whitman, W. B. (2001) Relationship of 16S rRNA sequence similarity to DNA hybridization in prokaryotes. *Int J Syst Evol Microbiol* **51:** 667–678.

Kim, S-G., Rhee, S-K., Ahn, C. Y., Ko, S-R., Choi, G-G., Bae, J-W., Park, Y-H., and Oh, H-M. (2006) Determination of cyanobacterial diversity during algal blooms in Daechung reservoir, Korea, on the basis of *cpcBA* intergenic spacer analysis. *Appl Environ Microbiol* **72:** 3252–3258.

Kimura, M. (1980) A simple method for estimating evolutionary rates of base substitutions through comparative studies of nucleotide sequences. *J Mol Evol* **16:** 111–120.

Kimura, M. (1983) The neutral theory of molecular evolution. In: M. Nei and R. K. Koehn (Eds.), *Evolution of Genes and Proteins*. Sinauer Associates Inc. Publishers, Sunderland, MA, USA. pp. 208–233.

King, J. L., and Jukes, T. H. (1969) Non-Darwinian evolution. *Science* **164:** 788–798.

Kitts, C. L. (2001) Terminal-restriction fragment patterns: a tool for comparing microbial communities and assessing community dynamics. *Curr Issues Intest Microbiol* **2:** 17–25.

Koizumi, Y., Kelly, J. J., Nakagawa, T., Urakawa, H., El-Fantroussi, S., Al-Muzaini, S., Fukui, M., Urushigawa,Y., and Stahl, D.A. (2002) Parallel characterization of anaerobic toluene- and ethylbenzene degrading microbial consortia by PCR-denaturing gradient gel electrophoresis, RNA-DNA membrane hybridization, and DNA microarray technology. *Appl Environ Microbiol* **68:** 3215–3225.

Komárek, J. (2003) Problem of the taxonomic category "species" in cyanobacteria. *Arch Hydrobiol Algol Stud* **109:** 281–297.

Komárek, J. (2005) The modern classification of Cyanoprokaryotes (cyanobacteria). *Oceanological and Hydrobiological Studies* Vol XXXIV, Supplement 3, pp. 5–17.

Komárek, J., and Anagnostidis, K. (1986) Modern approach to the classification system of Cyanophytes. 2: Chroococcales. *Arch Hydrobiol Algol Stud* **43:** 157–226.

Komárek, J., and Anagnostidis, K. (1989). Modern approach to the classification system of Cyanophytes 4. Nostocales. *Arch Hydrobiol Suppl* **82:** 247–345

Komárek, J., and Kovacik, L. (1989) Trichome structure of four *Aphanizomenon* taxa (Cyanophyceae) from Czechoslovakia, with notes on the taxonomy and delimitation of the genus. *Plant Syst Evol* **164:** 47–64.

Komárek, J., and Komárkova, J. (2006) Diversity of *Aphanizomenon*-like cyanobacteria. *Czech Phycology, Olomouc* **6:** 1–32.

Konstantinidis, K. T., and Tiedje, J. M. (2005) Genomic insights that advance the species definition for prokaryotes. *Proc Natl Acad Sci USA* **102:** 2567–2572. doi: 10.1073/pnas. 0409727102

Konstantinidis, K. T., Ramette, A., and Tiedje, J. M. (2006) The bacterial species definition in the genomic era. *Phil Trans R Soc* B **361:** 1929–1940. doi: 1098/rstb.2006.1920.

Konstantinidis, K. T., and Tiedje, J. M. (2007) Prokaryotic taxonomy and phylogeny in the genomic era: advancements and challenges ahead. *Curr Opin Microbiol* **10:** 504–509.

Koonin, E. V. (2009) Darwinian evolution in the light of genomics. *Nucleic Acids Res* **37:** 1011–1034.

Kumari, N., Srivastava, A. K., Bhargava, P., and Rai, L. C. (2009) Molecular approaches towards assessment of cyanobacterial diversity. *Afr J Biotechnol* **8:** 4284–4298. http: //academicjournals.org/AJB

Kurmayer, R., Dittmann, E., Fastner, J., and Chorus, I. E. (2002) Diversity of microcystin genes within a population of the toxic cyanobacterium *Microcystis* spp. in Lake Wannsee (Berlin, Germany). *Microb Ecol* **43:** 107–118.

Kwon, S. W., Park, J.-Y., Kim, J.-S., Kang, J.-W., Cho, Y.-H., Lim, C.-K., Parker, M. A., and Lee, G.-B.. (2005) Phylogenetic analysis of the genera *Bradyrhizobium*, *Mesorhizobium*, *Rhizobium*, and *Sinorhizobium* on the basis of 16S rRNA gene and internally transcribed spacer region sequences. *Int J Syst Evol Microbiol* **55** : 263–270.

Laamanen, M. J., Forsström, L., and Sivonen, K. (2002) Diversity of *Aphanizomenon flos-aquae* (cyanobacterium) populations along a Baltic Sea salinity gradient. *Appl Environ Microbiol* **68:** 5296–5303.

Lachance, M. (1981) Genetic relatedness of heterocystous cyanobacteria by deoxyribonucleic acid-deoxyribonucleic acid reassociation. *Int J Syst Evol Microbiol* **31:** 139–147.

Laguerre, G., Allard, M., Revoy, F., and Amarger, N. (1994) Rapid identification of rhizobia by restriction fragment length polymorphism analysis of PCR-amplified 16S rRNA genes. *Appl Environ Microbiol* **60:** 56–63.

Laloui, W., Palinska, K.A., Rippka, R., Partensky, F., Tandeau de Marsac, N., Herdman, M., and Iteman, I. (2002) Genotyping of axenic and non-axenic isolates of the genus *Prochlorococcus* and the OMF-'*Synechococcus*' clade by size, sequence analysis or RFLP of the internal transcribed spacer of the ribosomal operon. *Microbiology* **148:** 453–465.

Lane, D.J., Pace, B., Olsen, G. J., Stahl, D. A., Sogin, M. L. and Pace, N. R. (1985) Rapid determination of 16S ribosomal RNA sequences for phylogenetic analyses. *Proc Natl Acad Sci USA* **82:** 6955–6959.

Larkin, M. A., Blackshields, G., Brown, N. P., Chenna, R., McGettigan, P. A., McWilliam, H., Valentin, F., Wallace, I. M., Wilm, A., Lopez, R., Thompson, J. D., Gibson, T. J. and Higgins, D. G. (2007) ClustalW and ClustalX version 2.0. *Bioinformatics* **23:** 2947–2948. doi.10: 1093/bioinformatics/blm/04

Lavin, P., Gómez, P., González, B., and Ulloa, O. (2008) Diversity of the marine picocyanobacteria *Prochlorococcus* and *Synechococcus* assessed by terminal restriction fragment length polymorphisms of 16S-23S rRNA internal transcribed spacer sequences. *Revista Chilena de Historia Natural* **81:** 515–531.

Lawrence, J. G., and Ochman, H. C. (1998) Molecular archaeology of *Escherichia coli* genome. *Proc Natl Acad Sci USA* **95:** 9413–9417.

Lebaron, P., Catala, P., Fajon, C., Joux, F., Baudart, J., and Bernard, L. (1997) A new sensitive, whole cell hybridization technique for detection of bacteria involving a biotinylated oligonucleotide probe targeting rRNA and tyramide signal amplification. *Appl Environ Microbiol* **63:** 3274–3278.

Lehtimäki, J., Lyra, C., Suomalainen, S., Sundman, P., Rouhiainen, L., Paulin, L., Salkinoja-Salonen, M., and Sivonen, K. (2000) Characterization of *Nodularia* strains, cyanobacteria from brackish waters, by genotypic and phenotypic methods. *Int J Syst Evol Microbiol* **50:** 1043–1053.

Li, R., Carmichael, W. W., Liu, Y., and Watanabe, M. M. (2000) Taxonomic revaluation of *Aphanizomenon flos-aquae* NH-5 based on morphology and 16S rRNA gene sequences. *Hydrobiologia* **438:** 99–105.

Ligon, P. J. B., Meyer, K. G., Martin, J. A., and Curtis, S. E. (1991) Nucleotide sequence of a 16S rRNA gene from *Anabaena* sp. strain PCC 7120. *Nucleic Acids Res* **19:** 4553.

Liu, W-T., Marsh, T. L., Cheng, H., and Forney, L. J. (1997) Characterization of microbial diversity by determining terminal restriction fragment length polymorphism of genes encoding 16S rRNA. *Appl Environ Microbiol* **63:** 4516–4522.

Liu, W.T., Mirzabekov, A. D., and Stahl, D. A. (2001) Optimization of an oligonucleotide microchip for microbial identification studies: a non-equilibrium dissociation approach. *Environ Microbiol* **3:** 619–29.

Liu, Z., DeSantis, T. Z., Andersen, G. L., and Knight, R. (2008) Accurate taxonomy assignments from 16S rRNA sequences produced by highly parallel pyrosequencers. *Nucleic Acids Res* **36:** 1–11. doi: 10, 1093/nar/gkn 491.

Loughney, K., Lund, E., and Dahlberg, J. E. (1982) tRNA genes are found between 16s and 23s rRNA genes in *Bacillus subtilis. Nucleic Acids Res* **10:** 1607–1624.

Lovell, C. R., Friez, J., Longshore, J. W., and Bagwell, C. E. (2001) Recovery and phylogenetic analysis of *nifH* sequences from diazotrophic bacteria associated with dead and above ground biomass of *Spartina alterniflora. Appl Environ Microbiol* **67:** 5308–5314.

Loy, A., Lehner, A., Lee, N., Adamczyk, J., Meier, H., Ernst, J., Schleifer, K. H., and Wagner, M. (2002) Oligonucleotide microarray for 16S rRNA gene-based detection of all recognized lineages of sulfate-reducing prokaryotes in the environment. *Appl Environ Microbiol* **68:** 5064–5081.

Lu, W. (1999) Application of molecular biological techniques to the identification of cyanobacteria. Master's thesis, Liverpool John Moores University, Liverpool, England.

Lu, W., Evans, E. H., McColl, S. K., and Saunders, V. A. (1997) Identification of cyanobacteria by polymorphisms of PCR-amplified ribosomal DNA spacer region. *FEMS Microbiol Lett* **153:** 141–149.

Ludwig, W., and Klenk, H. (2001) Overview: a phylogenetic backbone and taxonomic framework for procaryotic systematics, In: Boone, D. R., and and Castenholz, R. W. (Eds.) *Bergey's Manual of Systematic Bacteriology*, vol. 1. Springer-Verlag, New York, NY, USA. pp. 49–65.

Ludwig, W., Strunk, O., Westram, R., Richter, L., Meier, H., Yadhukumar, Buchner, A., Lai, T., Steppi, S., Jobb, G., Förster, W., Brettske, I., Gerber, S., Ginhart, A. W., Gross, O., Grumann, S., Hermann, S., Jost, R., König, A., Liss, T., Lüβmann, R., May, M., Nonhoff, B., Reichel, B., Strehlow, R., Stamatakis, A., Stuckmann, N., Vilbig, A., Lenke, M., Ludwig, T., Bode, A., and Schleifer, K-H. (2004) ARB: a software environment for sequence data. *Nucleic Acid Res* **32:** 1363–1371.

Lyra, C., Hantula, J., Vainio, E., Rapala, J., Rouhiainen, L., and Sivonen, K. (1997) Characterization of cyanobacteria by SDS-PAGE of whole-cell proteins and PCR-RFLP of the 16S rRNA gene. *Arch Microbiol* **168:** 176–184.

Lyra, C., Suomalainen, S., Gugger, M., Vezie, C., Sundman, P., Paulin, L., and Sivonen, K. (2001) Molecular characterization of planktic cyanobacteria of *Anabaena, Aphanizomenon, Microcystis* and *Planktothrix* genera. *Int J Syst Evol Microbiol* **51:** 513–526.

Maidak, B. L., Cole, J. R., Lilburn, T. G., Parker, C. T., Jr., Saxman, P. R., Farris, R. J., Garrity, G. M., Olsen, G. J., Schmidt, T. M. and Tiedje J. M. (2001) The RDP-II (Ribosomal Database Project). *Nucleic Acids Res* **29:** 173–174.

Maiden, M. C., Bygraves, J. A., Feil, E., Morelli, G, Russell, J. E., Urwin, R., Zhang, Q., Zhou, J., Zurth, K., Caugant, D. A., Feavers, I. M., Achtman, M., and Spratt, B. G. (1998) Multilocus sequence typing: a portable approach to the identification of clones within populations of pathogenic microorganisms. *Proc Natl Acad Sci USA* **95:** 3140–3145.

Majer, D., Lewis, B. G., and Mithen, R. (1998) Genetic variation among field isolates of *Pyrenopeziza brassicae. Plant Pathol* **47:** 22–28.

Majer, D., Mithen, R., Lewis, B. G., Vos, P., and Oliver, R. P. (1996) The use of AFLP fingerprinting for the detection of genetic variation in fungi. *Mycol Res* **100:** 1107–1111.

Makarova, K. S., Ponomarev, V. A. and Koonin, E. V. (2001) Two C or not two C: recurrent disruption of Zn-ribbons, gene duplication, lineage-specific gene loss, and horizontal gene transfer in evolution of bacterial ribosomal proteins. *Genome Biol* **2:** RESEARCH 0033.

Malone, T. C., Ducklow, H. W., Peele, E. R., and Pike. S. E. (1991) Picoplankton carbon flux in Chesapeake Bay. *Mar Ecol Prog Ser* **78:** 11–22.

Manichanh, C., Chapple, C. E., Frangeul, L., Gloux, K., Guigo, R., and Dore, J. (2008) A comparison of random sequence reads versus 16S rDNA sequences for estimating the biodiversity of a metagenomic library. *Nucleic Acids Res* **36:** 5180–5188.

Margulies, M., Egholm, M., Altman, W. E., Attiya, S., Bader, J. S., Bemben, L. A., Berka, J., Braverman, M. S., Chen, Y.J., Chen, Z., Dewell, S. B., Du, L., Fierro, J. M., Gomes, X. V., Godwin, B. C., He, W., Helgesen, S., Ho, C. H., Irzyk, G. P., Jando, S. C., Alenquer, M. L. I., Jarvie, T. P., Jirage, K. B., Kim, J-B., Knight, J. R., Lanza, J. R., Leamon, J. H., Lefkowitz, S. M., Lei, M., Li, J., Lohman, K. L., Lu, H., Makhijani, V. B., McDade, K. E., McKenna, M. P., Myers, E. W., Nickerson, E., Nobile, J. R., Plant, R., Puc, B. P., Ronan, M. T., Roth, G. T., Sarkis, G. J., Simons, J. F., Simpson, J. W., Srinivasan, M., Tartaro, K. R., Tomasz, A., Vogt, K. A., Volkmer, G. A., Wang, S. H., Wang, Y., Weiner, M. P., Yu, P., Begley, R. F., and Rothberg, J. M. (2005) Genome sequencing in micro-fabricated high-density picolitre reactors. *Nature* (London) **437:** 376–380. doi: 10.1038/nature03959

Marsh, T. L., Saxman, P., Cole, J., and Tiedje, J. (2000) Terminal restriction fragment length polymorphism analysis Program, a web-based research tool for microbial community analysis. *Appl Environ Microbiol* **66:** 3616–3620.

Martin, W., and Koonin, E.V. (2006) Introns and the origin of nucleus-cytosol compartmentation. *Nature* (London) **440:** 41–45.

Martin, W., and Muller, M. (1998) The hydrogen hypothesis for the first eukaryote. *Nature* (London) **392:** 37–41.

Martin, W., Rujan, T., Richly, E., Hansen, A., Cornelsen, S., Lins, T., Leister, D., Stoebe, B., Hasegawa, M., and Penny, D. (2002) Evolutionary analysis of *Arabidopsis*, cyanobacterial, and chloroplast genomes reveals plastid phylogeny and thousands of cyanobacterial genes in the nucleus. *Proc Natl Acad Sci USA* **99:** 12246–12251.

Martínez-Murcia, A. J., Acinas, S. G., and Rodriguez-Valera, F. (1995) Evaluation of prokaryotic diversity by restrictase digestion of 16S rDNA directly amplified from hypersaline environments. *FEMS Microbiol Ecol* **17:** 247–255.

Martiny, J. G. H., Bohannan, B. J. M., Brown, J. H., Colwell, R. K., Fuhrman, J. A., Green, J. L., Horner-Devine, M. C., Kane, M., Krumins, J. A., Kuske, C. R., Morin, P. J., Naeem, S., Ørveås, L., Reysenbach, A-L., Smith, Val, H., and Staley, J. T. (2006) Microbial biogeography: putting microbes on the map. *Nat Rev Microbiol* **4:** 102–112. doi: 10.1038/nrmicro1341

Massol-Deya, A. A., Odelson, D. A., Hickey, R. F., and Tiedje, J. M. (1995) Bacterial community fingerprinting of amplified 16S and 16-23S ribosomal DNA gene sequences and restriction endonuclease analysis (ARDRA). In: Akkermans, A. D. L., Elsas, J. D. V., and Bruin, F. J. D. (Eds.) *Molecular Microbial Ecology Manual*. Kluwer Academic Publishers, Dordrecht, Netherlands.

Mayr, E. (1997) *Systematics and the Origin of Species from the Viewpoint of a Zoologist*. Columbia University Press. New York, NY, USA.

Mazard, S., Fuller, N. J., Orcutt, K. M., Bridle, O., and Scanlan, D. J. (2004) PCR analysis and distribution of unicellular cyanobacterial diazotrophs in the Arabian Sea. *Appl Environ Microbiol* **70:** 7355–7364.

Mazel, D., Houmard, J., Castets, A. M., and Tandeau de Marsac, N. (1990) Highly repetitive DNA sequences in cyanobacterial genomes. *J Bacteriol* **172:** 2755–2761.

Miller, S. R., Augustine, S., Olson, T. L., Blankenship, R. E., Selker, J., and Wood, A. M. (2005) Discovery of a free-living chlorophyll *d*-producing cyanobacterium with a hybrid proteobacterial/cyanobacterial small-subunit rRNA gene. *Proc Natl Acad Sci USA* **102:** 850–855.

Moeseneder, M. M., Arrieta, J. M., Muyzer, G., Winter, C., and Herndl, G. J. (1999) Optimization of terminal restriction fragment length polymorphism analysis for complex marine bacterioplankton communities and comparison with denaturing gradient gel electrophoresis. *Appl Environ Microbiol* **65:** 3518–3525.

Moeseneder, M. M., Winter, C., and Herndl, G. J. (2001) Horizontal and vertical complexity of attached and free-living bacteria of the Eastern Mediterranean Sea determined by 16S rDNA and 16S rRNA fingerprints. *Limnol Oceanogr* **46:** 95–107.

Moffitt, M. C., Blackburn, S. I., and Neilan, B. A. (2001) rRNA sequences reflect the ecophysiology and define the toxic cyanobacteria of the genus *Nodularia*. *Int J Syst Evol Microbiol* **51:** 505–512.

Moore, L. R., and Chisholm, S. W. (1999) Photophysiology of the marine cyanobacterium *Prochlorococcus:* ecotypic differences among cultured isolates. *Limnol Oceanogr* **44:** 628–638.

Moore, L. R., Rocap, G., and Chisholm, S. W. (1998) Physiology and molecular phylogeny of coexisting *Prochlorococcus* ecotypes. *Nature* (London) **393:** 464–467.

Moreira, D., and Philippe, H. (2000) Molecular phylogeny: pitfalls and progress. *J Theor Biol* **3:** 9–16.

Morgan, E. A., Ikemura, T., and Nomura, M. (1977) Identification of spacer tRNA genes in individual ribosomal RNA transcription units of *Escherichia coli*. *Proc Natl Acad Sci USA* **74:** 2710–2714.

Moyer, C. L., Dobbs, F. C., and Karl, D. M. (1994) Estimation of diversity and community structure through restriction fragment length polymorphism distribution analysis of bacterial 16S rRNA genes from a microbial mat at an active, hydrothermal vent system, Loihi seamount, Hawaii. *Appl Environ Microbiol* **60:** 871–879.

Murray, R. G. E., and Schleifer, K. H. (1994) Taxonomic notes: a proposal for recording the properties of putative taxa of procaryotes. *Int J Syst Bacteriol* **44:** 174–176.

Muyzer, G. (1999) DGGE/TGGE: a method for identifying genes from natural ecosystems. *Curr Opin Microbiol* **2:** 317–322.

Muyzer, G., de Waal, E. C., and Uitterlinden, A. G. (1993) Profiling of complex microbial populations by denaturing gradient gel electrophoresis analysis of polymerase chain reaction-amplified genes coding for 16S rRNA. *Appl Environ Microbiol* **59:** 695–700.

Muyzer, G., Teske, A., and Wirsen, C. O (1995) Phylogenetic relationships of *Thiomicrospira* species and their identification in deep-sea hydrothermal vent samples by denaturing gradient gel electrophoresis of 16S rDNA fragments. *Arch Microbiol* **164:** 165–172.

Myers, J. L., Sekar, R., and Richardson, L. L. (2007) Molecular detection and ecological significance of the cyanobacterial genera *Geitlerinema* and *Leptolyngbya* in Blackband disease of corals. *Appl Environ Microbiol* **73:** 5173–5182.

Nakamura, Y., Itoh, T., Matsuda, H., and Gojobori, T. (2004) Biased biological functions of horizontally transferred genes in prokaryotic genomes. *Nat Genet* **36:** 760–766.

Nakatsu, C. H. (2007) Soil microbial community analysis using denaturing gradient gel electrophoresis. *Soil Sci Soc Am J* **71:** 562–571.

Navarro, E., Simonet, P., Normand, P., and Bardin, R. (1992) Characterization of natural populations of *Nitrobacter* spp. using PCR/RFLP analysis of the ribosomal intergenic spacer. *Arch Microbiol* **157:** 107–115.

Nei, M. (1983) Genetic polymorphism and the role of mutation in evolution. In: Koehn, P. K., and Nei, M. (Eds.) *Evolution of Genes and Proteins*. Sinauer Associates Inc. Publishers, Sunderland, MA, USA. pp. 165–190.

Nei, M. (1984) Genetic polymorphism and neomutationism. In: Mani, G. S. (Ed.) *Evolutionary Dynamics of Genetic Diversity*. Springer-Verlag, New York, NY, USA. pp. 214–241.

Nei, M. (1987) *Molecular Evolutionary Genetics*. Columbia University Press, New York., NY, USA

Nei, M. (2005) Selectionism and neutralism in molecular evolution. *Mol Biol* **22:** 2318–2342. doi: 10. 1093/molbev/msi242

Nei, M., and Kumar, S. (2000) *Molecular Evolution and Phylogenetics*. Oxford University Press, New York, NY, USA. pp. 333

Nei, M., and Rooney, A. P. (2005) Concerted and birth-and-death evolution of multigene families. *Annu Rev Genet* **39:** 121–152.

Neilan, B. A., Jacobs, D., and Goodman, A. E. (1995) Genetic diversity and phylogeny of toxic cyanobacteria determined by DNA polymorphisms within the phycocyanin locus. *Appl Environ Microbiol* **61:** 3875–3883.

Neilan, B. A., Jacobs, D., Deldot, T., Blackall, L. L., Hawkins, P. R., Cox, P. T., and Goodman, A. E. (1997) Ribosomal-RNA sequences and evolutionary relationships among toxic and nontoxic cyanobacteria of the genus *Microcystis*. *Int J Syst Bacteriol* **47:** 693–697.

Nelissen, B., De Baere, R., Wilmotte, A., and De Wachter, R. (1996) Phylogenetic relationships of nonaxenic filamentous cyanobacterial strains based on 16S rRNA sequence analysis. *J Mol Evol* **42:** 194–200.

Nelissen, B., Wilmotte, A., Neefs, J.-M., and De Wachter, R. (1994) Phylogenetic relationships among filamentous helical cyanobacteria investigated on the basis of 16S ribosomal RNA gene sequence analysis. *Syst Appl Microbiol* **17:** 206–210.

Nichols, J. M., Foulds, I. J., Crouch, D. H., and Carr, N. G. (1982) The diversity of cyanobacterial genomes with respect to ribosomal RNA cistrons. *J Gen Microbiol* **128:** 2739–2746.

Nishihara, H., Miwa, H., Watanabe, M., Nagashima, M., Yagi, O., Takamura, Y. (1997) Random amplified polymorphic DNA (RAPD) analysis for discriminating genotypes of *Microcystis* cyanobacteria. *Biosci Biotechnol Biochem* **61:** 1067–1072.

Nishizawa, T., Asayama, M., Fujii, K., Harada, K., and Shirai, M. (1999) Genetic analysis of the peptide synthetase genes for a cyclic heptapeptide microcystin in *Microcystis* spp. *J Biochem* (Tokyo) **126:** 520–529.

Normand, P., Ponsonnet, C., Nesme, X., Neyra, M., and Simonet, P. (1996) ITS analysis of prokaryotes. In: Akkermans, A. D. L., van Elsas, J. D., and de Bruijn, F. J. (Eds.) *Molecular Microbial Ecology Manual*, vol. 3.4.5. Kluwer Academic Publishers, Dordrecht, Netherlands. p. 1–12.

Nübel, U., De Baere, R., Wilmotte, A., and De Wachter, R. (1996) Phylogenetic relationships of nonaxenic filamentous cyanobacterial strains based on the 16S rRNA sequence analysis. *J Mol Evol* **42:** 194–200.

Nübel, U., Garcia-Pichel F. , and Muyzer, G. (1997) PCR primers to amplify 16S rRNA genes from cyanobacteria. *Appl Environ Microbiol* **63:** 3327–3332.

Oberholster, P. J., Botha, A-M., Muller, K., and Cloete, T. E. (2005) Assessment of the genetic diversity of geographically unrelated *Microcystis aeruginosa* strains using amplified fragment length polymorphisms (AFLPs). *Afr J Biotechnol* **4:** 389–399.

Ochman, H., and Moran, N. A. (2001) Genes lost and genes found: evolution of bacterial pathogenesis and symbiosis. *Science* **292:** 1096–1099.

Ochman, H., Lawrence, J. G., and Groisman, E. A. (2000) Lateral gene transfer and the nature of bacterial innovation. *Nature* (London) **405:** 299–304.

Ojaimi, C., Davidson, B. E., Saint Girons, I., and Old, I. G. (1994) Conservation of gene arrangement and an unusual organization of rRNA genes in the linear chromosomes of the Lyme disease spirchaetes *Borrelia burgdorferi, B. garinii,* and *B. afzelii*. *Microbiology* **140:** 2931–2940.

Olsen, G. J., Matsuda, H., Hagstrom, R. and Overbeek, R. (1994) FastDNAml: a tool for construction of phylogenetic trees of DNA sequences using maximum likelihood. *Comput Appl Biosci* **10:** 41–48.

Orcutt, K. M., Rasmussen, U., Webb, E. A., Waterbury, J. B., Gundersen, K., and Bergman, B. (2002) Characterization of *Trichodesmium* spp. by genetic techniques. *Appl Environ Microbiol* **68:** 2236–2245.

Oren, A. (2004) A proposal for further integration of the cyanobacteria under the bacteriological code. *Int J Sys Evol Microbiol* **54:** 1895–1902.
Oren, A. (2011) Cyanobacterial systematic and nomenclature as featured in the *International Bulletin of Bacteriological Nomenclature and Taxonomy/International Journal of Systematic Bacteriology/International Journal of Systematic and Evolutionary Microbiology. Int J Syst Evol Microbiol* **61:** 10–15.
Oren, A., and Tindall, B. J. (2005) Nomenclature of the cyanophyta/cyanobacteria/cyanoprokaryotes under the international code of nomenclature of prokaryotes. *Arch Hydrobiol Algol Stud* **117:** 39–52.
Osborn, A. M., Moore, E. R. B., and Timmis, K. N. (2000) An evaluation of terminal-restriction fragment length polymorphism (T-RFLP) analysis for the study of microbial structure and dynamics. *Environ Microbiol* **2:** 39–50.
Otsuka, S., Suda, S., Li, R., Matsumoto, S., and Watanabe, M. M. (2000) Morphological variability of colonies of *Microcystis* morphospecies in culture. *J Gen Appl Microbiol* **46:** 39–50.
Otsuka, S., Suda, S., Li, R., Watanabe, M., Oyaizu, H., Matsumoto, S., and Watanabe, M. M. (1998) 16S rDNA sequences and phylogenetic analyses of *Microcystis* strains with and without phycoerythrin. *FEMS Microbiol Lett* **164:** 119–124.
Otsuka, S., Suda, S., Li, R., Watanabe, M., Oyaizu, H., Matsumoto, S., and Watanabe, M. M. (1999) Phylogenetic relationships between toxic and non-toxic strains of the genus *Microcystis* based on 16S to 23S internal transcribed spacer sequence. *FEMS Microbiol Lett* **172:** 15–21.
Otsuka, S., Suda, S., Shibata, S., Oyaizu, H., Matsumoto, S., and Watanabe, M. M. (2001) A proposal for the unification of five species of the cyanobacterial genus *Microcystis* Kützing ex Lemmermann 1907 under the Rules of the Bacteriological Code. *Int J Syst Evol Microbiol* **51:** 873–879.
Page, R. W., Jackson, M. J., and Krassay, A. A. (2000) Constraining sequence stratigraphy in north Australian basins: SHRIMP U–Pb zircon geochronology between Mt Isa and McArthur River. *Aus J Earth Sci* **47:** 431–459.
Palenik, B., and Swift, H. (1996) Cyanobacterial evolution and prochlorophyte diversity as seen in DNA-dependent RNA polymerase gene sequences. *J Phycol* **32:** 638–646.
Palinska, K. A., Liesack, W., Rhiel, E. and Krumbein, W. E. (1996) Phenotype variability of identical genotypes: the need for a combined approach in cyanobacterial taxonomy demonstrated on *Merismopedia*-like isolates. *Arch Microbiol* **166:** 224–233.
Palinska, K. A., Thomasius, C. F., Marquardt, J., and Golubic, S. (2006) Phylogenetic evaluation of cyanobacteria preserved as historic herbarium exsiccata. *Int J Syst Evol Microbiol* **56:** 2253–2263.
Palmer, C., Bik, E. M., Eisen, M. B., Eckburg, P. B., Sana, T. R., Wolber, P. K., Relman, D. A., and Brown, P. O. (2006) Rapid quantitative profiling of complex microbial populations. *Nucleic Acids Res* **34:** e5 doi: 10.1093/nar/gnj007
Pedrós-Alió, C. (2006) Marine microbial diversity: can it be determined? *Trends Microbiol* **14:** 257–263.
Pei, A., Nossa, C. W., Chokshi, P., Blaser, M. J., Yang, L., Rosmarin, D. M., and Pei, Z. (2009) Diversity of 23S rRNA genes within individual prokaryotic genomes. *PLoS ONE* **4:** e5437. doi: 10.1371/journal.pone.0005437
Pei, A. Y., Oberdorf, W. E., Nossam C. W., Agarwal, A., Chokshi, P., Gerz, E. A., Lee, P., Yang, L., Poles, M., Brown, S. M., Sotero, S., DeSantis, T., Brodie, E., Nelson, K., and Pei, Z. (2010) Diversity of 16S rRNA genes within individual prokaryotic genomes. *Appl Environ Microbiol* **76:** 3886–3897. doi: 10.1128/AEM.02953-09
Perna, N. T., Plunkett, G. 3rd, Burland, V., Mau, B., Glasner, J. D., Rose, D. J., Mayhew, G. F., Evans, P. S., Gregor, J., Kirkpatrick, H. A. *et al.* (2001) Genome sequence of enterohaemorrhagic *Escherichia coli* O157: H7. *Nature* (London) **409:** 529–533.
Persson, P.-E. (1996) Cyanobacteria and off-flavours. *Phycologia* **35:** 168–171.
Perumal, G. M., Ganesan, V., and Anand, N. (2009) Identification and phylogenetic analysis of filamentous cyanobacteria using random amplified polymorphic DNA (RAPD) finger printing. *Afr J Biotechnol* **8:** 974–978.
Pinevich, A. V. (2008) The paradoxes of biodiversity, phylogeny and taxonomy of cyanobacteria. *Moscow Univ Biol Sci Bull* **63:** 21–24.
Postius, C., and Ernst, A. (1999) Mechanisms of dominance: coexistence of picocyanobacterial genotypes in a freshwater ecosystem. *Arch Microbiol* **172:** 69–75.
Postius, C., Ernst, A., Kenter, U., and Böger, P. (1996) Persistence and genetic diversity among strains of phycoerythrin-rich cyanobacteria from the picoplankton of Lake Constance. *J Plankton Res* **18:** 1159–1166.
Prabina, B. J., Kumar, K., and Kannaiyan, S. (2005) DNA amplification fingerprinting as a tool for checking genetic purity of strains in the cyanobacterial inoculum. *World J Microbiol Biotechnol* **21:** 629–634. doi:10.1007/s11274-004-3566-5
Rajaniemi, P., Hrouzek, P., Kaštovská, K., Willame, R., Rantala, A., Hoffmann, L., Komárek, J., and Sivonen, K. (2005) Phylogenetic and morphological evaluation of genera *Anabaena, Aphanizomenon, Trichormus* and *Nostoc* (Nostocales, Cyanobacteria). *Int J Syst Evol Microbiol* **55:** 11–26.
Ramakrishnan, B., Lueders, T., Conrad, R., and Friedrich, M. (2000) Effect of soil aggregate size on methanogenesis and archaeal community structure in anoxic rice field soil. *FEMS Microbiol Ecol* **32:** 261–270.
Rasmussen, U., and Svenning, M. M. (1998) Fingerprinting of cyanobacteria based on PCR with primers derived from short and long tandemly repeated repetitive sequences. *Appl Environ Microbiol* **64:** 265–272.

Rastogi, R., Wu, M., Das Gupta, I., and Fox, G. E. (2009) Visualization of ribosomal RNA operon copy number distribution. *BMC Microbiol* **9:** 208. doi: 10.1186/1471-2180-9-208
Rippka, R. (1988) Recognition and identification of cyanobacteria. *Methods Enzymol* **167:** 28–67.
Rippka, R., and Cohen-Bazire, G. (1983) The Cyanobacteriales: a legitimate order based on the type strain *Cyanobacterium stanieri*? *Ann Microbiol* (Paris) **134:** 21–36.
Rippka, R., and Herdman, H. (1992) *Pasteur Culture Collection of Cyanobacteria: Catalogue and Taxonomic Handbook. I. Catalogue of Strains*. Institut Pasteur, Paris.
Rippka, R., Deruelles, J., Waterbury, J. B., Herdman, M., and Stanier, R. Y. (1979) Generic assignments, strain histories and properties of pure cultures of cyanobacteria. *J Gen Microbiol* **111:** 1–61.
Robertson, B. R., Tezuka, N., and Watanabe, M. M. (2001) Phylogenetic analyses of *Synechococcus* strains (cyanobacteria) using sequences of 16S rDNA and part of the phycocyanin operon reveal multiple evolutionary lines and reflect phycobilin content. *Int J Syst Evol Microbiol* **51:** 861–871.
Robinson, N. J., Robinson, P. J., Gupta, A., Bleasby, A. J., Whitton, B. A., and Morby, A. P. (1995) Singular over-representation of an octameric palindrome, HIP1, in DNA from many cyanobacteria. *Nucleic Acids Res* **23:** 729–735.
Rocap, G., Moore, L. R., and Chisholm, S. W. (1999) Molecular phylogeny of *Prochlorococcus* ecotypes, In: Charpy, L., and Larkum, A. W. D. (Eds.) *Marine Cyanobacteria*. Bulletin de l'Institut Océanographique, Monaco, Special Issue 19. pp. 107–116.
Rocap, G., Distel, D. L., Waterbury, J. B., and Chisholm, S. W. (2002) Resolution of *Prochlorococcus* and *Synechococcus* ecotypes by using 16S–23S ribosomal DNA internal transcribed spacer sequences. *Appl Environ Microbiol* **68:** 1180–1191.
Rosselló-Móra, R., and Amann, R. (2001) The species concept for prokaryotes. *FEMS Microbiol Rev* **25:** 39–67. doi: 10.1111/j.1574-6976.2001.tb00571.x
Rost, B., and Sander, C. (1996) Bridging the protein sequence-structure gap by structure predictions. *Annu Rev Biophys Biomol Struct* **25:** 113–136.
Rouhiainen, L., Sivonen, K., Buikema, W., and Haselkorn, R. (1995) Characterization of toxin-producing cyanobacteria by using an oligonucleotide probe containing a tandemly repeated heptamer. *J Bacteriol* **177:** 6021–6026.
Rudi, K., Skulberg, O. M., and Jakobsen, K. S. (1998) Evolution of cyanobacteria by exchange of genetic material among phyletically related strains. *J Bacteriol* **180:** 3453–3461.
Rudi, K., Skulberg, O. M., Skulberg, R., and Jakobsen, K. S. (2000) Application of sequence-specific labeled 16S rRNA gene oligonucleotide probes for genetic profiling of cyanobacterial abundance and diversity by array hybridization. *Appl Environ Microbiol* **66:** 4004–4011.
Saitou, W., and Nei, M. (1987) The neighbor-joining method: A new method for reconstructing phylogenetic trees. *Mol Biol Evol* **4:** 406–425.
Saker, M. L., and Neilan, B. A. (2001) Varied diazotrophies, morphologies and toxicities of genetically similar isolates of *Cylindrospermopsis raciborskii* (Nostocales, Cyanophyceae) from Northern Australia. *Appl Environ Microbiol* **67:** 1839–1845.
Sarma, T. A. (1981) Life cycle of *Stichosiphon indica* in cultures. *New Phytol* **88:** 103–110.
Scheldeman, P. B., Baurain, D., Bouhy, R., Scott, M., Muhling, M., Whitton, B. A., Belay, A., and Wilmotte, A. (1999) *Arthrospira* ('*Spirulina*') strains from four continents are resolved into only two clusters, based on amplified ribosomal DNA restriction analysis of the internally transcribed spacer. *FEMS Microbiol Lett* **172:** 213–222.
Schmidt, T. M., DeLong, E. F., and Pace, N. R. (1991) Analysis of marine picoplankton community by 16S rRNA gene cloning and sequencing. *J Bacteriol* **173:** 4371–4378.
Schönhuber, W., Fuchs, B., Juretschko, S., and Amann, R. (1997) Improved sensitivity of whole-cell hybridization by the combination of horseradish peroxidase-labeled oligonucleotides and tyramide signal amplification. *Appl Environ Microbiol* **63:** 3268–3273.
Schönhuber, W., Zarda, B., Eix, S., Rippka, R., Herdman, M., Ludwig, W., and Amann, R. (1999) *In situ* identification of cyanobacteria with Horseradish peroxidase-labeled, rRNA-targeted oligonucleotide probes. *Appl Environ Microbiol* **65:** 1259–1267.
Schopf, J. W. (1968) Microflora of the Bitter Springs Formation, Late Precambrian, Central Australia. *J Paleontol* **42:** 651–688.
Schopf, J. W. (1993) Microfossils of the Early Archaean Apex Chert: New Evidence of the Antiquity of Life. *Science* **260:** 640–646.
Schopf, J. W. and Walter, M. R. (1983) Archean microfossils, new evidence of ancient microbes In: Schopf, J. W. (Ed.) *Earth's Earliest Biosphere: its Origin and Evolution*. Princeton University Press, Princeton, NJ, USA. pp 214–239.
Schopf, J. W., Kudryavtsev, A. B., Agresti, D. G., Wdowiak, T. J., and Czaja, A. D. (2002) Laser-Raman imagery of Earth's earliest fossils. *Nature* (London) **416:** 73–76
Sen, S., and Nilsson, L. (1998) Molecular dynamics of duplex systems involving PNA: structural and dynamical consequences of the nucleic acid backbone. *J Am Chem Soc* **120:** 619–631.
Sergeev, V. N. (2009) The distribution of microfossil assemblages in Proterozoic rocks. *Precambrian Res* **173:** 212–222.

Seo, P.-S., and Yokota, A. (2003) The phylogenetic relationships of cyanobacteria inferred from 16S rRNA, *gyrB*, *rpoC1* and *rpoD1* gene sequences. *J Gen Appl Microbiol* **49:** 191–203.

Sessitsch, A., Weilharter, A., Gerzabek, M. H., Kirchmann, H., and Kandeler, E. (2001) Microbial population structures in soil particle size fractions of a long-term fertilizer field experiment. *Appl Environ Microbiol* **67:** 4215–4224.

Sharma, M., and Shukla, Y. (2009) The evolution and distribution of life in the Precambrian eon-Global perspective of the Indian record. *J Biosci* **34:** 765–776. http: / /www.ias,ac.in/jbiosci

Sharples, G. J., and Lloyd, R. G. (1990) A novel repeated sequence located in the intergenic regions of bacterial chromosomes. *Nucleic Acids Res* **18:** 6503–6508.

Sheldon, P. R. (1987) Parallel gradualistic evolution of Ordovician trilobites. *Nature* (London) **330:** 561–563.

Sigler, W.V., Bachofen, R., and Zeger, J. (2003) Molecular characterization of endolithic cyanobacteria inhabiting exposed dolomite in central Switzerland. *Environ Microbiol* **5:** 618–627.

Skulberg O. M., and Skulberg, R. (1985) Planktic species of *Oscillatoria* (cyanophyceae) from Norway. *Arch Hydrobiol [Suppl 71, 1/2] Algol Stud* **38/39:** 157–174.

Small, J., Call, D. R., Brockman, F. J., Straub, T. M., and Chandler, D. P. (2001) Direct detection of 16S rRNA in soil extracts by using oligonucleotide microarrays. *Appl Environ Microbiol* **67:** 4708–4716.

Smith, J. K., Parry, J. D., Day, J. G., and Smith, R. J. (1998) A PCR technique based on HIP1 interspersed repetitive sequence distinguishes cyanobacterial species and strains. *Microbiology* **144:** 2791–2801.

Sompong, U., Anuntalabhochai, S., Cutler, R. W., Castenholz, R. W., and Peerapornpisal, Y. (2008) Morphological and phylogenetic diversity of cyanobacterial populations in six hot springs of Thailand. *ScienceAsia* **34:** 153–162.

Song, J., Lee, S.-C., Kang, J., W., Baek, H.-J., and Suh, J.-W. (2004) Phylogenetic analysis of *Streptomyces* spp. isolated from potato scab lesions in Korea on the basis of 16S rRNA gene and 16S-23S rDNA internally transcribed spacer sequences. *Int J Syst Evol Microbiol* **54:** 203–209.

Song, T., Martensson, L., Eriksson, T., Zheng, W., and Rasmussen, V. (2005) Biodiversity and seasonal variation of the cyanobacterial assemblage in a rice paddy field in Fujian, China. *FEMS Microbiol Ecol* **54:** 131–140.

Sorek R, Zhu, Y., Creevey, C. J., Francino, M. P., Bork, P., and Rubin, E. M. (2007) Genome-wide experimental determination of barriers to horizontal gene transfer. *Science* **318:** 1449–1452.

Soria-Carrasco, V., and Castresana, J. (2008) Estimation of phylogenetic inconsistencies in the three domains of life. *Mol Biol Evol* **25 :** 2319–2329.

Sournia, A. (1968) La cyanophycee *Oscillatoria* (*Trichodesmium*) dans le plancton marin. *Nova Hedwigia* **15:** 1–12.

Srivastava, A. K., and Schlessinger, D. (1990) Mechanism and regulation of bacterial ribosomal RNA processing. *Annu Rev Microbiol* **44:** 105–129.

Srivastava, A. K., Bhargava, P., Kumar, A., Rai, L. C., and Neilan, B. A. (2009) Molecular characterization and the effect of salinity on cyanobacterial diversity in the ricefields of Eastern Uttar Pradesh. *Saline Systems* 5: 4 doi: 10.1186/1746-1448-5-4; http: / /www.salinesystems.org/content/5/1/4.

Srivastava, P. (2005) Vindhyan akinetes: An indicator of Mesoproterozoic biospheric evolution. *Orig Life Evol Biosph* **33:** 175–183.

Stackebrandt, E., and Goebel, B. M. (1994) Taxonomic note: a place for DNA-DNA re-association and 16S rRNA sequence analysis in the present species definition in Bacteriology. *Int J Syst Bacteriol* **44:** 846–849.

Stackebrandt, E., and Ebers J. (2006) Taxonomic parameters revisited: tarnished gold standards. *Microbiol Today* **8:** 6–9.

Staley, J. T. (1999) Bacterial biodiversity: a time for place. *ASM News* **10:** 1–6.

Staley, J. T. (2006) The bacterial species dilemma and the genomic-phylogenetic species concept. *Phil Trans R Soc* B **361:** 1899–1909.

Staley, J. T., and Gosink, J. J. (1999) Poles apart: biodiversity and biogeography of polar sea ice bacteria. *Annu Rev Microbiol* **53:** 189–215. doi: 10.1146/annurev.micro.53. 1.189

Stanier, R.Y., Sistrom, W. R., Hansen, T. A., Whitton, B. A., Castenholz, R. W., Pfennig, N., Gorlenko, V. N., Kondratieva, E. N., Eimhjellen, K. E., Whittenbury, R., Gherna, R. L., and Trüper, H. G. (1978) Proposal to place nomenclature of the cyanobacteria (blue-green algae) under the rules of the International Code of Nomenclature of Bacteria. *Int J Syst Bacteriol* **28:** 335–336.

Stoesser, G., Baker, W., van den Broek, A., Camon, E., Garcia-Pastor, M., Kanz, C., Kulikova, T., Leinonen, R., Lin, Q., Lombard, V., Lopez, R., Parkinson, H. *et al.* (2002) The EMBL nucleotide sequence database. *Nucleic Acids Res* **30:** 21–26.

Stulp, B. K. (1983) Morphological and molecular approaches to the taxonomy of the genus *Anabaena* (Cyanophyceae, Cyanobacteria). Drukkerij van Denderen B. V., Groningen, Germany.

Suda, S., Watanabe, M. M., Otsuka, S., Mahakahant, A., Yongmanitchai, W., Nopartnaraporn, N., Liu, Y., and Day, J. G. (2002) Taxonomic revision of waterbloom forming species of Oscillatorioid cyanobacteria. *Int J Syst Evol Microbiol* **52:** 1577–1595.

Swofford, D. L., Olsen, G. J., Waddell, P. J., and Hillis, D. M. (1996) Phylogenetic inference.In: Hillis, D. M., Moritz, C., and Mable, B. K. (Eds.) *Molecular Systematics*. Sinauer, Sunderland, Massachussetts, USA. pp 407–509.
Tanabe, Y., Kasai, F., and Watanabe, M. M. (2007) Multilocus sequence typing (MLST) reveals high genetic diversity and clonal population structure of the toxic cyanobacterium *Microcystis aeruginosa*. *Microbiology* **153:** 3695–3703.
Taton, A., Grubisic, S., Brambilla, E., De Wit, R., and Wilmotte, A. (2003) Cyanobacterial diversity in natural and artificial microbial mats of Lake Fryxell (McMurdo Dry Valleys, Antarctica): a morphological and molecular approach. *Appl Environ Microbiol* **69:** 5157–5169.
Tettelin, H., Masignani, V., Cieslewicz, M. J., Donati, C., Medini, D., Ward, N. L., Angiuoli, S. V., Crabtree, J., Jones, A. L., Durkin, A. S., DeBoy, R. T., Davidsen, T. M., Mora, M., Scarselli, M., Ros, I. M., Peterson, J. D., Hauser, C. R., Sundaram, J. P., Nelson, W. C., Madupu, R., Brinkae, L. M., Dodson, R. J., Rosovitz, M. J., Sullivan, S. A., Daugherty, S. C., Haft, D. H., Selengut, J., Gwinn, G. L., Zhou, L., Zafar, N., Khouri, H., Radune, D., Dimitrov, G., Watkins, K., O'Connor, K. J. B., Smith, S., Utterback, T. R., White, O., Rubens, C. E., Grandi, G., Madoff, L. C., Kasper, D. L., Telford, J. L., Wessels, M. R., Rappuoli, R., and Fraser, C. M. (2005) Genome analysis of multiple pathogenic isolates of *Streptococcus agalactiae*: implications for the microbial 'pangenome'. *Proc Natl Acad Sci USA* **102:** 13950–13955.
Thacker, R. W., and Paul, V. J. (2004) Morphological, chemical and genetic diversity of tropical marine cyanobacteria *Lyngbya* spp. and *Symploca* spp. (Oscillatoriales) *Appl Environ Microbiol* **70:** 3305–3312.
Thompson, J. D., Higgins, D. G. and Gibson, D. J. (1994) CLUSTAL W: improving the sensitivity of progressive multiple sequence alignment. *Comput Appl Biosci* **8:** 189–191.
Thuret, G. (1875) Essay de classification des Nostochinées. *Ann Sci Nat* **6:** 372–382.
Tiedje, J. M., Asuming-Brempong, S., Nusslein, K., Marsh, T. L., and Flynn, S. J. (1999) Opening the black box of soil microbial diversity. *Appl Soil Ecol* **13:** 109–122.
Tillett, D., Parker, D. L., and Neilan, B. A. (2001) Detection of toxigenicity by a probe for the microcystin synthetase a gene (*mcyA*) of the cyanobacterial genus *Microcystis*: comparison of toxicities with 16S rRNA and phycocyanin operon (phycocyanin intergenic spacer) phylogenies. *Appl Environ Microbiol* **67:** 2810–2818.
Tindall, B. J., Rosselló-Móra, R., Busse, H.-J., Ludwig, W., and Kämpfer, P. (2010). Notes on the characterization of prokaryote strains for taxonomic purposes. *Int J Syst Evol Microbiol* **60:** 249–266.
Tomioka, N., and Sugiura, M. (1984) Nucleotide sequence of the 16S-23S spacer region in the *rrn*A operon from a blue-green alga, *Anacystis nidulans*. *Mol Gen Genet* **193:** 427–430.
Tomioka, N., Shinozaki, K. and Sugiura, M. (1981) Molecular cloning and characterization of ribosomal RNA genes from a blue-green algae, *Anacystis nidulans*. *Mol Gen Genet* **184:** 359–363.
Tomitani, A., Knoll, A. H., Cavanaugh, C. M., and Ohno, T. (2006) The evolutionary diversification of cyanobacteria: Molecular phylogenetic and palaeontological perspectives. *Proc Natl Acad Sci USA* **103:** 5442–5447.
Tringe, S., von Mering, C., Kobayashi, A., Salamov, A. A., Chen, K., Chang, H. W., Podar, M., Short, J. M., Mathur, A. J., Detter, J. C., Bork, P., Hugenholtz, P., and Rubin, E. M. (2005) Comparative metagenomics of microbial communities. *Science* **308:** 554–557.
Turner, S. (1997) Molecular systematics of oxygenic photosynthetic bacteria. *Plant Syst Evol* **11:** 13–52.
Turner, S., Huang, T.-C., and Chaw, S.-M. (2001) Molecular phylogeny of nitrogen fixing unicellular forms. *Bot Bull Acad Sin* **42:** 181–186.
Turner, S., Pryer, K.M., Miao, V.P., and Palmer, J.D. (1999) Investigating deep phylogenetic relationships among cyanobacteria and plastids by small subunit rRNA sequence analysis. *J Eukaryot Microbiol* **46:** 327–338.
Tyson, G. W., Chapman, J., Hugenholtz, P., Allen, E. E., Ram, R. J., Richardson, P. M., Solovyev, V. V., Rubin, E. M., Rokhsar, D. S., and Banfield, J. F. (2004) Community structure and metabolism through reconstruction of microbial genomes from the environment. *Nature* (London) **428:** 37–43.
Urbach, E., Scanlan, D. J., Distel, D. L., Waterbury, J. B., and Chisholm, S. W. (1998) Rapid diversification of marine picophytoplankton with dissimilar light-harvesting structures inferred from sequences of *Prochlorococcus* and *Synechococcus* (cyanobacteria). *J Mol Evol* **46:** 188–201.
Valério, E., Chambel, L., Paulino, S., Fario, N., Pereira, P., and Tenreiro, R. (2009) Molecular identification, typing and traceability of cyanobacteria from freshwater reservoir. *Microbiology* **155:** 642–656.
Van Gijlswijk, R. P. M., Wiegant, J.,Raap, A. K., and Tanke, H. J. (1996) Improved localization of fluorescent tyramides for *in situ* fluorescence hybridization using dextran sulfate and polyvinyl alcohol. *J Histochem Cytochem* **44:** 389–392.
Venter, J. C., Remington, K., Heidelberg, J. F., Halpern, A. L., Rusch, D., Eisen, J. A., Wu, D., Paulsen, I., Nelson, K. E., Nelson, W., Fouts, D. E., Levy, S., Knap, A. H., Lomas, M. W., Nealson, K., White, O., Peterson, J., Hoffman, J., Parsons, R., Baden-Tillson, H., Pfannkoch, C., Rogers, Y.-H., and Smith, H. O. (2004) Environmental genome shotgun sequencing of the Sargasso Sea. *Science* **304:** 66–74.
Vincent, W. F., Bowman, J. P., Rankin, L. M., and McMeekin, T. A. (1999) Phylogenetic diversity of picocyanobacteria in Arctic and Antarctic ecosystems. In: Bell, C. R., Brylinsky, M., and Johnson-Green, P. (Eds.) *Microbial Biosystems: New Frontiers*: Atlantic Canada Society for Microbial Ecology, Halifax, Nova Scotia, Canada.

Vinuesa, P., Rademaker, J. L. W., De Bruijn, F. J., and Werner, D. (1998) Genotypic characterization of *Bradyrhizobium* strains nodulating endemic woody legumes of the Canary Islands by PCR-restriction fragment length polymorphism analysis of genes encoding 16S rRNA (16S rDNA) and 16S–23S rDNA intergenic spacers, repetitive extragenic palindromic PCR genomic fingerprinting, and partial 16S rDNA sequencing. *Appl Environ Microbiol* **64:** 2096–2104.

Vos, P., Hogers, R., Bleeker, M., Reijans, M., Van de Lee, T., Hornes, M., Frijters, A., Pot, J., Peleman, J., Kuiper, M., and Zabaeu, M. (1995) AFLP: a new technique for DNA fingerprinting. *Nucleic Acids Res* **23:** 4405–4414.

Wang, R. F., Beggs, M. L., Robertson L. H. and Cerniglia, C. E. (2002) Design and evaluation of oligonucleotide-microarray method for the detection of human intestinal bacteria in fecal samples. *FEMS Microbiol Lett* **213:** 175–182.

Wang, R. F., Beggs, M. L., Erickson, B. D. and Cerniglia, C. E. (2004) DNA microarray analysis of predominant human intestinal bacteria in fecal samples. *Mol Cell Probes* **18:** 223–234.

Ward, D. M., Ferris, M. J., Nold, S. C., Bateson, M. M., Kopkczynski, E. D., and Ruff-Roberts, A. L. (1994) Species diversity in hot spring microbial mats as revealed by both molecular and enrichment culture approach-relationship between biodiversity and community structure. In: Stal, L. J., and Caumette, P. (Eds.) *Microbial Mats, Structure, Development and Environmental Significance* (NATO ASI Series, G.35), Springer-Verlag, Heidelberg, Germany pp. 33–44.

Watanabe, M. (1996) Isolation, cultivation, and classification of bloom-forming *Microcystis* in Japan. In: Watanabe, M. F., Harada, K., Carmichael, W. W., and Fukui, H. (Eds.) *Toxic Microcystis*. CRC Press, Boca Raton, FL, USA pp. 13–34.

Wayne, L. G., Brenner, D. J., Colwell, R. R., Grimont, P. A. D., Kandler, O. , Krichevsky, M. I., Moore, L. H., Moore, W. E. C., Murray, R.G.E., Stackebrand, E., Starr, M. P., and Trüper, H.G. (1987) Report of the ad- hoc committee on reconciliation of approaches to bacterial systematics. *Int J Syst Bacteriol* **37:** 463–464.

Waterbury, J. B., and Rippka, R. (1989) The order *Chroococcales*. In: Krieg. N. R., and Holt, J. B. (Eds.) *Bergey's Manual of Systematic Bacteriology*, vol. 3. Williams & Wilkins, Baltimore, MD, USA. pp. 1728– 1746.

Weiwen, Z., Tieying, S., Xiaodong B., Birgitta, B., and Ulla, R. (2002) High cyanobacterial diversity in coralloid roots of cycads revealed by PCR fingerprinting. *FEMS Microbiol Ecol* **40:** 215–222.

Welsh, J., and McClelland, M. (1990) Finger printing of genomes using PCR with arbitrary primers. *Nucleic Acids Res* **18:** 7213–7218.

West, N. J., and Adams, D. G. (1997) Phenotypic and genotypic comparison of symbiotic and free-living cyanobacteria from a single field site. *Appl Environ Micriobiol* **63:** 4479–4484.

West, N. J., Schönhuber, W. A., Fuller, N. J., Amann, R. I., Rippka, R., Post, A. F., and Scanlan, D. J. (2001) Closely related *Prochlorococcus* genotypes show remarkably different depth distributions in two oceanic regions as revealed by *in situ* hybridization using 16S rRNA-targeted oligonucleotides. *Microbiology* **147:** 1731–1744.

Whitman, W. B., Coleman, D. C., and Wiebe, W. J. (1998) Prokaryotes: the unseen majority. *Proc Natl Acad Aci USA* **95:** 6578–6583.

Whitaker, R. J. (2006) Allopatric origins of microbial species. *Phil Trans R Soc* B **361:** 1975–1984.

Whitaker, R. J., Grogan, D. W., and Taylor, J. W. (2003) Geographic barriers isolated endemic population of hyperthermophilic archaea. *Science* **301:** 976–978. doi: 10.1126/science.1086909.

Whitton, B. A., and Potts, M. (Eds.) (2000) Introduction to the Cyanobacteria. In: Whitton, B. A., and Potts, M. (Eds.) *The Ecology of Cyanobacteria: Their Diversity in Time and Space*. Kluwer Academic Publishers, Dordrecht, Netherlands. pp.1–11.

Wille, N. (1904) Die Schizophyceen der Plankton Expedition. In: Hensen, V. (Ed.) Ergebnisse der Plankton-Expedition der Humbolt-Stiftung. Lipsius and Tischer, Kiel, Germany.

Willame, R., Boute, C., Grubisic, S., Komárek, J., Hoffmann, L., and Wilmotte, A. (2006) Morphological and molecular characterisation of planktonic cyanobacteria from Belgium and Luxembourg. *J Phycol* **42:** 1312–1332.

Wilmotte, A. (1994) Molecular evolution and taxonomy of the cyanobacteria. In: Bryant, D. A. (Ed.) *The Molecular Biology of Cyanobacteria*. Kluwer Academic Publishers, Dordrecht, Netherlands. pp. 1–25.

Wilmotte, A., and Golubic, S. (1991) Morphological and genetic criteria in the taxonomy of Cyanophyta/Cyanobacteria. *Arch Hydrobiol* **64:** 1–24.

Wilmotte, A., and Herdman, M. (2001) Phylogenetic relationships among the cyanobacteria based on 16S rRNA sequences. In: Garrity, G. M. (Ed.) *Bergey's Manual of Systematic Bacteriology*. Springer-Verlag, New York, NY, USA. pp. 487–493.

Wilmotte, A. M. R., and Stam, W. T. (1984) Genetic relationships among cyanobacterial strains originally designated as `*Anacystis nidulans*' and some other *Synechococcus* strains. *J Gen Microbiol* **130:** 2737–2740.

Wilmotte, A., Neefs, J. M., and De Wachter, R. (1994) Evolutionary affiliation of the marine nitrogen-fixing cyanobacterium *Trichodesmium* sp. strain NIBB 1067, derived by 16S ribosomal RNA sequence analysis. *Microbiology* **140:** 2195–2164.

Wilmotte, A., Turner, S., van der Peer, Y., and Pace, N. R. (1992) Taxonomic study of marine Oscillatorian strains (Cyanobacteria) with narrow trichomes. II. Nucleotide sequence analysis of the 16S ribosomal RNA. *J Phycol* **28:** 828–838.

Wilson, K. M., Schembri, M. A., Baker, P. D., and Saint, C. P. (2000) Molecular characterization of the toxic cyanobacterium *Cylindrospermopsis raciborskii* and design of a species-specific PCR. *Appl Environ Microbiol* **66:** 332–338.

Wilson, L. A., and Sharp, M. P. (2006) Enterobacterial repetitive intergenic consensus (ERIC) sequences in *Escherichia coli*: evolution and implications for ERIC-PCR. *Mol Biol Evol* **23:** 1156–1168.

Woese, C. R. (1987) Bacterial evolution. *Microbiol Rev* **51:** 221–271.
Woese, C. R., Kandler, O., and Wheelis, M. (1990) Towards a natural system of organisms. Proposal for the domains Archaea, Bacteria and Eucarya. *Proc Natl Acad Sci USA* **87:** 4576–4579.
Wolf, Y. I., Rogozin, I. B., Grishin, N. V., and Koonin, E. V. (2002) Genome trees and the trees of life. *Trends Genet* **18:** 472–479.
Worden, A. Z., Chisholm, S. W., and Binder, B. J. (2000) *In situ* hybridization of *Prochlorococcus* and *Synechococcus* (Marine cyanobacteria) spp. with rRNA-targeted peptide nucleic acid probes. *Appl Environ Microbiol* **66:** 284–289.
Wright, S. (1931) Evolution in Mendelian populations. *Genetics* **16:** 97–159.
Wu, Z., Shi, J., and Li, R. (2010) Unraveling molecular diversity and phylogeny of *Aphanizomenon* (Nostocales, Cyanobacteria) strains isolated from China. *J Phycol* **46:** 1048–1058.
Wuyts, J., De Rijk, P., Van de Peer, Y., Winkelmans, T. and De Wachter, R. (2001) The European Large Subunit Ribosomal RNA Database. *Nucleic Acids Res* **29:** 175–177.
Wuyts, J., Van de Peer, Y., Winkelmans, T. and De Wachter, R. (2002) The European database on small subunit ribosomal RNA. *Nucleic Acids Res* **30:** 183–185.
Xu, D., and Cote, J.-C. (2003) Phylogenetic relationships between *Bacillus* species and related genera inferred from comparison of 3'-end 16S rDNA and 5'-end 16S-23S ITS nucleotide sequences. *Int J Syst Evol Microbiol* **53:** 695–704.
Yap, W.H., Zhang, Z., and Wang, Y. (1999) Distinct types of rRNA operons exist in the genome of the actinomycete *Thermomonospora chromogena* and evidence for horizontal transfer of an entire rRNA operon. *J Bacteriol* **181:** 5201–5209.
Yeager, C. M., Kornosky, J. L., Morgan, R. E., Cain, E. C., Garcia-Pichel, F., Housman, D. C., Belnap, J., and Kuske, C. R. (2007) Three distinct clades of cultured heterocystous cyanobacteria constitute the dominant N_2-fixing members of biological soil crusts of the Colorado Plateau, USA. *FEMS Microbiol Ecol* **60:** 85–97.
Zehr, J. P., Limburger, R. J., Ohki, K., and Fujita, Y. (1990) Antiserum to nitrogenase generated from an amplified DNA fragment from natural populations of *Trichodesmium* spp. *Appl Environ Microbiol* **56:** 3527–3531.
Zehr, J. P., Mellon, M. T., and Hiorns, W. D. (1997) Phylogeny of cyanobacterial *nifH* genes: evolutionary implications and potential applications to natural assemblages. *Microbiology* (Reading) **143:** 1443–1450.
Zeigler, D. R. (2003) Gene sequences useful for predicting relatedness of whole genomes in bacteria. *Int J Syst Evol Microbiol* **53:** 1893–1900. doi: 10.1099/ijs.0.02713-0
Zhaxybayeva, O., Lapierre, P., and Gogarten, J. P. (2004) Genome mosaicism and organismal lineages. *Trends Genet* **20:** 254–260.
Zhaxybayeva, O., Gogarten, J. P., Charlebois, R. L., Doolittle, W. F., and Papke, R. T. (2006) Phylogenetic analyses of cyanobacterial genomes: quantification of horizontal transfer events. *Genome Res* **16:** 1099–1108. doi:10.1101/gt.5322306
Zhaxybayeva, O., Doolittle, W. F., Papke, R. T., and Gogarten, J. P. (2009) Intertwined evolutionary histories of marine *Synechococcus* and *Prochlorococcus marinus*. *Genome Biol Evol* **1:** 325–339. doi: 10.1093/gbe/evp032
Zuckerkandl, E., and Pauling, L. (1965) Evolutionary divergence and convergence in proteins. In: Bryson, V., and Vogel, H. J. (Eds.) *Evolving Genes and Proteins*. Academic Press, New York, NY, USA. pp. 97–166.

CHAPTER 2

GENOMES

Microbial genetics has witnessed rapid and very dynamic changes during the past two decades. Starting from genes to clones, a number of individual genes in diverse microbial species have been identified and genetic engineering techniques have enabled plant and animal scientists to produce genetically modified organisms. The emergence of 16S rRNA as a molecular marker for classification and identification of microorganisms ushered in a new phase to establish phylogenetic relationships. Due to a number of inadequacies, pitfalls and unresolved taxonomic puzzles, the choice fell on a group of housekeeping genes. Soon after, the whole genome sequencing projects added a new dimension to our understanding of genomes of microorganisms and genome-based phylogeny came into existence. This is known as 'Phylogenomics'. The first landmark is the publication of complete genome sequence of *Haemophilus influenzae* in 1995 and this event transformed biology into a throughput endeavour (Fleischmann *et al.*, 1995). The current rate of sequencing of prokaryotic genomes is 20 sequenced genomes per every month. By the end of March, 2009 a total of 1775 prokaryotic genomes have been sequenced that are available at National Center for Biotechnology Information (NCBI) genome database (Ahmed, 2009). Thus the study of genomes of microorganisms (genomics in the case of microbial species or metagenomics in the case of whole communities) can be complemented with the study of mRNAs (also known as transcriptome, transcriptomics) or proteins (proteome, proteomics). A study of these three areas requires a large sequencing capability and very powerful bioinformatics tools to organize and study the massive amount of information. Undoubtedly, the massive synergistic efforts put all over the world for human genome sequencing project left behind sequencing capabilities and bioinformatics resources that are readily available by the click of the mouse.

The discovery that greatly influenced the field of microbiology is the cultivation-independent approach for the study of microbial diversity with the help of identifiable environmental DNA sequences. In other words, it is the application of genomic sciences to understand the structure and function of ecosystems. This subject is known as 'metagenomics' and studies in this direction have yielded 1.2 million new genes. With this, a change in focus has taken place from an organism-based study to the environment-based approach (Bertin *et al.*, 2008; Frias-Lopez *et al.*, 2008; Ward *et al.*, 2008; Biers *et al.*, 2009). A correlation of genomic features found at a distinct sampling site with physical, chemical and biotic information led to the identification of organism-specific adaptations. Of all the ecosystems, understanding the structure and function of marine ecosystems attracted the attention of scientists. The importance given to this field can be gauged by the number of projects that focus their attention on microorganisms of medical and biotechnological interest. Nearly 130 marine isolates are being sequenced (Moore Foundation; http://www.moore.org). The sequencing of whole genome of *H. influenzae* was followed by the sequencing of two other bacterial genomes, i.e. *Mycoplasma genitalium* and *Methanococcus jannaschii* sequenced by TIGR (The Institute for Genomic Research, Rockville, Maryland, USA which is now a part of J. Craig Venter Institute, JCVI). Kaneko *et al.* (1996) for the first time sequenced the genome of *Synechocystis* sp. strain PCC 6803 at Kazusa DNA Research Institute, Japan. This constitutes the first fully sequenced genome of a cyanobacterium as well as an autotroph. This was followed by the completion of genome sequence of *Escherichia coli* K-12 (Blattner *et al.*, 1997). Due to the availability of complete genome sequence of *Synechocystis* sp. strain PCC 6803 along with complete set of genes provided a platform to the cyanobacteriologists for furthering research in this direction. A collective effort by Satoshi Tabata (Kazusa), Nori Murata (Okazaki) and Minoru Kanehisa (Kyoto) resulted into a cyanobacterial DNA chip consortium for the analysis of gene expression profiles by using *Synechocystis* DNA microarrays. The consortium with 25 members facilitated collaborative research between molecular biologists and bioinformaticians. The genome entries made by the sequencing teams at the primary repositories like GenBank (produced at NCBI; http://www.ncbi.nlm.nih.gov.Genbank/), European Molecular Biology Laboratory (EMBL; http://www.embl.de/) and DNA Data Bank of Japan (DDBJ; http://www.ddbj.nig.ac.jp/) have to be updated by the sequencing teams themselves. Due to lack of proper maintenance at these sites and as the databases such as Swiss-Prot (http://www.ebi.ac.uk.swissprot/; initially founded at Swiss Institute of Bioinformatics but later European Bioinformatics Institute and Georgetown University became collaborating partners with UniProt/Swiss-Prot manually annotated and UniProtKB/TrEMBL automatically annotated protein databases) and KEGG (Kyoto Encyclopedia of Genes and Genomes, Japan; http://www.genome.jp/kegg/) are unable to keep pace with the rapidly increasing data on annotated sequences, the bioinformatics centre of Kyoto University (the home of KEGG) developed a community database system known as CYORF. The collaboration of scientists at Kazusa, biologists at the DNA chip consortium and Bioinformaticians at Kyoto University resulted in CYORF database release 1.0. Nakamura *et al.* (2000) created an online genome database for *Synechocystis* sp. strain PCC 6803 termed as CyanoBase that provides annotations for each of the 3168 protein-coding genes deduced from the entire nucleotide sequence of the genome. It can be accessed at http://www.kazusa.or.jp:8080/cyano. They have also developed another site known as CyanoMutants as a repository database for the storage and distribution of information about *Synechocystis* sp. strain PCC 6803 and its mutants.

Complete genome sequences give us the opportunity to seek answers to unresolved questions provided powerful comparative methods of analysis are employed. The *E. coli* K-12 genome is 4,639,221 bp long and possesses 4,288 predicted genes. The genome of *Synechocystis* sp. strain PCC 6803 constitutes the second largest set after *E. coli* with 3,168 proteins encoded. Many cyanobacterial proteins have homologues among bacterial and/or chloroplast proteins. By the application of a combination of state-of-the-art computer methods it has been possible to identify homologous genes from the databases and at the same time predict the proteins with probable function. It is the non-homologous portions that constitute the greatest challenge for gene identity and the proteins encoded by these sequences. By January, 2011, 59 cyanobacterial genomes have been either fully or partially sequenced. Amongst these, 39 sequenced genomes are available through scientific publications and 20 are available on the websites of the sequencing Institutes. The 39 fully sequenced cyanobacterial strains fall into various taxonomic groups (Table 1). The unicellular forms comprise *Acaryochloris marina* MBIC11017 (Swingley *et al.*, 2008), *Cyanothece* sp. ATCC 51142 (Welsh *et al.*, 2008),

Table 1: Fully sequenced cyanobacterial members with their genome sizes, habitats and accession numbers.

S.no.	Cyanobacterium	Abbreviation	Genome size (MB)	Habitat	Genome Accession no.
1	*Acaryochloris marina* MBIC 11017	Ama-MBIC11017	6.50	M	embl: CP000828 Swingley *et al.* (2008)
2.	*Anabaena* sp. strain PCC 7120	Ana-PCC 7120	6.41	FW	embl: BA000019 Kaneko *et al.* (2001)
3.	*Anabaena variabilis* ATCC 29413	Ava-ATCC 29413	6.37	S	embl: CP000117
4.	*Arthrospira platensis* NIES-39	Asp-NIES-39	6.80	SL	GenBank/EMBL/DDBJ/APO11615 Fujisawa *et al.* (2010)
5.	*Arthrospira* sp. PCC 8005	Asp-PCC 8005	6.27	SL	DDBJ/EMBL/GenBank ADDH00000000 Janssen *et al.* (2010)
6.	*Cyanothece* sp. ATCC 51142	Cth-ATCC 51142	4.93	M	embl: CP000806 Welsh *et al.* (2008)
7.	*Cyanothece* sp. CCY0110	Cth-CCY0110	5.99	M	gb: 1101676644636–1101676644658
8.	*Crocosphaera watsonii* WH8501	Cwa-WH8501	6.20	M	embl: AADV02000100
9	*Cylinsrospermopsis raciborskii* CS-105	CS-105	3.89	FW	ncbi:ACYA00000000 Stucken *et al.* (2010)
10.	*Gloeobacter violaceus* PCC 7421	Gvi-PCC 7421	4.66	R	embl: BA000045 Nakamura *et al.* (2003)
11.	*Lyngbya aestuari* PCC 8106	Lae-PCC 8106	7.08	M	gb: 1099428180563–1099428180584
12.	*Lyngbya majuscula* 3L	Lma-PCC 7120	~8.5	M	DDBJ/EMBL/GenBank AEPQ00000000 Jones *et al.* (2011)
13.	*Microcoleus chthonoplastes* PCC 7420	Mch-PCC 7420	8.68	M	gb:1103659003780–1103659003836
14.	*Microcystis aeruginosa* NIES-843	Mic-NIES843	5.84	FW	embl: AP009552 Kaneko *et al.* (2007)
15.	*Microcystis aeruginosa* PCC 7806	Mic-PCC 7806	5.17	FW	embl: AM778843–AM778958 Frangeul *et al.* (2008)
16.	*Nostoc azollae* 0708	NoAz0708	5.84	Sy	Ran *et al.* (2010)
17.	*Nodularia spumigena* CCY9414	Nsp-CCY9414	5.34	M	gb:1099428179735–1099428179797
18.	*Nostoc punctiforme* PCC 73102	Npu-PCC 73102	8.20	Sy	Meeks *et al.* (2001)

Table 1: contd....

Table 1: contd....

S.no.	Cyanobacterium	Abbreviation	Genome size (MB)	Habitat	Genome Accession no.
19.	*Oscillatoria* sp. PCC 6506	Osc-PCC 6506	6.70	FW	DDBJ/EMBL/GenBank CACA01000001 to CACA01000377 Méjean *et al.* (2010)
20.	*Prochlorococcus marinus* SS120	Pro-SS120	1.75	M	embl: AE017126 Dufresne *et al.* (2003)
21.	*Prochlorococcus marinus* AS9601	Pro-AS9601	1.67	M	embl: CP000551 Kettler *et al.* (2007)
22.	*Prochlorococcus marinus* MED4	Pro-MED4	1.68	M	embl: BX548174 Rocap *et al.* (2003)
23.	*Prochlorococcus marinus* MIT9211	Pro-MIT9211	1.69	M	embl: AALP01000001 Kettler *et al.* (2007)
24.	*Prochlorococcus marinus* MIT9215	Pro-MIT9215	1.73	M	embl: CP000825 Kettler *et al.* (2007)
25.	*Prochlorococcus marinus* MIT9301	Pro-MIT9301	1.64	M	embl: CP000576 Kettler *et al.* (2007)
26.	*Prochlorococcus marinus* MIT9303	Pro-MIT9303	2.68	M	embl: CP000554 Kettler *et al.* (2007)
27.	*Prochlorococcus marinus* MIT9312	Pro-MIT9312	1.70	M	embl: CP000111 Coleman *et al.* (2006)
28.	*Prochlorococcus marinus* MIT9313	Pro-MIT9313	2.41	M	embl: BX572095 Rocap *et al.* (2003)
29.	*Prochlorococcus marinus* MIT9515	Pro-MIT9515	1.70	M	embl: CP000552 Kettler *et al.* (2007)
30.	*Prochlorococcus marinus* NATL1A	Pro-NATL1A	1.86	M	embl:CP000553 Kettler *et al.* (2007)
31.	*Prochlorococcus marinus* NATL2A	Pro-NATL2A	1.84	M	embl: CP000095
32.	*Raphidiopsis brookii* D9	D9	3.20	FW	ncbi:ACYB00000000 Stucken *et al.* (2010)
33.	*Synechococcus* sp. PCC 7335	Syn-PCC 7335	5.96	M	gb: 1103496006889–1103496006899
34.	*Synechococcus* sp. BL107	Syn-BL107	2.28	M	gb: 1099739244347 Dufresne *et al.* (2008)
35.	*Synechococcus* sp. CC9311	Syn-CC9311	2.61	M	embl: CP000435 Palenik *et al.* (2003)
36.	*Synechococcus* sp. CC9605	Syn-CC9605	3.51	M	embl: CP000110 Dufresne *et al.* (2008)
37.	*Synechococcus* sp. CC9902	Syn-CC9902	2.23	M	embl: CP000097 Dufresne *et al.* (2008)
38.	*Synechococcus elongatus* PCC 6301	Syn-PCC 6301	2.69	FW	embl: AP008231 Sugita *et al.* (2006)
39.	*Synechococcus* sp. PCC 7002	Syn-PCC 7002	3.01	HS	embl: CP000951
40.	*Synechococcus elongatus* PCC 7942	Syn-PCC 7942	2.69	FW	embl: CP000100
41.	*Synechococcus* sp. RCC307	Syn-RCC307	2.22	M	embl: CT978603 Dufresne *et al.* (2008)

Table 1: contd....

Table 1: contd....

S.no.	Cyanobacterium	Abbreviation	Genome size (MB)	Habitat	Genome Accession no.
42.	*Synechococcus* sp. RS9916	Syn-RS9916	2.66	M	gb: 1100013018508 Dufresne *et al.* (2008)
43.	*Synechococcus* sp. RS9917	Syn-RS9917	2.58	M	gb: 1099465004208 Dufresne *et al.* (2008)
44.	*Synechococcus* sp. WH5701	Syn-WH5701	2.86	M	gb: 1099465003749–1099465003864 Dufresne *et al.* (2008)
45.	*Synechococcus* sp. WH7803	Syn-WH7803	2.37	M	embl: CT971583 Dufresne *et al.* (2008)
46.	*Synechococcus* sp. WH7805	Syn-WH7805	2.62	M	gb: 1099646010155–1099646010157 Dufresne *et al.* (2008)
47.	*Synechococcus* sp. WH8102	Syn-WH8102	2.43	M	gb: BX548020 Palenik *et al.* (2003)
48.	*Synechocystis* sp. PCC 6803	Syn-PCC 6803	3.57	FW	embl: BA000022 Kaneko *et al.* (1996)
49.	*Thermosynechococcus elongatus* BP-1	Tel-BP1	2.59	HS	embl: BA000039 Nakamura *et al.* (2002)
50.	*Trichodesmium erythreum* IMS101	Ter-IMS101	7.80	M	embl: CP000393

Note: Genome details of nine strains (Ava-ATCC 29413; Cth-CCY0110; CWa-WH8501; Mch-PCC 7420; Pro-NATL2A; Syn-PCC 7002, Syn-PCC 7942, Syn-PCC 7335 and Ter-IMS101) can be found on the website of JGI and those of two others (Lae-PCC 8106 and Nsp-CCY9414) on the website of JCVI. The rest of the 39 fully sequenced cyanobacterial genomes are described in the text. This Table is based on the list of strains with their genome sizes provided in Frangeul *et al.* (2008) *BMC Genomics* **9**:274; doi:10.1186/1471-2164-9-274]. Abbreviations: FW=Freshwater; HS=Hot Spring; M=Marine; R=Rock; S=Soil; SL=Salt Lakes; Sy=Symbiotic.

Gloeobacter violaceus PCC 7421 (cyanobacterium lacking thylakoids; Nakamura *et al.*, 2003), *Microcystis aeruginosa* NIES-843 (Kaneko *et al.*, 2007), *M. aeruginosoa* PCC 7806 (Franguel *et al.*, 2008), 12 strains of *Prochlorococcus marinus* of which six strains (NATL1A, SS120, MIT 9211, MIT 9303, MIT 9313) are low-light (LL-) and six others (MIT9215, MIT9301, MIT9312, AS9601, MIT9515, MED4) are high-light (HL-) adapted strains (Rocap *et al.*, 2003; Dufresne *et al.*, 2005; Kettler *et al.*, 2007), *Synechococcus elongatus* PCC 6301 (Sugita *et al.*, 2006), and 11 strains of *Synechococcus* (WH8102, CC9311, Palenik *et al.*, 2003, 2006; BL107, CC9605, CC9902, RCC307, RS9916, RS9917, WH5701, WH7803, WH7805; Dufresne *et al.*, 2008), *Synechocystis* sp. strain PCC 6803 (Kaneko *et al.*, 1996), and *Thermosynechococcus elongatus* strain BP-1 (Nakamura *et al.*, 2002). The non-heterocystous filamentous forms, whose genomes are sequenced, belong to *Arthrospira (Spirulina) platensis* NIES-39 (Fujisawa *et al.*, 2010), *Arthrospira* sp. PCC 8005 (Janssen *et al.*, 2010), *Oscillatoria* sp. PCC 6506 (Méjean *et al.*, 2010b), *Raphidiopsis brookii* D9 (a strain that is defective in heterocyst formation; Stucken *et al.*, 2010), and *Lyngbya majuscula* 3L (Jones *et al.*, 2011). The heterocystous filamentous forms, whose genomes have been sequenced, are *Anabaena* PCC 7120 (Kaneko *et al.*, 2001), *Cylindrospermopsis raciborskii* CS-505 (Stucken *et al.*, 2010), *Nostoc azollae* 0708 (Ran *et al.*, 2010) and *Nostoc punctiforme* ATCC 29133 (PCC 73102; Meeks *et al.*, 2001). The draft sequences of 17 cyanobacterial genomes [*Arthrospira maxima* CS-328, *Crocosphaera watsonii* WH8501 (syn. *Synechocystis* sp. WH8501), six strains of *Cyanothece* sp. (strains PCC 7424, PCC 7425, PCC 7822, PCC 8801, PCC 8802, CCY 0110), *P. marinus* NATL2A, *Synechococcus elongatus* PCC 7942, *Synechococcus* sp. PCC 7002 (ATCC 27264; synonym of *Agmenellum quadruplicatum* PR-6); *Synechococcus* sp. PCC 7535, *Synechococcus* sp. strain JA-3-3A'b (also designated as Cyanobacteria bacterium Yellow Stone A-Prime), *Synechococcus* sp. strain JA-2-3 B'a (2-13) (also known as Cyanobacteria bacterium Yellow Stone B-Prime), *Trichodesmium erythraeum* IMS101, *Anabaena variabilis* ATCC 29413 and *Cyanobacterium* UCYN-A] are available on the website of DOE Joint Genome Institute (JGI; http://www.jgi.doe.gov/) supported by US Department of Energy, Office of Science, Walnut Creek, California and the genomes of three other cyanobacteria (*Microcoleus chthonoplastes* PCC 7420, *L. aesturii* PCC 8106 and *Nodularia spumigena* CCY9414) are put on the website of JCVI. The particulars of these 20 genomes will only be mentioned as given in the published literature for comparative purposes only. Of the 59 strains of cyanobacteria, *P. marinus* (12 strains), 13 strains of *Synechococcus* (WH8102, CC9311, BL107, CC 9605, CC9902, RCC307, RS9916, RS9917, WH5701, WH7803, WH7805, PCC 7002 and PCC 7335), *Cyanothece* sp. (CCY0110, ATCC 51142), *C. watsonii* WH8501, *T. erythraeum* IMS101, *Lyngbya majuscula*, *Lyngbya* sp. PCC 8106 and *No. spumigena* CCY9414 are marine in their habitat where as the rest of the members are from other aquatic (freshwater or hot-spring) and terrestrial (soil, rock, or symbiotic) habitats. Unicellular strains *C. watsonii* WH8501, seven strains of *Cyanothece* sp. (CCY0110, ATCC 51142, PCC 7424, PCC 7425, PCC 7822, PCC 8801 and PCC 8802), *Oscillatoria* PCC 6506, three strains of *Synechococcus* [PCC 7335, JA-3-3A'b and JA-2-3B'a (2-13)], *T. erythraeum* IMS101 and the heterocystous filamentous forms listed above fix nitrogen. However, in these three unicellular genera there is a temporal separation of photosynthesis and nitrogen fixation, the former occurring in the day and the latter during the night time. A phylogenetic analysis of 38 sequenced cyanobacterial genomes, based on concatenated 23S and 16S rRNA sequences, revealed well defined unicellular, filamentous non-heterocystous and heterocystous clusters. The smallest genome is of 1.55 Mb (*Synechococcus* sp. strain RS9916) and the largest one (8.68 Mb) is represented by *M. chthonoplastes* PCC 7420 (Fig. 1; Frangeul *et al.*, 2008). Fujisawa *et al.* (2010) divided the sequenced genomes of cyanobacteria into six groups based on unicellular and filamentous habit with or without the capacity to fix nitrogen: (i) unicellular forms (ii) unicellular forms that fix nitrogen, (iii) unicellular or short filamentous, (iv) filamentous types, (v) filamentous, nitrogen-fixing but non-heterocystous forms and (vi) filamentous, nitrogen-fixing and heterocystous members. The third group consists of a single representative *Synechococcus* sp. strain PCC 7002 that exhibits a filamentous tendency and marine in its habitat. The genome sequencing of *A. marina* CCMEE 5410, *Calothrix* sp. SCO1, *Gloeothece* sp. PCC 6909/1, *Prochlorococcus* sp. UH18301 and six *Synechococcus* strains (UW1, UW69, UW90, UW92, CB0101 and CB0205) is in progress. One can access the complete microbial genomes at http://www.ncbi.nlm.nih.gov/genomes/lproks.cgi, the microbial genome sequencing projects in progress at http://www.ncbi.nlm.nih.gov/genomes/lproks.cgi and the eukaryotic genome sequencing projects at http://www.ncbi.nlm.nih.gov/genomes/leuks.cgi.

I. METHODS

Sequencing refers to the recognition of the order in which nucleotides are arranged in a molecule of DNA. A number of sequencing methods have been developed paving the way for automation as a result of which the process is accomplished with ease and greater speed. Maxam-Gilbert sequencing method (developed by A. Maxam and W. Gilbert during 1976-77) and chain termination method discovered by Frederick Sanger and commonly known as Sanger method were the two earliest sequencing methods. The latter method gained popularity due to the requirement of few toxic chemicals and lower amounts of radioactivity. Sanger's method requires a single-stranded DNA template, a DNA primer, a DNA polymerase, radioactively or fluorescently-labelled nucleotides and modified nucleotides known as dideoxynucleotides. The addition of the each one of the four

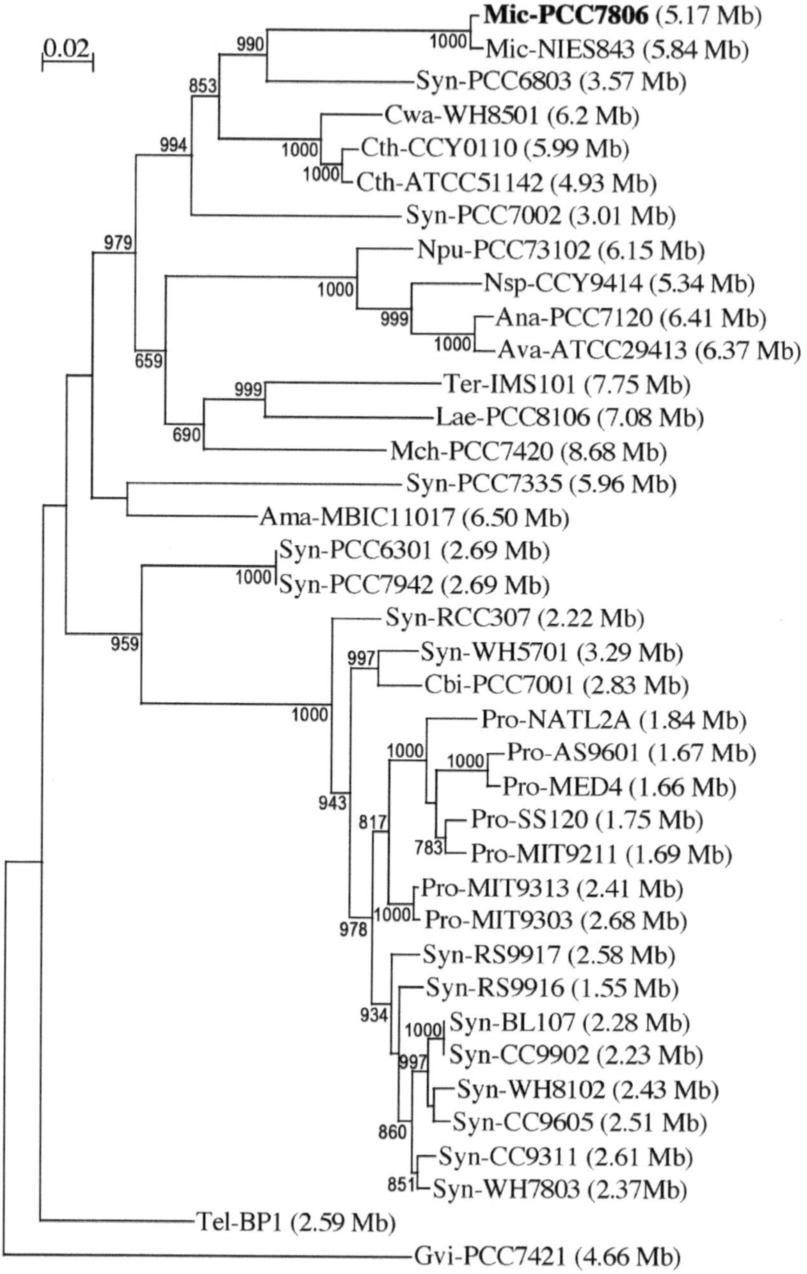

Figure 1: Phylogenetic maximum likelihood (ML) tree based on the concatenated 23S-16S rDNA sequences of diverse cyanobacterial lineages taken from public databases. The scale bar represents the average number of nucleotide substitutions per site. Genome sizes in megabases (Mb) are mentioned in parentheses. Trees were constructed using three methods (ML, Neighbor Joining and Maximum Parsimony). ML bootstrap values are indicated only if the bootstrap values obtained with the three methods are > 500 (1000 resamplings). See Table 1 for strain designations. With the kind permission of N. Tandeau de Marsac, Institut Pasteur, Unité des Cyanobactéries; CNRS, URA2172, F-75015, Paris, France.
[Frangeul *et al.* (2008) *BMC Genomics* **9:** 274; doi:10.1186/1471-2164-9-274]

dideoxynucleotides in four separate sequencing reactions leads to termination of chain elongation leading to DNA strands of varying lengths. These are separated by denaturing polyacrylamide urea gels with each of the four reactions run in one of the four individual lanes (lanes A, T, G, C). Visualization of the DNA bands either by autoradiography or ultraviolet light facilitates reading of DNA sequence from the X-ray film or gel image. The relative positions of the different bands among the four lanes are then used to infer the DNA sequence (from bottom to top). These methods are suitable to sequence relatively short fragments of DNA ~300 to 1000 nucleotides long. Out of a number of modified Sanger methods, fluorescent labelling of primers at 5′-end and the dideoxynucleotides has gained popularity and facilitated the development of automated and high-throughput DNA sequence analyzers. Vast majority of the sequencing projects employ this method. Chain-termination-based commercial kits are now available. Figure 2 elaborates the organism-based sequencing and

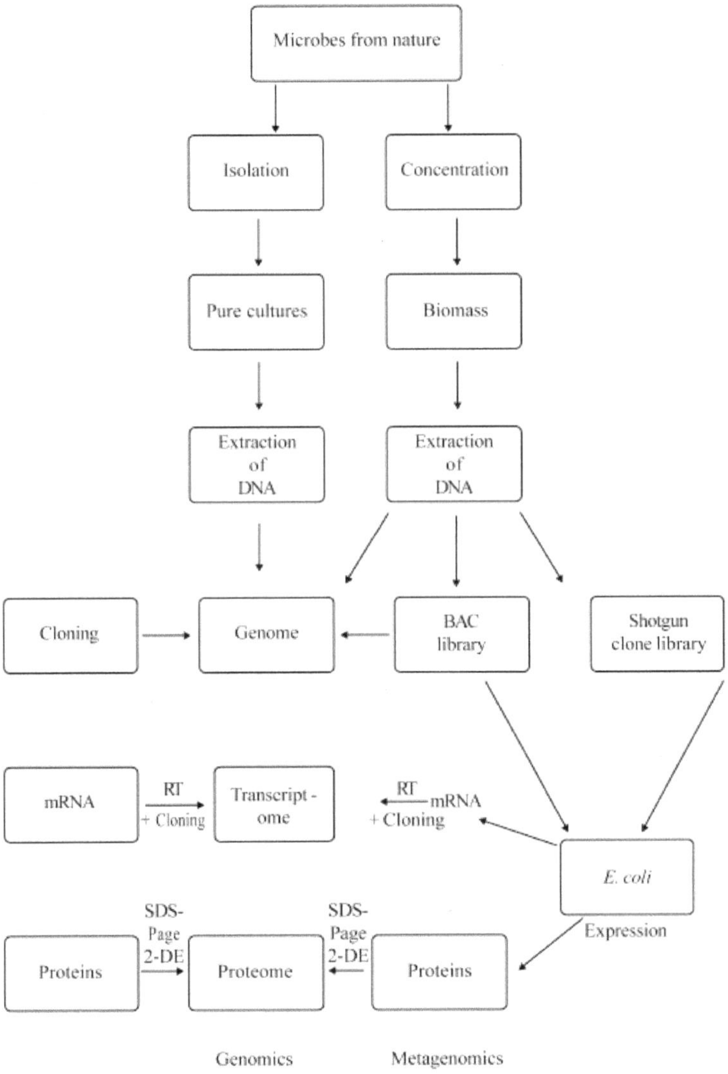

Figure 2: Schematic representation of steps involved in genomics and metagenomics, transcriptomics and proteomics.

metagenomics-based sequencing techniques. Metagenomic data obtained on sequencing of DNA, mRNA or proteins isolated from natural samples is compared with the genomic or metabolic data obtained from cultured strains that is available on the net (Fig. 3). Large scale DNA sequencing strategies in which DNA extracted from cells/environmental samples is cloned into appropriate vectors such as phagemids or bacterial artificial chromosome (BAC) and the libraries constructed are used in sequencing (Fig. 4). These large fragments of DNA (up to 100 kb) contain several genes arranged in a precise order in which they are found in the genomes they come from and in determining the exact order of sequence of genes it is quite necessary to clone overlapping fragments (Fig. 5). Alternatively, the extracted DNA either from the cells or environmental sequences is fragmented into small pieces (approximately 3 kb) and is cloned into conventional vectors in a procedure called "shotgun cloning". These are subjcted to sequencing techniques. The sequences are then assembled by using overlapping DNA regions. A number of new sequencing methods have been developed that parallelly conduct the sequencing process leading to the production of thousands or millions of sequences at once.

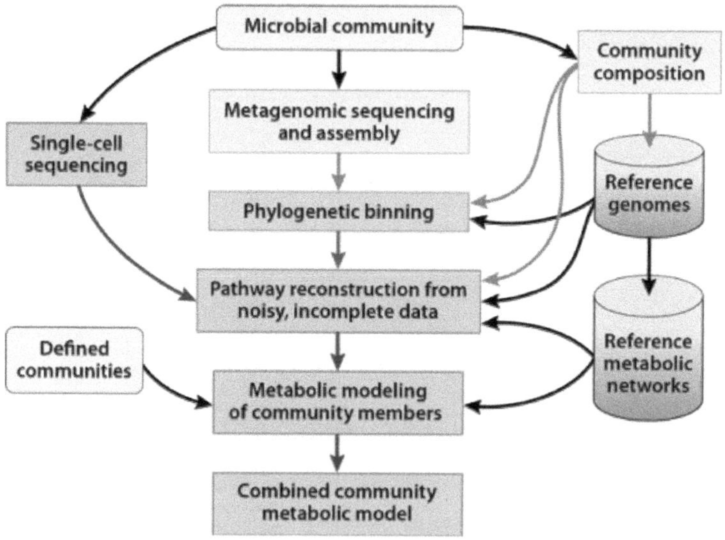

Figure 3: Work flow for reconstructing metabolic models from metagenomic data. Courtesy U.S. Department of Energy Genomic Science program and the website http://genomicscience.energy.gov.

Majority of the prokaryotic genomes have been sequenced at JCVI (http://www.jcvi.org, founded by J. Craig Venter who spearheaded privately the sequencing of human genome project that was accomplished in 2000 and discovered the shotgun sequencing method). JCVI coordinates researches at four organizations, the Center for Advancement of Genomics, The Institute for Genome Research, The Institute for Biological Energy Alternatives and the J. Craig Venter Science Foundation Joint Technology Center and the JCVI has facilities in Rockville, Maryland and La Jolla, California. The other sequencing Institutes are DOE-JGI (that unites the expertise of five national laboratories: Lawrence Berkeley, Lawrence Livemore, Los Alamos, Oak Ridge along with Hudson Alpha Institute for Biotechnology and is operated by the University of California for the U.S. Department of Energy), Gordon and Betty Moore Foundation Microbial Genome Sequencing Project (http://www.moore.org), Kazusa DNA Research Institute, Kisarazu, Japan (http://www.kazusa.org) and

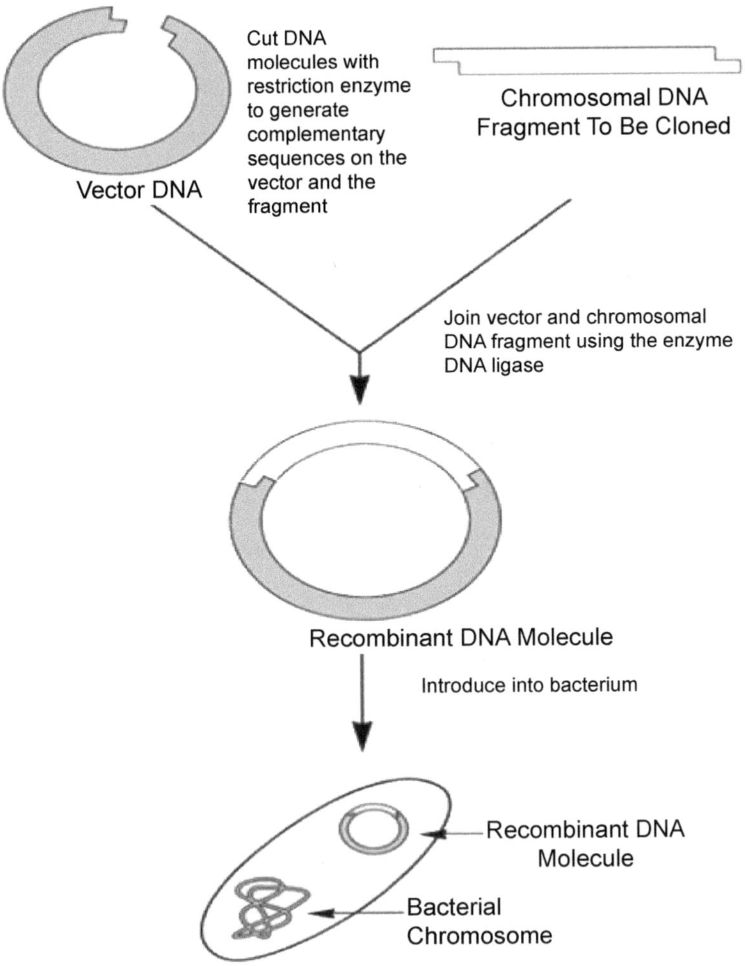

Figure 4: Cloning DNA in plasmids. Courtesy U.S. Department of Energy Genomic Science program and the website http://genomicscience.energy.gov.

many more individual Universities and Research Centres all over the world. Various aspects of computer methods in sequence analysis from database searching, performing multiple sequence alignment, construction of phylogenetic trees to predicting protein structure and function have been reviewed (Doolittle, 1996). The first such search tool was developed by Altschul *et al.* (1990) known as basic local alignment search tool (BLAST). A number of new generation, powerful and sensitive bioinformatics tools are made available subsequently (Henikoff and Henikoff, 1997; Neuwald *et al.*, 1997; Altschul *et al.*, 1997). The BLAST programme is available for public use at NCBI website http://blast.ncbi.nlm.nih.gov/Blast.cgi that helps in the identification of nucleotide sequences (BLASTn), protein sequences (BLASTp) and compares translational products of a nucleotide sequence of both strands against a protein database (BLASTx), searches a translational nucleotide database using a protein query (TLASTn) and converts nucleotide query sequence into protein sequences (TBLASTx). Other programmes used are based on interpolated Markov models (Salzberg *et al.*, 1998) and coding

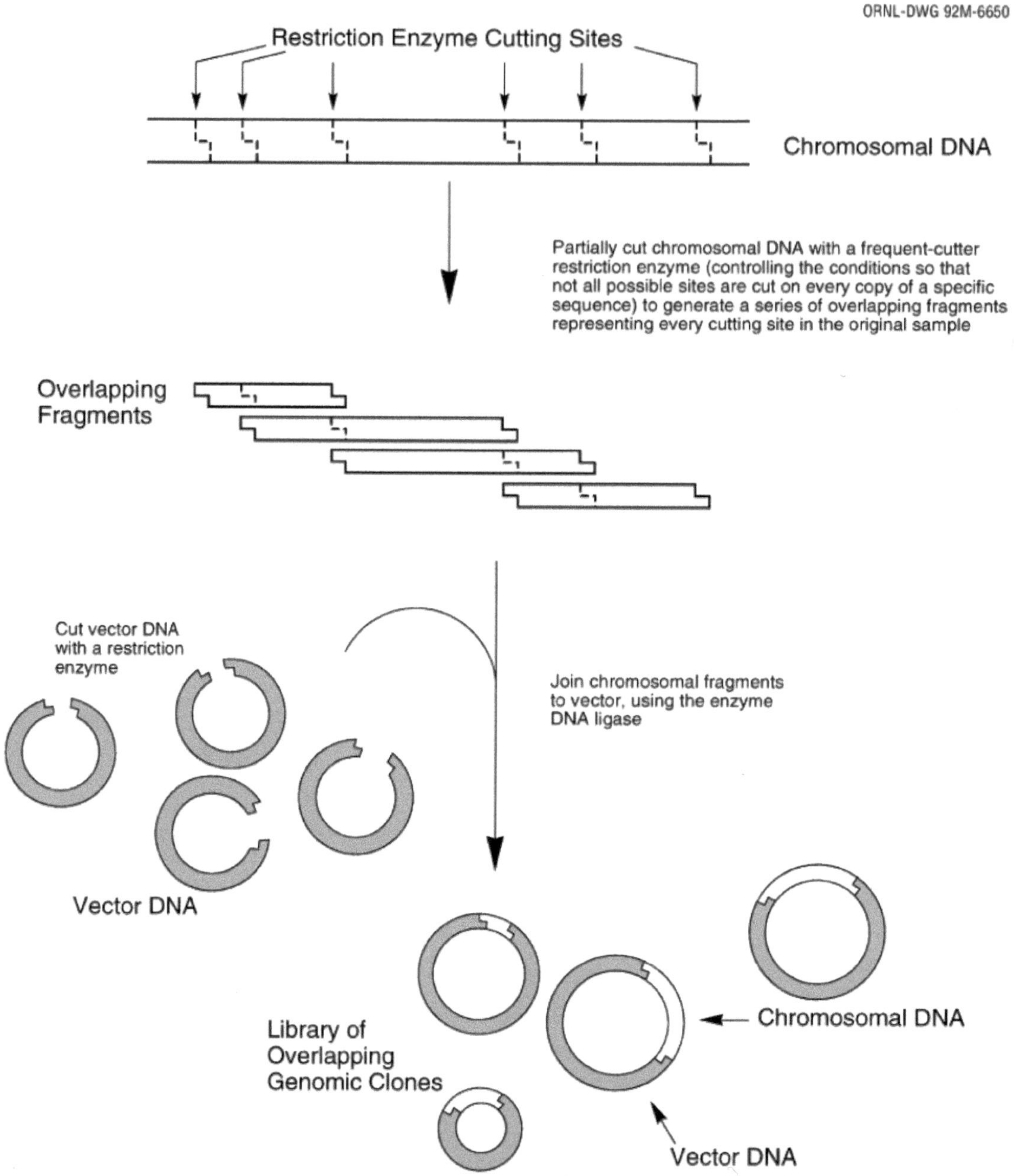

Figure 5: Construction of overlapping clone library. Courtesy U.S. Department of Energy Genomic Science program and the website http://genomicscience.energy.gov.

region identification based on CRITICA (Badger and Olsen, 1999). Gapped BLAST is an extension of the BLAST method that incorporates statistical analysis of alignments with gaps leading to high search sensitivity. PSI-BLAST (position-specific iterative BLAST) enables one to identify similarities between distant protein families (Altschul *et al.*, 1997; Altschul and Koonin, 1998). Gene searching models such as Critica, Glimmer or Generation are employed during the genome annotation process to identify sequences coding for proteins or various types of RNA (tRNA, rRNA), ribosome-binding sites, terminators, insertion sequences, promoter regions, laterally transferred genes and non-coding regions or genes showing relatively weak sequence similarities. It is the latter regions designated as 'twilight zone' that pose a real challenge for the automatic sequence annotation software programmes developed (Koonin and Galperin, 1997). Some of these are GeneQuiz (a workbench for sequence analysis; Scharf *et al.*, 1994) and MAGPIE system architecture (a fully automated genome analysis programme; Gaasterland and Sensen, 1996). The former is an annotation platform that serves both prokaryotic and eukaryotic sequence annotation, while the latter is meant exclusively for prokaryotic genome annotation. Other prokaryotic annotation platforms are Imagene (Medigue *et al.*, 1999), ATUGC (Bazzan *et al.*, 2003), GenDB (Meyer *et al.*, 2003), SABIA (Almeida *et al.*, 2004), MaGe (Vallenet *et al.*, 2006) and AGMIAL (Bryson *et al.*, 2006). Lombardot *et al.* (2006) created a site, Megx.net-a database resource for marine ecological genomics, useful for genomic and metagenomic data through which genome browsing, environmentally relevant protein families and group specific genes can be successfully identified. Additionally, it is possible to identify laterally transferred genes, or transposase and phage insertions by the TETRA software tool that computes tetranucleotide usage patterns. To address to the problems created by wrong annotation and provide a means of expert review, Markowitz *et al.* (2009) created a website known as IMGER that helps in the systematic and efficient revision of microbial genome annotation.

From the known gene sequence, the amino acid sequence of the putative protein is deduced and functional annotation is carried out, which is a challenging task because one has to take into account the molecular, cellular and phenotypic functions of the particular protein. At molecular level whether it is an enzyme, transporter, repressor or a structural protein has to be determined. At cellular level the role of the protein in a particular metabolic pathway or signalling cascade has to be assessed. Finally, the effect of the protein on general properties of the organism such as gliding motion or other cellular appendages or sporulation has to be determined. A quite useful approach in genome annotation is to identify gene pair-based close bidirectional best hits (BeTs) across two genomes by taking into account conserved gene clusters between them (Overbeek *et al.*, 1999). Genes that are co-transcribed generally are associated with the same function or participate in the same metabolic pathway. Identification of operons not only helps in understanding gene regulation but also provides important information on genome annotation. One can take into account the presence of gene clusters and their order in well studied prokaryotic systems such as *E. coli* by using intergenic distance distributions and of functional relationships between them. Methods so developed have been useful in predicting the existence of operons with a maximum accuracy of 88% in *E. coli* chromosome (Salgado *et al.*, 2000). Prediction of operons by a computational method is based on the presence of conserved gene pairs and the frequencies of their occurrence in various bacterial and archaeal genomes, the critical points of evaluation being the maintenance of a certain intergenic distance and that all genes in an operon are on the same strand. This method, however, does not take into account either the functions of the genes or their promoters and terminators (Ermolaeva *et al.*, 2001).

Another statistical method of predicting operons is based on the probabilistic approach by taking into account all possible parameters such as intergenic distance distribution, promoters, terminators, gene expression profiles and functional annotation (Craven *et al.*, 2000). Co-expression pattern from DNA microarray experiments has been taken as a tool supported by a Bayesian classification for prediction of operons (Sabatti *et al.*, 2002). A computational method developed by Zhang *et al.* (2007) considered the fact that the genes of an operon tend to encode enzymes that catalyze successive reactions in metabolic pathways. Dam *et al.* (2007) presented a logistic function-based classifier to predict the existence of operons in a genome which could detect operons with accuracy levels of 84.6% and 83.3% for *E. coli* and *B. subtilis*, respectively. Based on phylogenetic information and comparative genomic data such as intergenic distances, Bergman *et al.* (2007) constructed a highly sensitive and accurate operon prediction algorithm that could predict operons in the genome of *Bacillus anthracis* with high efficiency. As a test case, they have identified an operon (BA1489-92), in the genome of this organism, consisting of four genes that are functionally not so related but which escaped detection by the other operon prediction methods developed earlier. The co-transcription of these four genes has been confirmed by RT-PCR experiments. Operon prediction, based on comparative genomics of cyanobacteria, has been attempted by Chen *et al.* (2004) who took into account *P. marinus* MED4 and *P. marinus* MIT9313 as the reference genomes to predict operons in the genome of *Synechococcus* sp. WH8102 that constituted the target genome. Pair-wise comparison of orthologous genes by the COGNITOR program me (described below) using BLASTp resulted in 1448 pairs of matching genes between *Synechococcus* sp. strain WH8102 and *P. marinus* MED4 on the one hand and 1976 pairs of genes between *Synechococcus* sp. strain WH8102 and *P. marinus* MIT9313 on the other. There were 1422 pairs of BeTs between the two reference genomes. By taking into 1500 bp as the intergenic distance threshold level it was possible to identify 446 candidate operons in the target genome, of which 206 operons are shared with both the two reference genomes and 185 operons are shared with *P. marinus* MIT9313 only. Among the 446 candidate operons in the target genome, 242 are located on the positive strand whereas the rest 204 are located on the negative strand (Chen *et al.*, 2004).

Another approach is based on orthologous genes. Orthologs are conserved genes in different species that have evolved from a common ancestor (Fitch, 1970). Tatusov *et al.* (1997) first conceived the idea of constructing Clusters of Orthologous Groups (COGs) of proteins and included in this exercise six prokaryotic (*H. influenzae, Mycoplasma genitalium, Synechocystis* sp. strain PCC 6803, *Methanococcus jannaschii, Mycoplasma pneumoniae, E. coli*) and one eukaryotic (yeast) sequenced genomes available at that time. Every sequence from one genome was compared with every sequence from all other genomes and the BeTs in each of the genomes were determined. Each COG comprises proteins that are considered to be orthologous that is they are descendents of a common ancestral gene. To make it more meaningful phylogenetically, they combined the genomes of *H. influenzae* and *E. coli* to represent the gram-negative bacterial clade. Likewise, a gram-positive clade was constituted by merging the genomes of *M. genitalium* and *M. pneumoniae*. The other three clades were cyanobacteria, archaea and fungi represented by *Synechocystis* sp. strain PCC 6803, *M. jannaschii* and yeast, respectively. All these five clades constituted the first such COG database covering nearly 37% of the total number of sequences. A revision of the COG data set was affected with the addition of the genome of *Helicobacter pylori* increasing the number of COGs from 720 to 860 (Koonin *et al.*, 1998). With the inclusion of as many as 21 more completely sequenced genomes by 2000, a second release of COG database comprising 2,091 COGs was made with improvements made to identify new COGs in a new COGNITOR program me (Tatusov *et al.*, 2000). Tatusov *et al.* (2001) published the new COG database consisting of 2,791 COGs with as many as 45,350 proteins from 30 completely sequenced genomes. In addition, they provided a supplement for the COGs from the genome sequences of multicellular prokaryotes (nematode *Caenorhabditis elegans* and the fruitfly *Drosophila melanogaster*) shared by prokaryotes or archaea. Furthermore, another updated version of the database contained euKaryotic Orthologous Groups of proteins, named as KOGs with 59, 838 proteins distributed into 4,852 KOGs representing 110,655 analyzed eukaryotic gene products from three animals (*C. elegans, D. melanogaster* and *Homo sapiens*), one plant (*Arabidopsis thaliana*) and two fungi (*Saccharomyces cerevisiae* and *Schizosaccharomyces pombe*) and the intracellular microsporidian parasite *Encephditozoon cuniculi* (Tatusov *et al.*, 2003). Due to these developments, the COG database has become the main platform for functional annotation of newly sequenced genomes and for drawing conclusions on genome evolution. The role of COG database in comparative and functional genomics has been elaborated by Kaufmann (2006). Functionally, the COGs have been classified into 18 broad categories. Once the category of COG is known, it is easier to predict the function of new protein sequence on the basis of known function of existing COGs because it is very likely that the new sequence also exhibits the same or relatively similar cellular function as exhibited by other members of the same COG. Besides the COGNITOR, other tools of COG database comprise phylogenetic patterns search, extended phylogenetic patterns search, phylogenetic COG ranking, gene or domain fusion and gene context tool. The COG database will help us in identifying the core genes or minimal genome (genes with orthologues), conserved hypothetical proteins, PACE proteins (proteins of archaea conserved in eukaryotes) and orphan ORFs (ORFans) that do not match with any known sequence. When two genomes are compared, clusters of orthologous groups (COGs) of proteins are identified by the BLASTp analysis against COG reference data set and the variable genes (genes without orthologues) in the two organisms/strains under comparison are separated. The updated version of the COGs for unicellular organisms and the eukaryotic KOGs are accessible at http:// www.ncbi.nlm.nih.gov/ COG/ and via ftp at ftp://ftp.ncbi.nih.gov/pub/COG/, respectively.

A similarity search of the total protein-encoding genes is conducted against databases and the genes are classified into different categories. These are: (i) protein-coding genes whose function can be predicted, (ii) protein-coding genes without function prediction, (iii) genes without function with similarity, (iv) genes without function without similarity, (v) protein-coding genes coding signal peptides and (vi) protein-coding genes coding transmembrane proteins. The overall percentage coding capacity of the genome is calculated based on the percentage of genes that encode proteins with probable function. The putative protein-encoding genes (or open reading frames, ORFs) are generally identified by the start codons such as ATG,GTG,TTG or ATT and these are then denoted by giving a serial number with three letters. The first of these letters represents species name, the second letter specifies the length of ORF (if longer than 100 codons it is denoted by 'l', if shorter than 100 codons then by ('s') and the third letter represents the reading direction on the circular map (if towards right it is denoted by 'r' and if it is towards left it is represented by 'l'). For example, in *Synechocystis* sp. strain PCC 6803 a putative protein encoding gene or ORF, *sll0163* suggests that the *Synechocystis* (*s*) gene is longer than 100 codons (*l*) and is read from left side (*l*) and 0163 is the serial number of that ORF. Depending on the total number of genes so identified one can calculate the gene density that is the presence of one gene in relation to number of bp in a particular genome. Specific gene locations on the chromosomes are designated by mentioning the coordinates of bp in which the gene or ORF is located. For example, four copies rRNA gene clusters in the genome of *Anabaena* sp. strain PCC 7120 occur in the order of 16S-23S-5S at coordinates 2,375,734-2,302,211; 2,500,525-2,505,531; 4,919,771-4,914,765 and 5,947,188-5,942,409, respectively. A convenient method to identify the site of origin and termination of replication is to find out a shift in GC content known as GC skew. The leading strand is generally found to contain more guanine than cytosine residues. This fact is used to predict the origin and terminus locations. It is represented by a sum of (G–C)/

(G+C) in adjacent regions of chromosome from an arbitrary start to a given point in a sequence. The value reaches its maximum at the termination point while it is at its minimum at replication origin. Thus a GC skew analysis is conducted all over the genome to identify the probable shift to locate the origin of replication of the chromosome (Grigoriev, 1998).

II. GENOMES SEQUENCED

A brief account of the basic features of 39 fully sequenced cyanobacterial genomes is presented here followed by a comparative account of the major groups of functional categories of genes.

i) Genome of A. marina MBIC 11017: A unicellular cyanobacterium was initially isolated as a minor symbiont from a colonial ascidian together with a major symbiont *Prochloron* by Miyashita *et al.* (1996) from Palau Island. The cells of this bacterium are spherical, ellipsoidal, sheathed, non-motile and devoid of gas vacuoles. Its growth in ascidians has been confirmed by the work of Miyashita *et al.* (2003) and that of Kühl *et al.* (2005). It is also reported to grow in association with eukaryotic macroalgae (Murakami *et al.*, 2004; Ohkubo *et al.*, 2006) and in microbial mats in the Salton Sea (a saline and highly eutrophic California Lake; Miller *et al.*, 2005). A characteristic feature of this bacterium is the presence of chlorophyll *d* (3-desvinyl-3-formyl chlorophyll *a*) as the major light-harvesting pigment which has a formyl group at 3rd position in ring A structure. The absorption maximum of this pigment lies at 700 nm in organic solvents and up to 715 nm *in vivo*. The pigment composition and the organization of photosynthetic reaction centers (Miyashita *et al.*, 1997; Akiyama *et al.*, 2001, 2002, 2004) have been determined and the taxonomic identification as *A. marina* was subsequently reported (Miyashita *et al.*, 2003). Swingley *et al.* (2008) reported a total genome size of 8.3 Mb that includes the main genome of *A. marina* MBIC 11017 (with 6,503,723 bp) and 9 circular plasmids (pREB1-pREB9) of variable sizes. The largest plasmid is pREB1 (with 374,161 bp) and the smallest being pREB9 (with 2,133 bp). The G+C content of the main genome is 47% and that of the plasmids ranged from 43% (pREB9) to 47% (pREB1). The total number of genes in the main circular genome is 6,342 of which 14 represent pseudogenes. The genome has a coding capacity of 85% and the average length of a gene is 867 bp. There are six rRNA genes with 69 tRNAs. The ORFs without similarities to known genes are 28% and the conserved hypothetical ORFs range up to 18%. The insertion elements in the genome number up to 285 and their number in the plasmids varies from one (as in pREB9) to 30 (as in pREB1). The number of ORFs in the plasmids is in decreasing order, i.e. 417, 392, 382, and 279 in case of pREB1, pREB2, pREB3 and pREB4 respectively. Likewise in plasmids pREB5, pREB6, pREB7, pREB8 and pREB9 there are 224, 192, 174, 120 and 4 ORFs respectively. Several of the plasmids share significant regions of homology and the genes encode key proteins governing important metabolic reactions.

ii) Genome of Cyanothece sp. ATCC 51142: The genome of this marine diazotrophic cyanobacterium has been completely sequenced by Welsh *et al.* (2008). The most unique feature of this genome is the existence of six separate elements, a circular chromosome, four plasmids and a linear chromosome. The circular chromosome is 4,934,271 bp long with a G+C content of 37.9%. The linear chromosome is 429,701 bp long with a G+C content of 38.6%. Put together the genome has a length of 5,460,377 bp with a G+C content of 37.9%. The predicted genes in the circular and linear chromosome have been determined to be 4,762 and 449, respectively making up a total of 5,304 genes in this organism. The genes with assigned function are 2,584 and 127, respectively in the circular and linear chromosomes making up a total of 2,735 genes (51.6%). The hypothetical genes and genes of unknown function are 1,710 and 468 in the circular chromosome and 298 and 24 in the linear chromosome, respectively.

There are two rRNA operons on the circular chromosome with none being represented on the linear chromosome. There are 43 tRNA genes on the circular chromosome. When compared to the circular chromosome which has 45.7% genes with no assigned function, the linear chromosome has a higher percentage (71.7%) of genes of unknown function. At least 50 of the protein-coding genes on the linear chromosome are functional. The linear chromosome has 38 genes whose corresponding copies are represented in the circular chromosome and another 42 genes on linear chromosome are unique that have homologues in other organisms. The length of the largest plasmid is 39,620 bp while the medium ones have 31,856 bp and 14,685 bp and the smallest one has 10,244 bp.

***iii) Genome of G. violaceus PCC 7421*:** Complete genome structure of *G. violaceus* PCC 7421 has been presented by Nakamura *et al.* (2003) and the details of individual genes are listed in Cyanobase at http://www.kazusa or jp./cyanobase. This cyanobacterium is distinctive in its cellular organization as it lacks photosynthetic thylakoids. The genome of this organism is a circular molecule of 4,659,019 bp with an average G+C content of 62%. No plasmids have been detected during this study. The number of potential protein-coding genes is 4,430. Of these 41% (1836 genes) showed sequence similarity to genes of known function and 37% (1635 genes) of them are hypothetical genes. The rest 22% (959 genes) do not bear any resemblance to the known and registered gene sequences at public databases. There is one copy of rRNA gene cluster in counter-clockwise direction in the order of 16S-trnI-trnA-23S-5S. A total of 45 tRNA genes (including one rRNA gene cluster) are present representing 44 tRNA species. *G. violaceus* is unique in possessing an intron-less trnL-UAA. A single gene for tm-(transfer-messenger) RNA is found that probably plays a role in the degradation of aberrantly synthesized proteins. It is interesting to note that genes showing sequence similarity to those of *E. coli* and *Streptomyces coelicolor* are present. After substraction of these genes, *G. violaceus* PCC 7421 possesses 610 genes unique to three other cyanobacteria (*Anabaena* sp. strain PCC 7120, *Synechocystis* sp. strain PCC 6803 and *T. elongatus* BP-1) and at least half of them are of known function (related to antenna components, chlorophyll synthesis, photosystems and carbon fixation). At least 995 genes of this cyanobacterium (23%) do not bear any resemblance to three genomes under comparison (see Table 5). Of these, 684 do not bear any resemblance to the registered genes.

***iv) Genome of M. aeruginosa*:** The genomes of two strains of this cyanobacterium *M. aeruginosa* NIES-843 and *M. aeruginosa* PCC 7806 have now been sequenced.

a) Genome of M. aeruginosa NIES-843: Kaneko *et al.* (2007) sequenced the genome of *M. aeruginosa* NIES-843. A single circular chromosome of 5,842,795 bp long with an average G+C content of 42.3% is present. The putative protein-coding genes are 6,312. Of these 45% of them show sequence similarity to genes of known function, 32% are similar to hypothetical genes and the remaining 23% bear no resemblance to any known genes in the databases. One unique feature is the presence of putative transposases (469 gene sequences amounting to 7.4%) and large repeats (11.7% of the genome). Genes involved in the biosynthesis of microcystins, cyanopeptolin and for modification of small polypeptides have been identified. A large number of genes belong to restriction modification system. There are two operons for rRNA synthesis and 42 tRNA genes representing 41 tRNA species. An 8 bp palindromic sequence (5'-GCCATCGC-3'), characteristic of many cyanobacterial genomes, known as HIP1 is present in 1821 copies in all in the genome of NIES-843 and the frequency of its occurrence is one copy/3209 bp which is quite lower than in others.

b) Genome of M. aeruginosa PCC 7806: It has a circular chromosome of 5,172,804 bp long with a G+C content of 42%. The putative protein-coding sequences are 5,292 (Frangeul *et al.*, 2008). The proteins of known function are found to be 3,322 (amounting to 63% of the coding capacity). The genome contains as many as 1971 genes representing clusters of atypical genes (CAGs) purported to have been acquired through LGT. There are 838 genes (16%) whose products are not similar to >40% of any proteins in *M. aeruginosa* NIES-843 and of these 644 of the proteins bear no apparent similarity to >40% of any proteins in the other cyanobacterial genomes sequenced so far. The putative transposases are encoded by 362 genes (representing 6.8%). The proportion of large repeats equals the number in *M. aeruginosa* NIES-843 (i.e. 11.7%) and the length of the repeats is >1000 bases and there is >90% identity in these sequences. There are two rRNA operons and 41tRNA loci on the chromosome. Three groups of proteins have been identified after BLASTp analysis of 5,292 genes, on the basis of a sequence identity of 40% between individual protein sequences. The first group designated as "Maeru40" group consists of 764 genes (14.4%) specific to M. *aeruginosa* strain PCC 7806 and not found in 15 other cyanobacterial genomes under comparison. Of these, 438 genes (8.3%) have no homologues in the databases. The second group ("core40" group) comprises 652 proteins (12.3%) that had high degree of similarity with at least one gene sequence in each of the 15 other genomes. The third group ("other40") has the rest of the genes bearing high degree of similarity to the coding sequences in the other genomes. Significantly, the existence of two genes, the first one (*mic5251*) that is similar to a gene sequence of *Synechocystis* sp. strain PCC 6803 responsible for perceiving osmotic stress and cold stress and the second one (*mic5237*) identical to *Anabaena* sp. strain PCC 7120 gene known as *orrA* gene help *M. aeruginosa* strain PCC 7806 to overcome stress due to lower temperature and high pressure of waters during winter months. Gene sequences bearing homology to eukaryotic genes *actM* (that encodes actin) and *pfnM* (that synthesizes actin cognate binding partner) have been discovered in this strain that enable it to resist hydrostatic and osmotic pressures. Cell wall and polysaccharide synthesizing genes that have a role in colony formation have been identified in this strain as also a cluster of 12 genes responsible for the formation of gas vesicles. *M. aeruginosa* PCC 7806 appeared to be closer to *Cyanothece* sp. ATCC 51142, *Cyanothece* sp. CCY0110 and *C. watsonii* WH8501 on the basis of comparison of coding sequences of 44 cyanobacterial genomes. On the other hand, on the basis of 16S-23S rDNA phylogeny *M. aeruginosa* PCC 7806 appeared closer to *Synechocystis* sp. strain PCC 6803. The number of orthologous genes shared among these three closely related cyanobacteria with *M. aeruginosa* PCC 7806 is 1789. Although there appeared to be a low synteny score in the cyanobacterial genomes when compared to bacterial genomes there are at least four gene clusters relating to ribosomal proteins that are syntenic in the genomes of *M. aeruginosa* strain PCC 7806, *C. watsonii* WH8501 and *Synechocystis* sp. strain PCC 6803. Phosphate transport system (6 genes), inorganic carbon concentrating mechanism (5 genes), ATP synthase (8 genes) and genes of unkown function (4 genes) are the other gene clusters identified to be syntenic.

***v) Genomes of Prochlorococcus strains*:** Rocap *et al.* (2003) sequenced and compared the genomes of HL-adapted *P. marinus* MED4 and LL-adapted *P. marinus* MIT9313. The genome of the former is the smallest (with 1,657,990 bp and a G+C content of 30.8%) among oxygenic photoautotrophs while the genome of the latter is significantly larger (with 2,410,873 bp with a G+C content of 50.7%). The number of protein-coding genes and those proteins with assigned function are 1,716 and 1,134 in MED4 and 2,278 and 1,366 in MIT9313, respectively. There are 1352 genes common to both strains. Interestingly, excepting 38 of the common genes the rest all are shared by *Synechococcus* sp. strain WH8102. Genes responsible for the synthesis of chlorophyll *a/b* binding protein, (divinyl), chlorophyll *b* (i.e. chlorophyll *a* oxygenase that can convert divinyl chlorophyll *a* into divinyl chlorophyll *b*) and α-carotene (lycopene-ε-cyclase) are characteristic of the genomes of *Prochlorococcus.* The genome of MED4 has 364 genes specific to it and the corresponding genes are absent in MIT9313. Likewise, MIT9313 has 923 genes specific to it but these are absent in MED4 genome. The smaller size of MED4 genome is attributed due to a loss of a number of genes and the strain-specific genes are suggested to have been acquired through phages. There is one rRNA operon in MED4 while there are two in case of MIT9313. The number of tRNA genes is 37 in MED4 when compared to 43 present in MIT9313. The identification of the gene for chlorophyll *a* oxygenase in the genome of *Prochlorococcus* has been confirmed by a comparative genomic approach (Satoh and Tanaka, 2006).

On the basis of 16S rRNA phylogeny it was suggested that *P. marinus* MIT9313 branches at the *Prochlorococcus* radiation coming closer to *Synechococcus* group (Fuller *et al.*, 2003). *P. marinus* SS120 comes closer to *P. marinus* MED4 in its genome size being larger by only about 90 kb. Dufresne *et al.* (2005) compared the genomes of *P. marinus* MED4, *Prochlorococcus* SS120 and *P. marinus* MIT9313 and *Synechococcus* sp. strain WH8102. The gene order in MED4 and SS120 is highly conserved with only five inversions larger than 20 kb having taken place in both these genomes whereas large number of inversions, translocations and extensive genome rearrangements have taken place in between *Synechococcus* and *Prochlorococcus*, on the one hand, and between MIT9313 and the two *Prochlorococcus* strains on the other. These workers are of the view that the divergence of the clade with MED4 and SS120 is a more recent one than that containing *P. marinus* MIT9313. They have identified 1306 orthologous genes that are highly conserved in all marine picocyanobacteria. Total percentages of protein-coding genes in these four organisms under comparison revealed that *P. marinus* MED4 has the highest percentage (73.2%) of such genes with lowest (51.1%) being represented in *Synechococcus* sp. strain WH8102. Thus *P. marinus* MED4 representing HL-clade appears to be a more recently evolved as it is very less diversified than other LL-clades and the smallest genome size of this strain is explained on the basis that extensive gene loss had occurred during its evolution from a *Prochlorococcus* ancestor with a larger genome. This has been substantiated on the basis of (i) absence of a number of DNA repair genes in *P. marinus* MED4 and (ii) the absence of *ada* gene that encodes 6-O-metrhylguanine-DNA methyltransferase (in both *Prochlorococcus* MED4 and *P. marinus* SS120) that repairs alkylated forms of guanine and thymine in DNA. Due to this, a mutational pressure to high A+T content exists in these two strains.

Kettler *et al.* (2007) sequenced the genomes of seven *P. marinus* strains (MIT9211, MIT 9215, MIT9515, NATL1A, MIT9301, MIT9303 and AS9601 have been sequenced at JCVI, Maryland, USA; NATL2A has been sequenced at Joint Genome Institute, USDOE, California) and presented a comparative account together with those of *P. marinus* strains MED4, MIT9313 (Rocap *et al.*, 2003), SS120 (Dufresne *et al.*, 2003) and MIT9312 (Coleman *et al.*, 2006) sequenced earlier. The essential features of these genomes are presented in Table 2. The smallest genome is that of *P. marinus* MED4 (with 1.6 Mbp) and the largest is that of MIT9303 (2.7 Mbp). A complete reannotation of the 12 genomes by a uniform method was carried out by employing gene prediction programmes like CRITICA and GLIMMER. As a result, there is a progressive reduction in the number of core genes to 1250 (in contrast to 1352 detected by Rocap *et al.*, 2003) amounting to 40% to 70% of the genes in any of the 12 strains. A comparison of these genomes with the genomes of *Synechococcus* sp. (strains CC9311, CC9902, WH8102 and CC 9605) revealed 33 genes specific to *Prochlorococcus* (in contrast to 38 noted by Rocap *et al.*, 2003) but absent from some or all *Synechococcus* strains. Of these 33 genes, 13 are exclusively represented in *Prochlorococcus* strains only and absent in *Synechococcus* strains. Eight of these genes whose putative function has been assigned relate to HL-inducible protein (to overcome high-light intensity stress), a possible sodium-solute transporter, an iron sulphur protein and a transcription factor (deoR-like). Significantly, all *Prochlorococcus* genomes lack genes for divinylprotochlorophyllide reductase (*dvr*), allophycocyanin (*apcABCDE*), some phycoerythrins, phycobilisome linkers and molybdoprotein biosynthesis (*mobA*, *moaABCDE* necessary for nitrate reductase).

Table 2: Properties of genomes of *P. marinus* strains.

Isolate/Strain	Light adaptation	Clade	Genome size (Mb)	No. of genes	tRNAs	%G+C
MED4 (CCMP1986)	HL	I	1.68	1,929	37	30.8
MIT9515	HL	I	1.70	1,908		30.8
MIT9301	HL	II	1.64	1,907		31.4
AS9601	HL	II	1.67	1,926		31.3
MIT9215	HL	II	1.73	1,989		31.1
MIT9312	HL	II	1.70	1,962		31.2
NATL1A	LL	I	1.86	2,201		35.1
NATL2A	LL	I	1.84	2,158		35.0
CCMP1375 (SS120)	LL	II	1.75	1,925		36.4
MIT9211	LL	III	1.69	1,855		38.0
MIT9303	LL	IV	2.68	3,022		50.1
MIT9313	LL	IV	2.41	2,843	43	50.7

An international joint venture between Joint Genome Institute, J. Craig Venter Institute and Genoscope led to the sequencing of 12 *Prochlorococcus* and 11 *Synechococcus* strains listed in Tables 2 and 3, respectively.

vi) Genomes of Synechococcus strains: The properties of eleven of the genomes of *Synechococcus* strains are summarized in Table 3. The genomes of *Synechococcus* sp. strain WH8102 (Palenik *et al.*, 2003) and *Synechococcus* sp. strain CC9311 (Palenik *et al.*, 2006) sequenced earlier have been compared by Dufresne *et al.* (2008) with the genomes of nine other strains of *Synechococcus* sequenced by them. These strains have been isolated from the Mediterranean Sea, the Red Sea, the Pacific and Atlantic oceans from varying depths of 5 m (*Synechococcus* sp. strain CC9902 from the Pacific Ocean) to 1000 m (*Synechococcus* sp. strain BL107 from the Mediterranean Sea). With the exception of *Synechococcus* sp. strain WH5701 and *Synechococcus* sp. strain RCC307 that belong to Marine sub-cluster 5.2 and 5.3 respectively, the rest of the strains belong to marine sub-cluster 5.1 and cover nine of the ten clades described by Fuller *et al.* (2003). [Cluster 5 was initially recognized by Herdman *et al.* (1979) by combining the the former MC-A and MC-B defined by Waterbury and Rippka (1989). In cluster 5 *Synechococcus* strains from coastal, euryhaline and strictly marine habitats are included. Strictly marine PE-containing strains have been grouped into sub-cluster 5.1, while euryhaline strains lacking PE were put in subcluster 5.2 including strains WH5701 and WH8007]. The smallest genome is that of *Synechococcus* sp. strain RCC307 with 2.22 Mbp while the largest genome is of *Synechococcus* sp. WH5701 with 2.86 Mbp. The G+C contents vary from 52.5% (*Synechococcus* sp. strain CC9311) to 66.0% (*Synechococcus* sp. strain WH5701). There are two rRNA genes in all strains except in case of

Table 3: Properties of genomes of *Synechococcus* strains (Modified from Dufresne *et al.*, 2008).

Strain	Clade	Pigment Type[a]	Genome Size (Mb)	No. of genes	rRNAs	%G+C
1) CC9311	I	3d (CA)	2.61	2,944	2	52.5
2) CC9605	II	3c	2.51	2,645	2	59.2
3) WH8102	III	3c	2.43	2,583	2	59.4
4) CC9902	IV	3d (CA)	2.23	2,358	2	54.2
5) BL107	IV	3d (CA)	2.28	2,553	2	54.2
6) WH7803	V	3a	2.37	2,586	2	60.2
7) WH7805	VI	2	2.62	2,934	2	57.5
8) RS9917	VIII	1	2.58	2,820	2	64.8
9) RS9916	IX	3d (CA)	2.66	3,009	2	59.8
10) WH5701	NA	1	2.86	3,129	2	66.0
11) RCC307	X	3b	2.22	2,583	1	60.8

a-The pigment types are as per the classification provided by Six *et al.* (2007); CA=Chromatic adapter; NA=Not applicable.

Synechococcus sp. strain RCC307 which has one rRNA gene. However, the number of tRNA genes is variable from 42 (BL107) to 46 (RS9917) or 48 (WH7805 and WH5701).

After a comparison of *Synechococcus* genomes, Dufresne *et al.* (2008) identified 1572 gene families that constitute a core genome in this genus. This represents 52% of the total genome of *Synechococcus* sp. strain WH5701 and 67% of the total genome of *Synechococcus* sp. strain CC9902. When a comparison is made with *Prochlorococcus*, the number of gene families in the core genome is further reduced to 1228 gene families. However, Palenik *et al.* (2003) identified 1314 genes common to *Synechococcus* sp. strain WH8102 and *Prochlorococcus* strain MED4 and MIT9313 and 736 genes specific to *Synechococcus* sp. strain WH8102.

There is fairly a constant number of genes (748±85) shared by 2–10 *Synechococcus* genomes. Significantly, four *Synechococcus* genomes (BL107, CC9311, CC9605 and CC9902) are characteristic in possessing *isiA* and *isiB* genes that encode PSI antenna protein CP43 and flavodoxin, respectively that are separated while in other freshwater cyanobacteria these two genes are present in the same operon. Additionally, the genomes of these four strains also showed a ferredoxin gene existing in four to five copies that makes them versatile in switching over from iron-depleted (flavodoxin) to iron-enriched habitats (ferredoxin). There is a considerable degree of variation in the presence of unique genes (from 91 to 845) depending on genome size. These are distributed in a region called as 'genomic islands' that are characterized by (i) a deviation in tetranucleotide frequency, (ii) longer intergenic sequences (~105 bp) than shorter intergenic regions (~50 bp) outside these islands, (iii) a correlation between the size of genomic islands and the length of the genome in general, (iv) a high degree of similarity in genes that are shared by 11 genomes and (v) the acquisition of these genes through LGT. Similarly, in the genome of *Synechococcus* sp. strain WH 8102 the genes governing utilization of broad range of nitrogenous substances, some having transport potential, cell envelope genes, genes governing sialic acid synthesis in regions of low G+C content point out to the group of recently acquired genes through LGT. The existence of 16 probable phage integrases (that encode site-specific recombinases) in the regions of low G+C of the genome of *Synechococcus* sp. strain WH 8102 prompted Palenik *et al.* (2003) to conclude that these are akin to pathogenicity islands transferred between strains of pathogenic bacteria.

***vii) Genome of S. elongatus* PCC 6301 *(formerly Anacystis nidulans Berkeley strain 6301)*:** The complete nucleotide sequence of this freshwater cyanobacterium was determined by Sugita *et al.* (2006) and is available at web database, CYORF (http://www.cyano.genome.jp/). The genome of this organism consists of a single circular chromosome of 2,696,255 bp long with a G+C content of 55.5%. The number of protein-coding genes assigned on the chromosome is 2,525. The potential protein-coding genes (56%) showed sequence similarities to known function. 35% of the genes showed sequence similarities to the hypothetical genes. The rest of the 9% genes have no significant similarities to any of the predicted proteins in the public DNA databases. Two copies of rRNA gene cluster (rrnA and rrnB) have been assigned on the chromosome at 1,050 kb and 1,650 kb positions in the order of 16S-23S-5S. These two operons are separated at a distance of 6,05,653 bp measured from 5′-ends of the clusters. A total of 42 tRNA genes representing 42 tRNA species have been identified on the chromosome.

***viii) Genome of Synechocystis* sp. *strain* PCC 6803:** Kaneko *et al.* (1995, 1996) determined the sequences of the entire genome and assigned the essential protein-coding regions. *Synechocystis* sp. strain PCC 6803 has a circular chromosome with a total of 3,573,470 bp and G+C content of 47.7%. The length of the genome closely corresponded to the estimation of genome size (3.6 Mb) determined on the basis of restriction and linking analysis performed earlier by Kotani *et al.* (1994). A total of 3168 potential protein-coding genes were allocated on the chromosome. But in a subsequent analysis conducted in 2002, the assignment of potential protein-coding regions has been revised to 3264 (http://www.kazusa.or.jp/cyanobase/synechocystis/). Of the allocated genes, 145 (4.6%) have already been reported. A sub-total of genes similar to genes of known function are 1623 and similar hypothetical genes are 1133. These two together form genes similar to registered genes and the sub-total is 2,756. The genes without similarity number up to 508. The essential coding capacity of the genome of this organism is 55% and the remaining 45% (1426 genes) showed no significant similarity to any registered genes. The overall gene density (described as the number of genes in relation to length of bases) has been found to be approximately one gene per 1.1 kb. The average length of encoded gene product has a length of 326 amino acids, the longest one being 4,199 amino acids long (*Slr 408*). Two copies of rRNA gene clusters with identical 5,028 bp sequences are present. However, these are arranged in reverse orientation with respect to one another at map positions 68% and 93% of the genome. There are 42 tRNA genes in the genome of *Synechocystis* sp. strain PCC 6803.

As many as seven plasmids are reported in this organism. Three of them are small plasmids (with 5.2 kb, 2.4 kb and 2.3 kb length) and the rest are large. The nucleotide sequencing of four large plasmids (pSYSM:120 kb; pSYSX:106 kb; pSYSA:103 kb and pSYSG:44 kb) of *Synechocystis* sp. strain PCC 6803 revealed a total of 397 potential protein-coding genes and 77% of these have unknown function. The presence of sequence HIP1: 5′-GCCATCGC-3′specific in these plasmids indicates that they may have originated from different cyanobacterial strains. Of the 397 genes, similarity search for proteins coded by these genes indicated 23% (92 genes) having sequence similarity to genes of known function, 36% of genes (142 genes) being identical to hypothetical genes and the remaining 41% (163 genes) were non-identical to genes of known function (Kaneko *et al.*, 2003).

***ix) Genome of T. elongatus BP-1*:** The genome of this thermophilic cyanobacterium is a circular molecule consisting of 2,593,857 bp with an average G+C content of 53.9% (Nakamura *et al.*, 2002). No plasmid sequences have been encountered. The potential protein-coding sequences are 2,743 of which total number of protein-coding sequences is 2,475. Of these, 1,569 (i.e. 63%) genes have matched with the gene sequences of *Synechocystis* sp. strain PCC 6803 and *Anabaena* sp. strain PCC 7120. Genes of unknown function are 522 in number. *T. elongatus* BP-1 genome shares 116 (3%) and 254 (10%) genes with *Synechocystis* sp. strain PCC 6803 and *Anabaena* sp. strain PCC 7120 genomes, respectively. The gene density is one gene per every 1,048 bp. The start codons identified are ATG, GTG, TTG or ATT. There is one copy of rRNA gene cluster in the order of 16S-23S-5S at the coordinates 2,330,963-2,336,733. There is a group I intron sequence of 745 bp length in the 23S RNA gene cluster that encodes a homing endonuclease. An eight base palindromic sequence known as highly iterated palindromic sequence 1 (HIP1) is represented by 3,681 copies in the genome of this cyanobacterium. The frequency of occurrence of HIP1 is much higher (one copy per 705 bp) than found in *Synechocystis* sp. strain PCC 6803 (one copy per 1131 bp) and *Anabaena* sp.strain PCC 7120 (one copy per 1219 bp). Moreover, the frequency of occurrence of HIP1 in the genome of *T. elongatus* BP-1 itself varied as it was highest in protein-coding regions (one copy per 672 bp), lower in intergenic regions (one copy per 1209 bp) and lowest in RNA-encoding regions (one copy per 4328 bp). On the contrary, only one copy of HIP1 has been encountered in 110,339 bp regions corresponding to mobile introns and insertion sequences. There are 42 tRNA genes representing 42 tRNA species widely distributed all over the genome.

***x) Genomes of Arthrospira*:** *(a) A. (Spirulina) platensis NIES-39:* Fujisawa *et al.* (2010) reported the complete genome sequence *A. platensis* NIES-39 based on optical mapping. A circular chromosome (of 6.8 Mb) possesses 6630 protein-coding sequences of which 5157 are orthologues (78%) of known function. A comparison of 38 cyanobacterial genomes brought out 694 COGs common to all six groups belonging to photosynthetic, housekeeing and cyanobacteria-specific genes. Among these, 938 genes are represented in this category in *A. platensis*. COGs specific for *A. platensis* are 1066 among which 2056 genes are present. Of these, 71 are common to group (iv) and group (v) recognized by them. Among the heterocystous members 223 gene clusters are common to them. Significant features of the genome of *A. platensis* are (i) 150 group II introns (amounting to 612 kb) are present of which 71 of them are responsible for reverse transcriptase/maturases, (ii) presence of as many as 22 adenyl cyclases of which 10 are membrane associated signal transduction proteins, (iii) two-component signal transduction systems comprise of 84 putative histidine kinases of which 33 encode hybrid histidine kinases and 65 putative genes for response regulators, (iv) additionally methyl-accepting chemoreceptor proteins (8 genes), Ser/Thr protein kinases (43 genes), response regulators (20 genes) with an N-terminal receiver domain and helix-turn-helix DNA-binding C-terminal region are present, (v) 66 putative genes for σ^{70} type transcription factors are represented. Another remarkable feature of the genome is the presence of genes for heterocyst differentiation (*hetU*, *hetR* and *hetF*) and nitrogen fixation while at the same time no nitrogenase genes are present. Likewise, genes for type IV pilus that confer twitching motility are present but with no obvious twitching motility. Comparative genomic analysis revealed seven COG clusters for filamentous organization including *Synechococcus* sp. strain PCC 7002, seven genes of which are present in the genome of *A. platensis*. That is why *Synechococcus* sp. strain PCC 7002 has been projected as a primitive filamentous type and included as the sole representative of group (iii) by Fujisawa *et al.* (2010).

b) Genome of Arthrospira sp. *PCC 8005*: Whole genome shotgun sequencing with pyrosequencer technology assisted by MaGe automated annotation brought out important features of the genome of *Arthrospira* sp. PCC 8005 (Janssen *et al.*, 2010). The genome (with 6,279,260 bases) has a G+C content of 44.7% and 5,856 protein-coding sequences and comes closer to the genomes of *Arthrospira maxima* CS-328 and *Lynbya* PCC 8106 in showing highest overall synteny. There are 176 genes encoding RNAs. Among the unique coding sequences mention may be made of (i) desaturases that are required for production of β-carotene and two important fatty acids linoleic acid and γ-linolenic acid; (ii) for utilizing nitriles as sole source of nitrogen (*nthPAB*) and (iii) a bidirectional hydrogenase (*hox* and *hyp* genes). The genome contains highly repetitive sequences of tandem repeats, short palindromic repeats and insertion elements (Janssen *et al.*, 2010).

xi) Genome of L. majuscula 3L: The draft genome of *L. majuscula* 3L has been published with partial annotation of genes governing secondary metabolite production, DNA replication, recombination and repair, cell wall biogenesis and signal transduction mechanisms (Jones *et al.*, 2011). The size of the genome (~8.5 Mb) comes closer to the genomes of *N. punctiforme* (8.2 Mb) and *T. erythraeum* (7.8 Mb). With a mol % G+C of 44, the genome has 7,479 protein-coding genes and this again is higher than the number of genes found in *N. punctiforme* (6,086 genes) and *T. erythraeum* (4,451 genes). There are two rRNA operons and 56 tRNAs. Most of the housekeeping genes (102) found in bacteria have been identified (101 genes) in *L. majuscula* 3L. The largest cluster of COGs identified belong to DNA replication, recombination and repair (9%) followed by those of cell wall biogenesis (8%) and signal transduction mechanisms (7%). When compared to the diversity of natural products produced by this organism, the number of genes identified (126 genes) appear to be low. Of these, 44% have been assigned to non-ribosomal peptide synthetase- and/or polyketide synthase related genes. There are eight clusters of genes assigned to the probable synthesis of natural products of which two of them encode curacin A (an anticancer agent and tubulin polymerization inhibitor) and barbamide (molluscicide). The genome of *L. majuscula* 3L contains 15 genes assigned to the synthesis of sigma factors. Type I and type II σ^{70} factors Sig A to Sig E, respectively are encoded by five genes each. The remaining 5 genes belong to type III transcription factors, of which two resemble Sig F of *Synechocystis* sp. strain PCC 6803. Three others are distinctive from all other known Type III sigma factors.

***xii) Genome of Oscillatoria* sp. *PCC 6506*:** The genome of *Oscillatoria* PCC 6506 is of 6.7 Mb. Automatic annotation of the sequenced genome of *Oscillatoria* sp. PCC 6506 by MaGe annotation software revealed 6,007 coding sequences among which genes encoding RNAs (84), photosynthesis and *nif* gene cluster (37) and transposases (51) are recognized (Méjean *et al.*, 2010b). Among the genes encoding RNA, 70 genes have been allocated for tRNAs, 10 for small non-coding RNAs and 4 for rRNAs. Overall 3.7% of the genes are responsible for the synthesis of secondary metabolites. The gene clusters responsible for the synthesis of anatoxin-a and homoanatoxin-a (Cadel-Six *et al.*, 2009; Méjean *et al.*, 2009, 2010a) and cylindrospermopsin (Mihali *et al.*, 2008; Mazmouz *et al.*, 2010) have been identified on the genome. In addition, four other gene clusters are dedicated for the synthesis of polyketide synthase and non-ribosomal peptide synthetases. The presence of transposase sequences close to the toxin encoding gene clusters denotes LGT events.

***xiii) Genomes of C. raciborskii CS-505 and R. brookii D9*:** Both *C. raciborskii* CS-505 and *R. brookii* D9 constitute components of freshwater blooms while the strains of the former produce cylindrospermopsin or paralytic shellfish poisons, the toxic strains of the latter produce cylindrospermopsin and/or deoxycylindrospermopsin and anatoxin-a. The genomes of *C. raciborskii* CS-105 (3.89 Mb) and *R. brookii* D9 (3.2 Mb) are smaller in size amongst the filamentous and/or filamentous, heterocystous forms comparable to the genome size of the unicellular *Synechocystis* sp.

Table 4: General features of the genomes of strains *Cylindrospermopsis raciborskii* CS-505 and *Raphidiopsis brookii* D9 in comparison with four other fully sequenced genomes of filamentous cyanobacteria.

	D9	CS-505	Avar	Anab	Tery	Npun
Genome size (Mb)	3.20	3.89	6.34	6.41	7.75	8.23
G+C content (%)	40	40.2	41	41	40.8	41
Genes	3,088	3,968	5,134	5,432	5,542	6,501
Total CDS	3,010	3,452	5,039	5,368	4,452	6,087
Function assigned*	1,979	1,922	3,799	3,892	2,729	0
Unclassified	1,031	1,530	1,244	1,474	2,347	6,087
rRNA genes	9	9	12	12	5	12
tRNA genes	42	42	47	48	38	98
Transposases	9	77	57	145	260	112
Phage integrases	-	2	10	-	3	22
Repeated regions	53	406				
Plasmids	?	?	3	6	-	5
Unique CDS	394	794				
Function assigned*	157	291				
Unclassified	237	503				

Abbreviations: Avar: *Anabaena variabilis* ATCC 29413; Anab: *Anabaena* sp. PCC 7120; Tery: *Trichodesmium erythraeum* IMS101; Npun: *Nostoc punctiforme* PCC 73102.
*Function assigned according to COGs.
doi:10.1371/journal.pone.0009235.t002

With the kind permission of Mónica Vásquez, Department of Molecular Genetic and Microbiology, Faculty of Biological Sciences, Pontificia Universidad Católica de Chile, Santiago, Chile & Millenium Nucleus EMBA, Santiago, Chile, [Stucken *et al.* (2010) *PLoS ONE* **5(2)**: e9235. doi:10.1371/journal.pone.0009235].

strain PCC 6803 (Table 4). With a G+C content of 40%, the number of genes in *C. raciborskii* CS-505 is 3,968 of which 3,452 constitute the coding sequences (Stucken *et al.*, 2010). In case of *R. brookii* D9 the total number of genes is 3,088 of which 3,010 comprise the coding sequences. With the number of genes for rRNA and tRNA being 9 and 42, respectively in both the organisms, the genome of *C. raciborskii* CS-505 additionally possesses large number of genes for transposases, phage integrases and repeated regions (Fig. 6). A comparison of the sequenced genomes of heterocystous cyanobacteria led to the identification of COGs for filament formation (32 genes), diazotrophic metabolism (49 genes) and heterocyst development (149 genes) and nitrogen fixation (49 genes). The difference in the number of genes required for filament formation though appeared to be small in between *C. racoborskii* CS-505 (23 genes) and *R. brookii* D9 (20 genes), the common set of genes required for filament formation has further been reduced to 10 genes by the inclusion of unfinished genomes of *Nostoc azollae* 078 and *M. chthonoplastes* PCC 7420 in the sequence comparison. The number of genes required for diazotrophy in *C. raciborskii* CS-505 is 38 but in *R. brookii* D9 only five of these are present. The core number of orthologues accounted for 10 genes. Likewise, the COGs for heterocyst development have been found to be 149, among these 58 unique genes have been found that are common to all heterocyst-forming species and this number is further reduced to 41 if *N. azollae* 078 is included for sequence comparison.

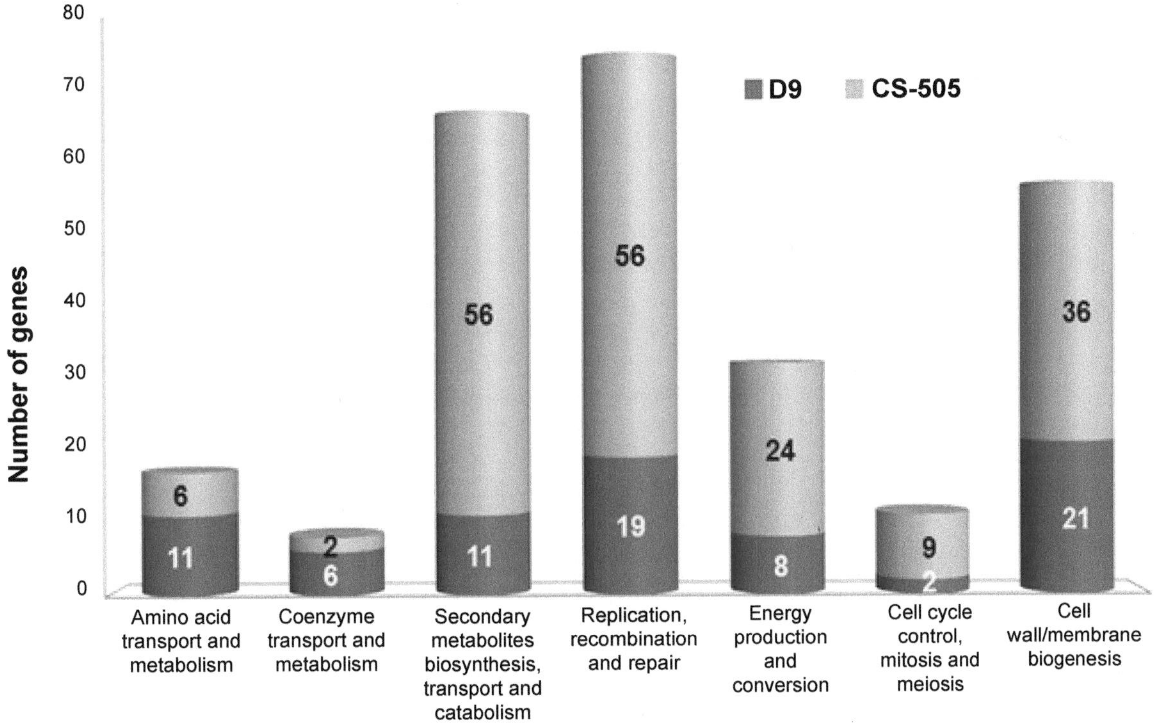

Figure 6: Distribution of the unique coding squences (CDS) of *Cylindrospermopsis raciborskii* CS-505 and *Raphidiopsis brookii* D9 into Cluster of Orthologous Groups (COGs). Only COG categories over-represented by CDS of *C. raciborskii* CS-505 or *R. brookii* D9 are shown. Unique CDS were obtained by a bidirectional best hits (BeTs) search between both genomes using a 30% cutoff. With the kind permission of M. Vásquez, Department of Molecular Genetic and Microbiology, Faculty of Biological Sciences, Pontificia Universidad Católica de Chile, Santiago, Chile & Millenium Nucleus EMBA, Santiago, Chile, [Stucken *et al.* (2010) *PloS ONE* **5(2):** e9235; doi:10.1371/journal.pone.0009235] doi:10.1371/journal.pone.0009235.g002

The number of genes for heterocyst development has been found to be 58 and 54 in *C. raciborskii* CS-505 and *R. brookii* D9, respectively. Though only four genes make a difference in the number of genes required for heterocyst development, *R. brookii* D9 does not form heterocysts. In *Anabaena* sp. strain PCC 7120 the number of genes required for heterocyst development have been found to be 77, of which 55 homologues are represented in *C. raciborskii* CS-505 without the gene sequences for *hetC*, *ccbP*, *hetL*, *hetN* and *patS*. On the other hand, *R. brookii* D9 lacks genes for nitrogen fixation (*nif* cluster-related genes, uptake hydrogenase gene cluster with *hupSL* and *hupW* and a set of genes involved in general metabolism) and heterocyst-specific glycolipid clusters (*hglEGDCA* and *hetM*). However, it possesses identical cluster of genes required for heterocyst envelope polysaccharides as present in *C. raciborskii* CS-505. The COGs identified for *nif* gene cluster in the heterocystous cyanobacteria on the basis of sequence comparisons is 49 which is the upper limit and 38 of them are noted in *C. raciborskii* CS-505. All the *nif* genes are present together in a 15 kb region in the genome of *C. raciborskii* CS-505 very much resembling the arrangement of genes in *nif2* gene cluster of *A. variabilis* ATCC 29413 (see for details Chapter 4). There is a greater degree of synteny in the genes for biosynthesis of cylindrospermopsin and saxitoxin in both the organisms (Stucken *et al.*, 2010).

***xiv) Genome of Anabaena* sp. *strain PCC 7120*:** The complete genomic sequence of *Anabaena* sp. strain PCC 7120 was reported by Kaneko *et al.* (2001). A single circular chromosome of 6,413,771 bp long with a G+C content of 41.3% is present in this organism. The total number of protein-coding genes is 5,368. The genome has a coding capacity of 45% and the encoded proteins showed similarity to known and predicted proteins of known function. Hypothetical gene products amount to 27% and the remaining (28%) genes lacked similarity to genes of known and predicted proteins in public databases. There are four sets of rRNA operons (rrnA-rrnD) in the sequence of 16S-23S-5S. The number of tRNA genes is 48 and 42 tRNA genes correspond to 42 tRNAs. Genes *fdxN*, *nifD* and *hupL* present in vegetative cells encode 57, 12 and 10 proteins, respectively including recombinases that help in excision and genome rearrangement during heterocyst differentiation.

Six plasmids designated as pCC7120α (with a length of 408,101 bp), pCC7120β (186,614 bp), pCC7120γ (101,965 bp), pCC7120δ (55,414 bp), pCC7120ε (40,340 bp) and pCC7120ζ (5,584 bp) have been characterized. From these plasmids, more than 100 genes have been cloned and sequenced much earlier to the sequencing of the genome. The distribution and functional identity of some of the genes is now known. A total of 764 genes have been assigned to the six plasmids. Excepting pCC7120ζ, the rest of them show genes for site specific recombinases, two of these are present on pCC 7120α while the others have one each. A gene (*all7280*) that encodes adenine-specific DNA methyltransferase and. ζ-carotene desaturase (*all7255*) which converts ζ-carotene to lycopene are located on plasmid pCC7120α. Of the 11 putative sigma factors three (*SigB*, *SigB3* and SigB4) have been assigned to the plasmids. Genes for DNA replication (i.e., DNA polymerase III β subunit, DNA polymerase γ and τ subunits, DNA polymerase δ′ subunit and single-strand DNA-binding proteins) are present on the plasmid pCC7120β. A cluster of three genes presumptively involved in encoding ABC phosphonate transporter is present on pCC7120γ. Sugaya *et al.* (2003) investigated the probable causes for the large genome size in *Anabaena* sp.strain PCC 7120 by comparing the gene-location distance and the distribution patterns of functional categories in the genomes of *Anabaena* sp. strain PCC 7120, *Synechocystis* sp. strain PCC 6803 and *T. elongatus* BP-1. They suggested that the present *Anabaena* sp. strain PCC 7120 genome might have resulted due to a whole genome duplication event during evolution. Subsequently, they further ruled out the sequential tandem duplications of genome segments and LGT events as the probable causes but attributed the large genome size due to whole genome duplication (Sugaya *et al.*, 2004).

***xv) Genome of Nostoc azollae 0708*:** Ran *et al.* (2010) published the complete genome sequence of the endosymbiont of *Azolla filiculoides* and reported complete genome erosion in *N. azollae* 0708. It has a circular chromosome of 5,846,145 bp with a low G+C content (38.3%) with two circular plasmids (Fig. 7). Of the 5,337 ORFs, the number of functional coding sequences is 3,668 while the rest of them are represented by pseudogenes. The proportion of pseudogenes to the number of predicted genes has been found to be highest in one of the plasmids P1 (63:51) followed by the chromosome (1623:3206) whereas in the second plasmid P2 it was 3:11. The accumulation of pseudogenes in certain important processes of DNA replication (*dnaA*) on the main chromosome and on the plasmid,

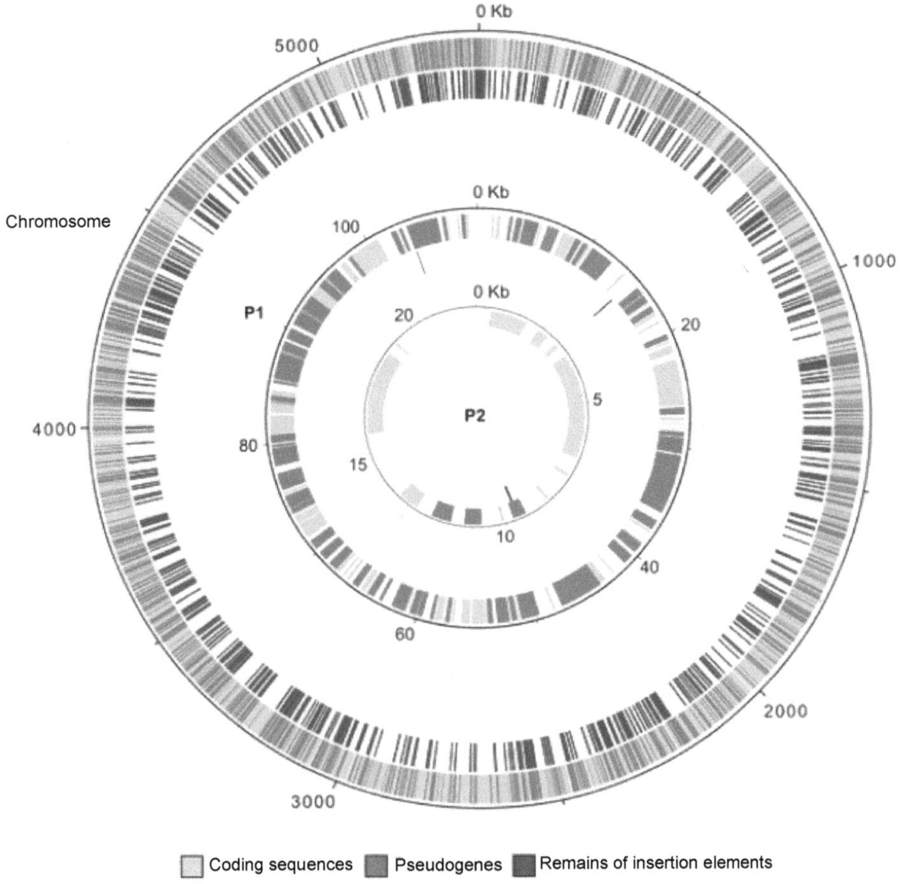

Figure 7: Map of the main chromosome, and plasmids (P1, P2) of the *'Nostoc azollae'* 0708 genome. The distribution of pseudogenes (red) and remains of insertion elements (blue) are indicated. Predicted genes are indicated by grey color. The highest level of gene erosion (number of pseudogenes:number of predicted genes) is found in the plasmid P1. Note that the occurrence of insertion elements appears to be correlated with the distribution of pseudogenes. The P1 and P2 plasmids only contain two and one remains of insertion elements, respectively. With the kind permission of B. Bergman, Department of Botany, Stockholm University, Stockholm, Sweden. [Ran *et al.* (2010) *PLoS ONE* **5(7):** e11486. doi:10.1371/journal.pone.0011486] doi:10.1371/journal.pone.0011486.g003.

Color image of this figure appears in the color plate section at the end of the book.

i.e. *dnaX, dnaN, holB* and *ssb* and DNA repair genes *recD, alkA* and *recQ* suggests that they are functionally redundant. Though genes for biosynthesis of photosynthetic pigments, proteins of PSI and PSII, nitrogen fixation and heterocyst differentiation are well conserved, phylogenetic analysis placed the genome of this organism (5.49 Mb) more closer to *C. raciborskii* CS-505 (3.88 Mb) and *R. brookii* D9 (3.19 Mb) rather than to the larger genomes of *Anabaena* sp. strain PCC 7120, *A. variabilis* ATCC 29413 and *Nostoc punctiforme* ATCC 73102 (Fig. 8). On the other hand, *N. azollae* 0708 shares highest number of COGs of proteins with *Anabaena* sp. strain PCC 7120, *A. variabilis* ATCC 29413 and *N. punctiforme* PCC 73102. In this respect, similar functional categories of COGs are shared in between *N. azollae* 0708 and *N. punctiforme* PCC 73102 in signal transduction mechanisms, cell membrane biogenesis, carbohydrate and amino acid metabolism and it is of special interest to know that the pseudogenes are under represented in these categories of genes in *N. azollae* 0708. Ran *et al.* (2010) determined the "core" and "shell" gene sets identified earlier for cyanobacterial genomes by Shi and

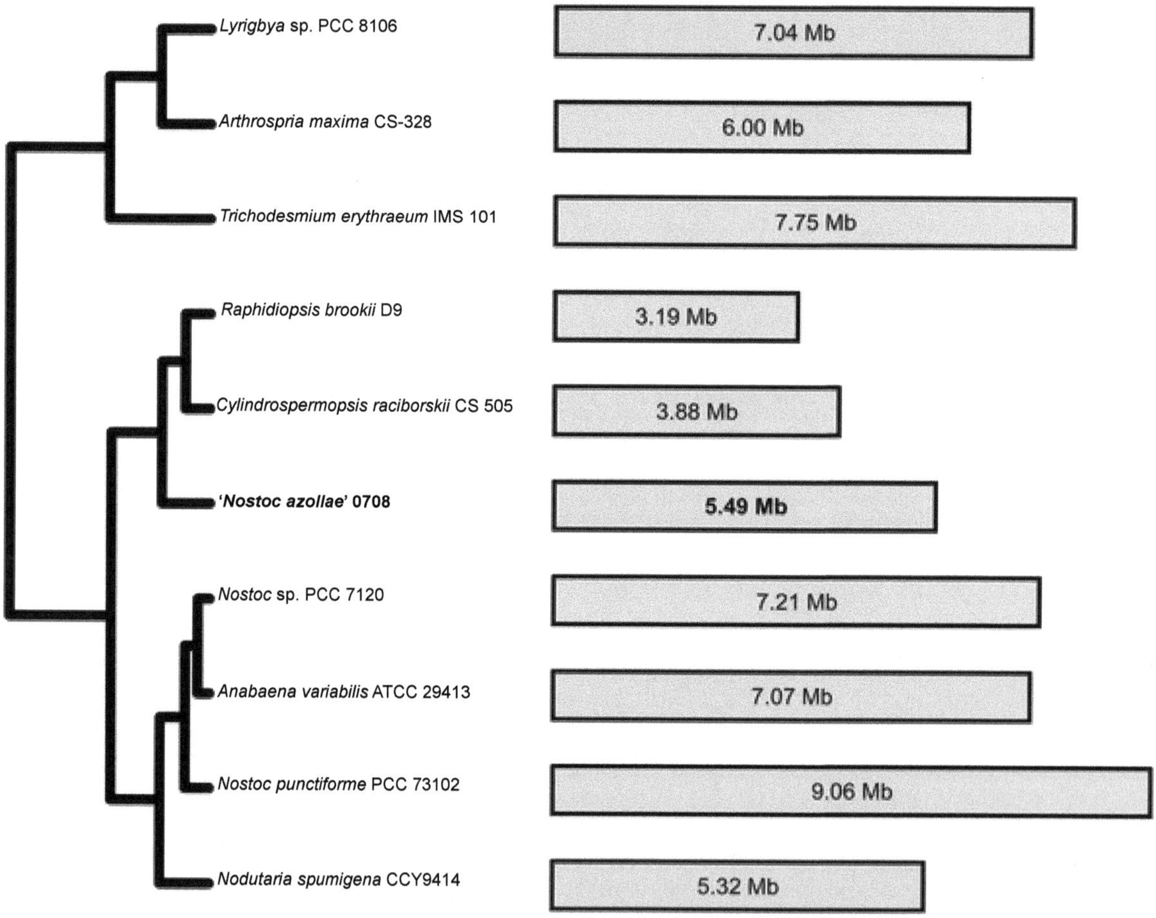

Figure 8: Phylogenetic tree and genome sizes for ten filamentous cyanobacterial species. The closest relatives to *'Nostoc azollae'* 0708 are *Raphidiopsis brookii* D9 and *Cylindrospermopsis raciborskii* CS-505, the two multicellular cyanobacteria with the smallest known genomes. The tree is a subclade from a maximum likelihood analysis of all cyanobacterial genomes available from NCBI and IMG/ER. With the kind permission of B. Bergman, Department of Botany, Stockholm University, Stockholm, Sweden [Ran *et al.* (2010) *PLoS ONE* **5(7):** e11486. doi:10.1371/journal.pone.0011486] doi:10.1371/journal.pone.0011486.g002.

Falkowski (2008) as 200 and 682, respectively for *N. azollae* 0708 and in this respect the endosymbiont resembles the free-living *Nostoc* and *Anabaena* species in retaining the important copies of the genes in both core and shell regions. However, it is quite apparent that the greater erosion in minimal set of genes governing glycolysis (*pfkA, gapA, pykA, gpmA, ldh*), nucleic acid replication, recombination and repair has taken place in the genome of the symbiont consistent with the needs of the host plant. The presence a phosphoenolpyruvate-dependent sugar phosphotransferase system in *N. azollae* 0708 akin to the major carbohydrate transport system in bacteria suggests that the cyanobacterium receives the required carbohydrates from the host plant. In return, the host plant receives fixed nitrogen from the symbiont as the symbiont has lost few genes in the areas of amino acid transport and metabolism, uptake of bicarbonate and phosphate and the ability to utilize alternative combined nitrogen sources thus reducing the symbiont to a nitrogen fixer.

***xvi) Genome of N. punctiforme ATCC 29133*:** The genome (8.2 Mb) has a mol% G+C of 41.5 and only 94% of the sequenced genome has been annotated and in this respect a preliminary analysis revealed 6,086 protein-coding ORFs of which 5314 are associated with previously recognized ORFs. The genes that encode proteins of known or probable function function are 3328 (amounting to 45%). The genes that encode conserved hypothetical and hypothetical proteins of unkown function are 1986 (constituting 27% of the total). Genes that do not bear resemblance to the previously known genes constitute 29% of the total. A comparison with *Anabaena* sp. strain PCC7120 genome revealed that *N. punctiforme* ATCC 29133 possesses 4814 (86%) of the ORFs of *Anabaena*. However, the number of ORFs of *N. punctiforme* in *Anabaena* is 5610 (Meeks *et al.*, 2001).

III. CATEGORIES OF GENES

According to the principles laid down by Riley (1993) the putative genes, whose function is known have been grouped into 14 categories. A comparative account of these groups is provided in Table 5 for four genera, i.e. *Synechocystis* sp.strain PCC 6803, *G. violaceus* PCC 7421, *T. elongatus* BP-1 and *Anabaena* sp. strain PCC 7120. The number of genes in each category is more in *Anabaena* sp. PCC 7120. It may be because of the fact that it is a nitrogen fixer. So genes associated with heterocyst differentiation and nitrogen fixation are additionally present in this organism. Genes related to regulatory functions are represented in large numbers in *Anabaena* sp. PCC 7120 (339) with minimum being represented in *T. elongatus* BP-1 (87).

***1) Biosynthesis of co-factors, prosthetic groups and carriers*:** *Synechococcus* sp. strain WH8102 possesses genes for the synthesis of plastocyanin (copper) for photosynthetic electron transport instead of ferredoxin and a cobalt-dependent ribonucleotide reductase (governed by *SYNW1692*; rather than iron-containing one as noted in many cyanobacteria) to overcome the iron deficiency. Genes for iron-dependent metalloenzymes (cytochrome P450 two additional cytochrome *c* molecules and one or two additional ferredoxins) are present in *Synechococcus* sp. strain CC9311 (Palenik *et al.*, 2006). Nicotinamide adenine dinucleotide (NAD) participates in a number of metabolic and regulatory processes. NAD(P) co-factors assume significance due to their role in photosynthesis and respiration. Taking into account *Synechocystis* sp. strain PCC 6803 as a model organism, Gerdes *et al.* (2006) compared the genomes of *E. coli* K12MG1655 and twelve other cyanobacteria (*A. variabilis* ATCC 29413, *N. punctiforme* PCC 73102, *Anabaena* sp. strain PCC 7120, *S. elongatus* PCC 7942, *P. marinus* MIT9313, *P. marinus* subsp. *marinus* strain CCMP1375 (MED4), *P. marinus* subsp. *pastoris* strain CCMP1986, *Synechococcus* sp. strain WH8102, *T. erythraeum* IMS101, *T. elongatus* BP-1, *C. watsonii* WH8501, *G. violaceus* PCC 7421). In all the cyanobacterial genomes examined including the model organism,

Table 5: Distribution of genes to the functional categories in four cyanobacterial genomes.

Functional Category	*Synechocystis* sp. strain PCC6803[a]	*G. violaceus* PCC7421[b]	*T. elongatus* BP-1[c]	*Anabaena* sp. strain PCC7120[d]
Amino acid biosynthesis	97	106	99	111
Biosynthesis of co-factors, Prosthetic groups and carriers	124	141	120	152
Cell envelope	67	61	56	80
Cellular processes	78	86	78	94
Central intermediary metabolism	31	30	23	70
Energy metabolism	93	102	72	98
Fatty acid, phospholipid and sterol metabolism	39	53	29	41
Photosynthesis and respiration	141	158	124	156
Purines, Pyrimidines, nucleosides, nucleotides	41	46	41	57
Regulatory functions	146	195	87	339
DNA replication, recombination, repair	60	68	62	93
Transcription	30	44	29	38
Translation	168	193	151	189
Transport and binding proteins	196	226	153	294
Other categories	312	327	266	584
Total number of genes of known function	2756	1836	1390	2396
Total number of genes	3264	4430	2475	5368

a (after Kaneko *et al.*, 1996); b (after Nakamura *et al.*, 2003); c (after Nakamura *et al.*, 2002); d (after Kaneko *et al.*, 2001).

a *de novo* pathway concerning the conversion of aspartate to NaMN is governed by a group of three genes (*nadB*, *nadA* and *nadC*) scattered all over the genomes. The presence of three other genes (*nadD*, *nadE* and *nadK*) that mediate conversion of NaMN to NADP are seen in all the genomes with the exception of *nadD* and *nadE* orthologues form a chromosomal cluster in all cyanobacteria except in the model organism where these occur in distal chromosomal loci.

***2) Cell envelope*:** Rocap *et al.* (2003) identified a 41.8 kb cluster of surface polysaccharide genes in a lower G+C (42%) region than the genome as a whole suggesting that it might have been acquired by *P. marinus* MIT9313 through LGT. In contrast, *P. marinus* MED4 has acquired a considerably large cluster (74.5 kb) consisting of 67 potential surface polysaccharide genes. *Synechococcus* sp. strain WH8102 possesses genes *SwmA* and *SwmB* for swimming motility. These genes are located in a low G+C region of the genome and of these *SwmB* encodes a large protein with 10,791 amino acids. Homologues of *pilB*, *-C*, *-D*, *-Q* and *-T* that govern twitching motility are represented in the genome of *Synechococcus* sp. strain WH8102 but orthologues of these genes are only represented in the genome of *P. marinus* MIT9313 but not in MED4 (Palenik *et al.*, 2003). Palenik *et al.* (2006) compared cell envelope of *Synechococcus* sp. strain CC9311 with that of *Synechococcus* sp. strain WH8102. There are a number of differences in the synthesis of LPS as *Synechococcus* sp. strain CC9311 lacks genes that encode KDO (a molecule necessary for biosynthesis of typical LPS) and for rhamnose biosynthesis (this sugar also is a potential component of LPS). It means that the type of LPS synthesized by *Synechococcus* sp. strain CC9311 is quite different in its properties thus affecting the sensitivity of this organism to certain phages. The presence of genes for pilin and pilin biosynthesis is indicative of pili-mediated twitching motility or DNA uptake in *Synechococcus* sp. strain CC9311. However, this strain lacks the two major surface proteins, *SwmA* and *SwmB* involved in swimming motility of *Synechococcus* sp. strain WH8102. A comparison of cell envelope related genes in the genomes of unicellular and filamentous cyanobacteria, *P. marinus* MED4, *Synechocystis* sp. strain PCC 6803, *T. erythraeum* IMS101, and *Anabaena* sp.strain PCC 7120 revealed 100, 186, 266 and 294 genes, respectively (Yang *et al.*, 2007). Thus the number of cell envelope related genes gradually increased from unicellular to filamentous cyanobacteria (Table 6). Minimum number of genes has been found for the biosynthesis of peptidoglycan, lipopolysaccharide, outermembrane proteins and genes of unknown function in *P. marinus* MED4 while there is large number of genes found in each category in *Anabaena* sp. strain PCC 7120.

Table 6: Cell envelope-related genes in cyanobacterial genomes.

Sr. No.	Cyanobacterium	Peptidoglycan	LPS	Exoplysaccharide	OM proteins	Unknown	Total
1.	*P. marinus* MED4	29	40	14	16	2	100
2.	*Synechocystis* sp. strain PCC 6803	37	73	28	40	8	186
3.	*T. erythraeum* IMS101	47	90	48	63	18	266
4.	*Anabaena* sp. strain PCC 7120	48	113	61	60	12	294

3) Energy metabolism: The number of genes allocated to this functional category seems to be lowest in *T. elongatus* BP-1 (72) and highest in *G. violaceus* PCC 7421 (102) (Table 5). Detailed explanation with reference to the processes governed is only available for *Cyanothece* sp. ATCC 51142. The genome of this organism contains all the genes related to fermentation processes (yielding end products like ethanol, lactate, acetate and hydrogen) that are known to occur in darkness coinciding with nitrogen fixation. A 20.2 kb cluster containing genes related to glucose and pyruvate metabolism has been identified on the linear chromosome while the genes related to carbohydrate and energy metabolism are distributed at multiple loci on the circular chromosome. The presence of lactate dehydrogenase (*ldh*) gene cluster on the linear chromosome that mediates the terminal step of lactate fermentation is unique to this organism. Another unique feature is the presence of a gene that encodes phosphoenolpyruvate carboxykinase that mediates the first step in gluconeogenesis required for dark metabolism.

4) Fatty acid, phospholipid and sterol metabolism: Genes related to fatty acid metabolism have been found to be lowest in *T. elongatus* BP-1(29), followed by *Synechocystis* sp. strain PCC 6803 (39) and *Anabaena* sp. strain PCC 7120 with maximum number of up to 53 found in *G. violaceus* PCC 7421 (Table 7). The standard nomenclature of fatty acids indicates the number of carbon atoms, the number of double bonds and the position of double bonds. For example stearic acid is a 18 carbon compound with no double bonds and designated as 18:0. If a double bond is introduced into stearic acid at carbon 9, it is converted to oleic acid, designated as 18:1, Δ^{9c}. That is one double is introduced between C9–C10. Saturated fats contain no double bonds-the more double bonds there are, the greater the degree of unsaturation. These are then converted to acyl-CoA esters. Subsequently, triacylglycerols are formed by step-wise acylation of glycerol-3-phosphate. Additional double bonds may be added typically to Δ^{12}, Δ^{15} and Δ^{6}. Fatty acid desaturases are the enzymes that introduce the double bonds at specific sites. Fatty acid desaturation of membrane lipids is shown to increase the fluidity of membrane lipids and is suggested to help to overcome the stress caused by high temperature and light intensity (Stubbs and Smith, 1984; Murata and Wada, 1995; Gombos *et al.*, 1997). In the thylakoid membranes of cyanobacteria monogalactosyldiacylglycerol (MGDG) comprises half of the components and plays an important role in the development of membrane system. Awai *et al.* (2006) compared the genomes of *Synechocystis* sp. strain PCC 6803 and *Anabaena* sp. strain PCC 7120 and identified genes encoding monoglucosyldiacylglycerol (MGlcDG) synthase that mediates galactolipid biosynthesis. The amino acid sequence of both the MGlcDG synthases showed significant

Table 7: Distribution of types and number of fatty acid desaturase genes in some cyanobacteria (modified after Chi *et al.*, 2008).

Sr. No.	Name of strain	Types of fatty acid desaturases	No. of copies	Total no.
1.	*Anabaena* sp. PCC 7120	Δ9, Δ12, Δ15	Δ9-two genes	4
2.	*A. variabilis* ATCC 29413	Δ9, Δ12, Δ15	Δ9-two genes	4
3.	*C. watsonii* WH8501	Δ9, Δ12, Δ15	Δ12-two genes	4
4.	*G. violaceus* PCC 7421	Δ9, Δ12	Δ9-six genes; Δ12-two genes	8
5.	*N. punctiforme* ATCC 29133 (PCC73102)	Δ9, Δ12, Δ15	Δ9-two genes	4
6.	*P. marinus* NATL1A	Δ9, Δ12	One gene each	2
7.	*P. marinus* NATL2A	Δ9, Δ12	One gene each	2
8.	*P. marinus* MIT9211	Δ9, Δ12	Δ12-two copies	3
9.	*P. marinus* MIT9301	Δ9, Δ12	Δ9-two genes	4
10.	*P. marinus* MIT9303	Δ9, Δ12	Δ9-two genes	4
11.	*P. marinus* MIT9312	Δ9, Δ12	Δ12-two genes	3
12.	*P. marinus* MIT9313	Δ9, Δ12	One gene each	2
13.	*P. marinus* AS9601	Δ9, Δ12	Δ12-two genes	3
14.	*P. marinus* MIT9515	Δ9, Δ12	One gene each	2
15.	*P. marinus* subsp. marinus CCMP1375 (SS120)	Δ9, Δ12	Δ12-two genes	3
16.	*P. marinus* subsp. *marinus* CCMP1375 (MED4)	Δ9, Δ12	Δ12-two genes	3
17.	*S. elongatus* PCC 7942	Δ9	One gene	1
18.	*S. elongatus* PCC 6301	Δ9	One gene	1
19.	*Synechococcus* sp. BL107	Δ9, Δ12	Δ9-two genes	3
20.	*Synechococcus* sp.CC9311	Δ9, Δ12	Δ9-two genes	3
21.	*Synechococcus* sp.CC9605	Δ9, Δ12	One gene each	2
22.	*Synechococcus* sp.CC9902	Δ9, Δ12	Δ9-two genes	3
23.	*Synechococcus sp. JA-2-3B'a(2-13)*	Δ9	One gene	1
24.	*Synechococcus* sp. JA-3-3A'b	Δ9	One gene	1
25.	*Synechococcus* sp. RCC307	Δ9, Δ12	Δ9-two genes	3
26.	*Synechococcus* sp. RS9916	Δ9	Δ9-two genes	2
27.	*Synechococcus* sp. RS9917	Δ9, Δ12	Δ9-two genes	3
28.	*Synechococcus* sp. WH5701	Δ9, Δ12	Two genes each	4
29.	*Synechococcus* sp. WH7803	Δ9, Δ12	Two genes each	4
30.	*Synechococcus* sp.WH7805	Δ9, Δ12	Two genes each	4
31.	*Synechococcus* sp. WH8102	Δ9, Δ12	Δ12-two genes	3
32.	*Synechocystis* sp. PCC 6803	Δ9, Δ12, Δ15, Δ6	One gene each	4
33.	*T. elongatus* strain BP-1	Δ9	Three genes	3
34.	*Trichodesmium erythraeum* IMS101	Δ9, Δ12, Δ15	One gene each	3
35.	*Lyngbya* sp. PCC 6106	Δ9, Δ12, Δ15, Δ6	One gene each	4
36.	*N. spumigena* CCY9914	Δ9, Δ12, Δ15, Δ6	Δ9-two genes	5
37.	*Cyanothece* sp. 0110	Δ9, Δ12, Δ15, Δ6	Δ9-two genes	5

similarities and the genes were identified from the category of genes of unknown function. Chi *et al.* (2008) presented a comparative analysis of fatty acid desaturase genes by comparing 37 cyanobacterial genomes. The cyanobacterial genomes sequenced by various workers have been downloaded and a comparison made. The types and number of fatty acid desaturase genes are listed in Table 7. In *Synechococcus* sp. strain JA-2-3B'a (2-13), *Synechococcus* sp. strain JA-3-3A'b, *S. elongatus* PCC 6301 and *S. elongatus* PCC 7942 one gene each of Δ^9 desaturase is found. In *Synechococcus* sp. RS9916 also a single type of desaturase (Δ^9) is present but there are two genes. Likewise, in *T. elongatus* BP-1, Δ^9

desaturase gene is present in three copies. In other nine strains of *Synechococcus* sp., Δ^9 and Δ^{12} type of desaturases are seen with a variation in the total number of genes from two to four. Likewise, in all *Prochlorococcus* strains Δ^9 and Δ^{12} type of desaturases have been encountered with a variation in the total number of genes from two to four. Exceptionally, in *G. violaceus* PCC 7421, Δ^9 and Δ^{12} types of desaturases are present with six genes of the former and two genes of the latter type. In heterocystous nitrogen-fixing forms (*Anabaena* sp. strain PCC 7120, *A. variabilis* ATCC 29413, *N. punctiforme* ATCC 29133) three types desaturases Δ^9, Δ^{12} and Δ^{15} have been detected with total number of genes being four to five. However, in *No. spumigena* CCY9914 besides Δ^9, Δ^{12}, Δ^{15} desaturases, an additional Δ^6 type of desaturase has also been met with. Likewise, *Cyanothece* sp. CCY0110, *Lyngbya* sp. PCC 8106 and *Synechocystis* sp. strain PCC 6803 also possess four types of fatty acid desaturases.

***5) Photosynthesis and respiration*:** Genes related to the synthesis of proteins of PSI are designated as Psa genes while those associated with PSII are named as Psb genes. The PSI is composed of 11–14 subunits designated as PsaA to PsaO. There is a great degree of similarity between the subunits of green plants, eukaryotic algae and cyanobacteria except that in cyanobacteria genes encoding PsaG and PsaH are absent (Xu *et al.*, 2001; Scheller *et al.*, 2001). Furthermore, cyanobacterial PSI does not contain the cluster of light-harvesting proteins (LHCI) characteristically found in the PSI of eukaryotic algae and higher plants. The cyanobacterial PSI complex exists in a trimeric form (Jordan *et al.*, 2001) whereas the PSI of higher plants, green algae and red algae exists in a monomeric state (Gardian *et al.*, 2007). Though the electron transport chain of PSI in eukaryotic algae and cyanobacteria is very similar, there exist certain differences in the number of chlorophyll molecules associated with PSI and PSII. The crystal structure of the different components of photosynthetic electron transport chain have been published that relate to PSI (Jordan *et al.*, 2001), cytochrome *b6/f* (Stroebel *et al.*, 2003), LHCI-PSI (Amunts *et al.*, 2007) and PSII (Umena *et al.*, 2011). Allen *et al.* (2011) integrated these and proposed a molecular model of the Z-scheme of photosynthesis. This gives an idea of the various component proteins of PSI and PSII and their arrangement in the thylakoid (Fig. 9) which makes it easier to understand the genes related to PSI and PSII of cyanobacteria presented below.

Genes allocated for photosynthesis and respiration are found to be 124 and 141 in case of *T. elongatus* BP-1 and *Synechocystis* sp. strain PCC 6803, respectively. In the other two cyanobacterial genomes, i.e. *G. violaceus* PCC 7421 (158 genes) and *Anabaena* sp. strain PCC 7120 (154 genes), almost same number of genes is found (Table 5). Kaneko *et al.* (1996) identified 124 genes in *Synechocystis* sp. strain PCC 6803 relating to PSI and PSII, phycobilisome formation, ATP synthesis, CO_2 fixation and electron transport systems. In the genome of *T. elongatus* BP-1, the gene *psbV* that encodes cytochrome C550 existed in a tandemly duplicated state. There is one copy of *psbA* that encodes reaction centre D1 complex of PSII and one copy of each gene of the cytochrome b6/f complex (*petA-petD*, *petG*, *petM* and *petN*). Complete sets of genes for phycobilisome components (*cpcA-cpcG* for phycocyanin and *apcA-apcF* for allophycocyanin) are present in this organism but the absence of genes *petL* (encoding a membrane spanning component) and *petE* (that encodes plastocyanin) is significant (Nakamura *et al.*, 2002). In the genome of *N. punctiforme* PCC 73102 genes *psbA* (four complete and three truncated copies), *psbD* (one complete and one truncated copy) and single copies of *psaA* and *psaB* encoding proteins of photosynthetic electron transport are present in a putative operon. Genes of the cytochrome b6/f complex *petB*(cytochromeB)-*petD*(subunit IV)-*petC*(Fe/S protein)-*petA*(Cytochrome f)-*petE*(plastocyanin) and *petJ* (cytochromec553) are present with an additional copy of the gene *petJ*. The synthesis of plastocyanin in presence of sufficient amounts of copper and cytochrome c553 in presence of iron are indicated as soluble electron carriers between the cytochrome b6/f and PSI complexes. Ferredoxin-NADP oxidoreductase and ferredoxin are encoded by a single gene (*petH*)

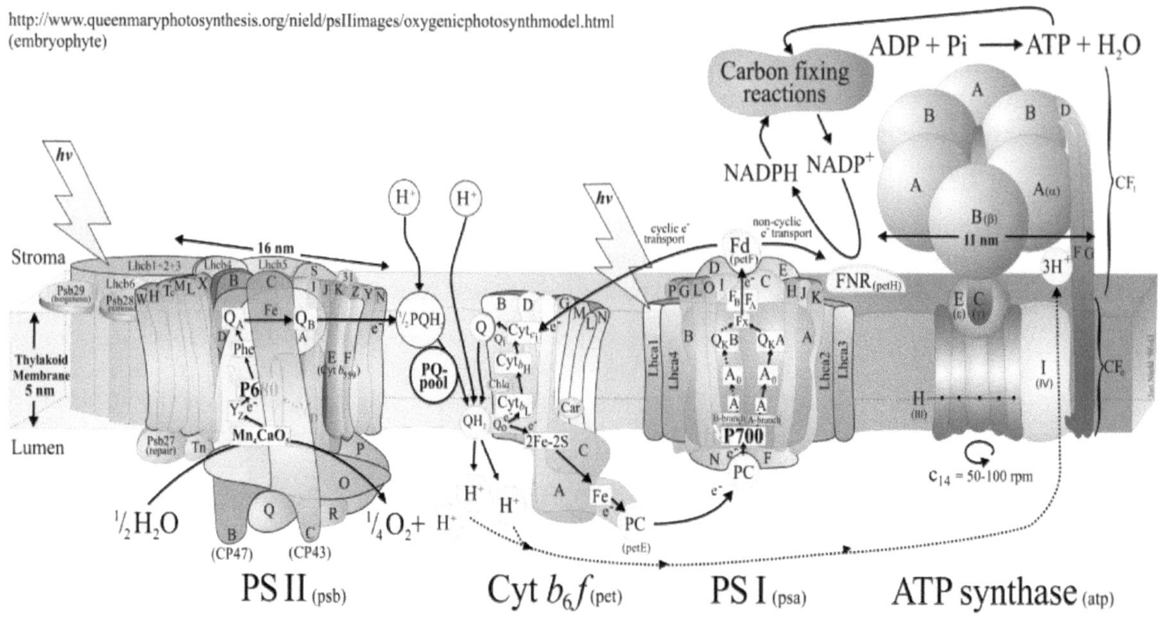

Figure 9: Model depicting the molecular machinery involved in the Z-scheme of oxygenic photosynthesis. Abbreviations: Lhca/*b* proteins: Light (*hv*) is captured and channelled to the P680 reaction centre of PSII, or the P700 reaction centre of PSI. PSII: In caps= *psb* gene products; P680= primary donor; Phe= pheophytin; Q_A,Q_B= secondary plastoquinone acceptors; PQ= plastoquinone. Cyt b_6f: caps= *pet* gene products; Cyt b_H= high potential; Cyt b_L= low potential; Note: total of 4 haem centres per monomer (Fe), including c_i= link for cyclic e^- transport; Q= quinone, Car= carotenoid, Chl*a*= Chlorophyll *a* molecule. PSI: In caps= *psa* gene products; P700= primary donor; PC= plastocyanin; A_0 (Chl*a*), A_1 (phylloquinone)= primary and intermediate e^- acceptors; F_A, F_B and F_X= bound FeS centres; Fd= ferredoxin. A, A' accessory chls are not shown. Approximate dimensions as indicated, in nm. Courtesy of Jon Nield, Queen Mary, University of London, UK; freely available from http://www.queenmaryphotosynthesis.org/nield/psIIimages/oxygenicphotosynthmodel.html.

Color image of this figure appears in the color plate section at the end of the book.

and 15 putative genes (9 encoding 2Fe-2S *petF* type and 6 encoding 4Fe-4S type), respectively (Meeks *et al.*, 2001). Rocap *et al.* (2003) observed that the genomes of *P. marinus* strains, MED4 and MIT9313 lacked phycobilisome structural proteins and enzymes involved in their biosynthesis. *P. marinus* MED4 has a single gene (*psbA*) where as *P. marinus* MIT9313 has two such genes. Both possess a single copy of the gene *psbD*. The *psbV* gene is absent in *P. marinus* MED4. As *P. marinus* MED4 is a HL-adapted strain, it has 22 genes that synthesize high-light inducible proteins (HLIPs). One copy of the gene encoding chlorophyll *a*/*b*-binding antenna protein is present in *P. marinus* MED4 whereas *P. marinus* MIT 9313 has two copies of this gene. Both of the strains lacked the typical cyanobacterial RUBP carboxylase/oxygenase (RuBisCO form IB) but instead possess genes for the synthesis of RuBisCO of form IA. These genes are located adjacent to carboxysome shell protein encoding genes. The similarity of these genes to those of γ-proteobacterium *Acidothiobacillus ferroxidans* suggests that they might have been acquired through LGT. Both these strains have genes for oligopeptide transporters and some sugar transporters which indicates that they can be partial heterotrophs if the need arises. Genes encoding some enzymes of TCA cycle such as 2-oxoglutarate dehydrogenase, succinyl-CoA synthase and succinyl-CoA-acetoacetate-CoA transferase are absent in both these strains. In a very exhaustive study, Dufresne *et al.* (2008) compared the genomes of all

11 strains of *Prochlorococcus* sequenced so far and confirmed the loss of all phycobilisome genes in these strains excepting those that encode phycoerythrin in LL strains and in a few HL-strains. The genome of *G. violaceus* PCC 7421 possesses *cpe* gene cluster that is related to phycoerythrin (*cpeR-cpeZ-cpeY-cpeA-cpeB-yef58-cpeF-cpeT*) and a rod-linker component gene *cpc*. A number of genes related to PSI (*PsaI,PsaJ, PsaK* and *PsaX*), PSII (*PsbY, PsbZ* and *Psb27*), a rod core linker peptide (*cpcG*) and degradation of phycobilisomes (*nblA*) are absent. The genes related to oxygen evolution (*PsaF, PsbO, PsbU* and *PsbV*) are poorly conserved (Nakamura *et al.*, 2003).

In cyanobacteria cytochrome b6/f complex is shared by both photosynthetic and respiratory electron transport systems. Accordingly, genes encoding mitochondrial type-1 multiprotein complex and bacterial type-2 FAD-containing NADH dehydrogenase (NADH:plastoquinoloxidoreductase) are present. In the genome of *N. punctiforme* ATCC 29133, a type-1 *ndhckj* contiguous cluster, four copies of large subunit type-2 dehydrogenase genes and possibly four copies of *ctaCDE* operon, governing the synthesis of cytochrome c oxidase are present. For carbon assimilation single copies of phosphoribulokinase and RuBisCO are present. The cluster of RuBisCO consists of seven genes in the order of *rbcL – rbcX – rbcS – OrfH1 –* $OrfH_2$ *–* $OrfH_3$*-rca*. A RuBisCO transcriptional regulator (rbcR or CbbR) is present near the cluster of seven genes that encode proteins of carbon concentrating mechanism (CCM). The order of these genes is *ccmK3 – ccmK2 – ccmL – ccmM – ccmN-fpg – ccmK1 – ccmL-rbcR* where *fpg* encodes putative formamidopyrimidine-DNA glycosylase. Genes *ccmK4* and *ccmK5* are present separately. Five copies of the gene encoding carbonic anhydrase are distributed solitarily through out the genome. An operon for glucose catabolism consists of genes in the order of *fbp – tal – zwf – opcA* that encode respectively fructose-1,6-biphosphatase, transaldolase, glucose-6-phosphate dehydrogenase (G6PD) and a protein allosteric effector of G6PD (Meeks *et al.*, 2001). *Anabaena* sp. strain PCC 7120 possesses the same arrangement of genes in this operon. In *Synechococcus* sp. strain WH8102 and *P. marinus* MED4 a *zwf-opcA* operon is present with other genes located elsewhere. In *S. elongatus* PCC 7942 the *tal* gene is located away from the cluster but in *Synechocystis* sp. strain PCC 6803 all the genes of the operon are dispersed all over the chromosome.

In cyanobacteria several genes have been identified whose products constitute members of the light-harvesting complex. Amongst these are high-light inducible proteins (HLIPs) or small cab-like proteins (Scp; Dolganov *et al.*, 1995; Funk and Vermaas, 1999). The expression of these genes has been observed under conditions of high light-stress, nitrogen starvation and low temperature. The genomes of marine [*P. marinus* MED4 (HL-adapted), *P. marinus* LL-strains MIT9313 and SS120; *Synechococcus* sp. strain WH8102] and freshwater (*Synechocystis* sp. strain PCC 6803, *N. punctiforme* ATCC 29133 and *Anabaena* sp. strain PCC 7120) cyanobacteria revealed a large number of *hli* genes in the former than in the latter. The highest number of such genes (22) has been detected in *P. marinus* MED4 whereas the two LL-strains have 9 (MIT9313) and 13 (SS120) and the *Synechococcus* sp. strain WH8102 has 8 *hli* genes. The number of *hli* genes in freshwater strains has been found to be 4, 8 and 9 genes, respectively in case of *Synechocystis* sp. strain PCC 6803, *Anabaena* sp. strain PCC 7120 and *N. punctiforme* ATCC 29133. The presence of significantly high number of *hli* genes in *P. marinus* strain MED4 signifies the occurrence of gene duplication events due to the high light growth conditions imposed upon the organism and the associated selection pressure. The distribution of the total 73 *hli* genes identified, separated into 24 clusters (although 11 of these represent single gene clusters, the largest of which has 11 *hli* genes), reveales a strong divergence between marine and freshwater species (Bhaya *et al.*, 2002). In two strains of *Synechococcus* sp. strain CC9311 and WH8102 the number of predicted genes for photosynthesis and light-harvesting are almost similar. However, there is a difference in the number of genes related to HLIP. There are 14 HLIP genes in *Synechococcus* sp. strain CC9311 where as there are 8 such genes in *Synechococcus* sp. strain WH8102. This suggests that *Synechococcus* sp. strain CC9311 is adapted to grow in HL habitats (Rocap *et al.*, 2003). In *S. elongatus* PCC 6301 there are 139 genes coding for photosynthesis-related components (Sugita *et al.*, 2006). In *Prochlorococcus* strains light regulation of photosynthetic CO_2 assimilation seems to have been lost as genes (*ndh4, ndhF4, CphX*) regulating low-affinity carbon transport are missing as well as genes encoding a thioredoxin and ferredoxin. These genes are characteristically represented in all genomes of *Synechococcus* sequenced to date (Dufresne *et al.*, 2008).

Photosynthetic pigments: Phycobilisomes (PBSs) are highly organized complexes of phycobiliproteins (PBPs) that contain allophycocyanins (APCs) as a central core to which the phycocyanins (PCs) and phycoerythrins (PEs) are attached to give rise to radiating arms or rods. All the PBPs in cyanobacteria consist of two distinct subunits α and β and are assembled into trimeric $(\alpha\beta)_3$ or mostly as hexameric $(\alpha\beta)_6$ complexes. In freshwater *Synechococcus* strains (*S. elongatus* PCC 7942 and *Synechocystis* sp. strain PCC 6803) while the core is composed of APC, the rods are entirely composed of PC. In PE-containing strains of *Synechococcus*, one basal disc in the rod is composed of PC and the rest is occupied by the red coloured phycoerythrobilin (PEB) at a molecular ratio of 1:2. In *Synechococcus* sp. strain WH7805 at the basal part of the rod instead of PC, an optically variant form of PC is present that is linked to PC and PEB at a molecular ratio of 2:1. However, in most PE-containing strains the distal part of the rods consists of two types of PE (PEI and PEII). PEII always binds with PEB and another orange coloured pigment known as phycourobilin (PUB; A_{max}=495 nm) whereas PEI binds only PEB or both PEB and PUB. Six *et al.* (2007) recognized three major pigment types in *Synechococcus* where type 1 is represented by forms with PC (strains RS9917 and WH5701), type 2 have PC and PEI (strains WH8018 and WH7805) and type 3 have PC, PEI and PEII. Type 3 has further been subdivided into four subtypes 3a-3d depending on the ratio of PEB and PUB bound to PEs. Strains that belong to subtype 3a have a low ratio of PEB and PUB (WH7803) whereas those included in 3b and 3c exhibit moderate and high ratios of PEB and PUB, respectively (strains CC9605, WH8102, RCC307, BL107 and CC9902). The fourth subtype 3d is characterized by the strains that possess variable amounts of PEB and PUB depending on the quality of light (white/green or blue) under which they are grown (strains CC9311 and RS9916). These show type IV chromatic adaptation. They proposed models for PBS structure for different *Synechococcus* pigment types and subtypes (Fig. 10). Besides the 11 strains of *Synechococcus* whose genomes have been sequenced (Table 3), Six *et al* (2007) included eight other strains that belonged to pigment type 2 (strain WH8018), pigment type 3a (strains AlmoO3 and RS9912), pigment type 3c (strains WH8103 and Oli31) and pigment type 3d (strains WH8020, MI I.I and M16.17strains). As in other cyanobacteria, the PBS genes are clustered together in these 19 strains of *Synechococcus*. The first small cluster is represented by four APC core genes *apcE-A-B-C* and two other APC core genes *apcD* and *apcF* are distributed away from PBS genes. Most of the rod genes exist in large clusters whose size depends on the complexity of pigment type, i.e. from 9–10 kbp in pigment type 1 to 27–28.5 kbp in pigment subtype 3d. The distribution and number of genes in the three pigment types is depicted in Table 8.

The assembly of PBS is mediated by linker polypeptides and each trimeric or hexameric subassembly contains at least one specific liker polypeptide. This in turn determines the type, location and aggregation state of the PBP within the rod and also modulates the spectroscopic properties. Glazer (1985) has provided a system of abbreviations for the linkers and their characterization for sequence and molecular weight determinations has been accomplished. Most of the linker polypeptides are associated with APC, PC or PE and so designated as APC-associated linkers (L_C, i.e. core linker), PC and PE associated linkers are (L_R for rod linker) where as L_{RC} is designated for PBSs

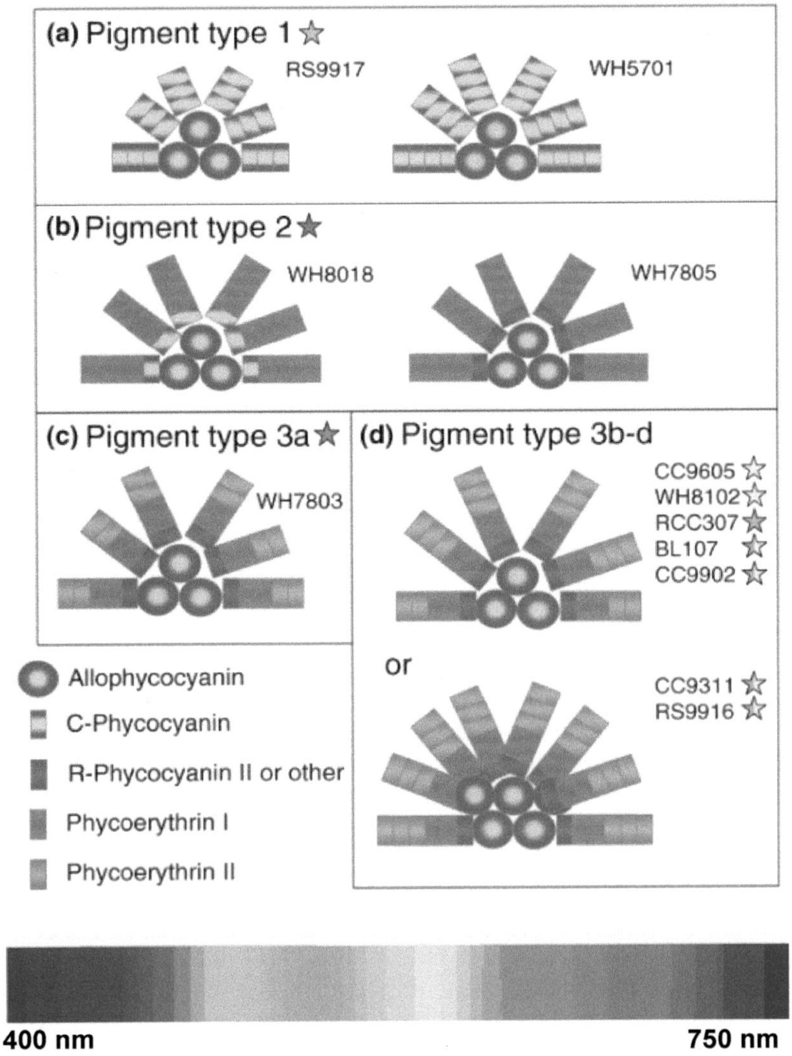

Figure 10: Proposed models of PBS structure for the different *Synechococcus* pigment types and subtypes. PBS cores are generally composed of three cylinders, but in some chromatic adapters possessing an extended L_{CM}, it is likely composed of two additional half cylinders (according to Ducret *et al.*, 1998). In pigment type 1, rods are composed of C-PC only; in pigment type 2, rods are composed of either C-PC, or R-PCIII and a PEI-like phycobiliprotein; in pigment type 3, rods comprise R-PC and two PE types (PEI and PEII). Cells of the latter pigment type bind PEB and PUB at a low (3a), medium (3b), high (3c) or variable (3d or type IV chromatic adapter) ratio. Colored stars indicate the pigment type of each strain. Lower block represents color code. With the kind permission of F. Partensky, UMR 7144 Université Paris VI and CNRS, Station Biologique, Groupe Plancton Océanique, F-29682 Roscoff cedex, France [Six *et al.* (2007) *Genome Biology* **8:** R259; doi:10.1186/gb-2007-8-12-r259].

Color image of this figure appears in the color plate section at the end of the book.

rod core-linker and L_{CM} denotes core-membrane linker of PBS. Ferredoxin-NADP oxidoreductase (FNR) transfers electrons from ferredoxin to $NADP^+$ to generate NADPH with an average value of 1.3 FNR per PBS. It has three functional domains. One of the domains at N-terminal end

Table 8: Phycobilisome rod genes in different pigment types of *Synechococcus* strains (Six *et al.*, 2007).

Pigment type	Strains	Total number of genes for PBP	Allocation for different components
Type 1	WH5701, RS9917	15	two *apcB-A* operons for C-PC α & β subunits; two rod core linker genes (*cpcGI* & *cpcGII*); two *cpcC* and *cpcD* rod linker genes; (an additional *cpcCIII* in WH5701); four genes encoding three different lyases (CpcE/F, CpcS and CpcT); one *pcy* gene for PCB: ferredoxin oxidoreductase.
Type 2	WH7805	19	one set of PE α- and β-subunit encoding genes three linker genes (*cpeC*, *cpeE* and *mpeD*-like gene); six putative lyase genes; two genes for PEP biosynthesis (*pebA* and *pebB)* and five unkown genes.
Type 3a	WH7803	6	additional genes including *unk 7/8* and *unk 9*
3d		12	additional genes (*mpeA, B, C, D, E, F* or *G, U, Y, Z* and *unk 7, 8* and *9*) of which those underlined are novel PII genes. These include PEB synthesis genes *PebA/B* and a number of lyase genes.

Unk=unknown

(of ~9 kDa) is generally found to be homologous to small PC rod-linker polypeptide CpcD. γ-subunits serve as PE linker polypeptides and contain two subtypes of covalently attached linear tetrapyrrole chromophores, phycoerythrobilin (PEB) and phycourobilin (PUB). Guan *et al.* (2007) downloaded (http://img.jgi.doe.gov/cgi-bin/pub/main.cgi) 25 genomes of cyanobacteria and compared the types and number of the linker polypeptides (Table 9). All *Prochlorococcus* strains revealed one FNR component each. However, three strains [NATLA, CCMP1375 (SS120) and MIT9211 all three LL-adapted strains) additionally possess a γ-subunit. In five *Synechococcus* strains (CC9311, WH8102, CC9605, CC9902 and WH5701) also a γ-subunit linker is present but in the rest of the cyanobacterial genomes examined the γ-subunit linker is absent. Homologues for genes *cpcC* and *cpcD* that encode PC-associated L_R linker polypeptides are absent in the genome of *Synechococcus* sp. strain WH8102. The absence of homologues for *nblA* and *nblB*, involved in the degradation of phycobilisomes during nutrient stress in cyanobacteria, is another significant feature of *Synechococcus* sp. strain WH8102 genome. Maximum number of linkers up to 13 is found in *C. watsonii* WH8501 (with 9 L_R and one each of the other four) and in *T. erythraeum* IMS101 (with 7 L_R, 2 L_{RC} and 2L_{CM} and one each of L_C and FNR). The total number of linker genes in the three heterocystous nitrogen-fixing species is found to be 10 with a variation of L_R linker genes up to five and two L_{RC} genes in *N. punctiforme* PCC 73102. In all 25 cyanobacteria put together, the number of linker polypeptides number up to 192. Of these, 167 genes encode PBS-associated linkers and 25 genes relate to the synthesis of FNR. Another important finding is that these genes, encoding PBS-associated linkers, PBSs as well as enzymes involved in biosynthesis, exist in clusters. In total 36 such gene clusters (1.5 kb to 13.2 kb) were found in the cyanobacterial genomes examined. FNR genes are widely distributed in the genomes but most of the PBS-linkers and PBSs are clustered and transcribed in the same direction.

Homologues of all chlorophyll biosynthesis genes have been identified in *A. marina* MBIC11017. The presence of genes governing the synthesis of protoporphyrin IX monomethyl ester oxidative cyclase (*AcsF*) and chlorophyll synthase (*ChlG*) that are responsible for the

Table 9: Distribution of genes for linker polypeptides in the various sequenced cyanobacterial genomes.

Sr. No.	Species	L_C	L_R	L_{RC}	L_{CM}	¥	FNR	No. of genes
1.	*P. marinus* sp. NATL2A	-	-	-	-	1	1	2
2.	*P. marinus* MIT9313	-	-	-	-	-	1	1
3.	*P. marinus* MIT9312	-	-	-	-	-	1	1
4.	*P. marinus* CCMP1986 (MED4)	-	-	-	-	-	1	1
5.	P. marinus CCMP1375 (SS120)	-	-	-	-	1	1	2
6.	*P. marinus* MIT9211	-	-	-	-	1	1	2
7.	*G. violaceus* PCC 7421	1	9	-	1	-	1	12
8.	*C. watsonii* WH8501	1	8	1	1	-	1	12
9.	*Synechocystis* sp. strain PCC 6803	1	3	2	1	-	1	8
10.	*Synechococcus* sp. CC9311	1	5	3	1	1	1	12
11.	*Synechococcus* sp. WH8102	1	4	2	1	1	1	10
12.	*Synechococcus* sp. CC9605	1	4	3	1	1	1	11
13.	*Synechococcus* sp. CC9902	1	5	3	1	1	1	12
14.	*Synechococcus* sp. PCC 6301	1	3	1	1	-	1	7
15.	*Synechococcus* sp. WH7805	1	3	2	1	-	1	8
16.	*Synechococcus* sp. WH5701	1	3	2	1	1	1	9
17.	*Synechococcus* sp. RS9917	1	2	2	1	-	1	7
18.	*Synechococcus* sp. JA-3-3A'b	1	4	1	2	-	1	9
19.	*Synechococcus* sp. JA-2-3B'a(2-13)	1	3	1	1	-	1	7
20.	*S. elongatus* PCC 7942	1	3	1	1	-	1	7
21.	*T. elongatus* BP-1	1	3	1	1	-	1	7
22.	*T. erythraeum*	1	7	2	2	-	1	13
23.	*Anabaena* sp. strain PCC 7120	1	3	4	1	-	1	10
24.	*A. variabilis* ATCC 29413	1	3	4	1	-	1	10
25.	*N. punctiforme* PCC 73102	1	5	2	1	-	1	10

synthesis of chlorophyll *a* from Protoporphyrin IX are noted in this organism. This signifies that chlorophyll *d* in this organism may be synthesized from chlorophyllide *a* or chlorophyll *a*. Another significant departure from the most of the cyanobacteria (where β-carotene is present) is the presence of α-carotene. In addition, zeaxanthin, an oxidative product of β-carotene has been identified as a major carotenoid. Eleven different proteins predicted to be associated with the biosynthesis of α-carotene and zeaxanthin are encoded by the genome of *A. marina* MBIC11017. These genes seem to be distributed through out the main genome without being clustered into operons. There are multiple copies of genes encoding PC core and linker proteins (*cpcA-cpcG*) with none for PE. Three copies of *cpcA* gene and one copy of *cpcB* gene are located on the main chromosome while other PBP-related genes are found in a number of large clusters on the plasmid pREB3. The gene *apcE* that encodes core-membrane linker peptide is absent in this organism. Typical PBS assembly is not seen but PBPs consist of four ring-shaped structures that are formed due to linking of PC and/APC subunits without core membrane linker required for PBS assemblies. Three copies each of *psbA* and *psbB* genes are present. There are eight genes that encode chlorophyll-binding proteins for additional light-harvesting. All genes required for the synthesis of cytochrome b6/f complex proteins are present in single copies with the exception of *petH* (present in three copies) and *petJ* (present in two copies) (Swingley *et al.*, 2008).

Sato (2003) described the utility of Gclust-a software programme to identify novel photosynthesis-related genes by clustering of 102,513 proteins from four eukaryotes (red alga, green plant, yeast, nematode) and eight cyanobacteria. Whole genome comparisons are based on either orthologues defined by bidirectional BeTs or by the use of homologous group of genes. Through this software, it was possible to identify 25 homologous genes involved in photosynthesis. A comparison of proteins encoded by 15 complete cyanobacterial genomes, including genomes of five strains of

P. marinus led to the identification of (i) minimal set of genes common for all the cyanobacteria, (ii) previously unidentified candidate photosynthetic genes and (iii) the first phototroph. Anoxygenic procyanobacteria have been considered as the first ancestral phototrophs from which through LGT events the evolution of forms like *Heliobacillus, Chlorobi, Chloroflexi,* purple bacteria and the extant cyanobacteria has been visualized. Massive LGT events encompassing >100 genes would have given rise to the present day cyanobacteria. It was also possible to show that cyanobacteria share a number of photosynthesis genes that are missing in genomes of other photoautotrophs (Mulkidjanian *et al.*, 2006). A clustering of proteins encoded by the genomes under comparison led to the identification of 3,188 protein families known as Cyanobacterial clusters of orthologous groups of proteins (CyOGs). In all the 15 genomes, 892 CyOGs are represented whereas 162 have been located in 14 of the 15 genomes. These together constitute the combined set of 1,054 CyOGs and considered as the core. Marine picocyanobacterial genomes such as those of *Prochlorococcus* have 52–66% of these CyOGs while the *Anabaena* sp. strain PCC 7120 genome has only 25% of these CyOGs. It was possible to identify 31 new proteins (so 31 new genes) previously listed as pseudogenes in the original submissions and the encoded proteins did not figure in the protein databases. Due to the reflection of majority of CyOGs groups from *Prochlorococcus* strains, certain important components of PSI and PSII are absent in the core set. For example, *psbU* and *psbV* genes are absent from the core set. On the contrary, PSI components PsaI, PsaJ, PsaK, and PSII components PsbZ are absent from the genome of *G. violaceus* PCC 7421 but are represented in the genomes of the rest. There are very few regulatory genes in the core set as these are lacking in *Prochlorococcus* genome. Among the core CyOGs, 84 of them are shared with higher plants (*Arabidopsis thaliana* and *Oryza sativa*), a red alga (*Cyanidioschyzon merolae*) and a diatom (*Thalassiosira pseudonana*). Approximately half of these relate to PSI, PSII, light-harvesting systems or HLIPs or early light-inducible protein superfamily. It is significant to note that 35 CyOGs represented in 14 cyanobacterial genomes have their counterparts in chloroplast-containing eukaryotic genomes but not met within the other (>350) prokaryotic and eukaryotic genomes. Anoxygenic phototrophic bacteria did not reveal any of the homologues of photosynthetic genes of cyanobacteria. However, there appears to be a difference in the number of core genes that have been identified for the genus *Prochlorococcus*. While describing gene order phylogeny of the genus *Prochlorococcus*, Luo *et al.* (2008) identified 1,131 orthologous genes shared by 13 genomes of which 12 belong to *Prochlorococcus* and one to the outgroup *Synechococcus* sp. strain WH8102. The phylogeny of these 12 *Prochlorococcus* genomes inferred from either 16S rRNA or random concatenation of 100 protein sequences sampled from each core genome appeared to be similar (Fig. 11a,b) and corresponded with the concensus tree of all core genes (Fig. 11c), constructed either by NJ or ML methods. The number of orthologous genes is lower than 1,273 reported earlier by Kettler *et al.* (2007) where outgroup has not been included. The 1,054 CyOGs forming the core described by Mulkidjanian *et al.* (2006) falls in between the two numbers.

As the cyanobacteria are the first organisms to have oxygenic photosynthesis, it is probable that they also must have developed protective mechanisms to scavenge superoxide radicals. Enzymes such as superoxide dismutases (SODs), catalases, peroxidases are important in the scavenging of superoxide radicals. Besides these, a group of enzymes known as peroxiredoxins (Prxs) constitute antioxidant enzymes known in a number of bacteria, archaea, yeast, algae, higher plants and animals. It is suggested that the Prxs function as regulators of redox-mediated signal transduction at least in some eukaryotes (Hoffmann *et al.*, 2002; Wood *et al.*, 2003a,b; Dietz, 2003; Veal *et al.*, 2004). The substrates for the Prxs include hydrogen peroxides, alkyl hydroperoxides and peroxynitrites. The catalytic mechanism of Prxs involves the oxidation of cysteine to sulphemic acid by the peroxide substrate. The genome of *Synechocystis* sp. strain PCC 6803 revealed the presence of one gene each

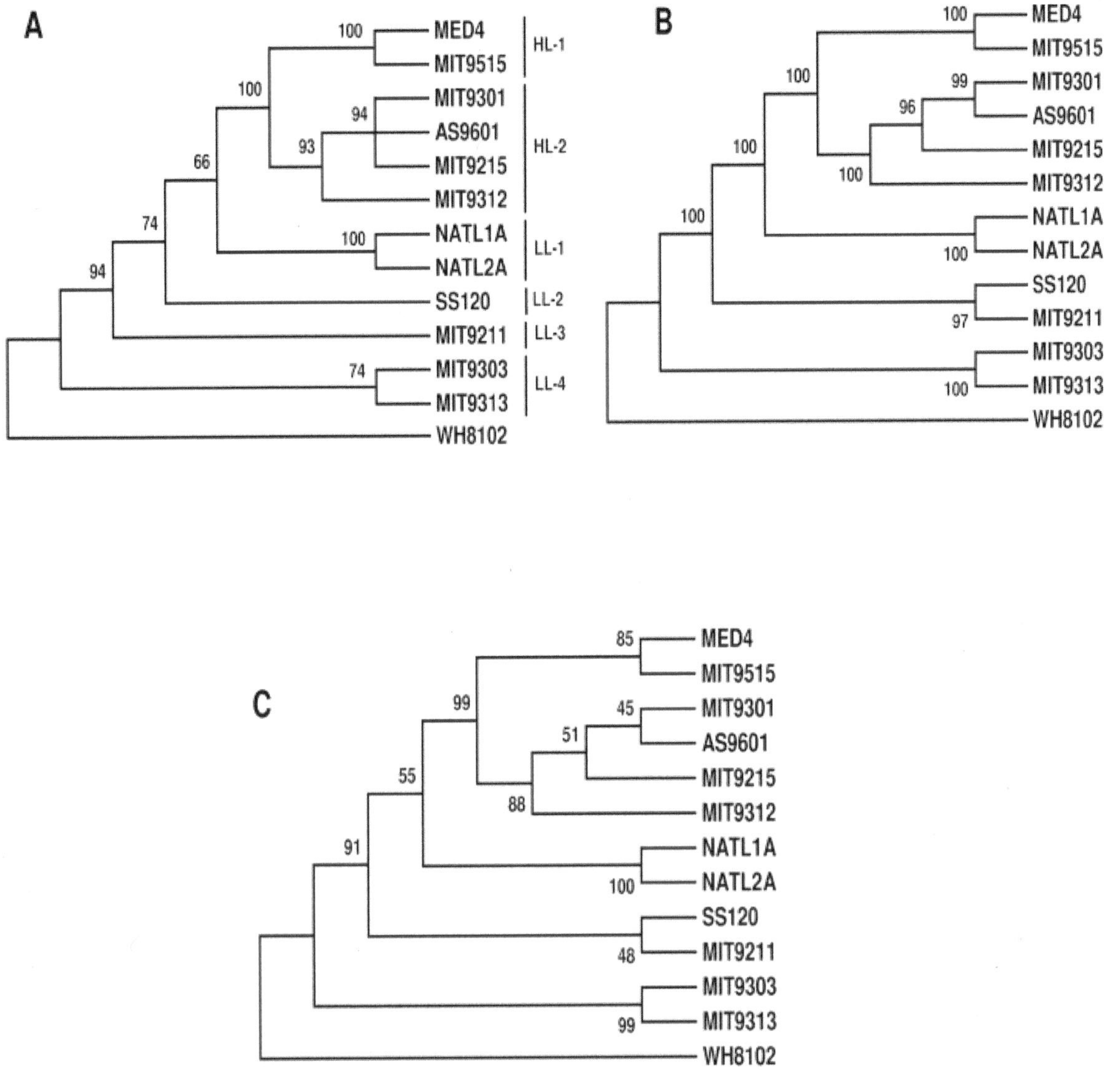

Figure 11: Phylogeny of 12 *Prochlorococcus* genomes inferred from (A) 16S rRNA, (B) random concatenation of 100 protein sequences sampled from core genome, and (C) concensus tree of all core genes. Trees were reconstructed by neighbor-joining (A), maximum parsimony (B), and maximum parsimony (C) methods. The trees are reprints from Kettler *et al.* (2007). HL, high light adapted strains; LL, low light adapted strains. With the kind permission of R. Friedman, Department of Biological Sciences, University of South Carolina, Columbia, South Carolina, USA. [Luo *et al.* (2008) *PLoS ONE* **3(12):** e3837. doi:10.1371/journal.pone.0003837] doi:10.1371/journal.pone.0003837.g001

for SOD, catalase and five genes that encode Prxs. A comparison of the genomes of *Synechocystis* sp. strain PCC 6803 and *S. elongatus* PCC 7942 by bioinformatic approach revealed five and six genes for Prxs, respectively. The expression of these genes was monitored under stress induced by light, hydrogen peroxide, salinity as well as nitrogen and iron deficiency (Stork *et al.*, 2005). In *T. elongatus* BP-1, two SOD genes *(tlr0036/SodM* and *tll1519/SodF*) and a thioredoxin peroxidase *(tll1454)* that help in scavenging superoxide radicals and hydrogen peroxide, respectively have been identified (Nakamura *et al.*, 2002). Due to the high concentrations of metal ions in the coastal environment, *Synechococcus* sp. strain CC9311 makes greater use of copper for purposes of making Cu-Zn SOD not seen in marine cyanobacteria. A multicopper oxidase characteristically present in this strain helps in the oxidation or detoxification of higher levels of reduced copper (Palenik *et al.*, 2006). It is suggested that oxidative stress in *M. aeruginosa* PCC 7806 leads to programmed cell death (PCD) akin to the capsase-like activity of eukaryotic cells that mediates PCD. Of the five genes identified in *M. aeruginosa* PCC 7806, three of them (*Mic0980, Mic3930* and *Mic4051*) showed similarity in sequences to those present in *M. aeruginosa* NIES-843. But due to lack of capsase-like activity in these gene products their role in processes other than PCD is indicated. The gene products of the other two genes *Mic1068* and *Mic5406* showed some resemblance to human capsase proteins (Frangeul *et al.*, 2008).

6) Regulatory functions: The genes allocated for regulatory functions in nine cyanobacteria have been listed in Table 10. Among these, genes encoding two-component regulatory systems, transcription factors and circadian clock proteins have been identified in different genomes of cyanobacteria. These are presented in the following sub-sections.

a) Two-component regulatory systems: The signal sensing and transmitting component consists of a sensor histidine kinase (or transmitter) which autophosphorylates an invariant histidine residue in an ATP-dependent mechanism in response to an environmental signal. The phosphorylated transmitter transfers the phosphate to an invariant aspartate residue in a cognate receiver protein called response regulator (or receiver). The response regulators often have an input domain that defines a DNA-binding motif through which the protein regulates transcription, although some have an output domain. There also exists a class of complex signal transducers that contain both transmitter and receiver domain. Genes for two-component systems identified in some cyanobacterial genomes are listed (Table 10). Although genes for biosynthesis of cAMP and cGMP have been identified in a number of cyanobacteria (Katayama and Ohmori, 1997; Kasahara and Ohmori, 1999; Terauchi and Ohmori, 1999; Ochoa de Alda and Houmard, 2000; Ochoa de Alda *et al.*, 2000), for the first time in *Synechocystis* sp.strain PCC 6803 two putative cyclic nucleotide phosphodiesterases (the enzymes that are involved in the degradation of cyclic nucleotides cAMP and cGMP) encoded by genes *sll1624* and *slr2100* have been identified. The presence of five proteins (probably cyclic nucleotide receptors)

Table 10: Distribution of genes related to regulatory functions in the genomes of some cyanobacteria.

No.	Name	Two-component systems			
		Sensory kinases	Response regulators	Hybrid type	Ser/Thr protein kinases
1.	*Anabaena* sp. strain PCC 7120	71	71	53	-
2.	*G. violaceus* PCC 7421	27	37	12	-
3.	*M. aeruginosa* NIES-843	22	23	-	-
4.	*N. punctiforme* PCC 73102	255			55
5.	*Synechocystis* sp. strain PCC 6803	26	38	17	-
6.	*Synechococcus* sp. WH8102	5	9	-	-
7.	*Synechococcus* sp. CC9311	11	17	-	-
8.	*S. elongatus* PCC 6301	13	21	3	-
9.	*T. elongatus* BP-1	17	27	6	11

(1) Kaneko *et al.* (2001); (2) Nakamura *et al.* (2003); (3) Kaneko *et al.* (2008); (4) Meeks *et al.* (2001); (5) Kaneko *et al.* (1996); (6) & (7) Palenik *et al.* (2003); (8) Sugita *et al.* (2006) and (9) Nakamura *et al.* (2002).

two of them being more specific to cGMP (*slr2104* that encodes an element of two-component regulatory systems and *slr1575* that is responsible for cyclic nucleotide-gated cation channel) and the rest three specific to cAMP (one of which *sll1371* that encodes a putative transcription factor, CRP-Syn) signifies the presence of cGMP and cAMP signalling pathways. CRP-Syn has been identified as a global regulator since it can bind several genes (encoding enzymes involved in transport and metabolism) upstream of *crp, Cya1* and *slr1575* (Ochoa de Alda *et al.*, 2000). *P. marinus* MIT9313 has very few genes for two-component sensor kinase systems and the number of such genes was even further reduced in *P. marinus* MED4 (Rocap *et al.*, 2003). Both these strains also lack the circadian clock genes *pex* and *kaiA*. In the absence of regulation by two-component regulatory systems and circadian proteins, the challenge of regulation of gene activity in this minimal photoautotroph seems to be fulfilled by the presence of non-coding RNAs (ncRNAs). These are functional RNA molecules mostly without a protein-coding function. These genes are normally located in the intergenic regions. Seven ncRNAs have been identified in cyanobacteria and were first described in *P. marinus* MED4. These are designated as Yfr1 to Yfr7 (representing cYanobacterial functional RNA; Axmann *et al.*, 2005). Of these, Yfr7 is considered to be homologous to the bacterial 6S RNA that possesses global regulatory functions. The second such molecule is Yfr1 which is required for growth of *S. elongatus* PCC 6301 under multiple stress conditions (Nakamura *et al.*, 2007). The predicted secondary structures of Yfr1 of *Prochlorococcus* strains MED4, MIT9313 and *Synechococcus* sp. strain WH8102 are presented in Fig. 12. Although only three Cis-acting anti-sense RNAs (asRNAs) have been reported earlier in cyanobacteria (Csiszar *et al.*, 1987; Duhring *et al.*, 2006; Hernandez *et al.*, 2006), none have been identified in *Prochlorococcus* genomes. The genome of *S. elongatus* PCC 6301 has the minimum number of genes (37) for two-component signal transduction system noted in freshwater cyanobacteria (Sugita *et al.*, 2006). Steglich *et al.* (2008) followed transcription of intergenic regions of *P. marinus* strain MED4 cells subjected to different types of stress conditions such as light qualities and quantities, phage infection and phosphate starvation. Analysis of microarray expression data from intergenic regions revealed the existence of as many as 276 novel transcriptional units. Among these were 12 new ncRNAs, 24 asRNAs as well as 113 short mRNAs. With two additional ncRNAs identified by homology, the number of ncRNAs totalled up to 14. In addition five of the seven ncRNAs have also been detected (Yfr2 and Yfr4-Yfr7). The distribution of various ncRNAs, asRNAs and short mRNAs in the genomes of various strains of *Prochlorococcus* (MED4, MIT9312, MIT9315, MIT9515, MIT9301, NATL1A and NATL2A) has been compared. In view of the occurrence of considerable number of ncRNAs equivalent to those found in bacterial cells it is suggested that these RNA regulators are likely to play a major regulatory role in this organism. Sugita *et al.* (2006) identified gene sequences similar to SRP RNA, tmRNA, 6SaRNA and RNase P subunit B in *S. elongatus* PCC 6301.

The genome of *N. punctiforme* PCC 73102 revealed unusually high number of (255) combined two-component signal transduction proteins. Majority (at least 153) of these transmitters are unlinked to a response regulator. Single domain sensor histidine kinases comprise only about 53% of this class of genes. Genes encoding transmitter-receivers and transmitter-receiver-transmitters constitute 47%. The remaining 102 response regulator genes consist of 36% encoding receivers with no apparent output domain while the remaing are helix-turn-helix DNA-binding output domains. Meeks *et al.* (2001) suggested that the presence of unusually high frequency of response regulators without output domains is indicative of the multiprotein phosphorelay signalling systems that help in integrating multiple environmental signals. Simple chemotaxis sensor histidine kinases and complex-chromophore-binding sensor histidine kinase proteins (complementary chromatic adaptation, circadian rhythm and phytochrome proteins) are included here. There are 55 ORFs

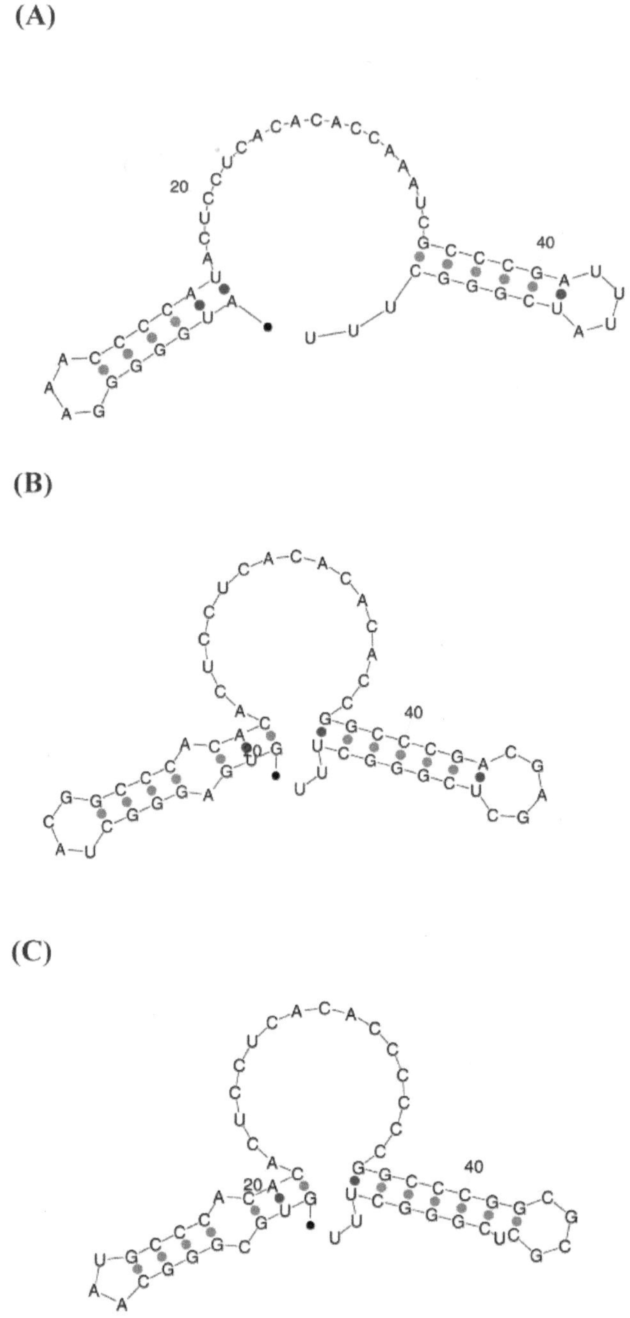

Figure 12: Predicted secondary structures of Yfr1 in *P. marinus* strains MED4 (A), MIT 9313 (B) and *Synechococcus* sp. strain WH 8102 (C) by MFOLD (as per Zuker, 2003). With the kind permission of W. R. Hess, University Freiburg, Institute of Biology II/Experimental Bioinformatics, Schänzlestrasse, Freiburg, Germany [Axmann *et al.* (2005) *Genome Biology* **6:** R73 doi:10.1186/gb-2005-6-9-r73].

Color image of this figure appears in the color plate section at the end of the book.

that are devoted to encode eukaryotic serine/threonine protein kinases whose function remains unclear. Homologs of *kaiABC* in a cluster with *cikA* exist elsewhere, although there is no evidence of a circadian system.

In *Synechocystis* sp. strain PCC 6803 the phosphorus sensing mechanism involves SphS (an orthologue of PhoR of *E. coli*) and the response regulator SphR (an orthologue of PhoB of *E. coli*), respectively. All phosphorus assimilation genes are regulated by these two gene products. As in *E. coli*, SphS senses inorganic phosphorus (P_i) levels in the environment and if it is low it gets autophosphorylated and activates SphR. Phosphorylation of SphR in turn leads to activation of genes connected with phosphorus assimilation. All such genes whose transcription is regulated by SphR are collectively called as Pho regulons. The phosphorylated SphR activates transcription of two operons (*sphX-pstS1-C1-A1-B1-B1′* and *pst2-C2-A2-B2*) that encode ABC-type high affinity P_i-uptake systems (Hirani *et al.*, 2001; Suzuki *et al.*, 2004). *Synechococcus* sp. strain WH8102 has only five sensor histidine kinases and nine response regulator genes of which one (SYNW1598) is a pseudogene (Palenik *et al.*, 2003). In *S. elongatus* PCC 6301 there are 13 and 21 potential genes for sensory histidine kinases and response regulators, respectively. There are another three genes of hybrid type. At least half of these relate to phosphate sensing *sphS* (*syc0424c*) and *sphR* (*syc0533d*) genes that encode the sensor kinase. A comparison of the genomes of *Synechococcus* sp. strain CC9311 and *Synechococcus* sp. strain WH8102 revealed 28 genes in the former (11 genes for histidine kinases and 17 genes for response regulators) and a minimum of 14 genes in the latter (five histidine kinase genes and 9 response regulator genes). The arguments put forward are that *Synechococcus* sp. strain CC9311 is a coastal strain and the environment being highly variable it requires greater adaptation whereas *Synechococcus* sp. strain WH8102 is an open ocean strain and grows almost in constant environmental conditions. That is why it possesses minimal regulatory functions. *Synechococcus* sp. strain CC9311 also apparently lacks a phosphate sensor–response regulator (seen in several cyanobacteria and bacteria) because of availability of phosphate at high concentrations in the coastal environment.

In *T. elongatus* BP-1 there are 17 and 27 potential genes for histidine kinases and response regulators, respectively. These relate to highly conserved genes for a drug sensor (*tlr0437*), a phosphonate sensor (*tll0925*), a KaiC-interacting protein (*tlr0029*), a phytochrome-like circadian input kinase (*tll0899*) and three motility related CheA-like proteins (*tlr0349, tll0568* and *tll1021*). A total of 11 genes for Ser/Thr protein kinases have been identified in *T. elongatus* BP-1. Genes for protein phosphatases in this organism include GlnB phosphatase (*tlr2243*) and two genes for adenyl cyclase (*tll2280, tll2410*) (Nakamura *et al.*, 2002). Nakamura *et al.* (2003) have assigned 76 genes for a two-component system in *G. violaceus* PCC 7421 that includes 27 genes for sensor histidine kinases, 37 response regulators and 12 for hybrids of sensor histidine kinases and response regulators. Of the 37 response regulators, 27 of them contained sequences for transcription factors. Except for the protein kinase domains of the serine/threonine protein kinases (15 of them) noted in this organism, these do not bear any resemblance to those known in other cyanobacteria such as *Synechocystis* sp. strain PCC 6803, *Anabaena* sp. strain PCC 7120 and *T. elongatus* BP-1. A total of 195 genes that encode two-component signal transduction systems have been identified in *Anabaena* sp. strain PCC 7120. Of these 71 genes each one has been allocated for sensory kinases and response regulators and 53 genes for hybrid sensory kinase and response regulator. Of these, 85 of them are arranged in clusters of 2 to 5 genes and 110 of them are present individually. The gene arrangement in a five gene cluster (*alr3155-alr3159*) is in the order of a sensory kinase-a response regulator-a phytochrome-like sensory kinase, a response regulator-a sensor/regulator hybrid. Thirteen of the sensory kinases bear a conserved domain that resembles serine/threonine kinase. It is interesting to note that genes for two-component system (two genes for sensory kinases paired with response regulators and four additional genes for response regulators) are also represented in the plasmids pCC7120 α, β, δ and ε in *Anabaena* sp. strain PCC 7120. Another unique feature of *Anabaena* sp. strain PCC 7120 genome is the presence of proteins with 4 or more repeating units of ~31 amino acids containing tryptophan and aspartic acid (WD). Proteins with WD repeats are known in eukaryotes (*Arabidopsis thaliana*-59 genes; *Caenorhabditis elegans*-88 genes; *Saccharomyces cerevisiae*-58 genes) but these are earlier reported only in case of *Synechocystis* sp. strain PCC 6803 (5 genes). There are 20 such genes on the chromosome of *Anabaena* sp. strain PCC 7120 and four on the plasmids. These have 4 to 15 WD repeats. The presence of conserved regions of serine/threonine kinases at the N-terminal regions of three genes (*all0438, alr3119* and *all3169*) and the gene products of *alr0029, alr2800* and *alr7129* exhibiting similarity to proteins of plant-disease resistant genes signifies that *Anabeana* sp. strain PCC 7120 shares a eukaryotic signalling pathway (Kaneko *et al.*, 2001). Zhang *et al.* (2007) conducted a genome-wide survey of Ser/Thr kinase genes in 21 species of sequenced cyanobacterial genomes. This constitutes the first detailed study on the diversity, conservation, domain structure and evolution of Ser/Thr kinases in cyanobacteria. The Ser/Thr kinase genes are absent in four of the *P. marinus* strains (CCMP1375, CCMP1986, MIT9312 and NTL2A) and one *Synechococcus* strain (WH8102). The number of Ser/Thr kinase genes varied greatly from lowest one (in *P. marinus* MIT9313 and three *Synechococcus* strains CC9311, CC9605 and CC9902) to 56 (*N. punctiforme* PCC 73102). A phylogenetic tree constructed on the basis of 16S rRNA sequence analysis and distribution of Ser/Thr genes is presented in Fig. 13. The 286 homologous Ser/Thr kinase genes have been divided into

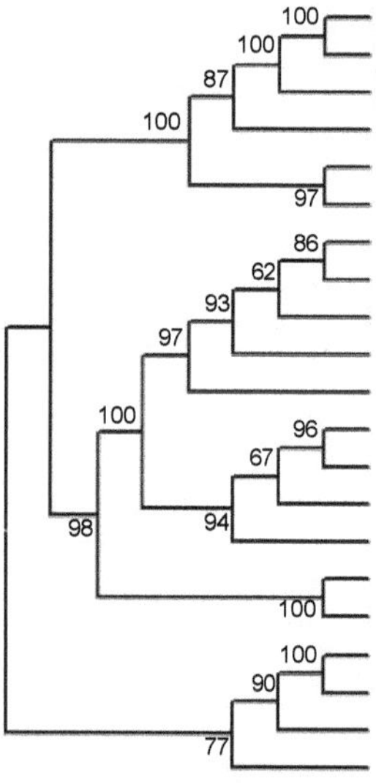

Strain	Key Feature	Acronym	Total Proteins	Total STKs	Additional domains
***Anabaena* PCC7120**	Filamentous freshwater	7120	6214	48(0.77%)	27(56.3%)
***Anabaena variabilis* ATCC 29413**	Filamentous soil	Ava	5746	53(0.92%)	25(47.2%)
***Nostoc punctiforme* PCC73102**	Filamentous symbiont	Npun	7364	56(0.76%)	27(48.2%)
Trichodesmium erythraeum IMS101	Filamentous marine	Tery	4759	40(0.84%)	31(77.5%)
Synechocystis sp. PCC6803	Unicellular freshwater	6803	3564	7(0.20%)	2(28.6%)
Crocosphaera watsonii WH8501	Unicellular marine	Cwat	5967	28(0.47%)	5(17.9%)
Prochlorococcus marinus CCMP1986	Unicellular marine	1986	1716	-	-
Prochlorococcus marinus MIT9312	Unicellular marine	9312	1808	-	-
Prochlorococcus NTL2A	Unicellular marine	NTL2A	1894	-	-
Prochlorococcus marinus CCMP1375	Unicellular marine	1375	1882	-	-
Prochlorococcus marinus MIT9313	Unicellular marine	9313	2273	1(0.04%)	-
Synechococcus sp. CC9605	Unicellular marine	9605	2695	1(0.04%)	-
Synechococcus sp. CC9902	Unicellular marine	9902	2321	1(0.04%)	-
Synechococcus sp. WH8102	Unicellular marine	8102	2526	-	-
Synechococcus sp. CC9311	Unicellular marine	9311	2892	1(0.03%)	-
Synechococcus elongatus PCC6301	Unicellular freshwater	6301	2527	5(0.20%)	1(20.0%)
Synechococcus elongatus PCC7942	Unicellular freshwater	7942	2662	5(0.20%)	1(20.0%)
Synechococcus sp. JA-2-3Ba(2-13)	Unicellular hot spring	CYB	2862	8(0.28%)	2(25.0%)
Synechococcus sp. JA-3-3Ab	Unicellular hot spring	CYA	2760	8(0.29%)	3(37.5%)
Gloeobacter violaceus PCC7421	Unicellular rock	7421	4430	14(0.32%)	3(21.4%)
Thermosynechococcus elongatus BP1	Unicellular hot spring	TBP1	2475	10(0.40%)	4(40.0%)

Figure 13: Phylogenetic tree of the sequenced cyanobacterial strains and STK information. A phylogenetic tree for 21sequenced cyanobacteria constructed based on 16s rRNA is shown here. Numbers appearing at the nodes corresponded to the values produced by bootstrap analysis (1000 replicates). Names of marine nitrogen-fixing strains are marked in grey boxes. Filamentous diazotrophic strains capable of heterocyst differentiation are marked in bold. Percentages in brackets represent total STKs as a percentage of total proteins and total additional domains as a percentage of total STKs. With the kind permission of S. Qin, Ocean University of China, Yushan Road, Qingdao, China [Zhang *et al.* (2007) *BMC Genomics* **8:** 395; doi:10.1186/1471-2164-8-395].

three families, cbSTKI, cbSTKII and cbSTKIII. Their structure and distribution among the sequenced genomes is presented in Fig. 14. In *M. aeruginosa* NIES-843 there are 22 genes for histidine kinases and 23 genes for response regulator. Two genes of histidine kinases (designated as MAE46010) and two genes for hybrid histidine kinases (designated as MAE21690 and MAE37480) are reported to be unique. Five sets of histidine kinases and response regulators have been found adjacent to each other while the rest of them are found scattered throughout the chromosome. According to Kaneko *et al.* (2008) five genes of histidine kinases (those found in *M. aeruginosa* NIES-843) designated as MAE03210 (*hik34*), MAE14410 (*hik2*), MAE36080 (*hik33*), MAE 52650 (*SphS*) and MAE 60820 (*sasA*) are well conserved in all cyanobacterial genomes sequenced to date. Frangeul *et al.* (2008) identified genes coding for a Ser/Thr kinase (*mic0129*) and a Ser/Thr phosphatase of the PPP family (*mic4622*) within two clusters that may be involved in cell wall synthesis in *M. aeruginosa* PCC 7806. Swingley *et al.* (2008) reported the existence of 170 genes in *A. marina* MBIC11017 assigned to two-component regulatory system consisting of sensory histidine kinases and response regulators.

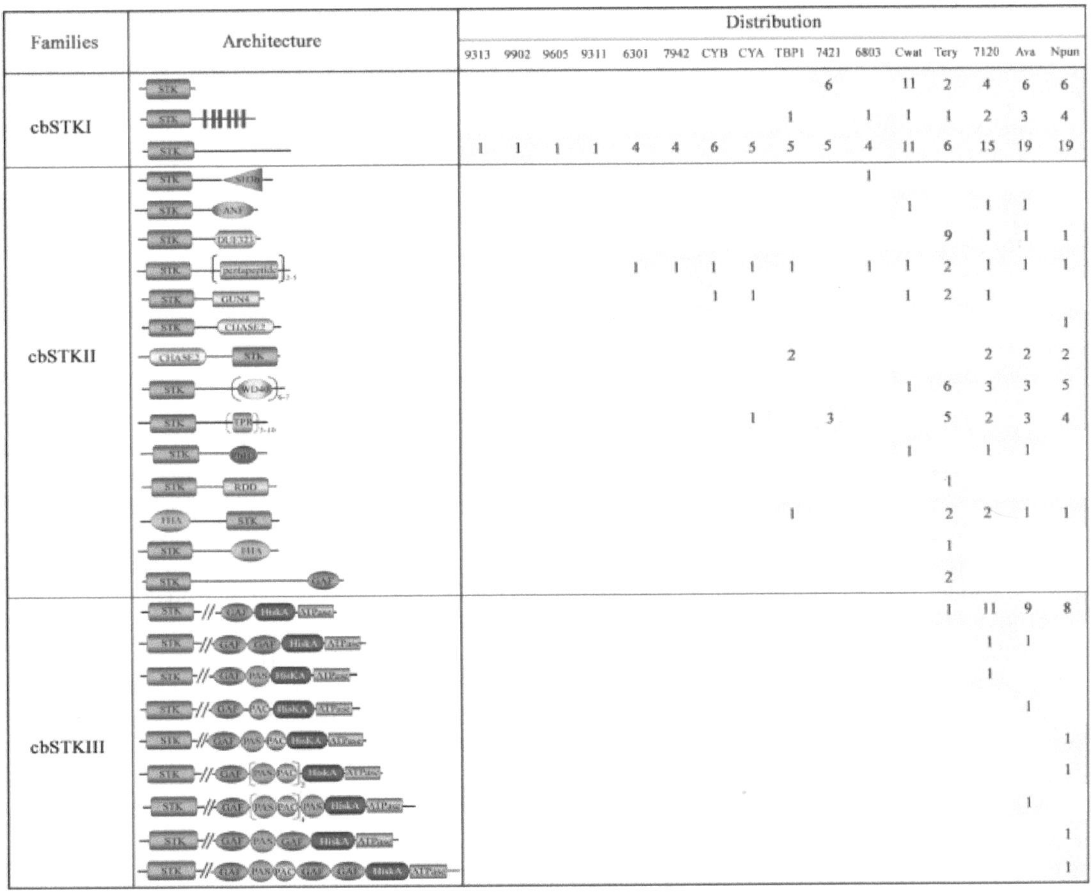

Figure 14: Schematic representation and distribution of putative cyanobacterial serine/threonine protein kinases. Strain names are as in Figure 13. With the kind permission of S. Qin, Ocean University of China, Yushan Road, Qingdao, China [Zhang X. *et al.* (2007) *BMC Genomics* **8:** 395; doi:10.1186/1471-2164-8-395].

Color image of this figure appears in the color plate section at the end of the book.

Su *et al.* (2007) conducted a computational prediction analysis of Pho regulons in 19 sequenced cyanobacterial genomes and found orthologues of SphR (sll0337) of *Synechocystis* sp. strain PCC 6803 in 16 genomes (*A. variabilis* ATCC 29413, *Anabaena* sp. strain PCC 7120 , *G. violaceus* PCC 7421, *P. marinus* MED4, MIT9312, MIT9313, NATL2A, *S. elongatus* PCC 7942, *S. elongatus* PCC 6301, *Synechococcus* sp. strain WH8102, CC9605, *Synechococcus* sp. strain JA-3-3-A'b, *Synechococcus* sp. strain JA-2-3B'a (2–13) *Synechocystis* sp. strain PCC 6803, *T. elonagtus* BF-1 and *T. erythraeum* IMS101) and absent in three of them (*P. marinus* CCMP1375, *Synechococcus* sp. strain CC9902 and CC9311). The *sphS* and *sphR* genes are predicted to be present in the same operon (in *A. variabilis* ATCC 29413, *Anabaena* sp. strain PCC 7120, *P. marinus* MED4, MIT9312, NATL2A and WH8102, *Synechococcus* sp. CC9605, *Synechococcus* sp. strain JA-3-3-A'b, *Synechococcus* sp. strain JA-2-3B'a (2–13), *S. elongatus* PCC 7942, *S. elongatus* PCC 6301) or split into two separate operons (as in *G. violaceus* PCC 7421, *T. elongatus* BF-1 and *Synechocystis* sp. strain PCC 6803). In the rest of the two cyanobacterial genomes the *sphS* gene is either not encoded (*T. erythraeum* IMS101) or has undergone frame-shift mutation (*P. marinus* MIT 9313). The activation of the genes of P_i-intake by SphR is due to its binding to at least three tandem repeats of 8 bp, i.e. CTTAACCT collectively designated as Pho-boxes (Suzuki *et al.*, 2004). With the help of the computational program so developed, Su *et al.* (2007) found Pho-boxes for genes transcribing for ABC-type P_i Pst transporter complex genes in most of these genomes. The components of P_i-transporter identified were a phosphate-binding subunit PstS, the transmembrane subunits PstC and PstA and the ATP-binding subunit PstB and the respective genes form an operon, *pstS-C-A-B*. However, in some of them *pstS* gene is present separately (as in *P. marinus* strains CCMP1375, MED4, MIT9312, MIT9313 and *Synechococcus* sp. CC9605). An additional phosphate-binding subunit SphX, encoded by *sphX*, is present in the Pst transporter complex in case of *A. variabilis* ATCC 29413, *Anabaena* sp. strain PCC 7120, *G. violaceus* PCC 7421, *Synechococcus* sp. strain WH8102, *S. elongatus* PCC 6301, *S. elongatus* PCC 7942, *Synechocystis* sp. strain PCC 6803, *T. elongatus* BF-1 and *T. erythraeum* IMS101. Phosphonate utilizing genes (of *Anabaena* sp. strain PCC 7120, *Synechococcus* sp. strain JA-2-3B'a (2–13) and T. *erythraeum* IMS101) and phosphatase A (*phoA*) genes (of *A. variabilis* ATCC 29413, *Anabaena* sp. strain PCC 7120, *G. violaceus* PCC 7421, *P. marinus* MED4, MIT9312, NATL2A, *S. elongatus* PCC 6301, *S. elongatus* PCC 7942, *Synechococcus* sp. strain WH8102, *Synechococcus* sp. strain JA-3-3-A'b, *Synechococcus* sp. strain JA-2-3B'a (2–13) and *Synechocystis* sp. strain PCC 6803) and extracellular nuclease (*nucH*) genes that can utilize the P_i moiety from nucleic acids in the environment (in case of *A. variabilis* ATCC 29413, *Anabaena* sp. strain PCC 7120, *S. elongatus* PCC 6301 and *S. elongatus* PCC 7942) all possess high scoring SphR-binding sites suggesting that they are under the regulation of SphR.

b) Transcription factors: Wu *et al.* (2007) created a database (cTF) containing 1288 putative transcription factors from 21 fully sequenced cyanobacterial genomes. This database enables one to identify the genes encoding transcription factors from any selected genome and is freely accessible at http://cegw.com/. *N. punctiforme* PCC 73102 genome contains 13 apparent alternative σ^{70} subunits in addition to the primary σ^{70} subunit. Campbell *et al.* (1998) indicated that it is due to mutation of only one alternative σ^{70} the phenotype of *N. punctiforme* PCC 73102 changed to a higher symbiotic interaction. The substantial capacity of *N. punctiforme* PCC 73102 for differential gene expression in response to a variety of environmental conditions is revealed by the presence of 57 ancillary transcriptional regulatory proteins. At least 67% of these bear resemblance to regulatory proteins such as TetR, XylR and ArsA. Besides, there are genes that encode proteins AraC (6 copies), LysR (8 copies) and MocR (two copies; Meeks *et al.*, 2001). Nakamura *et al.* (2002) allocated 21 genes for transcription factors in *T. elongatus* BP-1which are categorized into LuxR family (4 genes), OmpR family (7 genes), LysR family (3 genes, *rbcR, ntcB* and *ndhR*), CRP family

(3 genes including *ntcA*), ARS family (2 genes), Fur family (3 genes) and single genes for heat-shock gene repressor *HrcA*. In addition, a complete set of genes for σ factors for RNA polymerase (*SigA-D, SigF* and *SigG1/G2*) except SigE homologue have been identified. *G. violaceus* PCC7421 possesses relatively large number of transcription factors in the LuxR, LysR, PadR, TetR and MarR families. There are 14 genes for σ factors that include one gene for SigA, five genes for group 2 σ, six genes for extra-cytoplasmic type σ and two genes for group 3 σ factors (Nakamura *et al.*, 2003). In *Synechococcus* sp. strain WH8102, in addition to the principal RNA polymerase σ factor (SigA *SYNW1783*) five types of type 2 σ factors are characteristically present as in other cyanobacteria. One homologue of type 3 σ factor is present in contrast to seven noted in other cyanobacterial genomes sequenced earlier (*Synechocystis* sp. strain PCC 6803, *Anabaena* sp. strain PCC 7120 and *T. elongatus* BP-1 (Palenik *et al.*, 2003). *S. elongatus* PCC 6301 has 36 genes assigned for transcription factors. Of these, nine of the encoded transcription factors belong to the response regulator OmpR family while the others fall into LysR family (Sugita *et al.*, 2006). In *M. aeruginosa* NIES-843, 43 genes encode transcription factors. A global regulator for nitrogen assimilation *NtcAI,* a regulator for nirate assimilation (NtcB), a regulator for RuBisCO (RbcR), a regulator for genes for subunits of NAD dehydrogenase (NdhR), a cyclic AMP receptor protein for cell motility (SyCrp1), an inhibitor for genes for GroESL (HrcA) and a regulator for zinc efflux system (ZinR) are common with other cyanobacteria (Kaneko *et al.*, 2008).

Cyclic AMP receptor protein (CRP) is an important transcriptional regulator in bacteria and the range of genes regulated by it in cyanobacteria has come out by the studies of Xu and Su (2009), who conducted a computational prediction of CRP-binding sites in 29 cyanobacterial genomes. *Synechocystis* sp. strain PCC 6803 genome has two genes *sll1371* and *sll1924* that encode SyCRP1 and SyCRP2, respectively. In case of *Anabaena* sp. strain PCC 7120 also there are two genes *alr0295* and *alr2325* that encode AnCRPA and AnCRPB, respectively. Orthologues of SyCRP1 one each have been detected in ten other cyanobacterial genomes [*A. marina* MBIC 11017, *A. variabils* ATCC 29413, strains of *P. marinus* MIT 9313 and MIT 9303, strains of *Synechococcus* sp. CC9311, CC9605, JA-3-3-A'b, JA-2-3-3B'a (2–13), *T. elongatus* BP-1 and *T. erythraeum* IMS101]. The rest of the 17 cyanobacterial strains [*G. violaceus* PCC 7421, *P. marinus* strains AS9601, MIT9211, MIT9215, MIT9301, MIT9312, MIT9515, NATL1A, NATL2A, CCMP1375, MED4, *S. elongatus* PCC 6301, *S. elongatus* PCC 7942, and strains of *Synechococcus* sp. strain PCC 9902, RCC307, WH7803, WH8102] do not have the *crp* gene in their genomes. There seems to be a great variation in the number and type of target genes in the 12 cyanobacterial strains that showed the *crp* gene. Majority of the CRP-regulated genes are distributed among five cyanobacterial genomes (*Synechocystis* sp. strain PCC 6803, *Anabaena* sp. strain PCC 7120, *A. variabilis* ATCC 29413, *A. marina* MBIC11017 and *T. erythraeum* IMS101) and the number of target genes in these five cyanobacteria differed from one another in the functional domains of photosynthesis, carbon fixation, nitrogen assimilation, transporters/porins, kinases and transcription factors. A comparison made on the basis of total number of genes, number of transcription units, number of CRP-regulated genes, number of CRP-regulated transcription units and the number of CRP-regulated genes shared with *E. coli* revealed that *Anabaena* sp. strain PCC 7120 ranks first followed by *Synechocystis* sp. strain PCC 6803 (Xu and Su, 2009).

In enteric bacteria such as *E. coli* and other proteobacteria, nitrogen control is mediated by NtrB-NtrC two-component system (Reitzer, 2003). In contrast, in cyanobacteria nitrogen control is governed by NtcA, a global transcriptional regulator that belongs to a class of CRP-family. This helps in the repression of pathways for assimilation of nitrogen when more readily assimilable form of nitrogen is available. The binding sites of NtcA on DNA are highly conserved and possess a sequence of GTAN8TAC. In addition, all NtcA-activated genes also contain a 10-mer sequence (Herrero *et al.*, 2001). In view of regulation by NtcA of many genes involved not only in nitrogen metabolism but also in photosynthesis, a phylogenetic footprinting is conducted to identify conserved motifs in intergenic regions of orthologous genes from closely related genomes. Such a study conducted by Su *et al.* (2005) on nine cyanobacterial genomes (*G. violaceus* PCC 7421, *Anabaena* sp. strain PCC 7120, *P. marinus* strains CCMP1375, MED4 and MIT9313, *S. elongatus* PCC 6301, *Synechococcus* sp. strain WH8102, *Synechocystis* sp. strain PCC 6803 and *T. elongatus* BP-1) revealed the existence of strong NtcA promoter regions (i) for the ammonium transporters of the *amt* family in all the nine cyanobacterial genomes, and (ii) in the regulatory regions of *nirA* in all genomes excepting in *Synechococcus* sp. strain CCMP1375 and *P. marinus* MED4. The clustering of genes of nitrate/nitrite transporters of the ABC-type transporters *nrtABCD* in *Anabaena* sp. strain PCC 7120, *G. violaceus* PCC 7421, *Synechocystis* sp. strain PCC 6803, *S. elongatus* PCC 6301 and *T. elongatus* BP-1 with nitrate reductase (*narB*) and nitrite reductase A (*nirA*) and regulator ntcB upstream of transporters signifies the regulatory control of all these genes by NtcA. A number of transcription factors belonging to the families LuxR, LysR, AraC and TetR are encoded by the genome of *A. marina* MBIC11017 (Swingley *et al.*, 2008).

LexA is a transcriptional regulator that exists as a dimer and regulates transcription of genes connected with SOS response in *E. coli* (Fernandez *et al.*, 2000). Under normal growth conditions, LexA prevents transcription of genes under its control by binding to their promoter regions but during SOS response the binding of RecA to the damaged DNA strand triggers autocleavage of LexA into monomers that no longer can bind to its binding sites. Due to this, all such genes under the regulation of LexA are induced and expressed in a co-ordinated manner (Michel, 2005). But the type of genes regulated by LexA in different cyanobacteria has been found to be different. Studies conducted on *Synechocystis* sp. strain PCC 6803 point out that LexA may regulate transcription of genes of carbon metabolism (Domain *et al.*, 2004), bidirectional hydrogenase genes (*hoxEFUYH*; Gutekunst *et al.*, 2005; Oliveira and Lindblad, 2005) and *crhR* gene that encodes a RNA helicase (Patterson-Fortin *et al.*, 2008). The binding of LexA to a 14 bp pseudo-palindromic sequence in *Anabaena* sp. strain PCC 7120 has been demonstrated (Mazon *et al.*, 2004). In this organism, the *hoxEFUYH* exists as two operons, i.e. *hoxEF* and *hoxUYH*. Sjoholm *et al.* (2007) demonstrated that LexA regulates transcription of these two operons by binding to the upstream of promoter regions of *hoxE* and *hoxU*. In *Synechocystis* also a LexA-binding region consists of a 13 bp pesudo-palindromic segment 5′-AGTAACTAGTTCG-3′ upstream of the promoter of *hoxE* (Gutekunst *et al.*, 2005). Li *et al.* (2010) conducted a computational analysis of genes regulated by LexA in 33 cyanobacterial genomes. Six of the cyanobacterial genomes (*G. violaceus* PCC 7421, *Synechococcus* sp. JA-3-3-A′b, *Synechococcus* sp. JA-2-3-3B′a (2–13), *S. elongatus* PCC 6301, *S. elongatus* PCC 7942, *T. elongatus* BP-1 and *T. erythraeum* IMS101) do not possess a *lexA* gene. The SOS response in these strains is suspected to be regulated by a transcriptional regulator other than LexA. The rest of the 27 genomes of cyanobacteria [*A. marina* MBIC11017, *Anabaena* sp. strain PCC 7120, *A. variabilis* ATCC 29413, *Cyanothece* sp. PCC 8801, *M. aeruginosa* NIES-843, *N. punctiforme* PCC 73102, 12 strains of *P. marinus* (AS9601, CCMP1375, MED4, MIT9211, MIT9215, MIT9301, MIT9303, MIT9312, MIT9313, MIT9515, NATL1A, NATL2A), 8 strains of *Synechococcus* sp. (CC9311, CC9605, CC9902, PCC 7002, RCC307, WH8102, WH7803 and WH8102) and *Synechocystis* sp. strain PCC 6803] revealed the presence of *lexA* gene. According to Li *et al.* (2010) LexA might still serve as a transcriptional regulator for the SOS response in many of these cyanobacterial strains as LexA-binding sites of the core SOS-responsive genes *recA* and *lexA* are the most conserved. However, the extent to which other SOS-responsive genes are regulated is found to be different in these strains. For example, *ruvB* gene that encodes DNA helicase B has the LexA-binding sites in the HL-adapted *P. marinus* strains (MIT9312, MIT9515, MIT9215, MED4 and AS9601). Of the four nucleotide excision repair genes *uvrA*, *uvrB*, *uvrC* and *uvrD*, known to be regulated by LexA in *E. coli*, only *uvrD* has the LexA-binding sites in *P. marinus* MIT 9312. Apart from the SOS-responsive genes, LexA is known to regulate other photosynthetic genes such as PSII reaction center proteins PsbY (in *A. variabilis* ATCC 29413, *A. marina* MBIC11017, strains of *P. marinus* AS9601, MIT9215, MIT9301, MIT9312 and MIT9515), CP47 (in *Synechocystis* sp. strain PCC 6803) and a protein involved in PSI (in *M. aeruginosa* NIES-843, *P. marinus* strains MIT9312, NATL2A, *Synechococcus* sp. PCC 7002). In addition, 20 genes encoding transporters have been predicted to possess LexA-binding sites (in case of *A. marina* MBIC 11017, *A. variabilis* ATCC 29413, *M. aeruginosa* NIES-843, *N. punctiforme* PCC 73102, *Anabaena* sp. strain PCC 7120, *P. marinus* strains MIT9215, MIT9303, MIT9515, *Synechococcus* sp. strain WH8102 and *Synechocystis* sp. strain PCC 6803). Among these, the toxin and antibiotic exporter genes of *A. variabilis* ATCC 29413 and the genes for major drug efflux transporter in *P. marinus* (strains AS9601, MIT9301, MIT9515) and *Synechococcus* sp. strain WH 8102 stand out as importantly LexA-regulated. The overall conclusion is that at least in *Synechocystis* sp. strain PCC 6803 LexA no longer regulates SOS response (Li *et al.*, 2010).

c) Circadian clock proteins: *S. elongatus* PCC 7942 is the model organism for the studies on the circadian clock (see Chapter 6 on Circadian rhythms). Nakamura *et al.* (2002) identified genes *kaiA* (*tlr0481*), *kaiB* (*tlr0482*), *kaiC* (*tlr0483*), *sasA* (*tlr0029*), *cikA* (tll0899), *pex* (*tlr1955*) and *cpmA* (*tll1189*) in *T. elongatus* BP-1. A homologue of *pixJ1* (tll0569; that encodes bacteriophytochrome) required for positive phototaxis, *cikA* (*tll0899*) gene, two genes for flavin-binding cytochrome-like proteins (*tll0552* and *tll0425*) and one phototropin gene (*tll1282* that encodes flavin-binding photoreceptor) are also present in the genome of *T. elongatus* BP-1. KaiA, KaiB and KaiC protein encoding genes (*kaiA*, *kaiB* and *kaiC*) are absent in *G. violaceus* PCC 7421 Of the known input modifiers *sasA*, *cikA*, *ldpA* and *pexA* only *ldpA* gene sequence is present. Two output modifier genes (*rpoD2* and *cpmA*) also have been identified (Nakamura *et al.*, 2003). In the genome of *Anabaena* sp. PCC7120, major genes encoding KaiA, B and C proteins, as also of the genes of the input pathway (*cikA* and *pex*) and output modifier proteins (*rpoD2* and *cpmA*) and activator of KaiABC expression (*sasA*) have been identified (Kaneko *et al.*, 2001). All genes governing circadian rhythms are present in *M. aeruginosa* PCC 7806. However, the existence of genes encoding light-regulated two-component system consisting of *cph1* (for phytochrome) and its response regulator (*Rcp1*) in this organism is a question that needs to be investigated as to which of these processes gains control over the other (Frangeul *et al.*, 2008).

***7) DNA replication, recombination and repair*:** Kaneko *et al.* (1996) identified inteins (intervening protein sequences that are excised during post-translational modification) in the genes governing DNA synthesis in *Synechocystis* sp. strain PCC 6803. Four intein sequences have been identified in the genes of DNA helicase (*dnaB*), a subunit of DNA polymerase III (*dnaX*), DNA gyrase B subunit (*gyrB*) and the alpha subunit of DNA polymerase III (*dnaE*). In *Anabaena* sp. strain PCC 7120, two copies of *dnaB* are present, distributed one each on the chromosome (*all0578*) and the plasmid pCC7120α (*all7274*). An intein of 429 amino acid residues is present only in the former. *dnaE* gene is split into two parts (*all3578* and *alr1054*) separated by a 3 Mb portion and these two encode N-terminal and C-terminal portions, respectively of DnaE protein (Kaneko *et al.*, 2001). In *T. elongatus* BP-1, two split *dnaE* genes (*tll2056* and *tll2069*) are separated by a 10.2 kb region in the chromosome that are suspected to encode N-terminal and C-terminal parts of the DnaE protein, respectively. In addition, genes for natural transformation (*comA*, *comE*, *comM*) and for genetic recombination (*recA*, *recF*, *recG*, *recJ* and *recQ*) have been identified in this organism (Nakamura *et al.*, 2002). *P. marinus* strain MED4 lacks genes for several DNA repair pathways including recombinational repair (*recJ* and *recQ*) and damage reversal gene (*mutT*). The loss of *mutY* gene (the product of which removes adenosines incorrectly paired with oxidatively damaged guanine residues) in this organism is suggested to be responsible for increased rate of G.C to T.A transversions (Rocap *et al.*, 2003). A comparison of the genomes of strains of *P. marinus* strain MED4 and SS120 revealed that besides the absence of *ada* gene (that encodes 6-O-methylguanine-DNA methyltransferase which repairs alkylated forms of guanine and thymine in DNA), a number of genes such as *mutY* (that encodes A/G specific DNA glycosylase), *recQ* (that encodes Superfamily II DNA helicase), *recJ* (that encodes single-stranded DNA-specific exonuclease), *exol/lxseA* (that is responsible for exonuclease VII large subunit) and *xseB* (that encodes exonuclease VII small subunit) are absent in MED4 whereas SS120 possesses *mutY, recJ, exol/lxseA* and *xseB*. On the contrary, the presence of all these genes including that of *ada* has been demonstrated in *Synechococcus* strains MIT9313 and WH8102 (Dufresne *et al.*, 2005). Heavy duplication of genes related to DNA repair and recombination (primarily *recA*) seems to be a characteristic feature of the genome of *A. marina* MBIC11017. The percentage of such portions in the genome seems to be higher (18.7%) than noted in case of *Synechocystis* sp. strain PCC 6803 (11.2%) and *Anabaena* sp. strain PCC 7120 (5.8%) (Swingley *et al.*, 2008).

8) DNA restriction and modification: In *M. aeruginosa* PCC 7806 there are 21 potential genes for restriction enzymes localized together where as *M. aeruginosa* NIES-843 has 17 genes for restriction enzymes. Of these, 14 genes are common to both these strains. Among these, seven and eight of the restriction enzymes seem to be specific for *M. aeruginosa* PCC 7806 and *M. aeruginosa* NIES-843, respectively. A comparison of the genomes of these two *Microcystis* strains with *C. watsonii* WH8501 and *Synechocystis* sp. strain PCC 6803 revealed that the presence of 6-mer sequences signifies the existence of restriction sites. As in case of *Synechocystis* sp. strain PCC 6803 the absence of 6-mer sequences corresponded with the absence of restriction enzymes. A large majority of these 6-mer sequences constitute palindromic sequences amounting to nearly 51% of the rarest 1% 6-mers in *M. aeruginosa* PCC 7806 (Frangeul *et al.*, 2008). The genome of *M. aeruginosa* NIES-843 has 62 putative restriction-modification genes belonging to Type I RMs (4) and Type II RMs (58). In addition, four potential RM-related loci are disrupted by insertion sequences (ISs) (Kaneko *et al.*, 2007).

9) Transport and binding proteins: In the genome of *N. punctiforme* ATCC 29133 there are 89 ORFs that encode ATPase domain of assigned and unassigned membrane-associated ATP-binding cassette transport systems (designated accordingly as ABC-transporters). Additionally, 48 permeases not associated with ABC-transporters for the transport of organic carbon and iron are present. Two complete ATP-dependent phosphate transport systems (each consisting of *pstS, pstC, pstA, pstB*) and a periplasmic protein for sulphate transport are present (Meeks *et al.*, 2001). A gene cluster comprising of *nirA* (nitrite reductase) – *nrtA* - *nrtB* – *nrtC* – *nrtD* (ABC transporter)—*narB* (nitrate reductase) is present in *Anabaena* sp. strain PCC 7120 for the uptake and reduction of nitrite and nitrate, respectively. However, there appear to be two independent nitrite/nitrate transport systems in *N. punctiforme* ATCC 29133. This is evident by the presence of *nirA* and *narB* with an intervening permease that is meant for the transport of nitrite/nitrate. Besides these, a cluster of four genes bearing 90% similarity to nrtABCD nitrate transporter of *Anabaena* sp. strain PCC 7120. *P. marinus* MIT9313 has lost a 25-gene cluster that governs nitrate/nitrite transporter and nitrate reductase but nitrite reductase gene has been retained. This gene is flanked by a proteobacterial type nitrite transporter rather than a typical cyanobacterial type nitrate/nitrite permease. This suggests that these genes have been acquired through LGT which reflects that these two strains are able to adjust to the particular environmental niche in which the specific nutrient is prevalent. Accordingly, *P. marinus* MED4 that has even lost the ability to utilize nitrite is adapted to grow at the surface waters where ammonia is available and high light conditions prevail. On the other hand, *P. marinus* MIT9313 which can utilize nitrite is adapted to grow at greater depths where nitrite is available and low light conditions prevail. Genes necessary for the transport and utilization of urea, cyanate and oligopeptides have been identified in *P. marinus* MED4 but with no transporters for amino acid uptake. *P. marinus* MIT9313 has transporters for urea, amino acids and oligopeptides but lacks the genes necessary for cyanate utilization. In both the genomes, genes required for ammonia transport (*amt1*) and its incorporation through GS-GOGAT pathway are present besides the gene (*ntcA*) that encodes the nitrogen transcriptional regulator. A high affinity phosphate transport system encoded by *pstS* and *pstABC* is present in the genomes of both the strains with an additional copy of the phosphate-binding component *pstS* for utilization of orthophosphate from deeper waters. *P. marinus* MED4 possesses P-related regulatory genes (*phoB* and *phoR*-two component system and the transcriptional activator *ptrA*) whereas this system seems to be non-functional in *P. marinus* MIT9313 as it is interrupted by two frame-shifts and further degeneration noted in *ptrA*. The presence of genes encoding flavodoxin (*isiB*) in both *P. marinus* MED4 and MIT9313 has been demonstrated. An Fe-induced transcriptional regulator in both strains include (*Fur* that represses iron uptake genes) and numerous genes involved in high-affinity iron scavenging system and absence of genes concerned with Fe-siderophore complex are the additional characteristics of the iron acquisition system (Rocap *et al.*, 2003). These iron-related genes are absent in *Synechococcus* sp. strain WH8102 and this explains its dominance in iron-limited equatorial Pacific. In the genome of *Synechococcus* sp. strain WH8102, 5 to 6% of the gene complement is devoted for transport phenomena. Nearly 60% of the ORFs encode ABC- transporters. P-type ATPase transporter is found with only one transporter for copper. The uptake of urea and phosphate is facilitated by the synthesis of solute-binding proteins. For deriving nitrogen, the transporters for amino acids and oligopeptides are present. This strain is additionally equipped with genes for cyanate utilization as a source of nitrogen. In the marine waters, phosphonates (compounds with C-P bonds) produced by eukaryotic phytoplanktons (Coccolithophorids of Chrysophyta) constitute alternative P-source. So *Synechococcus* sp. strain WH8102 is equipped with genes for transport of phosphonates and the presence of genes for multiple phosphatases (*SYNW0120*, *SYNW0196*, *SYNW2391* and *SYNW2390*) explains the ability to obtain P from such environments. The presence of efflux transporters in the ABC family and efflux pumps for metals (*SYNW1472* and *SYNW0900*), chromate (*SYNW1323*) and arsenite (*SYNW1039*) are the additional features (Palenik *et al.*, 2003). *Synechococcus* sp. strain CC9311 is an inhabitant of coastal environment and thus is adapted to grow in higher phosphate concentrations. Accordingly, this strain lacks genes for a phosphate sensor-response regulator commonly seen in other cyanobacteria. In addition, it lacks several alkaline phosphatases that are present in *Synechococcus* sp. WH8102, which is an inhabitant of the open oceans. Metal transporters like FeOA/B transporter for iron that is absent in *Synechococcus* sp. strain WH8102 is seen in the coastal *Synechococcus* sp. strain CC9311 to meet the higher concentrations of iron in the coastal environment. In order to adjust to the osmotic changes in the coastal environment, *Synechococcus* sp. strain CC9311 is equipped with a metallothionein gene (*SmtA*) and five and two mechanosensitive ion channels of the type *MseS* and *MseL*, respectively compared with only two of *MesS* in *Synechococcus* sp. strain WH8102 (Palenik *et al.*, 2006). In *M. aeruginosa* NIES-843 synthesis of proteins required for phosphate acquisition is regulated by a two component system Hik7 (SphS) and Rre29 (SphR) that regulates the expression of genes, i.e. ABC-type of phosphate transporter. Three operons that encode subunits for ABC-type phosphate transporters (MAE18310-MAE18280, MAE18380-MAE18340 and MAE09280-MAE09250), three monocistronic genes for phosphate-binding periplasmic proteins (MAE18390, MAE32380, MAE38290) and two genes for alkaline phosphatases (MAE50240 and MAE16640) are present in the genome of this cyanobacterium (Kaneko *et al.*, 2007).

10) Insertion sequences: ISs constitute small mobile DNA elements capable of transposition that is mediated by a self-encoded transposase. These are classified into various families or groups on the basis of similarity, types of transposases and length of inverted repeats generated by insertion.. Their number varies from genus to genus. Kaneko *et al.* (1996) identified 990 ORFs bearing significant similarities to bacterial transposases spread over the entire genome of *Synechocystis* sp. strain PCC 6803. Only 26 of these encode functional transposases. The remaining ORFs seem to be disrupted by mutations such as frame-shift and deletions and also insertion of other ISs suggest that a lot of genomic rearrangements took place in this organism. In *N. punctiforme* ATCC 29133 there exist 150 ORFs that regulate the synthesis of transposases. STRR1 and STRR2 also occur frequently in this organism but STRR3 and LTRR (37-bp repeat) as noted in *Anabaena* sp. strain PCC 7120 are not found in *N. punctiforme* ATCC 29133 (Meeks *et al.*, 2001). In *Anabaena* sp. strain PCC 7120 as many as 145 genes encode transposases. The distribution of these genes on the chromosome and plasmids have been found to be 86 and 59, respectively. The plasmid pCC7120α contains 44 of these genes. The presence of majority of these genes in regions characteristic of IS-like elements, i.e. inverse repeats and/or duplications at both termini is significant. In *Synechocystis* sp. strain PCC 6803, *Anabaena* sp. strain PCC 7120, *T. elongatus* BP-1 and *G. violaceus* PCC 7421 there are 73 copies (in six families), 65 copies (in seven families), 52 copies (in five families) and 22 copies (in four families), respectively. A specific region designated as 'cold spot' in the genome of *T. elongatus* BP-1 occupies a 300 kb region and possesses high numbers of ISs and mobile introns. This region overlaps with a 'hot spot' region consisting of all the conserved genes among cyanobacterial species. At least 74% of the genes of *T. elongatus* BP-1 are common with the genes of *Synechocystis* sp. strain PCC 6803 and *Anabaena* sp. strain PCC 7120 (Nakamura *et al.*, 2002). In the genome of *M. aeruginosa* NIES-843 there are a total of 452 copies of ISs. However, in the genome of *M. aeruginosa* PCC 7806 there are only four putative ISs. Put together thus in *Microcystis* genome there are 456 copies grouped into 37 groups (ISMae 1 to ISMae37). ISMae1 and ISMae4 are common to both the strains but in *M. aeruginosa* NIES-843 there are 10 copies of ISMae1 and 33 copies of ISMae4. A second type of insertion sequences, termed as <u>m</u>iniature <u>i</u>nvert-repeat <u>t</u>ransposable <u>e</u>lements (MITEs) are characteristic in not having an encoded transposase. As many as 517 copies of MITEs ranging in size from 15 bp to 435 bp are present in the genome of *M. aeruginosa* NIES-843 and are classified into eight groups (*MaeMITEa* to *MaeMITEh*) (Kaneko *et al.*, 2008). The size of ISs constitute 11.8% of the genome in this organism with ISs (10% of the genome equal to 583 kb) and MITEs occupying the rest (1.8% of the genome of 105 kb). Importantly, the presence of ISs and MITEs around the *mcy* gene cluster signifies that transposition of this cluster between the individuals may be more common and thus contributes to a lot of genetic diversity in the natural populations of this toxic cyanobacterium. Most of the Group II introns are found in the ISs or intergenic regions. Group II introns are characteristic in coding for self-splicing ribozyme and with the help of the maturase or reverse transcriptase activity these function as retromobile genetic elements. In *T. elongatus* BP-1, 28 copies of Group II introns are present with high sequence similarity (with 87.2 to 100% for the type TEII3 and 85.3% to 100% for the TEII4) (Nakamura *et al.*, 2002). In *G. violaceus* PCC 7421 one copy of Group II intron has been identified at the map position 168,850-171,364 bp (Nakamura *et al.*, 2003). An eight base palindromic sequence, known as high-iterated palindrome (HIP1) sequence is present in the genomes of many cyanobacteria. But the number of copies of this sequence and its distribution varies in different genomes of cyanobacteria. As for example, the genome of *T. elongatus* BP-1 possesses as many as 3681 copies and the average frequency of occurrence is one copy/705 bp which is much higher than that present in *Synechocystis* sp. strain PCC 6803 (one copy/1131 bp). In *T. elongatus* BP-1 the frequency of occurrence of HIP1 is more in protein-coding regions (one copy/672 bp) than in RNA-encoding regions (one copy/4328 bp) and intergenic regions (one copy/1209 bp). The presence of one copy of HIP1 in the regions of mobile introns signifies that the HIP1 sequences are characteristic to the genomes of cyanobacteria. In the genome of *N. punctiforme* ATCC 29133, multiple copies of HIP1 sequences are known to occur as octanucleotide sequence (GCGATCGC). The frequency of such HIP1 is one sequence every 1200 bp (Meeks *et al.*, 2001). Elhai *et al.* (2008) identified mobile repeated elements known as small dispersed repeats (SDRs) having 21 to 27 nucleotides and have been grouped into 8 families. Three of the families represented by SDR4, SDR5 and SDR6 share a common predicted secondary structure. It is significant to note the distribution of SDR5 family within the octamer, HIP1.

***11) Toxic and bioactive compounds*:** It is intriguing to find out 62 ORFs encoding proteins involved in the synthesis of microcystins in the genome of *N. punctiforme* ATCC 29133 (Meeks *et al.*, 2001). *Microcystis* produces microcystins, cyanopeptolins and many other unidentified bioactive substances. In *M. aeruginosa* NIES-84, the biosynthesis of microcystins is mediated by a non-ribosomal peptide synthetase (NRPS) gene cluster (*mcyA-J*) and cyanopeptolins by cyanopeptolin synthase gene cluster (*mcnA-C* and *mcnE-G*). A third novel unknown NRPS gene cluster is also located on the chromosome of this organism. The organization of *mcy* gene cluster is similar to the other microcystin-producing cyanobacteria in having two sub-clusters transcribed in opposite directions. The first sub-cluster encodes NRPSs with accessory domains (*mcyA*, *mcyB* and *mcyC*) and the second encodes polyketide synthase (PKS) (*mcyD*) and mixed PKS-NRPS (*mcyE* and *mcyG*) and a racemase gene (*mcyF*). There are three additional genes located downstream this sub-cluster, i.e. two genes involved in the biosynthesis of microcystin (*mcyI* and *mcyJ*) and the third gene (*mcyH*) that is helpful in the transport of microcystin. A thioesterase gene (*mcyT*) that is implicated in the induction of microcystin is located within *mcy*-gene cluster in *Planktothrix agardhii* but this is located outside *mcy* gene cluster in *M. aeruginosa* NIES-843 (Kaneko *et al.*, 2007). Frangeul *et al.* (2008) compared the two strains of *M. aeruginosa* NIES-843 and PCC 7806 and confirmed the existence of three large gene clusters for NRPS/PKS. One of the clusters (PKSI) seems to govern the synthesis of aeruginoside and a protein similar to enedyine-type antitumor antibiotic (characteristically produced in some actinobacteria). The second (PKSII) cluster is responsible for the production of a number of tailoring enzymes and a third PKS cluster synthesizes compounds of chalcone/Stilbene family. One of the three clusters is suggested to govern the production of microcystins, cyanopeptolins and aeruginosins. The genes for cyanopeptolin (*mcn* cluster) bear resemblance to the anabaenopeptolide genes of *Anabaena* strain 90 and to the cyanopeptolin genes of *M. wesenbergii*. However, additionally the genome of *M. aeruginosa* PCC 7806 revealed the presence of a gene cluster for the synthesis of patellamide, a substance that is characteristically produced by the symbiotic cyanobacteria present in Ascidians. In contrast, *M. aeruginosa* NIES-843 produces microcyclamide that resembles patellamide. It is interesting to note the presence of transporter genes for these secondary metabolites along with those that confer resistance in the immediate vicinity of genes governing their synthesis. Thus the cells are able to survive in presence of these substances when they are released to outside by the efflux transporters (those of the uncharacterized PKSI/PKSIII hybrid compound) present (Frangeul *et al.*, 2008).

***12) Nif genes*:** The nitrogen-fixing non-heterocystous forms whose genomes have been sequenced include *Cyanothece* sp. ATCC 51142, *C. watsonii* WH8501, *T. erythraeum* IMS101, *Synechococcus* RF-1 and *Synechococcus* sp. JA-3-3A,b. The heterocystous cyanobacteria whose genomes have been sequenced are *Anabaena* sp. strain PCC 7120, *A. variabilis* ATCC 29413, *N. punctiforme* ATCC 29133 and *No. spumigena* CCY9414. In all these organisms, all nitrogen fixation genes (*nif* genes) are located in a contiguous cluster of 34 genes in a 28-kb region. Other structural genes in this region include *nifD* and *K* (that encode dinitrogenase, Mo-Fe protein) and *nifH* (that encodes dinitrogenase reductase, Fe-protein). In the above heterocystous forms (excepting in *No. spumigena*) and probably in *Synechococcus* RF-1 also the order of *nif* genes is *nifB – fdxn – nifS – nifU –nifH – nifD – nifK – Orf – nifE – nifN – nifX – Orf – nifW – hesA- hesB – fdxH*. The location and size of the excision element during heterocyst differentiation in *N. punctiforme* ATCC 29133 and *Anabaena* sp. strain PCC 7120 are found to be different. In the former the excision element is located at the 3′-end of the *nifD* gene and is of 24-kb whereas in the latter, *nifD* gene is interrupted by a 11-kb excision element. However, the two excision elements show a similarity in possessing a highly conserved excision gene (*xisA*) present at the beginning of the elements. Other unique features of *N. punctiforme* ATCC 29133 genome are (i) the absence of 55 kb excision element in *fdxN* gene (that is present in *Anabaena* sp. strain PCC 7120), (ii) the presence of a haemoglobin-like gene upstream of *nifH* as in *N. commune*, (iii) the synthesis of molybdenum-dependent nitrogenase in the heterocysts (in contrast to the presence of an alternative vanadium-dependent nitrogenase synthesized in the absence of molybdenum as noted in *A. variabilis* ATCC 29413 but not in *Anabaena* sp. strain PCC 7120), and (iv) the presence of additional copies of *nifH* (two copies), *nifE* and *nifN* (one additional copy each). The other genes governing the biosynthesis of Mo-Fe factors (*nifB, fdxN, nifS, nifU* and *nifV*), Mo-Fe factor assembly (*nifE* and *nifN*), iron uptake (*feoA* and *feoB*) and genes of unknown function too are represented in this cluster. This single largest cluster consists of two adjacent regulons on opposite strands with *modB* gene (that encodes molybdate ATP-binding cassette transporter permease) occurring at the end. In the heterocystous cyanobacteria, on the other hand, the corresponding clusters are interrupted by insertion elements (Welsh *et al.*, 2008). Genes for heterocyst initiation (*ntcA, hetR, hetN* and *hetF*), pattern formation (*patA, patS* and *patN*) and transition to non-dividing state (*hetC* as well as *hetP*) have been recognized in the genome of *Anabaena* sp. strain PCC 7120 on the basis of orthologues present in the related genus *N. punctiforme* ATCC 29133. Many of the genes (*ntcA, hamA, devA, hetR, patB, devR, devBCA*) encode proteins which have 90% sequence similarity among the two organisms. The gene products that share 60–70% amino acid identity between these two cyanobacteria are (i) a positive regulator for heterocyst differentiation (*hetF*), (ii) a sensor histidine kinase of a two-component regulatory system (*hepA* and *hepK*), (iii) pattern formation (*hglK* as well as *patA*), and (iv) a polyketide synthase (*hetM*). Other genes of *N. punctiforme* ATCC 29133 that show weak similarity with *Anabaena* sp. strain PCC 7120 genes are *hepC* (required for heterocyst envelope synthesis), *hetP* and *hetC* (similar to ABC protein exporters and required early in heterocyst differentiation). However, homologues of *hetP* have 70% similarity where as the homologues of *hetC* have 66% similarity in amino acid sequence. The gene product of *patS* is a peptide of 13 amino acids in *N. punctiforme* ATCC 29133 while it may be of 13 or 17 amino acids long in case of *Anabaena* sp. strain PCC 7120. One gene cluster consisting of *hglE*-(two unknown genes)-*hglD-hglC*-(one unknown gene)-*hetM-hetN-hetI-hglK* present on the chromosome of *Anabaena* sp. strain PCC 7120 is suggested to play a role in the synthesis and deposition of heterocyst glycolipid layer. The organization of *nif*-gene cluster in the genomes of *C. raciborskii* CS-505, *R. brookii* D9 (Stucken *et al.*, 2010) and *N. azollae* 0708 (Ran *et al.*, 2010) has already been presented in the discussion on their genomes. A comprehensive account on this subject is presented in Chapter 4.

IV. PHYLOGENOMICS

Besides the sequence-based phylogenies (single gene such as 16S rRNA gene, MLST analysis and whole genome), the utility of concatenation of protein-encoding genes or their products and the gene order in the construction of phylogenetic trees has been tested during recent years. Random regrouping (concatenation) of genes or their products much depends on the genes so selected whether or not such genes are liable to be transferred through LGT and on the nature of gene product. When gene order phylogenies are chosen, two important points are computed. These are breakpoint and inversion distances. Breakpoint distances denote the number of gene adjacencies that are present in one genome and absent in the other. So it reveals the dissimilarity of the gene order between two genomes. An inversion distance is indicative of the minimum number of inversion events that is required to convert one genome into another (Raubeson *et al.*, 1992; Olmstead and Palmer, 1994; Blanchette *et al.*, 1999; Belda *et al.*, 2005; Wang *et al.*, 2006). Kettler *et al.* (2007) identified a set of core genes common to 12 strains of *Prochlorococcus* and selected 100 such genes for drawing phylogenetic relationships. The phylogenetic trees of all these genes are congruent with the 16S rRNA and 16S–23S rRNA ITS phylogenies. Random regroupings of 100 such protein sequences (repeated 100 times) also yielded the same phylogeny and agreed with the topology derived from 16S rRNA gene. However, the position of two LL-strains of *P. marinus* MIT9211 and SS120 differed. Likewise, phylogenomics of 11 marine strains of *Synechococcus* based on concatenated alignments of 1,129 core proteins revealed phylogenetic trees consistent with 16S rRNA phylogeny. Two sub-groups have been recognized within the sub-cluster 5.1. In one of the sub-groups strains of *Synechococcus* WH8102, CC0605, CC9902 and BL107 are clustered together while in the second sub-group other strains of *Synechococcus* WH7803, WH7805, CC9311, RS9916, and RS9917 are grouped together. In contrast, in the 16S rRNA tree the position of *Synechococcus* strains RS9916 and RS9917 remained uncertain. Due to the fact that *Synechococcus* sp. strain RCC307 remained outside the sub-cluster 5.1, these workers preferred to create a new sub-cluster for this strain as 5.3 that has been consistent with 16S rRNA phylogeny. They further concluded that (i) the ancestor for sub-cluster 5.3 diverged even before the split between sub-cluster 5.1 and all other *Prochlorococcus* strains, and (ii) the rapid diversification process envisaged by Urbach *et al.* (1998) for marine picocyanobacteria seems to hold good in explaining the dominance of *Prochlorococcus* and sub-cluster 5.1 of *Synechococcus* in the marine ecosystem. Extending the criteria of average nucleotide identity (ANI) value of more than 95% (in relation to RB of 70% of DNA:DNA hybridization) for defining a species put forward by Konstantinidis and Tiedje (2005), these workers found that ANI appeared to be a better marker to assign the strains to a particular clade. However, the average ANI values worked out for marine picocyanobacteria centered around 87 to 91%, which is far below the level suggested by Konstantinidis and Tiedje (2005) for proteobacteria. Dufresne *et al.* (2008) suggested that both rRNA gene identity and ANI have fallen short to resolve the differences between the two *Prochlorococcus* and *Synechococcus* strains. Luo *et al.* (2008) subjected 12 genomes of *P. marinus* strains (MED4, MIT9515, MIT9312, AS9601, MIT9301, MIT9215-HL strains; SS120, MIT9211, NATL2A, NATL1A, MIT9303, MIT9313-LL) and one of *Synechococcus* strain (WH8102 as an outgroup) to BLASTCLUST analysis (a software that is helpful in clustering sequences based upon at least 30% similarity over a minimum of their 50% lengths). This revealed the presence of 1131 orthologous genes shared by all the 13 genomes. However, due to the existence of LGT and non-availability of any orhologous gene that is not liable to LGT, these workers preferred to construct phylogenies based on the gene order. The breakpoint and inversion distance-based phylogenies have been found to be consistent with the sequence-based and gene content-based trees in showing all six HL-strains in a monophyletic cluster with two clades (*P. marinus* MED4 and MIT9515 in the first clade and the other four strains of *P. marinus* MIT9312, AS9601, MIT9301 and MIT9215 in the second clade). However, in case of LL-strains, the inversion and breakpoint distance-based phylogenies showed the existence of two clusters (one for *P. marinus* NATL1A and NATL2A; and the other for *P. marinus* MIT9303 and MIT9313) consistent with sequence-based trees. At the same time, the two gene order phylogenies also gave support to the formation of a separate cluster for the two unresolved strains of *P. marinus*, i.e. MIT9211 and

SS120. This clade is closer to 6 HL-strains rather than to *P. marinus* NATL1A and NATL2A clade. On the other hand, sequence-based trees showed that *P. marinus* strains NATL1A and NATL2A are closer to the HL-strains. They have further integrated the gene order data as well as concatenated alignments and proposed that the situation can be resolved by proposing two alternative phylogenies. In the first, the clade with *P. marinus* strains MIT9211 and SS120 is closer to HL-strains where as in the second the clade with *P. marinus* strains NATL1A and NATL2A is closer to HL-strains. Johnson and Chisholm (2004) identified overlapping genes that are adjacent and located on either of DNA strands in the genomes of many microbes. These share one or more nucleotides in their coding sequences. A comparison of such overlapping genes among 1352 orthologous genes in two closely related strains of *P. marinus* (MIT9313 and MED4) revealed that there are 422 and 330 orthologues that are part of an overlapping gene pair in *P. marinus* MIT9313 and MED4, respectively. Among these orthologues, 69% and 83% in case of *P. marinus* MIT9313 and MED4, respectively have members of the overlap pair as orthologues. They concluded that overlapping genes are not directly related to either GC content or reduction in genome size but are more likely to be conserved. The order of genes in chromosomes and nucleotide sequences appeared to be similar in 23 cyanobacteria whose genomes have been compared by quantitative methods of estimation of gene order similarity. The reconstructions of phylogenies based on the similarity of gene orders led to the suggestion that genome rearrangements in marine picocyanobacteria are fixed at a low rate while in other groups the gene order can change very rapidly. The closest relatives of chloroplasts had been the strains of *Synechococcus* from hotsprings and the existence of *Synechococcus* strains in different environments (marine and freshwaters and hotsprings) is indicative of evolutionarily distant lines (Markov and Zakharov, 2009).

Molecular synapomorphies are protein signature sequences that represent conserved portions or indels that are specific to a group of taxa on the basis of which one can construct phylogenetic trees and compare these with the whole genome-based or single gene-based (16S rRNA) phylogenetic trees. Gupta (2009) compared phylogenetic trees constructed for 34 cyanobacterial sequenced genomes based on concatenated sequences for 45 conserved proteins and also the 16S rRNA gene. Out of more than 40 molecular signatures, 15 cyanobacteria-specific signature sequences have been identified that are helpful in delimiting several important taxonomic clades of cyanobacteria.

LITERATURE CITED

Ahmed, N. (2009) A flood of microbial genomes-do we need more? *PLoS ONE* **4:** e5831. doi:10.1371/journal.pone.0005831

Akiyama, M., Gotoh, T., Kise, H., Miyashita, H., Mimuro, M., and Kobayashi, M. (2004) Stoichiometries of chlorophyll *d*/PSI and chlorophyll *a*/PSII in a chlorophyll *d*-dominated cyanobacterium *Acaryochloris marina. Jpn J Phycol* **52:** 67–72.

Akiyama, M., Miyashita, H., Kise, H., Watanabe, T., Miyachi, S., and Kobayashi, M. (2001) Detection of chlorophyll *d* and pheophytin *a* in a chlorophyll *d*-dominating oxygenic photosynthetic prokaryote *Acaryochloris marina. Anal Sci* **17:** 205–208.

Akiyama, M., Miyashita, H., Kise, H., Watanabe, T., Mimuro, M., Miyachi, S., and Kobayashi, M. (2002) Quest for minor but key chlorophyll molecules in photosynthetic reaction centers-unusual pigment composition in the reaction centers of the chlorophyll *d*-dominated cyanobacterium *Acaryochloris marina. Photosynth Res* **74:** 97–107.

Allen, J. F., de Paula, W. B. M., Puthiyaveetil, S., and Nield, J. (2011) A structural phylogenetic map for chloroplast photosynthesis. *Trends Plant Sci* **16:** 645–655.

Almeida, L. G. P., Paixao, R., Souza, R.C., da Costa, G. C., Barrientos, F. J., dos Santos, M. T., de Almeida, D.F., and Vasconcelos, A. T. R. (2004) A system for automated bacterial (genome) integrated annotation-SABIA. *Bioinformatics* **20:** 2832–2833.

Altschul, S. F., and Koonin, E. V. (1998) Iterated profile searches with PSI-BLAST: a tool for discovery in protein databases. *Trends Biochem Sci* **23:** 494–497.

Altschul, S. F., Gish, W., Miller, W., Myers, E. W., and Lipman, J. (1990) Basic local alignment tool. *J Mol Biol* **215:** 403–410. doi: 10.1006/mbi.1990.9999

Altschul, S. F., Madden, T. L., Schaffer, A. A., Zhang, J., Zheng, Z., Miller, W., and Lipman, D. J. (1997) Gapped BLAST and PSI-BLAST—a new generation of protein database search programs. *Nucleic Acids Res* **25:** 3389–3402.
Amunts, A., Drory, O., and Nelson, N. (2007) The structure of a plant photosystem I supercomplex at 3.4 Å resolution. *Nature* (London) **447:** 58–63.
Awai, K., Kakimoto, T., Awai, C., Kaneko, T., Nakamura, Y., Takamiya, K-i., Wada, H., and Ohta, H. (2006) Comparative genomic analysis revealed a gene for monoglucosyldiacylglycerol synthase, an enzyme for photosynthetic membrane lipid synthesis in cyanobacteria. *Plant Physiol* **141:** 1120–1127.
Axmann, I. M., Kensche, P., Vogel, J., Kohl, S., Herzel, H., and Hess, W. R. (2005) Identification of cyanobacterial non-coding RNAs by comparative genome analysis. *Genome Biol* **6:** R73. doi:10.1186/gb-2005-6-9-r73
Badger, J. H., and Olsen, G. J. (1999) CRITICA: Coding region identification tool invoking comparative analysis. *Mol Biol Evol* **16:** 512–524.
Bazzan, A. L. C., Duarte, R., Pitinga, A. N., Schroeder, L. F., Souto, F. D. A., and Ceroni da Silva, S. (2003) ATUGC—an agent-based environment for automatic annotation of genomes. *Int J Comp Inform Syst* **12:** 241–273.
Belda, E., Moya, A., and Silva, F. J. (2005) Genome rearrangement distances and gene order phylogeny in {gamma} proteobacteria. *Mol Biol Evol* **22:** 1456–1467.
Bergman, N. H., Passalacqua, K. D., Hanna, P. C., and Qin, Z. S. (2007) Operon prediction for sequenced bacterial genomes without experimental information. *Appl Environ Microbiol* **73:** 846–854. doi: 10.1128/AEM.01686-06
Bhaya, D., Dufresne, A., Vaulot, D., and Grossman, A. (2002) Analysis of *hli* gene family in marine and freshwater cyanobacteria. *FEMS Microbiol* **215:** 209–219.
Biers, E. J., Sun, S., and Howard, E. C. (2009) Prokaryotic genomes and diversity in surface ocean waters: Integrating the global ocean sampling metagenome. *Appl Environ Microbiol* **75:** 2221–2229. doi:10.1128/AEM02118-08
Blanchette, M., Kunisawa, T., and Sankoff, D. (1999) Gene order break point evidence in animal mitochondrial phylogeny. *J Mol Evol* **49:** 193–203.
Blattner, F. R., Plunkett, G. III., Bloch, C. A., Perna, N. T., Burland, V., Riley, M., Collado-Vides, J., Glasner, J. D., Rode, C. K., Mayhew, G., Gregor, J., Davis, N. W., Kirkpatrick, H. A., Goeden, M. A., Rose, D. J., Man., B., and Shao, Y. (1997) The complete genome sequence of *Escherichia coli* K-12. *Science* **277:** 1453–1462.
Bryson, K., Loux, V., Bossy, R., Nicolas, P., Chaillou, S., van de Guchte, M., Penaud, S., Maguin, E., Hoebeke, M., Bessières, P., and Gibrat, J-F. (2006) AGMIAL: Implementing an annotation strategy for prokaryotic genomes as a distributed system. *Nucleic Acids Res* **34:** 3533–3545. doi: 10.1093/nar/gkl471
Campbell, E. L., Brahamsha, B., and Meeks, J. C. (1998) Mutation of an alternative sigma factor in the cyanobacterium *Nostoc punctiforme* results in increased infection of its symbiotic plant partner, *Anthoceros punctatus*. *J Bacteriol* **180:** 4938–4941.
Cadel-Six, S., Iteman, I., Peyraud-Thomas, C., Mann, S., Ploux, O., and Méjean, A. (2009) Identification of a polyketide synthase coding sequence specific for anatoxin-a-producing *Oscillatoria* cyanobacteria. *Appl Environ Microbiol* **75:** 4909–4912.
Chen, X., Su, Z., Dam, P., Palenik, P., Xu, Y., and Jiang, T. (2004) Operon prediction by comparative genomics: An application to the *Synechococcus* sp. WH8102 genome. *Nucleic Acids Res* **32:** 2147–2157. doi: 10.1093/nar/gkh510
Chi, X., Yang, Q., Zhao, E., Qin, S., Yang, Y., Shen, J., and Lin, H. (2008) Comparative analysis of fatty acid desaturases in cyanobacterial genomes. *Comp Funct Genomics* **2008**. Article Id:284508. doi:10.1155/2008/284508
Coleman, M., Sullivan, M. B., Martiny, A. C., Steglich, C., Barry, K., DeLong, E. F., and Chisholm, S. W.(2006) Genomic islands and the ecology and evolution of *Prochlorococcus*. *Science* **311:** 1768–1770. doi:10.1126/science.1122050
Craven, M., Page, D., Shavlik, J., Bockhorst, J., and Glasner, J. (2000) A probabilistic learning approach to whole-genome operon prediction. *Proc 8th Internl Conf on Intelligent Systems for Molecular Biology*, AAAI Press, San Diego, CA, USA pp. 116–127.
Csiszar, K., Houmard, J., Damerval, T., and Tandeau de Marsac, N. (1987) Transcriptional analysis of the cyanobacterial *gvpABC* operon in differentiated cells: Occurrence of an antisense RNA complementary to three overlapping transcripts. *Gene* **60:** 29–37.
Dam, P., Olman, V., Harris, K., Su, Z., and Xu, Y. (2007) Operon prediction using both genome-specific and general genome information. *Nucleic Acids Res* **35:** 288–298. doi:10.1093/nar/gkl1018
Dietz, K. J. (2003) Plant peroxiredoxins. *Annu Rev Plant Biol* **54:** 93–107.
Dolganov, N. A. M., Bhaya, D., and Grossman, A. R. (1995) Cyanobacterial protein with similarity to the chlorophyll a/b-binding proteins of higher plants: Evolution and regulation. *Proc Natl Acad Sci USA* **92:** 636–640.
Domain, F., Houot, L., Chauvat, F., and Cassier-Chauvat, C. (2004) Function and regulation of the cyanobacterial genes *lexA*, *recA* and *ruvB*: LexA is critical to the survival of cells facing inorganic carbon starvation. *Mol Microbiol* **53:** 65–80.
Doolittle, R. F (Ed.) (1996) *Computer Methods for Macromolecular Sequence Analysis. Methods Enzymol* **266**. Academic Press, San Diego, CA, USA.
Ducret, A., Muller, S. A., Goldie, K. N., Hefti, A., Sidler, W. A., Zuber, H., and Engel, A. (1998) Reconstitution, characterisation and mass analysis of the pentacylindrical allophycocyanin core complex from the cyanobacterium *Anabaena* sp. PCC 7120. *J Mol Biol* **278:** 369–388.

Dufresne, A., Garczarek, L., and Partensky, F. (2005) Accelerated evolution associated with genome reduction in a free-living prokaryote. *Genome Biol* **6:** R14. http://genomebiology.com/2005/6/2/R14
Dufresne, A., Salanoubat, M., Partensky, F., Artiguenave, F., Axmann, I. M., Barbe, V., Duprat, S., Galperin, M. Y., Koonin, E. V., Le Gall, F., Makarova, K. S., Ostrowski, M., Oztas, S., Robert, C., Rogozin, I. B., Scanlan, D. J., Tandeau de Marsac, N., Weissenbach, J., Wincker, P., Wolf, Y. I., and Hess, W. R. (2003) Genome sequence of the cyanobacterium *Prochlorococcus marinus* SS120, a nearly minimal oxyphototrophic genome. *Proc Natl Acad Sci USA* **100:** 10020–10025.
Dufresne, A., Ostrowski, M., Scanlan, D. J., Garczarek, L., Mazard, S., Palenik, B. P., Paulsen, I. T., Tandeau de Marsac, N., Wincker, P., Dossat, C., Ferriera, S., Johnson, J., Post, A. F., Hess, W. R., and Pratensky, F. (2008) Unraveling the genome mosaic of a ubiquitous genes of marine cyanobacteria. *Genome Biol* **9:** R90. doi:10.1186/gb-2008-9-5-r90
Duhring, U., Axmann, I. M., Hess, W. R., and Wilde, A. (2006) An internal antisense RNA regulates expression of the photosynthesis gene *isiA*. *Proc Natl Acad Sci USA* **103:** 7054–7058.
Elhai, J., Kato, M., Cousins, S., Lindblad, P., and Costa, J. L. (2008) Very small mobile repeated elements in cyanobacterial genomes. *Genome Res* **18:** 1484–1499. doi:10.1101/gr.074336.107
Ermolaeva, M. D., White, O., and Salzberg, S. L. (2001) Prediction of operons in microbial genomes. *Nucleic Acids Res* **29:** 1216–1221.
Fernandez De Henestrosa, A. R., Ogi, T., Aoyagi, S., Chafin, D., Hayes, J. J., Ohmori, H., and Woodgate, R. (2000) Identification of additional genes belonging to the LexA regulon in *Escherichia coli*. *Mol Microbiol* **35:** 1560–1572.
Fitch, W. M. (1970) Distinguishing homologous from analogous proteins. *Syst Zool* **19:** 99-113.
Fleischmann, R. D., Adams, M. D., White, O., Clayton, R. A., Kirkness, E. F., Kerlavage, A. R., Bult, C. J., Tomb, B. A., Dougherty, B. A., Merrik, J., McKenney, K., Sutton, G., FitzHugh, W., Fields, C., Gocayne, D., Scott, J., Shirley, R., Liu, L.-I., Glodek, A., Kelley, J. M., Weidman, J. F., Phillips, C. A., Spriggs, T., Hedblorn, E., Cotton, M. D., Utterback, T. R., Hanna, M. C., Nguyen, D. T., Saudek, D. M., Brandon, R. C., Fine, L. D., Fritchman, J. L., Fuhrmann, J. L., Geoghagen, N. S. M., Gnehm, C. L., McDonald, L. A., Small, K. V., Fraser, C. M., Smith, H. O., and Venter, J. C. (1995) Whole-genome random sequencing and assembly of *Haemophilus influenzae* Rd. *Science* **269:** 496–512.
Frangeul, L., Quillardet, P., Castets, A. M., Humbert, J-F., Matthijs, H. C. P., Cortez, D., Tolonen, A., Zhang, G-C., Gribaldo, S., Kehr, J-C., Zilliges, Y., Ziemert, N., Becker, S., Talla, E., Latifi, A., Billault, A., Lepelletier, A., Dittmann, E., Bouchier, C., and Tandeau de Marsac, N. (2008) Highly plastic genome of *Microcystis aeruginosa* PCC 7806, a ubiquitous toxic freshwater cyanobacterium. *BMC Genomics* **9:** 274 doi:10.1186/1471-2164-9-274
Frias-Lopez, J., Shi, Y., Tyson, G. W., Coleman, M. L., Schuster, S. C., Chisholm, S. W., and De Long, E. F. (2008) Microbial community gene expression in ocean surface waters. *Proc Natl Acad Sci USA* **105:** 3805–3810. doi:10.1073/pnas.0708897105
Fujisawa, T., Narikawa, R., Okamoto, S., Ehira, S., Yoshimura, H., Suzuki, I., Masuda, T., Mochimaru, M., Takaichi, S., Awai, K., Sekine, M., Horikawa, H., Yashiro, I., Omata, S., Takarada, H., Katano, Y., Kosugi, H., Tanikawa, S., Ohmori, K., Sato, N., Ikeuchi, M., Fujita, N., and Ohmori, M. (2010) Genomic structure of an economically important cyanobacterium, *Arthrospira* (*Spirulina*) *platensis* NIES-39. *DNA Res* **17:** 85–103. doi:10.1093/dnares/dsq004
Fuller, N. J., Marie, D., Partensky, F., Vaulot, D., Post, A. F., and Scanlan, D. J. (2003) Clade-specific 16S ribosomal DNA oligonucleotides reveal the predominance of a single marine *Synechococcus* clade throughout a stratified water column in the Red Sea. *Appl Environ Microbiol* **69:** 2430–2443.
Funk, C., and Vermaas, W. (1999) A cyanobacterial gene family coding for single-helix proteins resembling part of the light-harvesting proteins from higher plants. *Biochemistry* **38:** 9397–9404.
Gaasterland, T., and Sensen, C. W. (1996) Fully automated genome analysis that reflects user needs and preferences. A detailed introduction to the MAGPIE system architecture. *Biochimie* **78:** 302–310.
Gardian, Z., Bumba, L., Schrofel, A., Herbstova, M., Nebesarova, J., and Vacha, F. (2007) Organization of photosystem I and photosystem II in red alga *Cyanidium caldarium* encounter of cyanobacterial and higher plant concepts. *Biochim Biophys Acta* **1767:** 725–731.
Gerdes, S. Y., Kurnasov, O. V., Shatalin, K., Polanuyer, B., Sloutsky, R., Vonstein, V., Overbeek, R., and Osterman, A. L. (2006) Comparative genomics of NAD biosynthesis in cyanobacteria. *J Bacteriol* **188:** 3012–3023. doi:10.1128/jb.188.8.3012-3023.2006
Glazer, A. N., Williams, R. C., Yeh, S. W., and Clark, J. H. (1985) Kinetics of energy flow in the phycobilisome core. *Science* **230:** 1051–1053.
Gombos, Z., Kanervo, E., Tsvetkova, N., Sakamoto, T., Aro, E.-M., and Murata, N. (1997) Genetic enhancement of the ability to tolerate photoinhibition by introduction of unsaturated bonds into membrane glycerolipids, *Plant Physiol* **115:** 551–559.
Grigoriev, A. (1998) Analyzing genomes with cumulative skew diagrams. *Nucleic Acids Res* **26:** 2286–2290.
Guan, X., Qin, S., Zhao, F., Zhang, X., and Tang, X. (2007) Phycobilisomes linker family in cyanobacterial genomes: Divergence and evolution. *Int J Biol Sci* **3:** 434–445. www.biolsci.org.

Gupta, R. S. (2009) Protein signatures (molecular synapomorphies) that are distinctive characteristics of the major cyanobacterial clades. *Int J Syst Evol Microbiol* **59:** 2510–2526. doi:10.1099/ijs.0.005678-0

Gutekunst, K., Phunpruch, S., Schwarz, C., Schuchardt, S., Schulz-Friedrich, R., and Appel, J. (2005) LexA regulates the bidirectional hydrogenase in the cyanobacterium *Synechocystis* sp. PCC 6803 as a transcription activator. *Mol Microbiol* **58:** 810–823.

Henikoff, S., and Henikoff, J. G. (1997) Embedding strategies for effective use of information from multiple sequence alignments. *Protein Sci* **6:** 698–705.

Herdman, M., Janvier, M., Rippka, R., and Stanier, R. Y. (1979) Genome size of cyanobacteria. *J Gen Microbiol* **111:** 73–85.

Hernandez, J. A., Muro-Pastor, A. M., Flores, E., Bes, M. T., Peleato, M. L., and Fillat, M. F. (2006) Identification of a *furA* cis antisense RNA in the cyanobacterium *Anabaena* sp. PCC 7120. *J Mol Biol* **355:** 325–334.

Herrero, A., Muro-Pastor, A. M. and Flores, E. (2001) Nitrogen control in cyanobacteria. *J Bacteriol* **183:** 411–425.

Hirani, T. A., Suzuki, I., Murata, N., Hayashi, H., and Eaton-Rye, J. J. (2001) Characterization of a two-component signal transduction system involved in the induction of alkaline phosphatase under phosphate-limiting conditions in *Synechocystis* sp. PCC 6803. *Plant Mol Biol* **45:** 133–144.

Hofmann, B., Hecht, H. J., and Flohe, L. (2002) Peroxiredoxins. *Biol Chem* **383:** 347–364.

Janssen, P. J., Morin, N., Mereay, M., Leroy, B., Wattiez, R., Vallaeys, T., Waleron, K., Waleron, M., Wilmotte, A., Quillardet, P., Tandeau de Marsac, N., Talla, E., Zhang, C.-C., and Leys, N. (2010) Genome sequence of the edible cyanobacterium *Arthrospira* sp. PCC 8005. *J Bacteriol* **192:** 2465–2466.

Johnson, Z. I., and Chisholm, S. W. (2004) Properties of overlapping genes are conserved across microbial genomes. *Genome Res* **14:** 2268–2272. doi:10.1101/gr.2433104.

Jones, A., Manroe, E. A., Podell, S., Hess, W. R., Klages, S., Esquenzi, E., Niessen, S., Hoover, H., Rothmann, M., Lasken, R. S., Yates III, J. R., Reinhardt, R., Kube, M., Burkart, M. D., Allen, E. E., Dorrestein, P. C., Gerwick, W. H., and Gerwick, L. (2011) Genomic insights into physiology and ecology of the marine filamentous cyanobacterium, *Lyngbya majuscula*. *Proc Natl Acad USA* **108:** 8815–8820. doi:10.1073/pnas.11011137108.

Jordan, P., Fromme, P., Witt, H. T., Klukas, O., Saenger, W., and Krauss, N. (2001) Three-dimensional structure of cyanobacterial photosystem I at 2.5 Åresolution. *Nature* (London) **411:** 909–917.

Kaneko, T., Nakajima, N., Okamoto, S., Suzuki, I., Tanabe, Y., Tamaoki, M., Nakamura, Y., Kasi, F., Watanabe, A., Kawashima, K., Kishida, Y., Ono, Y., Shimizu, Y., Takahashi, C., Minami, C., Fujishiro, T., Kohara, M., Katoh, M., Nakazaki, N., Nakayama, S., Yamada, M., Tabata, S., and Watanabe, M. M. (2007) Complete genomic structure of the bloom-forming toxic cyanobacterium *Microcystis aeruginosa* NIES-843. *DNA Res* **14:** 247–256.

Kaneko, T., Nakamura, Y., Sasamoto, S., Watanabe, A., Kohara, M., Matsumoto, M., Shimpo, S., Yamada, M., and Tabata, S. (2003) Structural analysis of four large plasmids harboring in a unicellular cyanobacterium *Synechocystis* sp. PCC 6803. *DNA Res* **10:** 221–228.

Kaneko, T., Nakamura, Y., Wolk, C. P., Sasamoto, S., Watanabe, A., Iriguchi, M., Ishikawa, A., Kawashima, K., Kimura, T., Kishida, Y., Kohara, M., Matsumoto, M., Matsuno, A., Muraki, A., Nakazaki, N., Shimpo, S., Sugimoto, M., Takazawa, M., Yamada, M., Yasuda, M., and Tabata, S. (2001) Complete genomic sequence of the filamentous nitrogen-fixing cyanobacterium *Anabaena* sp. strain PCC 7120. *DNA Res* **8:** 205–213.

Kaneko, T., Sato, S., Kotani, H., Tanaka, A., Asamizu, E., Nakamura, Y., Miyajima, N., Hirosawa, M., Sugiura, M., Sasamoto, S., Kimura, T., Hosouchi, T., Matsuno, A., Muraki, A., Nakazaki, N., Naruo, K., Okumura, S., Shimpo, S., Takeuchi, C., Wada, T., Watanabe, A., Yamada, M. Yasuda, M., and Tabata, S. (1996) Sequence analysis of the genome of the unicellular cyanobacterium *Synechocystis* sp. strain PCC 6803. II. Sequence determination of the entire genome and assignment of potential protein-coding regions. *DNA Res* **3:** 109–136.

Kaneko, T., Tanaka, A., Sato, S., Kotani, H., Sazuka, T., Miyajima, N., Sugiura, M., and Tabata, S. (1995) Sequence analysis of the genome of the unicellular cyanobacterium *Synechocystis* sp. strain PCC 6803. I. Sequence features in the 1 Mb region from map positions 64% to 92% of the genome. *DNA Res* **2:** 153–166.

Kasahara, M., and Ohmori, M. (1999) Activation of a cyanobacterial adenylate cyclase, CyaC, by autophosphorylation and a subsequent phosphotransfer reaction. *J Biol Chem* **274:** 15167–15172.

Katayama, M., and Ohmori, M. (1997) Isolation and characterization of multiple adenylate cyclase genes from the cyanobacterium *Anabaena* sp. strain PCC 7120. *J Bacteriol* **179:** 3588–3593.

Kaufmann, M. (2006) The role of COG database in comparative and functional genomics. *Curr Bioinformatics* **1:** 291–300.

Kettler, J. C., Martiny, A. C., Huang, K., Zucker, J., Coleman, M. L., Rodrigue, S., Chen, F., Lapidus, A., Ferriera, S., Johnson, J., Steglich, C., Church, G. M., Richarson, P., and Chisholm, S. W. (2007) Patterns and implications of gene gain and loss in the evolution of *Prochlorococcus*. *PLoS Genet* **3:** e231. doi:10.1371/journal.pgen-0030231

Kotani, H., Kaneko, T., Matsubayashi, T., Sato, S., Sugiura, M., and Tabata, S. (1994) A physical map of the genome of a unicellular cyanobacterium *Synechocystis* sp. strain PCC 6803. *DNA Res* **1:** 303–307.

Koonin, E. V., and Galperin, M. Y. (1997) Prokaryotic genomes: The emerging paradigm of genome-based microbiology. *Curr Opin Genet Develop* **7:** 757–763.

Koonin, E. V., Tatusov, R. L., and Galperin, M. Y. (1998) Beyond complete genomes: From sequence to structure and function. *Curr Opin Struct Biol* **8:** 355–63.

Konstantinidis, K. T., and Tiedje, J. M. (2005) Genomic insights that advance the species definition for prokaryotes. *Proc Natl Acad Sci USA* **102:** 2567–2572.

Kühl, M., Chen, M., Ralph, P. J., Schreiber, U., and Larkum, A. W. D. (2005) A niche for cyanobacteria containing chlorophyll *d*. *Nature* (London) **433:** 820

Li, S., Xu, M., and Su, Z. (2010) Computational analysis of LexA regulons in cyanobacteria. *BMC Genomics* **11:** 527. doi:10.1186/1471-2164-11-527

Lombardot, T., Kottmann, R., Pfeffer, H., Richter, M., Teeling, H., Quast, C., and Glöckner, F. O. (2006) Megx.net-database resources for marine ecological genomics. *Nucleic Acids Res* **34:** D390-D393 doi: 10.1093/nar/gkj070

Luo, H., Shi, J., Arndt, W., Tang, J., and Friedman, R. (2008) Gene order phylogeny of the genus *Prochlorococcus*. *PLoS ONE* **3:** e3837.

Markov, A. V., and Zakharov, I. A. (2009) Evolution of gene orders in genomes of cyanobacteria. *Russian J Genet* **45:** 906–916.

Markowitz, V. M., Mavromatis, K., Ivanova, N. N., Chen, I-M. A., Chu, K., and Kyrpides, N. C. (2009) IMGER: A system for microbial genome annotation expert review and curation. *Bioinformatics* **25:** 2271–2278.

Mazmouz, R., Chapuis-Hugon, F., Mann, S., Pichon, V., Méjean, A., and Ploux, O. (2010) Biosynthesis of cylindrospermopsin and 7-epicylindrospermopsin in *Oscillatoria* PCC 6506; identification of the gene cluster and toxin analysis. *Appl Environ Microbiol* **76:** 4943–4949.

Mazon, G., Erill, I., Campoy, S., Cortes, P., Forano, E., and Barbe, J. (2004) Reconstruction of the evolutionary history of the LexA-binding sequence. *Microbiology* **150:** 3783–3795.

Medigue, C., Rechenmann, F., Danchin, A., and Viari, A. (1999) Imagene: An integrated computer environment for sequence annotation and analysis. *Bioinformatics* **15:** 2–15.

Meeks, J. C., Elhai, J., Thiel, T., Potts, M., Larimer, F., Lamerdin, J., Predki, P., and Atlas, R. (2001) An overview of the genome of *Nostoc punctiforme*, a multicellular symbiotic cyanobacterium. *Photosynth Res* **70:** 85–106.

Méjean, A., Mann, S., Maldiney, T., Vassiliadis, G., Lequin, O., and Ploux, O. (2009) Evidence that biosynthesis of the neurotoxic alkaloids anatoxin-a and homoanatoxin-a in the cyanobacterium *Oscillatoria* PCC 6506 occurs on a modular polyketide synthase initiated by L-proline. *J Am Chem Soc* **131:** 7512–7513.

Méjean, A., Mann, S., Vassiliadis, G., Lombard, B., Loew, D., and Ploux, O. (2010a) *In vitro* reconstitution of the first steps of anatoxin-a biosynthesis in *Oscillatoria* PCC 6506: From free L-proline to acyl carrier protein bound dehydroproline. *Biochemistry* **49:** 103–113.

Méjean, A., Mazmouz, R., Mann, S., Calteau, A., Médigue, C., and Ploux, O. (2010b) The genome sequence of the cyanobacterium *Oscillatoria* sp. PCC 6506 reveals several gene clusters responsible for the biosynthesis of toxins and secondary metabolites. *J Bacteriol* **192:** 5264–5265. doi: 10.1123/JB.00704-10.

Meyer, F., Goesmann, A., McHardy, A.C., Bartels, D., Bekel, T., Clausen, J., Kalinowski, J., Linke, B., Rupp, O., and Giegerich, R. (2003) GenDB-an open source genome annotation system for prokaryote genomes. *Nucleic Acids Res* **31:** 2187–2195.

Michel, B. (2005) After 30 years of study, the bacterial SOS response still surprises us. *PLoS Biol* **3:** e255. doi:10.137/journal.pbio.0030255

Mihali, T. K., Kellmann, R., Muenchhoff, J., Barrow, K. D., and Neilan, B. A. (2008) Characterization of the gene cluster responsible for cylindrospermopsin biosynthesis. *Appl Environ Microbiol* **74:** 716–722.

Miller, S. R., Augustine, S., Olson, T. L., Blankenship, R. E., Selker, J., and Wood, A. M. (2005) Discovery of a free-living chlorophyll *d*-producing cyanobacterium with a hybrid proteobacterial/cyanobacterial small-subunit rRNA gene. *Proc Natl Acad Sci USA* **102:** 850–855.

Miyashita, H., Adachi, K., Kurano, N., Ikemoto, H., Chihara, M., and Miyachi, S. (1997) Pigment composition of a novel oxygenic photosynthetic prokaryote containing chlorophyll *d* as the major chlorophyll. *Plant Cell Physiol* **38:** 274–281.

Miyashita, H., Adachi, K., Kurano, N., Ikemoto, H., Chihara, M., and Miyachi, S. (2003) *Acaryochloris marina* Gen. et Sp. nov. (Cyanobacteria), an oxygenic photosynthetic prokaryote containing Chl *d* as a major pigment. *J Phycol* **39:** 1247–1253.

Miyashita, H., Ikemoto, H., Kurano, N., Adachi, K., Chihara, M., and Miyachi, S. (1996) Chlorophyll *d* as a major pigment. *Nature* (London) **383:** 402.

Mulkidjanian, A. Y., Koonin, E. V., Makarova, K. S., Mekhedov, S. L., Sorokin, A., Wolf, Y. I., Dufresne, A., Partensky, F., Burd, H., Kanadzey, D., Haselkorn, R., and Galperin, M. Y. (2006) The cyanobacterial genome core and the origin of photosynthesis. *Proc Natl Acad Sci USA* **103:** 13126–13131. doi: 10.1073/pnas.065709103

Murakami, A., Miyashita, H., Iseki, M., Adachi, K., and Mimuro, M. (2004) Chlorophyll *d* in an epiphytic cyanobacterium of red algae. *Science* **303:** 1633.

Murata, N., and Wada, H. (1995) Acyl-lipid desaturases and their importance in the tolerance and acclimatization to cold of cyanobacteria. *Biochem J* **308:** 1–8.

Nakamura, T., Naito, K., Yokota, N., Sugita, C., and Sugita, M. (2007) A cyanobacterial non-coding RNA, Yfr1, is required for growth under multiple stress conditions. *Plant Cell Physiol* **48:** 1309–1318.

Nakamura, Y., Kaneko, T., and Tabata, S. (2000) CyanoBase, the genome database for *Synechocystis* sp. PCC 6803: Status for the year 2000. *Nucleic Acids Res* **28:** 72.

Nakamura, Y., Kaneko, T., Sato, S., Ikeuchi, M., Katoh, H., Sasamoto, S., Watanabe, A., Iriguchi, M., Kawashima, K., Kimura, T., Yishida, Y., Kiyokawa, M., Matsumoto, M., Matsuno, A., Nakazaki, N., Shimpo, S., Sugimoto, M., Takeuchi, C., Yamada, M., and Tabata, S. (2002) Complete genome structure of the thermophilic cyanobacterium *Thermosynechococcus elongatus* BP-1. *DNA Res* **9:** 123–130.

Nakamura, Y., Kaneko, T., Sato, S., Minuro, M., Miyashita, H., Tsuchiya, T., Sasamoto, S., Watanabe, A., Kawashima, K., Kishida, Y., Kiyokawa, C., Kohara, M., Matsumoto, M., Matsuno, A., Nakazaki, N., Shimpo, S., Takeuchi, C., Yamada, M., and Tabata, S. (2003) Complete genome structure of *Gloeobacter violaceus* PCC 7421, a cyanobacterium that lacks thylakoids. *DNA Res* **10:** 137–145.

Neuwald, A. F., Liu, J. S., Lipman, D. J., and Lawrence, C. E. (1997) Extracting protein alignment models from the sequence database. *Nucleic Acids Res* **25:** 1665–1677.

Ochoa de Alda, J. A. G., and Houmard, J. (2000) Genomic survey of cAMP and cGMP signaling components in the cyanobacterium *Synechocystis* PCC 6803. *Microbiology* **146:** 3183–3194.

Ochoa de Alda, J. A. G., Ajlani, G., and Houmard, J. (2000). The *Synechocystis* PCC 6803 *cya2*, a prokaryotic gene that encodes a guanylyl cyclase. *J Bacteriol* **182:** 3839–3842.

Ohkubo, S., Miyashita, H., Murakami, A., Takeyama, H., Tsuchiya, T., and Mimuro, M. (2006). Molecular detection of epiphytic *Acaryochloris* spp. on marine macroalgae. *Appl Environ Microbiol* **72:** 7912–7915.

Oliveira, P., and Lindblad, P. (2005) LexA, a transcription regulator binding in the promoter region of the bidirectional hydrogenase in the cyanobacterium *Synechocystis* sp. PCC 6803. *FEMS Microbiol Lett* **251:** 59–66.

Olmstead, R. G., and Palmer, J. D. (1994) Chloroplast DNA systematics: A review of methods and data analysis. *Am J Bot* **81:** 1205–1224.

Overbeek, R., Fonstein, M., D'Souza, M., Pusch, G. D., and Maltsev, N. (1999) The use of gene clusters to infer functional coupling. *Proc Natl Acad Sci USA* **96:** 2896–2901.

Palenik, B., Brahamsha, B., Larimer, F. W., Land, M., Hauser, L., Chain, P., Lamerdin, J., Regala, W., Allen, E. E., NcCarren, J., Paulsen, I., Dufresne, A., Pratensky, F., Webb, E. G., and Waterbury, J. (2003) The genome of a motile marine *Synechococcus*. *Nature* (London) **424:** 1037–1042.

Palenik, B., Ren, Q., Dupont, C. L., Myers, G. S., Heidelberg, J. F., Badger, J. H., Madupu, R., Nelson, W. L., Brinkac, L. M., Dadson, R. L., Durkin, A. S., Daugherty, S. C., Sullivan, S. A., Khouri, H., Mohamoud, Y., Halpin, R., and Paulsen, I. T. (2006) Genome sequence of *Synechococcus* CC9311: Insights into adaptation to a coastal environment. *Proc Natl Acad Sci USA* **103:** 13555–13559.

Patterson-Fortin, L. M., and Owttrim, G. W. (2008) A *Synechocystis* LexA-orthologue binds direct repeats in target genes. *FEBS Lett* **582:** 2424–2430.

Ran, L., Larsson, J., Vigil-Stenman, T., Nylander, J. A. A., Ininbergs, K., Zheng, W-W., Lapidus, A., Lowry, S., Haselkorn, R., and Berman, B. (2010) Genome erosion in a nitrogen-fixing vertically transmitted endosymbiotic multicellular cyanobacterium. *PLoS ONE* **5:** e11486. doi:10.1371/journal.pone.0011486

Raubeson, L. A., and Jansen, R. K. (1992) Chloroplast DNA: Evidence on the ancient evolutionary split in vascular land plants. *Science* **255:** 1697–1699.

Reitzer, L. (2003) Nitrogen assimilation and global regulation in *Escherichia coli*. *Annu Rev Microbiol* **57:** 155–176.

Riley, M. (1993) Functions of the gene products of *Escherichia coli*. *Microbiol Rev* **57:** 862–952.

Rocap, G., Larimer, F. W., Lamerdin, J., Malfatti, S., Chain, P., Ahlgren, N. A., Arellano, A., Coleman, M., Hauser, L., Hess, W. R., Johnson, Z. I., Land, M., Lindell, D., Post, A. F., Regala, W., Shal, M., Shaw, S. L., Steglich, C., Sullivan, M. B., Ting, C. S., Tolonen, A., Webb, E. A., Zinser, E. R., and Chisholm, S. W. (2003) Genome divergence in two *Prochlorococcus* ecotypes reflects oceanic niche differentiation. *Nature* (London) **424:** 1042–1047.

Sabatti, C., Rohlin, L., Oh, M. K., and Liao, J. C. (2002) Co-expression pattern from DNA microarray experiments as a tool for operon prediction. *Nucleic Acids Res* **30:** 2886–2893.

Salgado, H., Moreno-Hagelsieb, G., Smith, T. F., Collado-Vides, J. (2000) Operons in *Escherichia coli*: Genomic analyses and predictions. *Proc Natl Acad Sci USA* **97:** 6652–6657.

Salzberg, S. L., Delcher, A. L., Kasif, S., and White, O. (1998) Microbial gene identification using interpolated Markov models. *Nucleic Acids Res* **26:** 544–548.

Sato, N. (2003) Gclust: Genome-wide clustering of protein sequences for identification of photosynthesis-related genes resulting from massive horizontal gene transfer. *Genome Informatics* **14:** 585–586.

Satoh, S., and Tanaka, A. (2006) Identification of chlorophyllide *a* oxygenase in the *Prochlorococcus* genome by a comparative genome approach. *Plant Cell Physiol* **47:** 1622–1629. doi:10.1093/pep/pcl026

Scharf, M., Schneider, R., Casari, G., Bork, P., Valencia, A., Ouzounis, C., and Sander, C. (1994) GeneQuiz: A workbench for sequence analysis. *ISMB* **2:** 348–353.

Scheller, H. V., Jensen, P. E., Haldrup, A., Lunde, C., and Knoetzel, J. (2001) Role of subunits in eukaryotic photosystem I. *Biochim Biophys Acta* **1507:** 41–60.

Shi, T., and Falkowski, P. G. (2008) Genome evolution in cyanobacteria: The stable core and the variable shell. *Proc Natl Acad Sci USA* **105:** 2510–2515.

Six, C., Thomas, J-C., Garczarek, L., Ostrowski, M., Dufresne, A., Blot, N., Scanlan, D. J., and Partensky, F. (2007) Diversity and evolution of phycobilisomes in marine *Synechococcus* sp.: A comparative genomics study. *Genome Biol* **8:** R259s. doi:10.1186/gb-2007-8-12-r259

Sjoholm, J., Oliveira, P., and Lindblad, P. (2007) Transcription and regulation of the bidirectional hydrogenase in the cyanobacterium *Nostoc* sp. strain PCC 7120. *Appl Environ Microbiol* **73:** 5435–5446.

Stork, T., Michel, K-P., Pistorius, E. K., and Dietz, K-J. (2005) Bioinformatic analysis of the genomes of the cyanobacteria *Synechocystis* sp. PCC 6803 and *Synechococcus elongatus* PCC 7942 for the presence of peroxiredoxins and their transcription regulation under stress. *J Exp Bot* **56:** 3193–3206. doi:10.1093/jxb/eri316

Steglich, C., Futschik, M. E., Lindell, D., Voss, B., Chisholm, S. W., and Hess, W. R. (2008) The challenge of regulation in a minimal photoautotroph: Non-coding RNAs in *Prochlorococcus*. *PLoS Genet* **4:** e1000173. doi:10.1371/journal.pgen.1000173

Stroebel, D., Choquet, Y., Popot, J. L., and Picot, D. (2003) An atypical haem in the cytochrome b(6)f complex. *Nature* (London) **426 :** 413–418.

Stubbs, C. D., and Smith, A. D. (1984) The modification of mammalian membrane polyunsaturated fatty acid composition in relation to membrane fluidity and function. *Biochim Biophys Acta* **779:** 89–137.

Stucken, K., John, U., Cembella, A., Murillo, A. A., Soto-Liebe, K., Fuentes-Valdés, J. J., Friedel, M., Plominsky, A. M., Vásquez, M., and Glöckner, G. (2010) The smallest known genomes of multicellular toxic cyanobacteria: Comparison of minimal gene sets for linked traits and the evolutionary implications. *PLoS ONE* **5:** e9235. doi: 10.1371/journal.pone.0009235

Su, Z., Olman, V., Mao, F., and Xu, Y. (2005) Comparative genomics analysis of *ntcA* regulons in cyanobacteria: Regulation of nitrogen assimilation and its coupling to photosynthesis. *Nucelic Acids Res* **33:** 5156–5171. doi:10.1093/nar/gki817

Su, Z., Olman, V., and Xu, Y. (2007) Computational prediction of Pho regulons in cyanobacteria. *BMC Genomics* **8:** 156. doi: 10.1186/1471-2164-8-156

Sugaya, N., Murakami, H., Satoh, M., Aburatani, S., and Horimoto, K. (2003) Gene distribution patterns on cyanobacterial genomes. *Genome Informatics* **14:** 561–562.

Sugaya, N., Sato, M., Murakami, H., Imaizumi, A., Aburatani, S., and Horimoto, K. (2004) Causes for the large genome size in a cyanobacterium *Anabaena* sp. PCC 7120. *Genome Informatics* **15:** 229–238.

Sugita, C., Ogata, K., Shikata, M., Jikuya, H., Takano, J., Furllmichi, M., Kanchisa, M., Omata, T., Sugiura, M., and Sugita, M. (2006) Complete nucleotide sequence of the freshwater unicellular cyanobacterium *Synechococcus elongatus* PCC 6301 chromosome: Gene content and organization. *Photosyntth Res* doi:10.1007/s11120-006-9122-4

Suzuki, S., Ferjani, A., Suzuki, I., and Murata, N. (2004) The SphS-SphR two-component system is the exclusive sensor for the induction of gene expression in response to phosphate limitation in *Synechocystis*. *J Biol Chem* **279:** 13234–13240.

Swingley, W.D., Chen, C., Chuung, P. C., Conrad, A. L., Dejesa, L. C., Hao, J., Honchak, B. M., Karbach,L. E., Kurdoglu, A., Lahiri, S., Mastrain, S. D., Miyashita, H., Page, L., Ramakrishna, P., Satoh, S., Sattley, W.M., Shimada, Y., Taylor, H. L., Tomo, T., Tsuchiya,T., Wang, W. T., Raymond, J., Mimuro, M., Blankenship, R. E., and Touchman, J. W. (2008) Niche adaptation and genome expansion in the chlorophyll *d*-producing cyanobacterium *Acaryochloris marina*. *Proc Natl Acad Sci USA* **105:** 2005–2010. doi:10.1073/pnas.0709772105

Tatusov, R. L., Koonin, E. V., and Lipman, D. J. (1997) A genomic perspective on protein families. *Science* **278:** 631–637.

Tatusov, R. L., Galperin, M. Y., Natale, D. A., and Koonin, E. V. (2000) The COG database: A tool for genome-scale analysis of protein functions and evolution. *Nucleic Acids Res* **28:** 33–36.

Tatusov, R. L., Natale, D. A., Garkavtsev, I. V., Tatusova, T. A., Shankavaram, V. T., Rao, B. S., Kiryutin, B., Galperin, M. Y., Fedorova, N. D., and Koonin, E. V. (2001) The COG database: new developments in phylogenetic classification of proteins from complete genomes. *Nucleic Acids Res* **29:** 22–28.

Tatusov, R. L., Fedorova, N. D., Jackson, J. D., Jacobs, A. R., Kiryutin, B., Koonin, E. V., Krylov, D. M., Mazumder, R., Mekhedov, S. L., Nikolskoya, A. N., Rao, B. S., Smirnov, S., Sverdlov, A. V., Vasudevan, S., Wolf, Y. I., Yin, J. J., and Natale, D. A. (2003) The COG database: an updated version includes eukaryotes. *BMC Bioinformatics* **4:** 41. doi:10.1186/1471-2105-4-41

Terauchi, K., and Ohmori, M. (1999) An adenylate cyclase, Cya1, regulates cell motility in the cyanobacterium *Synechocystis* sp. PCC 6803. *Plant Cell Physiol* **40:** 248–251.

Umena, Y., Kawakami, K., Shen, J-R., and Kamiya, N. (2011) Crystal structure of oxygen-evolving photosystem II at a resolution of 1.9 angstrom. *Nature* (London) **473:** 55–60.doi: 10.1038/nature09913

Urbach, E., Scanlan, D. J., Distel, D. L., Waterbury, J.B., and Chisholm, S. W. (1998) Rapid diversification of marine picophytoplankton with dissimilar light-harvesting structures inferred from sequences of *Prochlorococcus* and *Synechococcus* (Cyanobacteria). *J Mol Evol* **46:** 188–201.

Vallenet, D., Labarre, L., Rouy, Z., Barbe, V., Bocs, S., Cruveiller, S., Lajus, A., Pascal, G., Scarpelli, C., and Medigue, C. (2006) MaGe: A microbial genome annotation system supported by synteny results. *Nucleic Acids Res* **34:** 53–65.

Veal, E. A., Findlay, V. J, Day, A. M, Bozonet, S. M., Evans, J. M., Quinn, J., Morgan, .BA. (2004) A 2-Cys peroxiredoxin regulates peroxide-induced oxidation and activation of a stress-activated MAP kinase. *Mol Cell* **15:** 129–139.

Wang, L-S., Warnow, T., Moret, B., Jansen, R., and Raubeson, L. (2006) Distance-based genome rearrangement phylogeny. *J Mol Evol* **63:** 473–483.

Ward, D. M., Cohan, F. M., Bhaya, D., Heidelberg, J. F., Kühl, M., and Grossman, A., (2008) Genomics, environmental genomics and the issue of microbial species. *Heredity* **100:** 207–219. doi:10.1038/sj.hdy.6801011

Waterbury, J.B. and Rippka, R. (1989) Order *Chroococcales* Wettstein 1924, emend. Rippka et al. 1979. In: Staley, J. T., Bryant, M. P., Pfenning, N., and Holt, J. G. (Eds.), *Bergey's Manual of Systematic Bacteriology*, vol. 3. Williams & Wilkins, Baltimore, USA pp. 1728–1746.

Welsh, E. A., Liberton, M., Stöckel, J., Loh, T., Elvitigalo, T., Wang, C., Wollman, A., Fulton, R. S., Clifton, S. W., Jacobs, J. M., Aurora, R., Ghosh, B. K., Sherman, L. A., Smith, R. D., Wilson, R. K., and Pakrasi, H. B. (2008) The genome of *Cyanothece* 51142, a unicellular diazotrophic cyanobacterium important in the marine nitrogen cycle. *Proc Natl Acad Sci USA* **105:** 15094–15099. doi:1073/pnas.0805418105

Wood, Z. A., Poole, L. B., and Karplus, P. A. (2003a) Peroxiredoxin evolution and the regulation of hydrogen peroxide signaling. *Science* **300:** 650–653.

Wood, Z. A., Schroder, E., Robin Harris, J., and Poole, L. B. (2003b) Structure, mechanism and regulation of peroxiredoxins. *Trends Biochem Sci* **28:** 32–40.

Wu, J., Zhao, F., Wang, S., Deng, G., Wang, J., Bai, J., Lu, J., Qu, J., and Bao, Q. (2007) cTF: A database for comparative genomics for transcription factors in cyanobacteria. *BMC Genomics* **8:** 104. doi:10.1186/1471-2164-8-104.

Xu, M., and Su, Z. (2009) Computational prediction of cAMP receptor (CRP) binding sites in cyanobacterial genomes. *BMC Genomics* **10:** 23. doi:10.1186/1471-2164-10-23

Yang, Y., Qin, S., Zhao, F., Chi, X., and Zhang, X. (2007) Comparison of envelope-related genes in unicellular and filamentous cyanobacteria. *Comp Funct Genomics* **2007**. doi:10.1155/2007/25751

Zhang, X., Zhao, F., Guan, X., Yang, Y., Liang, C., and Qin, S. (2007) Genome-wide survey of putative serine/threonine protein kinases in cyanobacteria. *BMC Genomics* **8:** 395. doi:10.1186/1471-2164-8-395

Zheng, Y., Szustakowski, J. D., Fortnow, L., Roberts, R. J., and Kasif, S. (2002) Computational identification of operons in microbial genomes. *Genome Res* **12:** 1221–1230.

Zuker, M. (2003) Mfold web server for nucleic acid folding and hybridization prediction. *Nucleic Acids Res* **31:** 3406–3415.

CHAPTER 3

AKINETES

Many filamentous cyanobacteria of the orders Nostocales and Stigonematales (corresponding t subsections IV and V, respectively of Rippka *et al.*, 1979) exhibit relatively simple morphology and th vegetative cells differentiate into heterocysts and akinetes during growth. Based on this taxonomi clustering and some biochemical and genetic similarities between these two cell types, akinete have been suggested to be the progenitors of heterocysts (Wolk *et al.*, 1994). Akinetes have been firs reported by Carter (1856) followed by other workers. The term 'akinetes' has been derived from th Greek word 'akinetos' which means 'motionless' and represent thick-walled, resting cells (Adam and Duggan, 1999). The discovery of Mesoproterozoic *Archaeoellipsoides*, microfossils of akinete (as long as 135 μm) of heterocystous cyanobacteria, indicated that the atmospheric oxygen level that persisted 1500 million years ago were well above 1% and these akinetes were comparable i morphology with the present day bloom-forming *Anabaena* sp. (Golubic *et al.*, 1995). *Archaeoellipsoide* has been reported from Newari locality of Mesoproterozoic Kheinjua Formation of the Sem Group, Vindhyan Supergroup in India (Srivastava, 2005). An important group of substances such a 2-methylhopane hydrocarbons serve as indicators of cyanobacteria and in other words the origin of oxygenic photosynthesis (Summons *et al.*, 1999). In this connection, the akinetes of *N. punctiform* are distinctive in possessing 2-methylbacteriohopanepolyol (2-MeBHP) lipids 34-fold higher tha vegetative cells and these are localized in the outer membranes. Since 2-MeBHPs are associated wit outer membranes of akinetes, functionally they can not be linked to oxygenic photosynthesis as ha been generally done to trace the origin of oxygenic photosynthesis in a geobiological perspectiv (Doughty *et al.*, 2009).

Akinetes of cyanobacteria have been shown to be tolerant to desiccation (Yamamoto, 1975; Sutherland *et al.*, 1979; Rai and Pandey, 1981; Sili *et al.*, 1994; Baker and Bellifemine, 2000) and seven year-old desiccated akinetes exhibited germination (Sili *et al.*, 1994). Sediments as old as 64 years revealed the presence of viable akinetes (Livingstone and Jaworskii, 1980). The thermal tolerance of akinetes, represented by their germination capacity after heat treatments, varied from species to species. The germination potential of *Anabaena vaginicola* akinetes was almost completely lost after heat treatment at 55°C for 8 min or 60°C for 2 min (Rai and Pandey, 1981) while the akinetes of *Anabaena circinalis* survived a heat treatment of 45°C for 60 min (Baker and Bellifemine, 2000). Akinetes of *Anabaena* dried in sunlight (10 h) and by heat treatment (60°C for 50 h) germinated only up to 40% and 15%, respectively (Hori *et al.*, 2003). Akinetes of *A. circinalis* have been reported to be susceptible to ultraviolet radiations (Fay, 1988). Dried akinetes of *Anabaena cylindrica* exposed to desiccation (28 d) and extreme low temperature (–80ºC) survived better than those subjected to vacuum (0.7 x 10–3 KPa) and high temperature (80ºC) under the laboratory conditions. Some proportion of the akinetes also survived for a period of 10 d in low Earth Orbit and also simulated conditions of Mars without UV radiations (28 d) (Olsson-Francis *et al.*, 2009).

Due to their resistance to various extremes in environmental variables, the akinetes of cyanobacteria have been equated with bacterial endospores (Nichols and Carr, 1978; Nichols and Adams, 1982). Since they are not heat-resistant, some others considered them equivalent to cysts of *Azotobacter* due to their metabolic activities and rapid growth on return to favourable environmental conditions (Sutherland *et al.*, 1979). A considerable wealth of information has accumulated on their ultrastructure, formation and metabolic activities and factors regulating their formation (Fogg *et al.*, 1973; Nichols and Carr, 1978; Adams and Carr, 1981; Pandey and Talpasayi, 1981; Nichols and Adams, 1982; Herdman, 1988; Sarma, 1993; Adams and Duggan, 1999).

I. AKINETE DIFFERENTIATION

***1) Genera studied*:** A number of cyanobacteria such as *Anabaena circinalis* (Fay *et al.*, 1984; van Dok and Hart, 1996), *Anabaena cylindrica* (Wolk, 1965; Simon, 1977); *Anabaena. doliolum* (Singh, 1967; Singh and Srivastava, 1968; Singh *et al.*, 1972b; Tyagi, 1974), *Anabaena fertilissima* (Reddy and Talpasayi, 1981), *Anabaena iyengarii* (Agrawal and Singh, 2000), *A. oscillarioides* (Rother and Fay, 1979), *Anabaena torulosa* (Sarma and Swarn Kanta, 1979; Sarma and Malhotra, 1989; Ahuja *et al.*, 2008), *Anabaena variabilis* (Singh *et al.*, 1972b), *Anabaenopsis arnoldii* (Reddy, 1983), *Aphanizomenon flos-aquae* (Wildman *et al.*, 1975; Thompson *et al.*, 2009), *Aph. ovalisporum* (Sukenik *et al.*, 2009), *Cylindrospermum licheniforme* (Fisher and Wolk, 1976; Hirosawa and Wolk, 1979a,b), *Cylindrospermum majus* (Singh and Sinha, 1965), *Cylindrospermopsis raciborskii* (Moore *et al.*, 2004, 2005), *Fischerella muscicola* (Singh *et al.*, 1972a), *Nodularia spumigena* (Pandey and Talpasayi, 1980) *Nostoc linckia* (Singh *et al.*, 1972b), *Nostoc* PCC 7524 (Sutherland *et al.*, 1979, 1985a,b); *Nostoc punctiforme* ATCC 29133 (Argueta and Summers, 2005), *Nostoc spongiaeforme* (Thiel and Wolk, 1983), *Nostoc* sp. (Ahluwalia and Kumar, 1983) and *Nostochopsis lobatus* (Agrawal and Singh, 2000), *Scytonema fritschii* (Sarma and Ghai, 1998) and *Westiellopsis prolifica* (Agrawal and Singh, 2000) have been employed for investigations on the pattern of akinete differentiation.

***2) Position of akinetes*:** The akinetes are formed either adjacent to or away from the heterocysts. In some species of *Anabaena* and *Nostoc* the initiation of akinete differentiation starts midway between two heterocysts and proceeds in both directions approaching the heterocysts (Fig. 1 A, B; Fig. 2 A, B). In species of *Cylindrospermum* (Fig. 3 A), *Gloeotrichia* (Fig. 2 D) and certain other species of *Anabaena* and *Nostoc* (Fig. 1A, B), cells adjacent to the heterocysts differentiate into akinetes. However, the

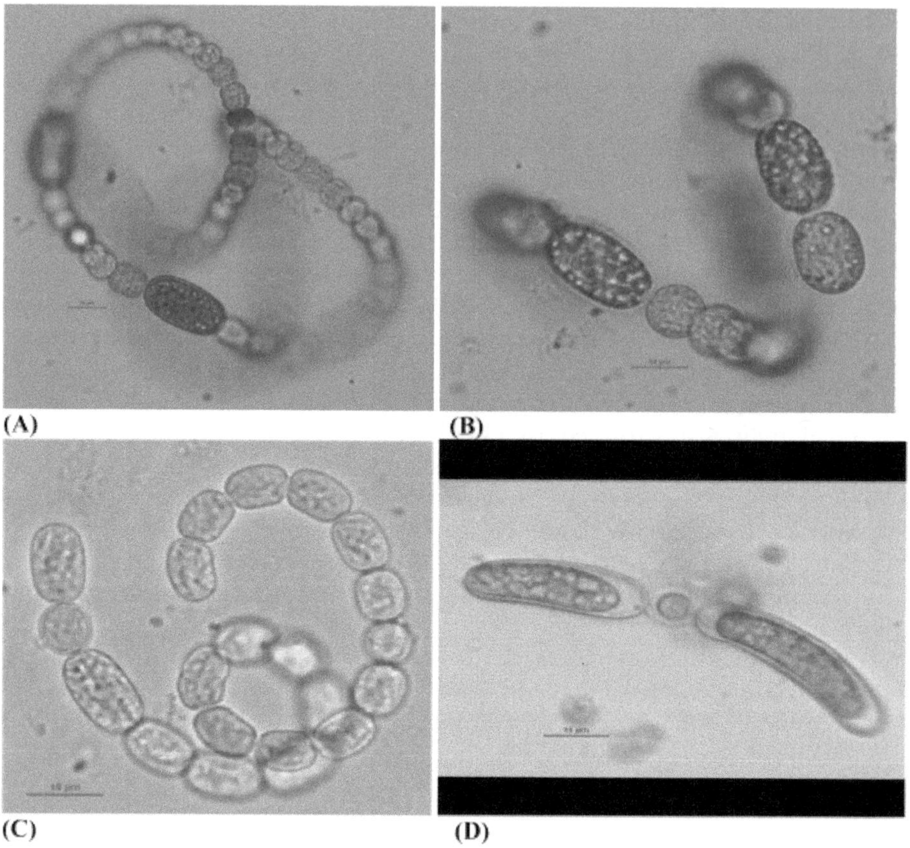

Figure 1: Development of akinetes in species of *Anabaena*. *A. circinalis* (A), *A. crassa* (B), *A. lamermannii* (C) and mature akinetes of *A. lamermannii* (D). In (A) and (B) development of akinetes a little away from heterocysts can be noted whereas in (C) and (D) the development of akinetes on either side of the heterocyst can be noticed. The magnification bar in the figures represents 10 µm. Pictures courtesy Mark T. Aubel, GreenWater Laboratories/CyanoLab, 205, Zeagler Drive, Suite 302, Palatka, Florida 32177, USA. (http://www.greenwaterlab.com/photo_algal.htm).

number of akinetes formed in the majority of genera is quite large as for example in certain species of *Anabaena*, akinetes are formed in strings or chains as all vegetative cells get transformed into akinetes (Fig. 2 C). In *A. lammermanni* the akinetes are formed in large numbers and they settle to the bottom of the water body (Fig. 4 A). However, few akinetes are formed in *Cylindrospermum* (subterminally beside a heterocyst on either side of filament) and *Gloeotrichia* (where only one subterminal akinete is present in each filament). In certain species of *Anabaena* (*A. circinalis*), *Anabaenopsis* (*A. raciborskii*) two akinetes differentiate in the middle of the trichome. The akinetes formed in case of *Anabaena* cf. *macrospora* appear to be the largest in size (Fig. 4 B).

3) *Structure*: In the light microscope, the akinetes appear as thick-walled and granulated structures very much larger than their antecedent vegetative cells. They appear as dark brown or blackish structures. The ultrastructure of akinetes revealed that photosynthetic lamellae remain intact and the presence of glycogen, cyanophycin, lipid granules and polyhedral bodies has been demonstrated.

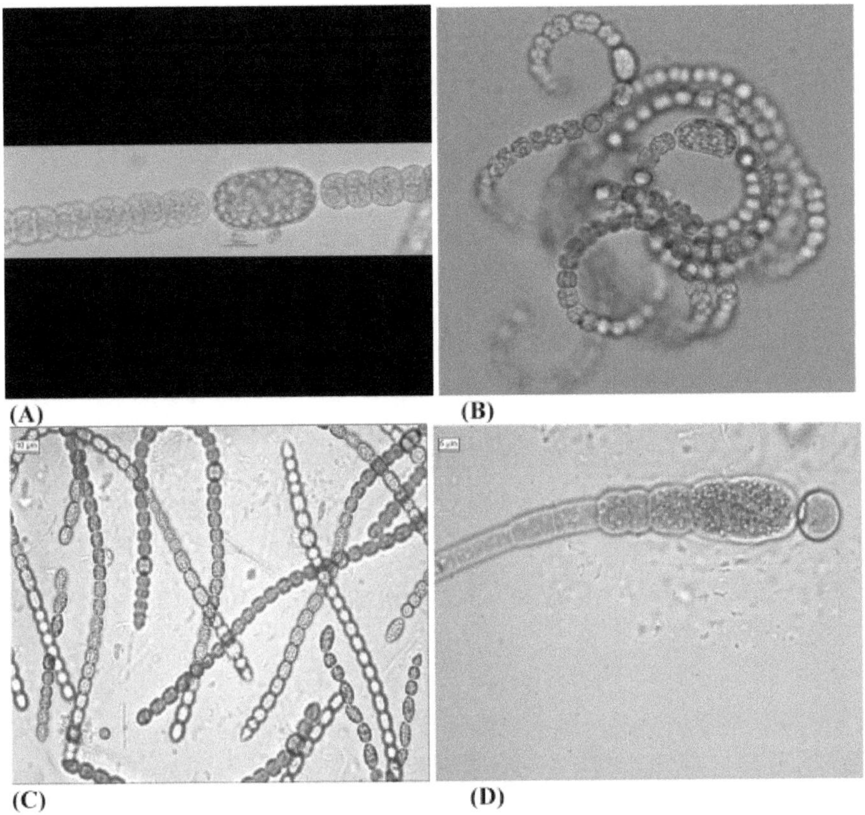

Figure 2: Development of akinetes in *Anabaena planctonica* (A), *A. flos-aquae* (B), *Anabaena* sp. (C) and *Gloeotrichia* sp. (D). The magnification bar in (D) represents 5 µm, in the rest it is 10 µm. Pictures (A) and (B) courtesy Mark T. Aubel, GreenWater Laboratories/CyanoLab, 205, Zeagler Drive, Suite 302, Palatka, Florida 32177,USA. (http://www.greenwaterlab.com/photo_algal.htm). Pictures (C) and (D) courtesy G. L. Tiwari, Department of Botany, University of Allahabad, Allahabad-211002, India.

However, polyphosphate granules are reported to be absent (Wilden and Mercer, 1963; Leak and Wilson, 1965; Miller and Lang, 1968; 1969; Jensen an Clark, 1969; Wildman *et al.*, 1975; Sutherland *et al.*, 1979; Grilli-Caiola and de-Vecchi, 1980). The thickened envelope of akinetes is composed of polysaccharides and glycolipids (Cardemil and Wolk, 1981).

4) Biochemical composition: The envelopes of akinetes of *A. cylindrica* consist 41% of carbohydrate, 24% amino compounds and 1% lipids on a dry weight basis (Dunn and Wolk, 1970). The envelope of akinetes of the same organism contained polysaccharides with subunits of glucose and mannose linked by 1–3 β-glycosidic bonds (Cardemil and Wolk, 1976, 1979). Akinetes of *A. fertlissima* have been shown to contain four glycolpids in contrast to five glycolipids detected in vegetative cells (Reddy, 1976). Many polar lipid components specific to akinetes of *A. torulosa*, *A. oscillarioides*, *Nostoc calcicola* and *N. ellipsosporum* have been recognized.

A change in the pigment composition of akinetes has been reported by some workers (Fay, 1969a; Wolk and Simon, 1969). Mature akinetes of *A. cylindrica* revealed the absence of phycocyanin with the replacement of chlorophyll by phaeophytin and a simultaneous decrease of β-carotene and

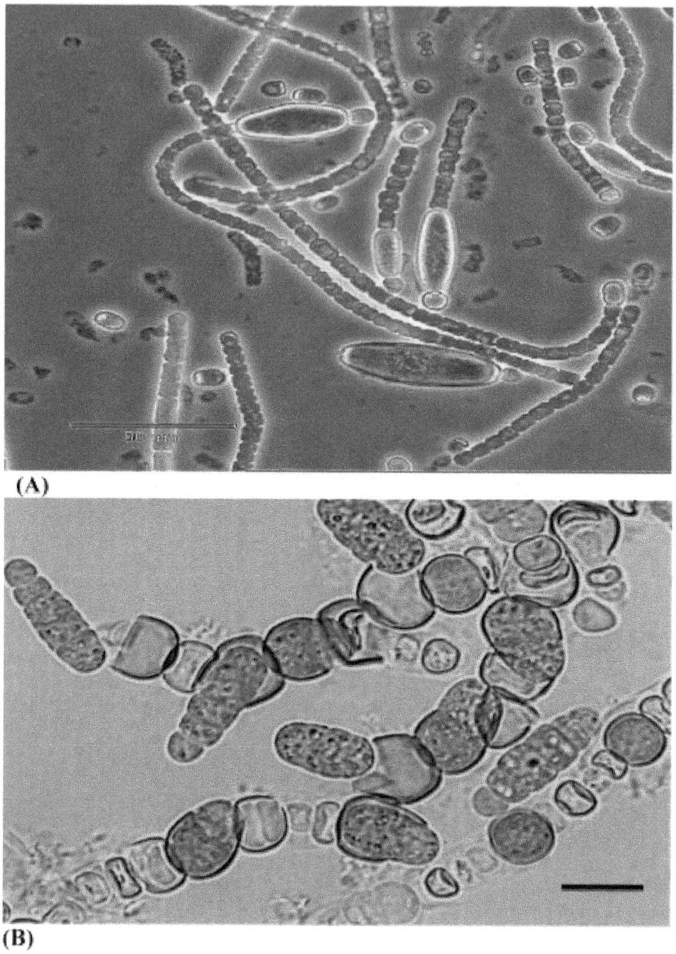

Figure 3: Akinetes of *Cylindrospermum* sp. strain PCC 7417 (A). Germinating akinetes of *Nodularia spumigena* (B). Magnification bar represents 50 μm (A) and 10 μm (B). Picture A courtesy of Rolf Schauder (University of Frankfurt), Mark Schneegurt (Wichita State University) and Cyanosite (www-cyanosite.bio.purdue.edu). Picture B courtesy Christina Esplund, Linnaeus University, School of Natural Sciences, Department of Marine Ecology, SE-391 82 Kalmar, Sweden.

increase of xanthophylls (Fay, 1969a). The amount of chlorophyll doubled where as the phycocyanin content was unchanged in *Nostoc* PCC 7524 (Sutherland *et al.*, 1979). Chauvat *et al.* (1982) reported a decrease in 90% of the photosynthetic and respiratory activities of akinetes in maturation phase where there was no loss of chlorophyll and phycocyanin contents. On the other hand, akinetes of *A. doliolum* lost all photosynthetic pigments at maturity (Rao *et al.*, 1984, 1987). Reddy (1983a) reported changes in phycobiliproteins during akinete formation in *A. fertilissima* whereas phycoerythrins were not detectable in akinetes of *No. spumigena* (Pandey and Talpasayi,1981).

Akinetes of *Cylindrospermum* possessed 30-times more DNA than vegetative cells (Ueda and Sawada, 1972) whereas DNA content was equal in akinetes and vegetative cells of *A. cylindrica* (Simon, 1977). However, the ratio of RNA:DNA in akinetes of *A. cylindrica* decreased (13:1) than in

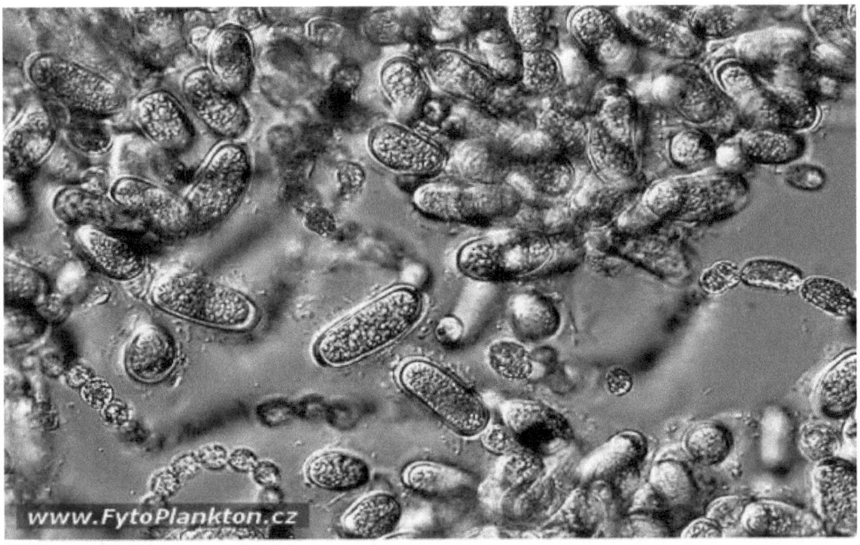

(A)

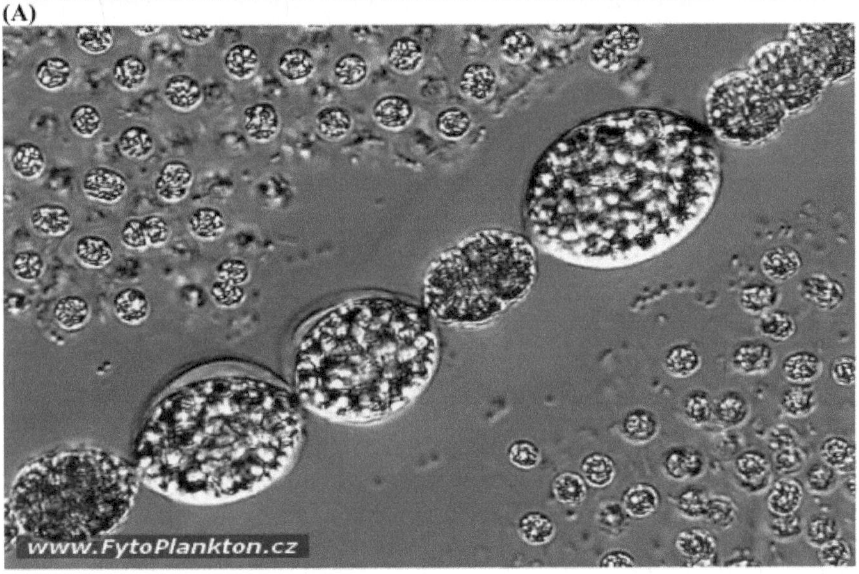

(B)

Figure 4: Mass of akinetes of *Anabaena lammermannii* (A) and *Anabaena* cf. *macrospora* (B) as seen through Nomarski contrast. Magnification of both (A) and (B) at x 400. Pictures courtesy P. Znachor, Laboratory of Phytoplankton Ecology, Institute of Hydrobiology, Biology Centre ASCR, Na Sádkách , 37005 České Budějovice, Czech Republic. (www.FytoPlankton.cz).

vegetative cells (19:1). The vegetative cells and akinetes of *Nostoc* PCC 7524 contained approximately the same amounts of DNA and RNA (Sutherland *et al.*, 1979).

***5) Metabolic activities*:** The metabolic activities of isolated akinetes of cyanobacteria differed from one another. In general, a lowering of photosynthetic activity was noted in the akinetes of *A. cylindrica* (Fay, 1969b), *Nostoc* PCC 7524 (Sutherland *et al.*, 1979; Chauvat *et al.*, 1982) and *A. torulosa*

(Sarma *et al.*, 2000). A high respiratory activity was noted in the akinetes of *A. cylindrica* (Fay 1969b; Herdman, 1987), *N. spongiaeforme* (Thiel and Wolk, 1983) and *A. torulosa* (Ahuja *et al.*, 2008) but it was reduced in case of the akinetes of *Nostoc* PCC 7524 (Chauvat *et al.*, 1982) and *A. doliolum* (Rao *et al.*, 1984). Nitrogenase activity was negligible in the akinetes of *A. cylindrica* (Fay, 1969b) and those of *S. fritschii* (Sarma and Ghai, 1998). A sharp decline in nitrogen fixation was noted in case of akinete development of *Anabaena* sp. (Ahluwalia and Kumar, 1982). Singh and Kashyap (1988) have detected low rates of respiration, evolution of O_2 and lack of photosynthetic pigments along with a lowering of enzymes of inorganic nitrogen metabolism in the mature akinetes of *F. muscicola*. Akinetes of *N. spongiaeforme* are metabolically active capable of synthesizing protein and lipid and respire in the dark and also evolve oxygen in light (Thiel and Wolk, 1983).

***6) Pattern of akinete differentiation*:** In diverse cyanobacteria akinetes may differentiate adjacent to heterocysts as in *A. cylindrica* (Wolk, 1965), *C. licheniforme* (Fisher and Wolk, 1976; Hirosawa and Wolk, 1979a) and other akinete-forming *Anabaena* strains (Rippka *et al.*, 1979) or midway between two heterocysts as noted in *A. doliolum* (Tyagi, 1974), certain *Nostoc* strains (Rippka *et al.*, 1979), *Nostoc* PCC 7524 (Sutherland *et al.*, 1979), *A. torulosa* (a halotolerant strain; Fernandes and Thomas, 1982) and *N. punctiforme* ATCC 29133 (Meeks *et al.*, 2002). After such initiation, cells on either side start developing into akinetes and the gradients of maturation reach the two heterocysts so that in a short time all the cells tend to get differentiated into strings of akinetes. In cultures grown in presence of nitrate or ammonia that lack heterocysts, the differentiation of akinetes occurred randomly in the filaments (Meeks *et al.*, 2002). The akinete formation in *A. torulosa* is initiated simultaneously in all vegetative cells and cells away from the heterocysts (5th to 6th cell in between two heterocysts in nitrate-free medium) always matured first. Thus the lag in the formation of akinetes is dependent on the lag in the maturation period (Sarma and Swarn Kanta, 1979) in contrast to the lag in the initiation of akinete formation in *A. cylindrica* (Simon, 1977). Wolk (1965, 1966) first advocated that heterocysts play a key role in akinete differentiation in *A. cylindrica* since vegetative cells adjacent to heterocysts always first differentiated into akinetes. A lack of such relationship between the two cell types was subsequently reported (Eberly, 1966; Hill, 1970; Burger, 1974; Wildman *et al.*, 1975; Sutherland *et al.*, 1980). Due to the fact that ceratin amino acid analogues such as 7-azatryptophan altered the pattern in the differentiation of heterocysts, some workers were interested to examine whether akinete differentiation depended on pattern of heterocysts or is independent of heterocysts. A common control mechanism for the differentiation of heterocysts and akinetes in *Nostoc* PCC 7524 has been proposed by Sutherland *et al.* (1979) on the basis of alteration in the position of akinetes due to the presence of 7-azatryptophan and canavine. However, in *Anabaena* sp. the position of akinetes has been altered in presence of 7-azatryptopohan and canavine. In *A. torulosa* the differentiation of akinetes always occurred away from the heterocysts in nitrate-free cultures. In a number of nutrients tested, the same pattern of akinete differentiation occurred as in nitrate-free medium but in nitrogen sources like nitrate, nitrite and carbon sources such as acetate or citrate the trichomes fragmented releasing heterocysts into medium and few-celled trichomes had undergone akinete differentiation in the absence of heterocysts at high frequency (Fig. 5; Sarma and Malhotra, 1989). In *S. fritschii* during akinete differentiation either in nitrate-free or nitrate medium a second round of heterocyst differentiation precedes akinete development and the cultures undergo a perceptible change of colour from blue-green to brown. Upon maturity, the akinetes acquired thick envelopes and were seen in elongated series interrupted by the dead necridia which resulted due to the crumpling of the newly developed heterocysts (Figs. 6, 7).

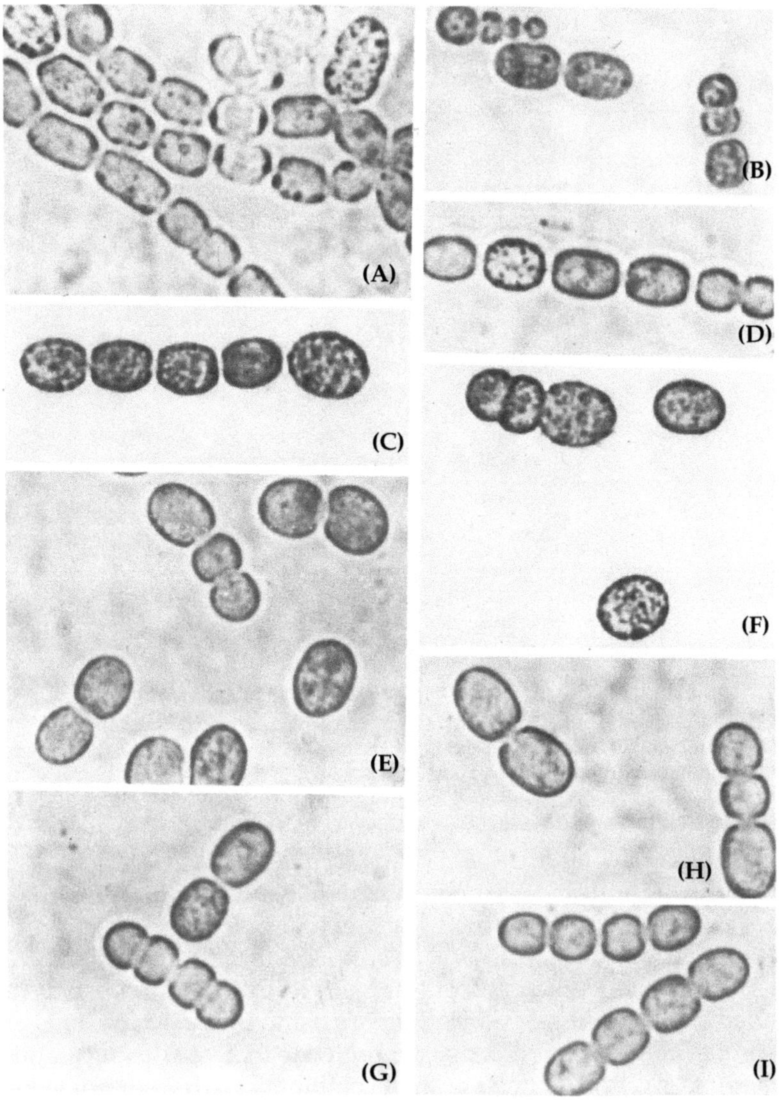

Figure 5: Akinete formation in *Anabaena torulosa*. (A) nitrate-free medium, 20th day; (B) germination of akinetes in nitrate-free medium, 2nd day; (C) a hormogone without heterocysts showing maturation of akinetes in nitrate medium, 8th day; (D) a hormogone showing maturation of akinete in nitrate medium with citrate, 6th day; (E) akinete in the process of germination in nitrate medium, 10th day; (F) akinetes from nitrate medium undergoing germination, 18th day; (G) two pairs of akinetes of which one pair has germinated giving rise two-celled germlings and of the second pair one has lost the granular contents from nitrate medium with citrate, 10th day; (H) akinetes from nitrite with citrate undergoing germination, 10th day; (I) hormogones in pairs from nitrite medium with citrate, 10th day. x 1,950 (From Sarma and Malhotra, 1989).

***7) Reserve products*:** Glycogen and cyanophycin granule polypeptide (CGP) constitute the major reserve products in cyanobacteria (Wolk, 1973; Shively, 1974). Electron microscope studies revealed the presence of glycogen granules in the interlamellar spaces of vegetative cells (Ris and Singh, 1961; Pankratz and Bowen, 1963; Giesey, 1964) and abundance of CGP granules in older cells or

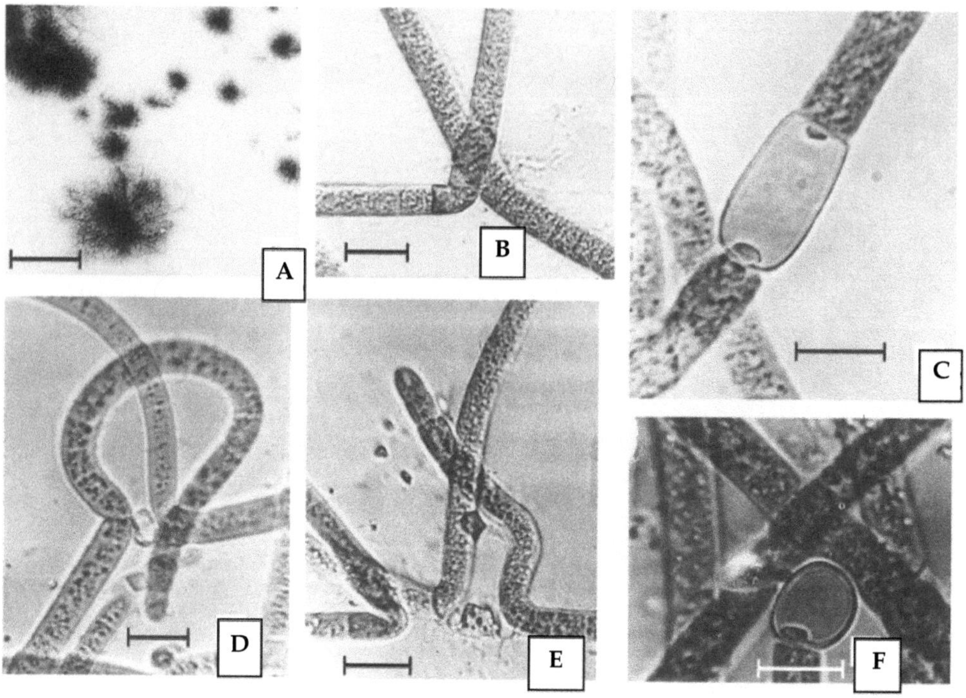

Figure 6: Morphological features of *Scytonema fritschii* in cultures. (A) colonies of *S. fritschii* on nitrate-free agar medium; (B), (C) and (D) types of false-branching in *S. fritschii*; (E) an intercalary heterocyst; (F) a terminal heterocyst. The magnification bar in all pictures represents 10 µm (From Sarma and Ghai, 1998).

akinetes (Lang, 1967). Most of the akinetes of cyanobacteria reveal the presence of gylcogen and CGP granules. Exceptionally the akinetes of *Cyanospira rippkae and C. capsulata* lack CGP granules (Sili *et al.*, 1994). The isolation of an arginine biosynthesis mutant of *N. ellipsosprum* was unable to accumulate CGP but even in its absence akinetes formed were devoid of CGP granules (Leganes *et al.*, 1998). Various investigations carried out on *A. torulosa* revealed that (i) glycogen accumulation during akinete formation is a light-dependent process and no glycogen formation was found in darkness or in cultures supplemented with DCMU in light (Sarma and Swarn Kanta, 1979); (ii) glycogen was maximum at the onset of akinete initiation that continued to increase during maturation period; (iii) inorganic nitrogen sources, carbon sources or their combinations that stimulated akinete formation also supported enhancement in the content of glycogen (Swarn Kanta and Sarma, 1980) and (iv) of the 21 amino acids L-methionine, L-tyrosine, L-glycine and L-histidine supported the formation of maxium glycogen content with concomitant formation of akinetes (Sarma and Swarn Kanta, 1980). The relative amounts of glycogen and CGP were compared in cells undergoing akinete formation in *A. torulosa* (Sarma and Khattar, 1986a,b). Sutherland *et al.* (1985b) suggested that internal reserves in akinetes of *Nostoc* PCC 7524 were not necessary for germination since carbon dioxide fixation commenced immediately in light. Since microcycle akinete formation bypasses vegetative growth of *A. torulosa*, it is likely to depend on the internal reserves of glycogen and CGP. This is evident by the decreased levels of the two reserves from akinete stage to microcycle akinete formation (Sarma and Malhotra, 1989).

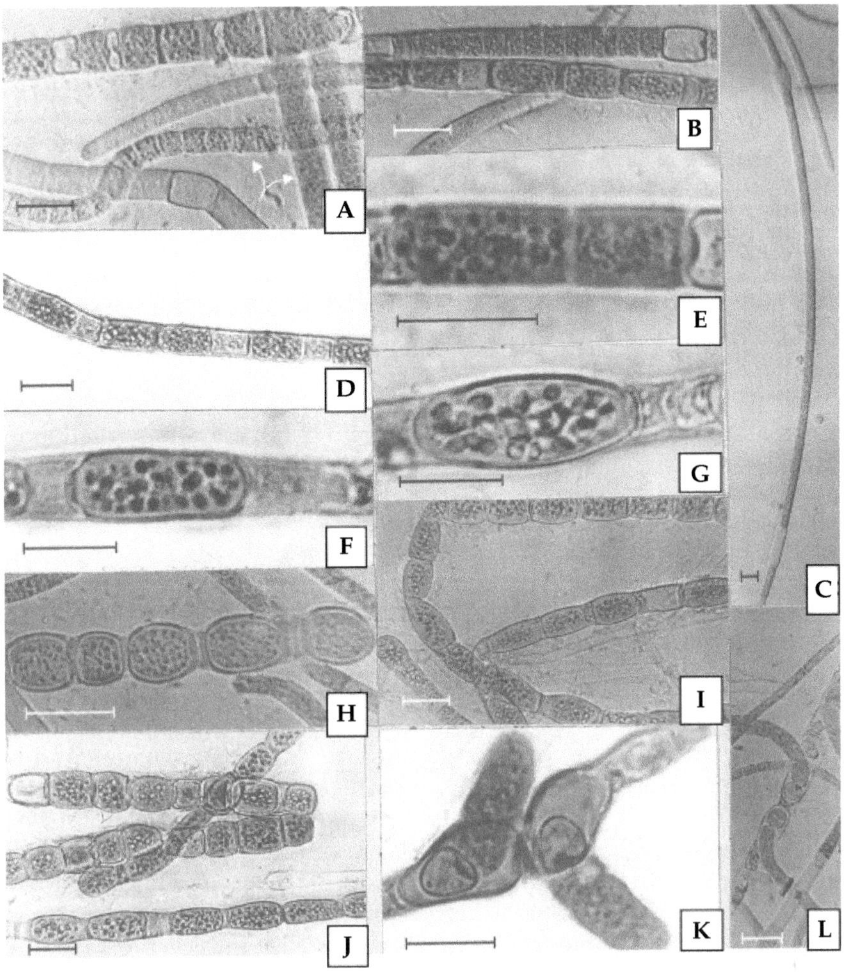

Figure 7: Pattern of akinete formation in *Scytonema fritschii.* (A) a vegetative filament with an intercalary heterocyst (*lower*) showing cells which are longer than wide and two filaments (*upper*, indicated by arrows exhibiting cell division from nitrate-free medium on 26th day; (B) two filaments from late exponential growth phase; of the two filaments, in the upper filament the interheterocyst distance increased to 16 cells while the lower filament shows maturing akinetes sandwiched by heterocysts; (C) a filament with two intercalary heterocysts showing an inter-heterocyst distance of 26 cells from nitrate-free medium on 6th day; (D) a filament undergoing second phase of heterocyst differentiation with potential akinetes in between from nitrate-free medium on 28th day; (E) magnified portion of a filament with two heterocysts on either side of two cells undergoing akinete differentiation from nitrate-free medium on 30th day; (F), (G) advanced stage of akinete differentiation showing deposition of cell wall layers revealing a thick exine and a smooth intine from nitrate-free medium on 42nd day; (H) a chain of mature akinetes from nitrate medium on 28th day; (I), (J) chains of maturing akinetes with heterocysts in between from nitrate medium on 28th day; (K) fully developed germlings being released laterally or terminally from nitrate-free medium on 3rd day of seeding; (L) advanced stage of akinete germination showing germlings still held by old walls of akinetes from nitrate-free medium on 5th day of inoculation. The magnification bar in all pictures represents 10 μm (From Sarma and Ghai, 1998).

8) Enzyme levels: The levels of enzymes such as α-amylase, catalase, peroxidase, glutamine synthetase (GS) have been compared in cells undergoing akinete differentiation in *A. torulosa* (Sarma *et al.*, 1977; Swarn Kanta *et al.*, 1978; Sarma and Swarn Kanta, 1982). Higher levels of catalase and peroxidase in the maturation phase of akinetes of *A. torulosa* suggested that the akinetes with fairly higher levels of peroxidase have fully operative H_2O_2 scavenging mechanisms. The increase of acid and alkaline phosphatases correlated well with the disappearance of polyphosphate granules in the akinetes that constitute the phosphate reserves of vegetative cells (Keck and Stich, 1957; Talpasayi, 1963; Wolk, 1973). Sarma and Mehta (1985) suggested that akinete formation in *A. torulosa* in presence of sodium β-glycerophosphate was due to the ability of the organism to solubilize such bound phosphate groups and utilized it for growth and akinete differentiation. GS has been recognized as the key enzyme for the incorporation of ammonia into the metabolic pool in cyanobacteria (Haselkorn, 1978). The activity of GS declined coinciding with maturation phase signifying that nitrogen fixation process slows down at the time of akinete differentiation. This is supported by the observations of Sutherland *et al.* (1979) who noticed a lowering of nitrogenase activity at the time of akinete differentiation in *Nostoc* PCC 7524. In addition, ATPase activity decreased from initiation through maturation phase to the mature akinete phase (Sarma and Swarn Kanta, 1982). A comparison of the activities of GS, glutamate pyruvate transaminase (GPT) and glutamate oxoglutarate transaminase (GOT) from different nutrient conditions inducing and inhibiting akinete differentiation revealed akinete formation occurred in nutrient combinations where GS activity decreased (Sarma and Swarn Kanta, 1982). Subsequent studies on the metabolic activities of *A. doliolum* (Rao *et al.*, 1984) and *F. muscicola* (Singh and Kashyap, 1988) confirmed the above observations.

II. FACTORS AFFECTING AKINETE DIFFERENTIATION

***1) Environmental factors*:** Light and temperature have been shown to regulate akinete development in cyanobacteria. Wolk (1965) found 80 ft. c. light intensity to be optimum for akinete formation in *A. cylindrica*. Increase of light intensity lowered akinete frequency in *Nostoc* PCC 7524. Akinete differentiation was reported in presence of light in *A. torulosa* and *No. spumigena* and dark incubated cultures of these cyanobacteria did not differentiate akinetes (Sarma and Swarn Kanta, 1979; Pandey and Talpasayi, 1980). The quality of light required for akinete differentiation in *No. spumigena* was noted to be red, yellow and blue regions of the spectra (Pandey and Talpasayi, 1980). Possible involvement of red light in the akinete differentiation of *A. torulosa* was reported by Fernandes and Thomas (1982). Akinete formation was highest in red light when compared to white and green lights but blue light reduced akinete production considerably in *Aph. flos-aquae* (Thompson *et al.*, 2009). On the other hand, light limitation due to increase in culture density and self-shading has been suggested to be the major factor for akinete formation (Fay, 1969b; Fay *et al.*, 1984; Wyman and Fay, 1986; Nichols and Adams, 1982; Herdman, 1987, 1988). Light limitation was found to induce the formation of akinetes in *A. cylindrica* (Nichols *et al.*, 1980), *A. circinalis* (Fay *et al.*, 1984) and *G. echinulata* (Wyman and Fay, 1986). Light limitation does not seem to be the main trigger in *A. circinalis* (van Dok and Hart, 1996).

Temperatures above 20°C (up to 30°C) enhanced akinete frequency in *A. cylindrica* (Wolk, 1965). Increase of temperature to 30–35°C induced akinete formation in *No. spumigena* in a minimal medium or in distilled water (Pandey and Taplpasayi, 1981). Incipient desiccation appeared to be the trigger for akinete formation in *A. variabilis* and *N. linckia* (Reddy, 1983b). In a number of planktonic cyanobacteria temperature seems the main trigger for akinete formation (Li *et al.*, 1997).

***2) Nutritional factors*:** Of the nutritional factors phosphate deficiency, carbon and nitrogen sources have received greater attention.

***(i) Phosphate (P) deficiency and nitrogen assimilation*:** P-deficiency has been recognized as the most significant factor governing akinete development in *A. cylindrica, A. doliolum, A. torulosa, A. circinalis,*

Cylindrospermum sp., and *F. muscicola* (Wolk, 1965; Kaushik *et al.*, 1971; Fisher and Wolk, 1976; Nichols and Adams, 1982; Herdman, 1987, 1988; Sarma and Khattar, 1992). However, in certain other forms such as *Aph. flos-aquae* and *Nostoc* PCC 7524, P-deficiency did not stimulate the formation of akinetes (Rother and Fay, 1979; Sutherland *et al.*, 1979). Studies on *A. torulosa* revealed that the induction of akinetes in P-deficient conditions depended greatly on the source of nitrogen as nitrate-free cultures were unable to fix nitrogen and the cultures did not differentiate akinetes. However, nitrate enriched cultures deficient in P alone supported the formation of akinetes (Fig. 8). Thus P-deficient nitrate enriched cultures supported enhanced frequency of akinetes with subsequent sub-culturing in a similar medium. These observations signify that (i) regulation of akinete differentiation in P-deficient cultures proceeds through the regulation of N-assimilating systems as nitrogen-fixing and nitrate utilizing cultures revealed no greater differences in P-uptake; and (ii) nitrate uptake by *A. torulosa* seems to be not dependent on P since nitrate uptake and GS activity did not differ in presence or absence of P. The most significant aspect of this study is the decrease of cellular P levels with a concomitant increase of extracellular P in the medium (Sarma and Khattar, 1992).

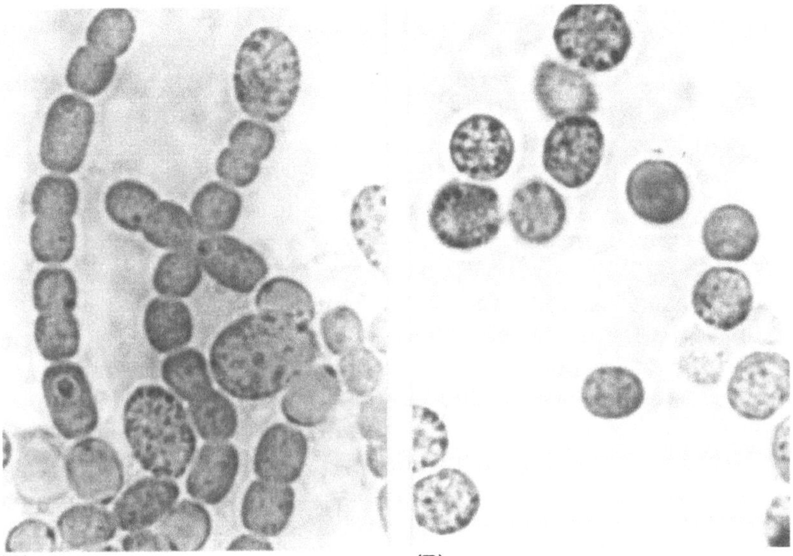

(A) (B)

Figure 8: Pattern of akinete differentiation in *Anabaena torulosa*. Akinetes from nitrate-free medium with phosphate on 20th day (A) and akinetes from nitrate medium without phosphate on 10th day (B). x1320 (From Sarma and Khattar, 1992).

ii) Nitrogen sources: Conflicting reports exist in literature on the role of nitrogen sources in inducing akinete differentiation in cyanobacteria. Nitrate nitrogen favoured akinete formation in several *Anabaena* spp. (Canabaeus, 1929) and *A. torulosa* (Sarma and Malhotra, 1989; Sarma and Khattar, 1994) whereas the incorporation of nitrogen sources did not affect akinete differentiation in *A. cylindrica* (Wolk, 1965), *Nostoc* PCC 7524 (Sutherland *et al.*, 1979) and other *Nostoc* strains (Rippka *et al.*, 1979). On the other hand, the frequency of akinetes decreased in *A. doliolum* in presence of nitrate and nitrite (Singh and Srivastava, 1968). However, nitrate enhanced early and en masse akinete differentiation in *N. linckia* while ammonium chloride delayed and reduced the frequency (Singh *et al.*, 1972b). In nitrogen sources such as nitrate and nitrite, *A. torulosa* experiences extensive fragmentation of the filaments soon followed by the initiation of akinete development on the 4th day and mature akinetes appeared with greater frequency (up to 30%) by 10th day thus hastening the akinete formation by 10 days period. In presence of ammonium chloride neither the period nor the frequency of akinetes seems to have been affected as the response resembled that of cultures from nitrate-free medium.

***iii) Carbon sources*:** Wolk (1965) reported that akinete formation in *A. cylindrica* was stimulated by acetate and calcium glucoronate. Glucose stimulated akinete production in *A. doliolum* (Tyagi, 1974). On the contrary, incorporation of sucrose (0.02%) in nutrient medium prolonged the exponential phase of cultures in *Nostoc* PCC 7524 and akinete differentiation was delayed (Sutherland *et al.*, 1979). Acetate and citrate induced high frequency of akinetes in *A. torulosa* (Sarma and Malhotra, 1989). Inorganic carbon sources, i.e. carbonates and bicarbonates of sodium, potassium and ammonium carbonate, induced early development of akinetes in *A. torulosa* (Sarma and Garg, 1985).

***iv) Other factors*:** Sarma *et al.* (1998) identified aeration as another factor that regulates akinete differentiation in *A. torulosa*. The effect of nutrients and aeration on O_2 evolution, photosynthetic pigments and the accumulation of CGP content of cells during akinete differentiation in the same organism were subsequently reported (Sarma *et al.*, 2000, 2004).

Deficiency of iron induced akinete formation in *A. circinalis* whereas a deficiency in other elements either decreased their frequency (as in case of Mg and Ca) or had no effect (Mo, SO_4) (Sinclair and Whitton, 1977). Depletion of iron has been found to be a trigger for akinete differentiation in *Anabaena* while at the same time omission of other elements such as phosphorus, magnesium, calcium, manganese, zinc, molybdenum, copper, boron and cobalt had not induced akinete formation (Hori *et al.*, 2003). Likewise, the limitation of iron and trace elements had no effect on akinete differentiation in *A. circinalis*.

Pandey and Kashyap (1987) reported that metabolic inhibitors like sodium azide, sodium fluoride and sodium arsenate stimulated akinete differentiation either in terms of the time required for differentiation or akinete frequency in *A. doliolum* (Ads-strain). Hydrogen ion concentration in the range of 7.9 to 9.0 was noted to be favourable in unbuffered media in case of *A. cylindrica* and organic buffers of the dipeptides of DL-alanyl glycine and DL-alanyl DL-alanine favoured akinte formation (Wolk, 1965). In case of *Aph. flos-aquae* HEPES (N-2-hydroxyethylpiperazine-N′-2-ethanesulfonic acid) buffer stimulated akinete production (Rother and Fay, 1979).

Culture filtrates of *C. licheniforme* in phosphate-free standard medium stimulated the formation of akinetes in a fresh inoculum (Fisher and Wolk, 1976). Hirosawa and Wolk (1979b) identified this substance as having two-fused five membered rings (one of which is a lactam and the other has a thio-keto group) that induced akinetes in phosphate enriched cultures as well. Other cyanobacteria like *Aph. flos-aquae* (Rother and Fay, 1979), *Nostoc* PCC 7524 (Sutherland *et al.*, 1979), *A. cylindrica* (Nichols and Adams, 1982) and *Nostoc* sp. (Ahluwalia and Kumar, 1983) did not respond to the addition of culture filtrates. Induction of proakinetes by neo-peptone within 48 h in nitrogen-free and ammonium grown cultures of *A. cylindrica* was reported (Sharma, 1984). Moreover, the inductive response of akinetes was found to be associated with some factors with a molecular weight of 10,000–20,000 and increased with the age of the culture.

***v) Carbon and nitrogen assimilation-an interaction*:** Four nutritional categories, i.e. photoautotrophs, phototrophs (growth in presence of reduced carbon sources in light), photoheterotrophs and chemoheterotrophs have been generally recognized in cyanobacteria. The differentiation of akinetes by *A. torulosa* in phototrophic, photoheterotrophic and heterotrophic conditions was supported by the presence fructose (Sarma and Khattar, 1993). Fructose incorporation in nitrogen-fixing photoautotrophic cultures enhanced nitrogenase activity by nearly 30 times and led to akinete differentiation within 10 days as it occurred in the presence of nitrate. However, when fructose was supplemented in nitrate medium there was no synergistic effect of these two nutrients. In cyanobacteria, the breakdown of fixed carbon sources is known to occur through oxidative pentose phosphate (OPP) pathway (Cheung and Gibbs, 1965; Pelroy and Bassham, 1972; Doolittle and Singer, 1974) with glucose-6-phosphate dehydrogenase (G6PDH) as the first enzyme involved in this route (Stewart, 1980; Smith, 1982). The levels of G6PDH in relation to nitrogen source and phase of akinete differentiaition revealed that nitrogen-fixing cultures exhibited a very high activity than in presence of nitrate in all the four nutritional conditions (Sarma and Khattar, 1993). The fact that nitrate is lowering the rate of fructose dissimilation through OPP-pathway suggests that fructose might be channeled to accumulate in the form of glycogen. Under all the four nutritional conditions, the presence of nitrate favoured the accumulation of more levels of glycogen when compared to nitrogen fixing conditions. These observations point out that (i) in nitrogen-fixing cultures fructose favoured nitrogen fixation due to generation of reducing power through the OPP-pathway (Neur *et al.*, 1983); (ii) in presence of nitrate a lowering of G6PDH suggests a slowing down of OPP-pathway and the extra carbon is channelized to form glycogen; (iii) nitrate uptake and reduction proceed under photoheterotrophic and chemoheterotrophic conditions leading to akinete differentiation and (iv) the cells of *A. torulosa* are committed to undergo akinete differentiation when a balance between carbon and nitrogen nutrition is maintained, suggesting that C:N ratios govern akinete differentiation in this organism (Sarma and Khattar, 1993). Rao *et al.* (1987) opined that a continuation of photosynthesis with a simultaneous cessation in nitrogen fixation led to an increase in C:N ratios during akinete differentiation in *A. doliolum*. But the findings of Sarma and Khattar (1993) on *A. torulosa* and those of Sarma and Ghai (1998) on *S. fritschii* revealed that in all nutrient conditions favouring akinete differentiation the C:N ratios decreased due to a decrease in photosynthesis and concomitant accumulation of nitrogen reserves. Evidences for the existence of an interplay between carbon and nitrogen metabolism during akinete differentiation have been presented by Ahuja *et al.* (2008) who compared enzymes of carbon metabolism (glycolytic cycle-pyruvate kinase; glyoxylate cycle -NAD^+ linked malate dehydrogenase and C_4 carbon fixation pathway-$NADP^+$ linked malate dehydrogenase) in nutrient-enriched and -depleted cultures under aerated and unaerated conditions. Alongside a comparison of respiratory O_2 uptake, nitrate uptake and nitrate reductase levels under the same conditions prompted them to conclude that the cells of *A. torulosa* are able to sense the nitrogen and carbon levels and thus are able to modulate the changes required to trigger either akinete development or continued vegetative growth (Fig. 9).

***vi) Mutational studies*:** Mutants defective in akinete differentiation arose spontaneously in *C. majus* (Singh and Sinha, 1965) and in *A. doliolum* following UV (Singh, 1967) or MNNG-treatment (Dhar, 1979) and in *G. ghosei* (Singh and Tiwari, 1970). The simultaneous mutational loss of the ability to produce both heterocysts and akinetes in *Cylindrospermum* sp. (Singh *et al.*, 1972b) and *A. variabilis* (Singh, 1978) suggests a common component of wall polysaccharides. The altered structure of the akinete envelope in a Het^- mutant of *Nostoc* PCC 7524 (Sutherland *et al.*, 1980) supports the existence of such a common subunit component in both the cell types. This is further substantiated by the molecular genetic analysis that has been presented in the following section. However, isolation of oligosporogenic and non-sporulating mutants in *A. doliolum* by acridine dyes indicated the probable involvement of an extrachromosomal determinant in this cyanobacterium (Dikshit *et al.*, 1981).

The mutants with the loss of akinete phenotype can be recognized into two categories. The first type does not differentiate akinetes at all (non-sporulating) as observed in *A. doliolum* (Singh, 1967; Singh and Dikshit, 1976; Dhar, 1979), *A. variabilis* (Singh, 1978), *N. linckia* (Singh *et al.*, 1972b)

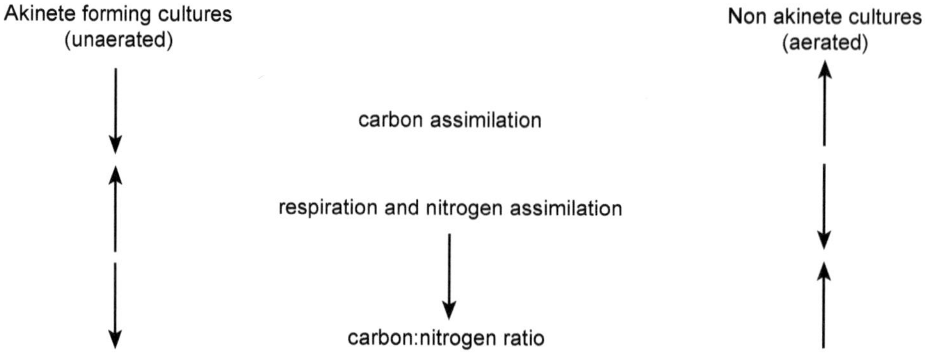

Figure 9: Interaction between carbon and nitrogen metabolism in *Anabaena torulosa* during akinete formation under unaerated and aerated cultures (From Ahuja *et al.*, 2008).

and *G. ghosei* (Singh *et al.*, 1972b). The second type of mutants differentiated akinetes with a lower frequency (oligosporogenic) as in *A. doliolum* (Dikshit *et al.*, 1981).

***vii) Genes involved in akinete differentiation*:** Early studies relate to the identification of common genes to heterocysts and akinetes that regulate synthesis of envelope polysaccharides. The *hepA* gene that encodes a polysaccharide layer in heterocysts (Wolk *et al.*, 1994) was also shown to be required for development of envelopes of akinetes in *A. variabilis* (Leganes, 1994). DevR is a response regulator component of a two-component regulatory system involved in heterocyst polysaccharide synthesis. This has caused increased akinete induction in *N. punctiforme* ATCC 29133 (Campbell *et al.*, 1996). Introduction of *devR* gene into cells of *Nostoc* sp. strain 29133 resulted in a stimulation of akinete formation in both ammonium-supplemented and N_2-fixing cultures. This prompted Campbell *et al.* (1996) to conclude that akinete and heterocyst differentiation is influenced by similar phosphorelay systems and there might be some interchange of information between the two cell types. Leganés *et al.* (1994) showed that mutations in two genes *hetR* and *hetP* that affect heterocyst differentiation have different effects on akinete differentiation in *N. ellipsopsorum*. In the *hetR* interrupted mutant neither heterocysts nor akinetes were produced there by showing that the transcriptional regulator protein HetR was essential for both heterocyst and akinete formation. In contrast, the *hetP* mutant formed akinetes but could not differentiate heterocysts. The observations of Wong and Meeks (2002) merit mention here who showed that a *hetR* mutant strain of *N. punctiforme* ATCC 29133 was able to form cold-resistant akinete-like structures devoid of typical granulation of akinetes.

Zhou and Wolk (2002) utilized *avak* gene of *A. variabilis* along with its promoter and fused it with GFP (green fluorescent protein) transcriptional reporter and introduced this construct in *A. cylindrica*. The expression of this gene in high levels in akinetes of *A. cylindrica* when compared to heterocysts and vegetative cells confirmed that this gene as an akinete marker gene. Likewise, a similar sequence to *avak* gene in *N. punctiforme* ATCC 29133 is also similarly regulated (Argueta *et al.*, 2004). But these studies have not been able to identify the function of these gene products during akinete differentiation. DNA microarray analysis of global gene expression profiles of *N. punctiforme* ATCC 29133 revealed transcription patterns of 6,893 genes. Of these, 497 genes were equally up-regulated and down-regulated during the time the cells entered into akinete phase. The down-regulated genes belonged to core functions that were associated with a non-growth state. There seems to be very little requirement of an adaptation to enter into akinete phase. A comparison of gene expression profiles (including those of *hetR*) of N_2-grown cultures and cells entering into akinete phase does not support the view that the akinetes serve as the progenitors of heterocysts as stated in the beginning of this Chapter but there seems to be considerable evolutionary divergence between the akinetes and heterocysts (Campbell *et al.*, 2007).

Despite the availability of a lot of literature on akinete differentiation on a number of cyanobacteria under varied environmental and nutritional conditions, a single trigger for akinete formation under the diverse conditions examined has not been identified.

III. AKINETE GERMINATION

The akinetes of certain cyanobacteria are reported to germinate too readily in the same medium of their formation (Fay, 1969b; Singh *et al.*, 1972b; Fogg *et al.*, 1973; Rother and Fay, 1977) or upon transfer to fresh medium (Miller and Lang, 1968; Stulp and Stam, 1982; Sutherland *et al.*, 1985a). In the akinetes of *A. variabilis* JS/07 an expanding electron dense layer, present in between the outer cell wall and the akinete coat, usually swells causing bursting of the spore coat during germination. Such akinetes usually release uniformly mostly single cells that undergo division into two cells, of which one cell differentiates into a heterocyst (Braune, 1980). Similarly, Skill and Smith (1987) noted synchronous akinete germination in *Anabaena* PCC 7937 and *Nostoc* PCC 6720 in which the germlings extrude from their envelopes as single cells. If germination is taking place in a medium free of combined nitrogen, the first heterocyst differentiates even when the germling is three cells long as in *Nostoc* PCC 7524 (Sutherland *et al.*, 1985b), *Anabaena* PCC 7937, *Nostoc* PCC 6720 (Skill and Smith, 1987) and *Cyanospira capsulata* (Sili et *al.*, 1994). During the first few hours of germination, the akinetes of *A. doliolum* undergo a change of colour from brown to blue-green and the contents withdraw from the spore wall. This single cell structure undergoes division leading to the formation of a four-celled germling, the terminal cell of which usually gets differentiated into a heterocyst (Rai *et al.*, 1988). Three modes of release of germlings from the germinating akinetes have been noticed. The most common mode is the formation of pore at one end through which the germling is released out, represented by a single cell or 2–3 celled-structure, while the spore coat remains intact (Skill and Smith, 1987; Sili *et al.*, 1994) This mode of germination of akinetes has been noted in *No. spumigena* (Fig. 3 B) and *S. fritschii* (Fig. 7 K, L; Sarma and Ghai, 1998). In the second mode of release the spore coat breaks into two equal halves at the mid point as noticed in case of *A. doliolum* (Singh and Srivastava, 1968; Rai *et al.*, 1988) and *G. ghosei* (Singh *et al.*, 1972b). Complete dissolution of the spore coat constitutes the third mode of release of germlings. The emergence of four-cell germlings due to a rupture of akinete envelope of *Cylindrospermopsis raciborskii* was preceded by an initiation phase of elongation followed by endospore separation from the akinete wall and division to give rise to a germling (Moore *et al.*, 2004).

A requirement of light for germination of akinetes has been found in majority of the cyanobacteria such as *A. circinalis* (van Dok and Hart, 1997; Thompson *et al.*, 2009), *A. cylindrica* (Yamamoto, 1976), *A. doliolum* (Singh and Sunita, 1974; Rai *et al.*, 1988), *A. iyengarii* (Agrawal and Singh, 2000), *A. torulosa* (Sarma and Malhotra, 1989), *A. variabilis* (Braune, 1979), *Anabaena verrucosa* and *Aph. flos-aquae* (Kim *et al.*, 2005), *Anabaena* PCC 7937 and *Nostoc* PCC 6720 (Skill and Smith, 1987) *Nostoc* PCC 7524 (Chauvat *et al.*, 1982; Sutherland *et al.*, 1985a,b), *No. spumigena* (Myers *et al.*, 2010), *N. lobatus* (Agrawal and Singh, 2000), *C. raciborskii* (Moore *et al.*, 2005) and *W. prolifica* (Agrawal and Singh, 2000) examined with the exception of akinetes of *Anabaena azollae* that germinated in dark in presence of fructose (Neely-Fisher *et al.*, 1989). A continuous exposure to light for 24 h has been found to be necessary for germination of *No. spumigena* akinetes with red light (620 to 665 nm) necessary for initiating germination. Mostly, light of low intensity (0.5 $\mu E\ m^{-2}\ s^{-1}$) has been found to be favourable for akinete germination (Huber, 1985). During the first 24 h of the germination process the energy required has been suggested to be derived from the aerobic oxidation of endogenous carbon reserves (Rai *et al.*, 1988). The energetic and metabolic requirements for the germination of akinetes of *Nostoc* PCC 7524 during the first 9 to 10 h could not be met either from cyclic photophosphorylation of PSI or respiration but for complete germination it required resumption of full photosynthetic activity with both photosystems functional (Chauvat *et al.*, 1982). Similarly, germination potential has been found to be highest when both photosystems are functional and if PSII inhibition is caused by DCMU, cyclic photophosphorylation needed the support of respiration to provide the required energy for the germination process (Herdman, 1987, 1988).

The nascent germlings of *A. doliolum* acquire first the capabilities of oxygenic photosynthesis and expression of nitrate reductase and GS activities prior to the establishment of a regular pattern of heterocyst differentiation which generally takes 24–72 h. During this period the presence of significant activities of GOT and GPT indicates a rapid turnover of the required amino acids for growth of the germlings (Rai *et al.*, 1988). Another alternative that has been suggested to meet the requirements of nitrogen before the establishment of nitrogen fixation capability is the dissolution of the thick peptidoglycan layer that facilitates the release of a lipopolysaccharide-like laminated layer in *Nostoc* PCC 7524 (Sutherland *et al.*, 1979) and *Cyanospira* sp. (Sili *et al.*, 1994). Neely-Fisher *et al.* (1989), who reported akinete germination of the symbionts of *Azolla pinnata* in presence of fructose in darkness, observed that the addition of nitrogen sources in presence of fructose did not cause any additional advantage. On the contrary, the addition of ammonium chloride (5 mM) in presence of fructose reduced the frequency of akinete germination in dark. Sutherland *et al.* (1985a,b) observed synchronized akinete germination in *Nostoc* PCC 7524 by dilution of akinetes into fresh medium with or without nitrate in light and the germlings in these media differentiated first heterocyst from the terminal cell of a germling around 19 h followed by the second heterocyst that appeared at the other terminus after few cell divisions. During the synchronized akinete germination, the synthesis of RNA and proteins began immediately and continued during the germination process but DNA synthesis started only after 90 h by which time the germling growth with ten vegetative cells and two terminal heterocysts would have taken place. The continued cell division in the absence of DNA synthesis has been explained by Adams and Duggan (1999) on the basis that the akinetes possess 10 genome equivalents of DNA so that these are redistributed to the ten vegetative cells during germling growth (Sutherland *et al.*, 1979).

Relatively well synchronized germination of akinetes (up to 90%) was noted after 43 h of transfer to fresh medium (Stulp and Stam, 1982). In certain members like *Nostoc* PCC 7524, Sutherland *et al.* (1985b) obtained nearly 100% synchronous germination in fresh medium within 24 h. In *A. torulosa* induction of akinete formation occurred in nitrogen (nitrate and nitrite) and carbon sources (acetate and citrate) and their combinations on the 10th day (Sarma and Malhotra, 1989). Akinetes in these nutrients had undergone germination too readily in the same medium with maximum percentage in presence of citrate with nitrite followed by acetate. The percentage of germination was found to be highest in nitrite followed by nitrate. The resulting 2–3 celled-germlings have immediately undergone a second round of akinete formation circumventing the process of vegetative growth. This has been designated as microcycle akinete formation. Maximum percentage of microcycle akinetes was observed in nitrite, citrate with nitrate and lowest in case of acetate in presence of the two nitrogen sources (Fig. 10; Sarma and Malhotra, 1989). Ammonium chloride inhibited the process of germination in case of *No. spumigena* (Huber, 1985), *A. circinalis* (van Dok and Hart, 1997) and *A. torulosa* (Sarma and Malhotra, 1989).

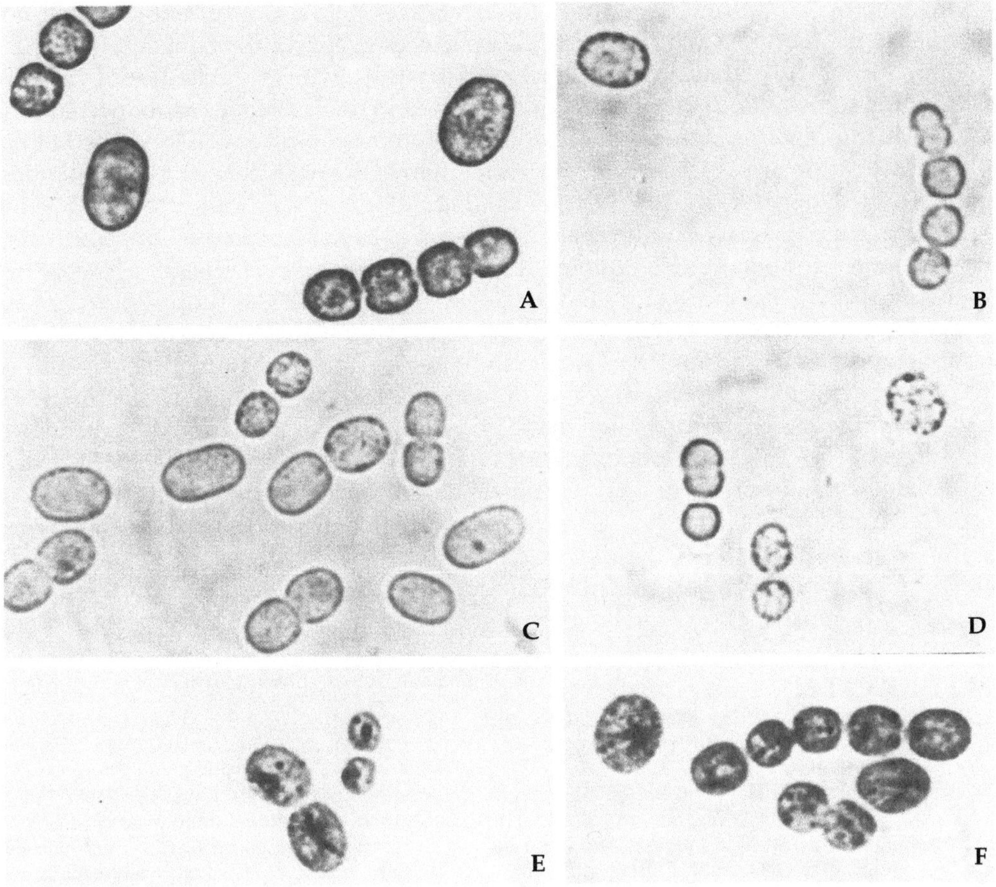

Figure 10: Microcycle akinete formation in *Anabaena torulosa*. Accumulation of granular contents in the hormogones leading to microcycle akinete formation circumventing vegetative growth in *A. torulosa*. (A), (B) in nitrate medium, 18th day; (C), (D) in nitrate with citrate, 20th day; (E), (F) in nitrate medium, 18th day. A-f x 1,950 (From Sarma and Malhotra, 1989).

IV. AKINETE GERMINATION IN RELATION TO BLOOM DEVELOPMENT

A number of investigations dealt on the role of akinetes as potential source of inoculum for development of harmful algal blooms. Akinetes of *A. fertilissima* and *Anabaenopsis arnoldii* from the sediments of Sambhar Salt Lake, the largest lake in India situated in east-central Rajasthan, exhibited 97% and 50% germination potential, respectively (Reddy, 1983b). The akinetes of *No. spumigena* in the sediments of Peel-Harvey estuarine system in western Australia constituted the chief source of inoculum for bloom formation (Huber, 1984). Cmiech *et al.* (1984) studied the emergence of blooms of *A. spiroides*, *A. flos-aquae*, *A. solitaria*, *G. echinulata* and *Aph. flos-aquae* and suggested that in the lake of York shire (New York), the surface populations of *A. solitaria* are generally removed by the germination of 40% of the akinetes immediately after their formation. Akinetes of *Aph. ovalisporum*, deposited in the sediments of Lake Kinneret, Israel, constituted a source of development for the ensuing year (Hadas *et al.*, 1999). Head *et al.* (1998, 1999) attributed the development of *A. solitaria* blooms to a small overwintering population of vegetative cells besides germinating akinetes from the sediment of a mesotrophic lake. Due to the absence of vegetative filaments of *A. flos-aquae* in the water column during winter, Baker (1999) concluded that the bloom originated from the germinating akinetes from the sediments. This was further confirmed in case of bloom formation by *A. circinalis* in the lower Murray River, Australia by Baker and Bellifemine (2000) who identified temperature, salinity and desiccation as parameters for potential management strategies to control germination of akinetes. The average density of *Anabaena ucrainica* akinetes was found to be 1.5 x 10^4 cm^{-3} in the sediments of a small agricultural reservoir situated in Shiga Prefecture, Japan and constituted an important source for bloom development as the high germination rates of akinetes was found in between 14 to 23°C (Tsujimura and Okubo, 2003). Tsujimura (2004) proposed sediment drying as a means of reducing the germination frequency to prevent recurrence of blooms of *A. ucrainica*. In the establishment of blooms of *G. echinulata* light, temperature, sediment mixing and sediment origin have been reported to be important factors that affected germination of akinetes. Akinetes from shallow sediments contributed more to the population rather than from deep water sediments (Karlsson-Elfgren *et al.*, 2004). A count of akinetes with empty envelopes along the littoral region as well as at the deep water portion of a pond Bugach near Krasnoyarsk, Russia revealed that it is the akinetes present in the littoral zone first germinated and contributed to the early bloom population of *A. flos-aquae* (Kravchuk *et al.*, 2002). Similar spatial and vertical distribution and seasonal dynamics in the bloom populations of *A. flos-aquae* due to akinetes in the sediments of littoral region and their germination has been found in a Siberian reservoir (Kravchuk *et al.*, 2011).

LITERATURE CITED

Adams, D. G., and Carr, N. G. (1981) The developmental biology of heterocyst and akinete formation in cyanobacteria. *Crit Rev Microbiol* **9:** 45–100.

Adams, D. G., and Duggan, P. S. (1999) Heterocyst and akinete differentiation in cyanobacteria. *New Phytol* **144:** 1–33.

Agrawal, S. C., and Singh, V. (2000) Vegetative survival, akinete formation and germination in three blue-green algae and one green alga in relation to light intensity, temperature, heat shock and UV exposure. *Folia Microbiol* **45:** 439–446.

Ahluwalia, A. S., and Kumar, H. D. (1982) Cellular differentiation and nitrogenase activity in the cyanobacterium *Anabaena*. *Beitr Biol Pflanzen* **57:** 459–467.

Ahluwalia, A. S., and Kumar, H. D. (1983) Pattern of akinete differentiation in the blue-green alga *Nostoc* sp. *Beitr Biol Pflanzen* **57:** 459–467.

Ahuja, G., Khattar, J. I. S., and Sarma, T. A. (2008) Interaction between carbon and nitrogen metabolism during akinete development in the cyanobacterium *Anabaena torulosa*. *J Basic Microbiol* **48:** 125–129.

Argueta, C., and Summers, M. L. (2005) Characterization of a model system for the study of *Nostoc punctiforme* akinetes. *Arch Microbiol* **183:** 338–346.

Argueta, C., Yuksek, K., and Summers, M. L. (2004) Construction and use of GFP reporter vectors for analysis of cell-type-specific gene expression in *Nostoc punctiforme*. *J Microbiol Methods* **59:** 181–188.

Argueta, C., Yuksek, K., Patel, R., and Summers, M. L. (2006) Identification of *Nostoc punctiforme* akinete-expressed genes using differential display. *Mol Microbiol* **61:** 748–757.

Baker, P. D. (1999) Role of akinetes in the development of cyanobacterial populations in the lower Murray River, Australia. *Mar Freshwater Res* **50:** 265–279.

Baker, P. D., and Bellifemine, D. (2000) Environmental influences on akinete germination of *Anabaena circinalis* and implications for management of cyanobacterial blooms. *Hydrobiologia* **427:** 65–73.

Braune, W. (1979) C-phycocyanin—the main photoreceptor in the light-dependent germination process *of Anabaena* akinetes. *Arch Microbiol* **122:** 289–295.

Braune, W. (1980) Structural aspects of akinete germination in the cyanobacterium *Anabaena variabilis*. *Arch Microbiol* **126:** 257–261.

Burger, J. (1974) A study of two populations of *Anabaena planctonica* Brunnth. (Cyanophyta) from Minnesota. *Phycologia* **13:** 125–129.

Campbell, E. L., Hagen, K. D., Cohen, M. F., Summers, M. L., and Meeks, J. C. (1996) The *devR* gene product is characteristic of receivers of two-component regulatory systems and is essential for heterocyst development in the filamentous cyanobacterium *Nostoc* sp. strain ATCC 29133. *J Bacteriol* **178:** 2037–2043.

Campbell, E. L., Summers, M. L., Christman, H., Martin, M. E., and Meeks, J. C. (2007) Global gene expression patterns of *N. punctiforme* in steady state dinitrogen grown heterocyst-containing cultures and at single time points during the differentiation of akinetes and hormogonia. *J Bacteriol* **189:** 5247–5256.

Canabaeus, L. (1929) Über die Heterocysten und Gasvakuolen der Blaualgen und ihre Beziehung zueinander. *Pflanzenforschung* **13:** 1–48.

Cardemil, L., and Wolk, C. P. (1976) The polysaccharides from heterocysts and spore envelopes of a blue-green alga. *J Biol Chem* **251:** 2967–2975.

Cardemil, L., and Wolk, C. P. (1979) The polysaccharides from heterocysts and spore envelopes of a blue-green alga. Structure of basic repeating subunit. *J Biol Chem* **254:** 736–741.

Cardemil, L., and Wolk, C. P. (1981) Polysaccharides from the envelopes of heterocysts and spores of the blue-green algae *Anabaena variabilis* and *Cylindrospermum licheniforme*. *J Phycol* **17:** 234–240.

Carter, H. J. (1856) Notes on the freshwater infusoria of the island of Bombay. No. 1. Organisation. *Ann Magazine Natural Hist* **18:** 115–132, 221–249.

Chauvat, F., Corre, B., Herdman, M., and Joset-Espardellier, F. (1981) A freezing method for the purification and storage of akinete suspensions of the cyanobacterium *Nostoc* PCC 7524. *FEMS Microbiol Lett* **10:** 319–321.

Chauvat, F., Corre, B., Herdman, M., and Joset-Espardellier, F. (1982) Energetic and metabolic requirements for the germination of akinetes of the cyanobacterium *Nostoc* PCC 7524. *Arch Microbiol* **133:** 44–49.

Cheung, W. Y., and Gibbs, M. (1965) Dark and photometabolism of sugars by a blue-green alga *Tolypothrix tenuis*. *Plant Physiol* **41:** 451–462.

Cmiech, H. A., Reynolds, C. S., and Leedale, G. F. (1984) Seasonal periodicity, heterocyst differentiation and sporulation of planktonic cyanophyceae in a shallow lake, with special reference to *Anabaena solitaria*. *Eur J Phycol* **19:** 245–257.

Dhar, B. (1979) Nutritional control of frequency of MNNG induced true branching habit in *Anabaena doliolum*. *Mol Gen Genet* **173:** 171–181.

Dikshit, K. L., Dikshit, R. P., and Verma, S. C. (1981) Induction of oligosporogenous mutants of a blue-green alga *Anabaena doliolum* by acriflavine and acridine orange. *Ann Bot* **48:** 235–242.

Doolittle, W.F., and Singer, R.A., (1974) Mutational analysis of dark endogenous metabolism in the blue-green bacterium *Anacystis nidulans*. *J Bacteriol* **119:** 677–683.

Doughty, D. M., Hunter, R. C., Summons, R. E., and Newman, D. K. (2009) 2-Methylhopanoids are maximally produced in akinetes of *Nostoc punctiforme*: Geobiological implications. *Geobiology* **7:** 524–532.

Dunn, J. H., and Wolk, C. P. (1970) Composition of cellular envelopes of *Anabaena cylindrica*. *J Bacteriol* **103:** 153–158.

Eberly, W. R. (1966) Notes on some new and rare Myxophyceae in laboratory culture. *Trans Am Microscop Soc* **85:** 130–138.

Fay, P. (1969a) Cell differentiation and pigment composition in *Anabaena cylindrica*. *Arch Microbiol* **67:** 62–70.

Fay, P. (1969b) Metabolic activities of isolated spores of *Anabaena cylindrica*. *J Exp Bot* **20:** 100–109.

Fay, P. (1988) Viability of akinetes of the planktonic cyanobacterium *Anabaen circinalis*. *Proc R Soc* B London **234:** 283–301.

Fay, P. Lynn, J. A., and Majer, S. C. (1984) Akinete development in the planktonic blue- green alga *Anabaena circinalis*. *British Phycol J* **19:** 163–173.

Fernandes, T. A., and Thomas, J. (1982) Control of sporulation in the filamentous cyanobacterium *Anabaena torulosa*. *J Biosci* **4:** 85–94.

Fisher, R. W., and Wolk, C. P. (1976) Substance stimulating the differentiation of spores of the blue-green alga, *Cylindrospermum licheniforme*. *Nature* (London) **259:** 394–395.

Fogg, G. E., Stewart, W. D. P., Fay, P., and Walsby, A. E. (1973) *The Blue-green Algae*. Academic Press, London. pp. 459.

Forchhammer, K., and Tandeau de Marsac, N. (1994) The PII protein in the cyanobacterium *Synechococcus* sp. strain PCC 7942 is modified by serine phosphorylation and signals the cellular N-status. *J Bacteriol* **176:** 84–91.

Fritsch, F. E. (1945) *The Structure and Reproduction of the Algae*, Vol. II, Cambridge University Press, Cambridge, England, UK pp. 796.

Giesy, R. M. (1964) A light and electron microscope study of intralamellar polyglucoside bodies in *Oscillatoria chalybea*. *Am J Bot* **51:** 388–396.

Golubic, S., Sergeev, V. N., and Knoll, A. H. (1995) Mesoproterozoic *Archaeoellipsoides*: Akinetes of heterocystous cyanobacteria. *Lethaia* **28:** 285–298.

Grilli-Caiola, M., and de-Vecchi, L. (1980) Akinete ultrastructure of *Nostoc* species isolated from cycad coralloid roots. *Can J Bot* **58:** 2513–2519.

Hadas, O., Pinkas, R., Delphine, E., Vardi, A., Kaplan, A. and Sukenik, A. (1999) Limnological and ecophysiological aspects of *Aphanizomenon ovalisporum* bloom in Lake Kinneret, Israel. *J. Plankton Res* **21:** 1439–1453.

Haselkorn, R. (1978) Heterocysts. *Annu Rev Plant Physiol* **29:** 319–344.

Head, R. M., Jones, R. I., and Bailey-Watts, A. E. (1998) Akinete germination and recruitment of planktonic cyanobacteria from lake sediments. *Verh Int Ver Limnol* **26:** 1711–1715.

Head, R. M., Jones, R. I. and Bailey-Watts, A. E. (1999) An assessment of the influence of recruitment from the sediment on the development of planktonic populations of cyanobacteria in a temperate mesotrophic lake. *Freshwater Biol* **41:** 759–769.

Herdman, M. (1987) Akinetes: Structure and function. In: Fay, P., and Van Baalen, C. (Eds.) *The Cyanobacteria*. Elsevier, Amsterdam, The Netherlands pp. 227–250.

Herdman, M. (1988) Cellular differentiation: Akinetes. *Methods Enzymol* **167:** 222–232.

Hill, H. (1970) A new form of *Raphidiopsis mediterranae* Skuja found in Minnesota lake. *Phycologia* **9:** 73–77.

Hirosawa, T., and Wolk, C. P. (1979a) Factors controlling the formation of akinetes adjacent to heterocysts in the cyanobacterium, *Cylindrospermum licheniforme* Kuetz. *J Gen Microbiol* **114:** 423–432.

Hirosawa, T., and Wolk, C. P. (1979b) Isolation and characterization of a substance which stimulates the formation of akinetes in the cyanobacterium *Cylindrospermum licheniforme* Kuetz. *J Gen Microbiol* **114:** 433–441.

Hori, K., Okamoto, J., Tanji, Y., and Unno, H. (2003) Formation, sedimentation and germination properties of *Anabaena* akinetes. *Biochem Eng J* **14:** 67–73.

Huber, A. L. (1984) *Nodularia* (Cyanobacteriaceae) akinetes in the sediments of the Peel-Harvey estuary western Australia: Potential inoculum source for *Nodularia* blooms. *Appl Environ Microbiol* **47:** 234–238.

Huber, A. L. (1985) Factors affecting germination of akinetes of *Nodularia spumigena* (Cyanobacteriaceae). *Appl Environ Microbiol* **49:** 73–78.

Jensen, T. E., and Clark, R. L. (1969) Cell wall and coat of the developing akinete of a *Cylindrospermum* species. *J Bacteriol* **97:** 1494–1495.

Karlsson-Elfgren, I., Rengefors, K., and Gustafsson, S. (2004) Factors regulating recruitment from the sediment to the water column in the bloom-forming cyanobacterium *Gloeotrichia echinulata. Freshwater Biol* **49:** 265–273.

Kaushik, M., and Kumar, H. D. (1970) The effects of light on growth and development of two nitrogen fixing blue-green algae. *Arch Microbiol* **79:** 52–57.

Kaushik, M., Kumar, H. D., and Singh, H. N. (1971) Studies on growth and development of two nitrogen fixing blue-green algae. I. Carbon and phosphorus nutrition. *Z Pflanzenphysiol* **65:** 432–442.

Keck, K., and Stich, H. (1957) The wide spread occurrence of polyphosphate in lower plants. *Ann Bot* **21:** 611–619.

Kim, B. H., Lee, W. S., Kim, Y.-O., Lee, H.-O., and Han, M.-S. (2005) Relationship between akinete germination and vegetative population of *Anabaena flos-aquae* (Nostocales, Cyanobacteria) in Seokchon reservoir (Seoul, Korea). *Arch Hydrobiol* **163:** 49–64.

Kravchuk, E. S., Ivanova, E. A., and Glayshev, M. I. (2002) Seasonal dynamics of the numbers of *Anabaena flos-aquae* (Lyngb.) Breb. akinetes in the surface layer of bottom sediments and bulk water. *Doklady Biol Sci* **384:** 233–234.

Kravchuk, E. S., Ivanova, E. A., and Gladyshev, M. I. (2011) Spatial distribution of resting stages (akinetes) of the cyanobacteria *Anabaena flos-aquae* in sediments and its influence in pelagic populations. *Mar Freshwater Res* **62:** 450–461. doi:10.1071/MF10256.

Lang, N. J. (1967) The fine structure of blue-green algae. *Annu Rev Microbiol* **22:** 15–46.

Leak, L. V., and Wilson, G. B. (1965) Electron microscopic observations on a blue-green alga, *Anabaena* sp. *Can J Genet Cytol* **7:** 237–249.

Leganés, F., Fernandez-Pinãs, F., and Wolk, C. P. (1994) Two mutations that block heterocyst differentiation have no effect on akinete differentiation in *Nostoc ellipsosporum*. *Mol Microbiol* **12:** 679–684.

Leganés, F., Fernandez-Pinãs, F., and Wolk, C. P. (1998) A transposition-induced mutant of *Nostoc ellipsosporum* implicates an arginine-biosynthetic gene in the formation of cyanophycin granules and of functional heterocysts and akinetes. *Microbiology* **144:** 1799–1805.

Li, R., Watanabe, M., and Watanabe, M. M. (1997) Akinete formation in planktonic *Anabaena* spp. (*Cyanobacteria*) by the treatment with low temperature. *J Phycol* **33:** 576–584.

Livingstone, D., and Jaworski, G. H. M. (1980) The viability of akinetes of blue-green algae recovered from the sediments of Rostherne Mere. *Br Phycol J* **15:** 357–364.

Miller, M. M., and Lang, N. J. (1968) The fine structure of akinete formation and germination in *Cylindrospermum*. *Arch Microbiol* **60:** 303–313.

Meeks, J. C., Campbell, E.L., Summers, M.L., and Wong, F. C. (2002) Cellular differentiation in the cyanobacterium *Nostoc punctiforme*. *Arch Microbiol* **178:** 395–403.

Miller, M. M., and Lang, N. J. (1968) The fine structure of akinete formation and germination in *Cylindrospermum*. *Arch Microbiol* **60:** 303–313.

Moore, D., McGregor, G. B., and Shaw, G. (2004) Morphological changes during akinete germination in *Cylindrospermopsis raciborskii* (Nostocales, Cyanobacteria). *J Phycol* **40:** 1098–1105.

Moore, D., O'Donohue, M., Garnett, C. M., Shaw, G. R., and Critchley, C. (2005) Factors affecting akinete differentiation in *Cylindrospermopsis raciborskii. Freshwater Biol* **50:** 345–352.

Myers, J. M., Beardall, J., Allinson, G., Salzman, S., and Gunthorpe, L. (2010) Environmental influences on akinete germination and development in *Nodularia spumigena* (Cyanobacteriaceae) isolated from Gippsland lakes, Victoria, Australia. *Hydrobiologia* **649:** 239–247.

Neely-Fisher, D. L., Bruce White, W., and Fisher, R. W. (1989) Fructose-induced dark germination of *Anabaena* akinetes. *Curr Microbiol* **19:** 139–142.

Neur, G., Papen, H., and Bothe, H. (1983) Heterocyst biochemistry and differentiation. In: Papargeorgiou, G. C., and Packer, L. (Eds.) *Photosynthetic Prokaryotes: Cell Differentiation and function* Elsevier, New York, NY, USA. pp. 219–242.

Nichols, J. M., and Adams, D. G. (1982) Akinetes. In: Carr, N. G., and Whitton, B. A. (Eds.) *The Biology of Cyanobacteria.* Blackwell Scientific Pulications, Oxford, UK pp. 387–412.

Nichols, J. M., and Carr, N. G. (1978) Akinetes of cyanobacteria. In: Chambers, G., and Vary, J. M. (Eds.) *Spores* Am Soc Microbiologists, Washington, pp. 335–343.

Nichols, J. M., Adams, D. G., and Carr, N. G. (1980) Effect of canavanine and other amino acid analogues on akinete formation in the cyanobacterium *Anabaena cylindrica. Arch Microbiol* **127:** 67–75.

Olsson, Francis, K., de La Torre, R., Towner, M. C., and Cockell, C. S. (2009) Survival of akinetes (resting state cells of cyanobacteria) in low Earth orbit and simulated extraterrestrial conditions. *Origins of Life and Evolution of Biospheres* **39:** 565–579.

Pandey, K. D., and Kashyap, A. K. (1987) Factors affecting formation of spores (akinetes) in cyanobacterium *Anabaena doliolum* (Ads. Strain). *J Plant Physiol* **127:** 123–134.

Pandey, R. K., and Talpasayi, E. R. S. (1980) Control of sporulation in a blue-green alga, *Nodularia spumigena* Mertens. *Indian J Bot* **3:** 128–133.

Pandey, R. K., and Talpasayi, E. R. S. (1981) Physiological and biochemical aspects of formation and germination of spores in blue-green algae. In: Bharadwaj, D.C. (Ed.) *Recent Advances in Cryptogamic Botany,* Palaeobotanical Society, Lucknow. pp. 1–14.

Pankratz, H. S., and Bowen, C. C. (1963) Cytology of blue-green algae. I. The cells of *Symploca mucorum. Am J Bot* **50:** 387–399.

Pelroy, R. A., and Bassham, J. A. (1972) Photosynthesis and dark carbon metabolism in unicellular blue-green algae. *Arch Microbiol* **86:** 25–28.

Rai, A. K., and Pandey, G. P. (1981) Influence of environmental stress on the germination of *Anabaena vaginicola* akinetes. *Ann Bot* **48:** 361–370.

Rai, A. N., Rao, V. V., and Singh, H. N. (1985) The biology of the cyanobacterial (blue-green-algal) akinetes (spores). *J Plant Sci Res* **1:** 1–20.

Rai, A. N., Rao, V. V., and Singh, H. N. (1988) Metabolic changes associated with akinete germination in the cyanobacterium *Anabaena doliolum. New Phytol* **109:** 133–138.

Rao, V. V., Ghosh, R., and Singh, H. N. (1987) Diazotrophic regulation of akinete development in the cyanobacterium *Anabaena doliolum. New Phytol* **106:** 161–168.

Rao, V. V., Rai, A. N., and Singh, H. N. (1984) Metabolic activities of akinetes of the cyanobacterium *Anabaena doliolum*: Oxygen exchange, photosynthetic pigments and enzymes of nitrogen metabolism. *J Gen Microbiol* **130:** 1299–1302.

Reddy, P. M. (1976) Physiological and biochemical studies on cellular differentiation with special reference to perennating structures of blue-green algae. Ph. D. Thesis in Botany, Banaras Hindu University, Varanasi, India.

Reddy, P. M., and Talpasayi, E. R. S. (1981) Some observations related to red-far red antagonism in germination of spores of the cyanobacterium *Anabaena fertilissima. Biochem Physiol Pflanzen* **176:** 105–107.

Reddy, P. M. (1983a) Changes in phycobiliproteins during spores (akinetes) differentiation in a cyanobacterium *Anabaena ferilissima. Biochim Biophys Acta* **761:** 191–195.

Reddy, P. M. (1983b) Effects of temperature pre-treatment on the viability of spores of halophilic blue-green algae. *Hydrobiologia* **106:** 235–240.

Rippka, R., Deruelles, J., Waterbury, J. B., Herdmann, M., and Stanier, R. Y. (1979) Generic assignments, strain histories and properties of pure cultures of cyanobacteria. *J Gen Microbial* **111:** 1–61.

Ris, H., and Singh, R. N. (1961) Electron microscope studies on blue-green algae. *J Biophys Biochem Cytol* **9:** 63–80.

Rother, J. A., and Fay, P. (1977) Sporulation and the development of planktonic blue-green algae in two Salopian meres. *Proc R Soc* B London **196:** 317–332.

Rother, J. A., and Fay, P. (1979) Blue-green algal growth and sporulation in response to simulated surface bloom conditions. *Br Phycol J* **14:** 59–68.

Sarma, T. A. (1993) Akinete differentiation in the cyanobacterium *Anabaena torulosa. Proc Natl Sem Cyanobacterial Research-Indian Scene,* NFMC, Bharathidasan University, Tirucirapally. pp. 50–62.

Sarma, T. A., and Garg, R. (1985) Effect of carbonates and bicarbonates on the growth and sporulation in two blue-green algae. *Proc Indian Acad Sci (Plant Sci)* **94:** 45–50.

Sarma, T. A., and Ghai, R. (1998) Pattern of akinete differentiation in the cyanobacterium *Scytonema fritschii. Folia Microbiol* **43:** 649–656
Sarma, T. A., and Khattar, J. I. S. (1986a) Accumulation of cyanophycin and glycogen during sporulation in the blue-green alga *Anabaena torulosa. Biochem Physiol Pflanzen* **181:** 155–164.
Sarma , T. A., and Khattar, J. I. S. (1986b) Accumulation of reserve products during sporulation in *Anabaena torulosa* under phototrophic, photoheterotrophic and heterotrophic conditions. *Biochem Physiol Pflanzen* **181:** 541–549.
Sarma, T. A., and Khattar, J. I. S. (1992) Phosphorus deficiency, nitrogen assimilation and akinete differentiation in the cyanobacterium *Anabaena torulosa. Folia Microbiol* **37:** 223–226.
Sarma, T. A., and Khattar, J. I. S. (1993) Akinete differentiation in photoheterotrophic and chemoheterotrophic conditions in *Anabaena torulosa. Folia Microbiol* **38:** 335–340.
Sarma, T. A., and Khattar, J. I. S. (1994) Photoheterotrophic and chemoheterotrophic dinitrogen fixation and nitrate utilization by the cyanobacterium *Anabaena torulosa. Folia Microbiol* **39:** 404–408.
Sarma, T. A., and Malhotra, J. (1989) Induction of the formation of akinetes and microcycle akinetes in the cyanobacterium *Anabaena torulosa. Biochem Physiol Pflanzen* **184:** 95–106.
Sarma, T. A., and Mehta, M. (1985) Phosphate utilization and the activity of phosphatases during sporulation in two blue-green algae. *Biochem Physiol Pflanzen* **180:** 401–405.
Sarma, T. A., and Swarn Kanta (1979) Biochemical studies on sporulation in blue green algae. 1. Glycogen accumulation. *Z Allg Mikrobiol* **19:** 571–575.
Sarma, T. A., and Swarn Kanta (1980) Biochemical studies on sporulation in blue-green algae. III. Effect of amino acids on glycogen accumulation. *Z Allg Mikrobiol* **20:** 647–650.
Sarma, T. A., and Swarn Kanta (1982) Variation in enzyme levels during sporulation in blue-green algae. *Biochem Physiol Pflanzen* **177:** 278–282.
Sarma, T. A., Ahuja, G., and Khattar, J. I. S. (2000) Effect of nutrients and aeration on O_2 evolution and photosynthetic pigments of *Anabaena torulosa. Folia Microbiol* **45:** 434–438.
Sarma, T. A., Ahuja, G. and Khattar, J. I. S. (2004) Nutrient stress causes akinete differentiation in cyanobacterium *Anabaena torulosa* with concomitant increase in nitrogen reserve substances. *Folia Microbiol* **49:** 557–562.
Sarma, T. A., Khattar, J. I. S., Ahuja, G., and Saini, V. (1998) Influence of aeration and nutrients on akinete differentiation in the cyanobacterium *Anabaena torulosa.* In: Subramanian, G., Kaushik, B. D., and Ventataraman, G. S. (Eds.) *Cyanobacterial Biotechnology.* Oxford and IBH Publishing Comp., New Delhi. pp. 99–102.
Sarma, T. A., Swarn Kanta, Ahuja, B. S., and Usha Kiran (1977) Amylase activity in a blue-green alga. *Curr Sci* **46:** 609–610.
Sharma, P. (1984) Heterocyst and akinete induction with altered pattern in *Anabaena cylindrica* caused by neopeptone. *Arch Microbiol* **139:** 196–201.
Shively, J. M. (1974) Inclusion bodies of prokaryotes. *Annu Rev Microbiol* **28:** 167–187.
Sili, C., Ena, A., Materassi, R., and Vincenzini, M. (1994) Germination of desiccated aged akinetes of alkaliphilic cyanobacteria. *Arch Microbiol* **162:** 20–25.
Simon, R. D. (1977) Sporulation in the filamentous cyanobacterium, *Anabaena cylindrica.* The course of spore formation. *Arch Microbiol* **111:** 283–288.
Sinclair, C., and Whitton, B. A. (1977) Influence of nitrogen source on morphology of Rivulariaceae (Cyanophyta). *J Phycol* **13:** 335–340.
Singh, H. N. (1967) Genetic control of sporulation in the blue-green alga *Anabaena doliolum.* Bharadwaja. *Planta* **75:** 33–38.
Singh, H. N., and Srivastava, B.S. (1968) Studies on morphogenesis in blue-green alga. I. Effect of inorganic nitrogen sources on developmental morphology of *Anabaena doliolum. Can J Microbiol* **14:** 1342–1348.
Singh, H. N., and Sunita, K. M. (1974) Biochemical study of spore germination in the blue-green alga *Anabaena doliolum. J Exp Bot* **25:** 837–845.
Singh, R. N., and Dikshit, R. P. (1976) Mutagenesis of blue-green alga, *Anabaena doliolum* Bharadwaja. *Mutat Res* **35:** 65–78.
Singh, R. N., and Sinha, R. (1965) Genetic recombination in a blue-green alga, *Cylindrospermum majus* Kuetz. *Nature* (London) **207:** 782–783.
Singh, R. N., and Tiwari, D. N. (1970) Frequent heterocyst germination in the blue-green alga, *Gloeotrichia ghosei* Singh. *J Phycol* **6:** 172–176.
Singh, R. N., Singh, S. P. and Singh, P. K. (1972a) Genetic regulation of nitrogen fixation in Blue-green Algae. In: Desikachary, T.V. (Ed.) *Taxonomy and Biology of Blue-green Algae,* University of Madras, Madras. pp. 263–268.
Singh, R. N., Tiwari, D. N., and Singh, V. P. (1972b) Genetic basis of cellular differentiation in blue-green algae. In: Desikachary, T.V. (Ed.) *Taxonomy and Biology of Blue-green Algae,* University of Madras, Madras. pp. 27–37.
Singh, S., and Kashyap, A. K. (1988) Metabolic characteristics of akinetes of the cyanobacterium *Fischerella muscicola. New Phytol* **110:** 97–100.
Singh, V. P. (1978) Mutations defective in nitrogen fixation, heterocyst and spore production in the blue-green alga, *Anabaena variabilis. Biochem Physiol Pflanzen* **174:** 76–79.

Simon, R. D. (1977) Sporulation in the filamentous cyanobacterium *Anabaena cylindrica*. The course of spore formation. *Arch Microbiol* **111:** 283–288.

Skill, S. C., and Smith, R. J. (1987) Synchronous akinete germination and heterocyst differentiation in *Anabaena* PCC 7937 and *Nostoc* PCC 6720. *J Gen Microbiol* **133:** 299–303.

Smith, A.J. (1982) Modes of cyanobacterial carbon metabolism. In: Carr, N. G., and Whitton, B. A. (Eds.) *The Biology of Cyanobacteria*, Blackwell Scientific Publications, Oxford, UK pp. 47–85.

Srivastava, P. (2005) Vindhyan akinetes: An indicator of Mesoproterozoic biospheric evolution. *Origin of Life and Evolution of Biospheres* **33:** 175–183.

Stewart, W. D. P. (1980) Some aspects of structure and function in N_2-fixing cyanobacteria. *Annu Rev Microbiol* **34:** 497–536.

Stulp, B. K., and Stam, W. T. (1982) General morphology and akinete germination of a number of *Anabaena* strains (Cyanophyceae) in culture. *Arch Hydrobiol* Suppl **63:** 35–52.

Sukenik, A., Hadas, O., Stojkovic, S., Malinsky-Rushansky, N., Vinter-Motzini, Y., and Beardall, J. (2009) Fluorescence microscopy reveals variations in cellular composition during formation of akinetes in the cyanobacterium *Aphanizomenon ovalisporum*. *Eur J Phycol* **44:** 309–317.

Summons, R. E., Jahnke, L. L., Hope, J. M., and Logan, G. A. (1999) 2-Methylhopanoids as biomarkers for cyanobacterial oxygenic photosynthesis. *Nature* (London) **400:** 554–557.

Sutherland, J. M., Herdman, M., and Stewart, W. D. P. (1979) Akinetes of the cyanobacterium *Nostoc* PCC 7524: Macromolecular composition, structure and control of differentiation. *J Gen Microbiol* **115:** 273–287.

Sutherland, J. M., Herdman, M., and Stewart, W. D. P. (1980) Structure and differentiation of akinetes and their relation to heterocysts of the cyanobacterium *Nostoc* PCC 7524. *Soc Gen Microbiol* **7:** 105.

Sutherland, J. M., Reaston, J., Stewart, W. D. P., and Herdman, M. (1985a) Akinetes of the cyanobacterium *Nostoc* PCC 7524: Macromolecular and biochemical changes during synchronous germination. *J Gen Microbiol* **131:** 2855–2863.

Sutherland, J. M., Stewart, W. D. P., and Herdman, M. (1985b) Akinetes of the cyanobacterium, *Nostoc* PCC 7524. Morphological changes during synchronous germination. *Arch Microbiol* **142:** 269–274.

Swarn Kanta, and Sarma, T. A. (1980) Biochemical studies on sporulation in blue-green algae. II. Factors affecting glycogen accumulation. *Z Allg Mikrobiol* **20:** 459–463.

Swarn Kanta, Sarma, T. A., and Ahuja, B. S. (1978) A kinetic study of amylase from a blue-green alga. *Indian J Exp Biol* **16:** 1211–1212.

Talpasayi, E. R. S. (1963) Polyphosphate containing particles of blue-green algae. *Cytologia* **28:** 76–80.

Tandeau De Marsac, N. and Houmard, J. (1996) Acclimation of the filamentous cyanobacteria *Calothrix* spp. to the environment: Physiological and molecular studies. In: Subramanian G., Kaushik, B. D., and Venkataraman, G. S. (Eds.) *Cyanobacterial Biotechnology* Oxford and IBH Publishing Co. Pvt. Ltd, New Delhi, India, pp. 215–224.

Thiel, T., and Wolk, C. P. (1983). Metabolic activities of spores of *Nostoc spongiaeforme*. *J Bacteriol* **156:** 369–374.

Thompson, P. A., Jameson, I., and Blackburn, S. I. (2009) The influence of light quality on akinete formation and germination in the toxic cyanobacterium *Anabaena circinalis*. *Harmful Algae* **8:** 504–512.

Tsujimura, S. (2004) Reduction of germination frequency in *Anabaena* akinetes by sediment drying: A possible method by which to inhibit bloom formation. *Water Res* **38:** 4361–4366.

Tsujimura, S., and Okubo, T. (2003) Development of *Anabaena* blooms in a small reservoir with dense sediment akinete population, with special reference to temperature and irradiance. *J Plankton Res* **25:** 1059–1067.

Tyagi, V. V. S. (1974) Some observations on the pattern of sporulation in the blue-green alga *Anabaena doliolum*. *Ann Bot* **38:** 1107–1111.

Ueda, K., and Swada, M. (1972) DNA content of akinete cells in *Cylindrospermum*. *Cytologia* **37:** 519–529.

van Dok, W., and Hart, B. T. (1996) Akinete differentiation in *Anabaena circinalis* (Cyanophyta) *J Phycol* **32:** 557–565.

van Dok, W., and Hart, B. T. (1997) Akinete germination in *Anabaena circinalis* (Cyanophyta). *J Phycol* **33:** 12–17.

Wilden, D. C., and Mercer, C. V. (1963) The ultrastructure of the heterocyst and akinete of the blue-green algae. *Arch Microbiol* **47:** 19–31.

Wildman, R. B., Loescher, J. H., and Winger, C. L. (1975) Development and germination of akinetes of *Aphanizomenon flos-aquae*. *J Physiol* **11:** 96–104.

Wolk, C.P. (1965) Control of sporulation in a blue-green alga. *Develop Biol* **12:** 15–35.

Wolk, C. P. (1966) Evidence of a role of heterocysts in the sporulation of a blue-green alga. *Am J Bot* **53:** 260–262.

Wolk, C. P. (1973) Physiology and cytological chemistry of blue-green algae. *Bacteriol Rev* **37:** 32–101.

Wolk, C. P., and Simon, R. D. (1969) Pigments and lipids of heterocysts. *Planta* **86:** 92–97.

Wolk, C. P., Ernst, A., and Elhai, J. (1994) Heterocyst metabolism and development. In: Bryant D. E. (Ed.) *The Molecular Biology of Cyanobacteria*. Kluwer Academic, Dordrecht, The Netherlands. pp. 769–823.

Wong, F. C., and Meeks, J. C. (2002) Establishment of a functional symbiosis between the cyanobacterium *Nostoc punctiforme* and the bryophyte hornwort *Anthoceros punctatus* requires genes involved in nitrogen control and initiation of heterocyst differentiation. *Microbiology* **148:** 315–323.

Wyman, M., and Fay, P. (1986) Interaction between light quality and nitrogen availability in the differentiation of akinetes in the planktonic cyanobacterium *Gloeotrichia echinulata*. *Br Phycol J* **21:** 147–154.
Yamamoto, Y. (1975) Effect of desiccation on the germination of akinetes of *Anabaena cylindrica*. *Plant Cell Physiol* **16:** 749–752.
Yamamoto, Y. (1976) Effect of some physical and chemical factors of the germination of akinetes of *Anabaena cylindrica*. *J Gen Appl Microbiol* **22:** 311–323.
Zhou, R., and Wolk, C. P. (2002) Identification of an akinete marker gene in *Anabaena variabilis*. *J Bacteriol* **184:** 2529–2532.

CHAPTER 4

HETEROCYST

Members of the orders Nostocales (except Oscillatoriaceae) and Stigonematales corresponding to Sections IV and V of Rippka *et al.* (1979) differentiate specialized cells known as heterocysts during nitrogen-deficient growth conditions. The importance of the presence or absence of heterocysts for taxonomic purposes was realized long time back by Bornet and Flahault (1886) who divided the order Hormogonales into Homocystaceae and Heterocystaceae. Although once considered to be a "botanical enigma" (Fritsch, 1951), the heterocysts have now emerged as a suitable prokaryotic model for understanding cellular differentiation and pattern formation in multicellular cyanobacteria and we are at a threshold of understanding the web of molecular events involved in their differentiation (Adams and Duggan, 1999; Adams, 2000; Wolk, 2000; Zhang *et al.*, 2006; Kumar *et al.*, 2010; Flores and Herrero, 2010). Thus the development of heterocyst from its antecedent vegetative cell proceeds through very important morphological, structural, biochemical and genomic changes culminating in the production of a slightly enlarged cell with microoxic internal atmosphere. Such a cell after formation can neither revert to a vegetative cell nor divide but can only fix nitrogen. That is why it is often designated as a terminally differentiated cell.

1) MORPHOLOGY

The heterocysts differ from vegetative cells in being slightly larger, double-walled, pale yellow or hyaline (colourless) with homogeneous cell contents and are connected to the vegetative cells by means of the glistening nodule/granule-like structures known as polar nodules/granules. The heterocysts are generally round (as in species of *Anabaena, Nostoc, Gloeotrichia* and *Rivularia*) or rectangular (as in species of *Aulosira, Scytonema* and *Hapalosiphon*). They differentiate in intercalary position at regularly spaced intervals all along the length of the filament, with the intervening number of vegetative cells being fairly constant. In general, the inter-heterocyst distance is occupied by 8 to 10 vegetative cells in species of *Anabaena* and *Nostoc* (Fig. 1A,B,E and F). Besides the intercalary heterocysts, the terminal cells of the filaments in these genera also differentiate into heterocysts. Thus there are intercalary and terminal heterocysts. However, in genera belonging to Rivulariaceae (*Rivularia* and *Gloeotrichia*) there is a single basal heterocyst (Fig. 1C). The terminal and basal heterocysts are connected with adjacent vegetative cells by a single polar nodule. Exceptionally, in certain species of *Anabaenopsis* (*A. circularis* and *A. arnoldii*) the intercalary heterocysts are paired (Fig. 1D). Species of the genus *Cylindrospermum* are characteristic in possessing terminal heterocysts, one on either side of the filament (Desikachary, 1959). In branched members of Stigonematales

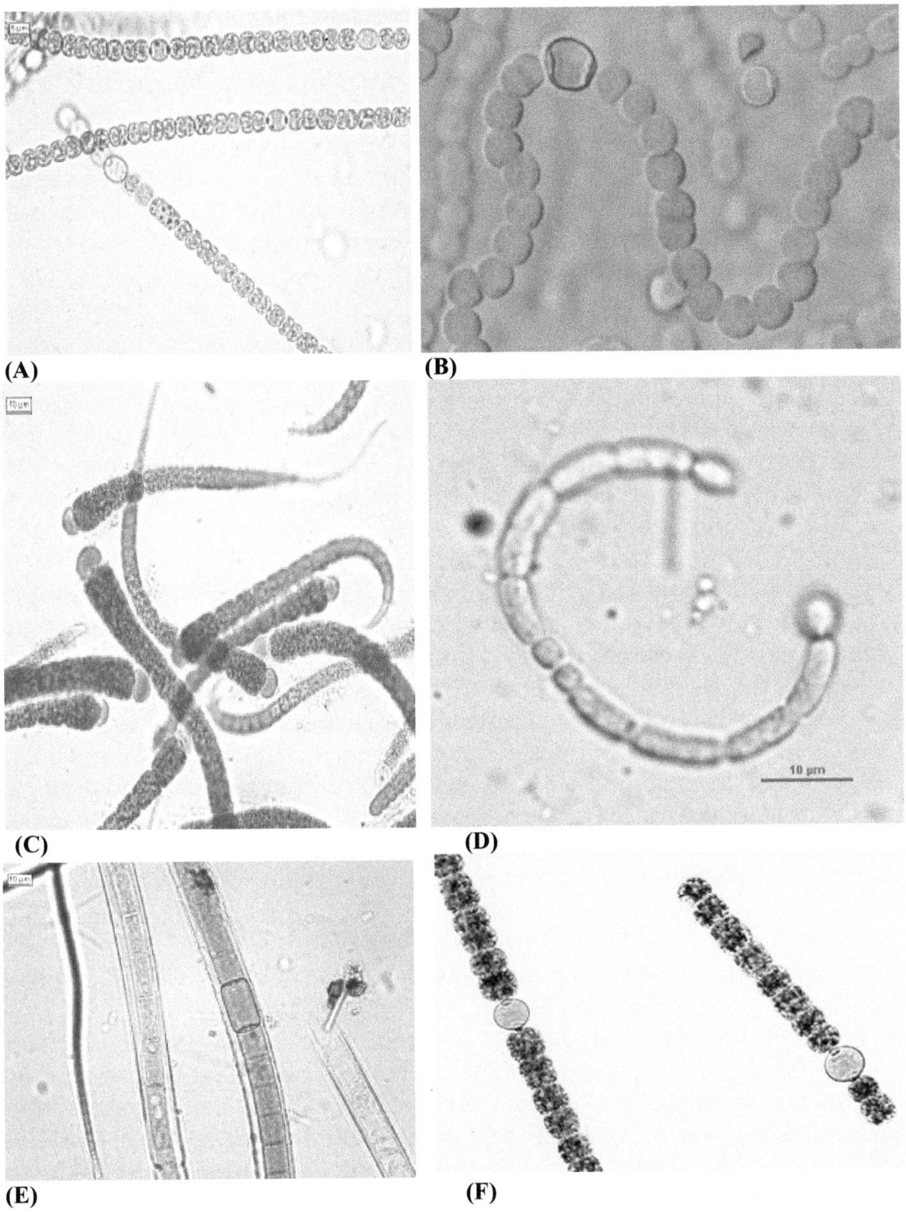

Figure 1: Light micrographs of *Anabaena* sp. (A), *Nostoc* sp. (B), *Calothrix* sp. (C), *Anabaenopsis circularis* (D), *Scytonema* sp. (E) and *Anabaena* sp. (F) showing intercalary heterocysts with two polar nodules (in A,B, D, E and F) and a single basal heterocyst with a single polar nodule (in B). The presence of a pair of heterocysts is characteristically seen in (D). The polar nodules can be seen clearly in (F). Magnification bar in the pictures represents 5 µm (A) and 10 µm (C, D and E). Pictures A,C and E courtesy G.L. Tiwari, Department of Botany, University of Allahabad, Allahabad-211002, India. Picture B courtesy Isao Inouye (Uinversity of Tsukuba), Mark Schneegurt (Wichita State University) and Cyanosite (www-cyanosite.bio.purdue.edu). Picture F courtesy A. L. Baker, Department of Biological Sciences, 124 Spaulding Life Sciences, Academic Way, University of New Hampshire, Durham, NH 03824 USA. http://cfb.unh.edu/phycokey/phycokey.htm.

(*Nostochopsis lobatus*, *Brachytrichia balani*, and *Mastigocladus laminosus*) the presence of intercalary heterocysts with two and three polar nodules has been demonstrated, besides lateral ones with single polar nodule (Iyengar and Desikachary, 1953; Venkataraman, 1957; Desikachary, 1959). Materials growing in nature or from nitrogen-deficient cultures show detached free heterocysts intermingled with normal filaments. The detachment of intercalary heterocysts generally leads to fragmentation of the filaments and they grow once again to exhibit the normal pattern of differentiation. Likewise, if terminal heterocysts get detached as in species of *Anabaena*, *Nostoc*, *Cylindrospermum* and in the Rivulariaceae, the terminal cells undergo differentiation replacing the older ones.

2) ULTRASTRUCTURE

Of the four methods employed, i.e. disruption by French Press, sonication, osmotic shock and lysozyme treatment for the isolation of intact structurally complete heterocysts of *Anabaena cylindrica*, Fay and Lang (1971) favoured the use of lysozyme treatment over the other three methods. Different degrees of damage involving disruption of the heterocyst cell wall and plasmolemma have been generally observed in the heterocysts derived from the other methods. Most of the ultrastuctural studies conducted on heterocysts of *Anabaena azollae* (Lang, 1965, 1968) and *A. cylindrica* Lemm. (Wildon and Mercer, 1963a; Lang and Fay, 1971; Fay and Lang, 1971; Winkenbach *et al.*, 1972) have clearly brought to light that of the two walls of the heterocyst seen in the light microscope, it is the outer wall that it is deposited external to the inner wall that is contiguous with the adjacent vegetative cells. These observations thus have cleared the misapprehensions held earlier that it is the outer wall that belongs to the vegetative cell and it is the inner wall that is secreted from cell membrane during the heterocyst development (Fritsch, 1945). The inner wall that is contiguous with the adjacent vegetative cells is composed of four layers (L_I–L_{IV}) and the outer wall is composed of three distinct layers, i.e. the outer most fibrous layer, middle homogeneous and the inner laminated layer in contact with L_{IV} of the inner wall. Extensive reorganization of thylakoids in the heterocyst takes place as revealed by the presence of coiled and densely layered structures. The deposition of an opaque, electron dense material near the narrow neck of the junction between vegetative cell and heterocyst results in the formation of a polar nodule. At this place, the thylakoid membranes undergo sufficient disintegration (Lang and Fay, 1971). Heterocysts are devoid of the granular inclusions such as carboxysomes, cyanophycin, glycogen and polyphosphate granules which are present in large numbers in vegetative cells (Wildon and Mercer, 1963b; Lang, 1968).

The three layers of the outer wall specific to an individual heterocyst are deposited one after another. These are absent at the narrow point of contact with the heterocyst and the adjacent vegetative cell. It is at this junction that the plasmolemma of vegetative cell and heterocyst are interconnected. These structures have been variously designated in the literature. For example, pore-like structures (Drawert and Metzner, 1956; Metzner, 1955), plasmodesmata (Hagedorn, 1960; Pankratz and Bowen, 1963) and microplasmodesmata (Lang and Fay, 1971; Giddings and Staehelin, 1978); of which, the last one is generally in use. The number of microplasmodesmata decreases three to five-fold (Fay, 1992) and these constitute the conduit for intercellular transport of metabolites.

Anabaena variabilis ATCC 29413, a cyanobacterium that exhibits heterotrophic growth (Wolk and Schaffer, 1976), differentiates heterocysts with thin wall layers and lacking distinct polar nodules although heterocyst-specific laminated layer is present (Haury and Spiller, 1981). It also formed larger, more oblong heterocysts (when compared to rectangular ones in control cultures) filled with glycogen granules and with a precocious maturation of the envelope layers in presence of fructose (40 mM) (Lang *et al.*, 1987). Heterocyst differentiation in the branched *M. laminosus* UTEX1931 occurred in both the wide (basal or prostrate) filaments and narrow branching filaments with a tendency of almost all the cells in the former type having a potential to develop into heterocysts. Mostly in these, continuous rows of four or more heterocysts are generally found. The fine structure details agreed with the heterocysts of *A. cylindrica* except that the thylakoid membranes are densely packed with occasional lamellar stacks embedded in granular cytoplasm (Nierzwicki-Bauer *et al.*, 1984a).

3) BIOCHEMICAL COMPOSITION

The differentiation of heterocysts in *A. cylindrica* was accompanied by a gradual loss of phycocyanin and the ability to fix CO_2. The presence of photosystem I (PSI) with a gradual increase in the content of β-carotene is a characteristic feature of heterocysts with a simultaneous decrease in the content of xanthophyll (Fay, 1969; Wolk and Simon, 1969). A relationship between the age of culture and the contents of the major phycobiliproteins in the heterocysts of *Anabaena* sp. L-31 has been established where heterocysts from 2-day old cultures exhibited very less amounts of these pigments whereas heterocysts from 5-day old cultures acquired more of the biliproteins very much similar to their contents in vegetative cells (Thomas, 1972). Reconstitution experiments with thylakoid membranes from heterocysts revealed that PSI activity was maximal when cytochrome *c-553* or plastocyanin were present. Hydrogen, NADH/NADPH or ascorbate or dichlorophenolindophenol couple served as electron donors (Hawkesford *et al.*, 1983). The presence of cytochromes *c-553*, *b-563* and *f-557* (corresponding to cytochrome *f* of higher plants and algae) and absence of cytochrome *b-559* in the isolated heterocysts of *N. muscorum* suggested the degradation of photosystem II (PSII; Almon and Böhme, 1980). However, the presence of phycobiliproteins in the heterocysts of *A. variabilis* (Peterson *et al.*, 1981; Ke *et al.*, 1983), an *Anabaena* sp. (Yamanaka and Glazer, 1983) and the endophyte of *Azolla*, *A. azollae* (Ke *et al.*, 1983; Tyagi *et al.*, 1981) has been unequivocally demonstrated based on microspectrophotometric and fluorescence action spectra. Further, the phycobiliproteins in the heterocysts of *A. variabilis* appeared to be organized into phycobilisomes (Ke *et al.*, 1983). During the thylakoid rearrangement, a 20 kDa polypeptide component of the heterocyst phycobiliprotein particle helps in attachment of the phycobiliproteins to the photosynthetic lamellae (Yamanaka and Glazer, 1983). The endosymbionts of liverworts and *Gunnera* lack the phycobiliproteins (Rodgers and Stewart, 1977; Silvester, 1976) but the heterocysts from the endophytes of *Azolla caroliniana* and *A. pinnata* (Kaplan and Peters 1981; Tyagi *et al.*, 1981; Kaplan *et al.*, 1986) not only possess the phycobiliproteins but they are also effective in harvesting light energy for production of O_2 evolution (Ray *et al.*, 1979) and PSI-linked acetylene reduction (Tyagi *et al.*, 1981). Peterson *et al.* (1981) observed fluorescence emission bands specific for phycocyanin and allophycocyanin in the heterocysts of *A. variabilis* and this has also been confirmed by the fluorescence microscopy of the individual heterocysts. The presence of PSI and absence of PSII was confirmed and the phycobiliproteins in heterocysts were effective in supporting light-dependent acetylene reduction. Kaplan *et al.* (1986) found that there is no appreciable difference in the composition of phycobiliproteins of vegetative cells and heterocysts of the endophyte of *A. caroliniana* when a comparison was made from the young apices to the mature portions of leaves. In contrast, filaments of *Nostoc* sp. and a free-living *Nostoc* sp. 7422 from *Cycas revoluta* revealed the localization of phycoerythrins in the vegetative cells of both strains but absent in the heterocysts (Lindblad and Bergman, 1989). Fluorescence emission and absorption spectra of single *Anabaena* sp. strain PCC 7120 cells at excitation and detection areas of less than 1.0 µm showed the existence of low PSII activity (Ying *et al.*, 2002).

The composition of heterocyst glycolipids (Hgls) and envelope polysaccharides has been investigated in great detail. Heterocysts contain polar glycerolipids that are common with vegetative cells. These include monogalactosyl diacylglycerol, digalactosyl diacylglycerol, phosphatidylglycerol and sulphoquinovosyl diacylglycerol which contain mainly C_{16} and C_{18} fatty acyl substitutes. On the other hand, the heterocyst glycolipids (Hgls) are localized in the laminated layer and they are suggested to provide an impervious layer for gases like O_2 (Winkenbach *et al.*, 1972). The major Hgl fractions contain glycosides of C_{26} and C_{28} polyhydroxyalkanes and glucose esters of C_{26} and C_{28} hydroxyfatty acids (Bryce *et al.*, 1972; Lambein and Wolk, 1973). Investigations on biosynthesis of Hgls in *A. cylindrica* by the incorporation of sodium acetate-[1-^{14}C] revealed the existence of the label in long chain mono-, di-, and trihydric alcohols with non-saponifiable glycosides (monohydric alcohols) occupying the major fraction in the young (60 h old) cultures. In the older cultures (234 h) the radioactivity was distributed among the three fractions by almost equal percentages (Abreu-Grobois *et al.*, 1977). Similar studies on *A. cylindrica* confirmed a simultaneous increase in the activity of key enzymes involved in the biosynthesis of polyhydroxyalkanes and hydroxyfatty acid moieties of Hgls during heterocyst differentiation (Krepski and Wolk, 1983). In the members of Nostocaceae, the presence of triols and the corresponding C-3 ketones as aglycones has been reported. Accordingly, the structure of the Hgls of the marine cyanobacterium *Nodularia harveyana*, established by spectroscopic and chemical methods, have been shown to be 1-(O-α-D-glucopyranosyl)-3R, 25R-hexacosanediol, 1-(O-α-D-glucopyranosyl)-3S, 25R-hexacosanediol and 1-(O-α-D-glucopyranosyl)-3-keto-25R-hexacosanol (Soriente *et al.*, 1992). However, the Hgls of *Cyanospira rippkae* have been found to be 1-(O-α-D-glucopyranosyl)-3R, 27R-octacosanediol and 1-(O-α-D-glucopyranosyl)-27 keto-3R-octacosanol (Soriente *et al.*, 1993). Gambacorta *et al.* (1998) reported the existence of tetrols in *Scytonema hofmanni*, *Calothrix desertica*, *Chlorogloeopsis fritschii* and *Fischerella muscicola*. Triols and the corresponding glycosides as in Nostocaceae were found only in *Microchate* sp. and *Tolypothrix tenuis*.

Bauersachs *et al.* (2009) examined the distribution of glycolipids in 34 axenic strains of cyanobacteria with the help of high performance liquid chromatography coupled to electrospray ionization tandem mass spectrometry and stated that the heterocystous members of the families Nostocaceae and Rivulariaceae are characteristic in possessing glycolipids with sugar moieties glycosidically bound to long chain diols, triols, keto-ols and keto-diols. Their absence in the unicellular and filamentous non-heterocystous cyanobacteria affords a chemotaxonomic criterion for distinguishing these members. With an increase in growth temperature, a decrease in keto-ols and keto-diols relative to the contents of diols and triols is suggestive of an adaptation for protecting nitrogenase from being inactivated while selectively permitting diffusion of molecular nitrogen into the heterocysts. The presence of heterocyst-specific glycolipids in the Pleistocene and Eocene Arctic sediments is of great ecological and evolutionary significance on the role played by these cyanobacteria in maintaining nitrogen balance in the past in these ecosystems (Bauersachs *et al.*, 2010). Awai *et al.* (2009) summarized the current status of our knowledge on the Hgls and the role played by them in the heterocystous cyanobacteria with an emphasis on the relationship of the enzymes mediating their synthesis with polyketide synthases.

The envelope polysaccharides account for 62% of the heterocyst wall in case of *A. cylindrica* and are composed of glucose (Glc), mannose (Man), galactose and xylose (Xyl) as the subunits with the first sugar as the major component (Dunn and Wolk, 1970). While the aminosugars are absent in the heterocyst envelope polysaccharides, isotopic labelling studies indicated that the content of murein (as assessed by the content of muramic acid) of the antecedent vegetative cell is retained during heterocyst differentiation (Dunn *et al.*, 1971). Studies of Cardemil and Wolk (1979) revealed that the heterocyst envelope polysaccharides of *A. cylindrica* contain repeating units of one mannose and three glucose residues linked by (β 1–3) glycosidic bonds with xylose and galactose residues being present as side branches. Digestion of the heterocyst envelope polysaccharides by treatment with (β 1–3) endoglucanase resulted in the release of glucose, a tri- (Man-Glc-Glc) and pentasaccharide (Man-Glc-Glc-Glc and Xyl as side branch) and the backbones terminate in the pentasaccharide subunit. Heterocyst envelope polysaccharides of *Cylindrospermum licheniforme* resemble those of *A. cylindrica* in possessing terminal xylose and galactose residues as side branches that are absent in the envelope polysaccharides of *A. variabilis*. However, the envelope polysaccharides from *A. variabilis* and *C. licheniforme* in addition possess terminal arabinose residues (Cardemil and Wolk, 1981a). Purified heterocyst preparations from *A. variabilis* synthesized envelope polysaccharides when incubated in presence of C^{14}-labelled mannose, arabinose and glucose residues and bacitracin inhibited the incorporation of the sugar residues into the polysaccharides (Cardemil and Wolk, 1981b).

4) PERMEABILITY TO GASES

The heterocysts should be permeable to N_2 and at the same time restrict the entry of O_2. What are the probable routes through which the gases make their entry into the heterocysts? In the intact filaments, two routes are considered important for the entry of gases, the surface layers and through the polar nodules and microplasmodesmata. During heterocyst differentiation the deposition of glycolipid layer as a laminated layer and the envelope polysaccharides (that have a different composition when compared to exopolysaccharides or sheaths) are the first important developmental protective strategies. Secondly, the absence of PSII remarbably decreases the generation of O_2 as a byproduct of photosynthesis. Thirdly, the enhanced respiratory activity shown by the heterocysts enables to scavenge O_2 from the site of nitrogen fixation. Since the Hgls and envelope polysaccharides appear to protect from the entry of gases, the microplasmodesmata that constitute the connecting link between the adjacent vegetative cells appear to be the appropriate alternative. The permeability of N_2 and O_2 to the heterocysts has been assessed in *Anabaena flos-aquae* by Walsby (1985). The heterocysts in this organism retain gas vacuoles and the collapse of gas vacuoles to 50% of their initial level by pressure diffusion of N_2 or O_2 was tested. The average permeability coefficient of the surface layers to N_2 and O_2 was found to be 0.4 μm s^{-1} which is within the range that could be provided by few glycolipid layers. On this basis, it was concluded that the main route for the entry of gases is through the surface layers, because the permeability of the pore region to gases is considerably less. A consideration of the metabolism of H_2, N_2 and O_2 provided ratios ranging from 0.9 to 1.2 for C_2H_4 to H_2 produced by the isolated heterocysts of *Anabaena* sp. strain CA-V and the H_2 evolution occurred in two phases one by a burst phase followed by a steady-state phase. But due to H_2-simulated C_2H_2 reduction, H_2 evolution was abolished. A persistent O_2 uptake (0.7 μmol of O_2 4 per mg dry weight per h) by the isolated heterocysts (35% of the whole filaments) suggested that O_2 impermeability is not a pre-requisite for nitrogen fixation by the heterocysts (Smith *et al.*, 1985). The first direct evidence that the Hgls and the Hep layer provide the requisite O_2 impermeability came from the studies of Murry and Wolk (1989) who characterized mutants EF114 and EF116 of *Anabaena* sp. strain PCC 7120 isolated by Wolk *et al.* (1988). EF114 deficient in Hgls reduced acetylene strictly under anaerobic conditions. EF116 that lost the cohesiveness of the envelope polysaccharide also lost the low affinity respiratory component associated with the heterocysts. The behavior of EF114 was also similar in this respect (Murry and Wolk, 1989). The other evidence in favour of Hgls and envelope polysaccharide providing the requisite O_2 impermeability was that of Kangatharalingam *et al.* (1992) who observed that the thickness of especially heterocyst glycolipids of *A. flos-aquae* gradually thickened with an increase in the exogenous pO_2 from 4 to 40 kPa. Cultures grown at these pO_2 when transferred to ambient pO_2 exhibited higher nitrogenase activity after 24 h of transfer. Walsby (2007) proposed that the main gas diffusion pathway into the heterocysts is through the polar nodules that are present at the junction of the heterocysts with vegetative cells. Here a third type of membrane (in addition to cytoplasmic membrane and thylakoids) is found forming a honey-comb like structure (Lang and Fay, 1971; Braun-Howland *et al.*, 1988) around the polar nodule with transmembrane proteins making pores through which gas exchange can take place. For scavenging O_2 that enters through these pores, the presence of respiratory enzymes in high concentration is a great possibility due to the oxidation of diaminobenzidine (Murry *et al.*, 1981).

5) RESPIRATORY ACTIVITY

According to Walsby (1985) O_2 enters into heterocysts more slowly than N_2 but upon entry it is immediately removed by a number of enzymes. The respiratory activity of heterocysts seems to serve two functions, the first is to scavenge O_2 from the site of nitrogen fixation and the second is to generate ATP in dark for supporting nitrogen fixation. Fay and Walsby (1966) measured the rate of O_2-uptake by isolated intact heterocysts of *A. cylindrica* that was 40% more than the intact filaments. However, Bradley and Carr (1976) reported much lesser rates of O_2-uptake. Nitrogen-fixing cyanobacteria show a high affinity and low affinity O_2 uptake systems. The high affinity O_2 uptake system is associated with growth in presence fixed nitrogen sources whereas the low affinity O_2 uptake system is operative during nitrogen fixation in dark. The low affinity O_2 uptake is associated with an oxidase operative in the heterocysts (Jensen and Cox, 1983). A characterization of mutants defective in heterocyst envelope polysaccharides showed that these mutants lacked the low affinity oxidase associated with the heterocysts (Murry and Wolk, 1989). The generation of reduced cytochrome *c-553* and donation of electrons from cytochrome *c-553* to either PSI or respiratory cytochrome oxidases in vegetative cells (Peschek, 1987) or heterocysts (Böhme and Almon, 1983; Böhme and Ernst, 1984; Houchins and Hind, 1983; Houchins, 1984 1985) has been demonstrated in *A. variabilis* ATCC 29413 or *Anabaena* sp. strain PCC 7120. Wastyn *et al.* (1988) reported that the rates of cytochrome *c-553* oxidation were far greater by the cytoplasmic membrane and thylakoids from heterocysts of *A. variabilis* ATCC 29413 than by the corresponding membranes from vegetative cells. Likewise, Houchins and Hind (1984) found that the level of cytochrome *aa3* in heterocyst membranes was 4 and 100 times higher than in the vegetative cells of *Anabaena* sp. strain PCC 7120. The observations of Wastyn (1988) on immunoblotting experiments as well as those of Häfele *et al.* (1988) on other biochemical properties have brought to light that the oxidase is cytochrome oxidase *aa3* type, a mitochondrial-type oxidase that is unique for cyanobacteria. An operon coding the subunits I, II and III of cytochrome *aa3* has been cloned from *Synechococcus vulcanus* (Tano *et al.*, 1991), *Synechocystis* sp. strain PCC 6803 (Alge and Peschek, 1993; Alge *et al.*, 1999) and *A. variabilis* ATCC 29413 (Schmetterer *et al.*, 2001). Jones and Haselkorn (2002) cloned and sequenced two operons of cytochrome oxidase (*cox*) of *aa3*-type, *coxBACI* and *coxBACII* encoding three subunits in each case. Of these, the latter was induced more than 20-fold at middle to late stages (12 to 14 h) of heterocyst differentiation in *Anabaena* sp. strain PCC 7120. While the expression of *coxBACII* was restricted to the proheterocysts and heterocysts, the expression of *coxBACI* was found in all cells when grown with nitrate or ammonium. Further, it nominally increased after a nitrogen step-down that was followed by a gradual decline afterwards. These observations were confirmed with the use of *gfp-met2* as a reporter gene with *coxBACII* and *coxBACI* promoters in the wild-type *Anabaena* sp. strain PCC 7120. The *coxBACI* operon of *Anabaena* sp. strain PCC 7120 is an orthologue of *A. variabilis* ATCC 29413 reported earlier by Schmetterer *et al.* (2001). A *coxAII* disruptant mutant that is deficient in subunit I of cytochrome *c* oxidase specific to the heterocysts and proheterocysts grew normally under dizotrophic growth conditions with the differentiation of normal functional heterocysts. This amply signifies that the loss of one *cox* gene is compensated by the presence of the other terminal oxidases present in the heterocysts. This is possible because *Anabaena* sp. strain PCC 7120 appears to possess four types of respiratory oxidase operons, i.e. *coxBACII, coxBACI, cyDAB* (a two-subunit operon encoding cytochrome D ubquinol oxidase, corresponding to genes *all4023* and *all4024*, respectively) and *coxBACIII*. The last two operons were assigned on the genome of *Anabaena* sp. strain PCC 7120 by Kaneko *et al* (2001).

The expression of *coxBACIII* operon in the developing heterocysts around 9 h after nitrogen step-down has been reported by Valladares *et al.* (2003). Inactivation of both *coxBII* and *coxAIII* greatly affected nitrogenase activity as well as diazotrophic growth under aerobic conditions but inactivation of either of these did not matter. This emphasizes the contribution of *coxBACII* and *coxBACIII* complexes to enhanced respiratory activity of the heterocysts to scavenge any oxygen that sneaks into these cells and the environment conducive for efficient functioning of nitrogenase is thus created in these cells (Valladares *et al.*, 2003). Heterocyst development in the double *cox* mutant (*coxBIIcoxAIII*; strain CSAV141) of *Anabaena* sp. strain PCC 7120 was greatly affected at the level of reorganization of thylakoid membranes and the formation of honey-comb like structure. The developing heterocysts in CSAV141 also accumulated considerable number of glycogen granules. Although the deposition of Hgls and the heterocyst envelope polysaccharides is normal and expression of heterocyst-specific *nifHDK* and *fdxH* genes took place, the nitrogenase activity under aerobic conditions was greatly repressed (Valladares *et al.*, 2007). Besides *coxI* that is expressed in vegetative cells of *A. variabilis* ATCC 29413, the expression of *coxII* was up-regulated after nitrogen step-down in this organism. Single (*coxI* or *coxII*) or double (*coxI* and *coxII*) mutants of *A. variabilis* ATCC 29413 exhibited normal diazotrophic growth (Pils *et al.*, 2004). The expression of rubrerythrin (RbrA, a gene product of *alr1174*) in the heterocysts of *Anabaena* sp. strain PCC 7120 that acts like an FNR-dependent peroxidase confers additional protection against O_2 damage to nitrogenase. A mutant of *rbrA* differentiated heterocysts but its ability to fix nitrogen was reduced to 8% to that of the fixation rate noted under anaerobic conditions. The heterocysts of the mutant revealed the presence of high internal concentrations of H_2O_2. Further, the expression of *rbrA* from two promoters and one of the promoters producing the smallest transcript being under the control of HetR favour its role in the early phases of nitrogenase induction in differentiating heterocysts. These observations are confirmed by the presence of very low concentrations of H_2O_2 in the heterocysts of a double mutant (*nifH*$^-$ *rbrA*$^-$) than in the vegetative cells (Zhao *et al.*, 2007).

In addition, the ability of isolated heterocysts to carry out oxidative phosphorylation (Tel-Or and Stewart, 1976; Peterson and Wolk, 1978), the dependence of nitrogen fixation in dark by respiratory O_2-uptake (Fay, 1976) and enhanced levels of dehydrogenases (Winkenbach and Wolk, 1973; Apte *et al.*, 1978) constitute evidences for high respiratory metabolism of heterocysts.

6) CYANOPHYCIN

Cyanophycin granule polypeptide [(CGP); multi-L-arginyl poly-(L-aspartic acid)] is a water insoluble polymer of aspartic acid and arginine (with a molecular weight of 25–100 kD) and serves as a nitrogen reserve of cyanobacteria (Simon, 1971; Allen, 1984). Lang *et al.* (1972) have demonstrated that the "structured granules" of *Anabaena* are composed of CGP. In *A. cylindrica* the CGP content of cells increased during stationary phase coinciding with a decrease in soluble protein content and the addition of chloramphenicol has considerably enhanced the CGP content (Simon, 1973a). CGP synthesis is mediated by a non-ribosomal synthetic process (Simon, 1973b) and the basic requirements of the biosynthetic reaction have been studied *in vitro* that required the two constituent amino acids, Mg-ATP, cyanophycin primer and the enzyme (Simon, 1976). Cultures of *Aphanocapsa* sp. 6308 (ATCC 27150) starved of light, CO_2, sulphur and phosphorus showed enhanced content of CGP when compared to control cultures. However, nitrogen limitation or lowering of growth temperature caused a decrease in CGP content (Allen *et al.*, 1980). A number of studies have indicated that the CGP in cyanobacterial cells is broken down during nitrogen deprivation and the synthesis of CGP ensues once nitrogen is replenished in the medium (Allen and Hutchinson, 1980; Allen and Hawley, 1983; Mackerras *et al.*, 1990). In view of this, it is suggested that CGP acts as a transient reservoir of nitrogenous material (Shively, 1974; Carr, 1988; Mackerras *et al.*, 1990; Li *et al.*, 2001). The synthesis of CGP depended on the degradation of cellular proteins as well as newly assimilated nitrogen in *A. cylindrica* (Simon, 1973b) and *Aphanocapsa* sp. 6308 (ATCC 27150) (Allen and Hawley, 1983). This has been confirmed in *Synechocystis* sp. strain PCC 6803 by the use of N^{15} labelling and H^1NMR spectroscopy (Allen *et al.*, 2005) and proton nuclear magnetic resonance spectroscopy (Kolodny *et al.*, 2006). The synthesis of CGP took place by the addition of monomers to the pre-existing polymer (Allen *et al.*, 2005). The non-template-dependent polymerization of CGP takes place in three steps of chain initiation, elongation and termination mediated by cyanophycin synthetase (CphA). The enzyme fom *A. variabilis* ATCC 29413 (Ziegler *et al.*, 1998), *Synechocystis* sp. strain PCC 6803 (Aboulmagd *et al.*, 2000, 2001) and the thermophilic *Synechococcus* sp. strain MA19 (Hai *et al.*, 1999, 2002) has been purified. The specific activity of the enzyme from *A. variabilis* ATCC 29413 is 2 to 5 fold higher than that of the enzyme from *Synechocystis* sp. strain PCC 6803 and the *cphA* gene from *A.variabilis* ATCC 29413 has been cloned and expressed in *E.coli*. CphA from *A. variabilis* ATCC 29413, with a molecular weight of 100 kD, appeared as a single band in sodium dodecyl sulphate/ polyacrylamide gel electrophoresis. Since the native enzyme has a molecular weight of 230 kDa, determined as per size-exclusion chromatography, it is suggested that the native enzyme exists as a dimer with two ATP-binding sites (Zeigler *et al.*, 1998). The *cphA* gene from *Anabaena* sp. strain PCC 7120 has been expressed in *E. coli* and the recombinant protein exhibited highest specific CphA activity by synthesizing CGP expressed as 6.7 nmol arginine min^{-1} mg $protein^{-1}$ (Voss *et al.*, 2004).

Gupta and Carr (1981b) have found that the enzymes involved in the synthesis (*cphA*) and breakdown (cyanphycinase, *cphB*) of CGP are more active in heterocysts than vegetative cells. The accumulation of higher levels of CGP in the heterocysts signifies that it may serve as a reserve material for biosynthetic reactions (Carr, 1988). The expression of *nifH* (nitrogenase protein) at the late stage of heterocyst differentiation in *Anabaena* sp. strain PCC 7120 is followed by the synthesis of CGP and its accumulation at the neck joining the heterocyst with the vegetative cells fixes the formation of polar nodule (Simon, 1987; Sherman *et al.*, 2000). During nitrogen deficiency, there is rapid mobilization of CGP in *A. cylindrica* and *Synechocystis* sp. strain 6803 (Mackerras *et al.*, 1990). The diazotrophic unicellular *Cyanothece* sp. ATCC 51142 showed accumulation of CGP after a peak in nitrogenase synthesis and the knock-out mutants (of *cphA* and *cphB*) of this organism had to depend on phycobilisome degradation in the absence of CGP synthesis (Li *et al.*, 2001). The growth properties of a mutant of *A. variabilis* ATCC 29413, isolated after insertional mutagenesis of *cphA* gene, revealed the non-essential nature of CGP metabolism for aerobic nitrogen fixation. The mutant in the absence of CGP synthesis and the polar nodule in the heterocysts exhibited a similar growth behaviour as the wild-type irrespective of the nitrogen source and fixed nitrogen under aerobic conditions (Ziegler *et al.*, 2001). Two gene clusters, *cph1* and *cph2* governing the synthesis and breakdown of CGP in the genome of *Anabaena* sp. strain PCC 7120 have been identified. The expression of *cph1* (*cphA1* and *cphB1*) and *cph2* (*cphA2* and *cphB2*) gene clusters although was noted in ammonium, nitrate and

N_2 grown cultures, the expression of *cph2* gene cluster was higher in the absence of ammonium in the medium. The contribution of *cph1* gene cluster for CGP accumulation in vegetative cells and heterocysts was more than that of *cph2* gene cluster. Mutational analysis revealed that the breakdown of CGP appeared to have more significant bearing on the diazotrophic growth potential rather than the synthesis of CGP. Though CGP synthesis in the heterocysts is a normal phenomenon it is not required for heterocyst function (Picossi *et al.* 2004).

7) DEVELOPMENT

Around the same time when Fritsch (1951) was discussing the enigmatic nature of heterocyst, Fogg (1942, 1944, 1949 and 1951) made important observations on the inhibition of heterocyst development in *A. cylindrica* due to the presence of combined nitrogen sources and an inverse relationship was noted by him between the frequency of heterocysts and concentration of nitrogen source in the medium. Certain sources like nitrate nitrogen, glycine and asparagine caused a decrease in the frequency of heterocysts and ammonium nitrogen totally suppressed differentiation. The induction of heterocysts in the nitrogen-deficient media and their repression in nitrate-, nitrite- or ammonium-containing media correlated their role in conferring the ability to grow in the absence of nitrogen and so their role in nitrogen fixation. This aspect received considerable attention by employing different heterocystous cyanobacteria. Species of *Anabaena, Nostoc* and *Nodularia* differentiate heterocysts in a regularly spaced pattern with almost constant number of vegetative cells (8–10) intervening two heterocysts. The frequency of heterocysts is usually represented as percentage, i.e. their number per 100 vegetative cells and thus 8–10% of cells in nitrogen-deficient cultures represent heterocysts. Species of other heterocystous cyanobacteria *Aulosira, Calothrix, Cylindrospermum, Gloeotrichia, Hapalosiphon, Mastigocladus, Scytonema, Stigonema* and *Tolypothrix* have been studied for the frequency of heterocysts in relation to the presence of nitrogen sources. Exceptionally, *Chlorogloea fritschii* initially described as a unicellular heterocystless nitrogen-fixing genus (Mitra, 1950; Fay and Fogg, 1962) later turned out to be a filamentous form producing heterocysts (Fay *et al.*, 1964; Stanier *et al.*, 1971). Thus heterocyst formation has been found to be inhibited in the presence of combined nitrogen (nitrate, nitrite and ammonium ions) sources in the medium in batch cultures of *Camptylonema lahorense* (Pandey and Mitra, 1962), *C. fritschii* (Fay *et al.*, 1964; Peat and Whitton, 1967), *Anabaena ambigua* Rao (Talpasayi and Kale, 1967), *A. flos-aquae* A-37 (Mickelson *et al.*, 1967), *A. variabilis* (Ogawa and Carr, 1969); *Anabaenopsis raciborskii* Wolosz, *C. majus* Kuetz, *Anabaena* sp. (Singh *et al.*, 1972) and *Anabaena doliolum* (Singh and Srivastava, 1968; Tyagi, 1973a). Of the three nitrogen sources ammonium repressed the formation of heterocysts and nitrogen fixation at even very low concentrations (Mickelson *et al.*, 1967). Thomas and David (1971) observed that the presence of nitrate nitrogen completely inhibited the induction of heterocysts in *Anabaena* sp. L-31 in batch cultures but in continuous cultures no inhibition was observed at high dilution rates with a consequent increase in heterocyst frequency. Kulasooriya *et al.* (1972) conducted a detailed study on the frequency of heterocysts and the rate of nitrogen fixation. Since growth of *A. cylindrica* in ammonium nitrogen caused complete repression of both heterocyst differentiation and nitrogen fixation, transfer of cultures into nitrogen-deficient medium resulted in a resumption of the two processes. They correlated the ability to differentiate heterocysts with an attainment of a C:N ratio from 4.5:1 to 6:1. Photoautotrophically grown cultures of *A. cylindrica* when starved of nitrogen (bubbled with argon +CO_2) formed higher frequency of heterocysts than when aerated with N_2+CO_2 in continuous cultures (Kulasooriya *et al.*, 1972). Heterocystless cultures derived from combined nitrogen sources when shifted to a medium deficient in nitrogen (nutrient shift or nitrogen step-down) undergo differentiation of heterocysts simultaneously at regularly spaced pattern resembling the pattern in nitrogen-free medium (Talpasayi and Kale, 1967; Kulasooriya *et al.*, 1972).

Heterocyst differentiation in *A. ambigua* has been reported to be stimulated by red light (600–700 nm) whereas green light (450–550 nm) has been found to be inhibitory. A linear increase in heterocyst frequency with increase in light intensity with a maximum frequency at 1300 lux was noted (Kale and Talpsayi, 1969; Kale, 1972). Kaushik and Kumar (1970) reported similar results on the intensity and quality of light required for heterocyst formation in *A. doliolum*. A photomorphogenetic control in heterocyst differentiation has largely been ruled out on the basis that the low light has not been stimulatory and the involvement of a phytochrome-like pigment in the differentiation process seems to be remote because of requirement of (i) high light intensity for longer duration, (ii) CO_2 in presence of light, (iii) red light for differentiation followed by yellow and blue lights and complete inhibition in green light and (iv) white light for attainment of maximum frequency of heterocysts. The recognition of incipient heterocyst-like structures in cultures of *Nostoc muscorum* and *C. fritschii* grown in dark with granular inclusions and packed thylakoids and their subsequent differentiation into mature heterocysts upon exposure to light suggests that light is required for completion of the differentiation process probably in order to meet the photoautotrophic requirements (Whitton and Peat, 1967; Ginsberg and Lazaroff, 1973).

The differentiation of heterocysts after nitrogen shift-down is generally completed within 12 h as in *A. doliolum* (Tyagi, 1973a), 16 h as in *A. cylindrica* (Wilcox *et al.*, 1973a) and about 24 h in *A. variabilis* (Ogawa and Carr, 1969), *Anabaena catenula* (Wilcox *et al.*, 1973a), 24–36 h in *Nostoc punctiforme* (Wong and Meeks, 2001) and 36 to 48 h in *Anabaena* sp. strain PCC 7120 (Hebbar and Curtis, 2000). During this period, a molecular network of events takes place in an orderly step-by-step programme. Cell division takes place preceding heterocyst development and the selection of one of the products of division by competition and commitment leads to first a proheterocyst development by the end of 6–12 h. The proheterocysts are recognizable by their pale colour and slightly enlarged size. These events are followed by middle and late developmental changes leading to a mature heterocyst. The early, middle and late phases are associated with the expression of a number of genes (or clusters of genes) and the interplay of the respective gene products that govern the development and maintenance of a pattern of heterocysts. The initiation of the events very much depends on the global nitrogen regulator NtcA which perceives the signals from the signalling molecule (2-oxoglutarate) regarding the nitrogen status of the cells. According to status of nitrogen sufficiency or deficiency being a transcriptional regulator NtcA regulates the expression of the concerned set of genes. Under nitrogen-deficient conditions, NtcA triggers the expression of the master regulator of heterocyst differentiation, HetR. The mutual interaction of these two regulators decides the future course of differentiation process. The laying down of outer wall layers over the pre-existing cell wall of the vegetative cell such as an outermost fibrous layer, middle homogeneous layer and the innermost laminated layer are the morphological changes associated with middle and late phases of heterocyst differentiation which generally occurs between 18 to 24 h of differentiation. The outermost fibrous and middle homogeneous wall layers are composed of polysaccharides and since these are in the form of an envelope it is known as heterocyst envelope polysaccharide (Hep) layer which is synthesized and deposited under the regulation of *hep* genes. The innermost laminated layer is composed of heterocyst-specific Hgls synthesized and transported by *hgl* genes. At the narrow point of contact between the vegetative cell and a heterocyst the walls get constricted and constriction-specific genes regulate this process. The junctional part of the septum of a vegetative cell with that of the heterocyst forms microplasmodesmata whose formation is regulated by the synthesis and their close integration brings into picture the means of transport channels between the two cells. The deposition of cyanophycin at the two poles for the formation of polar nodules makes the differentiation complete. However, it must be stated that the functional nature of heterocysts would only be reflected by the expression of nitrogen fixation genes. All these processes are supported by signal transduction proteins and to some extent sigma factor proteins. So as to keep the internal atmosphere of the heterocyst to nearly microoxic conditions a simultaneous down-regulation of PSII and an up-regulation in respiratory activity takes place. The following account summarizes the initial events of phycobiliprotein degradation and the gene expression profiles during differentiation of heterocyst.

Soon after nitrogen step-down, there is a transient degradation of phycobiliproteins in all the vegetative cells and after the differentiation of heterocysts the level of these pigments is regained to the normal levels in *A. cylindrica*. The development of proheterocysts occurred within a period of 4.5 h and mature heterocysts appeared by about 14 h that coincided with the development of nitrogenase activity around 13.5 h (Bradley and Carr, 1976). In case of unicellular cyanobacteria phycobiliprotein degradation in response to nitrogen deprivation leads to chlorosis (Boersch, 1910; Allen and Smith, 1969). The process of chlorosis has been very intensively investigated in these forms and the complementation of a non-bleaching mutant led to the identification of a gene *nblA* of *S. elongatus* PCC 7942. NblA, the protein encoded by *nblA*, is responsible for bringing about the degradation of phycobiliproteins *in vivo* (Collier and Grossman, 1994). While *S. elongatus* PCC 7942 has one *nblA* gene, the genome of *Synechocystis* sp. strain PCC 6803 has two homologues of *nblA*, *nblA1* and *nblA2* whose corresponding proteins are required for phycobiliprotein degradation *in vivo* (Baier *et al.*, 2001). NblA is a small protein (6 kDa) of 65 amino acids arranged into two α-helices assembled at an angle of 37° in an antiparallel V-shaped arrangement. Strauss *et al.* (2002) cloned the *nblA* gene from *Anabaena* sp. strain PCC 7120 and the recombinant NblA produced in *Escherichia coli* though existed as a dimer but formed trimers in solution. The crystal structure of NblA from the same organism revealed it to be a dimer consisting of four-helices formed by two dimers (Fig. 2). Site-directed mutagenesis helped in the identification of Leu51 and Lys53 at the C-terminal region responsible for binding to the α-chains of phycocyanin (at Gln16-Leu39) and phycoerythrin (at Leu19-Ser45) (Bienert *et al.*, 2005). The importance of other amino acid residues in NblA such as Ser9, Arg22, Arg56 and Lys33 has been recognized for interaction with phycobilisomes (Dines *et al.*, 2008). *Anabaena* sp. strain PCC 7120 has two *nblA* genes, one on the chromosome designated as *nblA* (*asr4517*) and the other on delta plasmid known as *nblA-p* (*asr8504*). *A. variabilis* ATCC 29413 has only one *nblA* gene. The amino acid sequence of NblA proteins of both these *Anabaena* strains is 100% identical but the gene sequences have 96% identity (Baier *et al.*, 2004). Baier *et al.* (2004) reported the essentiality of NblA for phycobiliprotein degradation *in vivo* but not for the development of fully functional heterocysts. The *nblA* gene disruptant mutants for *nblA* and *nblA-p* of *Anabaena* sp. strain PCC 7120 accumulated high levels of phycobiliproteins not only in the vegetative cells but also in the fully developed heterocysts. Localization of the phycobiliproteins by the immunogold particle technique revealed more phycocyanin in the heterocysts. The presence of *nblAI* gene just upstream of *cpeBA* (that encodes phycoerythrin apoprotein), regulation of its expression and interactions of NblAI with phycobilisomes has been reported (Luque *et al.*, 2003). The localization of NblA in the differentiated heterocysts of *Tolypothrix* sp. PCC 7601 pointed out that NblA is a necessary co-factor for degradation of phycobilisomes but not the triggering factor (Ochoa de Alda *et al.*, 2004). In this connection, the interaction of NblA with ClpC, a heat shock protein (HSP100) belonging to cyanobacterial Clp protease, assumes significance (Karradt *et al.*, 2008).

The basic studies on protein turnover in *N. muscorum* cells subjected to nitrogen step-down revealed that several sets of proteins are synthesized in vegetative cells but some are preferentially allowed to be continued in proheterocysts. There existed a qualitative difference in the sets of proteins

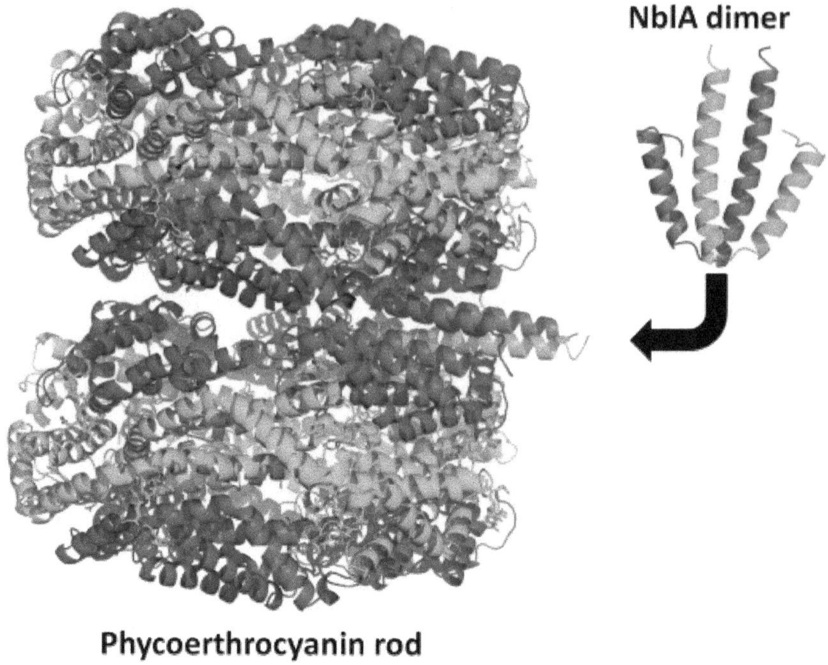

Figure 2: Free NblA dimer (from the crystal structure) and how it could penetrate into the gap formed between two hexamers of phycoerythrocyanin rod of *Anabaena* sp. strain PCC 7120. Picture courtesy of Noam Adir, Schulich Faculty of Chemistry, Technion-Israel Institute of Technology, Technion City, Haifa 32000, Israel.

Color image of this figure appears in the color plate section at the end of the book.

expressed in proheterocysts which was apparent after 9 h of nitrogen limitation (Fleming and Haselkorn, 1974). In order to understand the number of genes that are expressed from proheterocyst to mature heterocyst formation in *A. variabilis* ATCC 29413, Lynn *et al.* (1986) conducted DNA:RNA hybridization studies. Three important aspects that emerged from these studies are: (i) the expression of the genome was higher in vegetative cells (65%) than in the heterocysts (45%); (ii) the number of heterocyst-specific mRNA transcripts was approximately equivalent to 1000 genes; (iii) a constitutive expression of about 900 to 1300 genes occurred in both vegetative cells and heterocysts and (iv) the frequency of abundance of transcripts in heterocysts was higher than those in the vegetative cells. Lynn and Ownby (1987) confirmed the high turnover of mRNA in the isolated heterocysts of *A. variabilis* ATCC 29413 that continued for 2 h but decreased thereafter. However, in heterocysts of intact filaments the transcript accumulation continued as noted earlier (Lynn *et al.*, 1986). The enhanced gene expression patterns of *A. variabilis* ATCC 29413 after 8.5 h of nitrogen starvation could be correlated with the increase of nitrogenase activity that continued for another 5 to 6 h coinciding with the appearance of mature heterocysts. Simultaneously, the levels of transcripts for phycocyanin and allophycocyanin decreased within the first hour of nitrogen starvation. This is because of the short half-lives (16 to 18 min) of the transcipts of both *cpc* and *apc* genes. This decrease was soon made up by a rapid breakdown of proteins followed by enhanced nitrogen fixation (Wealand *et al.*, 1989). Synthesis of dinitrogenase reductase and its even distribution in the newly differentiated heterocysts of the symbionts of *Azolla filiculoides*, *A. caroliniana*, and *A. pinnata* coincided with the formation of contorted membranes and polar nodules (Braun-Howland *et al.*, 1988). But it was with the application of DNA microarray techniques that a wealth of information on gene expression profiles has been generated. Ehira *et al* (2003) elegantly demonstrated that the large-scale expression of genes during heterocyst differentiation in *Anabaena* sp. strain PCC 7120 is physically clustered in such a manner that they could be termed as 'expression islands'. They first synthesized cDNA in presence of Cy3-dUTP label or Cy5-dUTP label by reverse transcriptase PCR (RT-PCR) from the different mRNAs isolated at early (1–3 h), middle (8 h) and late stages (24 h) of heterocyst differentiation. The respective single-stranded cDNAs were pooled. The double-stranded fluorescently-labelled cDNA so synthesized was denatured at 95°C for 2 min, cooled and then applied to a DNA microarray on a glass slide consisting of 2407 clones of genomic sequences of 3 kbp each with 10 ORFs. There was sufficient overlapping of the ORFs in the selected clones so as to allow hybridization a very successful event. A procedure of normalization of fluorescence intensity was followed to determine the level of expression of genes. A comparison of such microarray from *Anabaena* grown on ammonium nitrogen was made with the one that has been subjected to nitrogen step-down. Transient gene expression was noted at very early stages of 1 to 3 h and middle stage (8 h). A co-ordinated up-regulation of continuous genomic regions, termed as expression of genomic islands, took place. The expression of *hetR* at 3rd h with a maximum of its accumulation by 8th h coincided with expression of *ntcA* which is consistent with the expression of *hetR* and *ntcA* as previously reported by Muro-Pastor *et al.* (2002). The expression islands pertained to 21 ORFs of carbohydrate metabolism probably related to heterocyst envelope polysaccharide, genes related to nitrogen fixation (expressed as operons *nifB-fdxN-nifS-nifU*, *nifHDK*, *hesAB-fdxH* and *nifENX* after the excision of necessary DNA elements) and heterocyst glycolipid synthetase genes (*hglE*, *hglD*, and *hglC*).

Another such a gene expression profile pertaining nitrogen fixation and heterocyst differentiation by a *Nostoc* strain 0102 symbiotic to *Gunnera megallanica* and *G. manicata* was constructed along a symbiotic profile from apex to mature parts of the leaf glands. The transcripts for *hetR*, *ntcA*, *glnB* and *nifH* were isolated and the expression patterns were determined as one-step RT-PCR along eight different positions in the leaf gland. The use of *hetR* and *glnB* as signalling molecules indicated an up-regulation of the former with a concomitant increase in the frequency of heterocysts and a down-regulation of the latter along the symbiotic profile. These results were corroborated with the respective proteins isolated and confirmed through Western blot analysis. The expression of *nif H* was quite high at the growing apices. Soon after infection, the heterocyst frequency increased to about 10–20% at the 3–4 mm from apex, and a little farther away (5–6 mm from apex) the heterocyst frequency reached 20–30% and 11–12 mm away from apex the heterocyst frequency attained a maximum of 50–60% (Wang *et al.*, 2004). Global gene expression profiles for *N. punctiforme*, another symbiotic cyanobacterium, were determined from steady-state cultures undergoing heterocyst differentiation. Three developmental states of hormogonium, heterocysts and akinetes were subjected for this analysis after a nitrogen step-down and compared with those of the ammonium-grown cultures. As many as 495 genes were expressed in the heterocyst stage, of which 373 were up-regulated (Campbell *et al.*, 2007). A significant discovery is the synthesis in large amounts of a nitrogen-stress induced non-coding RNA, designated as NsiR1 of 60 nucleotides long, in *Anabaena* sp. strain PCC 7120 that triggers heterocyst differentiation leading to nitrogen fixation. That is NsiR1 acts as a switch from nitrogen-replete conditions to nitrogen-limited cultures. Thus its expression is restricted to developing heterocysts and requires the participation of HetR and NtcA. The location of the gene for NsiR1 upstream of *hetF* and individual transcripts with cell-type specific promoter and a Rho-dependent terminator make it a unique non-coding RNA transcribed in high amounts after a nitrogen step-down (Ionescu *et al.*, 2010).

The discovery of a number of regulatory genes (*ntcA, hetR, hetA, hetP, hetC, hetF, hetL, hetY, hetZ* and *asr1734*), development genes (*devBCA* operon, *devH, devR* and *devT*), those of Hgls and Hep layer and pattern formation genes (*patA, patB, hetN* and *patS*) has been made by the characterization of different classes of mutants that are deficient in heterocyst development and nitrogen fixation. Moreover, the participation of genes for sigma factors and signal transduction in expression of the above genes or clusters of genes has also been indicated. A redeeming feature is that most of the knowledge in these areas has been generated by the use of mostly a single cyanobacterium, *Anabaena* sp. strain PCC 7120. A number of molecular tools and processes have been developed to pursue the gene expression profiles. The following account highlights the developments in these areas.

8) MUTANTS DEFECTIVE IN HETEROCYST DEVELOPMENT AND NITROGEN FIXATION

N-methyl-N′-nitro-N-nitrosoguanidine (NTG-)-induced mutants of *N. muscorum* deficient in heterocyst differentiation and nitrogen fixation have been put to use in genetic recombination experiments (Stewart and Singh, 1975). NTG-induced mutants of *N. muscorum* and *N. linckia* have been characterized with respect to het$^-$, nif$^-$ (heterocyst-less, non-nitrogen fixing), het$^+$, nif$^-$ (heterocystous and non-nitrogen-fixing) and M-het$^+$, nif$^+$ (multiple heterocystous) phenotypes. The involvement of a common genetic determinant(s) for heterocyst differentiation and the expression of *nif* genes in both the organisms and for heterocyst spacing pattern in *N. linckia* have been suggested (Singh *et al.*, 1977). A large number of mutants (65) of *A. variabilis* ATCC 29413 that required inorganic fixed nitrogen sources for growth have been isolated after NTG-mutagenesis. These showed altered pattern of heterocyst differentiation (with low and high frequency of heterocysts), with thick or incompletely developed heterocyst envelopes and protoplasts separated from heterocysts (Currier *et al.*, 1977). Some of these mutants of *A. variabilis* ATCC 29413 deficient in Hgls showed very low or no nitrogenase activity when grown under aerobic conditions suggesting a direct correlation between the absence of Hgls in the laminated layer and O_2-induced damage. The revertants of this class of mutants showed normal synthesis of Hgls and aerobic nitrogen fixation. Other classes of mutants in which Hgls were normal either required a supply of reductant in the form of dithionite or microaerophilic conditions for nitrogen fixation (Haury and Wolk, 1978). A short-trichome mutant of *N. muscorum* exhibiting het$^-$, nif$^-$ phenotype with reduced growth rate than the wild-type even in nitrate-medium has been reported that reverted to wild-type with a frequency of 2.62×10^{-7} (Padhy and Singh, 1978). Species of *Nostoc* (*N. linckia*), *Anabaena* (*Anabaena* strain CA) and *Gloeotrichia* (*G. ghosei*) have been put to use for the isolation of such mutants that showed altered pattern of nitrogen fixation and required the exogenous supply of fixed nitrogen sources for their survival (Tiwari, 1977; Tiwari *et al.*, 1979; Gotto *et al.*, 1979; Grillo *et al.*, 1979). Some of the NTG-induced mutants of *G. ghosei* showed the phenotypic features with loss of both polarity and heterocyst differentiation, longer filaments with normal heterocysts and chains of heterocysts (Mishra and Tiwari, 1986). Heterocystless mutants of *Anabaena* sp. strain PCC 7120 exhibited whole cell acetylene reduction activity under aerobic or microaerophilic conditions but exhibited similar growth pattern in the fixed nitrogen sources as well as no alteration in either glutamine synthetase [GS; L-glutamate:ammonia ligase (ADP-forming); EC 6.3.1.2] and glutamate synthase [GOGAT; L-glutamate:oxoglutarate amidotransferase; EC 1.4.7.1] activities (Spence and Stewart, 1987). Although a physiological and biochemical analysis of the above mutants is available, information on genes that regulate either the formation of heterocysts or causes for deficiency in nitrogen fixation is not available.

Anabaena sp. strain PCC 7120 possesses three restriction enzymes Asp7120I, Asp7120II and Asp7120III (corresponding to AvaI, AvaII and AvaIII of *A. variabilis*) that act as isoschizomers on the corresponding restriction sites at *Asp7120I*, *Asp7120II* (Duyvesteyn *et al.*, 1983) and *Asp7120III* (Elhai *et al.*, 1997) on the plasmids introduced into its cells. In order to overcome the activity of isoschizomers *in vivo*, the restriction sites on the plasmids are removed. In this way it has been possible to develop shuttle vectors capable of getting transferred in between *E. coli* and *Anabaena* sp. strain PCC 7120 (Wolk *et al.*, 1984). Wolk *et al.* (1988) for the first time utilized molecular tools for the identification of genes governing the synthesis of heterocyst envelope polysaccharides of *Anabaena* sp. strain PCC 7120. In doing so, they took advantage of the RP-4 (as helper plasmid)-mediated conjugative transfer of pBR322-based shuttle plasmids from *E. coli* to *Anabaena* sp. strain PCC 7120 (Wolk *et al.*, 1984). A pRL25C cosmid vector that is equipped with the capacity to replicate in *E. coli* as well as *Anabaena* sp. strain PCC 7120 was used to generate a library of 1,054 cosmid clones consisting of size-fractionated (about 40 kb) genomic DNA of *Anabaena* sp. strain PCC 7120 and RP-4 was transferred to this library by conjugation. A number of mutants unable to grow aerobically in a nitrogen-free medium but required fixed nitrogen sources for their survival were isolated after UV-mutagenesis. Among these, EF116 which showed defects in heterocyst envelope polysaccharide (which lost cohesiveness when compared to the wild-type) was chosen and a complementation of the mutant EF116 by transfer of genomic DNA fom the cosmid library through triparental conjugation enabled them to identify a 2.8 kb fragment of DNA derived fom one of the cosmids. Another mutant EF113 was complemented by a single cosmid bearing a 4.8 kb genomic DNA. Other mutants EF114 and EF122 are unique for the lacking the Hgl layer whereas EF104 showed the absence of nitrogen fixation even under anaerobic conditions. This technique of conjugative transfer of genes into *Anabaena* sp. strain PCC 7120 has been followed invariably by most of the investigators all world over and has become the focal point for transfer of reporter genes to identify various functions during heteocyst differentiation (Elhai and Wolk, 1988). Elhai *et al.* (1997) suggested that the success of triparental conjugation very much depended on the number of restriction sites (*Asp7120I*, *Asp7120II* or *Asp7120III*) carried on the plasmids intended to be transferred into *Anabaena* cells and the number of corresponding restriction enzymes present in *Anabaena* cells. Plasmids with their restriction sites (*Asp7120I*, *Asp7120II* or *Ava7120III*) modified by methylation in *E. coli* have a greater efficiency of being transferred as the restriction sites are protected against the activity of restriction enzymes. Another technique that has become handy is the transposon Tn5-mediated mutagenesis. Tn5 has four *AvaI* restriction sites and when transferred through pBR322bla::Tn5 construct into *Anabaena* cells, it might be cleaved by the Asp7120I restriction enzyme. In order to protect Tn5 from being cleaved, in the first instance pBR322bla::Tn5 construct is used to transform *E. coli* HB101 cells containing AvaI and AvaII methylase activity. That is *AvaI* restriction sites on Tn5 get methylated so that upon entry into *Anabaena* cells the Asp7120I restriction enzyme is unable to act on these sites. The pBR322::Tn5 is then transferred into *Anabena* sp. strain PCC 7120 cells through triparental conjugation. Tn5 has been found to get transposed into a number of sites on the genome of *Anabaena* sp. strain PCC 7120. Such cells with Tn5 can be screened and selected by the antibiotic resistance cassette present in Tn5 sequence. Out of a total of 400 colonies isolated from four conjugation plates, one isolate T-123 was unable to grow in a medium without combined nitrogen. That is it is unable to fix nitrogen (Fix^-) although it formed the heterocysts that resemble in every respect with those of the wild-type strain (Borthakur and Haselkorn, 1989).

A comprehensive account on the diversity of mutants related to the structure and function of heterocysts was published by Ernst *et al.* (1992). Their study is most significant for two reasons. The first is that they provided a classification of the mutants and the second is that the differentiation of heterocysts is not exclusively under environmental control but a certain degree of development must precede the synthesis of nitrogenase. The different phenotypic features and their combinations in different mutants of *Anabaena* sp. strain PCC 7120 isolated after Tn5 insertional mutagenesis have been presented. They are as follows: (i) Fox^- mutants are those that are incapable of fixing nitrogen in the presence of O_2; (ii) Fix^- mutants are those Fox^- mutants that expressed very low C_2H_2 reducing activity under any experimental condition; (iii) Het^- mutants are unable to differentiate heterocysts even after prolonged nitrogen deprivation, of these some showed fragmenting phenotype and others non-fragmenting phenotype; (iv) Hen^- mutants are those that are unable to synthesize heterocyst envelope; (v) Hgl^- represents the inability to synthesize Hgls and (vi) Dab^- phenotypic feature is denoted by the inability to oxidize diaminobenzidine (DAB) at the polar nodules that represents a deficiency in the respiratory enzymes and the fully developed normal heterocysts oxidize DAB as first reported by Murry and Wolk (1989). Mutants with a combination of the above phenotypic features are classified into different classes: (i) $Fox^-Het^-Fix^-$ fragmenting; (ii) $Fox^-Het^-Fix^-$ non-fragmenting; (iii) $Fox^-Het^+Dab^-Hen^-Fix^-$; (vi) Fox^- $Het^+Dab^+Hen^-Fix^+$; (v) $Fox^-Het^+Dab^+Hen^+Fix^+$; (vi) $Fox^-Het^+Dab^+Hen^+Fix^-$ and (vii) $Hen^-Hgl^-Hep^-$. The properties of the above classes of mutants point out to the existence of different developmental stages through which the differentiation of heterocyst proceeds with a concomitant expression of the concerned genes or clusters of genes.

The discovery of triparental conjugation and the use of Tn5 to obtain gene disruptant mutants of cyanobacteria paved the way for the elucidation of gene expression profiles in association with specific reporter genes. The specific molecular tools designed for this purpose involved localization of transcription using fusions with *lac*, *lux*, and *gfp* as reporter genes (Wolk, 1996). In *E. coli* lactose is hydrolysed by the activity of β-galactosidase (β-gal) encoded by the *lacZ* gene. Besides lactose, β-gal can also hydrolyze other chromogen substrates such as O-nitrophenol β-D-galactopyranoside (DNPG), 5-bromo-4-chloro-3-indolyl-β-D galactopyranoside (X gal) and 3,4 cyc lohexenoesculetin-β-D-galactopyranoside (S-gal) which produce yellow, blue and black products/ precipitates, respectively. There are two major limitations for using *lacZ* as the reporter gene. Firstly, it involves the usage of very costly chemicals and secondly the cells have to be lysed for the assay of β-gal activity. A number of vector systems such as multicopy (plasmid) and sigle copy (bacteriophage) vectors have been used to construct *lacZ* gene fusions by either *in vivo* or *in vitro* manipulations (Silhavy and Beckwith, 1985). Insertion of an RNaseIII cleavage site in between a polylinker cloning site and the promoterless *lacZ* gene facilitated the generation of an unchanging 5′-end for the *lacZ* mRNA. This led to the independent translation of *lacZ* mRNA and the β-gal level reflects the expression of gene in question (Linn and Pierre, 1990). Lammers *et al.* (1986) used *lacZ* as a reporter for finding out the region in 11-kb element of *nifD* responsible for excision during heterocyst differentiation in *Anabaena* sp. strain PCC 7120. This was based on an assay developed in *E. coli* by transposing a *mini-Mu-lac* gene fusion in *nifD* element. The excision of *nifD* element also resulted in the loss of *lacZ* gene and the colonies could be easily recognized by the absence of blue colour on X-gal plates. The firefly luciferase gene (*lux*) has been cloned (de Wet *et al.*, 1985) and it is a reporter with high sensitivity and could be easily identified by its luminescence that requires no post-translational modifications and is active immediately after translation. Luciferin, the substrate required for monitoring the assay of firefly luciferase is very costly and the use of tetracycline (an antibiotic that is photoinactivated) as a selectable marker and the necessity to permeabilize the cells during assay of the luminescence are the other disadvantages. So *luxABCDE* operon from the bacterium *Vibrio fischeri* has been generally used as reporter gene. The bacterial luciferase encoded by *luxAB* genes requires no addition of substrates but the gene products of *luxC*, *D* and *E* serve as substrates *in vivo*. In addition, bacterial luciferase oxidizes *n*-decanal, an aldehyde in presence of

$FMNH_2$ and light. The expression of *luxAB* as reporter gene ceases once cell death ensues. That is why a promoterless *luxAB* was used as a reporter gene to give rise to transcriptional fusions with the desired promoter of the gene to be expressed. The expression of the transcripts in response to nitrogen step-down or various environmental shifts could be identified with increased or decreased luminescence. Since the plasmid used for the transfer lacks the restriction sites for AvaIII and AvaII enzymes, it is not subjected to the degradative action of the corresponding enzymes inside *Anabaena* sp. strain PCC 7120 cells upon transfer (Elhai and Wolk, 1990; Wolk *et al.*, 1991; Ernst *et al.*, 1992; Maldener *et al.*, 1994; Cai and Wolk, 1997). Elhai and Wolk (1990) successfully demonstrated the expression of large and small subunits of RuBisCO in all vegetative cells of *Anabaena* sp. strain PCC 7120 when *luxAB* was introduced downstream of P_{rbcLS}. Likewise, the expression of *glnA* (GS gene) was found in vegetative cells as well as heterocysts but the expression of nitrogenase genes *nifHDK* was restricted to heterocysts. *Anabaena* sp. PCC 7118, a heterocystless mutant strain, fixed nitrogen under anaerobic conditions and transcriptional fusions of *luxAB* with P_{nifHDK} showed expression in specialized cells that looked different (lacking phycocyanin and occurring at different lengths all over the filament) from vegetative cells. These results emphasize that *Anabaena* sp. PCC 7118 though lacked full development of heterocysts has retained the pattern (Elhai and Wolk, 1990). A Tn5 derivative, Tn5-1063 with *luxAB* as reporter gene has advantages of locating the sites of gene insertions, mapping their position and subsequent cloning. By sequencing the identity of the gene is known (Wolk *et al.*, 1991).

The *gfp* gene of jellyfish *Aequoria victoria*, which encodes green fluorescent protein (GFP), has been cloned after two decades of its intial discovery. GFP as a reporter can be used in the hosts across all kingdoms and different colour variants like blue, yellow and red can be used together in systems to find out interplay of genes and their products. Upon excitation by UV-light (395 nm) the inert *gfp* reporter gene emits green light (509 nm) that can be easily visualized by using epifluorescence microscopy. Due to its fairly longer half-life extending many bacterial generations it sometimes creates problem as the dead cells also continue to show luminescence unlike *luxAB* reporter. In view of this, construction of *gfp* variants with shorter half-lives of 40, 60 and 110 min was feasible with the addition of degradative tag to the C-terminal end of wild-type *gfp* (Andersen *et al.*, 1998). The expression of a number of early genes after nitrogen step-down in *Anabaena* sp. strain PCC 7120 has been studied with the help of *gfp* transcriptional fusions (Khudyakov and Golden, 2004; Wang and Xu, 2005; Zhao *et al.*, 2005; Zhang *et al.*, 2009; Mella-Herrera, 2010). Cell-type specific gene expression in *N. punctiforme* ATCC 29133 was analysed by the construction of two transcriptional shuttle vectors using *gfp* as the reporter gene. Fluorescence of GFP was mainly restricted to the heterocysts when P_{hetR}-*gfp* transcriptional fusion was introduced into *N. punctiforme* ATCC 29133 (Argueta *et al.*, 2004).

9) GENES ESSENTIAL DURING DEVELOPMENT

A) *Early genes*: *Regulatory genes*

***i) ntcA*:** Chastain *et al.* (1990) identified a DNA-binding factor (VF1) from *Anabaena* sp. strain PCC 7120 cells that binds to three adjacent sites in the *xisA* upstream region. Vega-Palas *et al.* (1990) first identified <u>*ntc*</u>*A* gene (for nitrogen control) in *Synechococcus* sp. PCC 7120 while characterizing pleiotropic mutants that were unable to utilize inorganic nitrogen in a form other than ammonium. The gene product of *ntcA* is a DNA-binding protein and a transcriptional activator that belongs to the CRP-family of bacterial transcriptional regulators (Vega-Palas *et al.*, 1992). The DNA-binding activity is mediated by the presence of a helix-turn-helix at its C-terminal end (Vega-Palas *et al.*,

Table 1: Regulatory genes expressed during early phases of heterocyst differentiation in *Anabaena* sp,. PCC 7120.

Gene	Gene product	Time (h) of induction after nitrogen-step-down	Mutant phenotype	No. of tsps in the promoter	References
ntcA (=BifA; VF1) (*alr4392*)	DNA-binding protein, transcriptional activator of CRP family; binds to consensus sequence -GTAN$_8$TAC-	Constitutive expression; peak in expression after 12 h	Het$^-$; Unable to grow in N_2 and NO_3^-, can grow only on NH_4^+	Three; –49bp,–136bp and –180bp	Chastain *et al.* (1990), Wei *et al.* (1993), Ramasubramanian *et al.* (1996), Muro-Pastor *et al.* (2002)
hetR Master Regulator (*alr2329*)	DNA-binding protein and a serine protease; autoregulatory (HetR from -271 tsp) and positive feed-back control	2 h	Het$^-$; Mch-phenotype if overexpressed	Four; –184 bp, –271 bp, –696 bp and –728bp	Buikema & Haselkorn (1991a,b, 2001), Black *et al.* (1993), Huang *et al.* (2004)
hepA (*hetA*) (*alr2835*)	ATP-binding cassette transporter; regulated by hepK	7 h	Defective in cohesiveness of Heps; Fox$^-$ and Fix$^+$		Holland and Wolk (1990), Wolk *et al.* (1993)
patS	Peptide of 17 amino acids; C-terminal RGSGR sequence required for activity	6 h	Non-expression leads to normal pattern; Mch-phenotype if overexpressed	tsp located at -314 bp	Yoon & Golden (1998, 2001)
hetC	ABC type protein transporter;	4–6 h	Het$^-$; expression pattern resembled that of hetR;	tsp-571 bp NtcA activated; tsp-293 bp NtcA activated but dependent on HetR	Khudyakov & Wolk (1997), Muro-Pastor *et al.* (1999, 2009)
hetF (*all1730*)	Capsase-hemoglobinase type protease; brings positive auturegulation of HetR		Het$^-$; Mch-phenotype on overexpression		Wong & Meeks (2001)*, Wolk *et al.* (2007), Risser & Callahan (2008)
hetL (*all3740*)	Possesses pentapeptide A(D/N)LX* repeats *represents polar amino acid	Expression not cell specific	Het$^+$; Mch-phenotype if overexpressed; inactivation of hetR/ hetC is epistatic to hetL overexpression		Liu & Golden (2002)
hetY	Could be an ATPase				Yoon *et al.* (2003)
hetP		6 h	Het$^-$? but Mch-phenotype on overexpression; Could even bypass HetR	Five tsps.; HetR binding in between –727 & –545 tsps	Fernandez-Pinas *et al.* (1994), Higa & Callahan (2010)
hetZ (*alr0099*)			Het$^-$;		Zhang W. *et al.* (2007)

*First described in *N. punctiforme* ATCC 29133 by Wong and Meeks (2001).

1992). The presence of *ntcA* in a number of nitrogen-fixing as well as non-nitrogen-fixing cyanobacteria has been demonstrated and NtcA from *Anabaena* sp. strain PCC 7120 shows 77% identity with that of *Synechococcus elongatus* PCC 7942 NtcA and the helix-turn-helix of the two species is identical (Frías *et al.*, 1993). Simultaneously, Wei *et al.* (1993) identified *bifA* gene from *Anabaena* sp. strain PCC 7120 that encodes a sequence-specific DNA-binding protein. The VF1 (Chastain *et al.*, 1990), BifA (Wei *et al.*, 1993) and NtcA (Frías *et al.*, 1993) are identical. The requirement of NtcA for the expression of nitrogen assimilation and heterocyst development genes in *Anabaena* sp. strain PCC 7120 was demonstrated by Frías *et al.* (1994). An *ntcA* disruptant mutant required ammonium for growth and was unable to grow in nitrate-enriched and nitrogen-deficient media because of lack of synthesis of enzymes of nitrate metabolism and nitrogen fixation, respectively. The genes of nitrogen assimilation (*nifHDK* encoding nitrogenase, *nir* for nitrite reductase and *glnA*) and heterocyst development (*hetR*, a regulatory gene) are not expressed in the *ntcA* mutant after nitrogen step-down. Wei *et al.* (1994) cloned *ntcA* (*bifA*) gene of *Anabaena* sp. strain PCC 7120, inactivated it by the insertion of a Ω spectinomycin (Sp) and streptomycin (Sm) resistance cassette, cloned it into a suicide plasmid (pAM1320) and introduced it into *Anabaena* sp. strain PCC 7120 by conjugation. Single recombinants carried the *ntcA* disrupted gene integrated into the chromosome as well as the wild-type *ntcA* gene whereas the double recombinants (AMC236) possessed only the disrupted gene. Complementation of AMC236 with wild-type gene on a shuttle vector restored the wild-type character (AMC273 and AMC274). AMC236 grew only on ammonium medium and was unable to grow and differentiate heterocysts in a nitrogen-free medium while the wild-type and AMC273 and AMC274 grew in all nitrogen-enriched as well as nitrogen-deficient media. AMC236 did not show rearrangement of *nifD* or *fdxN* elements after nitrogen step-down. The molecular mechanism by which NtcA exerts such a regulation is explained by the presence of a consensus NtcA-binding sequence in the promoter regions of genes under its control. The binding of BifA (NtcA) to the promoter regions of *xisA* (that encodes a site-specific recombinase), *glnA* and *rbcL* and *nifH* genes has been reported by Ramasubramanian *et al.* (1994b). The presence of NtcA-binding sites upstream of the promoter regions in case of *glnA* (between –125 and –148 bp) and *rbcL* (between -12 and +12 bp and -43 to -54 bp) have been recognized. Luque *et al.* (1994) reported the presence of a palindromic DNA sequence $GTAN_8TAC$ as the target site for NtcA-binding in the promoter regions of nitrogen-regulated genes of *S. elongatus* PCC 7942. Frías *et al.* (1997) identified and cloned an operon of nitrite and nitrate assimilation genes consisting of *nir-nrtABC* (elements of nitrate permease system)-*narB* (nitrate reductase structural gene). In presence of ammonium, the repression of this operon takes place by the interaction of NtcA with 460 bp region upstream of the start of the *nir* gene and in presence of nitrate the induction of this operon takes place. So in this respect, NtcA acts as a repressor as well as an activator of enzymes of nitrogen assimilation. The NtcA-dependent promoters thus represent a different class of promoters that possess a -10 box (TAN_3T) and an NtcA-binding site characterized by the presence of the signature sequence $GTAN_8TAC$ in lieu of a –35 box. The regulation of NtcA of *Anabaena* sp. strain PCC 7120 by a redox-dependent mechanism has been reported by Jiang *et al.* (1997) in which the cysteine residues of NtcA are expected to play a key role. This is based on the interaction of NtcA with the promoter region of *gor* gene that encodes glutathione reductase. Further, the binding of NtcA with the promoter regions of not only the genes involved in nitrogen assimilation but also of carbon fixation (*rbcLS*), *xisA* (that encodes a site-specific recombinase expressed during heterocyst differentiation) and *ntcA* itself assumes significance and underlines the role of NtcA as a global regulatory protein. Further evidences that cellular redox status influences the expression of *ntcA* gene expression in *Synechocystis* sp. strain PCC 6803 were presented by Alfonso *et al.* (2001). Two transcripts of *ntcA*, one shorter (0.8 kb) and the other longer (1.2 kb) have been detected. The shorter transcript is light- and nitrogen-dependent whereas the longer transcript is synthesized constitutively. A concomitant increase of NtcA protein inside the cells coincided with the accumulation of the shorter transcript. The inhibition of NtcA accumulation in presence of electron transport inhibitors and the accumulation of the two transcripts inside glucose-grown cells in darkness emphasizes that it is electron transport and not light per se that regulates the expression of *ntcA*. Primer extension analysis and gel mobility shift assays revealed that NtcA is bound to the shorter transcript and not to the longer transcript. But the presence of three transcription start points (tsps) at locations –49, –136 and –180 have been identified in the promoter region of *ntcA* and these have been designated as P1, P2 and P3, respectively (Ramasubramanian *et al.*, 1996; Muro-Pastor *et al.*, 2002). Two putative NtcA-binding sites are located at -143.5 (GTAN$_8$AAC) and –103.5 (GTAN$_8$TAC) in relation to tsp of *ntcA* and these are referred as 'distal' and 'proximal' NtcA-binding sites, respectively. The distal NtcA-binding site overlapps with –10 mer sequence of the P2 promoter region of *ntcA* (Muro-Pastor *et al.*, 2002). *In vitro* binding assays performed with DNase protection method revealed that NtcA is bound to the distal site (Ramasubramanian *et al.*, 1996). The expression of *ntcA* in relation to exogenous nitrogen source and heterocyst differentiation has been studied. Transcripts of *ntcA* from P1 and P2 are produced in all nitrogen sources. The transcripts from P1 are subjected to increase 6–12 h after a nitrogen step-down and are localized very strongly in the heterocysts and those from P2 are constitutively expressed. The transcripts from P3 appear after 6–12 h of nitrogen step-down but the expression of *ntcA* from this tsp is very transient. However, the expression of *ntcA* from P1 and P3 is dependent on NtcA and HetR (Muro-Pastor *et al.*, 2002). To identify which of these transcripts first appears in the proheterocysts, Olmedo-Verd *et al.* (2006) made three deletions from 5′-end of the promoter region of *ntcA* gene (involving the three tsps) and introduced the altered versions of the *ntcA* promoter into the wild-type at a neutral site (the *nuiA-nucA* region) of *Anabaena* α-megaplasmid. The integration of the respective altered promoter constructs of *ntcA* by homologous recombination also altered the expression of wild-type *ntcA* gene. The expression of *ntcA* from P1 and P3 was localized in the proheterocysts and heterocysts and these observations have been additionally confirmed by the expression of *ntcA-gfp* translational fusion. Olmedo-Verd *et al.* (2008) studied the binding of NtcA to wild-type and mutated proximal and distal NtcA-binding sites of the P2 promoter region of *ntcA* of *Anabaena* sp. strain PCC 7120. Mutation of the proximal NtcA-binding site of *ntcA* from GTAN$_8$TAC to CTAN$_8$TAC abolished binding of NtcA to the two sites but the mutation of distal NtcA-binding site from GTAN$_8$TAC to CTAN$_8$AAC did not abolish NtcA binding to this site though more NtcA concentration was required than when compared to the wild-type. However, mutation of both the proximal and distal NtcA-binding sites still resulted in a lower NtcA-binding. These results are explained as follows. In nitrate-replete medium, *Anabaena* sp. strain PCC 7120 shows expression of NtcA from P2 and as well as P1 that would not require the presence of either NtcA or HetR (Muro-Pastor *et al.*, 2002). When cells are subjected to nitrogen deprivation, the low levels of NtcA present in the cells bind to the proximal NtcA-binding site thus increasing the production of transcripts from P1. Then in presence of HetR and other HetR-dependent elements, the transient expression from P3 takes place that in turn is required for the localized induction of NtcA from P1 site in the prospective proheterocysts and heterocysts. So the most important event during the nitrogen step-down is the binding of NtcA to the proximal NtcA-binding site because this determines which of the promoters are to be utilized for transcription of *ntcA* (Olmedo-Verd *et al.*, 2008). As cyanobacteria perceive nitrogen status by sensing intracellular 2-oxoglutarate levels (Muro-Pastor *et al.*, 2001), Valladares *et al.* (2008) reported transcription activation by NtcA and 2-oxoglutarate of three genes *hetC*, *nrrA* and *devB* involved in different stages of heterocyst differentiation.

Thiel and Pratte (2001) showed that MM3, an *ntcA* mutant of *A. variabilis* ATCC 29413, failed to grow in nitrate-enriched medium as well as differentiate heterocysts after a nitrogen step-down. Su *et al.* (2005) employed a bioinformatics approach to find out NtcA-binding sites in the genes of a genome and concurrently in the orthologues of the other genomes. As *glnA* is known to be transcribed from different tsps, it is of interest to know which one of the four transcripts is synthesized in the proheterocysts and heterocysts of *Anabaena* sp. strain PCC 7120. By the use of *lacZ* fusion constructs of the four tsps in the upstream region of the promoter of *glnA*, Valladares *et al.* (2004) showed that *glnA* is expressed from the P1 region in the proheterocysts and heterocysts of *Anabaena* sp. strain PCC 7120 after a nitrogen step-down. The mutation of NtcA-binding site in the P1 region and the binding of NtcA to the P1 site appear to govern the expression of *glnA* in vegetative cells and heterocysts. A comparison of the sequenced genomes (*Gloeobacter violaceus* PCC 7421, *Anabaena* sp. strain PCC 7120, *P. marinus* CCMP1375, *P. marinus* MED4, *P. marinus* MIT9313, *Synechococcus elongatus* PCC 6301, *Synechococcus* sp. strain WH8102, *Synechocystis* sp. strain PCC 6803 and *Thermosynechococcus elongatus* BP-1) revealed that NtcA-regulated promoters are found not only for genes of nitrogen assimilation but they are coupled to genes involved in various stages of photosynthesis thus highlighting the importance of these genes as regulatory points in the two major metabolic processes. Besides regulating the expression of genes of nitrogen and carbon metabolism, NtcA also is reported to control the expression of genes related to the maintenance of iron homeostasis. As cyanobacteria require iron for the synthesis of ferredoxins and other haem-related proteins, the expression of *furA* during heterocyst differentiation has received attention. López-Gomollón *et al.* (2007) found that the levels of *furA* transcripts and the FurA protein increased in proheterocysts and mature heterocysts as a result of nitrogen shift-down in the wild-type strain of *Anabaena* sp. strain PCC 7120. The expression of *furA* in *ntcA* mutant of *Anabaena* sp. strain PCC 7120 was depressed and consequently the levels of mRNA of *furA* and FurA protein decreased. These differences were attributable to the binding of NtcA to the promoter region of *furA* in the wild-type and the lack of it in the *ntcA* mutant.

***ii) hetR*:** Buikema and Haselkorn (1991a) characterized 140 Fox$^-$ mutants of *Anabaena* sp. strain PCC 7120 isolated after diethyl sulphate mutagenesis followed by penicillin enrichment. Out of these, seven mutants exhibited morphologically abnormal phenotypes. Mutant 216, originally suspected to be Fox$^-$ and Het$^-$ phenotype, was unable to fix nitrogen even under anaerobic conditions and it turned out to be Fix$^-$ and Het$^-$ while the rest of the six had Fox$^-$ and Het$^-$ phenotype. A cosmid library of the genome of *Anabena* sp. strain PCC 7120 was put to use for complementation and a cosmid with 9.5 kb region of the genome could complement mutant 216 as well as two other mutants. Subcloning of this fragment pinpointed a 2.4 kb region specific for complementing mutant 216 and sequencing of this fragment revealed a single ORF designated by them as *hetR* (Buikema and Haselkorn, 19991b), a regulatory gene that is essential and controls heterocyst differentiation. The gene product, HetR (consisting of 299-amino acid residues) is a DNA-binding protein (Huang *et al.*, 2004) but it is a unique protease whose Ser179 is converted to Asn in mutant 216. Directed mutagenesis of *hetR* in the wild-type *Anabaena* sp. strain PCC 7120 was achieved by transferring *hetR* interrupted gene [with a neomycin (Nm) resistance cassette carried by plasmid pWB216S2.4] sequence through conjugation with *E. coli* MC1061. This has led to the isolation of single (Nm-resistant) and double (Nm- and sucrose-resistant, 5%) recombinants with a phenotype similar to that of mutant 216. The presence of *hetR* in extra copies in the wild-type resulted in multiple-contiguous heterocyst (Mch) phenotype with enhanced (10–25%) heterocyst frequency than the wild-type (5 to 10%) in a nitrogen deficient medium and 5% of cells even differentiated into heterocyst-like cells under repressing (nitrate-enriched) conditions. But these appeared to be non-functional. The expression of *hetR* at very early phase during heterocyst differentiation (within 6 h) was supported by the increased (3 to 5-fold) transcript levels of *hetR* when *Anabaena* sp. strain PCC 7120 was subjected to a nitrogen step-down compared to very low levels of its expression in the organism grown in nitrogen-replete medium. Mainly two major transcripts of *hetR* gene (of 1.4 kb and 1.9 kb long) have been detected in the wild-type. These results emphasize that heterocyst differentiation requires a functional *hetR* gene in *Anabaena* sp. strain PCC 7120.

Cloning and sequencing of *hetR* gene from *A. variabilis* ATCC 29413 and the characteristics of a *hetR* disruptant mutant of this organism have been reported by Schiefer *et al.* (2002). The *hetR* mutant of *A. variabilis* ATCC 29413 exhibited a Fox$^-$ and Het$^-$ phenotype but under anaerobic conditions it showed Fix$^+$ and Het$^-$ character. The vegetative cells of *hetR* mutant showed *nif2* enzyme after 4 h of nitrogen step-down. Though the distribution of *hetR* gene has been reported to be restricted to the heterocystous cyanobacterial species such as *Anabaena, Nostoc* and *Calothrix* (Buikema and Haselkorn, 1991a), the presence of *hetR* gene sequences in non-heterocystous, filamentous cyanobacteria such as *Symploca* sp. PCC 8002 and *Trichodesmium* spp. (which fix nitrogen aerobically in a light/dark cycle) and *Leptolyngbya* sp. PCC 7310 (which fixes nitrogen anaerobically) has been reported by Janson *et al.* (1998). But the induction of *hetR* in *Symploca* sp. PCC 8002 after nitrogen deprivation did not occur. However, in some of the non-heterocystous, non-diazotrophic, filamentous forms (*Schizothrix calcicola, Oscillatoria lutea, Lyngbya* sp., *Phormidium mucicola* and *A. platensis*) not only the occurrence of *hetR* and *patS* gene sequences but also their functional nature has been demonstrated (Zhang *et al.*, 2009). An interesting case is represented by *Trichodesmium* which fixes nitrogen aerobically in the absence of heterocyst differentiation. Certain morphological and physiological changes in the trichomes of *Trichodesmium erythraeum* IMS101 lead to the formation of groups of cells known as diazocytes. Depending on the length of the trichome, the regions of diazocyte development range up to a maxium of four. Such diazocyte development has been detected in natural populations of *Trichodesmium* as well. Diazocytes show decreased synthesis of cyanophycin granules and increased synthesis of additional membranes (Fredriksson and Bergman, 1995), GS (Carpenter *et al.*, 1992) and cytochrome oxidase levels (Bergman *et al.*, 1993). The requirement of *ntcA* and *hetR* genes for the development of diazocytes in *T. erythraeum* IMS101, their constitutive expression during 12h/12h light/dark cycle and a diurnal cycle between the expression of *hetR* (peaking in dark) and *nifH* (peaking in light) have been substantiated by El-Shehawy *et al.* (2003).

Black *et al.* (1993) found expression of *hetR* gene at very low levels as reflected by the feeble fluorescence of $P_{hetR\text{-}luxAB}$ transcriptional construct in *Anabaena* sp. strain PCC 7120 in all cells derived from nitrogen-replete conditions and localized to proheterocysts or heterocysts in the organism subjected to a nitrogen step-down. Induction of *hetR* begins within 2 h after nitrogen shift-down in all the cells but it increased to 20-fold in the heterocyst-forming cells. They suggested that since induction of *hetR* requires a functional *hetR* gene, the autoregulatory nature of this gene is explained, meaning there by that *hetR* gene is under the control of positive feed-back. Besides its autoregulatory nature, HetR exhibits autodegradable property as evidenced from the recombinant HetR protein overproduced in *E. coli* BL21 (DE3). The presence of serine-protease inhibitors such as phenylmethanesulphonyl fluoride (PMSF) and Dansyl fluoride (DnsF) prevented the protease activity and autodegradation of HetR. That HetR is an unusual serine-type protease is also proved by the absence of autodegradation of recombinant HetR protein of mutant 216 where Ser179, essential for proteolytic activity of HetR, was modified to Asn179, though both inhibitors PMSF and DnsF could label the Ser179Asn-recombinant HetR (Zhou *et al.*, 1998). The requirement of HetR for heterocyst differentiation and Ser152 as the active site for this function has been reported by Dong *et al.* (2000). Of the two mutants isolated by site-directed mutagenesis, $HetR_{Ser142Ala}$ and $HetR_{Ser152Ala}$ of *Anabaena* sp. strain PCC 7120 the former formed heterocysts like the wild-type but the latter did not form heterocysts after nitrogen step-down. However, the mutant synthesized the $HetR_{Ser152Ala}$ protein after a nitrogen step-down very much in a similar manner to the mutant 216 that synthesized $HetR_{Ser179Asn}$ even in the absence of heterocyst differentiation. When compared to wild-type which supported maximum levels of HetR synthesis during early hours (within 6 h), the two mutants continued to synthesize mutant proteins ($HetR_{Ser152Ala}$ and $HetR_{Ser179Asn}$) for 24 h after nitrogen step-down.

Buikema and Haselkorn (2001) turned their attention to studies on expression of *hetR* in *Anabaena* sp. strain PCC 7120. The presence of four tsps in the promoter region of *hetR* located at –184, –271, –696 and –728 positions relative to the start codon of HetR has been demonstrated. In the *hetR* mutant 216, the 1.4 kb and 1.9 kb transcripts were produced at considerably lower levels with apparently no transcripts of 1.5 kb corresponding to the tsp located at -271. HetR produced from -271 tsp appears to be responsible for autoregulation of *hetR* in *Anabaena* sp. strain PCC 7120. They have chosen the *petE* gene that encodes a copper protein, plastocyanin in *S. elongatus* PCC 7942 under inducible conditions in presence of copper for the expression of *hetR* (Ghassemian *et al.*, 1994). Introduction of *hetR* gene under the regulation of *petE* promoter on a multicopy plasmid into *Anabaena* resulted in the production of Mch-phenotype in presence of copper (due to induction of *petE* gene) but in absence of copper (due to repression *petE* gene) the transformant produced normal pattern of heterocyst differentiation. Since *patA* gene regulates the normal pattern of heterocysts and its deletion caused the production of only terminal heterocysts, they introduced a *hetR-gfp* fusion construct into wild-type and *patA* deletion mutant. After a nitrogen step-down, the GFP fluorescence appeared mostly in all cells but later appeared to be restricted to the proheterocysts and heterocysts in the wild-type whereas it is restricted to only in the terminal heterocysts without any fluorescence being detected in vegetative cells in the *patA* deletion mutant. This shows that *hetR* expression is primarily in the differentiating cells leading to heterocyst formation and for localization of HetR protein in the proheterocysts or heterocysts it requires the presence of PatA.

Of the four tsps identified for *hetR*, the tsps located at –728 and –271 were found to be under the influence of NtcA, the regulator for nitrogen metabolism (Muro-Pastor *et al.*, 2002) It would be worthwhile to recall here that the tsp located at –271 is also dependent on HetR, through which autoregulation of *hetR* is reported to occur (Buikema and Haselkorn, 2001). Expression of NtcA was found to be higher in cells subjected to nitrogen step-down than in cells grown in presence of nitrate or ammonium. For this purpose, the tsps located at positions -180 and –49 bp in the promoter region of *ntcA* are put to use. In *ntcA* (CSE2 reported by Frías *et al.*, 1994) or *hetR* (DR884a reported by Black *et al.*, 1993) insertional mutants, the tsps at positions –180 and –49 of *ntcA* are not utilized. Muro-Pastor *et al* (2002) thus concluded that there is a mutual dependence of the two regulatory genes *ntcA* and *hetR*. This has been explained by the timing of expression of the two genes. In the wild-type, *hetR* expression begins in just less than 3 h and assumes maximum levels by 4.5 to 9 h whereas *ntcA* transcripts begin to appear only after 4.5 and 6 h after nitrogen deprivation. However, in *ntcA* mutant the *hetR* transcripts did not increase in response to nitrogen deprivation. Since *ntcA* expression continues even in organisms grown on fixed nitrogen sources such as nitrate or ammonia, the basal levels of NtcA present in cells after nitrogen step-down suffice the purpose for induction of *hetR* gene. Higher levels of *ntcA* transcripts required for heterocyst differentiation are induced in turn by *hetR* gene, a situation that is completely lacking in *hetR* mutant Thus the mutual dependence of NtcA and HetR is confirmed. Rajagopalan and Callahan (2010) analysed the utility of the four tsps of *hetR* by separating the promoter region into five segments and each of these having one or more tsp fused to *gfp* was individually introduced into wild-type *Anabaena* sp. strain PCC 7120. Of the four tsps, tsp located at –271 appeared to be tightly regulated and more widely used during early phases of heterocyst differentiation. This is evident by the expression of *–271 tsp-gfp* fusion construct in the wild-type during the first 4 h in spatially patterned cells destined to differentiate into heterocysts after a nitrogen step-down. The wild-type strain bearing the *–728/-696 tsp-gfp* fusion constructs showed significant levels of fluorescence in all cells in nitrogen-replete and -starved conditions. Similarly, *–184 tsp–gfp* fusion construct exhibited fluorescence uniformly in all cells. In a *patA* deletion mutant the expression of *271-lacZ* fusion construct was down-regulated which correlated with reduced heterocyst frequency where only terminal heterocysts differentiated but the transcripts from -271 tsp of *hetR* were completely lacking in strains overexpressing HetN and PatS, the two negative regulators of heterocyst differentiation. Important features of heterocyst differentiation such as commitment of cells to differentiate, the number of cells that undergo differentiation and the timing of these all depended on –271 tsp of *hetR* as its deletion caused a delay in all these parameters.

Structural studies on HetR of *Anabaena* sp. strain PCC 7120 showed that it exists *in vivo* as a homodimer by the formation of a disulphide bond between Cys48 residues of the two monomers. A mutant of *hetR* with a change in the Cys48 to Ala48, isolated after site-directed mutagenesis, lost the ability for HetR dimerization as well as heterocyst differentiation. So HetR dimerization is required for heterocyst differentiation. The mutant protein $HetR_{Cys48Ala}$ retained its active Ser residue and so its proteolytic activity. In order to show the importance of HetR homodimer formation for heterocyst differentiation, Huang *et al.* (2004) introduced the mutant $hetR_{Cys48Ala}$ gene into a $hetR^-$ mutant 884a of *Anabaena* sp. strain PCC 7120 reported earlier by Black *et al.* (1993) to generate a mutant C48. Mutant C48 produced $HetR_{Cys48Ala}$ protein that did not form a dimer *in vivo*. The isolated $HetR_{Cys48Ala}$ protein monomers did not interact with each other as well as with wild-type HetR monomers *in vitro*. Upon nitrogen step-down, mutant C48 failed to form heterocysts. When the $hetR_{Cys48Ala}$ gene construct was introduced into wild-type on a multi-copy plasmid, the resulting strain differentiated heterocysts with a normal pattern suggesting that the expression of mutant gene did not interfere with normal HetR function. Huang *et al.* (2004) also adduced evidences to show that HetR homodimer is a DNA-binding protein and regulates the activity of *hetR*, *hepA* and *patS* by binding to their promoter regions during early phases of heterocyst differentiation. The up-regulation in the activity of *hetR* required for heterocyst differentiation is prevented by the C-terminal pentapeptide (RGSGR for Arg-Gly-Ser-Gly-Arg) sequence of PatS due to its inhibition of the DNA-binding activity of HetR. This is how PatS or the synthetic RGSGR pentapeptide (at 1 μM) inhibits the formation of heterocysts in *Anabaena* sp. strain PCC 7120 as reported by Yoon and Golden (1998). For example, *patS* deletion causes an aberrant pattern with Mch-phenotype after nitrogen step-down (Yoon and Golden, 1998). Like PatS, another negative regulator of heterocyst differentiation is HetN. When *hetN* expression is switched off due to non-expression of $P_{petE\text{-}hetN}$ in the absence of copper (a situation resembling *hetN* deletion) it caused the formation of first a normal pattern followed by Mch-phenotype during the subsequent rounds of heterocyst differentiation (Callahan and Buikema, 2001). Both PatS and HetN act independently but have complementary function and help in the establishment and maintenance of the pattern. Their non-expression leads to an aberrant pattern as well as Mch-phenotype. In this background, the characterization of a mutant S2-45, isolated upon Tn5-1058 insertion in *Anabaena* sp. strain PCC 7120 genome, showed that it formed very long, irregularly spaced strings of heterocysts. Since attempts for the reconstruction of the mutant by introducing the Tn5 inserted region of S2-45 into wild-type failed, Khudyakov and Golden (2004) cured the plasmid from S2-45 by growing it in the absence of the antibiotics and the resulting strain (AMC1285) showed the original Mch-phenotype. In order to make conjugative transfer possible through different pDU-1-based plasmids carrying Nm^r marker, the original Nm-Bm-Sm resistance genes in the chromosome of S2-45 mutant have been replaced by Sp-Sm resistance cassette. The resulting strain AMC1286 exhibited Mch-phenotype. Since extra copies of *hetR* in the wild-type cause the formation of Mch-phenotype, it would be interesting to know the response of an already Mch-phenotype (AMC1286) to the expression of *hetR* on a multicopy plasmid in it. The introduction of plasmid pWB216S2.4 with Nm resistance marker and a wild-type *hetR* gene into AMC1286 resulted in the formation of a regular pattern of heterocyst differentiation. This led them to the conclusion that the mutation in the original S2-45 had been a recessive one in *hetR* gene. The cloning and sequencing of the *hetR* gene from AMC1285 (the cured strain) revealed a mutation at position 223 amino acid residue from Arg to Trp due to a C toT transition at 667 nucleotide in the *hetR* gene. Khudyakov and Golden (2004) cloned the mutant gene without its upstream regulator region and introduced it through a suicide plasmid into wild-type thus enabling it to undergo homologous recombination and locate itself into *hetR* locus, producing a merodiploid (AMC1287). The AMC1286 and AMC1287 strains have been shown to be insensitive to the signals of PatS and HetN because overexpression of *patS* did not cause any inhibition in heterocyst differentiation. The expression of *patS* gene exclusively in the heterocysts of AMC1286 and AMC1287 as revealed by the use of *gfp* as reporter gene suggests that the mutant $\text{HetR}_{\text{Arg223Trp}}$ protein did not interfere with *patS* expression. Likewise, *hetN* overexpression in AMC1287 failed to suppress heterocyst differentiation in nitrogen-free medium. On top of all the above, Khudyakov and Golden (2004) generated a mutant AMC1288 by the overexpression of $hetR_{Arg223Trp}$ in $hetR_{Arg223Trp}$ background (in AMC1286). After nitrogen step-down, AMC1288 differentiated all its vegetative cells in to heterocysts thus forming 100% heterocysts where signalling mechanism of *hetR* to and from *patS* and *hetN* was completely lost. Thus the above studies tell us that HetR is the master regulator of heterocyst differentiation and has different functions that can be separated by mutation (Khudyakov and Golden, 2004). It autoregulates its own levels, it induces the expression of *ntcA*, it activates the transcription of early genes and sends and receives signals to and from PatS and HetN.

In view of the interdependence of *ntcA* and *hetR* on the expression of one another during heterocyst differentiation in *Anabaena* sp. strain PCC 7120 (Muro-Pastor *et al.*, 2002), the effects of overexpression of *ntcA* were examined by transferring *ntcA* gene into the wild-type *Anabaena* sp. strain PCC 7120 and its *hetR* mutant (isolated by Buikema and Haselorn, 1991b). The two new strains generated have been designated as CSEL1 and CSEL2, respectively and constitutive overexpression of *ntcA* was found in the filaments grown in ammonium and subjected to nitrogen step-down. The content of NtcA protein in strains CSEL1 and CSEL2 was at least 25 times the amount of NtcA protein in the wild-type grown on ammonium medium. When CSEL1 was subjected to nitrogen step-down, heterocyst differentiation occurred with the expression of NtcA-dependent genes such as *devBCA* operon, or *coxII* operon and *xisA* gene (for excision of 11 kb intervening fragment in *nifD*) but the organism could not grow in nitrogen-deficient medium. On the other hand, CSEL2 strain exhibited the expression of *devBCA* operon and excision of 11 kb element even in presence of ammonium medium. These results thus emphasize that some of the NtcA-dependent genes can be expressed despite the absence of *hetR* gene (Olmedo-Verd *et al.*, 2005). In this connection, it would be appropriate to mention here that HetR induces the expression of six regulatory genes in response to growth in nitrogen-deficient medium. These are *hetP* (Fernandez-Pinas *et al.*, 1994), *hetC* (Khudyakov and Wolk, 1997), *devA* (Cai and Wolk, 1997), *ntcA* (Muro-Pastor *et al.*, 2002), *hetR* (Huang *et al.*, 2004) and *patS* (Huang *et al.*, 2004). The mutual regulation of *ntcA* and *hetR* during heterocyst differentiation required the presence of two similar PP2C-type protein phosphatases. PrpJ encoded by *prpJ* (*all1731*) described by Jang *et al.* (2007) constitutes a new control point for heterocyst maturation in *Anabaena* sp. strain PCC 7120, since a disruptant mutant (S20) of *prpJ* failed to develop the glycolipid laminated layer. Jang *et al.* (2009) identified another ORF (*all2470*) and redesignated the two ORFs *all1731* and *all2470* as *prpJ1* and *prpJ2*, respectively and their corresponding gene products PrpJ1 and PrpJ2 together are required for mutual regulation of *ntcA* and *hetR*. Although *prpJ2* is nonessential for diazotrophic growth, a double mutant for *prpJ1prpJ2* was unable to grow diazotrophically and resembled in its properties with the S20 mutant of *prpJ1*. The double mutant showed decreased transcripts for *ntcA* and *hetR* and accordingly the proteins NtcA and HetR when compared to the wild-type after a nitrogen step-down. A consensus NtcA-binding site (GTAGCCACAGGTAC) upstream of tsp in the promoter region of *prpJ2* has been recognized and a change in this sequence to CCGGCCACAGGCA resulted in a loss of NtcA-binding. This emphasizes the role of NtcA in regulating the expression of *prpJ2*. Another significant feature is the absence of induction of *prpJ2* in a *hetR* mutant. A nitrogen responsive regulator, NrrA encoded by *nrrA* (ORF *all4312*) has been described by Ehira and Ohmori (2006a). The expression of *nrrA* in the *prpJ1prpJ2* double mutant was unaffected so also the expression of *prpJ2* remained unaffected in a *nrrA* mutant. Taking these into consideration, Jang *et al.* (2009) concluded that for interaction of NtcA and HetR, the presence of PrpJ1 and PrpJ2 is required.

Direct regulation of heterocyst differentiation by HetR in *Anabaena* sp. strain PCC 7120 has been also determined by an epistasis analysis conducted by Orozco *et al.* (2006). They created a *hetR* null mutant (UHM103) that showed a Het$^-$ phenotype under all conditions tested and UHM103 is identical to the previously described *hetR*$^-$ mutants (Buikema and Haselkorn, 1991b; Black *et al.*, 1993; Dong *et al.*, 2000). By utilizing UHM103, they produced double mutants, i.e. *hetR*$^-$*patA*$^-$, *hetR*$^-$*hetN*$^-$ and *hetR*$^-$*patS*$^-$ which also showed a Het$^-$ phenotype identical to UHM103. This confirms that HetR directly regulates heterocyst differentiation and inactivation of *hetR* leads to epistatic effect on the genes tested. Risser and Callahan (2007) employed error-prone PCR amplification procedure for introducing mutations in the promoter and coding sequence of *hetR* of *Anabaena* sp. strain PCC 7120 and introduced the mutated *hetR* gene sequences into UHM103 in a het$^-$ background by conjugation. The exconjugants selected were further characterized for their growth and heterocyst differentiation in a nitrogen-deficient medium. Thirty-three mutants having amino acid substitutions from residue 33 to 150 at N-terminus of the HetR were unable to survive nitrogen deprivation and differed in heterocyst frequencies to less than 2%. Out of three amino acid substitutions in *hetR* gene, i.e $hetR_{Cys48Ala}$, $hetR_{Ser152Ala}$ and $hetR_{Ser179Asn}$, Risser and Callahan (2007) showed that $hetR_{Ser152Ala}$ and $hetR_{Cys48Ala}$ gene sequences on a plasmid could complement the mutant UHM103, restoring the heterocyst frequencies almost at par with that of the wild-type. This shows that these two amino acid residues Cys48 and Ser152 are not essential for heterocyst differentiation, in contrast to their essentiality reported earlier (Dong *et al.*, 2000; Huang *et al.*, 2004). Similarly, mutant 216 of *Anabaena* sp. strain PCC 7120 that had a mutation in $hetR_{Ser179Asn}$ (Dong *et al.*, 2000; Huang *et al.*, 2004) did not differentiate heterocysts and in the experiments of Risser and Callahan (2007) $hetR_{Ser179Asn}$ did not complement UHM103. Mutants with amino acid substitutions $hetR_{Asp17Glu}$ and $hetR_{Gly36Ala}$ caused a Het$^-$ phenotype whereas $hetR_{His69Tyr}$ substitution resulted in heterocyst differentiation to less than 0.1% of cells. These three amino acid substitutions were found to be essential for normal heterocyst differentiation and levels of HetR turnover. The other important findings are that (i) Residue Asp17 of HetR is important for DNA-binding activity but not Cys48 of HetR as has been reported earlier for this activity by Huang *et al.* (2004); (ii) the amino acid substitution in $hetR_{Cya48Ala}$ prevented the formation of HetR homodimer formation, thus confirming the earlier observations of Huang *et al.* (2004) and HetR homodiner formation has been observed in the wild-type, and other HetR amino acid substitutions such as $HetR_{Asp17Glu}$, $HetR_{Gly36Ala}$ and $HetR_{His69Tyr}$; (iii) replacement of 2 bp in the *hetR* coding region in the wild-type of *Anabaena* sp. strain PCC 7120 to cause a Cys48Ala substitution did not affect heterocyst differentiation; (iv) the wild-type HetR protein and the recombinant proteins produced (in respect of the amino acid substitutions in *hetR* mentioned in this study) in the experiments of Risser and Callahan (2007) were all devoid of the protease activity and accordingly their autodegradation *in vitro* was absent; and (v) lastly, according to Risser and Callahan (2007) we are presently not in a position to comprehend the intricate relationship between HetR protein turn over and protease activity on the one hand and protease activity of HetR and heterocyst differentiation on the other.

***iii) hetP*:** Mutation in P6 of *Anabaena* sp. strain PCC 7120 caused a Het$^-$ phenotype (Ernst *et al.*, 1992). Cloning and sequencing of the corresponding wild-type ORF located at approximately 3.44 Mb on the physical map of the chromosome revealed it to be an ORF of 474 or 525 bp encoding a protein of 159- or 175- amino acid residues, respectively that has no match in the databases. Fernandez-Pinas *et al.* (1994) named this ORF in P6 as *hetP*. Although reconstruction of the mutation has been successful, the wild-type *hetP* sequence could not complement the mutant until it is associated with an additional ORF situated at the 3′-end of *hetP*. It could not be conclusively ascertained whether the Het$^-$ phenotype of P6 has been caused due to loss of *hetP* as such or due to a polar effect of the Tn5 insertion or both. When *luxAB* has been used as the reporter gene to study transcription, mRNA of *hetP* accumulated 2.5 fold higher in a double recombinant within 6 h when compared to 12 h taken by the single recombinant. However, transformation of P6 or wild-type strain with a plasmid (pDU1) bearing *hetP* caused Mch-phenotype in both the transformants. In this respect, the effect of ectopic expression of *hetP* resembled the effects of such an expression of *hetR*, the master regulator of heterocyst differentiation. Indeed, Higa and Callahan (2010) demonstrated that the expression of *hetP* on a replicative plasmid could even bypass the requirement for *hetR*, when a *hetR* deletion mutant was transformed with a plasmid bearing *hetP* gene sequence. Such a transformant differentiated heterocysts that appeared normal but possessed nitrogenase activity under anaerobic conditions despite the production of Hgls and envelope polysaccharides. A *hetP* deletion mutant differentiated very few heterocysts and they further showed that there are two homologues of *hetP* represented by ORFs *asl1930* and *alr2902* that could readily complement the *hetP* deletion mutant. This is in sharp contrast to the failure of wild-type *hetP* gene sequence to complement *hetP* disruptant mutant (Fernandez-Pinas *et al.*, 1994). The use of *hetP-gfp* as a reporter gene in strains lacking HetR, PatA or HetF showed GFP fluorescence typical of HetR fluorescence after a nitrogen shift-down. The presence of five tsps located at –727, –545, –208, –177 and –12 bp in the promoter region of *hetP* and a probable HetR-binding region of a seven base pair invert repeat between –727 and-545 tsps confirmed the regulation of *hetP* by HetR. Haselkorn (2010) examined the role of *hetP* as a new player in the cascade controlling heterocyst differentiation mediated by HetR.

***iv) hetC*:** Khudyakov and Wolk (1997) identified *hetC* gene while characterizing the transposon Tn5-1058 induced mutant C3 of *Anabaena* sp. strain PCC 7120 that did not differentiate heterocysts after a nitrogen step-down. The transposition of Tn5-1058 in an ORF designated by them as *hetC* encodes a protein of 1,044 amino acid residues. It is located 1,165 bp 5′-end of *hetP*. The HetC protein exhibited a similarity to Hly family of bacterial protein exporters in its two-thirds C-terminal sequence while the rest of the N-terminal portion has no similarity to any protein in the database. Functional copies of *hetC* supplied in trans position in C3 mutant on a multicopy plasmid complemented *hetC* and restored the Fox$^+$ and Het$^+$ phenotype of the wild-type. The participation of *hetC* in the regulation of heterocyst differentiation has been confirmed by its expression during the first 4 h after nitrogen step-down. The presence of a canonical NtcA-binding sequence (GTAACATGAGATAC) upstream of the tsp located at -571 bp in the promoter region of *hetC* lends support to the suggestion that *hetC* is another gene that is activated by NtcA (Muro-Pastor *et al.*, 1999). Further, in an *ntcA* interrupted mutant (CSE2) the expression of *hetC* has been impaired. In a *hetC* disruptant mutant (DR1653), the induction of *hetR* by NtcA is not disturbed and is similar to wild-type cells indicating that *hetR* induction by NtcA is independent of the expression of *hetC*. The accumulation of *hetC* transcripts in the wild-type was evident by 6 h after nitrogen step-down which reached its peak by 24 h by which time mature heterocysts begun to appear. Since HetC is a protein of the ABC type transporters, it has been suggested to be involved in the export of an inhibitor of heterocyst differentiation as originally suggested by Khudyakov and Wolk (1997). Xu and Wolk (2001) showed that *hetC* gene is very strongly expressed in the small cells that resolve themselves into proheterocysts as reflected by the GFP fluorescence when *hetC-gfp* was used as a reporter gene. This is akin to the expression of *hetR-gfp* in the small cells. The importance of *hetC* can be assessed by the number of genes up-regulated by it after a nitrogen step-down in *Anabaena* sp. PCC 7120. Wang and Xu (2005) constructed vectors with the promoter regions of *hetR, hetC, hepB, hepK, devB. hglD, hglE* (*alr5351*), *hetN, patS, patA, patB, xisA, nifB, coxBII* and *rbcL* with *gfp* as the reporter gene and transformed the wild-type and *hetC* disruptant mutant of *Anabaena* sp. strain PCC 7120. Except for *rbcL* that is down-regulated, the rest of all other genes under study were up-regulated specially in developing proheterocysts and heterocysts in the wild-type. The up-regulation of *hepB, hglD, xisA, nifB* and *patA* was reported here for the first time. In the *hetC* disruptant mutant, the up-regulation of *hepB, hepK devB, patS, hetN* and *coxBII* was noted in the proheterocysts while *hglD, hglE, patB, nifB* and *xisA* were not expressed in the the smaller dividing cells that tend to resolve into proheterocysts. However, *patA* is the only gene that is down-regulated in all cells of the *hetC* disruptant mutant. There are subtle differences in the genes expressed in the *hetC* mutant, while the expression of *hepB* and *devB* is delayed, the expression of *hglD, hglE, patB, xisA* and *nifB* depended on *hetC*. In this connection, it may be noted that the expression of *hepA* also depended on *hetC* (Xu and Wolk, 2001). Other pattern formation or maintenance genes, *patS* and *hetN* are expressed in *hetC* mutant to the same extent as they are expressed in the wild-type. Another significant aspect of *hetC* regulation is the expression of a number of cell division genes (*ftsE, alr1706; ftsW, all0154, ftsX, all1757, ftsY, all1759, ftsZ, alr3858; minCDE, alr3455-alr3456-asr3457; sulA, all2390* and two other ORFs *all2033* and *all2797*) in the wild-type after 24 h of nitrogen step-down. Using *gfp* as reporter gene, the expression of all these cell division genes in *hetC* mutant showed that the *ftsZ* gene is up-regulated in proheterocysts that continued to divide more actively than vegetative cells. But the expression of *sulA* appears to be not affected in the *hetC* mutant suggesting that it is not under the regulation of *hetC*. Muro-Pastor *et al.* (2009) identified a second tsp located at –293 in the promoter region of *hetC* that is NtcA activated but dependent on HetR. The first tsp located at –571 in the promoter region of *hetC* has already been described that is NtcA activated but independent of HetR (Muro-Pastor *et al.*, 1999). Similarly, the promoter region of *devBCA* operon, expressed during maturation phase of heterocyst differentiation, also possesses a second tsp located at -454 bp that is NtcA activated and dependent on HetR. The first tsp located at -704 bp in the promoter region of *devBCA* operon has been described earlier (Cai and Wolk, 1997; Fiedler *et al.*, 2001). Thus it is interesting to know that both *hetC* and *devBCA* operon, the former exerting its effects during early phases and the latter exerting its effects in late phases of heterocyst differentiation possess similar tandem promoter arrangement.

v) hetF: During the isolation and characterization of transposon (Tn5-1063) insertion Fox^- mutants of *N. punctiforme* strain ATCC 29133, out of 69 short-listed mutants with a probable Fox^- phenotype, six have been selected for their inability to grow in a nitrogen-deficient medium and differentiate heterocysts. In one of these mutants (UCD415), the transposon insertion has been found in an ORF bearing 92% similarity to *hetR* gene of *Anabaena* sp. strain PCC 7120 but in the rest five of them the transposon insertion occurred in a new gene sequence that has been named by Wong and Meeks (2001) as *hetF*. The *hetF* gene sequence is 2.4 kbp long and the deduced amino acid sequence of the

HetF protein resembled a capsase-hemoglobinase type of protease (Aravind and Koonin, 2002). The wild-type and *hetF* disruptant mutant (UCD416) have been transformed with a multicopy plasmid (pSCR60) carrying *hetF* gene to generate strains UCD487 and UCD488, respectively. The presence of the multicopy plasmid with functional *hetF* gene not only restored the ability of UCD416 to differentiate heterocysts but also resulted in Mch-phenotype upon a nitrogen step-down. In this respect, the effects of inactivation of *hetF* and the presence of *hetF* in multiple copies resembled the effects of *hetR*. The absence of transcripts of *hetF* in UCD416 cells after nitrogen step-down as evidenced by the unchanged luciferase activity when *luxAB* gene was used as a reporter gene indicates that *hetF* is constitutively expressed. Moreover, the *hetR* gene expression, which is found in the wild-type within 3 h of nitrogen step-down with a maxiumum at 12 h, has been found to be delayed in UCD416 till 6 to 12 h and the maximum levels of *hetR* transcripts appeared only after 24 to 36 h. Further, transcriptional fusion of *hetR* with *gfp* showed GFP fluorescence mainly in the heterocysts after a nitrogen step-down in wild-type but in UCD416 when *hetR-gfp* was expressed on a multicopy plasmid, the HetR-GFP fusion protein appeared in all cells under nitrogen enriched medium. These results signify that *hetF* cooperates with *hetR* in a positive regulatory pathway of heterocyst differentiation in *N. punctiforme* strain ATCC 29133 (Wong and Meeks, 2001). The necessity of *hetF* homologue in *Anabaena* sp. strain PCC 7120 for nitrogen fixation has been demonstrated (Wolk *et al.*, 2007). PatS controls the formation of intercalary heterocysts and so the pattern formation in *Anabaena* sp. strain PCC 7120. Since PatS also regulates the transcription and HetR activity much in a similar manner as *hetF*, the combined effects of *hetF* and *patS* on the *hetR* transcription and activity of HetR has been examined by Risser and Callahan (2008). The main findings of their study are summarized here: (i) the *hetF* gene (*all1730*) of *Anabaena* sp. strain PCC 7120 is similar to *hetF* from *N. punctiforme* ATCC 29133; (ii) deletion of *hetF* in *Anabaena* sp. strain ATCC 7120 resulted in a Het^- phenotype (UHM130); (iii) since HetF belongs to CHF class of cysteine proteases, the active site His201 and Cys246 residues at the respective codons have been mutated to the codons of Tyr and Ala, respectively and the mutated *hetF* gene sequences have been introduced into UHM130. Due to failure of the two mutations to complement UHM130, the protease nature of HetF has been confirmed; (iv) HetR protein levels in *hetF* and *PatA* single and double mutants were found to be higher than in the wild-type of *Anabaena* sp. strain PCC 7120; (v) the increased levels of HetR in *hetF* and *patA* deletion single and double mutants has been found to be a post-transcriptional event and (vi) The necessity of *hetF* for transcription of *hetR* from –271 tsp indicates that *hetF* is required for positive autoregulation of *hetR* and its localization in cells that differentiate into heterocysts. Risser and Callahan (2008) confirmed this by changing the expression profile of *hetR* from its native promoter into a transposition carried on a multicopy plasmid with two independent inducible promoters of *petE* (inducible in presence of copper) and *nirA* (inducible in presence of nitrate). The corresponding constructs were then introduced into strains having either *hetF*, *patA* or *hetF* and *patA* deleted and also *hetR* deleted strains. The levels of HetR in wild-type and *hetR* deletion mutant increased due to expression under inductive conditions in presence of copper and nitrate and this in turn caused Mch-phenotype as reported earlier (Buikema and Haselkorn, 2001). Likewise, the levels of HetR in *hetF* and/or *patA* deletion mutants also increased suddenly under both the inductive conditions. Since transcription of *patS* and *hetR* is dependent on *hetR*, the effects of deletion of *hetF* on the patterned induction of *patS* and *hetR* was examined by introducing transcriptive fusions of *gfp* with the promoters of *patS* and *hetR* in *hetR* deletion, *hetR* mutant (Ser179Asn) and *hetF* deletion strains. The appearance of green fluorescence due to expression of $P_{hetR\text{-}gfp}$ fusion after nitrogen step-down was restricted to only heterocysts in the wild-type whereas it was uniformly seen in all vegetative cells in the other three strains. Likewise, $P_{patS\text{-}gfp}$ fusion was also expressed in a patterned manner that is restricted to heterocysts in the wild-type but not in the other three strains where it was uniformly seen in all vegetative cells. The above discussion amply brings into focus the positive regulation of *hetR* by HetF and its localization in the differentiating cells leading to heterocyst formation.

***vi) hetL*:** The pattern of heterocyst differentiation in *Anabaena* sp. strain PCC 7120 seems to be regulated by PatS which diffuses as an inhibitor forming a gradient from the heterocysts/proheterocysts and is involved in cell to cell signalling. The presence of a C-terminal pentapeptide (RGSGR) sequence and the role of synthetic pentapeptides in inhibition of heterocyst pattern formation have been emphasized (see under section PatS). While deciphering the sequence of molecular events that are involved in PatS signalling pathway, Liu and Golden (2002) identified a new gene *hetL* (*all3740*) that is suggested to play a non-essential role because a *hetL* null mutant differentiated heterocysts normally and exhibited diazotrophic growth. HetL exclusively possesses pentapeptide repeats with a consensus sequence of Ala(Asp/Asn)LeuX* (*representing a polar amino acid). Wild-type *Anabaena* sp. strain PCC 7120 transformed with plasmid pAM2065 carrying *hetL* gene resulted in Mch-phenotype (strain LD115). The frequency of heterocysts in strain LD115 was 18% in nitrate-enriched medium that is twice that of the wild-type in a nitrogen-deficient medium. Strain LD115 differentiated 20% and 34% heterocysts after 24 and 48 h, respectively after nitrogen step-down. Studies on *hetL* overexpression in *hetR* (strain 216), *ntcA* (AMC236) and *hetC* (DR1653) deletion strains showed that inactivation of *hetR* and *hetC* is epistatic to *hetL* overexpression with complete absence of heterocyst differentiation. In AMC236, *hetL* overexpression forced heterocyst differentiation to the level of proheterocyst formation. The presence of ammonium or PatS-5 pentapeptide in medium could not repress the formation of heterocysts when *hetL* is overexpresssed. The expression of *hetL-gfp* translational fusion in *Anabaena* sp. strain PCC 7120 showed fluorescence of GFP in vegetative cells as well as heterocysts. So the expression of *hetL* is not cell-specific.

***vii) hetY*:** Another gene that influences heterocyst development, designated as *hetY* (*alr2300*) which is adjacent to *patS* gene (*asl2301*) in *Anabaena* sp. strain PCC 7120 has been discovered by Yoon *et al.* (2003). Overexpression of *hetY* caused partial suppression of heterocyst formation which led to increase in the heterocyst spacing and disturbance in the original pattern. The *hetY* gene encodes a protein that could be an ATPase or kinase. Yoon *et al.* (2003) suggested that HetY could be an ATPase that is based on its phylogenetic relationship with YjeE ATPase from *Haemophilus influenzae*. HetY and YjeE exhibit 31.6% identity and 62.4% similarity and both of these proteins possess a glycine-rich P-loop region characteristic of ATPases. YjeE homologues in many bacterial genomes are adjacent to *amib*, a gene that is involved in cell wall synthesis. Based on these circumstantial evidences, the authors concluded that HetY could be involved in cell wall synthesis since *hetY* inactivated strains required more time for maturation of heterocysts and showed abnormalities in heterocyst morphology.

***viii) hetZ*:** Another gene cluster that regulates heterocyst differentiation and pattern formation in *Anabaena* sp. strain PCC 7120 has been discovered (Zhang W. *et al.*, 2007). Gene *hetZ* (*alr0099*) is located at the 5′-end of a pair of genes, *patU5* and *patU3*. In fact *patU5* (*asr0100*) and *patU3* (*alr0101*) of *Anabaena* sp. strain PCC 7120 represent 5′-and 3′-end portions of *patU* of *N. punctiforme*. The homologues of the above three genes are found in other filamentous heterocystous (*A. variabilis* ATCC 29413, *N. punctiforme* ATCC 29133, *Nodularia spumigena* CCY 9414) and filamentous non-heterocystous cyanobacteria that fix nitrogen under aerobic (*T. erythraeum* IMS 101) and microaeropihilic (*Lyngbya* sp. PCC 8106) conditions. A re-examination of the mutant 1801 (in which the transposition of Tn5-10876 occurred at 373 bp from the 5′-end of ORF *alr0099*) described by Ning and Xu (2004b) as a

Het⁻ phenotype now produced only 2–8% proheterocysts without shaking the cultures but upon shaking the cultures it neither produced heterocysts nor a pattern of phycobilisome fluorescence. Likewise, in a second mutant DRHB2.14 (in which the Tn903 got inserted at 615 bp from 5′-end of ORF *alr0099*) reported by Ning and Xu (2004b) and in the mutant *hetZ::C.K2* the differentiation of heterocysts proceeded with a normal pattern after nitrogen step-down but mature heterocysts appeared less than 1%. Similarly, FQ422 mutant (with insertion of transposon at 58 bp upstream of the 3′ end of the ORF *alr0099*) described by Fan *et al.* (2005) also had a similar phenotype. Because of the essential nature of *alr0099* for heterocyst differentiation it was designated as *hetZ* by Zhang *et al.* (2007). The up-regulation of *hetZ, patU5* and *patU3* was noted in proheterocysts and heterocysts. A mutant of *patU3* showed Mch-phenotype but when introduced into a *patA⁻* mutant it restored the formation of intercalary heterocysts. The phenotype of *patU3* mutant resembles the phenotype of a *patS* deletion mutant. But PatU3 does not resemble either PatS (Yoon and Golden, 1998) or for that matter HetN (Li *et al.*, 2002) in having the pentapeptide RGSGR. A very weak expression pattern of *hetC* in mutants *hetZ::C.K.2* and *hetZ::Tn5-10876* and a very strong expression of hetR in the cells of the latter mutant without any regular pattern are suggestive of an important role for the gene cluster *hetZ-patU5-patU3.*

***ix) asr1734*:** Gene *asr1734* of *Anabaena* sp. strain PCC 7120 encodes a protein Asr1734 of 93 amino acid residues that negatively regulates heterocyst differentiation Wu *et al.* (2007) . Asr1734 inhibited heterocyst formation when present in extra copies or when it is overexpressed in wild-type or in strains AMC451 (mutant of *patS* with *asr1734* overexpressed) and AMC1286 (mutant of $hetR_{Arg229Trp}$ with *asr1734* overexpressed) that exhibited Mch-phenotype. But in the wild-type overexpression of *asr1734* brought about degradation of phycobiliproteins in the vegetative cells. The expression of $P_{asr1734\text{-}gfp}$ as a reporter in the wild-type after a nitrogen step-down was restricted to the developing proheterocysts or heterocysts. Deletion of *asr1734* from the chromosome by the introduction of a suicide plasmid pAM3234 by homologous recombination resulted in a strain designated as AMC1252. This strain showed a constitutive formation of heterocysts (2%) in nitrate medium but differentiated high frequency of heterocysts (15%) after a nitrogen step-down. Ammonium grown cultures of AMC1252 when shifted to nitrate medium produced 5% heterocysts. Transcripts for *ntcA* and *hetR* increased in AMC1252 and in an *asr1734* overexpression strain after 6 h of nitrogen step-down. It is thus concluded that Asr1734 inhibits heterocyst differentiation downstream of *hetR*. Asr1734 has been predicted to have α-helical structure and the C-terminal region contains several basic amino acids. Mutation of Arg84 and Arg86 resulted in a double mutant Arg84Gly/Arg86Gly and such mutated *asr1734* when introduced into wild-type there was no inhibition of heterocyst differentiation showing there by the requirement of C-terminal region for stability or function of Asr1734 Wu *et al.* (2007). It is to be noted that homologues of *asr1734* are not present in unicellular cyanobacteria or the non-heterocystous filamentous *T. erythraeum* but database searches revealed its orthologues to be present in only heterocystous forms like *A. variabilis* ATCC 29413 and *N. punctiforme* PCC 73102. However, in *N. punctiforme* PCC 73102 it is considered to be an orphan gene and the predicted crystal structure of the gene product Npun_R1517 revealed an extensively interlocked homodimer structure with a number of clefts and cavities (Ni *et al.*, 2009).

B) Late genes

i) ***Development genes*:** The genes that are expressed considerably at late stages (around 12 to 14 h after nitrogen step-down) of heterocyst development and maturation are classified as development genes. In this category are included genes that are required in the synthesis and transport of Hgls and envelope polysaccharides, which are deposited one after another in the form of a laminated layer and heterocyst envelope, respectively at the time of maturation of heterocyst. The characterization of a Tn5-1063 insertion mutant, M7 (Fox^-, Het^+ and Hen^-) revealed that it is deficient in aerobic nitrogen fixation though differentiates heterocysts that are devoid of both normal polysaccharide layer and heterocyst-specific glycolipids (Ernst *et al.*, 1992). Maldener *et al.* (1994) recovered the Tn5-inserted chromosomal segment from M7 and found that due to its insertion a 732-bp ORF has been inactivated. This has been designated by them as *devA* (for development). Wild-type *Anabaena* sp. strain PCC 7120 subjected to directed mutagenesis of *devA* gene also yielded a mutant phenotype that resembled M7. The wild-type gene sequence of *devA* on a shuttle vector could also complement the M7 mutant. Sequencing of *devA* and the deduced amino acid sequence of DevA protein (244 amino acids) showed it to be a component of a periplasmic permease required for heterocyst maturation. DevA protein showed a strong similarity to ATP-binding subunit of the traffic ATPases. Two additional genes, *devC* and *devB* are present upstream of *devA*, and constitute a *devBCA* operon (Fiedler *et al.*, 1998). *devC* (earlier designated by Maldener *et al.*, 1994 as *OrfA*) occurs 1800 bp upstream of *devA* and is 1154 bp long that encodes a membrane-spanning protein (of 384 amino acids long) of ABC transporters. *devB* is 1425 bp long and present upstream of *devC* separated by 41 bp that encodes a membrane fusion protein (of 474 amino acids length) of ABC exporters. Although M7 synthesizes heterocyst-specific glycolipids, these are not transported to the envelope to form the laminated

Table 2: Development genes expressed at late stages of heterocyst differentiation in *Anabaena* sp. strain PCC 7120.

Gene	Gene product and time of appearance	Role	Mutant phenotype	Reference(s)
devBCA operon (dev for development) *devB* (*alr3710*) 1425 bp long	Membrane fusion protein of 474 amino acids long; 12 to 14 h of N step-down	Heterocyst maturation	Fox^-, Het^+, Hen^-	Fiedler *et al.* (1998)
*devC** (*alr3711*) 1154 bp long	Membrane spanning protein of ABC transporters of 384 amino acids long; 12 to 14 h of N step-down	Heterocyst maturation	Fox^-, Het^+, Hen^-	Fiedler *et al.* (1998)
devA (*alr3712*) 732 bp long	Periplasmic permease of 244 amino acids long; 12–14 h of N step-down	Heterocyst maturation	Fox^-, Het^+, Hen^-	Maldener *et al.* (1994)
devH (*alr3592*)	DNA binding protein/ transcriptional regulator; 29 kDa protein bears resemblance to cAMP receptor protein and nitrogen regulator NtcA; at 24 h a major transcript of 1.0 kb and a minor transcript of 1.25 kb appear at 48 h after N step-down	Regulates expression of *hgl* genes *hglC*, *hglE1*, *hglB*, *hglD* and *hglE2*	Fox^-, defective in the formation of Hgls	Hebbar and Curtis (2000), Ramirez *et al.* (2005)
*devR*** (*alr0442*) 405 bp long	Receiver domain of response regulator of 135 amino acids long	Responsible for a sensory pathway that is not responsive to cellular N status	Fox^-	Zhu *et al.* (1998), Zhou and Wolk (2003)

*The former gene designation of *devC* was *OrfA* described by Maldenet *et al.* (1994).
*******devR* was initially described by Campbell *et al.* (1996) in *N. punctiforme* ATCC 29133.

layer due to a deficiency in any one of the three genes of *devBCA* operon. Ultrastructural studies confirmed the absence of heterocyst glycolipid laminated layer (Fiedler *et al.*, 1998). Fiedler *et al.* (2001) found that the promoter regions of *devBCA* operon have specific NtcA-binding sites at tsp positions located at -762 bp in the *devBCA* operon of *A. variabilis* ATCC 29413 and at –704 bp position in the *devBCA* operon in case of *Anabaena* sp. strain PCC 7120 suggesting that the *devBCA* operon is activated by NtcA. Another gene, *devH*, encoding a putative DNA-binding protein required for heterocyst function, has been detected during the characterization of mutant AD239 of *Anabaena* sp. strain PCC 7120. DevH protein bears resemblance to the cyanobacterial proteins of the cAMP receptor (CRP) proteins to which NtcA, the global nitrogen regulator belongs. Although DevH lacks the five conserved glycine residues in the N-terminal region necessary for regulatory properties of the proteins of the CRP family, it possesses the helix-turn-helix motif in the C-terminal region implicated in DNA binding (Hebbar and Curtis, 2000). A Nm resistance cassette has been introduced in the sequence of *devH* and its transfer via plasmid pBN1-239B into *Anabaena* sp. strain PCC 7120 through conjugation from *E. coli* resulted in inactivation of the *devH* gene. The disruption of *devH* has been further confirmed by Southern analysis. The *devH* mutant (A57) phenotypically is Fox^-, differentiates heterocysts without a laminated layer and so is unable to fix nitrogen aerobically. Two transcripts of *devH*, the shorter (1.0 kb) more abundant one and the longer (1.25kb) less abundant one in a ratio of 5:1 have been induced in the wild-type after 24 and 48 h after nitrogen step-down, respectively and these are absent in A57 (Hebbar and Curtis, 2000). The requirement of *devH* for the synthesis of heterocyst glycolipid layer has further been demonstrated by Ramírez *et al.* (2005). *E. coli* BL21 (DE3) was transformed with plasmid pET23930a carrying *devH* gene sequence and the recombinant DevH protein has been utilized for production of polyclonal antibodies. The presence of DevH protein (29 kDa) in the cell-free extracts of *Anabaena* sp. strain PCC 7120, after nitrogen step-down, has been demonstrated by cross reaction with the antiserum of recombinant DevH. Mutant A57 though produced DevH protein it has been non-functional due to the presence of extra 58 amino acid residues in between the helix-turn-helix and the C-terminal portion. The mutant A57 also showed rearrangements in the *nifHDK* operon but the corresponding transcripts are not detectable in the extracts grown under aerobic conditions. But under anoxic conditions mutant A57 exhibited Fix^+ phenotype signifying the synthesis of nitrogenase. Importantly, the expression of genes specific for heterocyst-specific glycolipid biosynthesis, i.e. *hglC* and *hglE1* is greatly reduced while those of others (*hglB*, *hglD* and *hglE2*) are not detectable in mutant A57. All these genes are of course expressed in the wild-type (Ramírez *et al.*, 2005).

A *devR* gene, essential for the development of mature heterocysts in *N. punctiforme* ATCC 29133, a symbiotic cyanobacterium of *A. punctatus* has been discovered by Campbell *et al.* (1996). The discovery of *devR* can be traced to a transposon-induced mutant UCD311 which is classified as a Fox^- mutant. The identity of DevR with the receiver domain of a response regulator of the two-component regulatory system emphasizes its role in the development of mature heterocysts. Though the heterocyst-specific glycolipids are synthesized in the cells they are not transported and deposited in the *devR*-disruptant mutant. The participation of DevR in a phosphorelay signal transduction pathway has been confirmed by the biochemical and genetic studies on DevR protein of *N. punctiforme* ATCC 29133 (Hagen and Meeks, 1999). A recombinant His-tagged protein of DevR, produced in *E. coli* BL21 (DE3), has been phosphorylated *in vitro* by a histidine kinase, EnvZ (of *E. coli* involved in osmoregulation). The phosphotransferase role for DevR has been confirmed by the isolation of mutants defective in phosphotransferase activity. Since aspartate (D) residue at position 53 constitutes the phosphorylation site, a change in codon 53 for aspartate GAT to CAA (for glutamine, Q) and GAA for glutamate (E) resulted in mutants UCD425 and UCD436, respectively.

The phenotypes of these two mutants resembled that of UCD311 in being Fox⁻ except that in presence of combined nitrogen extensive fragmentation of the filaments resulted in short 2–3 celled fragments (Hagen and Meeks, 1999).

In *Anabaena* sp. strain PCC 7120, *alr4674* encodes DevT that resembles a Ser/Thr protein phosphatase. Three important features of *devT* are that (i) it is not under the regulation of NtcA, (ii) its expression is to some extent inhibited by HetR and (iii) it accumulates specifically in mature heterocysts, though initially it is constitutively expressed in all cells. In view of these, Espinosa *et al.* (2010) concluded that DevT plays an unknown role in heterocyst maturation.

ii) Genes for Heterocyst-specific Glycolipids: Amongst a number of mutants of *Anabaena* sp. strain PCC 7120 isolated by Buikema and Haselkorn (1991b), the characterization of strain 543 led to the identification of a gene, *hglK* responsible for the transport of heterocyst glycolipids outside the cells and their assembly into the laminated layer (Black *et al.*, 1995). Strain 543 differentiated heterocysts without the laminated layer of glycolipids though the Hgls are synthesized by the cells. The N-terminal portion of HglK protein is organized into four membrane spanning domains which help the protein to anchor to the cell membrane and the C-terminal portion consists of 36 pentapeptide repeats. The first and third amino acid residues of the pentapeptide repeats, Ala and Leu, respectively are highly conserved while the rest three are variable, i.e. AXLXX. Cloning and sequencing of the PCR-generated fragments of the strain 543 genomic DNA revealed that a base change at the amino acid residue 496 resulted in a stop codon just before the pentapeptide repeat due to which the C-terminal portion got deleted. Accordingly, a mutant, constructed by the insertion of Sp-Sm resistance cassette just at the beginning of the *hglK* gene, also showed a similar phenotype as that of strain 543. The *hglK* gene seems to be highly conserved and restricted to the heterocystous cyanobacteria in its distribution (Black *et al.*, 1995). Bauer *et al.* (1997) described the presence of *hglB* (*alr5357*), *hglC* (*alr5355*) and *hglD* (*alr5354*) upstream of *hetN* in *Anabaena* sp. strain PCC 7120. These three genes are shown to be essential for Hgl synthesis as null mutations in any one of the three genes resulted in strains that are capable of differentiating heterocysts devoid of a Hgl laminated layer. Cohen *et al.* (1994) isolated a Tn5-1063 transposon mutant of *N. punctiforme* ATCC 29133 with *luxAB* as reporter gene (strain 307) that failed to develop mature heterocysts with glycolipid laminated layer but fixed nitrogen under anoxic conditions. Cloning and sequence analysis of the mutant genome revealed it to be an ORF of 3,159 bp and this has been designated as *hglE* by Campbell *et al.* (1997). The putative gene product of *hglE* is a protein of 115.4 kDa. HglE protein consists of active sites characteristic of polyketide synthases (PKS) with a consensus motif GPX_5TACSS and also possesses 16 highly conserved amino acids within a region of 200 amino acids that constitutes the PKS domain. The presence of *hglE* gene sequence exclusively in heterocystous cyanobacteria (*A. variabilis* ATCC 29413, *Anabaena* sp. strain PCC 7120, *Nostoc* sp. strain ATCC 27896, *Nostoc* sp. strain MAC, *N. punctiforme* ATCC 29133, *Nostoc* sp. strain ATCC 29106, and other free-living and symbiotic *Nostoc* strains) and the inactivation of which resulted in the failure to form Hgl layer led them to believe that the synthesis of Hgls is mediated through the PKS pathway (Campbell *et al.*, 1997). The upregulation of *hglE* (*alr5351*) and another ORF *alr5347* by sodium chloride in *Anabaena* sp. strain PCC 7120 and its absence in the adenyl cyclase C (CyaC; this is an important enzyme which mediates the synthesis of cAMP) disruptant mutant demonstrates that sodium chloride-induced enhancement in gene expression leads to the deposition of Hgls (Imashimizu *et al.*, 2005).

Maldener *et al.* (2003) characterized five mutants of *Anabaena* sp. strain PCC 7120, of which T22 and P2 are important to be noted in connection with Hgls. In case of mutant T22, the ORF *alr2887* has been shown to be responsible for export of Hgls. This mutant formed heterocyst envelope layer of polysaccharides but could not transport Hgls from within the cell to form the laminated layer. Mutant P2 could not synthesize the Hgls but formed the Hep layer. The particular ORF has been identified to be *alr5357* which happens to be *hglB* (formerly *hetM*).

So far we described the importance of *hglK, hglB, hglC, hglD* and *hglE* in the deposition of Hgls. A genome-wide expression of genes during nitrogen step-down of *Anabaena* sp. strain PCC 7120 revealed that many of the genes are physically clustered and expressed together as 'islands" on the chromosome. As many as 18 ORFs (*all5343* to *alr5360* including *hglE, hglD, hglC, hetM, hetN* and *hetI* genes; the three latter genes encode a polyketide synthase, ketoacyl reductase and phosphopantethienyl transferase, respectively) are expressed together (Ehira *et al.*, 2003). Such an expression of genes for synthesis and transport of Hgls has been confirmed subsequently by Fan *et al.* (2005). On the basis of a comprehensive analysis of 71 Fox$^-$ mutants described earlier (Wolk *et al.*, 1991), involving as many as 106 loci of insertions of transposon over 29 kb chromosomal region of *Anabaena* sp. strain PCC 7120, Fan *et al.* (2005) identified a cluster of genes required for the synthesis and deposition of Hgls. In this connection, ORFs *alr5351* to *alr5357* are shown to be required for the biosynthesis of Hgls and ORFs *all5345* to *all5347* for their deposition. Of these ORFs, the identity of at least four of them has already been established, i.e. *alr5351* (*hglE*), *alr5354* (*hglD*), *alr5355* (*hglC*) and *alr5357* (*hglB*).

Apart from the participation of the above mentioned genes in the synthesis and transport of Hgls, the requirement of two more genes seems to be essential for the formation of the laminated layer comprising of Hgls. A protein phosphatase J encoded by *prpJ* (*all1731*) of *Anabaena* sp. strain PCC 7120 appears to govern the synthesis of the more abundant HGL, i.e. 1-(α-D-glucopyranosyl)-3,2,5-hexacosanediol. A mutant of *prpJ*, (S20) obtained by targeted inactivation (by the introduction of a Sp-Sm resistance cassette, corresponding to the catalytic domain of PrpJ) could not synthesize the more abundant Hgl resulting in the absence of laminated layer. Thus in mutant S20 the expression of *hglE* and *nifH* are affected. PrpJ (consisting of 758 amino acid residues with a molecular mass of 83.7 kDa) possesses three domains: an N-terminal domain (of 255 amino acid residues) of unknown function, a central domain (comprising of amino acid residues 256 to 534) of catalytic activity specific for PP2C-type protein phosphatases and a C-terminal domain (constituting amino acid residues 535 to 758) of unknown function. It is important to note here that amino acid residues 583 to 605 of the C-terminal domain are hydrophobic and could form a speculated transmembrane domain for its being located on the plasma membrane. Thus PrpJ is another important new control point for heterocyst maturation (Jang *et al.*, 2007). Another trimeric pore-forming outer membrane β-barrel protein encoded by *hgdD* (heterocyst glycolipid deposition protein; *alr2887*) of *Anabaena* sp. strain PCC 7120 has been reported to be essential for diazotrophic growth especially for heterocyst maturation (Moslavac *et al.*, 2007). Mutants of *hgdD*, although showed a decreased synthesis of Hgl1 and Hgl2, could not transport the Hgls to their site of deposition. Ultrastructural studies of the developing heterocysts of *hgdD* mutants showed the absence of laminated layer, rearrangements of thylakoids, honey-comb like structures and polar nodules, suggesting that the heterocyst maturation has been arrested (Moslavac *et al.*, 2007). At this juncture, it is important to recall here a similar phenotype of mutants of *devBCA* operon (Fiedler *et al.*, 1998, 1999) where mutants of *devA* (*alr3712*), *devB* (*alr3710*) or *devC* (*alr3711*) could not deposit the laminated layer despite the synthesis of Hgls by them. Stucken *et al.* (2010) identified the *hgl* gene cluster comprising of *hglE, hglG, hglD, hglC, hglA* and *hetM* along with a *devBCA* operon in the toxic *Cylindrospermopsis raciborskii* DS-505 and a *devBCA* operon in the non-nitrogen fixing *Raphidiopsis brookii* D9.

iii) Genes for heterocyst envelope polysaccharides: During heterocyst maturation, after the deposition of Hgl laminated layer, the formation of a new layer of heterocyst envelope polysaccharide layer (Hep) appears to be a pre-requisite for acquisition of Fox^+ and Fix^+ phenotype. Any deficiency in the formation of Hep layer leads to Fox^- phenotype. The genes involved in the regulation of Hep layer are classified under *hep* genes. Holland and Wolk (1990) recognized *hetA* gene (2,555 bp) that encodes a protein (of 601 amino acid residues) necessary for improving the cohesiveness of the heterocyst envelope polysaccharides. Mutant EF116 of *Anabaena* sp. strain PCC 7120 defective in *hetA* showed a reduction in the cohesiveness of the polysaccharides. The expression of *hetA* has been found to be prominent around 7 h after nitrogen step-down. An unexplained but significant feature of this gene is the occurrence of four identical repeats of the sequence 5′-TTCAAAA-3′ situated at the 3′-end and 12 identical repeats of the sequence 5′-CCCCAAT-3′ that extend into 5′-end of a second ORF located downstream of *hetA*. Such direct repeats were reported earlier in between *petA* and *petC* of *Nostoc* sp. strain PCC 7906 (Kallas *et al.*, 1988). The gene identity of *hetA* was shown to be in fact *hepA* subsequently.

Zhu *et al.* (1998) identified the essential nature of *hepA* gene (*alr2835*) of *Anabaena* sp. strain PCC 7120 for the deposition of Hep layer. The gene product of *hepA* is a member of the family of ATP-binding proteins of ATP-binding cassette transporters. The expression of *hepA* in response to nitrogen deprivation has been reported (Holland and Wolk, 1990; Wolk *et al.*, 1993; Ehira *et al.*, 2003). In an attempt to find out the genes that regulate the expression of *hepA*, Zhu *et al.* (2001) subjected strain DR 1069 (that possesses a chromosomal *hepA*::*luxAB* fusion construct) to Tn5-1058 induced mutagenesis. The inactivation of a gene designated as *hcwA* in this mutant HNL3 resulted in the down-regulation of *hepA* but the expression of *hepA* was restored to normal levels only after complementation of the mutant with wild-type *hcwA* gene. The gene product of *hcwA* resembles the N-acetylmuramoyl-1-alanine amidase that has degradative effects on the cell wall. It was suggested by Zhu *et al.* (2001) that degradation of peptidoglycan wall layer is a prerequisite for heterocyst differentiation. Another gene, *hepC* (*alr2834*) is localized upstream of *hepA*. The predicted product of *hepC* is a UDP-galactose-lipid carrier transferase. In the intervening region of *hepC* and *hepA*, DNA sequences required for induction of *hepA* are localized in between –574 and –440 bp and –340 and –169 bp relative to the tsp of *hepA*. The regulation of *hepA* by *hepK* (*all4496*) has been found to be significant for two reasons. The first one is that inactivation of *hepK* resulted in repression of *hepA* and secondly the synthesis of Hep layer also is inhibited. Functionally, HepK is predicted to be a sensor protein-histidine kinase belonging to two-component regulatory systems and DevR protein constitutes the response regulator (receiver domain) described earlier by Campbell *et al.* (1996) in *N. punctiforme* ATCC 29133. However, the corresponding ortholog of *devR* of *Anabaena* $devR_A$, located 1.4 Mb away from *hepK* has been identified in the genome *Anabaena* sp. strain PCC 7120 (Kaneko *et al.*, 2001). Zhou and Wolk (2003) cloned and sequenced $devR_A$ (*alr0442*) and conducted *in vitro* studies on recombinant proteins produced in *E. coli* BL21(DE3) that have been tagged with a hexahistidine tag (H_6) at their N-terminal end. In presence of ATP, autophosphorylation of cytoplasmic portion of HepK (a truncated portion of HepK with amino acid residues 267 to 575) took place but not of HepK and in presence of ADP dephosphorylation of phosphorylated cytoplasmic part of HepK took place. The phosphoryl group from cytoplasmic part of HepK has been readily transferred to the receiver domains H_6DevR or H_6DevR_A. The phosphorylation site on HepK has been found to be His348 and the receiver of phosphoryl group on $DevR_A$ has been detected to be Asp53 residue. Site-directed mutagenesis of His348 to Ala348 of HepK resulted in a loss of *in vitro* autophosphorylation activity and caused the production of heterocysts without the Hep layer but with HGL laminated layer. So this constitutes the first two-component regulatory system known to regulate heterocyst development. Another two-component histidine kinase encoding gene, *alr0117* of *Anabaena* sp. strain PCC 7120 has been identified by Ning and Xu (2004a) that has been suggested to control the formation of Hep layer. Maldener *et al.* (2003) identified two ORFs *alr3698* (*hepB*) and *all4388* that are important in the synthesis and deposition of Hep layer, respectively in *Anabaena* sp. strain PCC 7120. Mutants M22 (*alr3698*) and α21 (*all4388*) were isolated by Tn5-1063 and Tn5-1065 insertions, respectively. The heterocysts formed in these mutants were devoid of the Hep layer though the Hgl synthesis and deposition continued. The expression of these genes was followed by transcriptional fusions with *luxAB* as the reporter gene.

Ehira *et al.* (2003) detected a 28 kb region, on the genome of *Anabaena* sp. strain PCC 7120 (precisely at 3.45 Mb position), consisting of 21 ORFs that is expressed after nitrogen step-down as "HEP island". All these genes (ORFs *alr2823* to *alr2843* including *hepA* and *hepC*) are presumed to be important in the carbohydrate metabolism and towards the formation of Hep layer during heterocyst maturation. Among these, the requirement of nine genes (*alr2825, alr2827, alr2831, alr2832, alr2833, alr2837, alr2839, alr2840* and *alr2841*) has been demonstrated for the formation of Hep layer that confers the Hep^+ and Fox^+ phenotype. Tn5-1063 transposon insertional mutagenesis in between ORFs *alr2825* to *alr2841* resulted in the generation of mutants (Hep^- and Fox^-) that lacked the Hep layer although the synthesis of Hgl laminated layer continued (Huang *et al.*, 2005). The existence of two genes ouside the "HEP island" has been found to be essential for the formation of Hep layer. These are ORF *all4160* that occurs at a position other than the "HEP island" and far from *hepB* and another ORF, *alr3699* that occurs just downstream of *hepB*. Thus we have hep genes distributed at three places in the genome of *Anabaena* sp. strain PCC 7120. The first is the "HEP island" of genes, second is the *hepB-alr3699* cluster and the third *all4160*. The expression of *all4160, alr3699* and *hepB* occurred in proheterocysts and heterocysts of *Anabaena* sp. strain PCC 7120 as revealed by the GFP fluorescence when *gfp* has been used as a reporter gene under the influence of the respective promoter regions. The up-regulation of *all4160, alr3699* and *hepB* not only occurred in response to nitrogen deprivation but also they are co-transcribed under the regulation of *hepK* and *hepN* as confirmed by their down-regulation in *hepK* and *hepN* disruption mutants (Wang *et al.*, 2007). Fan *et al.* (2006) described the properties of mutants that regulate the Hep layer. Insertional mutants of *henR* (*alr1086*) are characterized by the presence of traces of heterocyst-specific glycolipids and are devoid of the Hgl layer as well as an hep layer. Because of its heterocyst envelope regulation, this gene was named as *henR*. The heterocysts stained by Alcian Blue were not positive suggesting the complete absence of the envelope. The constriction specific regulatory gene, *conR* (*all0187*) when inactivated though produced Hep^+ and Hgl^+ phenotype these layers did not join at the narrow point of contact between the vegetative cells and heterocysts. As the constriction did not get deepen it resulted in a loose arrangement permitting the entry of oxygen and so a loss of nitrogen-fixing ability. So *conR* gene is constriction-specific rather than heterocyst-specific and it is regulatory in nature. Mutants of *hepS* (*all2760*) are Hep^- and Hgl^+. It is a putative Ser/Thr kinase and the nature and type of proteins phosphorylated by it are yet to be discovered. Mutants defective in *hepN* (*alr0177*) showed Hep^- phenotype that lost quickly the Hgls by fragmenting into small pieces which could not be recovered for analysis but could be only visualized in scanning electron micrographs (Fan *et al.*, 2006). Mutations in four regulary genes, i.e. *hepK, hepN, henR* and *hepS* have interrelated effects in *Anabaena* sp. strain PCC 7120 (Lechno-Yossef *et al.*, 2006). Mutants FQ671 (*hepN*), FQ1487 (*hepS*), Y7 (*hepK*) and FQ1227 (*henR*) were subjected to nitrogen step-down with wild-type *Anabaena* sp. strain PCC 7120 as control. Except *henR* mutant that showed Hep^- and Hgl^- phenotype the rest three exhibited Hep^- and Hgl^+ phenotype. At 14 h after nitrogen step-down, mRNA was isolated from the cultures of the wild-type and the four mutants and the corresponding cDNA was synthesized. Oligonucleotide probes were designed for

6126 of 6129 ORFs of *Anabaena* sp. strain PCC 7120 genome and hybridization was conducted. In terms of cluster analysis and functional categories of genes 329 ORFs were differentially expressed. Seven clusters of ORFs have been recognized. In cluster 1, most of the genes were down-regulated (mentioned in perentheses) in *hepK* (167), *hepN* (117), *hepS* (65) and *henR* (53) mutants and most of the ORFs in *hepK* and *hepN* mutants pertained to '*nif*' and '*pol*' (polysaccharide envelope) islands. In the cluster 2, 18 genes were all down-regulated in the *hepN* mutant and most of these (12) pertained to 'Hep' island and 10 of these were down-regulated in *hepS* mutant as well. In cluster 3, in the *henR* and *hepS* mtants 13 and 10 genes respectively mostly belonging to 'Hgl' island were down-regulated. In cluster 4, 9 genes were down-regulated in the *hepK* mutant. Cluster 5 contains 70 ORFs that were up-regulated mostly in *hepK* and/or *hepN* mutant. Cluster 6 consists of seven ORFs that were all up-regulated in the *hepK* mutant, of these up-regulation of *furA* is noteworthy. In cluster 7, there were 23 ORFs up-regulated in *hepN* mutant. HepS and HenR appear to belong to a common regulatory pathway because of the regulation of common set of genes under their control. The expression of *patA* gene in all the four mutants except in *hepK* mutant signifies that *patA* expression is inhibited by products related to heterocyst development.

C) Genes for sigma factors

Sigma factors of RNA polymerase help in transcription of genes by binding to appropriate recognition sequences in the promoter region and facilitate open complex formation. There are two families of sigma factors, i.e. σ^{70} and σ^{54}. The σ^{70} family has been divided into three groups. Group 1 sigma factors or primary sigma factors that control transcription of housekeeping genes and are essential for survival. Group 2 sigma factors are very much similar to group 1 but are non-essential. Group 3 sigma factors are also known as alternative sigma factors and they are specific for transcribing mRNAs for production of proteins in specialized organs (of motility) or processes such as stress responses, endospore formation and extra-cytoplasmic functions. Sigma factors of the family σ^{54} are not represented in cyanobacteria. *Anabaena* sp. strain PCC 7120 possesses 12 putative genes for sigma factors, nine of them are located on the chromosome (group 1: *sigA, all5263*; group 2: *sigB2, alr3800; sigC, all1692; sigD, alr3810*; and *sigE, alr4249*; group 3: *sigF, all3853; sigG, alr3280; sigI, all2193* and *sigJ, alr0277*) and three of them are on the plasmids (*sigB, all7615; sigB3, all7608; sigB4, all7179*). Yoshimura *et al.* (2007) presented a phylogenetic analysis of sigma factors of four (*Anabaena* sp. strain PCC 7120, *A. variabilis* ATCC 29413, *Synechocystis* sp. 6803 and *S. elongatus* PCC 7942) cyanobacteria. The importance of group 2 sigma factors during a change in nutrients, light and temperature conditions in cyanobacteria has been highlighted (Osanai *et al.*, 2008). A primary sigma factor encoding gene *sigA* has been isolated and characterized by Brahamsha and Haselkorn (1991). Two transcripts of 1.7 kb and 2.2 kb are encoded by the monocistronic *sigA* gene with the former remaining constant under nitrogen-enriched as well as nitrogen-depleted conditions while the latter showed enhaced levels of its production after nitrogen step-down in *Anabaena* sp. strain PCC 7120. Studies on cloning, expression and inactivation of *sigB* and *sigC* genes of *Anabaena* sp. strain PCC 7120 revealed that SigB (with 332 amino acid residues) and SigC (with 416 amino acid residues) are produced only under nitrogen-limited conditions but these are not required either for nitrogen fixation or heterocyst differentiation (Brahamsha and Haselkorn, 1992). During a search of DNA-binding proteins that are able to bind to the promoter region of *rbcL* gene, Ramasubramanian *et al.* (1994a) isolated *sigA* gene of *Anabaena* sp. strain PCC 7120. A mutation in *sigH* of *N. punctiforme*, a symbiont of *Anthoceros punctatus*, enhanced the infection potential by 6-fold with a simultaneous increase in nitrogenase activity without any apparent increase in the heterocyst frequency (Campbell *et al.*, 1998). Khudyakov and Golden (2001) identified genes *sigD*, *sigE* and *sigF* in *Anabaena* sp. strain PCC 7120. Insertional inactivation of the three sigma factor genes has been achieved by the transfer of suicide plasmids pAM2178 (carrying Ω Sp-Sm resistance cassette) or pAM2179 (with Cm^r Em^r cassette) fused with *lacZ* as the reporter gene through conjugation with *E. coli* that permitted blue-white colony screening on X-gal plates. Inactivation of any of these genes did not significantly affect the growth of the mutants in nitrate nitrogen but mutants of *sigD* and *sigE* did not perform well under diazotrophic growth conditions. A *sigDsigE* double mutant differentiated proheterocysts after nitrogen step-down but had undergone extensive fragmentation. Likewise, another *sigEsigF* double mutant also established poorly under nitrogen limitation and showed extensive bleaching. However, it is significant to note that the expression of the three sigma factor genes was quite feeble after a nitrogen step-down and heterocyst differentiation in *Anabaena* sp. strain PCC 7120 as evidenced by the luminescence from *luxAB* reporter (Khudyakov and Golden, 2001). A group 2 sigma factor gene, *sigC* (*sll0184*) has been shown to control the expression of *glnB* (that encodes PII protein) in the stationary phase nitrogen-limited cultures of *Synechocystis* sp. strain PCC 6803 (Asayama *et al.*, 2004). Aldea *et al.* (2007) identified the probable role of eight sigma factor genes, four each belonging to group 2 (*sigB2, sigC, sigD* and *sigE*), and group 3 (*sigF, sigG* and *sigI, sigJ*) during heterocyst differentiation. The up-regulation of *sigC, sigE* (previously *sigF*) and *sigG* during heterocyst differentiation of *Anabaena* sp. strain PCC 7120 at 4 h, 16 h and 9 h, respectively has been demonstrated by using transcriptional fusions with *gfp* as the reporter gene. Five other sigma factor genes *sigB2, sigD, sigI* and *sigJ* were expressed in vegetative cells of nitrate-grown filaments but after nitrogen step-down their expression was found in both vegetative cells as well as heterocysts. Transcriptional fusions of *gfp* with the promoters of *sigC, sigE* and *sigG* revealed important information on the expression of genes in the development of heterocyst. The expression of *sigC* coincided with the expression of important heterocyst-specific early genes *hetC, hetP, devH, patS, patA* and *patB*; *hep* genes (*hepA, hepB, hepC,* and *hepK*) and Hgl genes (*hglC, hglD, hglE, hglK, hglB, hetN* and *hetI*). The genes governing the synthesis of envelope layers and respiratory terminal oxidases are expressed around 9 h corresponding with the expression of *sigG*. The expression of *sigE* around 16 to 17 h of nitrogen step-down suggests its involvement in the transcription of *nif* genes that are expressed at late stages of heterocyst formation.

D) Signal transduction genes

Two-component, one-component and Ser/Thr and Tyr protein kinases constitute three major signal transduction systems. Of these three types, the former two have been well characterized in the prokaryotes while the third Ser/Thr or Tyr kinases, widely distributed in eukaryotes, are the newly identified systems in prokaryotes. In the first type, there are two components; one is a sensor kinase that transfers phosphate from a His residue to an Asp residue in the response regulator (Stock *et al.*, 1990; Parkinson, 1993). One-component systems have known input and ouput domains but lack histidine kinase and the receiver domains. These are the major signal transduction systems identified in prokaryotes (Ulrich *et al.*, 2005). The current status on the purported role of Ser/Thr protein kinases and phosphatases in changing growth conditions, cell metabolism, photosynthesis and stress responses in cyanobacteria has been reviewed (Zhang, C-C. *et al.*, 2005; Zhang, X. *et al.*, 2007). Among the genomes so far sequenced, except for some unicellular forms (four strains of *Prochlorococcus* and one marine *Synechococcus* WH8102) Ser/Thr protein kinases are present in the rest of them but abundantly represented in filamentous, nitrogen-fixing forms. In *Anabaena* sp. strain PCC 7120 five Ser/Thr protein kinases have been characterized. These are PknA (alr4366; Zhang, 1993),

Pkn D (alr4368; Zhang and Libs, 1998), PknE (alr3732; Zhang *et al.*, 1998), PknC (all4813; Gonzalez *et al.*, 2001) and Pkn22 (alr2502; Xu *et al.*, 2003). Of these, PknC has broad substrate specificity.

A protein phosphatase 1/2A/2B homologue (*prpA*) has been found 301 bp downstream of *pknE* gene in *Anabaena* sp. PCC 7120. Recombinant proteins produced in *E. coli* have been used to raise polyclonal antibodies against these and are used for immunodetection of the corresponding proteins under various growth conditions. Though both these genes are closely linked but they are not expressed together (Zhang *et al.*, 1998). PrpA(95 kDa) protein has been found to be expressed in wild-type cells after nitrogen step-down whereas expression of PknE (53 kDa) protein has been found be high in nitrate-enriched medium but it decreased on nitrogen step-down up to 3 h after which it increased on long time exposure in nitrogen-free medium. An Nm resistance cassette has been used to generate disruptant mutants. ARAs revealed that aerobic and anaerobic nitrogen fixation rates have decreased considerably with abnormal heterocyst structure, although there was no difference in glycolipids of heterocysts (Zhang *et al.*, 1998). The involvement of two genes encoding protein kinases of the HstK family, *pkn44* (*all1625*) and *pkn30* (*all3691*) in the synthesis of minor heterocyst-specific glycolipid in *Anabaena* sp. strain PCC 7120 has been reported by Shi *et al.* (2007). The double mutant (D4.3) strain revealed a Fox^- and Hgl^- phenotype due to which aerobic nitrogen fixation and a minor heterocyst glycolipid are absent. D4.3 strain showed a delay of 24 h in heterocyst formation that was arrested at an early stage due to the absence of minor heterocyst glycolipid component. They concluded that the synthesis of major and minor Hgls is regulated by PrpJ and *pkn44/pkn30* regulatory pathways, respectively.

PatA encoded by *patA* gene of *Anabaena* sp. PCC 7120 contains 379 amino acids and the carboxy-terminal domain bears similarity to CheY of *E. coli* and other response regulators of the two-component regulatory systems. A mutant of *patA* differentiates only terminal heterocysts and exhibits poor growth in nitrogen-free medium compared to wild-type (Liang *et al.*, 1992). Cohen *et al.* (1994) isolated and characterized 14 Fox^- and Sym^+ mutants isolated after transposon Tn5-1063 insertion mutagenesis of *N. punctiforme* ATCC 29133, a symbiont of *A. punctatus*. One of the mutants of this group, UCD311 failed to develop mature heterocysts and fixed nitrogen under anoxic conditions though it was capable of synthesizing the heteocyst envelope. Sequence analysis of the interrupted gene revealed an ORF which has been designated as *devR* (for developmental regulation). The deduced amino acid sequence of DevR revealed it to be a receiver domain of two-component regulatory system of *N. punctiforme* ATCC 29133. DevR bears similarity to CheY of *E. coli* and SpoOF of *Bacillus subtilis* and possesses Asp57 as the phosphorylation site (Campbell *et al.*, 1996). The synthesis of heterocyst envelope polysaccharide of *Anabaena* sp. strain PCC 7120 has been reported to be under the regulation of *hepK* (*all4496*) that encodes a sensory protein histidine kinase and $devR_A$ (*devR* of *Anabaena*; *alr0442*) that encodes its corresponding cognate response regulator (Zhu *et al.*, 1998; Zhou and Wolk, 2003). Mutation of *hepK* prevented the induction of *hepA* (*alr2835*) as well as envelope polysaccharide. The gene product, Alr2835 resembles a member of the family of ABC transporters and is essential for the synthesis of Hep layer (Zhu *et al.*, 1998). *In vitro* autophosphorylation [in presence of (γ-^{32}P)] and dephosphorylation (in presence of ADP) of the cytoplasmic portion of HepK and the transfer of phosphoryl group to DevR (of *N. punctiforme* ATCC 29133) and $DevR_A$ have been demonstrated using hexa-histidine tagged to N-terminus of the recombinant proteins produced in *E. coli* (Zhou and Wolk, 2003). Ning and Xu (2004a) identified a two-component histidine kinase gene (*alr0117*) that may control heterocyst envelope polysaccharide layer and certain late maturation stages in heterocyst development of *Anabaena* sp. strain PCC 7120. In a transposon (Tn5-1087b) generated mutant, with its insertion in between nucleotides between 342 and 343 of *alr0117*, the expression of *hepA* was completely blocked while *patB* expression occurred constitutively at low levels. The expression of three other genes tested (*hetM, hglE* and *hetN*) was unaffected. Targeted inactivation of the gene in the wild-type resulted in the same mutation and complementation of the mutant with wild-type copy restored the phenotype. Fan *et al.* (2006) identified three additional signal transduction genes required for heterocyst maturation in *Anabaena* sp. strain PCC 7120 during characterization of 1076 Fox$^-$ mutants generated by transposon (Tn5-1063) insertion mutagenesis. These are a transcriptional regulator, *conR* (*all0187*), a response regulator, *henR* (*alr1086*) and a Ser/Thr kinase, *hepS* (*all2760*). They also confirmed the observations of Ning and Xu (2004a) regarding the gene *alr0117* and in consultation with them named this gene as *hepN*. The regulatory nature of HenR is exemplified by the nature of the pleiotropic effects it exerted on the envelope polysaccharide and heterocyst glycolipid layers. So it is suggested that HenR is a part of master system that regulates synthesis of heterocyst envelope that may be bifurcated into one controlling deposition of glycolipid layer and the other regulating the deposition of polysachharide layer. (The properties of the mutant phenotypes of these four regulatory genes are described under section late genes or maturation genes).

patB gene of *Anabaena* sp. strain PCC 7120 encodes a predicted protein of 529 amino acid residues and its N-terminal region bears similarity to Fe4-S4 bacterial-type ferredoxins while the C-terminal region is in the form of helix-turn-helix, typical of transcriptional regulators. In PAT-2 mutant, heterocysts appeared after a delay 42 h when subjected to nitrogen-step -down. *patB* sequences are found in other species of *Anabaena, Nostoc* and *Calothrix* and absent in unicellular cyanobacteria and other filamentous cyanobacteria that fix nitrogen under anaerobic conditions (Liang *et al.*, 1993). There are as many as 118 genes that probably encode transcriptional regulators in *Anabaena* sp. strain PCC 7120 (Ohmori *et al.*, 2001). Koksharova and Wolk (2002) reported four more genes *abp1, abp2, abp3* and *abp4* that encode DNA-binding proteins. The existence of *hetR* as homodimer, its DNA-binding activity and its inhibition by PatS have been reported by Huang *et al.* (2004). Ehira and Ohmori (2006a) employed 5336 oligonucleotide probes in a DNA microarray analysis and found up-regulation in the transcript levels of 410 genes at one or more time points (3, 8 and 24 h) after nitrogen step-down of *Anabaena* sp. strain PCC 7120. Of these, transcripts of ten genes from among the 108 presumptive transcriptional regulators were up-regulated prominently with transcripts of *all4312* superceding the rest. The rest nine genes up-regulated have been found to be transcriptional regulators (*alr2178, alr2208, alr2597, alr2625* and *all4925*), two-component response regulator (*all3788*), *hetR* (*alr2329*), *devH* (*alr3592*; a transcriptional regulator) and *ntcA* (*alr4392*). The ORF *all4312* has been designated as *nrrA* (nitrogen responsive regulator) that encodes a response regulator, of the OmpR family. These results have further been confirmed by conducting real-time reverse transcription PCR. The presence of specific NtcA-binding region in the promoter region of *nrrA* indicates that this gene is regulated by NtcA. The NrrA levels were enhanced within 3 h of nitrogen step-down. Heterocyst differentiation was delayed in a *nrrA* deletion mutant up to 24 h of nitrogen step-down due to lower levels of *hetR* induction (Ehira and Ohmori, 2006a). Simultaneouly, Muro-Pastor *et al.* (2006) published similar results on the role of gene product of *all4312*. Their investigation adduced additional evidences for the suggested role of All4312. Expression of *all4312* occurred in all cells including proheterocysts and heterocysts as evidenced by the fluorescence of GFP when *gfp* was used as the reporter gene with promoter region of *all4312*. In the *hetR* mutant strain, the expression of *all4312* was not altered. NtcA binding to the NtcA-binding region located in the promoter region of *all4312* at 41.5 nucleotides with respect to tsp is enhanced in presence of 2-oxoglutarate. Ehira and Ohmori (2006b) showed that NrrA directly regulates expression of *hetR* during heterocyst differentiation in *Anabaena* sp. strain PCC 7120 as demonstrated by its binding *in vitro* to the nitrogen responsive tsps located at -728 and -696 positions. This suggests that transcription from these sites is directly regulated by NrrA. In the *nrrA* mutant, the transcripts from the above corresponding tsps are either absent or reduced after nitrogen step-down. Since transcription of *hetR* at tsp at -728bp is dependent on NtcA, it may be that NtcA binding to this tsp is mediated by NrrA. Further, when NrrA is overexpressed, it leads to up-regulation of HetR synthesis. Ehira and Ohmori (2006b) explained the sequence of events as follows: nitrogen limitation causes increased production of NrrA stimulated by NtcA and in turn NrrA induces the expression of *hetR* resulting in heterocyst differentiation.

10) SIGNALLING MOLECULES

Two signalling molecules calcium and 2-oxoglutarate (2-OG) have been identified that transmit signals for the need to differentiate heterocysts to fulfill diazotrophic growth.

***i) Calcium*:** Calcium is required for a number of physiological processes in cyanobacteria. These include the dependence of PSII activities of *Anacystis nidulans in vitro* (England and Evans, 1983) and *in vivo* (Becker and Brand, 1985), phosphate uptake by *A. nidulans* and *Oscillatoria limnetica* (Kerson *et al.*, 1984) and in the protection of nitrogenase in *Gleocapsa* sp. 1430/3 (Falah Hamadi and Gallon, 1981) and *Gloeothece* sp. ATCC 27152 from O_2 damage (Gallon and Falah Hamadi, 1984). The requirement of calcium as a macronutrient for aerobic nitrogen fixation by two species of *Anabaena* (*Anabaena* sp. and *Anabaena* sp. ATCC 33047) and its requirement in traces when grown in microaerophilc conditions, prompted Rodríguez *et al.* (1990) to suggest that calcium has role in conferring protection to nitrogenase present in heterocysts. A calcium-mediated regulation of heterocyst frequency and nitrogen fixation has been demonstrated in *Nostoc* sp. PCC 6720 (ATCC 27895) by Smith *et al.* (1987) who showed that a decrease of calcium concentration (0.01 mM to 0.1 mM) stimulated heterocyst production from 5% to 9% whereas at calcium concentrations normally employed in growth media (0.1 to 1.0 mM) the heterocyst frequency did not fluctuate (remained at 5%). In presence of calcium ionophore, compound A23187 (calcimycin; in between 0.01 μM to 0.5 μM), that facilitates calcium uptake, and 0.1 to 1.0 mM Ca^{2+}, the frequency of heterocysts increased from 7.0% to 10.5%. This signifies that Ca^{2+} and A23187 showed synergistic effect. A reduction in the frequency of heterocysts has been noted by the inclusion of calcium channel inhibitor, lanthanium (La^{3+}). $^{45}Ca^{2+}$-uptake experiments showed that an increase in intracellular Ca^{2+} occurred under low illumination probably due to a decreased rate of efflux of Ca^{2+} and in presence of A23187, the intracellular Ca^{2+} increased. Intracellular $^{45}Ca^{2+}$ and heterocyst frequency in *Nostoc* sp. PCC 6720 have been shown to be negatively correlated with incident light (Smith and Wilkins, 1988).

The cellular Ca^{2+} concentration is modulated by a special calcium-binding protein known as calmodulin (calcium modulated protein) which interacts with cellular proteins that are to be activated by the release of calcium. Calmodulin and calcium-mediated regulation of cellular processes has thus received attention not only in eukaryotes but in prokaryotes as well (Onek and Smith, 1992). Isolation and characterization of calmodulin-like proteins from cyanobacteria have been successful. Pettersson and Bergman (1989) put forward evidences for the presence of calmodulin-like protein in vegetative cells and heterocysts of three species of *Anabaena*. The immunogold particles are more densely distributed in the central portion of vegetative cells while in heterocysts they are evenly distributed. These studies revealed the presence of a polypeptide of about 17 kDa that interacted with the antiserum raised against spinach calmodulin. Onek *et al.* (1992) studied the properties of calmodulin-like protein from *Nostoc* sp. PCC 6720 and showed that the 21 kDa protein possessed characteristics of eukaryotic calmodulin and activated pea NAD kinase *in vitro* in a Ca^{2+}-requiring reaction. The intracellular free Ca^{2+} ($[Ca^{2+}]_i$), in response to increasing extracellular Ca^{2+} during heat and cold shock, exhibited a transient burst in $[Ca^{2+}]_i$ in a recombinant strain of *Anabaena* sp. strain PCC 7120 that expresses constitutively apoaequorin (a calcium-binding sensitive luminescent protein). The luminescence of the cells, measured in a luminometer after treatment with coelenterazine, depended on the concentration of $[Ca^{2+}]_i$ (Torrecilla *et al.*, 2000). These studies paved the way for the identification of a specific calcium signature during heterocyst differentiation in the recombinant strain of *Anabaena* sp. strain PCC 7120 for apoaequorin and a *hetR* mutant strain expressing apoaequorin after a nitrogen step-down. The fluctuation in the calcium signature (increase, decrease or suppression) due to the addition of compound A23187 (5 μM), trifluoperazine (an inhibitor of eukaryotic calmodulin, 5 μM) and BAPTA-AM (an intracellular calcium chelator; 1,2-bis(2 aminophenoxy) ethane N,N,N′,N′ tetraacetic acid tetrakis(acetoxy-methyl ester at 300 μM) also prevented heterocyst differentiation. The generation of calcium signature in *hetR* mutant strain signifies that the calcium signal during heterocyst differentiation may be an early step preceding expression of *hetR* gene (Torrecilla *et al.*, 2004).

A cyanobacterial calcium-binding protein (CcbP), encoded by the gene *alr1010* of *Anabaena* sp. strain PCC 7120, overexpressed in *E. coli* has molecular mass of 21 kDa with 126 amino acid residues and no known motifs or domains to bind Ca^{2+}. An insertion mutant of CcbP (CCBP-M) and a transformant with P*petE*$^{-ccbP}$ that had *ccbP* gene of *Anabaena* sp. strain PCC 7120 under the control of *petE* promoter (CCBP-IE) have been subjected to nitrogen step-down experiments to follow heterocyst differentiation. CCBP-M showed Mch-phenotype with an increase in heterocyst frequency than that of the wild-type. On the other hand, CCBP-IE formed low frequency of heterocysts in the absence of copper (no expression of *ccbP* gene) but in presence of copper the expression of *ccbP* led to complete suppression of heterocyst differentiation. These results thus show that CcbP negatively regulates heterocyst differentiation. It has been further confirmed by the expression of rat CaM gene (*cam*) in the wild-type and CCBP-M of *Anabaena* sp. strain PCC7120 under the regulation of *petE* promoter. The failure of wild-type carrying *cam* gene to differentiate heterocysts and the suppression of Mch-phenotype in CCBP-M carrying *cam* gene in presence of copper confirmed that CcbP negatively regulates heterocyst differentiation. Additionally, *A. variabilis* ATCC 29413 transformed with a plasmid $P_{petE\text{-cam}}$ did not differentiate heterocysts after nitrogen step-down in presence of copper. Since *patS* deletion mutant exhibited Mch-phenotype, its transformation with plasmid $P_{petE\text{-ccbP (eryR)}}$ resulted in suppression of heterocysts when induced with copper. Likewise, the *hetR* overexpression strain when transformed with plasmid $P_{petE\text{-ccbP(eryR)}}$, instead of Mch-phenotype heterocyst differentiation was suppressed completely in presence of copper. The expression of *ccbP* gene in wild-type strain along with GFP as reporter gene ($P_{ccbP\text{-gfp}}$) resulted in strong fluorescence in vegetative cells and a weak fluorescence in the heterocysts. It is expected that due to down-regulation of *ccbP* in the heterocysts the $[Ca^{2+}]_i$ in the heterocysts is higher than in the vegetative cells (Zhao *et al.*, 2005).The increase of $[Ca^{2+}]_i$ and down-regulation of *ccbP* in the developing heterocysts and mature heterocysts have been linked to HetR and NtcA mediated reactions, respectively. The increase of $[Ca^{2+}]_i$ has been shown to be due to the degradation of CcbP by the proteolytic activity of HetR that requires Ca^{2+}. During the degradation process, two Ca^{2+} held by each molecule of CcbP are released. The proteolytic activity of HetR (besides the capability to autodegrade itself) seems to be specific for CcbP as it did not degrade other proteins such as NtcA of *Anabaena* sp. strain PCC 7120 and bovine serum albumin. As HetR is a serine-type protease, the active serine residues at 152 and 179 positions when mutated to Ala and Asn respectively, the mutant forms of HetR did not bring about the degradation of CcbP. However, another mutant form, $HetR_{C48A}$ that cannot exist as a dimer but possesses proteolytic activity, could digest CcbP. The down-regulation of *ccbP* brought about by NtcA is attributed to the presence of potential NtcA-binding site (GTTCTGAGTGGTCACA) in the promoter region of *ccbP* as noted earlier by Herrero *et al.* (2004). This is further confirmed by the *ccbP* mRNA levels in the wild-type that remained fairly constant during the first 3 h of nitrogen step-down and NtcA deletion mutant which showed 50% increase of *ccbP* mRNA levels during the same period. Interestingly, 2-OG promoted interaction of NtcA with its binding sequence in the promoter of *ccbP* (Shi *et al.*, 2006).

ii) 2-OG: The first indication that cyanobacteria perceive nitrogen status by sensing intracellular 2-OG levels came from the studies of Muro-Pastor *et al.* (2001) who monitored the levels of 19 different amino acids and 2-OG in mutants of *Synechocystis* sp. strain PCC 6803 with altered levels of GS-activity. Irrespective of the levels of glutamine and glutamate, the two key amino acids synthesized during the ammonia assimilation (GS-GOGAT) pathway, it is the levels of 2-OG that are perceived by cyanobacteria to sense combined nitrogen limitation. Another observation that lends support to 2-OG as the signalling molecule is the increased *in vitro* binding affinity of NtcA to the promoter region for *S. elongatus* PCC 7942 *glnA* in presence of 2-OG as well as its transcription rate (Vazquez-Bermudez *et al.*, 2002; Tanigawa *et al.*, 2002). Another important protein that regulates nitrogen metabolism in unicellular cyanobacteria is PII, a product of *glnB*. (There are three sub-families of PII proteins. $NifI_1$ and $NifI_2$ constitute two sub-families while the third sub-family comprises of all other PII proteins including products of *glnB* and *glnK*). The binding of 2-OG to PII not only stimulates phosphorylation of PII (at Ser49) in *S. elongatus* PCC 7942 (Forchhammer and Tandeau de Marsac, 1994, 1995) but also inhibits its dephosphorylation reaction (Irmler *et al.*, 1997; Ruppert *et al.*, 2002). Under nitrogen-limiting conditions, NtcA also regulates *glnB* at the transcriptional and post-translational levels (Lee *et al.*, 1999; Sauer *et al.*, 1999). In turn it has also been observed that PII stimulates and inhibits NtcA activity under nitrogen deprivation and in presence of nitrate enriched medium, respectively (Aldehni *et al.*, 2003; Paz-Yepes *et al.*, 2003; Aldehni and Forchhammer, 2006). The discovery of a transcriptional activator, PipX in *S. elongatus* PCC 7942 provides a connecting link between NtcA and PII which are the two known regulators of nitrogen metabolism in cyanobacteria. 2-OG favours interaction between NtcA and PipX under nitrogen deprivation but impairs binding to PII, the deciding factor being the intracellular concentration of 2-OG (Espinosa *et al.*, 2006). But all the above studies conducted mostly on unicellular cyanobacteria clearly emphasize that in perceiving nitrogen status the most important interactions occur between NtcA, PII and 2-OG. However, it may be noted that PII has not been well characterized in the filamentous heterocystous cyanobacteria.

In order to study the effect of enhanced intracellular levels of 2-OG, Li *et al.* (2003) constructed a recombinant strain of *Anabaena* sp. strain PCC 7120 with *E. coli* ketoglutarate permease gene (*kgtP*) under the influence of *petE* promoter that is expressed in presence of copper. The *kgtP* gene construct, carried by a shuttle vector (pRL25c), was transferred through conjugation and the exconjugants (KGTP strains) were selected in presence of 50 μg Nm ml^{-1}. Uptake kinetics of [1-^{14}C]2-OG by the KGTP strain showed that in presence of copper the *kgtP* gene was expressed and significant amounts of 2-OG were taken up by the cells. In nitrate medium, the addition of 2-OG (1.0 mM) resulted in heterocyst differentiation with a normal pattern. When PatS inactivated strain (AMC451; isolated by Yoon and Golden, 1998) was grown in presence of 2-OG (25 mM), the interval between two heterocysts was reduced to four cells (from the interval of 8 cells observed in the absence of 2-OG in the *patS* mutant). Another significant feature is the occurrence of heterocysts in pairs that constituted 71% of all the heterocysts when compared to 59% paired heterocysts in the absence of 2-OG. Ammonia repressed heterocyst differentiation in the KGTP strain but in presence of 2-OG the time taken for heterocyst differentiation was considerably reduced to 6 h after which ammonia could no longer repress heterocyst differentiation. This signifies that the presence high intracellular concentrations of 2-OG in the cells, the commitment of cells for heterocyst differentiation is reached very early than in its absence (Li *et al.*, 2003). The same group of workers extended these studies by synthesizing a non-metabolizable fluorinated 2-OG derivative, 2,2-difluoropentanedioic acid (DFPA) that could be used as a signalling molecule for KGPT strain. DFPA (at 0.5 mM) triggered the formation of heterocysts (5.5%) in KGPT strain even in presence of ammonium when compared to 7.1% observed in the wild-type in a nitrogen-free medium. Both 2-OG and DFPA enhanced the binding of NtcA to *ntcA* promoter regions having consensus binding sequences $GTAN_8TAC$ and $GTAN_8AAC$ at positions –110 and –150 bp, respectively (Laurent *et al.*, 2005). Studies on crystal structures of NtcA of *Anabaena* sp. strain PCC 7120 revealed three forms, one apoform and two other ligand-bound forms that can undergo complex formation with either 2-OG or DFPA. Though the binding of NtcA in its apoform to DNA is favoured, the presence of 2-OG enhances this binding activity as NtcA undergoes a conformational change (Zhao *et al.*, 2010). Chen *et al.* (2011) characterized two important residues in the effector-binding domain of NtcA of *Anabaena* sp. strain PCC 7120 and showed that any change in these two residues leads to loss of 2-OG recognition of the effector-binding domain and eventually to a failure of signal transmission and heterocyst differentiation.

In vitro transcription of three NtcA-activated genes at different times during heterocyst differentiation in *Anabaena* sp. strain PCC 7120, i.e. *hetC*, *nrrA* and *devB* could occur in the absence of any activators but the presence of NtcA and 2-OG was sufficient for open complex formation and transcript production (Valladares *et al.*, 2008). Studies on PII (*glnB*) mutants of *Anabaena* sp. strain PCC 7120 showed that the mutant MP2α grew slowly in all nitrogen enriched media and exhibited low nitrogenase activity but formed heterocysts with a normal pattern. Mass spectrometric analysis of PII from MP2α under nitrogen-fixing conditions showed nitration of Tyr51 with absence of phosphorylation of Ser49. This suggests that the PII is required as a regulator of nitrogen metabolism but not essential for heterocyst differentiation. Additionally, PII mutants of *Anabaena* sp. strain PCC 7120 for Ser49 to Ala49 and Tyr51 to Phe51 also grew slowly in nitrogen-fixing medium and mutant Ser49Ala showed impaired nitrate uptake but differentiated heterocysts. In both these mutants the up-regulation of *ntcA* was absent (Zhang, Y. *et al.*, 2007).

11) CELL DIVISION

Wilcox *et al.* (1973a) first suggested that in the heterocyst pattern formation in species of *Anabaena* (*A. catenula* and *A. cylindrica*) one of the cells midway between two heterocysts undergoes an asymmetrical division giving rise to a group of four or more cells and a smaller cell of this group becomes a proheterocyst which develops into to a mature one later. Mechanical breakage of filaments near proheterocysts caused a regression of the proheterocyst into vegetative cell that continued to divide. Adams and Carr (1981) confirmed the observations of Wilcox *et al.* (1973a), in so far as the occurrence of asymmetric cell division is concerned as the frequency of asymmetric divisions increased more than three fold causing the formation of heterocysts in pairs, when the light intensity was increased by nearly 10-fold (500 $\mu E\ cm^{-2}\ s^{-1}$). The frequency of double heterocysts was directly proportional to the period of incubation at high light intensity implying that at high light intensity the cells divided symmetrically resulting in pairs of heterocysts. A period of 8 h incubation in normal light was sufficient to restore the frequency to control values. Two important things emerge from these studies: (1) preceding heterocyst differentiation asymmetrical cell division leads to the formation of a small cell intitial destined to become a proheterocyst, and (2) this cell loses the capacity to divide in due course of time. However, Asai *et al.* (2009) demonstrated that division of a mother cell is not a pre-requisite for heterocyst development after nitrogen step-down. In specially developed silicon mould microchambers (200 x 200 x 8 μm) the growth of single filaments of *Anabaena* sp. strain PCC 7120 was traced in a nitrate-replete medium followed by a nitrogen step-down both in liquid and solid media. Heterocyst differentiation started after 30 h and 10 h in liquid and solid media, respectively. The initial distributions of *hetR-gfp* expression and phycobilisome fluorescence signals could not be correlated to the subsequent pattern of heterocyst development (Figs. 3 and 4). This is in agreement with the results of Toyoshima *et al.* (2010) who observed that about 10% heterocysts were formed without prior division when *Anabaena* sp. strain PCC 7120 was subjected to nitrogen step-down and the growth of individual filaments was followed on agar plates. A *hetR-gfp* transformant showed quartlet of cells from which the selection for proheterocyst rested in 75% of cases on either of the outer cells. According to them it is still not clear as to what are the molecular events that distinguish the inner and outer cells of the quartlet (Toyoshima *et al.*, 2010). Filamentous mutants of *S. elongatus* PCC 7942 formed the starting point for recognition of cell division genes such as *flm* (Dolganov and Grossman, 1993), *ftn2* and *ftn6* (Koksharova and Wolk, 2002). Cells of mutants *ftn2* and *ftn6* are 100- and 20-fold longer than the wild-type, respectively. Other cell division genes have been identified on the basis of their orthologues present in *E. coli* and *B. subtilis*. Previously identified genes (*ftsZ, minC, minD, minE* and *sulA*) and four newly identified genes *ftsE, ftsI, ftsQ* and *ftsW* from the genome of *S. elongatus* PCC 7942 have been characterized by comparative Tn5-692 insertional mutagenesis. Five additional cell division genes *cdv1, cdv2, cdv3, ftn6* and *cikK* have also been reported (Miyagishima *et al.*, 2005). Studies on cell division gene *ftsZ* in relation to heterocyst differentiation are only available. Bacterial *ftsZ* (for filamenting temperature-senstive mutant Z) encodes a protein FtsZ that possesses GTPase activity and due to this it polymerizes into a ring-like structure at the mid-point of the cell where it recruits other proteins to form a septum. A homologue of the *ftsZ* has been cloned and sequenced along with its flanking regions from *Anabaena* sp. strain PCC 7120. The deduced amino acid sequence of FtsZ from *Anabaena* sp. strain PCC 7120 ($FtsZ_{Ana}$ with 379 amino acid residues) showed an identity of 49% to *E. coli* FtsZ. Another ORF downstream of *ftsZ* encodes glutathione synthetase that is transcribed in the opposite direction (Doherty and Adams, 1995). The essential nature of *ftsZ* and the glutathione synthetase gene for cell growth in all nitrogen sources has been established (Zhang *et al.*, 1995). $FtsZ_{Ana}$ overexpressed in *E. coli* has been purified (47 kDa) and rabbit polyclonal antibodies raised against this recombinant protein have been used in an immunodetection assay. The $FtsZ_{Ana}$ with the same molecular mass has been detected from only vegetative cell preparations and heterocyst preparations are devoid of this protein (Kuhn *et al.*, 2000). The localization and formation of a ring-like structure at the site of septum formation by $FtsZ_{Ana}$ in *Anabaena* sp. PCC 7120 has been demonstrated by the expression of *ftsZ* gene under the influence of *petE* promoter fused to *gfp* ($P_{petE\text{-}ftsZ\text{-}gfp}$). Since SulA encoded by *sulA* gene from *E. coli* inhibits GTPase activity of $FtsZ_{Ana}$ *in vitro*, introduction of *sulA* under the influence of *petE* promoter into the strain with $P_{petE\text{-}ftsZ\text{-}gfp}$ inhibited ring formation by $FtsZ_{Ana}$ and heterocyst differentiation in presence of copper. These results suggest that (i) heterocyst differentiation and cell division are coupled processes and (ii) the arrest of cell division has no relation to the levels of 2-OG as the signalling molecule (Sakr *et al.*, 2006).

12) PATTERN FORMATION

The reappearance of the same pattern of heterocysts simultaneously from heterocystsless cultures subjected to nitrogen stress is explained by the existence of incipient or proheterocysts at the same loci in the filaments of *A. ambigua* and *A. cylindrica* (Talpasayi and Kale, 1967; Talpasayi and Bahal,

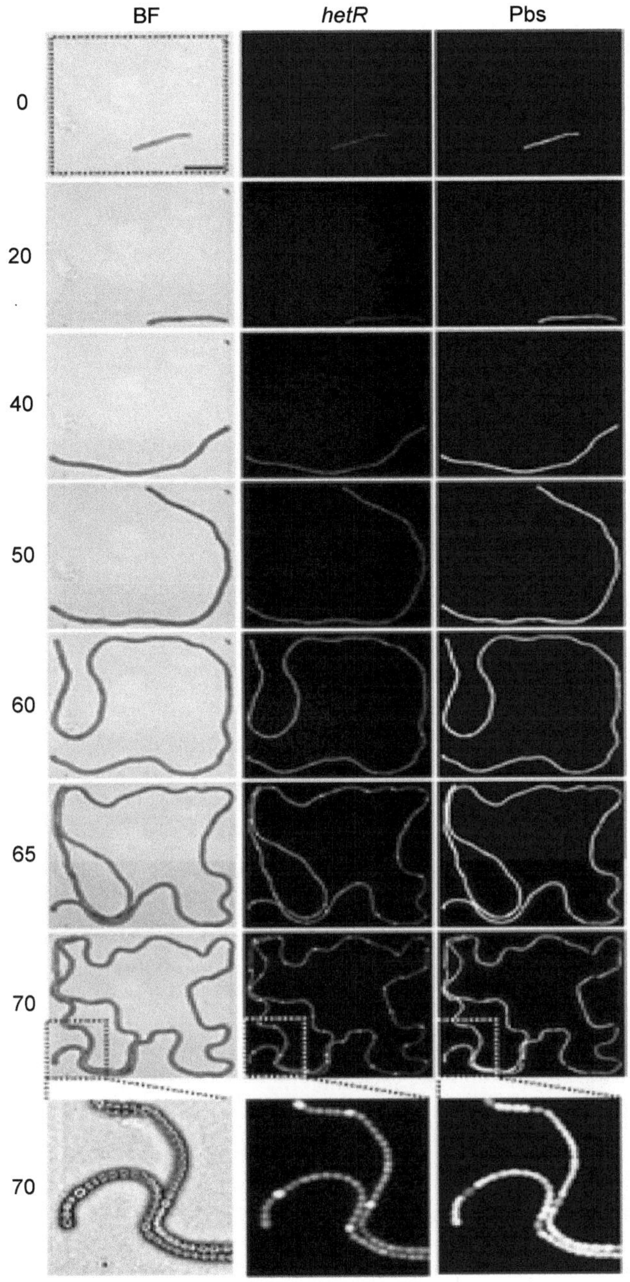

Figure 3: Heterocyst differentiation in microelectromechanical system-assisted liquid microchambers. Spatiotemporal dynamics in morphological changes (BF, bright field), *hetR* promoter (P*hetR*) activity monitored with a transcriptional *gfp* fusion reporter (*hetR*), and phycobilisome fluorescence (Pbs) in an individual *Anabaena* filament at the indicated times after nitrogen step-down. The dashed square at the top indicates the space of a microchamber. The bottom panel (taken at 70 h) shows a magnification of part of the filament. With the kind permission of H. Iwasaki, Department of Electrical Engineering and Biological Science, Waseda University (TWIns), Tokyo, Japan & PRESTO, Japan Science and Technology Agency (JST), Tokyo, Japan [Asai *et al.* (2009) *PLoS ONE* **4(10):** e7371. doi:10.1371/journal.pone.0007371.g001].

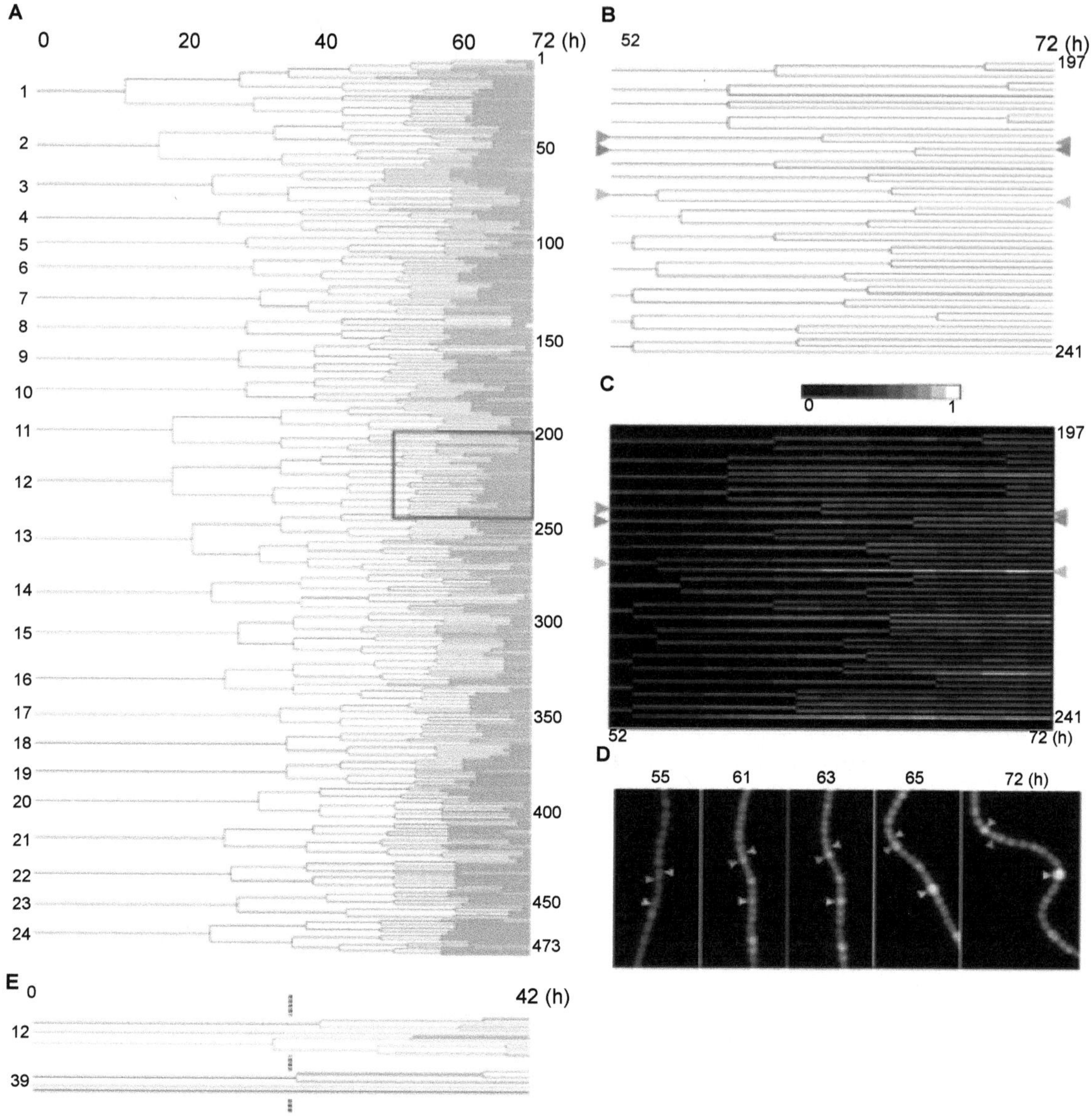

Figure 4: Cell lineage of an individual *Anabaena* filament. (A) The cell lineage analysis representing spatiotemporal profiles of cell division (branches) and heterocyst differentiation (red) in the bacterial filament. The horizontal axis shows time (h) after nitrogen step-down. (B, C) Magnification of a part of the *Anabaena* cell lineage shown in the blue square in panel A (B) and the same lineage superimposed with the spatiotemporal dynamics of P*hetR::gfp* signals (C). Arrowheads indicate some cells showing different *hetR* expression dynamics (D) Micrographs of P*hetR::gfp* expression patterns from the corresponding part of the bacterial filament. (E) Cells that differentiated into heterocysts without cell divisions (numbers 12 and 39). With the kind permission of H. Iwasaki, Department of Electrical Engineering and Biological Science, Waseda University (TWIns), Tokyo, Japan, & PRESTO, Japan Science and Technology Agency (JST), Tokyo, Japan [Asai *et al.* (2009) *PLoS ONE* **4(10):** e7371. doi:10.1371/journal.pone.0007371.g002].

Color image of this figure appears in the color plate section at the end of the book.

1967; Wilcox, 1970). Proheterocysts are slightly enlarged cells that are similar to vegetative cells in their shape and pigmentation, filled with granular inclusions and in some cases even gas vacuoles. Wilcox *et al.* (1973b) have identified seven stages of development of mature heterocysts from proheterocysts in case of *A. cylindrica* and *A. catenula* supported by ultrastructural studies. The first signs of development are the laying down of the outer fibrous layer of the heterocyst envelope that constitutes stage I. At stage II, the fibrous layer is completely laid down with the junction between the developing heterocyst and the vegetative cells gets demarcated. This progressively increases through stages III and IV at which time the disorganization of photosynthetic lamellae takes place with increased intracellular spaces. The deposition of inner laminated layer of the envelope becomes apparent by stage V and with this the developing heterocysts appear to be quite distinct from the vegetative cells. The condensation of the contorted lamellae to the two poles and the formation of polar plug at the two junctions with vegetative cells constitute the stage VI. Heterocyst maturation and aging constitutes the stage VII. The regression of proheterocysts into vegetative cells is quite possible up to stage V but not those which have crossed this stage and entered stage VI. Once the filaments complete the same pattern of heterocyst development after nitrogen shift-down, futher growth of the organism in nitrogen-deficient medium requires the continuation of the same pattern. Two important things that have been taken into consideration are cell division leading to increased filament length in between two heterocysts and the localization of the new site of proheterocyst in between the existing mature heterocysts, establishing a regular spacing of the new heterocysts. What triggers cell division before a proheterocyst differentiates? This has been discussed at length by various hypotheses put forward from time to time. Fogg (1949) put forward the view that there exist nutritional gradients in the filaments caused by the decreasing concentration of specific inhibitory substance probably ammonia that falls below a critical level and such a cell is triggered to undergo development into a proheterocyst, thus maintaining the same distance between two heteocysts. The hypothesis of Fritsch (1951) though on similar lines envisages the existence of growth-promoting substances probably derivatives of ammonia contribute to the growth of the intervening portion of vegetative cells in between two heterocysts and where the concentration of this growth-promoting substance falls below a critical level such cells differentiate into proheterocysts and then they mature into heterocysts. It means that the inhibitory substances or the growth-promoting substances released by the heterocysts prevent the adjacent vegetative cells from getting differentiated into heterocysts. Wolk (1967) demonstrated that the filaments of *A. cylindrica* when fragmented show increased heterocyst frequency due probably to the rate of cell division and not to any of the stimulatory substances or as a result of injury. But he suggested that the normal pattern shown by the filaments in nitrogen-free medium is due to the inhibitory substance released by the heterocyst itself. Diffusion of this compound away from the heterocyst, its loss or degradation from vegetative cells would cause a gradient of concentration and where the concentration of the inhibitor falls below a threshold level only then another proheterocyst would develop (Wolk, 1975, 1989). Wilcox *et al.* (1973a) advanced and extended the same theory and in support of this they showed that as the gradient of the inhibitory substance decreases and reaches a certain critical or threshold level, such cells undergo an asymmetric division producing few cells and there exists a competition in between the new products of division and only the smallest daughter cell is resolved to differentiate into a proheterocyst (Mitchison and Wilcox, 1972). Adams and Carr (1981) also found the existence of asymmetric cell divisions but observed the production of double heterocysts in *A. cylindrica* due to an increase in the intensity of illumination to 500 $\mu E\ cm^{-2}\ s^{-1}$, 10-fold higher than normal intensity. This is explained by the occurrence of symmetric cell division due to 12 h at high light intensity and it took nearly 8 h at normal light intensity to return to normal pattern. This brings out two issues, one is that the proheterocysts inhibit adjacent vegetative cells from differentiation and the second one is that it should also inhibit its own development. Wilcox *et al.* (1973a,b) provided evidences for these points by breaking points of contact between the proheterocysts and adjacent vegetative cells in such a manner that proheterocysts gradually are left with one, two or more adjacent vegetative cells. In cases where single cells are left attached with heterocysts, the proheterocysts had undergone regression into vegetative state in more than 75% such cases. Gradually, the percentage of regression decreased to zero as the number of adjacent vegetative cells increased in number suggesting that the regression of proheterocysts very much depended upon an inhibitory substance produced by it. When percentage of regression is high it means the proheterocysts accumulated the inhibitory substance beyond the threshold level and so further process of differentiation is discontinued and regression is the outcome. It means the inhibitor has to be destroyed in the proheterocyst or it should be lost to the adjacent vegetative cells in order to establish a gradient so that again the same pattern of proheterocyst formation is continued. So it is possible that the inhibitory zone extends by 4-5 cells on either side of the proheterocyst and that the number of intervening vegetative cells always are kept constant (Mitchison *et al.*, 1976). That specific gradients exist under normal growth conditions in heterocyst formation was also confirmed in the genus *Cylindrospermum* sp. which possesses two terminal heterocysts one on each side of the filament. Breakage of heterocysts on either side of the filament leads to formation of a new heterocyst in its place only after the required cell divisions have taken place. If both heterocysts are detached then a heterocyst would be formed first at one of the ends that is farthest from the heterocyst in the original filament and the second heterocyst would be formed after a certain vegetative growth, keeping the interheterocyst distance constant (Reddy and Taplasayi, 1974). Synchronous induction of heterocyst formation occurred after 24 h of nitrogen step-down and heterocyst-specific characteristics were discernible following filament fragmentation in *C. licheniforme* (Van De Water and Simon, 1982). The pattern of basal heterocyst differentiation and polarity in *Calthrix brevissima* are maintained during diazotrophic growth conditions but in presence of combined nitrogen sources heterocyst differentiation is completely suppressed with retention of polarity (in nitrate medium) and complete loss of polarity and development of false branches (in ammonium medium). After a nitrogen shift-down from ammonium medium, the differentiation of heterocysts occurred at regularly spaced intervals with breakage of filaments at the site of differentiation with restoration of polarity (Rai *et al.*, 1978). In symbiotic cyanobacteria such as *N. punctiforme* the normal pattern seems to be disturbed as the heterocyst frequency considerably increases from 25% to 60% of the total cells and the interheterocyst distance also decreases to a great extent (Meeks *et al.*, 2002). Nierzwicki-Bauer *et al.* (1984b) proposed a method for the identification of proheterocysts and heterocysts from complex branched filamentous types such as *M. laminosus* by staining with toluidine blue. Fixed and embedded materials that are sectioned could be readily examined under the light as well as electron microscope.

***Modifiers of the pattern*:** In order to further emphasize their point that proheterocyst regression is dependent on the concentration of the inhibitor, Mitchison and Wilcox (1973) showed that incorporation of 7-azatryptophan (AT), an analogue of tryptophan, significantly reduced the proheterocyst regression and caused differentiation of pairs of heterocysts in *A. catenula* and *A. cylindrica*. Agrawal and Kumar (1978) reported that in *A. doliolum* the addition of AT resulted in proheterocyst regression with a frequency of 100%, 35% and 10% in case of ammonium-, nitrite- and nitrate-containing media, respectively. A screening of twelve amino acid analogues and related compounds by Rogerson (1979) revealed that AT, MSX and β-2-thionyl-DL-alanine considerably reduced the interheterocyst distance and enhanced the frequency of heterocysts in *A. variabilis*.

Since the effects of AT are inhibited in presence of chloramphenicol, it was suggested that AT might be incorporated into proteins in place of tryptophan in case of *Anabaena* sp. CA due to a feedback inhibition caused in tryptophan biosynthesis at the level of the first enzyme, anthranilate synthetase (Bottomley *et al.*, 1980). Another explanation provided for the mode of action of AT is that at concentrations at which AT triggered heterocyst formation, it caused inhibition of GOGAT in *Anabaen* sp. strain CA, the second enzyme involved in ammonia assimilation. In addition, AT caused a decline in intracellular glutamate pool, decreased C-phycocyanin content and inhibited nitrogenase activity leading to nitrogen starvation and triggering more heterocyst differentiation (Chen *et al.*, 1987). Triggering of heterocyst differentiation in presence of MSX is due to a direct or indirect inhibition of GS activity in *Anabaena* spp. (Stewart and Rowell, 1975). Stacey *et al.* (1979) confirmed that AT can relieve the repressive effects of ammonium nitrate in medium by enhancing heterocyst frequency and nitrogenase synthesis and does not inhibit GS activity as MSX does. So GS is not solely responsible for the induction of heterocysts and nitrogenase in cyanobacteria. Adams (1992) also observed the formation of heterocysts in pairs in presence of AT in *A. cylindrica*. According to him though asymmetrical cell divisions occur in case of *A. cylindrica* during heterocyst differentiation at least 10% of these produce daughter cells of the same size and the presence of AT does not alter the situation and so double heterocyst formation results. The decrease of interheterocyst interval depended on the duration of incubation in AT and the decrease was directly proportional to the concentration resulting in multiple heterocysts.

A number of metabolic inhibitors such as thiol inhibitors (mercaptoethanol, p-mercuribenzoate, iodoacetate and ethyl maleimide) sodium azide and glycine inhibited heterocyst differentiation in *A. ambigua* with a reduction in their frequency (Bahal and Talpasayi, 1970a,b). Inhibitors of photosynthesis (DCMU), respiration (2,4 dinitrophenol, sodium azide and iodoacetic acid) effectively inhibited heterocyst differentiation in *A. doliolum*. Glucose reversed the effects of DCMU but not those of 2,4-dinitrophenol (Tyagi, 1973a). Antibiotics such as chloramphenicol (Talpasayi and Kale, 1967) and streptomycin (Kumar and Kaushik, 1971) inhibited heterocyst differentiation suggesting the requirement of protein synthesis during heterocyst differentiation. Actinomycin D (an inhibitor of DNA-dependent RNA synthesis) and mitomycin C (an inhibitor of DNA-dependent DNA synthesis) failed to inhibit heterocyst differentiation in *A. doliolum* (Tyagi, 1973b, 1975). The presence of rifampicin, chloramphenicol and mitomycin C at concentrations 0.2, 0.4 and 1.0 $\mu g\ ml^{-1}$, respectively inhibited proheterocyst and mature heterocyst formation in *A. cylindrica*. Concentrations lower than these permitted the development of proheterocysts but inhibited the development of mature heterocysts. However, the production of heterocysts in chains, disturbing the normal pattern, in *Nostoc linckia* due to the presence of rifampicin (0.1 $\mu g\ ml^{-1}$) alone or in combination with 2-4-dichlorophenoxyacetic acid (100 $\mu g\ ml^{-1}$) has also been reported (Tiwari *et al.*, 1981). The commitment point is the point at which the commitment of a cell to differentiate into a heterocyst can not be reversed by the addition of fixed nitrogen sources or by incubating in darkness. The commitment points for the nitrogen sources, darkness, rifampicin, proflavin and fluorouracil were found to be different (Adams and Carr, 1989). Heterocysts in *A. azollae* developed only up to proheterocyst level in presence of tunicamycin (beyond 0.2 $\mu g\ ml^{-1}$) and though heterocyst-specific glycolipids are synthesized, they are not transported to form the laminated layer (Reddy *et al.*, 1989). The induction of double and multiple heterocysts with altered spacing and pattern formation by neopeptone (0.4 $\mu g\ ml^{-1}$) has been noted in *A. cylindrica* (Sharma, 1984). Ultrastructural studies on heterocysts developed in presence of neopeptone revealed normal structural features with well developed polar nodules (Sharma and von Hofsten, 1987).

13) GENES REGULATING THE PATTERN

i) patA (all0521): During isolation and characterization of Tn5-induced mutants, Liang *et al.* (1992) recognized a PAT-1 mutant that differentiated heterocysts at the terminal positions. The cloned fragment did not contain Tn5 but had a frame-shift mutation due to introduction of a single adenine in a stretch of eight adenine residues at positions 316–323. Due to this, there was termination in the reading frame, 40 nucleotides downstream of the shift point. To recognize the particular gene sequence from *Anabaena* genome, a cosmid library of genes of *Anabaena* sp. strain PCC 7120 was used for complementation. The complementing gene sequence (1.1kb) restored the wild-type phenotype and was designated as *patA* (for pattern). *patA* was then disrupted by the introduction of a Sp-Sm resistance cassette and then transferred into *Anabaena* sp. strain PCC 7120. Single and double recombinants were obtained. Cloning and sequence analysis of *patA* gene and its deduced amino acid sequence revealed it to be a protein with 379 amino acids belonging to the group of response regulators. The C-terminal region of PatA is a phosphoacceptor (receiver) domain bearing resemblance to CheY of *E. coli*. *patA* gene is required for the formation of intercalary heterocysts and in the *patA* mutants the pattern of heterocyst formation is thus disturbed because of the formation of only terminal heterocysts. *patA* mutation also suppresses Mch-phenotype when *hetR* gene is overexpressed (Liang *et al.*, 1992). Orozco *et al.* (2006) created a *patA* null mutant (UHM101) that differentiates only terminal heterocysts resembling the *patA* mutant phenotype described previously (Liang *et al.*, 1992). A *patA hetN* double mutant (UHM112, *patA* deletion mutant with *hetN* expression under *petE* promoter) exhibited delayed Mch-phenotype. Unlike the *patA* deletion mutant which differentiated only terminal heterocysts, under inductive conditions in presence of copper UHM112 differentiated multiple heterocysts at terminal positions. In addition, they created a double deletion mutant (*patA*$^-$ and *patS*$^-$) that had the expression of *hetN* under the *petE* promoter. The Mch-phenotype of this mutant is distinctive for the reason that in addition it differentiated multiple terminal heterocysts thus decreasing sufficiently the interheterocyst distance. This suggests that the inactivation of *patS* is mainly responsible for such a phenotype. Makarova *et al.* (2006) reported a highly conserved N-terminal domain in PatA protein and termed it as PATAN (after PatA N-terminus) and a potential helix-turn-helix domain, characteristic of proteins that bring about protein-protein interactions associated with development and differentiation in cyanobacteria. Studies on transcriptional regulation of *patA* gene revealed the presence of three tsps at –305, –614 and –645 nucleotide positions. Transcription from tsp position -305 occurred in all cells of *Anabaena* sp. strain PCC 7120 after nitrogen step-down whereas transcription from tsps –614 and –645 positions occurred in proheterocysts. Introduction of $P_{patA\text{-}gfp}$ construct showed GFP fluorescence at the site of cell division and overexpression of *patA* resulted in the formation of enlarged cells. It means that the N-terminal and C-terminal domains of PatA function independently and *patA* gene has a role in cell division (Young-Robbins *et al.*, 2010). The expression of *patA* in the heterocysts after 3 to 6 h of nitrogen step-down of *Anabaena* sp. strain PCC 7120 is probably because of the presence of two conservative NtcA-binding sites in the promoter region of *patA* and its interaction with NtcA (Bastet *et al.*, 2010).

ii) patB: Another suspected Fix$^-$ mutant, isolated and characterized by Liang *et al.* (1993) by Tn5-induced mutagenesis of *Anabaena* sp. strain PCC 7120, PAT-2 exhibited a delay of ~21 h in heterocyst formation with a decreased interheterocyst distance of 3 to 8 cells in the older cultures resulting in increased heterocyst frequency. The cloned fragment from PAT-2 mutant did not reveal the interposed Tn5 but contained a deletion of G at nucleotide 1342 position resulting in a frame-shift mutation. This caused the deletion of 62 amino acid C-terminal helix-turn-helix portion from the protein. Studies on complementation of the gene from the cosmid lirary of *Anabaena* sp. strain

PCC 7120 genome, led to the identification of a 1.6 kb ORF named as *patB*. The encoded protein is predicted to contain a Fe_4S_4 bacterial ferredoxin sequence at its N-terminus and a helix-turn-helix at its C-terminus, characteristic of a transcriptional regulator. The essential nature of *patB* gene has been demonstrated by Jones *et al.* (2003) who showed that *patB* deletion mutants were unable to grow in a nitrogen-deficient medium. Mutants of *patB* with defect in the N-terminal portion or the original PAT-2 grew slowly and revealed Mch-phenotype after nitrogen shift-down. Expression of $P_{patB\text{-}gfp}$ gene construct showed green fluorescence exclusively in heterocysts.

iii) hetN: From amongst the repertoire of mutants isolated by Ernst *et al.* (1992), mutant N10 with a Het^- phenotype has been further characterized by Black and Wolk (1994). During reconstruction of a series of Tn5 insertion mutants at positions adjacent to the original site of insertion, one of the mutants that exhibited Mch-phenotype had Tn5 insertion nearest to the 5′-end of an ORF. This gene with 915 bp has been termed *hetN* whose product is similar to ketoacyl reductases. Two other genes, one upstream of *hetN* (at a distance of 556 bp) designated as *hetM* (1,518 bp) and another downstream of *hetN*, designated as *hetI* (711 bp) but present on the opposite strand, have been discovered. The N-terminal end of HetM resembles an acyl-carrier protein with its central portion similar to keto reductases whereas *hetI* encodes products similar to biosynthesis and export of cyclic peptides. When *hetN* is introduced into *Anabaena* sp. strain PCC 7120 on a replicating plasmid it completely suppressed heterocyst differentiation. But when *hetN* is introduced along with *hetI* on a replicating plasmid heterocyst formation is not suppressed. Attempts to inactivate *hetI* failed as fully segregated double recombinants could not be obtained. An intact copy of wild-type *hetN* gene could not complement mutant N10 to regain the wild-type character. Inactivation of *hetN* resulted in Mch-phenotype but on repeated subculturing the mutant changed to Het^- phenotype. Coincidentally, Bauer *et al.* (1997) also identified similar sequence of genes in the order *hetM* (*hglB*)-*hetN*-*hetI*. The gene *hetM* has been redesignated by them as *hglB* responsible for Hgl synthesis. The role of *hetN* in the maintenance of heterocyst pattern in *Anabaena* sp. strain PCC 7120 has further been investigated by Callahan and Buikema (2001) who replaced chromosomal *hetN* promoter region with the *petE* (that encodes plastocyanin) promoter (inducible in presence of copper) of *Anabaena* sp. strain PCC 7120. The resultant strain 7120PN, showed expression of *hetN* in presence of copper that suppressed heterocyst differentiation. This is in agreement with earlier observations (Black and Wolk, 1994; Bauer *et al.*, 1997). In the absence copper, as there was no expression of *hetN* (this situation almost equals to creating a null mutant for *hetN* and also obviates the ill effects of transposon mutagenesis), the organism showed a normal wild-type pattern of heterocyst differentiation up to 48 h but afterwards the Mch-phenotype prevailed. This has been attributed to the complete lack of HetN protein. Since the presence of *hetI* along with *hetN* on a replicating plasmid did not cause any inhibition in heterocyst differentiation and to rule out the possibility of any cis effects of *hetI* on *hetN* (Black and Wolk, 1994), they introduced $P_{petE\text{-}hetN}$ on a replicating plasmid (pSMC15) into wild-type. Complete suppression of heterocyst differentiation occurred as noted in case of 7120PN. *hetN* also prevented the patterned expression of *hetR*, the master regulator. This has been confirmed by introducing a $P_{hetR\text{-}gfp}$ fusion construct in plasmid pSMC127 into wild-type and 7120PN. The wild-type showed increased GFP fluorescence characteristically in groups of cells and to a lower level in all other cells in presence of copper but 7120PN exhibited decreased fluorescence of *hetR* uniformly in all cells. As it is reported earlier that ectopic overexpression of *hetR* leads to Mch-phenotype, $P_{petE\text{-}hetR}$ translational fusion was introduced into wild-type. In presence of copper overexpression of *hetR* led to Mch-phenotype in nitrogen-free medium as well as nitrate enriched medium. When a plasmid with $P_{petE\text{-}hetR}$ was introduced into 7120PN strain, overexpression of both *hetR* and *hetN* took place in presence of copper and the Mch-phenotype was suppressed and with only 1% heterocysts in nitrogen-free medium and none in presence of combined nitrogen. A heterocyst suppression signal by *hetN* has been suggested to originate from the heterocysts as is evident by the localized fluorescence of GFP in the heterocysts. They further showed that *hetN* is not required for Hgl synthesis (Callahan and Buikema, 2001). However, Li *et al.* (2002) suggested that HetN plays a role in fatty acid synthesis. Overexpression of *hetN* from *Anabaena* sp. strain PCC 7120 in *E. coli* caused the accumulation of HetN as inclusion bodies. Purified recombinant protein was used to raise polyclonal antibodies. The presence of HetN protein, in the heterocyst preparations of *Anabaena* sp. strain PCC 7120 subjected to nitrogen step-down, was demonstrated by immunoblotting and the concentration of HetN was 0.5 to 1.0% of the total protein. The association of HetN in higher concentrations (3.5 times) on the thylakoid membranes than that with the cytoplasmic membrane prompted them to conclude that HetN participates in fatty acid synthesis. They confirmed the observations of Callahan and Buikema (2001) on the effect of overexpression of HetN on *hetR*. The C-terminal pentapeptide RGSRG motif, that is characteristic of *patS* which suppressed heterocyst differentiation, has also been found in the gene sequence of *hetN*. If the suppression of heterocyst differentiation by *hetN* is due to the presence of this sequence, any mutation caused in these bases should not suppress heterocyst differentiation. Mutant copies of *hetN* (*Arg132Lys*, *Gly134Ser*, *Ser135Asp* and *Arg136Leu*) when introduced into wild-type on a multicopy plasmid (pRL25C), also suppressed heterocyst differentiation ruling out the possibility of the functional nature of the pentapeptide sequence (Li *et al.*, 2002). The biochemical properties of HetN revealed that it could hydrolyze ATP or GTP *in vitro* in presence of magnesium ions and the ketoacyl reductase activity has been predicted to reside in the conserved triad (Ser142-Tyr155-Lys159) of amino acids. Purification of recombinant proteins of HetN produced in *E. coli* has been feasible with the introduction of an eight amino acid fusion tag (Trp-Ser-His-Pro-Gln-Phe-Glu-Lys) at the C-terminus, known as Strep Tag II. Mutations in the conserved triad (*hetNSer142Asp*, *hetNTyr155Val* and *hetNLys159Glu*) did not affect the activity of ATP hydrolysis indicating there by that the catalytic mechanism of ATP/GTP hydrolysis of HetN is different from that of the reductase activity. The heterocyst suppressing activity of HetN has been tested by introducing mutant alleles of *hetN*, *hetNSer142Asp*, *hetNTyr155Val* and *hetNLys159Glu* on a multicopy plasmid (pRL25T) into *Anabaena* sp. strain PCC 7120. Mutant alleles, *hetNSer142Asp* and *hetNTyr155Val* suppressed heterocyst differentiation whereas mutant *hetNLys159Glu* did not suppress heterocyst differentiation. This is suggestive of the fact that the heterocyst suppression activity of *hetN* resides in Lys159 (Liu and Chen, 2009).

iv) patS: Originally identified as a missense mutation in a fragment of DNA (3.3 kb), carried by the conjugal plasmid and when introduced into *Anabeana* sp. strain PCC 7120, it completely suppressed the heterocyst formation. Yoon and Golden (1998) identified a short ORF of 54 bp from the above fragment that encodes a small peptide of 17 amino acids with its N-terminal part containing five hydrophobic amino acids and 5 to 8 amino acids at the C-terminal position that are important for its activity. The ORF has been named as *patS* (for pattern suppression). Overexpression studies were conducted by using *patS* gene under the control of *glnA* promoter and it resulted in complete suppression of heterocysts. When *patS* was put under the influence of promoter of *petE*, in the absence of copper since there was no expression of *patS*, the organism exhibited a normal pattern of heterocyst differentiation but in presence of copper due to induction of *patS* a decrease in heterocyst frequency was noted. Deletion of *patS* gene (strain AMC451) resulted in Mch-phenotype. The fluorescence of GFP has been restricted to proheterocysts during first 6 h after nitrogen step-down when $P_{patS\text{-}gfp}$ was used as a reporter gene. A synthetic pentapeptide PatS-5 (RGSGR), of the C-terminal amino acids, caused inhibition of heterocyst differentiation at very low concentrations. However, PatS-4 (GSGR) or PatS-4 (GSSR) exhibited reduced activity. The Mch-phenotype of AMC451 has been suppressed by the exogenously added PatS-5. These workers suggested that the PatS synthesized in proheterocysts diffuses along the filaments forming a gradient of inhibitor concentration on both sides of the proheterocyst and wherever the concentration is low such cells are destined to undergo differentiation into heterocyst (Yoon and Golden, 1998). Yoon and Golden (2001) recognized a group of few cells which express the GFP fluorescence in between two heterocysts when *patS-gfp* was used as a reporter gene and these resolve amongst themselves for differentiation into a proheterocyst. It appears PatS signalling is involved in the resolution of these clusters. Different transcripts of *patS* gene have been detected in vegetative cells grown on nitrate and after nitrogen step-down, where tsp located at –314 bp from the initiation codon was utilized in the former cells and this was replaced by a transcript synthesized from the tsp localized at –39 bp in the latter cells. Since *patS* mutant formed heterocysts without nitrogenase activity and a regular pattern in a nitrate medium, they suggested that PatS and the nitrogen fixation products together constitute the signals for heterocyst differentiation.

Synthetic *patS* minigenes encoding the 4 to 8 amino acids of the C-terminus of PatS have been expressed under different promoters such as P_{petE}, P_{patS} or P_{rbcL} in *Anabaena* sp. strain PCC 7120. Heterocyst formation was inhibited in case of *patS-5* to *patS-8* constructs (*patS-5*, RGSGR; *patS-6*, Glu-RGSGR; *patS-7*, Asp-Glu-RGSGR and *patS-8*, Cys-Asp-Glu-RGSGR) but not in case of *patS-4* (Gly-Ser-Gly-Arg). Although *patS-5* inhibited heterocyst differentiation as reported earlier (Yoon and Golden, 1998), it could not complement the AMC451 mutant. The site of action of PatS seems to be localized in the cytoplasm as fusion proteins of GFP with PatS-5 and a full length of PatS with a hexahistidine tag at its C-terminus both inhibited heterocyst differentiation. But expression of full length *patS* under the influence of *hepA* promoter ($P_{hepA\text{-}patS17}$) that is specific for proheterocysts in a *patS*-null mutant (AMC451) complemented it, restored the wild-type character and suppressed the Mch-phenotype. These observations emphasize that for cell-cell signalling full length PatS of the wild-type is functional. Two unrelated genes for heterocyst differentiation, i.e. *all3290* and orf77 (an unannotated ORF) that possess the pentapeptode sequence RGSGR motif when overexpressed under the control of *petE* promoter, also suppressed heterocyst differentiation (Wu *et al.*, 2004).

Besides *patS*, another gene that suppresses heterocyst formation is *hetN*. Both appear to have independent pathways for heterocyst suppression in the absence of ammonia. Inactivation of these two genes in *Anabaena* sp. strain PCC 7120 caused almost every cell to get differentiated into a heterocyst and the differentiation even took place in presence of nitrate, in contrast to wild-type that did not form heterocysts in presence of nitrate (Borthakur *et al.*, 2005). Orozco *et al.* (2006) created a *patS* null mutant (UHM114) that exhibited Mch-phenotype and resembled previously reported *patS* replacement mutants (Yoon and Golden, 1998, 2001). By utilizing a *patA* null mutant (UHM101), they studied the effect of inactivation of *patS* by generating a *patS*$^-$ and *patA*$^-$ double mutant (UHM113) which differentiated single heterocysts at terminal and intercalary positions separated by stretches of vegetative cells. The phenotype of UHM113 resembled the phenotype of *patS* null mutant (UHM114). A functional *patS* gene seems to be necessary for the formation of a *patA* phenotype. The second aspect of epistasis anslysis of *hetR* and *patS* genes is that when both these genes are overexpressed together, the site of action of *patS* appeared to be downstream of *hetR* transcription in the molecular events leading to heterocyst differentiation (Orozco *et al.*, 2006).

The presence of *hetR* and *patS* gene sequences in five non-heterocystous, non-diazotrophic, filamentous cyanobacteria (*A. platensis*, *Lyngbya* sp. FACHB 722, *Oscillatoria lutea* FACHB 406, *Phormidium mucicola* FACHB 723, *Schizothrix calcicola* FACHB 404) has been demonstrated by Zhang *et al.* (2009). Polyclonal antibodies raised against HetR protein of *Anabaena* sp. strain PCC 7120 cross reacted with the HetR protein present in the above five cyanobacteria. *het R* and *patS* are present close to each other and *patS* gene is located upstream of *hetR*. In *A. platensis*, *patS* encodes a large protein of 84 amino acids with RGSGR pentapeptide sequence at its C-terminus. The functional nature of *hetR* and *patS* gene sequences of *A. platensis* ($hetR_{ar}$ and $patS_{ar}$) has been tested by introducing the promoter regions of these genes along with *gfp* as reporter gene into *Anabaena* sp. strain PCC 7120 through conjugation. The exconjugants displayed GFP fluorescence typical of the expression of these genes in *Anabaena* sp. strain PCC 7120, with $hetR_{ar}$ expression not restricted to heterocysts but $patS_{ar}$ expression restricted to heterocysts after a nitrogen step-down. These observations amply signify that these gene sequences have evolved much before the appearance of heterocyst and their presence in non-heterocystous members indicates much wider function for HetR and PatS (Zhang *et al.*, 2009).

***v) Nature of the inhibitor*:** According to Fogg (1949) the nature of the inhibitor that forms a gradient on either side of a heterocyst is ammonia or a derivative of it and when the concentration of this substance falls below a certain critical level such cells tend to differentiate into heterocysts. Wolk (1967) first suggested that it is the heterocysts that produce the inhibitor that prevented the neighbouring cells to differentiate into heterocysts. Diffusion of this compound away from heterocysts or its loss or degradation within vegetative cells results in the establishment of a gradient in the vegetative cells on either side of the heterocyst. This was explained on the basis that when proheterocysts are formed they further produce the inhibitory substance and establish gradient on either side of it (Wolk and Quine, 1975). Since glutamine is the end product of nitrogen fixation and is transported into adjacent vegetative cells, glutamine or a derivative of it may be responsible for the pattern formation (Thomas *et al.*, 1977). Glutamine or a derivative of it could play such a role has been emphasized by Wolk (1979, 1982, 1989, 1991). Since proheterocysts are not equipped with the capacity to fix nitrogen to establish such a gradient of glutamine, it was suggested that the source of glutamine is from extensive proteolysis that undergoes in the vegetative cell before becoming a proheterocyst (Fleming and Haselkorn, 1974). Thiel and Leone (1986) challenged this on the basis that (i) mutants of *A. variabilis* with enhanced glutamine uptake showed conversion of most of the glutamine to glutamate; (ii) the intracellular concentrations of glutamine, glutamate, arginine, ornithine and citrulline are higher in glutamine grown cells rather than those cells grown in ammonium nitrogen or N_2; (iii) differentiation of heterocyst occurred in glutamine-containing media; (iv) due to high intracellular concentration of glutamine there is no question of a gradient being established and even then heterocyst differentiation took place. The effects of the amino acid analogue, AT in inducing pairs of heterocysts and multiple contiguous heterocysts are explained on the basis of inhibition of GOGAT activity and the interconnected events leading to a general nitrogen-deficiency (Chen *et al.*, 1987). The observations of Bottomley *et al.* (1980) on the feedback inhibition by AT of the first enzyme in the tryptophan biosynthesis pathway and the possibility of AT being incorporated into proteins that have a regulatory role in heterocyst differentiation point out to the involvement of tryptophan metabolism in the heterocyst differentiation (Adams, 1992; Bottomley *et al.*, 1980; Chen *et al.*, 1987). Adams (1992) proposed a modification of inhibitor/diffusion model in which all vegetative cells produce constitutively an inactive inhibitor that is supposed to be activated by a co-inhibitor produced by a proheterocyst or mature heterocyst. In presence of combined nitrogen sources like ammonium, the inhibitor is active in all cells and so no differentiation of heterocysts takes place but in nitrogen-limited cultures, the co-inhibitor produced by proheterocyst or heterocyst will interact with the inactive inhibitor present in adjacent vegetative cells and activate it so that the adjacent cells are inhibited.

The controversy with regard to the chemical nature of the inhibitor(s) seems to have been set at rest with the discovery of a number of genes and the interaction of their products. The discovery of *hetR* (Buikema and Haselkorn, 1991b), *patA* (Liang *et al.*, 1992), *patB* (Liang *et al.*, 1993), *patS* (Yoon and Golden, 1998), *hetN* (Ernst *et al.*, 1992) and *hetC* (Xu and Wolk, 2001) have heralded a new phase in understanding pattern formation. The pattern suppression gene, *patS* that encodes a polypeptide of 15–17 amino acids with a C-terminal pentapeptide (RGSGR) sequence is responsible for pattern formation and synthetic RGSGR sequence at 1 μM inhibits the formation of heterocysts (Yoon and Golden, 1998). PatS, synthesized in the mature heterocysts, is the main inhibitor that is supposed to form the gradient in the vegetative cells. Yoon and Golden (2001) elegantly demonstrated this by making transcriptional fusions between $P_{patS\text{-}gfp}$ and showed that at the time the products of asymmetric cell division are to be resolved for proheterocyst differentiation it is the level of PatS that deterimes the course of events. Thus PatS controls the pattern of intercalary heterocysts. Like PatS, HetN (a gene product of *hetN*, Ernst *et al.*, 1992) is another negative regulator of heterocyst differentiation. PatS and HetN act independently and have complementary functions and help in the establishment and maintenance of the pattern. Complete inactivation of both *patS* and *hetN* leads to an aberrant multiple-chain heterocyst phenotype (Borthakur *et al.*, 2005). A *patA* null mutant formed terminal heterocysts only (Orozoco *et al.*, 2006) and a mutation in *patB* caused a delay in heterocyst formation. HetC regulates a number of cell division genes (Xu and Wolk, 2001).

HetR is the master regulator of heterocyst differentiation and its up-regulation is very important for heterocyst differentiation to be initiated. That is HetR protein activates its own transcription and *patS* gene expression. PatS inhibits the upregulation of *hetR* by preventing the DNA-binding activity of HetR (Black *et al.*, 1993; Golden and Yoon, 2003; Huang *et al.*, 2004). Signalling between HetR, PatS and HetN is important for normal pattern formation and maintenance but when it is lost it leads to loss of heterocyst differentiation (Khudyakov and Golden, 2004). To explain the role of HetR, PatS and HetN in pattern formation of heterocysts four models have been presented. The first one proposed by Allard *et al.* (2007) visualized pattern formation without patterning proteins, HetR, PatS and HetN. These workers laid emphasis on cell growth and division together with the dynamics of fixed nitrogen. Three important testable hypotheses put forward by them are (i) the initial heterocyst positions can be correlated to the fast growing and or small vegetative cells; (ii) the release of fixed nitrogen albeit at small rates 10% or so will trigger PatS-induced resolution of proheterocysts and (iii) fixed nitrogen efflux pumps operating in vegetative cells cause the gradients in fixed nitrogen and leads to the resolution of proheterocysts. The rest of the three models are on similar lines. Risser and Callahan (2009) envisaged an activator-inhibitor model. HetR is the activator that forms gradients in the vegetative cells adjacent to the heterocysts and PatS and HetN are the two inhibitors originating from heterocysts also form gradients in vegetative cells. One of the inhibitors is overexpressed leading to activator decay committing the cell to differentiate into a proheterocyst. Gerdtzen *et al.* (2009) not only took into account the gene interactions between *ntcA*, *hetR* and *patS* involved in heterocyst differentiation but also the influence of knock-out or overexpression of the individual elements. Primarily based on the one-dimensional activator-inhibition model, this model highlighted the role of HetR as an activator and PatS as the inhibitor that can inhibit the production of HetR (Figs. 5 and 6). The model presented by Zhu *et al.* (2010) is a mathematical model incorporating the molecular interactions that regulate heterocyst pattern formation and maintenance in *Anabaena* sp. strain PCC 7120. This model is based on the concept of a non-diffusing activator (HetR) and two inhibitors (PatS and HetN) that show different diffusion rates and temporal expression patterns.

14) NITROGEN FIXATION

The overall reaction of nitrogen fixation and the role of gene products have been depicted in Fig. 7. The arrangement of nitrogen fixation (*nif*) genes in diazotrophic cyanobacteria such as *Anabaena*

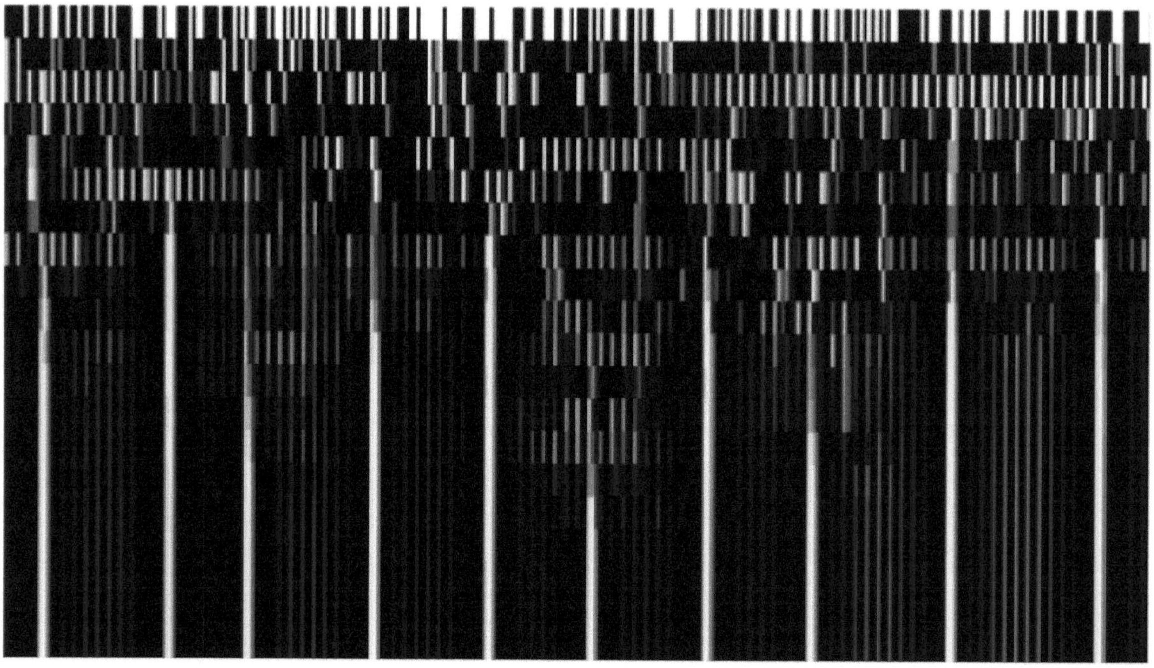

Figure 5: Convergence plot of the network for an array of 100 cells starting from random uniformly distributed binary initial conditions, 0/black or 1/white for *ntcA*, *hetR* and *patS*. Each new row represents the expression levels after one random iteration starting from the state of the previous row. The system converges to one attractor where only some cells differentiate to heterocysts, represented by high values in HetR (the stable vertical lines formed in the figure). The average distance between heterocysts is approximately 10 with a transport factor *D*=0.767. With the kind permission of Z. P. Gerdtzen, Centre for Biochemical Engineering and Biotechnology, Department of Chemical Engineering and Biotechnology, University of Chile, Av. Beauchef 850, Santiago 837-0448, Chile. [Gerdtzen *et al.* (2009) *BMC Bioinformatics* **10**(Suppl 6): S16 doi:10.1186/1471-2105-10-S6-S16].

sp. strain PCC 7120, into a *nif* gene cluster, has been largely understood from the studies of Mazur *et al.* (1980), Rice *et al.* (1982) and Golden *et al.* (1985). There are seventeen genes associated with *nif* gene cluster organized into several operons and attain a length of 23 kb. These are expressed as one of the "genomic islands" after a nitrogen step-down at late stages of heterocyst differentiation (Golden *et al.*, 1991; Ehira *et al.*, 2003). The *nif* operons consist of *nifB-fdxN-nifS-nifU*, *nifH-nifD-nifK*, *nifE-nifN-nifX-nifW* and *hesAB-fdxH- feoA*. The transcripts of *fdxH* also appear as monocistronic entities. Two more genes *xisA* and *xisF* that encode site-specific recombinases are present as the interrupting DNA elements in the *nifD* and *fdxN* genes, respectively (Golden *et al.*, 1987, 1988, 1991; Mulligan and Haselkorn, 1989; Ehira *et al.*, 2003; Fig. 8). *A. variabilis* ATCC 29413 is a heterotrophic cyanobacterium that differentiates heterocysts in aerobic conditions and under strictly anaerobic conditions the vegetative cells also exhibit nitrogenase activity. This organism contains three *nif* gene clusters, i.e. *nif1*, *nif2* and *vnf*. The nitrogenases encoded by *nif1* and *nif2* clusters contain

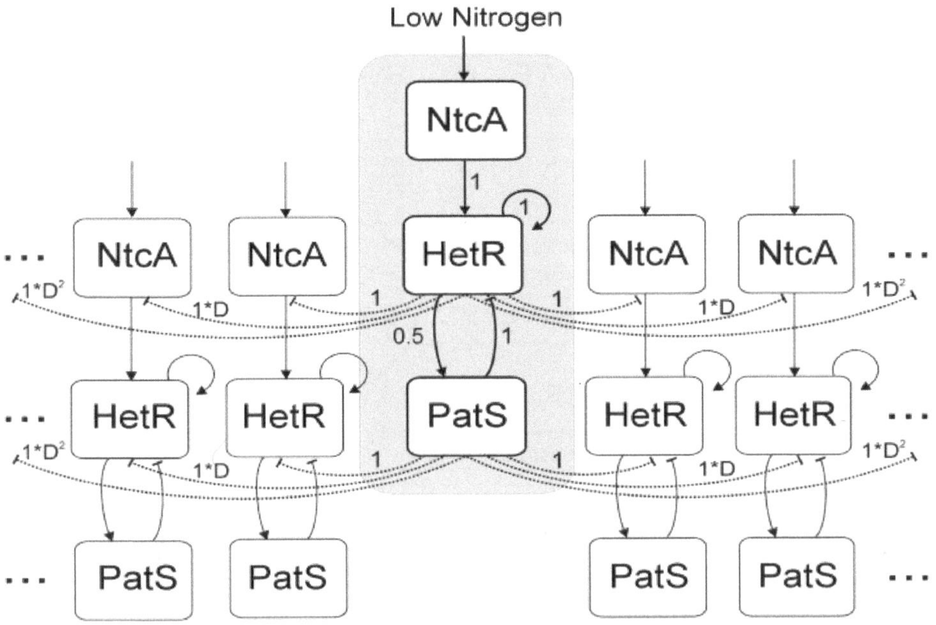

Figure 6: Diagram of the network considered. Cells are organized in a cyclical manner. Direct interactions are represented by solid lines and indirect interactions by dashed lines. Arrow heads indicate activation and vertical lines indicate inhibition. Numbers indicate the strength of the interactions considered among the elements of the network. With the kind permission of Z. P. Gerdtzen, Centre for Biochemical Engineering and Biotechnology, Department of Chemical Engineering and Biotechnology, University of Chile, Av. Beauchef 850, Santiago 837-0448, Chile [Gerdtzen *et al.* (2009) *BMC Bioinformatics* **10** (Suppl 6): S16 doi:10.1186/1471-2105-10-S6-S16].

Mo-Fe co-factor whereas the nitrogenase from *vnf* gene cluster contains a vanadium-Fe co-factor. The arrangement of genes in *nif1* gene cluster is almost the same as in *Anabaena* sp. strain PCC 7120, except that it does not contain the 55-kb element in *fdxN* gene and is expressed only in the heterocysts with Mo-dependent nitrogenase produced under aerobic conditions (Thiel *et al.*, 1995). In the *nif2* gene cluster that lacks the *fdxN* gene, the homologues of *nifE1* and *nifN1* are fused into a single ORF *nifEN2*. The *nif2* encoded Mo-nitrogenase is operative in vegetative cells and heterocysts under strictly anaerobic conditions (Fig. 9; Thiel *et al.*, 1995, 1997). The *vnf* cluster is expressed in the heterocysts (Thiel, 1993, 1996). Although there are evidences for greater genome erosion of the endosymbiont *Nostoc azollae* 0708, all the genes of the *nif1* gene cluster are intact and are flanked on either side by the presence of transposase genes (Ran *et al.*, 2010). There is a greater degree of similarity in the arrangement of genes of *nif1* gene cluster between *N. azollae* 0708 and *Anabaena* sp. strain PCC 7120 on the one hand and *N. punctiforme* ATCC 29133 (PCC 73102) on the other that also has *nif1* gene cluster (Fig. 8). In the toxic *Cylindrospermopsis raciborskii* CS-505 *nif* gene cluster comprises of a 15 kb region with contiguous *nifHDK* operon and 13 other genes related to nitrogen fixation. This cluster is similar to the *nif1* gene cluster of *Anabaena* sp. strain PCC 7120 and *A. variabilis* ATCC 29413 described above but the gene organization is comparable to the *nif2* gene cluster of *A. variabilis* ATCC 29413 and *N. azollae* 0708 (Fig. 8). In *C. raciborskii* CS-505 a *vnf* gene cluster is also present at a different locus (Stucken *et al.*, 2010). *Raphidiopsis brookii* D9, an isolate from Brazil, does not fix nitrogen and so did not reveal a *nif* gene cluster in its genome but the presence of coding sequences for coenzymes and amino acid transport are present (Figs. 9 and 10; Stucken *et al.*, 2010).

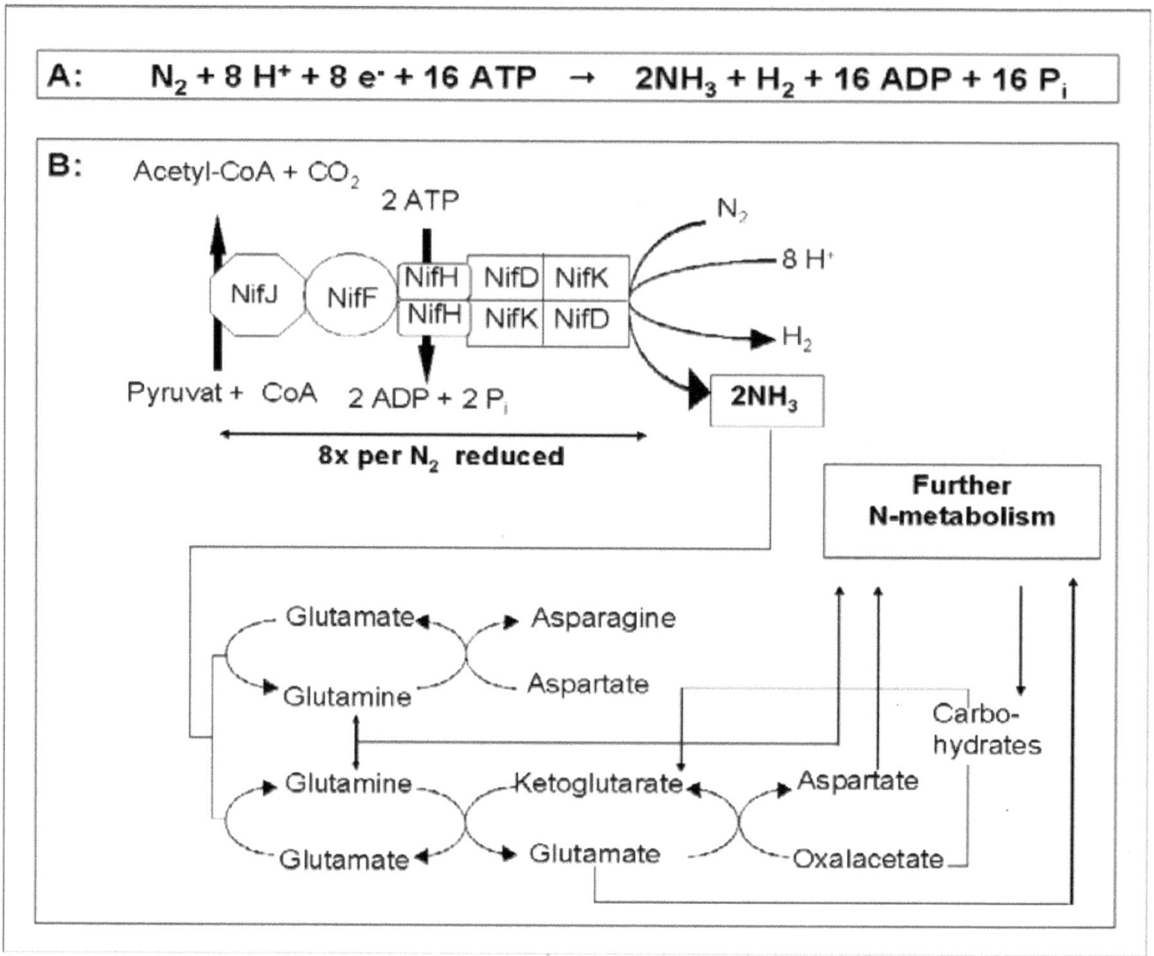

Figure 7: Reaction and molecular mechanism of biological nitrogen fixation. (A) General reaction of molecular nitrogen fixation B. Schematic structure and operation of the nitrogenase enzyme complex and subsequent metabolism of nitrogen. Electrons are transferred from reduced ferredoxin (or flavodoxin) via azoferredoxin to molybdoferredoxin. Each mol of fixed nitrogen requires 16 mol ATP hydrolyzed by the NifH protein. The NH_3 produced is utilised in the synthesis of glutamine or glutamate, respectively, for N-metabolism. NifJ: pyruvate flavodoxin/ferrodoxin oxidoreductase, NifF: Flavodoxin/ Ferredoxin). With the kind permission of C. Kneip, Department of Cell Biology, Philipps-University Marburg, Marburg, Germany & Department of Molecular Biology, Max-Planck-Institute for Infection Biology, Berlin, Germany. [Kneip *et al.* (2007) *BMC Evolutionary Biology* **7:** 55 doi: 10.1186/1471-2148-7-55].

***i) Genome rearrangement*:** The interrupting element in the *nifD* gene is of 11 kb and possesses 11 bp direct repeats (GGATTACTCCG) at each of its ends and contains the *xisA* gene. The second interrupting element is of 55 kb and interrupts the *fdxN* gene in the *nifB-fdxN-nifS-nifU* operon and possesses 5 bp direct repeats (TATTC) at its ends. This large interrupting element contains *xisF* gene. During heterocyst differentiation, both these interrupting elements (11 and 55 kb) are excised at the direct repeats by the activity of the corresponding recombinase proteins and the gene continuities are established. The excised elements remain as circular DNA molecules inside the cytoplasm of the differentiating heterocysts. The expression of all the genes in the *nif* gene cluster enables the

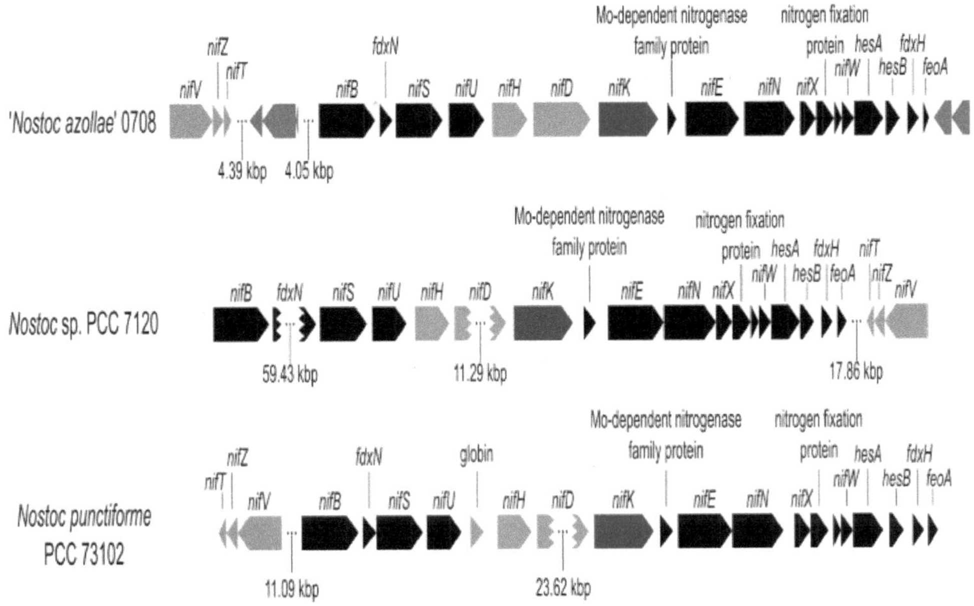

Figure 8: Illustration of genes related to N_2-fixation, a highly conserved gene cluster in cyanobacteria. The structural genes for the nitrogenase enzyme (*nifHDK*) are highlighted in color for clarity. Also, genes which differ in terms of occurrence and/or organization are indicated in grey. The nitrogenase enzyme catalyzes the fixation of atmospheric dinitrogen gas. Transposases are indicated in red. Three dots indicate gaps and incision elements, with the length of the omitted sequence given. With the kind permission of B. Bergman, Department of Botany, Stockholm University, Stockholm, Sweden [Ran *et al.* (2010) *PLoS ONE* **5(7)**: e11486. doi:10.1371/journal.pone.0011486] doi:10.1371/journal.pone.0011486.g007.

Color image of this figure appears in the color plate section at the end of the book.

organism to complete the differentiatiation of heterocysts and fix nitrogen under aerobic conditions. Golden *et al.* (1985) first reported the excision of the 11-kb element flanking the *nifK* and *nifD* genes precisely around 18 h after nitrogen step-down during heterocyst differentiation in *Anabaena* sp. strain PCC 7120. This results in juxtaposition of the two genes and continuity of the *nifD* protein-coding sequence coming under the transcriptional control of *nifH* promoter region (Haselkorn *et al.*, 1983, 1986). In many of the non-heterocystous diazotrophic cyanobacteria that have been examined there is no 11-kb element in the *nifD* gene and the *nifKDH* is a contiguous region. Examples include *Plectonema* (Barnum and Gendel, 1985), *Cyanothece* sp. and *Synechococcus* sp. (Kallas *et al.*, 1985) and *Pseudoanabaena* ATCC 29210 and *Oscillatoria tenuis* UTEX 1566 (Saville *et al.*, 1987) that fix nitrogen under microaerophilic or anaerobic conditions. Non-nitrogen fixing unicellular cyanobacteria like *Cocchochloris peniocystis* ATCC 27147, *Aphanocapsa* sp. ATCC 27178 and *Synechocystis* sp. ATCC 29109 neither contain the *nifKDH* nor the 11-kb sequences in their genomes (Saville *et al.*, 1987). The genomes of many heterocystous cyanobacteria such as strains of *Anabaena, Nostoc* and *Calothrix* contain the 11 kb element. However, exceptions do exist that do not contain the 11 kb element in their genomes such as *Fischerella* sp. ATCC 27929 (Saville *et al.*, 1987) and certain strains of *Anabaena* symbiotic in *A. caroliniana* and *Anthoceros punctatus* (Franche and Cohen-Bazire, 1985, 1987; Nierzwicki-Bauer and Haselkorn, 1986; Meeks *et al.*, 1988), *N. azollae* 0708 endosymbiont of *A. filiculoides* (Ran *et al.*, 2010) and *C. raciborskii* CS-505 (Stucken *et al.*, 2010). However, differences in the freshly isolated and cultured strains of the symbionts of *Azolla* and *A. punctatus* do exist. For example, in the

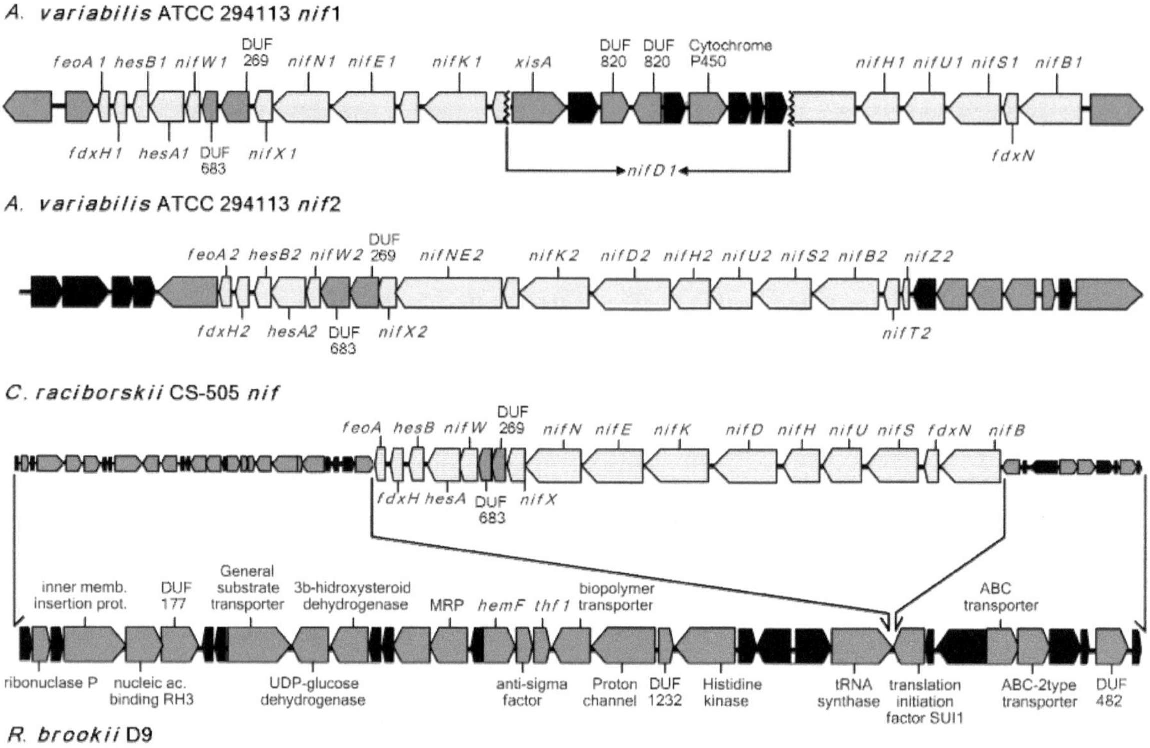

Figure 9: Schematic representation of the synteny within the vicinity of the *nif* gene clusters. The scheme represents the 15 kb gene cluster containing the *nifHDK* and the other 13 nitrogen fixation related genes in *Cylindrospermopsis raciborskii* CS-505 compared with the *nif1* and *nif2* gene clusters of *Anabaena variabilis* ATCC 29413 and the synteny regions between *C. raciborskii* CS-505 and *Raphidiopsis brookii* D9. The synteny regions between *C. raciborskii* CS-505 and *R. brookii* D9 are delimited by the arrows. *nif* genes are represented by light grey and dashed lines. Genes in black correspond to hypothetical proteins and grey genes to proteins with assigned function. With the kind permission of Mónica Vásquez, Department of Molecular Genetic and Microbiology, Faculty of Biological Sciences, Pontificia Universidad Católica de Chile, Santiago, Chile & Millenium Nucleus EMBA, Santiago, Chile. [Stucken *et al.* (2010) *PLoS ONE* **5(2):** e9235. doi:10.1371/journal.pone.0009235] doi:10.1371/journal.pone.0009235.g003.

freshly isolated symbionts of the former host from Senegal the 11 kb element was absent while the symbionts of the latter (freshly isolated as well as the cultured strains) showed the presence of *xisA* gene (Meeks *et al.*, 1988). There is considerable variation in the length of *nifD* element in the different cyanobacteria examined so far. The shortest *nifD* element is found in *Anabaena* sp. strain ATCC 33047 (4 kb; Henson *et al.*, 2005) and the largest (24 kb) in *N. punctiforme* ATCC 29133 (Meeks *et al.*, 2001) with the intermediate size of 11 kb detected in *Anabaena* sp. strain PCC 7120 (Golden *et al.*, 1985; 1988) and *A. variabilis* ATCC 29413 (Brusca *et al.*, 1989). In all the *nifD* elements of these cyanobacteria the *xisA* gene is located closest to the 3′-end of *nifD*. *Anabaena* sp. strain ATCC 33047 is an exception in possessing a changed 11-bp repeat nearest to the *xisA* by one nucleotide (CGGAGTAATTC) while the direct repeat farthest from *xisA* is identical (CGGAGTAATCC) to the direct repeats from all others (Henson *et al.*, 2005). There appears to be considerable sequence conservation in all the

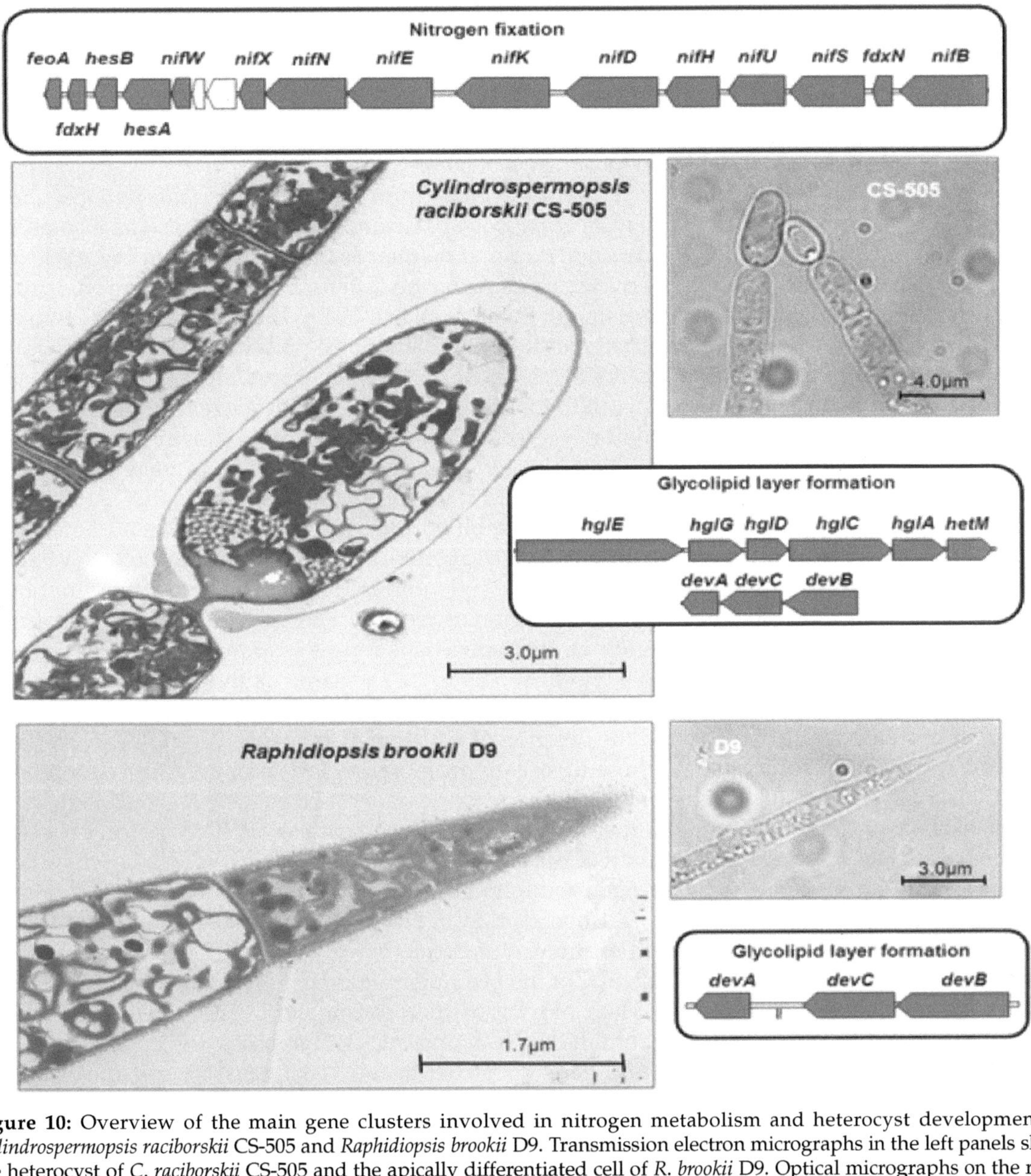

Figure 10: Overview of the main gene clusters involved in nitrogen metabolism and heterocyst development in *Cylindrospermopsis raciborskii* CS-505 and *Raphidiopsis brookii* D9. Transmission electron micrographs in the left panels show the heterocyst of *C. raciborskii* CS-505 and the apically differentiated cell of *R. brookii* D9. Optical micrographs on the right panels exhibit the Alcian blue staining characteristic of polysaccharides in the heterocyst. With the kind permission of Mónica Vásquez, Department of Molecular Genetic and Microbiology, Faculty of Biological Sciences, Pontificia Universidad Católica de Chile, Santiago, Chile & Millenium Nucleus EMBA, Santiago, Chile. [Stucken *et al.* (2010) *PLoS ONE* **5(2):** e9235. doi:10.1371/journal.pone.0009235] doi:10.1371/journal.pone.0009235.g001.

nifD elements examined. The upstream region of *xisA* (129 bp), *xisA* itself and a small ORF (of ≈500 nucleotides) are well conserved regions in the *nifD* elements (Henson *et al.*, 2005). The importance of a 6-bp region within the 11-bp direct repeats in excision and recombination has been stressed by the studies conducted on the ability of XisA to excise substrate plasmids with the *nifD* flanking regions in *E. coli*. Any change in the nucleotide sequence from a pyrimidine to a purine by mutation in the 11-bp repeats affected recombination (Henson *et al.*, 2008).

Brusca *et al.* (1989) demonstrated that the 11-kb element present in *A. variabilis* had a sequence similarity of 96% with the 11-kb element of *Anabaena* sp. strain PCC 7120 and it was also excised during heterocyst differentiation very much in a similar manner as the 11-kb element of *Anabaena* sp. strain PCC 7120 is excised. Though the two cyanobacteria are different in their diazotrophic growth potential, the organization of the 11-kb element, the presence of 11-bp direct repeats at the two ends of the 11-kb element, its excision during heterocyst differentiation, the presence of *xisA* gene encoding the site-specific recombinase and the ability of *xisA* gene of *A. variabilis* to complement the *xisA* gene of *Anabaena* sp. strain PCC 7120 all point out to the similarities in the genome rearrangement during heterocyst differentiation. The 5′-regulatory region of *xisA* gene has been shown to be important for excision of 11-kb element as its deletion blocked the expression of *xisA* in vegetative cells of *Anabaena* sp. strain PCC 7120 but not in *E. coli* cells. Mutants of *Anabaena* sp. strain PCC 7120 lacking the *nifD* element grew normally in presence of combined nitrogen and differentiated heterocysts in a nitrogen-free medium (Brusca *et al.* 1990). A vegetative cell DNA-binding factor (VF1) has been reported to govern the expression of *xisA* in cells undergoing heterocyst differentiation in *Anabaena* sp. strain PCC 7120. The 5′-upstream region of *xisA* has a 66-bp region consisting of three binding sites for VF1 adjacent to one another with consensus repeat sequences (ACATT) probably required for VF1-DNA interaction. This region seems to be the same as the regulatory region of *xisA* described by Brusca *et al.* (1990). Step-by-step deletion of 16 and 18 bp from the 66-bp region resulted in decreased VF1-binding and the deletion of a further 12 bp caused complete elimination of VF1-binding. Certain other genes whose upstream promoter regions bear an affinity for binding by VF1 are *rbcL* and *glnA* (Chastain *et al.*, 1990).

The excision of 11-kb element and the transcription of genes of the *nifHDK* operon seem to be independent of one another. A *xisA* mutant of *Anabaena* sp. strain PCC 7120 (DW12-2.2), which could not excise the 11-kb element and fix nitrogen, exhibited transcription of *nifH* and *nifD* genes without gene rearrangement. The absence of *nifK* transcription in DW12-2.2 suggests that its expression is dependent on the *nifH* promoter. Another mutant of *Anabaena* sp. strain PCC 7120 (LW1), with a deletion of 1.68-kb DNA fragment consisting of the promoter region of *nifH*, formed heterocysts but could not grow diazotrophically due to lack of transcription through *nifD*. However, LW1 showed normal excision of the 11-kb element from the *nifHDK* operon (Golden *et al.*, 1991).

The transposition of *mini-Mu-Kan-lac* gene construct in the 11-kb element and the fragment carrying the *nifD* gene was utilized in the transformation of *E. coli*. Such transformants of *E. coli* showed excision of the 11-kb element from the *nifD* gene and along with this the loss of *lac* gene facilitated the recognition of the white colonies (with a frequency of 3%) amongst the vast majority of the blue-colony formers on X-gal plates (Lammers *et al.* 1986). The sequencing of the *xisA* gene revealed that it runs counter to the direction of *nif* genes beginning 240 bp from the recombination site. The cloning of the 11-kb element into a plasmid vector and its propagation in *E. coli* resulted in a decreased frequency of the rearrangement. A second rearrangement of genes involving the excision of an interrupting sequence in *nifS* gene takes place at the same time of the excision of the *nifD* element but mediated by a site-specific recombinase other than XisA protein (Golden *et al.*, 1987). Golden *et al.* (1988) reported the excision of the third DNA element from the *fdxN* gene of

Anabaena sp. strain PCC 7120 that involves the removal of the 55-kb DNA element, positioned on the chromosome between the genes encoding the large and small subunits of ribulose-biphosphate carboxylase (*rbcL* and *rbcS*) and the *nifS* gene, by a site-specific recombinase encoded by *xisF*. Under microaerobic heterocyst-inducing conditions, the excision of 55-kb element takes place while the excision of 11-kb element excision does not occur. Of the cyanobacteria examined for the presence of the 55-kb element, *A. variabilis* ATCC 29413 (Herrero *et al.*, 2001), *N. azollae* 0708 (Ran *et al.*, 2010) and *C. raciborskii* CS-505 (Stucken *et al.*, 2010) lacked it while *Anabaena* sp. strain PCC 7120, *A. cylindrica* and *Nostoc* strain MAC possess the 55 kb-element in their genomes (Meeks *et al.*, 1988; Brusca *et al.*, 1989; Haselkorn, 1992). The requirement of two more genes that are present downstream of *xisF*, *xisH* and *xisI* for the excision of the 55-kb element has been shown by Ramaswamy *et al.* (1997). Deletion of the 3.2 kb region consisting of *xisH* and *xisI* blocked the excision of the *fdxN* element. The presence of *xisH* and *xisI* in extracopies on a replicating plasmid in *Anabaena* sp. PCC 7120 resulted in independent excision of the 55-kb element in vegetative cells itself. In view of these observations, these two genes are suggested to regulate the *xisF* function. Further, the inactivation of *xisA* gene by transposon mutagenesis containing a Sp- and Sm-resistance gene cassette inhibited the excision of 11 kb element as well as nitrogen fixation in *Anabaena* sp. strain PCC 7120. However, other geneome rearrangements involving the excision of the 55 kb element from the *fdxN* gene continued uninhibited. Thus these two excision events appear to be not linked to one another and occur independently of one another (Golden and Wiest, 1988; Carrasco *et al.*, 1994).

Besides the just described three gene rearrangements, i.e. excision of *nifD*, *nifS* and *fdxN* elements to give rise to continuity of gene sequences and expression of these genes in the heterocyst for nitrogen fixation to continue therein, another gene rearrangement involving the excision of a 10.5 kb DNA element from *hupL* gene (encoding a large subunit of uptake hydrogenase) has been described by Carrasco *et al.* (1995). The *hupL* element (9,435 bp) is characteristic in possessing 16-bp direct repeats (CACAGCAGTTATATGG) at each of its ends and a site-specific recombinase gene *xisC* towards one of its ends. The XisC of *Anabaena* sp. strain PCC 7120 is homologous to the XisA site-specific recombinase of the same organism and both XisA and XisC bear no sequence similarity to any other site-specific recombinase known so far. The excision of *hupL* element, located >700 kb from *nifD* and *fdxN* elements, takes place exactly coinciding with the excision of *nifD* and *fdxN* elements at about 18 h during, heterocyst differentiation. A *xisC* (*alr0677*) mutant of *Anabaena* sp. strain PCC 7120 exhibited no excision of *hupL* element and the presence of extra copies of *xisC* in wild-type on a replicating plasmid (overexpression) caused excision of the *hupL* element even under nitrogen-enriched conditions. Functionally, XisC has been shown to be similar to the phage integrases (Carrasco *et al.*, 2005).

ii) Requirements of nitrogen fixation: Either for aerobic or anaerobic nitrogen fixation the requirements are the expression of *nif* genes, a reductant and a source of ATP. The *nifHDK* operon becomes functional when *nifD* element is excised and the operon continuity is maintained during heterocyst differentiation. The products of *nifHDK* operon collectively are designated as the enzyme nitrogenase which consists of dintrogenase (Mo-Fe protein or protein I) and dinitrogenase reductase (Fe-protein or protein II). Dinitrogenase (220 to 240 kDa) is a heterotetramer of ($\alpha_2\beta_2$) polypeptides and the α, β subunits are encoded by *nifD* (Lammers and Haselkorn, 1983; Golden *et al.*, 1985) and *nifK* (Mazur and Chui, 1982), respectively. Dinitrogenase reductase (60 to 70 kDa), a homodimer is the gene product of *nifH*, supplies electrons to dintrogenase. Dinitrogenase possesses four 4Fe-4S centres arranged in two clusters and two copies of iron-molybdenum co-factor. On the other hand, dinitrogenase reductase has a single 4Fe-4S cluster. The O_2-sensitivity of the enzyme nitrogenase is due to the irreversible oxidation of 4Fe-4S clusters and the co-factors. Cyanobacterial nitrogenases exhibit considerable degree of similarity in the sequence of amino acids, three-dimensional structure and other enzymatic properties with nitrogenases from *Azotobacter vinelandii* or *Klebsiella pneumoniae* (Wolk *et al.*, 1994). There are two other proteins synthesized during *nif* gene expression in two different cyanobacteria. The first one is in *Nostoc commune* that synthesizes a hemoprotein, encoded by *glbN* gene, located in between *nifU* and *nifH*, during prolonged incubation under microaerophilic diazotrophic conditions (Potts *et al.*, 1992). The second one is in case of *Anabaena* sp. strain PCC 7120, GroEL, a multifunctional chaperonin, is synthesized in vegetative cells and heterocysts during *nif* gene expression (Jäger and Bergman, 1990) that probably helps in proper folding of nitrogenase (Govezensky *et al.*, 1991).

The reductive nature of heterocysts has been highlighted by their ability to reduce 2,3,5-triphenyl tetrazolium chloride (TTC; 0.01% w/v) and the red formazan crystals can be readily seen in the heterocysts of *A. variabilis*, *C. licheniforme*, *Nostoc* sp. *Calothrix* sp., and *Scytonema* sp. under the light microscope (Drawert and Metzner, 1956; Tischer, 1957). Their reductive nature and nitrogenase activity in heterocysts has further been confirmed (Talpasayi and Kale, 1967; Fay *et al.*, 1968; Stewart *et al.*, 1969). A gradient in the formation of formazan in heterocysts and the intervening vegetative cells does not result after treatment with TTC due to competition for electrons between TTC and O_2. On the other hand, treatment with 3,3′,-(3,3′-dimethoxy-4,4′-biphenylene)-bis-(2-p-nitrophenyl)-5-phenyl-2H tetrazolium chloride (NBT; 0.01% w/v) resulted in a gradient of the blue colour formation highest (reducing power) in heterocysts and proheterocysts (of *A. cylindrica*, *A. inaequalis*, *A. variabilis* and *Anabaenopsis circularis*) followed by vegetative cells. Nitrogen fixation by the heterocysts has been found to be adversely affected by the reduction of TTC or NBT as well as the detachment of heterocysts with the adjacent vegetative cells (Fay and Kulasooriya, 1972). It means that the attachment of heterocysts with the adjacent vegetative cells is important for nitrogen fixation to proceed. In this connection, the results of Weare and Benemann (1973) are important as breakage of filaments by blending at the junctions of heterocysts with vegetative cells caused sufficient decrease in the capacity of nitrogen fixation by *A. cylindrica*. Direct demonstration that heterocysts fix nitrogen came from studies on the incorporation of ^{15}N (Ohmori and Hattori, 1971) and ^{13}N (Wolk *et al.*, 1974) by the intact filaments of *A. cylindrica*. Inhibition of ^{13}N fixation into heterocysts was caused by carbon monoxide, hydrogen gas or prior growth of the cyanobacterium in media enriched with combined nitrogen. The principal products of ^{13}N-fixation in a number of cyanobacteria examined (*A. cylindrica*, *A. variabilis* ATCC 29413, *C. licheniforme* ATCC 29412, *Gloeothece* sp. 6909, ATCC 27512 and *Plectonema boryanum* UTCC 594) have been found to be [^{13}N]-NH_4 and [^{13}N]-glutamine (Thomas *et al.*, 1977; Meeks *et al.*, 1978). A decrease in the label of glutamine and a corresponding increase in the label in glutamate suggested the mediation of GS-GOGAT pathway for ammonia assimilation after nitrogen fixation or exogenous ammonia assimilation (Meeks *et al.*, 1978; Haselkorn, 1978). The above observations establish clearly the role of heterocysts in nitrogen fixation under aerobic conditions but under specialized microaerophilic or anaerobic conditions the induction of nitrogenase in the vegetative cells of *A. cylindrica* has also been demonstrated (Smith and Evans, 1970, 1971; Haystead *et al.*, 1970; Van Gorkom and Donze, 1971). The ability of isolated heterocysts to perform nitrogen fixation has largely been shown by acetylene reduction assay (Stewart *et al.*, 1968).

Isolated heterocysts of *A. cylindrica* Lemm or *A. variabilis* ATCC 29413 reduced acetylene very much like the intact filaments in light and did not require any additional co-factors (Wolk and Wojciuch, 1971a,b; Peterson and Burris, 1976; Jensen *et al.*, 1986). The requirement of light for nitrogen fixation by the isolated heterocysts of *A. cylindrica* and its inhibition by O_2 concentrations greater than atmospheric levels and by dark exposure all go in favour of the heterocyst being the site for nitrogen fixation (Weare and Banemann, 1973). There are evidences for the synthesis of nitrogenase in the heterocysts of *Nostoc muscorum* as revealed by the isolation and separation of nitrogenase proteins labelled with $^{34}SO_4^-$ (Fleming and Haselkorn, 1973). At least 10 different sets of proteins are synthesized after nitrogen step-down of *N. muscorum* and only some of these including nitrogenase are allowed to continue in the proheterocysts. A rapid turnover of protein seems to take place once the cell is committed to differentiate into a proheterocyst and the protein profile of proheterocysts differ greatly from those of vegetative cells (Fleming and Haselkorn, 1974). Since incorporation of ^{15}N-label by the heterocysts is no greater than vegetative cells (Ohmori and Hattori, 1971), Wolk *et al.* (1974) incubated intact filaments of *A. cylindrica* in presence of ^{13}N gas and found the label of fixed ^{13}N in the heterocysts by autoradiography. Heterocysts contributed to about 90% of the total nitrogenase activity of *A. cylindrica* grown under aerobic conditions based on acetylene reduction assays and possessed PSI as well as oxidative phosphorylation activities. They are unable to carry out PSII activities and fix CO_2. Photosynthetic electron transport and ATP generation supported nitrogenase to function in the absence of O_2 evolution (Tel-Or and Stewart, 1976, 1977). In contrast, Thomas *et al.* (1977) reported 34% of the acetylene-reducing activity of whole filaments by the isolated heterocysts of *A. cylindrica*. These workers put forward direct evidences for the reduction of dinitrogen in the heterocysts as well as the formation of glutamine in the heterocysts, its transportation into the adjacent vegetative cells and conversion of glutamine into glutamate supported by isotopically labelled substrates. The presence of GOGAT in the heterocysts of *Anabaena* sp. strain PCC 7120 was demonstrated by the formation of [^{14}C] glutamate from [^{14}C] glutamine in presence of α-oxoglutarate (Gupta and Carr, 1981a). Isolated heterocysts of *A. variabilis* contributed to 60% of the nitrogenase activity of the whole filaments and heterocysts derived from ^{55}Fe enriched media yielded dinitrogenase and dinitrogenase reductase proteins to 91% and 69%, respectively of the whole filaments (Peterson and Wolk, 1978).

A comparative study of the whole filaments and isolated heterocysts of a mutant strain of *Anabaena* sp. strain CA revealed that the heterocysts possessed chlorophyll *a* to the same extent as the vegetative cells and half the amounts of C-phycocyanin and allophycocyanin. A light-dependent reduction of acetylene was stimulated by the presence of 5% H_2 under argon that continued for more than 3 h (Kumar *et al.*, 1982). Murry *et al.* (1984) demonstrated the presence of dinitrogenase that is restricted to the heterocysts by immunoferritin labelling of Mo-Fe protein in cultures of *A. cylindrica* treated with DCMU and allowed to undergo differentiation of heterocysts in presence of argon. Heterocysts of *N. muscorum* are devoid of nitrate reductase apoprotein but when nitrate reductase apoprotein was added to heterocyst preparations, the reduction of nitrate to nitrite occurred that is suggestive of the presence of Mo-co-factor (Kumar *et al.*, 1985). Ammonium ions inhibited nitrogen fixation in *No. spumigena* strain AV1 with a corresponding suppression of *nifH* as reflected by the transcript and dinitrogenase reductase protein levels. But the expression of *ntcA* and *hetR* genes resulted in the differentiation of heterocysts despite the presence of ammonium ions (Vintila and El-Shehawy, 2007).

***iii) Supply of Reductant*:** The heterocystous cyanobacteria possess two types of ferredoxins, PetF in the vegetative cells encoded by *petF* and the second *fdxH* expressed in the heterocysts. The role of the latter, [2Fe-2S]-type of ferredoxin as an electron carrier for supplying electrons to nitrogenase (Bothe, 1970; Smith *et al.*, 1971), the generation of reduced ferredoxin from the electrons supplied by NADPH (Bothe, 1970; Smith *et al.*, 1971; Apte *et al.*, 1978), the mediation of ferredoxin-$NADP^+$ reductase (FNR; EC 1.18.1.2) in the reduction of ferredoxin and the operation of pentose phosphate pathway for the supply of NADPH in the heterocysts (Stanier and Cohen-Bazire, 1977; Haselkorn, 1978; Stewart, 1980; Bothe, 1982) have been demonstrated. Privalle and Burris (1984) reported the stimulation of acetylene reduction by intermediates of glycolysis and TCA cycle. Cell-free extracts from purified heterocysts of *A. variabilis* ATCC 29413 have been employed for *in vitro* demonstration of electron donation to nitrogenase in light and dark. Glycolytic cycle supported nitrogen fixation in dark with intermediates of glycolysis like glucose-6-phosphate (G6P), fructose-1,6-biphosphate and dihydroxyacetone phosphate each with NAD or NADP constituting the effective co-factors. NADH (and hydrogen) contributed electrons for nitrogenase via PSI in the light which is regulated by the presence of NADP that competed for electrons with nitrogenase. However, NADPH served as a source of electrons to nitrogenase in dark but this was inhibited by the presence of NADP. Since this inhibition was not reverted with time, it was suggested that FNR mediates the electron flow. In dark this enzyme directly mediates the reduction of ferredoxin that in turn supplies electrons to nitrogenase while in light the electrons fom NADPH have to first pass through PSI before reducing ferredoxin (Schrautemeier *et al.*, 1984). In addition to the intermediates of glycolysis mentioned above and confirming the role of NADPH and NADH in light and dark, respectively, Neur and Bothe (1985) reported that unphosphorylated sugars like glucose, fructose and erythrose also served as electron donors. Pyruvate with coenzyme A supported C_2H_2 reduction by the heterocysts of *A. cylindrica* or *A. variabilis* ATCC 29413. The cell-free extracts from heterocysts of these cyanobacteria could bring about reduction of ferredoxin in presence of pyruvate and coenzyme A in light as well as dark suggesting the mediation of pyruvate:ferredoxin oxidoreductase (PFO) in heterocysts (Neur and Bothe, 1985). The gene encoding PFO of *A. variabils* ATCC 29413 and *Anabaena* sp. PCC 7119 has been identified as *nifJ* by Schmitz *et al.* (1993).

Lyons and Thiel (1995) cloned and sequenced the operon *nifB-fdxN-nifS* and *nifU* of *A. variabilis* ATCC 29413. The requirement of NifB for both Mo-dependent and V-dependent nitrogenases and the non-essential nature of gene products of both *nifS* and *nifU* were some important features of nitrogen fixation by this organism. The existence of a [2Fe-2S]-type of ferredoxin in the heterocysts different from the one present in vegetative cells of *A. variabilis* ATCC 29413 that could utilize the reducing power either generated by H_2/hydrogenase (from *Clostridium pasteurianum*) or by G6P and FNR (from *A. variabilis*) has been demonstrated in *A. variabilis* ATCC 29413 by Schrautemeier and Böhme (1985). This has further been confirmed by a comparison of the amino acid composition, molecular weight and the redox potential of the heterocyst and vegetative cell ferredoxins from *A. variabilis* ATCC 29413 (Böhme and Scrautemeier, 1987). Schrautemeier *et al.* (1995) identified the two ferredoxins of *A. variabilis* ATCC 29413 as FdxH1 and FdxH2 encoded by the genes *fdxH1* and *fdxH2*, belonging to two nif clusters *nif1* and *nif2*, respectively. FdxH1 resembles the [2Fe-2S] ferredoxin from *Anabaena* sp. strain PCC 7120 whereas FdxH2 resembles the FdxH from *P. boryanum* PCC 73110. The expression of *fdxH2* and *nif2* gene system occurred in both *A. variabilis* ATCC 29413 and *P. boryanum* PCC 73110 afte nitrogen deprivation under anaerobic conditions whereas the expression of *fdxH1* and *nif1* gene system occurred under aerobic conditions in *A. variabilis* ATCC 29413. Evolutionarily it is suggested that the *nif2* gene system operates in the vegetative cells of non-heterocystous cyanobacteria and *nif1* gene system has branched off from *nif2* gene system for its expression exclusively in heterocysts under aerobic conditions. Thus in *A. variabilis* ATCC 29413 both *nif1* and *nif2* gene systems are operative depending on the oxygen relations of the environment.

Thiel and Pratte (2001) provided evidences in favour of independent expression profiles of *nif1* and *nif2* in *A. variabilis* ATCC 29413. The growth properties and expression of *nif1* and *nif2* under aerobic and anaerobic conditions of three mutants, i.e. (i) *ntcA* mutant (MM3) which could not grow in presence of nitrate as sole source of nitrogen, (ii) *nif1* mutant that produces heterocysts but lacks *nif1* nitrogenase (JE994), and (iii) a mutant that is Het^- and $Nif1^-$ (NF76) have brought to light that cyanobacteria sense nitrogen sufficieny or deficiency at the filament level as a whole but not at the level of individual cells as conceived earlier in the heterocyst pattern formation. *In situ* localization of *nif2* was monitored in a strain (JE35) with *nif2::lacZ* fusion. After allowing normal heterocyst pattern formation under aerobic conditions for 48 h, strain JE35 was incubated under anoxic conditions. The existence of fluorescence in the cells adjacent to heterocysts expressing *nif2* makes it clear that the pattern of hetrocyst formation has no relation to the expression of *nif2*. The cross-functionality of nitrogenase components of NifH1 and VnfH of *A. variabilis* ATCC 29413 was established by Pratte *et al.* (2006) who isolated single (*nifH1* or *vnfH*) and double mutants (*nifH1* and *vnfH*) of *A. variabilis* ATCC 29413. The *nifH1* mutant could grow diazotrophically in the absence of molybdate because of the expression and substitution by *vnfH*. Likewise, in the *vnfH* mutant the vanadium nitrogenase could be functional because of the expression and substitution by *nifH1*, it means the absence of V-Fe co-factor is fulfilled by the presence of Mo-Fe co-factor. On the other hand, the double mutant of *nifH1* and *vnfH* could not grow diazotrophically due to the absence of both Mo-Fe and V-Fe co-factors with or without molybdate or vanadate (Pratte *et al.*, 2006). Transcriptional pattern of *nifH1* and *vnfH* in *A. variabilis* ATCC 29413 revealed that the previously identified tsps of both *nifH1* and *vnfH1* did not drive the expression of the *lacZ* gene as the reporter gene. A further search upstream the promoter regions resulted in the recognition of a promoter within *nifV1* and a promoter upstream of *nifB1* contributed to the expression of *nifHDK1* where the contribution of *nifB1* promoter to the transcript pool was considerably larger (Ungerer *et al.*, 2010).

Molecular cloning and sequence analysis of the gene (*fdxH*) encoding heterocyst ferredoxin of *Anabaena* sp. strain PCC 7120 have been performed. Two major transcripts of *fdxH* (0.59 and 1.85 kb) have been detected at late stages of heterocyst differentiation with a tsp located at 132 bp upstream of *fdxH* (Böhme and Haselkorn, 1988). By site-directed mutagenesis of the positively charged amino acid residues present on the surface of FdxH, Schmitz *et al.* (1993) identified two lysine residues located at positions 10 and 11 of FdxH of *Anabaena* sp. strain PCC 7120 that interact with dinitrogenase reductase (*nifH*). Exchange of the two lysine residues of FdxH by Glu10 and Ala11 of PetF in a chimeric protein PetF:FdxH with N-terminal half (58-amino acid residues) of the former and C-terminal half (40-amino acid residues) of the latter resulted in a decreased affinity to nitrogenase. Three dimensional structure of [2Fe-2S] ferredoxin of *Anabaena* sp. strain PCC 7120 is very similar to the plant-type ferredoxins with the iron sulphur cluster positioned at the outer edge of the molecule. The secondary structure revealed the presence of seven strands of β-pleated sheet, two α-helices and seven type 1 turns (Jacobson *et al.*, 1993). Razquin *et al.* (1995) compared the catalytic potential of FdxH, PetE and flavodoxin from *Anabaena* sp. strain PCC 7120 in transferring electrons to nitrogenase. Reconstitution experiments with nitrogenase showed the effectiveness of only FdxH in transferring electrons to nitrogenase. FdxH and PetE could effectively transfer electrons to nitrogenase in reconstituted experiments from heterocyst preparations under anaerobic conditions. Flavodoxin proved to be important under iron-limited conditions and could effectively mediate electron transfer from PSI to $NADP^+$. *petH*, located away from the *nif* gene cluster on the genome map of *Anabaena* sp. strain PCC 7120, encodes FNR in the heterocysts after a nitrogen step-down. The FNR protein increased by 10-fold in the heterocysts when compared to vegetative cells as revealed by immunoquantification experiments. However, the purified FNR from heterocysts did not differ from the FNR from vegetative cells (Razquin *et al.*, 1996). Schrautemeier *et al.* (1994) characterized the [2Fe-2S]-type of ferredoxin encoded by *fdxH* gene as well as a new 2[4Fe-4S]-type of ferredoxin encoded by another *fdxH* gene, located immediately downstream of the former, in the non-heterocystous cyanobacterium, *Plectonema boryanum* PCC 73110. The latter type of ferredoxin has been reported for the first time in cyanobacteria. When *P. boryanum* PCC 73110 was cultured in presence of combined nitrogen, it expressed a [2Fe-2S]-type of ferredoxin (type-1) encoded by *petF* gene. The presence of *fdxH* gene in a non-heteocystous cyanobacterium is distinctive and the recombinant protein produced in *E. coli* has a molecular mass of 10.9 kDa with 98 amino acid residues. The presence of Lys10 and Lys11 that interact with dinitrogenase reductase has already been reported by Schmitz *et al.* (1993). Masephol *et al.* (1997) constructed three mutant strains of *Anabaena* sp. strain PCC 7120 for the *fdxH* gene. The performance of the mutants showed that though FdxH is very necessary for optimum diazotrophic growth but it is not essential for nitrogen fixation.

iv) Assimilation: Isotopic labelling studies (Ohmori and Hattori, 1971; Wolk *et al.*, 1974; Thomas *et al.*, 1977) and enzymatic composition of heterocysts and vegetative cells (Fleming and Haselkorn, 1973; Thomas *et al.*, 1977; Peterson and Wolk, 1978; Haselkorn, 1978; Murry *et al.*, 1984; Wolk, 1982, 1994) are unequivocal in suggesting the operation of GS-GOGAT pathway in cyanobacteria for assimilation of ammonium generated through nitrogen fixation or taken up exogenously. The nitrogen fixed in the heterocysts in the form of glutamine is transported into the adjacent vegetative cells where it is converted to two molecules of glutamate by the activity of GOGAT activity. The possible exchanges of metabolites from vegetative cells and heterocysts and vice-versa have been depicted in the model presented in Fig. 10 of Chapter 7. It has been proposed that one molecule of glutamate is retained by the vegetative cells and the second molecule of glutamate is transported back to the heterocyst for further amination (Haselkorn, 1978). This is dependent mostly on the GS and GOGAT levels. The presence of GS activity in higher levels under nitrogen-limited conditions in several cyanobacteria established that it is the primary ammonia assimilating enzyme (Dharmawardene *et al.*, 1973; Stewart and Rowell, 1975; Wolk *et al.*, 1976; Meeks *et al.*, 1977, 1978; Rowell *et al.*, 1977; Thomas *et al.*, 1977). The occurrence of almost equal levels of GS activity in both vegetative cells and heterocysts of *Nostoc* sp. symbiotic to *Nephroma* (Bergman and Rai, 1989) and *Anthoceros* (Rai *et al.*, 1989) and in *A. cylindrica* (Renström-Kellner *et al.*, 1990) has been reported. Due to the absence of GOGAT activity in the heterocysts of *Anabaena*, Thomas *et al.* (1977) suggested that the vegetative cells support this function since they contained higher GOGAT activity. There are contradicting reports on the presence of GOGAT activity in the heterocyst preparations. In case of *Anabaena* sp. strain PCC 7120, Gupta and Carr (1981a) reported GOGAT activity from the heterocyst preparations where as the GOGAT activity observed in the heterocyst extracts of *A. variabilis* ATCC 29413 was ascribed to glutaminase activity (Rai *et al.*, 1982). Wolk *et al.* (1994) agreed with the findings of Rai *et al.* (1982).

Studies on the regulation of GS activity in cyanobacteria and gene expression studies during heterocyst differentiation after nitrogen step-down have brought in new information in our understanding of the above issues. Three types of GSs are known. GSI, encoded by *glnA* gene, is present in most of the enterobacteria and in cyanobacteria. Rhizobiaceae, Frankiaceae and Streptomycetaceae possess GSI as well as GSII. GSIII, a gene product of *glnN*, has been detected in *Bacteroides fragilis* and *Butyrivibrio fibrisolvens* (two obligate anaerobes present in mammal intestines). GSIII has been first described from *Synechocystis* sp. strain PCC 6803 by Reyes *et al.* (1994) in addition to GSI but *glnN* sequence has not been observed from the filamentous heterocystous cyanobacterial species examined (*A. variabilis* ATCC 29413, *Anabaena* sp. strain PCC 7120, *Nostoc* sp. strains PCC 6720, PCC 7413, PCC 6705, *Calothrix* sp. strain PCC 7601 and *Fischerella* sp. strain UTEX 1829). While the transcription of *glnA* is regulated by NtcA, the transcription of *glnN* gene in *Synechocystis* sp. strain PCC 6803 is under the regulation of some other unidentified additional factors (Reyes *et al.*, 1997). The *glnN* gene product GSIII protein helps *S. elongatus* PCC 7942 to overcome from prolonged nitrogen chlorosis (Sauer *et al.*, 2000). *Pseudoanabaena* sp. PCC 6903 is an exception in cyanobacteria in possessing only *glnN* gene to meet its requirements (Crespo *et al.*, 1998). The properties of the three GS types are found to be different. Most importantly, GSI is a dodecameric protein with 12 identical subunits (unless specially mentioned GS activity in the following account specially refers to GSI activity), GSII is an octameric protein while GSIII is a hexameric protein.

The regulation of GS activity in cyanobacteria has been found to be different from that of the process observed in bacteria. In bacteria like *E. coli*, GS is subjected to cumulative feedback inhibition by the end products of glutamine metabolism, ADP and AMP and other nucleotides. In presence of the enzyme adenyltransferase (EC2.7.7.42) adenylation (attachment of an adenyl group to Tyr397 to each of the twelve identical subunits) of GS takes place. The extent of adenylation of different Tyr residues may be different in different molecules and so accordingly the GS activity may be inversely proportional to the number of adenylated subunits. The process of deadenylation also is mediated by adenyltransferase and so the GS activity is restored to normal levels. The cyanobacterial GS activity is not subjected to adenylation-deadenylation processes (Dharamwardene *et al.*, 1973; Rowell *et al.*, 1977, 1979; Stacey *et al.*, 1979; Fisher *et al.*, 1981). In *Anabaena* sp. strain PCC 7120 the levels of GS activity are lower in ammonium-grown cells than the cells grown in presence of nitrate or molecular nitrogen. The GS in this organism also is neither adenylated nor subjected to feedback inhibition (Orr and Haselkorn, 1981, 1982). The regulation of GS activity in cyanobacteria has been found to be at two levels, i.e. first at the transcriptional level and the second at post-transcriptional level depending on the availability of carbon and nitrogen sources. At transcriptional level NtcA exerts its regulation on the type of transcript(s) to be produced in vegetative cells and heterocysts and at post-transcriptional level the interaction of two GS-inactivating factors inhibits the GS activity (Muro-Pastor *et al.*, 2005). The interaction of two inactivating factors (IFs), IF7 and IF17 with GS leads to its complete inactivation *in vivo* (García-Domínguez *et al.*, 1999). In *Synechocystis* sp. PCC 6803 *gifA* and *gifB* genes encode IF7 and IF17, respectively. These two genes are under the transcriptional regulation of NtcA with NtcA-binding sites located at tsps 7.5 bp and 30.5 bp upstream of promoters of *gifA* and *gifB*, respectively. Thus NtcA represses the transcription of *gifA* and *gifB* genes and at the same time stimulates the transcription of *glnA* gene when GS activity has to be maintained at optimum levels, especially under nitrogen deprivation (García-Domínguez *et al.*, 2000). In case of *Anabaena* sp. strain PCC 7120 only one GS-inactivating factor IF7A encoded by *gifA* (*asl2339*) has been found. Galmozzi *et al.* (2010) have investigated the regulatory mechanism of *Anabaena* GS by IF7A. The important findings are that (i) *Anabaena* GS was not inactivated by *Synechocystis* IF7 and IF17 whereas *Synechocystis* GS was inactivated by IF7A; (ii) the presence of two consensus NtcA-binding sites at 28.5 bp and 77.5 bp in the promoter region upstream of the tsp of *gifA* and *in vitro* binding by NtcA to these two sites and the presence of 2-OG enhances this binding; (iii) the transcripts for GS decreased in ammonium grown cells of *Anabaena* sp. strain PCC 7120 with a corresponding increase in the levels of mRNA of *gifA* and so also the protein levels of GS and IF7A showed similar variation; (iv) a nitrogen step-down resulted in enhanced transcript and protein levels of *glnA* and a corresponding decrease in the transcript and protein levels of *gifA*; (v) in an *ntcA* mutant (CSE2) of *Anabaena* sp. strain PCC 7120 the induction of GS was impaired after a nitrogen step-down but the transcript and protein levels of *gifA* in CSE2 were unaltered irrespective of nitrogen source; (vi) a chimeric gene construct of *gifA* of *Anabaena* sp. strain PCC 7120 and *gifB* of *Synechocystis* sp. strain PCC 6803 when introduced into wild-type *Anabaena* sp. strain PCC 7120 at a neutral *nucA-nuiA* region of α-megaplasmid, in the resulting strain (ACHI) IF7A protein and the chimeric protein IF7A/IF17 exhibited a decrease upon ammonia addition and increase after a nitrogen step-down. The GS activity in wild-type showed a typical slow inactivation but in the ACHI the GS activity reached about 40% of the initial activity; (vii) a recombinant chimeric protein of IF7A and IF17 produced in *E. coli* was less effective than IF7A in GS inactivation; (viii) when *gfp* was used as the reporter gene and $P_{gifA\text{-}gfp}$ was introduced into *Anabaena* by conjugation and integrated into *gifA* locus, the expression of

GFP after a transfer to ammonium medium was very much higher in the vegetative cells than that observed in heterocysts after 4.5 to 6.5 h. These results thus confirm that there was no inactivation of GS in the heterocysts and the *gifA* gene expression in vegetative cells and heterocysts is differential, i.e. induced in the former and repressed in the latter (Galmozzi *et al.*, 2010).

The expression of *glnA* in both vegetative cells and heterocysts of *Anabaena* sp. strain PCC 7120 has been demonstrated by Elhai and Wolk (1990) when *luxAB* was used as the reporter gene. Tumer *et al.* (1983) observed that *glnA* is transcribed from different promoters depending on the nitrogen source but a single *glnA* promoter seems to operate under nitrogen-deficient or nitrogen-fixing conditions. The presence of two NtcA-binding regions located at 125 bp and 148 bp upstream of the tsps in the promoter region of *glnA* of *Anabaena* sp. strain PCC 7120 have bee dectected (Ramasubramanian *et al.*, 1994b). The order of binding of NtcA to the promoters of a number of genes under its transcriptional control has been reported to be *glnA*>*xisA*>*rbcLS*>*nifH*. Thus *glnA* occupies the first position in the sequence of genes under the transcriptional control of NtcA and is expressed at higher levels under nitrogen-limited conditions (Herrero *et al.*, 2001). However, the expression of *glnA* gene from multiple promoters with as many as five tsps in *Anabaena* sp. strain PCC 7120 has been reported by Valladares *et al.* (2004). In the vegetative cells and heterocysts the transcript from promoter 1 is synthesized that has a canonical NtcA-binding site, located at 93 bp upstream of start of *glnA*. Mérida *et al.* (1991) while studying the regulation of GS in *Synechocystis* sp. strain PCC 6803 found a parallelism in GS activity and the internal concentration of α-ketoglutarate (or 2-oxoglutarate; 2-OG) and indicated that it could play an important positive role of an effector of GS activity. The *in vitro* binding affinity of NtcA to the *glnA* promoter of *S. elongatus* PCC 7942 in presence of 2-OG and enhancement in its transcription rate constitute additional evidences for the role of 2-OG (Vazquez-Bermudez *et al.*, 2002; Tanigawa *et al.*, 2002). An analysis of *glnA* and *gif* mutants of *Synechocystis* sp. PCC 6803 also strongly indicated that cyanobacteria perceive nitrogen status by sensing intracellular 2-OG levels (Muro-Pastor *et al.*, 2001). Another new player in the nitrogen control of cyanobacteria is the product of *glnB* gene, popularly known as PII protein (Arcindéguy *et al.*, 2001). The interactions of GS, 2-OG and PII are discussed under section on Signalling molecules.

The structural, mechanistic and regulatory properties of GOGAT enzymes and their role in the amino acid metabolism of plants have been summarized. Depending on the electron donor for the conversion of L-glutamine to glutamate and 2-OG, three forms of GOGAT enzymes have been recognized, i.e. NADPH-GOGAT (EC1.4.1.13), NADH-GOGAT (EC 1.4.1.14) and reduced ferredoxin (Fd)-GOGAT [(EC 1.4.7.1); (Suzuki and Knaff, 2005)]. Studies conducted on unicellular cyanobacteria *Synechococcus* sp. PCC 6301 (Marqués *et al.*, 1992) and *Synechocystis* sp. strain PCC 6803 (Navarro *et al.*, 1995) showed the presence of Fd-GOGAT but the genome of latter organism contained a gene sequence similar to NADPH-GOGAT of *E. coli* (Kaneko *et al.*, 1996). *P. boryanum* IAM-M101 strain dg5 (isolated by Fujita *et al.*, 1996) has three genes *glsF* and *gltB* and *gltD* that encode respectively, Fd-GOGAT, and two subunits of NADH-GOGAT (Okuhara *et al.*, 1999). The *gltB* encodes a large subunit (of 1530 amino acids) and *gltD* encodes a small subunit (of 492aa) of NADH-GOGAT that bears sufficient homology to the NADPH-GOGAT of *E. coli*. The three genes have been cloned and the characterization of insertional mutants (by the introduction of a kanamycin-resistance gene, *neo*) highlighted the importance of Fd-GOGAT enzyme in the assimilation of ammonia in light when CO_2 was supplied. The *glsF* mutant had 2.5–3.3-fold decreased levels of chlorophyll, phycocyanin and allophycocyanin when compared to unaltered pigment ratios in the *gltB* and *gltD* mutants and the wild-type (Okuhara *et al.*, 1999). The Fd-GOGAT, the only GOGAT enzyme present in *Anabaena* sp. strain PCC 7120, is not operative in the heterocysts thus setting aside the controversy over the existence of GOGAT activity mentioned earlier (Martín-Figueroa *et al.*, 2000). This has been supported by the uniform levels of the transcripts of *glsF* and the activity of Fd-GOGAT in all the vegetative cells grown in all nitrogen sources but its absence in the heterocysts has been supported by the results from enzyme activity, Western and Northern blot analyses. The presence of NADP-isocitrate dehydrogenase activity in the heterocysts is suggested to be responsible for the generation of 2-OG and NADPH and the latter is utilized to reduce ferredoxin. In the model presented by them they envisaged that the 2-OG is transported to the adjacent vegetative cells and in exchange the heterocysts receive a glutamate molecule. Furthermore, they suggested that the transport of fixed nitrogenous substance from heterocysts could as well be glutamine or arginine because the heterocysts have full potential to synthesize arginine which could be incorporated into cyanophycin. The breakdown of cyanophycin could serve as a rich source of arginine as the fixed product that could be transported into vegetative cells depending on the need (Martín-Figueroa *et al.*, 2000).

Now we shall discuss the transport of metabolites from heterocysts to vegetative cells and vice-versa. Although it is mostly agreed that glutamine is the end product of ammonia assimilation, whether the transported nitrogenous substance is glutamine as such or a derivative of it is not yet clear. Another likely alternative molecule projected is Ala. Two investigations support this alternative. Nitrogen starved cells of *A. cylindrica* (F) when exposed to $^{14}CO_2$ the label most prominently accumulated in glutamine when incubated under ammonium and Ala was the most labelled product under N_2 (Lawrie *et al.*, 1976). The second investigation is that of Jüttner (1983) who similarly found good amount of labelled Ala from the extracts of heterocysts when whole filaments of *A. cylindrica* (F) were exposed to $^{14}CO_2$ for very brief periods (20 s). Ala can be formed through a transamination reaction from glutamate (Wolk *et al.*, 1976; Meeks *et al.*, 1978). Furthermore, the existence of five types of amino acid transport systems viz., high affinity system for basic amino acids, neutral amino acid transport systems (N-I and N-II), low affinity, passive transport system for basic amino acids and low affinity system for the uptake of acidic amino acids is known in cyanobacteria. The repression of N-I transport system that is required for optimum diazotrophic growth in the heterocysts, the expression of BgtA, a high affinity basic amino acid uptake transporter and N-II in vegetative cells as well as heterocysts, the specific release of Ala by mutants impaired in *natB* and *natCA* operon and the requirement of functional heterocysts for maximum release of alanine all point out to the likely transport of Ala from vegetative cells to heterocysts (Montesinos *et al.*, 1995, 1997; Picossi *et al.*, 2005; Pernil *et al.*, 2008). Moreover, the catabolic function of a compartmentalized alanine dehydrogenase has been demonstrated by the expression of *ald* (*alr2355*) gene fused with *gfp* as a reporter gene inside the heterocysts of *Anabaena* sp. strain PCC 7120. Thus Ala catabolism appears to be essential for diazotrophic growth (see Amino acid transport and diazotrophic growth).

With regard to carbon metabolism occurring in the heterocysts, the operation of a sucrose cycle of synthesis (mediated by the expression of *spsB* and *sppA*) and catabolism (mediated by *invB*) and its linking to glycogen turnover during nitrogen-fixing conditions has come to light (Cumino *et al.*, 2007). Further, the regulation of genes of sucrose biosynthesis by NtcA amply signifies that besides exercising nitrogen control, it also regulates carbon metabolism (Marcozzi *et al.*, 2009). All these have been dealt in detail under the section on Carbon metabolism. The above aspects discussed make it clear that heterocysts have the potential to build up the required carbon and nitrogen intermediates required for diazotrophic growth, at the same time depending on the requirement carbon as well as nitrogen inputs are received from vegetative cells. But exactly what is the nature of these intermediates awaits further research. An attempt made in this direction by Popa *et al.* (2007) merits mention. Carbon and nitrogen fixation and the probable metabolite exchange between heterocysts and vegetative cells of *A. oscillarioides* after enrichments with $^{13}CO_3^-$ and $^{15}N_2$ revealed that newly fixed nitrogen is rapidly exported from heterocysts and located in the vegetative cells. There was a difference in the allocation of C and N in between the older and newly formed heterocysts as the latter had higher enrichments. The neck region of heterocysts, cell division septae and cyanophycin granules represented sites most active assimilation and biosynthesis (Popa *et al.*, 2007).

v) Transport: The important proteins that are essential for maintaining the integrity of filaments have largely been identified by the work on fragmentation mutants. This led to the the researches on the probable modes of transport of substances in between vegetative cells and heterocysts. First we describe the fragmentation mutants, the proteins identified followed by means of transport between the cells of a filament of *Anabaena* sp. strain PCC 7120. Though mutants of *Anabaena* sp. strain PCC 7120 that have a tendency to fragment into few cells or even single cells under nitrogen-deficient conditions have earlier been isolated, the genes regulating them or complementing them have not been identified (Buikema and Haselkorn, 1991a; Khudyakov and Pinevich, 1992). Bauer *et al.* (1995) characterized three such mutants, strain 129, strain 415 and a third strain from among the lot isolated by their group (Buikema and Haselkorn, 1991a). Besides fragmenting into a single-celled state, strain 129 shows signs of proheterocyst differentiation before fragmenting but is not capable of performing aerobic nitrogen fixation due to the presence of very few mature heterocysts. The gene, identified by complementation of wild-type DNA, *fraC* encodes a phenylalanine-rich protein of 179 amino acid residues and this protein seems to be essential for the integrity of cell junctions. A comparison of the *fraC* gene sequences of strain 129 and the wild-type showed two base changes, one at position 1108 (from T to G in the threonine codon that is a silent mutation) and the other at position 770 (a frame-shift caused due to deletion of a single base). Strain 415 also fragments extensively after nitrogen step-down, differentiates proheterocysts and few heterocysts but is unable to perform aerobic nitogen fixation. Since *fraC* could not complement the mutation in strain 415, it appeared to have arisen due to a mutation in a gene other than *fraC*. The identification of another gene *fraH* has been possible by the isolation of mRNA and corresponding cDNA synthesis from a third mutant at 30 h after nitrogen step-down. Insertional inactivation of *fraH* gene sequence in the wild-type resulted in a mutant phenotype resembling the original phenotype in showing slightly longer fragments with proheterocyst and heterocyst formation in higher frequency than the other two mutants described above. The wild-type showed the mRNA transcripts of *fraH* after 12 h of transfer to a nitrogen-free medium. Nayar *et al.* (2007) identified a gene upstream of *hetR* of *Anabaena* sp. strain PCC 7120 and designated it as *fraG* (*alr2338*) on the basis of the phenotype of the mutants that undergo fragmentation and are deficient in the synthesis of Hgls. The gene *alr2338* has been disrupted at positions +124 bp, +158 bp, +176 bp and +1770 bp relative to the translational start site. In addition, a 592 bp portion of the *fraG* gene was replaced in the wild-type by the introduction of Sp-Sm resistance cassette through a suicide plasmid (pAN120). The resultant mutant (UHM127) and the *fraG* disruptant mutants exhibited the same phenotype, i.e. fragmentation of the filaments to the extent of being in a single celled stage within 24 h after nitrogen step-down and in their inability to synthesize Hgls. The wild-type synthesizes heterocyst glycolipids after 15 h of nitrogen step-down but by this time UHM127 did not synthesize the Hgls and did not undergo fragmentation. A gene construct carrying promoter of *fraG* fused to *gfp* when introduced into wild-type through a shuttle vector, the expression of GFP was found to be higher in the developing and mature heterocysts. The gene product, FraG has been predicted to be an integral membrane protein (with 751 amino acid residues) essential for maintaining filament integrity and heterocyst maturation.

A few investigations have been devoted to understand what constitutes the continuity between vegetative cells and heterocysts or vice-versa, the periplasm of the two cells or the microplasmodesmata present in between cell junctions. Contradicting evidences have been presented with reference to periplasm as the route for transport of substances. Mariscal *et al.* (2007) favour the view based on *patS* expression in proheterocysts and mature heterocysts. When *patS* promoter was fused with genetically engineered *gfp* [that has the ability to move from cytoplasm into the periplasm of proheterocysts and heterocysts due to the presence of twin-arginine signal sequence (ss)] as the reporter gene and introduced into *Anabaena* sp. PCC 7120, fluorescence could be detected in the periplasm of not only the proheterocysts and heterocysts but also in the periplasm of adjacent vegetative cells. According to these workers, periplasm continuity can form a suitable conduit for the transport of substances into heterocysts and vegetative cells and vice-versa (Mariscal *et al.*, 2007). On the other hand, Zhang *et al.* (2008) put forward the view that there exists a periplasmic barrier for GFP to be transported from cell-to-cell in *Anabaena* sp. strain PCC 7120. They chose three genes, two of them (*hepA* expressed around 5–10 h and *patB* expressed around 18 h) exclusively expressed in proheterocysts and heterocysts and the third gene *rbcL* expressed in vegetative cells (but turned off in heterocysts at late maturation phase) and the respective plasmids with gene constructs $P_{hepA\text{-}ssgfp}$, $P_{patB\text{-}ssgfp}$ and $P_{rbcL\text{-}ssgfp}$ were introduced separately into wild-type *Anabaena* sp. strain PCC 7120. As controls similar gene constructs without *ss* for *gfp* have been used. Through techniques such as FRAP (fluorescence recovery after photobleaching) and FLIP (fluorescence loss in photobleaching), the green fluorescence has been observed only in proheterocysts or heterocysts and no intercellular transfer through periplasm could take place. Likewise, in vegetative cells green fluorescence protein could not move into proheterocysts or heterocysts despite possessing signal sequence (Zhang *et al.*, 2008). To understand the mechanism of intercellular molecular exchange, Mullineaux *et al.* (2008) employed non-fluorescent acetoxymethylester derivative, calcein that reacts with endogenous esterases to form a fluorescent hydrophilic product. Once taken up by the cells if calcein is not exchanged in between the cells, the fluorescence of calcein should remain constant but if there is a rapid transfer of calcein from one cell to the other the decrease and increase of fluorescence could be observed. For determining the rate constants of calcein exchange, they employed *Anabaena* sp. strain PCC 7120 cells grown in nitrate medium and after nitrogen step-down. Visualization of calcein fluorescence by FRAP followed by their imaging through a confocal microscope enabled them to suggest that (i) there exists a rapid exchange of calcein in between vegetative cells; (ii) the fluorescence in vegetative cells adjacent to heterocysts remained fairly constant due to the presence of polar nodules; (iii) there is no calcein fluorescence in the periplasm of the cells; (iv) the exchange of calcein only takes place in between the vegetative cells of heterocystous cyanobacteria but not in between vegetative cells of non-heterocystous cyanobacteria like *Oscillatoria*; (v) *A. variabilis* mutant lacking cyanophycin synthase does not form polar nodules in its heterocysts and calcein exchange in between vegetative cells and heterocysts has been found to be much faster than in wild-type; (vi) they suggested that the likely candidate suitable for forming pore-like structures at cell junctions occupied by microplasmodesmata is FraG (or SepJ the product of gene *alr2338*), the channel-forming protein; (vii) mutants of FraG showed negligible exchange of calcein fluorescence in between the vegetative cells; and (viii) FraG constitutes part of molecular machinery required for intercellular molecular exchange (Mullineaux *et al.*, 2008). Flores *et al.* (2007) characterized a *sepJ* insertion mutant that exhibited extensive fragmentation without heterocyst development and nitrogen fixation even under anoxic conditions. The SepJ protein is thus shown to be necessary for filament integrity and its localization at the intercellular septa has been confirmed by the expression of *gfp* as a reporter gene in the wild-type. Summing up cell-cell communication in filamentous cyanobacteria, Haselkorn (2008) cautioned that while the evidences putforward by Zhang *et al.* (2008) and Mullineaux *et al.* (2008) appear to be convincing we should wait more for confirmation from structural studies combined with fluorescence involving FraG and other subunits that might constitute the microplasmodesmata. The other component proteins that help FraG in getting organized into microplasomdesmata have now been identified by Merino-Puerto *et al.* (2010). Of the four Fra proteins identified, i.e. FraC (*alr2392*), FraD (*alr1603*), FraE (*alr2394*), FraH (*alr1603*), the first three of them are encoded by the respective genes as a single operon constitutively and *fraH* is induced under nitrogen deprivation. The localization of FraC and FraD at the intercellular septa has been confirmed by the *gfp* transcriptional fusions with the promoter regions of *fraC* and *fraD*. So the three proteins FraG, FraC and FraD help in organization of the intercellular septa. The observations of Mariscal *et al.* (2011) assume significance in defining the role of FraG as an intercellular conduit for molecular exchange. The recognition of three domain structure for FraG (or SepJ; an N-terminal coiled-coil domain, a central linker and a C-terminal permease domain) and deletion mutants of the three domains focused on the importance of coiled-coil domain in the localization of FraG protein at the intercellular septa and for maintenance of the integrity of the filaments. Expression of *conR* (*all0187*), up-regulated in proheterocysts and heterocysts at 9 h after nitrogen step- down of *Anabaena* sp. strain PCC 7120, with $P_{conR\text{-}gfp}$ as reporter gene revealed that this gene plays an important role in septum formation. Mutants of ConR have been found to be defective in the septum formation and differentiated longer heterocysts with their polar nodules partially open. Though the mutant possessed nitogenase and fixed nitrogen to the extent of 70% of the wild-type, the deficiency of diazotrophic growth has been due to lack transportation of fixed nitrogen to the vegetative cells (Mella-Herrera *et al.*, 2010).

***vi) Carbon metabolism*:** The activities of G6P-dehydrogenase and 6-phosphogluconate hydrogenase were (6 to 8-fold) higher in isolated heterocysts of *A. cylindrica* (Winkenbach and Wolk, 1973; Lex and Carr, 1974). Due to this reason heterocysts are suggested to receive a source of reductant in the form of NADPH for nitrogenase to function in light as well as dark (Bothe, 1970). G6P-dehydrogenase is encoded by *zwf* gene and its inactivation in *Nostoc* sp. strain ATCC 29133 caused failure of diazotrophic growth in light or in dark in presence of organic carbon sources. The *zwf* transcript was expressed as an operon and contained sequences for genes encoding fructose-1,6-biphosphatase, transaldolase as well as an undesignated gene known as *opcA*. Complementation of the wild-type genes restored the diazotrophic growth potential signifying that in light OPP pathway provides the requisite reductant for nitrogen fixation and respiration in the heterocysts (Summers *et al.*, 1995). Enzymes of glycolysis and part of TCA cycle occur in heterocysts (Bothe, 1982). In addition, the presence of pyruvate-ferredoxin oxidoreductase (Neur and Bothe, 1982), isocitrate dehydrogenase (Papen *et al.*, 1983) and glutamate-oxoglutarate amidotransferase (Häger *et al.*, 1983) has been demonstrated. Due to the presence of all the enzymes for the conversion of G6P to oxoglutarate, heterocysts are not dependent on vegetative cells for a supply of glutamate (Neuer and Bothe, 1983; Papen *et al.*, 1986).

CO_2-fixation by the isolated heterocysts of *A. cylindrica* is negligible due to lack or low levels of RUBP-carboxylase (Fay and Walsby, 1966; Winkenbach and Wolk, 1973; Lex and Carr, 1974; Codd *et al.*, 1980) and as such in the absence of CO_2-fixation, the required reduced carbon compounds are supplied by the adjacent vegetative cells. Wolk (1968) demonstrated that when *A. cylindrica* filaments were exposed to radioactive carbon, the movement of carbon label moved from vegetative cells to the heterocysts. But the nature of carbon compounds that move from vegetative cells into the heterocysts to support nitrogen fixation remained sketchy for a long time. However, persistent efforts were made in this direction. $^{14}CO_2$-labelling of the filaments of *A. cylindrica* revealed that the heterocysts isolated after short intervals of 20 s showed labelled alanine, glutamate and glutamine and G6P. When alanine was supplied to the isolated heterocysts it got readily converted to glutamate and glutamine under N_2-H_2 atmosphere. It was thus concluded that alanine serves as a precursor for reducing equivalents required by the heterocysts (Jüttner, 1983). A survey of 33 organic compounds that included organic acids, polyhydric alcohols, sugars and sugar phosphates identified D-erythrose as the best source for supporting acetylene reduction by the isolated heterocysts of *Anabaena* sp. strain PCC 7120. While D-erythrose enhanced acetylene reduction by 10 times the rate of control, the supply of D-glucose, G6P, fructose-1,6-biphosphate and sucrose supported acetylene reduction by 4-fold (Privalle and Burris, 1984). However, a lot of literature has accumulated on the role of sucrose synthesis and degradation in the heterocysts to support nitrogen fixation. The role of sucrose metabolism in relation to heterocyst differentiation and nitrogen fixation is presented below.

Sucrose biosynthesis is known to occur in plants and cyanobacteria. Three enzymes are essential in this respect collectively known as sucrose biosynthesis-related proteins. A two-step biosynthesis of sucrose is mediated by sucrose phosphate synthase (SPS; UDP-glucose:D-fructose-6-phosphate 2-α-D-glucosyltransferase; EC 2.4.1.14) and sucrose phosphate phosphatase (SPP; Sucrose-6-phosphate-phosphohydrolase; EC 3.1.3.24). In the first step, the biosynthesis of sucrose-6-phosphate takes place in presence of precursors, fructose-6-phosphate and UDP-glucose, energy for which is derived from the cleavage of uridine diphosphate. In the second terminal step, the phosphate group from sucrose-6-phosphate is hydrolysed releasing free sucrose. The third enzyme sucrose synthase (SuS; UDP:glucose-D-fructose-2-α-D-glucosyltransferase; EC 2.4.1.13) synthesizes sucrose in a single step but this very enzyme can break down sucrose into its subunits. Alkaline/neutral invertases (A/N-Invs) irreversibly hydrolyze sucrose when there is a high demand for hexoses. Schilling and Ehrnsperger (1985) were the first to have reported SuS activity in cell-free extracts from vegetative cells of *A. variabilis*. Sucrose biosynthesis was mediated by SuS in vegetative cells and sucrose degradation was attributed to the existence of an A/N-Inv activity in the heterocyst preparations. A study of the biochemical properties of SuS from *Anabaena* sp. strain PCC 7119 led to the identification of two isoforms (Sus-I and SuS-II). SuS-II, a tetramer with a molecular mass of 92 kDa, differed from SuSs of higher plants in substrate specificity, regulation by metal ions and ATP and in its N-terminal sequence (Porchia *et al.*, 1999). The *susA* gene, encoding Sus-II isoform of SuS, has been cloned from *Anabaena* sp. strain PCC 7119 and the deduced amino acid identity with higher plant enzyme is of the order of 30–40%. The *susA* gene was expressed in *E. coli* and the recombinant protein produced was identical in its biochemical properties with the native enzyme. It is of interest to know that SuS has a taxonomic significance because of its occurrence in several nitrogen-fixing cyanobacterial species and its absence in two of the unicellular non-diazotrophic species investigated (Curatti *et al.*, 2000). The involvement of sucrose in the diazotrophic metabolism of *Anabaena* sp. strain PCC 7119 has been demonstrated by Curatti *et al.* (2002) who cloned the *sus* gene sequences from *Anabaena* sp. strain PCC 7119 and *Anabaena* sp. strain PCC 7120 and designated them as *susA* and *susB*, respectively. The *susA* gene expression was measured by the transcript abundance in terms of competitive RT-PCR and the SuSA protein (formerly SS-II) levels. A *susA* gene disruptant mutant, LC30 neither showed SuSA protein nor its activity. LC60 strain, an overexpression mutant of *susA*, showed enhanced (10-fold) SuSA protein and its activity. There was accumulation of sucrose in the LC30 cells while it did not accumulate in the LC60 cells suggesting that SuSA is involved in the cleavage of sucrose *in vivo*. The wild-type strain, on the other hand, showed no accumulation of sucrose but showed high invertase activity when compared with the cells grown in presence of nitrate than in presence of ammonium. Both the mutants performed equally well in nitrate and ammonium supplemented cultures but in nitrate-free medium LC60 grew poorly despite the differentiation of heterocysts. Thus a role for SuSA has been ascribed in the diazotrophic metabolism of *Anabaena* sp. strain PCC 7119. Curatti *et al.* (2006) showed that *Anabaena* sp. strain PCC 7120 possesses *susA* sequence besides the *susB* sequence reported by them earlier (Curatti *et al.*, 2002). SuSA contributes to the major extractable SuS activity. Both *susA* and *susB* promoter sequences of *Anabaena* sp. strain PCC 7120 revealed sequences similar to those present in *rbcLS* promoter sequences signifying their ammonium-upregulated nature. Further, it was shown that the expression of *susA* and *rbcLS* is developmentally restricted to the vegetative cells under nitrogen-fixing conditions showing there by that nitrogen fixation, carbon fixation and sucrose metabolism are under the same regulatory net work (Curatti *et al.*, 2006). The probable involvement of *susA* of *Anabaena* sp. strain PCC 7119 in the synthesis of polysaccharides from sucrose has been investigated by Curatti *et al.* (2008). Nitrogen and carbon sources and light affected the expression of *susA* gene as they affected the accumulation of glycogen. A comparison of the wild-type and a *susA* disruptant mutant revealed that the latter accumulated lesser amounts of glycogen under nitrogen-fixing conditions. A strain of *Anabaena* sp. strain PCC 7119 constitutively expressing *susA* showed higher levels of glycogen. Based on these observations they concluded that SuSA plays an important role in recognizing environmental and nutritional signals and modulate the expression of *susA* gene.

Porchia and Salerno (1996) reported for the first time the activities of SPS and SPP in *Anabaena* sp. strain PCC 7119. Two species of SPSs, SPS-I and SPS-II have been separated, purified and their properties studied. These enzymes considerably differed in their properties from those of the higher plants in having lower molecular mass, broad substrate specificity, requirement of Mg^{2+} and Mn^{2+} and in being monomeric in composition. Both SPS-I and SPS-II can utilize fructose-6-phosphate as their substrate besides UDP-glucose, while the former additionally accepts GDP-glucose and TDP-glucose as substrates the latter can accept ADP-glucose. Curatti *et al.* (1998) identified the gene *spsA* encoding SPS of *Synechocystis* sp. strain PCC 6803, a non-nitrogen fixing unicellular form. Cumino *et al.* (2001) compared the biochemical properties of the native SPP from *Anabaena* sp. strain PCC 7120 and the recombinant SPP produced in *E. coli* and these were shown to be identical. SPP of *Anabaena* sp. strain PCC 7120 differed with the corresponding enzymes from higher plants in its C-terminal sequence, molecular mass and subunit composition. Sucrose synthesis in the heterocysts of *Anabaena* sp. strain PCC 7120 and its probable role in nitrogen fixation have been elucidated by Cumino *et al.* (2007). Their observations confirmed the operation of a cycle of sucrose synthesis and degradation and its linking to glycogen accumulation during nitrogen fixation. The basis of their conclusions are that (i) sucrose synthesis occurs in light as well as in dark incubated cultures; (ii) total activities of SPSs and SPPs are higher in nitrogen-fixing conditions rather than in presence of combined nitrogen sources; (iii) the increase in the levels of transcripts for *spsA* and *spsB* corresponded with increase in the levels of SPS-A and SPS-B proteins; (iv) the expression of *spsA* in the vegetative cells and *spsB* and *sppA* in the heterocysts and vegetative cells contributed to sucrose synthesis; (v) the preferred glucosyl donor for SPS-A is UDP-glucose and for SPS-B it is either UDP-glucose or ADP-glucose; (vi) transcriptional fusions of optimized version of *gfp-mut2* with the promoters of *spsA*, *spsB* and *sppA* upstream of their tsps at 950 bp, 977 bp and 559 bp respectively, showed the fluorescence of GFP in vegetative cells ($P_{\textit{-spsA-gfp}}$) and vegetative cells and heterocysts ($P_{\textit{-spsB-gfp}}$ and $P_{\textit{-sppA-gfp}}$); (vii) two tsps at –54 bp and –93 bp have been observed for *spsB* resulting in the formation of RNA_{I} and RNA_{II} in nitrate-free, nitrate and ammonium grown cultures but RNA_{II} transcripts were 4-fold more abundant than that of RNA_{I} in nitrogen-fixing cultures; (viii) the consensus NtcA-binding site GTA(N_8)ACA was found at –39 bp with respect to tspII at –93 bp. The regulation of sucrose metabolism genes by NtcA has been examined by Marcozzi *et al.* (2009) who showed that induction of *spsB* (sucrose synthesis) gene after nitrogen step-down in *Anabaena* sp. strain PCC 7120 was abolished in an *ntcA*-disruptant mutant with an increase in the activity of SuSA (sucrose cleavage) activity. Both *spsB* and *susA* promoter regions showed interaction with NtcA *in vitro* at the consensus NtcA-binding sequences. These observations confirm that sucrose metabolism and nitrogen metabolism are transcriptionally under the control of NtcA. Fieulaine *et al.* (2005) reported the crystal structure of SPP from *Synechocystis* sp. strain PCC 6803 that revealed the presence of a core domain consisting of the active site and a small cap having a glucose-binding site. There are two flexible hinge tongs that link the two regions. The expression pattern of *spsA* and *sppA* genes of *Synechococcus* sp. strain PCC 7002, a marine cyanobacterium, has been reported for the first time and revealed cotranscription of the genes (Cumino *et al.*, 2010).

Vargas *et al.* (2003) reported the occurrence of two A/N-Invs in heterocystous cyanobacteria and one A/N-Inv in the unicellular forms based on a search of the fully sequenced cyanobacterial genomes. *Anabaena* sp. strain PCC 7120 possesses two genes, *invA* and *invB* that encode corresponding proteins with a molecular mass of 53–55 kDa. Lopez-Igual *et al.* (2010) demonstrated that the expression of *invA* and *invB* occurred under diazotrophic conditions but the abundance of the transcript and corresponding protein was higher for latter gene. The expression of *invA* gene was found uniformly in all vegetative cells as reflected by the GFP fluorescence when a transcriptional fusion construct $P_{\text{-}invA\text{-}gfp}$ was introduced into *Anabaena* sp. strain PCC 7120. The expression of *invB* was restricted to the developing heterocysts or mature heterocysts suggesting that *invB* plays an important role in regulating diazotrophic growth. This is confirmed by the characteristics of *hetR*, *invA* and *invB* mutants. The expression of *invB* gene was found all along the length of the filament in the the *hetR* mutant. In the *invA* mutant the expression of *nifHDK* was impaired associated with higher frequency of heterocysts. Loss of Inv-B function in the *invB* mutant led to impaired diazotrophic growth.

Thus there are a number of evidences in favour of the existence of intricate relationship between sucrose synthesis and breakdown, and conversion of sucrose to glycogen and its utilization during diazotrophic growth in cyanobacteria. Earlier studies of Ernst and Böger (1985) highlighted the importance of glycogen supported nitrogen fixation. When *A. variabilis* ATCC 29413 was subjected to nitrogen step-down, heterocyst differentiation was preceded by a round of fast accumulation of glycogen. There is a temporal separation of glycogen accumulation and nitrogen fixation, the former occurring in the light and the latter taking place in the dark. The observations of Jensen *et al.* (1986) on the high and sustained nitrogen-fixing capacity of isolated heterocysts, derived from fructose-grown *A. variabilis* ATCC 29413, was due to the presence of glycogen granules inside them. Ernst *et al.* (1990) reported that a modification of nitrogenase induced by C-limitation could be overcome due to endogenously stored glycogen and the presence of fructose in presence of air. A positive regulation of sugar catabolic pathways by a group 2 σ factor SigE (sll1689) in *Synechocystis* sp. strain PCC 6803 involving glycolysis, oxidative pentose phosphate pathway and glycogen catabolism may explain its transcriptional control over these processes (Osnai *et al.*, 2005).

Cumino *et al.* (2002) conducted a phylogenetic analysis of sucrose biosynthesis-related proteins from cyanobacteria (*Anabaena* sp. strain PCC 7120, *Anabaena* sp. strain PCC 7119; *N. punctiforme*, *Prochlorococcus marinus* MED4, *P. marinus* MIT 9313, *Synechococcus* sp. strain WH 8102 and *Synechocystis* sp. strain PCC 6803) and higher plants (rice, maize, potato, *Arabidopsis thaliana* and *Medicago truncatula*) and indicated that *Anabaena* SPS maintains a glucosyltransferase domain that constitutes the minimal catalytic unit of all SPPs and some SPSs. Lunn (2002) traced the origin of sucrose synthesis to the proteobacteria or a common ancestor of the proteobacteria and cyanobacteria. Vargas *et al.* (2003) conducted a phylogenetic analysis of A/N-Invs of cyanobacteria and suggested that the origin of higher plant A/N-Invs could be from an orthologous ancestral gene of a cyanobacteria as a result of endosymbiotic event.

vii) Amino acid transport and diazotrophic growth: A number of amino acids are utilized as alternative nitrogen sources by cyanobacteria by causing a repression in nitrogenase activity (Neilson and Larsson, 1980; Vaishampayan, 1982; Rawson, 1985; Spence and Stewart, 1986). However, transport systems for the uptake of amino acids have been reported in a few cases. The kinetics of transport for Glu in *A. cylindrica* (Rowell *et al.*, 1977), Pro (Spence and Stewart, 1986) and Gln/Glu in *Anabaena* sp. strain PCC 7120 (Flores and Muro-Pastor, 1988) have been investigated. Strasser and Falkner (1986) reported the existence of a highly functional transport system for Glu/Asp uptake by a *Nostoc* sp., an endosymbiont of *Geosiphon pyriforme*. High affinity and low affinity transport systems for Glu and Gln have been reported by Chapman and Meeks (1983) in *A. variabilis* ATCC 29413. The K_m values for high affinity system were found to be 13.8 and 100 µM for Gln and Glu, respectively whereas the low affinity system showed K_m values to be 1.1 and 1.4 mM, respectively for Gln and Glu. A single transport system for Leu (with a K_m of 10.8 µM) in the same organism was reported by Thiel (1988) that was inhibited by Ala, Gly, Val and Met. Herrero and Flores (1990) showed that the uptake of L-Arg is governed by an energy requiring high affinity process and a low affinity uptake system and the cells accumulated the substrate in the former but not in the latter. Two such uptake systems have been noted for the uptake of L-Lys also. Montesinos *et al.* (1995, 1997) characterized a number of amino acid analogue resistant mutants of *Anabena* sp. strain PCC 7120. Mutants developed in presence of L-canavine sulfate (50 µM), δ-hydroxylysine (50 µM), 4-aza-DL-leucine (100 µM), MSX (50 µM) and azaserine (50 µM) were characterized for the transport of different amino acids, intracellular concentrations and the type of amino acid released by them. On the basis of transport kinetics of various amino acids they suggested that five different types of transport systems are operative in *Anabaena* sp. strain PCC 7120. These are: (i) a high affinity system for basic amino acids (Arg and Lys; which could transport His and Orn as well); (ii) two neutral amino acid transport systems N-I and N-II (which could transport Ala, Asn, Gln, Met, Ser and Thr) and in addition the former could transport Gly, Leu, Phe and Pro; (iii) a low affinity, passive amino acid transport system for basic amino acids and (iv) a low affinity system for the uptake of acidic amino acids. In addition to *Anabaena* sp. strain PCC 7120, Montesinos *et al.* (1997) included eight representative strains belonging to unicellular, filamentous, heterocystous unbranched and branched members in their study for determining the uptake of thirteen ^{14}C-labelled amino acids. The presence of two genes *natA* and *natB* that encode a conserved component and a periplasmic binding protein, respectively of an ABC-type permease has been demonstrated in all the cyanobacteria. Gene disruptant mutants for *natA* and *natB* of *Anabaena* sp. strain PCC 7120 released Ala, Val, Phe, Ile and Leu into the medium. Significantly, the diazotrophic growth potential as well as heterocyst differentiation was affected in these mutants. The same group of workers further identified all the subunits of N-I neutral amino acid permease. These consist of two ATPases [encoded by *natA* (*all1046*) and *natE* (*all2912*)], two transmembrane proteins [encoded by *natC* (*all1047*) and *natD* (*all1284*)] and a periplasmic binding protein [(encoded by *natB* (*alr1834*)]. Two of these genes *natA* and *natC* are present as an operon. The expression of *natCA* operon and *natB* constitutively in the vegetative cells of *Anabaena* sp. strain PCC 7120 but not in the heterocysts was supported by the corresponding increase in the transcript levels and also by the use of $P_{natC\text{-}lacZ}$ fusion construct as a reporter gene. Ala was released at the highest level by *natB* disruptant mutants (CSS6) that exhibited 75% of the diazotrophic growth rate of the wild-type. The growth rate of *natCA* mutants was reduced to 50% of that of the wild-type. The release of Ala was affected in a mutant of *natB* that was unable to differentiate heterocysts. It means for highest release of Ala the presence of functional heterocysts is required, meaning thereby that L-Ala is transported into heterocysts from vegetative cells (Picossi *et al.*, 2005). The component proteins of two ABC-type amino acid transporters Bgt and N-II have been identified in *Anabaena* sp. strain PCC 7120 by Pernil *et al.* (2008). Bgt constitutes a high affinity basic amino acid transporter with BgtA and BgtB as components. BgtA is an ATPase subunit encoded by *alr4167* and BgtB is a composite protein that has subunits for a periplasmic-binding and transmembrane domains encoded by *alr3187*. N-II constitutes the second transporter for acidic and neutral polar amino acids. NatF, a periplasmic substrate-binding protein and NatG and NatH are transmembrane proteins of N-II transporter. The ATPase subunit BgtA serves both the transporters. Bgt and N-II are expressed both in the vegetative cells and heterocysts. Since Bgt is dispensable for diazotrophic growth, N-I that is exclusively expressed in vegetative cells (Picossi *et al.*, 2005) together with N-II constitute the major transport systems for 98% of the amino acids and are important for diazotrophic growth.

The main conclusions of the above investigations are that: (i) heterocysts export amino acids to the periplasmic space from where they would diffuse out before being taken up by vegetative cells; (ii) through N-I transporter the vegetative cells take up a number of amino acids and recapture a number of hydrophobic amino acids leaked out from the cells; (iii) there is a differential expression of the *nat* genes in the vegetative cells and heterocysts; (iv) since continuity of periplasmic space in between vegetative cells and heterocysts of a filament is maintained, there would be an exchange of these molecules between the two cell types; (iv) though Gln is projected as the product of nitrogen fixation that is transferred to vegetative cells, there exists a possibility for Ala being considered as a candidate because synthesis of Ala in the heterocysts has been demonstrated (Lawrie, 1976; Wolk *et al.*, 1976; Jüttner, 1983); (v) the release of Ala by *natB* mutants of *Anabaena* sp. strain PCC 7120 requires functional heterocysts and (vi) Ala can be a good substrate for alanine dehydrogenase and a number of other tranaminases.

Alanine dehydrogenase (Ald; EC 1.4.1.1) is an NAD-dependent enzyme that catalyzes the reversible deamination of L-Ala to pyruvate, ammonia, NADH and H^+. Ald is a hexameric protein and the molecular mass ranges from 270 kDa as in *A. cylindrica* (Rowell and stewart, 1975) to 240 kDa as in *Phormidium lapideum* (Sawa *et al.*, 1994). The molecular mass of the subunits accordingly is 43 kDa in the former and 41 kDa in the latter. Ald from *P. lapideum* does not possess any sequence similarity with other dehydrogenases but it is closely related to the family of 2-hydroxyacid dehydrogenases. The subunits of Ald get folded into two identical domains that are separated by a cleft. A region of (120) amino acids that is highly conserved is clustered around the cleft that might contain the active site of the enzyme (Sedelnikova *et al.*, 1998; Baker *et al.*, 1998). Ald activity is required for phycobilisome degradation in *S. elongatus* PCC 7942 when subjected to nitrogen starvation. Inactivation of *ald* gene of *S. elongatus* 7942 by insertion of a Sp^r cassette resulted in an Ald-deficient mutant that exhibited down-regulation of a number of nitrogen stress-related genes, foremost of which are *nblA* and *nblC*. The wild-type cells showed high Ald activity upon nitrogen starvation to maintain low cellular levels of L-Ala while the Ald-mutant accumulated higher levels of L-Ala and in the absence of induction of *nblA* and *nblC* phycobilisome degradation did not occur despite nitrogen deficiency (Lahmi *et al.*, 2006). An L-amino acid oxidase has been identified associated with thylakoid membranes of *Synechocystis* sp. strain PCC 6803. It can oxidize a number of basic L-amino acids with L-Cys and L-Arg serving as preferred substrates (with 150% and 100% activity when L-Arg is taken as standard) from among a number of others, the oxidation of L-Ala being 37%. This amino acid oxidase can also function as L-Arg dehydrogenase mediating electron transfer from L-Arg to the respiratory electron transport (Schriek *et al.*, 2009). Such amino acid oxidases with dual function are yet to be demonstrated in the heterocystous cyanobacteria. However, Ala catabolism seems to be essential for the diazotrophic growth of *Anabaena* sp. strain PCC 7120 because in an *ald* (ORF alr2355) inactivated mutant the diazotrophic growth potential decreased by 50%. A strain with *ald-gfp* gene construct showed the expression of *ald* in the developing and mature heterocysts (Pernil *et al.*, 2010).

viii) Hydrogen metabolism: Reduction of dinitrogen to ammonia mediated by nitrogenase is associated with concomitant evolution of hydrogen. At least 25% of the electrons during nitrogen fixation process are allocated for an obligatory reduction of protons (H^+) to form H_2 (Tamagnini *et al.*, 2002). However, according to Simpson and Burris (1984) nitrogenase generates one mol of H_2 for every mol of N_2 reduced. Hydrogen evolution has been viewed as an inbuilt trait of diazotrophs. Photobiological production of hydrogen has been receiving attention since last three decades, since it can be used as a clean, storable, transportable and renewable energy source. Many nitrogen-fixing bacteria (mainly species of *Clostridium*; Calusinska *et al.*, 2010), cyanobacteria (Tamagnini *et al.*, 2007) and certain green algae (*Chlamydomonas reinhardtii, Scenedesmus obliquus, Chlorella fusca* and *Nephroselmis olivacea*; Melis and Happe, 2001) are being exploited for ameliorating the situation in energy consumption and saving spheres (Lopes Pinto *et al.*, 2002; Tamagnini *et al.*, 2007). H_2 evolution and its subsequent consumption provide a means of protection against O_2 damage to nitrogenase. The oxidation of hydrogen (the oxyhydrogen) in a reaction termed as Knallgas reaction ($2H_2 + O_2 \rightarrow 2H_2O$) facilitates the removal O_2 from the site of nitrogenase. Walsby (1985) suggested that the molar flux of O_2 into the heterocysts is half of that of N_2. Besides the high respiratory activity, the Knallgas reaction provides additional protection. The organism saves part of its reductant as well as ATP for activity of nitrogenase if H_2 evolved is not allowed to be diffused out. So the hydrogen evolved is immediately consumed that is catalyzed by an uptake hydrogenase. There is a strong correlation between the activity of uptake hydrogenase and nitrogen fixation in a number of heterocystous cyanobacteria (Lambert and Smith, 1981; Houchins, 1984; Wolk *et al.*, 1994; Oxelfelt and *et al.*, 1995; Tamagnini *et al.*, 2002, 2007). Uptake hydrogenases function in the direction of H_2 uptake and recycle or reutilize the H_2/electrons so as to facilitate metabolism. This is supported by the induction of uptake hydrogenase activity in heterocystous cyanobacteria by H_2 (Tel-Or *et al.*, 1977), production of ATP via respiratory chain by coupling to O_2 (Peterson and Burris, 1978; Houchins and Burris, 1981a) and acetylene reduction by the supply of electrons (Bothe *et al.*, 1977). Stimulation in uptake hydrogenase activity of *Anabaena* sp. strain PCC 7120 was observed by the inclusion of H_2 or removal of O_2 in the gas phase of the cultures (Houchins and Burris, 1981b). It seems mere exogenous supply of H_2 is not sufficient to maintain uptake hydrogenase activity but it is dependent on the extent of induction of nitrogenase activity *in vivo* (Lambert and Smith, 1981; Oxelfelt *et al.*, 1995).

A comparative study of nitrogen fixation and H_2 uptake by four cyanobacteria (*A. variabilis, Nostoc spongiaeforme, Westiellopsis prolifica* and *Nostoc* sp.) revealed that (i) a concentration of up to 20% H_2 enhanced nitrogenase activity; (ii) *N. spongiaeforme* showed higher H_2 uptake activity under both aerobic and anaerobic conditions and (iii) addition of DCMU did not inhibit H_2 uptake in case of *N. spongiaeforme* (Vyas and Kumar, 1995). In case of *A. variabilis* ATCC 29413 exogenous H_2 has little effect on the stimulation of *in vivo* H_2 uptake during early stages of nitrogenase induction (Troshina *et al.*, 1996). Photoautotrophically grown cultures of *Anabaena cycadae* showed enhancement in the activities of nitrogenase as well as uptake hydrogenase but dark incubation in presence of glucose caused a reduction in nitrogenase activity to half of that observed in light with out uptake hydrogenase activity. Fresh isolates of *A. cycadae* from its host tissue exhibited higher nitrogenase activity with no apparent uptake hydrogenase activity (Kumar *et al.*, 1986). Nitrogen-fixing cultures of *A. variabilis* ATCC 29413 showed fructose-dependent H_2 evolution but not those grown in presence of ammonium. Fructose (up to 10 mM) was dissimilated into H_2 and CO_2 but the presence of DCMU inhibited H_2 evolution (Reddy *et al.*, 1996). Houchins and Burris (1981b,c) demonstrated the presence of an O_2-sensitive reversible hydrogenase in *Anabaena* sp. strain PCC 7120 that was independent of nitrogen-fixing capacity of the organism. The activity of reversible hydrogenase was enhanced 2–3 fold under microaerophilic conditions but remained unaffected by the addition of H_2. Due to their functional ability of catalyzing the production and consumption of H_2 both *in vitro* as well as *in vivo*, the reversible enzymes have been redesignated as bidirectional enzymes in the literature. Based on the metallocenters, three groups of hydrogenases, i.e. [NiFe] hydrogenases, [FeFe] hydrogenases and [Fe] hydrogenases have been recognized in prokaryotes. The existence of latter type of hydrogenases has been described in methane bacteria (Pilak *et al.*, 2006). All cyanobacterial hydrogenases are of [NiFe]-type (Tamagnini *et al.*, 2000, 2002, 2005). Nitrogen-fixing cyanobacteria possess uptake hydrogenase with the exception of *Synechococcus* sp. BG 043511. In addition, they may also have a bidirectional hydrogenase and in this respect *Nostoc* sp. strain PCC 73102 is an exception (Tamagnini *et al.*, 1997). The non-nitrogen-fixing unicellular cyanobacteria generally possess the bidirectional enzyme but exceptionally *Gloeobacter violaceus* PCC 7421 does not possess this enzyme (Nakamura *et al.*, 2003; Ludwig *et al.*, 2006). Min and Sherman (2010) reported the presence of an uptake hydrogenase as well as a bidirectional hydrogenase in the versatile unicellular nitrogen-fixing cyanobacterium *Cyanothece* sp. strain ATCC 51142 that showed maximum rates of H_2 production under photosynthetic conditions. A phylogenetic analysis of hydrogenase genes from five groups of photosynthetic bacteria including cyanobacteria revelaed the green sulfur bacterium *Chloroflexus aurantiacus* as the closest ancestor for vertical transmission of genes into cyanobacteria (Ludwig *et al.*, 2006).

Cyanobacterial uptake and bidirectional hydrogenases are nickel-dependent enzymes (Zhang *et al.*, 1984; Almon and Böger, 1984; Pederson *et al.*, 1986; Hausinger, 1987). The effect of environmental factors (light/dark or continuous light regimes, aerobic/anaerobic conditions) and nutrients such as Ni^{2+} on H_2 formation in *A. cylindrica* has been investigated (Daday *et al.*, 1977; Daday and Smith, 1983). Ni^{2+} (at 0.68 μM) supported maximum (175 nmol h^{-1} mg^{-1}) H_2-uptake activity when compared to Ni^{2+} depleted cultures after 5 d growth of *A. cylindrica* (Daday *et al.*, 1985). Xiankong *et al.* (1984) suggested that nickel may be required for either activation of an uptake hydrogenase or H_2 formation or in the synthesis of another protein involved in H_2 uptake. On the basis of the presence of a dark uptake hydrogenase activity and a light-dependent uptake in presence of DCMU at low O_2 concentration in *Anabaena* strains CA and IF, these workers further concluded that H_2 may serve as electron donor to both photosynthetic and respiratory electron transport chains. Kumar and Kumar (1990) studied H_2 evolution by *Anabaena* sp. strain CA in presence of fructose and erythrose and photosynthetic inhibitors and similarly concluded that the all electrons irrespective of source must pass through plastoquinone pool before reducing ferredoxin. Photoheterotrophic cultures (with glucose and DCMU in light) of *Nostoc* sp. strain Cc (symbiont of *Cycas circinalis*) when supplemented with Ni consumed 10 times more H_2 than control cultures (with DCMU). The biomass increased with improved nitrogen status underlining the importance of the availability of Ni as an important factor in establishing symbiosis (Tredici *et al.*, 1990). The localization of uptake hydrogenase in the membranes of heterocysts of *Anabaena* sp. strain PCC 7120 (Peterson and Burris, 1978; Houchins and Burris, 1981b) and other cyanobacteria (Eisbrenner *et al.*, 1978; Tel-Or *et al.*, 1978) has been demonstrated. However, the localization of uptake hydrogenase in both vegetative cells and heterocysts of *N. punctiforme* ATCC 29133 and some other symbiotic strains has been shown to be membrane bound by immunological techniques (Lindblad and Sellstedt, 1990; Rai *et al.*, 1992). Seabra *et al.* (2009) conducted immunolocalization based on antisera prepared from cyanobacterial hydrogenase enzymes and showed that in *Anabaena* sp. strain PCC 7120 the uptake hydrogenase is restricted to the heterocysts while in *N. punctiforme* ATCC 29133 the uptake hydrogenase was found in both the vegetative cells and heterocysts. In the non-heterocystous, filamentous *Lyngbya majuscula* the uptake hydrogenase was found to be membrane bound. The bidirectional hydrogenase has been shown to be present in both vegetative cells and heterocysts of *Anabaena* sp. strain PCC 7120 in a soluble or loosely membrane bound form (Houchins and Burris, 1981b).

Genome sequencing of a number of cyanobacteria has facilitated the recognition of genes governing the synthesis of hydrogenases. The hydrogen uptake (*hup*) genes encode the uptake hydrogenases while hydrogen oxidation (*hox*) genes encode the bidirectional hydrogenases. Post-translational steps lead to maturation of uptake and bidirectional hydrogenases. These involve the insertion of the metal ions into the active site(s) of hydrogenases and attachment of ligands to the Fe-atom. A set of six genes, *hypABCDEF* (*hyp* for hydrogenase pleiotropic) encode the proteins necessary for this function. The *hyp*-gene cluster of *Anabaena* sp. strain PCC 7120 is depicted in Fig. 11A. A final step in the maturation process involves the proteolytic cleavage of approximately 30 amino acid residues at the C-terminal end of the large subunit of the two hydrogenases mediated by *hupW* (for uptake hydrogenase) and *hoxW* (for bidirectional hydrogenase) genes, the products of which are known as C-terminal endopeptidases. In *S. elongatus* PCC 7942, *Synechococcus* sp. (strains WH8102, CC 9902, CC 9905), *Gloeobacter violaceus* PCC 7421 and strains of *P. marinus* (MIT 9312, MIT 9313, CCMP 1378, NATL 2A, sub. sp. *marinus* CCMP 1375) the *hup, hox* and *hyp* genes are absent. Certain other strains of *Synechococcus* sp. (PCC 7942, PCC 6308, PCC 6301 and PCC 7002) possess *hox* and *hyp* genes while *hox* genes alone are represented in *Synechocystis* sp. strain PCC 6803, *A. platensis, Spirulina subsala* and *Prochlorothrix hollandica*. In the heterocystous forms *N. punctiforme* and *Nostoc* Mitsui 38901 are exception in having *hup* and *hyp* genes whereas in other heterocystous members (*A. variabilis* ATCC 29413, *Anabaena* sp. strain PCC 7120, *Anabaena siamensis* TISTR 8012) *hup, hox* and *hyp* genes are represented. In *Lyngbya majuscula*, a non-heterocystous cyanobacterium that fixes nitrogen under microaerophilic conditions, the *hup, hox* and *hyp* genes are present (Tamagnini *et al.*, 2002, 2005, 2007).

Uptake hydrogenases of cyanobacteria are heterodimeric proteins consisting of a large subunit (α subunit) encoded by *hupL* and a small subunit (β subunit) encoded by *hupS*. HupL (60 kDa) possesses the active site with four cysteine residues two of which join the Fe and Ni atoms. HupS (35 kDa) contains the Fe-S clusters, important in passing electrons from the active site to the electron acceptor in the respiratory chain to produce ATP. The bidirectional hydrogenase of cyanobacteria is encoded by *hoxEFUYH* genes. The properties of the bidirectional enzyme have been understood from several cyanobacterial species (Hallenbeck and Benemann, 1978; Llama *et al.*, 1979; Asada *et al.*, 1987; Schmitz *et al.*, 1995, 2002). The bidirectional enzyme is sensitive to O_2, thermotolerant and has high affinity to H_2. Unlike the uptake hydrogenase, the bidirectional hydrogenase does not require the biosynthesis of nitrogenase as a pre-requisite for its biosynthesis as has been shown in some unicellular strains (Howarth and Codd, 1985). Schmitz *et al.* (2002) showed that the enzyme from *Synechocystis* sp. strain PCC 6803 consists of a large subunit HoxH with the active site consisting of six cysteine residues involved in the binding of nickel. The small subunit HoxY contains four cysteine residues that are supposedly involved in coordinating the putative [4Fe-4S] cluster. HoxHY together constitute the hydrogenase component while HoxEFU (encoded by *hoxEFU*) is the diaphorase component (dihydrolipoamide:NAD oxidoreductase) that transfers electrons to $NAD(P)^+$. The HoxU and HoxF proteins are the small and large subunits of the diaphorase moiety. The HoxEFUYH subunits are of 20, 61, 28, 24 and 49 kDa molecular weight, respectively with the dimeric enzyme complex of bidirectional hydrogenase Hox $(EFUYH)_2$, attaining a molecular weight of 375 kDa. Comparison of gene sequences of *hoxEFUYH* or the deduced amino acid sequences of the corresponding proteins of *A. variabilis* ATCC 29413, *A. variabilis* IAM M58, *Anabaena* sp. strain PCC 7120, *Synechococcus* sp. strain PCC 6301 and *Synechocystis* sp. strain PCC 6803 revealed a high degree of homology with the identity in case of *A. variabilis* ATCC 29413 and *Anabaena* sp. strain PCC 7120 reaching as high as 95% (Tamagnini *et al.*, 2002, 2007).

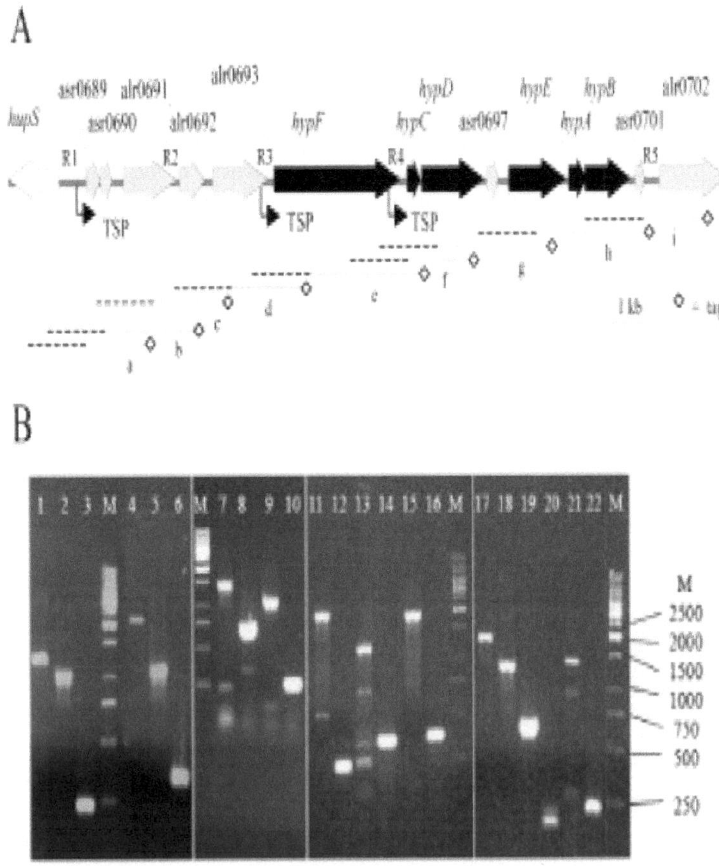

C

cDNA	PCR product	PCR primers	Size (bp)
a	1	asr0689+tag	1263
a	2	asr0690+tag	1053
a	3	alr0691+tag	238
b	4	asr0690+tag	1900
b	5	alr0691+tag	1085
b	6	alr0692+tag	339
c	7	alr0691+tag	1474
c	8	alr0692+tag	728
d	9	alr0693+tag	1086
d	10	*hypF.1*+tag	252
e	11	*hypF.2*+tag	2059
e	12	*hypC*+tag	221
f	13	*hypC*+tag	1237
f	14	*hypD*+tag	344
g	15	*hypD*+tag	1958
g	16	*hypE*+tag	371
h	17	*hypE*+tag	1945
h	18	*hypA*+tag	1409
h	19	*hypB*+tag	684
h	20	asr0701+tag	166
i	21	asr0701+tag	1457
i	22	alr0702+tag	202

Figure 11: (A) Physical map of the extended hyp-operon, covering a region of almost 14 kb, in *Anabaena* sp. strain PCC 7120. Apart from the *hyp*-genes (depicted in black) the region also includes five genes upstream of *hypF* (depicted in grey), the two genes downstream of *hypB* (depicted in grey), and *asr0697* (between *hypD* and *hypE*, depicted in grey). The overlapping cDNA of the region are marked a-i. Identified transcriptional start points, TSP, of the extended *hyp*-operon located upstream of *asr0689*, *hypF* and *hypC* are marked with arrows. Number R1–R6 represents the intergenic regions where conserved sequences were found in the extended *hyp*-operon. (B) Agarose gel showing the overlapping amplified PCR products, 1–22, of the RT-PCR experiments, a-i. (C) Primer pairs (1–22) used in the mapping of the extended *hyp*-operon consisting of overlapping cDNA (a-i). The size indicates the expected length of the PCR-product of the forward primer together with the reverse tag-primer. With the kind permission of P. Lindblad, Department of Photochemistry and Molecular Science, Uppsala University, P.O. Box 523, S 75120, Uppsala, Sweden [Agervald *et al.* (2008) *BMC Microbiology* **8**:69; doi:10.1186/1471-2180-8-69].

The arrangement of genes for the small and large subunit of uptake hydrogenases in most of the cyanobacteria examined is almost contiguous with the *hupS* being present upstream of the *hupL* gene. The 10.5 kb element present in *hupL* gene is excised during heterocyst differentiation in *Anabaena* sp. strain PCC 7120 which has already been described under section dealing with Genome Rearrangement. The intergenic region of *hupSL* in heterocystous cyanobacteria is variable (such as 75 bp in *A. variabilis* ATCC 29413, 68 bp in *Anabaena* sp. strain PCC 7120, and *N. muscorum* CCAP 1453/12, 144 bp in *Nostoc* sp. strain PCC 7422 and 192 bp *N. punctiforme* ATCC 29133) with prominent feature being the presence of short tandemly repeated repetitive sequences (Tamagnini *et al.*, 2002, 2007). Carrasco *et al.* (1995) first demonstrated that the *hupSL* transcripts appear during heterocyst differentiation in *Anabena* sp. strain PCC 7120 after a nitrogen shift-down. Subsequently, this has been confirmed in case of *N. muscorum* and *A. variabilis* ATCC 29413 where the expression of *hupSL* occurred about 24 h after nitrogen step-down and it has been suggested to be regulated at the transcriptional level (Axelsson *et al.*, 1999; Happe *et al.*, 2000). A tsp is located 259 bp upstream of *hupS* start codon of *N. muscorum* (Happe *et al.*, 2000; Lindberg *et al.*, 2000). *A. variabilis* ATCC 29413 lacks the 10.5 kb element interrupting *hupL* gene and the *hupSL* genes were transcribed as a single operon. The presence of a tsp at 103 bp upstream of 5′-end of the transcript and a Fnr (fumarate nitrate reductase regulator) binding site are the characteristic features of transcripts. The functional nature of the HupSL has been assessed by the isolation of a *hupSL* disruptant mutant that could not reoxidize the H_2 evolved during nitrogen fixation, as a result of which the mutant evolved three times more H_2 than the wild-type. A transcriptional analysis of hydrogenase genes in *A. variabilis* ATCC 29413 revealed that the *hupL* gene transcripts appeared both in heterocysts and vegetative cells and in vegetative cells of the ammonium grown organism (Boison *et al.*, 2000). Axelsson and Lindblad (2002) found that the transcriptional activity of *hupSL* genes in case of *N. muscorum* and *N. punctiforme* ATCC 29133 were enhanced in response to the addition of 9% H_2 in air that coincided with enhanced uptake hydrogenase activity. A low level of oxygen and addition of nickel also enhanced the *hupSL* transcript levels thus confirming the earlier studies made on uptake hydrogenase activity in case of *Anabaena* sp. strains CA and IF (Xiankong *et al.*, 1984), *A. cylindrica* (Houchins and Burris, 1981b; Daday *et al.*, 1985), *Oscillatoria subbrevis* strain 111 (Kumar and Polasa, 1991) and *N. punctiforme* ATCC 29133 (Oxelfelt *et al.*, 1995). In aerobically grown *L. majuscula* CCAP 1446/4 nitrogen fixation with concomitant synthesis of HupL protein occurred in the dark phase. There was an apparent HupL turnover with its degradation in the light phase and synthesis in the dark phase emphasizing a correlation between nitrogen fixation and uptake hydrogenase activity (Leitão *et al.*, 2005). The presence of putative NtcA-binding sites in the *hupSL* promoter of the heterocystous cyanobacteria such as *Nostoc* PCC 7422 (Yoshino *et al.*, 2007), *A. variabilis* ATCC 29413 (427 bp upstream of tsp of hupS; Weyman *et al.*, 2008) and in the non-heterocystous cyanobacteria *L. majuscula* CCP 1446/4 (in between 240 and 227 bp upstream of tsp of hupS; Leitão *et al.*, 2005) and *Gloeothece* sp. ATCC 27152 (238 bp upstream of tsp of *hupS*; Oliveira *et al.*, 2004) has been demonstrated. Holmqvist *et al.* (2009) identified the NtcA-binding site at –258.5 bp (upstream) to tsp in the promoter of *hupSL* of *N. punctiforme* ATCC 29133. The removal of NtcA-binding site has no effect on the expression of *hupSL* gene as truncated versions of the *hupSL* promoter without the NtcA-binding site as short as –57 bp to tsp could elicit the expression of either *gfp* or *luxAB* as reporter genes. They suggested that heterocyst-specific expression of the *hupSL* genes reside in this short promoter region or in the downstream untranslated leader sequence. The *hupSL* genes with the probable location of the maturation genes in relation to *hyp*-gene cluster in various filamentous heterocystous and non-heterocystous cyanobacteria have been presented (Fig. 12).

Transcriptional analysis of *hoxEFUYH* genes, encoding the bidirectional hydrogenase, in *Anabaena* sp. strain PCC 7120 showed that the transcripts appeared as two units because *hoxEF* and *hoxUYH* are separated by a distance of 8.8 kb. However, along with *hoxEF* an ORF *alr0750* also was transcribed and similarly, *hoxUYH* transcript consisted of two other ORFs, *alr0763* and *alr0765*. Two tsps, one occurring at 66 bp upstream of *alr0750* and the other at 57 bp upstream of *hoxU* have been identified. Specific interactions between LexA and NtcA lead to a transcriptional regulation of *hox* genes and the binding of these transcriptional regulators to the *hox* promoter region has been confirmed by electrophoretic mobility shift assays (Sjoholm *et al.*, 2007).

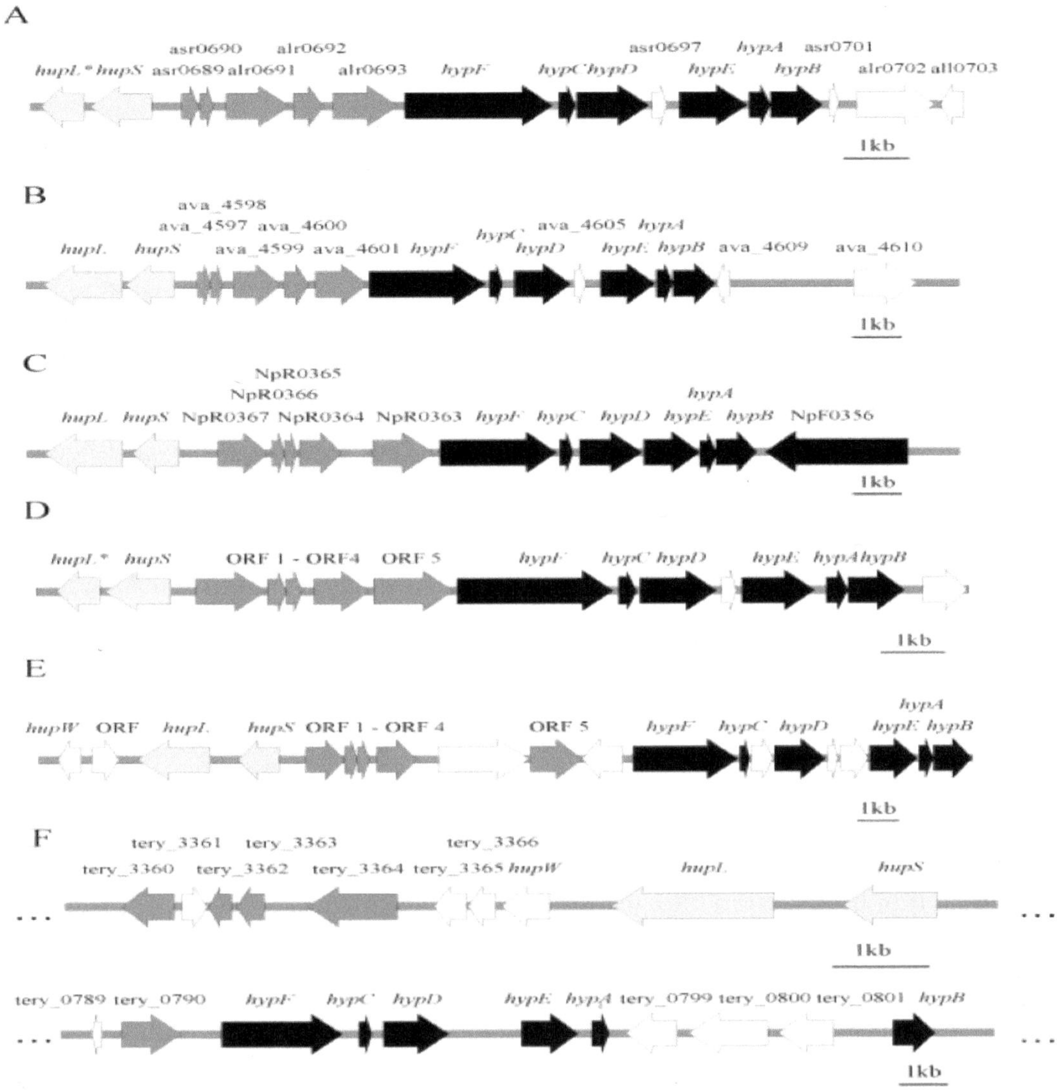

Figure 12: Physical map of the genomic arrangement of the structural hydrogenase genes, *hupSL* (depicted in light grey), the putative maturation genes of the small subunit of the uptake hydrogenase (dark grey), and *hyp* genes (black) of filamentous nitrogen-fixing cyanobacteria. (A) *Anabaena* sp. strain PCC 7120, (B) *Anabaena variabilis* ATCC 29413, (C) *Nostoc punctiforme* PCC 73102, (D) *Nodularia spumigena* CCY9414, (E) *Lyngbya majuscula* CCAP 1446/4, and (F) *Trichodesmium erythraeum* IMS101. The putative maturation genes of the small subunit of the uptake hydrogenase are located in close vicinity to the structural genes, *hupSL*, and often in between *hupSL* and the *hyp*-genes. In *T. erythraeum* IMS101 *hupSL* and the putative maturation genes are separated from the *hyp*-gene cluster by approximately 3.8 Mb. *Indicates the N end fragment of the *hupL* (*hupL5'*), as annotated in vegetative cells. With the kind permission of P. Lindblad, Department of Photochemistry and Molecular Science, Uppsala University, P.O. Box 523, S 75120, Uppsala, Sweden [Agervald *et al.* (2008) *BMC Microbiology* **8:** 69; doi:10.1186/1471-2180-8-69].

A survey of the genomes of *Anabaena* sp. strain PCC 7120, *N. punctiforme* ATCC 29133/PCC 73102 and *Synechocystis* sp. strain PCC 6803 for genes encoding C-terminal endopeptidases revealed *hupW* and *hoxW* genes, for the uptake and bidirectional hydrogenases, respectively. *Anabaena* sp. strain PCC 7120 has a *hoxW* gene (*all0770*) and a *hupW* (*alr1423*) gene for the bidirectional and uptake hydrogenases, respectively synthesized by it. In *N. punctiforme* ATCC 29133 a *hupW* gene (*c509/r320*) is present for the only uptake hydrogenase present while *Synechocystis* sp. strain PCC 6803 showed a *hoxW* (*slr1876*) for the only bidirectional hydrogenase produced by it (Wünschiers *et al.*, 2003). The *hupW* gene is present downstream of the large cluster of genes that encode Mo-nitrogenase. The unicellular *Gloeothece* sp. ATCC 27152 that fixes nitrogen aerobically showed the presence of *hupW*, 184 bp upstream of *hupL*. RT-PCR experiments revealed that *hupSLW* genes are co-transcribed under nitrogen-fixing conditions but not in presence of nitrate nitrogen. The presence of a putative NtcA-binding site in the promoter region of *hupS* around –41.5 bp with respect to tsp is another significant feature highlighting the role of NtcA in the transcriptional regulation of *hupSLW* (Oliveira *et al.*, 2004). The *hypABCDEF* genes are present as an operon as in *Anabaena* sp. strain PCC 7120 and *A. variabilis* ATCC 29413 but they are spread out on the genome of *Synechocystis* sp. strain PCC 6803. The role of *hyp* genes in the latter organism has been identified by the characterization of deletion and disruptant mutants that did not exhibit any hydrogenase activity (Hoffmann *et al.*, 2006). Significant levels of transcripts for *hypFCDEAB* with *hup* and *hox* genes of hydrogenases occurred concomitantly under nitrogen-fixing conditions than under non-nitrogen-fixing conditions in *L. majuscula* CCAP 1446/4. The involvement of NtcA and LexA in the transcriptional regulation of *hypFCDEAB* genes has been demonstrated by electrophoretic mobility shift assays (Ferreira *et al.*, 2007). Agervald *et al.* (2008) demonstrated that the *hypABCDEF* genes in *Anabaena* sp. strain PCC 7120 are transcribed as a single operon with tsps at 280 bp and 445 bp upstream of *hypF* and *hypC*, respectively. This 14-kb region with the tsps of the different *hyp*-genes, the amplified products in RT-PCR experiments, and the respective primers employed are depicted in Fig. 11 B and C.

15) OTHER FUNCTIONS

Now that the role of heterocyst in nitrogen fixation has been unequivocally proved, according to Haselkorn (2010) it is futile to attribute any function to the heterocyst other than nitrogen fixation. Probably these 'other functions' listed below have been ascribed to the heterocyst at a time when the main role in nitrogen fixation remained obscure. Because the evidences presented in favour of the germination of heterocyst and its role in akinete formation cannot be completely disregarded these are considered here. There are at least four other functions attributed to heterocysts. The heterocysts serve as (i) archaic reproductive structures; (ii) storage cells; (iii) attachment organs and (iv) structures that regulate akinete formation.

***i) Reproductive structures*:** Instances of heterocyst germination have been recorded from specimens collected from natural habitats as well from pure cultures of certain members. Some of the materials reported with germinating heterocysts are those of *Nostoc ellipsosprum* (Bornet and Thuret, 1880), *Anabaena cycadeae* Reinke (Spratt, 1911), *Calothrix* sp. (Steinecke, 1932) and *Rivularia* sp. (Desikachary, 1946). Wolk (1965) reported a germination frequency of 10% in case of *Anabaena cylindrica* grown presence of glucose and ammonium chloride. Treatment of *Nostoc linckia* with ultraviolet light induced heterocyst germination (up to 5%) and the percentage of germination increased to about 24% in a mutant of this organism. Exceptionally a very high percentage (83%) of heterocyst germination has been recorded in the wild- and its non-sporulating mutant type of *Gloeotrichia ghosei* (Singh and

Tiwari, 1970; Singh *et al.*, 1972). Wilcox *et al.* (1973a,b) suggested that regression of proheterocysts into vegetative cells could be possible only up to stage V in differentiation process.

ii) Storage cells: The heterocysts have been considered as storage organs for a number of enzymes (Canabaeus, 1929), cyanophycin and growth-promoting substances (Fritsch, 1940).

iii) Attachment organs: Species of *Anabaena, Rivularia, Gloeotrichia* and *Scytonema* growing on terrestrial substrata produce hormogones and with the help of terminal or intercalary heterocysts they get attached to the substratum and develop.

iv) Role in akinete formation: In traditional taxonomic literature, distinction of species of *Gloeotrichia* and *Cylindrospermum* is generally made on the morphology of akinetes. Because of the differentiation akinetes just adjacent to the heterocysts in *A. cylindrica*, Wolk (1965) first advocated that heterocysts play a role in akinete formation. The intiation of akinetes midway between two heterocysts and progressive maturation towards heterocyst has also been noted in some species of *Anabaena* and *Nostoc*. This aspect has been disciussed in the Chapter 3.

LITERATURE CITED

Aboulmagd, E., Oppermann-Sanio, F. B., and Steinbüchel, A. (2000) Molecular characterization of the cyanophycin synthetase from *Synechocystis* sp. strain PCC 6803. *Arch Microbiol* **174:** 297–306.

Aboulmagd, E., Oppermann-Sanio, F. B., and Steinbüchel, A. (2001) Purification of *Synechocystis* sp. strain PCC 6803 cyanophycin synthetase and its characterization with respect to substrate and primer specificity. *Appl Environ Microbiol* **67:** 2176–2182. Doi:10.1128/AEM.6752176-2182.2001

Abreu-Grobois, F. A., Billyard, T. C., and Walton, T. J. (1977) Biosynthesis of heterocyst glycolipids of *Anabaena cylindrica. Phytochemistry* **16:** 351–354.

Adams, D. G. (1992) The effect of 7-azatryptophan on heterocyst development in the cyanobacterium *Anabaena cylindrica. J Gen Microbiol* **138:** 335–362.

Adams, D. G. (2000) Heterocyst formation in cyanobacteria. *Curr Opin Microbiol* **3:** 618–624.

Adams, D. G., and Carr, N. G. (1981) Heterocyst differentiation and cell division in the cyanobacterium *Anabaena cylindrica.* Effect of high light intensity. *J Cell Sci* **49:** 341–352.

Adams, D. G., and Carr, N. G. (1989) Control of heterocyst development in the cyanobacterium *Anabaena cylindrica. J Gen Microbiol* **135:** 839–849.

Adams, D. G., and Duggan, P. S. (1999) Heterocyst and akinete differentiation in cyanobacteria. Tansley Review No. 107. *New Phytol* **144:** 3–33.

Agervald, Å., Stensjö, K., Holmqvist, M., and Lindblad, P. (2008) Transcription of the extended *hyp*-operon in *Nostoc* PCC 7120. *BMC Microbiol* **8:** 69. doi:10.1186/1471-2180-8-69

Agrawal, M., and Kumar, H. D. (1978) Effect of 7-azatryptophan on heterocyst differentiation in *Anabaena doliolum* Bharadwaja. *Proc Indian Acad Sci* **87:** 31–39.

Aldea, M. R., Mella-Herrera, R. A., and Golden, J. W. (2007) Sigma factor genes sigC, sigE and sigG are upregulated in heterocysts of the cyanobacterium *Anabaena* sp. strain PCC 7120. *J Bacteriol* **189:** 8392–8396.

Aldehni, M. F., and Forchhammer, K. (2006) Analysis of a non-canonical NtcA-dependent promoter in *Synechococcus elongatus* and its regulation by NtcA and PII. *Arch Microbiol* **184:** 378–386.

Aldehni, M. F., Sauer, J., Spielhaupter, C., Schmid, R., and Forchhammer, K. (2003) Signal transduction protein P(II) is required for NtcA-regulated gene expression during nitrogen deprivation in the cyanobacterium *Synechococcus elongatus* strain PCC 7942. *J Bacteriol* **185:** 2582–2591.

Alfonso, M., Perewoska, I., and Kirilovsky, D. (2001) Redox control of *ntcA* gene expression in *Synechocystis* sp. strain PCC 6803. Nitrogen availability and electron transport regulate the levels of the NtcA protein. *Plant Physiol* **125:** 969–981.

Alge, D., and Peschek, G. A. (1993) Identification and characterization of the *ctaC* (*coxB*) gene as part of an operon encoding subunits I, II, and III of the cytochrome *c* oxidase (cytochrome *aa3*) in the cyanobacterium *Synechocystis* PCC 6803. *Biochem Biophys Res Commun* **191:** 9–17.

Alge, D., Wastyn, M., Mayer, C., Jungwirth, C., Zimmermann, U., Zoder, R., Fromwald, S., and Peschek, G. A. (1999) Allosteric properties of cyanobacterial cytochrome *c* oxidase: Inhibition of the coupled enzyme by ATP and stimulation by ADP. *IUBMB Life* **48:** 187–197.

Allard, J. F., Hill, A. L., and Rutenberg, A. D. (2007) Heterocyst patterns without patterning proteins in cyanobacterial filaments. *Develop Biol* **312:** 427–434.

Allen, M. M. (1984) Cyanobacterial cell inclusions. *Annu Rev Microbiol* **38:** 1–25.

Allen, M. M., and Hawley, M. A. (1983) Protein degradation and synthesis of cyanophycin granule polypeptide in *Anabaena* sp. *J Bacteriol* **154:** 1480–1484.

Allen, M. M., and Hutchison, F. (1980) Nitrogen limitation and recovery in the cyanobacterium *Aphanocapsa* 6308. *Arch Microbiol* **128:** 1–7.

Allen, M. M., and Smith, A. J. (1969) Nitrogen chlorosis in blue-green algae. *Arch Microbiol* **69:** 114–120.

Allen, M. M., Hutchinson, F., and Weathers, P. J. (1980) Cyanophycin granule polypeptide formation and degradation in the cyanobacterium *Aphanocapsa* 6308. *J Bacteriol* **141:** 687–693.

Allen, M. M., Yuen, C., Medeiros, L., Zizlsperger, N., Farooq, M., and Koloday, N. H. (2005) Effect of light and chloramphenicol stress incorporation of nitrogen into cyanophycin in *Synechocystis* sp. strain PCC 6803. *Biochim Biophys Acta* **1725:** 241–246.

Almon, H., and Böhme, H. (1980) Components and activity of photosynthetic electron transport system of intact heterocysts isolated from the blue-green alga *Nostoc muscorum*. *Biochim Biophys Acta* **592:** 113–120.

Almon, H., and Böger, P. (1984) Ni-dependent uptake hydrogenase activity in the blue-green alga *Anabaena variabilis*. *Z Naturforsch* **39:** 90–92.

Andersen, J. B., Sternberg, C., Poulsen, L. K., Bjorn, S. P., Givskov, M., and Molin, S. (1998) New unstable variants of green fluorescent protein for studies of transient gene expression in bacteria. *Appl Environ Microbiol* **64:** 2240–2246.

Apte, S. K., Rowell, P., and Stewart, W. D. P. (1978) Electron donation to ferredoxin in heterocysts of the N_2-fixing alga *Anabaena cylindrica*. *Proc R Soc* B London **200:** 1–25.

Aravind, L., and Koonin, E. V. (2002) Classification of the capsase-hemoglobinase fold: Detection of new families and implications for the origin of the eukaryotic separins. *Proteins* **46:** 355–387.

Arcondéguy, T., Jack, R., and Merrick, M. (2001) PII signal transduction proteins, pivotal players in microbial nitrogen control. *Microbiol Mol Biol Rev* **65:** 80–105.

Argueta, C., Yukesk, K., and Summers, M. (2004) Construction and use of GFP reporter vectors for analysis of cell-type-specific gene expression in *Nostoc punctiforme*. *J Microbiol Methods* **59:** 181–188.

Asada, Y., Kawamura, S., and Ho, K.-K. (1987) Hydrogenase from the unicellular cyanobacterium, *Microcystis aeruginosa*. *Phytochemistry* **26:** 637–640.

Asai, H., Iwamori, S., Kawai, K., Ehira, S., Ishihara, J-i., Aihara, K., Shoji, S., and Iwasaki, H. (2009) Cyanobacterial cell lineage analysis of the spatio-temporal *hetR* expression profile during heterocyst pattern formation in *Anabaena* sp. PCC 7120. *PLoS ONE* **4**: e7371. doi:10.1371/journal.pone.007371.

Asayama, M., Imamura, S., Yoshihara, S., Miyazaki, A., Yoshida, N., Sazuka, T., Kaneko, T., Ohara, O., Tabata, S., Osanai, T., Tanaka, K., Takahashi, H., and Shirai, M. (2004) SigC, the group 2 sigma factor of RNA polymerase contributes to the late-stage gene expression and nitrogen promoter recognition in the cyanobacterium *Synechocystis* sp. strain PCC 6803. *Biosci Biotechnol Biochem* **68:** 477–487.

Awai, K., Lechno-Yossef, S., and Wolk, C. P. (2009) Heterocyst glycolipids. In: *Lipids in Photosynthesis. Adv Photosynth Resp* **30:** 179–202.

Axelsson, R., and Lindblad, P. (2002) Transcriptional regulation of *Nostoc* hydrogenases: Effects of oxygen, hydrogen and nickel. *Appl Environ Microbiol* **68:** 444–447.

Axelsson, R., Oxelfelt, F., and Lindblad, P. (1999) Transcriptional regulation of *Nostoc* uptake hydrogenase. *FEMS Microbiol Lett* **170:** 77–81.

Bahal, M., and Talpasayi, E. R. S. (1970a) Heterocyst differentiation in *Anabaena ambigua* Rao. III. Effect of Thiol inhibitors. *Indian Biologist* **2:** 56–65.

Bahal, M., and Talpasayi, E. R. S. (1970b) Effect of azide and glycine on heterocyst development in *Anabaena ambigua* Rao. *Indian J Exp Biol* **8:** 237–238.

Baier, K., Lehmann, H., Stephan, D. P., and Lockau, W. (2004) NblA is essential for phycobilisome degradation in *Anabaena* sp. strain PCC 7120 but not for development of functional heterocysts. *Microbiology* **150:** 2739–2749.

Baier, K., Nicklisch, S., Grundner, C., Reinecke, J., and Lockau, W. (2001) Expression of two *nblA*-homologous genes is required for phycobilisome degradation in nitrogen-starved *Synechocystis* sp. PCC 6803. *FEMS Microbiol Lett* **195:** 35–39.

Baker, P. J., Sawa, Y., Shibata, H., Sedelnikova, S. E., and Rice, D. W. (1998) Analysis of the structure and substrate binding of *Phormidium lapideum* alanine dehydrogenase. *Nat Struct Biol* **5:** 561–567.

Barnum, S. R., and Gendel, S. M. (1985) Organization of nitrogen fixation genes in a nonheterocystous, filamentous cyanobacterium. *FEMS Microbiol Lett* **29:** 339–342.

Bastet, L., Boileau, C., Bédu, S., Janicki, A., Latifi, A., and Zhang, C.-C. (2010) NtcA regulates *patA* expression in *Anabaena* sp. strain PCC 7120. *J Bacteriol* **192:** 5257–5259.

Bauer, C. C., Buikema, W. J., Black, K., and Haselkorn, R. (1995) A short-filament mutant of *Anabaena* sp. strain PCC 7120 that fragments in nitrogen-deficient medium. *J Bacteriol* **177:** 1520–1526.

Bauer, C. C., Ramaswamy, K. S., Endley, S., Scappino, L. A., Golden, J. W., and Haselkorn, R. (1997) Suppression of heterocyst differentiation in *Anabaena* PCC 7120 by a cosmid carrying wild-type genes encoding enzymes for fatty acid synthesis. *FEMS Microbiol Lett* **151:** 23–30.

Bauersachs, T., Compaoré, J., Hopmans, E. C., Stal, L. J., Schouten, S., and Sinninghe Damsté, S. (2009) Distribution of heterocyst glycolipids in cyanobacteria. *Phytochemistry* **70:** 2034–2039.

Bauersachs, T., Speelman, E. N., Hopmans, E. C., Reichart, G. J., Schouten, S., and Sinninghe Damsté, J. S. (2010) Fossilized glycolipids reveal past ocanic N_2 fixation by heterocystous cyanobacteria. *Proc Natl Acad Sci USA* **107:** 19190–19194.

Becker, D. W., and Brand, J. J. (1985) *Anacystis nidulans* demonstrates a photosystem II cation requirement satisfied only by Ca^{2+}or Na^{+} *Plant Physiol* **79:** 552 558.

Bergman, B., and Rai, A. (1989) The *Nostoc-Nephroma* symbiosis: Localization, distribution pattern and levels of key proteins involved in nitrogen and carbon metabolism of the cyanobiont. *Physiol Plant* **77:** 216–224.

Bergman, B., Siddiqui, P. J. A., Carpenter, E. J., and Peschek, G. A. (1993) Cytochrome oxidase: Subcellular distribution and relation to nitrogenase expression in the nonheterocystous marine cyanobacterium *Trichodesmium thiebautii*. *Appl Environ Microbiol* **59:** 3239–3244.

Bienert, R., Baier, K., Volkmer, R., Lockau, W., and Heinemann, U. (2005) Crystal structure of NblA from *Anabaena* sp. PCC 7120, a small protein playing a key role in phycobilisome degradation. *J Biol Chem* **281:** 5216–5223.

Black, K., Buikema, W. J., and Haselkorn, R. (1995) The *hglK* gene is required for localization of heterocyst-specific glycolipids in the cyanobacterium *Anabaena* sp. strain PCC 7120. *J Bacteriol* **177:** 6440–6448.

Black, T. A., and Wolk, C. P. (1994) Analysis of a Het^- mutation in *Anabaena* sp. strain PCC 7120 implicates a secondary metabolite in the regulation of heterocyst spacing. *J Bacteriol* **176:** 2282–2292.

Black, T. A., Cai, Y., and Wolk, C. P. (1993) Spatial expression and autoregulation of *hetR*, a gene involved in the control of heterocyst development in *Anabaena*. *Mol Microbiol* **9:** 77–84.

Boison, G., Böthe, H., and Schmitz, D. (2000) Transcriptional analysis of hydrogenase genes in the cyanobacteria *Anacystis nidulans* and *Anabaena variabilis* monitored by RT-PCR. *Curr Microbiol* **40:** 315–321.

Böhme, H. (1987) Regulation of electron flow to nitrogenase in a cell free system from heterocysts of *Anabaena variabilis*. *Biochim Biophys Acta* **891:** 121–128.

Böhme, H., and Almon, H. (1983) Reactions of hydrogen and oxygen with photosystem I of isolated heterocysts from *Anabaena variabilis* (ATCC 29413). *Biochim Biophys Acta* **722:** 401–407.

Böhme, H., and Ernst, A. (1984) Redox reactions of cytochrome *b*-564 and *f*-556 in isolated heterocysts of *Anabaena variabilis* (ATCC 29413). *Adv Photosynth Res* **2:** 639–642.

Böhme, H., and Haselkorn, R. (1988) Molecular cloning and nucleotide sequence analysis of the gene coding for heterocyst ferredoxin from the cyanobacterium *Anabaena* sp. strain PCC 7120. *Mol Gen Genetics* **214:** 278–285.

Böhme, H., and Schrautemeier, B. (1987) Comparative characterization of ferredoxins from heterocysts and vegetative cells of *Anabaena variabilis*. *Biochim Biophys Acta* **891:** 1–7.

Boresch, K. (1910) Zur Physiologie der Blaualgenfarbstoffe. *Lotos* (Prague) **58:** 344–345.

Bornet, E., and Flahault, C. (1886) *Révision des Nostocacées*. Reprint 1959. Cramer, Weinheim.

Bornet, E., and Thuret, G. (1880) *Notes Algologiques*. II. G. Masson, Paris.

Borthakur, D., and Haselkorn, R. (1989) The mutagenesis of *Anabaena* sp. strain PCC 7120: Isolation of a new mutant unable to grow without combined nitrogen. *J Bacteriol* **171:** 5759–5761.

Borthakur, P. B., Orozco, C. C., Young-Robbins, S. S., Haselkorn, R., and Callahan, S. M. (2005) Inactivation of *patS* and *hetN* causes lethal levels of heterocyst differentiation in the filamentous cyanobacterium *Anabaena* sp. PCC 7120. *Mol Microbiol* **57:** 111–123.

Bothe, H. (1970) Photosynthetischie Stickstoffixierung mit einem zellfreien Extrakt aus der Blaualge *Anabaena cylindrica*. *Ber Dtsch Bot Ges* **83:** 421–432.

Bothe, H. (1982) Nitrogen fixation In: Carr, N. G. and Whitton, B. A. (Eds.) *The Biology of Cyanobacteria*. Blackwell Scientific Publications, Oxford, UK pp. 87–104.

Bothe, H., and Eisbrenner, G. (1977). Effect of 7-azatryptophan on nitrogen fixation and heterocyst formation in the blue-green alga *Anabaena cylindrica*. *Biochem Physiol Pflanzen* **171:** 322–332.

Bothe, H., Neuer, G., Kalbe, I., and Eisbrenner, G. (1980) Electron donors and hydrogenase in nitrogen fixing microorganisms. In: Stewart, W. D. P. and Gallon, J. R. (Eds.) *Nitrogen Fixation*. Academic Press, New York, NY, *USA* pp. 83–112.

Bothe, H., Tennigkeit, J., and Eisbrenner, G. (1977) The utilization of molecular hydrogen by the blue-green alga *Anabaena cylindrica*. *Arch Microbiol* **114:** 43–49.

Bottomley, P. J., Van Baalen, C., and Tabita, F. R. (1980) Heterocyst differentiation and tryptophan metabolism in the cyanobacterium *Anabaena* sp. CA. *Arch Biochem Biophys* **203:** 204–213.

Bradley, S., and Carr, N. G. (1976) Heterocyst and nitrogenase development in *Anabaena cylindrica*. *J Gen Microbiol* **96:** 175–184.

Brahamsha, B., and Haselkorn, R. (1991) Isolation and characterization of the gene encoding the principal sigma factor of the vegetative cell RNA polymerase from the cyanobacterium *Anabaena* sp. strain PCC 7120. *J Bacteriol* **173:** 2442–2450.

Brahamsha, B., and Haselkorn, R. (1992) Identification of multiple RNA polymerase sigma factor homologs in the cyanobacterium *Anabaena* sp. strain PCC 7120: Cloning, expression, and inactivation of the sigB and sigC genes. *J Bacteriol* **174:** 7273–7282.

Braun-Howland, E. B., Lindblad, P., Nierzwicki-Bauer, S. A., and Bergman, B. (1988) Dinitrogenase reductase (Fe-protein) of nitrogenase in the cyanobacterial symbionts of three *Azolla* species: Localization and sequence of appearance during heterocyst differentiation. *Planta* **176:** 319–322.

Brusca, J. S., Chastain, C. J., and Golden, J. W. (1990) Expression of the *Anabaena* sp. strain PCC 7120 *xisA* gene from a heterologous promoter results in excision of the *nifD* element. *J Bacteriol* **172:** 3925–3931.

Brusca, J. S., Hale, M. A., Carrasco, C. D., and Golden, J. W. (1989) Excision of an 11-kilobase-pair DNA element from within the *nifD* gene in *Anabaena variabilis* heterocysts. *J Bacteriol* **171:** 4138–4145.

Bryce, T. A., Weln, D., Walsby, A. E., and Nichols, B. W. (1972) Monohexoside derivatives of long-chain polyhydroxyalcohols: a novel class of glycolipids specific to heterocystous algae. *Phytochemistry* **11:** 295–302.

Buikema, W. J., and Haselkorn, R. (1991a) Isolation and complementation of nitrogen fixation mutants of the cyanobacterium *Anabaena* sp. strain PCC 7120. *J Bacteriol* **173:** 1879–1885.

Buikema, W. J., and Haselkorn, R. (1991b) Characterization of a gene controlling heterocyst differentiation in the cyanobacterium *Anabaena* 7120. *Genes Dev* **5:** 321–330.

Buikema, W. J., and Haselkorn, R. (2001) Expression of the *Anabaena hetR* gene from a copper-regulated promoter leads to heterocyst differentiation under repressing conditions. *Proc Natl Acad Sci USA* **98:** 2729–2734.

Cai, Y., and Wolk, C. P. (1997) *Anabaena* sp. strain PCC 7120 responds to nitrogen deprivation with a cascade-like sequence of transcriptional activations. *J Bacteriol* **179:** 267–271.

Callahan, S. M., and Buikema, W. J. (2001) The role of HetN in maintenance of the heterocyst pattern in *Anabaena* sp. PCC 7120. *Mol Microbiol* **40:** 941–950.

Calusinska, M., Happe, T., Joris, B., and Wilmotte, A. (2010) The surprising diversity of clostridial hydrogenases: A comparative genomic perspective. *Microbiology* **156:** 1575–1588.

Campbell, E. L., Brahamsha, B., and Meeks, J. C. (1998) Mutation of an alternative sigma factor in the cyanobacterium *Nostoc punctiforme* results in an increased infection of its symbiotic plant partner, *Anthoceros punctatus*. *J Bacteriol* **180:** 4938–4941.

Campbell, E. L., Cohen, M. F., and Meeks, J. C. (1997) A polyketide-synthase-like gene is involved in the synthesis of heterocyst glycolipids in *Nostoc punctiforme* strain ATCC 29133. *Arch Microbiol* **167:** 251–258.

Campbell, E. L., Hagen, K. D., Cohen, M. F., Summers, M. L., and Meeks, J. C. (1996) The *devR* gene product is characteristic of receivers of two-component regulatory systems and is essential for heterocyst development in the filamentous cyanobacterium *Nostoc* sp. strain PCC 29133. *J Bacteriol* **178:** 2037–2043.

Campbell, E. L., Summers, M. L., Christman, H., Martin, M. E., and Meeks, J. E. (2007) Global gene expression patterns of *Nostoc punctiforme* in steady-state dinitrogen-grown heterocyst containing cultures and at single time points during differentiation of akinetes and harmogonia. *J Bacteriol* **189:** 5247–5256.

Canabaeus, L. (1929) Über die Heterocysten und Gasvaakuolen der Blaualgen und ihre Beziehungen zueinander. In: Kolkowitz, R. (Ed.) *Pflanzenforschung*. Vol 13, Jena, Fischer. pp. 1–48.

Cardemil, L., and Wolk, C. P. (1979) The polysaccharides from heterocyst and spore envelopes of a blue-green alga: Structure of the basic repeating unit. *J Biol Chem* **254:** 736–741.

Cardemil, L., and Wolk, C. P. (1981a) Polysaccharides from the envelopes of heterocysts and spores of the blue-green algae *Anabaena variabilis* and *Cylindrospermum licheniforme*. *J Phycol* **17:** 234–240.

Cardemil, L., and Wolk, C. P. (1981b) Isolated heterocysts of *Anabaena variabilis* synthesize envelope polysaccharide. *Biochim Biophys Acta* **674:** 265–276.

Carpenter, E. J., Bergman, B., Dawson, R., Siddiqui, P. J., Soderback, E., and Capone, D. G. (1992) Glutamine synthetase and nitrogen cycling in colonies of the marine diazotrophic cyanobacteria *Trichodesmium* spp. *Appl Environ Microbiol* **58:** 3122–3129.

Carr, N. G. (1988) Nitrogen reserves and dynamic reservoirs in cyanobacteria. In: Rogers, L. J., and Gallon, J. R. (Eds.) *Biochemistry of the Algae and Cyanobacteria*. Clarendon Press, Oxford, U.K. pp. 13–39.

Carrasco, C. D., Buettner, J. A., and Golden, J. W. (1995) Programmed DNA rearrangement of a cyanobacterial *hupL* gene in heterocysts. *Proc Natl Acad Sci USA* **92:** 791–795.

Carrasco, C. D., Holliday, S. D., Hansel, A., Lindblad, P., and Golden, J. W. (2005) Heterocyst-specific excision of the *Anabaena* sp. strain PCC 7120 *hupL* element requires *xisC*. *J Bacteriol* **187:** 6031–6038.

Carrasco, C. D., Ramaswamy, K. S., Ramasubramanian, T. S., and Golden, J. W. (1994) *Anabaena xisF* gene encodes a developmentally regulated site-specific recombinase. *Genes Dev* **8:** 74–83.

Chapman, J. S., and Meeks, J. C. (1983) Glutamine and glutamate transport by *Anabaena variabilis*. *J Bacteriol* **156:** 122–129.

Chastain, C. J., Brusca, J. S., Ramasubramanian, T. S., Wei, T. F., and Golden, J. W. (1990) A sequence-specific DNA-binding factor (VF1) from *Anabaena* sp. strain PCC 7120 vegetative cells binds to three adjacent sites in the *xisA* upstream region. *J Bacteriol* **172:** 5044–5051.

Chen, C., Van Baalen, C., and Tabita, F. R. (1987) Nitrogen starvation mediated by DL-7-azatryptophan in the cyanobacterium *Anabaena* sp. strain CA. *J Bacteriol* **169:** 1107–1113.

Chen, Y-F., Motteux, O., Bédu, S., Li, Y.-Z., and Zhang, C. C. (2011) Characterization of two critical residues in the effector domain of NtcA in the cyanobacterium *Anabaena* sp. strain PCC 7120. *Mol Microbiol* **63:** 32–38.

Codd, G. A., Okabe, K., and Stewart, W. D. P. (1980) Cellular compartmentation of photosynthetic and photorespiratory enzymes in the heterocystous cyanobacterium *Anabaena cylindrica*. *Arch Microbiol* **124:** 149–154.

Cohen, M. F., Wallis, J. G., Campbell, E. L., and Meeks, J. C. (1994) Transposon mutagenesis of *Nostoc* sp. strain ATCC 29133, a filamentous cyanobacterium with multiple cellular differentiation alternatives. *Microbiology* **140:** 3233–3240.

Collier, J. L., and Grossman, A. R. (1994) A small polypeptide triggers complete degradation of light-harvesting phycobiliproteins in nutrient-deprived cyanobacteria. *EMBO J* **13:** 1039–1047.

Crespo, J. L., García-Domínguez, M., and Florencio, F. J. (1998) Nitrogen control of the *glnN* gene that codes for GS type III, the only glutamine synthetase in the cyanobacterium *Pseudoanabaena* sp. PCC 6903. *Mol Microbiol* **30:** 1101–1112.

Cumino, A., Curatti, L., Giarrocco, L., and Salerno, G. L. (2002) Sucrose metabolism: *Anabaena* sucrose-phosphate synthase and sucrose-phosphate phosphatase define minimal functional domains shuffled during evolution. *FEBS Lett* **517:** 19–23.

Cumino, A., Ekeroth, C., and Salerno, G. L. (2001) Sucrose-phosphate phosphatase from *Anabaena* sp. strain PCC 7120: Isolation of the protein and gene revealed significant structural differences from the higher-plant enzyme. *Planta* **214:** 250–256.

Cumino, A., Marcozzi, C., Barreiro, R., and Salerno, G. (2007) Carbon cycling in *Anabaena* sp. PCC 7120. Sucrose synthesis in the heterocysts and possible role in nitrogen fixation. *Plant Physiol* **143:** 1385–1397.

Cumino, A. C., Perez-Cenci, M., Giarrocco, L. E., and Salerno, G. L. (2010) The proteins involved in sucrose synthesis in the marine cyanobacterium *Synechococcus* sp. PCC 7002 are encoded by two genes transcribed from a gene cluster. *FEBS Lett* **584:** 4655–4660.

Curatti, L., Flores, E., and Salerno, G. (2002) Sucrose is involved in the diazotrophic metabolism of the heterocyst-forming cyanobacterium *Anabaena* sp. *FEBS Lett* **513:** 175–178.

Curatti, L., Folco, E., Desplats, P., Abratti, G., Limones, V., Herrera-Estrella, L., and Salerno, G. L. (1998) Sucrose-phosphate synthase from *Synechocystis* sp. PCC 6803: Identification of the *spsA* gene and characterization of the enzyme expressed in *E. coli*. *J Bacteriol* **180:** 6776–6779.

Curatti, L., Giarrocco, L., and Salerno, G. L. (2006) Sucrose synthase and RuBisCO expression is similarly regulated by the nitrogen source in the nitrogen fixing cyanobacterium *Anabaena* sp. *Planta* **223:** 891–900.

Curatti, L., Giarrocco, L. E., Cumino, A. C., and Salerno, G. L. (2008) Conversion of sucrose to polysaccharides in filamentous nitrogen fixing cyanobacteria. *Planta* **228:** 617–625.

Curatti, L., Porchia, A. C., Herrera-Estrella, L., and Salerno, G. L. (2000) A prokaryotic sucrose synthase gene (*susA*) isolated from a filamentous nitrogen fixing cyanobacterium encodes a protein similar to those of plants. *Planta* **211:** 729–735.

Currier, T. C., Huary, J. F., and Wolk, C. P. (1977) Isolation and preliminary characterization of auxotrophs of a filamentous cyanobacterium. *J Bacteriol* **129:** 1556–1562.

Daday, A., and Smith, G. D. (1983) The effect of nickel on the hydrogen metabolism of the cyanobacterium *Anabaena cylindrica*. *FEMS Microbiol Lett* **20:** 327–330.

Daday, A., Mackerras, A. H., and Smith, G. D. (1985) The effect of nickel on hydrogen metabolism and nitrogen fixation in the cyanobacterium *Anabaena cylindrica*. *J Gen Microbiol* **131:** 231–238.

Daday, A., Platz, R., and Smith, G. D. (1977) Anaerobic and aerobic hydrogen gas formation by the blue-green alga *Anabaena cylindrica*. *Appl Environ Microbiol* **34:** 478–483.

Desikachary, T. V. (1946) Germination of the heterocysts in two members of the Rivulariaceae. *J Indian Bot Soc* **25:** 11–17.

Desikachary, T. V. (1959) *Cyanophyta*. Indian Council of Agricultural Research, New Delhi.

de Wet, J. R., Wood, K. V., Helinski, D. R., and DeLuca, M. (1985) Cloning of firefly luciferase cDNA and the expression of active luciferase in *Escherichia coli*. *Proc Natl Acad Sci USA* **82:** 7870–7873.

Dharmawardene, M. W. N., Haystead, A., and Stewart, W. D. P. (1973) Glutamine synthetase of the nitrogen-fixing alga *Anabaena cylindrica*. *Arch Microbiol* **90:** 281–295.

Dines, M., Sendersky, E., David, L., Schwarz, R., and Adir, N. (2008) Structural, functional and mutational analysis of the NblA protein provides insight into possible modes of interaction with the phycobilisome. *J Biol Chem* **283:** 30330–30340.

Doherty, H. M., and Adams, D. G. (1995) Cloning and sequence of *ftsZ* and flanking regions from the cyanobacterium *Anabaena* PCC 7120. *Gene* **163:** 93–96.

Dolganov, N., and Grossman, A. R. (1993) Insertional inactivation of genes to isolate mutants of *Synechococcus* sp. strain PCC 7942: Isolation of filamentous strains. *J Bacteriol* **175:** 7644–7651.

Dong, Y., Huang, X., Wu, X. Y., and Zhang, J. (2000) Identification of active site of HetR protease and its requirement for heterocyst differentiation in the cyanobacterium *Anabaena* sp. strain PCC 7120. *J Bacteriol* **182:** 1575–1579.

Drawert, H., and Metzner, I. (1956) Fluoreszenz und electronen mikroskopische Beobachtungen an *Cylindrospermum* und anderan Cyanophyceen. *Ber Beutesch Bot Ges* **69:** 291–301.

Dunn, J. H., and Wolk, C. P. (1970) Composition of cellular envelopes of *Anabaena cylindrica. J Bacteriol* **103:** 153–158.

Dunn, J. H., Simon, R. D., and Wolk, C. P. (1971) Incorporation of amino sugars into walls during heterocyst differentiation. *Develop Biol* **26:** 159–164.

Duyvesteyn, M. G. C., Korsuize, J., de Waard, A., Vonshak, A., and Wolk, C. P. (1983) Sequence-specific endonucleases in strains of *Anabaena* and *Nostoc. Arch Microbiol* **134:** 276–281.

Ehira, S., and Ohmori, M. (2006a) NrrA, a nitrogen-responsive regulator facilitates heterocyst development in the cyanobacterium *Anabaena* sp. strain PCC 7120. *Mol Microbiol* **59:** 1692–1703.

Ehira, S., and Ohmori, M. (2006b) NrrA directly regulates expression of *hetR* during heterocyst differentiation in the cyanobacterium *Anabaena* sp. strain PCC 7120. *J Bacteriol* **188:** 8520–8525.

Ehira, S., Ohmori, M., and Sato, N. (2003) Genome-wide expression analysis of the responses to nitrogen deprivation in the heterocyst-forming cyanobacterium *Anabaena* sp. strain PCC 7120. *DNA Res* **10:** 97–113.

Eisbrenner, G., Distler, E., Floener, L., and Böthe, H. (1978) The occurrence of the hydrogenase in some blue-green algae. *Arch Microbiol* **118:** 177–184.

Elhai, J., and Wolk, C. P. (1988) Conjugal transfer of DNA to cyanobacteria. *Methods Enzymol* **167:** 747–754.

Elhai, J., and Wolk, C. P. (1990) Developmental regulation and spatial pattern of expression of structural genes for nitrogenase in the cyanobacterium *Anabaena*. *EMBO J* **9:** 3379–3388.

Elhai, J., Vepritskiy, A., Muro-Pastor, A. M., Flores, E., and Wolk, C. P. (1997) Reduction of conjugal transfer efficiency by three restriction activities of *Anabaena* sp. strain PCC 7120. *J Bacteriol* **179:** 1998–2005.

El-Shehawy, R., Lugomela, C., Ernst, A., and Bergman, B. (2003) Diurnal expression of *hetR* and diazocyte development in the filamentous non-heterocystous cyanobacterium *Trichodesmium erythraeum. Microbiology* **149:** 1139–1146.

Ernst, A., and Böger, P. (1985) Glycogen accumulation and the induction of nitrogenase activity in the heterocyst-forming cyanobacterium *Anabaena variabilis. J Gen Microbiol* **131:** 3147–3153.

Ernst, A., Reich, S., and Böger, P. (1990) Modification of dinitrogenase reductase in the cyanobacterium *Anabaena variabilis* due to C starvation and ammonia. *J Bacteriol* **172:** 748–755.

Ernst, A., Black, T., Cai, Y., Panoff, J.-M., Tiwari, D. N., and Wolk, C. P. (1992) Synthesis of nitrogenase in mutants of the cyanobacterium *Anabaena* sp. strain PCC 7120 affected in heterocyst development or metabolism. *J Bacteriol* **174:** 6025–6032.

England, R. R., and Evans, E. H. (1983) A requirement for Ca^{2+} in O_2-evolving photosystem 2 preparations from the cyanobacterium, *Anacystis nidulans. Biochem J* **201:** 473–476.

Espinosa, J., Brunner, T., Fiedler, N., Forchhammer, K., Muro-Pastor, A. M., and Maldener, I. (2010) DevT (Alr4674), resembling a Ser/Thr protein phosphatase, is essential for heterocyst function in the cyanobacterium *Anabaena* sp. PCC 7120. *Microbiology* **156:** 3544–3555.

Espinosa, J., Forchhammer, K., Burillo, S., and Contreras, A. (2006) Interactive network in cyanobacterial nitrogen regulation: PipX, a protein that interacts in a 2-oxoglutarate dependent manner with PII and NtcA. *Mol Microbiol* **61:** 457–469. doi:10.1111j.1365-2598.2006.05231

Falah Hamadi, A., and Gallon, J. R. (1981) Calcium ions, oxygen and acetylene reduction (nitrogen fixation) in *Gloeocapsa* sp. 1430/3. *J Gen Microbiol* **125:** 391–398.

Fan, O., Huang, G., Lechno-Yossef, S., Wolk, C. P., Kaneko, T., and Tabata, S. (2005) Clustered genes required for synthesis and deposition of envelope glycolipids in *Anabaena* sp. strain PCC 7120. *Mol Microbiol* **58:** 227–243.

Fan, O., Lechno-Yossef, S., Ehira, S., Kaneko, T., Ohmori, M., Sato, N., Tabata, S., and Wolk, C. P. (2006) Signal transduction genes required for heterocyst maturation in *Anabaena* sp. strain PCC 7120. *J Bacteriol* **188:** 6688–6693.

Fay, P. (1969) Cellular differentiation and pigment composition in *Anabaena cylindrica. Arch Microbiol* **67:** 62–70.

Fay, P. (1973) The heterocyst. In: Carr, N. G., and Whitton, B. A. (Eds.) *The Biology of Blue-green Algae*. Blackwell Scientific Publications, Oxford, UK pp. 238–259.

Fay, P. (1976) Factors influencing dark nitrogen fixation in a blue-green alga. *Appl Environ Microbiol* **31:** 376–379.

Fay, P., and Fogg, G. E. (1962) Studies on nitrogen fixation by blue-green algae. III. Growth and nitrogen fixation in *Chlorogloea fritschii* Mitra. *Arch Microbiol* **42:** 310–321.

Fay, P., and Kulasooriya, S. A. (1972) Tetrazolium reduction and nitrogenase activity in heterocystous blue-green algae. *Arch Microbiol* **87:** 341–352.

Fay, P., and Lang, N. J. (1971) The heterocysts of blue-green algae. I. Ultrastructural integrity after isolation. *Proc R Soc* B London **178:** 185–192.

Fay, P., and Walsby, A. E. (1966) Metabolic activities of isolated heterocysts of the blue-green alga *Anabaena cylindrica*. *Nature* (London) **209:** 94–95.

Fay, P., Kumar, H. D., and Fogg, G. E. (1964) Cellular factors affecting nitrogen fixation in the blue-green alga *Chlorogloea fritschii*. *J Gen Microbiol* **35:** 351–360.

Fay, P., Stewart, W. D. P., Walsby, A. E., and Fogg, G. E. (1968) Is the heterocyst the site of nitrogen fixation in blue-green algae? *Nature* (London) **220:** 810–812.

Fernandez-Pinas, F., Leganes, F., and Wolk, C. P. (1994) A third genetic locus required for the formation of heterocysts in *Anabaena* sp. strain PCC 7120. *J Bacteriol* **176:** 5277–5283.

Ferreira, D., Leitão, E., Sjöholm, J., Oliveira, P., Lindblad, P., Moradas-Ferreira, P. and Tamagnini, P. (2007) Transcription and regulation of the hydrogenase(s) accessory genes, *hypFCDEAB*, in the cyanobacterium *Lyngbya majuscula* CCAP 1446/4. *Arch Microbiol* **188:** 609–617.

Fiedler, G., Arnold, M., Hannus, S., and Maldener, I. (1998) The *devBCA* exporter is essential for envelope formation in heterocysts of the cyanobacterium *Anabaena* sp. strain PCC 7120. *Mol Microbiol* **27:** 1193–1202.

Fiedler, G., Arnold, M., Hannus, S., and Maldener, I. (1999) An ABC exporter is essential for the localisation of envelope material in heterocysts of cyanobacteria. In: Peschek, G. A., Loeffelhardt, W., and Schmetterer, G. (Eds.) *The Phototrophic Prokaryotes*. Plenum Publishing Corporation, New York, N.Y. USA pp. 529–537.

Fiedler, G., Muro-Pastor, A. M., Flores, E., and Maldener, I. (2001) NtcA-dependent expression of the *devBCA* operon, encoding a heterocyst-specific ATP-binding cassette transporter in *Anabaena* sp. *J Bacteriol* **183:** 3795–3799.

Fieulaine, S., Lunn, J. E., Borel, F., and Ferrer, J.-L. (2005) The structure of a cyanobacterial sucrose-phosphate reveals two sugar tongs that release free sucrose in the cell. *Plant Cell* **17:** 2049–2058.

Fisher, R., Tuli, R., and Haselkorn, R. (1981) A cloned cyanobacterial gene for glutamine synthetase functions in *Escherichia coli*, but the enzyme is not adenylylated. *Proc Natl Acad Sci USA* **78:** 3393–3397.

Fleming, H., and Haselkorn, R. (1973) Differentiation in *Nostoc muscorum*. Nitrogenase is synthesized in heterocysts. *Proc Natl Acad Sci USA* **70:** 2727–2731.

Fleming, H., and Haselkorn, R. (1974) The program of protein synthesis during heterocyst differentiation in nitrogen-fixing blue-green algae. *Cell* **3:** 159–170.

Flores, E., and Herrero, A. (2010) Compartmentalized function through cell differentiation in filamentous cyanobacteria. *Nat Rev Microbiol* **8:** 39–50.

Flores, E., and Muro-Pastor, M. I. (1988) Uptake of glutamine and glutamate by the dinitrogen fixing cyanobacterium *Anabaena* sp. PCC 7120. *FEMS Microbiol Lett* **56:** 127–130.

Flores, E., Pernil, R., Muro-Pastor, A. M., Mariscal, V., Maldener, I., Lechno-Yossef, S., Fan, Q., Wolk, C. P., and Herrero, A. (2007) Septum localized protein required for filament integrity and diazotrophy in the heterocyst-forming cyanobacteria. *J Bacteriol* **189:** 3884–3890.

Fogg, G. E. (1942) Studies on nitrogen fixation by blue-green algae. I. Nitrogen fixation by *Anabaena cylindrica* Lemm. *J Exp Biol* **19:** 78–87.

Fogg, G. E. (1944) Growth and heterocyst production in *Anabaena cylindrica* Lemm. *New Phytol* **43:** 164–175.

Fogg, G. E. (1949) Growth and heterocyst production in *Anabaena cylindrica* Lemm. II. In relation to carbon and nitrogen metabolism. *Ann Bot* **13:** 240–259.

Fogg, G. E. (1951) Growth and heterocyst production in *Anabaena cylindrica* Lemm. III. The cytology of heterocysts. *Ann Bot* **15:** 23–35.

Forchhammer, K., and Tandeau de Marsac, N. (1994) The PII protein in the cyanobacterium *Synechococcus* sp. strain PCC 7942 is modified by serine phosphorylation and signals the cellular N-status. *J Bacteriol* **176:** 84–91.

Forchhammer, K., and Tandeau de Marsac, N. (1995) Phosphorylation of the PII protein (*glnB* gene product) in the cyanobacterium *Synechococcus* sp. strain PCC 7942: Analysis of the *in vitro* kinase activity. *J Bacteriol* **177:** 5812–5817.

Forsberg, A. J., Pavitt, G. D., and Higgins, C. F. (1994) Use of transcriptional fusions to monitor gene expression: A cautionary tale. *J Bacteriol* **176:** 2128–2132.

Fredriksson, C., and Bergman, B. (1995) Nitrogenase varies diurnally in a subset of cells within colonies of the non-heterocystous cyanobacteria *Trichodesmium* spp. *Microbiology* **141:** 2471–2478.

Franche, C., and Cohen-Bazire, G. (1985) The structural *nif* genes of four symbiotic *Anabaena azollae* show a highly conserved physical arrangement. *Plant Sci* **39:** 125–131.

Franche, C., and Cohen-Bazire, G. (1987) Evolutionary divergence in the *nifK, D, H* gene region among nine symbiotic *Anabaena azollae* and between *Anabaena azollae* and some free-living heterocystous cyanobacteria. *Symbiosis* **3:** 159–178.

Frías, J. E., Flores, E., and Herrero, A. (1994) Requirement of the regulatory protein NtcA for the expression of nitrogen assimilation and heterocyst development genes in the cyanobacterium *Anabaena* sp. PCC 7120. *Mol Microbiol* **14:** 823–832.

Frías, J. E., Flores, E., and Herrero, A. (1997) Nitrate assimilation gene cluster from the heterocyst forming cyanobacterium *Anabaena* sp. strain PCC 7120. *J Bacteriol* **179:** 477–486.

Frías, J. E., Mérida, A., Herrero, A., Martín-Nieto, J., and Flores, E. (1993) General distribution of the nitrogen control gene *ntcA* in cyanobacteria. *J Bacteriol* **175:** 5710–5713.
Fritsch, F. E. (1940) Some points in the structure of an *Anabaena*. *New Phytol* **3:** 85–96.
Fritsch, F. E. (1945) *The Structure and Reproduction of the Algae.* Vol. II. Cambridge University Press, Cambridge, UK pp. 796–800.
Fritsch, F. E. (1951) The Heterocyst: A botanical enigma. *Proc Linn Soc* London **162:** 194–211.
Fujita, Y., Takagi, H., and Hase, T. (1996) Identification of the *chlB* gene and the gene product essential for light-independent chlorophyll biosynthesis in the cyanobacterium *Plectonema boryanum*. *Plant Cell Physiol* **37:** 313–323
Gallon, J. R., and Falah Hamidi, A. (1984) Studies on the effects of oxygen on acetylene reduction (nitrogen fixation) in *Gloeothece* sp. ATCC 27152 *J Gen Microbiol* **130:** 495–503.
Galmozzi, C. V., Saelices, L., Florencio, F. J., and Muro-Pastor, M. I. (2010) Post-transcriptional regulation of glutamine synthetase in the filamentous cyanobacterium *Anabaena* sp. strain PCC 7120: Differential expression between vegetative cells and heterocysts. *J Bacteriol* **192:** 4701–4711.
Gambacorta, A., Pagnotta, E., Romano, I., Sodano, G., and Trincone, A. (1998) Heterocyst glycolipids from nitrogen fixing cyanobacteria other than Nostocaceae. *Phytochemistry* **48:** 801–805.
García-Domínguez, M., Reyes, J. C., and Florencio, F. J. (1999) Glutamine synthetase inactivation by protein-protein interaction. *Proc Natl Acad Sci USA* **96:** 7161–7166.
García-Domínguez, M., Reyes, J. C., and Florencio, F. J. (2000) NtcA represses transcription of *gifA* and *gifB*, genes that encode inhibitors of glutamine synthetase type 1 from *Synechocystis* sp. strain PCC 6803. *Mol Microbiol* **35:** 1192–1201.
Gerdtzen, Z. P., Salgado, J. C., Osses, A., Asenjo, J. A., Rapaport, I., and Andrews, B. (2009) Modeling heterocyst pattern formation in cyanobacteria. *BMC Bioinformatics* **10:** S16. doi:10.1186/1471-2105-10-S6-S16.
Ghassemian, M., Wong, B., Ferreira, F., Markley, J. L., and Straus, N. A. (1994) Cloning, sequencing and transcriptional studies of the genes for cytochrome c-553 and plastocycanin from *Anabaena* sp. PCC 7120. *Microbiology* **140:** 1151–1159.
Giddings, T. H., and Staehelin, L. A. (1978) Plasma membrane architecture of *Anabaena cylindrica*: occurrence of microplasmodesmata and changes associated with heterocyst development and the cell cycle. *Cytobiologie* **16:** 236–249.
Ginsberg, R., and Lazaroff, N. (1973) Ultrastructural development of *Nostoc muscorum*. *J Gen Microbiol* **75:** 1–9.
Golden, J. W., and Wiest, D. R. (1988) Genome rearrangement and nitrogen fixation in *Anabaena* blocked by inactivation of *xisA* gene. *Science* **242:** 1421–1423.
Golden, J. W., and Yoon, H. S. (2003) Heterocyst development in *Anabaena*. *Curr Opin Microbiol* **6:** 557–563.
Golden, J. W., Carrasco, C. D., Mulligan, M. E., Schneider, G. J., and Haselkorn, R. (1988) Deletion of a 55-kilobase-pair DNA element from the chromosome during heterocyst differentiation of *Anabaena* sp. strain PCC 7120. *J Bacteriol* **170:** 5034–5041.
Golden, J. W., Mulligan, M. E., and Haselkorn, R. (1987) Different recombination site specificity of two developmentally regulated genome rearrangements. *Nature* (London) **327:** 526–529.
Golden, J. W., Robinson, S. J., and Haselkorn, R. (1985) Rearrangement of nitrogen fixation genes during heterocyst differentiation in the cyanobacterium *Anabaena*. *Nature* (London) **314:** 419–423.
Golden, J. W., Whorff, L. L., and Wiest, D. R. (1991) Independent regulation of *nifHDK* operon transcription and DNA rearrangement during heterocyst differentiation in the cyanobacterium *Anabaena* sp. strain PCC 7120. *J Bacteriol* **173:** 7098–7105.
Golden, S. S., Brusslan, J., and Haselkorn, R. (1987) Genetic engineering of the cyanobacterial chromosome. *Methods Enzymol* **153:** 215–231.
Gonzalez, L., Phalip, V., and Zhang, C.-C. (2001) Characterization of PknC, a Ser/Thr kinase with broad substrate specificity from the cyanobacterium *Anabaena* sp. strain PCC 7120. *Eur J Biochem* **268:** 1869–1875.
Gotto, J. W., Tabita, F. R., and Van Baalen, C. (1979) Mutants of *Anabaena* strain CA altered in their ability to grow under nitrogen-fixing conditions. *J Bacteriol* **140:** 327–332.
Govezensky, D., Greener, T., Segal, G., and Zamir, A. (1991) Involvement of GroEL in *nif* gene regulation and nitrogenase assembly. *J Bacteriol* **173:** 6339–6346.
Grillo, J. F., Bottomley, P. J., Van Baalen, C., and Tabita, F. R. (1979) A mutant of *Anabaena* sp. CA with oxygen-sensitive nitrogenase activity. *Biochem Biophys Res Commun* **89:** 685–693.
Gupta, M., and Carr, N. G. (1981a) Detection of glutamate synthase in heterocysts of *Anabaena*. *J Bacteriol* **148:** 980–982.
Gupta, M., and Carr, N. G. (1981b) Enzymology of arginine metabolism in heterocyst-forming cyanobacteria. *FEMS Microbiol Lett* **12:** 179–181.
Hagedorn, H. (1960) Electronemikroskopische Untersuchungen an Blaualgen. *Naturwissenschaften* **47:** 430.
Hagen, K. D., and Meeks, J. H. (1999) Biochemical and genetic evidence for participation of DevR in a phosphorelay signal transduction pathway essential for heterocyst maturation in *Nostoc punctiforme* ATCC 29133. *J Bacteriol* **181:** 4430–4434.

Hai, T., Oppermann-Sanio, F. B., and Steinbüchel, A. (1999) Purification and characterization of cyanophycin synthetase from the thermophilic *Synechococcus* sp. strain MA19. *FEMS Microbiol Lett* **181:** 229–236.

Hai, T., Oppermann-Sanio, F. B., and Steinbüchel, A. (2002) Molecular characterization of a thermostable cyanophycin synthetase from the thermophilic cyanobacterium *Synechococcus* sp. strain MA19 and *in vitro* synthesis of cyanophycin and related polyamines. *Appl Environ Microbiol* **68:** 93–101. doi:10.1128/AEM.68.1.93-101.202

Hallenbeck, P. C., and Benemann, J. R. (1978) Characterization and partial purification of the reversible hydrogenase of *Anabaena cylindrica*. *FEBS Lett* **94:** 261–264.

Happe, T., Schütz, K., and Böhme, H. (2000) Transcriptional and mutational analysis of the uptake hydrogenase of the filamentous cyanobacterium *Anabaena variabilis* ATCC 29413. *J Bacteriol* **182:** 1624–1631.

Haury, S. F., and Wolk, C. P. (1978) Classes of *Anabaena* mutants with oxygen-sensitive nitrogenase activity. *J Bacteriol* **136:** 688–692.

Herrero, A., and Wolk, C. P. (1986) Genetic mapping of the chromosome of the cyanobacterium, *Anabaena variabilis*. *J Biol Chem* **261:** 7748–7754

Häfele, U., Scherer, S., and Böger, P. (1988) Cytochrome aa_3 from heterocysts of the cyanobacterium *Anabaena variabilis*: Isolation and spectral characterization. *Biochim Biophys Acta* **934:** 186–190.

Häger, K. P., Danneberg, G., and Böthe, H. (1983) The glutamate synthase in heterocysts of *Nostoc muscorum*. *FEMS Microbiol Lett* **17:** 179–183.

Hansel, A., Axelsson, R., Lindberg, P., Troshina, O. Y., Wünschiers, R., and Lindblad, P. (2001) Cloning and characterisation of a *hyp* gene cluster in the filamentous cyanobacterium *Nostoc* sp. strain PCC 73102. *FEMS Microbiol Lett* **201:** 59–64.

Haselkorn, R. (1978) Heterocysts. *Annu Rev Plant Physiol* **29:** 319–344.

Haselkorn, R. (1992) Developmentally regulated gene rearragements in prokaryotes. *Annu Rev Genet* **26:** 113–130.

Haselkorn, R. (2008) Cell-cell communication in filamentous cyanobacteria. *Mol Microbiol* **70:** 783–785.

Haselkorn, R. (2010) A new player in the regulatory cascade controlling heterocyst differentiation. *Mol Microbiol* **77:** 537–539.

Haselkorn, R., Golden, J. W., Lammers, P. J., and Mulligan, M. E. (1986) Developmental rearrangement of cyanobacterial nitrogen fixation genes; *Trends Genet* **2:** 255–259.

Haselkorn, R., Rice, D., Curtis, S. E., and Robinson, S. J., (1983) Organization and transcription of genes important in *Anabaena* heterocyst differentiation. *Ann Microbiol* (Inst. Pasteur) **134B:** 181–193.

Haury, J. F., and Spiller, H. (1981) Fructose uptake and influence on growth of and nitrogen fixation by *Anabaena variabilis*. *J Bacteriol* **147:** 227–235.

Hausinger, R. P. (1987) Nickel utilization by microorganisms. *Microbiol Rev* **51:** 22–42.

Hawkesford, M. J., Houchins, J. P., and Hind, G. (1983) Reconstitution of photosynthetic electron transfer in cyanobacterial heterocyst membranes. *FEBS Lett* **159:** 262–266.

Haystead, A., Robinson, R., and Stewart, W. D. P. (1970) Nitrogenase activity in extracts of heterocystous and non-heterocystous blue-green algae. *Arch Microbiol* **74:** 235–243.

Hebbar, P. B., and Curtis, S. E. (2000) Characterization of *devH*, a gene encoding a putative DNA binding protein required for heterocyst function in *Anabaena* sp. strain PCC 7120. *J Bacteriol* **182:** 3572–3581.

Henson, B. J., Watson, L. E., and Barnum, S. R. (2005) Chacterization of a 4 kb variant of the *nifD* element in *Anabaena* sp. strain ATCC 33047. *Curr Microbiol* **50:** 129–132.

Henson, B. J., Pennigton, L. E., Watson, L. E., and Barnum, S. R. (2008) Excision of the *nifD* element in the heterocystous cyanobacteria. *Arch Microbiol* **189:** 357–366.

Herrero, A., and Flores, E. (1990) Transport of basic amino acids by the dinitrogen-fixing cyanobacterium *Anabaena* PCC 7120. *J Biol Chem* **265:** 3931–3935.

Herrero, A., Muro-Pastor, A. M., and Flores, E. (2001) Nitrogen control in cyanobacteria. *J Bacteriol* **183:** 411–425.

Higa, K. C., and Callahan, S. M. (2010) Ectopic expression of *hetP* can partially bypass the need for *hetR* in heterocyst differentiation by *Anabaena* sp. strain PCC 7120. *Mol Microbiol* **77:** 562–574.

Hoffmann, D., Gutekunst, K., Klissenbauer, M., Schluz-Friedrich, R., and Appel, J. (2006) Mutagenesis of hydrogenase accessory genes of *Synechocystis* sp. PCC 6803. Additional homologues of *hypA* and *hypB* are not active in hydrogenase maturation. *FEBS J* **273:** 4516–4527.

Holland, D., and Wolk, C. P. (1990) Identification and characterization of *hetA*, a gene that acts early in the process of morphological differentiation of heterocysts. *J Bacteriol* **172:** 3131–3137.

Holmqvist, M., Stensjo, K., Oleviera, P., Lindberg, P., and Lindblad, P. (2009) Characterization of *hupSL* promoter activity in *Nostoc punctiforme* ATCC 29133. *BMC Microbiol* **9:** 54 doi: 10.1186/1471-2180-9-54.

Houchins, J. P. (1984) The physiology and biochemistry of hydrogen metabolism in cyanobacteria. *Biochim Biophys Acta* **768:** 227–255.

Houchins, J. P. (1985) Electron transfer chains of cyanobacterial heterocysts. In: Ludden, P. W., and Burris, J. E. (Eds.) *Nitrogen Fixation and CO_2 Metabolism*. Elsevier Science Publishers, Amsterdam. The Netherlands. pp. 261–268.

Houchins, J. P., and Burris, R. H. (1981a) Light and dark reactions of the uptake hydrogenase in *Anabaena* 7120. *Plant Physiol* **68:** 712–716.

Houchins, J. P., and Burris, R. H. (1981b) Physiological reactions of the reversible hydrogenase from *Anabaena* 7120. *Plant Physiol* **68:** 717–721.

Houchins, J. P., and Burris, R. H. (1981c) Occurrence and localization of two distinct hydrogenases in the cyanobacterium *Anabaena* sp. strain 7120. *J Bacteriol* **146:** 209–214.

Houchins, J. P., and Hind, G. (1983) Flash spectroscopic characterization of photosynthetic electron transport in isolated heterocysts. *Arch Biochem Biophys* **725:** 138–145.

Houchins, J. P., and Hind, G. (1984) Concentration and function of membrane-bound cytochromes in cyanobacterial heterocysts. *Plant Physiol* **76:** 456–460.

Howarth, D. C., and Codd, G. A. (1985) The uptake and production of molecular hydrogen by unicellular cyanobacteria. *J Gen Microbiol* **131:** 1561–1569.

Huang, G., Fan, Q., Lechno-Yossef, S., Wojciuch, E., Wolk, C. P., Kaneko, T., and Tabata, S. (2005) Clustered genes required for the synthesis of heterocyst envelope polysaccharide in *Anabaena* sp. strain PCC 7120. *J Bacteriol* **187:** 1114–1123.

Huang, X., Dong, Y., and Zhao, J. (2004) HetR homodimer is a DNA-binding protein required for heterocyst differentiation, and the DNA-binding activity is inhibited by PatS. *Proc Natl Acad Sci USA* **101:** 4848–4853.

Imashimizu, M., Yoshimura, H., Katoh, H., Ehira, S., and Ohmori, M. (2005) NaCl enhances cellular cAMP and upregulation of genes related to heterocyst development in the cyanobacterium *Anabaena* sp. strain PCC 7120. *FEMS Microbiol Lett* **252:** 97–103.

Ionescu, D., Voss, B., Oren, A., Hess, W. R., and Muro-Pastor, A. M. (2010) Heterocyst-specific transcription of NsiR1, a non-coding RNA encoded in a tandem array of direct repeats in cyanobacteria. *J Mol Biol* **398:** 177–188.

Irmler, A., Sanner, S., Dierks, H., and Forchhammer, K. (1997) Dephosphorylation of the phosphoprotein PII in *Synechococcus* PCC 7942: Identification of an ATP and 2-oxoglutarate-regulated phosphatase activity. *Mol Microbiol* **26:** 81–90.

Iyengar, M. O. P., and Desikachary, T. V. (1953) Occurrence of three-pored heterocysts in *Brachytrichia balani* (Lloyd) Born. & Flah. *Curr Sci* **22:** 180–181.

Jacobson, B. L., Chae, Y. K., Markley, J. L., Rayment, I., and Holden, H. M. (1993) Molecular structure of the oxidized, recombinant heterocyst [2Fe-2S] ferredoxin from *Anabaena* 7120 determined to 1.7 Å resolution. *Biochemistry* **32:** 6788–6793.

Jäger, K. M., and Bergman, B. (1990) Localization of a multifunctional chaperonin (GroEL protein) in nitrogen fixing *Anabaena* PCC 7120. Presence in vegetative cells and heterocysts. *Planta* **183:** 120–125.

Jang, J., Shi, L., Tan, H., Janicki, A., and Zhang, C.-C. (2009) Mutual regulation of *ntcA* and *hetR* during heterocyst differentiation requires two similar PP2C-type protein phosphatases, PrpJ1 and PrpJ2, in *Anabaena* sp. strain PCC 7120. *J Bacteriol* **191:** 6059–6066.

Jang, J., Wang, L., Jeanjean, R., and Zhang, C. C. (2007) PrpJ, a PP2C-type protein phosphatase located on the plasma membrane, is involved in heterocyst maturation in the cyanobacterium *Anabaena* sp. PCC 7120. *Mol Microbiol* **64:** 347–358.

Janson, S., Matveyev, A., and Bergman, B. (1998) The presence and expression of *hetR* in the non-heterocystous cyanobacterium *Symploca* PCC 8002. *FEMS Microbiol Lett* **168:** 173–179.

Jensen, B. B., and Cox, R. P. (1983) Effect of oxygen concentration on dark nitrogen fixation and respiration in cyanobacteria. *Arch Microbiol* **135:** 287–292.

Jensen, B. B., Cox, R. P., and Burris, R. H. (1986) Isolation of cyanobacterial heterocysts with high and sustained dinitrogen-fixation capacity supported by endogenous reductants. *Arch Microbiol* **145:** 241–247.

Jiang, F., Mannervik, B., and Bergman, B. (1997) Evidence for redox regulation of the transcription factor NtcA, acting both as an activator and a repressor, in the cyanobacterium *Anabaena* PCC 7120. *Biochem J* **327:** 513–517.

Jones, K. M., and Haselkorn, R. (2002) Newly identified cytochrome *c* oxidase operon in the nitrogen-fixing cyanobacterium *Anabaena* sp. strain PCC 7120 specifically induced in heterocysts. *J Bacteriol* **184:** 2491–2499.

Jones, K. M., Buikema, W. J., and Haselkorn, R. (2003) Heterocyst-specific expression of *patB*, a gene required for nitrogen fixation in *Anabaena* sp. strain PCC 7120. *J Bacteriol* **185:** 2306–2314.

Jüttner, F. (1983) ^{14}C-labeled metabolites in heterocysts and vegetative cells of *Anabaena cylindrica* filaments and their presumptive function as transport vehicles of organic carbon and nitrogen. *J Bacteriol* **155:** 628–633.

Kale, S. R. (1972) Effect of light on the development of heterocysts in *Anabaena ambigua* Rao. In: Desikachary, T. V. (Ed.) *The Biology of Blue-green Algae*. University of Madras, Madras pp. 203–208.

Kale, S. R., and Talpasayi, E. R. S. (1969) Heterocysts: A review. *Indian Biologist* **1:** 19–29.

Kallas, T., Coursin, T., and Rippka, R. (1985) Different organization of *nif* genes in non-heterocystous and heterocystous cyanobacteria. *Plant Mol Biol* **5:** 321–329.

Kallas, T., Spiller, S., and Malkin, R. (1988) Primary structure of cotranscribed genes encoding the Rieske Fe-S and cytochrome *f* proteins of the cyanobacterium *Nostoc* PCC 7906. *Proc Natl Acad Sci USA* **85:** 5794–5798.

Kaneko, T., Sato, S., Kotani, H., Tanaka, A., Asamizu, E., Nakamura, Y., Miyajima, N., Hirosawa, M., Sugiura, M., Sakamoto, S., Kimura, T., Hosouchi, T., Matsuno, A., Muraki, A., Nakazaki, A., Naruo, K., Okumura, S., Shimpo, S., Takeuchi, C., Wada, T., Watanabe, A., Yamada, M., Yasuda, M., and Tabata, S. (1996) Sequence analysis of the genome of the unicellular cyanobacterium *Synechocystis* sp. strain PCC 6803. II. Sequence determination of the entire genome and assignment of potential protein-coding regions. *DNA Res* **3:** 109–136.

Kaneko, T., Nakamura, Y., Wolk, C. P., Kuritz, T., Sasamoto, S., Watanabe, A., Iriguchi, M., Ishikawa, A., Kawashima, K., Kimura, T., Kishida, Y., Kohara, M., Matsumoto, M., Matsuno, A., Muraki, A., Nakazaki, N., Shimpo, S., Sugimoto, M., Takazawa, M., Yamada, M., Yasuda, M., and Tabata, S. (2001) Complete genomic sequence of the filamentous nitrogen-fixing cyanobacterium *Anabaena* sp. strain PCC 7120. *DNA Res* **8:** 205–213.

Kaplan, D., and Peters, G. A. (1981) The *Azolla-Anabaena* relationship X. $^{15}N_2$ fixation and transport in main stem axes. *New Phytol* **89:** 337–346.

Kaplan, D., Calvert, H. E., and Peters, G. A. (1986) The *Azolla-Anabaena* relationship. XII. Nitrogenase activity and phycobiliproteins of the endophyte as a function of leaf age and cell type. *Plant Physiol* **80:** 884–890.

Kangatharalingam, N., Priscu, J. C., and Paerl, H. W. (1992) Heterocyst envelope thickness, heterocyst frequency and nitrogenase activity in *Anabaena flos-aquae*: Influence of exogenous oxygen tension. *J Gen Microbiol* **138:** 2673–2678.

Karradt, A., Sobanski, J., Mattow, J., Lockau, W., and Baier, K. (2008) NblA, a key protein of phycobilisome degradation interacts with ClpC, a HSP100 chaperone partner of a cyanobacterial Clp protease. *J Biol Chem* **283:** 32394–32403.

Kaushik, M., and Kumar, H. D. (1970) The effect of light on growth and development of two blue-green algae. *Arch Microbiol* **74:** 52–57.

Ke, B., Fang, Z. X., Lu, R. Z., Calvert, H. E., and Dolan, E. (1983) The presence of phycobilisomes in heterocysts of *Anabaena variabilis*. *Photobiochem Photobiophys* **6:** 25–31.

Khudyakov, I., and Wolk, C. P. (1997) *hetC*, a gene coding for a protein similar to bacterial ABC protein exporters, is involved in early regulation of heterocyst differentiation. *J Bacteriol* **179:** 6971–6978.

Khudyakov, I. Y., and Golden, J. W. (2001) Identification and inactivation of three group 2 sigma factor genes in *Anabaena* sp. strain PCC 7120. *J Bacteriol* **183:** 6667–6675.

Khudyakov, I. Y., and Golden, J. W. (2004) Different functions of HetR, the master regulator of heterocyst differentiation in *Anabaena* sp. PCC 7120, can be separated by mutation. *Proc Natl Acad Sci USA* **101:** 16040–16045.

Khudyakov, I. Y., and Pinevich, A. V. (1992) Unicellular mutant of the filamentous cyanobacterium *Anabaena* sp. PCC 7118. *Mikrobiologiya* **60:** 704–708.

Kerson, G. W., Miernyk, J. A., and Budd, K. (1984). Evidence for the occurrence of, and possible physiological role for, cyanobacterial calmodulin. *Plant Physiol* **75:** 222–224.

Kneip, C., Lockhart, P., Voβ, C.,and Maier, U. G. (2007) Nitrogen fixation in eukaryotes. New models for symbiosis. *Evol Biol* **7:** 55. doi:10.1186/1471-2148/7/55

Koksharova, O. A., and Wolk, C.P. (2002) A novel gene that bears a DnaJ motif influences cyanobacterial cell division. *J Bacteriol* **184:** 5524–5528.

Kolodny, N. H., Bauer, D., Bryce, K., Klucevsek, K., Lane, A., Medeiros, L., Mercer, W., Moin, S., Park, D., Petersen, J., Wright, J., Yuen, C., Wolfson, A. J., and Allen, M. M. (2006) Effect of nitrogen source on cyanophycin synthesis in *Synechocystis* sp. strain PCC 6803. *J Bacteriol* **188:** 934–940.

Krepski, W. J., and Wolk, C. P. (1983) Biosynthesis of the envelope glycolipids during heterocyst development in *Anabaena cylindrica*. *J Gen Microbiol* **129:** 105–110.

Kuhn, I., Peng, L., Bedu, S., and Zhang, C.-C. (2000) Developmental regulation of the cell division protein FtsZ in *Anabaena* sp. strain PCC 7120, a cyanobacterium capable of terminal differentiation. *J Bacteriol* **182:** 4640–4643.

Kulasooriya, S. A., Lang, N., and Fay, P. (1972) The heterocysts of blue-green algae. III. Differentiation and nitrogenase activity. *Proc R Soc* B London **181:** 199–209.

Kumar, A., and Kumar, H. D. (1980) Differential effects of amino acid analogs on growth and heterocyst differentiation in two nitrogen fixing blue-green algae. *Curr Microbiol* **3:** 213–218.

Kumar, A., Tabita, F., and Van Baalen, C. (1982) Isolation and characterization of heterocysts from *Anabaena* sp. strain CA. *Arch Microbiol* **133:** 103–109.

Kumar, A. P., Perraju, B. T. V. V., and Singh, H. N. (1986) Carbon nutrition and regulation of uptake hydrogenase activity in free-living and symbiotic *Anabaena cycadae*. *New Phytol* **104:** 115–120.

Kumar, A. P., Rai, A. N., and Singh, H. N. (1985) Nitrate reductase activity in isolated heterocysts of the cyanobacterium *Nostoc muscorum*. *FEBS Lett* **179:** 125–128.

Kumar, D., and Kumar, H. D. (1990) Effect of some inhibitors and carbon sources on acetylene reduction and hydrogen production of isolated heterocysts of *Anabaena* sp. (strain CA). *Eur J Phycol* **25:** 375–379.

Kumar, H. D., and Kaushik, M. (1971) Studies on growth and development of two nitrogen-fixing blue-green algae. *Zeitschrift Pflanzenphysiol* **65:** 443–453.

Kumar, K., Mella-Herrera, R. A., and Golden, J. W. (2010) Cyanobacterial heterocysts. *Cold Spring Harb Perspect Biol* **2:** a000315. Epub.

Kumar, S., and Polasa, H. (1991) Influence of nickel and copper on photobiological hydrogen production and uptake in *Oscillatoria subbrevis* strain 111. *Proc Indian Nat Sci Acad* **57:** 281–285.

Lahmi, R., Sendersky, E., Perelman, A., Hagemann, M., Forchhammer, K., and Schwarz, R. (2006) Alanine dehydrogenase activity is required for adequate progression of phycobilisome degradation during nitrogen starvation in *Synechococcus elongatus* PCC 7942. *J Bacteriol* **188:** 5258–5265.

Lambein, F., and Wolk, C. P. (1973) Structural studies on the glycolipids from the envelope of the heterocyst of *Anabaena cylindrica*. *Biochemistry* **12:** 791–798.

Lambert, G. R., and Smith, G. D. (1981) The hydrogen metabolism of cyanobacteria (blue-green algae). *Biol Rev* **56:** 589–660.

Lammers, P. J., and Haselkorn, R. (1983) Sequence of the *nifD* gene coding for the α subunit of dinitrogenase from the cyanobacterium *Anabaena*. *Proc Natl Acad Sci USA* **80** 4723–4727.

Lammers, P. J., Golden, J. W., and Haselkorn, R. (1986) Identification and sequence of a gene required for developmentally regulated DNA excision in *Anabaena*. *Cell* **44:** 905–911.

Lang, N. J. (1965) Electron microscopic study of heterocyst development in *Anabaena azollae* Strasburger. *J Phycol* **1:** 127–134.

Lang, N. J. (1968) The fine structure of blue-green algae. *Annu Rev Microbiol* **22:** 15–46.

Lang, N. J., and Fay, P. (1971) The heterocysts of blue-green algae. II. Details of ultrastructure. *Proc R Soc* B London **178:** 193–203.

Lang, N. J., Krupp, J. M., and Koller, A. L. (1987) Morphological and ultrastructural changes in vegetative cells and heterocysts of *Anabaena variabilis* grown with fructose. *J Bacteriol* **169:** 920–923.

Lang, N. J., Simon, R. D., and Wolk, C. P. (1972) Correspondence of cyanophycin granules with structured granules in *Anabaena cylindrica*. *Arch Microbiol* **83:** 313–320.

Laurent, S., Chen, H., Bédu, S., Ziarelli, F., Peng, L., and Zhang, C.-C. (2005) Nonmetabolizable analogue of 2-oxoglutarate elicits heterocyst differentiation under repressive conditions in *Anabaena* sp. PCC 7120. *Proc Natl Acad Sci USA* **102:** 9907–9912.

Lawrie, A. C., Codd, G. A., and Stewart, W. D. P. (1976) The incorporation of nitrogen into products of recent photosynthesis in *Anabaena cylindrica* Lemm. *Arch Microbiol* **107:** 15–24.

Lechno-Yossef, S., Fan, Q., Ehira, S., Sato, N., and Wolk, C. P. (2006) Mutations in four regulatory genes have interrelated effects on heterocyst maturation in *Anabaena* sp. strain PCC 7120. *J Bacteriol* **188:** 7387–7395.

Lee, H.-M., Vazquez-Bermudez, M. F., and Tandeau de Marsac, N. (1999) The global nitrogen regulator NtcA regulates transcription of the signal transducer PII (GlnB) and influences its phosphorylation level in response to nitrogen and carbon supplies in the cyanobacterium *Synechococcus* sp. strain PCC 7942. *J Bacteriol* **181:** 2697–2702.

Leganes, F., Fernandes-Pinas, F., and Wolk, C. P. (1998) A transposon-induced mutant of *Nostoc ellipsoporum* implicates an arginine-biosynthetic gene in the formation of cyanophycin granules and of functional heterocysts and akinites. *Microbiology* **144:** 1799–1805.

Leitão, E., Oxelfelt, F., Oliveira, P., Moradas-Ferreira, P., and Tamagnini, P. (2005) Analysis of the *hupSL* operon of the nonheterocystous cyanobacterium *Lyngbya majuscula* CCAP 1446/4: Regulation of transcription and expression under a light-dark regime. *Appl Environ Microbiol* **71:** 4567–4576.

Lex, M., and Carr, N. G. (1974) The metabolism of glucose by heterocysts and vegetative cells of *Anabaena cylindrica*. *Arch Microbiol* **101:** 161–167.

Li, B., Huang, X., and Zhao, J. (2002) Expression of *hetN* during heterocyst differentiation and its inhibition of *hetR* up-regulation in the cyanobacterium *Anabaena* sp. strain PCC 7120. *FEBS Lett* **517:** 87–91.

Li, H., Sherman, D. M., Bao, S., and Sherman, L. A. (2001) Pattern of cyanophycin accumulation in nitrogen-fixing and non-nitrogen-fixing cyanobacteria. *Arch Microbiol* **176:** 9–18.

Li, J. H., Laurent, S., Konde, V., Bédu, S., and Zhang, C.-C. (2003) An increase in the level of 2-oxoglutarate promotes heterocyst development in the cyanobacterium *Anabaena* sp. strain PCC7120. *Microbiology* **149:** 3257–3263. doi: 10.1099/mic.0.26462-0.

Liang, J., Scappino, L., and Haselkorn, R. (1992) The *patA* gene product, which contains a region similar to CheY of *Escherichia coli*, controls heterocyst pattern formation in the cyanobacterium *Anabaena* 7120. *Proc Natl Acad Sci USA* **89:** 5655–5659.

Liang, J., Scappino, L., and Haselkorn, R. (1993) The *patB* gene product, required for growth of the cyanobacterium *Anabaena* sp. strain PCC 7120 under nitrogen-limiting conditions, contains ferredoxin and helix-turn-helix domains. *J Bacteriol* **175:** 1697–1704.

Lindblad, P., and Bergman, B. (1989) Occurrence and localization of phycoerythrin in symbiotic *Nostoc* of *Cycas revoluta* and in the free-living isolated *Nostoc* 7422. *Plant Physiol* **89:** 783–785.

Lindblad, P., and Sellstedt, A. (1990) Occurrence and localization of an uptake hydrogenase in the filamentous heterocystous cyanobacterium *Nostoc* PCC 73102. *Protoplasma* **159:** 9–15.
Lindberg, P., Hansel, A., and Lindblad, P. (2000) *hupS* and *hupL* constitute a transcription unit in the cyanobacterium *Nostoc* sp. PCC 73102. *Arch Microbiol* **174:** 129–133.
Linn, T., and Pierre, R. St. (1990) Improved vector system for constructing transcriptional fusions that ensures independent translation of *lacZ*. *J Bacteriol* **172:** 1077–1084.
Liu, D., and Golden, J. W. (2002) *hetL* overexpression stimulates heterocyst formation in *Anabaena* sp. strain PCC 7120. *J Bacteriol* **184:** 6873–6881.
Liu, J., and Chen, W.-L. (2009) Characterization of HetN, a protein involved in heterocyst differentiation in the cyanobacterium *Anabaena* sp. strain PCC 7120. *FEMS Microbiol Lett* **297:** 17–23.
Llama, M. J., Serra, J. L., Rao, K. K., and Hall, D. O. (1979) Isolation and characterization of the hydrogenase activity from the nonheterocystous cyanobacterium *Spirulina maxima*. *FEBS Lett* **98:** 342–346.
Lockau, W., Peterson, R. B., Wolk, C. P., and Burris, R. H. (1978) Modes of reduction of nitrogenase in heterocysts isolated from *Anabaena* species. *Biochim Biophys Acta* **502:** 298–308.
Lopes Pinto, F. A., Troshina, O., and Lindblad, P. (2002) A brief look at three decades of research on cyanobacterial hydrogen evolution. *Internl J Hydrogen Energy* **27:** 1209–1215.
López-Gomollón, S., Hernández, J. A., Wolk, C. P., Peleato, M. L., and Fillat, M. F. (2007) Expression of *furA* is modulated by NtcA and strongly enhanced in heterocysts of *Anabaena* sp. PCC 7120. *Microbiology* **153:** 42–50.
López-Igual, R., Flores, E., and Herrero, A. (2010) Inactivation of a heterocyst-specific invertase indicates a principal role of sucrose catabolism in heterocysts of *Anabaena* sp. *J Bacteriol* **192:** 5526–5533.
Ludwig, M., Schulz-Friedrich, R., and Appel, J. (2006) Occurrence of hydrogenases in cyanobacteria and anoxygenic photosynthetic bacteria: Implications for the phylogenetic origin of cyanobacterial and algal hydrogenases. *J Mol Evol* **63:** 758–768.
Luque, I., Flores, E., and Herrero, A. (1994) Molecular mechanism for the operation of nitrogen control in cyanobacteria. *EMBO J* **13:** 2862–2869.
Luque, I., Ochoa de Alda, J. A. G., Richaud, C., Zabulon, G., Thomas, J.-C., and Houmard, J. (2003) The NblAI protein from the filamentous cyanobacterium *Tolypothrix* PCC 7601: Regulation of its expression and interactions with phycobilisome components. *Mol Microbiol* **50:** 1043–1054.
Lunn, J. E. (2002) Evolution of sucrose synthesis. *Plant Physiol* **128:** 1490–1500.
Lynn, M. E., and Ownby, J. D. (1987) Transcriptional activity of heterocysts isolated from *Anabaena variabilis*. *Arch Microbiol* **148:** 115–120.
Lynn, M. E., Bantle, J. A., and Ownby, J. D. (1986) Estimation of gene expression in heterocysts of *Anabaena variabilis* by using DNA-RNA hybridization. *J Bacteriol* **167:** 940–946.
Lyons, F. M., and Thiel, T. (1995) Characterization of *nifB*, *nifS* and *nifU* genes in the cyanobacterium *Anabaena variabilis*. NifB is required for the vanadium-dependent nitrogenase. *J Bacteriol* **177:** 1570–1575.
Mackerras, A. H., De Chazal, N. M., and Smith, G. D. (1990) Transient accumulations of cyanophycin in *Anabaena cylindrica* and *Synechocystis* 6803. *J Gen Microbiol* **136:** 2057–2065.
Makarova, K. S., Koonin, E. V., Haselkorn, R., and Galperin, M. Y. (2006) Cyanobacterial response regulator PatA contains a conserved N-terminal domain (PATAN) with an alpha-helical insertion. *Bioinformatics* **22:** 1297–1301.
Maldener, I., Fiedler, G., Ernst, A., Fernández-Piñas, F., and Wolk, C. P. (1994) Characterization of *devA*, a gene required for the maturation of proheterocysts in the cyanobacterium *Anabaena* sp. strain PCC 7120. *J Bacteriol* **176:** 7543–7549.
Maldener, I., Hannus, S., and Kammerer, M. (2003) Description of five mutants of the cyanobacterium *Anabaena* sp. strain PCC 7120 affected in heterocyst differentiation and identification of transposon-tagged genes. *FEMS Microbiol Lett* **224:** 205–213.
Marcozzi, C., Cumino, A. C., and Salerno, G. L. (2009) Role of NtcA, a cyanobacterial global nitrogen regulator, in the regulation of sucrose metabolism gene expression in *Anabaena* sp. PCC 7120. *Arch Microbiol* **191:** 255–263.
Mariscal, V., Herrero, A., and Flores, E. (2007) Continuous periplasm in a filamentous, heterocyst-forming cyanobacterium. *Mol Microbiol* **65:** 1139–1145.
Mariscal, V., Herrero, A., Nenninger, A., Mullineaux, C. W., and Flores, E. (2011) Functional dissection of three domain SepJ protein joining the cells in cyanobacterial trichomes. *Mol Microbiol* **79:** 1077–1088.
Marqués, S., Florencio, F. J., and Candau, P. (1992) Purification and characterization of the ferredoxin-glutamate synthase from the unicellular cyanobacterium *Synechococcus* sp. PCC 6301. *Eur J Biochem* **296:** 69–77.
Martín-Figueroa, E., Navarro, F., and Florencio, F. J. (2000) The GS-GOGAT pathway is not operative in the heterocysts. Cloning and expression of *glsF* gene from the cyanobacterium *Anabaena* sp. PCC 7120. *FEBS Lett* **476:** 282–286.
Masephol, B., Schölisch, K., Görlitz, K., Kutzki, C., and Böhme, H. (1997) The heterocyst-specific *fdxH* gene product of the cyanobacterium *Anabaena* sp. PCC 7120 is important but not essential for nitrogen fixation. *Mol Gen Genet* **253:** 770–776.

Mazur, B. J., and Chui, C-F. (1982) Sequence of the gene coding for the β-subunit of dinitrogenase from the blue-green alga *Anabaena*. *Proc Natl Acad Sci USA* **79:** 6782–6786.

Mazur, B. J., Rice, D., and Haselkorn, R. (1980) Identification of blue-green algal nitrogen fixation genes by using heterologous DNA hybridization probes. *Proc Natl Acad Sci USA* **77:** 186–190.

Meeks, J. C., Campbell, E. L., Summers, M. L., and Wong, F. C. (2002) Cellular differentiation in the cyanobacterium *Nostoc punctiforme*. *Arch Microbiol* **178:** 395–403.

Meeks, J. C., Joseph, C. M., and Haselkorn, R. (1988) Organization of the *nif* genes in cyanobacteria in symbiotic association with *Azolla* and *Anthoceros*. *Arch Microbiol* **150:** 61–71.

Meeks, J. C., Elhai, J., Thiel, T., Potts, M., Larimer, F., Lamerdin, J., Predki, P., and Atlas, R. (2001) An overview of the genome of *Nostoc punctiforme*, a multicellular, symbiotic cyanobacterium. *Photosynth Res* **70:** 85–106.

Meeks, J. C., Wolk, C. P., Lockau, W., Schilling, N., Shaffer, W., and Chieu, W. S. (1978) Pathways of assimilation of $[^{13}N]N_2$ and $^{13}NH_4$-by cyanobacteria with and without heterocysts. *J Bacteriol* **134:** 125–130.

Meeks, J. C., Wolk, C. P., Thomas, J., Lockau, W., Shaffer, P. W., Austin, S. M., Chien, W.-S., and Galonsky, A. (1977) The pathways of assimilation of $^{13}NH_4'$ by the cyanobacterium, *Anabaena cylindrica*. *J Biol Chem* **252:** 7894–7900.

Melis, A., and Happe, T. (2001) Hydrogen production: Green algae as a source of energy. *Plant Physiol* **127:** 740–748.

Mella-Herrera, R. A., Neunuebel, M. R., and Golden, J. W. (2010) *Anabaena* sp. strain PCC 7120 conR contains a LytR-CpsA-Psr domain, is developmentally regulated, and is essential for diazotrophic growth and heterocyst morphogenesis. *Microbiology* **157:** 617–626. doi:10.1099/mic.0.046128-0

Mérida, A., Candau, P., and Florencio, F J. (1991) Regulation of glutamine synthetase activity in the unicellular cyanobacterium *Synechocystis* sp. strain PCC 6803 by the nitrogen source: Effect of ammonium. *J Bacteriol* **173:** 4095–4100.

Merino-Puerto, V., Mariscal, V., Mullineaux, C. W., Herrero, A., and Flores, E. (2010) Fra proteins influencing filament integrity, diazotrophy and localization of septal protein SepJ in the heterocyst forming cyanobacterium *Anabaena* sp. *Mol Microbiol* **75:** 1159–1170.

Metzner, I. (1955) Zur chemie und zum submikroskopischen aufbau der zellwainde, Scheiden und Gallerten von Cyanophyceen. *Arch Mikrobiol* **22:** 45–77.

Mickelson, J. C., Davis, E. B., and Tischer, R. G. (1967) The effect of various nitrogen sources upon heterocyst formation in *Anabaena flos-aquae* A-37. *J Exp Bot* **18:** 397–405.

Min, H., and Sherman, L. A. (2010) Hydrogen production by the unicellular diazotrophic cyanobacterium *Cyanothece* sp. strain ATCC 51142 under continuous light. *Appl Environ Microbiol* **76:** 4293–4301. doi: 10.1128/AEM.00146-10.

Mishra, A. K., and Tiwari, D. N. (1986) Mutagenesis and isolation of morphological mutants impaired in nitrogen-fixing capacity from a cyanobacterium *Gloeotrichia ghosei*. *New Phytol* **103:** 69–77.

Mitchison, G. J., and Wilcox, M. (1972) Rule governing cell division in *Anabaena*. *Nature* (London) **239:** 110–111.

Mitchison, G. J., and Wilcox, M. (1973) Alteration in heterocyst pattern of *Anabaena* produced by 7-azatryptophan. *Nat New Biol* **246:** 229–233.

Mitchison, G. J., Wilcox, M., and Smith, R. J. (1976) Measurement of an inhibitory zone. *Science* **191:** 866–868.

Mitra, A. K. (1950) Two new algae from Indian soils. *Ann Bot* **14:** 457–464.

Miyagishima, S-y., Wolk, C. P., and Osteryoung, K. W. (2005) Identification of cyanobacterial cell division genes by comparative and mutational analyses. *Mol Microbiol* **56:** 126–143.doi: 10.1111/j.1365-2958.2005.04548.x

Montesinos, M. L., Herrero, A., and Flores, E. (1995) Amino acid transport systems required for diazotrophic growth in the cyanobacterium *Anabaena* sp. strain PCC 7120. *J Bacteriol* **177:** 3150–3157.

Montesinos, M. L., Herrero, A., and Flores, E. (1997) Amino acid transport in taxonomically diverse cyanobacteria and identification of two genes encoding elements of a neutral amino acid permease putatively involved in recapture of leaked hydrophobic amino acids. *J. Bacteriol.* **179:** 853-862.

Moslavac, S., Nicolaisen, K., Mirus, O., Dehni, F. A., Pernil, R., Flores, E., Maldener, I., and Schleiff, E. (2007) A TolC-like protein is required for heterocyst development in *Anabaena* sp. strain PCC 7120. *J Bacteriol* **189:** 7887–7895.

Mulligan, M. E., and Haselkorn, R. (1989) Nitrogen fixation (*nif*) genes of the cyanobacterium *Anabaena* sp. strain PCC 7120. *J Biol Chem* **264:** 19200–19207.

Mullineaux, C. W., Mariscal, V., Nenninger, A., Khanum, H., Herrero, A., Flores, E., and Adams, D. G. (2008) Mechanism of intercellular molecular exchange in heterocyst-forming cyanobacteria. *EMBO J* **27:** 1299–1308.

Muro-Pastor, A. M., Flores, E., and Herrero, A. (2009) NtcA-regulated heterocyst differentiation genes *hetC* and *devB* from *Anabaena* sp. strain PCC 7120 exhibit a similar tandem promoter arrangement. *J Bacteriol* **191:** 5765–5774.

Muro-Pastor, A. M., Olmedo-Verd, F., and Flores, E. (2006) *All4312*, NtcA-regulated two-component response regulator in *Anabaena* sp. strain PCC 7120. *FEMS Microbiol Lett* **256:** 171–177.

Muro-Pastor, A. M., Valladares, A., Flores, E., and Herrero, A. (1999) The *hetC* gene is a direct target of the NtcA transcriptional regulator in cyanobacterial heterocyst development. *J Bacteriol* **181:** 6664–6669.

Muro-Pastor, A. M., Valladares, A., Flores, E., and Herrero, A. (2002) Mutual dependence of the expression of the cell differentiation regulatory protein HetR and the global nitrogen-regulator NtcA during heterocyst development. *Mol Microbiol* **44:** 1377–1385.

Muro-Pastor, M. I., Reyes, J. C., and Florencio, F. J. (2001) Cyanobacteria perceive nitrogen status by sensing intracellular 2-oxoglutarate levels. *J Biol Chem* **276:** 38320–38328.

Muro-Pastor, M. I., Reyes, J. C., and Florencio, F. J. (2005) Ammonium assimilation in cyanobacteria. *Photosynth Res* **83:** 135–150.

Murry, M. A., and Wolk, C. P. (1989) Evidence that the barrier to the penetration of oxygen into heterocysts depends upon two layers of the cell envelope. *Arch Microbiol* **151:** 469–474.

Murry, M. A., Hallenbeck, P. C., and Banemann, J. R. (1984) Immunochemical evidence that nitrogenase is restricted to heterocysts in *Anabaena cylindrica*. *Arch Microbiol* **137:** 194–199.

Murry, M. A., Olafsen, A. G., and Banemann, J. R. (1981) Oxidation of diaminobenzidine in the heterocysts of *Anabaena cylindrica*. *Curr Microbiol* **6:** 201–206.

Nakamura, Y., Kaneko, T., Sato, S., Minuro, M., Miyashita, H., Tsuchiya, T., Sasamoto, S., Watanabe, A., Kawashima, K., Kishida, Y., Kiyokawa, C., Kohara, M., Matsumoto, M., Matsuno, A., Nakazaki, N., Shimpo, S., Takeuchi, C., Yamada, M., and Tabata, S. (2003) Complete genome structure of *Gloeobacter violaceus* PCC 7421, a cyanobacterium that lacks thylakoids. *DNA Res* **10:** 137–145.

Navarro, F., Chávez, S., Candau, P., and Florencio, F. J. (1995) Existence of two ferredoxin-glutamate synthases in the cyanobacterium *Synechocystis* sp. PCC 6803: isolation and insertional inactivation of *gltB* and *gltS* genes. *Plant Mol Biol* **27:** 753–767.

Nayar, A. S., Yamamura, H., Rajagopalan, R., Risser, D. D., and Callahan, S. M. (2007) FraG is necessary for filament integrity and heterocyst maturation in the cyanobacterium *Anabaena* sp.strain PCC 7120. *Microbiology* **153:** 601–607. doi:10.1099/mic.0.2006/002535-0

Neilson, A. H., and Larsson, T. (1980) The utilization of organic nitrogen for growth of algae: Physiolosical aspects. *Physiol Plant* **48:** 542–553.

Neur, G., and Bothe, H. (1982) The pyruvate-ferredoxin oxidoreductase in heterocysts of the cyanobacterium *Anabaena cylindrica*. *Biochim Biophys Acta* **716:** 358–365.

Neur, G., and Bothe, H. (1983) Anaplerotic reactions in *Anabaena cylindrica*. *FEBS Lett* **158:** 79–83.

Neur, G., and Bothe, H. (1985) Electron donation to nitrogenase. *Arch Microbiol* **143:** 185–191.

Ni, S., Benning, M. M., Smola, M. J., Feldmann, E. A., and Kennedy, M. A. (2009) Crystal structure of Nupn_R1517, a putative negative regulator of heterocyst differentiation from *Nostoc punctiforme* PCC 73102. *Proteins* **74:** 794–798. doi:10.1002/prot.22308

Nierzwicki-Bauer, S. A., and Haselkorn, R. (1986) Differences in mRNA levels in *Anabaena* living freely or in symbiotic association with *Azolla*. *EMBO J* **5:** 29–35.

Nierzwicki-Bauer, S. A., Balkwill, D. L., Stevens, Jr., S. E. (1984a) Heterocyst differentiation in the cyanobacterium *Mastigocladus laminosus*. *J Bacteriol* **157:** 514–525.

Nierzwicki-Bauer, S. A., Balkwill, D. L., Stevens, Jr., S. E. (1984b) A new method for identification of heterocysts and proheterocysts in morphologically complex cyanobacteria. *Biotechnic Histochem* **59:** 163–170.

Ning, D., and Xu, X. (2004a) *alr0117*, a two-component histidine kinase gene, is involved in heterocyst development in *Anabaena* sp. PCC 7120. *Microbiology* **150:** 447–453.

Ning, D., and Xu, X. (2004b) A novel region involved in heterocyst development and pattern formation in the genome of *Anabaena* sp. PCC 7120. *Shui Sheng Sheng Wu Xue Bao (Acta Hydrobiologica Sinica)* **28:** 1–5 (in Chinese). Cited from Zhang *et al*. (2007).

Ochoa de Alda, J. A. G., Lichtle, C., Thomas, J. C., and Houmard, J. C. (2004) Immunolocalization of NblA, a protein involved in phycobilisome turnover during heterocyst differentiation in cyanobacteria. *Microbiology* **150:** 1377–1384. doi:10.1099/mic.0.26992-0

Ogawa, R. E., and Carr, N. G. (1969) The influence of nitrogen on heterocyst production in blue-green algae. *Limnol Oceanogr* **14:** 342–350.

Ohmori, M., and Hattori, A. (1971) Nitrogen fixation and heterocysts in the blue-green alga *Anabaena cylindrica*. *Plant Cell Physiol* **12:** 961–967.

Ohmori, M., Ikeuchi, M., Sato, N., Wolk, C. P., Kaneko, T., Ogawa, T., Kanehisa, M., Goto, S., Kawashima, S., Okamoto, S., Yoshimura, H., Katoh, H., Fujisawa, T., Ehira, S., Kamei, A., Yoshihara, S., Narikawa, R., and Tabata, S. (2001) Characterization of genes encoding multi-domain proteins in the genome of the filamentous nitrogen-fixing cyanobacterium *Anabaena* sp. strain PCC 7120. *DNA Res* **8:** 271–284.

Okuhara, H., Matsumara, T., Fujita, Y., and Hase, T. (1999) Cloning and inactivation of genes encoding ferredoxin- and NADH-dependent glutamate synthase in the cyanobacterium *Plectonema boryanum*. Imbalances in nitrogen and carbon assimilations caused by deficiency of the ferredoxin-dependent enzyme. *Plant Physiol* **20:** 33–41.

Oliveira, P., Leitão, E., Tamagnini, P., Moradas-Ferreira, P., and Oxelfelt, F. (2004) Characterization and transcriptional analysis of *hupSLW* in *Gloeothece* sp. ATCC 27152: An uptake hydrogenase from a unicellular cyanobacterium. *Microbiology* **150:** 3647–3655.

Olmedo-Verd, E., Flores, E., Herrero, A., and Muro-Pastor, A. M. (2005) HetR-dependent and -independent expression of heterocyst related genes in an *Anabaena* strain overproducing the NtcA transcription factor. *J Bacteriol* **187:** 1985–1991.

Olmedo-Verd, E., Muro-Pastor, A. M., Flores, E., and Herrero, A. (2006) Localized induction of *ntcA* regulatory gene in developing heterocysts of *Anabaena* sp. strain PCC 7120. *J Bacteriol* **188:** 6694–6699.

Olmedo-Verd, E., Valladares, A., Flores, E., Herrero, A., and Muro-Pastor, A. M. (2008) Role of two NtcA-binding sites in the complex *ntcA* gene promoter of the heterocyst-forming cyanobacterium *Anabaena* sp. strain PCC 7120. *J Bacteriol* **190:** 7584–7590.

Onek, L. A., and Smith, R. J. (1992) Calmodulin and calcium-mediated regulation in prokaryotes. *J Gen Microbiol* **138:** 1039–1049.

Onek, L. A., Lea, P. J., and Smith, R. J. (1992) Isolation and characterization of calmodulin-like protein from the cyanobacterium *Nostoc* sp. PCC 6720. *Arch Microbiol* **161:** 352–358.

Orozco, C. C., Risser, D. D., and Callahan, S. M. (2006) Epistasis analysis of four genes from *Anabaena* sp. strain PCC 7120 suggests a connection between PatA and PatS in heterocyst pattern formation. *J Bacteriol* **188:** 1808–1816.

Orr, J., and Haselkorn, R. (1981) Kinetic and inhibition studies of glutamine synthetase from the cyanobacterium *Anabaena* 7120. *J Biol Chem* **256:** 13099–13104.

Orr, J., and Haselkorn, R. (1982) Regulation of glutamine synthetase activity and synthesis in free-living and symbiotic *Anabaena* spp. *J Bacteriol* **152:** 626–635.

Osanai, T., Ikeuchi, M., and Tanaka, K. (2008) Group 2 sigma factors in cyanobacteria. *Plant Physiol* **133:** 490–506.

Osnai, T., Kanesaki, T., Nakano, T., Takahashi, H., Asayama, M., Shirai, M., Kanehisa, M., Suzuki, I., Murata, N., and Tanaka, K. (2005) Positive regulation of sugar catabolic pathways in the cyanobacterium *Synechocystis* sp. PCC 6803 by the group 2 σ factor Sig E. *J Biol Chem* **280:** 39653–30659. http://www.jbc.org

Oxelfelt, F., Tamagnini, P., Salema, R., and Lindblad, P. (1995) Hydrogen uptake in *Nostoc* strain PCC 73102: Effects of nickel, hydrogen, carbon and nitrogen. *Plant Physiol Biochem* **33:** 617–623.

Padhy, R. N., and Singh, P. K. (1978) Genetical studies on the heterocyst and nitrogen fixation of the blue-green alga *Nostoc muscorum*. *Mol Gen Genet* **162:** 203–211.

Pandey, D. C ., and Mitra, A. K. (1962) Production of heterocysts in certain Myxophyceae as influenced by the concentration of nitrogen in the medium. *Naturwissenschaften* **49:** 89–91.

Pankratz, H. S., and Bowen, C. C. (1963) Cytology of blue-green algae. I. The cells of *Symploca muscorum*. *Am J Bot* **50:** 387–399.

Papen, H., Neur, G., Refaian, M., and Bothe, H. (1983) The isocitrate dehydrogenase from cyanobacteria. *Arch Microbiol* **134:** 73–79.

Papen, H., Neur, G., Sauer,A., and Bothe, H. (1986) Properties of the glyceraldehyde-3-P dehydrogenase in heterocysts and vegetative cells of cyanobacteria. *FEMS Microbiol Lett* **36:** 201–206.

Parkinson, J. S. (1993) Signal transduction schemes of bacteria. *Cell* **73:** 857–871.

Paz-Yepes, J., Flores, E., and Herrero, A. (2003) Transcriptional effects of the signal transduction protein P(II) (*glnB* gene product) on NtcA-dependent genes in *Synechococcus* sp. PCC 7942. *FEBS Lett* **543:** 42–46.

Peat, A., and Whitton, B. A. (1967) Environmental effects on the structure of the blue-green alga *Chlorogloea fritschii*. *Arch Microbiol* **57:** 153–180.

Pederson, D. M., Daday, A., and Smith, G. D. (1986) The use of nickel to probe the role of hydrogen metabolism in cyanobacterial nitrogen fixation. *Biochemie* **68:** 113–120.

Pernil, R., Herrero, A., and Flores, E. (2010) Catabolic function of compartmentalized alanine dehydrogenase in the heterocyst-forming cyanobacterium *Anabaena* sp. strain PCC 7120. *J Bacteriol* **192:** 5165–5172.

Pernil, R., Picossi, S., Mariscal, V., Herrero, A., and Flores, E. (2008) ABC-type amino acid uptake transporter Bgt and N-II of *Anabaena* sp. strain PCC 7120 share an ATPase unit and are expressed in vegetative cells and heterocysts. *Mol Microbiol* **67:** 1067–1080.

Peschek, G. A. (1987) Respiratory electron transport. In: Fay, P. and Van Baalen, C. (Eds.) *The Cyanobacteria*. Elsevier Science Publishers, Amsterdam. The Netherlands. pp. 119–169.

Peterson, R. B., and Burris, R. H. (1976) Properties of heterocysts isolated with colloidal silica. *Arch Microbiol* **108:** 35–40.

Peterson, R. B., and Burris, R. H. (1978) Hydrogen metabolism in isolated heterocysts of *Anabaena* 7120. *Arch Microbiol* **116:** 125–132.

Peterson, R. B., and Wolk, C. P. (1978) High recovery of nitrogenase activity of ^{55}Fe-labeled nitrogenase in heterocysts isolated from *Anabaena variabilis*. *Proc Natl Acad Sci USA* **75:** 6271–6275.

Peterson, R. B., Dolan, E., Calvert, H. E., and Ke, B. (1981) Energy transfer from phycobiliproteins to photosystem I in vegetative cells and heterocysts of *Anabaena variabilis*. *Biochem Biophys Acta* **634:** 237–248.

Pettersson, A., and Bergman, B. (1989) Calmodulin in heterocystous cyanobacteria: Biochemical and immunological evidence. *FEMS Microbiol Lett* **60:** 95–99.

Picossi, S., Montesinos, M. L., Pernil, R., Lichtlé, C., Herrero, A., and Flores, E. (2005) ABC-type neutral amino acid permease N-1 is required for optimal diazotrophic growth and is repressed in the heterocysts of *Anabaena* sp. strain PCC 7120. *Mol Microbiol* **57:** 1582–1592.

Picossi, S., Valladares, A., Flores, E., and Herrero, A. (2004) Nitrogen-regulated genes for the metabolism of cyanophycin, a bacterial nitrogen reserve polymer: Expression and mutational analysis of two cyanophycin synthetase and cyanophycinase gene clusters in heterocyst-forming cyanobacterium *Anabaena* sp. PCC 7120. *J Biol Chem* **279:** 11582–11592.

Pilak, O., Mamat, B., Vogt, S., Hagemeier, C. H., Thauer, R. K., Shima, S., Vonrhein, C., Warkentin, E., and Ermler, U. (2006). The crystal structure of the apoenzyme of the iron-sulfur cluster-free hydrogenase. *J Mol Biol* **358:** 798–809.

Pils D., Wilken, C., Valladares, A., Flores, E., and Schmetterer, G. (2004) Respiratory terminal oxidases in the facultative chemoheterotrophic and dinitrogen fixing cyanobacterium *Anabaena variabilis* strain ATCC 29413: Characterization of the *cox2* locus. *Biochim Biophys Acta* **1659:** 32–45.

Popa, R., Weber, P. K., Pett-Ridge, J., Finzi, J. A., Fallon, S. J., Hutcheon, I. D., Nealson, K. H., and Capone, D. G. (2007) Carbon and nitrogen fixation and metabolite exchange in and between individual cells of *Anabaena oscillariodes. ISMEJ* 1–7. www.nature.com/ismej.

Porchia, A. C., and Salerno, G. L. (1996) Sucrose biosynthesis in a prokaryotic organism: Presence of two sucrose-phosphate synthases in *Anabaena* with remarkable differences compared with the plant enzymes. *Proc Natl Acad Sci USA* **93:** 13600–13604.

Porchia, A. C., Curatti, L., and Salerno, G. L. (1999) Sucrose metabolism in cyanobacteria: Sucrose synthase from *Anabaena* sp. strain PCC 7119 is remarkably different from the plant enzymes with respect to substrate affinity and amino-terminal sequence. *Planta* **210:** 34–40.

Potts, M., Angeloni, S. V., Ebel, R. E., and Bassam, D. (1992) Myoglobin in a cyanobacterium. *Science* **256:** 1690–1692.

Pratte, B. S., Eplin, K., and Thiel, T. (2006) Cross-functionality of nitrogenase components NifH1 and VnfH in *Anabaena variabilis*. *J Bacteriol* **188:** 5806–5811.

Privalle, L. S., and Burris, R. H. (1984) D-erythrose supports nitrogenase activity in isolated *Anabaena* sp. strain 7120 heterocysts. *J Bacteriol* **157:** 350–356.

Rai, A. K., Pandey, K. D., and Kashyap, A. K. (1978) Heterocyst differentiation and polarity in *Calothrix*. *New Phytol* **81:** 647–651.

Rai, A. N., Borthakur, M., Singh, S., and Bergman, B. (1989) *Anthoceros-Nostoc* symbiosis: Immunoelectronmicroscopic localization of nitrogenase, glutamine synthetase, phycoerythrin and ribulose-1,5-biphosphate carboxylase/oxygenase in the cyanobiont and the cultured (free-living) isolate *Nostoc* 7801. *J Gen Microbiol* **135:** 385–395.

Rai, A. N., Borthakur, M., Söderbäck, E., and Bergman, B. (1992) Immunogold localization of hydrogenase in the cyanobacterial-plant symbioses *Peltigera canina, Anthoceros punctatus* and *Gunnera magellanica*. *Symbiosis* **12:** 131–144.

Rai, A. N., Rowell, P., and Stewart, W. D. P. (1982) Glutamate synthase activity of heterocysts and vegetative cells of the cyanobacterium *Anabaena variabilis* Kütz. *J Gen Microbiol* **128:** 2203–2205.

Rajagopalan, R., and Callahan, C. M. (2010) Temporal and spatial regulation of the four transcription start sites of *hetR* from *Anabaena* sp. strain PCC 7120. *J Bacteriol* **192:** 1088–1096.

Ramasubramanian, T. S., Pu, F., and Golden, J. W. (1994a) Isolation of *Anabaena* sp. strain PCC 7120 *sigA* gene in a transcriptional interface selection. *J Bacteriol* **177:** 6676–6678.

Ramasubramanian, T. S., Wei, T.-F., and Golden, J. W. (1994b) Two *Anabaena* sp. strain PCC 7120 DNA-binding factors interact with vegetative cell- and heterocyst specific genes. *J Bacteriol* **176:** 1214–1223.

Ramasubramanian, T. S., Wei, T.-F., Oldham, A. K., and Golden, J.W. (1996) Transcription of the *Anabaena* sp. strain PCC 7120 *ntcA* gene: Multiple transcripts and NtcA binding. *J Bacteriol* **178:** 922–926.

Ramaswamy, K. S., Carrasco, C. D., Fatma, T., and Golden, J. W. (1997) Cell-type specificity of the *Anabaena fdxN*-element rearrangement requires *xisH* and *xisI*. *Mol Microbiol* **23:** 1241–1249.

Ramírez, M. E., Hebbar, P. B., Zhou, R., Wolk, C. P., and Curtis, S. E. (2005) *Anabaena* sp. strain PCC 7120 gene *devH* is required for synthesis of heterocyst glycolipid layer. *J Bacteriol* **187:** 2326–2331.

Ran, L., Larsson, J., Vigil-Stenman, T., Nylander, J. A. A., Ininbergs, K., Zheng, W.-W., Lapidus, A., Lowry, S., Haselkorn, R., and Berman, B. (2010) Genome erosion in a nitrogen-fixing vertically transmitted endosymbiotic multicellular cyanobacterium. *PLoS ONE* **5:** e11486. doi:10.1371/journal.pone.0011486

Rawson, D. M. (1985) The effects of exogenous amino acids on growth and nitrogenase activity in the cyanobacterium *Anabaena cylindrica* PCC 7122. *J Gen Microbiol* **131:** 2549–2554.

Ray, T. B., Mayne, B. C., Tola, Jr., R. E., and Peters, G. A. (1979) *Azolla-Anabaena* relationship. VIII. Photosynthetic characterization of the association and individual partners. *Plant Physiol* **64:** 791–795.

Razquin, P., Fillat, M. F., Schmitz, S., Stricker, O., Böhme, H., Gómez-Moreno, C., and Peleato, M. L. (1996) Expression of ferredoxin-NADP+ reductase in heterocysts of *Anabaena* sp. *Biochem J* **316:** 157–160.

Razquin, P., Schmitz, S., Peleato, M. L., Fillat, M. F., Gómez-Moreno, C., and Böhme, H. (1995) Differential activities of heterocyst ferredoxin, vegetative cell ferredoxin, and flavodoxin as electron carriers in nitrogen fixation and photosynthesis in *Anabaena* sp. *Photosynth Res* **43:** 35–40.

Reddy, P. M., and Talpasayi, E. R. S. (1974) Heterocyst formation in a blue-green alga, *Cylindrospermum. Nature* (London) **248:** 493–494.

Reddy, P. M., Conway, C. M., and Fisher, R. W. (1989) Tunicamycin suppresses differentiation and stimulates dedifferentiation of heterocysts in the cyanobacterium *Anabaena azollae* Straus. *Curr Microbiol* **19:** 109–114.

Reddy, P. M., Spiller, H., Albrecht, S. L., and Shanmugam, K. T. (1996) Photodissimilation of fructose to H_2 and CO_2 by a dinitrogen fixing cyanobacterium *Anabaena variabilis. Appl Environ Microbiol* **62:** 1220–1226.

Renström-Kellner, E., Rai, A. N., and Bergman, B. (1990) Correlation between nitrogenase and glutamine synthetase expression in the cyanobacterium *Anabaena cylindrica. Physiol Plant* **80:** 12–19.

Reyes, J. C., and Florencio, F. J. (1994) A new type of glutamine synthetase in cyanobacteria: The protein encoded by the *glnN* gene supports nitrogen assimilation in *Synechocystis* sp. strain PCC 6803. *J Bacteriol* **176:** 1260–1267.

Reyes, J. C., Muro-Pastor, M. I., and Florencio, F. J. (1997) Transcription of glutamine synthetase genes (*glnA* and *glnN*) from the cyanobacterium *Synechocystis* sp. strain PCC 6803 is differently regulated in response to nitrogen availability. *J Bacteriol* **179:** 2678–2689.

Rice, D., Mazur, B. J., and Haselkorn, R. (1982) Isolation and physical mapping of nitrogen fixation genes from the cyanobacterium *Anabaena* 7120; *J. Biol. Chem.* **257:** 13157–13163.

Rippka, R., Deruelles, J., Waterbury, J. B., Herdman, M., and Stanier, R. Y. (1979) Generic assignments, strain histories and properties of pure cultures of cyanobacteria. *J Gen Microbiol* **111:** 1-61.

Risser, D. D., and Callahan, S. M. (2007) Mutagenesis of *hetR* reveals amino acids necessary for HetR function in the heterocystous cyanobacterium *Anabaena* sp. strain PCC 7120. *J Bacteriol* **189:** 2460–2467.

Risser, D. D., and Callahan, S. M. (2008) HetF and PatA control levels of HetR in *Anabaena* sp. strain PCC 7120. *J Bacteriol* **190:** 7645–7654.

Risser, D. D., and Callahan, S. M. (2009) Genetic and cytological evidence that heterocyst patterning is regulated by inhibitor gradients that promote activator decay. *Proc Natl Acad Sci USA* **106:** 19884–19888.

Rodgers, G. A., and Stewart, W. D. P. (1977) The cyanophyte-hepatic symbiosis. I. Morphology and physiology. *New Phytol* **78:** 441–458.

Rodríguez, H., Rivas, J., Gurrero, M. G., and Losada, M. (1990) Ca^{2+} requirement for aerobic nitrogen fixation by heterocystous blue-green algae. *Plant Physiol* **92:** 886–890.

Rogerson, A. C. (1979) Modifiers of heterocyst repression and spacing and formation of heterocysts without nitrogenase in the cyanobacterium *Anabaena variabilis. J Bacteriol* **140:** 213-219.

Rowell, P., and Stewart, W. D. P. (1975) Alanine dehydrogenase of N_2-fixing blue-green alga, *Anabaena cylindrica. Arch Microbiol* **107:** 115–124.

Rowell, P., Enticott, S., Stewart, W. D. P. (1977) Glutamine synthetase and nitrogenase activity in the blue-green alga *Anabaena cylindrica. New Phytol* **79:** 41–54.

Rowell, P., Sampaio, M. J. A. M., Ladha, J. K., and Stewart, W. D. P. (1979) Alteration of cyanobacterial glutamine synthetase activity *in vivo* in response to light and NH_4^+. *Arch Microbiol* **120:** 195–200.

Ruppert, U., Irmler, A., Kloft, N., and Forchhammer, K. (2002) The novel protein phosphatase PphA from *Synechocystis* PCC 6803 controls dephosphorylation of the signaling protein PII. *Mol Microbiol* **44:** 855–864.

Sakr, S., Jeanjean, R., Zhang, C.-C., and Arcondeguy, T. (2006) Inhibition of cell division suppresses heterocyst development in *Anabaena* sp. strain PCC 7120. *J Bacteriol* **188:** 1396–1404.

Sauer, J., Dirmeier, U., and Forchhammer, K. (2000) The *Synechococcus* strain PCC 7942 *glnN* product (glutamine synthetase III) helps in the recovery from prolonged nitrogen chlorosis. *J Bacteriol* **182:** 5615–5619.

Sauer, J., Gorl, M., and Forchhammer, K. (1999) Nitrogen starvation in *Synechococcus* PCC 7942: Involvement of glutamine synthetase and NtcA in phycobiliprotein degradation and survival. *Arch Microbiol* **172:** 247–255.

Saville, B., Straus, N., and Coleman, J. R. (1987) Contiguous organization of nitrogenase genes in a heterocystous cyanobacterium. *Plant Physiol* **85:** 26–29.

Sawa, Y., Tani, M., Murata, K., Shibata, H., and Ochiai, H. (1994) Purification and characterization of alanine dehydrogenase from a cyanobacterium, *Phormidium lapideum. J Biol Chem* **116:** 995–1000.

Schiefer, W., Schütz, K., Hachtel, W., and Happe, T. (2002) Molecular cloning and characterization of *hetR* genes from filamentous cyanobacteria. *Biochim Biophys Acta* **1577:** 139–143.

Schilling, N., and Ehrnsperger, K. (1985) Cellular differentiation of sucrose metabolism in *Anabaena variabilis. Z Naturforsch* **40:** 776–779

Schmetterer, G., Valladares, A., Pils, D., Steinbach, S., Pacher, M., Muro-Pastor, A. M., Flores, E., and Herrero, A. (2001) The *coxBAC* operon encodes a cytochrome *c* oxidase required for heterotrophic growth in the cyanobacterium *Anabaena variabilis* strain ATCC 29413. *J Bacteriol* **183:** 6429–6434.

Schmitz, O., Boison, G., Hilscher, R., Hundeshagen, B., Zimmer, W., Lottspeich, F., and Böthe, H. (1995) Molecular biological analysis of a bidirectional hydrogenase from cyanobacteria. *Eur J Biochem* **233:** 266–276.

Schmitz, O., Boison, G., Salzmann, H., Bothe, H., Schütz, K.,Wang, S., and Happe, T. (2002) HoxE-a subunit specific for the pentameric bidirectional hydrogenase complex (HoxEFUYH) of cyanobacteria. *Biochim Biophys Acta* **1554:** 66–74.

Schmitz, O., Kentenich, T., Zimmer, W., Hundeshagen, B., and Böthe, H. (1993) Identification of the *nifJ* gene encoding for pyruvate: ferredoxin oxidoreductase in dinitrogen fixing cyanobacteria. *Arch Microbiol* **160:** 62–67.

Schmitz, S., Scrautemeier, B., and Böhme, H. (1993) Evidence from directed mutagenesis that positively charged amino acids are necessary for interaction of nitrogenase with the [2Fe-2S] heterocyst ferredoxin (FdxH) from the cyanobacterium *Anabaena* sp. strain PCC 7120. *Mol Gen Genet* **240:** 455–460.

Schrautemeier, B., and Böhme, H. (1985) A distinct ferredoxin for nitrogen fixation isolated from heterocysts of the cyanobacterium *Anabaena variabilis*. *FEBS Lett* **184:** 304–308.

Schrautemeier, B., Böhme, H., and Böger, P. (1984) *In vitro* studies on pathways and regulation of electron transport to nitrogenase with a cell-free extract from heterocysts of *Anabaena variabilis*. *Arch Microbiol* **137:** 14–20.

Schrautemeier, B., Cassing, A., and Böhme, H. (1994) Characterization of the genome region encoding an fdxH-type ferredoxin and a new 2[4Fe-4S] ferredoxin from the non-heterocystous, nitrogen fixing cyanobacterium *Plectonema boryanum*. *J Bacteriol* **176:** 1037–1046.

Schrautemeier, B., Neveling, U., and Schmitz, S. (1995) Distinct and differently regulated Mo-dependent nitrogen fixing systems evolved for heterocysts and vegetative cells of *Anabaena variabilis* ATCC 29413: Characterization of the *fdxH1/2* gene regions as part of *nif1/2* gene clusters. *Mol Microbiol* **18:** 357–369.

Schriek, S., Kahmann, V., Staiger, D., Pistorius, E. K., and Michel, P. (2009) Detection of an L-amino acid dehydrogenase activity in *Synechocystis* sp. PCC 6803. *J Exp Bot* **60:** 1035–1046. doi:10.1093/jxb/ern352

Seabra, R., Santos, R., Moradas-Ferreira, P., and Tamagnini, P. (2009) Immunolocalization of the uptake hydrogenase in the marine cyanobacterium *Lyngbya majuscula* CCAP 1446/4 and two *Nostoc* strains. *FEMS Microbiol Lett* **292:** 57–62.

Sedelnikova, S., Rice, D. W., Shibata, H., Sawa, Y., and Baker, P. J. (1998) Crystallization of alanine dehydrogenase from *Phormidium lapideum*. *Acta Cryst* **54:** 407–408.

Sharma, P. (1984) Heterocyst and akinete induction with altered pattern in *Anabaena cylindrica*, caused by neopeptone. *Arch Microbiol* **139:** 196–201.

Sharma, P., and von Hofsten, A. (1987) Ultrastructural studies of heterocyst induction by neopeptone in *Anabaena cylindrica*. *FEMS Microbiol Lett* **48:** 127–131.

Sherman, D. M., Tucker, D., and Sherman, L. A. (2000) Heterocyst development and localization of cyanophycin in N_2-fixing cultures of *Anabaena* sp. PCC 7120 (Cyanobacteria). *J Phycol* **36:** 932–941.

Shi, L., Li, J-H., Cheng, Y., Wang, L., Chen, W-L., and Zhang, C-C. (2007) Two genes encoding protein kinases of the HstK family are involved in the synthesis of the minor heterocyst-specific glycolipid in the cyanobacterium *Anabaena* sp. strain PCC 7120. *J Bacteriol* **189:** 5075–5081. doi:10.1128/JB.00323-07

Shi, Y., Zhao, W., Zhang, W., Ye, Z., and Zhao, J. (2006) Regulation of intracellular free calcium concentration during heterocyst differentiation by HetR and NtcA in *Anabaena* sp. PCC 7120. *Proc Natl Acad Sci USA* **103:** 11334–11339. doi:10.1073/pnas.0602839103

Shively, J. M. (1974) Inclusion bodies of prokaryotes. *Annu Rev Microbiol* **28:** 167–187.

Silhavy, T. J., and Beckwith, J. R. (1985) Use of *lac* fusions for the study of biological problems. *Microbiol Rev* **49:** 398–418.

Silvester, W. B. (1976) Endophyte adaption in *Gunnera-Nostoc* symbiosis. In: Nutman, P. S. (Ed.) *Symbiotic Nitrogen Fixation in Plants*, Vol 7. Cambridge University Press, Cambridge, UK. pp. 521–538.

Simon, R. D. (1971) Cyanophycin granules from the blue-green alga *Anabaena cylindrica*: A reserve material consisting of copolymers of aspartic acid and arginine. *Proc Natl Acad Sci USA* **68:** 265–267.

Simon, R. D. (1973a) The effect of chloramphenicol on the production of cyanophycin granule polypeptide in the blue-green alga *Anabaena cylindrica*. *Arch Microbiol* **92:** 115–122.

Simon, R. D. (1973b) Measurement of the cyanophycin granule peptide contained in the blue-green alga *Anabaena cylindrica*. *J Bacteriol* **114:** 1213–1216.

Simon, R. D. (1976) The biosynthesis of multi-L-arginyl poly (L-aspartic acid) in the filamentous cyanobacterium *Anabaena cylindrica*. *Biochim Biophys Acta* **422:** 407–418.

Simon, R. D. (1987) Inclusion bodies in the cyanobacteria: Cyanophycin, polyphosphate, polyhedral bodies. In: Fay, P., and Van Baalen, C. (Eds.) *The Cyanobacteria*. Elseiver Publishers Amsterdam, The Netherlands. pp. 199–225.

Simon, R. D., and Weathers, J. P. (1976) Determination of the structure of the novel polypeptide containing aspartic acid and arginine which is found in cyanobacteria. *Biochim Biophys Acta* **420:** 166–176.

Simpson, F. B., and Burris, R. H. (1984) A nitrogen pressure of 50 atmospheres does not prevent evolution of hydrogen by nitrogenase. *Science* **224:** 1095–1097.
Singh, H. N., Ladha, J. K., and Kumar, H. D. (1977) Genetic control of heterocyst formation in the blue-green algae *Nostoc muscorum* and *Nostoc linckia*. *Arch Microbiol* **114:** 155–159.
Singh, H. N., and Srivastava, B. S. (1968) Studies on morphogenesis in a blue-green alga. I. Effect of inorganic nitrogen sources on developmental morphology of *Anabaena doliolum*. *Can J Microbiol* **14:** 1341–1346.
Singh R. N., and Tiwari, D. N. (1970) Frequent heterocyst germination in the blue-green alga *Gloeotrichia ghosei* Singh. *J Phycol* **6:** 172–176.
Singh, R. N., Tiwari, D. N., and Singh, V. P. (1972) Genetic basis of cellular differentiation in blue-green algae. In: Desikachary, T. V. (Ed.) *The Biology of Blue-green Algae*. University of Madras, Madras. pp. 27–37.
Sjoholm, J., Oliveira, P., and Lindblad, P. (2007) Transcriptional regulation of the bidirectional hydrogenase in the cyanobacterium *Nostoc* sp. strain PCC 7120. *Appl Environ Microbiol* **73:** 5435–5446.
Smith, R. J., and Wilkins, A. (1988) A correlation between intracellular calcium and incident irradiance in *Nostoc* 6720. *New Phytol* **109:** 157–161.
Smith, R. J., Hobson, S., and Ellis, I. R. (1987) Evidence for calcium-mediated regulation of heterocyst frequency and nitrogenase activity in *Nostoc* 6720. *New Phytol* **105:** 531–541.
Smith, R. L., Kumar, D., Xiankong, Z., Tabita, F. R., and Van Baalen, C. (1985) H_2, N_2 and O_2 metabolism by isolated heterocysts from *Anabaena* sp. strain CA. *J Bacteriol* **162:** 565–570.
Smith, R. V., and Evans, M. C. W. (1970) Soluble nitrogenase from vegetative cells of the blue-green alga *Anabaena cylindrica*. *Nature* (London) **225:** 1253–1254.
Smith, R. V., and Evans, M. C. W. (1971) Nitrogenase activity in cell-free extracts of the blue-green alga, *Anabaena cylindrica*. *J Bacteriol* **105:** 913–917.
Smith, R. V., Noy, R. J., and Evans, M. C. W. (1971) Physiological electron donor to the nitrogenase of the blue-green alga *Anabaena cylindrica*. *Biochim Biophys Acta* **253:** 104–109.
Spence, D. W., and Stewart, W. D. P. (1986) Proline inhibits N_2-fixation in *Anabaena* 7120. *Biochem Biophys Res Commun* **139:** 940–946.
Soriente, A., Gambacorta, A., Trincone, A., Sili, C., Vincenzini, M., and Sudano, G. (1993) Heterocyst glycolipids of the cyanobacterium *Cyanospira rippkae*. *Phytochemistry* **33:** 393–396.
Soriente, A., Sodano, G., Gambacorta, A., and Trincone, A. (1992) Structure of the heterocyst "glycolipids" of the marine cyanobacterium *Nodularia harveyana*. *Tetrahedron* **48:** 5375–5384.
Spence, D. W., and Stewart, W. D. P. (1986) Proline inhibits N_2-fixation in *Anabaena* 7120. *Biochem Biophys Res Commun* **139:** 940–946.
Spence, D. W., and Stewart, W. D. P. (1987) Heterocystless mutants of *Anabaena* PCC 7120 with nitrogenase activity. *FEMS Microbiol Lett* **40:** 119–122.
Spratt, E. R. (1911) Some observations of the life-history of *Anabaena cycadeae*. *Ann Bot* **25:** 369–380.
Stacey, G., Bottomley, P. J., Van Baalen, C., and Tabita, F. R. (1979) Control of heterocyst and nitrogenase synthesis in cyanobacteria. *J Bacteriol* **137:** 321–326.
Stanier, R. Y., and Cohen-Bazire, G. (1977) The photosynthetic prokaryotes: the cyanobacteria. *Annu Rev Microbiol* **31:** 225–274.
Stanier, R. Y., Kunisawa, R., Mandel, M., and Cohen-Bazire, G. (1971) Purification and properties of unicellular blue-green algae (order Chroococcales). *Bacteriol Rev* **35:** 171–205.
Steincke, F. (1932) Das Auskeimen alter Heterocysten bei *Calothrix weberi*. (cited from Desikachary, 1959).
Stewart, W. D. P. (1964) The effect of available nitrate and ammonium nitrogen on the growth of two nitrogen-fixing blue-green algae. *J Exp Bot* **15:** 138–145.
Stewart, W. D. P. (1980) Some aspects of structure and function in nitrogen fixing cyanobacteria. *Annu Rev Microbiol* **34:** 497–536.
Stewart, W. D. P., and Rowell, P. (1975) Effects of L-methionine-DL-sulphoximine on the assimilation of newly fixed ammonia, acetylene reduction and heterocyst formation in *Anabaena cylindrica*. *Biochem Biophys Res Commun* **65:** 846–856.
Stewart, W. D. P., and Singh, H. N. (1975). Transfer of nitrogen-fixing *(nif)* genes in the blue-green alga *Nostoc mucorum*. *Biochem Biophys Res Commun* **62:** 62–69.
Stewart, W. D. P., Fitzgerald, G. P., and Burris, R. H. (1968) Acetylene reduction by nitrogen-fixing blue-green algae. *Arch Microbiol* **62:** 336–348.
Stewart, W. D. P., Haystead, A., and Pearson, H. W. (1969) Nitrogenase activity in heterocysts of blue-green algae. *Nature* (London) **224:** 226–228.
Stock, J. B, Stock, A. M., and Motten, J. M. (1990) Signal transduction in bacteria. *Nature* (London) **344:** 395–400.
Strasser, P., and Falkner, G. (1986) Characterization of the glutamate/aspartate transport system in a symbiotic *Nostoc* sp. *Planta* **168:** 381–385.

Strauss, H., Misselwitz, R., Labudde, D., Nicklisch, S., and Baier, K. (2002) NblA from *Anabaena* sp. PCC 7120 is a mostly α-helical protein undergoing reversible trimerization in solution. *Eur J Biochem* **269:** 4617–4624. doi:10.1046/j.1432-1033.2002.03161.x

Stucken, K., John, U., Cembella, A., Murillo, A. A., Soto-Liebe, K., Fuentes-Valdés, J. J., Friedel, M., Plominsky, A. M., Vásquez, M., and Glöckner, G. (2010) The smallest known genomes of multicellular toxic cyanobacteria: Comparison of minimal gene sets for linked traits and the evolutionary implications. *PLoS ONE* **5:** e9235. doi: 10.1371/journal.pone.0009235

Su, Z., Olman, V., Mao, F., and Xu, Y. (2005) Comparative genomic analysis of NtcA regulons in cyanobacteria: regulation of nitrogen assimilation and its coupling to photosynthesis. *Nucleic Acids Res* **33:** 5156–5171.

Summers, M. L., Wallis, J. G., Campbell, E. L., and Meeks, J. C. (1995) Genetic evidence of a major role for glucose-6-phosphate dehydrogenase in nitrogen fixation and dark growth of the cyanobacterium *Nostoc* sp. strain ATCC 29133. *J Bacteriol* **177:** 6184–6194.

Suzuki, A., and Knaff, D. B. (2005) Glutamate synthase: Structural, mechanistic and regulatory properties and role in the amino acid metabolism. *Photosynth Res* **83:** 193–217.

Talpasayi, E. R. S. and Bahal, M. (1967) Cellular differentiation in *Anabaena cylindrica. Zeitschrift Pflanzenphysiol* **56:** 100–107

Talpasayi, E. R. S., and Kale, K S. (1967) Induction of heterocysts in the blue-green alga *Anabaena ambigua. Curr Sci* **36:** 218–219

Tamagnini, P., Axelsson, R., Lindberg, P., Oxelfelt, F., Wünschiers, R., and Lindblad, P. (2002) Hydrogenases and hydrogen metabolism of cyanobacteria. *Microbiol Mol Biol Rev* **66:** 1–20.

Tamagnini, P., Costa, J.-L., Almeida, L., Oliveira, M.-J., Salema, R., and Lindblad, P. (2000) Diversity of cyanobacterial hydrogenases, a molecular approach. *Curr Microbiol* **40:** 356–361.

Tamagnini, P., Leitão, E., and Oxelfelt, F. (2005) Uptake hydrogenase in cyanobacteria: Novel input from non-heterocystous strains. *Biochem Soc Trans* **33:** 67–69.

Tamagnini, P., Leitão, E., Oliveira, P., Ferriera, D., Pinto, F., Harris, D., Heidorn, T., and Lindblad, P. (2007) Cyanobacterial hydrogenases: Diversity, regulation and applications. *FEMS Microbiol Rev* **31:** 692–720.

Tamagnini, P., Oxelfelt, F., Salema, R., and Lindblad, P. (1995) Immunological characterization of hydrogenases in the nitrogen-fixing cyanobacterium *Nostoc* sp. strain PCC 73102. *Curr Microbiol* **31:** 102–107.

Tamagnini, P., Troshina, O., Oxelfelt, F., Salema, R., and Lindblad, P. (1997) Hydrogenases in *Nostoc* sp. strain PCC 73102, a strain lacking a bidirectional enzyme. *Appl Environ Microbiol* **63:** 1801–1807.

Tanigawa, R., Shirokane, M., Maeda Si, S., Omata, T., Tanaka, K., and Takahashi, H. (2002). Transcriptional activation of NtcA-dependent promoters of *Synechococcus* sp. PCC 7942 by 2-oxoglutarate *in vitro*. *Proc Natl Acad Sci USA* **99:** 4251–4255.

Tano, H., Ishizuka, M., and Sone, N. (1991) The cytochrome *c* oxidase genes in blue-green algae and characteristics of the deduced protein sequence for subunit II of the thermophilic cyanobacterium *Synechococcus vulcanus. Biochem Biophys Res Commun* **181:** 437–442.

Tel-Or, E., and Stewart, W. D. P. (1976) Photosynthetic electron transport, ATP synthesis and nitrogenase activity in isolated heterocysts of *Anabaena cylindrica. Biochim Biophys Acta* **423:** 189–195.

Tel-Or, E., and Stewart, W. D. P. (1977) Photosynthetic components and activities of nitrogen-fixing isolated heterocysts of *Anabaena cylindrica. Proc R Soc* B London **198:** 61–86.

Tel-Or, E., Luijk, L. W., and Packer, L. (1977) An inducible hydrogenase in cyanobacteria enhances N_2 fixation. *FEBS Lett* **78:** 49–52.

Tel-Or, E., Luijk, L. W., and Packer, L. (1978) Hydrogenase in N_2-fixing cyanobacteria. *Arch Biochem Biophys* **185:**185–194.

Thiel, T. (1988) Transport of leucine in the cyanobacterium *Anabaena variabilis. Arch Microbiol* **149:** 466–470.

Thiel, T. (1993) Characterization of genes for an alternative nitrogenase in the cyanobacterium *Anabaena variabilis. J Bacteriol* **175:** 6276–6286.

Thiel, T. (1994) Genetic analysis of cyanobacteria. In: Bryant, D. A. (Ed.) *The Molecular Biology of Cyanobacteria.* Kluwer Dordrecht, The Netherlands. pp. 581–611.

Thiel, T. (1996) Isolation and characterization of the *nifEN* genes of the cyanobacterium *Anabaena variabilis. J Bacteriol* **178:** 4493–4499.

Thiel, T., and Leone, M. (1986) Effect of glutamine on growth and heterocyst differentiation in the cyanobacterium *Anabaena variabilis. J Bacteriol* **168:** 769–774.

Thiel, T., and Pratte, B. (2001) Effect on heterocyst differentiation of nitrogen fixtion in vegetative cells of the cyanobacterium *Anabaena variabilis* ATCC 29413. *J Bacteriol* **183:** 280–286.

Thiel, T., and Wolk, C. P. (1983) Metabolic activities of isolated akinetes of the cyanobacterium *Nostoc spongiaeforme. J Bacteriol* **156:** 369–374.

Thiel, T., Lyons, E. M., and Erker, J. C. (1997) Characterization of genes for a second Mo-dependent nitrogenase in the cyanobacterium *Anabaena variabilis. J Bacteriol* **179:** 5222–5225.

Thiel, T., Lyons E. M., Erker J. C., and Ernst, A. (1995) A second nitrogenase in vegetative cells of a heterocyst-forming cyanobacterium. *Proc Natl Acad Sci USA* **92:** 9358–9362.

Thiel, T., Lyons, E. M., and Thielemeier, J. (1999) Organization and regulation of two clusters of *nif* genes in the cyanobacterium *Anabaena variabilis*. In: Peschek, G.A., Lossffelhardt, W., and Schmetterer, G. (Eds.) *The Phototrophic Prokaryotes*. Kluwer, Plenum. New York, USA. pp. 517–521.

Thomas, J. (1970) Absence of the pigments of photosystem II of photosynthesis in heterocysts of a blue-green alga. *Nature* (London) **228:** 181–183.

Thomas, J. (1972) Relationship between age of culture and occurrence of the pigments of photosystem II of photosynthesis in heterocysts of a blue-green alga. *J Bacteriol* **110:** 92–95.

Thomas, J., and David, K. A. V. (1971) Studies on the physiology of heterocyst production in nitrogen-fixing blue-green alga *Anabaena* sp. L-31 in continuous culture. *J Gen Microbiol* **66:** 127–131.

Thomas, J., Meeks, J. C., Wolk, C. P., Shaffer, P. W., Austin, S. M., and Chien, W. S. (1977) Formation of glutamine from [^{13}N] ammonia, [^{13}N] dinitrogen and [^{14}C] glutamate by heterocysts isolated from *Anabaena cylindrica*. *J Bacteriol* **129:** 1545–1555.

Tiwari, D. N. (1977) Pattern of heterocyst differentiation as a function of nitrogen fixing ability in blue-green alga *Nostoc linckia*. *Beitr Biol Pfianzen* **53:** 177–186.

Tiwari, D. N., Pandey, A. K., and Mishra, A. K. (1981) Action of 2,4-dichlorophenoxyacetic acid and rifampicin on heterocyst differentiation in the blue-green alga, *Nostoc linckia*. *J Biosci* **3:** 33–39.

Tiwari, D. N.,Yadava, P. K., and Mishra, A. K. (1979) Developmental behaviour in mutants of *Gloeotrichia ghosei* defective in nitrogen fixation. *Microbios Lett* **12:** 121–131.

Torrecilla, I., Leganés, F., Bonilla, I., and Fernández-Piñas, F. (2000) Use of recombinant aequorin to study calcium homeostasis and monitor calcium transients in response to heat and cold shock in cyanobacteria. *Plant Physiol* **123:** 161–175.

Torrecilla, I., Leganés, F., Bonilla, I., and Fernández-Piñas, F. (2004) A calcium signal is involved in heterocyst differentiation in the cyanobacterium *Anabaena* sp. PCC 7120. *Microbiology* **150:** 3731–3739. doi:10.1099/mic.0.27403-0

Toyoshima, M., Sasaki, N. V., Fujiwara, M., Ehira, S., Ohmori, M., and Sato, N. (2010) Early candidacy for heterocyst differentiation into heterocysts in the filamentous cyanobacterium *Anabaena* sp. PCC 7120. *Arch Microbiol* **192:** 23–31.

Tredici, M. R., Margheri, M. C., De Philippis, R., and Materassi, R. (1990) The role of hydrogen metabolism in photoheterotrophic cultures of the cyanobacterium *Nostoc* sp. strain Cc isolated from *Cycas circinalis* L. *J Gen Microbiol* **136:** 1009–1015.

Troshina, O. Y., Serebrykova, L. T., and Lindblad, P. (1996) Induction of H_2-uptake and nitrogenase activities in the cyanobacterium *Anabaena variabilis* ATCC 29413: Effects of hydrogen and organic substrate. *Curr Microbiol* **33:** 11–15.

Tumer, N. E., Robinson, S. J., and Haselkorn, R. (1983) Different promoters for the *Anabaena* glutamine synthetase gene during growth using molecular or fixed nitrogen. *Nature* (London) **306:** 337–342.

Tyagi, V. V. S. (1973a) Effects of some metabolic inhibitors on heterocyst formation in the blue-green alga, *Anabaena doliolum*. *Ann Bot* **37:** 361–368.

Tyagi, V. V. S. (1973b) Effects of actinomycin D on growth, heterocyst formation, and nucleic acid synthesis in the blue-green alga, *Anabaena doliolum*. *Curr Sci* **42:** 343–346.

Tyagi, V. V. S. (1975) The heterocysts of blue-green algae (Myxophyceae). *Biol Rev* **50:** 247–284.

Tyagi, V. V. S., Ray, T. B., Mayne, B. C., and Peters, G. A. (1981) The *Azolla- Anabaena* relationship. XI. Phycobiliproteins in the action spectrum for nitrogenase-catalyzed acetylene reduction. *Plant Physiol* **68:** 1479–1484.

Ulrich, L. E, Koonin, E. V, and Zhulin, I. B. (2005) One-component systems dominate signal transduction in prokaryotes. *Trends Microbiol* **13:** 52–56.

Ungerer, J. L., Pratte, B. S., and Thiel, T. (2010) RNA processing of nitrogenase transcripts in the cyanobacterium *Anabaena variabilis*. *Arch Microbiol* **192:** 3311–3320.

Vaishampayan, A. (1982) Amino acid nutrition in the blue-green alga *Nostoc muscorum*. *New Phytol* **90:** 545–549.

Valladares, A., Flores, E., and Herrero, A. (2008) Transcription activation by NtcA and 2-oxoglutarate of three genes involved in heterocyst differentiation in the cyanobacterium *Anabaena* sp. strain PCC 7120. *J Bacteriol* **190:** 6126–6133.

Valladares, A., Herrero, A., Pils, D., Schmetterer, G., and Flores, E. (2003) Cytochrome *c* oxidase genes required for nitrogenase activity and diazotrophic growth in *Anabaena* sp. PCC 7120. *Mol Microbiol* **47:** 1239–1249.

Valladares, A., Maldener, I., Muro-Pastor, A. M., Flores, E., and Herrero, A. (2007) Heterocyst development and diazotrophic metabolism in terminal respiratory oxidase mutants of the cyanobacterium *Anabaena* sp. strain PCC 7120. *J Bacteriol* doi:10.1128/JB.00220-07

Valladares, A., Muro-Pastor, A. M., Herrero, A., and Flores, A. (2004) The NtcA-dependent P1 promoter is utilized for *glnA* expression in N_2-fixing heterocysts of *Anabaena* sp. strain PCC 7120. *J Bacteriol* **186:** 7337–7343.

Van De Water, S. D., and Simon, R. D. (1982) Induction and differentiation of heterocysts in the filamentous cyanobacterium *Cylindrospermum licheniforme*. *J Gen Microbiol* **128:** 917–925.

Van Gorkom, H. J., and Donze, M. (1971) Localization of nitrogen fixation in *Anabaena*. *Nature* (London) **234:** 231–232.

Vargas, W., Cumino, A., and Salerno, G. L. (2003) Cyanobacterial alkaline/neutral invertases. Origin of sucrose hydrolysis in the plant cytosol. *Planta* **216:** 951–960.

Vazquez-Bermudez, M. F., Herrero, A., and Flores, E. (2002). 2- oxoglutarate increases the binding affinity of the *ntcA* (nitrogen control) transcription factor for the *Synechococcus glnA* promoter. *FEBS Lett* **512:** 71–74.

Vega-Palas, M. A., Flores, E., and Herrero, A. (1992) NtcA, a global nitrogen regulator from the cyanobacterium *Synechococcus* that belongs to the Crp family of bacterial regulators. *Mol Microbiol* **6:** 1853–1859.

Vega-Palas, M. A., Madueno, F., Herrero, A., and Flores, E. (1990) Identification and cloning of a regulatory gene for nitrogen assimilation in the cyanobacterium *Synechococcus* sp. strain PCC 7942. *J Bacteriol* **172:** 643–647.

Venkataraman, G. S. (1957) A note on the occurrence of three-pored heterocysts in *Mastigocladus laminosus* Cohn. *Curr Sci* **26:** 254.

Vintila, S., and El-Shehawy, R. (2007) Ammonium ions inhibit nitrogen fixation but does not affect heterocyst frequency in the bloom-forming *Nodularia spumigena* strain AV1. *Microbiology* **153:** 3704–3712.

Voss, I., Diniz, S. C., Aboulmagd, E., and Steinbuchel, A. (2004) Identification of the *Anabaena* sp. strain PCC 7120 cyanophycin synthetase as a suitable enzyme for production of cyanophycin in Gram-negative bacteria like *Pseudomonas putida* and *Ralstonia eutropha*. *Biomolecules* **5:** 1588–1595.

Vyas, D., and Kumar, H. D. (1995) Nitrogen fixation and hydrogen uptake in four cyanobacteria. *Int J Hydrogen Energy* **20:** 163–168.

Walsby, A. E. (1985) The permeability of heterocysts to the gases nitrogen and oxygen. *Proc R Soc* B London **226:** 345–366.

Walsby, A. E. (2007) Cyanobacterial heterocysts: Terminal pores proposed as sites of gas exchange. *Trends in Microbiol* **15:** 340–349.

Walsby, A. E., and Nichols, B. W. (1969) Lipid composition of heterocysts. *Nature* (London) **221:** 673–674.

Wang, C. M., Ekman, M., and Bergman, B. (2004) Expression of cyanobacterial genes involved in heterocyst differentiation and dinitrogen fixation along a plant symbiosis development profile. *Mol Plant-Microbe Interactions* **17:** 438–443.

Wang, Y., and Xu, X. (2005) Regulation by *hetC* of genes required for heterocyst differentiation and cell division in *Anabaena* sp. strain PCC 7120. *J Bacteriol* **187:** 8489–8493.

Wang, Y., Lechno-Yossef, S., Gong, Y., Fan, Q., Wolk, C. P., and Xu, X. (2007) Predicted glycosyl transferase genes located outside the HEP island are required for formation of heterocyst envelope polysaccharide in *Anabaena* sp. strain PCC 7120. *J Bacteriol* **189:** 5372–5378.

Watsyn, M., Achatz, A., Molitor, V., and Peschek, G. A. (1988) Respiratory activities and aa_3-type cytochrome oxidase in plasma and thylakoid membranes from vegetative cells and heterocysts of the cyanobacterium *Anabaena* ATCC 29413. *Biochim Biophys Acta* **935:** 217–224.

Wealand, J. L., Myers, J. A., and Hirschberg, R. (1989) Changes in gene expression during nitrogen starvation in *Anabaena variabilis*. *J Bacteriol* **171:** 1309–1313.

Weare, N. M., and Benemann, J. R. (1973) Nitrogen fixation by *Anabaena cylindrica*. I. Localization of nitrogen fixation in the heterocysts. *Arch Microbiol* **90:** 323–332.

Wei, T.-F., Ramasubramanian, T. S., and Golden, J. W. (1994) *Anabaena* sp. strain PCC 7120 *ntcA* gene required for growth on nitrate and heterocyst development. *J Bacteriol* **176:** 4473–4482.

Wei, T.-F., Ramasubramanian, T. S., Pu, F., and Golden, J. W. (1993) *Anabaena* sp. strain PCC 7120 *bifA* gene encoding a sequence-specific DNA-binding protein cloned by *in vivo* transcriptional interference selection. *J Bacteriol* **175:** 4025–4035.

Weyman, P. D., Pratte, B., and Thiel, T. (2008) Transcription of *hupSL* in *Anabaena variabilis* ATCC 29413 is regulated by NtcA and not by hydrogen. *Appl Environ Microbiol* **74:** 2103–2110.

Whitton, B. A., and Peat, A. (1967) Heterocyst structure in *Chlorogloea fritschii*. *Arch Microbiol* **58:** 324–338.

Wilcox, M., Mitchison, G. J., and Smith, R. J. (1973a) Pattern formation in the blue-green alga, *Anabaena*. I. Basic mechanisms. *J Cell Sci* **13:** 707–723.

Wilcox, M., Mitchison, G. J., and Smith, R. J. (1973b) Pattern formation in the blue-green alga, *Anabaena*. II. Controlled proheterocyst regression. *J Cell Sci* **13:** 637–649.

Wildon, D. C., and Mercer, F. V. (1963a) The ultrastructure of the heterocyst and akinete of the blue-green algae. *Arch Microbiol* **47:** 19–31.

Wildon, D. C., and Mercer, F. V. (1963b) The ultrastructure of the vegetative cell of the blue-green algae. *Aus J Biol Sci* **16:** 585–596.

Wilson, G. A., Bertrand, N., Patel, Y., Hughes, J. B., Feil, E. J., and Field, D. (2005) Orphans as taxonomically restricted and ecologically important genes. *Microbiology* **151:** 2499–2501.

Winkenbach, F., and Wolk, C. P. (1973) Activities of enzymes of oxidative and reductive pentose phosphate pathways in heterocysts of a blue-green alga. *Plant Physiol* **52:** 480–483.

Winkenbach, F., Wolk, C. P. and Jost, M. (1972) Lipids of membranes and the cell envelope in heterocysts of a blue-green alga. *Planta* **107:** 69–80.

Wolk, C. P. (1965) Control of sporulation in a blue-green alga. *Develop Biol* **12:** 15–35.

Wolk, C. P. (1967) Physiological basis of the pattern of vegetative growth of a blue-green alga. *Proc Natl Acad Sci USA* **57:** 1246–1251.

Wolk, C. P. (1968) Movement of carbon from vegetative cells to heterocysts in *Anabaena cylindrica. J Bacteriol* **96:** 2138–2143.

Wolk, C. P. (1975) Differentiation and pattern formation in filamentous blue-green algae. In: Gerhardt, P., Costilow, R. N., and Sadoff. H. L. (Eds.) *Spores* IV. American Sciety for Microbiology, Washington, DC., USA. pp. 85–96.

Wolk, C. P. (1979) Intercellular interactions and pattern formation in filamentous cyanobacteria,. In: S. Subtelny, S., and Konigsberg, I. R. (Eds.) *Determinants of Spatial Organization*. Academic Press, Inc., New York, N.Y., USA. pp. 247–266.

Wolk, C. P. (1982) Heterocysts. In: Carr, N. G., and Whitton, B. A. (Eds.) *The Biology of the Cyanobacteria*. Blackwell Scientific Publications, Ltd., Oxford, UK. pp. 359–386.

Wolk, C. P. (1989) Alternative models for the development of the pattern of spaced heterocysts in *Anabaena* (Cyanophyta). *Plant Syst Evol* **164:** 27–31.

Wolk, C. P. (1996) Heterocyst formation. *Annu Rev Genet* **30:** 59–78.

Wolk, C.P. (2000) Heterocyst formation in *Anabaena*. In:. Brun, Y.V., and Shimkets, L.J. (Eds) *Prokaryotic Development*. American Society for Microbiology, Washington, DC., USA. pp. 83–104.

Wolk, C. P., and Schaffer, P. W. (1976) Heterotrophic micro- and macro-cultures of a nitrogen-fixing cyanobacterium. *Arch Microbiol* **110:** 145–147.

Wolk, C. P., and Simon, R. D. (1969) Pigments and lipids of heterocysts. *Planta* **86:** 92–97.

Wolk, C. P., and Quine, M. P. (1975) Formation of one dimensional patterns by stochastic processes and by filamentous blue-green algae. *Dev Biol* **46:** 370–382.

Wolk, C. P., and Wojciuch, E. (1971a) Photoreduction of acetylene by heterocysts. *Planta* **97:** 126–134.

Wolk, C. P., and Wojciuch, E. (1971b) Biphasic time course of solubilization of nitrogenase during cavitation of aerobically grown *Anabaena cylindrica. J Phycol* **7:** 339–344.

Wolk, C. P., Austin, S. M., Bortins, J., and Galonsky, A. (1974) Autoradigraphic localization of ^{13}N after fixation of ^{13}N-labelled nitrogen gas by a heterocyst forming blue-green alga. *J Cell Biol* **61:** 440–453.

Wolk, C. P., Cai, Y., and Panoff, J.-M. (1991) Use of transposon with luciferase as a reporter to identify environmentally responsive genes in a cyanobacterium. *Proc Natl Acad Sci USA* **88:** 5355–5359.

Wolk, C. P., Cai, Y., Cardemil, L., Flores, E., Hohn, B., Murry, M., Schmetterer, G., Schrautemeier, B., and Wilson, R. C. (1988) Isolation and complementation of mutants of *Anabaena* sp. strain PCC 7120 unable to grow aerobically on dinitrogen. *J Bacteriol* **170:** 1239–1244.

Wolk, C. P., Elhai, J., Kuritz, T., and Holland, D. (1993) Amplified expression of a transcriptional pattern formed during development of *Anabaena. Mol Microbiol* **7:** 441–445.

Wolk, C. P., Ernst, A., and Elhai, J. (1994) Heterocyst metabolism and development. In: *The Molecular Biology of Cyanobacteria*, 2nd edn. Bryant, D. A. (Ed.) Kluwer, Dordrecht, The Netherlands. pp. 769–823.

Wolk, C. P., Fan, Q., Zhou, R., Huang, G., Lechno-Yossef, S., Kuritz, T., and Wojciuch, E. (2007) Paired cloning vectors for complementation of mutations in the cyanobacterium *Anabaena* sp. strain PCC 7120. *Arch Microbiol* **188:** 551–563.

Wolk, C. P., Thomas, J., Shaffer, P. W., Austin, S. M., and Galonsky, A. (1976) Pathway of nitrogen metabolism after fixation of ^{13}N-labeled nitrogen gas by the cyanobacterium *Anabaena cylindrica. J Biol Chem* **251:** 5027–5034.

Wolk, C. P., Vonshak, A., Kehoe, P., and Elhai, J. (1984) Construction of shuttle vectors capable to conjugative transfer from *Escherichia coli* to nitrogen-fixing filamentous cyanobacteria. *Proc Natl Acad Sci USA* **81:** 1561–1565.

Wong, F. C., and Meeks, J. C. (2001) The *hetF* gene product is essential to heterocyst differentiation and affects HetR function in the cyanobacterium *Nostoc punctiforme. J Bacteriol* **183:** 2654–2661.

Wu, X., Lee, D. W., Mella, R. A., and Golden, J. W. (2007) The *Anabaena* sp. strain PCC 7120 *asr1734* gene encodes a negative regulator of heterocyst development. *Mol Microbiol* **64:** 782–794. doi: 10.1111/j.1365-2958.2007.05698.x

Wu, X., Liu, D., Lee, M. H., and Golden, J. W. (2004) *patS* minigenes inhibit heterocyst development of *Anabaena* sp. strain PCC 7120. *J Bacteriol* **186:** 6422–6429.

Wünschiers, R., Batur, M., and Lindblad, P. (2003) Presence and expression of hydrogenase specific C-terminal endopeptidases in cyanobacteria. *BMC Microbiol* **3:** 8.

Xiankong, Z., Tabita, F. R., and Van Baalen, C. (1984) Nickel control of hydrogen production and uptake in *Anabaena* spp. strains CA and 1F. *J Gen Microbiol* **130:** 1815–1818.

Xu, W. L., Jeanjean, R., Liu, Y. D., and Zhang, C. C. (2003) Pkn22 (*alr2502*) encoding a putative Ser/Thr kinase in the cyanobacterium *Anabaena* sp. PCC 7120 is induced by both iron starvation and oxidative stress and regulates the expression of *isiA*. *FEBS Lett* **553:** 179–182.

Xu, X., and Wolk, C. P. (2001) Role for *hetC* in the transition to a non-dividing state during heterocyst differentiation in *Anabaena* sp. *J Bacteriol* **183:** 393–396.

Yamanaka, G., and Glazer, A. N. (1983) Phycobiliproteins in *Anabaena* 7119 heterocysts. In: Papageorgiou, G. C., and Packer, L. (Eds.) *Photosynthetic Prokaryotes: Cell Differentiation and Function*. Elsevier, New York, N.Y., USA. pp. 69–90.

Ying, L., Huang, X., Huang, B., Xie, J., Zhao, J., and Zhao, X. S. (2002) Fluorescence emission and absorption spectra of single *Anabaena* sp. strain PCC 7120 cells. *Photochem Photobiol* **76:** 310–313.

Yoon, H.-S., and Golden, J. W. (1998). Heterocyst pattern formation controlled by a diffusible peptide. *Science* **282:** 935–938.

Yoon, H.-S., and Golden, J. W. (2001) PatS and products of nitrogen fixation control heterocyst pattern. *J Bacteriol* **183:** 2605–2613.

Yoon, H.-S., Lee, M. H., and Golden, J. W. (2003) *Anabaena* sp. strain PCC 7120 *hetY* gene influences heterocyst development. *J Bacteriol* **185:** 6995–7000.

Yoshimura, H., Okamoto, S., Tsumuraya, Y., and Ohmori, M. (2007) Group 3 sigma factor gene, *sigJ*, a key regulator of desiccation tolerance, regulates the synthesis of extracellular polysaccharide in cyanobacterium *Anabaena* sp. strain PCC 7120. *DNA Res* **14:** 13–24.

Yoshino, F., Ikeda, H., Masukawa, H., and Sakurai, H. (2007) High photobiological hydrogen production activity of a *Nostoc* sp. PCC 7422 uptake hydrogenase-deficient mutant with high nitrogenase activity. *Marine Biotechnol* **9:** 101–112.

Young-Robbins, S. S., Risser, D. D., Moran J. R., Haselkorn, R., and Callahan, S. M. (2010) Transcriptional regulation of the heterocyst patterning gene *patA* from *Anabaena* sp. strain PCC 7120. *J Bacteriol* **192:** 4732–4740. doi:10.1128/JB.00577-10

Zhang, C.-C. (1993) A gene encoding a protein related to eukaryotic protein kinases from the filamentous heterocystous cyanobacterium *Anabaena* PCC 7120. *Proc Natl Acad Sci USA* **90:** 11840–11844.

Zhang, C.-C., and Libs, L. (1998) Cloning and characterisation of the *pknD* gene encoding a eukaryotic-type protein kinase in the cyanobacterium *Anabaena* sp. PCC7120. *Mol Gen Genet* **258:** 26–33.

Zhang, C.-C., Friry, A., and Peng, L. (1998) Molecular and genetic analysis of two closely linked genes that encode, respectively, a protein phosphatase 1/2A/2B homolog and a protein kinase homolog in the cyanobacterium *Anabaena* sp. strain PCC 7120. *J Bacteriol* **180:** 2616–2622.

Zhang, C.-C., Huguenin, S., and Friry, A. (1995) Analysis of genes encoding the cell division protein FtsZ and a glutathione synthetase homologue in the cyanobacterium *Anabaena* sp. PCC 7120. *Res Microbiol* **146:** 445–455.

Zhang, C-C., Jang, J., Sakr, S. and Wang, L. (2005) Protein phosphorylation on Ser, Thr and Tyr residues in cyanobacteria. *J Mol Biol Biotechnol* **9:** 154–166. doi:19.1159/000089644

Zhang, C-C., Laurent, S., Sakr, S., Peng, L., and Bédu, S. (2006) Heterocyst differentiation and pattern formation in cyanobacteria: A chorus of signals. *Mol Microbiol* **59:** 367–375. doi.10.1111/j.1365-2958.2005.04979.x

Zhang, J. Y., Chen, W. L., and Zhang, C.-C. (2009) *hetR* and *patS*, two genes necessary for heterocyst pattern formation are wide spread in filamentous non-heterocyst-forming cyanobacteria. *Microbiology* **155:** 1418–1426. doi:10.1099/mic.0.027540-0

Zhang, L.-C., Chen, Y.-F., Chen, W.-L., and Zhang, C.-C. (2008) Existence of periplasmic barriers preventing green fluorescent protein diffusion from cell to cell in the cyanobacterium *Anabaena* sp. strain PCC 7120. *Mol Microbiol* **70:** 814–823.

Zhang, W., Du, Y., Khudyakov, I., Fan, Q., Gao, H., Ning, D., Wolk, C. P., and Xu, X. (2007) A gene cluster that regulates both heterocyst differentiation and pattern formation in *Anabaena* sp. strain PCC 7120. *Mol Microbiol* **66:** 1429–1443. doi: 1111/j.1365-2598.2007.05997.x

Zhang, X., Tabita, F. R., and Van Baalen, C. (1984) Nickel control of hydrogen production and uptake in *Anabaena* spp. strains CA and IF. *J Gen Microbiol* **130:** 1815–1818.

Zhang, X., Zhao, F., Guan, X., Yang, Y., Liang, C., and Qin, S. (2007) Genome-wide survey of putative serine/threonine protein kinases in cyanobacteria. *BMC Genomics* **8:** 395. doi: 10.1186/1471-2164-8-395

Zhang, Y., Pu, H., Wang, Q., Cheng, S., Zhao, W., Zhang, Y., and Zhao, J. (2007) PII is important in regulation of nitrogen metabolism but not required for heterocyst formation in the cyanobacterium *Anabaena* sp. PCC 7120. *J Biol Chem* **282:** 33641–33648.

Zhao, M.-X., Jiang, Y. L., He, Y.-X., Chen, Y.-F., Teng, Y.-B., Chen, Y., Zhang, C.-C., and Zhou, C.-Z. (2010) Structural basis for the allosteric control of the global transcription factor NtcA by the nitrogen starvation signal 2-oxoglutarate. *Proc Natl Acad Sci USA* **107:** 12487–12492. doi: 10.1073/pnas.100 1556107

Zhao, W., Ye, Z., and Zhao, J. (2007) RbrA, a cyanobacterial rubrerythrin, functions as a FNR-dependent peroxidase in heterocysts in protection of nitrogenase from damage by hydrogen peroxide in *Anabaena* sp. PCC 7120. *Mol Microbiol* **66:** 1219–1230. doi.10.1111/j.1365.2958.2007.05994.x

Zhao, Y., Shi, Y., Zhao, W., Huang, X., Wang, D., Brown, N., and Brand, J. (2005) CcbP, a calcium-binding protein from *Anabaena* sp. PCC 7120, provides evidence that calcium ions regulate heterocyst differentiation. *Proc Natl Acad Sci USA* **102:** 5744–5748. doi./10:1973/pnas.051782102

Zhou, R., and Wolk, C. P. (2003) A two-component system mediates developmental regulation of biosynthesis of a heterocyst polysaccharide. *J Biol Chem* **278:** 19939–19946.

Zhou, R., Wei, X., Jiang, N., Li, H., Dong, Y., His, K.-L., and Zhao, J. (1998) Evidence that HetR protein is an unusual serine-type protease. *Proc Natl Acad Sci USA* **95:** 4959–4963.

Zhu, J., Kong, R., and Wolk, C. P. (1998) Regulation of *hepA* of *Anabaena* sp. strain PCC 7120 by elements 5′ from the gene and by *hepK*. *J Bacteriol* **180:** 4233–4242.

Zhu, J., Jäger, K., Black, T., Zarka, K., Koksharova, O., and Wolk, C. P. (2001) HcwA, an autolysin, is required for heterocyst maturation in *Anabaena* sp. strain PCC 7120. *J Bacteriol* **183:** 6841–6851.

Zhu, M., Callahan, S. M., and Allen, J. S. (2010) Maintenance of heterocyst patterning in a filamentous cyanobacterium. *J Biol Dyn* **4:** 621–633.

Ziegler, K., Diener, A., Herpin, C., Richter, R., Deutzmann, R., and Lockau, W. (1998) Molecular characterization of cyanophycin synthetase, the enzyme catalyzing the biosynthesis of the cyanobacterial reserve material multi-L-arginyl-poly-L-aspartate (cyanophycin). *Eur J Biochem* **254:** 154–159.

Ziegler, K., Stephan, D. P., Pistorius, E. K., Ruppel, H. G., and Lockau, W. (2001) A mutant of the cyanobacterium *Anabaena variabilis* ATCC 29413 lacking cyanophycin synthetase. *FEMS Microbiol Lett* **196:** 13–18.

CHAPTER 5

MOVEMENTS

Motility confers on the microorganisms the ability to survive in their natural habitats. In order to cope up with environmental conditions of excess or deficiency of light/nutrients (or chemicals) the microorganisms exhibit necessary adjusting movements. The first organ of motility that attracted the attention of bacterial motility is flagella. Ever since the discovery of bacteria it has served as an important taxonomic trait to distinguish bacterial species. Motility due to flagella is present in both eubacteria and archaebacteria. The bacterial flagella are composed of three parts, a basal body, the hook and the filament. The basal body acts as the motor, the hook joins the filament to the basal body and the filament acts as the propeller. The filament is composed of many thousands of molecules of a single protein known as flagellin and assumes a diameter of 20 nm. There are certain subtle differences between the organization of the proteins of the motor of eubacteria and archaebacteria. Moreover, the polar and lateral flagella differ in their structure while the former is sheathed and thicker the latter is unsheathed and thinner. In spirochetes the flagella are unusually located internally in the periplasmic space. Type IV pili (Tfp), junctional pore complex (JPC), ratchet structure and contractile cytoskeleton are the other motility structures present in eubacteria (Bardy *et al.*, 2003). The movements associated with Tfp are known as twitching movements. *Pseudomonas aeruginosa* and *Neisseria gonorrhoeae* are the classical examples that exhibit twitching movements. The myxobacterium *Myxococcus xanthus* exhibits two types of gliding motion, i.e. social gliding and adventurous gliding. Social gliding is exhibited by groups of cells and adventurous gliding by single cells. It is interesting to note that social gliding is mediated by Tfp. Besides helically arranged protein fibrils, JPC is another organelle that helps *M. xanthus* in its adventurous gliding (Mattick, 2002; Nudleman and Kaiser, 2004).

A number of eubacteria and certain filamentous cyanobacteria exhibit surface-linked movements known as gliding movements (Castenholz, 1973, 1982; Hoiczyk, 2000; McBride, 2000, 2001; Adams, 2001). In the latter, helically arranged protein fibrils and JPC help in the gliding process by secreting mucilage (Wolgemuth *et al.*, 2002; Wolgemuth and Oster, 2004). Ratchet structure is a characteristic feature of *Cytophaga-Flavobacterium* group of bacteria that show gliding due to the movement of cell surface components. In these organisms, specific motility proteins are anchored to the cytoplasmic and outer membrane. Movements of the cytoplasmic proteins may be driven by the proton motive force and their interaction with the outer membrane proteins may propel the cells forward (Bardy *et al.*, 2003). Contractile cytoskeleton is a peculiar structure seen in *Spiroplasma melliferum* which exhibits motility in the absence of flagella and genes that regulate gliding. The cytoskeleton consists of an unique 59 kDa protein which is supposed to act as the linear motor that is internally attached to the cytoplasmic membrane.

Movements in response to light are classified into phototaxis, photokinesis and photophobic responses. Phototaxis is described as movement directed to the orientation of incident light while photokinesis is the speed of movement regulated by total light intensity. Photophobic response is the reversal of the direction of movement (Diehn *et al.*, 1979). Likewise, movements in response to chemical environment are known as chemotactic movements while those in relation to water are termed as hydrotactic movements. How do cyanobacteria sense and respond to light, their phototactic behaviour and how does light play a role in Tfp-dependent motility are the areas which received attention during the past two decades (Häder, 1987a,b; Bhaya *et al.*, 2001; Mullineaux, 2001; Armitage and Hellingwerf, 2003; Nakasugi *et al.*, 2006, 2007).

Bacterial chemotaxis has been investigated in very great detail with *Escherichia coli* emerging as the best example. There are three basic steps in bacterial chemotaxis. The first is signal perception by bacterial chemoreceptors located in the membrane. The second is signal transduction from chemoreceptors to the motor of the flagella and the third is the adaptation of the signal to desensitize the initial signal output (Lux and Shi, 2004). A chemoreceptor is a transmembrane protein that is divided into three portions, i.e. an extra-cytoplasmic ligand-binding portion and a transmembrane portion that connects to a cytoplasmic portion consisting of the signalling and methyl-accepting domains. That is why these are called as methyl-accepting chemoreceptor proteins (MCPs) or methyl-accepting chemotaxis proteins (Kort *et al.*, 1975). Located at the poles of the bacterial cell, MCPs are constitutive and are not involved in transport or metabolism. These are very sensitive detection devices that can detect concentrations of specific ligands at concentrations ranging from μM to nM range (Clarke and Koshland, 1979; Biemann and Koshland, 1994; Lin *et al.*, 1994). This is possible because the MCPs function as signalling lattices by forming initially stable homodimers and later these tend to become trimers of dimers (Lux and Shi, 2004). Signals perceived by MCPs are transmitted across the membrane which triggers an excitation response that is conveyed to two sensory proteins, i.e. CheA (a histidine kinase) and CheY (a response regulator). CheA interacts with the cytoplasmic signalling domain of MCP in association with CheW. As a result of which a tight ternary complex is produced. CheY, a response regulator phosphorylated by CheA, can bind to the motor of the flagella and affect a change in the direction of rotation and thus in the swimming direction (Armitage and Hellingwerf, 2003; Lux and Shi, 2004). These transmembrane chemoreceptors constitute nearly 90% of the total number of chemotaxis transducer molecues in *E. coli* (Hazelbauer and Harayama, 1983; Alexandre and Zhulin, 2001) suggesting that there are many ways to sense chemicals in bacteria.

I. TYPES OF MOVEMENTS

Three types of motility are noted in cyanobacteria, i.e. (A) Gliding (B) Swimming and (C) Twitching.

A) Gliding motility

i) Occurrence: Unicellular cyanobacteria are generally immotile with the exception of certain marine *Synechococcus* strains (Castenholz, 1973; Brock, 1978) and *Synechococcus* sp. strain PCC 6910 (Waterbury and Rippka, 1989) which exhibit gliding motility. Members of Oscillatoriaceae (species of *Oscillatoria*, *Lyngbya*, *Phormidium* and *Spirulina*) and Nostocaceae (*Anabaena* sp.) show gliding motion. Gliding is a slow uniform and highly coordinated movement that involves revolution around the long axis of the filaments during which the filaments do not show contractions or any sort of alterations in morphology. However, *Anabaena* sp. shows lateral and bending movements. The speed at which these organisms travel has been noted to be 10 μm s^{-1}. Another characteristic feature of gliding is that the trichomes leave behind tracts of mucilage as they move ahead (Hoiczyk, 2000).

ii) Structures associated with gliding

a) S-layer and Fibrils: Haften and Castenholz (1971) showed the existence of a "locomotory machinery" in *Oscillatoria princeps*. This is assembled in parallel fibrils with diameters of 5 to 8 nm which are aligned in a helical array just beneath the outer membrane. The arrangement of the fibrils and helical path of the filaments are together taken as an indication that these are associated with motility (Halfen, 1973). Studies on the envelope structure of four gliding filamentous cyanobacteria by freeze substitution, freeze fracturing and negative staining revealed a complex external surface layer consisting of two structural elements, i.e. a tetragonal crystalline S-layer attached to the outer membrane and a group of parallel, helically arranged surface fibrils (of 8 to 12 nm in diameter; with serrated appearance in cross section) spaced at 14 nm with each other on the top of the S-layer (Hoiczyk and Baumeister, 1995). Of the four cyanobacteria studied, the revolution of fibrils is in a clockwise direction in the two *Phormidium* sp. while they are in a counter-clock direction in *O. princeps* and *Lyngbya aeruginosa* (Hoiczyk and Baumeister, 1995). The serrated fibrils appear to correspond to the fibrils described earlier except for their location (Halfen, 1973, 1979). In contrast, Adams *et al.* (1999) reported the existence of a complex array of fibrils located between the peptidoglycan layer and the outer membrane of several *Oscillatoria* spp. The arrangement of each fibril in relation to the outer membrane and the peptidoglycan layer is such that the outer membrane makes contact with the peptidoglycan layer between each fibril (Adams *et al.*, 1999). Both the S-layer and fibrillar layer present above the outer membrane reported in the studies of Hoiczyk and Baumeister (1995) were not observed by Adams *et al.* (1999) in the *Oscillatoria* strains examined by them.

Considerable differences exist in literature on the diameter of the fibrils observed by various workers. The fibrils located on the surface of peptidoglycan layer reported by Halfen and Castenholz (1971) measured 5 to 8 nm in diameter (Halfen, 1973, 1979) whereas those observed by Adams *et al.* (1999) were 25 to 30 nm in diameter. However, the observations of Hoiczyk and Baumeister (1995) closely corresponded with those of Halfen and Castenholz (1971) who reported a diameter of 8 to

12 nm. The angle between the fibrils and the long axis of the filaments in all the above studies was noted to be 25 to 30°. It is suggested that the fibrils help the filaments in rotating them about their long axis as they glide. The motor or driving force for gliding is presumed to be provided by the angle of the fibrils and the rotation of the filaments and the direction of rotation is characteristic of the species.

The fibrils are made up of a glycoprotein known as oscillin consisting of 646 amino acid residues. The N-terminal region (of 554 amino acid residues) comprises of 46 repeats of a calcium-binding motif followed by a 92 residue C-terminal domain. They suggested that oscillin is essential for gliding on the basis of helically arranged fibrils, an S-layer underlying the oscillin fibrils and the secreted slime. The thrust is said to be generated by an interaction between the extruded slime, the surface of the cell and the solid surface (Hoiczyk and Baumeister, 1995). However, it is pertinent to point out that oscillin does not show any similarity with any other known motor proteins but shows similarity with SwmA protein identified on the surface of a swimming *Synechococcus* and HylA, another glycoprotein identified from non-motile filamentous cyanobacterium *Anabaena* sp. strain PCC 7120 (Brahamsha, 1996).

***b) Structure of JPCs*:** The existence of fine pores of 14–16 nm in diameter grouped in a circumference around the septa of at least two dozen species of cyanobacteria has been reported earlier (Guglielmi and Cohen-Bazire, 1982). Due to the fact that these pores did not traverse through the cyanobacterial walls, no definite role in gliding could be ascribed to them. Hoiczyk and Baumeister (1998) unequivocally demonstrated through light and electron microscopy that these pores are in fact part of a larger structure designated as JPC. The JPC is an organelle having a diameter of 70–80 nm and a length of 32 nm that spans the peptidiglycan and outer membrane. The JPC is cylindrical with a central bulge of 14 nm in diameter and two pores of 8 nm in diameter on either side. The JPCs were further shown to encircle the cell and are present near the septa in both *Phormidium uncinatum* and *Anabaena variabilis.* However, in the former the pores are aligned in a single row while in the latter several rows of pores are present on either side of the septum. Another important feature is the angle with which the JPCs are inclined relative to the cell axis. It is suggested that 30–40° angle provides the necessary direction to the slime extrusion which helps in propelling the cells forward.

***iii) Mechanism of gliding*:** Two models were put forward to explain the mechanism of gilding. *(a) Surface waves*: Halfen and Castenholz (1970, 1971) proposed that the driving force for gliding may be provided by the rhythmic distortion of the protein fibrils present in the cell wall leading to the formation of a wave over the surface of the filament that extends from one end to the other. The waves transmitted through the outer membrane interact with the substratum and thus move the filament forward, i.e. in the opposite direction. The propagation of the wave in the reverse direction results in a movement to the backward direction.

b) Extrusion of slime: The JPCs steadily secrete mucilage which flows in tight contact with the surface of the filament that makes it adhere to the surface. This adhesion and continuous secretion of mucilage finally causes locomotion of the filament. The change in the direction of slime extrusion explains the reversal in the direction of gliding. The helically arranged oscillin fibrils help in rotating the filament and the direction of rotation is determined by the orientation of the fibrils. Hoiczyk and Baumeister (1955, 1998) put forward a hypothesis that the surface striations formed by the protein fibrils act as channels for the extruded slime to flow along the filament.

This model has raised more questions than has provided answers for gliding motion. The following questions require consideration: whether (i) there is any synchronization in the JPCs of a filament? (ii) secretion of mucilage by JPCs is so forceful enough that it can provide sufficient force to move a filament? and (iii) slime secretion is the basis for movement or is it extruded as a consequence of motility? However, most of the corollary evidences point towards the following: (i) because of slime extrusion the gliding motion occurs as supported by the relationship between the speed at which the secretion of mucilage in *Synechococcus* takes place; (ii) the direction of mucilage secretion and locomotion are opposite to each other and (iii) mucilage secretion is able to provide necessary power for gliding. The mechanism of gliding by bacterial and cyanobacterial cells by slime extrusion model was explained by Jeon and Dobrynin (2005) who performed molecular dynamics simulations by compressing polymer chains through a molecular nozzle. The force with which the polymer chains are extruded through the molecular nozzle provides the force sufficient for propulsion.

B) Swimming

Waterbury *et al.* (1985) discovered that a marine species of *Synechococcus* sp. strain WH8102 could swim in the absence of any specific organs of motility unlike other eubacteria which are able to swim by means of flagella. The swimming speeds ranged from 5 to 25 μm s^{-1} (Waterbury *et al.*, 1985; Willey, 1988). During swimming the cells rotate about their longitudinal axis as they move and so generate both torque and thrust. Willey (1988) further observed that the swimming activity slows down and the cells become immotile due to the increase in the viscosity of medium. A number of techniques (transmission electron microscopy, freeze-fracture, freeze-etching, high intensity dark-field microscopy and motility-dependent amplitude spectra) employed were not helpful in unraveling any of the structures associated with swimming in *Synechococcus* sp. strain WH8102 (Waterbury *et al.*, 1985; Willey, 1988).

The swimming motility has been attributed due to the presence of a glycoprotein situated on the cell surface (constituting the S-layer) which is not an integral outer membrane protein but can easily be detached from cells by treatment with EDTA. This glycoprotein, designated as SwmA, has a molecular weight of 130 kDa (Brahamsha, 1996). SwmA is a glycosylated polypeptide consisting of 835 amino acid residues and contains repeats of Gly and Asp that serve as calcium-binding motifs (Brahamsha, 1999). There are a number of similarities between SwmA, oscillin and HylA (Brahamsha, 1996) in having multiple repeats of calcium-binding motifs in the N-terminal region and are homologous to the 47 amino acid domain at the C-terminal region without calcium-binding motifs (Hoiczyk and Baumeister, 1997). A functional relationship between SwmA and oscillin has also been indicated. While oscillin is shown to be associated with generating thrust for gliding with the help of slime produced on solid surface, SwmA alters the cell's surface characteristics or shape in such a way that rotation of cell results in thrust. The mechanism of swimming by *Synechococcus* is not clearly understood due to lack of knowledge on: (1) the number of genes and their gene products, (2) the presence of proteins other than SwmA and (3) the factors that generate torque and the signal transduction pathway. A beginning in this direction has been made by the isolation of a mutant strain, i.e. *Synechococcus* sp. strain S1A1, after insertional inactivation of *swmA* gene. It possesses major and minor cell surface polypeptides found in the wild-type and is shown to be deficient only in SwmA protein due to which it lost motility along with the outer S-layer (McCarren *et al.*, 2005). In addition, they noted fibrillar structures intervening the region between the S-layer and outer membrane. The presence of S-layers is also reported in other gliding cyanobacteria as discussed above (Hoiczyk and Baumeister, 1995; Hoiczyk and Hansel, 2000; Smarda *et al.*, 2002). The fibrillar structures that characterize the space in between the S-layer and the outer membrane in *Synechococcus* sp. strain WH8102 (McCarren *et al.*, 2005) also are reported to be present in marine *Synechococcus* isolates

(P-11-16 and P-11-17) (Perkins *et al.*, 1981) and a freshwater strain of *Synechococcus* (Schultze-Lam and Beveridge, 1994) which lack motility.

It is significant to note that Samuel *et al.* (2001) employed cryopreservation and freeze-substitution techniques to study the envelope structure of *Synechococcus* sp. strain WH8113. The ultrastructure studies of the cell envelope revealed an S-layer (of approximately 35 nm thick), outer membrane (10 nm thick), a peptidoglycan layer (15 nm thick) and cell membrane (10 nm thick). The presence of profusely large number of spicules (as long as 150 nm at a distance of 12 to 24 nm), with a uniform thickness of 5 nm, have been demonstrated by them (Fig. 1a,b). The radius of the spicules is found to be approximately 30 nm. The spicules are suggested to extend through channels in the surface layer and pass through the outer membrane, peptidoglycan layer and the cell membrane. The deprivation of calcium ions caused a disruption not only in the integrity of the spicules but also resulted in a separation of the thylakoid membrane (Fig. 2). The structural arrangement of spicules has been correlated to their function. Thus *Synechococcus* sp. strain WH8113 could swim by oscillating its surface (obviously by the rowing motion caused by the spicules) in the form of a travelling wave. The power for motility is derived from the ion-motive force across the cell membrane as suggested by Willey *et al.* (1987).

The genome of marine *Synechococcus* sp. strain WH8102 has been sequenced and analysis of the 2.4 mega base genome revealed three major portions, i.e. (i) a highly conserved core (consisting of nearly half the genome) associated with regions or gene sequences of marine adaptations (i.e. Ni-dependent superoxide dismutase, carboxysomes and DNA repair); (ii) marine adaptations (Na-dependent transporters) and (iii) a portion of genome unique to strain WH8102 consisting of gene sequences governing motility, phage-related regions, efflux transporters, glycosyltransferases, nitrate reduction, phycobilisomes and carbonic anhydrase (Palenik *et al.*, 2003). Although *Synechococcus* sp strain WH8102 does not possess pili or surface associated twitching motility, it is interesting to note that the genome contains six pil-like genes. Of these at least *pilB*, *pilC* and *pilD* have been identified. However, these do not encode the full complement of proteins required for pilus assembly and function (Palenik *et al.*, 2003). Genes required for motility are found in at least two widely separated regions, i.e. *Swm A* and *SwmB*. Of these two, the latter is found to be a very large ORF for 10,791 amino acids that is the longest ORF ever reported and constitutes more than 1% of the genome size.

***i) Mechanism of swimming*:** There are two important steps in swimming. First is generation of thrust and second is generation of torque. SwmA is shown to be required for the generation of thrust but not torque (Brahamsha, 1996). The coupling of the generation of thrust and torque is required for swimming but torque generation still remains unanswered. Currently two models are in vogue to describe the mechanism of swimming. These are (a) self-electrophoresis model and (b) wave generation model (Brahamsha, 1999).

a) *Self-electrophoresis model*: The cells are supposed to carry a fixed charge on their surface that is shielded by counter ions in the medium. If the cells are able to pump in ions at one end and pump them out at the other end, an electric force will be set up in the external medium. The external fluid layer containing the counter ions are driven over the surface of cells just like the treads of a tank that enables the cell to move in the opposite direction to the movement of external layer. Pitta and Berg (1995) ruled out such a mechanism as the electrophoretic mobility of *Synechococcus* cells is essentially zero.

b) *Wave generation model*: Ehlers *et al.* (1996) proposed this model by envisaging the generation of longitudinal or transverse waves of 0.2 μm long, 0.02 μm of amplitude and travelling at 160 μm s^{-1}

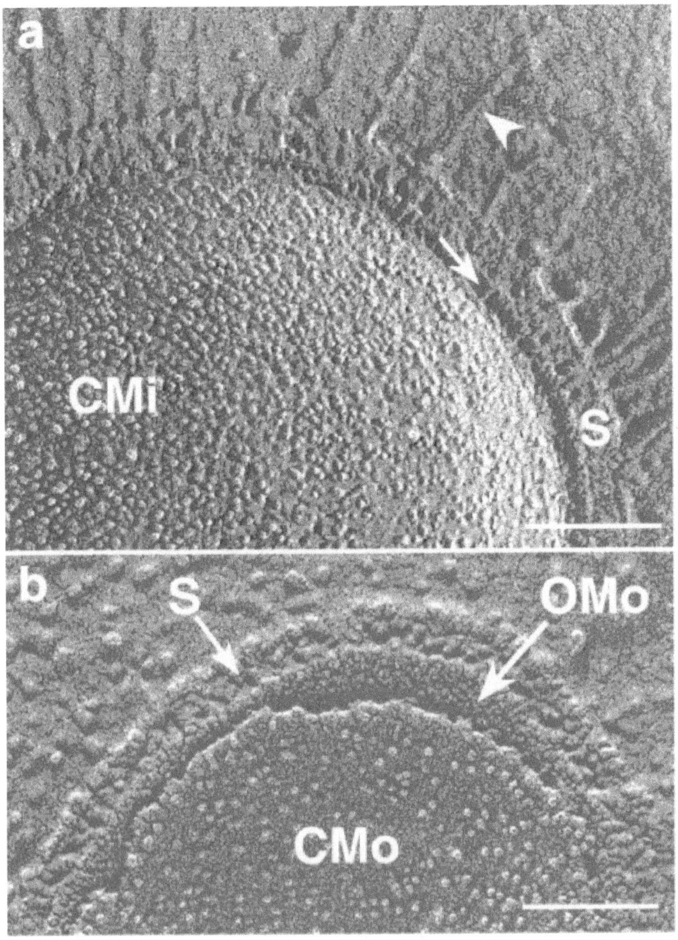

Figure 1: (a) Membrane fracture uncovering the inner cell membrane (CMi of *Synechococcus* sp. strain WH8113) showing a dense distribution of intramembrane particles. Fracture plane then crosses the surface layer (S) into the surrounding medium showing spicules (arrowhead) that extend from the cell surface. Arrow shows fiber extending from outer membrane to cell membrane. Scale bar, 100 nm. (b) *Synechococcus* (strain WH8113). Complementary fracture plane showing the outer leaflet of the cell membrane (CMo) which has fewer intramembrane particles than the inner leaflet. The fracture then crosses to the outer leaflet of the outer membrane (OMo), and then turns to fracture across the surface layer (S). Scale bar, 100 nm. With the kind permission of A.D.T. Samuel, Rowland Institute for Science, Cambridge, Massachusetts, USA and Department of Molecular & Cell Biology, Harvard University, Cambridge, Massachusetts, USA [Samuel et al. (2001) *BMC Microbiol.* **1:** 4. doi: 10.1186/1471-2180-1-4].

would allow the cell to swim at speeds of 25 µm s^{-1}. Wave generation is explained by the expansion and contraction of the local regions of the outer membrane producing thrust in such a manner that the cell moves in the direction of the wave (Brahamsha, 1999). There are no experimental evidences presented in favour of this model. However, the energy required for swimming has been suggested to be available either through sodium pump operating in the membrane or due to the existence of calcium channels. Willey *et al.* (1987) reported a definite requirement of sodium for swimming in *Synechococcus* sp. strain WH8113 as the swimming was directly proportional to the concentration of sodium in the medium and it was shown that concentrations below 10 mM, the organism was immotile. Sodium motive force exhibited a direct correlation with external concentration of sodium while proton motive force, electrical potential, the proton diffusion gradient and sodium diffusion

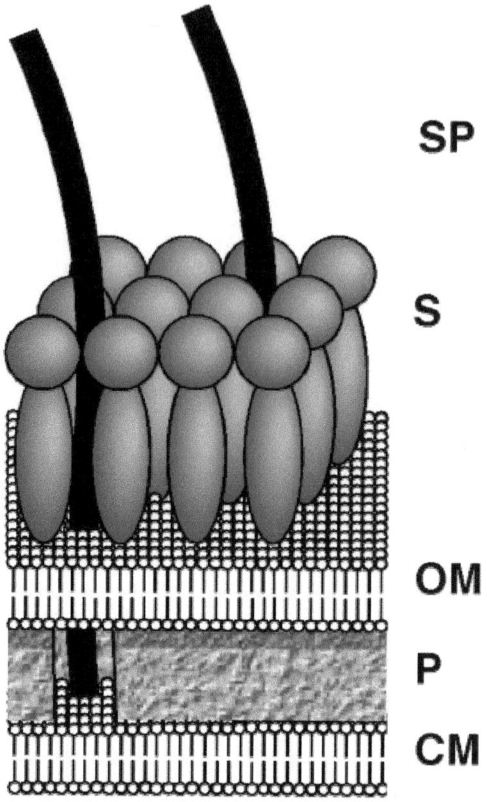

Figure 2: Diagram imaging the present knowledge of the cell envelope structure of *Synechococcus*. Present observations taken together suggest that the spicules (SP) extend through the surface layer (S) and outer membrane (OM) to contact the cell membrane (CM) (as shown in the cutaway of the peptidoglycan layer (P)).With the kind permission of A.D.T. Samuel, Rowland Institute for Science, Cambridge, Massachusetts, USA and Department of Molecular & Cell Biology, Harvard University, Cambridge, Massachusetts, USA [Samuel *et al.* (2001) *BMC Microbiol.* **1:** 4. doi: 10.1186/1471-2180-1-4].

gradient did not show such a correlation suggesting that the sodium motive force provides the direct source of energy for swimming in this organism. On the basis of the requirement of calcium for swimming, Pitta *et al.* (1997) concluded that the cell surface experiences longitudinal or transverse waves and the wave motion is suggested to be coordinated or driven by the changes in calcium concentration. Calcium depolarization is suggested to cause a local swelling of the cell surface which is responsible for the production of waves (Ehlers *et al.*, 1996). Whether it is the sodium motive force or calcium that provides the energy for thrust, according to Samuel *et al.* (2001) the spicules noted by them in the cell envelope of *Synechococcus* sp. strain WH8113 are well positioned to transduce energy at the cell membrane into mechanical work at the cell surface.

C) Twitching movements

The presence of pili or fimbriae on the cell surface is a characteristic feature of a number of eubacteria. The pili present on *E. coli* have been classified into four categories, i.e. Type I, II, III and IV on the basis of their diameter, number and distribution (Hayes, 1968). In cyanobacteria, the presence of pili-like structures on the cell surfaces of certain unicellular and filamentous forms has been demonstrated

(Lounatmaa *et al.*, 1980; Dick and Stewart, 1980; Vaara, 1982; Vaara *et al.*, 1984; Vaara and Vaara, 1988). But their formation, function and mechanisms of cellular and environmental controls have not been clearly understood. That is why a few of the cyanobacteria that exhibit gliding motility have been re-examined for the presence of flagella or pili. Amongst these, *P. uncinatum* and the marine *Synechococcus* species did not reveal any such appendages (Waterbury *et al.*, 1985: Hoiczyk and Baumeister, 1995, 1997).

Synechocystis sp. strain PCC 6803 possesses peritrichous pili of two distinct types, i.e. thick pili, (i.e. Tfp) having a diameter of 6–8 A° and a variable length of 2.5 µm, are required for twitching and thin pili (with a diameter of 3–4 A°), shorter than Tfp and covering the entire cell surface, are not required for motility (Bhaya *et al.*, 1999; Fig. 3a,b). The twitching motility of *Synechocystis* sp. strain PCC 6803 is akin to the social gliding shown by *M. xanthus*. When plated on agar plates, *Synechocystis* sp. PCC 6803 forms irregular, flat and spreading colonies and the cells exhibit phototactic movement en masse, whereas the mutant cells of this organism form dome-shaped and circular colonies that remain at the same place where they are seeded.

Tfp are composed of pilin sub-units (Sastry *et al.*, 1983). Each sub-unit is synthesized as a precursor which undergoes cleavage by a specific pre-pilin peptidase that cuts off the N-terminal part. This part of the mature pilin is highly conserved while the C-terminal part is not (Sastry *et al.*, 1985). Although the arrangement of pilin sub-units of Tfp in *Synechocystis* sp. strain PCC 6803 is not known, in *N. gonorrhoeae* it is composed of five pilin monomers per helical turn with a rise of approximately 41 Å per turn (Mattick, 2002). Calcium-dependent motility of *Synechocystis* sp. strain PCC 6803 was visualized by computer-aided video-microscopy (Moon *et al.*, 2004). On the basis of chelation studies and the use of calcium channel inhibitor (pimozide), calcium efflux blocker (orthovanadate), calcium ionophore (A23187) and calmodulin antagonists (trifluoroperazine and chlorpromazine), they concluded that Ca^{2+} plays a significant role in regulating photomovement.

Twitching movements take place in short intermittent jerks of up to several micrometers. The cells require a very moist surface and also cells have to exist in groups. Twitching is due to the presence of Tfp. The retraction of the pili provides the necessary driving force for twitching. Besides twitching, Tfp also are helpful in transformation, infection and adhesion (McBride, 2000, 2001). At least 12 Tfp components have been shown to affect transformation of *Synechocystis* sp. strain PCC 6803. Of these, PilA1 and PilT1 exert profound influence since *pilA1* mutants are deficient in Tfp on their cell surface while those of *pilT1* are hyperpiliated (Yoshihara *et al.*, 2001; Okamoto and Ohmori, 2002; Nakasugi and Neilan, 2005; Nakasugi *et al.*, 2006). Gene *slr0388* of *Synechocystis* sp. strain PCC 6803, annotated earlier under hypothetical proteins, has now been identified as a *comF* gene that regulates natural transformation, phototactic motility and piliation. This constitutes the second such gene that regulates transformation competency in *Synechocystis* sp. strain PCC 6803, the first one being *comA* (Yoshihara *et al.*, 2001). Gene disruption mutants of *comF* synthesized abundant PilA1 protein while at the same time the transcript and corresponding protein levels of PilT1 and ComA remained unaffected. The mutants exhibited a clumping phenotype and lost phototactic motility as well as transformation potential (Nakasugi *et al.*, 2006). However, the essentiality of another gene (*ssr3341*) for motility and natural transforming ability of *Synechocystis* sp. stain PCC 6803 has been demonstrated. This gene encodes an RNA chaperone (Hfq) which is reported to have multiregulatory roles in improving functional efficiency and stability of a number of small regulatory RNAs and overall translational performance. Inactivation of this gene caused loss of both phototactic and transformation abilities of the mutant and DNA microarray analysis of this mutant showed that the expression of a set of genes (*slr1667*, *slr1668*, *slr2015*, *slr2016* and *slr2018*), involved in motility, is greatly affected. All these genes in turn appear to be controlled by the cAMP-receptor protein (CRP) of *Synechocystis* sp. strain PCC 6803, SYCRP1 (Dienst *et al.*, 2008).

CRPs are global transcriptional regulatory proteins that bind to sequence-specific promoter regions of the genes under their control. Hedger *et al.* (2009) identified additional target genes for SYCRP1 of *Synechocystis* sp. strain PCC 6803 and the transcription of these genes, i.e. *slr1351* (mufF), *sll1874* ($ChlA_{II}$), *sll1708* (NarL) and *slr0442* is stimulated by illumination. According to them the binding of SYCRP1 to the promoter regions is similar to the binding of CRPs to Class I and Class II promoter regions of *E. coli* and so a similar mechanism of transcriptional activation is envisaged. *P. aeruginosa* is the best studied system to understand twitching motility mediated by Tfp. Genetic studies have established that there are at least 40 genes involved (McBride, 2001). The genes concerned with pilus biogenesis are designated as *pil* genes. Gene products PilU and PilT1 provide energy for retraction of Tfp while PilB provides energy essential for assembly of pilin sub-units. The pilin protein PilA1 which forms the basic structure of Tfp is synthesized by *pilA1* gene. PilD is required for processing of pre-pilin which serves as a bifunctional enzyme for N-terminal processing of prepilin and for methylation of N-terminal amino acid of the mature protein. PilE, PilV, FimT and FimU are other pre-pilin like proteins which are also processed by PilD. The functions of these prepilins are not known. PilQ is required to allow Tfp to cross the outer membrane. This protein forms highly stable complexes of 12–14 subunits with central channels that range from 5 to 10 nm in diameter.

***i) cAMP and twitching motility*:** cAMP is known as an important signalling molecule in prokaryotes as well as eukaryotes. Cellular cAMP levels are known to change in response to changes in environmental conditions such as light-dark, low pH-high pH and oxic-anoxic states (Ohmori *et al.*, 1988; Ohmori, 1989) and in case of nitrogen deficiency and sufficiency (Hood *et al.*, 1979). The addition of cAMP to cells of *Spirulina platensis* stimulated respiration and gliding movement (Ohmori *et al.*, 1992). The biosynthesis of cAMP is mediated by the enzyme adenylcyclase (Cya). Of the six classes of Cyas known, only Class III type is the universal class widely distributed among prokaryotes and eukaryotes. Terauchi and Ohmori (1999) identified two genes, *Cya1* (*slr1991*) and *Cya2* (sll0646) that encode Cyas in *Synechocystis* sp. strain PCC 6803. These bear homology to the eukaryotic Cyas and belong to the Class III . The disruption of *Cya1* gene resulted in a simultaneous reduction in cAMP levels with simultaneous loss of motility. The addition of extracellular cAMP regained motility of *Synechocystis* sp. strain PCC 6803. SYCRP1 binds to the promoter regions of genes under its control in presence of cAMP in *Synechocystis* sp. strain PCC 6803 (Yoshimura *et al.*, 2000). Disruption of the gene for cAMP receptor protein (SYCRP1, *sll1371) in Synechocystis* sp. strain PCC 6803 resulted in mutants that are non-motile with reduced Tfp on the cell surface signifying that cAMP-SYCRP1 complex controls the biogenesis of Tfp (Yoshimura *et al.*, 2002). On the contrary, a re-examination of the motility behaviour of the *cya1* and *sycrp1* gene disruptant mutants of *Synechocystis* sp. strain PCC 6803 suggested that the mutants exhibited motility and phototactic behavior. But these mutants seem to be impaired in one particular phase of phototaxis (Bhaya *et al.*, 2006). Blue light (450 nm) markedly increased cAMP content in the cells of *Synechocystis* sp. strain PCC 6803 while red (630 nm) and far red (720 nm) lights did not. Disruption of *cya1* gene encoding adenylycyclase resulted in no appreciable effect on cellular cAMP level even in blue light. Accordingly, blue light stimulated motility in wild-type while in *cya1* mutant cells blue light did not stimulate motility (Terauchi and Ohmori, 2004). They thus concluded that a blue light-cAMP signal transduction system is operative in the motility of *Synechocystis* sp. strain PCC 6803.

***ii) pil* Genes:** Genome sequencing of *Synechocystis* sp. strain PCC 6803 (Kaneko *et al.*, 1996) facilitated the identification of a number of genes involved in motility and phototaxis. By undertaking

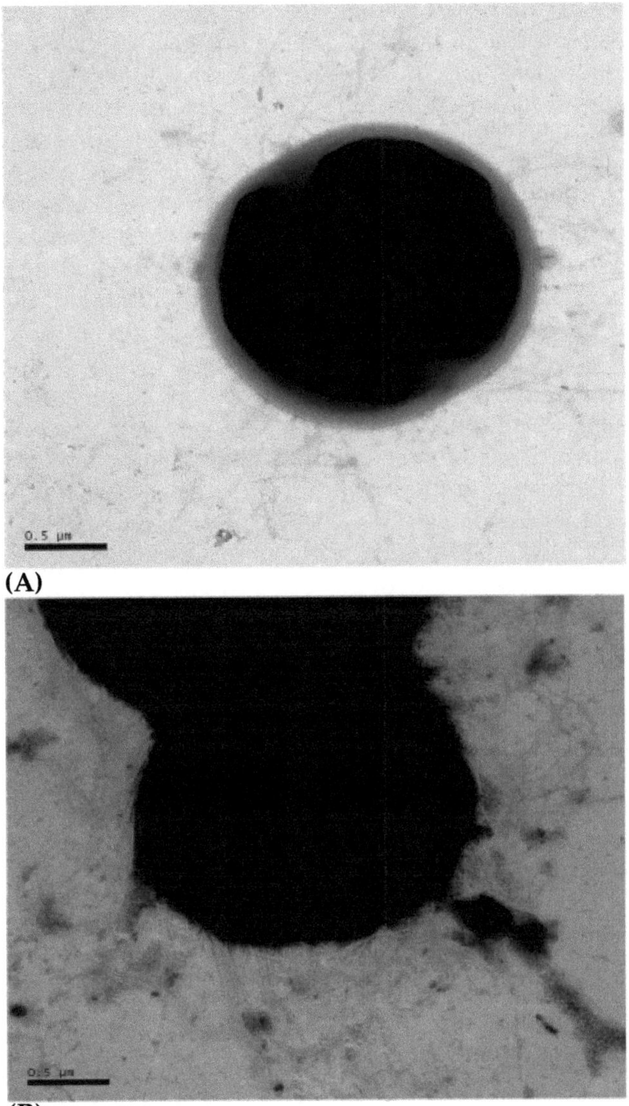

(A)

(B)

Figure 3: Type IV pili *of Synechocystis* sp. strain PCC 6803. Pictures kind courtesy of Professor D. Dienst, Humboldt University Berlin, Institute of Biology/Plant Biochemistry, Chausseestr. 117, 10115 Berlin/Germany.

mutational analysis of genes governing pilus biogenesis, motility and transformation, a number of workers identified their role. Okamoto and Ohmori (1999) presented evidences for the requirement of *pilT/pilC* gene cluster for motility and transformation competency in *Synechocystis* sp. strain PCC 6803. The role of an alternative sigma factor (SigF) in motility and pilus formation in *Synechocystis* sp. strain PCC 6803 was studied by Bhaya *et al.* (1999). Mutation of SigF conferred a pleiotropic effect on the cells such as (i) accumulation of extracellular pigments and proteins, (ii) the loss of pili leading to loss of motility and so phototaxis and (iii) a significant reduction in the levels of mRNA of *pilA1*

A2 genes governing subunits of pili. Amongst extracellular proteins, a glycoprotein designated as *HylA,* similar to motility related proteins such as oscillin and SwmA, has been identified. Of the six genes governing prepilins, at least *pilA1* and *pilA2* are known to encode proteins with greatest similarity to bacterial (*Pseudomonas* and *Myxococcus*) PilA proteins. It was concluded that (i) the directional motility of *Synechocystis* depends upon the activity of Tfp; (ii) the cells of *Synechocystis* possess different types of pili based on the presence of several components and homologues between Type II and Type IV pili and (iii) SigF controls the *pil* genes. These workers envisaged a larger role for SigF in the adaptation of cyanobacteria to various environmental stresses by its control of cell surface characteristics and pili-dependent motility. Bhaya *et al.* (2000) identified the genes involved in Tfp biogenesis in *Synechocystis sp.* strain PCC 6803. The major subunit of thick pilus is encoded by gene *pilA1* as evidenced by the fact that mutants of *pilA1* are devoid of Tfp or the thick pili. Genes *pilC* and *pilD* govern pilus biogenesis as mutants of these lack both thick and thin pili and are immotile. This signifies that both *pilC* and *pilD* genes are required for assembly of both types of pili. Gene *pilT1* appears to regulate pilus retraction while *pilT2* appears to be involved in phototaxis. This is evident from the mutants of *pilT1* which are immotile but show hyperpiliation and those of *pilT2* exhibit motility but are defective in phototaxis. Yoshihara *et al.* (2001) identified genes *pilA1, pilB1, pilM, pilN, pilO* and *pilQ* that are required together for motility and transformation competency in *Synechocystis* sp. strain PCC 6803. DNA microarray analysis of the target gene for SYCRP1 of *Synechocystis* sp. strain PCC 6803 revealed that of the four ORFs (*slr2015, slr2016, slr2017*and *slr2018*) only the gene products of the first three exhibited homology to the Tfp pre-pilin from *P. aeruginosa* and *M. xanthus* (Yoshimura *et al.*, 2002). On the basis these studies, ORFs *slr2015, slr2016* and *slr2017* were designated as *pilA9, pilA10* and *pilA11*, respectively. It is to be noted that *pilA1* to *pilA8* genes were assigned previously by Yoshihara *et al.* (2001). On the other hand, disruptants for ORF of *slr2018* exhibited a non-motile phenotype similar to *pil10* and *pil11* disruptants reported by Bhaya *et al.* (2001). Yoshimura *et al.* (2002) concluded that not only *pilA1* but also other putative *pil* operon products or their combination may be required for the structural components of Tfp and cell motility.

Proteomic characterization of membrane proteins of *Synechocystis* sp. strain PCC 6803 revealed that PilQ protein (*Slr1277*) is located in the outer membrane while PilM, PilN, PilO and PilA1 proteins are located in the plasma membrane (Norling *et al.*, 1998; Huang *et al.*, 2002, 2004). PilQ belongs to a secretin family which is responsible for the formation of a pore for the transport of the pilus across the membrane (Yoshihara and Ikeuchi, 2004).

Okamoto and Ohmori (2002) determined the biochemical properties of pilT protein after its purification from *E. coli* cells in which cyanobacterial *pilT* gene was expressed. The *pilT* gene product also exhibited ATPase activity signifying that *pilT* is responsible for the generation of the thrust required for motility and the competence in transformation. A mutant of *Synechocystis* deficient in *pilT* exhibited a pleiotropic behaviour by the loss of cell motility, increased length of surface pili and the loss of competence in transformation.

Nakasugi and Neilan (2005) detected the presence of Tfp on the cell surface of the toxic *Microcystis aeruginosa* PCC 7806 who identified four genes (*pilA, pilB, pilC* and *pilT*) governing their formation. These genes are similar to the already known sequences found in *Synechocystis* sp. strain PCC 6803 and other bacteria. Due to the presence of *pilT* genes sequences in the non-toxic strains of *M. aeruginosa,* they concluded that lateral gene transfer might be taking place from toxigenic strains to non-toxic strains through Tfp-mediated transformation.

To sum up, motility genes in *Synechocystis* sp. strain PCC 6803 can be grouped into three types: (i) primary motility genes known as *pil* genes that are essential for pilus biogenesis and assembly; (ii) secondary genes that are involved in motility and (iii) *pix* genes that govern positive phototaxis

(Yoshihara *et al.*, 2002). The primary motility genes are, i.e. *pilA1, pilB1, pilC, pilD, pilM, pilN, pilO, pilQ* and *pilT* which are also required for transformation competency (Bhaya *et al.*, 2000; Yoshihara *et al.*, 2001). There are eleven *pilA*-like genes that encode pre-pilins. Thick and thin pilus biogenesis is governed by *pilA1*. *PilA9, pilA10* and *pilA11* are responsible for cell motility. The other *pilA*-like genes are not essential for motility or phototaxis (Yoshihara *et al.*, 2001). Except in case of *pilT1* the rest of the mutants completely lost the pili from their surface. Exceptionally *pilT1* mutant exhibited enhanced assembly of the pili (Okamoto and Ohmori, 1999; Bhaya *et al.*, 2000). The expression of *pilA, pilB, pilC* and *pilT* genes of *M. aeruginosa* PCC 7806 has been studied and purified PilT protein showed ATPase activity. Heterologous expression of *pilT* gene, from *M. aeruginosa* PCC 7806, could complement the *pilT* mutant phenotype of *Pseudomonas aeruginosa* but not that of *Synechocystis* sp. strain PCC 6803 (Nakasugi *et al.*, 2007).

The secondary genes are identified to be Ser/Thr protein phosphatase genes and Ser/Thr protein kinase genes that are not linked to pilus assembly. Kamei *et al.* (1998) identified a novel gene involved in high-light resistance in the cyanobacterium *Synechocystis* sp. strain PCC 6803 and showed that this gene sequence *slr2031* plays a crucial role in the motility of this organism. A eukaryote-type protein kinase designated as SpkA (Synechocystis protein kinase) gene is shown to regulate motility in *Synechocystis* via phosphorylation of the membrane proteins. This is evidenced by the isolation of the SpkA-disrupted mutant lacking motility (Kamei *et al.*, 2001). The molecular mechanism by which the protein phosphorylation regulates motility in *Synechocystis* sp. strain PCC 6803 is not yet clear. The third type of genes governing positive phototaxis include a gene cluster designated as *pix* genes, i.e. *pixGHIJ1J2L* (detailed below in phototaxis; Yoshihara *et al.*, 2000; Bhaya *et al.*, 2001).

II. PHOTOTAXIS

Phototactic movements carried out by gliding, swimming or twitching are helpful to the cyanobacteria in enabling them to absorb maximum light for efficient photosynthesis. Positive phototaxis is directed towards the source of illumination while negative phototaxis is elicited in the opposite direction away from the incident light.

A) Action spectra and nature of photoreceptors

The action spectra for positive as well as negative phototaxis for a few of the cyanobacteria have been determined. Early studies on the action spectra for phototaxis of some filamentous cyanobacteria revealed that photosynthetic pigments were involved in the photorception (Nultsch, 1961; 1962; Tyagi, 1976). However, inhibitors of the photosynthetic electron transport chain were not effective to inhibit phototaxis (Tyagi, 1976). Since action spectrum for negative phototaxis in *A. variabilis* was in far red light (730 nm), so it was concluded that it is not the photosynthetic pigments that are the photoreceptors for phototaxis but some other photoreceptor must be involved (Nultsch and Schuchart, 1985). Certain of the unicellular cyanobacteria such as *S. elongatus* and *Synechocystis* sp. strain PCC 6803 also exhibit phototactic movements (Stanier *et al.*, 1971; Castes *et al.*, 1986; Ramsing *et al.*, 1997 and Choi *et al.*, 1999). The action spectra for positive phototaxis appeared to be species specific as in *P. uncinatum* (390, 480, 560 nm), *Cylindrospermum alatosporum* (450 and 640 nm), *A. variabilis* (550 and 730 nm) and in the thermophilic *Thermosynechococcus elongatus* (640 nm) light of different wave lengths (mentioned in parentheses) is utilized (Hader, 1987a; Kondou *et al.*, 2001). *T. elongatus* showed several action peaks at 530, 570, 640 and 680 nm at a fluence rate of 10 μmol m^{-2} s^{-1} but at higher fluence rates the red action peaks (640 and 680 nm) disappeared and far-red action peaks (at 720 and 740 nm) appeared. The motility rates of the cells were found to be 0.1 to 0.3 μm s^{-1}. However, in *Synechocystis* sp. strain PCC 6803 the absorption spectrum for positive phototaxis has been reported to be between 560 to 720 nm (Choi *et al.*, 1999; Ng *et al.*, 2003). Likewise, the action spectra for negative phototaxis of *Synechocystis* sp. strain PCC 6803 has been found to be in the longer ultraviolet (360 nm), high intensity blue light (470 nm) and red light regions (600 to 700 nm) while in *A. variabilis* it lies between 500-560 nm and 700nm (Nultsch *et al.*, 1983; Choi *et al.*, 1999). The influence of light and temperature on the phototaxis of *Synechococcus* isolates from hot springs of Octopus Yellow Stone National Park was studied (Ramsing *et al.*, 1997). The involvement of a phytochrome-like receptor in the phototaxis of *Synechocystis* sp. strain PCC 6803 was predicted based on the wave length dependency and phototactic speed (Choi *et al.*, 1999).

There appears to be a great diversity in the type and functional role of the photoreceptor pigments in cyanobacteria. The first phytochrome-like protein was discovered in *Fremyella diplosiphon* that regulates complementary chromatic adaptation (Kehoe and Grossman, 1996). Cyanobacterial phytochromes (Cph1 and Cph2) are akin to plant-type phytochromes and have been characterized in *Synechocystis* sp. strain PCC 6803. The chromophore in plant phytochromes is known as phytochromobilin wheras in Cph1 and Cph2 the chromophore is phycocyanobilin. In general, the N-terminal part of the protein consists of the photosensory core and the C-terminal region consists of a histidine kinase that is regulatory in nature. GAF domain is the photosensory domain in most of the phytochromes and phytochrome-related proteins such as Cph1 and Cph2 (Rockwell *et al.*, 2006). Cph1 is a light-regulated biliprotein kinase and the *cph1* gene is located upstream of a gene that encodes a response regulator to Cph1 in the genome of *Synechocystis* sp. strain PCC 6803. Typical of the plant-type phytochromes, Cph1 undergoes the classical red [P(r)] and far-red light [P(fr)] shifts and these chromophore structures are well conserved (Hughes *et al.*, 1997; Yeh *et al.*, 1997). Cph2 of *Synechocystis* sp. strain PCC 6803 differs from Cph1 in lacking the histidine kinase domain (Park *et al.*, 2000) but contains GGDEF and EAL output domains (Montgomery and Lagarias, 2002). Rockwell *et al.* (2009) identified that the P(fr) states of bacteriophytochromes and Cph1 are structurally dissimilar and it is the propionate side chains that perform different functions in these two classes. The diversity of photoreceptor pigments in cyanobacteria is known collectively as cyanobacteriochromes (CBCRs) that have absorption maxima spanning the entire visible spectrum (Rockwell *et al.*, 2010). Though the CBCRs differ in their spectral properties they share a common mechanism of light sensing by the 15/16 double-bond of the bilin chromophore. One of the subfamilies of CBCRs is distinctive in showing blue/green light sensing and these have evolved a thioether linkage to a second cysteine within the bilin-binding GAF domain (Rockwell *et al.*, 2011). They have in addition characterized the dual cysteine photosensors from *N. punctiforme* ATCC 29133. Mutants of *Synechocystis* sp. strain PCC 6803, isolated after site-directed mutagenesis of Cph1, acquired the potential to sense light from near UV light to infrared range (Rockwell *et al.*, 2011). Furthermore, it has been demonstrated that the second cysteine residue in the bilin-binding GAF domain is well conserved in the DXCF motif (Asp-Xaa-Cys-Phe). Such CBCRs perform blue/green light shifts by exhibiting isomerization of phycocyanobilin (PCB) into phycoviolobin (PVB), respectively. More than ten phycoviolobin variants of the CBCR subfamily from *N. punctiforme* ATCC 29133 have been produced in *E. coli* by genetic engineering techniques and the recombinant proteins exhibited well resolved blue/green light sensing states (Rockwell *et al.*, 2012). Moon *et al* (2010) reported the role of cyanopterin in UV/blue light signal transduction of *Synechocystis* sp. strain PCC 6803 phototaxis. Based on the properties of pterine glycosyltransferase gene (*pgtA*)-inactivated mutants of *Synechocystis* sp. PCC 6803 and inhibitory studies on pterin biosynthesis in the wld-type they suggested that cyanopterin is involved in the inhibition of negative phototaxis of the wild-type by sensing UV-A.

Fiedler *et al.* (2004) compared the growth performance of Cph1 and Cph2 mutants of *Synechocystis* sp. strain PCC 6803 under different light qualities and quantities. Cph1 mutant showed a reduced growth under far-red light whereas the growth of Cph2 mutant strains was inhibited by red light. On the other hand, Cph1 and Cph2 double mutants exhibited impaired growth under high light conditions. Cph2 possesses the conserved Cys residues at positions Cys 129 and Cys 1022 that may be involved in sensing red/far red and blue light, respectively. Fiedler *et al.* (2005) demonstrated that the Cph2 mutants in which Cys1022 has been replaced with Val1022 showed photomovement of the cells towards blue light. Furthermore, the importance of Cys129 in the N-terminal chromophore-binding domain of Cph2 has been revealed through the studies on mutants lacking Cys129. Cph2 could perform its function independently and the presence of other photoreceptos such as TaxD1 protein (similar in sequence to bacteriophytochrome and MCPs that regulated phototxis; Yoshihara *et al.*, 2000; Bhaya *et al.*, 2001a) as well as a BLUF protein (product of slr 1694 gene) of *Synechocystis* sp. strain PCC 6803 did not interfere with Cph2 function.

The mechanism of phototaxis in *Synechocystis* sp. strain PCC 6803 is explained on the basis of existence of certain regulatory elements homologous to those described in bacterial chemotaxis (Bhaya *et al.*, 2001a). Three genetic loci specific to phototaxis (*tax1*, *tax2* and *tax3*) were identified on the basis of the isolation and characterization of 300 transposon-tagged mutants of *Synechocystis* sp. strain PCC 6803. Of these 90% of them were immotile and the rest exhibited negative phototaxis. The products of *tax1* and *tax3* are homologous to bacterial chemotaxis proteins and are related to Tfp-mediated phototaxis (Bhaya *et al.*, 2001b). They concluded that a single transposon insertion in each mutant could be traced at different positions in the same genetic locus. Secondly, it was easy to map the position of the inserted transposon because of the known genome sequence of *Synechocystis* sp. strain PCC 6803. This was confirmed by the similar phenotype of these mutants with mutants isolated after gene inactivation experiments (Bhaya *et al.*, 1999, 2000, 2001b). TaxD1 has been identified as the photoreceptor in phototaxis of *Synechocystis* sp. strain PCC 6803 (Yoshihara *et al.*, 2000; Bhaya *et al.*, 2001a). TaxD1 has a sequence similarity to both bacteriophytochromes and MCPs. It perceives light signals and triggers a chain of phosphorylation reactions involving histidine kinase TaxAY1, which is classified as a hybrid sensor histidine kinase has a fused CheY-like regulator domain at its C-terminal end. This in turn modulates the activity of motility motor (Bhaya *et al.*, 2001b). The *taxD3* locus (*taxD3/ctr1/pilJ*) has been identified to encode a chemo-receptor-like protein and a CheA-like histidine kinase (encoded by *taxAY3*). Both *taxAY3* and *taxD3* disruptant mutants though yielded non-motile phenotypes with complete loss of Tfp, the presence of thin pili on the cell surface still could be demonstrated by electron microscopy (Bhaya *et al.*, 2001b; Chung *et al.*, 2001; Yoshihara *et al.*, 2002). TaxD3 lacks a sensing domain but possesses a tetratricopeptide (TPR) domain at the N-terminus. Several of the mutants were affected in the structure of specialized proteins with coiled-coil domains, or with TPR domains or certain others with tandem pentapeptide repeat domains (Bhaya *et al.*, 2001a). Interestingly, *taxD2* locus which contains all the components of a chemotaxis-like system does not appear to be involved in motility (Yoshihara *et al.*, 2001). Bhaya (2004) summarized the current status on phototaxis in cyanobacteria and highlighted the necessary gaps to be filled in this area. The wild-type cells of *Synechocystis* sp. strain PCC 6803 showed red light-dominated action spectra (645 and 704 nm) for positive phototaxis whereas a mutant of the same organism required blue light for negative phototaxis (Ng *et al.*, 2003). It is concluded that while positive phototaxis is controlled by TaxD1 protein, negative phototaxis is mediated by certain other blue-light photoreceptors. It is important to note here that the 645 nm action spectrum reported for positive phototaxis is the typical action maxima for phytochrome mediated response in plants. Ng *et al.* (2003) further emphasized that TaxD1 protein is an atypical phytochrome or the features and activities of the phytochrome-like receptor are modified by the association of TaxD1 *in vivo*.

Spectroscopic analysis of heterologously expressed *Synechocystis* PixD exhibited blue/green light shifted forms characteristic of flavin-bound BLUF domain (Masuda *et al.*, 2004). A new dimension to the regulation of phototaxis was added by the studies of Okajima *et al.* (2005) who demonstrated that there can exist two or more photoreceptors (other than phytochrome-like receptor) that mediate phototaxis in cyanobacteria. A comparative study of *Synechocystis* sp. strain PCC 6803 and *S. elongatus* for a signalling mechanism that regulates phototaxis revealed BLUF (a sensor of Blue-Light Using FAD) as the photorceptor protein. While phytochrome-like photoreceptor and BLUF perceive blue light, they even predicted that the existence of a probable photoreceptor that is active in orange-red light. These studies are consistent with the action spectra for phototaxis because monochromatic light of 500–700 nm was effective while light below 500 nm was ineffective. BLUF is the gene product of *pixD*. Gene disruption studies invoving *pixJ1* and *pixD* showed that the mutants exhibited negative phototaxis toward monochromatic light of 500–700 nm. The crystal structure of BLUF protein of *T. elongatus* BP-1 revealed it to be a decameric assembly and the monomers (encoded by *Tll0078*) consist of a BLUF domain at its N-terminal region. The overall structure of the BLUF domain consisted of five-stranded β-sheet with two α-helices running parallel with it (Kita *et al.*, 2005). The intramolecular hydrogen bonded FAD chromophore exhibited photochemical reaction and the role of conserved Gln50 and Tyr8 has been detected to be critical (Kita *et al.*, 2005; Okajima *et al.*, 2006). The structure of *Synechocystis* sp. strain PCC 6803 PixD (a product of *slr1694*) has been subsequently reported by Yuan *et al.* (2006) where the BLUF domain of *Synechocystis* resembled the BLUF domain of Tll0078 of *T. elongatus* BP-1. The role of the conserved Gln50 and Tyr8 has been confirmed by them in *Synechocystis* sp. strain PCC 6803 as reported earlier by Okajima *et al.* (2006). Additionally, the role of Trp91 in interacting with flavin to bring conformational changes in response to light absorption has been highlighted (Yuan *et al.*, 2006). Masuda *et al.* (2008) identified the crucial role of the conserved Met93 in the BLUF protein (PixD/*slr1694*) of *Synechocystis* sp. strain PCC 6803 for light-dependent signal transduction and suggested that the Met93 residue is highly conserved in all BLUF domains. Fukushima *et al.* (2008) studied the properties of mutants of *T. elongatus* BP-1 where the conserved amino acid residues Tyr8 and Gln50 of the BLUF protein were modified to Phe, Ala and Asn (Tyr8Phe, Tyr8Ala, Gln50Asn and Gln50Ala) by site-directed mutagenesis. All the mutants lacked the typical response of red shift upon blue-light illumination exhibited by the wild-type. Tanaka *et al.* (2009) reported oligomeric-state-dependent conformational changes of BLUF protein of T. *elongatus* (TePixD/Tll0078) and the protein existed in pentameric and decameric states and it is the latter that brought about the observed conformational change in the BLUF protein and so is functionally important.

B) Pix-genes and mechanism of phototaxis

Signal transduction pathway for phototactic motility in *Synechocystis* sp. strain PCC 6803 has been predicted to be on similar lines as that establihed for bacterial chemotaxis. Accordingly, the gene cluster governing positive phototaxis *pixGHIJ1J2L* releases products similar to PatA, CheY, CheW, MCPs and CheA, respectively (Yoshihara *et al.*, 2000; Bhaya *et al.*, 2001). Thus it was suggested that the gene cluster *pixGHIJ1J2L* forms the regulatory machinery for light perception and signal transduction pathway for phototactic movement in *Synechocystis* sp. strain PCC 6803. The gene for CheA in *Synechocystis* sp. strain PCC 6803 is present as a split gene at separate loci on the bacterial chromosome as *pix L-N* and *pix L-C*. (Yoshihara *et al.*, 2001). PatA is a cyanobacteria-specific response regulator which contains a region similar to CheY of *E. coli* that was reported in *Anabaena* sp. strain 7120, known to control heterocyst pattern formation (Liang *et al.*, 1992). CheY homologue interacts with either pilT or pilU proteins that supply energy for retraction (Yoshihara *et al.*, 2000; Bhaya *et al.*, 2001).

On the other hand, the photoreceptor protein for phototaxis in *S. elongatus* is localized at the two poles of the cells. This photoreceptor protein is a product of the gene *SepixJ* (a constituent of *Synechococcus elongatus pix*-gene cluster) that is homologous to *pixJ* of *Synechocystis* sp. strain PCC 6803. Furthermore, the *SepixJ* is an important constituent of the *Sepix*G gene cluster having *SepixGHIJL* that is homologous to *pix G* gene cluster (*pixGHIJ1J2L*) of *Synechocystis* sp. strain PCC 6803 (Kondou *et al.*, 2002). Ishizuka *et al.* (2006) studied the characteristics of PixJ of *T. elongatus* BP-1 and this CBCR exhibited photoconversion between blue light (433 nm) and green light (531 nm) absorbing forms. On the basis of its expression studies in *Synechocystis* sp. strain PCC 6803 it was concluded that PVB acts as the chromophore in contrast to PCB of *cph1*.

Okajima *et al.* (2005) suggested that the phytochrome-like photoreceptor and BLUF act as master switch between positive and negative phototaxis. If in fact BLUF is the photoreceptor, then it should be able to interact with PatA-like response regulator. Since sequence analysis of the genome of *Synechocystis* sp. strain PCC 6803 revealed six homologous *patA*-like genes (Kaneko *et al.*, 1996), Okajima *et al.* (2005) conducted yeast two-hybrid screening that confirmed the product of *pixD* (i.e. BLUF) self-interacted and interacted with *pixE* proteins strongly suggesting the existence of a possible signal transduction pathway. They speculated that the blue-light photoreceptors might be active for switching on negative phototaxis to avoid photoinhibition although photosynthetic light is available. It is clear that there are many gaps in understanding the mechanism of phototaxis. Though the structures associated with gliding, i.e. fibrillar apparatus, JPC and the spicules are described, how these perform their functions is yet to be identified. It is not enough to know the presence of gene sequences homologous to known sequences found in *Synechocystis* sp. strain PCC 6803. For example, genes essential for motility and positive phototaxis of *Synechocystis* sp. strain PCC 6803 have been found in *Anabaena* sp. strain PCC 7120 (Kaneko *et al.*, 2001) and *Nostoc punctiforme* (Meeks *et al.*, 2002). However, the former does not develop harmogonia while the latter forms hormogonia which exhibit gliding. Further, it is of interest to know that the motile harmogonia produced by *Calothrix* sp. PCC 7601 differentiates pili-like structures (Damerval *et al.*, 1991). The role of such pili in gliding is yet to be worked out.

LITERATURE CITED

Adams, D. G. (2001) How do cyanobacteria glide? *Microbiology Today* **28:** 131–133.

Adams, D. G., Ashworth, D., and Nelmes, B. (1999) Fibrillar array in the cell wall of a gliding filamentous cyanobacterium. *J Bacteriol* **181:** 884–892.

Alexandre, G., and Zhulin, I. B. (2001) More than one way to sense chemicals. *J Bacteriol* **18:** 4681–4686.

Armitage, J. P., and Hellingwerf, K. J. (2003) Light-induced behavioral responses ('phototaxis') in prokaryotes. *Photosynth Res* **76:** 145–155.

Bardy, S. L., Ng, Y. M. S., and Jarrell, K. F. (2003) Prokaryotic motility structures. *Microbiology* **149:** 295–304.

Bhaya, D. (2004) Light matters: phototaxis and signal transduction in unicellular cyanobacteria. *Mol Microbiol* **53:** 745–754. doi:10.1111/j.1365-2598.2004.04160.x

Bhaya, D., Bianco, N.R., Bryant, D., and Grossman, A. (2000) Type IV pilus biogenesis and motility in the cyanobacterium *Synechocystis* sp. PCC 6803. *Mol Microbiol* **37:** 941–951.

Bhaya, D., Nakasugi, K., Fazeli, F., and Burriesci, M. S. (2006) Phototaxis and impaired motility of *Synechocystis* sp. strain PCC 6803. *J Bacteriol* **188:** 7306–7310. doi:10.1128/JB.00573-06.

Bhaya, D., Takahashi, A., Shahi, P., and Grossman, A. R. (2001a) Novel motility mutants of *Synechocystis* strain PCC 6803 generated by *in vitro* transposon mutagenesis. *J Bacteriol* **183:** 6140–6143.

Bhaya, D., Takahashi, A., and Grossman, A. R. (2001b) Light regulation of type IV pilus-dependent motility by chemosensor-like elements in *Synechocystis* PCC 6803. *Proc Natl Acad Sci USA* **98:** 7540–7545.

Bhaya, D., Watanabe, N., Ogawa, T., and Grossman, A. R. (1999) The role of an alternative sigma factor in motility and pilus formation in the cyanobacterium *Synechocystis* sp. strain PCC 6803. *Proc Natl Acad Sci USA* **96:** 3188–3193.

Biemann, H. P., and Koshland, D. E. Jr. (1994) Aspartate receptors of *Escherichia coli* and *Salomnella typhimurium* bind ligand with negative and half-of-the-sites cooperativity. *Biochemistry* **33:** 629–634.

Brahamsha, B. (1996) An abundant cell-surface polypeptide is required for swimming by the non-flagellated marine cyanobacterium *Synechococcus. Proc Natl Acad Sci USA* **93:** 6504–6509.

Brahamsha, B. (1999) Non-flagellar swimming in marine *Synechococcus. J Mol Microbiol Biotechnol* **1:** 59–62.

Brock, T. D. (1978) *Thermophilic Microorganisms and Life at High Temperature.* Springer-Verlag, New York, N.Y, USA.

Castenholz, R. W. (1973) Movements. In: Carr, N.G. and Whitton, B.A. (Eds.) *The Biology of Blue-green Algae.* Blackwell Scientific Publications, Oxford, U.K. pp. 320–339.

Castenholz, R. W. (1982) Motility and taxes. In: Carr, N. G. and Whitton, B.A. (Eds.) *The Biology of Cyanobacteria.* Blackwell Scientific Publications, Oxford, U.K. pp. 413–439.

Castes, A.-M., Houmard, J., and Tandeau de Marsac, N. (1986) Is cell motility a plasmid-encoded function in the cyanobacterium *Synechocystis* 6803? *FEMS Microbiol Lett* **37:** 277–281.

Choi, J. S., Chung, Y. H., Moon, Y. J., Kim, C., Watanabe, M., Song, P. S., Joe, C. O., Bogorad, L., and Park, Y. M. (1999) Photomovement of the gliding cyanobacterium *Synechocystis* PCC 6803. *Photochem Photobiol* **70:** 95–102.

Clarke, S., and Koshland D. E. Jr. (1979) Membrane receptors for aspartate and serine in bacterial chemotaxis. *J Biol Chem* **254:** 9695–9702.

Damerval, T., Guglielmi, G., Houmard, J., and Tandeau de Marsac, N. (1991) Harmogonium differentiation in the cyanobacterium *Calothrix* : a photoregulated developmental process. *Plant Cell* **3:** 191–201.

Dick, H., and Stewart, W. D. P. (1980) The occurrence of fimbriae on a N_2-fixing cyanobacterium which occurs in lichen symbiosis. *Arch Microbiol* **109:** 107–109.

Diehn, B., Feinleib, M., Haupt, W., Hildebrand, E., Lenci, F., and Nultsch, W. (1979) Terminology of behavioral responses of motile microorganisms. *Photochem Photobiol* **26:** 559–560.

Dienst, D., Dühring, U., Mollenkopt, H-J., Vogel, J., Golecki, J., Hess, W. R., and Wilde, A. (2008) The cyanobacterial homologue of the RNA chaperone Hfq is essential for motility of *Synechocystis* sp. PCC 6803. *Microbiology* **154:** 3134–3143.

Guglielmi, G., and Cohen-Bazire, G. (1982) Structure et distribution des pores et des perforations de l'enveloppe de peptidoglycane chez quelques cyanobacteries. *Protistologica* **18:** 3923–3932.

Ehlers, K. M., Samuel, A. D. T., Berg, H. C., and Montgomery, R. (1996) Do cyanobacteria swim using traveling surface waves? *Proc Natl Acad Sci USA* **93:** 6340–8343.

Fiedler, B., Borner, T., and Wilde, A. (2005) Phototaxis in the cyanobacterium *Synechocystis* sp. PCC 6803: role of different photoreceptors. *Photochem Photobiol* **81:** 1481–1488

Fiedler, B., Broc, D., Schubert, H., Rediger, A., Börner, T., and Wilde, A. (2004) Involvement of cyanobacterial phytochromes in growth under different light qualities and quantities. *Photochem Photobiol* **79:** 551–555.

Fukushima, Y., Murai, Y., Okajima, K., Ikeuchi, M., and Itoh, S. (2008) Photoreactions of Tyr8- and Gln50-mutated BLUF domains of the PixD protein of *Thermosynechococcus elongatus* BP-1. Photoconversion at low temperature without Tyr8. *Biochemistry* **47:** 660–669.

Häder, D. P. (1987a) Photosensory behavior in prokaryotes. *Microbiol Rev* **51:** 1–21.

Häder, D. P. (1987b) Photomovement In: Fay, P., and van Baalen, C. (Eds.) The *Cyanobacteria.* Elsevier, New York, pp. 325–346.

Halfen, L. N. (1973) Gliding motility in *Oscillatoria*: ultrastructure and chemical characterization of the fibrillar layer. *J Phycol* **9:** 248–253.

Halfen, L. N. (1979) Gliding movements. In: Haupt, W., and Feinleib, M.E. (Eds.) *Encyclopedia of Plant Physiology.* Vol 7. Physiology of movements. Springer, Berlin Heidelberg, New York, pp. 250–267.

Halfen, L. N., and Castenholz, R. W. (1970) Gliding in a blue-green alga: a possible mechanism. *Nature* (London) **225:** 1163–1165.

Halfen, L.N., and Castenholz, R. W. (1971) Gliding in a blue-green alga, *Oscillatoria princeps. J Phycol* **7:** 133–145.

Hayes, W. (1968) *The Genetics of Bacteria and their Viruses.* Blackwell Scientific Publications, Oxford, Edinburgh, UK. 2nd edn. 925 pp.

Hedger, J., Holmquist, P. C., Leigh, K. A., Saraff, K., Pomykal, C., and Summers, M. L. (2009) Illumination stimulates cAMP-receptor protein dependent transcriptional activation from regulatory regions containing class I and class II promoter elements in *Synechocystis* sp. PCC 6803. *Microbiology* **155:** 2994–3004. doi: 10.1099/mic0.028035-0.

Hoiczyk, E. (2000) Gliding motility in cyanobacteria: observations and possible explanations. *Arch Microbiol* **174:** 11–17.

Hoiczyk, E., and Baumeister, W. (1995) Envelope structure of four gilding filamentous cyanobacteria. *J Bacteriol* **177:** 2387–2395.

Hoiczyk, E., and Baumeister, W. (1997) Oscillin, an extracellular, Ca^{2+}-binding glycoprotein essential for gliding motility of cyanobacteria. *Mol Microbiol* **26:** 699–708.

Hoiczyk, E., and Baumeister, W. (1998) The junctional pore complex, a prokaryotic secretion organelle, is the molecular motor underlying gliding motility in cyanobacteria. *Curr Biol* **8:** 1161–1168.

Hoiczyk, E., and Hansel, A. (2000) Cyanobacterial cell walls: News from an unusual prokaryotic envelope. *J Bacteriol* **182:** 1191–1199.

Hazelbauer, G. L., and Harayama, S. (1983) Sensory transduction in bacterial chemotaxis. *Int Rev Cytol* **81:** 33–70.

Hood, E. E., Armour, S., Ownby, J. D., Handa, A. K., and Bressan, R. A. (1979) Effect of nitrogen starvation on the level of adenosine 3′,5′-monophosphate in *Anabaena variabilis*. *Biochim Biophys Acta* **588:** 193–200.

Huang, F., Hedman, E., Funk, C., Kieselbach, T., Schroder, W. P., and Norling, B. (2004) Isolation of outer membrane of *Synechocystis* sp. PCC 6803 and its proteomic characterization. *Mol Cell Proteomics* **3:** 586–595.

Huang, F., Parmryd, I., Nilsson, F., Persson, A. L., Pakrasi, H. B., Andersson, B., and Norling, B. (2002) Proteomics of *Synechocystis* sp. strain PCC 6803: Identification of plasma membrane proteins. *Mol Cell Proteomics* **1:** 956–966.

Hughes, J., Lamparter, T., Mittmann, F., Hartmann, E., Gartner, W., Wilde, A., and Borner, T. (1997) A prokaryotic phytochrome. *Nature* (London) **386:** 663.

Ishizuka, T., Shimada, T., Okajima, K., Yoshihara, S., Ochiai, Y., Katayama, M., and Ikeuchi, M. (2006) Characterization of cyanobacteriochrome TePixJ from a thermophilic cyanobacterium *Thermosynechococcus elongatus* strain BP-1. *Plant Cell Physiol* **47:** 1251–1261.

Jeon, J., and Dobryin, A.V. (2005) Polymer containment and bacterial gliding motility. *Eur Phy J* (E) **17:** 361–372.

Kamei, A., Ogawa, T., and Ikeuchi, M. (1998) Identification of a novel gene (*slr2301*) involved in high-light resistance in the cyanobacterium *Synechocystis* sp. PCC 6803. In: Garab, G. (Ed.) *Photosynthesis : Mechanisms and Effects*. Kluwer Academic Publishers, Dordrecht, The Netherlands. pp. 2901–2905.

Kamei, A., Yuasa, T., Orikawa, K., Geng, X. X., and Ikeuchi, M. (2001) A eukaryote type protein kinase, SpkA, is required for normal motility of the unicellular cyanobacterium *Synechocystis* sp. strain PCC 6803. *J Bacteriol* **183:** 1505–1510.

Kaneko, T., Nakamura, Y., Wolk, C. P., Kuritz, T., Sasamoto, S., Watanabe, A., Iriguchi, M., Ishikawa, A., Kawashima, K., Kimura, T., Kishida, Y., Kohara, M., Matsunoto, M., Matsuno, A., Muraki, A., Nakazaki, N., Shimpo, S., Sugimoto, M., Takazawa, M., Yamada, M., Yasuda, M., and Tabata, S. (2001) Complete genome sequence of the filamentous nitrogen-fixing cyanobacterium *Anabaena* sp. strain PCC 7120. *DNA Res* **8:** 205–213; 227–253.

Kaneko, T., Sato, S., Kotani, H., Tanaka, A., Asamizu, E., Nakamura, Y., Miyajima, N., Hirosawa, M., Sugira, M., Sasamoto, S., Kimura, T., Hosouchi, T., Matsuno, A., Muraki, A., Nakazaki, N., Naruo, K., Okumura, S., Shimpo, S., Takeuchi, C., Wada, T., Watanabe, A., Yamada, M., Yasuda, M., and Tabata, S. (1996) Sequence analysis of the unicellular cyanobacterium *Synechocystis* sp. strain PCC 6803. II. Sequence determination of the entire genome and assignment of potential protein-coding regions. *DNA Res* **3:** 109–136.

Kasahara, M. and Ohmori, M. (1999) Activation of a cyanobacterial adenylate cyclase, Cya C, by autophosphorylation and subsequent phosphotransfer reaction. *J Biol Chem* **274:** 15167–15172.

Kehoe, D. M., and Grossman, A.R. (1996) Similarity of a chromatic adaptation sensor to phytochrome and ethylene receptors. *Science* **273:** 1409–1412.

Kita, A., Okajima, K., Morimoto, Y., Ikeuchi, M., and Miki, K. (2005) Structure of a cyanobacterial BLUF protein, Tll0078, containing a novel FAD-binding blue light sensor domain. *J Mol Biol* **349:** 1–9.

Kondou, Y., Mogami, N., Hoshi, F., Kutsuna, S., Nakazawa, M., Sakurai, T., Matsui, M., Kaneko, T., Tabata, S., Tanaka, I., and Manabe, K. (2002) Bipolar localization of putative photoreceptor protein for phototaxis in thermophilic cyanobacterium *Synechococcus elongatus*. *Plant Cell Physiol* **43:** 1585–1588.

Kondou, Y., Nakazawa, M., Higashi, S., Watanabe, M., and Manabe, K. (2001) Equal-quantum action spectra indicate fluence-rate-selective action of multiple photoreceptors for photomovement of the thermophilic cyanobacterium *Synechococcus elongatus*. *Photochem Photobiol* **73:** 90–95.

Kort, E. N., Goy, M. F., Larsen, S. H., and Adler, J. (1975) Methylation of a membrane protein involved in bacterial chemotaxis. *Proc Natl Acad Sci USA* **72:** 3939–3943.

Liang, J., Scappino, L., and Haselkorn, R. (1992) The *patA* gene product, which contains a region similar to CheY of *Escherichia coli*, controls heterocyst pattern formation in the cyanobacterium *Anabaena* 7120. *Proc Natl Acad Sci USA* **89:** 5655–5659.

Lin, L.N., Li, J., Brandts, J.F., and Weis, R.M. (1994) The serine receptor of bacterial chemotaxis exhibits half-site saturation for serine binding. *Biochemistry* **33:** 6564–6570.

Lounatmaa, K., Vaara, T., Osterlund, K., and Vaara, M. (1980) Ultrastructure of the cell wall of a *Synechocystis* strain. *Rev Can Microbiol* **26:** 204–208.

Lux, R., and Shi, W. (2004). Chemotaxis-guided movements in bacteria. *Cri Rev Oral Biol Med* **15:** 207–220.

Masuda, S., Hasegawa, K., Ishii, A., and Ono, T. A. (2004) Light-induced structural changes in a putative blue-light receptor with a novel FAD binding fold sensor of blue-light using FAD (BLUF); Slr1694 of *Synechocystis* sp. PCC6803. *Biochemistry* **43:** 5304–5313.

Masuda, S., Hasegawa, K., Ohta, H., and Ono, T-a. (2008) Crucial role in light signal transduction for the conserved Met93 of the BLUF protein PixD/Slr1694. *Plant Cell Physiol* **49:** 1600–1606.

Mattick, J. S. (2002) Type IV pili and twitching motility. *Annu Rev Microbiol* **56:** 289–314.

McBride, M. (2000) Bacterial gliding motility: mechanisms and mysteries. *Am Soc Microbiol News* **66:** 203–210.

McBride, M.J. (2001) Bacterial gliding motility: Multiple mechanisms for cell movement over surfaces. *Annu Rev Microbiol* **55:** 49–75.

McCarren, J., Heuser, J., Roth R., Yamada, N., Martone, M., and Brahamsha, B. (2005) Inactivation of *SwmA* results in the loss of an outer cell layer in a swimming *Synechococcus* strain. *J Bacteriol* **187:** 224–230.

Meeks, J. C., Campbell, E. L., Summers, M. L., and Wong, F. C. (2002) Cellular differentiation in the cyanobacterium in *Nostoc punctiforme*. *Arch Microbiol* **178:** 395–403.

Merz, A.J., and Sheetz, S. M. (2000) Pilus retraction powers bacterial twitching motility. *Nature* (London) **407:** 98–102.

Moon, Y.-J., Park, Y. M., Chung Y.-H., and Chol, J. S. (2004) Calcium is involved in photomovement of cyanobacterium *Synechocystis* sp. PCC 6803. *Photochem Photobiol* **79:** 114–119.

Moon-Y. J., Lee, E. M., Park, Y. M., Park, Y. S., Chung, W-I., and Chung, Y-H. (2010) The role of cyanopterin in UV/blue light signal transduction of cyanobacterium *Synechocystis* PCC 6803 phototaxis. *Plant Cell Physiol* **51:** 969–980.

Montgomery, B. L., and Lagarias, J. C. (2002) Phytochrome ancestry: sensors of bilins and light. *Trends Plant Sci* **7:** 357–366.

Mullineaux, C. W. (2001) How do cyanobacteria sense and respond to light. *Mol Microbiol* **41:** 965–971.

Nakasugi, K., and Neilan, B.A. (2005) Identification of pilus-like structures and genes in *Microcystis aeruginosa* PCC 7806. *Appl Environ Microbiol* **71:** 7621–7625. doi: 10.1128/AEM.71.11.7621-7625.2005

Nakasugi, K., Alexova, R., Svenson, C. J., and Neilan, B. A. (2007) Functional analysis of PilT from the toxic cyanobacterium *Microcystis aeruginosa* PCC 7806. *J Bacteriol* **189:** 1689–1697.

Nakasugi, K., Svenson, C. J., and Neilan, B. A. (2006) The competence gene, comF, from *Synechocystis* sp. strain PCC 6803 is involved in natural transformation, phototactic motility and piliation. *Microbiology* **152:** 3623–3631. doi:10.1099/mic.0-29189-0

Ng, W. O., Grossman, A. R., and Bhaya, D. (2003) Multiple light inputs control phototaxis in *Synechocystis* sp. strain PCC 6803. *J Bacteriol* **185:** 1599–1607.

Norling, B., Zak, E., Andersson, B., and Pakrasi, H. (1998) 2D-isolation of pure plasma and thylakoid membranes from the cyanobacterium *Synechocystis* PCC 6803. *FEBS Lett* **436:** 189–192.

Nudleman, E., and Kaiser, D. (2004) Pulling together with Type IV pili. *J Mol Microbiol Biotechnol* **7:** 52–62.

Nultsch, W. (1961) Der Einfluss des Lichtes auf die Bewegung der Cyanophyceen. *Planta* **56:** 632–647.

Nultsch, W. (1962) Phototaktische Aktionsspektren von Cyanophyceen. *Ber Deutsch Bot Ges* **75:** 443–453.

Nultsch, W., and Schuchart, H. (1985) A model of the phototactic reaction chain of the cyanobacterium *Anabaena variabilis. Arch Microbiol* **142:** 180–184.

Nultsch,W., Schuchart, H., and Koenig, F. (1983) Effects of sodium azide on phototaxis of the blue-green alga *Anabaena variabilis* and consequences to the two-photoreceptor systems-hypothesis. *Arch Microbiol* **134:** 33–37.

Ohmori, M. (1989) cAMP in *Anabaena cylindrica*: rapid changes in cellular levels in response to changes in extracellular environments. *Plant Cell Physiol* **30:** 911–914.

Ohmori, M., and Okamoto, S. (2004) Photoresponsive cAMP signal transduction in cyanobacteria. *Photochem Photobiol Sci* **3:** 503–511.

Ohmori, M., Ohmori, K., and Hasunuma, K. (1988) Rapid change in cyclic-3′,5′-AMP concentration triggered by a light-off or light-on signal in *Anabaena cylindrica. Arch Microbiol* **150:** 203–204.

Ohmori, M., Terauchi, K., Okamoto, S., and Watanabe, M. (1992) Regulation of cAMP-mediated photosignaling by a phytochrome in the cyanobacterium *Anabaena cylindrica. Photochem Photobiol* **75:** 675–679.

Okajima, K., Fukushima, Y., Suzuki, H., Kita, A., Ochiai, Y., Katayama, M., Shibata, Y., Miki, K., Noguchi, T., Itoh, S., and Ikeuchi, M. (2006) Fate determination of the flavin photoreceptions in the cyanobacterial blue light receptor TePixD (Tll0078). *J Mol Biol* **363:** 10–18.

Okajima, K., Yoshihara, S., Fukushima, Y., Geng, X., Katayama, M., Higashi, S., Watanabe, M., Sato, S., Tabata, S., Shibata, Y., Itoh, S., and Ikeuchi, M. (2005) Biochemical and functional characterization of BLUF-type flavin-binding proteins of two species of cyanobacteria. *J Biochem* **137:** 741–750.

Okamoto, S., and Ohmori, M. (1999) Analysis of cyanobacterial motility in *Synechocystis* sp. PCC 6803. *Plant Cell Physiol* **40** Suppl. 135S.

Okamoto, S., and Ohmori, M. (2002) The cyanobacterial pil-T protein is responsible for cell motility and transformation hydrolyzes ATP. *Plant Cell Physiol* **43:** 1127–1136.

Palenik, B., Brahamsha, B., Larimer, F.W., Land, M., Hauser, L.,Chain, P., Lamerdin, J., Regala, W., Allen, E.E., McCarren, J., Paulsen, I., Dufresne A., Partensky, F., Webb, A., and Waterbury, J. (2003) The genome of a motile marine *Synechococcus*. *Nature* (London) **424:** 1037–1042.

Park, C. M., Kim, J. I., Yang, S. S., Kang, J. G., Kang, J. H., Shim, J. Y., Chung, Y-H., Park, Y-M., and Song, P. S.(2000) A second photochromic bacteriophytochrome from *Synechocystis* sp. PCC 6803: spectral analysis and down-regulation by light. *Biochemistry* **39:** 10840–10847.

Perkins, F., Haas, L., Phillips, D., and Webb, K. (1981) Ultrastructure of a marine *Synechococcus* possessing spinae. *Can J Microbiol* **27:** 318–329.

Pitta, T. P., and Berg, H. C. (1995) Self-electrophoresis is not the mechanism for motility in swimming cyanobacteria. *J Bacteriol* **177:** 5701–5703.

Pitta, T. P., Sherwood, E. E., Kobel, A. M., and Berg, H. C. (1997) Calcium is required for swimming by the non-flagellated cyanobacterium *Synechococcus* strain WH8113. *J Bacteriol* **179:** 2524–2528.

Ramsing, N. B., Ferris, J. M., and Ward, D. M. (1997) Light induced motility of thermophilic *Synechococcus* isolates from Octopus Spring, Yellowstone National Park. *Appl Environ Microbiol* **63:** 2347–2354.

Rockwell, N. C., and Lagarias, J. C. (2010) A brief history of phytochromes. *ChemPhysChem* **11:** 1172–1180. doi: 10.1000/cphc.200999894

Rockwell, N. C., Martin, S. S., Feoktistova, K., and Lagarias, J. C. (2011) Diverse two-cysteine photocycles in phytochromes and cyanobacteriochromes. *Proc Natl Acad Sci USA* **108:** 11854–11859.

Rockwell, N. C., Martin, S. S., Gulevich, A. G., and Lagarias, J. C. (2012) Phycoviolobilin formation and spectral tuning in the DXCF cyanobacteriochrome subfamily. *Biochemistry* **51:** 1449–1463.

Rockwell., N. C., Shang, L., Martin, S. S., and Largaias, J. C., (2009) Distinct classes of red/far-red photochemistry within the phytochrome superfamily. *Proc Natl Acad Sci USA* **106:** 6123–6127.

Rockwell, N. C., Su, Y-S., and Lagarias, J. C. (2006) Phytochrome structure and signaling mechanisms. *Ann Rev Plant Biol* **57:** 837–858.

Samuel, A. D. T., Petersoen, J. D., and Reese, T. (2001) Envelope structure of *Synechococcus* sp.WH8113, a non-flagellated swimming cyanobacterium. *BMC Microbiol* **1:** 4. doi: 10.1186/1471-2180-1.4.

Sastry, P. A., Finlay, B. B., Pasloske, B. L., Paranchych, W., Pearlstone, J. R., and Smille, L. B. (1985) Comparative studies of the amino acid and nucleotide sequences of pilin derived from *Pseudomonas aeruginosa* PAK and PAO. *J Bacteriol* **164:** 571–577.

Sastry, P. A., Pearlstone, J. R., Smillie, L. B., and Paranchych, W. (1983) Amino acid sequence of pilin isolated from *Pseudomonas aeruginosa*. *FEBS Lett* **151:** 253–256.

Schultze-Lam, S., and Beveridge, T. J. (1994) Physico-chemical characteristics of the mineral-forming S-layer from the cyanobacterium *Synechococcus* strain GL24. *Can J Microbiol* **40:** 216–223.

Smarda, J. D., Smajs, D., Komrska, J., and Krzyzanek, V. (2002) S-layers on cell walls of cyanobacteria. *Micron* **33:** 257–277.

Stanier, R. Y., Kunisawa, R., Mandel, M., and Cohen-Bazire, G. (1971) Purification and properties of unicellular blue-green algae (Chroococcales). *Bacteriol Rev* **35:** 171–205.

Stock, J. B., Stock, A. M., and Motten, J. M. (1990) Signal transduction in bacteria. *Nature* (London) **344:** 395–400.

Tanaka, K., Nakasone, Y., Okajima, K., Ikeuchi, M., Tokutomi, S., and Terazima, M. (2009) Oligomeric-state-dependent conformational change of the BLUF protein TePixD (Tll0078). *J Mol Biol* **386:** 1290–1300.

Terauchi, K., and Ohmori, M. (1999) An adenylate cyclase, Cya 1, regulates cell motility in the cyanobacterium *Synechocystis* sp. PCC 6803. *Plant Cell Physiol* **40:** 248–251.

Terauchi, K., and Ohmori, M. (2004) Blue-light stimulates cyanobacterial motility via a cAMP signal transduction system. *Mol Microbiol* **52:** 303–309.

Tyagi, V. V. S. (1976) Studies on phototaxis in the blue-green alga, *Cylindrospermum*. *Biochem Physiol Pflanzen* **170:** 17–22.

Vaara, T. (1982) The outermost surface structures in Chroococcalean cyanobacteria. *Rev Can Microbiol* **28:** 929–941.

Vaara,T., and Vaara, M. (1988) Cyanobacterial fimbriae. *Methods Enzymol* **167:** 189–195.

Vaara, T., Ranta, H., Lounatmaa, K., and Korhonen, T. K. (1984) Isolation and characterization of pili (fimbriae) from *Synechocystis* CB3. *FEMS Microbiol Lett* **21:** 329–334.

Waterbury, J. B., and Rippka, R. (1989) Order Chroococcales Wettstein 1924. Emend. Rippka *et al.*, 1979. *Bergey's Manual of Systematic Bacteriology* Williams and Wilkins, Baltimore, USA. pp. 1728–1746.

Waterbury, J. B., Willey, J. M., Franks, D. G., Valois, F. W., and Watson, S. W. (1985) A cyanobacterium capable of swimming motility. *Science* **230:** 74–75.

Willey, J. M. (1988) Characterization of swimming motility in a marine unicellular cyanobacterium. Ph.D. Thesis. Woods Hole Oceanographic Institution and Massachusetts Institute of Technology, Cambridge, MA, USA.

Willey, J. M., Waterbury J. B., and Greenberg, E. P. (1987) Sodium-coupled motility in a swimming cyanobacterium. *J Bacteriol* **169:** 3429–3434.

Wolgemuth, C. W., and Oster, G. (2004) The junctional pore complex and the propulsion of bacterial cells. *J Mol Microbiol Biotechnol* **7:** 72–77.
Wolgemuth C., Hoiczyk, E., Kaiser, D., and Oster, G. (2002) How myxobacteria glide. *Curr Biol* **12:** 369–377.
Yeh, K. C., Wu, S. H., Murphy, J. T., and Lagarias, J. C. (1997) A cyanobacterial phytochrome two-component light sensory system. *Science* **277:** 1505–1508.
Yoshihara, S., and Ikeuchi, M. (2004) Phototactic motility in the unicellular cyanobacterium *Synechocystis* sp. PCC 6803. *Photochem Photobiol Sci* **3:** 512–518.
Yoshihara, S., Geng, X., Okamoto, S., Yura, K., Murata, T., Go, M., Ohmori, M., and Ikeuchi, M. (2001) Mutational analysis of genes involved in pilus structure, motility and transformation competency in the unicellular motile cyanobacterium *Synechocystis* sp. PCC 6803. *Plant Cell Physiol* **42:** 63–73.
Yoshihara, S., Geng, X., and Ikeuchi, M. (2002) *pilG* Gene cluster and split *pilL* genes involved in pilus biogenesis, motility and genetic transformation in the cyanobacterium *Synechocystis* sp. PCC 6803. *Plant Cell Physiol* **43:** 513–521.
Yoshihara, S., Katayama, M., Geng, X., and Ikeuchi, M. (2004) Cyanobacterial phytochrome-like PixJ1 holoprotein shows novel reversible photoconversion between blue- and green-absorbing forms. *Plant Cell Physiol* **45:** 1729–1737.
Yoshihara, S., Suzuki, F., Fujita H., Geng, X. X., and Ikeuchi, M. (2000) Novel putative photoreceptor and regulatory genes required for the positive phototactic movement of the unicellular motile cyanobacterium *Synechocystis* sp. PCC 6803. *Plant Cell Physiol* **41:** 1299–1304.
Yoshimura H., Hisabori, T., Yanagisawa, S., and Ohmori, M. (2000) Identification and characterization of a novel cAMP receptor protein in the cyanobacterium *Synechocystis* sp. PCC 6803. *J Biol Chem* **275:** 6241–6245.
Yoshimura, H., Yoshihara, S., Okamoto, S., Ikeuchi, M., and Ohmori, M. (2002) A cAMP receptor protein, SYCRP1, is responsible for the cell motility of *Synechocystis* sp. PCC 6803. *Plant Cell Physiol* **43:** 400–463.
Yuan, H., Anderson, S., Masuda, S., Dragnea, V., Moffat, K., and Bauer, C. (2006) Crystal structures of the *Synechocystis* photoreceptor Slr1694 reveal distinct structural states related to signaling. *Biochemistry* **45:** 12687–12694.

CHAPTER 6

CIRCADIAN RHYTHMS

The rotation of the earth about its axis relative to Sun causes the formation of day/night cycles with a regular periodicity. Accordingly, all living organisms adjust to this external clock by choosing their respective environmental niches. They do so by expressing endogenous rhythms that can measure the passage of time on an approximate 24 h scale. Hence these rhythms are known as circadian (circa= approximately; dies= a day) rhythms. The underlying machinery responsible for the circadian rhythms is known as circadian clock or biological clock that has three important components, i.e.

input pathway, central oscillator and output pathway. The input pathway resets the phase of the oscillation by responding to external/environmental signals. Of these, the light/dark cycle is the most predominant external cycle and a major environmental stimulus for the synchronization of circadian clock. The central oscillator (also designated as central pacemaker) receives signals through the input pathway and generates the oscillation. The output pathway translates the oscillation into behavioural and physiological rhythms. The circadian rhythms are exhibited by almost all groups of organisms from prokaryotes to eukaryotes. These daily rhythms regulate a wide variety of biological activities that help the organisms to adapt to daily changes in the environmental conditions.

Three important criteria are characteristic of circadian rhythms. Firstly, these persist in constant environmental conditions of temperature, light or darkness with a period of 24 h. That is they have an inherent periodicity of 24 h. Secondly, these rhythms are temperature compensated suggesting that these remain constant over a permissive range of temperatures. Thirdly, these endogenous rhythms can be tuned to oscillate with exactly 24 h period. This is also known as entrainment that can be achieved by the variation of light/dark and/or temperature cycles. This enables the living organisms to keep track of time in their local environment to exactly 24 h daily cycles.

Circadian rhythms were for the first time detected in 1729 by a French astronomer, Jean Jacques d'Ortous de Marian who observed that daily leaf movements of heliotrope persisted even when the plant was kept in dark for several days. That is the leaves of the plant continued to open during the day and close at night despite the absence of sunlight. But it was Erwin Bünning's work, on leaf movements in *Phaseolus* that formed the foundation for circadian (biological) clock research (Bünning, 1973). Recognizing the importance of circadian clock in the biological world, many experimental systems have been developed for elucidating the biochemical, molecular and genetic basis of the circadian rhythms. Although mutational studies on eukaryotic groups of organisms such as algae (*Chlamydomonas reinhardtii*), fungi (*Neurospora crassa*), insects (*Drosophila melanogaster*), fishes (hamster), mammals (mouse) and plants (*Arabidopsis thaliana*) have been made, only *D. melanogaster* and *N. crassa* have been amenable for a mutational analysis of the genes governing circadian rhythms (Golden *et al.*, 1997). Thus in general, these studies led every one to believe that the circadian rhythms are characteristic of eukaryotes alone. Moreover, it was generally thought that the prokaryotes are entirely devoid of these rhythms. The idea that the prokaryotes either unicellular or multicellular do not show circadian rhythms has gained strength to such an extent that many workers held it as a dogma and excluded prokaryotes while proposing models for circadian mechanisms (Johnson *et al.*, 1996).

The variations in response to light and dark cycles can be plotted into phase response curves. The magnitude of phase shift as a result of the application of a given stimulus is plotted at different points of time during the entire period of circadian cycle. The phase response curves show peaks alternating with troughs coinciding with the alternating light:dark (L:D) cycles during the entrainment period. These persist even after the cells are withdrawn and are subjected to a continuous light (LL) or continuous dark (DD) period. This period through which the rhythms continue to be expressed is known as free-running period. These are helpful in understanding the circadian oscillator.

Despite many odds, the persistent quest for identifying circadian rhythms in prokaryotic systems continued. It is in this context that the cyanobacteria emerged as suitable experimental systems for unraveling the molecular mechanism of circadian rhythms. However, the complete genome sequencing of *Synechocystis* sp. strain PCC 6803 (Kaneko *et al.*, 1996) paved the way for the identification of *Synechococcus* sp. strain PCC 7942 (with a genome size of 2.7 mb almost equivalent to the genome size of *Escherichia coli*) as the novel experimental system for the study of circadian rhythms.

I. EARLY STUDIES

Gallon *et al.* (1974) reported that nitrogen fixation and photosynthesis in cultures of *Gloeocapsa* depended on the age of cultures and there was a rise and fall in their activities in continuous light. However, these variations have not been correlated to daily cyclic variation. Nitrogen fixation by *Gloeocapsa* grown in 12:12 light and dark cycles was observed only during dark period (Millineaux *et al.*, 1981). However, in case of *Anabaena cylindrica* where nitrogenase is restricted to heterocysts and PSII activity is confined to vegetative cells, nitrogen fixation preferentially occurred during light period. The circadian control of nitrogen fixation in *Gloeocapsa* was apparent only when the cultures were subjected to 16:8 light and dark cycles where nitrogenase activity was maximal after 8 h of the onset of darkness. Likewise, a temporal separation of nitrogen fixation and oxygen production in *Plectonema boryanum* was also reported (Weare and Benemann, 1994).

Stal and Krumbein (1985) demonstrated that nitrogenase activity peaked during darkness when the cells of *Oscillatoria* sp. were subjected to LD cycle of 16:8 h. Besides, they observed that nitrogen fixation as well as photosynthetic activity reciprocally reached maximum activities that persisted even after the cells adapted to LD cycle are shifted to LL condition. Likewise, Mitsui *et al.* (1986) reported a temporal separation of nitrogen fixation and photosynthesis in two species of *Synechococcus* (Miami BG 43511 and 43522) where the pattern (peaking of the activities once daily) was established after the cells were transferred to LL for 20 h.

Though these studies have not been correlated with the circadian rhythms but unequivocally pointed towards a temporal separation of nitrogen fixation (occurring at night) from photosynthesis (taking place in the day) while at the same time dependent on LD cycles. The first report on the existence of circadian rhythms in cyanobacteria was made in *Synechococcus* sp. strain RF-1 that fixed nitrogen only during dark phase while the fixation varied at different levels in continuous light (Grobbelaar *et al.*, 1986; Huang and Chow, 1986). These workers identified that the pattern seen in LL after an adaptive period in LD to be an endogenous rhythm. Chen *et al.* (1991) also reported rhythmic amino acid uptake by *Synechococcus* RF-1. In a series of communications, Huang and coworkers have confirmed the existence of circadian clock in cyanobacteria. This was substantiated by the isolation of mutants of *Synechococcus* RF-1 that showed alteration in the rhythm as well (Huang *et al.*, 1990, 1993, 1994; Huang and Chou, 1991; Huang and Grobbelaar, 1995; Chen *et al.*, 1993,1998). Sweeney and Borgese (1989) reported the operation of temperature compensated cell division cycle in *Synechooccus* sp. strain WH 7803 in a 24 h period.

The above studies thus dispelled the dogma regarding the absence of circadian rhythms in prokaryotes. Huang and Grobbelaar (1995) studied several aspects of endogenous rhythm in *Synechococcus* RF-1 for nitrogenase activity and suggested that: (i) the rhythms were temperature-compensated showing approximate 24 h periods in constant conditions between 22°C and 33°C (Huang *et al.*, 1990); (ii) the endogenous nitrogenase rhythm in RF-1 is independent of nitrogen fixation *per se* as evidenced by the accumulation of *nifH* transcripts under only nitrogen-fixing conditions; (iii) RF-1 also showed rhythmic uptake of eight amino acids. Furthermore, Schneegurt *et al.* (1994) demonstrated the persistent 24 h alteration in photosynthesis and nitrogen fixation in *Cyanothece* sp. strain ATCC 51142 with a rhythmic accumulation of stored carbohydrate.

II. CHOICE OF EXPERIMENTAL ORGANISM

Convincing evidences for the existence of circadian rhythms in cyanobacteria have been provided with the studies on *Synechococcus* sp. strain PCC 7942 (identified initially as *Anacystis nidulans* and later came to be known as *Synechococcus* R-2), an amenable experimental material for a mutational analysis of genes governing circadian rhythms. The choice for such studies was made on the basis of its (i) natural transformability, (ii) ability to receive DNA by conjugation from *E. coli* at high efficiency and (iii) potential to express reporter genes. *Synechococcus* sp. strain PCC 7942 which is not diazotrophic meets all the above requirements and was chosen as the experimental system (Golden *et al.*, 1997). In many of the communications that followed, this organism has been referred to as *Synechococcus* sp. strain PCC 7942 but a redesignation of this strain as *S. elongatus* PCC 7942 settled the ambiguity.

Li and Golden (1993) employed a promoterless *luxAB* gene set that encodes the luciferase from *Vibrio harveyi* and fused it to the promoter of a PSII gene, *psbAI* (that encodes the photosystem II reaction center D1 protein). This fusion product has a strong expression in *S. elongatus* PCC 7924 because of its integration at a site in the chromosome that has been used previously for harbouring heterologous genes (Kondo *et al.*, 1993). The resulting strain, designated as AMC149, emitted light, showed a persistent 24 h rhythm of bioluminescence when the cells are shifted from LD to LL. By using this strain it was possible to demonstrate the phase resetting cues at different times during circadian cycle that is characteristic of circadian clock controlled phenomena as seen in eukaryotes. Exposure of cultures to opposite LD cycles resulted in opposite peaks of light production after transfer to LL. Furthermore, single pulses of 4 h of darkness in otherwise LL conditions shifted the peaks sooner or later. Another important feature is that the period remained close to 24 h when measured at 25°C, 30°C and 35°C signifying that the bioluminescence rhythm is also temperature compensated (Kondo *et al.*, 1993). The rhythmic expression of bioluminescence from *luxAB* reporter has also been demonstrated in two other cyanobacteria (*Synechocystis* sp. strain PCC 6803 and *Anabaena* sp. strain PCC 7120) that have been widely manipulated at the level of molecular genetics.

Liu *et al.* (1995) concluded that bioluminescence is an accurate parameter to measure the activity of *psbAI* gene in AMC149 because of rhythmic variation in mRNA of *psbAI* gene and the luciferase enzyme. The bacterial luciferase enzyme uses oxygen and reduced flavin mononucleotide as substrate and these are expected to vary in a circadian fashion as a consequence of rhythmic photosynthesis. A long chain aldehyde, N-decanal has been used as a third substrate for luciferase at 3% (v/v) dissolved in vacuum pump oil. Colonies of AMC149 were formed in 10 days in LL condition at 30°C. After subjecting the plates for 12 h darkness to reset the circadian clocks, the bioluminescence was monitored at various times during LL by exposing the colonies to decanal vapours. Bioluminescence reached a stable level within 15 min and continued longer than a week.

After having chosen the ideal experimental organism as *S. elongatus* PCC 7942 and its genetically transformed strain (with bioluminescence *luxAB* reporter gene) AMC149, the next hurdle to overcome is the rapid screening of thousands of colonies for circadian phenotypes. Kondo and Ishiura (1994) developed a device that can screen thousands of colonies with the help of a charge-coupled device (CCD) camera and a photomultiplier tube device. A turntable device that can hold a dozen petri-plates (each with as many as 1000 widely separated colonies) helped in screening the individual colonies for the light emission. Circadian profiles of bioluminescence of individual colonies could be recorded after several days of monitoring with this automated screening. Genetic analysis of 800 insertion mutant strains revealed that none of them was clearly arrhythmic. On the basis of waveforms of bioluminescence rhythms, the genes were grouped into five classes. Class 1 to class 3 rhythms though possess symmetrical waveforms but peaked in different phases. Strains grouped in class 4 rhythms are with asymmetric, saw-toothed waveforms whereas those grouped in class 5 show binodal peaks per cycle. These point towards the existence of several pathways of output from the circadian oscillator in *S. elongatus* PCC 7942 (Liu *et al.*, 1995).

In order to determine the number of genes that control the circadian clock in *S. elongatus* PCC 7942, Liu *et al.* (1995) made several fusions of the regulatory regions of known genes to *luxAB*. All such resulting mutants showed typical bioluminescence characteristic of the *S. elongatus* PCC 7942 but at lower amplitude. This confirmed that all the known genes are under the regulatory control of circadian clock. In an alternative approach, to assess the percentage of genes under circadian control throughout the genome, Liu *et al.* (1995) performed a random fusion of a promoterless *luxAB* gene set throughout the chromosome. This was feasible because *luxAB* can undergo homologous recombination between a segment on the chromosome and a cloned copy of the same region on a non-replicating strand. As a result of the single crossover event, the transforming plasmid gets incorporated into the chromosome at the site of recombination. Rhythmic transcription can easily be measured by following the luminescence (Andersson *et al.*, 2000; Mackey *et al.*, 2007). Most of the *S. elongatus* promoters drive *luxAB* with a bioluminescence with peaks at dusk (in a light-dark cycle) or subjective dusk (corresponding to dusk in continuous light) or a trough at dawn or subjective dawn. As a matter of convention, subjective dusk is considered as incubation period of 12 h, 36 h and so on in LL after a 12-h pulse of darkness. Likewise, subjective dawn is around 24 h, 48 h and so on in LL after a 12-h pulse of darkness. It has thus been possible to identify dusk peaking genes (*psbAI, psbAII, psbAIII, KaiA, KaiBC* and *rmA*) in *S. elongatus* PCC 7942 with Class I expression patterns (Liu *et al.*, 1995; Ishiura *et al.*, 1998). Plasmid pBR322 carries a kanamycin resistance marker and thus it is easy to score recombinants for kanamycin resistance. Secondly, due to the presence of *bom* site, it can easily be mobilized in a conjugational event from *E. coli* to *S. elongatus* PCC 7942. This was considered to be the most efficient way of transfer rather than by transformation (Tsinoremas *et al.*, 1994). Nearly 80 per cent of the recombinant clones showed a circadian rhythm with characteristic bioluminescence, divisible into two classes. Strains that showed peaks during LL have been included in Class I. In contrast, those strains with peaks of bioluminescence at other times quite opposite that of AMC149 have been grouped in Class II. AMC287 is a strain in which *luxAB* gene set has been integrated into *purF* region governing a key enzyme in purine biosynthesis. It was also found that *purF* region lies immediately adjacent to *purL* region in *S. elongatus* PCC 7942 and is co-transcribed with *purL* region, the difference being the bioluminescence pattern resembled that of Class 1 group of strains. Liu *et al.* (1996) found that a second promoter that is responsible for Class 2 type of bioluminescence of *purF* appears to lie within the C-terminal coding region of *purL*. A Tn5-*luxAB* transposon construct developed by Wolk *et al.* (1991) has been used for creating random *luxAB* fusions throughout *S. elongatus* PCC 7942 genome. A new circadian class 2 gene designated as *opcA* has been identified on the basis of transposition of a modified Tn5 element conferring kanamycin resistance into *S. elongatus* by triparental conjugation. Although the biochemical properties of the gene product of *opcA* are not known, it is very much required for the functioning of glucose-6-phosphate dehydrogenase (encoded by *Zwf*) and for the oxidative pentose phosphate pathway. A mutant, isolated by the insertion of Tn5 into *Zwf* upstream of *opcA*, was able to transcribe *opcA* independently of *Zwf* as a circadian class 2 gene. However, inactivation of *opcA* gene affected reductant production at night and caused a limitation in the supply of substrate for luciferase (Min and Golden, 2000).

The main conclusions drawn from the above studies are that (i) there is a global circadian control of gene expression in S. *elongatus* PCC 7942, (ii) there is some individuality in the expression of genes and this can be directly correlated to temporal separation of biochemical processes, and (iii) the circadian regulation imposes rhythmicity of transcription via a general mechanism that is not gene specific.

A bacterial luciferase gene set derived from *Xenorhabdus luminescens* (Xl *luxAB*) was used as a reporter gene and fused with the promoter region of *psbA1* gene to identify the expression of *psbA1* gene in *Thermosynechococcus elongatus* BP-1, a thermophilic cyanobacterium. Rhythmic expression of the gene, as measured by the luminescence with an automated monitoring apparatus, occurred with a period close to 25 h for at least 10 days in constant light and temperature. The rhythm was temperature compensated over a wide range of temperatures (30 to 60°C) (Onai *et al.*, 2004).

III. CIRCADIAN CLOCK

Three essential components have been identified, i.e. the circadian clock also known as the oscillator, a photoreceptor for setting the phase of the clock and a receptor for perceiving environmental stimuli such as temperature (input pathway) and a means of relaying clock information to the various behaviours that are under circadian control (output pathway). The bioluminescence system with reporter *luxAB* marker developed for S. *elongatus* PCC 7942 fulfills all these requirements for explaining cyanobacterial circadian clock. Knowledge that has been generated on the circadian clock in cyanobacteria can be useful to understand circadian clocks in higher organisms, including mammals (Ditty *et al.*, 2003; Mackey *et al.*, 2008; Eguchi *et al.*, 2008; Johnson *et al.*, 2008a,b; Johnson, 2010).

A) ***The oscillator in Synechococcus*:** The oscillator should be able to control rhythmicity and period of all downstream behaviour. In order to examine this, mutants affected in clock phenotype would be best suited for the purpose. Realizing this, Kondo *et al.* (1994) isolated mutants of strain AMC149 after mutagenesis with methylmethane sulfonate and screened over 1,50,000 clones for their bioluminescence. Twelve mutants were further characterized with altered bioluminescence rhythms whose periods ranged from 16 h to 60 h. Certain of the mutants exhibited bioluminescence at such a low amplitude that it can even be considered as having no rhythms (arrhythmia). They also demonstrated that it is possible to clone mutant genes by complementation involving single recombination strategy similar to that used for random insertion of *luxAB* genes throughout the genome. In this regard, they identified a period extender gene (*pex*) in *S. elongatus* PCC 7942 that has a 22-h period and designated it as SP22. Kutsuna *et al.* (1998) showed by sequence analysis that SP22 did not have a mutation in the genomic DNA segment carried on pS1K1 (and that *sp22* mutation was later found in a recently cloned new clock gene, *KaiC).* As a matter of fact, *pex* gene that was carried on *pS1K1* was a suppressor gene for the *sp22* mutation. *Pex* gene encodes a protein of 148 amino acid residues that is thought to modulate function of the central clock oscillator. Genetic analysis of such mutants offered the possibility that the circadian clock (oscillator) in *Synechococcus* is under the control of three ORFs that may form one operon. X-ray crystallography and biochemical characterization of Pex from *S. elongatus* PCC 7942 revealed that the molecule has a $\alpha+\beta$ structure with a winged helix motif and is expected to function as a dimer. By the winged portion, Pex is able to recognize dsDNA. The DNA-binding ability of Pex has further been substantiated by the isolation of a *pex* mutant in which Pex fails to bind to DNA. Due to this, the period-extension activity of *pex* gene is lost in the mutant. So Pex is suggested to be a DNA-binding transcription factor (Arita *et al.*, 2007).

B) ***Genes that regulate the clock*:** Kondo *et al.* (1994) identified the above three ORFs by screening as many as 500,000 clones of *S. elongatus* PCC 7942 obtained after mutagenesis of AMC149 strain by EMS and characterized at least 100 mutants with altered circadian phenotypes. These mutants did not show any other altered phenotype except for the differences in the period of the rhythms. The recognition of the mutants and their isolation was possible because of the addition of wild-type *Synechococcus* DNA. The DNA retrieved from such mutants showed that the wild-type DNA could complement the mutant phenotype. From these studies they concluded that a cluster of three adjacent genes designated as *kaiA*, *kaiB* and *kaiC* regulate the circadian clock in *S. elongatus* PCC 7942. The meaning of '*Kai*' in Japanese means "rotation" or "cycle". It is further interesting to know that these three genes could be rescued by a plasmid carrying the entire *kaiABC* cluster. Mapping studies by DNA sequencing revealed that at least 19 mutations could be assigned to the three *kai* genes. All these mutations were suggested to be mis-sense mutations due to single nucleotide exchanges. Further analysis revealed that (i) the three genes had two mutations mapped to each one of them, (ii) all possible clock phenotypes belonged to *kaiC*, (iii) there is no significant similarity in between the *kai* genes or any other known gene sequence from prokaryotes or eukaryotes except that gene sequences similar to *kaiC* have been noted in certain archaebacteria and (iv) the *kai* group of genes appear to be clock-specific because deletion of either one of the genes or the entire cluster does not affect the viability of the strain but results in arhythmicity.

Ishiura *et al.* (1998) proposed a feedback model to explain the working of circadian oscillator. A promoter situated upstream of *kaiA* is responsible for the production of monocistronic *kaiA* mRNA where as the promoter located upstream of *kaiBC* cistrons produces a dicistronic transcript for *kaiB* and *kaiC* genes. It has also been observed that the transcripts of *kaiA* and *kaiBC* show rhythmicity. Kutsuna *et al.* (2005) studied the transcriptional regulation of *kaiBC* and identified a region of 56 bp sequence from transcription start site (upstream of the operon that is located from –55 to +1). This region has been designated as the sufficient promoter region (SPR). Of this, especially the sequence from –52 to –28 bp was found to be essential for *kaiBC* transcription. Deletion of SPR from upstream or downstream sequences almost completely eliminated promoter activity. At the same time, a constitutive negative regulatory region upstream of SPR (base pairs –897 to –56) was also identified that extended into the coding region of *kaiA*.

Inactivation of one or the other of the *Kai* genes abolishes the rhythms and lowers *kaiBC* promoter activity. This has been termed as negative feed back where as *kaiA* overexpression enhanced *kaiBC* promoter activity and this has been termed as positive feed back. *kaiC* overexpression resets the phase of the rhythms. Accordingly, the *kaiC* expression is directly linked to the phase of oscillation. Nishimura *et al.* (2002) introduced mutations in *kaiA* region by PCR mutagenesis, a technique that was earlier reported in case of bacteria (Diaz *et al.*, 1991) and *Synechocystis* sp. strain PCC 6803 (Narusaka *et al.*, 1999).

C) *Properties of Kai proteins*: KaiA possesses an amino-terminal domain (KaiA 135N) that has no function in the autophosphorylation and on the basis of NMR structure determination it has been designated as a pseudo-receiver domain. KaiA protein exists at least as long and short types. Amino acid sequence of long type consists of approximately 300 amino acid residues. This has been confirmed in case of *S. elongatus* PCC 7942, *Synechocystis* sp. strain PCC 6803 and *Synechococcus* sp. strain WH 8102. The conservation in the amino acid sequence in the amino-terminal region (200 amino acids) of the long type of KaiA protein is very limited whereas a high degree of conservation persists in the 100 amino acid residues of the carboxy-terminal region. The short type of KaiA protein (from *Anabaena* sp. strain PCC 7120 and *Nostoc punctiforme* ATCC 29133) is independent of carboxy-terminal domains and these lack the amino-terminal regions. Proteolytic studies of KaiA (284 amino acid residues) protein yielded three major protein species with molecular weights of 22kDa, 17kDa and 13kDa consisting of 1–189, 1–154 and 164–284 amino acid residues, respectively. The C-terminal domain of KaiA consisting nearly of 100 amino acid residues is sufficient for binding to KaiC. Crystal and solution structures of KaiA domains reveal it to be a dimer with tight interactions (Williams *et al.*, 2002; Uzumaki *et al.*, 2004; Ye *et al.*, 2004; Garces *et al.*, 2004; Vakonakis *et al.*, 2004; Fig. 1A).

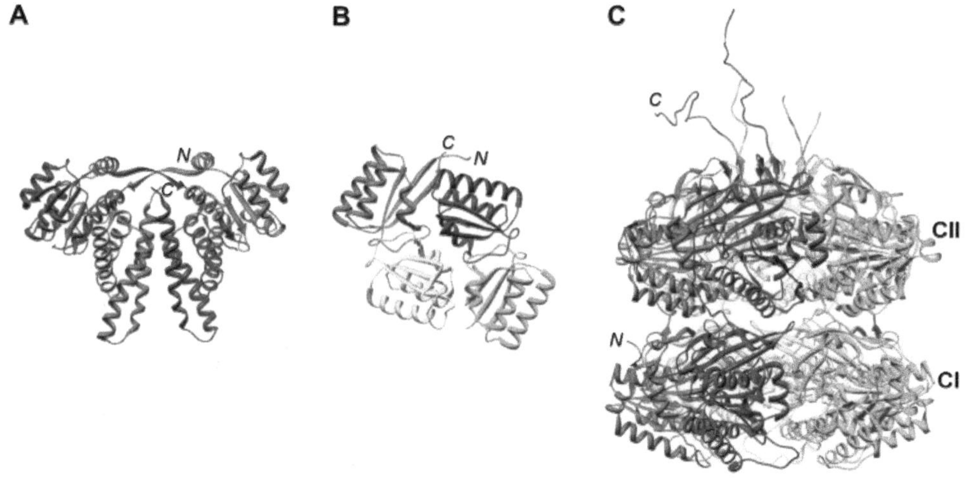

Figure 1: Three-dimensional structures of the cyanobacterial clock proteins KaiA, KaiB, and KaiC. (A) Crystal structure of the domain-swapped KaiA dimer from *S. elongatus*, PDB-ID 1R8J, http://www.rcsb.org (Ye *et al.*, 2004). (B) Crystal structure of the KaiB tetramer from *Synechocystis*, PDB-ID 1WWJ (Hitomi *et al.* 2005). (C) Crystal structure of the KaiC hexamer from *S. elongatus*, PDB-ID 2GBL (Pattanayek *et al.*, 2006). Subunits are colored differently and N- and C-terminal ends are labelled. ATP molecules bound between subunits in the N-terminal CI and C-terminal CII halves of KaiC were omitted. The images were produced with the UCSF Chimera package (Pettersen *et al.*, 2004). Illustration courtesy Martin Egli, Department of Biological Sciences, Vanderbilt University, Nashville, Tennessee, USA.

Color image of this figure appears in the color plate section at the end of the book.

KaiB consists of 108 amino acid residues and has a molecular weight of 12 kDa (Iwase *et al.*, 2004). NMR structure analysis of KaiB from *Synechocystis* sp. strain PCC 6803 revealed it to be a tetramer (Fig. 1B). This homotetrameric structure of KaiB forms an open square that is suitable for protein-protein interaction with positively charged residues being located around the perimeter. KaiB is localized on the phospholipid-rich membrane and gets translocated into the cytosol to interact with other Kai components, KaiA and KaiC. Further, the sensory domain of KaiB resembles thioredoxin family of proteins with β-α-β topology similar to the sensory domain of SasA (Hitomi *et al.*, 2005). However, studies on X-ray crystal structure (at a resolution of 2.6 A°) of KaiB from *Thermosynechococcus elongatus* BP-1 revealed an overall shape of the tetramer to be an elongated hexagonal plate with a single positively charged cleft flanked by two negatively charged ridges. The biological function of KaiB thus seems to be dependent on the positively charged cleft flanked by negatively charged ridges. This has been further confirmed by the site-directed mutagenesis of *kaiB* where substitution of residues Lys-11 or Lys-13 by Ala in the cleft region or deletion of C-terminal residues from 95–108 forming the ridge, dampens the *in vivo* circadian rhythms (Iwase *et al.*, 2005).

KaiC belongs to the family of DNA-binding ATPase superfamily of proteins, ATPase (DnaB) and a DNA recombinase (RecA; Leipe *et al.*, 2000). It forms a hexamer *in vitro* (Mori *et al.*, 2002; Hayashi *et al.*, 2003). Mori *et al.* (2002) reported a weak DNA-binding activity of KaiC protein. KaiC consists of 519 amino acid residues and has a duplicated tandem structure, i.e. the first and second halves of the KaiC amino acid sequences are similar. These two regions are designated as CI and CII domains that possess Walker's motifs (also called as P-loops). Two KaiA-binding sites are present, one in the middle (at approximately 240 to 260 aa residues) and the second at 500 to 519 aa residues. Three dimensional structure of KaiC hexamers has been studied by electron cryomicroscopy. Each KaiC monomer is a dumb-bell-shaped structure composed of two domains divided by relatively narrow portion. These domains probably correspond to the N-(amino acid residues 1–268) and C-terminal halves (amino acid residues 269–518) which have a duplicate amino acid sequence (Hayashi *et al.*, 2003). Pattanayek *et al.* (2004) studied the crystal structure of KaiC that revealed a doughnut-shaped structure with a central pore/channel. The hexameric structure showed an ATP-binding and a scaffold portion for Kai-protein complex formation (Fig. 1C). The central channel runs through that is wide open at N-terminal portion and narrows down at C-terminal region. The channel in CII half represents the portion where it binds to DNA.

KaiC is a kinase that autophosphorylates its own Ser and Thr residues *in vitro* (Nishiwaki *et al.*, 2000) and shows a clear circadian phosphorylation rhythm *in vivo* (Iwasaki *et al.*, 2002). Mutation of either of these residues nullified circadian rhythms in KaiBC gene expression. Two autophosphorylation sites on KaiC have been identified by Nishiwaki *et al.* (2004) by using mass spectrometry. These are at Ser431 and Thr432. By site directed mutagenesis, these amino acids were substituted by Ala to obtain single and double mutants. The phosphorylation of KaiC was reduced in the former while in the latter it was completely abolished suggesting that these are the sites that are phosphorylated *in vivo* as well. With loss of phosphorylation, the mutants lost circadian rhythm. These proteins were expressed in *E. coli* BL21, purified and *in vitro* complex formation studies were conducted further to know the interactions. Though KaiC was able to form hexamers *in vitro* it was unable to form complexes with KaiA, KaiB and SasA. These observations further signify that KaiC phosphorylation regulates its own transcriptional repression activity by controlling its binding affinity to other clock proteins. On the other hand, Xu *et al.* (2004) identified three potential phosphorylation sites at Thr432, Ser431 and Thr426 residues in the KaiCII domains by crystallographic and mutagenic analysis. Substitution of these residues by Ala singly or in combination altered phosphorylation of KaiC with concomitant abolition of circadian rhythms. Analysis of the mutants suggested that Ser431 and Thr426 may share a phosphate that can shuttle between these two residues there by modulating the KaiC activity. So phosphorylation of Thr432 and Ser431/Thr426 appear to be the key component for circadian clock to function in cyanobacteria. McClung (2007) suggested that KaiC possesses at least three enzyme activities, i.e. an autokinase, autophosphatase and ATPase and these three activities mutually influence each other. Evidences in favour of a weak and stable ATPase activity of KaiC are presented by Terauchi *et al.* (2007). In presence of KaiA and KaiB the ATPase activity oscillated with a circadian period *in vitro*. At different temperatures too the ATPase activity of KaiC was not variable suggesting that temperature compensation of circadian period seems to be mainly dependent on this reaction. They thus concluded that KaiC ATPase activity is the most fundamental reaction underlying circadian periodicity in cyanobacteria.

D) *Interaction between Kai proteins*: Two types of interactions are known, i.e. homotypic and heterotypic interactions (Iwasaki *et al.*, 1999). Based on studies on resonance energy transfer to assay protein-protein interactions, Xu *et al.* (1999) demonstrated that KaiB proteins interact among themselves (homotypic interaction). Based on the studies on a long period mutant that exhibits heterotypic interaction between KaiA and KaiB, it was suggested that interaction between Kai proteins is important for clock mechanism (heterotypic interaction). Further a negative feed back control of KaiC expression by KaiC protein to induce the formation of circadian oscillation in cyanobacteria. Two important points that emerge are that (i) protein-protein interactions are necessary for clock expression and (ii) the oscillation is sustained by *KaiA* by enhancing *KaiC* expression.

The phosphorylation of KaiC constitutes a potential signal for two important processes. Firstly, ectopic overexpression of KaiA leads to increased *kaiBC* transcription that in turn increases *kaiC* gene expression patterns. Secondly, ectopic overproduction of KaiC represses *kaiC* gene expression patterns. Thus KaiA and KaiC are suggested to be positive and negative regulators of *kaiBC* expression (Ishiura *et al.*, 1998). In both situations, i.e. due to elevated *kaiC* gene expression and its repression, the cells experience an arrhythmic state. This pointed toward the existence of a feed back regulatory mechanism required for a cyclic pattern of gene expression regulated by the circadian clock.

Iwasaki *et al.* (1999) reported a yeast two-hybrid system that was useful for demonstrating the interaction of KaiA with KaiC both *in vitro* as well as in *Synechococcus* cells. Two KaiA-binding domains (C_{KABD1} and C_{KABD2}) in KaiC have been identified by Taniguchi *et al.* (2001) by the use of yeast two-hybrid system and further confirmed by *in vitro* interactions. These two domains, i.e. C_{KABD1} and C_{KABD2} were found on corresponding C-terminal regions CI and CII, respectively. Many of the mutations reported earlier by Ishiura *et al.* (1998) in these two domains of KaiC exhibited altered interaction of KaiC with KaiA. Similar studies were made by Vakonakis *et al.* (2004a) and Vakonakis and LiWang (2004) on *T. elongatus* BP-1. NMR structural studies revealed that the C-terminal domain of KaiA that interacts with KaiC is a novel α-helical homodimeric structure. This is shown to interact with the linker region connecting the two globular KaiC domains. A KaiC region of ~12 kDa interacting with C-terminal domain of KaiA (residues 180–283; ThKai A180C) from *T. elongatus* BP-1 has also been identified by Vakonakis *et al.* (2004b). Vakonakis and LiWang (2004) identified a KaiC derived peptide (residues 488–518, CII_{ABD}) that interacts with terminal domain of KaiA (residues 180–230; ThKai A180C). Studies on NMR structure of the complex (~32kDa) revealed that CII_{ABD} binding to ThKai A180C alters the ThKai A180C dimerization angle suggesting that KaiA-KaiC affinity can be modulated by changes in dimerization geometry of the KaiA C-terminal domain. These protein-protein interactions have been explained in relation to the crystal structures of the three Kai proteins (Egli *et al.*, 2007).

Williams *et al.* (2002) explained the functional aspects of the three proteins of the circadian oscillator, probable role of *cikA* gene product in signal perception and the perception of the signal by KaiA protein that triggers autophosphorylation of KaiC protein. Additionally, KaiC phosphorylation rate determines (i) chromosome condensation (Mori and Johnson, 2001a); (ii) its aggregation state into homo and heterotypic interactions (Iwasaki *et al.*, 1999; Nishiwaki *et al.*, 2000) and (iii) its interaction with regulatory proteins such as SasA protein kinase (Iwasaki *et al.*, 2000) (Fig. 2). Another similarity of cyanobacterial circadian clock to post-transcriptional control of eukaryotic clock is the existence of KaiC in phosphorylated forms *in vivo* (Iwasaki *et al.*, 2002). KaiA is known to stabilize KaiC in its phosphorylated form and KaiB antagonizes the effect of KaiA (Iwasaki *et al.*, 2002; Williams *et al.*, 2002; Kitayama *et al.*, 2003; Ye *et al.*, 2004). Phosphorylated KaiC forms a tight complex with KaiA that is found at maximum level at clock time of 16 to 18 h (Kageyama *et al.*, 2003; Kitayama *et al.*, 2003). The ratio of phosphorylated to unphosphorylated KaiC is correlated with the period at which the clock runs (Ye *et al.*, 2004). Kitayama *et al.* (2003) estimated that KaiB and KaiC oscillate around an average of approximately 10,000 molecules per cell, while KaiA is present at an almost constant level of nearly 500 molecules per cell throughout the circadian cycle. While KaiC is present in the cytosol (near to the nucleoid region), KaiB exists in a bounded state to the membrane as well as occurs freely in the cytosol. Around clock time of 20 h, KaiB seems to be released from membrane into the cytosol there by its cytosolic concentration increases. The released KaiB attaches to KaiA-KaiC complex to form large clock complex (Kitayama *et al.*, 2003). Regulation of KaiC phosphorylation is proposed to be achieved due to the existence of clock proteins KaiA and KaiB in multiple states. This has been explained on the basis that even when *kaiBC* gene is regulated by KaiC protein, sustained oscillation of gene expression occurred and sustained oscillation of KaiC phosphorylation continued in the absence of transcription and translation processes (Kurosawa *et al.*, 2006a). A novel mutant of KaiC that was unable to reset the circadian clock has been isolated after mutagenesis of *S. elongatus*

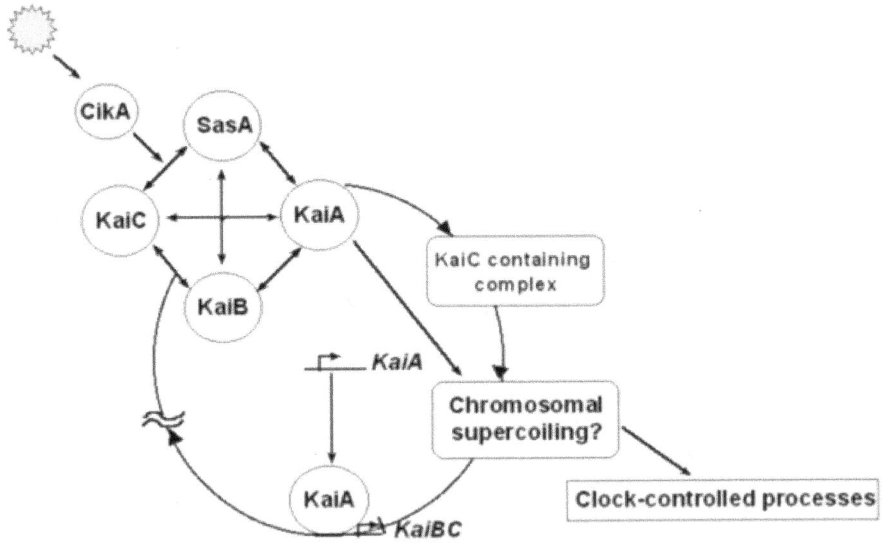

Figure 2: Molecular feedback loops of cyanobacteria. A cluster of *KaiABC* genes controls circadian rhythms in cyanobacteria. *KaiA* gene product acts as a positive regulator for *KaiBC* transcription, while *KaiBC* products along with other proteins inhibit their own transcription. With kind permission of V. K. Sharma, Chronobiology Laboratory, Evolutionary and Organismal Biology Unit, Jawaharlal Nehru Centre for Advanced Scientific Research, Jakkur, PO Box 6436, Bangalore 560 064, Karnataka, India [Paranjpe, D. A. and Sharma, V. K. (2005) *J Circadian Rhythms* **3:** 7 doi:10.1186/1740-3391-3-7].

PCC 7942 with EMS. This mutant designated as *pr1* (for KaiC phase response 1) exhibited normal growth as well as circadian rhythm similar to the wild-type. Interestingly, however, the rhythmicity in KaiC levels and its phosphorylation rates were greatly affected due to the mutation. Additionally, KaiC phosphorylation state was reduced along with decreased autoregulatory activity of KaiC. These studies emphasize the importance of protein phosphorylation cycle (PPC) of KaiC in resetting the cyanobacterial clock. In support of their above studies, Kurosawa *et al.* (2006b) presented a mathematical model for gene protein dynamics that can generate a sustained oscillation in both DD and LL conditions. According to them, KaiC phosphorylation seems to be the basis for the sustained oscillations in the absence of transcriptional or translational regulation.

Negative feed back of KaiC on its own promoter is another similarity that the cyanobacterial circadian clock shares with the eukaryotic clock proteins. Continuous over expression of KaiC represses *kaiBC* promoter (P*kaiBC*) and represses the rhythms of all promoters in the *S. elongatus* PCC 7942 genome. Two classes of *kaiC* repression have been identified by Nakahira *et al.* (2004). The first class termed as "high amplitude" expression, exhibited by 5–10% promoters including P*kaiBC*, is clock-dominated. The second type of response, exhibited by 90–95% of the promoters, has been termed as "low amplitude". This is a clock-modulated response where the genes have significant basal activity and are rhythmically modulated by KaiABC oscillator. KaiC thus coordinates the genome-wide gene expression as suggested by Johnson (2004). Genes that encode the proteins intrinsically associated with the circadian clock are associated with the second class. When KaiC was kept under the control of a synthetic promoter P*trc* from *E. coli*, KaiB and KaiC could restore the complete circadian rhythm in KaiBC inactivated arrhythmic strain of *S. elongatus* PCC 7942.

In contrast to autoregulatory nature of the circadian feed back loops in eukaryotes, in cyanobacteria the Kai promoters can be functionally replaceable. The replacement of P*kaiBC* has been reported first by Xu *et* al. (2003) followed by the replacement of P*kaiA* and P*kaiBC* by Nakahira *et al.* (2004). These observations signify that the circadian feedback loop in cyanobacteria does not require negative feed back of clock proteins upon specific clock promoters. The expression of an appropriate level of Kai proteins suffices the purpose for the oscillation of the circadian clock. Mori and Johnson (2001a) suggested that KaiC might mediate both its negative feedback regulation and global regulation of the cyanobacterial genome by bringing oscillations in the condensation and/ or supercoiling status of the entire cyanobacterial chromosome. Subsequent observations of Xu *et al.* (2003), Min *et al.* (2004) and Nakahira *et al.* (2004) added support to the observations of Mori and Johnson (2001a) that the cyanobacterial genome becomes an "oscillating nucleoid" or in other words an "oscilloid". This oscilloid model proposes that the global regulation of gene expression is in other words dependent on the rhythmic changes of chromosome topology (Woelfle and Johnson, 2006). They also emphasized the role of '*cis*' elements that not only globally mediate rhythmic gene expression governed by the chromosomal status but also on a promoter-by-promoter basis by specific '*trans*' acting factors. Woelfle *et al.* (2007) put this model to test by assaying changes in the topology of a non-essential cryptic plasmid (pANS) in *S. elongatus* PCC 7942. This plasmid was chosen as a reporter to identify the superhelical state of DNA in cyanobacteria that can change in a circadian fashion *in vivo*. They found that the plasmid topology changed in a circadian manner with rhythmic KaiC phosphorylation. When other cyanobacterial promoters were removed from their native chromosomal surroundings and introduced into this plasmid these also exhibited rhythmicity in expression that were dependent on KaiC and were repressed by *kaiC* overexpression. There is a possibility that KaiC-containing protein complexes either directly or indirectly modulate chromosome structure leading to global regulation of promoter activities in cyanobacteria.

Kucho *et al.* (2004a) demonstrated that DNA microarrays with unmodified oligonucleotide probes have been helpful in the detection of gene transcripts with greater sensitivity and specificity. With the application of such probes, they conducted a global analysis of circadian expression of genes in *T. elongatus* BP-1 and identified 2397 genes that accounted for 95% of genes in this organism. A comparison of the gene expression profiles at two different circadian times, i.e. LL2 (early subjective day) and LL14 (early subjective night) revealed that (i) the expression levels of 143 genes were significantly different at the two circadian times; (ii) the expression of 69 genes was enhanced at LL14 that were responsible for NADH dehydrogenase complex and other respiratory genes had peak expressions, and (iii) the expression of the remaining 74 genes was enhanced in the early subjective day corresponding to photosynthesis related genes (light-harvesting phycobilisome complex, chlorophyll synthesis, ferredoxin, chaperonin DnaK). The expression levels of genes involved in transcription and translation encoding elongation factor (EF-G) and eight ribosomal proteins were highest in between LL2 and LL14 (Kucho *et al.*, 2004b). In order to understand cellular physiology and the mechanism of clock-controlled genes, Kucho *et al.* (2005) used DNA microarrays to identify the cycling genes that are under circadian control. The expression of most of the genes peaked around the time of transition from day to night so as to enable the organism to adjust to the night condition. These genes related to a wide variety of metabolic pathways, membrane transport and signal transduction. Genes involved in respiration and poly-3-hydroxyalkonoate synthesis showed coordinated circadian expression suggesting that the regulation is important for the supply of energy and carbon source in the night. Genes involved in transcription and translation also followed a circadian cycling. Stoeckel *et al.* (2008) identified genes encoding different enzymes of central metabolic pathways such as glycolysis, oxidative pentose phosphate pathway and glycogen metabolism that showed oscillating transcript levels during diurnal cycles in *Cyanothece* sp. ATCC 51142. Since in this organism photosynthesis and nitrogen fixation are temporally separated, increased transcriptional activity at night coincided with the demands of nitrogen fixation.

Although the data obtained overwhelmingly support the idea that the regulation of clock genes in *S. elongatus* PCC 7942 resembles autoregulatory feed back models put forward for eukaryotes but the recent work focuses on the aspect that the mechanism seems to be more complex (Iwasaki *et al.*, 2002; Xu *et al.*, 2003; Tomita *et al.*, 2005). Ditty *et al.* (2005) detected that the rhythmic expression pattern of *kaiA* or *kaiBC* operon in certain of the mutants was altered but the rhythms persisted. Even if expression of *kaiA* or *kaiBC* occurred 12 h out of phase from the normal (and thus 12 h out of phase from other Kai locus) it did not affect the time required for one cycle. These results signify that the *Cis* elements within the promoters of the kai genes are not necessary to sustain clock functions.

On the basis of enhancement in KaiA-KaiB interaction *in vitro* by KaiC, Iwasaki *et al.* (1999) suggested the formation of a heteromultimeric complex containing all the three Kai proteins. Further, immunoprecipitation analysis at two points of the day revealed the existence of interactions between KaiC-SasA and KaiB-SasA (Iwasaki *et al.*, 2000). By studying a number of arrhythmic mutants of *S. elongatus* PCC 7942 that lack each functional Kai protein or SasA, Kageyama *et al.* (2003) demonstrated that KaiC forms multimeric protein complexes with other Kai proteins and SasA in a circadian fashion. That is the size of the complex varies from ~350 and 400–600 kDa during the subjective day and night respectively. This signifies that KaiC functions as a scaffold protein.

i) *Phosphorylation Cycle*: As phosphorylation of KaiC reaches its maximum rate at late night, it coincides with the association of all the component proteins (KaiA, KaiB, KaiC and SasA) into a periodosome. This assembly and disassembly of the component proteins coincides with the 24 h cycle and also with the rhythmicity in the phosphorylation of KaiC. When key residues in KaiC are mutated no phosphorylation takes place. Due to this the periodosome formation does not occur and the rhythms are abolished. Golden (2004) put forward a simple model explaining these facts.

Mori *et al.* (2002) included KaiC in the superfamily of RecA/DnaB of proteins that has a bearing on its enzymatic activity and its functional role in bringing global changes in gene expression patterns. These observations were based on analytical and ultracentrifugational analyses of the association of KaiC molecules into hexameric rings in presence of ATP. Hayashi *et al.* (2003) reported Mg-ATP induced hexameric ring structure of KaiC whereas in the absence of Mg-ATP KaiC remained as a monomer. SDS-PAGE analysis revealed a ladder of six bands at 56, 120, 173, 232, 287 and 340 kDa corresponding to cross-linked species ranging from monomer to hexamer. KaiC existed as a monomer below 1μM concentration of Mg-ATP and with the increase of concentration above this hexameric ring structure was favoured with no monomer existing at concentrations of 100 μM. This was further confirmed by the studies on single and double mutations in the Walker's motif A. Each single mutation resulted in a reduced affinity for ATP to different extents and greatly reduced hexamerization. The KaiC hexamer has the shape of a hexagonal pot with a large internal cavity. The diameter and the height of the hexameric KaiC are equal, i.e. 100 Å. It has only one opening and the cavity has a depth of 73 Å. However, the cavity has a width of 34 Å at its widest part and its entry is narrowed to 18 Å. Two molecules of KaiA can interact with one molecule of KaiC hexamer. However, the interaction of one molecule of KaiA with KaiC hexamer will suffice to enhance KaiC phosphorylation (Hayashi *et al.*, 2006). Of the two ATP-binding sites of KaiC, one at the N-terminal end is a high affinity site where as the C-terminal ATP-motif is a low affinity site. It is the N-terminal motifs that are responsible for hexamerization while the C-terminal motifs are responsible for both stabilization and phosphorylation of the KaiC hexamer. These studies are supported by ATPχS filter-binding assay of KaiC derived from *T. elongatus* BP-1 and its mutants (Hayashi *et al.*, 2004). KaiC phosphorylation can be reconstituted *in vitro* by using purified KaiA, KaiB, KaiC proteins with ATP. The phosphorylation and dephosphorylation cycles of KaiC *in vitro* were self-sustained and temperature compensated also. Due to the replication of these parameters in some of the KaiC mutants *in vivo*, Nakajima *et al.* (2005) proposed that the primary circadian oscillator KaiC in cyanobacteria is autonomous and not dependent on transcription and translation feedback loop of KaiBC. This chemical oscillator could reproduce temperature compensated rhythms and altered phenotypes in mutants (Naef, 2005).

Emberly and Wingreen (2006) observed a perfect synchronization between the release of monomers from KaiC hexamers during the day and formation of clusters of KaiC hexamers in the night. They also highlighted the importance of collective assembly and disassembly of proteins in biochemical networks in designing novel protein-based oscillators. The *in vitro* KaiC phosphorylation cycle and the changes in Kai protein complexes were also confirmed by Kageyama *et al.* (2006). Each phosphorylation cycle seems to be promoted by active and repeated association of KaiA with KaiC. High phosphorylation rates of KaiC induced the association of the KaiC hexamer with KaiB and inactivated KaiA leading to a dephosphorylation cycle, coinciding with formation of monomeric KaiC subunits from KaiC hexamers. Reduced phosphorylation led to the dissociation of KaiB from KaiC reactivating KaiA in the process. They are of the view that similar association and dissociation kinetics may be occurring *in vivo*. Mori *et al.* (2007) elucidated the ticking of an *in vitro* circadian clock by visualizing the rhythmicity of the KaiABC complex formation and its dissociation though electron microscopy and also quantifying by two-dimensional blue-native/ SDS-PAGE electrophoresis. Fluorescence resonance energy transfer with two populations of fluorescently labelled KaiC hexamers confirmed that monomer exchange among hexamers occurs (Fig. 3). According to them, this monomer exchange might be to maintain synchrony among KaiC hexamers in the reaction that helps in maintaining a high amplitude oscillation for several days (Fig. 4). *In vitro* oscillator has been subjected to entrainment by temperature pulses and could be reset. Based on computer modelling of interactions, Yoda *et al.* (2007) confirmed the experimental findings of Kageyama *et al.* (2006) for the existence of allosteric transition of KaiC hexamers as well as monomer shuffling. Furthermore, Miyoshi *et al.* (2007) presented a mathematical model for the *in vitro* functionality of KaiC phosphorylation cycle as well as its coupling to *in vivo* transcriptional/ translational feedback under LL conditions. According to van Zon *et al.* (2007) the synchronization of KaiC phosphorylation and dephosphorylation cycles very much depends on the existence of two types of hexamers of KaiC molecules, one the slowest and the other fastest. Due to a differential affinity, the former outcompetes the latter for KaiA molecules and in this process the fastest ones are slowed down making the entire population of KaiC molecules to oscillate in phase. Eguchi *et al.* (2008) explained the robustness of circadian oscillation of KaiC phosphorylation *in vitro* by putting forward a mathematical model and suggested that it is resilient to changes in the concentration of Kai proteins as well as a change in the system size.

Most of the mathematical models so far proposed took into consideration the equivalence of the three phosphorylation sites at Ser431, Thr432 and Thr 426 on KaiC molecule (Emberly and Wingreen, 2006; Kurosawa *et al.*, 2006; Mehra *et al.*, 2006; Takigawa-Imamura and Mochizuki, 2006; Clodong *et al.*, 2007; Miyoshi *et al.*, 2007; Mori *et al.*, 2007; van Zon *et al.*, 2007; Yoda *et al.*, 2007). However, the work of Rust *et al.* (2007) brought to light that there exist four forms of KaiC molecules *in vivo* that undergo phosphorylation differentially and each of these can be separated by means of SDS-PAGE and could be quantified by mass spectrometry. According to them, there are four forms of KaiC, i.e. unphosphorylated KaiC, KaiC phosphorylated at Ser431 (S-KaiC), Thr432 (T-KaiC) and KaiC phosphorylated both at Ser431 and Thr432 (ST-KaiC) which oscillated with a circadian period but with different phases. When unphosphorylated KaiC was incubated with KaiA, phosphorylation first occurred at Thr432 resulting in the formation of T-KaiC followed by the phosphorylation of Ser431.

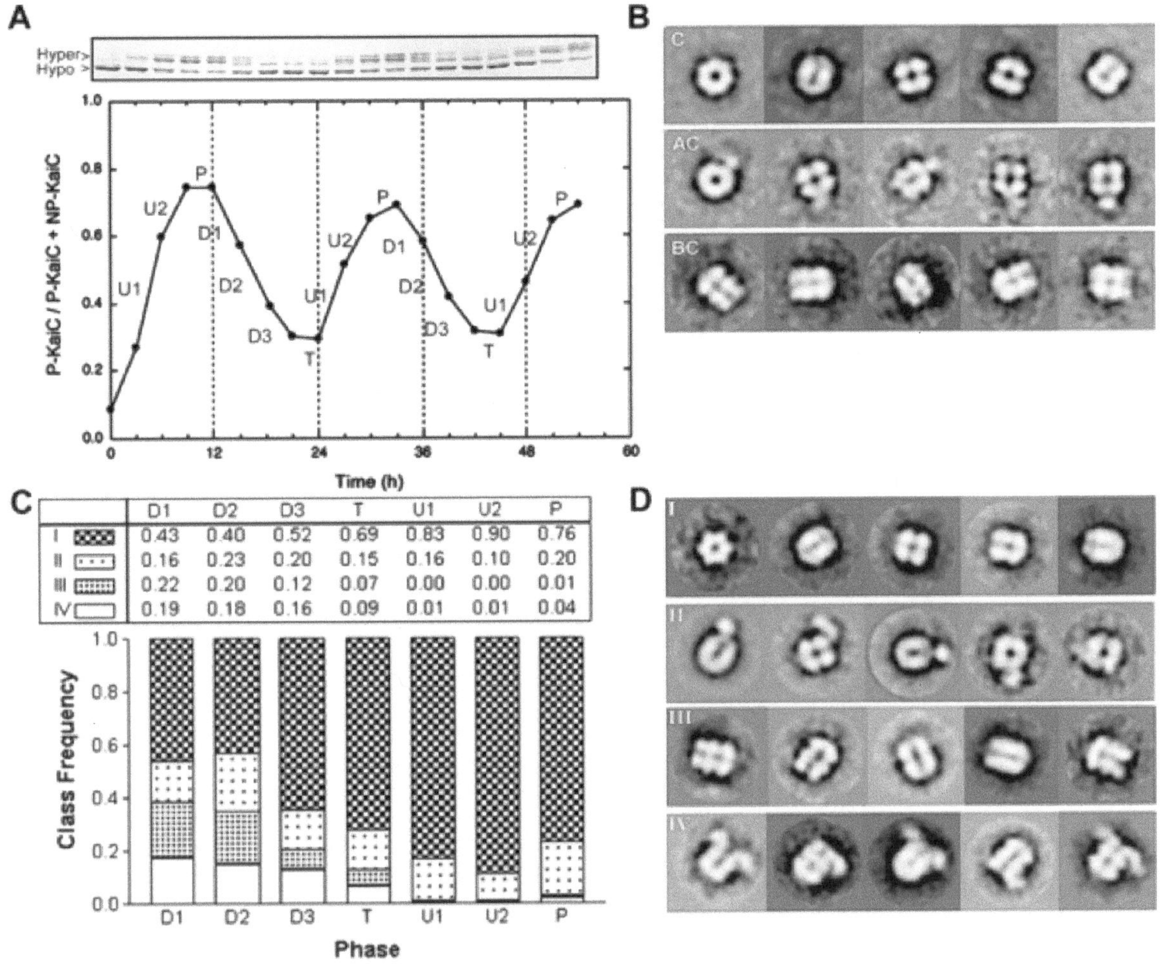

	D1	D2	D3	T	U1	U2	P
I	0.43	0.40	0.52	0.69	0.83	0.90	0.76
II	0.16	0.23	0.20	0.15	0.16	0.10	0.20
III	0.22	0.20	0.12	0.07	0.00	0.00	0.01
IV	0.19	0.18	0.16	0.09	0.01	0.01	0.04

Figure 3: EM Analysis of the Temporal Sequence of KaiABC Complex Formation.
(A) The *in vitro* oscillation of KaiC phosphorylation (upper panel), in which the lower band is hypophosphorylated KaiC (NP-KaiC) and the upper bands are different forms of hyperphosphorylated KaiC (P-KaiC). The lower panel is a quantification of the ratio of P-KaiC to total KaiC (P-KaiCþNP-KaiC) as a function of time. The oscillation cycle is divided into seven phases for the negative-stain EM analyses in (C) and (D).
(B) Representative EM class-sum images of KaiC alone (labeled C), KaiA-KaiC complexes (AC), and KaiB_KaiC complexes (BC).
(C) Determination of the relative frequency of the four meta classes at each phase of the KaiABC cycling reaction. The table reports the meta-class frequencies (top), and is plotted in the histogram (bottom) for each phase (as defined in [A]).
(D) Representative EM class-sum images generated for various phases of the KaiABC cycling reaction. The class-sum images were assigned to one of four structurally distinct meta classes (I–IV). With the kind permission of C. H. Johnson, Department of Biological Sciences, Vanderbilt University, Nashville, Tennessee, USA [Mori *et al.* (2007) *PLoS Biol* **5(4):** e93. doi:10.1371/journal.pbio.0050093] doi:10.1371/journal.pbio.0050093.g001.

This caused the production of second form of KaiC, i.e. ST-KaiC. Dephosphorylation of ST-KaiC at its Thr432 favoured the production of S-KaiC but due to the presence of KaiA rephosphorylation of S-KaiC at Thr432 resulted in ST-KaiC production and the concentration of S-KaiC remained low. But once the activity of KaiA was reduced, dephosphorylation of ST-KaiC was gradually favoured and

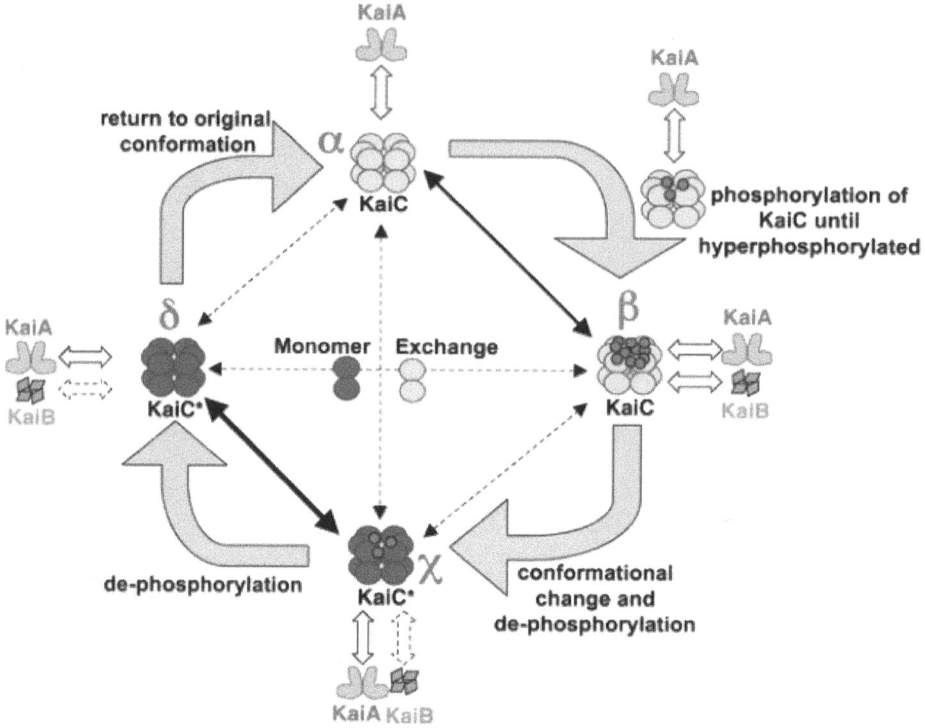

Figure 4: Depicted is a representation of the mathematical model for phosphorylation cycle of a KaiC hexamer and its association with KaiA and KaiB. A KaiC monomer is shown as a double circle that forms a hexamer that can associate/dissociate with KaiA and/or KaiB. KaiC hexamers are depicted in both their default status (light blue color) and in an altered KaiC* state that has undergone a conformational change (darker blue color). Red dots are phosphates attached to phosphorylation sites on KaiC (residues T426, S431 and T432). KaiC hexamers can exchange monomers between hexamers in any of the states, depicted by the double-headed arrows in the central region of the figure. The relative rates of exchange between these states are illustrated by the type of double-headed arrows (thick line = high rate; thin line = medium rate; and dashed line = low rate). With the kind permission of C. H. Johnson, Department of Biological Sciences, Vanderbilt University, Nashville, Tennessee, USA [Mori *et al.* (2007) *PLoS Biol* **5(4):** e93. doi:10.1371/journal.pbio.0050093] doi:10.1371/journal.pbio.0050093.g003.

Color image of this figure appears in the color plate section at the end of the book.

the built up of S-KaiC caused complete inactivation of KaiA. This led to dephosphorylation of T-KaiC and ST-KaiC due to which S-KaiC remained as the dominant form. Lastly, when dephosphorylation of enough S-KaiC had occurred then the activity of KaiA returned to normal and the cyclical formation of the four forms of KaiC took place (Rust *et al.*, 2007). Additional support for this has been provided by Li *et al.* (2009) who advocated that Thr432 constitutes the regulator for the oscillation amplitude while Ser431 serves as the major phase regulator. These dual phosphorylation sites on KaiC thus coordinate and control phosphorylation period. On the other hand, the importance of Thr426 to control KaiC phosphorylation status *in vitro* and *in vivo* has been emphasized by Xu *et al.* (2009) who adduced mutational and biochemical evidences in support of their conclusions. A change in Ser431 and Thr432 by way of site-directed mutagenesis or due to an inhibition of phosphorylation then Thr426 is liable to undergo phosphorylation. Kim *et al.* (2008) studied the nature of KaiA binding sites on KaiC (C_{KABD1} and C_{KABD2}) in the C-terminal loops (termed by them as A-loops).

They suggested that a conformational change in KaiC molecule makes the A-loops to be in an exposed or buried state. When A-loops in KaiC are exposed, the autokinase activity is favoured due to its interaction with KaiA at its binding site, moving ATP closer to the site. This results in a stabilization of the A-loops and phosphorylation continues. The autophosphatase activity of KaiC is prevalent when A-loops are in a buried state thereby disallowing the interaction of A-loops with KaiA. These conformational changes in KaiC molecule are suggested to be brought about by KaiB which interacts with KaiC at the A-loops. The type of conformational change KaiC undergoes during autophosphorylation and dephosphorylation cycles at Ser431 and Thr432 is explained by Chang *et al.* (2011). According to them autophosphorylation at Ser431 stabilizes the CII ring making it rigid but autophosphorylation at Thr432 relieves this rigidity to some extent. In presence of KaiA and KaiB the rigid and flexible states of CII ring of KaiC are rhythmically controlled and at the same time this sets the rhythm for the ATPase activity of the CI ring too. The evidences for such dynamic interactions of KaiA, KaiB and KaiC have been provided by Pattanayek *et al.* (2008) on the basis of X-ray crystallography, cryo-electron microscopy and gel-electrophoresis studies. The simultaneous binding of KaiA and KaiB to KaiC explains the role of KaiB in bringing conformational changes in KaiC and its role in phosphorylation and dephosphorylation of KaiC. Qin *et al.* (2010) studied the intermolecular associations between wild-type and mutant Kai proteins. They used KaiC mutant proteins in a hyperphosphorylated (with a change in Ser431Glu and Thr432Glu or Ser431Asp and Thr432) and hypophosphorylated (Ser431Ala and Thr432Ala or Ser431Ala and Thr432) state. The KaiB proteins with a change in Arg22 to Cys22 and Arg74 to Cys74 were used. The important findings are that: (i) KaiA and KaiB proteins rhythmically interact with KaiC poteins; (ii) KaiA has a higher affinity to hyperphosphorylated KaiC; (iii) hyperphosphorylated KaiC mutant proteins formed stable complexes with the three proteins; (iv) in the absence of KaiB, hyperphosphrylated KaiC could not form stable complex with KaiA; (v) mutant forms of KaiB interacted with KaiC to form KaiB.KaiC complexes or in presence of KaiA formed KaiA.KaiB.KaiC complexes very much in a similar fashion as those of the wild-type proteins; (vi) free and associated forms of the three proteins oscillated in phase and anti-phase, respectively and in this way the observations of van Zon *et al.* (2007) are confirmed. The KaiC phosphorylation rhythm has been confirmed *in vitro* at varying concentrations of Kai proteins and the regulation of phosphorylation cycle by KaiA and KaiB proteins has been termed as "parameter-tuning" and "state-switching", respectively by Nakajima *et al.* (2010). They suggested that the clock proteins oscillate *in vivo* in a circadian fashion and a possible entrainment mechanism is operative. A mathematical model presented by Brettschneider *et al.* (2010) could reproduce the phosphorylation dynamics of KaiC *in vitro* and identified KaiA inactivation as the primary event in the phosphorylation cycle. The inactivation of KaiA has been suggested to be due to its sequestration by KaiBC complex and evidences for a possible temperature entrainment mechanism have been presented with a temperature-dependent oscillation of KaiAC and KaiBC complexes.

ii) *Relationship between PPC and transcriptional-translational feed-back loop (TTFL)*: When *S. elongatus* PCC 7942 was grown in LL, it exhibited a turnover of KaiC protein in a circadian phase-dependent manner (Imai *et al.*, 2004). Proper circadian changes in KaiC accumulation involved transcriptional, translational and post-transcriptional processes. Inhibition of translation led to KaiC degradation and phosphorylation proceeded within at least 4 h in a circadian phase-dependent manner. Due to the reconstitution of a self-sustained circadian oscillator *in vitro*, Nakajima *et al.* (2005) put forward the idea that the circadian oscillator *in vivo* also is an autonomous KaiC phosphorylation unit that is not dependent on transcription and translation feedback loop. In fact some of the mathematical models also supported this idea (Kurosawa *et al.*, 2006b). However, convincing evidences for the operation of two KaiC-based oscillators, i.e. the first one by the PPC and the second one by TTFL have been put forward by Kitayama *et al.* (2008). The KaiA overexpression mutants (*ox-kaiA*) of *S. elongatus* PCC 7942 showed normal temperature-compensated circadian oscillation, despite KaiC being present in a phosphorylated but in an arrested state. It means even when the PPC is not allowed to operate, there is an additional input of KaiC proteins through TTFL. Kitayama *et al.* (2008) used three types of KaiC mutants to demonstrate the existence of TTFL in the absence of PPC. The first category of KaiC mutants are those in which Ser431 and Thr432 have been substituted by Ala (isolated by Nishiwaki *et al.*, 2004) and in this respect KaiC in these mutants constitutively exists always in an hypophosphorylated state but *kaiBC* expression in these mutants fluctuated arrhythmically. The second category of KaiC mutants are those in which Ser431 and Thr432 have been substituted by Glu. Since Glu mimics phosphorylation at these sites and so KaiC is said to exist in a hyperphosphorylated state. These mutants showed a dampened but a clear rhythm. The levels of KaiB and KaiC proteins also fluctuated rhythmically. An autokinase deficient mutant with a change in Lys294 to His294 (isolated by Hayashi *et al.*, 2004) produced unphosphorylated KaiC. All these point to the functioning of a second KaiC oscillator dependent on TTFL that is also equally important for the maintenance of circadian rhythms in cyanobacteria. Qin *et al.* (2010) preferred to call the PPC-based KaiC oscillator as the post-translational oscillator (PTO) and considered it as the core pacemaker and the TTFL as the slave oscillator. Based on the mathematical modelling and biochemical evidences they concluded that: (i) the core oscillator can function independently of TTFL; (ii) any inhibition of PTO/PPC would adversely affect TTFL; (iii) the slave oscillator can make additional inputs of KaiC protein into the core oscillator; and (iv) the core oscillator is coupled to the slave oscillator. The importance of TTFL has been emphasized when cell doubling times are faster and fall below 24 h. In such cells the PTO is unable to meet the demand to maintain circadian rhythms but a feedback of KaiC proteins from TTFL helps to cope up the situation (Fig. 5). Zwicker *et al.* (2010) used stochastic simulations to show that PTO and TTFL are intertwined and work in tandem to maintain robust circadian rhythms in cyanobacteria experiencing high growth conditions. These developments have been summarized (Johnson *et al.*, 2008, 2011).

IV. INPUT PATHWAYS

Phase resetting studies on mutant AMC149 led to the alternation of light and dark phases (Kondo *et al.*, 1993). Apart from light, other environmental signals like temperature, pH, osmotic environment may also play a significant role in phase setting input to the circadian clock in *S. elongatus* PCC 7942. However, the participation of photoreceptors (phytochome or other photoreceptor) cannot be ruled out. In this connection, the observations of Schmitz *et al.* (2000) are noteworthy who identified *cikA* (circadian input kinase) gene sequence in *S. elongatus* PCC 7942 that encodes a bacteriophytochome related kinase. This has been identified as a key component of the input pathway that supplies environmental information to the circadian oscillator in *S. elongatus* PCC 7942. It possesses three domains a GAF, a histidine protein kinase (HPK) and a pseudo-receiver domain (PsR) similar to those of response regulator family (Schmitz *et al.*, 2000). Further it was observed that inactivation of CikA abolishes the ability of the clock to be reset by a dark pulse. The possibility of CikA binding to the chromophore has been ruled out due to the absence of the conserved amino acid residues in the chromophore binding domain and *in vivo* binding experiments were negative (Matsuda *et al.*, 2003). The manner in which CikA functions and the type of environmental signal it perceives has come to light by the studies of Ivleva *et al.* (2006) who presented evidences that it senses not light but the

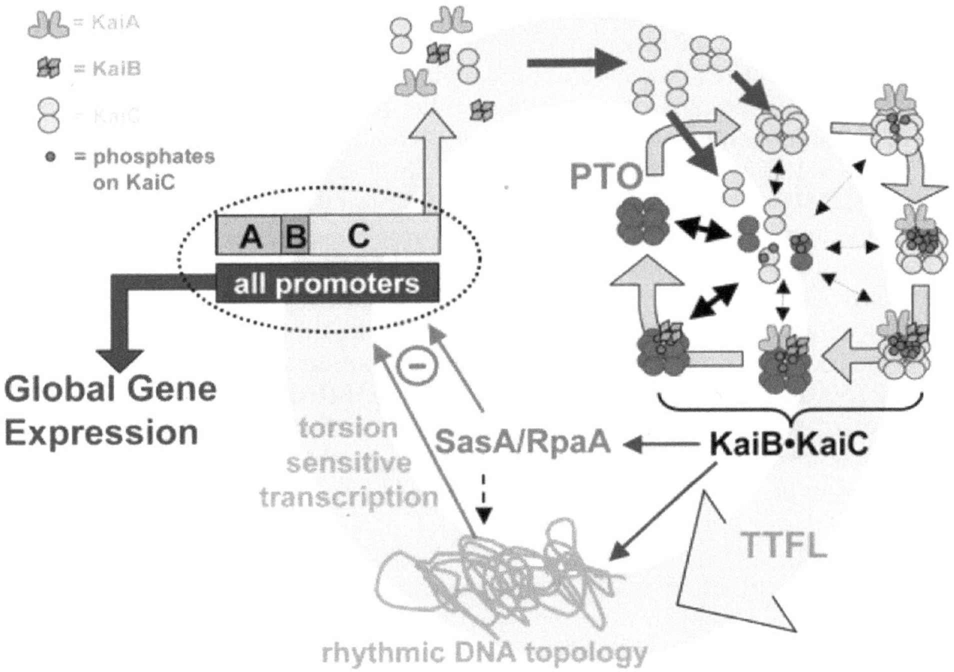

Figure 5: The core PTO is embedded in a larger TTFL. The PTO is linked to the damped TTFL (indicated by the pink background circle) by transcription and translation of the kaiABC cluster. Global gene expression is mediated by rhythmic modulation of the activity of all promoters, including those driving the expression of the central clock gene cluster, kaiABC (= ABC in figure). Rhythmic DNA torsion and/or transcriptional factor activity (e.g. RpaA/SasA) modulate global promoter activities. Cyclic changes in the phosphorylation status of KaiC that mediate the formation of the KaiBNKaiC complex regulate DNA topology/transcriptional factors. The PTO (cycle connected by lavender arrows in upper right quadrant) is determined by KaiC phosphorylation as regulated by interactions with KaiA and KaiB. Robustness is maintained by synchronization of KaiC hexameric status via monomer exchange (Ito *et al.*, 2007; Mori *et al.*, 2007). Monomer exchange is depicted in the figure by "dumbbell" KaiC monomers exchanging with KaiC hexamers in the middle of the PTO cycle; phase-dependent rate of monomer exchange is indicated by the thickness of the double-headed black arrows. The shade of KaiC hexamers (dark versus light blue) denotes conformational changes that roughly equate to kinase versus phosphatase forms. New synthesis of KaiC feeds into the KaiABC oscillator as non-phosphorylated hexamers or as monomers that exchange into pre-existing hexamers. If the new synthesis of KaiC occurs at a phase when hexamers are predominantly hypo-phosphorylated, the oscillation of KaiC phosphorylation is reinforced (enhanced amplitude). If on the other hand, new synthesis of unphosphorylated KaiC happens at a phase when hexamers are predominantly hyper-phosphorylated, this leads to an overall decrease in the KaiC phosphorylation status, thereby altering the phase of the KaiABC oscillator (phase shift) and/or reducing its amplitude. With the kind permission of C. H. Johnson, Department of Biological Sciences, Vanderbilt University, Nashville, Tennessee, USA [Qin *et al.* (2010) *PLoS Biol* **8(6)**: e1000394. doi:10.1371/journal.pbio.1000394] doi:10.1371/journal.pbio.1000394.g007.

Color image of this figure appears in the color plate section at the end of the book.

redox state of the plastoquinone pool inside the cells. The ligand-binding role of the pseudoreceiver domain of CikA has been demonstrated by its binding to quinone directly. As already mentioned, CikA associates with Kai proteins and influences the phosphorylation state of KaiC during the resetting of circadian phase by a dark pulse. Thus it seems that the resetting of clock in *S. elongatus* PCC 7942 is metabolism-dependent and the oscillator KaiC responds to the signals transmitted by CikA. The three domains, i.e. GAF, HPK and PsR are shown to be necessary for the functioning of CikA whereas N-terminal portion seems to be not necessary *in vivo*. The localization of CikA near the pole in presence of PsR domain, as revealed by green fluorescent reporter, suggests that it interacts with other components near the cell pole. But in the absence of PsR, CikA cannot localize at cell poles but gets uniformly distributed though out the cell. These observations signify that it is the PsR component that mediates interaction with other components of the input pathway (Zhang *et al.*, 2006). Four component proteins of the input pathway working in tandem with CikA have been identified by Mackey *et al.* (2008). They have identified four new genes *nhtA* (orf2160), *prkE* (orf0600), *ircA* (orf2387) and *cdpA* (orf1604) that encode aminotransferase-1, phase-resetting kinase, input-related cytochrome and cell division and phase protein, respectively. The involvement of these four proteins in the CikA input pathway has been confirmed by the isolation of mutants inactivated in the respective genes by the insertion of an antibiotic resistance cassette. *S. elongatus* PCC 7942 with P*kaiBC*::*luxAB* luciferase as reporter gene was transformed with the inactivated *nhtA*/*prkE*/*ircA*/*cdpA* gene and allowed to segregate so that mutant strains possessed corresponding inactivated genes in question. The *nhtA* null mutant behaved very much in a similar fashion as that of the wild-type but the *prkE* null mutant exhibited an unpredictable behavior with either the phase being advanced or delayed by about 6 to 11 h. The *cikAprkE* double mutant resembled a *cikA* null mutant in showing a non-resetting phenotype. The *ircA* and *cdpA* genes could not be inactivated completely as revealed by existence of wild-type copies of these genes by PCR analysis. Overexpression mutants of these genes against wild-type and *cikA* background have been isolated and their analysis revealed that the corresponding proteins are involved in either circadian period regulation, phase resetting and cell division. Mackey *et al.* (2008) are of the opinion that these four proteins function through protein phosphorylation, iron-cluster biosynthesis and redox regulation.

SasA (*Synechococcus* adaptive sensor A) is a histidine kinase (Iwasaki *et al.*, 2000). This is a member of the bacterial two-component regulatory system. It senses specific external stimuli or ligands and activates its autophosphorylation at a conserved His-residue and then transfers the phosphoryl group to an Asp residue in the cognate response regulator. SasA binds to KaiC to form a complex. This regulator drives physiological responses to the stimulus or signal by activating downstream genes or enzyme activities. The sensory domain of SasA has an amino acid sequence similar to that of full length KaiB and is sufficiently similar to bind to KaiC (Iwasaki *et al.* 2000). In *sasA*-inactivated strains, *kaiBC* expression is dramatically lowered with a concomitant lowering of the amplitude in rhythm and shortening of the period. SasA overexpression also strongly interfered with the rhythmicity. Although SasA is not required to drive a basic oscillation, it forms a secondary loop to stabilize Kai-based timing loop by coupling with KaiC. SasA levels essentially remain constant throughout the circadian cycle. SasA is important for growth in LD cycle conditions. The mutants devoid of SasA did not show any growth in LL. Iwasaki and Kondo (2004) concluded that SasA is interesting as a multifunctional regulator affecting the clock system, photoresponses and metabolism. Recently, Takai *et al.* (2006) identified a DNA-binding response regulator, RpaA (Regulator of phycobilisome associated) as a part of SasA. Mutants of *S. elongatus* PCC 7942 defective in SasA and RpaA showed a severe inhibition in circadian transcription. This two-component regulatory system is responsible for receiving the circadian time keeping signal from the oscillator to the transcription machinery. Moreover, autophosphorylation of SasA is enhanced in response to KaiC binding (Smith and Williams, 2006; Takai *et al.*, 2006) and this phosphate group is then transferred to RpaA (Takai *et al.*, 2006). According to Takai *et al.* (2006) the KaiC-SasA-RpaA pathway is likely to be the major pathway for circadian gene expression. This has further been confirmed by Taniguchi *et al.* (2007) who reported a novel gene *labA* (low amplitude and bright) that plays a role in negative feedback regulation of KaiC. The overexpression of *labA* significantly lowered circadian gene expression whereas its disruption abolished transcriptional repression caused by overexpression of KaiC.

They further suggested that *labA* and *sasA* function in parallel to regulate KaiBC expression. Upon interaction with KaiC, SasA gets autophosphorylated and activates RpaA which in turn controls circadian gene expression. RpaA is considered to be a transcription factor that controls global gene expression through basic transcription machinery. The SasA-RpaA pathway seems to be activated by Kai proteins during the subjective day. It is further suggested that KaiC represses RpaA function through the LabA-dependent pathway that is operative in the subjective night phase.

Katayama *et al.* (2003) identified *ldpA* (light-dependent period A) gene that encodes a new component of the input pathway. In contrast to *cikA*-deficient mutants, inactivation or deficiency of *ldpA* still can reset the clock in response to a dark pulse. Based on electron paramagnetic resonance spectra, the existence of two $Fe_4 S_4$ clusters have been shown in the gene product of *ldpA* suggesting the involvement of LdpA in the regulation of the circadian period by sensing light-dependent changes in electron transport. Further the existence of two redox-active centers consistent with two $[Fe_4S_4]^{2+/1+}$ clusters in LdpA and the requirement of LdpA for light-dependent modulation of *cikA* have a bearing in sensing redox reactions. Apart from this, LdpA copurifies with Kai proteins that are integral part of the circadian clock.

V. OUTPUT PATHWAYS

Tsinoremas *et al.* (1994) identified a gene that is part of the specific output pathway of the clock in *S. elongatus* PCC 7942. By following the procedure of insertional inactivation of genes developed by Dolganov and Grossman (1993), a number of mutants that showed short period (M2), arrhythmic (M11) and low amplitude oscillation of bioluminescence (M16) have been subsequently isolated and characterized (Tsinoremas *et al.*, 1994). Further studies revealed that these mutants not only exhibited defects in the genes governing circadian period (since they resembled mutants developed after methylmethane sulfonate mutagenesis described above) but also were shown to be defective in newly identified clock ORFs. Tsinoremas *et al.* (1996) reported that a mutation in the *rpoD2* gene that encodes a group 2 σ^{70}–like transcription factor causes a low amplitude phenotype in the *psbAI* gene expression rhythm. These workers proposed a model in which temporal incorporation of the RpoD2 sigma factor into RNA polymerase decreases the expression of *psbAI* during subjective night and generates a high amplitude circadian oscillation of *psbAI* expression. Further the specificity of the mutation for a subset of genes indicates that the *rpoD2* gene is a component of the output pathway in the *Synechococcus* circadian clock. Another gene *ndhD* regulates a similar phenotype but the expression of rhythms by several other genes was not affected by the mutation. Inactivation of *rpoD2* decreases the amplitude of the expression rhythm not only of *ndhD* but also that of *psbAI*. However, it had no affect on the expression rhythm of *purF* that encodes the oxygen-sensitive enzyme glutamine phosphoribosyl pyrophosphate amidotransferase (glutamine PRPP amidotransferase), a key regulatory enzyme in the purine biosynthetic pathway. Katayama *et al.* (1999) postulated that the activities of enzymes that are specifically expressed in the night, like *purF* are inhibited by the approach of the peak of a subset of class 1 rhythms of photosynthetic genes *psbAI* and *psbAII* into their phase.

The genetically engineered strain of *S. elongatus* PCC 7942 possessing *psbAI*-promoter belongs to Class 1` gene family exhibiting peaks at dusk period and troughs at dawn. In contrast Class 2 genes such as *purF* shows peaks and troughs at dawn and dusk, respectively. Johnson and Golden (1999) are of the view that the Class 2 gene expression pattern of *purF* gene is related to oxygen sensitivity of glutamine PRPP amidotransferase and it is similar to the temporal separation of nitrogenase in certain nitrogen-fixing cyanobacteria described earlier (Liu *et al.*, 1996; Johnson *et al.*, 1998).

Liu *et al.* (1995) proposed a model for explaining non-specific circadian control and a circadian regulation by specific *trans* factors. It has been suggested that Class 1 specific *Cis*-elements are turned on during the day by a Class 1-specific *trans* factor. Likewise, Class 2 specific *Cis* elements would be turned on at night by a Class 2-specific *trans* factor. Though it is difficult to imagine the existence of *trans* factors for all the large number of genes under the circadian control, it is suggested that there might be involvement of some global factors. Moreover, the discovery of a gene whose altered expression significantly lowered the amplitude of the luminescence rhythm driven by some promoters such as *psbAI* but not of luminescence rhythm driven by other promoters such as that of *purF*. These observations point towards the existence of subsets of clock-controlled genes in *S. elongatus* PCC 7942.

Katayama *et al.* (1999) identified a gene involved in the output pathway of *S. elongatus* PCC 7942 while examining a transposon Tn5-generated mutant *tnp6* that is affected in both amplitude and phasing of the *psbAI:luxAB* circadian expression rhythm. They introduced a derivative of TN5 into the chromosomes of reporter strains in which cyanobacterial promoters drive the *Vibrio harveyi luxAB* genes. As a result of which the oscillation of bioluminescence could be measured as a function of circadian gene expression. This mutant gene has been designated as *cpmA* (circadian phase modifier) and is shown to change the circadian phasing of promoter activity for one of the genes that encodes a central clock component (*kaiA::luxAB*) but it had little effect on the other two clock genes (*kaiB::luxAB*). They further concluded that the coordinated expression of *Kai* genes is not essential for the circadian time keeping in *Synechococcus*.

i) *Role of light and dark*: Entrainment is generally defined as the matching of the period of biological clock exactly equal to the environmental cycle. The primary signals of entrainment used are light and dark cycles that can set the phase of the circadian clock. Though phytochromes, rhodopsins or cryptochromes are the photopigments involved in circadian entrainment in other organisms, in case of cyanobacteria the photopigments involved in the process are poorly understood (Johnson, 1995; Roenneberg and Foster, 1997; Johnson and Golden, 1999). Blue and red lights are most effective in setting the phase of the cyanobacterial clock. However, it was not possible to reverse the phase of the clock in red or far-red lights, respectively. The action spectrum does not coincide with either that of photosynthesis or phytochrome. The available evidences suggest that there are certain specific unknown pigments that perceive the signals in the input pathway of the circadian clock of *S. elongatus* PCC 7942 and *Synechococcus* RF-1.

In cyanobacteria, the rhythmicity of the circadian cycle has been tested invariably in LL condition. The next question that emerges is whether light is required to run the circadian clock? Kondo *et al.* (1994) used light pulses that can reset the phase of the clock in DD in the photoautotrophic *S. elongatus* PCC 7942. They observed that circadian clock continued to function even in DD. Some other studies examined whether light is necessary for the continued functioning of the circadian clock or only certain metabolic rate has to be maintained for the clock to express. The presence of circadian rhythms in the cyanobacterium *Synechocystis* sp. strain PCC 6803 was shown by Aoki *et al.* (1995) by the rhythmic expression of *dnaK* gene (DnaK is a member of well-conserved heat shock proteins that plays a protective role in supporting growth at high temperatures beyond the normal physiological range). They fused a promoterless *luxAB* gene set downstream the promoter segment of the *Synechocystis dnaK* gene and introduced it into a specific site of the *Synechocystis* chromosome. Aoki *et al.* (1997) further showed that the circadian rhythms persisted even in DD. They selected *Synechocystis* sp. strain PCC 6803 and *Cyanothece* sp. ATCC 51142 that can grow heterotrophically on glucose and glycerol, respectively for such studies. *Synechocystis* can grow heterotrophically on glucose in darkness when the cells are exposed to daily brief light pulses. This has been termed as light-activated heterotrophic growth. With the help of *dnaK::luxAB* reporter strain of *Synechocystis*, it was demonstrated that the luminescence rhythm persisted for many cycles in DD. Likewise, Schneegurt *et al.* (1997) demonstrated that in *Cyanothece* sp. ATCC 51142 the rhythms of nitrogenase activity and carbohydrate content persisted for 4 days when grown on glycerol. These studies thus supported the idea that the circadian clock can run in DD if the specific metabolic rate is maintained and that the clock is not dependent on light. To sum up, it can be stated that the photoautotrophic cyanobacteria continue to operate the circadian clock even in DD as also the heterotrophic cyanobacteria that exhibit the circadian rhythms in DD. It means as light serves as a signal in the input pathway, so also DD serves as a signal for the circadian clock.

Cervený and Nedbal (2009) established a close relationship between respiratory activity and glycogen degradation in the long day (LD cycle of 16:8) condition coinciding with a transition from aerobic photosynthetic metabolism and dark respiratory phase at the dusk period. They extrapolated these findings to the accumulation of KaiB tetramer formation and the apparent link with circadian mechanism and the dark respiration in *Cyanothece* sp. ATCC 51142.

ii) *Cell division and the rhythms*: Early studies indicated that there existed a bi-directional interdependent linkage between circadian rhythms and cell division cycle (CDC) (Ehet and Wille, 1970; Klevecz, 1976; Edmunds, 1988). This is especially true for unicellular eukaryotic organisms where the timing of CDC is determined by circadian oscillator. It means that it is the circadian oscillator that specifies the time for cell division to occur even though the cells have attained the maturity to undergo the division process. An alternative suggested is that circadian clock mechanism oscillates independently of CDC but the CDC is dependent on the phase of circadian clock in a manner that is gated to occur coinciding with circadian phases (Ehet and Wille, 1970; Edmunds, 1988; Goto and Johnson, 1995). Mori *et al.* (1996) examined the rates of cell division and DNA content in rapidly growing cells of *S. elongatus* PCC 7942 and its mutant strain AMC149 that had *psbAI::luxAB* reporter gene set in its genome and two other mutant strains C22a and C27a (previously designated as SP22 and LP27, respectively, isolated after chemical mutagenesis of AMC149). Cells dividing at rates as fast as 10 h were controlled by circadian oscillator for cell division and gene expression as well as DNA replication occurred continuously throughout the circadian cycle within the population of dividing cells but the cytokinesis is gated by the circadian clock such that division is forbidden in the early subjective night. So it was felt necessary to examine whether genes related to cytokinesis (or septum formation) play a role in the circadian gating of cell division. A gene designated as *ftsZ* that is ubiquitously present in bacteria and chloroplasts encodes FtsZ protein. This protein assembles in the form of a Z ring in the middle of the cell that helps in the formation of a septum (Corton *et al.*, 1987; Bi and Lutkenhaus, 1991). An *ftsZ* gene identified from *Anabaena* was amplified by PCR and the PCR product was sequenced. Due to its similarity to regions of known *ftsZ* genes, it was introduced with its promoter along with a bacterial luciferase gene set into *S. elongatus* PCC 7942. It was found that the activity of *ftsZ* was highest when cells in the population were not dividing. Overexpression of *ftsZ* halted cell division and caused cells to become filamentous. In such mutants, robust circadian fluctuations of other genes such as *psbA1*, *KaiBC* and *ftsZ* persisted for at least 4 to 5 days that suggested regulation of gene expression and gating of cell division is not affected by halting cell division (Mori *et al.*, 1996).

In the colonies of unicellular cyanobacteria, when the products of cell division get separated and move away from one another do the rhythms exhibited by certain cells influence other cells or not? Mihalceseu *et al.* (2004) examined this question by inserting *psbAI::luxAB* and *psbAI::luxCDE* and measured the rhythms exhibited by the individual cells when seeded on agar plates and followed them up after every cell division by capturing their images division after division. Such individual cells of *S. elongatus* PCC 7942 exhibited robust and resilient circadian rhythms by the expression of the luciferase gene. The perturbations caused due to high intracellular noise and frequent cell divisions did not affect the individual rhythms and the cell to cell interactions did not exist. This confirms that the circadian clock of individual cyanobacterial cells is quite stable and there is no intercellular coupling effect (Amdaoud *et al.*, 2007).

iii) *Adaptive fitness*: It is generally accepted that circadian rhythms help the organism for better adaptation to the environmental conditions but convincing tests to prove this point have come up only in case of the cyanobacterium *S. elongatus* PCC 7942 (Kondo *et al.*, 1994; Ouyang *et al.*, 1998; Ishiura *et al.*, 1998). Wild-type and mutant strains of this organism that exhibited differences in the circadian period were grown in single strain and in mixed cultures. For example, mutant strains C22a and C28a showed periods of 23 h and 30 h, respectively whereas wild-type had a period of 25 h. There was no difference in the performance of the mutant strains in LL and in LD cycles when grown in single strain cultures suggesting that the difference in circadian periods did not contribute to any added advantage or disadvantage to the respective mutant strain. When mutant and wild-type strains were allowed to compete with each other in mixed cultures in different LD regimes such as 22 h cycle (LD 11:11h), a 24 h cycle (LD 12:12 h) and 30 h cycle (LD 15:15 h), the particular mutant whose period most closely matched that of LD-cycle emerged out successfully and eliminated other strains including the wild-type (Fig. 6). Woelfe *et al.* (2004) emphasized the adaptive value of circadian clocks in cyclic environments alone. A comparison of clock-disrupted cyanobacterial strains with those having a functional biological clock revealed that the latter defeated the former in rhythmic environments (Fig. 7). But this inherent advantage of the functional clock seems to disappear in constant environments.

VI. *KAI* GENES IN OTHER CYANOBACTERIA

Apart from *S. elongatus* PCC 7942, *kai* genes have now been identified in more than forty cyanobacteria belonging to diverse taxonomic groups (Lorne *et al.*, 2000). In all the six cyanobacteria (*S. elongatus* PCC 7942, *Synechocystis* sp. strain PCC 6803, *Anabaena* sp. strain PCC 7120, *Prochlorococcus marinus* MED4, *P. marinus* MIT 9313, *Synechococcus* sp. strain WH 8102 and *N. punctiforme* ATCC 29133) in which these genes have now been sequenced, the ORFs resemble only two *kai* genes, i.e. *kaiB* and *kaiC*. In addition, *kaiA* gene sequence is diversified. Furthermore, the genome of *Synechocystis* sp. strain PCC 6803 contains multiple gene sequences of *kaiB* and *kaiC* but it possesses only one *kaiA* gene. Dvornyk *et al.* (2002) identified a genetic polymorphism in *kaiABC* gene family in *Nostoc muscorum* permanently exposed to harmful UV radiations at two sites (Lower Nahal Oven, Mount Carmel and Lower Nahal Keziv, western upper Galilee) in Israel designated as "Evolution Canyons" (EC I and II). Five distinct subfamilies, *kai* I to *kai* V consisting of 20 functional genes and pseudogenes have been identified with variation in the number of member genes in each subfamily. Subfamilies I and II had 5 member genes each with 2 member genes in subfamilies IV and V while subfamily III had 6 member genes. Subfamilies IV and V occurred at EC I and EC II, respectively. Exceptionally, *kai* IV subfamily genes were found only as a single copy halotypes whereas genes of the *kai* V subfamily were observed in multiple copy halotypes. This polymorphism appeared to be evolutionarily recent as such radiation of genes was absent in other closely related genera such as *Nodularia*, *Cylindrospermum* and *N. punctiforme* ATCC 29133.

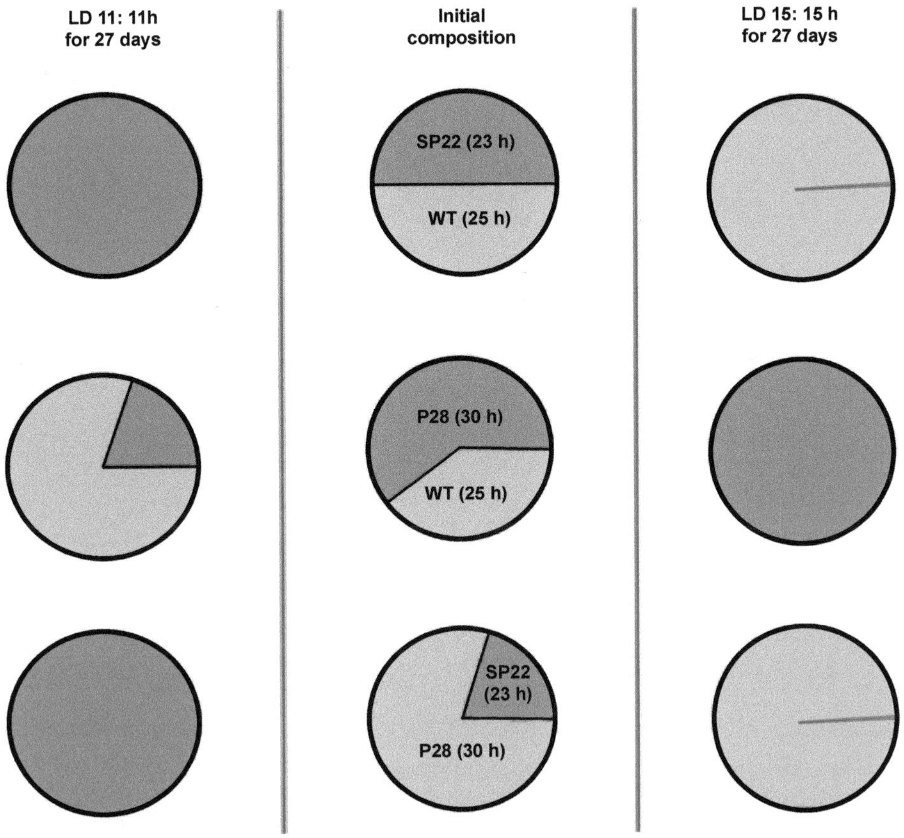

Figure 6: Circadian resonance in cyanobacteria. Rhythmic strains having different free-running periods were competed under LD cycles of different lengths. Strains whose free-running period matched that of LD cycles out-competed those with deviant periods. Middle panels represent initial composition of the competing strains. Values in the parenthesis indicate the free-running period of the cyanobacterial strains. (Figure modified after Ouyang *et al.*, 1998) With kind permission of V. K. Sharma, Chronobiology Laboratory, Evolutionary and Organismal Biology Unit, Jawaharlal Nehru Centre for Advanced Scientific Research, Jakkur, PO Box 6436, Bangalore 560 064, Karnataka, India [Paranjpe, D. A. and Sharma, V. K. (2005) *J Circadian Rhythms* **3**:7 doi:10.1186/1740-3391-3-7].

VII. ORIGIN AND EVOLUTION OF KAI GENES AND OTHER CLOCK COMPONENTS

Dvornyk *et al.* (2003) compared complete sequence data of 70 prokaryotic organisms and traced the origin and evolution of *kai* gene family. The *kaiA* and *kaiB* genes originated in cyanobacteria and the formation of the three gene cluster, *kaiABC* evolved about 1,000 My. The *kai* gene sequences were also found in certain Archaebacteria and Proteobacteria besides of course in cyanobacteria. Phylogenetic analysis of *kaiB* and *kaiC* genes suggested that cyanobacteria have probably two types of circadian system; one that is based on KaiABC- and the second KaiBC-based. The *kaiA* sequence appeared to be the youngest while *kaiC* appeared to be evolutionarily the oldest. They concluded that the evolution of prokaryotic circadian pacemakers have evolved parallely with geological history. In *P. marinus* MED4 a component of KaiABC oscillator has been lost during evolution and

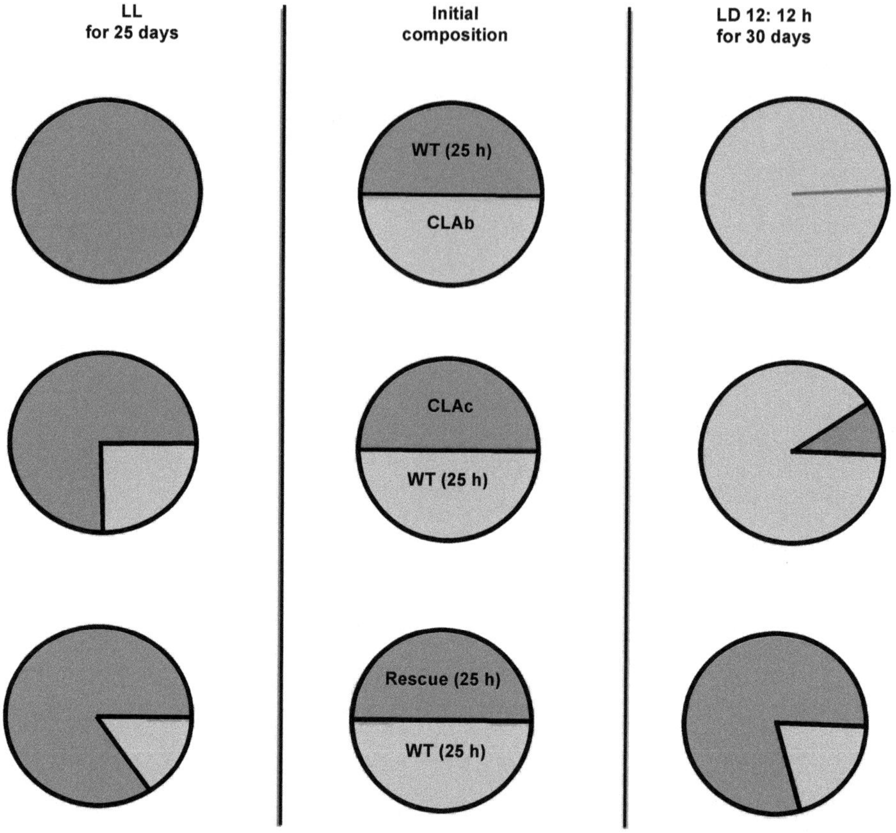

Figure 7: Competition between rhythmic and arrhythmic strains of cyanobacteria. Mutant strains with arrhythmic (CLAc), or dampened (CLAb) bioluminescence rhythm, as well as the rescued strains were competed against wild type strain under periodic and constant environments (LD cycles and LL, respectively). Rhythmic strains out competed the wild type strain under LD cycles, but the arrhythmic strains out competed rhythmic strains under LL. Middle panels represent initial composition of the competing strains. Values in the parenthesis indicate the free-running period of the cyanobacterial strains. (Figure modified after Woelfle *et al.*, 2004). With kind permission of V. K. Sharma, Chronobiology Laboratory, Evolutionary and Organismal Biology Unit, Jawaharlal Nehru Centre for Advanced Scientific Research, Jakkur, PO Box 6436, Bangalore 560 064, Karnataka, India [Paranjpe, D. A. and Sharma, V. K. (2005) *J Circadian Rhythms* **3**: 7 doi:10.1186/1740-3391-3-7].

only *kaiBC* sequences are present in this organism and KaiC exists in a hyperphosphorylated state. It is suggested that this evolutionary reduction in the clock components would suffice the timing mechanism in view of the stable environment in which the organism grows (Axmann *et al.*, 2009). The genome streamlining of *P. marinus* PCC 9511 resulted in loss of robustness of the circadian clock. Holtzendorff *et al.* (2008) demonstrated that though there is a diurnal variation in the mRNA levels of *kaiBC* operon and certain other selected genes such as *psbA* expression rhythms, under LL condition the robustness of the rhythm is lost. Though *kaiBC* gene sequences are present and the essential properties of KaiC are well conserved in *Legionella pneumophila*, an opportunistic human pathogen, their functional organization is yet to be uncovered (Lozza-Correa *et al.*, 2010).

The structure and molecular phylogeny of *sasA* genes in cyanobacteria revealed that *sasA* occurred in a single copy. Two Clades, Clade A and Clade B have been recognized. Clade A has been shown to be associated with species possessing all the three *kai* genes while Clade B was represented in *Synechococcus* sp. strain WH 8102 with the three clock oscillators or *Prochlorococcus* strains having only *kaiB* and *kaiC*. It is interesting to note that the trees of the *kaiBC* cluster and *sasA* genes from the same species have similar topologies. One of the possibilities for evolution of *sasA* gene sequence is by lateral gene transfer of *kaiB* and its fusion with the two component histidine kinase in other Proteobacteria to form *sasA* and its subsequent transfer back to cyanobacteria. This has been ruled out as the triple-domain structure homolog of *sasA* gene with KaiB-like sensor domain does not occur in other prokaryotes. As *kaiB* genes originated only in cyanobacteria, it is logical to conclude that *sasA* gene has originated in cyanobacteria apparently through fusion of two-component histidine kinase and ancestral *kaiB* gene to form the currently observed triple-domain structure (Dvornyk *et al.*, 2004).

According to Dvornyk *et al.* (2004) the ancestral circadian system consisted of *kaiB* and *kaiC* genes which were not in a cluster. To enhance the performance of this system *sasA* gene supposedly evolved as a universal input-output regulator. Due to the fusion of *kaiB* and *kaiC*, *sasA* must have become an indispensable part of *kaiBC-sasA* circadian system. Subsequent emergence of *kaiA* gene resulted in the formation of *kaiABC-sasA* system. Dvornyk (2005) conducted a phylogenetic analysis of *ldpA* gene from the fully sequenced genomes of prokaryotes and found that the topology of the phylogenetic tree is consistent with the phylogenetic analyses of SasA gene (Dvornyl *et al.*, 2004) and *kaiBC* operon (Dvornyk *et al.*, 2003) conducted earlier by his group. These results further lend support to the existence of KaiABC- and KaiBC-based circadian systems in prokaryotes.

The structure, polymorphism, mutation rates and selective constraints of *cpmA* genes have been compared among cyanobacteria. The *cpmA* gene sequences have also been found in some other bacteria and archaea. However, no *cpmA* gene sequences have been found in photosynthetic α-proteobacteria (*Rhodobacter, Rhodospirillum* and *Rhodopseudomonas*) or in *Chloroflexus* all of which have been reported to have *kai* genes and/or *kaiBC* operon laterally transferred from cyanobacteria. Exceptionally, *Trichodesmium erythaeum* possessed two copies of *cpmA* gene sequences while the rest of the cyanobacteria examined had only one *cpmA* gene (Dvornyk *et al.*, 2006). Based on the evolutionary analyses of the key components (*kaiB*, *kaiC* and *sasA* genes), it was hypothesized that each type of cyanobacterial circadian system (*kaiABC* cluster or *kaiBC*) has specific functional and selective constraints largely because of the emergence of *kaiA* gene (Dvornyk *et al.*, 2003, 2004, 2006).

LITERATURE CITED

Amdaoud, M., Vallade, M., Weiss-Schaber, C., and Mihalcescu, I. (2007) Cyanobacterial clock, a stable phase oscillator with negligible intercellular coupling. *Proc Natl Acad Sci USA* **104:** 7051–7056.

Andersson, C. R., Tsinoremas, N. F., Shelton, J., Lebedeva, N. V., Yarrow, J., Min, H., and Golden S. S. (2000) Application of bioluminescence to the study of circadian rhythms in cyanobacteria. *Methods Enzymol* **305:** 527–542.

Aoki, S., Kondo, T., and Ishiura, M. (1995) Circadian expression of the *dnaK* gene in the cyanobacterium *Synechocystis* sp. strain PCC 6803. *J Bacteriol* **177:** 5606–5611.

Aoki, S., Kondo, T., Wada, H., and Ishiura, M. (1997) Circadian rhythm of the cyanobacterium *Synechocystis* sp. strain PCC 6803 in the dark. *J Bacteriol* **79:** 5751–5755.

Arita, K., Hashimoto, H., Igari, K., Akaboshi, M., Kutsuna, S., Sato, M., and Shimizu, T. (2007) Structural and biochemical characterization of a cyanobacterium circadian clock-modifier protein. *J Biol Chem* **292:** 1128–1135.

Axmann, I. M., Dühring, U., Seeliger, L., Arnold, A., Vanselow, T., Kramer, A., and Wilde, A. (2009) Biochemical evidence for a timing mechanism in *Prochlorococcus*. *J Bacteriol* **191:** 5342–5347. doi:10.1128/JB.00419-09

Bell-Pederson, D., Cassone, V. M., Earnest, D. J., Golden, S. S., Hardin, P. E., Thomas, T. L., and Zoran, M. J. (2005) Circadian rhythms from multiple oscillators: Lessons from diverse organisms. *Nat Rev Genet* **6:** 544–556.

Bi, E. F., and Lutkenhaus, J. (1991) FtsZ ring structure associated with division in *Escherichia coli*. *Nature* (London) **354:** 161–164.

Brettschneider, C., Rose, R. J., Hertel, S., Axmann, I. M., Heck, A. J. R., and Kollmann, M. (2010) A sequestration feedback governs dynamics and temperature entrainment of the KaiABC circadian clock. *Mol Syst Biol* **6:** 389. doi.1038/msb.2010.44

Bünning, E. (1973) *The Physiological Clock* (3rd edn.) Springer-Verlag, New york, NY. USA.

Cervený, J., and Nedbal, L. (2009) Metabolic rhythms of the cyanobacterium *Cyanothece* sp. ATCC 51142 correlate with modeled dynamics of the circadian clock. *J Biol Rhythms* **24:** 295–303.

Chang, Y-G., Kuo, N-W., Tseng, R., and LiWang, A. (2011) Flexibility of the C-terminal, or CII, ring of KaiC governs the rhythm of the circadian clock of cyanobacteria. *Proc Natl Acad Sci USA* **108:** 14431–14436. www.pnas.org/cgi/doi/10.1073/pnas.1104221108

Chen, T.-H., Chen, T.-L., Huang, L.-M., and Huang, T.-C. (1991) Circadian rhythm in amino acid uptake by *Synechococcus* RF-1. *Plant Physiol* **97:** 55–59.

Chen, T-H., Pen, S-Y., and Huang, T-C. (1993) Induction of nitrogen-fixing rhythm *Synechococcus* RF-1 by light signals. *Plant Sci* **92:** 179–82.

Chen, Y-B., Dominic, B., Mellon, M. T., and Zehr, J. P. (1998) Circadian rhythm of nitrogenase gene expression in the diazotrophic filamentous nonheterocystous cyanobacterium *Trichodesmium* sp. strain IMS101. *J Bacteriol* **180:** 3598–3605.

Clodong, S., Dühring, U., Kronk, L., Wilde, A., Axmann, I., Herzel, H., and Kollmann, M. (2007) Functioning and robustness of a bacterial circadian clock. *Mol Syst Biol* **3:** 90.

Corton, J. C., Ward, Jr., J. E., and Lutkenhaus, J. (1987) Analysis of cell division gene *ftsZ* from gram–ve and gram+ve bacteria. *J Bacteriol* **169:** 1–7.

Diaz, J. J., Rhoads, D. D., and Roufa, D. J. (1991). PCR-mediated chemical mutagenesis of cloned duplex DNAs. *Biotechniques* 11, 204–206, 208, 210–211.

Ditty, J. L., Canales, S. R., Anderson, B. E., Williams, S. B., and Golden, S. S. (2005) Stability of the *Synechococcus elongatus* PCC 7942 circadian clock under directed anti-phase expression of *Kai* genes. *Microbiology* **151:** 2605–2613.

Ditty, J. L., Williams, S. B., and Golden, S. S. (2003) A cyanobacterial circadian timing mechanism. *Annu Rev Genet* **37:** 513–543. doi:10.1146/annurev.genet.37.110801.142716

Dolganov, N., and Grossman, A. R. (1993) Insertional inactivation of genes to isolate mutants of *Synechococcus* sp. strain PCC 7942: Isolation of filamentous mutants. *J Bacteriol* **175:** 7644–7651.

Dvornyk, V. (2005) Molecular evolution of *ldpA*, a gene mediating the circadian input signal in cyanobacteria. *J Mol Evol* **60:** 105–112.

Dvornyk, V. (2006) Subfamilies of *cpmA*, a gene involved in circadian output, have different evolutionary histories in cyanobacteria. *Microbiology* **152:** 75–84.

Dvornyk, V., Deng, H.-W., and Nevo, E. (2004) Structure and molecular phylogeny of *SasA* genes in cyanobacteria: Insights into evolution of the prokaryotic circadian system. *Mol Biol Evol* **21:** 1468–1476.

Dvornyk, V., Vinogradova, O., and Nevo, E. (2002) Long term microclimatic stress causes rapid adaptive radiation of KaiABC clock gene family in a cyanobacterium, *Nostoc linckia* from "Evolution Canyons" I and II, Israel. *Proc Natl Acad Sci USA* **99:** 2082–2087.

Dvornyk, V., Vinogradova, O., and Nevo, E. (2003) Origin and evolution of circadian clock genes in prokaryotes. *Proc Natl Acad Sci USA* **100:** 2495–2500.

Edmunds, L. N., Jr. (1988) *Cellular and Molecular Bases of Biological Clocks*. Springer-Verlag, New York, N. Y. USA.

Egli, M., Pattanayek, R., and Pattanayek, S. (2007) Protein-protein interactions in the cyanobacterial KaiABC circadian clock. In: Boeyens, J. C. A., and Ogilvie, J. F. (Eds.) *Models, Mysteries and Magic of Molecules.* Springer. Dordrecht, The Netherlands. pp. 283–289.

Eguchi, K., Yoda, M., Terada, T. P., and Sasai, M. (2008) Mechanism of robust circadian oscillation of KaiC phosphorylation *in vitro*. *Biophys J* **95:** 1773–1784.

Ehet, C. F., and Wille, J. J. (1970) The photobiology of circadian rhythms in protozoa and other eukaryotic microorganisms. In: Halldal, P. (Ed.) *Photobiology of Microorganisms*. Wiley, London, UK. pp. 369–416.

Emberly, E., and Wingreen, N. S. (2006) Hourglass model of a protein based circadian oscillator. *Phys Rev Lett* **96:** 038303.

Gallon, J. R., LaRue, T. A., and Kurz, W. G. W. (1974) Photosynthesis and nitrogenase activity in the blue-green alga, *Gloeocapsa*. *Can J Microbiol* **20:** 1633–1637.

Garces, R. G., Wu, N., and Gillon, W., and Pai, E. F. (2004) *Anabaena* circadian clock proteins KaiA and KaiB reveal potential common binding site to their partner KaiC. *EMBO J.* **23:** 1688–1698.

Golden, S. S. (2004) Meshing the gears of the cyanobacterial circadian clock. *Proc Natl Acad Sci USA* **101:** 13697–13698.

Golden, S. S., Ishiura, M., Johnson, C. H., and Kondo, T. (1997) Cyanobacterial circadian rhythms. *Annu Rev Plant Physiol Plant Mol Biol* **48:** 327–354.

Goto, K., and Johnson, C. H. (1995) Is the cell division cycle gated by a circadian clock? The case of *Chlamydomonas reinhardtii*. *J Cell Biol* **129:** 1061–1069.

Grobbelaar, N., Huang, T.-C, Lin, H.-Y., and Chow, T.-J. (1986) Dinitrogen-fixing endogenous rhythm in *Synechococcus* RF-1. *FEMS Microbiol Lett* **37:** 173–177.

Hayashi, F., Itoh, N., Uzumaki, T., Iwase, R., Tsuchiya, Y., Yamakawa, H., Morishita, M., Onai, K., Itoh, S., and Ishiura, M. (2004) Roles of two ATPase motif containing domains in cyanobacterial clock protein KaiC. *J Biol Chem* **279:** 52321–52337.

Hayashi, F., Iwase, R., Uzumaki, T., and Ishiura, M. (2006) Hexamerization by the N-terminal domain and intersubunit phosphorylation by the C-terminal domain of cyanobacterial circadian clock protein KaiC. *Biochem Biophys Res Commun* **318:** 864–872.

Hayashi, F., Suzuki, H., Iwase, R., Uzumaki, T., Miyake, A., Shen, J.-R., Imada, K., Furukawa, Y., Yonemura, K., Namba, K., and Ishiura, M. (2003) ATP-induced hexameric ring structure of the cyanobacterial circadian clock protein KaiC. *Genes to Cells* **8:** 287–296.

Hitomi, K., Oyama, T., Han, S., Arvai, A. S., and Getzoff, E. D. (2005) Tetrameric architecture of the circadian clock protein KaiB. *J Biol Chem* **280:** 19127–19135.

Holtzendorff, J., Partensky, F., Mella, D., Lennon, T.-F., Hess, W. R., and Garczarek, L. (2008) Genome streamlining results in loss of robustness of the circadian clock in the marine cyanobacterium *Prochlorococcus marinus* PCC 9511. *J Biol Rhythms* **23:** 187–199.

Huang, T.-C., and Chow, T.-J. (1986) New type of N_2 fixing unicellular cyanobacterium (blue-green alga). *FEMS Microbiol Lett* **36:** 109–110.

Huang, T.-C., and Grobbelaar, N. (1995) The circadian clock in the prokaryote *Synechococcus* RF-1. *Microbiology* **141:** 535–540.

Huang, T.-C., and Chou, W.-M. (1991) Setting of the circadian N_2-fixing rhythm of the prokaryotic *Synechococcus* sp. RF-1 while its *nif* gene is repressed. *Plant Physiol* **96:** 324–326.

Huang, T.-C., Chen, H.-M., Pen, S.-Y., and Chen, T.-H. (1994) Biological clock in the prokaryote *Synechococcus* RF-1. *Planta* **193:** 131–136.

Huang, T.-C., Tu, J., Chow, T.-J., and Chen, T.-H. (1990) Circadian rhythm of the prokaryote *Synechococcus* sp. RF-1. *Plant Physiol* **92:** 531–533.

Huang, T.-C., Wang, S.-T., and Grobbelaar, N. (1993) Circadian rhythm mutants of the prokaryotic *Synechococcus* RF-1. *Curr Microbiol* **27:** 249–254.

Imai, K., Nishiwaki, T., Kondo, T., and Iwasaki, H. (2004) Circadian rhythms in the synthesis and degradation of master protein KaiC in the cyanobacteria. *J Biol Chem* **279:** 36534–36539.

Ishiura, M., Kutsuna, S., Aoki, S., Iwasaki, H., Andersson C. R., Tanabe, A., Golden S. S., Johnson, C. H., and Kondo, T. (1998) Expression of a gene cluster KaiABC as a circadian feedback process in cyanobacteria. *Science* **281:** 1519–1523.

Ito, H., Kageyama, H., Mutsuda, M., Nakajima, M., Oyama, T., and Kondo, T. (2007) Autonomous synchronization of the circadian KaiC phosphorylation rhythm. *Nat Struct Mol Biol* **14:** 1084–1088.

Ivleva, N. B., Bramlett, M. R., Lindahl, P. A., and Golden, S. S. (2005) LdpA: a component of the circadian clock senses redox state of the cell. *EMBO J* **24:** 1202–1210.

Ivleva, N. B., Gao, T., LiWag, A. C., and Golden, S. S. (2006) Quinone sensing by the circadian input kinase of the cyanobacterial circadian clock. *Proc Natl Acad Sci USA* **103:** 17468–17473.

Iwasaki, H., and Kondo, T. (2004) Circadain timing mechanism in the prokaryotic clock system of cyanobacteria. *J Biol Rhythms* **19:** 436–444.

Iwasaki, H., Taniguchi, Y., Ishiura, M., and Kondo, T. (1999) Physical interactions among circadian clock proteins, KaiA, KaiB and KaiC, in cyanobacteria. *EMBO J* **18:** 1137–1145.

Iwasaki, H., Williams, S. B., Kitayama, Y., Ishiura, M., Golden, S. S., and Kondo, T. (2000) A KaiC-interacting sensory histidine kinase SasA, necessary to sustain robust circadian oscillation in cyanobacteria. *Cell* **101:** 223–233.

Iwasaki, H., Nishiwaki, T., Kitayama, Y., Kakajima, M., and Kondo, T. (2002) KaiA-stimulated KaiC phosphorylation in circadian timing loops in cyanobacteria. *Proc Natl Acad Sci USA* **99:** 15788–15793.

Iwase, R., Imada, K., Hayashi, F., Uzumaki, T., Namba, K., and Ishiura, M. (2004) Crystallization and preliminary crystallographic analysis of the circadian clock protein KaiB from the thermophilic cyanobacterium *Thermosyechococcus elongatus* BP-1. *Acta Crystallogr* **D60:** 727–729.

Iwase, R., Imada, K., Hayashi, F., Uzumaki, T., Morishita, M., Onai, K., Furukawa, Y., Namba, K., and Ishiura, M. (2005) Functionally important substructures of circadian protein KaiB in a unique tetramer complex. *J Biol Chem* **280:** 43141–43149.

Johnson, C. H. (1995) Photobiology of circadian rhythms. In: Horspool, W. M., and Song, P.-S. (Eds.) *CRC Handbook of Organic Photochemistry and Photobiology*. CRC Press, Boca Raton, Florida, FL, USA. pp. 1602–1610.

Johnson, C. H. (2004) Global orchestration of gene expression by the biological clock of cyanobacteria. *Genome Biology* **5:** 217. http://genomebiology.com/2004/5/217

Johnson, C. H. (2004) Precise circadian clocks in cyanobacteria. *Curr Issues Mol Biol* **6:** 103–110.

Johnson, C. H. (2010) Circadian clocks and cell division. What's the pacemaker? *Cell Cycle* **9:** 3864–3873. doi:10.4161/cc.9.19.13205

Johnson, C. H., and Golden, S. S. (1999) Circadian programs in cyanobacteria: Adaptiveness and mechanism. *Annu Rev Microbiol* **53:** 389–409.
Johnson, C. H., Egli, M., and Stewart, P. C. (2008a) Structural insights into a circadian oscillator. *Science* **322:** 697–701.
Johnson, C. H., Golden, S. S., and Kondo, T. (1998) Adaptive significance of circadian programs in cyanobacteria. *Trends Microbiol* **6:** 407–410.
Johnson, C. H., Golden, S. S., Ishiura, M., and Kondo, T. (1996) Circadian clocks in prokaryotes. *Mol Microbiol* **21:** 5–11.
Johnson, C. H., Mori, T., and Xu, Y. (2008b) A cyanobacterial circadian clockwork. *Curr Biol* **18:** 816–825. doi:10.1016/j.cub.2008.07.012
Johnson, C. H., Stewart, P. L., and Egli, M. (2011) The cyanobacterial circadian system: From biophysics to bioevolution. *Annu Rev Biophys* **40:** 143–167.
Kageyama, H., Kondo, T., and Iwasaki, H. (2003) Circadian formation of clock protein complexes by KaiA, KaiB, KaiC and SasA in cyanobacteria. *J Biol Chem* **278:** 2388–2395.
Kageyama, H., Nishiwaki, T., Nakajima, M., Iwasaki, H., Oyama, T., and Kondo, T. (2006) Cyanobacterial circadian pacemaker: Kai protein complex dynamics in the KaiC phosphorylation cycle *in vitro*. *Mol Cell* **23:** 161–171. doi:10.1016/j.molcel.2006.05.039
Kaneko, T., Sato, S., Kotani, H., Tanaka, A., Asamizu, E., Nakamura, Y., Miyajima, N., Hirosawa, M., Sigiura, M., Sasamoto, S., Kimura, T., Hosouchi, T., Matsuno, A., Muraki, A., Nakazaki, N., Naruo, K., Okumura, S., Shimpo, S., Takeuchi, C., Wada, T., Watanabe, A., Yamada, M., Yasuda, M., and Tabata, S. (1996) Sequence analysis of the genome of the unicellular cyanobacterium *Synechocystis* sp. strain PCC 6803. II. Sequence determination of the entire genome and assignment of potential protein-coding regions. *DNA Res* **3:** 109–136.
Katayama, M., Kondo, T., Xiong, J., and Golden, S. S. (2003) *ldpA* Encodes an iron-sulfur protein involved in light-dependent modulation of the circadian period in the cyanobacterium *Synechococcus elongatus* PCC 7942. *J Bacteriol* **185:** 1415–1422.
Katayama, M., Tsinoremas, N. F., Kondo, T., and Golden, S. S. (1999) *cpmA*, a gene involved in an output pathway of the cyanobacterial circadian system. *J Bacteriol* **181:** 3516–3524.
Kim, Y.-I., Dong, G., Carruthers Jr., C. W., Golden, S. S., and Liwang, A. (2008) The day/night switch in KaiC, a central oscillator component of the circadian clock of cyanobacteria. *Proc Natl Acad Sci USA* **105:** 12825–12830.
Kitayama, Y., Iwasaki, H., Nishiwaki, T., and Kondo, T. (2003) KaiB functions as an attenuator of KaiC phosphorylation in the cyanobacterial clock system. *EMBO J* **22:** 2127–2134.
Kitayama, Y., Nishiwaki, T., Terauchi, K., and Kondo, T. (2008) Dual KaiC-based oscillations constitute the circadian system of cyanobacteria. *Genes Dev* **22:** 1513–1521.
Kiyohara, Y. B., Katayama, M., and Kondo, T. (2005) A novel mutation of KaiC affects resetting of the cyanobacterial circadian clock. *J Bacteriol* **187:** 2559–2564.
Klevecz, R. R. (1976) Quantized generation time in mammalian cells as an expression of the cellular clock. *Proc Natl Acad Sci USA* **73:** 4012–4016.
Kondo, T., and Ishiura, M. (1994) Circadian rhythms of cyanobacteria: monitoring the biological clocks of individual colonies by bioluminescence. *J Bacteriol* **176:** 1881–1885.
Kondo, T., Strayer, C. A., Kukarni,. D., Taylor, W., Ishiura, M., Golden, S. S., and Johnson, C. H. (1993) Circadian rhythms in prokaryotes: luciferase as a reporter of circadian gene expression in cyanobacteria. *Proc Natl Acad Sci USA* **90:** 5672–5676.
Kondo, T., Tsinoremas, N. F., Golden, S. S., Johnson, C. H., Kutsuna, S., and Ishiura, M. (1994) Circadian clock mutants of cyanobacteria. *Science* **266:** 1233–1236.
Kucho, K., Tsuchiya, Y., Okumoto, Y., Harada, M., Yamada, M., and Iushiura, M. (2004b) Construction of unmodified oligonucleotide-based microarrays in the thermophilic cyanobacterium *Thermosynechococcus elongatus* BP-1: screening of the candidates for circadianly expressed genes. *Genes Genet Syst* **79:** 319–329.
Kucho, K., Okamoto, K., Tsuchiya, Y., Nomura, S., Nango, M., Kanehisa, M., and Ishiura, M. (2005) Global analysis of circadian expression in the cyanobacterium *Synechocystis* sp. strain PCC6803. *J Bacteriol* **187:** 2190–2199.
Kucho, K., Yoneda, H., Harada, M., and Ishiura, M. (2004a) Determinants of sensitivity and specificity in spotted DNA microarrays with unmodified oligonucleotides. *Genes Genet Syst* **79:** 189–197.
Kurosawa, G., Aihara, K., and Iwasa, Y. (2006) A model for the circadian rhythm of cyanobacteria that maintains oscillation without gene expression. *Biophys J* **91:** 2015-2-23.
Kutsuna, S., Kondo, T., Aoki, S., and Ishiura, M. (1998) A period-extender gene, *pex*, that extends the period of the circadian clock in the cyanobacterium *Synechococcus* sp. strain PCC 7942. *J Bacteriol* **180:** 2167–2174.
Kutsuna, S., Nakahira, Y., Katayama, M., Ishiura, M., and Kondo, T. (2005) Transcriptional regulation of the circadian clock operon *KaiBC* by upstream regions in cyanobacteria. *Mol Microbiol* **57:** 1474–1484.
Leipe, D.D., Aravind, L., Grishin, N.V. and Koonin, E.V. (2000) The bacterial replicative helicase DnaB evolved from a RecA duplication. *Genome Res* **10:** 5–16.

Li, C., Chen, X., Wang, P., and Wang, W. (2009) Circadian KaiC phosphorylation. A multilayer network. *PLoS Comput Biol* **5:** e1000568. doi: 10.1371/journal.pcbi. 1000568.

Li, R., and Golden, S. S. (1993) Enhancer activity of light-responsive regulatory elements in the untranslated leader regions of cyanobacterial *psbA* genes. *Proc Natl Acad Sci USA* **90:** 11678–11682.

Liu, Y., Golden, S. S., Kondo, T., Ishiura, M., and Johnson, C. H. (1995a) Bacterial luciferase as a reporter gene expression in cyanobacteria. *J Bacteriol* **177:** 2080–2086.

Liu, Y., Tsinoremas, N. F., Johnson, C. H., Lebedeva, N. V., Golden, S. S., Ishiura, M., and Kondo, T. (1995b) Circadian orchestration of gene expression in cyanobacteria. *Genes Dev* **9:** 1469–1478.

Liu, Y., Tsinoremas, N. F., Golden, S. S., Kondo, T., and Johnson, C. H. (1996) Circadian expression of genes involved in the purine biosynthetic pathway of the cyanobacterium *Synechococcus* sp. strain PCC 7942. *Mol Microbiol* **20:** 1071–1081.

Lorne, J., Scheffer, J., Lee, A., Painter, M., and Miao, V. P. W. (2000) Genes controlling circadian rhythm are widely distributed in cyanobacteria. *FEMS Microbiol Lett* **189:** 129–133.

Lozza-Correa, M., Gomez-Valero, L., and Buchrieser, C. (2010) Circadian clock proteins in prokaryotes: hidden rhythms? *Frontiers Microbiol* doi: 10.3388/frmicb.2010.00130

Mackey, S. R., Choi, J.-S., Kitayama, Y., Iwasaki, H., Dong, G., and Golden, S. S. (2008) Proteins found in CikA interaction assay link the circadian clock, metabolism and cell division in *Synechococcus elongatus*. *J Bacteriol* **190:** 3738–3746.

Mackey, S. R., Ditty, J. L., Clerico, E. M., and Golden, S. S. (2005) Detection of rhythmic bioluminescence from luciferase reporters in cyanobacteria. In: Rosato, E. (Ed.) *Circadian Rhythms: Methods and Protocols. Methods in Molecular Biology* **328:** 115–129. Humana Press, Totowa, N. J., USA. doi:10.1007/978-1-59745-257-1_8

Matsuda, M., Michel, K.-P., Zhang, X., Montgomery, B. L., and Golden, S. S. (2003) Biochemical properties of CikA an unusual phytochome-like histidine protein kinase that resets the circadian clock in *Synechococcus elongatus* PCC 7942. *J Biol Chem* **278:** 19102–19110.

McClung, C. R. (2007) The cyanobacterial circadian clock is based on the intrinsic ATPase activity. *Proc Natl Acad Sci USA* **104:** 16727–16728.

Mehra, A., Hong, C, I., Shi, M., Loros, J. J., Dunlap, J. C. and, Ruoff, P. (2006) Circadian rhythmicity by autocatalysis. *PLoS Comput Biol* **2:** e96

Millineaux, P. M., Gallon, J. R., and Chaplan, A. E. (1981) Acetylene reduction (nitrogen fixation) by cyanobacteria grown under alternating light-dark cycles. *FEMS Microbiol Lett* **10:** 245–247.

Mihalcescu, I., Hsing, W., and Leibler, S. (2004) Resilient circadian oscillator revealed in individual cyanobacteria. *Nature* (London) **430:** 81–85.

Min, H., and Golden, S. S. (2000) A new circadian Class-2 gene, opcA, whose product is important for reductant production at night in *Synechococcus elongatus* PCC 7942. *J Bacteriol* **182:** 6214–6221.

Min, H., Liu, Y., Johnson, C. H., and Golden, S. S. (2004) Phase determination of circadian gene expression in *Synechococcus elongatus* PCC 7942. *J Biol Rhythms* **19:** 103–112.

Mitsui, A., Kumazawa, S., Takahashi, A., Ikemoto, H., and Arai, T. (1986) Strategy by which nitrogen-fixing unicellular cyanobacteria grow photoautotrophically. *Nature* (London) **323:** 720–722.

Miyoshi, F., Nakayama, Y., Kaizu, K., Iwasaki, H., and Tomita, M. (2007) A mathematical model for the Kai-protein-based chemical oscillator and clock gene expression rhythms in cyanobacteria. *J Biol Rhythms* **22:** 69–80. doi:10.1177/0748730406295749.

Mackey, S. R., Ditty, J. L., Clerico, E. M., and Golden, S. S. (2007) Detection of rhythmic luminescence from luciferase reporters in cyanobacteria. In: Rosato, E. (Ed.) *Circadian Rhythms: Methods and Protocols. Methods in Molecular Biology* **362.** Humana Press, Totowa, N. J. USA.

Mori, T., and Johnson, C. H. (2001a) Circadian programming in cyanobacteria. *Seminar Cell Dev Biol* **18:** 7235–7242.

Mori, T., and Johnson, C. H. (2001b) Independence of circadian timing from cell division in cyanobacteria. *J Bacteriol* **183:** 2439–2444.

Mori, T., Binder, B., and Johnson, C. H. (1996) Circadian gating of cell division in cyanobacteria growing with average doubling times of less than 24 h. *Proc Natl Acad Sci USA* **93:** 10183–10188.

Mori, T., Saveliev, S. V., Xu, Y., Stafford, W. F. Cox, M. M., Inman, R. B., and Johnson, C. H. (2002) Circadian clock protein KaiC forms ATP-dependent hexameric rings and binds DNA. *Proc Natl Acad Sci USA* **99:** 17203–17208.

Mori, T., Wiliams, D. R., Byrne, M. O., Qin, X., Egli, M., Machab, H. S. Stewart, P. L., and Johnson, C. H. (2007) Elucidating the ticking of an *in vitro* circadian clock work. *PLoS Biol* **5:** e93. doi:10.1371/journal.pbio.0050093.

Naef, F. (2005) Circadian clocks go *in vitro*: Purely post-translational oscillators in cyanobacteria. *Mol Syst Biol* doi:10.1038/msb4100027-E1-E5.

Nair, U., Ditty, J. L., Min, H., and Golden, S. S. (2002) Roles for sigma factors in global circadian regulation of the cyanobacterial genome. *J Bacteriol* **184:** 3530–3538.

Nakajima, M., Imai, K., Ito, H., Nishiwaki, T., Murayama, Y., Iwasaki, H., Oyama, T., and Kondo, T. (2005) Reconstitution of circadian oscillation of cyanobacterial KaiC phosphorylation *in vitro*. *Science* **308:** 414–415.

Nakajima, M., Ito, H., and Kondo, T. (2010) *In vitro* regulation of circadian phosphorylation rhythm of cyanobacterial clock protein KaiC by KaiA and KaiB. *FEBS Lett* **584:** 898–902.

Nakahira, Y., Katayama, M., Miyashita, H., Kutsuna, S., Iwasaki, H., Oyama, T., and Kondo, T. (2004) Global gene expression by KaiC as a master process of prokaryotic circadian system. *Proc Natl Acad Sci USA* **101:** 881–885.

Narusaka, Y., Narusaka, M., Satoh, K., and Kobayashi, H. (1999) *In vitro* random mutagenesis of the D1 protein of the photosystem II reaction center confers phototolerance on the cyanobacterium *Synechocystis* sp. PCC 6803. *J Biol Chem* **274:** 23270–23275.

Nishimura, H., Nakahira, Y., Imai, K., Tsuruhara, A., Kondo, H., Hayashi, H., Hirai, M., Saito, H., and Kondo, T. (2002) Mutations in KaiA, a clock protein, extend the period of circadian rhythm in the cyanobacterium *Synechococcus elongatus* PCC 7942. *Microbiology* **148:** 2903–2909.

Nishiwaki, T., Iwasaki, H., Ishiura, M., and Kondo, T. (2000) Nucleotide binding and autophosphorylation of clock protein KaiC as a circadian timing process of cyanobacteria. *Proc Natl Acad Sci USA* **97:** 495–499.

Nishiwaki, T., Satomi, Y., Nakajima, M., Lee, C., Kiyohara, R., Kageyama, H., Kitayama, Y., Temamoto, M., Yamaguchi, A., Hajikata, A., Go, M., Iwasaki, H., Takao, T., and Kondo, T. (2004) Role of KaiC phosphorylation in the circadian clock system of *Synechococcus elongatus* PCC 7942. *Proc Natl Acad Sci USA* **101:** 13927–13932.

Onai, K., Morishita, M., Itoh, S., Okamoto, K., and Ishiura, M. (2004) Circadian rhythms in the thermophilic cyanobacterium *Thermosynechococcus elongatus*, compensation of period length over a wide temperature range. *J Bacteriol* **186:** 4972–4977.

Ouyang, Y., Andersson, C. R., Kondo, T., Golden, S. S., and Johnson, C. H. (1998) Resonating circadian clocks enhance fitness in cyanobacteria. *Proc Natl Acad Sci USA* **95:** 8660–8664.

Paranjpe, D. A., and Sharma, V. K. (2005) Evolution of temporal order in living organisms. *J Circadian Rhythms* **3:** 7. doi:10.1186.1740-3391-3-7

Pattanayek, R., Wang, J., Mori, T., Xu, Y., Johnson, C. H., and Egli, M. (2004) Circadian clock protein: Crystal structure of KaiC protein and functional insights. *Mol Cell* **15:** 375–388.

Pattanayek, R., Williams, D. R., Pattanayek, S., Mori, T., Johnson, C. H., Stewart, P. L., and Egli, M. (2008) Structural model of the circadian clock KaiB-KaiC complex and mechanism for modulation of KaiC phosphorylation. *EMBO J* **27:** 1767–1778.

Pettersen, E. F., Goddard, T. D., Huang, C. C., Couch, G. S., Greenblatt, D. M., Meng, E. C., and Ferrin, T. E. (2004) UCFS–Chimera —a visualization system for exploratory research and analysis. *J Comput Chem* **13:** 1605–1612.

Qin, X., Byrne, M., Xu, Y., Mori, T., and Johnson, C. H. (2010) Coupling of a core post-translational pacemaker to a slave transcription/translation feedback loop in a circadian system. *PLoS Biol* **8:** e1000394.

Qin, X., Byrne, M., Mori, T., Zon, P., Williams, D. R., Mchaourab, H., and Johnson, C. H. (2010) Intermolecular associations determine the dynamics of the circadian oscillations *in vitro*. *Proc Natl Acad Sci USA* **107:** 14805–14810.

Roenneberg, T., and Foster, R.G. (1997) Twilight times: light and circadian system. *Photochem Photobiol* **66:** 549–561.

Rust, M. J., Markson, J. S., Lane, W. S., Fisher, D. S., and O'Shea, E. K. (2007) Ordered phosphorylation governs oscillation of a three-protein circadian clock. *Science* **318:** 809–812.

Schmitz, O., Williams, S. B., Katayama, M., Kondo, T., and Golden, S. S. (2000) CikA, a bacteriophytochrome that resets the cyanobacterial circadian clock. *Science* **289:** 765–768.

Schneegurt, M. A., Sherman, D. M., Nayar, S., and Sherman, L. A. (1994) Oscillating behaviour of carbohydrate granule formation and dinitrogen fixation in the cyanobacterium *Cyanothece* sp. strain ATCC 51142. *J Bacteriol* **176:** 1586–1597.

Schneegurt, M. A., Sherman, D. M., and Sherman, L. A. (1997) Growth, physiology, and ultrastructure of a diazotrophic cyanobacterium, *Cyanothece* sp. strain ATCC 51142, in mixotrophic and chemoheterotrophic cultures. *J Phycol* **33:** 632–642.

Smith, R. M., and Williams, S. B. (2006) Circadian rhythms in gene transcription imparted by chromosome compaction in the cyanobacterium *Synechococcus elongatus*. *Proc Natl Acad Sci USA* **103:** 8564–8569. doi/10.1073/pnas.0508696103

Stal, L. J., and Krumbein, W. E. (1985) Nitrogenase activity in the non-heterocystous cyanobacterium *Oscillatoria* sp. grown under alternating light-dark cycles. *Arch Microbiol* **143:** 67–71.

Stoeckel, J., Welsh, E. A., Liberton, M., Kunnvakkam, R., Aurora, R., and Pakrasi, H. B. (2008) Global transcriptomic analysis of *Cyanothece* 51142 reveals robust diurnal oscillation of central metabolic processes. *Proc Natl Acad Sci USA* **105:** 6156–6161.

Sweeney, B. M., and Borgese, M. B. (1989) A circadian rhythm in cell division in a prokaryote, the cyanobacterium *Synechococcus* WH7803. *J Phycol* **25:** 183–186.

Takai, N., Nakajima, M., Oyama, T., Kito, R., Sugita, C., Sugita, M., Kondo, T., and Iwasaki, H. (2006) A KaiC associating SasA-RpaA two component regulatory system as a major circadian timing mediator in cyanobacteria. *Proc Natl Acad Sci USA* **103:** 12109–12114.

Takigawa-Imamura, H., and Mochizuki, A. (2006) Predicting regulation of the phosphorylation cycle of KaiC clock protein using mathematical analysis. *J Biol Rhythms* **21:** 405–416.

Taniguchi, Y., Katayama, M., Ito, R., Takai, N., Kondo, T., and Oyama, T. (2007) *labA* : a novel gene required for negative feed back regulation of the cyanobacterial circadian clock protein KaiC. *Genes Dev* **21:** 60–70.

Terauchi, K., Kitayama, Y., Nishiwaki, T., Miwa, K., Murayama, Y., Oyama, T., and Kondo, T. (2007) ATPase activity of KaiC determines the basic timing for circadian clock of cyanobacteria. *Proc Natl Acad Sci USA* **104:** 16377–16381.

Taniguchi, Y., Yamaguchi, A., Hijikata, A., Iwasaki, H., Kamagata, K., Ishiura, M., Go, M., and Kondo, T. (2001) Two KaiA binding domains of cyanobacterial circadian clock protein KaiC. *FEBS Lett* **496:** 86–90.

Tomita, J., Nakajima, M., Kondo, T., and Iwasaki, H. (2005) No transcription-translation feedback in circadian rhythm of KaiC phosphorylation. *Science* **307:**251–254.

Tsinoremas, N. F., Ishiura, M., Kondo, T., Andersson, C. R., and Tanaka, K. (1996) A sigma factor that modifies the circadian expression of a subset of genes in cyanobacteria. *EMBO J* **15:** 2488–2495.

Tsinoremas, N. F., Schaefer, M. R., and Golden, S. S. (1994) Blue and red light reversibly control psbAI expression in the cyanobacterium *Synechococcus* sp. strain PCC 7942. *J Biol Chem* **269:** 16143–16147.

Uzumaki, T., Fujita, M., Nakatsu, T., Hayashi, F., Shibata, H., Itoh, N., Kato, H., and Ishiura, M. (2004) Crystal structure of the C-terminal clock-oscillator domain of the cyanobacterial KaiA protein. *Nat Struct Mol Biol* **11:** 623–631.

Vakonakis, I., and LiWang, A. C. (2004) Structure of the C-terminal domain of the clock protein KaiA in complex with a KaiC-derived peptide: Implications for KaiC regulation. *Proc Natl Acad Sci USA* **101:** 10925–10930.

Vakonakis, I., Klewer, D. A., Williams, S. B., Golden, S. S., and LiWang, A. (2004a) Structure of the N-terminal domain of the circadian clock-associated histidine kinase SasA. *J Mol Biol* **342:** 9–17.

Vakonakis, I., Sun, J., Wu, T., Holzenburg, A., Golden, S. S., and LiWang, A.C. (2004b) NMR-structure of the Kai C-terminal domain of KaiA, a circadian clock protein: Implications for KaiA-KaiC interaction. *Proc Natl Acad Sci USA* **101:** 1479–1484.

van Zon, J. S., Lubensky, D. K., Altena, P. R. H., Rein ten Wolde, P. (2007) An allosteric model of circadian KaiC phosphorylation. *Proc Natl Acad Sci* USA **104:** 7420–7425. doi:10.1073/pnas.0608665104

Weare, N. M., and Benemann, J. R., (1974) Nitrogenase activity and photosynthesis in *Plectonema boryanum. J Bacteriol* **119:** 258–265.

Williams, S. B., Vakonakis, I., Golden, S. S., and Li Wang, A. C. (2002) Structure and function from the circadian clock protein KaiA of *Synechococcus elongatus:* A potential clock input mechanism. *Proc Natl Acad Sci USA* **99:** 15357–15362.

Woelfle, M. A., and Johnson, C. H. (2006) No promoter left behind: Global circadian gene expression in cyanobacteria. *J Biol Rhythms* **21:** 419-431. doi:10.1127/0748730406294418

Woelfle, M. A., Ouyang, Y., Phanvijhitsiri, K., and Johnson, C. H. (2004) The adaptive value of circadian clocks: An experimental assessment in cyanobacteria. *Curr Biol* **14:** 1481–1486.

Woelfle, M. A., Xu, Y., Qin, X., and Johnson, C. H. (2007) Circadian rhythms of superhelical status of DNA in cyanobacteria. *Proc Natl Acad Sci USA* **104:** 18819–18824.

Wolk, C. P., Cai, Y., and Panoff, J-M. (1991) Use of transposon with luciferase as a reporter to identify environmentally responsive genes in a cyanobacterium. *Proc Natl Acad Sci USA* **88:** 5355–5359.

Xu, Y., Mori, T., and Johnson, C. H. (2000) Circadian clock-protein expression in cyanobacteria: Rhythms and phase setting. *EMBO J* **19:** 3349–3357.

Xu, Y., Mori, T., and Johnson, C. H. (2003) Cyanobacterial circadian clockwork: Roles of KaiA, KaiB, and the *KaiBC* promoter in regulating KaiC. *EMBO J* **22:** 2117–2126.

Xu, Y., Mori, T., Pattanayek, R., Pattanayek, S., Egli, M., and Johnson, C. H. (2004) Identification of key phosphorylation sites in the circadian clock protein KaiC by crystallographic and mutagenic analysis. *Proc Natl Acad Sci USA* **101:** 13933–13938.

Xu, Y., Mori, T., Qin, X., Yan, H., Egli, M., and Johnson, C. H. (2009) Intramolecular regulation of phosphorylation status of the circadian clock protein KaiC. PLoS ONE 4: e7509. doi:10.137/journal.pone.0007509

Xu, Y., Piston, D., and Johnson, C. H. (1999) A bioluminescence resonance energy transfer (BRET) system—application to interacting circadian clock proteins. *Proc Natl Acad Sci USA* **96:** 15–16.

Ye, S., Vakonakis, I., Ioerger, T. R., LiWang, A. C., and Sacchettini, J. C. (2004) Crystal structure of circadian clock protein KaiA from *Synechococcus elongatus*. *J Biol Chem* **279:** 20511–20518.

Yoda, M., Eguchi, K., Terada, T. P., and Sasai, M. (2007) Monomer-shuffling and allosteric transition in KaiC circadian oscillation. *PLoS ONE* **2:** e408. doi:10.1371/journal.pone.0000408

Zhang, X., Dong, G., and Golden, S. S. (2006) The pseudo-receiver domain of CikA regulates the cyanobactertial circadian input pathway. *Mol Microbiol* **60:** 658–668.

Zwicker, D., Lubensky, D. K., and Rein ten Wolde, P. (2010) Robust circadian clocks from coupled protein-modification and transcription-translation cycles. *Proc Natl Acad Sci USA* Early Edition doi: 10. 1073/pnas.1007613107

CHAPTER 7

SYMBIOSIS

The term symbiosis was coined by Frank in 1877 and was described by Anton de Bary (1879) as living together of two different organisms in a permanent or long lasting association. On the basis of how each partner is benefited in the association, three broad categories of symbiotic relationships are recognized, i.e. commensalism, parasitism and mutualism. In commensalism, one member is helped and the other member is neither helped nor harmed. In parasitism one partner is benefited and the other partner is harmed. In mutualism both partners help each other. Symbiotic relationships encompass all major groups of living world. There is a great variation in the size of individuals that take part in the symbiotic association. The terminology in describing symbiotic relationships has now been defined by the International Society of Endocytobiology (Schenk, http://www.endocytobiology.org). In relation to cyanobacteria, the terms cyanobiont, cyanelle and cyanome need to be elaborated. A cyanobiont is defined as a cyanobacterial partner in a symbiotic relationship. Cyanelle is a genetically autonomous symbiotic endophyte. Cyanome is a symbiotic consortium of host and intracellular cyanobacterium.

Amongst the plant-microbe associations, those formed by cyanobacteria have attracted worldwide attention. Most symbiotic relationships of cyanobacteria are facultative suggesting that upon isolation each partner can exist independently and proliferate. However, when the two partners enter into symbiosis the whole unit (cyanome) proliferates. Each partner thus contributes to the nutrition of the other without causing any harm. Thus it can be safely stated that the symbioses formed by cyanobacteria are mutualistic. Cyanobacterial symbioses are widespread in aquatic and terrestrial environments. Of the aquatic bodies, in marine waters their symbiotic association has been reported with sponges, ascidians (sea squirts), echiuroid worms, diatoms, dinoflagellates and a protozoan in the plankton (Carpenter, 2002). In terrestrial environments, cyanobacteria are known to enter into symbiotic associations with almost all groups of plants. Cyanobacterial symbioses with fungi, bryophytes (hepatics, hornworts and mosses), pteridophytes (water fern *Azolla*), cycads and one angiosperm (*Gunnera*) have been recorded. Due to a difference in size of plants being invaded by the cyanobacteria, the larger partner is thus called as the host and the cyanobacteria are referred to as cyanobionts. Except that the association with fungi leads to the formation of a specialized lichen thallus, in the rest of the associations with plants the general structure in which the cyanobiont is housed is also formed in the absence of it. The cyanobiont fixes nitrogen and transfers the fixed nitrogen to the host plant and in return the latter provides fixed carbon to the cyanobiont. Thus in these mutualistic associations the cyanobacteria constitute a driving force in the evolution of their hosts (Usher *et al.*, 2007).

I. SPONGES

The existence of sponges (Phylum-Porifera) can be dated back to Precambrian. As many as 9000 living sponge species are distributed on tropical reefs from lower to higher latitudes (Brusca and Brusca, 1990). Sponges are known as filter feeders and pump large volumes of water through their canal system (Reiswig, 1971, 1974; Pile *et al.*, 1996). With reference to their nutrition and other features they very much resemble Protozoa, since a number of amoeboid cells move freely in the sponge matrix. Sponges harbour a diversity of prokaryotic and eukaryotic symbionts and they account for 40% of their biomass (Vacelet, 1975; Vacelet and Donadey, 1977). The range of these symbionts includes archaebacteria, heterotrophic bacteria, cyanobacteria, green algae, red algae, dinoflagellates, cryptophytes and diatoms (Larkum *et al.*, 1987; Santavy *et al.*, 1990; Duglas, 1994; Preston *et al.*, 1996). It is also true for a single given species of sponge. As for example, *Theonella swinhoei* possesses heterotrophic bacteria, unicellular cyanobacteria and filamentous heterotrophic bacteria (Bewley *et al.*, 1996). On the other hand, sponges of *Aplysina* spp. show different bacterial genera such as *Bacillus* sp., *Micrococcus* sp., *Arthrobacter* sp., *Vibrio* sp., *Pseudoalteromonas* sp. (Hentschel *et al.*, 2001). Likewise, sponge *Rhopaloeides odorabile* is a good habitat for β-proteoacteria, γ-proteobacteria, cytophagas, actinobacteria and green sulphur bacteria (Webster *et al.*, 2001).

Cyanobacterial symbionts are known to occur in many sponge genera. Their location can be extracellular in sponge tissue or intracellular in specialized vacuoles. Due to their growth, the symbionts enable the sponges to compete for substrata with algae and corals in illuminated areas (Wilkinson, 1983; Hinrichsen, 1997). Most common cyanobacterial symbionts belong to *Aphanocapsa feldmannii* (Fremy) group (Feldman, 1933) which is present in the surface tissues of as many as 60 sponge species (belonging to 13 orders). Other cyanobacteria such as *Synechocystis* (Larkum *et al.*, 1988), *Oscillatoria* (Wilkinson, 1992), *Anabaena* (Larkum, 1999), *Cyanobacterium* (Webb and Mass, 2002) and *Synechococcus* (Hentschel *et al.*, 2002; Usher *et al.*, 2004a) have been reported from different sponge genera. According to one estimate, the proportion of cyanobacterial biomass is equal to that of sponge cells by meeting 50% of the sponge's energy budget and 80% of the sponge's carbon budget through photosynthesis of cyanobionts (Wilkinson, 1983; Cheshire *et al.*, 1997). However, of the above cyanobionts, only two of them have been frequently reported from many sponge genera. These are *Oscillatoria spongeliae* and *Synechococcus spongiarum*. The former has been recorded from three species of sponges belonging to Dysideidae (Order Dictyoceratida), i.e. *Lamellodysidea* (formerly *Dysidea*) *herbacea*, *L. chlorea* and *L. granulosa* (Larkum *et al.*, 1987; Hinde *et al.*, 1994; Thacker and Starnes, 2003) whereas the latter has been reported from a wider variety of sponges. These are sponge genera *Xestospongia muta* (Petrosiidae, Haplosclerida; Gómez *et al.*, 2002), *Aplysina aerophora* (Aplysinidae, Verongida; Hentschel *et al.*, 2002) and *Chondrilla nucula* (Chondrillidae, Chondrosida; Usher *et al.*, 2004b). Usher *et al.* (2004a) distinguished the *Synechococcus* spp. associated with sponges to be different from the planktonic *Synechococcus* though as yet there are no evidences for the symbionts to be host-specific.

Growth of sponges is generally measured in terms of dry weight and the area under cover occupied by the sponges at the beginning of the experiment. *L. chlorea* lost both dry weight and nearly 40% of the area under shading after 15 days of incubation where as in *X. exigua* the loss was insignificant. However, the biomass of the cyanobiont in *L. chlorea* (as measured in terms of chlorophyll content) did not differ very much from controls. These studies help us in understanding the relationship between the symbionts. The cyanobiont *O. spongeliae* contributes to the overall development of its host in a mutualistic association whereas *S. spongiarum* may be commensals that draw resources from its host without significantly affecting sponge mass. But when *X. exigua* is under shade it is likely that *S. spongiarum* is consumed by its host (Thacker, 2005).

Sponges growing in Great Barrier Reef of Australia have been examined for the nature of cyanobionts, ultrastructure and pigment composition. *O. spongeliae* has been detected in *Dysidea herbacea* and in many sponge species. Two more unidentified *Oscillatoria* species have been found in an unidentified sponge species and the ascidian *Trididemnum miniatum*. Along with *Oscillatoria* sp., the latter contained *Prochloron* as well. All the three *Oscillatoria* species could be distinguished on the basis of thylakoid arrangement (Larkum *et al.*, 1987). Usher *et al.* (2006) characterized the unicellular cyanobionts of marine sponges from Australia and the Mediterranean by transmission electron microscopy, cell shape and size and thylakoid arrangement. The cells of *S. spongiarum* are oval and the turns of thylakoids increase from 1 to 5. The cells are located in the outer wall of sponges *C. nucula*, *C. australiensis* and *Ircinia variabilis*. Cells of *A. feldmannii* that occupy the matrix of sponges *I. variabilis* and *Petrosia ficiformis* are almost spherical and the turns of thylakoids increase from 2.5 to 6. The cells of *Synechococcus* spp. are oval and have a spiral thylakoid with 2–3 turns and the cells occupy top few mm of sponge *Haliclona* sp. Thus the cell size and shape of *S. spongiarum* and *Synechococcus* spp. are similar suggesting that these cyanobionts are morphologically indistinguishable though derived from different geographic locations or hosts. They could only be identified on the basis of number of turns of the thylakoids. However, sponges *Cymbastela marshae* and *I. variabilis* revealed the presence of symbionts *Oscillatoria* sp. and *Aphanocapsa raspagaigellae*, respectively which could readily be identified both by size and ultrastructural features. One interesting feature noticed is that in *C. australiensis*, the cyanobiont is reported to be transmitted vertically, i.e. through sponge eggs (Usher *et al.*, 2001). Moreover, attempts to culture sponge-associated cyanobacteria have not been successful. In addition, the absence of these species in a free state in water samples suggests that the cyanobionts may not survive outside their hosts (Usher *et al.*, 2004b). Histological and ultrastructural studies on sponge *Tethya orphei* (Demospongiae), collected from Arì Athol coral stones of Maldives Islands, revealed the presence of *O. spongeliae* in the cortical region and penetrated deeply inside the choanosomal region overlapping with siliceous spicule bundles. The proliferation of the cyanobiont was so extensive that it could be vertically transmitted from sponge to sponge which confirms that the association is mutualistic (Gaino *et al.*, 2006).

The molecular marker that has revolutionized the understanding of microbial ecology is 16S rRNA gene amplification of microbial samples. Besides understanding the phylogenetic relationships, distribution patterns and diversity of microbes in the environment, this approach has an added advantage in bypassing the requirement of culturing the microbes. A number of workers have utilized this molecular marker for understanding the distribution and specificity of the symbionts that occur in various sponge species. Hentschel *et al.* (2002) investigated the diversity of symbiotic microbial communities of 190 sponge species collected from all over the world by 16S rDNA sequences. A total of 14 monophyletic sponge-specific clusters belonging to different bacterial divisions have been observed of which seven sequences from cyanobacteria inhabiting sponges *Aplysina aerophoba* and *T. swinhoei* could be divided into two clades, i.e. *Synechococcus*/*Prochlorococcus* and *Pleurocapsa*. Webb and Mass (2002) found coccoid cyanobacteria in *Mycale* (*Carmia*) *hentscheli* by epifluorescence microscopy. The amplification of 16S rRNA genes of these organisms revealed the presence of four closely related clones which had a high (8%) sequence divergence. The clones exhibited closest similarity to *Cyanobacterium stanieri* followed by *Prochloron* sp. and *Synechocystis* sp. In order to find out the host specificity of a particular cyanobiont in certain species of sponges or population of sponges, Thacker and Starnes (2003) subjected the DNA samples of cyanobionts of *L. (Dysidea) herbacea* 1A and 1B and *L. granulosa* for amplification and sequence determination of 16S rDNA. These three sponge species exhibited a high degree of specificity for a particular cyanobacterial group emphasizing the probability of coevolution of both the host and cyanobiont. Usher *et al.* (2004a) studied the distribution and phylogeny of unicellular cyanobacterial symbionts of selected marine sponges (*A. aerophoba, I. variabilis* and *P. ficiformis*) from the Mediterranean, Australia (four *Chondrilla* species) and *Haliclona* sp. from both the regions by direct 16S rDNA sequencing. The cyanobionts comprise at least four closely related species of *Synechococcus* that included *A. feldmannii* from *P. ficiformis* and *C. nucula*. A hitherto undescribed symbiont of sponges related to *Oscillatoria rosea* has been recorded from *Cymbastela marshae* from Australia. Diverse sponge genera from the Mediterranean, Indian, Pacific and Southern oceans showed the existence of *S. spongiarum*. Four dictyoceratid marine sponges (*L. herbacea, L. chlorea, Lendenfeldia chondrodes* and *Phyllospongia papyracea*) from reef sites of Republic of Palau showed the presence of symbionts belonging to α-proteobacteria group as revealed by 16S rRNA gene analysis (sequences that fit into *Rhodobacter* sp.) besides harbouring *O. spongeliae*. Interestingly, *L. chondrodes* alone showed additionally the presence of *Synechocystis* sp. in both surface (pinacoderm) and internal mesohyl whereas *O. spongeliae* was restricted to mesohyl. Specific location of these cyanobionts was confirmed by the fluorescence *in situ* hybridization experiments. *P. papyracea* contained significant number of γ-proteobacteria (Ridley *et al.*, 2005). Steindler *et al.* (2005) traced the 16S rRNA phylogeny of sponge-associated cyanobacteria. The amplification of 16S rRNA was carried out with primers 361F(5′-GAATTTTCCGCAATGGGC-3′) and 1459 R (5′-GGTAAYGACTTCGGGCRT-3′). Most of the sequences matched with species of *Synechococcus, Prochlorococcus* and members of Oscillatoriales suggesting their polyphyletic origin and that these represent multiple independent symbiotic events. Taylor *et al.* (2005) studied the biogeography of bacteria associated with the marine sponge *Cymbastela concentrica* by using 16S rDNA-DGGE (denaturing gradient gel electrophoresis). The DGGE banding patterns indicated different bacterial communities in this sponge from tropical versus temperate Australia. The tropical forms of *C. concentrica* showed the similar cyanobionts reported earlier in *Mycale hentscheli* from New Zealand by Webb and Mass (2002).

Wilkinson and Fay (1979) reported fixation of nitrogen by sponges growing in coral reef of Red Sea. Only those sponges gave positive acetylene reduction assay (ARA) that harbour cyanobionts and those that lacked cyanobionts were reported negative for ARA. It is further suggested that these cyanobionts play an important role in maintaining the nitrogen balance in marine environments with low available N where the sponges grow. A variety of functions have been attributed to these symbionts, of which mention may be made of nutrient acquisition and growth of the sponges (Wilkinson and Vacelet, 1979; Frost and Williamson, 1980; Wilkinson, 1992; Vacelet *et al.*, 1995; Hill, 1996), stabilization of sponge skeleton (Rützler, 1985), processing of metabolic waste (Beer and Ilan, 1998) and secondary metabolite production (Unson *et al.*, 1994; Bewley *et al.*, 1996; Flowers *et al.*, 1998; Schmidt *et al.*, 2000). The secondary metabolites consist of inhibitors of cell division and various enzymes, that affect multiplication of viruses, fungi and a number of microbes. Some other secondary metabolites exhibit anti-inflammatory, antitumor, antiviral properties and toxicity to cellular and cardiovascular systems (Munro *et al.*, 1999; Lee *et al.*, 2001). Other functions performed by these metabolites are predator and competitor deterrence (Pawlik *et al.*, 1995; Thacker *et al.*, 1998; Engel and Pawlik, 2000) and resistance to malignant microbial infections (Garson, 2001; Thakur *et al.*, 2003).

The existence of *O. spongeliae* in the tropical marine sponge *L. herbacea* has been identified on the basis of flow cytometry separation of the cyanobiont from sponge cells. The role of the cyanobiont in the production of polychlorinated compounds has also been emphasized. Thus this constitutes the first report on the production of secondary metabolites by a cyanobiont from sponges. Faulkner *et al.* (1994) separated cells of sponge *L. herbacea* and its symbiont *O. spongeliae* on the basis of fluorescence using a cell sorter (Becton-Dickson FAC Star Plus) to obtain 10 million non-fluorescent sponge cells and 2 million fluorescent cyanobacterial filaments and showed by NMR spectroscopy that the cyanobacterial fraction is associated with signals due to 13-demethylisodysidenin as the major chlorinated metabolite. On the other hand, the sponge cell fraction unambiguously showed sesquiterpenes herbadysidolide and spirodysin by GC-MS analysis. Similarly, the presence of 2-(2′,4′-dibromophenyl)-4,6-dibromophenol as the major metabolite of *L. herbacea* (from a shallow lagoon near Hotel Nikko in Palau) up to 6% dry weight originated from the cyanobiont. This was confirmed by flow cytometry and NMR spectroscopy that the metabolite was from the fraction of cyanobacterial symbiont. It was further suggested that these compounds serve the role of chemical defense of the sponge against predators and bacterial invasion.

II. ECHIUROID WORMS

Two worms, *Ikedosoma gogoshimense* and *Bonellia fuliginosa* that grow in the muddy sand at low tide levels and coral reefs, respectively possess cyanobionts in their subepidermal connective tissues. No details of their nature, type of symbionts and their interaction are known (Carpenter, 2002).

III. CORALS

Lesser *et al.* (2004) identified unicellular, symbiotic cyanobacteria in the host cells of the coral *Montastraea cavernosa* collected from the Caribbean Islands. These authors suggest that the cyanobionts coexist with symbiotic dinoflagellates (zooxanthellae) of the coral and form long term association within the host cells.

IV. ASCIDIANS

These are classified as Chordates as their larvae possess a notochord. These live as permanently attached or buried in the sand or mud. In the family Didemnidae of ascidians, five genera form symbiotic association with two cyanobionts *Synechocystis* and *Prochloron*.

The presence of photosynthetic organisms and the ability of ascidians to evolve oxygen can be dated back to 1935. Initially identified as *Synechocystis didemni*, it was later named as *Prochloron didemni* (Lewin, 1975; Newcomb and Pugh, 1975). *P. didemni* is characteristic in being a prokaryote that possesses chlorophyll *a* and *b* but lacks phycobiliproteins (Lewin and Withers, 1975). Chlorophyll *b* is present in a bound form to a protein that differs from Cab protein of plastids of green algae and higher plants (La Roche *et al.*, 1996). The 16S rRNA sequencing revealed closer resemblances to cyanobacterial lineages. The understanding of microenvironment of *P. didemni* in ascidians remained incomplete and attempts to cultivate the symbiont have not been successful despite unconfirmed reports that exist in literature (Kühl and Larkum, 2002).

The cells of *Prochloron* reside in cloacae or embedded in the folds of gelatinous matrix extracellular to the host that forms a transparent upper tunic/test (Lewin and Cheng, 1989; Hirose *et al.*, 1996, 1998). Alberte (1987) reported that up to 60% of carbon demand of ascidians is met by the transfer of photosynthates from *Prochloron*. The contribution of *Prochloron* to the carbon demand of ascidians differs from species to species (Koike and Suzuki, 1996) as exemplified by *Dimemnum molle* which cannot depend solely on photosynthates from its symbiont where as *Lissoclinum voeltzkowi* the carbon demand can be fully met by *Prochloron* (Koike *et al.*, 1993).

V. ALGAE

A) Diatoms: Cyanobacteria form three types of symbioses with diatoms. These are formation of microbial spheres, epiphytic and intracellular associations.

i) Microbial spheres: Brehm *et al.* (2003) observed the formation of loose associations of bacteria, cyanobacteria and diatoms leading to the formation of spherical objects known as microbial spheres. In the first phase, bacteria and diatoms come together and are held in a matrix. In the second phase, cyanobacteria penetrate these spheres and arrange themselves on the surface. The formation of such spheres and their proliferation in non-axenic cultures of *Phormidium* from North Sea microbial mats by the entrapment of phototrophic bacteria and diatoms (especially *Navicula*) was observed. Chemotactic responses are indicated for such an association and possible nutritional interactions are indicated.

In another loose association, the presence of a number of diatoms (of the genera *Amphora, Berkeleya, Cymbella, Entomoneis, Epithemia, Lunella, Mastogloia, Nitzschia* and *Rhopalodia*) deep inside the colonies of *Rivularia* growing in brackish waters of Baltic Sea has been noted. The advantages of such association for diatoms can be protection from grazing, free mobility in the secreted mucilage and supply of organic and inorganic nutrients (Snoeijis and Murasi, 2004).

ii) Epiphytic and endophytic associations: These two types are considered together here because of their interchangeability from one type to the other and the early events associated with the discovery of this symbiosis. *Richelia* is a short filamentous, heterocystous cyanobacterium with 4–10 vegetative cells. A single terminal heterocyst is present that is slightly smaller in diameter than the vegetative cells. The filaments are slightly tapered and do not possess gas vesicles (Janson *et al.*, 1995). Due to these morphological features it is considered closer to *Calothrix* and in fact Lemmermann (1899) initially identified the epiphyte of *Rhizosolenia* as *Calothrix rhizosoleniae*. Subsequently, the endosymbiont of *Rhizosolenia* was given the name of *Richelia intracellularis* (Schmidt, 1901). The endosymbiotic nature of *R. intracellularis* was confirmed by Lemmermann (1905) not only in *Rhizosolenia* but also in *Hemiaulus* and its occurrence as epiphyte on *Chaetoceros*. Generic names *Calothrix* and *Richelia*, have been used by Carpenter (2002) for the epiphyte and the endosymbiont, respectively. Reports on the epiphytic nature of *R. intracellularis* growing on *Chaetoceros* (Janson *et al.*, 1999) and *Bacteriastrum* (Villareal, 1992; Rai *et al.*, 2000; Carpenter, 2002) exist in literature. *R. intracellularis* establishes as an epiphyte by attaching to the spaces in between the cells in chains of diatom colonies of *Chaetoceros*. The endosymbiotic nature of *R. intracellularis* in diatom *Hemiaulus* spp. (Kimor *et al.*, 1978; 1992; Heinbokel, 1986; Villareal, 1994) and *Rhizosolenia clevi* var. *communis* (Sundström, 1984) has been described with two in the former and 2–4 filaments inside the cells of the latter. *Rhizosolenia* does not fix nitrogen in the absence of its endosymbiont *R. intracellularis* (Villareal, 1987). Nitrogen fixation by *R. intracellularis* can fully support the needs of its host (Villareal, 1990, 1991). The division cycle of *Rhizosolenia*-*R. intracellularis* symbiosis (Villareal, 1989) and a preliminary characterization of this symbiosis *in vitro* (Villareal, 1990) have been reported.

The justification for designating the epiphyte and endosymbiont by different generic names has been examined by studying the genetic diversity of *R. intracellularis*. Janson *et al.* (1999) used PCR amplification of *hetR* gene sequence unique for heterocystous cyanobacteria for this purpose. The host for epiphytic *R. intracellularis* was *Chaetoceros* sp. (two species collected from the Pacific Ocean). Species of *Hemiaulus* showed variable number of cells in their filaments as well as in the number of endosymbiotic *R. intracellularis* filaments. For example *H. hauckii* (with 5–6 cells per filament from the Pacific Ocean) contained two *R. intracellularis* filaments in each cell while *H. membranaceous* (with 10 cells per filament from Atlantic and Pacific Oceans) showed the presence of two *R. intracellularis* filaments in each cell and *R. clevei* var. *communis* (from Pacific Ocean) contained four *R. intracellularis* filaments in each cell. The amplification products of *hetR* gene sequences of the epiphytic and endosymbiotic *R. intracellularis* all belonged to the same group and there was no difference in the sequences of *R. intracellularis* growing intracellularly in *H. membranaceous* either from Pacific or Atlantic Oceans. A high degree of host specificity was also deduced based on the divergence of the sequences between symbionts from different genera.

Gómez *et al.* (2005) studied the distribution pattern of *R. intracellularis* as an epiphyte on *Chaetoceros compressus* from the Pacific Ocean. Colonies of *C. compressus* consisted of 10–16 cells and 1–9 epiphytic filaments of *R. intracellularis*. Variables of environment (temperature and salinity) and nutrients (nitrate and phosphate) have been taken into account to assess the distribution of *Rhizosolenia* + *Richella* consortia. These studies have been extended to various depths (5 to 200 m) and latitudes (at nine stations in between 30° 30′ to 34° 15′ N in May and at 10 stations in between 30° to 34° 20′ N during July) of Pacific Ocean at Kurashio, Oyashia currents surrounding Japan and the Celebes, Sulu and South China Sea. The occurrence of *R. intracellularis* as endosymbiont of *Rhizosolenia clevei* and free-living *R. intracellularis* in the samples prompted the authors to suggest that the free-living *R. intracellularis* filaments might have originated due to their release from the surface of the symbiotic diatom plasmolemma that lacked the frustule. Such nascently released *R. intracellularis* filaments could colonize the scenescent cells in colonies of *C. compressus* lacking the epiphytic *R. intracellularis*. Once *R. intracellularis* establishes on a single cell, it then spreads to other diatom cells in the colony. These events have been correlated with the distribution patterns of the two diatoms and *R. intracellularis*. The occurrence of *C. compressus* with its epiphyte is restricted to the Indian and western Pacific Oceans whereas the endosymbiont *R. intracellularis* in other diatoms is ubiquitous in warm oceans. The occurrence of *Richelia-Chaetoceros* consortia exclusively in the periphery of the geographic proliferations of *C. compressus* coincided with the overlapping area of the populations of asymbiotic *C. compressus* and *R. intracellularis* as an endosymbiont in *R. clevei*. Further support for their findings is derived from the work of Janson *et al.* (1999) who reported a similarity of *hetR* gene sequences of endosymbiotic *R. intracellularis* in *R. clevei* and the epiphytically growing *R. intracellularis* on *Chaetoceros*. Further, as free-living *R. intracellularis* needs a support (due to lack of gas vacuoles in its cells) and as *Chaetoceros* cannot survive in oligotrophic waters both find it mutually beneficial leading to the formation of *Richelia-Chaetoceros* consortia. Foster and Zehr (2006) compared the utility of the molecular markers *nifH*, *hetR* and 16S rRNA sequences in the characterization of diatom-diazotroph associations (DDAs).

***iii) Role of DDAs in the open oceans*:** The nitrogen cycle has been recognized as an important intrinsic component of the ocean ecosystem and is supposed to play a greater role in the response of oceans to global environmental changes (Zehr and Kudela, 2011). DDAs and *Trichodesmium* constitute the important components of the upper euphotic zone flora (White *et al.*, 2007a,b; Dore *et al.*, 2008). The contribution of *R. intracellularis* to the enrichment of nitrogen status of tropical marine waters as an epiphyte and endophyte of diatoms has been highlighted (Venrick, 1974; Carpenter *et al.*, 1999; Scharek *et al.*, 1999a,b; Capone, 2001). According to one estimate, based on ARAs of samples assayed from SW N Atlantic (27° N 50° W), the contribution of *Richelia* as an endosymbiont of the diatom *Hemiaulus* has been noted on the average to the tune of 3110±1315 µmol N m^{-2} d^{-1} (Carpenter *et al.*, 1999). The rates of nitrogen fixation by the DDAs assume significant proportions both in the open oceans (Fong *et al.*, 2008; Zeev *et al.*, 2008; Kitajima *et al.*, 2009; Foster *et al.*, 2009, 2011; Turk *et al.*, 2011; Villareal *et al.*, 2011) and nearshore systems (White *et al.*, 2007a,b; Subramanian *et al.*, 2008). However, along a Mediterranean transect the dominance of rhizobia has been reported with lower concentrations of unicellular diazotrophic cyanobacteria in the western Mediterranean Sea and *Richelia* in the eastern basin. (Le Moal *et al.*, 2011). *R. intracellularis* contributes to the extent of 35% to 48% of nitrogen demand in the Gulfs of California at Guaymas and Carmen basins (White *et al.* 2007a).

Zeev *et al.* (2008) estimated low rates of nitrogen fixation (~1.1 nmol N L^{-1} day^{1}) by *R. intracellularis* from ultraoligotrophic waters of Levantine basin of the eastern Mediterranean Sea. The microscopic identification has been correlated with the reverse transcribed PCR amplification of the *nifH* gene that showed 98.8% identity with the known *nifH* sequence of *R. intracellularis*. The wide spread occurrence of *Hemiaulus-R. intracellularis* symbiosis in southwest North Atlantic Ocean has been reported by Villareal(1994). The influence of Amazon river plume on the distribution of free-living and symbiotic cyanobacteria in the western tropical north Atlantic has been reported (Foster *et al.*, 2007). By using *nifH* gene amplification by quantitative PCR method, Foster *et al.* (2009) showed the gene abundance and gene expression of diazotrophic populations from the Eastern Equatorial Atlantic. *H. hauckii-R. intracellularis* association has been found to be abundant (>10^4 *nifH* copies L^{-1}) in the north-west of the Congo River plume. Although *Calothrix-Chaetoceros* association is poorly represented at the surface, its abundance (3.7 x 10^4 *nifH* copies L^{-1}) at a depth of 40 m in the equatorial upwelling region has been noted. In case of *Rhizosolenia-R.intracellularis* association though the number of gene copies is lowest, the transcript abundance has been found to be high (9.4 x 10^1 to 1.8 x 10^4). These observations are supported by the nanometer scale secondary ion mass spectrometry approach to measure nitrogen fixation rates and the release of fixed products into the natural waters. *Richelia* and *Calothrix* symbionts fixed 171–420 times more nitrogen than the free cells. The cell specific rates (1.15–7.5 fmol N $cell^{-1}$ h^{-1}) among the symbionts resembled each other and the fixed nitrogen was rapidly transferred. On the average *Richelia* as an endosymbiont fixed 81–744% more nitrogen than what is needed for its growth and transferred nearly 97.3% of the fixed nitrogen to the diatom cell (Foster *et al.*, 2011). In the tropical waters of eastern North Atlantic, in the vicinity of Cape Verde Islands, the nitrogen fixation rates >6 nmol N L^{-1} h^{-1} have been found. The amplification of *nifH* transcripts by RT-PCR resulted in 605 *nifH* transcripts of which 76% belonged to six operational diazotrophic populations. The contribution of unicellular diazotrophic cyanobacteria appeared to be significant in both the coastal and open ocean waters.

Studies on the DDAs of oligotrophic waters of North Pacific Ocean, specially northwest of Hawaii have been conducted mostly near station ALOHA (22° 45′ N, 158° 00′ W), recognized as one of the Hawaii Ocean Time Series (HOT) programme (Heinbokel, 1986; Scharek *et al.*, 1999a,b; White *et al.*, 2007a,b; Fong *et al.*, 2008). The other stations include the CLIMAX station (Vanrick, 1974) and some to farther east (Mague *et al.*, 1974; Wilson *et al.*, 2008). The distribution and significance of *Rhizosolenia-R. intracellularis* association has been for the first time reported by Vanrick (1974). Nitrogen transport by the vertically migrating diatom mats in the North Pacific Ocean seems to be the regulating feature for the occurrence of the blooms in this oligotrophic water body (Villareal *et al.*, 1993). The temporal variation in the abundance of *H. hauckii-R. intracellularis* association has been recorded by Scharek *et al.* (1999a,b). Quantitative RT-PCR determination of *nifH* gene copies revealed the abundance of *Trichodesmium, Crocosphaera watsonii* and *R. intracellularis* but the active nitrogen-fixing member among these has been found to be *Trichodesmium* (Fong *et al.*, 2008). Through ocean colour satellite data, Wilson *et al.* (2008) detected *Rhizosolenia-R. intracellularis* association in the late summer blooms of eastern Pacific Ocean (30° N, 156° W) and that the fixed nitrogen by this association seems to propel chlorophyll blooms.

While summarizing the occurrence of DDAs in marine waters, Monteiro *et al.* (2010) proposed a model simulating the global distribution of these symbioses. According to them the largest distribution of the DDAs centered in the Indian Ocean and the subtropical Atlantic Ocean but predicted inconsistent distribution of DDAs with smaller biomass values in the Pacific Ocean. Moreover, the conspicuous absence of DDAs northwest of Hawaii in their model is in sharp contrast to the distribution patterns of DDAs in this region described above. Other reports pertain to the abundance of *Rhizosolenia-R. intracellularis* and *Hemiaulus-R. intracellularis* associations from the Bay of Bengal (Kulkarni *et al.*, 2010) and the occurrence of five species of *Rhizosolenia* and two species of *Hemiaulus* with *R. intracellularis* as an endosymbiont in the tropical western Indian Ocean waters (at Zanzibar and Dar es Salam; Lyimo, 2011).

iv) *Other symbiotic associations of diatoms*: Intracellular inclusions in the form of "sphaeroid bodies" in certain diatoms such as *Epithemia* and *Rhopalodia* have been identified. These bodies are surrounded by a double membrane as confirmed by ultrastructural studies (Drum and Pankratz, 1965; Geitler, 1977). These are considered as unique organelles or obligate endosymbiotic intracellular organisms (Fig. 1). The ability of *Rhopalodia gibba* to fix nitrogen (Floener and Bothe, 1980) lent support for further identification of enzymatic machinery for nitrogen fixation of cyanobacterial origin by Prechtl *et al.* (2004) who employed molecular markers such as 16S rRNA analysis and *nifD* genes. The ability of *R. gibba* to fix nitrogen in light and the gram-negative nature of the thylakoids of the sphaeroid bodies and the sequence similarities (of 16S rRNA and *nifD* genes) suggested that the genome is closely related to *Cyanothece* sp. ATCC 51142. They further concluded that the sphaeroid bodies represent vertically transmitted structures that were integrated into host cell from a free-living state.

A symbiotic association between three partners involving a diatom, a protozoan and a cyanobacterium has been described where the diatom frustules are inhabited by the protozoan *Solenicola* having coccoid cyanobacteria as its endosymbiont (Buck and Bentham, 1998). Carpenter (2002) also mentions the endosymbiotic nature of certain coccoid cyanobacteria in *Coscinodiscus* cells collected from the Indian Ocean near Zanzibar but details of this symbiotic association are not available. Endosymbiotic coccoid *Cyanothece*-like cells numbering up to 20–30 per cell in chain-forming diatom *Climacodium frauenfeldianum* have been identified with the help of 16S rDNA analysis (Carpenter and Janson, 2000).

B) Dinoflagellates: Commonly referred to as "zooxanthellae", the dinoflagellates constitute endosymbionts in marine invertebrates and protists (Banaszak *et al.*, 1993; Trench, 1997) and are ubiquitous in their distribution in coral reef ecosystems (Taylor, 1974; Trench, 1993; Rowan, 1998; Baker, 2003). These are nutritionally versatile organisms exhibiting autotrophic (in symbiotic association), mixotrophic and heterotrophic modes of nutrition. Some even are parasitic. The presence of certain coloured bodies termed as "phaeosomes' (Schütt, 1895) in certain dinoflagellates (*Omnithoceros, Histioneis, Citharistes* and *Amphisolenia*) has been confirmed by ultrastructural studies (Lucas, 1991). On the basis of thylakoid arrangement, location of carboxysomes and shape and size of cells four distinct forms of cyanobacteria have been identified which are larger than the planktonic free-living *Synechococcus*. In *Omnithoceros*, the symbionts are held externally between the upper and

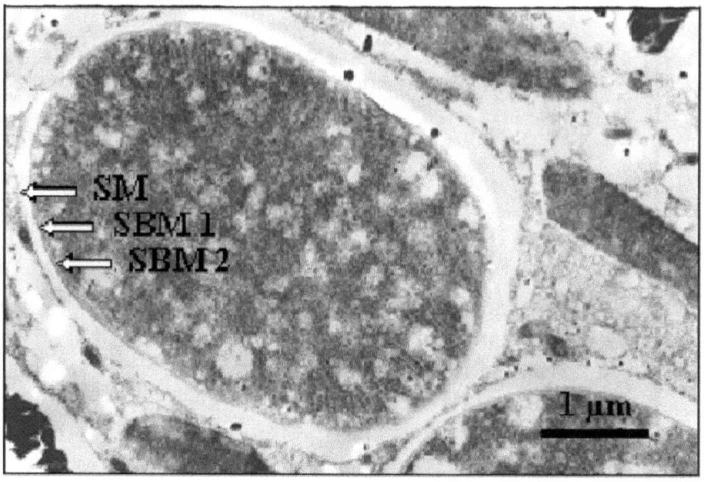

Figure 1: A spheroid body of the diatom *Rhopalodia gibba*. SM: Symbiontophoric membrane; SBM: Spheroid body membrane. With the kind permission of C. Kneip, Department of Cell Biology, Philipps-University Marburg, Marburg, Germany & Department of Molecular Biology, Max-Planck-Institute for Infection Biology, Berlin, Germany [Kneip *et al.* (2007) *BMC Evolutionary Biology* **7:** 55 doi:10.1186/1471-2148-7-55].

lower cingular list whereas a pocket formed by singular groove holds the symbionts in *Parahistoneis*. A small chamber in the girdle floor (as in *Histoneis*) and the chamber opening to the girdle reduced to a small hole (as in *Citharistes*) are the other devices developed to hold the symbionts. While in all the above four dinoflagellates, the cyanobionts are external to the protoplast, in *Amphisolenia* the symbionts are very much entrenched in the cytoplasm of the host cell as revealed by transmission electron microscopy.

VI. FUNGI

Endocyanosis and the formation of lichen thallus are the two modes of symbiotic associations of cyanobacteria with fungi.

A) Endocyanosis: This involves the formation of a bladder by the mycobiont *Geosiphon pyriformis* (Kütz.) Von Wettstein, as the cyanobiont *Nostoc punctiforme* is engulfed. Evidences in support of *G. pyriformis* belonging to the order Glomales, Glomomycetes (formerly placed in Zygomycetes) have been presented by Gehrig *et al.* (1996) by the sequencing of 16S rRNA bringing it closer to other fungi of this order that form arbuscular mycorrhizal associations. *G. pyriformis* is a soil inhabiting species that occurs in larger abundance near Spessart Mountains, Bibergemünd, Germany (Kluge, 2002). The successful symbiotic association of *G. pyriformis* with the cyanobiont results in the formation of a bladder of up to 2 mm long and 5 mm in diameter (Kluge, 2002). Initially both partners exist in soil leading an independent life. The fungus is coenocytic and grows extensively below the surface of soil. The cyanobacterium also exists freely in soil. In order to reach a successful association *N. punctiforme* has to get transformed into a non-motile primordial stage in its life cycle that is formed from the pre-existing motile hormogonium. Although fluorescence-labelled lectin-specific sugar was identified as mannose, due to its presence in hormogonia and late primordial stages this does not provide unequivocal evidence in favour of being considered as the signalling molecule for recognition. Schüßler *et al.* (1997) studied the ability of hormogonia, primordial and vegetative colonies to bind fluorescein isothiocyanate (FITC)-conjugated lectins with sugar specificity to [α-D-mannose/α-D-glucose], N-acetyl-β-D-glucosamine oligomers, α-L-fucose, β-D-galactose, α-D-galactose, N-acetyl-α-D-glucosamine and salic acid. The presence of large amounts of α-D-mannosyl or α-D-glucosyl residues in the extracellular slime of *N. punctiforme* correlated with the primordial stage. So it is suggested to play a role in specific recognition between the two partners. However, studies of Wolf (2003) have identified β-1,4-linked galactose isomers might play a role in the recognition process. These results further emphasize the involvement of a lectin-mediated mechanism for specific partner recognition in *Geosiphon* symbiosis.

Once recognition is made, the fungus starts making an invagination so that the cyanobiont is engulfed and taken in. As soon as this is completed, the bladder formation is initiated and achieved. It is thus possible for the fungus to form a bladder at each successful contact with the cyanobiont. The cyanobiont in the bladder initially suffers a shock as evidenced by a transient loss of photosynthetic pigments. But soon it recovers, multiplies rapidly, differentiates heterocysts and regains its photosynthetic activity. Thus the cyanobiont gets established in the mature bladder. It is demarcated from the contents of the bladder by the membrane surrounding the filament of the cyanobiont that is known as symbiosome membrane. Experiments with ^{14}C revealed that the cyanobiont is photosynthetically active but the nutrient exchange between the two partners is not properly understood. A beginning in this direction seems to have been made by Schüßler *et al.* (2006) who discovered a monosaccharide transporter gene named as glomeromycotan monosaccharide transporter (GP MST1) in *G. pyriformis*. This transporter has a highest affinity for glucose followed by mannose, galactose and fructose.

B) Lichens: Lichen is a Latin word that carries the meaning of "tree moss". Lichens represent symbiotic association between a fungal and algal partner to form a thallus which otherwise could not be formed by either of them alone. This association is purely ectosymbiotic interrelationship that bestows the thallus with a unique morphology and extraordinary physiological properties. Simon Schwendener first described this symbiotic association in 1867 (Honegger, 2000).

***i) Occurrence*:** Majority of lichens grow in very diverse habitats in all continents. These include forests, deserts, tundras, grasslands and thickly populated areas such as towns and cities. Some of the lichens are very specialized as they are restricted to a particular environmental condition, climate, geography or substrate. Lichens grow on a variety of substrata such as bark, wood, rocks, mosses, soil, dead vegetation, leaves, pinecones, and manmade objects. While many of the lichens grow on such generalized substrata, some of them are specific to a type of rock (calcareous, siliceous, sandstone, granite etc.) or the bark of a certain trees (conifers and oaks etc.).

***ii) The two partners*:** As many as 15,000 species of lichens have been reported from all over the world. Nearly 1/5th of all known fungal species participate in lichenization (Lutzoni *et al.*, 2001; Kirk *et al.*, 2001). In North America and New Foundland, 3600 and 1000 species of lichens, respectively are reported to occur. The fungal partner is designated as mycobiont and may belong to either Ascomycota or Basidiomycota. Lichenized species of the former represent 42% (i.e. nearly 30,000 species) and accordingly these lichen thalli are known as ascolichens. Members belonging to three major classes of Ascomycota, i.e. Lichinomycetes, Eurotinomycetes and Lecanoromycetes are involved in the formation of lichen thalli. The fructifications of ascolichens are typically those of Ascomycetes, i.e. apothecia, perithecia or pseudothecia and other ascocarps. In case of Basidiomycota members of the class Agaricomycetes (exclusively Agaricales) enter into lichen symbiosis (Büdel, 1992; Rambold *et al.*, 1998; Rikkinen *et al.*, 2002; Herrera-Campos *et al.*, 2005; Lücking, 2008). In basidiolichens, basidiocarps represent the fruiting bodies. Lichens are named according to the fungal partner.

The algal partner is known as photobiont or phycobiont. It may be either a green alga and/ or a cyanobacterium. In the vast majority of lichens four genera of algae are known to participate in the symbiosis. Of these, two are green algae (*Trebouxia* sp. and *Trentepohlia* sp.) and two belong to cyanobacteria, i.e. *Nostoc* sp. and *Scytonema* sp. (Ahmadjian, 1967; Büdel and Hensen, 1983; Tschermak-Woess, 1988; Nyati *et al.*, 2007). Lichens with a single photobiont (either a green alga or cyanobacterium) are known as bipartite lichens. In some lichens, the green alga constitutes the main photobiont and the cyanobacterium is restricted to reproductive structures known as cephalodia. Such lichens are known as tripartite lichens. According to Laundon (1995) when the nature of photobiont is not known it is designated as photomorph. The unidentified green and blue-green algae are thus called as chloromorph and cyanomorph, respectively.

Exceptionally, *Petroderma maculiforme* is the only brown alga that enters into symbiotic association with an undescribed species of ascomycetous fungus *Verrucaria* (Wynne, 1969). This lichen was collected from intertidal rock surfaces in northern California and the fungal species was identified as *V. tavaresiae* (Moe, 1997). The thallus organization of this lichen with the nature of symbiont interaction (Sanders *et al.*, 2004) and ultrastructural studies on the photobiont in free-living and in lichen symbiosis (Sanders *et al.*, 2005) have been described. Two photobionts belonging to Xanthophyta have also been reported in some other lichen thalli (Tschermak-Woess, 1988).

Cyanolichens , i.e. lichens possessing cyanobacteria constitute 10% of the lichen species. In the order Lecanorales of lichens, the most commom photobiont is *Nostoc* (Friedl and Büdel, 1996). About half of the cyanolichens have green algae as the main photobiont. A few other filamentous, heterocystous cyanobacterial photobionts belong to the genera *Calothrix, Dichothrix, Fischerella, Stigonema* and *Tolypothrix* (Tschermak-Woess, 1988; Oksanen, 2004). Cyanolichens are generally restricted to or are most abundant in old growth and mature forests. Some of the examples of cyanolichens belong to the species of *Coccocarpia, Lobaria, Leptogium, Nephroma, Peltigera, Pseudocyphellaria* and *Sticta* (Figs. 2 to 5). The dominance of lichens in a wide variety of habitats reflects on their ability to tolerate extremes of environmental conditions such as cold, desiccation, heat, UV radiations and visible light and other harsh environmental conditions. High levels of UV radiations are frequently met within the Polar and higher mountain regions and so the lichens occurring in these regions must be able to cope up with these radiations. The differences in the physiology of green algae and cyanobacteria may be useful in explaining the differences in the physiology of the photobionts of lichens. Cyanolichens differ from green algal lichens in their photochemical apparatus in at least two important aspects that make them more susceptible to high light levels. Cyanobacteria lack zeaxanthin-violoxanthin cycle that acts as a screen, for high light intensity, present in green algae (Demming-Adams, 1990). Secondly, PSII reaction center protein D1 has a lower resistance to photoinhibition in cyanobacteria than green algal D1 protein (Clarke *et al.*, 1993). Thus the cyanobionts of cyanolichens are weaker for defense against excess light. Further, cyanobacteria and cyanolichens are not able to photosynthesize without liquid water (Lange *et al.*, 1986). Accordingly, light and availability of liquid water appear to be crucial factors for the viability of cyanolichens. Combination of high wind speed and high irradiance levels cause higher rates of evaporation leading to thallus desiccation. Combined effects of desiccation and high irradiance will also increase risk for thallus damages (Demming-Adams, 1990; Gauslaa and Solhaug, 1996). However, the generation of UV absorbing compounds, quenching of toxic intermediates and damage repair are some of the strategies evolved by the lichens. Increased production of lichen phenolics by the mycobiont in *Cetraria islandica* in response to UV-B radiation exhibited no perceptible changes in chlorophyll and carotenoid concentrations (Bachereau and Asta, 1997). Although a negative correlation between UV-B radiation and phenolics has been made in certain lichens (Swanson and Fahselt, 1997), due to the absorption of UV-B and UV-C bands by

Figure 2: Some common cyanolichens. (A) *Lobaria pulmonaria*, on *Acer saccharum*. From the North Shore of Lake Superior, Ontario. (B) *Lobaria scrobiculata* (damp), on *Acer macrophyllum*. From the central Coast Range of California. (C) *Leptogium cyanescens* on mossy *Acer saccharum* bark. From Acadia National Park, Maine. (D) *Peltigera aphthosa*, on mossy soil. From the Cascade foothills, western Oregon. Photographs courtesy Stephen Sharnoff, Missouri Botanical Garden, University of California, Berkeley, CA, USA (http://www.sharnoffphotos.com/).

Color image of this figure appears in the color plate section at the end of the book.

secondary lichen products a protective role has been assigned to these substances (Quilhot *et al.*, 1995, 2002; Holder *et al.*, 2000). We are presently at the threshold to understand these mechanisms and whether these are specially adopted either by the mycobiont or the photobiont is the question that has been addressed to during recent years. In the Antarctic epilithic lichen *Xanthoria elegans*, the presence of parietin and β-carotene as the UV-protective compounds in upper cortex of mycobiont has been demonstrated (Wynn-Williams, 2000). It is interesting to note that certain cyanobacteria (Garcia-Pichel and Castenholz, 1993) produce mycosporine-like amino acids that act as intracellular, cytoplasmic sunscreen pigments.

iii) Form of lichen thallus: The lichen thalli assume a variety of colours and shapes. They appear greyish-green, white, orange, yellow, yellowish-green, brown or black. Traditionally, the lichens are divided into three types based on the form of thallus. The first type is crustose thallus which forms a crust over the substratum. The second type is a foliaceous thallus which mostly resembles a leaf-like structure and dried up thallus of a liverwort with lobed and irregular margins. The lichens that assume a brush-like structure and are branched are included in the third type known as fruticose lichens. These grow upright or pendant attached to bark of trees or rocks.

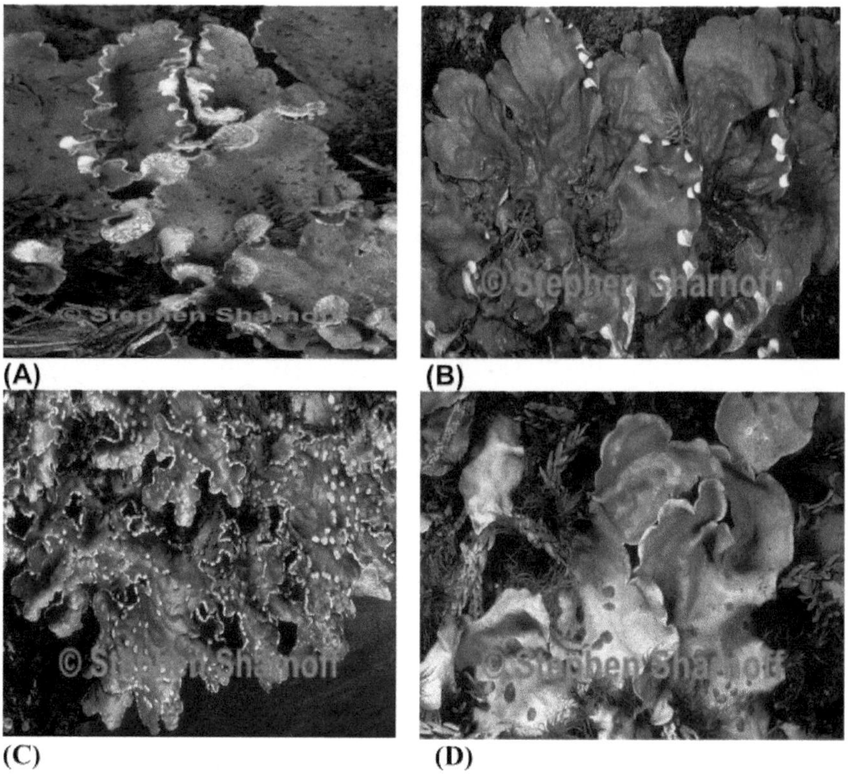

Figure 3: Some common cyanolichens. (A) *Peltigera leucophlebia*, on sandy soil. From northern interior British Columbia. (B) *Peltigera neopolydactyla*, on mossy dead *Picea*. From Montague Island, Prince William Sound, Alaska. (C) *Pseudocyphellaria crocata*, on *Arctostaphylos*. From Oregon Dunes National Recreation Area, coastal Oregon. (D) *Nephroma arcticum*, on mossy soil. From Southeast Alaska. Photographs courtesy Stephen Sharnoff, Missouri Botanical Garden, University of California, Berkeley, CA, USA (http://www.sharnoffphotos.com/).

Color image of this figure appears in the color plate section at the end of the book.

iv) *Internal structure*: Structurally, the foliose type of lichen shows greater internal tissue organization. A transverse section of foliose lichen reveals upper and lower cortex. Immediately beneath the upper cortex there is an algal zone in which the cells of the photobiont are present. The distribution of photobiont in the lichen thallus may be random through out the vegetative body of the thallus. Such thalli are known as homoiomerous thalli. In majority of lichen thalli, the photobiont is distributed just below the upper surface forming a well developed stratified layer. Such thalli are known as heteromerous thalli. The cells of photobiont are surrounded by thin-walled loosely packed hyphae and at points of contact are shown to be intimately surrounded by the mycobiont. This is followed by the central region known as medulla which is occupied by loosely entangled hyphal threads. The medulla merges with the lower epidermis from which a number of rhizoids develop that anchor the thallus to the substratum and also help in absorption. In case of crustose lichens the upper cortex is composed of hyphal layer which is either rudimentary or highly developed. Beneath this layer an algal layer is present in close association with the upper layer. A loosely entangled mass of hyphae constitutes the medulla which then merges with the lower epidermis. The lower epimdermis may be well developed or completely absent. In fruticose lichens there is no such differentiation as an upper and lower surface but the thallus is composed of centrally grouped, compact hyphae that form

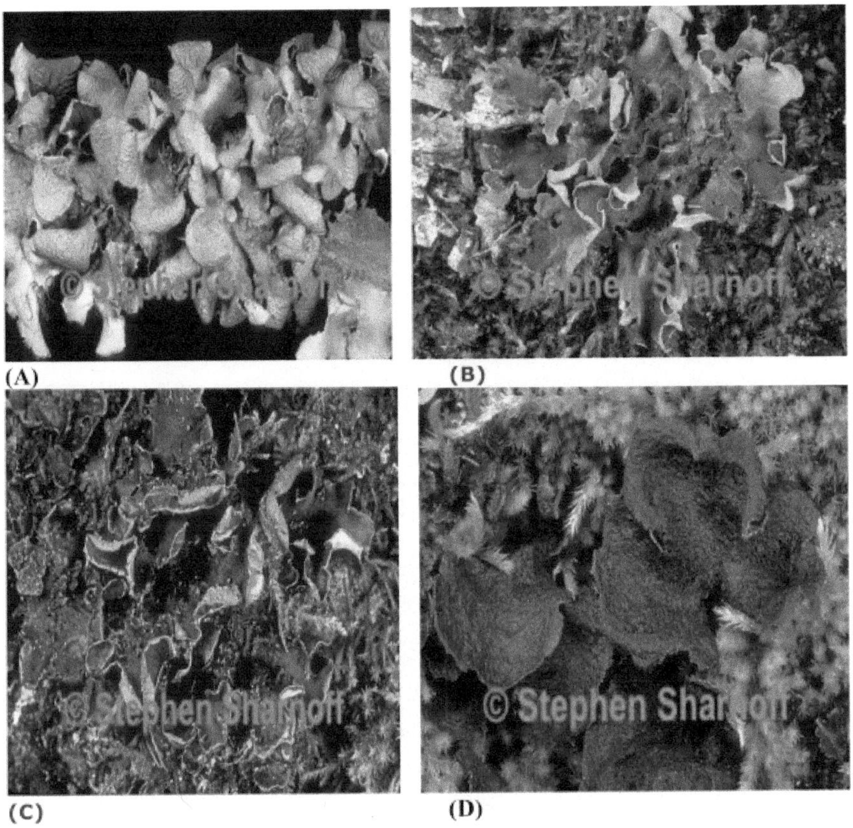

Figure 4: Some common cyanolichens. (A) *Nephroma bellum*, on mossy soil. From northern interior British Columbia. (B) *Nephroma resupinatum* on mossy *Quercus*. From the northern Sierra Nevada, California (C) *Nephroma helveticum* subsp. *helveticum*,on mossy rock. From the North Shore of Lake Superior, Ontario. (D) *Sticta fuliginosa*, on mossy rock. From Shasta-Trinity National Forest, northwestern California. Photographs courtesy Stephen Sharnoff, Missouri Botanical Garden, University of California, Berkeley, CA, USA (http://www.sharnoffphotos.com/).

Color image of this figure appears in the color plate section at the end of the book.

the cylindrical, upright branched portion. This is the medulla and around is the lower cortex. This is followed by the algal zone and upper cortex. The medullary group of hyphae and the thickened cortical hyphal threads provide the requisite mechanical strength to the thallus. The central threads form an attachment organ at the base.

v) *Reproduction*: The photobiont reproduces asexually within the lichen thallus itself. Even otherwise, the lichen thallus exhibits both asexual and sexual reproduction. During asexual reproduction, the mycobiont and the photobiont separate from the lichen thallus and grow out to form a new lichen thallus. The structures formed may be called as soredia, blastidia, isidia and lobules. These are scattered by wind, animals, birds, mites, ticks and rain to new locations where they develop into new lichen thalli. In addition to these structures, the mycobiont forms specialized structures known as pycnidia which release conidia. These are carried out by wind currents and when they fall on suitable substrata by the side of the photobiont, they germinate and envelope the photobiont to again form a lichen thallus. Sexual reproduction is exhibited by the mycobiont leading to the

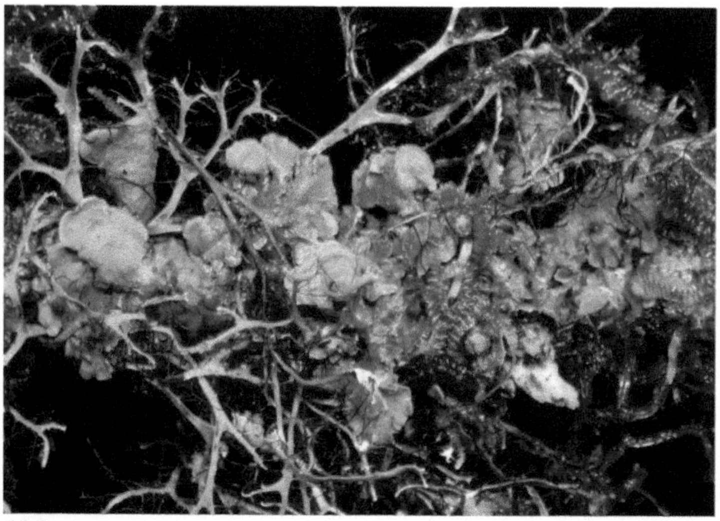

(A)

(B)

Figure 5: Cyanolichens from Kenya (A) *Coccocarpia erythroxyli.* (B) *Coccocarpia palmicola.* Photographs courtesy J. Rikkinen, Department of Biological and Environmental Sciences, PO Box 65, University of Helsinki, FIN-00014, Helsinki, Finland.

Color image of this figure appears in the color plate section at the end of the book.

formation of fruiting bodies. The conidia released from pycnidia recognize a tiny thread (trichogyne) on a lichen surface and attaches itself. Both these fuse to form a diploid cell. At this place a fruiting body develops. Meiotic division is followed by a mitotic division, that results in the formation of eight ascospores. The haploid ascospores during germination on suitable substrata find a suitable photobiont to form a new lichen thallus. Apothecia and perithecia are the two fruiting bodies that are generally formed. Apothecia are disk or cup-shaped, macroscopic and form spores centrally whereas perithecia are flask-shaped and are located on the upper surface opening by means of a pore. Both apothecia and perithecia form asci with ascospores.

vi) *Fossil records*: Hallbauer *et al.* (1977) described *Thuchomyces lichenoides* consisting of only mycobiont that has been assigned to Precambrian period. *Pelicothallos* reported from Tertiary period was in fact only a photobiont and the mycobiont was lacking in this 'lichen' (Sherwood-Pike, 1985). Ziegler (1992) described apothecia-like structures on a lichen thallus from Triassic sediments. *Spongiophyton* is another lichen thallus described from Lower Devonian period of North America (Stein *et al.*, 1993), that did not reveal the presence of either mycobiont or photobiont. All these fossil records fall short of the description of true lichen symbiosis due to the absence of one or the other partners in the thallus (Taylor and Taylor, 1993). The first convincing unequivocal evidence for the presence of a lichen symbiosis (mycobiont and photobiont together) in a new fossil genus *Winfrenatia*, a cyanolichen from Rhynie Chert of the Lower Devonian, was described by Taylor *et al.* (1997). The mycobiont was composed of superimposed layers of coenocytic hyphae present as a network. The cells of a coccoid cyanobacterium with rings of mucilage are entangled in the spaces of the net. The division of the photobiont in three planes characteristic of coccoid cyanobacteria resulted in cell clusters presumably up to 64 cells. The presence of endospores and soredia as the reproductive structures was also indicated. Phylogenetically, the mycobiont is suggested to be closer to Zygomycetes while the photobiont is almost similar to *Gloeocapsa* and *Chroococcidiopsis.*

vii) *Nature of symbiosis*: Lichen symbiosis has been traditionally considered to be a mutualistic association (Nash, 1996). The mycobiont is benefited to a large extent as it can meet nearly 70–80% of its carbon requirements from the photobiont (Smith, 1980; Tapper, 1981). Reports on the supply of nutrients from the mycobiont to the photobiont do not exist in literature (Hill, 1994). The argument that the photobiont is benefited by getting a shelter in an otherwise hostile outer environment seems no longer acceptable because the photobiont's growth and multiplication are very much restricted in the lichen symbiosis than when it exists freely in nature (Honegger, 1993; Ahmadjian, 1993). These observations generated interest in defining the role of mycobiont as nothing but a parasite and the concept that lichen symbiosis is a form of controlled parasitism gained strength.

Comparative phylogenetic analysis has been carried out by matching the sequences of 16S rRNA and 23S rRNA genes of lichenized and non-lichenized fungi of Ascomycota (representing ~75% species). These results suggest that the evolution of free-living non-lichenized Ascomycota had taken place due to losses of the lichen symbiosis. The genera *Penicillium* and *Aspergillus* must have originated from such lichenized mycobionts (Lutzoni *et al.*, 2001). By combining Bayesian phylogenetic tree sampling methodology and a statistical model of trait evolution, Lutzoni *et al.* (2001) estimated the rates of gains or losses of lichenization on each of the 19,900 phylogenetic trees and suggested that during evolution of Ascomycota there have been at least 1.5 times as many losses of the lichen symbiotic state than gains. The rates of loss of lichen symbiotic habit led to the development of non-lichenized fungi called as lichenicolous fungi (that can dwell on or in lichens as parasites, commensals or saprophytes).

On the basis of frequency of heterocysts in cyanobionts, relative proportion of green algal cells vs cyanobacterial cells per unit of fungus as variables, a theoretical model has been predicted to explain the cost of lichen symbiosis. The model explains that the mycobiont derives maximum benefit by altering the role of the photobiont in tripartite lichens. The mycobiont meets the negligible cost of differentiating cephalodia (to accommodate the cyanobiont), thereby allowing the increase in frequency of heterocysts so as to restrict its growth. The reported heterocyst frequencies in bipartite (2 to 8%) and tripartite lichens (10 to 55%) reported in literature support the above predictions. The mycobiont thus derives maximum benefit from the cyanobiont by meeting its nitrogen requirements whereas the green alga meets the carbon requirements (Hyvärinen *et al.*, 2002). These observations further lend support to the concept that lichen symbiosis is not a mutualistic association but a predominantly of commensalism or even emphasizes the role of a parasite to the mycobiont. Joneson *et al.* (2011) identified the expression of 41 and 33 candidate genes by the fungal and algal partners, respectively in *Cladonia grayi* lichen symbiosis. Proteins involved in self and non-self recognition, lipid metabolism and negative regulation of glucose repressible genes are highly expressed in the mycobiont whereas the phycobiont *Asterochloris* showed the up-regulation of chitinase-like protein, an amino acid metabolism protein and a protein arginine methyltransferase.

viii) *Specificity of the partners*: A number of investigators contributed to our understanding of cyanobiont specificity in cyanolichens. Traditional taxonomy helped in the identification of cyanobiont to be a species of *Nostoc*. The questions that have been addressed are whether: (i) there is any diversity in the *Nostoc* strains that form symbiotic association or a single strain is invariably represented in all cyanolichens, (ii) the same *Nostoc* strain forms association in a particular species of cyanolichen in geographically distant regions, (iii) there is any diversity in the mycobiont that forms symbiosis and (iv) morphological diversity of the lichen thallus is imposed by the photobiont or mycobiont. To find out strain differences in *Nostoc*, molecular markers have been of great help in providing requisite answers.

The most widely used molecular markers for resolving strain differences of *Nostoc* and in determining specificity are $tRNA^{Leu}$ (UAA) intron and 16S rDNA sequences. The fungal-specific markers are internal transcribed spacer (ITS) of the 5.8S gene of nuclear rDNA coding for RNAs of the small and large subunits of the ribosome and intergeneric spacer (IGS) separating two consecutive repeats. The 5.8S gene is homologous to a portion of the 23S gene of prokaryotes. It is highly conserved, small in size and provides characters that are helpful in resolving differences in between taxa (Hillis and Dixon, 1991). A brief account on the types of introns, their properties is presented here.

Introns are defined as sequences of DNA that interrupt coding sequences of many genes and also referred to as intervening sequences (Lewin, 2008). Depending on their structural and functional features and nature of splicing mechanisms these are classified into four groups: (i) spliceosomal introns, (ii) group I introns, (iii) group II introns and (iv) archaeal introns (Belfort *et al.*, 1995; Nilsen, 2003). (i) Spliceosomal introns are the conventional introns present in eukaryotic cells. Along with coding sequences, these are also transcribed into mRNA but are excised during the processing of mRNA by spliceosomes (Logsdon, 1998). (ii) Group I introns can catalyze their own excision. They are widely distributed in the genes of mitochondria, plastids, nuclear rRNA genes, bacterial tRNA genes, genes of eukaryotes and viruses (Saldanha *et al.*, 1993). (iii) Group II introns differ from group I introns in their splicing mechanism. Though they can catalyze their own excision, the splicing mechanism resembles those of eukaryotes. They are exclusively found in genes coding for proteins, tRNA and rRNA genes of organelles (Michel *et al.*, 1989; Saldanha *et al.*, 1993). (iv) Archaeal introns are present in archaebacteria and these do not have the self-splicing mechanism of group I and group II introns. These are spliced by an endonuclease that cuts at the exon-intron junction (Kjems and Garrett, 1991; Lykke-Andersen *et al.*, 1997).

Xu *et al.* (1990) for the first time reported a group I self-splicing intron in the gene for leucine transfer RNA [$tRNA^{Leu}$ (UAA)] in two species of *Anabaena*, *Anabaena* sp. strain PCC 7120 and *Anabaena azollae*. It is of 249 bp in the former and 291 bp in the latter. It is interesting to know that an intron of similar nature and in the same identical position is present in the same tRNA gene of chloroplasts of higher plants. This further strengthens the concept of endosymbiotic origin of chloroplasts. The variability of intron in the $tRNA^{Leu}$ (UAA) gene in filamentous heterocystous cyanobacteria has been investigated further. Kuhsel *et al.* (1990) sequenced tRNA^{Leu} (UAA) introns in *Scytonema* strain PCC 7110, *Anacystis* strain R2 and *Phormidium* strain N182. The presence of group I introns in bacteriophages, eubacteria, mitochondria, chloroplasts and nuclear genetic systems of eukaryotes has been subsequently demonstrated. Studies of Paquin *et al.* (1997) revealed that in some cyanobacteria a group I intron is present in tRNA^{fMet} gene that is absent in plastids. On the basis of phylogenetic analysis of group I intron in tRNA^{Leu} (UAA) and tRNA^{fMet} genes, they suggested that (i) tRNA^{Leu} (UAA) intron may be absent or present in closely related species, (ii) tRNA^{fMet} intron is of more recent origin in cyanobacteria and (iii) the presence of tRNA^{Leu} (UAA) intron in both cyanobacteria and plastids suggests their acquisition from a common ancestor.

Thus tRNA^{Leu} (UAA) intron sequencing has been used for differentiating closely related organisms. Due to difficulties encountered in the cultivation of cyanobionts of lichens and the probability of growth of contaminating organisms along with the isolated cyanobionts, group I intron of tRNA^{Leu} (UAA) gene has been the choice for establishing the identity of the cyanobionts in lichen symbiosis as also in the cyanobionts entering into symbiotic relationships with bryophytes. Paulsrud and Lindblad (1998) studied the genetic diversity of photobionts in four lichen species (*Nephroma arcticum, Peltigera aphthosa, P. membranacea* and *P. canina*). Of these, *N. arcticum* and *P. aphthosa* are tripartite lichens where the former shows internal cephalodia and the latter forms external cephalodia. The other two species of *Peltigera* are bipartite lichens. The intron sequence of tRNA^{Leu} (UAA) gene was analyzed for the photobionts of the above lichens and compared with laboratory cultures of *Nostoc punctiforme* PCC 73102, *Nostoc muscorum* CCAP1453/12 and *Anabaena* sp. strain PCC 7120. The interesting findings are that (i) all the photobionts showed the presence of introns consistently with a length of around 300 bp, (ii) all the photobionts were identified as one *Nostoc* strain, (iii) the existence of one intron type in a thallus indicates that the same *Nostoc* strain colonized each lichen thallus, (iv) geographically separated lichen thalli also revealed the same intron sequence suggesting the high degree of specificity and (v) similarity in sequences of introns matched with two free-living *Nostoc* strains (*Nostoc muscorum* CCAP 1453/12 and *N. punctiforme* PCC 73102).

The photobionts of bipartite and tripartite lichens differed with respect to the variability of the sequences of the introns corresponding to bases 99 to 143 in *Anabaena* sp. strain PCC 7120. The photobionts of bipartite lichens (*P. membranacea* and *P. canina*) showed similarity in the sequences of intron tRNA^{Leu} (UAA) gene with those of *N. muscorum* whereas photobionts from tripartite lichens (*N. arcticum* and *P. aphthosa*) exhibited sequence similarity in intron closer to *N. punctiforme* PCC 73102. In bipartite lichens the photobiont mainly is suggested to carry on photosynthesis and in tripartite lichens the photobiont is restricted to the cephalodia where it performs nitrogen fixation. Whether this dual functional difference can be correlated with the variability in the sequences of the introns or not needs further investigation.

The ability of the mycobionts of tripartite lichens to enter into symbioses with either only a green alga or a cyanobacterium leads to the development of different morphotypes (Armaleo and Clerc, 1991; Goffinet and Bayer, 1997). The morphotypes of tripartite lichens are designated as cyanosymbiodemes for those with a cyanobacterium and chlorosymbiodemes for those with a green alga (Renner and Galloway, 1982). Such morphotypes may lead an independent life or at times come together to live side by side. The latter association is termed as photosymbiodemes. These have been reported in *Peltigera* (Brodo and Richardson, 1979), *Nephroma* and *Lobaria* (Tønsberg and Holtan-Hartwig, 1983) and *P. venosa* (Ott, 1988). The morphotypes may exhibit differences in growth form as in *Lobaria* (Jordan, 1972) and *Sticta* (Galloway, 1994), their anatomy as in *P. venosa* (Ott, 1988), chemistry as in *N. arcticum* (Renner and Galloway, 1982; Tønsberg and Holtan-Hartwig,

1983), habitat preferences (White and James, 1988) and morphology also as in *P. aphthosa* group (Holtan-Hartwig, 1993; Vitikainen, 1994; Goward *et al.*, 1995) and *Pseudocyphellaria* (Renner and Galloway, 1982; Galloway, 1988). The generic identity and the specificity of the cyanobacterial strains present in the morphotypes of tripartite lichens have been investigated. Miao *et al.* (1997) reported the existence of different *Nostoc* strains in the morphotypes of *P. membranacea*. On the other hand, Paulsrud *et al.* (1998) detected that there exists specificity and the same strain of *Nostoc* is represented in the morphotypes examined from geographically distant areas in central Sweden and collection sites in Sweden and Finland. The identity of *Nostoc* strain was based on similarity in the sequences of intron of $tRNA^{Leu}$ (UAA) gene of the photobionts. The bipartite and tripartite lichens of *P. aphthosa* harboured the same *Nostoc* strain as revealed by the matching of intron sequences of the photosymbiodeme. Similarly, intron sequences of photobiont of one specimen of bipartite lichen *P. neopolydactyla* (from central Finland) were similar to the sequences of photobionts of tripartite lichens collected from Finland and Sweden. It was thus concluded that the apparent diversity in *Nostoc* strains associated with *P. neopolydactyla* infact is dependent on the particular fungal chemotype that establishes the association.

The diversity of photobionts in lichens from geographically distant regions has been studied by Paulsrud *et al.* (2000). The photobionts of *P. membranacea* collected from Oregon (USA) and in Sweden showed identical sequences of the intron of $tRNA^{Leu}$ (UAA) gene. Similarly, *Nephroma resupinatum* thalli inhabiting Oregon and Finland showed similar sequences of the intron. These results indicate that the same photobiont is present in particular species of thallus irrespective of its place of collection. On the contrary, the cyanobionts of *P. neopolydactyla* collected from Oregon and Washington revealed intron sequences different from the sequences of cyanobionts present in the thalli of *P. neopolydactyla* collected from central Finland. At least two different *Nostoc* strains have been identified in the materials from USA thus confirming their earlier observations on *P. neoploydactyla* (Paulsrud *et al.*, 1998). Further, two different *Nostoc* strains were represented in different samples of *P. brittanica* and five different strains seem to be associated with six specimens from Oregon and Washington based on the intron sequences. The diversity of *Nostoc* strains in populations of *P. neopolydactyla* is whether due to the particular chemotype of the fungus involved in the association or due to the existence of several morphological and chemical races in this species of *Peltigera* remains to be elucidated. To establish specificity of the cyanobiont, populations of *P. aphthosa* growing in field were subjected to asceptic removal of their cephalodia (containing the cyanobiont *Nostoc*) and seven axenic cultures of *Nostoc* (five isolates from lichen thalli: two strains from *P. aphthosa-Nostoc* Pa-1 and *Nostoc* Pa-2; one each from *P. membranacea*, *P. canina* and *N. resupinatum*, i.e. *Nostoc*-Pm, *Nostoc*-Pc and *Nostoc*-Nr, respectively; *N. punctiforme* PCC 73102 and *Anabaena* sp. strain PCC 7120) were inoculated on the surface of *P. aphthosa* thalli. After the development of new cephalodia, 80 such cephalodial cyanobionts were analyzed for $tRNA^{Leu}$ (UAA) intron sequences and compared with the sequences of seven axenic cultures as well as those of cyanobionts of thalli occurring in nature. Interestingly, none of the inoculated strains appeared in the newly generated cephalodia but all the 80 cephalodia contained the same sequences of $tRNA^{Leu}$ (UAA) intron that were originally present in the cephalodia of the thalli at the site. Two of the inoculated strains survived as epiphytes on the same thalli and they belonged to the isolates from bipartite *Peltigera* species. These results suggest that cyanobacterial association and lichen-forming fungi can be specific and stable (Paulsrud *et al.*, 2001).

Rikkinen *et al.* (2002) subjected cyanobionts of cyanolichens from northern Europe, western north America and central China for 16S rDNA and $tRNA^{Leu}$ (UAA) intron sequence analysis. The former helped in resolving phylogenetic relationships while the latter enabled in the identification of *Nostoc* strains associated with lichen symbiosis. All *Nostoc* strains formed a monophyletic group as revealed by 16S rDNA sequencing. Further, *Nostoc* strains were divided into two sub-groups, cyanobionts of epiphytic lichens designated as *Nephroma* guild and the rest as *Peltigera* guild, irrespective of geographical origin or generic identity of the lichen species. According to them, the cyanolichens show specificity to a cyanobiont on a community scale, in other words suggesting that lichens of a particular habitat exhibit specificity for a group of cyanobacterial strains. Low cyanobiont selectivity was reported in lichen specimens from Antarctica. Free-living cyanobacteria and cyanobionts from five lichen species (*Massalongia carnosa, Leptogium puberulum, Psoroma cinnamomeum, Placopsis parellina* and *Placopsis contortuplicata*; the first two being bipartite and the rest tripartite) collected from Livington Island (maritime Antarctica) were analyzed for $tRNA^{Leu}$ (UAA) intron as a genetic marker for the identification of cyanobacterial strains. All the lichen species examined shared the same *Nostoc* strain with an additional *Nostoc* strain in two of the lichens. There was no difference in the *Nostoc* strains in between bi- and tripartite lichens. This has been explained as a selection pressure in the harsh environment and that the mycobionts in order to survive in the extreme climatic condition have no greater choice for the selection of their cyanobionts (Wirtz *et al.*, 2003).

The identity and specificity of major photobionts of *Pseudocyphellaria* have been examined by comparing 16S rRNA gene sequences. On this basis, cyanobacterial and green algal isolates could easily be distinguished one from the other. With the help of both 16S rRNA gene and $tRNA^{Leu}$ (UAA) intron sequences of isolates it was possible to identify the individual photobionts. The genetic diversity of cyanobiont and mycobionts was investigated using $tRNA^{Leu}$ (UAA) intron sequences and ITS sequences (of 5.8S gene), respectively. Two *Nostoc* strains have been identified as species-specific. On the other hand, 5.8S ITS sequences did not show much variation in the mycobionts of *P. crocata* and *P. neglecta*. Further, the two symbionts have been shown to be specific for all samples (Summerfield *et al.*, 2002). The $tRNA^{Leu}$ (UAA) intron sequences of a number of symbiotic strains (54 of them) of *Nostoc* that are derived from lichens [species of *Peltigera* (18), *Nephroma* (7)], bryophytes [*Blasia* (6), *Anthoceros fusiformis* (4)] and gymnosperms (*Cycas circinalis, C. rumphii, Encephalartos lebomboensis* and *Zamia pumila*) have been compared with the sequences of diverse free-living cyanobacteria belonging to all five taxonomic subsections and the evolutionary patterns deduced. The $tRNA^{Leu}$ (UAA) intron sequences in various strains of *Nostoc* exhibited high similarity and shared a highly conserved intron sequence with few variable positions. These differences have been found in one stem-loop (P6b) of the $tRNA^{Leu}$ (UAA) intron. Degenerate heptanucleotide repeats are characteristic of this region that fold into a hairpin structure. All *Nostoc* strains exhibited differences in the number of heptanucleotide repeats thus causing size variations and also by the presence of other sequences not having the heptanucleotide repeats (Costa *et al.*, 2002). Further, the regions that flank these sequences contained the same or similar heptanucleotide repeats. The different groups of degenerate heptanucleotide repeats could be distinguished into two classes from the P6b stem-loop of the intron, i.e. *N. punctiforme* PCC 73102 (two groups of repeats one with a consensus sequence 5′-TDNGATT-3′ and the other its pairing repeat with 3′-AATYHAA-5′) and *Nostoc commune* (two groups of repeats with a consensus sequence 5′-NNTGAGT-3′ and its base pairing repeat 3′-AACTCHN-5′). The cause of variations in the introns has been attributed due to slipped strand mispairing during replication and homologous recombination among different loci in the genome.

An important aspect of group I introns pertains to their mobility. Group I introns of $tRNA^{Leu}$ (UAA) gene are once considered to be immobile and are of ancient origin. It was presumed that these introns are older than the divergence of cyanobacteria and chloroplasts (Kuhsel *et al.*, 1990; Xu *et al.*, 1990; Delwiche and Palmer, 1997). Group I introns have now been shown to be mobile. They can insert themselves into intronless genes. This process has been termed as homing (Dujon,

1989). Their ability to get excised and get inserted at new locations is based on the presence of ORF governing the production of DNA endonucleases. At least 30% of the group I introns possess such ORFs that encode site specific DNA endonucleases. Although the location of such ORFs varies in the conserved secondary RNA structure, it does not interfere with the folding of the catalytic core (Lambowitz *et al.*, 1999).

On the contrary, according to Rudi *et al.* (2002) the evolutionary pattern of $tRNA^{Leu}$ (UAA) intron is involved with lateral gene transfer (LGT) in cyanobacteria. Evidences adduced in support of this are a higher sequence similarity with introns in $tRNA^{Ile}$ (CAU) and $tRNA^{Arg}$ (CCU) genes of α- and β-proteobacteria and sporadic distribution of $tRNA^{Leu}$ (UAA) intron in *Nostoc* and *Microcystis* radiations. Further, the sequences of tRNA gene along with flanking regions and its intron have provided sufficient support for LGT as the means of distribution and evolution of $tRNA^{Leu}$ (UAA) intron. Intronless strains showed the absence of tRNA gene and flanking regions and strains with introns showed the presence of tRNA gene and flanking regions. It is not due to intron mobility but rather due to its instability that the sporadic distribution of this intron is seen in genus *Microcystis*.

The validity of $tRNA^{Leu}$ (UAA) intron sequences as a molecular marker for the identification and measuring taxonomic relationships in cyanobacteria has been questioned by Oksanen *et al.* (2004) based on the comparison of the molecular phylogeny deduced from 16S rRNA gene sequences. The two classes of heptanucleotide repeats in the P6b stem-loop described by Costa *et al.* (2002) have been found among distant relatives whereas some close relatives harboured different repeat classes with a high sequence difference. Symbiont specificity in bipartite lichens *P. crocata*, *P. neglecta* and *P. perpetua* from Northern and Southern Hemispheres has been investigated based on $tRNA^{Leu}$ (UAA) intron sequences where the fungal partner was identified on the basis of 5.8S ITS of the nuclear encoded ribosomal repeat unit and a part of the gene encoding β-tubulin. Both 5.8S ITS and β-tubulin gene sequence analyses have confirmed that all the three species of *Pseudocyphellaria* examined actually represent morphotypes of the same phylogenetic fungal species. Five cyanobionts have been identified from a total of 36 specimens of the above three species collected from various geographical regions of the two Hemispheres. Of these, two strains of *Nostoc* are represented in a number of specimens while three have been restricted to one lichen thallus each of *P. crocata* from Australia and two specimens from Chile (Summerfield and Eaton-Rye, 2006).

The term 'selectivity' has been used instead of 'specificity' by some investigators to describe the choice of a partner during symbiosis. According to Galun and Bubrik (1984), selectivity means 'preferential interaction between organisms'. Thus the two terms, specificity and selectivity have been used synonymously (Beck *et al.*, 2002). The foregoing account on the selectivity of a cyanobiont by an ascomycetous fungus during lichen symbiosis is either based on 16S rRNA gene and/or $tRNA^{Leu}$ (UAA) intron sequence analysis. According to few workers, the former marker is too conservative to be relied upon to distinguish between species or strains of *Nostoc* whereas the latter falls short of explaining the presence or absence of heptanucleotide repeats in the P6b stem-loop (Fox *et al.*, 1992; Stenroos *et al.*, 2006). Besides, $tRNA^{Leu}$ (UAA) intron sequencing has generated a controversy about its suitability to be employed for distinguishing between strains of *Nostoc* (Rudi *et al.*, 2002; Oksanen *et al.*, 2004). Another important aspect that has been over-looked is the lack of correlation of *Nostoc* clades with fungal taxa. Stenroos *et al.* (2006) selected 16S rDNA, $tRNA^{Leu}$ (UAA) intron, partial *rbcL* and *rbcX* genes of the *rbcLX* gene cluster as molecular markers to investigate selectivity of lichen mycobionts and cyanobionts (*Nostoc*). Such an analysis of 122 new sequences generated from 45 lichen collections from various geographical regions revealed that lichens *Pseudocyphellaria*, *Sticta*, *Collema*, and *Leptogium* appear to prefer certain *Nostoc* strains whereas lichen genera *Peltigera*, *Lobaria*,

Nephroma and *Stereocaulon* are able to accept a broader spectrum of *Nostoc* strains as photobionts. Specimens collected from different geographical regions showed similar *Nostoc* strains suggesting that *Nostoc* taxa are very widely distributed. In other words, *Nostoc* strains diversified (as the other *Nostoc* strains that associate with *Blasia* and *Macrozamia*) and are versatile in forming bipartite (*Pseudocyphellaria crocata*) or tripartite (*Lobaria pulmonaria*) lichens. Strains of *Nostoc* present in lichens *Stereocaulon* spp. and *N. arcticum* (with external cephalodia) do not appear in the '*Pseudocyphellaria* clade' as also certain of the free-living *Nostoc* spp. They concluded that some of the *Nostoc* taxa are specialized in leading a symbiotic life with only lichen-forming fungi.

Cyanobiont selectivity of eight epiphytic lichens (*Lobaria pulmonaria*, tripartite lichen with internal cephalodia; *Nephroma bellum*, *N. laevigatum*, *N. parile*, *N. resupinatum*, *P. triptophylla*, *P. leucophlebia*, tripartite lichen with external cephalodia and *P. praetextata*) of an old growth forest area in Finland was examined by sequencing of three gene loci (partial 16S rDNA, partial *rbcL* and complete *rbcX* gene of the *rbcLX* gene cluster). The above lichen species growing on same old aspen (*Populus tremula*) and adjacent trees in the same stand were compared to know whether they share the same cyanobiont or harbour different strains specific to the lichen. The main findings can be summarized as follows: (i) all the lichen species showed the sequence similarities of the cyanobiont to be *Nostoc*, (ii) there is no correlation of the geographic origin of the samples, (iii) the sequences of the cyanobionts in all the epiphytic species are distributed into two major clades, (iv) cyanobionts of *P. leucophlebia* and *P. praetextata* group together in clade I with four *P. leucophlebia* samples from lithophytic habitats, (v) clade II consisted of two subgroups, subgroup II*a* having the two cyanobionts from *N. laevigatum* together with cyanobionts of *Degelia plumbea*, *L. scrobiculata*, *N. helveticum*, *N. tangeriense* and *Pseudocyphellaria crocata* and subgroup IIb consisting of all the cyanobionts of *N. bellum*, *N. parile*, *N. resupinatum* and *P. triptophylla* irrespective of their geographical origin together with three cyanobionts of *L. pulmonaria* and (vi) the cyanobionts of *L. pulmonaria* revealed sequence similarities to the clade II signifying that *L. pulmonaria* is more versatile and can form symbioses with a wide range of *Nostoc* strains. This could also lead to different *Nostoc* strains in a single lichen thallus (Myllys *et al.*, 2007). In this context, it is interesting to note that the thalli of *L. pulmonaria* do not always contain cephalodia (Zoller *et al.*, 1999).

The clades I and II identified by these workers corresponded to those already described earlier by Rikkinen *et al.* (2002), Lohtander *et al.* (2003) Rikkinen (2003, 2004) and Oksanen *et al.* (2004). Moreover, following Rikkinen *et al.* (2002) the clades I and II were designated as "*Peltigera* guild" and "*Nephroma* guild", respectively based on the ecology of the lichens. Though "*Peltigera* guild" includes the terricolous lichens, the "*Nephroma* guild" corresponded only to epiphytic lichens of old growth forests. However, "*Nephroma* guild" does not constitute a homogeneous group of epiphytic lichens since it consisted of certain other cyanobionts collected from terrestrial and lithophytic habitats (Stenroos *et al.*, 2006).

The variability in the morphotypes of Peltigerineae is whether due to differences in the photobionts or genetically distinct mycobionts has been investigated. Though the first such molecular study was conducted by Armaleo and Clerc (1991), the results did not have sufficient resolution to distinguish the presence of two species. DePriest and Been (1992) used ITS region of 5.8S gene of lichenized fungus *Cladonia chlorophaea* for such studies. The fungus specific primers designed for mycorrhizal fungi by Egger (1995) were put to test for the amplification of the 5.8S ITS region of mycobionts using total lichen DNA as template (Goffinet and Bayer, 1997). Further, these workers also used ITS of 5.8S gene sequences of mycobiont members of photomorph pairs in *P. aphthosa*, *P. brittannica*, *P. leucophlebia*, *N. expallidum* and *N. arcticum*. One possibility suggested for the development of a tripartite lichen thallus in the genera studied is the fusion of the chloromorph and cyanomorph lobes. The supporting evidences cited in favour of this are soredial fusion in the asexual life cycle of *P. didactyla* (Stocker-Wörgötter and Türk, 1990) and the presence of chemical hybrids between two varieties of this species (Goffinet and Hastings, 1995). The existence of similar 5.8S ITS sequences of the mycobionts of fused cyanomorph-chloromorph structures of *Peltigera* and *Nephroma* prompted them to conclude that the same species of mycobiont is involved in the symbiosis. Phylogenetic patterns among *Nostoc* cyanobionts within bi- and tripartite lichens of the genus *Pannaria* have been deduced based on 16S rRNA gene sequences. The cyanobionts have been derived from 21 *Pannaria* species from both Northern and Southern hemispheres and compared with 69 free-living and symbiotic strains. Most of the *Nostoc* sequences from *Pannaria* are distributed among the "*Nephroma*-guild" and within two subgroups of "*Peltigera*-guild". Cyanobionts from several tripartite lichens of *Pannaria* are grouped together with the *Nostoc* sequences of cyanobionts of corticolous bipartite lichens from both the hemispheres (Arve *et al.*, 2008). The morphotypes of the photobionts of tropical lichen genera *Dictyonema, Acantholichen* and *Coccocarpia* and some other *Stereocaulon* lichens have been identified microscopically under *Scytonema* and a unicellular member *Chroococcus*. Phylogenetic analysis of the photobionts of the above lichens revealed the presence of a previously unrecognized lineage of filamentous cyanobacteria clustered into a single clade known as *Rhizonema*. But the *Rhizonema* clade did not cluster with *Scytonema* (Lücking *et al.*, 2009). Otálora *et al.* (2010) investigated the symbiotic specificity at an intercontinental spatial-scale among gelatinous lichens belonging to Collemataceae (Lecanoromycetes). They conducted a phylogenetic study on the basis of *rbcLXS* sequences of *Nostoc* strains sampled from 79 thalli belonging to 24 species of Collemataceae. Most of the fungal species belonging to Collemataceae showed a generalist pattern of association with *Nostoc* strains. However, only in case of five species a one-one specificity has been noted. According to them, during the course of evolution these five mycobiont species represent independent transitions from a generalized pattern of photobiont association.

ix) *Role of lectins in lichen symbiosis*: Lectins are a class of proteins or glycoproteins which specifically bind to cell surface carbohydrate moieties. Their presence has been detected in animals, plants, fungi and bacteria. In most of the plants and legumes lectins have been detected mostly from seeds. However, in a number of cases their presence has been demonstrated in other vegetative parts such as leaves, stems, barks and roots. The important properties of lectins are that they: (i) are of non-immune origin; (ii) are soluble or membrane bound; (iii) agglutinate erythrocytes with a high specificity (so also called as haemagglutinins); and (iv) stimulate resting lymphocytes to actively divide. The role of lectins in rhizobium-legume symbiosis, in the formation of arbuscular endomycorrhiza and parasitism has been clearly established (Albrecht *et al.*, 1999; Hirsch, 2004; Brewin, 2004; Tikhonovich *et al.*, 2004). The mitogenic, antitumor, antiviral and immunity stimulating potential of mushroom lectins have been summarized (Singh *et al.*, 2010).

In lichen symbiosis the mycobiont first recuits a photobiont. In doing so the initial interaction of the mycobiont with the photobiont is through the secretion of lectins that bind to the cell surface carbohydrate moieties of the phycobiont. Two classes of lectins with different roles have been detected. The first type is the secreted lectins which promote the recruitment of algal cells to the neighbourhood of the mycobiont. These intial interactions lead to the binding of the cells of the photobiont to the mycobiont and also prevent the loss of the cells of photobiont. The second type is the algal binding proteins (ABP) that have enzyme activity and promote the physical interaction between the fungus and its specific algal partner. These two types are described here seaparately.

A number of phytohaemagglutinins have been isolated and characterized from various lichen species. Haemagglutinins isolated from the cyanolichen thalli *P. canina* and *P. polydactyla* could bind to the cell walls of the cyanobiont *Nostoc* sp. (Lockhart *et al.*, 1978). Likewise, Ingram (1982) conducted a detailed survey of haemagglutinins and haemolysins from the extracts of 36 species of lichens from 19 genera. The assays involved 20 types of erythrocytes and the extracts of *Usnea fragilescens*, *Parmelia caperata* and *Lepraria incana* exhibited haemagglutinin and haemolysin activities to over 13 and 10 types of erythrocytes, respectively. Proteins extracted from the lichen thalli of *Xanthoria parietina* could bind to the cell surface of the cultured material of the phycobiont but not to the freshly isolated phycobionts. Using fluorescamine-labelled proteins from *X. parietina*, Bubrick and Galun (1980) demonstrated strong binding to the cultured phycobionts obtained from *X. parietina*, *Caloplaca auriantia* and *C. citrine* but not to the cultured or freshly isolated phycobionts from the thalli of *Cladonia convoluta*, *Ramalina duriaei* and *R. pollinaria*. The binding very much depended on the cell wall chemistry of the phycobiont and specially has been correlated with the presence of high levels of acidic polysaccharides. Further, the binding of fluorescamine-labelled protein could also bind to the isolated cell walls of the phycobionts. Lectins from the thalli of *Peltigera horizantalis* have been found to bind to the cell surface of cultured phycobionts but not to the cell surface of freshly isolated phycobionts (Petit, 1982). Another purified phytolectin from *Peltigera canina* var *canina* has been found to be thermostable with a molecular weight of 80,000–90,000 and it could be used as cytochemical marker in tissue (Petit *et al.*, 1983). In order to visualize the cell surface receptors of mycobionts and phycobionts through fluorescence microscopy, Marx and Peveling (1983) employed a number of purified lectins. The mycobionts showed more affinity to most of the lectins than the phycobionts. However, all investigated species of *Trebouxia* could bind lectin ConA to their cell surface but not to the cell surface of the two species of *Pseudotrebouxia*.

Cultures of the mycobiont from the thallus of *Xanthoria* produced ABP that is restricted to the cell walls of the mycobiont and has a molecular weight of 12 kDa (Bubrick *et al.*, 1981, 1985). A similar ABP from cyanolichen *Nephroma laevigatum* revealed a heterodimeric structure with 52 kDa and 55 kDa subunits (Kardish *et al.*, 1991). Characterization of lectins with homodimer structure from the thalli of *Peltigera membranacea* (Lehr *et al.*, 1995) and *P. aphthosa* (Lehr *et al.*, 2000) has been reported. Similarly, a homodimeric (each subunit with 16.5 kDa) lectin with haemagglutinating activity has been purified from the lichen *Dictyonema glabratum* which harbours *Scytonema* as a photobiont. Structurally, it has been shown to be a glycoprotein with neutral monosaccharides galactose, xylose, glucose, mannose, in addition to glucosamine. The haemagglutinating activity of the lectin has been inhibited by N-acetylgalactosamine (Elifio *et al.*, 2000). A glycoprotein lectin with D-galactose specificity has been reported from the tripartite lichen *P. aphthosa* (Feoktistov *et al.*, 2009).

The ABPs from *X. parietina*, *Evernia prunastri* and *P. canina* have been characterized. The ABP from *X. parietina* has been shown to be a lectin exhibiting arginase activity that can hydrolyze arginine to ornithine and urea (Molina *et al.*, 1993). While ABP is retained by the mycobiont, *X. parietina* also secreted a lectin that too possessed the arginase activity. This is designated as sectered arginase of *Xanthoria* (SAX). Both of them differed in the glycosyl moieties.The glycosyl moiety of SAX is composed of galactose and glucose while the endogenous arginase contained equimolar amounts of N-acetyl-D-glucosamine and glucose (Molina and Vicente, 1995, 1996, 2000). The SAX and the ABP have the same molecular mass and the same qualitative amino acid composition. That is why these two glycoproteins have been considered as isolectins (Molina and Vicente, 2000). When purified ABP from *X. parietina* was administered to the cultures of phycobiont of the same thallus, in the absence of the cell wall receptor for the ABP, cellular disorganization took place (Molina and Vicente, 1996). This has been attributed to an increase in the concentration of algal putrescine followed by apparent loss of chlorophyll. The loss of chlorophyll has been suggested to be due to putrescine-activated glucanse activity that led to the disruption of the membranes (Molina *et al.*, 1998a). Ferritin- or fluorescein-labelled ABP from *X. parietina* could bind to the cell walls of the phycobiont only when the phycobiont possessed urease activity in its cell walls. Due to the binding there is an inhibition in both the arginase activity of the ABP and the urease activity of phycobiont cell wall. Moreover, when purified ABP is added to the cultures of the phycobiont with urease activity, no cellular disorganization occurred (Molina *et al.*, 1996, 1998b).

E. prunastri (a lichen belonging to Parmeliaceae) has been shown to produce both ABP and SAE (a secreted arginase from *Evernia*). But the polysaccharide moiety of the latter is composed of fructose, mannose and glucose (Planelles and Legaz, 1987). The arginase activity of both SAE and SAX has been shown to be dependent on Mn^{2+}. Although Ca^{2+} could not replace Mn^{2+}, its addition in presence of Mn^{2+} significantly inhibited arginase activity. The lectin function of both SAE and SAX has been demonstrated by their binding to the polysaccharide moiety of the urease on the cell walls of the phycobionts. However, binding of SAE to the cell walls of the phycobiont of *E. prunastri* has been found to be specific as it could not bind to the cell walls of the phycobiont of *X. parietina*. On the other hand, SAX is non-specific in its binding activity as it could bind to the cell walls of the phycobionts of both *E. prunastri* and *X. parietina*. The polysaccharide moiety of the urease significantly contained α-D-galactose which serves as a ligand for lectin binding. So the important requirement for binding of ABP to the cell walls of homologous and heterologous phycobionts is the presence of the galactose residues as the ligand molecule. This indeed has been demonstrated to be the mechanism for ensuring specificity of the phycobiont in the formation of lichen symbiosis (Legaz *et al.*, 2004). Hydrolysis of the α-1,4-polygalactoside moiety of urease located in the cell walls of the phycobiont with α-1,4-galactosidase caused the release of high amounts of D-galactose. This resulted in an inhibition of lectin binding due to the absence of the ligand for binding but if the hydrolysis is performed with β-4-galactosidase there was no effect on binding as very low amounts of D-galactose could be released (Sacristán *et al.*, 2006). Partial amino acid sequences of SAE and SAX have been determined. The former showed an undecapeptide (a peptide with 11 amino acid residues) which is homologous to the Mn^{2+}-binding site where as the latter contained a heptapeptide and an undecapeptide (Sacristán *et al.*, 2008). The cell recognition model of chlorolichens has now been extended to cyanolichens. *Leptogium corniculatum* is shown to secrete an arginase that could bind to the cell wall ligand of the cyanobiont *Nostoc* strain. Furthermore, the secreted arginase could also bind to some extent to the cell walls of *Trebouxia* from *E. prunastri* (Sacristán *et al.*, 2007; Vivas *et al.*, 2010). *P. canina*, a cyanolichen, is shown to produce and secrete an arginase that acts as a lectin in binding to the cell surface urease present on the cell wall of the cyanobiont *Nostoc* (Díaz *et al.*, 2009). The secreted lectin from the thallus of *P. canina* has been suggested to act as a chemoattractant and induces chemotropism in compatible *Nostoc* cells by contriction-relaxation pulses similar to that of myosin II (Díaz *et al.*, 2011).

x) *Role of hydrophobins in symbiosis*: Proteins that occur on the surface of mycelial (asmomycetous and basidiomycetous) fungi are known as hydrophobins (Wösten, 2001). These confer water repellent properties to conidia, hyphae and multicellular structures. As many as 50 hydrophobins have so far been described from a wide variety of fungi including saprophytic moulds, edible mushrooms, plant pathogens and fungi forming mycorrhizal associations (Whiteford and Spanu, 2002). They enable the fungi to escape their aquatic environment and come in contact with air and play a greater role in fungal pathogenesis (Carpenter *et al.*, 1992; Zhang *et al.*, 1994; Spanu, 1997; Arntz and Tudzynski, 1997), ectomycorrhizal associations (Tagu *et al.*, 1996) and symbiotic association with green algae and/or cyanobacteria (Honegger, 1991; Scherrer *et al.*, 2000). Hydrophobins thus constitute a group of morphogenetic proteins that can be classified into two groups, i.e. class I and class II based on their hydropathy patterns and soluble characteristics. The former are present in both ascomycetous and basidiomycetous fungi whereas the latter have been detected only in ascomycetous fungi. Class II hydrophobins are soluble in hot SDS solution and those of class I dissociate into monomers upon incubation in trifluoroacetic acid and performic acid (de Vries *et al.*, 1993; Wessels *et al.*, 1991).

Structurally, hydrophobins of both the classes are about 100 amino acids long and contain eight conserved cysteine residues (Wessels, 1994, 1997). The cysteine residues are involved in intramolecular cross-linkages by forming disulphide bonds with each other. Studies on cerato-ulmin indicated that these disulphide bridges are formed with probable linkage of Cys1-Cys2, Cys3-Cys4, Cys5-Cys6 and Cys7-Cys-8. Hydrophobins once produced in the fungal cell are secreted to outside via ER-Golgi pathway. These contain at their N-terminal part a signal peptide consisting of a few amino acids. It is speculated that this signal peptide serves as an anchor to the hydrophobins to the outside of the fungal cell walls. The property of self assembly is retained even after the disulphide bridges are reduced and the sulphydryl groups are blocked with iodoacetamide, although the resulting proteins were unable to re-form disulphide linkages. It is suggested that the disulphide bonds inhibit premature self-assembly of the hydrophobins prior to their secretion into the wall (deVocht *et al.*, 2000). The polyhydrophobins of *Claviceps* spp. are 400 amino acids long. Those from *C. fusiformis*, CFTH1 are tripartite hydrophobins, i.e. three class II bimodular hydrophobins are encoded as a single protein by a multimeric gene. However, the physical and chemical properties of CFTH1 resemble those of class II hydrophobins. A penta hydrophobin has also been described from *C. purpurea* (Whiteford and Spanu, 2002).

Hydrophobins of class I from *Schizophyllum commune* are the best studied and the growing fungal tips secrete SC3 monomers that assemble spontaneously at hydrophilic/hydrophobic interfaces, i.e. between water and air, water and oil, or water and hydrophobic solid like teflon into an amphipathic film/membrane (Wösten *et al.*, 1993, 1994a,b, 1995). The hydrophilic side of the membrane orients and attaches itself to the cell wall while the hydrophobic side gets exposed to the hydrophobic environment. Thus aerial hyphae and spores tend to become aerophobic and hyphae that grow over hydrophilic substrata get themselves attached to the surface (Wösten, 2001).

The capacity of lichens to withstand repeated cycles of desiccation and flooding enables them to colonize extreme habitats. The inner cavities of lichens are lined with hydrophobic membrane. This is also known as rodlet layer. The isolation and cloning of genes and localization of hydrophobins has now been achieved both from ascomycetous and basidiomycetous fungi that form the lichen thalli. Honegger (1991) proposed two major functions to the rodlet layer. The first is that at the immediate contact site of the growing hyphae with the cell wall surface of algal cells, the hydrophobins diffuse and spread over the surface of the photobiont. Thus a continuous surface hydrophobic layer is formed at wall-air interfaces of both the partners. Secondly, it prevents accumulation of water in the interior of the thallus and helps in the supply of optimal flow of water and solutes from the exterior to the algal layer and vice versa. This has also received much support from other workers (Wessels, 2000; Dyer, 2002).

The first report on the occurrence of hydrophobins in lichen-forming (ascomycetous) fungi *Xanthoria parietina* and *X. ectaneoides* was made by Scherrer *et al.* (2000) who demonstrated that hydrophobins of class I designated as XPH1 and XEH1, respectively produced by these fungi are of the size of 10 kDa. These assembled *in vitro* as shown by transmission electron microscope studies into a rodlet layer with individual rodlets of ~10nm. This was further confirmed by the fact that the antibodies raised against hydrophobins are bound to this structure (Scherrer *et al.*, 2002; Trembley *et al.*, 2002a). A common feature of *X. parietina* and *Dictyonema glabratum* is the presence of gas-filled spaces in the photobiont layers. Despite the fact that the thalli of *D. glabratum* daily undergo the cycles of hydration and desiccation, the walls of the air cavities are lined by the fungal hyphae that are covered by the rodlet layer indicating the presence of hydrophobins. Three genes *DGH1*, *DGH2* and *DGH3* governing the production of hydrophobins DGH1, DGH2, and DGH3, respectively have been identified in *D. glabratum*. Of these three hydrophobins, DGH1 is of 14-kDa and most abundant protein. The N-terminal sequence of this protein has been used to carry out cDNA cloning by RT-PCR. The co-amplification of cDNA fragments that encode hydrophobins DGH2 and DGH3 was also possible when the cDNA encoding the signal peptide was cloned by RACE-PCR. These three hydrophobins also share 54–66% amino acid identity (Trembley *et al.*, 2002a). The differential expression of these three genes depended on age and location of the particular hydrophobin in the thallus.

xi) *Co-speciation versus algal switching*: Ahmadjian (1987) advocated the hypothesis of co-evolution for lichens especially when one or both symbionts appear obligate and specialized. However, this hypothesis has not been put to test. Thompson (1994) suggested that co-evolution directly requires an assessment of increased fitness resulting from genetic change. Another alternative suggested for co-evolution is to indirectly demonstrate parallel cladogenesis or co-speciation of symbiont lineages (Page and Hafner, 1996). This can be an acceptable proposition if there are highly specific associations between the algal and fungal partners to the extent the algal partner is vertically transmitted throughout the same fungal lineage. If horizantal transfer of algal partners through a fungal lineage takes place then the concept of cladogenesis or co-speciation could be rejected. Evidences in support of horizantal transfer of algal partners through fungal lineages was also designated as "algal switching" by Piercey-Normore and DePriest (2001) who tested this by using nuclear ITS phylogenies of algal and fungal partners from 33 natural lichen associations of Cladoniaceae predominantly harbouring the green alga *Asterochloris*. Random amplified polymorphic DNA analysis with 23 primers showed little polymorphism among *Asterochloris* assemblages suggesting that a low level of variation exists across the entire algal genome. This clade, designated as clade I comprising *Asterochloris* assemblages also overlaps with the sequences of *Trebouxia glomerata*, *T. irregularis* and *T. pyriformis* along with 33 natural lichen-forming algae. It was also shown that clade I genotypes associate with lichen-forming fungi of different orders such as *Stereocaulon* (Stereocaulaceae, Lecanorales), *Pycnothelia papillaria* (Cladoniaceae, Lecanorales) and *Anzia carnionivea* (Trapeliaceae, Agyriales). On this basis they rejected parallel cladogenesis and co-speciation and proposed that switching of highly selected algal genotypes occurs repeatedly among lichen symbioses.

xii) *Diversity of cyanolichens*: McCune (1993) has developed a conceptual framework for understanding the distribution of epiphytic lichens along with three gradients such as time, height and moisture. This is known as 'similar gradient hypothesis' that predicts a special sequence of successional events in lichen flora as the particular stand matures. Chlorolichens and electorioid lichens are present in young regenerating stands and as the stand matures these are replaced by cyanolichens followed by bryophytes. Cyanolichens constitute 42% of the lichen community of Northwest old-growth *Pseudotsuga-Tsuga* forests. Cyanolichens include all macrolichens, mainly populations of *Lobaria oregana* that are concentrated in the "light transition zone". This extends from about 13 to 37 m height in an overall canopy height of 50–60 m, as revealed by the vertical stratification studies of these forests (McCune *et al.*, 1997). Other cyanolichen representatives that inhabit these areas are other species of *Lobaria*, *Nephroma*, *Peltigera* and *Pseudocyphellaria*. Some of these are indicators of acidic deposition (Denison *et al.*, 1977; James *et al.*, 1977; Gauslaa, 1995) and ecological continuity (Rose, 1976, 1988; Goward, 1994). Cyanolichens are considered to be important source of nitrogen for forest ecosystems because of their ability to fix atmospheric nitrogen through their cyanobionts (Pike, 1978; Antoine, 2004) and more particularly in old-growth temperate forests of North America (Sollins *et al.*, 1980). Old-growth forests are considered to be important natural havens for many epiphytic macrolichen communities. The biomass of old-growth associated lichens increases slowly in the old-growth conifer forests of Pacific North West of North America where the cyanolichens can exceed 1 T ha^{-1} of biomass (McCune, 1993; Sillett, 1995; Neitlich and McCune, 1997). Besides fixing nitrogen, the cyanolichens also adsorb nitrogen from the atmosphere and ultimately release it into their immediate environment. Thus they are considered to contribute significantly to the nitrogen budgets of some ecosystems in different geographical areas such as Sweden (Kallio, 1974; Huss-Danell, 1977), North Carolina (Becker *et al.*, 1977; Becker, 1980), Columbia (Forman, 1975), Chile (Godoy *et al.*, 2001) and British Columbia (Campbell and Fredeen, 2004). Epiphytic cyanolichens probably make significant contribution specially in some coniferous forests where nitrogen is limiting (Rhoades, 1995; Nash, 1996). Goward and Arsenault (2000a) identified 31 epiphytic (tree-dwelling) cyanolichens colonizing conifers in the intermontane forests of British Columbia. Of these, at least 12 species are considered to be rare. Maximum diversity of cyanolichens was encountered in lowland old-growth rain forests specially the wettest subzones of the Interior Ceder-Hemlock zone. A significant component of lichen diversity in the humid coastal forests of Nova Scotia is represented by the cyanolichens (Casselman and Hill, 1995; Seaward *et al.*, 1997). Most common species represented are those of *Lobaria* as well as *Collema subflaccidum, Leptogium cyanescens* and *Parmeliella triptophylla*. The rare species are represented by *Coccocarpia palmicola, Degelia plumbea, Erioderma pedicellatum, Leptogium corticola, L. laceroides, L. saturinum, Nephroma laevigatum, N. helveticum, Pannaria conoplea, Pseudocyphellaria perpetua* and *Sticta fuliginosa*. Studies on total epiphyte cover, spatial distribution and succession of epiphytes of *Tsuga heterophylla* (western hemlock) in an old-growth Douglas-fir forest brought to light that foliose macrolichens *Lobaria, Pseudocyphellaria* and *Sticta* were most abundant in the lower and mid canopy regions (Lyons *et al.*, 2000). The abundance of *Lobaria pulmonaria* in the Interior Cedar-Hemlock forests of east-central British Columbia correlated very well as an indicator species to assess the diversity with stand age and the functional role of dominant species in the ecosystem. The increase in the density of *L. pulmonaria* populations along with other cyanolichens is expected to play a predominant role in maintaining the nitrogen budget where atmospheric nitrogen deposition is relatively low in this region (Campbell and Fredeen, 2004).

Cyanolichens are late colonists in the sequence of stand development and they are found on the relatively older lower branches and inner branch regions. This is supported by the total biomass that is greatest in the mid canopy where large branches are present. Canopy lichen abundance of regenerating hemlock forests (*Tsuga heterophylla*) in wet temperate rain forests of central-interior British Columbia has been surveyed. Cyanolichen taxa such as *Nephroma helveticum, Sticta fulginosa* and *Pseudocyphellaria anomala* reached their abundance at mid-canopy (12–24 m) positions. The abundance of smaller cyanolichen thalli in regenerating hemlock forests casts doubt as to whether these represent cyanolichen thalli facing severe growth constraints or were simply thalli that experienced greater fragmentation. It is concluded that the 120-140 year old hemlock forests have not yet attained sufficient old-growth characteristics to support the growth of cyanolichens (Radies and Coxson, 2004).

One of the world's tallest and most massive forests containing as many as 20 conifer species exceeding in 60 m in height are situated in the west coast of North America (Van Pelt, 2001). These constitute the temperate rain forests that at one time extended from Alaska to California. A number of workers have explored lichen biogeography in the region of California forests covering wilderness areas, national and state parks, watersheds and broader geographic regions. One of the tallest rain forest species is *Sequoia sempervirens* (coast redwood) besides *Picea sitchensis* (Sitka spruce) that support the growth of a number of epiphytic lichens. Epiphytic cyanolichens (*Erioderma sorediatum, Lobaria oregana, L. pulmonaria, L. scrobiculata, Nephroma bellum, N. helveticum, N. laevigatum, Pseudocyphellaria anomala, P. antraspis, P. crocata* and *Sticta limbata*) contributed less to overall epiphytic biomass than bryophytes and ferns but they were more abundant than chlorolichens. Of all the cyanolichens *L. pulmonaria* was the most abundantly represented in upper crowns with nearly three times as much biomass as all cyanolichens put together. *L. oregana* is reported to grow in the middle to upper crown. *L. scrobiculata, P. anomala, S. limbata* and *P. crocata* occurred nearly to the treetops categorized under high exposue cyanolichens (Ellyson and Sillett, 2003).

However, studies on floristic and community composition of lichens in relation to environment variables such as climate, geography and stand structure are lacking. Jovan (2002) first conducted landscape level of epiphytic lichen diversity of northern and central California. This was followed by another such study highlighting regional variations in epiphytic marcolichen communities of the same region depended on temperature, elevation and moisture variables. Cyanolichens constituted the Greater Central Valley group dominated by other nitrophilous group of lichens. Representatives of cyanolichens recorded were species of *Leptogium* and *Collema*. In the Northwest Coast Group, cyanolichens had the highest species richness, diversity and abundance while nitrophiles are rare (Jovan and McCune, 2004). Berryman and McCune (2006) compared epiphytic macrolichen communities among forest types in the Blue River watershed of western Oregon. The cyanolichens were largely limited to lower elevation forests (470–950 m) of the western hemlock series. The diversity of cyanolichens was more in old-growth and mature stands at low elevations.

Phytogeographic and taxonomic studies on cyanolichens from central Europe revealed the presence of *Anema nodulosum, A. prodigulum, Lempholemma intricatum, Leptogium ferax, Porocyphus rehmicus* and *Zahlbrucknerella calcarea* (from Slovak Republic for the first time), *Leptogium biatorinum* and *L. magnussonii* (from Hungary) and *A. prodigulum, Heppia adglutinata, L. biatorinum* and *Psorotrichia taurica* (from Czech Republic) (Czeika *et al.*, 2004). The probability of detecting five rare epiphytic macrolichens can be improved by adopting model-based stratification studies that enabled the detection of cyanolichens *Nephroma laevigatum, N. occultum, N. parile, Lobaria scrobiculata* and *Pseudocyphellaria rainierensis* in the Pacific North West. Tree models constructed by using topographic and bio-climatic variables have revealed a direct correlation to the presence of common lichens such as *L. oregana, L. pulmonaria, P. anomala* and *P. anthraspis* (Edwards *et al.*, 2004).

xiii) *Endangered cyanolichens*: Epiphyte abundance between stands may vary in age. Management strategies have established a strong relationship between stand age and epiphytic biomass, particularly for cyanolichens that are dominant in old-growth forests but nearly absent from young stands (McCune, 1993). Thus the cyanolichens are generally restricted to or most abundant in old-growth and mature forests. Genera such as *Erioderma, Leptogium, Lobaria, Nephroma, Pannaria, Parmeliella, Peltigera, Pseudocyphellaria, Solorina* and *Sticta* are abundantly represented in these habitats. At one time very well documented on the deciduous and coniferous trees as epiphytes in humid regions throughout Northern Hemisphere, cyanolichens are dwindling in their numbers due to ecological disturbances such as acid rain, acidification of the substratum and low buffering capacity of coniferous bark (Nieboer *et al.*, 1984; Farmer *et al.*, 1992; Gauslaa and Holien, 1998). It is also true for populations of cyanolichens, especially on conifers, in areas subject to acid rain in North America and western and central Europe (Goward and Arsenault, 2002a). Similarly, the prospects of finding the boreal felt lichen, *Erioderma pedicellatum* that typically colonizes coniferous trees in Atlantic Canada are rare (Maass, 1980, 1983; Robertson, 1998; Sipman, 2002). *E. pedicellatum* is a large- to medium-sized foliose cyanolichen. It grows on and attached to trees in cool moist areas. It is typically found on the tree trunks of *Abies balsamea* (Balsam fir) in Canada and at times on *Picea mariana* (black spruce) and *P. glauca* (white spruce) and rarely on *Acer rubrum* (red maple). It is listed in the endangered wildlife in Canada and also critically endangered globally and listed on IUCN Red List of Threatened Species. The natural disturbances that affect its population are stand scenescence, forest blow down, insect outbreaks and grazing by invertebrates. Along with any one of these, anthropogenic activities/disturbances such as wood harvesting, urban development, fire, air pollution, pesticides and climatic changes may act together. A five-year management plan (2006-2011) for the boreal felt lichen in Newfoundland and Labrador has been prepared by the Wild Life Division, Department of Environment and Conservation in collaboration with Boreal Felt Lichen Working Group. The possible management strategies for the preservation of cyanolichens in northeastern North America have been highlighted (Richardson and Cameron, 2004).

Goward and Arsenault (2000b) proposed the existence of a "dripzone effect" by which nutrients that leach from the upper branches of *Populus* enhance the pH of the neighbouring conifers. Due to this *Lobaria pulmonaria* and other epiphytic cyanolichens attain their maximum development over the lower branches of conifers. This phenomenon seems to be prevalent in the forests of humid regions of south-central Columbia that are not subjected to acid rain. Arsenault and Goward (2000) further emphasized that aspen and cottonwood should be considered trees of keystone nature as they efficiently pump nutrients from soil to the canopy and later release these to nearby pines altering the acidic pH of the bark of pines that is otherwise unsuitable for the growth of cyanolichens. In support of this they indicated that the copious presence of cyanolichens over the bark of pines, spruces, hemlocks and other members of the Pinaceae in some portions of North America. Though the proposed hypothesis of "dripzone effect" needs experimental evidence, it may derive logical support from the fact that decreasing numbers of aspens contribute to the general loss of epiphytic macrolichen diversity on conifers.

In a conservation strategy, transplantation of over 1000 mature thalli of *Lobaria oregana* and *Pseudocyphellaria rainierensis* has been done into the crowns of old-growth (400–700 year), mature (140–150 year), young (30–40 year) and recent clear-cut areas of Oregon Cascades. Since the growth rates were very much lower in clear-cuts due to high (50–90%) mortality rates, it is suggested to include young forests in conservation strategies (Sillett and McCune, 1998). In order to increase biodiversity of lichens in managed forests in Pacific North West, Sillett and Goslin (1999) advocated the retention of remnant trees, maintenance of hardwoods followed by longer rotation periods to conserve old-growth associated lichens. Alternative silvicultural practices are receiving attention in view of decreasing populations of aspen (*Populus tremula*) that supports growth of epiphytic lichens. The rate of removal of aspen was nearly up to 50%. Cyanolichens *Collema curtisporum, C. furfuraceum* and *Leptogium saturinum* can be suitably preserved if selective thinning of aspens is conducted (Hedenås and Ericsona, 2003).

VII. BRYOPHYTES

Of the 340 known genera of liverworts, only two of them form symbioses with cyanobacteria. These are *Blasia pusilla* and *Cavicularia densa* belonging to Blasiales. Among the six hornworts, five of them, i.e. *Anthoceros, Dendroceros, Notothylas, Phaeoceros* and *Leiosporoceros* form symbiotic association with cyanobacteria (Rai *et al.*, 2000; Adams, 2000; Villarreal and Renzaglia, 2006; Adams and Duggan, 2008). In most of these, the symbiont is a species of *Nostoc*. Granhall and Hofsten (1976) reported intracellular localization of *Nostoc* filaments in a species of *Sphagnum* from Swedish mires. *Pleurozium schreberi*, a species of feather moss forms symbiotic association with cyanobacteria (DeLuca *et al.*, 2002) that has great impact on nitrogen cycle in mature boreal forests (Zackrisson *et al.*, 2004).

The gametophytes of liverworts and hornworts are prostrate, few cm long and attached to the substrata by rhizoids. The location of the cyanobionts varies in the host plants. In case of the widely distributed *B. pusilla*, *Nostoc* is restricted to specialized auricles that develop on the ventral, three-celled slime hairs. *Nostoc* is attracted. The symbiotic association of *Nostoc* with *Blasia* is worth noting because of the two types of asexual propagules produced by it. On the dorsal side of the thallus, stellate gemmae are produced while small, ovoid gemmae are produced in long-necked flask-shaped receptacles. The stellate gemmae are of particular interest because these are equipped with two lobes and two auricles containing symbiotic *Nostoc* colonies (Renzaglia, 1982; Duckett and Renzaglia, 1993). It is a miniature *Blasia* gametophyte as upon dispersal it germinates to give rise to a new thallus with the cyanobiont. In contrast, *C. densa* is a rare plant endemic to Japan and produces gemmae in crescent-shaped receptacles near the mature lobes. Two types of gemmae are produced by the thallus. The first type is small and pale gemmae which upon dispersal get infected with endosymbiont. The second type is the larger, stellate gemmae which get infected by the endosymbiont even while attached to the plant. The latter infact represent miniature thalli and are equipped with two symbiotic auricles and it is into these auricles the symbiont enters (Fig. 6 A-D).

In hornworts, the invasion of cyanobacteria takes place through the ventral mucilage clefts that are formed throughout the life of the plants. The middle lamella between internal cells separates to form a schizogenous space and *Nostoc* gets established as a globose colony. In *Anthoceros* and *Phaeoceros* slime cavities within the thallus open to the ventral surface by means of pores. The only exception in hornworts is *Leiosporoceros dussii*, a monotypic genus considered to be the most genetically and morphologically divergent hornwort (Duff *et al.*, 2004; Villarreal and Renzaglia, 2006). The plants form large rosettes. The male and female plants grow intermixed. The female gametophytes are strap-shaped and fleshy with dorsal outgrowths. In transverse section the thallus is 9–15 celled-thick in the middle and taper to margins with 2–4 cells. Multiple schizogenous cavities (2–6) are some times present. This hornwort is distinguishable from others by the absence of ventral clefts and the cyanobacteria are present in ventrally developed schizogenous canals that are dichotomously branched and run parallel to the main axis of the thallus (Villarreal and Renzaglia 2006). Exceptionally, in case of *P. screberi* besides *Nostoc* sp. a species each of *Calothrix* and *Stigonema* are also associated. Villarreal and Renzaglia (2006) described the ultrastructure and development of schizogenous canals in *L. dussii* and the ultrastructure of *Nostoc* filaments. Young plants (80–90 day old) develop mucilage clefts that are randomly scattered along the flattened anterior rim of lobes. Small intercellular spaces or cavities that extend to several cell lengths are scattered throughout the interior of the thallus. These irregular cavities are schizogenous, i.e. formed by gradual separation of middle lamella. These cavities contain mucilage, often interconnected and are in close proximity to apical cells. The mucilage strands elongate and branch synchronously with apical growth of the thallus. Although a single invasion of *Nostoc* is required for strand production, non-specificity is suggested on the basis of the presence of two distinct *Nostoc* species.

Campbell and Meeks (1989) identified early events in the initiation of *Anthoceros punctatus* and *Nostoc* symbiosis. A hormogonium-inducing factor (HIF) is released by *A. punctatus* in a nitrogen-limited or growth conditioned medium devised for it. HIF transforms *Nostoc* filaments into hormogonia at a high frequency. Since gametophytic tissue of *A. punctatus* requires a source of fixed nitrogen for its continuous growth, the release of HIF and the subsequent hormogonia so produced at high frequency would colonize the slime cavities in the newly formed gametophytic tissue. The HIF is suggested to be heat labile substance, 12–14 kD in size and complexed by polyvinylpyrrolidone.

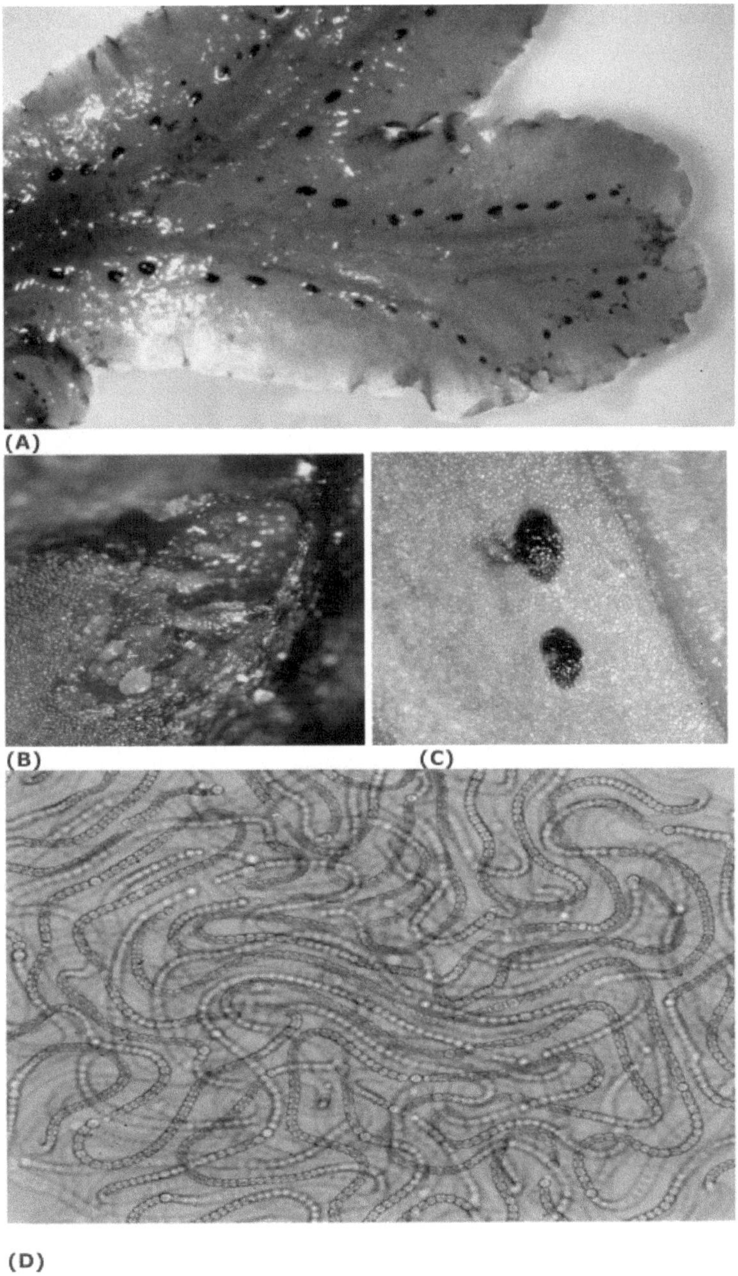

Figure 6: Cyanobacterial symbiosis with *Cavicularia densa*. (A) *Cavicularia* thallus lobe with dark *Nostoc* colonies. (B) Symbiotic (*Nostoc*-infected) and non-symbiotic gemmae emerging from a semi-closed receptacle on the thallus surface. (C) Auricles dorsal view with endosymbiotic *Nostoc* colonies. (D) Cyanobiont *Nostoc* sp. isolated from symbiotic gemmae of *C. densa*. Photographs courtesy J. Rikkinen, Department of Biological and Environmental Sciences, PO Box 65, University of Helsinki, FIN-00014, Helsinki, Finland.

Color image of this figure appears in the color plate section at the end of the book.

Certain chemoattractants released by the host plant enable the motile hormogonia to glide efficiently towards the gametophytic tissue and colonize the slime cavities of *A. punctatus* (Knight and Adams, 1996). The association of type IV pili on the surface of the hormogonia has been shown to be essential for symbiotic competence, since mutation of pil-like genes resulted in altered surface piliation and reduced symbiotic competence (Duggan *et al.*, 2007). DNA microarray analysis of 6,893 genes of *N. punctiforme* revealed that as many as 1827 genes are differentially transcribed in 24-h hormogonia, majority of which were associated with genes encoding proteins for signal transduction and transcriptional regulation (Campbell *et al.*, 2007). The formation of hormogonia greatly correlated with the loss of nitrogen fixation and decrease of photosynthetic carbon fixation and ammonia assimilation by 30 and 40%, respectively. However, the ability of nitrogen and carbon fixation was restored back within 96 h of infection by hormogonia.

i) *Genetic diversity of Nostoc strains*: A comparison of 31 cyanobionts of hornwort *Phaeoceros* (isolated from several closely spaced locations) by PCR amplification of short arbitrary primers or primers specific for the regions flanking the 16S–23S rRNA internal transcribed spacer revealed that (i) a diversity of symbiotic cyanobacteria infect *Phaeoceros*, (ii) the same thallus could be infected with many different cyanobacterial strains, (iii) identical symbiotic strains found in different thalli at the same place were never found to be in a free-living state and (iv) one of the cyanobionts was a species of *Calothrix* while the rest belonged to *Nostoc* (West and Adams, 1997). Studies on the genetic diversity of *Nostoc* strains, based on $tRNA^{Leu}$ (UAA) intron sequence, revealed that many different *Nostoc* strains are involved in the symbiotic associations with *Anthoceros fusiformis* and *B. pusilla* (Costa *et al.*, 2001). A phylogenetic analysis of $tRNA^{Leu}$ (UAA) intron sequences of cyanobionts derived from *B. pusilla* (78 sequences) and *C. densa* (12 sequences) revealed a great homogeneity in the symbiotic *Nostoc* strains. These belonged to a specific group of symbiotic strains that bear closer resemblance to those of hornworts and cycads on the one hand and terricolous cyanolichens on the other (Rikkinen and Virtanen, 2008).

ii) *Reconstitution of symbiosis in vitro*: Rodgers and Stewart (1977) first provided information on the morphological and physiological features of the symbiosis of *A. punctatus* and *B. pusilla* with *N. sphaericum*. Reconstitution of the symbioses by a range of symbiotic and free-living *Nostoc* spp. in case of *Blasia* (Rodgers and Stewart, 1977; Meeks, 1998; Wong and Meeks, 2002) and *Anthoceros* (Enderlin and Meeks, 1983; Kimura and Nakano, 1990) has been successful. In the studies on *Blasia*, a number of free-living and symbiotic strains of *Nostoc*, *Chlorogloea fritschii* and *Fischerella ambigua* were unable to reconstitute the symbiosis. Likewise, free-living and symbiotic strains of *Anabaena* and *Nostoc* were unable to reconstitute the symbiosis with *Anthoceros*. Reconstitution experiments with *Phaeoceros*, on the other hand, revealed that not only free-living and symbiotic species of *Nostoc* but also a species of *Calothrix* and *Chlorogloeopsis* could establish symbioses, although *Chlorogloeopsis* has never been found to be a natural symbiont of any liverwort or hornwort (West and Adams, 1997). *A. punctatus*-*Nostoc* (*N. punctiforme*) interactions can be defined in two stages. The first is the infective stage during which the pre-existing cavities in *A. punctatus* are invaded by motile hormogonia leading to the symbiotic association. The stimulus is provided by the HIF and chemoattractants released by the host. Once inside the mucilaginous cavity, the cyanobiont is probably regulated by the host to the extent further hormogonia are no longer formed. In the second stage, the hormogonium gets deeply entrenched and divides to form filaments with high frequency of heterocysts (Enderlin and Meeks, 1983), equipped with higher rates of nitrogen fixation (Steinberg and Meeks, 1989) and releases the fixed nitrogen to the host (Meeks *et al.*, 1985; Joseph and Meeks, 1987). Once hormogonia are formed and infection of the gametophytic tissue takes place, there should be a mechanism to repress the formation of further cycles of hormogonia so that the cyanobiont can differentiate heterocysts and take up the function of nitrogen fixation. To examine this aspect, Cohen *et al.* (1994) characterized a transposon-induced mutant of *N. punctiforme* that is more (50-fold) infective than its wild type culture. Two genes have been identified that have bearing on the symbiotic potential of *N. punctiforme*. These are *hrmA* (which is an unique sequence not represented in any database of gene sequences) and *hrmU* that has similarity to the sequences of a NAD(P)-H-dependent oxidoreductases. Reporter gene constructs of *hrmA* and *hrmU* have been induced by aqueous extract of *A. punctatus*. Since the aqueous extract of *A. punctatus* eliminates the HIF-stimulated hormogonium formation, it is likely that gene products of *hrmAU* operon may block hormogonium formation by producing an inhibitor or catabolizing an activator (Meeks, 1998). The increased infectivity of the mutant is explained on the basis that it undergoes hormogonium cycle even in presence of inhibitor/repressor of HIF (Cohen and Meeks, 1997). Campbell *et al.* (2003) reported the presence of a repressor gene (*hrmR*) within the *hrm* locus. Sequence analysis of *hrm* locus of *N. punctiforme* revealed the existence of four more genes besides *hrmU* and *hrmA*. These are *hrmI, hrmR, hrmK* and *hrmE* with two unknown short ORFs in between *hrmK* and *hrmE* that code unknown proteins. The gene sequences of *hrmR* and *hrmI* are transcribed in the same direction as 5′ of *hrmU*. The ORFs *hrmK* and *hrmE* are transcribed in the opposite direction from the *hrmRIUA* cluster. The four additional genes (*hrmI, hrmR, hrmK* and *hrmE*) identified have similarity to genes encoding enzymes of carbon metabolism. The protein transcripts of *hrmA* and *hrmU* genes have shown similarities to hexuronic acid metabolism of certain heterotrophic bacteria. Thus the gene products of the four genes, i.e. *hrmE, hrmK, hrmR* and *hrmI* have 55%, 49%, 39% and 36% similarities to aldehyde reductase, gluconate kinase, transcriptional receptors *LacI/GalR* family, uronate isomerase, respectively. The gene product of *hrmU* has 57% and 55% similarities to 2-keto-3-deoxygluconate and mannonate oxidoreductase, respectively. The product of *hrmR* gene has been shown to be a DNA-binding protein that regulates the transcription of its own gene and a near by gene *hrmE*. The activity of the repressor was inhibited by galacturonate or the lysate from induced *N. punctiforme* cells suggesting that the binding of repressor is modulated by a sugar molecule. A model for explaining the regulation of genes in the *hrm* locus by plant signals has been proposed by Campbell *et al.* (2003).

Mutants of *N. punctiforme* unable to differentiate heterocysts (*Het*$^-$) were tested for their ability to establish a symbiotic association with *A. punctatus* in cultures. Three mutants tested were defective in *ntcA, hetR* and *hetF* regions. *NtcA* controls the transcription of a number of genes in the course of heterocyst differentiation (Herrero *et al.*, 2001; Fiedler *et al.*, 2001). *HetR* is the first heterocyst-specific gene and is considered to be the primary activator of heterocyst differentiation and is indirectly dependent on *ntcA* (Wolk, 2000). *HetF* is a positive activator of heterocyst differentiation that helps in the enhanced transcription of *hetR* (Wong and Meeks, 2001). Among these, mutants defective in *hetF* and *hetR* regions infected *A. punctatus* with similar frequency as wild type but did not support the diazotrophic growth of the plant partner. Thus functional *hetF* and *hetR* regions along with certain heterocyst regulatory elements are required for heterocyst differentiation and nitrogenase expression in both free-living and symbiotic states. A mutant defective in *ntcA* region was unable to infect *A. punctatus* though produced hormogonia at a low frequency. The ability for production of hormogonia and infection of the host matched that of wild-type when mutants defective in *ntcA* were complemented with functional copies of *ntcA*. Thus it is clear that a functional *ntcA* is essential for hormogonia formation, infection and a stable symbiosis to ensue. Chapman *et al.* (2008) reported that muation of the *cyaC* gene that encodes the multidomain adenyl cyclase enzyme had differential symbiotic competence in *N. punctiforme*. An omega neomycin phosphotransferase gene with pSCR19 was ligated to the 3′- and 5′-ends of the *cyaC* gene to obtain mutants C3212 and C1068, respectively.

Although C3212 and C1068 exhibited similar growth potential in a nitrogen-deficient medium as the wild-type but formed hormogonia 12 h earlier than the wild-type. When co-cultured with *B. pusilla* tissue, C3212 showed higher symbiotic competence (67.12% infected colonies in 28 days) compared to the reduced symbiotic competence of C1068 (8.5% infected colonies in 28 days).

Flavonoids, secreted by legumes, have been suggested to be the signalling molecules for the rhizobia (Fisher and Long, 1992) and these substances also are suspected to play a role in the symbiotic association of mycorrhizal fungi (Xie *et al.*, 1995; Shirley, 1996; Stafford, 1997). In an attempt to find out the universality of the communication/signalling mechanism in between host and its symbiont, Rasmussen *et al.* (1996) reported the induction of *nod* genes in *Rhizobium* by the seed exudates of *Gunnera*, a host that harbours the cyanobiont *Nostoc* sp. Alternatively, the flavonoid naringin induced the expression of *hrmA* gene in *N. punctiforme* (Cohen and Yamasaki, 2000). Transcription of specific sets of genes in bacteria is regulated by alteration of the sigma subunit of RNA polymerase. Campbell *et al.* (1998) examined the role of alternative group 2 sigma factor in the development and symbiotic interaction of *N. punctiforme* with *A. punctatus*. Thus the induction of transcription of *SigH* was noted within 1.5 h of exposure of *N. punctiforme* with *A. punctatus*. This is the second genetic target in *N. punctiforme* that responds to chemical signals from *A. punctatus*.

iii) ***Structural and physiological changes in the symbiont***: Once the symbiont enters the mucilaginous cavity of the host, it is subjected to a number of structural and physiological changes. In general, it has been observed that the growth rate of the symbiont is slowed down but how this is regulated is not known. In addition, the cell size of the symbiont also increases. One important structural change is the differentiation of heterocysts with high frequency. The physiological changes mainly relate to photosynthetic and nitrogen metabolism. An eight-fold decrease in light-dependent CO_2 fixation and RuBisCO specific activity in reconstituted symbiotic tissues of *A. punctatus* with *Nostoc* sp. strain UCD 7801 has been reported. Studies with enzyme-linked immunosorbent assays with polyclonal antibodies against RuBisCO revealed that the regulation of RuBisCO activity in the symbiotic state is by a posttranslational mechanism rather than by an alteration in RuBisCO protein synthesis (Steinberg and Meeks, 1989). The RuBisCO protein and phycobiliprotein content of vegetative cells of the symbiont and the free-living cyanobacterium were found to be the same (Steinberg and Meeks, 1989; Rai *et al.*, 1989; Meeks, 1990). The heterocyst frequency increased to 43–45% with a concomitant enhancement in nitrogenase activity to 23.5–185.7 (nmol. min. mg^{-1}). Nearly 80% of the fixed nitrogen is released as ammonia by the symbiont. The activity of GS is reduced to 15% to that of cultures of the symbiont (Hill, 1975; Rodgers and Stewart, 1977; Meeks *et al.*, 1985). However, the studies of Joseph and Meeks (1987) revealed a reduction of three to four-fold in GS activity in *Nostoc* sp. strain 7801 grown in symbiotic association with *A. punctatus*. A correlation between the level of GS expression and the extent of symbiotic heterocyst differentiation has not been noticed. According to these workers, the regulation of GS takes place by a posttranslational mechanism in *A. punctatus* associated *Nostoc* sp. strain 7801. Steinberg and Meeks (1991) demonstrated that symbiotic association of *Nostoc* with *A. punctatus* derived at least one-third of the reductant required for nitrogenase activity by the photosynthates of the symbiont. However, steady state levels of nitrogenase activity of symbiotic *Nostoc* depended largely on the endogenous carbohydrate reserves of either *A. punctatus* or *Nostoc* strain.

The explanation given to the differentiation of heterocysts at a high frequency by the symbiont is that it has evolved a different regulatory mechanism for the differentiation of heterocysts even in presence of ammonia or fixed nitrogen *in situ*. A symbiotic sensing and signalling pathway for the differentiation of heterocysts in symbionts (different from the one that regulates heterocyst differentiation in free-living cyanobacteria on nitrogen starvation) has been proposed by Meeks et al. (1988). The evidences in favour of this hypothesis are that (i) vegetative cells of symbiotic cyanobacteria (growing in association with *A. punctatus, Azolla caroliniana, Zamia skinneri* and *Gunnera manicata*) possess cyanophycin and phycobiliproteins signifying that the presence of these nitrogen reserves does not in any way cause a nitrogen deficiency; (ii) the heterocysts of cyanobacteria in association with *A. punctatus* and *Azolla* sp. contain phycobiliproteins, a feature that is entirely lacking in the heterocysts of free-living cyanobacteria, (iii) the end product of nitrogen fixation translocated in the symbiotic forms is ammonia whereas in free-living cyanobacteria glutamine is translocated to adjacent vegetative cells and (iv) the symbiotically associated *Nostoc* continues to differentiate heterocysts at a high frequency even though the vegetative cells are present in exogenous ammonia secreted by it. In this respect the symbiont is suggested to possess a different regulatory mechanism from the free-living cyanobacterial sensing and signalling pathway to nitrogen deprivation. Evidences in support of such a mechanism are the isolation of mutants of *N. punctiforme* defective in nitrate assimilation. Nitrate neither supported growth nor repressed heterocyst differentiation. Symbiotic tissues of *A. punctatus* established with the above mutants responded differently in presence of nitrate by the repression of nitrogenase (probably heterocyst differentiation as well). This has been explained on the basis that the plant partner ceases to produce signals controlling cellular differentiation in the symbiont. On this basis, Meeks (1998) suggested the existence of a combined nitrogen-independent sensing and signalling pathway in the heterocyst differentiation of the symbiont. Further, it is suspected that the symbiotic sensing and signalling pathway overtakes the combined nitrogen deprivation sensing and signalling system (that operates in the free-living growth state) so that heterocysts continue to differentiate even in presence of nitrogen derived ammonium. Strains (mutants or natural isolates) of *Nostoc* that lack symbiotic sensing and signalling pathway, however, fail to form a successful nitrogen-fixing symbiotic association with *A. punctatus* because of their failure to respond to the plant signals that initiate enhanced level of heterocyst differentiation.

The organization of *nif* genes in *Nostoc* sp. 7801 in symbiotic association with *A. punctatus* as revealed by restriction fragments is consistent with a contiguous *nifHDK* operon where as *nifD* and *nifK* genes are nominally separated by an unknown length of intervening sequence in cultured isolates and *Anabaena* sp. strain PCC 7120 (Meeks *et al.*, 1988).

The importance of nitrogen fixation by mosses has been emphasized. Nitrogen fixation reported from Swedish mires was shown to be due to endophytic *Nostoc* sp. present in the hyaline cells of leaves of *Sphagnum* sp. (Granhall and Hofsten, 1976; Solheim and Zielke, 2002). This is the only report on the probable existence of a cyanobiont in a moss species. No further details are available about this association. In all the other reports on nitrogen fixation by mosses, the symbionts are reported to be epiphytic (Solheim and Zielke, 2002). *Sphagnum riparium* and *Drepanocladus exannulatus* exhibited peaks in light-dependent nitrogen fixation around 16°C and 11°C (Basilier *et al.*, 1978). *Pleurozium schreberi*, a feather moss ubiquitous in boreal forests that accounts for 90% of the total cover, has been shown to be associated with *Nostoc* sp. (DeLuca *et al.*, 2002). Zackrisson *et al.* (2004) emphasized that this association exerts a great impact on the nitrogen cycle of mature boreal forests. The epiphytic association of a number of cyanobacteria with many moss genera has been examined by the use of a variety of microscopic techniques (confocal laser scanning microscopy, epifluorescence and scanning microscopy). By confocal laser scanning microscopy of moss genera *Sanionia uncinata, Calliargon richardsonii* and *Bryum pseudotriquetrum* (from Norway) and *Hyloconium splendens* (from Sweden), it was possible to see deeper layers and create three-dimensional images. In *S. uncinata* the leaves have formed grooves completely enclosing the cyanobacteria. The spaces between stem and leaves of *C. richardsonii* serve as good sites for the colonization of the cyanobacteria. The prevalence of

Nostoc sp. over *Microchate* sp., *Gloeocapsa* sp. and *Asterocapsa* sp. has been noted with other fungal mycelia and heterotrophic bacteria (Spaink *et al.*, 2004). The association of *Nostoc* sp., *Calothrix* sp. and *Stigonema* sp. with *P. schreberi* in contributing towards nitrogen fixation in boreal forests of northern Sweden has been recognized (Gentili *et al.*, 2005). *Nostoc* sp. and *Calothrix* sp. have been successfully isolated and reconstitution experiments with feather moss shoots yielded potential shoots that effectively fixed nitrogen. The temperature optima for nitrogen fixation by *Nostoc* sp. (5°C) and *Calothrix* sp. (above 30°C) suggests that during winter *Nostoc* sp. fixes high levels of nitrogen while *Calothrix* sp. did not. At 30°C, nitrogen fixation by *Calothrix* sp. was three-fold greater than *Nostoc* sp. The ecological implications are that these two genera are active nitrogen fixers depending on the season. *Nostoc* sp. with low temperature optimum for nitrogen fixation is active during winter months whereas *Calothrix* sp. fixes nitrogen in mid-summer. Although colonization of *P. schereberi* with *Nostoc* sp. and *Stigonema* sp. has been predominantly observed, *Stigonema* sp. could not be isolated into pure cultures.

VIII. *AZOLLA*

The genus *Azolla* is a heterosporous floating aquatic fern (of the family Salviniaceae) that is distributed in various types of aquatic bodies such as freshwater ponds, lakes or streams. Species of *Azolla* are native to Asia (e.g. China and Vietnam), Africa (e.g. Senegal, Zaire, Sierra Leone), N. and S. America and the two poles. It is widely distributed in tropical, sub-tropical and temperate regions of the world. *Azolla-Anabaena* symbiotic association is distinctive for the fact that *Anabaena azollae* is known to persist throughout the reproductive cycle of the plant. It is transmitted from generation to generation (i.e. vertically) through the megasporophyte (sporocarp) of *Azolla*.

The classification of *Azolla* is based on morphological and reproductive features and accordingly seven species of *Azolla* have been recognized (Svenson, 1944). These are *A. caroliniana*, *A. filiculoides*, *A. mexicana*, *A. microphylla*, *A. rubra*, *A. pinnata* and *A. nilotica*. Based on the presence of number of floats on megaspores, these seven species have been distributed into two sections (or subgenera). *Azolla* (or *Euazolla*) is characteristic in possessing three floats on the megaspores to which the first five species belong and *Rhizosperma*, possesses nine floats on the megaspore to which the last two mentioned species, i.e. *A. pinnata* (with two varieties *A. pinnata* var. *imbricata* and *A. pinnata* var. *pinnata*) and *A. nilotica* belong. Of these species, the neotropical species *A. caroliniana*, *A. mexicana* and *A. microphylla* exhibit closer resemblances in vegetative, ecophysiological and genetical features and are distinct from other taxa (Lumpkin *et al.*, 1991). In addition, physiological and biochemical features have also been taken into account to confirm the traditional classification. However, except for the taxonomic status of *A. rubra* in *Euazolla* (some consider it as a subspecies of *A. filiculoides*) the rest of the species have closer affinities. This has been confirmed by DNA:DNA hybridization studies and the generation of hybrids between *A. mexicana-A. caroliniana* and *A. microphylla* (Zimmerman *et al.*, 1991).

Unlike other symbiotic associations of cyanobacteria, *Azolla-Anabaena* association has been effectively put to use as a biofertilizer in rice cultivation in South and Southeast Asia. The biomass of *Azolla* under optimal conditions gets doubled in less than 48 h with the addition of nitrogen by its potential to fix nitrogen. *Azolla* thus can accumulate 5–6.5% N on dry weight basis. Apart from serving as manure, fertilization of rice fields with *Azolla* has two other advantages. It can be used as a fodder and a weed suppressor (Moore, 1969).

In *Azolla-Anabaena* symbiosis the cyanobiont is transferred from one generation to another through megasporophyte. The development of megasporocarp and microsporocarp is identical and *Anabaena* is packaged into both the structures (Perkins and Peters, 1992). Due to rupture of microsporocarp, the cyanobiont is released to outside environment and does not play any role in the continuity of the symbiont (Calvert *et al.*, 1983; Peters and Calvert, 1983). On the contrary, the symbiont is located inside the indusium at the tip and above the megaspore. It is during the germination of the megaspore within the megasporophyte and during embryogenesis that the *Anabaena* cells are able to re-establish the symbiosis. A number of workers have studied events leading to the inoculation process and suggested processes for re-establishment of the symbiosis (Calvert *et al.*, 1983; Becking, 1987). But none of these studies have focussed attention on the probable role of epidermal trichomes in establishing symbiosis. Although studies of Campbell (1893), Dunham and Fowler (1987) mention the existence of subepidermal trichomes associated with *Anabaena* but no definite role has been assigned to them. Calvert *et al.* (1985) were the first to have demonstrated that the first cells to contact the symbiont during the re-establishment of symbiosis are the epidermal trichomes present on the developing embryo.

Establishment of symbiosis between *A. mexicana-A. azollae* has been investigated by Peters and Perkins (1993) following gametogenesis and embryogenesis. By employing light, scanning electron and transmission electron microscopy it was shown that (i) the cyanobiont is located in the space between the indusium and apical membrane that is designated as inoculation chamber; (ii) *Anabaena* cells that exist in the inoculation chamber are in the form of akinetes; (iii) embryo develops cotyledonary leaf primordium possessing epidermal hairs during which the akinetes germinate to give rise to germlings; (iv) the rapidly developing cotyledonary leaf with its four seriate trichomes engulfs the akinetes, geminating akinetes and germlings and (v) the cells of the trichomes exhibit transfer cell ultrastructure (TCU). The manner in which the indusial cap is generated is explained as follows.

Morphologically, the cotyledonary leaf is unlobed and vase shaped. By its faster uneven growth rate, it rapidly surrounds the entire meristem. As the emerging sporophyte releases the indusial cap it is completely surrounded by the cotyledonary leaf. *Anabaena* cells in the first instance are associated with hairs of cotyledonary leaf and later their association is shifted to the hairs of developing modified true leaf. These modified true leaves differ from the leaves of the mature vegetative sporophyte by the absence of bilobed character. These develop a cavity-like structure that extends from the dorsal side to the ventral achlorophyllous portion. This cavity-like structure contains mostly the akinetes of *Anabaena*. The inoculum for the first true leaf cavities is provided by the cells of *Anabaena* in the form of undifferentiated trichomes entrapped in the region of apical meristem of the cotyledonary leaf, modified true leaves and their associated hairs (Figs. 7 and 8). Thus these epidermal hairs help in redistributing these filaments into developing symbiotic cavities (Calvert *et al.*, 1985). It is interesting to note that the nature and number of hairs associated with the different kinds of leaves are different. For example, the cotyledonary leaf hairs are uniseriate and four in number whereas the first modified true leaf possesses two branched hairs each of which consists of six to seven cells. However, the second, third and fourth modified true leaves contain a single multibranched hair consisting of six to eight cells.

A correlation of leaf development in *Azolla* with nitrogen fixation by the symbiont reveals that the fixed nitrogen content increases with maturity of the leaf. The role of hormogones for causing fresh infection has been envisaged during the differentiation of leaf cavities in the developing leaves from the apical meristem and it is at this juncture that specialized epidermal partitioning trichomes play a role in the distribution of the symbiont. The concentration of HIF seems to decrease from apex downwards where as the HRF seems to increase in its concentration from base to apex of the stem axis. In other words, the repression in the formation of hormogonia is brought about by *Azolla* in its leaf cavities enabling the symbiont to differentiate heterocysts at a high frequency. The role of HRF has been discussed in case of *A. punctatus-Nostoc* symbiosis. In case of *Azolla*, the fronds

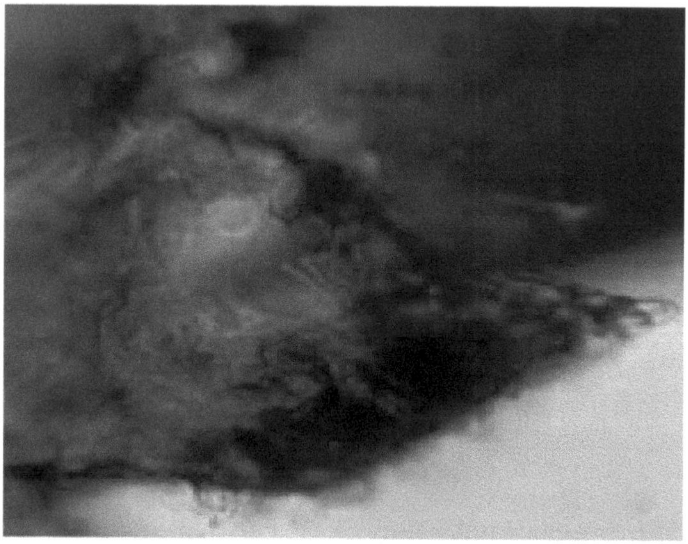

Figure 7: Confocal laser scanning electron micrograph of leaf cavity of *Azolla filiculoides* with *Anabaena* inside. Photograph courtesy E. S. Pierson[1], L. Masselink[2] and M. M. L. van Kempen[2], Departments of General Instruments[1] and Aquatic Ecology & Environmental Biology[2], Faculty of Science, Mathematics and Computing Science, Radboud University, Huygensgebouw, Heyendaalseweg 135, NL-6525, AJ Nijmegen, The Netherlands

Color image of this figure appears in the color plate section at the end of the book.

assume a reddish colour due to enhanced accumulation of deoxyanthocyanins (Pieterse *et al.*, 1977; Wagner, 1997). It is significant that aqueous extracts of *A. pinnata* and *A. filiculoides* induced *hrmA-luxB* genes in *N. punctiforme* strain UCD 328 releasing the HRF. This was compared with the levels of deoxyanthocyanins that repressed hormogonia formation in *Nostoc* sp. Thus this constitutes the first report on deoxyanthocyanins serving as molecules of plant recogntion by symbionts (Cohen *et al.*, 2002). Moreover, the induction of *hrmA-luxAB* genes by *Azolla* spp. is due to a synergistic action of deoxyanthocyanins and other plant-derived components and the percentage of induction by these substances is far more higher than the flavonoid naringin (Cohen and Yamasaki, 2000) and by extracts of *A. punctatus* (Cohen and Meeks, 1997).

i) *Isolation and cultivation of the symbiont*: A number of workers have reported the isolation and cultivation of the symbiont from *Azolla* (Newton and Herman, 1979; Gates, 1980; Tel-Or *et al.*, 1983; Subramanian and Malliga, 1988). The inoculum in these studies has been derived from the microdissection or homogenization of surface sterilized megasporocarps of *Azolla* or *Azolla* fronds or algal packets isolated from the fronds. Cultures thus propagated exhibited morphological resemblances to species of *Anabaena* or *Nostoc* but one distinctive feature that has been noted is that the cultured isolates are morphologically distinct from the filaments of symbionts freshly isolated from *Azolla*.

This has raised a number of questions. The first is that whether the isolated strains do really represent the symbiont located in the leaf cavities of *Azolla* or represent some other strains that are epiphytic. Secondly, is it possible for the existence of more than one strain in the leaf cavities of *Azolla* and if so is it that one strain is variable which is facultative and amenable for cultivation and not the other strain that is obligate in nature. This obligate symbiont is metabolically deficient and could not be cultured so far. Thirdly, the taxonomic identity of the symbiont whether it is a species of *Anabaena* or *Nostoc* has been questioned. The following studies have provided answers to these questions.

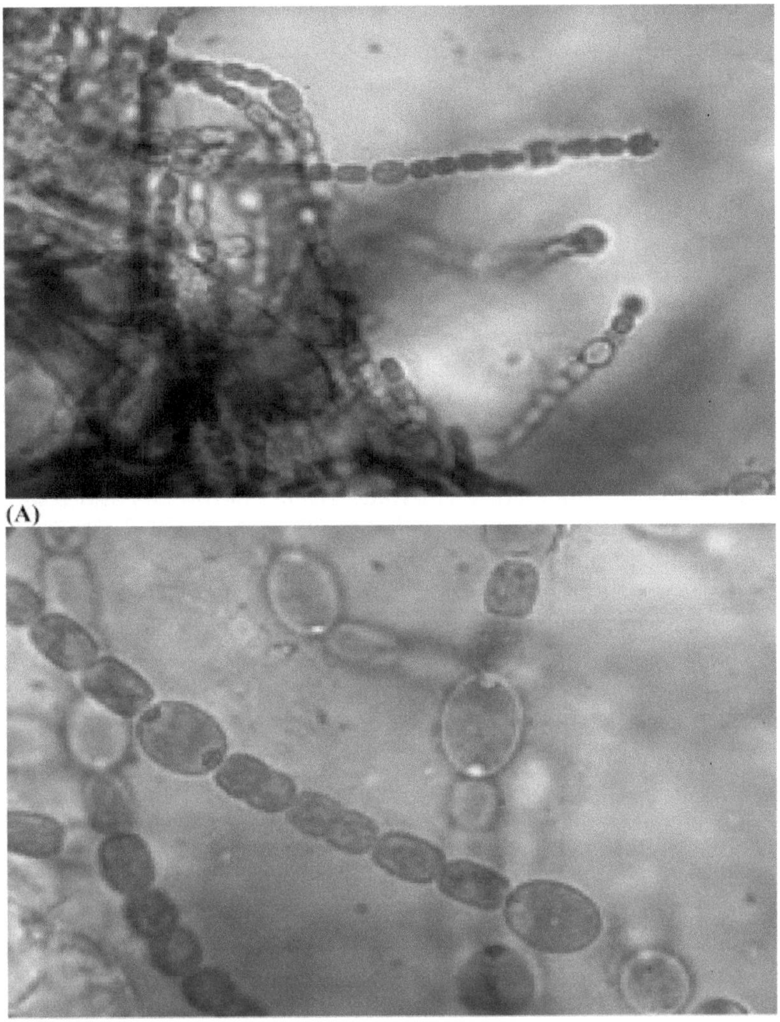

Figure 8: (A) Photomicrograph of tichomes of *Azolla* with those of *Anabaena* surrounding them. (B) Tichomes of *Anabaena* from *Azolla*. Photograph courtesy E. S. Pierson[1], L. Masselink[2] and M. M. L. van Kempen[2], Departments of General Instruments[1] and Aquatic Ecology & Environmental Biology[2], Faculty of Science, Mathematics and Computing Science, Radboud University, Huygensgebouw, Heyendaalseweg 135, NL-6525, AJ Nijmegen, The Netherlands.

Differences in morphological and physiological characteristics between cultures of *Azolla* symbionts and the fresh isolates from *Azolla* plants have been studied by the application of immunological and lectin hemagglutination studies (Newton and Herman, 1979; Gates, 1980; Ladha and Watanabe, 1982; McCowen *et al.*, 1987). The role of lectins in establishing *Azolla-Anabaena* symbiosis has been examined (Mellor *et al.*, 1981). Agglutination of human erythrocytes was caused by extracts from *A. caroliniana-Anabaena* symbiosis and *Anabaena*-free *Azolla* plants whereas extracts of fresh symbionts from *A. caroliniana* or *A. azollae* cultures did not cause agglutination of human erythrocytes. Kobiler *et al.* (1981) observed haemagglutination activity in extracts of free-living *A. azollae* and very low activity in those from *A. filiculoides-Anabaena* symbiosis. Ladha and Watanabe (1982) reported that there exists a high antigenic similarity between *Anabaena* freshly isolated from

Azolla plants and the cultures of the symbionts from all over the world. They further concluded that (i) at least two cultures are not true representatives of the symbiosis with *Azolla* or their antigenic properties might have undergone a change during isolation and cultivation; (ii) the presence of cross-reactive antigens between *Azolla* leaves and surface of *A. azollae*. By the application of indirect immunofluorescence technique, Ladha and Watanabe (1984) demonstrated agglutination of human and rabbit erythrocytes by the extracts of *Azolla-Anabaena* symbiosis and *Anabaena*-free *Azolla* plants. Such agglutination was not observed in case of *A. azollae* freshly separated from *Azolla* plants or free-living *A. azollae*.

Additional evidences in support of these differences have been put forward by restriction fragment length polymorphism (RFLP) as well. Frenche and Cohen-Bazire (1985, 1987) demonstrated that the cultures of the symbiont from *A. filiculoides* did not reveal any common hybridization bands with freshly isolated cyanobionts from *Azolla* spp. including *A. filiculoides*. Likewise, cultures of *Anabaena* from *A. caroliniana* and its freshly isolated symbiont exhibited differences in the RFLP patterns of genes sequences *glnA*, *psbA*, *rbcS* and *nifH* (Nierzwicki-Bauer and Haselkorn, 1986). Gebhardt *et al.* (1991) identified a common cyanobacterial symbiont in *A. mexicana* and *A. pinnata* by RFLP analyses with both single copy *glnA* and *rbcS* gene probes and a multicopy *psbA* gene probe. In these studies *nifD* excision probe and a *xisA* gene probe of *Anabaena* sp. strain PCC 7120 were employed for comparing the sequences in DNA extracted from free-living isolates and freshly isolated cyanobionts from the two species of *Azolla*. The sequences homologous to these probes from free-living isolates were found to be homologous to *Anabaena* sp. strain PCC 7120 sequences whereas the DNA from freshly isolated symbionts did not reveal any homology. From these studies Gebhardt *et al.* (1991) concluded that (i) the isolates were different from the major cyanobiont that resides in the leaf cavities of *Azolla* spp., (ii) these isolates are ubiquitously present as a culturable minor cyanobacterial symbiont in at least three species of *Azolla*, i.e. *A. caroliniana*, *A. filiculoides* and *A. pinnata*.

Taxonomically the cyanobionts from other symbiotic associations have been identified as species of *Nostoc* except in case of *Azolla* where the cyanobiont has been accorded the taxonomic assignment as *Anabaena* (*A. azollae*). Few workers preferred to designate the symbiont of *Azolla* as a species of *Nostoc* instead of *Anabaena* on the basis of formation of hormogonia in the former according to the classification of Rippka *et al.* (1979). Meeks *et al.* (1988) have assigned the status of *Nostoc* sp. to the cyanobiont of *Azolla* where as Grilli-Caiola *et al.* (1992) have considered the cyanobiont as a species of *Trichormus*. Ran *et al.* (2010) confirmed the symbiont of *A. filiculoides* to be a species of *Nostoc* and designated it as *Nostoc azollae* 0708. The partners of *A. filiculoides* symbiosis are depicted in Fig. 9.

A comparison of the symbionts from different hosts suggested that those from *Azolla* are markedly very close to each other on the basis of morphological, physiological properties and phycobiliprotein content and at the same time different from the symbionts of other hosts. Besides, the property of marked tendency to produce hormogonia has been noted in *Nostoc* species symbiotic with other hosts, a tendency that is lacking in *A. azollae* (Vagnoli *et al.*, 1992). Furthermore, strains of the cyanobiont from different *Azolla* species share a high degree of morphological identity but have been shown to differ at molecular level (Frenche and Cohen-Bazire, 1987; McCowen *et al.*, 1987; Plazinski *et al.*, 1988). Zimmerman *et al.* (1989) conducted a detailed study of the several cyanobionts from *Azolla*. Out of 10 secondary (minor) cyanobionts isolated from *Azolla* (*A. caroliniana*, *A. filiculoides*, *A. mexicana* and *A. pinnata*), six of them were identified as *Anabaena* and four of them as *Nostoc* on the basis of morphological features. All six strains of *Anabaena* are unique as they resembled one another and did not bind the lectins. Of these, five of them (two isolates of *Anabaena* from *A. caroliniana*, isolated by I. Newton, USA and R. Caudales, Rutgers University, USA; two isolates from *A. pinnata* isolated

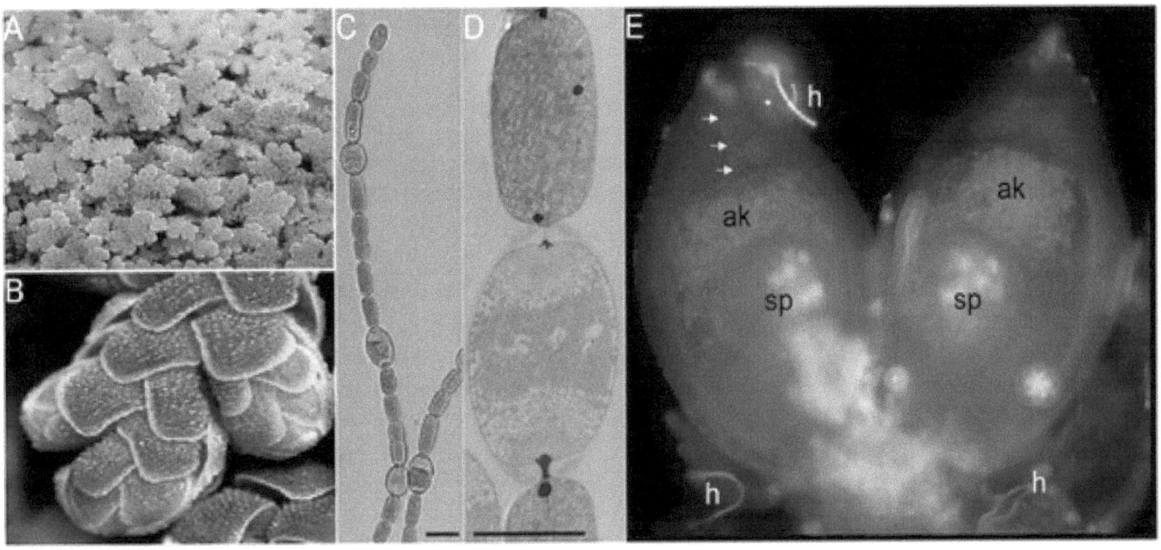

Figure 9: The partners in the *Azolla* symbiosis.
A) Fronds of the *Azolla filiculoides* Lam. plant. B) Close up of an *Azolla* branch showing the apex and the alternating 'stacked' dorsal leaves, each containing a cavity in which the cyanobiont (*Nostoc azollae* 0708) filaments reside. C) Light micrograph of the cyanobiont. The larger cells in the vegetative filaments represent the nitrogen-fixing heterocysts. Scale bar=5 μm. D) Transmission electronmicrograph of the cyanobiont. Note the thicker cell-walls and the electron dense polar nodules of the heterocyst (middle cell) at the interface to flanking vegetative cells, which function as combined N storage structures (cyanophycin). Scale bar=5 μm. E) A snap-shot in the vertical transmission process of the cyanobiont between *Azolla* plant generations, using fluorescence microscopy. Pairs of megasporocarps (blue) develop at the underside of the cyanobacterial colonized *Azolla* leaves. Filaments of the motile cyanobacterial cell stage (red), the hormogonia (h), are attracted to the sporocarps, gather at the base and subsequently move towards the tip, before entering the sporocarps via channels (white arrows). Once inside the sporocarp the hormogonia differentiate into individual thick walled resting spores (or akinetes; ak), seen as the intensively red fluorescing small inoculum on top of the megaspores (sp). With the kind permission of B. Bergman, Department of Botany, Stockholm University, Stockholm, Sweden. [Ran *et al.* (2010) *PLoS ONE* **5(7):** e11486. doi:10.1371/journal.pone.0011486] doi:10.1371/journal.pone.0011486.g001.

Color image of this figure appears in the color plate section at the end of the book.

by S. Nierzwicki-Bauer, USA and R. Fisher, USA and one isolate from *A. filiculoides* isolated by E. Tel-Or, Israel) showed closer resemblances to *A. variabilis* ATCC 29413 at the level of electrophoretic enzymatic analysis of nearly 12 important enzymes. On the other hand, all the *Nostoc* strains were distinct from one another morphologically but unrelated enzymatically.

The symbionts from *Azolla* formed hormogonia only upon transfer to fresh medium while in the symbionts from other hosts hormogonia are generally present in their populations. Secondly, *Anabaena* strains from *Azolla* species do not produce typical *punctiforme* stage, a feature that is characteristic of *Nostoc* symbiont. Moreover, these *Anabaena* strains prefer fructose as the carbon source (Vagnoli *et al.*, 1992). Four isolates of *A. azollae* from different species of *Azolla* that showed hormogonia production have been assigned to the genus *Nostoc* though they do not possess *punctiforme*-stage (Tomaselli *et al.*, 1988). However, it is suggested that *Azolla* possesses a *Nostoc* species as the major symbiont. This symbiont has not so far been isolated as it is not amenable for laboratory cultivation because of its obligate nature or certain deficiency in its metabolism. Besides this *Nostoc* sp., minor symbionts capable of free-living growth are also suggested to be present. It is these minor symbionts that have been readily isolated into cultures (Peters, 1991). Ran *et al.* (2010) have conducted phylogenetic analysis of the symbiont *N. azollae* 0708 which is clustered with *Cylindrospermopsis raciborskii* CS-505 and *Raphidiopsis brookii* D9 rather than with other *Anabaena* species.

ii) *Exchange of nutrients during symbiosis*: Initial characterization of *Azolla-A. azollae* symbiosis revealed that *A. caroliniana* plants free of the symbiont could be obtained by treatment with combination of antibiotics (aureomycin, penicillinG, streptomycin, bacitracin, polymyxin-B-sulphate). Such *Azolla* plants required combined nitrogen for their growth (Peters and Mayne, 1974a). This substantiates that the symbiont fixes nitrogen and meets the nitrogen requirements of the host as well as its own. The nitrogenase activity as determined by ARAs has indicated that *A. azollae* is responsible for the fixation of nitrogen and there exists a considerable exchange of metabolites between the fern and symbiont (Peters and Mayne, 1974b). Hill (1975) for the first time documented that the symbiont differentiates heterocysts at high frequency from near zero at the growing point to nearly 20–30% of all cells in mature leaves. The vegetative cells also undergo an increase in their cell size. The developmental pattern of *Anabaena* could be correlated with that of the fern as the cells of the symbiont at the apex remain small and devoid of heterocysts and cannot fix nitrogen. Soon after the colonization of leaf cavities, the symbiont differentiates heterocysts at high frequency and fixes nitrogen (Hill, 1977). N^{15}-labelling experiments assisted by ARAs confirmed that *A. azollae* in symbiosis with *Azolla* fixes nitrogen and supports growth of the host plant in a nitrogen deficient medium (Peters *et al.*, 1977). Further, [$^{13}N_2$] N_2 incorporation studies of Meeks *et al.* (1987) have confirmed that the symbiont fixes nitrogen.

Newton and Cavins (1976) detected high levels of free ammonia in the intracellular nitrogenous pool of *Azolla* plants growing in N_2-free media with the symbiont. Thus this free ammonia seems to serve as the major source of nitrogen for the symbiotic relationship. Symbiont-free *Azolla* plants or *Azolla* plants with symbionts grown in nitrate media did not contain free ammonia to that level indicating that the free ammonia has been derived out of nitrogenase of the algal cells.

The distribution of ammonia assimilating enzymes such as glutamate dehydrogenase and glutamine synthetase in between the host plant and symbiont revealed that the former contributed to nearly 75% of the enzyme pool suggesting that both the partners in this association have the capacity to synthesize glutamate either through GDH or GOGAT pathways (Ray *et al.*, 1978).

A definite developmental role for ensuring *Azolla-Anabaena* symbiosis has been assigned to the hair cells residing in the cavity of the host plant. These hairs have been shown to establish elaborate cell wall in growths that are characteristic of transfer cells (Duckett *et al.*, 1975). Peters *et al.* (1978) suggested that the hairs in mature and immature cavities might be involved in nitrogen assimilation and the release of fixed nitrogen respectively. Scanning and transmission electron microscopy of *A. azollae* symbiotic in *A. pinnata* revealed that the filaments are loosely entangled and are seen adhering to the protruding hairs and folded cell walls of the cavities. The presence of frequent invaginations or unbranched two-celled hairs and the cytoplasm of the mature hair cells contain a transparent network with blebs and vesicles. These structures appear to help in exchange of metabolites between the symbiotic partners (Neumüller and Bergman, 1981). These hair cells are postulated to play an important role in transferring fixed nitrogen to the host due to the presence of high activity of ammonia-assimilating enzymes (Uheda, 1986).

Two major isoforms of glutamine synthestase (GS) are known in higher plants, i.e. GS1 and GS2. The former is known to occur in the cytoplasm while the latter is in the plastids/chloroplasts. Differential roles for these two isoforms have been indicated. GS1 in roots helps in the assimilation of ammonia derived from nitrate reduction while GS2 helps in the assimilation of ammonia produced during photorespiration (Cren and Hirel, 1999; Ireland and Lea, 1999; Lancien *et al.*, 2000; Tobin and Yamaya, 2001). Uheda *et al.* (2004) demonstrated the location of GS2 isoform of the enzyme in mature leaves of *A. filiculoides* by the use of immunoelectron microscopy and immunogold labelling. Most of the label was found in chloroplasts but in hair cells abundant label was noted both in chloroplasts and in the cytoplasm. The existence of weak label in the hairs of the cyanobiont-free plants points towards the important role of hair cells in the assimilation of ammonia.

Peters and Calvert (1983) suggested that the regulation of GS in the association is at the level of function rather than synthesis. According to Orr and Haselkorn (1982) the GS activity in the symbiotic association was barely detectable either by applying enzyme assays or by radioimmuno assay. These are 5–10% lower in the symbiont *in situ* than in the free-living form. Molecular mechanisms governing gene activation and gene repression during heterocyst differentiation in cyanobacteria indicate that regulation is affected at the level of transcription (see chapter 4). Differences in mRNA levels of genes for nitrogenase and GS have been studied in the endosymbiont of *Azolla* by northern hybridization by using the corresponding cloned genes from *Anabaena* sp. strain PCC 7120. The *nif* genes (*nifH*, *nifD*, and *nifK*) are transcribed from a single *nif HDK* operon. A lowering of GS transcript levels up to 10% was found in the endosymbiont compared with the levels in free-living *A. azollae* (Nierzwicki-Bauer and Haselkorn, 1986). This leads to the conclusion that the host plant some how regulates transcription of the *Anabaena glnA* gene. In bacteria *glnA* transcription is regulated by the ratio of glutamine to ketoglutarate, a high ratio of these leads to repression. If the same mechanism is operative in the *Azolla-Anabaena* symbiosis, in that case it is of no use for the fern to transport glutamine to the symbiont as that will eventually be metabolized. Nierzwicki-Bauer and Haselkorn (1986) concluded that the fern might produce a non-metabolizable analog of glutamine that may act as a co-repressor of the *Anabaena glnA* gene but which does not affect the eukaryotic gene.

Photosynthetic efficiency of the symbiont of *Azolla* has been studied. *A. azollae* accounted for 7.5 to 15% of the total chlorophyll content of the symbiosis (Peters and Mayne, 1974a). Symbionts isolated from *Azolla* could fix CO_2 (Peters, 1975). Photosynthetic characterization of the symbiotic association and that of individual partners indicated that the fixation of CO_2 occurs via the Calvin cycle but the role of the symbiont contributing to the photosynthesis in the intact association was not apparent in the action spectrum (Ray *et al.*, 1979). A comparison of photosynthetic rates of *Azolla-Anabaena* symbiosis and *Anabaena*-free *Azolla* plants suggested that *Anabaena* contributes very little to carbon fixation. The symbiont is dependent on the host for its carbon that is supplied in the form of sucrose. Due to a five-fold decrease in the level of transcription of RuBisCO in the symbiont of *Azolla*, Nierzwicki-Bauer and Haselkorn (1986) concluded that the regulation of transcription of *rbcL-rbcS* operon may be at the level of repression by the host through a diffusible carbohydrate. ATP necessary to perform nitrogen fixation is synthesized photosynthetically in the heterocysts of the symbiont (Kaplan and Peters, 1981). Further, the heterocysts of the symbionts also are shown to possess RuBisCO, the activity of which is generally absent in the heterocysts but present in vegetative cells of free-living cyanobacteria. Thus it has been possible to detect the presence of *rbcL* and *rbcS* genes in heterocysts and vegetative cells of *Anabaena* spp. in association with or isolated from *Azolla-Anabaena* symbiosis by the application of *in situ* hybridaztion studies. However, the functional role of RuBisCO in heterocysts of the symbionts of *Azolla* remains to be elucidated (Madan and Nierwicki-Bauer, 1993).

Studies on genome erosion of the symbiont *N. azollae* 0708 brought out a number of new observations that explain the symbiosis in *Azolla*. Due to the accumulation of large number of pseudogenes (31.2%) and the presence of more than 600 insertion sequences in many of the genes, regulating important functional areas of DNA replication, repair, glycolysis and biosynthesis of co-factors of the endosymbiont, have become redundant. Furthermore, other processes that are impaired relate to the uptake of bicarbonate, phosphate, nitrate and urea. However, due to the presence of the *'nif'*-gene cluster intact along with 22 genes governing heterocst differentiation, it has become an obligatory confinement for the endosymbiont to fix nitrogen and release it to the host. It is also unable to reassimilate the released nitrogen due to the absence of transport mechanisms. So the host environment becomes very restrictive and the host supplies only limited supply of carbohydrates and phosphate so that the growth of the symbiont is slowed down (Ran *et al.*, 2010). The affected metabolic processes in the symbiont *N. azollae* 0708 have been depicted in Fig. 10.

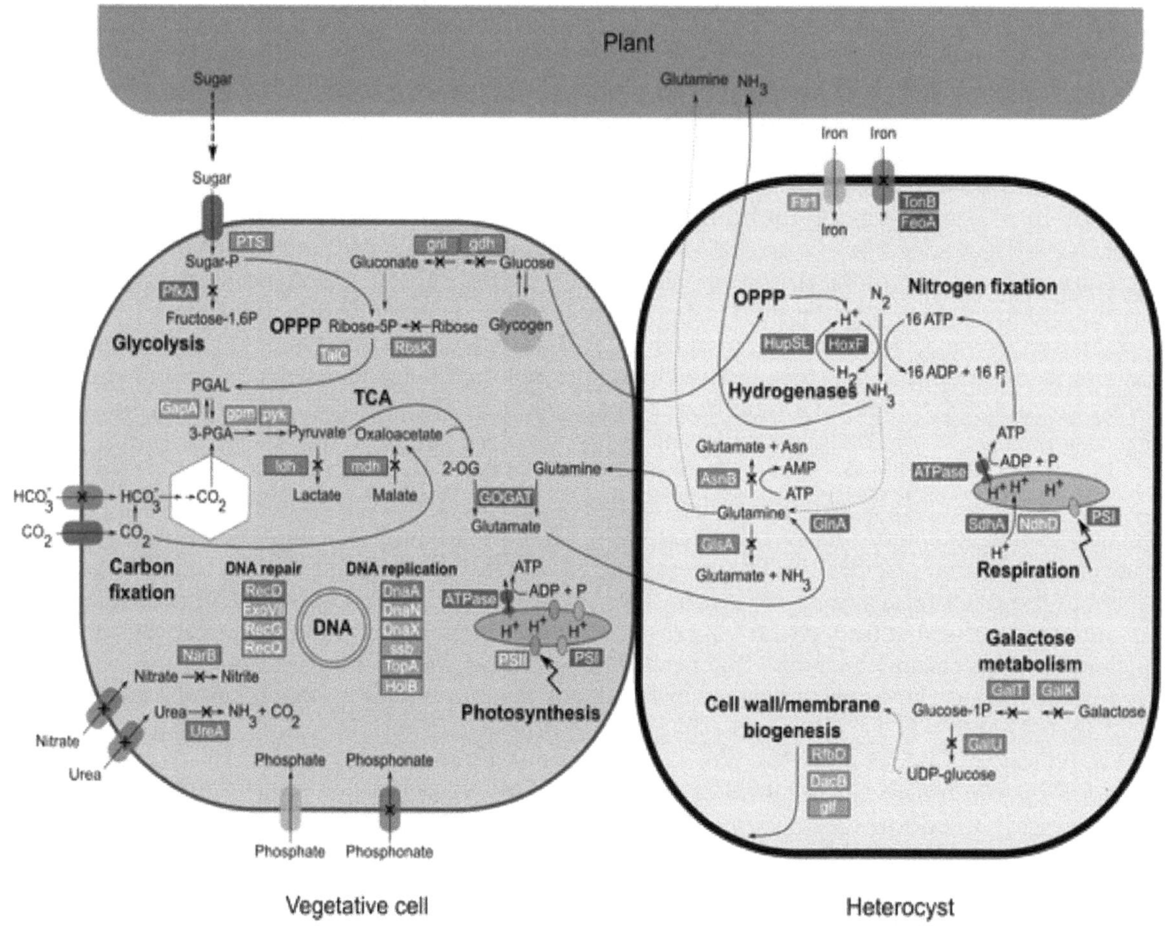

Figure 10: Schematic illustration of important metabolic and genetic information pathways in *Nostoc azollae*. The left cell represents a vegetative cell while the right a nitrogen-fixing heterocyst. Red color indicates pseudogenes lacking a functional counterpart in the *N. azollae* genome. Orange indicates pseudogenes where a functional counterpart is present elsewhere in the genome. Fully functional gene(s) are illustrated (blue) only if their function is linked to other processes in the figure. The localization of pathways in vegetative cells or heterocysts is representative only for nitrogen fixation (heterocysts) and PSII activity (vegetative cells). Note that only a minor part of the nitrogen fixed in heterocysts is incorporated using the GS-GOGAT pathway and used for synthesis of amino acids, while most is exported to the plant as NH_3. Sugar is provided by the plant in an as yet unknown form; putatively imported via the sugar phosphotransferase system. Function has been lost in the glycolytic pathway as the *pfkA* gene, encoding 6-phosphofructokinase, is a pseudogene and sugar metabolism in the *Azolla* cyanobiont probably proceeds via the oxidative pentose phosphate pathway. Extensive loss of function is evident among genes involved in uptake and transport of nutrients and *N. azollae* has lost the capacity to both import and metabolise alternative nitrogen sources. With the kind permission of B. Bergman [Ran *et al.* (2010) *PLoS ONE* **5(7):** e11486. doi:10.1371/journal.pone.0011486] doi:10.1371/journal.pone.0011486.g006.

IX. CYCADS

The order Cycadales is an ancient group of Gymnosperms that has existed for nearly 300 million years. Commonly known as cycads, these are characteristic in possessing pinnately compound leaves and giant cones. The major reasons for their long survival history are (i) their ability to survive drought and fire, (ii) their resistance to pathogens and predators and (iii) the production of protective, secondary metabolites. Modern cycads comprise of 11 genera (Brenner *et al.*, 2003).

Reinke in 1872 for the first time reported the existence of endosymbionts in cycad roots. The symbiotic association of cycads with nitrogen-fixing cyanobacteria is restricted to the coralloid roots. These roots first begin to make their appearance as apogeotropic roots with papillose sheath. These are also termed as precoralloid roots. The precoralloid roots are adventitious in nature and emerge from the hypocotyls below the cotyledonary petiole. As the precoralloid roots mature, they acquire the cyanobiont that generally occupies the cortical zone. Some other changes such as replacement of sheath layer by a dermal layer with scattered lenticels are characteristic of the maturation process.

i) *Cyanobionts*: Most commonly observed cyanobionts are species of *Nostoc* or *Anabaena* but other members such as species of *Spirulina, Oscillatoria, Rivularia* and *Calothrix* have also been reported. Species of *Nostoc* are mostly found in the coralloid roots of many cycads such as *Bowenia, Cycas, Dioon, Encephalartos, Macrozamia, Stangeria* and *Zamia* (Figs. 11 to 13). Species of *Anabaena*, i.e. *Anabaena cycadeae* and *A. circinalis* and *N. ellipsosporum* are present in the coralloid roots of *Cycas* (*C. revoluta*). Additionally, *Nostoc* sp. is characteristic of the coralloid roots of *Macrozamia lucida* whereas species of *Calothrix* can infect the coralloid roots of *Encephalartos hildebrandtii*. Norstog and Nicholls (1997) also mention the names of *Anabaena macrozamiae, Nostoc cycadeae, N. punctiforme* and *N. commune* as endophytes of cycads.

Endosymbionts of *Zamia* were first recorded by Schacht in 1853 in *Z. pumila*. Since then coralloid roots or roots with nodules harbouring cyanobionts have been reported in various species of *Zamia* such as *Z. floridana* (Watanabe, 1924; Chaudhuri and Akthar, 1931; Stewart and Rodgers, 1977; Lindblad, 1984), *Z. furfuracea* (Neumann, 1977), *Z. lindenii* (Spratt, 1915) and *Z. skinneri* (Lindblad *et al.*, 1985).

ii) *Infection process*: The invasion of coralloid roots by the cyanobiont can occur at the base, the apex or in an intercalary position. The exact mode of entry is not known but it is generally believed that it is through wounds or natural openings that the hormogonia invade the roots (Nathanielsz and Staff, 1975a; Ow *et al.*, 1999). Nathanielsz and Staff (1975b) observed that the cyanobionts *Nostoc* or *Anabaena* occur both intercellularly and intracellularly in the cells of inner and outer cortex of apogeotropic roots of *Macrozamia communis*. Algal invasion is preceded by mucus secretion by cells of algal zone in the cortex and that is deposited in the intercellular spaces of cortical parenchyma. During the invasion, an actual algal invasion front of finely granular mucus material bypasses mucus already deposited in the intercellular spaces filling much of the cell cavity. This is followed by the entry of large number of algal symbionts. The entry of the cyanobiont is facilitated by the contribution of other microorganisms in the infection process. It has been noted that the primary infection of the coralloid roots of cycads by *Bacillus radicola* stimulated their development. This is followed by the entry of *Azotobacter* and finally *Anabaena* (Spratt, 1915). McLuckie (1922) concluded that infection by soil bacteria contributed to the growth of coralloid roots in *Macrozamia spiralis*. The distortions caused by bacteria or bacteroids in the developing coralloid roots facilitated the entry of *Anabaena*. On the contrary, the presence of other microorganisms had a negative effect on the establishment of symbiosis in cycads (Wittman *et al.*, 1965; Grilli-Caiola, 1975a).

Figure 11: Common cycads that show the development of coralloid roots. (A) *Cycas revoluta*; (B) *Dioon edule*; (C) *Encephalartos hildenbrandtii*; (D) *Macrozamia dyeri*. Photographs courtesy and copyright Jody L. Haynes, The Cycad Society, Inc. (www.cycad.org)

Color image of this figure appears in the color plate section at the end of the book.

Figure 12: Common cycads that show the development of coralloid roots (A) *Macrozamia macdonnellii;* (B) *Zamia floridana;* (C) *Zamia cremnophila;* (D) *Zamia splendens*. Photographs courtesy and copyright Jody L. Haynes, The Cycad Society, Inc. (www.cycad.org)

Color image of this figure appears in the color plate section at the end of the book.

(A)

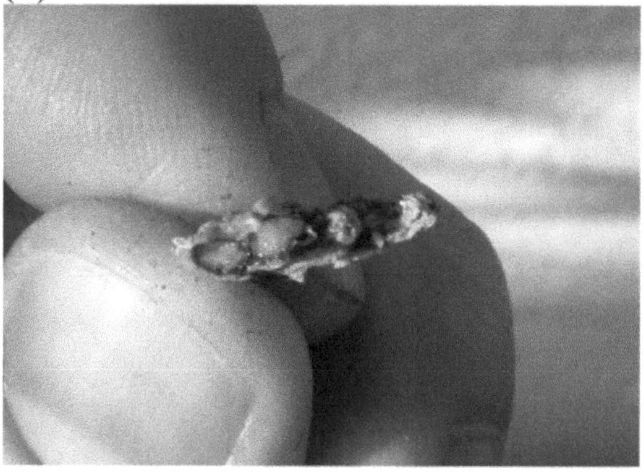

(B)

Figure 13: (A) A mass coralloid roots of *Cycas revoluta;* (B) A coralloid root magnified. Photographs courtesy and copyright Jody L. Haynes, The Cycad Society, Inc. (www.cycad.org)

Color image of this figure appears in the color plate section at the end of the book.

Wittmann *et al.* (1965) have observed that the coralloid roots of *M. communis* are bacteria-free. The presence of other eukaryotic algae, fungi and bacteria has never been observed in the coralloid roots along with cyanobionts (Grilli-Caiola, 1980). Obukowicz *et al.* (1981) demonstrated the presence of phenolic deposits in the mucilaginous material of algal zone by ferric chloride ultrahistochemistry of *Cycas revoluta-Anabaena* symbiosis. The presence of excessive phenolic deposits (in adjacent cells of the plasma membrane, endoplasmic reticulum within the plastids and at the periphery) in the surrounding host tissues facilitates the exclusion of other microorganisms and permits only *Anabaena* into the coralloid roots. They noted intracellular localization of cyanobionts confirming the earlier observations of Grilli (1963) and Nathanielsz and Staff (1975b).

Studies on the ultrastructure and developmental cycle of symbionts living in the coralloid roots (of *C. circinalis, C. revoluta, C. rumphii, E. altensteinii, E. lebomboensis, E. lehmannii, E. longifolius, E. natalensis, E. umbeluxiensis, E. villosus, M. communis* and *Dioon edule*) revealed that all the symbionts have characteristics of *Nostoc* with vegetative cells and enlarged heterocysts but without akinetes. However, symbionts showed a gradual and continuous increase in the frequency of heterocysts from apex to the base of the coralloid roots (Grilli-Caiola, 1980). The cyanobiont of *Z. skinneri* does not show any sub-cellular modifications in the coralloid roots and healthy vegetative cells of the symbiont show thylakoids with phycobilisomes, carboxysomes, cyanophycin and glycogen granules. Furthermore, the ultrastructural details of the coralloid roots of *Z. skinneri* revealed the presence of characteristic transfer cells that play a role in the transfer of solutes between the symbiont and the host (Lindblad *et al.*, 1985). The presence of akinetes as noted by Lindblad *et al.* (1985) in *Z. skinneri* has also been substantiated earlier in *C. revoluta* (Zhu, 1982), *Dioon edule* (Grilli-Caiola, 1975b) and *M. communis* (Grilli-Caiola, 1974). Baulina and Lobakova (2003) observed vegetative cells with reduced cell walls in the cyanobionts of *C. circinalis*, *Ceratozamia mexicana* and *Encephalartos villosus* growing in the intercellular spaces of cyanobacterial zone of cortex. These cells resembled protoplasts and sphaeroplasts and overproduced mucilage-like polysaccharide- and protein-like substances that hastend the death of the cells.

Morphological and physiological properties of the symbionts residing in the coralloid roots of cycads are shown to be different with those isolated and grown in culture (Grilli, 1963; Grilli-Caiola, 1974, 1975a,b). One of the most distinguishing features of the symbionts is the continuous increase in the frequency of heterocysts from apex to the base of the coralloid roots (Grilli-Caiola, 1980). Besides the increase in the percentage of heterocysts from the growing tip of the coralloid roots (as noted in *Zamia*) towards basal, older parts, there was increase in the frequency of multiple (double, triple and quadruple) heterocysts (Lindblad *et al.*, 1985). This feature was also noted in other cycad species (Spratt, 1911; Grilli-Caiola, 1975a, 1980; Zhu, 1982).

An inverse relationship has been noted between ARAs and the total heterocyst frequency suggesting that it is likely that nitrogenase is active only in single heterocysts and in one of the heterocysts occurring in a group (Lindblad, 1984; Lindblad *et al.*, 1985).

Lindblad and Bergman (1989) showed the localization of phycoerythrin associated with thylakoids of vegetative cells of the symbiotic *Nostoc* growing in the coralloid roots of *C. revoluta* with the help of immunocytochemistry but phycoerythrin was absent in the heterocysts. Nitrogenase activity assayed by ARAs and ^{15}N fixation by cyanobionts freshly isolated from coralloid roots of *Macrozamia riedlei* showed a direct correlation with the frequency of heterocysts and these depended mostly on the substrates supplied by the host (Lindblad *et al.*, 1991).

iii) *Diversity of Nostoc strains*: A detailed study on the diversity of cyanobionts has been conducted by Grobbelaar *et al.* (1987) on the basis of light microscopy. They identified *N. commune*, *N. punctiforme*, *N. ellipsosporum*, *N. paludosum*, *N. muscorum* and *Calothrix* sp. as the cyanobionts from the coralloid roots of 31 species of *Encephalartos* indigenous to S. Africa. Subsequent studies on the genetic diversity of *Nostoc* strains occurring in the coralloid roots of cycads have been made by the application of molecular techniques such as RFLP, PCR fingerprinting and sequencing of $tRNA^{Leu}$ (UAA) intron but the results obtained in each of these are not comparable. Lindblad *et al.* (1989) used Southern blot technique and cloned *Anabaena* sp. strain PCC 7120 *nif*K and *gln*A genes as probes and compared RFLPs of cyanobionts freshly isolated from coralloid roots of *Ceratozamia mexicana*, *C. robusta*, *Dioon spinulosum*, *Z. furfuracea* and *Z. skinneri*. Differences in the sizes of their DNA fragments hybridizing with both probes indicated that different cyanobacterial species and/or strains were associated with the symbiosis. Alternatively, when symbionts freshly isolated from roots of *E. altensteinii* and three independently isolated strains from the same coralloid root were compared, they turned out to be the same organism. It is thus likely that a mixture of *Nostoc* strains can be associated with a single cycad species (Lindblad *et al.*, 1989).

PCR fingerprinting with primers derived from short tandemly repeated repetitive sequences (STRR) of cyanobionts from the coralloid roots of a number of cycads revealed the existence of numerous cyanobacteria in a single coralloid root. The diversity of cyanobionts was noted in the apical, middle and basal regions of a single cluster (Zheng *et al.*, 2002). Contrary to the above findings, Costa *et al.* (2004) concluded that a single *Nostoc* strain is associated in individual coralloid roots of cycads on the basis of $tRNA^{Leu}$ (UAA) intron sequencing. Natural populations of *M. riedlei* and those growing in greenhouses of Perth (Australia) have been selected for the isolation of cyanobionts from the coralloid roots. Several *Nostoc* strains appeared to be involved in this symbiosis. However, when single coralloid roots or when all coralloid roots from the same plant were examined only one *Nostoc* strain appeared to be associated with the symbiosis. On the basis of these findings, Costa *et al.* (2004) contested the findings of Zheng *et al.* (2002). Analysis of cyanobiont diversity from 31 species of Australian cycad genus *Macrozamia* based on sequencing of 16S rDNA revealed that there is negligible host specialization. From among the 56 cyanobionts isolated into cultures 22 have been identified to belong to *Nostoc* sp. and one to *Calothrix* sp. The former cyanobiont seems to be widely represented within 14 species of *Macrozamia* (Gehringer *et al.*, 2010).

X. *GUNNERA*

The genus *Gunnera* with about 40 species belongs to the monogeneric family Gunneraceae of angiosperms. However, many workers preferred to classify this genus in the family Haloragaceae (Moore *et al.*, 2010). The species of this genus are distributed almost exclusively throughout the Southern Hemisphere. Shindler's work (1905) has been recognized as the most complete morphological revision of the genus. He has recognized five subgenera viz., *Perpensum*, *Pseudogunnera*, *Milligania*, *Misandra* and *Panke* based on their morphology, respective modes of propagation and their geographic distribution. A new South American species *G. herteri* has been discovered in 1933 that has been placed in a new subgenus *Ostenigunnera* (Mattfeld, 1933).

According to Bergman *et al.* (1992a) the subgenera *Pseudogunnera* (*G. macrophylla*), *Perpensum* (*G. perpensa*) and *Ostenigunnera* (*G. herteri*) have one species each whereas subgenus *Misandra* has two species (*G. lobata* and *G. magellanica*, the latter is a small, stoloniferous species). In the subgenus *Milligania* six species have been included. By far the largest subgenus *Panke* is represented by the gigantic forms with thick and fleshy stems. The largest and most famous species of this subgenus are *G. manicata* (giant ornamental rhubarb attaining a height of 6–8 ft and a native of Brazil; Fig. 14 A) and *G. tinctoria* (Chilean rhubarb; Fig. 14 B). These are very attractive ornamental plants. Many species of *Gunnera* show preference for high altitudes. *G. tinctoria* is a fully naturalized exotic plant of New Zealand but in view of its weedy potential it has been placed in National Pest Plant Accord List.

Species of *Gunnera* are perennial herbs. There is great variation in the size of the plant. The upright stem portion attains a height of 2 cm as in subgenus *Milligania* and up to 6 m as in the subgenus *Panke* with intermediate range of height plants found in the subgenus *Misandra*. Species belonging to the subgenus *Panke* are the largest herbs known on earth. Other important features are slender or fleshy stems; leaves possess long petioles (rhubarb-like); flowers in compound spikes, uni- or bisexual without petals. Anatomically polystelic nature provides evidence of their aquatic ancestry.

i) *Cyanobiont*: The cyanobiont in *Gunnera* has initially been classified as *Scytonema* or *Anabaena* but now it is generally accepted that the cyanobiont is a species of *Nostoc*. Though initially identified as

Figure 14: Plants of *Gunnera manicata* (A) and *Gunnera tinctoria* (B) in the fore-ground. Pictures downloaded with the kind permission of Gerd Seehawer, Schneverdingen, Germany. (http://www.dendroimage.de/galleng.htm).

N. gunnerae because of its close resemblance to free-living *N. punctiforme* the cyanobiont has been redesignated as *N. punctiforme* (Harder, 1917; Winter, 1935).

ii) *Nature of symbiosis*: The nature of *Gunnera-Nostoc* symbiosis is facultative. The two partners upon separation can lead an independent existence. Reinke made the initial observations on the *Gunnera-Nostoc* symbiosis. The true nature of the symbiosis was for the first time confirmed by Silverster and Smith (1969). All the species of *Gunnera* form the glands at the base of their leaves and possess the endosymbiont *Nostoc*. Even with the cyanobiont *Nostoc*, the symbiosis is relatively non-specific as a number of strains and species can establish symbiosis with *Gunnera* (Bonnett and Silvester, 1981). Of the many cyanobacterial symbioses, *Gunnera-Nostoc* symbiosis represents the most advanced type because of the cyanobiont being in the intracellular state (Bergman *et al.*, 1992b).

iii) *The glands*: The host plant possesses peculiar glands at the base of each leaf stalk. Reinke (1873) first described the glands containing algae in *Gunnera* and their development has been described subsequently in *G. macrophylla* (Miehe, 1924). The glands have also been designated as "stem glands" by Merker (Solereder, 1908), "rosette organs" (Batham, 1943), "transformed root primordia" (Schaede, 1951) and nodules (Silvester and McNamara, 1976) in the literature.

The formation of the glands under sterile conditions in the absence of the symbiont signifies that their development is governed by the plant genes. However, it is universally accepted that (i) for the initiation and further maintenance of the symbiosis the glands are required and (ii) symbiosis-specific genes may also operate in *Gunnera-Nostoc* symbiosis but these are suggested to be required for establishing contact between the two partners but are not involved in the formation of the gland.

iv) *Morphology and development of the gland*: The general structure and development of the stem glands in *Gunnera* spp. have been studied by a number of workers. Bonnett (1981) provided information on the development of glands in *G. chilensis*. The development of two unequal meristematic masses at the base of the developing cotyledons constitutes the first step. These soon assume the shape of papillae. There are 6–9 papillae of which 5–8 papillae are arranged in a circle and one is located in the center. The outer papillae in the circle are slightly bent outwards whereas the central one is straight. The papillae grow and undergo division perpendicular to the hypocotyls. These are visible to outside as hemispherical protuberances on the outside of the stem. Due to dissolution of the middle lamellae of the cell walls running perpendicular to the hypocotyl, a number of channels are formed between the papillae. At the bottom of these channels, the cells get separated forming a cavity. Due to the production copious mucilage, the epidermis ruptures and the mucus flows out through the channels. The chemical nature of mucus reveals the presence of carbohydrates, tannic acids and other polyphenols. It is acidic and believed to be antifungal in nature.

New glands continue to be differentiated at the base of each leaf primordium under nitrogen-limited conditions (Chiu *et al.*, 2005). The nutritional factors that govern the development of the glands *in vitro* have been studied by Chiu *et al.* (2005) who suggested that (i) nitrogen status of the plant is the main determinant for gland development; (ii) the presence of exogenous carbon sources like sucrose accelerated gland development under nitrogen-limited conditions; (iii) under nitrogen replete conditions gland development did not take place; and (iv) a high concentration of sucrose under nitrogen enriched conditions stimulated the development of a callus-like outgrowth instead of a gland. After sometime, growth and cell divisions are slowed down with a cessation in the production of mucilage. Due to the formation of new cell layers, the glands turn brownish and develop a cork-like surface. At this stage, infection is no longer possible. In the absence of infection, disintegration of glands ensues afterwards. So during the production of copious mucilage infection is a probable event.

v) *Infection process*: The infection of each gland represents an independent event and everytime *Nostoc* filaments growing outside need to gain entry into the new differentiating gland. The manner in which the symbiont is brought nearer to the gland is not clearly known. It is likely that during the emergence of the seedling of *Gunnera, Nostoc* filaments present in the soil may get entrapped in the mucilage or filaments of the symbiont may stick to the trichomes/hairs located on the gland surface. The filaments of *Nostoc* may be attracted towards gland by the chemoattractants present in the mucilage. The presence of fimbriare on the cell surface of the symbiont may also be helpful in the attachment to the gland surfaces (Dick and Stewart, 1980; Lindblad and Bergman, 1990). Thus mucilage serves as an important communication channel between the symbionts prior to infection. How the symbiont survives in the acidic pH of the mucilage secreted by the glands of *Gunnera* (in case of *G. manicata* the pH is 5.0; and in case of *G. chilensis* the pH is 4–5) is not understood. But it has been confirmed by many workers that the symbiont grows rapidly in the mucilage and forms a thin film over the gland surface. Two observations of Rasmussen *et al.* (1994) that (i) mucilage secreted from the glands induces hormogonia formation in vegetative *Nostoc* filaments and (ii) mucilage contains certain factors that stimulate growth and induce protein synthesis in compatible *Nostoc* strains are worth noting. Molecular characterization of the mucilage secreted by *Gunnera* stem gland revealed arabinose and galactose in molar proportions of 1.00:0.25 with an additional 0.13 parts of glucoronic acid. Dot blot analysis confirmed the existence of arabinogalactan proteins (Rasmussen *et al.*, 1996).

Schaede (1951, 1962) suggested that the symbiont penetrates into the root cells where the formation of canals allows free passsage for the algal filaments. This mode of entry of the symbiont has not received much support. However, it is now generally accepted that the entry of hormogonia into the interior of the gland is affected via the mucilage channels (Batham, 1943; Silvester and McNamara, 1976; Towata, 1985). In order to reach the interior and finally the cavity, the hormogonia have to travel against the flow of mucilage and indeed it requires a great force. This may be achieved by the chemotaxis of the hormogonia due to the presence of chemoattractants in high concentration in the cavity of the gland. In otherwords, the cyanobiont responds to the signals of the host by showing chemotaxis. Indeed chemotaxis has been shown to be possible in *N. punctiforme* PCC 73102 with exudates or crushed extracts of *G. manicata* and a number of other hosts (*Cycas revoluta* and *Blasia pusila*) and non-host plants (*Trifolium repens, Arabidopsis thaliana* and *Oryza sativa*) as well (Nilsson *et al.*, 2006). In other cyanobacterial symbioses such as lichens, lectins are known to play a role in the recognition process (described above in detail under the section on lichens) but in case of *Gunnera-Nostoc* symbiosis the role of lectins in the recognition of partners is yet to be established. However, Khamar *et al.* (2010) identified multiple role of soluble sugars in the establishment of *Gunnera-Nostoc* endosymbiosis. Prior to the establishment of the cyanobiont, the mucilage contained higher levels of galactose and arabinose that had little effect on hormogonia formation. After the establishment of the cyanobiont, the mature glands accumulated higher concentrations of glucose and fructose that favoured the vegetative growth of the cyanobiont. This is supported by the accumulation of starch in the cortical cells of the nitrogen-starved plants and simultaneously the expression of host genes encoding enzymes involved in starch hydrolysis took place.

It is well known that *nod* genes of *Rhizobium* are involved in the production of Nod-factors, chitin oligomers with an acyl chain at the non-reducing end that have numerous effects on the host plant. Although common *nod* gene sequences (*nod ABC* or the regulatory gene *nod D* or the host specific gene *nod L*) are not present in *Nostoc* but Nod-D binding part of the *nod* promoter and *nod* box showed hybridization with specific *nod* genes such as *nodEF*. The presence of gene sequences homologous to *nodMN* in *Nostoc* genome is of interest because *NodM* has been reported to be involved in the synthesis of glucosamine that acts like a plant defence elicitor. So it is likely that *Nostoc* can also produce a factor containing glucosamine that induces cell divisions in *Gunnera* gland channels. Nod-factors activate cytokinin pathway, which include cell division and nodule development (Frugier *et al.*, 2008). Exopolysaccharide governing *exo* gene sequences of *Rhizobium*, important for establishing successful nodulation of legumes, are found in *Nostoc*.

vi) *Intracellular localization*: This is the only cyanobacterial symbiosis in which the cyanobiont becomes intracellular. In order to enter the cells, the cyanobiont has to penetrate the cell walls. The infection process and ultrastructure of *Gunnera-Nostoc* symbiosis have been investigated by Silvester and McNamara (1976) who highlighted that the cyanobiont upon reaching the cavity, first penetrates the thin-walled meristematic cells situated at the base of the gland. This has also been confirmed by the work of Bonnett (1990). The cells of the symbiont become intracellular and are surrounded by a host membrane. The question that has to be answered is whether the cyanobiont gains entry by sheer mechanical force or through the invaginations/cell foldings or enzymic degradation of cell wall. One of the theories put forward is that the cyanobiont gains entry through the cell wall foldings/invaginations. It has been observed that the cell wall foldings/invaginations take place independently even in the absence of the symbiont (Schmidt, 1901). During entry into the cell the symbiont acquires a host membrane that separates it from the cytoplasm of the host cell (Von Neumann *et al.*, 1970; Silvester and McNamara, 1976). There are no evidences for the production of pectolytic or cellulolytic enzymes or the release of IAA by the cyanobiont that can trigger cellulase activity. The second theory putforward is the cell wall dissolution theory. The observations of Towata (1985) that the bacteria associated with the cyanobiont help in the degradation of cell wall also received attention. The presence of large populations of other microorganisms in the fresh mucilage secretions besides the symbiont has also been confirmed (Flowers, 1998). Towata (1985) further mentions that fungi such as *Penicillium*, *Fusarium* and *Alternaria* have been isolated from the mucilage of *G. kaalensis*. The pectolytic and cellulolytic enzymes contributed by these fungi have been suggested to bring about the dissolution of the host cell wall and/or middle lamellae. However, the establishment of *Gunnera-Nostoc* symbiosis *in vitro* in the absence of the said fungi rules out the probable role of these fungi (Chiu *et al.*, 2005). After the dissolution of cell wall, the cyanobiont is taken in by the folding of the plasmolemma. This theory has received much support from the works of Jönsson (1894), Miehe (1924), Schaede (1951) and Johansson and Bergman (1992). The absence of wall rupture after the infection by the cyanobiont suggests that the initial dissolution must have been repaired. Observations of Silvester and McNamara (1976) also confirm this theory but in their reconstitution experiments they did not come across any bacteria or other microorganisms that could contribute to the dissolution of the cell wall. The presence of infoldings has finally been suggested as individual variations that can be noted depending on the species of *Gunnera* used in the study.

The infected cell soon gets filled with the ensuing growth of the endosymbiont. The cytoplasm of the cell gets occupied to a peripheral portion. The endosymbiont is shared between the daughter cells as cell organelles during cell division and wall penetration does not seem to be a requirement subsequently. However, the host plasma membrane around the symbiont serves as a selective barrier. The symbiont exhibits an altered morphology as the cells are spherical rather than cylindrical, aseriate or in short filaments with altered distribution and shape of the thylakoids in the cells and with a high heterocyst frequency.

vii) *Specificity*: It is generally accepted that all the species of *Gunnera* have *Nostoc* spp. as endosymbiont. In order to find out specificity of the *Nostoc* spp. a number of workers have established symbiosis *in vitro* by a range of *Nostoc* spp. and closely related *Anabaena* spp. Reconstitution experiments have been successful in *G. chilensis* with *N. punctiforme* (Reinke, 1873; Jönsson, 1894) or *Nostoc* symbionts from *G. chilensis* or *G. arenaria* (Johansson and Bergman, 1992). *Nostoc* symbionts from *G. arenaria, C. revoluta, Anthoceros* spp, *P. polydactyla* and *Nostoc commune* reconstituted symbiosis in *G. manicata* (Bonnett and Silvester, 1981). Successful establishment of symbiosis was monitored by microscopic examination, nitrogenase activity and total nitrogen content of the established symbiosis. Symbionts of *Nostoc* from *G. arenaria* (New Zealand), *C. revoluta* (Green house, New Zealand), *Anthoceros* sp. (New Zealand), *P. polydactyla* and a soil cyanobacterium *N. commune* (Netherlands) entered into successful symbiosis with *G. manicata*. Except the symbiont from *P. ploydactyla*, the rest of the four associations performed well with reference to the nitrogenase activity, nitrogen per plant, number of infected glands per plant and nitrogen content per gland. Strains incapable of producing hormogonia did not infect the glands. Moreover, *Nostoc* symbionts from *Macrozamia lucida, Anabaena azollae* (from *Azolla*) and two another *Anabaena* species (*A. oscillarioides* and *A. flos-aquae*) did not establish symbiosis with *G. manicata* (Bonnett and Silvester, 1981). This clearly demonstrates that motility is esstential for infection to take place.

Zimmerman and Bergman (1990) studied the diversity of cyanobionts and correlated it with the occurrence of *Gunnera* species and their habitat. For this purpose they had chosen twelve cyanobionts isolated from *Gunnera* plants growing in Sweden (*G. manicata* 1 and 2 from two sites, *G. tinctoria* and *Gunnera* sp.), New Zealand [*G. arenaria* (2), *G. chilensis* (3) from five different sites] and USA (*G. arenaria, G. kaalensis, G. killipiana* and *Gunnera* sp.) and compared their protein profiles and RFLP polymorphisms (by hybridizations with heterologous *nif*H and *gln*A probes). These results point towards the identical nature of the cyanobionts from different *Gunnera* species growing at the same site in Sweden. However, a different cyanobiont was detected in *G. manicata* growing at a different site in Sweden. On the other hand, of the five cyanobionts from two species of *Gunnera* collected in the same location in New Zealand three subgroups of cyanobionts have been dectected. The cyanobionts from three different species of *Gunnera* grown in different localities were found to be different. These results emphasize that there is no great critical selective factor required by *Gunnera*.

In case of heterocystous cyanobacteria a specific family of STRR sequences have been described (Mazel, 1990). The number of copies of such sequences was estimated to be about 100 per genome in *Calothrix* species in which these sequences were initially discovered. The long tandemly repeated repetitive (LTRR) sequences are 37-bp long and have been identified in *Anabaena* sp. strain PCC 7120. STRR sequences have been put to use as a valuable tool for the identification and characterization of cyanobacteria, specially the toxin-producing strains (Rouhiainen *et al.*, 1995). Thus the conserved status of these sequences makes them suitable for identification of cyanobacteria. A fingerprint method consisting of STRR and LTRR sequences was developed for the recognition of 35 symbiotic isolates of *Nostoc* from *Gunnera* spp. The results showed a high degree of heterogeneity among the isolates from the same *Gunnera* species as well as different species (Rasmussen and Svenning, 1998). Rasmussen and Svenning (2001) have drawn similar conclusions based on RFLP-patterns of amplified 16S rRNA gene and 16S-23S ITS region as well as DGGE of PCR products of *hetR* gene. Similarly, cyanobionts isolated from 11 different geographical areas have been subjected for PCR fingerprinting analysis with STRR sequences as primers. These studies revealed that (i) all the 45 cyanobionts isolated could be divided into ten groups of which five cyanobiont isolates were found to be unique, (ii) most of the groups are restricted in their distribution to one geographical area and (iii) a low cyanobiont specificity has been indicated because many cyanobiont strains have been detected within and between 11 different *Gunnera* species. Further, more than one cyanobiont has been associated with symbiosis within the same plant as well as within the same stem gland (Nilsson *et al.*, 2000). The diversity of *Nostoc* strains occurring in three populations of *G. tinctoria* from Chile has been studied by PCR amplification of STRR sequences of the symbiotic tissue of *Gunnera*. There is great variation in the number of *Nostoc* strains establishing symbiosis both within and between *Gunnera* populations. However, one *Nostoc* strain or closely related strains have been found within an individual host plant (Guevara *et al.*, 2002).

16S rDNA sequence analysis has been regarded as one of the most valid criteria for unraveling taxonomic relationships between closely related species or genera (Weisburg *et al.*, 1991). The diversity of *Nostoc* strains establishing symbiosis with *Gunnera* has been analysed by this marker. Svenning *et al.* (2005) conducted a phylogenetic analysis of symbiotic *Nostoc* strains establishing symbiosis with species of *Gunnera* by RFLP of the 16S rDNA sequences and 16S-23S ITS regions. Cyanobionts from *Gunnera* have been identified as a distinct clade. They are of the opinion that it is not justified to assign all symbiotic *Nostoc* species to *N. punctiforme*.

viii) *Nutrient exchange*: The cyanobiont experiences a sudden change in its photosynthetic and nitrogen fixing capabilities. Although the cyanobiont possesses the pigments, RuBisCO enzyme and other enzymes of the Calvin's cycle in the same proportion as that of the free-living cyanobacterium, its autotrophic metabolism is shifted towards a heterotrophic metabolism. Alongside, the frequency of heterocysts increases to 30% contributing towards an increase in nitrogen fixation. The host meets the carbon requirements of the cyanobiont that is transported in the form of sucrose. The transport of carbon source from the host to the cyanobiont and the transport of fixed nitrogen from the cyanobiont to the apex of the plant body seem to be through the conducting elements. All these aspects have been investigated in greater detail.

The following observations signify that the cyanobiont leads a heterotrophic existence. The proportion of pigments such as chlorophyll *a* (Söderback *et al.*, 1990), phycobiliproteins and the carboxylating enzyme (Söderback and Bergman, 1992) of the cyanobionts of *G. megallanica* are similar to the free-living cyanobacteria. Söderback and Bergman (1993) observed that the photosynthetic capabilities of *G. chilensis* and *G. megallanica* are better than their counterparts isolated into cultures. The most important aspect is the high RuBisCO activity associated with low *in vivo* CO_2 fixation. $^{14}CO_2$-incorporation revealed a fixed carbon translocation from the leaves to the symbiotic tissue resulting in high rates of nitrogen fixation. Additionally, a major reduction in PSII activity results when free-living *Nostoc* enters into symbiosis with *Gunnera*. This is evidenced by a smaller pool size of electron acceptors (QA) and plastoquinone and a reduction in the ability to utilize light by PSII units. These changes have been correlated with partial degradation of DI protein. It is not that the PSII is entirely absent but the PSII efficiency has been largely reduced which signals a shift from autotrophic to a heterotrophic mode of nutrition (Black and Osborne, 2004). While the photosynthetic ability of the cyanobiont is thus limited, the nitrogen fixation potential is enhanced due to the differentiation of heterocysts at a high frequency, enhancement in nitrogenase activity and respiratory electron transport. Freshly isolated *Nostoc* clusters from the glands of *G. megallanica* performed high rates of nitrogen fixation in the presence of glucose, fructose and sucrose. When nitrogenase activity of the symbiont was inhibited *in vivo* by high O_2 levels, the accumulation of sucrose, glucose and fructose occurred (Parsons, 2002). Uptake of glucose analogue, 3-[^{14}C]-O-methyl-glucose (^{14}C-OMG) by symbiotic and free-living *Nostoc* revealed that the uptake is mediated by a hexose transporter and that the uptake slowed down with the increase of heterocyst frequency in the symbiont signifying that the uptake process was specifically associated with vegetative cells. Further, glucose also is shown to be metabolized through glycolysis as well as incomplete citric acid cycle in symbiotic cells (Black *et al.*, 2002).

The cyanobiont can support the growth and can meet the nitrogen demand of the host as evidenced by its growth in nitrogen-deficient media. The rates of nitrogen fixation by the *Gunnera-Nostoc* symbiosis have been reported to be of the order of 72 kg N ha^{-1} a^{-1} (Silvester and Smith, 1969). However, some investigators concluded that this symbiosis has a decreased capacity for nitrogen fixation due to the existence of low frequencies of heterocysts in *G. tinctoria* (Silvester, 1975) or the presence of large number of degenerate cells (Towata, 1985). Osborne *et al.* (1992) showed that nitrogen fixed by the cyanobiont could meet the complete nitrogen requirements of *G. tinctoria*. The cyanobiont occupies less than 1% of the total biomass of the host and cyanobiont put together. The ability of the host to utilize nitrate or ammonium has been shown to be very much limited.

Nitrogen fixation in relation to gland development has been studied in *G. megallanica* by the assay of total and specific nitrogenase activity and the frequency of heterocysts both in intact glands and sections of the gland apex downwards. The apical portion of the gland consists of vegetative cells of the cyanobiont loaded with high density of storage granules. Gradually downwards the cyanobiont differentiated heterocysts where the frequency ranged from zero to 30% with toal and specific nitrogenase activity being maximum and nitrogenase protein was localized only in the heterocysts. Progressively down the gland although the heterocyst frequency increased to as high as 60%, the nitrogenase activity was lower. However, the decrease in nitrogenase activity may be due to certain other factors as the nitrogenase protein remained at the same level (Söderback *et al.*, 1990). Transcription and protein profiles of genes related to heterocyst differentiation and dinitrogen fixation have been examined in *Nostoc* strain 0102 (isolated from Gunnera) under simulated symbiotic conditions in *G. megallanica* and *G. manicata*. The expression of *hetR* gene correlated positively with high frequency of heterocysts. The expression of genes *ntcA* and *nifH* was also high whereas the expression of *gln*B showed decreased expression. These studies have been successfully conducted by using RT-PCR and Western blot analysis (Wang *et al.*, 2004). A comparison of protein expression profiles of freshly isolated cyanobiont from the glands of *G. manicata* with cultures of the same strain has revealed that a significant number of proteins are abundant whereas certain other proteins are either down-regulated or entirely missing in a symbiotic state. For example, nitrogenase and enzymes of OPP pathway are highly expressed while enzymes of Calvin's cycle are down-regulated (Ekman *et al.*, 2006).

By the use of ^{15}N, Silvester *et al.* (1996) confirmed that only 12% of the fixed nitrogen is retained by the freshly isolated cyanobiont from *G. megallanica* whereas the rest 88% is released as NH_3 outside the cells. Within the intact glands, the cyanobiont retained only 2–5% of the fixed N_2 and up to 30% of the extracellular N is in the form of asparagine after one hour. The release of ammonia by the excised glands suggests that the excised glands because of their surrounding *Gunnera* envelope/membrane also are akin to the intact glands and in these cases rates of recovery of NH_3 corresponded well with C_2H_2 reduction rates. The overall evidences suggest that *Gunnera* regulates the activity of GS of the cyanobiont but not at the level of its synthesis as noted in case of *Anthoceros-Nostoc* symbiosis (Joseph and Meeks, 1987). A metabolic model has been proposed that envisages the nitrogen fixation by the cyanobiont supported by either light or carbohydrates derived from the host. The assimilation of fixed nitrogen seems to require the input of 1 mol oxaloacetate for the export of 1 mol of asparagine (Silvester *et al.*, 1996).

In root nodule symbiosis, there is ample evidence for the transport of nitrogen compounds through xylem from the nodule. The importance of phloem has been indicated in the transport of carbon and also in the retranslocation of nitrogen when it reaches the leaves. Nitrogen fixation by the cyanobionts in *Gunnera* seems to be supported by the translocation of carbon from the host presumably by way of phloem. *G. monoica* is a stoloniferous plant with 4–6 leaves at each node and is amenable for assay of nitrogenase activity by ARAs conducted by the incubation of intact stolons and nodes in an atmosphere of 10% v/v of acetylene in air. Such experiments have been helpful in assessing ^{15}N as percentage of all N atoms, µg ^{15}N g^{-1} dry weight of plant tissue for understanding the distribution of the isotope and ^{15}N% as a percentage of all the ^{15}N taken up in a particular organ. Rapid translocation of recently fixed nitrogen from mature regions to the apex of *G. monoica* has been demonstrated by ^{15}N pulse-chase experiments. Additionally, stem-girdling experiments provided evidence for the translocation of recently fixed nitrogen through phloem. It means that the flow of fixed nitrogen via phloem must be counter to the flow of carbohydrates into the glands (Stock and Silvester, 1994).

XI. PHYLOGENY OF *NOSTOC* STRAINS ACROSS ALL SYMBIOSES

In most of the cyanobacterial symbioses the cyanobiont is a strain of *Nostoc*. In the different symbioses described in this Chapter, the diversity and phylogeny of *Nostoc* strains have been discussed. The molecular markers selected for this purpose varied greatly as also the hosts from which the cyanobionts have been derived. These include comparison of the sequences of $tRNA^{Leu}$ (UAA) intron (Lindblad *et al.*, 1989; Paulsrud and Lindblad, 1998; Paulsrud *et al.*, 1998, 2000; Costa *et al.*, 2001, 2002, 2004; Wirtz *et al.*, 2003; Summerfield and Eaton-Rye, 2006; Rikkinen and Virtanen, 2008), 16S rDNA (Oksanen *et al.*, 2004; Myllys *et al.*, 2007; Gehringer *et al.*, 2010), 16S rRNA-23S rRNA ITS region (West and Adams, 1997), 16S rDNA and $tRNA^{Leu}$ (UAA) intron (Rikkinen *et al.*, 2002; Summerfiedld *et al.*, 2002; Stenroos *et al.*, 2006) and gene locus *rbcLXS* (O'Brien *et al.*, 2005; Stenroos *et al.*, 2006; Myllys *et al.*, 2007; Otálora *et al.*, 2010). The other molecular markers include PCR fingerprinting of STRR sequences (Zheng *et al.*, 2002; Guevara *et al.*, 2002; Costa *et al.*, 2004) STRR and LTRR sequences (Rasmussen and Svenning, 1998; Nilsson *et al.*, 2000) and RFLP patterns of *nifK* and *glnA* (Lindblad *et al.*, 1989) and *nifH* and *glnA* genes (Zimmerman and Bergman, 1990). Either a high degree of specificity or a great diversity in the strains of *Nostoc* has been reported. The study of Costa *et al.* (2002) consisted of 54 symbiotic strains of *Nostoc* derived from different hosts such as *Peltigera* (18 species), *Nephroma* (7 species), *Blasia* (6 species) *A. fusiformis* (4 species) and few cycads (*C. circinalis*, *C. rumphii*, *E. lebomboensis* and *Z. pumila*). They compared the stem-loop (P6b) of $tRNA^{Leu}$ (UAA) intron sequences that possesses degenerate heptapeptide repeats. According to them there is a high degree of similarity and the *Nostoc* strains shared high degree of conserved intron sequence. Oksanen *et al.* (2004) questioned the validity of $tRNA^{Leu}$ (UAA) intron sequence comparisons for deriving the phylogeny of *Nostoc* strains. According to few workers the locus of *rbcLXS* is quite suitable and reliable molecular marker that can provide the degree of variation needed to unravel the specificity of *Nostoc* strains. Moreover, the results from *rbcLXS* locus are very much comparable to the multilocus sequence typing approach (O'Brien *et al.*, 2005; Stenroos *et al.*, 2006; Myllys *et al.*, 2007; Otálora *et al.*, 2010) but which is not the case with 16S rDNA sequences (Costa *et al.*, 2002; Oksanen *et al.*, 2004; Rikkinen *et al.*, 2004; Stenroos *et al.*, 2006). The observations of Rikkinen *et al.* (2002) revealed (i) the presence of *Nostoc* strains that are specific to the species of *Peltigera* ("Peltigera guild") and *Nephroma* ("*Nephroma* guild") corresponding to clade I and clade II, respectively; and (ii) the cyanobiont selection very much depended on a community scale depending on the habitat. The existence of the two clades, clade I and clade II has further been substantiated by other workers (Lohtander *et al.*, 2003; Rikkinen *et al.*, 2003, 2004; Oksanen *et al.*, 2004; Stenroos *et al.*, 2006; Myllys *et al.*, 2007). Though clade I consisted of a homogeneous sequences of *Nostoc* strains from terricolous lichens, the clade II has been found be heterogeneous with sequences of *Nostoc* strains from epiphytic lichens as well as cyanobionts collected from terrestrial and lithophytic habitats (Stenroos *et al.*, 2006).

Papaefthimiou *et al.* (2008) conducted a preliminary clustering of *Nostoc* strains (cyanobionts from bryophytes, cycads and *Gunnera* spp.) with the help of amplified rDNA restriction analysis (ARDRA) and the phylogeny has been reconstructed on the basis of 16S rRNA gene sequences coupled with morphological characterization. They also confirmed the existence of all cyanobionts from the above hosts in two clades. Together with these several free-living *Nostoc* strains of the species of *N. muscorum*, *N. calcicola*, *N. edaphium*, *N. ellipsosporum* and strains related to *N. commune* are clustered together. A phylogenetic study conducted by Otálora *et al.* (2010), based on *rbcLXS* sequences sampled from 79 lichen thalli of Collemataceae with 163 *Nostoc* sequences from GenBank, recognized two major clades. Clade I included both free-living and symbiotic *Nostoc* strains. Clade II comprised of both free-living and symbiotic strains. However, clade II could be further resolved into three sub-clades. In sub-clade I besides free-living *Nostoc* strains, cyanobionts from North American (*Peltigera dactyla*), South American (*Leptogium azureum*, *L. cyanescens*) and a European (*Fuscopannaria leucophaea*) lichen species are clustered together. Cyanobionts from other lichen thalli (*Leptogium lichenoides*, *Sticta hypochroa*, *S. gaudichaldia*, *Peltigera malacea* and *Protopannaria pezizoides*) are present in the sub-clade II. Large majority of the symbiotic *Nostoc* strains (from *G. pyriforme*, vast majority of lichens including members of Collemataceae and *C. circinalis*, *G. manicata* and *Stangeria paradoxa*) and some free-living ones are clustered together. In this respect, the phylogenetic relationships drawn by others (Wirtz *et al.*, 2003; O'Brien *et al.*, 2005; Stenroos *et al.*, 2006; Myllys *et al.*, 2007) have been confirmed by them.

LITERATURE CITED

Adams, D. G. (2000) Symbiotic interactions. In: Whitton, B. A., and Potts, M. (Eds.) *Ecology of Cyanobacteria: Their Diversity in Time and Space*. Kluwer Academic, Dordrecht, The Netherlands. pp. 523–561.

Adams, D. G., and Duggan, P. S. (2008) Cyanobacteria-bryophyte symbioses. *J Exp Bot* **59:** 1047–1058. doi:10.1093/jxb/em.005.

Ahmadjian, V. (1967) A guide to the algae occurring as lichen symbionts: Isolation, culture, cultural physiology, and identification. *Phycologia* **6:** 127–160.

Ahmadjian, V. (1987) Coevolution in lichens. *Ann New York Acad Sci* **503:** 307–315.

Ahmadjian, V. (1989) Studies on the isolation and synthesis of bionts of the cyanolichen *Peltigera canina* (Peltigeraceae). *Plant Syst Evol* **165:** 29–38.

Ahmadjian, V. (1993) *The Lichen Symbiosis*. John Wiley & Sons, New York, NY, USA.

Alberte, R. S., Cheng, L., and Lewin, R. A. (1987) Characteristics of *Prochloron*-ascidian symbioses. 2. Photosynthesis irradiance relationships and carbon balance of association from Palau Micronesia. *Symbiosis* **4:** 147–170.

Albrecht, C., Geurts, R., and Bisseling, T. (1999) Legume nodulation and mycorrhizae formation: Two extremes in host specificity meet. *EMBO J* **18:** 281–288.

Antoine, M. E. (2004) An ecophysiological approach for quantifying nitrogen fixation by *Lobaria oregana*. *Bryologist* **107:** 82–87.

Armaleo, D., and Clerc, P. (1991) Lichen Chimeras: DNA analysis suggests that one fungus forms two morphotypes. *Exptl Mycol* **15:** 1–10.

Arntz, C., and Tudzynski, P. (1997) Indentification of genes induced in alkaloid-producing cultures of *Claviceps* sp. *Curr Genet* **31:** 357–360.

Arsenault, A., and Goward, T. (2000) The drip zone effect: New insights into the distribution of rare lichens. In: Darling, L. M. (Ed.) *Proc Biology and Management of Species and Habitats at Risk* **2:** 767–768. Ministry of Environment, Lands and Parks, Victoria, B. C. and University College of the Cariboo, Kamloops, B. C.

Arve, E., Dimitra, P., Eli Helene, R., and Anton, L. (2008) Phylogenetic patterns among *Nostoc* cyanobionts within bi- and tripartite lichens of the genus *Pannaria*. *J Phycol* **44:** 1049–1059.

Bachereau, F., and Asta, J. (1997) Effects of solar ultraviolet radiation at high altitude on the physiology and the biochemistry of a terricolous lichen (*Cetraria islandica* [L] Ach.). *Symbiosis* **23:** 197–217.

Baker, A. C. (2003) Flexibility and specificity in coral-algal symbiosis: Diversity, ecology and biogeography of *Symbiodinium*. *Annu Rev Ecol Evol Syst* **34:** 661–689.
Banaszak, A. T., Iglesias-Prieto, R., and Trench, R. K. (1993) *Scrippsiella velellae* sp. nov. (Peridiniales) and *Gloeodinium viscum* sp. nov. (Phytodiniales), dinoflagellate symbionts of two hydrozoans (Cnidaria). *J Phycol* **29:** 517–528.
Basilier, K., Granhall, U., and Stenström, T.-A. (1978) Nitrogen fixation in wet minerotrophic moss communities of a sub-arctic mire. *Oikos* **31:** 236–246.
Batham, E. J. (1943) Vascular anatomy of New Zealand species of *Gunnera*. *Trans R Soc New Zealand* **73:** 209–216.
Baulina, O. I., and Lobakova, E. S. (2003) A typical cell forms overproducing extracellular substances in populations of cycad cyanobionts. *J Microbiol* **72:** 701–712.
Beck, A., Friedl, T., and Rambold, G. (1998) Selectivity of photobiont choice in a defined lichen community: Inferences from cultural and molecular studies. *New Phytol* **139:** 709–720.
Beck, A., Kasalicky, T., and Rambold, G. (2002) Myco-photobiontal selection in a Mediterranean cryptogam community with *Fulgensia fulgida*. *New Phytol* **153:** 317–326.
Becker, V. E. (1980) Nitrogen fixing lichens in forests of the Southern Appalachian Mountains of North Carolina. *Bryologist* **83:** 29–39.
Becker, V. E., Reeder, J. E., and Stetler, R. (1977) Biomass and habitat of nitrogen-fixing lichens in an oak forest in the North Carolina Piedmont. *Bryologist* **80:** 93–99.
Becking, J. H. (1987) Endophyte transmission and activity in the *Anabaena-Azolla* association. *Plant Soil* **100:** 183–212.
Beer, S., and Ilan, M. (1998) *In situ* measurements of photosynthetic irradiance responses of two Red Sea sponges growing under dim light conditions. *Mar Biol* **131:** 613–617.
Belfort, M., Reaban, M. E., Coetzee, T., and Dalgaard, J. Z. (1995) Prokaryotic introns and inteins: A panoply of form and function. *J Bacteriol* **177:** 3897–3903.
Bergman, B., Johansson, C., and Söderback, E. (1992a) The *Nostoc-Gunnera* symbiosis. *New Phytol* **122:** 379–400.
Bergman, B., Johansson, C., and Söderback, E. (1992b) Cyanobacterial-Plant symbioses. *Symbiosis* **14:** 61–81.
Berryman, S., and McCune, B. (2006) Epiphytic lichens along gradients in topography and stand structure in western Oregon, USA. *Pacific Northwest Fungi* **1:** 1–38.
Bewley, C. A., Holland, N. D., and Faulkner, D. J. (1996) Two classes of metabolites from *Theonella swinhoei* are localized in distinct populations of bacterial symbionts. *Experientia* **52:** 716–722.
Black, K., and Osborne, B. (2004) An assessment of photosynthetic downregulation in cyanobacteria from *Gunnera-Nostoc* symbiosis. *New Phytol* **162:** 125–132.
Black, K. G., Parsons, R., and Osborne, B. A. (2002) Uptake and metabolism of glucose in the *Nostoc-Gunnera* symbiosis. *New Phytol* **153:** 297–305.
Bonnett, H. T., and Silvester, W. B. (1981) Specificity in the *Gunnera-Nostoc* endosymbiosis. *New Phytol* **89:** 121–128.
Brehm, U., Krumbein, W. E., and Palinska, K. A. (2003) Microbial spheres: A novel cyanobacterial-diatom symbiosis. *Naturwissenschaften* **90:** 136–140.
Brenner, E. D., Stevenson, D. W., and Twigg, R. W. (2003) Cycads: evolutionary innovations and the role of plant-derived toxins. *Trends Plant Sci* **8:** 1360–1385.
Brewin, N. J. (2004) Plant cell wall remodeling in the *Rhizobium*-legume symbiosis. *Critical Rev Plant Sci* **23:** 293–316.
Brodo, I. M., and Richardson, D. H. S. (1979) Chimeroid associations in the genus *Peltigera*. *Lichenologist* **10:** 157–170.
Brusca, R. C., and Brusca, G. J. (1990) Phylum Porifera: The sponges. In: Sinauer, A. D. (Ed.) *Invertebrates*, Sinauer Press, Sunderland, Mass, USA pp. 181–210.
Bubrick, P., and Galun, M. (1980) Proteins from the lichen *Xanthoria parietina* which bind to phycobiont cell walls. Correlation between binding patterns and cell wall chemistry. *Protoplasma* **104:** 167–173.
Bubrick, P., Frensdorff, A., and Galun, M. (1985) Proteins from the lichen *Xanthoria parietina* which bind to phycobiont cell wall. Isolation and partial purification of an algal-binding protein. *Symbiosis* **1:** 85–95.
Bubrick P., Galun, M., and Fresnsdorff, A. (1981) Proteins from the lichen *Xanthoria parietina* which bind to phycobiont cell wall. Localization in the intact lichen and cultured mycobiont. *Protoplasma* **105:** 207–211.
Buck, K., and Bentham, W. N. (1998) A novel symbiosis between a cyanobacterium, *Synechococcus* sp., an aplastidic protist, *Solenicola setigera*, and a diatom, *Leptocylindrus mediterraneus*, in the open ocean. *Mar Biol* **132:** 349–355.
Büdel, B. (1992) Taxonomy of lichenized procaryotic blue-green algae. In: Reisser, W. (Ed.) *Algae and Symbioses*. Biopress, Bristol, UK. pp. 301–324.
Büdel, B. (1999) Ecology and diversity of rock-inhabiting cyanobacteria in tropical regions. *Eur J Phycol* **34:** 361–370.
Büdel, B., and Henssen, A. (1983) *Chroococcidiopsis* (Cyanophyceae), a phycobiont in the lichen family Lichinaceae. *Phycologia* **22:** 367–375.
Calvert, H. E., Perkins, S. K., and Peters, G. A. (1983) Sporocarp structure in the heterosporous water fern *Azolla mexicana* Presl. *Scanning Electron Microscopy* **3:** 1499–1510.

Calvert, H. E., Pence, M. K., and Peters, G. A. (1985) Ultrastructural ontogeny of leaf cavity trichomes in *Azolla* implies a functional role in metabolite exchange. *Protoplasma* **129:** 10–27.

Calvert, H. E., Perkins, S. K., and Peters, G. A. (1985) Involvement of epidermal trichomes in the continuity of the *Azolla-Anabaena* symbiosis through the *Azolla* life cycle. *Am J Bot* **72:** 808 (Abstr.).

Cameron, R. P., and Richardson, D. H. S. (2006) Occurrence and abundance of epiphytic cyanolichens in protected ares of Nova Scotia, Canada. *Opuscula Philolichemnum* **3:** 5–14.

Campbell, D. H. (1893) On the development of *Azolla filiculoides* Lam. *Ann Bot* **7:** 155–187.

Campbell, E. L., and Meeks, J. C. (1989) Characteristics of hormogonia formation by symbiotic *Nostoc* spp. in response to the presence of *Anthoceros punctatus* or its extracellular products. *Appl Environ Microbiol* **55:** 125–131.

Campbell, E. L., and Meeks, J. C. (1992) Evidence for plant-mediated regulation of nitrogenase expression in the *Anthoceros-Nostoc* symbiotic association. *J Gen Microbiol* **138:** 473–480.

Campbell, E. L., Brahamsha, B., and Meeks, J. C. (1998) Mutation of an alternative sigma factor in the cyanobacterium *Nostoc punctiforme* results in increased infection of its symbiotic plant partner, *Anthoceros punctatus*. *J Bacteriol* **180:** 4938–4941.

Campbell, E. L., Wong, F. C. Y., and Meeks, J. C. (2003) DNA binding properties of the HrmR protein of *Nostoc punctiforme* responsible for transcriptional regulation of genes involved in the differentiation of hormogonia. *Mol Microbiol* **47:** 573–582.

Campbell, E. L., Summers, M. L., Christman, H., Martin, M. E., and Meeks, J. C. (2007) Golbal gene expression patterns of *Nostoc punctiforme* in steady-state dinitrogen-grown heterocyst-containing cultures and at single time points during the differentiation of akinetes and hormogonia. *J Bacteriol* **189:** 5247–5256.

Campbell, J., and Fredeen, A. L. (2004) *Lobaria pulmonaria* abundance as an indicator of macrolichen diversity in Interior Cedar-Hemlock forests of east-central British Columbia. *Can J Bot* **82:** 970–982.

Capone, D. G. (2001) Marine nitrogen fixation: What's the fuss? Commentary. *Curr Opinion Microbiol* **4:** 341–348.

Carpenter, C. E., Mueller, R. J., Kazmierczak, P., Zhang, L., Villalon, D. K., and Van Alfen, N. K. (1992) Effect of a virus on accumulation of a tissue-specific cell surface protein of the fungus *Cryphonectria* (*Endothia*) *parasitica*. *Mol Plant-Microbe Interact* **4:** 55–61.

Carpenter, E. J. (2002) Marine cyanobacterial symbiosis. *Proc R Irish Acad* **102**B: 15–18.

Carpenter, E. J., and Janson, S. (2000) Intracellular cyanobacterial symbionts in the marine diatom *Climacodium frauenfeldianum* Grunow. *J Phycol* **36:** 540–544.

Carpenter, E. J., Montoya, J. P., Burns, J., Mulholland, M., Subramaniam, A., and Capone, D. G. (1999) Extensive bloom of a N_2 fixing symbiotic association in the tropical Atlantic Ocean. *Mar Ecol Prog Ser* **188:** 273–283.

Casselman, K. L., and Hill, M. J. (1995) Lichens as a monitoring tool: A Pictou County (Nova Scotia) perspective. In: Herman, T. B., Bondrup-Nielson, S., Willison, J. H. M., and Munro, N. W. P. (Eds.) *Ecosystem Monitoring in and Protected Areas*. Proc II Internl Conf *Science and the Management of Protected Areas*. Dalhousie University, Halifax, Nova Scotia, Canada. pp. 237–244.

Chapman, K. E., Duggan, P. S., Billington, N. A., and Adams, D. G. (2008) Mutation of different sites in the *Nostoc punctiforme cyaC* gene, cncoding the multidomain enzyme adenyl cyclase, results in different levels of infection of the host plant *Blasia pusilla*. *J Bacteriol* **190:** 1843–1847.

Chaudhuri, H., and Akhtar, A. R. (1931) The coral-like roots of *Cycas revoluta*, *Cycas circinalis* and *Zamia floridana* and the alga inhabiting them. *J Indian Bot Soc* **10:** 43–59.

Cheshire, A. C., Wilkinson, C. R., Seddon, S., and Wetsphalen, G. (1997) Bathymetric and seasonal changes in photosynthesis and respiration of the phototrophic sponge *Phyllospongia lamellosa* in comparison with respiration by the heterotrophic sponge *Ianthella basta* on Davies Reef, Great Barrier Reef. *Mar Freshwater Res* **48:** 589–599.

Chiu, W.-L., Peters, G. A., Levieille, G., Still, P. C., Cousins, S., Osborne, B., and Elhai, J. (2005) Nitrogen deprivation stimulates symbiotic gland development in *Gunnera manicata*. *Plant Physiol* **139:** 224–230.

Clarke, A. K., Soitamo, A., Gustafsson, P., and Oquist, G. (1993) Rapid interchange between two distinct forms of cyanobacterial photosystem II reaction center protein D1 in response to photoinhibition. *Proc Natl Acad Sci USA* **90:** 9973–9977.

Cohen, M. F., and Meeks, J. C. (1997) A hormogonium regulating locus, hrmUA, of the cyanobacterium *Nostoc punctiforme* strain ATCC 29133 and its response to an extract of a symbiotic plant partner *Anthoceros punctatus*. *Mol Plant-Microbe Interact* **10:** 280–289.

Cohen, M. F., and Yamasaki, H. (2000) Flavonoid-induced expression of a symbiosis-related gene in the cyanobacterium *Nostoc punctiforme*. *J Bacteriol* **182:** 4644–4646.

Cohen, M. F., Sakihama, Y., Takagi, Y. C., Ichiba, T., and Yamasaki, H. (2002) Synergistic effect of deoxyanthocyanins from symbiotic fern *Azolla* spp. on *hrmA* gene induction in the cyanobacterium *Nostoc punctiforme*. *Mol Plant-Microbe Interact* **15:** 875–882.

Cohen, M. F., Wallis, J. G., Campbell, E. L., and Meeks, J. C. (1994) Transposon mutagenesis of *Nostoc* sp. strain 29133, a filamentous cyanobacterium with multiple cellular differentiation alternatives. *Microbiology* **140:** 3233–3240.

Costa, J.-L., Lindblad, P., and Martinez Romero, E. (2004) Sequence based data supports a single *Nostoc* strain in individual coralloid roots of cycads. *FEMS Microbiol Ecol* **49:** 481–487.
Costa, J.-L., Paulsrud, P., and Lindblad, P. (2002) The cyanobacterial tRNA^{Leu} (UAA) intron: Evolutionary patterns in a genetic marker. *Mol Biol Evol* **19:** 850–857.
Costa, J.-L., Paulsrud, P., Rikkinen, J., and Lindblad, P. (2001) Genetic diversity of *Nostoc* symbionts endophytically associated with two bryophyte species. *Appl Environ Microbiol* **67:** 4393–4396.
Cren, M., and Hirel, B. (1999) Glutamine synthetase in higher plants: Regulation of gene and protein expression from the organ to the cell. *Plant Cell Physiol* **40:** 1187–1193.
Czeika, H., Czeika, G., Guttová, A., Farkas, E., Lõkös, L., and Halda, J. (2004) Phytogeographic and taxonomic remarks on eleven species of cyanophilic lichens from central Europe. *Preslia* **76:** 183–192.
de Bary, A. (1879) *Die Erscheinung der Symbiose*. Strasburg, Germany.
DeLuca, T. H., Zackrisson, O., Nilsson, M. C., and Sellstedt, A. (2002) Quantifying nitrogen-fixation in feather moss carpets of boreal forests. *Nature* (London) **419:** 917–920.
Delwiche, C. F., and Palmer, J. D. (1997) The origin of plastids and their spread via secondary symbiosis. *Plant Syst Evol* Suppl **11:** 53–86.
Demming-Adams, B. (1990) Carotenoids and photoprotection in plants: a role for the xanthophyll zeaxanthin. *Biochim Biophys Acta* **1020:** 1–24.
Denison, R., Caldwell, B., Bormann, B., Eldred, L., Swanberg, C., and Anderson, S. (1977) The effects of acid rain on nitrogen fixation in western Washington coniferous forests. *Water Air and Soil Pollution* **8:** 21–34.
DePriest, P. T., and Been, M. D. (1992) Numerous group I introns with variable distributions in the ribosomal DNA of a lichen fungus. *J Mol Biol* **228:** 315–321.
de Vocht, M. L., Reviakine, I., Wösten, H. A. B., Brisson, A. D. R., Wessels, J. G. H., and Robillard, G. T. (2000) Structural and functional role of the disulfide bridges in the hydrophobin SC3. *J Biol Chem* **275:** 28428–28432.
de Vries, O. M. H., Fekkes, M. P., Wösten, H. A. B., and Wessels, J. G. H. (1993) Insoluble hydrophobin complexes in the walls of *Schizophyllum commune* and other filamentous fungi. *Arch Microbiol* **159:** 330–335.
Díaz, E.-M., Sacristán, M., Legaz, M.-E., and Vicente, C. (2009) Isolation and characterization of a cyanobacterium-binding protein and its cell wall receptor in the lichen *Peltigera canina*. *Plant Signal Behav* **4:** 598–603.
Díaz, E.-M., Vicente-Manzanares, M., Sacristán, M., Vicente, C., and Legaz, M.-E. (2011) Fungal lectin of *Peltigera canina* induces chemotropism of compatible *Nostoc* cells by constriction-relaxation pulses of cyanobiont cytoskeleton. *Plant Signal Behav* **6:** 1525–1536.
Dick, H., and Stewart, W. D. P. (1980) The occurrence of fimbriae on a N_2-fixing cyanobacterium which occurs in lichen symbioses. *Arch Microbiol* **124:** 107–109.
Dore, J. E., Letelier, R. M., Church, M. J., and Karl, D. M. (2008) Summer phytoplankton blooms in the oligotrophic North Pacific subtropical gyre: Historical perspective and recent observations. *Prog Oceanogr* **76:** 2–38. doi:10.1016/j.pocean.2007.10.002
Drum, R. W., and Pankratz, S. (1965) Fine structure of an unusual cytoplasmic inclusion in the diatom genus *Rhopalodia*. *Protoplasma* **60:** 141–149.
Duckett, J. G., and Renzaglia, K. S. (1993) The reproductive biology of the liverwort *Blasia pusilla* L. *J Bryol* **17:** 541–552.
Duckett, J. G., Prasad, A. K. S. K., Davies, D. A., and Walker, S. (1977) A cytological analysis of the *Nostoc*-bryophyte relationship. *New Phytol* **79:** 349–362.
Duckett, J. G., Toth, R., and Soni, S. L. (1975) An ultrastructural study of the *Azolla, Anabaena azollae* relationship. *New Phytol* **75:** 111–118.
Duff, R. J., Cargill, D. C., Villarreal, J. C., and Renzaglia, K. S. (2004) Phylogenetic relationships of the hornworts based on *rbcL* sequence data: Novel relationships and new insights. In: Goffinet, B., Hollowell, V., and Magill, R. (Eds.) *Molecular Systematics of Bryophytes. Monographs in Systematic Botany from the Missouri Botanical Garden* **98:** 41–58.
Duggan, P. S., Gottardello, P., and Adams, D. G. (2007) Molecular analysis of genes in *Nostoc punctiforme* involved in pilus biogenesis and plant infection. *J Bacteriol* **189:** 4547–4551.
Duglas, A. E. (1994) *Symbiotic Interactions*. Oxford University Press, Oxford, UK. pp. 148.
Dujon B. (1989) Group I introns as mobile genetic elements: Facts and mechanistic speculations—A review. *Gene* **82:** 91–114.
Dunham, D. G., and Fowler, K. (1987) Megaspore germination, embryo development and maintenance of the symbiotic association in *Azolla filiculoides* Lam. *Bot J Linnean Soc* **95:** 43–53.
Dyer, P. S. (2002) Hydrophobins in the lichen symbiosis. *New Phytol* **154:** 1–4.
Edwards, T. C., Jr., Cutler, R., Geiser, L., Alegria, J., and McKenzie, D. (2004) Assessing rarity and seral stage association of species with low detectability: Lichens in western Oregon and Washington forests. *Ecol Applications* **14:** 414–424.
Egger, K. N. (1995) Molecular analysis of ectomycorrhizal fungal communities. *Can J Bot* **73** (Suppl): S1415–S1422.

Ekman, M., Tollback, P., Klintt, J., and Bergman, B. (2006) Protein expression profiles in an endosymbiotic cyanobacterium revealed by a proteomic approach. *Mol Plant-Microbe Interact* **19:** 1251–1261.
Elifio, S. L., Da Silva, M., De Laourdes, C. C., Iacomini, M., and Gorin, P. A. J. (2000) A lectin from lichenized basidiomycete *Dictyonema glabratum*. *New Phytol* **148:** 327–334.
Ellyson, W. J., and Sillett, S. C. (2003) Epiphyte communities on Sitka Spruce in an old-growth redwood forest. *Bryologist* **106:** 197–211.
Enderlin, C. S., and Meeks, J. C. (1983) Pure culture and reconstitution of the *Anabaena-Azolla* symbiotic association. *Planta* **158:** 157–165.
Engel, S., and Pawlik, J. R. (2000) Allelopathic activities of sponge extracts. *Mar Ecol Prog Ser* **207:** 273–281.
Farmer, A.M., Bates, J. F., and Bell, J. N. B. (1992) Ecophysiological effects of acid rain on bryophytes and lichens. In: Bates, J. W., and Farmer, A. M. (Eds.) *Bryophytes and Lichens in a Changing Environment*, Chapter 11. Oxford University Press, Oxford, UK. pp. 284–313.
Faulkner, D. J., Unson, M. D., and Bewley, C. A. (1994) The chemistry of some sponges and their symbionts. *Pure Appl Chem* **66:** 1883–1890.
Feldman, J. (1933) Sur quelques cyanphycées vivant dans le tissu des éponges de banyules. *Arch Zool Exp Gén* **75:** 331–404.
Feoktistov, A. S., Kitashov, A. V., and Lobakova, E. S. (2009) The characterization of lectins from the tripartite lichen *Peltigera aphthosa* (L.) Willd. *Moscow Univ Biol Sci Bull* **64:** 23–27.
Fiedler, G., Muro-Pastor, A., Flores, E., and Maldener, I. (2001) Ntc-dependent expression of the *devBCA* operon, encoding a heterocyst-specific ATP-binding cassette transporter in *Anabaena* spp. *J Bacteriol* **183:** 3795–3799.
Fisher, R., and Long. S. R. (1992) Rhizobium-plant signal exchange. *Nature* (London) **357:** 655–660.
Floener, L., and Bothe, H. (1980) Nitrogen fixation in *Rhopalodia gibba*, a diatom containing blue-green inclusions symbiotically. In: Schwemmler, W., and Schenk, H. E. A. (Eds.) *Endocytobiology, Endosymbiosis and Cell Biology*. Walter de Gruyter & Co., Berlin, Germany. pp. 541–552.
Flowers, A. E., Garson, M. J., Webb, R. I., Dumdei, E. J., and Charan, R. D. (1998) Cellular origin of chlorinated diketopiperazines in the dictyoceratid sponge *Dysidea herbacea* (Keller). *Cell Tissue Res* **292:** 597–607.
Fong, A. A., Karl, D., Lukas, R., Letelier, R. M., Zehr, J. P., Church, M. J. (2008) Nitrogen fixation in an anticyclonic eddy in the oligotrophic North Pacific Ocean. *ISME J* **2:** 663–676.
Forman, R. T. T. (1975) Canopy lichens with blue-green algae: A nitrogen source in a Columbian rain forest. *Ecology* **56:** 1176–1184.
Foster, R. A., and Zehr, J. P. (2006) Characterization of diatom-cyanobacteria symbioses on the basis of *nifH*, *hetR*, and 16S rRNA sequences. *Environ Microbiol* **8:** 1913–1925.
Foster, R. A., Kuypers, M. M. M., Vagner, T., Paerl, R. W., Musat, N., and Zehr, J. P. (2011) Nitrogen fixation and transfer in the open ocean diatom cyanobacterial symbioses. *ISME J* **5:** 1484–1493. doi:10.1038/ismej.204.26
Foster, R. A., Subramaniam, A., and Zehr, J. P. (2009) Distribution and activity of diazotrophs in the Eastern Equatorial Atlantic. *Environ Microbiol* **11:** 741–750.
Foster, R. A., Subramaniam, A., Mahaffey, C., Carpenter, E. J., Capone, D. G., and Zehr, J. P. (2007) Influence of the Amazon River plume on distributions of free-living and symbiotic cyanobacteria in the western tropical north Atlantic Ocean. *Limnol Oceanogr* **52:** 517–532.
Fox, G. E., Wisotzkey, J. D., and Jurtshuk, P. Jr. (1992) How close is close: 16S rRNA sequence identity may not be sufficient to guarantee species identity. *Int J Syst Bacteriol* **42:** 166–170.
Franche, C., and Cohen-Bazire, G. (1985) The structural *nif* genes of four symbiotic *Anabaena azollae* show a highly conserved physical arrangement. *Plant Sci* **39:** 125–131.
Franche, C., and Cohen-Bazire, G. (1987) Evolutionary divergence in the *nif* K, D, H genes region among nine symbiotic *Anabaena azollae* and between *Anabaena azollae* and some free-living heterocystous cyanobacteria. *Symbiosis* **3:** 159–178.
Frank, A. B. (1877) Ueber die biologischen Verhältnisse des Thallus einiger Krustenflechten. *Beitr Biol Pflanzen* **2:** 123–200.
Friedl, T., and Büdel, B. (1996) Photobionts. In: Nash III, T.H., (Ed.) *Lichen Biology*. Cambridge University Press, Cambridge, UK. pp. 217–239.
Frost, T. M., and Williamson, C. E. (1980) *In situ* determination of the effect of symbiotic algae on the growth of freshwater sponge *Spongilla lacustris*. *Ecology* **61:** 1361–1370.
Frugier, F., Kosuta, S., Murray, J., Crespi, M., and Szczyglowszki, K. (2008) Cytokinin: A secret agent of symbiosis. *Trends Plant Sci* **13:** 115–120.
Gaino, E., Sciscioli, M., Lepore, E., Rebora, M., and Corriero, G. (2006) Association of the sponge *Tethya orphei* (Porifera, Desmospongiae) with filamentous cyanobacteria. *Invertebrate Biol* **125:** 281–287.
Galloway, D. J. (1988) Nomenclatural notes on *Pseudocyphellaria* VI. Two endemic Australian taxa. *Lichenologist* **29:** 599–601.

Galun, M., and Bubrik, P. (1984) Physiological interactions between the partners of the lichen symbiosis. In: Linskens, H. F., and Heslop-Harrison, J. (Eds.) *Encyclopedia of Plant Physiology: Cellular Interactions*. Springer, Berlin, Germany. pp. 362–401.

Garcia-Pichel, F., and Castenholz, R. W. (1993) Occurrence of UV-absorbing, mycosporine-like compounds among cyanobacterial isolates and an estimate of their screening capacity. *Appl Environ Microbiol* **59:** 163–169.

Garson, M. J. (2001) Ecological perspectives on marine natural product biosynthesis. In: McClintock, J. B., and Baker, B. J. (Eds.) *Marine Chemical Ecology*. CRC Press, Boca Raton, Florida, FL, USA. pp. 71–114.

Gates, J. E., Fisher, R. W., Goggin, T. W., and Azrolan, N. I. (1980) Antigenic differences between *Anabaena azollae* fresh from the *Azolla* fern leaf cavity and the free-living cyanobacteria. *Arch Microbiol* **128:** 126–129.

Gauslaa, Y. (1995) *Lobarion*, an epiphytic community of ancient forests, threatened by acid rain. *Lichenologist* **27:** 59–76.

Gauslaa, Y., and Solhaug, K. A. (1996) Differences in the susceptibility to light stress between epiphytic lichens of ancient and young boreal forest stands. *Functional Ecology* (British Ecological Society) **10:** 344–354. doi:10.2307/2390282

Gebhardt, J. S., and Nierzwicki-Bauer, S. A. (1991) Identification of a common cyanobacterial symbiont associated with *Azolla* spp. through molecular and morphological characterization of free-living and symbiotic cyanobacteria. *Appl Environ Microbiol* **57:** 2141–2146.

Gehrig, H., Schüßler, A., and Kluge, M. (1996) *Geosiphon pyriformis*, a fungus forming endocytobiosis with *Nostoc* (cyanobacteria), is an ancestral member of the Glomales: Evidence by SSU rRNA analysis. *J Mol Evol* **43:** 71–81.

Gehringer, M. M., Pengelly, J. J. L., Cuddy, W. S., Fieker, C., Forster, P. I., and Neilan, B. A. (2010) Host selection of symbiotic cyanobacteria in 31 species of the Australian cycad genus: *Macrozamia* (Zamiaceae). *Mol Plant-Microbe Interact* **23:** 811–822. doi: 10. 1094/MPMI-23-6-0811

Geitler, L. (1977) Zur Entwicklungsgeschichte der Epithemiaceen *Epithemia*, *Rhopalodia* and *Denticula* (Diatomophyceae) und ihre vermulich symbintischen Sphäroidkörper. *Plant Syst Evol* **128:** 265–275.

Gentili, F., Nilsson, M.-C., Zackrisson, O., DeLuca, T. H., and Sellstedt, A. (2005) Physiological and molecular diversity of feather moss associative N_2-fixing cyanobacteria. *J Exp Bot* **56:** 3121–3127.

Godoy, R., Oyarún, C. E., and Gerding, V. (2001) Precipitation chemistry in deciduous and evergreen Nothofagus forests of southern Chile under low-deposition climate. *Basic Appl Ecol* **2:** 65–72.

Goffinet, B., and Bayer, R. J. (1997) Characterization of mycobionts of photomorph pairs in the Peltigerineae (lichenized ascomycetes) based on internal transcribed spacer sequences of the nuclear ribosomal DNA. *Fungal Genet Biol* **21:** 228–237.

Goffinet, B., and Hastings, R. I. (1995) Two new sorediate taxa of *Peltigera*. *Lichenologist* **27:** 43–58.

Gómez, R., Erpenbeck, D., van Dijk, T., Richelle-Maurer, E., Devijver, C., Brackman, J. C., Woldringh, C., and van Soest, R. W. M. (2002) Identity of cyanobacterial symbiont of *Xestospongia muta*. *Boll Mus Ist Biol Univ Genova* **66–67:** 82–83.

Gómez, F., Furuya, K., and Takeda, S. (2005) Distribution of the cyanobacterium *Richelia intracellularis* as an epiphyte of the diatom *Chaetoceros compressus* in the western Pacific Ocean. *J Plankton Res* **27:** 323–330.

Goward, T. (1994) Notes on old growth dependent epiphytic macrolichens in inland British Columbia, Canada. *Acta Bot Fennica* **150:** 31–38.

Goward, T., and Arsenault, A. (2000a) Inland old-growth rain forests: Safe havens for rare lichens? In: Darling, L. M. (Ed.) *Proc Biology and Management of Species and Habitats at Risk*, **2:** 759–766. Ministry of Environment, Lands and Parks, Victoria, B. C. and University College of the Cariboo, Kamloops, B. C.

Goward, T., and Arsenault, A. (2000b) Cyanolichen distribution in young unmanaged forests: A dripzone effect? *Bryologist* **103:** 28–37.

Goward, T., Goffinet, B., and Vitikainen, O. (1995) Synopsis of the genus *Peltigera* (lichenized Ascomycetes) in British Columbia, with a key to the North American species. *Can J Bot* **73:** 91–111.

Granhall, U., and Hofsten, A. (1976) Nitrogenase activity in relation to intracellular organisms in *Sphagnum* mosses. *Physiol Plant* **36:** 88–94.

Grilli, M. (1963) Observazioni sulle strutture e le infrastrutture delle cellule dei tubercoli radicali di *Cycas revoluta*. *Annali della Faoltà di Agraria* **3:** 467–481.

Grilli-Caiola, M. (1974) A light and electron microscope study of the blue-green algae living either in the coralloid roots of *Macrozamia communis* or isolated in culture. *Giornale Botanica Italiano* **108:** 161–173.

Grilli-Caiola, M. (1975a) A light and electron microscope study of the blue-green algae living in the coralloid roots of *Encephalartos altensteinii* and in culture. *Phycologia* **14:** 25–33.

Grilli-Caiola, M. (1975b) Structural and ultrastructural aspects of blue-green algae growing in the coralloid roots of *Dioon edule* and in culture. *Phykos* **14:** 29–34.

Grilli-Caiola, M. (1980) On the phycobionts of cycad coralloid roots. *New Phytol* **85:** 537–544.

Grilli-Caiola, M., Forni, C., and Castagnola, M. (1992) *Anabaena azollae* akinetes in the sporocarps of *Azolla filiculoides* Lam. *Symbiosis* **14:** 247–264.

Grobbelaar, N., Scott, W. E., Hattingh, W., and Marshall, J. (1987) The identification of the coralloid endophytes of the southern African cycads and the ability of the isolates to fix dinitrogen. *S Afr J Bot* **53:** 111–118.

Guevara, R., Armesto, J. J., and Caru, M. (2002) Genetic diversity of *Nostoc* microsymbionts from *Gunnera tinctoria* revealed by PCR-STRR fingerprinting. *Microb Ecol* **44:** 127–136.

Hall, J. W., and Swanson N. P. (1968) Studies on fossil *Azolla*: *Azolla montana*, a Cretaceous megaspore with many small floats. *Am J Bot* **55:** 1055–1061.

Hallbauer, D. K., Jahns, H. M., and Beltmann, H. A. (1977) Morphological and anatomical observations on some Precambrian plants from the Witwaterstand, South Africa. *Geol Rundschau* **66:** 477–491.

Harder, R. (1917) Ernahrungsphysiologische Untersuchungen an Cyanophyceen, hauptsächlich dem endophytischen *Nostoc punctiforme*. *Zeitschrift für Botanik* **9:** 143–242.

Hedenås, H., and Ericsona, L. (2003) Response of epiphytic lichens on *Populus tremula* in a selective cutting experiment. *Ecological Applications* **13:** 1124–1134.

Heinbokel, J. F. (1986) Occurrence of *Richelia intracellularis* (Cyanophyta) within diatoms *Hemiaulus hauckii* and *H. membranaceus* off Hawaii. *J Phycol* **22:** 399–403.

Hentschel, U., Hopke, J., Horn, M., Friedrich, A. B., Wagner, M., Hackr, J., and Moore, B. S. (2002) Molecular evidence for a uniform microbial community in sponges from different oceans. *Appl Environ Microbiol* **68:** 4431–4440.

Hentschel, U., Schmid, M., Wagner, M., Fieseler, L., Gernert, C., and Hacker, J. (2001) Isolation and phylogenetic analysis of bacteria with antimicrobial activities from the Mediterranean sponges, *Aplysina aerophoba* and *Aplysina cavernicola*. *FEMS Microbiol Ecol* **35:** 305–312.

Herrera-Campos, M. A., Huhndorf, S., and Lücking, R. (2005) The folicolous lichen flora of Mexico IV: A new, folicolous species of *Pyrenothrix* (Chaetothyriales: Pyrenotrichaceae). *Mycologia* **97:** 356–361.

Herrero, A., Muro-Pastor, A. M., and Flores, E. (2001) Nitrogen control in cyanobacteria. *J Bacteriol* **183:** 411–425.

Hill, D. J. (1975) The pattern of development of *Anabaena* in the *Azolla-Anabaena* symbiosis. *Planta* **122:** 179–184.

Hill, D. J. (1977) The role of *Anabaena* in the *Azolla-Anabaena* symbiosis. *New Phytol* **78:** 611–616.

Hill, D. J. (1994) The nature of symbiotic relationship in lichens. *Endeavour* **18:** 96–103.

Hillis, D. M., and Dixon, M. T. (1991) Ribosomal DNA: Molecular evolution and phylogenetic inference. *Quart Rev Biol* **66:** 411–453.

Hill, M. S. (1996) Symbiotic zooxanthellae enhance boring and growth rates of the tropical sponge. *Anthosigmella varians* forma *varians*. *Mar Biol* **125:** 649–654.

Hinde, R., Pironit, F., and Borowitzka, M. A. (1994) Isolation of *Oscillatoria spongeliae,* the filamentous cyanobacterial symbiont of the marine sponge *Dysidea herbacea*. *Mar Biol* **119:** 99–104.

Hinrichsen, D. (1997) Coral reefs in crisis. *BioScience* **47:** 554–558.

Hirose, E., Maruyama, T., Cheng, L., and Lewin, R. A. (1996) Intracellular symbiosis of a photosynthetic prokaryote *Prochloron* sp., in a colonial ascidian. *Invertebrate Biol* **115:** 343–348.

Hirsch, A. M. (2004) Plant microbe symbioses: A continuum from commensalisms to parasitism. *Symbiosis* **37:**

Holder, J. M., Wynn-Williams, D. D., Rull Perez, F., and Edwards, H. G.M, (2000) Raman spectroscopy of pigments and oxalates *in situ* within epilithic lichens: *Acarospora* from the Antarctic and Mediterranean. *New Phytol* **145:** 271–280.

Holtan-Hartwig, J. (1993) The lichen genus *Peltigera* exclusive of the *P. canina* group, in Norway. *Sommerfeltia* **15:** 1–77.

Honegger, R. (1991) Functional aspects of the lichen symbiosis. *Annu Rev Plant Physiol Plant Mol Biol* **42:** 553–578.

Honegger, R. (1993) Developmental biology of lichens. *New Phytol* **125:** 659–677.

Honegger, R. (2000) Great Discoveries in Bryology and Lichenology: Simon Schwendener (1829–1919) and the dual hypothesis of lichens. *Bryologist* **103:** 307–313.

Honegger, R., and Hatsch, A. (2001) Immunocytochemical location of the (1→3) (1→4)-β-glucan lichenin in the lichen-forming ascomycete *Cetraria islandica* (Icelandic moss). *New Phytol* **150:** 739–746.

Honegger, R., Peter, M., and Scherrer, S. (1996) Drought-induced structural alterations at the mycobiont-photobiont interface in a range of foliose macrolichens. *J Protoplasma* **190:** 221–232.

Huss-Danell, K. (1977) Nitrogenase activity in the lichen *Stereocaulon paschele*: Recovery after dry storage. *Physiol Plant* **41:** 158–161.

Hyvärinen, M., Härdling, R., and Tuomi, J. (2002) Cyanobacterial lichen symbiosis: the fungal partner as an optimal harvester. *Oikos* **98:** 498–504.

Ingram, G. A. (1982) Haemagglutinins and haemolysins in selected lichen species. *Bryologist* **85:** 389–393.

Ireland, R. J., and Lea, P. J. (1999) The enzymes of glutamine, glutamate, asparagine, and aspartate metabolism. In: Singh, B. K. (Ed.) *Plant Amino Acids, Biochemistry and Biotechnology*. Marcel Dekker, New York, NY, USA. pp. 49–109.

James, P. W., Hawksworth, D. L., and Rose, F. (1977) Lichen communities in the British Isles: A preliminary conspectus. In: Seaward, M. R. D. (Ed.) *Lichen Ecology*. Academic Press, London, England, UK. pp. 295–413.

Janson, S., Rai, A. N., and Bergman, B. (1995) Intracellular cyanobiont *Richelia intracellularis*. Ultrastructure and immunolocalization of phycoerythrin, nitrogenase, Rubisco and glutamine synthetase. *Mar Biol* **124:** 1–8.

Janson, S., Wouters, J., Bergman, B., and Carpenter, E. J. (1999) Host specificity in the *Richelia*-diatom symbiosis revealed by *hetR* gene sequence analysis. *Environ Microbiol* **1:** 431–438.

Johansson, C., and Bergman, B. (1992) Early events during the establishment of *Nostoc-Gunnera* symbiosis. *Planta* **188:** 403–413.

Johansson, C., and Bergman, B. (1994) Reconstitution of symbiosis of *Gunnera manicata* Linden: Cyanobacterial specificity. *New Phytol* **126:** 643–652.

Joneson, S., Armaleo, D., and Lutzoni, F. (2011) Fungal and algal gene expression in early developmental stages of lichen-symbiosis. *Mycologia* **103:** 291–306. doi:10.3852/10-064

Jönsson, B. (1894) Studier öfver algparasitism hos *Gunnera* L. *Botaniska Notiser* 1–20.

Jordan, W. P. (1972) Erumpent cephalodia, an apparent case of phycobial influence on lichen morphology. *J Phycol* **8:** 112–117.

Joseph, C. M., and Meeks, J. C. (1987) Regulation of expression of glutamine synthetase in a symbiotic *Nostoc* strain associated with *Anthoceros punctatus*. *J Bacteriol* **169:** 2471–2475.

Jovan, S. (2002) A landscape-level analysis of epiphytic lichen diversity in northern and central California: Environmental predictors of species richness and potential observer effects. *Bull Calif Lichen Soc* **9:** 1–7.

Jovan, S., and McCune, B. (2004) Regional variation in epiphytic macrolichen communities in northern and central California forests. *Bryologist* **107:** 328–339.

Kallio, P. (1974) Nitrogen fixation in subarctic lichens. *Oikos* **25:** 194–198.

Kaplan, D., and Peters, G. A. (1981) The *Azolla-Anabaena azollae* relationship. X. $^{15}N_2$ fixation and transport in main stem axes. *New Phytol* **89:** 337–346.

Kardish, N., Silberstein, L., Flemminger, N., and Galun, M. (1991) Lectin from the lichen *Nephroma laevigatum*. Localization and function. *Symbiosis* **11:** 47–62.

Khamar, H. J., Breathwaite, E. K., Prasse, C. E., Fraley, E. R., Secor, C. R., Chibane, F. L., Elhai, J., and Chiu, W.-L. (2010) Multiple role of soluble sugars in the establishment of *Gunnera-Nostoc* endosymbiosis. *Plant Physiol* **154:** 1381–1389.

Kimor, B., Gordon, N., and Neori, A. (1992) Symbiotic associations among the microplankton in oligotrophic marine environments, with special reference to Gulf of Aqaba, Red Sea. *J Plankton Res* **14:** 1217–1231.

Kimor, B., Reid, F. M., and Jordan, J. B. (1978) An unusual occurrence of *Hemiaulus membranaceus* Cleve (Bacillariophyceae) with *Richelia intracellularis* Schmidt (Cyanophyceae) off the coast of Southern California in October 1976. *Phycologia* **17:** 162–166.

Kimura, J., and Nakano, T. (1990) Reconstitution of *Blasia-Nostoc* symbiotic association under axenic conditions. *Nova Hedwigia* **50:** 181–200.

Kirk, P. M, Cannon, P. F., David, J. C., and Stalpers, J. A. (2001) Ainsworth and Bisby's Dictionary of the Fungi, 9th edn. CAB International, Egham, UK.

Kitajima, S., Furuya, K., Hashihama, F., Takeda, S., and Kanda, J. (2009) Latitudinal distribution of diazotrophs and their nitrogen fixation in the tropical and subtropical western North Pacific. *Limnol Oceanogr* **54:** 537–547. doi:10.4319/lo.2009.54.2.0537

Kjems, J., and Garrett, R. A. (1991) Ribosomal RNA introns in archaea and evidence for RNA conformational changes associated with splicing. *Proc Natl Acad Sci USA* **88:** 439–443.

Kluge, M. (2002) A fungus eats a cyanobacterium: The story of the *Geosiphon pyriformis* endocyanosis. *Proc R Irish Acad* **102B:** 11–14.

Kneip, C., Lockhart, P., Voβ, C., and Maier, U. G. (2007) Nitrogen fixation in eukaryotes. New models for symbiosis. *Evol Biol* **7:** 55. doi:10.1186/1471-2148/7/55

Knight, C. D., and Adams, D. G. (1996) A method for studying chemotaxis in nitrogen fixing cyanobacterium-plant symbiosis. *Physiol Mol Plant Path* **49:** 73–77.

Kobiler, D., Cohen-Sharon, A., and Tel-Or, E. (1981) Recognition between the N_2-fixing *Anabaena* and the water fern *Azolla*. *FEBS Lett* **133:** 157–160.

Koike, I., and Suzuki, T. (1996) Nutritional diversity of symbiotic ascidians in a Fijian seagrass meadow. *Ecol Res* **11:** 381–386.

Koike, I., Yamamuro, M., and Pollard, P. C. (1993) Carbon and nitrogen budgets of two ascidians and their symbiont, *Prochloron*, in a tropical seagrass meadow. *Aus J Mar Freshwater Res* **44:** 173–182.

Kühl, M., and Larkum, A. W. D. (2002) The microenvironment and photosynthetic performance of *Prochloron* sp. in symbiosis with didemnid Ascidians. In: Seckbach, J. (Ed.) *Cellular Origin in Extreme Habitats*. Vol. 3. *Symbiosis, Mechanisms and Model Systems*. Kluwer Academic Publishers Dordrecht, The Netherlands. pp. 273–290.

Kuhsel, M. G., Strickland, R., and Palmer, J. D. (1990) An ancient group I intron shared by eubacteria and chloroplasts. *Science* **250:** 1570–1573.

Kulkarni, V. V., Chitari, R. R., Narle, D. D., Patil, J. S., and Anil, A. C. (2010) Occurrence of cyanobacteria-diatom symbioses in the Bay of Bengal: Implications in biogeochemistry. *Curr Sci* **99:** 736–737.

Ladha, J. K., and Watanabe, I. (1982) Antigenic similarity among *Anabaena azollae* separated from different species of *Azolla*. *Biochem Biophys Res Comm* **109:** 675–682.
Ladha, J. K., and Watanabe, I. (1984) Antigenic analysis of *Anabaena azollae* and the role of lectin in the *Azolla-Anabaena* symbiosis. *New Phytol* **98:** 295–300.
Lambowitz, A. M. (1989) Infectious introns. *Cell* **56:** 323–326.
Lambowitz, A. M., Caprara, M. G., Zimmerly, S., and Perlman, P. S. (1999) Group I and group II ribozymes as RNPs: Clues to the past and guides to the future. In: Gesteland, R. F., Cech, T. R., and Atkins, J. F. (Eds.) *The RNA World*. Cold Spring Harbor Laboratory Press, Cold Spring Harbor, New York, NY, USA. pp. 451–485.
Lancien, M., Gadal, P., and Hodges, M. (2000) Enzyme redundancy and the importance of 2-oxoglutarate in higher plant ammonium assimilation. *Plant Physiol* **123:** 817–824.
Lange, O. L., Kilian, E., and Ziegler, H. (1986) Water vapor uptake and photosynthesis in lichens: performance differences in species with green and blue-green algae as phycobionts. *Oecologia* **71:** 104–110.
Lange, O. L., Green, T. G. A., and Ziegler, H. (1988) Water related photosynthesis and carbon isotope discrimination in species of taxonomic distribution and existence of regions of homology with symbiotic genes of *Rhizobium*. *Can J Microbiol* **37:** 171–181.
Larkum, A. W. D. (1999) The cyanobacteria of coral reefs. Marine cyanobacteria. In: Charpy, L., and Larkum, A. W. D. (Eds.) *Bull de Institut Oceanographique Monaco Numero special 19*. Monaco Musee Oceanographique. pp. 149–167.
Larkum, A. W. D., Cox, G. C., and Dibrayawan, T. P. (1988) Prokaryotic algal symbionts of coral reef songes. *Proc 6th Int Coral Reef Symp* 1983. **3:** 163–169.
Larkum, A. W. D., Cox, G. C., Hiller, R. G., Parry, D. L., and Dibbayawan, T. P. (1987) Filamentous cyanophytes containing phycourobilin and in symbiosis with sponges and an ascidian of coral reefs. *Marine Biol* **95:** 1–13.
La Roche, J., van der Staay, G. W. M., Partensky, F., Ducret, A., Aebersold, R., Li, R., Golden, S. S., Hiller, R. G., Wrench, P. M., Larkum, A. D. M., and Green, B. R. (1996) Independent evolution of the Prochlorophyta and green plant chlorophyll a/b light-harvesting proteins. *Proc Natl Acad Sci USA* **93:** 15244–15248.
Laundon, J. R. (1995) On the classification of lichen photomorphs. *Taxon* **44:** 387–389.
Lee, Y. K., Lee, J.-H., and Lee, H. K. (2001) Microbial symbiosis in marine sponges. *J Microbiol* (Korea) **39:** 254–264.
Legaz, M. E., Fontaniella, B., Millanes, A. M., and Vicente, C. (2004) Secreted arginases from phylogenetically far-related lichen act as cross-recognition factor for two different algal cells. *Eur J Cell Biol* **83:** 1–12.
Lehr, H., Fleminger, G., and Galun, M. (1995) Lectin from the lichen *Peltigera membranacea* (Ach.) Nyl.: Characterization and function. *Symbiosis* **18:** 1–13.
Lehr, H., Galum, M., Ott, S., Jahns, H. M., and Fleminger, G. (2000) Cephalodia of the lichen *Peltigera aphthosa*. Specific recognition of the compatible photobiont. *Symbiosis* **29:** 357–365.
Lemmermann, E. (1899) Ergebnisse einer Reise nach dem Pacific (H. Scauinsaland 1896/97). *Planktanolgen, Abhandl Aus Naturw Ver Bremen* **16:** 313–398.
Lemmermann, E. (1905) Die Algenflora der Sandwich-Inseln. Ergebnisse einer nach dem Pacific, H. Schauinsland 1896/97. *Engler's Bot Jahrbücher Syst, Pflanzengeschichte Pflanzengeographie* **34:** 607–663.
Le Moal, M., Collin, H., and Biegala, I. C. (2011) Intriguing diversity among diazotrophic picoplankton along a Mediterranean transect: A dominance of rhizobia. *Biogeosci* **8:** 827–840. doi:10. 5194/bg-8-827-2011.
Lesser, M. P., Mazel, C. H., Gorbunov, M. Y., and Falkowski, P. G. (2004) Discovery of symbiotic nitrogen-fixing cyanobacteria in corals. *Science* **305:** 997–1000.
Lewin, B. (2008) *Genes IX*. Jones and Bartlett Publishers, Boston and London.
Lewin, R. A. (1975) A marine *Synechocystis* (Cyanophyta, Chroococcales) epizoic on ascidians. *Phycologia* **14:** 153–160.
Lewin, R. A., and Cheng, L. (Eds.) (1989) *Prochloron—a microbial enigma*. Chapman & Hall, New York, NY, USA.
Lewin, R. A., and Withers, N. W. (1975) Extraordinary pigment composition of a prokaryotic alga. *Nature* (London) **256:** 735–737.
Lin, C., and Watanabe, I. (1988) Study on the association between *Anabaena azollae* and *Azolla microphylla* during the germination of megasporocarps. *Symbiosis* **5:** 199–208.
Lindblad, P. (1984) Diversion between C_2H_2-reduction and heterocyst frequency in a cycad root. In: Veeger, C. V., and Newton, W. E. (Eds.) *Advances in Nitrogen Fixation Research*. Nühoff/Junk, Publishers, The Hague. The Netherlands. pp. 511.
Lindblad, P., Atkins, P., and Pate, J. S. (1991) N_2-fixation by freshly isolated *Nostoc* from coralloid roots of the cycad *Macrozamia riedlei* (Fisch-Gaud) Gardn. *Plant Physiol* **95:** 753–759.
Lindblad, P., and Bergman, B. (1989) Occurrence and localization of phycoerythrin in symbiotic *Nostoc* of *Cycas revoluta* and in the free-living isolated *Nostoc* 7422. *Plant Physiol* **89:** 783–785.
Lindblad, P., and Bergman, B. (1990) The cycad-cyanobacterial symbiosis. In : Rai, A. N. (Ed.) *Handbook of Symbiotic Cyanobacteria*. CRC Press, Boca Raton, Florida, FL, USA. pp. 137–159.
Lindblad, P., Bergman, B., Hofsten, A. V., Hällbom, L., and Nylund, J. E. (1985) The cyanobacterium-*Zamia* symbiosis: An ultrastructural study. *New Phytol* **101:** 707–716.

Lindblad, P., Hällbom, L., and Bergman, B. (1985) The cyanobacterium-*Zamia* symbiosis: C_2H_2-reduction and heterocyst frequency. *Symbiosis* **1:** 19–28.
Lindblad, P., Haselkorn, R., Bergman, B., Nierzwicki-Bauer, S. A. (1989) Comparison of DNA restriction fragment length polymorphisms of *Nostoc* strains in and from cycads. *Arch Microbiol* **152:** 20–24.
Lockhart, C. M., Rowell, P., and Stewart, W. D. P. (1978) Phytohaemagglutinins from the nitrogen-fixing lichens *Peltigera canina* and *P. polydactyla*. *FEMS Microbiol Lett* **3:** 127–130.
Logsdon, J. M. Jr. (1998) The recent origin of spliceosomal introns revisited. *Curr opin Genet Dev* **8:** 637–648.
Lohtander, K., Oksanen, I., Rikkinen, J. (2003) Genetic diversity of green algal and cyanobacterial photobionts in *Nephroma* (Peltigerales). *Lichenologist* **35:** 325–339.
Lucas, I. A. N. (1991) Symbionts of the tropical Dinophysiales (Dinophyceae). *Ophelia* **33:** 213–224.
Lücking, R. (2008) Folicolous lichenized fungi. *Flora Neotropica Monograph* **103:** 1–867.
Lücking, R., Lowrey, J. D., Sikaroodi, M., Gillevet, P. M., Chaves, J. L., Sipman, H. J. M., and Bungartz, F. (2009) Do lichens domesticate photobionts like farmers domesticate crops? Evidence from a previously unrecognized lineage of filamentous cyanobacteria. *Am J Bot* **96:** 1409–1418.
Lumpkin, T. A., Zimmerman, W. J., and Watanabe, I. (1991) The *Anabaena-Azolla* symbiosis: Diversity and relatedness of neotropical host taxa. *Plant Soil* **137:** 167–170.
Lutzoni, F., Pagel, M., and Reeb, V. (2001) Major fungal lineages are derived from lichen symbiotic ancestors. *Nature* (London) **411:** 937–940.
Lyimo, T. J. (2011) Distribution and abundance of the cyanobacterium *Richelia intracellularis* in the coastal waters of Tanzania. *J Ecol Nat Environ* **3:** 85–94.
Lykke-Andersen, J., Aagaard, C., Semionenkov, M., and Garrett, R. A. (1997) Archaeal introns: Splicing, intercellular mobility and evolution. *Trends Biochem Sci* **22:** 326–331.
Lyons, B., Nadkarni, N. M., and North, M. P. (2000) Spatial distribution and succession of epiphytes of *Tsuga heterophylla* (western hemlock) in an old-growth Douglas-fir forest. *Can J Bot* **78:** 957–968.
Maass, W. S. G. (1980) *Erioderma pedicellatum* in North America: A case study of a rare and endangered lichen. *Proc N S Inst Sci* **30:** 69–87.
Maass, W. S. G. (1983) New Observations on *Erioderma* in North America. *Nordic J Bot* **3:** 567–575.
Madan, A. P., and Nierzwicki-Bauer, S. A. (1993) *In situ* detection of transcripts for ribulose-1,5-biphosphate carboxylase in cyanobacterial heterocysts. *J Bacteriol* **175:** 7301–7306.
Mague, T., Weare, N., and Holm-Hansen, O. (1974) Nitrogen fixation in the North Pacific Ocean. *Mar Biol* **24:** 109–119.
Marx, M., and Peveling, E. (1983) Surface receptors in lichen symbionts visualized by fluorescence microscopy after use of lichens. *Protoplasma* **114:** 52–61.
Mattfeld, J. (1933) Weiteres zur Kenntnis der *Gunnera herteri* Osten. Ostenia (Coleccion de Trabajos Botanicos), Montevideo, pp. 102–118.
Mazel, D., Houmard, J., Castets, A. M., Tandeau de Marsac, N. (1990) Highly repetitive DNA sequences in cyanobacterial genomes. *J Bacteriol* **172:** 2755–2761.
McCowen, S. M., MacArthur, L., and Gates, J. E. (1987) *Azolla* fern lectins that specifically recognize endosymbiotic cyanobacteria. *Curr Microbiol* **14:** 329–333.
McCune, B. (1993) Gradients in epiphyte biomass in three *Pseudotsuga-Tsuga* forests of different ages in western Oregon and Washington. *Bryologist* **96:** 405–411.
McCune, B. (1994) Using epiphyte litter to estimate epiphyte biomass.*Bryologist* **97:** 396–401.
McCune, B., Amsberry, K. A., Camacho, F. J., Clery, S., Cole, C., Emerson, C., Felder, G., French, P., Greene, D., Harris, R., Hutten, M., Larson, B., Lesko, M., Majors, S., Markwell, T., Parker, G. G., Pendergrass, K., Peterson, E. B., Peterson, E. T., Platt, J., Proctor, J., Rambo, T., Rosso, A., Shaw, D., Turner, R., and Widmer, M. (1997) Vertical profile of epiphytes in a Pacific old-growth forest. *Northwest Science* **71:** 145–152.
McLuckie, J. (1922) Studies in symbiosis. The apogeotropic roots of *Macrozamia spiralis* and their physiological significance. *Proc Linn Soc New South Wales* **47:** 319–328.
Meeks, J. C. (1990) Cyanobacterial-bryophyte associations. In: Rai, A. N. (Ed.) *Handbook of Symbiotic Cyanobacteria*. CRC Press, Boca Raton, Florida, FL, USA. pp. 43–63.
Meeks, J. C. (1998) Symbiosis between nitrogen-fixing cyanobacteria and plants. *BioScience* **48:** 266–276.
Meeks, J. C., Joseph, C. M., and Haselkorn, R. (1988) Organization of the *nif* genes in cyanobacteria in symbiotic association with *Azolla* and *Anthoceros*. *Arch Microbiol* **150:** 61–71.
Meeks, J. C., Enderlin, C. S., Joseph, C. M., Chapman, J. S., and Lollar, M. W. L. (1985) Fixation of $[^{13}N]N_2$ and transfer of fixed nitrogen in the *Anthoceros-Nostoc* symbiotic association. *Planta* **164:** 406–414.
Meeks, J. C., Steinberg, N. A., Enderlin, C. S., Joseph, C. M., and Peters, G. A. (1987) *Azolla-Anabaena* relationship. XIII. Fixation of $[^{13}N2]N_2$. *Plant Physiol* **84:** 883–886.

Mellor, R. B., Gadd, G. M., Rowell, P., and Stewart, W. D. P. (1981) A phytohaemagglutinin from the *Anabaena azollae* symbiosis. *Biochem Biophys Res Comm* **99:** 1348–1353.
Miao, V. P. W., Rabenau, A., and Lee, A. (1997) Cultural and molecular characterization of photobionts of *Peltigera membranacea*. *Lichenologist* **29:** 571–586.
Michel, F., Umesono, K., and Ozeki, H. (1989) Comparative and functional anatomy of group II catalytic introns—a review. *Gene* **82:** 5–30.
Miehe, H. (1924) Entwickliungsgeschichtliche untersurbung der algensymbiose bei *Gunnera macrophylla* Bl. *Flora* **117:** 1–15.
Moe, R. (1997) *Verrucaria taveresiae* sp. nov., a marine lichen with a brown algal photobiont. *Bull Calif Lichen Soc* **4:** 7–11.
Molina, M. C., and Vicente, C. (1993) Loss of photoergonic condition of *Xanthoria parietina* photobiont effected by an algal-binding protein isolated from the same lichen species. In: Sato, V., Ishida, S., and Ishikawa, H. (Eds.) *Endocytobiology*. Tübingen University Press, Tübingen, Germany. pp. 69–74.
Molina, M. C., and Vicente, C. (1995) Correlationships between enzymatic activity of lectins, putrescine content and chloroplast damage in *Xanthoria parietina* photobionts. *Cell Adhesion Commun* **3:** 1–12.
Molina, M. C., and Vicente, C. (1996) Ultrastructural deterioration of *Xanthoria parietina* (L.) Th. Fr. phycobiont induced by a *Xanthoria* lectin. *Phyton* (Austria) **36:** 197–208.
Molina, M. C., and Vicente, C. (2000) Purification and characterization of two isolectins with arginase activity from the lichen *Xanthoria parietina*. *J Biochem Mol Biol* **33:** 300–307.
Molina, M. C., Bajon, C., Sauvanet, A., Robert, D., and Vicente, C. (1998b) Detection of polysaccharides and ultrastructural modification of the photobiont cell wall produced by two arginase isolectins from *Xanthoria parietina*. *J Plant Res.* **111:** 191–197.
Molina, M. C., Muñiz, E., and Vicente, C. (1993) Enzymatic activities of algal-binding protein and its algal cell wall receptor in the lichen *Xanthoria parietina*. An approach to the parasitic basis of mutualism. *Plant Physiol Biochem* **31:** 131–142.
Molina, M. C., Stocker-Wörgötter, E., Türk, R., and Vicente, C. (1998a) Secreted, glycosylated arginase from *Xanthoria parietina* thallus induces loss of cytoplasmic material from *Xanthoria* photobionts. *Cell Adhesion Commun* **6:** 481–490.
Molina, M. C., Vicente, C., Pedrosa, M. M., and Muniz, E. (1996) Binding of a labeled lectin from the lichen *Xanthoria parietina* to its own phycobiont and analysis of its enzymatic activity. *Phyton* (Austria) **36:** 145–158.
Monteiro, F. M., Follows, M. J., and Dutkiewicz, S. (2010) Distribution of diverse nitrogen fixers in the global ocean. *Global Biogeochem Cycles* **24:** GB3017. doi:10.1029/2009GB003731
Moore, A. W. (1969) *Azolla*: Biology and agronomic significance. *Bot Rev* **35:** 17–34.
Moore, M. J., Soltis, P. S., Bell, C. D., Burleigh, J. D., and Soltis, D. E. (2010) Phylogenetic analysis of 83 plastid genes further resolves the early diversification of eudicots. *Proc Natl Acad Sci USA* **107:** 4623–4628.
Munro, M.H.G., Blunt, J.W., Dumdei, E.J., Hickford, S.J.H., Lill, R.E., Li, S., Battershill, C.N., and Duckworth, A.R., (1999) The discovery and development of marine compounds with pharmaceutical potential. *J Biotechnol* **70:** 15–25.
Myllys, L. M., Stenroos, S., Thell, A., and Kuusinen, Y. M. (2007) High cyanobiont selectivity of epiphytic lichens in old growth boreal forest of Finland. *New Phytol* **173:** 621–629.
Nash, III, T. H. (1996) Nitrogen, its metabolism and potential contribution of ecosystems. In: Nash, III, T. H. (Ed.) *Lichen Biology*. Cambridge University Press, Cambridge, UK. pp. 121–135.
Nash, III, T. H. (Ed.) (1996) *Lichen Biology*. Cambridge University Press, Cambridge, UK.
Nathanielsz, C. P., and Staff, I. A. (1975a) A mode of entry of blue-green algae into the apogeotropic roots of *Macrozamia communis*. *Am J Bot* **62:** 232–235.
Nathanielsz, C. P., and Staff, I. A. (1975b) On the occurrence of intracellular blue-green algae in cortical cells of the apogeotropic roots of *Macrozamia communis* L. Johnson. *Ann Bot* **39:** 363–368.
Neitlich, P. N., and McCune, B. (1997) Hotspots of epiphytic lichen diversity in two young managed forests. *Conser Biol* **11:** 172–182.
Neuman, D. (1977) Ultrastrukturelle Untersuchungen zur Symbiose von Cyanophyceen mit Cycadeen (*Cycas circinnalis* L., *Zamia furfuraceae* L.). *Biochem Physiol Pflanzen* **171:** 313–322.
Neumüller, M., and Bergman, B. (1981) The ultrastructure of *Anabaena azollae* in *Azolla pinnata*. *Physiol Plant* **51:** 69–76.
Newcomb, E. H., and Pugh, T. D. (1975) Blue-green algae associated with ascidians of the Great Barrier Reef. *Nature* (London) **253:** 533–534.
Newton, J. W., and Cavins, J. F. (1976) Altered nitrogenous pools induced by the *Azolla-Anabaena azollae* symbiosis. *Plant Physiol* **58:** 798–799.
Newton, J. W., and Herman, A. I. (1979) Isolation of cyanobacteria from the aquatic fern, *Azolla*. *Arch Microbiol* **120:** 161–165.
Nieboer, E., McFarlane, J. D., and Richardson, D. H. S. (1984) Modification of plant cell buffering capacities by gaseous air pollutants. In: Koziol, M., and Whatley, F. R. (Eds.) *Gaseous Air Pollutants and Plant Metabolites*, Butterworths, London, UK. pp. 313–333.

Nierzwicki-Bauer, S. A., and Haselkorn, R. (1986) Differences in mRNA levels in *Anabaena* living freely or in symbiotic association with *Azolla*. *EMBO J* **5:** 29–35.

Nilsen, T. W. (2003) The spliceosome: The most complex macromolecular machine in the cell? *Bioessays* **25:** 1147–1149.

Nilsson, M., Bergman, B., and Rasmussen, U. (2000) Cyanobacterial diversity in geographically related and distant host plants of the genus *Gunnera*. *Arch Microbiol* **173:** 97–102.

Nilsson, M., Rasmussen, U., and Bergman, B. (2006) Cyanobacterial chemotaxis to extracts of host and nonhost plants. *FEMS Microbiol Ecol* **55:** 382–390.

Norstog, K. J., and Nicholls, T. J. (1997) *The Biology of the Cycads*. Cornell University Press, Ithaca, NY, USA.

Nyati, S, Beck, A., and Honegger, R. (2007) Fine structure and phylogeny of green algal photobionts in the microfilamentous genus *Psoroglaena* (Verrucariaceae, lichen-forming ascomycetes). *Plant Biol* **9:** 390–399.

O'Brien, H. E., Miadlikowska, J., and Lutzoni, F. (2005) Assessing host specialization in symbiotic cyanobacteria associated with four closely related species of the lichen fungus *Peltigera*. *Eur J Phycol* **40:** 363–378.

Obukowicz, M., Schaller, M., and Kennedy, G. S. (1981) Ultrastructure and phenolic histochemistry of the *Cycas revoluta-Anabaena* symbiosis. *New Phytol* **87:** 751–759.

Oksanen, I. (2004) Specificity in cyanobacterial lichen symbiosis and bioactive metabolites (ethylene and microcystins). Dissertat Biocentri Viikki Univ. Helsingiensis **23:** 1–76.

Oksanen, I., Lohtander, K., Sivonen, K., and Rikkinen, J. (2004) Repeat-type distribution in *trnL* intron does not correspond with species phylogeny: Comparison of the genetic markers 16S rRNA and *trnL* intron in heterocystous cyanobacteria. *Int J Syst Evol Microbiol* **54:** 765–772.

Orr, J., and Haselkorn, R. (1982) Regulation of glutamine synthetase activity and synthesis in free-living and symbiotic *Anabaena* spp. *J Bacteriol* **152:** 626–635.

Osborne, B. A., Cullen, A., Jones, P. W., and Campbell, G. J. (1992) Use of nitrogen by the *Nostoc-Gunnera tinctoria* (Molina) Mirbel symbiosis. *New Phytol* **120:** 481–487.

Otálora, M. A. G., Martínez, I., O'Brien, H., Molina, M. C., and Aragóu, G. (2010) Multiple origins of high reciprocal symbiotic specificity at an intercontinental spatial scale among gelatinous lichens (Collemataceae; Lecanoromycetes). *Mol Phyl Evol* **56:** 1089–1095.

Ott, S. (1988) Photosymbiodemes and their development in *Peltigera venosa*. *Lichenologist* **20:** 361–368.

Ow, M. C., Gantar, M., and Elhai, J. (1999) Reconstitution of a cycad-cyanobacterial association. *Symbiosis* **27:** 125–134.

Page, R. D., and Hafner, M. S. (1996) Molecular phylogenies and host-parasite cospeciation: Gophers and lice as a model system. In: Harvey, P. H., Leigh, A. J., Brown. J., Smith, M., and Nee, S. (Eds.) *New Uses for New Phylogenies*. Oxford University Press, New York, NY, USA. pp. 255–272.

Pandey, K. D., Kashyap, A. K., and Gupta, R. K. (1992) Nitrogen fixation by cyanobacteria associated with moss communities in Schimacher oasis, Antarctica. *Israel J Bot* **41:** 187–198.

Papaefthimiou, D., Hrouzek, P., Mugnai, M. A., Lukesova, A., Turicchia, S., Rasmussen, U., and Ventura, S. (2008) Differential patterns of evolution and distribution of symbiotic behaviour in nostocacean cyanobacteria. *Syst Evol Microbiol* **58:** 553–564.

Paquin, B., Kathe, S. D., Nierzwicki-Bauer, S. A., and Shub, D. A. (1997) Origin and evolution of group I introns in cyanobacterial tRNA genes. *J Bacteriol* **179:** 6798–6806.

Parsons, R. (2002) Nodule function and regulation in the *Gunnera-Nostoc* symbioses. *Proc R Irish Acad* **102:** 41–43.

Paulsrud, P., and Lindblad, P. (1998) Sequence variation of the $tRNA^{Leu}$ (UAA) intron as a marker for genetic diversity and specificity of symbiotic cyanobacteria in some lichens. *Appl Environ Microbiol* **64:** 310–315.

Paulsrud, P., Rikkinen, J., and Lindblad, P. (1998) Cyanobiont specificity in some *Nostoc*-containing lichens and in a *Peltigera aphthosa* photosymbiodemes. *New Phytol* **139:** 517–524.

Paulsrud, P., Rikkinen, J., and Lindblad, P. (2000) Spatial patterns of photobiont diversity in some *Nostoc*-containing lichens. *New Phytol* **146:** 291–299.

Paulsrud, P., Rikkinen, J., and Lindblad, P. (2001) Field investigations on cyanobacterial specificity in *Peltigera aphthosa*. *New Phytol* **152:** 117–123.

Pawlik, J. R., Chanas, B., Toonen, R. J., and Fenical, W. (1995) Defenses of Carribean sponges against predatory reef fish. I. Chemical deterrency. *Mar Ecol Prog Ser* **127:** 183–194.

Perkins, S.K., and Peters, G. A. (1992) The *Azolla-Anabaena* symbiosis: Endophyte continuity in the *Azolla* life-cycle is facilitated by epidermal trichomes. I. Partitioning of the endophytic *Anabaena* into developing sporocarps. *New Phytol* **123:** 53–64.

Peters, G. A. (1975) The *Azolla-Anabaena* relatioship III. Studies of metabolic capabilities and a further characterization of the symbiont. *Arch Microbiol* **103:** 113–122.

Peters, G. A. (1991) *Azolla* and other plant-cyanobacteria symbioses: Aspects of form and function. *Plant Soil* **137:** 25–36.

Peters, G. A., and Calvert, H. E. (1983) The *Azolla-Anabaena* symbiosis. In: Gof, T. L. J. (Ed.) *Algal Symbiosis*. Cambridge University Press, New York, NY, USA. pp. 109–145.

Peters, G. A., and Mayne, B. C. (1974a) The *Azolla, Anabaena azollae* relationship. I. Initial characterization of the association. *Plant Physiol* **53:** 813–819.

Peters, G. A., and Mayne, B. C. (1974b) The *Azolla, Anabaena azollae* relationship. II. Localization of nitrogenase activity as assayed by acetylene reduction. *Plant Physiol* **53:** 820–824.

Peters, G. A., and Perkins, S. K. (1993) The *Azolla-Anabaena* symbiosis: Endophyte continuity in the *Azolla*-life cycle is facilitated by epidermal trichomes. II. Re-establishment of the symbiosis following gametogenesis and embryogenesis. *New Phytol* **123:** 65–75.

Peters, G. A., Toia, R. E. Jr., and Lough, S. M. (1977) *Azolla-Anabaena azollae* relationship. V. $^{15}N_2$ fixation, acetylene reduction, and H_2 production. *Plant Physiol* **59:** 1021–1025.

Peters, G. A., Toia, R. E. Jr., Levine, N. J., and Raveed, D. (1978) *Azolla-Anabaena azollae* relationship. VI. Morphological aspects of the association. *New Phytol* **80:** 583–593.

Petit, P. (1982) Phytolectins from the nitrogen-fixing lichen *Peltigera horizantalis*: The binding pattern of protein extract. *New Phytol* **91:** 705–710.

Petit, P., Lallemant, R. and Savoye, D. (1983) Purified phytolectin from the lichen *Peltigera canina* var. *canina* which binds to the phycobiont cell walls and its use as a cytochemical marker *in situ*. *New Phytol* **94:** 103–110.

Piercey-Normore, M. D., and DePriest, P. T. (2001) Algal switching among lichen symbioses. *Am J Bot* **88:** 1490–1498.

Pieterse. A. H., de Lange, and van Vliet, J. P. (1977) A comparative study of *Azolla* in the Netherlands. *Acta Bot Neerl* **26:** 433–449.

Pike, L. H. (1978) The importance of epiphytic lichens in mineral cycling. *Bryologist* **81:** 247–257.

Pile, A. J., Patterson, M. R., and Witman, J. D. (1996) *In situ* grazing on plankton <10 µm by the boreal sponge *Mycale lingua*. *Mar Ecol Progr Ser* **141:** 95–102.

Planelles, V., and Legaz, M. E. (1987) Purification and some properties of the secreted arginase of the lichen *Evernia prunastri* and its regulation by usnic acid. *Plant Sci* **51:** 9–16.

Plazinski, J., Franche, C., Liu, C.-C., Lin, T., Shaw, W., Gunning, B. E. S., and Rolfe, B. G. (1988) Taxonomic status of *Anabaena azollae*: An overview. *Plant Soil* **108:** 185–190.

Prechtl, J., Kneip, C., Lockhart, P., Wenderoth, K., and Maier, U.-G. (2004) Intracellular sphaeroid bodies of *Rhopalodia gibba* have nitrogen-fixing apparatus of cyanobacterial origin. *Mol Biol Evol* **21:** 1477–1481.

Preston, C. M., Wu, K. Y., Molinski, T. F., and DeLong, E. F. (1996) A psychrophilic crenarchaeon inhabits a marine sponge: *Crenarchaeon symbiosum* gen. nov., sp. nov. *Proc Natl Acad Sci USA* **93:** 6241–6246.

Quilhot, W., Piovano, M., Chamy, M. C., and Garbarino, J. A. (1995) Studies on Chilean lichens. XXII. Secondary products from *Degelia* Gayana. *Lichenologist* **27:** 405–407.

Quilhot, W., Rubio, C., Fernandez, E., and Hidalgo, M. E. (2002) Efectos de la radiacion UV solar en la acumulacion de 1′-cloropanarina en *Erioderma leylandii* (Pannariaceae, Ascomycotina liquenizado), Laguna San Rafael, Aisen, Chile [Effects of UV solar radiation on the accumulation of 1′-chloropannarin in *Erioderma leylandii* (Pannariaceae, lichenized Ascomycotina) Laguna San Rafael, Aisen, Chile]. *Boletin del Museo Nacional de Historia Natural, Chile* **51:** 75–80.

Radies, D. N., and Coxson, D. S. (2004) Macrolichen colonization on 120–140 year old *Tsuga heterophylla* in wet temperate rainforests of central-interior British Columbia: A comparison of lichen response to even-aged versus old-growth stand structures. *Lichenologist* **36:** 235–247.

Rai, A. N. (1990) *CRC Handbook of Symbiotic Cyanobacteria*. CRC Press, Boca Raton, Florida, FL, USA.

Rai, A. N., and Bergman, B. (2002) Cyanolichens. *Proc R Irish Acad* **102B:** 19–22.

Rai, A. N., Söderbäck, E., and Bergman, B. (2000) Cyanobacterium-plant symbioses. *New Phytol* **147:** 449–481.

Rai, A. N., Borthakur, M., Singh, S., and Bergman, B. (1989) *Anthoceros-Nostoc* symbiosis: Immunoelectronmicroscopic localization of nitrogenase, glutamine synthetase, phycoerythrin and ribulose-1,5-biphosphate carboxylase/oxygenase in the cyanobiont and the cultured (free-living) isolate *Nostoc* 7801. *J Gen Microbiol* **135:** 385–395.

Rambold, G., Friedl, T., and Beck, A. (1998) Photobionts in lichens: Possible indicators of phylogenetic relationships? *Bryologist* **101:** 392–397.

Ran, L., Larsson, J., Vigil-Stenman, T., Nylander, J. A. A., Ininbergs, K., Zheng, W-W., Lapidus, A., Lowry, S., Haselkorn, R., and Berman, B. (2010) Genome erosion in a nitrogen-fixing vertically transmitted endosymbiotic multicellular cyanobacterium. *PLoS ONE* **5:** e11486. doi:10.1371/journal.pone.0011486

Rasmussen, U., and Svenning, M. M. (1998) Fingerprinting of cyanobacteria based on PCR with primers derived from short and long tandemly repeated repetitive sequences. *Appl Environ Microbiol* **64:** 265–272.

Rasmussen, U., and Svennig, M. (2001) Characterisation by genotypic methods of symbiotic *Nostoc* strains isolated from five species of *Gunnera*. *Arch Microbiol* **176:** 204–210.

Rasmussen, U., Johansson, C., and Bergman, B. (1994) Early communication in the *Gunnera-Nostoc* symbiosis: Plant-induced cell differentiation and protein synthesis in the cyanobacterium. *Mol Plant-Microbe Interact* **7:** 696–702.

Rasmussen, U., Johansson, C., Renglin, A., Peterson, C., and Bergman, B. (1996) A molecular characterization of the *Gunnera-Nostoc* symbiosis: Comparison with the *Rhizobium*-plant and *Agrobacterium*-plant interactions. *New Phytol* **133:** 391–398.

Ray, T. B., Mayne, B. C., Toia, R. E. Jr., and Peters, G. A. (1979) *Azolla-Anabaena* relationship VIII. Photosynthetic characterization of the association and individual partners. *Plant Physiol* **64:** 791–795.

Ray, T. B., Peters, G. A., Toia, R. E. Jr., and Mayne, B. C. (1978) *Azolla-Anabaena* relationship VII. Distribution of ammonia-assimilating enzymes, protein, and chlorophyll between host and symbiont. *Plant Physiol* **62:** 463–467.

Reinke, J. (1872) Parasitiscbe *Anabaena* in Wurzelen der Cycadeen. *Gott Nachr* 107.

Reiswig, H. M. (1971) Particle feeding in natural populations of three marine demosponges. *Biol Bull* **141:** 568–591.

Reiswig, H. (1974) Water transport, respiration and energetics of three tropical marine sponges. *J Exp Mar Biol Ecol* **14:** 231–249.

Renner, B., and Galloway, D. J. (1982) Photosymbiodemes in *Pseudocyphellaria* in New Zealand. *Mycotaxon* **16:** 197–231.

Renzaglia, K. S. (1982) A comparative developmental investigation of the gametophyte generation in the Metzgeriales (Hepatophyta). *Bryophyt Bibl* **24:** 1–253.

Renzaglia, K. S., Duff, R. J., Nickrent, D. L., and Garbary, D. (2000) Vegetative and reproductive innovations of early land plants: Implications for a unified phylogeny. *Trans R Soc* B London **355:** 769–793.

Rhoades, F. R. (1995) In: Lowman, N. D., and Nadkarni, N. M. (Eds.) *Forest Canopies*. Academic Press, Toronto, Ontario, Canada. pp. 353–408.

Richardson, D. H. S., and Cameron, R. P. (2004) Cyanolichens: Their response to pollution and possible management strategies for their conservation in northeastern *North American Naturalist* **11:** 1–22.

Ridley, C. P., Faulkner, D. J., and Haygood, M. G. (2005) Investigation of *Oscillatoria spongeliae*-dominated bacterial communities in four dictyoceratid sponges. *Appl Environ Microbiol* **71:** 7366–7375.

Rikkinen, J. (2003) Ecological and evolutionary role of photobiont-mediated guilds in lichens. *Symbiosis* **34:** 99–110.

Rikkinen J. (2004) Ordination analysis of $tRNA^{Leu}$(UAA) intron sequences from lichen-forming *Nostoc* strains and other cyanobacteria. *Symbolae Botanicae Upsalienses* **34:** 377–391.

Rikkinen, J., and Virtanen, V. (2008) Genetic diversity in cyanobacterial symbionts of thalloid bryophytes. *J Exp Bot* **59:** 1013–1021. doi: 10.1093/jxb/em003.

Rikkinen, J., Oksanen, I., and Lohtander, K. (2002) Lichen guilds share related cyanobacterial symbionts. *Science* **297:** 357.

Rippka, R., Deruelles, J., Waterbury, J. B., Herdman, M., and Stanier, R. Y. (1979) Generic assignments, strain histories and properties of pure cultures of cyanobacteria. *J Gen Microbiol* **111:** 1–61.

Robertson, A. (1998) The boreal felt lichen, *Erioderma pedicellatum* (Hue) P. M. Jorg, in Newfoundland, Report to Department of Forest Recources and Agrifoods, Government of Newfoundland and Labrador, St. Johns. pp. 63.

Rodgers, G. A., and Stewart, W. D. P. (1977) The cyanophyte-hepatic symbiosis. I. Morphology and physiology. *New Phytol* **78:** 441–458.

Rose, F. (1976) Lichenological indicators of age and environmental continuity in woodlands. In: Brown, D. H., Hawksworth, D. L., and Bailey, R. H. (Eds.) *Lichenology: Progress and Problems*. Academic Press, London, UK. pp. 279–307.

Rose, F. (1988) Phytogeographical and ecological aspects of Lobarian communities in Europe. *Bot J Linn Soc* **96:** 69–79.

Rossbach, M., and Lambrecht, S. (2006) Lichens as biomonitors: Global, regional and local aspects. *Croatica Chemica Acta* **79:** 119–124.

Rouhiainen, L., Sivonen, K., Buikema, W. J., and Haselkorn, R. (1995) Characterization of toxin-producing cyanobacteria by using an oligonucleotide probe containing a tandemly repeated heptamer. *J Bacteriol* **177:** 6021–6026.

Rowan, R. (1998) Diversity and ecology of zooxanthellae on coral reefs. *J Phycol* **34:** 407–417.

Rudi, K., and Jakobsen, K. S. (1997) Cyanobacterial $tRNA^{Leu}$ (UAA) group I introns have polyphyletic origin. *FEMS Microbiol Lett* **156:** 293–298.

Rudi, K., and Jakobsen, K. S. (1999) Complex evolutionary patterns of $tRNA^{Leu}$ (UAA) group I introns in cyanobacterial radiation. *J Bacteriol* **181:** 3445–3451.

Rudi, K., Fossheim, T., and Jakobsen, K. S. (2002) Nested evolution of a $tRNA^{Leu}$ (UAA) group I intron by both horizantal intron transfer and recombination of the entire tRNA locus. *J Bacteriol* **184:** 666–671.

Rützler, K. (1985) Association between Caribbean sponges and photosynthetic organisms. In : Rützler, K. (Ed.) *New Perspectives in Sponge Biology*, Smithsonian Institution Press, Washington, D. C. USA. pp. 455–466.

Sacristán, M., Millanes, A. M., Legaz, M. E., and Vicente, C. (2006) A lichen lectin specifically binds to the α-1,4-polygalactoside moiety of urease located in the cell wall of homologous algae. *Plant Signal Behav* **1:** 23–27.

Sacristán, M., Millanes, A. M., Fontaniella, B., Legaz, M. E., and Vicente, C. (2008) A first attempt to elucidate amino acid sequence of some lichen lectins. *Phyton* (Austria) **77:** 253–262.

Sacristán, M., Vivas, M., Millanes, A. M., Fontaniella, B., Vicente, C., and Legaz, M. E. (2007) The recognition pattern of green algae by lichenized fungi can be extended to lichens containing cyanobacterium as photobiont. In: Mendez-Vilas, A. (Ed.) *Communicating Current Research and Educational Topics and Trends in Applied Microbiology*. Formatex Research Center, Badajoz, Spain pp. 213–219.

Sanders, W. B., Moe, R. L., and Ascaso, C. (2004) The intertidal marine lichen formed by the pyrenomycetous fungus *Verrucaria tavaresiae* and the brown alga *Pteroderma maculiforme* (Phaeophyceae): Thallus organization and symbiont interaction. *Am J Bot* **91:** 511–522.

Sanders, W. B., Moe, R. L., and Ascaso, C. (2005) Ultrastructural study of the brown alga *Pteroderma maculiforme* (Phaeophyceae) in the free-living state and in lichen symbiosis with the intertidal marine fungus *Verrucaria tavaresiae* (Ascomycotina). *Eur J Phycol* **40:** 353–361.

Saldanha, R., Mohr, G., Belfort, M., and Lambowitz, A. M. (1993) Group I and group II introns. *FASEB J* **7:** 15–24.

Santavy, D. L., Willenz, P., and Colwell, R. R. (1990) Phenotypic study of bacteria associated with the Caribbean sclerosponge *Ceratoporella nicholsoni. Appl Environ Microbiol* **56:** 1750–1762.

Schacht, H. (1853) Beitrag zur Entwicklungs-Geschichte der Wurzel. *Flora* **17:** 257–266.

Scharek, R., Latasa, M., Karl, D. M., and Bidigare, R. R. (1999a) Temporal variations in diatom abundance and downward vertical flux in the oligotrophic North Pacific gyre. *Deep Sea Res* **46:** 1051–1075. doi:10.1016/S0967-0637(98)00102-2

Scharek, R., Tupas, L. M., and Karl, D. M. (1999b) Diatom fluxes to the deep sea in the oligotrophic North Pacific gyre at Station ALOHA. *Mar Ecol Prog Ser* **182:** 55–67. doi:10.3354/meps182055

Schaede, R. (1951) Über die Blaualgensymbiose von *Gunnera. Planta* **39:** 154–170.

Schaede, R. (1962) *Die Pflanzlichen Symbiosen*. Gustav Fischer Verlag, Stuttgart, Germany.

Scherrer, S., and Honegger, R. (2003) Inter- and intraspecific variation of homologous hydrophobin (H1) gene sequences among *Xanthoria* spp. (lichen-forming ascomycetes). *New Phytol* **158:** 375–389.

Scherrer, S., De Vries, O. M. H., Dudler, R., Wessels, J. G. H., and Honegger, R. (2000) Interfacial self-assembly of fungal hydrophobins in the lichen-forming ascomycete *Xanthoria parietina* and *X. ectaneoides. Fungal Genet Biol* **30:** 81–93.

Scherrer, S., Haisch, A., and Honegger, R. (2002) Characterization and expression of *XPH1*, the hydrophobin gene of the lichen-forming ascomycete *Xanthoria parietina. New Phytol* **154:** 175–184.

Schindler, A.K. (1905) Haloragaceae. In: Engler, A. *Das Pflanzenreich* **4:** 225.

Schmidt, E. W., Obraztova, A.Y., Davidson, S. K., Faulkner, D. J., and Haygood, M. G. (2000) Indentification of the antifungal peptide-containing symbiont of the marine sponge *Theonella swinhoei* a novel Delta-Proteobacterium Candidatus *Entotheonella palauensis. Mar Biol* **136:** 969–977.

Schmidt, J. (1901) Ueber *Richelia intracellularis*, eine neue in Plankton-Diatomeen lebende Alge. *Hedwigia* **40:** 112–115.

Schüßler, A., Meyer, T., Gehrig, H., and Kluge, M. (1997) Variations of lectin binding sites in extracellular glycoconjugates during the life cycle of *Nostoc punctiforme*, a potentially endosymbiotic cyanobacterium. *Eur J Phycol* **32:** 233–239.

Schüßler, A., Martin, H., Cohen, D., Fitz, M., and Wipf, D. (2006) Characterization of a carbohydrate transporter from symbiotic glomeromycotan fungi. *Nature* (London) **444:** 933–936.

Schütt, F. (1895) Peridineen der Plankton-Expedition. *Ergebnisse der Plankton-Expedition der Humboldt Stiftung* **4:** 1–170.

Seaward, M. R. D., Lynds, A., and Richardson, D. H. S. (1997) Lichens of Beaverbrook, Nova Scotia. *Proc Nova Scotia Institute of Science* **41:** 93–103.

Sherwood-Pike, M. A. (1985) *Pelicothallos* Dilcher, an overlooked fossil lichen. *Lichenologist* **17:** 114–115.

Shirley, B. W. (1996) Flavonoid biosynthesis: 'new' functions for an 'old' pathway. *Trends Plant Sci* **1:** 377–382.

Silvester, W. B. (1975) Endophyte adaptation in *Gunnera-Nostoc* symbiosis. In: Nutman, P. S. (Ed.) *Symbiotic Nitrogen Fixation in Plants*. Cambridge University Press, New York, NY, USA. pp. 521–538.

Silvester, W. B., and McNamara, P. J. (1976) The infection process and ultrastructure of the *Gunnera-Nostoc* symbiosis. *New Phytol* **77:** 135–141.

Silvester, W. B., and Smith, D. R. (1969) Nitrogen fixation by *Gunnera-Nostoc* symbiosis. *Nature* (London) **224:** 1231.

Silvester, W. B., Parsons, R., and Watt, P. W. (1996) Direct measurement of release and assimilation of ammonia in the *Gunnera-Nostoc* symbiosis. *New Phytol* **132:** 617–625.

Sillett, S. C. (1995) Branch epiphyte assemblages in the forest interior and on the clearcut edge of 700-year old forest canopy in western Oregon. *Bryologist* **98:** 301–312.

Sillett, S. C., and Goslin, M. N. (1999) Distribution of epiphytic macrolichens in relation to remnant trees in multiple-age Douglas-fir forest. *Can J For Res* **29:** 1204–1215.

Sillett, S. C., and Mc Cune, B. (1998) Survival and growth of cyanolichen transplants in Douglas-fir forest canopies. *Bryologist* **101:** 20–31.

Singh, R. S., Bhari, R., and Kaur, H. P. (2010) Mushroom lectins: Current status and future perspectives. *Critical Rev Biotechnol* **30:** 99–126. doi:10.3109/07388550903365048

Sipman, H. J. M. (2002) The significance of the northern Andes for lichens. *Bot Rev* **68:** 88–99.

Smith, D. C. (1980) Mechanisms of nutrient movement between lichen symbionts. In: Cook, C. B., Pappas, P. W., and Rudolph, E. D. (Eds.) *Cellular Interactions in Symbiosis and Parasitism*. Ohio State University Press, Ohio, USA. pp. 197–227.

Snoeijs, P., and Murasi, L. W. (2004) Symbiosis between diatoms and cyanobacterial colonies. *Vie Milieu* **54:** 163–170.

Söderback, E., and Bergman, B. (1992) The *Nostoc-Gunnera megallanica* symbiosis: Phycobiliproteins, carboxysomes and Rubisco in the cyanobiont. *Physiol Plant* **84:** 425–432.

Söderback, E., and Bergman, B. (1993) The *Nostoc-Gunnera* symbiosis: Carbon fixation and translocation. *Physiol Plant* **89:** 125–132.

Söderback, E., Lindblad, P., and Bergman, B. (1990) Developmental patterns related to nitrogen fixation in the *Nostoc-Gunnera megallanica* Lam. symbiosis. *Planta* **182:** 355–362.

Solereder, H. (1908) *Systemic Anatomy of the Dicotyledons.* Vol I. Revised D. H. Scott (1908) Claredon Press, Oxford, UK.

Solheim, B., and Zielke, M. (2002) Associations between cyanobacteria and mosses. In: Rai, A. N., Bergman, B., and Rasmussen, U. (Eds.) *Cyanobacteria in Symbiosis.* Kluwer Academic Publishers, Dordrecht, The Netherlands. pp. 137–152.

Sollins, P., Grier, C. C., McCorison, F. M., Cromack Jr., K., Fogel, R., and Fredriksen, R. L. (1980) The internal element cycles of an old-growth Douglas-fir ecosystem in western Oregon. *Ecological Monographs* **50:** 261–285.

Spaink, H. P., Solheim, B., Wiggen, H., Röberg, S. (2004) Associations between arctic cyanobacteria and mosses. *Symbiosis* **37:** 169–187.

Spanu, P. (1997) HCF-1, a hydrophobin from the tomato pathogen *Cladosporium fulvum. Gene* **193:** 89–96.

Spratt, E. R. (1911) Some observations on the life history of *Anabaena cycadeae. Ann Bot* **25:** 369–380.

Spratt, E. R. (1915) The root-nodules of the Cycadaceae. *Ann Bot* **29:** 619–626.

Stafford, H. A. (1997) Roles of flavonoids in symbiotic and defense functions in legume roots. *Bot Rev* **63:** 27–39.

Stein, W. E., Harmon, G. D. and Hueber, F. M. (1993) *Spongiophyton* from the Lower Devonian of North America reinterpreted as a lichen. *Am J Bot* **80:** 93.

Stenroos, S., Högnabba, F., Myllys, L., Hyvönen, J., and Thell, A. (2006) High selectivity in symbiotic associations of lichenized ascomycetes and cyanobacteria. *Clastidics* **22:** 230–238.

Steinberg, N. A., and Meeks, J. C. (1989) Photosynthetic CO_2 fixation and ribulose biphosphate carboxylase/oxygenase activity of *Nostoc* sp. strain UCD 7801 in symbiotic association with *Anthoceros punctatus. J Bacteriol* **171:** 6227–6233.

Steinberg, N. A., and Meeks, J. C. (1991) Physiological sources of reductant for nitrogen fixation activity in *Nostoc* sp. strain UCD 7801 in symbiotic association with *Anthoceros punctatus. J Bacteriol* **173:** 7324–7329.

Steindler, L., Huchon, D., Avni, A., and Ilan, M. (2005) 16S rRNA Phylogeny of sponge-associated cyanobacteria. *Appl Environ Microbiol* **71:** 4127–4131.

Stewart, W. D. P., and Rodgers, G. A. (1977) The cyanophyte-hepatic symbiosis. *New Phytol* **78:** 459–471.

Stock, P. A., and Silvester, W. B. (1994) Phloem transport of recently fixed nitrogen in the *Gunnera-Nostoc* symbiosis. *New Phytol* **126:** 259–266.

Stocker-Wörgötter, E. (1995) Experimental cultivation of lichens and lichen symbionts. *Can J Bot* **73:** S579–S589.

Stocker-Wörgötter, E., and Türk, R. (1990) Thallus formation of the cyanobacterial lichen *Peltigera didactyla* from soredia under laboratory conditions. *Bot Acta* **103:** 315–321.

Stocker-Wörgötter, E., and Türk, R. (1994) Artificial resynthesis of the photosymbiodemes *Peltigera leucophlebia* under laboratory conditions. *Cryptogamic Bot* **4:** 300–308.

Strasburger, E. (1873) Ueber *Azolla.* Verlag von Ambr Abel., Leipzig, Germany.

Subramaniam, A., Yager, P. L., Carpenter, E. J., Mahaffey, C., Bjorkman, K., Cooley, S., Kustka, A. B., Montoya, J. P., Sañudawilhelmy, S. A., Shipe, R., and Capone, D. G. (2008) Amazon River enhances diazotrophy and carbon sequestration in the tropical North Atlantic Ocean. *Proc Natl Acad Sci USA* **105:** 10460–10465.

Subramanian, G., and Malliga, P. (1988) Isolation of *Anabaena azollae* from the megasporocarp of *Azolla. Curr Sci* **57:** 1352–1353.

Summerfield, T. C., and Eaton-Rye, J. J. (2006) *Pseudocyphellaria crocata, P. neglecta* and *P. perpetua* from Northern and Southern Hemipsheres are a phylogenetic species and share cyanobionts. *New Phytol* **170:** 597–607.

Summerfield, T. C., Galloway, D. J., and Eaton-Rye, J. J. (2002) Species of cyanolichens from *Pseudocyphellaria* with indistinguishable ITS sequences have different photobionts. *New Phytol* **155:** 121–129.

Sundström, B. G. (1984) Observations on *Rhizosolenia clevei* Ostenfeld (Bacillariophyceae) and *Richelia intracellularis* Schmidt (Cyanophyceae). *Bot Mar* **27:** 345–355.

Svenson, H. K. (1944) The new world species of *Azolla. Am Fern J* **34:** 69–84.

Svenning, M. M., Eriksson, T., and Rasmussen, U. (2005) Phylogeny of symbiotic cyanobacteria within the genus *Nostoc* based on 16S rDNA sequence analysis. *Arch Microbiol* **183:** 19–26.

Swanson, A., and Fahselt, D. (1997) Effects of ultraviolet on polyphenolics of *Umbilicaria americana. Can J Bot* **75:** 284–289.

Tagu, D., Nasse, B., and Martin, F. (1996) Cloning and characterization of hydrophobin-encoding cDNAs from the ectomycorrhizal basidiomycete *Pisolithus tinctorius. Gene* **168:** 93–97.

Tapper, R. (1981) Direct measurement of translocation of carbohydrate in the lichen *Cladonia convoluta,* by quantitative autoradiography. *New Phytol* **89:** 429–437.

Taylor, D. L. (1974) Symbiotic marine algae: Taxonomy and biological fitness. In: Vernberg, C. B. W. (Ed.) *Symbiosis in the Sea.* University of South Carolina Press, Columbia, USA. pp. 245–262.

Taylor, M. W., Schupp, P. J., de Nys, R., Kjelleberg, S., and Steinberg, P. D. (2005) Biogeography of bacteria associated with the marine sponge *Cymbastella concentrica. Environ Microbiol* **7:** 419–433.

Taylor, T. N., and Taylor, E. L. (1993) *The Biology and Evolution of Fossil Plants*. Prentice-Hall, Englewood Cliffs, NJ, USA.

Taylor, T. N., Hass, H., and Kerp, H. (1997) A cyanolichen from the Lower Devonian Rhynie Chert. *Am J Bot* **84:** 992–1004.

Tel-Or, E., Sandovsky, T., Kobiler, D., Arad, C., and Weinberg, R. (1983) The unique symbiotic properties of *Anabaena* in the water fern *Azolla*. In: Papegeorgiou, G. C. and Packer, L. (Eds.) *Photosyntheic Prokaryotes: Cell Differentiation and Function*. Elsevier Science, New York, NY, USA. pp. 303–314.

Thacker, R. W. (2005) Impacts of shading on sponge-cyanobacteria symbioses: A comparison between host-specific and generalist associations. *Integr Comp Biol* **45:** 369–376.

Thacker, R. W., and Starnes, S. (2003) Host specificity of the symbiotic cyanobacterium *Oscillatoria spongeliae* in marine sponges, *Dysidea* spp. *Marine Biol* **142:** 643–648.

Thacker, R. W., Becerro, M. A., Lumbang, W. A. and Paul, V. J. (1998) Allelopathic interactions between sponges on a tropical reef. *Ecology* **79:** 1740–1750.

Thakur, N. L., Hentschel, U., Krasko, A., Pabel, C. T., Anil, A. C., and Müller, W. E. G. (2003) Antibacterial activity of the sponge *Suberites domuncula* and its primmorphs: Potential basis for epibacterial chemical defense. *Aquatic Microbial Ecol* **31:** 77–83.

Thompson, J. N. (1994) *The Coevolutionary Process*. University of Chicago Press, Chicago, Illinois, USA.

Tikhonovich, I., Lutenberg, B., and Prvorov, N. (2004) *Biology of Molecular Plant-Microbe Interactions*. Vol. 4, CPL Scientific Publishers Ltd. Berkshire, UK. pp. 633.

Tobin, A. K., and Yamaya, T. (2001) Cellular compartmentation of ammonium assimilation in rice and barley. *J Exp Bot* **52:** 591–604.

Tomaselli, L., Margheri, M. C., Giovannetti, L., Sili, C., and Carlozzi, P. (1988) The taxonomy of *Azolla* spp. cyanobionts. *Annali di Microbiologia* **38:** 157–161.

Tønsberg, T., and Holtan-Hartwig, J. (1983) Phycotype pairs in *Nephroma*, *Peltigera* and *Lobaria* in Norway. *Nordic J Bot* **3:** 681–688. doi:10.1111/j.1756-1051.1983.tb.01479.x

Towata, E. M. (1985) Mucilage glands and cyanobacterial colonization in *Gunnera kaalensis* (Haloragaceae). *Bot Gaz* **146:** 56–62.

Trembley, M. L., Ringli, C., and Honegger, R. (2002) Differential expression of hydrophobins *DGH1*, *DGH2*, and *DGH3* and immunolocalization of DGH1 in strata of the lichenized basidiocarp of *Dictyonema glabratum*. *New Phytol* **154:** 185–195.

Trembley, M. L., Ringli, C., and Honegger, R. (2002a) Hydrophobins DGH1, DGH2, and DGH3 in the lichen-forming basidiomycete *Dictyonema glabratum*. *Fungal Genet Biol* **35:** 247–259.

Trembley, M. L., Ringli, C., and Honegger, R. (2002b) Morphological and molecular analysis of early stages in the resynthesis of the lichen *Baeomyces rufus*. *Mycol Res* **106:** 768–776.

Trench, R. K. (1993) Microalgal–invertebrate symbioses: a review. *Endocytobiosis Cell Res.* **9:** 135–175.

Trench, R. K. (1997) Diversity of symbiotic dinoflagellates and the evolution of microalgal invertebrate symbiosis. *Proc Int Coral Reef Symp* 8th Panama **2:** 1275–1286.

Tschermak-Woess, E. (1988) The algal partner. In: Galun, M. (Ed.) *CRC Handbook of Lichenology*, Vol. I. CRC Press, Boca Raton, Florida, FL, USA. pp. 39–92.

Turk, K. A., Rees, A. P., Zehr, J. P., Pereiro, N., Shelley, R., Lohan, M., Malcolm, E., Woodward, E. M. S., and Gilbert, J. (2011) Nitrogen fxation and nitrogenase (*nifH*) expression in tropical waters of the eastern North Atlantic. *ISME J* **5:** 1201–1212. doi:10.1038/ismej.2010.205.

Tumer, N. E., Robinson, S. J., and Haselkorn, R. (1983) Different promoters for the *Anabaena* glutamine synthetase gene during growth using molecular or fixed nitrogen. *Nature* (London) **306:** 337–342.

Uheda, E. (1986) Isolation of hair cells from *Azolla filiculoides* var. *japonica* leaves. *Plant Cell Physiol* **27:** 1255–1261.

Uheda, E., Maejima, K., and Shiomi, N. (2004) Localization of glutamine synthetase isoforms in hair cells of *Azolla* leaves. *Plant Cell Physiol* **45:** 1087–1092.

Unson, M. D., and Faulkner, D. J. (1993) Cyanobacterial symbiont biosynthesis of chlorinated metabolites from *Dysidea herbacea* (Porifera). *Cell Mol Life Sci* **49:** 349–353.

Unson, M. D., Holland, N. D., and Faulkner, D. J. (1994) A brominated secondary metabolite synthesized by the cyanobacterial symbiont of a marine sponge and accumulation of the crystalline metabolite in the sponge tissue. *Mar Biol* **119:** 1–11.

Usher, K. M., Bergman, B., and Raven, J. A. (2007) Exploring cyanobacterial mutualisms. *Annu Rev Ecol Evol Sys* **38:** 255–273. doi: 10.1146/annu rev ecolo sys38. 091 206.095 641.

Usher, K. M., Fromont, J., Sutton, D. C., and Toze, S. (2004a) The biogeography and phylogeny of unicellular cyanobacterial symbionts in sponges from Australia and the Mediterranean. *Microb Ecol* **48:** 167–177.

Usher, K. M., Kuo, J., Fromont, J., Kuo, J., and Sutton, D. C. (2001) Vertical transmission of cyanobacterial symbionts in the marine sponge *Chondrilla australiensis* (Desmospongiae). *Hydrobiologia* **461:** 15–23.

Usher, K. M., Kuo, J., Fromont, J., Toze, S., and Sutton, D. C. (2006) Comparative morphology of five species of symbiotic and non-symbiotic coccoid cyanobacteria. *Eur J Phycol* **41:** 179–188.

Usher, K. M., Toze, S., Fromont, J., Kuo, J., and Sutton, D. C. (2004b) A new species of cyanobacterial symbiont from the marine sponge *Chondrilla nucula*. *Symbiosis* **36:** 183–192.

Vacelet, J. (1975) Etude en microscopie dectronique de l'associat~one ntre bacteries et spongiaires du genre *Verongia* (Dictyoceratida). *J Microsc Biol Cell* **23:** 271–288.

Vacelet, J., and Donadey, C. (1977) Electron microscope study of the association between some sponges and bacteria. *J Exp Mar Ecol* **30:** 301–314.

Vacelet, J., Boury-Esnault, N., Fiala-Medioni, A., and Fisher, C. R. (1995) A methanotrophic carnivorous sponge. *Nature* (London) **377:** 296.

Vagnoli, L., Margheri, M. C., Allotta, G., and Materassi, R. (1992) Morphological and physiological properties of symbiotic cyanobacteria. *New Phytol* **120:** 243–249.

Van Pelt, R. (2001) *Forest giants of the Pacific Coast.* Global Forest Society and University of Washington Press, Seattle, WA, USA.

Venrick, E. (1974) The distribution and significance of *Richelia intracellularis* Schmidt in the North Pacific central gyre. *Limnol Oceanogr* **19:** 437–445.

Villarreal, J. C. A., and Renzaglia, S. (2006) Structure and development of *Nostoc* strands in *Leiosporoceros dussii* (Anthoceratophyta): A novel symbiosis in land plants. *Am J Bot* **93:** 693–705.

Villareal, T. A. (1987) Evaluation of nitrogen fixation in the diatom genus *Rhizosolenia* Ehr. in the absence of its cyanobacterial symbiont *Richelia intracellularis* Schimidt. *J Plankton Res* **9:** 965–971.

Villareal, T. A. (1989) Division cycles in the nitrogen-fixing *Rhizosolenia* (Bacillariophyceae) *Richelia* (Nostocaceae) symbiosis. *Br Phycol J* **24:** 357–365.

Villareal, T. A. (1990) Laboratory culture and preliminary characterization of the nitrogen-fixing *Rhizosolenia-Richelia* symbiosis. *Mar Ecol* **11:** 117–132.

Villareal, T. A. (1991) Nitrogen-fixation by the cyanobacterial symbiont of the diatom genus *Hemiaulus*. *Mar Ecol Prog Ser* **76:** 201–204.

Villareal, T. A. (1992) Marine nitrogen-fixing diatom-cyanobacteria symbioses. In: Carpenter, E. J., Capone, D. G., and Rueter, J. G. (Eds.) *Marine Pelagic Cyanobacteria: Trichodesmium and Other Diazotrophs*. Kluwer Academic Publishers, Amsterdam, The Netherlands. pp. 163–175.

Villareal, T. A. (1994) Widespread occurrence of the *Hemiaulus*-cyanobacteria symbiosis in the southwest North Atlantic Ocean. *Bull Mar Sci* **54:** 1–7.

Villareal, T. A., Adornato, L., Wilson, C., and Schoenbaechler, C. A. (2011) Summer blooms of diatom-diazotroph assemblages and surface chlorophyll in the North Pacific gyre: A disconnect. *J Geophys Res* **116:** doi:10.1029/2010JC006268, 2011

Villareal. T. A., Altabet, M. A., and Culver-Rymzsa, K. (1993) Nitrogen transport by vertically migrating diatom mats in the North Pacific Ocean. *Nature* (London) **363:** 709–712.

Vitikainen, O. (1994) Taxonomic revision of *Peltigera* (lichenized Ascomycotina) in Europe. *Acta Bot Fennica* **152:** 1–96.

Vivas, M., Sacristán, M., Legaz, M. E., and Vicente, C. (2010) The cell recognition model in chlorolichens involving a fungal lectin binding to an algal ligand can be extended to cyanolichens. *Plant Biol* **12:** 615–621.

Von Neumann, D., Ackerman, M., and Jacob, F. (1970) Cited from Silvester, W. B. and McNamara, P. J. (1976).

Wagner, G. M. (1987) *Azolla*: A review of its biology and utilization. *Bot Rev* **63:** 1–26.

Wang, C. M., Ekman, M., and Bergman, B., (2004) Expression of cyanobacterial genes involved in heterocyst differentiation and dinitrogen fixation along a plant symbiosis development profile. *Mol Plant-Microbe Interactions* **17:** 438–443.

Watanabe, K. (1924) Studien über die Koralloide von Cycas revolute. Botanical Magazine **38:** 165–187.

Webb, V. L., and Mass, E. W. (2002) Sequence analysis of 16S rRNA gene of cyanobacteria associated with the marine sponge *Mycale* (*Carmia*) *hentscheli*. *FEMS Microbiol Lett* **207:** 43–47.

Webster, N. S., Wilson, K. J., Blackall, L. L., and Hill, R. T. (2001) Phylogenetic diversity of bacteria associated with the marine sponge *Rhopaloeides odorabile*. *Appl Environ Microbiol* **67:** 434–444.

Weisburg, W. G., Barns, S. M., Pelletier, D. A., and Lane, D. J. (1991) 16S ribosomal DNA amplification for phylogenetic study. *J Bacteriol* **173:** 697–703.

Wessels, J. G. H. (1994) Developmental regulation of fungal cell wall formation. *Annu Rev Phytopathol* **32:** 413–437.

Wessels, J. G. H. (1997) Hydrophobins: Proteins that change the nature of the fungal surface. *Adv Microb Physiol* **38:** 1–45.

Wessels, J. G. H. (2000) Hydrophobins, unique fungal proteins. *Mycologist* **14:** 153–159.

Wessels, J. G. H., de Vries, O. M. H., Ásgeirsdóttir, S. A., and Schuren, F. H. J. (1991) Hydrophobin genes involved in formation of aerial hyphae and fruit bodies in *Schizophyllum commune*. *Plant Cell* **3:** 793–799.

West, N. J., and Adams, D. G. (1997) Phenotypic and genotypic comparison of symbiotic and free-living cyanobacteria from a single field site. *Appl Environ Microbiol* **63:** 4479–4484.

White, A. E., Prahl, F. G., Letelier, R. M., and Popp, B. N. (2007a) Summer surface waters in the Gulf of California: Prime habitat for biological N_2 fixation, *Global Biogeochem Cycles* **21:** GB2017, doi:10.1029/ 2006GB002779

White, A. E., Spitz, Y. H., and Letelier, R. M. (2007b) What factors are driving summer phytoplankton blooms in the North Pacific subtropical gyre? *J Geophys Res* **112:** C12006. doi:10.1029/2007JC004129

White, F. J., and James, P. W. (1988) Studies on the genus *Nephroma* II. The Southern temperate species. *Lichenologist* **20:** 103–166.

Whiteford, J. R., and Spanu, P. D. (2002) Hydrophobins and the interactions between fungi and plants. *Mol Plant Pathol* **3:** 391–400.

Wilkinson, C. R. (1983) Net primary productivity in coral reef sponges. *Science* **219:** 410–412.

Wilkinson, C. R. (1992) Symbiotic interactions between marine sponges and algae. In: Reisser, W. (Ed.) *Algae and Symbioses.* Bioprocess, Bristol, England, UK. pp. 112–151.

Wilkinson, C. R., and Fay, P. (1979) Nitrogen fixation in coral reef sponges with symbiotic cyanobacteria. *Nature* (London) **279:** 527–529.

Wilkinson, C. R., and Vacelet, J. (1979) Transplantation of marine sponges to different conditions of light and current. *J Exp Mar Biol Ecol* **37:** 91–104.

Wilson, C., Villareal, T. A., Maximenko, N., Bograd, S. J., Montoya, J. P., and Schoenbaechler, C. A. (2008) Biological and physical forcings of late summer chlorophyll blooms at 30°N in the oligotrophic Pacific. *J Mar Syst* **69:** 164–176. doi:10.1016/j.jmarsys.2005.09.018

Winter, G. (1935) Über die Assimilation des Luftstichstoffs durch endophytische Blaualgen. *Beitr Biol Pflanzen* **23:** 295–305.

Wirtz, N., Lumbsch, H. T., Green, T. G. A., Turk, R., Pintado, A., Sancho, L., and Schroeter, B. (2003) Lichen fungi have low cyanobiont selectivity in maritime Antarctica. *New Phytol* **160:** 177–183.

Wittman, W., Bergersen, F. J., and Kennedy, G. S. (1965) The coralloid roots of *Macrozamia communis* L.Johnson. *Aus J Biol Sci* **18:** 1129–1134.

Wolf, E. (2003) Partnererkennung und Inkorporation bei der Pilz-Endocyanose *Geosiphon pyriformis*. Technische Universität Darmstadt, Hessische Landes-und Hochschulbibliothek.

Wolk, C. P. (2000) Heterocyst formation in *Anabaena*. In: Brun, Y. V., and Shimkers, L. J. (Eds.) *Prokaryotic Development*. American Society for Microbiology, Washington, D. C. USA. pp. 83–104.

Woodson, S. A., and Cech, T. R. (1989) Reverse splicing of the *Tetrahymena* group I introns: Implication for the directionality of splicing and for intron transposition. *Cell* **57:** 335–345.

Wong, F. C. Y., and Meeks, J. C. (2001) The *hetF* gene product is essential to heterocyst differentiation and affects HetR function in the cyanobacterium *Nostoc punctiforme. J Bacteriol* **183:** 2654–2661.

Wong, F. C. Y., and Meeks, J. C. (2002) Establishment of a functional symbiosis between the cyanobacterium *Nostoc punctiforme* and the bryophyte *Anthoceros punctatus* requires genes involved in nitrogen control and initiation of heterocyst differentiation. *Microbiology* **148:** 315–323.

Wösten, H. A. B. (2001) Hydrophobins: Multipurpose proteins. *Annu Rev Microbiol* **55:** 625–646.

Wösten, H . A. B., Ásgeirsdóttir, S. A., Krook, J. H., Drenth, J. H. H., and Wessels, J. G. H. (1994) The SC3 hydrophobin self-assemles at the surface of aerial hyphae as a protein membrane constituting the hydrophobic rodlet layer. *Eur J Cell Biol* **63:** 122–129.

Wösten, H. A. B., De Vries, O. M. H., and Wessels, J. G. H. (1993) Interfacial self-assembly of a fungal hydrophobic rodlet layer. *Plant Cell* **5:** 1567–1574.

Wösten, H. A. B., Ruardy, T. G., van der Mei, H. C., Busscher, H. J., and Wessels, J. G. H. (1995) Interfacial self-assembly of a *Schizophyllum commune* hydrophobin into an insoluble amphipathic membrane depends on surface hydrophobicity. *Colloids Surf* **5:** 189–195.

Wösten, H. A. B., Schuren, F. H. J., and Wessels, J. G. H. (1994) Interfacial self-assembly of a hydrophobin into an amphipathic membrane mediates fungal attachment to hydrophobic surfaces. *EMBO J* **13:** 5848–5854.

Wynne, M. J. (1969) Life history and systematic studies of Pacific North American Phaeophyceae (brown algae). *Univ Calif Publi Bot* **50:** 1–88.

Wynn-Williams, D. D., (2000) Cyanobacteria in the deserts—life at the limit?. In: Whitton, B.A., Potts, M. (Eds.), *The Ecology of Cyanobacteria*. Kluwer Academic Publishers, Dordrecht, The Netherlands. pp. 341–366.

Xie, Z.-P., Stachelibn, C., Vierheilig, H., Wiemken, A., Jabbouri, S., Broughton, W. J., Vögeli-Lange, R., and Boller, T. (1995) Rhizobial nodulation factors stimulate mycorrhizal colonization of nodulating and nonnodulating soybeans. *Plant Physiol* **108:** 1519–1525.

Xu, M. Q., Kathe, S. D., Goodrichblair, H., Nierzwicki-Bauer, S. A., and Shub, D. A. (1990) Bacterial origin of a chloroplast intron-conserved self-splicing group I introns in cyanobacteria. *Science* **250:** 1566–1570.

Yoshimura, I., Kurokawa, T., and Kinoshita, Y. (1993) Development of lichen thalli *in vitro. Bryologist* **96:** 412–421.

Yoshimura, I., Kurokawa, T., Yamamoto, Y., and Kinoshita, Y. (1994) *In vitro* development of the lichen thallus of some species of *Peltigera*. *Cryptogamic Botany* **4:** 314–319.

Zackrisson, O., DeLuca, T. H., Nilsson, M.-C., Sellstedt, A., and Berglund, L. (2004) Nitrogen fixation increases with successional age in boreal forests. *Ecology* **85:** 3327–3334.

Zeev, E. B., Yogev, T., Man-Ahronovich, D., Kress, N., Hernt, B., Bêjá, O., and Berman-Frank, I. (2008) Seasonal dynamics of the endosymbiotic nitrogen-fixing cyanobacterium *Richelia intracellularis* in the eastern Mediterranean Sea. *ISME J* **2:** 911–923.

Zehr, J. P., and Kudela, R. M. (2011) Nitrogen cycle of the open ocean: From genes to ecosystems. *Annu Rev Mar Sci* **3:** 197–225.

Zhang, L., Villalon, D., Sun, Y., Kazmierczak, P., and Van Alfen, N. K. (1994) Virus-associated down-regulation of the gene encoding cryparin, an abundant cell-surface protein from the chestnut blight fungus, *Cryphonectria parasitica*. *Gene* **139:** 59–64.

Zheng, W., China, P. R., Song, T., Bao, X., Bergman, B., and Rasmussen, U. (2002) High cyanobacterial diversity in coralloid roots of cycads revealed by PCR fingerprinting. *FEMS Microbiol Ecol* **40:** 215–222.

Zhu, C. (1982) Fine structure of blue-green algae and the cells lined along the endophyte cavity in the coralloid roots of *Cycas*. *Acta Bot Sinica* **24:** 109–114 (in Chinese).

Ziegler, R. (1992) Komplex-thallöse, fossil Organismen mit blattflechtenartigen Bau aus dem mittleren Keuper (Trias, Karn) Unterfrankens. In: Kovar-Eder (Ed.) *Palaeovegetational development in Europe and regions relevant to its palaeofloristic evolution*. Museum of Natural History, Vienna. pp. 341–349.

Zimmerman, R. A., and Dahlberg, A. E. (Eds.) (1996) *Ribosomal RNA: Structure, Evolution, Processing and Function in Protein Biosynthesis*. CRC Press, Boca Raton, Florida, FL, USA.

Zimmerman, W. J., and Bergman, B. (1990) The *Gunnera* symbiosis: DNA restriction fragment length polymorphism and protein comparisons of *Nostoc* symbionts. *Microb Ecol* **19:** 291–302.

Zimmerman, W. J., Rosen, B. H., and Lumpkin, T. A. (1989) Enzymatic lectin and morphological characterization and classification of presumptive cyanobionts from *Azolla* Lam. *New Phytol* **113:** 497–503.

Zimmerman, W. J., Watanabe, I., Ventura, T., Payawal, P., and Lumpkin, T. A. (1991) Aspects of the genetic and botanical status of neotropical *Azolla* species. *New Phytol* **119:** 561–566.

Zoller, S., Lutzoni, F., and Scheidegger, C. (1999) Genetic variation within and among populations of the threatened lichen *Lobaria pulmonaria*. Switzerland and implications for its conservation. *Mol Ecol* **8:** 2049–2059.

CHAPTER 8

CYANOPHAGES

The discovery of viruses infecting bacteria independently by F.W. Twort and Felix d'Herelle in 1915 and 1917, respectively laid a foundation for a study of these viruses. These have been termed as bacteriophages or briefly as phages. Researches on the biology of phages can be divided into three phases. In the first phase (1915–1970), basic studies on phages contributed to our understanding of phage morphology, structure and multiplication cycles. The second phase (1970–1990) dealt with genetics of phages that enabled us to have a better understanding of basic biological processes. The major breakthroughs in phage research ushered in new areas of study in molecular biology that had instant ramifications into genetic engineering and biotechnology. The third phase of phage research, started in 1990s, pertains to the extensive studies on the diversity of marine viruses including phages determined on the basis of identification of gene sequences from marine waters. Thus the study of uncultured phage communities along with uncultured microbial gene sequences has given rise to a new branch of genetics known as metagenomics (Handelsman, 2004; Schloss and Handelsman, 2005; Zinger *et al.*, 2011). When compared to cellular metagenomic surveys, 65 to 95% of marine viral metagenomic sequences had no counterparts in previously described sequences. In this connection, the uncultured phage communities appear to be most diverse as 3000 to 7000 viral types could be analysed in a 200 litre sample of seawater (Breitbart *et al.*, 2002). The acquisition of host genes by some of these phages and their transfer back into hosts has provided means for the understanding of not only lateral gene transfer (LGT) in the marine ecosystem but also the evolution of phage genomes along with the genomes of their hosts (Breitbart and Rohwer, 2005; Stern and Sorek, 2010). In addition, studies on viral mortality rates of marine heterotrophic bacteria and cyanobacterial picoplanktons have highlighted the role of marine viruses in biogeochemical cycling and the probable ecological effects (Fuhrman, 1999; Wommack and Colwell, 2000; Storrie-Lombardi and Pinkart, 2007; DeCorte *et al.*, 2010; Breitbart, 2011; Danovaro *et al.*, 2011). Rightly, this third phase also termed as 'third age of phages' by Mann (2005) has diverted the attention of every one from the labaoratory to the deep seas.

It was almost fifty years after the discovery of bacteriophages that the isolation of first virus, LPP-1 active against three genera of cyanobacteria, i.e. *Lyngbya*, *Phormidium* and *Plectonema boryanum* was made (Safferman and Morris 1963, 1964a). This generated a lot of enthusiasm for two reasons. Firstly, it was thought that the discovery of these viruses would help in the study of genetics of cyanobacteria and secondly in the control of algal blooms. Thus a number of workers reported the isolation of cyanophages from all over the world. But the euphoria generated by the discovery of these viruses died down without meeting the two expectations. Freshwater cyanophages infective against various unicellular and filamentous cyanobacteria are summarized along with their sensitive hosts and country of origin (Tables 1 and 2).

Table 1: Freshwater cyanophages that infect unicellular cyanobacteria.

Sr. No.	Cyanophage	Host(s)	Country	Authors
1	AS-1	*Anacystis nidulans*	USA	Safferman *et al.* (1972)
2	AS-1M	*Synechococcus cedrorum*	USA	Sherman and Connelly (1976)
3	AS-1K	*Synechococcus* sp.	Russia	Goryushin and Shainskaya (1984)
4	AC-1	*A. nidulans* *Chroococcus minor*	India	Venkataraman *et al.* (1973)
5	AM-1	*A. nidulans* *Microcystis aeruginosa*	Russia	Sirenko *et al.* (1976)
6	Aph-1	*Aphanothece stagnina*	Scotland	Barnet *et al.* (1981)
7	SM-1	*Synechococcus elongatus & M. aeruginosa*	USA	Safferman *et al.* (1969b)
8	SM-2	*S. elongatus & M. aeruginosa*	USA	Fox *et al.* (1976)
9	S-1	*Synechococcus* sp.	USA	Adolph and Haselkorn (1973a)
10	S-2L	*Synechococcus* sp.	Russia	Khudyakov (1977)
11	S-3L	*S. schmidlea*	Russia	Koz'yakov (1981, 1982)
12	S-4L	*S. elongatus*	Russia	Khudyakov and Matev (1982)
13	S-8K	*Synechococcus* sp.	Russia	Goryushin and Averikiev (1982)
14	Ma-LBP	*M. aeruginosa*	Australia	Tucker and Pollard (2005)

I. NOMENCLATURE AND CLASSIFICATION

To remove the ambiguity that persisted in the nomenclature and classification of viruses, International Committee on Taxonomy of viruses (ICTV) constituted subcommittees on vertebrate, invertebrate, bacterial, plant, protozoal and fungal viruses for receiving taxonomic proposals and framing guidelines for the classification of viruses. The ICTV in its IXth Report recommended the inclusion of six orders, 87 families, 19 subfamilies and 348 genera of viruses (Ackermann, 2011). In case of bacteriophages, Bradley's (1967) classification formed the basis and ICTV took into account 70 different properties of bacterial viruses to classify them. On the basis of phage morphology, they have recommended the division of bacteriophages into different families by adding a suffix–viridae to the names of families of phages. There are a total of ten families and the first three families belonging to the Order Caudovirales (tailed phages), i.e. Myoviridae, Siphoviridae (also termed as Styloviridae) and Podoviridae are widely represented. The largest among these is the family Siphoviridae represented by 3262 members followed by Myoviridae (with 1312 members) and Podoviridae (771 members; Ackermann, 2007). The members belonging to these three families contain double-stranded (ds) DNA as genetic material. The properties of all the families of phages with the type of nucleic acid, examples and total members in each family are depicted in Fig. 1. Representatives of T-even phages are grouped in Myoviridae. Thus the T-series of phages with even numbers, i.e. T2, T4 and T6 are morphologically, genetically and serologically related amongst themselves and different from the rest. These are autonomous virulent and do not require presence of bacterial chromosome for their multiplication, so it is destroyed soon after infection. Moreover, the DNA of these phages possesses 5-hydroxymethylcytosine in place of normal cytosine. These exhibit icosahedral symmetry where the heads are either isometric (P2) or elongated (T2) and possess a contractile tail with tail plate and tail fibers. The archetype of Myoviridae is phage T4. On the basis of phylogenetic relationships based on head and tail gene sequences four sub-groups are recognized, i.e. (i) T-evens: members of which are closely related to T4. (ii) Pseudo-T-evens: These are phylogenetically divergent but morphologically similar. Examples are RB49, RB42 that infect *E. coli.* (iii) Schizo-T-evens: These have more elongated

Table 2: Freshwater cyanophages infective against filamentous cyanobacteria.

Sr. No.	Cyanophage	Hosts	Country	Authors
1	LPP-1	*Lyngbya* sp. *Phormidium* sp., and *Plectonema boryanum*	USA	Safferman and Morris (1963)
2	LPP-1G (G-III)	*P. boryanum*	Israel	Padan *et al.* (1967)
3	P-2, P-3 and P-4	*P. boryanum*	India	Singh and Singh (1967)
4	LPP-DUN (D-1)	*P. boryanum*	Scotland	Daft *et al.* (1970)
5	LPP-2	*Lyngbya* sp., *Phormidium* sp., and *P. boryanum*	USA	Safferman *et al.* (1969a)
6	LPP-3	*P. boryanum*	Russia	Mendzhul *et al.* (1988)
7	P-5 and P-6	*P. boryanum*	India	Singh *et al.* (1972)
8	LPP-2N	*P. boryanum*	USA	Booth (1975)
9	AR-1 and AR-2	*Anabaenopsis raciborskii* and *Raphidiopsis indica*	India	Singh and Singh (1967)
10	A-1	*Anabaena cylindrica*	India	Singh *et al.* (1972)
11	A-2	*Anabaena variabilis*	Sweden	Granhall and Hofsten (1969)
12	A-4L	*A.variabilis*	Russia	Khudyakov and Gromov (1973)
13	C-1	*Cylindrospermum* sp.	India	Singh and Singh (1967)
14	N-1	*Nostoc muscorum*	USA	Adolph and Haselkorn (1971)
15	N-2 N-L	*Nostoc linckia* *N. linckia*	Russia Russia	Mendzhul *et al.* (1983)
16	TAuHN-1	*Tolypothrix tenuis, Aulosira fertilissima, N. muscorum, Hapalosiphon intricatus*	India	Kaushik and Venkataraman (1973)
17	O-1	*Oscillatoria chlorina*	India	Trivedi and Oza (1979)
18	A-1 (L) to App (L)*	*A. variabilis*	Russia	Koz'yakov (1972, 1977)
19	AN-Series 10-26**	*A. variabilis,* *N. linckia,* *Nostoc* sp. ATCC 29131-c, *Nostoc* sp. ATCC 29131-b, *Nostoc* sp. ATCC 29105-a, *Nostoc* sp. ATCC 29151	USA	Hu *et al.* (1981)
20	NP-1T	*Nostoc* sp., *Plectonema* sp.	Russaia	Muradov *et al.* (1990a)

*Of these cyanophages all but A-1(L) and A-4(L) have been lost (See Hu *et al.*, 1981).
**AN-10 was isolated as a contaminating phage in the stock of A-4(L) received from USSR by Hu *et al.* (1981). These sixteen new isolates exhibited variations in sensitizing the 8 strains of *Anabaena variabilis* and other *Nostoc* strains (See Hu *et al.*, 1981).

heads than the first two. *Aeromonas* phages such as nt-1, kvp20, 65 and Aer 1 belong to this category (Tétart *et al.*, 2001; Desplats *et al.*, 2002) and (iv) Exo-T-evens: Marine cyanophages such as S-PM2 that have diverged from T4 and unique for carrying "photosynthetic" genes that encode homologues of D1 and D2 proteins of the host PSII are included here (Hambly *et al.*, 2001).

Siphoviridae possess phage heads with icosahedral symmetry with long flexible or rigid non-contractile tails with or without tail fibers. Representatives of this family are phage lambda, T1 and T5. Phage T5 that occupies an intermediate position in between the T-even and T-odd phages, interestingly is autonomous virulent like T-even phages but retains ordinary cytosine in its DNA

Morphology	Nucleic acid	Family(-viridae)	Example	Members	Characteristics
Tailed	DNA, ds, L	*Myo-*	T4	1312	Tail long, contractile
	ds, L	*Sipho-*	λ	3262	Tail long, non-contractile
	ds, L	*Podo-*	T7	771	Tail short
Polyhedral	DNA, ss, C	*Micro-*	ϕX174	38	
	ds, C, T	*Cortico-*	PM2	3?	Complex capsid, lipids
	ds, L	*Tecti-*	PRD1	19	Internal membrane
	RNA, ss, L	*Levi-*	MS2	38	Envelope, lipids
	ds, L, S	*Cysto-*	ϕ6	3	Filaments or rods
Filamentous	DNA, ss, C	*Ino-*	fd	66	Envelope, lipids
Pleomorphic	DNA, ds, C, T	*Plasma-*	L2	5	Spindle-shaped, no capsid

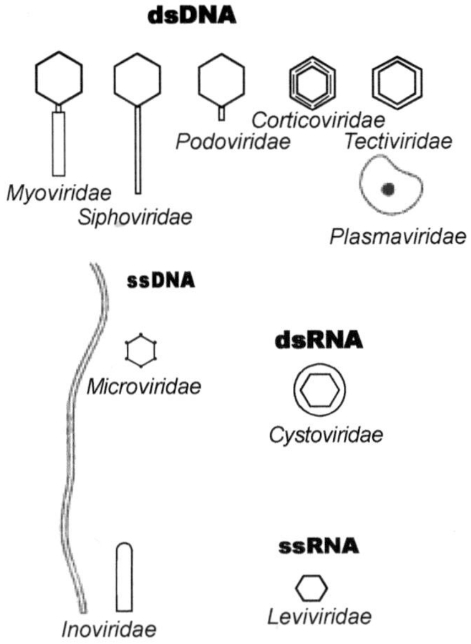

Figure 1: Major bacteriophage groups. C, circular; L, linear; S, segmented; T, superhelical; ds, double-stranded; ss, single-stranded. Modified with the permission of H-W. Ackermann, Department of Microbiology, Immunology and Infectiology, Faculty of Medicine, Laval University, Vandry Pavilion, Avenue de la Médecine, Quebec, QC, G1K 7P4, Canada. (Ackermann, 2011).

like the other T-odd phages. Podoviridae have heads with short non-contractile tail that is shorter than head. Examples are T3 and T7. These are morphologically, genetically and serologically are unrelated amongst each other and with the rest. These T-odd phages are dependent virulent as throughout their replication bacterial chromosome remains intact. The DNA of these phages contains normal cytosine.

Cyanophages have been designated by the first letters of the sensitive hosts followed by Arabic numerals in order of their discovery. Most of the cyanophages discovered till 1990 have been designated like this. But with the discovery of marine cyanophages and the adoption of nomenclature norms along with the first letter(s) of the name of host, the first letter of the area/locality is added from which collection is made followed by the letter representing the morphotype of the virus. Accordingly, S-SSM2 phage (described by Sullivan *et al.*, 2003) the first letter S denotes the host *Synechococcus*,

SS represents Sargasso Sea from which collection was made and M represents a myovirus and 2 represents the second virus in the series. Safferman *et al.* (1983) proposed a classification for cyanophages in accordance with ICTV guidelines for bacteriphage classification. The cyanophages whose morphology and ultrastructure have been described till then have been included and classified into three families, i.e. Myoviridae, Siphoviridae (or Styloviridae) and Podoviridae and recognized representative genera of cyanophages as Cyanomyovirus, Cyanosiphovirus and Cyanopodovirus, respectively. A cyanomyovirus is a cyanophage with contractile ('myos' in Greek means muscle) tail. AS-1 is recognized as the type species. Other cyanophages such as AS-1M, N-1, A-1(L) and A-2 having similar morphology are included here. Likewise, Cyanosiphovirus is a cyanophage with non-contractile tail but the tail extends as a pillar ('sipho' in Greek means a 'siphon' or tube-like). The type species is S-1. Other members are S-2L and SM-2. The cyanopodovirus is characterized by the presence of a short tail that is smaller than the length of the head. Because this short tail assumes the shape of a foot ('podos' in Greek means foot) it is named as a cyanopodovirus. LPP-1 is the type species. Other members are LPP-1G, LPP-2, SM-1, A-4(L) and AC-1.

II. FRESHWATER CYANOMYOVIRUSES

The type species of cyanomyoviruses AS-1 that infects *Anacystis nidulans* and *Synechococcus cedrorum* has been isolated from a waste stabilization pond located at Sun City, Lake Centre, Florida (Safferman *et al.*, 1972). The AS-1 particle contains dsDNA and appears polyhedral in structure with a long contractile tail. The edge to edge diameter of its head is 90 nm and the tail measures 243.5 nm long and 22.5 nm in diameter. Magnesium ions are not required for stability of virus, while NaCl increases the adsorption rate by 34%. The virus has greatest stability between pH 4 and 10. The other cyanomyoviruses isolated and characterized were AS-1M, A-2, A-1(L), Ma-LMM01, N-1 and S-3L. Cyanophage AS-1M is morphologically identical to AS-1 in having edge-edge diameter of the head of 90 nm and a contractile tail of 240 nm long and 22.0 nm in diameter. It also infects *A. nidulans* and *S. cedrorum* but exhibits rapid multiplication cycle (8 h) than AS-1 and completely lyses the host within 12 h after infection (Sherman and Connelly, 1976). Cyanophage A-2 active against *Anabaena variabilis* (Granhall and Hofsten, 1969) was reported from Sweden. A-1(L) through A-9 (L) series of cyanophages have been isolated from Leningrad (Kozyakov, 1977). A-1(L) and A-4(L) which sensitize *A. variabilis* have been examined in detail and the rest of them have been lost (Hu *et al.*, 1981; Gromov, 1983). The diameter of the capsid of A-1(L) is 60 nm and has a contractile tail of 83 nm long. Other cyanomyoviruses described infect various strains of *Anabaena* and *Nostoc* among which AN-10 and AN-15 are morphologically alike with phage heads (60 and 63 nm, respectively) and contractile tails (100 nm; Hu *et al.*, 1981).

A cyanophage (Ma-LMM01), a myovirus with a head (86 nm dia.) and tail (209 nm length and 24 nm width) possessing a central tube of 9 nm, has been isolated from Lake Mikata in Fukui Prefecture, Japan that infects the toxic strain of *Microcystis aeruginosa* NIES298 (Yoshida *et al.*, 2006). This cyanophage with a latent period of 6 to 12 h has a burst size of 50 to 120 particles per cell. The phage possesses dsDNA as its genome with a size of ca. 160 kbp and a low level of methylation. Phylogenetic analysis of the deduced amino acid sequence of the putative sheath protein revealed that it forms a sister group with freshwater and marine cyanophages although it possesses morphological similarities with T4 like-phages. In view of the fact that *M. aeruginosa* NRC-1 that was used as a host strain for the isolation of other cyanophages (SM-1 by Safferman *et al.*, 1969; SM-2 by Fox *et al.*, 1976) was later shown to be a strain of *Synechococcus* sp. (Suttle, 2000), the report by Yoshida *et al.* (2006) assumes siginificance and seems to constitute the first report on the isolation of a cyanophage infectious against a toxic strain of *M. aeruginosa*. Cyanophages phylogenetically related to Ma-LMM01 infecting *M. aeruginosa* have been analysed by amplifying by real-time PCR the ribonucleotide reductase α-subunit, ribonucleotide reductase β-subunit and sheath protein genes (ORFs 090, g91 and ORF092) with primers specially developed for these genes. The prevalence of the phage sequences relating to cyanophages Ma-LMM01, Ma-LMM02, Ma-LMM03 and Ma-HPM05 were noted when the blooms of *Microcystis* were represented in Hirosawanoike Pond and Lake Mikata. However, no lytic phages active against *M. aeruginosa* strain NIES298 could be isolated (Takashima *et al.*, 2007). One of the probable reasons suggested is that Ma-LMM01 may enter into a lysogenic cycle. A supporting evidence in favour of this is the recognition of Ma-LMM01 phage-specific insertion sequence elements (IS607-cp) in four strains of *M. aeruginosa* (NIES90, NIES112, NIES604 and RM6) and in the genome of *Cyanothece* sp. The insertion element IS607-cp is present in multiple copies in the four strains of *M. aeruginosa* and depending on the location on their genomes three groups of IS607-cp have been recognized. While IS607-cp-1 is present in three strains (NIES90, NIES112 and NIES604), IS607-cp-2 and IS607-cp-3 are found specific to NIES90 and RM6, respectively (Kuno *et al.*, 2010).

Adolph and Haselkorn (1971) isolated cyanophage N-1 that infects the nitrogen-fixing cyanobacterium *Nostoc muscorum*. The viral particles are polyhedral with a contractile tail attached to the head by a tail capital similar to LPP-1. The diameter of the head is 55 nm which contains DNA with a molecular weight of 38 x 10^6 daltons and G+C content of 41%. In mature N-1 virion a total of 19 structural proteins were resolved with a molecular weight of 37,000 and 14,000 daltons. It would require about 34% of the coding capacity of N-1 DNA which is similar to LPP-1 (Sherman and Haselkorn, 1970c; Adolph and Haselkorn, 1973b). A cyanophage that infects a *Synechococcus* sp. has been isolated along with its host from a reservoir in Korea bearing morphological resemblances to AS-1 (Kim and Choi, 1994). Koz'yakov (1981) described a new cyanophage S-3L active against *Synechococcus schmidlea*. The phage head has a diameter of 81.6 nm and a tail of 16.3 nm long with a contractile sheath. The phage DNA contains 46.8 to 47.6 mol% G+C. The adsorption time was 60 minutes with a latent period and eclipse period of 11 and 4 h, respectively. The burst size was approximately 35 particles per infected cell.

III. FRESHWATER CYANOSIPHOVIRUSES

S-1, the type species of the genus Cyanosiphovirus, infecting *Synechococcus* has been isolated by Adolph and Haselkorn (1973a). This phage has an isometric head with a diameter of 50 nm and a rigid non-contractile tail of 140 nm long. The genome of the phage is dsDNA with a molecular weight of 23–26 x 10^6 and a G+C content of 70–74%. There are 13 structural proteins of which the major structural proteins have molecular weights of 39,000, 11,000 and 10,000. Another *Synechococcus* phage S-(2)L, isolated from Russia by Khudyakov (1977), also possesses an isometric head that is slightly larger (56 nm in diameter) with a non-contractile tail (120 nm long) but possesses several short thin fibers at the distal end. The molecular weight of dsDNA is 26 x 10^6. Khudyakov *et al.* (1978) reported that adenine in this phage DNA is substituted by 2,6-diaminopurine. Other cyanosiphoviruses that infect *Synechococcus* are S-4(L) and S-5(L) which have been reported by Khudyakov and Matveyev (1981, 1982). Another related member of this group is SM-2 that infects *Synechococcus* and *Microcystis*. It also has an isometric head that is 50–55 nm in diameter with a flexible non-contractile tail of 130–140 nm long (Fox *et al.*, 1976; Leach *et al.*, 1980). A cyanosiphovirus with a capsid diameter of 70 nm and a long non-contractlie tail, infecting the bloom-forming toxic cyanobacterium *Cylindrospermopsis raciborskii*, has been reported from Australian freshwater lakes (Pollard and Young, 2009). The latent period of this has been reported to be 21 h with a burst size of 64 phage particles per cell.

IV. FRESHWATER CYANOPODOVIRUSES

This group includes LPP-1 initially described by Safferman and Morris (1963) and G-III or LPP-1G (Padan *et al.*, 1967; Ginzberg *et al.*, 1968; Fig. 2) and D-1 (Daft *et al.*, 1970) strains. LPP-1G and D-1 are serologically related to LPP-1 but cross reaction between LPP-1G and D-1 has not been tested.

LPP-1 has an icosahedral head (with an edge to edge distance of 60 nm) and a very short tail. The tail is 10–20 nm long and 15 nm in diameter with an outer sheath and a tail capital attached to one of the vertices of the hexagonal capsid (Luftig and Haselkorn, 1967). The viral nucleic acid is DNA with a contour length of 13.2 ± 0.5 μm. The amino acid content of the virus accounted for 52% of the particle weight. Ten principal structural proteins have been identified in mature LPP-1 virion (Luftig and Haselkorn, 1968). According to Sherman and Haselkorn (1970c) the major head proteins are 39,000 and 13,000 molecular weight species while the tail protein has a molecular weight of 80,000. This would amount to a 35% coding capacity of LPP-1 DNA.

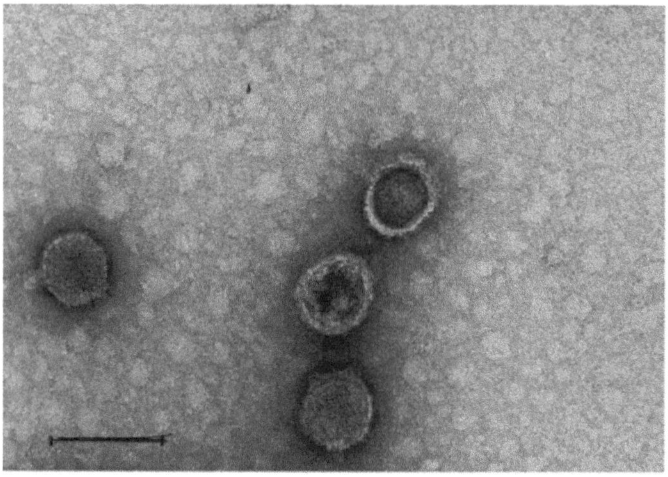

Figure 2: Negative stained LPP-1G phage particles. The magnification bar represents 100 nm. Picture courtesy M. Kessel, Department of Membrane and Ultrastructural Research, The Hebrew University-Hadassah Medical School, Jerusalem, Israel.

The differences in serological relationships have led to further classification of LPP group of viruses into LPP-1 and LPP-2 types (Safferman *et al.*, 1969a). The LPP-2 group of cyanophages is similar in morphology to LPP-1 with an average head to head distance of LPP-2 virus being 57.3 ± 0.27 nm. The molecular weight is similar to LPP-1, differing only by 2% in values of G+C content. Sherman and Haselkorn (1970a,b) reported major differences between the structural proteins. LPP-2 has 14 structural proteins with a molecuar weight of 74,9000 Da and would require 50% coding capacity of LPP-2 (Table 3). Cyanophage N-2 that is morphologically similar to LPP-1 but specific to *Nostoc linckia* has been described by Mendzhul *et al.* (1983). So it is important to note that N-2 is cyanopodovirus and differs from N-1 which is a cyanomyovirus.

The cyanophage SM-1 infects *Synechococcus elongatus* and *M. aeruginosa*. Electron microscopy of purified virus shadowed with chromium revealed polyhedral particles with an average diameter of 88 nm (Safferman *et al.*, 1969b) with five or six knob like projections but with no obvious tail. Mackenzie and Haselkorn (1972a), however, have observed exceedingly short tail. A

Table 3: Comparative account of LPP-1 group of cyanopodoviruses.

Sr. No.	Property	LPP-1	LPP-1G	D-1
1	Morphology	Icosahedral head with tail	Icosahedral head with tail	Icosahedral head with tail
2	Size of Particle	Head 60 nm dia. Tail 15 nm dia. 15–20 nm long	Head 50–60 nm dia. Tail 28–37 nm long	Head 58.6 nm dia. Tail 20 nm long
3	Nucleic acid	DNA	DNA	DNA
4	Plaque size	0.1–1.0 mm	Nt	Nt
5	Adsorption period	60 min.	60 min.	30 min.
6	Latent period	8 h	6–7 h	40 h
7	Temperature of Inactivation	60 min at 55°C	Nt	Nt
8	Stability	Stable at 4°C for 3 h	Nt	Nt
9	Requirement of Mg^{2+}	Require Mg^{2+}	Nt	Nt
10	Serological relationship	Neutralization with D-1 antiserum	Neutralization with LPP-1 antiserum; Neutralization with D-1 antiserum not tested	Neutralization with LPP-1 antiserum; Neutralization with LPP-1G antiserum not tested

Nt: Not tested

dozen proteins have been identified in the virion and the dsDNA has a G+C content of 66–67 mol%, identical to host. The infection process of SM-1 has been described by Mackenzie and Haselkorn (1972b).

Venkataraman *et al.* (1973) isolated a cyanophage that infects unicellular *A. nidulans* and the colonial form *Chroococcus minor* from a waste stabilization pond inside IARI, New Delhi. This cyanophage has a polyhedral head of 62.5 nm in diameter with a non-contractile tail of 25 nm (Sharma and Venkataraman, 1977). AC-1 is considered close to SM-1 in its morphology. The differences between SM-1, AC-1 and LPP-1 are given in Table 4.

Phage AN-22 described by Hu *et al.* (1981) is a cyanopodovirus with a edge-edge diameter of the head of 54 nm and an exceedingly short tail. From freshwater and seawater bodies of Romania, Moisa *et al.* (1981a) isolated 31 strains of cyanophages active against *Phormidium* sp. Of these, three types of virus particles differing in tail length and with a morphology similar to that of T-odd

Table 4: Comparative account of freshwater cyanopodoviruses.

Property	SM-1	AC-1	LPP-1
Shape and size of particles	Polyhedron with no obvious tail; 88 nm dia.	Polyhedron with tail; 25 nm dia.	Polyhedron with tail; 15–20 nm
Stability	Stable in H_2O	Stable in its lysate	Unstable; requires Mg^{2+} ions
Adsorption	----	40 min.	60 min.
Latent period	30 h	30 h	8 h
Plaques	Less than 1–2 mm	4–6 mm after 10 days	Less than 1 mm
Temperature of inactivation	60 min at 55°C	Nt	60 min at 55°C
Serological relationship	No neutralization with LPP-1, AS-1 antiserum	Nt	No neutralization with SM-1 & AS-1 antiserum
Host range	*S. elongatus* *M. aeruginosa*	*A. nidulans* *C. minor*	*Lyngbya* sp., *Phormidium* sp., *P. boryanum*

Nt: Not tested

coli-phages are described (Moisa *et al.*, 1981b). A temperate cyanophage NP-1T, discovered by Muradov *et al.* (1990a,b) from Russia, is unique because it infects and lysogenizes heterocystous nitrogen-fixing cyanobacterium *Nostoc* on the one hand and non-heterocystous filamentous *Plectonema* on the other. The phage capsid has a diameter of 78 nm and lacks a distinct tail. DNA of the phage has a mean contour length of 14.4 μm and molecular mass of 28MDa. The specificity of the phage was confirmed from the data on restriction analysis using the EcoRV enzyme that hydrolyses DNA of cyanophages to yield different number of fragments. Ma-LBP, a cyanophage infecting *M. aeruginosa* has been isolated from Lake Baroon, Sunshine Coast, Queensland, Australia by Tucker and Pollard (2005). Two strains of the virus of Ma-LBP (a podovirus), resembling T7 in their morphology, have been detected with short stubby tails but the head diameters varied by 10 nm. In one of the strains the head diameter was 42 nm whereas in the second it was 52 nm. The virus had a latent period of 11.2 h with an average burst size of 28 virus particles per host cell. The titer of the virus from lake waters was found to be 5.6 x 10^4 ml^{-1} that represents 0.23% of the natural viral population of the Lake (2.46 x 10^7 ml^{-1}). In view of this the authors concluded that the virus controls the populations of *Microcystis*.

V. CYANOPHAGES OF UNCERTAIN IDENTITY

A number of cyanophages isolated on *P. boryanum* such as P-2, P-3, P-4 (Singh and Singh, 1967), P-5, and P-6 (Singh *et al.*, 1972) are included here primarily because they sensitize the same host strain (*P. boryanum*) with minor differences in plaque morphology, latent period etc. The cyanophages P-2 to P-6 have not been either morphologically or serologically characterized for relationships amongst themselves or with those of others, described above infecting *Plectonema*.

The cyanophage C-1 sensitizing *Cylindrospermum* sp. (Singh and Singh, 1967), AR-1 attacking *Anabaenopsis raciborskii*, *A. circularis* and *Raphidiopsis indica* (Singh and Singh 1967), AR-2 that infects *A. raciborskii* (Singh *et al.*, 1972) and A-1 (Singh *et al.*, 1972) are the other cyanophages discovered causing lysis of the nitrogen fixing cyanobacteria from India. Of these cyanophages, A-1 which sensitizes *Anabaena cylindrica* is of special interest because of its unique morphology amongst the hither to discovered freshwater cyanophages. The phage possesses a sheath of 20 nm broad and 100 nm long. A tail or wick of 130–300 nm long is attached to sheath. These particles are headless. It is suspected that they contain RNA, and their morphology brings them closer to raphidosomes reported in *Saprospira grandis* (Lewin, 1963) and *Actinomyces streptomycini* (Rautenstein *et al.*, 1966).

Kaushik and Venkataramn (1973) described a cyanophage TAuHN-1 infecting *Tolypothrix tenuis*, *Aulosira fertilissima*, *N. muscorum* and *Hapalosiphon intricatus*. A series of viruses belonging to A-4L and N-L infecting *A. variabilis*, *Nostoc linckia* and S-8K and S-2L infecting *Synechococcus*, respectively have been isolated from Russia (Gromov, 1983). Trivedi and Oza (1979) reported the isolation of cyanophage O-1 active against *Oscillatoria chlorina* from Ahmedabad. The phage is a slender rod having a diameter of 25 nm and a length of 300 nm.

Phlips *et al.* (1990) reported the lysis of *M. aeruginosa* by a cyanophage designated as MA 1, though they were not able to isolate and characterize the particles. Manage *et al.* (1999) investigated the seasonal changes in titres of cyanophages in Furuike Pond located in Sancho, Matsuyama City (Japan). The titres of cyanophages infecting *M. aeruginosa* were at 2 x 10^2 and 4.2 x 10^4 PFU ml^{-1} during the months of June and September, respectively and this coincided with a decrease in the population of the host cells from their initial densities of 1.8 x 10^4 and 9.4 x 10^5 cells ml^{-1} in the corresponding months. During winter season, however, the cyanophages were not detectable and the host was represented at low levels. Honjo *et al.* (2006) studied the diversity of virus-like agents killing

M. aeruginosa in Furuike pond that represents a hypereutrophic pond. Although filtered pond waters caused lysis of the host cells in cultures, it was not possible to isolate the viral particles. Selective amplification of the genetic material of the agents followed by pulse-field gel electrophoresis (PFGE) demonstrated the presence of viral communities differing in genome sizes.

VI. STABILITY

Safferman and Morris (1964a) reported that LPP-1 retains about 70% of its initial activity when stored in its own lysate at 4°C while the requirement of Mg^{2+} for its infectivity was subsequently demonstrated (Schneider *et al.*, 1964; Goldstein *et al.*, 1967). Cyanophages SM-1, AS-1 and S-2L are stable in distilled water and Mg^{2+} is not required for stability or infectivity (Safferman *et al.*, 1969b; 1972; Khudyakov, 1977). Safferman *et al.* (1969b) noted no significant loss in the viability of SM-1 phage, when stored at 4°C. Moreover, SM-1 is identical to LPP-1 in its requirement to pH and thermal stability. Cyanophages AC-1 and TAuHN-1 have been reported to be stable in their own lysates at 4°C (Venkataraman *et al.*, 1973; Kaushik and Venkataraman, 1973). Amla (1978) reported slight inactivation of AS-1 by EDTA and sodium citrate that has been found to be concentration-dependent. Studies on the effects of metallic ions on the stability of N-1 indicated that the growth medium of the host (modified Chu-10 medium) and saline magnesium chloride conferred total stability to the virus N-1 (Padhy and Singh, 1978a).

VII. ADSORPTION

The time taken for nearly 95% of the cyanophage particles to get adsorbed to the host cell wall varied from cyanophage to cyanophage. Although LPP-1 is stable in its own lysate, it needs Mg^{2+} for its biological activity (Schneider *et al.*, 1964; Goldstein *et al.*, 1967). This apparently reflects the requirement of Mg^{2+} for adsorption. Moreover, SM-1 and AS-1 do not require Mg^{2+} for infectivity (Safferman *et al.*, 1969a, 1972). No special requirements for adsorption of cyanophages LPP-2 and D-1 have been reported (Safferman *et al.*, 1969b; Daft *et al.*, 1970). The effect of physical and chemical agents on adsorption of N-1 phage has been studied (Padhy and Singh, 1977a).The effect of temperature on the adsorption and one-step growth of N-1 was described by Padhy and Singh (1977b). The age of host, ions and pH have been shown to influence adsorption of N-1 to the cells of *N. muscorum* (Padhy and Singh, 1978b). The rate of adsorption of N-1 was reduced with the aging of algal cultures and complete loss of adsorption occurred when dead algal cells were used. Further, N-1 did not require the presence of saline-magnesium chloride, L-tryptophan and L-phenylalanine for adsorption and optimum pH for adsorption of N-1 was noted to be 7.6–8.1 (Padhy and Singh, 1978b). The involvement of host polysaccharides in AS-1 adsorption has been indicated (Samimi and Drews, 1978). The adsorption of AS-1 was enhanced in dark by increasing the concentration of sodium (Cseke and Farkas, 1979).

Bobrovnik *et al.* (1978) studied the kinetics of adsorption of cyanophage A-1 and suggested an enzymatic character of virus adsorption. Illumination and concentration of sodium in the external medium enhanced the adsorption of phage AS-1 to *A. nidulans* while illumination alone doubled phage adsorption at normal sodium concentration (Na = 11.7 mM). An increase of sodium concentration (Na = 0.11 mM) resulted in a 10-fold increase in phage adsorption in darkness (compared to that found in light) (Cseke and Farkas, 1979). Olivera *et al.* (1982) reported the adsorption of AS-1 cyanophage to vesicles of known lipid composition especially phosphatidecholine (from soybean) and cholesterol but not the algal lipids.

Light exerted a strong influence on the adsorption of AS-1 as revealed by the studies of Kao *et al.* (2005) who compared the infection process of this phage on parent *S. elongatus* PCC 7942 and its mutants defective in *kaiA* and *kaiB* genes. The circadian regulatory proteins had only minor effects on phage AS-1 infection as reflected by the phage yield. Significant lowering in phage production was noted as a result of low light.

VIII. PHAGE DEVELOPMENT

The general pattern of multiplication of cyanophages is identical to that of virulent bacteriophages. Adsorption, penetration, intracellular development, maturation and lysis are the general steps recognized in the lytic cycle. Electron microscopy of thin sections of *P. boryanum* cells infected with LPP-1 reveals that the first sign of infection is invagination of photosynthetic lamellae (Brown *et al.*, 1966; Smith *et al.*, 1966). The space between the folded lamellae and plasma membrane in which the virus particles later appear has been designated as virogenic stroma. Studies on internal development of cyanophage G-III revealed that as maturation of virus particles progressed they appear to be attached to the photosynthetic lamellae and considerable reorganization in photosynthetic lamellae occurred leading to the formation of vesicle-like bodies in the cells of the host (Fig. 3; Padan *et al.*, 1967). It is probably to create more internal space to facilitate the accommodation of large number of maturing virions inside the cells. This has been confirmed by the studies of Luftig and Haselkorn (1967) with LPP-1 infected cells of *P. boryanum*. Cyanophages belonging to LPP-2 group also exhibit similar multiplication cycle (Safferman *et al.*, 1969a).

The multiplication rate of D-1 is reported to be fast where the titre increased from 10^3 to 10^{12} PFU/ml within 40 h when compared to LPP-1 where the rates of increase were noted to be from 20 to 10^4 PFU/ml in 36 h (Daft *et al.*, 1970). Sherman and Haselkorn (1970a) confirmed the findings of Brown *et al.* (1966) and Smith *et al.* (1966) on LPP-1 infected cells of *P. boryanum* who suggested that

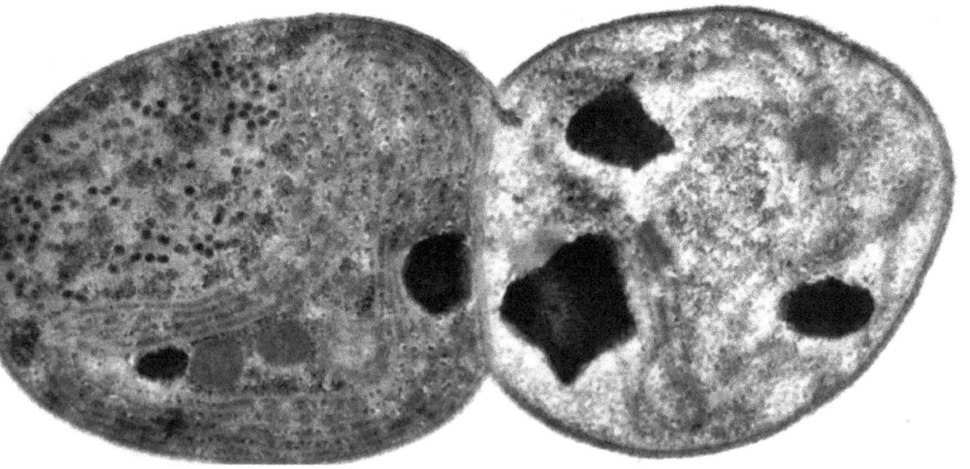

Figure 3: Electron micrograph of *Plectonema boryanum* cell depicting invaginated photosynthetic lamellae and the viriogenic stroma containing the viral heads inside the cell 7 h after infection with LPP-1G. Cells were fixed in 4% glutaraldehyde in 0.1 M cacodylate buffer at pH 6.9 and post-fixed in 1% osmium tetraoxide in the same buffer; scale represents 1 μm. Picture courtesy M. Kessel, Department of Membrane and Ultrastructural Research, The Hebrew University-Hadassah Medical School, Jerusalem, Israel.

the invagination occurred first at 3 h after infection and was distinct by the 4th h corresponding to the end of eclipse period. By the end of latent period almost all the cells showed invaginated lamellae and viriogenic stroma containing phage particles. The breakdown of host DNA has also been noted in LPP-1 infected cells of *P. boryanum* coinciding with LPP-1 synthesis (Sherman and Haselkorn, 1970b). A characterization of virus-induced proteins and the follow up of the course of their synthesis during LPP-1 infection in the cells of *P. boryanum* has been reported (Sherman and Haselkorn, 1970c). These workers identified 3 classes of viral proteins: three proteins belonging to the earliest class begin to appear soon after infection and their synthesis completed by 4th h. This is followed by the synthesis of early proteins (another 3 proteins) by the 2nd h and continues to be synthesized until lysis. The late class of proteins belonged to viral structural proteins. A total of 17 distinct virus induced proteins were detected after infection. They accounted for 65% of the coding capacity of viral genome. Sherman *et al.* (1976) conducted detailed investigations on the ultrastructure of infection and assembly of phage AS-1M. The first noticeable evidence of phage infection was a degradation of the nucleoplasm of the host cell. Precisely after 6 h of infection many phage precursors in the process of assembly could be seen. Based on these studies, they have speculated that empty phage heads are formed first and then filled with DNA. Three classes of proteins are synthesized after AS-1M phage infection in the cells of *S. cedrorum*. These are classified as early proteins which are produced until 4 h after infection, middle proteins which are produced until 3 h after infection until lysis; and late proteins, i.e. the phage structural proteins which are synthesized until 4 h after infection (Sherman and Pauw, 1976). They further suggested that a breakdown of the host DNA begins immediately after infection that is optimal by 3 h. This process is in response to a phage-dependent nuclease activity that has the characteristics of an early protein. Phage DNA synthesis begins at about 3 h after infection and the enzymes responsible for this appear to be middle proteins and require post-infection RNA transcription.

Transmission electron micrographs of cyanophage SM-2 infected cells of *S. elongatus* revealed virion heads 10 h after infection correlating well with 6–8 h eclipse period for the virus (Leach *et al.*, 1980). One of the significant features of the development of the cyanophage SM-2 is that photosynthetic lamellae remain intact throughout the infection cycle even upon accumulation of additional virion heads.

The latent period of cyanophages of LPP-1 group has been detected to be 8 h, 6–7 h and 40 h in LPP-1, LPP-1G and D-1, respectively (Safferman and Morris, 1964a; Padan *et al.*, 1967; Daft *et al.*, 1970). Similar to LPP cyanophages, N-1 has a latent period of 7 h (Adolph and Haselkorn, 1972). The burst size of N-1 is 100 PFU/infected cell. AS-1 has a latent period of 8.5 h and an average burst size of 50 PFU/infected cell (Safferman *et al.*, 1972). SM-2 has a latent period of 30 h with a rise period of 20 h and an average burst size of 100 PFU/infected cell (Mackenzie and Haselkorn, 1972b). The latent period of AC-1 is reported to be 30 h (Sharma and Venkataraman, 1977).

i) *Dependency of replication cycle on photosynthesis*: LPP-1 infected cells of *P. boryanum* showed an inhibition of photoassimilation of carbon dioxide (Ginzberg *et al.*, 1968). The dependeny of LPP-1G viral multiplication on photosynthesis is limited to the first 3 h of eclipse period and for ATP production (Padan *et al.*, 1970). In the presence of CCCP (inhibitor of cyclic photophosphorylation), no virus is produced whereas in the presence of DCMU and anaerobiosis (presence of cyclic photophosphorylation) the viral reproductive cycle is completed. Thus ATP for the reproductive cycle is provided by photosystem I (PSI) through cyclic photophosphorylation. Wu *et al.* (1968) demonstrated that energy for viral development is provided by either oxidative phosphorylation or photophosphorylation and when both systems are operative, infection is faster than with either system alone. In the absence of both systems, development is completely inhibited.

A decrease in CO_2 fixation rate was noted in *Plectonema* as a result of LPP-1 infection and this closely corresponded with the invagination of the photosynthetic lamellae. It is in this area that the newly synthesized virions accumulated (Sherman and Haselkon, 1970a,b). Sherman and Haselkorn (1971) observed that DCMU had little effect on LPP-1 multiplication except that it decreased the burst size. This modest decrease in burst size was attributed to the lowering of cyclic photophosphorylation relative to non-cyclic type. Padan *et al.* (1971b) showed that LPP-1G is a thermosensitive cyanophage, since the development is inhibited at 35°C. This is explained due to the cessation of CO_2 photoassimilation as well as invagination of photosynthetic lamellae.

Mackenzie and Haselkorn (1972c) have shown that for the synthesis of SM-1 phage, photosynthetic metabolism of the host is essential since withdrawl of CO_2 or of light completely suppressed viral growth. Moreover, the incorporation of DCMU or CCCP also suppressed viral growth. Adolph and Haselkorn (1972) reported similar observations after N-1 infection.

The development of phage progeny of AS-1M in *S. cedrorum* requires the photosynthetic machinery intact, since phage development did not occur when photosynthesis was inhibited by DCMU, CCCP, valinomycin, nigericin or lack of CO_2. However, supplementation of glucose in dark or in light with DCMU (10^{-5}M) supported phage production to about 10% of control (Sherman, 1976). Synthesis of cyanophages A-1(L) and A-4(L) did not occur in darkness but illumination, though for brief periods, supported reduced phage burst. The presence of DCMU and dinitrophenol did not inhibit their development in light (Al-Musavi, 1977).

Amla (1979) compared intracellular development of phage AS-1 in two pigment mutants, i.e. blue and yellow of *A. nidulans*. Phage AS-1 development was subjected to DCMU-mediated inhibition in the cells of blue mutant similar to that of parent strain. Whereas in the cells of the yellow mutant, phage multiplication was slow with reduced yield. Cyanophage AS-1 infection caused significant changes in the structure of photosynthetic apparatus. PSI reaction centre was the most resistant to degeneration (Gusev *et al.*, 1980a). Gusev *et al.* (1980b) employed ESR, luminescence and polarography methods to identify the specific reaction centre in the photosynthetic apparatus of *A. variabilis* infected by cyanophage A-1(L). The virus first showed disturbance in the functioning of PSII followed by alterations in the thylakoid arrangement. Mamkayeva *et al.* (1980) observed the maturation of A-1(L) cyanophages in *A. variabilis* in an electron transparent local zone near or between thylakoids in cells exposed to DCMU (5×10^{-6} M). Photosynthetic inhibitors and removal of light suppressed the replication of SM-2 suggesting that it is dependent on host's photosynthetic metabolism for energy required for replication (Benson and Martin, 1981). Cyanophage SM-2 has also been reported to be highly dependent for its development on photosynthetic metabolism of the host cells (Benson and Martin, 1981). Amla and Saxena (1983) described the metabolic aspects of cyanophage AS-1 replication in *A. nidulans*. Viral multiplication did not occur in dark and the latent period was prolonged when the infected cells of *A. nidulans* were in dark for 6 h followed by illumination in presence of CCCP and DCMU. Phage growth was abolished due to chloramphenicol treatment whereas treatment with rifampicin and actinomycin-D before eclipse period abolished AS-1 development. While majority of the evidences presented above substantiate photosynthesis-dependent cyanophage multiplication cycle, Bisen *et al.* (1988) reported that LPP-1 multiplication cycle in *Phormidium uncinatum* does not depend on photosynthesis.

Cyanophage N-1 infected cells of *N. muscorum* supported virus development in the absence of PSII activity and energy for phage multiplication is provided by cyclic photophosphorylation as well as via a degradation of carbohydrate reserves (Kashyap and Singh, 1989). Cyanophage infection not only inhibited PSII activity at the level of secondary acceptor, Q(B) of PSII but also D1 protein in *Synechococcus* sp. strain PCC 6301 cells that is comparablable to DCMU-induced protection of D1 protein turnover, in AS-1 infected cells (Teklemariam *et al.*, 1990).

ii) *Nucleic acid metabolism*: A ribonuclease for cyanophage AS-1 has been detected in the infected cells of *A. nidulans*. Further purification and its characterization have shown a relative specificity of uridylic acid (Lehmann *et al.*, 1979). Cyanophage AS-1 specific endonuclease has been purified (molecular weight 40,000 ± 4000) from *A. nidulans* and it is further reported to bring about breakdown of host DNA (Szekeres, 1981).

The DNA metabolism of cyanophage AS-1 infected cells of *A. nidulans* has been investigated in detail. Hwang-Lee *et al.* (1982a) identified an unique species of phage-induced light DNA (PIL-DNA) in the infected cells accounting for as much as 50% of total DNA at ~5 h post-infected period and the same was found in negligible amounts after 9 h during a 16 h infective cycle. This has been further confirmed by a differential labelling experiment in which C^{14}-incorporation begins around 2 h of post-infection and is maximum by 4–6 h followed by a decrease thereafter. Whereas H^3-adenine accumulates preferentially in AS-1 DNA by the middle of the lytic cycle at which time no H^3-adenine incorporation took place in PIL-DNA (Hwang-Lee *et al.*, 1982b). The observation that PIL-DNA could be differentially labelled has further been extended to other phage-host systems, i.e. AS-1, AS-1M, AS-1 lab stock with *A. nidulans* 625 and *S. cedrorum* 1191 by using precursors, i.e. H^3-uracil, H^3-thymidine and C^{14}-glucose-6-phosphate (Blashka *et al.*, 1983).

Hwang-Lee *et al.* (1983a) demonstrated that both phage and host DNA are involved in the production of PIL-DNA and its production is more dependent on the integrity of the phage DNA than that of the host. Subsequently, biophysical characterization of PIL-DNA revealed that it is modified by the presence of 5-methylcytosine and resembles AS-1 DNA in nucleotide sequence (Hwang-Lee *et al.*, 1983b). Bancroft and Smith (1988a,b) divided cyanophages A-4L, AN-13, AN-23, A-1L and AN-10 into two classes on the basis of analysis of restriction endonuclease sites. Cyanophages A1-L and AN-10 possess adenosine methylated GATC sequences (as Sau3AI cleaves but MboI does not) and are thus tolerant to this dam-like methylation. It is interesting to note that DNA of the host *Anabaena* sp. strain PCC 7120 also appears to be modified at the same base sequences. But AN-10 showed considerably less modification. The rest of the three cyanophages neither contained adenosine methylated or unmodified GATC sequences (as both Sau 3AI and MboI fail to cleave) and so are intolerant to such methylation. These results suggest that DNA methylation as well as restriction may be detrimental for phage reproduction if phage DNA is correspondingly not methylated. Cyanophage A-4(L) represents an extreme example of sequence counter-selection as it has eliminated from its genome many or all of the numerous sequences modified in *Anabaena* DNA (Lambert and Carr, 1984; Bancroft and Smith, 1988a,b). Bancroft and Smith (1989) further mapped the average sites of several restriction endonucleases in DNA isolated from five cyanophages, i.e. A-1(L), A-4(L), AN-10, AN-13 and AN-23 which were replicated in *Anabaena* sp. strain PCC 7120. It is appropriate to mention here that the cyanophages A-1(L) and A-4(L) were isolated by Koz'yakov (1977) and the rest three were isolated by Hu *et al.* (1981). AN-10, though isolated by Hu *et al.* (1981) as a contaminant from A-4L, differed in the restriction map but appeared to be related to A-1(L). On the basis of restriction maps the cyanophages AN-13 and AN-23 are related. These observations are further supported by the fact that the genomes of these two phages are 46 kbp long and hybridization between their DNAs has also been possible (Hu *et al.*, 1981). The restriction maps of A-1L and AN-10 appeared to be circular while the maps of other phages are linear.

Schneider and Haselkorn (1988) identified restriction fragments of cyanophage N-1 DNA which contains genes that transcribe early proteins in infected cells of *Anabaena*. The sequence of two promoter regions to each other in the –35 and –10 regions were nearly found to be identical to consensus sequence for promoters in *E. coli*. Muradov *et al.* (1990a,b) detected sequences homologous to nitrogenase structural genes (*nifD* and *K*) of *Klebsiella pneumoniae* in the DNA of various moderate NP-1T cyanophage strains infecting *Nostoc* sp. N-39 lysogenized by these cyanophages. The *nif* genes mentioned have not been detected in DNA of *Nostoc* sp. N-39 primary culture.

Szekers *et al.* (1983) showed that the endonuclease splits host DNA but not AS-1 phage DNA. Further, AS-1 phage DNA was shown to be not only resistant to AS-1 endonuclease but also to a number of restriction endonucleases, the recognition site of which contains a central G-dc dinucleotide. The sites attacked preferentially by the AS-1 endonuclease are specially protected on AS-1 DNA molecule. However, AS-1 DNA when inserted into plasmid pBR322 and cloned into *E. coli* was found to be susceptible to AS-1 endonuclease.

iii) *Disturbances in other metabolic features*: Mann *et al.* (1975) for the first time identified that accumulation of guanosine-3 diphosphate-5-diphosphate (ppGpp) occurred during energy limitation in the cells of *A. nidulans*. Nitrogen starved (MSX-treated) and energy depleted cells of *A. nidulans* have been shown to accumulate large amounts of ppGpp. However, this accumulation of ppGpp does not occur in the host cells infected with cyanophage AS-1 and also in *E. coli* cells infected with T4 suggesting that it is the difference in control of nutritional or energy metabolism rather than differences in ability to synthesize ppGpp (Borbely *et al.*, 1980). Nitrogen starvation in *A. nidulans* led to the accumulation of ppGpp (Friga *et al.*, 1981). Now it is being increasingly realized that ppGpp is centrally involved in the adaptation of cyanobacterial cells to a wide range of environmental stresses. The regulatory function of ppGpp lies in the fact that it redirects transcription to overcome starvation (Magnusson *et al.*, 2005; Braeken *et al.*, 2006). Studies on bacteria revealed that MazG gene encodes enzymes that hydrolyze the harmful non-canonical nucleotides such as ppGpp (Galperin *et al.*, 2006). Thus these enzymes have been designated as "house-cleaning" enzymes. Other functions of MazG are (i) it is implicated as a regulator of programmed cell death in *E. coli* and (ii) it prevents the accumulation of ppGpp in response to amino acid starvation (Gross *et al.*, 2006). It has been suggested that the phage encoded MazG operates to reduce the pool of ppGpp in *Synechococcus* cells (Clokie and Mann, 2006). Mann *et al.* (2005) sequenced the cyanophage S-PM2 genome and identified 239 ORFs. Of these, ORF136 encodes a homologue of the bacterial MazG. The probability of MazG altering the physiology of phage infected cells has been tested by Bryan *et al.* (2008). By designing primers internally to the ORF136 gene (a total length of 438 bp), PCR amplification from the genome of cyanophage S-PM2 resulted in a product of 279 bp (64% of the gene sequence). By employing these primers, the genomes of 41 marine cyanophages (isolated from different marine waters) have been amplified by PCR. After conducting the sequence analysis of this product, computational analysis of data was performed by BlastN and protein:protein searches. The sequences of 15 of the cyanophages (S-Bn1, S-IO41, IO50, S-BP3, S-MM1, S-MM5, RS-9, RS-11, RS18, RS23, RS26, RS-27, RS37, RS56, FWP) showed 99% identity to the reference phage S-PM2. A phylogenetic analysis of the 16 phage sequences (15 + reference phage S-PM2) with all known bacterial sequences included in Clade A, Clade B and other sequences available in database of NCBI revealed that all 16 cyanophages are closeted into a single group, so classified as Group I. The translated sequences also exhibited a high degree of similarity to bacterial MazG protein from the NCBI database. Group II consisted of 26 cyanophages of which only 9 showed (IO15, IO18, IO17, IO39, IO41, RS38, RS-39, RS-85, S-IO80) similarity to the MazG sequence of S-PM2. MazG gene is highly conserved in *Synechococcus* myoviruses and podoviruses. Because of the fact *Prochlorococcus* and *Synechococcus* phages' MazG genes do not cluster with their host's genes it seems that they have not acquired it from their hosts but acquired them by LGT from other sources. It seems S-PM2 might have acquired this gene from an ancestor like *Chloroflexus aurantiacus* rather than from its host *Synechococcus*.

Other metabolic changes relate to the activity of enzymes of carbon and nitrogen metabolism. The infection of phage AS-1 was shown to be responsible for transformation of a low activity form of glucose-6-phosphate dehydrogenase to a hyperactive form (Cseke *et al.*, 1981). Blashka *et al.* (1982) studied the incorporation of 22 radioactive metabolites and identified three classes of substances, i.e. (i) substances exclusively incorporated in the infected state, (ii) compounds incorporated at a 2- to 3-fold rate in infected cells and (iii) compounds incorporated relatively equally by infected and non-infected cells. Studies on the metabolic changes associated with cyanophage N-1 infection of *N. muscorum* cells revealed an increase in the level of glucose-6-phosphate dehydrogenase, and a lowering of glutamine synthetase activity. However, nitrogenase activity remained high throughout eclipse period which decreased rapidly soon after 5 h (Amla *et al.*, 1987). Nitrate reductase activity of phage N-1 infected cells of *N. muscorum* was elevated (Bisen *et al.*, 1986a) and this was suggested to be at the level of synthesis of new enzyme by phage N-1 mediated molybdenum co-factor synthesis (Bagchi *et al.*, 1987). Phage N-1 was shown to require products of glutamine synthetase activity, i.e. glutamine and its further metabolic products for its replication while the inhibition of glutamate-oxaloacetate transaminase and significant increase of glutamate-pyruvate transaminase has been suggested to involve a shift in nitrogen metabolism in favour of N-1 multiplication (Kashyap *et al.*, 1988). Mendzhul *et al.* (1988) observed changes in the specific activity of glucose-6-phosphate dehydrogenase in cyanophage LPP-3 infected cells of *P. boryanum* that were shown to be dependent on the illumination. Low illumination (400 Lux) that supported optimum virus development enhanced the activity of the enzyme by 75% while high illumination (2000 Lux) lowered the specific activity of the enzyme by 62%.

IX. ROLE OF FRESHWATER CYANOPHAGES IN THE CONTROL OF CYANOBACTERIAL BLOOMS

The rapid disappearance of cyanobacterial blooms under natural conditions suggested the presence of a factor that is generally associated with degeneration. Krauss (1961) first suspected that the sudden disappearance of cyanobacterial blooms is due to the presence of viral agents. Safferman and Morris (1964b) demonstrated that the control of cyanobacteria with virus by artificially producing algal blooms in stainless-steel tanks containing 112 litres of tap water supplemented with inorganic nutrients. The temperature of this pond was maintained at 32°C. The viral agent was added after considerable growth of *P. boryanum* and degeneration of the organism was assessed by counting the fragments of filaments. Complete lysis or disappearance of algal cells occurred about 7 days after viral treatment.

Upon a close examination of three hosts sensitive to LPP-1, it was suggested that these three genera *Lyngbya*, *Phormidium* and *P. boryanum* have close morphological characteristics (Safferman and Morris, 1964b). These observations were further confirmed and in a re-classification of the members of Oscillatoriaceae, Drouet (1963) showed that all the three members belonged to a single species of the genus *Schizothrix*, i.e. *S. calcicola*. This has been considered as one of the most widely distributed and most frequently encountered cyanobacterium on earth. Safferman and Morris (1967) thus speculated that LPP-1 had a lot of ecological significance in naturally controlling the populations of *Plectonema* and related organisms since none of the cyanobacterial strains were found dominantly in 11 of the 12 waste stabilization ponds surveyed. Moreover, the dominant strains in these ponds are reported to be not susceptible to LPP viruses. The incidence and magnitude of LPP-group indicated that several ponds (Arkansas, California, Florida, Indiana, Missouri, South Dakota and Texas) supported high cyanophage titres. New Hampshire pond failed to reveal the presence of the virus. The distribution of cyanophages in fish-ponds of Israel and the prevalence of these phages in high titre has been reported (Padan *et al.*, 1967; Padan and Shilo, 1969).

Shane (1971) surveyed the distribution of LPP-group of viruses in various types of natural waters like rivers, streams, lakes, farms and residential ponds, oxidization and stabilization lagoons and industrial storage waters. The virus was found during all seasons of the year and throughout a wide range of pH. The virus was consistently absent from the headwaters of the three rivers. However, the incidence of the virus was noted to increase when rivers flowed through more populated areas. So a relationship seems to exist between domestic and industrial pollution and with the ecology of these viruses. Singh (1973) studied the occurrence and distribution of cyanophages in ponds, sewage water and rice fields of Cuttack, India. Two types of cyanophages, i.e. clear plaque-forming (virulent) and turbid plaque-forming (lysogenic) strains were observed in natural habitats. This constitutes the first report on the occurrence of cyanophages from the rice fields of India. Seasonal periodicity of cyanophages AC-1, infecting *A. nidulans* ad *C. minor*, was studied by Sharma and Venkataraman (1978). These workers selected three ponds, a waste stabilization pond each at IARI, New Delhi, Nagpur (Maharashtra) and Ahmedabad (Gujarat) as well as freshwaters from River Ganges at Haridwar (Uttarakhand). AC-1 has not been found at Nagpur and Ahmedabad or in Ganges water. The cyanophage has been found in high titres in the ponds at New Delhi, which showed a maximum in May-June, while it was lowest during winter (November to January).

Desjardins and Olson (1983) demonstrated some control of bloom concentrations of *P. boryanum* in outdoor pond facilities with cyanophage LPP-1. This cyanophage was most effective when present before the bloom developed indicating that cyanophages are likely to be useful in preventing a cyanobacterial bloom rather than destroying a bloom already in progress. The different degrees of genetic variability of cyanophages attacking cyanobacteria have been shown to be due to degree of virus-induced inhibition of photoautotrophic energy metabolism and to the degree of modification of anabolic functions and subcellular structures of the cell (Mendzhul *et al.*, 1982). This has further been experimentally substantiated by the distribution of cyanophages. The interaction of the cyanophages LPP-1 and Aph-1 with their hosts, i.e. *P. boryanum* and *Aphanothece stagnina*, respectively in chemostat cultures revealed fluctuations in number of host cells and phage particles (Barnet *et al.*, 1981). The fluctuation in the phage-host cell numbers has been shown to be due to evolution of new phage-resistant strains of the hosts. PR-1 was resistant to LPP-DUN1 while it was susceptible to a mutant phage. However, PR-2 was unaffected by this phage. PR-2 grew more slowly than wild-type. Barnet *et al.* (1984) investigated the effect of suspended particulate material on cyanophages-cyanobacteria interactions in liquid cultures. These studies revealed that (i) the suspended silt protected cyanobacteria from lysis by the phage and (ii) the number of cyanophages and host cells oscillated in a reciprocal manner.

Studies on ecological dynamics of the toxic bloom-forming cyanobacterium *M. aeruginosa* and its cyanophages in Lake Mikata (Japan) have been made by the quantification of phycocyanin intergeneric spacer (PC-IGS) sequences in case of the former and the PCR amplification of sheath protein gene (g91) in case of the latter (Yoshida *et al.*, 2008). Simultaneously, the relative abundance of microcystin-producing sub-population of the species of *M. aeruginosa* has been determined by the amplification of *mcy* gene. The population of *M. aeruginosa* was represented by two peaks as reflected by copies of PC-IGS sequences to maximum of 3×10^3 copies ml^{-1} in the month of May and 2 x 10^5 copies in the month of July. These numbers are considered to be higher by about 200 times the cell number determined by microscopy. This has been explained on the basis that the cells of *Microcystis* contain several copies of *mcyA* gene (up to a maximum of 11) during the transition from the logarithmic growth phase to the stationary phase. However, a positive correlation existed between the total microcystin concentration and the copy number of the *M. aeruginosa mcyA* gene (Kurmayer and Kutzenberg, 2003; Yoshida *et al.*, 2005, 2007). A change in the toxin-producing sub-population of *M. aeruginosa* was noted over a period of time with the growth of non-toxic strains outcompeting in summer. The cyanophage abundance was negatively correlated with the abundance of *M. aeruginosa*. During summer, when cyanophage abundance increased, a simultaneous decline in the host population was noted. So a seasonal shift in the composition of the different sub-populations of *M. aeruginosa* was evident. On the basis of phylogenetic classification of the *g91* sequence analysis, a number of genotypes (a to j) of cyanophages have been noted during the period of study with genotype 'a' being most frequently represented in Lake Mikata. This sequence appeared to be identical to the sequences of Ma-LMM01, Ma-LMM02, Ma-LMM03 and Ma-LMM05 that are specific to only microcystin-producing populations of *M. aeruginosa* (Yoshida *et al.*, 2006; Takashima *et al.*, 2007; Yoshida *et al.*, 2008). Although genotypes 'b' to 'j' were represented only once, they had 98.9% identity to genotype 'a'. It is concluded that the cyanophage assemblage has the ability to kill only a small proportion of *M. aeruginosa* population paving the way for the replacement of phage-sensitive population by phage-insensitive population. The present understanding is that viruses will infect the most abundant host species because of high encounter rates and "killing the winner" viruses maintain the coexistence of competing species, a hypothesis proposed earlier by Thingstad and Lignell (1997). To quantify cyanophage Ma-LMM01 infected cells in the natural populations of *M. aeruginosa*, Yoshida *et al.* (2010) detected *g91* (RNA polymerase B) gene transcripts. Based on the relative abundance of *g91* gene transcripts before onset of release of virus progeny in the culture based experiments, it was feasible to assay of transcripts of *g91* in the infected cells under natural conditions by real-time reverse transcription PCR. The levels of expression of *g91* (10^{-2} per *rnpB* transcript) in natural waters corresponded with the expression levels in cultures.

X. MUTATIONAL STUDIES ON FRESHWATER CYANOPHAGES

Plaque-type, host-range and conditional lethal mutants have been isolated and characterized. Spontaneous mutants of LPP-1 exhibiting unstable rapid lysis (*r*) character were described by Safferman and Morris (1964b). Singh *et al.* (1972) characterized stable *r* mutants of LPP-1 in addition to minute, host-range (*h*) and acriflavine-resistant mutants. Rimon and Oppenheim (1975) isolated nitrosoguanidine (MNNG)-induced temperature-sensitive (*ts*) mutants of LPP-2 SP-1 who showed complementation and recombination among these mutants. Singh and Chaubey (1976) reported the induction r-mutants with hydroxylamine. Singh and Kashyap (1976) conducted a detailed study on induction of *h* and *ts* markers in LPP-1 after mutagenesis with MNNG and 2-aminopurine and suggested that these represented point mutations. Amla (1979, 1981) reported *h* and minute plaque-forming mutants of phage AS-1 whereas AS-1 resistant mutants of *A. nidulans* were reported by Bisen *et al.* (1985). Cyanophage LPP-1 resistant mutants of *P. uncinatum* were subsequently isolated and characterized (Bisen *et al.*, 1986b). The isolation and characterization of host-range mutants (Sarma and Kaur, 1993, 1997) and temperature-sensitive mutants of cyanophage N-1 (Sarma and Singh, 1994, 1995) were reported for the first time. The development of cyanophage N-1 has been studied under the stress of heavy metals (Sarma and Kaur, 2005).

XI. MARINE CYANOPHAGES

Viruses constitute the most abundant biological entities in marine waters exceeding the numbers of their hosts and as per present estimates there are nearly 10^{27} to 10^{31} viruses in the world's Oceans (Wilhelm and Suttle, 1999; Suttle, 2005; Breitbart and Rohwer, 2005) and these amount to ~270 Mt of C. If these viruses are laid end to end, they would span between Earth and the Sun 10^{13} times (Hendrix, 2003). A number of workers have demonstrated the existence of viruses in high concentrations (ranging from 10^5 to 10^8 particles ml^{-1} seawater) with transmission electron microscopy, after concentrating either by filtration or through centrifugation (Bergh *et al.*, 1989; Børsheim *et al.*, 1990; Bratbak *et al.*, 1990, 1992; Hara *et al.*, 1991; Heldal and Bratbak, 1991; Proctor and Fuhrman, 1990, 1991, 1992; Suttle *et al.*, 1990, 1991). The titres of viruses or virus-like particles (VLPs) represented in different environmental samples are presented in Table 5. Attempts to isolate cyanophages from seawaters were very few and sporadic (Moisa *et al.*, 1981; Suttle *et al.*, 1990) but it was only in 1993 that the marine cyanophages have been well characterized (Suttle and Chan, 1993; Waterbury and Valois, 1993; Wilson *et al.*, 1993).

Table 5: Virus-like particles in different environmental samples

Sr. No.	Type of water body	Titre ml^{-1}water/ g^{-1} soil	Authors
1	Surface seawater	10^7	Fuhrman, 1999; Wommack and Colwell, 2000
2	Marine sediment	10^8	Danovaro *et al.*, 2001
3	Soil	~10^8	Williamson *et al.*, 2003
4	Freshwater	10^5 to 10^9	Kepner *et al.*, 1998; Chen *et al.*, 2001; Vrede *et al.*, 2003
5	Hotsprings	10^5 to 10^9	Breitbart *et al.*, 2004
6	Hypersaline waters	10^6 to 10^9	Oren *et al.*, 1997; Dyall-Smith *et al.*, 2003
7	Solar salterns	~10^{10}	Guixa-Boixareu *et al.*, 1996; Pedros-Alio *et al.*, 2000a,b

Studies on marine cyanophages have dealt upon the isolation and morphological characterization of cyanophages infecting *Synechococcus* and *Prochlorococcus* from marine waters of different geographical locations. Phage titres have been assayed on the basis of culture-based plaque assay and the phage diversity has been determined by a number of techniques developed for the purpose. Culture-independent assays of phage diversity included the assay of phage-specific genes (encoding different proteins of capsid, sheath or tail) such as *g20*, *g23* or *g91* or some of the host genes identified in the marine cyanophages such as *psbA* and *psbD* gene sequences or other allied photosynthesis genes. Another important characteristic of viruses that can be readily determined is the genome size. The genome sizes range from few kb to several hundred thousand kb. This chraracteristic of phages has been utilized for determining the frequency distribution of genome sizes. As conventional electrophoresis does not provide necessary resolution, pulsed-field gel electrophoresis (PFGE) or denaturing gradient gel electrophoresis (DGGE) have been employed. This has revolutionized marine viral enumeration studies. The role of marine cyanophages in the control of prokaryotic marine phytoplanktons has received much attention. All these aspects are discussed in detail in the following sections.

i) *Isolation and morphological characterization*: Waterbury and Valois (1993) isolated marine cyanophages infecting marine *Synechococcus* strains belonging to two clusters. Cluster A is characterized by strains (WH8012, WH8103, WH8108, WH8018, WH8109 and WH8017) that occur in open ocean and coastal waters, possess phycoerythrin as the primary light-harvesting pigment and require high salt concentration for growth while Cluster B is composed of strains (WH7803 and WH8101) that occur in coastal waters, possess phycocyanin as the primary light-harvesting pigment and do not require elevated salt concentration but at the same time are halophilic. Among the phages isolated by them (75 in total) representatives from each of the three families of tailed phages are included. Phage 9 infected as many as 10 of 13 strains tested from Marine Cluster (MC)-A. Phages 5 and 7 infected only the host strain used for isolation. Phage 20 not only infected strains of MC-A but also infected strain WH8101 from MC-B. Phage 1 isolated on WH8101 from MC-B also infected 5 of 13 strains belonging to MC-A. All the cyanophages isolated are virulent phages. Attempts to induce lysogenic cells to produce phage by using temperature shock, light shifts, UV or X-ray irradiation and treatment with mitomycin C have not been successful. According to them, these lytic cyanophages have negligible effect in regulating the populations of *Synechococcus* in marine waters. They based their conclusions on the loss of infectivity by cyanophages due to solar radiation and viral decay rates (Heldal and Bratbak, 1991; Suttle and Chan, 1992). Further, they suggested that evolution of phage-resistant strains and their coexistence with phages is the likely final outcome. The emergence of phage-sensitive cells from the phage-resistant populations enables the viruses to maintain the titres up to 10^3 to 10^4 cyanophages ml^{-1} in both inshore and offshore waters. Garza and Suttle (1998) studied the rates of mortality of *Synechococcus* and the evolution of UV-resistant viral communities by incubating two cyanophage isolates S-PWM1 and S-PWM3 either with or without host cells of *Synechococcus* sp in UV-transparent bags at the surface and at different depths of seawater at Port Aransas, Texas. Likewise, natural populations of *Synechococcus* sp. and cyanophages were (collected, filtered to remove zooplankton grazers, bacteria and host cells) put in UV-transprent bags and were held at the surface and various depths of the experimental site. The mortality rates of the host in two offshore locations have been noted to be 1 and 8%. In contrast, nearshore incubations of cyanophages were responsible for removing only ≤1% of *Synechococcus* cells on a daily basis during winter and spring. Natural cyanophage communities tolerate damage by solar radiation better in summer than in winter. Net decay rates of cyanophage infectivity in sunlight were similar whether host cells were present or not indicating that detectable cyanophage production did not occur during day time *in situ* incubations. Natural viral communities appeared to be more resistant to DNA damage than cultured cyanophage Syn-M3 infecting *Synechococcus* sp. DC2 (WH7803) and the marine *Vibrio* phage PWH3a-P1 when incubated from dawn to dusk in offshore and coastal waters of western Gulf of Mexico. The DNA damage was measured by estimating the number of sun-light induced formation of cyclobutane pyrimidine dimmers (CPDs) and pyrimidine pyrimidone photoproducts (PPs) by a highly specific radioimmunoassay (Weinbauer *et al.*, 1999).

Wilson *et al.* (1993) isolated cyanomyoviruses (S-BM1, S-PM1, S-PM2 and S-WHM1) and cyanosiphoviruses (S-BS1 and S-PS1) infecting marine *Synechococcus*. The diameter of the head of S-PM2 is 90 nm with a tail of 165 nm long and 20 nm wide (Fig. 4 A and B). Cyanophages S-BM1 and S-PM1 had a similar morphology in having isometric heads with a diameter ~88 nm, extended tails ~57 nm long and 31 nm wide (at the base plate) which are connected to the head with a collar, 8nm long and 14 nm wide and tail plates ~45 nm wide. Cyanophage S-WHM1 had an isometric head of approximately the same size as those of S-BM1 and S-PM1 and a longer tail ~108 nm long and 23 nm wide. Cyanosiphoviruses (S-PS1 and S-BS1) have isometric heads with a diameter of 90 nm and had rigid tails ~165 nm long and 20 nm wide. A base plate of 43 nm wide was connected to the tail. Protruding from each plate corner was a spike ~45 nm long and tail fibres that are 1.5 times the length of the tail (Table 6). The genome size of cyanomyoviruses isolated were ca. 80–85 kb and those of cyanosiphoviruses were ~90–100 kb. The restriction patterns of cyanomyoviruses from Woods Hole appeared to be unique from others. The restriction patterns of cyanomyoviruses

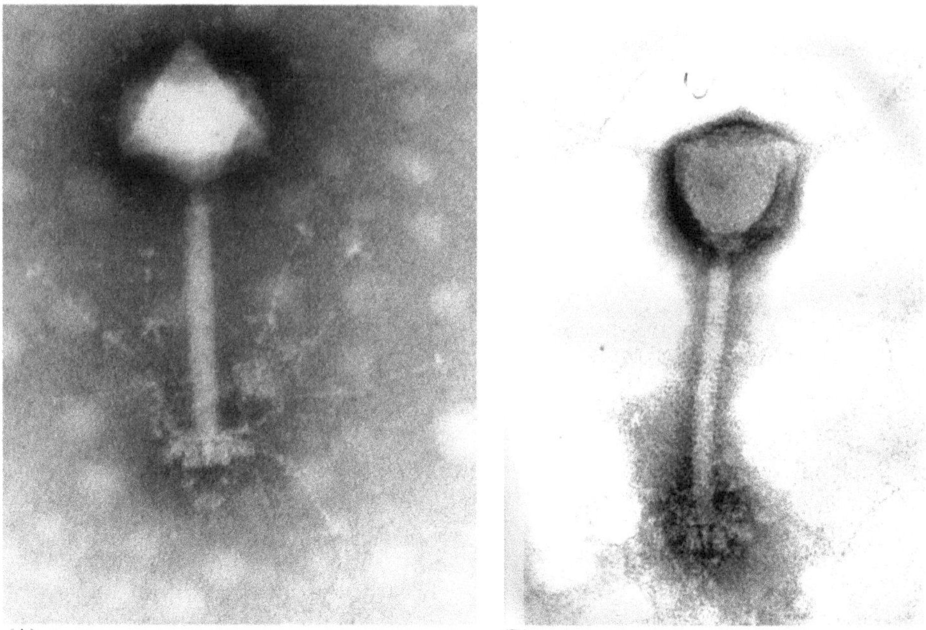

(A) (B)

Figure 4: (A) Transmission electron microscope image of the *Synechococcus* Phage S-PM2. (B) Negative (phosphotungstate-) stained particle of S-PM2. x 297,000. Picture (A) with the kind permission of N. H. Mann, Department of Biological Sciences, University of Warwick, Coventry, UK. [Mann, N. H. (2005) *PLoS Biol* **3(5):** e182. doi:10.1371/journal.pbio.0030182] Doi:10.1371/journal.pbio.0030182.g002.
Images (A) and (B) courtesy H.-W. Ackermann, Department of Microbiology, Immunology and Infectiology, Faculty of Medicine, Laval University, Vandry Pavilion, Avenue de la Médecine, Quebec, QC, G1K 7P4, Canada.

and cyanosiphoviruses (from Bermuda and English Channel) resembled each other. There is a limited degree of homology among all cyanophage DNAs as revealed by Southern hybridization. The structural protein in each case was ~53 to 54 kDa.

Lu *et al.* (2001) isolated and characterized a number of cyanophages from three Georgia's coastal river (Savannah, Altamaha and Satilla) estuaries by selecting four phycoerythrin (PE)-containing strains (WH7803, WH7805, WH8108, and WH81103) and seven phycocyanin (PC)-containing (WH8101, WH8007, WH5701, CCMP1628, CCMP1629, CCMP1630 and CCMP1632) strains of *Synechococcus* sp. as the hosts for isolation. Cyanophages P3 (isolated on WH7805), P5 and P16 (isolated on WH 7803) cross-infected majority of PE- and PC-containing strains and so had broad host range. Cyanophages P68 and P76 (isolated on PC-containing strain WH 8007) exhibited a very narrow host range as the former cross infected only WH 8101 while the latter did not cross infect any other strain. Cyanophages (P2, P39, P53, P56, P71, P73, P79, P81, P82 and P83) isolated on WH 7805 cross infected other PE strains such as WH7803 and WH8108. Of the cyanophages P8, P12 and P14 isolated on PC-containing strain WH8101, only P8 and P14 cross-infected WH7803 and WH 7805 whereas P12 did not cross infect any other strain. Cyanophages P71, P72, P73, P79, P81, P82 and P83 isolated on PE-containing strain WH7805, all cross infected WH7803 and WH8108. Morphologically 80% of the cyanophages belonged to the family Myoviridae (Table 8).

Northern waters of Moreton Bay, Queensland, Australia supported expansive bloom of *Lyngbya majuscula* that produces toxic substances responsible for dermatitis and asthma-like symptoms.

Table 6: Marine cyanophages isolated on *Synechococcus* strains.

Location	Family	Cyanophage	Head Dia. (nm)	Tail L X W (nm)	Authors
English Channel	M	S-PM2	90	165 x 20	Wilson *et al.* (1993)
Gulf Stream	M	Syn-14	93	136 x 21	Waterbury and Valois (1993)
		Syn-33	-	-	
NE Providence Channel	M	Syn-26	-	-	Waterbury and Valois (1993)
Sargasso Sea	M	S-SSM1	95	150 x 25	Sullivan *et al.* (2003)
		S-SSM2	80	225 x 22	Waterbury and Valois (1993)
		Syn-2	66	149 x 17	
		Syn-10	100	145 x 19	
		Ø2	-	-	
Shelf	M	S-ShM1	80	120 x 25	Sullivan *et al.* (2003)
		S-ShM2	70	100 x 30	
Woods Hole	M	S-WHM1	88	108 x 23	Wilson *et al.* (1993)
		Syn-1	-	-	Waterbury and Valois (1993)
		Syn-9	87	153 x 19	
		Ø9	-	-	
Slope	M	S-SM1	90	80 x 25	Sullivan *et al.* (2003)
Bermuda	M	S-BM1	88	57 x 31	Wilson *et al.* (1993)
		S-BM4	-	-	
		S-BM5	-	-	
Plymouth		S-BM6	-	-	
		S-PM1	88	108 x 23	
		S-PM2	-	-	
Bermuda	S	S-BS1	90	165 x 43	Wilson *et al.* (1993)
Plymouth	S	S-PS1	90	165 x 43	
Bermuda	P	S-BP1	-	-	Fuller *et al.* (1998)
		S-BP2	-	-	
		S-BP3	-	-	
Red Sea	M	S-RSM1	-	-	Wilson (1994)
		S-RSM2	-	-	Millard *et al.* (2004)
		S-RSM28	-	-	
		S-RSM88	-	-	
Miami, Fla	M	S-MM1	-	-	Wilson (1994)
		S-MM2	-	-	
		S-MM3	-	-	
		S-MM4	-	-	
		S-MM5	-	-	
		S-MM7	-	-	
Bergen, Norway	M	S-BnM1	-	-	Wilson (1994)
		S-Bn3	-	-	Fuller *et al.* (1998)
Gulf of Mexico	M	S-PWM1	-	-	Suttle and Chan (1993)
		S-PW3	-	-	
Gulf Stream		Ø14, Ø33	-	-	Waterbury and Valois (1993)
		Ø34			
		Ø12	-	-	Wilson (1994)
Rhode Island	M	S-RIM1 to S-RIM40	-	-	Marston and Sallee (2003)
Kagoshima Bay	M	S-KM1	-	-	Okunishi *et al.* (2003)
Baltimore Inner Harbor and Chesapeake Bay	M	S-CBM2	90 ± 5	-	Wang and Chen (2007)
Baltimore Inner Harbor and Chesapeake Bay	S	S-CBS2	90 ± 3	-	Wang and Chen (2007)
		S-CBS3	56 ± 4	-	
		S-CBS4	72 ± 4	-	
Baltimore Inner Harbor and Chesapeake Bay	P	S-CBP1	52 ± 3	-	Wang and Chen (2007)
		S-CBP2	55 ± 5	-	
		S-CBP3	55 ± 4	-	

The sudden disappearance of this bloom was attributed to the presence of virus-like particles akin to cyanosiphoviruses. Electron microscopy revealed the existence of these particles inside the cells of *L. majuscula* and caused a decrease in photosynthetic efficiency and electron transport (Hewson *et al.*, 2001). A cyanomyovirus S-KM1 infecting *Synechococcus* sp. (PE-rich strain KFM001) with a capsid diameter of 83 nm and a contractile tail of 166 nm long has been isolated from Kagoshima Bay by Okunishi *et al.* (2002). Phage titers of S-KM1, at the Kagoshima Bay, have been found to be almost undetectable but a corresponding increase in its titer (7.5 x 10^3 ml^{-1}) occurred about one week after the onset of the bloom of the PE-rich strain of *Synechococcus*. They provided estimates of host population lysed per day from 0.0055% to 2.1% suggesting that the lytic phages control the populations of their respective hosts.

Cyanophages infecting *Prochlorococcus* and *Synechococcus* (44 of them) have been characterized with reference to morphology and host-range (20 different host strains belonging to both of the genera; Tables 6 and 7). Of these, 31 cyanophages have been isolated from Red Sea, Gulf Stream and BATS by Sullivan *et al.* (2003) and these workers included in their study two cyanophages reported earlier by Wilson *et al.* (1993) from English Channel and Woods Hole and 11 cyanophages isolated by Waterbury and Valois (1993) from different locations such as Sargasso Sea, Woods Hole, Gulf Stream and English Channel. This report constitutes the first comprehensive account of the marine

Table 7: Marine cyanophages isolated on *Prochlorococcus*.

Location	Family	Cyanophage(s)	Head Dia. (nm)	Tail L x W (nm)	Authors
BATS	P	P-SSP1	45	-	Sullivan *et al.* (2003)
BATS	P	P-SSP2	55	-	Sullivan *et al.* (2003)
BATS	P	P-SSP3 P-SSP4 P-SSP7 P-SSP8	50	-	Sullivan *et al.* (2003)
BATS	P	P-SSP5 P-SSP6	55	6 x 10	Sullivan *et al.* (2003)
Gulf Stream	P	P-GSP1 Syn12	45 45	- 8 x 10	Sullivan *et al.* (2003)
Red Sea	P P	P-RSP1 P-RSP2 P-RSP3	55 40 50	6 x 10 - -	Sullivan *et al.* (2003)
Sargasso Sea	P	Syn-5	-	-	Waterbury and Valois (1993)
Slope	S P	P-SS1 P-SS2 P-SP1	45 x 90 50 x 100 60	280 x 15 260 x 12 8 x 12	Sullivan *et al.* (2003)
BATS	M	P-SSM1 P-SSM2 P-SSM3 P-SSM4 P-SSM5 P-SSM6	60 85 95 80 60 90	160 x 20 110 x 25 105 x 25 160 x 20 20 x 80 150 x 28	Sullivan *et al.* (2003)
Red Sea	M	P-RSM1 P-RSM2 P-RSM3	80 75 80	110 x 27 170 x 23 175 x 23	Sullivan *et al.* (2003)
Shelf	M	P-ShM1 P-ShM2	75 -	120 x 18 -	Sullivan *et al.* (2003)

Note: M = Myoviridae; P = Podoviridae; S = Siphoviridae

Table 8: Cyanophages isolated[a] on PE and PC-containing strains of *Synechococcus*[b] and the strains on which cross-infection was possible (after Lu *et al.*, 2001).

Cyanophage(s)	Host strain on which isolated	Strains on which cross-infection was possible
P2, P39, P53, P56, P71, P72, P73, P79, P81, P82 and P83	PE-containing strain WH 7805	PE-containing strains WH 7803 and WH 8108
P3	PE-containing strain WH 7805	Majority of PE- and PC-containing strains
P5 and P16	PE-containing strain WH 7803	Majority of PE- and PC-containing strains
P8 and P14	PC-containing strain WH 8101	PE-containing strains WH 7803 and WH 7805
P12	PC-containing strain WH 8101	Did not cross-infect on any other strain
P68	PC-containing strain WH 8007	PC-containing strain WH 8101
P76	PC-containing strain WH 8007	Did not cross-infect on any other strain

[a]From the Georgia coastal river (Savannah, Altamaha and Satilla) estuaries.
[b]Four PE-containing strains (WH7803, WH 7805, WH 8103 and WH 8108) and seven PC-containing strains (WH 8101, WH 8007, WH 5701, CCMP 1628, CCMP 1629, CCMP 1630 and CCMP 1632) of *Synechococcus* have been used by Lu *et al.* (2001) in this study. PE-Phycoerythrin; PC-Phycocyanin

cyanophages that were isolated from different depths and locations. The diversity of phages was represented by all the three groups of tailed phages. Cyanomyoviruses are quite unique in having a broader host range as they not only infected and cross-infected *Prochlorococcus* strains among each other but also did so between *Prochlorococcus* and *Synechococcus* strains. Those cyanophages isolated on *Synechococcus* exhibited broader host range as they infected low-light (LL-) adapted strains of *Prochlorococcus* as well. This was explained on the basis that the LL-adapted *Prochlorococcus* strains are less diverged from *Synechococcus* and they in fact are closely related with each other on the basis of rDNA sequence homology. But none of these phage strains have infected marine bacterial isolates. Cyanosiphoviruses have been represented by two cyanophages (P-SS1 and P-SS2) that were specific to their respective hosts. In contrast, cyanopodoviruses were extremely host-specific with ony two cross infections out of a possible 300. High light-(HL-) adapted strains of *Prochlorococcus* are wide spread in marine waters and these supported the multiplication of cyanopodoviruses that have a restricted host-range. Titres of cyanophages infecting strains of *Synechococcus* were relatively high (10^6 phages ml^{-1}) when compared to those sensitizing *Prochlorococcus* strains. Certain of the strains of *Prochlorococcus* (MIT 9303, MIT 9313 and SS120) had low or no cyanophage strains infecting them while on other hosts the cyanophage titres were generally lower (350 phages ml^{-1}) (Table 9). However, representatives of cyanophages infecting the respective *Prochlorococcus* strains could be found to depths of 70 m to 100 m. These studies were extended to include surface coastal waters (mesotrophic) to open ocean sites (oligotrophic) of the Atlantic Ocean in relation to the natural abundance of the respective host strains. Nearly 98% representative morphotypes of cyanophages infecting HL-adapted *Prochlorococcus* were of cyanopodoviruses while the rest were cyanomyoviruses. LL-adapted *Prochlorococcus* strains were infected by cyanomyoviruses (65%) and the rest belonged to cyanopodoviruses. On the other hand, maximum infection of *Synechococcus* was by cyanomyoviruses (93%) and the rest belonged to the other two categories.

A similar study conducted by McDaniel *et al.* (2006) at the Gulf of Mexico led to the characterization of 35 marine cyanophages isolated on *Synechococcus* sp. strain WH7803 as the host. Morphologically, 94% of these were found to be myoviruses and the rest of them belonged to podovirus category. The cross infectivity of these strains was tested on 24 strains of *Synechococcus* isolated from the surface waters and extending up to a depth of 75 m. Some of the cyanophages (SPGM-01, SPGM-02 and

Table 9: *Prochlorococcus* cyanophages with their host of isolation and cross-infectivity between strains of the same genus and those of *Synechococcus** (after Sullivan *et al.*, 2003).

Cyanophage(s)	Host on which isolated	Hosts on which cross-infection was possible
Syn1	*Synechococcus* WH8101	*Synechococcus* WH7803, WH8017 and WH8018
S-SM1, S-ShM1 and S-SSM1	*Synechococcus* WH6501	*Synechococcus* WH8102
Syn9	*Synechococcus* WH8109	Broad host range; infected HL-adapted MIT 9515 and MED4 strains of *Prochlorococcus*
P-RSP3, P-SSP8	LL-adapted strain NATL2A of *Prochlorococcus*	Did not cross infect any other strain of *Prochlorococcus* both HL-adapted as well as LL-adapted strains and of *Synechococcus* strains
P-RSM3	LL-adapted strain NATL2A of *Prochlorococcus*	Cross infected only LL-adapted MIT9211, NATL1A of *Prochlorococcus*
P-SSM3, P-SSM4, P-SSM5, P-SSM6, P-RSM2	LL-adapted strain NATL2A of *Prochlorococcus*	Cross infected other LL-adapted strains of *Prochlorococcus*
P-RSP2 P-SSP2, P-SSP4	HL-adapted strain of *Prochlorococcus* MIT9302 HL-adapted strain of *Prochlorococcus* MIT9312	Did not cross infect other HL-adapted strains or LL-adapted strains of *Prochlorococcus* or other strains of marine *Synechococcus* strains
P-SSP3	HL-adapted strain of *Prochlorococcus* MIT9312	Cross-infected only high-light adapted strain of *Prochlorococcus* MIT9515

**Synechococcus* strain WH8020 did not support the growth of any of the 44 phage strains tested (see text). HL-high light; LL-low light.

SPGM-03) had a very restricted host range as they infected only one strain of the host whereas majority of the cyanophage isolates (SPGM-04, SPGM-11 to SPGM-35) exhibited a broad host range as they infected 10 to 17 strains of the host. The rest of the cyanophages (SPGM-05 to SPGM-10) infected 3 to 8 host strains and so had an intermediate host range (McDaniel *et al.*, 2006). Virus isolation studies conducted from Indian Ocean by plaque assay method on *Synechococcus* WH7803 revealed that cyanophages were abundant from the surface waters to a depth of 10 m with a maximum at 1.00 am after which there was a decrease in their numbers. Nearly 50 such cyanophages (S-IO1 to S-IO50) have been isolated whose morphotype could not be established (Clokie *et al.*, 2006). Tables 6 and 7 summarize some of the marine cyanophages isolated and characterized against marine *Synechococcus* and *Prochlorococcus* strains. A probable explanation that is given for the cross-infectivity of *Prochlorococcus* and *Synechococcus* strains by the marine cyanomyoviruses is the existence of additional sets of tRNA genes in these viral genomes accommodating AT-rich codons. These additional tRNA gene sets enable these viruses to improve their overall translational efficiency and infect *Synechococcus* strains which have generally a high mol% G+C content (Enav *et al.*, 2012).

The marine cyanophages described so far infect the coastal and oceanic *Synechococcus* strains. In an attempt to find out the cyanophages, from surface waters of Baltimore Inner Harbor and Chesapeake Bay, infecting the estuarine strains of the same genus, Wang and Chen (2007) for the first time described one cyanomyovirus (S-CBM2 on host strain CB0101), three cyanosiphoviruses (S-CBS2, S-CBS3 and S-CBS4 isolated on host strains CB0204, CB0202 and CB0101, respectively) and three cyanopodoviruses (S-CBP1, S-CBP2 and S-CBP3 isolated on CB0101, CB0208 and CB0101, respectively) against four strains of estuarine *Synechococcus* belonging to MC-B isolated earlier by Chen *et al.* (2004) from Chespeake Bay.

While most of the marine cyanophages isolated were against the two unicellular marine phytoplankton components (*Synechococcus* and *Prochlorococcus* strains), Jenkins and Hayes (2006) investigated the diversity of cyanophages infecting *Nodularia spumigena* from the brackishwaters of Baltic Sea. The existence of four cyanomyoviruses (N-BM1, N-BM2, N-BM3 and N-BM4), five cyanosiphoviruses (N-BS1, N-BS2, N-BS3, N-BS4 and N-BS5) and seven unidentified morphotypes has been demonstrated. Cross-infectivity studies on 13 genotypes of *Nodularia* revealed that N-BM1 infected three other strains besides the original strain on which it was isolated. Isolate 12, an unidentified morphotype of the cyanophages, also cross-infected many strains. The dimensions of the siphoviruses seemed to represent the largest particle sizes reported hitherto for marine cyanophages. The diameter of heads of siphoviruses N-BS1 and N-BS4 were of the order of 137 x 120 nm and 127 x 122 nm with a tail length of 834 nm and 888 nm, respectively.

***ii) Genomic sequences of marine cyanophages*:** Database on phage genomic sequences shows 430 completed dsDNA phage genomes. Of these, only eight belong to marine bacteriophages. These include six vibrio phages (vpV262, KVP40, VP16T, VP16C, K139, and VHML), one roseophage SI01 and one *Pseudomonas* phage (PM2). By now complete genome sequences of eleven cyanophages have been published. Of these, five are cyanomyoviruses that are T4-like (P-SSM2, P-SSM4, S-PM2, Syn5 and Ma-LMM01), one unclassified virus (Syn 9) and five cyanopodoviruses (P-SSP7, Syn5, P60, PaV-LD and Pf-WMP4). The details are as follows.

a) Phage P60: Lu *et al.* (2001) first reported the isolation of cyanophage P60 (P for Phi), a podovirus infecting *Synechococcus* sp. strain WH7803 from one of Georgia's coastal rivers, Satilla River. Genome sequencing of this phage revealed interesting features, i.e. (i) the phage genome has as many as 80 potential ORFs, (ii) the length of phage genome is 47,872 bp and consists gene sequences similar to other lytic phages such as T7, Phi-YeO3-12 and SI01 and (iii) similarities in coding sequences of nucleotide metabolism of phage and marine unicellular cyanobacteria exist (Chen and Lu, 2002).

b) Prochlorococcus viruses: The myoviruses of *Prochlorococcus*, P-SSM2 and P-SSM4 exhibit T-4 like capsid, base-plate and tail structure and are similar in morphology except that the heads of the former are slightly larger (85 nm) than the latter (80 nm). However, the tails of P-SSM4 are slightly longer (160 x 20 nm) than in case of P-SSM2 (110 x 25 nm). In both the cases, morphotypes with and without contractile tails have been reported (Fig. 5). The sequencing of P-SSM2 and P-SSM4 and P-SSP7 cyanophage genomes infecting strains of *Prochlorococcus* revealed the presence of an integrase gene similar to that of T7 gene in cyanophage P-SSP7 alone. Gene sequences of unknown function accounted for 60% and 58% in phages P-SSM2 and P-SSM4, respectively. Among the identified genes, genes similar to phage T4 were of the order of 14% and 21% in P-SSM2 and P-SSM4, respectively (Fig. 6). Table 10 summarizes genome characteristics of these phages. Of the two myoviruses, P-SSM2 possesses two large clusters of 24 genes related to sugar epimerase, transferase and synthase genes probably involved in lipopolysaccharide synthesis. The genomes of all the three phages contained photosynthesis genes (*psbA*, *hlip*) and a gene from aldolase family (*talC*) that helps in alternative routes of carbon metabolism during phage infection. It is also significant to note that P-SSM2 and P-SSM4 also contained phosphate-inducible genes (*phoH* and *pstS*) that are helpful in phosphate stress. Certain of these gene sequences have been suggested to represent "signature" sequences of cyanophage genes (Sulliivan *et al.*, 2005; Table 10). P-SSP7 is a cyanopodovirus (Fig. 7 A) and its

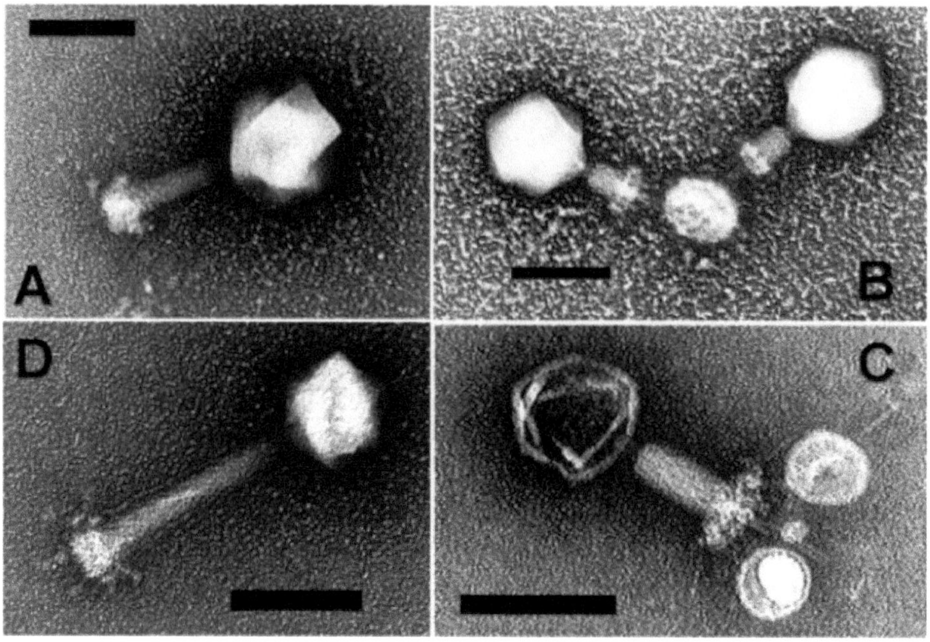

Figure 5: Electron micrograph of negative-stained *Prochlorococcus* Myoviruses P-SSM2 and P-SSM4. Myovirus P-SSM2 with (A) non-contracted tail and (B) contracted tail, and myovirus P-SSM4 with (C) contracted tail and (D) non-contracted tail. Note the T4-like capsid, base-plate, and tail structure in both myoviruses. Scale bars indicate 100 nm. With the kind permission of S. W. Chisholm, Department of Biology, Massachusetts Institute of Technology, Cambridge, Massachusetts, USA. [Sullivan *et al.* (2005) *PLoS Biol* **3(5):** e144. doi:10.1371/journal.pbio.0030144] doi:10.1371/journal.pbio.0030144.g002.

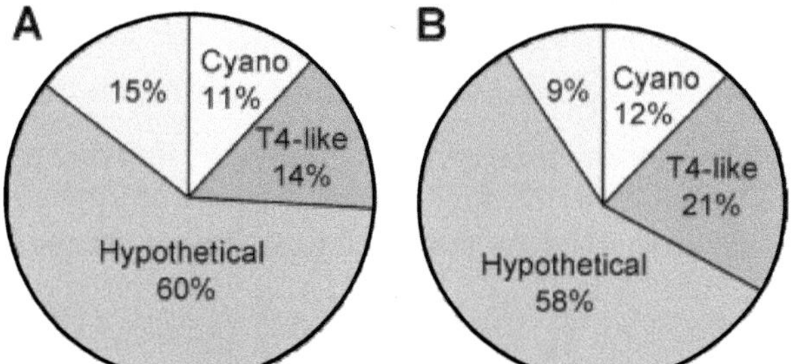

Figure 6: Taxonomy of best BLASTp hits for P-SSM2 and P-SSM4. Each predicted coding sequence from both phage genomes was used as a query against the nonredundant database to identify the taxon of the best hit. Blue slices indicate phage hits, while yellow slices indicate cellular hits. With the kind permission of S. W. Chisholm, Department of Biology, Massachusetts Institute of Technology, Cambridge, Massachusetts, USA. [Sullivan *et al.* (2005) *PLoS Biol* **3(5):** e144. doi:10.1371/journal.pbio.0030144] doi:10.1371/journal.pbio.0030144.g005.

Color image of this figure appears in the color plate section at the end of the book.

genome has 540 ORFs, nearly half of which could not be assigned any function (nearly 54%). This podophage possesses 31% of the translated ORFs similar to T7 phage encoded proteins and 4% typical for it. There are 11% of the proteins synthesized that are similar to cyanobacterial (7%) and bacterial proteins (4%) (Fig. 7 B). P-SSP7 has 16 genes common with the 26 core genes identified for T7. These relate to shut-down of host transcription, initiation of phage gene transcription, degradation of host

Table 10: Genome characteristics of *Prochlorococcus* cyanophages (after Sullivan *et al.* (2005)).

Cyanophages	Length (bp)	G+C content	ORFs	Functional genes
P-60	47,872 bp		80	Genes of nucleotide metabolism similar to cyanobacteria
S-PM2	196,280 bp	37.8%	239	*petE, petF*
P-SSP7	~45 kb	38.7%	54	Of the 26 core phage T7 genes 15 are present; integrase gene present
P-SSM2	252 kb	35.5%	327	Phage T4-like genes 14% present; genes of unkown function 60%; integrase gene lacking; two large clusters of genes (24 in total) present relating to lipopolysaccaride synthesis; phosphate inducible genes (*phoH* and *pstS*) helpful in phosphate stress are present
P-SSM4	178 kb	36.7%	198	Phage T4-like genes 21% present; genes of unkown function are 58%; integrase gene lacking; two large clusters of genes (24 in total) absent; phosphate inducible genes (*phoH* and *pstS*) helpful in phosphate stress are present
Pf-WMP4	40,938 bp		45	
Syn-9	162,109 bp		184	*nblA* gene that degrades phycobilisomes; integrase (recombinase) present, 44 ORFs with known function and the rest of unknown function

DNA, DNA replication, DNA packaging and virion formation are important among others. Some of the T7 functions absent in P-SSP7 relate to ligation of DNA fragments, inhibition of host RNAP, lysis events and interactions specific to host cell envelope during virion formation are important. There seems to be a functional *int* gene in P-SSP7 genome that has a 42-bp matching sequence downstream of $tRNA^{Leu}$ gene of the host *P. marinus* MED4 suggesting the probability of P-SSP7 undergoing a lysogenic event (Sullivan *et al.*, 2005).

c) *Phage S-PM2*: The genome of S-PM2 is a dsDNA molecule that is circularly permuted and consists of 196,280 bp with mol% G+C of 37.8. The total ORFs represented are 239 and possess sequences for 25 tRNAs. Of these, 40 ORFs have been organized into discrete clusters that are associated with virus morphogenesis, nucleotide metabolism, gene regulation and DNA replication and repair. ORFs of unkown function suspected to be involved in the adaptation of the metabolism of host cell to the requirements of phage replication are present. Twenty ORFs have been shown to be identical to the cyanobacterial hosts (Mann *et al.*, 2005). It may be mentioned that S-PM2 is the first "photosynthetic" marine cyanophage that has been reported to carry PSII reaction center genes, *psbD* and an interrupted *psbA* gene sequences in its genome (Mann *et al.*, 2003). Two photosynthesis electron transport genes encoding plastocyanin (*petE*) and ferredoxin (*petF*) that are present in the *Prochlorococcus* phage P-SSM2 are not represented in S-PM2 genome (Mann *et al.*, 2005). Though the adsorption of S-PM2 to the cells of *Synechococcus* sp. strain WH7803 was light-dependent, neither photosynthesis is required for adsorption nor does it make any difference of the presence or absence of PSII reaction center genes in its genome (Jia *et al.*, 2010).

d) *Phage Syn5*: This phage, isolated by Waterbury and Valois (1993) from Sargasso Sea, infects *Synechococcus* sp. strain WH8109 belonging to MC-A, Calde II. It exhibited a restricted host range as it did not infect 19 other strains of *Synechococcus* and *Prochlorococcus*. Pope *et al.* (2007) investigated the structural properties, capsid organization and genome sequence of this cyanophage. Cryo-electronmicroscopy revealed that the virus particles have an isometric icosahedral head of ~60 nm in diameter with a short tail of 25 nm length. One of the most distinguishing features of this

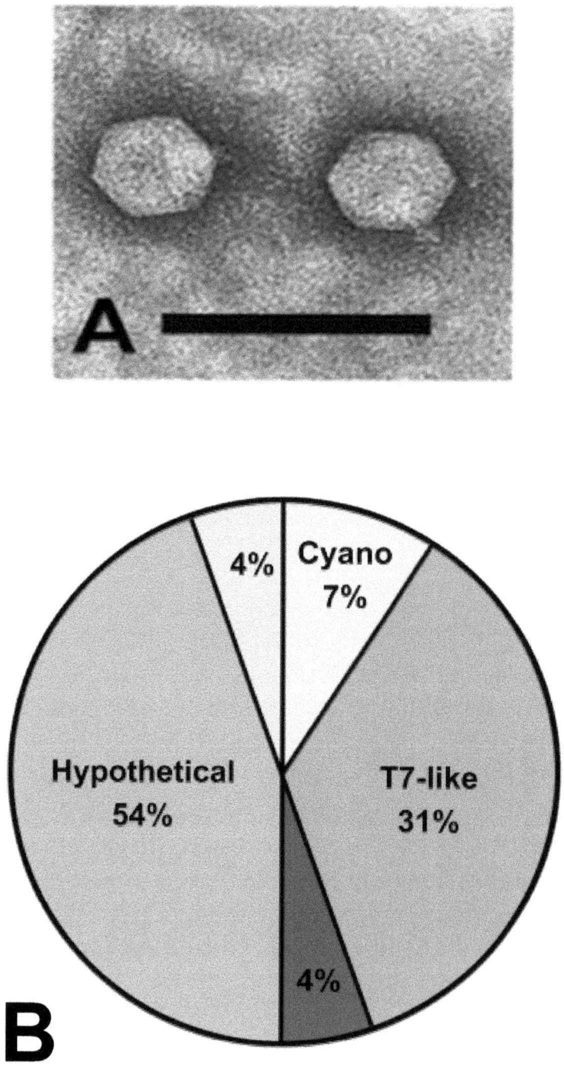

Figure 7: Features of the *Prochlorococcus* Podovirus P-SSP7 (A) Electron micrograph of negative-stained podovirus P-SSP7. Note the distinct T7-like capsid and tail structure. Scale bar indicates 100 nm. (B) Taxonomy of best BLASTp hits for P-SSP7. Each predicted coding sequence from the phage genomes was used as a query against the nonredundant database to identify the taxon of the best hit. Blue slices indicate phage hits, while yellow slices indicate cellular hits. With the kind permission of S. W. Chisholm, Department of Biology, Massachusetts Institute of Technology, Cambridge, Massachusetts, USA. [Sullivan *et al.* (2005) *PLoS Biol* 3**(5):** e144. doi:10.1371/journal.pbio.0030144] doi:10.1371/journal.pbio.0030144.g001.

Color image of this figure appears in the color plate section at the end of the book.

phage is the presence of a "horn" which is a slender elongated fibrous protrusion of approximately 50 nm long and 10 nm wide, tapering to 2–5 nm at the tip. TEM of infected cells revealed the eclipse period to be of 1 h. By 2 h, 30% of the infected cells showed five or more phage heads in the cell. Cell lysis ensued after 10 h of infection. The genome of Syn5 is 46, 214 bp long with a 237 bp terminal direct repeat. There are 61 ORFs and some resemble T7 phage gene sequences and so Syn5 shares some of the features of T7. The existence of an integrase gene signifies that the phage is a temperate one. The structural proteins of the phage are encoded by eight of the 11 ORFs assigned to structural proteins. The amino acid sequence of these proteins has similarity to the proteins of enteric phages. The occurrence of genes is without identifiable promoter regions and Shine-Dalgarno sequences. The presence of overlapping start and stop codons signifies a translational coupling by which the translation of the downstream genes is dependent on the translation of the preceding genes as has been identified in T4. As per these criteria, thirty five of the Syn5 ORFs could be translationally coupled. The arrangement of genes in Syn5 is very similar to that found in other cyanophages such as P60 and P-SSP7. PSI and PSII reaction center genes are absent in the genome of Syn5 such as those reported in other cyanophages (Mann *et al.*, 2003; Sullivan *et al.*, 2005; Weigele *et al.*, 2007; Sharon *et al.*, 2009). Raytcheva *et al.* (2010) reported an eclipse period of ~45 min and a latent period of ~60 min with a burst size of 20–30 particles per cell, in contrast to the findings of Pope *et al.* (2007). The intracellular assembly of Syn5, as observed by them through cryo-electron images, showed the presence of protein capsids filled with scaffolding protein, though this protein is not associated with mature free virion particles. According to them the presence of the scaffolding protein in virion heads even before packaging of DNA into the heads is akin to the intracellular development of enteric dsDNA phages. Since cyanobacteria are the earliest prokaryotes this type of intracellular development of cyanophages seems to have originated with these organisms.

e) *Cyanophage Pf-WMP4:* This phage infects *Phormidium foveolarum* and resembles T7. The genome of this phage is 40,938 bp long and consists of 45 ORFs, nine of which are similar to T7. *In vitro* transcription led to the identification of seven promoters to the left (Liu *et al.*, 2007).

f) *Cyanophage Ma-LMM01*: The phage genome is 162,109 bp long with 184 ORFs. Database search revealed that there are 44 (24%) ORFs whose functional identity could be assigned and the rest of the 140 ORFs have no detectable homologues. The most significant feature is the presence of *nblA* gene that mediates degradation of phycobilisome complex. A site-specific recombinase or integrase (*int-* gene used by temperate phages) and two prophage anti-repressor genes (that mediate integration into host genome) are located within 7.5 kb with a typically low G+C content. Similar to the other cyanomyoviruses or phage T4, Ma-LMM01 genome has 4.3% and 5.4% of transmembrane proteins and proteins with signal peptides, respectively. Only four ORFs-ORF1, ORF41, ORF108 and ORF109 had their best sequence homologies in other viruses. The virion head consists of two major head proteins resembling phage lambda. ORF19 encodes tail sheath protein (Yoshida *et al.*, 2008).

g) *Cyanophage syn9*: This cyanophage has a contractile tail but this has been treated as an unclassified virus (Yoshida *et al.*, 2008). The genome of this phage is 177,300 bp long with 226 putative ORFs and six tRNA genes. The phage has 19% of the genes that bear resemblance to phage T4. The assembly of capsid proteins resembles those of and bears structural and topological relationships with phage SPO1 and herpes virus. Evidences for different patterns of DNA replication have been presented (Weigele *et al.*, 2007).

h) *P-SS2*: This is the first marine cyanosiphovirus whose genome has been sequenced. This phage has been isolated against *P. marinus* MIT9313 as the sensitive host from 83 m deep

Atlantic Ocean slope waters (Sullivan *et al.*, 2009). The particle size of this phage is the largest (with a head of 75 nm broad and 140 nm long and a 325 nm long tail of 12 nm diameter) among 17 other similar phages that infect species of enteric bacteria, mycobacteria, *Streptomyces*, *Methanobacterium* and *Lactococcus*. The genome of P-SS2 is ~108 kb long, surpassed by the T5 genome (122 kb), and encodes 131 predicted proteins. Of these, six structural proteins have been identified on the basis of sequence comparisons while 35 proteins during infection cycle. Some of the distinctive features of the genome of P-SS2 are that (i) it lacks the photosynthesis genes that are widely represented in the marine cyanomyoviruses; (ii) the phage P-SS2 may be a temperate phage due to the presence of *int*, *bet*, *exo* and a 53 bp attachment site; and (iii) the host attachment site is located near an insertion sequence activity which could mobilize a gene involved in nitrogen stress response. The proteome analysis revealed that the 35 proteins identified during infection cycle corresponded to structural proteins. The structural genes were clustered into "modules", one of the largest such module consists of 28 structural genes including six structural genes for tail fibers (ORFs 067, 072, 073, 074, 077, 091). Of these six tail fiber genes, ORF 073 encodes a protein of 1627-amino acid residues. This has been found to be similar to the tail fiber protein of P-SSM2, a myovirus. The low mol% G+C of this gene signifies that it might have been acquired through LGT (Sullivan *et al.*, 2009).

i) *PaV-LD*: A tailless *Planktothrix agardhii* virus from Lake Donghu, China has been reported. The genome of this phage is unique for carrying NblA gene (ORF 022L) that encodes non-bleaching protein involved in phycobilisome degradation. Electron microscopy and spectroscopic analysis of the PaV-LD infected cells revealed a significant decline of phycobilin content. The expression of the NblA gene during infection cycle has been confirmed by Western blot analysis. Another significant feature is the presence of a linear chromosome (95,299 bp) with 142 potential genes. Protein analysis of the phage revealed 13 major structural proteins (27 kDa to 172 kDa). Genes that encode tail proteins are absent in the genome (Gan *et al.*, 2011).

Besides the individual genomes of the cyanophages presented above, comprehensive accounts on genome comparison of *Prochlorococcus* and *Synechococcus* myoviruses, recombination studies on Rhode Island myovirus (RIM) strains and annotation of the myoviral genomes as per the standards laid down by the Genome Standards Consortium (GSC) are available. Sullivan *et al.* (2010) reported genome analysis of marine cyanophage myoviruses (16 of them) and compared these with the genomes of T4-like myoviruses (10) from different hosts and environments. The 16 marine cyanophage myoviruses included six of those (P-SSM2, P-SSM4, P-HM1, P-HM2, P-RSM4 and P-SSM7) isolated on *Prochlorococcus* and ten of those (S-PM2, Syn9, Syn19, Syn33, Syn1, S-ShM2, S-SM2, S-SSM7, S-SSM5 and S-SM1) isolated against *Synechococcus* strains. Except for the sizes of the genomes of S-SSM7 (232 kb) and P-SSM2 (252 kb), the rest of the genomes ranged in between 174 kb to 196 kb. The mol% G+C for *Prochlorococcus* myoviruses was lower (37± 1%) than the mol% G+C of *Synechococcus* myoviruses (40.1 + 1.0%). All the 26 genomes consisted of a T4-like core of 38 genes that encode capsid proteins and DNA replication genes. In addition, there are 25 genes that form cyanophage-core genes which are shared by all the 16 cyanophage myoviruses. The total number of core genes for cyanophage myoviruses thus comes to 63 genes. The T4-like genes shared among the non-cyanophage myoviruses are 32 genes. Of the 25 cyanophage-core genes, the cyanophage myoviral genomes consisted of six host genes (*psbA*, *mazG*, *phoH* gene that encodes a phosphate-starvation inducible protein; *hsp20*, *hli03* and *cobS*, cobalamine synthetase gene), two virion structural genes and 17 hypothetical genes. The *Synechococcus* myoviral genomes consisted of different gene subsets such as *petE*, *zwf*, *gnd*, *prnA* and *cpeT* that are not found in the genomes of *Prochlorococcus* myoviruses. Marston and Amrich (2009) studied recombination and microdiversity among five RIM populations of RIM2, RIM17, RIM24, RIM49/Syn9 and RIM50. These five RIM strains co-occur in the Rhode Island's coastal waters and have been frequently collected repeatedly over years. They chose multilocus sequence analysis by selecting four core phage-specific genes (*g20*, *g23*, *g43*, DNA polymerase gene; and cyanophage Syn9-g101 gene specific for tail fiber protein) and four phage-encoded host-derived genes (*psbA*, *psbD*, *cobS*, and *phoH*). The development of suitable primers for the PCR amplification of the selected genes and the sequence determinations for 60 concatenated sequences led to the identification of single nucleotide polymorphisms (SNPs) in these eight loci. Intragenic recombination has been observed in two phage-encoded host genes (*psbA* and *psbD*) and to a limited extent in the major capsid protein gene (*g23*). In most cases such SNPs are neutral but in some others they resulted in amino acid substitutions. So they concluded that there is every possibility for changes in the population dynamics of host-virus interactions. Metagenomic analysis of the five RIM strains revealed an average nucleotide identity of 99.3–99.8%. The GSC has developed a checklist for annotation of Genome/Metagenome sequences. Following these guidelines, Duhaime *et al.* (2011) have enriched the public databases by annotating 30 genomes of marine bacteriophages publicly available. Among these are included P-SS2, *Prochlorococcus* phages (PM-2, P-SSP7, P-SSM2 and P-SSM4) and *Synechococcus* phages (P60, S-PM2, S-RSM4, Syn5 and Syn9).

iii) *Enumeration techniques for marine cyanophages*: *(a) Transmission electron microscopy (TEM), epifluorescence microscopy and staining techniques*: The advantages of positive and negative staining of aquatic viruses for the basic studies with electron microscopy have been highlighted by Ackermann and Haldal (2010). The general stains employed for staining of aquatic viruses are phosphotungstate and uranylacetate. The former confers only negative staining whereas the latter gives both positive and negative staining. The probable artefacts encountered in staining, electron microscopy, counting and size determinations of viruses have been discussed at length and refinements in these techniques suggested. TEM has been the best choice for viewing viruses directly because of the best resolution of up to few nanometers and at the same time it permits the determination of shape of the virus. Sieburth (1979) conducted first studies on marine viruses by employing TEM. Torella and Morita (1979) reported a titer of 10^4 viruses from coastal Oregon waters by using TEM. These studies were followed by the investigations of a number of workers (Bergh *et al.*, 1989; Børsheim *et al.*, 1990; Proctor and Fuhrman, 1990; Wommack *et al.*, 1992; Bratbak and Heldal, 1993 Wilson *et al.*, 1998). Enumeration of marine viruses through TEM requires the transfer of viruses either directly or after concentration by ultrafiltration onto the electron microscopy grids. As it is a time consuming process (and the equipment being bulky and expensive), it is not suited for field studies (Paul *et al.*, 1991; Wommack *et al.*, 1992; Bratbak and Heldal, 1993). The second approach is the use of epifluorescence microscopy combined with fluorescent staining of viruses. In this technique, marine virus particles are collected on 0.02 µm pore-size filters and then stained with a DNA-specific fluorescent dye such as DAPI (4′, 6′-diamino-2-phenylindole) and observed under the epifluorescnce microscopy (Sieburth *et al.*, 1988). DAPI was the first fluorochrome used for labelling and counting VLPs by epifluorescence microscopy (Proctor and Fuhrman, 1990; Hara *et al.*, 1991; Fuhrman and Suttle, 1993; Suttle, 1993). Other fluorescent dyes used were YO-PRO-1 {4-[3-methyl-2,3-dihydro-(benzo-1,3-oxazole)-2 methylmethyledene]-1-3, trimethylammoniumpropyl)-quinilinium diiodide}, YOYO-1, POPO-1 or SBYR Green-1 that are reported to possess very high binding coefficients for nucleic acids and such stained viruses have been viewed under epifluorescence microscope (Hirons *et al.*, 1994; Hennes and Suttle, 1995). SBYR Green-stained samples of sea water when viewed through epifluorescence microscope revealed images of bacterial cells as bigger rounded structures and viral particles as pin-head structures (Fig. 8). In most of the isolation protocols described, the use

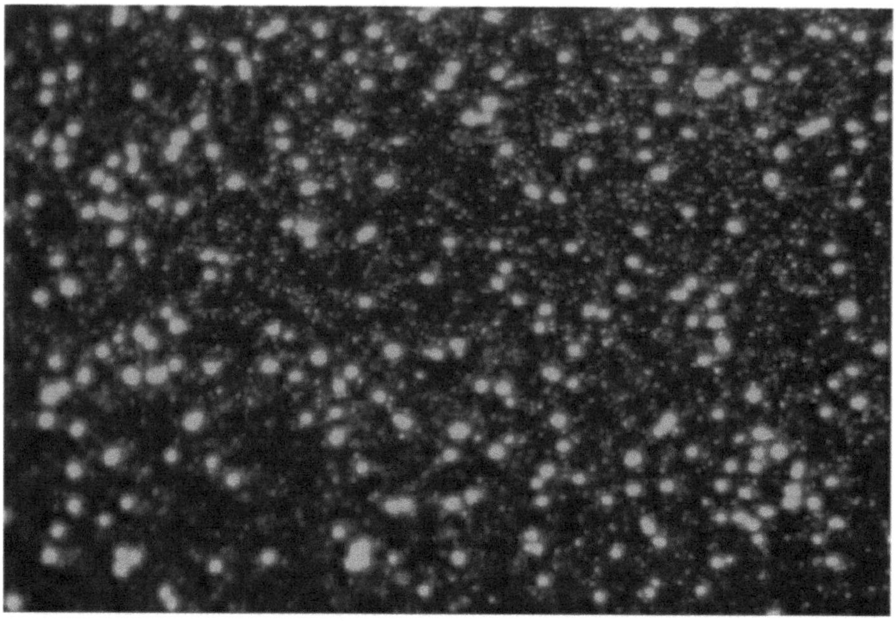

Figure 8: An image from an epifluorescence microscope of seawater stained with the dye SYBR Green to reveal the bacterial cells and smaller viral particles. With the kind permission of N. H. Mann Department of Biological Sciences, University of Warwick, Coventry, UK. [Mann, N. H. (2005) *PLoS Biol* **3(5):** e182. doi:10.1371/journal.pbio.0030182]. doi: 10.1371/journal.pbio.0030182.g001. Image courtesy Jed Fuhrman, Department of Biological Sciences, University of Southern California, MC0371, Los Angeles, CA 90089, USA.

Color image of this figure appears in the color plate section at the end of the book.

of 25 mm Anodisc™ membranes with a built-in support disc has been practised. But Budinoff *et al.* (2011) advocated the use of 13 mm Anodisc™ or Nucleopore™ membranes (with pore sizes 0.015–0.030 µm) without a support disc. Hennes *et al.* (1995) used fluorescently stained viruses as probes to detect specific bacteria and cyanobacteria. Bettarel *et al.* (2000) compared five different methods for the enumeration of marine viruses. These are TEM, and epifluorescence microscopy with different fluorochromes such as DAPI, YO-PRO-1, SYBR Green I. Of these the use of YO-PRO-1 in combination with epifluorescence microscopy proved to be the best for the enumeration of T7 virus particles from cultures as well as viruses from different freshwater ecosystems that differed in their trophic levels. Marie *et al.* (1999) enumerated a marine virus PpV-01 in culture infecting *Phaeocystis pouchetii* as well as natural samples through flow cytometry, a technique that has been used for counting the number of bacterial or cyanobacterial (*Prochlorococcus* or *Synechococcus*) cells from marine environments (Fig. 9). After fixing the samples with 0.5 % glutaraldehyde, the samples were deep-frozen and in presence of SYBR Green I, the enumeration of viruses has been done by flow cytometry (Fig. 10). However, the titres of phages were found to be slightly on higher scale when compared to epifluorescence microscopy or by TEM. They also found that the viruses displayed a depth-dependent profile from the Mediterranean Sea as that of the cells of the host. A disadvantage with SBYR Green I is that its fluorescence fades within 30 seconds under some conditions. As a result of which one is bound to use high concentrations of SBYR Green I and also employ an anti-fading mixture (Noble and Fuhrman, 1998). That is why some workers preferred to use SBYR Gold as the fluorochrome instead of SBYR Green I. In this connection, the work of Chen *et al.* (2001) merits mention because they employed digital image analysis and flow cytometry to enumerate marine

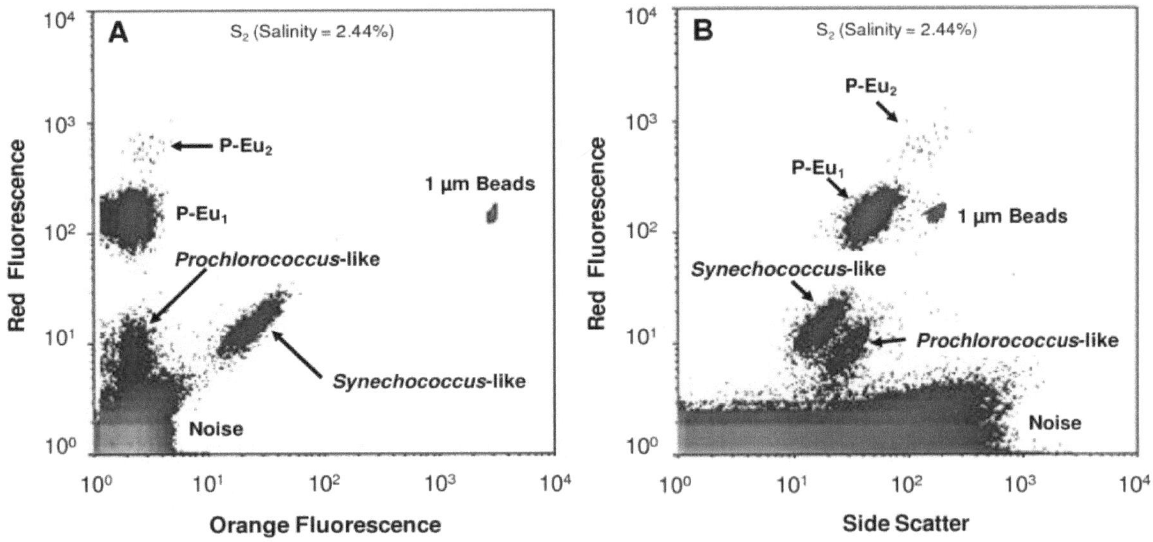

Figure 9: Cytometric differentiation of picophytoplankton populations. These different sub-populations were identified according to their differences in side scatter and both red and orange fluorescence; *Synechococcus* emit a strong orange fluorescence whereas *Prochlorococcus* emit only red fluorescence when excited by blue light (Goericke and Repeta, 1992). In addition, *Synechococcus* cells are significantly larger than *Prochlorococcus* cells (ca. 1–0.6 μm in diameter respectively) (Chisholm, 1992). Pico-eukaryotes were identified by their larger size (SSC) and higher red fluorescence. Scatter plot of orange versus red fluorescence (A) and side scatter versus red fluorescence (B) observed on Station S2 (i.e. Salinity = 2.44%), showing: (i) 2 populations of pico-cyanobacteria, one exhibiting fluorescence and side-scatter characteristics of *Prochlorococcus* sp. (referred as *Prochlorococcus*-like) and the second exhibiting fluorescence and side-scatter characteristics of *Synechococcus* (referred as *Synechococcus*-like) and (ii) 2 populations of pico-eukaryotes (P-Eu$_1$ and P-Eu$_2$).With the kind permission of M. Schapira, School of Biological Sciences, Flinders University, GPO Box 2100, Adelaide SA 5001, Australia & Southern Ocean Group & Ifremer-LERN, Avenue du Général de Gaulle, 14520 Port en Bessin, France. [Schapira *et al. Saline Systems* (2010) **6:** 2; doi:10.1186/1746-1448-6-2].

viruses that were stained with SBYR Gold. A comparison of the estimates made with SYBR Gold-stained particles using digitized images gave 1.3 times higher titre in comparison to those based on direct counting with epifluorescence microscopy. Direct epifluorescence counting of particles stained with fluorochrome SBYR Gold yielded 1.34 times higher values than those estimated by TEM. Brussaard (2004) optimized procedures for the enumeration of viruses through flow cytometry where it was recommended fixation of samples with glutaraldehyde (0.5% final concentration) for 15–30 min followed by storing of samples in liquid nitrogen at –80°C. Samples when required were thawed, diluted in Tris-EDTA buffer (pH 8.0) and stained with SYBR Green I for 10 min in the dark at 80°C and cooled before analysis. The counts of various viruses made as per this technique yielded better results and the use of low storage temperatures prevented loss of up to 50 to 80% total virus abundance (Fig. 10). The disadvantages of prefixation with glutaraldehyde or formaldehyde of the marine viral samples for examination through epifluorescence microscopy have been highlighted by Wen *et al.* (2004). Prolongation of post-fixation period resulted in a decrease of viral abundance by about 72% in 16 days. They recommended the use of unfixed samples stained with YO-PRO-1, SBYR Green I or SBYR Gold for enumeration through epifluorescence microscopy. Prolonged incubation periods at –20°C for at least 2 weeks or in case of YO-PRO-1 even for one year did not affect the cyanophage numbers. Other techniques include most probable number assays (Waterbury and Valois, 1993), radiotracer analysis (Fuhrman and Noble, 1995; Steward *et al.*, 1992a,b, 1996), inhibition

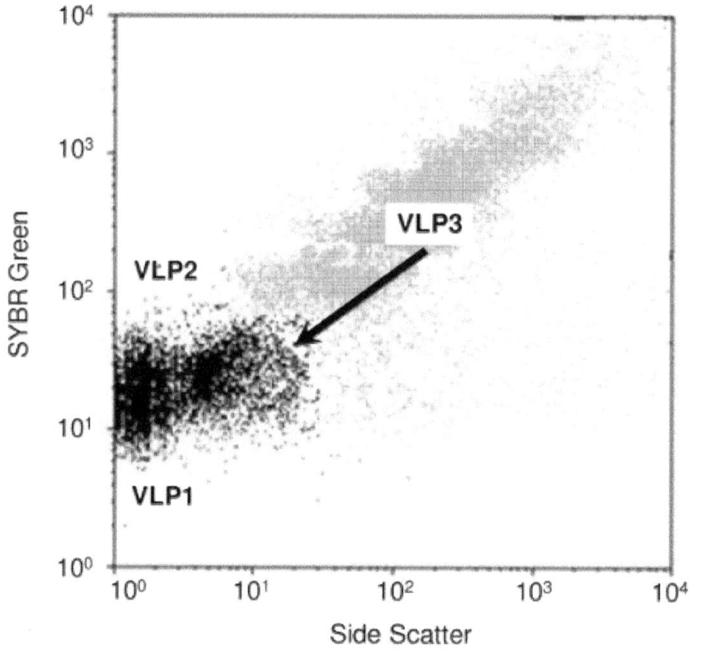

Figure 10: Cytometric differentiation of virus-like particles. Scatter plot of side scatter (SSC) versus green fluorescence (SYBR Green) showing 3 viral sub-populations: VLP1, VLP2 and VLP3. Sub-populations of virus-like particles (VLP) were discriminated based on their differences in SYBR Green fluorescence and SSC, as non-overlapping classes of size and green fluorescence according to Brussaard (2004). VLP1 and VLP2 correspond to populations observed previously in seawater samples and described as bacteriophages (Marie *et al.*, 1999). The sub-population VLP3 exhibited the same SYBR Green fluorescence level than VLP2 but was characterized by higher SSC. Recent work have indicated that viral population presenting a cytometric signature similar to the one observed here for VLP3, were likely to be phytoplankton viruses (Buussard *et al.*, 2005, 2008). With the kind permission of M. Schapira, School of Biological Sciences, Flinders University, GPO Box 2100, Adelaide SA 5001, Australia & Ifremer-LERN, Avenue du Général de Gaulle, 14520 Port en Bessin, France. [Schapira *et al.* (2010) *Saline Systems* **6**: 2; doi:10.1186/1746-1448-6-2].

experiments (Bratbak *et al.*, 1992; Heldal and Bratbak, 1991) and measurements of dissolved esterase activity (Brussaard *et al.*, 1995; van Boekel *et al.* 1992). These studies provided concentrations of marine viruses ranging from 10^6 ml^{-1} to 10^8 ml^{-1} thus confirming the earlier observations where the marine viral concentrations ranged in between 10^7 ml^{-1} to 10^8 ml^{-1} for a variety of marine waters (Bergh *et al.*, 1989; Proctor and Fuhrman, 1990). A number of workers reported similar results from several locations and depths.

b) *Molecular techniques*: *(i) T4 gene 20*: T4 gene20 (*g20*) encodes a protein, gp20 also known as portal protein that is essential in the morphogenesis of phage head. It helps in the initiation of head assembly by the formation of proximal vertex, for DNA packaging and for binding with the head-tail junction. Sequence analysis of cyanophage genomes of S-PM2, S-WHM1 and S-BnM1 revealed a high degree of similarity of a particular ORF. This ORF from these phages bears 36% identity to the T4 *g20* sequence. The ORF encodes a polypeptide of 550 amino acids long (Fuller *et al.*, 1998). They employed two cyanophage-specific (CPS) primers CPS-1 and CPS-2 to amplify the *g20* sequence by PCR from a number of cyanophages belonging to the family Myoviridae previously isolated from different geographical locations. They suggested that by employing the CPS-1 and CPS-2 primers it would be possible to identify the marine cyanophage assemblages quantitatively (by cPCR) and qualitatively following phylogenetic analysis of amplified products. The amplification of cyanophages belonging to Podoviridae (S-BP1, S-BP2, S-BP3 and phi12) by primers CPS1 and CPS2 was not possible. These primers were subsequently employed in PCR-DGGE to study cyanophage diversity along a transect of the Atlantic Ocean (Wilson *et al.*, 1999, 2000). Using a myoviral-specific PCR-sequencing approach, with *g20* as the marker gene, Jameson *et al.* (2011) reported a high diversity of myoviruses along a transect of North-South Atlantic Ocean. According to them the myoviruses are distributed over large areas of Atlantic Ocean irrespective of the abundance and diversity of the co-occurring host genera, *Prochlorococcus* and *Synechococcus*. DGGE was combined with PCR to "fingerprint" and compare algal-viral communities both spatially and temporally (Short and Suttle, 2000, 2002, 2003). Zhong *et al.* (2002) studied the phylogenetic diversity of marine cyanophage isolates and those of natural assemblages by using CPS1 and CPS8 as primers for *g20* gene sequence amplification. A portion of *g20* corresponding to ~592 bp was subjected to cloning and sequence analysis. These studies revealed the presence of nine phylogenetic groups belonging to a total of 114 different *g20* homologue sequences from estuarine and oligotrophic offshore environments. A great divergence in cyanophage community structure was found in estuaries and open-ocean samples. Of the nine groups only three of them consisted of cyanophages and the rest of the unidentified groups are suspected to represent either bacteriophages that are infective on cyanobacteria or cyanophages infective on other unidentified members of cyanobacteria.

Instead of directly assessing cyanophage diversity by the amplification of *g20* gene sequence and subjecting to DGGE analysis, Marston and Sallee (2003) first identified 36 different cyanomyoviruses infecting the different strains of *Synechococcus* over a three-year sampling survey conducted at Rhode Island's coastal waters. Then the *g20* gene sequences from natural waters have been amplified by using a different set of PCR primers such as CPS1/CPS8, CPS1/CPS2 and CPS1/CPS4 for different cyanophages. Most of the phage diversity has been represented by cyanomyoviruses, however, at least 50% of the samples containing cyanophage sequences could not be amplified by the primers employed. This reflects that other morphotypes belonging to podovirus and siphovirus categories may also be important components of the cyanophage community in the area studied by them. The cyanophage diversity and community structure is also shown to be directly dependent on the physical structure of water column as indicated by temperature and salinity. This was substantiated by PCR amplification of *g20* gene fragments followed by DGGE which suggests that certain cyanophages are wide spread whereas certain others are specially distributed at greater depths in three inlets of British Columbia (Frederickson *et al.*, 2003). Mann (2003) highlighted the factors affecting the diversity of cyanophages, their interactions with their hosts and selective pressures in the marine environment.

Studies on the genetic diversity and population dynamics of cyanophage communities in the Chesapeake Bay revealed unique *g20* clonal sequences different from open-ocean waters (Wang and Chen, 2004). Terminal-restriction fragment length polymorphism (T-RFLP) based on the *g20* was developed by which 15 *g20* genotypes have been identified. Of these, only one was closely related to the sequences from open ocean waters. The T-RFLP profiles of the *g20* gene demonstrated that the myovirus populations exhibited great change seasonally (where phage titres increased from 2.2 PFU ml^{-1} in April to a high of 559.6 PFU ml^{-1} in September) rather than spatially. The seasonal variations appeared to be dependent on the variations of the host cell densities (*Synechococcus* sp. CB 0101 ranging from $2.6 \pm 0.5 \times 10^2$ cells ml^{-1} in February to $8.1 \pm 1.0 \times 10^4$ cells ml^{-1} in July). Dorigo *et al.*, (2004) analysed cyanophage diversity of Lake Bourget (France) by the amplification of *g20* gene with the help of PCR primers CSP-1 and CSP-8 followed by DGGE. Out of 47 sequences analysed,

35 distinct myovirus g20 genotypes were identified. Upon phylogenetic analysis, these could be divided into seven distinct taxonomic units. The interesting feature is that some of these freshwater cyanophage sequences have been found to be similar to marine cyanophage sequences than to other freashwater cyanophage sequences.

Short and Suttle (2005) reported the occurrence of *g20* sequences from samples collected from wide range of freshwaters (Lake Constance, Germany; Chilliwack Lake and Cultus Lake from Canada; a Catfish production pond located at Limco, Ic. in Crowley, La) as well as marine waters (Gulf of Mexico, the Arctic, an Arctic cyanobacterial mat, Southern, Northeast and Southeast Pacific Oceans and Chuckchi Sea). After DNA extraction and PCR amplification, the separation was done by DGGE and then the *g20* gene was amplified with CPS-4 (nondegenerate upstream) and G20-2 (degenerate downstream) primers. The *g20* gene sequences showed nearly 99% identity from environments that differed greatly in temperature, salinity and location. They concluded that it is very unlikely that the viruses and their hosts occur in both freshwaters and marine waters. Moreover, the CPS primers might not be specific for cyanophages due to the detection of *g20* sequences from a depth of 3,246 m in the Chuckchi Sea. A phylogenetic analysis revealed the presence of four new *g20* clusters of which two have been entirely found in freshwaters. The genetic diversity of cyanophages in the Gulf of Aquaba has been studied by Mühling *et al.* (2005) who simultaneously determined the genetic diversity of *Synechococcus* sp. and correlated the prevalence of a particular host strain with a particular phage genotype. The genetic diversity of *Synechococcus* was assayed with RNA polymerase core subunit gene *rpoc1* and the phage diversity infecting the particular genotype of *Synechococcus* was identified by PCR amplification of *g20* gene sequence and subsequent analysis by DGGE. The main findings are that (i) *Synechococcus* abundance was represented at a maximum cell density of 3.4 x 10^4 cells ml^{-1}, though the preponderance of *Prochlorococcus* was greater by 12-folds, (ii) the sudden disappearance of *Synechococcus* was attributed to depleted levels of phosphate concentration, (iii) plaque assay studies, conducted on four host strains of *Synechococcus* (WH7803, WH8103, RS9906 and RS9911), revealed clear-plaque formation more readily on WH7803 rather than on other hosts, (iv) PCR amplification of *rpoc1* gene from the samples taken from a depth of 10 m resulted in eight clone libraries. Thirty six clones from each of the clone libraries were analysed by RFLP to reveal genetic diversity of *Synechococcus*. Season-specific prevalence of the genotypes of *Synechococcus* was noted where a number of genotypes disappeared with the onset of summer. During autumn RFLP types S19 and S12 were predominantly represented and only RFLP type S17 could be detected during spring. Genotype S19 accounted for 31 of the 36 clones analyzed during July and 18 of the 36 clones analysed during October. Likewise, the cyanophage population was dominated by only two genotypes (P26 and P27) during the summer months. As it was not possible to cultivate the dominant *Synechococcus* RFLP types (S19 and S12), the representative phage genotype infecting the host strains could not be ascertained. They concluded that viral infection plays a major role in controlling the populations of *Synechococcus* in the Gulf of Aquaba where grazing is insignificant. The interaction with the co-occurring phages results in a qualitative and quantitative change in the genetic diversity of *Synechococcus* strains. They did not agree with the suggestion of Waterbury and Valois (1993) that cyanophage-resistant mutants are dominantly represented in the oligotropic regions of oceans. Their observations also derive support from the findings of Suttle and Chan (1994) who demonstrated lack of such resistance in the host cells in offshore environments. Of the 35 marine cyanophages isolated and characterized from the Gulf of Mexico, the *g20* gene sequence could be amplified from the 20 of the phages suggesting that only 63% of them are myoviruses (McDaniel *et al.*, 2006). PFGE of viral community structure of Norwegian coastal waters revealed the presence of 29 different viral populations with a genome size variation ranging from 26 to 500 kb. The cyanophages infecting *Synechococcus* (represented to 7.3 x 10^4 ml^{-1}) have been detected at a concentration of 7.2 x 10^3 particles ml^{-1}. The genetic diversity of cyanophages has been determined by the amplification of *g20* gene sequence (by CPS1 and CPS2 primers) and the homology of all viruses belonged to cyanomyoviruses. Seasonal shifts in the populations of cyanophages were noted with one community dominantly represented in spring and summer and a different one in fall (Sandaa and Larsen, 2006). Isothermal nucleic acid amplification assay, i.e. signal-mediated amplification of RNA technology (SMART) has been used for the first time by Hall *et al.* (2002) to detect DNA extracted from marine cyanophages. This technique is based on the target-dependent production of multiple copies of an RNA signal that is measured by an enzyme linked oligosorbent assay. Cyanophage S-PM2 and S-BnM1 specific signals for *g20* gene sequences could be identified very easily. Extensive purification of phage DNA is not required as signals could be obtained from crude phage lysates as well. So this technique does not appear to be best suited for identifying phage DNA sequences but helps in detecting RNA transcripts. By using this technique it has been possible to differentiate the S-PM2 phage-infected and non-infected cells of *Synechococcus* sp. strain WH7803 and the expression of *g20* mRNA peaked at 10 h after infection coinciding with the onset of lysis (Fig. 11; Wharam *et al.*, 2007).

From the above studies, it has been possible to (i) determine diversity of marine cyanophages as measured by variability in the sequence of *g20* across gradients in time and space from different environments, (ii) the *g20* diversity was as great within the sample as between Oceans, (iii) the phage diversity increased as *Synechococcus* abundance increased, and (iv) temporal studies correlated cyanophage diversity with *Synechococcus* diversity. The marine cyanophage diversity described by various workers on the basis of *g20* sequences could be grouped into nine phylogenetic clusters and with the *g20* sequences originated from freshwater environments constituted the tenth cluster. Of the nine marine clusters, three are those from the cultured representatives of marine cyanophages whereas the remaining six marine clusters (Clusters A to F) and the freshwater cluster represent environmental samples. Sullivan *et al.* (2008) are of the opinion that the *g20* marker falls short of explaining the diversity of marine cyanophages. This conclusion was based on a comprehensive phylogenetic study of (i) all the marine cyanophages isolated by other workers to date, (ii) their own cyanophage isolates obtained after enlarging the range of host strains during isolation, (iii) the assemblages identified after metagenomic analysis of marine waters from different oceans and (iv) also the data from Global Ocean Sampling expedition (GOS). Comeau *et al.* (2010) conducted a comprehensive metagenomic analysis of more than 300 scaffolds of T4-type cyanophages from GOS project. They designated the T4-type cyanophages as Cyano-T4s and compared the gene order with those of the cultured representatives of Cyano-T4s. There is a closer resemblance in the synteny between the gene sequences of uncultured Cyano-T4s with the core genome of known and cultured Cyano-T4s. According to them the cultured Cyano-T4s represent the great diversity of such phages from marine surface waters.

In order to find out a diagnostic marker for detecting freshwater cyanophages, Baker *et al.* (2006) have tried to amplify the *g20* sequence by the PCR primers developed earlier (CPS1 and CPS2 primers developed by Fuller *et al.*, 1998; CPS1 and CPS8 primers described by Zhong *et al.*, 2002 and *g20* fD2 and CPS2 primers developed by Hambly, 2002). But as the *g20* gene sequences could not be amplified from freshwater cyanophages, they developed special primers for the amplification of phage AN-15 gene that encodes major capsid protein (MCP). First they identified the MCP by subjecting the purified viral particles to SDS-PAGE analysis. A dominant protein at approximately 37 kDa was supposed to represent the MCP. N-terminal amino acid analysis of this protein revealed the presence of amino acids Ala-Leu-Thr-Thr-Leu-Thr-Pro. A degenerate primer PAN 15 F1 (5′-ACN ACN YTN CAN CC-3′) was designed based on the amino acid sequence that helped in the

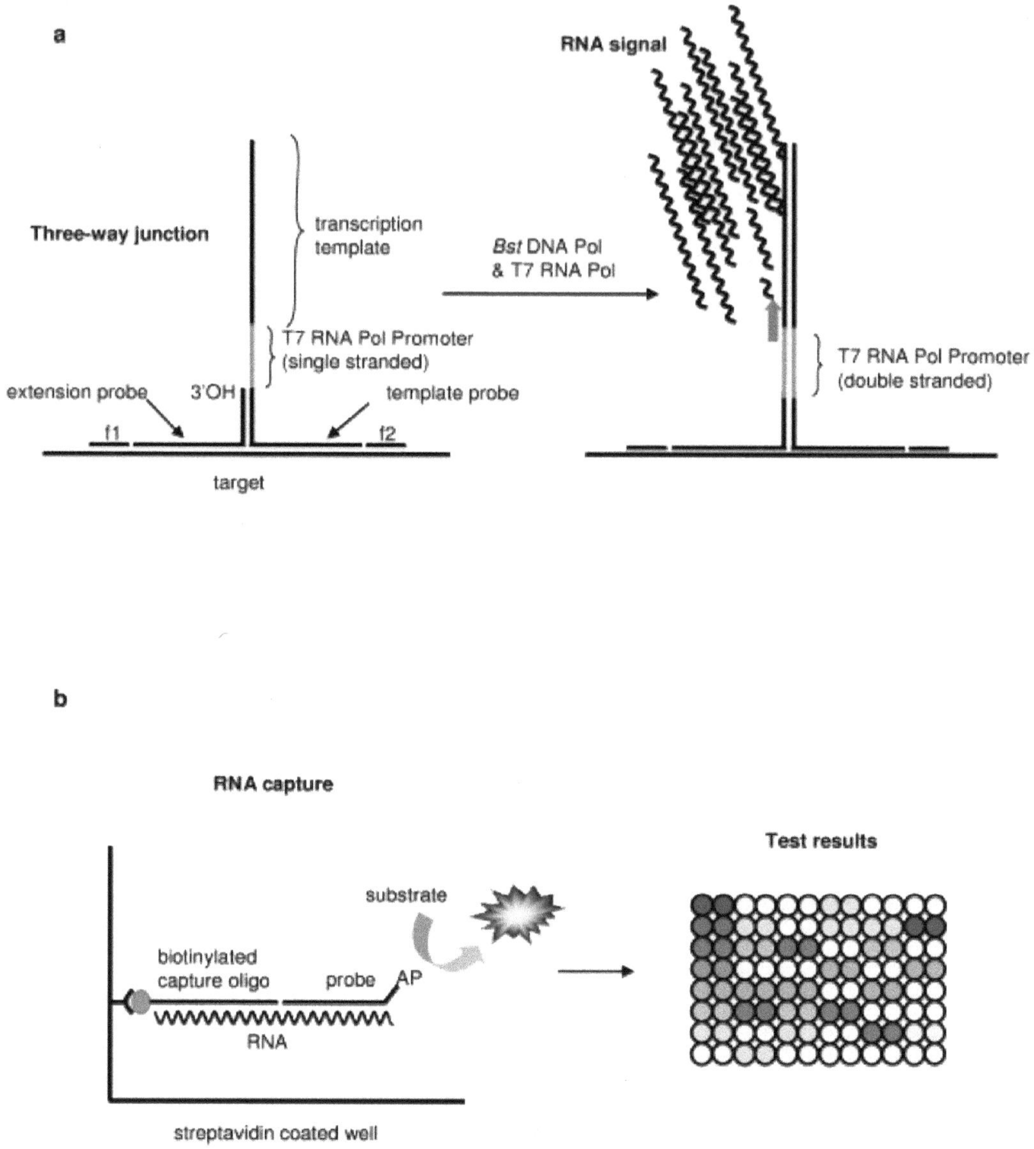

Figure 11: The SMART assay. (a) Specific probes hybridise with the target to form a three-way junction (3WJ), assisted by facilitator probes (f1 & f2). The 3WJ initially contains a single-stranded, inactive T7 RNA polymerase promoter sequence. The promoter is made double-stranded (active) by extension (by *Bst* DNA polymerase) off the 3′ of the extension probe, leading to the generation of large amounts of RNA signal (by T7 RNA polymerase), which may itself be amplified if required. (b) Detection of RNA signal by ELOSA (Enzyme Linked OligoSorbant Assay). The assay uses 2 specific probes: a biotinylated capture probe and enzyme (Alkaline phosphatase, AP) linked detection probe. Non-specific nucleic acid and 3WJ probes are removed, following binding in a streptavidin coated well, and RNA signal is detected via a colour change. Quantification of signal takes place in a 96 well plate, allowing multiple samples to be analysed simultaneously. With the kind permission of S.D. Wharam, Bigelow Laboratory for Ocean Sciences, 180 McKown Point, West Boothbay Harbor, Maine 04575, USA. [Wharam *et al.* (2007) *Virology Journal* **4:** 52. doi:10.1186/1743-422X-4-52].

amplification of a 150 bp fragment. Sequence analysis suggested that the 150 bp fragment matched with N-terminal region of the MCP. As the sequence match confirmed that it is MCP gene sequence, a pair of new primers MCPF5 and MCPR5, were designed. A search of the database revealed that there are no known sequences either at nucleotide level or protein level. MCP gene sequences from viruses AN-15, N-1 and A-1(L) have been amplified by PCR using MCPF5/R5 and PCR products approximately 350 bp were obtained. Environmental samples of cyanophages from freshwater lakes of United Kingdom (Priest Pot, Esthwaite Water and Burrate Reservoir) subjected to PCR amplification of MCP gene sequence revealed a total of 26 sequences that were almost identical. Based on the differences at the nucleotide level in these sequences, these were grouped into 11 different genotypes (A to K). Of these, A to F have been most frequently encountered. Genotypes identical to that of AN-15 and A-1(L) have been detected in samples collected from Burrator Reservoir and Esthwaite Water. Genotype A was found only in Priest Pot, an eutrophic pond. This constitutes the first report on the identification of natural viral assemblages that infect filamentous cyanobacteria inhabiting freshwater habitats. The generation of strongest PCR products from natural waters with the sudden disappearance of the filamentous bloom-forming cyanobacterium at Esthwaite water strongly suggests the involvement of cyanophages in controlling bloom populations. Wilhelm *et al.* (2006) explained an interesting case relating to the occurrence of *g20* gene sequences in freshwaters in the absence of a host. The distribution and diversity of cyanophages occurring in Lake Erie has been studied by *g20* sequence analyses after PCR amplification by primers CPS1 and CPS8 to amplify the region from ~826 to 1376 of the *g20* capsid assembly protein. The *g20* sequence homologues present in the natural water samples as well as ballast of commercial ships were related to marine cyanophages. It is significant to note that the cyanophages present in the freshwaters of this Laurentian Great Lake infected only the marine *Synechococcus* sp. strain WH7803 that is not an inhabitant of these waters, out of a number of other strains (WH5701, WH8007, WH8101, WH8102, PCC 7002, PCC 7942, PCC 6803 and BO 8807) of the same genus and *M. aeruginosa* LE3, *Microcystis* sp. UTEX 2386, *Anabaena* sp. UTEX 2558, *Anabaena* sp. UTEX 2576 and *Oscillatoria luteus* tested for host-range. It is suggested that the ships' ballast water may be responsible for the transfer of marine cyanophages and other marine microbes that can survive their movement into freshwater. The purified cyanophage samples viewed through TEM revealed capsids with a mean diameter of 69.6 nm and a mean tail length of ~147.4 nm (Wilhelm *et al.*, 2006). Thus ship's ballast water seems to be an important conduit for the bioinvasion of microorganisms (of the order of 6.8×10^{19} viruses and 3.9×10^{18} bacterial cells per annum) at the Port of Hampton Roads in Lower Chesapeake Bay (Drake *et al.*, 2007). Employing CPS1 and CPS8 primers, the amplification of *g20* sequences by PCR from the flood waters of a Japanese paddy field, Wang *et al.* (2010) found a great diversity in cyanophages and several unique groups that were more closely related to those cyanophages from freshwaters than to those from marine waters.

ii) *g23*: Genes *g18* to *g23* of phage T4 encode different proteins of the virion. Tail sheath (*g18*), tail tube (*g19*), head portal protein (*g20*), prohead core protein (*g21*), a scaffold protein (*g22*) and the major capsid protein (*g23*) are encoded by the respective genes mentioned in the parentheses. Hambly *et al.* (2001) identified conserved sequence in the genome of marine cyanophage S-PM2 that is homologous to the gene sequences (*g18* to *g23*) of T4. In the conserved sequences of cyanophage DNA, there are certain insertions and deletions of small ORFs of unknown function. The gene sequences identified relate to the structural components of the phage head and contractile tail. In view of the structural differences in the morphology of these phages, considerable divergence in amino acid sequence of the respective proteins may exceed uniformly by about 50%. Short fragments (450 bp) of the gene encoding major capsid protein (*g23*) have been amplified to assess the diversity of marine cyanophages from the brackish Baltic Sea infecting *No. spumigena*. The encoded protein has been shown to be identical to the *No. spumigena*-specific cyanophages described above (Jenkins and Hayes, 2006). Jia *et al.* (2006) reported molecular characterization of T4-type bacteriophages from rice straw and a surface soil in a Japanese rice field by the amplification of *g23* gene sequence. They detected great diversity in T4-type cyanophages distinctive from those found in the marine environments. But interestingly, the phylogenetic analysis showed that two novel sub-groups of the T4-type phages were more closely related to the marine cyanophage isolates of exo-T-evens described ealier by Hambly *et al.* (2001).

iii) *phoH*: The *pho* regulon regulates phosphate uptake and metabolism under low phosphate conditions. A bioinformatics study of the fully sequenced marine phage genomes available with GenBank revealed the presence of *pho* regulon genes in nearly 40% of them. Of the genes of the *pho* regulon, *phoH* from marine viruses appeared as a distinct cluster, different from the *phoH* sequences of the bacterial hosts. With the help of the PCR amplification of the *phoH* sequences, Goldsmith *et al.* (2011) determined the diversty of the *phoH* sequences from different depth profiles of Sargasso Sea and six geographically distinct locations. Due to the presence of unique *phoH* clusters at different depths and geographic location, *phoH* gene sequence has been projected as a molecular marker to assess marine phage diversity.

iv) *"Photosynthetic" phages*: Photosynthesis genes *psbA* and *psbD* encode PSII core reaction centre proteins D1 and D2, respectively. The *hli* gene encodes the high-light inducible proteins (HLIPs). HLIPs are suggested to protect photosynthetic apparatus from excess excitation energy during stressful conditions in cyanobacteria (He *et al.*, 2001). Two other related genes *petE* and *petF* encode plastocyanin and ferredoxin, respectively. Mann *et al.* (2003) reported for the first time the existence of *psbD* and an interrupted *psbA* gene sequences in the genome of cyanophage S-PM2 that infects *Synechococcus* sp. strain WH7803. Lindell *et al.* (2004) investigated the different aspects of transfer of photosynthesis genes to and from *Prochlorococcus* viruses (Table 11). Upon a close examination of the photosynthetic gene sequences of selected cyanophages P-SSP7, P-SSM2 and P-SSM4 and those of their hosts, it is concluded that the phage photosynthesis genes are (i) highly conserved sequences, (ii) show high degree of homology to the host gene sequences (85% and 95% in nucleotide and amino acid identities, respectively), (iii) arranged together suggesting that they might be expressed simultaneously during infection (Miller *et al.*, 2003), (iv) the presence of a 7-amino acid sequence at the C-terminal portion of phage D1 protein similar to the cyanobacterial D1 proteins as well as in non-green algal plastids, (v) the presence of a 7-amino acid sequence in the centre of the D2 protein

Table 11: Photosynthetic genes in certain cyanophages (based on the data from Lindell *et al.*, 2004).

Cyanophage	Characteristics	Host strain	Photosynthetic Genes
P-SSP7	T7-like Podovirus Capsid ~50 nm; Non-contractile tail; genome 45 kb	High-light adapted *Prochlorococcus* Strain MED4	D1 and one HLIP
P-SSM2	T4-like Myovirus Capsid~85 nm; long contractile tails	Low-light adapted Strains of *Prochlorococcus* NATL1A, NATL2A and MIT9211	D1, six HLIPs, ferredoxin and plastocyanin
P-SSM4	T4-like Myovirus Capsid ~80 nm Long contractile tails	Two low-light adapted strains (NATL1A, NATL2A) and two high-light adapted strains (MED4 and MIT9215) of *Prochlorococcus*	D1, D2 and four HLIPs

that is also found characteristically in the *Prochlorococcus* strains MED4 and SS120 (but not in other cyanobacterial or eukaryotic D2 proteins) and (vi) the absence of additional amino acid sequences in the D2 proteins of *Synechococcus* sp. strain WH8102 as well as the *Synechococcus* phage S-PM2. When cyanophage S-PM2 infects *Synechococcus* cells it encodes D1 and D2 proteins that enable the host cells to overcome photoinhibition. It means the phage is able to get energy supply for its replication in an otherwise inhospitable host environment (Bailey *et al.*, 2004). Phage-mediated genes synthesized hemoxygenase (*ho1*), 15,16-dihydrobiliverdin:ferredoxin oxidoreductase (*pebA*) in case of P-SSM2 and phycobilin:ferredoxin oxidoreductase (*pcyA*) gene in case of P-SSM4 during their replication in *Prochlorococcus* thus enhancing their fitness in the host that is otherwise inherently incompetent to synthesize phycobiliproteins (Dammeyer *et al.*, 2008).

Phylogenetic analysis of phage *psbA* gene sequences revealed that the *psbA* genes fall into a clade of *Synechococcus* or *Prochlorococcus* hosts. The cyanophages used in this study are S-PM2, S-WHM1 and S-BM4, S-RSM2 (isolated by Wilson *et al.*, 1993) and S-RSM28 and S-RSM88 isolated from the Gulf of Aquaba by Millard *et al.* (2004). By the use of degenerate PCR primers and Southern blotting, cyanomyovirus populations screened from Red Sea showed the presence of a copy of the *psbA* gene among 37 of the 68 isolates. In addition to *psbA* gene, all phages showed the presence of *psbD* gene. The DNA sequences of phages S-PM2 and S-RSM88 are identical to the extent of the presence of 212 nucleotides comprising group I intron sequences in between codons 334 and 335. However, in phages S-RSM2, S-BM4 and S-WHM1 and the *psbA* and the *psbD* genes are adjacent to each other but differ in the nucleotide sequences from each other (Millard *et al.*, 2004). Lindell *et al.* (2005) investigated the functional nature of the "photosynthetic" genes carried by the cyanophages infecting *Synechococcus* and *Prochlorococcus* who assayed the D1 protein levels. When cellular *psbA* gene stopped producing D1 protein, the phage-encoded D1 protein increased gradually to cope up with the deficiency. These observations suggest that the phage replication cycle is dependent on photosynthesis and the phage-encoded proteins sustained photosynthesis of the host ensuring its replication. Zeidner *et al.* (2005) analysed environmental viral samples and prophage populations for the *psbA* and *psbD* gene sequences and observed significant differences from those of the host genes. On the basis of probable codon usage it is suggested that photosynthesis genes carried by the cyanophages are evolving and are under levels of purifying selection that make them indistinguishable from the sequences of their hosts. There is every possibility of these gene sequences being exchanged and reshuffled between *Synechococcus* and *Prochlorococcus* via the phage intermediates (Zeidner *et al.*, 2005). Studies on evolutionary aspects of core PSII genes (*psbA* and *psbD*) in marine cyanophages and their hosts have been conducted by sequence analysis of 33 cultured marine cyanophages (belonging to the three groups of tailed phages) and viral DNA samples collected from field samples. Nearly 88% of the phage genomes contained *psbA* gene sequence while 50% of the phage genomes possessed both *psbA* and *psbD* genes. The existence of *psbA* gene in all cyanomyoviruses and *Prochlorococcus* podoviruses has been demonstrated but this gene sequence could not be amplified from the siphoviruses of *Prochlorococcus* or podoviruses infecting *Synechococcus*. Phylogenetic analysis of *psbA* and *psbD* gene sequences from cultured cyanobacteria and cyanophages segregated into distinct clusters (Figs. 12 and 13). Furthermore, *psbA* and *psbD* genes were involved in a transfer from the hosts to phages in at least four and two discrete steps, respectively. This was followed by horizontal and vertical gene transfer between cyanophages (Sullivan *et al.*, 2006). The existence of *psbA* gene sequences in two of the three cyanopodoviruses (S-CBP1 and S-CBP3) and one cyanomyovirus (S-CBM2)) infecting the estuarine *Synechococcus* strains has been demonstrated subsequently (Wang and Chen, 2007). The sequences from the two podoviruses very much resembled ten environmental *psbA* sequences reported earlier from Chesapeake Bay (Bench *et al.*, 2007), Mediterranean waters and

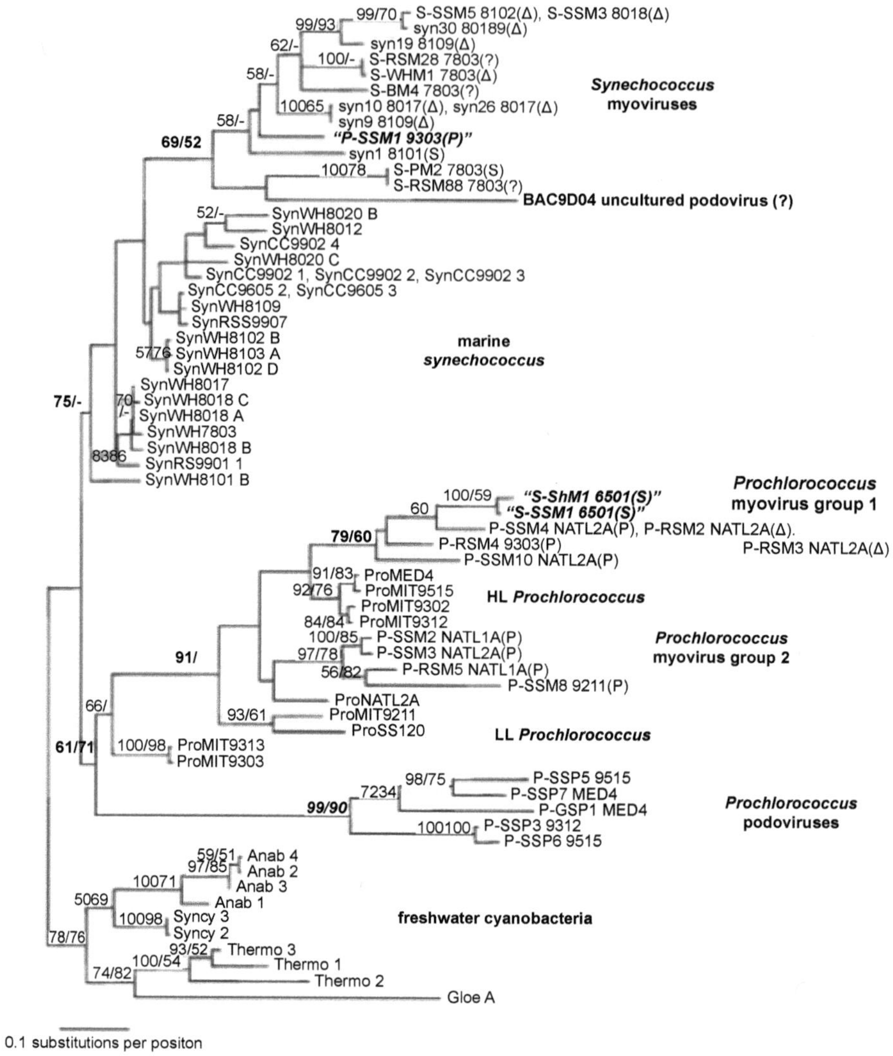

Figure 12: Phylogenetic tree of *psbA* gene sequences from cultured cyanobacteria and cyanophages. Phages are listed by their name, followed by their original host. Phages that are known to infect both *Prochlorococcus* and *Synechococcus* hosts are indicated with a "Δ"; those that infect only one genus are labeled either P (infect only *Prochlorococcus* hosts) or S (infect only *Synechococcus* hosts), while those that are unknown are designated with a "?". Phages shown in italics and bracketed with "**" were isolated on hosts that do not belong to the same cluster and are thus exceptions to the general clustering pattern. Taxa are color coded according to the following biological groupings: myoviruses (red), podoviruses (black), marine *Synechococcus* hosts (light blue), marine *Prochlorococcus* hosts (dark green, LL; light green, HL), freshwater cyanobacteria (dark blue). The tree topology was estimated by LogDet analysis of 1st and 2nd codon positions. Sequences where intragenic recombination was detected using other methods (see Materials and Methods) were not included in these phylogenetic analyses. Branch lengths were estimated by maximum likelihood under a model with nonstationary nucleotide frequencies. Numbers at the nodes represent neighbor-joining bootstrapping and maximum likelihood puzzling support. Anab, *Anabaena;* Gloe, Gleobacter; HL, high-light adapted; LL, low-light adapted; Syncy, *Synechocystis;* Thermo, *Thermosynechococcus*. With the kind permission of S. W. Chisholm, Department of Biology, Massachusetts Institute of Technology, Cambridge, Massachusetts, USA. [Sullivan *et al.* (2006) *PLoS Biol* **4(8):** e234. doi:10.1371/journal.pbio.0040234] doi:10.1371/journal.pbio.0040234.g001.

Color image of this figure appears in the color plate section at the end of the book.

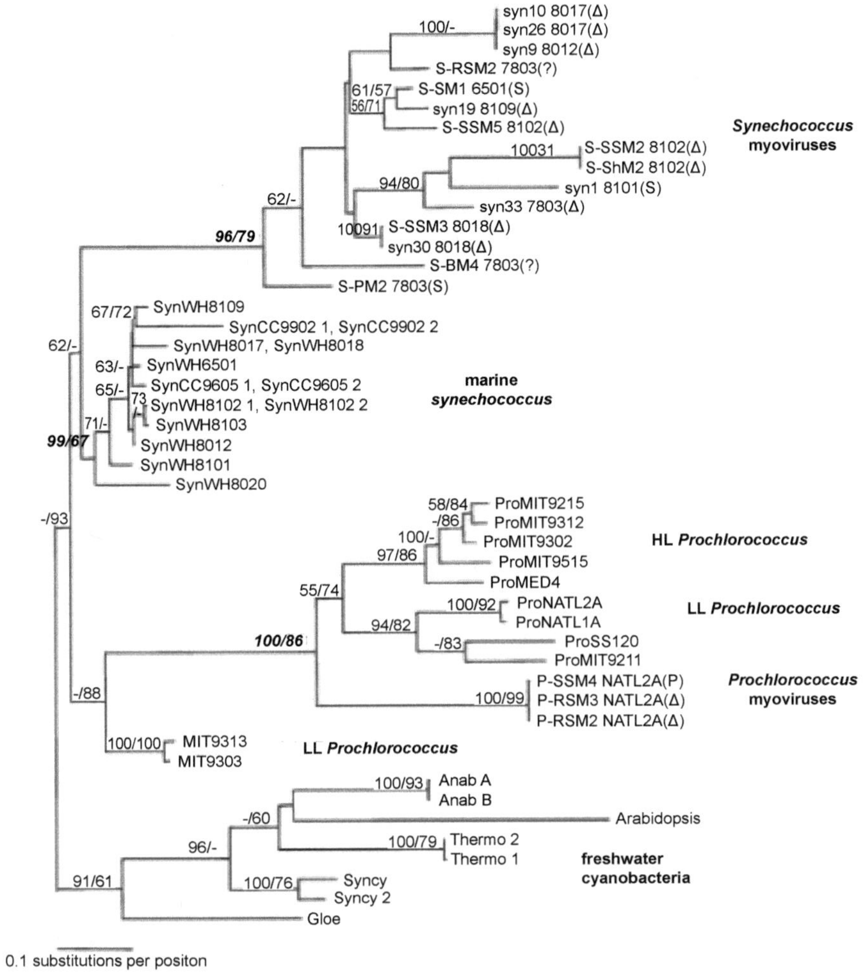

Figure 13: Phylogenetic tree of *psbD* gene sequences from cultured cyanobacteria and cyanophages. Details as in Figure 10. Sequences where intragenic recombination was detected using other methods (e.g. P-SSM1) were not included in these phylogenetic analyses. With the kind permission of S.W. Chisholm, Department of Biology, Massachusetts Institute of Technology, Cambridge, Massachusetts, USA. [Sullivan *et al.* (2006) *PLoS Biol* **4(8):** e234. doi:10.1371/journal.pbio.0040234] doi:10.1371/journal.pbio.0040234.g002

Color image of this figure appears in the color plate section at the end of the book.

Red Sea (Zeidner *et al.*, 2005) and from Sorcerer II Global Ocean Sampling Project (Rusch *et al.*, 2007) but these are distantly related to the *psbA* sequences from marine *Synechococcus*, *Prochlorococcus* and their viruses. Moreover, on the basis of average GC% , 3rd codon GC% and the triplet sequences from D1 protein motifs of partial *psbA* genes from marine picocyanobacteria and their cyanophages along with environmental *psbA* sequences reported so far, Wang and Chen (2007) identified five clusters of *psbA* sequences, i.e. (i) marine *Synechococcus* strains (ii) *Synechococcus* myoviruses (iii) *Synechococcus* podoviruses and environmental viral sequences (iv) *Prochlorococcus* podoviruses and (v) *Prochlorococcus* strains and their myoviruses. The absence of *psbD* gene sequence from the five marine cyanophages described by Wang and Chen (2007) is consistent with the observations of Sullivan *et al.* (2006) who suggested the absence of *psbD* gene sequence from cyanophages with restricted host range. None of the cyanophages belonging to Podoviridae and Styloviridae tested so far contained the *psbD* gene. Wang and Chen (2008) investigated the prevalence of highly host-specific podoviruses and siphoviruses infecting estuarine *Synechococcus* strains. Of these, three podovirusres, three siphoviruses and one myovirus have been characterized with reference to their morphology, host-range, growth and genetic features. While *psbA* gene sequences have not been found in the siphoviruses, the *psbA* gene sequences found in the two podoviruses clustered with some environmental *psbA* sequences forming a unique cluster that is distantly related to the already known *psbA* clusters. Another significant feature is the presence of highly conserved DNA polymerase (*pol*) gene in these estuarine podoviruses that makes them unique to develop a host-phage relationship different from what has been observed in waters of open Oceans.

Sharon *et al.* (2007) conducted a global survey of the prevalence of phage photosynthesis genes and their relationship with cyanobacterial photosynthesis genes from diverse oceanic environments. They concluded that the cyanophage genomes have a long term effect on the evolution of cyanobacterial photosynthetic apparatus by shifting these genes from one host to another and the viral *psbA* genes are actively expressed in the host cells in the marine environment to prolong photosynthesis albeit for short periods of lytic cycle. Paul and Sullivan (2005) are of the opinion that *psbA* gene is transferred from host to phage multiple times while *psbD* was transferred a single time from host to phage but these have not gone back from phage to host. However, *hlip* gene acquired from host by the phage has undergone some evolutionary changes and in some cases might have been returned to the host. Although it is difficult to explain these hypotheses with the limited phylogenetic information now available, it seems probable that the phage genomes have evolved by acquiring genes of interest from their hosts that increase phage fitness and potential for replication. The range of genetic diversification in *psbA* and *psbD* gene sequences discovered in cyanophages might be useful for the evolution of the host genomes as well. Thus the cyanophage-encoded photosynthesis genes represent one of the best examples for the occurrence of LGT in marine microbial communities and this will no doubt continue to drive evolutionary forces to shape the biosphere of the future (Hambly and Suttle, 2005). Summarizing the role of light in the marine cyanophage-host interactions, Clokie and Mann (2006) suggested that it is desirable to find out (i) the fitness benefits of acquiring host genes by marine cyanophages, (ii) the probable effects of diurnal cycle on the infection cycle of phages infecting marine cyanobacteria and (iii) as to how the LGT continues to be an important component of the arms race between the marine cyanophages and their hosts.

Photosynthesis gene *psbA* has been detected in the viral populations of Norwegian coastal waters by Sandaa *et al.* (2008) who separated viral genomes ranging in size from 31 to 380 kb by PFGE followed by PCR combined with cloning and sequencing. The contribution of *psbA* gene carried by the phage and that of the host has been examined by modelling experiments during lytic cycle of the phage P-SSP7 in the cells of *P. marinus* MED4. As phage genome replication proceeds consequent upon the degradation of host genome and also D1 protein degradation takes place due to phage infection, there is a shortfall in the supply of D1 protein. Since the infected cells are no longer in a position to supply D1 protein, phage encoded *psbA* gene fulfills the role and gives advantage to the phage for completing its life cycle (Fig. 14; Bragg and Chisholm, 2008). Lytic cycle of P-SSP7 in the host *P. marinus* MED4 has been followed for genome-wide expression dynamics of both host and phage. Four phage-encoded host metabolism genes *psbA*, *hli*, *talC* and ribonucleotide reductase (*nrd*) are transcribed together as a functional unit. These genes help the phage to meet its energy requirements and in the deoxynucleotide production for phage replication in oligotropic marine waters. The presence of such genes in marine waters leads to the formation of genome islands and

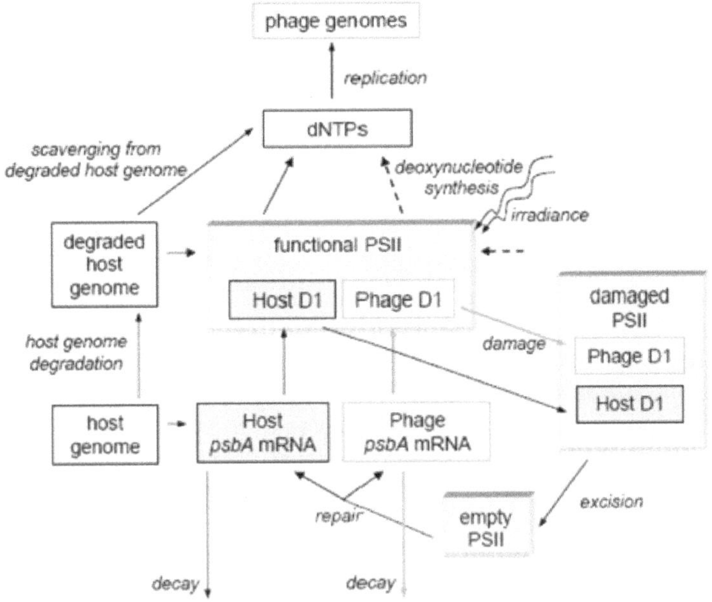

Figure 14: Schematic diagram of model, phage genomes are made using dNTPs from two possible sources. First, dNTPs can be made by scavenging deoxynucleotides from the host genome. This process can occur in the dark, but is bolstered by photosynthesis. Second, dNTPs can be newly synthesized by a process that is dependent on the products of photosynthesis (dashed lines). Photosynthesis is dependent on functional PSII subunits, which contain the D1 protein. During exposure to light, D1 proteins can become damaged, and are excised from PSII subunits, and replaced with D1 proteins from either host or phage encoded *psbA* mRNAs. With the kind permission of J. G. Bragg, Department of Civil and Environmental Engineering, Massachusetts Institute of Technology, Cambridge, Massachusetts, USA [Bragg, J. G. and Chisholm, S. W. (2008) *PLoS ONE* **3(10):** e3550. doi:10.1371/journal.pone.0003550]. doi:10.1371/journal.pone.0003550.g001

the presence of their homologues in cyanophages suggests that phage-mediated transfer of host genes has a major impact on microbial evolution. The ability of the phage to use up-regulated host genes is one of the starting points for their stable incorporation into phage genomes and their subsequent transfer back to genome islands. The activation of host genes during infection may be directing the co-evolution of gene content in both host and phage genomes (Lindell *et al.*, 2007). Chénard and Suttle (2008) conducted a phylogenetic analysis of 45 *psbA* gene sequences generated from a number of environmental viral samples collected from freshwaters (Lake Contance, West Germany, and Experimental Lakes Area of Ontario, Canada) and marine waters (collected from Arctic Ocean, Gulf of Mexico, NE Pacific Ocean). The important findings are that (i) the sequences from freshwaters have an evolutionary history distinct from those of marine waters, and (ii) *psbA* sequences from the same geographical area clustered into different clades suggesting that the area supported the presence of different types of phages. However, other findings such as the differences in *psbA* sequences from *Prochlorococcus* and *Synechococcus* viral genomes and as also between the sequences of myoviruses and podoviruses agreed with the observations of other workers reported earlier. The observations of Mann *et al.* (2003) on the interrupted nature of *psbA* gene in the genome of S-PM2 due to the presence of a group I intron has been confirmed by other workers but its expression during the lytic cycle of S-PM2 has been studied by Millard *et al.* (2010). The group I intron is shown to be spliced during the entire lytic cycle thus providing a continuous supply of D1 protein but due to the appearance of both spliced and unspliced *psbA* transcripts they predicted a regulatory role for intron splicing.

They found the presence of a homing endonuclease (HE; *FCphl*) gene adjacent to *psbA* gene sequence in the genome of S-PM2. Furthermore, the intergenic region between *psbA* and *FCphl* is occupied by antisense RNA gene (asRNA). Millard *et al.* (2010) are of the opinion that asRNA is co-regulated with *psbA* and *F-Cphl* genes and is involved with their expression.This appears to be the first report on the presence of asRNA in the genomes of bacteriophages, a property that is widely represented in eukaryotes and to certain extent in prokaryotes.

Besides the core PSII genes in marine cyanophages, Sharon *et al.* (2009) demonstrated the presence of core PSI genes (*PsaA, B, C, D, E, K* and a unique *J* and *F* fusion gene) in the genomes of marine cyanophage sequences collected from metagenomic data of the GOS expedition. The gene products of these seven genes have been suggested to be sufficient to form an intact monomeric PSI complex and thus the marine cyanophages are able to funnel reducing power from respiratory and other electron transfer chains to the PSI complex. Furthermore, the presence of gene sequences of electron transfer protein genes like NADP(H) dehydrogenase (NDH-1) in the marine cyanophage genomes signifies that the marine cyaophages tinker with the photosynthetic electron flow of their hosts and can derive benefits of ATP and NADPH from PSII or by the operation of PSI obtaining only ATP, depending on the metabolic state of the host cell (Alperovitch-Lavy *et al.*, 2011; Philosof *et al.*, 2011).

Other host-related genes that play an important role in the multiplication of marine cyanophages relate to the biosynthesis of polyamines. Polyamines influence the structure and oxygen evolution rate of the PSII reaction centre (Bograh *et al.*, 1997). These are suggested to stabilize *psbA*-2 mRNA (Mulo *et al.*, 1998). Sequencing of marine cyanophage genomes revealed the presence of *speD* gene in S-PM2 (Mann *et al.*, 2005) and P-SSM4 (Sullivan *et al.*, 2005) that encodes S-adenosylmethionine a key enzyme in the biosynthesis of polyamines spermine and spermidine. During infection these phages may derive either of the functions mentioned above to prolong the functional integrity of photosystem.

Metagenomic analysis of Chesapeake Bay waters led to the identification of two myoviruses P-SSM2 and P-SSM4 represented by 878 and 723 homologous sequences, respectively. The podoviruses have been represented by P-SSP7 and P60 with homologous sequences of 543 and 368, respectively. Thus the cyanomyoviruses and cyanopodoviruses are equally well represented in the Chesapeake Bay metagenome. Cyanophage P60 appears to be the most commonly detected phage based on the second version of phage proteomic tree (Rohwer and Edwards, 2002; Edwards and Rohwer, 2005). The cyanophage homologous sequences accounted for 14% of the total metagenomic sequences specific for phages. A total of 17,124 bp of *psbA* homologous sequences have been noted that constitute 5% of the 341 kb of cyanophage homologues identified in the library. On the basis of sequence length and percentage of total genome, *psbA* gene sequences have been found in the majority of cyanophages. Thus it has been concluded that cyanomyoviruses with larger genomes carried both *psbA* and *psbD* genes while cyanopodoviruses and cyanosiphoviruses that exhibited narrow host range contained only *psbA* gene. Another interesting feature is the existence of sequences homologous to ORFs of the genome of cyanomyovirus P-SSM2 in the metagenomic library of Chesapeake Bay (Bench *et al.*, 2007). Metagenomic analysis of the sequences from the GOS expedition (conducted at five sampling sites GS19, GS20, GS26, GS34 and GS51) revealed a high concentration of viral sequences pertaining to 1,54, 662 viral genes encoding various metabolic and cellular functions. The acquisition of host genes that are environmentally important for the viral population seems to be a more wide-spread phenomenon than previously envisaged. The most predominant sequences relate to tailed bacteriophage sequences. Of the 27 complete aquatic viral genomes analysed only one reference bacteriophage genome has been abundantly represented. This sequence seems to be closely related to the marine cyanophage P-SSM4 but it is not identical to it. This cyanophage seems to influence the distribution, abundance and diversity of its host identified to be closer to *P. marinus* MIT9312 (Williamson *et al.*, 2008).

v) *Genes related to pigment biosynthesis*: Of the two oceanic cyanobacteria *Synechococcus* and *Prochlorococcus*, the former possesses phycobilisome rods that contain phycoerythrin (Waterbury and Rippka, 1989) which helps in the transfer of harvested light energy to PSI and PSII. The latter does not possess phycobilisome rods but possesses alternative antenna based on chlorophyll (Goericke and Repeta, 1992; Lichtle *et al.*, 1995; La Roche *et al.*, 1996; Ting *et al.*, 2002). Some strains of *Prochlorococcus* express genes that are remnants of the ancestral *Synechococcus* phycobilisomes (Hess *et al.*, 1996). Genes involved in the biosynthesis of phycoerythrobilin (PEB) and phycocyanobilin (PCB) are designated as *pebA* and *pcyA*. The sequence of reactions involved in the biosynthesis of PEB and PCB is described here. Heme in presence of heme oxygenase (HO-1) is converted to an open chain product by the addition of seven electrons in presence of oxygen into biliverdin 1x alpha (BV) that constitutes the substrate for various enzymes of the FDBR family. Two sequential two electron reductions catalysed by the enzymes PebA and PebB give rise to PEB. Phycoerythrobilin synthase (PebS) is a product of the gene *pebS* that catalyses a unique four electron reduction of BVIx alpha to PEB. BV is reduced to PCB in presence of the enzyme PcyA with the addition of 4 electrons. All cyanobacteria possess the enzymes PebA and PebB. Enzymes HO-1 and PcyA are found both in cyanobacterial cells and marine cyanophages. The PebS enzyme is found only in marine cyanophage populations. Exceptionally, marine cyanophages P-SSP7 (a *Prochlorococcus* podovirus), Syn5 and P-60 (*Synechococcus* podoviruses) do not possess any of the bilin biosynthesis genes. Of the myoviruses, *Prochlorococcus* myovirus P-SSM2 contains *pebS*, *ho1* and *petF* genes where as P-SSM4 consists of only one *pcyA* gene (Sullivan *et al.*, 2005; Mann *et al.*, 2005; Weigele *et al.*, 2007). *Synechococcus* myoviruses S-PM2 and Syn9 possess *cpeT* gene (Mann *et al.*, 2005). *Synechococcus* sp. strain PCC 7002 possesses a *cpcT* gene that is paralogous to *cpeT* gene and the *cpcT* gene encodes a lyase that attaches the phycobilin chromophore to the β-subunit of phycocyanin (Shen *et al.*, 2006). Of the cyanobacterial strains, six of the HL-*Prochlorococcus* strains (MED4, MIT9215, MIT9301, MIT9312, MIT9515 and AS9601) investigated possess *pebA*, *pebB*, *pcyA*, *ho1*, and *petF* genes while six of the LL-*Prochlorococcus* strains (NATL1A, NATL2A, SS120, MIT9211,, MIT9303 and MIT9313) showed in addition to these, the gene *cpeT*. *Synechococcus* strains (CC9311, CC9605, CC9902, WH8102, JA-2-3B'a (2-13) and JA-3-3Ab) also contained the six genes as seen in LL-*Prochlorococcus* strains. Dammeyer *et al.* (2008) investigated the functional aspects of the genes *ho1*, *pebA*, *petF* in P-SSM2 and a *pcyA* homolog in the phage P-SSM4. Cyanophage P-SSM2 *pebA* encodes PebS via an intermediate 15, 16 dihydrobiliverdin and in this sense *pebA* of the phage carries out a reaction in a single step that is performed in two steps in cyanobacterial cells and these genes are expressed during the first hour of infection in the cells. On the basis of identification of bilin biosynthesis genes of the host and phage genomes from oceanic communities of marine cyanophages and the subsequent phylogenetic analysis revealed that PebA, PebS and HO-1 are grouped into a monophyletic cluster and have been acquired from marine cyanobacteria only once. Whereas *pcyA* phage sequences group into two clusters and may have been obtained in two steps.

Millard *et al.* (2009) conducted a comparative genomic analysis of five marine cyanomyoviruses including cyanophage S-RSM4 and detected a hypervariable region in between genes *g15-g18* characterized by the presence of *ptoX* (plastoquinol terminal oxidase), *gnd* (6-phosphogluconate dehydrogenase), *zwf* (glucose-6-phosphate dehydrogenase) and *petE* (plastocyanin) genes. The acquisition of these host genes by the cyanomyoviruses might have taken place as independent events and has contributed to the evolution of these cyanomyoviruses. A bioinformatics analysis of GOS expedition viral metagenomes in combination with the genomes of viruses and microorganisms enabled Sharon *et al.* (2011) to identify the presence of hitherto undiscovered 34 microbial gene families in the 452 viral scaffolds of the metagenome. These are related to energy metabolism (electron transport genes), PSI and PSII, translation and post-translation related genes. This amply signifies that the marine viruses control translation and post-translational events in their host cells besides the already known transcriptional control.

XII. LYSOGENY

The best example for explaining lysogenic behaviour is phage λ and *E. coli* system. This has been used as a tool in gene clong experiments as well. A number of workers reported lysogenic behaviour among cyanobacterial cells and their induction by various agents. However, the nature of temperate phages, the molecular biology of their integration and excision events and the type of transductional events mediated by these phages still remains obscure. The following account summarizes the nature of lysogenic behaviour and of temperate cyanophages of both freshwater and marine origin.

***i) Lysogeny in freshwater cyanophages*:** A number of workers have reported the lysogenic behavior of *P. boryanum* cells consequent upon infection by LPP-1 and these reports focussed their attention on a variety of conditions that are helpful in inducing a lytic cycle. Singh and Singh (1972) not only observed lysogeny in *P. boryanum* but also demonstrated the transfer of antibiotic resistance marker from resistant cells of *P. boryanum* to the antibiotic-sensitive cells through transduction. Lysogeny was for the first time reported due to infection of LPP-1D (Delaware strain) and LPP-2 that are temperate viruses on *P. boryanum* (Cannon *et al.*, 1971). Cannon and Shane (1972) reported the effects of antibiotic stress on protein synthesis in the establishment of lysogeny in *P. boryanum* cells. Padan *et al.* (1972) isolated a temperate cyanophage LPP2-SPI which was able to enter into lysogenic cycle in *P. boryanum.* Studies on stable lysogeny by SPI cts1, a ts-mutant of SPI and its induction by heat (treatment at 40°C) of host cells of *P. boryanum* revealed that shortly after infection a leakage of macromolecules from the host cells was observed with a simultaneous inhibition in CO_2 fixation. However, induced prophage neither caused cellular leakage of macromolecules nor CO_2 fixation was inhibited (Rimon and Oppenheim, 1975). Cannon *et al.* (1976) cultured *P. boryanum* IU594 and *P. boryanum*-co-resistant to LPP-1, LPP-1D and LPP-2 in chemostats to evaluate the development of phage-resistant strains and the development of lysogeny. After three oscillations in the virus titer and cell numbers both of them stabilized and due to development of LPP 1-resistant strains of *P. boryanum* the phage was eliminated. However, the existence of phage titres of LPP-1D and LPP-2 throughout the growth in chemostat signifies the induction of lysogenic cells. *P. uncinatum* cells lysogenic for cyanophage LPP-1 produced phage at the rate of 3×10^3 PFU ml^{-1} and treatment with mitomycin C enhanced phage production by 5-fold. Treatments with UV-light and heat did not lead to induction of lysogenic cells (Bisen *et al.*, 1986b). A temperate cyanopodovirus N(S)1 that formed turbid plaques on a species of *Anabaena* 77S15 has been described (Franche, 1987). Sode *et al.* (1994) isolated a marine temperate cyanophage ms-1 infecting *Synechococcus* sp. NKBGO42902, a strain that is known to produce cytokinin-like compounds. The induction of the temperate cyanophage ms-1 in the lysogenic cells by treatment with copper ions (3.1×10^{-4} M of copper sulphate) appeared to be specific for Cu^{2+} as other metallic ions Cr, Pb, Co and Zn were not able to induce the lytic cycle. Another cyanobacterium that has been reported to be lysogenic is *Phormidium persicinum*. The phage belongs to the family of Siphoviridae and has a genome size of 50 kbp. Induction of lysogenic cells into lytic cycle in presence of mitomycin C

(30 μg ml^{-1}) was followed by electron microscopy (Ohki and Fujita, 1996). A temperate phage-specific to *Trichodesmium* sp. has been isolated by Ohki (1999) and it is speculated that the phage had a specific role in the regulation of *Trichodesmium* biomass.

***ii) Lysogeny in marine cyanophages*:** Phosphate availability has been shown to affect kinetics of cyanophage infection and also is probably responsible for induction of lysogenic *Synechococcus* cells (Wilson *et al.*, 1996, 1998). A study conducted over a period of one year (during October, 1999 to October, 2000) at Tampa Bay, Florida revealed prophage induction by *Synechococcus* cells after treatment with mitomycin C. Due to this, there was appreciable increase in the titre of virus particles that was inversely correlated with cyanobacterial abundance. Induction of prophage was predominantly noted during summer months. The limited lysogeny noted during summer months has been explained on the basis that exposure to UV or high temperature, lysogenic *Synechococcus* naturally had undergone induction thereby decreasing the population density. It was concluded that lysogenization is not a random event but a seasonal pattern that is consistent with the occurrence of lysogeny when host population was poorly represented in nature (McDaniel *et al.*, 2002). During a *Synechococcus* bloom in a pristine fjord in British Columbia, Canada, it was noted that 0.6% of the cells are lysogens as they could be induced by the addition of mitomycin C (Ortmann *et al.*, 2002). McDaniel and Paul (2005) demonstrated induction of lysogenic cells of marine *Synechococcus* collected from ten locations of Gulf of Mexico when amended with either nutrients (such as nitrate, ammonium, urea or phosphate) or mitomycin C. In five of the nine experiments induction occurred only in the latter. Phosphate enrichment led to induction in only one Tampa Bay sample. Nutrient enrichment, on the contrary, resulted in a decrease of the lytic phage production.

Put together, the number of *Synechococcus* cells occurring at any given time and the cyanophage titers, lysogeny in *Synechococcus* is a most likely event taking place during low ambient cyanophage abundance in marine waters. Similar results have been reported from the Gulf of Mexico and Mississippi River plume by Long *et al.* (2008). The extent to which the strains of *Synechococcus* enter into lysogeny or liable to be infected by lytic phages has been investigated by McDaniel *et al.* (2006). Of the 25 strains of *Synechococcus* that occur in the Gulf of Mexico, 11 strains showed induction after treatment with mitomycin C and the released particles have been stained with SBYR Gold and viewed through epifluorescence microscopy. One of the lysogenic strains appeared to be unique because of its induction not only by continuous light but also by the release of particles that contained single-stranded DNA as genome.

XIII. ROLE OF CYANOPHAGES IN MARINE ECOSYSTEM

The virtual abundance of viruses in marine waters superceding the numbers of marine phytoplanktons has led to the concept of marine viruses being treated as virioplankton. Viral abundance in relation to environmental parameters, the probable impact of viral-mediated lysis of marine phytoplankton, the concept of microbial loop and the role of viruses in maintaining host community diversity have received attention (Fig. 15; Wommack and Colwell, 2000). Two genera of cyanobacteria that are important components of marine phytoplankton are *Prochlorococcus* and *Synechococcus*. Of these two, *Prochlorococcus* is numerically the dominant prokaryote inhabiting the temperate and tropical oceans (Partensky *et al.*, 1999). It is less than one micron in diameter and contributes significantly to global photosynthesis (Liu *et al.*, 1997, 1998). In marine waters the organism is represented to a density of 10^5 cells ml^{-1}. This cyanobacterium is adapted to live in oligotrophic waters and is represented by two ecotypes adapted to grow under HL- and LL-conditions and are physiologically and genetically

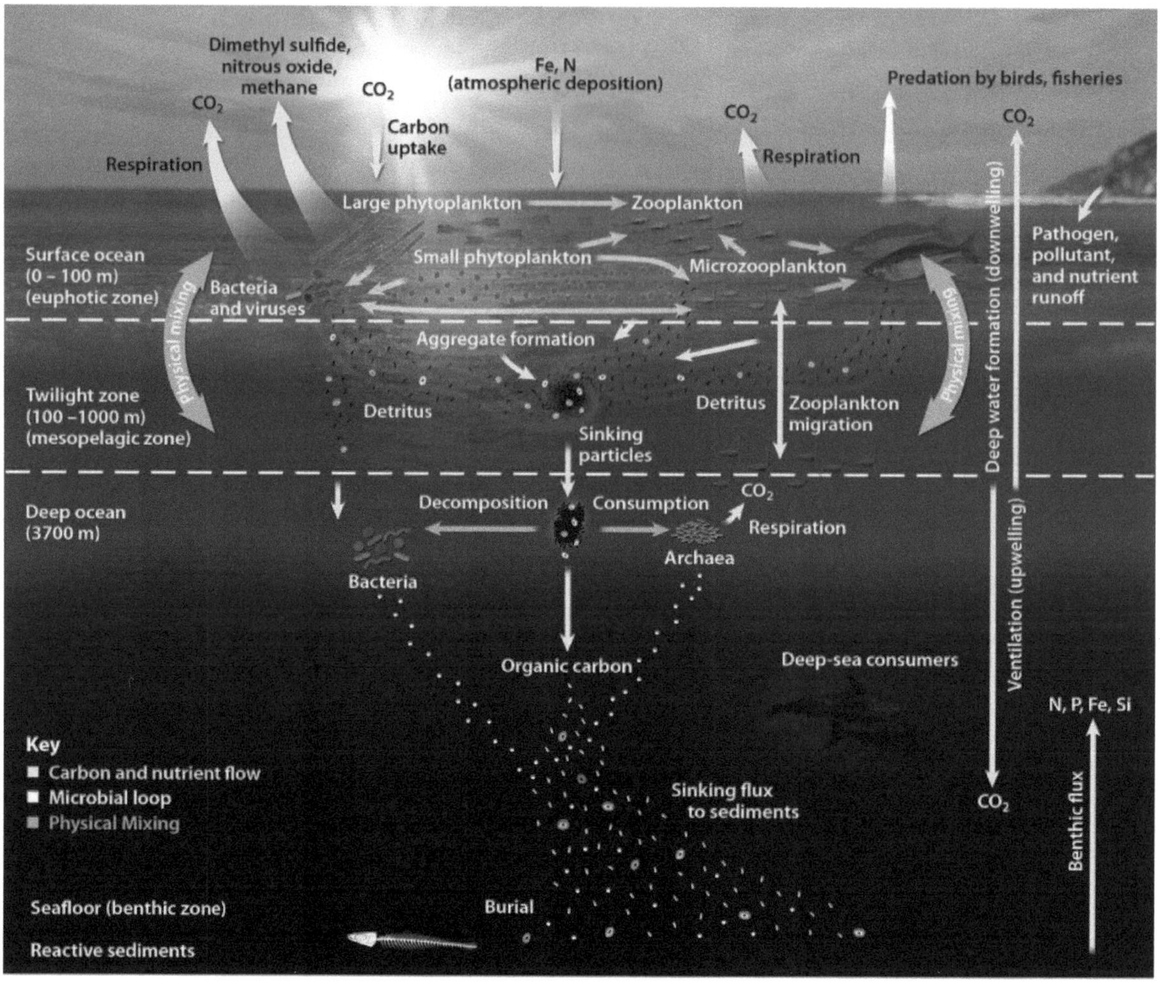

Figure 15: Microbial loop and the role of virioplankton in causing the consequential carbon and nutrient cycling in the ocean ecosystem. Picture courtesy U.S. Department of Energy Genomic Science program and the website http://genomicscience.energy.gov.

Color image of this figure appears in the color plate section at the end of the book.

diverse (Moore *et al.*, 1998). These two ecotypes also exhibit differences in nitrogen and phosphorus utilization and copper (Mann *et al.*, 2002) and also phage sensitivity (Sullivan *et al.*, 2003). As already mentioned, marine *Synechococcus* is represented by many strains that are grouped into MC-A and MC-B. The distribution of these strains in relation to depth, salinity and nutrient concentrations has been investigated in detail. *Prochlorococcus* seems to be abundant than *Synechococcus*, as the former accounted to nearly 31% of the bacteria-sized organisms in the oligotrophic waters of North Pacific (Waterbury *et al.*, 1979; Campbell *et al.*, 1994). It was only through the use of flow cytometry and epifluorescence microscopy that *Prochlorococcus* emerged as the most important component of the marine phytoplankton. These two cyanobacteria are suggested to fix as much as 20–80% of the carbon in these environments (Joint and Pomroy, 1983; Li *et al.*, 1983; Waterbury *et al.*, 1986; Liu *et al.*, 1997). Although the existence of VLPs in the cells of marine phytoplankton was demonstrated by a number of workers, these observations, however, have not been correlated with the existence of marine phages that were specific for the respective marine planktons (Fuhrman and Suttle, 1993). Mayer and Taylor (1979) first isolated a large polyhedral virus measuring 130–135 nm in diameter that was specific for *Micromonas pusilla*, a marine Chrysophyte. In the Gulf of Mexico, the concentrations of cyanophages infecting a single strain of *Synechococcus* ranged in between 10^3 L^{-1} to 10^8 L^{-1} (Suttle *et al.*, 1993; Suttle and Chan, 1993). The concentrations of bacteriophages infecting various bacterial hosts were found to be of the order of 10^4 L^{-1} to 3.7×10^7 phage particles L^{-1}. However, great variability was noted in the titres of viruses that infect eukaryotic photosynthetic flagellates like *M. pusilla* (Cottrell and Suttle, 1991). Primary productivity was reduced by as much as 78% due to infection of phytoplankton by the respective viruses of diatoms, Cryptophytes, Prasinophytes and chroococcoid cyanobacteria in marine waters. So in addition to grazing and nutrient limitation, phage infection seems to play a regulating factor in the development of the phytoplankton communities (Suttle *et al.*, 1990). Proctor and Fuhrman (1990) suggested that up to 7% of the heterotrophic bacteria and 5% of the cyanobacteria from diverse marine locations contained mature phage and this indicates that up to 70% of the prokaryotes could be infected. This led to the understanding that a new pathway of carbon and nitrogen cycling in marine food webs with probable influence on LGT in marine microorganisms. Viral-mediated release of dissolved organic matter (DOM) can be a good source of major nutrients (C, N, and P) and trace elements (e.g. Fe) for other photosynthetic and heterotrophic microorganisms (Middelboe *et al.*, 1996; Göbler *et al.*, 1997; Middelboe *et al.*, 2003). This was shown to influence bacteria and phytoplankton species composition and succession (Göbler *et al.*, 1997; Brussaard *et al.*, 2004). Another area of global significance influenced by viruses is the augmentation in production of dimethyl sulfide (DMS) that helps in cloud formation (Charlson *et al.*, 1987). The production of dimethylsulfoniopropionate (DMSP) by many phytoplanktons and its subsequent cleavage by algal lyases and/or other lyases of other organisms into DMS and acrylic acid has been demonstrated. The build up of DMSP in nutrient media as a result of viral lysis of *M. pusilla*, *P. pouchetii* and *Emiliana huxleyi* has been confirmed (Hill *et al.*, 1998; Malin *et al.*, 1998; Wilson *et al.*, 1998). It seems, therefore, viral lysis of phytoplankton may constitute an important source for production of DMSP in the marine environment.

Suttle and Chan (1993) first re-emphasized the role of marine cyanophages in controlling cyanobacterial community structure in marine waters. The dynamics and distribution of marine cyanophages infecting *Synechococcus* DC2 and SYN48 along the transect of Western Gulf of Mexico in Aransas Pass, Texas revealed that the cyanophage concentrations ranged from few hundred ml^{-1} at a depth of 97 m to 4×10^5 ml^{-1} at the surface waters (83 km offshore). The highest concentrations of cyanophages at the surface waters coincided with *Synechococcus* density. As the density of the host increased beyond 10^3 ml^{-1}, then the titers of cyanophages increased suggesting that a minimum concentration of host is required for infection to occur. They estimated that on daily basis about 5 to 7% of *Synechococcus* cells are lysed. Cyanophages thus seem to constitute a dynamic marine planktonic communitiy and are responsible for causing significant losses of *Synechococcus* population on a daily basis. A comparison of viral production rates and the contact rates between cyanophage particles and the host cells at the farthest offshore waters and nearshore waters brought to light that most of the *Synechococcus* cells in the former locality are susceptible to infection while in the latter only 80% host cells are liable to be infected. This was explained on the basis that most of the cells are probably resistant as reported earlier by Waterbury and Valois (1993). This aspect of study has received attention during recent years. Stoddard *et al.* (2007) isolated 23 phage-resistant mutants by employing four strains of *Synechococcus* sp. (WH7803, WH8018, WH8012 and WH8101) challenged by 32 previously isolated cyanomyophages. The important findings are that (i) resistance is more likely due to a change in phage receptor sites, (ii) selection for resistance to one phage simultaneously conferred resistance to other phages, and (iii) there is no significant genetic similarity of the phages as revealed by the analysis of *g20* gene sequences. Thus the evolution of multiple phage resistance seems to influence population dynamics of the host. Mutation from phage sensitivity to phage resistance is known to lead to a configurational change in phage receptor sites (Inoue *et al.*, 1995; Bohannan and Lanski, 2000). Experimental evidences suggest that the phage-resistant mutants are less efficient than their sensitive counterparts in phage-free environments (Buckling and Rainey, 2002) and there exists a fitness cost for the evolution and maintenance of virus resistance (Bohannan and Lenski, 2000). The cost of maintaining virus resistance by *Synechococcus* has been investigated by Lennon *et al.* (2007) who conducted experiments with phylogenetically distinct strains of the host to resist 32 cyanomyophages (S-RIM1 to S-RIM13, S-RIM15 to S-RIM21, S-RIM23 to S-RIM36) collected from Mount Hope Bay, Rhode Island. The cost of resistance (COR) hypothesis was tested by comparing growth data and their resistance patterns. The growth rates have been determined by conducting co-culture experiments with a parent strain, its phage-resistant strain along with a reference strain. The reference strain (WH8101 lacks phycoerythin but instead contains phycocyanin and is distinguishable from others (WH7803, WH8018 and WH8012) that possess phycoerythrin. The quantification of cell densities of these strains was followed in nutrient media at a constant nitrogen concentration but the phosphate concentration was varied. The growth rate of virus-resistant mutant of *Synechococcus* (WH7803R8) was found to be inferior to that its parent strain (WH7803) and there appeared to be on the average a 20% reduction in relative fitness, irrespective of phosphorus availability. Likewise, the performance of phage-resistant mutant *Synechococcus* WH 8108 was significantly lower leading to an almost 18% reduction in relative fitness. Further, phage resistance among the 22 virus-resistant strains of *Synechococcus* (seven isolated from WH7803, four from WH8101, three from WH8012 and eight from WH8018) has been tested to find out the breadth of resistance (BOR). The BOR profile for each of the resistant mutant is represented by total resistance (summed up number of viruses to which a *Synechococcus* strain was resistant) and compositional resistance (that refers to the multivariate similarity of BOR profiles among different phage resistant strains of *Synechococcus*). Competition experiments revealed that *Synechococcus* growth rates were not affected by total resistance but its growth rates were affected by compositional resistance. The strains that were sensitive to S-RIM4 and S-RIM11 showed low fitness where as strains resistant to S-RIM7 and S-RIM20 showed relatively high fitness. So it depends on the fitness cost that the viruses impose on their hosts which matters in determining community structure of hosts. If the COR is small then resistant bacteria may be more common with concomitant lower titre of viruses and lower viral-induced mortality. On the other hand, if COR is high then phage-resistant cyanobacteria may be relatively rare leading to high titer of viruses and high virus-induced mortality. Stern and Sorek (2011) summarized bacterial and phage defensive and offensive strategies eventually leading to a phage-host arms race and the evolution of bacterial genomes.

A number of models have been presented to explain the role of virioplankton in maintaining microbial food web. It is generally assumed that the cell debris from lysis and the released viruses returns to feed bacteria in a closed loop where the oxidation of carbon compounds is followed by a regeneration of N and P (Bratbak *et al.*, 1990; Proctor and Fuhrman, 1991; Fuhrman, 1992). The viruses cause a general shift in the food web activity (mostly production and respiration) from larger organisms to bacteria (Fuhrman, 1999; Wilhelm and Suttle, 1999) and reduced protist production (Fuhrman, 1999). The first mathematical model based on empirical data confirmed that viral-mediated mortality of *Phaeocystis globosa* is an essential regulating factor for the nutrient cycling (Ruardij *et al.*, 2005). Other aspects that merit mention are (i) infection of primary producers (10%

of the phytoplankton mortality), (ii) consumption of viruses by nanozooplankton (13% of the virus production), (iii) increased bacterial production and respiration rates (33% above the level without the viruses) and (iv) a loss nearly of 20% in nanozooplankton production.

Hannen *et al.* (1999) developed two artificial laboratory-scale enclosures filled with lake water (130 litres each) to simulate the conditions of Lake Loosdrecht, The Netherlands and the succession of microorganisms was followed consequent upon the lysis of filamentous cyanobacteria by the cyanophages. The cyanobacterial populations consisted of *Oscillatoria* c.f. *limnetica* and *Prochlorothix hollandica*. Electron microscopic observations revealed the presence of lambda-like cyanophage particles inside the cells of the above cyanobacteria as well as attached to their filaments. However, the titre of free viral particles has not been assayed. The microbial communities were dominated by heterotrophic bacteria belonging to the Cytophagales and actinomycetes, that are known to degrade complex organic substances. These were followed by the appearance of rotifers that are known to connect the microbial loop with higher trophic levels. DGGE was chosen to follow the bacterial and eukaryotic communities and were analysed by the amplification of the sequences of 16S and 18S rDNA fragments. These observations confirm and strongly support the idea that the marine viruses in general can structure microbial communities.

Sano *et al.* (2004) repeated the experiments of Wilcox and Fuhrman (1994) and confirmed the finding that marine viruses need a minimal concentration (>3% of natural concentration) in order to find hosts. They demonstrated further that viruses from different habitats such as near-shore marine sites, lakewater, marine sediments and soil could be propagated on microbial flora from marine water sample. This led to the idea that viruses can move between different ecosystems and propagate and that the possibility of LGT between microorganisms of different biomes exists. Breitbart *et al.* (2004) found the distribution of podophage T7-like DNA polymerase sequences in major biomes including marine, freshwater, sediment, terrestrial, extreme and metazoan-associated environments. A wealth of knowledge has accumulated on various aspects of abundance and importance of viruses in the environment, culture-based and culture-independent studies of viral diversity, the movement of viruses between biomes, metagenomic studies of viral diversity, the role of mobile genetic elements and LGT (Breitbart and Rohwer, 2005; Breitbart *et al.*, 2007). Current aspects of viral genomics that remain to be elucidated are as to whether (i) the key "signature" cyanophage sequences confer any benefit on the phages and (ii) the role of such host genes acquired by the phage in phage ecology and host range. The marine phages have developed adaptive mechanisms to infect their hosts growing in oligotropic environments of marine waters (O'Connell, 2005). Benthic viruses (vibriobenthos) accounted for 10^7 to 10^{10} particles g^{-1} of benthos of fresh and marine water sediments. As revealed by PFGE and analysis of their genomes, vibriobenthic assemblages are highly diverse and distinct from virioplankton (Danovaro *et al.*, 2008). Breitbart (2012) highlighted many of the "established truths" about marine viruses and few of the challenges that lie ahead for the future researchers in this field.

XIV. METAGENOMICS

Our understanding of microbial physiology, genetics and ecology is mainly based on only 1% of the microbes obtained in pure cultures. So the vast majority of the microbes remain unexplored (Schloss and Handelsman, 2005). The development of culture-independent approaches for the identification of microorganisms from their natural habitats with the analysis of 5S and 16S rRNA gene sequences paved the way for the recognition of new gene sequences as well as new microbes (Lane *et al.*, 1985; Giovannoni *et al.*, 1990; Schmidt *et al.*, 1991; Stahl *et al.*, 1985; Barns *et al.*,1994; Eden *et al.*, 1999).

By these efforts it has now been possible to identify a number of heterotrophic bacteria that have not been cultured hitherto. In the study of uncultured microorganisms, metagenomics occupies a central place. The term 'Metagenomics', was coined by Handelsman *et al.* (1998), to represent the analysis of gene sequences of uncultured microorganisms. This involves direct isolation of DNA sample from the environment, cloning of it into a suitable vector, followed by the transformation of host bacterium like *E. coli* by this vector and screening the transformant for the respective trait. The transformed clones of *E. coli* can be maintained as a metagenomic library. Metagenomics has also been alternatively called as community genomics, environmental genomics and population genomics.

Two types of approaches have been followed in the study of such cloned genes, i.e. (i) function-driven analysis and (ii) sequence-driven analysis. Function-driven analysis involves heterologous gene expression and identifying the protein synthesized from such a gene sequence. In sequence-driven analysis cloned DNA preparation is identified by sequence analysis of phylogenetic markers such as 'anchor' sequences (16S rRNA and *recA*) or for other conserved genes by hybridization or by PCR or random sequencing. With the help of sequence based screening for small molecules it has now been possible to discover new antibiotics (such as turbomycin A, turbomycin B, violacein, deoxyviolacein, terragine A, indirubin and a number of long-chain N-acyl amino acid antibiotics). This subject has been reviewed (Handelsman, 2004; Schloss and Handelsman, 2005). The diversity of viruses in a particular sample can be estimated by either employing the amplification and sequencing of conserved genes or partially sequencing shotgun libraries. Metagenomic analysis deals with the discovery of the uncultured marine viral communities by the latter technique (Breitbart *et al.*, 2002; Hendrix, 2003; Pedulla *et al.*, 2003) and those of marine cyanophages by the former. The sequencing of shotgun libraries of 168 viral assemblages collected from four oceanic regions revealed that several hundred thousand distinct viral species are dispersed in these waters. Most of these sequences do not resemble any of the previously reported sequences suggesting that much of the viral diversity remains unexplored and uncharacterized. From these studies it is concluded that it is through the selective pressure prevailing in a particular environment that selects a certain viral type to be an effective infectious agent (Angly *et al.*, 2006). The diversity of marine cyanophages has been determined with the help of the amplification of conserved *psbA* and *psbD* genes from diverse marine habitats (Venter *et al.*, 2004; Angly *et al.*, 2006; Zeidner *et al.*, 2005; DeLong *et al.*, 2006; Sharon *et al.*, 2007; Bench *et al.*, 2007) and *g20* gene sequence (Wilson *et al.*, 1999, 2000; Zhong *et al.*, 2002; Marston and Sallee, 2003; Frederickson *et al.*, 2003; Dorigo *et al.*, 2004; Wang and Chen, 2004; Mühling *et al.*, 2005; McDaniel *et al.*, 2006; Baker *et al.*, 2006; Wilhelm *et al.*, 2006).

LITERATURE CITED

Ackermann, H.-W. (2007) 5500 Phages examined in the electron microscope. *Arch Virol* **152:** 277–243.

Ackermann, H.-W. (2011) Bacteriophage taxonomy. *Microbiology Australia*. May 2011. pp. 90–94.

Ackermann, H.-W. and Haldal, M. (2010) Basic electron microscopy of aquatic viruses. Chapter 18. In: *Manual of Aquatic Viral Ecology*. The Americal Society of Limnology and Oceanography. pp. 132–192.

Ackerman, H.-W., Audurier, A., Berthiaume, L., Jones, L. A., Mayo, J. A., and Vidaver, A. K. (1978) Guidelines for bacteriophage characterization. *Adv Virus Res* **23:** 1–24.

Adolph, K. W., and Haselkorn, R. H. (1971) Isolation and characterization of a virus infecting the blue-green alga *Nostoc muscorum*. *Virology* **46:** 200–208.

Adolph, K. W., and Haselkorn, R. H. (1972) Photosynthesis and the development of blue-green algal virus N-1. *Virology* **47:** 370–374.

Adolph, K. W., and Haselkorn, R. (1973a) Isolation and characterization of a virus infecting a blue-green alga of the genus *Synechococcus*. *Virology* **54:** 230–236.

Adolph, K. W., and Haselkorn, R. (1973b) Blue-green algal virus N-1: Physical properties and disassembly into structural parts. *Virology* **53:** 427–440.

Al-Musavi, R. A. (1977) Effect of photosynthesis and respiration on growth of cyanophages of *Anabaena variabilis*. *Mikcrobiologiya* **46:** 725–729.

Alperovitch-Lavy, A., Sharon, I., Rohwer, F., Aro, E.-M., Glaser, F., Milo, R., Nelson, N., and Béjà, O. (2011) Reconstructing a puzzle: Existence of cyanophages containing both photosystem-I and photosystem-II gene suites inferred from oceanic metagenomic datasets. *Environ Microbiol* **13:** 24–32. doi:10.1111/j.1462-2920.2010.02304.x

Amla, D. V. (1978) Stability, adsorption and growth characteristics of cyanophage AS-1. In: S. P. Singh, Tiwari, D. N., Kashyap A. K., and Yadav, P. K. (Eds.) *Advances in Cyanophyte Research*. Banaras Hindu University, Varanasi. pp. 131–139.

Amla, D. V. (1979) Characteristics of pigment mutants of *Anacystis nidulans*: UV sensitivity and multiplication of cyanophage AS-1. *Biochem Physiol Pflanzen* **174:** 678–684.

Amla,D.V. (1981) Isolation of characteristics of minute plaque forming mutant of cyanophage AS-1. *Biochem Physiol Pflanzen* **176:** 83–89.

Amla, D. V., and Saxena, P. N. (1983) Metabolic aspects of cyanophage AS-1 replication and reproduction in cyanobacterium *Anacystis nidulans*. *Biochem Physiol Pflanzen* **178:** 225–236.

Amla, D. V., Rowell, P., and Stewart, W. D. P. (1987) Metabolic changes associated with cyanophage N-1 infection of cyanobacterium *Nostoc muscorum*. *Arch Microbiol* **148:** 321–327.

Angly, F., Felts, B., Breitbart, M., Salamon, P., Edwards, R., Carlson, C., Chan, A. M., Haynes, M., Kelley, S., Liu, H., Mahaffy, J. M., Mueller, J. E., Nulton, J., Olson, R., Parsons, R., Rayhawk, S., Suttle, C. A., and Rohwer, F. (2006) The marine viriomes of four oceanic regions. *PLoS Biol* **4:** 2121–2131.

Bagchi, S. N., Kaloya, P., and Bisen, P. S. (1987) Effect of cyanophage N-1 infection on the synthesis and stability of *Nostoc muscorum* nitrate reductase. *Curr Microbiol* **15:** 61–65.

Bailey, S., Clokie, M. R. J., Millard, A., and Mann, N. H. (2004) Cyanophage infection and photoinhibition in marine cyanobacteria. *Res Microbiol* **155:** 720–725.

Baker, A. C., Goddard, V. J., Davy, J., Scroeder, D. C., Adams, D. G., and Wilson, W. H. (2006) Identification of a diagnostic marker to detect freshwater cyanophages of filamentous cyanobacteria. *Appl Environ Microbiol* **72:** 5713–5719.

Bancroft, I., and Smith, R. J. (1988a) The isolation of genomic DNA from cyanophages infecting *Nostoc* and *Anabaena* species of cyanobacteria. *New Phytol* **110:** 233–239.

Bancroft, I., and Smith, R. J. (1988b) Analysis of restriction endonuclease sites in cyanophages infecting the heterocystous cyanobacteria. *J Gen Virol* **69:** 739–743.

Bancroft, I., and Smith, R. J. (1989) Restriction mapping of genomic DNA from five cyanophages infecting the heterocystous cyanobacteria *Nostoc* and *Anabaena*. *New Phytol* **113:** 161–166.

Barnet, Y. M., Daft, M. J., and Stewart, W. D. P. (1981) Cyanobacteria-cyanophage interactions in continuous culture. *J Appl Bacteriol* **51:** 541–552.

Barnet, Y. M., Daft, M. J., and Stewart, W. D. P. (1984) The effect of suspended particulate material on cyanobacteria-cyanophage interactions in liquid culture. *J Appl Microbiol* **56:** 109–116.

Barns, S. M., Fundyga, R. E., Jeffries, M. W., and Pace, N. R. (1994) Remarkable archaeal diversity detected in a Yellowstone National Park hot spring environment. *Proc Natl Acad Sci USA* **91:** 1609–1613.

Bench, S. R., Hanson, T. E., Williamson, K. E., Ghosh, D., Radosovich, M., Wang, K., and Wommack, K. E. (2007) Metagenomic characterization of Chesapeake Bay virioplankton. *Appl Environ Microbiol* **73:** 7629–7641.

Benson, R., and Martin, E. (1981) Effect of photosynthetic inhibitors and light dark regimes on the replication of cyanophage SM-2. *Arch Microbiol* **129:** 165–167.

Bergh, O., Børsheim, K. Y., Bratbak, G., and Heldal, M. (1989) High abundance of viruses found in aquatic environments. *Nature* (London) **340:** 467–468.

Bettarel, Y., Sime-Ngando, T., Amblard, C., and Laveran, H. (2000) A comparison of methods for counting viruses in aquatic systems. *Appl Environ Microbiol* **66:** 2283–2289.

Bisen, P. S., Audholia, S., and Bhatnagar, A. K. (1985) Mutation to resistance for virus AS-1 in the cyanobacterium *Anacystis nidulans*. *Microbios Lett* **29:** 7–13.

Bisen, P. S., Audholia, S., Bhatnagar, A. K., and Bagchi, S. N. (1986b) Evidence for lysogeny and viral resistance in the cyanobacterium *Phormidium uncinatum*. *FEMS Microbiol Lett* **52:** 225–228.

Bisen, P. S., Audholia, S., Shukla, H. O., Gupta, A., and Singh, D. P. (1988) Evidence for photosynthetic independence of viral multiplication of cyanophage LPP-1 infected cyanobacterium *Phormidium uncinatum*. *FEMS Microbiol Lett* **52:** 225–228.

Bisen, P. S., Bagchi, S. N., and Audholia, S. (1986a) Nitrate reductase activity of a cyanobacterium *Phormidium uncinatum* after cyanophage LPP-1 infection. *FEMS Microbiol Lett* **33:** 69–72.

Blashka, K. H., Hwang-Lee, L., Cohn, G., Blamire, J., and McGowan, R. E. (1982) Altered metabolite incorporation into cyanobacterium *Anacystis nidulans* as a result of cyanophage AS-1 infection. *Microbios* **34:** 141–152.

Blashka, K. H., Hwang-Lee, L., McGowan, R. E., and Blamire, J. (1983) DNA metabolism during infection of *Anacystis nidulans* by cyanophage AS-1: 3-alternative precursors and host/phage systems for PIL-DNA. *Microbios* **36:** 93–100.
Bobrovnik, S. A., Mendzhul, M. I., and Lysenko, T. G. (1978) Kinetics mechanism and thermodynamics of cyanophage A-1 adsorption on the cells of algae-host. *Biofizika* **23:** 489–493.
Bograh, A., Gingras, Y., Tajmir-Riahi, H. A., and Carpentier, R. (1997) The effects of spermine and spermidine on the structure of photosystem II proteins in relation to inhibition of electron transport. *FEBS Lett* **402:** 41–44.
Bohannan, B. J. M., and Lanski, R. E. (2000) Linking genetic change to community evolution: Insights from studies of bacteria and bacteriophage. *Ecol Lett* **3:** 362–377.
Booth, S. J. (1975) Isolation, identification and partial characterization of cyanophage LPP-2N. Ph. D. Thesis, University of Nebraska, NE, USA.
Borbely, G., Kaki, C., Gulyas, A., and Farkas, G. L. (1980) Bacteriophage infection interferes with guanosine 3′-diphosphate-5′-diphosphate accumulation induced by energy and nitrogen starvation in the cyanobacterium *Anacystis nidulans*. *J Bacteriol* **144:** 859–864.
Børsheim, K. Y., Bratbak, G., and Heldal, M. (1990) Enumeration and biomass estimation of planktonic bacteria and viuses by transmission electron microscopy. *Appl Environ Microbiol* **56:** 352–356.
Bradley, D. E. (1967) Ultrastructure of bacteriophage and bacteriocins. *Bacteriol Rev* **31:** 230–314.
Braeken, K., Moris, M., Daniels, R., Vanderleyden, J., and Michiels, J. (2006) New horizons for (p)ppGpp in bacterial and plant physiology. *Trends Microbiol* **14:** 45–54.
Bragg, J. G., and Chisholm, S. W. (2008) Modeling the fitness consequences of a cyanophage encoded photosynthetic gene. *PLoS ONE* **3:** e3550. doi:10.1371/journal.pone. 0003550
Bratbak, G., and Heldal, M. (1993) Total count of viruses in aquatic environments. In: Kemp, P. F., Sherr, B. F., and Cole, J. J. (Eds.) *Current Methods in Aquatic Microbial Ecology*. Lewis Publishers, London, UK. pp. 135–138.
Bratbak, G., Heldal, M., Norland, S., and Thingstad, T. F. (1990) Viruses as partners in spring bloom microbial trophodynamics. *Appl Environ Microbiol* **56:** 1400–1405.
Bratbak, G., Heldal, M., Thingstad, T. F., Rieman, B., and Haslund, O. H. (1992) Incorporation of viruses into the budget of microbial C-transfer. A first approach. *Mar Ecol Prog Ser* **83:** 273–280.
Breitbart, M. (2012) Marine viruses: Truth or dare. *Annu Rev Mar Sci* **4:** 425–448. doi:10.1146/annurev-marine120709–142805
Breitbart, M., and Rohwer, F. (2005) Here a virus, there a virus, everywhere the same virus? *Trends Microbiol* **13:** 278–284.
Breitbart, M., Miyake, J. H., and Rohwer, F. (2004) Global distribution of identical phage-encoded DNA sequences. *FEMS Microbiol Lett* **236:** 244–256.
Breitbart, M., Salamon, P., Andresen, B., Mahaffy, J. M., Segall, A. M., Mead, D., Azam, F., and Rohwer, F. (2002) Genomic analysis of uncultured marine viral communities. *Proc Natl Acad Sci USA* **99:** 14250–14255.
Breitbart, M., Thompson, L. R., Suttle, C. S., and Sullivan, M. B. (2007) Exploring the vast diversity of marine viruses. *Oceanography* **20:** 135–139.
Breitbart, M., Wegley, L., Leeds, S., Schoenfeld, T., and Rohwer, F. (2004) Phage community dynamics in hot springs. *Appl Environ Microbiol* **70:** 1633–1640.
Brown, R. M., Jr., Smith, K. M., and Walne, P. L. (1966) Replication cycle of the blue-green algal virus LPP-1. *Nature* (London) **212:** 729–730.
Brussaard, C. P. D. (2004) Optimization of procedures for counting viruses by flow cytometry. *Appl Environ Microbiol* **70:** 1506–1513.
Brussaard, C. P. D., Kuipers, B., and Veldhuis, M. J. W. (2005) A mesocosm study of *Phaeocystis globosa* population dynamics. I. Regulatory role of viruses in bloom control. *Harmful Algae* **4:** 859–874.
Brussaard, C. P. D., Riegman, R., Noordeloos, A. A. M., Codee, G. C., Witte, H., Kop, A. J., Nieuwland, G., van Duyl, F. C., and Bak, R. P. M. (1995) Effects of grazing sedimentation and phytoplankton cell lysis on the structure of a coastal pelagic food web. *Mar Ecol Prog Ser* **123:** 259–271.
Brussaard, C. P. D., Timmermans, K. R., Uitz, J., and Veldhuis, M. J. W. (2008) Virioplankton dynamics and virally induced phytoplankton lysis versus microzooplankton grazing southeast of the Kerguelen (Southern Ocean). *Deep-Sea Res II* **55:** 752–765.
Bryan, M. J., Burroughs, N. J., Spence, E. M., Clokie, M. R. J., Mann, N. H., and Bryan, S. J. (2008) Evidence for the intense exchange of MazG in marine cyanophages by horizontal gene transfer. *PLoS ONE* **3:** e2048. doi:10.1371/journal.pone. 0002048
Buckling, A., and Rainey, P. B. (2002) Antagonistic co-evolution between a bacterium and a bacteriophage. *Proc R Soc* B London **269:** 1385–1391.
Budinoff, C. R., Loar, S. N., LeCleir, G. R., Wilhelm, S. W., and Buchan, A. (2011) A protocol for the enumeration of aquatic viruses by epifluorescence microscopy using Anodisc™ membranes. *BMC Microbiol* **11:** 168. doi:10.1186/1471-2-180-11-168.

Campbell, L., Nolla, H. A., and Vaulot, D. (1994) The importance of *Prochlorococcus* to community structure in the central Pacific Ocean. *Limnol Oceanogr* **39:** 954–961.

Cannon, R. E., and Shane, M. S. (1972) The effect of antibiotic stress on protein synthesis in the establishment of lysogeny of *Plectonema boryanum*. *Virology* **49:** 130–133.

Cannon, R. E., Shane, M. S., and Bush, V. N. (1971) Lysogeny of a blue-green alga, *Plectonema boryanum*. *Virology* **45:** 149–153.

Cannon, R. E., Shane, M. S., and Whitaker, J. M. (1976) Interaction of *Plectonema boryanum* (Cyanophyceae) and the LPP-cyanophages in continuous culture. *J Phycol* **12:** 418–421.

Charlson, R. J., Lovelock, J. E., Andreae, M. O., and Warren, S. G. (1987) Oceanic phytoplankton, atmospheric sulphur, cloud albedo and climate. *Nature* (London) **326:** 655–661.

Chen, F., and Lu, J. (2002) Genomic sequence and evolution of marine cyanophage P60: A new insight on lytic and lysogenic phages. *Appl Environ Microbiol* **68:** 2589–2594.

Chen, F., Lu, J.-R., Binder, B. J., Liu, Y.-C., and Hodson, R. E. (2001) Application of digital image analysis and flow cytometry to enumerate marine viruses stained with SBYR Gold. *Appl Environ Microbiol* **67:** 539–545.

Chen, F., Wang, K., Kan, J., Bachoon, D. S., Lu, J., Lau, S., and Campbell, L. (2004) Phylogenetic diversity of *Synechococcus* in the Chesapeake Bay revealed by ribulose-1,5-biphosphate carboxylase-oxygenase (RuBisCO) large subunit gene (rbcL) sequences. *Aquat Microb Ecol* **136:** 153–164.

Chénard, C., and Suttle, C. A. (2008) Phylogenetic diversity of sequences of cyanophage photosynthetic gene *psbA* in marine and freshwaters. *Appl Environ Microbiol* **74:** 5317–5324.

Chisholm, S. W. (1992) Phytoplankton size. In: Falkowski, P. G., and Woodhead, A. D. (Eds.) *Primary Productivity and Biogeochemical Cycles in the Sea*. Plenum Press, New York, NY, USA. pp. 213–237.

Clokie, M. R. J., and Mann, N. H. (2006) Marine cyanophages and light. *Environ Microbiol* **8:** 2074–2082.

Clokie, M. R. J., Millard, A. D., Mehta, J. Y., and Mann, N. H. (2006) Virus isolation studies suggest short-term variations in abundance in natural cyanophage populations of the Indian Ocean. *J Mar Biol Ass* UK **86:** 499–505.

Comeau, A. M., Arbiol, C., and Krisch, H. M. (2010) Gene network visualization and quantitative synteny analysis of more than 300 marine T4-like phage scaffolds from the GOS metagenome. *Mol Biol Evol* **27:** 1935–1944. doi:10.1093/molbev/msq076

Cottrell, M. T., and Suttle, C. A. (1991) Wide-spread occurrence and clonal variation in viruses which cause lysis of a cosmopolitan, eukaryotic marine plankter, *Micromonas pusilla*. *Mar Ecol Prog Ser* **78:** 1–9.

Cseke, C. A., and Farkas, G. L. (1979) Effect of light on the attachment of cyanophage AS-1 to *Anacystis nidulans*. *J Bacteriol* **137:** 667–669.

Cseke, C. A., Balogh, A., and Farkas, G. L. (1981) Redox modulation of glucose-6-phosphate dehydrogenase in *Anacystis nidulans* and its 'uncoupling' by phage infection. *FEBS Lett* **126:** 85–88.

Daft, M. J., Begg, J., and Stewart, W. D. P. (1970) A virus of blue-green algae from freshwater habitats in Scotland. *New Phytol* **69:** 1029–1038.

Dammeyer, T., Bagby, S. C., Sullivan, M. B., Chisholm, S. W., and Frankenberg-Dinkel, N. (2008) Efficient phage-mediated pigment biosynthesis in oceanic cyanobacteria. *Curr Biol* **18:** 442–448.

Danovaro, R., and Serresi, M. (2000) Viral density and Virus-to-bacterium ratio in deep-sea sediments of the Eastern Mediterranean. *Appl Environ Microbiol* **66:** 1857–1861.

Danovaro, R., Corinaldesi, C., Dell'Anno, A., Fuhrman, J. A., Middelburg, J. J., Noble, R. T., and Suttle, C. A. (2011) Marine viruses and global climate change. *FEMS Microbiol Rev* **35:** 993–1034.

Danovaro, R., Corinaldesi, C., Filippini, M., Fischer, U. R., Gessner, M. O., Jacquet,S., Magagnini, M., and Velimirov, B. (2008) Viriobenthos in freshwater and marine sediments. *Freshwater Biol* **53:** 1186–1213.

Danovaro, R., Dell'Anno, A., Trucco, A., Serresi, M., and Vanucci, S. (2001) Determination of virus abundance in marine sediments. *Appl Environ Microbiol* **67:** 1384–1387.

DeCorte, D., Sintes, E., Winter, C., Yokokawa, T., Reinthaler, T., and Herndl, G. J. (2010) Links between viral and prokaryotic communities throughout the water column in the (sub)tropical Atlantic Ocean. *ISME J* **4:** 1431–1442. doi:10.1038/ismej.2010.65

d'Herelle, F. (1917) Sur un microbe invisible antagoniste des bacillus dysentériques. *C.R Hebd Seances Acad Sci D* **165:** 373–375.

DeLong, E. F., Preston, C. M., Mincer, T., Rich, V., Hallam, S. J., Frigaard, N.-U., Martinez, A., Sullivan, M. B., Edwards, R., Brito, B. R., Chisholm, S. W., and Karl, D. M. (2006) Community genomics among stratified microbial assemblages in the Ocean's interior. *Science*: **311:** 496–503.

Desjardins, P. R., and Olson, O. B. (1983) Viral control of nuisance cyanobacteria. *California Water Resources Contribution No. 185*. pp. 1–35.

Desplats, C., Dez, C., Tétart, F., Eleaume, H., and Krisch, H. M. (2002) Snapshot of the genome of the pseudo-T-even bacteriophage RB49. *J Bacteriol* **184:** 2789–804.

Diener, T. O. (1963) Physiology of virus-infected plants. *Annu Rev Phytopathol* **1:** 197–218.
Dorigo, U., Jacquet, S., and Humbert, J. F. (2004) Cyanophage diversity, inferred from *g20* gene analyses, in the largest natural lake in France, Lake Bourget. *Appl Environ Microbiol* **70:** 1017–1022.
Drake, L. A., Doblin, M. A., and Dobbs, F. C. (2007) Potential microbial bioinvasions via ship's ballast water, sediment and biofilm. *Mar Pollu Bull* **55:** 333–341.
Drouet, F. (1963) Ecophenes of *Schizothrix calcicola* (Oscillatoriaceae). *Proc Natl Acad Sci USA* **115:** 261–281.
Duhaime, B. M., Kottmann, R., Field, D., and Glöckner, F. O. (2011) Enriching public descriptions of marine phages using Genomic Standards Consortium MIGS standard. *Stand Genomic Sci* **4:** 1 doi:10.4056/sigs.621069
Dyall-Smith, M., Tang, S., and Bath, C. (2003) Haloarchael viruses: How diverse are they? *Res Microbiol* **154:** 309–313.
Eden, P. A., Schmidt, T. M., Blakemore, R. P., and Pace, N. R. (1991) Phylogenetic analysis of *Aquaspirillum magnetotacticum* using polymerase chain reaction-amplified 16S rRNA-specific DNA. *Int J Syst Bacteriol* **41:** 324–325.
Edwards, R. A., and Rohwer, F. (2005) Viral metagenomics. *Nat Rev Microbiol* **3:** 504–510. www. nature.com/reviews/micro.
Enav, H., Bèjá, O., and Mandel-Gutfreund, Y. (2012) Cyanophage tRNAs may have a role in cross-infectivity of oceanic *Prochlorococcus* and *Synechococcus* hosts. *ISME J* **6:** 619–628. doi:10.1038/ismej.2011.146
Fox, J. A., Booth, S. J., and Martin, E. L. (1976) Cyanophage SM-2: A blue-green algal virus. *Virology* **73:** 557–560.
Franche, C. (1987) Isolation and characterization of a temperate cyanophage for tropical *Anabaena* strain. *Arch Microbiol* **148:** 172–177.
Frederickson, C. M., Short, S. M., and Suttle, C. A. (2003) The physical environment affects cyanophage communities in British Columbia inlets. *Microb Ecol* **46:** 348–357.
Friga, G. M., Borbely, G., and Farkas, G. L. (1981) Accumulation of guanosine tetraphosphate (ppGpp) under nitrogen starvation in *Anacystis nidulans*, a cyanobacterium. *Arch Microbiol* **129:** 341–343.
Fuhrman, J. A. (1992) Bacterioplankton roles in cycling of organic matter: The microbial food web. In: Falkowski, P. G., and Woodhead, A. D. (Eds.) *Primary Productivity and Biogeochemical Cycles in the Sea*. Plenum Press, New York, NY, USA. pp. 361–383.
Fuhrman, J. A. (1999) Marine viruses and their biogeochemical and ecological effects. *Nature* (London) **399:** 541–548.
Fuhrman, J. A., and Noble, R. T. (1995) Viruses and protists cause similar bacterial mortality in coastal seawater. *Limnol Oceanogr* **40:** 1236–1242.
Fuhrman, J. A., and Suttle, C. A. (1993) Viruses in marine planktonic systems. *Oceanography* **6:** 51–63.
Fuller, N. J., Wilson, W. H., Joint, I. R., and Mann, N. H. (1998) Occurrence of a sequence in marine cyanophages similar to that of T4 g20 and its application to PCR-based detection and quantification techniques. *Appl Environ Microbiol* **64:** 2051–2060.
Galperin, M. Y., Moroz, O. V. , Wilson, K. S., and Murzin, A. G. (2006) House cleaning, a part of good housekeeping. *Mol Microbiol* **59:** 5–19.
Gan, E. B., Gui, J. F., and Zhang, Q.-Y. (2011) A novel cyanophage with cyanobacterial non-bleaching protein A in the genome. *J Virol* **85:** 305–314. doi:10.1128/VI.06282-11JVI06282-11
Garza, D. R., and Suttle, C. A. (1998) The effect of cyanophages on the mortality of *Synechococcus* and selection for UV-resistant viral communities. *Microb Ecol* **36:** 281–292.
Ginzberg, D., Padan, E., and Shilo, M. (1968) Effect of cyanophage infection on CO_2 photoassimilation in *Plectonema boryanum*. *J Virol* **2:** 695–701.
Giovannoni,, S. J., Britschgi, T. B., Moyer, C. L., and Field, K. G. (1990) Genetic diversity in Sargasso Sea bacterioplankton. *Nature* (London) **345:** 60–63.
Göbler, C. J., Hutchins, D. A., Fisher, N. S., Cosper, E. M., and Sanudo-Wilhelmy, S. A. (1997) Release and bioavailability of C, N, P, Se and Fe following viral lysis of a marine chrysophyte. *Limnol Oceanogr* **42:** 1492–1504.
Goericke, R., and Repeta, D. J. (1992) The pigments of *Prochlorococcus marinus*: The presence of divinyl chlorophyll a and b in a marine prokaryote. *Limnol Oceanogr* **37:** 425–433.
Goldsmith, D. B., Crosti, G., Dwivedi, B., McDaniel, L. D., Varsani, A., Suttle, C. A., Weinbauer, M. G., Sandaa, R. A., and Breitbart, M. (2011) Development of *phoH* as a novel signature gene for assessing marine phage diversity. *Appl Environ Microbiol* **77:** 7730–7739.
Goldstein, D. A., Bendet, I. J., Lauffer, M. A., and Smith, K. M. (1967) Some biological and physicochemical properties of blue-green algal virus LPP-1. *Virology* **32:** 601–613.
Goryushin, V. A., and Shainskaya, O. A. (1984) Isolation and characterization of a strain of the cyanophage AS-1: AS-1K. *Microbiologiya* **46:** 70–75.
Granhall, U., and Hofsten, A. V. (1969) The ultrastructure of a cyanophage attack on *Anabaena variabilis*. *Physiol Plant* **22:** 713–722.
Gromov, B. V. (1983) Cyanophages. *Ann Microbiol* (*Inst. Pasteur*) **134:** 43–59.

Gromov, B. V., and Koz'yakov, S. Y. (1970) Study of the properties of the interrelationship between populations of the blue-green alga *Plectonema boryanum* Gom, and the cyanophages LPP-1. *Vestn Leningrad Univ Ser Biol* **1:** 128–135.

Gross, M., Marianovsky, I., and Glaser, G. (2006) MazG-a regulator of programmed cell death in *Escherichia coli*. *Mol Microbiol* **59:** 590–601.

Guixa-Boixareu, N., Calderon-Paz, J., Heldal, M., Bratbak, G., and Pedros-Alio, C. (1996) Viral lysis and bacteriovory as prokaryotic loss factors along a salinity gradient. *Aquat Microb Ecol* **11:** 215–227.

Gusev, M. V., Zhivepistseva, I. V., Matorin, D. N., Nikitino, K. A., and Timofeev, K. N. (1980a) Degradation of *Anacystis nidulans* cells during treatment with cyanophage AS-1. *Vopr Virusol* **3:** 330–333.

Gusev, M. V., Zhivepistseva, I. V., Matorin, D. N., Nikitino, K. A., and Timofeev, K. N. (1980b) Sequence of degradation process in *Anabaena variabilis* photosynthetic apparatus affected by cyanophage A-1L. *Fiziol Rast* (Mosc) **27:** 470–476.

Hall, M. J., Wharam, S. D., Weston, A., Cardy, D. L. N., and Wilson, W. H. (2002) Use of signal-mediated amplification of RNA technology (SMART) to detect marine cyanophage DNA. *BioTechniques* **32:** 604–611.

Hambly, E. (2002) *Molecular Studies of Marine Cyanophages.* Ph.D. Thesis. University of Warwick, Coventry, UK.

Hambly, E., and Suttle, C. A. (2005) The viriosphere, diversity, and genetic exchange within phage communities. *Curr Opin Microbiol* **8:** 444–450.

Hambly, E., Tétart, F., Desplats, C., Wilson, W. H., Krisch, H. M., and Mann, N. H. (2001) A conserved genetic module that encodes the major virion components in both coliphage T4 and the marine cyanophages S-PM2. *Proc Natl Acad Sci USA* **98:** 11411–11416.

Handelsman, J. (2004) Metagenomics: Application of genomics to uncultured organisms. *Microbiol Mol Biol Rev* **69:** 669–685.

Handelsman, J., Rondon, R., Brady, S. F., Clardy, J., and Goodman, R. M. (1998) Molecular biological access to the chemistry of unknown soil microbes: A new frontier for natural products. *Chem Biol* **5:** R245–R249.

Hannen, E. J., Zwart, G., van Agterveld, M. P., Gons, H. J., Ebert, J., and Laanbrock, H. J. (1999) Bacterial and eukaryotic community structure after mass lysis of filamentous cyanobacteria associated with viruses. *Appl Environ Microbiol* **65:** 795–801.

Hara, S.,Terauchi, K., and Koike, I. (1991) Abundance of viruses in marine waters: Assessment of epifluorescence and transmission electron microscopy. *Appl Environ Microbiol* **57:** 2731–2734.

He, Q., Dolganov, N., Björkman, O., and Grossman, A. R. (2001) The high light inducible polypeptides in *Synechocystis* PCC6803: expression and function in high light. *J Biol Chem* **276:** 306–314.

Heldal, M., and Bratbak, G. (1991) Production and decay of viruses in aquatic environments. *Mar Ecol Prog Ser* **72:** 205–212.

Hendrix, R. W. (2003) Bacteriophage genomics. *Curr Opin Microbiol* **6:** 506–511.

Hennes, K. P., and Suttle, C. A. (1995) Direct counts of viruses in natural sea water and laboratory cultures by epifluorescence microscopy. *Limnol Oceanogr* **40:** 1054–1059.

Hennes, K. P., Suttle, C. A., and Chan, A. M. (1995) Fluorescently labeled virus probes show that natural virus populations can control the structure of marine microbial communities. *Appl Environ Microbiol* **61:** 3623–3627.

Hess, W. R., Partensky, F., Van Der Staay, G. W. M., Garcia-Fernandez, J. M., Borner, T., and Vaulot, D. (1996). Coexistence of phycoerythrin and a chlorophyll a/b antenna in a marine prokaryote. *Proc.Natl Acad Sci USA* **93:** 11126–11130.

Hewson, I., O'Neil, J. M., and Dennison, W. C. (2001) Virus-like particles associated with *Lyngbya majuscula* (Cyanophyta, Oscillatoriaceae) bloom decline in Moreton Bay, Australia. *Aquat Microb Ecol* **25:** 207–213.

Hill, R. W., Wb, A., Cottrell, M. T., and Dacey, J. W. H. (1998) Virus-mediated total release of dimethylsulfoniopropionate from marine phytoplankton: a potential climate process. *Aquat Microb Ecol* **14:** 1–6.

Hirons, G. T., Fawcett, J. J., and Crissman, H. A. (1994) TOTO and YOYO: New very bright fluorochromes for DNA content analysis by flow cytometry. *Cytometry* **15:** 129–140.

Honjo, M., Matsui, K., Ueki, M., Nakamura, R., Fuhrman, J. A., and Kawabata, Z. (2006) Diversity of virus-like agents killing *Microcystis aeruginosa* in a hypereutrophic pond. *J Plankton Res* **28:** 407–412.

Hu, N.-T., Thiel, T., Giddings, T. H., and Wolk, C. P. (1981) New *Anabaena* and *Nostoc* cyanophages from sewage settling ponds. *Virology* **114:** 236–246.

Hwang-Lee, L., Blashka, K. H., Blamire, J., and McGowan, R. E. (1982a) DNA metabolism during infection of *Anacystis nidulans* by cyanophage AS-1: 1. Identification of a unique species of DNA. *Microbios* **35:** 49–62.

Hwang-Lee, L., Blashka, K. H., McGowan, R. E., and Blamire, J. (1982b) DNA metabolism during infection of *Anacystis nidulans* by cyanophage AS-1: 2. Analysis of DNA synthesis patterns and discrimination of precursor utilization. *Microbios* **36:** 111–118.

Hwang-Lee, L., Cohn, G., Blashka, K. H., Blamire, J., and McGowan, R. E. (1983a) DNA metabolism during infection of *Anacystis nidulans* by cyanophage AS-1: 4. Studies on source of PIL-DNA. *Microbios* **36:** 101–112.

Hwang-Lee, L., Cohn, G., Blashka, K. H., McGowan, R. E., and Blamire, J. (1983b) DNA metabolism during infection of *Anacystis nidulans* by cyanophage AS-1: 5. Biophysical characterization of phage-induced light DNA. *Microbios* **36:** 191–208.

Inoue, T., Matsuzaki, S., and Tanaka, S. (1995) A 26 kDa outer membrane protein, OmpK, common to *Vibrio* species is the receptor for broad-host-range vibriophage, KVP40. *FEMS Microbiol Lett* **125:** 101–105.
Jameson, E., Mann, N. H., Joint, H., Sambles, C., and Mühling, M. (2011) The diversity of cyanomyovirus populations along a North-South Atlantic Ocean transect. *ISME J* **5:** 1713–1721. doi:10.1038/ismej.2011.54
Jenkins, C. A., and Hayes, P. K. (2006) Diversity of cyanophages infecting the heterocystous filamentous cyanobacterium *Nodularia* isolated from the brackish Baltic Sea. *J Mar Biol Ass* UK **86:** 529–536.
Jia, Y., Ying, S., Shan, J., Millard, A., Clokie, M. R., and Mann, N. H. (2010) Light-dependent adsorption of photosynthetic cyanophages to *Synechococcus* sp. WH7803. *FEMS Microbiol Lett* **310:** 120–126.
Jia, Z., Ishihara, R., Nakajima, Y., Asakawa, S., and Kimura, M. (2006) Molecular characterization of T4-type bacteriophages in a rice field. *Environ Microbiol* **9:** 1091–1096. doi:10.1111/j.1462-2920.2006.01207.x
Jiang, S., Steward, G., Jellison, R., Chu, W., and Choi, S. (2004) Abundance, distribution and diversity of viruses in alkaline, hypersaline Mono Lake, California. *Microb Ecol* **47:** 30–40.
Joint, I. R., and Pomroy, A. J. (1983) Production of picoplankton and small nanoplankton in the Celtic Sea. *Mar Biol* **77:** 19–27.
Kao, C. C., Green, S., Stein, B., and Golden, S. S. (2005) Diel infection of a cyanobacterium by a contractile bacteriophage. *Appl Environ Microbiol* **71:** 4276–4279.
Kashyap, A. K., and Singh, S. (1989) Changes in photoelectron transport activity in cyanophage N-1 infected cells of *Nostoc muscorum. Current Microbiol* **18:** 151–155.
Kashyap, A. K., Rai, A. N., and Singh, S. (1988) Effect of cyanophage N-1 development on nitrogen metabolism of cyanobacterium *Nostoc muscorum. FEMS Microbiol Lett* **51:** 145–148.
Kaushik, B. D., and Venkataraman, G. S. (1973) Isolation of a new cyanophage TAuHN-1. *Curr Sci* **42:** 395–396.
Kepner, R. J., Wharton, R. J., and Suttle, C. (1998) Viruses in Antarctic lakes. *Limnol Oceanogr* **42:** 1754–1761.
Khudyakov, I. Ya. (1977) Characteristics of a new cyanophage S-2(L) lysing the unicellular cyanobacterium belonging to *Synechococcus* genus. *Mikrobiologiya.* **46:** 547–553.
Khudyakov, I. Ya., and Gromov, B. V. (1973) The temperate cyanophage A-4 (L) of the blue-green alga *Anabaena variabilis. Mikrobiologiya* **42:** 904–907.
Khudyakov, I. Ya., and Matveyev, A. V. (1981) Comparative characteristics of cyanophages of unicellular cyanobacterium *Synechococcus* sp. 698. In: *Viruses of Microorganisms.* Academy of Nauk SSSR, Pushchino. pp. 104–106.
Khudyakov, I. Ya., and Matveyev, A. V. (1982) New cyanophages S-4(L) and S-5(L) causing lysis of the cyanobacterium *Synechococcus elongatus. Mikrobiologiya* **51:** 102–106.
Khudyakov, I. Ya., Kirnos, M. D., Alexandrushkina, N. I., and Vanyushin, B. F. (1978) Cyanophage S-2L contains DNA with 2,6-diaminopurine substituted for adenine. *Virology* **88:** 8–18.
Kim, M., and Choi, Y.-K. (1994) A new *Synechococcus* cyanophage from a reservoir in Korea. *Virology* **204:** 338–342.
Koz'yakov, S. Ya. (1977) Cyanophages of the series A(L) specific for the blue-green alga, *Anabaena variabilis.* Eksperimental' naya al'gologiya. *Trudi Petergof Biol Inst* (Leningrad State University) **25:** 151–175.
Koz'yakov, S. Ya. (1981) Characteristics of a new cyanophage for the cyanobacterium *Synechococcus schmidlea. Mikrobiologiya.* **50:** 543–549.
Koz'yakov, S.Ya. (1982). Peculiarities of a new cyanophage specific for the cyanobacterium *Synechococcus schmidlea. Microbiology* (New York) **50:** 395–401.
Koz'yakov S. Ya., Gromov B. V., Khudyakov I. Y. (1972) A-1(L) cyanophage of the blue-green alga *Anabaena variabilis. Mikrobiologiya* **41:** 555–559.
Krauss, R. W. (1961) Algae and Metropolitan Wastes. Trans of 1960 Seminar. Tech. Report W60-3, Robert A. Taft Engineering Corporation. US Public Health Service, Cincinnati, Ohio, USA. p. 40.
Kurmayer, R., and Kutzenberg, T. (2003) Application of real-time PCR for quantification of microcystin genotypes in a population of the toxic cyanobacterium *Microcystis* sp. *Appl Environ Microbiol* **69:** 6723–6730.
Kuno, S., Yoshida, T., Kamikawa, R., Hosada, N., and Sako, Y. (2010) The distribution of a phage-related insertion sequence element in the cyanobacterium, *Microcystis aeruginosa. Microbes Environ* **25:** 295–301. doi:10.1264/jsme2.ME.10125
La Roche, J., Van Der Staay, G. W. M., Partensky, F., Ducret, A., Aebersold, R., Li, R., Golden, S. S., Hiller, R. G., Wrench, P. M., Larkum, A. W. D., and Green, B. R. (1996). Independent evolution of the prochlorophyte and green plant chlorophyll *a/b* light-harvesting proteins. *Proc. Natl. Acad. Sci. USA* **93:** 15244–15248.
Lambert, G. R., and Carr, N. G. (1984) Resistance of DNA from filamentous and unicellular cyanobacteria to restriction endonuclease cleavage. *Biochem Biophys Acta* **781:** 45–55.
Lane, D. J., Pace, B., Olsen, G. J., Stahl, D. J., Sogin, M. L., and Pace, N. R. (1985) Rapid determination of 16S ribosomal RNA sequences for phylogenetic analyses. *Proc Natl Acad Sci USA* **82:** 6955–6959.
Leach, J. E., Lee, K. W., Benson, R. L., and Martin, E. L. (1980) Ultrastructure of the infection cycle of cyanophage SM-2 in *Synechococcus elongatus* (Cyanophyceae). *J Phycol* **16:** 307–310.

Lehmann, J., Volkl, W., Udvardy, J., Borbely, G., Sivok, S., and Farkas, G. L. (1979) Characterization of a ribonuclease from *Anacystis nidulans* infected with cyanophage AS-1. *Phytochemistry* **18:** 541–544.

Lennon, J. T., Khatana, S. A. M., Marston, M. F., and Martiny, J. B. H. (2007) Is there a cost of virus resistance in marine cyanobacteria? *ISME J* **1:** 300–312. www.nature.com/ismej

Lewin, R. A. (1963) Rod-shaped particles of *Saprospira*. *Nature* (London) **198:** 103–104.

Li, W. K. W., Subba Rao, D. V., Harrison W. G., Smith, J. C., Cullen, J. J., Irwin, B., and Platt, T. (1983) Autographic picoplankton in the tropical ocean. *Science* **219:** 292–295.

Lichtle, C., Thomas, J. C., Spilar, A., and Partensky, F. (1995). Immunological and ultrastructural characterization of the photosynthetic complexes of the prochlorophyte *Prochlorococcus* (oxychlorobacteria) 1. *J Phycol* **31:** 934–941.

Lindell, D., Jaffe, J. D., Johnson, Z. I., Church, G. M., and Chisholm, S. W. (2005) Photosynthesis genes in marine viruses yield proteins during host infection. *Nature* (London) **438:** 86–89.

Lindell, D., Jaffe, J. D., Coleman, M. L., Futschik, M. E., Axman, I. M., Rector, T., Kettler, G., Sullivan, M. B., Steen, R., Hess, W. R., Church, G. M., and Chisholm, S. W. (2007) Genome wide expression dynamics of a marine virus and host reveal features of co-evolution. *Nature* (London) **449:** 83–86.

Lindell, D., Sullivan, M. B., Johnson, Z. I., Tolonen, A. C., Rohwer, F., and Chisholm, S. W. (2004) Transfer of photosynthesis genes to and from *Prochlorococcus* viruses. *Proc Natl Acad Sci USA* **101:** 11013–11018.

Liu, H., Campbell, L., Landry, M. R., Nolla, H. A., Brown, S. L., and Constantinou, J. (1998) *Prochlorococcus* and *Synechococcus* growth rates and contributions to production in the Arabian Sea during the 1995 Southwest and Northeast monsoons. *Deep Sea Res* II **45:** 2327–2352.

Liu, H., Nolla, H. A., and Campbell, L. (1997) *Prochlorococcus* growth rate and contribution to primary production in the equatorial and subtropical North Pacific Ocean. *Aquat Microb Ecol* **12:** 39–47.

Liu, X., Shi, M., Kong, S., Gao, Y., and An, C. (2007) Cyanophage Pf-WMP4, a T7 like phage infecting the freshwater cyanobacterium *Phormidium foveolarum*: Complete genome sequence and DNA translation. *Virology* **366:** 28–39.

Long, A., McDaniel, L. D., Mobberley, J., and Paul, J. H. (2008) Comparison of lysogeny (prophage induction) in heterotrophic bacterial and *Synechococcus* populations in the Gulf of Mexico and Mississippi river plume. *ISME J* **2:** 132–144.

Lu, J., Chen, F., and Hodson, R. E. (2001) Distribution, isolation, host specificity and diversity of cyanophages infecting marine *Synechococcus* spp. in river estuaries. *Appl Environ Microbiol* **67:** 3285–3290.

Luftig, R., and Haselkorn, R. (1967) Morphology of a virus of blue-green algae and properties of its deoxyribonucleic acid. *J Virol* **1:** 344–361.

Luftig, R., and Haselkorn, R. (1968a) Studies on the structure of blue-green algae virus LPP-1. *Virology* **34:** 644–674.

Luftig, R., and Haselkorn, R. (1968b) Comparison of blue-green algae virus LPP-1 and the morphologically related viruses GIII and coliphage T7. *Virology* **34:** 675–678.

Mackenzie, J. J., and Haselkorn, R. (1972a) Physical properties of blue-green algal virus SM-1 and its DNA. *Virology* **49:** 497–504.

Mackenzie, J. J., and Haselkorn, R. (1972b) An electron microscope study of infection by the blue-green algal virus SM-1. *Virology* **49:** 505–516.

Mackenzie, J. J., and Haselkorn, R. (1972c) Photosynthesis and the development of blue-green algal virus SM-1. *Virology* **49:** 517–521.

Magnusson, L. U., Farewell, A., and Nystrom, T. (2005) ppGpp: A global regulator in *Escherichia coli*. *Trends Microbiol* **13:** 236–242.

Malin, G., Wilson, W. H., Bratbak, G., Liss, P. S., and Mann, N. H. (1998) Elevated production of dimethylsulfide resulting from viral infection of cultures of *Phaeocystis pouchetii*. *Limnol Oceanogr* **43:** 1389–1393.

Mamkayeva, K. A., Al-Musavi, R. A., and Gromov, B. V. (1980) The effect of photosynthesis in the ultrastructure of a cyanophage during its intracellular growth. *Mikrobiologiya* **49:** 631–633.

Manage, P. M., Kawabata, Z., and Nakano, S.-I. (1999) Seasonal changes in densities of cyanophages infectious to *Microcystis aeruginosa* in hypereutrophic pond. *Hydrobiologia* **411:** 211–216.

Mann, N. H. (2003) Phages of marine cyanobacterial picoplankton. *FEMS Microbiol Rev* **27:** 17–34.

Mann, N. H. (2005) The third age of phage. *PLos Biol* **3:** e182. doi:10.1371/journal.pbio.0030182

Mann, N. H., Ahlgren, N., Moffett, J. W., and Chisholm, S. W. (2002) Copper toxicity and cyanobacteria ecology in the Sargasso Sea. *Limnol Oceanogr* **47:** 976–988.

Mann, N., Carr, N. G., and Midgley, J. E. (1975) RNA synthesis and the accumulation of guanine nucleotides during growth shift down in the blue-green alga *Anacystis nidulans*. *Biochim Biophys Acta* **402:** 41–50.

Mann, N. H., Clokie, M. R. J., Millard, A. J., Cook, A., Wilson, W. H., Wheatley, P. J., Letarov, A., and Krisch, H. M. (2005) The genome of S-PM2, a "photosynthetic" T4-type bacteriophage that infects marine *Synechococcus* strains. *J Bacteriol* **187:** 3188–3200.

Mann, N. H., Cook, A., Millard, A., Bailey, S., and Clokie, M. (2003) Marine ecosystems: Bacterial photosynthesis genes in a virus. *Nature* (London) **424:** 741.

Marie, D., Brussaard, C. P. D., Thyrhaug, R., Bratbak, G., and Vaulot, D. (1999) Enumeration of marine viruses in culture and natural samples by flow cytometry. *Appl Environ Microbiol* **65:** 45–52.

Marston, M., and Amrich, C. G. (2009) Recombination and microdiversity in coastal marine cyanophages. *Environ Microbiol* **11:** 2893–2903.

Marston, M. F., and Sallee, J. L. (2003) Genetic diversity and temporal variation in the cyanophage community infecting marine *Synechococcus* species in Rhode Island's coastal waters. *Appl Environ Microbiol* **69:** 4639–4647.

Mathews, R. E. F. (1979) Classification and nomenclature of viruses. Third report of the International Committee on Taxonomy of viruses. *Intervirology* **12:** 129–296.

Mayer, J. A., and Taylor, F. J. R. (1979) A virus which lyses the marine nanoflagellate *Micromonas pusilla*. *Nature* (London) **281:** 299–301.

McDaniel, L., and Paul, J. H. (2005) Effect of nutrient addition and environmental factors on prophage induction in natural populations of marine *Synechococcus* species. *Appl Environ Microbiol* **71:** 842–850.

McDaniel, L., dela Rosa, M., and Paul, J. H. (2006) Temperate and lytic cyanophages from the Gulf of Mexico. *J Mar Biol Ass* UK **86:** 517–527.

McDaniel, L., Houchin, L. A., Williamson, S. J., and Paul, J. H. (2002) Lysogeny in marine *Synechococcus*. *Nature* (London) **415:** 496. www.nature.com

Mendzhul, M. I., Karnaukhov, V. N., Martsenyuk, P. P., and Grigoryas, V. A. (1982) Features of interaction between virus and cell in cyanophage-cyanobacteria system. *Mikrobiol* ZH (Kiev) **44:** 63–70.

Mendzhul, M. I., Lysenko, T. G., and Semenova, G. A. (1983) Biological and physicochemical properties of N-2. *Mikrobiol* ZH (Kiev) **45:** 60–66.

Mendzhul, M. I., Lysenko, T. G., Shainkskaya, O. A., and Gerashchenko, I. V. (1988) Changes in the specific activity of glucose-6-phosphate dehydrogenase in cyanophage LPP-3-cyanobacterium *Plectonema boryanum*. *Mikrobiol* ZH (Kiev) **50:** 88–89.

Middelboe, M., Jorgensen, N. O. G., and Kroer, N. (1996) Effects of viruses on nutrient turnover and growth efficiency of non-infected marine bacterioplankton. *Appl Environ Microbiol* **62:** 1991–1997.

Middelboe, M., Riemann, L., Steward, G. F., Hansen, V., and Nybroe, O. (2003) Virus-induced transfer of organic carbon between marine bacteria in a model community. *Aquat Microb Ecol* **33:** 1–10.

Millard, A., Clokie, M. R., Shub, D. A., and Mann, N. H. (2004) Genetic organization of the *psbAD* region in phages infecting marine *Synechococcus* strains. *Proc Natl Acad Sci USA* **101:** 11007–11012.

Millard, A. D., Gierga, G., Clokie, M. R., Evans, D. J., Hess, W. R., and Scanlan, D. J. (2010) An antisense RNA in a lytic cyanophage links *psbA* to a gene encoding a homing endonuclease. ISME J **4:** 1121–1135. doi: 10.1038/ismej.2010.43

Millard, A. D., Zwirglanaier, K., Downey, M. J., Mann, N. H., and Scanlan, D. J. (2009) Comparative genomics of marine cyanomyoviruses reveals the widespread occurrence of *Synechococcus* host genes localized to a hyperplastic region: Implications for mechanisms of cyanophage evolution. *Environ Microbiol* **11:** 2370–2387.

Miller, E. S., Heidelberg, J. F., Eisen, J. A., Nelson, W. C., Durkin, A. S., Ciecko, A., Feldblyum, T. V., White, O., Paulsen, I. T., Nierman, W. C., Lee, J., Szcypinski, B., and Fraser, C. M. (2003) Complete genome sequence of the broad-host-range vibriophage KVP40: Comparative genomics of a T4-related bacteriophage. *J Bacteriol* **185:** 5220–5223.

Moisa, I., Sotropa, E., and Velehorschi, V. (1981a) Investigations on the presence of cyanophages in fresh and seawaters of Romania. *Rev Roum Med Virol* **32:** 127–132.

Moisa, I., Sotropa, E., and Velehorschi, V. (1981b) Sequence of morphological alterations in blue-green algae in the course of cyanophage infection. *Virologie* **32:** 133–137.

Moore, L. R., Rocap, G., and Chisholm, S. W. (1998) Physiology and molecular phylogeny of co-existing *Prochlorococcus* ecotypes. *Nature* (London) **393:** 464–467.

Mühling, M., Fuller, N. J., Millard, A., Somerfield, P. J., Marie, D., Wilson, W. H., Scanlan, D. J., Post, A. F., Joints, I., and Mann, N. H. (2005) Genetic diversity of marine *Synechococcus* and co-occurring cyanophage communities. Evidence for viral control of phytoplankton. *Environ Microbiol* **7:** 499–508.

Mulo, P., Eloranta, T., Aro, E. M., and Maenpaa, P. (1998) Disruption of a spe-like open reading frame alters polyamine content and *psbA*-2 mRNA stability in the cyanobacterium *Synechocystis* sp. PCC 6803. *Bot Acta* **111:** 71–76.

Muradov, M., Cherkasov, G. V., Akhmedova, D. U., and Khalmuradov, A. G. (1990a) A new temperate cyanophage NP-1T lysogenizing cyanobacterial cultures belonging to the genera *Nostoc* and *Plectonema*. *Microbiologya* **59:** 1038–1045.

Muradov, M., Cherkasov, G. V., Akhmedova, D. U., Kamilova, F. D., Mukhamedov, R. S., Abdukarimov, A. A., and Khalmuradov, A. G. (1990b) Comparative study of NP-1T cyanophage, which lysogenize nitrogen-fixing bacteria of the genera *Nostoc* and *Plectonema*. *Microbiologiya* **59:** 558–563.

Noble, R. T., and Fuhrman, J. A. (1998) Use of SYBR Green I for rapid epifluorescence counts of marine viruses and bacteria. *Aquat Microb Ecol* **14:** 113–118.

O'Connell, D. (2005) Viral genomics: Small is beautiful. *Nat Rev Microbiol* **3:** 520.

Ohki, K. (1999) A possible role of temperate phage in the regulation of *Trichodesmium* biomass. *Bull Inst Oceanogr Monaco* **19:** 287–292.

Ohki, K., and Fujita, Y. (1996) Occurrence of a temperate cyanophage lysogenizing the marine cyanophyte *Phormidium persicinum*. *J Phycol* **32:** 365–370.

Okunishi, S., Yoshimura, N., Maeda, H., Yoshikawa, T., and Sakata, T. (2002) Isolation and characterization of a cyanophage infecting cyanobacterial picoplankton, *Synechococcus* sp. from Kagoshima Bay. *Microbes Environ* **17:** 148–152.

Okunishi, S., Yoshimura, N., Maeda, H., Yoshikawa, T., and Sakata, T. (2003) A virulent cyanophage affects the seasonal abundance of cyanobacterial picoplankton (*Synechococcus* sp.) in Kagoshima Bay. *Microbes Environ* **18:** 10–15.

Olivera, A. R., Mudd, J. B., and Desjardins, P. R. (1982) AS-1 Cyanophage adsorption to liposomes. *J Gen Virol* **61:** 153–156.

Oren, A., Bratba, G., and Heldal, M. (1997) Occurrence of virus-like particles in the Dead Sea. *Extremophiles* **1:** 143–149.

Ortmann, A. C., Lawrence, J. E., and Suttle, C. A. (2002) Lysogeny and lytic viral production during a bloom of the cyanobacterium *Synechococcus* spp. *Microb Ecol* **43:** 225–231.

Padan, E., and Shilo, M. (1969) Distribution of cyanophages in natural habitats. *Verh Int Verein Limnol* **17:** 747–751.

Padan, E., and Shilo, M. (1973) Cyanophages-Viruses attacking blue-green algae. *Bacteriol Rev* **37:** 343–370.

Padan, E., Ginzburg, D., and Shilo, M. (1970) The reproductive cycle of cyanophage LPP1-G in *Plectonema boryanum* and its dependence on photosynthetic and respiratory system. *Virology* **40:** 514–521.

Padan, E., Raboy, B., and Shilo, M. (1971a) Endogenous dark respiration of the blue-green alga, *Plectonema boryanum. J Bacteriol* **106:** 45–50.

Padan, E., Rimon, A., Ginzberg, D., and Shilo, M. (1971b) A thermosensitive cyanophage (LPP1-G) attacking the blue-green alga *Plectonema boryanum*. *Virology* **45:** 773–776.

Padan, E., Shilo, M., and Kislev, N. (1967) Isolation of "cyanophages" from freshwater ponds and their interaction with *Plectonema boryanum*. *Virology* **32:** 234–246.

Padan, E., Shilo, M., and Oppenheim, A. B. (1972) Lysogeny of the blue-green alga *Plectonema boryanum* by LPP2-SPI cyanophage. *Virology* **47:** 525–526.

Padhy, R. N., and Singh, P. K. (1977a) Effects of physical and chemical agents on the blue-green algal virus N-1. *Acta Virol* **21:** 264–267.

Padhy, R. N., and Singh, P. K. (1977b) Effect of temperature on adsorption and one-step growth of the *Nostoc* virus N-1. *Arch Microbiol* **155:** 163–167.

Padhy, R. N., and Singh, P. K. (1978a) Stabilizing effects of metallic ions in the blue-green algal virus N-1. *Biochem Physiol Pflanzen* **173:** 188–192.

Padhy, R. N., and Singh, P. K. (1978b) Effect of host aging, ions and pH on the adsorption of cyanovirus N-1 to *Nostoc muscorum*. *Arch Microbiol* **156:** 289–292.

Partensky, F., Hess, W. R., and Vaulot, D. (1999) *Prochlorococcus*, a marine photosynthetic prokaryote of global significance. *Microbiol Mol Biol Rev* **63:** 106–127.

Paul, J. H., and Sullivan, M. B. (2005) Marine phage genomics: What we learned? *Curr Opin Biotechnol* **16:** 299–307.

Paul, J. H., Jiang, S. C., and Rose, J. B. (1991) Concentration of viruses and dissolved DNA from aquatic environments by vortex flow filtration. *Appl Environ Microbiol* **57:** 2197–2204.

Pedros-Alio, C., Calderon, Paz, J., MacLean, M., Medina, G., Marrase, C., Gasol, J., and Boixereu, N. (2000a) The microbial food web along salinity gradients. *FEMS Microbiol Ecol* **32:** 143–155.

Pedros-Alio, C., Calderon-Paz, J., and Gasol, J. (2000b) Comparative analysis shows that bacteriovory, not viral lysis, controls the abundance of heterotrophic prokaryotic plankton. *FEMS Microbiol Ecol* **32:** 157–165.

Pedulla, M. L., Ford, M. E., Houtz, J. M., Karthikeyan, T., Wadsworth, C., Lewis, J. A., Jacobs-Sera, D., Falbo, J., Gross, J., Pannunzio, N. R., Brucker, W., Kumar, V., Kandasamy, J., Keenan, L., Bardarov, S., Kirakov, J., Lawrence, J. G., Jacobs, W. R. Jr., Hendrix, R. W., and Hatfull, G. F. (2003) Origins of highly mosaic mycobacteriophage genomes. *Cell* **113:** 171–182.

Philosof, A., Battchikova, N., Aro, E-M., and Béjà, O. (2011) Marine cyanophages: Tinkering with the electron transport chain. *ISME J* **5:** 1568–1570.

Phlips, E. J., Monegue, R. L., and Aldridge, F. J. (1990) Cyanophages which impact bloom-forming cyanobacteria. *J Aquat Plant Manag* **28:** 92–97.

Pollard, P. C., and Young, L. M. (2009) Lake viruses lyse cyanobacteria, *Cylindrospermopsis raciborskii*, enhances filamentous-host dispersal in Australia. *Acta Oecologica* **36:** 114–119.

Pope, W. H., Weigele, P. R., Chang, J., Pedulla, M. L., Ford, M. E., Houtz, J. M., Jiang, W., Chiu, W., Hatfull, G. F., Hendrix, R. W., and King, J. (2007). Genome sequence, structural proteins, and capsid organization of the cyanophages Syn5: A "horned" bacteriophage of marine *Synechococcus*. *J Mol Biol* **368:** 966–981.

Proctor, L. M., and Fuhrman, J. A. (1990) Viral mortality of marine bacteria and cyanobacteria. *Nature* (London) **343:** 60–62.

Proctor, L. M., and Fuhrman, J. A. (1991) Roles of virus infection in organic particle flux. *Mar Ecol Prog Ser* **69:** 133–142.

Proctor, L. M., and Fuhrman, J. A. (1992) Mortality of marine bacteria in response to enrichments of the virus size fraction from seawater. *Mar Ecol Prog Ser* **87:** 283–293.

Rautenstein, Y. J., Stefanov, S. B., Smiernova, E. I., and Khadjimurti, L. N. (1966) On defective lysogeny in *Actinomyces streptomycini* Kras. *Microbiology* **30:** 860–863.

Raytcheva, D. A., Haase-Pettingell, C., Piret, J. H., and King, J. (2010) Intracellular assembly of cyanophage Syn5 proceeds through a scaffold-containing procapsid. *J Virol* doi:10.1128/JVI.01601-10.

Rimon, A., and Oppenheim, A. B. (1975) Heat induction of the blue-green alga *Plectonema boryanum* lysogenic for cyanophages SPI cts1. *Virology* **64:** 454–463.

Rocap, G., Larimer, F. W., Lamerdin, J., Malfatti, S., Chain, P., Ahlgren, N. A., Arellano, A., Coleman, M., Hauser, L., Hess, W. R., Johnson, Z. I., Land, M., Lindell, D., Post, A. F., Regala, W., Shah, M., Shaw, S. L., Steglich, C., Sullivan, M., Ting, C. S., Tolonen, A., Webb, E. A., Zinser, E. R., and Chisholm, S. W. (2003) Genome divergence in two *Prochlorococcus* ecotypes reflects oceanic niche differentiation. *Nature* (London) **424:** 1042–1047.

Rohwer, F., and Edwards, R. (2002) The phage proteomic tree: Genome based taxonomy for phage. *J Bacteriol* **184:** 4529–4535.

Ruardij, P., Veldhuis, M. J. W., and Brussaard, C. P. D. (2005) Modeling the bloom dynamics of the polymorphic plankter *Phaeocystis globosa*: Impact of grazers and viruses. *Harmful Algae* **4:** 941–963.

Rusch, D. B., Halpem, A. L., Sutton, G., Heidelberg, K. B., Williamson, S., Yooseph, S., Wu, D., Eisen, J. A., Hoffman, J. M., Remington, K., Beeson, K., Tran, B., Smith, H., Baden-Tillson, H., Stewart, C., Thorpe, J., Freeman, J., Andrews-Pfankoch, C., Venter, J. E., Li, K., Kravitz, S., Heidelberg, J. F., Utterback, T., Rogers, Y.-H., Falcon, L. I., Souza, V., Bonilla-Rosso, G., Eguiarte, C. E., Karl, D. M., Sathyendranath, S., Platt, T., Bermingham, E., Gallardo, V., Tomayo-Castillo, G., Ferrari, M. R., Strausberg, R. L., Nealson, K., Friedman, R., Frazier, M., and Venter, J. C. (2007) The Sorcerer II global ocean sampling expedition: Northwest Atlantic through eastern tropical Pacific. *PLoS Biol* **5:** e77.

Safferman, R. S., and Morris, M. E. (1963) Algal virus: Isolation. *Science* **140:** 679–680.

Safferman, R. S., and Morris, M. E. (1964a) Growth characteristics of the blue-green algal virus LPP-1. *J Bacteriol* **88:** 771–775.

Safferman, R. S., and Morris, M. E. (1964b) Control of algae with viruses. *J Am Water Works Ass* **56:** 1217–1224.

Safferman, R. S., and Morris, M. E. (1967) Observations on the occurrence, distribution, and seasonal incidence of blue-green algal viruses. *Appl Microbiol* **15:** 1219–1222.

Safferman, R. S., Cannon, R. F., Desjardins, P. R., Gromov, B. V., Haselkorn, R., Sherman, L. A., and Shilo, M. (1983) Classification and nomenclature of viruses of cyanobacteria. *Intervirology* **19:** 61–66.

Safferman, R. S., Diener, T. O., Desjardins, P. R., and Morris, M. E. (1972) Isolation and characterization of AS-1, a phycovirus infecting the blue-green algae *Anacystis nidulans* and *Synechococcus cedrorum*. *Virology* **47:** 105–113.

Safferman, R. S., Schneider, I. R., Steere, R. L., Morris, M.E., and Diener, T. O. (1969a) Phycovirus SM-1: A virus infecting unicellular blue-green algae. *Virology* **37:** 386–395.

Safferman, R. S., Morris, M. E., Sherman, L. A., and Haselkorn, R. (1969b) Serological and electron microscopic characterization of a new group of blue-green algal viruses (LPP-2). *Virology* **39:** 775–780.

Samami, B., and Drews, G. (1978) Adsorption of cyanophage AS-1 to unicellular cyanobacteria and isolation of receptor material from *Anacystis nidulans*. *J Virol* **25:** 164–174.

Sandaa, R.-A., and Larsen, A. (2006) Seasonal variations in virus-host populations in Norwegian coastal waters: Focusing on the cyanophage community infecting marine *Synechococcus* spp. *Appl Environ Microbiol* **72:** 4610–4618.

Sandaa, R. A., Clokie, M., and Mann, N. H. (2008) Photosynthetic genes in viral populations with a large genome size range from Norwegian coastal waters. *FEMS Microbiol Ecol* **63:** 2–11.

Sano, E., Carlson, S., Wegley, L., and Rohwer, F. (2004) Movement of viruses between biomes. *Appl Environ Microbiol* **70:** 5842–5846.

Sarma, T. A., and Kaur, B. (1993) Spontaneous and induced host-range mutants of cyanophage N-1. *Arch Virol* **130:** 195–200.

Sarma, T. A., and Kaur, B. (1997) Characterization of host-range mutants of cyanophage N-1. *Acta Virol* **41:** 245–250.

Sarma, T. A., and Kaur, S. P. (2005) Development of cyanophage N-1 under heavy metal stress. *Acta Viriol* **49:** 23–28.

Sarma, T. A., and Singh, R. (1994) Isolation and characterization of temperature-sensitive mutants of cyanophage N-1. *Acta Virol* **38 :** 11–16.

Sarma, T. A., and Singh, R. (1995) Characterization of ts-mutants of cyanophage N-1 by their inactivation by physical and chemical agents. *Acta Virol* **39:** 65–68.

Schloss, P. D., and Handelsman, J. (2005) Metagenomics for studying unculturable microorganisms: Cutting the Gorgian knot. *Genome Biol* **6:** 229. http://genomebiology.com/2005/6/8/229

Schmidt, T. M., DeLong, E. F., and Pace, N. R. (1991) Analysis of a marine picoplankton community by 16S rRNA gene cloning and sequencing. *J Bacteriol* **173:** 4371–4378.

Schneider, G. J., and Haselkorn, R. (1988) Characterization of two early promoters of cyanophage N-1. *Virology* **167:** 150–155.

Schneider, I. R., Diener, T. O., and Safferman, R. S. (1964) Blue-green algal virus LPP-1: purification and partial characterization. *Science* **144:** 1127–1130.

Shane, M. S. (1971) Distribution of blue-green algal viruses in various types of natural waters. *Water Res* **5:** 711–716.

Sharma, C. R., and Venkataraman, G. S. (1977) Cyanophage AC-1 infecting blue-green alga *Anacystis nidulans*. *Curr Sci* **46:** 496–497.

Sharma, C. R., and Venkataraman, G. S. (1978) Seasonal periodicity of cyanophage AC-1. *Curr Sci* **47:** 914–915.

Sharon, I., Alperovitch, A., Rohwer, F., Haynes, M., Glaser, F., Atamna-Ismaeel, N., Pinter, R. Y., Partensky, F., Koonin, E. V., Wolf, Y. I., Nelson, N., and Béjà, O. (2009) Photosystem I gene cassettes are present in marine virus genomes. *Nature* (London) **461:** 258–261. doi:10.1038/nature08284

Sharon, I., Battchikova, N., Aro, E-M., Giglione, C., Meinnel, T., Glaser, F., Pinter, R. Y., Breitbart, M., Rohwer, F., and Béjà, O. (2011) Comparative metagenomics of microbial traits within oceanic viral communities. *ISME J* **5:** 1178–1190. doi:10.1038/ismej.2011.2

Sharon, I., Tzahor, S., Williamson, S., Shmoish, M., Man-Aharonovich, D., Rusch, D. B., Yooseph, S., Zeidner, G., Golden, S. S., Mackey, R., Adir, N., Weingart, U., Horn, D., Venter, J. C., Mandel-Gutfreund, Y., and Béjà, O. (2007) Viral photosynthetic reaction center genes and transcripts in the marine environment. *ISME J* **1:** 492–501.

Shen, G., Saunée, N. A., Williams, S. R., Gallo, E. F., Schluchhter, W. M., and Bryant D. A. (2006) Identification and characterization of a new class of bilin lyase. The *cpcT* gene encodes a bilin lyase responsible for attachment of phycobilin to Cys-153 on the β-subunit of phycocyanin in *Synechococcus* sp. PCC 7002. *J Biol Chem* **281:** 17768–17778.

Sherman, L. A. (1976) Infection of *Synechococcus cedrorum* by cyanophage AS-1M. III. Cellular development and phage metabolism. *Virology* **71:** 199–206.

Sherman, L. A., and Connelly, M. (1976) Isolation and characterization of a cyanophage infecting the unicellular blue-green algae *Anacystis nidulans* and *Synechococcus cedrorum*. *Virology* **72:** 540–544.

Sherman, L. A., and Haselkorn, R. (1970a) LPP-1 infection of the blue-green alga *Plectonema boryanum*. I. Electron microscopy. *J Virol* **6:** 820–833.

Sherman, L. A., and Haselkorn, R. (1970b) LPP-1 infection of the blue-green alga *Plectonema boryanum*. II. Viral deoxyribonucleic acid synthesis and host deoxyribonucleic acid breakdown. *J Virol* **6:** 834–840.

Sherman, L. A., and Haselkorn, R. (1970c) LPP-1 infection of the blue-green alga *Plectonema boryanum*. III. Protein synthesis. *J Virol* **6** :841–846.

Sherman, L. A., and Haselkorn, R. (1971) Growth of the blue-green algae virus LPP-1 under conditions which impair photosynthesis. *Virology* **45:** 739–746.

Sherman, L. A., and Pauw, P. (1976) Infection of *Synechococcus cedrorum* by AS-1M. II. Protein and DNA synthesis. *Virology* **71:** 17–27.

Sherman, L. A., Connelly, M., and Sherman, D. (1976) Infection of *Synechococcus cedrorum* by the cyanophage AS-1M. 1. Ultrastructure of infection and phage assembly. *Virology* **71:** 1–16.

Short, S. M., and Suttle, C. A. (2000) Denaturing gradient gel electrophoresis resolves virus sequences amplified with degenerate primers. *BioTechniques* **28:** 20–22.

Short, S. M., and Suttle, C. A. (2002) Sequence analysis of marine virus communities reveals that groups of related algal viruses are widely distributed in nature. *Appl Environ Microbiol* **68:** 1290–1296.

Short, S. M., and Suttle, C. A. (2003) Temporal dynamics of natural communities of marine algal viruses and eukaryotes. *Aquat Microb Ecol* **32:** 107–119.

Short, C. M., and Suttle, C. A. (2005) Nearly identical bacteriophage structural gene sequences are widely distributed in both marine and freshwater environments. *Appl Environ Microbiol* **71:** 480–486.

Sieburth, J. M. (1979) *Sea Microbes*. Oxford University Press, New York, NY, USA. 491 pp.

Sieburth, J. M., Johnson, P. W., and Hargraves, P. E. (1988) Ultrastructure and ecology of *Aureococcus anophagefferens* gen. et sp. nov. (Chrysophyceae) the dominant picoplankter during a bloom in Narragansett Bay, Rhode Island USA Summer. *J Phycol* **24:** 416–425.

Singh, P. K. (1973) Occurrence and distribution of cyanophages in ponds, sewage and rice fields. *Arch Mikrobiol* **89:** 169–72.

Singh, R. N., and Chaubev, I. J. (1976) The genetics of cyanophyceae and cyanophages: problems and prospects. *J. Cytol. Genet.* **11:** 116–121.

Singh, R. N., and Kashyap, A. K. (1977) Isolation and characterization of temperature- sensitive mutants of cyanophage LPP-1. *Mol Gen Genet* **154:** 31–34.

Singh, R. N., and Singh, P. K. (1967) Isolation of cyanophages from India. *Nature* (London) **216:** 1020–1021.

Singh, R. N., and Singh, P. K. (1972) Transduction and lysogeny in blue-green algae. In: Desikachary,T. V. (Ed.) *Taxonomy and Biology of Blue-Green Algae*. University of Madras, Madras, India. pp. 258–262.

Singh, R. N., Singh, P. K., Kashyap, A. K., Sarma, T. A., Dhar, B., Chaubey, I. J., and Choudhury, I. D. (1972) Isolation and characterization of new cyanophages and mutations of LPP-1 and host alga *Plectonema boryanum*. In: Desikachary, T. V. (Ed) Taxonomy and Biology of Blue-green Algae, University of Madras, Madras. pp. 585–591.

Sirenko, L. A., Myslovich, V. O., Goryushin, V. A., Mikhailyuk, D. P. (1976) Effect of infection with cyanophage AM-1 on the metabolism of the blue-green alga. *Fiziol. Rast.* **23:** 1214–1218.

Smith, K. M., Brown, Jr., R. M., Goldstein, D. A., and Walne, P.L. (1966) Culture methods for the blue-green alga *Plectonema boryanum* and its virus, with an electron microscope study of virus-infected cells. *Virology* **28:** 580–591.

Sode, K., Oonari, R., and Oozeki, M. (1997) Induction of a temperate marine cyanophage by heavy metal. *J Mar Biotechnol* **5:** 178–180.

Sode, K., Oozeki, M., Asakawa, K., Burgess, J. G., and Matsunaga, T. (1994) Isolation of marine cyanophages infecting the marine unicellular cyanobacterium *Synechococcus* sp. NKBG042902. *J Mar Biotechnol* **1:** 189–192.

Stahl, D. A., Lane, D. J., Olsen, G. J., and Pace, N. R. (1985) Characterization of a Yellowstone hot spring microbial community by 5S rRNA sequences. *Appl Environ Microbiol* **49:** 1379–1384.

Stern, A., and Sorek, R. (2011) The phage-host arms race: Shaping the evolution of microbes. *Bioessays.* **33:** 43–51. doi:10.1002/bies.201000071

Steward, G. F., Smith, D. C., and Azam, F. (1996) Abundance and production of bacteria and viruses in the Bering and Chukehi seas. *Mar Ecol Prog Ser* **131:** 287–300.

Steward, G. F., Winker, J., Cochlan, W. P., Smith, D. C., and Azam, F. (1992a) Estimation of virus production in the sea. II. Field results. *Mar Microbial Food Webs* **6:** 79–90.

Steward, G. F., Winker, J., Smith, D. C., Cochlan, W. P., and Azam, F. (1992b) Estimation of virus production in the sea. I. Method development. *Mar Microbial Food Webs* **6:** 57–78.

Stoddard, L. I., Martiny, J. B., and Marston, M. F. (2007) Selection and characterization of cyanophage resistance in marine *Synechococcus* strains. *Appl Environ Microbiol* **73:** 5516–5522.

Storrie-Lombardi, M. C., and Pinkart, H. C. (2007) Co-evolution of cyanophage and cyanobacteria in Antarctic lakes: Adaptive responses to high UV flux and global warming. In: Hoover, R. B., Levin, G. V., and Rozanov, A. (Eds.) *Astrobiology and Planetary Missions*. X. SPIE, Bellingham, WA, USA. pp. 1–13

Sullivan, M. B., Coleman, M. L., Weigele, P., Rohwer, F., and Chisholm, S. W. (2005) Three *Prochlorococcus* cyanophage genomes: Signature features and ecological interpretations. *PLoS Biol* **3:** e144 0790-0806. www.plosbiology.org

Sullivan, M. B., Coleman, M. L., Quinlivan, V., Rosenkrautz, J. E., DeFrancesco, A. S., Tan, G., Fu, R., Lee, J. A. Waterbury, J. B., Bielawski, J. P., and Chisholm, S. W. (2008) Portal protein diversity and phage ecology. *Environ Microbiol* **10:** 2810–2823. doi.10.1111/S. 1462–2920. www. blackwell. synergy.com

Sullivan, M. B., Huang, K. H., Ignacio-Espinoza, J. C., Berlin, A. M., Kelly, L., Weigele, P. R., DeFrancesco, A. S., Kern, S. E., Thompson, L. R., Young, S., Yandava, C., Fu, R., Krastins, B., Chase, M., Sarracino, D., Osburne, M. S., Henn, M. R., and Chisholm, S. W. (2010) Genome analysis of oceanic cyanobacterial myoviruses compared with T-4 like myoviruses from diverse hosts and environments. *Environ Microbiol* **12:** 3035–3056. doi:10.1111/j.1462-2920.2919.02280.x

Sullivan, M. B., Krastins, B., Hughes, J. L., Kelly, L., Chase, M., Sarracino, D., and Chishom, S. W. (2009) The genome and structural proteome of an ocean siphovirus: A new window into the cyanobacterial "mobilome". *Environ Microbiol* **11:** 2935–2951. doi: 1111/j.1462-2920.2009 02081.x

Sullivan, M. B., Lindell, L., Lee, J.A., Thompson, L. R., Bielawski, J. P., and Chisholm, S. W. (2006) Prevalence and evolution of core photosystem II genes in marine cyanobacterial viruses and their hosts. *PLoS Biol* **4:** e234. www.plosbiology.org

Sullivan, M. B., Waterbury, J. B., and Chisholm, S. W. (2003) Cyanophages infecting the oceanic cyanobacterium *Prochlorococcus*. *Nature* (London) **424:** 1047–1050.

Suttle, C. A. (1993) Enumeration and isolation of virus. In: Kemp, P. F., Sherr, B. F., Sherr, E. B., and Cole, J. J. (Eds.) *Handbook of Methods in Aquatic Microbial Ecology* **28:** 237–243.

Suttle, C. A. (2000) Cyanophages and their role in the ecology of cyanobacteria. In: Whitton, B. A., and Potts, M. (Eds.) *The Ecology of Cyanobacteria*. Kluwer Academic Publishers, Dordrecht, The Netherlands. pp. 563–589.

Suttle, C. A. (2005) Viruses in the sea. *Nature* (London) **437:** 356–361.

Suttle, C. A., and Chan, A. M. (1993) Marine cyanophages infecting oceanic and coastal strains of *Synechococcus*: Abundance, morphology, cross-infectivity and growth characteristics. *Mar Ecol Prog Ser* **92:** 99–109.

Suttle, C. A., and Chan, A. M. (1994) Dynamics and distribution of cyanophages and their effect on marine *Synechococcus* spp. *Appl Environ Microbiol* **60:** 3167–3174.

Suttle, C. A., and Chen, F. (1992) Mechanisms and rates of decay of marine viruses in seawater. *Appl Environ Microbiol* **58:** 3721–3729.

Suttle, C. A., Chan, A. M., and Cottrell, M. T. (1990) Infection of phytoplankton by viruses and reduction of primary productivity. *Nature* (London) **347:** 467–469.

Suttle, C. A., Chan, A. M., and Cottrell, M. T. (1991) Use of ultrafiltration to isolate viruses from seawater which are pathogens of marine phytoplankton. *Appl Environ Microbiol* **57:** 721–726.

Szekers, M. (1981) Phage induced development of site specific endonuclease in *Anacystis nidulans*, a cyanobacterium. *Virology* **111:** 1–10.

Szekers, M., Szmidt, A. E., and Torok, I. (1983) Evidence for restriction modification-like system in *Anacystis nidulans* infected by cyanophage AS-1. *Eur J Biochem* **131:** 137–141.

Takashima, Y., Yoshida, T., Yoshida, M., Shirai, Y., Tomaru, Y., Takao, Y., Hiroishi, S., and Nagasaki, K. (2007) Development and application of quantitative detection of cyanophages phylogenetically related to cyanophages Ma-LMM01 infecting *Microcystis aeruginosa* in freshwater. *Microbes Environ* **22:** 207–213.

Teklemariam, T. A., Demeter, S., Dea, Z., Suranyi, G., and Borbely, G. (1990) AS-1 cyanophage infection inhibits the photosynthetic electron flow of photosystem II in *Synechococcus* sp. PCC 6301, a cyanobacterium. *FEBS Lett* **270:** 211–215.

Tétart, F., Desplats, C.,Kutateladze, M., Monod, C., Ackermann, H-W., and Krisch, H. M. (2001) Phylogeny of the major head and tail genes of the wide-ranging T4 bacteriophages. *J Bacteriol* **183:** 358–366.

Thingstad, T. F., and Lignell, R. (1997) A theoretical approach to the question of how trophic interactions control carbon demand, growth rate, abundance and diversity. *Aquat Microb Ecol* **13:**19–27.

Ting, C.S., Rocap, G., King, J., and Chisholm, S.W. (2002) Cyanobacterial photosynthesis in the oceans: The origins and significance of divergent light-harvesting strategies. *Trends Microbiol* **10:** 134–142.

Torrella, F., and Morita, R. Y. (1979) Evidence by electron micrographs for a high incidence of bacteriophage particles in the waters of Yaquina Bay, Oregon: Ecological and taxonomical implications. *Appl Environ Microbiol* **37:** 774–778.

Trivedi, J. P., and Oza, P. P. (1979) A lytic agent for blue-green alga *Oscillatoria chlorina* from cooling towers. *Comp Physiol Ecol* **4:** 207–212.

Tucker, S., and Pollard, P. (2005) Identification of cyanophages Ma-LBP and infection of the cyanobacterium *Microcystis aeruginosa* from an Australian subtropical Lake by the virus. *Appl Environ Microbiol* **71:** 629–655.

Twort, F.W. (1915) An investigation on the nature of ultra-microscopic viruses. *Lancet ii* : 1241–1243.

van Boekel, W. H. M., Hansen, F. C., Riegman, R., and Bak, R. P. M. (1992) Lysis-induced decline of a *Phaeocystis* spring bloom coupling with the microbial food web. *Mar Ecol Prog Ser* **81:** 269–276.

Venkataraman, G. S., Kaushik, B. D., Subramanian, G., Shanmugasundaram, G., and Govindarajan, A. (1973) Cyanophage AC-1: A phage infecting unicellular and colonial blue-green algae. *Curr Sci* **42:** 104–105.

Venter, J. C., Remington, K., Heidelberg, J., Halpern, A. L., Rusch, D., and Eisen, J. A. (2004) Environmental genome shotgun sequencing of the Sargasso Sea. *Science* **304:** 66–74.

Vrede, K., Stensdotter, U., and Lindstrom, E. (2003) Viral and bacterioplankton dynamics in two lakes with different humic contents. *Microb Ecol* **46:** 406–415.

Wang, G., Murase, J., Asakawa, S., and Kimura, M. (2010) Unique viral capsid assembly protein gene (*g20*) of cyanophages in the floodwater of a Japanese paddy field. *Biol Fertil Soils* **46:** 93–102. doi:10.1007/s00374-009-0410-y.

Wang, K., and Chen, F. (2004) Genetic diversity and population dynamics of cyanophage communities in the Chesapeake Bay. *Aquat Microb Ecol* **34:** 105–116.

Wang, K., and Chen, F. (2008) Prevalence of highly host-specific cyanophages in the estuarine environment. *Environ Microbiol* **10:** 300–312.

Waterbury, J. B., and Rippka, R. M. (1989) Subsection I. Order Chroococcales Wettstein 1924, emend. Rippka *et al.*, 1979, p. 1728–1729. In: Krieg, N. R., and Holt, J. B. (Eds.) *Bergey's Manual of Systematic Bacteriology*. Vol. 3. Williams and Wilkins, Baltimore, USA.

Waterbury, J. B., and Valois, F. W. (1993) Resistance to co-occurring phages enables marine *Synechococcus* communities to coexist with cyanophages abundant in seawater. *Appl Environ Microbiol* **59:** 3393–3399.

Waterbury, J. B., Watson, S. W., Guillard, R. R. L., and Brand, L. E. (1979) Widespread occurrence of a unicellular marine planktonic cyanobacterium. *Nature* (London) **277:** 293–294.

Waterbury, J. B., Watson, S. W., Valois, F. W., and Franks, D. G. (1986) Biological and ecological characterization of the marine unicellular cyanobacterium *Synechococcus*. *Can Bull Fish Aquat Sci* **214:** 71–120.

Weigele, P. R., Pope, W. H., Pedulla, M. L., Houtz, J. M., Smith, A. L., Conway, J. F., King, J., Hatfull, G. F., Lawrence, J. G., and Hendrix, R. W. (2007) Genomic and structural analysis of Syn9, a cyanophage infecting marine *Prochlorococcus* and *Synechococcus*. *Environ Microbiol* **9:** 1675–1695.

Weinbauer, M. G., Wilhelm, S. W., Suttle, C. A., Pledger, R. J., and Mitchell, D. L. (1999) Sunlight-induced DNA damage and resistance in natural viral communities. *Aquat Microb Ecol* **17:** 111–120.

Wen, K., Ortamnn, A. C., and Suttle, C. A. (2004) Accurate estimation of viral abundance by epifluorescence microscopy. *Appl Environ Microbiol* **70:** 3862–3867.

Wharam, S. D., Hall, M. J., and Wilson, W. H. (2007) Detection of virus mRNA within infected host cells using an isothermal nucleic acid amplification assay: Marine cyanophage gene expression within *Synechococcus* sp. *Virology J* **4:** 52 doi:10.1186/1743-422X-4-52

Wilcox, R. M., and Fuhrman, J. A. (1994) Bacterial viruses in coastal seawater: Lytic rather than lysogenic production. *Mar Ecol Prog Ser* **114:** 35–45.

Wilhelm, S. W., and Suttle, C. A. (1999) Viruses and nutrient cycles in the sea. *BioScience* **49:** 781–788.

Wilhelm, S. W., Carberry, M. J., Eldridge, M. L., Poorvin, L., Saxton, M. A., and Doblin, M. A. (2006) Marine and freshwater cyanophages in a Laurentian Great Lake: Evidence from infectivity assays and molecular analyses of *g20* genes. *Appl Environ Microbiol* **72:** 4957–4963.

Williamson, K. E., Wommack, K. E., and Rodosevich, M. (2003) Sampling natural viral communities from soil for culture-independent analyses. *Appl Environ Microbiol* **69:** 6628–6633.

Williamson, S. J., Rusch, D. B., Yooseph, S., Halpern, A. L., Heidelberg, K. B., Glass, J. I., Andrews-Pfannkoch, C., Fadrosh, D., Miller, C. S., Sutton, G., Frazier, M., and Venter, J. C. (2008) The Sorcerer II Global Ocean Sampling expedition: Metagenomic characterization of viruses within aquatic microbial samples. *PLoS ONE* **3:** e1456. doi:10.1371/journal.pone.0001456.

Wilson, W. H. (1994) Characterization of viruses which infect marine phytoplankton. Ph.D. thesis. University of Warwick, Coventry, United Kingdom.

Wilson, W. H., Carr, N. G., and Mann, N. H. (1996) The effect of phosphate status on the kinetics of cyanophage infection in the Oceanic cyanobacterium *Synechococcus* sp. WH 7803. *J Phycol* **32:** 506–516.

Wilson, W. H., Joint, I. R., Carr, N. G., and Mann, N. H. (1993) Isolation and molecular characterization of five marine cyanophages propagated on *Synechococcus* sp. strain WH 7803. *Appl Environ Microbiol* **59:** 3736–3743.

Wilson, W. H., Nicholas, J. F., Joint, I. R., and Mann, N. H. (1999) Analysis of cyanophage diversity and population structure in a south-north transect of the Atlantic Oean. *Bull Inst Oceanogr* (Monaco) **19:** 209–26.

Wilson, W. H., Nicholas, J. F., Joint, I. R., and Mann, N. H. (2000) Analysis of cyanophage diversity in the marine environment using denaturing gradient gel electrophoresis. In: Bell, C. R., Bryliksky, M., and Johnson-Green, P. (Eds.) *Microbial Biosystems: a New Frontier*. Proc 8th Intl Symp on Microbial Ecology, Halifax, Nova Scotia, Canada. pp. 565–570.

Wilson, W. H., Turner, S., and Mann, N. H. (1998) Population dynamics of phytoplankton and viruses in a phosphate limited mesocosm and their effect on DMSP and DMS production. *Eustuarine Coastal Shelf Sci* **46:** 49–59.

Wommack, K. E., and Colwell, R. R. (2000) Virioplankton: Viruses in aquatic ecosystems. *Microbiol Mol Biol Rev* **64:** 69–114.

Wommack, K. E., Hill, R. T., Kellel, M., Russek-Cohen, E., and Colwell, R. A. (1992) Distribution of viruses in the Chesapeake Bay. *Appl Environ Microbiol* **58:** 2965–2970.

Wu, J. H., Choules, G. L., and Lewin, R. A. (1968) Early stage of the infection process in blue-green algal virus system, as affected by KCN and light. In: *Biochemical Regulation in Diseased Plants or Injury*. Photopathological Society of Japan, Tokyo. pp. 153–160.

Yoshida, M., Yoshida, T., Takashima, Y., Kondo, R., and Hiroishi, S. (2005) Genetic diversity of the toxic cyanobacterium *Microcystis* in Lake Mikata. *Environ Toxicol* **20:** 229–234.

Yoshida, M., Yoshida, T., Takashima, Y., Hosoda, N., and Hiroishi, S. (2007) Dynamics of microcystin-producing and non-microcystin-producing *Microcystis* population is correlated with nitrate concentration in a Japanese Lake. *FEMS Microbiol Lett* **266:** 49–53.

Yoshida, M., Yoshida, T., Kashima, A., Takashima, Y., Hosoda, N., Nagasaki, K., and Hiroishi, S. (2008) Ecological dynamics of the toxic bloom-forming cyanobacterium *Microcystis aeruginosa* and its cyanophages in freshwater. *Appl Environ Microbiol* **74:** 3269–3273.

Yoshida, M., Yoshida, T., Yoshida-Takashima, Y., Kashima, A., and Hiroishi, S. (2010) Real-time PCR detection of host-mediated cyanophage gene transcripts during infection of a natural *Microcystis aeruginosa* population. *Microbes Environ* **25:** 211–215. doi:10.1264/jsme2-ME10117

Yoshida, T., Nagasaki, K., Takashima, Y., Shirai, Y., Tomaru, Y., Takao, Y., Sakamoto, S., Hiroishi, S., and Ogata, H. (2008) Ma-LMM01 infecting toxic *Microcystis aeruginosa* illuminates diverse cyanophages genome strategies. *J Bacteriol* **190:** 1762–1772.

Yoshida, T., Takashima, Y., Tomaru, Y., Shirai, Y., Takao, Y., Hiroishi, S., and Nagasaki, K. (2006) Isolation and characterization of a cyanophage infecting the toxic cyanobacterium *Microcystis aeruginosa*. *Appl Environ Microbiol* **72:** 1239–1247.

Zeidner, G., Bielawski, J. P., Shmoish, M., Scanlan, D. J., Sabehi, G., and Béjà, O. (2005) Potential photosynthesis gene recombination between *Prochlorococcus* and *Synechococcus* via viral intermediates. *Environ Microbiol* **7:** 1505–1513.

Zhong, Y., Chen, F., Wilhelm, S. W., Poorvin, L., and Hodson, R. E. (2002) Phylogenetic diversity of marine cyanophage isolates and natural virus communities as revealed by sequences of viral capsid assembly protein *g20*. *Appl Environ Microbiol* **68:** 1576–1584.

Zinger, L., Gobet, A., and Pommiers, T. (2011) Two decades of describing unseen majority of aquatic microbial diversity. *Mol Ecol* doi:10.1111/j.1365.294x.2011.05362.x

CHAPTER 9

CYANOBACTERIAL TOXINS

Any substance that is poisonous in nature is termed as a toxin. Toxins produced by cyanobacteria are known as cyanobacterial toxins or cyanotoxins. Of the 150 known genera of cyanobacteria, 40 species of cyanobacteria have been implicated in the formation of toxic blooms in fresh, marine and estuarine waters (Carmichael, 1997; Saker *et al.*, 1999). The bloom algae are characterized by the presence of reddish brown, refractile structures inside their cells known as gas vacuoles that help them to keep afloat in water. Due to mixing of agricultural and industrial waste-water, the nutrient status of water bodies undergoes a change. In this context, nitrogen and phosphorus enrichment leads to the formation of nutrient saturated water bodies defined as eutrophic waters. The growth of bloom algae is thus dependent on variations in the nutrients especially of nitrogen and phosphorus, light intensity, temperature, pH and other factors. Prepas and Charette (2003) summarized the factors responsible for eutrophication of water bodies worldwide and suggested methods of control. The earlier concept that a low ratio of nitrogen to phosphorus (N:P) is responsible for the predominance of bloom algae has now been questioned. Downing *et al.* (2001) who analyzed data from 99 temperate lakes suggested that it is the variations in the total N and P contents in waters more important rather than their ratio. Due to their excessive growth, the waters emanate foul smell that is obnoxious in nature and the waters are unfit for use in any manner either for animal or human consumption or for recreational purposes.

World over the drinking water requirements are generally met from surface waters, since the groundwater supplies are unable to meet the demand of the ever increasing population. It is estimated that at least 48% of the lakes of North America are eutrophic suggesting that they are rich in nutrients and low in oxygen favouring the growth of cyanobacterial blooms posing problems of water quality in the United States (Reynolds, 2004). Thus the bloom-forming cyanobacteria have received a worldwide attention to identify the nature of toxins produced by them. The problem with cyanobacterial toxins is that the waters are unfit even for irrigation purposes as these toxins persist in waters even after the disappearance or death of the bloom algae. Thus the cyanobacterial toxins tend to accumulate (bioaccumulation) at different trophic levels leading to biomagnification. World Health Organization (WHO) recommended that the focus be made on cyanobacterial toxins by considering cyanobacteria in drinking water as a chemical problem (Chorus and Bartram, 1999). On the basis of the characteristics of bioaccumulation of cyanobacterial toxins, their bioavailability for exerting the toxicity, exposure and uptake routes a theoretical predictive model has been developed that enables to evaluate the risk to aquatic organisms and human beings (White *et al.*, 2005).

In general, three important features of cyanobacterial toxins are worth considering. Firstly, the cyanobacterial toxins are colourless, odourless and low molecular weight compounds that are formed as secondary metabolites. These do not play any role in growth and primary metabolism of the cyanobacterium. Secondly, these compounds are highly toxic to all living organisms such as fishes, animals and human beings. Thirdly, the cyanobacterial toxins have proved to be highly poisonous than the mycotoxins, and toxins of higher plant origin as well as shellfish toxins (Codd, 1995; Carmichael, 1997; Falconer, 1998; Sivonen and Jones, 1999; Metcalf and Codd, 2004).

Toxic bloom-forming cyanobacteria belong to the genera *Anabaena*, *Anabaenopsis*, *Aphanizomenon*, *Cylindrospermopsis*, *Microcystis*, *Oscillatoria*, *Planktothrix*, *Phormidium*, *Nodularia* and *Nostoc*. Table 1 summarizes the species of the respective genera with their geographic location. Amongst these, except *Nodularia* that is known to inhabit marine and brackish waters, the rest of the cyanobacteria are freshwater bloom-formers (Sivonen and Jones, 1999). *Hapalosiphon hibernicus* strain B2-3-1 has been reported to be toxic but it is not clear whether it has been isolated from terrestrial or freshwater mud-habitats (Princep *et al.*, 1992). Exceptionally, a lichen-associated *Nostoc* sp. strain IO-102I is reported to be toxic (Oksanen *et al.*, 2004). *Nostoc* sp. strain-152, a freshwater cyanobacterium has been reported to be toxic (Sivonen *et al.*, 1990a). Oksanen *et al.* (2004) concluded that *Nostoc* sp. strain-IO-102I is distantly related to *Nostoc* sp. strain-152 but closely related to *Nostoc punctiforme* PCC 73102 and other symbiotic strains. It is quite intriguing to note that 95% of the genera of cyanobacteria belonging to the five taxonomic groups revealed a common neurotoxic amino acid termed as β-N-methylamino-L-alanine (Cox *et al.*, 2005).

Classification: The cyanobacterial toxins are classified either on the basis of their chemical properties or symptoms of toxicity (Gorham and Carmichael, 1988; Codd, 1995, 2000; Briand *et al.*, 2003; Haider *et al.*, 2003). On the basis of their chemical structure cyanobacterial toxins are divided into (1) cyclic peptides (2) alkaloids and (3) lipopolysaccharides in order of their decreasing toxicity. On the basis of their toxicity symptoms, the cyanobacterial toxins are classified as (1) hepatotoxins (2) neurotoxins and (3) dermatotoxins. The latter classification is followed in this text.

1) Hepatotoxins: Microcystins (MCs), nodularin and cylindrospermopsin (CYN) are the three hepatotoxins produced by different cyanobacteria. MCs are produced by the freshwater bloom-forming, unicellular and colonial *Microcystis* species, mainly *M. aeruginosa* that is worldwide in its distribution (Fig. 1 A). Three other species of *Microcystis* that produce MCs are *M. botrys*, *M. viridis* and *M. wesenbergii* (Fig. 1 B, C and D). Filamentous forms of *Anabaena* (Fig. 1 E), *Anabaenopsis*, *Aphanizomenon* (Fig. 2 A-E), *Phormidium*, *Planktothrix* (Fig. 1 F) and *Nostoc* also produce MCs, besides the terrestrial genus *Hapalosiphon*. As many as 90 variants of MCs have so far been recorded which show variations in the degree of methylation, hydroxylation, epimerization, peptide sequence and toxicity (Rinehart *et al.*, 1988; Codd, 1995, 2000; Sivonen, 1996; Dow and Swoboda, 2000; Kaeberinck and Neilan, 2001; Pearson *et al.*, 2010). Nodularins are produced by *Nodularia spumigena* (Fig. 3) whereas CYN is produced by *Cylindrospermopsis* (Fig. 4 A), *Raphidiopsis* (Fig. 4B) and certain other filamentous forms. MCs are cyclic heptapeptides while nodularins are cyclic pentapeptides. CYN is an alkaloid.

The target tissue in animals including man is liver. The symptoms of poisoning by MCs in human beings range from weakness, loss of appetite, vomiting, diarrhea and cancer. It is because of their carcinogenic nature, a lot of attention has been given to MC related research (Ueno *et al.*, 1996; Zhou *et al.*, 2002).

2) Neurotoxins: Anatoxins and STXs that are alkaloid in nature are included here. Several freshwater bloom-forming algae *Anabaena flos-aquae* (anatoxins a,b,c) and *Aphanizomenon flos-aquae* (Fig. 2 B), *Oscillatoria* (*O. mougeotii*) and *Cylindrospermopsis* are known to produce anatoxins (Mitrovic *et al.*, 2004). STXs are toxins generally found in the dinoflagellates that cause paralytic shellfish poisoning and these toxins find their way into human beings when the shellfish is consumed. More than 30 STXs are produced by the freshwater cyanobacteria. Other cyanobacteria such as *Lyngbya majuscula* and *O. mougeotii* are also known to produce STXs (Ferreira *et al.*, 2001). Anatoxins and STXs mainly act upon the nervous system and also affect skin, liver and gastro-intestinal region. The occurrence of cyanobacterial neurotoxins, their chemical properties, mode of action and biosynthetic pathways have been reviewed (Aráoz *et al.*, 2010).

3) Dermatotoxins: Species of *Lyngbya* (Fig. 5), *Oscillatoria* and *Schizothrix* have been reported to be responsible for the commonly observed severe dermatitis in swimmers. Severe oral and gastrointestinal inflammation also has been noted in certain cases of ingestion. The toxins of *Lyngbya*, i.e. aplysiotoxins and debromoaplysiotoxins are tumor promoters. *O. nigroviridis* and *Schizothrix calcicola* also produce debromoaplysiotoxins along with other two such toxins (Chorus and Bartram, 1999).

Table 1: Toxic cyanobacterial genera with types of toxins produced from different countries.

Country	Toxic Species	Cyanotoxin	Authors
Australia	*A. circinalis*	STXs	Carmichael (1992); Humpage *et al.* (1994)
	Aph. ovalisporum	Cylindrospermopsin	Humpage *et al.* (1994); Negri *et al.* (1995, 1997); Banker *et al.* (1997)
	M. flos-aquae	Microcystins	
	N. spumigena	Nodularins	Baker and Humpage (1994); Jones *et al.* (1994)
	Aphanothece sp., *Microcystis* sp.	Microcystins	Dasey *et al.* (2005)
	Microcystis sp., *Anabaena*, sp., *Aphanizomenon* sp., *Nodularia* sp.	Microcystins Nodularins	Kemp and John (2006)
	Nodularia spumigena	Nodularins	Jones *et al.* (2006)
Belgium	*M. aeruginosa*	Microcystins	Wirsing *et al.* (1998)
	Microcystis sp., *Planktothrix* sp., *Woronichinia* sp., *Aphanizomenon* sp.	Microcystins	Williame *et al.* (2005)
Brazil	*Microcystis* sp., *Anabaena* sp.	Microcystins	Jochimsen *et al.* (1998); Pouria *et al.* (1998); Lagos *et al.* (1999); Hirooka *et al.* (1999); Matthiensen *et al.* (2000)
	C. raciborskii	Saxitoxins	
	Microcystis sp.	Microcystins	
	M. panniformis, M. protocystis, M. novacekii, Aph. gracile, Aph. cf. *manguinii, Aph.* cf. *issastschenkoi*	Microcystins	Costa *et al.* (2006)
	C. raciborskii	STX	
	Aphanocapsa cumulus	Microcystins	
Canada	*Anabaena flos-aquae*	Anatoxin-a	Carmichael *et al.* (1975); Devlin *et al.* (1977) Matsunaga *et al.* (1989); Mahmood and Carmichael (1987)
		Anatoxin-a(s)	Krishnamurthy *et al.* (1986); Harada *et al.* (1991a)
	Anabaena sp., *Aphanizomenon* sp., *Microcystis* sp., *Oscillatoria* sp.,	Microcystins	Hrudey *et al.* (1994, 1999); Zurawell *et al.* (1999)
	M. aeruginosa	Microcystins	Gupta *et al.* (2001)
	Microcystis sp., *Anabaena* sp.	Microcystins	Giani *et al.* (2005); Rolland *et al.* (2005)
China	*Anabaena* spp., *Oscillatoria* spp.	Microcystins	Xu *et al.* (2000)
Czech Republic	*Microcystis* spp., *P. agardhii, Aph. flos-aquae*	Microcystins	Marsalek *et al.* (2001)
	Microcystis spp.	Microcystins	Znachor *et al.* (2006)
Denmark	*A. lammermannii*	Anatoxin-a(s)	Henriksen *et al.* (1997); Onodera *et al.* (1997)
	M. botrys	Microcystins	
	P. agardhii, P. mougeotii	Microcystins	Henriksen *et al.* (1996)
	Anabaena spp., *P. agardhii, Aph. flos-aquae*	Microcystins	Henriksen (2001)
	M. aeruginosa	Microcystins	Albay *et al.* (2003, 2005)
Egypt	*Oscillatoria tenuis*	Microcystins	Brittain *et al.* (2000)
England	*Nostoc* sp.	Microcystins	Beattie *et al.* (1998)

Table 1: contd....

Table 1: contd....

Country	Toxic Species	Cyanotoxin	Authors
Finland	*A. flos-aquae, A. lammermannii, A. circinalis* *Anabaena* sp., *Aphanizomenon* sp.	Microcystins Anatoxin-a	Sivonen *et al.* (1990b, 1992a) Sivonen *et al.* (1989a)
France	*A. circinalis* *Aph. flos-aquae*	Microcystins	Vezie *et al.* (1998) Maatouk *et al.* (2002)
Germany	*Anabaena* sp. *Aphanizomenon* sp. *Microcystis* sp. *Planktothrix* sp., *Microcystis* sp. *Anabaena* sp., *Aphanizomenon* sp.	Anatoxin-a Microcystins Microcystins Microcystins	Bumke-Vogt *et al.* (1999) Fromme *et al.* (2000) Hummert *et al.* (2001); Wiedner *et al.* (2001); Frank (2002)
Greece	*Anabaenopsis millerii* *Microcystis* sp., *Anabaena* sp., *Anabaenopsis* sp., *Aphanizomenon* sp., *Cylindrospermopsis* sp. *Aph. ovalisporum*	Microcystins Microcystins Microcystins	Lanaras and Cook (1994) Cook *et al.* (2004) Gkelis *et al.* (2005)
Holland	*P. agardhii*	Microcystins	Janse *et al.* (2005)
Hungary	*C. raciborskii*	Cylindrospermopsin	Törökné (1997)
Ireland	*Oscillatoria* sp. *Anabaena* sp.	Anatoxin-a	Edwards *et al.* (1992) James *et al.* (1997)
Israel	*Aph. ovalisporum*	Cylindrospermopsin	Banker *et al.* (1997); Shaw *et al.* (1999)
Italy	*A. planktonica* bloom	Anatoxin-a	Bruno *et al.* (1994)
Japan	*M. viridis* *Microcystis* blooms	Microcystins Microcystins	Kusumi *et al.* (1987); Watanabe *et al.* (1986); Tsuji *et al.* (1996); Park *et al.* (1998); Matsunaga *et al.* (1999)
New Zealand	*No. spumigena*	Nodularins	Carmichael *et al.* (1988a); Rinehart *et al.* (1988)
Norway	*A. flos-aquae* *P. agardhii* *P. formosa* *Anabaena* sp., *Microcystis* sp., *Oscillatoria* sp.	Microcystins Microcystins Homoanatoxin-a Microcystins	Sivonen *et al.* (1992a); Sivonen (1990b); Luukkuainen *et al.* (1993) Skulberg *et al.* (1992) Utkilen *et al.* (2001)
Portugal	*M. aeruginosa, M. wesenbergi, A. flos-aquae, A. scheremetievi, Aph. flos-aquae*	Microcystins	Vasconcelos (1993), Vasconcelos *et al.* (1996, 2001); Vasconcelos and Pereira (2001)
Scotland	*Oscillatoria* sp. Benthic *M. aeruginosa, A. flos-aquae* *N. spumigena*	Anatoxin-a Microcystins Nodularins	Edwards *et al.* (1992); Codd *et al.* (1995) Dow and Swoboda (2000) Dahlmann *et al.* (2005)
Slovak Republic	*M. aeruginosa*	Microcystins	Sedmak and Kosi (1997)
Spain	*Microcystis* sp., *Oscillatoria* sp., *Aphanizomenon* sp., *Anabaena* sp., *Aph. ovalisporum*	Microcystins	Moreno *et al.* (2004)

Table 1: contd....

Table 1: contd....

Country	Toxic Species	Cyanotoxin	Authors
Sweden	*M. aeruginosa, M. viridis, P. prolifica, P. agardhii*	Microcystins	WHO (2003)
Switzerland	*O. limosa*	Microcystins	Mez *et al.* (1996)
	Oscillatoria spp., *Phormidium* spp.	Hepatotoxins & neurotoxins	Mez *et al.* (1997)
Turkey	*M.aeruginosa*	Microcystins	Albay *et al.* (2005)
USA	*A. circinalis*	Anatoxin-a	Schwimmer and Schwimmer (1964); Gibson and Smith (1982);
	Hapalosiphon hibernicus	Microcystins	Princep *et al.* (1992); Puschner *et al.* (1998); Jacoby *et al.* (2000);
	M. aeruginosa	Microcytins	Rinkta-Kanto *et al.* (2005); Oberholster *et al.* (2006)

I. HEPATOTOXINS

Structurally MC and nodularin are cyclic peptides, the former being a heptapeptide and the latter a pentapeptide. Both MCs and nodularins possess a unique hydrophobic amino acid known as Adda (Eriksson *et al.*, 1987). The toxicity of MCs and nodularins is due to the presence of Adda chain and any change in the Adda chain reduces the toxicity of these toxins (Carmichael, 1992; Stotts *et al.*, 1993; Dow and Swoboda, 2000; Duy *et al.*, 2000). On the other hand, CYN is a cyclic guanidine alkaloid. The information on their structure and biosynthesis is presented here.

A) Microcystins: These are the most widely distributed because a number of cyanobacteria produce these toxins. MCs are cyclic heptapeptides with a molecular weight ranging from 909 to 1,115 D. Because initial isolation of this toxin was done from *M. aeruginosa*, the toxin was named as microcystin (Botes *et al.*, 1985). Structural analysis of MCs from *M. aeruginosa* and *A. flos-aquae* revealed equimolar amounts of glutamic acid, alanine, arginine and leucine besides β-methylaspartic acid and the peptides appeared to be cyclic (Krishnamurthy *et al.*, 1986). The general structure of MC is cyclo (D-Ala^1-L-Xaa^2-D-erythro-β-methyl-D-iso-Asp^3-L-Zaa^4-Adda^5-D-isoGlu^6-N-methyldehydro-Ala^7 (Carmichael *et al.*, 1988; Krishnamurthy *et al.*, 1989). Adda (2S, 3S, 8S, 9S)-3-amino-9-methoxy-2, 6, 8-trimethyl-10 phenyldeca-4-6-dienoic acid) is the novel β-amino acid that is responsible for the toxicity of MC. The toxin consists of two protein and five non-protein amino acids. The variability in the different forms of MCs is due to the variability in the former while the latter appear to be relatively constant. The first letter code of the variable L-amino acids at position 2 (Xaa) and 4 (Zaa) is taken to designate the different forms of MCs. The most widely studied variant of MC is MC-LR with L-leucine and L-arginine at Xaa and Zaa positions (Fig. 6). MC-LR is produced by *M. aeruginosa* (Botes *et al.*, 1985), *M. aeruginosa* UTEX 2388 (Oh *et al.*, 2000), *M. aeruginosa* strain B 2666 (Diehnelt *et al.*, 2006), *M. viridis* (Watanabe *et al.*, 1988), *A. flos-aquae* (Rinehart *et al.*, 1988; Krishnamurthy *et al.*, 1989; Sivonen *et al.*, 1992a), *Anabaena* sp. strain 90, *Anabaena* sp. strain 141 (Sivonen *et al.*, 1992a) and Antarctic cyanobacterial communities (Jungblut *et al.*, 2006). Botes *et al.* (1984) characterized MCs from *M. aeruginosa* in which Xaa and Zaa residues identified were L-leucine and L-alanine (MC-LA) and L-tyrosine and L-methionine (MC-YM), respectively. Three additional forms of MCs have been identified by Botes *et al.* (1985) that consisted of L-tyrosine at Xaa and variable aminoacids at Zaa positions i.e., MC-YR(L-tyrosine and L-arginine), MC-YA (L-tyrosine and L-alanine) and MC-YM(O) (L-tyrosine and methionine S-oxide). MC-YR production by *M. viridis* (Watanabe *et al.*, 1988) and *M. aeruginosa* UTEX 2388 (Oh *et al.*, 2000) has also been reported subsequently. Gathercole and

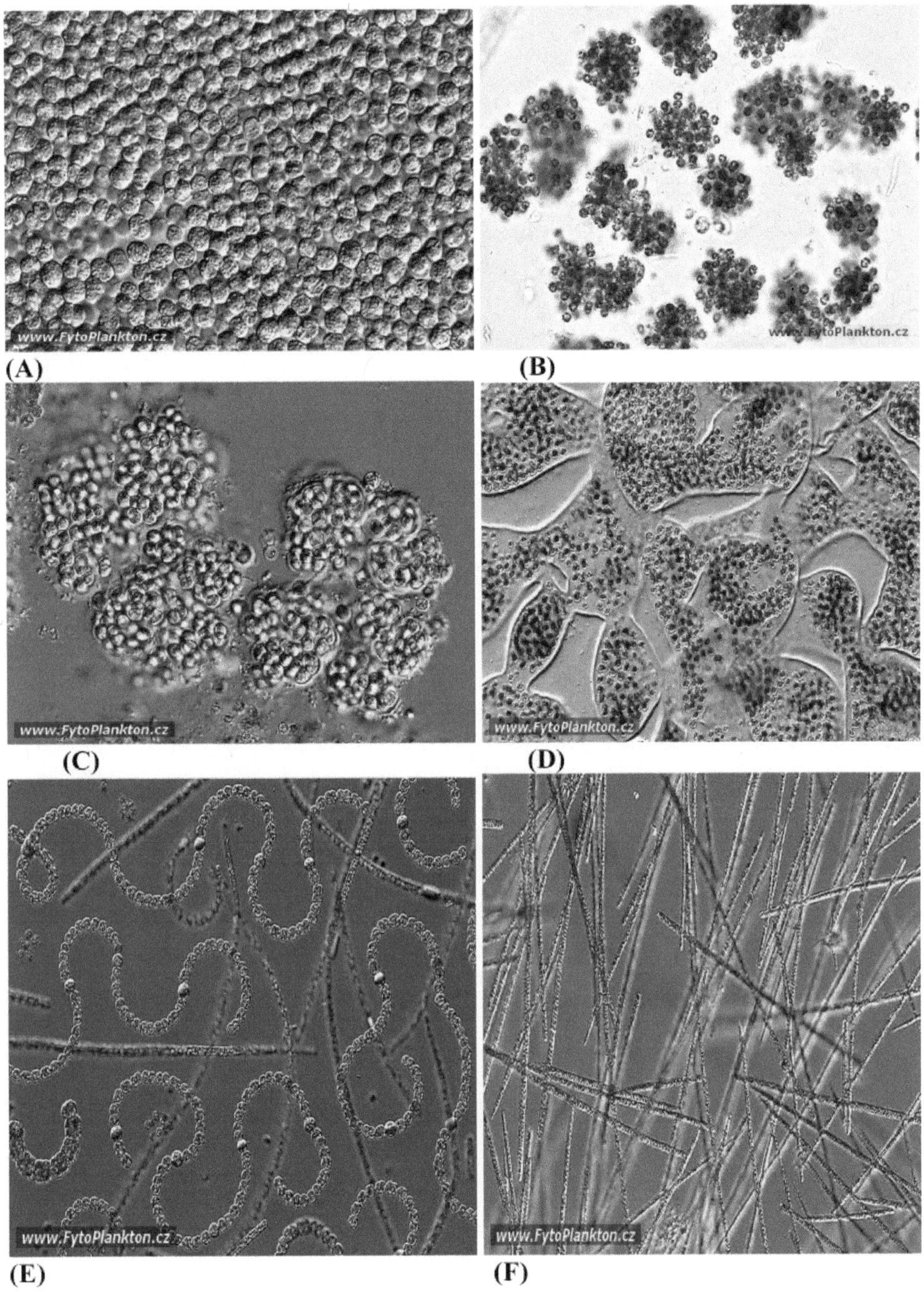

Figure 1: Representatives of planktonic cyanobacteria from different water bodies of Czech Republic. *Microcystis aeruginosa* (A), *M. botrys* (B), *M. viridis* (C), *M. wesenbergii* (D), *Anabaena flos-aquae* (E), and *Planktothrix agardhii* (F). Bright field (B) and differential interface contrast microscopic (DIP) or Nomarski contrast images (the rest) with a magnification of x 300 (E, F), x 400 (B, C and D) and x 600 (A). Pictures courtesy P. Znachor, Laboratory of Phytoplankton Ecology, Institute of Hydrobiology, Biology Centre ASCR, Na Sádkách , 37005 České Budějovice, Czech Republic (www.FytoPlankton.cz).

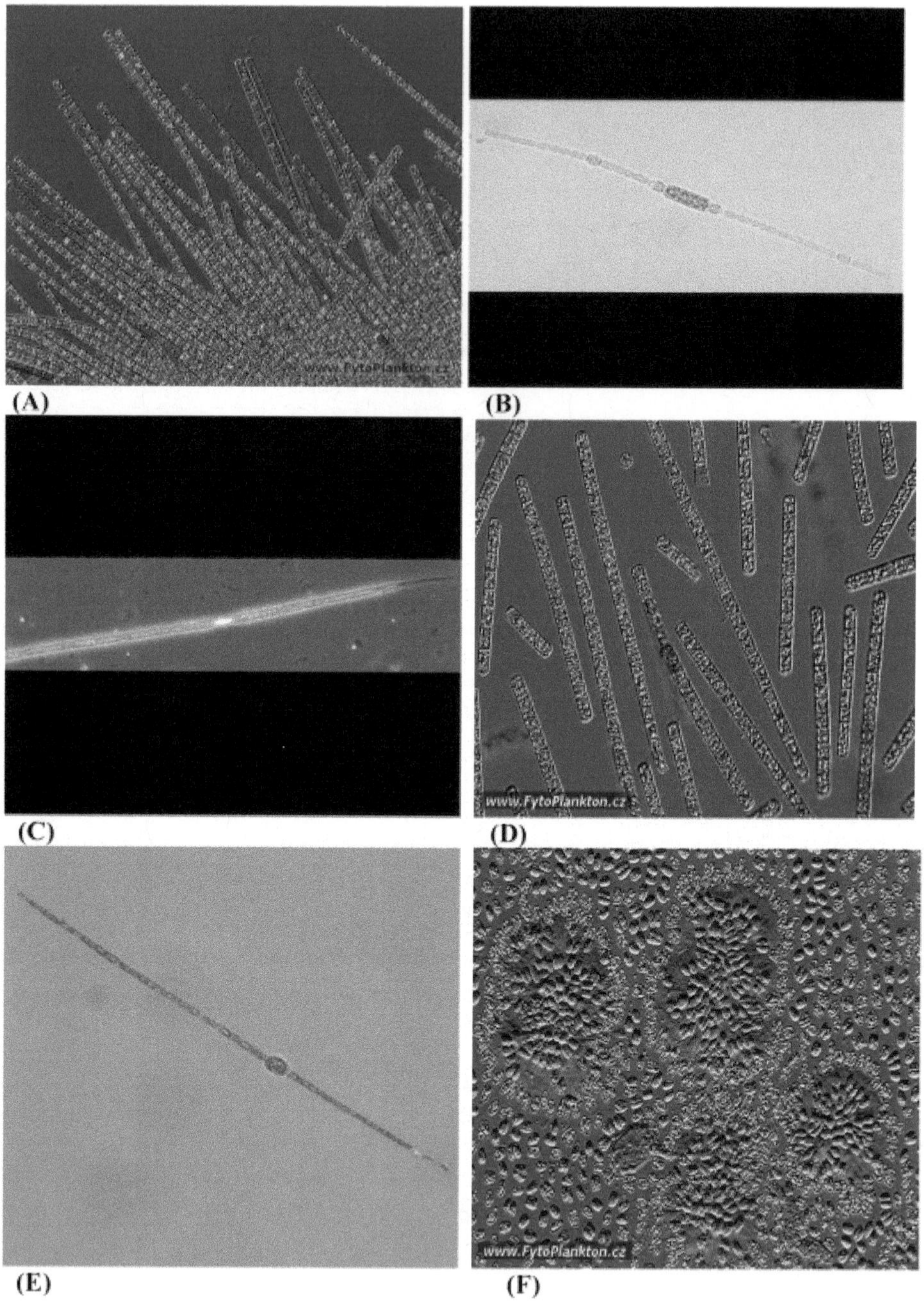

Figure 2: Toxic bloom-forming cyanobacteria. *Aphanizomenon flos-aquae* x 400 (A), *Aphanizomenon gracile* (B), *Aphanizomenon issatschenkoi* (C), *Aphanizomenon klebhanii* x 600 (D), *Aphanizomenon ovalisporum* (E), *Woronichinia naegeliana* x 400 (F). Pictures A, D and F are Nomarski contrast images courtesy P. Znachor, Laboratory of Phytoplankton Ecology, Institute of Hydrobiology, Biology Centre ASCR, Na Sádkách , 37005 České Budějovice, Czech Republic. (www.FytoPlankton.cz). Pictures B, C and E courtesy Mark T. Aubel, GreenWater Laboratories/CyanoLab, 205, Zeagler Drive, Suite 302, Palatka, Florida 32177, USA, (http://www.greenwaterlab.com/photo_algal.htm).

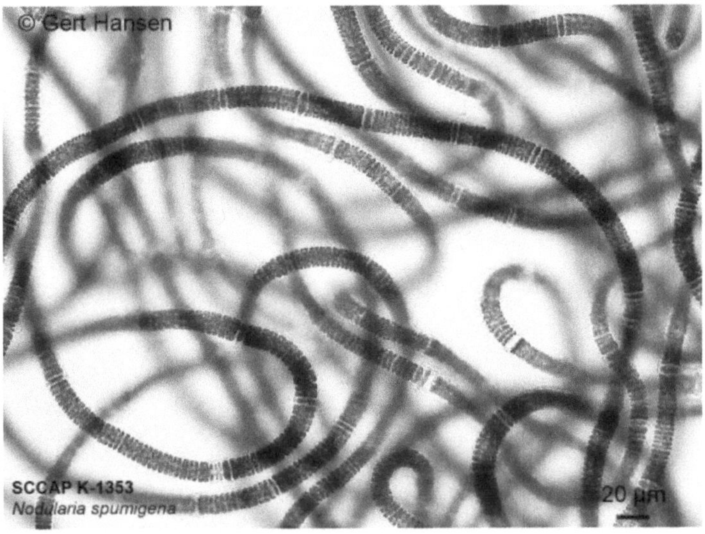

Figure 3: Filaments of *Nodularia spumigena* from Scandinavian Culture Collection for Algae and Protozoa (SCCAP K-1353). Picture courtesy Gert Hansen, **S**candinavian **C**ulture **C**ollection of **A**lgae & **P**rotozoa, Department of Biology, University of Copenhagen, Oester Farimagsgade 2D, DK-1353, Copenhagen, Denmark (http://www.sccap.dk/).

Thiel (1987) reported the presence of L-leucine and aminoisobutyric acid at Xaa and Zaa positions in a strain of *M. aeruginosa*. Watanabe *et al.* (1988) compared the MC content of different strains of *Microcystis* (belonging to *M. aeruginosa*, *M. viridis* and *M. wesenbergii*). One strain of *M. aeruginosa* showed high amounts of MC-YR and a lesser amount of MC-LR. However, three toxins, i.e. MC-RR, MC-YR and MC-LR were detected in two other strains of *M. aeruginosa* and four of M. *viridis*. The main component of toxin in these species was MC-RR while *M. wesenbergii* did not show the presence of any of the three MCs observed in the rest of the two species. Another variant, MC-LY with L-leucine and L-tyrosine at Xaa and Zaa positions has been reported from a strain of *M. aeruginosa* (Stoner *et al.*, 1989). MC-HtyR is a rare variant of MC produced by *A. flos-aquae*, *M. aeruginosa*, *M. aeruginosa* UTEX and *M. viridis* with homo-tyrosine and arginine at Xaa and Zaa positions (Botes *et al.*, 1985; Watanabe *et al.*, 1988; Harada *et al.*, 1991a; Oh *et al.*, 2000). Another such variant containing 1,2,3,4-tetrahydrotyrosine and arginine, designated as MC-(H_4)YR, from *Microcystis* spp. was isolated by Harada *et al.* (1991a). MC-RR has also been recorded from *A. flos-aquae*, *Anabaena* sp. strain 90, and *Anabaena* sp. strain 141 (Sivonen *et al.*, 1992a). Bloom samples of *Microcystis* spp. produced MC-AR, MC-FR (phenylalanine and arginine), MC-M(O)R and MC-WR (1-tryptophan and arginine) (Namikoshi *et al.*, 1992a). L-Glutamate and its δ-methylester [E(Ome)] are reported to exist at Xaa and Zaa positions in MC-L-Glu[E(Ome)] in the MC produced by *Anabaena* sp. strain 186 (Namikoshi *et al.*, 1998). Seven new MCs containing dehydroalanine (Dha) or L-serine (L-Ser) instead of N-methyldehydroalanine unit and the L-Glu and/its δ-methyl ester [E (Ome) in place of the two variable L-amino acids at Xaa and Zaa positions have been isolated from *Anabaena* sp. strain 186 (Namikoshi *et al.*, 1998). *M. aeruginosa* UTEX B 2666 produced MC-LF (with leucine and phenylalanine at Xaa and Zaa positions) besides MC-LA, MC-LAba and MC-LR (Diehnelt *et al.*, 2006). Other MCs such as MC-VF, MC-HilR, MC-LL, MC-LF and MC-LW respectively contain valine, phenylalanine; homoisoleucine, arginine; leucine, leucine; leucine, phenylalanine and leucine, 1-tryptophan at Xaa and Zaa positions. These have been described from cultures of *M. aeruginosa*

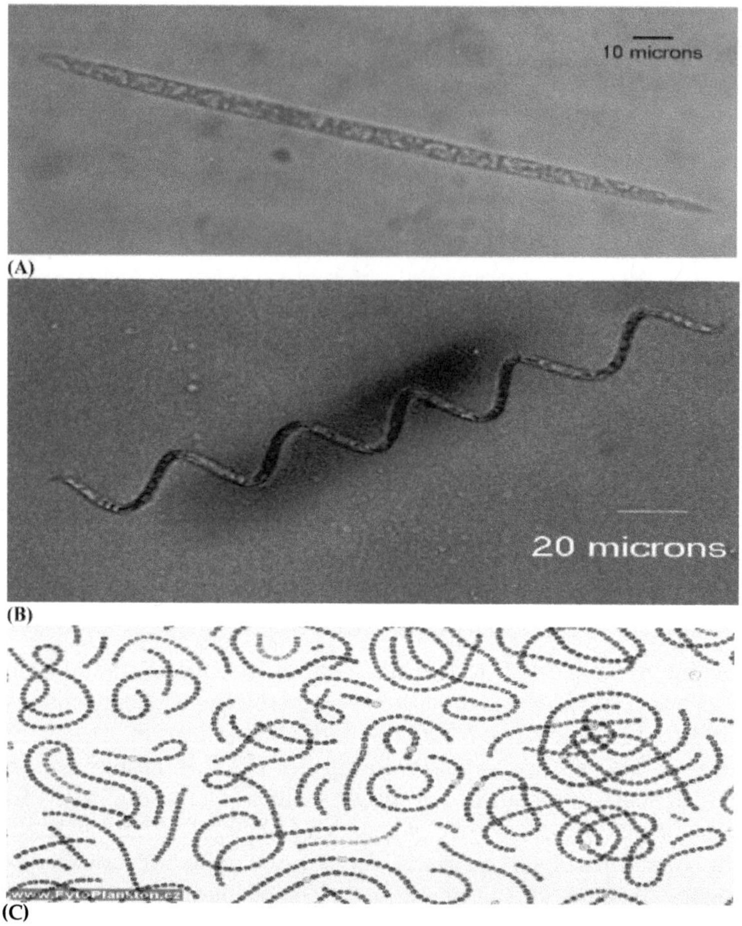

Figure 4: *Cylindrospermopsis raciborskii*, straight morphotype (A), *Raphidiopsis curvata* (B) and *Anabaena circinalis* (C). Pictures A and B courtesy Roger Burks (University of California at Riverside), Mark Schneegurt (Wichita State University) and Cyanosite (www-cyanosite.bio.purdue.edu). Picture C courtesy P. Znachor, Laboratory of Phytoplankton Ecology, Institute of Hydrobiology, Biology Centre ASCR, Na Sádkách, 37005 České Budějovice, Czech Republic (www.FytoPlankton.cz).

(Craig *et al.*, 1993; Azevedo *et al.*, 1994; Bateman *et al.*, 1995; Namikoshi *et al.*, 1995; Diehnelt *et al.*, 2006). Table 2 summarizes the various MCs that show variations at Xaa and Zaa positions.

Besides changes in L-amino acids at Xaa and Zaa positions, the variation in MCs is also due to demethylation in amino acids 3 and/or 7, modification of Adda and changes in amino acids at 1, 6 and 7 positions (Rinehart *et al.*, 1994; Sivonen, 1996). These variants of MCs are described here and presented in Table 3. The substitution of β-methylaspartic acid and N-methylhydroalanine by isoaspartic acid and dehydroalanine, respectively in *O. agardhii* resulted in the formation of [D-Asp^{3}, Dha^{7}] MC-RR (Krishnamurthy *et al.*, 1989; Luukkainen *et al.*, 1993). The same MC variant was also reported in *Anabaena* sp. strain 202A1 (Sivonen *et al.*, 1992a). New variants of MCs that have a modified Adda, i.e. DMAdda (*O*-demythylAdda) and ADMAdda (*O*-acetyl-*O*-demythylAdda) have further been isolated and characterized (Namikoshi *et al.*, 1991). Thus [DM Adda^{5}] MC-LR was reported from *Nostoc* sp. strain 152 (Sivonen *et al.*, 1990a) and *Nostoc* sp. strain IO-102-I (Oksanen *et al.*, 2004). Likewise, [ADM Adda^{5}] MC-LR was detected in *Microcystis* spp. (Namikoshi *et al.*, 1992a), *Nostoc* sp. strain 152 (Sivonen *et al.*, 1990, 1992b) and *Nostoc* sp. strain IO-102-I (Oksanen *et al.*, 2004).

Figure 5: *Lyngbya wollei* mats (A), a single filament of *L. wollei* (B), filaments of *Lyngbya* sp. stained with 5-Sytox green under epifluorescence microscope (C) and same as in (C) high resolution (D). Pictures A and B courtesy Mark T. Aubel, GreenWater Laboratories/CyanoLab, 205, Zeagler Drive, Suite 302, Palatka, Florida 32177, USA (http://www.greenwaterlab.com/photo_algal.htm). Pictures C and D courtesy Barry H. Rosen, USGS, USGS-Office of the Regional Executive-SE Area 12703 Research Parkway, Orlando, FL 32826, USA (http://microbiology.usgs.gov/image_gallery_cyanobacteria_lyngbya_epifluorescence_sytox.html).

Color image of this figure appears in the color plate section at the end of the book.

The substitution by D-serine in place of D-alanine in *Nostoc* sp. strain 152 resulted in the formation of [D-Ser^1, ADM Adda^5] MC-LR (Sivonen *et al.*, 1992b). The presence of N-methylserine (Mser) in place of N-methyldehydroalanine in *Microcystis* spp. formed [Mser^7] MC-LR (Namikoshi *et al.*, 1992a, 1995). This has also been reported subsequently in *M. aeruginosa* strain B2666 (Diehnelt *et al.*, 2006). The replacement of D-Glu by D-Glu (OC_3H_7O) in *Microcystis* spp. formed [D-Glu (OC_3H_7O)^6] MC-LR (Namikoshi *et al.*, 1992a). In *Nostoc* sp., L-arginine is replaced by L-homoarginine resulting in the formation of [D-Asp^3, ADM Adda^5] MC-LHar (Namikoshi *et al.*, 1991). The same species of MC was also noted in *Nostoc* sp. strain 152 (Sivonen *et al.*, 1991, 1992b). In *A. flos-aquae*, L-tyrosine is substituted by L-homotyrosine (Hty) forming a modified form of MC-YR that is known as [D-Asp^3] MC-HtyR (Harada *et al.*, 1991a). [D-Asp^3] MC-LR was reported from *A. flos-aquae* and *M. viridis* (Krishnamurthy *et al.*, 1989; Harada *et al.*, 1991a,), *O. agardhii* (Luukkainen *et al.*, 1993) and Antarctic cyanobacterial communities (Jungblut *et al.*, 2006). Likewise, [Dha^7] MC-LR is known from *M. aeruginosa* (Harada *et al.*, 1991b), *Anabaena* sp. strain 202A1 and 202A2 (Sivonen *et al.*, 1992a) and *O. agardhii* (Luukkainen *et al.*, 1993). Another variant of MC, [D-Asp^3, Dha^7] MC-LR was reported

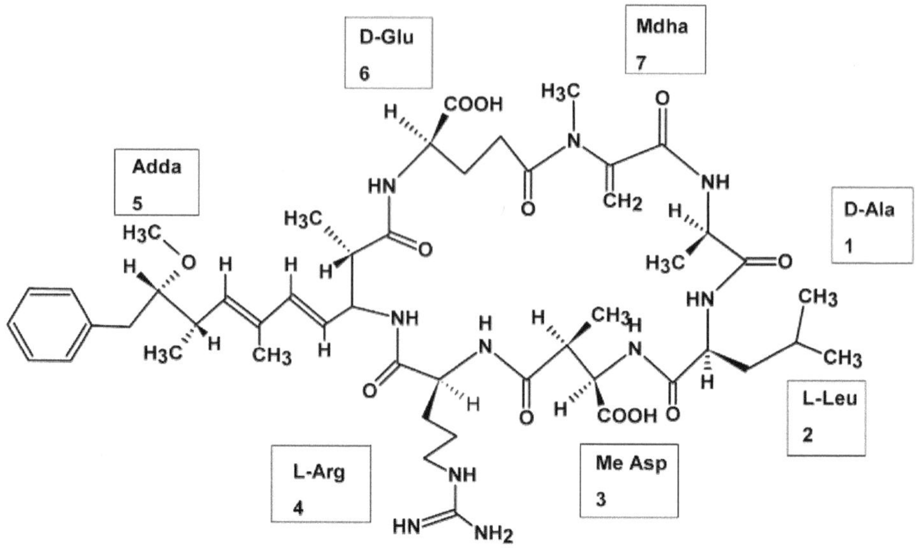

Figure 6: Structure of MC-LR.

from *M. aeruginosa* (Botes *et al.*, 1985) and *Anabaena* sp. strain 202A1 and 202A2 (Sivonen *et al.*, 1992a). [D-Asp[3]] MC-RR is produced by *O. agardhii* (Luukkainen *et al.*, 1994), *Anabaena* sp. (Sivonen *et al.*, 1992a) and from the bloom samples of *P. rubescens* (Briand *et al.*, 2005). [Dha[7]] MC-RR is produced by *M. aeruginosa, O. agardhii* (Luukkainen *et al.*, 1993) and *Anabaena* sp. (Sivonen *et al.*, 1992a). The bloom samples of *P. rubescens* showed the presence of [D-Asp[3], Dhb[7]] MC-RR (Briand *et al.*, 2005). Dienhelt *et al.* (2006) reported the occurrence of [Asp[3]] MC-LA and [Asp[3]] MC-LL in *M. aeruginosa* strain B2666. Sivonen *et al.* (1992a) isolated and characterized MC-LR, [D-Asp[3]] MC-LR, [Dha[7]] MC-LR, [D-Asp[3], Dha[7]] MC-LR, MC-RR, [D-Asp[3]] MC-RR, [Dha[7]] MC-RR and [D-Asp[3], Dha[7]] MC-RR from seven freshwater strains of *Anabaena* spp.

Quantitative differences between toxins within and between strains were found and the most abundant MCs found were either MC-LR or MC-RR or their demethyl derivatives in a strain. Thirteen (two green- and the rest red-pigmented) strains of *O. agardhii* were analyzed for MC content and the species of MCs. Each strain produced 2 to 5 variants of MC, i.e. [D-Asp[3]] MC-LR, [Dha[7]] MC-LR, [D- Asp[3], Dha[7]] MC-RR, [D-Asp[3]] MC-RR and [Dha[7]] MC-RR. Of these, [D-Asp[3]] MC-RR was most frequently found up to 80–99% of total MC content in each of the nine strains. However, [Dha[7]] MC-RR was the principal MC species in the rest of the four strains of which two were red-pigmented (Luukkainen *et al.*, 1993). Three new variants of MCs were also isolated but only one could be identified as [D-Asp[3], Mser[7]] MC-RR that has N-methyl-serine in place of Mdha. Such sustitution of N-methyl-serine has been described in *Microcystis* spp. as [Mser[7]] MC-LR (Namikoshi *et al.*, 1992c), in *Nostoc* sp. strain 152 as [ADM Adda[5], Mser[7]] MC-LR (Sivonen *et al.*, 1992b) and L-serine variants in two *Anabaena* sp. strain 202 are [L-Ser[7]] MC-LR and [L-Ser[7]] MC-RR and in *Anabaena* sp. strain 66 as [L-Ser[7]] MC-HtR (Namikoshi *et al.*, 1992b).

The blooms of *Microcystis* and *Planktothrix* were predominantly represented in lakes of Belgium and Luxemburg followed by those of *Anabaena* and *Woronichinia* and *Aphanizomenon*. Six variants of MCs were present in most of the blooms where MC-LR was predominantly represented in 64% of analyzed blooms, followed by MC-YR (55%). MC-LY has not been detected (Willame *et al.*, 2005).

Table 2: MC-variants due to change in positions of L-amino acids at Xaa(2) and Zaa(4) positions in the structure of microcystins.

MC-Variant	Xaa(2)	Zaa(4)	Organism	Reference(s)
MC-LA	Leucine	Alanine	*M. aeruginosa* *M. aeruginosa* UTEX B 2666 *M. viridis*	Botes *et al.* (1984); Diehnelt *et al.* (2006); Kaya and Watanabe (1990)
MC-Laba	Leucine	Aminoisobutyric acid	*M. aeruginosa* *M. aeruginosa* UTEX B 2666	Gathercole and Thiel (1987); Diehnelt *et al.* (2006)
MC-AR	Alanine	Arginine	*Microcystis* spp.*	Namikoshi *et al.*(1992a)
MC-YM	Tyrosine	Methionine	*M. aeruginosa*	Botes *et al.* (1984); Elleman *et al.* (1978)
MC-YA	Tyrosine	Alanine	*M. aeruginosa*	Botes *et al.* (1985)
MC-LR	Leucine	Arginine	*M. aeruginosa* *A. variabilis* *M. viridis** *Anabaena* strain 90; *Anabaena* spp. *Anabaena* sp. strain 141; *A. flos-aquae* *M. aeruginosas* train B2666 *M. aeruginosa* UTEX 2388 Antarctic* cyanobacteria	Botes *et al.* (1985) Krishnamurthy *et al.* (1989) Rinehart *et al.* (1988); Watanabe *et al.* (1988) Sivonen *et al.* (1992a) Diehnelt *et al.* (2006) Oh *et al.* (2000) Jungblut *et al.* (2006)
MC-LY	Leucine	Tyrosine	*M. aeruginosa*	Stoner *et al.* (1989)
MC-FR	Pheneylalanine	Arginine	*Microcystis* spp.*	Namikoshi *et al.* (1992a)
MC-M(O)R	Methionine S-Oxide	Arginine	*Microcystis* spp.*	Namikoshi *et al.* (1992a)
MC-YM(O)	Tyrosine	Methionine S-Oxide	*M. aeruginosa*	Elleman *et al.* (1978); Botes *et al.* (1985)
MC-RR	Arginine	Arginine	*M. aeruginosa* *M. viridis** *Anabaena* sp. strain 60 *A. flos-aquae* *Anabaena* sp strain 90; sp. strain 141 *Anabaena* sp. strain 90; sp. strain 202A1 *M. aeruginosa* UTEX 2388	Kusumi *et al.* (1987) Painuly *et al.* (1988) Watanabe *et al.* (1988) Sivonen *et al.* (1992a) Rapala *et al.* (1997) Oh *et al.* (2000)
MC-YR	Tyrosine	Arginine	*M. aeruginosa* *M.viridis** *M.aeruginosa* UTEX 2388	Botes *et al.* (1985) Watanabe *et al.* (1988) Oh *et al.* (2000)
MC-HtyR	Homotyrosine	Arginine	*A. flos-aquae*	Harada *et al.* (1991a)
MC-(H_4) YR	1,2,3,4,-tetrahydrotyrosine	Arginine	*Microcystis* spp.	Harada *et al.* (1991a)

Table 2: contd....

Table 2: contd....

MC-Variant	Xaa(2)	Zaa(4)	Organism	Reference(s)
MC-WR	1-tryptophane	Arginine	*Microcystis* spp.*	Namikoshi *et al.* (1992a)
MC-LW	Leucine	1-tryptophane	*M. aeruginosa*	Bateman *et al.* (1995)
MC-L-Glu [E(ome)]	Glutamate	δ-methylester [E(ome)]	*Anabaena* sp. strain 186	Namikoshi *et al.* (1998)
MC-LF	Leucine	Phenylalanine	*M. aeruginosa* strain B2666	Azevedo *et al.* (1994); Diehnelt *et al.* (2006)
MC-LL	Leucine	Leucine	*M. aeruginosa*	Craig *et al.* (1993)
MC-HilR	Homo-isoleucine	Arginine	*M. aeruginosa*	Namikoshi *et al.* (1995)
MC-VF	Valine	Phenylalanine	*M. aeruginosa*	Bateman *et al.* (1995)

*Represents natural bloom samples/natural materials and the rest are from axenic cultures of the respective cyanobacteria.

Table 3: MC-Variants of cyanobacteria with substitutions in 1st to 7th positions of the general structure of microcystin.

MC-Variant[a]	Organism	Reference(s)
[D-Asp[3], Dha[7]] MC-LR	*M. aeruginosa* *Anabaena* sp. strain 202A1; 202A2	Botes *et al.* (1985); Harada *et al.* (1991b) Sivonen *et al.* (1992a)
[D-Asp[3]] MC-LR	*A. flos-aquae*; *M. viridis** *Anabaena* sp. strain 60; *Anabaena* sp. strain 90, 141; *A. flos-aquae* *O. agardhii* Antarctic cyanobacterial comunities*	Krishnamurthy *et al.* (1989); Harada *et al.* (1990b, 1991a) Sivonen *et al.* (1992a) Luukkainen *et al.* (1993) Jungblut *et al.* (2006)
[Dha[7]] MC-LR	*M. aeruginosa* *Anabaena* sp. strain 202A1; strain 202A2 *O. agardhii*	Harada *et al.* (1991b) Sivonen *et al.* (1992a) Luukkainen *et al.* (1993)
[D-Asp[3], ADM Adda[5]] MC-LR	*Nostoc* sp. *Nostoc* sp. strain 152 *Nostoc* sp. strain IO-102-I	Namikoshi *et al.* (1990) Sivonen *et al.* (1990a) Oksanen *et al.* (2004)
[D-Asp[3], Dha[7]] MC-RR	*O. agardhii** *Anabaena* strain 202A1	Krishnamurthy *et al.* (1989); Luukkainen *et al.* (1993, 1994); Sivonen *et al.* (1992a)
[L-Ser[7]] MC-LR	*Microcystis* spp.* *M. aeruginosa* strain B2666	Namikoshi *et al.* (1992c) Diehnelt *et al.* (2006)
[ADM Adda[5]] MC-LR	*Nostoc* sp. *Nostoc* sp. strain 152 *Nostoc* sp. strain IO-102-I	Namikoshi *et al.* (1990a) Sivonen *et al.* (1990a, 1992b) Oksanen *et al.* (2004)
[D-Asp[3], ADM Adda[5]] MC-LHar	*Nostoc* sp. strain 152	Sivonen *et al.* (1992b)
[D-Asp[3]] MC-RR	*O. agardhii* *Anabaena* sp. strain 60; sp. strain 141; *A. flos-aquae* *P. rubescens**	Meriluto *et al.* (1989); Luukkainen *et al.* (1993) Sivonen *et al.* (1992a) Briand *et al.* (2005)
[ADM Adda[5]] MC-LHar	*Nostoc* sp. *Nostoc* sp. strain 152	Namikoshi *et al.* (1990) Sivonen *et al.* (1990a, 1992b)
[D-Ser[1], ADM Adda[5]] MC-LR	*Nostoc* sp. strain 152	Sivonen *et al.* (1992b)
[ADM Adda[5], MeSer[7]] MC-LR	*Nostoc* sp. strain 152	Sivonen *et al.* (1992b)

Table 3: contd....

Table 3: contd....

MC-Variant[a]	Organism	Reference(s)
[D-Asp[3]] MC-HtyR	*A. flos-aquae*	Harada *et al.* (1991a)
[D-Glu-OC_2H_3OH)[6]] MC-LR	*Microcystis* spp.*	Namikoshi *et al.* (1992a)
[Dha[7]] MC-E(OMe) E(OMe)	*Anabaena* sp. strain 186	Namikoshi *et al.* (1998)
[D-Asp[3], Dha[7]] MC-E(OMe) E(OMe)	*Anabaena* sp. strain 186	Namikoshi *et al.* (1998)
[L-Ser[7]] MC-E(OMe) E(OMe)	*Anabaena* sp. strain 186	Namikoshi *et al.* (1998)
[D-Asp[3], L-Ser[7]] MC-E(Ome) E(OMe)	*Anabaena* sp. strain 186	Namikoshi *et al.* (1998)
[Dha[7]] MC-EE(OMe)	*Anabaena* sp. strain 186	Namikoshi *et al.* (1998)
[D-Asp[3], Dha[7]] MC-EE(OMe)	*Anabaena* sp. strain 186	Namikoshi *et al.* (1998)
[L-Ser[7]] MC-EE(OMe)	*Anabaena* sp. strain 186	Namikoshi *et al.* (1998)
[DM Adda[5]] MC-LR	*Nostoc* sp. strain 152 *Microcystis* spp.* *Nostoc* sp. strain IO-102-I	Sivonen *et al.* (1992b) Namikoshi *et al.* (1992a) Oksanen *et al.* (2004)
[D-Asp[3], Dhb[7]] MC-RR	*P. rubescens**	Briand *et al.* (2005)
[Asp[3]] MC-LA	*M. aeruginosa* strain B2666	Dienhelt *et al.* (2006)
[Asp[3]] MC-LL	*M. aeruginosa* strain B2666	Dienhelt *et al.* (2006)
[D-Asp[3], D-Glu(OCH_3)[6] MC-LR	*A. flos-aquae*	Sivonen *et al.* (1992d)
[(6Z)-Adda[5]] MC-LR	*M. viridis**	Harada *et al.* (1990a,b)
[L-Ser[7]] MC-LR	*Anabaena* sp.	Namikoshi *et al.* (1992c)
[D-Asp[3], ADM Adda[5], Dhb[7]] MC-RR	*Nostoc* sp.	Beattie *et al.* (1998)
[Dha[7]] MC-FR	*Microcystis* sp.	Luukkainen *et al.* (1994)
[Dha[7]]MC-HphR	*Anabaena* spp.	Namikoshi *et al.* (1992b)
[D-Asp[3], Dha[7]] MC-HtyR	*Anabaena* spp.	Namikoshi *et al.* (1992b)
[Dha[7]] MC-YR	*M. aeruginosa*	Sivonen *et al.*(1992c)
[D-Asp[3]] MC-YR	*Microcystis* spp.*	Namikoshi *et al.* (1992d)
[L-Ser[7]] MC-RR	*Anabaena* spp. *M. aeruginosa* (cultures as well as bloom samples)	Namikoshi *et al.* (1992c) Luukkainen *et al.* (1993)
[(6Z)-Adda[5]] MC-RR	*M. viridis*	Harada *et al.* (1990a, b)
[D-Asp[3], Me Ser[7]] MC-RR	*O. agardhii*	Luukkainen *et al.* (1993)
[Dha[7]] MC-HtyR	*Anabaena* spp.	Namikoshi *et al.* (1992b)
[D-Glu-$OC_2H_3(CH_3)OH$[6]] MC-LR	*Microcystis* spp.*	Namikoshi *et al.* (1992a)
[D-Asp[3], ADMAdda[5], Dhb[7]] MC-RR	*Nostoc* sp.	Beattie *et al.* (1998)
[L-Ser[7]] MC-HtyR	*Anabaena* sp.	Namikoshi *et al.* (1992b)
[D-Asp[3], ADM Adda[5], Dhb[7]] MC-HtyR	*Nostoc* sp.	Beattie *et al.* (1998)
[L-MeLan[7]] MC-LR	*Microcystis* spp.*	Namikoshi *et al.* (1995)

[a]Aba, aminoisobutyric acid; D-Asp, D-erythro-β-methylaspartic acid; Dha, dehydroalanine; Dhb, N-methyldehydrobutyrine; DMAdda, O-demethylAdda; ADMAdda, O-acetyl-O-demethylAdda; [E(Ome)], δ-methylester of L-Glu; Mser, N-methylserine; Har, homoarginine; Hph, Homophenylalanine; Hty, Homotyrosine; MeLan, N-methylanthionine; M(O), methionine S-oxide; MeSer, N-Methylserine.

*Denotes natural bloom samples/or natural materials and the rest are axenic cultures.

Studies on the development of cyanobacterial blooms in eight lakes of northern Poland revealed the existence of oscillatorian and nostocacean species reaching an average chlorophyll *a* concentration of 10 $\mu g\ L^{-1}$ corresponding to MC levels of 4–5 $\mu g\ L^{-1}$. Four variants of MCs, i.e. dm MC-RR, MC-RR, MC-YR and MC-LR were detected. Briand *et al.* (2005) detected two main variants of MCs, [D-Asp^3] MC-RR and [D-Asp^3, Dhb^7] MC-RR, from natural bloom populations of *P. rubescens* with high concentrations (6.7 $\mu g\ L^{-1}$) during August-December every year. The cellular concentration of MCs ranged from 0.1 to 0.3 pg $cell^{-1}$. Interestingly, laboratory cultures of the organism only produced [D-Asp^3] MC-RR with a cellular quota of MC of 0.3 to 0.7 pg $cell^{-1}$. Species of *Aphanizomenon*, including a Scandinavian species, *Aph. skujae* (Skuja) Kom.-Legn. & Cromb, are predominantly noted suggesting that the future toxin detection programmes should also focus attention on the evaluation of impending potential danger from neurotoxins as well (Mankiewicz *et al.*, 2005).

i) Biosynthesis: Multimodular enzyme complexes known as peptide synthetases mediate synthesis of small peptides through a thiotemplating mechanism rather than by mRNA. Cyclosporin (an immunosuppressant), β-lactam group of antibiotics like penicillins, cephalosporins and certain others such as gramicidin S, tyrocidin A and surfactins are synthesized through this route (Kleinkauf and von Doehren, 1996; Neilan *et al.*, 1999). These enzyme complexes consist of polyketide synthases (PKSs) and non-ribosomal peptide synthetases (NRPSs). Peptide synthetases possess a modular structure containing specific functional domains and these domains exhibit a high degree of sequence conservation (Stachelhaus *et al.*, 1995). The various functional domains for recognition, aminoacyl adenylation and thioesterification of its amino acid substrate as well as for the elongation of the growing peptide are present (Kleinkauf and von Doehren, 1996). MCs are synthesized through the pathway of mixed PKSs and NRPSs (Arment and Carmichael, 1996; Dittmann *et al.*, 1997). This is also known as combinatorial biosynthesis (McDaniel *et al.*, 1993; Stachelhaus *et al.*, 1995). Welker and von Doehren (2006) reviewed the synthesis of cyanobacterial peptides. A screening of the PKS and NRPS genes, from axenic cultures of freshwater and marine cyanobacteria and 14 sequenced genomes brought to light that their distribution is more common in the filamentous forms and the degenerate primers used in the identification of these modules would be helpful in the synthesis of natural products from cyanobacteria (Ehrenreich *et al.*, 2005). The genetic basis of cyanobacterial toxin production has been discussed (Kurmayer and Christiansen, 2009; Pearson *et al.*, 2010).

The origin of carbon atoms in the unusual amino acid, Adda and methylaspartic acid has been traced by C^{14}-incorporation. The methyl groups on C_6 and C_8 of Adda are methionine derived while Me group of C_2 of Adda might be probably derived through methionine or acetate (or possibly from propionate in the absence of acetate). Further it was shown that C_3–C_8 segment of Adda is acetate derived or likely from propionate in the absence of acetate. The remaining carbons in the Adda are phenylalanine derived. Phenylacetyl-CoA has been suggested to be the most probable initiator for Adda biosynthesis. The biosynthesis of Masp takes place as follows: A condensation reaction involving Acetyl-CoA with pyruvic acid results in the formation of 2-hydroxy-2-methylsuccinic acid. This gets converted to 2-hydroxy-3-methylsuccinic acid that upon oxidation gives rise to 2-oxo-3-methylsuccinic acid. A transamination reaction finally yields Masp (Moore *et al.*, 1991).

With the help of conserved gene sequences in bacteria and fungi for NRPSs, it was possible to identify the genes in *Microcystis* and *Anabaena* (Dittmann *et al.*, 1996). These genes were designated as *mcy* genes. To confirm this, insertional inactivation of these genes was done in the toxic *M. aeruginosa* PCC 7806 that yielded non-toxic transformants. Thus it was proposed that *mcyB* gene encodes MC synthetase (Dittmann *et al.*, 1997). The detection and characterization of MC and peptide synthetase genes in strains of *Anabaena*, *Aphanizomenon*, *Cylindrospermopsis*, *Microcystis*, *Nodularia*,

Nostoc, Oscillatoria, Plectonema and *Pseudoanabaena* was described by Neilan *et al.* (1999) who not only demonstrated their existence in cultures of these cyanobacteria but also from naturally occurring bloom populations.

The precursors required for synthesis are phenylacetate, malonyl coenzyme A, S-adenyl-L-methionine, glutamate, serine, alanine, leucine, D-methyl-isoaspartate and arginine. The gene cluster responsible for the synthesis of MC has now been sequenced (Nishizawa *et al.*, 2000; Tillett *et al.*, 2000). The role of PKS has been identified in the production of fatty acid side chain of Adda. The biosynthetic pathway of MC-LR starting with the biosynthesis of Adda is presented. Tillett *et al.* (2000) recognized two operons (from a large gene cluster) for *mcyABC* (peptide synthetase) and *mcyDE* (hybrid polyketide-peptide synthetase), as confirmed by the cloning and sequencing of MC synthetase gene cluster in two different strains of *M. aeruginosa*. The two operons recognised are *mcyABC* and *mcyDEFGHIJ* consisting of 10 genes. The genes *mcyA*, *mcyB* and *mcyC* have been identified as responsible for activation and incorporation of five amino acid constituents of the heptapeptide (Nishizawa *et al.*, 1999). Thus *mcyA1* is shown to be responsible for N-methyl-dehydroalanine; *mcyA2* for D-alanine; *mcyB1* for Xaa (at position 2); *mcyB2* for D-erythro-β-methyl-iso-aspartic acid; *mcyC1* for Zaa (at position 4). The second operon *mcyD-J* is responsible for Adda and D-glutamate synthesis (Nishizawa *et al.*, 2000; Tillett *et al.*, 2000). The two operons *mcyABC* and *mcyDEFGHIJ* are transcribed in opposite directions. Apart from the six genes (*mcyA* to *mcyE* and *mcyG*) responsible for the incorporation of precursors, the rest of the four genes were shown to encode monofunctional proteins that possess tailoring functions. Thus these four genes, i.e. *mcyJ*, *mcyF*, *mcyI* and *mcyH* help in O-methylation, epimerization, dehydration and cellular localization, respectively (Kaebernick *et al.*, 2002).

There are three D-amino acids in MCs, i.e. D-glutamate, D-alanine and D-Me-Asp/D-aspartate but there are only two racemase activities present in the gene cluster. An integral epimerase domain present in the NRPS module is responsible for the incorporation of D-alanine into heptapeptide backbone. Since an integrated epimerase domain for incorporating D-glutamate and D-MeAsp/D-aspartate is lacking, it is suggested that these two amino acids might be converted by racemases either before or after their incorporation. Nishizawa *et al.* (2001) showed that *mcyF* mediates synthesis of D-glutamate since gene disruption studies abolished MC production in *M. aeruginosa* K-139 but did not affect cell growth. Further, *mcyF* supported D-Glu-independent growth of an auxotroph of *E. coli* for D-Glu. Thus these workers reported for the first time the existence of the racemase in prokaryotic NRPS genes. On the contrary, Sielaff *et al.* (2003) presented evidences for the synthesis of aspartate racemases rather than glutamate racemases by *mcyF* gene and also detected a gene for L-glutamate specific racemase that is located outside the *mcy* gene cluster.

It has now been possible to identify the gene clusters for heptapeptide biosynthesis in *Planktothrix* and *Anabaena* strain 90. There are certain differences in the organization of the *mcy* gene cluster among the three MC producers (Table 4; Fig. 7). The significant features of MC biosynthesis in *Planktothrix* CYA 126/8 are that (i) sequence analysis revealed *mcy* region to consist of 55-kb cluster having 9 genes, (ii) of these, 8 genes (*mcyA, B, C, D, E, G, H* and *J*) are similar to *M. aeruginosa* genes governing peptide synthetases, (iii) *mcyT*, an additional gene encoding thioesterases, is present which is lacking in *M. aeruginosa*, (iv) the racemase genes *mcyF* and *mcyI* (the products of which are similar to 3-phosphoglycerate dehydrogenase genes) present in *M. aeruginosa* are reported to be absent in *P. agardhii*, (v) while *mcyA-C* and *mcyD-J* constitute two operons that are transcribed bidirectionally in *M. aeruginosa*, the whole of *mcy* gene cluster except *mcyT* represents a single operon and (vi) the gene *mcyJ* synthesizes O-methyltransferase. The probability of lateral gene transfer (LGT) of complete *mcy* gene clusters between two different genera has been ruled out (Christiansen *et al.*, 2003). In support of this, Rantala *et al.* (2004) presented phylogenetic evidences for the early evolution of MC synthesis. Three regions of *mcy* gene cluster (a fragment of *mcyA* gene, *mcyD* and *mcyE*) were amplified and sequenced in thirty-six strains of *Anabaena, Nodularia, Nostoc, Planktothrix* and *Microcystis*. To investigate LGT in the evolution of MC biosynthetic gene cluster, the amplification and sequencing of 16S rRNA and *rpoC1* genes was done. They concluded that (i) the ability to produce MC has been repeatedly lost in the derived lineages of cyanobacteria and (ii) the genes encoding nodularin synthetase are recently derived from those encoding MC synthetase.

Rouhiainen *et al.* (2004) sequenced and characterized the *mcy* gene cluster in *Anabaena* strain 90. The main findings are as follows. (i) The total size of *mcy* region is 55.4 kb. (ii) The *mcy* region consists of 3 operons, i.e. genes *mcyA, mcyB, mcyC* constitute the first operon that is transcribed in the opposite direction to the second. (iii) Genes *mcyG, mcyD, mcyI, mcyE, mcyF, mcyJ* form the second operon and gene *mcyH* represents the third operon. (iv) The *mcyA, mcyB, mcyC* are responsible for NRPSs where as *mcyD* is a PKS gene. Genes *mcyG* and *mcyE* show functions of peptide synthetase and PKS mixed genes. (v) Genes *mcyJ, mcyF* and *mcyI* encode a methyltransferase, an aspartate racemase and a D-3-phosphoglycerate dehydrogenase, respectively. Besides exhibiting differences in the arrangement of genes in *mcy* gene cluster (from those of *M. aeruginosa* and *P. agardhii*), the *Anabaena* strain 90 MC genes exhibited an average sequence match of 74% with the other two genera.

The knowledge that has been generated on the cloning and sequencing of the *mcy* gene cluster falls short of explaining as to how transcription of the essential mRNAs takes place. The work of Kaebernick *et al.* (2000, 2002) is significant in this direction. Northern blotting has been the traditional choice for transcriptional analysis of gene clusters while radioactive primer extensions or S1 nuclease mapping are useful in the recognition of transcription start sites (Domanski *et al.*, 1997). Since these methods were not successful in the transcriptional analysis of *mcyABCDEFGHIJ* genes, Kaebernick *et al.* (2002) employed reverse transcription (RT)-PCR to detect mRNA transcripts. To map the transcription initiation points, rapid amplification of C-DNA ends (RACE; also known as ligation-anchored PCA) was used by them. Due to this, a complete transcriptional analysis of *mcy*-gene

Table 4: Differences between *mcy* gene clusters of three major MC producers.

mcy gene cluster	*M. aeruginosa* PCC 7806 (1)	*Planktothrix* CYA 126/8 (2)	*Anabaena* strain 90 (3)
Length	55 kb	55 kb	55.4 kb
No. of operons	2	1	4
No. of genes	10	8	10
Organization of operons	*mcyA-mcyB-mcyC* *mcyD-mcyJ*	*mcyA-mcyB-mcyC-mcyD-mcyE-mcyG-mcyH-mcyJ*; *mcyT* additional gene; *mcyF* and *mcyI* absent	*mcyA-mcyB-mcyC* *mcyG-mcyD-mcyI-mcyE-mcyF-mcyJ* *mcyH*
Functions	*mcyA-mcyB-mcyC* NRPS; activation and incorporation of 5 amino acids *mcyG*-mcyE-*mcyD*-PKS; synthesize the Adda and D-glutamate	*mcyA-mcyB-mcyC* NRPS similar function *mcyD-mcyE-mcyG* PKS. similar function	*mcyA-mcyB-mcyC* NRPS genes *mcyE-mcyD-mcyG* PKS; similar function
References	Tillett *et al.* (2000)	Christiansen *et al.* (2003)	Rouhiainen *et al.* (2004)

Note: In (1) *mcyJ, mcyF, mcyI* and *mcyH* encode enzymes of O-methylation, epimerization, dehydration and cellular localization, respectively. In (2) the gene products of *mcyF* and *mcyI* that bear resemblance to 3-phosphoglycerate dehydrogenase are absent. The additional gene *mcyT* encodes a thioesterase and *mcyJ* encodes O-methyltransferase. In (3) *mcyI* encodes 3-phosphoglycerarate dehydrogenase as in (1) while *mcyF* encodes aspartate racemase and *mcyJ* is responsible for synthesis of O-methyltransferase.

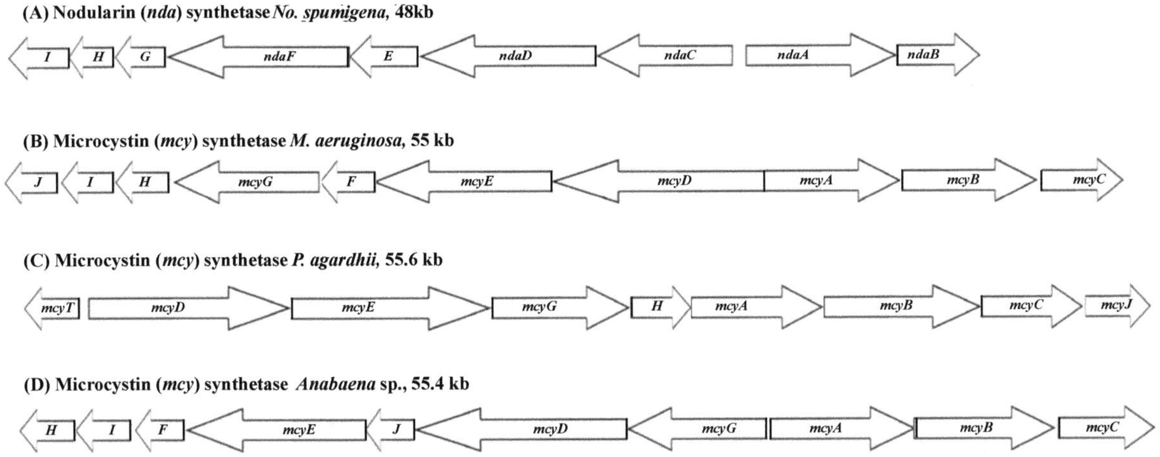

Figure 7: Hepatotoxin gene clusters from various cyanobacteria. Structures of the nodularin gene cluster of *No. spumigena* (A) and microcystin gene clusters of *M. aeruginosa* (B), *P. agardhii* (C) and *Anabaena* sp. 90 (D). Diagram redrawn and not to scale. With the kind permission of B. A. Neilan, Cyanobacteria and Astrobiology Research Laboratory, School of Biotechnology and Biomolecular Sciences, The University of New South Wales, Sydney 2052, NSW, Australia [Pearson *et al.* (2010) *Mar Drugs* **8:** 1650–1680; doi:10.3390/md8051650].

cluster could be possible. The important transcriptional events are as follows. (i) The gene cluster is transcribed as two polycistronic transcripts in two opposite directions from a central promoter located in between *mcyABC* and *mcyDEFGHIJ*. (ii) Two transcription start sites have been identified for both *mcyA* and *mcyD* that are dependent on high (68 µmol of photons $m^{-2} s^{-1}$) and low (16 µmol of photons $m^{-2} s^{-1}$) intensities of light, respectively. (iii) The polyketide and tailoring genes *mcyE* to *mcyJ* appear to possess individual promoters. (iv) The inter-cistronic start sites for genes *mcyE* to *mcyJ* suggests that the gene regulation is very complex with multiple and alternate messages.

***ii) Genetic diversity of MC producers*:** Mikalsen *et al.* (2003) identified two groups of genetic variants in closely related *Microcystis* strains (10 strains of *M. aeruginosa*, 2 strains of *M. viridis*, 2 strains of *M. botrys* and 2 *M.* spp., *M. cf. flos-aquae*, *M. cf. wesenbergii*, *M. cf. ichthyoblabe*) on the basis of genetic characterization of a region of adenylation domain in module *mcyB1*. They suggested that recombinational events in between *mcyB1* and the corresponding domains in *mcyC* are responsible for the observed genetic variation in MC production. This is substantiated by the following: (i) the strains containing *mcyABC* gene cluster produced detectable levels of MC; (ii) B-type variants of the *mcyB1* produced MC-LR isoforms and one strain (N-C 161/1) produced MC-LR as well as MC-YR. Another strain N-C 31 produced only MC-LR. N-C 118/2 and PCC 7806 produced large amounts of [Asp^3] MC-LR and MC-LR and (iii) all strains with C-type variants of the *mcyB1* module produced MC-RR and other members of sub-groups synthesized MC-LR in addition to MC-RR. Mikalsen *et al.* (2003) further concluded that: (i) the *mcyABC* gene cluster is conserved in all strains of *Microcystis*, (ii) the same gene arrangement in the synthetase-negative strains suggests that *mcyA* and *mcyB* genes have been lost in some strains, (iii) the MC synthesis is probably an ancestral feature of the genus *Microcystis* and (iv) LGT appears to be the mechanism for the evolution of structurally related peptides in diverse microorganisms as indicated earlier by Kleinkauf and van Doehren (1996).

Kurmayer *et al.* (2004) compared two populations of *Planktothrix*, i.e. *P. agardhii* and *P. rubescens* to find out the abundance of active and inactive MC genotypes. *P. agardhii*, a green-pigmented organism inhabiting shallow, mesotrophic or hypertrophic water bodies, showed two genotypes: those possessing *mcyA* gene and those lacking *mcyA* whereas all the 49 strains of *P. rubescens*, a red-pigmented bacterium growing in deep stratified oligo- or mesotrophic waters, showed the *mcyA* gene. Although one strain of *P. agardhii* and eight strains of *P. rubescens* exhibited MC synthetase genes but they did not produce MC. These results further demonstrate that the populations of *P. agardhii* do not show proper distribution of *mcy* genes suggesting that the absence/inactivation of *mcy* genes. This results in the development of non-toxic strains of *P. agardhii* that co-exist with the toxin-producing strains. Kurmayer *et al.* (2005) identified MC-producing ecotypes of *P. agardhii* or *P. rubescens* that differed in production of specific MC forms and correlated the genetic variation within adenylation domains (*Ad*). Thus the sequences of *mcyA Ad1*, *mcyB Ad1* and *mcyC Ad*1 were analysed in 17 *Planktothrix* strains. Two distinct *Ad*-genotypes were noted in the sequences of *mcyA Ad1* that differed in base pair composition and insertion of an N-methyltransferase (NMT) domain. The genotype *mcyA Ad1* possessing NMT, synthesized N-methyl-dehydroalanine while *mcyA Ad1* genotype without NMT produced dehydrobutyrine in position 7. The sequence of *mcyB Ad1* exhibited a lower variation where one *Ad*-genotype showed the formation homotyrosine and another *Ad*-genotype synthesized arginine at position 2. Significantly, the sequences of *mcyC Ad1* were highly similar and all these formed arginine in position 4.

Mbedi *et al.* (2005) studied the variability of *mcy* gene cluster (8 genes) in 46 strains of *Planktothrix*. The presence of PCR amplification products for two *mcy* regions in non-toxic strains was noticed whereas the MC-producing strains contained two variable (in sequence and length) *mcy* regions and four regions that were highly conserved and specific. In order to find out the development of non-toxic strains (deficient in MC production) in natural populations of *Planktothrix*, Christiansen *et al.* (2006) analyzed 29 strains of *P. rubescens* and found that some of the strains were in fact represent mutant strains of the organism due to either deletions or insertions in the *mcy* gene cluster. The deletions are of the order of 400 bp long in *mcyB* region in one strain and 1869 bp long in *mcyHA* region of three strains. Insertions of 1429 bp in *mcyD* region (in eight strains) and 1433 bp each in the regions of *mcyEG* (in one strain) and *mcyA* (in one strain) have been identified. The mosaic nature of *mcyABC* gene cluster in naturally occurring strains of *Microcystis* has largely been understood by the detection of point mutations, multiple recombination events, insertion element(s). Using probes for *dnaN*, *uma1*, *mcyA* (from *M. aeruginosa* PCC 7806), Tooming-Klunderud *et al.* (2008) found point mutations in the NMT domain of *mcyA* and multiple recombination events gave rise to 'phylogenetic mosaics' in *mcyA* (NMT domain) and the adenylation (A) domain sequences of *mcyB* and *mcyC*.

***iii) Other toxins of Microcystis*:** Besides various forms of MCs, *M. aeruginosa* also produces certain other toxins, i.e. micropeptins A and B that inhibit plasmin and trypsin (Okino *et al.*, 1993a), microginin that inhibits angiotensin-converting enzyme (Okino *et al.*, 1993b) and microviridins that inhibit elastase enzyme (Okino *et al.*, 1995). Microginins are linear tetra and pentapeptides in which chlorination occurs at N-terminal aliphatic moiety (Ishida *et al.*, 1998) whereas in aeruginosins and cyanopeptolins chlorination occurs at aromatic moieties. Agrawal *et al.* (2001) showed the presence of a protease inhibitor in *M. aeruginosa* whose identity has not been determined. It means that *M. aeruginosa* is a potential source of a number of MCs as well as enzyme inhibitors posing a great challenge for the herbivores with complex heterogeneous mixture of inhibitors and toxins (Fastner *et al.*, 2001; Welker *et al.*, 2003, 2004a,b). Cyanopeptolins possess an Arg or Lys residue at the N-terminal of the modified amino acid 3-amino-6-hydroxy-piperidone. These were identified as inhibitors of trypsin-like activity in *Daphnia* whereas peptides of microviridin class were moderately active. Microginins and MCs did not inhibit trypsin-like activity (Czarnecki *et al.*, 2006). There is a lot of structural diversity in the aeuginosins, microginins, cyanopeptolins and anabaenopeptins because of variability in the amino acid residues and modifications like glycosylation, sulfatation, methylation or halogenation (Ishida *et al.*, 1999, 2007; Rouhiainen *et al.*, 2000; von Elert *et al.*, 2005; Cadel-Six *et al.*, 2008). The gene clusters for the biosynthesis of cyanopeptolins, aeruginosins and microcyclamide have now been identified and the organization of genes in these is depicted in Fig. 8 A. Moreover, 22 structural variants of cyanopeptolins have been identified with 13 such structural variants produced by a single strain of *M. aeruginosa*. Trichamide, a cyclic peptide from *Trichodesmium erythraeum* predicted on the basis of its genome sequence, bears close resemblance to patellamide produced by cyanobacterial symbionts of ascidians. The genes of trichamide biosynthetic pathway include a precursor peptide gene, a putative heterocyclization gene, an oxidase, two proteases and hypothetical genes (Sudek *et al.*, 2006).

An operon for cyanopeptolin-984 (*mcn*) has been identied in *Microcystis* c.f. *wesenbergii* NIVA-CYA 172/5 which consists of four genes encoding peptide synthetases and one halogenase gene. Two additional ORFs are present in the 5′-flanking region that transcribe in the opposite direction, of which one encodes an ABC-transporter. The *mcn* operon bears resemblance to anabaenopeptilide synthetase operon (*apd*) from *Anabaena* strain 90 (Tooming-Klunderud *et al.*, 2007). The existence of a halogenase gene in 17 out of 28 axenic strains of *Microcystis* sp. that mediates chlorination of cyanopeptolins and aeruginosins has been demonstrated. It is located in between two genes coding for NRPS modules and is present only in the toxic strains of the genus and absent in the non-toxic strains (Cadel-Six *et al.*, 2008). A cyanopeptolin gene cluster has been identified in *Planktothrix* strain NIVA CYA 116, *Microcystis* sp. and *Anabaena* strain. While the latter two members possessed a halogenase gene, it is absent in the former member. Sequencing of this gene cluster showed that both *Microcystis* sp. and *Planktothrix* are related to each other rather than *Anabaena*. Because there is no clear evidence of both recombination and LGT between cyanopeptolin and MC gene sequences, it is suggested that the cyanopeptolin gene clusters in the three genera have evolved independently (Rounge *et al.*, 2007). In the aeruginosin gene cluster (*aer*) of *M. aeruginosa* (strains PCC 7806, NIES-98 and NIES-843) the genes encoding the first three NRPS modules exhibit a high degree of similarity with the exception of the genes encoding tailoring enzymes such as the halogenases and sulfotransferases. This is largely responsible for the diversity of aeruginosin congeners produced by the three strains (Ishida *et al.*, 2009).

M. aeruginosa NIES-298 produces a cytotoxic cyclic hexapeptide known as microcyclamide that is synthesized by ribosomal pathway. The biosynthetic pathway of microcyclamide is akin to the synthesis of patellamide of *Prochloron didemni*. From structural analysis of microcyclamide to genes and vice-versa revealed the mediation of two subtisilin-type proteases, a heterocyclization enzyme and six other gene products of which two bear no resemblance to patellamide synthetic intermediates (Ziemert *et al.*, 2008). A cluster of nine such genes for cyclamide synthesis in *M. aeruginosa* PCC 7806 revealed the presence of *patA* to *patG*, these seven genes are present in two groups *patA* to *patE* with intervening *mic3999* and *mic3998* and then *patF* and *patG* (Fig. 8 B; Frangeul *et al.*, 2008).

***iv) Factors affecting MC biosynthesis*:** The factors affecting biosynthesis of MCs in all the organisms known to produce this toxin are discussed in this section collectively. The effects of environmental factors such as light intensity, temperature and nutrients (nitrogen, phosphorus and trace elements) have been investigated in a number of MC-producing cyanobacteria. The results from these studies are not comparable with each other as the parameters taken for expressing MC content of cultures varied greatly.

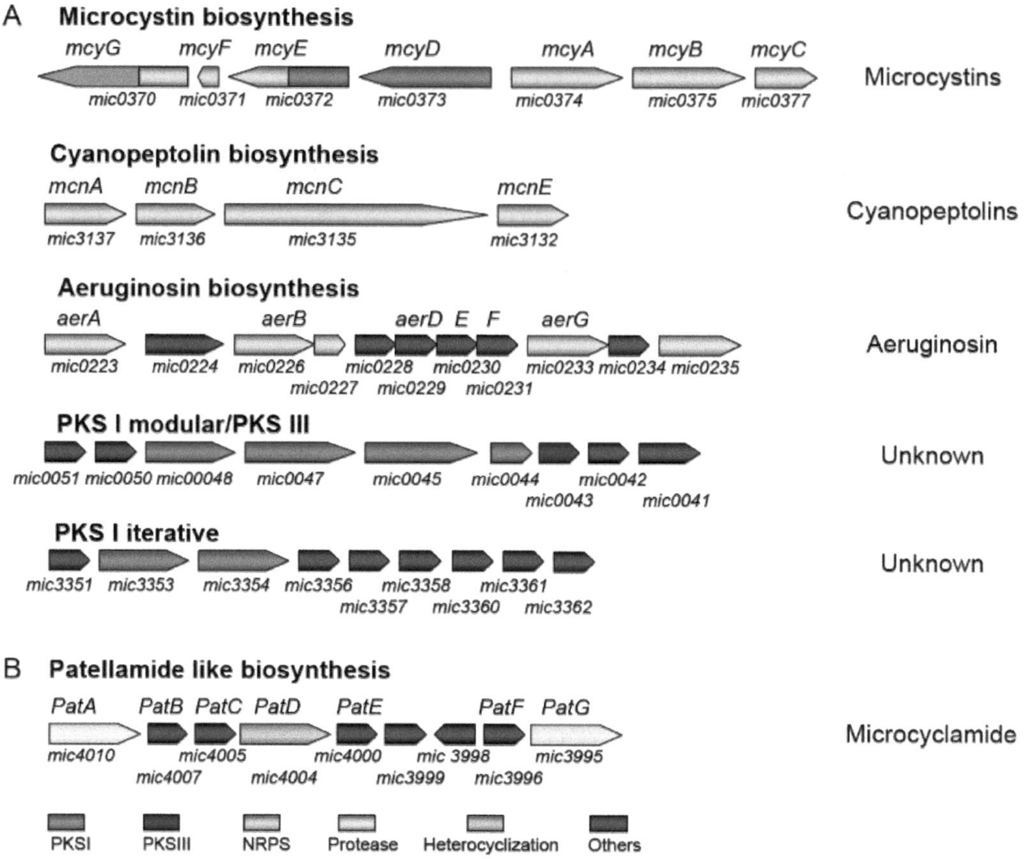

Figure 8: Schematic representation of secondary metabolite gene clusters in *Microcystis aeruginosa* PCC 7806. (A) Gene clusters encoding non-ribosomal peptide synthetetases (NRPS) and polyketide synthases (PKS). The names assigned to individual genes in *M. aeruginosa* PCC 7806, or to genes that were characterized in other cyanobacterial strains are indicated above the arrows. Products assigned to the respective pathways are shown on the right. (B) Gene cluster encoding enzymes potentially involved in a patellamide-like pathway. Names of patellamide biosynthesis genes are indicated above the arrows. Gene identifiers in the *M. aeruginosa* PCC 7806 genome are indicated below the arrows. With the kind permission of Tandeau de Marsac, Institut Pasteur, Unité des Cyanobactéries; CNRS, URA2172, F-75015, Paris, France [Frangeul *et al.* (2008) *BMC Genomics* **9:** 274; doi:10.1186/1471-2164-9-274].

Color image of this figure appears in the color plate section at the end of the book.

Effect of light on MC production has been studied by various workers (Gorham, 1964; Watanabe and Oishi, 1985; van der Westhuizen *et al.*, 1986; Codd and Poon, 1988). Van der Westhuizen and Eloff (1985) reported that increase of light intensity enhanced toxin production by *M. aeruginosa*. Four-fold increase in the MC content of *M. aeruginosa* in batch cultures was observed with an increase in light intensity from 7.53 microeinsteins m^{-2} s^{-1} to 30.1 microeinsteins m^{-2} s^{-1} (Watanabe and Oishi, 1985). The toxicity in these studies was reported to be independent of light intensity and toxicity was assayed by determining 50% lethal dose in a mouse bioassay. On the contrary, Utkilen and Gjolme (1992) reported a 2.5 fold increase in toxicity of *M. aeruginosa* CYA 228/1 in continuous cultures when the light intensity was increased from 20 to 40 microeinsteins m^{-2} s^{-1} but further increase in light intensity did not enhance the toxicity. These observations cannot be compared with the earlier studies as the content of MC was expressed in proportion to the increase in biomass or protein content of cultures. Red and green lights supported higher ratio of toxin to protein content than in white light (Utkilen and Gjolme, 1992). Five strains of *O. agardhii* produced demethyl-3-MC-RR at high nitrogen concentrations (84 mg^{-1} L), low light intensity (12 to 95 microeinsteins m^{-2} s^{-1}), phosphorus concentration (0.4 mg^{-1} L) and temperature (25°C) (Sivonen, 1990).

Photosynthetically active radiation (PAR) in between 10 and 403 μmol of photons m^{-2} s^{-1} with a light-dark (12h:12h) cycles in continuous cultures of *M. aeruginosa* strain PCC 7806 influenced the content of MC in between 34.5 and 81.4 fg $cell^{-1}$. This was significantly positively correlated with growth rate under PAR-limited growth but not under PAR-saturated growth. Intracellular MC content increased to 10 to 20% higher at a PAR of 10 and 40 μmol of photons m^{-2} s^{-1} than at the end of the dark period while at higher light intensities the MC content was inhibited (Wiedner *et al.*, 2003). Kaebernick *et al.* (2000) have chosen *mcyB* (for peptide synthetase) and *mcyD* (for PKS) genes for demonstrating the effect of light intensity/quality and oxidative stress on the transcriptional regulation of these genes for MC biosynthesis in *M. aeruginosa* PCC 7806. Under high light intensities and red light, the transcript levels of *mcyB* and *mcyD* increased while blue light and ceratin stress factors (such as methylviologen and sodium chloride) reduced the level of transcripts. It may be added that MC synthetase production is initiated at certain threshold light intensities such as between dark and low light and medium and high light. The influence of high and low light intensities on transcriptional regulation of *mcy* gene cluster has already been described earlier (Kaebernick *et al.*, 2002). Total MC content and selective production of a particular variant of MC in *P. agardhii* depended on light intensity in the range of 60–100 μmol m^{-2} s^{-1}. The transcripts of the *mcyA* gene and total MC production rate started increasing up to 60 μmol m^{-2} s^{-1} light intensity but the levels of these decreased with irradiance greater than 100 μmol m^{-2} s^{-1}. Increase of light intensity decreased the content of MC-deRR but with a concomitant three-fold increase in the content of MC-deLR signifying that *P. agardhii* becomes more toxic at high light intensities (Tonk *et al.*, 2005).

The rest of the studies are devoted to understand MC production in light, temperature and/or nitrogen/phosphorus enrichment or the supply of iron. *Anabaena* sp. cultured under light limitation at different temperatures has been assessed for production of MCs and neurotoxins (Rapala and Sivonen, 1998). Growth and MC content of different strains of *M. aeruginosa* have been compared under different light intensities and various nutrients (Hesse and Kohl, 2001). According to some workers, variation in light intensity had no significant effect on MC content of both *Planktothrix agardhii* and *M. aeruginosa* grown in turbidostats (Boettcher *et al.*, 2001). A study on the influence of trace metals such as Zn and Fe revealed that Zn enhanced growth and MC production while low Fe concentrations decreased growth but supported higher MC formation (Lukac and Aegerter, 1993). Iron uptake is shown to be light-dependent and the decrease in iron concentration resulted in a decline in the ratio of toxin to both dry weight as well as protein content. The effects of light intensity on toxin production reported earlier (Watanabe and Oishi, 1985; Van der Westhuizen and Eloff, 1985; Sivonen, 1990; Utkilen and Gjolme, 1992) has been suggested to be due to the effect of light intensity on iron uptake (Utkilen and Gjolme, 1995).

Utkilen and Gjolme (1995) observed that nitrate- and phosphate-limited conditions decreased the ratio of toxin content to dry weight in *M. aeruginosa* CYA 228/1. This decrease was not due to a decrease in MC production but was due to enhanced carbohydrate content under low nitrate and phosphate concentrations. However, the MC content remained constant when expressed in relation to protein content of the cells that appeared to be a better parameter for expression of MC content than dry weight (Utkilen and Gjolme, 1995). Phosphorus-limited chemostat cultures of *M. aeruginosa* UTEX 2388 showed a specific growth rate constant (μ) of 0.1 to 0.6 day^{-1} and was a function of cellular

P content under P-limitation. Both carbon fixation and the rate at which MC was produced in cells were higher at lower μ. Under P-limited conditions MC-LR was the dominant form produced (Oh *et al.*, 2000). On the other hand, N-limited continuous cultures of *M. aeruginosa* NASH 01-A19 exhibited a positive linear relationship between MC content per cell, protein, chlorophyll *a*, cell dry weight and cell volume. MC content of fast growing smaller cells was higher than in slow growing cells (Long *et al.*, 2001). Rapala *et al.* (1997) found that the toxin content in the cells of *Anabaena* sp. strain 90 and 202A1 increased with phosphate concentration in medium while at high temperature (25 to 30°C) and light intensities (2 to 100 μ mol m^{-2} s^{-1}) the toxin content decreased. Moreover, the production of different variants of MC varied with growth stimuli where MC-LR was shown to be associated with temperatures below 25°C whereas MC-RR was produced at higher temperature (35°C).

Variations in nitrogen and phosphorus concentrations influenced the MC production in *Microcystis* strains (Vézie *et al.*, 2002). Growth rates of *Microcystis* sp. (Utkilen and Gjolme, 1995; Rapala and Sivonen, 1998; Kotak *et al.*, 2000; Oh *et al.*, 2000) and MC content of *M. aeruginosa* are directly proportional to the phosphorus concentration (Jacoby *et al.*, 2000; Kotak *et al.*, 2000). On the other hand, a variation in MC content of *M. aeruginosa* was suggested to be in relation to N:P ratio and growth phase (Liu *et al.*, 2000). Nutrient (nitrogen, phosphorus or iron) and light limitation influence the energetic state of the cyanobacterial cells by causing a stress condition and eventually affect protein synthesis of the cells but not the synthesis of MC (Bickel and Lyck, 2001). A comparison of growth in different media revealed that *A. flos-aquae* UTEX 2383 supported maximum growth in terms of cell number, chlorophyll *a* content and specific growth rate along with toxicity of the cultures in a mouse bioassay (Gupta *et al.*, 2002).

B) Nodularins: *No. spumigena* was the first cyanobacterium that has been reported to be toxic to livestock (Francis, 1878). As a matter of fact nodularin was for the first time isolated from a species of *Anabaena* that was known to produce both MCs and nodularins (Carmichael, 1986, 1988). Toxic *Nodularia* spp. have been recorded from Germany (Kalbe and Tiess, 1964; Gussmann *et al.*, 1985), Denmark (Lindstrom, 1976), Australia (Main *et al.*, 1977), Finland (Persson *et al.*, 1984; Eriksson *et al.*, 1988) and Sweden (Edler *et al.*, 1985).

***i) Structure*:** The structure of nodularin was for the first time elucidated by Rinehart *et al.* (1988) who suggested that the amino acid composition of the pentapeptide is β-methylaspartic acid, glutamic acid, arginine, dehydrobutyric acid and Adda (Fig. 9). It is cyclo (-D-MeAsp^1-L-Arg^2-Adda^3-D-Glu^4-Mdhb^5). The molecular formula of nodularin is $C_{41}H_6N_8O_{10}$. It has a molecular weight of 824D (Rinehart *et al.*, 1988; Eriksson *et al.*, 1988; Runnegar *et al.*, 1988; Carmichael, 1988; Duy *et al.*, 2000). Because of the presence of Adda in its structure, nodularins possess the same toxicity as of MCs and cause liver damage leading to death of human beings. Nodularins possess carcinogenic activity (not shown by MCs) but both are found to be potent tumor promoters (Humpage *et al.*, 2000; Saker *et al.*, 2003).

An analogue of nodularin known as motuporin, found in the marine sponge *Theonella swinhoei*, differs from nodularin only by one amino acid. It possesses hydrophobic L-Val in place of L-Arg in nodularin (de Silva *et al.*, 1992). Namikoshi *et al.* (1994) described variants of nodularin where [DMAdda^3]-nodularin, [6(z)-Adda^3]-nodularin was found in natural sample while [d-Asp^1]-nodularin was detected in cultures. [L-Har^2]-nodularin was produced by a freshwater non-gas vacuolate strain of *Nodularia* PCC 7804 (Beattie *et al.*, 2000; Saito *et al.*, 2001). Mazur-Marzec *et al.* (2006) characterized eight nodularin variants. Besides the unmodified nodularin with an arginine residue (Nod-R), three variants are demethylated at different sites, i.e. on aspartic acid [D-Asp^1]-nodularin, the Adda residue [DMAdda^3]-nodularin and dehydrobutyric acid [dhb^5]-nodularin. Two other nodularin variants have an additional methyl group located in the Adda [MeAdda]-nodularin and Glu [Glu^4(Ome]-

nodularin. The linear nodularin and geometrical isomer of Nod-R have also been isolated. Of the eight variants [dhb^5]-nodularin and [MeAdda]-nodularin are the new variants. Of these, seven are reported to be produced by marine forms of *No. spumigena* while one nodularin has its origin from a freshwater strain *Nodularia* PCC 7804.

(ii) Biosynthesis: Labelling studies have shown that nodularin biosynthesis is a mixed PKS-NRPS pathway (Moore *et al.*, 1991; Rinehart *et al.*, 1994). Based on these results and those of *mcy* gene cluster (Tillett *et al.*, 2000) it has now been possible to predict the biosynthetic pathway of nodularin. Peptide synthetase and PKS gene homologues were identified in various strains of *Nodularia* isolated from globally diverse locations by degenerate PCR technique (Moffitt and Neilan, 2001). Various strains (21) of *No. spumigena*, *No. harveyana* and *No. sphaerocarpa* were subjected to 16S rDNA sequence analysis to find out their interrelationships. *No. spumigena* represented a distinct cluster producing nodularin while the other two species formed a second cluster representing the benthic non-toxic types (Moffitt *et al.*, 2001). Nodularin synthetase gene cluster of *No. spumigena* SOR 10 is a 48-kb cluster that consists of 9 genes, *nda A* to *nda I* and it is transcribed bidirectionally from a regulatory promoter

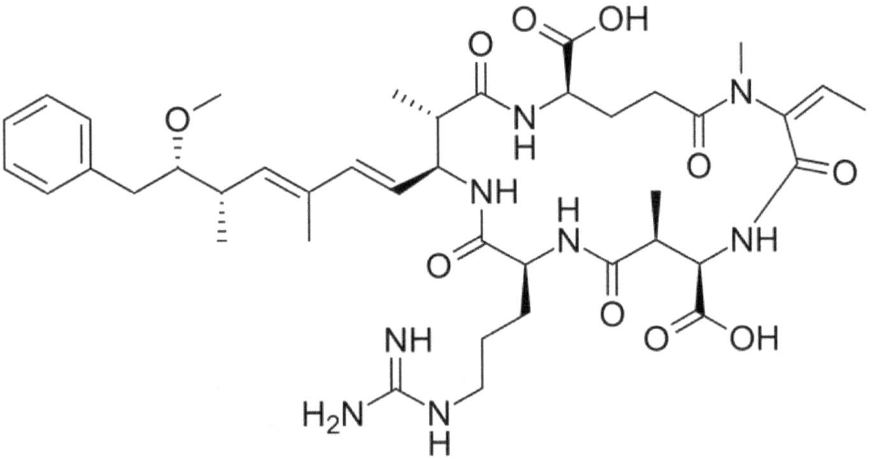

Figure 9: Structure of nodularin. Nodularin is a cyclic pentapeptide (MW 619). The L-Arg residue of nodularin may be replaced with a homoarginine (nodularin-Har) or valine residue (motuporin).

region located in between *nda A* and *nda C* (Fig. 7A). NRPS, PKS modules and tailoring enzymes are synthesized by this gene cluster. The presence of transposases flanking the *nda* gene cluster is akin to the *mcy* gene cluster. Comparative and phylogenetic analysis of *mcy* and *nda* gene clusters revealed that nodularin gene cluster might have evolved from *mcy* gene cluster (Moffitt and Neilan, 2004). The probable functions of the enzymes encoded by the *nda* gene cluster were compared to those of the homologous proteins of *mcy* gene cluster to assign their functions and catalytic activities. ORFs *nda CDEF* encode enzymes for biosynthesis of Adda while *nda GHAB* encode enzymes responsible for peptide synthesis, cyclization and transport (Moffitt and Neilan, 2004). There are four NRPS modules identified for nodularin. The active sites of D-Glu and D-MeAsp A domains are identical to MC synthetases. The active sites of *nda A* activates and methylates L-Thr as a precursor of Medhb and has little homology to the active site of *mcy A*. Conversion of Me-Thr to Medhb occurs via *nda H* enzyme similar to the conversion of Me-Ser to MeDha (Moffitt and Neilan, 2004).

iii) Factors Affecting biosynthesis of nodularins: High light intensity (105–155 µmol^{-2}S^{-1}), salinity (15%), phosphate concentration (1700–5500 µg L^{-1}) and temperature (19°C) supported increased formation of nodularin in *No. spumigena* with prolonged period of incubation (Lehtimaeki *et al.*, 1997). Extracellular nodularin concentration increased with period of incubation suggesting that cell lysis caused the release of the toxin. A comparative study in batch (phosphate 55–5,500 µg L^{-1}; nitrate 100–30,000 µg L^{-1}) and chemostat cultures (phosphate 5–315 µg L^{-1}) of *Nodularia* strain GR8b revealed that nodularin content per dry weight, nodularin to protein ratio and extracellular nodularin were highest after 25 days period but were not influenced by the nutrient concentration (Repka *et al.*, 2001). Nodularin concentration per cell was constant under different concentrations of nutrients.

C) Cylindrospermopsin (CYN): CYN is mainly produced by *Cylindrospermopsis raciborskii*. Besides CYN, *C. raciborskii* is also known to produce saxitoxins, anatoxin-a and PSPs. The genus *Cylindrospermopsis* was created by separating a species of *Anabaenopsis raciborskii* on the basis of presence of gas vacuoles and pattern of heterocyst differentiation (Seenayya and Subba Raju, 1972) that has been supported by other workers (Horecka and Komarek, 1979; Hindak, 1988). So far eight species of *Cylindrospermopsis*, i.e. *C. africana*, *C. cuspis*, *C. philippinensis* and *C. raciborskii* from Africa (Komárek and Kling, 1991); *C. allantoidispora*, *C. catemaco*, *C. tavernae* from Brazil (Komarkova, 1988; Komarkova *et al.*, 1999) and *C. curvispora* from Japan (Watanabe, 1995) have been described. Except the last mentioned species, the rest of them do not differentiate akinetes. Of the remaining seven species, *C. raciborskii* is the most frequently reported species and is known to produce the hepatotoxic alkaloid CYN. The occurrence of *C. raciborskii* in Waikato lakes of New Zealand has been reported (Ryan *et al.*, 2003). In Australia, *C. raciborskii* was responsible for Palm Island mystery disease (Byth, 1980; Bourke *et al.*, 1983). The only toxin produced by Australian strains is CYN and its hepatotoxicity has been subsequently established (Hawkins *et al.*, 1985; Hayman, 1992; Ohtani *et al.*, 1992; Saker *et al.*, 1999). *C. raciborskii* does not form typically surface blooms. Maximum densities of *C. raciborskii* occur at 2–3 m below the surface of waters. Secondly, taste and odour problems are not associated with blooms of *C. raciborskii* as it does not produce volatile organic compounds such as geosmin that is commonly associated with HABs (Saker and Griffiths, 2001). Other filamentous cyanobacteria that are reported to produce CYN are *Anabaena lapponica* from boreal environments (Spoof *et al.*, 2006), *Aphanizomenon ovalisporum* (Fig. 2 E; Banker *et al.*, 1997), *Aph. flos-aquae* (Preussel *et al.*, 2006), *Lyngbya wollei* (Fig. 5 A, B; Seifert *et al.*, 2007), strains of *Oscillatoria* (PCC 6407, PCC 6506, PCC 6602, PCC 7926 and PCC 10702; Mazmouz *et al.*, 2010), *Raphidiopsis curvata* (Fig. 4 B; Li *et al.*, 2001) and *Umezakia natans* (Harada *et al.*, 1994).

i) Structure: CYN is a sulfate ester of tricyclic guanidine (mol. wt. 415D) substituted with a hydroxymethyl uracil (Fig. 10; Ohtani *et al.*, 1992). It is a stable compound that is not removed by boiling (Norris *et al.*, 1999). As compared to MCs and nodularins, CYN acts slowly but it exerts its effects on a number body parts, e.g. kidneys, intestines, lungs though liver is the main organ affected (Hawkins *et al.*, 1985; Terao *et al.*, 1994; Falconer *et al.*, 1999; Seawright *et al.*, 1999; Shaw *et al.*, 2000). Besides CYN, *Cylindrospermopsis* is known to produce STX, a type of paralytic shellfish-poisoning toxins (Lagos *et al.*, 1999). CYN-producing cyanobacteria have now been detected in many European, Central Asian and American water bodies (Falconer, 2005).

Several analogues of CYN have been identified including those of 7-epi-CYN and deoxy-CYN. The toxicity of CYN has been largely due to the pyrimidine ring, the presence of a hydroxyl group on the uracil bridge and the ketoenol status of uracil (Mazur-Marzec, 2006). The LD_{50} (i.p. for 5 days; estimated by mouse bioassay) value of 7-epi-CYN of *Aph. ovalisporum* has been found to be 200 µg kg^{-1} body weight, a value that is similar to CYN. Its subsequent treatment with chlorine solution or chlorine-related oxidants yielded 5-chloro-CYN and cylindrospermic acid and both of these derivatives proved to be non-toxic even at doses 50 times higher (10 mg kg^{-1} mouse body weight) than LD_{50} of CYN. These results emphasize that the pyrimidine ring of uracil is essential for toxicity of CYN (Banker *et al.*, 2000, 2001).

ii) Biosynthesis: The biosynthetic pathway of CYN suggests that glycine is the precursor and later guanidinoacetic acid serves as the starter unit for the polyketide chain. C1 and C2 of glycine serve as C14 and C15 of CYN. Carbon atoms 4 to 13 arise from contiguous acetate units attached head to tail. The origin of NH-CO-NH segment in the uracil ring is not known (Bourgoyne *et al.*, 2000). The involvement of three genes *aoaA, aoaB* and *aoaC* that encode amidinotransferase, a hybrid NRPS/PKS and PKS, respectively have been reported to mediate the biosynthesis of CYN in *Aph. ovalisporum* (Shalev-Alon *et al.*, 2002). The synthesis of guanidinoacetic acid is mediated by the enzyme amidinotransferase which is then passed on to the hybrid NRPS/PKS and PKS enzyme systems for further synthesis. The characterization of amidinotransferase from *C. raciborskii* AWT205 has been reported (Kellmann *et al.*, 2006). A *cyr* gene cluster (*cyrA* to *cyrO*) for biosynthesis of CYN in *C. raciborskii* AWT205 has been sequenced and characterized. A 43 kb fragment of DNA consists of 15 ORFs that mediate biosynthesis, regulation and export of the toxin. The flanking regions on either side of the *cyr* cluster consists of *hyp* genes, the gene products of which act as molecular chaperones for the maturation step of hydrogenases. The precursor molecule glycine accepts an amino group via an amidinotransfer reaction that constitutes the first step. This is soon followed by five polyketide extensions and subsequent reduction reactions (Mihali *et al.*, 2008). Stucken *et al.* (2010) sequenced the genomes of *C. raciborskii* CS-505 and *Raphidiopsis brookii* D9. In the former the *cyr* gene cluster is spread over 41.6 kb region and organized into 16 ORFs that bear a high degree of synteny with those of the genes in the *cyr* gene cluster of *C. raciborskii* AWT205. This resemblance in the two gene clusters is not only limited to the arrangement of genes but also to the flanking regions in having *hyp* genes at both ends of the *cyr* gene cluster. In addition to *cyrL* and *cyrM*, *C. raciborskii* CS-505 *cyr* gene cluster revealed a 219 bp transposase element that is located between *cyrC* and *cyrE* (Fig. 11). This particular region with the transposase elements is also reflected in the CYN biosynthesis gene cluster of *R. brookii* D9 indicating that rearrangements have occurred in this section of the genome. The announcement of draft genome sequence of *Oscillatoria* PCC 6506 with several gene clusters responsible for toxins and secondary metabolites (Méjean *et al.*, 2010) and subsequent identification of complete gene cluster for CYN and 7-epi-CYN biosynthesis paved the way for understanding the complete set of biosynthetic reactions (Mazmouz *et al.*, 2010). Though this gene cluster is homologous with the gene cluster of *C. raciborskii* AWT205 there exist some differences in the number of genes and their arrangement. Instead of 15 *cyr* genes as in *C. raciborskii*,

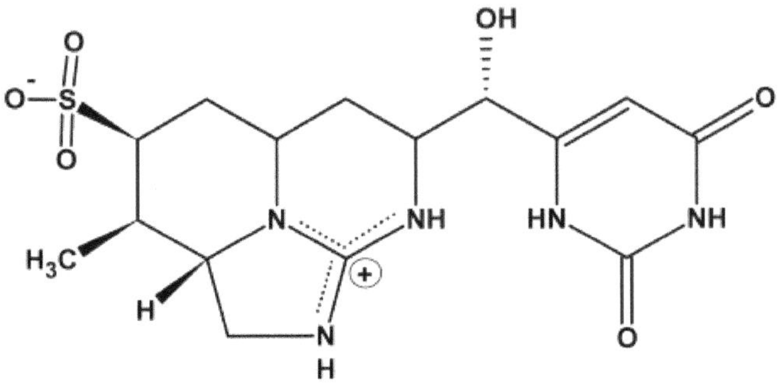

Figure 10: Structure of CYN

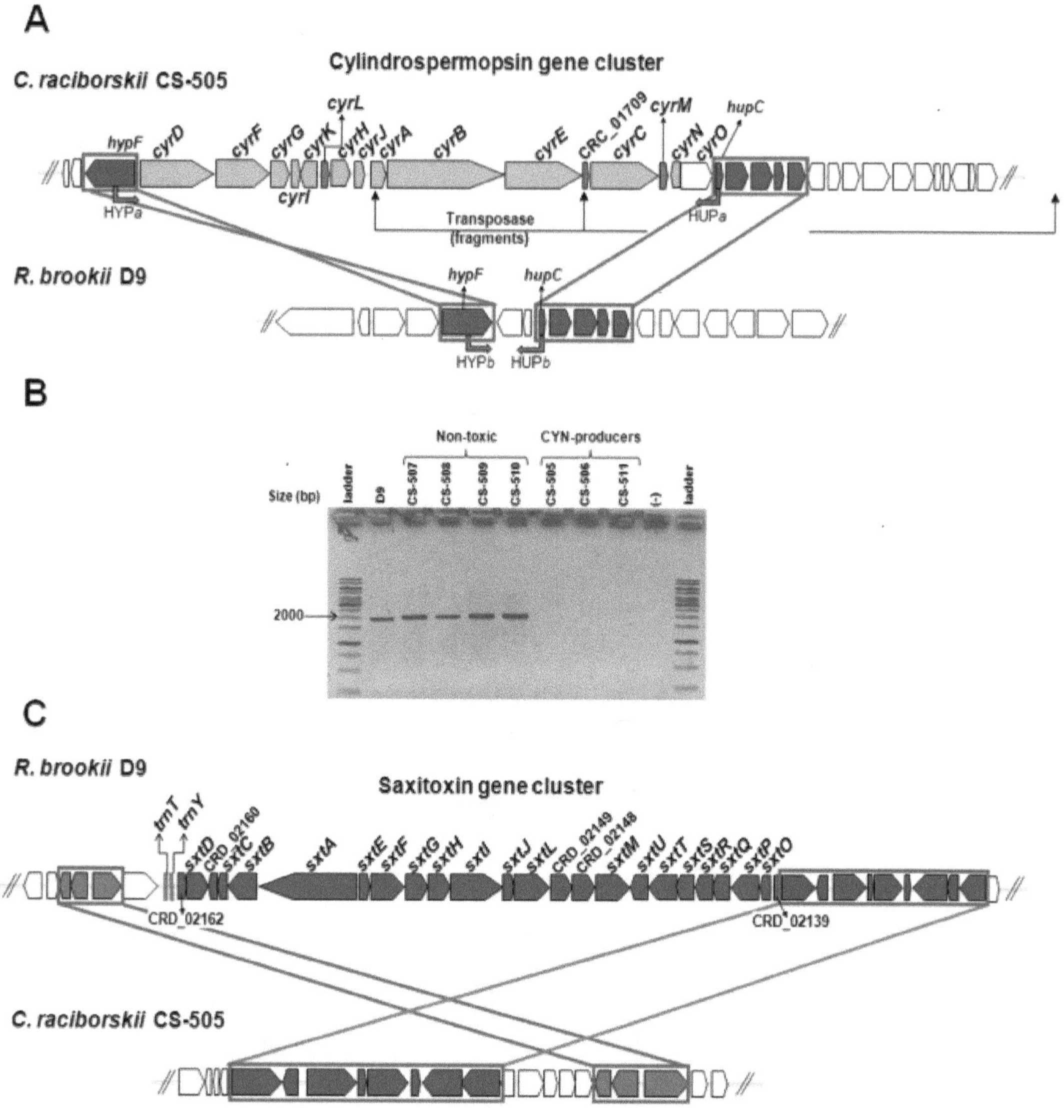

Figure 11: Schematic representation of the syntenic regions within the toxin gene clusters in *C. raciborskii* CS-505 and *R. brookii* D9. (A). Location of the CYN gene cluster of *C. raciborskii* CS-505 compared with the syntenic genomic region in *R. brookii* D9. (B). Gel electrophoresis of the PCR products from the *hypF/hupC* amplification in *R. brookii* D9 and in the strains of *C. raciborskii* non-toxic: CS-507, CS-508, CS-509 and CS-510. Producers of CYN: CS-505, CS-506 and CS-511 do not present amplification of the *hypF/hupC* region. C. Location of the STX gene cluster of *R. brookii* D9 compared with the syntenic genomic region in *C. raciborskii* CS-505. Genes participating in syntenic regions are depicted in blue and highlighted in the green boxes within the arrows; genes outside the syntenic regions are depicted in white. tRNAs and transposases are shown in red. The grey arrows show the position of the primer pairs HYPa/HUPa and HYPb/HUPb used to amplify the region between *hypF* and *hupC* genes in different strains of *C. raciborskii* and in *R. brookii* D9, respectively. Ladder: GeneRuler 1 kb DNA ladder (Fermentas, Ontario, Canada). With the kind permission of Moníca Vásquez, Department of Molecular Genetics and Microbiology, Faculty of Biological Sciences, Pontificia Universidad Católica de Chile, Santiago, Chile & Millenium Nucleus EMBA, Santiago, Chile [Stucken *et al.* (2010) *PLoS ONE* **5(2):** e9235; doi:10.1371/journal.pone.0009235] doi:10.1371/journal.pone.0009235.g004.

Color image of this figure appears in the color plate section at the end of the book.

there are 11 genes common (*cyrA* to *cyrK*) interspersed with three others like *Orf1* (that encodes an ATP-grasp protein) and *trp1* and *trp2* (that encode transposases) in *Oscillatoria* PCC 6506. The *cyrL* to *cyrO* genes are not represented in the *cyr* gene cluster of *Oscillatoria* PCC 6506. The *Oscillatoria* PCC 6506 *cyrA*, *cyrB* and *cyrC* gene products exhibited a high degree of sequence similarity with the gene products of *aoaA*, *aoaB* and *aoaC* of *Aph. ovalisporum*. Determination of the species of CYN produced by *Oscillatoria* PCC 6506 during biosynthesis led to the identification of 7-epi-CYN (68.6%) followed by CYN (30.2%) and traces of 7-deoxy-CYN (Mazmouz *et al.*, 2010).

***iii) CYN toxicity*:** As the total chemical synthesis has been achieved (Xie *et al.*, 2000), it is possible to test the analogues for determining the specific chemical groups responsible for transport and toxicity of CYN. Accordingly, Runnegar *et al.* (2002) tested the analogues CYN-DIOL (synthetic CYN lacking sulfate group) and the derivative of C7 epimer of CYN, EPI-DIOL (having H in place of OH at position 7) for transport and toxicity. Both these substances exhibited no difference in uptake and toxicity suggesting that neither sulphate group nor hydroxyl group in 7-epi-CYN contributes to the toxicity of CYN. Further, the synthesis of 7-deoxy-CYN was achieved and it was suggested that C7 oxygenation was not required for the inhibition of protein synthesis (Looper *et al.*, 2005).

***iv) Genetic diversity of CYN producers*:** Saker *et al.* (1999) described two morphological types of *C. raciborskii*, i.e. straight and coiled trichomes which exhibited CYN content ranging from 0.14 to 0.2%. Special morphological features of *C. raciborskii* in batch and continuous cultures (Shafik *et al.*, 2003) and the morphological changes that occur in bloom development were reported (Padisak, 2003). Sixteen strains of *Cylindrospermopsis* isolated from freshwater bodies (lakes and reservoirs) in Africa, America, Australia and Europe were subjected to 16S tRNA-23S rRNA ITS region. These strains are thus grouped into three clusters, i.e. (i) American strains, (ii) strains of Europe and (iii) African and Australian strains (Gugger *et al.*, 2005). A recent colonization of Australian waters by African strains could be a possible event.

Of the seven strains of *C. raciborskii* isolated from Australia, only five of them (CR1, CR2, CR3, CR4 and CR5) produced detectable levels of CYN in their cells in batch cultures. No correlation between the morphological type (straight or coiled trichomes) and toxin production could be observed. The addition of nitrate or ammonium compounds to medium did not increase the content of CYN but the strains responded more positively for CYN production in a growth medium without nitrogen sources (Saker and Neilan, 2001).

The occurrence of *C. raciborskii* from a Canadian mesotrophic to eutrophic hardwater lakes (Constance Lake, Ottowa, Ontario) has been reported (Hamilton *et al.*, 2004). Though initially known from tropical waters (Hawkins *et al.*, 1985), *C. raciborskii* has now been collected from sub-tropical and temperate regions of the world. Studies on bloom development in 39 Brazilian tropical reservoirs located in the semi-arid region of northeastern Brazil revealed that 80% of the total phytoplankton density was due to *C. raciborskii*. Low ammonia concentration, high temperature and pH values corresponded with high percentage growth of *Cylindrospermopsis* (Bouvy *et al.*, 2000). Bernard *et al.* (2003) presented a comparative account of the toxicity of *C. raciborskii* strains collected from different geographical regions. The Australian strains as also strains from Florida (USA) and Thailand produced CYN and caused liver damage associated with effects on the kidneys, adrenal glands, small intestine, lungs, thymus and heart whereas Brazilian isolates produced toxicity symptoms of STXs. Analysis by liquid chromatography with fluorescence detection revealed the presence of saxitoxin (STX), neoSTX and decarbamoyl STX. On the other hand, French isolates of *C. raciborskii* though were responsible for symptoms akin to CYN (moderate, multifocal hepatocellular necrosis combined with disorganization of parenchyma and congestion of inner sinusoid) did not produce CYN in culture extracts. Bouaicha and Nasri (2004) reported for the first time *C. raciborskii* from Algerian freshwaters. The diversity studies dealing with the *cyrB* gene adenylation domain sequences in 17 lakes of Florida revealed that the majority (92%) of the amplicons resembled those of *Aph. ovalisporum* with 99% identity. Curiously enough, *Aph. ovalisporum* is the only CYN-producing cyanobacterium distributed throughout the United States except in Florida where the occurrence of *C. raciborskii* has been reported (Yilmaz and Phlips, 2011).

D) Stability of Hepatotoxins: It is important to know the stability of MCs, nodularins and CYN either in natural waters or when stored in the laboratory for experiments related to toxicity studies. In natural fresh or marine waters exposure to light and interaction with dissolved organic matter may affect the toxicity. Very few studies are devoted towards understanding the stability of these toxins. Both MCs and nodularins are chemically stable (Harada, 1995; Harada *et al.*, 1996b). MCs released by *M. aeruginosa* after algicide treatment retained their toxicity for nearly 10 days, afterwards a decrease in MC content was noted (Jones and Orr, 1994). Either due to cell death during bloom senescence or during water treatment processes, the toxins are released into the surrounding waters and these persist for several days even after the blooms have disappeared (Lahti *et al.*, 1997). A decrease of MC content in natural waters has been attributed to (i) dilution of the toxin containing water with non-contaminated water; (ii) adsorption on particulate material; (iii) temperature- and pH-dependent decomposition, and (iv) photolysis and biological degradation (Harada, 1995). Nodularin is stable for nearly 10 days in aqueous solutions irrespective of light condition, i.e. light/dark or sunlight (Twist and Codd, 1997).

The degradation of MC-LR occurred in less than a week in reservoir water (Cousins *et al.*, 1996) when compared to sterilized reservoir water (12 days) and de-ionized water (27 days). The authors suggested that decomposition of hepatotoxins might be taking place by the disruption of the conjugated diene system in the Adda. The ability of *Pseudomonas aeruginosa* to degrade MC-LR through the activity of an alkaline protease is worth mentioning (Takenaka and Watanabe, 1997). The decrease in the concentrations of these toxins in non-sterile seawater suggested that certain bacteria might be responsible for degrading these toxins and that nodularin is more resistant to biodegradation than MCs.

Jones *et al.* (1994) reported the degradation of MCs by the bacterium *Sphingomonas* sp. strain MJ-PV isolated from Australian natural waters. Bourne *et al.* (1996) observed that *Sphingomonas* could breakdown MC-LR by ring opening resulting in an open chain linear compound that is 200 times less toxic. Hydrolysis of Arg-Adda and then Ala-Leu peptide bonds involving enzymatic degradation resulted in a linearized MC-LR (H-Adda-Glu-Mdha-Ala-Leu-MeAsp-Arg-OH) and then a tetrapeptide (H-Adda-Glu-Mdha-Arg-OH) (Bourne *et al.*, 1996). Species of *Sphingomonas* have now been isolated from Japanese lakes as well that have the potential to degrade MCs (Park *et al.*, 2001; Saitou *et al.*, 2003; Ishii *et al.*, 2004). However, Imanishi *et al.* (2005) reported another bacterium, B-9, isolated from a Japanese Lake Tsukui, that degraded MC-LR and nodularin. The degradation of MC to non-toxic Adda by B-9 involves sequential enzymatic hydrolysis of Arg-Adda, Ala-Leu and then Adda-Glu peptide bonds resulting in two unknown non-toxic intermediates (Imanishi *et al.*, 2005). Bourne *et al.* (2001) characterized the gene cluster for the bacterial degradation of MCs. A new rod-shaped gram-negative, single polarly flagellated bacterial isolate identified as *Paucibacter toxinivorans* gen. nov et sp. nov. capable of degrading MCs and nodularins was isolated from lake sediments of Finland and physiological properties of the bacterium have now been characterized (Rapala *et al.*, 2005).

E) Mode of action of Hepatotoxins: Signal transduction pathways are mediated by enzymes that catalyze phosphorylation and dephosphorylation of proteins at specific serine, threonine and tyrosine sites by protein kinases and protein phosphatases, respectively (Cohen, 1989). Substrate specificity, dependence on metal ions and sensitivity to inhibitory agents are taken into consideration for classifying protein phosphatases (PPs). Ser/Thr protein phosphatases are mainly divided into two categories. Type-1 phosphatases (PP-1) that are inhibited by two heat-stable proteins termed as inhibitor 1 and inhibitor 2 and preferentially dephosphorylate the β-subunit of phosphorylase kinase. Type-2 phosphatases (PP-2) are not inhibited by the heat-stable inhibitors and show preference to the α-subunit of phosphorylase kinase (Cohen, 1989). PP-2 are subdivided into spontaneously active (PP-2A), Ca^{2+}-dependent (PP-2B) and Mg^{2+}-dependent (PP-2C) classes. Thus PP-1 and PP2-A enzymes show metal-independent activities. Wera and Hemmings (1995) described the structure of each Ser/Thr protein phosphatase, its regulation and probable physiological role. By now as many as 40 tyrosine phosphatases have been characterized.

Okadaic acid was the first specific inhibitor of PPs and subsequently other inhibitors such as tautomycin, dinophysistoxin, calyculin, MCs and nodularins were discovered. The activities of PP-1 and PP-2A were inhibited by MC-LR at 1.7 nM and 0.04 nM concentrations (IC_{50}), respectively. Thus inhibition of PP-2A was greater than PP-1 and the inhibition of the activities of both PP-1 and PP-2A was 10-folds more than that of okadaic acid (Honakanen *et al.*, 1990). The Ki value for inhibition of PP-1 and PP-2A by MC-LR was found to be below 0.1 nM. However, the inhibition of PP-2B was 1000-folds less potent. Further, okadaic acid prevented the binding of MC-LR to PP-2A while protein inhibitors 1 and 2 prevented the binding of MC-LR to PP-1 and PP-2A. The potential value of MC-LR in the detection and analysis of protein kinases and phosphatases was highlighted (MacKintosh *et al.*, 1990).

MCs and nodularins in solution assume a chemical shape that is similar especially in the Adda-glutamate part of the cyanotoxin molecule (Rudolph-Boehner *et al.*, 1994; Annila *et al.*, 1996). Thus this region is important for the interaction with PPs and constitutes the essential part for exerting the toxicity (Barford and Keller, 1994; Goldberg *et al.*, 1995). The molecular basis of interactions of MCs with PPs has been explained (Bagu *et al.*, 1997; Dawson and Holmes, 1999). MCs interact covalently with PP-1c and PP-2Ac but nodularins do not bind covalently to PP-1c and PP-2Ac. MCs form this covalent link between the Mdha residue of MCs and Cys-273 of PP-1c bringing them closer to catalytic center of PP-1c. MCs exert little effect upon PP-2B. The difference in position of N-methyldehydrobutyrine residue in nodularin relative to N–methyldehydroalanine residue of MC-LR is responsible for the nature of binding. Further, Bagu *et al.* (1997) suggested that both okadaic acid and calyculinA are similar to MCs and nodularins in their tertiary structure and binding domain to PP1c. Besides these, X-ray crystal structures (Gauss *et al.*, 1997) and mutational analysis (Zhang *et al.*, 1994) provide additional evidences for the nature of interaction of MCs and nodularins with PP1. In this regard, Maynes *et al.* (2006) elucidated the crystal structures of nodularin-V (motuporin) and MC-LA bound to human protein phosphatase-1c (gamma isoform). This is another additional evidence for the covalent binding of MCs to PP1c but not nodularins to an active site cysteine residue (Cys-273). MC-HtyR, a dehydrobutyrine containing MC variant [Asp^3-ADM $Adda^5$-Dhb^7], isolated from *Nostoc* sp., showed a non-covalent interaction with PPs. It could potently inhibit PP1, PP2A, PP4 and PP5 with IC_{50} values similar to MC-LR. Besides, a new cyclic peptide containing 3-amino-6-hydroxy-2-piperidone named as nostocyclin exhibited 500-folds less potency for inhibition of PP1, PP2A, PP4 and PP5 (Hastie *et al.*, 2005).

CYN caused nearly 50% decrease in glutathione levels of cultured rat hepatocytes and this makes the hepatocytes more vulnerable for CYN toxicity (Runnegar *et al.*, 1994). The lowering of glutathione levels has been suggested to be due to an inhibition in its synthesis. Cytochrome P450 appears to be involved in the metabolism of CYN and the metabolites formed might even be more toxic and potential inhibitors of glutathione synthesis (Runnegar *et al.*, 1995).

In relation to toxicity three important points are worth considering. Firstly, the linearization of MCs or nodularin abolishes the toxic effects. Secondly, the overall shape of these molecules in relation to Adda side chain is critical for interaction with PPs. Modifications in the Adda such as demethylation of the C9 of Adda side chain have little effect on toxicity but the geometrical isomers in which C6 diene is changed from *trans* E configuration to the *cis* Z configuration possess decreased toxicity by 100-folds. Accordingly, the IC_{50} of such variants towards PPs is increased (Nishiwaki-Matsushima *et al.*, 1991). Thirdly, the presence of carboxylic acid on the glutamic acid residue to MCs is important to toxicity. MC variants with alterations in Xaa and Zaa at 2nd and 4th positions all possess almost the same toxicity as MC-LR as evidenced by the IC_{50} values (Craig *et al.*, 1993; Rinehart *et al.*, 1994).

F) Toxicity studies with Hepatotoxins: Cyanobacteria have been known to cause animal and human poisoning in different parts of the world for over last 100 years. Toxicity studies on plants and animals (mussels, crayfish and fishes used for human consumption) have shown that MCs accumulate in body parts and are harmful to human beings on consumption (De Figueiredo *et al.*, 2004). Toxicity in relation to animal testing data is generally expressed as LD_{50} value of the toxin that is required to kill 50% of the population of the tested animals. In growth measurements, concentration of the toxin inhibiting growth by 50% is expressed as EC_{50} whereas concentration of the toxin inhibiting enzyme activity or lethal to animalicules in solution by 50% is expressed as IC_{50} (Table 5).

i) Zooplanktons: The blooms of *M. aeruginosa* become predominant in natural waters by inhibiting the growth of other microalgae which serve as food for the zooplankton. This results in an unfavourable nutritive condition for the zooplanktons (Kurmayer and Juettner, 1999). Cyanobacterial toxicity to *Daphnia* has been reported as early as 1980s (Lampert, 1981; Fulton and Pearl, 1987; Benndorf and Henning, 1989). Biomass of *M. aeruginosa* or *Aph. flos-aquae* exerted considerable toxic effects on the life cycle (Reinikainen *et al.*, 1994), movements and feeding appendages of *Daphnia* (Haney *et al.*, 1995). Two alternatives have been suggested to explain the grazer resistance shown by certain zooplanktons. The first is the evolution of phenotypes with plasticity in feeding behavior through natural selection (Hariston *et al.*, 2001) and the second is the availability of phosphorus-deficient diets (DeMott and Dhawale, 1995).

Toxic cyanobacteria inhibit the feeding of zooplanktons and increase their mortality (Lampert, 1987; De Bernardi and Giussani, 1988, 1990). DeMott *et al.* (1991) compared the feeding habits of four species of zooplanktons that differed in their physiological sensitivities to MC-LR, nodularin and feeding behaviour in presence of toxic and non-toxic bloom algae. *Daphnia pulcaria* was less sensitive to these toxins (LD_{50} 21.4 µg ml^{-1} for 48 h) and stopped feeding in presence of toxic cells. On the contrary, *D. pulex* exhibited greater sensitivity (LD_{50} 9.6 µg ml^{-1} for 48 h) but showed uninhibited feeding on toxic cells. The copepod *Diaptomus birgei* was highly sensitive (LD_{50} 0.45 to 1.0 µg ml^{-1} for 48 h) but could survive better than the earlier two because of its capability in selecting its food and uninterrupted feeding. *Daphnia hyalina* (LD_{50} 11.6 µg ml^{-1} for 48 h) is moderately sensitive but showed rapid inhibition in feeding in presence of toxic cyanobacteria. It means in the absence of alternative food, the zooplanktons may become selective (so stop feeding) or may not be able to differentiate between toxic and non-toxic food materials (so show continued feeding behaviour). Kurmayer and Juettner (1999) confirmed the findings of DeMott *et al.* (1991) and suggested that copepods avoid toxic cyanobacteria and are highly selective while daphnid cladocerans are less selective. It is because of this reason they ingest both toxic and non-toxic strains of *Microcystis* (Rohrlack *et al.*,

Table 5: Toxicity assay systems with LD_{50} values of different cyanobacterial toxins.

Toxin	Assay organism	LD_{50} Value	Authors
MC-LR	*Daphnia pulcaria* *D. pulex* *D. hyaline* *Diaptomus birgei* Mouse bioassay** Pigs**	21.4 μg ml^{-1} 9.6 μg ml^{-1} 11.6 μg ml^{-1} 0.45 μg ml^{-1} 36–122 nasal 5–10 oral 25–150 50 72	DeMott *et al.* (1991) Dawson (1998) Ito *et al.* (1997) Dawson (1998) WHO (1998) Botes *et al.* (1985); Rinehart *et al.* (1988); Krishnamurthy *et al.* (1989); Watanabe *et al.* (1988) Beasley *et al.* (2000)
MC-LA	Mouse bioassay	50	Botes *et al.* (1984); Kaya and Watanabe (1990)
MC-AR	Mouse bioassay	250	Namikoshi *et al.* (1992a)
[D-Asp³] MC-LR	Mouse bioassay	160–300	Krishnamurthy *et al.* (1989); Harada *et al.* (1989, 1990b, 1991a); Sivonen *et al.* (1992a); Luukkainen *et al.* (1993)
[Dha⁷] MC-LR	Mouse bioassay	250	Harada *et al.* (1991b); Sivonen *et al.* (1992a); Luukkainen *et al.* (1993)
[DMAdda⁵] MC-LR	Mouse bioassay	90–100	Namikoshi *et al.* (1992a); Sivonen *et al.* (1992b)
[(6Z)-Adda⁵] MC-LR	Mouse bioassay	>1200	Harada *et al.* (1990a)
MC-LY	Mouse bioassay	90	Stoner *et al.* (1989)
MC-HilR	Mouse bioassay	100	Namikoshi *et al.* (1995)
[D-Asp³, ADMAdda⁵] MC-LR	Mouse bioassay	160	Sivonen *et al.* (1990a); Namikoshi *et al.* (1990)
[L-MeSer⁷] MC-LR	Mouse bioassay	150	Namikoshi *et al.* (1992a, 1995)
[D-Glu(OCH_3)⁶] MC-LR	Mouse bioassay	>1000	Sivonen *et al.* (1992d); Rinehart *et al.* (1994); Bateman *et al.* (1995)
[ADMAdda⁵] MC-LR	Mouse bioassay	60	Sivonen *et al.* (1990a); Namikoshi *et al.* (1990)
MC-M(O)R	Mouse bioassay	700–800	Namikoshi *et al.* (1992a)
MC-FR	Mouse bioassay	250	Namikoshi *et al.* (1992a)
MC-YM(O)	Mouse bioassay	56	Elleman (1978); Botes *et al.* (1985)
[ADMAdda⁵] MC-LHar	Mouse bioassay	60	Sivonen *et al.* (1990a); Namikoshi *et al.* (1990)
MC-RR	Mouse bioassay	600	Kusumi *et al.* (1987); Painuly *et al.* (1988); Watanabe *et al.* (1988); Sivonen *et al.* (1992a)

Table 5: contd....

Table 5: contd....

Toxin	Assay organism	LD_{50} Value	Authors
[(6Z)-Adda[5]] MC-RR	Mouse bioassay	>1200	Harada *et al.* (1990a)
MC-YR	Mouse bioassay Pigs**	70 1000	Botes *et al.* (1985); Watanabe *et al.* (1988) Falconer *et al.* (1999)
[D-Asp[3]] MC-HtyR	Mouse bioassay	160–300	Harada *et al.* (1991a)
[D-Glu-$OC_2H_3(CH_3)^6$] MC-LR	Mouse bioassay	>1000	Namikoshi *et al.* (1992a)
MC-HtyR	Mouse bioassay	80–100	Harada *et al.* (1991a)
MC-WR	Mouse bioassay	150–200	Namikoshi *et al.* (1992a)
[L-MeLan[7]] MC-LR	Mouse bioassay	1000	Namikoshi *et al.* (1995)
Nodularin	Mouse bioassay	50	Chorus and Bartram (1999)
CYN	Mouse bioassay	200–2000	
Anatoxin-a	Mouse bioassay	375 <100 i.v. 5000 oral 2000 nasal	Fitzepore *et al.* (1994)
Homoanatoxin	Mouse bioassay	250	Kupier-Goodman *et al.* (1999)
Anatoxin-a(s)	Mouse bioassay	50	Kupier-Goodman *et al.* (1999)
STXs	Mouse bioassay Male Mice Female mice	 10 3.4 i.v. 263 oral 8.0	 Wiberg and Stephensen (1960); Kupier-Goodman *et al.* (1999)
	Rats** 24 hrs old Rats 21 day old Rats Adults (60–70 days)	5.5 64 oral 8.3 270 oral 10 531 oral	Watts *et al.* (1966); Kupier-Goodman *et al.* (1999)
Antillatoxin*	Mouse bioassay	20.1±6.4 nM	Berman *et al.* (1999)
Kalkitoxin*	Mouse bioassay Goldfish (*Carassius auratus*) Brine shrimp (*Artemia salina*)	3.8±1.91 nM 700 nM 170 nM	Berman *et al.* (1999) Wu *et al.* (2000)

*The values represent LC_{50}.
**Until unless specially stated the toxins were given through i.p. route and the respective doses were in μg kg^{-1} body weight.

1999) and accumulate MCs (Mohamed, 2001). Thus this constitutes the first step in the food chain for biomagnification. Cultures of calanoid copepods *Acartia tonsa* and *Eurytemora affinis* and an oligotrich ciliate *Strombidium sulcatum* took up measurable amounts of tritiated nodularin when exposed to ^{3}H-dihydronodularin suggesting that dissolved toxins can get accumulated at higher trophic levels even if planktonic grazers would avoid feeding on toxic cyanobacteria (Karjalainen *et al.*, 2003).

ii) Fishes: Species of mussel (*Mytilus edulis* and *M. galloprovincialis*) accumulated MCs upon feeding toxic bloom of *M. aeruginosa* (Williams *et al.*, 1997; Amorim and Vasconcelos, 1999). Fish treated with purified MCs or fed on toxic cyanobacterial material exhibited pathological changes analogous to those reported for mammals. The studies on rainbow trout, *Salmo gairdneri* (Phillips *et al.*, 1985), common carp, *Cyprinus carpio* (Rabergh *et al.*, 1991) and rainbow trout *Oncorhynchus mykiss* (Tencalla and Dietrich, 1997) highlight such pathological changes. *O. mykiss* that has been fed with acutely toxic doses (5700 μg MC kg^{-1} body weight) of lyophilized culture of *M. aeruginosa* revealed that the toxin accumulated in the liver in a time-dependent manner. Hepatocyte necrosis has been suggested to be due to reversible and irreversible interaction of MC-LR with PP1 whereas apoptotic cell death after 48 h was primarily due to the irreversible inhibition of PP2A (Fischer *et al.*, 2000). An increase in liver mass (depicted by hepatosomatic index; HSI) was detected in *O. mykiss* exposed to *Microcystis* cells or its extracts with heterotrophic bacterial lipopolysaccharide (Best *et al.*, 2003). Together with an increase of HSI, the fish accumulated water in the gut. Cell contents of *Microcystis* stimulated a drinking response that led to fluid in the gut. Crayfish *Procambarus clarkii* has been reported to accumulate MCs in the intestine and hepatopancreas (Vasconcelos *et al.*, 2001). In European carp, *Cyprinus carpio* low concentrations of MCs cause damage to hepatopancreas and kidney (Fischer and Dietrich, 2000). The targets of MC-LR toxicity in embryos and larvae of freshwater fish, *Misguruns mizolepis* are liver and heart (Liu *et al.*, 2002). Developing embryos of fish are known to be more sensitive than juvinile stages and adults to MCs (Oberemm, 2001a; Wiegand and Pflugmacher, 2001). *Oreochromis nilotieus*, a freshwater fish accumulated MCs in the gut, liver and kidney (Mohamed *et al.*, 2003). Fishes that accumulate MCs in their muscle tissue tend to be dangerous to human beings when consumed (Magalhaes *et al.*, 2001). Employing *Thamnocephalus platyurus* bioassay, Blom *et al.* (2001) determined that [D-Asp^3, (E)-Dhb^7] MC-RR, produced by *Planktothrix rubescens*, has higher grazer toxicity compared to MC-LR, MC-RR, MC-YR and nodularin.

A number of investigators considered oxidative stress as the marker for assessing the toxicity of MCs. The variations in the levels of reduced gultathione (GSH) and malondialdehyde (MDA) have been selected as biomarkers for assessing the toxicity of MCs in three groups of silver carp (*Hypopththalmichthys molitrix* Val.) that were reared in aquarium, grown in store pond without cyanobacterial bloom and fishes exposed for 25 days to natural bloom populations of *M. ichthyoblabe* (60% by composition) and *M. aeruginosa* (40%). The levels of GSH were significantly elevated in fish exposed to toxic blooms (MC concentration of nearly 513 μg g^{-1} dry weight) than the other two groups of fishes. The increase of MDA levels was slightly higher in the fishes exposed to toxic blooms. This signifies that fishes exposed to MCs experience oxidative stress that is a fundamental initiating factor to disease development in living organisms (Blaha *et al.*, 2004). The levels of lipid peroxidation (LPO), superoxide dismutase (SOD), glutathione peroxidase (GPx) and glutathione reductase (GR) increased in a time-dependent manner in tilapia fish (*Oreochromis* sp.) exposed to 60 μg MC-LR $fish^{-1}$ day^{-1}. Significant increase of LPO level in liver, kidney and gills of fish treated with MC-LR for 21 days suggests that oxidative stress plays an important role in the toxicity to tilapia fish (Jos *et al.*, 2005). Tilapia fish, administered with a single dose of MCs (500 μg of MC-LR or MC-RR kg^{-1} body weight) showed enhanced actvities of SOD, catalase and LPO that is suggestive of an adaptive response of the fish to the exposure of MCs. Since these enzymes are considered as good oxidative stress biomarkers, they can be taken as valuable tools for the assessment of early responses of fish to these toxins (Prieto *et al.*, 2006). The toxicity of *P. rubescens* was studied by oral feeding of the cultures (80,000 cells 120 μl^{-1} equivalent to 0.6–0.9 μg MC-LR kg^{-1} body weight) in European whitefish (*Coregonus lavaretus*). Histological studies revealed tissue damage to liver, gastro-intestinal tract and kidney due to accumulation of MC (Ernst *et al.*, 2006).

The accumulation of MCs in freshwater mussel (*Hyridella menziesi*) and rainbow trout in lakes of Rotoiti and Rotoehu (New Zealand) has been investigated along with variations in concentrations of MCs in the bloom algae. Phytoplankton samples revealed 760 µg L^{-1} in Lake Rotoiti. Freshwater mussels suspended in cages in lake waters revealed variants of MC-LR, MC-YR, MC-RR, MC-AR, MC-FR, MC-LA and MC-WR but these were not detected in muscle or liver tissues of rainbow trout (Wood *et al.*, 2006). During a bloom of *No. spumigena* in Gippsland Lakes of Southern Victoria, nodularin concentration reached alarming levels in mussels and in prawn viscera at cell counts nearly 30,000 cells ml^{-1} (Van Buynder *et al.*, 2001). Freshwater mussel *Anodonta cygnea* accumulated CYN up to 2.52 µg g^{-1} tissue dry weight after being exposed to *C. raciborskii* cultures for 16 days. At the end of two-week period, CYN distribution in different parts of the body was 68.1% in haemolymph, 23.3% in viscera, 7.7% in foot and gonad and 0.9% in mantle (Saker *et al.*, 2004). A link in the food web of Baltic Sea for biomagnification of nodularin has been established. Eiders (a northern sea duck; *Somateria mollissima*) that fed on mussels in the Baltic Sea showed accumulation of nodularin in their liver from 3 to 180 µg kg^{-1} body weight and 0.1 to 5.8 µg nodularin per liver (dry weight) (Sipia *et al.*, 2005).

iii) Amphibians: Embryos of *Xenopus laevis* (African clawed frog) were exposed to biomass of *M. aeruginosa* (with MC-LR) and *M. wesenbergii* (without MC-LR) and purified MC-LR (0–250 µg L^{-1} and 500 µg L^{-1}) in a 96 h teratogenesis assay (FETAX). MC-LR and biomass of *M. aeruginosa* caused significant lethality up to 30% and 50%, respectively. The LC_{25} concentrations were 380 µg L^{-1} for MC-LR and 300 mg L^{-1} (=250 µg L^{-1} MC-LR) for biomass of *M. aeruginosa*. At lower concentrations of MC-LR (25 µg L^{-1}) there was increase in number of malformations. However, surviving embryos (53%) from among those treated with 250 µg L^{-1} of MC-LR and biomass of *M. wesenbergii* showed malformations (Dvorakova *et al.*, 2002).

iv) Mammals: In mammals when administered intravenously (i.v.) or intraperitoneally (i.p.), MCs get localized in the liver (Falconer *et al.*, 1986; Nishiwaki *et al.*, 1994). Extensive haemorrhage, accumulation of yellow fluid in body cavities and acute hepatic necrosis have been noted in sheep dosed with water containing bloom of *No. spumigena* and guinea pigs developed periacinar liver necrosis leading to death (Main *et al.*, 1977). Cows developed signs of anorexia, mental derangement, dehydration, recumbency and ruminal atony upon ingestion of *Microcystis* bloom water. Increase of liver associated enzyme activities (alkaline phosphatase, γ-glutamyltransferase, aspartate transaminase and lactate dehydrogenase) was noted after a week of bloom ingestion. Hepatic enlargement associated with hepatic dissociation, degeneration and necrosis was common in cows administered with bloom alga intraruminally. When orally ingested, MC-LR is transported across the ileum into blood stream via bile-acid transporter (also known as the multi-specific organic ion transport system) present in the hepatocytes and cells lining the small intestines (Runnegar *et al.*, 1991; Falconer *et al.*, 1992). The other variants of MCs are more hydrophobic than MC-LR and may cross cell membranes by other mechanisms including diffusion (Kuiper-Goodman *et al.*, 1999). The localization of MCs in liver appears to be due to the specificity in their binding to hepatocytes (Dawson, 1998; Chorus and Bartram, 1999; Dow and Swoboda, 2000).

Isotopically labelled MCs when administered either i.v. or i.p. in mice or rats, the accumulation of nearly 70% of the label was found in the liver (Falconer *et al.*, 1986; Runnegar *et al.*, 1986; Brooks and Codd, 1987; Robinson *et al.*, 1989, 1991; Meriluto *et al.*, 1990; Lin and Chu, 1994a; Nishiwaki *et al.*, 1994). Uptake studies on ^{3}H-dihydromicrocystin-LR by rat hepatocytes revealed that the uptake is a specific feature of hepatocytes whereas uptake by other cell lines (human hepatocarcinoma cell line HepG2 as well as mouse fibroblast cell line NIH-3T3 and human neuroblastoma cell line SH-

SY5Y) is negligible. Inhibition of the uptake of ^{3}H-dihydromicrocystin-LR by bile acid transport inhibitors (antamanide, 5 μM; sulfobromophthalein 50 μM and rifampicin 30 μM) indicates that the MC-LR uptake is through multispecific transport system for bile acids (Eriksson *et al.*, 1990a). In mice or rats the cause of death due to MC-LR was due to haemorrhagic shock secondary to massive hepatocellular necrosis and collapse of hepatic parenchyma (Theiss *et al.*, 1988). The hepatocyte deformation by MC-LR, 7-desmethyl-MC-RR and nodularin has been attributed to the increase in overall level of protein phosphorylation (Eriksson *et al.*, 1990b). Robinson *et al.* (1991) conducted a detailed study on the concentration of MC-LR given i.v. in liver. Though MC-LR concentration did not change in liver but some portion of the toxin was excreted via urine and the faecal route. This signifies the role of liver in detoxifying MC-LR through the microsomal enzyme inducers (Brooks and Codd, 1987). Kondo *et al.* (1996) identified three metabolic products, i.e. a glutathione conjugate, a cysteine conjugate and a conjugate with the oxidized Adda diene.

Sub-lethal doses of MC-LR (45 μg kg^{-1} body weight) were given to Bab/C mice and examined after 2, 4, 12 and 24 h of exposure. Hepatocellular hypertrophy, eosinophilic cytoplasmic condensation around the outer nuclear membrane and apoptosis of few hepatocytes were noted. Animals that received two daily doses showed marked increase in apoptotic hepatocytes and those with four and seven doses of the toxin showed hepatocytomegaly and karyomegaly (Guzman and Solter, 2002). Nodularin selectively induced left lobe atrophy in livers of F344 rats but simulataneously right and middle lobes increased in their weight. However, after three weeks of nodularin administration, when diethylnitrosamine (DEN) was given for a total period of 10 weeks, DEN-altered hepatocytes regained regenerating power with concomitant restoration of phosphatase activity (Lim *et al.*, 2001).

The LD_{50} values of MC, nodularin and CYN have been determined by mouse bioassay. Most of the structural variants of MCs and nodularins were toxic in the range 50–300 μg kg^{-1} body weight (Table 3). Structural modifications of the Adda-glutamate region such as changes in isomerization of Adda-diene (6(E) to 6(Z) or acylation of the glutamate yielded non-toxic derivatives of MCs and nodularins (Harada *et al.*, 1990; Rinehart *et al.*, 1994). Accordingly, the LD_{50} values for such variants were found to be very high, i.e. 1000 to 1200 μg kg^{-1} body weight. The doses of MCs required i.p. were 25–150 μg kg^{-1} body weight (Dawson, 1998; WHO, 1998). Concentrations for oral ingestions were much higher, i.e. 5 to 10 mg kg^{-1} body weight (Ito *et al.*, 1997; WHO, 1998). On the other hand, the inhalation toxicity of MC-LR is very high and the concentrations almost correspond to i.p. dose, i.e. 36–122 μg kg^{-1} body weight (Dawson, 1998). The concentration of nodularin required for LD_{50} is 50 μg kg^{-1} body weight given i.p. In case of CYN, the concentrations reported for LD_{50} were 200–2000 μg kg^{-1} body weight administered i.p. (Chorus and Bartram, 1999). The concentrations of MCs with lowest observed adverse effect level (LOAEL) and no observed adverse effect level (NOAEL) are 100 and 40 μg kg^{-1} body weight, respectively.

Oral administration of MC-LR (0, 40, 200 or 1000 μg kg^{-1} body weight for 13 weeks) in a good laboratory practice on 15 male and female mice revealed NOAEL of 40 μg kg^{-1} body weight per day (Fawell *et al.*, 1994). The LD_{50} values for other forms of MCs are known to be 70 μg kg^{-1} body weight (i.p. in mice) for MC-YR whereas MC-RR has an LD_{50} (i.p. in mice) of 300–600 μg kg^{-1} body weight (WHO, 2003). *M. aeruginosa* extract (containing at least seven variants of MCs of which MC-YR as a major constituent) given to pigs i.p. at doses of 280, 800, 1310 μg kg^{-1} body weight per day in a 44 day study revealed LD_{50} of nearly 100 μg. The LOAEL was found to be 280 μg kg^{-1} body weight per day (Falconer *et al.*, 1994). On the other hand, a lethal dose of MC-LR for pigs was 72 μg kg^{-1} body weight given i.p. (Beasley *et al.*, 2000). Toxicity studies with MC-LR on embryonic and foetal developmental studies in mice revealed a NOAEL of 600 μg kg^{-1} body weight per day (Fawell *et al.*, 1994). According to them MC-LR is not a selective developmental toxicant in mice.

Incidences of human poisoning relate both to the recreational and drinking waters. Toxicity due to blooms of *Anabaena* and *Microcystis* in recreational waters was reported from North America, United Kingdom, the Netherlands, and Australia. Though no fatalities were reported from these countries, illness due to cyanobacterial poisoning in Saskatchewan (Canada) developed stomach cramps, vomiting, diarrhoea, fever, headache, pains in muscles and joints and weakness (Dillenberg and Dehnel, 1960). Ingestion of lake water that developed toxic blooms of *M. aeruginosa* resulted in hospitalization of two military recruits with atypical pneumonia in United Kingdom but yet majority of them developed abdominal pain, nausea, vomiting and diarrhoea (Turner *et al.*, 1990). Besides these symptoms, dermal irritation, sore red eyes, sore throat and allergic responses have also been noted in some patients in Australia (Ressom *et al.*, 1994). Characteristic symptoms of MC-poisoning in human beings upon drinking water loaded with cyanobacterial toxins are weakness, anorexia, cold extremities, pallor, apathy, respiratory problems, gastroenteritis, vomiting and diarrhoea. These symptoms are associated with necrosis of the liver leading to death by either haemorrhagic shock or liver failure (Gorham and Carmichael, 1988; Codd, 1995, 2000; Dow and Swoboda, 2000; de Figueiredo *et al.*, 2004). The outbreak of Palm Island mystery disease in Australia (Byth, 1980; Hawkins *et al.*, 1985) was in fact associated with the development of hepatoenteritis in 140 school children and 10 adults due to intake of lake water treated with copper sulphate that suffered the bloom of *C. raciborskii* (Chorus and Bartram, 1999). It was suggested that the addition of copper sulphate to control the bloom alga caused lysis of the cyanobacterial cells releasing more toxin into the waters. Due to drinking of lakewater that developed immense cyanobacterial bloom, more than 2000 cases of gastroenteritis in children were reported from Brazil in 1988, of these 88 of them succumbed to death (Teixera *et al.*, 1993). The death of 76 hemodialysis patients in a dialysis center at Caruaru, Brazil is a glaring example of human misery suffered due to hepatotoxins. This was due to use of dialysate contaminated with cyanobacterial cells and toxins. This was designated as 'Caruaru syndrome' (Pouria *et al.*, 1998; Jochimsen *et al.*, 1998). Subsequent studies on the examination of liver tissues by light microscopy and electron microscopy revealed a number of histopathological changes. The characteristics of 'Caruaru syndrome' are painful, extreme hepatomegaly, jaundice, a bleeding diathesis, disruption of liver plates, liver cell deformity, necrosis, apoptosis and injuries to cell organelles. The serum samples from diseased patients and the carbon filters of the dialysis unit revealed the existence of variants of MCs, i.e. MC-YR, MC-LR and MC-AR. The concentration of MCs in liver samples from 39 patients, collected after the death of 52 patients in between February to December, 1996, was found to be 223 ng g^{-1} liver tissue (Carmichael *et al.*, 2001).

v) Carcinogenicity: The hepatocellular specificity of MC-LR has been confirmed by immunohistochemical studies in rats, mice and trout (Solter *et al.*, 1998; Yoshida *et al.*, 1998; Fischer *et al.*, 2000). Upon chronic administration (20 kg^{-1} body weight, 100 times over 28 weeks), MCs promoted the formation of multiple neoplastic nodules in the liver of mice (Ito *et al.*, 1997). Other pathological symptoms reported in mice for chronic administration are increased liver weight and hepatohistological damage (Heinze, 1999) and kidney damage in rats (Milutinovi *et al.*, 2002). The carcinogenicity of MCs has been tested *in vivo* animal experiments. These involve testing DNA-damaging chemicals that can initiate cancer development while other classes of chemicals are able to promote cancer (Boutwell, 1974; Yamasaki, 1988; Fitzgerald and Yamasaki, 1990). The potential of MCs as cancer promoting substances was thus tested *in vivo* on skin, liver and intestines of mice. Initial application of 7,12-dimethyl benzanthracene (DMBA) as a single dose (500 μg) on dorsal skin of mice followed by *Microcystis* extract (80 mg MC-YM L^{-1}) increased the frequency of skin papillomas (Falconer, 1991). In another carcinogenic bioassy, DEN was given i.p. at a dose of 200 mg kg^{-1} body weight followed by partial hepatectomy. MC-LR doses (1 and 10 μg kg^{-1} body weight; 10, 25 and 50 μg kg^{-1} body weight; twice weekly) were subsequently administered i.p. for 8 weeks. This resulted in three-fold increase of positive pre-neoplastic liver loci with a total increase in the area by 11-folds. MC-LR did not induce such a response when administered alone. N-methyl-N-nitrosourea was chosen as tumor initiator (at two oral doses of 40 μg kg^{-1} body weight) followed by drinking water (containing MCs equivalent to 1.2 or 4.2 mg kg^{-1} body weight per day for 22 weeks). MCs did not promote either lymphoid or duodenal tumors (Falconer and Humpage, 1996).

Male Fischer 344 rats were injected i.p. with DEN as tumor promoter (200 mg kg^{-1} body weight) and two weeks after initiation it was followed by nodularin injections i.p. (25 μg kg^{-1} body weight twice a week until 12 weeks). The production of four types of nodules with eosinophilic, clear, mixed and basophilic cells was noted. Two weeks after nodularin administration was stopped, a staining reaction (GST-P immunostaining, i.e. glutathione S-transferase placental form positive) revealed dense nodules that stained homogeneously and pale nodules that stained heterogeneously (Song *et al.*, 1999). Thus it was concluded that (i) nodularin has a promoting effect by inducing hepatocyte proliferation; (ii) proliferation of hepatocytes is a temporary phenomenon and (iii) enzyme altered hepatocyte dense nodules are progressive and pale nodules are regressive. Since MCs and nodularins are potent inhibitors of PP1 and PP2A, these are considered to be non-phorbol ester (12-O-tetradecanoyl phorbol 13-acetate, TPA)-type promoters. Okadaic acid, tautomycin and calyculin are the other such substances (Fujiki and Suganuma, 1993). Protein phosphatase inhibition leads to a shift in higher phosphorylation of certain target proteins such as tumor suppressor proteins. This is considered to be a major post-translational modification resulting in excessive signalling. This leads to cell proliferation, cell transformation and tumor promotion (Fujiki and Suganuma, 1993; Ohta *et al.*, 1994).

Cell death occurs either by programmed cell death (apoptosis) or necrosis. It is pertinent here to point out the differences between these two processes. Apoptosis is a normal physiological process that takes place to maintain cell homeostasis. A number of morphological and physiological changes occur that involve disintegration of cytoskeleton, cell shrinkage, chromatin condensation, and activation of specific proteases known as capsases (Kerr *et al.*, 1972; Wyllie *et al.*, 1980; Nicholson and Thornberry, 1997). In the early events, reversal in position of phosphatidyl serine takes place, i.e. phosphatidyl serine located on the inner face of plasma membrane becomes exposed on the outer surface. This acts as a recognition signal for phagocytosis. A number of external stimuli including receptor ligation and toxins are responsible for apoptosis. On the other hand, necrotic cell death results due to mechanical or toxic cell injury. A number of morphological changes occur in necrotic cells such as swelling, plasma membrane rupture, organelle breakdown and ultimately lysis followed by release of cell contents and an inflammatory response (Wyllie *et al.*, 1980). Characteristic centrilobular hepatocyte rounding, dissociation, necrosis and breakdown of sinusoidal epithelium leads to severe interhepatic haemorrhage. Acute toxicity leads to cytoskeletal collapse due to aggregation of actin filaments, microtubules and intermediate filaments within hepatocytes (Hooser *et al.*, 1991; Khan *et al.*, 1996; Wickstrom *et al.*, 1995). Rapid hyperphosphorylation of cytoskeletal proteins such as cytokeratin seems to be responsible for the pathogenesis secondary to the inhibition of PP1 and PP2A (Falconer and Yeung, 1992; Ohta *et al.*, 1992; Runnegar *et al.*, 1993).

Severe inhibition of PP1 and PP2A by MCs and nodularins leads to an imbalance in normal cellular processes such as cell proliferation and results in cancer production or apoptosis (Boe *et al.*, 1991; Fujiki and Suganuma, 1993; Toivola *et al.*, 1994; Fladmark *et al.*, 1998; Mankiewicz *et al.*, 2001). Stimulation in cytokinesis and decrease in the rate of apoptosis were observed when MC-LR was administered as a single dose of 1 nM or multiple doses of MC-LR at pM concentrations (Humpage and Falconer, 1999). This could be due to the tumor-producing activity of MC-LR. On the other hand, inhibition of cytokinesis at higher concentrations (10 nM) of MC-LR induced apoptosis (Humpage and Falconer, 1999). Nodularin is reported to activate several proto-oncogenes of the *fos* and *Jun* family that are considered to play a role in tumor promotion (Ohta *et al.*, 1994).

***vi) Toxicity studies in vitro*:** The viability of mouse B- and T-lymphocytes decreased by 23% *in vitro* treated with MC-LR after 24 h. Apoptotic cell death occurred in B-cells probably through B-cell antigen receptor and mitochondrial pathway while T-cells were unaffected (Teneva *et al.*, 2005a). Isolated rat hepatocytes treated with MC-LR showed cell shape changes (blebbing) associated with remarkable reorganization of microfilaments. These changes are specific to MC-LR because microfilament-modifying drugs like cytochalasin D and phalloidin could not bring out such changes (Eriksson *et al.*, 1989). Falconer and Yeung (1992) described cytoskeletal changes in cultured rat hepatocytes exposed to *Microcystis* toxins. The mechanism of *Microcystis* toxicity is due to cytoskeletal damage leading to loss of cell morphology, cell adhesion and finally cellular necrosis.

Besides hepatocytes, MCs and nodularins can also affect other types of cells. Fladmark *et al.* (1999) observed apoptosis in human embryo kidney HEK 293 cells, Swiss 3T3 mouse embryo fibroblasts, breast carcinoma cell line MCF-7 and in rat promyelocytic IPC-81 leukemia cells. Various other cellular deformities observed were membrane budding, cell shrinkage and organelle redistribution in human fibroblasts, endothelial cells, epithelial cells, lymphocytes and in rat promyelocytes (McDermott *et al.*, 1998; Mankiewicz *et al.*, 2000, 2001). Dose-related increase in chromosomal breakage has been observed in human lymphocytes (Repavich *et al.*, 1990; Premazzi and Volterra, 1993). Treatment of human and chicken peripheral blood lymphocytes with MC-LR (for 72 h in a proliferation assay) decreased T-cell proliferation at the highest dose whereas the proliferation of B-lymphocytes was inhibited at all doses. Apoptotic and necrotic cells increased with dose and time of exposure (Lankoff *et al.*, 2004). MC-LR- and MC-RL-containing extracts from bloom material of *M. aeruginosa* (growing in Sulejow Reservoir that serves as a source of drinking water) at acceptable dose of MC (WHO) caused activation of cytochrome *c* oxidase complex in mitochondria and it is suggested to serve as an indicator for assessing the toxicity *in vivo* (Majsterek *et al.*, 2004). Benthic marine cyanobacterial species from Portuguese coast (not associated with apparent bloom formation) produced novel toxins that induced apoptosis in SH-SY5Y-neuroblastoma cells, NRK kidney cells or fibroblasts. Aqueous extracts of the cyanobacteria inhibited thrombin-induced blood platelet activation, with decreased P-selection expression, platelet aggregation and shedding of platelet-derived microvesicles. Two of the cyanobacteria constitute important source for novel toxins that may be useful in mammalian signalling pathways of apoptosis and platelet formation (Selheim *et al.*, 2005).

***vii) Genotoxicity studies*:** Ames *Salmonella* assay (strains TA98, TA100, TA102) and *Bacillus subtilis* multigene sporulation test (using 168 and *hcr-9* strains) were used to detect mutagenic response for purified toxins that proved to be negative (Grabow *et al.*, 1982; Repavich *et al.*, 1990). However, point mutations were induced by MC-LR in four strains of *Salmonella typhimurium* by Ames test (Ding *et al.*, 1999). A genotoxic effect by MC-LR has been shown in *E. coli* (strain PQ37) (Mankiewicz *et al.*, 2002). Though point mutations were not induced, there was loss of heterozygosity when human lymphoblastoid TK6 cells were treated with MC-LR (Zhan *et al.*, 2004). MC-LR caused DNA damage in mouse embryo fibroblasts and BHK-21 cell lines (Rao *et al.*, 1998) as well as in cultured rat hepatocytes (Ding *et al.*, 1999). By employing Comet assay, Zegura *et al.* (2003) demonstrated MC-LR-induced DNA strand breaks in human hepatoma HepG2 cells. But these DNA strand breaks have now been attributed to early stages of apoptosis and not due to MC-LR-induced genotoxicity (Lankoff *et al.*, 2004). Moreover, Lankoff *et al.* (2003) reported that MC-LR did not cause any chromosomal damage in CHO-K1 cells as well as in human lymphocytes. Another explanation given to the DNA strand breaks is the probable role of MC-LR in inhibiting the DNA repair enzymes (Lankoff *et al.*, 2004). DNA-dependent protein kinase is the key enzyme identified in the repair of radiation-induced DNA breaks by mediating the non-homologous end joining (Douglas *et al.*, 2001). Lankoff *et al.* (2006) demonstrated that pre-treatment of human lymphocytes and human glioblastoma cell lines MO59J and MO59K with MC-LR (0.5 mg ml^{-1}) not only inhibited repair of radiation-induced damage but also enhanced the frequencies of chromosomal aberrations including dicentric chromosomes.

***viii) Protection against toxicity*:** The role of membrane bound vitamin E in protecting against MC-LR toxicity was investigated in mice fed with lower or higher vitamin E supplements, 8.33 or 33.3 units $mouse^{-1}$ day^{-1} for 4 weeks, respectively. The evaluation of vitamin E protection was performed by determining the levels of LPO, alanine transaminase and glutathione S-transferase. Some protection was noticed at lower vitamin E concentration by limiting toxin-induced leakage of alanine transaminase and glutathione-S-transferase. A higher dose of vitamin E significantly increased the rate of mortality and percentage weight loss of liver of mice when challenged with lethal dose of MC-LR (Gehringer *et al.*, 2003b). BALB/c mice pre-treated with selenium (sodium selenite administered at the rate of 1.5 µg $mouse^{-1}$ day^{-1} for two weeks) reduced the liver damage caused by MC-LR and decreased alanine trasaminase and LPO levels. However, the increase of liver glutathione peroxidase activity indicated the possible route of selenium protection in mice (Gehringer *et al.*, 2003a). In an attempt to find out a suitable chemoprotectant against the toxicity of MC-LR, Rao *et al.* (2004) screened a variety of substances either by way of pre-treatment, co-treatment or post-treatment against the lethal dose of MC-LR (100 µg kg^{-1} body weight) in a mouse (Swiss albino female) bioassay. Though some of them offered no protection (D-glucose, mannitol, L-cysteine, naringin and amifostine) but extended the survival time. Some others like N-acetylcysteine, glutathione and Trolox® partially protected on either pre- or co-treatment. On the other hand, pre-treatment with rifampin (25 mg kg^{-1}), cyclosporin (10 mg kg^{-1}) and silymarin (400 mg kg^{-1}) gave complete protection. Of these three agents, the first two gave complete protection when co-administered with MC-LR. However, in post-treatment rifampin was effective when administered after 15 minutes. Any chemoprotectant either should possess the ability to inhibit the interaction of MCs with PPs or inhibit the uptake of the toxin into hepatocytes. It was further suggested that possibly rifampin, cyclosporin and silymarin conferred protection by inhibiting intracellular transport of the toxin into the liver (Rao *et al.*, 2004). Oxidative damage caused by nodularin is alleviated in presence of melatonin. The activities of SOD, catalase, glutathione peroxidase in mouse exhibited an increase when co-treatment of nodularin (1, 5 µg kg^{-1} body weight) was given along with melatonin (5, 10, 15 mg kg^{-1} body weight) for seven days. A treatment with nodularin (5 µg kg^{-1} body weight for seven days) followed by a post-treatment with melatonin (5, 10, 15 mg kg^{-1} body weight) also protected from the oxidative damage when compared to nodularin treatment alone (Lankoff *et al.*, 2002).

II. NEUROTOXINS

These are one of the most toxic compounds produced by cyanobacteria that interfere with the functioning of neuromuscular system. Species of *Nostoc* (Davidson, 1959), *Anabaena* (Carmichael *et al.*, 1990), *A. planctonica* (Bruno *et al.*, 1994), *A. flos-aquae* (Devlin *et al.*, 1977; Carmichael *et al.*, 1975, 1979), *Oscillatoria* (Sivonen *et al.*, 1989a; Edwards *et al.*, 1992; Skulberg *et al.*, 1992) are reported to produce four neurotoxins, i.e. anatoxin-a, anatoxin-a(s), STX and neo-STX (Ressom *et al.*, 1994;

Chorus and Bartram, 1999). Paralysis of skeletal muscles and respiratory muscles leads to death of the individual within minutes (Ressom *et al.*, 1994; Singh *et al.*, 1999).

A) Anatoxin-a: It was first isolated from *A. flos-aquae*. Some species of *Anabaena, Planktothrix, Aphanizomenon* and *Cylindrospermum* (Sivonen and Jones, 1999; Ballot *et al.*, 2010), *M. aeruginosa* (Park *et al.*, 1993) *Phormidium favosum* (Gugger *et al.*, 2005) and *Oscillatoria* PCC 6506 (Méjean *et al.*, 2009) also are reported to produce this toxin.

Structurally, anatoxin-a is an alkaloid considered to be a secondary amine, 2-acetyl-9 azabicyclo(4-2-1) non-ene (Fig. 12; Huber, 1972). It has a molecular weight of 165 D (Carmichael, 1992). It is a structural analogue of cocaine and neurotransmitter acetylcholine (Mazur-Marzec, 2006). The biosynthesis of anatoxin-a has been shown to be analogous to tropane group of alkaloids found in higher plants (Gallon *et al.*, 1990, 1994). This was countered by Hemscheidt *et al.* (1995) based on ^{14}C-labelling and enzymatic experiments. They suggested that pyrroline, a precursor of anatoxin-a, is synthesized in *A. flos-aquae* starting from ornithine or arginine via putrescine. The derivation of carbon skeletons of anatoxin-a from either acetate or glutamate has been confirmed by ^{14}C-labelling experiments and the retention of glutamic acid residue during transformation of anatoxin-a further does not lend support to the biosynthesis of anatoxin-a similar to tropane group of alkaloids (Hemscheidt *et al.*, 1995). However, the biosynthesis of anatoxin-a and homoanatoxin-a has been reported to be mediated by the gene products of eight contiguous genes (*anaA* to *anaH*) in *Oscillatoria* PCC 6506. It is initiated by the loading of 1-proline on acyl-carrier protein where its ring structure gets oxidized into pyrroline oxidation state. This is soon followed by three successive PKS modules for elongation, reduction and cyclization and methylation. The loading of 1-proline in the first reaction has been confirmed by the mediation of purified protein of AnaC. The existence of gene sequences *anaC, anaE, anaF* and *anaG* in all anatoxin-a producing toxic strains and their absence from non-toxic strains of cyanobacteria has also been demonstrated (Méjean *et al.*, 2009).

The stability of anatoxin-a has been studied. Photolysis of anatoxin-a was dependent on the intensity of sunlight and pH of the medium with a half-life of 1–2 h whereas in absence of sunlight or metal ions such as copper and iron, the half- life was several days. The breakdown products of anatoxin-a exhibited no toxicity in mouse bioassay (Stevens and Krieger, 1991).

The mode of action of anatoxin-a is similar to acetylcholine and acts as a stereo-selective agonist of the acetylcholine receptors in both neurons and endplates with 8-10 fold higher potency (Carmichael *et al.*, 1975; Spivak *et al.*, 1980; Ressom *et al.*, 1994; Dow and Sowboda, 2000). Under normal conditions, acetylcholine released by neurons binds to acetylcholine receptor results in opening up of sodium channels. Continuous flow of sodium ions to muscle cells activates them. The degradation of acetylcholine by acetylcholinesterase leads to closure of sodium channel that makes

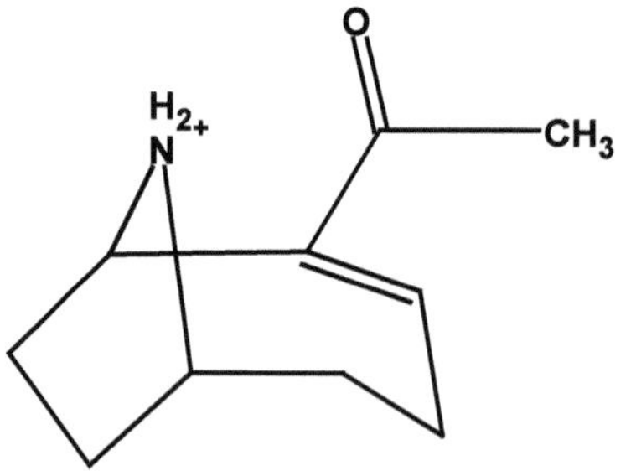

Figure 12: Structure of anatoxin-(a).

the muscles rest. Due to irreversible binding of anatoxin-a to acetylcholine receptor, constant flow of sodium ions to muscle cells results in membrane depolarization. This causes to open up voltage sensitive Ca^{2+} and Na^{2+} channels. The latter may then amplify the response activating further electrical transmission (Soliakov *et al.*, 1995) leading to overstimulation of muscle cells. When respiratory muscles are involved it results in lack of O_2 to brain leading to convulsions, suffocation and finally death (Mazur-Marzec, 2006).

Lowest dose of anatoxin-a causing death in mice (i.p; LD_{10}) was 200–250 µg kg^{-1} body weight (Stevens and Krieger, 1989, 1991) and LD_{50} of anatoxin-a was 375 µg kg^{-1} body weight (Fitzgeorge *et al.*, 1994). The i.v. LD_{50} values were less than 100 µg kg^{-1} body weight. LD_{50} values for oral administration and intranasal LD_{50} in mice were 5000 and 2000 µg kg^{-1} body weight, respectively (Fitzgeorge *et al.*, 1994). Although there are very few studies on toxicity of anatoxin-a, it is worthwhile to mention the studies of Astrachan *et al.* (1980) who suggested that doses of 200 or 125 µg kg^{-1} body weight of anatoxin-a given thrice a day caused foetal malformation in hamsters.

The binding of anatoxin-a to the acetylcholine receptor sites with greater specificity has been demonstrated by the competitive inhibition of $[H^3]$-acetylcholine and d-$[H^3]$-turbocurarine binding to the receptor site. The binding of anatoxin-a to receptor site of acetylcholine brings about conformational changes in the receptor-ion channel complexes. These conformational changes can be detected by the binding of certain compounds that block neuromuscular transmission without interacting with receptor site. The binding of certain ion channel blockers such as tritiated perhydrohistrionicotoxin (H_{12}-HTX) and phenylcyclidine (PCP) to the receptor-ion channel complex is stimulated by anatoxin-a. Aronstam and Witkop (1981) demonstrated that the binding ion-channel blockers such as H_{12}-HTX, PCP and phenylcylcidine methiodide is stimulated in presence of anatoxin-a to the acetylcholine receptor of *Torpedo* electric tissue.

Two nicotinic acetylcholine-binding sites were evaluated for binding by 18 analogues of anatoxin-a that differed in side chain structure. Anatoxin-a is the most potent competitor for ligand binding at $[H^3]$ nicotine and $[^{125}I]\alpha$-bungarotoxin-binding sites. All analogues have higher affinity at $[H^3]$ nicotine site compared to $[^{125}I]\alpha$-bungarotoxin-binding site (Wonnacott *et al.*, 1991).

The toxicity of anatoxin-a is 10-fold more potent than its optically inactive isomer in increasing mean arterial blood pressure and heart rate, pO_2 and pCO_2 of blood and pH. Severe hypoxemia and hypercapnia accompanied by acidosis were induced by anatoxin-a by 6 and 4 fold greater than its optical isomer (Adeyemo and Siren, 1992). A comparison of the effects of anatoxin-a in four different preparations revealed that anatoxin-a was 50 times more potent than nicotine and 20 times more potent than acetylcholine (Thomas *et al.*, 1993). These studies are supported by the potency of anatoxin-a in evoking 1A or type 2 current responses in rat hippocampal neurones far greater than nicotine or acetylcholine (Atkondon and Abulquerque, 1995). Similar to nicotine, anatoxin-a shows more potency than noradrenaline in releasing dopamine for strial nerve terminals from rat superfused hippocampal synaptosomes (Clarke and Reuben, 1996). Studies on genotoxic potential of anatoxin-a are not avilable. A benthic species of *Oscillatoria* that produced anatoxin-a was shown to be responsible for dog deaths, after consumption of water on the shores of Loch Insh, Scotland. The presence of the cyanobacterium in the stomach contents, their positive neurotic response in bioassay studies and *in vitro* production of anatoxin-a by the cultures of *Oscillatoria* emphasized the role of anatoxin-a (Edwards *et al.*, 1992). Developmental toxicity of anatoxin-a has been studied in pregnant mice by administering anatoxin-a i.p. (125 or 200 µg kg^{-1} body weight) on gestation days (GD), i.e. 8–12 or 13–17 days. Decreased motor activity was observed at 200 µg kg^{-1} body weight at both time periods whereas pup viability and weight on postnatal day remained unaffected. GD8 mouse embryos exposed to 0.1–0.25 μg anatoxin-a for 26–28 h showed no embryonic dysmorphology but there were perturbations in mouse yolk sac vasculature. Synergistic toxicity of MC-LR and anatoxin-a has been studied by orally administering MC-LR first (500, 1000 μg kg^{-1} body weight) followed by anatoxin-a (500, 1000 or 2500 μg kg^{-1} body weight). Neither death nor definite signs of intoxication were noted even at high concentrations. However, in amphibian toxicity assays involving exposure of toad embryos (stages 17 and 25 of *Bufo arenarum*) to anatoxin-a (0.3 to 30 mg L^{-1} for 10 days) resulted in 100% lethality at high dose in groups 6–13 days post-exposure (Rogers *et al.*, 2005).

The effects of anatoxin-a on non-neurone cells have been studied. In this connection, the studies of Rao *et al.* (2002) on anatoxin-a induced apoptosis in cultured rat thymocytes and African green monkey kidney cells are significant. Plasma membrane blebbing, condensed chromatin, nuclear fragmentation and formation of apoptotic bodies are the characteristics of anatoxin-a-induced apoptosis. The involvement of capsases (capsase-3 activity) and reactive oxygen species in apoptosis was also indicated.

Homoanatoxin-a (mol wt. 179 D) is structurally similar to anatoxin-a and is produced by an *Oscillatoria formosa (P. formosum)* strain (Fig. 13). Méjean *et al.* (2009) reported the production of anatoxin-a and homoanatoxin-a by *Oscillatoria* PCC 6506. A propinyl group at C-2 is characteristic instead of the acetyl group in anatoxin-a (Skulberg *et al.*, 1992). The LD_{50} of homoanatoxin-a was 288–578 μg kg^{-1} body weight. The LD_{50} value of homoanatoxin-a (i.p.) was 250 μg kg^{-1} body weight. Toxicity due to lethal doses leads to body paralysis, convulsions and death by respiratory arrest in 7–12 minutes (Kupier-Goodman *et al.*, 1999). The physiological effects of homoanatoxin-a are shown to be due to d-tubocurarine (Skulberg *et al.*, 1992). It also enhances the influx of Ca^{2+} ions in cholinergic nerve terminals (Aas *et al.*, 1996).

In an attempt to detect toxins from health food supplements (tablets and capsulses of *Spirulina*) marketed in Rome (Italy), Draisci *et al.* (2001) reported the existence of epoxyanatoxin-a and its analogues (18 to 19 μg g^{-1} in only two of the five samples tested) and dihydrohomoanatoxin-a (10 μg g^{-1} in three of the five samples). Thus these authors suggested that a potential human health hazard could be associated with the consumption of such food supplements. It is interesting to note that anatoxin-a and homoanatoxin-a have been isolated from *Oscillatoria* sp. strain 193 PCC 9240 and *O. formosa* NIVA CYA-92 PCC 10111, respectively. Other strains of *Oscillatoria* such as PCC 6407, PCC 6412 and PCC 9107 produced exclusively anatoxin-a where as strains PCC 9029 and PCC 6506 synthesized both anatoxin-a and homoanatoxin-a (Aráoz *et al.*, 2005).

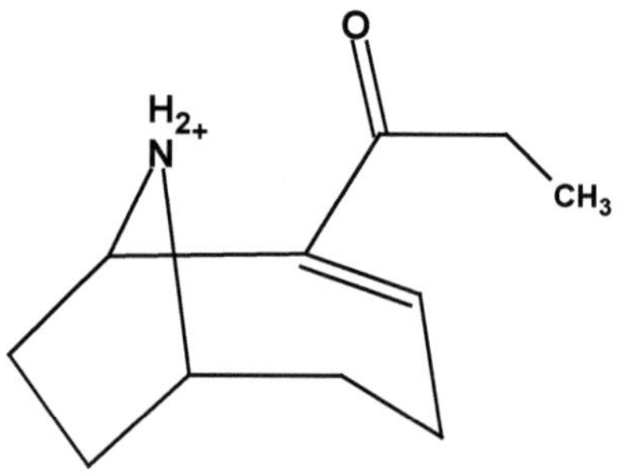

Figure 13: Structure of homoanatoxin-a.

B) Anatoxin-a(s): Anatoxin-a(s) is a unique N-hydroxyguanidine methylphosphatase ester with a molecular weight of 252 D (Fig. 14; Matsunaga *et al.*, 1989; Ressom *et al.*, 1994). It has been isolated from *A. flos-aquae* and *A. lammermanni*. Mahmood and Carmichael (1986a) reported the production of anatoxin-a(s) by *A. flos-aquae* NRC 525-17. One of the important syndromes of anatoxin-a(s) is salivation in vertebraes (Carmichael and Falconer, 1993). That is why the letter 's' is added as a suffix in parenthesis.

The biosynthesis of anatoxin-a(s) has been studied in *O. formosa* and all carbon atoms are suggested to be derived from amino acids. L-Methionine is suggested to contribute the methyl carbons or other donors to the C-1 pool whereas L-arginine accounts for C-2, C-4, C-5 and C-6 carbons of the toxin (Moore *et al.*, 1992, 1993). Purified anatoxin-a(s) has a LD_{50} value of 50 µg kg^{-1} body weight in a mouse bioassay. In acute toxicity tests of mice and rats, the signs of poisoning included excessive cholinergic stimulation. In contrast to anatoxin-a, anatoxin-a(s) inhibits the activity acetycholineterase. Complete inactivation of serum cholinesterase of rats was noted at doses of 350 and 600 µg kg^{-1} body weight. Anatoxin-a(s) reacts with cholinesterase phosphorylating it and forms a complex. As the enzyme is unable to degrade acetylcholine, overstimulation of muscle cells occurs (Carmichael, 1997; Singh *et al.*, 1999).

C) STXs: These are the most widely known group of neurotoxins mainly produced by marine dinoflagellate species of *Alexandrium* (Fig. 15 A), *Pyrodinium* and *Gymnodinium* (Fig. 15 B; Oshima *et al.*, 1993) These toxins are chemically most diverse comprising of macrolides, cyclic polyethers, sprolides and purine alkaloids. Their biosynthetic pathways and molecular genetics have been summarized (Kellmann *et al.*, 2010). The generic name 'saxitoxins' for the paralytic shellfish toxins (PSTs) produced by freshwater cyanobacteria was proposed by Sivonen and Jones (1999). STXs are tricyclic guanidium alkaloids (Fig. 16). There are more than 30 naturally occurring STX analogues. STXs accumulated by shellfish when consumed by human beings results in PSP. Cyanobacteria that are known producers of STXs are *A. circinalis* (Fig. 4C; Humpage *et al.*, 1994), *Planktothrix* sp. FP1 (Pomati *et al.*, 2000) from Australia, *Aph. flos-aquae* from USA (Alam *et al.*, 1973; Mahmood and Carmichael, 1986b) and Portugal (Pereira *et al.*, 2000), *Aph. gracile* (Fig. 2 C) from northeast Germany (Ballot *et al.*, 2010) and France (Ledereux *et al.*, 2010), *Aphanizomenon* sp. DC-1 from China (Liu *et al.*, 2006), *C. raciborskii* (Sawyer *et al.*, 1968; Lagos *et al.*, 1999; Bouvy *et al.*, 1999; Bernard *et al.*, 2003) and *R. brookii* D9 (Stucken *et al.*, 2010) from Brazil and *L. wollei* (Carmichael *et al.*, 1997) from USA are known producers of PSTs. There appears to be a geographical segregation of strains of *A. circinalis* since Australian isolates are known to produce PSTs whereas in other continents the production of other neurotoxins such as anatoxin-a and anatoxin-a(s) has been reported (Sivonen and Jones, 1999).

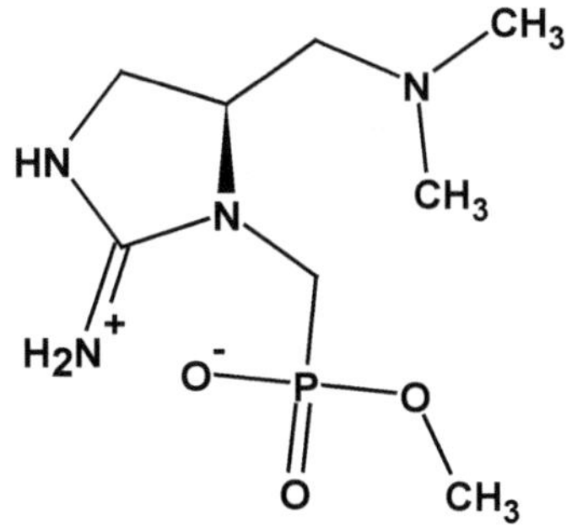

Figure 14: Structure of anatoxin-a(s).

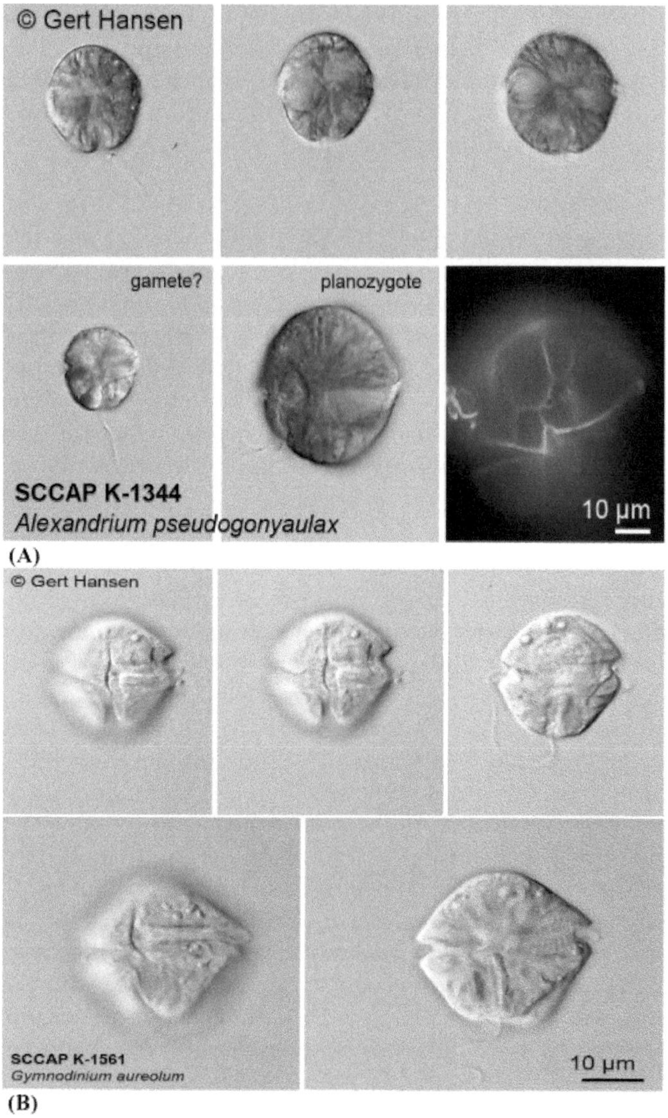

Figure 15: Saxitoxin-producing species of dinoflagellates *Alexandrium pseudogonyaulax* (A) and *Gymnodinium aureolum* from Scandinavian Culture Collection for Algae and Protozoa in different phases and views. Pictures courtesy Gert Hansen, Scandinavian Culture Collection of Algae & Protozoa, Department of Biology, University of Copenhagen, Oester Farimagsgade 2D, DK-1353, Copenhagen, Denmark (http://www.sccap.dk/).

Color image of this figure appears in the color plate section at the end of the book.

Structurally STXs are divided into two series, saxitoxins and neosaxitoxins. The difference in between the two is that in the latter a hydroxyl group is added at N-1 position. In each of these series the structural variants are synthesized by the presence and differential stereochemistry of sulphate groups at C-1 and the occurrence of N-sulphate carbamoyl groups at C-17. The diversity of PSTs produced by *L. wollei* appears to be greater than in any of the PST-producing cyanobacterial

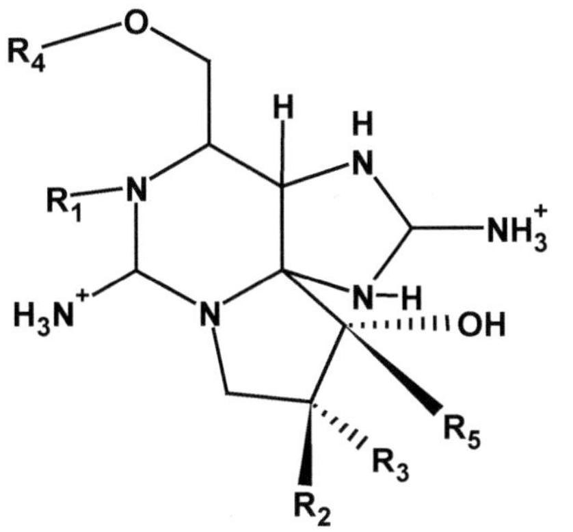

Figure 16: Structure of saxitoxin.

species. Besides six novel structural analogues, it is reported to produce decarbomyol STXs, (dcSTXs), decarbamyol gonyautoxins such as dcGTX2 and dcGTX3 (Fig. 17; Mihali *et al.*, 2011). The major PSTs produced by *Aph. flos-aquae* and *Aphanizomenon* sp. NH-5 have been reported to be STX and neoSTX. In case of *A. circinalis* STX and GTXs and C-1 and C-2 variants are the major STXs synthesized (Baker and Humpage, 1994; Humpage *et al.*, 1994). On the other hand, 14 strains of *Aph. gracile* have been reported to produce GTX5, dcSTX, STX and neoSTX (Ballot *et al.*, 2010).

STX production by *C. raciborskii* T3 exhibited a correlation with intracellular Na^+ levels. This was confirmed by supplying Na^+ channel-blocking agents like amiloride and lidocaine in presence of Na^+. Amiloride (1 mM) lowered cellular Na^+ and STX levels whereas lidocaine (1 μM) enhanced total cellular Na^+ and toxin levels. It was suggested that the metabolism of STX or toxin itself might be associated with maintenance of cellular homeostasis (Pomati *et al.*, 2004). Carneiro *et al.* (2009) investigated the effects of light intensity and quality on the production of STX and neoSTX by *C. raciborskii* in batch cultures. The production of neoSTX was higher in cultres under light/dark cycles at 100 μmol photons $m^{-2}\ s^{-1}$ than at 50 and 150 μmol photons $m^{-2}\ s^{-1}$ light intensity. The rhythmicity in the production of toxin was lost under red light. Transcription profiles of PST-producing and non-toxic strains of *A. circinalis* were compared by DNA microarray approach. Exposure of *A. circinalis* to lidocaine hydrochloride (1μM for 2 hrs) led to the expression of certain putative toxin-strain distinctive DNA fragments. On the other hand, there was no change in the expression of the same genes after exposure to 1μM lidocaine in STX-producing *C. raciborskii*. However, exposure to lidocaine enhanced transcription of genes involved in physiological adaptive responses and bloom formation in cyanobacteria such as gas vesicle structural protein A and phycocyanin (Pomati *et al.*, 2006).

i) Biosynthesis: A hypothetical biosynthesis pathway of STX consisting of nine reactions (Shimizu *et al.*, 1984) formed the basis for the identification of the role of enzymes (S-adenylhomocysteine hydrolase, methionine aminopeptidase, sulfotransferase, a Na^+-dependent transporter, aminotransferase and O-carbamoyl transferase) for its *in vitro* synthesis (Kellmann and Neilan, 2007). The identification of *sxtI* (that encodes O-methylcarbamoyltransferase) gene in *C. raciborskii* T3 as a marker gene for biosynthesis of STX and the presence of its homologue in two other PST-producing cyanobacteria paved the way for the sequence analysis of the entire gene cluster in *C. raciborskii* T3 (Kellmann *et al.*, 2008), *A. circinalis* AWQC131C and *Aphanizomenon* sp. NH-5 (Fig. 18; Mihali *et al.*, 2009). Due to the presence of *sxtI* gene in a truncated state in *L. wollei*, 400 bp amplicons of *sxtT* and *sxtH* (gene products responsible for hydroxylation at C-12 position) were helful in identifying and further

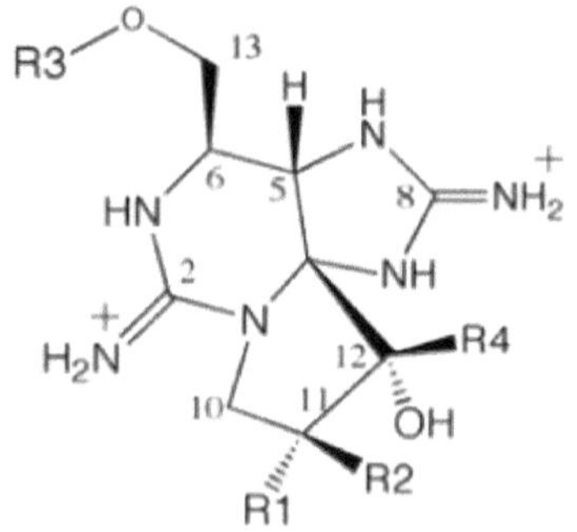

	R1	R2	R3	R4
dcSTX	H	H	H	OH
dcGTX2	OSO_3^-	H	H	OH
dcGTX3	H	OSO_3^-	H	OH
1	H	OSO_3^-	$COCH_3$	H
2	H	OSO_3^-	$COCH_3$	OH
3	OSO_3^-	H	$COCH_3$	OH
4	H	H	H	H
5	H	H	$COCH_3$	OH
6	H	H	$COCH_3$	H

Figure 17: Structure of the saxitoxin analogs identified in *Lyngbya wollei*. With the kind permission of B. A. Neilan, Cyanobacteria and Astrobiology Research Laboratory, School of Biotechnology and Biomolecular Sciences, The University of New South Wales, Sydney 2052, NSW, Australia. [Mihali *et al.* (2011) *PLoS ONE* **6(2):** e14657. doi:10.1371/journal.pone.0014657] doi:10.1371/journal.pone.0014657.g001.

sequencing of *sxt* gene cluster in this organism as well (Fig. 19; Mihali *et al.*, 2011). The *sxt* gene cluster of *C. raciborskii* T3 reveals 26 candidate STX genes (*sxtA* to *sxtZ*). The biosynthesis of STX is initiated by the gene product of *sxtA* which consists of four catalytic domains known as SxtA1 (methyltransferase), SxtA2 (GCN5-related N-acetyltransferase, GNAT domain), SxtA3 (an acyl carrier protein, ACP domain) and SxtA4 (aminotransferase domain). The precursors for STX biosynthesis arginine, S-adenosylmethionine and acetate are acted upon by SxtA. The first reaction in biosynthesis of STX is initiated by the loading of ACP (SxtA3) with acetate by SxtA1 forming propionyl ACP which then undergoes a Claisen condensation reaction (SxtA4) with arginine resulting in the formation of 4-amino-3-oxo-guanidinoheptane. This is followed by three heterocyclizations resulting in STX and the activity of tailoring enzymes facilitates the formation of numerous isoforms of the toxin (Kellmann *et al.*, 2008). The general pattern of biosynthesis in the four PST-producing cyanobacteria is almost the same with differences in the species of STXs produced. However, there are a number of differences in the structural organization of the *sxt* gene cluster. These are depicted in Fig. 20 and summarized (Table 6). It is pertinent to point out here that *A. circinalis* AWQC131C does not produce neoSTX because it is unable to convert STX into neoSTX by introducing a hydroxyl group at N-1 position due to the absence of *sxtX* gene in its cluster. The biosynthetic pathway for STX is represented in Fig. 14. Moustafa *et al.* (2009) who traced the origin of STX genes in cyanobacteria suggested that nine putative STX genes were horizontally transferred from non-cyanobacterial sources and 13 belonged to cyanobacteria while four others are shared by toxic as well as non- toxic strains. A comparison of gene clusters involved in the biosynthesis of STX in *R. brookii* D9 and CYN in *C. raciborskii* CS-505 revealed a great deal of synteny. The STX gene cluster of the former spans 25.7 kb region in its genome and is organized into 24 ORFs. Of these, 20 ORFs bear resemblance with those of *C. raciborskii* T3. Since 19 genes from among the STX gene clusters of *R. brookii* D9 and *C. raciborskii* T3 share 100% similarity, it has been suggested that this is the minimum number of genes required for STX production (Stucken *et al.*, 2010). Due to non-availability of molecular markers, so far it has not been possible to undertake studies on genetic diversity of PST-producing cyanobacteria. However, the sequencing of the *sxt* gene cluster in four cyanobacteria as described above seems to have overcome this problem. Ballot *et al.* (2010) employed *sxtA* gene as a marker for identifying PST-producing cyanobacteria from Lakes Melang and Scharmützel (northeast Germany)

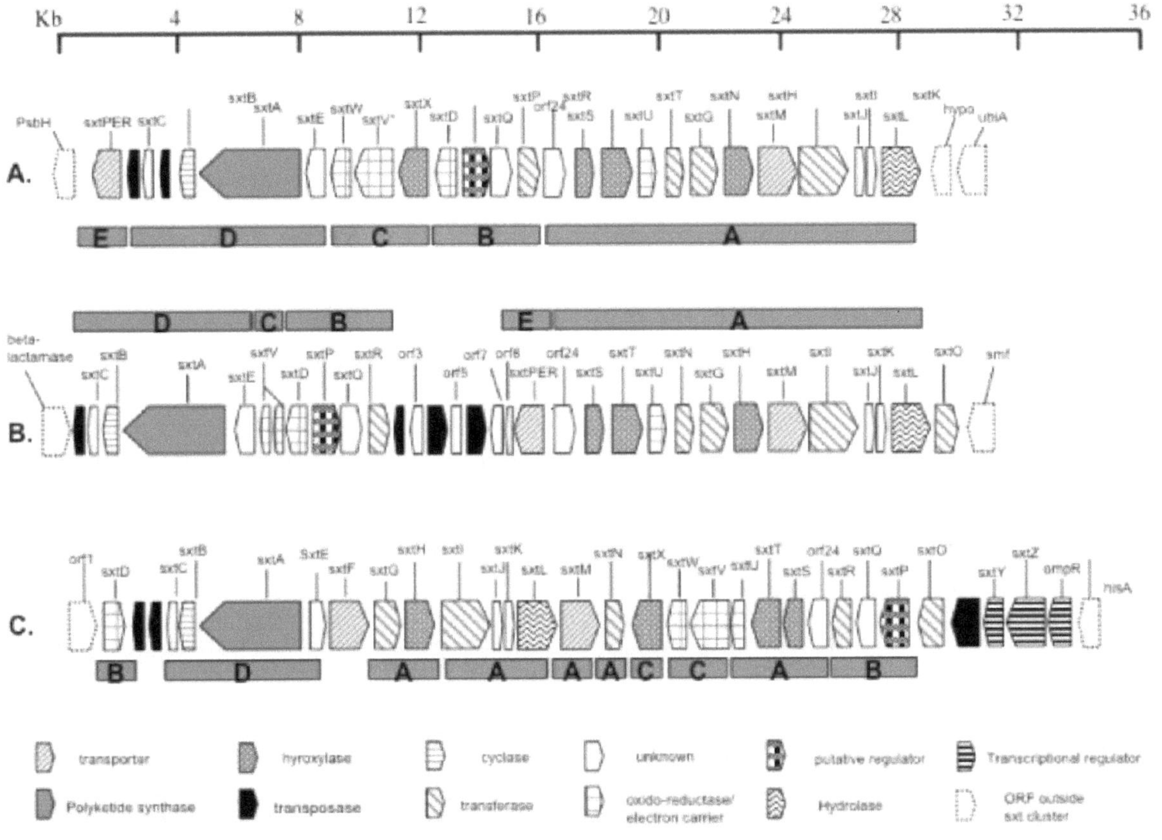

Figure 18: Structure of the paralytic shellfish toxin biosynthesis cluster identified in (A). *Aphanizomenon. sp.* NH-5, (B) *Anabaena circinalis* AWQC131C, C. *Cylindrospermopsis raciborskii* T3. The gene cluster schematic for *C. raciborskii* T3 has been adapted from kellmann *et al.* 2008 [29]. Segments A-E denote cluster fragments homologous in the three strains. The scale indicates length in thousand base pairs. ompR, transcriptional regulator of *ompR* family. With the kind permission of B. A. Neilan, Cyanobacteria and Astrobiology Research Laboratory, School of Biotechnology and Biomolecular Sciences, The University of New South Wales, Sydney 2052, NSW, Australia [Mihali *et al.* (2009) *BMC Biochemistry* **10:** 8;doi:10.1186/1471-2091-10-8].

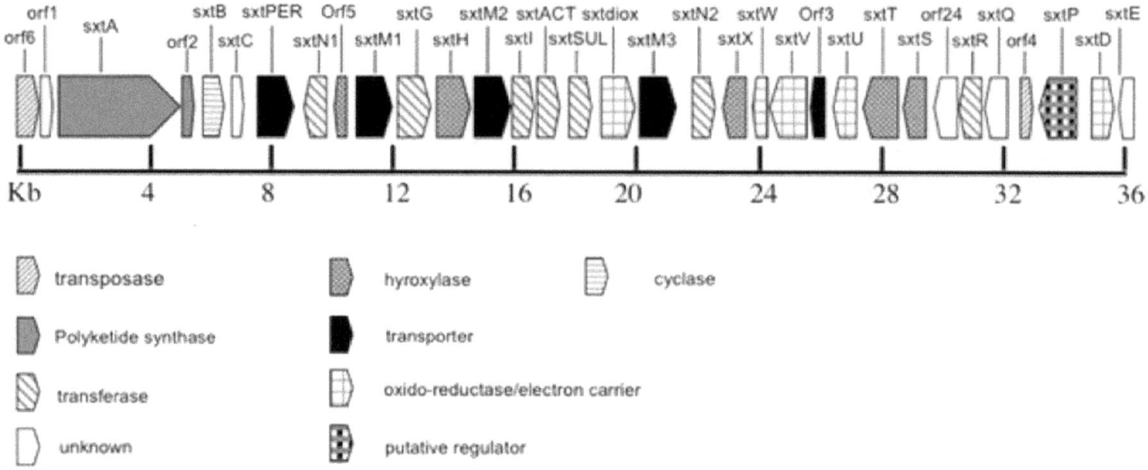

Figure 19: Organization of the PST biosynthesis cluster in *Lyngbya wollei*, scale indicates gene size in base pairs, direction of arrowed boxes indicates direction of transcription. With the kind permission of B. A. Neilan, Cyanobacteria and Astrobiology Research Laboratory, School of Biotechnology and Biomolecular Sciences, The University of New South Wales, Sydney 2052, NSW, Australia [Mihali *et al.* (2011) *PLoS ONE* **6(2):** e14657. doi:10.1371/journal.pone.0014657] doi:10.1371/journal.pone.0014657.g002.

and reported the presence of 14 strains of *Aph. gracile*. These strains produced STX, dcSTX, neoSTX and GTX5. Another study pertained to the monitoring of STX-producing cyanobacteria from the lakes, dams and rivers of Australia. Al-Tebrineh *et al.* (2010) presented a quantitative estimation of cell numbers of *A. circinalis* versus STX concentration and this correlated very well with the amplicon numbers of *sxtA* gene.

***ii) Toxicity studies with STXs*:** Of all the PSTs, toxicity studies only on STXs are available due to availability of this toxin in sufficient quantities when compared to the rest. The LD_{50} of STX for male mouse through different routes was 3.4 µg kg^{-1} body weight (i.v.), 10.0 µg kg^{-1} body weight (i.p.) and 263 µg kg^{-1} body weight (oral) whereas for female mouse LD_{50} of STX was 8.0 µg kg^{-1} body weight i.p. (Wiberg and Stephenson, 1960; Kupier-Goodman *et al.*, 1999). The LD_{50} of STX is 64 µg kg^{-1} body weight for 24 h old rats through oral ingestion while it is 5.5 µg kg^{-1} body weight through i.p. route. Wealings (rats of 21 days old) and adults (60–70 days) required LD_{50} of 270 and 531 µg kg^{-1} body weight through oral ingestion, respectively whereas through i.p. LD_{50} was 8.3 and 10 µg kg^{-1} body weight (Watts *et al.*, 1966; Kupier-Goodman *et al.*, 1999). The lethal dose of STX in human beings is 1–4 mg equivalents (Evans *et al.*, 1978). Clinical symptoms of PSP in human beings appear when 0.72 mg of STX is ingested whereas acute toxicity symptoms appear when 0.9 to 3.6 mg of STX is consumed (Levin, 1992).

Their toxicity is due to the presence of carbamoyl and hydroxyl groups in the molecule. These inhibit nerve conduction by blocking sodium channels causing a disturbance in the propagation of action potential to muscle cells. Sodium channels consist of a pore-forming α-subunit (of 260 kDa) associated with auxiliary β-subunit (of 33–36 kDa). There are six distinct receptor sites on sodium channel α-subunit and a number of neurotoxins such as tetrodotoxin, conotoxins, sea anemone toxins, brevetoxins and ciguatoxin all bind with high affinity and specificity to these six receptor sites. Characteristic alterations in the permeation and gating of sodium channels are brought out by these toxins. These have been helpful in understanding the structure and function of voltage-

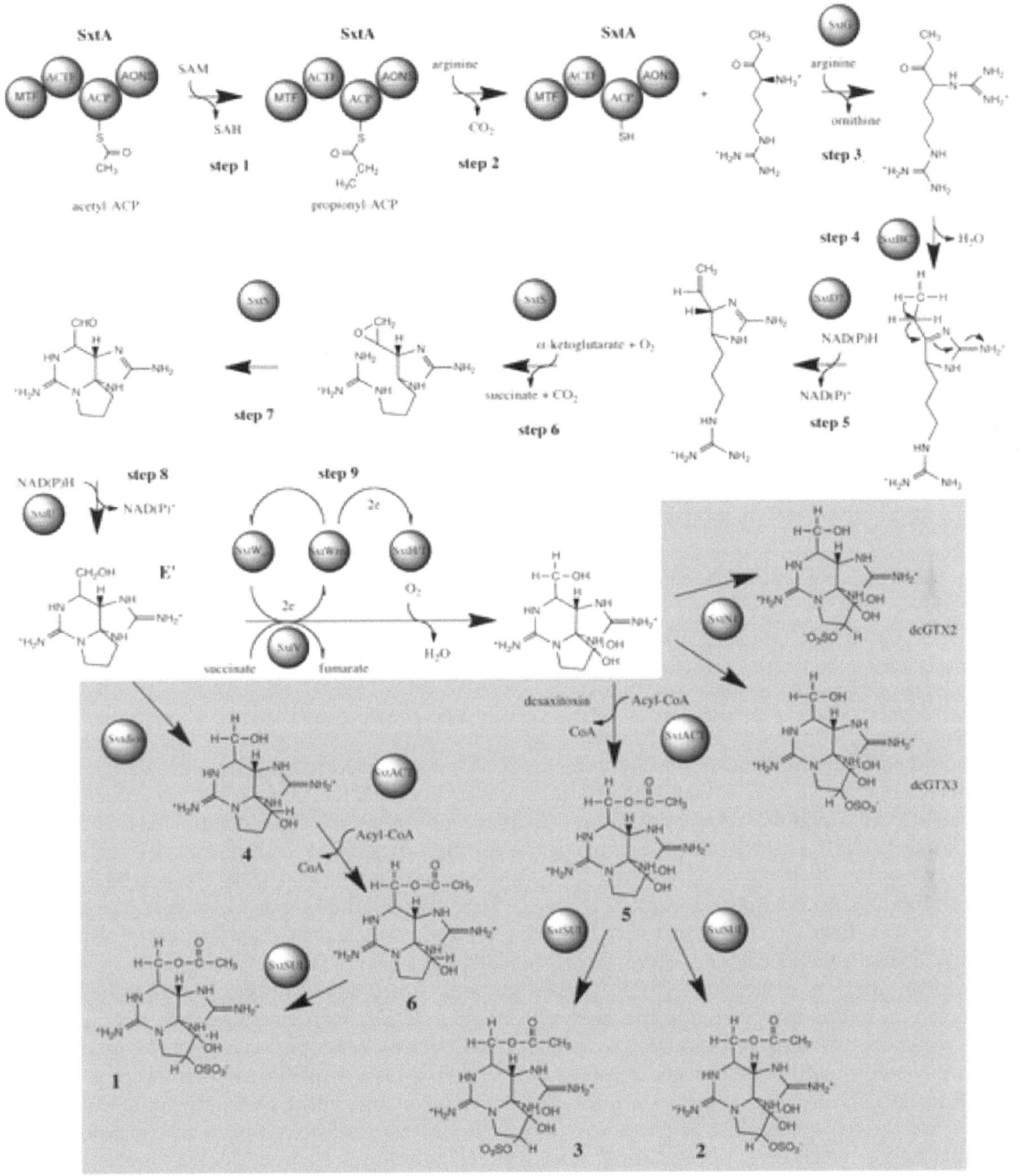

Figure 20: Proposed biosynthetic pathway for the PSTs produced by *Lyngbya wollei*. The gray shaded area highlights the steps that are unique to *L. wollei*. With the kind permission of B. A. Neilan, Cyanobacteria and Astrobiology Research Laboratory, School of Biotechnology and Biomolecular Sciences, The University of New South Wales, Sydney 2052, NSW, Australia [Mihali *et al*. (2011) *PLoS ONE* **6(2):** e14657. doi:10.1371/journal.pone.0014657] doi:10.1371/journal.pone.0014657.g004.

Table 6: Differences between *sxt* gene clusters of four PST-producing cyanobacteria.

sxt gene cluster	*C. raciborskii* T3 (1)	*Aphanizomenon* sp. NH-5 (2)	*A. circinalis* AWQC131C (3)	*L. wollei* (4)
Length	>35 kb	27.5 kb	29 kb	36 kb
Flanking regions	5′-*orf1* 3′-*orf34*	5′-*psbH* 3′-*ublA*	5′-β-lactamase 3′-*smf* gene	5′-*orf6* 3′unknown
Number	26 genes	22 genes	21 genes	31 genes
Genes absent	-	*sxtF, sxtO, sxtY, sxtZ, ompR*	*sxtF,sxtY,sxtZ, ompR,sxtX,sxtW*	*sxtL*
Additional known genes	-	*sxtPER*	*sxtPER*	*sxtdiox, sxtN2, sxtSUL sxtM1, sxtM2, sxtACT*
Additional unkown genes	-	-	*orf3,orf5, orf7* truncated intersected by transposases	*orf1, orf2, orf3, orf4, orf5, orf6, orf24*
PSTs	STX, neoSTX, dcSTX, GTX5	STX, neoSTX	STX, dcSTX, GTX2, GTX3, dcGTX3, dcGTX2, O- and N-sulfurylated C-1 and C-2 variants	STXdcSTX, dcGTX2, dcGTX3, 6 novel variants known as *L. wollei* toxins
References	Kellmann *et al.* (2008)	Mihali *et al.* (2009)	Mihali *et al.* (2009)	Mihali *et al.* (2011)

Note: Flanking regions: The 5′-*orf1* and 3′-*orf34* in (1) are of unknown function. The 5′-*psbH* and 3-*ublA* genes in (2) encode a photosystem II reaction center protein and a prenyltransferase involved in the biosynthesis of ubiquinone, respectively. The 3′-*smf* gene in (3) is presumably involved in DNA uptake. The 5′-*orf6* and 3′-flanking gene in (4) are of unknown function. Genes absent: The *sxtY*, *sxtZ* and *ompR* genes are involved in signal transduction and transcriptional regulation of PST production in (1). *sxtF* is involved in PST transport; sxtO encodes an adenylsulfate kinase; the gene product of *sxtW* resembles a ferredoxin; *sxtX* is responsible for the production of neoSTX so that is why (3) does not produce neoSTX. *sxtL* encodes GDSL lipase-like enzyme. Additional known genes: *sxtPER* belongs to drug and metabolite transport family of proteins probably fulfills the role of *sxtY*, *sxtZ* and *ompR* in both (2) and (3). *sxtdiox* in (4) encodes a dioxygenase; gene products of *sxtN2* and *sxtSUL* represent sulfotransferases; *sxtM1* and *sxtM2* encode exporters; *sxtACT* represents novel acyltransferase.

dependent sodium channels. A single molecule of STX is reported to interact with a sigle sodium channel (Catterall *et al.*, 1979). Since the guanidinium group of STX is positively charged it interacts with negatively charged carboxyl groups at the opening of the pore of the sodium channel (Cestele and Catterall, 2000). The flow of potassium or the resting potential of the membrane or membrane resistance are not affected (Adelman *et al.*, 1982; Gorham and Carmichael, 1988). Due to inhibition of sodium channels, acetylcholine cannot be released. Symptoms of tingling and numbness around lips are seen at low doses of STXs but at very high doses paralysis and death results due to respiratory failure (Carmichael and Falconer, 1993; Duy *et al.*, 2000).

In 1994, after an outbreak of PSP in Alaska, patients admitted to hospital with acute toxicity could recover in less than 24 h and the toxin was mostly cleared via urine (Gessner *et al.*, 1997). In North America, eating of contaminated mussels, clams, cockles, scallops, oysters, crabs and lobsters commonly leads to PSP from east coast (New Foundland to Massachusetts) to the west coast (Alaska to California). Though currently no antidote for PSP is available, anti-STX monoclonal antibodies show some protection against STX binding and reduction in peripheral nerve action potential in rat tibial nerve (Huot *et al.*, 1989; Chanh *et al.*, 1991). Potassium channel blocker, 4-amonipurine significantly reversed the effects of STX in rats suggesting that it can be used as an antidote for PSP (Chen *et al.*, 1996).

D) β-N-methylamino-L-alanine (BMAA): BMAA is a non-protein amino acid (Fig. 21) first detected in the cycad seeds where it is suggested to give chemical protection to the plant against herbivore consumption. However, BMAA was also found in the cyanobacterial symbiont in the coralloid roots of cycas. In the Guam ecosystem, biomagnification of BMAA has assumed alarming proportions (Cox *et al.*, 2003). BMAA from the cyanobacterium *Nostoc* gets accumulated in the seeds of cycads. The seeds are in turn eaten by flying foxes. The flying foxes are a delicacy for the Chamarro people of Guam. Apart from flying foxes, Chamarros also use cycad seed flour. At each trophic level, BMAA gets magnified 100-folds in its concentration. There is a 10,000-fold increase in the concentration of BMAA from the first trophic level (*Nostoc*) to flying foxes. Thus Chamarros accumulate BMAA in their brain tissue that leads to degenerative diseases such as amotrophic lateral sclerosis/Parkinsonism-dementia complex (ALS/PDC). The incidence of these diseases among Chamarro people of Guam is 50–100 times higher than at other places. BMAA is bound to proteins in the body and it is released slowly. As it is released, it gets incorporated into the amino acid sequences of proteins.

Cox *et al.* (2005) reported that BMAA is produced by 95% of the cyanobacteria tested belonging to all major taxonomic groups. BMAA is also found in other cyanobacterial-plant symbioses such as *Azolla filiculoides* and *Gunnera kauaiensis*. Cox *et al.* (2005) concluded that (i) the ability to produce BMAA by most of the cyanobacterial species examined suggests that this is a very highly conserved feature, (ii) the production of BMAA is not a consistent feature and it is a function of growth conditions and/or life cycle stages, (iii) symbiosis is not a pre-requisite for BMAA production, (iv) the magnitude to which human beings are exposed to BMAA can be gauged by the occurrence of marine blooms of *Trichodesmium thiebautii* and *T. erythraeum* in the Baltic Sea and other oceans; and the occurrence of *Prochlorococcus* in tropical and subtropical waters, and (v) in all toxicity studies, the assay of BMAA would be helpful in understanding its role in pathological symptoms.

E) Toxins from *L. majuscula*: *L. majuscula* is a good source of secondary metabolites. As many as 100 such substances have been identified from different isolates of this organism distributed all over the world. Their production cannot be correlated to any specific geogaphical location (Gerwick *et al.*, 2001) and samples of *L. majuscula* from Guam have yielded diverse metabolites such as indanone metabolites (Nagle *et al.*, 1996), lyngbyastatins (Harrington *et al.*, 1998; Luesch *et al.*, 2001; Williams *et al.*, 2003), malyngolide (Cardellina *et al.*, 1979), majusculamides (Marner and Moore, 1977), pitiamide (Nagle *et al.*, 1996) and peptolides (Luesch *et al.*, 2001). Toxicity studies on these metabolites are not available. A Jamaican strain of *L. majuscula* is known to produce jamaicamide and two regulatory proteins have been suggested to be involved in secondary metabolism and complementary chromatic adaptation (Jones *et al.*, 2009).

Tropical marine species of *Lyngbya* and *Symploca* have been compared for diversity of their morphological, chemical and genetic features. Characterization based on 16S rDNA analysis did not reveal any relationship between morphological and chemical divergence on the one hand and

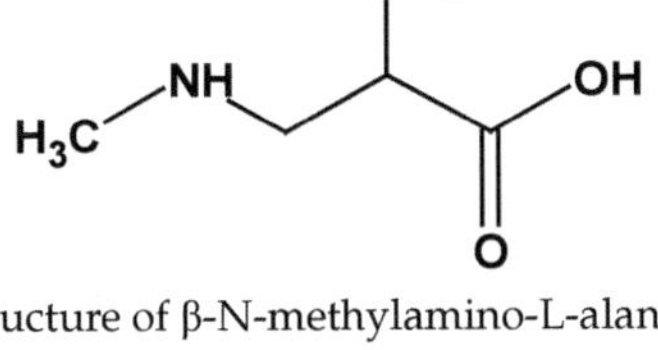

Figure 21: Structure of β-N-methylamino-L-alanine (BMAA).

genetic relatedness on the other. It is suggested that LGT and interaction of the specific genotype with environmental conditions play a predominant role in the production of variety of secondary metabolites (Thacker and Paul, 2004).

Unusual lipopeptides produced as secondary metabolites are characteristic of *L. majuscula*. Amongst these, antillatoxin (ATX) and kalkitoxin (designated as 1) are neurotoxins (Fig. 22; Orjala *et al.*, 1995; Berman *et al.*, 1999; Yokokawa *et al.*, 1999, 2000a,b; Wu *et al.*, 2000). ATX is toxic to fishes, so termed as ichthyotoxin. It is a cyclic lipopeptide whose total chemical synthesis has been achieved (Yokokawa *et al.*, 1999) but the synthetic sample is not identical in its toxicity to the natural one (Yokokawa *et al.*, 2000a,b). Along with ATX and kalkitoxin, curacin-A has also been identified as a natural product of *L. majuscula*. ATX and kalkitoxin produced a concentration-dependent neurotoxicity in rat cerebellar granule neurones whereas curacin-A was non-toxic. ATX caused acute neurotoxicity in short time whereas kalkitoxin produced a delayed neurotoxic response. LC_{50} values of 20.1 ± 6.4 and 3.8 ± 1.91 nM were observed for ATX and kalkitoxin, respectively (Berman *et al.*, 1999).

L. majuscula from Carribean collections produced kalkitoxin. Complete stereostructure, synthesis and biological properties of kalkitoxin have been described (Wu *et al.*, 2000). It is toxic to common gold fish (*Carassius auratus*, LC_{50}, 700 nM), potently toxic to brine shrimp (*Artemia salina*; LC_{50} 170 nM) and also potently inhibits cell division in fertilized sea urchin embryo assay (IC_{50} ≅25 nM). Synthetic kalkitoxin also is toxic to the same extent (LC_{50} 170nM) in brine shrimp assay. Natural kalkitoxin showed exceptional neurotoxicity in cell cultures of rat neurons with LC_{50} of 3.86 nM. Besides, it is also shown to be a potent blocker of the voltage senstive Na^+ channel in mouse neuro-2 a cells (EC_{50}= 1 nM). The biosnthesis of kalkitoxin is through polyketide/nonribosomal peptide synthetase pathway (Wu *et al.*, 2000).

In cerebellar granule cells exposed to ATX (100nM), a rapid increase in Ca^{2+} was antagonized by tetrodotoxin in a concentration-dependent manner. Further, ATX induced allosteric enhancement of $[H^3]$ batrachotoxin binding to site 2 of α-subunit. ATX also produced a strong synergistic binding in combination with brevetoxin for site 5 on the voltage gated Na^+ channel. Thus these results suggest that ATX is an activator of voltage-gated sodium channels (Li *et al.*, 2001).

III. DERMATOTOXINS

Aplysiatoxins (ATs), debromoaplysiatoxins (DATs) and lyngbyatoxins (LA-A, B, C) produced by species of benthic cyanobacteria like *Oscillatoria* (*O. nigroviridis*), *Lyngbya* (*L. majuscula*) and *Schizothrix* (*S. calcicola*) and *Planktothrix* are included here (Moore, 1977; Mynderse *et al.*, 1977; Chorus and Bartram, 1999). ATs and DATs are tumor promoters and activate protein C kinase (Mynderse *et al.*, 1977; Fujiki *et al.*, 1990). ATs and DATs cause dermatitis but LAs besides causing dermatitis are also responsible for oral or gastrointestinal inflammation (Cardellina *et al.*, 1979). These disorders were reported in swimmers in tropical and sub-tropical waters caused by *L. majuscula* (Osborne *et al.*, 2001). On Winward beaches of Hawaii in 1958 and on beaches in Okinawa (Japan) in 1968 cases of acute dermatitis (also known as "swimmers's itch") were reported (Moikeha and Chu, 1971; Hoshimoto *et al.*, 1976; Burke and Tester, 2002). A subcutaneous inflammation with signs of erythema, blisters and desquamation occurred within 12 h of exposure to the cyanobacterium. ATs are toxic to mice at a dose of 0.3 mg kg^{-1} body weight (Moore, 1977).

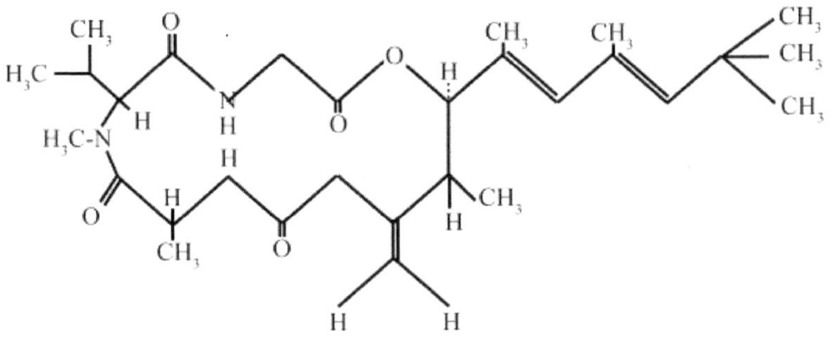

Antillatoxin A

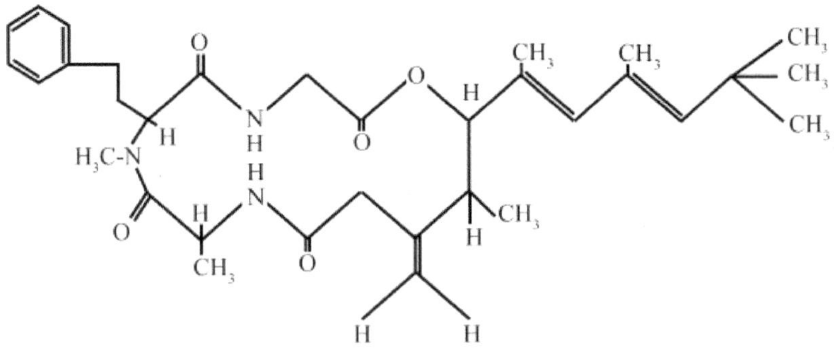

Antillatoxin B

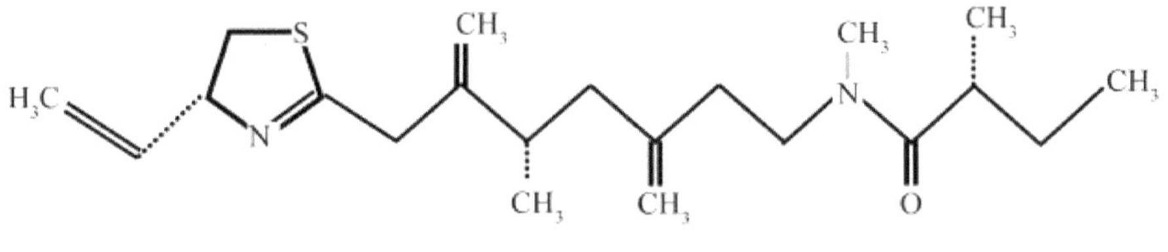

Kalkitoxin

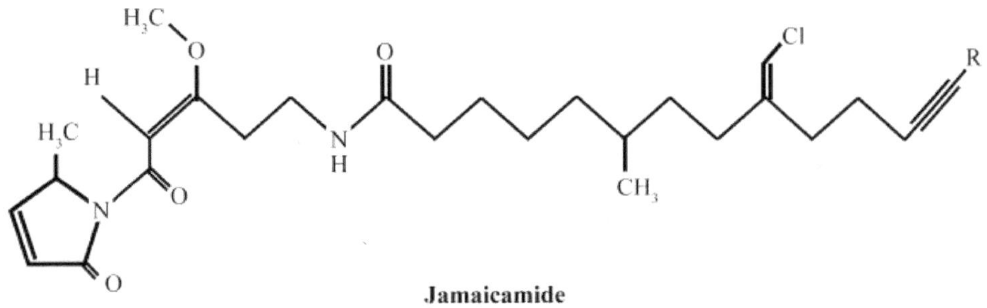

Jamaicamide

Figure 22: Structure of *Lyngbya* toxins.

IV. CYANOBACTERIAL LIPOPOLYSACCHARIDES (LPS)

Cyanobacterial LPS are reported to be responsible for an outbreak of gastro-enteritis in Sewickley, Pennsylvania (Lippy and Erb, 1976; Keleti *et al.*, 1979). These are considered to be 10-times less toxic than LPS of *Salmonella* (Keleti and Sykora, 1982; Raziuddin *et al.*, 1983; Bell and Codd, 1996). LPS molecules consist of O antigens, core polysaccharides and lipid A moieties. Lipid A region is responsible for its activity leading to symptoms such as fever, diarrhoea, vomiting and hypotension (Metcalf and Codd, 2004). The toxicity caused by LPS endotoxins remains largely unknown (Duy *et al.*, 2000). Stewart *et al.* (2006) reviewed the current knowledge on cyanobacterial LPS. These are weakly toxic when compared to LPS from the Enterobacteriaceae. Moreover, the toxic moiety of LPS, lipid A has not so far been detected in any cyanobacterial wall component. So it is highly improper to assign the above symptoms to the cyanobacterial LPS without any scientific basis.

V. OTHER MISCELLANEOUS TOXINS

Cyanobacterin has been isolated from *Scytonema hofmanii* (UTEX1581) that is reported to inhibit growth of various algae by exerting its action on cell membrane and thylakoids (Mason *et al.*, 1982). Fischerellin A is produced by *Fischerella muscicola* and has a minimum inhibitory concentration of 14 nM against *Synechococcus* PCC 6911 and inhibits photosystem II (Hagmann and Juettner, 1996). These two substances are suggested to be responsible for chemical defence. Kasumigamide, an antialgal tetrapeptide has been detected in cells of *M. aeruginosa* NIES-87. This peptide has a molecular formula of $C_{40}H_{50}N_8O_9$ and is a linear tetrapeptide containing N-terminal-α-hydroxyacid. Erythro-β-phenyl-D-serine, D-arginine, β-alanine and D-Pla are the four sub-units. It inhibited the growth of *Chlamydomonas neglecta* NIES-439 (Ishida and Murakami, 2000).

Anabaenopeptilide 90A and 90B are seven residue depsipeptides isolated from *Anabaena* strain 90, besides three six-residue peptides called anabaenopeptins. As yet their function has not been clearly understood, although four genes governing anabaenopeptilide synthetase domains *apdA*, *apdB* and *apdD* containing two, four and one module, respectively have been identified. The fourth gene *apdC*, present between modules six and seven, is similar to halogenase genes (Rouhiainen *et al.*, 2000). Besides the production of anabaenopeptilide 90A and 90B, *Anabaena* strain 90 also produces anabaenopeptins and both of these are serine protease inhibitors (Namikoshi and Rinehart, 1996; Weckesser *et al.*, 1996). Mutation of anabaenopeptilide gene in *Anabaena* strain 90 resulted in the enhanced production of anabaenopeptins (Repka *et al.*, 2004). Spiroidesin, a D-amino acid-containing lipopeptide was isolated and characterized from *Anabaena spiroides* that inhibited cell growth of *M. aeruginosa* at 1.6 μM (IC_{50}) (Kaya *et al.*, 2002). Lobocyclamide B, a cyclododecapeptide containing five β-hydroxy-α-amino acid residues was isolated from *L. confervoides*. This report constitutes first such occurrence of χ-hydroxythreonine in a natural peptide (MacMillan and Molinski, 2002). Aeruginosin 98-B from *M. aeruginosa* as the lead molecule, Radau *et al.* (2003) designed, synthesized and conducted inhibition tests selective against serine proteases. Cyanostatin A and B have been isolated from *Microcystis* sp. at Loch Rescobic in Scotland. These two are lipopeptides that contain 3-amino-2-hydroxydecadonic acid and weakly inhibit PP2A. However, these strongly inhibited the activity of leucine aminopeptidase M with IC_{50} values of 40 and 12 nM ml^{-1}, respectively (Sano *et al.*, 2005). Nostopeptolides (*nos*) A1 and A2, isolated from *Nostoc* sp. GSV224, are cyclic peptides containing nine amino acid residues, a butyric acid group and an internal acetate-derived unit linked by peptide and ester bonds. A gene cluster of eight ORFs encoding the biosynthesis and transport of this group of nonapeptides has also been characterized (Hoffmann *et al.*, 2003). A mixed NRPS-PKS seems to be operative in the biosynthesis of *nos*. Optimization of extraction parameters for anabaenopeptins were described where the use of aqueous methanol was found to be much advantageous during extraction and further handling with plastic containers (Morrison *et al.*, 2006).

VI. DETECTION AND ANALYSIS

In detection and analysis of cyanobacterial toxins, the factors governing their stability (in natural and laboratory conditions), their intra and extracellular concentrations, possible degradation by bacteria and the usage of glass vessels instead of plastic containers have to be taken into consideration (Bell and Codd, 1996). Broadly, four important steps are recognized. These are (i) extraction procedures, (ii) concentration/cleanup procedures, (iii) sample preservation and (iv) determination.

i) Extraction: A number of chemical and physical extraction procedures have been suggested. The cyanobacterial toxins generally are present inside the cells and only are liable to be released into external waters upon lysis/death of cyanobacterial blooms or artificially upon the application of algicides like copper sulphate. Extraction of toxins from lysed cells has been practiced by the use of a number of solvents such as acetic acid/water, alcohol/water and methanol/water mixtures. Boiling in hot water bath or using a microwave oven in water as solvent proved to be beneficial in the extraction of a number MC varants including MC-LW and MC-LF which are hydrophobic (Metcalf and Codd, 2000). Sonication and freezing-thawing are also practiced prior to solvent extraction. The method of Lawton *et al.* (1994) seems to be quite suited for extraction of intracellular toxin content. The bloom sample is filtered and subjected to freezing-thawing followed by extraction with 75% methanol.

ii) Concentration/cleanup procedures: The most widely preferred concentration step is by solid-phase extraction in which water containing the toxins is passed through an octadecyl (C18) coated silica adsorption catridge. The toxins are eluted into methanol and the volume is reduced by evaporation of methanol. Although these catridges are helpful in the extraction of a number of variants of MCs (except MC-LW whose recovery is poor) and nodularins with high efficiency, the only disadvantage lies in the fact that along with the toxins naturally occurring organic matter present in water is also concentrated. This further leads to interference during determination of the species of the toxin and its concentration. The use of C18 catridges with endcapped silanol groups has also been advocated for the efficient recovery of MC-RR (Ojanpera *et al.*, 1995). Aqueous or acidified methanol instead of pure methanol has been suggested as a better eluant at a pH in between 4 and 8 (Tsuji *et al.*, 1994a).

***iii) Sample preservation*:** Pure MCs and nodularins are relatively quite stable in sterile water. However, the only constraint is if the sample also contains pigments, photochemical degradation can occur quickly (Tsuji *et al.*, 1994a). So such toxin samples containing cellular material are better protected from light by preserving them in dark. Since the MCs and nodularins are also liable to microbial degradation, necessary steps are to be taken to deep freeze the samples.

iv) Detection: The methods can be broadly classified into three categories, i.e. biological, instrumental and molecular methods. Biological methods include both bioassays and biochemical techniques including enzyme assays. The protocols for obtaining purified toxins followed by their determination to picogram levels by the use of sophisticated instruments (HPLC, MS) have been perfected. The detection of cyanobacterial hepatotoxins by various methods has been reviewed (McElhiney and Lawton, 2005). The recognition of variants of cyanobacterial toxins and other secondary metabolites of

L. majuscula has been possible by the application of such physicochemical techniques. The availability of a number of molecular markers has enabled the identification of not only the toxic bloom algae from laboratory cultures but also from natural bloom samples to the species level.

1) Biological methods: ***(i) Bioassays*:** The most widely used conventional assay for determining the toxicity of cyanobacterial toxins is the mouse bioassay. The toxins are administered i.p. or orally to the inbred lines of mouse and the concentration of the toxin lethal to 50% of the population (LD_{50}) is calculated. In general, the toxicity of bloom material or extracts is tested and by inferring the toxic response, the identity of the class of toxin can be determined. Though this method is the officially approved method by the Association of Official Analytical Chemists (under EC Law) for determining marine STXs, it has certain limitations. The foremost is it is unethical and secondly this bioassay is only a qualitative test for detecting the presence or absence of a toxin but does not help in the chemical identity of the toxin. The MCs and nodularins cause death within 4 h with symptoms of liver haemorrhage. The neurotoxins cause death more quickly without any apparent tissue damage, i.e. within 15 minutes (Falconer, 1993). There are a number of limitations, i.e. (i) since this assay can potentially be calibrated against MC-LR, the toxicity of any toxin can only be expressed in terms of MC-LR equivalents, (ii) this test does not have the sensitivity to be applicable to water samples and (iii) water samples with 1–2 µg L^{-1} MC-LR cannot be used directly as it requires a considerable concentration of the sample. This is exemplified by the fact that for MC-LR with LD_{50} of 50 µg kg^{-1}, the lethal dose is about 1 µg for a 20 gm mouse which requires 1 L sample to be administered intraperitoneously which is not practicable.

A number of other bioassays have been proposed from time to time. Though some of these proved to be sensitive, these are neither specific to MCs nor applicable for measuring the levels of toxins in waters. Some of these are Microtox test involving luminescent bacteria (Lawton *et al.*, 1990; Volterra *et al.*, 1992), *Artemia* bioassay (Kiviranta *et al.*, 1991; Campbell *et al.*, 1994; Lahti *et al.*, 1995), assay with plant growth of *Sinapis alba* L. (Kos *et al.*, 1995), assay with African locust (Hiripi *et al.*, 1998), assay with larvae of crustacean *Thamnocephalus platyurus* (Törökné *et al.*, 2000) and toxicity test with crustacean *Daphnia* (Kyselkova and Marsalek, 2000). Cyanobacterial toxicity was assessed, using the crude extracts obtained from bloom samples dominated by *Microcystis* spp., by comparing the sensitivity of 17 acute bioassays. The bioassay employing the crustacean *T. platyurus* was highly sensitive followed by those of *Drosophila melanogaster*, protozoans *Spirostomum ambiguum* and *Tetrahymena termophyla* and the crustacean *Daphnia pulex* (Marsalek and Blaha, 2004).

As more sophisticated alternative *in vitro* biochemical methods have been developed, these have gained more prominence over mouse bioassay. For example, the inhibition of protein phosphatases (for the assay of MCs and nodularins), acetylcholinesterase [for the assay of anatoxin-a (s)] and protein synthesis (for the assay of CYN) have been developed. For the detection and quantification of STXs, STX-binding assays have been put to use. These methods proved to be highly sensitive, reproducible and can be performed at ease.

***ii) Immunoassays*:** Another important *in vitro* test system is immunoassay. The cyanobacterial toxins do not possess antigencity, i.e. these do not act as antigens to evoke an immune response for generation of antibodies specific against them. Such molecules that do not show antigenicity are known as heptens. That is why these are linked to other larger molecules to make a complex antigen that can generate the formation of specific antibodies. The immunoassay provides two advantages, i.e. this method is highly specific, sensitive and can detect the presence of toxin as low as 1 µg l^{-1} of MC-LR (a concentration considered as a guideline value according to WHO). However, the cross-reaction of cyanobacterial toxins with antibodies raised against a particular toxin variant very much depends on differences in the degree of concentration between variants.

Immunoassays are also combined with other enzyme-linked immunosorbent (ELISA) assays. Monoclonal antibodies (Kfir *et al.*, 1986) and polyclonal antibodies (Brooks and Codd, 1988) isolated against MC-LA were found to be specific to MC-LA. MC-LR-polylysine or MC-LR-ethylenediamine conjugated with bovine serum albumin was used to raise antibodies. Such antibodies raised after 4 weeks of immunization of rabbits were characterized in a radioimmunoassay (RIA), ELISA and indirect competitive ELISA. Of the two antigens MC-LR-ethylenediamine conjugated with bovine serum albumin proved to be better. RIA and ELISA revealed that the antibodies cross-reacted with other variants such as MC-RR, MC-YR and nodularin but had low reactivities with MC-LY and MC-LA (Chu *et al.*, 1989). Nagata *et al.* (1995) demonstrated protective effects of specific monoclonal antibodies on MC-LR-induced cell damages as assessed by morphological changes, lactate dehydrogenase release into medium and cell viability measured by using a tetrazolium dye in primary rat hepatocyte cultures. Weller *et al.* (2001) developed a direct competitive immunoassay of broad specificity that could be accomplished in a shortest time of 2 h and with a detection limit of 50 ng L^{-1}. Indirect immunoassay (12 h), antiimmune complex immunoassay (48 h), and immunoassay based on anti-idiotype antibody (12 h) exhibited lower cross-reactivities whereas direct immunoassay had high cross-reactivity for nodularin as well. The percentage cross reactivities ranged from 100% for MC-LR to a maximum of 156% in case of [Asp^3, Dhb^7]MC-RR with intermediate percentages of cross-reactivities for MC-RR (114%), MC-LA (115), MC-YR (118%), MC-LY (122%), MC-WR (147%), MC-LF (123%), MC-LW (104%), [Asp^3]MC-HtyR (101%), [Asp^3]MC-LR (114%) and nodularin (102%).

Specific mono- and polyclonal antibodies against MCs conjugated with N-methyldehydroalanine that is chemically modified (aminoethylation) have been developed. Such monoclonal antibodies were used in ELISA assays that could detect MCs as low as 1 ng ml^{-1}. A good correlation between ELISA assays and HPLC was established. This is suggested to be extremely sensitive analytical tool for direct measurement of toxins in cyanobacteria and water samples (Mikhailov *et al.*, 2001).

Polyclonal antibodies raised against MC-LR when employed in ELISA-assays showed cross reaction with MC-LR and other variants of MC (Chu *et al.*, 1990). On the other hand, monoclonal antibodies produced against MC-LR showed cross reaction with a number of MC variants and nodularins in an ELISA format (Usleber *et al.*, 1995). ELISA requires long analysis time and needs appropriate training (McDermott *et al.*, 1995; Pyo *et al.*, 2005). Fluorescence immunochromatography method and HPLC have been used for detection of MCs. The advantages of the former method are very short detection time and low detection limit while the latter method has an advantage of quantification of individual variants of MCs (Pyo *et al.*, 2005). Likewise, ELISA and HPLC have been compared for the detection of MCs in environmental samples. Commercially available ELISA kit was modified suitably for quantitative determination of MCs within 2 to 3 days of sampling and the assays were comparable to HPLC analysis. Modified ELISA is more simple and highly sensitive approach. Five species of *Phormidium*, i.e. *P. bijugatum*, *P. molle*, *P. papyraceum*, *P. uncinatum* and *P. autumnale* hitherto not listed amongst toxic cyanobacteria caused neuro/hepatotoxic symptoms in mice. Of the five species, *P. bijugatum* extracts even caused death. The toxicity of the *Phormidium* spp. was ascribed to STXs and MCs as confirmed by ELISA (Teneva *et al.*, 2005b). The analysis of serum samples of patients exposed to MCs during 1996 outbreak in Brazil by ELISA (19.9 ng ml^{-1}) was almost comparable to the detection of these by LC/MS (21.2ng ml^{-1}). This suggests that ELISA is a simple, low-cost, accessible method to screen human serum for evidence of MC exposure (Hilborn *et al.*, 2005).

During recent years, recombinant antibody technology has replaced the conventional methods of production of polyclonal and monoclonal antibodies specific for MCs. This method is very simpler and cost effective than the conventional methods used for antibody production. The existence of antibody variable domains in filamentous phages proved to be advantageous in the selection and enrichment of recombinant antibody fragments against a range of heptens including MCs from large antibody libraries (McCafferty *et al.*, 1990; McElhiney *et al.*, 2000). Antibody fragments so selected can be expressed in a bacterial host such as *E. coli* so that large amounts of antibodies are produced than those of existing methods. This has been termed as phage display technology. By the application of this technique, recombinant phage antibody clones have been isolated against MC-LR from a naïve human semi-synthetic phage display library (McElhiney *et al.*, 2000). Such selected phage antibodies were expressed as soluble single-chain antibody fragments (scAbs). Characterization by competitive ELISA, cross-reactivity of the most sensitive scAb clone as determined by ELISA of related toxins and the quantitative determination of the contents of MCs in cyanobacterial extracts enabled the detection of MC-LR, MC-RR, MC-LW and MC-LF in micromolar range. The isolation of scAb with the help of Griffin 1 phage display library proved to be 800 times more sensitive for MC-LR since it could detect the toxin below 1 $\mu g\ L^{-1}$ concentration. Further, immobilized anti-MC-LR scAb was used in immunoaffinity columns for concentrating MC-LR from large volumes of water for subsequent analysis by HPLC. The recovery rate was found to be 94% (McElhiney *et al.*, 2002). Thus this technology offers a large scale, low-cost, rapid, specific and sensitive assay method.

***iii) PP Inhibition*:** PP1 and PP2A are severely inhibited by MCs. The activity of PPs can be determined by the release of phosphate from phosphorylated protein substrates. Since MCs inhibit phosphatase activity, the extent of release of phosphate can be measured by this assay. Assay systems involving P^{32}-labelled substrate (MacKintosh *et al.*, 1990; Yoshizawa *et al.*, 1990; Honakanen *et al.*, 1990, 1994; Holmes and Boland, 1993; MacKintosh and MacKintosh, 1994; Chaivimol *et al.*, 1994; Ash *et al.*, 1995) and P^{33}-labelled substrate (Sahin *et al.*, 1995) have been described that measure the radiolabelled phosphate produced. MCs from environmental samples have been determined by using P^{32} -labelled substrate (Sim and Mudge, 1993; Jones and Orr, 1994; Lambert *et al.*, 1994; Lam *et al.*, 1995). Though the detection limits are around 0.1 $\mu g\ L^{-1}$ or lower than this (Lambert *et al.*, 1994; Xu *et al.*, 2000) due to cumbersome experimental procedures, costly nature of chemicals and precautions to be taken in handling the chemicals, it did not find favour in routine analysis. Alternatively, a colorimetric assay using p-nitrophenol (An and Carmichael, 1994; Ward *et al.*, 1997; Liu *et al.*, 2002; Akin-Oriola and Lawton, 2005) was developed. PP-inhibition assay is generally carried out by using a modified colorimetric method described by these workers in which the substrate p-nitrophenylphosphate (uncoloured) is allowed to react with PP1 (purified) and the resultant product p-nitrophenol (yellow colour) formed is assayed using a microtitre plate reader (Dynex MRX II with Revelation software version 4.02). Though not specific for MCs, PP inhibition would greatly help as a convincing screen for the presence of hepatotoxins in any given sample. One important precaution to be kept in view is the endogenous phosphatase activity of the extracts.

A colorimetric immuno-protein phosphatase inhibition assay was developed for the identification of MCs and nodularin from laboratory cultures and cyanobacterial bloom samples. Polyclonal antibodies raised against MC-LR were used in conjunction with protein phosphatase inhibition which was possible to distinguish seven variants of MCs (MC-LR, D-Asp³-RR, -LA, -LF, -LY, -LW and –YR) one from the other as well as nodularin on the one hand and okadaic acid, calyculin A and tautomycin on the other (Metcalf *et al.*, 2001). This technique is based on the ability of MC-LR antiserum to protect PP1 from the inhibitory action of cyanobacterial toxin. MC-LR antiserum was allowed to interact with MC-LR and after a certain period of incubation (15 min at 37°C), the colorimetric assay of PP inhibition was performed. In order to interpret the percentage of inhibition of PP by the toxin after interaction with MC-LR antiserum, a protective index (PI) is deduced by applying a formula PI = ($\%Act^{AS}$–$\%Act^{NS}$)/$\%Inhib^{NS}$. $\%Act^{AS}$ is the rate of p-nitrophenol production in the protein phosphatase inhibition assay for PP1 inhibitors after incubation with MC-LR antiserum, expressed as a percentage of the MC-free control value while $\%Act^{NS}$ is the rate of p-nitrophenol production in the protein phosphatase inhibition assay for PP1 inhibitors after incubation with null serum substracted from the rate of p-nitrophenol production by the MC-free control value. A PI value of nearly 1.00 was obtained for assays with MC-LY and nodularin where as the PI was 0.94 and 0.90 for MC-LA and MC-LR, respectively. In case of the rest, MC-Asp^3-RR, MC-LF, MC-LW and MC-YR the PI value ranged from 0.80 to 0.86. In case of non-cyanobacterial toxins such as tautomycin, calyculin A and okadaic acid the PI value was lowest at 0.18, 0.07 and 0.02, respectively suggesting that the MC-LR antiserum did not give protection. This method has an added advantage as concentrations as low as 10 nM could be detected.

Further, a new PP-inhibition assay that enables the recognition of variants of MCs and nodularin has been developed. In the first step, PP-inhibition by different variants of MCs and nodularin revealed IC_{50} values of 2.2, 1.8, 9 and 175 nM for MC-LR, nodularin, MC-YR and MC-RR, respectively. In the second step, taking MC-LR as standard ELISA assay was conducted that showed equivalence in toxin responses. When these toxins were determined through PP inhibition assay by taking MC-LR as standard, the ratio of values determined by PP2A inhibition to ELISA decreased in the order: nodularin (2.23), MC-LR (1.1)>MC-YR (0.63)>Mc-RR (0.06). The differential sensitivity of PP2A assay to various standards enabled the development of an indicative toxicity ranking (ITR) where a ranking of 1 was assigned to ratios > or =0.8 or greater and 3 (the lowest) to values < or =0.2. By this new method, analysis of samples and extracts of cyanobacterial mats from ponds and streams in cold temperate locations has been conducted. The disadvantage with LC-MS is its dependence on MC-LR as the standard commercial variable for the determination of all MC-variants whereas PP inhibition and ELISA not only help in the determination of the variant of MC but also its stoichiometry (Mountfort *et al.*, 2005).

A rapid and sensitive detection method for MCs in water samples was developed based on a new fluorescence immunochromatography assay system employing monoclonal antibodies of MC-LR and compared with cyanobacterial toxins analyzed through HPLC. MC-LR was conjugated with bovine serum albumin to produce monoclonal antibodies against MC-LR in BALB/c mice by an initial injection of 0.2 ml of the conjugate and incomplete Freund's adjuvant followed by booster injections of a similar kind. Hybridoma cell line (SP_2/O-Ag14) was screened for production of anti-MC-LR antibodies using an indirect ELISA. The positive hybridomas were cloned several times. Antibodies so produced were purified by membrane ultrafiltration, ammonium sulfate precipitation with final step protein G column. Fluorescence coated toxin from water sample was mixed with monoclonal antibodies against MC-LR. The intensity of fluorescence of the conjugates was detected on the detection zone that could be scanned by a Laser fluorescence scanner. The concentration of MC-LR in unknown sample was calculated from a standard curve. The monoclonal antibodies produced showed fairly good binding ability with MC-RR, MC-YR as well as MC-LR. This is the reason why fluorescence immunochromatography could detect individual amounts of MC variants but helped in the estimation of total MCs in a given sample. The performance of fluorescence immunochromatography for the detection of MCs compared very well with HPLC (with a correlation coefficient of 0.9929) (Pyo *et al.*, 2005). A minimum detection limit (MDL) of 11 pg filament^{-1} of single filaments of *Planktothrix* (*P. agardhii* and *P. rubescens*) was feasible by ELISA using antibodies raised against MC-LR. A comparison of the amount of MC-LR content from single filaments isolated from natural samples and those of laboratory cultures revealed that the content of MC-LR was below MDL in majority of the cases. Of the two species, *P. agardhii* from Bassenthwaite Lake (England) exhibited lowest MC concentration (0.7 fg μm^{-3}) whereas *P. rubescens* from Iznik Lake showed highest MC content (2.9 fg μm^{-3}). *P. agardhii* showed a better positive correlation between filament biovolume and MC content than *P. rubescens* from environmental samples. On the other hand, *P. rubescens* showed a good relationship between filament biovolume and MC content under cultural conditions (Akcaalan *et al.*, 2006).

***iv) Inhibition of acetylcholinesterase activity*:** Anatoxin-a(s) can be detected by estimating the activity of acetylcholinesterase. Since acetycholinesterase degrades acetylcholine into acetyl group and choline, the acetyl group can be assayed colorimetrically by complexing with dithiobisnitrobenzoic acid. A reduction in the absorbance at 410 nm over a period of 5 minutes would indicate the presence of anatoxin-a(s). However, it is not a specific inhibition assay for anatoxin-a(s) because organophosphorus pesticides also inhibit acetycholinesterase activity. If this assay is employed, it has to be substantiated by other confirmatory tests for toxins of cyanobacterial origin.

2) Instrumental methods of Analysis: ***(i) High Precision Liquid Chromatography (HPLC)*:** Routine analysis of peptide toxins has been traditionally carried out by HPLC (Harada *et al.*, 1996a,b; Meriluto, 1997; Meriluto *et al.*, 1996). Sample separation is followed by sample detection. Sample separation is achieved by a reversed phase C18 packed column (Krishnamurthy *et al.*, 1986; Harada *et al.*, 1988a,b; Flett and Nicholson, 1991; Rositano and Nicholson, 1998; Lawton *et al.*, 1994; Wirsing *et al.*, 1999), amide C16 column (Spoof *et al.*, 2001), internal surface reversed phase column (Meriluto and Eriksson, 1988) or an ion exchange column (Gathercole and Thiel, 1987) and an aqueous mobile phase containing methanol or acetonitrile. The use of an internal surface-reversed-phase column minimized the time compared to others (Meriluto and Eriksson, 1988; Meriluto *et al.*, 1990; Watanabe *et al.*, 1988; Kondo *et al.*, 1992b). Good resolution of the toxins from one another and other co-extracted compounds depends on the mobile phase used. As for example, MC-LR and MC-YR can be better resolved from one another with methanol-based mobile phase rather than with acetonitrile/ ammonium acetate mobile phase (Harada *et al.*, 1988a).

Detection of MCs has been traditionally carried out with UV-absorbance (Lawton *et al.*, 1994, 1995; ISO, 2005). Most of the MCs and nodularins have UV-absorption maxima at 238 nm. However, those MCs with aromatic amino acids such as tryptophan (MC-LW) have absorption maxima at lower wavelengths, i.e. 222 nm (Lawton *et al.*, 1994, 1995). The problems arising out of the co-eluted compounds and the use of plasticware have to be kept in view. As MC-LR is the standard available, the concentration of the rest of the MCs could only be expressed in terms of MC-LR equivalents. However, the total MC content can be determined with high degree of accuracy by UV-detection. Alternatively, the use of a photo-diode array (PDA) detection not only helps in the detection of a MC but also provides its absorption spectrum thus indicating the presence of MC in a given sample with confirmation. PDA detection also allows the determination of concentrations of MCs lower than 1µg L^{-1}. Semi-quantitative analyses of benthic cyanobacterial communities from Fresh Pond, McMurdo Ice Shelf of Antarctica by using HPLC-DAD and PP-inhibition assay revealed the presence of MC-LR and [D-Asp^3] MC-LR equivalent to 11.4 ng MC-LR mg^{-1} dry weight (Jungblut *et al.*, 2006). A survey of 40 ponds, lakes and hydroterrestrial environments in Bratina Island and four dry valleys of Antarctica by PCR amplification of 16S rRNA gene and *mcyE* gene brought to light that a *Nostoc* sp. is responsible for the production of novel MCs. Detected by ELISA, LC-MS and PP assay, the MCs consisted of [D-Asp^3] MC-RR, [D-Asp^3] MC-LR with Gly at position 1 instead of Ala and combinations of homoarginine [hAr^2] or [ADM Adda^5] substitution (Wood *et al.*, 2008).

The determination of CYN by HPLC has been described (Harada *et al.*, 1994) with detection at 262 nm but without the requirement of its extraction and concentration steps. Likewise, the determination of CYN to 1 μg L^{-1} with the help of LC/MS/MS did not require sample cleanup, extraction and concentration steps (Eaglesham *et al.*, 1999). Lyophilized cells of *Cylindrospermopsis* from Europe, Israel and Australia were analyzed for CYN content using HPLC. Extraction of CYN was optimized with 5% formic acid to prevent interference in chromatograms by contaminating compounds. Isocratic conditions of 5% (v/v) aqueous methanol plus 0.1% (v/v) trifluoroacetic acid as the mobile phase were used (Törökne *et al.*, 2004).

***ii) Mass spectrometric detection*:** During recent years, there has been a shift towards detection of MCs by mass spectrometry (MS) following HPLC separation. The detection of MCs is unequivocal because MCs produce characteristic ions in the mass spectra (Namikoshi *et al.*, 1992a,b; Kondo *et al.*, 1992b; Rinehart *et al.*, 1994; Lawton *et al.*, 1995; Yuan *et al.*, 1998, 1999a,b). MCs could be detected by LC/MS/MS system at concentrations as low as 0.1 μg L^{-1} that is ten times below the WHO guidelines for MC-LR concentration in natural waters (Poon *et al.*, 1993; Lawton *et al.*, 1995; Kondo *et al.*, 1992b, 1995; Tsuji *et al.*, 1994 b; Bateman *et al.*,1995; Hormazabal *et al.*, 2000; Ells *et al.*, 2000; Hummert *et al.*, 2000; Zweigenbaum *et al.*, 2000; Spoof *et al.*, 2003; Dahlmann and Luckas, 2005). By employing LC-MS, six MC variants have been identified (Lawton *et al.*, 1995). Tsuji *et al.*(1994b) detected individual MCs to the level of 0.02 μg L^{-1} by MS. Based on the mass spectrum of an available standard, the identification of a MC is possible. These data have to be additionally supplemented by toxicity studies so as to enable the determination of toxicity equivalents. Unknown MCs could also be identified on the basis of fragmentation patterns by a variation of MS known as MS/MS detection (Bateman *et al.*, 1995; Lawton *et al.*, 1995; Yuan *et al.*, 1998, 1999a,b). Matrix-Assisted Laser Desorption Ionization Time of Flight Mass Spectrometry (MALDI-TOF MS) was used to identify toxins (MCs, micropeptins and anabaenopeptolin) from intact microorganisms as well. Typing of the blooms into toxic and non-toxic phenotypes can be achieved within few minutes (Erhard *et al.*, 1997). Besides providing information on molecular mass of all peptides, MALDI-TOF-MS also helped in the detection of individual MC variants in cultures of *M. aeruginosa* (Robillot *et al.*, 2000). Analysis of intact colonies of *Microcystis* spp. from bloom populations by MALDI-TOF-MS revealed the identification of a total of 46 individual peptides (21 of which have not been described previously). In most colonies 2–10 individual peptides were detected. The principal component analysis revealed MC-LR, MC-RR, MC-YR, anabaenopeptins B and E/F, a microviridin and a new cyanopeptolin. Colonies of *M. aeruginosa* predominantly produced MCs whereas those of *M. ichthyoblable* were negative for MCs but produced anabaenopeptins. No peptide metabolites could be detected in 16 of the 19 colonies of *M. wesenbergii* (Welker *et al.*, 2004b). Twenty six strains of *M. aeruginosa* collected from potable water bodies of northern Portugal were isolated into pure cultures and the content of MCs was determined using MALDI-TOF MS, ELISA and PCR procedure targeting genes of *mcy* gene cluster. Values of MCs ranged from 0.02 to 0.53% dry weight of cultures as revealed by ELISA. MALDI-TOF-MS analysis demonstrated the presence of several chemically distinct variants of MCs along with aeruginosins, anabaenopeptins and several other unidentified peptide-like compounds (Saker *et al.*, 2005). *M. aeruginosa* B 2666, a strain that has not been investigated previously, released into medium eight variants of MCs, i.e. MC-LR, MC-LA, [MeSer^7] MC-LR, MC-LL, MC-LF, MC-L(Aba), [Asp^3] MC-LA and [Asp^3] MC-LL. Of these, the last two MCs were reported for the first time separated by reversed-phase microbore liquid chromatography and introduced directly into a hybrid linear ion-trap-Fourier transform ion cyclotron resonance mass spectrometer with electrospray ionization. This technique has the following advantages. (1) It is fairly rapid. (2) It provides reliable identification in the absence of desired standards. (3) The work can be carried out even though one does not possess additional information about structural variants from other analytical techniques (Diehnelt *et al.*, 2006).

Isolation and analysis of anatoxin-a and anatoxin-a(s) have been achieved by a four-step process involving extraction with 0.05 M acetic acid followed by a reversed phase, ODS (octadecylsianized) and a cation exchanger (COOH) organosilans bonded to silica gels were applied as a cleanup step. Separation and identification is achieved by TLC and HPLC chromatographies (Harada *et al.*, 1989). Thermospray-LC/MS was used to detect anatoxin-a that proved to be advantageous over the conventional HPLC/UV. This method was found to be more suitable for detection of anatoxin-a from cultures and naturally occurring bloom samples where the toxin could be detected at as low concentrations as 500 pg (Harada *et al.*, 1993). Further improvement in the detection of anatoxin-a was made by the solid-phase extraction of the toxin that enabled a recovery of 75.7 ± 7.2% at a concentration of 20 ng L^{-1}. The detection of trace amounts of anatoxin-a from natural waters by an automated on-line derivatization-LC-electrospray MS was reported where concentrations as low as 2.1 ng L^{-1} were detected (Takino *et al.*, 1999). A new LC/MS method has been developed for determination of anatoxin-a and homoanatoxin-a. The LC was coupled via an electron spray ionisation (ESI) source to an ion-trap MS in positive mode. The detection limit for anatoxin was 0.03 ng (on-column) corresponding to 0.6 µg L^{-1} (Furey *et al.*, 2003a). Employing post-synaptic membrane fractions of *Torpedo* electric tissue, the inhibition of binding of radioactively labelled bungarotoxin by anatoxin-a and homoanatoxin-a (purified from *Oscillatoria* sp. strain 193 PCC 9240 and *O. formosa* NIVA CYA-92 PCC 101111, respectively) to acetylcholine receptors was studied. This technique has been termed as receptor radioligand-binding assay and has a high sensitivity with a dectetion limit of about 1×10^{-8} M which is equivalent to 1 µg L^{-1} ($\approx 6 \times 10^{-9}$ M) proposed for anatoxin-a in drinking water. A concentration-dependent inhibition was observed with K_i of $5.4 \pm 1.1 \times 10^{-8}$ M for anatoxin-a and $7.4 \pm 0.9 \times 10^{-8}$ M for homoanatoxin-a (Aráoz *et al.*, 2005). Anatoxin-a produced by *P. favosum* was identified by HPLC with a UV-detector and by electrospray-Quadruple-Time of flight mass spectrometry and further confirmed by tandem mass spectrometry (Gugger *et al.*, 2005).

Solid-phase microextraction of anatoxin-a using three forms of polyaniline films (PANI) and a single form of polypyrrole led to the identification of leucoemeraldine form of PANI as best suited for extraction. Optimization parameters for better extraction were 32 µm thickness, a salt concentration of 10% (w/v), an extraction time of 30 min and a stirring rate of 100 ppm (Ghassempour *et al.*, 2005).

iii) Capillary Zone Electrophoresis (CZE): Though capillary electrophoresis has been suggested for the separation of MCs and nodularins (Boland *et al.*, 1993; Onyewuenyi and Hawkins, 1996; Bouaicha *et al.*, 1996; Bateman *et al.*, 1995; John *et al.*, 1997; Siren *et al.*, 1999), it is not as sensitive as HPLC. Water bloom samples and crude cyanobacterial extracts were subjected to CZE and micellar electrokinetic chromatography (MEKC) that enabled simultaneous separation of anatoxin-a, MC-LR and CYN. The detection limit of these toxins was in between 1–4 µg ml^{-1}. Both CZE and MEKC were sequentially applied for confirming the results obtained by conventional techniques. Polyimide-coated fused-silica capillary of 48.5 cm x 50 µm ID (40cm effective length) were used by hydrodynamic sample introduction for injecting samples. The voltage applied was +25 kv at a temperature of 25°C. The measurements were made spectrophotometrically at 270 nm where diode array detection permitted UV spectra to be recorded at 0.2 s intervals across the electropherogram. The separation of toxins showed a dependence on the concentration of SDS in the running electrolyte. The application of

MEKC has proved beneficial as it can overcome the problems of interference by high protein content of the samples during CZE separation. MEKC proved more advantageous for the analysis of CYN because of the retention of proteinaceous material by anionic micelles. Thus CZE and MEKC are useful, rapid and sensitive physicochemical assays for simultaneous determination of MC-LR, CYN and anatoxin-a (Vaas *et al.*, 2004).

Analysis of STXs, produced by *A. circinalis* and other cyanobacteria, in the range of 10–70 ppb was demonstrated by a single run HPLC method by employing ion exchange as a separation mode with a novel mobile phase system coupled to post column oxidation and fluorescence detection. During the analysis of STXs either the presence of cyanobacterial matrix or natural water constituents did not interfere in any manner (Papageorgiou *et al.*, 2005).

3) Molecular Probes: A number of molecular probes have been identified during recent years to identify the toxic blooms both from laboratory cultures and natural samples. The utility of such molecular probes is unequivocal in the identification of a particular toxic cyanobacterial species but these require trained personnel and also involve lot of expenditure with reference to sophisticated instruments and very costly chemicals that an ordinary public health institution can afford. Nevertheless these techniques are greatly helpful in the identification of toxic cyanobacteria up to species level. These are discussed here.

***i) PC-IGS sequences*:** The utility of intergeneric spacer (IGS) between the bilin subunit genes, i.e. β (*cpcB*) and α (*cpcA*) of the phycocyanin (PC) operon has been examined for the identification of cyanobacteria to the strain level. Thus PC-IGS sequence was amplified by PCR from DNA extracted from the environmental samples. Restriction endonuclease digestion of the PCR amplified products yielded many DNA profiles specific to different cyanobacteria. As for example, restriction enzyme profiles for *MspI, RsaI* and *Taq1* could be assigned to eight genera (40 strains). Genus *Microcystis* could be identified with restriction enzymes *AluI, CfoI* and *HaeIII* whereas *Anabaena* was identified with digestion profiles provided by *AluI, CfoI* and *ScrFI* (Neilan *et al.*, 1995).

***ii) RAPD and PCR of 10-mer sequences*:** Random oligonucleotide primers of 10 nucleotides long combined with PCR have given a molecular tool that is useful for recognition of DNA from various living organisms. The randomly amplified polymorphic DNA (RAPD) technique in combination with PCR helped in identifying the DNA profiles specific to the toxic cyanobacteria tested to the level of genus, species and strain. The genera *Anabaena* and *Microcystis* could be easily distinguished (Neilan, 1995).

***iii) Heptamer repeats*:** The genomes of filamentous, heterocystous cyanobacteria possess short tandemly repeated repititive (STRR) DNA sequences between or within genes. The genome of *Anabaena* sp. strain PCC 7120 consists of such heptamer repeats at many sites, i.e. in and around *PC* operon (Belknap and Haselkorn, 1987), after the *atpBE* operon (Curtis, 1987), after the *psbB* gene (Lang and Haselkorn, 1989), within the *nifB* operon (Mulligan and Haselkorn, 1989) and within *nifJ* gene (Bauer *et al.*, 1993). The heptamer repeat sequence CCCCAGT found in *Anabaena* sp. strain PCC 7120 is common with *Anabaena* and *Nostoc* strains. Genomes of non-heterocystous *Oscillatoria* sp. and *Microcystis* sp. are devoid of such heptamer sequences. It is concluded that this sequence is mobile between *Anabaena* and *Nostoc* strains (Bauer *et al.*, 1993). Rouhiainen *et al.* (1995) used RFLP analysis using STRR to distinguish *Anabaena* isolates producing hepatotoxins from those of neurotoxin-producing strains and from *Nostoc* spp.

***iv) 16S and 23S rRNA ITS sequences*:** 16S-23S rDNA ITS has helped in the recognition of three clusters among 47 strains of the genus of *Microcystis*. In the first cluster both toxic and non-toxic strains are included while the second consisted of only toxic strains and the third cluster comprised of non-toxic strains. Phylogenetic analysis based on the above studies enables us in understanding relationships amongst closely related species and strains (Otsuka *et al*., 1999b).

Evolutionary relationships among toxic and non-toxic strains of the genus *Microcystis* were traced with the help of 16S rRNA gene sequences. Strains of *Microcystis* with the ability to produce MCs formed a distinct cluster consistent with cell morphology, gas vacuolation and low G + C content. This cluster was shown to be quite distinct from the *Synechocystis* and *Nostoc* clusters. *Microcystis*-specific 16S rRNA sequence signatures were used to develop primers that enabled identification of potentially MC-producing strains of *Microcystis* via DNA amplification. It was possible to selectively amplify DNAs of all the strains belonging to *Microcystis* cluster by employing one of the primers located at 179 to 209 of the gene. The 209 F (forward) with a sequence of 5′-ATGTGCCGCGAGGTGAAACCTAAT-3′ was amplified alongwith 409 R (reverse) primer with a sequence of 5′-TTACAA(C/T)CCAA(G/A)(G/A)(G/A)CCTTCCTCCC-3′. Thus *M. aeruginosa* PCC 7806, PCC 7941, AWT 139, NIE 389 and PCC 7820, *M. viridis* NIES 102 and *M. wesenbergii* NIE 3107 were all producers of MCs as substantiated by LD_{50} in mouse bioassay and its associated histological studies.

***v) DNA-dependent RNA Polymerase gene*:** DNA-dependent RNA polymerase gene (*rpoC1*) has been recognized as a more discriminatory marker (Palenik and Haselkorn, 1992) than 16S rRNA that has been used more extensively for the identification of cyanobacteria (Lyra *et al*., 1997; Nuebel *et al*., 1997; Rudi *et al*., 1998). A 609-bp region of *rpoC1* was amplified by PCR with *C. raciborskii*-specific primers. Molecular characterization of *C. raciborskii* isolates (19 of them) from different geographical regions of Australia representing straight and coiled trichomes suggested that all isolates belonged to the same species. The utility of RAPD and STRR sequence profiles to recognise *C. raciborskii* isolates was of a limited value than the *rpoC1* gene analysis. The specific PCR targetting a specific region of *rpoC1* developed for *C. raciborskii* was found to be unique for this organism as it enabled identification of DNA from natural populations (Wilson *et al*., 2000). *A. circinalis* occurs in the blooms along with morphologically similar coiled *Anabaena* sp. in Murray-Darling Basin of Australia. Because it produces PSPs in this region of the world, *A. circinalis*-specific PCR assay of *rpoC1* gene has been developed to detect this orgainsm from the environmental samples (Fergusson and Saint, 2000).

***vi) mcyB and mcyA markers*:** The gene *mcyB* encodes a peptide synthetase of 242,334 Da protein containing two modules each possessing adenylation, thiolation and condensation domains. The *mcyB* may be involved with the activation of "variable L-amino acids". Thus genetic probes from *mcy* gene cluster, i.e. *mcyB* gene and an adenylation domain within this gene were used to detect MC-producing cyanobacteria (Neilan *et al*., 1999; Nishizawa *et al*., 1999). Although these two probes showed good correlation to toxin production, due to the sequence similarity of *mcyB* region with other peptide synthetase loci these genetic loci did not find favour for further application (Dittmann *et al*., 1997; Tillett *et al*., 1999). By using oligonucleotide primers for *mcyB* gene of the operon MC synthetase gene, PCR amplification of 60 *Microcystis* spp. strains from 15 reservoirs distributed in Brazil was performed. A correlation between the presence of *mcyB* gene and the presence of total MC as determined by ELISA revealed that a unique amplified product of approximately 780 bp was present in 18 strains of *Microcystis*. The lake waters contained a mixture of toxic and non-toxic strains of *Microcystis* (Bittencourt-Oliveira, 2003). A set of oligonucleotide primers was designed according to the squence of the peptide synthetase gene *mapep1* (module 3 of *mcyB*). These are forward primer mapep-2B with a sequence of 5′-GATTGAACGGATGGCAGCAC-3′ and mapep

2A reverse primer with a sequence of 5′GGCCGCTCTAGAAACAGGTA-3′ and for module 1 of *mcyA* gene M1F2 forward primer with a sequence of 5′-CAGTGACGAGTAATCCCGAT-3′ and M1R2 reverse primer with a sequence of 5′-GCCATTCTCGCATCTGAGGA-3′. PCR amplification of 20 toxic strains and 18 non-toxic strains of *Microcystis* revealed that PCR amplified products could be seen only in case of toxic strains. It was possible to enumerate the number of toxic cells (10^3–10^5 cells ml^{-1}) on the basis number of amplified products. A positive correlation existed in the toxin concentration and density of toxic cells in the environmental samples from Lakes of Mikata and Suigetsu (Yoshida *et al.*, 2003).

N-methyltransferase (NMT) domain encoded by *mcyA* gene has been chosen as a molecular probe to identify toxic *Microcystis*. NMT region was specifically present in 18 toxigenic strains of *Microcystis* but was not detected in non-toxic strains (17) of this genus. NMT-specific DNA was detected in natural populations and correlated well with PP-inhibition assays. Further, these results have been compared with those of 16S rRNA gene sequences and PC-IGS sequences. The strains of *Microcystis* formed a monophyletic group based on 16S rRNA gene sequence analysis while PC-IGS enabled the recognition of two groups, i.e. both toxic and non-toxic strains of the genus were randomly distributed. Interestingly, *mcyA* gene marker helped in the recognition of two coherent groups. In all NMT-positive strains a gene of unknown function (*umaI*) existed at a distance from *mcyC* gene. Though gene *umaI* also was detected in non-toxic strains it was not transcribed along with *mcyABC* (Tillett *et al.*, 2001). Analysis of naturally occurring strains of *Anabaena* sp. revealed a 1236 bp in-frame deletion that corresponded to the loss of the entire NMT domain but the synthesis of MCs continued in nine of these strains as they incorporated dehydroalanine in place of N-methyldehydroalanine and also elevated levels of L-Ser (Fewer *et al.*, 2008). However, majority of the non-toxic strains of *Planktothrix* sp. from nine European countries have lost the large *mcy* gene cluster but still retained *mcyT* gene along with flanking regions of *mcy* gene cluster. The involvement of *mcyT* in the synthesis of MCs has been confirmed by the isolation of *mcyT* inactivation mutant of a MC-producing strain that showed a reduction of 94 ± 2% in MC synthesis (Christiansen *et al.*, 2008).

The distribution of MC-producing and non-MC-producing genotypes of *M. aeruginosa* and *M. botrys* in European freshwater bodies was studied by PCR amplification of gene regions *mcyA* and *mcyB*. Individual colonies of the two species were estimated for the presence of the two MC-synthesizing genes, the type of MC-variants and total MC content. More than 75% of the colonies of the two species contained *mcy* genes whereas only 20% of the colonies identified as *M. ichthyoblabe* and *M. viridis* were positive for the amplification of *mcy* genes. However, colonies of *M. wesenbergi* did not give a PCR product for either of the *mcy* genes. A relationship between 'morphospecies' and MC genotype as well as 'morphospecies' and MC production have also been established (Via-Ordorika *et al.*, 2004).

The amplification of six characteristic segments of *mcy* gene cluster, three corresponding to NRPS genes (*mcyA*, *mcyB*, and *mcyC*) and three to PKS genes (*mcyD*, *mcyE* and *mcyG*), helped in the distinction of MC-producing and non-producing strains of *Microcystis*. These studies revealed that (i) all the MC-producing strains could be identified with help of specific amplicons and (ii) simultaneous amplification of several specific gene regions is feasible with cultured cells and field samples. The authors further suggested that the potential toxicity of a field sample could easily be determined on the basis of amplification of these two sets of genes from field samples (Ouahid *et al.*, 2005). The detection of MC-producing strains of cyanobacteria by the amplification of *mcyF* gene by PCR enabled Rantala *et al.* (2006) to identify *Microcystis* sp. (70%), *Planktothrix* sp. (63%) and *Anabaena* sp. (37%) with variable population densities in 70 Finnish Lakes The proportion of the three genera in the lakes very much depended on the trophic status of the lakes and eutrophic lakes supported the growth of the three bloom formers. With the help of real-time PCR of *mcyA* and phycocyanin markers it was possible to survey the fluctuations in the populations of MC-producing and non-MC-producing *Planktothrix agardhii* in a eutrophic French Lake. The proportion of MC-producing genotype fluctuated between 30–80% of the poupulation (Briand *et al.*, 2008).

vii) Taq nuclease assay or 5′-nuclease PCR: *TaqMan* (TM) PCR or *Taq* nuclease assay was first developed to enumerate specific genotypes of picocyanobacteria (Becker *et al.*, 2000). This was extended for simultaneous monitoring of two microbes (*E. coli* and MC-producing strains of *Microcystis*) by conducting PCR reaction sequence of specific probe dually labelled with fluorescent dye that is also known as *TaqMan* probe (Foulds *et al.*, 2002). The rate of exponential accumulation of the amplificon is monitored by the hydrolysis of TM probe in which it generates a fluorescent signal during the amplification process. To detect *E. coli lacZ* TM probe with a sequence of 5′-ATTCGCCATTCAGGCTGCGCAA-3′ and for the identification of a MC TM probe having a sequence of 5′-TTAAATCGGAAAT TATCCCAGAAAATGCCGT-3′ were selected. These dual labelled probes were modified with 5′-6-carboxyfluorescein reporter dye and a 3′ 6-carboxy N, N, N′, N′ tetramethylrhodamine quencher dye. LacZ primers were employed to amplify a region of *lacZ* gene in *E. coli* (180 bp amplificon) for the identification and enumeration of *E. coli* cells contaminating drinking water. In a similar manner, a region of *mcyA* gene from MC synthetase operon (1220 bp amplificon) was identified to be specific for toxin-producing strains of *M. aeruginosa*. 5′-Nuclease PCR reactions were performed in which a standard curve based on pre-determined cell or DNA concentrations was compared to assay the threshold cycle of unknown samples. The concentration of microbes in water samples was quantified by knowing standard units of gene copies of *lacZ* gene and *mcyA* gene μL^{-1}. Using the above methods, it was possible to detect three copies of target samples within 2 h after DNA extraction from both microbes (Foulds *et al.*, 2002).

Seasonal development of bloom populations of *Microcystis* was monitored along with toxic MC-producing genotype in Lake Wannsee (Berlin, Germany). *Taq* nuclease assay was performed for two gene regions, i.e. *PC-IGS* sequence for the estimation of total population and *mcyB* gene of *mcy* gene cluster. The development of threshold cycle of PCR enabled to infer the number of cells. These showed a correlation with microscopically determined cell numbers from logirthmic cultures. Identical amplification efficiencies for both genes were detected in all 10 strains of *Microcystis* cultures examined. The application of *Taq* nuclease assay for field samples revealed that the mean number of *mcy* genotype is stable from winter to summer and in Lake Wannsee its cell numbers could be enumerated. Further, the *mcy* genotypes formed a smaller part of the PC genotypes ranging from 1 to 38% and a correlation of 1:1 between cell number and the two selected markers existed. A parallel relationship between cell numbers estimated through inverted microscope counts and *Taq* nuclease assay was also established (Kurmayer and Kutzenberger, 2003).

VII. OCCURRENCE OF HARMFUL ALGAL BLOOMS (HABs)

Significant human health risk from HABs has been predicted on the basis of worldwide surveys on mass development of toxic cyanobacteria in potable and recreational waters. The percentage of toxic samples accounted for 60% of the blooms formed world over. In general, hepatotoxin producers are more common than the neurotoxin-producing ones. However, severe animal poisonings have been reported from North America, Europe and Australia due to neurotoxins rather than by hepatotoxins. Blooms producing CYN have been reported from Australia, Hungary, Japan, Israel and Germany.

Human health risk assessment could be possible by determining the toxin levels in potable and recreational water bodies. Knowledge on the nutrient status, periodicity in the development of HABs, the prevalence of toxic cyanobacterial species, nature and types of toxins produced and the level of toxins in these waters is very important. It is difficult to predict the concentration of toxins in the cells or in water during a bloom because of differences in the proportion of toxin-producers and non-toxin-producers within cyanobacterial populations. Despite this, the cellular contents of cyanobacterial toxins visa-vis the level of toxins present in water bodies per unit volume have been determined. Of the cyanobacterial toxins, nodularins were present to a maximum of 18 mg g^{-1} dry weight (Sivonen and Jones, 1999). On the basis of the number of cyanobacterial cells ml^{-1} water, three levels of adverse health effects have been recognised. Relatively low adverse health effects can be expected if the number of cyanobacterial cells ml^{-1} are about 20,000 or in otherwords on the basis of chlorophyll content if it is nearly 10 μg chlorophyll *a* L^{-1}. Moderate adverse health effects can result when the cell number is around 100,000 cyanobacterial cells ml^{-1} or equivalent to 50 μg chlorophyll *a* L^{-1}. High adverse health effects can arise due to scum formation by HABs. If in potable or recreational water bodies, a situation favouring moderate risk prevails and if bloom samples are predominantly represeted by *Microcystis* spp., then the MC level is likely to be 20 μg MC L^{-1} with average cellular MC content of 0.2 pg $cell^{-1}$ or 0.4 μg MC μg chlorophyll a^{-1}. If the bloom sample contains *P. agardhii*, then the levels of MCs can be 1–2 μg MC μg chlorophyll a^{-1} attaining concentrations of 200–400 μg MC L^{-1}. As per WHO (1998) guidelines the value of permissible MC-LR concentration in drinking water is 1 μg MC-LR L^{-1}. Moreover, the transition from moderate risk to high risk level due to the formation of scums is found to be achieved very swiftly in a short period of few h. Thus it becomes more obligatory to monitor the levels of toxic cyanobacteria on a day to day basis as the toxin concentrations may be enhanced 1000 times or more than the guidelines prescribed by WHO.

The conventional water treatment processes employ chlorination or ozonation and also use ferric chloride and potassium permanganate and it is very likely the finished waters may contain increased concentrations of MCs due to lysis of cyanobacterial cells (Lam *et al.*, 1995). The application of copper sulphate or the use of organic copper-chelated algicide caused lysis of bloom samples increasing concentrations of MCs from 4.7 μg L^{-1} to 1110 μg L^{-1} within 4 h post-treatment (Jones and Orr, 1994). Human illness attributed to toxic cyanobacterial lysis following copper sulphate treatment of drinking water sources has been reported from Charleston, West Virginia (Tisdale, 1931), Palm Island, Queensland, Australia (Bourke *et al.*, 1983) and Armidale, New South Wales, Australia (Falconer *et al.*, 1983).

***Occurrence of toxic blooms*:** Toxic blooms have been reported from almost all quarters of the world.

***Africa*:** The river Nile (used as drinking water source) at Sohag Province of Egypt supported the development of blooms of *O. tenuis* which is a known producer of MCs (Brittain *et al.*, 2000). Many fresh water bodies of Morocco developed toxic cyanobacterial blooms of the genera *Microcystis*, *Synechocystis*, *Pseudoanabaena* and *Oscillatoria* (Oudra *et al.*, 2002; Sabour *et al.*, 2002). Drinking water contamination of toxic *M. aeruginosa* led to the poisoning of livestock in S. Africa (Van Halderen *et al.*, 1995).

***America*:** The development of toxic blooms in Canada, North America and South America is described here. Majority of the Lakes and rivers in Northern America develop extensive blooms. The magnitude of the development of a mixed bloom of *M. aeruginosa* and *A. circinalis* in St. Johns

River can be seen in Fig. 23. Toxic blooms of *Anabaena, Aphanizomenon, Microcystis, Oscillatoria* and *Nodularia* have been reported from Canada (Carmichael, 1992). Detectable limits of MC-LR greater than 1 µg g^{-1} dry weight were found in 70% of the samples collected from 19 lakes of Alberta (Hrudey *et al.*, 1994). However, MC-LR concentrations of phytoplankton samples from lakes and dugouts ranged from 4 to 605 µg g^{-1} dry weight (Kotak *et al.*, 1993). Maximum values of MC-LR exceeding 1500 µg g^{-1} dry weight were also reported (Hrudey *et al.*, 1994; Zurawell *et al.*, 1999). The occurrence of blooms of *M. aeruginosa* in drinking water lakes of Canada with toxin levels much higher than permissible limits in both natural and treated waters has been reported (Gupta *et al.*, 2001). Cyanobacterial abundance and toxicity were predicted on the basis of studies conducted on 22 lakes of southern Qubec, Canada. Three important findings emerged out of these studies. Firstly, the biomass of the phytoplankton increased linearly with increase in total phosphorus concentration. Secondly, the biomass of the toxic genera, i.e. *Microcystis* and *Anabaena* correlated with the MC concentration. Thirdly, the most important factor appeared to be epilimnetic nitrogen that is responsible for the biomass of toxic species. Moreover, the level of MC per unit biomass did not vary significantly among the lakes (Giani *et al.*, 2005). Further studies on four eutrophic lakes in the eastern parts of Qubec (Canada) were chosen for a study of seasonal changes in the composition of cyanobacterial community. MC content was determined using PP-inhibition assay by using extracts of lyophilized plankton. Three lakes showed maximum toxicity in summer while the fourth showed maximum toxicity during spring due to the abundance of genera *Microcystis* and *Anabaena* (Rolland *et al.*, 2005). In a study on Lake Erie, Rinta-Kanto (2005) observed the persistence of toxic blooms of *Microcystis* responsible for MC-LR equivalents exceeding 1 µg L^{-1}, the safety limit set by WHO. By applying a combination of molecular probes and techniques of real-time PCR assay utilizing specific Primer-Taq Man probe, *Microcystis*-specific 16S rDNA fragment and a microcystin toxin synthetase

Figure 23: Bloom of *Microcystis aeruginosa* and *Anabaena circinalis* on the St. Johns River, Florida, USA. Picture courtesy John Burns (Cyanolab), Mark Schneegurt (Wichita State University) and Cyanosite (www.cyanosite.bio.purdue.edu).

Color image of this figure appears in the color plate section at the end of the book.

mcyD gene, the number of toxic cell numbers of *Microcystis* were determined to be in the range of 4×10^8 to 2×10^3 cells L^{-1}. By using sequence analysis of *mcyA* gene fragments, the occurrence of toxic *Microcystis* and *Planktothrix* was documented in the previously undetected localities in the western basin of Lake Erie (Rinta-Kanto and Wilhelm, 2006).

In as many as 27 states of USA, HABs have been reported (Codd *et al.*, 2005). Based on studies on the occurrence of toxic cyanobacterial blooms in surface drinking waters in the state of Florida, the incidence of primary hepatocellular carcinoma was diagnosed in patients in between 1981 and 1998 (Fleming *et al.*, 2002). *Lyngbya* dominated blooms in a North Louisiana reservoir were investigated to know their potential toxicity to the survival, reproduction and behaviour of *Daphnia magna* and *Pimephales promealis*. Risk evaluation experiments recommended the application of a herbicide whose active ingredient is copper for effective control of *Lyngbya* blooms (Mastin *et al.*, 2002). Physicochemical data from 241 lakes in Missouri, Iowa, northeastern Kansas and southern Minnesota revealed the total phosphorus and total nitrogen to be in the range of 2–995 and 90–15870 µg L^{-1}, respectively. The MC content of lake waters increased with increased latitutes from undetectable to 4500 ng L^{-1}. For effective lake management and minimization of human health risks, the authors suggested that the empirical relationships between environmental variables and MC values have to be taken into consideration (Graham *et al.*, 2004). A study on the genetic variation within and among populations of *M. aeruginosa* in southern Michigan Lakes revealed that 53 of the 67 isolates were genetically distinct as determined on the basis of *mcyA* gene marker. Besides studying the various limnological characteristics of these lakes, a relationship between survivorship of the isolate and composition of the medium has been established. For example, a low rate of survival of isolates from oligotrophic lakes was noted when cultured in nutrient-rich algal medium. Additionally, four strains of *M. aeruginosa* from three freshwater lakes of North America, Africa and Europe (isolated and maintained in pure cultures) were genetically analyzed and compared with those of natural populations. These four strains are not only related to each other but also exhibited similarity to most of the strains of Bear Lake and Spring Lake suggesting that a similarity of nutrient levels in these bodies and nutrient medium employed (Wilson *et al.*, 2005). Out of a total 182 lakes and reservoirs of Indiana surveyed for HABs, 19 of them showed detectable growth of *C. raciborskii* during late summer months of August and September of 2002 and 2003. Both straight and curled morphotypes have been detected. The bloom has been widely represented in 35% of the impoundments sampled (n=40) than in natural lakes. Though bottom temperatures of shallow lakes reached 22–23.5°C, the mean tempearure in the deepest waters of lakes in which *C. raciborskii* was detected was 18.2°C and with surface temperatures of 26.5°C. High epilimnetic phosphorus content is characteristic of these lakes (Jones ans Sauter, 2005). Sheldon Lake of Colorado, a public recreational focal point, supported the development of toxic *Microcystis* blooms as confirmed by the use of molecular markers of the *mcy* gene cluster (Oberholster *et al.*, 2006). Water samples (247) collected from the Salton Sea (a saline endorheic rift lake that is largest in California), dominated by *Synechococcus* sp. and benthic filamentous *Oscillatoria* sp., yielded the presence of MCs. Pure cultures of both the dominant organisms produced MC-LR and MC-YR (Carmichael *et al.*, 2006).

Extensive blooms of *Anabaenopsis* have been reported in Bedetti Lake, Argentina (Fig. 24). Brazil attracted the attention of the whole world in a glaring instance of the death of 60 patients of a kidney dialysis centre due to supply of drinking water from a lake that suffered from the blooms of *Anabaena* and *Microcystis* (Jochimsen *et al.*, 1998; Pouria *et al.*, 1998). Toxic cyanobacterial blooms have been recorded from various habitats of Brazil such as estuaries (Yunes *et al.*, 1996), lagoons (Azevedo *et al.*, 1994; Azevedo 1996; Porfirio *et al.*, 1999; Lagos *et al.*, 1999; Magalhaes *et al.*, 2001) and reservoirs (Bouvy *et al.*, 1999; Chelleppa *et al.*, 2000; Costa *et al.*, 2001; Molica *et al.*, 2002). Lakes in the state of Parana' of

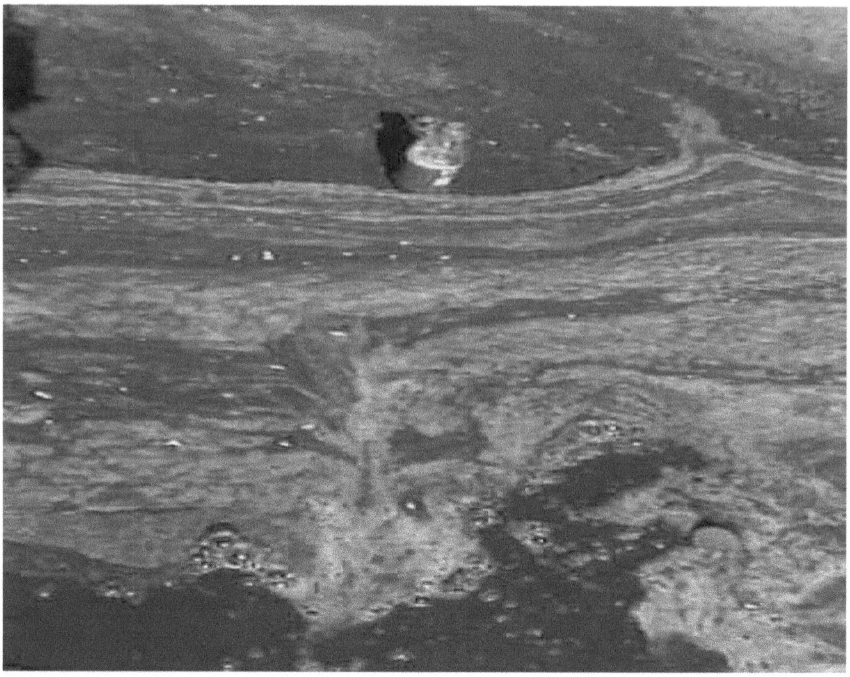

Figure 24: *Anabaenopsis* bloom in Bedetti Lake, Santo Tome, Santa Fe, Argentina. Picture courtesy Frederico Emiliani (Universidad Nacional del Litoral), Mark Schneegurt (Wichita State University) and Cyanosite (www.cyanosite.bio.purdue.edu).

Color image of this figure appears in the color plate section at the end of the book.

Brazil and Patos Lagoon estuary of Rio Grande do Sul of southern Brazil developed blooms of toxic *M. aeruginosa* (Hirooka *et al.*, 1999; Matthiensen *et al.*, 2000). Certain picpoplanktonic cyanobacteria from Caruaru reservoirs produced MCs in the concentrations ranging from 0.08 to 3.7 ng mg^{-1} dry weight suggesting that these organisms might have contributed to human poisoning in Caruaru for long time (Domingos *et al.*, 1998). In Chile also the lakes supported the growth of *Microcystis* spp. producing various MCs (Campos *et al.*, 1999; Neumann *et al.*, 2000). The development of massive blooms of *Pseudoanabaena schmdlei* along with *M. aeruginosa* in the man-made reservoir of Northeast Brazil resulted in the death of almost the whole fish population (Chellappa *et al.*, 2000). Substantial genetic diversity in *M. aeruginosa* was reported within and among lakes of Brazilian reservoirs using phycocyanin intergeneric spacer and flanking regions (*cpcBA*). Nine distinct genotypes were collected from the four sites with the presence of as many as six genotypes from a single site and one specific genotype existing at two sites. Moreover, a single genotype was represented along the depth gradient repeatedly (Bittencourt-Oliveira *et al.*, 2001). The occurrence of mixed cyanobacterial blooms comprising of *C. raciborskii, M. panniformis, M. protocystis, M. novacekii, Aph. gracile, Aph.* c.f. *manguinii, Aph.* c.f. *issastschenkoi* (Fig. 2C) was recorded in Armanado Ribeiro Goncalves reservoir of Brazil that serves as drinking water resource. *C. raciborskii* was predominantly represented during rainy season with levels of STXs at 3.14 µg L^{-1}. This was followed by the development of blooms of species of *Microcystis* with MCs at 8.8 µg L^{-1} in the transition period whereas in the dry period co-dominance of *C. raciborskii, Microcystis* spp. and *Aphanizomenon* spp. was noted with low levels of STXs (Costa *et al.*, 2006). By targetting *mcyA* as marker gene, analysis of shoreline distribution of cyanobacterial DNA from Lake Ontario (bordering Ontario, Canada and New York, USA) revealed the existence of *Microcystis*, *Planktothrix* and *Anabaena* sp. at 50% of the sampling stations. With the predominance of *Microcystis* sp. the MC concentration of the waters exceeded the WHO guidelines (Hotto *et al.*, 2007). A metagenomic approach by the isolation of 36 kb DNA fragments containing *mcyA2* and *mcyB1* genes from natural water samples and a correlation of these loci from individual cyanobacterial cells improved the analysis of MC-producing strains in the Laurentian Great Lakes (Allender *et al.*, 2009).

Asia: Amongst Asian countries, blooms of toxic cyanobacteria have been reported from South Korea, China, Japan, Phillippines and Africa. Species of *Microcystis*, *Anabaena*, and *Planktothrix*/*Oscillatoria* producing MCs and anatoxin-a have been predominantly represented from South Korean drinking water resources (Oh *et al.*, 2001; Park, 2001). Yu (1989) conducted an epidemiological study of human primary liver cancer in Qidong county in China and observed a correlation on the incidence of liver cancer about eight times in people who used pond and ditch water for drinking purposes. MC-producing species of *Anabaena* and *Oscillatoria* were detected in the ponds and ditches in Haimen City (Jian-Su province) and Fusui county (Guangxi province) in China and a significant correlation between the occurrence of these bloom species and the incidence of primary liver cancer has been established (Ueno *et al.*, 1996; Xu *et al.*, 2000). Lake Dianchi suffered due to development of toxic cyanobacterial blooms during recent years and the fate of MCs in this aquatic ecosystem has been studied. The concentrations of MCs ranged from 0.17 to 0.82 $\mu g\ L^{-1}$ that is below the permissible limits set by WHO. Of the three processes, i.e. absorption, photodegradation and biodegradation, photodegradation mainly is shown to be responsible for the elimination of MCs (Zhang *et al.*, 2004). In Japan cyanobacterial blooms of the species of *Microcystis*, i.e. *M. aeruginosa*, *M. flos-aquae*, *M. ichthyoblabe*, *M. novacekii*, *M. viridis* and *M. wesenbergii* have been reported on the basis of morphological characteristics (Komarek and Kling, 1991). Amongst these six species, some strains of *M. aeruginosa*, *M. ichthyoblabe* and *M. viridis* are known to be toxic (Watanabe *et al.*, 1989). Further, the presence of *Microcystis* blooms producing MC (Tsuji *et al.*, 1996; Park *et al.*, 1998; Matsunaga *et al.*, 1999) in water bodies and death of dozens of ducks (Matsunaga *et al.*, 1999) were reported. Five new protease inhibitors were isolated from blooms of Lake Teganuma (Japan) that characteristically inhibited at least one or two proteases in assays for trypsin, plasmin, chymotrypsin, leucine aminopeptidase, carboxypeptidase A and angiotensin converting enzyme (Kodania *et al.*, 1999). Blooms of toxic *M. aeruginosa* have been represented in the water bodies from Taiwan (Lee *et al.*, 1998) and Phillippines (Cuvin-Aralar *et al.*, 2002). The emergence of non-toxic strains of *Microcystis* from the populations of MC-producing *Microcystis* strains KLL MG-K and KLL MB-K, isolated from Lake Kinneret, Israel was studied. Non-toxic strains MG-J and MB-J spontaneously succeeded as evidenced by the loss of atleast 34 kb of the *mcy* region governing MC synthetase. Laboratory and field experiments highlighted the advantages of the toxic strain over non-toxic counterparts underlining the possible role of MC under natural conditions for suppressing the growth of non-toxic strains (Schatz *et al.*, 2005). The recurrence of toxic blooms in Taihu Lake, China and the risk of MC-LR exposure to human beings has been predicted on the basis of tolerable daily intake levels, although short-term genotoxicity assays revealed negative results (Wu *et al.*, 2006).

Australia and New Zealand: Newly constructed shallow lakes in Queensland, Australia supported the development blooms of *Aph. ovalisporum* that produces CYN. Distinctive nutrient parameters such as ready availability of nutrients, high chloride and hardness levels triggered the bloom development that constitutes the first report from Australia. Based on 16S rRNA gene sequence *Aph. ovalisporum* was shown to be identical to *Aph. ovalisporum* isolated from Lake Kinneret, Israel. The homology of this strain was much lower to other species of this genus, i.e. *Aph. flos-aquae*, *Aph. gracile* or *C. raciborskii* that are known to produce CYN (Shaw *et al.*, 1999). The toxic cyanobacterial blooms of serious nature in Australia are *M. aeruginosa*, *No. spumigena*, *A. circinalis* (known to produce PSPs) and *C. raciborskii* and the management strategies include the ability to monitor the growth of these HABs in catchments, source waters and the distribution systems (Steffenson *et al.*, 1999). Blooms of *M. aeruginosa* (Atkins *et al.*, 2001) and *Cylindrospermopsis* (Fitzgerald, 2001) have frequently been reported from Australia. Members of the order Chroococcales such as two forms of *Aphanothece* and non-gas vacuolated *Microcystis* species forming benthic communities of embayments in the Myall Lake of Australia proved to be toxic in mouse bioassay, PP inhibition and ELISA for MCs (Dasey *et al.*, 2005). Orielton Lagoon in Tasmania suffered due to the development of a bloom of *No. spumigena* due to lowering of salinity (15–20 g kg^{-1}) during December and January with concentrations of nodularin reaching as high as 2000–3500 μg g^{-1} dry weight (Jones *et al.*, 2006). Toxic cyanobacterial blooms of *Microcystis* and *Anabaena* are of wide spread occurrence in the Lakes of Victoria (Fig. 25). Blooms of *Aphanizomenon* and *Nodularia* occur in the southwest wetlands on the Swan coastal plain of western Australia. Out of a total of 32 natural bloom samples, 28 proved to be toxic with MC concentrations ranging from 1645 to 8428.6 μg L^{-1}. The distribution of MC variants revealed 100% by MC-LR followed by MC-YR (71.4%) and MC-RR (60.7%). The presence of nodularins (1664 μg L^{-1}) along with MCs constitutes the first report (Kemp and John, 2006).

Europe: Toxic blooms of *M. aeruginosa* have been reported from lakes and reservoirs of North and central Portugal as demonstrated by mouse bioassay with LD_{50} of 15 to 75 mg kg^{-1} body weight (Vasconcelos, 1993). Subsequent studies conducted on 30 lakes, rivers and reservoirs revealed toxic blooms from 18 of these water bodies with a predominance of *M. aeruginosa*, *M. wesenbergi*, *A. flos-*

Figure 25: *Microcystis aeruginosa* bloom, Lake Mokoan, Victoria, Australia. Picture courtesy Ben Long (University of Surrey), Mark Schneegurt (Wichita State University) and Cyanosite (www.cyanosite.bio.purdue.edu).

Color image of this figure appears in the color plate section at the end of the book.

aquae, *A. scheremetievi* and *Aph. flos-aquae* (Vasconcelos, 1994). The predominance of hepatotoxic water blooms known to produce diverse MCs (MC-LR, MC-RR and MC-YR and others) is shown to be associated with the dominance of *M. aeruginosa* in Portuguese freshwater bodies such as lakes, reservoirs and rivers (Vasconcelos *et al.*, 1996, 2001; Vasconcelos and Pereira, 2001). Vasconcelos (1999) reviewed the status of cyanobacterial toxins in Portugal and indicated the possible accumulation of MC-LR by edible molluscs, fishes, crayfishes and mussels and risk to human beings upon consumption. Blooms of *M. aeruginosa* and *Aph. flos-aquae* from Lake Grand-Lieu (Vezie *et al.*, 1998) and Saint-Caprais reservoir (Maatouk *et al.*, 2002) respectively, have been reported from France. The development of blooms of *M. aeruginosa* in Belgium near Liege in three adjacent ponds led to the death of birds (Wirsing *et al.*, 1998). Certain of the toxic cyanobacteria such as *Oscillatoria*, *Phormidium* and *Planktothrix* never form scums but the formation of dense mats of *Oscillatoria* and *Phormidium* in oligotrophic, alpine waters of Switzerland with hepatotoxic and neurotoxic effects have been reported (Mez *et al.*, 1997). The occurrence of HABs of *Planktothrix*, *Microcystis*, *Anabaena* and *Aphanizomenon*, known producers of MCs and anatoxin-a, have been detected in many of the water bodies in Germany (Hummert *et al.*, 2001; Wiedner *et al.*, 2001). Czech recreational and drinking water bodies supported the development of HABs containing *Microcystis* spp., *P. agardhii* and *Aph. flos-aquae* that produced MCs (Marsalek *et al.*, 2001). Three eutrophic Latvian lakes (Lakes Mazais, Lielais Balterzers and Sekitis) suffered the development of toxic *M. aeruginosa*, *Aph. flos-aquae* and *Anabaena flos-aquae* resulting in the deterioration of water quality (Eynard *et al.*, 2000). Likewise, in northeastern Slovene freshwaters the occurrence of *M. aeruginosa* blooms producing MCs has been reported (Sedmak and Kosi, 1997). In Scotland, Lake Loch Leven developed hepatotoxic blooms of *M. aeruginosa* and *A. flos-aquae* persistently year after year leading to the death of several hundreds of fish that accumulated on the shores (Codd *et al.*, 1995). The death of sheep and dogs and reports of human illness due to toxic blooms of *Microcystis* in various water bodies of UK have come to light (Dow and Swoboda, 2000; WHO, 2003). The severity of the bloom of *Aph. flos-aquae* in a lake of Surrey is depicted in Fig. 26. Northwest Atlantic, Orkney Islands, east coast of Scotland and the North and Baltic Sea developed HABs of *No. spumigena*. However, coastal zones supported the development other toxic cyanobcterial communities that influenced the toxin profiles and toxicity (Dahlmann *et al.*, 2005). Halinen *et al.* (2007) provided evidence for the occurrence of late summer blooms of *No. spumigena* and *Aph. flos-aquae* in the Baltic Sea. Toxic strains isolated from open-sea sampling stations and coastal Gulf of Finland revealed great genetic heterogeneity based on the sequencing of 16S rRNA. The blooms supported the production of many MC variants, i.e. MC-LR, [D-Asp^3] MC-LR, [D-Asp^3] MC-HtyR,, MC-HtyR, [D-Asp^3, Dha^7] MC-HtyR and [Dha^7] MC-HtyR (Halinen *et al.* 2007). A quantitative real-time PCR was applied to 120 samples from the Baltic Sea to amplify *ndaF* gene that encodes subunit F of nodularin synthetase of *No. spumigena*. The distribution of the cyanobacterium was found horizontally as well as vertically extending up to a depth of 30 m. The frequency distribution reflected by 30 *ndaF* gene copies ml^{-1} corresponding to 1–3 pg nodularin ml^{-1} water (Koskenniemi *et al.*, 2007).

Studies conducted in Sweden (over a period of six years from 1991 to 1997) revealed the development of HABs of *M. aeruginosa*, *M. viridis* and *P. prolifica* (Willen *et al.*, 2000). The development of intoxication in pets and in 121 human beings due to *P. agardhii* bloom by the production of MCs has been reported (WHO, 2003). HABs of species of *Anabaena*, *Microcystis* and *Oscillatoria* producing MCs have been recorded from Norwich water bodies (Utkilen *et al.*, 2001). Majority of the water bodies of Denmark also supported extensive development of HABs comprising of MC producers i.e., *Microcystis* spp., *Anabaena* spp., *P. agardhii* and *Aph. flos-aquae* (Henriksen, 2001). Hepatotoxic

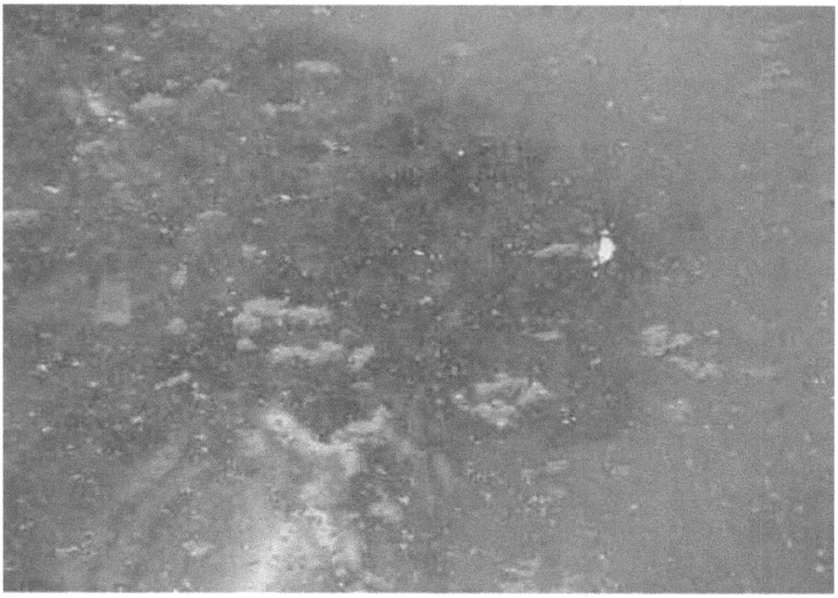

Figure 26: Bloom of *Aphanizomenon flos-aquae* in a lake in Surrey, UK. Picture courtesy John Burns (Cyanolab), Mark Schneegurt (Wichita State University) and Cyanosite (www.cyanosite.bio.purdue.edu).

Color image of this figure appears in the color plate section at the end of the book.

blooms of *A. flos-aquae* and *P. agardhii* were shown to be responsible for the death of fish and a cow, respectively in Denmark (Henriksen, 2001).

Three Turkish lakes Sapanca, Iznik and Taskisi were analysed for the contents of MCs in relation to the depth profiles. MCs were detected at depths of 15 and 25 m at concentrations of 3.65 μg L^{-1} MC-LR equivalents in Lake Sapanca where the surface waters did not contain any MCs. On the other hand, Lake Taskisi water samples revealed two unidentified MCs equal to a concentration of 2.43 μg L^{-1} MC-LR equivalents (Albay *et al.*, 2003). The occurrence of toxic blooms of *M. aeruginosa* in the Kucukcekmece lagoon (Istanbul, Turkey), the concentration of MC in waters and biomass of the bloom, factors affecting bloom of *M. aeruginosa* have been described for the first time over a period of three years (Albay *et al.*, 2005). The MC content of filtered water samples ranged between 0.06 and 24.2 μg L^{-1} MC-LR equivalents. A temperature of 24°C and 28.5°C favoured the highest MC concentration (>3 μg L^{-1}) and the highest biomass (>30 mg L^{-1}), respectively corresponding with a total N:P ratio of 7:1. The highest concentration of *M. aeruginosa* biomass was 173 mg L^{-1} and MC content of 24.2 μg L^{-1} MC-LR equivalents and highest salinity (8.8%) were measured concurrently in the lagoon. Of the 79 pond waters tested, 39 of them showed the growth of HABs with varying abundance of *Microcystis, Planktothrix* and *Anabaena*. The total MC concentrations ranged between <0.1 μg L^{-1} and up to 1000 μg L^{-1}. More than half samples contained MC levels of >10 μg L^{-1} that is ten times more than the permissible limit prescribed by WHO posing imminent danger for human health.

Comparative population dynamics of two *Planktothrix*-dominated Dutch freshwater lakes, i.e. Lake Tjeukemeer and Lake Klinckenberg revealed that in the former the concentration of MCs persisted throughout summer (with maximum concentrations of up to 30 μg L^{-1}) with *P. agardhii* as the dominant species whereas in the latter the maximum MC content of 140 μg L^{-1} persisted for a short duration with dominance of *P. rubescens* (Janse *et al.*, 2005). Spree, Dahme and Havel are three lowland rivers of Main Berlin waterways that are connected to some shallow lakes. The surface waters of this ecosystem are measured for concentrations of MCs (both cellular and extracellular), as measured by ELISA assay, varied between 0.14 and 119 μg L^{-1}. Nearly 74% of the samples showed MC values above 1 μg L^{-1} whereas 38% of the samples revealed MC content of 10 μg L^{-1}. Nearly 2% of the samples exhibited MC values of 100 μg L^{-1}. On the other hand, anatoxin-a levels of two Berlin Lakes Unterhavel and Mueggelsee were 0.01 μg L^{-1} (Fromme *et al.*, 2000).

The periodicity, species composition and MC concentration in 18 eutrophic Czech reservoirs were studied over a period of two years. The prevalence of toxic cyanobacterial blooms increased in the month of August with a shift in species predominance to *Microcystis* spp. 90% of the samples contained MCs and three MC variants (MC-LR, MC-RR and MC-YR) have been detected. The concentration of MCs was positively correlated with chlorophyll *a* and cyanobacterial biovolume. This study signifies the potential health risk to human population and the need to regularly monitor the development of toxic blooms (Znachor *et al.*, 2006). A survey conducted over a period of 13 years (1987 to 2000) in the Greek freshwater lakes, toxic blooms of *Microcystis, Anabaena, Anabaenopsis, Aphanizomenon* and *Cylindrospermopsis* were represented in 9 out of 33 lakes. The concentration of MC ranged from 50.3 to 1638 ± 464 μg g^{-1}. Lake Kastorea was chosen for studying seasonal variations of the cyanobacteria and the biovolume of the cyanobacteria was quite high (11 μL L^{-1}) throughout the year that was in excess of the levels proposed by WHO for recreational and drinking waters. Intracellular MC-LR concentrations of the blooms reached as high as 3186 μg L^{-1} during September to November thus posing a very high risk to human beings (Cook *et al.*, 2004). Lakes Lysimachia and Trichoonis in Greece supported the development of toxic blooms of *Aph. ovalisporum* constituting 99 and 58% of the total biomass, respectively. The concentration of MC-LR in Lake Lysimachia was 0.9 μg L^{-1} (Gkelis *et al.*, 2005). Strains of *Aph. ovalisporum* from Israel and Australia are reported to produce CYN (Banker *et al.*, 1997; Shaw *et al.*, 1999). MC content of *Microcystis* blooms from Belgium was 556 μg g^{-1} dry weight (Wirsing *et al.*, 1998). A taxonomic and morphological description of 21 cyanobacterial blooms from Belgium were presented along with their geographical distribution (Willame and Hoffmann, 1999). Blooms of *Microcystis* and *Planktothrix* were represented in almost all the 64 samples collected from southern part of Belgium and in Luxemberg as well as in bordering northeastern France followed by the genera *Anabaena, Woronichinia* and *Aphanizomenon*. MCs were found in 53% of the blooms and were mainly due to the preponderance of *Microcystis* but highest concentration of 2331 μg g^{-1} dry weight was found in a sample dominated by *Woronichinia naegeliana* (Fig. 2F; Willame *et al.*, 2005).

The genetic diversity of 107 *Microcystis* colonies (seven morphospecies) from 15 European Lakes was characterized on the basis of 16S rRNA gene ITS diversity. These studies eanabled the recognition of 59 distinct classes and suggested that there exists a significant genetic variation within and across habitats. This is supported by the fact that 93% of the population of *M. aeruginosa* consisted of two distinct classes whereas 24% of the population exhibited 14 classes found in two different lakes with one group being represented in lake across Czech Republic, Germany, Italy and Scotland. Thus the genetic heterogeniety of *M. aeruginosa* may have major ecological impact (Janse *et al.*, 2004).

Toxic cyanobacterial blooms of *Microcystis, Aphanizomenon* and *Anabaena* have been recorded along the Guadiana River (southwestern Spain) at different sites where the MC concentrations ranged from 0.10–21.86 MC-LR equivalent L^{-1}. Lakes Lysimachia and Trichonis in Greece supported the development of toxic blooms of *Aph. ovalisporum* constituting 99 and 58% of the total biomass, respectively. The concentration of MC-LR in Lake Lysimachia was 0.9 μg MC-LR equivalent L^{-1} in a site (Valdelacalzada) and 0.1 to 11.3 μg MC-LR equivalent L^{-1} at another site (Vitonogales). These results have been corraborated by the production of MC by the cultured isolates. *M. aeruginosa* thus exhibited variability in MC content either from different blooms or from the same bloom sample (Moreno *et al.*, 2004). Toxic blooms have been frequently reported from Portugal particularly in some of the rivers that flow from Spain (Ferreira *et al.*, 2001; Vasconcelos, 1993; Rocha *et al.*, 2002). Different strains of *Oscillatoria* and *Phormidium* producing anatoxin-a or homoanatoxin-a caused dog deaths (during 2002, 2003 and 2005) on the banks of Tarn River, France and pure cultures of eight strains produced these neurotoxins (Cadel-Six *et al.*, 2007).

A. circinalis is a potential health hazard because of its distribution in freshwaters across Europe, North America, Asia, South Africa, Japan, New Zealand and Australia (Baker, 1992). Moreover, there is a geographical segregation in the type of toxins produced by this bacterium. For example, strains from America and Europe produce only anatoxin-a (Schwimmer and Schwimmer, 1964; Gibson and Smith, 1982) whereas Australian strains exclusively produce STXs (Carmichael, 1992; Humpage *et al.*, 1994).

No. spumigena is wide spread in its occurrence within estuaries and coastal lagoons of Australia (Jones *et al.*, 1994; Heresztyn and Nicholson, 1997), Baltic Sea (Sivonen *et al.*, 1989b), German North Sea Coast (Nehring, 1993), New Zealand (Carmichael *et al.*, 1988b) and North America (Galat *et al.*, 1990). Death of domestic animals upon consumption of bloom material of *No. spumigena* due to massive liver failure has been reported from Australia, the Baltic Sea and New Zealand (Francis, 1878; Carmichael *et al.*, 1988b; Nehring, 1993; Ressom *et al.*, 1994).

VIII. METHODS OF CYANOBACTERIAL TOXIN REMOVAL

Cyanobacterial toxins are generally intracellular but due to environmental conditions or lysis of cyanobacterial cells, the toxins tend to become extracellular and thus pose a great health risk to human beings. For example, CYN can exist in extracellular state up to six weeks even after the complete disappearance of the bloom of *C. raciborskii* (McGregor and Fabbro, 2000). A number of methods have been proposed for intracellular and extracellular toxin removal. This subject has been reviewed (Hitzfeld *et al.*, 2000).

***i) Intracelluar toxin removal*:** There are two methods that are generally employed. The first is slow sand filtration and the second is the membrane filtration. Both these methods are practised to remove toxic cells from the waters by gently subjecting them to be separated from the water either by sand filtration or membrane filtration. Slow sand filtration is cheap and can be used in large scale. This generally enables filtration to be achieved gradually and slowly without subjecting any pressure on the cells. A sand bed develops a biofilm that allows biological degradation of dissolved substances. At times overloading of cells can limit the effectiveness of sand filtration (Chorus and Bartram, 1999). Instead of sand filtration, riverbank filtration has been advocated in which the lake or river water is made to pass across the soil on the bank. This has been termed as bank filtration (Miller and Fallowfield, 2001). This is a cost effective and low maintenance technique. Batch adsorption experiments with clay, sand and loam soils revealed more adsorption of MC-LR and nodularin in clay soils. The soils with highest organic carbon content (2.9%) and the highest clay content (16.1%) were found to be most effective in removing these toxins completely within 10–16 days in 2/3 soils incubated in the dark at 20°C (Miller *et al.*, 2001a,b; Miller and Fallowfield, 2001). The degradation has been largely attributed to both adsorption as well as degradation brought about by bacteria. The German Federal Environmental Agency (UBA) conducted laboratory and field experiments as well as bank filtration studies. These infiltration and bank filtration processes are normally considered as safe treatments. However, UBA identified a number of limitations. These are lack of sufficient light and/nutrients, low temperature, low clay or silt, anoxic conditions and absene of clogging layer, i.e. indicative of the absence of bacteria that bring out decomposition (Gruetzmacher, 2003). Membrane filtration involved physical removal of bloom material via a semi-permeable membrane. It may cause some damage to cells leading to release of toxins into waters. Although the content of toxins in waters is not increased (Chorus and Bartram, 1999), filtration can not remove extracellular toxins. Coagulation/flocculation and clarification are the other conventional treatment methods that are practised to remove cyanobacterial cells but are ineffective in removing extracellular toxins. Ferric chloride and aluminium sulphate are used as coagulating/flocculating agents. Optimization of doses and the pH are important (James and Fawell, 1991; Mouchet and Bonnelye, 1998). It has generally been observed that at times coagulation may also cause the release of toxins from the cyanobacterial cells (James and Fawell, 1991). The introduction of air bubbles into the system following a flocculation stage makes the flocs to float to the surface. This is termed as dissolved air floatation (DAF). This has an added avantage because cells can be removed easily and there is no danger of the cells getting lysed. Different cyanobacterial species behave differently for DAF treatment. As for example, *Microcystis* was removed by 40–80%, *Anabaena* by 90–100% but *Planktothrix* only by 30% in a Belgian DAF plant (Drikas and Hrudey, 1994). A combination of these methods with other techniques seems to be important in the conventional treatments.

***ii) Extracellular toxin removal*:** A number of physicochemical methods have been proposed for the removal of extracellular toxins from time to time. Adsorption to activated carbon, electrochemical degradation, photolysis, reverse osmosis and ultrasonication constitute the physical methods. Chemical treatment methods involve the use of chlorination, ozonation, permanganate and hydrogen peroxide.

A) Physical treatment methods for the removal of toxins: ***(i) Activated carbon*:** Carbon industry utilizes a number of carbonaceous precursors such as peat, bituminous and lignite coal, wood and coconut shell for preparing activated carbon (Wigmans, 1989). Apart from these, some other precursors include almond shells, olive stones, apricot stones, rice husks, cedar nutshells and corncobs. The purpose of activation is to create a porous structure that can readily adsorb pollutants. Three types of pores are generally recognized on the basis of their diameter, i.e. micropores (<20 Å), mesopores (20–500 Å) and macropores (>500 Å). Activated carbon is prepared by either physical activation or chemical activation. Physical activation involves two steps. In the first step pyrolysis is allowed in an inert environment at elevated temperatures of 650 to 850°C followed by oxidation with steam and/or carbon dioxide at silimar temperatures (Wigmans, 1989; Mazyck and Cannon, 2000). According to Wigmans (1989), carbon dioxide increases mesopores whereas steam creates a microporous structure. On the other hand, chemical activation does not require intensive heat and energy treatment but activation is achieved by treatment with zinc chloride, phosphoric acid or potassium hydroxide. Though chemical activation renders impressive porous structures with large surface areas, the resulting carbon has an acidic surface that hinders adsorption (Karanfil *et al.*, 1999). Powdered activated carbon (PAC) and granulated activated carbon (GAC) are the two sources generally employed.

In order to create a desired adsorbent, carbon activation is tailored by adding a catalyst like calcium. But in the long run, the deposition of calcium within the pores leads to a detrimental effect during reactivation procedures by leading to pore widening. To overcome the problems of calcium deposition, a reactivation process by the use of steam at 375°C followed by ramped-nitrogen treatment was advocatd by Mazyck and Cannon (2000, 2002) that can yield reactivated GAC similar to that of virgin GAC. PAC is quite costly and additional cost requirements for its removal and disposal add to its cost. The beneficial aspects of GAC are its long life, higher adsorptive capacity, the ease of operation, more efficient use of carbon, and the ability to regenerate the carbon for reuse.

PAC and GAC from different starting materials *viz.*, wood, coal, peat, coconut have been tested for toxin removal studies. PAC is useful in targeted treatments whereas GAC can be used continuously (Chorus and Bartram, 1999; Griffiths and Saker, 2003; Svreck and Smith, 2004). Of a number of carbon sources tested, wood-based GAC and PAC were more effective as MC adsorbents. Extracts from *M. aeruginosa* and *O. tenuis*, containing 2 mg L^{-1} MC were allowed to react with activated carbon source (100–500 mg L^{-1}) for seven days at 22 ± 1°C. PAC was more efficient than GAC in removing MCs. MC variants MC-RR, MC-YR and MC-WR from *M. aeruginosa* extracts adsorbed more significantly than MC-LR and MC-LHar from *O. tenuis* (Mohamed *et al.*, 1998). The surface characteristics of the carbon source appear to be more important in sorption performance rather than the source from which the PAC and GAC are derived. The removal of MCs is limited to a small percentage by many of the PACs used at a dosing rate of 1 kg m^{-3} with at least 50 kg m^{-3} required for effective removal of the toxins. Depending on the physical characteristics of PAC, the adsorption levels of MC-LR is reported to be between 20 to 280 µg mg^{-1} (Falconer *et al.*, 1983). The dominant factor influencing adsorption appears to be the mesopore range. Adsorption of MC-LR was found to be greatest at estimated diameter of 1.2 to 2.6 nm. Available surface area and presence of organic matter in the waters to be treated are the two important factors governing toxin removal. Toxin removal from pure water and water loaded with organic matter very much differed due to competition between binding sites.

The use of GAC filtration beds is increasingly becoming more common in modern treatment methods. However, the forbidding cost and the efforts to recharge are becoming impediments for the extensive use of GAC beds. GAC columns with 70 g of GAC (from different sources) could remove toxins to the extent of 50% only after passing 10 dm^3 water. The adsorption capacity of used GAC was considerably reduced when compared to unused GAC of the same type.

Another important aspect of cleaning of drinking waters is the treatment for removal of taste and odour problems created by the presence of compounds 2-methylisoborneol (2-MIB) and geosmin that are prevalent in HABs. The removal of these substances either by oxidation (treatment with chlorine or ozone) or by adsorption (PAC filter) has been attempted. 2-MIB has an effective molecular diameter of 6 Å and micropores of activated carbon are the best sites for effective adsorption as the compound is able to create more permanent adsorption sites (Pendleton *et al.*, 1997; Newcombe *et al.*, 1997). Of the two substances, the adsorption efficiency of geosmin to PAC was superior to 2-MIB. The removal efficiencies by PAC filter/adsorber varied from 25.7% to 88.4%. The remaining concentrations of these substances in the finished waters were below 30 ng L^{-1}. However, removal by oxidation was very weak but higher doses of ozone (3.8 mg L^{-1}) showed higher removal efficiency (84.8% for 2-MIB) with a contact time of 6.4 min (Jung *et al.*, 2005).

***ii) Electrochemical degradation*:** The treatment of various industrial wastewaters by electrochemical process for cleaning up has generated lot of interst. The electrochemical approach has been applied to dyeing wastewater (Vlyssides and Israilides, 1998), wastewater containing phenol (Chin and Cheng, 1985; Smith and Watkinson, 1981) or benzene derivatives (Comninellis and Plattnet, 1995). The effective removal of algal cells from pond water by electrochemical method (Feng *et al.*, 2003), prompted Feng *et al.* (2005) to apply this method for the degradation of MC-LR. The oxidation rate of MC-LR increased with the current by the hydroxyl radicals produced and the results are comparable to photocatalytic oxidation with no byproducts detected by HPLC. MC-LR (1.0 mg L^{-1}) was oxidized completely at an oxidation rate of 0.219 min^{-1} at an applied current of 100 mA with a half-life period of 2.5 min. Thus electrochemical method appears to be a promising approach.

iii) Photolysis: MCs are very stable under natural sunlight (Tsuji *et al.*, 1994a). On the other hand, Mazur and Plinski (2001) observed that the concentration of MC-LR, MC-RR and nodularin decreased by 30% over a period of 21 days when incubated at 17 ± 1°C in light (40 µmol photon $m^{-2} s^{-1}$). The decomposition or degradation of hepatotoxins by treatment with UV has been studied. Upon UV-irradiation, the decomposition of MCs proceeded by isomerization and the products tended to be non-toxic (Tsuji *et al.*, 1995). Photolytic and photocatalytic processes are recognized. Photolytic processes involve treatment with ozone and/or hydrogen peroxide in presence of UV light during which slow and reactive hydroxyl radicals are generated. Photocatalytic processes comprise oxidation in presence of oxides with semi-conductive properties such as TiO_2, ZnO, Fe_2O_3, WO_3 and CdSe (Fox and Dulay, 1993; Legrini *et al.*, 1993). Of these, TiO_2 has been widely used because it is chemically passive and stable over wide range of pH values. Radiations within range of 300 to 400 nm are absorbed by TiO_2 leading to the generation of hydroxyl radicals that have the highest oxidation potential than ozone or chlorine. This has opened up another area known as Advanced Oxidation Technologies (AOTs) and Membrane Separation Technologies (MSTs) that received great attention for environmental remediation. AOTs are highly effective both functionally and cost-wise (Fujishima *et al.*, 1999). The illumination (330–450 nm from a xenon UV lamp) of MC-LR (in 50 to 200 µM range) in presence of a catalyst like TiO_2 (1% m/v solution) caused photocatalytic destruction within a period of 10–40 minutes depending on the initial concentration. This method seems to be quite useful for rapid elimination of hepatotoxins from potable waters (Robertson *et al.*, 1997). A thin layer of TiO_2 deposited on glass not only caused the destruction of extracellular toxins but also proved to be lethal to toxic bacterial cells (*E. coli* within 2 h) and inhibited the growth of green algal cells (to the extent of 60% of growth of *Oedogonium*) (Makowski and Wardas, 2001). Although photocatalytic process is quite effective in eliminating extremely high concentrations of MC-LR, a variety of byproducts are generated. Thus Liu *et al.* (2002) suggested that TiO_2 photocatalysis has to be combined with treatment with H_2O_2. Since TiO_2 particles in suspension are difficult to be removed after their application, attempts have been made to prepare thin films and membranes by immobilizing TiO_2 (Balasubramanian *et al.*, 2003). Kinetic modelling studies of Feitz and Waite (2003), on TiO_2-catalyzed photodegradation, revealed that the strong adsorption of MC-LR to semiconductor surface sites is followed by a reaction between surface located organic radicals leading to the degradation of the toxin. The formation of organic radicals is explained on the basis of interaction of the desorbed surface bound superoxide with organic substances generating organic peroxy radicals. Further, the inhibition of the formation of organic radicals at alkaline pH due to the formation and deposition of carbonate on the surface prevents degradation of MC-LR. Choi *et al.* (2006) fabricated nanostructured TiO_2 films and membranes that have multiple applications such as decomposition of recalcitrant organic pollutants, destruction of biological toxins, inactivation and killing of pathogenic microorganisms, physical separation of contaminants and intermediate products and self-antibiofouling action.

iv) Reverse osmosis: A process that is generally used in desalination of brackish or estuarine waters, reverse osmosis has also been put to use for the removal of toxins. Reverse osmosis membranes gave a retention rate of 95% for MC-LR and MC-RR. These were also effective in retaining the nodularins The retention of both salt and the toxins from waters by reverse osmosis renders almost clean water for drinking purposes. Reverse osmosis does not help in the destruction or decomposition of toxins but only helps in their separation from water. The retentate has to be disposed off very carefully as it might contaminate other sources of water.

***v) Ultrasonication*:** Ultrasonic radiation inhibited growth of *Microcystis* and also did not contibute to the increase of the extracellular MC concentration. This suggests that ultrasonic irradiation shorter than 5 min did not lead to lysis of cells. Further, effective degradation of MCs occurred at 150 KHz and 30 W for 20 min sonication resulting in a removal rate of 70% (Ma *et al.*, 2005). However, it appears for complete degradation higher frequencies of sonication are required as exemplified by the studies of Song *et al.* (2005). Rapid degradation of MC-LR by ultrasonic irradiation at 640 KHz resulted in degradation products that did not exhibit any biological activity as assessed by PP1 inhibition assay. However, a concentration range of 0.03 to 3.0 µM and the presence of hydroxy radicals appeared to be important factors for degradation besides other processes of hydrolysis and pyrolysis (Song *et al.*, 2005).

(B) Chemical treatment methods for toxin removal: The utility of a number of oxidizing agents have been tried for the removal of cyanobacterial toxins. The chemical oxidants can be placed in increasing order of their oxidizing potential as: chlorine dioxide> chlorine>hypochlorous acid>permanganate>perhydroxy radical>hydrogen peroxide>ozone> hydroxy radical. Thus the hydroxy radical has the highest oxidizing potential of 2.80 Volts while chlorine dioxide has the lowest of 1.28 Volts. Although treatment time, concentration, pH and removal efficiency of various oxidizing chemicals have been worked out, the lacuna lies in the lack of characterization of decomposition products. The toxicity testing has also been limited to mouse bioassay. Since most of the oxidizing agents react with unsaturated bonds present in Adda moiety of MC, any change in its characteristics will lead to a lowering of its UV absorption at 238 nm leading to a notion that the toxin has been decomposed. In the absence of characterization of degradation products with reference to their quantitative determination and toxicity testing no oxidizing agent seems to serve the purpose. However, a lot of knowledge has been generated on the possible removal of MCs by chlorination, hydrogen peroxide, potassium permanganate and ozonation.

***i) Chlorination*:** Chlorine readily dissolves in water by forming hypochlorous acid. Depending on pH, the hypochlorous acid dissociates forming hypochlorite ions. This reaction is initiated at a pH of 5.0 and 100% dissociation takes place above pH 10.0. The efficacy of chlorination thus depends on the formation of hypochlorous acid. It largely depends on the chloride compound and concentration used. Water is exposed to varying doses of chlorine for 30 min at an elevated pH. More than 95% of MCs and nodularin can be removed by concentrations greater than 1 mg L^{-1} of aqueous chlorine and calcium hypochlorite whereas sodium hypochlorite and chloramines at the same dose can remove 40–80% of these toxins (Nicholson *et al.*, 1993, 1994). It is effective in removing most of the cyanobacterial toxins except anatoxin-a. On the other hand, CYN (at concentrations 20–24 µg L^{-1}) was effectively oxidized by 4 mg L^{-1} chlorine at pH 7.2–7.4 (Nicholson *et al.*, 1993). Though chlorination is practised very widely in USA for removing dissoloved algal toxins, chlorination byproducts have not been characterized which may also be toxic (Svreck and Smith, 2004). Chlorination can cause lysis of cyanobacterial cells and there is a chance of the toxin content increasing if applied in the initial stages. It is advisable that the cells are first removed by coagulation or sedimentation or flocculation. Chlorination thus can be best applied as a terminal step of the treatment process and this will also reduce the demand for higher levels of chlorine (Chorus and Bartram, 1999). Oxidation of MC-LR into non-toxic dihydroxyisomers of MC-LR by treatment with chlorine dioxide (ClO_2) was monitored by UV-spectroscopy and HPLC. The characterization of reaction products was performed with mass spectrometry analysis where toxicity was determined by PP1A inhibition assay. But the overall rate constant (k) for the reaction suggested that ClO_2 is not a suitable oxidant for the degradation of MCs in drinking water treatment processes (Kull *et al.*, 2004). Subsequently, they have suggested that the oxidation of MCs resulted in the permitted concentrations as per guidelines of WHO and also chloride at its normal concentration could be effectively used for the detoxification of MCs. Thus, the oxidation of MC-LR (at 10 μg L^{-1}) by ClO_2 (1 mg L^{-1}) led to the formation of non-toxic oxidation products, dihydroxyisomers of MC-LR. The presence of organic matter enabled ClO_2 to be consumed more rapidly by fulvic and humic acids, leaving a residual ClO_2 that could oxidize MC-LR. PP1 inhibition was observed by intact MC-LR while the oxidation products did not inhibit the activity of PP1 (Kull *et al.*, 2006). Electrogenerated active chlorine efficiently decomposed MC-RR and MC-LR within 15 min electrolysis under a current density of 8.89 mA cm^{-2} at 20°C and pH 7.00. The removal rates of MCs increased with increasing concentrations of chloride. The inactivation kinetics of MC-LR by free chlorine revealed that MC-LR could be inactivated more readily at pH 6.0 which decreased with increase of pH to 9.0 (Xagoraraki *et al.*, 2006).

ii) Hydrogen peroxide: Though hydrogen peroxide has been applied for the oxidation of phenolic wastewater and for the treatment of organic wastewater, its utility for treatment of toxin removal has been very restricted. Irradiation with UV light increases the efficacy of hydrogen peroxide since it dissociates to form two hydroxy radicals. However, there are no reports on treatment of hydrogen peroxide in combination with UV light for the removal of MCs from drinking water supplies. Treatment with hydrogen peroxide is not very effective as 17% of MC-LR was decomposed in 60 minutes with a 20 mg dm^{-3} solution of hydrogen peroxide (Drikas, 1994). Rositano *et al.* (1998) used 2 mg dm^{-3} hydrogen peroxide to treat 1 mg dm^{-3} solution of MC-LR. Though no toxin decomposition was noted within the first 10 min, combination with ozone, however, abolished the toxicity within 30 s.

iii) Permanganate: It generally attacks functional groups with multiple bonds. So it can be effective in decomposing MCs by attacking the unsaturated bonds in the Adda moiety. Rositano *et al.* (1998) reported that 1 mg dm^{-3} of potassium permanganate removed 95% of of 200 μg dm^{-3} of MC-LR in 30 min. It compared very well with the oxidation potential of chlorine but due to its higher oxidation potential it removed MCs much more rapidly than chlorine. Much more needs to be done with regard to its decomposition products, their toxicity and optimization parameters for MC destruction.

iv) Ozonation: A process that has generally been used for disinfecting purposes or to remove colour and/or odour of the waters in Europe and North America, ozonation has occupied a significant place in water treatment plants for the removal of cyanobacterial toxins. A two-stage ozonation water treatment is suggested that can be either with pre- and inter-ozonation, inter- and post-ozonation, or pre- and post-ozonation (Langlais *et al.*, 1991). Though early studies suggested the removal of MCs by a 2 s ozonation of waters, details regarding concentrations of MCs or ozone employed are not available (James and Fawell, 1991). Subsequent studies revealed that 800 μg L^{-1} of MC-LR could be oxidized to below the detectable limits of HPLC by concentrations of ozone lower than 0.2 mg L^{-1} within seconds to minutes whereas 88 μg L^{-1} nodularin was completely inactivated by 0.05 mg L^{-1} of ozone within 15 s (Nicholson *et al.*, 1993; Rositano and Nicholson, 1994; Rositano *et al.*, 1998). These studies indicate that the removal of MCs is directly proportional to the concentration of ozone and there exists a residual concentration of ozone after its demand has been met. Further, the efficacy of ozone is reduced by the presence of organic matter and alkalinity. Hitzfeld *et al.* (2000) have summarized the observations of various workers. In these studies the concentration of

MC-LR ranged in between 9 to 500 μg L^{-1} that required a dose of 0.2 to 1.5 mg L^{-1} ozone. However, it must be stated that the duration also ranged in between 5 to 30 min. Hart and Stott (1993) used 1 mg L^{-1} of ozone and a contact of 5 min to remove 9 μg L^{-1} of MC-LR with 50% efficiency in presence of organic matter. Bernazeau (1994) demonstrated that the demand for ozone is reduced in the absence of organic matter (0.2 mg L^{-1} ozone removed 500 μg L^{-1} of MC-LR with a contact time of 4 min with 99% efficiency) whereas in presence of organic matter at the same concentration of MC-LR the concentration of ozone (0.5 mg L^{-1}) and contact time (10 min) were increased with a reduction in efficiency by nearly 50%. The removal of MC-LR (nearly 200 μg L^{-1}) to the extent of 100% was also reported in two separate studies where ozone at a dose of 1.0 mg L^{-1} required a contact time of 5 min (Nicholson *et al.*, 1993; Rositano *et al.*, 1998). The experimental materials differed in various studies. Natural cyanobacterial bloom material or their extracts or lyophilized materials were used in toxin removal experiments. The removal of toxins was much easier with ozone treatment in the first two cases rather than with lyophilized materials. Pre-ozonation caused improved coagulation of material and has been thus widely used (Langlais *et al.*, 1991). Besides the inherent dangers of cell lysis, pre-ozonation does not meet the complete requirements since a second post-ozonation step is required with a higher dose of ozone to oxidize the excess of organic matter and toxins present. Thus post-ozonation is recommended as an advanced treatment process. Since ozone also causes lysis of cyanobacterial cells it is more important to practise ozonation at the end of treatment process (Chorus and Bartram, 1999; Haider *et al.*, 2003; Svreck and Smith, 2004).

Studies involving oxidation of anatoxin-a, anatoxin-a(s) and STXs by ozone are few. The concentration of ozone required was very much higher in raw water (4.5 mg L^{-1}) with no residual ozone but resulted in a decrease of anatoxin-a from 2.4 μg L^{-1} to 0.6 μg L^{-1} whereas in filtered water in the absence of organic material the dose of ozone decreased to 2.2 mg L^{-1} with the removal of anatoxin level to 0.3 μg L^{-1} without any residual ozone. Even higher doses of ozone (4.2 mg $min.^{-1}$ for 15 min.) were required for reducing the toxicity of *A. circinalis* extract to near the lethal threshold concentration (Rositano *et al.*, 1998). The oxidation of other PSTs such as GTX2, dcGTX2, dcGTX3, C1 and C2 by ozone is possible, yet the details pertaining to optimization parameters for destroying the neurotoxins have to be worked out (Nicholson and Rositano, 1997).

The oxidation efficiency of ozone is reduced at pH values greater than 7.5 due to a lower oxidation potential of ozone under these conditions. Knowledge on the byproducts of ozonation and their probable toxicity are yet to be assessed. In general, mouse bioassay was chosen in assessing the toxicity of residual toxins after ozonation and a more reliable sensitive assay such as PP-inhibition should be the choice in future studies involving the evaluation of toxicity of byproducts of ozonation.

IX. GUIDELINES FOR SAFER LIMITS OF CYANOTOXINS

In view of wide ranging genotoxic effects and carcinogenicity to human beings exerted by MCs, it was considered appropriate to take into consideration LOAEL or NOAEL from most suitable chronic or subchronic studies. A tolerable daily intake (TDI) for MC-LR is derived on the basis of body weight and the NOAEL for liver changes after a 13-week mouse bioassay conducted at the Water Research Centre in the United Kingdom (Fawell *et al.*, 1994). The TDI was thus fixed at 0.04 μg kg^{-1} body weight per day.

The maximum acceptable concentration (MAC) for MC-LR is calculated from the TDI and it was arrived at 1.5 μg L^{-1} per day.

Risk assessment of cyanobacterial toxins to human beings was considered in guidelines (WHO, 1998) but due to insufficient data on cyanobacterial toxins, guideline values for all these toxins could not be derived. However, a total guideline value for MC-LR (free plus cell bound) of 1 μg L^{-1} was set as a safer MAC (WHO, 2003). Based on the studies available on HABs in different countries, the levels of cyanobacterial toxins in potable waters and the overall risk assessment to human beings, now draft guidelines have been prepared by each country (Chorus, 2005). These guidelines cover safer limits of cyanobacterial toxins for potable and recreational waters. Variations from the general limit of 1 μg L^{-1} MC-LR equivalents have been proposed taking into consideration the average weight of the citizen and the proportion of TDI attributed to the consumption of water. In Australia, the guideline value was set at 1.3 μg L^{-1} MC-LR equivalents considering the average weight of a citizen as 60 kg and a proportion of daily intake at 0.9. In the absence of toxin monitoring data it was suggested to take into account the cell numbers of HABs. If toxic *M. aeruginosa* occurs in potable waters, 6,500 cells ml^{-1} is considered equivalent to 1.3 μg L^{-1} MC-LR equivalents (giving a toxic cell quota of 0.2 pg total MCs $cell^{-1}$). Potential health risk from recreational waters is considered a possibility when the cell number of toxic blooms would correspond to 20,000 cells ml^{-1}. Likewise, increasing risk to livestock health is likely when the concentration of MCs exceeds 2.3 μg L^{-1} MC-LR equivalents and a corresponding cell number of *Microcystis* population estimated to be 11,500 cells ml^{-1}. Although guidelines for fish and shellfish in Australia are not available, health alert levels for toxins in fishes, prawns and mussels in the state of Victoria, Australia have been proposed (Van Buynder *et al.*, 2001). The critical concentrations in seafood are determined for fishes (250 μg kg^{-1}), prawns (1,100 μg kg^{-1}) and mussels (1,500 μg kg^{-1}) taking into account TDI level for adults. Brazil adopted the standard value of 1 μg L^{-1} for MCs as mandatory and values of 3 μg L^{-1} for STXs and 15 μg L^{-1} for CYN were recommended guidelines. Besides these, a general alert is given when the toxic cyanobacterial cells exceed 20, 000 cell ml^{-1} and the application of algicides during water purification process is forbidden. A MAC of 1.5 μg L^{-1} for MC-LR in drinking water in Canada was adopted and a comprehensive plan is under preparation for including other MC variants. In New Zealand, the guideline values for MCs, nodularins and CYN are 1 μg L^{-1} whereas 6, 1, 2, and 3 μg L^{-1} have been prescribed for anatoxin-a, anatoxin-a(s), homoanatoxin and STXs, respectively. It is to be noted that except for MCs, WHO (2004) does not have guideline values for other cyanobacterial toxins. Countries such as Czech Republic, Finland, France, Germany, Hungary, Poland, South Africa and Spain adopted the WHO guideline value of 1 μg L^{-1} for MC-LR. Although there are no official guidelines for cyanobacterial toxins in countries like Belgium and Luxembourg, Denmark, Greece and Italy, depending on the requirements the WHO guideline of 1 μg L^{-1} for MC-LR is adopted. There are no federal regulatory guidelines for either cyanobacteria or their toxins in the USA but they have been included in the US Environmental Protection Agency "*Contaminant candidate list*" for further research to clarify the need for further regulation. However, several state or local authorities adopted a regulatory limit of 1 μg L^{-1} MC-LR equivalents.

X. PROTOCOLS FOR WATER TREATMENT

In order to provide potable water to the people every regulatory authority takes enough precautions during water treatment in treatment plants. Conventional water treatment plants during the 1960s and 1970s mainly relied on pre-chlorination and post-chlorination steps combined with the intervening steps of coagulation, flocculation, settling and sand filtration (Fig. 27). The inherent danger invoved in such treatment is that the pre-chlorination may cause lysis of the toxic blooms leading to an increase in the toxin content beyond the acceptable guidelines of WHO. In the absence of regular monitoring of surface waters for the development of blooms and if the number of toxic cyanobacterial cells exceeds the moderate adverse health effects to high risk level, then such treatments do not meet the standards and there is a risk of the population being exposed to the danger of these toxins. In order to overcome any such eventuality, water treatment lines advocated during 1990s relied mostly on pre-ozonation and post-ozonation steps followed by terminal disinfection by chlorine or chlorine dioxide with intervening steps of coagulation, flocculation, PAC filtration (Fig. 28). Sand filtration, disinfection by ozone, and a second stage filtration on GAC beds would be additional advantage for effective removal of the toxins. Even if the toxic cells are prone for lysis during preozonation, the removal of the toxins during PAC and sand filtration and the post-ozonation and final chlorination steps ensure complete removal of any traces of the toxins present.

Studies on the removal of cyanobacterial toxins from drinking waters for household purposes and small community systems have been made. Lawton *et al.* (1998) tested three domestic jug filtration units based on activated carbon and ion exchange resin. Approximately 60% of the filamentous cyanobacteria and 10% of the single cells of *Microcystis* could be removed with toxin removal rates ranging from 32 to 57% and the efficiency was increased to 88% by repeatedly passing the same water through the same filter. The variants of MCs tested were MC-LR, MC-LY and MC-LW. Laboratory-scale batch reactors were developed very much in a similar manner to the modern water works. Toxic cyanobacterial cells (cultures of *M. aeruginosa* or extracts of *P. rubescens* bloom samples) were

Raw Water

⇩

Prechlorination

⇩

Other Chemicals:

Coagulants

Flocculants (optional)

pH adjustment (when necessary)

Powdered Activated Carbon (Optional)

⇩

Coagulant-Flocculation-Settling

⇩

Sand Filtration

⇩

Post-chlorination

pH Adjustmeent

Remineralization (Optional)

⇩

To supply

Figure 27: Conventional treatment scheme applied to surface waters in the 60s and 70s

Raw Water

⇩

Preozonation

⇩

Remineralization (when necessary)

⇩

Other chemicals:

-Coagulants (all cases)

-Flocculants (optional)

-pH adjustment

Powdered Activated Carbon (optional)

⇩

Coagulant-Flocculation settlig or Flotation

⇩

First stage filtration on sand (or Dual-Media filter)

⇩

Disinfection by means of O_3

Oxidation by free radicals

⇩

Second stage Filtration on GAC

⇩

Final disinfection by Cl_2 or ClO_2

pH Adjustment (Optional)

Figure 28: Typical water treatment line for surface waters in the 90s.

subjected to ozonation followed by carbon and sand filtration steps. Ozone concentrations of 1.5 mg L^{-1} were required to destroy 5 x 10^5 *Microcystis* cells ml^{-1} (equivalent to total organic carbon of 1.56 mg L^{-1}) (Hoeger *et al.*, 2002). They suggested that for effective treatment of surface waters containing cyanobacterial blooms the regular monitoring of total organic carbon with cyanobacterial cell densities, ozonation combined with efficient filtration steps are required to provide safe drinking water. An immunoaffinity column (IAC) for the determination of MCs to the level of pg to µg L^{-1} in tap waters has been designed by Tsutsumi *et al.* (2000). The mean recovery level as determined by HPLC with UV detection or ELISA revealed MC-LR, MC-RR and MC-YR to be 91.8%, 77.3% and 86.4% respectively, in the range of 2.5–100 µg L^{-1}. The tap water cleaned up with IAC rendered effective elimination of the toxins when compared to octadecyl silanized catridge.

LITERATURE CITED

Aas, P., Eriksen, S., Kolderup, J., Lundy, P., Haugen, J.-E., Skulberg, O. M., and Fonnum, F. (1996) Enhancement of acetylcholine release by homoanatoxin-a from *Oscillatoria formosa*. *Environ Toxicol Pharmacol* **2:** 223–232.

Adelman, W. J., Fohlmeister, J. F., Sasner, J. J., and Ikawa, M. (1982) Sodium channels blocked by aphanotoxin obtained from the blue-green alga *Aphanizomenon flos-aquae*. *Toxicon* **20:** 513–516.

Adeyemo, O. M., and Siren, A.-L. (1992) Cardiorespiratory changes and mortality in the conscious rat induced by (+) and (±)-anatoxin-a. *Toxicon* **30:** 899–905.

Agrawal, M. K., Bagchi, D., and Bagchi, S. N. (2001) Acute inhibition of protease and suppression of growth in zooplankter, *Monia macrocopa*, by *Microcystis* blooms collected in Central India. *Hydrobiologia* **464:** 37–44.

Akcaalan, R., Young, F. M., Metcalf, J. S., Morrison, L. F., Albay, M., and Codd, G. A. (2006) Microcystin analysis in single filaments of *Planktothrix* spp. in laboratory cultures and environmental samples. *Water Res* **40:** 1583–1590.

Akin-Oriola, G. A., and Lawton, L. A. (2005) Detection and quantification of toxins in cultures of *Microcystis aeruginosa* (PCC 7820) by HPLC and protein phosphatase inhibition assay effect of blending various collectors at bulk. *Afr J Sci Technol* **6:** 1–10.

Alam, M., Ikawa, M., Sasner, J. J., Jr., and Sawyer, P. J. (1973) Purification of *Aphanizomenon flos-aquae* toxin and its chemical and physiological properties. *Toxicon* **11:** 65–72.

Albay, M., Akcaalan, R., Tufekci, H., Metcalf, J. S., Beattie, K. A., and Codd, G. A. (2003) Depth profiles of cyanobacterial hepatotoxins (microcystins) in three Turkish freshwater lakes. *Hydrobiologia* **505:** 89–95.

Albay, M., Matthiensen, A., and Codd, G. A. (2005) Occurrence of toxic blue-green algae in the Kuckcekmece lagoon (Istanbul, Turkey). *Environ Toxicol* **20:** 277–284.

Allender, C. J., LeCleir, G. R., Rinta-Kanto, J. M., Small, R. L., Stachwell, M. F., Boyer, G. L., and Wilhelm, S. W. (2009) Identifying the source of unkown microcystin genes and predicting microcystin variants by comparing genes with uncultured cyanobacterial cells. *Appl Environ Microbiol* **75:** 3598–3604.

Al-Tebrineh, J., Mihali, T. K., Pomati, T. and Neilan, B. A. (2010) Detection of saxitoxin-producing cyanobacteria and *Anabaena circinalis* in environmental water blooms by quantitative PCR. *Appl Environ Microbiol* **76:** 7836–7842. doi.10.1128/AEM.00174-10

Amorim, A., and Vasconcelos, V. (1999) Dynamics of microcystins in the mussel *Mytilus galloprovincialis*. *Toxicon* **37:** 1181–1185.

An, J., and Carmichael, W. W. (1994) Use of a colorimetric protein phosphatase inhibition assay and enzyme linked immunosorbent assay for the study of microcystins and nodularins. *Toxicon* **32:** 1495–1507.

Annila, A., Lehtimaeki, J., Manila, K., Eriksson, J. E., Sivonen, K., Rantala, T. T., and Drakenberg, T. (1996) Solution structure of nodularin an inhibitor of serine/threonine specific protein phosphatases. *J Biol Chem* **271:** 16695–16702.

Aráoz, R., Nghiem, H-O., Rippka, R., Palibroda, N., Tandeau de Marsac, N., and Herdman, M. (2005) Neurotoxins in axenic oscillatorian cyanobacteria: Coexistence of anatoxin-a and homoanatoxin-a determined by ligand-binding assay and GC/MS. *Microbiology* **151:** 1263–1273.

Aráoz, R., Molgó, J., and Tandeau de Marsac, N. (2010) Neurotoxic cyanobacterial toxins. *Toxicon* **56:** 813–828. doi:10.1016/j.toxicon.2009.07.036

Arment, A. R., and Carmichael, W. W. (1996) Evidence that microcystin is a thiotemplate product. *J Phycol* **32:** 591–597.

Aronstam, R. S., and Witkop, B. (1981) Anatoxin-a interactions with cholinergic synaptic molecules. *Proc Natl Acad Sci* USA **78:** 4639–4643.

Ash, C., MacKintosh, C., and MacKintosh (1995) Use of a protein phosphatase inhibition test for the detection of cyanobacterial toxins in water. *Water Sci Technol* **31 :** 51–53.

Astrachan, N. B., Archer, B. G., and Hilbelink, D. R. (1980) Evaluation of the sub-acute toxicity and teratogenicity of anatoxin-a. *Toxicon* **18:** 684–688.

Atkins, R., Rose, T., Brown, R. S., and Robb, M. (2001) The *Microcystis* cyanobacteria bloom in the Swan River-February 2000. *Water Sci Technol* **43:** 107–114.

Atkondon, M., and Abulquerque, E. X. (1995) Diversity of nicotinic acetylcholine receptors in rat hippocampal neurons. III. Agonist actions of the novel alkaloid epibatidine and analysis of type II current. *J Pharmacol Exp Ther* **274:** 771–782.

Azevedo, S. M. F. O. (1996) Current studies on toxic cyanobacteria in Brazil. *Reunião Anual da Federação de sociedades de Biologia Experimental*, 12 Caxambu, MG, 40p.

Azevedo, S. M. F. O., Evans, W. R., Carmichael, W. W., and Namikoshi, M. (1994) First report of *Microcystis* from a Brazilian isolate of the cyanobacterium *Microcystis aeruginosa*. *J Appl Phycol* **6:** 261–265.

Bagu, J. R., Skyes, B. D., Craig, M. M., and Holmes, C. F. B. (1997) A molecular basis for different interactions of marine toxins with protein phosphatase-1. *J Biol Chem* **272:** 5087–5097.

Baker, P. (1992) *Anabaena circinalis*. In: Tyler, P. (Ed.) *Identification of Common Noxious Cyanobacteria*. Part I. *Nostocales*. Melbourne Water Corporation, Melbourne, Australia.

Baker, P. D., and Humpage, A. R. (1994) Toxicity associated with commonly occurring cyanobacteria in surface waters of Murray-Darling basin, Australia. *Aus J Mar Freshwat Res* **45:** 773–786.

Balasubramanian, G., Dionysiou, D. D., Suidan, M. T., Subramanian, V., Baudin, I., and Laine, J. M. (2003) Titania powder modified sol-gel process for photocatalytic applications. *J Mat Sci* **38:** 823.

Ballot, A., Fastner, J., and Wiedner, C. (2010a) Paralytic shellfish poisoning toxin-producing cyanobacterium *Aphanizomenon gracile* in northeast Germany. *Appl Environ Microbiol* **76:** 1173–1180. doi:10.1128/AEM.02285-09

Ballot, A., Fastner, J., Lentz, M., and Wiedner, C. (2010b) First report of anatoxin-a producing cyanobacterium *Aphanizomenon issatschenkoi* in northeastern Germany. *Toxicon* **56:** 964–971.

Banker, R., Carmeli, S., and Hadas, O. (1997) Identification of cylindrospermopsin in *Aphanizomenon ovalisporum* (Cyanophyceae) isolated from Lake Kinneret, Israel. *J Phycol* **33:** 613–616.

Banker, R., Carmeli, S., Werman, M., Teltsch, B., Porat, R., and Sukenik, A. (2001) Uracil moiety is required for toxicity of the cyanobacterial hepatotoxin cylindrospermopsin. *J Toxicol Environ Health* **62:** 281–288.

Banker, R., Teltsch, B., Sukenik, A., and Carmeli, S. (2000) 7-Epicylindrospermopsin, a toxic minor metabolite of the cyanobacterium *Aphanizomenon ovalisporum* from lake Kinneret, Israel. *J Nat Prod* **63:** 387–389.

Barford, D., and Keller, J. (1994) Co-crystallization of the catalytic subunit of the serine/threonine specific protein phosphatase 1 from human in complex with microcystin-LR. *J Mol Biol* **235:** 763–766.

Bateman, K. P., Thibault, P., Douglas, D. J., and White, R. L. (1995) Mass spectral analyses of microcystins from toxic cyanobacteria using on-line chromatographic and electrophoretic separations. *J Chromatogr* **712:** 253–268.

Bauer, C., Scappino, L., and Haselkorn, R. (1993) Growth of the cyanobacterium *Anabaena* on molecular nitrogen: NifJ is required when iron is limited. *Proc Natl Acad Sci* USA **90:** 8812–8816.

Beasley, V. R., Lovell, R. A., Holmes, K. R., Walcott, H. E., Schaeffer, D. J., Hoffman, W. E., and Carmichael, W. W. (2000) Microcystin-LR decreases hepatic and renal perfusion, and causes circulatory shock, severe hypoglycemia, and terminal hyperkalemia in intravascular dosed swine. *J Toxicol Environ Health* **61:** 281–303.

Beattie, K. A., Kaya, K., and Codd, G. A. (2000) The cyanobacterium *Nodularia* PCC 7804, of freshwater origin, produces [L-Har^2] nodularin. *Phytochemistry* **54:** 57–61.

Beattie, K. A., Kaya, K., Sano, T., and Codd, G. A. (1998) Three dehydrobutyrine (Dhb)-containing microcystins from the cyanobacterium *Nostoc* sp. *Phytochemistry* **47:** 1289–1292.

Becker, S., Boeger, P., Oehlmann, R., and Ernst, A. (2000) PCR bias in ecological analysis: A case study for quantitative Taq nuclease assays in analyses of microbial communities. *Appl Environ Microbiol* **66:** 4486–4494.

Belknap, W. R., and Haselkorn, R. (1987) Cloning and light regulation of expression of the phycocyanin operon of the cyanobacterium *Anabaena*. *EMBO J* **6:** 871–884.

Bell, S. G., and Codd, G. A. (1996) Detection, analysis and risk assessment of cyanobacterial toxins. In: Hester, R. E., and Harrison, R. M. (Eds.) *Agricultural Chemicals and the Environment*. Issues in Environmental Science and Technology. No. 5. Royal Society of Chemistry, Cambridge, UK. pp. 109–122.

Benndorf, J., and Henning, M. (1989) *Daphnia* and toxic blooms of *Microcystis aeruginosa* in Bautzen reservoir (GDR). *Int Rev Hydrobiol* **74:** 233–248.

Berman, F. W., Gerwick, W. H., and Murray, T. F. (1999) Antillatoxin and kalkitoxin, ichthyotoxins from the tropical cyanobacterium *Lyngbya majuscula*, induce distinct temporal patterns of NMDA receptor-mediated neurotoxicity. *Toxicon* **37:** 1645–1648.

Bernard, C., Harvey, M., Briand, J. F., Bire, R., Krys, S., and Fontaine, J. J. (2003) Toxicological comparison of diverse *Cylindrospermopsis raciborskii* strains: Evidence of liver damage caused by a French *C. raciborskii* strain. *Environ Toxicol* **18:** 176–186.

Bernazeau, F. (1994) Can microcystins enter drinking water distribution systems? In: Steffensen, D., and Nicholson, B. (Eds.) *Toxic Cyanobacteria. Current Status and Research Management*. American Water Works Association Research Foundation, Denver, CO, USA. pp. 115–118.

Best, J. H., Eddy, F. B., and Codd, G.A. (2003) Effects of *Microcystis* cells, cell extracts and lipopolysaccharide on drinking and liver function in rainbow trout *Oncorhynchus mykiss* Walbaum. *Aquat Toxicol* **64:** 419–426.

Bickel, H., and Lyck, S. (2001) Importance of energy change for microcystin production. In: Chorus, I. (Ed.) *Cyanotoxins-Occurrence, Causes, Consequences*. Springer Verlag, KG, Berlin. pp. 133–141.

Bittencourt-Oliveira, M. C. (2003) Detection of potential microcystin-producing cyanobacteria in Brazilian reservoirs with *mcyB* molecular marker. *Harmful Algae* **2:** 51–60.

Bittencourt-Oliveira, M. C., Oliveira, M. C., and Bolch, C. J. S. (2001) Genetic variability of some Brazilian strains of *Microcystis aeruginosa* complex (Cyanophyceae/Cyanobacteria) using the nucleotide sequence analysis of the intergenic spacer and flanking regions from *cpc*BA-phycocyanin operon. *J Phycol* **37:** 810–818.

Blaha, L., Kopp, R., Simkova, K., and Mares, J. (2004) Oxidative stress biomarkers are modulated in silver carp (*Hypophthalmichthys molitrix* Val.) exposed to microcystin-producing cyanobacterial water bloom. *Acta Vet Brno* **73:** 477–482.

Blom, J. F., Robinson, J. A., and Juettner, F. (2001) Higher grazer toxicity of [D-Asp[3], (E)-Dhb[7]]microcystin-RR of *Planktothrix rubescens* as compared to different microcystins. *Toxicon* **39:** 1923–1932.

Boe, R., Gjersten, B. T., Vintermyr, O. K., Houge, G., Lanotte, M., and Doskeland, S. O. (1991) The protein phosphatase inhibitor okadaic acid induces morphological changes typical of apoptosis in mammalian cells. *Exptl Cell Res* **195:** 237–246.

Boettcher, G., Chorus, I., Ewald, S., Hintze, T., and Walz, N. (2001) Light limited growth and microcystin content of *Planktothrix agardhii* and *Microcystis aeruginosa* in turbidostats. In: Chorus, I. (Ed.) *Cyanotoxins-Occurrence, Causes, Consequences.* Springer Verlag, KG, Berlin, Germany, pp. 115–133.

Boland, M. P., Smillie, Chen, D. Z. X., and Holmes, C. F. B. (1993) A unified bioscreen for the detection of diarrhetic shellfish toxins and microcystins in marine and freshwater environments. *Toxicon* **31:** 1393–1405.

Botes, D. P., Tuinman, A. A., Wessels, P. L., Viljoen, C. C., Kruger, H., Williams, D. H., Santikarn, S., Smith, R. J., and Hammond, S. J. (1984) The structure of cyanoginosins-LA, a cyclic heptapeptide toxin from the cyanobacterium *Microcystis aeruginosa. J Chem Soc Perkin Trans* **I:** 2311–2318.

Botes, D. P., Wessels, P. L., Kruger, H., Runnegar, M. T. C., Santikarn, S., Smith, R. J., Barna, J. C. J., and Williams, D. H. (1985) Structural studies on cyanoginosins-LR, -YR, -YA, and-M, peptide toxins from *Microcystis aeruginosa. J Chem Soc Perkin Trans* **1:** 2747–2748.

Bouaicha, N., and Nasri, A.-B. (2004) First report of cyanobacterium *Cylindrospermopsis raciborskii* from Algerain waters. *Environ Toxicol* **19:** 541–543.

Bouaicha, N., Rivasseau, C., Hennion, M. C., and Sandra, P. (1996) Detection of cyanobacterial toxins (microcystins) in cell extracts by micellar electrokinetic chromatography. *J Chromatogr* **685:** 53–57.

Bourgoyne, D. L., Hemscheidt, T. K., Moore, R. E., and Runnegar, M. T. C. (2000) Biosynthesis of cylindrospermopsin. *J Org Chem* **65:** 152–156.

Bourke, A. T. C., Hawes, R. B., Neilan, A., and Stallman, N. D. (1983) An outbreak of hepato-enteritis (the Palm Island mystery disease) possibly caused by algal intoxication. *Toxicon* **3:** 45–48.

Bourne, D. G., Blakeley, R. L., Riddles, P., and Jones, G. J. (2005) Development of polymerase chain reaction and fluoresent *in situ* hybridization techniques for the detection of a bacterial strain that degrades the cyanobacterial toxin microcystin LR. *Mar Freshwater Res* **56:** 1127–1135.

Bourne, D. G., Jones, G. J., Blakeleyy, R. L., Jones, A., Negri, A. P., and Riddles, P. (1996) Enzymatic pathway for the bacterial degradation of the cyanobacterial cyclic peptide toxin microcystin LR. *Appl Environ Microbiol* **62:** 4086–4094.

Bourne, D. G., Riddles, P., Jones, G. J., Smith, W., and Blakeley, R. L. (2001) Characterization of a gene cluster involved in bacterial degradation of the cyanobacterial toxin microcystin-LR. *Environ Toxicol* **16:** 523–534.

Boutwell, R. K. (1974) The function and mechanism of promoters in carcinogenesis. *CRC Crit Rev Toxicol* **2:** 419–443.

Bouvy, M., Falcaeo, D., Marinho, M., Pagano, M., and Moura, A. (2000) Occurrence of *Cylindrospermopsis* (Cyanobacteria) in 39 Brazilian tropical reservoirs during the 1998 drought. *Aquat Microb Ecol* **23:** 13–27.

Bouvy, M., Molica, R., Oliveira, S., Marinho, M., and Beker, B. (1999) Dynamics of a toxic cyanobacterial bloom (*Cylindrospermopsis raciborskii*) in a shallow reservoir in the semi-arid region of Northeast Brazil. *Aquatic Microb Ecol* **20:** 285–297.

Briand, E., Gugger, M., Francois, J. C., Bernard, C., Humbert, J. F., and Quiblier, C. (2008) Temporal variations in the dynamics of potentially microcystin-producing strains in a bloom-forming *Planktothrix agardhii* (cyanobacterium) population. *Appl Environ Microbiol* **74:** 3839–3848.

Briand, J.-F., Jacquet, S., Bernard, C., and Humbert, J. F. (2003) Health hazards for terrestrial vertebrates from toxic cyanobacteria in surface water ecosystems. *Vet Res* **34:** 361–377.

Briand, J. F., Jacquet, S., Flinois, C., Avois-Jacquet, C., Maisonnette, C., Leberre, B., and Humbert, J. F. (2005) Variations in the microcystin production of *Planktothrix rubescens* (cyanobacteria) assessed from a four-year survey of Lac du Bourget (France) and from laboratory experiments. *Microb Ecol* **50:** 418–428.

Brittain, S., Mohamed, Z. A., Wang, J., Lehmann, V. K. B., Carmichael, W. W., and Rinehart, K. L. (2000) Isolation and characterization of microcystins from a River Nile strain of *Oscillatoria tenuis* Agardh ex Gomont. *Toxicon* **38:** 1759–1771.

Brooks, W. P., and Codd, G. A. (1987) Distribution of *Microcystis aeruginosa* peptide toxin and interactions with hepatic microsomes in mice. *Pharmacol Toxicol* **60:** 187–191.

Brooks, W. P., and Codd, G. A. (1988) Immunoassay of hepatotoxic cultures and water blooms of cyanobacteria using *Microcystis aeruginosa* peptide toxin polyclonal antibodies. *Environ Technol Lett* **9:** 1343–1348.

Bruno, M., Barbini, D. A., Pierdominici, E., Serse, A. P., and Ioppolo, A. (1994) Anatoxin-a and a previously unknown toxin in *Anabaena planctonica* from blooms found in lake Mulargia (Italy). *Toxicon* **32:** 369–373.

Bumke-Vogt, C., Mailahn, W., and Chorus, I. (1999) Anatoxin-a and neurotoxic cyanobacteria in German lakes and reservoirs. *Environ Toxicol* **14:** 117–125.

Burgoyne, D. L., Hemscheidt, T. K., Moore, R. E., and Runnegar, M. T. (2000) Biosynthesis of cylindrospermopsin. *J Org Chem* **65:** 152–156.

Burke, W. A., and Tester, P. A. (2002) Skin problems related to non-infectious coastal organisms. *Dermatologic Therapy* **15:** 10–17.

Byth, S. (1980) Palm Island mystery disease. *Med J Aus* **2:** 40–42.

Cadel-Six, S., Dauga, C., Castets, A. M., Rippka, R., Bouchier, C., Tandeau de Marsac, N., and Welker, M. (2008) Halogenase genes in non-ribosomal peptide synthetase gene clusters of *Microcystis* (cyanobacteria): Sporadic distribution and evolution. *Mol Biol Evol* **25:** 2031–2041.

Cadel-Six, S., Peyraud-Thomas, C., Brient, L., Tandeau de Marsac, N., Rippka, R., and Méjean, A. (2007) Different genotypes of anatoxin-producing cyanobacteria coexist in Tarn River, France. *Appl Environ Microbiol* **73:** 7605–7614.

Campbell, D. L., Lawton, L. A., Beattie, K. A., and Codd, G. A. (1994) Comparative assessment of the specificity of the brine shrimp and Microtox assays to hepatotoxic (microcystin-LR containing) cyanobacteria. *Environ Toxicol Water Qual* **9:** 71–77.

Campos, V., Cantarero, S., Urrutia, H., Heinze, R., Wirsing, B., Neumann, U., and Weckesser, J. (1999) Microcystin in cyanobacterial blooms in Chilean lake. *Syst Appl Microbiol* **22:** 169–173.

Cardellina, J. H., II, Moore, R. E., Arnold, E. V., and Clardy, J. (1979) Structure and absolute configuration of malyngolide, an antibiotic from the marine blue-green alga *Lyngbya majuscula* Gomont. *J Org Chem* **44:** 4039–4042.

Carmichael, W. W. (1986) Algal toxins. In: Callow, J. A. (Ed.) *Advances in Botanical Research*. Academic Press, London, UK. pp. 47–101.

Carmichael, W. W. (1988) Toxins of freshwater algae. In: Tu, A. T. (Ed.) *Marine Toxins and Venoms, Handbook of Natural Toxins*. Marcel Deckker, New York, NY, USA. pp. 121–147.

Carmichael, W. W. (1992) Cyanobacteria secondary metabolites-the cyanotoxins. *J Appl Bacteriol* **72:** 460–466.

Carmichael, W. W. (1994) The toxins of cyanobacteria. *Scientific American* **170:** 78–86.

Carmichael, W. W. (1997) The cyanotoxins. In: Callow, J. (Ed.) *Advances in Botanical Research*, Vol. 27, Academic Press, London, UK pp. 211–256.

Carmichael, W. W. (2001) Health effects of toxin-producing cyanobacteria: "The cyanoHABs". *Human and Ecological Risk Assessment* **7:** 1393–1407.

Carmichael, W. W., and Falconer, I. R. (1993) Disease related to freshwater algal blooms. In: Callow, J. A. (Ed.) *Advances in Botanical Research*, Academic Press, London, UK. pp. 187–210.

Carmichael, W. W., and Li, R. H. (2006) Cyanobacteria toxins in the Salton Sea. *Saline Systems* **2:** 5. doi: 10.1186/1746-1448-2-5

Carmichael, W. W., Biggs, D., and German, P. (1975) Toxicology and pharmacological action of *Anabaena flos-aquae* toxin. *Science* **187:** 542–544.

Carmichael, W. W., Biggs, D., and Petterson, M. A. (1979) Pharmacology of anatoxin-a, produced by freshwater cyanophyte *Anabaena flos-aquae* NRC-44-1. *Toxicon* **17:** 229–236.

Carmichael, W. W., Azevedo, S. M. F. O., An, J. S., Molica, R. J. R., Jochimsen, E. M., Lau, S., Rinehart, K. L., Shaw, G. R., and Eaglesham, G. K. (2001) Human fatalities from cyanobacteria: Chemical and biological evidence for cyanotoxins. *Environ Health Perspect* **109:** 663–668.

Carmichael, W. W., Beasley, V. R., Bunner, D. L., Eloff, J. N., Falconer, I. R., Gorham, P., Harada, K. I., Krishnamurthy, T., Yu, M.J., Moore, R., Rinehart, K., Runnegar, M., Skulberg, O. M., and Watanabe, M. (1988a) Naming of cyclic heptapeptide toxins of cyanobacteria (blue-green algae). *Toxicon* **26:** 971–973.

Carmichael, W. W., Eschedor, J. T., Patterson, G. M. L., and Moore, R. E. (1988b) Toxicity and partial structure of a hepatotoxic peptide produced by the cyanobacterium *Nodularia spumigena* Mertens emend L575 from New Zealand. *Appl Environ Microbiol* **54:** 2257–2263.

Carmichael, W. W., Evans, W. R., Yin, Q. Q., Bell, P., and Moczydloowsky, E. (1997) Evidence of paralytic shellfish poisons in the freshwater cyanobacterium *Lyngbya wollei* (Farlow ex Gomont) comb. nov. *Appl Environ Microbiol* **63:** 3104–3110.

Carmichael, W. W., Mahmood, N. A., and Hyde, E. G. (1990) Marine toxins: Origins, structure and molecular pharmacology. In: Hall, S., and Strichartz, G. (Eds.) *Natural Toxins from Cyanobacteria (Blue-Green Algae)*. American Chemical Society, Washington, DC, USA. pp. 87–106.

Carneiro, R. L., Elisângela, M., dos Santos, V., Furlanetto Pacheco, A. B., and de Oliveira e Azevedo, S. M. F. (2009) Effects of light intensity and light quality on growth and circadian rhythm of STX production in *Cylindrospermopsis raciborskii* (cyanobacteria). *J Plankton Res* **31:** 481–488.

Catterall, W. A., Morrow, G. S., and Hartshame, R. P. (1979) Neurotoxin binding to receptor sites associated with the voltage sensitive sodium channels in intact, lysed, and detergent solubilized brain membranes. *J Biol Chem* **254:** 11379–11387.

Cestele, S., and Catterall, W. A. (2000) Molecular mechanisms of neurotoxin action on voltage-gated sodium channels. *Biochemie* **82:** 883–892.

Chaivimol, J., Swoboda, U. K., and Dow, C. S. (1994) In: Codd, G. A., Jefferies, T. M., Keevil, C. W., and Potter E. (Eds.) *Detection Methods for Cyanobacterial Toxins*. Royal Society of Chemistry, Cambridge, UK. p. 172.

Chanh, T. C., Siwak, E. B., and Hewetson, J. F. (1991) Anti-idio-based vaccines against biological weapons. *Toxicol Appl Pharmacol* **108:** 183–193.

Chellappa, N. T., Costa, M. A. A. M., and Marinho, I. R. (2000) Harmful cyanobacterial blooms from semi-arid freshwater ecosystems of Northeast Brazil. *Aus Soc Limnol* (Newsletter) **38:** 45–49.

Chen, H.-M., Lin, C.-H., and Wang, T. M. (1996) Effects of 4-aminopurine on STX intoxication. *Toxicol Appl Pharmacol* **141:** 44–48.

Chin, D. T., and Cheng, C. Y. (1985) Oxidation of phenol with AC electrolysis. *Electrochem Sci Technol* **132:** 2605–2611.

Choi, H., Stathatos, E., Dinonysiou, D. D. (2006) Photocatalytic TiO_2 films and membranes for the development of efficient wastewater treatment and reuse systems. *Desalination* **202:** 199–206.

Chorus, I. (2005) *Current Approaches to Cyanotoxin Risk Assessment, Risk Management and Regulations in Different Countries.* Federal Environmental Agency, Berlin, Germany.

Chorus, I., and Bartram, J. (1999) *Toxic Cyanobacteria in Water: A Guide to Public Health Consequences, Monitoring and Management.* E & FN Sport, London, UK.

Christiansen, G., Fastner, J., Erhard, M., Borner, T., and Dittmann, E. (2003) Microcystin biosynthesis in *Planktothrix*: Genes, evolution, and manipulation. *J Bacteriol* **185:** 564–572.

Christiansen, G., Kurmayer, R., Liu, Q., and Boerner, T. (2006) Transposons inactivate biosynthesis of nonribosomal peptide microcystin in naturally occurring *Planktothrix* spp. *Appl Environ Microbiol* **72:** 117–123.

Christiansen, G., Molitor, C., Philmus, B., and Kurmayer, R. (2008) Nontoxic strains of cyanobacteria are the result of major gene deletion events induced by a transposable element. *Mol Biol Evol* **25:** 1695–1704.

Chu, F. S., Huang, X., and Wei, R. D. (1990) Enzyme-linked immunosorbent assay for microcystins in blue-green algal blooms. *J Assoc Anal Chem* **73:** 451–456.

Chu, F. S., Huang, X., Wei, R. D., and Carmichael, W. W. (1989) Production and characterization of antibodies against microcystins. *Appl Environ Microbiol* **55:** 1928–1933.

Clarke, P. B. S., and Reuben, M. (1996) Release of [^{3}H]-noradrenaline from rat hippocampal synaptosomes by nicotine: Mediation of different nicotinic receptor subtypes from striatal [^{3}H]-dopamine release. *Br J Pharmacol* **117:** 595–606.

Codd, G. A. (1995) Cyanobacterial toxins: occurrence, properties and biological significance. *Water Sci Technol* **32:** 149–156.

Codd, G. A. (2000) Cyanobacterial toxin, the perception of water quality, and the prioritisation of eutrophication control. *Ecol Eng* **16:** 51–60.

Codd, G. A., and Poon, G. K. (1988) Cyanobacterial toxins. *Proc Phytochem Soc Eur* **28:** 283–296.

Codd, G. A., Bell, S. G., and Brooks, W. P. (1989) Cyanobacterial toxins in water. *Water Sci Technol* **21:** 1–13.

Codd, G. A., Bell, S. G., Kaya, K., Ward, C. J., Beattie, K. A., and Metcalf, J. S. (1999) Cyanobacterial toxins, exposure routes and human health. *Eur J Phycol* **34:** 405–415.

Codd, G. A., Edwards, C., Beattie, K. A., Lawton, W. A., Campbell, D. L., and Bell, S. G. (1995) Toxins from cyanobacteria (blue-green algae). In Wiessner, W., Schnepf, E., and Starr, R. C. (Eds.) *Algae, Environment and Human Affairs.* Bioprocess Ltd., Bristol, England, UK. pp. 1–7.

Codd, G. A., Morrison, L. F. and Metcalf, J. S. (2005) Cyanobacterial toxins: Risk management and health protection. *Toxicol Appl Pharmacol* **203:** 264–272.

Cohen, P. (1989) The structure and regulation of protein phosphatases. *Annu Rev Biochem* **58:** 453–508.

Comninellis, C. H., and Plattnet, E. (1995) Electrochemical wastewater treatment. *Chimia* **42:** 250–252.

Cook, C. M., Vardaka, E., and Lanaras, T. (2004) Toxic cyanobacteria in Greek waters, 1987–2000: Occurrence, toxicity, and impacts in the Mediterranean region. *Acta Hydrochim Hydrobiol* **32:** 107–124.

Costa, I. A. S., Araujo, M. F. F., and Chellappa, N. T. (2001) Contribution to our knowledge of cyanobacteria from Brazil. Toxin producing species from a eutrophicated reservoir of Rio Grande do Norte State. *Proc Plankton Symp Espinho-Portugal.* p. 1999.

Costa, I. A. S., Azevedo, S. M. F. O., Senna, P. A. C., Bernardo, R. R., Costa, S. M., and Chellappa, N. T. (2006) Occurrence of toxin-producing cyanobacteria blooms in a Brazilian semiarid reservoir. *Braz J Biol* **66:** 211–219.

Cousins, I. T., Bealing, D. J., James, H. A., and Sutton, A. (1996) Biodegradation of microcystin-LR by indigenous mixed bacterial populations. *Water Res* **30:** 481–485.

Cox, P. A., Banack, S. A., and Murch, S. J. (2003) Biomagnification of cyanobacterial neurotoxins and neurodegenerative disease among Chamarro people of Guam. *Proc. Natl Acad Sci USA* **100:** 13330–13383.

Cox, P. A., Banack, S. A., Murch, S. J., Rasmussen, U., Tien, G., Bidigare, R. R., Metcalf, J. S., Morrison, L. F., Codd, G. A., and Bergman, B. (2005) Diverse taxa of cyanobacteria produce β-N-methylamino-L-alanine, a neurotoxic amino acid. *Proc Natl Acad Sci USA* **102:** 5074–5078.

Craig, M., Luu, H. A., McCready, T. L., Williams, D., Anderson, R. J., and Holmes, C. F. B. (1996) Molecular mechanisms underlying the interaction of motuporin and microcystins with type-1 and type-2A protein phosphatases. *Biochem Cell Biol* **74:** 569–578.

Craig, M., McCready, L., Luu, H. A., Smillie, M. A., Dubord, P., and Holmes, C. F. B. (1993) Identification and characterization of hydrophobic microcystins in Canadian freshwater cyanobacteria. *Toxicon* **31:** 1541–1549.

Cremer, J., and Henning, K. (1991) Application of reversed-phase medium-pressure liquid chromatography to the isolation, separation and amino acid analysis of two closely related peptide toxins of the cyanobacterium *Microcystis aeruginosa* strain PCC 7806. *J Chromatogr* **587:** 71–80.

Curtis, S. E. (1987) Genes encoding the beta and epsilon subunits of the proton-translocating ATPase from *Anabaena* sp. strain PCC 7120. *J Bacteriol* **169:** 80–86.

Cuvin-Aralar, M. L., Fastner, J., Focken, U., Becker, K., and Aralar, E. V. (2002) Microcystins in natural blooms and laboratory cultured *Microcystis aeruginosa* from Laguna de Bay, Phillippines. *Syst Appl Microbiol* **25:** 179–182.

Czarnecki, O., Henning, M., Lippert, I., and Welker, M. (2006) Identification of peptide metabolites of *Microcystis* (Cyanobacteria) that inhibit trypsin-like activity in planktonic herbivorous *Daphnia* (Cladocera). *Environ Microbiol* **8:** 77–87.

Dahlmann, E., Gerdts, W., and Hummert, H. P. (2005) Overview of key phytoplankton toxins and their recent occurrence in the North and Baltic Seas. *Environ Toxicol* **20:** 1–17.

Dahlmann, J., and Luckas, B. (2005) Microcystin analysis in drinking and surface water using a highly sensitive LC/MS/MS direct injection technique. *www.appliedbiosystems. com.*

Dasey, M., Ryan, N., Wilson, J., McGregor, G., Fabbro, L., Neilan, B. A., Burns, B. P., Kankaanpaeae, H., Morrison, L. F., Codd, G. A., Rissik, D., and Bowling, L. (2005) Investigations into the taxonomy, toxicity and ecology of benthic cyanobacterial accumulations in Myall Lake, Australia. *Marine Freshwater Res* **56:** 45–55.

Davidson, F. F. (1959) Poisoning of wild and domestic animals by a toxic water bloom of *Nostoc rivulare* Kuetz. *J Am Water Works Assoc* **51:**1277–1287.

Dawson, J. F., and Holmes, C. F. B. (1999) Molecular mechanisms underlying inhibition of protein phosphatases by marine toxins. *Frontiers in Bioscience* **4:** 646–658.

Dawson, R. M. (1998) The toxicology of microcystins. *Toxicon* **36:** 953–962.

De Bernardi, R., and Giussani, G. (1988) Discrimination between algae and artificial particles by freshwater and marine copepods. *Limnol Oceanogr* **33:** 397–408.

De Bernardi, R., and Giussani, G. (1990) Are blue-green algae a suitable food for zooplankton? An overview. *Hydrobiologia* **200/201:** 29–41.

De Figueiredo, D. R., Azeiteiro, U. M., Esteves, S. M., Goncalves, F. J. M., and Pereira, M. J. (2004) Microcystin-producing blooms-a serious global public health issue. *Ecotoxicol Environ Saf* **59:** 151–163.

DeMott, W. R., and Dhawale, S. (1995) Inhibition of *in vitro* protein phosphatase activity in three zooplankton species by microcystin-LR, a toxin from cyanobacteria. *Arch Hydrobiol* **134:** 417–424.

DeMott, W. R., Zhang, O.-X., and Carmichael, W. (1991) Effects of toxic cyanobacteria and purified toxins on the survival and feeding of a copepod and three species of *Daphnia*. *Limnol Oceanogr* **36:** 1346–1357.

de Silva, E. D., Williams, D. E., Anderson, R. J., Klix, H., Holmes, C. F. B., and Allen, T. M. (1992) Motuporin, a potent protein phosphatases inhibitor isolated from the Papua New Guinea sponge *Theonella swinhoei* Gray. *Tetrahedron Lett* **33:** 1561–1564.

Devlin, J. P., Edwards, O. E., Gorham, P. R., Hunter, N. R., Pike, R. K., and Stavric, B. (1977) Anatoxin-a, a toxic alkaloid Z from *Anabaena flos-aquae* NRC-44-1. *Can J Chem* **55:** 1367–1371.

Diehnelt, C. W., Dugan, N. R., Peterman, S. M., and Budde, W. L. (2006) Identification of microcystin toxins from a strain of *Microcystis aeruginosa* by liquid chromatography introduction into a hybrid linear ion trap-Fourier transform ion cyclotron resonance mass spectrometer. *Anal Chem* **78:** 501–512.

Dillenberg, H. O., and Dehnel, M. K. (1960) Toxic water bloom in Saskatchewan, 1959. *Can Med Assoc J* **83:** 1151–1154.

Ding, W. X., Shen, H.M., Zhu, H. G., Lee, B. L., and Ong, C. N. (1999) Genotoxicity of *Microcystis* cyanobacteria extract of a water source in China. *Mutat Res* **442:** 69–74.

Dittmann, E., Meissner, K., and Boerner, T. (1996) Conserved sequences of peptide synthetase genes in the cyanobacterium *Microcystis aeruginosa*. *Phycologia* **35** (Suppl 6): 62–67.

Dittmann, E., Neilan, B. A., Erhard, M., von Doehren, H., and Boerner, T. (1997) Insertional mutagenesis of a peptide synthetase gene that is responsible for hepatotoxin production in the cyanobacterium *Microcystis aeruginosa* PCC 7806. *Mol Microbiol* **26:** 779–787.

Domanski, T. L., de Jonge, B. L. M., and Bayles, K. W. (1997) Transcriptional analysis of the *Staphylococcus aureus* gene encoding penicillin-binding protein 4. *J Bacteriol* **179:** 2651–2657.

Domingos, P., Rubin, T. K., Molica, R. J. R., Azevedo, S. M. F. O., and Carmichael, W. W. (1998) First report of microcystin production by picoplanktonic cyanobacteria isolated from northeast Brazilian drinking water supply. *Environ Toxicol* **14:** 31–35.

Douglas, P., Moorhead, G. B. G., Ye, R., and Lees-Millers, S. P. (2001) Protein phosphatases regulate DNA-dependent protein kinase activity. *J Biol Chem* **22:** 18992–18998.

Dow, C. S., and Swoboda, U. K. (2000) Cyanotoxins. In: Whitton, B. A., and Potts, M. (Eds) *The Ecology of Cyanobacteria*. Kluwer Academic Publishers, Dordrecht, The Netherlands. pp. 614–632.

Downing, J. A., Watson, S. B., and Mc Cauley, E. (2001) Predicting cyanobacteria dominance in lakes. *Can J Fish Aquat Sci* **58:** 1905–1908.

Draisci, R., Ferretti, E., Palleschi, L., and Marchiafava, C. (2001) Identification of anatoxins in blue-green algae food supplements using liquid chromatography-tandem mass spectrometry. *Food Additives and Contaminanats* **18:** 525–531.

Drikas, M. (1994) Control and/or removal of toxins. In: Steffensen, D. A., and Nicholson, B. C. (Eds.) *Toxic Cyanobacteria: Current Status of Research and Management*. Australian center for water quality, Salisbury, Australia. p. 93

Drikas, M., and Hrudey, S. (1994) Control and removal of toxins: Summary of discussions. In: Steffensen, D., and Nicholson, B. C. (Eds.) *Toxic Cyanobacteria. Current status of Research and Management*. American Water Works Association Research Foundation, Denver, CO, USA.

Duy, T. N., Lam, P. K. S., Shaw, G. R., and Connell, D. W. (2000) Toxicology and risk assessment of freshwater cyanobacterial (Blue-green algal) toxins in water. *Rev Environ Contam Toxicol* **163:** 113–186.

Dvorakova, D., Dvorakova, K., Blaha, L., Marsalek, B., and Knotkova, Z. (2002) Effects of cyanobacterial biomass and purified microcystins on malformations in *Xenopus laevis*: Teratogenesis assay (FETAX). *Environ Toxicol* **17:** 547–555.

Eaglesham, G. K., Norris, R. L., Shaw, G. R., Smith, M. J., Chiswell, R. K., Davis, R. K., Neville, G. R., Seawright, A. A., and Moore, M. R. (1999) Use of HPLC-MS/MS to monitor cylindrospermopsin, a blue-green algal toxin, for public health purposes. *Environ Toxicol* **14:** 151–154.

Edler, L., Ferno, S., Lind, M. G., Lundberg, R., and Nilsson, P. O. (1985) Mortality of dogs associated with bloom of the cyanobacterium *Nodularia spumigena* in the Baltic Sea. *Ophelia* **24:** 103–109.

Edwards, C., Beattie, K. A., Scrimgeour, C. H., and Codd, G. A. (1992) Identification of anatoxin-a in benthic cyanobacteria (blue-green algae) and in associated dog poisonings at Loch Insh, Scotland. *Toxicon* **30:** 1165–1175.

Ehrenreich, I. M., Waterbury, J. B., and Webb, E. A. (2005) Distribution and diversity of natural product genes in marine and freshwater cyanobacterial cultures and genomes. *Appl Environ Microbiol* **71:** 7401–7413.

Ells, B., Froese, K., Hrudey, S. E., Purves, R. W., Guevremont, R., and Barnett, D. A. (2000) Detection of microcystins using electrospray ionization high-field asymmetric waveform ion mobility mass spectrometry/mass spectrometry. *Rapid Commun Mass Spectrom* **14:** 1538–1542.

Elleman, T. C., Falconer, I. R., Kackson, A. R. B., and Runnegar, M. T. (1978) Isolation, characterization and pathology of the toxin from a *Microcystis aeruginosa* (=*Anacystis cyanea*) bloom. *Aus J Biol Sci* **31:** 209–218.

Erhard, M., von Doehren, H., and Jungblut, P. (1997) Rapid typing and elucidation of new secondary metabolites of intact cyanobacteria using MALDI-TOF mass spectrometry. *Nat Biotechnol* **15:** 906–909.

Eriksson, J. E., Hagerstrand, H., and Isomaa, B. (1987) Cell selective cytotoxicity of peptide toxin fron the cyanobacterium *Microcystis aeruginosa*. *Biochim Biophys Acta* **930:** 304–310.

Eriksson, J. E., Meriluoto, J. A., Kujari, H. P., Osterlund, K., Fagerlund, K., and Hallbom, L. (1988) Preliminary characterization of a toxin isolated from the cyanobacterium *Nodularia spumigena*. *Toxicon* **26:** 161–166.

Eriksson, J. E., Groenberg, L., Nygard, S., Slotte, J. R., and Meriluto, J. A. O. (1990a) Hepatocellular uptake of ^{3}H-dihydroxymicrocystin LR, a cyclic peptide toxin. *Biochim Biophys Acta* **1025:** 60–66.

Eriksson, J. E., Paatero, G. I. L., Meriluoto, J. A. O., Codd, G. A., Kass, G. E. N., Nicotera, P., and Orrenius, S. (1989) Rapid micro filament reorganization induced in isolated rat hepatocytes by microcystin-LR, a cyclic peptide toxin. *Exp Cell Res* **185:** 86–100.

Erikson, J. E., Toivola, D., Meriluto, J. A. O., Karaki, H., Han, Y.-G., and Hartshorne, D. (1990b) Hepatocyte deformation induced by cyanobacterial toxins reflects inhibition of protein phosphatases. *Biochem Biophys Res Commun* **173:** 1347–1353.

Ernst, B., Hoeger, S. J., O'brien, E., and Dietrich, D. R. (2006) Oral toxicity of microcystin-containing cyanobacterium *Planktothrix rubescens* in European whitefish (*Coregonus lavaretus*). *Aqat Toxicol* **79:** 31–40.

Evans, T. M., Schllllnger, J. E., and Stuart, D. G. (1978) Rapid determination of bacteriological water quality by using *Limulus* lysate. *Appl Environ Microbiol* **35:** 376–382.

Eynard, F., Mez, K., Walther, J.-L. (2000) Risk of cyanobacterial toxins in Riga waters. *Water Res* **34:** 2979–2988.

Falconer, I. R. (1991) Tumor promotion and liver injury caused by oral consumption of cyanobacteria. *Environ Toxicol Water Qual* **6:** 177–184.

Falconer, I. R. (1993) Mechanism of toxicity of cyclic peptide toxins from blue-green algae. In: Falconer, I. R. (Ed.) *Algal Toxins in Seafood and Drinking Water*. Academic Press, London, UK. pp. 1538–1542.

Falconer, I. R. (1998) Algal toxins and human health. In: Hrubec, J. (Ed.) *The Handbook of Environmental Chemistry*, Vol. 5 part C, Springer-Verlag, Berlin, Germany. pp. 53–82.

Falconer, I. R. (2005) Cyanobacterial poisoning of livestock and people. In: Falconer, I. R. (Ed.) *Cyanobacterial Toxins of Drinking Water Supplies. Cylindrospermopsins and Microcystins*. CRC Press, Boca Raton, Florida, FL, USA.

Falconer, I. R., and Humpage, A. R. (1996) Tumor promotion by cyanobacterial toxins. *Phycologia* **35** (Suppl. 6): 74–79.

Falconer, I. R., and Yeung, D. S. (1992) Cytoskeletal changes in hepatocytes induced by *Microcystis* toxins and their relation to hyperphosphorylation of cell proteins. *Chem Biol Interact* **81:** 181–196.

Falconer, I. R., Beresford, A. M., and Runnegar, M. T. C. (1983) Evidence of liver damage by toxin from a bloom of the blue-green alga, *Microcystis aeruginosa. Med J Aus* **1:** 511–514.

Falconer, I. R., Buckley, T., and Runnegar, M. T. (1986) Biological half-life, organ distribution and excretion of ^{125}I-labelled toxic peptide from the blue-green alga *Microcystis aeruginosa. Aus J Biol Sci* **39:** 17–21.

Falconer, I. R., Burch, M.D., Steffensen, D. A., Choice, M., and Coverdale, O. R. (1994) Toxicity of blue-green alga (cyanobacterium) *Microcystis aeruginosa* in drinking water to growing pigs, as an animal model for human injury and risk assessment. *J Environ Toxicol Water Qual* **9:** 131–139.

Falconer, I. R., Choice, A., and Hosja, W. (1992) Toxicity of edible mussels (*Mytilus edulis*) growing naturally in an estuary during a water bloom of the blue-green alga *Nodularia spumigena. J Environ Toxicol Water Qual* **9:** 131–139.

Falconer, I. R., Hardy, S. J., Humpage, A. R., Froscio, S. M., Tozer, G. J., and Hawkins, P. R. (1999) Hepatic and renal toxicity of the blue-green alga (cyanobacterium) *Cylindrospermopsis raciborskii* in male Swiss albino mice. *Environ Toxicol* **14:** 143–150.

Falconer, I. R., Runnegar, M. T. C., and Huynh, V. L. (1983) Effectiveness of activated carbon in the removal of algal toxin from potable water supplies: A pilot investigation. In: Tenth Federal Convention of Australian Water and Wastewater Association, Sydney, Australia. pp. 1–8.

Fastner, J., Erhard, M., and von Doehren, H. (2001) Determination of oligopeptide diversity within a natural population of *Microcystis* spp. (Cyanobacteria) by typing single colonies of matrix-assisted laser desorption ionization-time of flight mass spectrometry. *Appl Environ Microbiol* **67:** 5069–5076.

Fawell, J. K., James, C. P., and James, H. A. (1994) Toxins from blue-green algae: Toxicological assessment of microcystin-LR and a method for its determination in water. Water Research Centre, Medmenham, UK. pp 1–16.

Feitz, A. J., and Waite, T. D. (2003) Kinetic modeling of TiO_2-catalyzed photodegradation of trace levels of microcystin-LR. *Environ Sci Technol* **37:** 561–568.

Feng, C., Sugiura, N., Shimada, S., and Maekewa, T. (2003) Development of a high performance electrochemical waster water treatment system. *J Hazard Matter* **103:** 65–78.

Feng, C., Sugiura, N., Masaaka, Y., and Maekawa, T. (2005) Electrochemical degradation of microcystin-LR. *J Environ Sci Health A* **40:** 453–465.

Fergusson, K. M., and Saint, C. P. (2000) Molecular phylogeny of *Anabaena circinalis* and its identification in environmental samples by PCR. *Appl Environ Microbiol* **66:** 4145–4148.

Ferreria, F. M. B., Soler, J. M., Fidalgo, M. L., and Vila, P. F. (2001) PSTs from *Aphanizomenon flos-aquae* (cyanobacteria) collected in the Crestuma-Lever reservoir (Duro River, northern Portugal). *Toxicon* **39:** 757–761.

Fewer, D. P., Tooming-Klunderud, A., Jokela, J., Wahlsten, M., Rouhiainen, L., Kristensen, T., Rohrlack, T., Jakobsen, K. S., and Sivonen, K. (2008) Natural occurrence of microcystin synthetase deletion mutants capable of producing microcystins in the genus *Anabaena* (cyanobacteria). *Microbiology* **154:** 1007–1014.

Fischer, W. J., and Dietrich, D. R. (2000) Pathological and biochemical characterization of microcystin-induced hepatopancreas and kidney damage in carp (*Cyprinus carpio*). *Toxicol Appl Pharmacol* **164:** 73–81.

Fischer, W. J., Hitzfeld, B. C., Tencalla, F., Eriksson, J. E., Mikhailov, A., and Dietrich, D. R. (2000) Microcystin LR toxicodynamics, induced pathology, and immunohistochemical localization in livers of blue-green algae exposed rainbow trout (*Oncorhynchus mykiss*). *Toxicol Sci* **54:** 365–373.

Fitzepore, R. B., Clark, S. A., and Keevil, C. W. (1994) Routes of intoxication. In: Codd, G. A., Jeffries, T. M., Keevil, C. W., and Potter, E. (Eds.) 1st International Symposium on *Detection methods for cyanobacterial (blue-green algal) toxins*. Royal Society of Chemistry, Cambridge, UK. pp. 69–74.

Fitzgerald, D. J. (2001) Cyanotoxins and human health-overview. In: Chorus, I. (Ed.) *Cyanotoxins-Occurrence, Causes and Consequences*. Springer, Berlin, Germany. pp. 179–190.

Fitzgerald, D. J., and Yamasaki, H. (1990) Tumor promotion: Models and assay systems. *Teratogen, Carcinogen, Mutagen* **10:** 89–102.

Fladmark, K. E., Brustugun, O. T., Hovald, R., Boe, R., Gjertsen, B. T., Zhivotovsky, B., and Doskeland, S. O. (1999) Ultrarapid scapase-3 dependent apoptosis induction by serine/threonine phosphatase inhibitors. *Cell Death and Differentiation* **6:** 1099–1108.

Fladmark, K. E., Serres, M. H., Lersen, N. L., Yasumoto, T., Aune, T., and Doskeland, S. O. (1998) Sensitive detection of apoptogenic toxins in suspension cultures of rat and salmon hepatocytes. *Toxicon* **36:** 1101–1114.

Fleming, L. E., Rivero, C., Burns, J., Williams, C., Bean, J. A., Shea, K. A., and Stinn, J. (2002) Blue-green algal (cyanobacterial) toxins, surface drinking water, and liver cancer in Florida. *Harmful Algae* **1:** 157–168.

Flett, D. J., and Nicholson, B. C. (1991) *Toxic Cyanobacteria in Water Supplies: Analytical Techniques.* Urban Water Research Association of Australia. 61 pp.

Foulds, I. V., Granacki, A., Xiao, C., Krull, U. J., Castle, A., and Horgen, P. A. (2002) Quantification of microcystin-producing cyanobacteria by 5′-nuclease PCR. *J Appl Microbiol* **93:** 825–834.
Fox, A. M., and Dulay, T. M. (1993) Heterogeneous photocatalysis. *Chem Rev* **93:** 341–357.
Francis, G. (1878) Poisonous Australian lake. *Nature* (London) **18:** 11–12.
Frangeul, L., Quillardet, P., Castets, A. M., Humbert, J-F., Matthijs, H. C. P., Cortez, D., Tolonen, A., Zhang, G-C., Gribaldo, S., Kehr, J-C., Zilliges, Y., Ziemert, N., Becker, S., Talla, E., Latifi, A., Billault, A., Lepelletier, A., Dittmann, E., Bouchier, C., and Tandeau de Marsac, N. (2008) Highly plastic genome of *Microcystis aeruginosa* PCC 7806, a ubiquitous toxic freshwater cyanobacterium. *BMC Genomics* **9:** 274 doi:10.1186/1471-2164-9-274.
Frank, C. A. (2002) Microcystin-producing cyanobacteria in recreational waters in southwestern Germany. *Environ Toxicol* **17:** 361–366.
Fromme, H., Koehler, A., Krause, R., and Fuehrling, D. (2000) Occurrence of cyanobacterial toxins-microcystins and anatoxin-a—in Berlin water bodies with implications to human health and regulations. *Environ Toxicol* **15:** 120–130.
Fujishima, A., Hashimoto, K., and Watanabe, T. (1999) *TiO_2 Photocatalysis-Fundamentals and Applications*. BKC, Tokyo, Japan.
Fujuki, H., and Suganuma, M. (1993) Tumor promotion by inhibitors of protein phosphatase 1 and 2A: The okadaic acid class of compounds. *Adv Cancer Res* **61:** 143–196.
Fujuki, H., Suganuma, M., Suguri, H., Yoshizawa, S., Takagi, K., Nakayasu, M., Ojika, M., Yamada, K., Yasumoto, T., Moore, R. E., and Sugimura, T. (1990) New tumor promoters from marine natural products. In: Hall, S., and Strichartz (Eds.) *Marine Toxins: Origin, Structure and Molecular Pharmacology*. American Chemical Society, Washington, D. C. USA. pp. 232–240.
Fulton, R. S. III, and Pearl, H. W. (1987) Toxic and inhibitory effects of the blue-green alga *Microcystis aeruginosa* on herbivorous zooplankton. *J Plankton Res* **9:** 837–855.
Furey, A., Crowley, J., Lehane, M., and James, K. J. (2003a) Liquid chromatography with electron spray ion-trap mass spectrometry for the determination of anatoxins in cyanobacteria and drinking water. *Rapid Commun Mass Spectrometry* **17:** 583–588.
Furey, A., Crowley, J., Shuilleabhain, A. N., Skulberg, O. M., and James, K. J. (2003b) The first identification of the rare cyanobacterial toxin, homoanatoxin-a, in Ireland. *Toxicon* **41:** 297–303.
Galat, D. L., Verdin, J. P., and Sims, L. L. (1990) Large-scale patterns of *Nodularia spumigena* blooms in Pyramid Lake, Neveda, determined from Landsat imagery: 1972–1986. *Hydrobiologia* **197:** 147–164.
Galey, F. D., Beasley, V. R., Carmichael, W. W., Kleppe, G., Hooser, S. B., and Haschek, W. M. (1987) Blue-green algae (*Microcystis aeruginosa*) hepatotoxicosis in dairy cows. *Am J Vet Res* **48:** 1415–1420.
Gallon, J. R., Chit, K. N., and Brown, E. G. (1990) Biosynthesis of the tropane-related cyanobacterial toxin anatoxin-a: Role of ornithine decarboxylase. *Phytochemistry* **29:** 1107–1111.
Gallon, J. R., Kittakoop, O., and Brown, E. G. (1994) Biosynthesis of anatoxin-a by *Anabaena flos-aquae*: Examination of primary enzymatic steps. *Phytochemistry* **35:** 1195–1203.
Gathercole, P. S., and Thiel, P. G. (1987) Liquid chromatographic determination of the cyanoginosins, toxins produced by the cyanobacterium *Microcystis aeruginosa*. *J Chromatogr* **408:** 435–440.
Gauss, C.-M., Sheppock, J. E., Narin, A. C., and Chamberlin, R. A. (1997) A molecular modeling analysis of the binding interactions between okadaic acid class of natural product inhibitors and the ser-thr phosphatases, PP1 and PP2A. *Bioorganic Medicinal Chem* **5:** 1751–1773.
Gehringer, M. M., Downs, K. S., Downing, T. G., Naude, R. J., and Shephard, E. G. (2003a) An investigation into effect of selemium supplementation on microcystin hepatotoxicity. *Toxicon* **41:** 451–458.
Gehringer, M. M., Govender, S., Shah, M., and Downing, T. G. (2003b) An investigation of the role of vitamin E in the protection of mice against microcystin toxicity. *Environ Toxicol* **18:** 142–148.
Gerwick, W. H., Tan, L. T., and Sitachitta, N. (2001) Nitrogen-containing metabolites from marine cyanobacteria. *Alkaloids Chem Biol* **57:** 75–184.
Gessner, B. D., Bell, P., Doucette, G. J., Moczydlowski, E., Pli, M. A., Van Dolah, F., and Hall, S. (1997) Hypertension and identification of toxin in human urine and serum following a cluster of mussel-associated paralytic shellfish poisoning outbreaks. *Toxicon* **35:** 711–722.
Ghassempour, A., Najafi, N. M., Mehdina, A., Davarani, S. S., Fallahi, M., and Nakhshab, M. (2005) Analysis of anatoxin-a using polyaniline as a sorbent in solid-phase microextraction coupled to gas chromatography-mass spectrometry. *J Chromatogr* **1078:** 120–127.
Giani, A., Bird, D. F., Prairie, Y. T., and Lawrence, J. F. (2005) Empirical study of cyanobacterial toxicity along a trophic gradient of lakes. *Can J Fish Acquat Sci* **62:** 2100–2109.
Gibson, C. E., and Smith, R. V. (1982) Freshwater plankton. In: Carr, N. G., and Whitton, B. A. (Eds.) *The Biology of Cyanobacteria*. Blackwell Scientific Publications, Oxford, UK. pp. 463–489.

Gkelis, S., Moustaka-Gouni, M., Sivonen, K., and Lanaras, T. (2005) First report of the cyanobacterium *Aphanizomenon ovalisporum* Forti in two Greek lakes and cyanotoxin occurrence. *J Plankton Res* **27:** 1295–1300.

Goldberg, J., Huang, H. B., Kwon, Y. G., Greengard, P., Narin, A. C., and Kuriyan, J. (1995) Three-dimensional structure of the catalytic subunit of protein serine/threonine phosphatase-1. *Nature* (London) **376:** 745–753.

Gorham, P. R. (1964) Toxic algae. In: Jackson, D. F. (Ed.) *Algae and Man*, Plenum Press, New York, NY, USA. pp. 307–336.

Gorham, P. R., and Carmichael, W. W. (1988) Hazards of freshwater blue-green algae (cyanobacteria). In: Lembi, C. A., and Waaland, J. R. (Eds.) *Algae and Human Affairs*. Cambridge University Press, Cambridge, UK. pp. 403–431.

Grabow, W. O., Du Randt, W. C., and Scott, W. E. (1982) *Microcystis aeruginosa* toxin: Cell culture toxicity, hemolysis and mutagenicity assays. *Appl Environ Microbiol* **43:** 1425–1433.

Graham, J. L., Jones, J. R., Jones, S. B., Downing, J. A., Clevenger, T. E. (2004) Environmental factors influencing microcystin distribution and concentration in the Midwestern United States. *Water Res* **38:** 4395–4404.

Griffiths, D. J., and Saker, M. L. (2003) The Palm Island Mystery Disease 20 years on: A review of research on the cyanotoxin cylindrospermopsin. *Environ Toxicol* **18:** 79–93.

Gugger, M., Lenoir, S., Berger, C., Ledreux, A., Druart, J. C., Humbert, J. F., Guette, C., and Bernard, C. (2005) First report in a river in France of the benthic cyanobacterium *Phormidium favosum* producing anatoxin-a associated with dog neurotoxicosis. *Toxicon* **45:** 919–928.

Gugger, M., Molica, R., Berre, B. L., Dufour, P., Bernard, C., and Humbert, J.-F. (2005) Genetic diversity of *Cylindrospermopsis* strains (cyanobacteria) isolated from four continents. *Appl Environ Microbiol* **71:** 1097–1100.

Gupta, N., Bhaskar, A. S. B., and Rao, P. V. L. (2002) Growth characteristics and toxin production in batch cultures of *Anabaena flos-aquae*: Effects of culture media and duration. *World J Microbiol Biotechnol* **18:** 29–35.

Gupta, S., Giddings, M., and Sheffer, M. (2001) Cyanobacterial toxins in drinking water: A Canadian perspective. In Chorus, I. (Ed.) *Cyanotoxins-Occurrence, Causes and Consequences*. Springer, Berlin, Germany. pp. 208–212.

Gussmann, H., Molzhan, M. J., and Bicks, B. (1985) Vergiftungen bei jungrindern durch die Blaualge *Nodularia spumigena*. *Montash Veterinaermed* **40:** 76–79. (In German) Cited from Sivonen, K., Kononen, K., Carmichael, W. W., Dahlem, A. M., Rinehart, K. L., Kiviranta, J., Niemela, S. I. (1989) Occurrence of the hepatotoxic cyanobacterium *Nodularia spumigena* in the Baltic Sea and structure of the toxin. *Appl Environ Microbiol* **55:** 1990–1995.

Guzman, R. E., and Solter, P. F. (2002) Characterization of sublethal microcystin-LR exposure in mice. *Vet Pathol* **39:** 17–26.

Hagmann, L., and Juettner, F. (1996) Fischerellin-A, a novel photosystem-II inhibiting allelochemical of the cyanobacterium *Fischerella muscicola* with antifungal and herbicidal activity. *Tetrahedron Lett* **37:** 6539-6542.

Haider, S., Naithani, V., Viswanathan, P. N., and Kakkar, P. (2003) Cyanobacterial toxins: A growing environmental concern. *Chemosphere* **52:** 1–21.

Halinen, K., Jokela, J., Fewer, D. P., Wahlsten, M., and Sivonen, K. (2007) Direct evidence that production of microcystins by *Anabaena* strains from the Baltic Sea. *Appl Environ Microbiol* **73:** 6543–6550.

Hamilton, P. B., Ley, L. M., Dean, S., and Pick, F. R. (2004) The occurrence of the cyanobacterium *Cylindrospermopsis raciborskii* in Constance Lake: An exotic cyanoprokaryote new to Canada. *Phycologia* **44:** 17–25.

Haney, J. F., Sasner, J. J., and Ikawa, M. (1995). Effects of products released by *Aphanizomenon flos-aquae* and purified STX on the movements of *Daphnia carinata* feeding appendages. *Limnol Oceanogr* **40:** 263–272.

Harada, K.-I., (1995) Chemistry and detection of microcystins. In: *Toxic Microcystis*. CRC Press, Boca Raton, New York, London, Tokyo. pp. 203–232.

Harada, K.-I., Matsuura, K., Suzuki, M., Oka, H., Watanabe, M. F., Oishi, S., Dahlem, A. M., and Beasley, V. R. (1988a) Analysis and purification of toxic peptides from cyanobacteria by reversed-phase high-performance liquid chromatography. *J Chromatogr* **448:** 275–283.

Harada, K.-I., Matsuura, K., Suzuki, M., Watanabe, M. F., Oishi, S., Dahlem, A. M., Beasley, V. R., and Carmichael, W. W. (1990a) Isolation and characterization of the minor components associated with microcystins LR and RR in the cyanobacterium (blue-green algae). *Toxicon* **28:** 55–64.

Harada, K.-I., Suzuki, M., Dahlem, A. M., Beasley, V. R., Carmichael, W. W., and Rinehart, K. L. (1988b) Improved method for purification of toxic peptides produced by cyanobacteria. *Toxicon* **26:** 433–439.

Harada, K.-I., Nagai, H., Kimura, Y., and Suzuki, M. (1993) Liquid chromatography/mass spectrometric detection of anatoxin-a, a neurotoxin from cyanobacteria. *Tetrahedron* **49:** 9251–9260.

Harada, K.-I., Ogawa, K., Kimura, Y., Murata, H., Suzuki, M., Thorn. P. M., Evans, W. R., and Carmichael, W. W. (1991a) Microcystins from *Anabaena flos-aquae* NRC 525-17. *Chem Res Toxicol* **4:** 535–540.

Harada, K. I., Ogawa, K., Matsuura, K., Murata, H., Suzuki, M., Watanabe, M. F., Itezono, Y., and Nakayama, N. (1990b) Structural determination of geometrical isomers of microcystins-LR and -RR from cyanobacteria by two-dimensional NMR spectroscopic techniques. *Chem Res Toxicol* **3:** 473–481.

Harada, K.-I., Ogawa, K., Matsuura, K., Nagai, H., Murata, H., Suzuki, M., Itezono, Y., Nakayama, N., Shirai, M., and Nakano, M. (1991b) Isolation of two toxic heptapeptide microcystins from an axenic strain of *Microcystis aeruginosa*, K-139. *Toxicon* **29:** 479–489.

Harada, K.-I., Ohtani, I., Iwamoto, K., Suzuki, M., Watanabe, M. F., Watanabe, M., and Terao, K. (1994) Isolation of cylindrospermopsin from a cyanobacterium *Umezakia natans* and its screening method. *Toxicon* **32:** 73–84.

Harada, K.-I., Murata, H., Qiang, Z., Suzuki, M., and Kondo, F. (1996a) Mass spectrometric screening method for microcystins in cyanobacteria. *Toxicon* **34:** 701–710.

Harada, K.-I., Tsuji, K., Watanabe, M. F., and Kondo, F. (1996b) Stability of microcystins from cyanobacteria-III. Effect of pH and temperature. *Phycologia* **35:** 83–88.

Harada, S., Kimura, Y., Ogawa, K., Suzuki, M., Dahlem, A. H., Beasley, V. R., and Carmichael, W. W. (1989) A new procedure for the analysis and purification of naturally occurring anatoxin-a from the blue-green alga *Anabaena flos-aquae*. *Toxicon* **27:** 1289–1296.

Harrington, G. G., Luesch, H., Yoshida, W. Y., Moore, R. E., Nagle, D. G., Park, P. U., Biggs, J., Paul, V. J., Mooberry, S. L., Corbett, T. H., and Valeriote, F. A. (1998) Isolation, structure determination and biological activity of dolastatin 12 and lyngbyastatin 1 from *Lyngbya majuscula/Schizothrix calcicola* cyanobacterial assemblages. *J Nat Prod* **61:** 1221–1225.

Hariston, N. G., Holtmeier, C. L., Lampert, W., Weider, L. J., Post, D. M., and Fischer, J. M. (2001) Natural selection for grazer resistance to toxic cyanobacteria: Evolution of phenotypic plasticity. *Evolution* **55:** 2203–2214.

Hart, J., and Stott, P. (1993) *Microcystin-LR removal from water*. Foundation for Water Research Report, FR0367, Swindon, UK.

Hastie, C. J., Borthwick, E. B., Morrison, L. F., Codd, G. A., and Cohen, P. T. (2005) Inhibition of several protein phosphatases by non-covalently interacting microcystin and a novel cyanobacterial peptide, nostocyclin. *Biochim Biophys Acta* **1726:** 187–193.

Hawkins, P. R., Chandrasena, N. R., Jones, G. J., Humpage, A. R., and Falconer, I. (1997) Isolation of *Cylinsrodpermopsis raciborskii* from an ornamental lake. *Toxicon* **35:** 341–346.

Hawkins, P. R., Runnegar, M. T. C., Jackson, A. R. B., and Falconer, I. R. (1985) Severe hepatotoxicity caused by the tropical cyanobacterium (Blue-green alga) *Cylindrospermopsis raciborskii* (Woloszynska) Seenaya and Subba Raju isolated from a domestic water supply reservoir. *Appl Environ Microbiol* **50:** 1292–1295.

Hayman, J. (1992) Beyond the Barcoo-probable human tropical cyanobacterium poisoning in outback Australia. *Med J Aus* **157:** 794–796.

Heinze, R. (1999) Toxicity of the cyanobacterial toxin microcystin-LR to rats after 28 days intake with drinking water. *Environ Toxicol* **14:** 57–60.

Hemscheidt, T., Rapala, J., Sivonen, K., and Skulberg, O. M. (1995) Biosynthesis of anatoxin-a in *Anabaena flos-aquae* and homoanatoxin-a in *Oscillatoria formosa*. *J Chem Soc Chemical Commun* 1361–1362.

Henriksen, P. (1996) Microcystin profiles and contents in Danish populations of cyanobacteria/blue-green algae as determined by HPLC. *Phycologia* (Supplement) **35:** 102–110.

Henriksen, P. (2001) Toxic freshwater cyanobacteria in Denmark. In: Chorus, I. (Ed.) *Cyanotoxins-Occurrence, Causes and Consequences*. Springer Verlag, KG, Berlin, Germany, pp. 49–56.

Henriksen, P., Carmichael, W. W., An, J., and Moestrup, Ø. (1997) Detection of an anatoxin-a(s)-like anticholineterase in natural blooms and cultures of cyanobacteria/blue-green algae from Danish lakes and in the stomach contents of poisoned birds. *Toxicon* **35:** 901–913.

Heresztyn, T., and Nicholson, B. C. (1997) Nodularin concentrations in Lakes Alexandria and Albert, South Australia, during a bloom of the cyanobacterium (blue-green alga) *Nodularia spumigena* and degradation of the toxin. *Environ Toxicol Water Quality* **12:** 273–282.

Hesse, K., and Kohl, J.-G. (2001) Effects of light and nutrient supply on growth and microcystin content of different strains of *Microcystis aeruginosa*. In: Chorus, I. (Ed.) *Cyanotoxins-Occurrence, Causes and Consequences*. Springer Verlag, KG, Berlin, Germany, pp. 152–158.

Hilborn, E. D., Carmichael, W. W., Wayne, W., Yuan, M., Azevedo, S. M. F. O. (2005) A simple colorimetric method to detect biological evidence of human exposure to microcystins. *Toxicon* **46:** 218–221.

Hindak, F. (1988) Planktonic species of two related genera, *Cylindrospermopsis* and *Anabaenopsis*, from western Slovakia. *Arch Hydrobiol* **80:** 283–302.

Hiripi, L., Nagy, L., Kalmar, T., Kovacs, A., and Voros, L. (1998) Insect (*Locusta migratoria migratorioides*) test monitoring the toxicity of cyanobacteria. *Neurotoxicol* **19:** 605–608.

Hirooka, E. Y., Pinotti, M. H., Tsutsumi, T., Yoshida, F., and Ueno, Y. (1999) Survey of microcystins in water between 1995 and 1996 in Parana, Brazil using ELISA. *Nat Toxins* **7:** 103–109.

Hitzfeld, B. C., Hoeger, S. J., and Dietrich, D. R. (2000) Cyanobacterial toxins: Removal during drinking water treatment, and human risk assessment. *Environ Health Perspect* **108** (Suppl): 113–122.

Hoeger, S. J., Dietrich, D. R., and Hitzfeld, B.C., (2002) Effect of ozonation on the removal of cyanobacterial toxins during drinking water treatment. *Environ Health Perspect* **110:** 1127-1132.

Hoffmann, D., Hevel, J. M., Moore, R. E., and Moore, B. S. (2003) Sequence analysis and biochemical characterization of the nostopeptolide A biosynthetic gene cluster from *Nostoc* sp. GSV 224. *Gene* **311:** 171–180.

Holmes, C. F. B., and Boland, M. P. (1993) Inhibitors of protein phosphatase-1 and -2A; two of the major serine/threonine protein phosphatases involved in cellular regulation. *Current Opin Struct Biol* **3:** 934–943.
Honakanen, R. E., Codispoti, B. A., Tse, K., and Boynton, A. L. (1994) Characterization of natural toxins with inhibitory activity against serine/threonine protein phosphatases. *Toxicon* **32:** 339–350.
Honakanen, R. E., Zwiller, J., Moore, R. E., Daily, S. L., Khatra, B. S., Dukelow, M., and Boynton, A. L. (1990) Characterization of microcystin-LR, a potent inhibitor of type 1 and type 2A protein phosphatase. *J Biol Chem* **265:** 19401–19404.
Hooser, S. B., Beasley, V. R., Lovell, R. A., Carmichael, W. W., and Haschek, W. M. (1991) Actin filament alterations in rat hepatocytes induced *in vivo* and *in vitro* by microcystin-LR, a hepatotoxin from the blue-green alga, *Microcystis aeruginosa. Vet Pathol* **28:** 259–266.
Horecka, M., and Komarek, J. (1979) Taxonomic position of three planktonic blue-green algae from the genera *Aphanizomenon* and *Cylindrospermopsis*. *Preslia* (Prague) **51:** 289–312.
Hormazabal, V., Ostensvik, O., Underdal, B., and Skulberg, O. M. (2000) Simultaneous determination of the cyanotoxins anatoxin A, microcystin desmethyl-3-RR, LR, RR and YR in water using liquid chromatography-mass spectrometry. *J Chromatogr Relat Technol* **23:** 3155–3164.
Hoshimoto, Y., Kamiya, H., Yamazato, K., and Nozawa, K. (1976) Occurrence of a toxic blue-green alga inducing skin dermatitis in Okinawa. In: Ohsaka, A., Hayashi, K., and Sawai, Y. (Eds.) *Animal, Plant and Microbial Toxins*. Plenum Publishing, New York, NY, USA. pp. 333–338.
Hotto, A. M., Satchwell, M. F., and Boyer, G. L. (2007) Molecular characterization of potential microcystin-producing cyanobacteria in Lake Ontario Embayments and nearshore waters. *Appl Environ Microbiol* **73:** 4570–4578.
Hrudey, S. E., Lambert, T. W., and Kenefick, S. L. (1994) Health risk assessment of microcystins in drinking water supplies. In: *Toxic Cyanobacteria-a Global Perspective*. Australian Centre for Water Quality Research, Adelaide, Australia. pp. 7–12.
Hrudey, S., Burch, S., Burch, M., Drikas, M., Greorgy, R., (1999) Remedial measures. In: Chorus, I., and Bartram, J. (Eds.), *Toxic Cyanobacteria in Water. A Guide to their Public Health Consequences, Monitoring and Management*. Routledge, London, pp. 275–312.
Huber, C. S. (1972) The crystal structure and absolute configuration of 2,9-diacetyl-9-azabicyclo[4,2,1]non-2,3-ene. *Acta Crystol* **238:** 2557–2582.
Hummert, C., Reichelt, M., and Luckas, B. (2000) New strategy for the determination of microcystins and diarrhetic shellfish poisoning (DSP) toxins, two potent phosphatases 1 and 2A inhibitors and tumor promoters. *J Anal Chem* **366:** 508–513.
Hummert, C., Reichelt, M., Weiss, J., Liebert, H. P., and Luckas, B. (2001) Identification of microcystins in cyanobacteria from the bleiloch former drinking-water reservoir (Thuringia, Germany). *Chemosphere* **44:** 1581–1588.
Humpage, A. R., and Falconer, I. R. (1999) Microcystin-LR and liver tumor promotion: Effects on cytokinesis, ploidy, and apoptosis in cultured hepatocytes. *Environ Toxicol Water Qual* **14:** 61–75.
Humpage, A. R., Hardy, S. J., Moore, E. J., Froscio, S. M., and Falconer, I. R. (2000) Microcystins (cyanobacterial toxins) in drinking water enhance the growth of aberrant crypt foci in the mouse colon. *J Toxicol Environ Health* **61:** 155–165.
Humpage, A. R., Rositano, J., Bretag, A. H., Brown, R., Baker, P. D., Nicholson, B. C., and Steffensen, D. A. (1994) Paralytic shellfish poisons from Australian cyanobacterial blooms. *Aus J Mar Freshwater Res* **45:** 761–771.
Huot, R. I., Armstrong, D. I., and Chanh, T. C. (1989) *In vitro* and *in situ* inhibition of the sodium channel blocker STX by monoclonal antibodies. *J Toxicol Environ Health* **27:** 381–393.
Imanishi, S., Kato, H., Mizuno, M., Tsuji, K., and Harada, K.-I. (2005) Bacterial degradation of microcystins and nodularin. *Chem Res Toxicol* **18:** 591–598.
Ishida, K., and Murakami, M. (2000) Kasumigamide, an antialgal peptide from the cyanobacterium *Microcystis aeruginosa. J Org Chem* **65:** 5898–5900.
Ishida, K., Christiansen, G., Yoshida, W. Y., Welker, M., Bonjoch, J., Hertweck, C., Börner, T., Hemscheidt, T. K., and Dittmann, E. (2007) Biosynthetic pathway and structure analysis of aeruginoside 126A and B, cyanobacterial peptides bearing an unusual 2-carboxy-6-hydroxyoctaindole moiety. *Chem Biol* **14:** 565–576.
Ishida, K., Matsuda, H., and Murakami, M. (1998) Four new microginins, linear peptides from the cyanobacterium *Microcystis aeruginosa. Tetrahedron.* **54:** 13475–13484.
Ishida, K., Okita, Y., Matsuda, H., Okino, T., and Murakami, M. (1999) Aeruginosins, protease inhibitors from the cyanobacterium *Microcystis aeruginosa. Tetrahedron.* **55:** 10971–10988.
Ishida, K., Welker, M., Christiansen, G., Cadel-Six, S., Bouchier, C., Dittmann, E., Hertweck, C., and Tandeau de Marsac, N. (2009) Plasticity and evolution of aeruginosin biosynthesis in cyanobacteria. *Appl Environ Microbiol* **75:** 2017–2026.
Ishii, H., Nishijima, M., and Abe, T. (2004) Characterization of degradation process of cyanobacterial hepatotoxins by a gram-negative aerobic bacterium. *Water Res* **38:** 2667–2676.
ISO 20179 (2005) Water Quality: Determination of microcystins-method by solid phase extraction (SPE) and high performance liquid chromatography (HPLC) with ultraviolet (UV) detection. ISO, Geneva, Switzerland.

Iteman, I., Rippka, R., Tandeau de Marsac, N., and Herdman, M. (2002) rDNA analyses of planktonic heterocystous cyanobacteria, including members of the genera *Anabaenopsis* and *Cyanospira*. *Microbiology* **148:** 481–496.

Ito, E., Kondo, F., Terao, K., and Harada, K.-I. (1997) Neoplastic nodular formation in mouse liver induced by repeated intraperitoneal injections of microcystin-LR. *Toxicon* **35:** 1453–1457.

Jacoby, J. M., Collier, D. C., Welch, E. B., Hardy, F. J., and Crayton, M. (2000) Environmental factors associated with a toxic bloom of *Microcystis aeruginosa*. *Can J Fish Aquat Sci* **57:** 231–240.

James, H., and Fawell, J. (1991) *Detection and Removal of Cyanobactrial Toxins from Freshwaters*. Report No. 0211. Foundation of Water Research, Murlow, Buckinghamshire, UK.

James, K. J., Sherlock, I. R., and Stack, M. A. (1997) Anatoxin-a in Irish freshwater and cyanobacteria, determined using a new fluorimetric liquid chromatographic method. *Toxicon* **35:** 963–971.

Jähnichen, S., Ihle, T., Petzoldt, T., and Benndorf, J. (2007) Impact of inorganic carbon availability on microcystin production by *Microcystis aeruginosa* PCC 7806. *Appl Environ Microbiol* **73:** 6994–7002.

Janse, I., Kardinaal, W. E. A., Meima, M., Fastner, J., Visser, P. M., and Zwart, G. (2004) Toxic and non-toxic *Microcystis* colonies in natural populations can be differentiated on the basis of rRNA gene internal transcribed spacer diversity. *Appl Environ Microbiol* **70:** 3979–3987.

Janse, I., Kardinaal, W. E. A., Agterveld, M. K-v., Meima, M., Visser, P. M., and Zwart, G. (2005) Contrasting microcystin production and cyanobacterial population dynamics in two *Planktothrix*-dominated freshwater lakes. *Environ Microbiol* **7:** 1514–1524.

Jochimsen, E. M., Carmichael, W. W., An, J., Cardo, D. M., Cookson, S. T., Holmes, C. E. M., Antunes, B. C., Liho, D., Lyra, T. M., Barreto, V. S. T., Azevedo, S. M. F. O., and Jarvis, W. R. (1998) Liver failure and death after exposure to microcystins at a haemodialysis center in Brazil. *New Eng J Med* **338:** 873–878.

John, W., Raynor, M. W., and Rae, B. (1997) Micellar electrokinetic capillary chromatography of algal toxins. *J High Res Chromatogr* **20:** 34–38.

Jones, A. C., Gerwick, L., Gonzalez, D., Dorrestein, P. C., and Gerwick, W. H. (2009) Transcriptional analysis of the jamaicamide gene cluster from the marine cyanobacterium *Lyngbya majuscula* and identification of possible regulatory proteins. *BMC Microbiol* **9:** 247. doi:1186/1471-2180-9-247.

Jones, G. J., and Orr, P. T. (1994) Release and degradation of microcystin following algicide treatment of a *Microcystis aeruginosa* bloom in a recreational lake, as determined by HPLC and protein phosphatase inhibition assay. *Water Res* **28:** 871–876.

Jones, G. J., Blackburn, S. I., and Parker, N. S. (2006) A toxic bloom of *Nodularia spumigena* Mertens in Orielton Lagoon, Tasmania. *Aus J Marine Freshwater Res* **45:** 787–800.

Jones, G. J., Bourne, D. G., Blakeley, R. L., and Doelle, H. (1994) Degradation of the cyanobacterial hepatotoxin microcystin by aquatic bacteria. *Nat Toxins* **2:** 228–235.

Jones, W. W., and Sauter, S. (2005) *Distribution and Abundance of Cylindrospermopsis raciborskii in Indiana Lakes and Reservoirs*. Office of Water Quality, Indiana Department of Environmental Management, Indianapolis, IN, USA.

Jos, A., Pichardo, S., Prieto, A. I., Repetto, G., Vazquez, C. M., Moreno, I., and Camean, A. M. (2005) Toxic cyanobacterial cells containing microcystins induce oxidative stress in exposed tilapia fish (*Oreochromis* sp.) under laboratory conditions. *Aquat Toxicol* **72:** 261–271.

Jung, S.-W., Baek, K.-H., and Yu, M.-J. (2005) Treatment of taste and odour material by oxidation and adsorption. Off-Flavours in the aquatic environment VI. *Water Sci Technol* **49:** 289–295.

Jungblut, A. D., Hoeger, S. J., Mountford, D., Hitzfeld, B. C., Dietrich, D. R., and Neilan, B. A. (2006) Characterization of microcystin production in an Antarctic cyanobacterial mat community. *Toxicon* **47:** 271–278.

Kaas, H., and Henriksen, P. (2000) STXs (PSTs) in Danish Lakes. *Water Res* **34:** 2089–2097.

Kaebernick, M., and Neilan, B. A. (2001)Ecological and molecular investigations of cyanotoxin production. *FEMS Microbiol Ecol* **35:** 1–9.

Kaebernick, M., Dittmann, E., Borner, T., and Neilan, B. A. (2002) Multiple alternate transcripts direct the biosynthesis of microcystin, a cyanobacterial nonribosomal peptide. *Appl Environ Microbiol* **68:** 449–455.

Kaebernick, M., Neilan, B. A., Boerner, T., and Dittsman, E. (2000) Light and the transcriptional response of the microcystin biosynthesis gene cluster. *Appl Environ Microbiol* **66:** 3387–3392.

Kaebernick, M., Rohrlack, T., Christoffersen, K., and Neilan, B. A. (2001) A spontaneous mutant of microcystin biosynthesis: Genetic characterization and effect on *Daphnia*. *Environ Microbiol* **3:** 669–679.

Kalbe, L., and Tiess, D. (1964) Entenmassensterben durch *Nodularia* wasserbute kleinen jasmundu dodden auf rugen. *Arch Exp Vet Med* **18:** 535–539. Cited from Carmichael, W. W., Eschedor, J. T., Patterson, G. M. L., and Moore, R. E. (1988) Toxicity and partial structure of a hepatotoxic peptide produced by the cyanobacterium *Nodularia spumigena* Mertens emend. L575 from New Zealand. *Appl Environ Microbiol* **54:** 2257–2263.

Karanfil, T., Kitis, M., Kilduff, J., and Wigton, A. (1999) Role of granular activated carbon surface chemistry on the adsorption of organic compounds. 2. Natural organic matter. *Environ Sci Technol* **33:** 3225–3233.

Karjalainen, M., Reinkainen, M., Lindvall, F., Spoof, L., and Meriluto, J. A. O. (2003) Uptake and accumulation of dissolved, radiolabeled nodularin in Baltic Sea zooplankton. *Environ Toxicol* **18:** 52–60.

Kaya, K., and Watanabe, M. M. (1990) Microcystin composition of an axenic clonal strain of *Microcystis viridis*-containing waterblooms in Papanese freshwaters. *J Appl Phycol* **2:** 173–178.

Kaya, K., Mahakhant, A., Keovara, L., Sano, T., Kubo, T., and Takagi, H. (2002) Spiroidesin, a novel lipopeptide from the cyanobacterium *Anabaena spiroides* that inhibits cell growth of the cyanobactrium *Microcystis aeruginosa. J Nat Prod* **65:** 920–921.

Kehr, J.-C., Zilliges, Y., Springer, A., Disney, M. D., Ratner, D. D., Bouchier, C., Seeberger, P. H., Tandeau de Marsac, N., and Dittmann, E. (2006) A mannan binding lectin is involved in cell-cell attachment in a toxic strain of *Microcystis aeruginosa. Mol Microbiol* **59:** 893–906.

Keleti, G., and Sykora, J. L. (1982) Production and properties of cyanobacterial endotoxins. *Appl Environ Microbiol* **43:** 104–109.

Keleti, G., Sykora, J. L., Libby, E. C., and Shapiro, M. A. (1979) Composition and biological properties of lipopolysaccharides isolated from *Schizothrix calcicola* (Ag.) Gomont (cyanobacteria). *Appl Environ Microbiol* **38:** 471–477.

Kellmann, R., and Neilan, B. A. (2007) Biochemical characterization of paralytic shellfish toxin biosynthesis *in vitro. J Phycol* **43:** 497–508.

Kellmann, R., Mihali, T. K., Jeon, Y. J., Pickford, R., Pomati, F., and Neilan, B. A. (2008) Biosynthetic intermediate analysis and functional homology reveal a putative saxitoxin gene cluster in cyanobacteria. *Appl Environ Microbiol* **74:** 4044–4053.

Kellmann, R., Mills, T., and Neilan, B. A. (2006) Functional modeling and phylogenetic distribution of putative cylindrospermopsin biosynthesis enzymes. *J Mol Evol* **62:** 267–280.

Kellmann, R., Stücken, A., Orr, R. J. S., Svendsen, H. M., and Jakobsen, K. S. (2010) Biosynthesis and molecular genetics of polyketides in marine dinoflagellates. *Mar Drugs* **8:** 1011-1-48. doi:10.3390/md8041011.

Kemp, A., and John, J. (2006) Microcystins associated with *Microcystis* dominated blooms in the Southwest wetlands, Western Australia. *Environ Toxicol* **21:** 125–130.

Kerr, J. F. R., Wyllie, A. H., and Currie, A. R. (1972) Apoptosis: A basic biological phenomenon with wide ranging implications in tissue kinetics. *Br J Cancer* **26:** 239–257.

Kfir, R., Johannsen, E., and Botes, D. P. (1986) Monoclonal antibody specific for cyanoginosin-LA: Preparation and characterization. *Toxicon* **24:** 543–552.

Khan, S. A., Wickstrom, M. L., Haschek, W. M., Schaeffer, D. J., Ghosh, S., and Beasley, V. R. (1996) Microcystin-LR and kinetics of cytoskeletal reorganization in hepatocytes, kidney cells, and fibroblasts. *Nat Toxins* **4:** 206–214.

Kiviranta, J., Sivonen, K., and Niemela, S. I. (1991) Detection of toxicity of cyanobacteria by *Artemia salina* bioassay. *Environ Toxicol Water Qual* **6:** 423–436.

Kleinkauf, H., and von Doehren, H. (1996) A nonribosomal system of peptide biosynthesis. *Eur J Biochem* **236:** 335–351.

Kodania, S., Suzuukia, S., Ishidaa, K., and Murakamia, M. (1999) Five new cyanobacterial peptides from water bloom materials of lake Teganuma (Japan). *FEMS Microbiol Lett* **178:** 343–348.

Komarek, J., and Kling, H. (1991) Variation in six planktonic cyanophyte genera in Lake Victoria (East Africa). *Arch Hydrobiol* **88:** 21–46.

Komarkova, J. (1998) The tropical planktonic genus *Cylindrospermopsis* (Cyanophytes, Cyanobacteria). In: Azevedo, S. M. F. O. (Ed.) Anais dos IV Congresso Latino-Americano de Ficologia. II. Reuniao lbero Americana de Ficologia e VII Reuniaeo Brasileira de Ficologia. Vol. I, Sao Paulo, Brazil. pp. 327–340.

Komarkova, J., Laudares-Silva, R., and Senna, P. A. C. (1999) Extreme morphology of *Cylindrospermopsis raciborskii* (Nostocales, Cyanobacteria) in the Lagoa do Peri, a freshwater coastal lagoon, Santa Catarina, Brazil. *Arch Hydrobiol* (Suppl. 129) *Algol Stud* **94:** 207–222.

Kondo, F., Ikai, Y., Oka, H., Okumura, M., Ishikawa, N., Harada, K.-I., Matsuura, K., Murata, H., and Suzuki, M. (1992a) Formation, characterization, and toxicity of the glutathione and cysteine conjugates of toxic heptapeptide microcystins. *Chem Rev Toxicol* **5:** 591–596.

Kondo, F., Ikai, Y., Oka, H., Ishikawa, N., Watanabe, M. F., Watanabe, M., Harada, K.-I., and Suzuki, M. (1992b) Separation and identification of microcystins in cyanobacteria by frit-fast atom bombardment liquid chromatography/mass spectrometry. *Toxicon* **30:** 227–237.

Kondo, F., Ikai, Y., Oka, H., Matsumoto, H., Yamada, S., Ishikawa, N., Tsuji, K., Harada, K.-I., Shimada, T., Oshikata, M., and Suzuki, M. (1995) Reliable and sensitive method for determination of microcystins in complicated matrices by frit-fast atom bombardment liquid chromatography/mass spectrometry. *Nat Toxins* **3:** 41–49.

Kondo, F., Matsumoto, H., Yamada, S., Ishikawa, N., Ito, E., Nagata, S., Ueno, Y., Suzuki, M., and Harada, K.-I. (1996) Detection and identification of metabolites of microcystins formed *in vivo* in mouse and rat livers. *Chem Res Toxicol* **9:** 1355–1359.

Kondo, R., Yoshida, T., Yuki, Y., and Shingo, H. (2000) DNA-DNA reassociation among a bloom-forming cyanobacterial genus, *Microcystis. Int J Syst Evol Microbiol* **50:** 767–770.

Kos, P., Gorzo, G., Suranyi, G., and Borbely, G. (1995) Simple and efficient method for isolation and measurement of cyanobacterial hepatotoxins by plant tests (*Sinapis alba* L.). *Anal Biochem* **225:** 49–53.

Koskenniemi, K., Lyra, C., Rajaniemi-Wacklin, P., Jokela, J., and Sivonen, K. (2007) Quantitative real-time PCR detection of toxic *Nodularia* cyanobacteria in the Baltic Sea. *Appl Environ Microbiol* **73:** 2173–2174.

Kotak, B. G., Kenefick, S. L., Fritz, D. L., Rousseaux, C. G., Prepas, E. E., and Hrudey, S. E. (1993) Occurrence and toxicological evaluation of cyanobacterial toxins in Alberta lakes and farm dugouts. *Water Res* **27:** 495–506.

Kotak, B.G., Lam, A. K.-Y., Prepas, E. E., Holmes, C. F. B., and Hrudey, S. E. (2000) Role of chemical and physical variables in regulating microcystin-LR concentration in phytoplankton of eutrophic lakes. *Can J Fish Aquat Sci* **57:** 1584–1593.

Krishnamurthy, T. Carmichael, W. W., and Sarver, E. W. (1986) Toxic peptides from freshwater cyanobacteria (Blue-green algae). I. Isolation, purification and characterization of peptides from *Microcystis aeruginosa* and *Anabaena flos-aquae. Toxicon* **24:** 865–873.

Krishnamurthy, T., Szafraniec, L., Hunt, D. F., Shabanowitz, J., Yates, J. R. III., Hauer, C. R., Carmichael, W. W., Skulberg, O., Codd, G. A., and Missler, S. (1989) Structural characterization of toxic cyclic peptides from blue-green algae by tandem mass spectrometry. *Proc Natl Acad Sci* USA **86:** 770–774.

Kuiper-Goodman, T., Falconer, I., and Fitzgerald, J. (1999) Human health aspects. In: Chorus, I., and Bartram, J. (Ed.) *Toxic Cyanobacteria in Water: A Guide to Public Health Consequences, Monitoring and Management*. WHO Publ., E. & F. N. Spon, London-New York. pp. 41–111.

Kull, T. P., Backlund, P. H., Karlsson, K. M., and Meriluto, J. A. (2004) Oxidation of the cyanobacterial hepatotoxin microcystin-LR by chlorine dioxide: Reaction kinetics, characterization, and toxicity of reaction products. *Environ Sci Toxicol* **38:** 6025–6031.

Kull, T. P., Sjovall, O. T., Tammenkoski, M. K., Backlund, P. H., and Meriluto, J. A. (2006) Oxidation of the cyanobacterial hepatotoxin microcystin-LR by chlorine dioxide: Influence of natural organic matter. *Environ Sci Technol* **40:** 1504–1510.

Kurmayer, R., and Christiansen, G. (2009) The genetic basis of toxin production in cyanobacteria. *Freshwater Rev* **2:** 31–50.

Kurmayer, R., and Juettner, F. (1999) Strategies for the co-existence of zooplankton with the toxic cyanobacterium *Planktothrix rubescens* in lake Zurich. *J Plankton Res* **21:** 659–683.

Kurmayer, R., and Kutzenberger, T. (2003) Application of real-time PCR for quantification of the toxic cyanobacterium *Microcystis* sp. *Appl Environ Microbiol* **69:** 6723–6730.

Kurmayer, R., Christiansen, G., Fastner, J., and Boemer, T. (2004) Abundance of active and inactive microcystin genotypes in populations of the toxic cyanobacterium *Planktothrix* spp. *Environ Microbiol* **6:** 831–841.

Kurmayer, R., Christiansen, G., Gumpenberger, M., and Fastner, J. (2005) Genetic identification of microcystin ecotypes in toxic cyanobacteria of the genus *Planktothrix*. *Microbiology* **151:** 1525–1533.

Kusumi, T., Ooi, T., Watanabe, M. M., Takahasi, H., and Kakisawa, H. (1987) Cyanoviridin RR, a toxin from the cyanobacterium (blue-green alga) *Microcystis viridis*. *Tetrahedron Lett* **28:** 4695–4698.

Kyselkova, I., and Marsalek, B. (2000) Using of *Daphnia pulex*, *Artemia salina* and *Tubifex tubifex* for cyanobacterial microcystins toxicity detection. *Biologia* **55:** 637–643.

Lagos, N., Onodera, H., Zagatto, P. A., Andrinolo, D., Azevedo, S. M., and Oshima, Y. (1999) The first evidence of paralytic shellfish toxins in the freshwater cyanobacterium *Cylindrospermopsis raciborskii*, isolated from Brazil. *Toxicon* **37:** 1359–1373.

Lahti, K., Ahtiainen, J., Rapala, J., Sivonen, K., and Niemela, S. I. (1995) Assessment of rapid bioassays for detecting cyanobacterial toxicity. *Lett Appl Microbiol* **21:** 109–114.

Lahti, K., Rapala, J., Faerdig, M., Niemela, M., and Sivonen, K. (1997) Presence of cyanobacterial hepatotoxin, microcystin-LR in particulate material and dissolved in lake water. *Water Res* **31:** 187–195.

Lam, A. K. Y., Prepas, E. E., Spink, D., and Hrudey, S. E. (1995) Chemical control of hepatotoxic phytoplankton blooms: implications for human health. *Water Res* **29:** 1845–1854.

Lambert, T. W., Boland, M. P., Holmes, C. F. B., and Hrudey, S. E. (1994) Quantitation of the microcystin hepatotoxins in water at environmentally relevant concentrations with the protein phosphatase bioassay. *Environ Sci Technol* **28:** 753–755.

Lampert, W. (1981) Inhibitory and toxic effects of blue-green algae on *Daphnia*. *Int Revue Ges Hydrobiol* **66:** 285–298.

Lampert, W. (1987) Laboratory studies on zooplankton-cyanobacteria interactions. *New Zealand J Mar Freshwater Res* **21:** 483–490.

Lanaras, T., and Cook, C. M. (1994) Toxin extraction from an *Anabaenopsis milleri*-dominated bloom. *Sci Technol Environ* **142:** 163–169.

Lang, J. D., and Haselkorn, R. (1989) Isolation, sequence and transcription of the gene encoding the photosystem II chlorophyll-binding protein phosphatase 1 and 2A from both mammals and higher plants. *Plant Mol Biol* **13:** 441–457.

Langlais, B., Reckhow, D. A., and Brink, D. R. (1991) *Ozone in Water Treatment. Application and Engineering*. CO/Chelesa, MI, American Water Works Association Research Foundation, Lewis Publishers.

Lankoff, A., and Banasik, A. (2001) The effect of cyanobacterial toxins on the stability of intracellular membranes. *Cell Biol Mol Lett* **6:** 410

Lankoff, A., Banasik, A., and Nowak, M. (2002) Protective effect of melatonin against nodularin-induced oxidative stress. *Arch Toxicol* **76:** 158–165.

Lankoff, A., Banasik, A., Obe, G., Deperas, M., Kuzminski, K., Tarczynska, M., Jurczak, T., and Wojcik, A. (2003). Effect of microcystin-LR and cyanobacterial extract from Polish reservoir of drinking water on cell cycle progression, mitotic spindle and apoptosis in CHO-K1 cells. *Toxicol Appl Pharmacol* **189:** 204–212.

Lankoff, A., Bialczyk, J., Dziga, D., Carmichael, W. W., Gradzka, I., Lisowska, H., Kuszewski, T., Gozdz, S., Piorun, I., and Wojcik, A. (2006) The repair of gamma-irradiation-induced DNA damage is inhibited by microcystin-LR, the PP1 and PP2A phosphatase inhibitor. *Mutagenesis* **21:** 83–90.

Lankoff, A., Carmichael, W. W., Grasman, K. A., and Yuan, M. (2004) The uptake kinetics and immunotoxic effects of microcystin-LR and chicken peripheral blood lymphocytes *in vitro*. *Toxicology* **204:** 23–40.

Lankoff, A., Krzowski, L., Glab, J., Banasik, A., Lisowska, H., Kuszewski, T., Gozdz, S., and Wojcik, A. (2004) DNA damage and repair in human peripheral blood lymphocytes following treatment with microcystin-LR. *Mutat Res* **559:** 131–142.

Lawton, L. A., Campbell, D. L., Beattie, K. A., and Codd, G. A. (1990) Use of a rapid bioluminescence assay for detecting cyanobacterial microcystin toxicity. *Lett Appl Microbiol* **11:** 205–207.

Lawton, L. A., Cornish, B. J. P. A., and MacDonald, A. W. R. (1998) Removal of cyanobacterial toxins (microcystins) and cyanobacterial cells from drinking water using domestic water filters. *Water Res* **32:** 633–638.

Lawton, L. A., Edwards, C., and Codd, G. A. (1994) Extraction and high-performance liquid chromatographic method for the determination of microcystins in raw and treated waters. *Analyst* **119:** 1525–1530.

Lawton, L. A., Edwards, C., Beattie, K. A., Pleasance, S., Dear, D. J., and Codd, G. A. (1995) Isolation and characterization of microcystins from laboratory cultures and environmental samples of *Microcystis aeruginosa* and from an associated animal toxicosis. *Nat Toxins* **3:** 50–57.

Leckoski, N., Fewer, D. P., and Sivonen, K. (2009) Wide spread occurrence and lateral gene transfer of the cyanobactin biosynthesis gene cluster in cyanobacteria. *Appl Environ Microbiol* **75:** 853–857.

Ledereux, A., Thomazeau, S., Catharine, A., Duval, C., Yéprémian, C., Marie, A., and Bernard, C. (2010) Evidence for saxitoxins production by the cyanobacterium *Aphanizomenon gracile* in a French recreational water body. *Harmful Algae* **10:** 88–97.

Lee, T. H., Chen, Y. M., and Chou, H. N. (1998) First report of microcystins in Taiwan. *Toxicon* **36:** 247–255.

Legrini, O., Oliveros, E., and Braun, A. M. (1993) Photochemical processes for water treatment. *Chem Rev* **93:** 671–698.

Lehtimaeki, J., Moisander, P., Sivonen, K., and Kononen, K. (1997) Growth, nitrogen fixation, and nodularin production by two Baltic Sea cyanobacteria. *Appl Environ Microbiol* **63:** 1647–1656.

Leusch, H., Pangilinan, R., Yoshida, W. Y., Moore, R. E., and Paul, V. J. (2001) Pitipeptolides A and B, new cyclodepsipeptides from the marine cyanobacterium *Lyngbya majuscula*. *J Nat Prod* **64:** 304–307.

Levin, R. E. (1992) Paralytic shellfish toxins: Their origin, characteristics, and methods of detection: A review. *J Food Biochem* **15:** 405–417.

Li, W. I., Berman, F. W., Yokokawa, F., Shioiri, T., Gerwick, W. H., and Murray, T. F. (2001) Antillatoxin is a marine cyanobacterial toxin that potently activates voltage-gated sodium channels. *Proc Natl Acad Aci USA* **98:** 7599–7604.

Lim, I. K., Park, T. J., Park, S. C., Yoon, G., Kwak, C. S., Lee, M. S., Song, K. Y., Choi, Y. K., and Hyun, B. H. (2001) Selective left-lobe atrophy by nodularin treatment accompanied by reduced protein phosphatase 1/2a and increased peroxisome proliferation in rat liver. *Int J Cancer* **91:** 32–40.

Lin, J. R., and Chu, F. S. (1994a) Kinetics of distribution of microcystin-LR in serum and liver cytosol of mice: An immunochemical analysis. *J Agric Food Chem* **42:** 1035–1040.

Lin, J. R., and Chu, F. S. (1994b) *In vitro* neutralization of the inhibitory effect of microcystin-LR to protein phosphatase 2A by antibody against the toxin. *Toxicon* **32:** 605–613.

Li, R., Carmichael, W. W., Brittain, S., Eaglesham, G. K., Shaw, G. R., Liu, Y., and Watanabe, M. M. (2001) First report of the cyanotoxins cylindrospermopsin and deoxycylindrospermopsin from *Raphidiopsis curvata* (cyanobacteria). *J Phycol.* **37:** 1121–1126.

Lindstrom, E. (1976) Et utbrud af algeforgiftning blandt hunde. *Dansk Veterinaertidsskr* **59:** 637–641 (In Danish) Cited from Sivonen, K., Kononen, K., Carmichael, W. W., Dahlem, A. M., Rinehart, K. L., Kiviranta, J., Niemela, S. I. (1989) Occurrence of the hepatotoxic cyanobacterium *Nodularia spumigena* in the Baltic Sea and structure of the toxin. *Appl Environ Microbiol* **55:** 1990–1995.

Lippy, E. C., and Erb, J. (1976) Gastrointestinal illness at Sewickley, PA. *J Am Water Works Assoc* **68:** 606–610.

Liu, B.-H., Yu, F.-Y., Huang, X., and Chu, F. F. (2000) Variation of microcystin content of *Microcystis aeruginosa* relative to medium N:P ratio and growth stage. *J Appl Microbiol* **89:** 323–329.

Liu, Y., Chen, W., Li, D., Shen, Y., Liu, Y., and Song, L. (2006) Analysis of paralytic shellfish toxins in *Aphanizomenon* DC-1 from Lake Dianchi, China. *Environ Toxicol* **21:** 289–295.

Liu, Y., Song, L., Li, X., Liu, T. (2002) The toxic effects of microcystin-LR on embryo-larval and juvenile development of loach, *Misguruns mizolepis* Gunthe. *Toxicon* **40:** 395–399.
Long, B. M., Jones, G. J., and Orr, P. T. (2001) Cellular microcystin content in N-limited *Microcystis aeruginosa* can be predicted from growth rate. *Appl Environ Microbiol* **67:** 278–283.
Looper, R. E., Runnegar, M. T. C., and Williams, R. W. (2005) Synthesis of putative structure of 7-deoxycylindrospermopsin: C7 oxygenation is not required for the inhibition of protein synthesis. *Angewandte Chemie* **44:** 3879–3881.
Luesch, H., Yoshida, W. Y., Moore, R. E., and Paul, V. J. (1999) Lybgbyastatin 2 and norlyngbyastatin 2, analogues of dolastatin G and nordolastatin G from the marine cyanobacterium *Lyngbya majuscula*. *J Nat Prod* **62:** 1702–1706.
Lukac, M., and Aegerter, R. (1993) Influence of trace metals on growth and toxin production of *Microcystis aeruginosa*. *Toxicon* **31:** 293–305.
Luukkainen, R., Namikoshi, M., Sivonen, K., Rinehart, K. L., and Niemela, S. I. (1994) Isolation and identification of 12 microcystins from four strains and two bloom samples of *Microcystis* spp.: Structure of a new hepatotoxin. *Toxicon* **32:** 133–139.
Luukkainen, R., Sivonen, K., Namikoshi, M., Faerdig, M., Rinehart, K. L., and Niemela, S. I. (1993) Isolation and identification of eight microcystins from thirteen *Oscillatoria* strains and structure of a new microcystin. *Appl Environ Microbiol* **59:** 2204–2209.
Lyra, C., Hantula, J., Vainio, E., Rapala, J., Rouhiainen, l., and Sivonen, K. (1997) Characterization of cyanobacteria by SDS-PAGE of whole-cell proteins and PCR/RFLP of the 16S rRNA gene. *Arch Microbiol* **168:** 176–184.
Lyra, C., Laamanen, M., Lehtimaeki, J. M., Surakka, A., and Sivonen, K. (2005) Benthic cyanobacteria of the genus *Nodularia* are non-toxic, without gas vaculoes, able to glide and genetically more diverse than planktonic *Nodularia*. *Int J Syst Evol Microbiol* **55:** 555–568.
Ma, B., Chen, Y., Hao, H., Wang, B., Lv, H., and Zhang, G. (2005) Influence of ultrasonic field on microcystins produced by bloom forming algae. *Colloids Surf Biointerfaces* **41:** 197–201.
Maatouk, I., Bouaicha, N., Fontan, D., and Levi, Y. (2002) Seasonal variation of microcystin concentration in Saint-Caprais reservoir (France) and their removal in a small full scale treatment plant. *Water Res* **36:** 2891–2897.
MacKintosh, C., and MacKintosh, R. W. (1994) The inhibition of protein phosphatases by toxins: implications for health and an extremely sensitive and rapid bioassay for toxin detection. In: Codd, G. A., Jefferies, T. M., Keevil, C. W., and Potter, E. (Eds.) *Detection Methods for Cyanobacterial Toxins*. Royal Society of Chemistry, Cambridge, UK. pp. 90–99.
MacKintosh, C., Beattie, K. A., Klumpp, S., Cohen, P., and Codd, G. A. (1990) Cyanobacterial microcystin-LR is a potent and specific inhibitor of protein phosphatases 1 and 2A from both mammals and higher plants. *FEBS Lett* **264:** 187–192.
MacKintosh, R. W., Dalby, K. N., Campbell, D. G., Cohen, P. T., Cohen, P., and MacKintosh, C. (1995) The cyanobacterial toxin microcystin binds covalently to cysteine-273 on protein phosphatase 1. *FEBS Lett* **371:** 236–240.
MacMillan, J. B., and Molinski, T. F. (2002) Lobocyclamide B from *Lyngbya confervoides*. Configuration and asymetric synthesis of beta-hydroxy-alpha-amino acids by (-)asparteine-mediated aldol addition. *Org Lett* **4:** 1883–1886.
Magalhaes, V. F., Soares, R. M., and Azevedo, S. M. F. O. (2001) Microcystin contamination in fish from the Jacarepagua Lagoon (Rio de Janeiro, Brazil): Ecological implication and human health risk. *Toxicon* **39:** 1077–1085.
Mahmood, N. A., and Carmichael, W. W. (1986a) The pharmacology of anatoxin-a(s), a neurotoxin produced by the freshwater cyanobacterium *Anabaena flos-aquae* NRC-17. *Toxicon* **24:** 425–434.
Mahmood, N. A., and Carmichael, W. W. (1986b) Paralytic shelfish poisons produced by the freshwater cyanobacterium *Aphanizomenon flos-aquae* NH5. *Toxicon* **24:** 175–186.
Mahmood, N. A., and Carmichael, W. W. (1987) Anatoxin-a(s), an anticholinesterase from the cyanobacterium *Anabaena flos-aquae* NCR-525-17. *Toxicon* **25:** 1221–1227.
Main, D. C., Berry, P. H., Peet, R. L., and Robertson, J. P. (1977) Sheep mortalities with the blue-green alga *Nodularia spumigena*. *Aus Vet J* **53:** 578–581.
Majsterek, I., Sicinska, P., Tarczynska, M., Zalewski, M., and Walter, Z. (2004) Toxicity of microcystin from cyanobacteria growing in a source of drinking water. *Toxicol Pharmacol* **139:** 175–179.
Makowski, A., and Wardas, W. (2001) Photocatalytic degradation of toxins secreted to water by cyanobacteria and unicellular algae and photocatalytic degradation of the cells of selected microorganisms. *Curr Topics Biochem* **25:** 19–25.
Mankiewicz, J., Komarkova, J., Izydorczyk, K., Jurczak, T., Tarczynska, M., and Zalewski, M. (2005) Hepatotoxic cyanobacterial blooms in the lakes of northern Poland. *Environ Toxicol* **20:** 499–506.
Mankiewicz, J., Walter, Z., Tarczynska, M., Fladmark, K. E., Doskeland, S. O., and Zalewski, M. (2000) Apoptotic effect of cyanobacterial blooms collected from water reservoirs. *Polish J Occupational Med Environ Health* **13:** 335–344.
Mankiewicz, J., Walter, Z., Tarczynska, M., Fladmark, K. E., Doskeland, S. O., and Zalewski, M. (2001) Apoptotic effect of cyanobacterial extract on rat hepatocytes and human lymphocytes. *Environ Toxicol Water Qual* **3:** 225–233.
Mankiewicz, J., Walter, Z., Tarczynska, M., Palyvoda, O., Wojtysiak-Staniaszczyk, M., and Zalewski, M. (2002) Genotoxicity of cyanobacterial extracts containing microcystins from Polish water reservoirs as determined by SOS chemostat and comet assay. *Environ Toxicol* **17:** 341–350.

Marner, F. J., and Moore, R. E. (1977) Majusculamides A and B, two epimeric lipopeptides from *Lyngbya majuscula* Gomont. *J Org Chem* **42:** 2815–2819.

Marsalek, B., and Blaha, L. (2004) Comparison of 17 biotests for detection of cyanobacterial toxicity. *Environ Toxicol* **19:** 310–317.

Marsalek, B., Blaha, L., Turanek, J., and Neca, J. (2001) Microcystin-LR and total microcystins in cyanobacterial blooms in the Czech Republic 1993–1998. In: Chorus, I. (Ed.). *Cyanotoxins-Occurrence, Causes and Consequences*. Springer, Berlin, Germany. pp. 56–62.

Mason, C. P., Edwards, K. R., Carlson, R. E., Pignatello, J., Gleason, P. K., and Wood, J. M. (1982) Isolation of chlorine-containing antibiotic from the freshwater cyanobacterium *Scytonema hofmanni*. *Science* **215:** 400–402.

Mastin, B. J., Rodgers, Jr., J. H., and Deardorff, T. L. (2002) Risk evaluation of cyanobacteria-dominated algal blooms in a North Lousiana reservoir. *J Aquat Ecosys Stress and Recovery* **9:** 103–114.

Mathys, W., and Surholt, B. (2004) Analysis of microcystins in freshwater samples using high performance liquid chromatography and an enzyme-linked immunosorbent assay. *Int J Hyg Environ Health* **207:** 601–605.

Matsunaga, H., Harada, K.-I., Senma, M., Ito, Y., Yasuda, N., Ushida, S., and Kimura, Y. (1999) Possible cause of unnatural mass death of wild birds in a pond in Nishinomiya, Japan: Sudden appearance of toxic cyanobacteria. *Nat Toxins* **7:** 81–84.

Matsunaga, S., Moore, R. E., Niemczura, W. P., and Carmichael, W. W. (1989) Anatoxin-a(s): A potent anticholinesterase from *Anabaena flos-aquae*. *J Am Chem Soc* **111:** 8021–8023.

Matthiensen, A., Beattie, K. A., Yunes, J. S., Kaya, K., and Codd, G. A. (2000) [D-Leu]microcystin-LR, from the cyanobacterium *Microcystis* RST 9501 and from a *Microcystis* bloom in the Patos Lagoon estuary, Brazil. *Phytochemistry* **55:** 383–387.

Maynes, J. T., Luu, H. A., Cherney, M. M., Andersen, R. J., Williams, D., Holmes, C. F., and James, M. N. (2006) Crystal structures of protein phosphatase-1 bound to motuporin and dihydromicrocystin-LA: Elucidation of the mechanism of enzyme inhibition by cyanobacterial toxins. *J Mol Biol* **356:** 111–120.

Mazmouz, R., Chapuis-Hugon, F., Mann, S., Pichon, V., Méjean, A., and Ploux, O. (2010) Biosynthesis of cylindrospermopsin and 7-epicylindrospermopsin in *Oscillatoria* sp. PCC 6506. Identification of the *cyr* gene cluster and toxin analysis. *Appl Environ Microbiol* **76:** 4943–4949. doi:10.1128/AEM.00717-10.

Mazur, H., and Plinski, M. (2001) Stability of cyanotoxins, microcystin-LR and nodularin in seawater and BG-11 medium of different salinity. *Oceanologia* **43:** 329–339.

Mazur-Marzec, H. (2006) Characterization of phycotoxins produced by cyanobacteria. *Oceanol Hydrobiol Stud* **35:** 85–109.

Mazur-Marzec, H., Meriluto, J., Plinski, M., and Szafranek, J. (2006) Characterization of nodularin variants in *Nodularia spumigena* from the Baltic Sea using liquid chromatography/mass spectrometry/mass spectrometry. *Rapid Commun Mass Spectrom* **20:** 2023–2032.

Mazyck, D., and Cannon, F. (2000) Overcoming calcium catalysis during the thermal reactivation of granular activated carbon. Part I. Steam-curing plus ramped-temperature N_2 treatment. *Carbon* **38:** 1785–1799.

Mazyck, D., and Cannon, F. (2002) Overcoming calcium catalysis during the thermal reactivation of granular activated carbon. Part II. Variation of steam curing reactivation parameters. *Carbon* **40:** 241–252.

Mbedi, S., Welker, M., Fastner, J., and Wiedner, C. (2005) Variability of the microcystin synthetase gene cluster in the genus *Planktothrix* (Oscillatoriales, Cyanobacteria). *FEMS Microbiol Lett* **245:** 299–306.

McCafferty, J., Griffiths, A. D., Winter, G., and Chiswell, D. J. (1990) Phage antibodies: Filamentous phage displaying antibody variable domains. *Nature* (London) **348:** 552–554.

McDaniel, R., Ebert-Khosla, S., Hopwood, D. A., and Khosla, C. (1993) Engineered biosynthesis of novel polyketides. *Science* **262:** 1546–1557.

McDermott, C. M., Nho, C. W., Howard, W., and Holton, B. (1998) The cyanobacterial toxin, microcystin-LR, can induce apoptosis in a variety of cell types. *Toxicon* **36:** 1981–1996.

McDermott, C. M., Feola, R., and Plude, J. (1995) Detection of cyanobacterial toxins (microcystins) in waters of northeastern Wisconsin by a new immunoassay technique. *Toxicon* **33:** 1433–1442.

McElhiney, J., and Lawton, L. A. (2005) Detection of the cyanobacterial hepatotoxins microcystins. *Toxicol Appl Pharmacol* **203:** 219–230.

McElhiney, J., Drever, M., Lawton, L. A., and Porter, A. J. (2002) Rapid isolation of a single-chain antibody against the cyanobacterial toxin microcystin-LR by phage display and its use in the immunoaffinity concentration of microcystins from water. *Appl Environ Microbiol* **68:** 5288–5295.

McElhiney, J., Lawton, L. A., and Porter, A. J. (2000) Detection and quantification of microcystins (cyanobacterial hepatotoxins) with recombinant antibody fragments isolated from a naïve human phage display library. *FEMS Microbiol Lett* **193:** 83–88.

McGregor, G. B., and Fabbro, L. D. (2000) Dominance of *Cylindrospermopsis raciborskii* (Nostocales, Cyanoprokaryota) in Queensland tropical and subtropical reservoirs: Implications for monitoring and management. *Lakes Reservoirs: Research Management* **5:** 195–205.

McGregor, D. P., Molloy, P. E., Cunnigham, C., and Harris, W. J. (1994) Spontaneous assembly of bivalent single-chain antibody fragments in *Escherichia coli*. *Mol Immunol* **31:** 219–226.

Méjean, A., Mann, S., Maldiney, T., Vassiliadis, G., Lequin, O., and Ploux, O. (2009) Evidence that biosynthesis of neurotoxic alkaloids anatoxin-a and homoanatoxin-a in the cyanobacterium *Oscillatoria* PCC 6506 occurs on a modular polyketide synthase initiated by 1-proline. *J Am Chem Soc* **131:** 7512–7513.

Méjean, A., Mazmouz, R., Mann, S., Calteau, A., Médigue, C., and Ploux, O. (2010) The genome sequence of the cyanobacterium *Oscillatoria* sp. PCC 6506 reveals several gene clusters responsible for the biosynthesis of toxins and secondary metabolites. *J Bacteriol* **192:** 5264–5265. doi:10.1128/JB.00704-10

Meriluto, J. (1997) Chromatography of microcystins. *Anal Chim Acta* **352:** 298.

Meriluto, J. A., and Eriksson, J. E. (1988) Rapid analysis of peptide toxins in cyanobacteria. *J Chromatogr* **438:** 93–99.

Meriluto, J. A., Harmala-Brasken, A.-S., Eriksson, J., Toivola, D., and Lindhol, T. (1996) Choosing analytical strategy for microcystins. *Phycologia* **35** (Suppl 6): 125–132.

Meriluto, J. A., Nygard, S. E., Dahlem, A. M., and Eriksson, J. E. (1990) Synthesis, organotropism and hepatocellular uptake of two tritium-labeled epimers of dihydromicrocystin-LR, a cyanobacterial peptide toxin analog. *Toxicon* **28:** 1439–1446.

Metcalf, J. S., and Codd, G. A. (2000) Microwave oven and boiling water bath extraction of hepatotoxins from cyanobacterial cells. *FEMS Microbiol Lett* **184:** 241–246.

Metcalf, J. S., and Codd, G. A. (2004) *Cyanobacterial Toxins in the Water Environment*. Foundation for Water Research, UK. pp. 1–36.

Metcalf, J. S., Bell, S. G., and Codd, G. A. (2001) Colorimetric immuno-protein phosphatase inhibition assay for specific detection of microcystins and nodularins of cyanobacteria. *Appl Environ Microbiol* **67:** 904–909.

Mez, K., Beattie, K. A., Codd, G. A., Hanselmann, K., Hauser, B., Naegeli, H., and Preisig, H. R. (1997) Identification of a microcystin in benthic cyanobacteria linked to cattle deaths on alpine pastures in Switzerland. *Eur J Phycol* **32:** 111–117.

Mez, K., Hanselmann, K., Naegeli, H., and Preisig, H. R. (1996) Protein-phosphatase-inhibiting activity in cyanobacteria from alpine lakes in Switzerland. *Phycologia* (Suppl) **35:** 133–139.

Mihali, T. K., Carmichael, W. W., and Neilan, B. A. (2011) A putative gene cluster from a *Lyngbya wollei* bloom that encodes paralytic shellfish toxin biosynthesis. *PLoS ONE* **6:** e14657. doi:10.1371/journal.pone.0014657

Mihali, T. K., Kellmann, R., Muenchhoff, J., Barrow, K. D., and Neilan, B. A. (2008) Characterization of the gene cluster responsible for cylindrospermopsin biosynthesis. *Appl Environ Microbiol* **74:** 716–722.

Mihali, T. K., Kellmann, R., and Neilan, B. A. (2009) Characterization of the paralytic shellfish toxin biosynthesis gene clusters in *Anabaena circinalis* AWQC131C and *Aphanizomenon* sp. NH-5. *BMC Biochemistry* **10:** 8.doi:10.1186/1471—2091-10-8

Mikalsen, B., Boison, G., Skulberg, O. M., Fastner, J., Davies, W., Gabrielsen, T. M., Rudi, K., and Jakobsen, K. S. (2003) Natural variation in the microcystin synthetase operon *mcyABC* and impact on microcystin production in *Microcystis* strains. *J Bacteriol* **185:** 2774–2785.

Mikhailov, A., Harmala-Brasken, A. S., Meriluto, J., Sorokina, Y., Dietrich, D., and Eriksson, J. E. (2001) Production and specificity of mono and polyclonal antibodies against microcystins conjugated through N-methylhydroalanine. *Toxicon* **39:** 477–483.

Miller, M. J., and Fallowfield, H. J. (2001) Degradation of cyanobacterial hepatotoxins in batch experiments. *Water Sci Technol* **43:** 229–232.

Miller, M. J., Critchley, M. M., Hutson, J., and Fallowfield, H. J. (2001a) The adsorption of hepatotoxins from water on to soil during batch studies. *Water Res* **35:** 1461–1468.

Miller, M. J., Hutson, J., and Fallowfield, H. J. (2001b) An overview of the potential for removal of cyanobacterial hepatotoxin from drinking water by riverbank filtration. *Environ Health* **1:** 82–93.

Milutinovi, A., Sedmak, B., Horvat-Znidarsic, I., and Suput, D. (2002) Renal injuries induced by chronic intoxication with microcystins. *Cell Mol Biol Lett* **7:** 139–141.

Mitrovic, S. M., Pflugmacher, S., James, K. J., Furey, A. (2004) Anatoxin-a elicits an increase in peroxidase and glutathione S-transferase activity in aquatic plants. *Aqua Toxicol* **68:** 185–192.

Moffitt, M. C., and Neilan, B. A. (2001) On the presence of peptide synthetase and polyketide synthase genes in the cyanobacterial genus *Nodularia*. *FEMS Microbiol Lett* **196:** 207–214.

Moffitt, M. C., and Neilan, B. A. (2004) Characterization of the nodularin synthetase gene cluster and proposed theory of evolution of cyanobacterial hepatotoxins. *Appl Environ Microbiol* **70:** 6353–6362.

Moffitt, M. C., Blackburn, S. I., and Neilan, B. A. (2001) rRNA sequences reflect the ecophysiology and define the toxic cyanobacteria of the genus *Nodularia*. *Int J Syst Evol Microbiol* **51:** 505–512.

Mohamed, Z. A. (2001) Accumulation of cyanobacterial hepatotoxins by *Daphnia* in some Egyptian irrigation canals. *Ecotoxicol Environ Safety* **50:** 4–8.

Mohamed, Z. A., Carmichael, W. W., and El-Sharouny, H. M. (1998) Activated carbon removal efficiency of microcystins in an aqueous cell extract of *Microcystis aeruginosa* and *Oscillatoria tenuis* strains isolated from Egyptian freshwaters. *Environ Toxicol* **14:** 197–201.

Mohamed, Z. A., Carmichael, W. W., and Hussein, A. A. (2003) Estimation of microcystins in the freshwater fish *Oreochromis niloticus* in an Egyptian fish farm containing a *Microcystis* bloom. *Environ Toxicol* **18:** 137–141.

Moikeha, S. N., and Chu, G. W. (1971) Dermatitis-producing alga *Lyngbya majuscula* Gomont in Hawaii. II. Biological properties of the toxic factor. *J Phycol* **7:** 8–13.

Molica, R., Onodera, H., Garcia, C., Rivas, M., Andrinolo, D., Nascimento, S., Meguro, H., Oshima, Y., Azevedo, S., and Lagos, N. (2002) Toxins in the freshwater cyanobacterium *Cylindrospermopsis raciborskii* (Cyanophyceae) isolated from Tabocas reservoir in Caruaru, Brazil, including demonstration of a new STX analogue. *Phycologia* **41:** 606–611.

Moore, B. S., Ohtani, I., de Koning, C. B., Moore, R. E., and Carmichael, W. W. (1992) Biosynthesis of anatoxin-a(s). Origin of the carbons. *Tetrahedron Lett* **33:** 6595–6598.

Moore, R. E. (1977) Toxins from blue-green algae. *BioScience* **27:** 797–802.

Moore, R. E. (1996) Cyclic peptides and depsipeptides from cyanobacteria: A review. *J Ind Microbiol* **16:** 134–143.

Moore, R. E., Chen, J. L., Moore, B. S., Patterson, G. M. L., and Carmichael, W. W. (1991) Biosynthesis of microcystin-LR. Origin of the carbons in the Adda and Masp units. *J Am Chem Soc* **113:** 5083.

Moore, R. E., Ohtani, I., Moore, B. S., de Koning, C. B., Yoshida, W. Y., Runnegar, M. T., and Carmichael, W. W. (1993) Cyanobacterial toxins. *Gazzatta Chimica Italiana* **123:** 329–336.

Moreno, I. M., Pereira, P., Franca, S., and Camean, A. (2004) Toxic cyanobacteria strains isolated from blooms in the Guadiana River (southwestern Spain). *Biol Res* **37:** 405–417.

Morrison, K. F., Parkin, G., and Codd, G. A. (2006) Optimization of anabaenopeptin extraction from cyanobacteria and the effect of methanol on laboratory manipulation. *Peptides* **27:** 10–17.

Mouchet, P., and Bonnelye, V. (1998) Solving algae problems: French expertise and world-wide applications. *J Water SRT – Aqua* **47:** 125–141.

Mountfort, D. O., Holland, P., and Sprosen, J. (2005) Method for detecting classes of microcystins by combination of protein phosphatase inhibition assay and ELISA: Comparison with LC-MS. *Toxicon* **45:** 199–206.

Moustafa, A., Loram, J. E., Hackett, J. D., Andersen, O. M., Plumley, F. G., and Bhattacharya, D. (2009) Origin of STX biosynthetic genes in cyanobacteria. *PLoS ONE* **4:** e5758. doi: 10.1371/journal.pone.0005758

Mulligan, M. E., and Haselkorn, R. (1989) Nitrogen fixation (*nif*) genes of the cyanobacterium *Anabaena* sp. strain PCC 7120: The *nifB-fdxN-nifS-nifU* operon. *J Biol Chem* **264:** 19200–19207.

Mynderse, J. S., Moore, R. E., Kashiwagi, M., and Norton, T. R. (1977) Antileukemia activity in the oscillatoriaceae: Isolation of debromoaplysiatoxin from *Lyngbya*. *Science* **196:** 538–540.

Nagata, S., Soutome, H., Tsutsumi, T., Hasegawa, A., Sekijima, M., Sugamata, M., Harada, K., Suganuma, M., and Ueno, Y. (1995) Novel monoclonal antibodies against microcystin and their protective activity for hepatotoxicity. *Nat Toxins* **3:** 78–86.

Nagle, D. G., Paul, V. J., and Roberts, M. A. (1996) Y paoamide, a new broadly acting feeding deterrent from the marine cyanobacterium *Lyngbya majuscula*. *Tetrahedron Lett* **37:** 6263–6266.

Namikoshi, M., and Rinehart, K. L. (1996) Bioactive compounds produced by cyanobacteria. *J Ind Microbiol* **17:** 373–384.

Namikoshi, M., Rinehart, K. L., Sakai, R., Sivonen, K., and Carmichael, W. W. (1990) Structure of three new cyclic hepatotoxins produced by the cyanobacterium (blue-green alga) *Nostoc* sp. strain 152. *J Org Chem* **55:** 6135–6139.

Namikoshi, M., Rinehart, K. L., Sakai, R., Stotts, R. R., Dahlem, A. M., Beasley, V. R., Carmichael, W. W., and Evans, W. R. (1992a) Identification of 12 hepatotoxins from a Homer lake bloom of the cyanobacteria *Microcystis aeruginosa*, *Microcystis viridis*, and *Microcystis wesenbergii*: Nine new microcystins. *J Org Chem* **57:** 866–872.

Namikoshi, M., Sivonen, K., Evans, W. R., Carmichael, W. W., Rouhiainen, L., Luukkainen, R., and Rinehart, K. L. (1992b) Structure of three new homotyrosine-containing microcystins and a new homophenylalanine variant from *Anabaena* sp. strain 66. *Chem Res Toxicol* **5:** 661–666.

Namikoshi, M., Sivonen, K., Evans, W. R., Carmichael, W. W., Sun, F., Rouhiainen, L., Luukkainen, R., and Rinehart, K. L. (1992c) Two new L-serine variants of microcystin-LR and -RR from *Anabaena* sp. strain 202 A1 and 202 A2. *Toxicon* **30:** 1457–1464.

Namikoshi, M., Sivonen, K., Evans, W. R., Sun, F., Carmichael, W. W., and Rinehart, K. L. (1992d) Isolation and structure of microcystins from a cyanobacterial water bloom (Finland). *Toxicon* **30:** 1473–1479.

Namikoshi, M., Choi, B. W., Sakai, R., Sun, F., Rinehart, K. L., Carmichael, W. W., Evans, W. R., Cruz, P., Munro, M. H. G., and Blunt, J. W. (1994) New nodularin: A general method for structure assignment. *J Org Chem* **59:** 2349–2357.

Namikoshi, M., Sun, F., Choi, B. W., Rinehart, K. L., Carmichael, W. W., Evans, W. R., and Beasley, V. R. (1995). Seven more microcystins from Homer lake cells: Application of the general method for structure assignement of peptides containing,-dehydroamino acid unit(s). *J Org Chem* **60:** 3671–3679.

Namikoshi, M., Yuan, M., Sivonen, K., Carmichael, W. W., Rinehart, K. L., Rouhiainen, L., Sun, S., Brittain, S., and Otsuki, A. (1998) Seven new microcystins possessing two L-glutamic acid units, isolated from *Anabaena* sp. strain 186. *Chem Res Toxicol* **11:** 143–149.

Negri, A. P., Jones, G. J., and Hindmarsh, M. (1995) Sheep mortality associated with shellfish poisoning toxins from the cyanobacterium *Anabaena circinalis*. *Toxicon* **33:** 1321–1329.

Negri, A. P., Jones, G. J., Blackburn, S. I., Oshima, Y., and Onodera, H. (1997) Effect of culture and bloom development and of sample storage on paralytic shellfish poisons in the cyanobacterium *Anabaena circinalis*. *J Phycol* **33:** 26–35.

Nehring, S. (1993) Mortality of dogs associated with a mass development of *Nodularia spumigena* (Cyanophyceae) in a brackish lake at the German North Sea coast. *J Plankton Res* **15:** 867–872.

Neilan, B. A. (1995) Identification and phylogenetic analysis of toxigenic cyanobacteria by multiplex randomly amplified polymorphic DNA PCR. *Appl Environ Microbiol* **61:** 2286–2291.

Neilan, B. A., Jacobs, D., and Goodman, A. E. (1995) Genetic diversity and phylogeny of toxic cyanobacteria determined by DNA polymorphisms within the phycocyanin locus. *Appl Environ Microbiol* **61:** 3875–3883.

Neilan, B. A., Jacobs, D., Dot, T. D., Blackall, L. L., Hawkins, P. R., Cox, P. T., and Goodman, A. E. (1997) rRNA sequences and evolutionary relationship among toxic and non-toxic cyanobacteria of the genus *Microcystis*. *Int J Syst Bacteriol* **47:** 693–697.

Neilan, B. A., Dittmann, E., Rouhiainen, L., Baas, R. A., Schaub, V., Sivonen, K., and Borner, T. (1999) Nonribosomal peptide synthesis and toxigenicity of cyanobacteria. *J Bacteriol* **181:** 4089–4097.

Neilan, B. A., Saker, M. L., Fastner, J., Torokne, A., and Burns, B. P. (2003) Phytogeography of the invasive cyanobacterium *Cylindrospermopsis raciborskii*. *Mol Ecol* **12:** 133.

Neumann, U., Campos, V., Cantarero, S., Urrutia, H., Heinze, R., Weckesser, J., and Erhard, M. (2000) Co-occurrence of non-toxic (Cyanopeptolin) and toxic (Microcystin) peptides in a bloom of *Microcystis* sp. from a Chilean lake. *Syst Appl Microbiol* **23:** 191–197.

Newcombe, G., Drikas, M., and Hayes, R. (1997) Influence of characterized natural organic material on activated carbon adsorption. II. Effect on pore volume distribution and adsorption of 2-methylisoborneol. *Water Res* **31:** 1065–1073.

Nicholson, B., and Rositano, J. (1997) Chemical methods for the destruction of cyanobacterial toxins. In: Workshop on cyanobacteria (blue-green algae) and their toxins, Brisbane, Australia, 1997. Australian Water and Wastewater Association, Sydney, Australia.

Nicholson, B., Rositano, J., Humpage, A., and Burch, M. (1993) Removal of algal toxins in water treatment processes. In: 15th AWWA Federal Convention, Gold Coast, Queensland, Australia, Australian Water and Wastewater Association Sydney, Auatralia pp. 327–331.

Nicholson, B., Rositano, J., and Burch, M. D. (1994) Destruction of cyanobacterial peptide hepatotoxins by chlorine and chloramines. *Water Res* **28:** 1297–1303.

Nicholson, D. W., and Thornberry, N. A. (1997) Capsases: killer proteases. *Trends Biochem Sci* **22:** 299–306.

Nishihara, H., Miwa, H., Watanabe, M., Nagashima, M., Yagi, O., and Takamura, Y. (1997) Random amplification polymorphic DNA (RAPD) analysis for discriminating genotypes of *Microcystis* cyanobacteria. *Biosci Biotech Biochem* **61:** 1067–1072.

Nishiwaki, R., Ohta, T., Sueoka, E., Suganuma, M., Harada, K., Watanabe, M. F., and Fujiki, H. (1994) Two significant aspects of microcystin-LR: Specific binding and liver specificity. *Cancer Lett* **83:** 283–289.

Nishiwaki-Matsushima, R., Nishiwaki, S., Ohta, T., Yoshizawa, S., Suganuma, M., Harada, K.-I., Watanabe, M. F., and Fujiki, H. (1991) Structure-function relationships of microcystins, liver tumor promoters, in interaction with protein phosphatase. *Jpn J Cancer Res* **82:** 993–996.

Nishizawa, T., Asayama, M., and Shirai, M. (2001) Cyclic heptapeptide microcystin biosynthesis requires the glutamate racemase gene. *Microbiology* **147:** 1235–1241.

Nishizawa, T., Asayama, M., Fujii, K., Harada, K.-I., and Shirai, M. (1999) Genetic analysis of the peptide synthetase genes for a cyclic heptapeptide microcystin in *Microcystis* spp. *J Biochem* **126:** 520–529.

Nishizawa, T., Ueda, A., Asayama, M., Harada, K., Ochi, K., and Shirai, M. (2000) Polykedtide synthase gene coupled to the peptide synthetase module involved in biosynthesis of the cyclic heptapeptide microcystin. *J Biochem* (Tokyo) **127:** 779–889.

Norris, R. L., Eaglesham, G. K., Pierens, G., Shaw, G. R., Smith, M. J., Chiswell, R. K., Seawright, A. A., and Moore, M. R. (1999) Stability of cylindrospermopsin, the toxin from the cyanobacterium *Cylindrospermopsis raciborskii*: Effect of pH, temperature, and sunlight on decomposition. *Environ Toxicol* **14:** 163–165.

Nuebel, U., Garcia-Pichel, F., and Muyzer, G. (1997) PCR primers to amplify 16S rRNA genes from cyanobacteria. *Appl Environ Microbiol* **63:** 3327–3332.

Oberemm, A. (2001a) Effects of cyanotoxins on early life stages of fish and amphibians. In: Chorus, I. (Ed.) *Cyanotoxins-Occurrence, Causes, Consequences*. Springer, Berlin, Germany. pp. 240–248.

Oberemm, A. (2001b) A fish-embryo bioassay for the assessment of cyanobacterial toxicity. In: Chorus, I. (Ed.) *Cyanotoxins-Occurrence, Causes, Consequences*. Springer, Berlin, Germany. pp. 339–343.

Oberholster, P. J., Botha, A-M., and Grobbelaar, J. U. (2004) *Microcystis aeruginosa*: Source of toxic microcystins in drinking water. *Afr J Biotechnol* **3:** 159–168.

Oberholster, P. J., Botha, A.-M., and Cloete, T. E. (2006) Toxic cyanobacterial blooms in shallow, artificially mixed urban lake in Colorado, USA. *Lakes & Reservoirs: Research and Management* **11:** 111–123.

Oh, H.-M., Lee, S. J., Jang, M.-H., and Yoon, B.-D. (2000) Microcystin production by *Microcystis aeruginosa* in a phosphorus-limited chemostat. *Appl Environ Microbiol* **66:** 176–179.

Oh, H.-M., Lee, S. J., Kim, J. H., Kim, H. S., and Yoon, B. D. (2001) Seasonal variation and indirect monitoring of microcystin concentrations in Daechung Reservoir, Korea. *Appl Environ Microbiol* **67:** 1484–1489.

Ohta, T., Nishiwaki, R., Yatsunami, J., Komori, A., Suganuma, M., and Fujiki, H. (1992) Hyperphosphorylation of cytokeratin 8 and 18 by microcystin-LR, a new liver tumor promoter, in primary cultured rat hepatocytes. *Carcinogenesis* **13:** 2443–2447.

Ohta, T., Sueoka, E., Iida, N., Komori, A., Suganuma, M., Nishiwaki, R., Tatematsu, M., Kim, S. J., Carmichael, W. W., and Fujiki, H. (1994) Nodularin, a potent inhibitor of protein phosphatase 1 and 2A, is a new environmental carcinogen in male F344 rat liver. *Cancer Res* **54:** 6402–6406.

Ohtani, I., Moore, R. E., and Runnegar, M. T. C. (1992) Cylindrospermopsin: A potent hepatotoxin from the blue-green alga *Cylindrospermopsis raciborskii*. *J Am Chem Soc* **114:** 7942–7944.

Ojanpera, I., Pelander, A., Vuori, E., Himberg, K., Waris, M., and Niinivara, K. (1995) Detection of cyanobacterial hepatotoxins by TLC. *J Planar Chromatogr* **8:** 69–72.

Okino, T., Murakami, M., Haraguchi, R., Munekata, H., and Matsuda, H. (1993a) Micropeptins A and B, plasmin and trypsin inhibitors from the blue-green alga *Microcystis aeruginosa*. *Tetrahedron Lett* **34:** 8131–8134.

Okino, T., Matsuda, H., Murakami, M., and Yamaguchi, K. (1993b) Microginin, an angiotensin-converting enzyme inhibitor from the blue-green alga *Microcystis aeruginosa*. *Tetrahedron Lett* **34:** 501–504.

Okino, T., Matsuda, H., Murakami, M., and Yamaguchi, K. (1995) New microviridins, elastase inhibitors from the blue-green alga *Microcystis aeruginosa*. *Tetrahedron* **51:** 10679–10686.

Oksanen, I., Jokela, J., Fewer, D. P., Wahlsten, M., Rikkinen, J., and Sivonen K. (2004) Discovery of rare and highly toxic microcystins from lichen-associated cyanobacterium *Nostoc* sp. strain IO-102-I. *Appl Environ Microbiol* **70:** 5756–5763.

Onodera, H., Oshima, Y., Henriksen, P., and Yasumoto, T. (1997) Confirmation of anatoxin-a(s) in the cyanobacterium *Anabaena lammermannii* as the cause of bird kills in Danish lakes. *Toxicon* **35:** 1645–1648.

Onyewuenyi, N., and Hawkins, P. (1996) Separation of toxic peptides (microcystins) in capillary electrophoresis, with the aid of organic mobile phase modifiers. *J Chromatogr* **749:** 271–278.

Orjala, J., Nagle, D. G., Hsu, V. L., and Gerwick, W. H. (1995) Antillatoxin-an exceptionally ichthyotoxic cyclic lipopeptide from the tropical cyanobacterium *Lyngbya majuscula*. *J Am Chem Soc* **117:** 8281–8282.

Osborne, N. J. T., Webb, P. M., and Shaw, G. R. (2001) The toxins of *Lyngbya majuscula* and their human and ecological health effects. *Environ Int* **27:** 381–392.

Oshima, Y., Blackburn, S. I., and Hallegraeff, G. M. (1993) Comparative study on paralytic shellfish toxin profiles of the dinoflagellate *Gymnodium catenatum* from three countries. *Mar Biol* **116:** 471–476.

Otsuka, S., Suda, S., Li, R. H., Watanabe, M., Oyaizu, H., Matsumoto, S., and Watanabe, M. M. (1998) 16S rDNA sequences and phylogenetic analysis of *Microcystis* strains with and without phycoerythrin. *FEMS Microbiol Lett* **164:** 119–124.

Otsuka, S., Suda, S., Li, R. H., Watanabe, M., Oyaizu, H., Matsumoto, S., and Watanabe, M. M. (1999a) Characterization of morphospecies and strains of the genus *Microcystis* (cyanobacteria) for a reconsideration of species classification. *Phycol Res* **47:** 189–197.

Otsuka, S., Suda, S., Li, R. H., Watanabe, M., Oyaizu, H., Matsumoto, S., and Watanabe, M. M. (1999b) Phylogenetic relationship between toxic and non-toxic strains of the genus *Microcystis* based on 16S to 23S internal transcribed spacer sequence. *FEMS Microbiol Lett* **172:** 15–21.

Ouahid, Y., Perez-Silva, G., and Del Campo, F. F. (2005) Identification of potentially toxic environmental *Microcystis* synthetase gene regions. *Environ Toxicol* **20:** 235–242.

Oudra, B., Loudiki, M., Vasconcelos, V., Sabour, B., Sbiyyaa, B., Oufdou, K., and Mezrioul, N. (2002) Detection and quantification of microcystins from cyanobacteria strains isolated from reservoirs and ponds in Morocco. *Environ Toxicol* **17:** 32–39.

Padisak, J. (2003) Estimation of minimum sedimentary inoculum (akinete) pool of *Cylindrospermopsis raciborskii*: A morphology and life cycle based method. *Hydrobiologia* **502:** 389–394.

Painuly, P., Perez, R., Fukai, T., and Shimizu, Y. (1988) The structure of a cyclic peptide toxin, cyanogenosin-RR, from *Microcystis aeruginosa*. *Tetrahedron Lett* **29:** 11–14.

Palenik, B., and Haselkorn, R. (1992) Multiple evolutionary origins of prochlorophytes, the chlorophyll-containing prokaryotes. *Nature* (London) **355:** 265–267.

Papageorgiou, J., Nicholson, B. C., Linke, T. A., and Kapralos, C. (2005) Analysis of cyanobacterial-derived STXs using high performance ion exchange chromatography with chemical oxidation/fluorescence detection. *Environ Toxicol* **5:** 549–559.

Park, H.-D. (2001) Cyanotoxins and cyanobacterial blooms in South Korean lakes. In Chorus, I. (Ed.) *Cyanotoxins-Occurrence, Causes, Consequences*. Springer, Berlin. pp. 68–75.

Park, H.-D., Sasaki, Y., Maruyama, T., Yanagisawa, E., Hiraishi, A., and Kato, K. (2001) Degradation of the cyanobacterial hepatotoxin microcystin by a new bacterium isolated from a hypertrophic lake. *Environ Toxicol* **16:** 337–343.

Park, H.-D., Twami, C., Watanabe, M. F., Harada, K.-I., Okino, T., and Hatashi, H. (1998) Temporal variabilities of the concentrations of intra- and extracellular microcystin and toxic *Microcystis* species in a hypertrophic Lake Suwa, Japan. *Environ Toxicol Water Qual* **13:** 61–72.

Park, H.-D., Watanabe, M. F., Harada, K., Hayashi, H., and Okino, T. (1993) Seasonal variation of *Microcystis* species and toxic heptapeptide microcystins in Lake Suwa. *Environ Toxicol Water Qual* **8:** 425–435.

Pearson, L., Mihali, T., Moffitt, M., Kellmann, R., and Neilan, B. (2010) On the chemistry, toxicology and genetics of the cyanobacterial toxins, microcystin, nodularin, STX, and cylindrospermopsin. *Mar Drugs* **8:** 1650–1680.

Pendleton, P., Wong, S., Schumann, R., Levay, G., Denoyel, R., and Rouquerol, J. (1997) Properties of activated carbon controlling 2-methylisoborneol adsorption. *Carbon* **35:** 1141–1149.

Pereira, P., Onodera, H., Andrinolo, D., Franca, S., Araujo, F., Lagos, N., and Oshima, Y. (2000) Paralytic shellfish toxins in the freshwater cyanobacterium *Aphanizomenon flos-aquae*, isolated from Montargil reservoir, Portugal. *Toxicon* **38:** 1689–1702.

Persson, P., Sivonen, K., Keto, K., Kononen, K., Niemi, M., and Viljamaa, H. (1984) Potentially toxic blue-green algae (cyanobacteria) in Finnish natural waters. *Aqua Fenn* **14:** 147–154.

Phillips, M. J., Roberts, R. J., Stewart, J. A., and Codd, G. A. (1985) The toxicity of the cyanobacterium *Microcystis aeruginosa* to rainbow trout, *Salmo gairdneri* Richardson. *J Fish Dis* **8:** 339–344.

Pomati, F., Kellmann, R., Cavalieri, R., Burns, B. P., and Neilan, B. A. (2006) Comparative gene expression of PSP-toxin-producing and non-toxic *Anabaena circinalis* strains. *Environ Internl* **32:** 743–748.

Pomati, F., Rossetti, C., Manarolla, G., Burns, B. P., and Neilan, B. A. (2004) Interactions between intracellular Na^+ levels and STX production in *Cylindrospermopsis raciborskii* T3. *Microbiology* **150:** 455–461.

Pomati, F., Sacchi, S., Rossetti, C., Giovannardi, S., Onodera, H., Oshima, Y., and Neilan, B. A. (2000) The freshwater cyanobacterium *Planktothrix* sp. FP1: Molecular identification and detection of paralytic shellfish poisoning toxins. *J Phycol* **36:** 553–562.

Poon, G. K., Griggs, L. J., Edwards, C., Beattie, K. A., and Codd, G. A. (1993) Liquid chromatography-electrospray ionization mass-spectrometry of cyanobacterial toxins. *J Chromatogr* **628:** 215–233.

Porfirio, Z., Ribeiro, P. M., Estevan, C. S., Ricardo, L. S., and Santana, A. E. G. (1999) Hepatosplenomegaly caused by an extract of cyanobacterium *Microcystis aeruginosa* bloom collected in the Manguaba Lagoon, Alagoas-Brazil. *Revista de Microbiol* **30:** 278–285.

Pouria, S., de Andrade, A., Barbosa, J., Cavalcanti, R. L., Barreto, V. T. S., Ward, C. J., Preiser, W., Poon, G. K., Neild, G. H., and Codd, G. A. (1998) Fatal microcystin intoxication in haemodialysis unit in Caruaru, Brazil. *The Lancet* **352:** 21–26.

Premazzi, G., and Volterra, L. (1993) *Microphyte Toxins: A Manual for Toxin Detection, Environmental Monitoring and Therapies to Counteract Intoxications*. JRC CEC, Luxembourg.

Prepas, E. E., and Charette, T. (2003) Worldwide eutrophication of water bodies: Causes, Concerns, Controls. In: Lokar, B. S. (Ed.) *Treatise on Geochemistry*. Vol. 9. Elsevier, Amsterdam, The Netherlands. pp. 311–331.

Preussel, K., Stuken, A., Wiedner, C., Chorus, I., and Fastner, J. (2006) First report on cylindrospermopsin producing *Aphanizomenon flos-aquae* (Cyanobacteria) isolated from two German lakes. *Toxicon* **47:** 156–162.

Prieto, A. I., Jos, A., Pichardo, S., Moreno, I., Camean, A. M., (2006) Differential oxidative stress responses to microcystin-LR and RR in intraperitoneally exposed tilapia fish (*Oreochromis* sp.). *Aquat Toxicol* **77:** 314–321.

Princep, M. R., Caplan, F. R., Moore, R. E., Patterson, G. M. L., Honakanen, R. E., and Boynton, A. L. (1992) Microcystin-LA from a blue-green alga belonging to the Stigonematales. *Phytochemistry* **31:** 1247–1248.

Puschner, B., Galey, F. D., Johnson, B., Dickie, C. W., Vondy, M., Francis, T., and Holstege, D. M. (1998) Blue-green algae toxicosis in cattle. *J Am Vet Med Assoc* **213:** 1605–1607.

Pyo, D., Park, G., Choi, J., and Oh, C. (2005) Microcystin detection characteristics of fluorescence immunochromatography and high performance liquid chromatography. *Bull Korean Chem Soc* **26:** 268–272.

Rabergh, C. M. I., Bylund, G., and Eriksson, J. E. (1991) Histopathological effects of microcystin-LR, a cyclic peptide toxin from the cyanobacterium (blue-green alga) *Microcystis aeruginosa*, on common carp (*Cyprinus carpio* L.). *Aquat Toxicol* **20:** 131–146.

Radau, G., Schermuly, S., and Fritsche, A. (2003) New cyanopeptide-derived low molecular weight inhibitors of trypsin-like serine proteases. *Arch Pharm* **336:** 300–309.

Rantala, A., Fewer, D. P., Hisbergues, M., Rouhiainen, L., Vaitomaa, J., Boerner, T., and Sivonen, K. (2004) Phylogenetic evidence for the early evolution of microcystin synthesis. *Proc Natl Acad Sci USA* **101:** 568–573.

Rantala, A., Rajaniemi-Wacklin, P., Lyra, C., Lepistó, L., Rintala, J., Mankiewicz-Boczek, J., and Sivonen, K. (2006) Detection of microcystin-producing cyanobacteria in Finnish Lakes with genus-specific microcystin synthetase gene E (*mcyE*) PCR and associations with environmental factors. *Appl Environ Microbiol* **72:** 6101–6110.

Rao, P. V. L., Battacharya, R., Parida, M. M., Jana, A. M., and Bhaskar, A. S. B. (1998) Freshwater cyanobacterium *Microcystis aeruginosa* (UTEX 2385) induced DNA damage *in vivo* and *in vitro*. *Environ Toxicol Pharmacol* **5:** 1–6.

Rao, P. V. L., Battacharya, R., Gupta, N., Parida, M. M., Bhaskar, A. S. B., and Dubey, R. (2002) Involvement of capsase and reactive oxygen species in cyanobacterial toxin anatoxin-a induced cytotoxicity and apoptosis in rat thymocytes and Vero cells. *Arch Toxicol* **76:** 227–235.

Rao, P. V. L., Gupta, N., and Jayaraj, R. (2004) Screening of chemoprotectants against cyclic peptide toxin microcystin-LR. *Inadian J Pharmacol* **36:** 87–92.

Rapala, J., and Sivonen, K. (1998) Assessment of environmental conditions that favour hepatotoxic and neurotoxic *Anabaena* spp. strains cultured under light limitation at different temperatures. *Microb Ecol* **36:** 181–192.

Rapala, J., Berg, K. A., Lyra, C., Niemi, R. M., Manz, W., Suomalainen, S., Paulin, L., and Lahti, K. (2005) *Paucibacter toxinivorans* gen nov., sp. nov., a bacterium that degrades cyclic cyanobacterial hepatotoxins microcystins and nodularin. *Int J Syst Evol Microbiol* **55:** 1563–1568.

Rapala, J., Sivonen, K., Lyra, C., and Niemela, S. I. (1997) Variation of microcystins, cyanobacterial hepatotoxins, in *Anabaena* spp. as a function of growth stimuli. *Appl Environ Microbiol* **63:** 2206–2212.

Raziuddin, S., Siegelman, H. W., and Tornabene, T. G. (1983) Lipopolysaccharides of the cyanobacterium *Microcystis aeruginosa*. *Eur J Biochem* **137:** 333–336.

Reinikainen, M., Ketola, M., and Walls, M. (1994) Effects of the concentrations of the toxic *Microcystis aeruginosa* and an alternative food on the survival of *Daphnia pulex*. *Limnol Oceanogr* **39:** 424–432.

Repavich, W. M., Sonzogni, W. C., Standridge, J. H., Wedepohl, R. E., and Meisner, L. F. (1990) Cyanobacteria (blue-green algae) in Wisconsin waters: Acute and chronic toxicity. *Water Res* **24:** 225–231.

Repka, S., Koivula, M., Jarjunpä, V., Rouhiainen, L., and Sivonen, K. (2004) Effects of phosphate and light on growth and bioactive peptide production by the cyanobacterium *Anabaena* strain 90 and its anabaenopeptilide mutant. *Appl Environ Microbiol* **70:** 4551–4560.

Repka, S., Mehtonen, J., Vaitomaa, J., Saari, I., and Sivonen, K. (2001) Effects of nutrients on growth and nodularin production of *Nodularia* strain GR8b. *Microb Ecol* **42:** 606–613.

Ressom, R., Soong, F. S., Fitzgerald, J., Turczynowicz, L., El Saadi, O., Roder, D., Maynard, T., and Falconer, I. (1994) *Health Effects of Toxic Cyanobacteria* (*Blue-green algae*). Australian National Health and Medical Research Council, Looking Glass Press, Australian Govt. Publication Service, Canberra, Australia. pp. 108.

Reynolds, K. A. (2004) Cyanobacteria: Natural organisms with toxic effects. *Water Conditiong Purif* **3:** 78–80.

Rinehart, K. L., Harada, K.-I., Namikoshi, M., Chen, C., Harvis, C. A., Munro, M. H. G., Blunt, J. W., Mulligan, P. E., Beasley, B. R., Dahlem, A. M., and Carmichael, W. W. (1988) Nodularin, microcystin, and the configuration of Adda. *J Am Chem Soc* **110:** 8557–8558.

Rinehart, K. L., Namikoshi, M., and Choi, B. W. (1994) Structure and biosynthesis of toxins from blue-green algae (cyanobacteria). *J Appl Phycol* **6:** 159–176.

Rinta-Kanto, J. M., and Wilhelm, S. W. (2006) Diversity of microcystin-producing cyanobacteria in spatially isolated regions of Lake Erie. *Appl Environ Microbiol* **72:** 5083–5085.

Rinta-Kanto, J. M., Ouellette, A. J. A., Boyer, G. L., Twiss, M. R., Bridgeman, T. B., and Wilhem, S. W. (2005) Quantification of toxic *Microcystis* spp. during the 2003 and 2004 blooms in Western Lake Erie using quantitative real-time PCR. *Environ Sci Technol* **39:** 4198–4205.

Roberstson, P. K. J., Lawton, L. A., Muench, B., and Rouzade, J. (1997) Process influencing the destruction of microcystin-LR by TiO_2 photocatalysis. *J Chem Soc Commun* **4:** 393–394.

Robillot, C., Vinh, J., Puiseux-Dao, S., and Hennion, M.-C. (2000) Hepatotoxin production kinetics of the cyanobacterium *Microcystis aeruginosa* PCC 7820 as determined by HPLC-Mass spectrometry and protein phosphatase bioassay. *Environ Sci Technol* **34:** 3372–3378.

Robinson, N. A., Pace, J. G., Matson, C. F., Dinterman, R. E., and Pace, J. G. (1989) Characterization of chemically tritiated microcystin-LR and its distribution in mice. *Toxicon* **27:** 1035–1042.

Robinson N. A., Pace, J. G., Matson, C. F., Miura, G. A., and Lawrence, W. B. (1991) Tissue distribution, excretion and hepatic biotransformation of microcystin-LR in mice. *J Pharmcol Exp Ther* **256:** 176–182.

Rocha, C., Galvao, H., and Barbosa, A. (2002) Role of transient silicon limitation in the development of cyanobacteria blooms in the Guadiana estuary, south western Siberia. *Mar Ecol Progr Ser* **228:** 35–45.

Rogers, E. H., Hunter, E. S. III., Moser, V. C., Phillips, P. M., Herkovits, J., Munoz, L., Hall, L. L., and Chernoff, N. (2005) Potential developmental toxicity of anatoxin-a, a cyanobacterial toxin. *J Appl Toxicol* **25:** 527–534.

Rohrlack, T., Dittmann, E., Henning, M., Borner, T., and Kohl, J.-G. (1999) Role of microcystins in poisoning and food ingestion inhibition of *Daphnia galeata* caused by the cyanobacterium *Microcystis aeruginosa*. *Appl Environ Microbiol* **65:** 737–739.

Rolland, A., Bird, D. F., and Giani, A. (2005) Seasonal changes in composition of the cyanobacterial community and the occurrence of hepatotoxic blooms in the eastern townships, Qubec, Canada. *J Plankton Res* **27:** 683–694.

Rositano, J., and Nocholson, B. C. (1994) *Water Treatment Techniques for the Removal of Cyanobacterial Toxins from Water.* Australian Centre for Water Treatment and Water Quality Research. Salisbury, South Australia.

Rositano, J., Nicholson, B., and Pieronne, P. (1998) Destruction of cyanobacterial toxins by ozone. *Ozone Sci Eng* **20:** 223–238.

Rouhiainen, L., Paulin, L., Suomalainen, S., Hyytiaenen, H., Buikema, W., Haselkorn, R., and Sivonen, K. (2000) Genes encoding synthetases of cyclic depsipeptides, anabaenopeptilides, in *Anabaena* strain 90. *Mol Microbiol* **37:** 156–167.

Rouhiainen, L., Sivonen, K., Buikema, W. J., and Haselkorn, R. (1995) Characterization of toxin-producing cyanobacteria by using oligonucleotide probe containing a tandemly repeated heptamer. *J Bacteriol* **177:** 6021–6026.

Rouhiainen, L., Vakkilainen, T., Siemer, B. L., Buikema, W., Haselkorn, R., and Sivonen, K. (2004) Genes coding for hepatotoxins (microcystins) in the cyanobacterium *Anabaena* strain 90. *Appl Environ Microbiol* **70:** 686–692.

Rounge, T. B., Rohrlack, T., Tooming-Klunderud, A., Kristensen, T., and Jakobsen, K. S. (2007) Comparison of cyanopeptolin genes in *Planktothrix, Microcystis,* and *Anabaena* strains: Evidence for independent evolution within each genus. *Appl Environ Microbiol* **73:** 7322–7330.

Rudi, K., Larsen, F., and Jakobsen, K. S. (1998) Detection of toxin producing cyanobacteria by use of paramagnetic beads for cell concentration and DNA purification. *Appl Environ Microbiol* **64:** 34–37.

Rudolph-Boehner, S., Mierke, D. F., and Morder, L. (1994) Molecular structure of cyanobacterial tumor-promoting microcystins. *FEBS Lett* **349:** 319–323.

Runnegar, M. T., Berndt, N., Kong, S. M., Lee, E. Y., and Zhang, L. (1995) *In vivo* and *in vitro* binding of microcystins to protein phosphatases 1 and 2A. *Biochem Biophys Res Commun* **216:** 162–169.

Runnegar, M. T. C., Falconer, I. R., and Silver, J. (1981) Deformation of isolated rat hepatocytes by a peptide hepatotoxin from the blue-green alga *Microcystis aeruginosa. Naunyn-Schmiedebergs Arch Pharmacol* **317:** 268–272.

Runnegar, M. T. C., Falconer, I. R., Buckley, T., and Jackson, A. R. (1986) Lethal potency and tissue distribution of ^{125}I-labelled toxic peptides from the blue-green alga *Microcystis aeruginosa. Toxicon* **24:** 506–509.

Runnegar, M. T. C., Falconer, I. R., Jackson, A. R. B., and McInnes, A. (1983) Toxin production by *Microcystis aeruginosa* cultures. *Toxicon* **3**(Suppl): 377–380.

Runnegar, M. T. C., Gerdes, R. G., and Falconer, I. R. (1991) The uptake of cyanobacterial hepatotoxin microcystin by isolated rat hepatocytes. *Toxicon* **29:** 43–51.

Runnegar, M. T. C., Jackson, A. R. B., and Falconer, I. R. (1988) Toxicity of the cyanobacterium *Nodularia spumigena* mertens. *Toxicon* **26:** 143–151.

Runnegar, M., Kong, S., and Berndt, N. (1993) Protein phosphatase inhibition and *in vivo* hepatotoxicity of microcystins. *Am J Physiol* **265:** G224–G230.

Runnegar, M. T., Kong, S. M., Zhong, Y. Z., and Lu, S. C. (1995) Inhibition of reduced glutathione synthesis by cyanobacterial alkaloid cylindrospermopsin in cultured rat hepatocytes. *Biochem Pharmacol* **49:** 219–225.

Runnegar, M. T., Kong, S. M., Zhong Y. Z., Ge, J. L., and Lu, S. C. (1994) The role of glutathione in the toxicity of a novel cyanobacterial alkaloid cylindrospermopsin in cultured rat hepatocytes. *Biochem Biophys Res Commun* **201:** 235–241.

Runnegar, M. T., Xie, C., Snider, B. B., Wallace, G. A., Weinreb, S. M., and Kuhlenkamp, J. (2002) *In vitro* hepatotoxicity of the cyanobacterial alkaloid cylindrospermopsin and related synthetic analogues. *Toxicol Sci* **67:** 81–87.

Ryan, E. F., Hamilton, D. P., and Barnes, G. E. (2003) Recent occurrence of *Cylindrospermopsis raciborskii* in Waikato lakes of New Zealand. *New Zealand J Mar Freshwater Res* **37:** 829–836.

Sabour, B., Loudiki, M., Oudra, B., Vasconcelos, V., Martins, R., Oubraim, S., and Fawzi, B. (2002) Toxicology of a *Microcystis ichtyoblabe* waterbloom from Lake Oued Mellah (Morocco). *Environ Toxicol* **17:** 24–31.

Sahin, A., Tencalla, F. G., Dietrich, D. R., Mez, K., and Naegeli, H. (1995) Enzymatic analysis of liver samples from rainbow trout for diagnosis of blue-green algae-induced toxicosis. *Am J Vet Res* **56:** 1110–1116.

Saito, K., Konno, A., Ishii, H., Saito, H., Nishida, F., Abe, T., and Chen, C. (2001) Nodularin-Har: A new nodularin from *Nodularia. J Nat Prod* **64:** 139–141.

Saitou, T., Sugiura, N., Itayama, T., Inamori, Y., and Matsumura, M. (2003) Degradation characteristics of microcystins by isolated bacteria from lake Kasumigaura. *J Water Sci Res* **52:** 13–18.

Saker, M. L., and Griffiths, D. J. (2001) Occurrence of blooms of the cyanobacterium *Cylindrospermopsis raciborskii* (Woloszynska) Seenayya and Subba Raju in a north Queensland domestic water supply. *Mar Freshwater Res* **52:** 907–915.

Saker, M. L., and Neilan, B. A. (2001) Varied diazotrophies, morphologies, and toxicities of genetically similar isolates of *Cylindrospermopsis raciborskii* (Nostocales, Cyanophyceae) from Northern Australia. *Appl Environ Microbiol* **67:** 1839–1845.

Saker, M. L., Fastner, J., Dittmann, E., Christiansen, G., and Vasconcelos, V. M. (2005) Variation between strains of the cyanobacterium *Microcystis aeruginosa* isolated from a Portuguese river. *J Appl Microbiol* **99:** 749–757.

Saker, M. L., Nogueria, I. C. G., Vasconcelos, V. M., Neilan, B. A., Eaglesham, G. K., and Pereira, P. (2003) Cyanobacterial toxins. In: Chorus, I., and Bartram, J. (Eds.) *Toxic Cyanobacteria in Water: A Guide to Their Public Health Consequences, Monitoring and Management*. E and FN Spoon, London, UK. pp. 41–111.

Saker, M. L., Metcalf, J. S., Codd, G. A., and Vasconcelos, V. M. (2004) Accumulation and depurination of the cyanobacterial toxin cylindrospermopsin in freshwater mussel *Anodonta cygnea*. *Toxicon* **43:** 185–194.

Saker, M. L., Neilan, B. A., and Griffiths, D. J. (1999a) Two morphological forms of *Cylindrospermopsis raciborskii* (cyanobacteria) isolated from Solomon dam, Palm Island, Queensland. *J Phycol* **35:** 599.

Saker, M. L., Thomas, A. D., and Norton, J. H. (1999b) Cattle mortality attributed to the cyanobacterium *Cylindrospermopsis raciborskii* in an outback region of north Queensland. *Environ Toxicol* **41:** 179–182.

Sano, T., Takagi, H., Morrison, L. F., Metcalf, J. S., Codd, G. A., and Kaya, K. (2005) Leucine aminopeptidase M inhibitors, cyanostatin A and B, isolated from cyanobacterial water blooms in Scotland. *Phytochemistry* **66:** 543–548.

Sawyer, P., Gentile, J., and Sasner, J. J. (1968) Demonstration of a toxin from *Aphanizomenon flos-aquae* (L) Ralfs. *Can J Microbiol* **14:** 1199–1204.

Schatz, D., Keren, Y., Hadas, O., Carmeli, S., Sukenik, A., and Kaplan, A. (2005) Ecological implications of the emergence of non-toxic subcultures from toxic *Microcystis* strains. *Environ Microbiol* **7:** 798–805.

Schwimmer, D., and Schwimmer, M. (1964). Algae and medicine. In: Jackson, D. F. (Ed.) *Algae and Man*. Plenum Press, New York, NY, USA. p. 434.

Seawright, A. A., Nolan, C. C., Shaw, G. R., Chiswell, R. K., Norris, R. L., Moore, M. R., and Smith, M. J. (1999) The oral toxicity for mice of the tropical cyanobacterium *Cylindrospermopsis raciborskii* (Woloszynska). *Environ Toxicol* **14:** 135–142.

Sedmak, B., and Kosi, G. (1997) Microcystins in Slovene freshwaters (central Europe)-first report. *Nat Toxins* **5:** 64–73.

Seenayya, G., and Subba Raju, S. N. (1972) On the ecology and systematics of the alga known as *Anabaenopsis raciborskii* (Wolosz.) Elenk. and a critical evaluation of the forms described under *Anabaenopsis*. In: Desikachary, T. V. (Ed.) *Taxonomy and Biology of Blue-green algae*. University of Madras, Madras pp. 52–57.

Seifert, M., McGregor, G., Eaglesham, G., Wickramasinghe, W., and Shaw, G. (2007) First evidence for the production of cylindrospermopsin and deoxycylindrospermopsin by the freshwater benthic cyanobacterium, *Lyngbya wollei* (Farlow ex Gomont) Speziale and Dyck. *Harmful Algae* **6:** 73–80.

Selheim, F., Herfindal, L., Martin, R., Vasconcelos, V., and Doskeland, S. O. (2005) Neuroapoptogenic and blood platelet targetting toxins in benthic marine cyanobacteria from Portuguese coast. *Aquat Toxicol* **74:** 294–306.

Shafik, H., Voeroes, L., Sprober, P., Presing, M., and Kovacs, A. W. (2003) Some special morphological features of *Cylindrospermopsis raciborskii* in batch and continuous cultures. *Hydrobiologia* **506–509:** 163–167.

Shalev-Alon, G., Sukenik, A., Livnah, O., Schwarz, R., and Kaplan, A. (2002) A novel gene encoding amidinotransferase in the cylindrospermopsin producing cyanobacterium *Aphanizomenon ovalisporum*. *FEMS Microbiol Lett* **209:** 87–91.

Shaw, G. R., Seawright, A. A., Moore, M. R., and Lam, P. K. (2000) Cylindrospermopsin, a cyanobacterial alkaloid: Evaluation of its toxigenic activity. *Ther Drug Monit* **22:** 89–92.

Shaw, G. R., Sukenik, A., Livne, A., Chiswell, R. K., Smith, M. J., Seawright, A. A., Norris, R. L., Eaglesham, G. K., and Moore, M. R. (1999) Blooms of the cylindrospermopsin containing cyanobacterium, *Aphanizomenon ovalisporum* (Forti) in newly constructed lakes, Queensland, Australia. *Environ Toxicol* **14:** 167–177.

Shimizu, Y., Norte, M., Hori, A., Genenah, A., and Kobayashi, M. (1984) Biosynthesis of saxitoxin analogues: The unexpected pathway. *J Am Chem Soc* **106:** 6433–6434.

Sielaff, H., Dittman, E., Tandeau De Marsac, N., Bouchier, C., von Doehren, H., Borner, T., and Schwecke, T. (2003) The *mcyF* gene of the microcystin biosynthetic gene cluster from *Microcystis aeruginosa* encodes an aspartate racemase. *Biochem J* **373:** 909–916.

Sim, A. T. R., and Mudge, L. M. (1993) Protein phosphatase activity in cyanobacteria-consequences for microcystin toxicity analysis. *Toxicon* **31:** 1179–1186.

Singh, D. P., Tyagi, M. B., and Kumar, A. (1999) Cyanobacterial toxins. In: Fatma, N. (Ed.) Cyanobacterial and Algal Metabolism and Environmental Biotechnology. Narosa Publishing House, New Delhi, India. pp. 61–72.

Sipia, V. O., Karlsson, K. M., Meriluto, J. A. O., and Kankaanpaa, H. T. (2005) Eiders (*Somateria mollissima*) to obtain nodularin, a cyanobacterial hepatotoxin, in Baltic Sea food web. *Environ Toxicol Chem* **23:** 1256–1260.

Siren, H., Jussila, M., Liu, H. W., Peltoniemi, S., Sivonen, K., and Riekkola, M. L. (1999) Separation, purity testing and identification of cyanobacterial hepatotoxins with capillary electrophoresis and electron spray mass spectrometry. *J Chromatogr A* **839:** 203–215.

Sivonen, K. (1990) Effect of light, temperature, nitrate, orthophosphate and bacteria on growth and hepatotoxin production by *Oscillatoria agardhii* strains. *Appl Environ Microbiol* **56:** 2658–2666.

Sivonen, K. (1996) Cyanobacterial toxins and toxin production. *Phycologia* **35:** 12–24.

Sivonen, K., and Jones, J. (1999) Cyanobacterial toxins. In: Chorus, I. and Bartram, J. (Eds.). *Toxic Cyanobacteria in Water: A Guide to Public Health Significance, Consequences, Monitoring and Management*. E & FN SPON, London. pp. 41–111.

Sivonen, K., Carmichael, W. W., Namikoshi, M., Rinehart, K. L., Dahlem, M., and Niemela, S. I. (1990a) Isolation and characterization of hepatotoxic microcystin homologues from the filamentous freshwater cyanobacterium *Nostoc* sp. strain 152. *Appl Environ Microbiol* **56:** 2650–2657.
Sivonen, K., Himberg, K., Luukkainan, R., Niemela, S. I., Poon, G. K., and Codd, G. A. (1989a) Preliminary characterization of neurotoxic cyanobacteria blooms and strains from Finland. *Toxic Assess* **4:** 339–352.
Sivonen, K., Kononen, K., Carmichael, W. W., Dahlem, A. M., Rinehart, K. M., Kiviranta, J., and Niemela, S. I. (1989b) Occurrence of the hepatotoxic cyanobacterium *Nodularia spumigena* in the Baltic Sea and the structure of the toxin. *Appl Environ Microbiol* **55:** 1990–1995.
Sivonen, K., Namikoshi, M., Evans, W. R., Carmichael, W. W., Sun, F., Rouhaiainen, L., Luukkainen, R., and Rinehart, K. L. (1992a) Isolation and characterization of a variety of microcystins from seven strains of the cyanobacterial genus *Anabaena*. *Appl Environ Microbiol* **58:** 2495–2500.
Sivonen, K., Namikoshi, M., Evans, W. R., Faerdig, M., Carmichael, W. W., and Rinehart, K. L. (1992b) Three new microcystins, cyclic heptapeptide hepatotoxins, from *Nostoc* sp. strain 152. *Chem Res Toxicol* **5:** 464–469.
Sivonen, K., Namikoshi, M., Evans, W. R., Gromov, B. V., Carmichael, W. W., and Rinehart, K. L. (1992c) Isolation and structures of five microcystins from a Russian *Microcystis aeruginosa* strain CALU 972. *Toxicon* **30:** 1481–1485.
Sivonen, K., Niemelä, S. I., Niemi, R. M., Lepistö, L., Luoma, T. H., and Räsänen, L. A. (1990b) Toxic cyanobacteria (blue-green algae) in Finnish fresh and coastal waters. *Hydrobiologia* **190:** 267–275.
Sivonen, K., Skulberg, O. M., Namikoshi, M., Evans, W. R., Carmichael, W. W., and Rinehart, K. L. (1992d) Two methyl ester derivatives of microcystins, cyclic heptapeptide hepatotoxins, isolated from *Anabaena flos-aquae* strain CYA 83/1. *Toxicon* **30:** 1465–1471.
Skulberg, O. M., Carmichael, W. W., Anderson, R. A., Matsunaga, S., Moore, R. E., Skulberg, R. (1992) Investigation of a neurotoxic Oscillatorialean strain (cyanophyceae) and its toxin. Isolation and characterization of homoanatoxin-a. *Environ Toxicol Chem* **11:** 321–329.
Smith, D. S., and Watkinson, A. (1981) Anodic oxidation of phenol for wastewater treatment. *Can J Chem Eng* **59:** 52–59.
Soliakov, L., Gallagher, T., and Wonnacott, S. (1995) Anatoxin-a evoked [^{3}H] dopamine release from rat striatal synaptosomes. *Neuropharmacol* **34:** 1535–1541.
Solter, P. F., Wollenberg, G. K., Huang, X., Chu, F. S., and Runnegar, M. T. (1998) Prolonged sublethal exposure to the protein phosphatase inhibitor microcystin-LR results in multiple dose-dependent hepatotoxic effects. *Toxicol Sci* **44:** 87–96.
Song, K. Y., Lim, I. K., Park, S. C., Lee, S.O., Park, H. S., Choi, K. Y., and Hyun, B. H. (1999) Effect of nodularin on the expression of glutathione S-transferase placental form and proliferating cell nuclear antigen in N-nitrosodiethylamine initiated hepatocarcinogenesis in the male Fischer 344 rat. *Carcinogenesis* **20:** 1541–1548.
Song, W., Teshiba, T., Rein, K., and O'Shea, K. E. (2005) Ultrasonically induced degradation and detoxification of microcystin-LR (cyanobacterial toxin). *Environ Sci Technol* **39:** 6300–6305.
Spivak, C. E., Witkop, B., and Albuquerque, E. X. (1980) Anatoxin-a: A novel, potent agonist at the nicotinic receptor. *Mol Pharmacol* **18:** 384–394.
Spoof, L., Karlsson, K., and Meriluto, J. (2001) High-performance liquid chromatographic separation of microcystins and nodularin, cyanobacterial peptide toxins, on C-18 and amide C-16 sorbents. *J Chromatogr* **909:** 225–236.
Spoof, L., Vesterkvist, P., Lindhol, T., and Meriluto, J. (2003) Screening for cyanobacterial hepatotoxins, microcystins and nodularins in environmental water samples by reversed-phase liquid chromatography-electron spray ionization mass spectrometry. *J Chromatogr* **1020:** 105–119.
Spoof, L., Berg, K. A., Rapala, J., Lahti, K., Lepist, L., Metcalf, J. S., Codd, G. A., and Meriluoto, J. (2006) First observation of cylindrospermopsin in *Anabaena lapponica* isolated from the boreal environment (Finland). *Environ Toxicol* **21:** 552–560.
Stachelhaus, T., Schneider, A., and Marahiel, M. A. (1995) Rational design of peptide antibiotics by targeted replacement of bacterial and fungal domains. *Science* **269:** 69–72.
Steffensen DA, Nicholson BC, Burch MD, Drikas M, Baker PD (1999) Ecology and management of blue–green algae (cyanobacteria) in Australia. *Environ Toxicol Water Qual* **14:** 183–195.
Stevens, D. K., and Krieger, R. I. (189) Application of a sensitive, GC-ECD analysis for anatoxin-a. *Toxicon* **27:** 79.
Stevens, D. K., and Krieger, R. I. (1991) Stability studies on the cyanobacterial nicotinic alkaloid anatoxin-a. *Toxicon* **29:** 167–169.
Stewart, I., Schluter, P. J., and Shaw, G. R. (2006) Cyanobacterial lipopolysaccharides and human health- a review. *Environ Health* **5:** 7. doi:10.1186/1476-069X-5-7
Stoner, R. D., Adams, W. H., Slatkin, D. N., and Siegelman, H. W. (1989) The effects of single L-amino acid substitutions on the lethal potencies of microcystins. *Toxicon* **27:** 825–828.
Stotts, R. R., Namikoshi, M., Haschek, W. M., Rinehart, K. L., Carmichael, W. W., Dahlem, A. M., and Beasley, V. R. (1993) Structural modifications imparting reduced toxicity in microcystins from *Microcystis* spp. *Toxicon* **31:** 783–789.

Stucken, K., John, U., Cembella, A., Murillo, A. A., Soto-Liebe, K., Fuentes-Valdés, J. J., Friedel, M., Plominsky, A. M., Vásquez, M., and Glöckner, G. (2010) The smallest known genomes of multicellular toxic cyanobacteria: Comparison minimal gene sets for linked traits and the evolutionary implications. *PLoS ONE* **5:** e9235. doi: 10.1371/journal.pone.0009235

Sudek, S., Haygood, M. G., Youssef, D. T. A., and Schmidt, E. W. (2006) Structure of trichamide, a cyclic peptide from the bloom-forming cyanobacterium *Trichdesmium erythraeum* predicted from the genome sequence. *Appl Environ Microbiol* **72:** 4382–4387.

Svreck, C., and Smith, W. (2004) Cyanobacterial toxins and the current state of knowledge on water treatment options: A review. *J Environ Eng Sci* **3:** 155–185.

Takenaka, S., and Watanabe, M. F. (1997) Microcystin-LR degradation by *Pseudomonas aeruginosa* alkaline protease. *Chemosphere* **34:** 749–757.

Takino, M., Daishima, S., and Yamaguchi, K. (1999) Analysis of anatoxin-a in freshwaters by automated on-line derivatization-liquid chromatography-electron spray mass spectrometry. *J Chromatogr* **862:** 191–197.

Teixera, M. G. L. C., Costa, M. C. N., Carvalho, V. L. P., Pereira, M. S., and Hage, E. (1993) Gastroenteritis epidemic in the area of the Itaparica Dam, Bahia, Brazil. *Bull Pan Am Health Org* **27:** 244–253.

Tencalla, F., and Dietrich, D. (1997) Biochemical characterization of microcystin toxicity in rainbow trout (*Oncorhynchus mykiss*). *Toxicon* **34:** 583–595.

Teneva, I., Dzhambazov, B., Koleva, L., Mladenov, R., and Schirmer, K. (2005b) Toxic potential of five freshwater *Phormidium* species (Cyanoprokaryota). *Toxicon* **45:** 711–725.

Teneva, I., Mladenov, R., Popov, N., and Dzhambazov, B. (2005a) Cytotoxicity and apoptotic effects of microcystin-LR and anatoxin-a in mouse lymphocytes. *Folia Biol* (Praha) **51:** 62–67.

Teneva, L., Visser, P. M., Christiansen, G., Dittman, E., Snelder, E. O., Wiedner, C., Mur, L. R., and Huisman, J. (2005) The microcystin composition of the cyanobacterium *Planktothrix agardhii* changes toward a more toxic variant with increasing light intensity. *Appl Environ Microbiol* **71:** 5177–5181.

Terao, K., Ohmori, S., Igarashi, K., Ohtani, I., Watanabe, M. F., Harada, K. I., Ito, E., and Watanabe, M. (1994) Electron microscopic studies on experimental poisoning in mice induced by cylindrospermopsin isolated from blue-green alga *Umezakia natans*. *Toxicon* **32:** 833–843.

Thacker, R. W., and Paul, V. J. (2004) Morphological, chemical, and genetic diversity of tropical marine cyanobacteria *Lyngbya* spp. and *Symploca* spp. (Oscillatoriales). *Appl Environ Microbiol* **70:** 3305–3312.

Thomas, P., Stephens, M., Wilkie, G., Amar, M., Lunt, G. G., Whiting, P., Gallagher, T., Pereira, E., Alkondon, M., and Alburquerque, E. X. (1993) (+) Anatoxin-a is a potent agonist at neural nicotinic acetylcholine receptors. *J Neurochem* **60:** 2308–2311.

Theiss, W. C., Carmichael, W. W., Wyman, J., and Bruner, R. (1988) Blood pressure and hepatocellular effects of the cyclic heptapeptide toxin produced by the freshwater cyanobacterium (blue-green alga) *Microcystis aeruginosa* strain PCC 7820. *Toxicon* **26:** 603–613.

Tillett, D., Dittmann, E., Erhard, M., von Döhren, H., Börner, T., and Neilan, B.A. (2000) Structural organization of microcystin biosynthesis in *Microcystis aeruginosa* PCC 7806: an integrated peptide-polyketide synthetase system. *Chem Biol* **7:** 753–764.

Tillett, D., Parker, D. L., and Neilan, B. A. (2001) Detection of toxigenicity by a probe for the microcystin A gene (*mcyA*) of the cyanobacterial genus *Microcystis*: Comparison of toxicities with 16S rRNA and phycocyanin operon (phycocyanin intergenic spacer) phylogenies. *Appl Environ Microbiol* **67:** 2810–2818.

Tisdale, E. S. (1931) Epidemic of intestinal disorders in Charleston W. Va., occurring simultaneously with unprecedented water supply conditions. *Am J Public Health* **21:** 198–200.

Toivola, D. M., Eriksson, J. E., and Brautigan, D. L. (1994) Identification of protein phosphatase 2A as the primary target for microcystin-LR in rat liver homogenates. *FEBS Lett* **344:** 175–180.

Tonk, L., Visser, P., Christiansen, G., Dittmann, E., Snelder, E. D. F. M., Wiedner, C., Mur, L., and Huisman, J. (2005) The microcystin composition of the cyanobacterium *Planktothrix agardhii* changes toward a more toxic variant with increasing light intensity. *Appl Environ Microbiol* **71:** 5177–5181.

Tooming-Klunderud, A., Mikalsen, B., Kristensen, T., and Jakobsen, K.S. (2008) The mosaic structure of the *mcyABC* operon in *Microcystis*. *Microbiology* **154:** 1886–1899.

Tooming-Klunderud, A., Rohrlack, T., Shalchian-Tabrizi, K., Kristensen, T., and Jakobsen, K. S. (2007) Structural analysis of a non-ribosomal halogenated cyclic peptide and its putative operon from *Microcystis*: Implications for evolution of cyanopeptolins. *Microbiology* **153:** 1382–1393.

Törökne, A. K. (1997) Interlaboratory trial using Thamnotox kit for detecting cyanobacterial toxins. VIII Int Conf on Harmful Algae. Vigo, Spain. p. 114.

Törökne, A. K., Laszlo, E., Chorus, I., Sivonen, K., and Barbosa, F. A. R. (2000) Cyanobacterial toxins detected by Thamnotox kit (A double blind experiment). *Environ Toxicol* **15:** 549–553.

Törökne, A., Asztalos, M., Bankine, M., Bickel, H., Borbely, G., Carmeli, S., Codd, G. A., Fastner, J., Huang, O., Humpage, A., Metcalf, J. S., Rabai, E., Sukenik, A., Suranyi, G., Vaas, G., and Weiszfeiler, V. (2004) Interlaboratory comparison trial on cylindrospermopsin measurement. *Anal Biochem* **332:** 280–284.

Tsuji, K., Naito, S., Kondo, F., Ishikawa, N., Watanabe, M. F., Suzuki, M., and Harada, K.-I. (1994a) Stability of microcystins from cyanobacteria: Effect of light on decomposition and isomerization. *Environ Sci Technol* **28:** 173–177.

Tsuji, K., Naito, S., Kondo, F., Watanabe, M. F., Suzuki, S., Nakazawa, H., Suzuli, M., Shimada, T., and Harada, K. (1994b) A clean-up method for analysis of trace amounts of microcystins in lake water. *Toxicon* **32:** 1251–1259.

Tsuji, K., Setsuda, S., Watanuki, T., Kondo, F., Nakazawa, H., Suzuki, M., and Harada, K.-I. (1996) Microcystin levels during 1992–95 for Lakes Sagami and Tsukui-Japan. *Nat Toxins* **4:** 189–194.

Tsuji, K., Watanuki, T., Kondo, F., Watanabe, M. F., Suzuki, S. Nakazawa, H., Suzuki, M., Uchida, H., and Harada, K.-I. (1995) Stability of microcystins from cyanobacteria-II. Effect of UV light on the decomposition and isomerization. *Toxicon* **33:** 1619–1631.

Tsutsumi, T., Nagata, S., Hasegawa, A., and Ueno, Y. (2000) Immunoaffinity column as clean-up tool for determination of trace amounts of microcystins in tap water. *Food Chem Technol* **38:** 593–597.

Turner, P. C., Gammie, A. J., Hollinrake, K., and Codd, G. A. (1990) Pneumonia associated with contact with cyanobacteria. *Br Med J* **300:** 1440–1441.

Twist, H., and Codd, G. A. (1997) Degradation of the cyanobacterial hepatotoxin, nodularin, under light and dark conditions. *FEMS Microbiol Lett* **151:** 83–88.

Utkilen, H., and Gjolme, N. (1992) Toxin production by *Microcystis aeruginosa* as a function of light in continuous cultures and its ecological significance. *Appl Environ Microbiol* **58:** 1321–1325.

Utkilen, H., and Gjolme, N. (1995) Iron-stimulated toxin production in *Microcystis aeruginosa. Appl Environ Microbiol* **61:** 797–800.

Utkilen, H., Skulberg, O. M., Skulberg, R., Gjolme, N., and Underdal, B. (2001) Toxic cyanobacterial blooms of inland waters in Southern Norway 1878–1998. In: Chorus, I. (Ed.) *Cyanotoxins-Occurrence, Causes and Consequences*. Springer, Berlin, Germany. pp. 46–49.

Ueno, Y., Nagata, S., Tsutsumi, T., Hasegawa, A., Watanabe, M. F., Park, H. D., Chen, G., and Yu, S. Z. (1996) Detection of microcystin, a blue-green algal hepatotoxin, in drinking water sampled in Haimen and Fusui, endemic areas of primary liver cancer in China, by highly sensitive immunoassay. *Carcinogenesis* **17:** 1317–1321.

Usleber, E., Scheider, E., Terplan, G., and Laycock,, M. V. (1995) Two formats of enzyme immunoassay for the detection of saxitoxin and other paralytic shellfish poisoning toxins. *Food Add Contam* **12:** 405–413.

Vaas, G., Gasper, A., Pager, C., Suranyl, G., Mathe, C., Hamvas, M. M., and Borbely, G. (2004) Analysis of cyanobacterial toxins (anatoxin-a, cylindrospermopsin, microcystin-LR) by capillary electrophoresis. *Electrophoresis* **25:** 108–115.

Vaitomaa, J., Rantala, A., Halinen, K., Rouhiainen, L., Tallberg, P., Mokelke, L., and Sivonen, K. (2003) Quantitative real-time PCR for determination of microcystin synthetase E copy numbers for *Microcystis* and *Anabaena* in Lakes. *Appl Environ Microbiol* **69:** 7289–7297.

Van der Westhuizen, A. J., and Eloff, J. N. (1985) Effect of temperature and light on the toxicity and growth of the blue-green alga *Microcystis aeruginosa* (UV-006). *Planta* **163:** 55–59.

Van der Westhuizen, A. J., Eloff, J. N., and Krueger, G. H. J. (1986) Effect of temperature and light (fluence rate) on the composition of the toxin of the cyanobacterium *Microcystis aeruginosa* (UV-006). *Arch Hydrobiol* **108:** 145–154.

Van Buynder, P. G., Oughtred, T., Kirby, B., Phillips, S., Eaglesham, G., Thomas, K., and Burch, M. (2001) Nodularin uptake by seafood during a cyanobacterial bloom. *Environ Toxicol* **16:** 468–471.

Van Halderen, A., Harding, W.R., Wessels, J. C., Schneider, D. J., Heine, E. W., Van der Merwe, J., and Fourie, J. M. (1995) Cyanobacterial (blue-green algae) poisoning of livestock in the western Cape province of South Africa. *J S Afr Vet Assoc* **66:** 260–264.

Vasconcelos, V. (1993) Toxicity of cyanobacteria in lakes of North and Central Portugal. Ecological implications. *Verh Internat Verein Limnol* **25:** 694–697.

Vasconcelos, V. (1994) Toxic cyanobacteria (blue-green algae) in Portuguese freshwaters. *Arch Hydrobiol* **130:**439–451.

Vasconcelos, V. (1999) Cyanobacterial toxins in Portugal: Effects on aquatic animals and risk for human health. *Braz J Med Biol Res* **32:** 249–254.

Vasconcelos, V., and Pereira (2001) Cyanobacteria diversity and toxicity in a wastewater treatment plant (Portugal). *Water Res* **35:** 1354–1357.

Vasconcelos, V., Oliveira, S., and Teles, F. O. (2001) Impact of a toxic and a non-toxic strain of *Microcystis aeruginosa* on the crayfish *Procambarus clarkii*. *Toxicon* **39:** 1461–1470.

Vasconcelos, V., Sivonen, K., Evans, W. R., Carmichael, W. W., and Namikoshi, M. (1996) Hepatotoxic microcystin diversity in cyanobacterial blooms collected in Portuguese waters. *Water Res* **30:** 2377–2384.

Vézie, C., Brient, L., Sivonen, K., Betru, G., Lefeuvre, J.-C., Salkinoja-Salonen, M. (1998) Variation of microcystin content of cyanobacterial blooms and isolated strains in Lake Grand-Lieu (France). *Microb Ecol* **35:** 126–135.

Vézie, C., Rapala, J., Vaitomaa, J., Seitsonen, J., and Sivonen, K. (2002) Effect of nitrogen and phosphorus on growth of toxic and non-toxic strains of *Microcystis* strains and on intracellular microcystin concentrations. *Microb Ecol* **43:** 443–454.

Via-Ordorika, L., Fastner, J., Kurmayer, R., Hisbergues, M., Dittmann, E., Komarek, J., Erhard, M., and Chorus, I. (2004) Distribution of microcystin-producing and non-microcystin-producing *Microcystis* sp. in European freshwater bodies: Detection of microcystins and microcystin genes in individual colonies. *System Appl Microbiol* **27:** 592–602.

Vlyssides, A. G., and Israilides, C. J. (1998) Electrochemical oxidation of a textile dye and finishing wastewater using a Pt/Ti electrode. *J Environ Sci Health* **33:** 847–862.

Volterra, L., Bruno, M., Gucci, P. M. B., and Pierdominici, E. (1992) Fast method for detecting toxic cyanophyte blooms. *Environ Toxicol Water Qual* **7:** 215–222.

von Elert, E., Oberer, L., Merkel, P., Huhn, T., and Blom, J. F. (2005) Cyanopeptolin 954, a chlorine-containing chymotrypsin inhibitor of *Microcystis aeruginosa* NIVA Cya 43. *J Nat Prod* **68:** 1324–1327.

Ward, C. J., Beattie, K. A., Lee, E. Y. C., and Codd, G. A. (1997) Colorimetric protein phosphatase inhibition assay of laboratory strains and natural blooms of cyanobacteria: Comparison of high-performance liquid chromatographic analysis for microcystins. *FEMS Microbiol Lett* **153:** 465–473.

Watanabe, M. (1995) Studies on planktonic blue-green algae. 5. A new species of *Cylindrospermopsis* (Nostocaceae) from Japan. *Bull Natl Sci Mus* **21:** 45–48.

Watanabe, M. F., and Oishi, S. (1985) Effects of environmental factors on toxicity of a cyanobacterium (*Microcystis aeruginosa*) under culture conditions. *Appl Environ Microbiol* **49:** 1342–1344.

Watanabe, M. F., Oishi, S., Harada, K., Matsuura, K., Kawai, H., and Suzuki, M. (1988) Toxins contained in *Microcystis* species of cyanobacteria (blue-green algae). *Toxicon* **26:** 1017–1025.

Watanabe, M. F., Oishi, S., Watanabe, Y., and Watanabe, M. (1986) Strong possibility of lethal toxicity in the blue-green alga *Microcystis viridis* Lemm. *J Phycol* **22:** 552–556.

Watanabe, M. F., Watanabe, M., Kato, T., Harada, K.-I., and Suzuki, M. (1989) Heptapeptide toxin production during the batch culture of two *Microcystis* species (cyanobacteria). *J Appl Phycol* **1:** 161–165.

Watts, J. S., Reilly, J., Dacosta, F., and Krop, S. (1966) Acute toxicity of paralytic shellfish poison in rats of different ages. *Toxicol Appl Pharmacol* **8:** 286–294.

Weckesser, J., Martin, C., and Jakobi, C. (1996) Cyanopeptolins, depsipeptides from cyanobacteria. *Syst Appl Microbiol* **19:** 133–138.

Welker, M., and Steinberg, C. H. (1999) Indirect photolysis of cyanotoxins: One possible mechanism for their low persistence. *Water Res* **33:** 1159–1164.

Welker, M., and von Doehren, H. (2006) Cyanobacterial peptides-Nature's own combinatorial biosynthesis. *FEMS Microbiol Rev* **30:** 530–563.

Welker, M., Christiansen, G., and von Dohren, H. (2004a) Diversity of coexisting *Planktothrix* (Cyanobacteria) chemotypes deduced by mass spectral analysis of microcystins and other oligopeptides. *Arch Microbiol* **182:** 288–298.

Welker, M., Brunke, M., Preussel, K., Lippert, I., and von Dohren, H. (2004b) Diversity and distribution of *Microcystis* (Cyanobacteria) oligopeptide chemotypes from natural communities studied by single colony mass spectrometry. *Microbiology* **150:** 1785–1796.

Welker, M., von Dohren, H., Tauscher, H., Steinberg, C. E. W., and Erhard, M. (2003) Toxic *Microcystis* in shallow lake Muggelsee (Germany)- dynamics, distribution, diversity. *Arch Hydrobiol* **157:** 227–248.

Weller, M. G., Zeck, A., Eikenberg, A., Nagata, S., Ueno, Y., and Niessner, R. (2001) Development of a direct competitive microcystin immunoassay of broad specificity. *Anal Sci* **17:** 1445–1448.

Welsh, J., and McClelland, M. (1990) Fingerprinting genomes using PCR with arbitrary primers. *Nucleic Acids Res* **18:** 7213–7218.

Wera, S., and Hemmings, B. A. (1995) Serine/threonine protein phosphatases. *Biochem J* **311:** 17–29.

White, S. H., Duivenvoorden, L. J., and Fabbro, L. D. (2005) A decision making framework for ecological impacts associated with the accumulation of cyanotoxins (cylindrospermopsin and microcystin). *Lakes Reservoirs: Research Management* **10:** 25–37. doi: 10.1111/j.1440-1770.2005.00258,x.

WHO (1998) *Guidelines for Drinking Water Quality*. 2nd edn. *Addendum to volume 2. Health Criteria and Other Supporting Information*. World Health Organization, Geneva, Switzerland.

WHO (2003) Algae and cyanobacteria in freshwater. In: *Guidelines for Safe Recreational Water Environments*. Volume 1: *Coastal and freshwaters*. World Health Organization, Geneva, Switzerland. pp. 136–158.

WHO (2004) *Guidelines for Drinking Water Quality*. 3rd Edition. World Health Organization, Geneva, Switzerland.

Wiberg, G. S., and Stephenson, N. R. (1960) Toxicological studies on paralytic shellfish poison. *Toxicol Appl Pharmacol* **2:** 607–615.

Wickstrom, M. L., Khan, S. A., Haschek, W. M., Wyman, J. F., Eriksson, J. E., Schaeffer, D. J., and Beasley, V. R. (1995) Alterations in microtubules, intermediate filaments, and microfilaments induced by microcystin-LR in cultured cells. *Toxicol Pathol* **23:** 326–337.

Wiedner, C., Chorus, I., and Fastner, J. (2001) The water bodies surveyed for cyanotoxins in Germany. In: Chorus, I. (Ed.) *Cyanotoxins-Occurrence, Causes, Consequences*. Springer, Berlin, Germany. pp. 6–21.

Wiedner, C., Visser, P. M., Fastner, J., Metcalf, J. S., Codd, G. A., and Mur, L. R. (2003) Effects of light on the microcystin content of *Microcystis* strain PCC 7806. *Appl Environ Microbiol* **69:** 1475–1481.

Wiegand, C., and Pflugmacher, S. (2001) Uptake of microcystin-LR in aquatic organisms. In: Chorus, I. (Ed.) *Cyanotoxins-Occurrence, Causes, Consequences*. Springer, Berlin, Germany. pp. 240–248.

Wigmans, T. (1989) Industrial aspects of production and use of activated carbon. *Carbon* **27:** 13–22.

Willen, E., Ahlgren, G., and Soederhielm, A.-C. (2000) Toxic cyanophytes in three Swedesh Lakes. *Verh Internat Verein Limnol* **27:** 560–564.

Williame, R., and Hoffmann, L. (1999) Bloom-forming blue-green algae in Belgium and Luxembourg. *Arch Hydrobiol* (Suppl) *Algol Stud* **94:** 365–376.

Williame, R., Jurczak, T., Iffly, J.-F., Kull, T., and Meriluto, J. (2005) Distribution of hepatotoxic cyanobacterial blooms in Belgium and Luxembourg. *Hydrobiologia* **551:** 99–117.

Williams, D. E., Dawe, S.C., Kent, M. L., Anderson, R. J., Craig, M., and Holmes, C. F. B. (1997) Bioaccumulation and clearance of microcystins from salt water mussels, *Mytilus edulis*, and *in vivo* evidence for covalently bound microcystins in mussel tissues. *Toxicon* **35:** 1617–1625.

Williams, J. G. K., Kubelik, A. R., Livak, K. J., Rafalski, J. A., and Tingey, S. V. (1990) DNA polymorphisms amplified by arbitrary primers are useful as genetic markers. *Nucleic Acids Res* **18:** 6531–6535.

Williams, P. G., Moore, R. E., and Paul, V. J. (2003) Isolation and structure determination of lyngbyastatin 3, a lyngbyastatin 1 homologue from the marine cyanobacterium *Lyngbya majuscula*. Determination of the configuration of the 4-amino-2, 2-dimethyl-3-oxopentanoic acid unit in majusculamide C, dolastatin 12, lyngbyastatin 1, and lyngbyastatin 3 from cyanobacteria. *J Nat Prod* **66:** 1356–1363.

Wilson, A. E., Sarnelle, O., Neilan, B. A., Salmon, T. P., Gehringer, M. M., and Hay, M. E. (2005) Genetic variation of the bloom-forming cyanobacterium *Microcystis aeruginosa* within and among lakes: Implications for harmful algal blooms. *Appl Environ Microbiol* **71:** 6126–6133.

Wilson, K. M., Schembri, M. A., Baker, P. D., and Saint, C. P. (2000) Molecular characterization of the toxic cyanobacterium *Cylindrospermopsis raciborskii* and design of a species-specific PCR. *Appl Environ Microbiol* **66:** 332–338.

Wirsing, B., Flury, T., Wiedner, C., Neumann, U., and Weckesser, J. (1999) Estimation of the microcystin content in cyanobacterial field samples from German lakes using the colorimetric protein-phosphatase inhibition assay and RP-HPLC. *Environ Toxicol* **14:** 23–29.

Wirsing, B., Hoffmann, L., Heinze, R., Klein, D., Daloze, D., Braekman, J. C., and Weckesser, J. (1998) First report on the identification of microcystin in a water bloom collected in Belgium. *Sys Appl Microbiol* **21:** 23–27.

Wonnacott, S., Jackman, S., Swanson, K. L., Rapoport, H., and Albuquerque, E. X. (1991) Nicotinic pharmacology of anatoxin analogs. II. Side chain structure-activity relationships at neuronal nicotinic ligand binding sites. *J Pharmacol Exptl Therapeutics* **259:** 387–391.

Wood, S. A., Briggs, L. R., Sporsen, J., Ruck, J.G., Wear, R. G., Holland, P. T., and Bloxham, M. (2006) Changes in concentrations of microcystins in rainbow trout, freshwater mussels and cyanobacteria in lakes Rotoiti and Rotoehu. *Environ Toxicol* **21:** 205–222.

Wood, S. A., Mountfort, D., Selwood, A. I., Holland, P. T., Puddick, J., and Cary, S. C. (2008) Widespread distribution and identification of eight novel microcystins in Antarctic cyanobacterial mats. *Appl Environ Microbiol* **74:** 7243–7251.

Wu, J.-Y., Xu, Q.-J., Gao, G., and Shen, J.-H. (2006) Evaluating genotoxicity associated with microcystin-LR and its risk to source water safety in Meiliang Bay, Taihu Lake. *Environ Toxicol* **21:** 250–255.

Wu, M., Okino, T., Nogle, L. M., Marquez, B. L., Williamson, R. T., Sitachitta, N., Berman, F. W., Murray, T. F., McGough, K., Jacobs, R., Colsen, K., Asano, T., Yokokawa, F., Shiori, T., and Gerwick, W. H. (2000) Structure, synthesis and biological properties of kalkitoxin, a novel neurotoxin from the marine cyanobacterium *Lyngbya majuscula*. *J Am Chem Soc* **122:** 12041–12042.

Wyllie, A. H., Kerr, J. F., and Currie, A. R. (1980) Cell death: The significance of apoptosis. *Int Rev Cytol* **68:** 251–306.

Xagoraraki, I., Harrington, G. W., Zulliger, K., Zeier, B., Krick, W., Karner, D. A., Standridge, J. H., and Westrick, J. (2006) Inactivation kinetics of the cyanobacterial toxin microcystin-LR by free chlorine. *J Environ Eng* **132:** 818–823.

Xie, C., Runnegar, M. T. C., and Snider, B. B. (2000) Total synthesis of (±) cylindrospermopsin. *J Am Chem Soc* **122:** 5017–5024.

Xu, L. H., Lam, P. K. S., Chen, J. P., Xu, J. M., Wong, B. S. F., Zhang, Y. Y., Wu, R. S. S., and Harada, K.-I. (2000) Use of protein phosphatase inhibition assay to detect microcystins in Donghu Lake and a fish pond in China. *Chemosphere* **41:** 53–58.

Xu, L. H., Lam, P. K., Chen, J. P., Xu, J. M., Wong, B. S., Zhang, Y. Y., Wu, R. S., and Harada, K.-I. (2002) Use of protein phosphatase inhibition assay to detect microcystins in Donghu Lake and a fish pond in China. *Chemosphere* **41:** 53–58.

Yamasaki, H. (1988) Multistage carcinogenesis: Implications for risk estimation. *Cancer Metastasis Rev* **7:** 5–18.

Yilmaz, M., and Phlips, E. J. (2011) Diversity and selection acting on cylindrospermopsin *cryB* gene adenylation domain sequences in Florida. *Appl Environ Microbiol* **77:** 2502-2507. doi:10.1128/AEM.02252-10.

Yokokawa, F., Fujiwara, H., and Shioiri, T. (1999) Total synthesis and revision of absolute configuration of antillatoxin, an ichthyotoxic cyclic lipopeptide of marine origin. *Tetrahedron Lett* **40:** 1915–1916.

Yokokawa, F., Fujiwara, H., and Shioiri, T. (2000a) Total synthesis and revision of absolute streochemistry of antillatoxin, an ichthyotoxic cyclic lipopeptide from marine cyanobacterium *Lyngbya majuscula*. *Tetrahedron* **56:** 1759–1775.

Yokokawa, F., Shioiri, T., and Gerwick, W. H. (2000b) Structure, synthesis and biological properties of kalkitoxin, a novel neurotoxin from the marine cyanobacterium *Lyngbya majuscula*. *J Am Chem Soc* **122:** 12041–12042.

Yoshida, T., Makita, Y., Tsutsumi, T., Nagata, S., Tashiro, F., Yoshida, F., Sekijima, M., Tamura, S.-I., Harada, T., Maita, K., and Ueno, Y. (1998) Immunohistochemical localization of microcystin-LR in the liver of mice: A study on the pathogenesis of microcystin-LR-induced hepatotoxicity. *Toxicol Pathol* **26:** 411–418.

Yoshida, T., Yuki, Y., Lei, S., Chinen, H., Yoshida, M., Kondo, R., and Hiroishi, S. (2003) Quantitative detection of toxic strains of the cyanobacterial genus *Microcystis* by competitive PCR. *Microbes Environ* **18:** 16–23.

Yoshizawa, S., Matsushima, R., Watanabe, M. F., Harada, K.-I., Ichihara, A., Carimachael, W. W., Fujiki, H. (1990) Inhibition of protein phosphatases by microcystins and nodularin associated with hepatotoxicity. *J Cancer Res Clin Oncol* **116:** 609–614.

Young, F. M., Thomson, C., Metcalf, J. S., Lucocq, J. M., and Codd, G. A. (2005) Immunogold localisation of microcystins in cryosectioned cells of *Microcystis*. *J Struct Biol* **151:** 208–214.

Yu, S.-Z. (1989) Drinking water and primary liver cancer. In: Tang, Z.-Y., Wu, M.-C., and Xia, S.-S. (Eds.) *Primary Liver Cancer*. Springer-Verlag, New York, NY, USA. pp. 30–37.

Yuan, M., Namikoshi, M., Otsuki, A., and Sivonen, K. (1998) Effect of amino acid-side chain on fragmentation of cyclic peptide ions: Differences of electron spray ionozation collision-induced decomposition mass spectra of toxic heptapeptide microcystins containing ADMAdda instead of Adda. *Eur Mass Spectrom* **4:** 287–298.

Yuan, M., Namikoshi, M., Otsuki, A., Rinehart, K. L., Sivonen, K., and Watanabe, M. F. (1999a) Low-energy collisionally activated decomposition and structural characterization of cyclic heptapeptide microcystins by electron spray ionization mass spectrometry. *J Mass Spectrom* **34:** 33–43.

Yuan, M., Namikoshi, M., Otsuki, A., Watanabe, M. F., and Rinehart, K. L. (1999b) Electrospray ionization mass spectrometric analysis of microcystins, cyclic heptapeptide hepatotoxins: Modulation of charge states and $[M+H]^+$ to $[M+Na]^+$ ratio. *J Am Soc Mass Spectrom* **10:** 1138–1151.

Yunes, J. S., Salomon, P. S., Matthiensen, A., Beattie, K. A., Raggett, S. L., and Codd, G. A. (1996) Blooms of cyanobacteria in the Patos Lagoon estuary, in Southern Brazil. *J Aquat Ecosys Health* **5:** 223–229.

Zegura, B., Sedmak, B., and Filipic, M. (2003) Microcystin-LR induces oxidative DNA damage in human hepatoma cell line HepG2. *Toxicon* **41:** 41–48.

Zhan, L., Sakamoto, H., Sakuraba, M., Wude, S., Zhang, L. S., Suzuki, T., Hayashi, M., and Honma, M. (2004) Genotoxicity of microcystin-LR in human lymphoblastoid TK6 cells. *Mutat Res* **557:** 1–6.

Zhang, L., Ping, X., and Yang, Z. (2004) Determination of microcystin-LR in surface water using high-performance liquid chromatography/tandem electrospray ionization mass detector. *Talanta* **62:** 193–200.

Zhang, Z., Zhao, S., Long, F., Zhang, L., Bai, G., Shima, H., Nagao, M., Lee, E. Y. C. (1994) A mutant of protein phosphatase-1 that exhibits altered toxin sensitivity. *J Biol Chem* **269:** 16997–17000.

Zhou, L., Yu, H., and Chen, K. (2002) Relationship between microcystin in drinking water and colorectal cancer. *Biomed Environ Sci* **15:** 166–171.

Ziemert, N., Ishida, K., Quillardet, P., Bouchier, C., Hertweck, C., Tndeau de Marsac, N., and Dittmann, E. (2008) Microcyclamide biosynthesis in two strains of *Microcystis aeruginosa*: From structure to genes and vice versa. *Appl Environ Microbiol* **74:** 1791–1797.

Znachor, P., Jurczak, T., Komarkova, J., Jezberova, J., Mankiewicz, J., Kastovska, K., and Zapomelova, E. (2006) Summer changes in cyanobacterial bloom composition and microcystin concentration in eutrophic Czech reservoirs. *Environ Toxicol* **21:** 236–243.

Zurawell, R. W., Kotak, B. G., and Prepas, E. E. (1999) Influence of lake trophic status on the occurrence of microcystin-LR in the tissue of pulmonate snails. *Freshwater Biol* **42:** 707–718.

Zweigenbaum, J. A., Henion, J. D., Beattie, K. A., Codd, G. A., and Poon, G. K. (2000) Direct analysis of microcystins by microbore liquid chromatography electrospray ionization ion-trap mass spectrometry. *J Pharm Biomed Anal* **23:** 723–733.

CHAPTER 10

STRESS RESPONSES

Microorganisms grow in the environments suitable for them and make a niche for themselves. Due to the fluctuations in the nutritional or environmental conditions they face various adverse situations. In order to adjust to such conditions, the microorganisms adapt to the new environment by altering their metabolic profile. Due to this plasticity, they are able to grow under the changing environment without detriment to growth and development. The availability of excess nutrients or lack of proper nutrients, light (high or low and UV-light), salinity, high solute concentrations (osmotic stress), temperature (high and low), hydrogen ion concentration (high or low pH) and lack of water (drought leading to desiccation) are the common abiotic stresses which microorganisms experience in nature.

The cyanobacteria grow in various types of habitats and gain a cosmopolitan distribution. Majority of them are aerobic photoautotrophs but some of them show the distinctive property of heterotrophic growth and survive for long periods in darkness. They grow in all types of terrestrial habitats where they are suggested to play important role in terrestrial ecosystems (Whitton, 1992; Whitton and Potts, 2000). They inhabit all types of water bodies that are salty, brackish or freshwater. The picoplanktonic cyanobacteria grow in all types of oceans from surface waters to greater depths (even extending up to 100 m). Freshwater bodies of differing trophic states support the growth of a number of cyanobacteria where bloom-forming and non-bloom-forming forms are prevalent. Their ability to survive in extremely high and low temperatures enables them to grow in hot springs, Arctic and Antarctic Lakes, snow and ice (Sabacka and Elster, 2006; Seckbach, 2007). Their growth under such harsh conditions is commonly associated with sharing their tasks and supporting each other by the formation of crusts or mats. These have been variously designated as "microbiotic soil crusts" (Elridge and Greene, 1994), "biological soil crusts" (also abbreviated to biocrusts; Belnap *et al.*, 2001a), "cryptogamic crusts" (Strandling *et al.*, 2002) and "cryptobiotic crusts" (Pócs, 2009). The biocrusts are dominated by either cyanobacteria or lichens or mosses (Fig. 1). Their symbiotic association from fungi to higher plants provides the best example of their existence in extremely desiccated state (Usher *et al.*, 2007). This is suggestive of the fact that the cyanobacteria can cope up with a wide spectrum of abiotic stresses such as heat and cold, desiccation, salinity, high light intensity, nutrient starvation, anaerobiosis and the presence of high concentrations of pesticides and heavy metals.

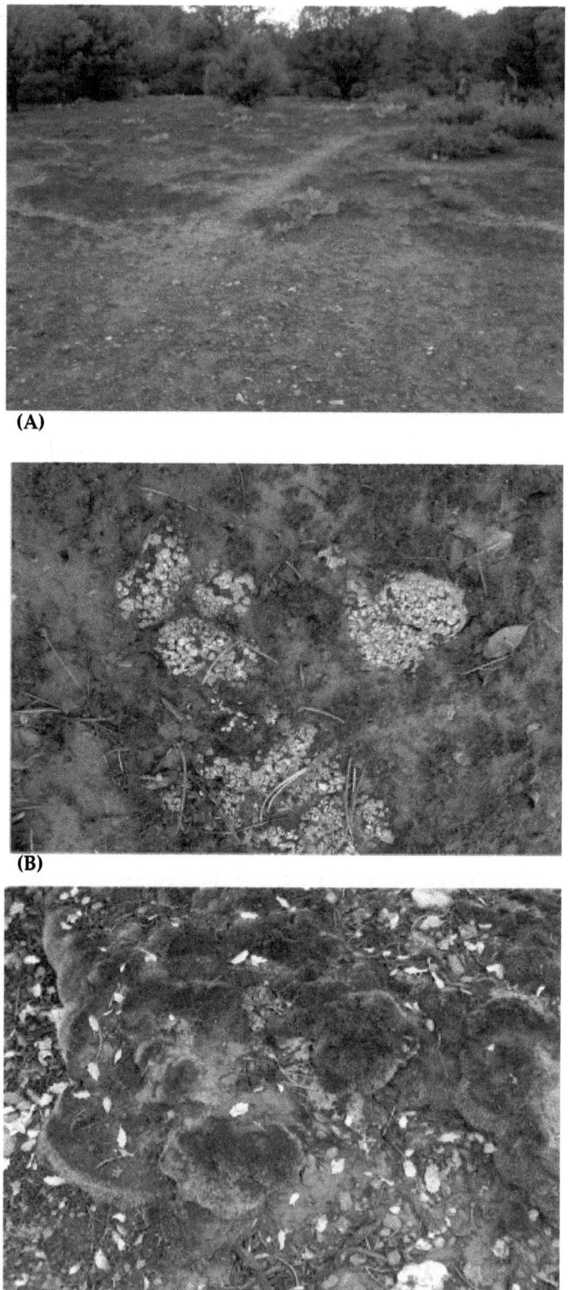

Figure 1: Biocrusts of Northern Arizona. Cyanobacteria dominated crusts that are quite sensitive to human interference (A); Lichen dominated biocrusts at a rest stop on the way to Phoenix, on a limestone type of soil (B); Moss dominated soil crusts (C). With the kind permission of T. R. Northen, Lawrence Berkeley National Laboratory 1, Life Sciences Division, 1 Cyclotron Road, Berkeley, CA 94720, USA; http://biocrust.posterous.com/.

Color image of this figure appears in the color plate section at the end of the book.

1) Habitat sensing: The microorganisms sense changes in the surrounding environment with the help of two-component regulatory systems. Okamoto *et al.* (2007) established a correlation between signal transduction domains and habitats in cyanobacteria by applying hmmpfam (Pfam-1s Version 14.0) program to the 14 cyanobacterial genomes [*Anabaena variabilis*, *Anabaena* sp. strain PCC 7120, *Gloeobacter violaceus* PCC 7421, *Nostoc punctiforme* ATCC 29133, *Synechocystis* sp. strain PCC 6803, *S. elongatus* PCC 6301, *S. elongatus* PCC 7942, *Thermosynechococcus elongatus* BP-1- all freshwater; *Crocosphaera watsoniii* WH8501, *Prochlorococcus marinus* MED4, *P. marinus* SS120, *P. marinus* MIT9313, *Synechococcus* sp. strain WH8102, and *Trichodesmium erythraeum*- all marine]. A positive correlation between habitats and functional domains of signalling pathways existed as freshwater species developed more signal transduction domains than the seawater species that had fewer domains. In view of the existence of the cyanobacteria since Precambrian period, these organisms must have faced very challenging situations to obtain sufficient levels of CO_2 from environments with wide fluctuations in O_2 levels ranging from anaerobiosis to saturation levels. Thus the cyanobacteria during the course of evolution have developed a CO_2 concentrating mechanism (CCM) along with the evolution of carbonic anhydrases (CAs) and the main carboxylating enzyme, RUBP carboxylase/ oxygenase (RuBisCO). Badger *et al.* (2006) reviewed the current status of the genomic diversity of the CCM among cyanobacterial species and the evolution of this mechanism to match with their habitat requirements. A comparison 18 genomes of cyanobacteria (besides the 14 genomes mentioned above, four others, i.e *Synechococcus* sp. strain PCC 7002, *Synechococcus* sp. strain CC9605, *Synechococcus* sp. strain CC9902 and *Synechococcus* sp. strain CC9311 are included in this study) with their habitats revealed a correlation between the type of carboxysome and the environment. Cyanobacterial species inhabiting open oceans are equipped with α-type of carboxysomes (so classified as α-cyanobacteria that possess form 1A type of RuBisCO to which *Prochlorococcus* and α-*Synechococcus* spp. belong) whereas freshwater and other marine species possess β-type of carboxysomes (so called as β-cyanobacteria that possess form 1B type of RuBisCO). Four CAs have been identified that are responsible for generation of CO_2 within the carboxysome and three of these (CcaA, CcmM and CsoS3) are related to carboxysome function. There are limited number of transporters for bicarbonate (SbtA- HCO_3^-, BicA- HCO_3^- and $NDH1_4$- CO_2) and CO_2 uptake in the open ocean species of *Prochlorococcus*. Species inhabiting coastal and estuarine environments with fluctuations in CO_2 and temperature are equipped with variable number of bicarbonate and CO_2 uptake systems. Likewise, species inhabiting symbiotic environments, thermal hot springs and calcareous rocks (where fluctuations in Ci, temperature, O_2 and nutrients exist) showed most number of composite Ci transporters (BCT- HCO_3^-, SbtA- HCO_3^-, BicA- HCO_3^-, NDH-1_3-CO_2 and NDH-1_4-CO_2; the former three being located on the cytoplasmic membrane and the latter two on the thylakoids).

The physiology of abiotic stress has been understood in plants and cyanobacteria. Mainly two types of damages are known to be caused by most of the abiotic stresses. The first is the water-deficit stress that is caused by a number of environmental conditions including drought, salinity, heat and cold. The second type of damage results due to the production of reactive oxygen species (ROS) that damage cellular constituents including proteins and nucleic acids. This is called as oxidative stress.

2) Water-deficit stress: The water status of a cell (or tissue) is defined by two parameters, i.e. water potential and relative water content. Water potential of a system is represented by Ψ and is measured in units of pressure (M Pascal or MPa). This is similar to electric potential, just as the current flows from a compartment of high potential to a compartment of low potential when a connection is made between them. Thus water is driven from a high potential to a low potential. When sufficient water is available, the internal cell environment is thus fully hydrated because of movement of water from outside (high water potential) to inside the cells (low water potential) and the cells appear fully turgid. However, water potential is governed by two factors. The first is the concentration of solutes that creates the solute- or osmotic potential and the pressure exerted by the pressure- or turgor potential. The osmotic potential is generally lower (more negative) than the water potential and the turgor potential is the difference between them. The soluble proteins and other molecules interact with water by their polar and charged groups that are often positioned around their external surfaces. This results in the formation of a highly ordered structure known as 'hydration shell' or 'water shell'. But in conditions of drought (a water-deficit stress), heat (at high temperature) or cold (due to chilling and freezing) and high salinity the water potential of the environment decreases as a result water flows out of the cells and the cells face water-deficit stress. As a result of this, the protective hydration shell around the macromolecules is disturbed leading to their denaturation and consequent loss of activity. In order to combat these different stress conditions, the microbial (/plant) cells show some common and some variable metabolic responses. Water-deficit in the environment leads to drought conditions and not all organisms can face this challenge but some of them do so by developing tolerance to drought. They do so by not desiccating themselves but by taking recourse to store water (as in desert cacti) or synthesize water (as in rodents) to cater to future needs. It is difficult to set internal limits of water content common to all organisms as especially it is difficult to identify a specific feature or trait common to them. The limits and frontiers of desiccation tolerant life and constraints of such tolerance have been reviewed (Alpert, 2005, 2006). Desiccation tolerance in prokaryotes has received greater attention quite early and the cyanobacteria have emerged as good experimental systems to study desiccation (Potts, 1994, 1999, 2001). A detailed account on desiccation has been presented later at the end of this chapter.

3) Oxidative stress: Atmospheric oxygen is generally non-reactive to most of the organic molecules because it possesses two electrons that have parallel spins, whereas the organic molecules have paired electrons with opposite spins. Oxygen can react with a divalent reductant provided it has two unpaired electrons that spin opposite to those of oxygen, Due to the spin restriction, oxygen can not be reduced in biochemical reactions but if oxygen is activated it can participate in reactions with organic molecules. This activation of oxygen may take place by two different mechanisms. The first is by the formation of singlet oxygen species ($^1O_2$*). This is possible by the absorption of sufficient energy to reverse the spin of one of the unpaired electrons to generate $^1O_2$* species. During PSI and PSII and respiration as well as during chemical or environmental stress conditions, the $^1O_2$* species is generated in plants and cyanobacteria. $^1O_2$* is more reactive towards organic molecules than oxygen and can transfer its excitation energy to other biological molecules or can react with them to produce endoperoxides or hydroperoxides (Haliwell and Gutteridge, 1989). In the second mechanism, the $^1O_2$* species is reduced step-wise to produce superoxide radicals ($O_2^{\bullet -}$), hydrogen peroxide (H_2O_2) and hydroxyl radicals (OH·). All these ($^1O_2$*, $O_2^{\bullet -}$, H_2O_2, OH·) are powerful oxidizing agents and are collectively known as ROS. Both $O_2^{\bullet -}$ and OH· have an unpaired electron that makes them highly reactive with organic molecules. The $^1O_2$* species has a very short half-life in cells (Gorman and Rodgers, 1992) and it can last for nearly 4 µs in water and 100 µs in a non-polar environment (Foyer and Harbinson, 1994). The $O_2^{\bullet -}$ anion is moderately reactive and has a half-life of 2–4 µs. Because of its negative charge it cannot diffuse through biological membranes and so dismutated readily to H_2O_2. It oxidizes the [4 Fe-4S]$^{2+}$ clusters to [3 Fe-4S]$^{1+}$ releasing iron (Fe^{2+}). Thus the activity of quinones and metal-containing enzyme complexes (Fe^{3+} and Cu^{2+}) gets affected. Further, $O_2^{\bullet -}$ can cause the formation of hydroperoxy radicals ($HO^{\bullet}_2$) by protonation that can cross biological membranes and remove hydrogen atoms from polyunsaturated fatty acids and lipid hydroperoxides leading to autooxidation of lipids (Haliwell and Gutteridge, 1989). Though H_2O_2 is relatively less reactive, it is fairly long-lived molecule with a half-life of 1 ms and can diffuse to some distances from its site of production. Inactivation of Calvin cycle and superoxide dismutase enzymes by the oxidation of their thiol groups by H_2O_2 has been reported (Charles and Haliwell, 1980; Bowler *et al.*, 1994). Superoxide anions are scavenged by SODs to produce H_2O_2 and O_2. H_2O_2 forms the substrate for a number of decomposing reactions catalyzed by catalases and peroxidases. Due to its high reactivity, H_2O_2 gets reduced by metal ions leading to the formation of highly reactive OH· radical. The OH· radical oxidizes organic molecules in two different ways. In the first type of reaction, the OH· radical is added to the organic molecule to form a hydroxylated compound or the formation of organic radical plus water. The organic radicals can react with oxygen to generate peroxyl radicals, which can in turn produce other organic radicals. In this way organic radicals are produced one after another in a chain reaction. Although H_2O_2 and $O_2^{\bullet-}$ do not directly target DNA but can still be considered as potential mutagens due to their ability to cause the formation of OH· radicals that can cause extensive lesions in DNA. In order to protect the cells from ROS there are certain natural antioxidants that remove these radicals directly. Some of these are L-ascorbic acid (vitamin C), glutathione (GSH; a tripeptide of Glu-Cys-Gly) and α-tocopherol (vitamin E). Oxidative stress causes extensive damage and its effects are reflected in many biochemical pathways leading to slow degeneration and death of cells (Vranová *et al.*, 2002; Imlay, 2003; Latifi *et al.*, 2009).

4) Osmoregulation: Water is the universal solvent and dissolved in it are other cosolvents (solutes) that significantly influence the behaviour of water as a solvent. The composition of cosolvents, ionic strength, osmotic pressure (OP) and pH contribute to the characteristics of a solvent. Each cosolvent also contributes to the osmolality of the solution. Osmolality is defined as the OP of a solution at a particular temperature expressed as moles of solute per kg of solvent. In contrast, osmolarity is an approximation for osmolality expressed as moles of solute per litre of solution (Wood, 1999). Osmolarity is calculated as the sum of concentration of osmotically active solutes in a solution. OP of the extracellular environment may increase (osmotic upshift) or decrease (osmotic downshift). An OP differential between the inside and outside puts pressure on the cell membrane. Furthermore, at high external OP, the cell would lose water if it did not increase its internal OP. Thus OP must be regulated. However, enzymes need an environment where the ionic strength does not vary too drastically. So a particular OP range for the bacterial media that supports growth can be identified (as osmotolerant) and the cells try to adapt to changes in the cosolvent concentration by adjusting their internal cosolvent levels to balance the outside. However cells need 100–150 mM K^+ and about the same concentration of metabolic intermediates so they can not reduce their internal cosolvent levels below a minimum of approximately 250 mM. Conversely, an increase of the internal salt concentration to too high levels inhibits many enzymes. Very high concentrations of metal ions damage proteins by altering their structural conformations. During osmotic upshift, water loss results in a decrease of cell volume and the turgor is lost. Turgor pressure (ΔP) is a hydrostatic difference which balances the OP difference between cell interior and exterior. It is the turgor pressure that renders the chemical potentials of intracellular and extracellular water equal at equilibrium. An osmoregulatory response is one in which the cell exhibits a physiological response that mitigates passive adjustments in cell structure caused by changes in the extracellular osmolality. Enzymes are not generally affected by OP per se, but it is the ionic environment that affects their structure. The secret of adapting to osmotic upshift is the use of "compatible solutes". These are special solutes which, even in high concentrations, do not inhibit enzyme function. The most common are: glutamate (L-α-glutamate and β-glutamate, ectoine, hydroxyectoine, trehalose, glycine betaine, glucosylglycerol and proline (Fig. 2). Hence increasing the concentration of trehalose, proline and other neutral molecules does not hurt enzymes much whereas increasing concentration of ions such as K^+ does. Glycine betaine (and to a lesser extent proline and trehalose) are not merely just "extra" osmotic pressure, they are actual osmoprotectants and positively counteract the effect of increased ionic strength. For *E. coli* growing in a typical mineral salts medium plus glucose, the internal osmolality (or OP) is ~300 milli-osmolar. The pressure outwards on the cell wall is approximately 3.5 atmospheres. *E.coli* can adapt to grow in media containing up to about 4% NaCl if provided with osmoprotectants. Marine bacteria live at much greater osmotic upshifts than this.

All the mesophilic prokaryotes generally accumulate the above compatible solutes under various stress conditions. In contrast, thermophilic and hyperthermophilic organisms generally accumulate very unusual compatible solutes such as di-myo-inositol-1-1-phosphate, di-mannosyl-di-myoinositol-1-1-phosphate, di-glycerol phosphate, mannosylglycerate and mannosylglycerimide (Fig. 3). The accumulation of these unusual compatible solutes has not been encountered in organisms inhabiting low or moderate temperatures. The protective roles of mannoglycerate and di-glycerol phosphate during thermal denaturation of enzymes *in vitro* is another interesting point that goes in their favour to confer protection from high temperature related metabolic adjustments (Santos and da Costa, 2002). The osmoadaptive mechanisms of prokaryotes have been summarized (Empadinhas and da Costa, 2008). Two types of strategies have been envisaged. Of these, the first one known as "salt-in" strategy involves an influx of ions from the surrounding environment. This strategy seems to be restricted to the extreme halophilic Archaea such as *Halomicrobium, Haloarcula, Haloquadratum, Halorhabdus, Natronobacterium* and *Natronococcus* (all belonging to the family Halobacteriaceae), the halophilic bacteria of the order Haloanaerobiales and the bacterium *Salinibacter*. The major ion that is accumulated by these bacteria is K^+. The second strategy involves accumulation of low molecular weight organic compatible solutes. Large majority of the microorganisms rely on this strategy to overcome different stresses and this constitutes a versatile mode of osmotic adaptation in an otherwise osmotically changing environment.

I. SALT STRESS

Knowledge that has emanated from the studies on salt stress in cyanobacteria can be studied under the following heads: (1) Biochenmical and physiological studies (Reed *et al.*, 1986; Schubert *et al.*, 1993; Jeanjean *et al.*, 1993; Hagemann *et al.*, 1994, 1999; Marin *et al.*, 2002; Singh *et al.*, 2002; Ferjani *et al.*, 2003); (2) Synthesis of compatible solutes; (3) Na^+/H^+ antiporters; (4) Salt intake and cell signalling (Elanskaya *et al.*, 2002; Shoumskaya *et al.*, 2005); (5) Gene level responses that deal with salt stress tolerance; (6) Genome- and proteome-based studies (Hagemann *et al.*, 1991; Fulda *et al.*, 2000, 2006; Huang *et al.*, 2006) including those of microarray analysis of genes and their products (Kanesaki *et al.*, 2002; Marin *et al.*, 2004; Hagemann, 2011). These are presented below.

***i) Biochemical and physiological studies*:** The requirement of sodium for cyanobacterial growth at high pH (Allen and Arnon, 1955; Espie *et al.*, 1988), uptake of several inorganic nutrients such as inorganic carbon, nitrate, phosphate (Lara *et al.*, 1993; Avendaño and Fernández.-valiente, 1994) and photosynthetic electron transport at the O_2 evolving complex (Zhao and Brand, 1988) has been reported. Freshwater cyanobacteria tolerate low levels of sodium chloride (0.5 M) whereas cyanobacteria from hypersaline environments can tolerate up to concentrations as high as 3 M. Apte *et al.* (1987) examined the relationship between sodium uptake and salt tolerance in *Anabaena*

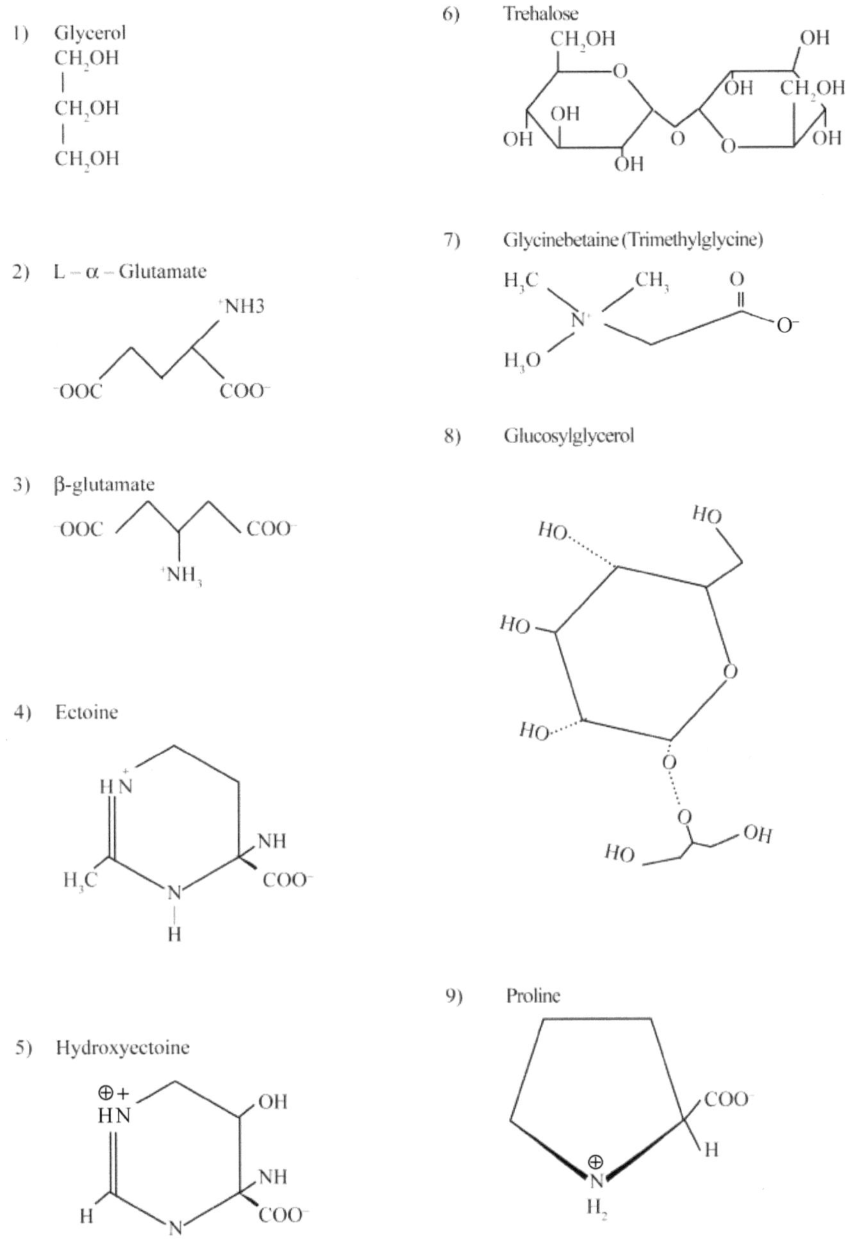

Figure 2: Common osmolytes synthesized by mesophilic prokaryotes in response to salt stress.

torulosa (an akinete-forming brackish water form that can tolerate 170 mM NaCl) and *Anabaena* sp. strain L-31 (a non-sporulating freshwater strain that can tolerate 55 mM NaCl). A negative correlation between Na^+ influx and salt tolerance was noted and inhibition of Na^+ influx (in presence of alkaline pH, K^+ above 25 mM, nitrate and ammonium ions) enhanced the salt tolerance of not only *A. torulosa* but also *Anabaena* sp. strain L-31. Apte and Bhagawat (1989) identified salinity stress-induced

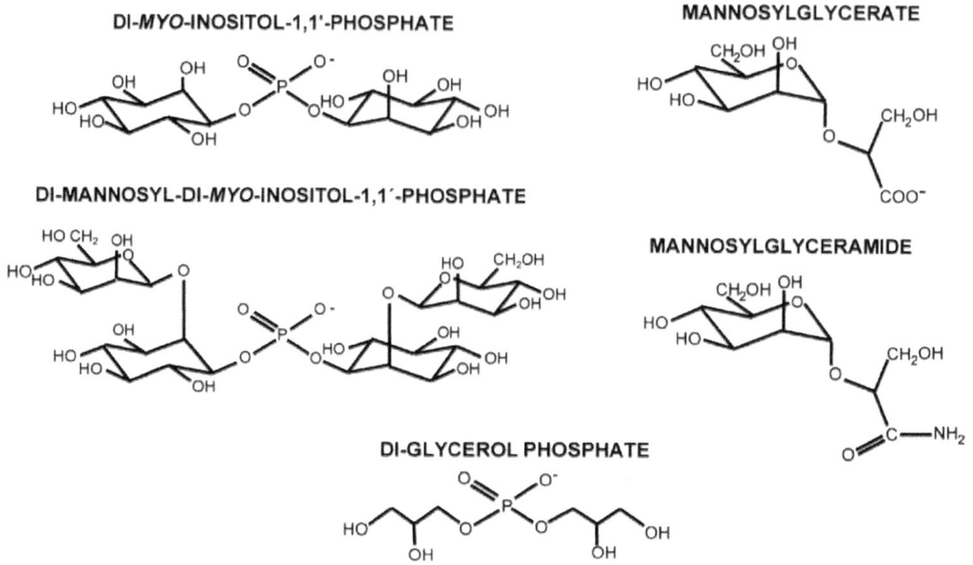

Figure 3: Special osmolytes synthesized by thermophilic and hyperthermophilic prokaryotes in response to salt stress.

proteins in *A. torulosa* and *Anabaena* sp. strain L-31. In the former a total of 11 salt stress proteins were induced within first 10 min with another protein (of 79 kD) induced after 30 min of exposure to 120 mM NaCl. All these proteins were found to be persistent and were synthesized at high rates excepting one of 29 kD protein. Of these, five salt stress proteins were found in the cytoplasm, three being found in the membrane and two appearing in both the fractions. In the latter, however, there appeared nine salt-stress proteins that were induced in the first 10 min followed by another protein after 2 h salt stress. But their synthesis appeared to be transient as only six out of the ten proteins were continuously synthesized that were localized in the cytoplasm and two were located in the membrane. Bhagawat and Apte (1989) presented a comparative analysis of proteins after heat shock, salinity and osmtic stress in *Anabaena* sp. strain L-31. By the use of inhibitors of transcription (rifampin 50 μg/ml) and translation (chloramphenicol 100 μg/ml), they identified the production of certain proteins common to all three types of stress conditions followed by the production of proteins specific to each stress. In response to salt stress at least the synthesis of four proteins was repressed with the induction or enhancement in the production of 13 other proteins of variable molecular weights. Salt tolerance of *A. torulosa* and *Anabaena* sp. strain L-31 was enhanced most effectively in presence of nitrogen sources such as nitrate, ammonium and glutamate followed by glutamine and aspartate. Competitive inhibition by nitrate and ammonium of the Na^+ influx seems to be the major mechanism for protection against salt stress (Reddy *et al.*, 1989). *A. torulosa* and *Anabaena* sp. strain L-31 displayed a differential response to salt and osmotic stresses. The former strain which is halotolerant (growth inhibition by 50% in presence of 250 mM NaCl) showed a osmosensitive behaviour (350 mM sucrose inhibited growth by 50%) where as the latter a freshwater strain that is sensitive to salt (inhibition of growth by 50% in presence of 100 and 125 mM NaCl) exhibited considerable osmotolerance (growth inhibition by 10 to 20% only in presence of 200 to 300 mM sucrose). The ionic component of salt stress inhibited nitrogenase activity in both strains but not by the osmotic stress. There was synthesis of common proteins in both of these stresses besides the appearance of unique proteins in salt-stress and osmotic stress (Fernandes *et al.*, 1993). Apte *et al.*

(1998) identified salt-responsive proteins in *A. torulosa* by subtractive RNA hybridization approach that provide inherent tolerance to many environmental stresses. Some of these proteins include KdpATPase (K^+ scavenging enzyme) and the molecular chaperone GroEL.

The initial damage of salt stress is perceived at the level of transient inactivation of proteins of photosystem I (PSI) and photosystem II (PSII) complexes (Allakhverdiev and Murata, 2008). The 'Z'-scheme of photosynthesis with the photosynthetic electron transport chain and the location of proteins up to the level of reduction of ferredoxin and generation of ATP is depicted in Fig. 4. Methyl viologen and p-benzoquinone constitute the *in vitro* electron acceptors for PSI and PSII, respectively whereas dichlorophenolindophenol and diphenylcarbazid serve as electron acceptors for PSI and PSII, respectively. Because salt stress due to NaCl has both osmotic and ionic effects (Allakhverdiev *et al.*, 2000a), the effects are perceived both at the level of cell membrane and thylakoid membranes. Across the cell membrane, the entry of Na^+ through Na^+/H^+ antiporter activity results in the accumulation of Na^+ at the PSII complex transiently inactivating the O_2-evolving machinery. This slow and irreversible inactivation of PSII has been attributed to be due to the destruction of Mn cluster that constitutes the catalytic center of O_2-evolving complex. In cyanobacteria, the O_2-evolving machinery is located on the luminal side of the thylakoid membrane that is stabilized by three extrinsic proteins, a 33-kD protein, cytochrome *c-550* and PsbU (Enami *et al.*, 1998; Shen *et al.*, 1998; Nishiyama *et al.*, 1999). Of these proteins, cytochrome *c-550* and PsbU are loosely bound to the donor side of the core complex of PSII (Nishiyama *et al.*, 1997, 1999). In presence of light, the accumulated

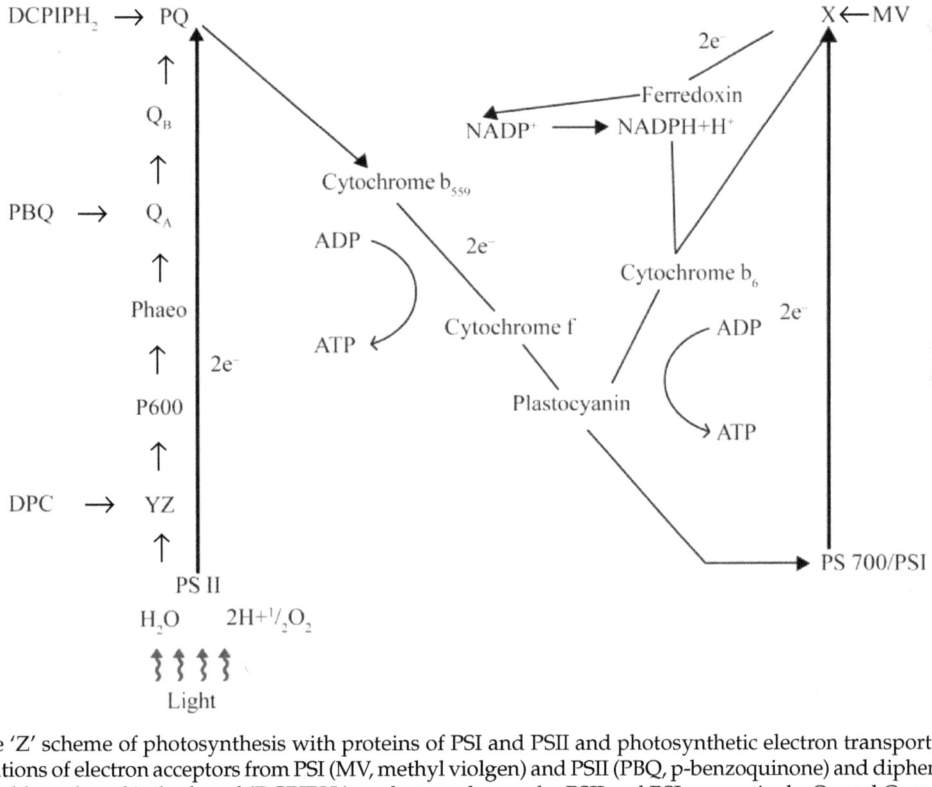

Figure 4: The 'Z' scheme of photosynthesis with proteins of PSI and PSII and photosynthetic electron transport chain with probable locations of electron acceptors from PSI (MV, methyl violgen) and PSII (PBQ, p-benzoquinone) and diphenylcarbazid (DPC) and dichlorophenol indophenol ($DCPIPH_2$) as electron donors for PSII and PSI, respectively. Q_A and Q_B are quinones, Phaeo=Phaeophytin, PQ=Plastoquinone; P_{680} and P_{700} are Chlorophyll *a* molecules. (Molecular machinery involved in the 'Z' scheme of photosynthesis is depicted in Figure 9 of Chapter 2).

Na^+ may be extruded out through the Na^+/H^+ antiporter activities due to the availability of energy from photosynthesis. In the absence of light due to the non-availability of energy, the Na^+/H^+ antiporter becomes inoperative resulting in an increase of Na^+. This leads to the accumulation of Na^+ at high concentrations in the intra-thylakoid space irreversibly inactivating O_2-evolving machinery. Allakhverdiev *et al.* (2001) presented evidences in favour of the existence of rapid and slow phases in NaCl-induced inactivation of Na^+/H^+ antiporters in *S. elongatus* PCC 7942. Light restoration of the activity of Na^+/H^+ antiporters constituted the slow phase. The use of uncouplers of photosynthetic electron transport such as carbonylcyanide m-chlorophenylhydrazone and carbonylcyanide p-trifluoro-methoxyhydrazone prevented the restoration and recovery of PSII and PSI suggesting that the ATP produced during photophosphorylation is important for tolerance of photosynthetic machinery to salt stress. Pomati *et al.* (2004) studied the effect of ion channel modulating agents on Na^+-K^+ fluxes of whole cells of *Cylindrospermopsis raciborskii* strain T3 and *C. racibosrskii* strain AWT 205 under salt stress (10 mM). An increase in extracellular pH to alkalinity was directly proportional to the increase in cellular Na^+ levels whereas at pH (8.0) that is optimum for growth, cellular K^+ levels predominantly increased. In presence of channel modulating agents, like lidocaine hydrochloride (1 μM) and veratridine (100 μM), the increase of Na^+ was coupled with K^+ but in presence of channel blockers such as aniloride (1 mM) and saxitoxin (1 μM) there was a decrease in the levels of both ions. Na^+/H^+ antiporter mediated H^+ accumulation caused the Na^+ efflux, energy for which may have been obtained through a proton motive force.

The presence of aquaporin proteins specific to water have been detected in all major groups of organisms. Although it has been suggested that water efflux from cells through the cytoplasmic membrane takes place due to higher osmotic stress, conclusive proof in favour of this has only been presented by Shapiguzov *et al.* (2005). The identification of an aquaporin (*aqpZ*) gene in *Synechocystis* sp. strain PCC 6803 and its mutation led to prevention of the water efflux due to osmotic stress of the mutant cells. PCR analysis showed that such mutant cells possessed a disrupted copy of the gene and the stress caused by 0.5 M sorbitol did not cause any shrinkage in mutant cells (due to prevention of water loss) but in the wild-type, the cell shrinkage was half of the original level (due to water loss). A comparison of the gene expression profiles, conducted after hyperosmotic stress for 15, 60 and 120 min, of the wild-type and the mutant revealed that the mRNA levels of nearly 150 genes showed more than two-fold increase in the wild-type cells at 15 min. However, in the mutant cells the expression of certain genes governing the synthesis of chaperones (*hspA, htpG*), proteases (*htrA. clpB, hhoA*), glucosylglycerol phosphate synthase (*ggps*) and many other proteins (26) of unknown function was significantly affected. On the other hand, at 60 min though similar sets of genes were induced in both wild-type and AqpZ mutant but it was less marked in the latter. This signifies that the disruption of *aqpZ* did not cause any direct influence on the expression of these genes. It was also noted that at both 15 min and 60 min the expression of *ggps* was markedly reduced. However, the essentiality of *aqpZ* gene for glucose metabolism during photomixotrophic growth of *Synechocystis* sp. strain PCC 6803 has been highlighted (Akai *et al.*, 2011).

Reports exist in literature on the involvement of lipids in conferring protection against salt stress (Huflejt *et al.*, 1990; Khamutov *et al.*, 1990; Ritter and Yopp, 1993). The presence of unsaturated fatty acids in the membrane lipids enhanced the tolerance of *Synechocystis* sp. strain PCC 6803 cells to salt stress. Tasaka *et al.* (1996) isolated mutants by targeted mutagenesis of the genes of fatty acid desaturases that produced low levels of unsaturated fatty acids. A comparison of desaturase (*des*)-deficient mutants of *Synechocystis* sp. strain PCC 6803 with that of wild-type suggested that unsaturated fatty acids are required for protection of photosynthetic machinery from salt stress. The cells deficient in *desA* and *desD* genes governing Δ^{12} and Δ^{6} desaturases, respectively contained monounsaturated fatty acids but not polyunsaturated fatty acids. A comparison of *desA*$^-$/*desD*$^-$ cells of *Synechocystis* sp. strain PCC 6803 to sodium-induced toxicity revealed that the recovery of these cells from salt stress is rather slow when compared to wild-type cells. Allakhverdiev *et al.* (1999) demonstrated that genetic engineering of the unsaturation of fatty acids in membrane lipids alters the tolerance of *Synechocystis* sp. strain PCC 6803 to salt stress. This is supported by inhibitory studies. Light reversal of NaCl-induced stress was eliminated by the presence of lincomycin, an inhibitor of photosynthesis. Likewise, the presence of N,N'-dicyclohexylcarbodiimide, an inhibitor of H^+-ATPase, enhanced the NaCl-induced toxicity suggesting the involvement of Na^+/H^+ antiporter in conferring salt tolerance. The presence of high degree of unsaturated fatty acids in the membrane lipids contributes to enhanced fluidity of membrane. The *desA*$^-$ and *desD*$^-$ deficient cells of *Synechocystis* sp. strain PCC 6803 exhibited more sensitive Na^+/H^+ antiporter to salt stress with a slow recovery than the wild-type cells. Transformation of *S. elongatus* PCC 7942 with *desA* gene from *Synechocystis* sp. strain PCC 6803 resulted in the synthesis of di-unsaturated fatty acids (Sakamoto *et al.*, 1994) which otherwise normally produced only saturated and monounsaturated fatty acids in its membrane lipids (Murata and Wada, 1995). Transformants of *S. elongatus* PCC 7942 with *desA* gene (Δ^{12}desaturase) exhibited differences over wild-type with reference to fatty acid synthesis (Allakhverdiev *et al.*, 2001). In wild-type cells, the synthesis of 16:0 (49%) and 16.1 (41%) is most abundant followed by low levels of 18:0, 18:1Δ^9 and 18:1Δ^{12}. On the other hand, in DesA$^+$ cells, the synthesis of 16:2$\Delta^{9,12}$ appeared (up to 15%) suggesting that certain amounts of 16:1Δ^9 has been desaturated. Likewise 18:2$\Delta^{9,12}$ also appeared accounting for 3% of the total fatty acids. The PSII of the DesA$^+$ cells of *S. elongatus* PCC 7942 appeared to be more resistant to salt stress. Light restoration of PSII activity occurred within 3 h to about 50% of the original level which remained constant for the next 10 h. The possible sites affected by the unsaturation of fatty acids in membrane lipids included water channels, K^+ (Na^+) channels, Na^+/H^+ antiporter system and H^+-ATPase(s) all of which are located on plasma membrane. These observations constitute direct evidences for the protection and maintenance of photosynthetic machinery under salt stress.

The inhibition in the repair of photodamaged PSII of *Synechocystis* sp. strain PCC 6803 has been identified by Allakhverdiev *et al.* (2002) to be due to suppression in the transcription and translation of *psbA* genes. An examination of the synergistic effects of light (250 to 2,000 $\mu E\ m^{-2}\ s^{-1}$) and salt stress (0.5 M) revealed that approximately 60% of the light-inducible genes were strongly inhibited by salt stress of which 20% were moderately suppressed. A proportion of 20% of the light-inducible genes were enhanced by 0.5 M NaCl. The levels of D1 protein that declined in high light intensity were returned to normal levels due to a repair in weak light as well as in low salt medium but in presence of high concentration of NaCl (1.0 M) there was no increase in the level of D1 protein. DNA microarray analysis revealed complete inhibition by NaCl (1.0 M) of light-induced transcripts of all light-inducible genes. These results have been found to be consistent with the results obtained with [S^{35}]-Met incorporation studies. However, the light-inducibility of some light-inducible genes was enhanced to some extent in low concentrations of NaCl (0.5 M). Allakhverdiev and Murata (2008) summarized recent progress on the effects of salt stress on photosynthesis in cyanobacteria and suggested that the damage caused to PSII and PSI occurs through rapid and slow changes. The first rapid changes are associated with a decrease of photosynthetic activity occuring within the first few min causing the efflux of water through water channels and influx of Na^+ into the cells. The slow changes that occur probably in few hours are associated with dissociation of extrinsic proteins from PSI and PSII. When salt-stressed (0.8 M NaCl) *Spirulina platensis* was incubated either in dark or under different light intensities the first phase was found to be independent of light that was characterized by a rapid decrease in PSII activity (20–50% in the first 15 min) followed by a subsequent recovery (86–92% of the initial level after 4 h). In the second phase, the decrease in PSII activity was directly proportional to the intensity of light without any decrease in the dark. However, the changes in O_2 evolution activity were independent of light intensity during both phases (Lu and Zhang, 2000). Cytochrome *c* oxidase (*ctaI*)-defective mutant of *Synechocystis* sp. strain PCC 6803 showed a substantial decrease in photosynthesis due to reduction in photochemical efficiency of PSII and of the chlorophyll in the reaction centre of the photooxidizable form of PSI (Ryu *et al.*, 2003). Incubation of *S. platensis* cells in sodium chloride (0.8 M) under moderate light intensity caused a decrease in PSII mediated oxygen evolution activity and increase in PSI activity along with increase of P700. Western blot analysis revealed a 40% loss in the thylakoid membrane protein D1 in the PSII reaction centre. SDS-PAGE analysis of cell-free extracts of cells treated with NaCl also showed alterations in thylakoid membrane proteins. Of these a 47-kDa chlorophyll protein and 94-kDa protein showed a decline in their amounts with an increase in 17-kDa protein. The diminishing of the former two proteins was associated with a decreased energy transfer from light-harvesting antenna to PSII in the extracts as well as alterations in chlorophyll fluorescence of whole cells and isolated thylakoids (Sudhir *et al.*, 2005). This has been further confirmed by Zhang *et al.* (2010) who reported a decrease in the excitation energy transfer from phycobilisomes to PSII with a concomitant increase in PSI electron transport activity in the salt-stressed cells of *S. platensis*. Salt stress in the same organism not only inhibited PSII activity but also caused alteration in the proteins of thylakoid membrane. The PsbO protein got dissociated and accumulated in the soluble fraction. This in turn resulted in the modification of Q_B niche from the acceptor side to an increase at the donor side (Gong *et al.*, 2008).

An osmotic stress protein of *Anabaena* sp. strain PCC 7120 that is immunologically related to plant dehydrins has been identified after an osmotic upshift due to the presence of sucrose, sorbitol and polyethylene glycol. This protein is a 40 kD osmotic-stress-induced protein when osmotic potential shift of 0.5 MPa (sucrose or PEG) or 1.2 MPa (sorbitol) was present in the external medium. The synthesis of plant dehydrin-like polypeptides was also noted in *Calothrix* sp. strain PCC 7601 and in *Nostoc* sp. strain PCC 7911 (Mac-R2) (Close and Lammers, 1993). The involvement of water channels in osmotic stress due to the presence of sorbitol (1.0 M) has been indicated in *Synechocystis* sp. strain PCC 6803 and the decrease in the cytoplasmic space by 50% suggested that water loss from cells constituted the primary event. Secondly, the O_2 evolution rate was reduced by 50% of the original level in presence of benzoquinone and DCBQ (an artificial electron acceptor). The transport of electrons by PSI from reduced DCPIP to MV decresed to 70% of the original level. This emphasizes that the activities of both PSII and PSI have declined in intact cells. The presence of p-chloromercuriphenyl-sulfonic acid (at100 μM), a water channel blocker, prevented the efflux of water and suppressed the inactivation of PSI and PSII and protected the cells against osmotic-stress induced by sorbitol. Further the inactivation of both PSII and PSI is suggested to be due to an increase in intrathylakoid concentration of K^+ ions that may have caused a dissociation of plastocyanin or cytochrome *c-553* from the PSI complex. The release of osmotic stress caused an inflow of water through water channels with a simultaneous efflux of K^+ ions and restoration of activities of PSI and PSII (Allakhverdiev *et al.*, 2000b).

***ii) Compatible solutes*:** The synthesis of compatible solutes is one of the mechanisms adopted by cyanobacteria in overcoming salt and osmotic stress (Empadinhas and da Costa, 2008; Klähn and Hagemann, 2011). In a comprehensive study on carbohydrate accumulation during salt stress, Reed *et al.* (1986) tested 71 strains of cyanobacteria from freshwater and marine environments. Of the 22 marine isolates, 41% accumulated glucosylglycerol [(2-O-α-D-glucopyranosyl)-glycerol; GG; principal among them being *Spirulina subsala, Synechococcus* sp. (4 strains)], 36% accumulated sucrose (*Anabaena* sp. CA, *Dermocarpa violacea*, *Nodularia harveyana*, two strains) and 23% accumulated trehalose. Of the 49 freshwater strains tested 76% accumulated sucrose in response to salt stress, 18% synthesized trehalose and 6% showed GG. There was absolutely no relationship between habitat or genus and the type of osmolyte synthesized. In another study, Stal and Reed (1987) tested 47 strains of cyanobacteria belonging to taxonomically different groups from marine microbial mats for osmolyte accumulation. There was no correlation between the chemical nature of the organic solute, the salt optimum or salt tolerance of the cyanobacterium examined. Sucrose was the main osmolyte accumulated in species of *Nostoc*/*Anabaena* group while trehalose accumulation was noted in marine heterocystous *Calothrix* sp. GG accumulation was noted characteristically in strains belonging to LPP-group. However, some members of this group accumulated sucrose or trehalose as well. There was variation between marine *Oscillatoria* spp. (trehalose accumulation) and freshwater *Oscillatoria* spp. with broad salinity tolerance (that accumulated sucrose). Samples from marine microbial mats revealed the presence of glycerol, GG, sucrose and trehalose. Of the total 130 cyanobacterial strains tested, half of them (66 strains) accumulated sucrose whereas 20, 24 and 18 strains accumulated trehalose, GG and glycine betaine, respectively (Erdmann, 1992; Reed *et al.*, 1986; Stal and Reed, 1987). The accumulation of the low molecular weight carbohydrates has been correlated with the habitat distribution of cyanobacteria in the marine environment. The synthesis of sucrose and trehalose was more rapid than that of GG. Strains that synthesized the former two compatible solutes inhabited brackish waters where as those that showed the synthesis of the latter commonly occurred in habitats of less variable salinity. Thus sucrose synthesis has been noted in case of *Anabaena variabilis* ATCC 29413, *Nostoc entophytum* CCAP1453/14 and *S. elongatus* PCC 6301 while tehalose synthesis was noted in case of *Microchate grisea* CCAP1451/1 and GG as well as sucrose was synthesized by *Synechococcus* sp. strain CCMP/SYN WH 5701 (Warr *et al.*, 1987). The synthesis of sucrose by salt-sensitive (freshwater) strains of cyanobacteria such as *Synechococcus* has been reported (Mackay *et al.*, 1984: Reed *et al.*, 1986; Joset *et al.*, 1996; Hagemann and Erdmann, 1997). The detection of sucrose synthase gene in some unicellular cyanobacteria (*Microcystis aeruginosa* PCC 7806, *G. violaceus* PCC 7421 and *T. elongatus* BP-1) and increase in its transcript levels in these organisms after salt treatment is suggestive of sucrose synthesis as an adaptive mechanism (Kolman *et al.*, 2012). The synthesis of GG has been first detected in *Synechococcus* sp. strain N100 (Borowitzka *et al.*, 1980). Subsequently, GG synthesis has been detected in the cyanobacterial strains that exhibited intermediary tolerance such as *Synechocystis* sp. strain PCC 6803 (Hagemann *et al.*, 1987; Erdmann *et al.*, 1992; Joset *et al.*, 1996; Hagemann and Erdmann, 1997). Salt-tolerant strains of cyanobacteria such as *Synechococcus* sp. strain PCC 7418 (later redesignated as *Aphanothece halophytica*) synthesize glycine betaine (Mackay *et al.*, 1984; Reed *et al.*, 1986).

Blumwald *et al.* (1983) have suggested that changes in cell volume caused by different osmotic environments could be monitored by electron spin resonance (ESR) methods. According to them these enable us to identify changes in internal cell volume by measuring ESR spectra with 2,2,6,6-tetramethyl-4-oxopiperidinoxy free radical (TEMPONE) as a spin probe. The accumulation of sucrose by *Synechococcus* sp. strain 6311 and *Nostoc muscorum* both freshwater strains is indicative of common osmoregulatory response where internal cellular sucrose concentrations reached up to 580 mM in presence of salt (0.6 M NaCl) for 40 h.

a) GG synthesis: Hagemann and Erdmann (1994) have suggested that the precursor for the synthesis of GG is ADP-Glucose (ADP-Glc) and not UDP-glucose as proposed earlier by Reed and Stewart (1985). ADP-Glc and glycerol-3-phosphate (G3P) react in presence of ADP-glucose: glycerol-3-phosphate 2-glycosyl transferase (GGP synthase) to produce ADP and GGP. The latter is further converted to ADP+GG+Pi in presence of glucosylglycerol-3-phosphate phosphohydrolase (GGP-phosphatase). These enzymes, i.e. GGP synthase and GGP-phosphatase are present in normal cells in a pre-formed inactive state and get activated *in vivo* when *Synechocystis* sp. strain PCC 6803 cells are put under salt stress or *in vitro* by the addition of NaCl to the homogenization buffer or assay buffer. The *in vivo* activation of GG synthesizing enzyme system in salt stressed cells of *Synechocystis* sp. strain PCC 6803 has been mainly attributed to the post-translational modification. The accumulation of GG in salt-stressed cells (684 mM NaCl) occurred linearly in the absence of fresh protein synthesis and is not affected by the presence of chloramphenicol (for at least in the first two h after salt-stress) and the type of proteins whose synthesis increased after salt stress were of low molecular weight. Such cells exhibited retarded growth in cultures (Hagemann *et al.*, 1990; Erdmann *et al.*, 1992). The expression of the gene *ggpS* has been compared between *Synechococcus* sp. strain PCC 7002 (a marine cyanobacterium) and *Synechocystis* sp. strain PCC 6803 (a freshwater isolate of Rippka *et al.*, 1979) to understand the regulatory differences. PCR amplification of *Synechococcus ggpS* gene resulted in a 2,030 bp DNA sequence and the protein deduced from it exhibited 85% similarity to the GGPS protein of *Synechocystis* sp. strain PCC 6803. A null mutant of *Synechococcus* sp. strain PCC 7002 for *ggpS* gene exhibited a salt-sensitive phenotype due to its inability to synthesize GG. By conducting Northern blot and immuno-blot experiments, the expression of the gene under salt shock revealed that while no expression of the gene occurred in low salt media, the levels of both mRNA of *ggpS* and protein GGPS increased by 100-fold in presence of high salt concentration. The fact that the synthesis of GGPS was inhibited in presence of chloramphenicol in case of *Synechococcus* but not in case of *Synechocystis* sp. strain PCC 7002, signifies that the regulation of the synthesis is at different levels. It seems that in case of *Synechococcus* sp. strain PCC 7002 it is at the level of transcription whereas in case of *Synechocystis* sp. strain PCC 6803 it is at post-translational level (Engelbrecht *et al.*, 1999).

A gene *stpA* (salt tolerance protein A) that encodes the enzyme GGP-phosphatase has been identified in *Synechocystis* sp. strain PCC 6803 and its mutation resulted in a salt-sensitive phenotype accompanied by the accumulation of GGP inside the cells. Specific ORFs were cloned into *aphII* gene catridge (aminoglycoside phosphotransferase II conferring kanamycin resistance) of *E. coli* plasmid pUC4K at selected unique restriction sites. The isolated plasmid DNA was used to transform *Synechocystis* sp. strain PCC 6803 cells. The transformants showing kanamycin resistance were selected and characterized. The levels of *stpA* mRNA increased in presence of NaCl concentrations above 170 mM. The *stpA* was cloned and its expression in *E. coli* cells produced a protein of 46-kDa that exhibited GGP-phosphatase activity. A consensus motif of acid phosphatases is a characteristic feature of GGP-phosphatase. The antibodies of *E. coli* protein neutralized the protein of similar size in the cell-free extracts of *Synechocystis* from salt-stressed cells. However, the protein produced in *E. coli* did not require the activation by NaCl the requirement of which was noted in case of cyanobacterial GGP-phosphatase. The basal levels of the enzymes of GG synthesis pathway noted in non-stressed cells or in low concentrations of NaCl might be due to the regulation at transcriptional level as the *stpA* mRNA levels in such cultures were very low. However, when the cells experience a salt-stress (684 mM NaCl) the *stpA* mRNA levels reached a maximum after 3 h of salt shock and levels of the enzyme reached a maximum around 5 h of salt shock. During this study, they also identified another ORF upstream of *stpA*, *OrfI* that showed similarities to bacterial two-component systems required for salt tolerance. Mutants impaired in both *stpA* and *OrfI* have also been isolated and characterized with reference to salt tolerance and synthesis of GG (Hagemann *et al.*, 1997). Similarly, the salt-dependent expression of the GGP synthase gene *ggpS* was examined in salt-stressed cells of *Synechocystis* sp. strain PCC 6803. Like the *stpA* gene product, very low levels of GGP synthase were noted in control cultures indicating that there is a constitutive expression of this gene. However, in salt-stressed cells of *Synechocystis* the levels of mRNA of *ggps*, that of protein GGPS and the synthesized GG linearly increased with the concentration of the salt (0, 171, 342, 513, 684, 855 and 1026 mM of NaCl). During the first 4 h of salt stress, a salt shock greater than 300 mM resulted in GG synthesis of approximately 20 μmol of GG $h^{-1}ml^{-1}$ $A750^{-1}$ when compared to the synthesis rate of 6.5 μmol of GG $h^{-1}ml^{-1}$ $A750^{-1}$ in low salt concentrations (171 mM NaCl) (Marin *et al.*, 2002).

When *ggpS*$^-$ mutant cells of *Synechocystis* sp. strain PCC 6803 were subjected to salt stress (of 450 mM NaCl) a cessation of cell division occurred suggesting that salt stress inhibited cell division. A comparison of salt stress with hyperosmotic stress (induced by sorbitol at 900 mM which was equivalent osmotic effect as that of 450 mM NaCl) revealed that while salt stress arrested cell division with an increase in cell size in case of *ggpS*$^-$ mutant but in case of hyperosmotic stress the inhibition of cell division was associated with decrease in cell size both in case of wild-type as well as the *ggpS*$^-$ mutant. The addition of GG to the medium resulted in an uptake of GG accompanied by renewed cell division as revealed by flow cytometry analysis. Although the mutant cells accumulated high levels of sucrose, it is not sufficient to overcome salt stress but the presence of GG is essential for overcoming salt stress. These results thus prompted Ferjani *et al.* (2003) to conclude that there is a qualitative difference in the protection given by sucrose and that of GG during salt stress.

A comparison of ionic osomotic stress caused by salt and nonionic osmotic stress induced by the presence of sorbitol and maltose revealed a differential response of *Synechocystis* sp. strain PCC 6803. The activation of the GG synthesizing enzymes in presence of salt resulted in GG synthesis and this was devoid of any cell volume changes. Although there was minor change in the cell volume in presence of sorbitol due to lack of activation of GG synthesizing enzyme system GG accumulation did not occur but instead sorbitol was accumulated by the cells. On the other hand, osmotic stress induced by maltose resulted in a decrease of cell volume due to efflux of water and in the absence of osmolyte accumulation the cells tended to be osmosensitive. Sorbitol was found to be toxic at a concentration of 700 mM while NaCl at a concentration of 1.2 M was toxic (that is equivalent to 2.4 M of sorbitol). In order to differentiate the osmotic and salt stress, the expression of the gene *ggpS* was followed by the cloning of this gene with *luxAB* as the reporter gene (RP_{ggpS}:: *luxAB*). These genes were integrated into the *Synechocystis* sp. strain PCC 6803 genome along with an antibiotic resistance gene catridge. The expression of the *ggpS* gene product was identified by the accumulation of LuxAB proteins. The addition of NaCl above 400 mM concentrations (500 and 600 mM) and sorbitol stress (at 600 mM) induced *luxAB* activity. However, in presence of maltose no such *luxAB* activity was noted indicating the absence of activation of *ggpS* promoter by the osmotic stress induced by maltose. In sharp contrast, the *ggpS* mRNA levels were found to be different. In case of salt stress, the *ggpS* mRNA reached a maximum after 0.5 h reaching a steady-state level at 8 h but in case of sorbitol-stressed cells, the mRNA levels were comparatively lower than NaCl-stressed cells (Marin *et al.*, 2006).

The presence of an active transport system for GG uptake was found in *Synechocystis* sp. strain PCC 6803. Uptake of C^{14}-labelled GG was enhanced with a gradual increase of NaCl (between 10 and 700 mM) with an optimal activation at 350 mM and a pH of 4 to 7 favoured a linear increase. Sucrose and trehalose competed well with GG transport. The uptake of GG in a salt-sensitive mutant conferred protection against salt stress thus enabling the mutant cells to grow in presence of salt. On the other hand, in wild-type cells the GG transporter most likely helped in preventing the cellular GG to leak out of the cells. A concentration-dependent inhibition in uptake of GG was noted in presence of uncouplers of photosynthetic and oxidative electron transport systems where as inhibitors of H^+-ATPases (localized in the cytoplasmic membrane) or cytochrome oxidase had no effect on the transport of GG. The kinetics of GG uptake resembled the sugar uptake systems noted in several cyanobacterial strains with a weak affinity (k_m value of 50 μM) to the substrate in contrast to high affinity system of glycine betaine transporter (a k_m of 2 μM) of *A. halophytica* (Mikkat *et al.*, 1996).

b) Glycine betaine: In *E. coli* the synthesis of glycine betaine takes place through a two-step reaction. The first step is mediated by a membrane-bound oxygen-dependent choline dehydrogenase that oxidizes choline to betaine aldehyde. In the second step, betaine aldehyde is catalyzed by a soluble NAD-dependent betaine aldehyde dehydrogenase to glycine betaine (Landfald and Strom, 1986; Andersen *et al.*, 1988). Moreover, the genes of this biosynthetic pathway and those of the choline uptake system, transport of glycine betaine and a regulatory protein are all clustered together (Lamark *et al.*, 1991). The expression of these genes is induced by the presence of choline and by the osmotic shock (Landfald and Strom, 1986). Nomura *et al.* (1995) reported for the first time the protective effects and acquisition of salinity tolerance by production of glycine betaine in *S. elongatus* PCC 7942 that was transformed with a shuttle plasmid carrying 9 kb fragment of *E. coli bet* gene cluster (consisting of *bet A, betB, betI,* and *betT* that encode choline dehydrogenase, betaine aldehyde dehydrogenase, a putative regulatory protein and choline transport system, respectively). *Bet*-transformants of *S. elongatus* PCC 7942 transported choline for the synthesis of glycine betaine under salt stress. The accumulation of glycine betaine up to 45 mM concentration inside the *bet*-containing cells led to a stabilization of the activities of PSI and PSII. The enhanced growth performance of *bet*-transformants when compared to wild-type suggested the gene cluster from *E. coli* was expressed and synthesis of functional proteins responsible for acquisition of salt resistance proceeded. Purified betaine aldehyde dehydrogenase (about 18-fold) from *A. halophytica* had a specific activity of 298.6 μmol min^{-1} mg^{-1} protein and the enzyme was a tetramer of 30-kDa identical subunits showing optimal activity at pH 7.5 and 25°C. The specific activity of the enzyme increased with an increase in external salinity (Incharoensakdi and Kum-Arb, 1998). Extending their studies on the protective role of glycine betaine during salt stress, Nomura *et al.* (1998) demonstrated that glycine betaine protects RuBisCO from inactivation in the *bet*-transformants of *Synechococcus*. A comparison of the wild-type and *bet*-transformants of *Synechococcus* for the levels of mRNA, protein and enzyme activity revealed that it is the activity of RuBisCO that is affected during salt stress in the wild-type cells whereas in the *bet*-transformants the activity was not affected due to the synthesis of glycine betaine. *In vitro* reactivation of RuBisCO by the presence of glycine betaine took place both in case of enzyme extracts of salt stressed cells as well as acid denatured enzyme. However, it is not understandable as to why in salt-stressed cells RuBisCO gets inactivated as NaCl helps in a spontaneous refolding of the dimeric RuBisCO from *Rhodospirillum rubrum* (Schmidt *et al.*, 1994) and prevents RuBisCO from getting dissociated into small subunit and large subunit from *A. halophytica* (Incharoensakdi *et al.*, 1986).

Another important biosynthetic pathway of glycine betaine is through methylation of glycine. Extremely halophilic bacteria like *Actinopolyspora halophila* (aerobic, heterotrophic eubacterium) and *Ectothiorhodospira halochloris* (anaerobic, phototrophic sulphur bacterium) synthesize glycine betaine from glycine through three step methylation reactions catalyzed by two methyltransferases. In the first step, glycine is methylated to sarcosine by glycine sarcosine methyltransferase (GSMT). The second step involving the methylation of sarcosine to dimethylglycine is again catalyzed by GSMT because of its broader substrate specificity. In the third step, dimethylglycine is methylated to form glycine betaine catalyzed by sarcosine dimethylglycine methyltransferase (SDMT). This path way is linked to the central metabolism as the precursor of glycine is 3-phosphoglycerate which is an intermediate of glycolysis. Glycine betaine accumulation was noted up to 33% of the cell dry weight in case of *A. halophila* (Yancey *et al.*, 1982) and up to 2.5 M in case of *E. halochloris* (Trüper and Galinski, 1990). Transformants of *E. coli* with the methyltransferase genes of *E. halochloris* produced glycine betaine and exhibited improved salt tolerance (Nyossöla *et al.*, 2000). Homologues of gene sequences of GSMT and SDMT of *E. halochloris* have been found in the halophilic cyanobacterium *A. halophytica* but with differing substrate specificity in case of the latter enzyme. Characaterization of N-methyltransferases from *A. halophytica* (ApGSMT and ApDMT) revealed a three step methylation reactions from glycine to glycine betaine similar to those described in the two distantly related eubacterial species. The accumulation of glycine betaine up to 2 M concentrations in the cells did not result in a feed back inhibition of the biosynthetic pathway. The rate of synthesis of these enzymes in *A. halophytica* cells linearly increased with an increase of external concentration of sodium chloride. The binding of methyl acceptors on to these enzymes has been identified due to the amino acids Arg-169 in ApGSMT and Pro-171 and Met-172 in ApDMT. A change in the amino acids from Arg-169 to Lys or Glu in ApGSMT and Pro-171 to Gln and/or Met-172 to Arg in ApDMT significantly decreased V_{max} and increased K_m for methyl acceptors (glycine, sarcosine and dimethylglycine). Transformants of *E. coli* cells with the two gene sequences for ApGSMT and ApDMT exhibited increased levels of accumulation of glycine betaine (Waditee *et al.*, 2003). Gene sequences of ApGSMT and ApDMT were ligated to *E. coli bet*-cluster (choline dehydrogenase and betaine aldehyde dehydrogenase) along with plasmid pUC303-Bm and used to transform freshwater *S. elongatus* PCC 7942 cells. The growth performance of the wild-type and the transformants under control conditions was similar but under salt stress (0.4 M NaCl) only transformants with ApGSMT and ApDMT genes exhibited growth but the wild-type and control cultures with *bet*-cluster could not grow. Likewise, transformants with the potentiality to accumulate glycine betaine could grow in higher concentrations of NaCl (0.5 M and 0.6 M) as well as in sea water from Mikaway Bay (of the Aichi Prefecture in Japan) with salinity units 30.4. The performance of the transformants was found to be better than *Synechococcus* cells with ApNhaP1gene both in 0.6 M NaCl medium and that of sea water. The accumulation of the protein levels of ApGSMT and ApDMT in the transformants of *Synechococcus* increased with the increasing levels of salinity and corresponded with the levels observed in case of *A. halophytica* cells. This is suggestive of the fact that the promoters of these genes from *A. halophytica* were recognized in *Synechococcus* cells. The levels of glycine betaine synthesized in the transformants of *Synechococcus* with ApGSMT and ApDMT genes was always higher (~5 fold higher in 0.5 mM NaCl) than those observed in case of *Synechococcus* cells with *E. coli bet*-gene cluster (at 0.3 M NaCl) (Waditee *et al.*, 2005). Since glycine is the precursor for the three step methylation pathway higher internal levels of glycine should be available to the organism to produce sufficient levels of glycine betaine under salt stress. As serine and glycine are readily interconvertible by serine hydroxymethyltransferase any attempt to enhance the levels of serine will contribute to the enhancement in the levels of glycine. Both serine and glycine constitute important intermediates in the photorespiratory pathway and these are recycled as 3-phosphoglycerate (3-PGA) to the Calvin Cycle. Serine is derived through the phosphorylation pathway from 3-PGA through the enzyme D-3-phosphoglycerate dehydrogenase (PGDH, EC 1.1.1.95) that constitutes the first step in this pathway. The identification of a single copy of the the gene encoding PGDH in the genome of *A. halophytica* (as revealed by shotgun cloning) led to a study of metabolic engineering of betaine accumulation in microbes and plants (Waditee *et al.*, 2007). The expression of ApPGDH in *E. coli* led to the production of enhanced levels of glycine betaine both from serine by choline oxidative pathway as well as glycine via glycine methylation and contributed to the increased salt tolerance. The levels of glycine betaine in the cells of *E. coli* overexpressing the genes ApPGDH, ApGSMT and ApDMT were about 1.2–1.8 fold higher than those cells expressing ApGSMT and ApDMT only. The levels of serine and glycine likewise were found to be enhanced to about 2.5–3.0 and 3–4 fold higher, respectively in the cells overexpressing the ApPGDH, ApGSMT and ApDMT than the control and the cells expressing ApGSMT and ApDMT. A measurement of intracellular levels of serine and glycine during salt stress of *A. halophytica* cells revealed that the levels of both these amino acids decreased. However an exogenous supply of these amino acids during salt stress contributed to the enhanced levels of glycine betaine synthesis and to the salt tolerance as well. The activity of ApPGDH was found to be enhanced by 1.5–2.3 fold upon increase of salinity from 0.5 to 2.5 M NaCl. A molecular characterization of the protein of ApPGDH suggested it be closer to PGDH from *Synechocystis* sp. strain PCC 6803 (~79% identity) with a sequence of 526 amino acid residues and a molecular weight of 56,410 Da. Besides possessing three domains such as nucleotide-binding, substrate-binding and regulatory domains, the ApPGDH protein consists of a C-terminal domain resembling the PGDHs of *Arabidopsis* and mammals.

A. halophytica contains a betaine transporter (BetT) that is active at alkaline pH and high salinity. BetT of *A. halophytica* (BetT $_{A.\ halophytica}$) belongs to the BCCT (betaine-choline-carnitive transporter) family and is distinctive in having an isoelectric pH in the acidic range (4.58) in contrast to the isolelectric pH in the basic range (9.54) of a typical BCCT transporter (opnD) like that of *Bacillus subtilis*. BetT $_{A.\ halophytica}$ is specific for the transport of betaine and its activity was enhanced in presence of sodium and thus it is suggested to be a Na^+-betaine symporter. Betaine uptake by the cells of *A. halophytica* was enhanced two and half times in presence of 2.0 M NaCl than in presence of 0.5 M NaCl. Maximum uptake of betaine occurred at a pH of 9.0. Betaine transporter gene from *A. halophytica* has been cloned and used in the transformation of *S. elongatus* PCC 7942 (that lacks a BetT). Transformants of *S. elongatus* PCC 7942 overexpressing this gene exhibited kinetic properties of BetT similar to that of *A. halophytica* cells. Studies on the overexpression of the BetT gene of *A. halophytica* in *E. coli* mutant MKH13 (deficient in *betT*, *putPA*, *proP* and *proU* genes) and its kinetic properties also confirmed that the BetT is a Na^+-betaine symporter (Laloknam *et al.*, 2006).

Another path way for the synthesis of glycine betaine is through oxidation of choline mediated by choline oxidase. This pathway is characteristic of the soil bacterium *Arthrobacter globiformis* that does not require any co-factors. The gene for this enzyme, *codA* has been cloned and used in the transformation of *S. elongatus* PCC 7942. The transformants exhibited higher levels of glycine betaine accumulation (60–80 mM) and improved tolerance to salt stress expressed in terms of growth, chlorophyll content and photosynthetic activity (Deshnium *et al.*, 1995).

The protective effects of glycine betaine reside in its dipole characteristics at physiological pH values and the presence of three hydrophobic methyl groups at the positively charged end of the molecule. The dipole seems to mask the Na^+ and Cl^- that enter the cells under salt stress and the hydrophobic methyl groups help in the stabilization of hydrophobic domains of proteins and thus prevent their distortion (Nomura *et al.*, 1998). The interaction of glycine betaine with hydrophilic and hydrophobic domains of macromolecules has been tested by examining the effects of salt stress and betaine synthesis on photoinhibition of PSII in transformants of *S. elongatus* PCC 7942 with *codA* gene (Deshnium *et al.*, 1995). The photoinhibition of PSII induced by strong light was enhanced due to salt stress (220 mM NaCl) and glycine betaine synthesized in the transformant cells protected PSII against photoinhibition. However, neither salt stress nor betaine synthesis affected photodamage to PSII. On the contrary, salt stress inhibited repair of photodamaged PSII and glycine betaine reversed this inhibitory effect of salt stress. Pulse-chase labelling experiments with [S^{35}]-Met revealed that under control conditions (20 mM NaCl) *de novo* synthesis of thylakoid membrane proteins was unaffected in presence of choline. In contrast, under salt stress (220 mM NaCl) the synthesis of these proteins decreased but their levels returned to control levels in presence of choline. Likewise, betaine did not affect the levels of radioactively labelled D1 protein in light under the non-stress condition (20 mM NaCl) but under salt stress (220 mM NaCl) there was significant inhibition in the synthesis of D1. Since it is already known that *S. elongatus* PCC 7942 genome consists of three *psbA* genes (*psbAI*, *psbAII* and *psbAIII*) and the expression of these genes is dependent on the intensity of light (the transcription of *psbAI* occurring under normal light conditions 70 µmol photons m^{-2} s^{-1} and the other two being expressed under strong illumination of 250 µmol photons m^{-2} s^{-1}; Golden *et al.*, 1986; Kulkarni *et al.*, 1992), Ohnishi and Murata (2006) observed the effects of salt stress and betaine on the level of *psbA* transcripts by Northern blotting. Due to the occurrence of such changes in the transcripts of *psbA* genes under all the conditions examined it was suggested that the inhibition and enhancement in the synthesis of D1 protein was neither due to salt stress nor betaine synthesis on transcription of *psbA* genes. Although the molecular mechanism responsible for the inhibition of degradation of D1 by salt stress is not known at present, it is suggested that salt stress might inhibit the interaction between the FtsH and DegP2 proteases and D1 or alternatively salt stress might inactivate these proteases.

The differential role of compatible solutes in countering the external salinity was examined by performing drying experiments with liposomes in presence of compatible solutes. Hincha and Hagemann (2004) observed that maximum protection was given by sucrose, trehalose and sorbitol from leakage of soluble marker from liposomes whereas GG and glycine betaine exerted hardly any effect. However, the performance of GG and glycine betaine was improved due to the presence of sucrose with them. By employing Fourier-transform infrared spectroscopy, phase transitions determined in the membranes due to the presence of solutes point out that all solutes decreased the phase transition temperatures corresponding closely with their ability to protect liposomes against leakage. GG protected the membranes under stress. This is akin to the property of glycerol on conferring membrane stability. Reports exist in literature on the effects of glycerol as a good protectant for biological (Lovelock, 1953; Hincha *et al.*, 1985; Santarius, 1992) and model membranes during freezing and thawing but not during drying and rehydration (Crowe *et al.*, 1990).

c) ***Trehalose***: A natural constituent of common food-stuffs (such as bread, wine, beer, vinegar and honey), trehalose is a multifunctional molecule (Elbein *et al.*, 2003). It is a non-reducing disaccharide that consists of two glucose units connected by 1α,-1α glycosidic linkage. Synthesized during osmotic stress as a compatible solute, trehalose is used as a carbon source by certain bacteria (Strøm *et al.*, 1998; Horlacher and Boss, 1997) and also constitutes the structural component of cell wall glycolipids (of *Mycobacteria*, *Nocardia*, *Rhodococcus* and *Corynebacterium*; Argülles, 2000; De Smet *et al.*, 2000; Richards *et al.*, 2002). In yeast, it is acculumated as a reserve compound and its synthesis during various abiotic stresses is considered as an adaptive response (Elbein, 1974; Thevelein, 1984; De Virgilio *et al.*, 1994; Hottiger *et al.*, 1994; Singer and Lindquist 1998a; Hounse *et al.*, 1998). The accumulation of trehalose in response to many abiotic stresses is known in various groups of organisms from bacteria to higher plants. Avonce *et al.* (2006) summarized five different pathways of trehalose biosynthesis in different groups of organisms (Fig. 5). The first one is a two-step biosynthetic pathway mediated by trehalose-6-phosphate synthase (TPS) and trehalose-6-phosphate phosphatase (TPP) and is known as TPS/TPP pathway. It occurs in eubacteria, archaea, fungi, insects and plants. A single step conversion of maltose mediated by trehalose synthase (TS) isomerizes the α-1- α-4 bond of maltose to α-1- α-1 bond forming trehalose. This is the second one known as TS pathway and has been reported in *Pimelobacter* sp. The third pathway is a two-step process that involves the conversion of maltooligosaccharides, glycogen or starch to trehalose. The first reaction is catalyzed by maltooligosyl trehalose synthase (TreY) that triggers the transglycosylation of the last glucose molecule of the maltooligosaccharides, glycogen or starch at their reducing end. This results in the conversion of α-1- α-4 bond to α-1- α-1 bond thus forming maltooligosyltrehalose which has trehalose at the end of the polymer. The second enzyme, maltooligosyltrehalose trehalohydolase (TreZ) catalyzes the hydrolytic release of trehalose. This pathway has been reported in the species of the thermophilic archaebacterium *Sulfolobus*. The fourth (TreP) and fifth (TreT) pathways are reversible pathways that involve the transfer of a glucose molecule to glucose-1-phosophate and ADP-glucose, respectively. The fourth pathway is mediated by trehalose phosphorylase (TreP) that has been detected in fungi but its *in vivo* participation in such a reaction is uncertain because it has been observed only *in vitro*. The fifth pathway is mediated by trehalose glycosyltransferring synthase (TreT) and it can also use UDP-glucose and GDP-glucose. This pathway has been reported in the hyperthermophilic archaebacterium *Thermococcus litoralis*.

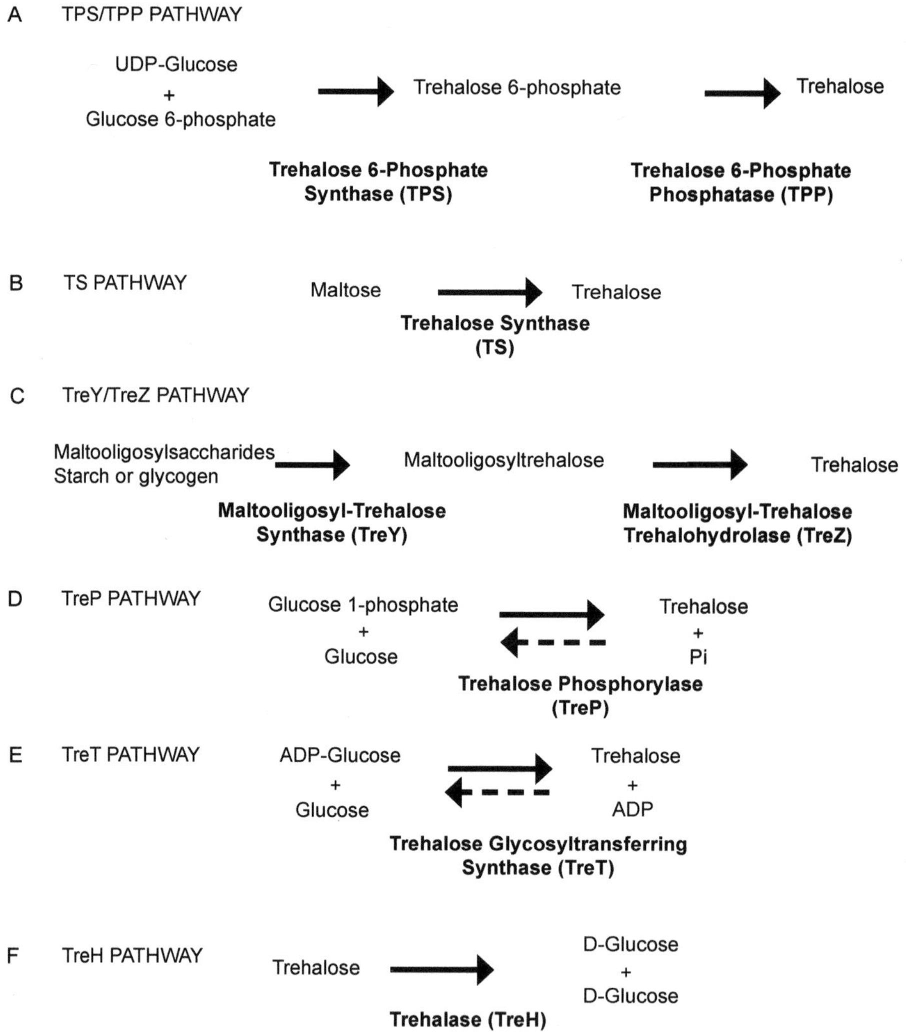

Figure 5: The trehalose biosynthetic pathways. The enzymes are indicated in bold. Dotted arrows in D and E pathways indicate reversible reaction. With the kind permission of G. Iturriaga, Instituto de Biotecnología-UNAM, Av. Universidad 2001, Col. Chamilpa, Cuernavaca 62210, Mexico. [Avonce *et al.* (2006) *BMC Evolutionary Biology* **6:** 109 doi:10.1186/1471-2148-6-109].

Studies on trehalose accumulation in cyanobacterial cells in response to salt stress and its biosynthesis in salt-stressed cyanobacterial cells are available. Trehalose biosynthesis in cyanobacteria occurs through the TreY-TreZ pathway. *S. platensis* tolerates salt concentrations up to 750 mM that is 1.5 times more than the concentration of sea water (Giovanna *et al.*, 1983; Vonshak *et al.*, 1988; Warr *et al.*, 1985; Zeng and Vonshak, 1998). Warr *et al.* (1985) reported that in addition to trehalose, *S. platensis* showed the accumulation of GG when exposed to salt stress. Trehalose accumulation along with sucrose in response to desiccating conditions has been reported by Higo *et al.* (2006) in *Anabaena* sp. strain PCC 7120. The expression of this gene in response to drought stress was determined using a DNA microarray technique (Katoh *et al.*, 2004; Higo *et al.*, 2006; Yoshimura *et al.*, 2006). Ohmori *et al.* (2009) observed that trehalose accumulation inside the cells of *S. platensis* was dependent on the presence of NaCl and its transport inside the cells because inhibition of the transport of Na^+ due to the addition of amiloride and monesin resulted in a decrease of trehalose. The synthesis of trehalose coincided with the increase of TreZ (or Mth), the enzyme that produces trehalose from maltooligosyl trehalose. The expression of the gene *mth* in response to salt stress and the increase of cellular Mth levels were noted in this organism. A comparative study of trehalose accumulation in a terrestrial cyanobacterium *Nostoc commune* (both cultures as well as naturally occurring colonies) in response to desiccation and salt stress revealed the existence of a correlation between loss of hydration and the level of accumulation of trehalose. This coincided with the loss of photosynthetic O_2 evolution of the cells in the dry state. As the colonies were rehydrated the photosynthetic O_2 evolution resumed with no detectable trehalose levels in the cells. Likewise, the presence of low concentrations of sodium chloride (0.2 M; non-ionic osmotic stress) inhibited photosynthetic O_2 evolution and induced the accumulation of trehalose. Thus trehalose accumulation seems to be an important feature in *N. commune* during desiccation (Sakamoto *et al.*, 2009).

The presence of mycosporine-like amino acids in unicellular cyanobacteria growing in a gypsum crust at the bottom of a hypersaline saltern pond has been demonstrated in concentrations as high as 98 mM. Significantly, two mycosprine-like amino acids with an absorption spectrum at 332 nm and 362 nm were present inside the cells. When the upper layers of the crusts were diluted, the mycosprine-like amino acids were rapidly leached out directly proportional to the dilution stress (Oren, 1997).

iii) Na^+/H^+ antiporters: The Na^+/H^+ antiporters are necessary to maintain ion homeostasis and regulate the exchange of Na^+ and H^+ across the membrane. So these will be helpful in overcoming salt stress by causing Na^+ efflux in exchange for H^+ thereby decreasing intracellular Na^+ ions. An electrochemical gradient of protons established by the respiratory chain or H^+-translocating ATPase drives this process. Thus the Na^+/H^+ antiporters play a very important role in Na^+ extrusion, pH homeostasis, cell volume regulation and establishment of electrical potential of Na^+ (Padan and Schuldiner, 1994, 1996). The Na^+/H^+ antiporter of *E. coli* is a high capacity Na^+ extrusion transporter responsible for the salt tolerance of the cells at alkaline pH (Padan and Schuldiner, 1994; Padan *et al.*, 2001). In cyanobacteria, the mechanism for salt tolerance is suggested to be due to the active efflux of Na^+ and the accumulation of K^+ (Reed and Stewart, 1985; Richtie, 1992). The active export of Na^+ across the membrane is suggested to be due to the action of primary Na^+-ATPase (Richtie, 1992) because the generation of substantial proton motive force is prevented due to adverse transmembrane pH gradient at alkaline pH. At this juncture, it may be noted that the optimum growth of cyanobacteria generally occurs at pH values ranging from 7.5 to 11.0.

Na^+/H^+ antiporters are generally classified as members of monovalent cation:proton antiporter (CPA) superfamily. These are subdivided into the Na^+ -transporting carboxylic acid decarboxylase

(NaT-DC) family found only in bacteria and CPA1 and CPA2 families found in many other kingdoms (Chang *et al.*, 2004). The genome of *S. elongatus* PCC 7942 (http://genome.jgi-psf.org/finished_microbes/synel/synel.home.html) revealed the presence of seven ORFs (*nha1* to *nha7*) that encode protein sequences of Na^+/H^+ antiporters. Of the seven antiporter proteins Nha3, Nha4 and Nha5 belong to the CPA2 family and the rest (Nha1, Nha2, Nha6 and Nha7) belong to the CPA1 family. The genome of *Synechocystis* sp. strain PCC 6803 has five genes that encode Na^+/H^+ antiporters as reported initially by Kaneko *et al.* (1996) and further studies conducted by Elanskaya *et al.* (2002). Of these five putative Na^+/H^+ antiporters, two are homologous to NhaP of *Pseudomonas aeruginosa* and the rest three are homologous to NapA type of antiporters of *Enterococcus hirae*. One of the NapA type antiporters is reported to be essential for the survival of this organism. However, six genes (slr1727, sll0273, sll0689, slr1595, slr0415 and sll0556) that govern the synthesis of Na^+/H^+ antiporters designated as NhaS1 to NhaS6 have been reported in *S. elongatus* PCC 7942 by Billini *et al.* (2008). There exists a sufficiently higher degree of homology between the amino acid sequences of antiporter proteins of *Synechococcus* and *Synechocystis*. There are two Na^+/H^+ antiporters in *A. halophytica* (ApNhaP and ApNhaPA-1) and the antiporter proteins bear resemblance to those of the above two organisms.

Waditee *et al.* (2002) demonstrated that the key factor in governing salt tolerance is the presence of an efficient Na^+/H^+ antiporter capable of pumping out intracellular of Na^+. *S. elongatus* PCC 7942, a freshwater cyanobacterium, was transformed with eight different types of expression plasmids each carrying a different gene (Na^+/H^+ antiporter, ApNhaP from *A. halophytica*, pUC303-*ApNhaP*; the molecular chaperone DnaK from *A. halophytica*, pUC-303 *dnaK*; catalase from *E. coli* pUC 303-*KatE*; betaine synthesis genes from *E. coli*, pUC303-*bet* and their combinations). Transformants of *Synechococcus* expressing ApNhaP exhibited growth in a medium containing 0.5 M NaCl as well as in sea water whereas those expressing betaine, catalase and chaperone could not grow in these media. Coexpression of ApNhaP with catalase or ApNhaP with catalase and betaine did not cause any additional benefit for salt tolerance than ApNhaP alone grown in presence of 0.5 M NaCl. However coexpression of ApNhaP with catalase resulted in enhanced salt tolerance of cells grown in sea water.

Na^+/H^+ antiporter from *A. halophytica*, ApNhaP, was found to be homologous to antiporters from plants, mammals and some bacteria (NhaP from *Pseudomonas* and SynNhaP from *Synechocystis*) and exhibited antiporter activity over a wide range of pH (5.0 to 9.0). The antiporter could complement the Na^+-sensitive phenotype of *E. coli* mutant (TO114) deficient in antiporter. ApNhaP showed high Ca^{2+}/H^+ antiporter activity at alkaline pH but no activity for Li^+/H^+ antiporter. The complementation of Ca^{2+}-sensitive phenotype of *E. coli* was also noted but not the Li^+-sensitive phenotype. The replacement of the C-terminal tail of ApNhaP with that of SynNhaP altered the ion specificity of the antiporter. Although both antiporters possess highest degree of homology, the C-terminal tail sequences appear to be distinct. C-terminal tail of ApNhaP possesses 22 basic and 14 acidic amino acids whereas that of SynNhaP consists of 15 basic and 24 acidic amino acids. In order to test the ion specificity of the C-terminal tails, chimeric antiporters were constructed that were designated as ASNhaP (ApNhaP with C-terminus of SynNhaP) and SANhaP (SynNhaP with C-terminus of ApNhaP). The exchange activities of chimeric antiporter ASNhaP with regard to Na^+/H^+ and Ca^{2+}/H^+ were comparable to that of parental ApNhaP but in addition showed Li^+/H^+ exchange activity that was not detectable in parental antiporter. In contrast, the chimeric SANhaP antiporter showed reduced exchange activities for Na^+/H^+ and Li^+/H^+ when compared to parental SynNhaP. The Ca^{2+}/H^+ exchange activity in the chimeric SANhaP was found to be pH-dependent while the parental SynNhaP showed pH-independent Ca^{2+}/H^+ activity (Waditee *et al.*, 2001). The isolation and functional characterization of Ca^{2+}/H^+ antiporter from *Synechocystis* sp. strain PCC 6803 and *A. halophytica* revealed that the antiporter proteins ApCAX and SynCAX are localized in the cytoplasmic membranes of these cyanobacteria and catalyzed Ca^{2+}/H^+ exchange reactions at alkaline pH. Mutant cells of *Synechocystis* disrupted in the *SynCAX* gene showed lower Ca^{2+} efflux activity with a salt-sensitive phenotype. There was gradual degradation of pigments in mutant cells at high pH where chlorophyll, phycocyanin and phycobiliprotein were 30%, 20% and 10% of the wild-type, respectively. Though the cellular K^+ levels were the same in the mutant and wild-type, the Na^+ content of the mutant cells was three times higher than the wild-type. *S. elongatus* PCC 7942 cells with *ApCAX* and *SynCAX* genes and their overexpression resulted in a salt-tolerant phenotype. Glu-74 and Glu-324 are the two conserved amino acid residues present in the transmembrane domain of the antiporter protein that is responsible for the Ca^{2+}/H^+ antiporter activity and for salt tolerance (Waditee *et al.*, 2004). Homologues of the NapA-type of Na^+/H^+ antiporters of *Synechocystis* sp. strain PCC 6803 have been characterized from *A. halophytica* with novel ion specificities that are involved in salt tolerance at alkaline pH. Two genes Ap-*napA1-1* and Ap-*napA1-2* that encode Ap-NapA1-1 and Ap-NapA1-2 polypeptides of the same size, respectively have been isolated and these antiporters complemented the salt-sensitive phenotype *E. coli*. The antiporter activities of Ap-NapA1-2 were significantly lower than the others but complemented K^+ uptake deficient mutant of *E. coli*. Amino acid residues Glu129, Asp225 and Asp226 of the transmembrane segment of the antiporter protein and Glu142 in the loop region are responsible for the activity. *S. elongatus* PCC 7942 exhibited enhanced salt tolerance due to overexpression of the gene Ap-*napA1-1* especially at the alkaline pH. Thus these two antiporters of *A. halophytica* exhibit different ion specificities (Wutipraditkul *et al.*, 2005). Furthermore, they identified the pH dependence of Ap-NapA1-1 and Ap-NapA1-2. However, AP-NhaP1 exhibited a high Na^+/H^+exchange activities at pH 6.0 to 9.0 whereas Ap-NapA1-1 had no Na^+/H^+ exchange activity at pH 6.0 but its activity increased with increasing pH with an optimum at pH 8.5. Likewise, pH dependence for Li^+/H^+ exchange activities by AP-NapA1-1 was demonstrated where as Ap-NhaP1 showed very little or no Li^+/H^+ exchange activity at all pHs examined. Genes encoding H^+-ATPase and Na^+-ATPase have been identified in *A. halophytica*. In case of the latter nine genes are organized into an operon. Na^+-ATPase conferred salt tolerance in a DKS mutant of *E. coli* deficient in ATPase. *S. elongatus* PCC 7942 acquired tolerance to salt stress up on the expression of *ApNa⁺-atp* operon and showed the localizationof Na^+-ATPase in its cytoplasmic membrane (Soontharapirakkul *et al.*, 2011).

The role of the genes *nhaS1* to *nhaS5* in governing the production of NhaS1 to NhaS5 in response to salt stress and the regulation of internal pH in the cells of *Synechocystis* sp. strain PCC 6803 has been identified by the generation of mutants. Single, double and triple mutants impaired in the respective genes of antiporters were generated by interposing genes conferring kanamycin resistance gene (from pUCΔk), chloramphenicol resistance gene (from pACyC184) or streptomycin resistance gene (from pBSL130) in ORFs cloned into *E. coli* vectors. Such plasmid DNA isolated from *E. coli* was used to transform the *Synechocystis* cells. Transformants for the respective antibiotics were selected. Completely segregated mutants revealed the inactivated genes for the respective antiporter genes. Double and triple mutants were generated by adding the required antiporter gene to the completely segregated inactivated genes with the Sm^r gene. The single, double and triple mutants of *Synechocystis* for *nhaS1*, *nhaS4* and *nhaS5* resembled wild-type cells in maintaining the same Na^+ and K^+ gradients at high salt concentrations suggesting that they are not required for adaptation of *Synechocystis* for salt stress. The activity of antiporter encoded by *nhaS3* seems to be sufficient for giving protection against salt stress (Elanskaya *et al.*, 2002).

Blanco-Rivero *et al.* (2005) isolated a transposon (Tn5-1063) generated mutant of *Anabaena* sp. strain PCC 7120 (designated as PHB11) that exhibited Na^+ sensitivity and lost the ability to grow at alkaline pH. The insertion of Tn5 inactivated a seven ORF operon that bears sequence resemblance to bacterial operons governing tolerance to high salt concentration and adaptation to alkaline pH. This operon, known as *mrp* operon (for multiple resistance and pH adaptation), comprised of *mrpA* that is responsible for Na^+ tolerance at high pH and together with *mrpB* and *mrpD* that encode protein sequences similar to hydrophobic subunits of the proton pumping NAD(P):ubiquinone oxidoreductase (Complex 1) found in mitochondria and eubacteria. PHB11 grew 20% less than that of wild-type and the cells lysed after 10-11 days growth at a pH of 10.5. During a shift of pH from 7.5 to 10.5 the inhibition of photosynthesis (measured as O_2 evolution) reached to a maximum of 95%. The *mrp* operon along with Tn5 sequence was recovered from PHB11 and used in the generation of single and double recombinant strains. The expression of this gene in these strains was monitored by the luminescence of *luxA* and *luxB* genes (from *Vibrio fischeri* as Tn5-1063 can make transcriptional fusions between these genes). It appeared to be similar in both the strains as the induction of *mrpA* was dependent on the concentration of Na^+ reaching its highest at 100 mM Na^+ in the medium. However, the response of the two strains to the combined effects of Na^+ concentration and elevated pH (10.5) was found to be different. In the double recombinant strain (which has lost wild copy of the *mrp* operon) though the induction of *mrpA* took place at relatively low concentration of Na^+ (3–25 mM), at elevated pH and high Na^+ concentrations the viability of this strain was affected similar to that of PHB11. In contrast, in the single recombinant strain the induction of *mrpA* was maximum at increasing concentration of Na^+ and alkaline pH suggesting that *mrpA* and for that matter the whole *mrp* operon confers tolerance to Na^+ at elevated pH.

Billini *et al.* (2008) showed that of the seven Na^+/H^+ antiporters of *S. elongatus* PCC 7942, Nha3 possessed the Na^+/H^+ antiporter activity because this antiporter complemented the salt-sensitive phenotype of *E. coli* TO114 strain that is is deficient in Na^+/H^+ transporter activity. The rest of the Na^+/H^+ antiporter proteins Nha1, Nha4, Nha6 and Nha7 exhibited low Na^+/H^+ antiporter activity and were not able to complement salt-sensitive phenotype of *E. coli* TO114 cells. The expression of the *nha* genes was monitored under salt stress and alkaline stress conditions (imposed by high pH) using real time reverse-transcription PCR. The expression of all the seven genes of the Na^+/H^+ antiporters under the stress conditions examined signifies that all these act synergistically to overcome salt and pH tolerance.

iv) Gene level responses: A five-fold increase in copy number of an endogenous plasmid (pSY10) of *Synechococcus* sp. NKBG 042902 has been reported in response to salt stress, a phenomenon that was specific to NaCl but not to either to KCl or sorbitol (Takeyama *et al.*, 1991). By employing random catridge mutagenesis, Hagemann and Zuther (1992) isolated three salt-sensitive mutants (143, 406 and 549) of *Synechocystis* sp. strain PCC 6803. The characterization of mutant 549 showed a deletion of 1.8 kb portion in the genome affecting four ORFs (a glycoprotease *gcp* homologue, the *psaFJ* gene and an unknown gene). The construction of single mutations and a study of their properties showed that only *gcp* gene affected the salt tolerance resulting in a salt-sensitive phenotype. Nitrogen starvation of the cells followed by changes in pigmentation (a decrease of phycobiliproteins and increase of carotenoids), and ultrastucture (higher accumulation of cyanphycin granules) are other properties observed in this mutant. The gene product of *gcp*, a glycoprotease seems to be related to cyanophycin degradation. Tha salt tolerance levels of wild-type, mutant 549 and the *gcp* single mutant varied. The former two tolerated up to 684 mM NaCl while the third showed reduced tolerance (550 mM NaCl). DNA:DNA hybridization studies indicated that recombination occurred via a double cross-over event with the replacement of wild-type alleles by the mutated copies in the *gcp* mutant (Zuther *et al.*, 1998).

Mutagenesis of *Anabaena* sp. strain PCC 7120 with Tn5 transposon carrying *luxAB* as the reporter gene resulted in the identification of a gene resembling in its sequence low-temperature inducible (*lti2*) gene sequence that has been identified to be a cold-inducible gene. This gene was inducible by low temperature, osmotic as well as salt stress. A second round of transposon mutagenesis of this mutant gave rise to another mutant in which the transcription of *lti2*-like gene was reduced and the insertion of the transposon was found to be in an ORF (known as *orrF*) adjacent to *lti2* gene that resembled in sequence with that of a response regulator. Complementation with the wild-type DNA fragment carrying the entire *orrF* sequence but not the truncated sequence restored the activity of response regulator (Schwartz *et al.*, 1998)

A gene (slr0897) of *Synechocystis* sp. strain PCC 6803, designated as *Ssglc*, encodes a β-1,4 glucanase-like protein (SsGlc). The deduced amino acid sequence of this protein showed a high degree of similarity to sequences of glycoside hydrolase (GH) family β-1,4-glucanases (cellulases) from various sources. Cloning and expression of this gene in *E. coli* resulted in the production of a protein that could hydrolyze barley β-glucan and lichenan (β-1,3,1,4-glucan) but not cellulose (β-1,4-glucan), curdlan (β-1,3 glucan) or laminarin (β-1,3,1,6- glucan). A mutant derived by the disruption of the *Ssglc* gene with kanamycin-resistance catridge gene exhibited similar growth performance as that of wild-type in continuous light (40 µmol of photons/m^2 per s), 12 h light (40 µmol of photons/m^2 per s) and 12 h dark cycle, cold stress (4°C), osmotic stress (100–400 mM sorbitol) and high light stress (200 µmol of photons/m^2 per s) but under salt stress (300–450 mM NaCl) the growth of the SsGlc mutant was significantly inhibited. In view of the above, Tamoi *et al.* (2007) concluded that SsGlc functions in salt stress tolerance.

***v) Genome- and proteome-based studies*:** The potential of the techniques like reverse transcriptase-polymerase chain reaction (RT-PCR) and DNA microarrays has been exploited for the identification of gene expression at genome-wide (global) level in cyanobacteria under various stress conditions. With the availability of complete genome sequence of *Synechocystis* sp. strain PCC 6803, it has now been possible to identify gene expression patterns of this organism under different stress conditions. Such expression studies have also been conducted in other unicellular and filamentous cyanobacteria. Proteome of an organism represents the total number of proteins encoded by its genome. The automated identification of proteins has been possible due to the MALDI-TOF analysis along with two-dimensional electrophoresis (2-DE). Thus these studies reveal knowledge on the level of protein expression, protein isoforms produced from each gene, the extent to which proteins are post-translationally modified and cellular and subcellular distribution of proteins (Figs. 6 and 7; Pandhal *et al.*, 2008; Castielli *et al.*, 2009). The cellular responses to varying degrees of salt stress are represented in Fig. 8 that incorporated the proteomic studies of Fulda *et al.* (1999, 2000, 2006) and Sudhir *et al.* (2005). The different cellular compartments that experience salt stress are noted to be cytoplasm, plasma membrane, periplasmic space and thylakoids. In the cytoplasm there is a general increase in salt stress-specific proteins with an increase in carbon metabolism. A change in the nature of lipids in the plsama membrane is associated with an increase in phosphate, nitrate and nitrite-binding proteins. The effect of salt stress on the periplasmic space is reflected at the level of synthesis of new extracellular layers. There is a structural reorganization in the proteins of the thylakoid membrane with repression in phycobilisome synthesis (Pandhal *et al.*, 2008). Kanesaki *et al.* (2002) performed a DNA microarray analysis of proteins in salt-stressed cells of *Synechocystis* sp. strain PCC 6803. Salt stress-induced proteins belonged to the class of ribosomal proteins and proteins of unknown function. Hyperosmotic stress and salt stress were perceived as different stimuli. Hyperosmotic stress caused the induction of certain specific gene products like 3-keto acyl-carrier protein reductase and rare lipoprotein A. However, there were some common gene products belonging to the category of heat shock proteins and the enzymes for the synthesis of GG induced in both hyperosmotic stress and salt-stress. Marin *et al.* (2004) investigated genome-wide expression of genes to salt stress by DNA microarray technique and compared these to changes in

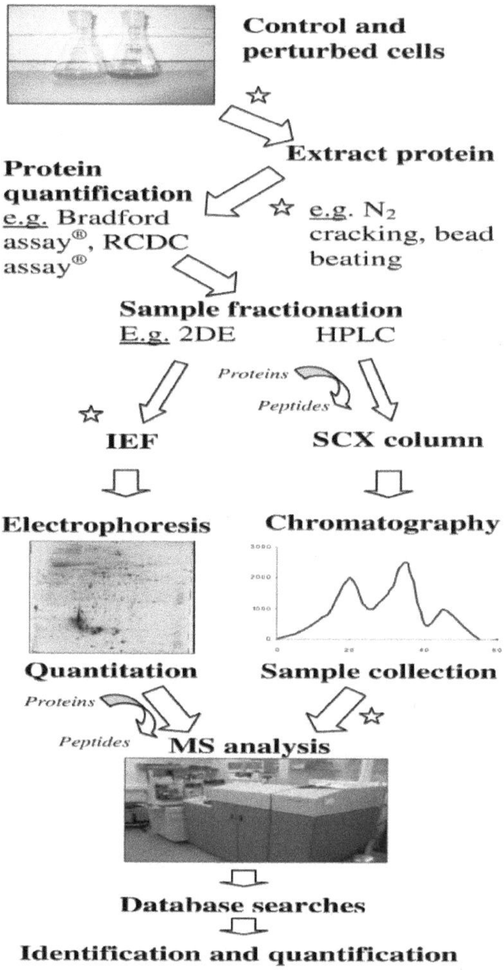

Figure 6: A general proteomic workflow. The main steps are, cell culture, protein extraction (liquid nitrogen freezing and cracking), sample fractionation (traditional 2DE and gel-free chromatography method), mass spectrometer analysis and protein quantitation and identification. Quantitation using 2DE is densitometry based and implemented with appropriate software (e.g. PD QUEST®, SameSpots®), but can also be undertaken on the mass spectrometer, relying on abundance of peptide peaks. Identification occurs with database search software (based on peptide mass fingerprint or sequence analysis). Proteins are commonly digested using trypsin: indicate essential salt removal stages. RCDC® – protein quantitation assay kit, IEF- isoelectric focusing, HPLC- High Performance Liquid Chromatography, SCX- Strong Cation Exchange. With the kind permission of C. A. Biggs, Biological and Environmental Systems Group, Department of Chemical and Process Engineering, The University of Sheffield, Mappin Street, Sheffield, S1 3JD, UK [Pandhal *et al.* (2008) *Saline Systems* **4:** 1; doi:10.1186/1746-1448-4-1].

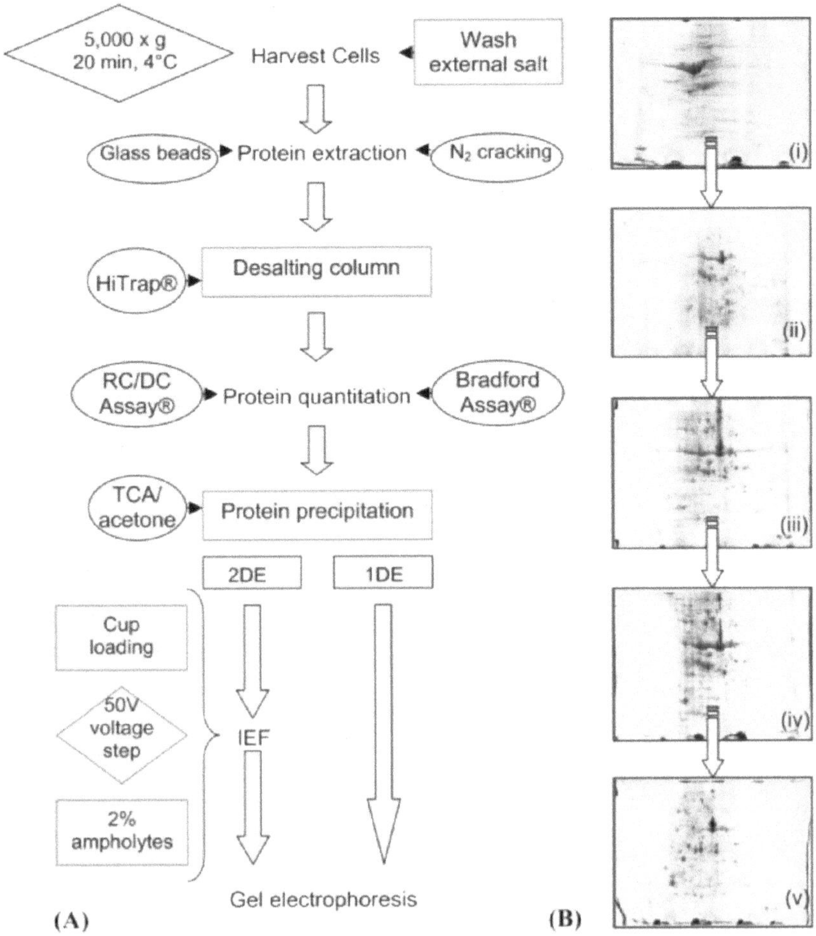

Figure 7: (A) A simple salt removal strategy for proteomic experiments using gel electrophoresis for protein fractionation. Desalting steps are in square boxes, example techniques in circular boxes and parameters in diamond boxes. (B) Euhalothece desalting optimisation tests. 7 cm 2 gels, pI 3–10. (i) No desalting steps (ii) TCA/acetone precipitation only (iii) Desalting column not used (iv) External salt not washed (v) Complete desalting procedure followed. With the kind permission of C. A. Biggs, Biological and Environmental Systems Group, Department of Chemical and Process Engineering, The University of Sheffield, Mappin Street, Sheffield, S1 3JD, UK [Pandhal *et al.* (2008) *Saline Systems* **4:** 1; doi:10.1186/1746-1448-4-1].

main physiological parameters. The expression of 240 genes was enhanced three-fold in the first 15 min while at the same time there was repression of 140 genes. After 24 h of salt stress, however, the activity of 39 genes remained significantly enhanced. This enhancement was supported by RNA isolation, cDNA synthesis from control cultures and DNA microarray hybridization and validation by Northern blotting experiments. Genes that are involved in the synthesis of proteins meant for salt acclimation are *ggps*, *stpA*, *glpK*, and *glpD* which mediate the synthesis of GG, ggABC genes for ABC-type translocator for compatible solutes, one of the six Na^+/H^+ antiporters, one of the nine mechanosensitive channel like proteins (*Slr0765*) and a probable DNA-binding protein (*Slr189*). The activity of certain genes was lower when salt stress continued even after 24 h. Some of these genes are that encode high-light induced proteins, heat shock protein HtpG, the RNA binding protein Rbp3

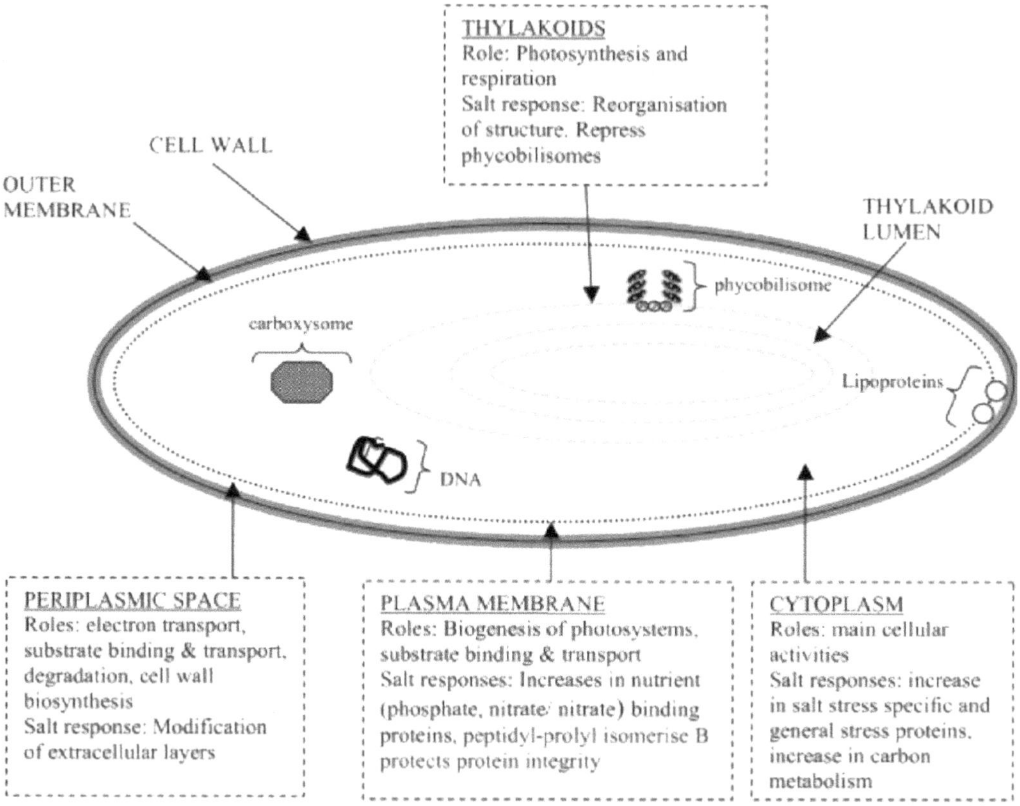

Figure 8: Salt responses at the protein level. A summary of responses to varying degrees of salt stress (~2–4% w/v NaCl) in the main cellular compartments of a unicellular cyanobacterium cell, as revealed by proteomics studies of Fulda *et al.* (1999, 2000, 2006) and Sudhir et al. (2005). With the kind permission of C. A. Biggs, Biological and Environmental Systems Group, Department of Chemical and Process Engineering, The University of Sheffield, Mappin Street, Sheffield, S1 3JD, UK [Pandhal *et al.* (2008) *Saline Systems* **4**:1; doi:10.1186/1746-1448-4-1].

and enzymes of carbohydrate metabolism such as phosphofructokinase, fructose-1,6-biphosphatase aldolase and pentose-5-phosphate-3-epimerase. However, the last three enzymes of carbohydrate metabolism showed unchanged expression after long term salt acclimation. Significantly, two genes that encode peptide deformylase and Met-aminopeptidase involved in polypeptide maturation meant for removal of N-formyl-Met and N-N-terminal Met, respectively showed increased level of mRNA after 24 h of salt stress.

Fulda *et al.* (2006) conducted a detailed proteome analysis of salt-shocked cells of *Synechocystis* sp. strain PCC 6803 by measuring S^{35}-methionine labelling combined with 2-DE protein separation and PMF for identification of the specific proteins. They identified 337 different protein species out of a total of 500 identified protein spots. These belonged to four functional groups (i) proteins specific to synthesis of GG, (ii) general stress proteins, (iii) enzymes of basic carbohydrate metabolism and (iv) hypothetical proteins. The first group of proteins is specific to salt stress and pertained to the synthesis of osmolyte GG and the expression of these genes was stably up-regulated whereas the second group of general stress proteins included the synthesis of GroEL1, DnaK2 and GrpE. Analysis of the total soluble fraction of *Synechocystis* cell extracts, 55 soluble proteins were found with either an expression level enhanced by salt shock (18 proteins) or even accumulated at high concentration after long term acclimation (37 proteins). However, for 40% of the proteins that were accumulated in salt acclimated *Synechocystis* cells, gene induction was not observed in the DNA microarray experiments, thereby indicating the involvement of post-transcriptional regulation in salt acclimation.

A comparison of the proteome with the transcriptome (mRNA levels of genes expressed after salt shock of earlier studies) revealed that 89% of the proteins so induced shortly after salt shock were also supported by the levels of mRNAs. The changes induced by salt stress in the proteins of plasma membranes were monitored by another proteomic screening performed by Huang *et al.* (2002, 2006). Differential and 2-DE protein profiles led to the identification of 109 proteins corresponding to 66 different gene products, of which the production of 20 of them was induced under salt stress. Half of these proteins related to periplasmic binding proteins belonging to the category of ABC-transporters or hypothetical proteins. Few of the gene products expressed at highest level are those of Fut A1 (an iron-binding protein) and Vipp[1] (for vesicle-inducing protein in plastids1) which have been suggested to be involved in protection of PSII and in thylakoid membrane formation, respectively. Other proteins identified in the plasma membrane of *Synechocystis* sp. strain PCC 6803 belong to the category of proteases (regulatory subunit of ATP-dependent Clp protease, the Hho-protease, a Hly protein), a protein belonging to CheY superfamily and two subunits of ATP synthase. At least there were 10 newly identified proteins belonging to hypothetical proteins that have no sequence similarities with any protein of known function. One of the most significant findings is the identification of a GG-binding protein of the ABC-type transporter located on plasma membrane. These observations are consistent with the earlier findings on salt-enhanced gene expression (Mikkat and Hagemann, 2000; Marin *et al.*, 2004).

The ability of many cyanobacterial strains to grow almost from neutral pH to alkaline range (of pH 10.0 to 11.0) makes them versatile not only to adjust to variable pH conditions but also to adjust to the concentrations of dissolved carbon dioxide/bicarbonate levels. Since the photoautotrophic growth of *Synechocystis* sp. strain PCC 6803 is similar at pH values ranging from 7.5 to 10.0, the probable changes that occur during a shift from pH 7.5 to pH 10.0 are worth following. Intracellular pH levels are also maintained consistent with the external pH. Accordingly, compartmentalization of pH in the vicinity of internal membranes and the cytosol has been observed. Thylakoid lumen has a pH that is ~2 units lower than the pH of the cytosol (Belkin and Packer, 1988). Any change in external pH by 2 units will increase internal pH by ~0.2 unit (Belkin and Packer, 1988; Ritchie, 1991). The isolation and characterization of a number of pH-sensitive PSII mutants of *Synechocystis* sp. strain PCC 6803 that were able to grow at alkaline pH (pH10.0) but not at pH 7.5 revealed that these mutants were deficient in either PsbO or PsbV lumen proteins (Eaton-Rye *et al.*, 2003; Summerfield *et al.*, 2005). Keeping these in view, Summerfield and Sherman (2008) studied global transcriptional response during a shift of *Synechocystis* cells from pH 7.5 to pH 10.0. Majority of the genes that were up-regulated during such a shift are those pertaining to the processes of photosynthesis and respiration. Other categories of proteins up-regulated pertain to the regulatory functions, transport and binding proteins. Similar studies on the proteomic analysis of plasma membranes of high pH-stressed cells of *Synechocystis* sp. strain PCC 6803 confirmed the up-regulation of transport and binding proteins of ABC transporters which included three phosphate transport proteins. Other categories of genes up-regulated during high pH stress belonged to cell division, signalling, photosynthesis and respiration (Zhang *et al.*, 2009). One of the probable mechanisms suggested for maintaining pH homeostasis is due to accumulation of acetolactate by the cells of *Synechocystis* sp. strain PCC 6803 under alkaline conditions (Maestri and Joset, 2000).

***vi) Salt intake and cell signalling*:** The major studies relate to the identification of genes governing histidine kinases (Hiks), the corresponding response regulators (Rres), Ser/Thr protein kinases (Spks) and transcription factors and their expression in response to salt stress. But how these coordinate with each other during salt-stress is yet to be studied. There appears to be a variation in the number of genes that encode histidine kinases (Hiks) in the genome of *Synechocystis*. Initially 42 Hiks have been identified (Mizuno *et al.*, 1996) but in subsequent studies we come across 43 Hiks (Marin *et al.*, 2003) and 47 Hiks (Murata and Suzuki, 2006). Some of these have been implicated in the acclimation to cold, phosphate starvation, light stress, osmotic and salt stress (Hirani *et al.*, 2001; Suzuki *et al.*, 2001, 2004; Mikami *et al.*, 2002; Marin *et al.*, 2003). To identify specific sensory Hiks that are involved in the perception of salt stress in *Synechocystis* sp. strain PCC 6803, a library of mutants defective in 43 Hiks have been generated and compared with wild-type cultures for the function of each Hik, the type and number of proteins synthesized during salt stress using DNA microarray analysis of genome-wide gene expression. Four Hiks, i.e. Hik16, Hik33, Hik34 and Hik41 were responsible for perception and transmission of signals for salt stress. Hik34 mutant cells were shown to be deficient in the expression of 28 genes related to salt stress and certain other general stress-related proteins such as chaperones (*dnaJ* for heat shock protein 40, *hsp17*, *dnaK* , the subunit of aprotein *clpB*) and superoxide dismutase (*SodB*). Some other genes (not generally related to stress response) that were not induced by salt stress and not expressed in Hik34 mutant cells were a gene for riboflavin kinase (*ribF*), a gene for hydrogenase expression/formation factor (*hypA*), a gene for group 2 σ factor (*SigB*) and 14 genes of unknown function. It is suspected that *SigB* might be responsible for transduction of salt signal perceived by Hik34 and in initiating the expression of a set of genes related to salt stress. Hik33 mutant cells were deficient in the expression of three high-light inducible (*hli*) genes, a *SigD* gene that encodes group 2 σ factor and a protein of unkown function. Some osmotic stress and cold-inducible genes were not expressed in Hik33 mutant cells confirming the earlier observations that Hik33 is a cold sensor (Suzuki *et al.*, 2001) and osmosensor (Mikami *et al.*, 2002). Under salt-stress, *hik33*-knockout mutants of *Synechocystis* sp. strain PCC 6803 exhibited reduced levels of cytoplasmic membrane proteins specially those of substrate-binding proteins of ABC-transporters and certain regulatory proteins including Rre13. Altogether 26 different proteins are differentially expressed (Li *et al.*, 2012). Close homologues of the gene governing Hik33 have been detected in *S. elongatus* PCC 7942, *Anabaena* sp. strain PCC 7120, *P. marinus* MED4 and *Bacillus subtilis*. Hik16 is considered to be a membrane bound sensor whereas Hik41 as the probable transducer of the signal and these two respond to only NaCl. However, it is interesting to know that the salt inducibility of genes governing the synthesis and uptake of GG, proteins involved in photosynthesis and certain other regulatory functions were unaffected in all the types of Hik mutants (Marin *et al.*, 2003). In addition to the above four Hiks (Hik16, Kik33, Hik34 and Hik41), the identification of Hik10 and the corresponding Rres Hik33-Rre31, Hik34-Rre1 and Hik10-Rre3 as well as Hik16-Hik41-Rre17 formed a system that regulated the transcription of a specific set of genes responsible for hyperosmotic stress. Hyperosmotic stress-inducible genes have been grouped into four classes based on an induction factor exceeding 4.0. Group I consisted of *fabG* and 10 other genes whose expression was considerably reduced due to mutation of Hik33. Thus this group of genes is regulated by Hik33-Rre31 system. There were 19 genes belonging to Group 2 that were affected in Hik34 mutant cells (compared to 28 genes in Hik34 mutant cells noted above). The genes whose expression was depressed in Hik34 mutant cells are *hspA* (for a small heat shock protein), *ClpB1* (for ClpB protease), *SodB* (for superoxide dismutase), *htpG* (for heat shock protein 90), *dnaK2* (for heat shock protein 70), *dnaJ* (for heat shock protein 40), *groEL2* (for chaperonin 60-2), *groEL1* (for chaperonin 60-1), *groES* (for chaperonin 10) and other genes of known and unkown function. Thus the expression of Group 2 genes was under the regulation of Hik34-Rre1 system. Genes of Group 3 essentially (those of unkown function *sll0939* and *slr* 0967) include those that were not affected due to mutation of Hik33 and Hik34 but were suppressed to basal levels of function in Hik16 and Hik41 mutant cells. The genes included in Group 4 are those whose induction of expression was unaffected by inactivation of any of the examined Hiks. Genes such as *rlpA* and *repA* (for rare lipoprotein A), *htrA* (for a serine protease) and 17 other genes belong to this category. A screening of gene knock-out libraries of *Synechocystis* sp. strain PCC 6803 for Hiks and Rres governing hyperosmotic stress revealed that 32 of the 52 hyperosmotic stress-inducible genes are under the control of Hik33, Hik34 or Hik16 and Hik41. However, the rest 20 genes induced by hyperosmotic stress are under the regulation of some other unknown mechanism(s). Moreover, the observations of Kanesaki *et al.* (2002) that there are special type of genes that are expressed under salt stress (*pbp* gene for penicillin-binding protein and *ctpA* gene for C-terminal processing protease) and hyperosmotic stress (*fabG, rlpA, repA*) are worth mentioning. Though there are similar kinds of Hiks operating under salt and hyperosmotic stress, the types of genes expressed under their control differ in the two stresses (Paithoonrangsarid *et al.*, 2004). This has further been confirmed by the screening of the gene knockout libraries of Rres by RNA slot-blot hybridization followed by a genome-wide DNA microarray (Shoumskaya *et al.*, 2005). In addition to Hik33, Hik34, Hik16 plus Hik41, Hik10 is also involved in the perception of salt stress and such signals are received by the Rre31, Rre17, Rre1 and Rre3. These thus regulate the expression of a large number of genes. The recognition of Hik33-Rre31, Hik34 or unidentified Hik-Rre1, Hik10-Rre3 and Hik16-Hik41-Rre17 systems function to perceive and transduce the signals of both salt and hyperosmotic stress as reported earlier (Paithoonrangsarid *et al.*, 2004). A single gene (*htrA* gene in case of Hik10-Rre3 system) or few genes (*sll0939, slr0967* and *sll0938* in case of Hik16-Hik41-Rre17 system) were induced in both salt and hyperosmotic stress. In contrast, in case of systems like Hik33-Rre31 and Hik34-Rre1 and unidentidfied Hik-Rre1 the expression of a group of genes was induced by salt-stress or hyperosmotic stress. Murata and Suzuki (2006) reviewed the utility of genome sequence of *Synechocystis* sp. strain PCC 6803 in understanding as to how cyanobacteria sense environmental stress. By performing targeted mutagenesis of 44 out of 47 Hiks and 42 out of 45 Rres, the mutant libraries were screened by genome-wide microarray analysis and slot-blot hybridization. They identified Hiks specific for low temperature, hyperosmotic and salt stress and manganese deficiency that perceive and transmit signals. Some of the genes for heat shock proteins (*hspA, dnaK2, dnaJ, groEL2, clpB1* and *clpC*), GG synthesis (*ggpS, glpD*), *sigD* and *hliA* appear to be induced commonly in response to salt-stress and osmotic stress.

Of the seven Spks identified in *Synechocystis* sp. strain PCC 6803 (SpkA to SpkG), the relative expression of SpkC, SpkD, SpkF and SpkG has been examined under seven stress conditions (Liang *et al.*, 2011). Significantly under high salt conditions SpkG gene is overexpressed when compared to control cultures and those derived from low and high temperatures and media deficient in carbon, nitrogen and phosphorus (Fig. 9). The transcriptional dynamics of these four genes based on semiquantitative RT-PCR showed that in the salt-stressed cells of *Synechocystis* sp. strain PCC 6803 the expression of SpkG is up-regulated while at the same time SpkC is down-regulated. However, in all other six conditions SpkG is down-regulated but SpkC is up-regulated in all conditions except in high salt condition. Construction of deletion mutants of SpkC and SpkG and their growth characteristics under high salt conditions have been helpful in identifying SpkG that plays an important role in perceiving high salt signal directly. Genome-wide patterns of transcription of genes in the wild-type and SpkG mutant have been compared by DNA microarray analysis (Fig. 10).

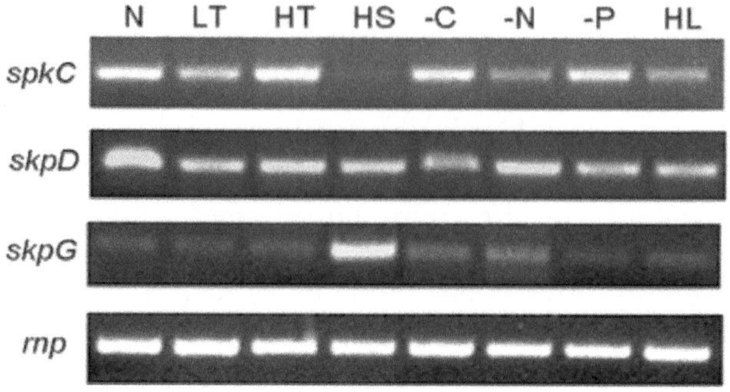

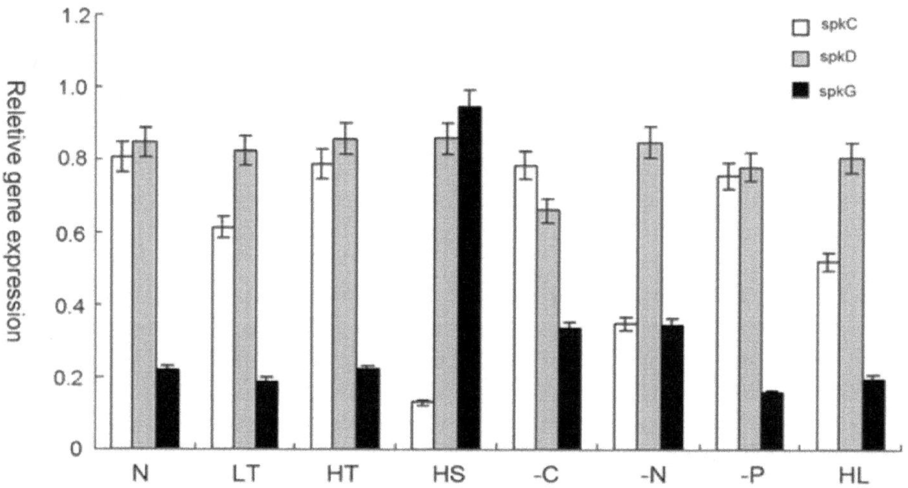

Figure 9: Semiquantitative RT-PCR reveals the transcriptional dynamics of spkC, spkD, spkF and spkG at different conditions. The transcript level of RNase P in each sample serves as a control. N, normal conditions; LT, low temperature; HT, high temperature; HS, high salt; -C, carbon-deficient; -N, nitrogen-deficient; -P, phosphorus-deficient. With the kind permission of S. Qin, Institute of Oceanology, Chinese Academy of Sciences, Qingdao, China. [Liang *et al.* (2011) *PLoS ONE* **6(5):** e18718. doi:10.1371/journal.pone.0018718] doi:10.1371/journal.pone.0018718.g001

Gene sequences governing the synthesis of Group 3 sigma factors SigH (sll0856), SigG (slr1545) and SigF (slr1564) and the regulatory protein RsbU (slr-2031) have been identified in the genome of *Synechocystis* sp. strain PCC 6803 and subjected to interposon mutagenesis to generate mutants to understand their role in stress tolerance. Mutant SigF showed a severe defect in the induction of salt-stress proteins. Though its ability to tolerate salt stress (up to 684 mM NaCl) was not apparently changed but its ability to tolerate still higher concentrations of salt was reduced. A comparison of wild-type and Sig mutants revealed no differences in their growth rates, pigment contents, photosynthetic and respiratory activities and the level of accumulation of GG (Huckauf *et al.*, 2000).

***vii) Common stress mechanisms*:** The genome of *A. halophytica* contains a gene *DnaK1* that encodes a polypeptide of 721 amino acid residues with a C-terminal tail composed of many acidic amino acids. In order to understand whether this protein has a chaperone like activity and the role of C-terminal

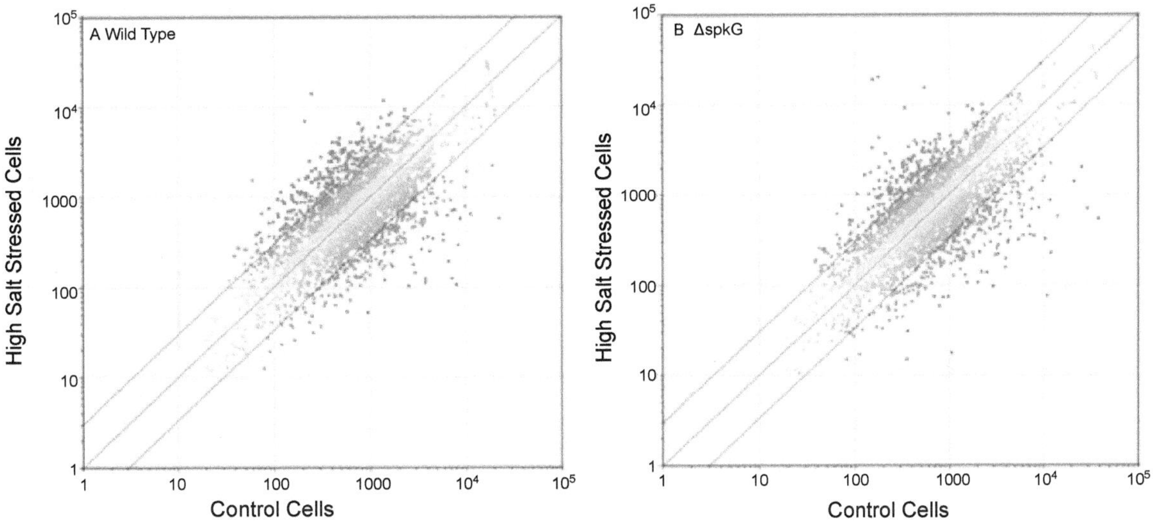

Figure 10: Genome-wide patterns of transcription of salt-stress regulated genes. (A). Gene expression in wild-type cells that had been exposed to 855 mM NaCl for 30 min was compared with that in unstressed cells. (B). Gene expression in mutant spkG cells that had been exposed to 855 mM NaCl for 30 min was compared with that in unstressed cells. Dots correspond to genes whose fold change >2 are beyond the reference lines. With the kind permission of S. Qin, Institute of Oceanology, Chinese Academy of Sciences, Qingdao, China. [Liang *et al.* (2011) *PLoS ONE* **6(5):** e18718. doi:10.1371/journal.pone.0018718] doi:10.1371/journal.pone.0018718.g004.

Color image of this figure appears in the color plate section at the end of the book.

tail, Hibino *et al.* (1999) conducted a comparative study of the activities of three different types of DnaK1 molecules (DnaK1 from *A. halophytica* and *S. elongatus* PCC 7942 and the C-terminal tail deleted DnaK1 from *A. halophytica*) produced in *E. coli* cells for their ATPase activity and refolding activity of the urea-denatured proteins under salinity stress. The ATPase activity of DnaK1 was very high at salinity levels of 1.0 M NaCl or KCl from *A. halophytica* but the ATPase activity of DnaK1 from *Synechococcus* decreased with increasing concentration of the salts. The deletion of C-terminal tail of DnaK1 of *A. halophytica* had no apparent effect on ATPase activity but the refolding activity of the DnaK1 protein is lost because it could not carry out the refolding of urea-denatured lactate dehydrogenase at high salinity. These results emphasize the role of DnaK1 and its C-terminal region in helping denatured proteins to overcome salt-stress (Hibino *et al.*, 1999).

Computational screening of intergenic regions of *S. elongatus* PCC 6301 and its comparison with three genomes of freshwater cyanobacteria brought to light the existence of small, regulatory, non-coding RNA (ncRNA) genes. These ncRNAs are shown to be required for growth under multiple stress conditions. One of these genes *Yfr1* is 65 nucleotides long and is located in between *gnaB* and *trxA* genes. This gene seems to be highly conserved among the genomes examined. Further, a mutant of *S. elongatus* PCC 6301 deficient in *Yfr1* was unable to tolerate various stress conditions such as oxidative stress and salt stress and showed accumulation of the transcripts of *sbtA* gene. *In vitro* interaction of Yfr1 RNA with *sbtA* mRNA suggests that the target of action of Yfr1 is *sbtA* mRNA (Nakamura *et al.*, 2007).

Gaber *et al.* (2007) identified two genes (slr1562 and ssr2061) in the genome of *Synechocystis* sp. strain PCC 6803 that encode two glutaredoxins (Grxs), Grx1 and Grx2 respectively. The deduced amino acid sequences of these are highly homologous to the Grxs from other organisms. Grxs are small, heat-stable oxidoreductases that function as alternative hydrogen donor in the reduction of intramolecular disulfide in ribonucleotide reductase, the essential enzyme for DNA synthesis. First discovered in *E. coli* (Grx1 and Grx3) and in yeast (Grx1 and Grx2), these three Grxs are small proteins with a molecular weight of 10 kDa and contain an active site of CPYC. A second group of Grxs comprising of Grx3, Grx4 and Grx5 with an active site of GFS corresponds to yeast. The Grxs found in *Synechocystis* share a high identity to the Grxs from other organisms. The mRNA levels of Grx2 of *Synechocystis* increased under oxidative stress conditions caused by high salinity, chilling or the addition of H_2O_2, methylviolgen or t-butylhydroperoxide. The overexpression of the gene (*ssr2061*) of *Synechocystis* in *E. coli* JM109 produced Grx2 that protected the cells of the latter to salt stess greater than 700 mM.

II. TEMPERATURE STRESS

The optimum growth temperature of mesophilic prokaryotes ranges between 25°C to 40°C. Beyond this the mesophilic prokaryotes are unable to grow and so this temperature seems to be demarcation point for mesophiles and those that are equipped to grow at temperatures above 40°C. These are known as thermophiles which have an optimum temperature range of growth between 50°C and 70°C while at the same time some are able to grow albeit slowly at 40°C as well. Examples of thermophiles are found among eubacteria, actinomycetes, fungi, protozoa and algae. Few other eubacteria and archaebacteria that are able to grow at still higher temperatures (between 80°C and 110°C) are known as hyperthermophiles. The hyperthermophiles are unable to survive below 60°C. Due to a "shift-down" or "shift-up" of the microbial cells from their optimum growth temperature, they experience a cold shock or heat shock, respectively.

A) Cold shock: A sudden shift-down in the growth temperature results in a cold shock leading to a transient cessation of growth. During this period, the general protein synthesis is severely inhibited and is accompanied by the expression of a group of well defined proteins. Among these, cold shock family of proteins (Csps), low temperature protein chaperones such as GroEL, GroES, caseinolytic proteases (Clps), RNA binding proteins (Rbps) and RNA helicases are important in bacteria (Weber and Marahiel, 2003; Gualerzi *et al.*, 2003; Phadtare and Severinov, 2010). Eventually, the synthesis of these proeins also decreases, followed by a phase of acclimatization to the low temperature and growth resumes. In contrast to heat shock, cold shock response appears to be predominantly post-transcriptional as no cold-specific σ factor has so far been identified. In *E. coli* many of the proteins that are up-regulated during cold shock are associated with the translational apparatus (Phadtare, 2004; Nierhaus and Wilson, 2004). The perceptible effects of cold shock are felt at different levels in the cells starting with a decrease in membrane fluidity (thus affecting active transport and protein secretion), stabilization of secondary structures of RNA and DNA (with a consequential decrease in translation and transcription) to slow or inefficient folding of proteins and finally to the cold-adapted functionality of the structure of ribosomes.

Studies on the responses of cyanobacteria to cold stress (Murata and Wada, 1995; Murata and Los, 1997; Los and Murata, 1998, 1999, 2000, 2004; Prakash *et al.*, 2010; Los *et al.*, 2010) have contributed to our understanding of the following six areas. These are as follows: (i) fatty acid desaturases that are responsible for adjustments in membrane fluidity; (ii) Rbps that probably function as RNA chaperones (similar to the Csps of *E. coli* and *Bacillus subtilis*); (iii) RNA helicases that help in the removal of secondary structures of mRNAs and improve efficiency in translation at the low temperature; (iv) adjustments in the structure of ribosomal proteins so that the translation efficiency is improved at the low temperature; (v) Clps that participate in the renewal of PSII and (vi) genome-wide gene expression patterns and the recognition of stress sensors and transducers.

i) Studies on the physiology of cold stress: The earliest studies on cold stress relate to the treatment of *Anacystis nidulans* cells to chilling temperatures (0°C–5°C) resulting in loss of viability (Rao *et al.*, 1977), a decrease in photosynthetic O_2 evolution, electron transport and phosphorylation (Forrest *et al.*, 1957; Jansz and Maclean, 1973; Rao *et al.*, 1977) and change in the light absorption spectrum of carotenoids (Brand, 1977). Observations on the release of certain amino acids such as glutamate and pteridines have also been reported (Forrest *et al.*, 1957; Jansz and Maclean, 1973). A chilling treatment (0°C for 10 min) of *A. nidulans* resulted in an inactivation of photosynthesis and Hill reaction with 1,4-benzoquinone and a modification of light absorption spectrum of carotenoids. These activities very much depended on the growth temperature and the temperature critical for chilling susceptibility. Accordingly, the mid point values for critical temperatures for chilling susceptibility were 4, 5 and 12°C for cells grown at 28, 33 and 38°C, respectively. One of the important parameters is the spectral change of carotenoids that represents a phase change in the cytoplasmic membrane due to chilling. Isolated membrane fragments and liposomes prepared from the lipids extracted from *A. nidulans* also showed similar alterations in spectral properties of carotenoids. According to Ono and Murata (1981a) besides the damage caused to photosynthetic system, the leakage of small solutes from the cytoplasmic membrane might be responsible for the loss of viability of *A. nidulans* subjected to chilling. Thus the chilling susceptibility has been correlated with the fluidity of membrane lipids in *A. nidulans*. Subsequently, Ono and Murata (1981b) reported that the loss of viability due to chilling has been due to a leakage of K^+ and Mg^{2+} and amino acids through the cytoplasmic membrane. The growth temperatures undertaken were 28 and 38°C for which the mid point values for critical temperature for chilling susceptibility were 5 and 14°C, respectively. Chilling-induced damage to photosynthetic electron transport and Hill reaction could be restored by suspending the shocked cells in presence of K^+ and Mg^{2+}.

Four abundant glycerolipids have been identified in the thylakoid membranes of chloroplasts and in the cells of cyanobacteria. These are monogalactosyl diacylglycerol (MGDG), digalactosyl diacylglycerol (DGDG), sulfoquinovosyl diacylglycerol (SQDG) and phosphatidylglycerol (PG; Fig. 11). These help in the maintenance of photosynthetic electron transport system. The association of SQDG and MGDG with ATP synthase and PSII reaction center complex, respectively have been known (Pick *et al.*, 1987; Murata and Siegenthaler (2004). Due to the presence of higher levels of 18:3 (9, 12, 15) fatty acids in the lipids of photosynthetic membranes of cyanobacteria, phase transition of these membranes occurs far below room temperature. Photosynthetic membranes of cyanobacteria also contain proportionately higher amounts of proteins and this leads to a higher phase-transition temperature of lipids (Ono and Murata, 1982; Omata and Murata, 1983). A comparison of phase-transition of cytoplasmic membranes with photosynthetic membranes of cyanobacteria revealed that the former exhibited a lower phase-transition temperature than the latter (Wada *et al.*, 1994). In support of the suggestion that the phase separation of lipids in the cytoplasmic membrane is indeed responsible for loss of photosynthetic activities and viability of cells, Ono and Murata (1981b) presented freeze-fracture electron micrographs of *A. nidulans* subjected to chilling injury. The existence of membrane lipids in liquid crystalline state above and in phase separation state below the critical temperatures for chilling susceptibility have been noted, i.e. at 14°C and 4°C for cells grown at 38°C and 28°C, respectively. When membrane lipids are in liquid-crystalline state, the intramembrane particles are distributed at random in the fracture faces but when membrane lipids are in the phase separation state the particles are displaced from the domain of gel phase. Accordingly, cells

1) Monogalactosyl diacylglycerol (MGDGI)

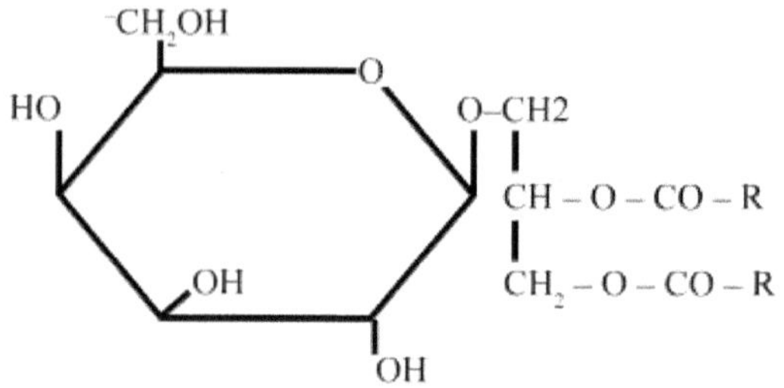

2) Digalactosyl diacylglycerol (DGDG)

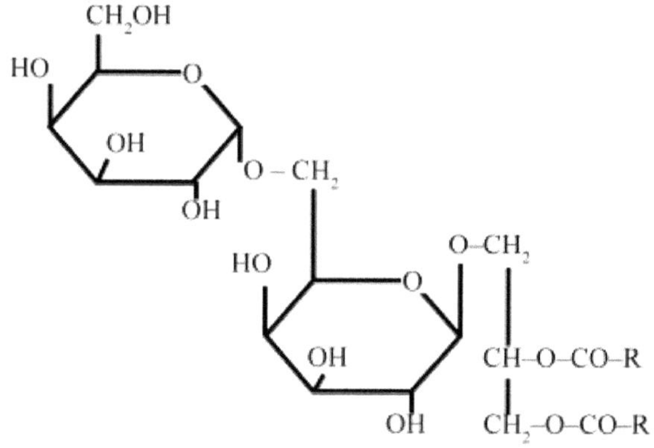

3) Sulfoquinovosyl diacylglycerol (SQDG)

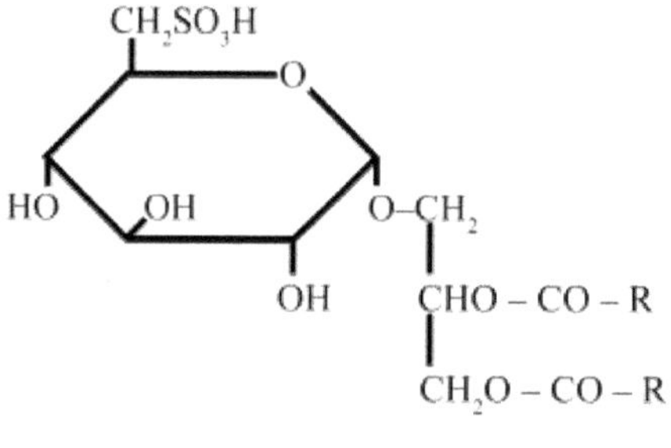

4. Phosphatidylglycerol (PG)

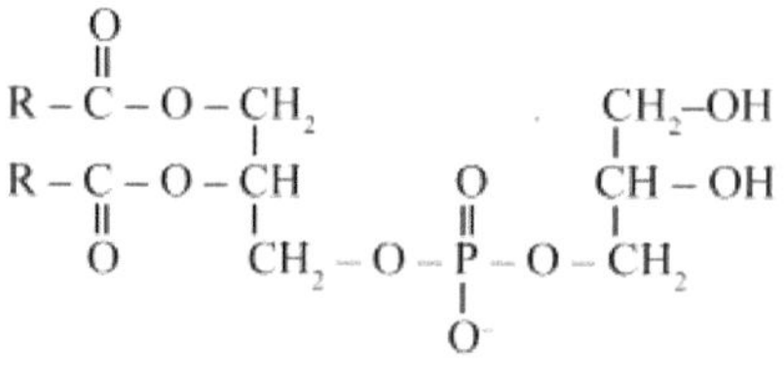

Figure 11: Molecular structure of the four major glycerolipids.

subjected to chilling injury revealed fracture faces containing particle-free and particle-containing regions (phase-separation state). The cells from normal growth temperatures revealed the existence of particles distributed at random (liquid-crystalline state). Moreover, *Anabaena variabilis*, a species that is not susceptible for chilling, revealed intramembrane particles distributed at random in cells derived from growth temperatures as well as chilling treatments. Further, the studies of Murata *et al.* (1979) and Sato *et al.* (1979) have confirmed that when growth temperatures are lowered a decrease in saturated fatty acids takes place with a concomitant increase in the unsaturated fatty acids. Ono and Murata (1982) observed that due to variation in growth temperature the fatty acid composition also varies and when double bonds in the fatty acids of lipids are introduced it leads to a decrease in the thermotrophic phase transition of the membranes. So the role of fatty acid desaturases in introducing double bonds in the fatty acids to produce more unsaturated fatty acids and their role in low temperature tolerance has received attention. A number of studies have correlated changes in fatty acid composition of the membrane lipids to the ambient growth temperature (Murata *et al.*, 1979, 1992a,b; Sato and Murata, 1980, 1981; Wada and Murata, 1990) and this has been regarded as an adaptive response (Murata and Nishida, 1987; Murata, 1989). Cossins (1994) termed this as homeoviscous adaptation.

ii) Role of fatty acid desaturases: Desaturases introduce double bonds into fatty acids and thus convert saturated fatty acids into unsaturated ones. Three types of desaturases, i.e. acyl-CoA desaturases, acyl-ACP (acyl-carrier protein) desaturases and acyl-lipid desaturases are known. The first type is present bound to the endoplasmic reticulum of animal, yeast and fungal cells. These introduce double bonds into fatty acids bound to coenzyme A. The second type, acyl-ACP desaturases, is present in the stroma of plant plastids and these can introduce double bonds into fatty acids that are bound to ACP. The third type acyl-lipid desaturases introduce double bonds into fatty acids of glycerolipds. They are bound to the endoplasmic reticulum and the chloroplast membranes of plant cells as well as the thylakoid membranes of cyanobacteria. It is the acyl-lipid desaturases that respond to a shift-down of growth temperature and convert saturated fatty acids into unsaturated fatty acids there by increasing the fluidity of the membranes. Murata *et al.* (1992) classified cyanobacterial strains into four groups based on the patterns of fatty acid desaturation. *S. elongatus* PCC 7942 (*A. nidulans* R2), *Synechococcus vulcanus* (a thermophilic cyanobacterium) and *Mastigocladus laminosus* belong to group 1 that can only introduce a double bond at Δ^9 position of fatty acids either at the *sn*-1 or *sn*-2 position during a 10–14 h temperature shift-down. In group 2, *Synechococcus* sp. strain PCC 7002, *A. variabilis* M3, *Plectonema boryanum* and *N. muscorum* are included. These strains can introduce double bonds at the Δ^9, Δ^{12} and Δ^{15} (ω^3) positions of the C18 acids at the *sn*-1 position and at the Δ^9, Δ^{12} positions of C16 acids at the *sn*-1 position. *Synechococcus* sp. strain PCC 7002 has three acyl-lipid desaturase genes (*desA*, Δ^{12} desaturase; *desB*, ω^3 or Δ^{15} desaturase and *desC*, Δ^9 desaturase). In case of *A. variabilis* M3, conversion of 16:0 to 16:1Δ^9 occurred in light as well as in dark due to cold shock and polyunsaturated fatty acids 16:2$\Delta^{9,12}$ and 18:3$\Delta^{9,12,15}$ also increased (Sato and Murata, 1980). Group 3 consists of *Synechocystis* sp. strain PCC 6714 and *S. platensis* that can introduce three double bonds at $\Delta^{6,9,12}$ positions of C18 acids at *sn*-1 position. *Synechocystis* sp. strain PCC 6803 and *Tolypothrix tenuis* belong to group 4 and these strains can introduce double bonds at the $\Delta^{6,9,12,15}$ positions of the C18 acids at the *sn*-1 position. *Synechocystis* sp. strain PCC 6803 contains four genes for acyl-lipid desaturases that have been cloned and their expression and activities noted in *E. coli*. These are *desA* (Wada *et al.*, 1993), *desB* (Sakamoto *et al.*, 1997a,b), *desC* (Sakamoto *et al.*, 1994) and *desD* (Reddy *et al.*, 1993) encoding acyl-lipid desaturases that introduce double bonds specifically at Δ^{12}, ω^3, Δ^9 and Δ^6 positions of C18 fatty acids, respectively. A temperature shift-down (38°C to 22°C) resulted in the desaturation at ω^3 position due to which the accumulation of 18:3$\Delta^{9,12,15}$ and 18:4$\Delta^{6,9,12,15}$ acids was noted (Wada and Murata, 1990).

Wada and Murata (1989) first isolated mutants of *Synechocystis* sp. strain PCC 6803 defective in fatty acid desaturases (FAD) at the 6th and 12th positions and designated them as Fad6 and Fad12, respectively. These mutants exhibited growth to the same extent as that of the wild-type at 34°C but they differed in their growth rate at 22°C. The growth rate of Fad12 was much lower than wild-type whereas Fad6 grew to the same rate as that of the wild-type. Fad6 mutant lacked polyunsaturated fatty acids 18:3$\Delta^{6,9,12}$ whereas Fad12 lacked 18:2$\Delta^{9,12}$ and 18:3$\Delta^{6,9,12}$ but contained low levels of 18:2$\Delta^{6,9}$ fatty acids. Since Fad6 mutant grew to the same extent as that of the wild-type at 22°C, the absence of 18:3 fatty acid did not make any difference but at the same time due to low levels of synthesis of 18:2 fatty acid in Fad12 the growth was lower than wild-type. So the importance of 18:2 for growth at low temperature has been recognized. The *desA* gene from *Synechocystis* sp. strain PCC 6803 has been cloned and used to transform *S. elongatus* PCC 7942 (which possesses only Δ^9 desaturase). The transformants acquired the ability to desaturate at Δ^{12} position also and synthesized 16:2$\Delta^{9,12}$ and 18:2$\Delta^{9,12}$ fatty acids and exhibited more cold tolerance (Wada *et al.*, 1990, 1994). The transformant *desA*$^+$ cells also appeared to be more tolerant to photoinhibition (Gombos *et al.*, 1997). The fact that *Synechocystis* sp. strain PCC 6803 can synthesize mono-, di- and triunsaturated fatty acids in its lipids (Wada and Murata, 1989) has been further exploited by the isolation of mutants deficient in desaturases (see Table 4, Chapter 2 for details of number and types of genes of desaturases in various cyanobacteria). Mutant Fad6, defective in desaturation of fatty acids at the Δ^6 position, synthesized only mono- and di-unsaturated fatty acids (Wada and Murata, 1989). This mutant was transformed by an interrupted gene sequence of *desA* with kanamycin resistance marker (*Km*^{r}) and the transformant was designated as Fad6/*desA*:*Km*^{r} (Wada *et al.*, 1992). The transformants defective in desaturation at the Δ^6 and Δ^{12} positions produced only monounsaturated fatty acids in its lipids. A comparison of the fatty acid profiles of wild-type, Fad6 and Fad6/*desA*:*Km*^{r} revealed that the content of saturated fatty acids (16:0) were 58, 59 and 57 mol% , respectively. While Fad6 showed 11 mol % of monounsaturated fatty acid 18:1Δ^9 and 25 mol% of diunsaturated fatty acid 18:2$\Delta^{9,12}$, the wild-type showed 7 mol% and 12 mol% of the respective mono- and disaturated fatty acids. The transformant Fad6/*desA*:*Km*^{r} produced higher levels of monounsaturated fatty acid up to 41 mol% and di- and polyunsaturated fatty acids were completely absent. At low temperature, Fad6 and transformant Fad6/*desA*:*Km*^{r} experienced severe impairment in growth due to a phenomenon termed as low temperature photoinhibition (Gombos *et al.*, 1992). Likewise, Wada *et al.* (1994) transformed *A. nidulans* R-2 SPc with *desA* gene from *Synechocystis* sp. strain PCC 6803 and demonstrated that the transformants of *A. nidulans* acquired the capability to introduce a second double bond into palmitoleic (16:1Δ^9) and oleic (18:1Δ^9) acids at the Δ^{12} position and at the same time showed tolerance to chilling stress.

Tasaka *et al.* (1996) isolated a series of mutants of *Synechocystis* sp. strain PCC 6803 defective in acyl-lipid desaturases and presented evidences in support for the role of polyunsaturated membrane lipids in growth, respiration and photosynthesis. Mutants deficient in Δ^{12} desaturase (*desA* gene) and Δ^6 desaturase (*desD* gene) showed increased levels of monounsaturated oleic acid at the expense of polyunsaturated fatty acids such as di-, tri- and tetraunsaturated fatty acids. The growth of these mutants was similar to that of the wild-type at 35°C but they did not grow well at 25°C. The *desA*$^-$/*desD*$^-$ mutants were unable to grow at 15°C probably due to their non-recovery from photo-induced damage to the PSII complex. Their inability to process precursor of D1 protein might be responsible for a short supply of mature D1 protein which is an essential component of the reaction center of the PSII complex. The transformants of *S. elongatus* PCC 7942 with *desA* gene from *Synechocystis* sp. strain PCC 6803 described above (Wada *et al.*, 1990, 1994), showed replacement of D1:1 isoform of D1 protein by D1:2 isoform within few h of cold acclimation (Campbell *et al.*, 1995). Thus it can be reasonably concluded that transformation of *S. elongatus* PCC 7942 with *desA* gene shifted the temperature critical for the replacement of D1:1 isoform by D1:2 isoform (Sippola *et al.*, 1998). Since it was not known as to whether the desaturases act on fatty acids in the glycerolipids of cytoplasmic membrane or thylakoid membranes or both, Mustardy *et al.* (1996) conducted immunochemical localization studies on acyl-lipid desaturases. A temperature shift-down resulted in the enhanced expression of desaturase genes followed by de novo synthesis of desaturases and the desaturases targeted to both the cytoplasmic as well as thylakoid membranes. So it means any qualitative or quantitative change that takes place in the fatty acid composition of glycerolipids due to a cold shock or chilling, it occurs both in the cytoplasmic as well as thylakoid membranes.

Sakamoto and Bryant (1998) demonstrated that it is not due to photoinhibition that the cyanobacterial cells show impairment in growth processes when subjected to low temperature. *S. elongatus* PCC 7002 exhibited chlorotic symptoms associated with a decrease in total amounts of phycobiliproteins and chlorophyll *a* when cultured at 15°C with nitrate as sole source of nitrogen. These symptoms were observed in cultures grown at low (50 µE m^{-2} s^{-1}) as well as high light (250 µE m^{-2} s^{-1}) intensities. When the chlorotic cultures were shifted from 15 to 38°C, the cultures quickly regained their blue-green colour. Moreover, due to the occurrence of exponential growth in presence of urea at 15°C, it was suggested that the chlorosis observed at the low temperature was due to a nitrogen limitation and not due to limited photosynthetic activity or photodamage caused to the photosynthetic apparatus. Extending their studies on this aspect, the growth performance, photosynthetic activity and nitrate assimilation of *S. elongatus* PCC 6301 were examined at different temperatures (15, 20 and 30°C). Growth rate decreased with decreasing temperature but it ceased completely at 15°C. Photosynthetic efficiency measured in terms of O_2 evolution rapidly declined at very high light intensity (3 mE m^{-2} s^{-1}) but at normal light intensity (250 µE m^{-2} s^{-1}) there was no inhibition in O_2 evolution. However, cells could not actively take up either nitrate or nitrite at 15°C, although nitrate and nitrite reductases were quite active inside the cells. So growth inhibition at the low temperatures is associated with inactivation of nitrate/nitrite transporters rather than due to photoinhibition (Sakamoto and Bryant, 1999).

Kiseleva *et al.* (2000) identified the presence of only one fatty acid desaturase gene (*desC* for Δ^9 desaturase) in the thermophilic cyanobacterium *S. vulcanus* which has an optimum growth temperature of 55°C at which the cells exhibited very low levels of mRNA of *desC* gene. After a temperature shift-down to 45°C, the level of the transcript increased by 10-fold accompanied by an increase in the enzyme. However, at 35°C the levels of both the transcript as well as the protein decreased by 3–4 times that was found at 45°C. But instead of a 10° shift, a sudden shift-down to 35°C was not accompanied by the accumulation of *desC* transcript or the enzyme. These workers suggested that the pre-existing enzyme was sufficient to bring about the required desaturation reaction and that the Δ^9 desaturase retains its activity at vey low temperatures. The cloning of the *desC* gene from *S. vulcanus* and its expression studies in *E. coli* demonstrated that the enzyme converted stearic acid into oleic acid. The protein has only 40–45% homology with known *desC* desaturases from other mesophilic cyanobacteria.

Extensive studies on the expression of *desA*, *desB* and *desD* genes of *Synechocystis* sp. strain PCC 6803 after a temperature shift-down from 34°C to 22°C have been conducted (Los and Murata, 1994; Los *et al.*, 1997). The level of transcripts for *desA* gene increased by 10-fold within 30 min of transfer to the low temperature but when the cultures were shifted back to 34°C the transcripts for *desA* gene disappeared rapidly. Northern and Western blotting analyses revealed that the level of *desD* mRNA increased 10-fold within 15 min of a temperature shift-down and the level of the Δ^6 desaturase enzyme doubled within 4 h of the start of cold treatment. The *desB* gene, that encodes ω^3 desaturase, responded more quickly to the temperature shift-down as the transcripts started appearing within 10 min and the mRNA levels increased 15-fold higher. This has been ascribed to be due to enhanced transcription as well as to the stabilization of the transcript. The *desB* mRNA levels gradually declined when the cells were shifted to high temperature (Los and Murata, 1994; Los *et al.*, 1997). Sakamoto and Bryant (1997) demonstrated cold-inducibility of the *desC* gene in *Synechococcus* sp. strain 7002. When the cultures of this cyanobacterium were shifted from 38°C to 22°C, the level of *desC* mRNA increased within 5 min. The *desC* gene, that encodes Δ^9 desaturase, in *S. elongatus* PCC 7942 has been shown to be induced within 30 min after a temperature shift from 36°C to 24°C (Ishizaki-Nishizawa *et al.*, 1996). Other important features of *des* gene expression in *Synechocystis* sp. strain PCC 6803 are: (i) transcription start sites for the *desA*, *desB* and *desD* genes are located at –114,–35, and –347 bp positions at both 34°C and 22°C; (ii) RNA polymerase utilizes the same promoters at both temperatures and (iii) the existence of a consensus sequence of 5′-GTTTGTTTT-3′ downstream the start sites for the cold-inducible desaturases.

The expression of three desaturase genes (*desA*, Δ^9 desaturase; *desB*, ω^3 or Δ^{15} desaturase and *desC*, Δ^9 desaturase) and the fatty acid composition were compared at two growth temperatures (38°C and 22°C) in two wild-type strains of the marine cyanobacterium *Synechococcus* sp. strain PCC 7002 (NIBB and PR6000). The levels of mRNA of the three genes increased within 15 min after a temperature shift-down to 22°C but these soon disappeared after a shift-up to 38°C. The overall fatty acid composition of the cells gown at 38°C was identical at low (50 $\mu E\ m^{-2}\ s^{-1}$) and high light (250 $\mu E\ m^{-2}\ s^{-1}$) intensities. However, ω^3 desaturation increased in cells grown at both light intensities at only 22°C. This signifies that fatty acid desaturation at the ω^3 position is regulated by the low temperature-dependent expression of *desB* gene. Based on the above results, Sakamoto *et al.* (1997) concluded that the rate-limiting step for increasing the desaturation of membrane lipids at the low temperature is post-transcriptional.

From the above discussion it is clear that a lowering of growth temperature increases the fluidity of the cytoplasmic membrane due to the synthesis of polyunsaturated fatty acids. What exactly is the threshold limit for perception of a decrease in temperature? A comparative study of *Synechocystis* sp. strain PCC 6803 grown at two different temperatures (36°C and 32°C) for fatty acid content and expression of *desA* gene revealed that the cells grown at 32°C contained more unsaturated fatty acids than the cells grown at 36°C. This suggests that the fluidity of membranes has increased at 32°C. When cells grown at 36°C were shifted to 28°C, the expression of *desA* gene was not noted whereas in the cells grown at 32°C, the expression of *desA* gene occurred at 26°C. It means the cells of *Synechocystis* sp. strain PCC 6803 were able to sense a change in temperature only when it exceeded 6°C (Los *et al.*, 1993). Vigh *et al.* (1993) examined the effects of hydrogenation of cells of *Synechocystis* sp. strain PCC 6803 grown at 36°C for 4 min. This treatment converted 5% of the unsaturated fatty acids to saturated fatty acids in the cell membrane but not in the thylakoid membranes. As a result, the membrane fluidity decreased and the *desA* gene expression took place at 30°C instead of 28°C. It means hydrogenation of lipids in plasma membrane increased the threshold temperature for the expression of *desA* gene by 2°C. Likewise, a temperature-dependent expression of the *desB* gene was compared between the wild-type and *desA*$^-$/*desD*$^-$ mutant of *Synechocystis* sp. strain PCC 6803. Wild-type cells grown at 36°C showed expression of *desB* gene at 28°C whereas *desB* gene was expressed in mutant *desA*$^-$/*desD*$^-$ cells at 32°C. Since the mutant contained only monounsaturated fatty acids, the fluidity of the cytoplasmic membrane decreased and the cold-induced expression of *desB* gene was shifted toward a lower temperature by 5°C (Los and Murata, 1999). The cold-inducible

Δ^9 desaturase gene (*desC*) also showed a temperature-dependent expression in the wild-type and *desA*$^+$ transformants of *S. elongatus* PCC 7942. Since the wild-type was able to synthesize only monounsaturated fatty acids, the *desC* gene expression occurred at 30°C. However, *desA*$^+$ transformant cells synthesized diunsaturated fatty acids at the expense of monounsaturated fatty acids and in these the induction of *desC* gene took place at 28°C. Due to the presence of diunsaturated fatty acids, an increase in the fluidity of membrane shifted the cold-inducible expression of *desC* gene towards higher temperatures by 2°C. The enhanced level of expression of desaturase genes and consequently elevated levels of unsaturated fatty acids at the low temperature has been explained to be due to negative temperature coefficients of the desaturases. That is why desaturases can be more active at lower temperatures (Murata and Wada, 1995).

The regulatory role of light has been emphasized in the induction of acyl-lipid desaturases in *Synechocystis* sp. strain PCC 6803 (Kis *et al.*, 1998). Cells grown photoautotrophically at 25°C in normal (70 μmol m^{-2} s^{-1}), moderately strong (500 μmol m^{-2} s^{-1}) and strong (2000 μmol m^{-2} s^{-1}) light did not show any difference in the level of fatty acid unsaturation. Light-activated heterotrophic growth (with 5 mM glucose in dark at 25°C and illuminated once daily for 10 min with a light intensity of 40 μmol m^{-2} s^{-1}) supported a 2-fold increase in monounsaturated fatty acids at the expense of polyunsaturated fatty acids. When compared to photoautotrophically grown cultures where the level of 18:1 fatty acid was 7%, in dark grown cultures these fatty acids increased to 17%. More significantly, in dark grown cultures, the level of α 18:3 and 18:4 fatty acid levels were reduced to traces. Temperature or light did not influence the level of *desC* gene transcripts for Δ^9 desaturase as it seems to be constitutively expressed at 25°C with or without illumination. The expression of *desB* gene in dark-grown cells was negligible but in light the mRNA of this gene increased by 10-fold reaching a maximum level within 15 min. The level of mRNAs for *desA* and *desD* were higher than the transcripts of *desB* but the expression of the former two in light increased by 10-fold. When photoaututrophically grown cultures at 35°C were transferred to 25°C in dark, the induction of *des* genes did not occur. It means the low-temperature-induction was blocked in the dark.

Studies of Yin *et al.* (2007) showed that the challenge from chill plus light stress could be overcome in *Synechocystis* sp. strain PCC 6803 by the presence of a gene *sll1242* named as *ccr–1* (cyanobacterial cold resistant) gene. *Synechocystis* sp. strain PCC 6803 can survive for >2 months at 5°C in the absence of light but in presence of light (100 μmol photons m^{-2} s^{-1}) at 5°C the cells completely lose their viability within 10 days. The tolerance of a strain to 5°C in presence (100 μmol photons m^{-2} s^{-1}) or absence of light is determined on the basis of its ability to reinitiate growth (ARG) after an exposure for 5 days. Photoautotrophic and mixotrophic (in presence of glucose) growth of such strains was assayed at a photosynthetic photon flux of 30 μmol photons m^{-2} s^{-1} light intensity at 30°C. From a random insertion mutant library (6000 clones) of *Synechocystis* sp. strain PCC 6803, 14 mutants have been isolated that exhibited increased chill-light sensitivity with an ARG of less than 20% of the wild-type. Inverse PCR and sequencing confirmed the presence of an interrupted gene sequences in six mutants, out of which mutant *sll1242* was selected for a further study. This mutant showed increased sensitivity to chill plus light and was unable to grow photoautotrophically and mixotrophically at 15°C whereas the wild-type could do so at different temperatures (30, 20 and 15°C). The photosynthetic efficiency of the mutant *sll1242* greatly decreased. Complementation of the wild-type gene sequence in the mutant *sll1242* confirmed the role of this gene in maintaining chill plus light tolerance. Another mutant with interrupted gene sequence for *desD* marker designated as *desD*:C.K 2 also showed increased sensitivity to chill-light stress. After a chill-light stress for 5 days, the ARG value of this mutant decreased to 5.6% of that of wild-type. Complementation of the mutant by *desD* gene increased the viability and the ARG reached up to 113.6%. This probably reflects on the role of

18:3 fatty acids in chill-light tolerance. The essentiality of α-tocopherol in tolerating chill-light stress has ben identified by subjecting the mutants (*slr1736, slr1737, slr0089, slr0090)* unable to synthesize α-tocopherol to chill-light stress. Acquired chill-light tolerance (ACLT) is a property that is exhibited by the wild-type when the cells are pre-conditioned at a temperature of 15°C. This pre-conditioning enhances considerably the cellular levels of α-tocopherol. The wild-type and the mutants exhibited almost the same growth pattern when subjected to chill in dark but when subjected to chill in light the mutants completely lost the ACLT. Complementation of the *slr0089* mutant with the wild-type gene regained the ACLT thus showing the essential nature of α-tocopherol in conferring chill-light tolerance to *Synechocystis* sp. strain PCC 6803 (Yang *et al.*, 2008).

***iii) Microarray studies*:** Suzuki *et al.* (2000 a,b) carried out a systematic disruption of putative genes for Hiks and identified Hik33 (*hik33*) as a regulator of the cold-inducible expression of *desB* gene in a strain that carried a reporter gene for bacterial luciferase. Genome-wide microarrays revealed that close to 50 genes are strongly induced in *Synechocystis* sp. strain PCC 6803 under cold stress. In addition to these, several other genes that are induced at low temperature belong to genes of known function, those responsible for oxidative stress and several others of unkown function. The important genes of known function that are induced at low temperature are genes for: (i) RNA polymerase (*rpo*); (ii) sigma factor D (*sigD*); (iii) elongation factor EF-G (*fus*); (iv) high-light-inducible proteins (*hliA, hliB, hliC*); (iv) subunit 4 of NADH dehydrogenase and (v) alternative form of cytochrome *c* (*cytM*) (Suzuki *et al.*, 2001). The number of genes under the control of Hik33 were found to be 23 out of 38 highly cold-inducible genes. It means the rest 15 of the 38 cold-inducible genes were not regulated by Hik33. So it was suggested that *Synechocystis* sp. strain PCC 6803 might possess another sensor or pathway for transduction of the low-temperature signal (Suzuki *et al.*, 2001). At this juncture, it is important to note that the Hik33 contains two transmembrane domains and several conserved domains, i.e. a HAMP-linker, a leucine zipper, a PAS domain and a histidine kinase domain (Williams and Stewart, 1999; Taylor and Zuhlin, 1999). The HAMP-linker possesses two helical regions that are responsible for transducing the stress signals via intramolecular structural changes between the two helical regions and leads to the intermolecular dimerization of the protein (Williams and Stewart, 1999; Arvind and Ponting, 1999; Arvind *et al.*, 2003). Inaba *et al.* (2003) conducted fourier transform infrared (FTIR) spectrometry of the cytoplasmic membranes of the double mutant *desA⁻/desD⁻* of *Synechocystis* sp. strain PCC 6803 (isolated by Tasaka *et al.*,1996) and showed that due to double mutation the cytoplasmic membrane acquired more rigidity than those from the wild-type cells at 22°C. Microarray analysis of cold-inducible genes in wild-type and the double mutant enabled them to recognize three groups of genes. The first group consisted of those genes that were induced by cold in the wild-type but were strongly cold-inducible in the *desA⁻/desD⁻* cells. Included in this group are certain heat shock genes (*hspA, dnaK2,* and *clpB2*), genes of the sulphate transport system (*sbpA, cysA, cysT* and *cysW*), RNA polymerase sigma factor (*sigB*), a histidine kinase (*hik34*) and a gene for a putative penicillin-binding protein (*psp*). The second group consisted of genes whose cold-inducibility is moderately enhanced by the double mutation and these are genes for high-light-inducible proteins (*hli* genes) and some genes of unknown function. The cold-inducibility of the third group of genes was unaffected by the double mutation and included here are cold shock genes for RNA-binding protein (such as *rbp1*) and cyanobacterial RNA helicase (*crhL*). Maximum induction of *hspA* and *dnaK2* was noted in *desA⁻/desD⁻* mutant cells at 24°C but wild-type cells did not show such induction. The cold-inducibility of *rbp1* (25°C) and *crhL* (~20°C) was unaffected in the double mutant. The role of *hik33* as a membrane sensor to detect a cold stress signal by sensing the rigidity of cytoplasmic membrane has been evaluated by the generation of another mutant by the inactivation of *hik33* gene in *desA*⁻/*desD*⁻ double mutant. Microarray analysis of *desA*⁻/*desD*⁻/*hik33*⁻ conducted after a cold shock revealed that the mutation of *hik33* abolished or reduced the cold-inducibility of 10 of the 17 genes in the second group (the expression of Hik33 regulated genes *hliA*, *hliB* and *sigD* was no longer inducible by cold) and 7 of the 25 genes in the third group. However, the inactivation of *hik33* did not affect significantly the cold-inducibility of 15 genes in the first group except for *hspA* and *cysA*. These results point out that most of the genes included in the group 2 and 3 are under the regulation of *hik33* and that very few genes of the first group are under the control of *hik33*. Los and Murata (2004) concluded that Hik33 perceives a decrease in membrane fluidity as the primary signal of cold stress. Since the transmembrane domains of Hik33 are associated with the lipid phase of the cytoplasmic membrane, it is likely that these two domains sense changes in the cytoplasmic membrane rigidity (Los *et al.*, 2008; Los and Zinchenko, 2009). To identify the corresponding Rre for Hik33, a Rre knock-out library (that contains all the Rre genes of *Synechocystis* sp. strain PCC 6803 in a disrupted state) was screened by RNA slot-blot hybridization. The cold-inducible genes whose expression is governed by Hik33 were chosen as the probes. In this process, Rre26 emerged as the candidate for the corresponding Rre for Hik33 and this constitutes the two-component (Hik33-Rre26) system for perceiving and transducing the cold stress signal (Murata and Los, 2006). The observation that Rre26 binds to the promoter region of *hliB* gene suggests that it may be involved in the transduction of the low-temperature signal (Kappel and van Waasbergen, 2007).

Studies conducted on gram-positive and gram-negative bacteria showed that the status of supercoiling of genomic DNA plays an important role in the regulation of gene expression in response to environmental stresses (Higgins, 1988; Wang and Lynch, 1993; Dorman, 2006). In cyanobacteria the temperature-dependent alterations in DNA supercoiling have been recognized as an important signal for gene expression in response to cold stress (Los, 2004; Prakash *et al.*, 2009). Inhibitors of DNA gyrase such as novobiocin have been used to study the changes in negative supercoiling as well as genome-wide expression of genes in *Synechocystis* sp. strain PCC 6803 during cold stress. An increase in the negative supercoiling of the promoter region of the the *desB* gene during cold stress controlled its expression at low temperatures. Since novobiocin inhibits the stress-induced changes in DNA supercoiling, there is a lack of transcription of many genes involved in cold stress. The expression of genes for Hik33, Hik34, *crhR* and *rbpA1* which are known to be obligatory for tolerating cold stress could be prevented by the inhibition of DNA gyrase (Prakash *et al.*, 2009).

iv) Cold shock proteins: The induction of a separate group of proteins known as Csps occurs during cold stress (Jones *et al.*, 1992, 1996; Von Bogelen and Neidhart, 1990). *E. coli* is the best example to understand the role of Csps. CspA is a small 5-kDa protein that is exclusively synthesized at low temperature. It binds to single-stranded DNA and is thought to function as either a general transcriptional activator of the cold shock regulon in *E. coli* (La Teana *et al.*, 1991) or an RNA chaperone (Jones and Inouye, 1994). Homologues of CspA have now been identified in many different eubacteria (Av-Gay *et al.*, 1992; Ray *et al.*, 1994; Willimsky *et al.*, 1992). Of the nine-member *E. coli* Csp family, CspA constitutes 10% of the total proteins during a cold shock (Jiang *et al.*, 1997). The three-dimensional structure of CspA reveals a five-stranded β-barrel structure (Newkirk *et al.*, 1994; Schindelin *et al.*, 1994) with two consensus RNA-binding motifs (RNAP1 and RNAP2). These help in the recognition and binding to RNA (Schroder *et al.*, 1995). Jiang *et al.* (1997) suggested that CspA prevents RNA secondary structure formation that helps in enhancing the translation at low temperatures. Nucleic acid-binding cold shock domain proteins (CSDs) are of wide spread occurrence in plants (Karlson and Imai, 2003). The CSDs encompass bacterial Csps and bind to single-stranded DNA/RNA and double-stranded DNA (Graumann and Marahiel, 1996). Graumann and Marahiel

(1998) speculated that the CSDs are ancient structures that existed even before the divergence of prokaryotes and eukaryotes. It is interesting to note that cyanobacteria do not possess CSDs but instead possess Rbps (Sato, 1995) that also possess a three-dimensional functional surface required for nucleic acid binding (Graumann and Marahiel, 1996). During the course of evolution, Rbps may have replaced CSDs in cyanobacteria (Graumann and Marahiel, 1998).

a) RNA-binding proteins: A gene (*rbpA*) that encodes an Rbp was for the first time reported by Sato (1994) in *A. variabilis* M3. This gene has been subsequently redesignated as *rbpA1* (Sato, 1995) and was shown to be organized in an operon with a ribosomal protein gene *rpsU*. A temperature shift-down from 38 to 22°C caused a drastic increase in the level of the transcript of *rbpA1* and it disappeared rapidly after a shift-up from 22°C to 38°C. The RbpA protein (it may be noted that the protein name Rbp is derived from the gene name '*rbp*' and it is not an abbreviation for 'RNA-binding protein' hence we can say Rbp protein) that was expressed in *E. coli* exhibited a strong affinity with RNA isolated from *Anabaena* cells. The RbpA protein contained only one RNA-recognition motif (RRM) with a short C-terminal glycine-rich sequence. This constitutes one of the smallest among known Rbps. Other cyanobacteria which are reported to possess Rbp genes are *S. elongatus* PCC 6301 (Sugita and Sugiura, 1994), *Chlorogloeopsis* sp. PCC 6912, *Anabaena* sp. strain PCC 7120 (Mulligan *et al.*, 1994), *Synechocystis* sp. strain PCC 6803 (Kaneko *et al.*, 1996), *S. elongatus* PCC 7942 and *Synechococcus* sp. strain PCC 7002 (Sato *et al.*, 1997a). In cyanobacteria there appears to be a correlation between the copy number of *rbp* genes and the size of the genome of the organism. *S. elongatus* PCC 6301 with a 2.6 Mb genome possesses two copies of *rbp* genes. As the genome size of *Synechocystis* sp. strain PCC 6803 is 3.6 Mb the number of copies of rbp genes increased to three (Kaneko *et al.*, 1996). In *A. variabils* M3, *Anabaena* sp. strain PCC 7118 and *Anabaena* sp. strain PCC 7120 with an estimated genome size of 6.4 Mb in each case, the number of *rbp* genes increased to eight (Bancroft *et al.*, 1989). Since the copy numbers of *rbp* genes seem to be consistent with the increase in genome size, the increase in copy number of *rbp* genes could have resulted due to duplication of the whole genome and not by simple gene duplication events. These genes encode small Rbps with one RRM with a short C-terminal glycine rich sequence bearing resemblance to the RbpA initially reported by Sato (1994). The Rbps seem to be unique for cyanobacteria as among the prokaryotes *Helicobacter pylori* (Tomb *et al.*, 1997) and *Treponema pallidum* (Fraser *et al.*, 1998) are the only other bacteria known to possess the *rbp* genes. It is interesting to note that the cyanobacterial RRM sequences are very much similar to chloroplast RRMs which supports the endosymbiotic origin of chloroplast (Sugita and Sugiura, 1994; Mulligan *et al.*, 1994).

Southern blot and immunoblot analyses revealed the existence of at least eight genes in *A. variabilis* M3 that encode Rbp proteins. Of the eight, the cloning of five of them (*rbpA1*, *rbpA3*, *rbpB*, *rbpC* and *rbpD*) was initially reported (Sato, 1994, 1995; Sato and Maruyama, 1997). The rest of the three genes (*rbpA2*, *rbpE* and *rbpF*) have been cloned subsequently (Maruyama *et al.*, 1999). Seven of the *A. variabilis* M3 Rbps possess a single RRM with a short C-terminal glycine rich sequence but RbpD protein lacks such a C-terminal glycine rich sequence. Moreover, exceptionally RbpD protein only contains a probable cAMP- and cGMP-dependent protein kinase phosphorylation site in the C-terminal region. Although Sato (1995) reported that the nucleotide sequence of *rbpA2* gene in *A. variabilis* M3 appeared to be identical to that of the *rbpA* gene of *Anabaena* sp. strain PCC 7120, Maruyama *et al.* (1999) observed small changes between the nucleotide positions 55, 299 and 300 from the initiation codon. These differences contributed to the variation in the sequences of amino acids at the corresponding positions. The amino acid sequences of these gene products were highly conserved, except that the RbpD protein lacked glycine-rich C-terminal domain present in all other known members of the gene family.

Gene expression studies on the eight genes of Rbps of *A. variabilis* M3 revealed that after a temperature shift from 38°C to 22°C, the transcripts of *rbpF* attained maximum levels at the first hour followed by the *rbpE* transcripts at the second hour. However, the mRNA levels of *rbpC*, *rbpA3*, *rbpE* and *rbpF* have been found to be transient. The gene transcripts of *rbpA1* and *rbpA2* were found to be high at steady state levels at the low temperature. However, mRNA levels of the former gene reached their maximum level within 3 h whereas the transcripts for the latter gene attained maximum levels within 2 h. The levels of these transcripts for these two genes declined gradually but when transferred back to 38°C these were barely detectable. There was a corresponding increase in the levels of Rbp proteins during cold shock. Two of the genes *rbpB* and *rbpD* were expressed more or less constitutively (Sato, 1994, 1995; Maruyama *et al.*, 1999). Sato and Murayama (1997) identified that the transcription of *rbpA3* is driven by two promoters. One of the promoters is active at high temperatures and its activity is suppressed at the low temperatures. In contrast, the activity of the other promoter increased after a temperature shift down from 38°C to 22°C. Further, to know the involvement of 5′-untranslated region (5′-UTR) in the cold regulation of the *rbpA1* gene, a series of deletions of the *rbpA1* gene were transcriptionally fused to *lacZ* promoter and then the constructs were introduced into *A. variabilis* M3. Additionally, a series of plasmids based on shuttle vector pRA101 were fused with short DNA fragments in the 5′-UTR region of the *rbpA1* gene and *lacZ* as reporter gene. These experiments revealed that short DNA fragments one from –139 to +201 and another from –139 to +149 both conferred cold regulation on the *lacZ* gene. Both transcripts of *lacZ* gene and the β-galactosidase activity increased at the low temperature. Sato and Nakamura (1998) concluded that DNA fragment from –139 to +140 conferred high constitutive expression of gene *rbpA1* at 38°C as well as 22°C. This indicates the constitutive nature of the promoter of the *rbpA1* gene and the 5′-UTR sequence is required to reduce transcription of the *rbpA1* gene at high temperature. At high temperatures the expression of this gene is repressed by the binding of repressor protein(s) to 5′-UTR region of the *rbpA1* gene. Gel mobility shift assays and affinity purification of DNA-binding proteins revealed the formation of two complexes. There are two polypeptides (75 and 32 kDa) that are involved in the formation of complex 1 and one polypeptide (72 kDa) is involved in the formation of complex 2 (Sato and Nakamura, 1998). Maruyama *et al.* (1999) compared the 5′-UTR of the *rbp* genes and found four conserved regions. The first one is the conserved ribosome-binding site (RBS; 5′-TTCGGAGA-3′) that is located about 6 bases upstream of the translation initiation codon. The conserved box I (5′-TCTCCGAA-3′), box II (5′-TTTGTTTNNAGT-3′) and box III ((5′-TTCGGYGA-3′) are located about 40, 90 and 110 bases upstream of the initiation codon, respectively. It is interesting to note that the RBS is complementary to box I and box III is complementary to box I. Phylogenetic analysis of Rbp proteins revealed that Rbps of cyanobacteria and glycine-rich proteins (GRPs) of eukaryotes do not belong to the same lineage but the Rbp proteins of cyanobacteria have been suggested to have diverged from the GRPs. The apparent similarities in both structure and regulation between the cyanobacterial Rbp proteins and eukaryotic GRPs may have been due to convergent evolution.

RNA-binding properties of Rbp proteins of *Anabaena* sp. strain PCC 7120 have been characterized by Hamano *et al.* (2004). Based on the comprehensive genomic data, the existence of two classes of Rbp proteins (Classes I and II) has been proposed that diverged before the diversification of cyanobacteria. Class I Rbp proteins are further classified into two types, i.e. with or without a C-terminal glycine-rich domain. RbpA1 belongs to the Class I Rbps with high affinity to C-rich and G-rich sequences. The Class II Rbps have low affinity to G-rich sequences. Site-specific mutagenesis analysis revealed that aromatic amino acids Tyr4 and Phe46 are very important in RNA-binding as well as secondary structure maintenance. However, it is also suggested that the C-terminal glycine-rich domain itself does not contribute to the RNA-binding but Arg83 located close to C-terminal end of RRM is important in RNA-binding. The cold regulation of the *rbp* genes in three cyanobacteria having different temperature optima for growth has ben compared. *Anabaena* sp. strain PCC 7120 (a mesophilic strain), *T. elongatus* BP-1 (a thermophilic strain) and *Oscillatoria* sp. SU1 (a psychrophilic strain) showed increase of gene transcripts and respective proteins at low temperatures. The enhancement in the level of gene transcripts and the Rbp proteins was found to be highest in case of *Oscillatoria* sp. SU1 which suggested that the mechanism of cold regulation of *rbp* genes is common in various cyanobacteria with different temperature optima (Ehira *et al.*, 2003). The effect of osmotic stress on the expression of *rbp* genes in *Anabaena* sp. strain PCC 7120 was followed by comparing the transcript levels and Rbp proteins of the genes *rbpA, rbpB, rbpC* and *rbpD* in cultures grown in source of combined nitrogen (ammonium ions) and nitrogen-fixing conditions. A transient increase in the levels of all the four gene products was noted under both conditions after a temperature shift from 30°C to 20°C. However, osmotic stress enhanced the levels of transcripts and the respective Rbp proteins in ammonium grown cultures. In the absence of combined nitrogen, osmotic stress repressed the short-term induction of *rbp* gene expression. These workers concluded that the nitrogen nutrition modulates the stress-responsive regulation of Rbp proteins that provides a mechanism to integrate environmental and developmental signals (Mori *et al.*, 2003).

b) RNA helicases: One of the most important emerging areas in nucleic acid metabolism is the recognition of RNA helicases that help in maintaining RNA secondary structure. Majority of RNA helicases belong to superfamily 2 (SF2) subclass of helicases and contain a conserved domain of eight or nine amino acids (Tanner and Linder, 2001; Tanner *et al.*, 2003; Rocak and Linder, 2004). Three subfamilies, i.e. DEAD, DEAH and DExH/D are recognized in SF2 on the basis of variations within a common DEAD motif (Asp-Glu-Ala-Asp) (Tanner and Linder, 2001). In addition, there are seven additional motifs whose sequences and spacing are highly conserved (Linder *et al.*, 1989; Schmid and Linder, 1992; Gorbalenya and Koonin, 1993; Pause and Sonenberg, 1993; Fuller-Pace *et al.*, 1993; Fuller-Pace, 1994, 2006; Tanner and Linder, 2001; Rocak and Linder, 2004). The specificity of RNA helicases to target RNA molecules is due to the amino acid sequences outside the conserved domain and these also help in protein-protein interactions. RNA helicases play a major role in all cellular processes that involve RNA maturation, functioning in ribosome biogenesis, RNA splicing, transport and turnover, transcription, translation initiation, RNAi, RNA editing and development. Moreover, due to the fact each RNA helicase performs a unique function their role is associated with diverse cellular functions. Now it is being realized that besides their participation in key housekeeping metabolic pathways, they also play a regulatory role in responding to changes in environmental variables such as temperature, light, oxygen and osmolarity. In cyanobacteria, three RNA helicases have been described with their specific role in a stress response. *Anabaena* sp. strain PCC 7120 genome possesses two genes designated as *crhC* (cyanobacterial RNA helicase cold) and *crhB* that encode CrhC and CrhB proteins, respectively. Both *crhC* and *crhB* genes belong to the DEAD box subfamily of RNA helicases. The deduced amino acid sequence of CrhC protein confirmed its identification as a novel RNA helicase that belongs to the DEAD box family of helicases (Gorbalenya and Koonin, 1993). These two genes are differentially expressed. The expression of *crhC* is limited to cold shock conditions whereas *crhB* is expressed under a vriety of conditions but its expression is enhanced in the cold (Chamot *et al.*, 1999). Other important features of CrhC are that (i) a FAT (Pelyalanine-Alanine-Threonine) box is present instead of the diagnostic SAT (Serine-Alanine-Threonine) box, (ii) the change in serine to phenylalanine might be due to a transition taking place in the TCC Ser-codon to a TTC Phe-codon and (iii) the ATPase activity and RNA helicase activity of CrhC is coupled by the SAT box. Studies on CrhC expression revealed that it is not a general stress protein but specifically induced by a temperature shift-down to 10°C and in this respect resembles the expression of *E. coli* RNA helicase gene *csdA*. The *crhC* promoter contains a σ^{70}-dependent –10 mer sequence (TAAGAT) located 9 nucleotides upstream the transcription start site but lacks the –35 mer like region characteristic of *E. coli* promoter region. An AT-rich sequence resembling an upstream enhancer element is located at positions +87 to +97. The *Anabaena* cold shock box 5′-TGACAGGCCGA-3′ matches with the *E. coli* cold shock box 5′-TGACGTACAGA-3′ at 7 of 11 positions. The regulation of cold shock-induced RNA helicase gene expression in *Anabaena* sp. strain PCC 7120, the cloning of *crhC* gene and its transcription analysis have been studied by Chamot and Owttrim (2000). The transcripts for *crhC* gene appeared only when the organism is cultivated below 25°C or upon a cold shock (due to a shift from 30°C to 20°C) with maximum level of transcripts appearing between 20°C and 15°C. The transcripts do not accumulate when the organism is grown at 25°C or higher than this (i.e. 30°C, 37°C and 43°C). Though light as such is not required for the accumulation of *crhC* gene transcript accumulation, light-derived metabolism seems to be essential for the transcript to appear during a cold shock in dark. The half-life of *crhC* transcript is longer (67 min) at 20°C than at 30°C (11 min). Thus mRNA stability seems to play an important role in the accumulation of *crhC* transcript accumulation. A time course of cold-induced CrhC protein accumulation in the cells was noted 15 min after transfer to 20°C from 30°C as supported by Wetsern blot analysis. Yu and Owttrim (2000) described the biochemical characteristics of CrhC of *Anabaena* sp. strain PCC 7120 expressed in *E. coli*. According to these workers, CrhC exhibits ATP-independent RNA-binding and RNA-dependent ATPase and (d)ATP-dependent RNA unwinding activities. The unwinding of double-stranded RNA duplexes occurs in 5′------3′direction and this indicates that an interaction with 5′ region is required for CrhC function *in vivo*. The RNA-binding by CrhC exhibited non-specific characteristics as it was bound to partially dsRNA substrate and ssRNA, similar to those of other helicases. Thus CrhC may help in unwinding RNA secondary structures at the 5′-end of the target mRNAs induced during cold stress and stabilize them at low temperature. This probably leads to an improvement in their translational efficiency there by removing the block in translation initiation induced by cold shock. The cold-induced transcripts are characteristic in possessing 5′-untranslated (5′-UTR) regions that are required for cold shock specific expression (Jones and Inouye, 1994). Consistent with all cold-induced transcripts, the *crhC* mRNA also possesses a highly structured 5′-UTR required for the expression of CrhC. The importance of 5′-UTR sequence of ChrC is recognized by the fact that a replacement of the *crhC* promoter and 5′-UTR sequences by heterlogous sequences required for expression in *E. coli* has abolished its cold-regualated expression. El-Fahmawi and Owttrim (2003) demonstrated that the CrhC is an integral membrane protein associated with cytoplasmic face of the cytoplasmic membrane. Immunoelectron microscopy identified the CrhC protein to exist in higher concentrations at the two poles of the cell. This polar biased localization of CrhC protein suggests that the role of CrhC RNA helicase in cold stress is associated with the cytoplasmic membrane and occurs at the poles of the cells.

The transcriptional response of cyanobacteria to light (Mohamed and Jansson, 1989; Golden, 1995; Reyes and Florencio, 1995; Richter *et al.*, 1998; Kis *et al.*, 1998) and a correlation of light-induced enhancement in transcript levels to the photosynthetic electron transport (Reyes and Florencio, 1995; Kis *et al.*, 1998) have been well documented. The role of light in influencing the gene expression post-transcriptionally by redox-mediated effects on mRNA stability added another dimension to understand the gene expression. In case of *Synechocystis* sp. strain PCC 6803, the transcripts for *psbA-2* and *psbA-3* were found to have increased stability in the dark controlled by the cellular redox potential (Mohamed *et al.*, 1993; Tyystjarvi *et al.*, 1998). Similarly, Kulkarni and Golden (1997) have noted that in case of *S. elongatus* PCC 7942 the 5′-UTR regions of the three *psbA* transcripts affect transcript turnover in response to light intensity most probably by the recruitment of RNA-binding proteins required for translation or degradation. Since the electron transport system for photosynthesis and respiration have a common point at the plastoquinone pool for the entry of electrons from PSII and NAD(P)H-dehydrogenase, the redox status of plastoquinone pool depends on both photosynthetic light as well as on the metabolism of endogenous respiratory substrates. It is now known that the redox status of electron transport chain carriers regulates the expression of certain genes such as *glnA* (glutamine synthetase; Reyes and Florencio, 1995), *glnB* (for PII protein; García-Domínguez and Florencio, 1997), *cpcBA* (for α- and β-subunits of phycocyanin; Li and Sherman, 2000), proteins of the photosynthetic electron transport chain (Alfonso *et al.*, 2000; Li and Sherman, 2000; El Bissati and Kirilovsky, 2001) and *ntcA* (a transcriptional regulator; Alfonso *et al.*, 2001). *Synechocystis* sp. strain PCC 6803 has a RNA helicase designated as CrhR (cyanobacterial RNA helicase redox) with a specific role in salt/cold shock (Vinnemeir and Hagemann, 1999). The role of CrhR in light stress specially in reactions of bidirectional RNA winding, RNA annealing, RNA strand exchange, redox-regulated expression has been defined (Kujat and Owittrim, 2000). The expression of CrhR in *Synechocystis* sp. strain PCC 6803 is regulated by light-driven changes in redox status of the electron transport chain between Q_A in PSII and Q_o in cytochrome b6/f. Kujat and Owittrim (2000) concluded that a redox-responsive RNA helicase may provide the photosynthetic cyanobacteria with an ability to regulate expression of redox-responsive genes at the transcriptional level. It is suggested that cyanobacteria may not sense light directly but rather they may perceive the light driven changes in redox potential of electeron carriers in the electron transport pathway. Chamot *et al.* (2005) studied the biochemical properties of CrhR and showed that like other RNA helicases, CrhR possesses RNA-stimulated ATPase and bidirectional ATP-stimulated RNA helicase activity. The unique feature of CrhR is the catalytic activity for RNA annealing and the occurrence of helicase and annealing reactions concurrently to promote RNA strand exchange through a branch migration mechanism. So the reactions catalyzed by CrhR are more diverse and RNA annealing at least is not a general property of the RNA helicases. It is interesting to note that CrhR comes closer to other proteins like RepA, DnaB and replicative DNA helicases and RecA in its biochemical properties. The role of RNA helicases in abiotic stress has been reviewed (Owttrim, 2006). The expression of *crhR* gene is shown to be regulated by the presence of LexA-related protein that acts as a repressor when CrhR is not required, i.e. under conditions which oxidize the electron transport chain. Patterson-Fortin *et al.* (2006) identified that the binding site of LexA-related protein is located downstream of the transcription start site of CrhR. This has been suggested to be consistent with the regulatory protein binding sites localized in other genes known to be regulated by either light or redox signals.

The functional role of CrhR in the acclimatization of *Synechocystis* sp. strain PCC 6803 to low temperature has been identified by the isolation of a mutant in which *crhR* gene has been interrupted with a spectinomycin-resistance gene cassette. DNA microarray analysis of genome-wide gene expression in wild-type and *crhR* mutant along with Northern and Western blotting analyses revealed that in the mutant the expression of genes encoding the heat shock proteins GroEL1 and GroEL2 was no longer inducible at the low temperature as supported by both transcript and protein levels. CrhR helps in the stabilization of the transcripts of the genes *groEL1* and *groEL2* during the first 5 to 60 min cold acclimatization and also helps in the enhancement of transcription of these genes during the later (3–5 h) part of the acclimatization process. CrhR thus regulates the low-temperature inducible expression of the above heat-shock proteins which in turn may be required for low-temperature acclimatization process (Prakash *et al.*, 2010).

c) The Clp proteases: A new family of molecular chaperones includes those of caseinolytic proteases (Clp/Heat shock protein, Hsp100) that consist of constitutive and stress-inducible representatives. These belong to a broader family of AAA+ proteins (ATPases associated with various cellular activities) that are known to play different roles such as protein folding, disassembly of protein complexes and translocation of proteins across membranes. Of the two classes of Clp proteins known, the first consists of relatively large proteins with two distinct ATP-binding domains (ClpA to -E and L), while the second has smaller proteins with only one such domain (ClpM, -N, -X, and -Y) (Thompson and Maurizi, 1994; Kessel *et al.*, 1995; Schirmer *et al.*, 1996). The spacer region between the two ATP-binding domains has been used to classify the Clp genes. ClpA family members have a spacer of 5 amino acids, ClpC members have 62 to 69 amino acids and ClpB members have the longest spacer of 123 and 131 amino acids. At least in ClpB, the spacer region of amino acids comprises 13.5% (Squires and Squires, 1992). ClpA, restricted to gram-negative bacteria like *E. coli*, is the best studied. It is a dimer of 84 kDa subunits and assembles into a hexamer in the presence of ATP (Maurizi, 1991). It associates with ClpP protein and brings down degradation of the selected polypeptide. In this process ClpA holds, unfolds and exposes the polypeptide for the proteolytic activity of ClpP (Hwang *et al.*, 1987; Katayama-Fujimara *et al.*, 1987; Thompson and Maurizi, 1994; Kessel *et al.*, 1995; Hoskins *et al.*, 1998; Kim *et al.*, 1998). Likewise, ClpC and ClpX (but not ClpB) facilitate the activity of ClpP. ClpA and ClpX exhibit chaperone activity independent of ClpP analogous to those of DnaK and DnaJ (Wickner *et al.*, 1994). ClpP proteolytic subunit exhibits low levels of proteolytic activity but when complexed with ClpA, ClpC or ClpX active holoenzymes are formed that are able to degrade denatured proteins (Katayama-Fujimara *et al.*, 1987; Wojtkowlak *et al.*, 1993; Shanklin *et al.*, 1995). The proteolytic subunit of ClpP forms a barrel-like structure consisting of two heptameric rings present opposite to each other. The proteolytc chamber thus formed possesses the catalytic triad of Ser-His-Asp amino acids. The capping of this chamber on one or both sides by either ClpA or ClpX hexameric rings takes place (Grimaud *et al.*, 1998). The interaction of ClpY with ClpQ to form a proteolytically active holoenzyme has also been reported (Missiakas *et al.*, 1996). ClpB is another well-characterized member of the Clp/Hsp100 family that is represented in most eubacteria and eukaryotes. This protein has a long intervening region between the two ATP-binding domains (Squires and Squires, 1992). Two forms of ClpB proteins are synthesized in eukaryotes and prokaryotes. In the former, cytosolic-nuclear (100–110 kDa) and mitochondrial (78 kDa) ClpB proteins are encoded by different genes (Leonhardt *et al.*, 1993; Sanchez and Lindquist, 1990) but in the latter a single gene encodes a large (94 kDa) and a small (78 kDa) ClpB protein from a second translational start site (Park *et al.*, 1993; Eriksson and Clarke, 1996). All ClpB proteins are induced by high temperatures and most of these are required for thermotolerance. Under normal conditions ClpB is a non-essential component of cells whose inactivation does not lead to any phenotypic changes (Sanchez and Lindquist, 1990; Eriksson and Clarke, 1996; Hong and Vierling, 2000). ClpB interacts with large protein aggregates that accumulate under heat stress and the ATP-induced structural changes enable it to bind to the protein aggregates for performing its function. In doing so, it cooperates with other Hsps such as DnaK-DnaJ-GrpE to suppress and remove aggregation of a protein substrate. This has been identified as a highly efficient multi-chaperone system in *E. coli* (Zolkiewski, 1999). The functional collaboration between ClpB and the DnaK system requires the ATP hydrolysis at the two binding sites of ClpB and these two chaperones act synergistically to remodel proteins and dissolve aggregates (Doyle *et al.*, 2007). ClpC is represented in cyanobacteria, plants and in most gram-positive eubacteria. ClpD is exclusively present in plants. Certain gram-positive eubacteria possess ClpE and ClpL. ClpB has a protein unfolding activity that is dependent on ATP hydrolysis. It promotes protein disaggregation by unfolding aggregated proteins and hands them over to DnaK chaperone system for refolding (Goloubinoff *et al.*, 1999; Lum *et al.*, 2005).

The genome of *Synechocystis* sp. strain PCC 6803 consists of four genes for ClpB, ClpC, ClpP and ClpX proteases. Moreover, up to four isozymes of ClpP are encoded by a multigene sub-family. *S. elongatus* PCC 7942 is a thermophilic cyanobacterium particularly sensitive to low temperatures (Siva *et al.*, 1977). Porankiewicz and Clarke (1997) examined the role of ClpB to understand whether this heat shock protein is also important during cold stress by the isolation of a *clpB* deletion strain (D*clpB*), They demonstrated that the deficient strain exhibited a retarded growth at 25°C with a generation time of 54.4 ± 7.5 h in contrast to the average generation time of 15.0 ± 0.9 h for the wild-type. However, at 37°C the generation time of the wild-type and mutant have been noted to be 7.5 ± 0.4 h and 8.6 ± 0.7 h, respectively. The photosynthetic activity of the wild-type and mutant, as expressed in terms of O_2 evolution, was almost the same at 37°C but when shifted to 25°C both wild-type and mutant suffered a loss of 50% of the photosynthetic activity. But in terms of recovery, the wild-type recovered faster by returning to 70% of control after 24 h at 25°C whereas the mutant recovered marginally to 43% of the control. A temperature shift-down from 37°C to 25°C induced the synthesis of ClpB protein (92 kDa) by 5-fold while the same could not be detected at the normal growth temperature (37°C). After 24 h acclimation at 25°C, a corresponding increase in shorter form ClpB′ (78 kDa) protein was noted that was synthesized from a second translational start site. But further decrease in temperature to 20°C and 15°C progressively decreased the induction of ClpB protein. These results signify that the ClpB induction contributes to the acclimation process to permissive low temperatures. On the other hand, ClpC is constitutively expressed in the wild-type as well as the *DclpB* mutant. A temperature shift from 37°C to 25°C resulted in a 3-fold increase in ClpC protein whereas in the wild-type a marginal increase was noted.

***d) Protein S21 of small subunit of ribosome*:** Sato (1994) identified a gene designated as *rpsU* for protein S21 which is a component of small subunit (SSU) of ribosomes that is located just downstream of the *rbpA1* gene in the genome of *A. variabilis* M3. During a temperature shift-down from 38°C to 22°C the combined transcripts of *rbpA1* and *rpsU* increased by 10-fold within 2.5 h. However, at 38°C the transcript of only *rpsU* was more abundant (Sato, 1994). Subsequently, it was demonstrated that S21 was present at an equimolar level relative to other ribosomal proteins at 22°C but the relative level of S21 decreased at high temperatures (Sato *et al.*, 1997b). However, cold-inducibility of S21 protein in *Synechocystis* sp. strain PCC 6803 has been explained due to the presence of *rpsU* downstream of the rRNA operon, away from the *rbpA1* gene. The presence or absence of S21 protein in a ribosome makes it translationally active or inactive. The pattern of cold-inducible accumulation of S21 is thus required for low temperature acclimatization of cyanobacteria for the translational apparatus.

The fore-going description of cold-induced physiological changes encompass (i) a transient cessation of growth due to an inhibition in photosynthesis and protein synthesis, (ii) a decrease in membrane fluidity followed by the expression of desaturase genes that bring about a qualitative and quantitative change in unsaturated fatty acids, (iii) the two-component system of Hik33-Rre26 for perception and transduction of cold-stress signals, (iv) the induction of rbp proteins and RNA helicases for improving the structure stabilization of mRNAs of cold-inducible gene transcripts thereby enhancing translational efficiency and (v) the production of ribosomes to function properly at the low temperature. It remains to be seen as to how all these processes are integrated and function in a co-ordinated manner to overcome the cold stress in cyanobacteria.

B) Heat shock: Generally, a short exposure of cells to heat shock reduces the synthesis of normal cellular proteins and at the same time induces the synthesis of a new set of proteins known as the heat shock proteins (Hsps). The Hsps help in the the folding of newly formed proteins and also help in the refolding of denatured proteins. The Hsps were initially discovered in *Drosophila melanogaster* larvae that were exposed to heat shock (Ritossa, 1962). Since then considerable wealth of information has accumulated on the physiology, biochemistry and functional interactions of the various classes of Hsps. These are thus designated as "molecular chaperones", and are recognized as a ubiquitous feature of model organisms undergoing experimental stress. The Hsps are classified on the basis of their molecular weight into chaperones having high and low molecular weight. These are designated as Hsp followed by the molecular weight in eukaryotes such as Hsp100, Hsp90, Hsp70 and Hsp60 which are the major chaperones. Correspondingly in prokaryotes such as *E. coli*, ClpB (96 kDa) and ClpA (84 kDa) belong to the Hsp100 family of chaperones. These are Clp proteases already described under cold shock response. ClpB exists as a hexa or heptameric ring and helps in the solubilization of protein aggregates. ClpA exists as a hexameric ring and helps in the unfolding of protein aggregates for solubilization. Heat shock protein, HtpG protein (71kDa) of *E. coli* belongs to the Hsp90 family and this protein exists as a dimer and helps in the prevention of protein aggregation. HtpG protein has the ability to recognize non-native proteins and can help refolding of proteins in presence of GroEL, GroES and ATP. Studies on *htpG* gene deletion or disrupted mutants of *E. coli* revealed the non-essential nature of this gene under both non-stress and stressed conditions. DnaK protein (69 kDa) belongs to the Hsp70 family and forms a substrate-binding channel and helps in co- and post-translational protein folding in association with DnaJ and GrpE acting as co-chaperones. DnaK binds to around 20% of newly synthesized polypeptides in *E. coli*, assisting in their folding and structural maintenance (Hartl and Hayer-Hartl, 2002). It also regulates the activity of key proteins involved in stress management and cell division (Parsell and Lindquist, 1993; McCarty and Walker, 1994; Tomoyasu *et al.*, 1998). The two co-chaperones, DnaJ and GrpE regulate the ATPase activity of *E. coli* DnaK. DnaJ enhances the substrate-binding activity of DnaK by stimulating its ATPase activity whereas GrpE is a nucleotide exchange factor, associating and dissociating with DnaK in presence or absence of ATP, respectively (Liberek *et al.*, 1991; Harrison, 2003). GroEL protein (57 kDa) corresponds to Hsp60. Many of the Hsp60 chaperones are also known as chaperonins. The term chaperonin was originally coined by Ellis (1990) to designate non-heat-induced Hsp60. GroEL exists as a two heptameric rings and in association with GroES (chaperonin-10) forms a functional complex that helps in post-translational protein-folding. One type of chaperonins, called chaperonin-60 (Cpn60) is homologous to GroEL protein of *E. coli*. Both DnaK (Hsp70) and GroEL (Hsp60) have been identified and characterized from various organisms and major cell organelles. Although Hsp70 proteins are synthesized under stress conditions but due to their constitutive expression under normal growth conditions these are considered as cellular thermometer (Craig and Gross, 1991). Thus these constitute the ubiquitous group of molecular chaperones that bring about protein-protein interactions and help in the folding and assembly of proteins. Single chain chaperonins are FtsH (HflB, 71 kDa), DegP (HtrA, 48 kDa) and Lon (87 kDa). FtsH forms a multimeric ring and helps in the folding and assembly of membrane proteins. HtrA exists as two hexameric rings and is the chaperone expressed at low temperatures. Lon exists as a tetramer and facilitates the substrate sequestration for proteolysis. Hsp33 (33 kDa) exists as a dimer and gives protection during oxidative stress. Ibp A and Ibp B of *E. coli* correspond to α-Hsp with a molecular weight of 16 kDa. This forms a sphere of 24 subunits that prevent protein aggregation. The heat inducibility of these chaperones and chaperonins is in the order of α-Hsp>DnaK>ClpB, HtpG, GroEL>Hsp33, Lon>ClpA.

A number of explanations have been put forward for the synthesis of Hsps in the stressed cells. According to the classical model, the accumulation of denatured proteins in the cytoplasm constitutes the major trigger (Craig and Gross, 1991; Bukau, 1993). Another model takes into account

DnaK protein acting as the direct thermometer for sensing heat (McCarty and Walker, 1991). The ribosome sensing hypothesis envisages regulation of stress genes at the level of translation process. The implication of Hsps in the acquisition of thermotolerance has been well established (Lehel *et al.*, 1993; Glatz *et al.*, 1997, 1999; Vigh *et al.*, 1998, 2007). Thus the Hsps protect and recover the stress damaged proteins in the cell. In *E. coli*, IbpA/B (small Hsps), GroEL, DnaK, ClpB constitute a sophisticated network of "holding" and "folding" chaperones that prevent protein aggregation during heat stress. They not only promote protein disaggregation during heat stress but also help in refolding the damaged proteins after the stress (Buchner, 1996; Mogk *et al.*, 1999, 2003). Initially it is the small Hsps that are bound to the denatured polypeptides which hand them over to DnaK and GroEL chaperones for refolding (Veinger *et al.*, 1998). Likewise, denatured protein aggregates are successfully resolubilized and refolded by the sequential action of ClpB and DnaK (Singer and Lindquist, 1998b; Goloubinoff *et al.*, 1999; Mogk *et al.*, 1999; Diamant *et al.*, 2000). Motohashi *et al.* (1999) reported the functional co-operation between ClpB and DnaK. The existence of DnaK and DnaJ as a stable trigonal ring complex in *Thermus thermophilus* (TDnaK.J complex) and the *dnaK* gene cluster also contains ClpB gene. Substrate proteins could be protected from aggregation and heat denaturation by TDnaK.J complex is dependent on ATP and TGrpE. Subsequent addition of ClpB at moderate temperature resulted in an efficient reactivation of the proteins. This study emphasizes the cooperation of TDnaK.J complex, TGrpE, TClpB and ATP for complete reactivation and TGroE/ES could not substitute TClpB. Diamant *et al.* (2001) demonstrated the regulation of molecular chaperones by the chemical chaperones *in vitro* as well as *in vivo* when *E. coli* cells were subjected to salt and heat stress simultaneously. Accordingly, the effects of four osmolytes (glycine betaine, glycerol, proline and trehalose) on the folding activities of individual and combinations of chaperones (GroEL, DnaK and ClpB) have been studied. Except trehalose, the low physiological concentrations of the rest of three osmolytes activated molecular chaperones in their local folding functions. Especially glycine betaine prevented protein aggregation during heat stress as well as provided maximum levels of stimulation for the chaperone-mediated protein folding. On the other hand, trehalose strongly inhibited DnaK-dependent chaperone networks such as DnaK+ GroEL and DnaK+ClpB probably due to high viscocity levels. Thus the levels of activity of molecular chaperones much depends on the level and nature of osmolytes that accumulate when the cells are simultaneously subjected to a salt and heat stress.

i) Synthesis of Hsps in response to heat stress: All cyanobacterial genomes sequenced so far possess two distinct hsp60 genes, a *groEL* gene as a part of *groESL* operon and a solitary *cpn60* gene. *Synechococcus* sp. strain PCC 6301 cells grown at 39°C when subjected to short-term exposure to 47°C synthesized a number of polypeptides that varied in their molecular weights (ranging from 91 kDa to 11.4 kDa) (Borbély *et al.*, 1985). Molecular cloning of genes *cpn60* and *dnaK* from *Synechocystis* sp. strain PCC 6803 revealed the existence of these genes in single copies and the transcripts were monocistronic. The transcript levels of *cpn60* and *dnaK* increased 30 and 25 fold, respectively after 90 min heat shock at 42°C. But after 2 h, the level of the transcripts of the two genes reached the pre-stress levels. Oxidative stress and UV-exposure also triggered the accumulation of the transcripts of these two genes. The deduced amino acid sequences of Cpn60 brings it closer to bacterial and plant Cpn60 proteins whereas DnaK protein sequence is homologous to its bacterial counterpart than to eukaryotic ones. Another siginificant feature is the presence of ORF60-5 that encodes a protein belonging to the peripheral membrane proteins involved in metabolite transport in bacteria (Chitnis and Nelson, 1991). When *Synechocystis* sp. strain PCC 6803 cells grown at 30°C were exposed to heat stress at 42°C, the induction of Hsps of different molecular sizes (79, 64, 15 and 14 kDa) was noted. Though the cells initially accumulated high levels of Hsp70 and Hsp15, the synthesis of Hsp64 and Hsp14 continued through 2 h. The Hsp70 and Hsp64 have been identified to be similar to DnaK and GroEL of *E. coli* (Lehel *et al.*, 1992). The cloning and characterization of the *groEL* gene of *Synechocystis* sp. strain PCC 6803 was performed by using *groEL* gene of *S. elongatus* PCC 7942 as a probe. A fragment of 3.25 kbp from the genome of *Synechocystis* sp. strain PCC 6803 revealed three ORFs. On the basis of sequencing of the three ORFs and deduced amino acid sequences, two of them corresponded to GroEL and GroES proteins (Lehel *et al.*, 1993). The third gene (ORF51), located from 460 to 612 bp of the sequence encodes a putative polypeptide of 51 amino acids, that did not bear any significant homology to any known sequence in the database. The GroES of *Synechocystis* sp. strain PCC 6803 showed 80% and 62–72% homology to GroES proteins of *S. elongatus* PCC 7942 and other bacterial species, respectively. In addition to *groESL* operon, *Synechocystis* genome also possesses two copies of *groEL* analogous genes. The transcript of the *groESL* operon was barely detectable at 30°C but after a heat shock at 42°C for 15 min the level of the transcript was enhanced by 100-fold, reaching a maximum in 14 h. The sequence of Cpn60 of *Synechocystis* sp. PCC 6803 (Chitnis and Nelson, 1991) showed same degree of similarity to the sequence of GroEL described by Lehel *et al.* (1993). However, the degree of homology of the sequence of GroEL of *Synechocystis* sp. PCC 6803 is higher to Cpn60 of *Bacillus subtils* than to its own sequence of Cpn60.

The role of HtpG, a prokaryotic homologue of Hsp90, in thermal stress management of cyanobacteria has been investigated. A single copy of the *htpG* gene sequence (ORF, *sll0430*) has been cloned from *S. elongatus* PCC 7942 and a comparison with the sequences in the database suggested that the deduced amino acid sequence bears closest homology (64% overall identity) to that of HtpG from *Synechocystis* sp. strain PCC 6803. Two unique features of HtpG are that a portion between 210 and 224 (numbered according to HtpG of *E. coli*) amino acid residues did not exist and residues at 14 and 33 in homologues from *S. elongatus* PCC 7942 and *Synechocystis* sp. strain PCC 6803, respectively were inserted between position numbers 409 and 410. Mutants of *S. elongatus* PCC 7942 in which *htpG* was inactivated by targeted mutagenesis exhibited similar growth pattern as that of wild-type at 30°C and 42°C but the cultures turned yellow. At 45°C, the *htpG* mutant grew with a lag period while the wild-type grew normally. After a heat shock (cells exposed to 50°C for 20 min) the survival of wild-type was 20% whereas the *htpG* mutant surived to 0.01 % only. Wild-type cells exhibited the ability of acquired thermotolerance (a pre-treatment for 60 min at 42°C before giving heat shock at 50°C) but the *htpG* mutant lost this ability. These results emphasize the essential nature of *htpG* gene in thermal stress management in cyanobacteria (Tanaka and Nakamoto, 1999). A *htpG* null mutant was generated by inserting a chloramphenicol resistance cassette in the *htpG* coding sequence of *Synechocystis* sp. strain PCC 6803. A comparison of *htpG* mutant with a mutant of *hsp16.6* and the double mutant (*htpG*$^-$:*hsp16.6*$^-$) for their growth, cell survival rate and rates of O_2 evolution revealed that all the three of them were sensitive to heat stress. Basal level of thermotolerance and acquired thermotolerance of the double mutant was lowest among the three when compared to wild-type (Fang and Barnum, 2003). Apart from the essential nature of HtpG during heat (Tanaka and Nakamoto, 1999), cold (Hossain and Nakamoto, 2002), salt (Huang *et al.*, 2002) and oxidative (Hossain and Nakamoto, 2003) stresses, other studies have indicated that *htpG* transcripts and HtpG protein accumulation occurred under high light intensity as well (Hihara *et al.*, 2001; Mary *et al.*, 2004).

Cloning and characterization of *groEL* gene and its contribution to thermotolerance of *Anabaena* sp. strain L-31 has been reported (Rajaram *et al.*, 2001; Rajaram and Apte, 2003). Cloning and characterization of *cpn60* gene from *Anabaena* sp. strain L-31 and its expression due to heat shock have been compared with that of *groEL* in response to nitrogen status. Severe inhibition of photosynthesis was associated with inhibition of nitrate reduction in nitrate supplemented cultures compared to nitrogen-fixing cultures. The expression of *groEL* was independent of the nitrogen status of the medium and readily expressed in response to heat shock but the expression of *cpn60* very much depended on the ability of the organism to utilize nitrate or ammonium ions (Rajaram and Apte, 2008).

The genome of *Synechocystis* sp. strain PCC 6803 has four *dnaJ* genes (*sll1666*, *sll1933*, *sll0093* and *slr0897*) whereas *P. marinus* MIT9313 and MED4 possess three and two *dnaJ* homologues, respectively. However, the homology and organization of the *dnaJ* genes differed in these organisms. While *dnaJ3* shared the strongest identity (about 60% identity) between *Synechocystis* sp. strain PCC 6803 and *P. marinus* MIT9313, *Prochlorococcus dnaJ1* is not well conserved (that showed only 34% identity). In both the *Prochlorococcus* strains *dnaJ1* is located close to *dnaK3* gene. Under high light exposure, the mRNA of *dnaJ3* dramatically increased in *Synechocystis* and *P. marinus* MIT9313 (Mary *et al.*, 2004).

The Hsps have been employed to measure evolutionary relationships among eubacterial groups. In this connection, sequence comparisons of GroEL from different bacterial species (resulted into phylogenetic trees based on distance or parsimony methods that were consistent with 16S rRNA trees) were resolved into different clusters of proteobacteria, chlamydiae, bacteroids, spirochaetes, firmicutes (gram-positive bacteria), cyanobacteria and chloroplasts. Detailed analyses suggested that the GroEL proteins serve as valuable evolutionary tools to draw phylogenetic relationships among eubacteria (Viale *et al.*, 1994). In a similar manner, coding region of of *cpn60* (GroEL or Hsp60) can be amplified with universal PCR primers. A cpn60 sequence database (consisting of more than 2000 such sequences from more than 240 genera of bacteria, eukaryotes and archaebacteria available at http://cpndb.cbr.nrc.ca) has been created that is useful in clinical, phylogenetic and microbial ecology studies (Hill *et al.*, 2004).

***ii) Expression of dnaK genes*:** Nimura *et al.* (1994a) identified three *dnaK* homologues *dnaK1*, *dnaK2* and *dnaK3* in *S. elongatus* PCC 7942. In contrast to DnaK1 and DnaK2 proteins that are present in the cytosol, DnaK3 protein is present in large amounts associated with thylakoid membranes. Characteristically, DnaK3 possessed a long C-terminal region, the deletion of which did not cause any alteration in the phenotype of the generated strain when compared to wild-type (Nimura *et al.*, 1994b, 1996). The genome of *Synechocystis* sp. strain PCC 6803 revealed the presence of four *dnaK* homologues (ORFs *sll0086*, *sll0170*, *sll1932* and sll0058; Kaneko *et al.*, 1996; Mary *et al.*, 2004) that showed high degree of homology to *S. elongatus* PCC 7942 *dnaK* homologues. DnaK3 protein of *Synechocystis* sp. strain PCC 6803 also showed a long C-terminal region resembling the DnaK3 of *S. elongatus* PCC 7942. The presence of three *dnaK* homologues has been demonstrated in the genome of *Anabaena* sp. strain PCC 7120 (http:www.kazusa.or.jp/cyano), *P. marinus* MIT9313 and MED4 strains (Mary *et al.*, 2004). The presence of three *dnaK* homologues is a rare feature among prokaryotes as only two *dnaK* homologues (*hsc62* and *hsc66*; of these the latter is induced by cold shock but not by heat shock) were reported in *E. coli* other than *dnaK* (Seaton and Vickery, 1994; Lelivelt and Kawula, 1995; Yoshimura *et al.*, 1998) and in *Borrelia burgdorferi* B21 two *dnaK* homologues are present (Fraser *et al.*, 1997).

Circadian control of *dnaK* gene expression in *Synechocystis* sp. strain PCC 6803 was monitored by using the promoterless bacterial luciferase genes (*luxAB* from *Vibrio harveyi*) as a reporter gene. The *luxAB* gene set was introduced downstream of the promoter region of *dnaK* gene of *Synechocystis*. The *dnaK* gene expression pattern oscillated with a period of 22 hr for at least 5 days in continuous light. The bioluminescence rhythm satisfied the three criteria set for circadian rhythms. Simultaneously, the mRNA levels of *dnaK* gene also oscillated in a rhythmic manner suggesting that the expression of *dnaK* gene is under circadian control (Aoki *et al.*, 1995). Isolation and characterization of *dnaK* locus from *A. halophytica* led to the identification of *grpE* and *dnaJ* genes in the order *grpE-dnaK1-dnaJ*. The mRNA levels for *dnaK1* gene increased upon subjecting the cells to heat stress. The functional aspect of the *dnaK1* locus has been tested by its co-expression with a plant plastocyanin gene in *E. coli* cells due to which there was a dramatic increase in the solubility of the plastocyanin protein. This signifies that the *dnaK1* produces a functional protein and it performs the functions assigned to DnaK/Hsp70 (Lee *et al.*, 1997).

Nimura *et al.* (2001) conducted gene disruption experiments in *S. elongatus* PCC 7942 by introducing kanamycin resistance marker in *dnaK1*, *dnaK2* and *dnaK3*. As *S. elongatus* PCC 7942 is known to possess multiple copies of the genome a successful transformation can lead to the disruption of all copies of the particular *dnaK* gene. Thus DnaK2 and DnaK3 were found to be essential as disruption of all copies of *dnaK2* and *dnaK3* was not possible but only in case of *dnaK1* successful transformants were obtained suggesting that all copies of this gene could be disrupted and the organism could lead a normal growth even in its absence. The expression of the three *dnaK* homologues in response to heat shock (a shift up from 30°C to 45°C) showed that only DnaK2 and GroEL proteins exhibited a typical heat shock response in the sense that the proteins increased in their level during the first 30 min and thereafter increased level was maintained for the entire period under study. The effects of overproduction of the three gene products were studied in *E. coli* by using an inducible expression system separately for each gene. Overproduction of DnaK1 and DnaK2 resulted in defective septation leading to the formation of long filamentous cells. Overproduction of DnaK3 resulted in cells that were swollen and twisted. As deletion of *dnaK* in *E. coli* causes a temperature-sensitive phenotype, a *dnaK756* deletion mutant of *E. coli* was tested for its ability to grow at non-permissive temperature when the three *dnaK* homologues of *S. elongatus* PCC 7942 were expressed. *dnaK2* could suppress growth deficiency at the non-permissive temperature while *dnaK1* and *dnaK3* could not suppress this phenotype. Overproduction of DnaK1 or DnaK3 in mutant *dnaK756* of *E. coli* resulted in growth inhibition at the permissive temperature. Furthermore, the levels of expression of DnaK2 and DnaK3 and GroEL proteins after heat shock in wild-type and *dnaK1* disruptant mutant cells (designated as DK 1KM) were found to be the same as revealed by Western blotting. The studies of Varvasovszki *et al.* (2003) further confirmed that in *Synechocystis* sp. strain PCC 6803, of the three *dnaK* homologues only *dnaK2* is transcriptionally active. The co-chaperones *dnaJ* and *grpE* were not inducible. Gene inactivation experiments with *dnaK2* yielded partial mutants that exhibited a temperature-sensitive phenotype with lower inducibility of GroEL and Hsp17.

Although DnaK1 of *A. halophytica* (Lee *et al.*, 1997) and DnaK1 of *S. elongatus* PCC 7942 (Nimura *et al.*, 2001) showed extensive sequence similarity and comparable *in vitro* chaperone capabilities, functionally they differed under *in vivo* conditions. Blanco-Rivero *et al.* (2005) compared the functional capabilities of DnaK1 from *A. halophytica* and *S. elongatus* PCC 7942 *in vivo* by expressing them in *E. coli* mutants defective in DnaK (*dnaK756*) or a null mutant. DnaK1 from *S. elongatus* PCC 7942 effectively replaced and complemented *E. coli* DnaK by performing all the major functions recognized for this chaperone. Some of the important functional differences in the two DnaK1 proteins are: (i) production of filamentous cells of *E. coli* due to overproduction of DnaK1 from *Synechococcus* and its absence due to overproduction of DnaK1 from *A. halophytica*; (ii) reduced levels of production of GroEL in *E. coli* cells when its own DnaK or *Synechococcus* DnaK1 are overproduced whereas there was no significant reduction in the levels of GroEL in *E. coli* when DnaK1 from *A. halophytica* was overproduced; (iii) Overproduction of DnaK1 from *Synechococcus* prevented aggregation of misfolded RuBisCO units in *E. coli* mutants while DnaK1 from *A. halophytica* could not prevent aggregation of RuBisCO subunits; and (iv) most importantly, DnaK from *E. coli* or DnaK1 from *Synechococcus* protected the *E. coli* null mutants (for *dnaK*) from heat stress by conferring thermotolerance but DnaK1 from *A. halophytica* could do so to much lower levels. By contrast, *dnaK* homologues of *Synechocystis* sp. strain PCC 6803 did not complement *dnaK* in *E. coli dnaK*-defective mutants. Rupprecht *et al.* (2007) demonstrated that all three homologues of *dnaK* are functional in *Synechocystis* sp. strain PCC 6803 and the three DnaK proteins share a high degree of homology to each other. However, they differ in their molecular masses, i.e. DnaK1, DnaK2 and DnaK3 possess 75 kDa, 67 kDa and 86 kDa molecular weights, respectively and share the common DnaK domain. All the three proteins showed C-terminal extensions and the observed differences are due to this. The expression of the three DnaK proteins in *Synechocystis* sp. strain PCC 6803 was demonstrated by Western blot analysis using antibodies specific to each one of the DnaK proteins. Specific physiological functions have been identified by the isolation of *dnaK* gene disruptant mutants by introducing kanamycin resistance marker. They confirmed the essentiality of *dnaK2* and *dnaK3* for survival of *Synechocystis* sp. strain PCC 6803 while deletion of *dnaK1* gene did not affect the survival as reported earlier in case of *S. elongatus* PCC 7942 (Nimura *et al.*, 2001). PCR and restriction analysis of the genomes of *dnaK2* and *dnaK3* disruptant mutants revealed the presence of wild-type genes suggesting that all the copies of the two genes could not be completely inactivated. DnaK1 protein was no longer produced in the *dnaK1* disruptant mutant but the levels of DnaK2 and DnaK3 in *dnaK2* and *dnaK3* disruptant mutants were almost comparable to the levels of these proteins in the wild-type. Deletion of *dnaK3* gene could only be possible when the *dnaK3* gene was expressed from a neutral site in the genome of *Synechocystis* sp. strain PCC 6803. The C-terminal region proved to be essential for proper functioning of DnaK3 protein, as a truncated *dnaK3* gene (without the C-terminal) region could not complement *dnaK3* when expressed from a different site on the chromosome of *Synechocystis*. By using an expression plasmid pMal-C2, all the three *dnaK* homologues have been individually cloned and *E. coli dnaK756* mutant was transformed. In none of the transformants, *dnaK* homologues of *Synechocystis* sp. strain PCC 6803 could complement the *dnaK* gene functions of *E. coli*. Although the three proteins increased in their levels after a heat shock, only the levels of DnaK2 protein were up-regulated consistent with the earlier observations on the same organsim (after a heat shock and other stresses like UV light, ethanol, salt and high light; Chitnis and Nelson, 1991; Varvasovszki *et al.*, 2003; Mary *et al.*, 2004; Fulda *et al.*, 2006) and on *S. elongatus* PCC 7942 (Nimura *et al.*, 2001).

A highly conserved N-terminal ATP-binding domain and a less conserved C-terminal peptide-binding domain are characteristically present in the DnaK proteins. It is through the peptide binding domain that DnaK3 is shown to be quantitatively associated with the cytosolic side of the thylakoid membranes in *S. elongatus* PCC 7942. It means that due to the absence of a thylakoid-targeting sequence, it was surmised that DnaK3 might require a receptor protein for its interaction with thylakoid membrane (Nimura *et al.*, 1996). Zhu *et al.* (1996) showed that the substrate-binding portions constitute β-sheet regions together with adjacent α-helical regions. In order to identify the role of DnaK3 proteins and their binding interactions with thylaloid membranes, Katano *et al.* (2006) isolated *dnaK3* mutants and their suppressors from *S. elongatus* PCC 7942. The mapping of the *dnaK3* regions led to the identification of T to C conversion in one of the mutants (NBC001) at 436th codon from TTG (Leu) to TCG (Ser) in the β-sheet region. The growth and O_2 evolved upon a temperature shift-up (from 30°C to 43°C) decreased in case of NBC001. Suppressor mutants were isolated by incubating NBC001 at the non-permissive temperature (43°C). Of the three suppressor mutants isolated, one of it mapped in *rpll24* (ribosomal protein large; *syc1876*) that encodes 50S ribosomal protein L24 whereas in the rest of the two suppressors the substrate-binding capacity was restored suggesting that these two were intragenic suppressors. Further, DnaK1 and DnaK2 proteins have been shown to be localized in the cytosol and partly in the membrane fractions in the wild-type cells grown at 30°C or 43°C. So also the same patterns of localization of these two proteins have been found in DnaK3 mutant and in its suppressor. On the other hand, the localization of DnaK3 protein in the membrane and membrane-bound fractions in the wild-type and its presence exclusively in the cytosol in NBC001 mutant at high temperature has been noted. But in another suppressor mutant NBC002, the pattern of localization of DnaK3 resembled the wild-type. A correlation of these observations with D1 protein suggests that in the wild-type and *dnaK3* mutant at both 30°C and 43°C the full length D1 protein with molecular weight of 24 kDa was present (that corresponded to the intermediate of D1 protein of *Synechocystis* sp. strain PCC 6803). The absence of such an intermediate D1 protein in the suppressor mutant at either of the temperatures led them to conclude that DnaK3 protein is involved in targeting the translational intermediates to the thylakoid membrane. Using the *lacZ* gene as a reporter, Sato *et al.* (2007) demonstrated that *dnaK2* is the only gene that is up-regulated in *S. elongatus* PCC 7942 during heat stress. Cells exposed to high light or high salt stress, *dnaK2-lacZ* expression was clearly inducible. On the other hand, *dnaK1* and *dnaK3* showed little response. In view of this, they postulated that DnaK2 might function as a general molecular chaperone, e.g. in processes of protein folding, oligomer assembly, and stabilization of the protein structure.

***iii) The role of proteases*:** The key enzymes of the protein degradation machinery are designated as peptidases or proteases. Peptidases catalyze the hydrolysis of peptides by digestion of the specific bonds inside the target molecule. Peptidases can be divided into two large groups according to their substrate specificities: endopeptidases and exopeptidases. Exopeptidases remove single or several amino acid residues, dipeptides or tripeptides, from N- or C-termini, and accordingly can be classified into mono-, di- and tripeptidases, respectively (Kenny, 1999; Rawlings and Barret, 1999). Endopeptidases can also remove single or several amino acid residues, but in contrast to exopeptidases, classification of endopeptidases is based on the active proteolytic residues of the enzymes, not on a substrate. Endopeptidases have been divided into four major groups: serine, cysteine, aspartic and metallo-peptidases (Barrett, 1994, 1995; Callis, 1995; Kenny, 1999).

Various studies have shown that cyanobacteria possess a set of proteases belonging to families such as Clp (Eriksson and Clarke, 1996; Porankiewicz *et al.*, 1998; Panichkin *et al.*, 2001), Deg (Sokolenko *et al.*, 2002), FtsH (Mann *et al.*, 2000; Bailey *et al.*, 2001), Ctp (Shestakov *et al.*, 1994; Ivleva *et al.*, 2002), Gsp (Zuther *et al.*, 1998) and SppA (Lensch *et al.*, 2001) that are also represented in non-photosynthetic prokaryotes.

***a) Clp proteases*:** There are four genes that encode Clp proteins (*ClpB*, *ClpC*, *ClpP* and *ClpX*) in *Synechocystis* sp. strain PCC 6803. Of these, *ClpC* and *ClpX* genes exist as single copies and *ClpB* and *ClpP* genes are present in multiple copies. Two *Prochlorococcus* strains (MED4 and MIT9313) also possess similar copies of Clp homologues. In all these three organisms, *ClpP* is variable as *ClpPI* appears to be monocistronic while *ClpPII* and *ClpPIII* are each part of bicistronic operons with *ClpX* and *ClpPIV*, respectively (Mary *et al.*, 2004). *S. elongatus* PCC 7942 has 10 distinct Clp proteins. Of these, four are Hsp100 chaperones (ClpB1-2, ClpC and ClpX), three Clp P proteins (ClpP1-3), ClpR (which is ClpP-like protein) and two adaptor proteins (ClpS1-2) (Schelin *et al.*, 2002; Clarke *et al.*, 2005). ClpR, having a protein sequence similar to ClpP, seems to lack the catalytic triad (Ser-His-Asp) characteristic of Ser-type proteases (Poranckiewicz *et al.*, 1999).

The cloning and sequencing of the single copy *ClpB* homologue from *S. elongatus* PCC 7942 was reported by Eriksson and Clarke (1996). *ClpB* gene consisted of 2,649 bp coding a polypeptide of 883 amino acids. The absence of –10 mer and –35 mer sequences near the promoter region of this gene signifies that either constitutive σ^{70} or the heat shock σ^{32} factors do not have a role in transcriptional events. The deduced protein sequence was most similar (70–75%) to the sequence of ClpB from bacteria and cytosolic form of ClpB from plants. Mutants isolated after gene inactivation experiments showed no synthesis of ClpB proteins whereas the wild-type produced long (93 kDa) and short forms (79 kDa) of ClpB after a heat shock.The level of ClpC remained unchanged in the wild-type but in the ClpB-defective mutant the level of ClpC doubled during heat stress. Taking photosynthesis as a metabolic indicator, a heat shock (shift-up from 37°C to 55°C for 10 min) caused a significant reduction in photosynthetic oxygen evolution in wild-type and the mutant. But a pre-treatment of cells at 50°C for 1.5 h resulted in a significant retention of the photosynthetic efficiency in the wild-type. Due to a reduction in the level of ClpB protein in the mutant, it lost the ability to tolerate the heat stress. These results thus signify the vital nature of the synthesis of ClpB protein for cell survival at high temperature and for acquiring thermotolerance.

Similarly, cloning and sequencing of ClpB gene from *Plectonema boryanum* (pClpB) showed it to consist of two highly conserved ATPase domains characteristic of other ClpB sequences examined. The accumulation of the transcripts of the small and long forms of pClpB and the respective proteins due to excess light at low temperature is another significant feature. Phylogenetic analysis of *ClpA*, *ClpB* and *ClpC* sequences from 27 different organisms (bacteria, yeast mitochondria and eukaryotes) revealed that the ClpB sequences are well resolved into bacterial and eukaryotic sequences with the pClpB sequences coming very closer to *Synechococcus* and other cyanobacterial ClpB sequences (Celerin *et al.*, 1998). Clarke and Eriksson (2000) demonstrated that the truncated form of ClpB (79 kDa protein) has a functional role in conferring thermotolerance in *S. elongatus* PCC 7942. Gene inactivation experiments led to the identification of two types of mutants: (i) in the first type *ClpBI-79* was inactivated so that it produced only ClpBI-93 protein and (ii) in the second type *ClpBI-93* was inactivated so that it produced only ClpB1-79. A temperature shift-up (37°C to 48.5°C) caused a six to seven fold induction of ClpBI-93 protein with a concomitant 2–3 fold increase in ClpBI-79 in the wild-type. However, in the two categories of mutants the levels of production of the respective ClpBI-93 or ClpBI-79 proteins gave thermal protection to a lower level than that of the wild-type. The overall data point out that the contribution of the two forms of ClpB to thermotolerance in wild-type was two-thirds and one-third by ClpBI-93 and ClpBI-79 proteins, respectively. In view of this, they questioned the suggestion that ClpBI-79 might be playing a regulatory role in the activity of full length ClpBI-93 as has been pointed out in case of *E. coli* (Park *et al.*, 1993). Eriksson *et al.* (2001) described the existence of a novel form of ClpB (ClpBII) in *S. elongatus* PCC 7942 that is not induced by heat or other stresses (light 1000 µmol photons m^{-2} s^{-1}; cold 25°C; salt 150 mM NaCl or H_2O_2 0.5 M). Though a constitutive protein, ClpBII was unable to complement ClpBI in its function for protecting the cells against heat stress as evidenced by the gene inactivation experiments. Thermotolerance assays (at 54°C for 15 min after a first pre-treatment at 48.5°C for 1.5 h) showed that the wild-type developed significant thermotolerance over ClpBI- and ClpBII-deficient mutants. The *ClpBII* gene is a single copy ORF (with 2,685 bp) without the typical –10 mer or –35 mer promoter sequences characteristic of *E. coli* genes. It possesses two ATP-binding domains (ATP1 and ATP2) that are separated by a spacer of 129 amino acids. Unlike *ClpBI* that produces a full length (93 kDa) and a truncated form of the protein (79kDa), *ClpBII* produced a single form of ClpBII (97 kDa) polypeptide that corresponded to the predicted size of ClpBII.

The multigene family of ClpP from *S. elongatus* PCC 7942 has been characterized. The three *ClpP* genes appeared to be different in their organization. *ClpPI* gene is monocistronic in contrast to *ClpPII* and *ClpPIII* genes that are part of bicistronic operons with *ClpX* and *ClpR*, respectively. Though not inducible by a heat (50°C for 2 h) or cold (25°C for 6 h) shock or high light (1000 µmol photons m^{-2} s^{-1}) or oxidation (H_2O_2 0.5 M), *ClpPII*, *ClpPIII*, *ClpX* and *ClpR* seemed to be essential for cell viability.

However, a gradual and slow accumulation of the respective proteins during high light stress and low temperature (except ClpP2 protein) occurred. Despite the occurrence of *ClpPII* and *ClpPIII* with *ClpX* and *ClpR* in bicistronic operons, the transcripts of these genes were produced as monocistronic constructs. Mutants defective in *ClpPI* produced lower levels of ClpP2 protein concomitant with increased levels of both ClpP3 and ClpR. The inactivation of *ClpPII* gene caused a rise in *ClpPI* transcripts but not the ClpP1 protein level. These results point out to the functional nature of the ClpP genes and there exists a regulatory complexity among these genes (Schelin *et al.*, 2002).

Andersson *et al.* (2006) described the chaperone activities of ClpC from *S. elongatus* PCC 7942 (SyClpC) overexpressed in *E. coli.* Such purified SyClpC exhibited *in vitro* chaperone activity both by preventing aggregation of unfolded polypeptides and by refolding the aggregated proteins into their native state. Gel filtration chromatography analysis revealed SyClpC to be a dimer but in presence of ATP and casein it forms a hexamer similar to other Hsp100 proteins. ATPase activity of SyClpC was found to be optimal between 37°C and 45°C within a pH range of 7.0 to 8.0. The presence of an adaptor protein MecA from *B. subtilis* enhanced the refolding activity of SyClpC by 3-fold. Since cyanobacteria lack MecA orthologs, the role of ClpS adaptor proteins (ClpS1 and ClpS2) in enhancing the ATPase activity of SyClpC was tested but SyClpS1 did not stimulate the ATPase activity of SyClpC. But it may be worthwile to mention that ClpS adaptor proteins bind directly to N-terminal destabilizing residues on the substrate protein, targeting them for degradation by ClpAP (Dougan *et al.*, 2002; Erbse *et al.*, 2006). Stanne *et al.* (2007) reported the existence of two mixed proteolytic cores. The first one consisted of ClpP1 and ClpP2 subunits associated with ClpX while the second one contained ClpP3 and ClpR that were bound to ClpC. Of these two, the latter was shown to be constitutively abundant and was required for cell viability and growth (Clarke and Eriksson, 1996; Schelin *et al.*, 2002). The localization of all Clp proteins in *S. elongatus* PCC 7942 was found to be in the soluble fraction except that a minor proportion (20–40%) of ClpC, ClpP1 and ClpR was also associated with membrane fraction. The levels of the Clp proteins in mutants defective in *ClpPI* and *ClpPII* were compared with the wild-type in relation to their localization. Consistent with the earlier observations of Schelin *et al.* (2002) mutant defective in ClpP1 showed a down-regulation of ClpP2 with a simultaneous increase in ClpP3 and ClpR. The localization of the latter protein increased in the membrane fraction in ClpP1 mutant when compared to wild-type although the former protein was associated in traces with the membrane fraction. The levels of ClpX were higher in the ClpP1-deficient mutant without any apparent change in the levels of ClpC, ClpS1 and ClpS2 relative to wild-type. In ClpP2-defective mutant except that the ClpP1 was associated with membranes in the absence of soluble ClpP2 protein, the levels of rest of the proteins resembled those of ClpP1-defective muant. Andersson *et al.* (2009) purified the ClpP3/ClpR complex and found it to consist of a heptameric complex with three and four units of ClpP3 and ClpR, respectively. This complex exhibited remarkable similarities to the eukaryotic 20 S proteosome.

Nitrogen straved cells of cyanobacteria generally exhibit chlorosis due to the degradation of phycobiliproteins. Although the detailed mechanism of degradation is not yet known, the role of "non-bleaching genes" has been identified after screening a number of non-bleaching (*nbl*) mutants of *S. elongatus* PCC 7942. Of these genes, *nblA* encodes a protein NblA that has been directly implicated in the degradative process. Crystal structure of NblA polypeptide from *Anabaena* sp. strain PCC 7120 revealed the existence of two α-helices, a shorter N-terminal and a longer C-terminal one. Through their C-terminal domains, the NblA molecules undergo dimerization leading to a homodimer formation (Bienert *et al.*, 2006). This has been confirmed from the crystallographic data from the NblA proteins of *Thermosynecococcus vulcanus* as well (Dines *et al.*, 2007). The interaction of NblA with the α-subunits of phycobiliproteins is mediated through the highly conserved amino acid residues at the

C-terminus (Bienert *et al.*, 2006). Karradt *et al.* (2008) showed that the NblA protein interacts with the ClpC from *Anabaena* sp. strain PCC 7120 with its highly conserved N-terminal helix to bring about the degradation. The gene encoding the Clp protein, identified as *alr2999*, initially was annotated as an "endopeptidase Clp ATP-binding chain ClpB" (bacteria.kazusa.or.jp/cyanobase/) but the BLAST search against the NCBI database (www.ncbi.nlm.nih.gov/BLAST/) revealed it to be identical (87%) to ClpC from *Synechocystis* sp. strain PCC 6803. Based on the site-directed mutagenesis at the highly conserved N-terminal domain and the C-terminal domain, the interactions of NblA protein with $ClpC_{Alr2999}$ protease and phycobiliproteins, respectively have been confirmed.

b) Deg Proteases: This family of ATP-dependent serine endopeptidases, wide-spread in bacteria and eukaryotes, is responsible for the degradation of misfolded and abnormal proteins of the cell membrane or periplasmic compartment (Strauch and Beckwith, 1988; Lipinska *et al.*, 1990; Kim *et al.*, 1999). This family is also known as Htr family of proteases and includes three related proteins DegP (HtrA), DegQ (HhoA) and Deg S (HhoB). These are very flexible with regard to their substrate specificity. Some of these cleave only one substrate but others act as general proteases on unfolded substrates. The proteolytic activity of these proteases is regulated by the PDZ domain [PDZ domain represents a common structural domain of 80–90 amino acids found generally in the signalling proteins of bacteria, yeasts, plants, viruses and animals. PDZ combines the first letters of three such proteins which were discovered for the first time, i.e. post-synaptic density protein, Drosophila disc large tumor suppressor (DlgA) and Zonula occludens-1 protein]. PDZ is also referred as DHR (Dlg homologous region) or GLGF (Glycine-Leucine-Glycine-Phenylalanine) domains. DegP of *E. coli* is a heat stress protease that functions in a homohexameric state in the periplasm (Kim *et al.*, 1999; Sassoon *et al.*, 1999) and is required for cell viability at elevated temperatures or under oxidative stress (Strauch *et al.*, 1989; Lipinska *et al.*, 1990; Skorko-Glonek *et al.*, 1997, 1999). At low temperatures DegP functions as a molecular chaperone while at high temperatures it develops proteolytic activity (Spiess *et al.*, 1999). Crystal structure of DegS of *E. coli* revealed that the PDZ domain is extended sideways to form a funnel shaped structure and the proteolytic centre existed in an inactive conformation. Specific interactions leading to a conformational shift and the activation of the proteolytic centre are required (Wilken *et al.*, 2004). In contrast, DegP of *E. coli* consists of two PDZ domains, PDZ1 and PDZ2. A cage-like hexameric complex is produced by the assembly of two trimers (Clausen *et al.*, 2002; Krojer *et al.*, 2002).

The genome of *Synechocystis* sp. strain PCC 6803 revealed the presence of three genes encoding HtrA (*htrA, slr1204*), HhoA (*hhoA, sll1679*) and HhoB (*hhoB, sll1427*) (Kaneko *et al.*, 1996). Though designated similar to Deg proteases of *E. coli*, these appear to be more similar to each other than to *E. coli* Deg proteases in their structure and function (Kieselbach and Funk, 2003; Huesgen *et al.*, 2005; Jansen *et al.*, 2005). HhoA is a soluble protein located in the periplasm and its association with plasma membrane has been demonstrated (Fulda *et al.*, 2000; Huang *et al.*, 2006). The role of Deg proteases in protecting the cells from high light-induced damage was examined in a glucose tolerant strain of *Synechocystis*. A comparison of a triple mutant (involving the step-wise inactivation of *hhoA*, then *hhoB* double mutant and finally *htrA*) with wild-type revealed that the amount of D1 in the mutant was estimated to be approximately 60% of the original level. But the amounts of FtsH protease in the wild-type and mutant were similar as found through immunodetection assy. These observations led them to conclude that these Deg proteases are involved in the resistance of *Synechocystis* to light stress and play either a direct or indirect role in the repair of PSII *in vivo* (Silva *et al.*, 2002). Since the triple mutant did not survive at elevated growth temperatures or high light intensities (Silva *et al.*, 2002; Barker *et al.*, 2006), it was concluded that that the Deg proteases protect *Synechocystis* during heat and light stresses. The essentiality of Deg proteases for PSII repair and D1 degradation was re-examined by constructing a triple mutant (*htrA⁻hhoA⁻hhoB⁻*) using the wild-type strain of *Synechocystis*. The wild-type, single and double mutants grew at 29°C or 37°C but the triple mutant was unable to grow at high temperature. Thus Barker *et al.* (2006) demonstrated unambiguously that although the Deg proteases are needed for photoprotection of *Synechocystis*, they do not play essential role in D1 protein turn over and PSII repair. However, their role in protecting the extra-cytoplasmic compartments from heat and light stress as well as from the oxidative damage caused by the ROS generated by photosynthetic electron transport has been stressed (Florian *et al.*, 2005).

The properties of HhoA protease from *Synechocystis* sp. strain PCC 6803 have been studied and the formation of hexameric complexes was shown to be regulated by the PDZ domain. *Synechocystis* HhoA protein consisted of a single PDZ domain whose activity increased with a rise in temperature and pH. The presence of Mg^{2+} or Ca^{2+} stimulated the proteolytic activity of HhoA (Huesgen *et al.*, 2007). Deg/HtrA proteases constitute important components of a network for PSII quality control in chloroplasts and cyanobacteria. The functional role of Deg proteases in maintaining protein quality control including PSII has been highlighted (Huesgen *et al.*, 2009).

c) FtsH proteases: This family is represented by bacterial FtsH protease that is a membrane integral ATP-dependent peptidase (Tomoyasu *et al.*, 1993). It participates in various degradative processes involving the degradation of translocase subunit SecY, heat stress transcription factor σ^{32} and transcriptional activator λCII (Herman *et al.*, 1995; Kihara *et al.*, 1995). It exists as a homodimeric or/and tetrameric structure (Akiyama *et al.*, 1995) and this multimeric structure is required for the catalytic activity (Karata *et al.*, 1999). Akiyama and Ito (2000) reported that the transmembrane organization is essential for the proteolytic activity against the integral membrane proteins. *Synechocystis* sp. strain PCC 6803 possesses four genes (*FtsH1*, *FtsH2*, *FtsH3* and *FtsH4*) that encode four respective FtsH proteases. Gene inactivation studies revealed that two mutations (FtsH1 and FtsH3) were lethal whereas the third (FtsH4) did not show any altered phenotype. The fourth mutation involving FtsH2 caused altered pigmentation leading to impaired PSI biogenesis (Mann *et al.*, 2000). FtsH2 of *Synechocystis* sp. strain PCC 6803, a homologue of *Arabidopsis* FtsH (Bailey *et al.*, 2001), is important in controlling the degradation of D1 protein during its repair from photoinhibition (Zach *et al.*, 2005). Mutants lacking FtsH2 (*slr0228*) of *Synechocystis* showed impaired rate of degradation of D1 (Silva *et al.*, 2003; Komenda *et al.*, 2006). It is not that the FtsH2 is involved in the removal of the breakdown products of D1 but its function is envisaged at an early stage of D1 degradation (Haussuhl *et al.*, 2001). D1 degradation is not completely blocked in mutants of *Synechocystis* that show extremely high rates of D1 degradation (Komenda *et al.*, 2010). These evidences put together with those of Barker *et al.* (2006) that selective turnover of D1 was still operational in triple mutants of Deg proteases, would suggest the possibility of other FtsH proteases being associated with this process. An *FtsH2* gene construct with a glutathione S-transferase tag (GST-tagged) was introduced into *Synechocystis* sp. strain PCC 6803 and when GST-tagged FtsH2 was purified, FtsH3 was also copurified along with it (Barker *et al.*, 2008). In addition, preliminary negative stain electron microscopy revealed FtsH2 to exist as hexameric complexes instead of being a dimer. So it means the *Synechocystis* FtsH proteases can exist as hetero-oligomeric complexes, consisting of two types of FtsH subunits.

The involvement of FtsH and Deg proteases in the degradation of UV-B- damaged PSII reaction center subunits D1 and D2 proteins of *Synechocystis* sp. strain PCC 6803 has been investigated. Gene inactivation experiments showed that it is the FtsH not Deg proteases that are responsible for the repair of PSII during and following UV-B irradiation. FtsH mutants (gene *slr0228*) exhibited increased sensitivity to UV-B radiation and the repair activity of the cells was impaired in visible light after UV-B exposure but Deg-G mutants resembled the wild-type in their inactivation and reactivation kinetics. Further, the degradation of UV-B damaged D1 and D2 proteins and their recovery during the repair period was retarded in FtsH mutants as revealed by immunoblot assays (Cheregi *et al.*, 2007).

d) Ctp proteases: This family of carboxy-terminal peptidases belongs to the C-terminal processing peptidases. Tsp peptidase of *E. coli* is involved in the degradation of mistranslated proteins that are targeted for degradation. Such proteins acquire *ssrA* sequence at their C-terminus which can be specifically recognized by the Tsp peptidase (Jentsch, 1996; Keiler *et al.*, 1996). This peptidase has a characteristic PDZ domain at its C-terminus that is required for substrate recognition (Ponting, 1997; Beebe *et al.*, 2000). *Synechocystis* sp. strain PCC 6803 has three genes that encode Ctp-peptidases, Ctp A, Ctp B and Ctp C that are homologous to Tsp peptidases of *E. coli* (Hara *et al.*, 1991; Silver *et al.*, 1992). The cyanobacterial Ctp peptidases reside in the periplasm (Fulda *et al.*, 2000; Zak *et al.*, 2001). The functional role of the *ctp* genes in *Synechocystis* has been examined in relation to processing of the precursor form of D1 (pD1) with its C-terminal extension. In *Synechocystis* the carboxy-terminal extension in pD1 is 16 amino acids long (Nixon *et al.*, 1992). For correct integration of D1 into tetramanganese cluster essential for catalysis of water oxidation reaction, the C-terminal extension has to be removed by the CtpA peptidase (Diner *et al.*, 1988). Shestakov *et al.* (1994) identified the role of CtpA peptidase in the biogenesis of photosynthetic machinery by the analysis of *Synechocystis* sp. strain PCC 6803 CtpA-deficient mutant. This mutant was unable to grow under photoautotrophic conditions and showed impaired C-terminal processing of pD1. By insertion of kanamycin resistance gene in the sequence of *ctpA* of *Synechocystis*, a mutant was isolated that showed D1 protein larger (by 2 kDa) than in the wild-type cells. Though the PSII complexes of the mutant exhibited functional reaction centres of PSII complex, it was unable to accept electrons from water (Anbudurai *et al.*, 1994). The amino acid residues Ser313 and Lys338 have been found to be critical for *in vivo* catalytic activity of CtpA as identified by X-ray crystal structure analysis of CtpA from *Scenedesmus obliquus* (Liao *et al.*, 2000). Detailed analysis of site specific mutations introduced at the conserved residues of CtpA in *Synechocystis* revealed three additional amino acid residues (Asp253, Arg255 and Glu316) important for the catalytic activity. These findings are supported by the photoautotrophic growth potential of the mutants along with their ability to process pD1 protein and the amount of pD1 versus mature D1 inside the cells (Inagaki *et al.*, 2001a,b). CtpC gene was found to be essential for cell viability as it was not possible to obtain complete gene-inactivated mutants (Ivleva *et al.*, 2002). Complete inactivation of CtpB did not lead to any visible phenotypic or physiological changes.

e) Spp family: This family is characterizd by the presence of two signal peptide peptidases differing in their molecular mass, with the shorter version representing the C-terminal domain of the longer one. *Synechocystis* sp. strain PCC 6803 has two genes *SppA1* and *SppA2* and a homologue of the former gene is found in *Arabidopsis thaliana* (Lensch *et al.*, 2001).

iv) Photosynthesis and heat shock: Due to heat shock, photosynthetic apparatus appears to be affected first before the impairment of any other metabolic process. Chlorophyll *a* fluorescence provides an important intrinsic probe for measuring the thermal stability of PSII-protein complexes. Heat inactivation of photosynthesis can as well be measured in terms of O_2 evolution. A disturbance in the integrity of thylakoid membranes causes damage to the integral proteins of the photosynthetic electron transport. Nishiyama *et al.* (1993) showed that thermal adaptation of photosynthetic apparatus very much depended on the growth temperature of *Synechococcus* sp. strain PCC 7002, a mesophilic cyanobacterium. The temperatures required for 50% inactivation of PSII activity were found to be 44°C and 48°C for cells grown at 25°C and 40°C, respectively. PSII activity thus appears to be most sensitive to heat and can adapt to high temperature. PSII-mediated electron transport showed greater thermal stability in thylakoid membranes isolated from cells which had adapted to high temperature than those from non-adapted cells. When the thylakoid membranes were isolated in presence of glycine betaine, such membranes were 40% more active than those prepared in its absence. Subsequent studies of these workers identified cytochrome *c-550* and PsbU (a constitutive extrinsic protein of PSII complex) as essential components of the PSII complex for protecting *Synechococcus* sp. strain PCC 7002 against heat-induced inactivation. These two protein molecules together provided higher thermal stability than either of these individually (Nishiyama *et al.*, 1994, 1997). Targeted mutagenesis of *Synechococcus* sp. strain PCC 7002 for *psbU* gene yielded mutants ($PsbU^-$) that were no longer able to adjust to a rise in growth temperature and lost the ability of thermal adaptation. The levels of heat shock proteins (Hsp70, Hsp60 and Hsp17) in wild-type and $PsbU^-$ cells remained the same suggesting that mutation in *psbU* did not affect the general heat shock response but it is the thermal acclimatization process that is affected. $PsbU^-$ cells grown at 25°C when exposed to 49°C for 10 min did not survive but when grown at 38°C only less than 10% of the $PsbU^-$ cells survived, in contrast to 80% survival exhibited by wild-type (Nishiyama *et al.*, 1999).

Light-mediated expression of heat shock genes (*htpG*, *groESL1*, *groEL2* and *hspA*) in *Synechocystis* sp. strain PCC 6803 has been studied by measuring the corresponding transcripts and protein levels in cells pre-treated (grown at 30°C in presence of 35 $\mu E\ m^2\ s^{-1}$) at 42°C for 1 hr in presence or absence of light and given a final shock at 48°C for 1 hr in presence (35 $\mu E\ m^2\ s^{-1}$) or absence of light. Light accelerated the expression of the genes whereas in dark except for *hspA* the expression of other genes was not as intense as in light. The kinetics of induction was different for the Hsp mRNAs. As for example, *htpG* gene showed the greatest level within 10–20 min after a temperature shift-up whereas *hspA* and *groEL2* mRNAs reached after 20 to 60 min and that of *groESL1* was slowest (Asaldulghani *et al.*, 2003).

Various proposals have been put forward to explain the synthesis of Hsps. One of the earliest models was based on the accumulation of denatured proteins inside cells that provides a sufficient signal for Hsp synthesis (Craig and Gross, 1991; Bukau, 1993). The second is the ribosome sensor hypothesis that transduces the signal of heat stress at the level of translational process to increase the production of Hsps. The third one was based on the extent of accumulation of ROS that could trigger the production of Hsps. Experimental evidences in support of any one of these is lacking. The first attempt in this direction was made by Horváth *et al.* (1998) who discovered a "fluidity" gene that encodes a small Hsp (sHsp), Hsp17. This protein (17 kDa), mostly associated with thylakoid membranes, acts as sensor for detecting thermal fluctuations and induced the synthesis of Hsps in *Synechocystis* sp. strain PCC 6803. Treatment of cells with benzyl alcohol (BA) caused a fluidizing effect on the membranes resulting in a reduction of thermotolerance. BA treated cells when subjected to heat stress the thermotolerance further decreased suggesting that both of these acted synergistically. This was substantiated by data on O_2 evolution and chlorophyll *a* fluorescence. Selective hydrogenation of cytoplasmic membrane without changing the fluidity of their thyakoid membranes was also chosen as a parameter to study the induction of Hsps. A Northern blot analysis of RNAs from cells exposed to heat stress (after acclimatizing to contrasting temperatures 22°C vs 36°C and treated with BA or catalytically saturated selectively) revealed that induction of *dnaK*, *groESL*, *cpn60* and *hsp17* genes occurred at significantly higher temperatures for cells grown at 36°C than those grown at 22°C. The threshold temperature was found to be 44°C for maximum induction for cells grown at 36°C but this decreased to 38–40°C for cells grown at 22°C. It means there is no particular temperature set point for the activation of heat shock genes. Moreover, unlike the induction of other Hsps, *hsp17* was induced at temperatures 44°C and above for samples grown at 36°C. A redox signalling pathway explains the diffusion rate of plastoquinone molecule and other mobile electron carriers also determine the thylakoid integrity and in turn the acquisition of thermotolerance (Huner *et al.*, 1996). In addition, the induction of *groESL* and *cpn60* were shown to be controlled by the redox state of the thylakoid membranes (Glatz *et al.*, 1997). Due to the absence of heat shock response when cytoplasmic membranes were selectively hydrogenated, these investigators concluded that the thylakoid membranes act as a cellular thermometer to transduce signals of heat stress. Török *et al.* (2001) substantiated the role of sHsp from *Synechocystis* sp. strain PCC 6803 and showed that Hsp17 plays a dual role in stabilizing heat-stressed membranes as well as transferring unfolded proteins to DnaK/DnaJ/GrpE and GroEL/ES chaperone network for subsequent refolding. The function of stabilizing the membranes depended on its preference for the liquid crystalline phase. So depending on whether Hsp17 exists in association with thylakoid membranes or cytosolic fraction, it may function as a membrane stabilizing factor or as a member of multi-chaperone network. These roles of Hsp17 were confirmed by studies on hsp17$^-$ deletion mutant. The sHsps (α-crystallin and Hsp17) from *Synechocystis* sp. strain PCC 6803 interacted with the polar head group of synthetic anionic membranes of dimyristoylphosphatidylglycerol and dimyristoylphosphatidylserine and strongly stabilized the liquid-crystalline state. Likewise, these two sHsps prevented the formation of inverted hexagonal structure and stabilized the bilayer liquid-crystalline state in membranes composed of nonbilayer lipid didaidophosphatidylethanolamine. Membranes consisting of MGDG and PG (enriched with unsaturated fatty acids and isolated from *Synechocystis* thylakoids) exhibited an increased molecular order in the fluid-like state after interaction with the sHsps. These results suggest that the sHsps can modulate membrane lipid polymorphism (Tsvetkova *et al.*, 2002). Giese and Vierling (2002) reported a deletion mutant (hsp16.6$^-$) of *Synechocystis* sp. strain PCC 6803 that showed a conditional lethal phenotype. The oligomeric nature (approximately 20 subunits) and its potential to bring about refolding of bound denatured proteins in association with DnaK/DnaJ chaperone system has also been reported (Mogk *et al.*, 2003). The role of sHsps in protecting wide range of cellular functions of *Synechocystis* sp. strain PCC 6803 during heat stress has come to light by employing immunoprecipitation and affinity chromatography. As many as 42 different proteins have been recovered which exhibited interactions with Hsp16.6 of *Synechocystis*. These were released by the ATP-dependent activity of DnaK and co-chaperones and all of them were found to be heat labile. Some of the proteins (13) identified by mass spectrometry related to metabolic processes ranging from transcription, translation, cell signalling and secondary metabolism. A comparative study of the wild-type and sHsp 16.6 deletion mutant showed that sHsp 16.6 contributed significantly to the acquisition of cellular thermotolerance (Basha *et al.*, 2004). The dynamic nature of the interactions between the subunits of sHsp oligomers has been investigated in a number of systems like wheat (Hsp 16.9), yeast (Hsp28), *Synechocystis* (Hsp16.6) and *Methanococcus tuberculosis* (Hsp16.3) and their reversible dissociation into subunits upon heating was observed (Haslbeck *et al.*, 1999; van Montfort *et al.*, 2001; Giese and Vierling, 2002; Gu *et al.*, 2002). In order to understand the interactions of the subunits of *Synechocystis* sp. strain PCC 6803 sHsp16.6 oligomers, Friedrich *et al.* (2004) designed the sHsp16.6 gene constructs with a C-terminal 8 amino acid tags (WSHPQFEK represnting Trp-Ser-His-Pro-Gln-Phe-Glu-Leu, respectively). *E.coli* (BL21) cells were transformed with such pJC2016.6 plasmid and the transformants produced the C-terminal tagged sHsp16.6. Separation of C-terminal tagged sHsp 16.6 by streptactin (resin) column chromatography and purification to >95% homogeneity was achieved, whereas the wild-type protein did not interact with streptactin-affinity resin. Such purified sHsp16.6 subunits formed oligomers containing 20 or 22 subunits *in vitro*. The addition of C-terminal tag did not cause any increase in the size of the oligomers and heat denatured luciferase used as substrate was reactivated *in vitro* by the interactions of sHsps and other molecular chaperones. In addition, strep-tagged Hsp18.1 from pea also exhibited similar *in vitro* interactions except that the strep-tagged and wild-type Hsp18.1 could associate to form dodecameric oligomers that were slightly larger than the wild-type Hsp18.1. Their interaction with the substrate molecules *in vitro* conform to the behaviour of their wild-type counterparts *in vivo*.

Balogi *et al.* (2005) identified a "heat-shock lipid" in the thylakoids that helps in the acclimatization of cyanobacteria during heat/light stress. This lipid, known as highly saturated monoglucosyldiacylglycerol (MGlcDG), increased in its level during stress. *In vitro* experiments carried out with thylakoid MGlcDG liposomes showed them to be quite stable even at extremely high temperatures. Moreover, of the five polar lipids tested, only MGlcDG strongly interacted with the sHsp (Hsp17) from *Synechocystis* sp. strain PCC 6803. Thus it was concluded that MGlcDG and Hsp17 play an important role in preserving functional integrity of thylakoid membranes and help in the development of acquired thermotolerance in light- or heat-stressed cyanobacteria.

S. vulcanus, a thermophilic cyanobacterium, constitutively expresses a sHsp that showed high homology to Hsp16.6 from the mesophilic cyanobacterium *Synechocystis* sp. strain PCC 6803. The gene *hspA* encoding this protein was cloned (Roy and Nakamoto, 1998). The gene product HspA showed homology to the α-crystallin-related sHsps from other organisms. HspA formed a large homo-oligomer consisting of 24 subunits that prevented thermal aggregation of model substrates such as porcine malic dehydrogenase at 45°C and 50°C and citrate synthase at 50°C (Roy *et al.*, 1999). In order to understand the role of sHsp in conferring cellular thermotolerance, Nakamoto *et al.* (2000) cloned *hspA* from *S. vulcanus* and used it to transform *S. elongatus* PCC 7942 with an expression vector p-ECT under the control of *tac* promoter. The transformant (ECT-16-1) constitutively expressed 16.6 kDa protein that accumulated in soluble and insoluble fractions. Viability of ECT-16-1 cells when treated at 50°C for 15 min was enhanced when compared to wild-type cells. Heat-adapted cells (45°C for 1 hour) when exposed to a temperature of 50°C for 60 min resulted in a 1000 and 10,000 fold increase in the survival at 50°C in case of wild-type and ECT-16-1 cells, respectively. HspA plays a role in preserving the integrity of thylakoid membranes under heat stress and its constitutive expression did not affect the expression of other Hsps such as GroEL and DnaK. However, *hspA* deletion mutant of *Synechocystis* sp. strain PCC 6803 constitutively expressed higher levels of GroESL1 and GroEL2 proteins under normal conditions (Asadulghani *et al.*, 2004). Nitta *et al.* (2005) showed that cultures of ECT-16-1 when subjected to heat shock (shifted to 45°C or 50°C for 15 min) or exposed to high light (1000 $\mu E.m^{-2}\ s^{-1}$ at 30°C), released HspA from thylakoids. This suggested that under normal conditions thylakoid membranes act as the reservoir for the HspA molecules. Under stressed conditions, the HspA protein molecules get dissociated from the thylakoids and interact with the non-native proteins as well as proteins from thylakoids and prevent them from aggregation. Immunocytochemical evidences confirmed the localization of the HspA protein molecules in the cytosol under stress conditions. This probably leads to the induction of GroES/GroEL and DnaK/DnaJ/GrpE chaperone networks. Additional evidences in favour of the sHsps to confer stress tolerance by protecting the integrity of thylakoid membranes were presented by Sakthivel *et al.* (2009). ECT-16-1 cells not only exhibited thermotolerance but also survived oxidative stress by exhibiting much better growth and viability in H_2O_2 than the wild-type strain. Moreover, all the pigment molecules (chlorophylls, carotenoids and phycocyanins) appeared normal in ECT-16-1 cells when compared to a continuous reduction in their content in the wild-type cells. Comparison of wild-type and *hspA* deletion mutant of *Synechocystis* sp. strain PCC 6803 also confirmed the role of sHsp in stabilizing the pigment molecules and thylakoid integrity. As *Synechocystis* constitutively expresses HspA protein it could overcome the oxidative stress but the *hsp⁻* mutant showed a considerable decrease in its viability and pigment content under oxidative stress. Nakamoto and Honma (2006) demonstrated that HspA from *S. elongatus* PCC 7942 interacted directly with PCs and suppressed their inactivation inder denaturing conditions. A molecular ratio of four HspA monomers per one PC (αβ) monomer gave maximum protection.

In order to understand the regulation of *hspA* gene, the *hspA* gene from *S. vulcanus* was overexpressed in *E. coli* under the control of *lac*-promoter. The transcription of *hspA* occurred constitutively at 30°C and 42°C in *E. coli* cells but the expression of *hspA* was low even in presence of the inducer Isopropyl βD-1 thiogalactopyranoside at 30°C. Heat induction of *hspA* occurred despite the replacement of native *hspA* promoter with *lacZ* promoter or by the addition of rifampin. These results suggest that the primary form of heat regulation is at the post-transcriptional level (Kojima and Nakamoto, 2005). Nakamoto and Vígh (2007) reviewed the current status of the sHsps in stress management and highlighted their role in membrane quality control and in maintaining the integrity of membranes under stress conditions.

To identify the components associated with thermal acclimatization of PSII in *Synechocystis* sp. strain PCC 6803 whole-genome microarrays have been conducted (Rowland *et al.*, 2010). The acclimatization of PSII to elevated temperatures lasted 480 min during which as many as 176 genes were expressed. These can be divided into seven distinct response profile groups of which early transient phase and sustained phase have been found to be important. The early transient phase is characterized by the expression of a number of stress response genes including Hsps and in the sustained phase genes involved in phycobilisome assembly and modification, photosynthesis, respiration, lipid metabolism and motility. Specific gene knock-outs related to heat shock response suggested that all mutants showed a lower PSII rates under normal growth conditions. Mutations in *clpB1*, *cpcC2*, *hspA*, *htpG* and *slr1674* (encoding a protein of unkown function) affected basal PSII thermotolerance. Long-term thermal acclimatization was affected by mutations in *cpcC2*, *hik34*, *hspA* and *hypA1*.

Apart from the induction of Hsps and their role in conferring thermotolerance in various cyanobacterial systems, other stress-inducible genes have been shown to play an important role. In this direction, iron-stress inducible (*isi*) genes, *isiA* (Laudenbach and Straus, 1988) and *isiB* (Laudenbach *et al.*, 1988) and the corresponding proteins protect cyanobacterial cells from oxidative and heat stresses. The *isiA* gene encodes IsiA or CP43′, a homologue of the CP43 chlorophyll-binding proteins (Burnap *et al.*, 1993) that act as antenna for photosystems and light energy dissipation in PSII (Park *et al.*, 1999; Sandström *et al.*, 2001; Cadoret *et al.*, 2004; Ihalainen *et al.*, 2005; Havaux *et al.*, 2005). Flavodoxin is the gene product of *isiB* and helps the organisms to substitute for ferredoxin in iron-limited habitats (Straus, 1994). These two genes exist as an operon in cyanobacteria like *Syenchocystis* sp. strain PCC 6803, *S. elongatus* PCC 7942 and *Synechococcus* sp. strain PCC 7002 (Laudenbach and Straus, 1988; Leonhardt and Straus, 1992; Burnap *et al.*, 1993; Straus, 1994) whereas in others they are either dispersed (*Anabaena* sp. strain PCC 7120; Leonhardt and Straus, 1994) or only one *isiA* homologue is present (*T. elongatus* BP-1) or both are absent (*Synechococcus* sp. strain WH8102). In addition to their induction in iron-limited habitats, induction of *isiA* and *isiB* genes under high salt, high temperature (Vinnemeier *et al.*, 1998), oxidative (Jeanjean *et al.*, 2003; Yousef *et al.*, 2003) and high-light stresses (Havaux *et al.*, 2005) has also been reported. Studies on *isiA* interrupted and *isiB* deletion mutants of *Synechocystis* sp. strain PCC 6803 suggested their involvement in countering heat and oxidative stresses. The presence of another gene *isiC* has been demonstrated downstream of *isiB* which was co-transcribed with *isiB*. The involvement of *isiB* and *isiC* in countering oxidative stress and the role of *isiA* and *isiB* in the management of heat and light stresses immediately after heat shock have been stressed. The optimization of light energy supply after a heat shock seems to be an essential step in preventing cellular damage as well as in the recovery from heat stress (Kojima *et al.*, 2006).

v) Regulation of Hsp synthesis: Heat shock responses involve protein unfolding and malfunction leading to cell death. The synthesis of Hsps counters this effect by refolding the denatured proteins and the proteases degrade those proteins which are not recoverable. The regulation of Hsp synthesis is well understood in case of *E. coli*. The Hsps in this organism nearly account to 2% and 20–25% of the total proteins when grown at 30°C and 46°C, respectively. This metabolic burden is maintained by the regulatory role of σ^{32} (a transcription factor) that is bound to the RNA polymerase (RNAP) which recognizes the promoters of the heat shock genes so that their transcription and translation can occur. It means the synthesis, activity and stability of σ^{32} constitute important events in signal perception and transduction of heat shock response. However, in cyanobacteria there are as yet no evidences for the involvement of any transcription factor that regulates the synthesis of Hsps. A sigma factor related stress response has been well established in case of *B. subtilis* also (Yura *et al.*, 2000; Dartigalongue *et al.*, 2001; Helmann *et al.*, 2001; El-samad *et al.*, 2005). In addition, several repressor proteins and their binding sites have been reported. In many of the gram-positive bacteria, HrcA repressor protein and its binding site CIRCE (for controlling inverted repeat for chaperone expression) have been reported (Narberhaus, 1999). CIRCE is an inverted repeat consisting of 9 bp separated by a 9 bp-spacer. The repressor protein HrcA negatively regulates the expression of *grpE-dnaK-dnaJ*- and *groESL* operons in *B. subtilis* (Hecker *et al.*, 1996; Narberhaus, 1999). The existence of a CIRCE element upstream of *hrcA* gene in *B. subtilis* signifies that the *hrcA* gene is subject to autoregulation. Another protein, HspR repressor and its binding site HAIR (a cis-element, HSPR-associated inverted repeat) have been found in *Streptomyces coelicolr* and *Mycobacterium tuberculosis* that regulate *dnaK* operon and *clpB* (Stewart *et al.*, 2002; Bucca *et al.*, 2003). In *B. subtilis*, a CtsR system (Derre *et al.*, 1999) and the RheR repressor in *Streptomyces albus* G (Servant *et al.*, 1999) have been reported.

Webb *et al.* (1990) first reported a CIRCE element in the 5′-UTR region of *groESL* operon in *S. elongatus* PCC 7942, but the corresponding repressor protein (HrcA) was not described. Nakamoto *et al.* (2003) conducted a search of 5′-upstream regions of a number of cyanobacterial *groE* genes and found the CIRCE sequence (5′-TTAGACTC-N9-GAGTGC-3′) with the exception of *S. vulcanus groEL2* (Furuki *et al.*, 1996). The transcriptional start sites of *groESL1* operons from *Synechocystis* sp. strain PCC 6803 and *S. elongatus* PCC 7942 have transcriptional start sites at the same nucleotide located within the CIRCE. A search of many cyanobacterial genomes for *hrcA* gene sequence (http://www.kazusa.or.jp./cyano/) revealed the presence of *hrcA* homologues in *Synechocystis* sp. strain PCC 6803, *T. elongatus* BP-1 and *Anabaena* sp. strain PCC 7120. The deduced amino acid sequence of HrcA was homologous to the other HrcA proteins. The absence of a CIRCE sequence upstream of the cyanobacterial *hrcA* genes indicates that the *hrcA* is not subject to autoregulation as reported in case of *B. subtilis*. In *Synechocystis* sp. strain PCC 6803, gene *sll1670* encodes HrcA and this regulates the *groESL1* operon and the *groEL2* gene. The *sll1670* gene inactivated mutants of *Synechocystis* sp. strain PCC 6803 showed a depression in the activity of *groESL1* and *groEL2* genes but the expression of other heat shock genes remained unaffected. The fact that the expression of *groESL* genes in the mutant is not completely repressed and the increase in their expression after a heat shock suggests that there are two different regulatory mechanisms. The first one is a negative regulation governed by HrcA and a second unknown positive one. Under normal growth conditions, the interaction of

HrcA with CIRCE repressed the transcription of *groEL1* and *groEL2*. But under heat stress conditions the inactivation of HrcA led to an induction of *groESL* genes. It is interesting to know that the *hrcA* mutant exhibited overexpression of GroEL that conferred improved thermotolerance associated with a reduced photobleaching of phycocyanin under heat stress. These workers also found consensus promoter sequences of the *E. coli* major sigma factor upstream of the CIRCE regions in all the *groESL* operons examined suggesting that *groESL* operon in cyanobacteria is also recognized by a major sigma factor that may confer a positive regulation.

Regulation of expression of *groESL* operon by two different small proteins in *S. elongatus* PCC 7942 and *Synechocystis* sp. strain PCC 6803 has been reported. Cloning and characterization of *orf 7.5* from *S. elongatus* PCC 7942 encoding a sHsp was reported (Nakamoto *et al.*, 2001). The transcripts of *orf 7.5* were barely detectable in cells grown at 30°C but a heat shock (40°C or 45°C) caused a transient increase in the levels of this transcript. Targeted gene disruption experiments led to the isolation of mutants that were unable to grow at 45°C and lost the ability for acquired thermotolerance at 50°C. The requirement of *orf 7.5* gene product in conferring thermotolerance was confirmed by complementing a wild-type *orf 7.5*. Due to a strong reduction in the levels of groESL transcript in the *orf 7.5* mutant, it was concluded that *orf 7.5* controls the expression of *groESL* operon. Sato *et al.* (2007) conducted primer extension analysis of *dnaK2* promoter region of *S. elongatus* PCC 7942 and recognized a 20 bp region designated as MARS (multi-stress associated regulatory sequence) that is found to be essential for stress induction of *dnaK2* gene. Although this region possesses inverse repeats that can act as probable recognition sites for a trans-factor, the probable interactions between these two are yet to be unfolded.

The regulation of sHsp genes in many bacterial species has been reported by the specific DNA sequences and their corresponding binding repressor proteins. The *ibpA* and *ibpB* sHSP homologs present in *E. coli* are regulated by σ^{32} (Allen *et al.*, 1992). In *S. albus* the sHSP gene *hsp18* is regulated by the RheA repressor that interacts with the inverted repeat present around the promoter region (Servant *et al.*, 1999, 2000). Another DNA element present at the 5′-UTR region of *hspA* genes in *Bradyrhizobium japonicum* is ROSE (repression of heat shock gene expression) that folds into stem-loop like structure in the corresponding mRNA and masks the ribosome-binding site to inhibit translation. Under heat stress, the melting of the secondary structure brings about the translation (Nocker *et al.*, 2001). In cyanobacteria also such sequences and the corresponding binding proteins in *hspA* gene regulation have been reported. A novel AT-rich imperfect inverted repeat (ACAAAgcAAA-TTTagTTGT) was detected at 5′UTR region of *hspA* gene sequence of *S. vulcanus*. The regulation of the *hspA* gene activity by a putative DNA binding-protein has been envisaged. A repressor protein from the unstressed cells of a closely related thermophilic *T. elongatus* BP-1 was isolated and its DNA-binding activity to the above sequence both *in vitro* as well as *in vivo* was lost at a heat shock temperature. The size of *hspA* mRNA (650 nucleotides) and the time course of its accumulation after a heat shock were similar in both the organisms and so the regulation of *hspA* gene expression appeared to be very similar (Kojima and Nakamoto, 2002).

Genome-wide expression of genes in wild-type and all *hik* mutants of *Synechocystis* sp. strain PCC 6803 (http://www.kazusa.or.jp/cyano/synechocystis/mutants; Suzuki *et al.*, 2000) was investigated by using DNA microarrays covering 3,075 of the 3,267 genes. *Synechocystis* grown at 34°C was given a heat shock for 20 and 60 min at 44°C and genome-wide transcription was assessed by isolating the mRNA. The level of transcripts for 59 genes rose by >3 fold whereas the activity of certain genes was suppressed to less than one-third (58 genes) and to less than half of their initial level (232 genes) in many others. Specially, the transcripts of genes *clpB1*, *hspA*, *groESL*, *htpG*, *dnaJ*, *dnaK2* rose by 6 fold during first 20 min at 44°C but the mRNA level of *hik34* gene increased by 19 fold. The level of *clpB1* and *dnaJ* decreased after 60 min of heat shock but the levels of *hspA*, *groEL1*and *groES* transcripts were still expressed at high levels after incubation of cells at 44°C for 60 min. On the other hand, the level of transcripts of most of the heat shock genes remained higher than in wild-type suggesting that Hik34 showed a negative effect on the expression of heat shock genes in the wild-type. Exceptionally, the level of transcripts of *hspA* gene continued to appear for a longer duration reaching a maximum after 40 min of heat shock. The identification of the respective proteins by 2-D electrophoresis and MALDI-TOF mass spectrometry revealed increased levels of the Hsps (GroEl, DnaK2, HspA, ClpB, HtpG and GroES) in *hik34* mutant cells. This has further been confirmed by the overexpression of *hik34* gene in *Synechocystis* introduced by using a multicopy plasmid pVZ321 under the promoter of *nrt* operon. Stable expression of the plasmid in the transformants led to depressed expression of Hsp genes (*htpG, hspA, groESL1, dnaK2* and *groEL2*). These results amply make it clear that it is not only the transcripts that increased but this is followed by the increase in corresponding proteins as well and that Hik34 negatively regulates the expression of certain Hsps related to thermotolerance. This has been explained on the basis that under non-stressed conditions Hik34 might be existing in an autophosphorylated state and the transfer of this phosphate group to a Rre might be repressing the expression of heat shock genes. Under heat shock conditions, autophosphorylation of Hik34 might not occur preventing the subsequent repression of gene expression. Of all Hik mutants, only *Hik34* mutant showed the expression of heat shock genes and allowed cells to survive an increase in growth temperature from 30°C to 48°C. Suzuki *et al.* (2006) have conducted a transcriptome analysis of *Synechocystis* sp. strain PCC 6803 at 34°C and after a heat shock at 44°C for 20 min and 60 min and proteome analysis after 60 min of heat shock. The levels of expression of mRNA at 44°C after 20 min increased by more than 8-fold in case of genes *groEL1, groEL2, hspA, htpG, dnaK2* and *clpB1* but some of the transcripts such as *groES, groEL2, hspA* and *clpB1* increased by 40-fold at 60 min. However, the level of transcripts of many of these genes decreased at 60 min point of time (compared to the level at 20 min) except for the levels of transcripts of *groEL1* and *clpC*. The proteome analysis conducted after a heat shock after 60 min exhibited increased levels of Hsps GroES, GroEL1, GroEL2, HspA, HtpG, DnaK2, ClpB1, ClpC and HhoA. In addition, they identified a number of enzymes, transcription factors and proteins whose level either increased or decreased during heat shock response and classified the different proteins into different classes.

A different approach of proteome analysis was undertaken by Hongsthong *et al.* (2009) to find out up-regulated and down-regulated proteins of plasma membrane (PM), thylakoid membrane (TM) and the cytosol in *S. platensis* after an increase in growth temperature from 35°C to 40°C. The identification of these proteins was done by 2-DE and MALDI-TOF mass spectrometry. On the basis of expression patterns of genes as defined by Lacerda *et al.* (2007), three categories of genes, i.e. resistance, adaptation and sustained tolerance have been identified during heat shock in the PM and soluble fraction. These were further classified into different clusters among these three categories (Figs. 12 and 13). Majority of the proteins belonged to the group of sustained tolerance and these were 46% each in case of PM and soluble fraction and 58% in the TM fraction. The up-regulated proteins belonged to two-component signal transduction (histidine kinase, Ser/Thr protein kinase and response regulators including Gly-Gly-Asp-Glu-Phe domains) and Hsps (GroEL, ClpB, DnaK/DnaJ). GroEL and ClpB were detected in the soluble fraction whereas DnaK in the TM fraction. Another up-regulated protein belonged to DEAD/DEAH box RNA helicases that are involved in RNA maturation, proof reading and enhancement of DNA-unwinding. The unique feature of this RNA helicase is its existence in a phosphorylated state suggesting that it might be involved in environmental sensing and signal transduction. Two down-regulated proteins were identified in the PM and soluble fractions while thirteen down-regulated proteins have been detected in the

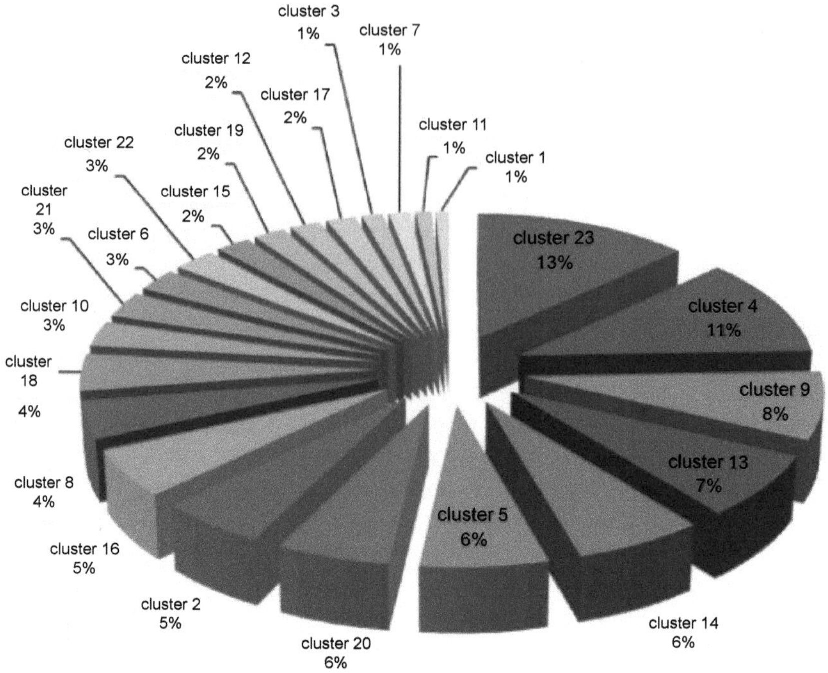

Figure 12: Pie charts representing percentage of each protein cluster identified in *Spirulina platensis* after a heat shock at 40°C. The protein clusters are classified by the expression pattern of all significant differentially expressed proteins in the plasma membrane fraction (clusters 7, 11, 12 and 21 are resistance proteins, cluster 22 is adaptation protein, clusters 1, 5, 6, 14, 15, 16, 18, 20 and 23 are sustained proteins, and clusters 2, 3, 4, 8, 9, 10, 13, 17 and 19 are undetermined protein trends). With the kind permission of A. Hongsthong, BEC Unit, National Center for Genetic Engineering and Biotechnology, 83 Moo8, Thakham, Bangkhuntien, Bangkok 10150, Thailand. [Hongsthong *et al.* (2009) *Proteome Science* **7**: 33; doi:10.1186/1477-5956-7-33].

Color image of this figure appears in the color plate section at the end of the book.

TM fraction. A decrease in the level of Δ^9-desaturase that catalyzes the first step in the fatty acid desaturation in the PM and TM of this cyanobacterium was noted. Another down-regulated protein belonged to uvrD/REP helicase that is known to be responsible for SOS response induced by UV light (Crowley and Hanawalt, 2001), indicating that the DNA damage caused due to heat stress is repaired by a repair system other than SOS repair.

III. OXIDATIVE STRESS

Of the ROS, the generation of $O_2^{\bullet-}$ and $^1O_2^*$ mainly occurs in the photosynthetic electron transport (PET) and during respiration in plants and cyanobacteria (Asada and Takahasi, 1987; Gutteriddge and Halliwell, 1990; Shiraishi *et al.*, 1994). These radicals subsequently give rise to H_2O_2 and OH·. The reaction of $O_2^{\bullet-}$ with H_2O_2 leads to the formation of OH· which is often catalyzed by metal ions, especially Fe^{3+}. This is known as Fenton or Haber-Weiss reaction (Haliwell and Gutteridge, 1986). As a consequence of this, all such enzymes and proteins in PET that possess iron will become the main targets of $O_2^{\bullet-}$ damage inside the cells (Haliwell and Gutteridge, 1986; Kuo *et al.*, 1987; Fridovich, 1989; Gutteridge and Haliwell, 1990; Gardner and Fridovich, 1991). The formation of $^1O_2^*$ in chlorophyll antenna is due to the transfer of energy from excited triplet-state chlorophyll molecules to molecules

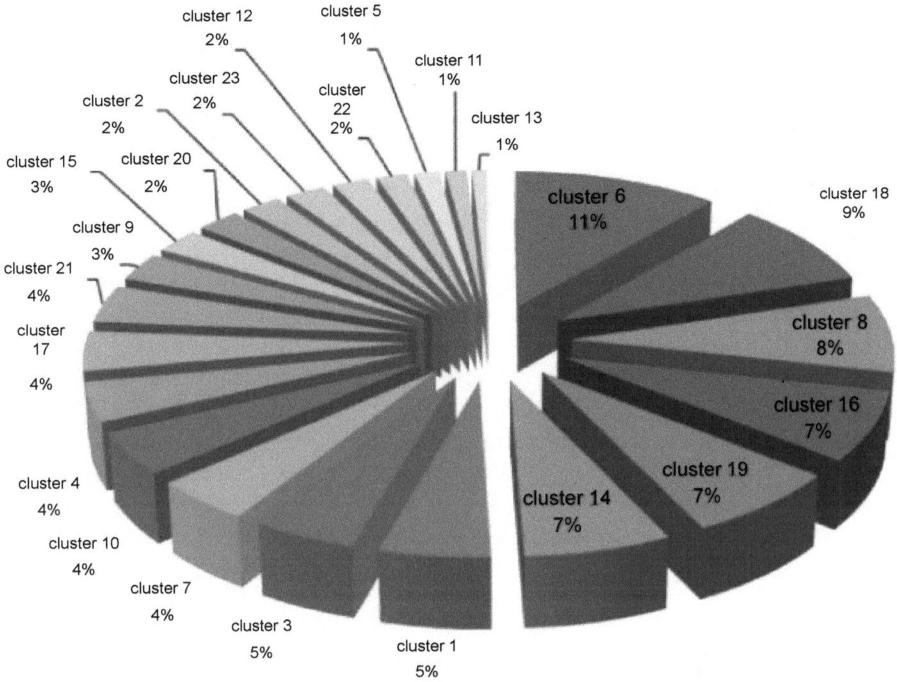

Figure 13: Pie charts representing percentage of each protein cluster identified in *Spirulina platensis* after a heat shock at 40°C. These are classified by the expression pattern of all significant differentially expressed proteins in the soluble fraction (clusters 9, 17 and 20 are resistance proteins, clusters 7, 12 and 21 are adaptation proteins, clusters 1, 2, 3, 6, 16, 18 and 19 are sustained proteins, and clusters 4, 5, 8, 10, 11, 13, 14, 15, 22 and 23 are undetermined protein trends). With the kind permission of A. Hongsthong, BEC Unit, National Center for Genetic Engineering and Biotechnology, 83 Moo8, Thakham, Bangkhuntien, Bangkok 10150, Thailand. [Hongsthong *et al.* (2009) *Proteome Science* **7:** 33; doi:10.1186/1477-5956-7-33].

Color image of this figure appears in the color plate section at the end of the book.

of oxygen. In turn ${}^1O_2{}^*$ can give rise to O_2^- anion (Asada and Takahasi, 1987; Symons, 1991). Thus the formation of both $O_2^{\bullet-}$ and ${}^1O_2{}^*$ by PET is generally favoured when normal metabolic pathways are slowed by physiological stress conditions. Under such conditions at the reducing site of PSI, NADPH utilization is suboptimal and due to lower NADP levels occasionally electrons are transferred to O_2 instead of NADP (Asada, 1994). This reaction is known as Mehler reaction and constitutes the major electron transfer route in presence of methyl viologen (MV; a herbicide also known as paraquat). MV accepts electrons from PSI and reduces O_2 to $O_2^{\bullet-}$. In addition, respiratory dehydrogenases of mitochondria and bacteria also have been shown to be important sources of $O_2^{\bullet-}$ and H_2O_2 (Salin, 1991). In order to overcome the oxidative stress and protect the cells from oxidative damage many enzymes are functional in the cells and mediate the removal of the ROS species. Among these SODs, monofunctional catalases, bifunctional catalases-peroxidases, peroxiredoxins, glutaredoxins and thioredoxins are important.

A) Superoxide dismutases (SODs; EC 1.15.1.1): Cu,Zn-SOD was the first SOD to be discovered and initially named as erythrocuprein or hemocuprein (McCord and Fridovich, 1969). Since then researches conducted in this area have unequivocally demonstrated the presence of SODs ubiquitously in all aerobic organisms and these enzymes help in dismutation of $O_2^{\bullet-}$ radicals. The

SODs thus scavenge $O_2^{\bullet -}$ radicals from the cells in a reaction that involves the formation of H_2O_2. It can be written as:

$2\,O_2^{\bullet -} + 2\,H^+ \longrightarrow H_2O_{2\,+}\,O_2$ (Reaction 1).

The rate of this reaction very much depends on the rate of diffusion of $O_2^{\bullet -}$ anion into the active site of the enzyme and the availability of hydrogen ions (Getzoff *et al.*, 1992). Depending on the type of prosthetic metal group present at the active site of the enzyme, SODs are classified into iron (Fe-SOD; *sodB*), manganese (Mn-SOD; *sodA*), copper,zinc (Cu,Zn-SOD; *sodC*) and nickel (Ni-SOD; *sodN*) containing enzymes. Among these, Fe-SOD is widely distributed in the algae, cyanobacteria and other prokaryotes and the stroma of chloroplasts. Though Mn-SOD is widely distributed in all prokaryotes and eukaryotes, its occurrence in cyanobacteria is restricted. After its first discovery in *Streptomyces* sp. with Ni as the catalytic metal (Youn *et al.*, 1996), Ni-SODs have been described subsequently in many groups of organisms. In 1997, Fridovich wrote in his review that research in this field has reached a happy state of maturity. Truely enough, we now know the crystal structures of all the SODs and also the lethal effects of the accumulated mutant proteins of the SODs (Miller, 2004).

The distribution of the different metal forms of SODs among cyanobacteria has been reported. A comparison of 64 cyanobacterial SOD sequences (including 24 completely sequenced genomes) revealed the occurrence of one *sodN* homologue mainly in *P. marinus* strains (AS9601, CCMP1986, CCMP1375, MIT9301, MIT9303, MIT9211, MIT9312, MIT9313, MIT9515, NATL1A, NATL2A), *Synechococcus* sp. (strains BL107, CC9605, CC9311, WH8102) *C. watsonii* WH8501 and *T. erythraeum* IMS101. Cu,Zn-SODs have been noted in strains of *Synechococcus* (BL107, CC9605, CC9311, CC9902, RS9916, WH7805) and *Lyngbya* sp. PCC 8106. The *sodA* gene sequences have been observed in *C. watsonii* WH8501, *Synechococcus* sp. (WH5701, RS9917), *T. elongatus* BP-1, *T. erythraeum* IMS101, *P. boryanum* UTEX485, *Leptolyngbya valderiana* BDU20041, *N. punctiforme* PCC 73102, and *A. variabilis* ATCC 29413. Among these, *P. boryanum* has three genes and *N. punctiforme* PCC 73102 and *A. variabilis* ATCC 29413 have two genes each while the others have one gene each encoding Mn-SOD. By far the largest distribution is that of Fe-SODs that are present in *Synechococcus* sp. (strains RS9916, RS9917, JA-3-3Ab, JA-2-3B'a, WH5701, WH7805; one gene each), *S. elongatus* PCC 6301 (two genes), *S. elongatus* PCC 7942 (two genes), *Synechocystis* sp. strain PCC 6803, *Cyanothece* sp. CCY0110, *T. elongatus* BP-1, *G. violaceus* PCC 7421, *Lyngbya* sp. PCC 8106, *S. platensis*, *P. boryanum* UTEX485 (one gene each), *Anabaena* (*Nostoc*) sp. PCC 7120 (two genes), *N. punctiforme* PCC 73102, *A. variabilis* ATCC 29413, *N. linckia*, *N. commune*, *N. commune* CHEN (one gene each; Regelsberger *et al.*, 2004; Priya *et al.*, 2007). This is in so far as the distribution of individual SODs in different cyanobacteria is concerned. But a close scrutiny reveals that while it is true that Ni-SOD is the only form found in many of the unicellular strains, a combination of Ni-SOD with either Cu,Zn-SOD (*Synechococcus* sp. strains BL107, CC9605, CC9311) or Mn-SOD (*C. watsonii* WH8501 and *T. erythraeum* IMS101) is also known. The occurrence of Cu,Zn-SOD is considerably rare but its occurrence in association with Fe-SOD is also seen (*Synechococcus* sp. strains RS9916, WH7805 and *Lyngbya* sp. PCC 8106). The presence of Fe- and Mn-SOD forms is noted in unicellular (*Synechococcus* sp. strains WH5701, RS9917, *T. elongatus* BP-1, *G. violaceus* PCC 7421), filamentous non-heterocystous (*P. boryanum* UTEX485) and heterocystous forms [*Anabaena* sp. strain PCC 7120, *A. variabilis* ATCC 29413 and *N. punctiforme* PCC 73102; Priya *et al.*, 2007]. However, Dupont *et al.* (2008) reported that *P. marinus* SS120 and *Synechococcus* sp. strain CC9902 also possess a *sodN* gene in each of them besides the *sodC* gene sequence in the latter. According to them *Synechococcus* sp. RS9917 has Fe and Cu,Zn-SODs but according to Priya *et al.* (2007) this strain possesses Fe- and Mn-SODs. Dupont *et al.* (2008) further showed the presence of Fe-SOD in *Synechococcus* sp. RCC307 and *Synechococcus* sp. strain WH7803 but the latter contained a gene encoding Cu,Zn-SOD as well.

The Fe- and Mn-SODs show high homology in their primary and three-dimensional structures. The crystal structures of Fe-SOD of *T. elongatus* BP-1 and Mn-SOD of *Anabaena* sp. strain PCC 7120 in monomeric and dimeric states with the catalytically active amino acid residues are depicted in Fig. 14. The number of amino acid residues and molecular weights differ in both cases. Cyanobacterial Fe-SODs possess 199–229 amino acid residues with a molecular weight ranging 21–25 kDa. The Mn-SODs have 200–316 amino acid residues with a molecular weight ranging 22–34 kDa. Both Fe- and Mn-SODs typically exist as homodimers or homotetramers. Topologically, the presence of α-N-terminal and a α/β-C-terminal domain is a characteristic feature of Fe- and Mn-SODs. In cyanobacteria, an intervening highly conserved sequence motif [Asp-Val-Trp-Glu-His-Ala-Tyr-Tyr-(Asp282-Tyr289)] is present that extends between the second α-helix and the first β-sheet of the C-terminal domain of both the SODs. The active site contains a single Fe-ion and the highly conserved Asp282 and His286 constitute the metal-binding amino acid residues (Fig. 15). The other amino acids of this highly conserved region are Glu285 and Tyr289 that form a dimer surface spanning the interface and bridging the active sites between the opposite halves of each subunit (Priya *et al.*, 2007). The Fe- and Mn-SODs differ in three important aspects. The first one pertains to the metal

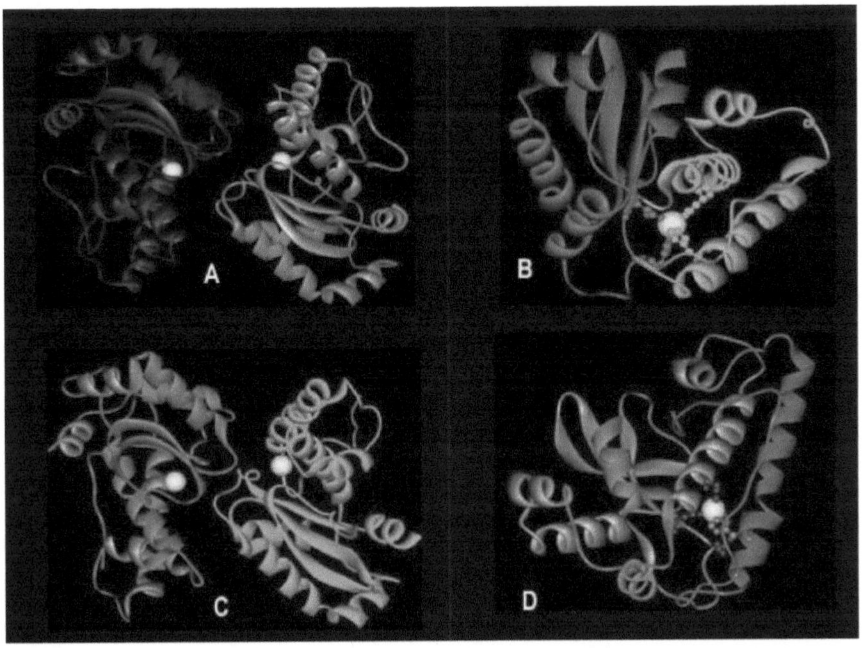

Figure 14: Structure of Fe and MnSOD. Structures are visualized using WebLab ViewerLite 4.2 software. Catalytically essential aspartate or histidine residues are represented in ball and stick mode binding the active metal (yellow) is shown to identify the location of the active site. Protein database codes are given in parentheses: (i) FeSOD (PDB 1gv3); (ii) MnSOD (PDB 1my6). (A) FeSOD of *Thermosynechococcus elongatus* BP-1 dimers are distinguished by colour (violet and slate), and structures are represented with the active site (yellow) of subunit. (B) Monomeric subunit of FeSOD represents an N-terminal (green) and a C-terminal (red). Similarly (C) represents dimer structure of *Anabaena* sp. MnSOD in pink and green with active site highlighted in yellow. (D) Monomeric MnSOD showing the N-terminal residues in blue and C-terminal in pink with metal binding ligands. The transmembrane hydrophobic pocket specific for MnSOD is highlighted in red (D). Picture with the kind permission of B. Priya, National Facility for Marine Cyanobacteria (Sponsored by Department of Biotechnology, Govt. of India), Department of Marine Biotechnology, School of Marine Sciences, Bharathidasan University, Tiruchirappalli, Tamil Nadu, India. [Priya *et al.* (2007) *BMC Genomics* **8:** 435. doi:10.1186/1471-2164-8-435].

Color image of this figure appears in the color plate section at the end of the book.

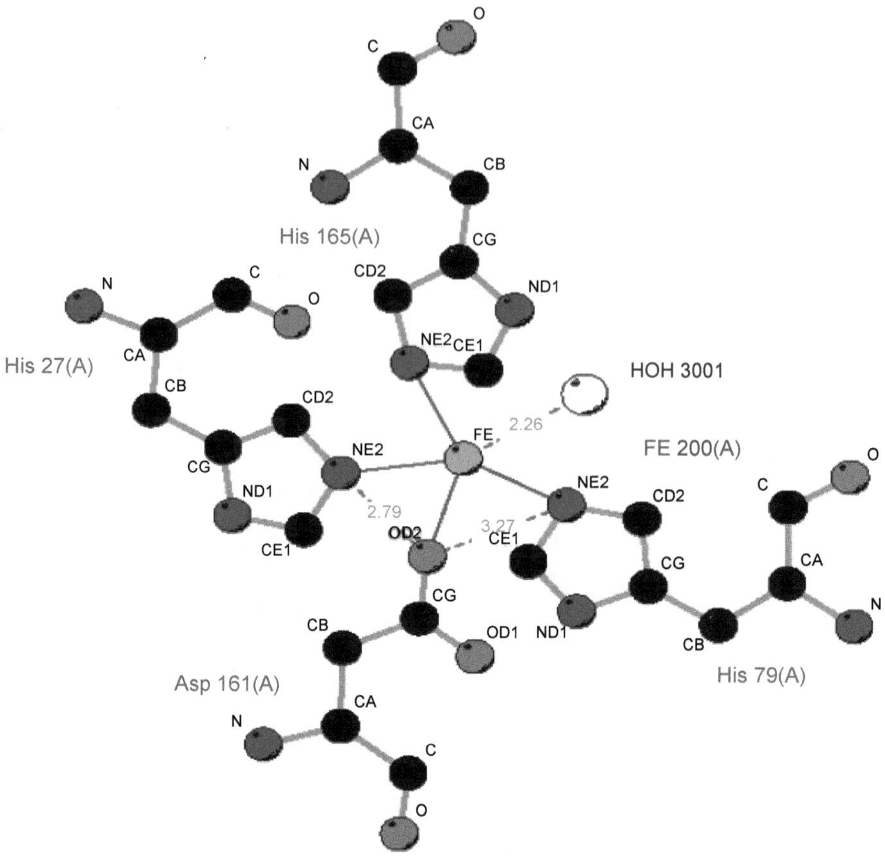

Figure 15: The active site residues of Fe Superoxide dismutase of *Thermosynechococcus elonagtus* BP-1. Picture with the kind permission of B. Priya, National Facility for Marine Cyanobacteria (Sponsored by Department of Biotechnology, Govt. of India), Department of Marine Biotechnology, School of Marine Sciences, Bharathidasan University, Tiruchirappalli, Tamil Nadu, India. [Priya *et al.* (2007) *BMC Genomics* **8:** 435; doi:10.1186/1471-2164-8-435].

Color image of this figure appears in the color plate section at the end of the book.

specific signature sequence which is Phe184X_3Ala188Gln189....Thr280....Phe/Tyr303 in case of Fe-SOD while in Mn-SOD it is Arg184X_3Gly188Gly189....Gly280....Trp303. Secondly, in Fe-SOD it is the Asp ligand that forms a hydrogen bond from the active site with the outer sphere residue of Trp243 whereas in Mn-SOD it is Gln262 that is involved in hydrogen bond formation. Thirdly, two Lys residues at positions 201 and 255 have been assigned a photosynthetic role only in Fe-SOD. The Mn-SOD is distinguishable from Fe-SOD by the presence of a transmembrane hydrophobic pocket. The 2.0Å resolution structure of the catalytic portion of the Mn-SOD from *Anabaena* sp. strain PCC 7120 showed that both the N-terminal helical hair pin (His94 and His145) and the C-terminal α/β-domain (His232 and Asp228) contribute ligands to the catalytic manganese site (Fig. 16). The ligands are tightly constrained by hydrogen bonding with surrounding residues either from the same monomer (Tyr98, Asn144, Trp194, Gln213, Val229 and Trp230) or from the neighbouring subunit (Glu231, Tyr235). Thus the symmetric dimeric structure of Mn-SOD is maintained by the

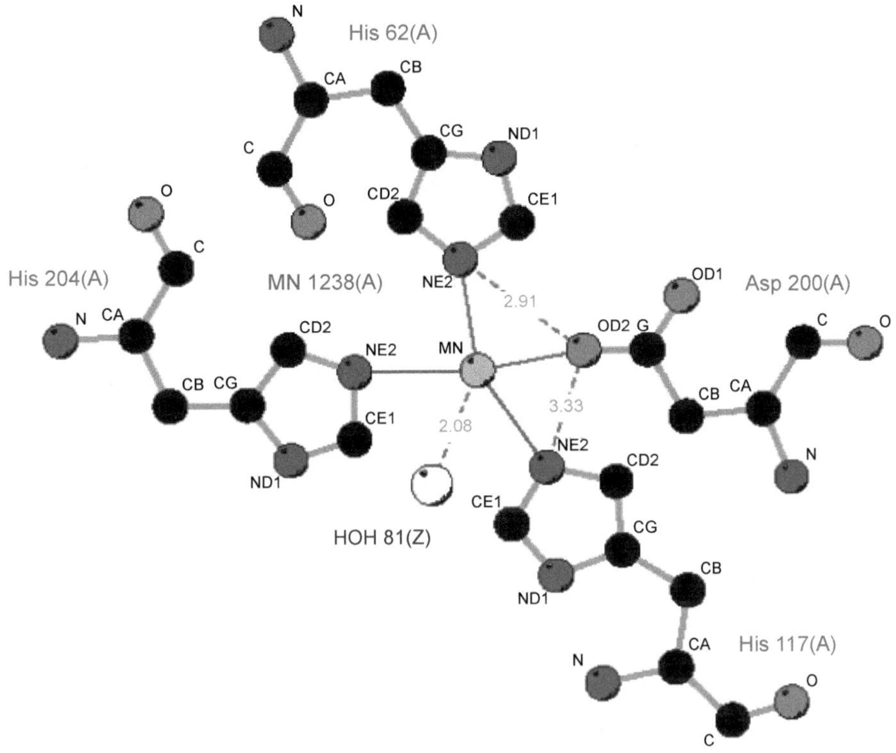

Figure 16: The active site residues of Mn Superoxide dismutase of *Anabaena* sp. strain PCC 7120. Picture with the kind permission of B. Priya, National Facility for Marine Cyanobacteria (Sponsored by Department of Biotechnology, Govt. of India), Department of Marine Biotechnology, School of Marine Sciences, Bharathidasan University, Tiruchirappalli, Tamil Nadu, India [Priya *et al.* (2007) *BMC Genomics* **8:** 435; doi:10.1186/1471-2164-8-435].

Color image of this figure appears in the color plate section at the end of the book.

contribution of both monomers to the active site and the substrate channel. The distance (18.4 Å) between the two Mn----Mn atoms is bridged by the hydrogen bonded His232 of one monomer with Glu231 of the other monomer (Atzenhofer *et al.*, 2002). The elucidated dimeric structure of Mn-SOD of *L. valderiana* BDU20041 and its catalytically essential Asp141 and His(4) residues are represented in ball and stick mode (Priya *et al.*, 2010; Figs. 17 and 18).

Cu, Zn-SODs exist as dimers and each monomer possesses ~174–233 amino acid residues rich in Gly (11–16%) that form eight β-sheets providing structural stability to the molecule. The primary and tertiary structure of Cu,Zn-SOD does not bear any resemblance to the Fe- and Mn-SODs and so the active site (Miller, 2004). Two highly conserved domains seem to be important in copper binding. The first one is Gly-Phe-His-(Ile-Leu-Val)-His-x-[Asn-Gly-Thr]-[Gly-Pro-Asp-Ala]-[Ser-Gln-Lys]-Cys and the second one is Gly-[Gly-Ala]-Gly-Gly-[Ala-Glu-Gly]=Arg[Phe-Ile-Leu]-[Asp-Gly]-Cys-Gly. The highly conserved His-residues in the former are important in the copper binding and the Cys residue in the latter are important in disulfide bonding. Four histidines coordinate the catalytic sphere of Cu^{2+} ion represented by His103, Nε2 of His105, His147 and His215 whereas His147, His157, His171 and Oδ1 of Asp174 form the catalytic domain of Zn^{2+} ion. The Ni-SODs possess a Ni-hook at the N-terminal region having a highly conserved region with (His-Cys-Asp-Gly-Pro-Cys-Val-Tyr-

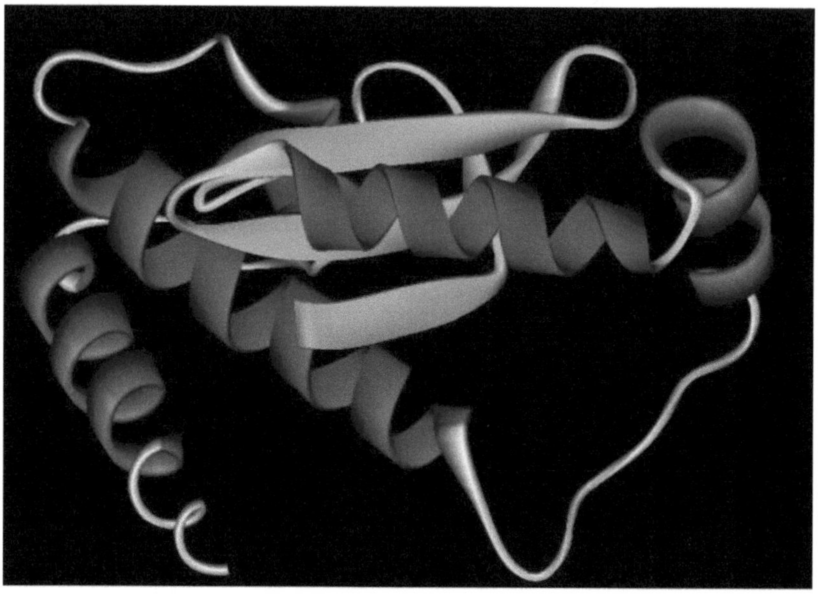

Figure 17: The elucidated structure of MnSOD of *Lyngbya valderiana* BDU20041 is a dimer with 6 helices and 3 sheets. Depicted above is a monomer showing helices (red) and strands (blue). Structures are visualized using WebLab ViewerLite 4.2 software [http://in.msi.com/]. Picture with the kind permission of D. Prabaharan, National Facility for Marine Cyanobacteria (Sponsored by Department of Biotechnology, Govt. of India), Department of Marine Biotechnology, School of Marine Sciences, Bharathidasan University, Tiruchirappalli, Tamil Nadu, India. [Priya *et al.* (2010) *Saline Systems* **6:** 6; doi:10.1186/1746-1448-6-6].

Color image of this figure appears in the color plate section at the end of the book.

Asp-Pro-Ala) that provides a unique identity to Ni-DODs. These essentially possess four α-helices held by a short connecting α-helix (Priya *et al.*, 2007).

The overall disproportionation reaction (Reaction 1) is accomplished by the SODs in two steps which are of the first order with respect to $O_2^{\bullet-}$ anion. This is explained by the 'ping-pong' mode of enzyme action. The two steps are sequential and the enzyme gets reduced (Reaction 2a) in the first step and this reduced enzyme participates in the second step by binding another $O_2^{\bullet-}$ anion with the generation of H_2O_2 and the release of the oxidized enzyme. These two steps can be represented as:

$Ez_{Ox} + O_2^{\bullet-} + H^+ \longleftrightarrow E_{red}(H^+) + O_2$ (Reaction 2a)

$Ez_{red}(H^+) + O_2^{\bullet-} + H^+ \longleftrightarrow Ez_{OX} + H_2O_2$ (Reaction 2b)

where Ez_{Ox} and Ez_{red} represent the oxidized and reduced states of the SODs. Accordingly, Fe^3 and Fe^{2+}, $Cu^{2+}Zn^{2+}$ and Cu^+Zn^{2+}, Mn^{3+} and Mn^{2+}, Ni^{3+} and Ni^{2+} represent the oxidized and reduced states respectively of the metal ions. The net positive electrostatic potential near the active site of the SODs and the negative charge of $O_2^{\bullet-}$ radical facilitates substrate binding. The ping-pong mechanism of action exhibited by SODs is explained on the basis that the reduced enzyme is still be able to bind the $O_2^{\bullet-}$ in the second step due to preservation of the electrostatic attraction of the positive charge upon metal ion reduction by the coupled uptake of a proton. It means that $O_2^{\bullet-}$ binding is favoured to two different versions of the same active site. The uptake of one proton in the first half of the reaction seems to provide sufficient driving force for substrate reduction (and binding) and the excess energy available is thereby utilized in the second step (Miller, 2004).

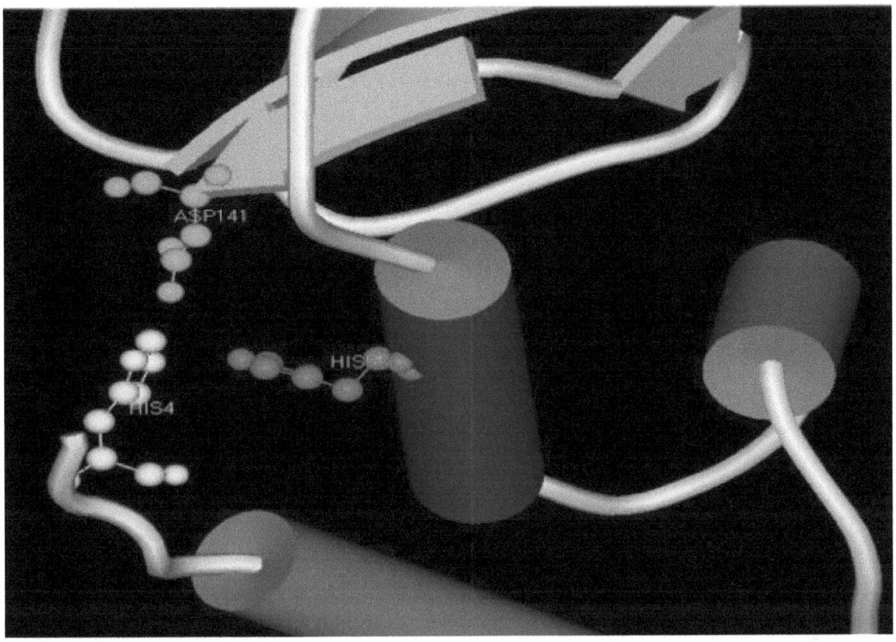

Figure 18: Metal binding sites of predicted MnSOD of *Lyngbya valderiana* BDU20041. Catalytically essential aspartate or histidine residues are represented in ball and stick mode. Structures are visualized using WebLab ViewerLite 4.2 software [http://in.msi.com/]. Picture with the kind permission of D. Prabaharan, National Facility for Marine Cyanobacteria (Sponsored by Department of Biotechnology, Govt. of India), Department of Marine Biotechnology, School of Marine Sciences, Bharathidasan University, Tiruchirappalli, Tamil Nadu, India. [Priya *et al.* (2010) *Saline Systems* **6:** 6; doi:10.1186/1746-1448-6-6].

Color image of this figure appears in the color plate section at the end of the book.

Anacystis nidulans (but now *S. elongatus* PCC 7942) possesses two genes that encode FeSOD. In order to study the role of Mn-SOD in conferring oxidative stress tolerance, Mn-SOD gene of *E. coli* has been subcloned into a *E. coli-A. nidulans* shuttle vector (pSG111) and used in the transformation of *E. coli* B101 and *A. nidulans* cells. The transformants of both the organisms exhibited increased tolerance to paraquat-induced oxidative stress suggesting that Mn-SOD provides protection against damage caused by $O_2^{\bullet-}$ anions (Gruber *et al.*, 1990). Initial studies were made on the isolation, purification and localization of Fe- and Mn-SODs in the vegetative cells and heterocysts of cyanobacteria. The localization of Fe-SOD in the soluble extracts of *A. cylindrica* has been demonstrated with the help of polyclonal antibodies raised against Fe-SOD. The presence of this enzyme in the heterocysts was shown with the help of immunogold labelling assisted with scanning electron microscopy (Canini *et al.*, 1992). A direct relationship between the levels of these enzymes and the ability to overcome the oxidative stress response has been established by the studies on the isolation of mutants in certain cyanobacteria like *P. boryanum* affecting Mn-SOD activity (Steinitz *et al.*, 1979) and *S. elongatus* PCC 7942 defective in Fe-SOD designated as *sodB*$^-$ (Laudenbach *et al.*, 1989; Herbert *et al.*, 1992). *SodB*$^-$ mutants of *S. elongatus* PCC 7942 showed no detectable synthesis of Fe-SOD and the *sodB* locus exists as a single copy in its genome transcribing a monocistronic mRNA (Laudenbach *et al.*, 1989). This is contrary to the two Fe-SOD gene sequences reported later (Priya *et al.*, 2007). Although the growth rates of both wild-type and the *sodB*$^-$ mutant were similar under normal growth conditions, the *sodB*$^-$ mutant exhibited increased sensitivity to oxidative stress caused by MV than the wild-type.

An interruption in the electron flow between the two (4Fe-4S) centers of PSI and cytochrome *f* was detected by measuring flash-induced absorption transient at 705 nm (Herbert *et al.*, 1992). The *sodB*$^-$ mutant of *S. elongatus* PCC 7942 showed similar photosynthetic pigments and catalase activity as the wild-type but the growth rate of the wild-type was faster with higher cyclic photosynthetic electron transport. Further, the *sodB*$^-$ mutant was more sensitive to oxidative stress induced by MV but not to the oxidative stress caused by norflurazon (an inhibitor that blocks the synthesis of phytoene desaturase and so the synthesis of β-carotene and other carotenoids; Thomas *et al.*, 1998). Based on the extensive studies made on *P. boryanum* (Asada *et al.*, 1975; Okada *et al.*, 1979; Steinitz *et al.*, 1979), Campbell and Laudenbach (1995) cloned the SOD genes in *P. boryanum* UTEX485 by employing a 23 base oligonucleotide primer (containing the sequence 5'-GATGTCTGGGAACACGCTTACTA-3') designed by the reverse translation of conserved region of amino acids at the C-terminal region of SOD. Of the four genes identified, *sodB* governs the synthesis of Fe-SOD enzyme whereas *sodA1*, *sodA2* and *sodA3* encoded Mn-SOD enzymes. In MV-resistant mutants *sodB* and *sodA1* genes were constitutively expressed. Under conditions of oxidative stress induced by MV, iron and nitrogen limitation, wild-type cells showed the accumulation of transcripts for *sodA2*. The MV-resistant mutant of *P. boryanum* showed high levels of *sodA3* transcipt accumulation. The genome of *Anabaena* sp. strain PCC 7120 consists of *sodB* and *sodA* genes encoding Fe- and Mn-SODs, respectively. The *sodA* gene exhibited a high homology to the *sodA2* gene from *P. boryanum* described earlier (Campbell and Laudenbach, 1995). The *sodB* and *sodA* genes of *Anabaena* sp. strain PCC 7120 were differentially expressed and differed in their localization patterns. Fe-SOD is present in the soluble fractions of vegetative cells and heterocysts whereas Mn-SOD is localized in the thylakoid membranes attached towards the luminal side by a motif at the N-terminus of the mature protein. The pre-protein of Mn-SOD also possesses a leader peptide which is shown to be essential for its activity as confirmed by its expression in *E. coli* cells. Though significant amounts of Fe-SOD have been detected in the isolated heterocysts, they did not contain any Mn-SOD. These observations emphasize the differential role of SODs in relation to the type of cells in which they are localized (Li *et al.*, 2002). Such a localization of Fe- and Mn-SODs has also been confirmed by the studies on *Anabaena* sp. strain PCC 7120. Alongside, studies on kinetics of enzyme activities of Fe- and Mn-SODs led to the determination of their catalytic rate constants (Regelsberger *et al.*, 2002, 2004). In order to find out the importance of N-terminal region and linker region of the Mn-SOD, two recombinant proteins were produced in *E. coli* BL21 (DE3) cells transformed with an expression vector containing Mn-Sod gene constructs which lacked either the N-terminal hydrophobic region (amino acids 2–29) or both the hydrophobic and the linker region (amino acids 2–61). The recombinant Mn-SODs produced by the transformants of *E. coli* were found to be homodimeric in organization and exhibited catalytic activity assayed by the conventional ferricytochrome *c* and by stopped flow spectroscopy. These results emphasize that the linker region is not involved in oligomerization and that the soluble catalytic portion is fully active (Regelsberger *et al.*, 2002). They confirmed the localization of Fe-SOD in the soluble fractions of vegetative cells and heterocysts of *Anabaena* sp. strain PCC 7120 with the help of activity, staining, inhibition patterns and immunogold labelling. *E. coli* BL21 (DE3) cells transformed with an expression vector, pET-28a carrying the cloned *sodB* gene from *Anabaena*, produced the recombinant enzyme. The purified enzyme (having a molecular mass of 42 kDa) was shown to exist in a homodimeric state (Regelsberger *et al.*, 2004). The Fe-SOD gene (*slr1516*) of *Synechocystis* sp. strain PCC 6803 has been cloned and its overexpression in *E. coli* BL21 (DE3) cells led to the production of recombinant protein at significantly higher levels. The recombinant protein was purified by gel filtration and ion-exchange chromatography to near homogeneity. The transformant *E. coli* cells tolerated the oxidative stress induced by MV (Bhattacharya *et al.*, 2004). The *sodA* gene inactivation in *Anabaena* sp.

strain PCC 7120 resulted in a strain (A17) that could not grow in the absence of combined nitrogen and was more sensitive to photoinhibition caused by increased light intensity. The restoration of wild-type character was possible by complementing the wild-type *sodA* gene in A17. The presence of Mn-SOD in vegetative cells and heterocysts was confirmed by immunoblotting. The inability of A17 to grow in a nitrogen-free medium was explained on the basis that the ROS produced in heterocysts inactivated nitrogenase in the absence of Mn-SOD. Thus the membrane associated Mn-SOD protects the photosynthetic apparatus as well as nitrogenase from oxidative damage (Zhao *et al.*, 2007). L. *valderiana* BDU20041, a marine isolate tolerant to acid black1, exhibited a 3-fold increase in the activity of Mn-SOD. The cloned gene (550 bp) exhibited 63.8% homology with other bacterial Mn-SODs and its deduced amino acid sequence revealed the active site residues to be His4, His58 and Asp141 along with highly conserved Mn-specific residues. The production of ROS in presence of acid black1 was confirmed by incubating the dye-treated cells with 2′,7′-dichlorofluorescein diacetate (5 µM) in dark at 27 ± 2°C for one hour and fluorescence measured in a spectrofluorimeter confirmed the presence of ROS (Priya *et al.*, 2010).

Phylogenetic analysis of various Fe- and Mn-SODs (83 SOD sequences of which 2 bacterial, three fungal, and six cyanobacterial Mn-SODs and eight cyanobacterial Fe-SODs included) revealed that: (i) the Fe-SODs were normally well separated from the Mn-SODs; (ii) combialistic SODs that can accept either Fe- or Mn- as co-factors are generally grouped with Fe-SODs; and (iii) the Fe-SODs of Archaea and thermophilic bacteria form a clade within the Mn-SOD cluster. The identity and per cent similarity of various cyanobacterial Mn-SODs (given in parentheses) are *Anabaena* sp. strain PCC 7120 —*N. punctiforme* M1 (57/71), *Anabaena* sp. strain PCC 7120—*N. punctiforme* M2 (68/78), *Anabaena* sp. strain PCC 7120—*P. boryanum* (56/70) and *Anabaena* sp. strain PCC 7120—*E. coli* (50/63) and that of *Anabaena* sp. strain PCC 7120 and *E. coli* Fe-SOD is 44/59. Phylogenetic analysis of Fe-SODs showed that the filamentous species formed a well separated cluster and the unicellular forms were well segregated into another cluster (Regelsberger *et al.*, 2002, 2004). All the four metalloforms of SODs formed distinct clusters in a radial Neighbour-Joining analysis of 105 cyanobacterial SOD sequences (of which 38 are Fe-SODs, 36 Mn-SODs, 13 Cu,Zn-SODs and 18 Ni-SODs; Priya *et al.*, 2010; Fig. 19).

An obligate requirement of nickel is shown by marine picoplanktonic cyanobacteria for the expression of *sodN* gene regardless of the presence of the type of nitrogen source in the medium. In all the strains of *Prochlorococcus* and some of the *Synechococcus* sp. Ni-SOD is alone represented. The presence of ORFs bearing significant similarity to Ni-SOD pre-proteins has been demonstrated in some of these cyanobacterial genomes sequenced (Dufresne *et al.*, 2003; Rocap *et al.*, 2003). The presence of an ORF that encodes a peptidase (*sodX*) downstream of the *sodN* gene in *P. marinus* strains (MED4, MIT9313, and SS120), *Synechococcus* sp. strain IMS101 and *C. watsonii* WH8501 has been detected (Eitinger, 2004). When oxygen-sentive *E. coli* mutant lacking the endogenous Fe- and Mn-SOD was transformed with *sodN* and *sodX* from *P. marinus* MIT9313, in the transformant co-expression of both the genes took place. The recombinant SOD of *E. coli* cells restored oxygen tolerance. It means for the production of catalytically active Ni-SOD, the co-expression of *sodX* with *sodN* is essential. A comparison of *Synechococcus* sp. strain CC9311 (a coastal isolate) and *Synechococcus* sp. strain WH8102 (an isolate from open oceans) for Ni-requirement showed that the former strain that possesses Cu,Zn-SOD besides Ni-SOD was able to withstand the Ni-deficiency because of its slow growth in presence of NH_4^+ and in the absence of Ni whereas the latter strain that contained only Ni-SOD could not grow in any nitrogenous medium devoid of Ni. It means *Synechococcus* strain CC9311 had a greater tolerance for Ni-starvation and that the Cu,Zn-SOD could not completely replace the Ni-SOD (Dupont *et al.*, 2008). Likewise, Ni-starvation inhibited growth of *Synechococcus*

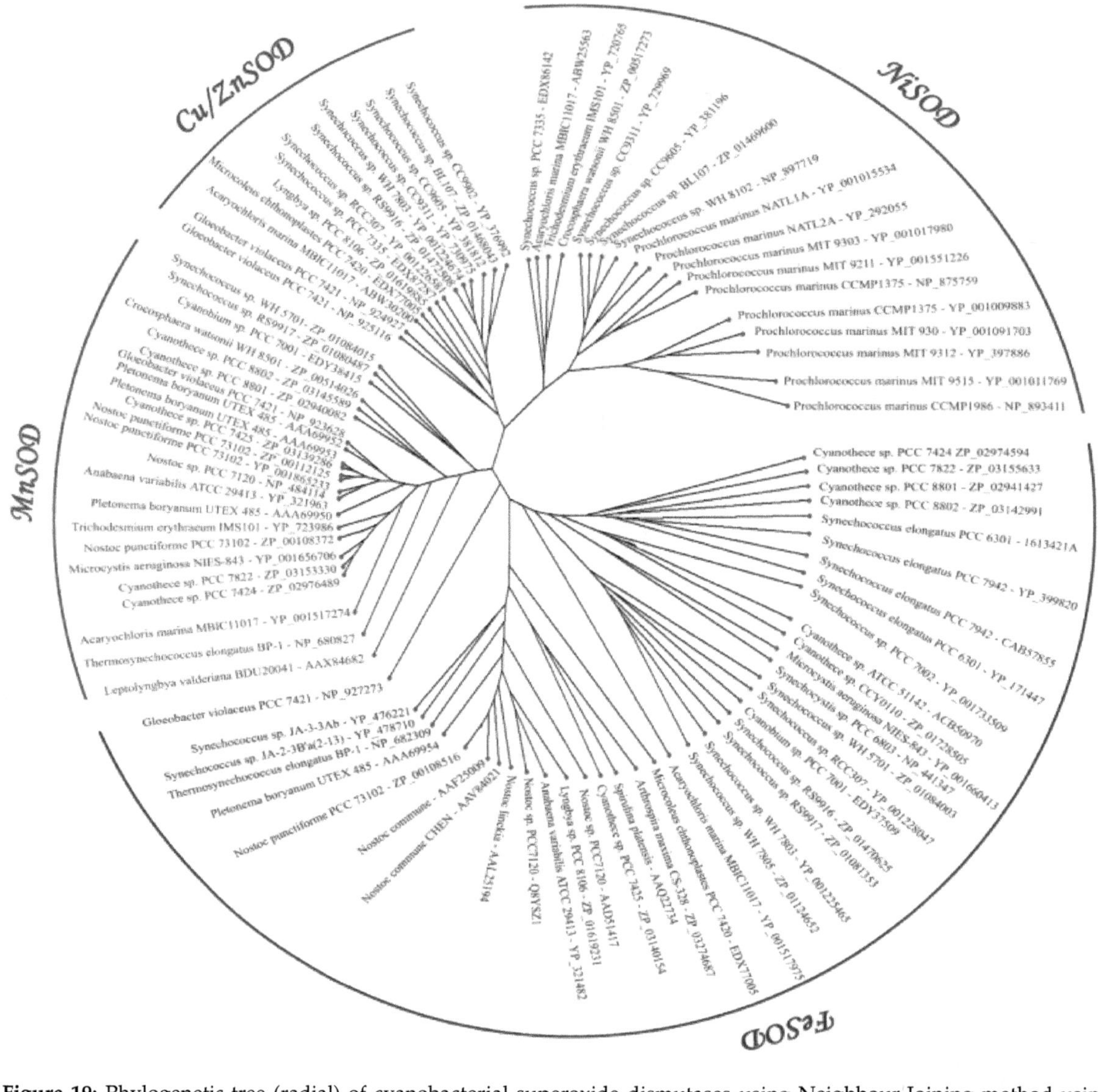

Figure 19: Phylogenetic tree (radial) of cyanobacterial superoxide dismutases using Neighbour-Joining method using Phylodraw (v.0.8) [http://pearl.cs.pusan.ac.kr/phylodraw/]. Picture with the kind permission of D. Prabaharan, National Facility for Marine Cyanobacteria (Sponsored by Department of Biotechnology, Govt. of India), Department of Marine Biotechnology, School of Marine Sciences, Bharathidasan University, Tiruchirappalli, Tamil Nadu, India. [Priya *et al.* (2010) *Saline Systems* **6:** 6; doi:10.1186/1746-1448-6-6].

Color image of this figure appears in the color plate section at the end of the book.

strain WH8102 (with Ni-SOD) and WH7803 (with Fe- and Cu,Zn-SOD) in a nitrate medium while the growth of *Synechococcus* sp. strains WH5701 (with an Fe- and Mn-SOD) and PCC 7002 (with an Fe-SOD gene) remained unaffected by the limitation of Ni. The total SOD activity decreased by 30% and 15% of the wild-type in *Synechococcus* sp. strains WH8102 and WH7803, respectively. The presence of an active SOD has been demonstrated by ^{63}Ni labelling studies and DGGE. Moreover, *Synechococcus* sp. strain WH7803 showed a requirement of Ni for its acclimatization in low-iron and high-light conditions of growth and depletion of Ni affected other metabolic pathways leading to a low SOD activity (Qiu and Price, 2009).

B) Monofunctional catalases (EC 1.11.1.6): Cyanobacteria in general lack genes for the synthesis of typical monofunctional catalases. The only complete gene sequence for a typical catalase is present in *N. punctiforme* PCC 73102 with a proximal haem ligand Tyr and the conserved distal residues His, Asn and Ser (Zámocký *et al.*, 2008). This enzyme contains haem *b* at the active site and uses NADPH as a redox-active co-factor. Truncated C-terminal gene sequences are present for typical catalases in *S. elongatus* PCC 7942 and *Cyanothece* sp. ATCC 51142. In *Anabaena* sp. strain PCC 7120 a fusion protein with a typical catalase-related domain was reported that bears closer resemblance to the fusion protein of allene oxide synthase from *Plexaura homomalla* and related corals (Oldham *et al.*, 2005). Some of the important properties of monofunctional catalases are that: (i) they can be inhibited by 3-amino-1,2,4-triazole, an inhibitor of eukaryotic catalases; (ii) their activity is constant over several pH units and (iii) they can not be reduced by dithionite. On the basis of their size, two subgroups are recognized, the small subunit (55 to 69 kDa) enzymes associated with haem *b* and the large subunit enzymes (75 to 84 kDa). The monofunctional catalase of *N. punctiforme* PCC 73102 belongs to the former group. The monofunctional catalases characterized in great detail are known to exist as homotetramers (Obinger *et al.*, 1997; Jakopitsch *et al.*, 1999).

Another minor gene family of catalases is represented by Mn-catalases (non-haem or dimanganese catalases or Mn-Cats). With the exception of *G. violaceus* PCC 7421, one ORF with similarity to Mn-Cat has been reported in the heterocystous, diazotrophic species. There are two [*N. punctiforme* PCC 73102 and *Anabaena* sp. strain PCC 7120] and three (*Cyanothece* sp. PCC 7424) paralogues for Mn-Cats in certain of them. At least ~50% of the investigated species contain neither a typical catalase nor bifunctional catalase-peroxidase or Mn-Cat. Studies on functional aspects of Mn-Cat genes (structural analysis, transcription or expression) do not exist in literature.

C) Bifunctional catalase-peroxidases (KatGs; EC 1.11.1.7): Initially ascorbate peroxidase (APx) activity was reported from crude extracts of cyanobacteria (Miyake *et al.*, 1991). So it was thought convenient to distinguish cyanobacteria into two groups, those possessing APx and those that do not. But analysis of 44 fully or partially sequenced cyanobacterial genomes did not reveal the presence of gene sequences for APx (Bernroitner *et al.*, 2009). KatGs are the only peroxidases that possess a very high catalase activity besides a substantial peroxidase activity and do not possess any sequence homology to the monofunctional catalases. But KatGs bear sufficient homology to plant APxs and yeast cytochrome *c* peroxidase (CcP) and that is why they are included in the Class I superfamily of peroxidases to which plant, fungal, protist and bacterial haem peroxidases belong (Welinder, 1992; Passardi *et al.*, 2007). The KatG encoding genes are present in nearly 30% of the cyanobacterial genomes with one *katG* gene in each of them. For example one KatG gene is found in *G. violaceus* PCC 7421, *S. elongatus* PCC 6301, *S. elongatus* PCC 7942, eight strains of *Synechococcus* sp. (strains WH7803, WH7805, CC9605, RS9916, RS9917, RCC307, WH5701, PCC7002), *Synechocystis* sp. strain PCC 6803, *A. marina* MBIC11017 and *Cyanothece* sp. CCY0110. Of these, only the last mentioned species is known to fix nitrogen and the rest are unable to fix nitrogen. In all *P. marinus* strains (AS9601, MIT9211, MIT9215, MIT9301, MIT9303, MIT9312, MIT9515, NATL1A, NATL2A, subsp. *marinus* strains CCMP1375, CCMP1986) some *Synechococcus* strains [BL107, CC9311, CC9902, JA-2-3B'a(2-13) and JA-3-3Ab], *M. aeruginosa* NIES-843, *T. elongatus* BP-1,, *C. watsonii* WH8501, strains of *Cyanothece* (ATCC 51142, PCC 7424, PCC 8801) *Lyngbya* sp. PCC 8106, *T. erythraeum* IMS101,

A. variabilis ATCC 29413, *No. spumigena* CCY9414, *N. punctiforme* PCC 73102 and *Anabaena* sp. strain PCC 7120 KatG gene sequences are absent (Bernroitner *et al.*, 2009). The cyanobacterial KatG enzyme is dimeric (Obinger *et al.*, 1997; Jakopitsch *et al.*, 1999) as also KatGs from *Bacillus stearothermophilus* (Loprasert *et al.*, 1989) and *Streptomyces* sp. (Youn *et al.*, 1995). Monomeric (*Halobacterium halobium*, Fukumori *et al.*, 1985) and tetrameric KatGs (*E. coli*; Claiborne and Fridovich, 1979; *Mycobacterium smegmatis*; Marcinkeviciene *et al.*, 1995 and *Rhodobacter capsulatus*; Hochman and Shemesh, 1987; Forkl *et al.*, 1993) are also known in the respective bacterial species mentioned in parentheses.

Cloning, nucleotide sequencing, expression of recombinant protein in *E. coli* of *katG* gene and KatG enzyme and its kinetic properties from *S. elongatus* PCC 7942 (Mutsuda *et al.*, 1996) and *Synechocystis* sp. strain PCC 6803 (Jakopitsch *et al.*, 1999; Regelsberger *et al.*, 1999) have been reported. The *S. elongatus* PCC 7942 KatG exhibits high homology with *Synechocystis* sp. strain PCC 6803 KatG enzyme. The ORF of *katG* of *S. elongatus* PCC 7942 is 2160 bp long that encodes a protein of 720 amino acids long with a molecular mass of ~79.9 kDa. The native enzyme purified to homogeneity also revealed a molecular mass of 150 kDa composed of two identical subunits with a mass of 79 kDa. The absorption spectrum revealed the presence of a typical of protohaem molecule per dimer. The native enzyme purified from the cells and the enzyme overexpressed in *E. coli* [BL21(DE3)pLyS] exhibited similar kinetic properties for catalase activity and the same ratio of catalase to peroxidase activity (Mutsuda *et al.*, 1996). The purification and kinetic properties of KatG enzyme from cytosolic extracts of *A. nidulans* (*S. elongatus* PCC 7942) revealed that the enzyme possesses two identical subunits of equal size (80.5 kDa). It exhibits both catalase and O-dianisidine peroxidase activity (as it accepts electrons from O-dianisidine but not from ascorbate, glutathione and NADH). The catalase activity has been found to be very efficient with a pH optimum between pH 6.5 and 7.5 (in contrast to broad pH range shown by monofunctional catalases) and a K_m for H_2O_2 of 4.3 mM with a calculated turnover number of 7200 s^{-1}. The KatG of *A. nidulans* is insensitive to the eukaryotic catalase inhibitor 3-amino-1,2,4-triazole (in contrast to the sensitive nature of monofunctional catalases) but sensitive to cyanide (K_i=27.2 μM; Obinger *et al.*, 1997). *Synechocystis* sp. strain PCC 6803 *katG* gene that encodes a bifunctional catalase-peroxidase has been cloned and the coding sequence was extended by attaching a hexahistidine tag at the C-terminus and the gene sequence was used to transform *E. coli* [BL 21(DE3)pLyS] strain using the pET-3a vector. The enzyme overexpressed in *E. coli* was purified to homogeneity and shown to exist in a homodimeric state with a molecular mass of 170 kDa. As typical of bifunctional catalase-peroxidase, the KatG enzyme showed a high catalase activity with a *Km* of 4.9 ± 0.25 mM for H_2O_2 and turnover number of 3500 s^{-1}. It also showed peroxidase activity with O-dianisidine, guaiacol and pyrogallol but not with NAD(P)H, ferricytochrome *c*, ascorbate or glutathione as electron donors (Jakopitsch *et al.*, 1999). The KatG enzyme from cytosolic extracts of *Synechocystis* sp. strain PCC 6803 purified to homogeneity by a six-step purification procedure also exhibited almost similar kinetic parameters described by Jakopitsch *et al.* (1999) for the recombinant enzyme produced in *E. coli*. Spectroscopy of the pyridine ferrochrome revealed iron protoporphyrin IX as the prosthetic group. Peptide mass mapping (by matrix-assisted laser desorption ionization time-of-flight mass spectrometry) established identity between the amino acid sequence of purified KatG from *Synechocystis* sp. strain PCC 6803 and the deduced amino acid sequence from *katG* gene sequence from the same organism (Regelsberger *et al.*, 1999).The *in vivo* role of KatG from *Synechocystis* sp. strain PCC 6803 was investigated by the isolation of a *katG* deletion mutant. Surprisingly the mutant showed resistance to H_2O_2 and MV-induced stress because of its residual H_2O_2 scavenging activity (30 times lower than that of the wild-type), though the mutant was indistinguishable from the wild-type in doubling time and other growth characteristics. The residual peroxidase activity seemed to be sufficient to protect the cells from oxidative stress and the protective role of KatG assumes significance in the survival of cells where H_2O_2 is generated by other algae or bacteria in the ecosystem. The residual peroxidase activity was shown to be light-dependent and due to the presence of a thioredoxin-specific peroxidase (Tichy and Vermaas, 1999).

In the first catalytic reaction cycle of catalases and peroxidases, the ferric enzyme is oxidized by H_2O_2 to the redox intermediate compound I with the release of a water molecule (Reaction 1). This has an oxoferryl center ($Fe^{IV}=O$) in combination with either a porphyrin π-cation radical or an amino acid radical ($R^{\bullet+}$). Catalse and peroxidase cycles differ in the use of second peroxide molecule. In catalase reaction, second peroxide molecule is used as a reducing agent for compound I thus regenerating the native enzyme with the release of O_2 (Reaction 2). This reaction is generally very faster than compound I formation. In the peroxidase reaction, compound I is transformed in the first one-electron reduction to compound II containing either oxoferryl ($Fe^{IV}=O$) center or an amino acid radical ($R^{\bullet+}$) in combination with FeIII (Reaction 3). Compound II is finally reduced back to the ferric peroxidase in a second one-electron reduction (Reaction 4). APxs and CcP cannot perform Reaction 2 and thus can reduce compound I via compound II exclusively thereby oxidizing their substrates ascorbate and cytochrome *c*, respectively. By contrast catalase-peroxidases perform both peroxidase (Reactions 1, 3 and 4) and catalytic cycle are active (Jakopitsch *et al.*, 2003c).

Peroxidases reduce peroxides by means of one-electron or one two-electron donors. One electron donors can be aromatic (phenols) or aliphatic (glutathione, NO^{-2} anions or metal cations Mn^{2+}) substances and the corresponding oxidation product eventually dimerizes. Two electron donors can be halides (chloride, bromide, iodide or thiocyanate) with the corresponding oxidation product being hypophalous acid or hypothiocyanate. That is why the peroxidases that accept electrons from halides are known as haloperoxidases (irrespective of the protein family that brings about this reaction). The active site of KatGs consists of a proximal triad (of Asp, Trp and His) and a distal triad (of Arg, Trp and His) of amino acids and this is similar to the active sites of APx and CcP. Despite this, the Class I peroxidases differ from one another in many respects. The KatGs have a predominant catalase activity whereas no such activity has been reported for either APx or CcP. Their reactivities toward H_2O_2, one electron donors and the spectral features of redox intermediates are found to be different. The KatGs are organized into a two domain monomeric structure with an N-terminal haem domain that is catalytic and a C-terminal domain that exhibits a high structural and sequence similarity with N-terminal domain but which is non-functional. But without the C-terminal domain, the N-terminal domain does not exhibit either catalase or peroxidase activity. The cloned *katG* of *Synechocystis* sp. strain PCC 6803 was subjected to PCR-based oligonucleotide site-directed mutagenesis to identify the role of distal amino acid triad (Arg-Trp-His). Overexpression of the respective six genes individually in *E. coli* [BL 21(DE3)pLySl] led to the production of six recombinant proteins where the distal triad amino acid residues were changed : Arg119Ala, Arg119Asn, Trp122Phe, Trp122Ala, His123Gln and His123Glu, i.e. Arg119 was exchanged with Ala or Asn; Trp122 was exchanged with Phe or Ala and His was exchanged with Gln or Glu. The Trp122 mutants completely lost their catalase activity but the turnover number of catalase activity of 0.02%, 0.03%, 0.5% and 14.6% has been found for His123Gln, His123Glu, Arg119Asn and Arg119Ala mutant proteins, respectively. The important role of Trp122 in catalase activity was identified by following the transition of the ferric enzyme to compound I formation by H_2O_2 spectroscopically. The distal His-Arg pair is found to be important in heterolytic cleavage of H_2O_2 where as the distal Trp is important in the compound I reduction by H_2O_2 (Regelsberger *et al.*, 2000). Thus the role of conserved Trp residues at the distal haem cavity site (Trp122) in the formation of compound I was followed by steady-state and stopped-flow spectroscopy. Mutants of *Synechocystis* sp. strain PCC 6803 Trp122Ala and Trp122Phe that completely lost their catalase activity were not affected in either the formation of compound I or its reduction to compound II in presence of O-dianisidine or pyrogallol. These observations signify that indole ring of distal tryptophan is essential for two-electron reduction of compound I by H_2O_2 and not for compound I formation (Regelsberger *et al.*, 2001).

The *Synechocystis* catalase-peroxidase enzyme also proved to be a good haloperoxidase capable of oxidizing chloride, bromide and iodide. This was tested by the halogenation of monochlorodimedon (MCD) by chloride, bromide and iodide by the wild-type and variant Arg119Ala, Trp122Phe and Trp122Ala enzymes. As mentioned earlier, the Trp122 mutated to Phe and Ala lost the catalase as well as halogenation activities but with an overall increase in peroxidase activity. This indicates that the same redox intermediate is involved in H_2O_2 and halide oxidation (Jakoptisch *et al.*, 2001). Recombinant wild-type KatG enzyme and the distal side variants (Arg119Ala, Arg119Asn, Trp122Phe, Trp122Ala, His123Gln and His123Glu) from *Synechocystis* sp. strain PCC 6803 have been subjected to UV-Vis and resonance Raman spectroscopy to find out distal cavity interactions. These involve H-bond networks that connect the distal side residues with proximal side of the haem pocket. The distal mutations affected the Fe haem coordination state and also the stability of the haem architecture (Heering *et al.*, 2002). Apart from the distal triad of the active site, the importance of distal Asp152 residue was recognized. The side chain carboxyl group of Asp152 is 7.8 Å away from the haem iron and is hydrogen bonded to two water molecules and a KatG-specific loop. Mutants of this distal Asp152 were generated by site-directed mutagenesis and Asp152Asn, Asp152Ser, and Asp152Trp variants showed a reduced catalase activity with the turnover number of 2.7%, 5.7% and 0.6% of the wild-type, respectively. The variants showed enhanced (2.7 times) peroxidase activity than wild-type and another variant Pro151Ala enzymes. The variant enzymes showed an altered pH profile and it was possible to determine spectroscopically the redox intermediate compound by H_2O_2. In the Asp152 variants the reactivity of compound toward aromatic one electron donors was enhanced with a decrease in the reactivity to H_2O_2. A mechanism of H_2O_2 oxidation different than that of monofunctional catalases involving the distal Trp122 and Asp152 has been predicted (Jakopitsch *et al.*, 2003a). Since the distal histidine (His123) forms a hydrogen bond with the adjacent asparagine (Asn153), the role of this hydrogen bond was investigated in the bifunctional activity of KatGs. Replacement of asparagine with either Ala (Asn153Ala) or aspartic acid (Asn153Asp) did not affect the peroxidase activity. However, the turnover number of catalase activity was found to be 16.5% and 6% of that of wild-type in Asn153Asp and Asn153Ala variants, respectively. This shows that Asn153 is an important part of the extended H-bond network of Kat G and it seems to provide optimal conditions for binding and oxidation of the second molecule of H_2O_2 necessary in the catalase reaction (Jakopitsch *et al.*, 2003b). Investigating the role of tyrosine, a distal side residue (Tyr249), Jakopitsch *et al.* (2003c) found that the bifunctional KatG completely lost its catalase activity with the replacement of tyrosine by phenylalanine (Tyr249Phe) but retained its peroxidase activity. Similar to the wild-type enzyme, Tyr249Phe variant enzyme could use a range of substrates (O-dianisidine, pyrogallol, guaiacol, tyrosine and ascorbate) for its peroxidase activity. The formation of redox intermediate compound I has been demonstrated spectroscopically by the addition of equimolar H_2O_2 by the variant enzyme. Further, for the first time the transition from compund I to an oxoferryl-like compund II was detected and the rate constants for the formation of both compounds determined by the sequential stopped-flow technique.

The formation of protein-based radicals was demonstrated in wild-type KatG enzyme of *Synechocystis* sp. strain PCC 6803 using a multifrequency EPR investigation (Ivancich *et al.*, 2003). EPR signal for the formation of compound I intermediate [FeIV=O por^{++}] and Tyr$^{\bullet}$ and Trp$^{\bullet}$ radicals was detected. The Trp$^{\bullet}$ radical formation was supposed to be from Trp106, which is located at a distance of 7–8 Å from haem propionate group. Further it has been suggested that the extensive

H-bond network involving Trp122, His123 and Arg119, the haem propionate group and Trp106 plays a key role on the formation of Trp• radical.

The existence of a covalent bond between Trp122, Tyr249 and Met275 and the formation of a novel covalent adduct has been demonstrated by the mass spectrometric analysis of the wild-type, recombinant and the corresponding variants (Trp122, Tyr249 and Met 275) of KatG enzymes from *Synechocystis* sp. strain PCC 6803 (Jakopitsch *et al.*, 2003d). As already mentioned the exchange of Trp122 with Phe or Ala (Regelsberger *et al.*, 2000) and Tyr249 with Phe (Jakopitsch *et al.*, 2003c) resulted in a complete loss of catalase activity but not peroxidase activity. The cross-linking of Trp122 and Tyr249 is thus important in catalase activity as variants of Met275 still allowed the covalent bond formation between Trp122 and Tyr249 (Jakpoitsch *et al.*, 2003d). The variants of Met275 showed a reduced catalase activity whereas the peroxidase activity is normal (Donald *et al.*, 2003). Further, the novel covalent adduct (of Trp122, Tyr249 and Met275) appears to be hydrogen bonded to a neighbouring Arg439 of *Synechosystis* KatG enzyme. In the KatG of *S. elongatus* PCC 7942 Arg409 is at H-bonding distance to both Tyr218 and Met244. A lower turnover number of catalase activity has been observed in case of mutant enzymes of KatG (Tyr249Phe, Met275Ile, Arg439Ala and Arg439Asn) of *Synechocystis* sp. strain PCC 6803 but the peroxidase activity remained unaffected. The typical KatG covalent adduct remained intact in the two Arg439 variants as in case of wild-type. In the Met275Ile variant the covalent link existed only between Tyr249 and Trp122 whereas in case of Tyr249Phe variant the link is absent. The radical species formed in the variants Tyr249Phe and Arg439Ala/Asn during their interaction with peroxoacetic acid was similar to wild-type and they are oxoferrylporphyrin radical, tryptophanyl radical and the tyrosyl radical. The Met275Ile variant did not form the Trp• species probably due to change in the haem distal site. The sequence of intermediates has been noted to be por$^{\bullet+}$, Trp• (or Trp$^{\bullet+}$) and Tyr• (Jakopitsch *et al.*, 2004). The reaction of KatG of *Synechocystis* sp. strain PCC 6803 with isoniazid resulted in the identification of the first compound I unequivocally as [Fe(IV)=O Por$^{\bullet+}$]. Subsequent to this, it was possible to identify two protein-based radicals Trp106• and a Tyr• formed by intramolecular electron transfer. Further, the reaction of isoniazid with KatG enzyme seemed to be very close to the haem distal side that could readily react with [Fe(IV)=O Por$^{\bullet+}$] species. However, the KatG from *M. tuberculosis* differed in its reaction with isoniazid and the site of formation of Trp• radical appeared to be different depending on the differences in substrate binding (Colin *et al.*, 2010).

Wild-type KatG enzyme from *Synechocystis* sp. strain PCC 6803 and its variants of proximal triad and distal side Asp152 and Asn153 residues have been studied in their ferric and ferrous state (at various pH values) by employing resonance Raman spectroscopy and electron absorption. A comparison of these with CcP revealed that a change in the proximal residues caused a weakening of haem binding to the protein and so these are important in conferring stability of the protein architecture. On the other hand, Asp152Ser variation caused an architectural change and the KatG very much appeared as a CcP. A change in Asn153 to Asp153 resulted in a disruption of the H-bond with distal His123. It has been further suggested that Asp152, important in H_2O_2 oxidation reaction, is expected to be H-bonded to nitrogen atom of Ile248. Any change in the Asp152 and Ile248 H-bond alters both the stability of haem architecture as well as the interaction between proximal His-Asp H-bond (Santoni *et al.*, 2004). Similar conclusions have been drawn by Heering *et al.* (2002) based on their studies on distal side variants already described earlier.

Jakopitsch *et al.* (2005) put forward a proposal for the catalytic mechanism of KatGs and reported the interconversion of ferrous enzymes, compound II and compound III by the use of wild-type and Tyr249Phe variants of KatG enzyme of *Synechocystis* sp. strain PCC 6803. The binding of O_2 to ferrous KatG and Tyr249Phe is reversible and the rate constants are comparable to those of horseradish peroxidase (HRP) but the dissociation constants of ferrous dioxygen are higher in case of Tyr249Phe (129µM) when compared to wild-type KatG (84 µM) and HRP (1.9 µM). The conversion of compound II to compound III is very much faster in case of variant Tyr249Phe than in case of HRP and other haem peroxidases.

In monofunctional catalases H_2O_2 oxidation and formation of O_2 occurs in two steps. In the first step, one molecule of H_2O_2 is oxidized by the catalase enzyme to form an intermediate compound I [oxoiron (IV) Porphyrin π-cation radical intermediate] which oxidizes another molecule of H_2O_2 to O_2 in the second step. The redox intermediates during catalase cycle of KatGs from different sources (*Synechocystis* sp. PCC 6803, *Burkholderia pseudomallei* and *Mycobacterium tuberculosis*) have been determined by the oxidation of of peroxoacetic acid and H_2O_2. The intermediates with different spectral features differed among the three KatGs (Jakopitsch *et al.*, 2007). Employing gas chromatography-mass spectrometry, O_2 evolved from a 50:50 mixture of $H_2{}^{18}O_2$ and $H_2{}^{16}O_2$ by KatGs from *Synechocystis* sp. strain PCC 6803 and *M. tuberculosis* was shown by two electron oxidation without breaking the O-O bond. The formation of $^{18}O_2$ and $^{16}O_2$ is indicative of two electron oxidation because O_2 evolution as $^{16}O\ ^{18}O$ has not been detected. This has been found to be the case in case of KatG variants (D152S, H123E, W122F, Y249F and R439A) that exhibited a very low catalase activity (Vlasits *et al.*, 2007). These results thus support the conclusions drawn earlier on monofunctional catalases (Jarnagin and Wang, 1958; Kato *et al.*, 2004). The kinetics of H_2O_2 oxidation by wild-type KatG and its variants (Trp122Phe, Asp152Ser and Glu253Gln) of *Synechocystis* sp. strain PCC 6803 have been probed by sequential-mixing stopped-flow spectroscopy (Vlasits *et al.*, 2010).

A molecular peculiarity of catalase-peroxidases is the presence of three large insertions in their catalytic domain, two of which are typical for KatGs. A gene duplication event in the ancestral hydroperoxidase gene may be responsible for the presence of an extra C-terminal copy. Both the N-terminal and C-terminal copies orient themselves and facilitate the functioning of both the haem group and essential amino acid residues at the active site (Zámocký *et al.*, 2001). Phylogenetic relationships in class I superfamily of bacterial, fungal and plant peroxidases (60 known sequences) possessing prosthetic haem group, peroxidase motif and peroxidatic activity, revealed the existence of four clearly separated clades, i.e. C-terminal and N-terminal domains of catalases-peroxidases, Apxs and CcPs. Evidences are presented in favour of the gene duplication event in catalse-peroxidases having occurred in the later phase of evolution after the formation of individual peroxidase families (Zámocký, 2004). An overview of the evolution of catalases from bacteria to humans covering their distribution, phylogeny structure and function of these enzymes has been presented. Their physiological response to oxidative stress and diseases resulting due to deficiency of catalases in human beings has been highlighted (Zámocký *et al.*, 2008). Evolution of structure and function of class I peroxidases has been studied by analyzing over 940 sequences in all available genomes of prokaryotes and eukaryotes. Phylogenetic analysis of 193 class I peroxidases showed the presence of two new well separated clades of hybrid peroxidases that might represent evolutionary bridges between catalase-peroxidases and CcPs (typeA) as well as between APxs (Zámocký *et al.*, 2010).

H_2O_2 as a signalling molecule: H_2O_2 has emerged as a signalling molecule that influences the transcription by changing the expression of hundreds of genes. Besides affecting the genes involved in ROS detoxification, the range of cellular processes influenced by H_2O_2 signalling encompass signal transduction, transcriptional regulation and the metabolism of proteins, carbohydrates and lipids (Stone and Yang, 2006; Vandenbroucke *et al.*, 2008). In *E. coli*, transcription factors SoxR and OxyR respond to $O_2^{\bullet -}$ and H_2O_2 stress, respectively. The reaction of OxyR with H_2O_2 causes the formation of an intramolecular disulfide bond that brings about a conformational change and makes it active

(Zheng *et al.*, 1998; Storz and Zheng, 2000; Pomposiello and Demple, 2001; Choi *et al.*, 2001). Another well studied repressor is PerR that was first discovered in *B. subtilis.* The PerR regulon includes a number of genes such as *katA* (for catalase), *ahpCF* (for alkyl hydroperoxide reductsae), *mrgA, hemAXCDBL, fur* and *perR.* PerR is a metalloprotein and its regulation involves both oxidation and metal ions (Herbig and Helmann, 2001; Fuangthong *et al.*, 2002; Mongkolsuk and Helmann, 2002; Helmann *et al.*, 2003). Three dimeric proteins Fur, Zur and PerR coordinate gene exression in *B. subtilis* in response to iron, zinc and H_2O_2 stress, respectively.

An ORF (*slr1738*) encoding PerR similar to that of *B. subtilis* has been identified in the genome of *Synechocystis* sp. strain PCC 6803 by employing a full genome cDNA microarray. Consistent with the known properties of PerR, its function as a repressor has been determined by the construction of a *perR* knock-out strain. It was possible to identify PerR-specific sequences upstream the promoters of the genes under its control. Another ORF (*sll1621*) that encodes a peroxiredoxin is regulated by PerR. The most important genes under its control are *isiA, sigD, sigB* and *mrgA* (*slr1894*). MrgA belongs to ferretin family of proteins that helps in the storage of iron and plays a key role in oxidative stress response. MrgA deletion mutants of *Synechocystis* sp. strain PCC 6803 have been found to be highly sensitive to peroxide.The down-regulated genes included those of chlorophyll biosynthesis pathway, regulatory genes and those of histidine kinases. A study of peroxide-induced gene expression in PerR and thioredoxin (TrxA1) mutants revealed that neither PerR nor TrxA1 is essential for peroxide protective response (Li *et al.*, 2004). Screening of a library of Hik mutants (44) of *Synechocystis* sp. strain PCC 6803 (by RNA slot blot hybridization and DNA microarray) revealed the involvement of four Hiks (Hik33, Hik34, Hik16 and Hik41) in H_2O_2 signal perception and transduction pathway. Of a total of 77 H_2O_2-inducible genes 26 were regulated by these four Hiks with an induction factor of 4.0. Hik33 was responsible for the induction of 22 of the 26 H_2O_2-inducible genes because all these genes were not expressed in Hik33 mutant cells in response to H_2O_2. The expression of only two genes was affected in case of Hik34, Hik16 or Hik41 mutants. H_2O_2-inducible genes are *hspA,* the *gifA* and *gifB* (for the subunits of glutamine synthetase inactivation factors), *hliA, hliB* and *hliC* genes, *nblA1* and *nblA2, dnaJ* and *perR.* Only nine H_2O_2-inducible genes are known to be under the regulation of PerR. Interestingly, PerR mutation abolished the induction of *nblA1, nblA2* and *ndh2* (which are under the control of Hik33) and six other genes (whose induction is not controlled by any of the four Hiks) suggesting that H_2O_2 signal reception requires the participation of both Hik33 and PerR. The induction of rest of the 45 H_2O_2-inducible genes was not affected either in wild-type or the Hik or PerR mutants implying the existence of another unknown H_2O_2 sensor for these genes (Kanesaki *et al.*, 2007). Miller *et al.* (2007) synthesized two new fluorescent probes, Peroxy Green (PG1) and Peroxy Crimson (PC1) that are helpful in visualizing endogenous H_2O_2 produced in living cells. This opens up a possibility for detailed studies on brain cell signalling in response to H_2O_2 and in understanding the diseases caused by a deficiency of catalases in human beings.

D) Peroxiredoxins (Prxs; EC 1.11.1.15): This family of enzymes was first described under the term 'peroxidoxin' (Chae *et al.*, 1994) but shortly afterwards the term 'peroxiredoxin' was suggested (Chae *et al.*, 1994) which has been widely accepted. The Prxs belong to members of the thioredoxin (Trx)-fold superfamily which have a thioredoxin-like fold and interact with either thiol- or disulfide-containing substrates. Six subclasses have now been recognized in Trx-fold superamily viz., glutathione peroxidases, glutathione-S-transferases, Trxs, glutaredoxins (Grx), DsbA (catalyzes disulfide formation) and Prxs (Schröder and Ponting, 1998). The structure, function and biology of Prxs has been investigated in detail (Stone and Yang, 2006; Veal *et al.*,2007; Winterbourn, 2008; Fourquest *et al.*, 2008; Hall *et al.*, 2009).

Prxs are generally divided into five subclasses based on their different subunit composition and the position and number of conserved cysteine residues. In the first subclass, 1-Cys prxs, one conserved cysteine residue is present in the N-terminal part of the enzyme whereas in the second subclass, 2-Cys Prxs, two cysteine residues are present with one each at the N- and C-terminal parts. Prx-Q is a chloroplastic enzyme which belongs to the third subclass, and is similar to *E.coli* bacterioferretin comigratory protein. Type II Prxs are atypical members that belong to the fourth subclass and possess only one conserved cysteine residue. These members can use Trx as well as Grx as electron donors and multiple isoforms of this enzyme are localized in many sub-cellular compartments. The last subclass is represented by NADPH-dependent glutathione peroxidase (GPx)-like proteins that carry out reduction of H_2O_2 and organic hydroperoxides to water and corresponding alcohols (Bernroitner *et al.*, 2009).

A complete survey of the 44 (fully or partially) sequenced genomes of cyanobacteria revealed the existence of gene sequences of all the five subclasses of Prxs mentioned above. Except the 12 strains of *P. marinus* and 60% of the *Synechococcus* strains, the rest of the genomes of cyanobacteria investigated possess one ORF with similarity to 1-Cys Prx. However, *A. marina* MBIC11017 contains two paralogues of 1-Cys Prx. The highly conserved sequence at the N-terminal domain is Val-Leu-Phe-Ser-His-Pro-X-Asp-Tyr-Thr-Pro-Val-<u>Cys</u>-Thr-Thr-Glu- that has one cysteine residue. The catalytic role of this enzyme has not yet been revealed in cyanobacteria. The presence of two conserved cysteine residues is a characteristic feature of 2-Cys Prxs. The N-terminal conserved domain possesses a sequence –Phe-Phe-Tyr-Pro-Leu-Asp-Phe-Thr-Phe-Thr=Phe-Val-<u>Cys</u>-Pro-Thr-Glu- and the C-terminal possesses a 'resolving' cysteine residue –Val-<u>Cys</u>-Pro-. These are dimeric enzymes and the two subunits are linked via a disulphide bond in the oxidized form. The presence of one ORF similar to 2-Cys Prx has been reported in all the genomes of cyanobacteria investigated. Type II Prxs are atypical Prxs in possessing one conserved cysteine residue. They utilize thioredoxin as electron donor to reduce peroxides and so are called as thioredoxin peroxidases. The presence of multiple isoforms of Type II Prxs has been demonstrated in the sub-cellular compartments of plants. A gene sequence of Type II Prxs is present in 15% of *P. marinus* and *Synechococcus* strains, *M. aeruginosa*, *Synechocystis* sp. strain PCC 6803, *A. marina* MBIC11017, *C. watsonii* WH8501, *Cyanothece* (strains ATCC 51142, CCY0110, PCC 7424 and PCC 8106), *T. erythraeum* IMS101, *A. variabilis* ATCC 29413, *No. spumigena* CCY9414, *N. punctiforme* PCC 73102 and *Anabaena* sp. strain PCC 7120. The cyanobacterial Type II Prxs contain a highly conserved sequence in the N-terminal region with one cysteine (-Leu-Pro-Gly-Ala-Phe-Thr-Pro-Thr-<u>Cys</u>-Ser-Ser-) residue whereas the second cysteine residue is not conserved in all species. It is absent in all *P. marinus* strains, in *M. aeruginosa*, *T. erythraeum* and all Nostocales. Type II Prx sequence is absent in 85% of *P. marinus* and *Synechococcus* strains, *G. violaceus* PCC 7421, *S. elongatus* PCC 6301, *S. elongatus* PCC 7942 and *T. elongatus* BP-1. The most widely represented Prx in cyanobacteria belongs to the PrxQ group with multiple isoforms in all cyanobacterial species investigated. Except 15% of *P. marinus* strains and 5% of *Synechococcus* strains that posess one isoform of Prx, the rest of them possess two (85% of *P. marinus* strains, 45% of *Synechococcus* strains, *Synechocystis* sp. strain PCC 6803, *C. watsonii* WH8501 and *Cyanothece* ATCC 51142), three [*M. aeruginosa* NIES-843, 45% of *Synechococcus* strains, *T. elongatus* BP-1, strains of *Cyanothece* (CCY0110 and PCC 8801), *Lyngbya* sp. PCC 8106, *T. erythraeum* IMS101] and four [*A. variabilis* ATCC 29413, *No. spumigena* CCY 9414, *N. punctiforme* PCC 73102 and *Anabaena* sp. strain PCC 7120] isoforms of PrxQ. Gene sequences of GPx are present in all *P. marinus* (with two such sequences in 85% and one in 15%) strains, *G. violaceus* PCC 7421, *S. elongatus* PCC 7942, *Synechococcus* (with two in 5% and one in 80%) strains, *Synechocystis* sp. strain PCC 6803 (two ORFs), *C. watsonii* WH8501, *N. punctiforme* PCC 73102 and *Cyanothece* PCC 8801. The presence of GPx sequence in *N. punctiforme*

PCC 73102 is exceptional because it is absent in most of the nitrogen-fixing species examined [*Cyanothece* CCY 0110, *T. erythraeum* IMS101, *A. variabilis* ATCC 29413, *No. spumigena* CCY 9414 and *Anabaena* sp. strain PCC 7120] as also in *M. aeruginosa* NIES-843, *T. elongatus* BP-1, *A. marina* MBIC11017, strains of *Cyanothece* sp. (ATCC 51142, PCC 7424) and *Lyngbya* sp. PCC 8106 (Bernroitner *et al.*, 2009).

The Prxs rely on the sulphur atom of a conserved cysteine residue termed as the peroxidative Cys (Cys_P) that helps in the cleavage of the peroxyl –O-OH bond. The second cysteine (if present) is known as the resolving cysteine (Cys_R) residue. It is on this basis that the Prxs were originally grouped into two subfamilies: 2-Cys Prx and 1-Cys Prx. On the basis of the catalytic mechanism, the 2-Cys Prx are divided into 'typical' which possess the Cys_P and Cys_R residues and form homodimers through the intersubunit disulfide bond and 'atypical' which possess one Cys_P and form intramolecular cysteine in the same polypeptide. The basic catalytic mechanism in all Prxs is suggested to take place in two steps. In the first step, the activated cysteine (Cys_P-SH) is oxidized by a peroxide substrate to sulphenic acid (Cys_P-SOH) with concurrent reduction of the peroxide (ROH). In the second step, the sulphenic acid then reacts with the sulphydryl group of Cys_R forming an intersubunit disulphide bond and simultaneously releasing water. The regeneration of oxidized catalytic thiol depends on glutathione (R'SH) or thioredoxin, glutaredoxin, cyclophilin or tryperodoxin (Wood *et al.*, 2003; Stork *et al.*, 2005). Prxs are efficient scavengers of very low concentrations of H_2O_2. Substrates include H_2O_2, alkyl hydroperoxides and peroxynitrite. Prxs are important components of antioxidant defence system and also play a significant role in redox-mediated signal transduction at least in some eukaryotes (Dietz *et al.*, 2003; Veal *et al.*, 2004; Dietz, 2011). These reactions have been best explained in case of 2-Cys Prxs which carry out a cycle of three distinct redox transformations within a single active site. Thus Cys_P goes through three different stages in the peroxidatic cycle: sulphydryl (-2) – sulphenic acid (O) - disulphide(-1) - sulphydryl (-2) (Aran *et al.*, 2009).

In some non-photosynthetic bacteria AhpC, the 2-Cys Prx gets reducing power from a flavoprotein AhpF, an NADPH disulphide oxidoreductase (Poole, 2005). In plants, the chloroplast 2-Cys Prx is reduced by electrons received from CDSP32, a protein containing two Trx domains in tandem (Tel-Or *et al.*, 1985). The homologues of AhpF and CDSP32 are absent in cyanobacterial genomes. The regeneration of oxidized form of 2-Cys Prx in eukaryotes is known to occur through an electron transfer from NADPH via NADPH: thioredoxin oxidoreductase (NTR) and a Trx together in a single protein (Pérez-Ruiz *et al.*, 2006). Only in case of *Anabaena* sp. strain PCC 7120 genome an NTR sequence has been reported (Florencio *et al.*, 2006). Cyanobacterial 2-Cys Prxs are also known as thioredoxin peroxidases because they use Trx as electron donor to reduce peroxides. The regeneration of oxidized form of 2-Cys Prx is suggested to be through the reduced ferredoxin and the enzyme ferredoxin: thioredoxin oxidoreductase plays a key role in supplying electrons to Trx (Meyer *et al.*, 2005; Buchanan and Balmer, 2005; Florencio *et al.*, 2006; Schürmann and Buchanan, 2008). Subsequently, the reduced Trx, catalyzes the reduction of a variety of target enzymes, including Prx, by disulphide/dithiol exchange (Buchanan and Balmer, 2005). In the thermophilic cyanobacterium *T. elongatus* BP-1, NADPH thioredoxin reductase C functions as an electron donor to 2-Cys Prx. An NAD(P)H oxidase complex induced by oxidative stress was shown to consist of NADPH thioredoxin reductase C and 2-Cys Prx. This provided evidence for their interaction *in vivo* and further *in vitro* NADPH-dependent H_2O_2 reduction assays required the presence of NADPH thioredoxin reductase C and 2-Cys Prx suggesting that the reductase supplies reducing power from NADPH to 2-Cys Prx. The NADPH thioredoxin reductase C is a unique enzyme that consists of a NADPH thioredoxin reductase domain and a thioredoxin-like domain with conserved –CXYC- motif in each of them. A point mutation in the conserved motif of the reductase domain resulted in the loss of NADPH oxidation whereas point mutation in the conserved motif of thioredoxin domain did not affect the NADPH oxidase activity (Sueoka *et al.*, 2009).

The genome of *Synechocystis* sp. strain PCC 6803 revealed the presence of five genes encoding Prxs, two genes of PrxQ (*sll0221* and *slr0242*) and one gene each of 2-Cys prx (*sll0755*), 1-Cys Prx (*slr1198*) and Type II Prx (*sll1621*) (Kaneko *et al.*, 1996; Kobayashi *et al.*, 2004; Hosoya-Matsuda *et al.*, *2005*). In case of *S. elongatus* PCC 7942 six putative genes of Prxs have been detected, i.e. one each of 1-Cys Prx (gene 915), 2-Cys Prx (gene 782) and four of PrxQ (genes 310, 439, 662 and 1668). The 2-Cys Prx gene (*sll0755*) of *Synechocystis* sp. strain PCC 6803 has been cloned and this gene was overexpressed in *E. coli*. This enzyme was able to reduce H_2O_2 and tertiary butyl hydroperoxide with thioredoxin from *E. coli* as electron donor (Yamamoto *et al.*, 1999). The purified protein (22.5 kDa) possesses two cysteine residues Cys50 and Cys172 required for the catalytic cycle. These are present in highly conserved motifs –FTFV<u>C</u>PTEI- (Phe-Thr-Phe-Val-Cys-Pro-Thr-Glu-Ile, the so-called F-motif) and EV<u>C</u>P (glu-Val-Cys-Pro). During the reaction cycle an intersubunit disulphide bond is formed between Cys50 of one subunit and Cys172 of the other. Significant conformational changes seem to be necessary during the catalytic cycle (Hirotsu *et al.*, 1999; Schröder *et al.*, 2000). So this novel antioxidant enzyme has been designated as thioredoxin peroxidase (TPx). *Synechocystis* cells were transformed with pY7Blue-T plasmid having TPx gene interrupted with spectinomycin/streptomycin resistance marker to generate *tpx*⁻ cells. The *tpx*⁻ mutant cells lost the peroxidase activity coupled with the electron flow in PSII. These results indicate that the TPx activity is coupled to photosynthetic electron transport system. Investigations on the *in vivo* role of 2-Cys Prx in *Synechocystis* sp. strain PCC 6803 and its homologue in *S. elongatus* PCC 7942 and the corresponding gene disruption mutants showed that 2-Cys Prx is important in protecting the cells against peroxide stress (Klughammer *et al.*, 1998; Yamamoto *et al.*, 1999). Moreover, studies on the 2-Cys Prx disruptant mutants of *Synechocystis* sp. strain PCC 6803 revealed that this Prx is important in photosynthetic adaptation as in the mutants the quantum yield decreased with a simultaneous saturation of electron transport rates at lower light intensity. These results thus suggest that the 2-Cys Prx plays an important protective role in photosynthesis (Klughammer *et al.*, 1998; Yamamoto *et al.*, 1999).

Perelman *et al.* (2003) identified an ORF downstream of *nblR* gene sequence (that encodes a response regulator NblR essential for cell survival under high light and nutrient starvation) of *S. elongatus* PCC 7942 that is homologous to thioredoxin-peroxidase-like (TplA). To understand the H_2O_2 detoxification mechanisms in this organism, they generated single and double mutants for *katG* and *tplA* genes. The double mutant surprisingly tolerated H_2O_2 stress and its ability to detoxify H_2O_2 compared very well with the wild-type and TplA mutant cells. This suggested that KatG activity was essential for this process and *tplA* gene was dispensable but when the cells experience oxidative stress under high light conditions TplA appears to be essential for growth. Cells lacking TplA could not compete with wild-type. These results emphasize different physiological roles for the enzymes depending on whether oxidative stress is caused by high light or H_2O_2.

A DNA microarray analysis of *Synechocystis* sp. strain PCC 6803 cells exposed to oxidative stress (MV for 15 min) showed significant induction of genes (*sll1621*, *slr1738*, *slr0074*, *slr0075* and *slr05890*) commonly under conditions of both normal and high light. When Fur-type transcriptional regulator encoded by *slr1738* (located near to *sll1621* gene) has been deleted, the deletion mutant under non-stressed conditions showed greatest depression in the expression of *sll1621*. The gene product of *sll1621*, a Type II Prx, has been shown to be essential for aerobic photoautotrophic growth both under high light (in liquid and solid media) and low light (solid media) conditions. The purified His-tagged recombinant protein SLR1738 produced in *E. coli* exhibited binding to the intergenic region between *sll1621* and *slr1738*. Thus the Fur homologue *slr1738* of *Synechocystis* plays an important regulatory role in the induction of antioxidant *sll1621* in response to oxidative stress (Kobayashi *et al.*, 2004).

Two ORFs, *sll1621* and *slr1198* of *Synechocystis* sp. PCC 6803 encode AphC/TSA family protein and antioxidant protein, respectively (Nakamura *et al.*, 1998). Hosoya-Matsuda *et al.* (2005) recognized these proteins by Trx-affinity chromatography and showed that these proteins are identical in sequence to Type II Prx (*sll1621*) and 1-Cys Prx. (*slr1198*). Recombinant SLL1621 protein expressed in *E. coli* when purified not only interacted with Trx but also exhibited glutathione-dependent peroxidase activity. Gene (*sll1621*) disruptant mutant was not viable even under very weak light conditions emphasizing the essential role of this Prx for photosynthetic growth. Although protein SLR1198 did not show peroxidase activity, the *slr1198* disruptant mutant exhibited poor growth. A bioinformatic analysis of the genomes of *Synechocystis* sp. strain PCC 6803 and *S. elongatus* PCC 7942 for Prxs and their transcript accumulation revealed that 1-Cys and 2-Cys Prxs of both organisms exhibited very high similarities of 90% and 88% with identities of 83% and 73%, respectively. The two PrxQ genes (*sll0221* and *sll0242*) of *Synechocystis* and one PrxQ gene (1668) of *S. elongatus* are atypical in possessing only one conserved Cys residue where as the other three PrxQ genes (310,439 and 662) of the latter organism possess two Cys residues in the conserved region that are spaced by only few amino acids similar to the eukaryotic PrxQs (Stork *et al.*, 2005). The mRNA levels of the five Prxs of *Synechocystis* and six Prxs of *S. elongatus* under standard growth and induced oxidative stress (H_2O_2 5 mM, MV 50 μM, light intensity 20, 200 and 800 μE $m^{-2} s^{-1}$ for 48 hrs, iron limitation and NaCl 0.6 M) conditions exhibited differences in between the two organisms. With respect to light intensity, the transcript levels of most of the Prx genes of *Synechocystis* (1-Cys Prx, 2-Cys Prx, Type II Prx and PrxQ-B1) accumulated to a higher level at medium and high light intensities whereas those (1-Cys Prx, 2-Cys Prx and PrxQ-A1) of *S. elongatus* were already up-regulated under low light intensity itself. The transcripts of Type II Prx showed highest up-regulation in response to H_2O_2 exposure than MV-treated cells consistent with earlier studies (Kobayashi *et al.*, 2004; Hosoya-Matsuda *et al.*, 2005). By contrast, in *S. elongatus* the transcript levels of 2-Cys Prx were strongly up-regulated in H_2O_2 as well as after MV-treatment confirming the earlier work of Perleman *et al.* (2003). In case of NaCl stress, most of the Prxs in both the organisms were up-regulated at different times, i.e. at 24 or 48 hrs (Stork *et al.*, 2005). Gene *alr2503* of *Anabaena* sp. strain PCC 7120 that encodes a putative PrxQ homologue is located in the same gene cluster as *pkn22* which ecncodes a Ser/Thr kinase. A knock-out mutant of *pkn22* (Mp22) was very sensitive to oxidative stress because of its failure to express *prxQ-A*. The mutant tolerated oxidative stress only after the introduction and expression of the *prxQ-A* gene in trans position. The properties of the recombinant PrxQ-A overexpressed in *E. coli* revealed that it (i) it protects DNA from being degraded by ROS; (ii) reduces H_2O_2 in presence of dithiothrietol and (iii) shows Trx-dependent peroxidase activity. The recognition of Cys-47 as the peroxidative residue and its replacement by Ser completely abolished the peroxidase activity. Type II Prx of *Anabaena* sp. strain PCC 7120 is a hybrid protein with a Grx domain fused at the C-terminal end that contains the conserved –CXXC- domain. This hybrid protein was expressed in *E. coli* and shown to have the highest peroxidase activity toward H_2O_2 using glutathione as electron donor. The existence of this protein in both vegetative cells and heterocysts has been demonstrated by immunoblot analysis (Hong *et al.*, 2008). Pérez-Pérez *et al.* (2009) conducted a comprehensive analysis of Prxs of *Synechocystis* sp. PCC 6803 and demonstrated that five Prxs (1-Cys Prx, 2-Cys Prx, Type II Prx and PrxQ1 and PrxQ2) of this organism could use TrxA, TrxB and TrxQ as electron donors. Glutathione or Grxs were inefficient as electron donors for the Prxs. Type II Prx and TrxQ combination proved to be the best that showed highest catalytic efficiency and this coincided with the maximum levels of expression of the corresponding gene transcripts and proteins under different stress conditions. The eukaryotic 2-Cys Prxs have been shown be sensitive to overoxidation due to the presence of structural Gly-Gly-Leu-Gly and Tyr-Phe motifs. The overoxidation of 2-Cys Prx results in its inactivation leading to a transient increase in H_2O_2 concentration. This results in H_2O_2-mediated signalling and is known as floodgate hypothesis. Thus this adaptation allows 2-Cys Prxs to act as floodgates, keeping resting levels of H_2O_2 low while permitting higher levels during signal transduction (Wood *et al.*, 2003). To know whether such an adaptation exists in cyanobacterial 2-Cys Prxs, Pascual *et al.* (2010) showed that several prokaryotic 2-Cys Prx sequences including those of cyanobacteria contained Gly-Gly(Leu-/Val-/Ile)-Gly and Try-Phe motifs characteristic of overoxidation sensitive ezymes. Further two cyanobacterial 2-Cys Prx enzymes (from *Anabaena* sp. strain PCC 7120 and *Synechocystis* sp. strain PCC 6803) have been tested for their sensitivity to overoxidation. The wild-type and mutant enzyme from *Anabaena* showed overoxidation at the Cys residue. In response to high light and H_2O_2, the 2-Cys Prx from *Anabaena* was reversibly overoxidized *in vitro*. These results emphasize that the two organisms adopt different strategies to cope up with H_2O_2. *Synechocystis* has low content of less sensitive 2-Cys Prx and high catalase activity whereas *Anabaena* possesses abundant, sensitive 2-Cys Prx but low catalase activity. In this respect the latter comes closer to the chloroplast system.

E) Thioredoxins (Trxs): Trxs are small proteins (of molecular weight ~12,000) ubiquitously present in all living organisms. The common active site sequence –Trp-Cys-Gly-Pro-Cys- is well conserved in all Trxs. These proteins share a common structure known as Trx motif consisting of four α-helices and five β-sheets (Eklund *et al.*, 1984; Capitani *et al.*, 2000). In photosynthetic organisms a number of processes are regulated by light and this can be due to light-driven changes in intracellular pH, ion concentration or reduced co-factors. The levels of dithiol form of Trxs are enhanced by light-driven enhancement in the levels of both NADPH and ferredoxin. Trxs in turn can function as reducing agents in processes such as sulphur metabolism and deoxynucleotide synthesis. Trxs mediate dithiol-disulphide exchange and thus regulate the activity of enzymes through the formation or the reduction of disulphide bridge in the target enzymes. This leads to a modulation of enzyme activity via a redox control. The known enzymes whose activity is controlled by Trxs belong to the Calvin cycle. The regulation of chloroplast NADP-dependent malate dehydrogenase by Trx *m* was reported (Scheibe, 1987) followed by the regulation of glyceraldehyde-3-phosphate dehydrogenase, fructose-1,6-biphosphatase, sedoheptulose-1,7-biphosphatase and phosphoribulokinase by Trx *f* and these are known as thiol enzymes in the chloroplasts (Marcus *et al.*, 1988; Porter *et al.*, 1988; Jacquot *et al.*, 1997). In prokaryotes, three phylogenetic groups of Trxs, *m*, *x* and *y* are known. Their corresponding homologues occur in chloroplasts and algae (Meyer *et al.*, 2005) and in cyanobacteria the genome of *Synechocystis* sp. strain PCC 6803 possesses one homologue from each of these families. These are TrxA (*slr0623*), TrxB (*slr1139*) and TrxQ (*slr0233*) belonging to *m*, *x* and *y* groups, respectively (Navarro and Florencio, 1996; Florencio *et al.*, 2006; Pérez-Pérez *et al.*, 2006).

A Trx from *Anabaena* sp. 7119 (Gleason *et al.*, 1985) and *A. nidulans* (*S. elongatus* PCC 7942; Muller and Buchanan, 1989) that was similar to *m*-type Trx in amino acid sequence and functional properties has been reported. Alam *et al.* (1989) identified two thioredoxin genes in *Anabaena* sp. strain PCC 7120. One of these encodes an unusual Trx that activates spinach fructose-1,6-biphosphatase and considered to be the counterpart of chloroplast Trx *f* (Gleason, 1990) but later designated as Trx 2 bearing no resemblance in amino acid sequence to known Trxs (Eklund *et al.*, 1991). Although the gene that encodes Trx 2 has been cloned from *Anabaena* sp. strain PCC 7120 and expressed in *E. coli*, its presence in the cells of *Anabaena* could not be demonstrated (Alam *et al.*, 1989). Western immunoblotting revealed the presence of Trx 2 in very low levels and it can be reduced by glutathione. The amino acid sequences of Trxs T1 and T2 from *Anabaena* sp. strain PCC 7120 showed the former protein to be similar to spinach chloroplast Trx *m* and the latter to a Grx. In this respect, cyanobacteria resemble plant chloroplasts in having two dissimilar Trxs- one that is related to more common protein found in prokaryotes and the other a Grx (Gleason, 1992).

The earliest studies on Trx gene disruption mutants of *S. elongatus* PCC 7942 (then described as *A. nidulans* R2) corresponding to genes *slr0623* and *sll1980* of *Synechocystis* sp. PCC 6803 that encode TrxA and a Trx-like protein respectively revealed that while the former is essential for photosynthetic growth (Muller and Buchanan, 1989), the latter was required for correct assembly of photosynthetic apparatus (Collier and Grossman, 1995). The first indication that Trx can serve as a source of reductant for Prxs came from the *katG* deletion mutant of *Synechocystis* that exhibited a light-dependent thiol-specific peroxidase activity (Tichy and Vermaas, 1999) already described earlier. Studies on Trx-deletion mutants of *Synechocystis* have shown that while TrxA is essential for the survival of the organism (as deletion mutants could not be obtained), deletion of TrxQ caused hypersensitivity to H_2O_2 suggesting that TrxQ plays a role in peroxide tolerance (Nishiyama *et al.*, 2001; Pérez-Pérez *et al.*, 2009).

In order to identify Trx-linked processes in cyanobacteria, Lindahl and Florencio (2003) screened cytosolic and peripheral membrane proteins by using an immobilized mutant form of Trx, TrxA (where Cys35 was exchanged with Ser by site-directed mutagenesis) as a bait. The formation of a covalent bond between the target proteins and TrxA was confirmed by non-reducing/reducing two-dimensional SDS/PAGE. Of the identified cytosolic (18) and membrane-associated (8) proteins, 20 of them are new candidate substrates for Trx. Phosphoglucomutase is an important enzyme whose activity was inhibited under oxidizing and activated under reducing (in presence of DTT) conditions *in vitro*. Light-induced redox regulation by Trx seems to be extensive as processes of assimilation and storage of carbon, sulphur and nitrogen, all enzymes of glycogen synthesis pathway (including the rate-limiting enzyme ADP-glucose pyrophosphorylase), cyanophycin synthesis, ammonia assimilation (GOGAT), carbon concentrating mechanism protein (CcmM) come under its influence. This shows that reduced Trx has a profound effect on all the anabolic processes and depending on the growth conditions the identities and relative abundance of Trx target proteins are likely to change. Cyanobacterial redox-mediated signalling and the regulation of transcription of key genes involved in acclimation response to a changing environment has remained obscure for a long time. In this direction Nakamura and Hihara (2006) first demonstrated that PedR, a LuxR-type regulator, gets activated and inactivated under low and high light conditions, respectively with a concomitant conformational change. Horiuchi *et al.* (2010) for the first time showed that PedR interacts with Trx to transmit the redox status of photosynthetic electron transport chain. This has been confirmed by the *in vitro* formation of intramolecular disulphide bond between Cys80 residue of the PedR homodimer and its reduction by TrxM1 and TrxX. Mutants of *Synechocystis* that lack ferredoxin-Trx reductase or NADPH-Trx reductase showed minimal conformational change in PedR or minimal transcriptional regulation of key genes under high light. So a transient inactivation of PedR results due to its reduction by Trx under high light. Trx-affinity chromatography is proving to be a useful method for understanding the Trx-network (Hisabori *et al.*, 2005).

The availability of reduced Trx very much depends on the reducing equivalents supplied by either NADPH or reduced ferredoxin. In presence of two Trx reductases, NADPH-Trx reductase (NTR) and ferredoxin-Trx reductase (FTR) the *in vitro* reduction of Trx has been demonstrated. Hishiya *et al.* (2008) identified the genes for these two Trx reductases and also four typical isoforms of Trx. Their studies on NTR and FTR disruptant mutants suggested that the NTR-Trx pathway is important for antioxidant system whereas FTR-Trx pathway may be beneficial in the control of cell growth rate. Of the four isoforms of Trx, the gene product of *slr0623*, the homologue of *m*-type Trx (i.e. TrxA) is the most abundant type and that accumulation of Trx isoforms is dependent on the expression of the redox-related proteins. A comparative analysis of NTR-2-Cys Prx system from plants and cyanobacteria has been presented and the gene sequences of NTR from *Anabaena* sp. strain PCC 7120 could complement the knock-out mutants for NTR gene of *Arabidopsis thaliana* but not those from *Synechocystis* sp. strain PCC 6803. So in this respect the *Anabaena* enzyme fulfills the role of plant enzyme (Pascual *et al.*, 2011).

F) Glutathione (GSH): GSH is a tripeptide of γ-L-glutamyl-L-Cysteinyl-Glycine and it is a non-protein thiol. It is widely distributed in some prokaryotes and in all eukaryotes with the exception of those that lack chloroplast or mitochondria. Among the prokaryotes, it is present in cyanobacteria, proteobacteria as well as in a few strains of gram-positive bacteria. The concentration of GSH ranges from 0.1 to 10 mM in bacteria (Fahey *et al.*, 1978; Fahey and Sundquist, 1991; Newton *et al.*, 1996). The thiol group of GSH is responsible for its antioxidant activity which forms a thiyl radical on oxidation, and then reacts with a second oxidized glutathione to form disulphide bonded dimeric form GSSG. The ratio of GSH:GSSG is carefully controlled in the cytoplasm and it mostly exists in its reduced state. GSH functions directly as a free radical scavenger by reacting chemically with singlet oxygen, superoxide and hydroxyl radicals. It plays a critical role in protection against oxidative stress caused by peroxides such as H_2O_2 or alkylhydroperoxides, and other environmental stresses such as osmotic shock, acidity and toxicity due to certain chemicals (methylglyoxal, hypochlorous acid and monochloramine). It also stabilizes membranes by removing acyl peroxides formed by lipid peroxidation reactions. Glutathione peroxidase uses GSH to reduce H_2O_2 producing GSSG. The redox potential of GSH is sufficient enough to reduce dehydroascorbate to ascorbic acid or to reduce disulphide bonds of proteins.

The reduction of GSSG to GSH is catalysed by the enzyme glutathione reductase (GR) the product of *gor* gene. Reducing equivalents from NADPH are used. Thus GR serves as the key link between the two redox couples [GSH/GSSG and NAD(P)H/NAD(P)] in the cell. However, GSH and NAD(P)H/NAD(P) do not exchange electrons directly. Multiple forms of GR exist in plants associated with different subcellular compartments. In cyanobacteria, for example in *Synechocystis* no gene encoding GR has been found. So how GSH could be regenerated from GSSG is not known (Li *et al.*, 2007). Under oxidative stress, GSH serves as a protein reductant either directly or through the reduction of Grxs. Glutathionylation is a process in which mixed disulphides are formed between GSH and the cysteines of proteins. At any given time ~2% of the total glutathione content in *E. coli* cells is in the form of protein disulphides (Miranda-Vizuete *et al.*, 1996). Protein glutathionylation seems to take place by direct oxidation of GSH and protein thiol followed by the formation of mixed disulphide bond (Thomas *et al.*, 1995; Hondrop and Matthews, 2004). This is a protective mechanism to prevent cellular proteins from being inactivated by oxidative stress, because overoxidation of protein thiols leads to the formation of sulfinic acid or higher oxidation states. Such proteins cannot be reduced by Trx or Grx and this leads to their irreversible inactivation. Glutathionylated proteins can be brought back into an active state by Grxs that mediate the reduction of disulphides, once the organism is brought back to normal environment (Ritz and Beckwith, 2001; Hondrop and Mathews, 2004). The many faces of glutathione in bacteria have thus been exposed (Masip *et al.*, 2006).

NADPH-dependent glutathione peroxidases (GPxs; EC 1.11.1.9 and EC 1.11.1.12) catalyze the reduction of H_2O_2 and organic hydroperoxides to water and corresponding alcohols using GSH as electron donor. Most of the cyanobacterial genomes (fully/partially) sequenced show the presence of an ORF with homology to plant GPxs (all strains of *P. marinus*, 11 strains of *Synechococcus*, *G. violaceus*

PCC 7421, *S. elongatus* PCC 6301, *S. elongatus* PCC 7942, *C. watsonii* WH8501, *Cyanothece* sp. PCC 8801 and *N. punctiforme* PCC 73102) except in case of *Synechocystis* sp. stain PCC 6803 and two strains of *Synechococcus* which have two paralogues. In the rest of the cyanobacterial genomes investigated [*M. aeruginosa* NIES-843, two strains of *Synechococcus*, *T. elongatus* BP-1, *A. marina* MBIC11017, strains of *Cyaonothece* (ATCC 51142, CCY0110 and PCC 7424), *Lyngbya* sp. PCC 8106, *T. erythraeum* IMS101, *A. variabilis* ATCC 29413, *No. spunimgena* CCY9414, *Anabaena* sp. strain PCC 7120] the GPx gene sequence is absent. The cynobacterial GPx have a conserved active site with three cysteine residues from N- to C-terminus, X<u>C</u>G, -P<u>C</u>N- and F<u>C</u>Y, respectively with the latter being absent in *Synechocystis* sp. strain PCC 6803 and *Synechococcus* PCC 7002 GPx proteins (Bernroitner *et al.*, 2009).

G) Glutaredoxins (Grxs): These are small glutathione-disulphide oxodoreductases belonging to the Trx-fold superfamily (Martin, 1995). These are present in bacteria, plants and mammals. The importance of Grx was recognized by Holmgren (1976) who first reported its requirement as an alternative hydrogen donor in the reduction of intramolecular disulphide in ribonucleotide reductase, an important enzyme for DNA synthesis. Three major groups of Grxs are known. The classical type Grx is a small 10 kDa protein with an active site of Cys-Pro-Tyr-Cys. *E. coli* (Grx1 and Grx3) and yeast (Grx1 and Grx2) type Grxs belong to this group. The second group possesses an active site of Cys-Gly-Phe-Ser and yeast Grxs (Grx3, Grx4 and Grx5) belong to this group (Rodriguez-Manzaneque *et al.*, 1999). *E. coli* Grx2 that is structurally related to glutathione-S-transferase belongs to the third group (Xia *et al.*, 2001). The Grxs catalyze reduction of protein disulphides through dithiol mechanism and GSH-protein mixed disulphides through monothiol mechanism in a coupled system with GSH, NADPH and GR and maintain the redox homoeostasis (Holmgren, 1989; Holmgren and Aslund, 1995). In dithiol mechanism both cysteines of the conserved motif are required for the dithiol redox reaction whereas in the monothiol mechanism only one cysteine of the active site is involved. Besides DNA synthesis, a number of cellular functions are influenced by Grxs such as regulation of transcription factors, reduction of dehydroascorbate, protein folding and sulphur metabolism (Wells *et al.*, 1993; Fernandes and Holmgren, 2004). The functions of Grxs stretch far beyond a simple Trx backup system as these enzymes are efficient in the reactivation of proteins inactivated during oxidative stress. Evidences in favour of the protection conferred by Grxs against oxidative stress have been put forward in the reactivation of HIV-1 protease (Davis *et al.*, 1997) and eukaryotic nuclear factor I (Bandyopadhyay *et al.*, 1998).

The protective role of Grxs against oxidative stress in cyanobacteria has been investigated. The completely sequenced genome of *Synechocystis* sp. strain PCC 6803 revealed the presence three probable genes of Grxs (Kaneko *et al.*, 1996). One of these, ORF *ssr2061* (of 264 codons) has been cloned and the deduced amino acid sequence of the Grx shares 29–42% identity in amino acid sequence of Grxs from *E. coli* (35%), yeast (31%), rice (42%) and human (29%). *E. coli* BL21 (DE3) cells transformed with pET-2061 vector overexpressed the recombinant protein (Grx2061; up to 20% of the total protein quantified by band scan) and the purified protein exhibited NADPH oxidation in the 2-hydroxyethyldisulphide assay. However, in the active site of Grx2061 between the active Cys residues, Tyr residue was replaced by Phe residue as observed in case of rice but in the rest the conserved active site is present. Higher Grx activity in the transformant enabled it to grow in presence of H_2O_2. The protective role of overexpressed Grx in *E. coli* cells against oxidative stress induced by H_2O_2 has been highlighted (Li *et al.*, 2005). Deletion of *ssr2061* was not lethal to *Synechocystis* sp. strain PCC 6803 but reduced its viability which indicates that *ssr2061* has a protective role against oxidative stress. A mixed disulphide affinity approach was undertaken to identify the target proteins interacting with Grx2061 *in vivo*. By employing mutated Grx2061 protein (at the second Cys residue to Ser of the active site; Cys-Pro-Phe-Cys to Cys-Pro-Phe-Ser) immobilized on to an affinity column, cytosolic proteins were passed through the column. Among the 42 proteins so identified 26 of them were novel to the cyanobacterium. These proteins are supposed to be involved in 12 cellular processes including oxidative stress response, Calvin's cycle, protein synthesis etc. Most importantly the interaction of Grx2061with Type II Prx, catalase-peroxidase, Trx-reductase and Grx1562 has been studied in detail. Loss of catalase activity (to nearly 30%) in presence of Grx suggested that catalase was Grx-regulated. Type II Prx acivity confirmed that it could also be regulated by Grx2061. So it means that Grx-dependent activation of Prx and deactivation of catalase can occur in presence of Grx2061. Another interesting feature is the interaction of Trx-reductase with Grx2061 that is reported for the first time from a photosynthetic organism. The interaction of Grx2061 with Grx1562 indicated that oxidized Grx2061 could keep redox equilibrium and can be reduced by Trx reductase. It means that Grx2061 can accept electrons from either glutathione or Trx-reductase. So the redox reactions pass through NADPH-TR-GSH-Grx and protein.

The progress made in redox biology with the help of proteomics using redox-affinity chromatography, two-dimensional electrophoresis and mass spectrometry has greatly improved our understanding of the intricate network of physiological processes at work in cyanobacteria experiencing oxidative stress (Nishiyama and Hisabori, 2009). Trxs and Grxs thus constitute unifying elements in redox biology. Both proteins act by changing the structure and activity of a broad spectrum of proteins, typically by modifying redox status. The range of target proteins in case of Trx and Grx is bound to increase as new types of orthologues of Trx and Grx are yet to be discovered through out the animal and plant kingdoms (Meyer *et al.*, 2009).

Oxidative stress in relation to iron limitation, high salinity, nitrogen metabolism and UV-irradiation has also received attention (Latifi *et al.*, 2009). A direct relationship between iron limitation and oxidative stress has been established by recent investigations. These will be presented along with observations that connect salinity and oxidative stress.

H) Iron deficiency and oxidative stress: The global transcriptional regulator, Fur has been identified to couple iron homeostasis and defence against ROS. Fur functions as a positive regulator of genes encoding iron storage proteins and SODs. This has led to the suggestion that lack of iron regulation might cause oxidative stress. Secondly, through Fenton reaction the reduced iron form can be used to convert less reactive H_2O_2 to more reactive oxygen species leading to oxidative damage. The induction of *isiA* and *isiB* genes has been noted under iron deficiency and also several other stresses including oxidative stress (Huvaux *et al.*, 2005; Kouril *et al.*, 2005). In response to iron limitation, some of the cyanobacterial species (*S. elongatus* PCC 7942, *Synechocystis* sp. strain PCC 6803 and *Anabaena* sp. strain PCC 7120) show the expression of *isiA* gene (Burnap *et al.*, 1993; Leonhardt and Straus, 1994; Xu *et al.*, 2003). The induction of *isiA* has also been observed under oxidative stress conditions (Jeanjean *et al.*, 2003; Yousef *et al.*, 2003; Li *et al.*, 2004; Singh *et al.*, 2004). Xu *et al.* (2003) reported that a putative Ser/Thr kinase gene *pkn22* (*alr2502*) in *Anabaena* sp. strain PCC 7120 is induced by both iron starvation and oxidative stress and regulates the expression of *isiA*. A *pkn22* insertion mutant (Mp22 obtained by the insertion of a spectinomycin/streptomycin resistance cassette) was unable to grow to the extent as that of wild-type under iron-limiting conditions and the gowth was completely arrested in iron-depleted conditions. The expression of genes *isiA*, *isiB*, *psbC* and *pkn22* was studied in presence of iron and under iron-deficient conditions both in wild-type and the Mp22 mutant. Additionally, MV-treated cells were also subjected to measure the transcripts for *pkn22*. In the wild-type, the trancripts for *pkn22*, *isiA*, *isiB* and *psbC* were noted in iron-limited conditions and in MV-stressed cells showed transcripts for *pkn22*. However, in case of Mp22 mutant the expression of *isiA* was completely abolished one hour after iron removal but there was no effect on the expression of either *isiB* or *psbC*. In the chromosome of *Anabaena* sp. strain PCC 7120, *alr2502* is located immediately upstream of a gene (*alr2503*) encoding a Prx. In the wild-type strain while *alr2503* is expressed under iron-limiting conditions, in case of Mp22 the expression of *alr2503* ia also abolished suggesting that *pkn22* regulates both *isiA* and *alr2503* induced by iron limitation. When *Anabaena* sp. strain PCC 7120 was subjected to iron-deficient conditions, the cells experienced a 100-fold increase in the amount of ROS compared to non-starved cells. Uncoupling of oxidative stress from iron deficiency by the use of a ROS quenching molecule tempol (4-hydroxy-2,2,6,6-tetramethylpiperidine 1-oxyl, a low molecular weight membrane permeable antioxidant) resulted in the non-induction of *isiA* gene (that encodes CP43′ protein) even under iron deficiency (Latifi *et al.*, 2005). Further, iron depletion also caused the induction of a PrxQ-A and a mutant deficient in the synthesis of this peroxidase was not viable under iron-deficent conditions.

Two ferretin-type storage complexes function in *Synechocystis* sp. strain PCC 6803 for the storage of iron inside the cells. These are bacterioferretin and MrgA [a member of the DPS for DNA-binding proteins from starved cells) proteins. Bacterioferretins (encoded by *bfr* genes) function as ferroxidases (BFR ferroxidases) that oxidize Fe^{2+} to Fe^{3+} while generating H_2O_2 and Fe^{3+} is stored as iron oxide in the cavity at the center of their 24-mer ultrastructure (Lewin *et al.*, 2005). Two bacterioferretin proteins one with a conserved haem ligand and the other with a conserved di-iron center are encoded by two *bfr* genes of *Synechocystis*. Mutant cells deficient in the production of either of these two proteins experienced a loss of nearly 50% of the cellular quota of iron even while growing under iron-sufficient conditions (Keren *et al.*, 2004). DPS proteins differ from other members of ferretin proteins in the absence of fifth C-terminal helix (Zeth *et al.*, 2004; Lewin *et al.*, 2005). Besides their main function of storage of iron, DPS proteins also act as DNA-binding proteins protecting DNA against oxidative stress, cold shock proteins, neutrophile activators or pili components (Zeth *et al.*, 2004). The DPS proteins form large (~150 kD) hexameric complexes that bind chromosomal DNA with little sequence specificity. As BFR ferroxidases generate H_2O_2, DPS proteins exhibit catalase activity donating the electron from Fe^{2+} to H_2O_2 (Li *et al.*, 2004). Site-directed mutants of *dpsA* of *S. elongatus* PCC 7942 exhibited very high sensitivity to photooxidative stress induced by high light and MV treatment (Dwivedi *et al.*, 1997). Inactivation mutants of *mrgA* of *Synechocystis* sp. strain PCC 6803 grew much slower than wild-type when transferred from iron replete to iron-depleted conditions. Moreover, when grown in iron-sufficient conditions the internal quota of iron of *mrgA* mutants very much resembled those of wild-type cells unlike the BFR mutants (Shcolnick *et al.*, 2007). Since in *Synechocystis* sp. strain PCC 6803, *mrgA* gene is a part of PerR regulon that is up-regulated during peroxide stress, an *mrgA* disrupted mutant was tested for its peroxide and light stresses. This mutant was found to be highly sensitive to very low peroxide levels but showed the up-regulation of a gene cluster (*sll1722-sll1726*) responsible for the synthesis of exopolysaccharide substance (EPS). Mutants of wild-type and *mrgA* mutant disrupted in EPS cluster were found highly sensitive to oxidative stress. In the absence of EPS the cells became more sensitive to peroxide and light. So the cells of *Synechocystis* sp. strain PCC 6803 are liable to be damaged more readily to external oxidative and light stresses in the absence of *mrgA* (Foster *et al.*, 2009). A comprehensive study by Shcolnick *et al.* (2009) revealed a direct relationship between iron homeostasis and oxidative stress in *Synechocystis* sp. strain PCC 6803. A comparative DNA microarray analysis of wild-type, wild-type cells treated with DFB (deferoxamine B; an iron chelating agent), *mrgA* and *perR* mutants (Li *et al.*, 2004), *mrgA* mutant treated with DFB, *bfrA+bfrB* double mutant (Keren *et al.*, 2004) and *bfrA+bfrB+mrgA* triple mutant (constructed by transformation of *bfrA+bfrB* mutant with *mrgA* disruption construct) cultured in iron-sufficient (10 µM) media was conducted. In addition, cells of wild-type and three disruption strains (*mrgA*, *bfrA+bfrB* and *bfrA+bfrB+mrgA*) were grown on iron-sufficient (10 µM Fe in an EDTA amended medium) and iron-limited (0.3 µM) media, washed free of iron, held in iron-free medium and exposed to different concentrations of H_2O_2 in darkness for 20 hrs. Such cells were also subjected to DNA microarray analysis. A number of genes [such as *isiAB* and iron transport components *futC* and *feoB*, the putative outer membrane transporter (*slr1406*) and the entire ABC-type iron transport genes (*slr1316* to *slr1319*)] were up-regulated in wild-type cells treated with DFB. Inactivation of *mrgA* showed significant changes in the pattern of expression of 255 genes about half of which were up-regulated (polysaccharide metabolism and all hypothetical genes) and half of them were down-regulated (HliA, *ssl2542* and HliB, *ssr2595*). By contrast, treatment of *mrgA* mutant cells with DFB affected the expression of 914 genes which is 4.8 times higher than that observed in wild-type. Direct comparison of DFB-treated wild-type cells and *mrgA* mutant cells yielded 779 differentially regulated genes. The down-regulated genes in *mrgA* mutant treated with DFB pertained to photosynthesis, respiration, ATP synthase, cytochrome oxidase, NADH dehydrogenase (associated with low affinity and high affinity CO_2 uptake mechanisms, 18 of 20 genes), phycobiliproteins, ferredoxin-nitrite reductase (*nir*, *slr0898*), nitrate/nitrite transporter subunit (*nrtD*, *sll1453*), glutamine synthetase (*glnA*, *slr1756*), the glutamate synthase large subunit (gltB, sll1502) and the entire nitrate/nitrite transporter system (*nrt*, *sll1450-sll1453*). The up-regulated genes belonged to *isiA* (*sll0247*, an 8-fold increase in transcript level; and all the genes in this operon), five detoxification genes (of which catalase, *sll1987* and SOD, *slr1516* increased by 2-fold) and Fur-like protein (only *slr 1738 perR* gene, not the other two Fur-like proteins, *sll0517* and *sll1937*). Virtually all genes under PerR regulon including *isiA* and *futA2* (*slr0513*) were up-regulated with a significant increase in the transcripts of *perR* gene (13.3 fold) in *mrgA* mutant treated with DFB than in wild-type. These results provide strong relationship between iron deficiency and oxidative stress. Iron-deficient wild-type cells could withstand H_2O_2 stress (up to 8 mM) 2-fold higher than iron-sufficient cells. Mutant *mrgA* was much more sensitive to H_2O_2 but *bfrA+bfrB* double mutant withstood exposure of up to 4 mM H_2O_2 and were not affected by iron availability. The triple mutant was much more sensitive to H_2O_2 when compared to *bfrA+bfrB* or wild-type. These results thus emphasize that the combined action of the two iron storage complexes coordinate accumulation and gradual release of iron for utilization minimizing the oxidative damage from its interactions with ROS produced in abundance during photosynthesis.

I) Salinity and oxidative stress: The expression of *isiA* and *isiB* genes of *Synechocystis* sp. strain PCC 6803 (Vinnenmeir *et al.*, 1998) and several genes of Prxs of *Synechocystis* and *S. elongatus* PCC 7942 (Stork *et al.*, 2005) were highly induced in response to salt stress. Bagchi *et al.* (2007) characterized a mutant of *S. elongatus* PCC 7942 that exhibited high tolerance to salinity as well as high constitutive expression of *isiA* gene. *S. elongatus* PCC 7942 transformed with *katE* from *E. coli* showed overexpression of catalase with a concomitant resistance to salinity (Kaku *et al.*, 2000). The observations on *A. doliolum* and *M. aeruginosa* point out that high salinity caused oxidative damage due to breakdown of ROS scavenging mechanism (Singh and Kshatriya, 2002) and induced the release of H_2O_2 (Ross *et al.*, 2006), respectively. Finally it is concluded that any physiological condition that decreases the balance of ATP to NADPH would result in ROS production and hence oxidative stress (Latifi *et al.*, 2009).

IV. DESICCATION

Desiccation can be defined as the ability to lose water (or dry up) to equilibrium with air and remain in a state of suspended metabolism for a very long time. Upon rehydration (or absorption of water) the cells regain the functions of active metabolism. The ability to tolerate desiccation is met within different groups of plants and animals. Among plants, bryophytes, pteridophytes and certain angiosperms are capable to lead a desiccated life. However, the most important stages are represented by the seeds, spores and pollen grains. Important animal species belonging to nematodes, rotifers and tardigrates exhibit desiccation tolerance. Desiccation can be complete or partial. Complete desiccation refers to drying to complete air dryness without any metabolism whereas during partial desiccation basal metabolism is still maintained. The term "anhydrobiosis" was coined to represent the state of complete desiccation (Crowe *et al.*, 1992). A quantitative definition of complete desiccation has also been proposed and widely accepted. Complete desiccation represents drying to <0.1 g H_2O g^{-1} of dry weight or mass that is to 10% water content or less than that. In other words, it is roughly equivalent to air-dryness at 50% relative humidity at 20°C and corresponds to a water potential of -100 MPa (Gaff, 1997; Haranczyk *et al.*, 1998; Proctor, 2003). The limits of complete desiccation in prokaryotes seem to be low as bacteria can survive drying up to 2% of water content (Potts, 1994). It means drying up to 10% or lower levels of water content would result in probably too little water left inside the cells to hydrate the macromolecules. There are many examples of surviving complete desiccation for considerable period of time. The longest duration of survival was represented by seeds of *Nelumbo nucifera* which germinated successfully after 1100 years (Shen-Miller *et al.*, 1995). Other examples are from mosses and liverworts which recovered after 20–25 years of complete desiccation. Many bacteria, fungi, lichens and microalgae are also endowed with the capacity to tolerate desiccation. The cyanobacterium *N. commune* survived complete desiccation without any oxidative damage to genomic DNA (due to a covalent modification) for 13 years (Shirkey *et al.*, 2003) and recovered after 13 years of herbarium storage (Shirkey *et al.*, 2000). Understanding the mechanism of desiccation tolerance would greatly help in identifying and developing techniques for the storage of tissues, cells and enzymes (Potts, 1994) and in the metabolic engineering of sensitive cells including those of humans (Potts, 2001; Potts *et al.*, 2005).

Desiccation is a very complex process that occurs very rapidly. The matric water potential or matric water pressure very much depends on the presence of particulate structures present in water viz., proteins, ribosomes and even bacterial cells. At the interface of these particulate structures, the water molecules tend to group themselves and have less tendency to react chemically. Thus these particulate structures and their interfaces tend to decrease the activity of water (a_w). The presence of solutes also lowers a_w. Depending on temperature, the rate at which water escapes as water vapour varies. For example, if the relative humidity is 100% at 20°C, it is decreased to 50% when the temperature is increased to 32°C at constant pressure. Likewise, water vaporization can also be influenced by the relative amounts of solutes present in a solution. In nature, bright sunny days result in completely dry atmosphere and the relative humidity drops nearly to 1 to 2%. The flora and fauna of that region start experiencing loss of internal water as vapour. In case of higher plants, the rate of transpiration will determine the rate at which water is lost as water vapour. In case of other pro- and lower-eukaryotes, there is no such regulation but the organisms tend to experience the loss of water as water vapour and start getting desiccated. But the formation of mist/dew in the early morning will increase the relative humidity of the atmosphere and the organisms start acquiring water or get rehydrated. The cycles of desiccation and rehydration assume greater significance in hot, arid and cold deserts. It is worth noting that as the organisms experience desiccation, they do so by simultaneously getting additional exposure to accompanying stress conditions such as freezing and/or halophilic or high light intensity conditions. As all these stress conditions produce ROS and oxidative stress is a consequence of desiccation as well as rehydration. The extent of water present inside the desiccated prokaryotic cells, the methods of water removal (preferential exclusion hypothesis in relation to replacement of water by the osmoprotectant substances, the differences between drying and salting or sugaring, the sensitivities of prokaryotes to air drying, the effects of freezing-thawing, the differences between freeze-drying and air drying) and the targets of desiccation damage caused in prokaryotes to proteins, nucleic acids and lipids and membranes are the important areas that received attention in studies on desiccation tolerance of prokaryotes (Potts, 1994, 1999, 2001; Potts *et al.*, 2005).

One of the desiccation tolerant and radioresistant bacterium isolated from irradiated meat more than 50 years ago is *Deinococcus radiodurans* (Deinococcaceae; Anderson *et al.*, 1956). More than 30 species of this genus have now been isolated from a variety of habitats, i.e. upper sand layers of deserts, Antarctic rocks, soil, water, faecal matter and hot springs (Ferreira *et al.*, 1997; de Groot *et al.*, 2005; Blasius *et al.*, 2008). Genome sequences of *D. radiodurans* (White *et al.*, 1999), *D. geothermalis* (Makarova *et al.*, 2007; Fig. 20 A) and *D. deserti* (deGroot *et al.*, 2009; Fig. 20 B and C) have already been published. Like *Deinococcus*, among the cyanobacteria the most dominant photosynthetic organism of the extreme arid and cold deserts is the coccoid *Chroococcidiopsis* (Fig. 21). Species of this genus have been found from diverse extreme habitats such as thermal springs (Vasishta, 1968; Anagnostidis and Pantazidou, 1988) , shady caves (Friedmann, 1961), rocks, walls and paddy fields (Tiwari, 1972), air spaces of porous rocks or underneath rock-soil surfaces of hot deserts (Stolz, 1990; Grilli-Caiola *et al.*, 1993; Billi *et al.*, 2001; Boison *et al.*, 2004; Wierzchos *et al.*, 2006) or ice-free Ross Deserts of Antarctica (Friedmann *et al.*, 1988; Friedmann, 1993; Nienow and Friedman, 1993) and hypersaline ponds and lakes (Dor and Paz, 1989; Dor *et al.*, 1991; Řezanka *et al.*, 2003). Endolithic cyanobacteria inhabiting harsh environments of cold and hot deserts are generally dominated by species of *Chroococcidiopsis*, *Myxosarcina* and *Gloeocapsa* (van Thielen and Garbary, 1999; Whitton and Potts, 2000; de la Torre *et al.*, 2003). However, endolithic cyanobacteria inhabiting non-desert environments like limestone of Niagara Escarpment and travertine deposits in Yellowstone National Park are dominated by both filamentous and unicellular types such as *Leptolyngbya*, *Nostoc* and *Synechocystis* (Gerrath *et al.*, 1995, 2000). Molecular characterization of endolithic cyanobacteria inhabiting exposed dolomite rocks in central Switzerland (accession No. AY153458) led to the identification of the organism as *Chroococcidiosps* sp. (91% identity; Sigler *et al.*, 2003). Endolithic biofilms colonizing Antarctic granite rocks also showed the presence of *Chroococcidiopsis*-like organism bearing resemblance (98% identity) to AY153458 (de los Ríos *et al.*, 2007). Cryptoendolithic cyanobacteria dominated by species of *Chroococcidiopsis* have been reported to cause chemical weathering in sandstones by bioalkalization in different parts of S. Africa, Australia, N. America and S. America (Büdel *et al.*, 2004). *Chroococcidiopsis* is morphologically simple but exhibits wide variability. It is a unique ancient type that resembles certain Proterozoic microfossils (Friedmann and Ocampo-Friedmann, 1995).

Ultrastructural studies on desiccated *Chroococcidiopsis* revealed the presence of multilayered envelopes and the ability to tolerate desiccation by this organism was correlated with the synthesis of these multilayered envelopes of extracellular polysaccharides (EPS) and production of spores. Moreover, the amounts of acid-, sulphate- and beta-linked polysaccharides increased in response to desiccation (Grilli-Caiola, 1993; 1996). Phylogenetic analysis of the desert strains of *Chroococcidiopsis* based on 16S rRNA suggested that four strains (057, 123, 171 and 029) were distinct from the type species *Chroococcidiopsis thermalis* PCC 7203 and one of the strains (029, from Negev Desert, Israel) has been identified as suitable organism for genetic studies on desiccation tolerance (Billi *et al.*, 2001).

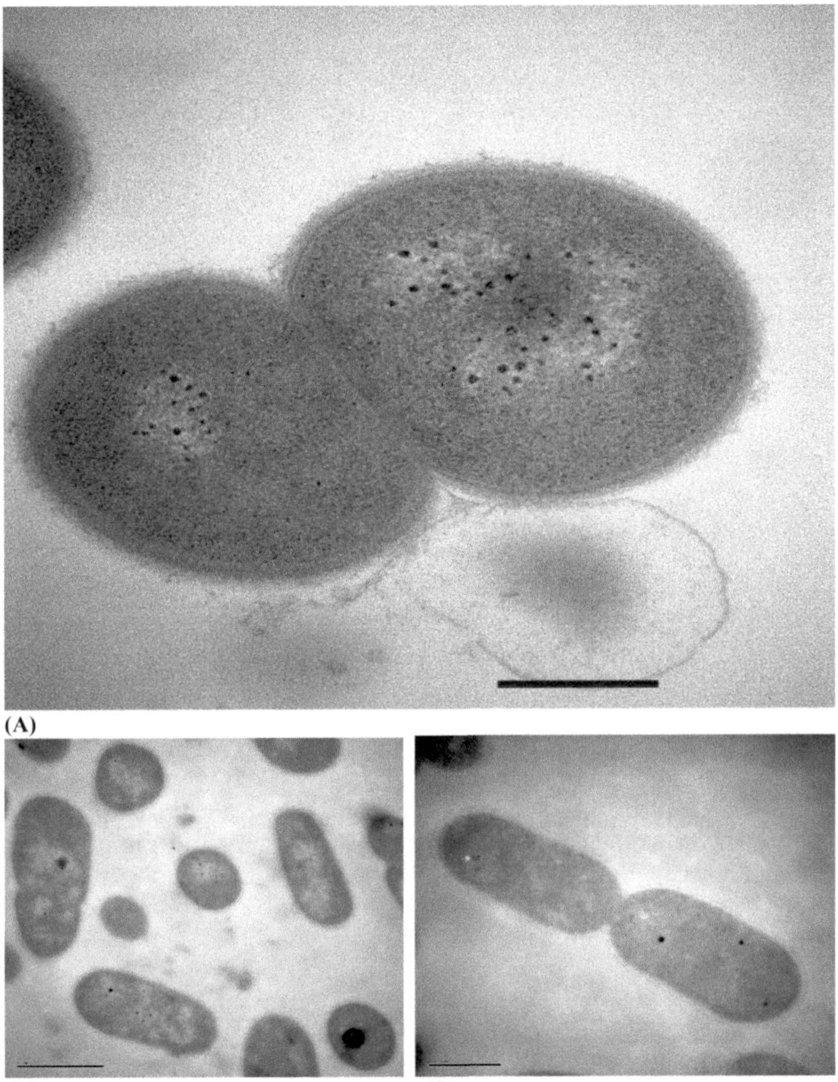

(A) (B) (C)

Figure 20: Electron micrographs of *Deinococcus geothermalis* with a scale bar, 0.5 µm (A) and *Deinococcus deserti* (B and C). Picture A courtesy Elena Gaidamakova, Dept of Pathology, USUHS 4301 Jones Bridge Rd, Bethesda, MD 20814, USA. Pictures B and C courtesy J.-P. Chauvin, Institut de Biologie du Développement de Marseille Luminy (IBDML), UMR 6216, Campus de Luminy Case 907, Marseille Cedex 9 F-13288, France, obtained through Arjan de Groot, CEA, DSV, IBEB, Laboratory of Microbial Ecology of the Rhizosphere and Extreme Environments (LEMiRE), Saint-Paul-lez-Durance, France.

Besides being desiccation tolerant, *Chroococcidiopsis* possesses a very good DNA repair system as severe damage caused by 2.5 and 5 kGy ionizing radiation could be repaired within 3 and 24 h, respectively. This reflects that this organism can withstand long periods of desiccation by its ability to repair DNA damages caused during desiccation (Billi *et al.*, 2000). Due to the above unique features *Chroococcidiopsis* has been projected as a suitable pioneer organism for the development of technologies for large-scale terrafarming on Mars (Friedmann and Ocampo-Friedmann, 1995).

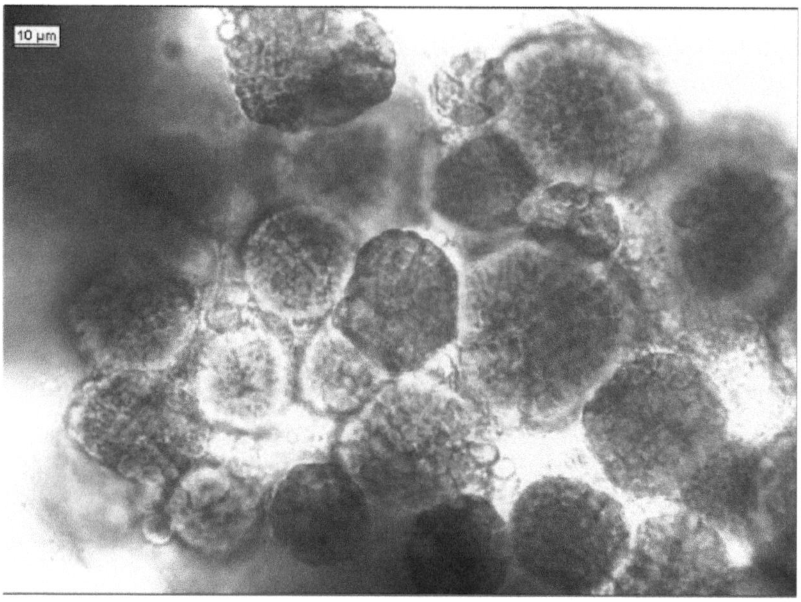

Figure 21: *Chroococcidiopsis* sp. Picture courtesy G. L. Tiwari, Department of Botany, University of Allahabad, Allahabad-211002, India.

Color image of this figure appears in the color plate section at the end of the book.

Dried monolayers of *Chroococcidiopsis* sp. 029 strain could not survive the harsh simulated martian-surface UV and visible light conditions as the cells were killed within 30 min but survived and exhibited slow growth albeit for short periods beneath 1 mm rocks if supplied with water and nutrients. Even after the cells had lost viability, enzymic activity and pigmentation there was no change in the morphology of the cells (Cockell *et al.*, 2005). Raman spectroscopic analysis of the cyanobacteria inhabiting the gypsum crystals from the Houghton Crater, Devon Island, Canadian High Arctic (the site of meteorite impact during the Miocene 23 My ago) revealed the existence of *Nostoc* and *Gloeocapsa*. Specific biosignatures developed for β-carotene and scytonemin in these crystals would be helpful in the detection of extinct or extant halotrophs inside sulphate crystals from Mars (Edwards *et al.*, 2005).

Epilithic and endolithic cyanobacteria growing on rocks from a limestone cliff in Beer, Devon, U.K. have been identified by culture-independent (16S rRNA sequencing) and culture-dependent isolation techniques. Among the 92 clones identified, the largest phylotype of 33 clones was distantly related to *Halothece* sp. from marine environments. Morphotypes bearing resemblance to *Leptolyngbya* (isolates OU-6, OU-8, OU-13 and OU-14), *Phormidium* (isolates OU-44 and OU-10) and a member of Pleurocapsales (isolates OU-11 and OU-12) have been isolated. With a view to select cyanobacteria for space missions, the rocks from the same site were exposed to the effects of low earth orbit at a height of 300 km for 10 days. A single cyanobacterial isolate (OU-20) bearing morphological resemblance to *Gloeocapsa* sp. survived the extreme radiation and desiccation conditions associated with space. It was possible to select and isolate the same organism (OU-20) by exposing the rocks from the same site to similar conditions of desiccation (0.7×10^{-3} kPa), vacuum and UV light (325–400 nm for different time intervals) under the laboratory conditions. Vacuum was found to be more detrimental than desiccation (Olsson-Francis *et al.*, 2010).

Other non-heterocystous and heterocystous cyanobacteria that tolerate extremes of desiccation have been reported from biological soil crusts all the world over. Generally, soil crusts dominated by cyanobacteria are dark coloured or black. The most common cyanobacterium that occurs world-wide in the soil crusts is *Microcoleus vaginatus*. The filaments are equipped with a thick external polysaccharide sheath (Fig. 22). This cyanobacterium generally occupies 0.2 to 0.5 mm beneath the soil surface where sufficient light is available for its growth and harmful effects of UV radiation are minimized. When sufficient moisture is available it migrates upwards and grows extensively and the populations of *Scytonema* and *Nostoc* with coloured sheaths make their appearance. When dry spells return, *M. vaginatus* again becomes subterranean (Belnap *et al.*, 2001b). The intertidal marine mats are dominated by the growth of *Microcoleus* sp. (Pentecost, 1985; Potts and Whitton, 1977, 1980) whereas terrestrial crusts are dominated by species of *Tolypothrix*, *Calothrix* and *Nostoc* (Whitton *et al.*, 1979; Potts and Whitton, 1980; Whitton, 1982). In the soil crusts of Sonoran hot

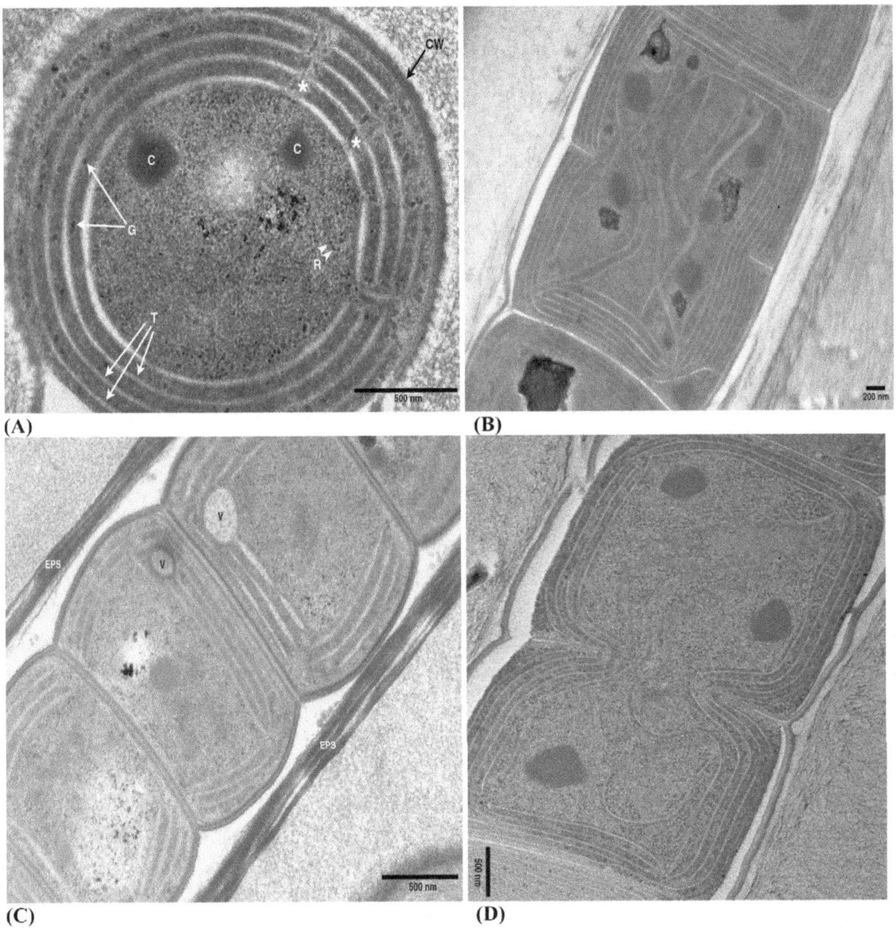

Figure 22: Transmission electron micrographs of cryo-fixed *Microcoleus* sp. in cross section (A) and longitudinal sections (B, C, and D). The arrangement of thylakoids and the the thick extracellular polysachharide mucilaginous sheath can be clearly seen. The magnification bars in represent 200 nm (B) and 500 nm (A, C and D). CW, cell wall; R, Ribosomes; G, polysaccharide (glycogen) granules; C, carboxysomes; asterisk, perforations; EPS, extracellular polysaccharides; V, vesicles. Pictures courtesy A. Kaplan, I. Ohad and R. Nevo, Department of Biologiclal Chemistry, Hebrew University, Jerusalem, Israel 972-2-658-5423.

desert in Arizona, N. America (that receives summer monsoons) the cyanobacteria are represented by species of *Lyngbya, Calothrix, Schizothrix* and *Nostoc*. In the cold deserts of N. America, the crusts are dominated by *M. vaginatus*. The cold deserts of N. America (Pacific North west where summers are hot and dry and winters are cool and moist) support the growth of either attached or vagrant forms of *Dermatocarpon miniatum, N. commune* and *N. flagelliforme*. The soil crusts in the regions of sub-Alpine, Alpine sods and Arctic tundras characteristically support the growth of *Nostoc* sp. The soil crusts of Californian Chaparrals (Mediterranean dry steppe) support the growth of *Schizothrix* and *Nostoc* (Belnap *et al.*, 2001b). The soil crusts from the cool and cold deserts of the Colorado Plateau are dominated by species of *Microcoleus* (*M. vaginatus* Gomont), *Phormidium* and *Schizothrix* (Garcia-Pichel *et al.*, 2001). Microbial mats from Hunt Ice Shelf, High Arctic, Canada (latitude 83° 04′N, longitude 74°25′W) are dominated by the growth of *Phormidium, Leptolyngbya, Nostoc* and *Gloeocapsa* among other heterotrophic bacteria, green microalgae, diatoms and microinvertebrates (Vincent *et al.*, 2000, 2004; Mueller *et al.*, 2005).

The tropical soil crusts of India are dominated by the growth of *Lyngbya arboricola, Plectonema notatum, Scytonema chiastum* and *S. ocellatum* among many other non-heterocystous and heterocystous cyanobacteria that appeared in cultures associated with these forms (Tirkey and Adhikary, 2005). Other terrestrial crusts grow on common building surfaces/roof tops or archaeological monuments. Species of *Gloeocapsa, Gloeothece, Aphanocapsa, Chroococcus, Scytonema* (*S. geitleri*) and *Tolypothrix* have been reported from building surfaces/roof tops (Tripathi *et al.*, 1991). *L. arboricola* has been reported from the bark surfaces of *Mangifera indica* as well (Tripathi and Srivastava, 2001). The growth of species of *Tolypothrix* (*T. scytonemoides*), *Calothrix* and *Nostoc* is a common feature on the surface of temple rocks in India (Amarpalli *et al.*, 1997; Tripathy *et al.*, 1997, 1999). Some of the dominant cyanobacterial species growing on the temple rocks of India are presented in Fig. 23. Deterioration of stone monuments of cultural heritage in European countries of the Mediterranean Basin caused by cyanobacteria and certain members of Chlorophyta has been reviewed. Lithophytic cyanobacterial species belonging to the genera *Gloeocapsa, Gloeothece, Chroococcus, Phormidium, Plectonema, Scytonema, Lyngbya* and *Microcoleus* have been recorded from six types of rocks (marble, limestone, travertine, dolomite, sandstone and granite) (Macedo *et al.*, 2009).

In studies dealing with desiccation, the dried mats of soil crusts are often used directly. Besides strains of *Chroococcidiopsis*, the most widely employed cyanobacterium to understand desiccation tolerance is *N. commune*. It is cosmopolitan in its distribution and occurs in terrestrial habitats such as rice fields (grows as sheres and the balls attain varying diameters up to 10 cm), nutrient poor soils and limestones (grows as parchament-like colonies) and assumes macroscopic proportions. An edible form of *Nostoc* (known as Ge-Xian-Mi in Chinese) identified as *N. flagelliforme* was reported from arid and semi-arid regions of China (Gao, 1998) and rice fields of Hefeng County of China (Qiu *et al.*, 2002). This form has been identified as *N. sphaeroides* (Huang *et al.*, 1998) and *N. commune* (Martinez and Querijero, 1986; Martinez, 1988; Potts, 2000). Colonies of *N. commune* either collected from the field or its clonal isolates are employed in studies devoted to desiccation tolerance [see Fig. 6 (D) of Chapter 1]. This organism has been reported to survive storage at –400 MPa (0% relative humidity) for centuries (Potts, 1994) and the desiccated samples for at least 64 years have been reported to be viable and cultures obtained there from (Shirkey *et al.*, 2003). The filaments are immobilized in the unusual EPS that contributes in many ways for its reported desiccation tolerance. The presence of two antibacterial metabolites (a terpenoid, anthroquinone) in *N. commune* (EAWAG 122b) and carotenoids, 2-hydroxymyxol-2′-fucoside, nostoxanthin and canthaxanthin in *N. commune* NIES (IAM M-13) make this organism unique (Jaki *et al.*, 2000; Takaichi *et al.*, 2009). Thus *N. commune* is a typical anhydrobiotic cell and has been developed as a prokaryotic model for answering fundamental

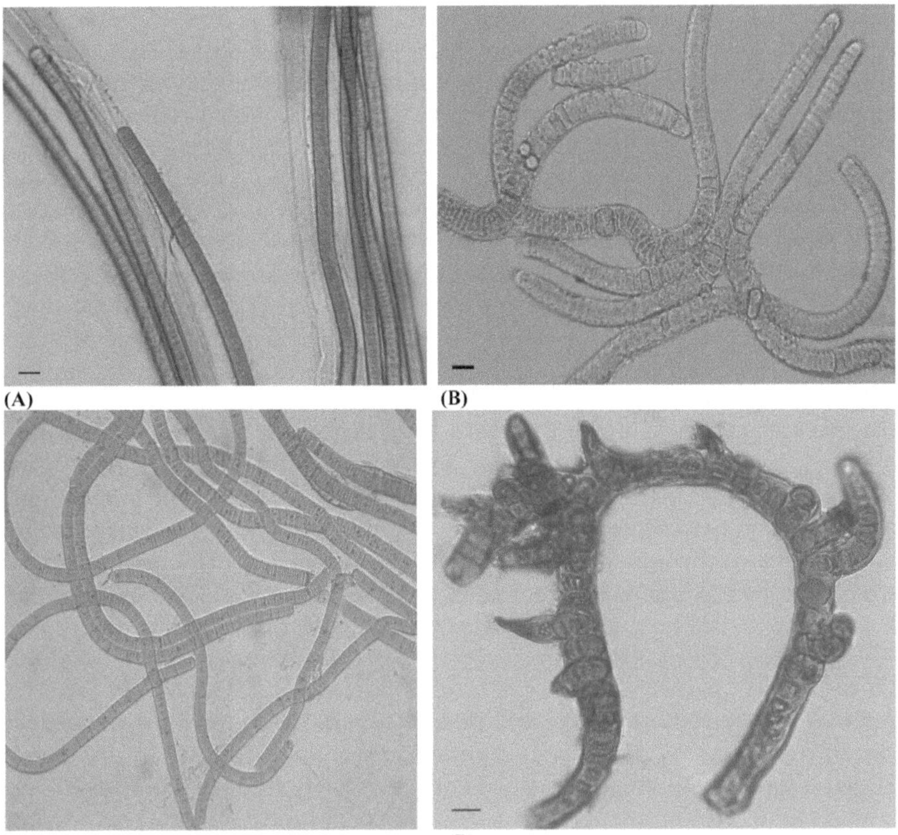

Figure 23: Cyanobacteria from temple monuments of Orissa, India. *Lyngbya* sp (A), *Tolypothrix* sp. (B), *Plectonema* sp. (C) and *Stigonema* sp (D). Magnification bar in all pictures represents 10 μm Pictures courtesy S.P. Adhikary, DBT-HRD Program, Centre for Biotechnology, Visva-Bharati, Santiniketan-731235, West Bengal, India.

questions in understanding desiccation tolerance (Potts, 1996, 2001, 2005). The synthesis and release of EPS is a common feature present in many cyanobacterial species as well as by the dominant cyanobacterial flora reported in the soil crusts. Other cyanobacteria like *Gloeocapsa* sp., *Anabaena* sp. strain PCC 7120, *A. variabilis* ATCC 29413 and *N. punctiforme* PCC 73102 have also been used for comparison. Knowledge that has accumulated on desiccation tolerance is presented in the following sections: (i) targets of desiccation damage in cyanobacterial cells, (ii) physiological studies dealing with photosynthesis and nitrogen fixation during desiccation and rehydration, (iii) membrane modification, (iv) synthesis and structure of EPS and its role in desiccation tolerance; (v) trehalose synthesis and desiccation tolerance, (vi) synthesis of mycosporine-like amino acids (MAAs) and (vii) the role of scytonemin in desiccation tolerance.

i) Targets of desiccation damage: The macromolecules proteins, DNA and lipids appear to be the main targets of damage during desiccation. When *N. commune* UTEX 584 cells grown on $Na_2{}^{35}SO_4$ were desiccated to a matric water potential of 99.5 MPa (a_w=0.5), the cells continued to incorporate S^{35} and maintained high levels of chloramphenicol-sensitive protein synthesis for more than 90 min. But long term desiccation (17 days) led to extensive proteolysis and the dissociation of phycobilisomes took place in presence of light or in its absence (Potts, 1985). Nitrogen fixation appeared to be quite sensitive even for short periods of drying (Potts *et al.*, 1984) but the enzyme nitrogenase reductase (*nifH* product; Fe-protein of nitrogenase) remained intact in immobilized, dehydrated and rehydrated cells of *N. commune* as revealed by immunoblotting (Peat *et al.*, 1988). During rehydration *N. muscorum* showed an instantaneous increase in the size of intracellular ATP pool which was inhibited by sodium azide (10 mmol/l) and carbonyl cyanide m-chlorophenylhydrazone (2 μmol/l) (Potts and Morrison, 1986). Although rapid cessation of nitrogen fixation occurred upon dehydration of *N. commune* cells (Potts *et al.*, 1984; Potts and Bowman, 1985), the fact that the presence of Fe-protein component of nitrogenase in a 10 year field-desiccated material of *N. commune* (Peat *et al.*, 1988) signifies that the enzyme components of nitrogen fixation appear to be intact during extended periods of desiccation. The resumption of the processes of photosythesis, respiration and nitrogen fixation was noted after specific lag periods during rehydration (Scherer *et al.*, 1984). The polysome turnover in the immobilized and desiccated cells (drying at –99.5 MPa for 2 h) of *N. commune* continued to persist but declined thereafter during long periods of desiccation (Angeloni and Potts, 1986). Thus the basic enzyme machinery inside the cells remains intact as evidenced by the quick regaining of vital metabolic processes upon rehydration. The stability of certain proteins even during extended periods of desiccation has been confirmed in the desiccated field materials of *N. commune* as well as in its clonal isolate *N. commune* UTEX 584 (Scherer and Potts, 1989; Hill *et al.*, 1994). The lipid biosynthesis in the rehydrated cells of *N. commune* was instantaneous with the appearance of steady-state radiolabel in PG and SQDG within the first few min whereas the abundant glycolipids MGDG and DGDG required 2 h time. The lag period was associated with the incorporation of $^{35}SO_3$ and the functional nature of the proteins required for uptake and phosphorylation of glycerol, uptake of sulphate and phosphate (Taranto *et al.*, 1993). The *rpoC1C2* gene transcripts governing the synthesis of the two subunits of RNA polymerase (γ and β′) readily disappeared in dehydrated cells of *N. commune* UTEX 584 but their levels attained to control levels within 60 min of rehydration. This was mainly due to the presence of RNA polymerase enzyme in a stable form during long-term desiccation as shown by immunoblotting analyses of protein extracts (Xie *et al.*, 1995). All the above studies point out that the proteins and metabolic potential of the cyanobacterial cells is retained during desiccation and the desiccated cells represent 'simply a collection of dried components' (Potts, 1994).

The damages caused to DNA during desiccation may be through chemical modifications (alkylation or oxidation), cross-linking or base removal such as depurination. Based on the corollary evidences of variations in the pH of internal cellular compartments and the rate of *in vitro* depurination of DNA, it was estimated that *Nostoc* DNA would achieve 1% depurination of the genome after storage at 37°C for 10 years (Potts, 1994). But in actual field conditions the organism faces extremely high temperatures and it is expected that the depurination rates must be higher than those mentioned. The presence of significant amounts of 6-methyladenine in the genome of *N. commune* has been reported but these residues are lost more readily than other purines (Jägger and Potts, 1988). The DNA of laboratory grown *N. commune* UTEX 584 showed light-dependent nicking for short periods when the cells were dried that lacked the normal capsular polysaccharide (Stulp and Potts, 1987). However, Shirkey *et al.* (2003) pointed out that the DNA from 13 year desiccated *N. commune* cells is devoid of the oxidatively modified DNA biomarkers such as 8-hydroxyguanine, 8-hydroxyadenine and 5-hydroxyuracil. Fifteen samples of form species of *N. commune* (obtained from different Herbaria in a desiccated state for varying periods, i.e. 26–149 years) were included in this study and five form species, i.e. WH002 (64 years dry), WH0014 (55 years dry), ALD 8122 (29 years dry), TOP 1993 (9 years dry) and ENG 1996 (6 years dry) showed viability. The genomic DNA derived from such cultures, rehydrated and desiccated materials was tested for the amplification of 10 gene loci. The genomic DNA existed in a covalently modified state and required pre-treatment with N-phenacylthiazolium bromide for amplification of gene loci (a reagent that is generally used to cleave DNA- and protein-linked advanced glycosylation end-products). *In vitro* desiccation of covalently closed circular DNA resulted in the loss of supercoiling, aggregation, transformation and transfection efficiency but these effects were minimized in presence of trehalose. It is expected that trehalose present in the *Nostoc* colonies provides the required milieu for protecting the cellular components including DNA from oxidative damage and degradation. The mechanism of damage due to desiccation and rehydration is mainly due to the production of ROS and the consequent oxidative stress induced in the model organisms under study. These have already been highlighted within the preceding section on oxidative stress.

ii) Physiological studies: During dehydration, photosynthetic systems undergo changes leading to cessation in photosynthetic activity. This is reflected at the level of (i) dissipation of absorbed light energy as heat, (ii) deactivation of PSI and PSII activities, (iii) delay in quenching of PSII fluoresecence, and (iv) inhibition of energy transfer from phycobilisomes to the anchor protein and the anchor protein to the PSII core complex. Inhibition of photosynthetic activity, in presence of high light intensity and temperature, leads to the formation of ROS and oxidative stress is inevitable. So it must be pretty harsh and very critical to survive under such conditions.

Most of the physiological studies conducted relate to understanding the recovery of photosynthesis, respiration and nitrogen fixation during rehydration. Photosynthetic rates of four strains of *Chroococcidiopsis* and one *Chroococcus* (all strains isolated from arid desert rocks) decreased with decreasing water potential caused by matric treatment when compared to osmotic control method. This was confirmed by $^{14}CO_2$-uptake studies (Potts and Friedmann, 1981). The effects of rehydration on the recovery of photosynthesis in *Scytonema geitleri* (growing on roof tops of buildings) indicated that photochemical reactions of photosynthesis recover first followed by increase in ATP pool size. Subsequently, CO_2 fixation and nitrogen fixation recovered. A reversal in the sensitivity of the processes was noted during dehydration (Tiwari and Tripathi, 1998). These results were confirmed by an assay of fluorescence measurements, ^{14}C-fixation, O_2 evolution/consumption and dichlorophenolindophenol reduction in *S. geitleri* where osmotic water potential (0 MPa) in presence of sucrose during rehydration showed more recovery of PSI than PSII than with matric treatments (Tiwari and Tripathi, 2001). The recovery of photosynthetic activity of *N. commune* during rehydration very much corresponded with the rate of absorption kinetics of water. This very much depended on the amount and structure of sheath and the extent of slime present outside the cells. The absorption kinetics of water uptake (at 1 min, 2 h and 9 h) by desiccated colonies of *N. commune* during rehydration was determined and the maximum absorption of water by a colony was 23-fold its initial dry weight after 26 h in dark. The fluorescence of phycobiliproteins and PSI complex were recovered within 1 min whereas it took 2 min for the recovery of PSI and cyclic electron flow around PSI. But the recovery of PSII occurred after a lag of 5 min and that too in two phases with half-life times of about 20 min and 2 h. Carbon fixation corresponded with the first phase of PSII recovery. Although the colonies absorbed 20 times water of their initial dry weight, water equivalent to twice to its initial dry weight was sufficient to maintain PSII activity (Satoh *et al.*, 2002). Activation of photosynthesis in the rehydrated sand crusts from Negev Desert, Isreal (mainly consisting of *Microcoleus* sp.) took place within less than 5 min which did not require *de novo* protein synthesis. During this time, over 50% of PSII activity, assembly of phycobilisomes and PSI antennae could be recovered. Another remarkable feature is that the soil crust algae did not show photoinhibition, a property generally exhibited by model organisms like *Synechocystis* sp. strain PCC 6803 when exposed to high light conditions. The rate of PSII repair was dependent on the intensity of light and its duration. So this explains how the crust organisms are able to readily adjust to the extreme environmental conditions of temperature, high light intensity and the daily desiccation and rehydration cycles prevalent in deserts (Harel *et al.*, 2004). Ohad *et al.* (2005) reported inactivation of photosynthetic electron flow during desiccation in native biological sand crusts and those crusts cultivated on nitrocellulose filters that contained filaments of *Microcoleus* sp. The morphological features of the cyanobacterium were confirmed by scanning-transmission and atomic force microscopy. Desiccation caused a complete loss of fluorescence in the native crusts accompanied by a decrease in 77 k PSII fluorescence emission relative to that of PSI. Further, a decrease of energy transfer from phycobilisomes to PSII was noted. The addition of trehalose (1.5 M) to the native crusts retrieved the loss of fluorescence by 50% but was accompanied by a decrease of 77 k PSI fluorescence induced by chlorophyll excitation. This is explained as a survival strategy of *Microcoleus* by showing alterations in energy transfer from antennae to reaction centres. Ohad *et al.* (2010) monitored high light-induced changes in *Mirocoleus*-dominated sand crusts maintained in the field (under natural light intensity of 200 µmol photons m^{-2} s^{-1}) in dry and wet experimental plots. A dramatic decline in chlorophyll fluorescence yield (50%) was observed. As it was not possible to measure the rates of uptake of CO_2 and release of O_2 in these natural crusts, nitrocellulose filter-grown *Microcoleus* was used to measure chlorophyll fluorescence at high light intensity (500 µmol photons m^{-2} s^{-1}). The loss of chlorophyll fluorescence was up to 18–23% of its initial value but the decrease in O_2 evolution was marginal suggesting an uncoupling between variable fluorescence and the photosynthetic electron transport. Laboratory grown liquid cultures of *Microcoleus* when exposed to very high light intensities (2000 µmol photons m^{-2} s^{-1}), the chlorophyll fluorescence and thermoluminescence emission were reduced to 85% and 90% of their initial values but O_2 evolution, O_2 uptake and PSI activity remained unaffected. It means that there was no inhibition of forward electron flow activity of *Microcoleus* PSII. As light-induced variable fluorescence was not inhibited by electron transfer inhibitors 3-(3,4-dichlorophenyl)-1,1-dimethylurea (DCMU), 2,5-dibromo-3-methyl-6-isopropylbenzoquinone (DMMIB) or the uncoupler carbonyl cyanide-p-trifluoromethoxyphenylhydrazone (FCCP), it ruled out the possibility of reduction of plastoquinone or O_2 or dissipation of excess light excitation as heat (via modulation of energy transfer from antenna to the reaction centres, known as non-photochemical quenching). So it was concluded that a greater back flow of electrons from Quinone A to phaeophytin took place, as the temperature required for thermoluminescence emissions was quite low when compared to other cyanobacteria used as models in photosynthesis experiments. *Microcoleus* thus has an efficient protective mechanism against light-induced oxidative stress and is able to thrive under the harsh desert conditions. When desiccated *N. flagelliforme* (for 2 years and 8 years dry storage) was rehydrated, recovery of photosynthesis and respiration in the 2-year sample was faster as it absorbed water much faster. But the 8-year sample failed to recover both the processes and showed PSI activity alone (Liu *et al.*, 2010). A dramatic increase in cAMP levels in the cells of *Anabaena* sp. strain PCC 7120 upon rehydration prompted Higo *et al.* (2008) to study the respiratory O_2 uptake and photosynthetic O_2 evolution in the wild-type and the adenyl cyclase gene (*cyaC*) disruptant mutant. During rehydration, the *cyaC* disruptant mutant showed enhanced O_2 uptake than wild-type which resulted in more oxidative stress as evidenced by lipid peroxidation and protein carbonylation. The *cyaC* disruptant mutant took more time for the recovery of photosynthesis as 70% of the O_2-evolving activity was reached after one day of rehydration. The growth of the *cyaC* mutant was perceptible only when it was able to recover from the damages caused by rehydration. It was suggested that cAMP regulates respiratory levels during rehydration in such a manner that the damages caused by oxidative stress are kept at minimum.

Cyanobacteria growing beneath pieces of gypsum rock, mainly composed of *Chroococcidiopsis* sp., after 2–3 days of rehydration exhibited nitrogenase activity aerobically in light. *Chroococcidiopsis* sp. strain PCC 7203, on the other hand, showed nitrogenase activity under microaerophilic and low light conditions. The presence of molybdenum nitrogenase 2 was confirmed by sequencing of *nifH* gene. It is suggested that the gypsum shards provide a good microhabitat for *Chroococcidiopsis* to fix nitrogen under low light and oxygen tensions (Boison *et al.*, 2004).

iii) EPS and its role in desiccation tolerance: Two types of cyanobacterial EPS can be recognized viz., the EPS associated with cell surface and the EPS released into the external medium and known as released polysaccharides (RPS). The EPS on cell surface are circumscribed in the form of sheaths, capsules and slimes depending on the thickness, consistency and appearance (De Philippis and Vincenzini, 1998, 2003). While the sheaths and capsules generally reflect the shape of the organism, the slimes form a dispersed material around the organism and do not reflect the shape of cells. Studies on chemical composition of EPS revealed that it can be composed of homoplysachharides or heteropolysaccharides. The number of monosaccharides in the latter range from six to twelve in nearly 75% of the cyanobacterial EPS. The hexoses (glucose, galactose, mannose and fructose), pentoses (ribose, xylose and arabinose), deoxyhexoses (fucose, rhamnose and methylrhamnose) and the acidic hexoses (glucoronic and galacturonic acids) constitute the subunits. Besides these, the presence of acetylated aminosugars and other non-carbohydrate constituents such as phosphate, lactate, acetate, sulphate and glycerol groups has been demonstrated in the cyanobacterial EPS. The presence of sulphate and uronic acids contribute to the overall negative charge of the EPS and makes it 'sticky' in nature (Pereiro *et al.*, 2009). Chi *et al.* (2007) for the first time reported the presence of a homopolysaccharide consisting of a single monosaccharide (α D-1,6-glucose) in *Cyanothece* sp. 113, a marine, diazotrophic cyanobacterium. In two strains of *Synechocystis*, the presence of large number of monosaccharides (11 to 12) has been reported. *Synechocystis* sp. strain PCC 6803 showed the presence of xylose, fucose, rhamnose, 3-O-methyl-deoxyhexose, mannose, glucose, galactose, 4-O-methylhexose and methylhexose in its EPS whereas *Synechocystis* sp. strain PCC 6714 contained arabinose and 3-O-methylpentose in addition to those present in the former strain. Uronic acids (16.4–16.7%), glucosamine and galactosamine (10–15%) and proteins (20–40% w/w) and sulphated residues are also present (Panoff *et al.* 1988).

The EPS of *N. commune* is a high molecular weight substance that accumulates to more than 60% of the dry weight of the colonies (Hill *et al.*, 1999).The biochemical and structural properties of the sheath of *N. commune* were examined in the field material as well as its isolate in cultures (*N. commune* DRH1). The sheath around the colonies possessed a pellicle consisting of Ca/Si rich material that acts as a barrier to the epiphytes. There appears to be a variation in the number of monosaccharides present in the EPS of species of *Nostoc* as for example three monosaccharide subunits (*N. flagelliforme* a desert species of China, Huang *et al.*, 1998), 6 monosaccharide subunits (freshwater species of *N. commune* from China, Brüll *et al.*, 2000) and seven monosaccharide subunits (*N. commune* from mountain area, Huang *et al.*, 1998; *N. commune* DIH1 from deserts of Mongolia, Helm *et al.*, 2000) have been reported in literature. Huang *et al.* (1998) compared hot-water soluble polysaccharides in the sheath material of field samples as well as RPS of laboratory grown cultures of *N. commune*, *N. flagelliforme* and *N. sphaeroides*. The sheath of the field samples consisted of three main monosacharides glucose, xylose and galactose in the ratio of 2:1:1. The presence of mannose to varying levels in the three species and of arabinose only in *N. flagelliforme* is significant. But the

RPS in cultures exhibited variations as for example *N. flagelliforme* possessed mannose, galactose, glucose and glucoronic acid but no arabonise whereas the RPS of *N. sphaeroides* and *N. commune* contained glucose (the major component), rhamnose, fucose, xylose, mannose and galactose. The linkage analysis of the EPS of the field samples showed that xylose, glucose and galactose are 1–4 linked, mannose, galactose and xylose are present as terminal residues and the branch points occurred in glucose (1-3,4 and 1-3,6 linkages) and xylose (1-3,4 linkage). On the other hand, Helm *et al.* (2000) reported the presence of glucose, galactose, xylose and nosturonic acid in the ratio of 2:1:1:1, respectively in the EPS of field sample of *N. commune* DRH1. Although the separation into individual monosaccharides was not possible, they suggested the presence of a pentamer as a repeating unit with a disaccharide (of glucose and xylose) and trisaccharide (β-D-nosturonic acid, β-D-Glucose and D-Galactose) present in 1:1 molar ratio. The glycan matrix is unique in possessing abundant water stress proteins (Wsp), a UV-A/B absorbing pigment, oxidized and reduced forms of the pigment scytonemin and two glycoproteins (75 kDa and 110 kDa). A carbohydrate-modifying enzyme (68 kDa) is the second abundant protein after Wsp (Hill *et al.*, 1994a,b). However, the glycan sheath of *N. commune* with its unique properties responsible for conferring tolerance to environmental stresses is also liable to bacterial degradation. A new species of *Paenibacillus* (*P. glycanilyticus*) which utilizes rhamnose and L-fucose has the capability to degrade the glycan of *N. commune* (Dasman *et al.*, 2002a). The purified enzyme (128.5 kDa) has been found to be unique as its N-terminal amino acid sequence did not bear any resemblance to hitherto known polysaccharide degrading enzymes. Complete degradation of *N. commune* polysaccharide into tetra and hexasaccharides took place at an optimum pH and temperature of 5.5 and 35°C, respectively (Dasman *et al.*, 2002b).

An unusual water flux in the EPS of the desiccated colonies of *N. commune* (collected from Topasail Island, North Carolina, U.S.A.) was noted. Desiccated colonies were rehydrated (in light of 150 µmol of photons m^{-2} s^{-1} at 25°C) for different intervals of time and the increase in weight was monitored till the time there was no further increase (equilibrium weight). During the first rehydration cycle it took 30 hr for attaining equilibrium weight. The same colonies were allowed to desiccate (at 24.5°C and 25% relative humidity) to attain an internal water cotent of 0.06 g H_2O g^{-1} dry weight of the colonies which closely corresponded with the initial dry weight of the colonies when they were collected. In the second and third rehydration cycles it required 12 h and 3 h for attaining the equilibrium weight, respectively. The increase in water content contributed to 20–40 fold increase in their weight as they attained equilibrium weight (Shaw *et al.*, 2003).

The accumulation of a novel Wsp inside the desiccated cells of *N. commune* (Scherer and Potts, 1989), its secretion into the EPS with UV-A/B absorbing pigments (Hill *et al.*, 1994) and the release of Fe-SOD upon rehydration of *N. commune* CHEN/1986 (Shirkey *et al.*, 2000) have been reported to play a role in desiccation tolerance. The existence of three of the Wsps (33, 37 and 39 kDa) with similar consensus N-terminal sequences (Ala-Leu-Tyr-Gly-Tyr-Thr-Ile-Gly-Glu) indicated that the variation in their molecular masses resulted due to proteolysis of the 39-kDa Wsp. The presence of the Wsp in the desiccated cells of the field materials of *N. commune* and the synthesis of Wsp (39 kDa) in the laboratory grown colonies subjected to drying go in favour of its suggested role in desiccation. The field materials, collected from China, Europe, North America, Aldabra Atoll (Indian Ocean) and Antarctica, desiccated (for years) revealed a similar pattern of Wsp on Western blots (Scherer and Potts, 1989). The presence of the N-terminal consensus sequence and another internal sequence (Glu-Ala-Arg-Val-Thr-Gly-Pro-Thr-Thr-Pro-Ile-Asp) in the Wsps bring them closer to carbohydrate-modifying enzymes. The presence of 1, 4-β-D-xylan-xylanohydrolase activity in the purified Wsps and the ionic interactions between the purified Wsp and and a xylan-containing UV-A/B absorbing pigment suggest that Wsp plays a role in the synthesis of the latter (Hill *et al.*, 1994).

This has been confirmed by Wright *et al.* (2005) who cloned the *wspA* gene that existed as a single copy in a monocistronic operon in the genome of *N. commune* DRH1. The transcription of the genes *wspA* and *sodF* and the synthesis and release of WspA into EPS have been induced upon desiccation or UV-A/B irradiation. The binding of the purified WspA protein, expressed in *E. coli*, to the UV-A/B absorbing pigment scytonemin emphasizes the predicted role of WspA in global stress response of *N. commune*. Morsay *et al.* (2008) identified two novel glycosidases in the EPS of *N. commune*. One of these proteins (20 kDa) with β-D-glucosidase activity has been identified as a cyanobacterial fasciclin proteins secreted on the cell surface of *N. punctiforme* PCC 73102, *A. variabilis* ATCC 29413 and *Anabaena* sp. strain PCC 7120. The second major water-soluble protein (36 kDa) was found to be identical to the Wsp that showed heat-resistant β-D-glucosidase activity. The presence of these two glycosidases has been suggested to dissolve the EPS material facilitating the dispersal of the filaments. The existence of stable SodF mRNA in abundance in the desiccated cells for prolonged periods (13 years) and its turnover within 15 min of rehydration followed by the release of Fe-SOD enzyme into the surrounding medium enabled *N. muscorum* to tolerate oxidative stress contributed by rehydration as well as UV-A/B exposure.

In agreement with the idea that desiccation-tolerant cyanobacteria also exhibit freezing-tolerance, desiccation-tolerant *N. commune* resisted freezing stress whereas desiccation-sensitive *Synechocystis* sp. strain PCC 6803 and *Fischerella muscicola* could not tolerate freezing-stress. This mostly depended on the ability of *N. commune* to convert the excess light energy absorbed by it to heat and the protection of the cells from photoinactivation (Lin *et al.*, 2004). The role of EPS in conferring desiccation and freezing tolerance was investigated by subjecting the EPS-depleted cells of *N. commune* to these stresses (Tamaru *et al.*, 2005) and by the addition of extracted EPS of cryptoendolithic *Nostoc* sp. to the cultures of *Chlorella* sp. and *Chroococcidiopsis* sp. (Knowles and Castenholz, 2008). Desiccated colonies did not possess the ability to evolve O_2, a property which they quickly regained by rehydration. However, in air-dried colonies (with 10% w/w water) and natural colonies of *N. muscorum* with EPS the capacity to evolve O_2 was not damaged. Removal of EPS by mechanical means of either the naturally occurring colonies or laboratory cultures of *N. commune* KU002 made them more desiccation- and heat- and freezing-sensitive (Tamaru *et al.*, 2005). The RPS of a cryptoendolithic *Nostoc* sp. CCMEE 6160 when added to the cultures of other naturally occurring algae in the community (*Chlorella* sp. CCMEE 6038 and *Chroococcidiopsis* sp. CCMEE 5056) protected them from desiccation tolerance. The staining of the cultures with SYTOX Green (a fluorescent stain) clearly differentiated the live (appear red due to autofluorescence) and dead cells (appear bright green due to the fluorescence of the dye). Protection from freezing stress by the addition of RPS to the two test cultures was not to the extent observed in desiccation stress. The differential staining by SYTOX Green of living and dead cells has been attributed to the integrity of the membranes, the former cells retained the integrity of the membranes and in the latter cells the integrity of membranes is damaged (Knowles and Castenholz, 2008).

iv) Membrane modification: During dehydration as the matric water potential inside the cells decreases, the hydration shell around the macromolecules is disturbed. At least in case of lipids the physical state of water molecules is not well understood but the loss of the water molecules influences greatly the physical state of the lipid molecules. The lipid bilayer of the cytoplasmic membrane as well as thylakoid membranes in cyanobacteria face the change in physical state of the lipid molecules at the time of desiccation as well as rehydration. During desiccation when water molecules around the lipid molecules are removed the packing density of the head groups increases in the lipid bilayer. Due to increase of packing density, there is every possibility for increase in vander waals interactions among the hydrocarbons. As a result of this, the temperature required to melt these chains (T_m) increases. For example, phosphatidylcholine of egg has a transition temperature of –7°C when hydrated and the T_m increases to 70°C when it is fully dehydrated. It means the phosphatidylcholine of egg is in gel phase at room temperature when it is dry but it will pass through a phase transition when rehydrated. When the lipid bilayers pass through such phase transitions during rehydration the membranes tend to become leaky. This is the real challenge faced by the cyanobacterial cells when desiccated cells are rehydrated. However, when phosphatidylcholine of egg is dehydrated in presence of disaccharides such as sucrose or trehalose, the T_m is considerably lowered to –20°C but at the same time when rehydrated it will not pass through a phase transition as it is already in a liquid crystalline phase.

Studies conducted on liposomes *in vitro* have been extended to intact cells (Crowe *et al.*, 1989; Hoekstra *et al.*, 1992; Leslie *et al.*, 1994, 1995; Crowe *et al.*, 2002). A number of studies conducted on sucrose, trehalose, glucose and the polymer hydroxyethyl starch (HES) established that these substances often form glasses (i.e. vitrify) at low water contents and their ability to hold the liposomes intact and protect them from leakage and fusion very much depended on the combination of two of these substances rather than each of them individually. Temperatures at which glass formation takes place, i.e. T_g is found to be directly proportional with the molecular weight and decreasing water content. For example, HES (molecular weight 500,000) though is a good glass former (T_g is more than 150°C) could not prevent leakage from unilamellar vesicles during drying or rehydration (because of its inability to interact with the polar head groups) but HES when combined with glucose which can directly interact with lipid head groups (in 1:1 ratio) effectively formed a stable glass that inhibited fusion and leakage (Crowe *et al.*, 1997). When we try to apply these to intact cells, the formation of stable carbohydrate glass by sucrose and trehalose may be expected as these are present in significantly high amounts in desiccated cells. As T_g in case of trehalose is higher than sucrose and also due to conversion of certain amounts of trehalose to its crystalline dihydrate form during absorption of moisture renders the glassy sugar to have exceptionally high T_g (Crowe *et al.*, 2002). Crowe *et al.* (2002) summarized evidences that go in favour of stabilizing effects of sucrose and trehalose in desiccated cells and that 'these sugars in the cytoplasm may be both necessary and sufficient for anhydrobiosis'. They highlighted the utility of these sugars in maintaing the mammalian cells in a dry state that has profound use in human clinical medicine. Increased levels of trehalose production during desiccation, heat shock, cold shock and oxidative stress and the ability of trehalose to protect proteins and cell membranes from the damages caused by the aforesaid stresses opened up new vistas to understand the multifunctional role of trehalose (Elbein *et al.*, 2003). Lipid soluble extracts of desiccated cells of *N. commune* revealed the presence of trehalose and sucrose and these were not detectable after rehydration. The presence of trehalose inside the cells was confirmed by immunoblotting (Hill *et al.*, 1994). Moreover, Hill *et al.* (1999) demonstrated that a low concentration of sucrose and trehalose mixture (equivalent to the concentrations present inside the cells of *N. commune* DRH1) with EPS in small amounts prevented the leakage of carboxyfluoroscein from membrane vesicles.

v) Trehalose synthesis: Avonce *et al.* (2006) outlined five main pathways for trehalose biosynthesis (see Fig. 5). The pathway of trehalose synthesis reported in thermophilic archaebacterium *Sulfolobus*, known as Tre Y/Tre Z pathway, involves two steps. In the first step, the terminal glucose subunit at the reduced end of starch or glycogen is transglycosylated from a α 1-α 4 to α 1-α 1 bond resulting in the formation of maltooligosyl trehalose mediated by the enzyme maltooligosyl trehalose synthase (Tre Y or Mts). In the second step, maltooligosyl trehalose with its terminal trehalose is hydrolysed to release trehalose mediated by maltooligosyl trehalose trehalose hydrolase (TreZ or Mth). This pathway is also operative in cyanobacteria. Katoh *et al.* (2004) identified a gene cluster in the genome of *Anabaena* sp. strain PCC 7120 consisting of trehalose metabolizing enzymes, i.e. Mts and Mth involved in the synthesis of trehalose and encoded by *mth* (*all0168*), *mts* (*all0167*) genes, respectively and a trehalase (TreH) encoded by *treh* (*all0166*) that is involved in trehalose hydrolysis. The expression of the three genes is up-regulated during dehydration. They also indicated that the genomes of *A. variabilis* ATCC 29413 and *N. punctiforme* PCC 73102 also possess the gene cluster mentioned above. A genome wide DNA microarray was conducted during early to late dehydration/desiccation. Of the seven categories of genes identified, four categories were up-regulated and three of them were down-regulated. Genes up-regulated belonged to the categories of cold shock, heat shock and osmotic (salt) stress. The genes of photosynthesis, nitrogen transport system, RNA polymerase and ribosomal proteins were down-regulated. Higo *et al.* (2006) demonstrated the expression of trehalose gene cluster in *Anabaena* sp. strain PCC 7120 during dehydration and showed that the amount of trehalose accumulated was small (0.05–0.1% of dry weight). Gene disruption (*mth*) mutants of *Anabaena* sp. strain PCC 7120 showed a simultaneous decrease in the amount of trehalose synthesized and the desiccation tolerance. But the disruption of *treH* resulted in an increase in the level of trehalose. A comparison of gene expression profiles of the wild-type and the gene disruption mutants of *Anabaena* by oligomeric DNA microarray revealed a correlation between trehalose content and the expression of two genes one of which is involved in the synthesis of a co-factor of DnaK. The accumulation of trehalose and the cessation of photosynthetic activity occurred during desiccation and salt stress (0.2 M) in *N. commune* (Sakamoto *et al.*, 2009). The rate of trehalose production exceeded that of trehalose hydrolysis under water stress. In this connection, TreH plays an important role in maintaining the increased levels of trehalose during desiccation of *N. punctiforme* IAM M-15. The absence of trehalose in fully hydrated cells and the induction of genes (*mth* and *mts*) during desiccation and salt stress resulted in the production of increased levels of trehalose (Yoshida and Sakamoto, 2009).

vi) MAAs: These are water soluble substituted cyclohexenes which are linked to amino acids and iminoalcohols. With absorption maxima between 310 and 360 nm, these absorb UV-A/B radiations and thus protect the cells from their harmful effects. By now the chemical structure of over 30 different MAAs have been elucidated (Bandaranayake, 1998). These serve as sunscreen compounds and their induction in response to wavelengths between 280 and 320 nm radiations suggest that they protect the cells. The presence of photoreceptors that sense the need for synthesis of MAAs has been reported in case of *N. commune* (Ehling-Schulz *et al.*, 1997) and *Chlorogloeopsis* sp. PCC 6912 (Portwich and Garcia-Pichel, 2000). Besides their role in giving protection to UV radiation, their role as multipurpose secondary metabolites has also been emphasized (Oren and Gunde-Cimerman, 2007). Intracellular accumulation of MAA with absorbance maximum at 320 nm in *Gloeocapsa* sp. strain Ca-90-Cal-G(2) correlated with increased resistance to UV radiation and there was 20 to 25% enhancement in desiccation tolerance (Garcia-Pichel *et al.*, 2001). The presence of MAAs in the glycan sheath of *N. commune* indicates that under the climatic conditions of hot and cold deserts the organism is able to resist the harmful UV radiations as well. The presence of sugar moieties in MAAs of *N. commune* varied from three to eight attaining a maximum molecular mass of 1801 Da. These pigments are released into the glycan sheath and they are attached non-covalently forming complexes with the glycan sheath (Böhm *et al.*, 1995). The induction of mycosporine by UV-B (<315 nm) in *N. commune* was associated with extracellular glycan synthesis (Ehling-Schulz *et al.*, 1997). The induction of shinorine in response to UV irradiation (280–320 nm) by the cyanobacteria *Chlorogloeopsis* sp. PCC 6912 (Portwich and Garcia-Pichel, 1999, 2000), *Anabaena* sp., *N. commune* and

Scytonema sp. (Sinha *et al.*, 2001), *Anabaena* sp. (Rozema *et al.*, 2002; Sinha *et al.*, 2003) and *Nodularia* sp. (Sinha *et al.*, 2003) has been reported. The synthesis of mycosporine-2-glycine (with maximum absorption at 331 nm) was reported for the first time in cultures of *Euhalothece* sp. (isolated from a gypsum crust in a hypersaline saltern pond in Israel) when grown at highlight intensities (Kedar *et al.*, 2002). A database on mycosporines and MAAs highlighting their molecular structures and biosynthetic routes has been created (Sinha *et al.*, 2007). Shinorine synthesis in *A. variabilis* ATCC 7937 was not induced by heat stress either in presence or absence of UV-irradiation but its synthesis in salt stress and ammonium treatment had synergistic effects with UV-irradiation. The induction of shinorine synthesis was dependent on the salt and ammonium concentration (Singh *et al.*, 2008). A combination of two genes for the biosynthesis of the common core (deoxygadusol) of all MAAs existed only in case of *A. variabilis* ATCC 29413 and the other three cyanobacteria tested (*Anabaena* sp. strain PCC 7120, *Synechocystis* sp. strain PCC 6803 and *S. elongatus* PCC 6301) did not reveal gene sequences governing the synthesis of MAAs.

vii) The role of scytonemin: The presence of scytonemin in the EPS of many cyanobacteria is known. Scytonemin (544 molecular weight) is a dimeric molecule consisting of indolic and phenolic subunits and its absorption is in between 325–425 nm with maximum at 370 nm. The 'scytonemin skeleton' has been suggested to result by the condensation of tryptophan- and phenylpropanoid-derived subunits (Proteau *et al.*, 1993). The synthesis of scytonemin in *N. commune* in response to UV-B irradiation was minimal but it was inducible in high amounts by UV-A irradiation (350–400 nm; Ehling-Schulz *et al.*, 1997). Fleming and Castenholz (2007) investigated the role of scytonemin synthesis in relation to UV-A irradiation and desiccation. In *Chroococcidiopsis* CCMEE 5056 and *N. punctiforme* PCC 73102, the synthesis of scytonemin occurred in large amounts when periodically desiccated (i.e. one day desiccated for every two days rehydrated) was irradiated with UV-A. Scytonemin so synthesized in *N. punctiforme* PCC 73102 remained quite stable in the EPS despite exposure to UV-A for two months. In *Chroococcidiopsis* CCMEE 246 that produces scytonemin constitutively in visible light, periodic desiccation partially inhibited scytonemin production.

(viii) Transcriptional regulation: Suzuki *et al.* (2004) identified two ORFs, *alr0295* and *alr2325* that encode proteins similar to cAMP receptor proteins (CRP) in *Anabaena* sp. strain PCC 7120 and designated these genes *ancrpA* and *ancrpB*. The corresponding proteins AnCrpA and AnCrpB showed binding to cAMP, the former was more specific to cAMP than the latter that showed weak binding to cAMP as well as cGMP. AnCrpA possesses a core of three amino acids (Arg-180, Glu-181 and Arg-185) as DNA-binding region specific to the five base pair sequence 5′-TGTGA-3′ of DNA just as in case of *E. coli*. AnCrpB lacks this DNA-binding motif and its C-terminal sequence does not bear any resemblance to *E. coli* CRP, SYCRP1 (*Synechocystis* cAMP receptor protein) and AnCrpA. The inability of AnCrpB to bind to DNA is due to difference in the sequence of amino acids at the DNA binding motif (Val-187, Glu-188 and Lys-192) of which Val-187 could not hydrogen bond with DNA. It is suggested that AnCrpB acts as a transcriptional regulator. AnCrpA has been suggested to regulate *nif*-related gene expression in *Anabaena* sp. strain PCC 7120 when grown on nitrate. The *ancrpA* disruptant mutant exhibited down-regulation of several genes of nitrogen fixation when grown on nitrate. It was possible to identify the specific genes by oligonucleotide microarray and qRT-PCR. The binding of AnCrp to 5′-upstream region of *nifB* gene was confirmed in presence of cAMP using electrophoretic mobility shift assays (Suzuki *et al.*, 2007). Higo *et al.* (2007) conducted a global transcriptional analysis in response to rehydration in *Anabaena* sp. strain PCC 7120. During dehydration, as many as 133 genes were down-regulated and 29 genes were up-regulated. The process of rehydration involved up-regulation of as many as 259 genes whereas 13 genes showed rehydration specific down-regulation. Among the 29 genes up-regulated during rehydration, eight genes were for transcriptional regulators. Six of these belonged to transcriptional repressor family and the rest two genes were *ancrpB* and ORF *alr0618*. Studies conducted on the gene disruptant mutants of *ancrpB* and *alr0618* conclusively showed that the former gene product AnCrpB is required for recovery of photosynthesis during late rehydration events and the latter gene product (Alr0618) is related to inorganic carbon uptake in particular. The cAMP-binding motif is absent in Alr0618 and so it does not bind to cAMP.

The contribution of EPS to desiccation tolerance in *Anabaena* sp. strain PCC 7120 has been further confirmed by the identification of *sigJ* gene (a group 3 sigma factor gene) that regulates synthesis of EPS. The expression of *sigJ* in terrestrial *Nostoc* HK-01 was significantly induced during mid-phase of desiccation. *Anabaena* sp. strain PCC 7120 is a freshwater strain and can not tolerate desiccation but a transformant of *Anabaena* (HE 0277) overexpressing the *alr0277* gene for *sigJ* acquired desiccation tolerance associated with 3.2 fold higher production of EPS. This has been further confirmed by oligonucleotide microarray. Alongwith *sigJ*, the expression of 112 genes was up-regulated and the expression of 42 genes was down-regulated in the HE 0277 cells during desiccation. Among the up-regulated genes more than 5% of them belonged to cell envelope genes and the other genes are for photosynthesis, respiration, transcription and regulatory functions (Yoshimura *et al.* 2007).

LITERATURE CITED

Akai, M., Oai, K., Kusano, M., Sato, M., Redestig, H., Toyooka, K., Morishita, M., Miyake, H., Hazama, A., Checchetto, K., Szabo, I., Matsuoka, K., Saito, K., Yasui, M., Ishiura, M., and Uozumi, N. (2011) Plasma membrane aquaporin AqpZ is essential for glucose metabolism during photomixotrophic growth of *Synechocystis* sp. PCC 6803. *J Biol Chem* **286:** 25224–25235. doi:10.1074/jbc.M111 236385

Akiyama, Y., and Ito, K. (2000) Roles of multimerization and membrane association in the proteolytic functions of FtsH (HflB). *EMBO J* **19:** 3888-3895.

Akiyama, Y., Yoshihisa, T., and Ito, K. (1995) FtsH, a membrane-bound ATPase, forms a complex in the cytoplasmic membrane of *Escherichia coli*. *J Biol Chem* **270:** 23485–23490.

Alam, J., Curtis, S., Gleason, F. K., Gerami-Nejad, M., and Fuchs, J. A. (1989) Isolation, sequence, and expression in *Escherichia coli* of an unusual thioredoxin gene from the cyanobacterium *Anabaena* sp. strain PCC 7120. *J Bacteriol* **171:** 162–171.

Alfonso, M., Perewoska, I., and Kirilovsky, D. (2000) Redox control of *psbA* gene expression in the cyanobacterium *Synechocystis* PCC 6803. Involvement of the cytochrome b6/f complex. *Plant Physiol* **122:** 505–515.

Alfonso, M., Perewoska, I., and Kirilovsky, D. (2001) Redox control of *ntcA* gene expression in *Synechocystis* sp. PCC 6803. Nitrogen availability and electron transport regulate the levels of the NtcA protein. *Plant Physiol* **125:** 969–981.

Allakhverdiev, S. I., and Murata, N. (2008) Salt stress inhibits photosystem II and I in cyanobacteria. *Photosynth Res* **98:** 529–539.doi:10.1007/s11 120-008-933-x

Allakhverdiev, S. I., Kinoshita, M., Inaba, M., Suzuki, I., and Murata, N. (2001) Unsaturated fatty acids in membrane lipids protects the photosynthetic machinery against salt induced damage in *Synechococcus*. *Plant Physiol* **125:** 1842–1853.

Allakhverdiev, S. I., Nishiyama, Y., Suzuki, I., Tasaka, Y., and Murata, N. (1999) Genetic engineering of the unsaturation of fatty acids in membrane lipids alters the tolerance of *Synechocystis* to salt stress. *Proc Natl Acad Sci USA* **96:** 5862–5867.

Allakhverdiev, S. I., Nishiyama, Y., Miyairi, S., Yamamoto, H., Inagaki, N., Kanesaki, Y., and Murata, N. (2002) Salt stress inhibits the repair of photodamaged photosystem II by suppressing the transcription and translation of *psbA* genes in *Synechocystis*. *Plant Physiol* **130:** 1443–1453.

Allakhverdiev, S. I., Sakamoto, A., Nishiyama, Y., Inaba, M., and Murata, N. (2000a) Ionic and osmotic effects of NaCl induced inactivation of photosystems I and II in *Synechococcus* sp. *Plant Physiol* **123:** 1047–1056.

Allakhverdiev, S. L., Sakamoto, A., Nishiyama, Y., and Murata, N. (2000b) Inactivation of photosystems I and II in response to osmotic stress in *Synechococcus*. Contribution of water channels. *Plant Physiol* **122:** 1201–1208.

Allen, M. B., and Arnon, D. I. (1955) Studies on nitrogen-fixing blue-green algae. I. Growth and nitrogen fixation by *Anabaena cylindrica*Lemm. *Plant Physiol* **30:** 366–372.

Allen, S. P., Polazzim, J. O., Giersem, J. K., and Easton, A. M. (1992) Two novel heat shock genes encoding proteins produced in response to heterologous protein expression in *Escherichia coli*. *J Bacteriol* **174:** 6938–6947.

Alpert, P. (2005) The limits and frontiers of desiccation tolerant life. *Integr Comp Biol* **45:** 685–595.

Alpert, P. (2006) Constraints of tolerance: Why are desiccation tolerant organisms so small or rare? *J Exp Biol* **209:** 1575–1584.

Amarpalli, R., Tripathy, P., and Adhikary, S. P. (1997) Epilithic blue-green algae (cyanobacteria) from temples of India and Nepal. Presence of UV sunscreen pigments. *Arch Hydrobiol Algol Stud* **86:** 147–161.

Anagnostidis, K., and Pantazidou, A. (1988) Endolithic cyanophytes from the saline thermal springs of Aedipsos, Hallas (Greece). *Arch Hydrobiol Algol Stud* **50–53** (Suppl 80): 555–559.

Anbudurai, P. R., Mor, T. S., Ohad, I., Shestakov, S. V., and Pakrasi, H. B. (1994) The *ctpA* gene encodes the C-terminal processing protease for the D1 protein of photosystem II reaction center complex. *Proc Natl Acad Sci USA* **91:** 8082–8086.

Andersen, P. A., Kaasen, I., Styrvold, O. B., Boulnois, G., and Strom, A.R. (1988) Molecular cloning, physical mapping and expression of the *bet* genes governing the osmoregulatory choline-glycine betaine pathway of *Escherichia coli. J Gen Microbiol* **134:** 1737–1746.

Anderson, A. W., Nordan, H. C., Cain, R. F., Parrish, G., and Duggan, D. (1956) Studies on a radio-resistant micrococcus. I. Isolation, morphology, cultural characteristics, and resistance to gamma radiation. *Food Technol* **10:** 575–578.

Andersson, F. I., Blankytny, R., Kirsten, J., Turgay, K., Bukau, B., Mogk, A., and Clarke, A. K. (2006) Cyanobacterial ClpC/Hsp100 protein displays intrinsic chaperone activity. *J Biol Chem* **281:** 5468–5475.

Andersson, F. I., Tryggvesson, A., Sharon, M., Diemand, A. V., Classen, M., Best, C., Schmidt, R., Schelin, J., Stanne, T. M., Bujau, B., Robisnson, C. V., Witt, S., Mogk, A., and Clarke, A. K. (2009) Structure and function of a novel type of ATP-dependent Clp protease. *J Biol Chem* **284:** 13519–13532.

Angeloni, S.V., and Potts, M. (1986) Polysome turnover in immobilized cells of *Nostoc commune* (cyanobacteria) exposed to water stress. *J Bacteriol* **168:** 1036–1039.

Aoki, S., Kondo, T., and Ishiura, M. (1995) Circadian expression of the *dnaK* gene in the cyanobacterium *Synechocystis* sp. strain PCC 6803. *J Bacteriol* **177:** 5606–5611.

Apte, S. K., and Bhagawat, A. A. (1989) Salinity-stress-induced proteins in two nitrogen-fixing *Anabaena* strains differentially tolerant to salt. *J Bacteriol* **171:** 909–915.

Apte, S. K., Reddy, B. P., and Thomas, J. (1987) Relationship between sodium influx and salt tolerance of nitrogen fixing cyanobacteria. *Appl Environ Microbiol* **53:** 1934–1939.

Apte, S. K., Fernandes, T., Bardan, H., and Ballal, A. (1998) Expression and possible role of stress-responsive proteins in *Anabaena. J Bio Sci* **23:** 399–406.

Aran, M., Ferrero, D. S., Pagano, E., and Wolosink, R. A. (2009) Typical 2-Cys peroxiredoxins-modulation by covalent transformations and noncovalent interactions. *FEBS J* **276:** 2478–2493.

Aravind, L., and Ponting, C. P. (1999) The cytoplasmic helical linker domain of receptor histidine kinase and methyl-accepting proteins is common to many prokaryotic signaling proteins. *FEMS Microbiol Lett* **176:** 111 –116.

Aravind, L., Anantharaman, V., and Iyer, L. M. (2003) Evolutionary connections between bacterial and eukaryotic signaling systems: A genomic perspective. *Curr Opin Microbiol* **6:** 490– 507.

Argüelles, J. C (2000) Physiological roles of trehalose in bacteria and yeast: A comparative analysis. *Arch Microbiol* **174:** 217–224.

Asada, K. (1994) Production and action of active oxygen species in photosynthesis tissues. In: Foyer, C. H. and Mullineaux, P. M. (Eds.), *Causes of Photooxidative Stress and Amelioration of Defense Systems in Plants.* CRC Press, Boca Raton, Florida, FL, USA. pp. 77–104.

Asada, K., and Takahashi, M. (1987) Production and scavenging of active oxygen photosynthesis. In: Kyle, D. J., Osmond, C. B., and Arntzen, C. J. (Eds.) *Photoinhibition,* Elsevier Science Publishers, Amsterdam, The Netherlands, pp. 227–287.

Asada, K., Yoshikawa, K., Takahashi, M., Maeda, Y., and Enmanji, K. (1975) Superoxide dismutases from a blue-green alga, *Plectonema boryanum. J Biol Chem* **250:** 2801–2807.

Asadulghani, Nitta, K., Kaneko, Y., Kojima, K., Fukuzawa, H., Kosaka, H., and Nakamoto, H. (2004) Comparative analysis of the *hspA* mutant and wild-type *Synechocystis* sp. strain PCC 6803 under salt stress: Evaluation of the role of *hspA* in salt-stress management. *Arch Microbiol* **182:** 487–497.

Asadulghani, Suzuki, Y., and Nakamoto, H. (2003) Light plays a role in the modulation of heat shock response in the cyanobacterium *Synechocystis* sp. PCC 6803. *Biochem Biophys Res Commun* **306:** 872–879.

Atzenhofer, W., Regelsberger, G., Jacob, U., Peschek, G. A., Furtmüller, P. G., and Obinger, C. (2002) The 2.0 Å resolution structure of the catalytic portion of a cyanobacterial membrane-bound manganese superoxide dismutase. *J Mol Biol* **321:** 479–489.

Avendaño, M. C., and Fernández-Valiente, E. (1994). Effect of sodium on phosphate uptake in unicellular and filamentous cyanobacteria. *Plant Cell Physiol* **35:** 1097–1101.

Av-Gay, Y., Aharonowitz, Y., and Cohen, G. (1992) *Streptomyces* contains a 7.0 kDa cold shock like protein. *Nucleic Acids Res.* **20:** 5478–5479.

Avonce, N., Mendoza-Vargas, A., Morett, E., and Iturrigate, G. (2006) Insights on the evolution of trehalose biosynthesis. *BMC Evol Biol* **6:** 109. doi: 10.1186/1471-2148-6-109

Badger, M. R., Price, G. D., Long, B. M., and Woodjer, F. J. (2006) The environmental plasticity and ecological genomics of the cyanobacterial CO_2 concentrating mechanism. *J Exp Bot* (*Phenotypic Plasticity and the Changing Environment*-Special Issue) **57:** 249–265.
Bagchi, S. N., Bitz, T., Pistorius, E. K., and Michel, K. P. (2007) A *Synechococcus elongatus* PCC 7942 mutant with a higher tolerance toward the herbicide bentazone also confers resistance to sodium chloride stress. *Photosynth Res* **92:** 87–101.
Bailey, S., Silva, P., Nixon, P., Mullineaux, C., Robinson, C., and Mann, N. (2001) Auxiliary functions in photosynthesis: The role of the FtsH peptidase. *Biochem Soc Trans* **29:** 455–459.
Balogi, Z., Török, Z., Balogh, G., Jósvay, K., Shigapova, N., Vierling, E., Vigh, L., and Horváth, I. (2005) "Heat shock lipid" in cyanobacteria during heat/light-acclimation. *Arch Biochem Biophys* **436:** 346–354.
Bancroft, I., and Smith, R. J. (1989) Restriction mapping of genomic DNA from five cyanophages infecting the heterocystous cyanobacteria *Nostoc* and *Anabaena*. *New Phytol* **113:** 161–166.
Bandaranayake, W. M. (1998) Mycosporines: are they nature's sunscreens? *Nat Prod Rep* **1998:** 159–171.
Bandyopadhyay, S., Starke, D. W., Mieyal, J. J., and Gronostajski, R. M. (1998) Thioltransferase (Glutaredoxin) reactivates the DNA-binding activity of oxidation inactivated nuclear factor I. *J Biol Chem* **273:** 392–397.
Barker, M., Boehm, M., Nixon, P. J., and Nield, J. (2008) Structural analysis of an FtsH2/FtsH3 complex isolated from *Synechocystis* sp. PCC 6803. In: Allen, J. F., Gantt, E., Golbeck, J. H., and Osmond, B. (Eds.) *Photosynthesis. Energy from the Sun*. 14th International Congress on Photosynthesis. Springer: The Netherlands, pp. 738–740.
Barker, M., de Vries, R., Nield, J., Komenda, J., and Nixon, P. J. (2006) The *deg* proteases protect *Synechocystis* sp. PCC 6803 during heat and light stresses but are not essential for removal of damaged D1 protein during the photosystem two repair cycle. *J Biol Chem* **281:** 30347–30355.
Barrett, A. J. (Ed.) (1994) *Proteolytic Enzymes: Serines and Cysteine Peptidase. Methods Enzymol* **244:** 765 pp. Academic Press Inc. San Diego, New York, Boston, London, Sydney, Tokyo and Toronto.
Barrett, A. J. (Ed.) (1995) *Proteolytic Enzymes: Aspartic and Metallopeptidase. Methods Enzymol* **248:** 870 pp. Academic Press Inc. San Diego, New York, Boston, London, Sydney, Tokyo and Toronto.
Basha, E., Lee, G. J., Breci, L. A., Hausrath, A. C., Buan, N. R., Giese, K. C., and Vierling, E. (2004) The identity of proteins associated with a small heat shock protein during heat stress *in vivo* indicates that these chaperones protect a wide range of cellular functions. *J Biol Chem* **279:** 7566–7575.
Beebe, K. D., Shin, J., Peng, J., Chaudhury, C., Khera, J., and Pei, D. (2000) Substrate recognition through a PDZ domain in tail-specific peptidase. *Biochemistry* **39:** 3149–3155.
Belkin, S., and Packer, L. (1988) Determination of pH gradients in intact cyanobacteria by electron spin resonance spectroscopy. *Methods Enzymol* **167:** 677–685.
Belnap, J., Büdel, B., and Lange, O. L. (2001a) Biological soil crusts: Characteristics and distribution. In: Belnap, J. and Lange, O. L. (Eds.) *Biological Soil Crusts: Structure, Function and Management. Ecol Studies* **15**, Springer, Berlin, Germany. pp. 3–30.
Belnap, J., Rosentreter, R., Leonard, S., Kaltenecker, J. H., Williams, J., and Elridge, D. (2001b) *Biological soil crusts: Ecology and Management*. Technical Reference 1730-2, US Department of Interior.
Bernroitner, M., Zámocký, M., Furtmüller, P. G., Peschek, G. A., and Obinger, C. (2009) Occurrence, phylogeny and function of catalases and peroxidases in cyanobacteria. *J Exp Bot* **60:** 423–440. doi:10.1093/jxb/em309
Bewley, J. D. (1979) Physiological aspects of desiccation tolerance. *Annu Rev Plant Physiol* **30:** 195–238.
Bhagawat, A. A., and Apte, S. K. (1989) Comparative analysis of heat shock, salinity and osmotic stress in the nitrogen fixing cyanobacterium *Anabaena* sp. strain L-31. *J Bacteriol* **171:** 5187–5189.
Bhattacharya, J., Dastidar, K. G., Chatterjee, A., Majee, M., and Majumdar, A. L. (2004) *Synechocystis* Fe superoxide dismutase gene confers oxidative stress tolerance to *Escherichia coli*. *Biochem Biophys Res Commun* **316:** 540–544.
Bienert, R., Baier, K., Volkmer, R., Lockau, W., and Heinemann, U. (2006) Crystal structure of NblA from *Anabaena* sp. PCC 7120, a small protein playing a key role in phycobilisome degradation. *J Biol Chem* **281:** 5216–5223.
Billi, D., Friedmann, E. I., Hofer, K. G., Grilli Caiola, M., and Ocampo-Friedmann, R. (2000) Ionizing-radiation resistance in the desiccation-tolerant cyanobacterium *Chroococcidiopsis*. *Appl Environ Microbiol* **66:** 1489–1492.
Billi, D., Friedmann, I., Helm, R. F., and Potts, M. (2001) Gene transfer to the desiccation tolerant cyanobacterium *Chroococcidiopsis*. *J Bacteriol* **183:** 2298–2305.
Billini, M., Stamatakis, K., and Sophianopoulou, V. (2008) Two members of a network of putative Na^+/H^+ antiporters are involved in salt and pH tolerance of the freshwater cyanobacterium *Synechococcus elongatus*. *J Bacteriol* **190:** 6318–6329.
Blanco-Rivero, A., Leganés, F., Fernández–Valiente, E., Calle, P., and Fernández-Piñas, P. (2005a) mrpA, a gene with roles in resistance to Na^+ and adaptation to alkaline pH in the cyanobacterium *Anabaena* sp. PCC 7120. *Microbiology* **151:** 1671–1682. doi:10.1099/mic.0.27848.0
Blanco-Rivero, M. C., Takabe, T., and Viale, A. M. (2005b) Functional differences between cyanobacterial DnaK1 chaperones from the halophyte *Aphanothece halophytica* and the freshwater species *Synechococcus elongatus* expressed in *Escherichia coli*. *Curr Microbiol* **51:** 164–170.

Blasius, M., Hubscher, U., and Sommer, S. (2008) *Deinococcus radiodurans*: What belongs to the survival kit? *Crit Rev Biochem Mol Biol* **43:** 221–238.

Blumwald, E., Mehlhorn, R. J., and Packer, L. (1983) Studies on osmoregulation in salt adaptation of cyanobacteria with ESR spin-probe techniques. *Proc Natl Acad Sci USA* **80:** 2599–2602.

Böhm, G. A., Pfleiderer, W., Böger, P., and Scherer, S. (1995) Structure of a novel oligosaccharide-mycosporine-amino acid ultraviolet A/B sunscreen pigment from the terrestrial cyanobacterium, *Nostoc commune*. *J Biol Chem* **270:** 8536–8539.

Borbély, G., Surányi, G., Korez, A., and Pálfi, Z. (1985) Effect of heat shock on protein synthesis in the cyanobacterium *Synechococcus* sp. strain PCC 6301. *J Bacteriol* **161:** 1125–1130.

Boison, G., Mergel, A., Jolkver, H., and Bothe, H. (2004) Bacterial life and dinitrogen fixation at a gypsum rock. *Appl Environ Microbiol* **70:** 7070–7077.

Borowitzka, I. L., Demmerle, S., Mackay, M. A., and Norton, R. (1980) Carbon-13 nuclear magnetic resonance study of osmoregulation in a blue-green alga. *Science* **210:** 650–651.

Bowler, C., Van Camp, W., Van Montagu, M., and Inzé, D. (1994) Superoxide dismutase in plants. *CRC Critical Rev Plant Sci* **13:** 199–218.

Brand, J. J. (1977) Spectral changes in *Anacystis nidulans* induced by chilling. *Plant Physiol* **59:** 970–973.

Brand, J. J., Kirchanski, S. J., and Ramirez-mitchell, R. (1979) Chill-induced morphological alterations in *Anacystis nidulans* as a function of growth temperature. *Planta* **145:** 63–68.

Brock, T. D. (1973) Lower pH limit for the existence of blue-green algae: evolutionary and ecological implications. *Science* **197:** 480–483.

Brüll, L. P., Huang, Z., Thomas-Oates, J. E., Paulsen, B. S., Cohen, E. H., and Michaelsen, T. E. (2000) Studies of polysaccharides from three edible species of *Nostoc* (cyanobacteria) with different colony morphologies: Structural characterization and effect on the complement system of polysaccharides from *Nostoc commune*. *J Phycol* **36:** 871–881.

Bucca, G., Brassington, A. M., Hotchkiss, G., Mersinias, V., and Smith, C. P. (2003) Negative feedback regulation of *dnaK*, *clpB*, and *lon* expression by the DnaK chaperone machine in *Streptomyces coelicolor*, identified by transcriptome and *in vivo* DnaK-depletion analysis. *Mol Microbiol* **50:** 153–166.

Buchanan, B. B., and Balmer, Y. (2005) Redox regulation: A broadening horizon. *Annu Rev Plant Biol* **56:** 187–220.

Büdel, B., Weber, B., Kühl, M., Pfanz, H., Sültemeyer, D., and Wessels, D. (2004) Reshaping of sandstone surfaces by cryptoendolithic cyanobacteria: bioalkalisation causes chemical weathering in arid landscapes. *Geobiology* **2:** 261–268.

Bukau, B. (1993) Regulation of the *Escherichia coli* heat-shock response. *Mol Microbiol* **9:** 671–680.

Burnap, R. L., Troyan, T., and Sherman, L. A. (1993) The highly abundant chlorophyll–protein complex of iron-deficient *Synechococcus* sp. PCC 7942 (CP43′) is encoded by the *isiA* gene. *Plant Physiol* **103:** 893–902.

Cadoret, J.-C., Demoulière, R., Lavaud, J., van Gorkom, H. J., Houmard, J., and Etienne, A.-L. (2004) Dissipation of excess energy triggered by blue light in cyanobacteria with CP43′ (*isiA*). *Biochim Biophys Acta* **1659:** 100–104.

Callis, J. (1995) Regulation of protein degradation. *Plant Cell* **7:** 845–857.

Campbell, D., Zhou, G., Gustafsson, P., Öquist, G., and Clarke, A. K. (1995) Electron transport regulates exchange of two forms of photosystem II D1 protein in the cyanobacterium *Synchococcus*. *EMBO J* **14:** 5457–5466.

Campbell, W. S., and Laudenbach, D. E. (1995) Characterization of four superoxide dismutase genes from a filamentous cyanobacterium. *J Bacteriol* **177:** 964–972.

Canini, A., Civitareale, P., Marini, S., Grilli-Caiola, M., and Rotilo, G. (1992) Purification of iron superoxide dismutase in the cyanobacterium *Anabaena cylindrica* Lemm and localization of the enzyme in heterocysts by immunogold labeling. *Planta* **187:** 438–444.

Capitani, G., Markovic-Housley, Z., DelVal, G., Morris, M., Jansonius, J. N., and Schürmann, P. (2000) Crystal structures of two functionally different thioredoxins in spinach chloroplasts. *J Mol Biol* **302:** 135–154.

Castielli, O., De La Cerda, B., Navarro, J. A., Hervás, M., and De La Rosa, M. A. (2009) Proteome analyses of the response of cyanobacteria to different salt conditions. *FEBS Lett* **583:** 1753–1758. doi:10.1016/j.febslet.2009.03.069

Celerin, M., Gilpin, A. A., Schisler, N. J., Ivanov, A. G., Miskiewicz, E., Krol, M., and Laudenbach, D. E. (1998) ClpB in a cyanobacterium: Predicted structure, phylogenetic relationships and regulation by light and temperature. *J Bacteriol* **180:** 5173–5182.

Cha, M. K., Hong, S. K., and Kim, I. H. (2007) Four thiol peroxidases contain a conserved GCT catalytic motif and act as a versatile array of lipid peroxidases in *Anabaena* sp. PCC 7120. *Free Radic Biol Med* **42:** 1736–1748.

Chae, H. Z., Chung, S. J., and Rhee, S.G. (1994) Thioredoxin-dependent peroxide reductase from yeast. *J Biol Chem* **269:** 27670–27678.

Chae, H. Z., Robison, K., Poole, L. B., Church, G., Storz, G., and Rhee, S. G. (1994) Cloning and sequencing of thiol-specific antioxidant from mammalian brain: Alkyl hydroperoxide reductase and thiol-specific antioxidant define a large family of antioxidant enzymes. *Proc Natl Acad Sci USA* **91:** 7017–7021.

Chamot, D., and Owttrim, G. W. (2000) Regulation of cold shock-induced RNA helicase gene expression in the cyanobacterium *Anabaena* sp. strain PCC 7120. *J Bacteriol* **182:** 1251–1256.

Chamot, D., Colvin, K. R., Kujat-Choy, S. L., and Owttrim, G. W. (2005) RNA structural rearrangement via unwinding and annealing by the cyanobacterial RNA helicase, CrhR. *J Biol Chem* **280:** 2036–2044.
Chamot, D., Magee, W. C., Yu, E., and Owttrim, G. W. (1999) A cold shock-induced cyanobacterial RNA helicase. *J Bacteriol* **181:** 1728–1732.
Chang, A., Lin, R., Studley, W. K., Tran, C., and Saier, Jr., M. (2004) Phylogeny as a guide to structure and function of membrane transport proteins. *Mol Membr Biol* **21:** 171–181.
Charles, S. A., and Haliwell, B. (1980) Effect of hydrogen peroxide on spinach (*Spinacia oleracea*) chloroplast fructose biphosphatase. *Biochem J* **189:** 373–376.
Cheregi, O., Sicora, C., Kos, P. B., Barker, M., Nixon, P. J., and Vass, I. (2007) The role of FtsH and Deg proteases in the repair of UV-B radiation-damaged photosystem II in the cyanobacterium *Synechocystis* PCC 6803. *Biochim Biophys Acta* **1767:** 820–828.
Chi, Z., Su, C. D., and Lu, W. D. (2007) A new exopolysaccharide produced by marine *Cyanothece* sp. 113. *Bioresource Technol* **98:** 1329–1332.
Chitnis, P. R., and Nelson, L. (1991) Molecular cloning of the genes encoding two chaperone proteins of the cyanobacterium *Synechocystis* sp. PCC 6803. *J Biol Chem* **266:** 58–65.
Choi, H., Kim, S., Mukhopadhyay, P., Cho, S., Woo, J., Storz, G., and Ryu, S. (2001) Structural basis of the redox switch in the OxyR transcription factor. *Cell* **105:** 103–113.
Claiborne, A., and Fridovich, I. (1979) Purification of the O-dianisidine peroxidase from *Escherichia coli* B. *J Biol Chem* **254:** 4245–4252.
Clarke, A. K., and Eriksson, M.-J. (1996) The cyanobacterium *Synechococcus* sp. PCC 7942 possesses a close homologue to the chloroplast ClpC protein of higher plants. *Plant Mol Biol* **31:** 721–730.
Clarke, A. K., and Eriksson, M. J. (2000) The truncated form of bacterial heat shock protein ClpB/Hsp100 contributes to the development of thermotolerance in the cyanobacterium *Synechococcus* sp. strain PCC 7942. *J Bacteriol* **182:** 7092–7096.
Clarke, A. K., MacDonald, T. M., and Sjögren, L. L. E. (2005) The ATP-dependent Clp protease in chloroplasts of higher plants. *Physiol Plant* **123:** 406–412.
Clausen, T., Southan, C., and Ehrmann, M. (2002) The HtrA family of proteases: Implications for protein composition and cell fate. *Mol Cell* **10:** 443– 455.
Close, T. J., and Lammers, P. J. (1993) An osmotic stress protein of cyanobacteria is immunologically related to plant dehydrins. *Plant Physiol* **101:** 773–779.
Cockell, C. S., Schuerger, A. C., Billi, B., Friedmann, E. I., and Panitz, C. (2005) Effects of a simulated martian UV flux on the cyanobacterium *Chroococcidiopsis* sp. 029. *Astrobiology* **5:** 127–140.
Colin, J., Jakopitsch, C., Obinger, C., and Ivancich, A. (2010) The reaction of *Synechocystis* catalase-peroxidase (KatG) with isoniazid investigated by multifrequency (9-285 GHz) EPR spectroscopy. *Appl Magn Reson* **37:** 267–277.
Collier, J. L., and Grossman, A. R. (1995) Disruption of a novel thioredoxin-like protein alters the cyanobacterial photosynthetic apparatus. *J Bacteriol* **177:** 3269–3276.
Cossins, A. R. (1994) Homeoviscous adaptation of biological membranes and its functional significance. In: Cossins, A. R. (Ed.) *Temperature Adaptation of Biological Membranes*, Portland, London, UK. pp. 63–76.
Craig, E. A., and Gross, C A. (1991) Is hsp70 the cellular thermometer? *Trends Biochem Sci* **16:** 135–140.
Crowe, J. H., Carpenter, J. F., Crowe, L. M., and Anchordoguy, T. J. (1990) Are freezing and dehydration similar stress vectors? A comparison of modes of interaction of stabilizing solutes with biomolecules. *Cryobiology* **27:** 219–231.
Crowe, J. H., Hoekstra, F. A., and Crowe, L. M. (1989) Membrane phase transitions are responsible for imbibitional damage in dry pollen. *Proc Natl Acad Sci USA* **86:** 520–523.
Crowe, J. H., Hoekstra, F. A., and Crowe, L. M. (1992) Anhydrobiosis. *Annu Rev Physiol* **54:** 579–599.
Crowe, J. H., Oliver, A. E., and Tablin, F. (2002) Is there a single biochemical adaptation to anhydrobiosis. *Integr Comp Biol* **42:** 497–503.
Crowe, J. H., Oliver, A. E., Hoekstra, F. A., and Crowe, L. M. (1997) Stabilization of dry membranes by mixtures of hydroxyethyl starch and glucose: The role of vitrification. *Cryobiology* **35:** 20–30.
Crowley, D. J., and Hanawalt, P. C (2001) The SOS-dependent upregulation of *uvrD* is not required for efficient nucleotide excision repair of ultraviolet light induced DNA photoproducts in *Escherichia coli*. *Mutat Res* **485:** 319–329.
Dartigalongue, C., Missiakas, D., and Raina, S. (2001) Characterization of the *Escherichia coli* sigma E regulon. *J Biol Chem* **276:** 20866–20875.
Dasman, Kajiyama, S., Kawasaki, H., Yagi, M., Seki, T., Fukusaki, E., and Kobayashi, A. (2002a) *Paenibacillus glycanilyticus* sp. nov. a novel species that degrades heteropolysacahride produced by the cyanobacterium *Nostoc commune*. *Int J Syst Evol Microbiol* **52:** 1669–1674. doi:10.1099/ijs.0.02171-0
Dasman, Kajiyama, S., Okazawa, A., Fukusaki, E., and Kobayashi, A. (2002b) Purification and properties of an enzyme capable of degrading the polysaccharide of the cyanobacterium *Nostoc commune*. *Z Natuforsch* **57:** 1042–1046.
Davis, D. A., Newcomb, F. M., Starke, D. W., Ott, D. E., Mieyal, J. J., and Yarchoan, R. (1997) Thioltransferase (glutaredoxin) is detected within HIV-1 and can regulate the activity of glutathionylated HIV-1 protease *in vitro*. *J Biol Chem* **272:** 25935–25940.

de Groot, A., Chapon, V., Servant, P., Christen, R., Fischer-Le Saux M., *et al.* (2005) *Deinococcus deserti* sp. nov., a gamma-radiation-tolerant bacterium isolated from the Sahara Desert. *Int J Syst Evol Microbiol* **55:** 2441–2446.

deGroot, A, Dulermo, R., Ortet, P., Blanchard, L., Guerin, P., Fernandez, B., Vacherie, B., Dossat, C., Jolivet, E., Siguier, P., Chandler, M.,Barakat, M., Dedieu, A., Barbe, V., Heulin, T., Sommer, S., Achouak, W., and Armengaud, J. (2009) Alliance of proteomics and genomics to unravel the specificities of Sahara bacterium *Deinococcus deserti. PLoS Genetics* **5:** e1000434. doi: 1371/journal.pgen.1000434

de los Ríos, A., Grube, M., Sancho, L. G., and Ascaso, C. (2007) Ultrastructural and genetic characteristics of endolithic biofilms colonizing Antarctic granite rocks. *FEMS Microbiol Ecol* **59:** 386–395.

de la Torre, J. R., Goebel, B. M., Friedmann, E. I., and Pace N. R. (2003) Microbial diversity of cryptoendolithic communities from the McMurdo Dry Valleys, Antarctica. *Appl Environ Microbiol* **69:** 3858–3867.

De Philippis, R., and Vincenzini, M. (1998) Exocellular polysaccharides from cyanobacteria and their possible applications. *FEMS Microbiol Rev* **22:** 151–175.

De Philippis, R., and Vincenzini, M. (2003) Outermost polysaccharidic investments of cyanobacteria: Nature, significance and possible applications. *Recent Res Dev Microbiol* **7:** 13–22.

Derre, I., Rapoport, G., and Msadek , T. (1999) CtsR, a novel regulator of stress and heat shock response, controls *clp* and molecular chaperone gene expression in gram-positive bacteria. *Mol Microbiol* **31:** 117–131.

Deshnium, P., Los, D. A., Hayashi, H., Mustardy, L., and Murata, N. (1995) Transformation of *Synechococcus* with a gene for choline oxidase enhances tolerance to salt. *Plant Mol Biol* **29:** 897–907.

De Smet, K. A., Weston, A., Brown, I. N., Young, D. B., and Robertson, B. D. (2000) Three pathways for trehalose biosynthesis in mycobacteria. *Microbiology* **146:** 199–208.

De Virgilio, C., Hottiger, T., Dominguez, J., Boller, T., and Wiemken, A. (1994) The role of trehalose synthesis for the acquisition of thermotolerance in yeast. I. Genetic evidence that trehalose is a thermoprotectant. *Eur J Biochem* **219:** 179–186.

Diamant, S., Ben-Zvi, A. P., Bukau, B., and Goloubinoff, P. (2000) Size-dependent disaggregation of stable protein aggregates by the DnaK chaperone machinery. *J Biol Chem* **275:** 21107–21113.

Diamant, S., Eliahu, N., Rosenthal, D., and Goloubinhoff, P. (2001) Chemical chaperones regulate molecular chaperones *in vitro* and in cells under combined salt and heat stress. *J Biol Chem* **276:** 39586–39591.

Dietz, K.-J. (2003) Plant peroxiredoxins. *Annu Rev Plant Biol* **54:** 93–107.

Dietz, K.-J. (2011) Peroxiredoxins in plants and cyanobacteria. *Antioxidants Redox Signal* **15:** 1129–1159. doi:10.1089/ars.2010.3657

Diner, B. A., Ries, D. F., Cohen, B. N., and Metz, J. G. (1988) COOH-terminal processing of polypeptide D1 of the photosystem II reaction center of *Scenedesmus obliquus* is necessary for the assembly of the oxygen-evolving complex. *J Biol Chem* **263:** 8972–8980.

Dines, M., Sendersky, E., Schwarz, R., and Adir, N. (2007) Crystallization of sparingly soluble stress-related proteins from cyanobacteria by controlled urea solublization. *J Struct Biol* **158:** 116–121.

Donald, L- J., Krokhin, O. V., Duckworth, H. W., Wiseman, B., Deemagaran, T., Singh, R., Switala, J., Carpena, X., Fita, I., and Loewen, P. C. C. (2003) Characterization of catalase-peroxidase KatG from *Burkholderia pseudomallei* by mass spectrometry. *J Biol Chem* **278:** 35687–35692.

Dor, I., and Paz, N. (1989) Temporal and spatial distribution of mat microalgae in the experimental solar ponds, Dead Sea area, Israel. In: Cohen, Y., and Rosenberg, E. (Eds.). *Microbial Mats: Physiological Ecology of Benthic Microbial Communities.* American Society for Microbiology, Washington, DC, USA. pp. 114–122.

Dor, I., Nehama, C., and Baldinger, I. (1991) Polymorphism and salinity tolerance as a criterion for differentiation of three new species of *Chroococcidiopsis* (Chroococcales) *Arch Hydrobiol Algol Stud* **64:** 411–421.

Dorman, C. J. (2006) DNA supercoiling and bacterial gene expression. *Sci Prog* **89:** 151–166.

Dougan, D. A., Reid, B. G., Horwich, A. L., and Bukau, B. (2002) ClpS, a substrate modulator of the ClpAP machine. *Mol Cell* **9:** 673–683.

Doyle, S. M., Hoskins, J. R., and Wickner, S. (2007) Collaboration between ClpB AAA$^+$ remodeling protein and the DnaK chaperone system. *Proc Natl Acad Sci USA* **104:** 11138–11144.

Dufresne, A., Salanoubat, M., Partensky, F., Artiguenave, F., Axmann, I. M., Barbe, V., Duprat, S., Galperin, M. Y., Koonin, E. V., Le Gall, F., Makarova, K. S., Ostrowski, M., Oztas, S., Robert, C., Rogozin, I. B., Scanlan, D. J., Tandeau de Marsac, N., Weissenbach, J., Wincker, P., Wolf, Y. I., and Hess, W. R. (2003) Genome sequence of the cyanobacterium *Prochlorococcus marinus* SS120, a nearly minimal oxyphototrophic genome. *Proc Natl Acad Sci USA* **100:** 10020–10025.

Dupont, C. L., Barbeau, K., and Palenik, B. (2008) Nickel uptake and limitation in marine *Synechococcus* strains. *Appl Environ Microbiol* **74:** 23–31.

Dwivedi, K., Sen A., and Bullerjahn, G. S. (1997) Expression and mutagenesis of *dpsA* gene of *Synechococcus* sp. PCC 7942, encoding a DNA-binding protein involved in oxidative stress protection. *FEMS Microbiol Lett* **155:** 85–91.

Eaton-Rye, J. J., Shand, J. A., and Nicoll, W. S. (2003) pH-dependent photoautotrophic growth of specific photosystem II mutants lacking lumenal extrinsic polypeptides in *Synechocystis* PCC 6803. *FEBS Lett* **543:** 148–153.

Edwards, H. G. M., Villar, S. E. J., Parnell, J., Cockell, C. S., and Lee, P. (2005) Raman spectroscopic analysis of cyanobacterial gypsum halotrophs and relevance for sulphate deposits on Mars. *Analyst* **130:** 917–923.

Ehira, S., Hamano, T., Hayashida, T., Kojima, K., Nakamoto, H., Hiyama, T., Ohmori, M., Shivaji, S., and Sato, N. (2003) Conserved temperature-dependent expression of RNA-binding proteins in the cyanobacteria with different temperature optima. *FEMS Microbiol Lett* **225:** 137–142.

Ehling-Schulz, M., Bilger, W., and Scherer, S. (1997) UV-B-induced synthesis of photoprotective pigments and extracellular polysaccharides in the terrestrial cyanobacterium *Nostoc commune. J Bacteriol* **179:** 1940–1945.

Eitinger, T. (2004) *In vivo* production of active nickel superoxide dismutase from *Prochlorococcus marinus* MIT9313 is dependent on its cognate peptidase. *J Bacteriol* **186:** 7821–7825.

Eklund, H., Cambillau, C., Sjoberg, B. M., Holmgren, A., Jornvall, H., Hoog, J. O., and Branden, C. I. (1984) Conformational and functional similarities between glutaredoxin and thioredoxins. *EMBO J* **3:** 1443–1449.

Eklund, H., Gleason, F. K., and Holmgren, A. (1991) Structural and functional relations among thioredoxins of different species. *Proteins Struct Funct Genet* **11:** 13–28.

Elanskaya, I. V., Karandashova, I. V., Bogachev, A. V., and Hagemann, M. (2002) Functional analysis of the Na^+/H^+ antiporter encoding genes of the cyanobacterium *Synechocystis* sp. PCC 6803. *Biochemistry* (Moscow) **67:** 432–440.

Elbein, A. D. (1974) The metabolism of α,α-trehalose. *Adv Carb Chem Biochem* **30:** 227–256.

Elbein, A. D., Pan, Y. T., Pastuszak, I., and Carroll, D. (2003) New insights on trehalose: a multifunctional molecule. *Glycobiology* doi:10.1093/glycob/cwg047

El Bissati, K., and Kirilovsky, D. (2001) Regulation of *psbA* and *psaE* expression by light quality in *Synechocystis* species PCC 6803. A redox control mechanism. *Plant Physiol* 125: 1988–2000.

El-Fahmawi, B., and Owttrim, G. W. (2003) Polar-biased localization of the cold stress-induced RNA helicase, CrhC, in the cyanobacterium *Anabaena* sp. strain PCC 7120. *Mol Microbiol* **50:** 1439–1448.

Ellis, R. J. (1990) The molecular chaperone concept. *Semin Cell Biol* **1:** 1–9.

Elridge, D. J., and Greene, R. S. B. (1994) Microbial soil crusts: A review of their roles in soil and ecological processes in the range lands of Australia. *J Soil Res* **32:** 389–415.

El-Samad, H., Kurata, H., Doyle, J. C., Gross, C. A., and Khammash, M. (2005) Surviving heat shock: control strategies for robustness and performance. *Proc Natl Acad Sci USA* **102:** 2736–2741.

Empadinhas, N., and da Costa, M. S. (2008) Osmoadaptation mechanisms in prokaryotes: distribution of compatible solutes. *Int Microbiol* **11:** 151–161. doi:10.2436/20.1501.01.55

Enami, I., Kikuchi, S., Fukuda, T., Ohta, H., and Shen, J. R. (1998) Binding and functional properties of four extrinsic proteins of photosystem II from a red alga *Cyanidium caldarium*, as studied by release-reconstitution experiments. *Biochemistry* **37:** 2787–2793.

Engelbrecht, F., Marin, K., and Hagemann, M. (1999) Expression of *ggps* gene, involved in osmolyte synthesis in the marine cyanobacterium *Synechococcus* sp. strain PCC 7002, revealed regulatory differences between this strain and the freshwater strain *Synechocystis* sp. strain PCC 6803. *Appl Environ Microbiol* **65:** 4822–4829.

Erbse, A., Schmidt, R., Bornemann, T., Schneider-Mergener, J., Mogk, A., Zahn, R., Dougan, D.A., and Bukau, B. (2006) ClpS is an essential component of the N-end rule pathway in *Escherichia coli. Nature* (London) **439:** 753–756.

Erdmann, N., Fulda, S., and Hagemann, M. (1992) Glucosylglycerol accumulation during salt acclimation of two unicellular cyanobacteria. *J Gen Microbiol* **138:** 363–368.

Eriksson, M.-J., and Clarke, A. K. (1996) The heat shock protein ClpB mediates the development of thermotolerance in the cyanobacterium *Synechococcus* sp. strain PCC 7942. *J Bacteriol* **178:** 4839–4846.

Eriksson, M. J., Schelin, J., Miskiewicz, E., and Clarke, A. K. (2001) Novel form of ClpB/Hsp 100 protein in the cyanobacterium *Synechococcus. J Bacteriol* **183:** 7392–7396.

Espie, G. S., Miller, A. G., and Canvin, D. T. (1988) Characterization of the Na^+ requirement in cyanobacterial photosynthesis. *Plant Physiol* **88:** 757–763.

Fahey, R. C., and Sundquist, A. R. (1991) Evolution of glutathione metabolism. *Adv Enzymol Relat Areas Mol Biol* **64:** 1–53.

Fahey, R. C., Brown, W. C., Adams, W. B., and Worsham, M. B. (1978) Occurrence of glutathione in bacteria. *J Bacteriol* **133:** 1126–1129.

Fang, S., and Barnum, S. P. (2003) The heat shock gene, *htpG*, and thermotolerance in the cyanobacterium *Synechocystis* sp. PCC 6803. *Curr Microbiol* **47:** 341–346.

Ferjani, A., Mustardy, L., Sulpice, R., Marin, K., Suzuki, I., Hagemann, M., and Murata, N. (2003) Glucosylglycerol, a compatible solute sustains cell division under salt stress. *Plant Physiol* **131:** 1628–1637.

Fernandes, A., and Holmgren, A. (2004) Glutaredoxins: Glutathione-dependent redox enzymes with functions far beyond a simple thioredoxin back up system. *Antioxd Redox Signal* **6:** 63–74.

Ferreira, A. C., Nobre, M. F., Rainey, F. A., Silva, M. T., Wait, R., *et al.* (1997) *Deinococcus geothermalis* sp. nov. and *Deinococcus murrayi* sp. nov., two extremely radiation resistant and slightly thermophilic species from hot springs. *Int J Syst Bacteriol* **47:** 939–947.

Fleming, E. D., and Castenholz, R.W. (2007). Effects of periodic desiccation on the synthesis of the UV-screening compound, scytonemin, in cyanobacteria. *Environ Microbiol* **9:** 1448–1455.

Florencio, F. J., Pérez-Pérez, M. E., Lopez-Maury, L., Mata-Cabana, A., and Lindahl, M. (2006) The diversity and complexity of the cyanobacterial thioredoxin systems. *Photosynth Res* **89:** 157–171.

Florian, H. P., Holger, S., and Iwona, A. (2005) The family of Deg proteases in cyanobacteria and chloroplasts of higher plants. *Physiol Plant* **123:** 413–420.

Forkl, H., Vandekerckhove, J., Drews, G., and Tadros, M. H. (1993) Molecular cloning, sequence analysis and expression of the gene for catalase-peroxidase (cpeA) from the photosynthetic bacterium *Rhodobacter capsulatus* B10. *Eur J Biochem* **214:** 251–258.

Forrest H. S., Van Baalen, C., and Myers, J. (1957) Occurrence of pteridines in a blue-green alga. *Science* **125:** 699–700.

Foster, J. S., Havemann, S. A., Singh, A. K., and Sherman, L. A. (2009) Role of *mrgA* in peroxide and light stress in the cyanobacterium *Synechocystis* sp. PCC 6803. *FEMS Microbiol Lett* **293:** 298–304(7).

Fourquet, S., Huang, M. E., D'Autreaux, B., and Toledano, M. B. (2008) The dual functions of thiol-based peroxidases in H_2O_2 scavenging and signaling. *Antioxid Redox Signal* **10:** 1565–1576.

Foyer, C. H., and Harbinson, J. (1994) Oxygen metabolism and the regulation of photosynthetic electron transport. In: Foyer, C. H., and Mullineaux, P. M. (Eds.) *Causes of Photooxidative Stress and Amelioration of Defense Systems in Plants*: CRC Press, Boca Raton , Florida, FL, USA. pp. 1–42.

Fraser, C. M., Casjens, S., Huang, W. M., Sutton, G. G., Rebecca Clayton, R., Lathigra, R., White, O. Ketchum, K. A., Dodson, R., Hickey, E. K., Gwinn, M., Dougherty, B., Tomb, J- F., Fleischmann, R. D., Richardson, D., Peterson, J., Kerlavage, A. R., Quackenbush, J., Salzberg, S., Hanson, M., van Vugt, R., Palmer, N., Adams, M. D., Gocayne, J., Weidman, J., Utterback, T., Watthey, L., McDonald, L., Artiach, P., Bowman, C., Garland, S., Fujii, C., Cotton, M. D., Horst, K., Roberts, K., Hatch, B., Smith, H. O., and Venter, J. C. (1997) Genomic sequence of a Lyme disease spirochaete, *Borrelia burgdorferi*. *Nature* (London) **390:** 580–586. doi:10.1038/37551

Fraser, C. M., Norris, S. J., Weinstock, G. M., White, O., Sutton, G. G., Dodson, R., Gwinn, M., Hickey, E. K., Clayton, R., Ketchum, K. A.,Sodergren, E., Hardham, J. M., McLeod, M. P., Salzberg, S., Peterson, J., Khalak, H., Richardson, D., Howell, J. K., Chidambaram, M., Utterback, T., McDonald, L., Artiach, P., Bowman, C., Cotton, M. D., Fujii, C., Garland, S., Hatch, B., Horst, K., Roberts, K., Sandusky,M., Weidman, J., Smith, H. O., and Venter, J. C. (1998) Complete Genome Sequence of *Treponema pallidum*, the syphilis spirochete. *Science* **281:** 375–388. doi:10.1126/science.281.5375.375

Fridovich, I. (1989) Superoxide dismutases, an adaptation to a paramagnetic gas. *J Biol Chem* **264:** 7761–7764.

Fridovich, I. (1995) Superoxide radical and superoxide dismutases. *Annu Rev Biochem* **64:** 97–112.

Fridovich, I. (1997) Superoxide anion radical ($O_2^{\bullet-}$). Superoxide dismutases and related matters. *J Biol Chem* **272:** 18515–18517.

Friedmann, E. I. (1961) *Chroococcidiopsis kashaii* sp. nov. and the genus *Chroococcidiopsis* (studies on cave algae from Israel III). *Österr Bot Z* **108:** 354–367.

Friedmann, E. I., and Ocampo-Friedmann, R. (1995) A primitive cyanobacterium as pioneer microorganism for terraforming Mars. *Adv Space Res* **15:** 243–246.

Friedmann, E.I., Hua, M., and Ocampo-Friedmann, O. (1988) Cryptoendolithic lichen and cyanobacterial communities of the Ross Desert, Antarctica. *Polarforschung* **58:** 251–159.

Friedmann, E. I., Kappen, L., Meyer, M. A., and Nienow, J. A. (1993) Long-term productivity in the cryptoendolithic microbial community of the Ross Desert, Antarctica. *Microb Ecol* **25:** 51–69.

Friedrich, K. L., Giese, K. C., Buan, N., and Vierling, E. (2004) Interactions between small heat shock protein subunits and substrate in small heat shock protein-substrate complexes. *J Biol Chem* **279:** 1080–1089.

Fukumori, Y., Fujiwara, T., Okada-Takahashi, Y., Mukohata, Y., and Yamanaka, T. (1985) Purification and properties of a peroxidase from *Halobacterium halobium* L-33. *J Biochem* **98:** 1055–1061.

Fuangthong, M., Herbig, A. F., Bsat, N., and Helmann, J. D. (2002) Regulation of the *Bacillus subtilis fur* and *perR* genes by PerR: not all members of the PerR regulon are peroxide inducible. *J Bacteriol* **184:** 3276–3286.

Fulda, S., Huang, F., Nilsson, F., Hagemann, M., and Norling, B. (2000) Proteomics of *Synechocystis* sp. strain PCC 6803. Identification of periplasmic proteins in cells grown at low and high salt concentrations. *Eur J Biochem* **267:** 5900–5907.

Fulda, S., Mikkat, S., Schroder, W., and Hagemann, M. (1999) Isolation of salt-induced periplasmic proteins from *Synechocystis* sp. strain PCC 6803. *Arch Microbiol* **171:** 214–217.

Fulda, S., Mikkat, S., Huang, F., Huckauf, J., Marin, K., Norling, B., and Hagemann, M. (2006) Proteome analysis of salt stress response in the cyanobacterium *Synechocystis* sp. strain PCC 6803. *Proteomics* **6:** 2733–2745.

Fuller-Pace, F. V. (1994) RNA helicases: modulators of RNA structure. *Trends Cell Biol* **4:** 271–274.

Fuller-Pace, F. V. (2006) DExD/H box RNA helicases: multifunctional proteins with important roles in transcriptional regulation. *Nucleic Acids Res* **34:** 4206–4215.

Fuller-Pace, F.V., Nicol, S. M., Reid, A. D., and Lane, D. P. (1993) DbpA: a DEAD box protein specifically activated by 23s rRNA. *EMBO J.* **12:** 3619–3626.

Furuki, M.,Tanaka, N., Hiyama, T., and Nakamoto, H. (1996) Cloning, characterization and functional analysis of *groEL*-like gene from thermophilic cyanobacterium *Synechococcus vulcanus*, which does not form an operon with *groES*. *Biochim Biophys Acta* **1294:** 106–110.

Gaber, A., El-Awady, M., Elarabi, N. I., and Soliman, M. H. (2007) Overexpression of glutaredoxin-2 from cyanobacterium *Synechocystis* PCC 6803 in *Escherichia coli* conferring enhanced salt tolerance. *Arab J Biotechnol* **10:** 13–22.

García-Domínguez, M., and Florencio, F. J. (1997) Nitrogen availability and electron transport control the expression of *glnB* gene (encoding PII protein) in the cyanobacterium *Synechocystis* sp. PCC 6803. *Plant Mol Biol* **35:** 723–734.

Garcia-Pichel, F., López-Cortés, A., and Nübel, U. (2001) Phylogenetic and morphological diversity of cyanobacteria in soil desert crusts from the Colorado plateau. *Appl Environ Microbiol* **67:** 1902–1910.

Gardner, P. R., and Fridovich, I. (1991) Superoxide sensitivity of the *Escherichia coli* 6-phosphogluconate dehydratase. *J Biol Chem* **266:** 1478–1483.

Gaff, D. F. (1997) Response of desiccation tolerant "resurrection" plants to water stress. *In*: Basra, A. S. and Basra, R. K. (Eds.) *Mechanisms of Environmental Stress Resistance in Plants*. Harwood Academic Publishers, London, UK. pp. 43–58.

Gao, K. S. (1998) Chinese studies on the edible blue-green alga, *Nostoc flagelliforme*: A review. *J Appl Phycol* **10:** 37–49.

Gerrath, J. F., Gerrath, J. A., and Larson, D. W. (1995) A preliminary account of the endolithic algae of limestone cliffs of the Niagara Escarpment. *Can J Bot* **73:** 788–793.

Gerrath, J. F., Gerrath, J. A., Matthes, U., and Larson, D. W. (2000) Endolithic algae and cyanobacteria from cliffs of the Niagara Escarpment, Ontario, Canada. *Can J Bot* **78:** 807–815.

Getzoff, E. D., Cabelli, D. E., Fisher, C. L., Parge, H. E., Viezzoli, S. M., Banci, L., and Hallewell, R. A. (1992) Faster superoxide dismutase mutants designed by enhancing electrostatic guidance. *Nature* (London) **358:** 347–351.

Giese, K. C., and Vierling, E. (2002) Changes in oligomerization are essential for the chaperone activity of a small heat shock protein *in vivo* and *in vitro*. *J Biol Chem* **277:** 46310–46318.

Giovanna, R., Cella, R., Camerino, G., and Ciferri, O. (1983) Resistance to azetidine-2-carboxylic acid and sodium chloride tolerance in carrot cell cultures and *Spirulina platensis*. *Plant Cell Physiol* **24:** 1073–1078.

Glatz, A., Horváth, I., Varvasovszki, V., Kovács, E., Török, Z., and Vígh, L. (1997) Chaperonin genes of the *Synechocystis* PCC 6803 are differentially regulated under light-dark transition during heat stress. *Biochem Biophys Res Commun* **239:** 291–297.

Glatz, A., Vass, I., Los, D. A., and Vígh, L. (1999) The *Synechocystis* model of stress: From molecular chaperones to membranes. *Plant Physiol Biochem* **37:** 1–12.

Gleason, F. K. (1990) Function of two dissimilar thioredoxins in the cyanobacterium, *Anabaena* sp. 7120. In: Baltscheffsky, M. (Ed.) *Current Research in Photosynthesis*, vol. 4, Kluwer Academic Publishers, Dordrecht, The Netherlands. pp. 175–178.

Gleason , F. K. (1992) Activities of two dissimilar thioredoxins from the cyanobacterium *Anabaena* sp. PCC 7120. *J Bacteriol* **174:** 2592–2598.

Gleason, F. K., Whittaker, M. M., Holmgren, A., and Jornvall, H. (1985) The primary structure of thioredoxin from the filamentous cyanobacterium *Anabaena* sp. 7119. *J Biol Chem* **260:** 9567–9573.

Golden, S. S. (1995) Light-responsive gene expression in cyanobacteria. *J Bacteriol* **177:** 1651–1654.

Golden, S. S., Brusslan, J., and Haselkorn, R. (1986) Expression of a family of *psbA* genes encoding a photosystem II polypeptide in the cyanobacterium *Anacystis nidulans* R2. *EMBO J* **5:** 2789–2798.

Goloubinoff, P., Mogk, A., Zvi, A. P., Tomoyasu, T., and Bukau, B. (1999) Sequential mechanism of solubilization and refolding of stable protein aggregates by a bichapherone network. *Proc Natl Acad Sci USA* **96:** 13732–13737.

Gombos, Z., Kanervo, E., Tsvetkova, N., Sakamoto, T., Aro, E.-M., and Murata, N. (1997) Genetic enhancement of the ability to tolerate photoinhibition by introduction of unsaturated bonds into membrane glycerolipids. *Plant Physiol* **115:** 551–559.

Gombos, Z., Wada, H., and Murata, N. (1992) Unsaturation of fatty acids in membrane lipids enhances tolerance of the cyanobacterium *Synechocystis* PCC 6803 to low temperature photoinhibiton. *Proc Natl Acad Sci USA* **99:** 9959–9963.

Gong, H., Tang, Y., Wang, J., Wen, X., Zhang, L., and Lu, C. (2008) Characterization of photosystem II in salt-stressed cyanobacterial *Spirulina platensis* cells. *Biochim Biophys Acta* **1777:** 488–495.

Gorbalenya, A. E., and Koonin, E. V. (1993) Helicases: amino acid sequence comparisons and structure-function relationships. *Curr Opin Struct Biol* **3:** 419–429.

Gorman, A. A., and Rodgers, M. A. (1992) Current perspectives of singlet oxygen detection in biological environments. *J Photochem Photobiol* **14:** 159–176.

Gottesman, S., and Maurizi, M. R. (1992) Regulation by proteolysis: Energy-dependent proteases and their targets. *Microbiol Rev* **56:** 592–621.

Graumann, P. L., and Marahiel, M. A. (1996) Some like it cold: Response of microorganisms to cold shock. *Arch Microbiol* **166:** 293–300.

Graumann, P. L., and Marahiel, M. A. (1998) A superfamily of proteins that contain the cold shock domain. *Trends Biochem Sci* **23:** 286–290.

Graumann, P. L., SchrÖder, K., Schmid, R., and Marahe, M. M. (1996) Cold shock stress-induced proteins in *Bacillus subtilis. J Bacteriol* **178:** 4611–4619.

Grilli Caiola, M., Billi, D., and Friedmann, E. I. (1996) Effect of desiccation on envelopes of the cyanobacterium *Chroococcidiopsis* sp. (Chroococcales). *Eur J Phycol* **31:** 97–105.

Grilli Caiola, M., Ocampo-Friedmann, R., and Friedmann, E. I. (1993) Cytology of long-term desiccation in the desert cyanobacterium *Chroococcidiopsis* sp. (Chroococcales). *Phycologia* **32:** 315–322.

Grimaud, R., Kessel, M., Beuron, F., Steven, A. C., and Maurizi M.R. (1998) Enzymatic and structural similarities between the *Escherichia coli* ATP-dependent proteases ClpXP and ClpAP. *J Biol Chem* **273:** 12476–12481.

Gruber, M. Y., Glick, B. R., and Thompson, J. E. (1990) Cloned manganese superoxide dismutase reduces oxidative stress in *Escherichia coli* and *Anacystis nidulans. Proc Natl Acad Sci USA* **87:** 2608–2612.

Gu, L., Abulimiti, A., Li, W., and Chang, Z. (2002) Monodisperse Hsp 16.3 nonamer exhibits dynamic association and reassociation, with the nonamer dissociation prerequisite for chaperone-like activity. *J Mol Biol* **319:** 517–526.

Gualerzi, C. O., Giuliodori, A. M., and Pon, C. L. (2003) Transcriptional and post-transcriptional control of cold shock genes. *J Mol Biol* **331:** 527–539.

Gutteridge, J. M. C., and Halliwell, B. (1990) The measurement and mechanism of lipid peroxidation in biological systems. *Trends Biochem Sci* **15:** 129–135.

Hagemann, M. (2011) Molecular biology of cyanobacterial salt acclimation. *FEMS Microbiol Rev* **35:** 87–123. doi: 10. 1111/j.1574-6976.2010.00234.x

Hagemann, M., and Erdmann, N. (1994) Activation and pathway of glucosylglycerol synthesis in the cyanobacterium *Synechocystis* sp. PCC 6803. *Microbiology* **140:** 1427–1431.

Hagemann, M., and Erdmann, N. (1997) Environmental stresses. In: Rai, A.K. (Ed.) *Cyanobacterial Nitrogen Metabolism and Environmental Biotechnology*. Springer-Verlag, Heidelberg; Narosa Publishing House, New Delhi, pp. 156–221.

Hagemann, M., and Zuther, E. (1992) Selection and characterization of mutants of the cyanobacterium *Synechocystis* sp. PCC 6803 unable to tolerate high salt concentrations. *Arch Microbiol* **158:** 429–434.

Hagemann, M., Erdmann, N., and Wittenburg, E. (1987) Synthesis of glucosylglycerol in salt-stressed cells of the cyanobacterium *Microcystis firma. Arch Microbiol* **148:** 275–279.

Hagemann, M., Schoor, A., Jeanjean, R., Zuther, E., and Joset, F. (1997) The *stpA* gene from *Synechocystis* sp. strain PCC 6803 encodes the glucosylglycerol-phosphate phosphatase involved in cyanobacterial osmotic response to salt shock. *J Bacteriol* **179:** 1727–1733.

Hagemann, M., Schoor, A.,Mikkat, S., Effmert, U., Zuther, E., Marin, K., Fulda, S., Vinnemeier, J., Kunert, A.,Milkowski, C., Probst, C., and Erdmann, N. (1999) The biochemistry and genetics of the synthesis of osmoprotective compounds in cyanobacteria. In: Oren, A. (Ed.) *Microbiology and Biogeochemistry of Hypersaline Environments.* CRC Press, Boca Raton, FL, USA. pp. 177–186.

Hagemann, M., Techel, D., and Rensing, L. (1991) Comparison of salt- and heat-induced alterations of protein synthesis in the cyanobacterium *Synechocystis* sp. PCC 6803. *Arch Microbiol* **155:** 587–592.

Hagemann, M., Wölfel, L., and Krüger, B. (1990) Alterations of protein synthesis in the cyanobacterium *Synechocystis* sp. PCC 6803 after a salt shock. *J Gen Microbiol* **136:** 1393–1399.

Halliwell, B., and Gutteridge, J. M. C. (1986) Oxygen free radicals and iron in relation to biology and medicine: Some problems and concepts. *Arch Biochem Biophys* **246:** 501–514.

Haliwell, B., and Gutteridge, J. M. C. (1989) *Free Radicals in Biology and Medicine.* Claredon Press, Oxford, UK.

Hall, A., Karplus, P. A., and Poole, L. B. (2009) Typical 2-Cys peroxiredoxins-structure, mechanisms and functions. *FEBS J* **276:** 2469–2477.

Hamano, T., Murakami, S., Takayama, K., Ehira, S., Maruyama, K., Kawakami, H., Morita, E. H., Hayashi, H., and Sato, N. (2004) Characterization of RNA-binding properties of three types of RNA-binding proteins in *Anabaena* sp. PCC 7120. *Cell Mol Biol* **50:** 613–624.

Hara, H., Yamamoto, Y., Higashitani, A., Suzuki, H., and Nishimura, Y. (1991) Cloning, mapping, and characterization of the *Escherichia coli prc* gene, which is involved in C-terminal processing of penicillin-binding protein 3. *J Bacteriol* **173:** 4799–4813.

Harel, Y., Ohad, I., and Kaplan, A. (2004) Activation of photosynthesis and resistance to photoinhibition in cyanobacteria within biological desert crust. *Plant Physiol* **136:** 3070–3079.

Haranczyk, H., Gazdzinski, S., and Olech, M. (1998) Initial stages of lichen hydration observed by proton magnetic relaxation. *New Phytol* **138:** 191–202.

Harrison, C. (2003) GrpE, a nucleotide exchange factor for DnaK. *Cell Stress Chaperones* **8:** 218–224.

Hartl, F. U., and Hayer-Hartl, M. (2002) Molecular chaperones in the cytosol: From nascent chain to folded protein. *Science* **295:** 1852– 1858.

Haslbeck, M., Walke, S., Stromer, T., Ehrnsperger, M., White, H. E., Chen, S., Saibil, H. R. and Buchner, J. (1999) Hsp26: a temperature-regulated chaperone. *EMBO J* **18:** 6744–6751.

Haussuhl, K., Andersson, B., and Adamska, I. (2001) A chloroplast DegP2 protease performs the primary cleavage of the photodamaged D1 protein in plant photosystem II. *EMBO J* **20:** 713–722.

Havaux, M., Guedeney, G., Hagemann, M., Yeremenko, N., Matthijs, H. C., and Jeanjean, R. (2005) The chlorophyll-binding protein IsiA is inducible by high light and protects the cyanobacterium *Synechocystis* PCC 6803 from photooxidative stress. *FEBS Lett* **579:** 2289–2293.

Hecker, M., Schumann, W., and Volker, U. (1996) Heat-shock and general stress response in *Bacillus subtilis. Mol Microbiol* **19:** 417–428.

Heering, H. A., Indiani, C., Regelsberger, G., Jalopitsch, C., Obinger, C., and Smulevich, G. (2002) New insights into the heme cavity structure of catalase-peroxidase, a spectroscopic approach to the recombinant *Synechocystis* enzyme and selected distal cavity mutants. *Biochemistry* **41:** 9237–9247.

Helm, R. F., Huang, Z., Edwards, D., Leeson, H., Peery, W., and Potts, M. (2000) Structural characterization of the released polysaccharide of desiccation-tolerant *Nostoc commune* DRH1. *J Bacteriol* **182:** 974–982.

Helmann, J. D., Wu, M. F., Kobel, P. A., Gamo, F. J., Wilson, M., Morshedi, M. M., Navre, M., and Paddon, C. (2001) Global transcriptional response of *Bacillus subtilis* to heat shock. *J Bacteriol* **183:** 7318–7328.

Helmann, J. D., Wu, M. F., Gaballa, A., Kobel, P. A., Morshedi, M. M., Fawcett, P., and Paddon, C. (2003) The global transcriptional response of *Bacillus subtilis* to peroxide stress is coordinated by three transcription factors. *J Bacteriol* **185:** 243–253.

Herbert, S. K., Samson, G., Fork, D. C., and Laudenbach, D. E. (1992) Characterization of damage to photosystems I and II in a cyanobacterium lacking detectable iron superoxide dismutase activity. *Proc Natl Acad Sci USA* **89:** 8716–8720.

Herbig, A. F., and Helmann, J. D. (2001) Roles of metal ions and hydrogen peroxide in modulating the interaction of the *Bacillus subtilis* PerR peroxide regulon repressor with operator DNA. *Mol Microbiol* **41:** 849–859.

Herman, C., Thevenet, D., D'Ari, R., and Bouloc, P. (1995) Degradation of sigma 32, the heat stress regulator in *Escherichia coli,* is governed by HflB. *Proc Natl Acad Sci USA* **92:** 3516–3520.

Herranen, M., Battchikova, N., Zhang, P., Graf, A., Sirpio, S., Paakkarinen, V., and Aro, E.-M. (2004) Towards functional proteomics of membrane protein complexes in *Synechocystis* sp. PCC 6803. *Plant Physiol* **134:** 470–481.

Hibino, T., Kaku, N., Yoshikawa, H., Takabe, T., and Takabe, T. (1999) Molecular characterization of DnaK from the halotolerant cyanobacterium *Aphanothece halophytica* for ATPase, protein folding and copper binding under various salinity conditions. *Plant Mol Biol* **40:** 409–418.

Higo, A., Ikeuchi, M., and Ohmori, M. (2008) cAMP regulates respiration and oxidative stress during rehydration in *Anabaena* sp. PCC 7120. *FEBS Lett* **582:** 1883–1888.

Higo, A., Katoh, H., Ohmori, K., Ikeuchi, M., and Ohmori, M. (2006) The role of a gene cluster for trehalose metabolism in dehydration tolerance of the filamentous cyanobacterium *Anabaena* sp. PCC 7120. *Microbiology* **152:** 979–987.

Higo, A., Suzuki, T., Ikeuchi, M., and Ohmori, M. (2007) Dynamic transcriptional changes in response to rehydration in *Anabaena* sp. PCC 7120. *Microbiology* **153:** 3685–3694. doi:10.1099/mic.0.2007/009223-0

Higgins, C. F., Dorman, C. J., Stirling, D. A., Waddell, L., Booth, I. R., May, G., and Bremer, E. A. (1988) Physiological role for DNA supercoiling in the osmotic regulation of gene expression in *S. typhimurium* and *E. coli. Cell* **52:** 569–584.

Hihara, Y., Kamei, A., Kanehisa, M., Kaplan, A., and Ikeuchi, M. (2001) DNA microarray analysis of cyanobacterial gene expression during acclimation to high light. *Plant Cell* **13:** 793–806.

Hill, D. R, Hladun, S. L., Scherer, S., and Potts, M. (1994a) Water stress proteins of *Nostoc commune* (Cyanobacteria) are secreted with UV-A/B-absorbing pigments and associate with 1,4-β-D-xylanxylanohydrolase activity. *J Biol Chem* **269:** 7726–7734.

Hill, D. R., Keenan, T. W., Helm, R. F., Potts, M., Crowe, L. M., and Crowe, J. H. (1999) Extracellular polysaccharide of *Nostoc commune* (cyanobacteria) inhibits fusion of membrane vesicles during desiccation. *J Appl Phycol* **9:** 237–248.

Hill, D. R., Peat, A., and Potts, M. (1994b) Biochemistry and structure of glycan secreted by desiccation-tolerant *Nostoc commune* (cyanobacteria). *Protoplasma* **182:** 126–148.

Hill, J. H., Penny, S. L., Crowell, K. G., Goh, S. H., and Hemmingsen, S. M. (2004) CpnDB: A chaperonin sequence database. *Genome Res* **14:** 1669–1675. www.genome.org

Hincha, D. K., and Hagemann, M. (2004) Stabilization of model membranes during drying by compatible solutes involved in the stress tolerance of plants and microorganisms. *Biochem J* **383:** 277–283.

Hincha, D. K., Heber, U., and Schmitt, J. M. (1985) Antibodies against individual thylakoid membrane proteins as molecular probes to study chemical and mechanical freezing damage *in vitro. Biochim Biophys Acta* **809:** 337–344.

Hirani, T. A., Suzuki, I., Murata, N., Hayashi, H., and Eaton-Rye, J. J. (2001) Characterization of a two-component signal transduction system involved in the induction of alkaline phosphatase under phosphate-limiting conditions in *Synechocystis* sp. PCC 6803. *Plant Mol Biol* **45:** 133–144.

Hirotsu, S., Abe, Y., Okada, K., Nagahara, N., Hori, H., Nishino, T., and Hakoshima, T. (1999) Crystal structure of a multifunctional 2-Cys peroxiredoxin heme-binding protein 23 kDa/proliferation-associated gene product. *Proc Natl Acad Sci USA* **96:** 12333–12338.

Hisabori, T., Hara, S., Fujii, T., Yamazaki, D., Hosoya-Matsuda, N., and Motohashi, K. (2005) Thioredoxin affinity chromatography: a useful method for further understanding the thioredoxin network. *J Exp Bot* **56:** 1463–1468.

Hishiya, S., Hatakeyama, W., Mizota, Y., Hosoya-Matsuda, N., Motohashi, K., Ikeuchi, M., and Hisabori, T. (2008) Binary reducing equivalents using NADPH: Thioredoxin reductase and ferredoxin-thioredoxin reductase in the cyanobacterium *Synechocystis* sp. strain PCC 6803. *Plant Cell Physiol* **49:** 11–18.

Hochman A., and Shemesh A. (1987) Purification and characterization of a catalase-peroxidase from the photosynthetic bacterium *Rhodopseudomonas capsulata*. *J Biol Chem* **262:** 6871–6876.

Hoekstra, F. A., Crowe, J. H., Crowe, L. M., van Roekel, T., and Vermeer, E. (1992) Do phospholipids and sucrose determine membrane phase transitions in dehydrating pollen species? *Plant Cell Environ* **15:** 601–606.

Holmgren, A. (1976) Hydrogen donor system for *Escherichia coli* ribonucleoside diphosphate reductase dependent upon glutathione. *Proc Natl Acad Sci USA* **73:** 2275–2279.

Holmgren, A. (1989) Thioredoxin and glutaredoxin systems. *Biol Chem* **264:** 13963–13966.

Holmgren, A., and Aslund, F. (1995) Glutaredoxin. *Methods Enzymol* **252:** 283–292.

Hondorp, E. R., and Matthews, R. G. (2004) Oxidative stress inactivates cobalamin-independent methionine synthase (MetE) in *Escherichia coli*. *PLoS Biol* **2:** 1738–1753.

Hong, S.-K., Cha, M.-K., and Kim, I.-H. (2008) A glutaredoxin-fused thiol peroxidase acts as an important player in hydrogen peroxide detoxification in late-phased growth of *Anabaena* sp. PCC7120. *Arch Biochem Biophys* **475:** 42–49.

Hong, S. W., and Vierling, E. (2000) Mutants of *Arabidopsis thaliana* defective in the acquisition of tolerance to high temperature stress. *Proc Natl Acad Sci USA* **97:** 4392–4397.

Hongsthong, A., Sirijuntarut, M., Yutthanasirikul, R., Senachak, J., Kurdrid, P., Chaevadhanarak, S., and Tanticharoen, M. (2009) Sub-cellular proteomic characterization of the high-temperature stress response of the cyanobacterium *Spirulina platensis*. *Proteome Sci* **7:** 33. doi:10.1186/1477-5966-7-33

Horiuchi, M., Nakamura, K., Koima, K., Nishiyama, Y., Hatakeyama, W., Hisabori, T., and Hihara, Y. (2010) The PedR transcriptional regulator interacts with thioredoxin to connect photosynthesis with gene expression in cyanobacteria. *Biochem J* **431:** 135–140.

Horlacher, R., and Boss, W. (1997) Characterization of TreR, the major regulator of the *Escherichia coli* trehalose system. *J Biol Chem* **272:** 13026–13032.

Horváth, I., Glatz, A., Varvasovszki, V., Török, Z., Páli, T., Balogh, G., Kovács, E., Nádasdi, L., Benkö, S., Joó, F., and Vígh, L. (1998) Membrane physical state controls the signaling mechanism of the heat shock response in *Synechocystis* PCC 6803: Identification of Hsp17 as a "fluidity gene". *Proc Natl Acad Sci USA* **95:** 3513–3518.

Hoskins, J. R. Pak, M., Maurizi, M. R., and Wickner, S. (1998) The role of ClpA chaperone in proteolysis by ClpAP. *Proc Natl Acad Sci USA* **95:** 12135–12140.

Hosoya-Matsuda, N., Motohashi, K., Yoshimura, H., Nozaki, A., Inoue, K., Ohmori M., and Hisabori, T. (2005) Anti-oxidative stress system in cyanobacteria. Significance of type II peroxiredoxin and the role of 1-Cys peroxiredoxin in *Synechocystis* sp. strain PCC 6803. *J Biol Chem* **280:** 840–846.

Hossain, M. M., and Nakamoto, H. (2002) HtpG plays a role in cold acclimation in cyanobacteria. *Curr Microbiol* **44:** 291–296.

Hossain, M. M., and Nakamoto, H. (2003) Role for the cyanobacterial HtpG in protection from oxidative stress. *Curr Microbiol* **46:** 70–76.

Hottiger, T., De Virgilio, C., Hall, N. M., Boller, T., and Wiemken, A. (1994) The role of trehalose synthesis for the acquisition of thermotolerance in yeast. II. Physiological concentrations of trehalose increase the thermal stability of proteins *in vitro*. *Eur J Biochem* **219:** 187–193.

Hounsa, C. G., Brandt, E.V., Thevelein, J., Hohmann, S., and Prior, B. A. (1998) Role of trehalose in survival of *Saccharomyces cerevisiae* under osmotic stress. *Microbiology* **144:** 671–680

Huang, F., Fulda, S., Hagemann, M., and Norling, B. (2006) Proteomic screening of salt-stress induced changes in plasma membranes of *Synechocystis* sp. strain PCC 6803. *Proteomics* **6:** 910–920.

Huang, L., McCluskey, M. P., Ni, H., and LaRossa. R. A. (2002) Global gene expression profiles of the cyanobacterium *Synechocystis* sp. strain PCC 6803 in response to irradiation with UV-B and white light. *J Bacteriol* **184:** 6845–6858.

Huang, Z. B., Liu, Y. D., Paulsen, B. S., and Klaveness, D. (1998) Studies on polysaccharides from three edible species of *Nostoc* (Cyanobacteria) with different colony morphologies: Comparison of monosaccharide compositions and viscosities of polysaccharides from field colonies and suspension cultures. *J Phycol* **34:** 962–968.

Huckauf, J., Nomura, C., Forchhammer, K., and Hagemann, M. (2000) Stress responses of *Synechocystis* sp. strain PCC 6803 mutants impaired in genes encoding putative alternative sigma factors. *Microbiology* **146:** 2877–2889.

Huesgen, P. F., Scholz, P., and Adamska, I. (2007) The serine protease HhoA from *Synechocystis* sp. strain PCC 6803: Substrate specificity and formation of a hexameric complex are regulated by the PDZ domain. *J Bacteriol* **189:** 6611–6618.

Huesgen, P. F., Schuhmann, H., and Adamska, I. (2005) The family of Deg proteases in cyanobacteria and chloroplasts of higher plants. *Physiol Plant* **123:** 413–420.

Huesgen, P. F., Schuman, H., and Adamska, I. (2009) Deg/HtrA proteases as components of a network for photosystem II quality control in chloroplasts and cyanobacteria. *Res Microbiol* **160:** 726–732. doi:10.1016/j.resmic.2009.08.005

Huflejt, M., Tremolieres, A., Pineau, B., Lang, J., Hatheway, J., and Packer, L. (1990) Changes in membrane lipid composition during saline growth of the freshwater cyanobacterium *Synechococcus* 6311. *Plant Physiol* **94:** 1512–1521.

Huner, N. P. A., Maxwell, D. P., Gray, G. R., Savitch, L. V., Krol, M., Ivanov, A. G., and Falk, S. (1996) Sensing environmental temperature change through imbalances between energy supply and energy consumption: Redox state of photosystem II. *Physiol Plant* **98:** 358–364.

Hwang, B. J., Park, W. J., Chung, C. H., and Goldberg, A. L. (1987) *Escherichia coli* contains a soluble ATP-dependent protease (Ti) distinct from protease La. *Proc Natl Acad Sci USA* **84:** 5550–5554.

Ihalainen, J. A., D'Haene, S., Yeremenko, N., van Roon, H., Arteni, A. A., Boekema, E. J., van Grondelle, R., Matthijs, H. C., and Dekker, J. P. (2005) Aggregates of the chlorophyll-binding protein IsiA (CP43′) dissipate energy in cyanobacteria. *Biochemistry* **44:** 10846–10853.

Imlay, J. A. (2003) Pathways of oxidative damage. *Annu Rev Microbiol* **57:** 395–418.

Inagaki, N., Maitra, R., Satoh, K., and Pakrasi, H. B. (2001a) Amino acid residues that are critical for *in vivo* catalytic activity of CtpA, the carboxy-terminal processing protease for the D1 protein of photosystem II. *J Biol Chem* **276:** 30099–30105.

Inagaki, N., Yamamoto, Y., and Satoh, K. (2001b) A sequential two-step proteolytic process in the carboxyl-terminal truncation of precursor D1 protein in *Synechocystis* sp. PCC 6803. *FEBS Lett* **509:** 197–201.

Incharoensakdi, A., and Kum-Arb, U. (1998) Betaine aldehyde dehydrogenase from a halotolerant cyanobacterium *Aphanothece halophytica*: Purification, properties and regulation by salinity. *J Sci Soc Thailand* **24:** 234–240.

Incharoensakdi, A., Takabe, T., and Akazawa, T. (1986) Effect of betaine on enzyme activity and subunit interaction of ribulose 1,5-bisphosphate carboxylase/oxygenase from *Aphanothece halophytica. Plant Physiol* **81:** 1044–1049.

Ishizaki-Nishizawa, O., Fujii, T., Azuma, M., Sekiguchi, K., Murata, N., Ohtani, T., and Toguri, T. (1996) Low-temperature resistance of higher plants is significantly enhanced by a nonspecific cyanobacterial desaturase. *Nat Biotechnol* **14:** 1003–1006.

Ivancich, A., Jakopitsch, C., Auer, M., Un, S., and Obinger, C. (2003) Protein-based radicals in catalase-peroxidase of *Synechocystis* sp. PCC 6803. A multifrequency EPR investigation of wild-type and variants on the environment of heme active site. *J Am Chem Soc* **125:** 14093–14102.

Ivleva, N., Sidoruk, K., Pakrasi, H. B., Shestakov, S.V. (2002) Study of functional roles of Ctp-family proteins in the cyanobacterium *Synechocystis* 6803. *Microbiology* **71:** 509–513.

Jacquot, J. P., Lancelin, J. M., and Meyer, Y. (1997) Thioredoxins: Structure and function in plant cells. *New Phytol* **136:** 543–570.

Jäger, K., and Potts, M. (1988) Distinct fractions of genomic DNA from cyanobacterium *Nostoc commune* that differ in the degree of methylation. *Gene* **74:** 197–201.

Jaki, B., Heilmann, J., and Sticher, O. (2000) New antibacterial metabolites from the cyanobacterium *Nostoc commune* (EAWAG 122b). *J Nat Prod* **63:** 1283–1285.

Jakopitsch, C., Auer, M., Regelsberger, G., Jantschko, W., Furtmüller, P. G., Rüker, F., and Obinger, C. (2003a) Distal site aspartate is essential in the catalase-peroxidase. *Biochemistry* **42:** 5292–5300.

Jakopitsch, C., Auer, M., Regelsberger, G., Jantschko, W., Furtmüller, P. G., Rücker, F., and Obinger, C. (2003b) The catalytic role of the distal site asparagine-histidine couple in catalase-peroxidases. *Eur J Biochem* **270:** 1006–1013.

Jakopitsch, C., Auer, M., Ivancich, A., Rücker, F., Furtmüller, P. G., and Obinger, C. (2003c) Total conversion of bifunctional catalase-peroxidase (KatG) to monofunctional peroxidase by exchange of a conserved distal side tyrosine. *J Biol Chem* **278:** 20185–20191.

Jakopitsch, C., Ivancich, A., Schmuckenschlager, F., Wanasinghe, A., Pöltl, G., Furtmüller, P. G., Rücker, F., and Obinger, C. (2004) Influence of the unusual covalent adduct on the kinetics and formation of radical intermediates in *Synechocystis* catalase-peroxidase: A stopped-flow and EPR characterization of the Met^{275}, Tyr^{249}, and Arg^{439} variants. *J Biol Chem* **279:** 46082–46095.

Jakopitsch, C., Kolarich, D., Petutschnig, G., Furtmuller, P. G., and Obinger, C. (2003d) Distal side tryptophan, tyrosine and methionine in catalase-peroxidases are covalently linked in solution. *FEBS Lett* **552:** 135–140.

Jakopitsch C., Regelsberger G., Furtmüller P. G, Rüker F., Peschek G. A., and Obinger C. (2001) Catalase-peroxidase from *Synechocystis* is capable of chlorination and bromination reactions. *Biochem Biophys Res Commun* **28:** 682–687.

Jakopitsch, C., Rüker, F., Regelsberger, G., Dockal, F., Peschek, G. A., and Obinger, C. (1999) Catalase-peroxidase from the cyanobacterium *Synechocystis* PCC 6803: Cloning, overexpression in *Escherichia coli*, and kinetic characterization. *Biol Chem* **380:** 1087–1096.

Jakopitsch, C., Vlasits, J., Wiseman, B., Loewen, P. C., and Obinger, C. (2007) Redox intermediates in the catalase cycle of catalase-peroxidases from *Synechocystis* sp. PCC 6803, *Burkholderia pseudomallei* and *Mycobacterium tuberculosis*. *Biochemistry* **46:** 1183–1193.

Jakopitsch, C., Wanasinghe, A., Jantschko, W., Furtmüller, P. G., and Obinger, C. (2005) Kinetics of interconversion of ferrous enzymes, compound II and compound III, wild-type *Synechocystis* catalase-peroxidase and Y249F. Proposal for the catalytic mechanism. *J Biol Chem* **280:** 9037–9042.

Jansen, T., Kidron, H., Taipaleenmaki, H., Salminen, T., and Maenpaa, P. (2005) Transcriptional profiles and structural models of the *Synechocystis* sp. PCC 6803 Deg proteases. *Photosynth Res* **84:** 57–63.

Jansz, E.R., and Maclean, F. I. (1973) The effect of cold shock on the blue-green alga *Anacystis nidulans. Can J Microbiol* **19:** 381–387.

Jarnagin, R. C., and Wang, J. H. (1958) Investigation of the catalytic mechanism of catalase and other ferric compounds with doubly ^{18}O-labeled hydrogen peroxide. *J Am Chem Soc* **80:** 786–787.

Jeanjean, R., Matthijs, H. C. P., Onana, B., Havaux, M., and Joset, F. (1993) Exposure of the cyanobacterium *Synechocystis* PCC 6803 to salt stress induces concerted changes in respiration and photosynthesis. *Plant Cell Physiol* **34:** 1073–1079.

Jeanjean, R., Zuther, E., Yeremenko, N., Havaux, M., Matthijs, H. C., and Hagemann, M. (2003) A photosystem 1 *psaFJ*-null mutant of the cyanobacterium *Synechocystis* PCC 6803 expresses the *isiAB* operon under iron replete conditions. *FEBS Lett* **549:** 52–56.

Jentsch, S. (1996) When proteins receive deadly messages at birth. *Science* **271:** 955–956.

Jiang, W., Hou, Y., and Inouye, M. (1997) CspA, the major cold-shock protein of *Escherichia coli*, is an RNA chaperone. *J Biol Chem* **272:** 196–202.

Jones, P. G., and Inouye, M. (1994) The cold-shock response-a hot topic. *Mol Microbiol* **11:** 811–818.

Jones, P. G., Cashel, M., Glaser, G., and Neidhardt, F. C. (1992) Function of a relaxed-like state following temperature downshifts in *Escherichia coli. J Bacteriol* **174:** 3909–3914.

Jones, P. G., Mitta, M., Kim, Y., Jiang, W., and Inouye, M. (1996) Cold shock induces a major ribosomal-associated protein that unwinds double-stranded RNA in *Escherichia coli. Proc Natl Acad Sci USA* **93:** 76–80.

Joset, F., Jeanjean, R., and Hagemann, M. (1996) Dynamics of the response of cyanobacteria to salt stress: Deciphering the molecular events. *Physiol Plant* **96:** 738–744.

Kaku, N., Hibino, T., Tanaka, Y., Ishikawa, H., Araki, E., Takabe, T., and Takabe, T. (2000) Effects of overexpression of *Escherichia coli* katE and *bet* genes on the tolerance for salt stress in a freshwater cyanobacterium *Synechococcus* sp. PCC 7942. *Plant Sci* **159:** 281–288

Kaneko, T., Sato, S., Kotani, H., Tanaka, A., Asamizu, E., Nakamura, Y., Miyajima, N., Hirosawa, M., Sugiura, M., Sasamoto, S., Kimura, T., Hosouchi, T., Matsuno, A., Muraki, A., Nakazaki, N., Naruo, K., Okumura, S., Shimpo,S., Takeuchi, C., Wada, T.,Watanabe, A., Yamada, M., Yasuda, M., and Tabata, S. (1996) Sequence analysis of the genome of the unicellular cyanobacterium *Synechocystis* sp. strain PCC6803. II. Sequence determination of the entire genome and assignment of potential protein-coding regions. *DNA Res* **3:** 109–136.

Kanesaki, Y., Suzuki, I., Allakhverdiev, S. I., Mikami, K., and Murata, N. (2002) Salt stress and hyperosmotic stress regulate the expression of different sets of genes in *Synechocystis* sp. strain PCC 6803. *Biochem Biophys Res Commun* **290:** 339–348.

Kanesaki, Y., Yamamoto, H., Paithoonrangsarid, K., Shoumskaya, M., Suzuki, I., Hayashi, H., and Murata, N. (2007) Histidine kinases play important roles in the perception and signal transduction of hydrogen peroxide in the cyanobacterium *Synechocystis* sp. PCC 6803. *Plant J* **49:** 313–324.

Kappell, A. D., and van Waasbergen, L. G. (2007) The response regulator RpaB binds the high light regulatory 1 sequence upstream of the high-light-inducible *hliB* gene from the cyanobacterium *Synechocystis* PCC 6803. *Arch Microbiol* **187:** 337–342.

Karata, K., Inagawa, T., Wilkinson, A. J., Tatsuta, T., and Ogura, T. (1999) Dissecting the role of a conserved motif (the second region of homologue) in the AAA family of ATPases. Site directed mutagenesis of the ATP-dependent peptidase FtsH. *J Biol Chem* **274:** 26225–26232.

Karlson, D., and Imai, R. (2003) Conservation of cold shock domain protein family in plants. *Plant Physiol* **131:** 12–15.

Karradt, A., Sobanski, J., Mattow, J., Lockau, W., and Baier, K. (2008) NblA, a key protein of phycobilisome degradation, interacts with ClpC, aHSP100 chaperone partner of a cyanobacterial Clp protease. *J Biol Chem* **283:** 32394–32403.

Katano, Y., Nimura-Matsune, K., and Yoshikawa, H. (2006) Involvement of DnaK3, one of the three DnaK proteins of cyanobacterium *Synechococcus* sp. PCC 7942, in translational process on the surface of thylakoid membrane. *Biosci Biotechnol Biochem* **70:** 1592–1598.

Katayama-Fujimura, Y. S., Gottesman, S., and Maurizi, M. R. (1987) A multiple-component, ATP-dependent protease from *Escherichia coli. J Biol Chem* **262:** 4477–4485.

Kato, S., Ueno, T., Fukuzumi, S., and Watanabe, Y. (2004) Catalase reaction by myoglobin mutants and native catalase: mechanistic investigation by kinetic isotope effect. *J Biol Chem* **279:** 52376–52381.

Katoh, H., Asthana, R. K., and Ohmori, M. (2004) Gene expression in the cyanobacterium *Anabaena* sp. PCC 7120 under desiccation. *Microb Ecol* **47:** 164–174.

Kedar, L., Kashman, Y., and Oren, A. (2002) Mycosporine-2-glycine is the major mycosporine-like amino acid in a unicellular cyanobacterium (*Euhalothece* sp.) isolated from a gypsum crust in a hypersaline saltern pond. *FEMS Microbiol Lett* **208:** 233–237.

Keiler, K. C., Waller, P. R., and Sauer, R. T. (1996) Role of a peptide tagging system in degradation of proteins synthesized from damaged messenger RNA. *Science* **271:** 990–993.

Kenny, A. J. (1999) Nomenclature and Classes of Peptidases. In: Sterchi, E. E., and Stöcker, W. (Eds.) *Proteolytic Enzymes. Tools and Targets*, Springer Verlag, Heidelberg, Germany. pp 1–8.

Keren, N., Aurora, R., and Pakrasi, H. B. (2004) Critical roles of bacterioferritins in iron storage and proliferation of cyanobacteria. *Plant Physiol* **135:** 1666–1673.

Kessel, M., Maurizi, M. R., Kim, B., Kocsis, E., Trus, B. L., Singh, S. K., and Steven, A. C. (1995) Homology in structural organization between *E. coli* ClpAP protease and the eukaryotic 26S proteosome. *J Mol Biol* **250:** 587–594.

Khamutov, G., Fry, I. V., Huflejt, M. E., and Packer, L. (1990) Membrane lipid composition, fluidity, and surface charge changes in response to growth of the freshwater cyanobacterium *Synechococcus* 6311 under high salinity. *Arch Biochem Biophys* **277:** 263–267.

Kieselbach, T., and Funk, C. (2003) The family of Deg/HtrA proteases: From *Escherichia coli* to *Arabidopsis*. *Physiol Plant* **119:** 337–346.

Kihara, A., Akiyama, Y., and Ito, K. (1995) FtsH is required for proteolytic elimination of uncomplexed forms of SecY, an essential protein translocase subunit. *Proc Natl Acad Sci USA* **92:** 4532–4536.

Kim, K. I., Park, S. C., Kang, S. H, Cheong, G. W., and Chung, C. H. (1999) Selective degradation of unfolded proteins by the self-compartmentalizing HtrA peptidase, a periplasmic heat stress protein in *Escherichia coli*. *J Mol Biol* **294:** 1363–1374.

Kim, S. W., Choi, I. H., Kim, S. N., Kim, Y. H., Pyo, S. N., and Rhee, D. K. (1998). Molecular cloning, expression, and characterization of *dnaK* in *Streptococcus pneumoniae*. *FEMS Microbiol Lett* **161:** 217–224.

Kis, M., Zsiros, O., Farkas, T., Wada, H. Nagy, F., and Gombos, Z. (1998) Light-induced expression of fatty acid desaturase genes. *Proc Natl Acad Sci USA* **95:** 4209–4214.

Kiseleva, L. L., Serebriiskaya, T. S., Horváth, I., Vigh, L., Lyukevich, A. A., and Losa, D. A. (2000) Expression of the gene for the Δ^9 acyl-lipid desaturase in the thermophilic cyanobacterium. *J Mol Microbiol Biotechnol* **2:** 331–338.

Klähn, S., and Hagemann, M. (2011) Compatible solute biosynthesis in cyanobacteria. *Environ Microbiol* **13:** 551–562. doi: 10. 1111/j.1462-2920.2

Klughammer, B., Baier, M., and Dietz, K. J. (1998) Inactivation by gene disruption of 2-cysteine-peroxiredoxin in *Synechocystis* sp. PCC 6803 leads to increased stress sensitivity. *Physiol Plant* **104:** 699–706.

Knowles, E. J., and Castenholz, R. W. (2008) Effect of exogenous extracellular polysaccharides on the desiccation and freezing tolerance of rock-inhabiting phototrophic microorganisms. *FEMS Microbiol Ecol* **66:** 261–270.

Kobayashi, M., Ishizuka, T., Katayama, M., Kanehisa, M., Bhattacharyya-Pakrasi, M., Pakrasi, H. B., and Ikeuchi, M. (2004) Response to oxidative stress involves a novel peroxiredoxin gene in the unicellular cyanobacterium *Synechocystis* sp. PCC 6803. *Plant Cell Physiol* **45:** 290–299.

Kojima, K., and Nakamoto, H. (2002) Specific binding of a protein to a novel DNA element in the cyanobacterial small heat-shock protein gene. *Biochem Biophys Res Commun* **297:** 616–624.

Kojima, K., and Nakamoto, H. (2005) Post-transcriptional control of the *hspA* heat-shock induction. *Biochem Biophys Res Commun* **331:** 583–588.

Kojima, K., Suzuki-Maenaka, T., Kikuchi, T., and Nakamoto, H. (2006) Roles of *isiABC* operon in protection from oxidative and heat stresses. *Physiol Plant* **128:** 507–519.

Kolman, M. A., Torres, L. L., Martin, M. L., and Salerno, G. L. (2012) Sucrose synthase in unicellular cyanobacteria and its relationship with salt and hypoxic stress. *Planta*: **235:** 955–964. doi:10. 1007/s 00425-011-1542-5

Komenda, J., Barker, M., Kuvikova, S., de Vries, R., Mullineaux, C. W., Tich, M., and Nixon, P. J. (2006) The FtsH protease slr0228 is important for quality control of photosystem II in the thylakoid membrane of *Synechocystis* sp PCC 6803. *J Biol Chem* **281:** 1145–1151.

Komenda, J., Knoppová, J., Krynická, V., Nixon, P. J., and Tichy, M. (2010) Role of FtsH2 in the repair of photosystem II in mutants of the cyanobacterium *Synechocystis* PCC 6803 with impaired assembly or stability of the $CaMn_4$ cluster. *Biochim Biophys Acta* **1797:** 566–575. doi:10.1016/ j.bbabio.2010.02.006

Kouril, R., Arteni, A. A., Lax, J., Yeremenko, N., D'Haene, S., Rögner, M., Matthijs, H. C., Dekker, J. P., and Boekema, E. J. (2005) Structure and functional role of supercomplexes of IsiA and Photosystem I in cyanobacterial photosynthesis. *FEBS Lett* **579:** 3253–3257.

Krojer, T., Garrido-Franco, M., Huber, R., Ehrmann, M., and Clausen, T. (2002) Crystal structure of DegP (HtrA) reveals a new protease-chaperone machine. *Nature* (London) **416:** 455–459.

Kujat, S. L., and Owttrim, G.W. (2000) Redox-regulated RNA helicase expression. *Plant Physiol* **124:** 703–713.

Kulkarni, R. D., and Golden, S. S. (1997) mRNA stability is regulated by a coding-region element and the unique 59-untranslated leader sequences of the three *Synechococcus psbA* transcripts. *Mol Microbiol* **24:** 1131–1142.

Kulkarni, R. D, Schaefer, M. R., and Golden, S. S. (1992) Transcriptional and posttranscriptional components of *psbA* response to high light intensity in *Synechococcus* sp. strain PCC 7942. *J Bacteriol* **174:** 3775–3781.

Kuo, C. F., Mashino, T., and Fridovich, I. (1987) α,β-Dihydroxyisovalerate dehydratase superoxide. A sensitive enzyme. *J Biol Chem* **262:** 4724–4727

Lacerda, C. M. R., Choe, L. H., and Reardon, K. F. (2007) Metaproteomics analysis of a bacterial community response to cadmium exposure. *J Proteome Res* **6:** 1145–1152.

Lamark, T., Kaasen, I., Eshoo, M. W., Falkenberg, P., McDougall, J., and Strom, A. R. (1991) DNA sequence and analysis of five *bet* genes encoding the osmoregulatory choline-glycine betaine pathway of *Escherichia coli. Mol Microbiol* **5:** 1049–1064.

Lara, C., Rodrŕguez, R., and Guerrero, M. G. (1993). Sodium-dependent nitrate transport and energetics of cyanobacteria. *J Phycol* **29:** 389–395.

Laloknam, S., Tanaka, K., Buaboocha, T., Waditee, R., Incharoensakdi, A., Hibino, T., Tanaka, Y., and Takabe, T. (2006) Halotolerant cyanobacterium *Aphanothece halophytica* contains a betaine transporter active at alkaline pH and high salinity. *Appl Environ Microbiol* **72:** 6018–6026.

Landfald, B., and Strom, A. R. (1986) Choline-glycine betaine pathway confers a high level of osmotic tolerance in *Eschcrichia coli. J Bacteriol* **165:** 849–855.

La Teana, A., Brandi, A., Falconi, M., Spurion, R., Pon, C., and Gualerzi, C. O. (1991) Identification of a cold shock transcriptional enhancer of the *Escherichia coli* gene encoding nucleoid protein H-NS. *Proc Natl Acad Sci USA* **88:** 10907–10911.

Latifi, A., Jeanjean, R., Lemeille, S., Havaux, M., and Zhang, C.-C. (2005) Iron starvation leads to oxidative stress in *Anabaena* sp. strain PCC 7120. *J Bacteriol* **187:** 6596–6598.

Latifi, A., Ruiz, M., and Zhang, C.-C. (2009) Oxidative stress in cyanobacteria. *FEMS Microbiol Rev* **33:** 258–278.

Latifi, A., Ruiz, M., Jeanjean, R., and Zhang, C. C. (2007) PrxQ-A, a member of the peroxiredoxin Q family, plays a major role in defense against oxidative stress in the cyanobacterium *Anabaena* sp. strain PCC7120. *Free Radic Biol Med* **42:** 424–431.

Laudenbach, D. E., and Straus, N. A. (1988) Characterization of a cyanobacterial iron stress-induced gene similar to *psbC. J Bacteriol* **170:** 5018–5026.

Laudenbach, D. E, Reith, M. E., and Straus, N. A. (1988) Isolation, sequence analysis, and transcriptional studies of the flavodoxin gene from *Anacystis nidulans* R2. *J Bacteriol* **170:** 258–265.

Laudenbach, D. E., Trick, C. G., and Straus, N. A. (1989) Cloning and characterization of an *Anacystis nidulans* R2 superoxide dismutase gene. *Mol Gen Genet* **216:** 455–461.

Lee, B. H., Hibino, T., Jo, J., Viale, A. M., and Takabe, T. (1997) Isolation and characterization of a *dnaK* genomic locus in a halotolerant cyanobacterium *Aphanothece halophytica. Plant Mol Biol* **35:** 763–775.

Lehel, C., Los, D., Wada, H., Györgyei, J., Horváth, I., Kovács, E., Murata, N., and Vigh, L. (1993) A second GroEL-like gene organized in a *groEL*-operon is present in the genome of *Synechocystis* sp. PCC 6803. *J Biol Chem* **263:** 1799–1804.

Lehel, C., Wada, H., Török, Z., Gombos, Z., Murata, N., and Vigh, L. (1992) Heat shock protein synthesis of the cyanobacterium *Synechocystis* PCC 6803. Purification of the GroEL-related chaperonin. *Plant Mol Biol* **18:** 327–336.

Lelivelt, M. J., and Kawula, T. H. (1995) Hsc66, an Hsp70 homolog in *Escherichia coli*, is induced by cold shock but not by heat shock. *J Bacteriol* **177:** 4900–4907.

Lensch, M., Herrmann, R. G., and Sokolenko, A. (2001) Identification and characterization of SppA, a novel light-inducible chloroplast peptidase complex associated with thylakoid membranes. *J Biol Chem* **276:** 33645–33651.

Leonhardt, K., and Straus, N. A. (1992) An iron stress operon involved in photosynthetic electron transport in the marine cyanobacterium *Synechococcus* sp. PCC 7002. *J Gen Microbiol* **138:** 1613–1621.

Leonhardt, K., and Straus, N. A. (1994) Photosystem II genes *isiA*, *psbDI* and *psbC* in *Anabaena* sp. PCC 7120: Cloning, sequencing and the transcriptional regulation in iron-stressed and iron-repleted cells. *Plant Mol Biol* **24:** 63–73.

Leonhardt, S. A., Fearon, K., Danese, P. N., and Mason, T. L. (1993) Hsp78 encodes a yeast mitochondrial heat shock protein in the Clp family of ATP-dependent proteases. *Mol Cell Biol* **13:** 6304–6313.

Leslie, S. B., Israeli, E., Lighthart, B., Crowe, J. H., and Crowe, L. M. (1995) Trehalose and sucrose protect both membranes and proteins in intact bacteria during drying. *Appl Environ Microbiol* **61:** 3592–3597.

Leslie, S. B., Teter, S. A., Crowe, L. M., and Crowe, J. H. (1994) Trehalose lowers membrane phase transitions in dry yeast cells. *Biochim Biophys Acta* **1192:** 7–13.

Lewin A, Moore GR, Le Brun NE (2005) Formation of protein-coated iron minerals. *Dalton Trans* 3597–3610.

Li, H., and Sherman, L. A. (2000) A redox-responsive regulator of photosynthesis gene expression in the cyanobacterium *Synechocystis* sp. strain PCC 6803. *J Bacteriol* **182:** 4268–4277.

Li, H., Singh, A. K., McIntyre, L. M., and Sherman, L. A. (2004) Differential gene expression in response to hydrogen peroxide and the putative PerR regulon of *Synechocystis* sp. strain PCC 6803. *J Bacteriol* **186:** 3331–3345.

Li, M., Huang, W., Yang, G., Liu, X., and Wu, Q. (2005) Expression and oxidative stress tolerance studies of glutaredoxin from cyanobacterium *Synechocystis* sp. PCC 6803 in *Escherichia coli. Protein Expr Purif* **42:** 85–91.

Li, M., Yang, Q., Zhang, L., Li, H., Cui, Y., and Wu, Q. (2007) Identification of novel targets of cyanobacterial glutaredoxin. *Arch Biochem Biophys* **458:** 220–228.

Li, T., Huang, X., Zhou, R., Liu, Y., Li, B., Nomura, C., and Zhao, J. (2002) Differential expression and localization of Mn and Fe superoxide dismutases in the heterocystous cyanobacterium *Anabaena* sp. strain PCC 7120. *J Bacteriol* **184:** 5096–5103.

Li, T., Yang, H. M., Cui, S. X., Suzuki, I., Zhang, L. F., Li, L., Bo, T., Wang, J., Murata, N., and Huang, F. (2012) Proteomic study of the impact of Hik33 mutation in *Synechocystis* sp. PCC 6803 under normal and salt stress conditions. *J Proteome Res* **11:** 502–514.

Liang, C., Zhang, X., Chi, X., Guan, X., Li, Y., Qin, S., and bo Shao, H. (2011) Serine/Threonine protein kinase SpkG is a candidate for high salt resistance in the unicellular cyanobacterium *Synechocystis* sp. PCC 6803. *PLoS ONE* **6:** e18718. doi:10.137/journal.pone.0018718

Liao, D. I., Qian, J., Chisholm, D. A., Jordan, D. B., and Diner, B. A. (2000) Crystal structures of the photosystem II D1 C-terminal processing protease. *Nat Struct Biol* **7:** 749–753.

Liberek, K., Marszalek, J., Ang, D., Georgopoulos, C., and Zylicz, M. (1991). *Escherichia coli* DnaJ and GrpE heat shock proteins jointly stimulate ATPase activity of DnaK *Proc Natl Acad Sci USA* **88:** 2874–2878.

Lin, Y., Hirai, M., Kashino, Y., Koike, H., Tuzi, S., and Satoh, K. (2004) Tolerance to freezing stress in cyanobacteria, *Nostoc commune* and some cyanobacteria with varying tolerances to drying stress. *Polar Biosci* **17:** 56–68.

Lindahl, M., and Florencio, F. J. (2003) Thioredoxin-linked processes in cyanobacteria are as numerous as in chloroplasts, but targets are different. *Proc Natl Acad Sci USA* **100:** 16107–16112.

Linder, E., Lasko, P. F., Ashburner, M., Leroy, P., Nielson, P. J., Nishi, K., Schnier, J., and Slonimski, P. P. (1989) Birth of the D – E – A – D box. *Nature* (London) **337:** 121–122.

Lipinska, B., Zylicz, M., and Georgopoulos, C. (1990) The HtrA (DegP) protein, essential for *Escherichia coli* survival at high temperatures, is an endopeptidase. *J Bacteriol* **172:** 1791–1797.

Liu, Y., Yu, L., Ke, W., Gao, X., and Qiu, B. (2010) Photosynthetic recovery of *Nostoc flagelliforme* upon rehydration after 2 years and 8 years dry storage. *Phycologia* **49:** 429–437. doi: 10.2216/09-011

Loprasert, S., Negoro, S. and Okada, H. (1989) Cloning, nucleotide sequence, and expression in *Escherichia coli* of the *Bacillus stearothermophilus* peroxidase gene (*perA*). *J Bacteriol* **171:** 4871–4875.

Los, D. A. (2004) The effect of low-temperature-induced DNA supercoiling on the expression of the desaturase genes in *Synechocystis*. *Cell Mol Biol* **50:** 605–612.

Los, D. A., and Murata, N. (1994) The low-temperature-induced accumulation of the *desA* transcript in *Synechocystis* PCC6803 is a result of both activation of transcription and maintenance of RNA stability. *Russian J Plant Physiol* **41:** 146–151.

Los, D. A., and Murata, N. (1998) Structure and expression of fatty acid desaturases. *Biochim Biophys Acta***1394:** 3–15.

Los, D. A., and Murata, N. (1999) Response to cold-shock in cyanobacteria. *J Mol Microbiol Biotechnol* **1:** 221–230.

Los, D. A., and Murata, N. (2000) Regulation of enzymatic activity and gene expression by membrane fluidity. *Science's Signal Transduction Knowledge Environment* (http://www.stke.org/cgi/content/full/OC_sigtrans; 2000/62/pe1).

Los, D. A., and Murata, N. (2004) Membrane fluidity and its roles in the perception of environment signals. *Biochim Biophys Acta* **1666:** 142–157.

Los, D. A., and Zinchenko, V. V. (2000) Regulatory role of membrane fluidity in gene expression. In: Wada, H., and Murata, N. (Eds) *Lipids in Photosynthesis. Structure, Function and Genetics*. Springer Science + Business Media, B.V. Berlin, Germany. pp. 329–348.

Los, D. A., Horvàth, I., Vigh, L., and Murata, N. (1993) The temperature-dependent expression of the desaturase gene *desA* in *Synechocystis* PCC 6803. *FEBS Lett* **318:** 57–60.

Los, D. A., Ray, M. K., and Murata, N. (1997) Differences in the control of the temperature-dependent expression of four genes for desaturases in *Synechocystis* sp. PCC 6803. *Mol Microbiol* **25:** 1167–1176.

Los, D. A., Suzuki, I., Zinchenko, V. V., and Murata, N. (2008) Stress responses in *Synechocystis*: regulated genes and regulatory systems. In: Herrero, A., and Flores, E. (Eds) *The Cyanobacteria: Molecular Biology, Genomics and Evolution*, Caister Academic Press: Norfolk, UK. pp. 117–157.

Los, D. A., Zorina, A., Sinetova, M., Kryazhov, S., Mironov, K., and Zinchenko, V. (2010) Stress sensors and signal transducers in cyanobacteria. *Sensors* **10:** 2386–2415. doi:10.3390/s100302386

Lovelock, J. E. (1953) Het (SicJ) mechanism of the protective action of glycerol against haemolysis by freezing and thawing. *Biochim Biophys Acta* **11:** 28–36.

Lu, C., and Zhang, J. (2000) Role of light in response of PSII photochemistry to salt stress in the cyanobacterium *Spirulina platensis*. *J Exp Bot* **51:** 911–917.

Macedo, M. F., Miller, A. Z., Dionísio, A., and Saiz-Jimenez, C. (2009) Biodiversity of cyanobacteria and green algae on monuments in the Mediterranean Basin: An overview. *Microbiology* **155:** 3476–3490.

Mackay, M. A., Horton, R. S., and Borowitzka, L. J. (1984) Organic osmoregulatory solutes in cyanobacteria. *J Gen Microbiol* **130:** 2177–2191.

Maestri, O., and Joset, F. (2000) Regulation by external pH and stationary growth phase of the acetolactate synthase from *Synechocystis* PCC 6803. *Mol Microbiol* **37:** 828–838.

Makarova, K. S., Omelchenko, M. V., Gaidamakova, E. K., Matrosova, V. Y., Vasilenko, A., Zhai, M., Lapidus, A., Copeland, A., Kim, E., Land, M., Mavromatis, K., Pitluck, S., Richardson, P. M., Detter, C., Brettin, T., Saunders, E., Lai, B., Ravel, B., Kemner, K. M., Wolf, Y. I., Sorokin, A., Gerasimova, A. V., Gelfand, M. S., Fredrickson, J. K., Koonin, E. V., and

Daly, M. J. (2007) *Deinococcus geothermalis*: the pool of extreme radiation resistance genes shrinks. *PLoS ONE* **2:** e955. doi:10.1371/journal.pone.0000955
Mann, N. H., Novac, N., Mullineaux, C. W., Newman, J., Bailey, S., and Robinson, C. (2000) Involvement of an FtsH homologue in the assembly of functional photosystem I in the cyanobacterium *Synechocystis* sp. PCC 6803. *FEBS* **479:** 72–77.
Marcinkeviciene, J. A., Magliozzo, R. S., and Blanchard, J. S. (1995) Purification and characterization of the *Mycobacterium smegmatis* catalase-peroxidase involved in isoniazid activation. *J Biol Chem* **270:** 22290–22295.
Marcus, F., Moberly, L., and Latshaw, S. P. (1988) Comparative amino acid sequence of fructose-1,6-bisphosphatase: Identification of a region unique to the light-regulated chloroplast enzyme. *Proc Natl Acad Sci USA* **85:** 5379–5383.
Marin, K., Huckauf, J., Fulda, S., and Hagemann, M. (2002) Salt-dependent expression of glucosylglycerol-phosphate synthase, involved in osmolyte synthesis in the cyanobacterium *Synechocystis* sp. strain PCC 6803. *J Bacteriol* **184:** 2870–2877.
Marin, K., Kanesaki, Y., Los, D. A., Murata, N., Suzuki, I., and Hagemann, M. (2004) Gene expression profiling reflects physiological processes in salt acclimation of *Synechocystis* sp. strain PCC 6803. *Plant Physiol* **136:** 3290–3300.
Marin, K., Strinberg, M., Eisenhut, M., Krämer, R., and Hagemann, M. (2006) Osmotic stress in *Synechocystis* sp. PCC 6803: Low tolerance towards nonionic osmotic stress results from lacking activation of glucosylglycerol accumulation. *Microbiology* **152:** 2023–2030.
Marin, K., Suzuki, I., Yamaguchi, K., Ribbeck, K., Yamamoto, H., Kanesaki, Y., Hagemann, M., and Murata, N. (2003) Identification of histidine kinases that act as sensors in the perception of salt stress in *Synechocystis* sp. strain PCC 6803. *Proc Natl Acad Sci USA* **100:** 9061–9066.
Martin, J. L. (1995) Thioredoxin-a fold for all reasons. *Structure* **3:** 245–250.
Martinez, M. R. (1988) *Nostoc commune* Vauch., a nitrogen-fixing blue-green alga, as source of food in the Philippines. *Phil Agric* **71:** 295–307.
Martinez, M. R., and Querijero, N. M. B. (1986) Macrocolony formation in the nitrogen-fixing blue-green alga, *Nostoc commune* Vauch. *Phil Agric* **69:** 547–565.
Maruyama, K., Sato, N., and Ohta. (1999) Conservation of structure and cold-regulation of RNA-binding proteins in cyanobacteria: Probable convergent evolution with eukaryotic glycine-rich RNA-binding proteins. *Nucleic Acids Res* **27:** 2029–2036.
Mary, I., Tu, C. J., Grossman, A., and Vaulot, D. (2004) Effects of high light on transcripts of stress-associated genes for the cyanobacteria *Synechocystis* sp. PCC 6803 and *Prochlorococcus* MED4 and MIT9313. *Microbiology* **150:** 1271–1281.
Masip, L., Veeravalli, K., and Georgiou, G. (2006) The many faces of glutathione in bacteria. *Antiox Redox Signal* **8:** 753–762.
Maurizi, M. R. (1991) ATP-promoted interaction between ClpA and ClpP protease in activation of Clp protease from *Escherichia coli*. *Biochem Soc Trans* **19:** 719–723.
McCarty, J. S., and Walker, G. C. (1991) DnaK as thermometer: Threonine-199 is site of autophosphorylation and is critical for ATPase activity. *Proc Natl Acad Sci USA* **88:** 9513–9517.
McCarty, J. S., and Walker, G. C. (1994) DnaK mutants defective in ATPase activity are defective in negative regulation of the heat shock response: Expression of mutant DnaK proteins results in filamentation. *J Bacteriol* **176:** 764–780.
McCord, J. M., and Fridovich, I. (1969) Superoxide dismutase an enzymic function for erythrocuprein (hemocuprein). *J Biol Chem* **244:** 6049–6055.
Meyer, Y., Reichheld, J. P., and Vignols, F. (2005) Thioredoxins in *Arabidopsis* and other plants. *Photosynth Res* **86:** 419–433.
Meyer, Y., Buchanan, B. B., Vignols, F., Reichheld, J. P. (2009) Thioredoxins and Glutaredoxins: Unifying elements in redox biology. *Annu Rev Genetics* **43:** 335–367.
Mikami, K., Kanesaki, Y., Suzuki, I., and Murata, N. (2002) The histidine kinase Hik33 perceives osmotic stress and low-temperature stress in *Synechocystis* sp. PCC 6803. *Mol Microbiol* **46:** 905–915.
Mikkat, S., Hagemann, M., and Schoor, A. (1996) Active transport of glucosylglycerol is involved in salt adaptation of the cyanobacterium *Synechocystis* sp. strain PCC 6803. *Microbiology* **142:** 1725–1732.
Miller, A-F. (2004) Superoxide dismutases active sites that save, but a protein that kills. *Curr Opin Chem Biol* **8:** 162–168.
Miller, E. W., Tulythan, O., Isacoff, E. Y., and Chang, C. (2007) Molecular imaging of hydrogen peroxide produced for cell signaling. *Nat Chemical Biol* **3:** 263–267. doi:10.1038/nchembio871
Miranda-Vizuete, A., Rodriguez-Ariza, A., Toribio, F., Holmgren, A., Lopez-Barea, J., and Pueyo, C. (1996) The levels of ribonucleotide reductase, thioredoxin, glutaredoxin 1, and GSH are balanced in *Escherichia coli* K12. *J Biol Chem* **271:** 19099–19103.
Missiakas, D., Schwager, F., Betton, J.-M., Georggopolous, C., and Raina, S. (1996) Isolation and characterization of HsIV HsIU (ClpQ ClpY) proteins involved in overall proteolysis of misfolded proteins in *E. coli*. *EMBO J* **15:** 6899–6909.
Miyake, C., Michihata, F., and Asada, K. (1991) Scavenging of hydrogen peroxide in prokaryotic and eukaryotic algae: acquisition of ascorbate peroxidase during the evolution of cyanobacteria. *Plant Cell Physiol* **32:** 33–43.
Mizuno, T., Kaneko, T., and Tabata, S. (1996) Compilation of all genes encoding bacterial two-component signal transducers in the genome of the cyanobacterium *Synechocystis* sp. strain PCC 6803. *DNA Res* **3:** 407–414.
Mogk, A., Deuerling, E., Vorderwulbecke, S., Vierling, E., and Bukau, B. (2003). Small heat shock proteins, ClpB and the DnaK system form a functional triade in reversing protein aggregation. *Mol Microbiol* **50:** 585–595.

Mogk, A., Tomoyasu, T., Goloubinoff, P., Rüdiger, S., Röder, D., Langen, H., and Bukau, B. (1999) Identification of thermolabile *Escherichia coli* proteins: prevention and reversion of aggregation by DnaK and ClpB. *EMBO J* **18:** 6934–6949.

Mohamed, A., and Jansson, C. (1989) Influence of light on accumulation of photosynthesis-specific transcripts in the cyanobacterium *Synechocystis*. *Plant Mol Biol* **13:** 693–700.

Mohamed, A., Eriksson, J., Osiewacz, H., and Jansson, C. (1993) Differential expression of the *psbA* genes in the cyanobacterium *Synechocystis*. *Mol Gen Genet* **238:** 161–168.

Mohamed, A., and Jansson, C. (1989) Influence of light on accumulation of photosynthesis-specific transcripts in the cyanobacterium *Synechocystis*. *Plant Mol Biol* **13:** 693–700.

Mongkolsuk, S., and Helmann, J. D. (2002) Regulation of inducible peroxide stress responses. *Mol Microbiol* **45:** 9–15.

Mori, S., Castoreno, A., Mulligan, M. E., and Lammers, P. J. (2003) Nitrogen status modulates the expression of RNA-binding proteins in cyanobacteria. *FEMS Microbiol Lett* **227:** 203–210.

Morsay, F. M., Kuzuha, S., Takani, Y., and Sakamoto, T. (2008) Novel thermostable glycosidases in the extracellular matrix of the terrestrial cyanobacterium *Nostoc commune*. *J Gen Appl Microbiol* **54:** 243–252.

Motohashi, K., Watanabe, Y., Yohda, M., and Yoshida, M. (1999) Heat-inactivated proteins are rescued by the DnaJ-GrpE set and ClpB chaperones. *Proc Natl Acad Sci USA* **96:** 7184–7189.

Müller, D. R., Vincent, W. F., Bonilla, S., and Laurion, I. (2005) Extremotrophs, extremophiles and broadband pigment strategies in a high arctic ice shelf system. *FEMS Microb Ecol* **53:** 73–87.

Muller, E. D. G., and Buchanan, B. B. (1989) Thioredoxin is essential for photosynthetic growth. *J Biol Chem* **264:** 4008–4014.

Mulligan, M. E., Jackman, D. M., and Murphy, S. T. (1994) Heterocyst forming filamentous cyanobacteria encode proteins that resemble eukaryotic RNA-binding proteins of the RNP family. *J Mol Biol* **235:** 1162–1170.

Murata, N. (1989) Low-temperature effects on cyanobacterial membranes. *J Bioenerg Biomembr* **21:** 61–75.

Murata, N., and Los, D. A. (1997) Membrane fluidity and temperature perception. *Plant Physiol* **115:** 875– 879.

Murata, N., and Los, D. A. (2006) Histidine kinase Hik33 is an important participant in cold signal transduction in cyanobacteria. *Physiol Plant* **126:** 17–27.

Murata, N., and Nishida, I. (1987) Lipids of blue-green algae (cyanobacteria). In: Stumpf, P. K. (Ed.) *The Biochemistry of Plants*, Vol. 9. Academic Press, San Diego, CA, USA. pp. 315–347.

Murata, N., and Siegenthaler, P. A. (2004) Lipids in photosynthesis: An overview. In: : Wada, H., and Murata, N. (Eds) *Lipids in Photosynthesis. Structure, Function and Genetics*. Springer Science + Business Media, B.V. Berlin, Germany. pp. 1–20.

Murata, N., and Suzuki, I. (2006) Exploitation of genomic sequences in a systematic analysis to access how cyanobacteria sense environmental stress. *J Exp Bot* **57:** 235–247.

Murata, N., and Wada, H. (1995) Acyl-lipid desaturases and their importance in the tolerance and acclimatization to cold of cyanobacteria. *Biochem J* **308:** 1–8.

Murata, N., Ishizaki-Nishizawa, O., Higashi, H., Hayashi, H., Tasaka, Y., and Nishida, I. (1992a) Genetically engineered alteration in the chilling sensitivity of plants. *Nature* (London) **356:** 710–713.

Murata, N., Ono, T., and Sato, N. (1979) Lipid phase of membrane and chilling injury in the blue-green alga *Anacystis nidulans*. In: Lyons, J. M., Graham, D., and Raison, J. K. (Eds.) *Low Temperature Stress in Crop Plants*. Academic Press, New York, NY, USA. pp. 337–345.

Murata, N., Wada, H., and Gombos, Z. (1992b) Modes of fatty-acid desaturation in cyanobacteria. *Plant Cell Physiol* **33:** 933–941.

Mustardy, L., Los, D. A., Gombos, Z., and Murata, N. (1996) Immunocytochemical localization of acyl-lipid desaturases in cyanobacterial cells: Evidence that both thylakoid membranes and cytoplasmic membranes are sites of lipid desaturation. *Proc Natl Acad Sci USA* **93:** 10524–10527.

Mutsuda, M., Ishikawa, T., Takeda, T., and Shigeoka, S. (1996) The catalase-peroxidase of *Synechococcus* PCC 7942: purification, nucleotide sequence analysis and expression in *Escherichia coli*. *Biochem J* **316:** 251–257.

Nakamoto, H., and Honma, D. (2006) Interaction of small heat-shock protein with light harvesting cyanobacterial phycocyanins under stress conditions. *FEBS Lett* **580:** 3029–3034.

Nakamoto, H., and Vígh, L. (2007) The small heat shock proteins and their clients. *Cell Mol Life Sci* **64:** 294–306.

Nakamoto, H., Suzuki, N., and Roy, S. K. (2000) Constitutive expression of a small heat-shock protein confers cellular thermotolerance and thermal properties to the photosynthetic apparatus in cyanobacteria. *FEBS Lett* **483:** 169–174.

Nakamoto, H., Suzuki, N., and Kojima, K. (2003) Targeted inactivation of the *hrcA* repressor gene in cyanobacteria. *FEBS Lett* **549:** 57–62.

Nakamoto, H., Tanaka, N., and Ishikawa, N. (2001) A novel heat shock protein plays an important role in thermal stress management in cyanobacteria. *J Biol Chem* **276:** 25088–25095.

Nakamura, K., and Hihara, Y. (2006) Photon flux density-dependent gene expression in *Synechocystis* sp. PCC 6803 is regulated by a small, redox-responsive, LuxR-type regulator. *J Biol Chem* **281:** 36758–36766.

Nakamura, T., Naito, K., Yokota, N., Sugita, C., and Sugita, M. (2007) A cyanobacterial non-coding RNA, Yfr1, is required for growth under multiple stress conditions. *Plant Cell Physiol* **48:** 1309–1318.

Nakamura, Y., Kaneko, T., Hirosawa, M., Miyajima, N., and Tabata, S. (1998) CyanoBase, a www database containing the complete nucleotide sequence of the genome of *Synechocystis* sp. strain PCC 6803. *Nucleic Acids Res* **26**: 63–67.
Narberhaus, F. (1999) Negative regulation of bacterial heat shock genes. *Mol Microbiol* **31**: 1–8.
Navarro, F., and Florencio, F. J. (1996) The cyanobacterial thioredoxin gene is required for both photoautotrophic and heterotrophic growth. *Plant Physiol* **111**: 1067–1075.
Newkirk, K., Feng, W., Jiang, W.,Tejero, R., Emerson, S. D., Inouye, M., and Montelione, G. T. (1994) Solution NMR structure of the major cold shock protein (CspA) from *Escherichila coli*: Identification of a binding epitope for DNA. *Proc Natl Acad Sci USA* **91**: 5114–5118.
Newton, G. L., Arnold, K., Price, M. S., Sherrill, C., Delcardayre, S. B., Aharonowitz, Y., Cohen, G., Davies, J., Fahey, R. C., and Davis, C. (1996) Distribution of thiols in microorganisms: mycothiol is a major thiol in most actinomycetes. *J Bacteriol* **178**: 1990–1995.
Nienow, J. A., and Friedmann, E. I. (1993) Terrestial lithophytic (rock) communities. In: Friedmann, E. J., and Thistle, A. B. (Eds.) *Antarctic Microbiology*. John Wiley & Sons, Inc., New York, NY, USA. pp. 343–412.
Nierhaus, K. H., and Wilson, D. N. (Eds.) (2004) *Protein Synthesis and Ribosome Structure. Translating the Genome.* Wiley-VCH Verlag GmbH & Co. KGaA, Weinheim, Germany.
Nimura, K., Takahashi, H., and Yoshikawa, H. (2001) Characterization of the *dnaK* multigene family in the cyanobacterium *Synechococcus* sp. strain PCC 7942. *J Bacteriol* **183**: 1320–1326.
Nimura, K., Yoshikawa, H., and Takahashi, H. (1994a) Identification of *dnaK* multigene family in *Synechococcus* sp. PCC 7942. *Biochem Biophys Res Commun* **201**: 466–471.
Nimura, K., Yoshikawa, H., and Takahashi, H. (1994b) Sequence analysis of the third *dnaK* homolog gene in *Synechococcus* sp. PCC 7942. *Biochem Biophys Res Commun* **201**: 848–854.
Nimura, K., Yoshikawa, H., and Takahashi, H. (1996) DnaK3, one of the three DnaK proteins of cyanobacterium *Synechococcus* sp. PCC 7942, is quantitatively detected in the thylakoid membrane. *Biochem Biophys Res Commun* **229**: 334–340.
Nishiyama, Y., and Hisabori, T. (2009) Physiological impact of thioredoxin- and glutaredoxin-mediated redox regulation in cyanobacteria. *Adv Bot Res* **52**: 187–205.
Nishiyama, Y., Hayashi, H., Watanabe, T. and Murata, N. (1994) Photosynthetic oxygen evolution is stabilized by cytochrome C_{550} against heat inactivation in *Synechococcus* sp. PCC 7002. *Plant Physiol* **105**: 1313–1317.
Nishiyama, Y., Kovács, E., Lee, C. B., Hayashi, H., Watanabe, T., and Murata, N. (1993) Photosynthetic adaptation to high temperature associated with thylakoid membranes of *Synechococcus* PCC 7002. *Plant Cell Physiol* **34**: 337–343.
Nishiyama, Y., Los, D.A., Hayashi, H., and Murata, N. (1997) Thermal protection of the oxygen-evolving machinery by PsbU, an extrinsic protein of photosystem II in *Synechococcus* sp. PCC 7002. *Plant Physiol* **115**: 1473–1480.
Nishiyama, Y., Los, D. A., and Murata, N. (1999) PsbU, a protein associated with photosystem II, is required for the acquisition of cellular thermotolerance in *Synechococcus* sp. PCC 7002. *Plant Physiol* **120**: 301–308.
Nishiyama, Y., Yamamoto, H., Allakhverdiev, S. I., Inaba, M., Yokota, A., and Murata, N. (2001) Oxidative stress inhibits the repair of photodamage to the photosynthetic machinery. *EMBO J* **20**: 5587–5594.
Nitta, K., Suzuki, N., Honma, D., Kaneko, Y., and Nakamoto, H. (2005) Ultrastructural stability under high temperature or intensive light stress conferred by a small heat shock protein in cyanobacteria. *FEBS Lett* **579**: 1235–1242.
Nixon, P. J., Barker, M., Boehm, M., de Vries, R., and Komenda, J. (2005) FtsH-mediated repair of the photosystem II complex in response to light stress. *J Exp Bot* **56**: 357–363.
Nixon, P. J., Trost, J. T., and Diner, B. A. (1992) Role of carboxyl-terminus of polypeptide D1 in the assembly of a functional water-oxidizing manganese cluster in photosystem II of the cyanobacterium *Synechocystis* sp. PCC 6803: Assembly requires a free carboxyl group at C-terminal position 344. *Biochemistry* **31**: 10859–10871.
Nocker, A., Hausherr, T., Balsiger, S., Krstulovic, N. P., Hennecke, H., and Narberhaus, F. (2001) A mRNA-based thermosensor controls expression of rhizobial heat shock genes. *Nucleic Acids Res* **29**: 4800–4807.
Nomura, M., Hibino, T., Takabe, T., Sugiyama, T., Yokota, A., Miyake, H., and Takabe, T. (1998) Transgenically produced glycine betaine protects ribulose 1,5-biphosphate carboxylase/oxygenase from inactivation in *Synechococcus* sp. PCC 7942 under salt stress. *Plant Cell Physiol* **39**: 425–432.
Nomura, M., Ishitani, M., Takabe, T., Rai, A. K., and Takabe, T. (1995) *Synchococcus* sp. PCC 7942 transformed with *Escherichia coli bet* genes produces glycine betaine from choline and acquires resistance to salt stress. *Plant Physiol* **107**: 703–708.
Nyossöla, A., Kerovuo, J., Kaukinen, P., Von Weymarn, N., and Reinikainen, T. (2000) Extreme halophiles synthesize betaine from glycine by methylation. *J Biol Chem* **275**: 22196–22201.
Obinger, C., Regelsberger, G., Strasser, G., Burner, U., and Peschek, G. A. (1997) Purification and characterization of a homodimeric catalase-peroxidase from the cyanobacterium *Anacystis nidulans*, *Biochem Biophys Res Commun* **235**: 545–552.
Ohad, I., Nevo, R., Brumfeld, V., Reich, Z., Tsur, T., Yair, M., and Kaplan, A. (2005) Inactivation of photosynthetic electron flow during desiccation of desert biological sand crusts and *Microcoleus* sp. enriched isolates. *Photochem Photobiol Sci* **4**: 977–982.

Ohad, I., Raanan, H., Keren, N., Tchernov, D., and Kaplan, A. (2010) Light-induced changes within photosystem II protects *Microcoleus* sp. biological desert sand crusts against excess light. *PLoS ONE* **5:** e11000. doi:10.1371/journal.pone.0011000

Ohnishi, N., and Murata, N. (2006) Glycine betaine counteracts the inhibitory effects of salt stress on the degradation and synthesis of D1 protein during photoinhibition in *Synechococcus* sp. PCC 7942. *Plant Physiol* **141:** 758–765.

Okada, S., Kanematsu, S., and Asada, K. (1979) Intracellular distribution of manganese and ferric superoxide dismutases in blue-green algae. *FEBS Lett* **103:** 106–110.

Okamoto, S., Yamanishi, Y., Ehira, S., Kawashima, S., Tonomura, K., and Kanehisa, M. (2007) Prediction of nitrogen metabolism-related genes in *Anabaena* by kernel-based network analysis. *Proteomics* **7:** 900–909. doi: 10.1002/pmic.200600862

Oldham, M. L., Brash, A. R, Newcomer, M. E. (2005) The structure of coral allene oxide synthase reveals a catalase adapted for metabolism of a fatty acid hydroperoxide. *Proc the Natl Acad Sci USA* **102:** 297–302.

Olsson-Francis, K., de la Torre, R., and Cockell, C. S. (2010) Isolation of novel extreme tolerant cyanobacteria from a rock-dwelling microbial community by using exposure to low earth orbit. *Appl Environ Microbiol* **76:** 2115–2121.

Omata, T., and Murata, N. (1983) Isolation and characterization of the cytoplasmic membranes from the blue-green alga (cyanobacterium) *Anacystis nidulans*. *Plant Cell Physiol* **24:** 1101–1112.

Ono, T.-A., and Murata, N. (1981a) Chilling susceptibility of the blue-green alga, *Anacystis nidulans* I. Effect of growth temperature. *Plant Physiol* **67:** 176–181.

Ono, T.-A., and Murata, N. (1981b) Chilling susceptibility of the blue-green alga, *Anacystis nidulans* II. Stimulation of the passive permeability of cytoplasmic membrane at chilling temperatures. *Plant Physiol* **67:** 182–187.

Ono, T.-A., and Murata, N. (1982) Chilling susceptibility of the blue-green alga, *Anacystis nidulans* III. Lipid phase of cytoplasmic membrane. *Plant Physiol* **69:** 125–129.

Oren, A. (1977) Mycosporine-like amino acids as osmotic solutes in a community of halophilic cyanobacteria. *Geomocrobiol J* **14:** 231–240.

Oren, A., and Gunde-Cimerman, N. (2007) Mycosporines and mycosporine-like amino acids: UV protectants or multipurpose secondary metabolites. *FEMS Microbiol Lett* **269:** 1–10.

Owttrim, G. W. (2006) RNA helicases and abiotic stress. *Nucleic Acids Res* **34:** 3220–3230.

Padan, E., and Schuldiner, S. (1994) Molecular physiology of the Na^+/H^+ antiporter in *Escherichia coli*. *J Exp Biol* **196:** 443–456.

Padan, E., and Schuldiner, S. (1996) Bacterial Na^+/H^+ antiporters-molecular biology, biochemistry and physiology. In: Konings, W. N., Kaback, H. R., and Lolkema, J. S. (Eds) *Handbook of Biological Physics*, Vol 2. Elsevier Science Publishers, Amsterdam, The Netherlands. pp. 501–531.

Padan, E., Venturi, M., Gerchman, Y., and Dover, N. (2001) Na^+/H^+ antiporters. *Biochim Biophys Acta* **1505:** 144–157.

Paithoonrangsarid, K., Shoumskaya, M. A., Kanesaki , Y., Satoh, S., Tabata, S., Los, D. A., Zinchenko, V. V., Hayashi, H., Tanticharoen, M., Suzuki, I., and Murata, N. (2004) Five histidine kinases perceive osmotic stress and regulate distinct sets of genes in *Synechocystis*. *J Biol Chem* **279:** 53078–53086.

Pandhal, J., Wright, P. C., and Biggs, C. A. (2008) Proteomics with a pinch of salt: A cyanobacterial perspective. *Saline systems* **4:** 1 doi:10.1186/1746-1448-4-1

Panichkin, V. B., Glazer, V. M., Zinchenko, V. V., Sokolenko, A., Herrmann, R. G., and Shestakov, S. V. (2001) *clpP2* gene encoding peptidase in cyanobacteria *Synechocystis* sp. PCC 6803 controls the sensitivity of cells to photoinhibition. *Izv Akad Nauk SSSR* Ser Biol **3:** 312–317.

Panoff, J. M., Priem, B., Morvan, H., and Joset, F. (1988) Sulphated exopolysaccharides produced by two unicellular strains of cyanobacteria, *Synechocystis* PCC 6803 and 6714. *Arch Microbiol* **150:** 558–563.

Park, S. K., Kim, K. I., Woo, K. M., Seol, J. H., Tanaka, K., Ichihara, A., Ha, D. B., and Chung, C. H. (1993) Site-directed mutagenesis of the dual translational initiation sites of the *clpB* gene in *Escherichia coli* and characterization of its gene product. *J Biol Chem* **268:** 20170–20174.

Park, Y.-I., Sandström, S., Gustafsson, P., and Öquist, G. (1999) Expression of the *isiA* gene is essential for the survival of the cyanobacterium *Synechococcus* sp. PCC 7942 by protecting photosystem II from excess light under iron limitation. *Mol Microbiol* **32:** 123–129.

Parsell, D. A., and Lindquist, S. (1993) The function of heat-shock proteins in stress tolerance: Degradation and reactivation of damaged proteins. *Annu Rev Genet* **27:** 437–496.

Pascual, M. B., Mata-Cabana, A., Florencio, F., Lindahl, M., and Cejudo, F. J. (2010) Overoxidation of 2-Cys peroxiredoxin in prokaryotes: Cyanobacterial 2-Cys peroxiredoxins sensitive to oxidative stress. *J Biol Chem* **285:** 34485–34492.

Pascual, M. B., Mata-Cabana, A., Florencio, F., Lindahl, M., and Cejudo, F. J. (2011) A comparative analysis of NADPH thioredoxin reductase C-2-Cys peroxiredoxin system from plants and cyanobacteria. *Plant Physiol* **155:** 1806–1816.

Passardi, F., Zamocky, M., Favet, J., Jakopitsch, C., Penel, C., Obinger, C., and Dunand, C. (2007) Phylogenetic distribution of catalase-peroxidases: are there patches of order in chaos? *Gene* **397:** 101–113.

Patterson-Fortin, L. M., Colvin, K. R., and Owttrim, G. W. (2006) A Lex-A related protein regulates redox-sensitive expression of the cyanobacterial RNA helicase, CrhR. *Nucleic Acids Res* **34:** 3446–3454.

Pause, A., and Sonenberg, N. (1993) Helicases and RNA unwinding in translation. *Curr Opin Struct Biol* **3:** 953–959.
Peat, A., Powell, N., and Potts, M. (1988) Ultrastructural analysis of the rehydration of desiccated *Nostoc commune* HUN (Cyanobacteria) with particular reference to the immunolabelling of NifH. *Protoplasma* **146:** 72–80.
Pentecost, A. (1985) Relationships between light, temperature and photosynthesis in a temperate *Microcoleus* (cyanobacterium) mat. *Microbios* **43:** 141–148.
Pereira, S., Zille, A., Micheletti, E., Moradas-Ferreira, P., DePhilippis, R., and Tamagnini, P. (2009) Complexity of cyanobacterial exopolysaccharides, composition, structures, inducing factors and putative genes involved in their biosynthesis and assembly. *FEMS Microbiol Rev* **33:** 917–941.
Perelman, A., Uzan, A., Hacohen, D., and Schwarz, R. (2003) Oxidative stress in *Synechococcus* sp. strain PCC 7942: various mechanisms for H_2O_2 detoxification with different physiological roles. *J Bacteriol* **185:** 3654–3660.
Pérez-Ruiz, J. M., Spinola, M. C., Kirchsteiger, K., Moreno, J., Sahrawy, M., and Cejudo, F. J. (2006) Rice NTRC is a high-efficiency redox system for chloroplast protection against oxidative damage. *Plant Cell* **18:** 2356–2368.
Pérez-Pérez, M. E., Florencio, F. J., and Lindahl, M. (2006) Selecting thioredoxins for disulphide proteomics: target proteomes of three thioredoxins from the cyanobacterium *Synechocystis* sp. PCC 6803. *Proteomics* **6** (Suppl. 1): S186–S195.
Pérez-Pérez, M. E., Mata-Cabana, A., Sánchez-Riego, M., Lindahl, M., and Florencio, F. J. (2009) A comprehensive analysis of the peroxiredoxin reductase system in the cyanobacterium *Synechocystis* sp. strain PCC 6803 reveals that all five peroxiredoxins are thioredoxin dependent. *J Bacteriol* **191:** 7477–7489.
Phadtare, S. (2004) Recent developments in bacterial cold-shock response. *Curr Issues Mol Biol* **6:** 125–136.
Phadtare, S., and Severinov, K. (2010) RNA modeling and gene regulation by cold shcok proteins. *RNA Biology* **7:** 788–795. doi:10.4161/rna.7.6.13482
Pick, U., Weiss, M., Gounaris, K., and Barber, J. (1987) The role of different thylakoid glycolipids in the function of reconstituted chloroplast ATP synthase. *Biochim Biophys Acta* **891:** 28–39.
Pócs, T. (2009) Cyanobacterial crust types, as strategies for survival in extreme habitats. *Acta Bot Hungarica* **51:** 147–178.
Pomati, F., Burns, B. P., and Neilan, B. A. (2004) Use of ion channel modulating agents to study cyanobacterial Na^+-K^+ fluxes. *Biol Proceed Online* **6:** 137–143. doi:10.1251bpo82
Pomposiello, P. J., and Demple, B. (2001) Redox-operated genetic switches: the SoxR and OxyR transcription factors. *Trends Biotechnol.* **19:** 109–114.
Ponting, C. P. (1997) Evidence for PDZ domains in bacteria, yeast, and plants. *Protein Sci* **6:** 464–468.
Portwich, A., and Garcia-Pichel, F. (1999) Ultraviolet and osmotic stresses induce and regulate the synthesis of mycosporines in the cyanobacterium *Chlorogloeopsis* PCC 6912. *Arch Microbiol* **172:** 187–192.
Portwich, A., and Garcia-Pichel, F. (2000) A novel prokaryotic UVB photoreceptor in the cyanobacterium *Chlorogloeopsis* PCC 6912. *Photochem Photobiol* **71:** 493–498.
Poole, L. B. (2005) Bacterial defenses against oxidants: mechanistic features of cysteine-based peroxidases and their flavoprotein reductases. *Arch Biochem Biophys* **433:** 240–254.
Porankiewicz, J., and Clarke, A. K. (1997) Induction of the heat shock protein ClpB affects cold acclimation in the cyanobacterium *Synechococcus* sp. strain PCC 7942. *J Bacteriol* **179:** 5111–5117.
Porankiewicz, J., Wang, J., and Clarke, A. K. (1999) New insights into the ATP-dependent Clp protease: *Escherichia coli* and beyond. *Mol Microbiol* **32:** 449–458.
Porankiewicz, J., Schelin, J., and Clarke A. K. (1998) The ATP-dependent Clp peptidase is essential for acclimation to UV-B and low temperature in the cyanobacterium *Synechococcus. Mol Microbiol* **29:** 275–283.
Porter, M. A., Stringer, C. D., and Hartman, F. C. (1988) Characterization of the regulatory thioredoxin site of phosphoribulokinase. *J Biol Chem* **263:** 123–129.
Potts, M. (1985) Protein synthesis and proteolysis in immobilized cells of the cyanobacterium *Nostoc muscorum* UTEX 584 exposed to matric water stress. *J Bacteriol* **164:** 1025–1031.
Potts, M. (1994) Desiccation tolerance of prokaryotes. *Microbiol Rev* **58:** 755–805.
Potts, M. (1996) The anhydrobiotic cyanobacterial cell. *Physiol Plant* **97:** 788–794.
Potts, M. (1999) Mechanisms of desiccation tolerance in cyanobacteria. *Eur J Phycol* **34:** 319–328.
Potts, M. (2000) *Nostoc. In*: Whitton B. A., and Potts M. (Eds.), *Ecology of Cyanobacteria: Their Diversity in Time and Space.* Kluwer Academic, Dordrecht, The Netherlands. pp. 465–504.
Potts, M. (2001) Desiccation tolerance: a simple process? *Trends Microbiol* **9:** 553–559.
Potts, M., and Bowman, M. A. (1985) Sensitivity of *Nostoc commune* UTEX 584 (cyanobacteria) to water stress. *Arch Microbiol* **141:** 51–56.
Potts, M., and Friedmann, E. I. (1981) Effects of water stress on cryptoendolithic cyanobacteria from hot desert rocks. *Arch Microbiol* **130:** 267–271.
Potts, M., and Morrison, N. S. (1986) Shifts in the intracellular ATP pools of immobilized *Nostoc* cells (cyanobacteria) induced by water stress. *Plant Soil* **90:** 211–221.
Potts, M., and Whitton, B. A. (1977). Nitrogen fixation by blue-green algal communities in the intertidal zone of the lagoon of Aldabra Atoll. *Oecologia (Berl.)* **27:** 275–283.

Potts, M., and Whitton, B. A. (1980). Vegetation of the intertidal zone of the lagoon of Aldabra, with particular reference to the photosynthetic prokaryotic communities. *Proc R Soc* B London **208:** 13–55.

Potts, M., Bowmann, M. A., and Morrison, N. S. (1984) The matric water potential (Ψ_m) in immobilized cultures of cyanobacteria. *FEMS Microbiol Lett* **24:** 193–196.

Potts, M., Slaughter, S. M., Hunneke, F-U., Garst, J. F., and Helm, R. F. (2005) Desiccation tolerance of prokaryotes: Application of principles to human cells. *Integr Comp Biol* **45:** 800–809.

Prakash, J. S. S., Krishna, P. S., Sirisha, K., Kanesaki, Y., Suzuki, I., Shivaji, S., and Murata, N. (2010) An RNA helicase, CrhR, regulates the low-temperature-inducible expression of heat-shock genes *groES*, *groEL* in *Synechocystis* sp. PCC 6803. *Microbiology* **156:** 442–451.

Prakash, J.S.S., Sinetova, M., Zorina, A., Kupriyanova, E., Suzuki, I., Murata, N., and Los, D. A. (2009) DNA supercoiling regulates the stress-inducible expression of genes in the cyanobacterium *Synechocystis*. *Mol Biosyst* **5:** 1904–1912.

Priya, B., Premanadh, J., Dhanalakshmi, R. T., Seethalakshmi, T., Uma, L., Prabaharan, D., and Subramanian, G. (2007) Comparative analysis of cyanobacterial superoxide dismutases to discriminate canonical forms. *BMC Genomics* **8:** 435. doi:10.1186/1471-2164-8-435.

Priya, B., Sivaprasanth, R. K., Jensi, V. D., Uma, L., Subramanian, G., and Prabaharan, D. (2010) Characterization of manganese superoxide dismutase from a marine cyanobacterium *Leptolyngbya valderiana* BDU 200041. *Saline Systems* **6**.6. doi:10.1186/1746-1448-6-6.

Proctor, M. C. F. (2003) Comparative ecophysiological measurements on the light responses, water relations and desiccation tolerance of the filmy ferns *Hymenophyllum wilsonii* Hook and *H. tunbrigense* (L.) Smith. *Ann Bot* **91:** 717–727.

Proteau, P. J., Gerwick, W. H., Garcia-Pichel, F., and Castenholz, R. (1993) The structure of scytonemin, an ultraviolet sunscreen pigment from the sheaths of cyanobacteria. *Experientia* **49:** 825–829.

Qiu, B., and Price, N. M. (2009) Different physiological response of four marine *Synechococcus* strains (Cyanophyceae) to nickel starvation under iron replete and iron-deplete conditions. *J Phycol* **45:** 1062–1071.

Qiu, B., Liu, J., Liu, Z., and Liu, S. (2002) Distribution and ecology of the edible cyanobacterium Ge-Xian-Mi (*Nostoc*) in rice fields of Hefeng County in China. *J Appl Phycol* **14:** 423–429.

Rajaram, H., and Apte, S. K. (2003) Heat-shock response and its contribution to thermotolerance of the nitrogen-fixing cyanobacterium *Anabaena* sp. strain L-31. *Arch Microbiol* **179:** 423–429.

Rajaram, H., and Apte, S. K (2008) Nitrogen status and heat-stress-dependent differential expression of the cpn60 chaperonin gene influences thermotolerance in the cyanobacterium *Anabaena*. *Microbiology* **154:** 317–325. doi: 10.1099/mic.0.2007/011064.0

Rajaram, H., Ballal, A. D., Apte, S. K., Wiegert, T., and Schumann, W. (2001) Cloning and characterization of the major *groESL* operon from a nitrogen-fixing cyanobacterium *Anabaena* sp. strain L-31. *Biochim Biophys Acta* **1519:** 143–146.

Rao, V. S. K., Brand, J. J., and Myers, J. (1977) Cold shock syndrome in *Anacystis nidulans*. *Plant Physiol* **59:** 965–969.

Rawlings, N. D., and Barrett, A. J. (1999) MEROPS: the peptide database. *Nucleic Acids Res.* **27:** 325–331.

Ray, M. K., Sitaramamma, T., Ghandhi, S., and Shivaji, S. (1994) Occurrence and expression of *cspA*, a cold shock gene, in Antarctic psychotrophic bacteria. *FEMS Microbiol Lett* **116:** 55–60.

Reddy, A. S., Nuccio,M. L., Gross,L. M., and Thomas,T. L. (1993) Isolation of Δ6-desaturase gene from the cyanobacterium *Synechocystis* sp. strain PCC 6803 by gain-of-function expression in *Anabaena* sp. strain PCC7120. *Plant Mol Biol* **27:** 283–300.

Reddy, B. M., Apte, S. K., and Thomas, J. (1989) Enhancement of cyanobacterial salt tolerance by combined nitrogen. *Plant Physiol* **89:** 204–210.

Reed, R. H., and Stewart, W. D. P. (1985) Osmotic adjustment and organic solute accumulation in unicellular cyanobacteria from freshwater and marine habitats. *Mar Biol* **88:** 1–9.

Reed, R. H., Borowitzka, L. J., Mackay, M. A., Chudek, J. A., Foster, R., Warr, S. R.C., Moore, D. J., and Stewart W. D. P. (1986) Organic solute accumulation in osmotically stressed cyanobacteria. *FEMS Microbiol Rev* **39:** 51–56.

Regelsberger, G., Atzenhofer, W., Rücker, F., Peschek, G. A., Jakopitsch, C., Paumann, M., Furtmüller, P. G., and Obinger, C. (2002) Biochemical characterization of a membrane bound manganese-containing superoxide dismutase from the cyanobacterium *Anabaena* sp. PCC 7120. *J Biol Chem* **277:** 43615–43622.

Regelsberger, G., Laaha, U., Dietmann, D., Rüker, F., Canini, A., Grilli-Caiola, M., Furtmüller, P. G., Kakopitsch, C., Peschek, G. A., and Obinger, C. (2004) The iron superoxide dismutase from the filamentous cyanobacterium *Nostoc* PCC 7120. *J Biol Chem* **279:** 44384–44393.

Regelsberger, G., Jakopitsch, C., Rüker, F., Krois, D., Peschek, G.A., and Obinger, C. (2000) Effect of distal cavity mutations on the formation of compound I in catalase-peroxidases, *J Biol Chem* **275:** 22854–22861.

Regelsberger, G., Jakopitsch, C., Furtmüller, P. G., Rüker, F., Switala, J., Loewen, P. C., and Obinger, C. (2001) The role of distal tryptophan in the bifunctional activity of catalase-peroxidases. *Biochem Soc Trans* **29:** 99–105.

Regelsberger, G., Obinger, C., Zoder, R., Altmann, F., and Peschek, G. A. (1999) Purification and characterization of a hydroperoxidase from the cyanobacterium *Synechocystis* PCC 6803: identification of its gene by peptide mass mapping using matrix assisted laser desorption ionization time-of-flight mass spectrometry. *FEMS Microbiol Lett* **170:** 1–12.

Reyes, J. C., and Florencio, F. J. (1995) Electron transport controls transcription of the glutamine synthetase gene (*glnA*) from the cyanobacterium *Synechocystis* sp. PCC 6803. *Plant Mol Biol* **27:** 789–799.
Řezanka, T., Viden, I., Go, J. V., Dor, I., and Dembitsky, V. M. (2003) Polar lipids and fatty acids of three wild cyanobacterial strains of the genus *Chroococcidiopsis*. *Folia Microbiol* **48:** 781–786.
Richards, A. B., Krakowka, S., Dexter, L. B., Schmid, H., Wolterbeek, A. P. M., Waalkens-Berendsen, D. H., Shigoyuki, A., and Kurimoto, M. (2002) Trehalose: A review of properties, history of use and human tolerance, and results of multiple safety studies. *Food Chem Toxicol* **40:** 871–898.
Ritchie, R. J. (1991) Membrane-potential and pH control in the cyanobacterium *Synechococcus* R-2 (*Anacystis nidulans*) PCC 7942. *J Plant Physiol* **137:** 409–418.
Ritchie, R. J. (1992) Sodium transport and the origin of the membrane potential in the cyanobacterium *Synechococcus* R-2 (*Anacystis nidulans*) PCC 7942. *J Plant Physiol* **139:** 320–330.
Ritchie, R. J. (1998) Bioenergetics of membrane transport in *Synechococcus* R-2 (*Anacystis nidulans, S. leopoliensis*) PCC 7942. *Can J Bot* **76:** 1127–1145.
Richter, S., Hagemann, M., and Messer, W. (1998) Transcription analysis and mutation of a *dnaA*-like gene in *Synechocystis* sp. strain PCC 6803. *J Bacteriol* **180:** 4946–4949.
Ritossa, F. (1962) A new puffing pattern induced by heat shock and DNP in *Drosophila*. *Experientia* **18:** 571–573.
Ritter, D., and Yopp, J. H. (1993) Plasma membrane lipid composition of the halophilic cyanobacterium *Aphanothece halophytica*. *Arch Microbiol* **159:** 435–439.
Ritz, D., and Beckwith, J. (2001) Roles of thiol-redox pathways in bacteria. *Annu Rev Microbiol* **55:** 21–48.
Rocak, S., and Linder, P. (2004) DEAD-box proteins: the driving forces behind RNA metabolism. *Nat Rev Mol Cell Biol* **5:** 232–241.
Rocap, G., Larimer, F. W., Lamerdin, J., Malfatti, S., Chain, P., Ahlgren, N. A., Arellano, A., Coleman, M., Hauser, L., Hess, W. R., Johnson, Z. I., Land, M., Lindell, D., Post, A. F., Regala, W., Shah, W. M., Shaw, S. L., Steglich, C., Sullivan, M. B., Ting, C. S., Tolonen, A., Webb, E. A., Zinser, E. R., and Chisholm, S. W. (2003) Genome divergence in two *Prochlorococcus* ecotypes reflects oceanic niche differentiation. *Nature* (London) **424:** 1042–1047.
Rodriguez-Manzaneque, M. T., Ros, Cabiscol, E., Sorribas, A., and Herrero, E. (1999) Grx5 glutaredoxin plays a central role in protection against protein oxidative damage in *Saccharomyces cerevisiae*. *Mol Cell Biol* **19:** 8180–8190.
Ross, C., Santiago-Vázquez, L., and Paul, V. (2006) Toxin release in response to oxidative stress and programmed cell death in the cyanobacterium *Microcystis aeruginosa*. *Aquat Toxicol* **78:** 66–73.
Rowland, J. G., Pang, X., Suzuki, I., Murata, N., Simon, W. J., and Slabas, A. R. (2010) Identification of components associated with thermal acclimation of photosystem II in *Synechocystis* sp. PCC 6803. *PLoS ONE* **5:** e10511. doi:10.10371/journal.pone.0010511
Roy, S. K., and Nakamoto, H. (1998) Cloning, characterization, and transcriptional analysis of a gene encoding an α-crystallin-related, small heat shock protein from the thermophilic cyanobacterium *Synechococcus vulcanus*. *J Bacteriol* **180:** 3997–4001.
Roy, S. K., Hiyama, T., and Nakamoto, H. (1999) Purification and characterization of the 16-kDa heat-shock-responsive protein from the thermophilic cyanobacterium *Synechococcus vulcanus*, which is an α-crystallin-related. *Eur J Biochem* **262:** 406–416.
Rozema, J., Björn, L. O., Bornman, J. F. *et al.* (2002) The role of UV-B radiation in aquatic and terrestrial ecosystems- an experimental and functional analysis of the evolution of UV-absorbing compounds. *J Photochem Photobiol* **66:** 2–12.
Rupprecht, E., Gathmann, S., Fuhrmann, E., and Schneider, D. (2007) Three different DnaK proteins are functionally expressed in the cyanobacterium *Synechocystis* sp. PCC 6803. *Microbiology* **153:** 1828–1841. doi:10.1099/mic0.2007/0058 76-0
Ryu, J.-Y., Suh, K.-H., Chung, Y.-H., Park, Y.-M., Chow, W. S., and Park, Y-II. (2003) Cytochrome *c* oxidase of the cyanobacterium *Synechocystis* sp. PCC 6803 protects photosynthesis from salt stress. *Mol cell* **16:** 74–77.
Sabacka, M., and Elster, J. (2006) Response of cyanobacteria and algae from Antarctic wetland habitats to freezing and desiccation stress. *Polar Biol* **30:** 31–37.
Sakamoto, T., and Bryant, D. A. (1997) Temperature-regulated mRNA accumulation and stabilization for fatty acid desaturase genes in the cyanobacterium *Synechococcus* sp. strain PCC 7002. *Mol Microbiol* **23:** 1281–1292.
Sakamoto, T., and Bryant, D. A. (1998) Growth at low temperature causes nitrogen limitation in the cyanobacterium *Synechococcus* sp. PCC 7002. *Arch Microbiol* **169:** 10–19.
Sakamoto, T., and Bryant, D. A. (1999) Nitrate transport and not photoinhibition limits growth of freshwater cyanobacterium *Synechococcus* species PCC 6301 at low temperature. *Plant Physiol* **119:** 785–794.
Sakamoto, T., Higashi, S., Wada, H., Murata, N., and Bryant, D. A. (1997a) Low-temperature-induced desaturation of fatty acids and expression of desaturase genes in the cyanobacterium *Synechococcus* sp. PCC 7002. *FEMS Microbiol Lett* **152:** 313–320.
Sakamoto, T., Stirewalt, V. L., and Bryant, D. A. (1997b) Two acyl-lipid Δ9 desaturase genes of the cyanobacterium *Synechococcus* sp. PCC 7002. In: Williams, J. P., Khan, M. U., and Lem, N. W. (Eds.) *Physiology, Biochemistry, and Molecular Biology of Plant Lipids* Kluwer, Dordrecht, The Netherlands. pp. 380–381.

Sakamoto, T., Wada, H., Nishida, I., Ohmori, M., and Murata, N. (1994) Δ^9 Acyl-lipid desaturases of cyanobacteria: Molecular cloning and substrate specificities in terms of fatty acids, *sn*-positions, and polar head groups. *J Biol Chem* **269:** 25576–25580.

Sakamoto, T., Yoshida, T., Arima, H., Hatanaka, Y., Takan, Y., and Tamaru, Y. (2009) Accumulation of trehalose in response to desiccation and salt stress in the terrestrial cyanobacterium *Nostoc commune*. *Phycological Res* **57:** 66–73.

Sakthivel, K., Watanabe, T., and Nakamoto, H. (2009) A small heat-shock protein confers stress tolerance and stabilizes thylakoid membrane proteins in cyanobacteria under oxidative stress. *Arch Microbiol* **191:** 319–328.

Salin, M. L. (1991). Chloroplast and mitochondrial mechanisms for protection against oxygen toxicity. *Free Rad Res Commun* **12–13:** 851–858.

Sanchez, Y., and Lindquist, S. L. (1990) HSP104 required for induced thermotolerance. *Science* **248:** 1112–1115.

Sandström, S., Park, Y-I., Öquist, G., and Gustafsson, P. (2001) CP43′, the *isiA* gene product, functions as an excitation energy dissipator in the cyanobacterium *Synechococcus* sp. PCC 7942. *Photochem Photobiol* **74:** 431–437.

Santarius, K. A. (1992) Freezing of isolated thylakoid membranes in complex media. VIII. Differential cryoprotection by sucrose, proline and glycerol. *Physiol Plant* **84:** 87–93.

Santoni, E., Jakopitsch, C., Obinger, C., and Smulevich, G. (2004) Comparison between catalase-peroxidase and cytochrome *c* peroxidase. The role of hydrogen bond networks for protein stability and catalysis. *Biochemistry* **43:** 5792–5802.

Santos, S., and da Costa, M. S. (2002) Compatible solutes of organisms that live in hot saline environments. *Environ Microbiol* **4:** 501–509.

Sassoon, N., Arie, J. P., and Betton, J. M. (1999) PDZ domains determine the native oligomeric structure of the DegP (HtrA) peptidase. *Mol Microbiol* **33:** 583–589.

Sato, M., Nimura-Matsune, K., Watanabe, S., Chibazakura, T., and Yoshikawa, H. (2007) Expression analysis of multiple *dnaK* genes in the cyanobacterium *Synechococcus elongatus* PCC 7942. *J Bacteriol* **189:** 3751–3758.

Sato, N. (1994) A cold-regulated cyanobacterial gene cluster encodes RNA-binding protein and ribosomal protein S21. *Plant Mol Biol* **18:** 165–170.

Sato, N. (1995) A family of cold-regulated RNA-binding protein genes in the cyanobacterium *Anabaena variabilis* M3. *Nucleic Acids Res* **23:** 2161–2167.

Sato, N., and Maruyama, K. (1997) Differential regulation by low temperature of the gene for an RNA-binding protein, *rbpA3*, in the cyanobacterium *Anabaena variabilis* strain M3. *Plant Cell Physiol* **38:** 81–86.

Sato, N., and Murata, N. (1980) Temperature shift-induced responses in lipids in the blue-green alga, *Anabaena variabilis*. The central role of diacylmonogalactosylglycerol in thermo-adaptation. *Biochim Biophys Acta* **619:** 353–366.

Sato, N., and Murata, N. (1981) Studies on the temperature shift induced desaturation of fatty acids in monogalactosyl diacylglycerol in the blue-green alga (cyanobacterium), *Anabaena variabilis*. *Plant Cell Physiol* **22:** 1043–1050.

Sato, N., and Nakamura, A. (1998) Involvement of the 5′-untranslated region in cold-regulated expression of the *rbpA*1 gene in the cyanobacterium *Anabaena variabilis* M3. *Nucleic Acids Res* **26:** 2192–2199.

Sato, N., Maruyama, K., Nishiyama, Y., and Murata, N. (1997a) Identification of a cold-regulated RNA binding protein from the marine cyanobacterium *Synechococcus* sp. PCC 7002 (Cyanophyta). *J Plant Res* **110:** 405–410.

Sato, N., Murata, N., Miura, Y., and Ueta, N. (1979) Effect of growth temperature on lipid and fatty acid compositions in the blue-green algae, *Anabaena variabilis* and *Anacystis nidulans*. *Biochim Biophys Acta* **572:** 19–28.

Sato, N. Tachikawa, T., Wada, A., and Tanaka, A. (1997b) Temperature-dependent regulation of the ribosomal small-subunit protein S21 in the cyanobacterium *Anabaena variabilis* M3. *J Bacteriol* **179:** 7063–7071.

Satoh, K., Hiral, M., Nishio, J., Yamaji, T., Kashino, Y., and Koike, H. (2002) Recovery of photosynthetic systems during rewetting is quite rapid in a terrestrial cyanobacterium *Nostoc commune*. *Plant Physiol* **43:** 170–176.

Scheibe, R. (1987) NADP$^+$-malate dehydrogenase in C3-plants: Regulation and role of a light-activated enzyme. *Physiol Plant* **71:** 393–400.

Schelin, J., Lindmark, F., and Clarke, A. K. (2002) The ClpP multigene family for the ATP-dependent Clp protease in the cyanobacterium *Synechococcus*. *Microbiology* **148:** 2255–2265.

Scherer, S., and Potts, M. (1989) Novel water stress protein from a desiccation-tolerant cyanobacterium: purification and partial characterization. *J Biol Chem* **264:** 12546–12553.

Scherer, S., Ernst, A., Chen, T.-W., and Böger, P. (1984). Rewetting of drought-resistant blue-green algae: time course of water uptake and reappearance of respiration, photosynthesis, and nitrogen fixation. *Oecologia* **62:** 418–423.

Schindelin, H., Jiang, W., Inouye, M., and Heinemann, U. (1994) Crystal structure of CspA, the major cold shock protein of *Escherichia coli*. *Proc Natl Acad Sci USA* **91:** 5119–5123.

Schmid, S. R., and Linder, P. (1992) D–E–A–D protein family of putative RNA helicases. *Mol Microbiol* **6:** 283–292.

Schmidt, M., Buchner, J., Todd, M. J., Lorimer, G. H., and Viitanen, P. V. (1994) On the role of GroES in the chaperonin-assisted folding reaction. *J Biol Chem* **269:** 10304–10311.

Schirmer, E. C., Glover, J. R., Singer, M. A., and Lindquist, S. (1996) HSP100/Clp proteins: a common mechanism explains diverse functions. *Trends Biol Sci* **21:** 289–296.

Schröder, E., and Ponting C. P. (1998) Evidence that peroxiredoxins are novel members of the thioredoxin fold superfamily. *Protein Sci* **7:** 2465–2468.

Schröder, E., Littlechild, J. A., Lebedev, A. A., Errington, N., Vagin, A. A., and Isupov, M. N. (2000) Crystal structure of decameric 2-Cys peroxiredoxin from human erythrocytes at 1.7 Å resolution. *Structure* **8:** 605–615.

Schroder, K., Graumann, P., Schnuchel, A., Holak, T. A., and Marahiel, M. A. (1995) Mutational analysis of the putative nucleic acid-binding surface of the cold-shock domain, cspb, revealed an essential role of aromatic and basic residues in binding of single-stranded-DNA containing the y-box motif. *Mol Microbiol* **16:** 699–708. doi: 10.1111/j.1365-2958.1995.tb02431.x

Schubert, H., Fulda, S., and Hagemann, M. (1993) Effects of adaptation to different salt concentrations on photosynthesis and pigmentation of the cyanobacterium *Synechocystis* sp. PCC 6803. *J Plant Physiol* **142:** 291–295.

Schürmann, P., and Buchanan, B. B. (2008) The ferredoxin/thioredoxin system of oxygenic photosynthesis. *Antioxid Redox Signal* **10:** 1235–1274.

Schwartz, S. H., Black, T. A., Jäger, K., Panoff, J.-M., and Wolk, C. P. (1998) Regulation of an osmoticum-responsive gene in *Anabaena* sp. strain PCC 7120. *J Bacteriol* **180:** 6332–6337.

Seaton, B. L., and Vickery, L. E. (1994) A gene encoding a DnaK/hsp70 homolog in *Escherichia coli. Proc Natl Acad Sci* USA **91:** 2066–2070.

Seckbach, J. (ed.) (2007) Algae and Cyanobacteria in Extreme Environments Vol 11 *Cellular Origin, Life in Extreme Habitats and Astrobiology*. Springer, Dordrecht, The Netherlands.

Servant, P., Grandvalet, C., and Mazodier, P. (2000) The RheA repressor is the thermosensor of the HSP18 heat shock response in *Streptomyces albus. Proc Natl Acad Sci USA* **97:** 3538–3543.

Servant, P., Rapoport, G., and Mazodier, P. (1999) RheA, the repressor of *hsp18* in *Streptomyces albus* G. *Microbiology* **145:** 2385–2391.

Shanklin, J., DeWitt, N. D., and Flanagan, J. M. (1995) The stroma of higher plant plastids contain ClpP and ClpC, functional homologs of *Escherichia coli* ClpP and ClpA: an archetypal two-component ATP-dependent protease. *Plant Cell* **7:** 1713–1722.

Shapiguzov, A., Lyukevich, A. A., Allakhverdiev, S. I., Sergeyenko, T. V., Suzuki, I., Murata, N., and Los, D. A. (2005) Osmotic shrinkage of cells of *Synechocystis* sp. PCC 6803 by water efflux via aquaporins regulates osmostress gene expression. *Microbiology* **151:** 447–455. doi:10.1099/mic.0.27530-0

Shaw, E., Hill, D. R., Brittain, N., Wright, W. J., Täuber, U., Marand, H., Helm, R. F., and Potts, M. (2003) Unusual water flux in the extracellular polysaccharide of the cyanobacterium *Nostoc commune. Appl Environ Microbiol* **69:** 5679–5684.

Shcolnick, S., Shaked, Y., and Keren, N. (2007) A role for mrgA, a DPS family protein, in the internal transport of Fe in the cyanobacterium *Synechocystis* sp. PCC 6803. *Biochim Biophys Acta* **1767:** 814–819.

Shcolnick, S., Summerfield, T. C., Reytman, L., Sherman, L. A., and Keren, N. (2009) The mechanism of iron homeostasis in the unicellular cyanobacterium *Synechocystis* sp. PCC 6803 and its relationship to oxidative stress. *Plant Physiol* **150:** 2045–2056.

Shen, J.-R., Qian, M., Inoue, Y., and Burnap, R. L. (1998) Functional characterization of *Synechocystis* sp. PCC 6803 DpsbU and DpsbV mutants reveals important roles of cytochrome *c-550* in cyanobacterial oxygen evolution. *Biochemistry* **37:** 1551–1558.

Shen-Miller, J., Mudgett, M. B., Schopf, J. W., Clarke, S., and Berger, R. (1995) Exceptional seed longevity and robust growth: Ancient sacred lotus from China. *Am J Bot* **82:** 1367–1380.

Shestakov, S. V., Anbudurai, P. R., Stanbekova, G. E., Gadzhiev, A., Lind, L. K., and Pakrasi, H. B. (1994) Molecular cloning and characterization of the *ctpA* gene encoding a carboxyl-terminal processing peptidase. Analysis of a spontaneous photosystem II-deficient mutant strain of the cyanobacterium *Synechocystis* sp. PCC 6803. *J Biol Chem* **269:** 19354–19359.

Shiraishi, T., Takahashi, M.-A., and Asada, K. (1994) Generation of O_2 anion radicals and hydroxyl radicals in chloroplast thylakoids. In: Asada, K., and Yoshikawa, T. (Eds.) *Frontiers of Reactive Oxygen Species in Biology and Medicine.* Excerpta Medica, Amsterdam, The Netherlands, pp. 31–32.

Shirkey, B., Kovarcik, D. P., Wright, D. J., Wilmoth, G., Prickett, T. F., Helm, R. F., Gregory, E. M. and Potts, M. (2000). Active Fe-containing superoxide dismutase and abundant sodF mRNA in *Nostoc commune* (Cyanobacteria) after years of desiccation. *J Bacteriol* **182:** 189–197.

Shirkey, B., McMaster, N. J., Smith, S. C., Wright, D. J., Rodriguez, H., Jaruga, P., Birincioglu, M., Helm, R. F., and Potts, M. (2003) Genomic DNA of *Nostoc commune* (cyanobacteria) becomes covalently modified during long-term (decades) desiccation but is protected from oxidative damage and degradation. *Nucleic Acids Res* **31:** 2995–3005.

Shoumskaya, M. A., Paithoonrangsarid, K., Kanesaki, Y., Los, D. A., Zinchenko, V. V., Tanticheran, M., Suzuki, I., and Murata, N. (2005) Identical Hik-Rre systems are involved in perception and transduction of salt signals and hypersmotic signals but regulate the expression of individual genes to different extents in *Synechocystis. J Biol Chem* **280:** 21531–21538.

Sigler, W. V, Bachofen, R., and Zeyer, J. (2003) Molecular characterization of endolithic cyanobacteria inhabiting exposed dolomite in central Switzerland. *Environ Microbiol* **5:** 618–627.

Silva, P., Choi, Y. J., Hassan, H. A., and Nixon, P. J. (2002) Involvement of the HtrA family of proteases in the protection of the cyanobacterium *Synechocystis* PCC 6803 from light stress and in the repair of photosystem II. *Phil Trans R Soc* B London **357:** 1461–1467.

Silva, P., Thompson, E., Bailey, S., Kruse, O., Mullineaux, C. W., Robinson, C., Mann, N. H., and Nixon, P. J. (2003) FtsH is involved in the early stages of repair of photosystem II in *Synechocystis* sp PCC 6803. *Plant Cell* **15:** 2152–2164.

Singer, M. A., and Lindquist, S. (1998a) Thermotolerance in *Saccharomyces cerevisiae*: the Yin and Yang of trehalose. *Trends Biotechnol* **16:** 460–468.

Singer, M. A., and Lindquist, S. (1998b) Multiple effects of trehalose on protein folding *in vitro* and *in vivo*. *Mol Cell* **1:** 639–648.

Singh, A. K., Li, H., and Sherman, L. A. (2004). Microarray analysis and redox control of gene expression in the cyanobacterium *Synechocystis* sp. PCC 6803. *Physiol Plant* **120:** 27–35.

Singh, D. P., and Kshatriya, K. (2002) NaCl-induced oxidative damage in the cyanobacterium *Anabaena doliolum*. *Curr Microbiol* **44:** 411–417.

Singh, S. C., Sinha, R. P., and Häder, D-P. (2002) Role of lipids and fatty acids in stress tolerance in cyanobacteria. *Acta Protozool* **41:** 297–308.

Singh, S. P., Klisch, M., Sinha, R. P., and Häder, D.-P. (2008) Effects of abiotic stressors on synthesis of the mycosporine-like amino acid shinorine in the cyanobacterium *Anabaena variabilis* PCC 7937. *Photochem Photobiol* **84:** 1500–1505.

Singh, S. P., Klisch, M., Sinha, R. P., and Häder, D.-P. (2009) Genome mining of mycosporine-like amino acid (MAA) synthesizing and non-synthesizing cyanobacteria. *Genomics* **95:** 120–128. doi: org/10.1016/j.ygeno.2009.10.002

Sinha, R. P., Ambasht, N. K., Sinha, J. P., Klisch, M., and Häder, D.-P. (2003) UV-B-induced synthesis of mycosporine-like amino acids in three strains of *Nodularia* (cyanobacteria). *J Photochem Photobiol B* **71:** 51–58.

Sinha, R. P., Klisch, M., Helbling, E. W., and Häder, D.-P. (2001) Induction of mycosporine-like amino acids (MAAs) in cyanobacteria by solar ultraviolet-B radiation. *J Photochem Photobiol B* **60:** 129–135.

Sinha, R. P., Singh, S. P., and Häder, D-P. (2007) Database on mycosporines and mycosporine-like amino acids (MAAs) in fungi, cyanobacteria, macroalgae, phytoplankton and animals. *J Photochem Photobiol* B *Biol* **89:** 29–35.

Sippola, K., Kanervo, E., Murata, N., and Aro, E.-M. (1998) A genetically engineered increase in fatty acid unsaturation in *Synechococcus* sp. PCC 7942 allows exchange of D1 protein forms and sustenance of photosystem II activity at low temperature. *Eur J Biochem* **251:** 641–648.

Siva, V., Rao, K., Brand, J. J., and Myers, J. (1977) Cold shock syndrome in *Anacystis nidulans*. *Plant Physiol* (Bethesda) **59:** 965–969.

Skorko-Glonek, J., Lipinska, B., Krzewski, K., Zolese, G., Bertoli, E., and Tanfani, F. (1997) HtrA heat stress peptidase interacts with phospholipid membranes and undergoes conformational changes. *J Biol Chem* **272:** 8974–8982.

Skorko-Glonek, J., Zurawa, D., Kuczwara, E., Wozniak, M., Wypych, Z., and Lipinska, B. (1999) The *Escherichia coli* heat stress peptidase HtrA participates in defense against oxidative stress. *Mol Gen Genet* **262:** 342–350.

Sokolenko, A., Pojidaeva, E., Zinchenko, V., Panichkin, V., Glaser, V. M., Herrmann, R. G., and Shestakov, S. V. (2002) The gene complement for proteolysis in the cyanobacterium *Synechocystis* sp. PCC 6803 and *Arabidopsis thaliana* chloroplasts. *Curr Genet* **41:** 291–310.

Soontharapirakkul, K., Promden, W., Yamada, N., Kageyama, H., Incharoensakdii, A., Iwamoto-Kihara, A., and Takabe, T. (2011) Halotolerant cyanobacterium *Aphanothece halophytica* contains an Na^+-dependent F1F0-ATP synthase with a potential role in salt stress tolerance. *J Biol Chem* **286:** 10169–10176.

Spiess, C., Beil, A., and Ehrmann, M. (1999) A temperature-dependent switch from chaperone to peptidase in a widely conserved heat stress protein. *Cell* **97:** 339–347.

Squires, C., and Squires, C. L. (1992) The Clp proteins: proteolysis regulators or molecular chaperones? *J Bacteriol* **174:** 1081–1085.

Stal, L. J., and Reed, R. H. (1987) Low molecular mass carbohydrate accumulation in cyanobacteria from marine microbial mat in response to salt. *FEMS Microbiol Lett* **45:** 305–312.

Stanne, T. M., Pojidaeva, E., Andersson, F. I., and Clarke, A. K. (2007) Distinctive types of ATP-dependent Clp proteases in cyanobacteria. *J Biol Chem* **282:** 14394–14402.

Steinitz, Y., Mazor, Z., and Shilo, M. (1979) A mutant of the cyanobacterium *Plectonema boryanum* resistant to photooxidation. *Plant Sci Lett* **16:** 327–335.

Stewart, G. R., Wernisch, L., Stabler, R., Mangan, J. A., Hinds, J., Laing, K. G., Young, D. B., and Butcher, P. D. (2002) Dissection of the heat-shock response in *Mycobacterium tuberculosis* using mutants and microarrays. *Microbiology* **148:** 3129–3138.

Stolz, J. F. (1990) Distribution of phototrophic microbes in the flat laminated microbial mat at Laguna Figueroa, Baja California, Mexico. *BioSystems* **23:** 345–357.

Stone, J. R., and Yang, S. (2006) Hydrogen peroxide: a signaling messenger. *Antioxid Redox Signal* **8:** 243–270. doi:10.1089/ars2006.8.243

Stork, T., Michel, K. P., Pistorius, E. K., and Dietz, K. J. (2005) Bioinformatic analysis of the genomes of the cyanobacteria *Synechocystis* sp. PCC 6803 and *Synechococcus elongatus* PCC 7942 for the presence of peroxiredoxins and their transcript regulation under stress. *J Exp Bot* **422:** 3193–3206.

Storz, G., and Zheng, M. (2000) Oxidative stress. In: Storz, G., and Hengge-Aronis, R. (Eds.) *Bacterial Stress Response*. ASM Press, Washington, D.C. USA. pp. 47–59.

Strandling, D. A., Thygerson, T., Walker, J. A., Smith, B. N. Hansen, L. D., Cridle, R. S., and Pendleton, R. L. (2002) Cryptogamic crust metabolism in response to temperature, water vapour and liquid water. *Thermochem Acta* **394:** 219–225.

Strauch, K. L., and Beckwith, J. (1988) An *Escherichia coli* mutation preventing degradation of abnormal periplasmic proteins. *Proc Natl Acad Sci USA* **85:** 1576–1580.

Strauch, K. L., Johnson, K., and Beckwith, J. (1989) Characterization of degP, a gene required for proteolysis in the cell envelope and essential for growth of *Escherichia coli* at high temperature. *J Bacteriol* **171:** 2689–2696.

Straus, N. A. (1994) Iron deprivation: physiology and gene regulation. In: Bryant, D.A. (Ed.) *The Molecular Biology of Cyanobacteria*. Kluwer Academic Publishers, Dordrecht, The Netherlands. pp. 731–75.

Strøm, A. R. (1998) Osmoregulation in the model organism *Escherichia coli* genes governing the synthesis of glycine betaine and trehalose and their use in metabolic engineering of stress tolerance. *J Biosci* **23:** 437–445. doi:10.1007/BF 02936137

Strøm, A. R., and Kaasen, I. (1993) Trehalose metabolism in *Escherichia coli* : stress protection and stress regulation of gene expression. *Mol Microbiol* **8:** 205–210.

Stulp, B. K., and Potts, M. (1987) Stability of nucleic acids in immobilized and desiccated *Nostoc commune* UTEX 584 (cyanobacteria). *FEMS Microbiol Lett* **41:** 241–245.

Sudhir, P.-R., Pogoryelov, D., Kovács, L., Garáb, G., and Murthy, S. D. S. (2005) The effects of salt stress on photosynthetic electron transport and thylakoid membrane proteins in the cyanobacterium *Spirulina platensis*. *J Biochem Mol Biol* **38:** 481–485.

Sueoka, K., Yamazaki, T., Hiyama, T., and Nakamoto, H. (2009) The NADPH thioredoxin reductase C functions as an electron donor to 2-Cys peroxiredoxin in a thermophilic cyanobacterium *Thermosynechococcus elongatus* BP-1. *Biochem Biophys Res Commun* **380:** 520–524. doi:10.1016/j.bbrc.2009.01.091

Sugita, M., and Sugiura, M. (1994) The existence of eukaryotic ribonucleoprotein consensus sequence-type RNA-binding proteins in a prokaryote, *Synechococcus* 6301. *Nucleic. Acids Res* **22:** 25–31.

Summerfield, T. C., and Sherman, L. A. (2008) Global transcriptome analysis of the alkali-tolerant cyanobacterium *Synechocystis* sp. strain PCC 6803. *Appl Environ Microbiol* **74:** 5276–5284.

Summerfield, T. C., Shand, J. A., Bentley, F. K., and Eaton-Rye, J. J. (2005) PsbQ (Sll1638) in *Synechocystis* sp. PCC 6803 is required for photosystem II activity in specific mutants and in nutrient-limiting conditions. *Biochemistry* **44:** 805–815.

Suzuki, I., Kanesaki, Y., Mikami, K., Kanehisa, M., and Murata, N. (2001) Cold-regulated genes under control of the cold sensor Hik33 in *Synechocystis*. *Mol Microbiol* **40:** 235–244.

Suzuki, I., Kanesaki, Y., Hayashi, H., Hall, J. J., Simon, W. J., Slabas, A. R., and Murata, N. (2005) The histidine kinase Hik34 is involved in thermotolerance by regulating expression of heat shock genes in *Synechocystis*. *Plant Physiol* **138:** 1409–1421.

Suzuki, I., Los, D. A., and Murata, N. (2000a) Perception and transduction of low temperature signals to induce desaturation of fatty acids. *Biochem Soc Trans* **28:** 628–630.

Suzuki, I., Los, D. A., Kanesaki, Y., Mikami, K., and Murata, N. (2000b) The pathway for perception and transduction of low-temperature signals in *Synechocystis*. *EMBO J* **19:** 1327–1334.

Suzuki, I., Simon, W. J., and Slabas, A. R. (2006) The heat shock response of *Synechocystis* sp. PCC 6803 analysed by transcriptomics and proteomics. *J Exp Bot* **57:** 1573–1578.

Suzuki, T., Yoshimura, H., Hisabori, T., and Ohmori, M. (2004) Two cAMP receptor proteins with different biochemical properties in the filamentous cyanobacterium *Anabaena* sp. PCC 7120. *FEBS Lett* **571:** 154–160.

Suzuki, T., Yoshimura, H., Ehira, S., Ikeuchi, M., and Ohmori, M. (2007) AnCrpA, a cAMP receptor protein, regulates *nif*-related gene expression in the cyanobacterium *Anabaena* sp. strain PCC 7120 grown with nitrate. *FEBS Lett* **581:** 21–28.

Symons, M. C. R. (1991) Free radicals in biological systems. In: Dreosti, I. E. (Ed.) *Trace Elements, Micronutrients and Free Radicals*. Humana Press, Totowa, NJ, USA. pp. 1–24.

Szalontai, B., Nishiyama, Y., Gombos, Z., and Murata, N. (2000) Membrane dynamics as seen by fourier transform infrared spectroscopy in a cyanobacterium, *Synechocystis* PCC 6803. The effects of lipid unsaturation and the protein-to-lipid ratio. *Biochim Biophys Acta* **1509:** 409–419.

Takaichi, S., Maoka, T., and Mochimaru, M. (2009) Unique carotenoids in the terrestrial cyanobacterium *Nostoc commune* NIES-24: 2-hydroxymyxol-2′-fucoside, nostoxanthin and canthaxanthin. *Curr Microbiol* **59:** 413–419.

Takeyama, H., Burgess, J. G., Sudo, H., Soda, K., and Matsunaga, T. (1991) Salinity dependent copy number increase of a marine cyanobacterial endogenous plasmid. *FEMS Microbiol Lett* **90:** 95–98.

Tamaru, Y., Takani, Y., Yoshida, T., and Sakamoto, T. (2005) Crucial role of extracellular polysaccharides in desiccation and freezing tolerance in the terrestrial cyanobacterium *Nostoc commune*. *Appl Environ Microbiol* **71:** 7327–7333.

Tamoi, M., Kurotaki, H., and Fukamizo, T. (2007) β-1,4-Glucanase-like protein from the cyanobacterium *Synechocystis* PCC 6803 is a β-1,3,4-glucanase and functions in salt stress tolerance. *Biochem J* **405:** 139–146.

Tanaka, N., and Nakamoto, H. (1999) HtpG is essential for the thermal stress management in cyanobacteria. *FEBS Lett* **458:** 117–123.

Tanner, N. K., and Linder, P. (2001) DExD/H box RNA helicases: from generic motors to specific dissociation functions. *Mol Cell* **8:** 251–262

Tanner, N. K., Cordin, O., Banroques, J., Doère, M., and Linder, P. (2003) The Q motif: a newly identified motif in DEAD box helicases may regulate ATP binding and hydrolysis. *Mol Cell* **11:** 127–138.

Taranto, P. A., Keenan, T. W., and Potts, M. (1993). Rehydration induces rapid onset of lipid biosynthesis in desiccated *Nostoc commune* (Cyanobacteria). *Biochim Biophys Acta* **1168:** 228–237.

Tasaka, Y., Gombos, Z., Nishiyama, Y., Mohanty, P., Ohba, T., Ohki, K., and Murata, N. (1996) Targeted mutagenesis of acyl lipid desaturases in *Synechocystis*: evidence for the important roles of polyunsaturated membrane lipids in growth, respiration and photosynthesis. *EMBO J* **15:** 6416–6425.

Taylor, B. L., and Zhulin, I. B. (1999) PAS domains: internal sensors of oxygen, redox potential, and light. *Microbiol Mol Biol Rev* **63:** 479–506.

Tel-Or, E., Huflejt, M., and Packer, L. (1985) The role of glutathione and ascorbate in hydroperoxide removal in cyanobacteria. *Biochem Biophys Res Commun* **132:** 533–539.

Thevelein, J. M. (1984) Regulation of trehalose mobilisation in fungi. *Microbiol Rev* **48:** 42–59.

Thomas, D. J., Avenson, T. J., Thomas, J. B., and Herbert, S. K. (1998) A cyanobacterium lacking iron superoxide dismutase is sensitized to oxidative stress induced with methyl viologen but is not sensitized to oxidative stress induced with norflurazon. *Plant Physiol* **116:** 1593–1602.

Thomas, J. A., Poland, B., and Honzatko, R. (1995) Protein sulfhydryls and their role in the antioxidant function of protein S-thiolation. *Arch Biochem Biophys* **319:** 1–9.

Thompson, M. W., and Maurizi, M. R. (1994) Activity and specificity of *Escherichia coli* ClpAP protease in cleaving model peptide substrates. *J Biol Chem* **269:** 18201–18208.

Tichy, M., and Vermaas, W. (1999) *In vivo* role of catalase-peroxidase in *Synechocystis* sp. strain PCC 6803. *J Bacteriol* **181:** 1875–1882.

Tirkey, J., and Adhikary, S. P. (2005) Cyanobacteria in biological soil crusts of India. *Curr Sci* **89:** 515–521.

Tiwari, B. S. and Tripathi, S. N. (1998) Effect of hydration and dehydration and dynamics of some physiological reactions in desiccation tolerant cyanobacterium *Scytonema geitleri*. *Indian J Biochem Biophys* **35:** 172–178.

Tiwari, B. S., and Tripathi, S. N. (2001) Photosynthetic activities of roof-top desiccation tolerant cyanobacterium *Scytonema geitleri* at varying hydration levels. *Algae* **16:** 445–455.

Tiwari, G. L. (1972) Morphology and life history of *Chroococcidiopsis* Geitler (Chroococcales). *Hydrobiologia* **40:** 177–182.

Tomb, J. F., White, O., Kerlavage, A. R., Clayton, R. A., Sutton, G. G., Fleischmann, R. D., Ketchum, K. A., Klenk, H. P., Gill, S., Dougherty, B. A., Nelson, K., Quackenbush, J., Zhou, L., Kirkness, E. F., Peterson, S., Loftus, B., Richardson, D., Dodson, R., Khalak, H. G., Glodek, A., McKenney, K., Fitzgerald, L. M., Lee, N., Adams, M. D., Hickey, E. K., Berg, D. E., Gocayne, J. D., Utterback, C. R., Peterson, J. D., Kelley, J. M., Cotton, M. D., Weidman, J. M., Fujii, C., Bowman, C., Watthey, L., Wallin, E., Hayes, W. S., Borodovsky, M., Karp, P. D., Smith, H. O., Fraser, C. M., and Venter, J. C. (1997) The complete genome sequence of the gastric pathogen *Helicobacter pylori*. *Nature* (London) **388:** 539–547.

Tomoyasu, T., Yamanaka, K., Murata, K., Suzaki, T., Bouloc, P., Kato, A., Niki, H., Hiraga, S., and Ogura, T. (1993) Topology and subcellular localization of FtsH protein in *Escherichia coli*. *J Bacteriol* **175:** 1352–1357.

Tomoyasu, T., Ogura, T., Tatsuta, T., and Bukau, B. (1998) Levels of DnaK and DnaJ provide tight control of heat shock gene expression and protein repair in *Escherichia coli*.*Mol Microbiol* **30:** 567–581.

Török, Z., Goloubinoff, P., Horváth, I., Tsvetkova, N. M., Glatz, A., Balogh, G., Varvasovszki, V., Los, D. A., Vierling, E., Crowe, J. H., and Vigh, L. (2001) *Synechocystis* Hsp17 is an amphitropic protein that stabilizes heat-stressed membranes and binds denatured proteins for subsequent chaperone-mediated refolding. *Proc Natl Acad Sci USA* **98:** 3098–3103.

Tripathi, S. N., and Srivastava, P. (2001) Presence of stable active scavenging enzymes superoxide dismutase, ascorbate peroxidase and catalase in a desiccation-tolerant cyanobacterium *Lyngbya arboricola* under dry state. *Curr Sci* **81:** 197–200.

Tripathi, S. N., Tiwari, B. S., and Talpasayi, E. R. S. (!991) Growth of cyanobacteria (blue-green algae) on urban buildings. *Energy and Buildings* **15:** 499–505.

Tripathy, P., Amarpalli, R., and Adhikary, S. P. (1997) Survey of epilithic bluegreen algae (cyanobacteria) from temples of India and Nepal. *Arch Hydrobiol Algol Stud* **87:** 43–57.

Tripathy, P., Roy, A., Anand, N., and Adhikary, S. P. (1999) Blue-green algal flora on the rock surface of temples and monuments of India. *Feddes Repertorium* **110:** 133–144.

Trüper, H. G., and Galinski, E. A. (1990) Biosynthesis and fate of compatible solutes in extremely halophilic phototrophic eubacteria. *FEMS Microbiol Rev* **75:** 247–254.

Tsvetkova, N. M., Horváth, I., Török, Z., Wolkers, W. F., Balogi, Z., Shigapova, N., Crowe, L. M., Tablin, F., Vierling, E., Crowe, J. H., and Vigh, L. (2002) Small heat-shock proteins regulate membrane lipid polymorphism. *Proc Natl Acad Sci USA* **99:** 13504–13509.

Tyystjarvi, T., Tyystjarvi, E., Ohad, I., and Aro, E. (1998) Exposure of *Synechocystis* 6803 cells to series of single turnover flashes increases the *psbA* transcript level by activating transcription and down-regulating *psbA* mRNA degradation. *FEBS Lett* **436:** 483–487.

Usher, K. M., Bergman, B., and Raven, R. A. (2007) Exploring cyanobacterial mutualisms. *Annu Rev Ecol Evol Syst* **38:** 255–273.

Van Bogelen, R., and Neidhardt, F. C. (1990) Ribosomes as sensors of heat and cold shock in *Escherichia coli*. *Proc Natl Acad Sci USA* **87:** 5589–5593.

Vandenbroucke, K., Robbens, S., Vandepoele, K., Inzé, D., Van de Peer, and Van Breusegen, F. (2008) Hydrogen peroxide-induced gene expression across Kingdoms: A comparative analysis. *Mol Biol Evol* **25:** 507–516.

van Montfort, R. L. M., Basha, E., Friedrich, K. L., Slingsby, C., and Vierling, E. (2001) Crystal structure and assembly of a eukaryotic small heat shock protein. *Nat Struct Biol* **8:** 1025–1030.

van Thielen, N., and Garbary, D. J. (1999) Life in the rocks: endolithic algae. In: Seckbach, J. (Ed.) *Enigmatic Microorganisms and Life in Extreme Environments*. Kluwer Academic Publishers, Dordrecht, The Netherlands. pp. 243–253

Varvasovszki, V., Glatz, A., Shigapova, N., Jósvay, K., Vigh, L., and Horváth, I. (2003) Only one *dnaK* homolog, *dnaK2* is active transcriptionally and is essential in *Synechocystis*. *Biochem Biophys Res Commun* **305:** 641–645.

Vasishta, P. C. (1968) Thermal cyanophyceae of India. *Phykos* **7:** 198–241.

Veal, E. A., Day, A. M., and Morgan, B. A. (2007) Hydrogen peroxide sensing and signaling. *Mol Cell* **26:** 1–14.

Veal, E. A., Findlay, V. J., Day, A. M., Bozonet, S. M., Evans, J. M., Quinn, J., and Morgan, B. A. (2004) A 2-Cys peroxiredoxin regulates peroxide-induced oxidation and activation of a stress-activated MAP kinase. *Mol Cell* **15:** 129–139.

Veinger, L., Diamant, S., Buchner, J., and Goloubinoff, P. (1998) The small heat-shock protein IbpB from *Escherichia coli* stabilizes stress-denatured proteins for subsequent refolding by a multichaperone network. *J Biol Chem* **273:** 11032–11037.

Venter, J. C., Remington, K., Heidelberg, J. F., Halpern, A. L., Rusch, D., Eisen, J. A., Wu, D., Paulsen, I., Nelson, K. E., Nelson, W., Fouts, D. E., Levy, S., Knap, A. H., Lomas, M. W., Nealson, K., White, O., Peterson, J., Hoffman, J. J., Parsons, R., Baden-Tillson, H., Pfannkoch, C., Rogers, Y. H., and Smith, H. O. (2004) Environmental genome shotgun sequencing of the Sargasso Sea. *Science* **304:** 66–74.

Viale, A. M., Arakaki, A. K., Soncini, F. C., and Ferreyra, R. G. (1994) Evolutionary relationships among eubacterial groups as inferred from GroEL (chaperonin) sequence comparisons. *Int J Syst Bacteriol* **44:** 527–533.

Vigh, L., Los, D. A., Horváth, I., and Murata, N. (1993) The primary signal in the biological perception of temperature: Pd-catalyzed hydrogenation of membrane lipids stimulated the expression of the *desA* gene in *Synechocystis* PCC 6803. *Proc Natl Acad Sci USA* **90:** 9090–9094.

Vigh, L., Maresca, B., and Harwood, J. L. (1998) Does the membrane's physical state control the expression of heat shock and other genes? *Trends Biochem Sci* **23:** 369–374.

Vigh, L. O., Nakamoto, H., Landry, J., Gomez-Munoz, A., Harwood, J. L., and Horvath, I. (2007) Membrane regulation of stress response from prokaryotic models to mammalian cells. *Ann New York Acad Sci* **1113:** 40–51.

Vincent, W. F., Gibson, J. A., Pienitz, R., Villeneuve, V., Broady, P. A., Hamilton, P. B., and Howard-Williams, C. (2000) Ice shelf microbial ecosystems in the High Arctic and implications for life on snowball Earth. *Naturwissenschaften* **87:** 137–141.

Vincent, W. F., Mueller, D. R., and Bonilla, S. (2004) Ecosystems on ice: the microbial ecology of Markham Ice Shelf in the high Arctic. *Cryobiology* **48:** 103–112.

Vinnemeier, J., and Hagemann, M. (1999) Identification of salt-regulated genes in the genome of the cyanobacterium *Synechocystis* sp. strain PCC 6803 by subtractive RNA hybridization. *Arch Microbiol* **172:** 377–386.

Vinnemeier, .J, Kunert, A., and Hagemann, M. (1998) Transcriptional analysis of the *isiAB* operon in salt-stressed cells of the cyanobacterium *Synechocystis* sp. PCC 6803. *FEMS Microbiol Lett* **169:** 323–330.

Vlasits, J., Furtmüller, P. G., Jakopitsch, C., Zámacký, M., and Obinger, C. (2010) Probing hydrogen peroxide oxidation kinetics of wild-type *Synechocystis* catalase-peroxidase (KatG) and selected variants. *Biochem Biophys Acta* **1804:** 799–805.

Vlasits, J., Jakopitsch, C., Schwanninger, M., Holubar, P., and Obinger, C. (2007) Hydrogen peroxide oxidation by catalase-peroxidase follows a non-scrambling mechanism. *FEBS Lett* **581:** 320–324.

Vonshak, A., Guy, R., and Guy, M. (1988) The response of the filamentous cyanobacterium *Spirulina platensis* to salt stress. *Arch Microbiol* **150:** 417–420.

Vranová, E., Inzé, D., and Van Breusegem, F. (2002) Signal transduction during oxidative stress. *J Exp Bot* **53:** 1227–1236.

Wada, H., and Murata, N. (1989) *Synechocystis* PCC 6803 mutants defective in desaturation of fatty acids. *Plant Cell Physiol* **30:** 971–978.

Wada, H., and Murata, N. (1990) Temperature-induced changes in the fatty acids composition of the cyanobacterium, *Synechocystis* PCC 6803. *Plant Physiol* **92:** 1062–1069.

Wada, H., Avelange-Macherel, M. H., and Murata, N. (1993) The *desA* gene of the cyanobacterium *Synechocystis* sp. strain PCC 6803 is the structural gene for Δ^{12} desaturase. *J Biol Chem* **175:** 6056–6058.

Wada, H., Gombos, Z., and Murata, N. (1990) Enhancement of chilling tolerance of a cyanobacterium by genetic manipulation of fatty acid desaturation. *Nature* (London) **347:** 200–203.

Wada, H., Gombos, Z., and Murata, N. (1994) Contribution of membrane lipids to the ability of the photosynthetic machinery to tolerate temperature stress. *Proc Natl Acad Sci USA* **91:** 4273–4277.

Wada, H., Gombos, Z., Sakamoto, T., and Murata, N. (1992) Genetic manipulation of the extent of desaturation of fatty acids in membrane lipids in the cyanobacterium *Synechocystis* PCC 6803. *Plant Cell Physiol* **33:** 535–540.

Waditee, R., Bhuiyan, Md. N. H., Rai, V., Aoki, K., Tanaka, Y., Hibino, T., Suzuki, S., Takano, J., Jagendorf, A. T., Takabe, T., and Takabe, T. (2005) Genes for direct methylation of glycine provide high levels of glycine betaine and abiotic stress tolerance in *Synechococcus* and *Arabidopsis*. *Proc Natl Acad Sci USA* **102:** 1318–1323.

Waditee, R., Bhuiyan, Md. N. H., Hirata, E., Hibino, T., Tanaka, Y., Shikata, M., and Takabe, T. (2007) Metabolic engineering of betaine accumulation in microbes and plants. *J Biol Chem* **282:** 34185–34193.

Waditee, R., Hibino, T., Tanaka, Y. Nakamura,T., Incharoensakdi, A., and Takabe, T. (2001) Halotolerant cyanobacterium *Aphanothece halophytica* contains an Na^+/H^+ antiporter, homologous to eukaryotic ones, with novel ion specificity affected by C-terminal tail. *J Biol Chem* **276:** 36931–36938.

Waditee, R., Hibino, T., Nakamura, T., Incharoensakdi, A., and Takabe, T. (2002) Overexpression of a Na^+/H^+ antiporter confers salt tolerance on a freshwater cyanobacterium, making it capable of growth in sea water. *Proc Natl Acad Sci USA* **99:** 4109–4114.

Waditee, R., Hossain, G. S., Tanaka, Y., Nakamura, T., Shikata, M., Takano, J., Takabe, T., and Takabe, T. (2004) Isolation and functional characterization of Ca^{2+}/H^+ antiporters from cyanobacteria. *J Biol Chem* **279:** 4330–4338.

Waditee, R., Tanaka, Y., Aoki, K., Hibino, T., Jikuya, H., Takano, J., Takabe, T., and Takabe, T. (2003) Isolation and functional characterization of N-methyltransferases that catalyze betaine synthesis from glycine in a halotolerant photosynthetic organism *Aphanothece halophytica*. *J Biol Chem* **278:** 4932–4942.

Wang, H. L., Postier, B. L., and Burnap, R. L. (2002) Polymerase chain reaction-based mutageneses identify key transporters belonging to multigene families involved in Na^+ and pH homeostasis of *Synechocystis* sp. PCC 6803. *Mol Microbiol* **44:** 1493–1506.

Wang, J. C., and Lynch, A. S. (1993) Transcription and DNA supercoiling. *Curr Opin Gene Dev* **3:** 764–768.

Warr, S. R. C., Reed, R. H., and Stewart, W. D. P. (1987) Low-molecular weight carbohydrate biosynthesis and the distribution of cyanobacteria (Blue-Green Algae) in marine environments. *Br Phycol J* **22:** 175–180.

Warr, S. R. C., Reed, R. H., Chudek, J. A., Foster, R., and Stewart, W. D. P. (1985) Osmotic adjustment in *Spirulina platensis*. *Planta* **163:** 424–429.

Webb, R., Reddy, K. J., and Sherman, L. A. (1990) Regulation and sequence of the *Synechococcus* sp. strain PCC 7942 *groESL* operon, encoding a cyanobacterial chaperonin. *J Bacteriol* **172:** 5079–5088.

Weber, M.H., and Marahiel, M.A. (2003) Bacterial cold shock responses. *Sci Prog* **86:** 9–75.

Welinder, K.G. (1992) Superfamily of plant, fungal and bacterial peroxidases, *Curr Opin Struct Biol* **2:** 388–393.

Wells, W. W., Yang, Y., and Diets, T. L. (1993) Thioltransferases. *Adv Enzymol* **66:** 149–201.

White, O., Eisen, J. A., Heidelberg, J. F, Hickey, E. K., Peterson, J. D., Dodson, R. J., Haft, D. H., Gwinn, M. L., Nelson, W. C., Richardson, D. L., Moffat, K. S., Qin, H., Jiang, L., Pamphile, W., Crosby, M., Shen, M., Vamathevan, J. J., Lam, P., McDonald, L., Utterback, T., Zalewski, C., Makarova, K. S., Aravind, L., Daly, M. J., Minton, K.W., Fleischmann, R. D., Ketchum, K. A., Nelson, K.E., Salzberg, S., Smith, H. O., Venter, J. C., and Fraser, C. M. (1999) Genome sequence of the radioresistant bacterium *Deinococcus radiodurans* R1. *Science* **286:** 1571–1577.

Whitton, B. A. (1982) *Rivers, Lakes and Marshes*. Hodder & Stoughton. London, UK. pp. 131.

Whitton, B. A. (1992) Diversity, ecology, and taxonomy of the cyanobacteria. *In*: Mann, N.H., and Carr, N.G., (Eds.) *Photosynthetic Prokaryotes*, Plenum Press, New York, NY, USA. pp. 1–51.

Whitton, B. A., and Potts, M. (2000) *Ecology of Cyanobacteria: Their Diversity in Time and Space*. Kluwer, Dordrecht, The Netherlands.

Whitton, B. A., Donaldson, A., and Potts, M. (1979) Nitrogen fixation by *Nostoc* colonies in terrestrial environments of Aldabra Atoll, Indian Ocean. *Phycologia* **18:** 278–287.

Wickner, S., Gottesman, S., Skowyra, D., Hoskins, J., McKenney, K., and Maurizi, M. R. (1994) A molecular chaperone, ClpA, functions like DnaK and DnaJ. *Proc Natl Acad Sci USA* **91:** 12218–12222.

Wierzchos, J., Ascaso, C., and Mckay, C. P. (2006) Endolithic cyanobacteria in halite rocks from hype rarid core of the Atacama Desert. *Astrobiology* **6:** 415–422.

Wilken, C., Kitzing, K., Kurzbauer, R., Ehrmann, M., and Clausen, T. (2004) Crystal structure of the DegS stress sensor: how a PDZ domain recognizes misfolded protein and activates a protease. *Cell* **117:** 483–494.

Williams, S. B., and Stewart, V. (1999) Functional similarities among two component sensors and methyl-accepting chemotaxis proteins suggest a role for linker region amphipathic helices in transmembrane signal transduction. *Mol Microbiol* **33:** 1093–1102.

Willimsky, G., Bang, H., Fischer, G., and Marahiel, M. A. (1992) Characterization of *cspB*, a *Bacillus subtilis* inducible cold shock gene affecting cell viability at low temperatures. *J Bacteriol* **174:** 6326–6335.

Wilson, N. W., and Nierhaus, K. H. (2004) The how and Y of cold shock. *Nat Struct Mol Biol* **11:** 1026–1028.

Winterbourn, C. C. (2008) Reconciling the chemistry and biology of reactive oxygen species. *Nat Chem Biol* **4:** 278–286

Wojtkowiak, D., Georgopoulos, C., and Zylicz, M. (1993) Isolation and characterization of ClpX, a new ATP-dependent specificity component of the Clp protease of *Escherichia coli*. *J Biol Chem* **268:** 22609–22617.

Wood, J. M. (1999) Osmosensing by bacteria: signals and membrane based sensors. *Microbiol Mol Biol Rev* **63:** 230–262.

Wood, Z. A., Schroder, E., Robin Harris, J., and Poole, L. B. (2003) Structure, mechanism and regulation of peroxiredoxins. *Trends Biochem Sci.* **28:** 32–40.

Wright, D. J., Smith, S. C., Joardar, V., Scherer, S., Jervis, J., Warren, A., Helm, R. F., and Potts, M. (2005) UV-irradiation and desiccation modulate the three-dimensional extracellular matrix of *Nostoc commune* (cyanobacteria). *J Biol Chem* **280:** 40271–40281.

Wutipraditkul, N., Waditee, R., Incharoensakdi, A., Hibino, T., Tanaka, Y., Nakamura,T., Shikata, M., Takabe, T., and Takabe, T. (2005) Halotolerant cyanobacterium *Aphanothece halophytica* contains NapA-type Na^+/H^+ antiporters with novel ion specificity that are involved in salt tolerance at alkaline pH. *Appl Environ Microbiol* **71:** 4176–4184.

Xia, B., Valmis-Gardikas, A., Holmgren, A., Wright, P. E., and Dyson, H. (2001) Solution structure of *Escherichia coli* glutaredoxin-2 shows similarity to mammalian glutathione S-transferases. *J Mol Biol* **310:** 907–918.

Xie, W., Tice, D., and Potts, M. (1995) Cell-water deficit regulates expression of *rpoC1C2* (RNA polymerase) at the level of mRNA in desiccation-tolerant *Nostoc commune* UTEX 584 (cyanobacteria). *FEMS Microbiol Lett* **126:** 159–164.

Xu, W. L., Jeanjean, R., Liu, Y. D., and Zhang, C. C. (2003) *pkn22* (*alr2502*) encoding a putative Ser/Thr kinase in the cyanobacterium *Anabaena* sp. PCC 7120 is induced by both iron starvation and oxidative stress and regulates the expression of *isiA*. *FEBS Lett.* **553:** 179–182.

Yamamoto, H., Miyake, C., Dietz, K. J., Tomizawa, K., Murata, N., and Yokota, A. (1999) Thioredoxin peroxidase in the cyanobacterium *Synechocytis* sp. PCC 6803. *FEBS Lett* **447:** 269–273.

Yancey, P. H., Clark, M. E., Hand, S. C., Bowlus, R. D., and Somero, G. N. (1982) Living with water stress: evolution of osmolyte systems. *Science* **217:** 1214–1222.

Yang, Y., Yin, C., Li, W., and Xu, X. (2008) α-Tocopherol is essential for acquired chill-light tolerance in the cyanobacterium *Synechocystis* sp. strain PCC 6803. *J Bacteriol* **190:** 1554–1560.

Yin, C., Li, W., Du, Y., Kong, R., and Xu, X. (2007) Identification of a gene, *ccr-1* (*sll1242*), required for chill-light tolerance and growth at 15°C in *Synechocystis* sp. PCC 6803. *Microbiology* **153:** 1261–1267.

Yoshida, T., and Sakamoto, T. (2009) Water-stress induced trehalose accumulation and control of trehalose in the cyanobacterium *Nostoc punctiforme* IAM M-15. *J Gen Appl Microbiol* **55:** 135–145.

Yoshimune, K., Yoshimura, T. and Esaki, N. (1998) Hsc62, a new DnaK homolog of *Escherichia coli*. *Biochem Biophys Res Commun* 250: 115–118.

Yoshimura, H., Ikeuchi, M., and Ohmori, M. (2006) Up-regulated gene expression during dehydration in a terrestrial cyanobacterium, *Nostoc* sp. strain HK-01. *Microbes Environ* **21:** 129–133.

Yoshimura, H., Okamoto, S., Tsumuraya, Y., and Ohmori, M. (2007) Group 3 sigma factor gene, *sigJ*, a key regulator of desiccation tolerance, regulates synthesis of extracellular polysaccharide in the cyanobacterium *Anabaena* sp. strain PCC 7120. *DNA Res* **14:** 13–24.

Youn, H.-D., Yim, Y.-I., Kim, K., Hah, Y. C., and Kang, S.-O. (1995) Spectral characterization and chemical modification of catalase-peroxidase from *Streptomyces* sp. *J Biol Chem* **270:** 13740–13747.

Youn, H. D., Youn, H., Lee, J. W., Yim, Y. I., Lee, J. H., Hah, Y. C., and Kang, S.-O. (1996) A novel nickel-containing superoxide dismutase from *Streptomyces* spp. *Biochem J* **318:** 889–896.

Yousef, N., Pistorius, E. K., and Michel, K.-P. (2003) Comparative analysis of *idiA* and *isiA* transcription under iron starvation and oxidative stress in *Synechococcus elongatus* PCC 7942 wild-type and selected mutants. *Arch Microbiol* **180:** 471–483.

Yu, E., and Owttrim, G. W. (2000) Characterization of the cold stress-induced cyanobacterial DEAD-box protein CrhC as an RNA helicase. *Nucleic Acids Res* **28:** 3926–3934.

Yura, T., Kanemori, M., and Morita, M. T. (2000) The heat shock response: regulation and function. In: Storz, G., and Hengge-Aronis, R. (Eds.), *Bacterial Stress Responses*. ASM Press, Washington, DC. USA. pp. 3–18.

Zach, A., Adi, I., Galit, S-V., and Wataru, S. (2005) FtsH proteases in chloroplasts and cyanobacteria. *Physiol Plant* **123:** 386–390.

Zak, E., Norling, B., Maitra, R., Huang, F., Andersson, B., and Pakrasi, H. B. (2001) The initial steps of biogenesis of cyanobacterial photosystems occur in plasma membranes. *Proc Natl Acad Sci USA* **98:** 13443–13448.

Zámocký, M. (2004) Phylogenetic relationships in class I of the superfamily of bacterial, fungal and plant peroxidases. *Eur J Biochem* **271:** 3297–3309.

Zámocký, M., Furtmüller, P. G., and Obinger, C. (2008) Evolution of catalases from bacteria to humans. *Antioxid Redox Signal* **10:** 1527–1547.

Zámocký, M., Furtmüller, P. G., and Obinger, C. (2010) Evolution of structure and function of class I peroxidases. *Arch Biochem Biophys* **500:** 45–57.
Zámocký, M., Regelsberger, G., Jakopitsch, C., and Obinger, C. (2001) The molecular peculiarities of catalase-peroxidases. *FEBS Lett* **492:** 177–182.
Zeng, M. T., and Vonshak, A. (1998) Adaptation of *Spirulina platensis* to salinity stress. *Comp Biochem Physiol* **120:** 113–118.
Zheng, M., Aslund, F., and Storz, G. (1998) Activation of the OxyR transcription factor by reversible disulfide bond formation. *Science* **279:** 1718–1721.
Zeth, K., Offermann, S., Essen, L. O., and Oesterhelt, D. (2004) Iron-oxo clusters biomineralizing on protein surfaces: structural analysis of *Halobacterium salinarum* DpsA in its low- and high-iron states. *Proc Natl Acad Sci USA* **101:** 13780–13785.
Zhang, L-F., Yang, H-M., Cui, S-X., Hu, J., Wang, J., Kuang, T-Y., Norling, B., and Huang, F. (2009) Proteomic analysis of plasma membranes of cyanobacterium *Synechocystis* sp. strain PCC 6803 in response to high pH stress. *J Proteome Res* **8:** 2892–2902. doi: 10. 1021/pr9000 24w
Zhang, T., Gong, H., Wen, X., and Lu, C. (2010) Salt stress induces a decrease in excitation energy transfer from phycobilisomes to photosystem II but an increase to photosystem I in the cyanobacterium *Spirulina platensis*. *J Plant Physiol* **167:**951–958.
Zhao, J. D., and Brand, J. J. (1988). Sequential effects of sodium depletion on photosystem II in *Synechocystis*. *Arch Biochem Biophys* **264:** 657–664.
Zhao, W., Guo, Q., and Zhao, J. (2007) A membrane-associated Mn-superoxide dismutase protects the photosynthetic apparatus and nitrogenase from oxidative damage in the cyanobacterium *Anabaena* sp. PCC 7120. *Plant Cell Physiol* **48:** 563–572.
Zhu, X., Zhao, X., Burkholder, W. F., Gragerov, A., Ogata, C. M., Gottesman, M. E., and Hendrickson, W. A. (1996) Structural analysis of substrate binding by the molecular chaperone DnaK. *Science* **272:** 1606–1614.
Zolkiewski, M. (1999) ClpB cooperates with DnaK, DnaJ and GrpE in suppressing protein aggregation. A novel multi-chaperone system from *Escherichia coli*. *J Biol Chem* **274:** 28083–28086.
Zuther, E., Scubert, H., and Hagemann, M. (1998) Mutation of a gene encoding a putative glycoprotease leads to reduced salt tolerance, altered pigmentation and cyanophycin accumulation in the cyanobacterium *Synechocystis* sp. strain PCC 6803. *J Bacteriol* **180:** 1715–1722.

INDEX

C

D

E

F

G

H

N

O

P

R

S

T

U

V

W

Color Plate Section

Chapter 2

Figure 7

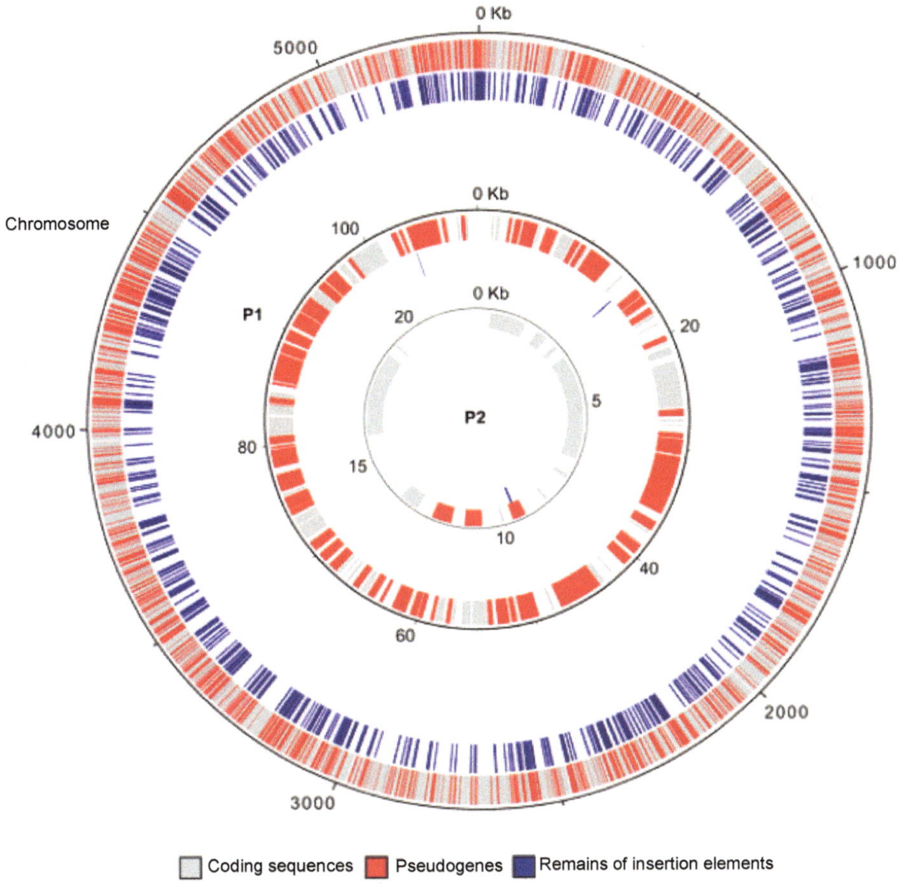

Figure 9

http://www.queenmaryphotosynthesis.org/nield/psIIimages/oxygenicphotosynthmodel.html
(embryophyte)

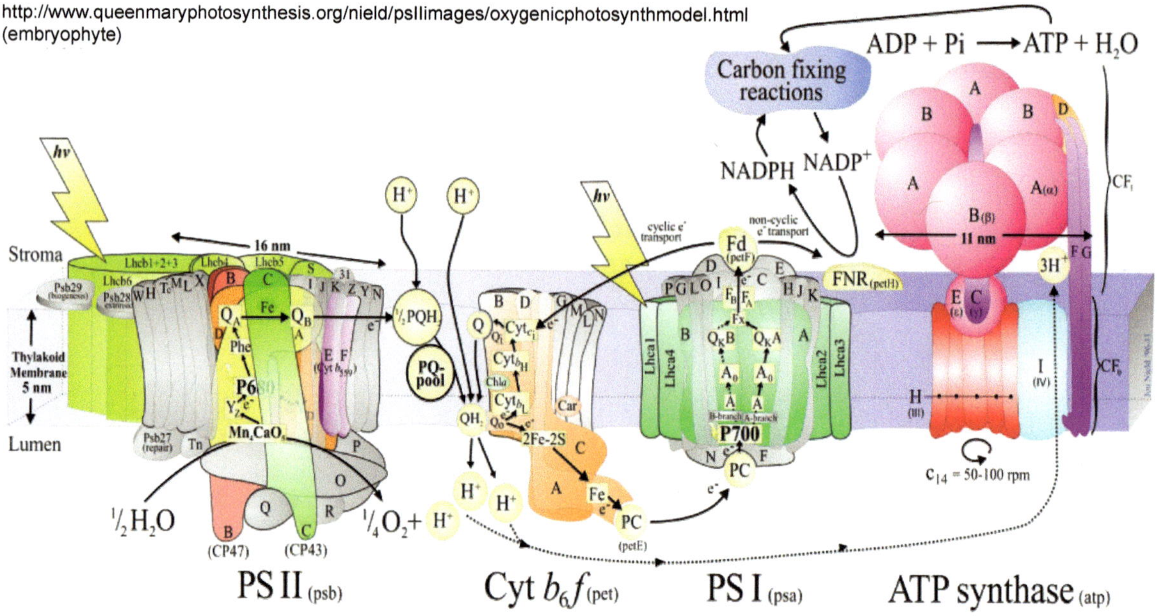

Figure 10

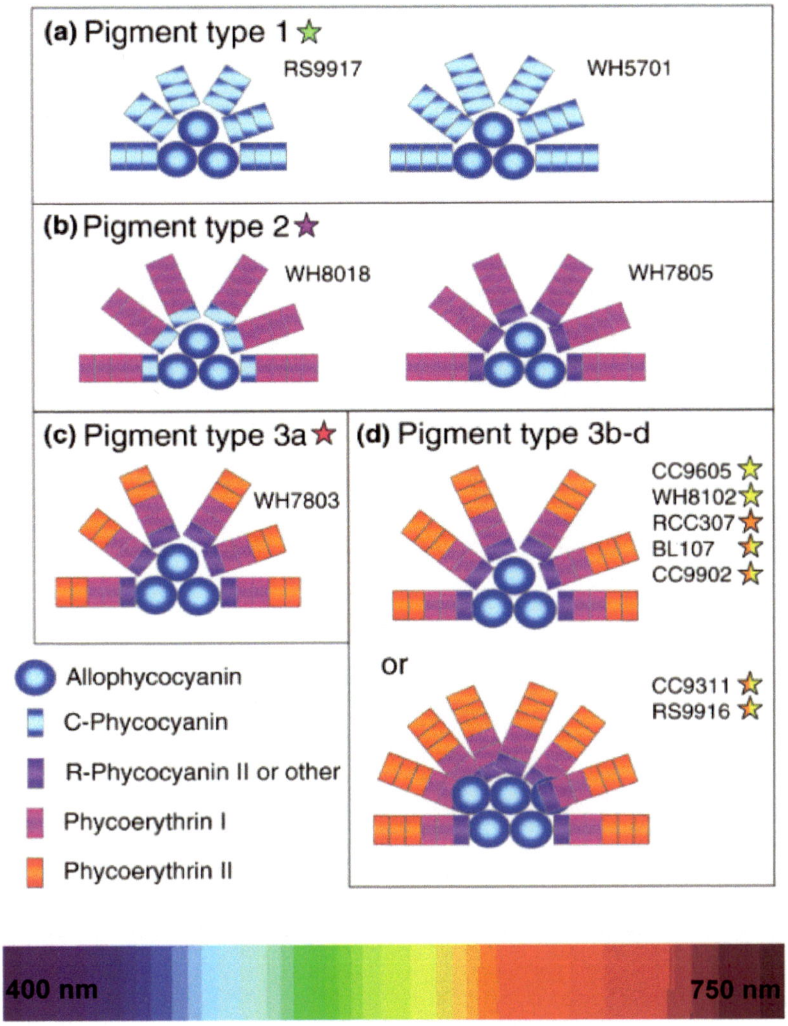
(a) Pigment type 1
RS9917
WH5701
(b) Pigment type 2
WH8018
WH7805
(c) Pigment type 3a
WH7803
(d) Pigment type 3b-d
CC9605
WH8102
RCC307
BL107
CC9902
or
CC9311
RS9916
Allophycocyanin
C-Phycocyanin
R-Phycocyanin II or other
Phycoerythrin I
Phycoerythrin II
400 nm
750 nm

Figure 12

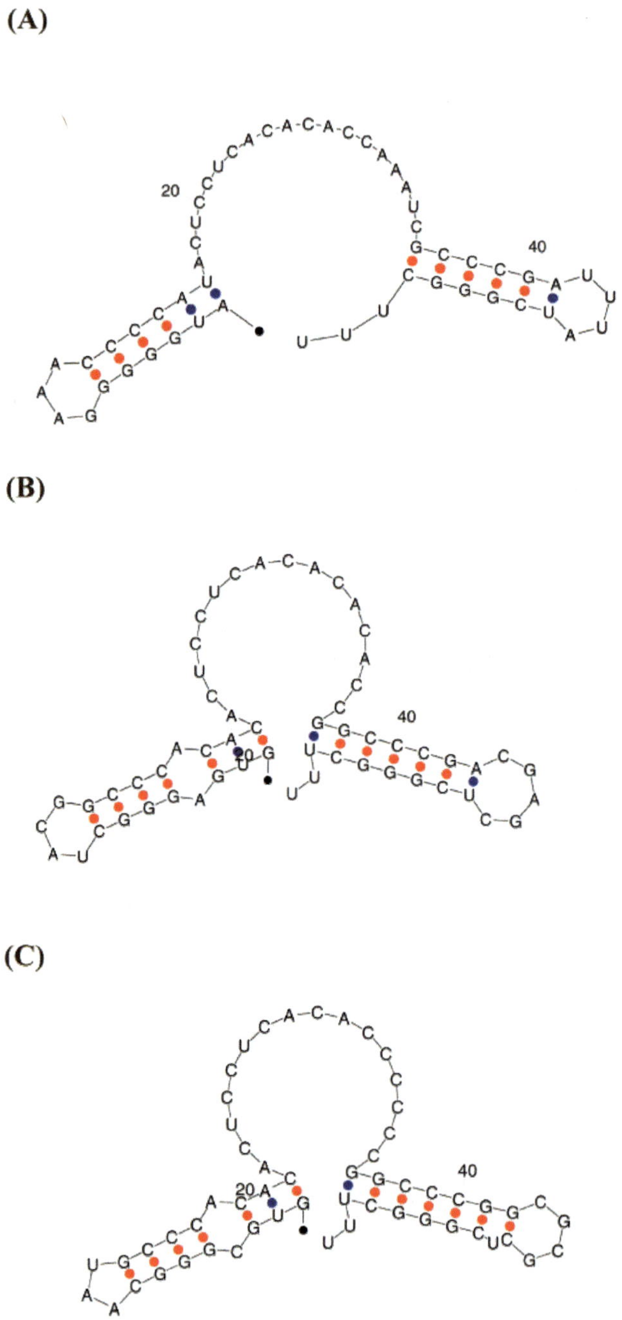

Figure 14

Families	Architecture	Distribution															
		9313	9902	9605	9311	6301	7942	CYB	CYA	TBP1	7421	6803	Cwat	Tery	7120	Ava	Npun
cbSTKI	STK										6		11	2	4	6	6
	STK – transmembrane helices									1		1	1	1	2	3	4
	STK ———	1	1	1	1	4	4	6	5	5	5	4	11	6	15	19	19
cbSTKII	STK – SH3b											1					
	STK – ANF												1		1	1	
	STK – DUF323													9	1	1	1
	STK – [pentapeptide]$_{2-5}$					1	1	1	1	1		1	1	2	1	1	1
	STK – GUN4							1	1				1	2	1		
	STK – CHASE2																1
	CHASE2 – STK									2					2	2	2
	STK – (WD40)$_{6-7}$												1	6	3	3	5
	STK – (TPR)$_{3-10}$								1		3			5	2	3	4
	STK – PbH1												1		1	1	
	STK – RDD													1			
	FHA – STK									1				2	2	1	1
	STK – FHA													1			
	STK ——— GAF													2			
cbSTKIII	STK // GAF HisKA ATPase													1	11	9	8
	STK // GAF GAF HisKA ATPase														1	1	
	STK // GAF PAS HisKA ATPase														1		
	STK // GAF PAC HisKA ATPase															1	
	STK // GAF PAS PAC HisKA ATPase																1
	STK // GAF [PAS PAC]$_2$ HisKA ATPase																1
	STK // GAF [PAS PAC]$_4$ PAS HisKA ATPase															1	
	STK // GAF PAS GAF HisKA ATPase																1
	STK // GAF PAS PAC GAF GAF HisKA ATPase																1

Chapter 4

Figure 2

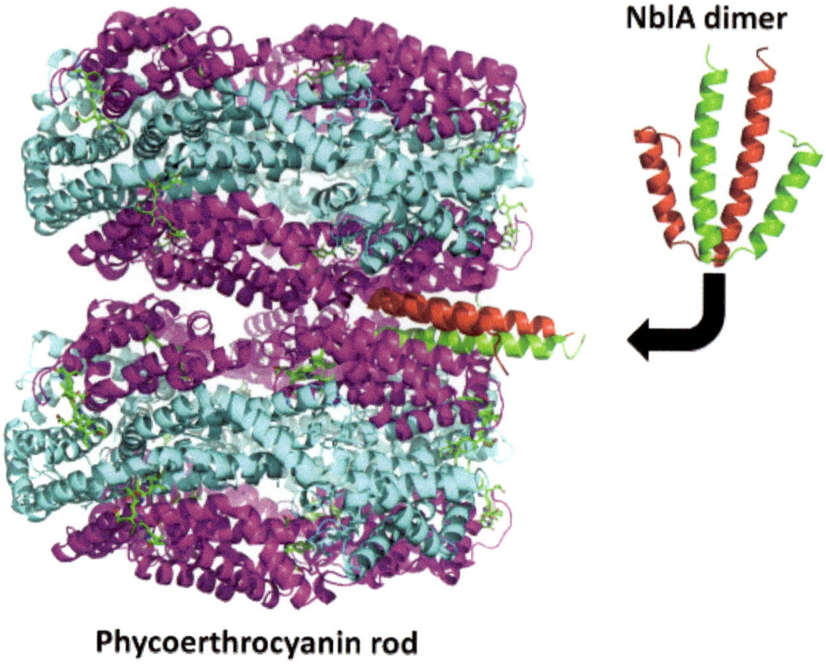

Figure 4

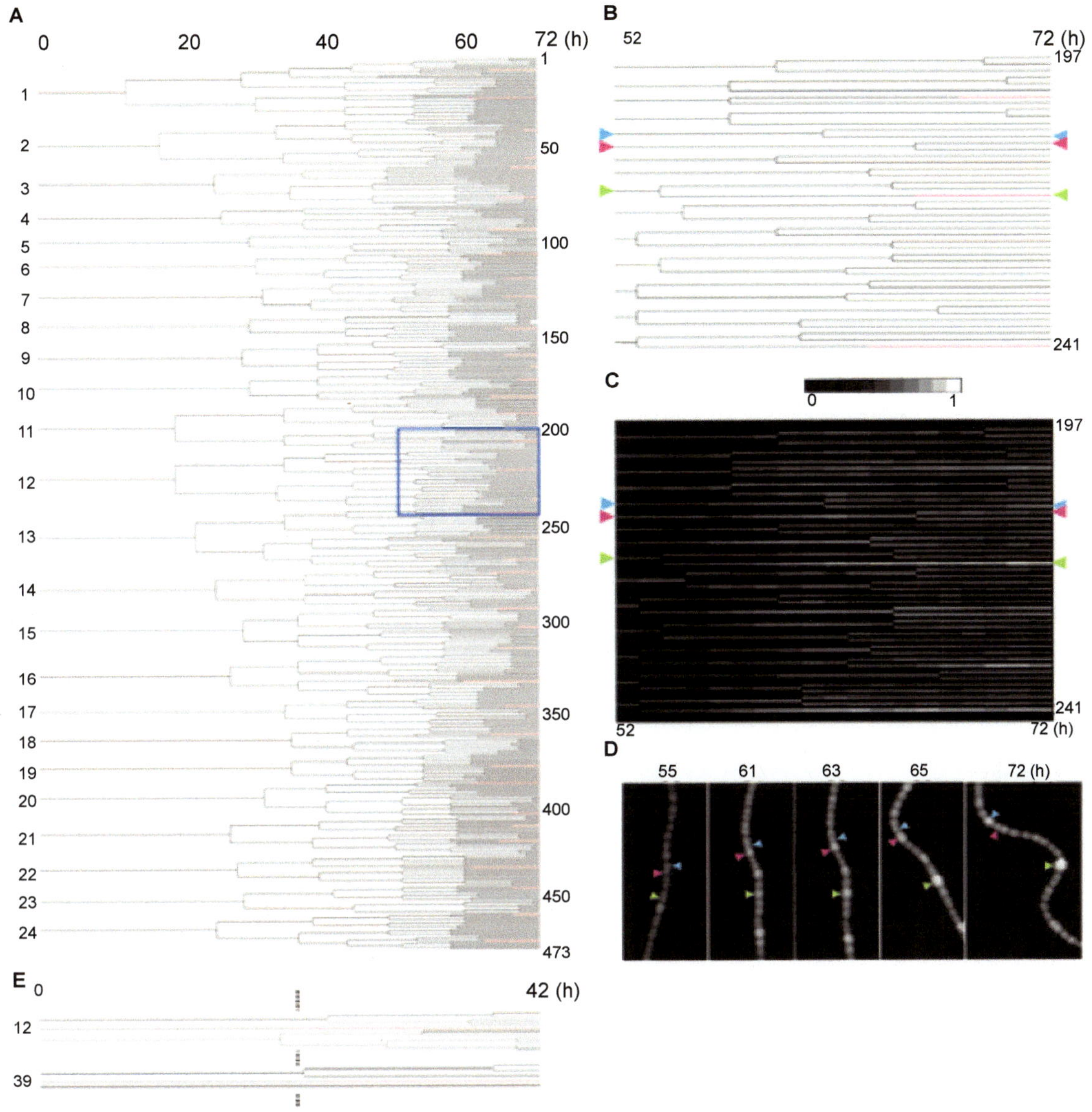
A
0
20
40
60
72 (h)
1
50
100
150
200
250
300
350
400
450
473
1
2
3
4
5
6
7
8
9
10
11
12
13
14
15
16
17
18
19
20
21
22
23
24
B
52
72 (h)
197
241
C
0
1
197
241
52
72 (h)
D
55
61
63
65
72 (h)
E
0
42 (h)
12
39

Figure 8

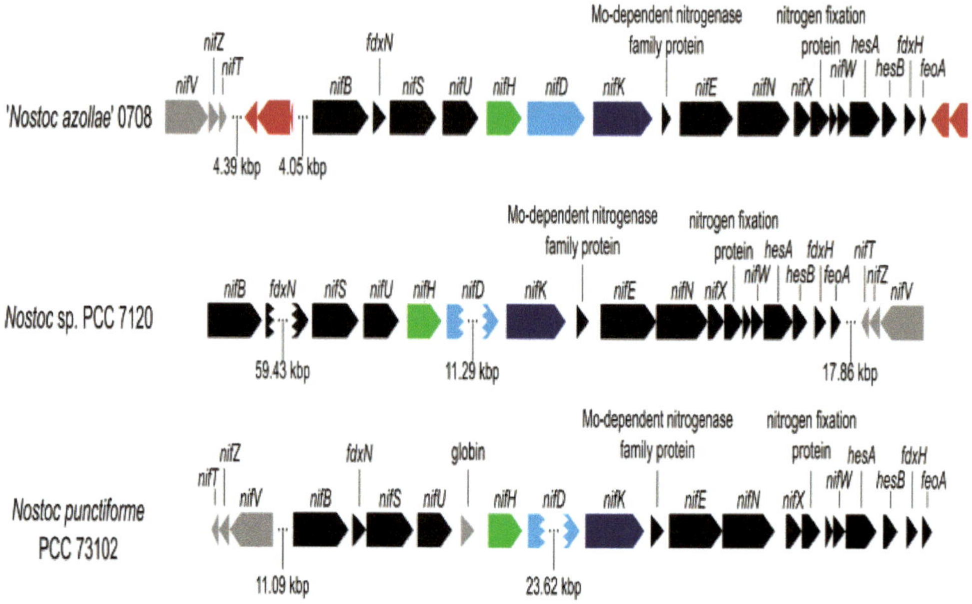
'Nostoc azollae' 0708
nifV
nifZ
nifT
4.39 kbp
4.05 kbp
nifB
fdxN
nifS
nifU
nifH
nifD
nifK
Mo-dependent nitrogenase family protein
nifE
nifN
nifX
nitrogen fixation protein
nifW
hesA
hesB
fdxH
feoA
Nostoc sp. PCC 7120
59.43 kbp
11.29 kbp
17.86 kbp
Nostoc punctiforme PCC 73102
globin
11.09 kbp
23.62 kbp

Chapter 6

Figure 1

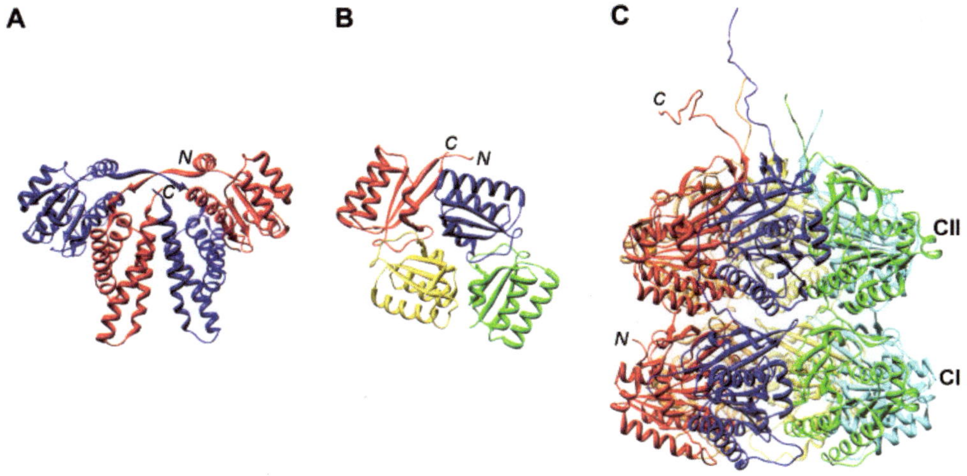

Figure 4

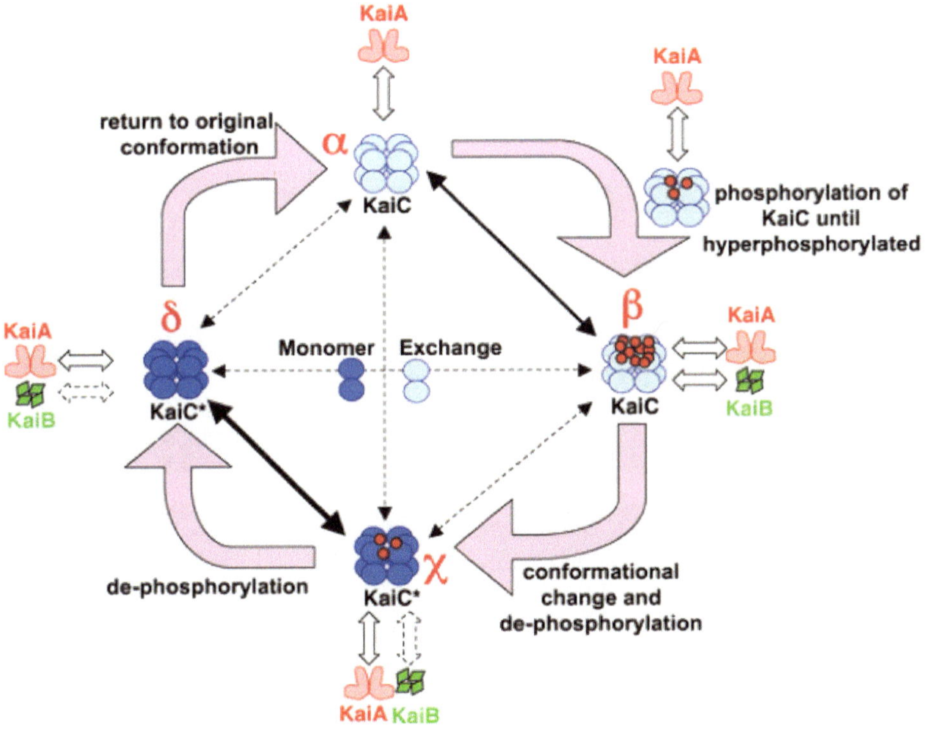

Figure 5

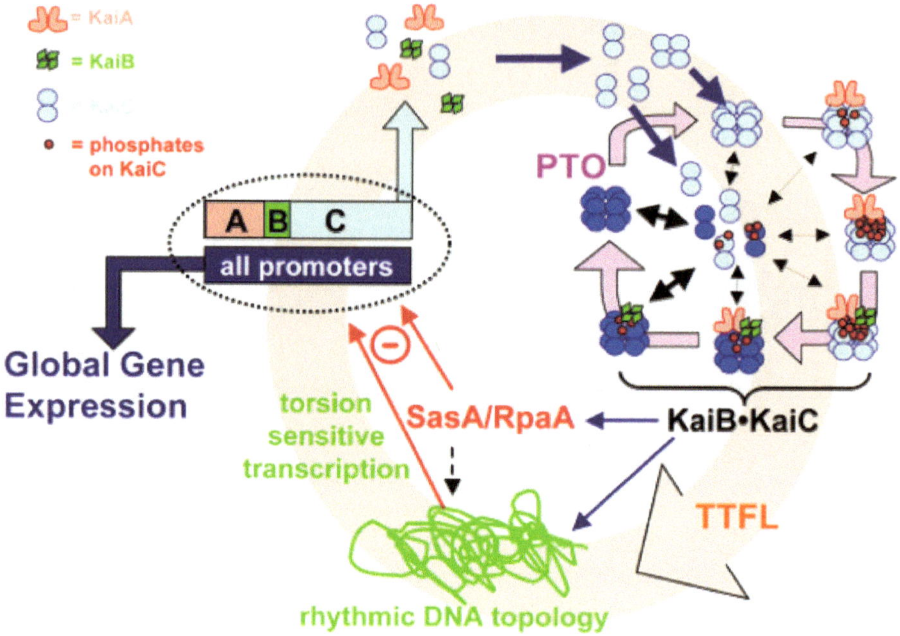

Chapter 7

Figure 2

Figure 3

(A) (B) (C) (D)

Figure 4

(A) (B) (C) (D)

Figure 5

(A)

(B)

Figure 6

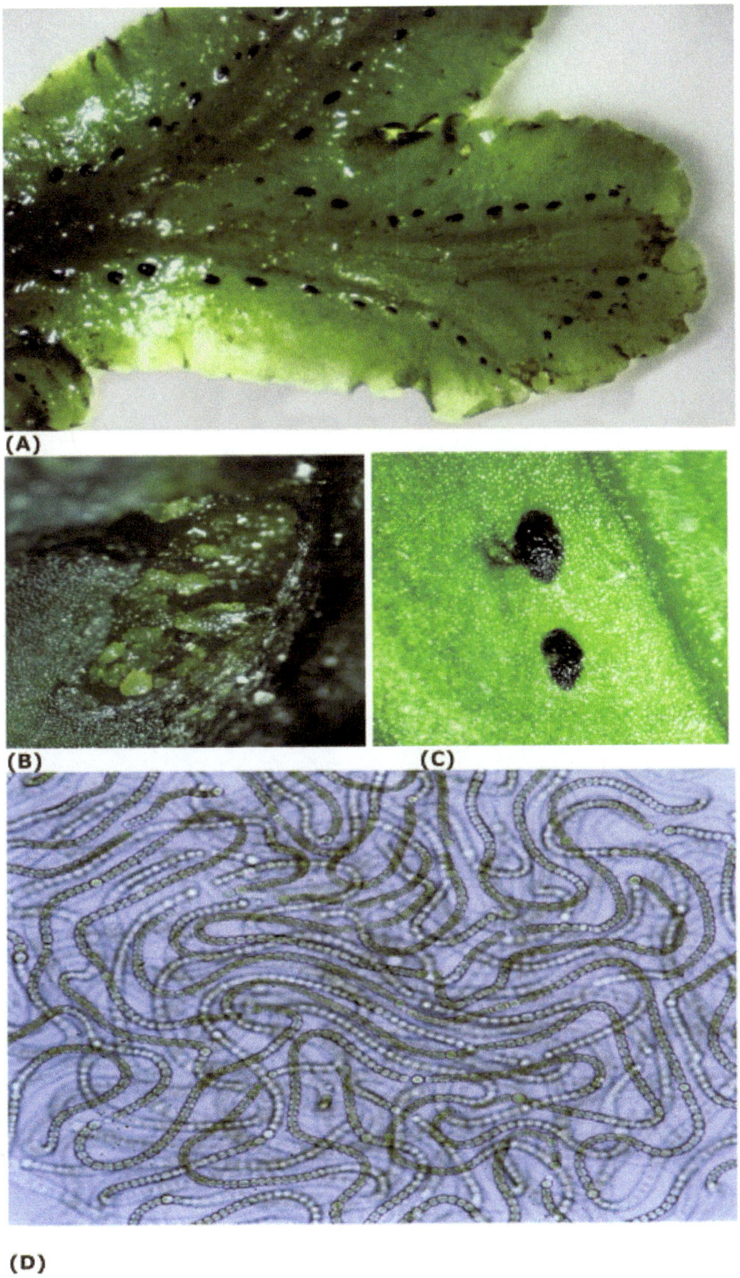

Figure 7

Figure 9

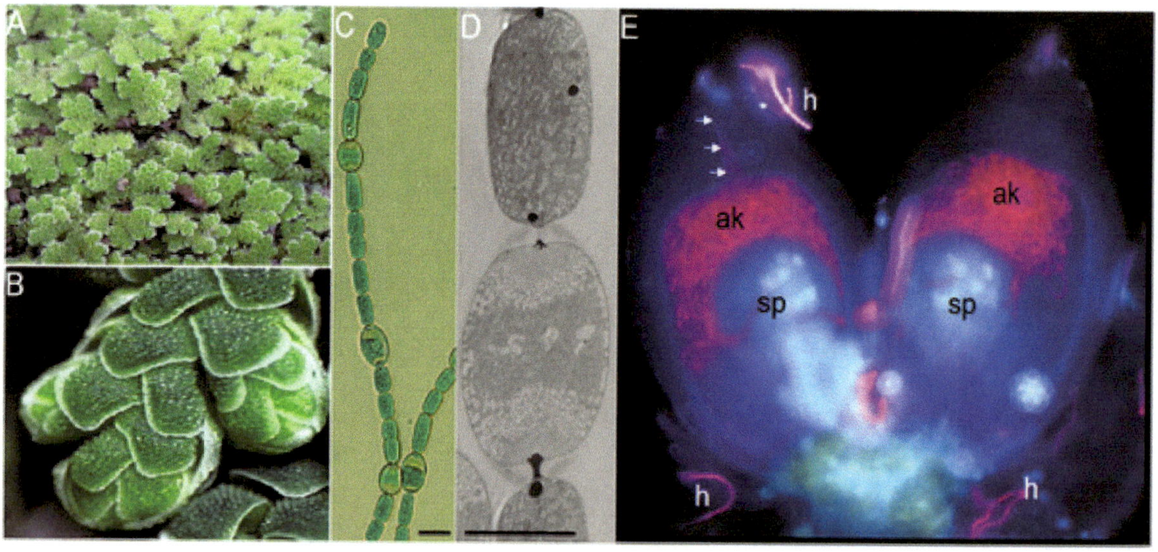

Figure 11

(A)

(B)

(C)

(D)

Figure 12

(A) (B) (C) (D)

Figure 13

(A)

(B)

Chapter 8

Figure 6

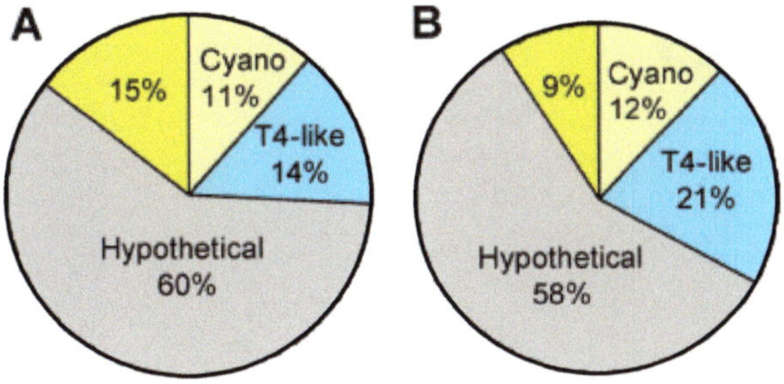

Figure 7

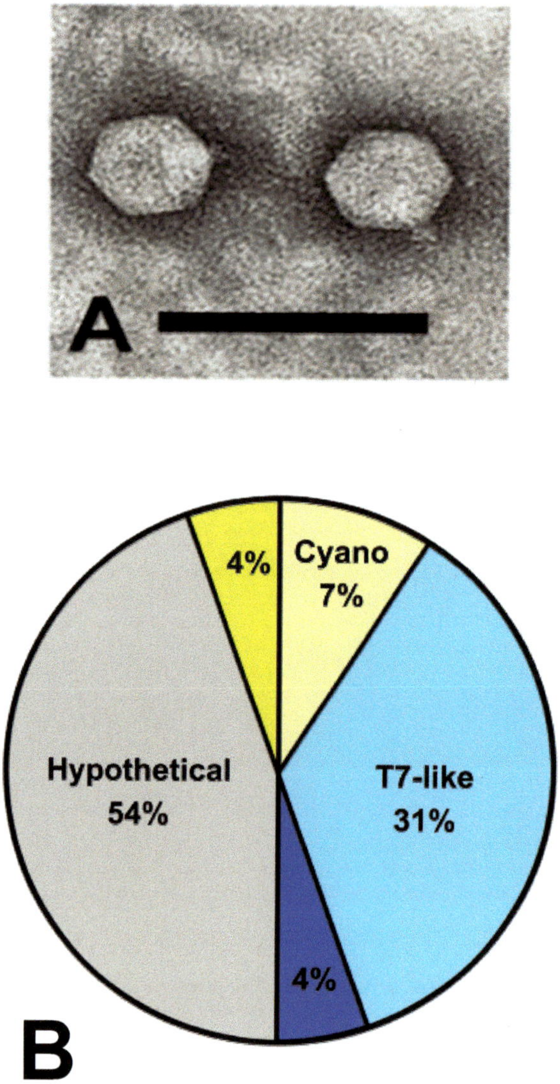

Figure 8

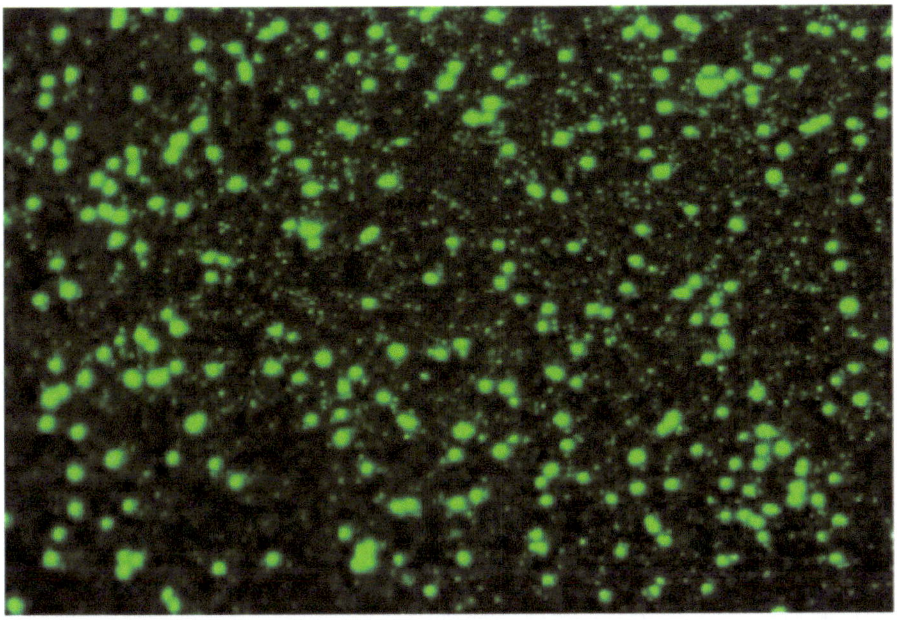

Figure 12

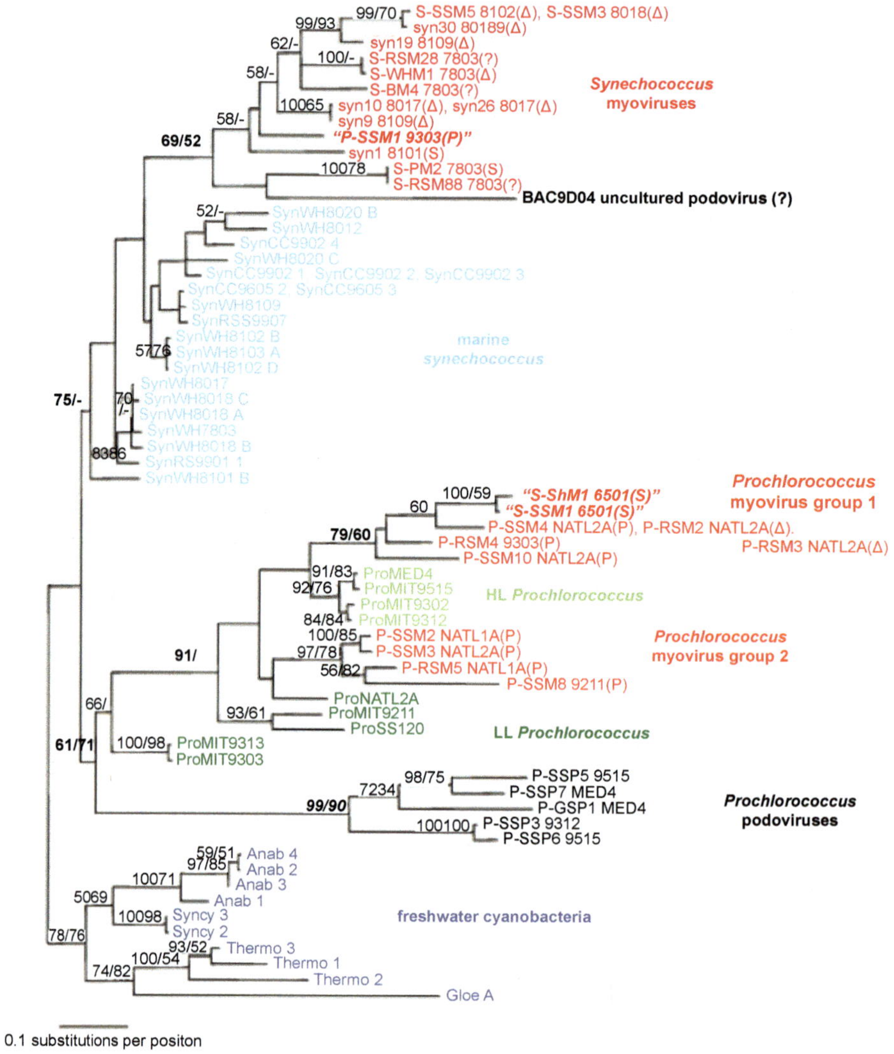

S-SSM5 8102(Δ), S-SSM3 8018(Δ)
syn30 80189(Δ)
syn19 8109(Δ)
S-RSM28 7803(?)
S-WHM1 7803(Δ)
S-BM4 7803(?)
syn10 8017(Δ), syn26 8017(Δ)
syn9 8109(Δ)
"P-SSM1 9303(P)"
syn1 8101(S)
S-PM2 7803(S)
S-RSM88 7803(?)
BAC9D04 uncultured podovirus (?)
Synechococcus
myoviruses
99/70
99/93
62/-
100/-
58/-
10065
58/-
69/52
10078
52/-
SynWH8020 B
SynWH8012
SynCC9902 4
SynWH8020 C
SynCC9902 1, SynCC9902 2, SynCC9902 3
SynCC9605 2, SynCC9605 3
SynWH8109
SynRSS9907
SynWH8102 B
SynWH8103 A
SynWH8102 D
SynWH8017
SynWH8018 C
SynWH8018 A
SynWH7803
SynWH8018 B
SynRS9901 1
SynWH8101 B
marine
synechococcus
5776
75/-
70
/-
8386
"S-ShM1 6501(S)"
"S-SSM1 6501(S)"
P-SSM4 NATL2A(P), P-RSM2 NATL2A(Δ),
P-RSM3 NATL2A(Δ)
P-RSM4 9303(P)
P-SSM10 NATL2A(P)
Prochlorococcus
myovirus group 1
100/59
60
79/60
91/83
ProMED4
ProMIT9515
92/76
ProMIT9302
84/84
ProMIT9312
HL Prochlorococcus
100/85
P-SSM2 NATL1A(P)
97/78
P-SSM3 NATL2A(P)
56/82
P-RSM5 NATL1A(P)
P-SSM8 9211(P)
Prochlorococcus
myovirus group 2
91/
ProNATL2A
93/61
ProMIT9211
ProSS120
LL Prochlorococcus
66/
61/71
100/98
ProMIT9313
ProMIT9303
98/75
P-SSP5 9515
7234
P-SSP7 MED4
P-GSP1 MED4
99/90
100100
P-SSP3 9312
P-SSP6 9515
Prochlorococcus
podoviruses
59/51
Anab 4
97/85
Anab 2
10071
Anab 3
Anab 1
5069
10098
Syncy 3
Syncy 2
freshwater cyanobacteria
78/76
93/52
Thermo 3
100/54
Thermo 1
74/82
Thermo 2
Gloe A
0.1 substitutions per positon

Figure 13

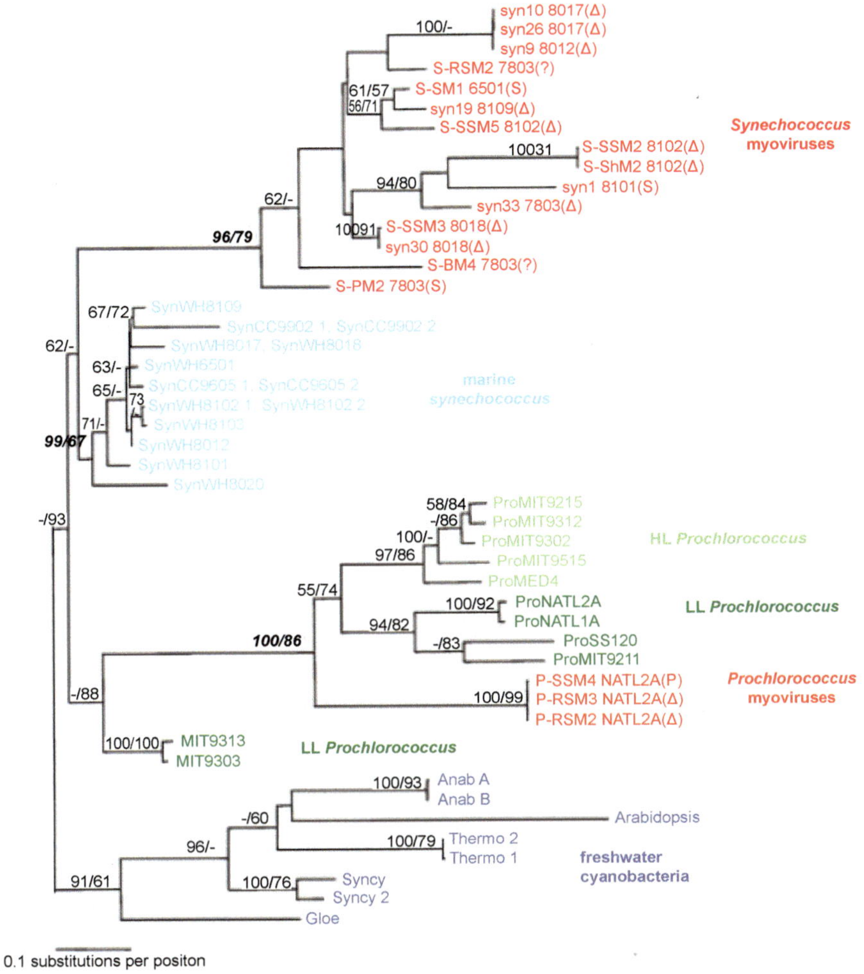
syn10 8017(Δ)
syn26 8017(Δ)
syn9 8012(Δ)
S-RSM2 7803(?)
S-SM1 6501(S)
syn19 8109(Δ)
S-SSM5 8102(Δ)
S-SSM2 8102(Δ)
S-ShM2 8102(Δ)
syn1 8101(S)
syn33 7803(Δ)
S-SSM3 8018(Δ)
syn30 8018(Δ)
S-BM4 7803(?)
S-PM2 7803(S)
Synechococcus myoviruses
SynWH8109
SynCC9902 1, SynCC9902 2
SynWH8017, SynWH8018
SynWH6501
SynCC9605 1, SynCC9605 2
SynWH8102 1, SynWH8102 2
SynWH8103
SynWH8012
SynWH8101
SynWH8020
marine synechococcus
ProMIT9215
ProMIT9312
ProMIT9302
ProMIT9515
ProMED4
HL Prochlorococcus
ProNATL2A
ProNATL1A
ProSS120
ProMIT9211
LL Prochlorococcus
P-SSM4 NATL2A(P)
P-RSM3 NATL2A(Δ)
P-RSM2 NATL2A(Δ)
Prochlorococcus myoviruses
MIT9313
MIT9303
LL Prochlorococcus
Anab A
Anab B
Arabidopsis
Thermo 2
Thermo 1
freshwater cyanobacteria
Syncy
Syncy 2
Gloe
0.1 substitutions per positon

Figure 15

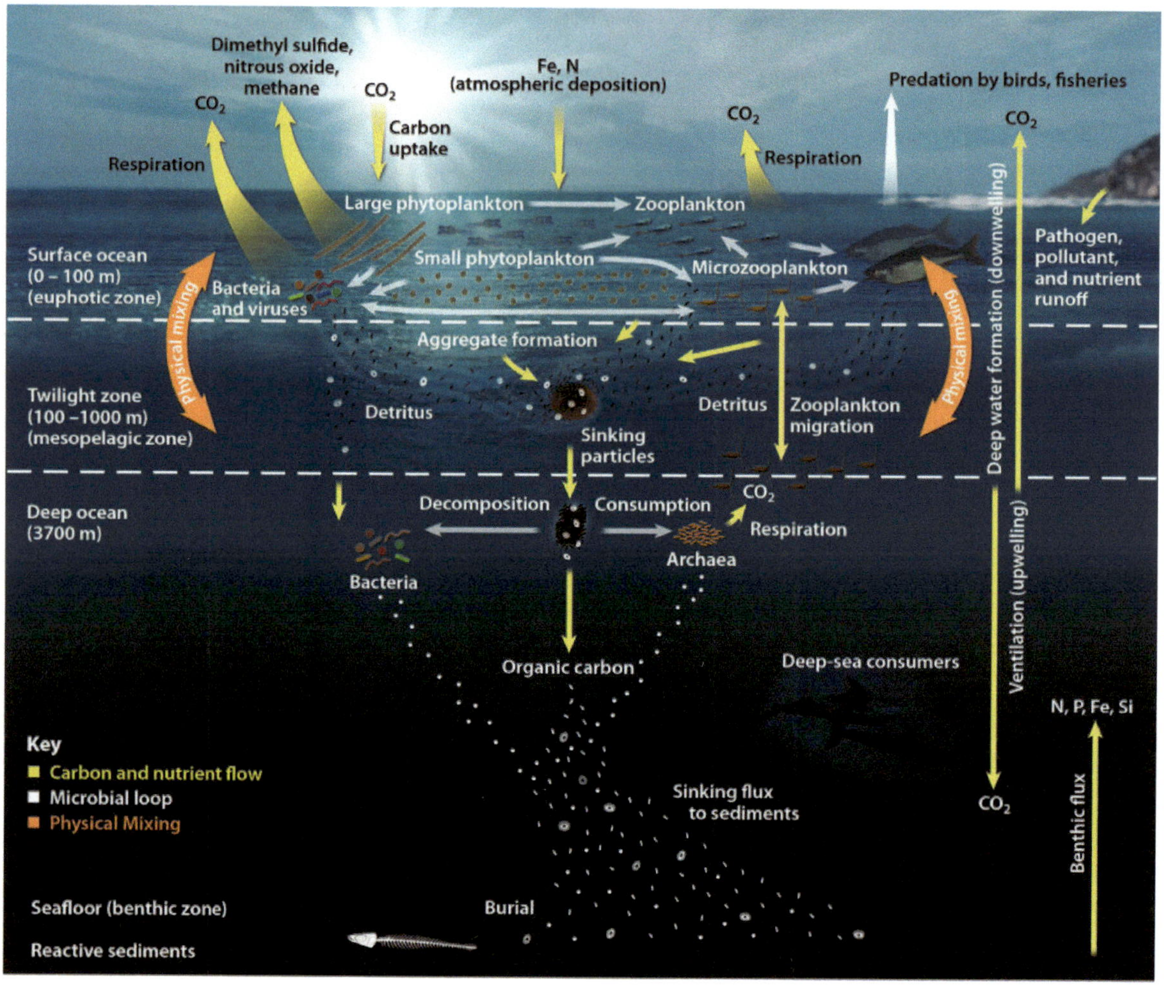
Dimethyl sulfide, nitrous oxide, methane
CO2
CO2
Carbon uptake
Fe, N (atmospheric deposition)
Predation by birds, fisheries
CO2
CO2
Respiration
Respiration
Large phytoplankton
Zooplankton
Small phytoplankton
Microzooplankton
Surface ocean (0 – 100 m) (euphotic zone)
Bacteria and viruses
Physical mixing
Pathogen, pollutant, and nutrient runoff
Aggregate formation
Twilight zone (100 –1000 m) (mesopelagic zone)
Detritus
Detritus
Zooplankton migration
Sinking particles
Physical mixing
Deep water formation (downwelling)
Deep ocean (3700 m)
Decomposition
Consumption
CO2
Respiration
Archaea
Bacteria
Ventilation (upwelling)
Organic carbon
Deep-sea consumers
N, P, Fe, Si
Key
Carbon and nutrient flow
Microbial loop
Physical Mixing
Sinking flux to sediments
CO2
Benthic flux
Seafloor (benthic zone)
Burial
Reactive sediments

Chapter 9

Figure 5

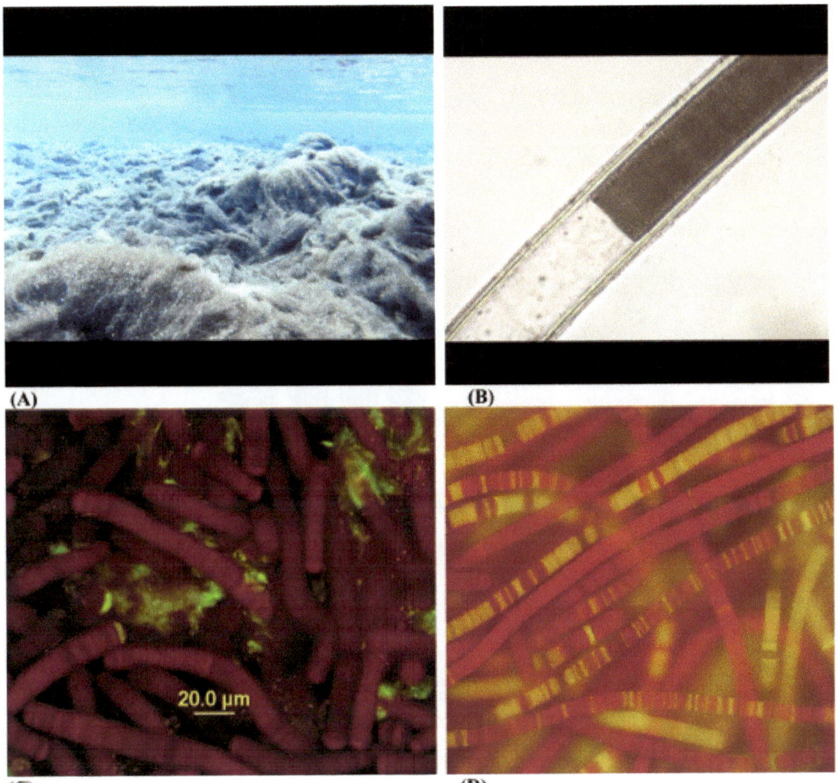

(A) (B) (C) (D)

Figure 8

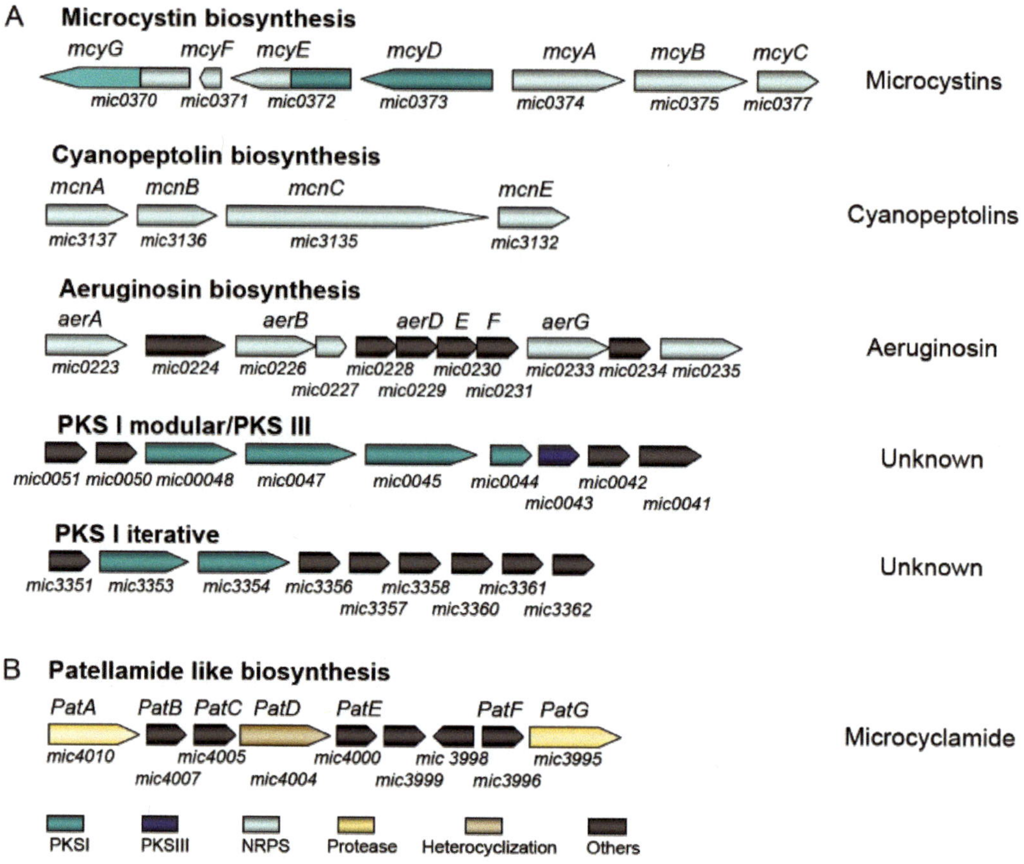
A Microcystin biosynthesis
mcyG mcyF mcyE mcyD mcyA mcyB mcyC
mic0370 mic0371 mic0372 mic0373 mic0374 mic0375 mic0377
Microcystins
Cyanopeptolin biosynthesis
mcnA mcnB mcnC mcnE
mic3137 mic3136 mic3135 mic3132
Cyanopeptolins
Aeruginosin biosynthesis
aerA aerB aerD E F aerG
mic0223 mic0224 mic0226 mic0228 mic0230 mic0233 mic0234 mic0235
mic0227 mic0229 mic0231
Aeruginosin
PKS I modular/PKS III
mic0051 mic0050 mic00048 mic0047 mic0045 mic0044 mic0042
mic0043 mic0041
Unknown
PKS I iterative
mic3351 mic3353 mic3354 mic3356 mic3358 mic3361
mic3357 mic3360 mic3362
Unknown
B Patellamide like biosynthesis
PatA PatB PatC PatD PatE PatF PatG
mic4010 mic4005 mic4000 mic 3998 mic3995
mic4007 mic4004 mic3999 mic3996
Microcyclamide
PKSI PKSIII NRPS Protease Heterocyclization Others

Figure 11

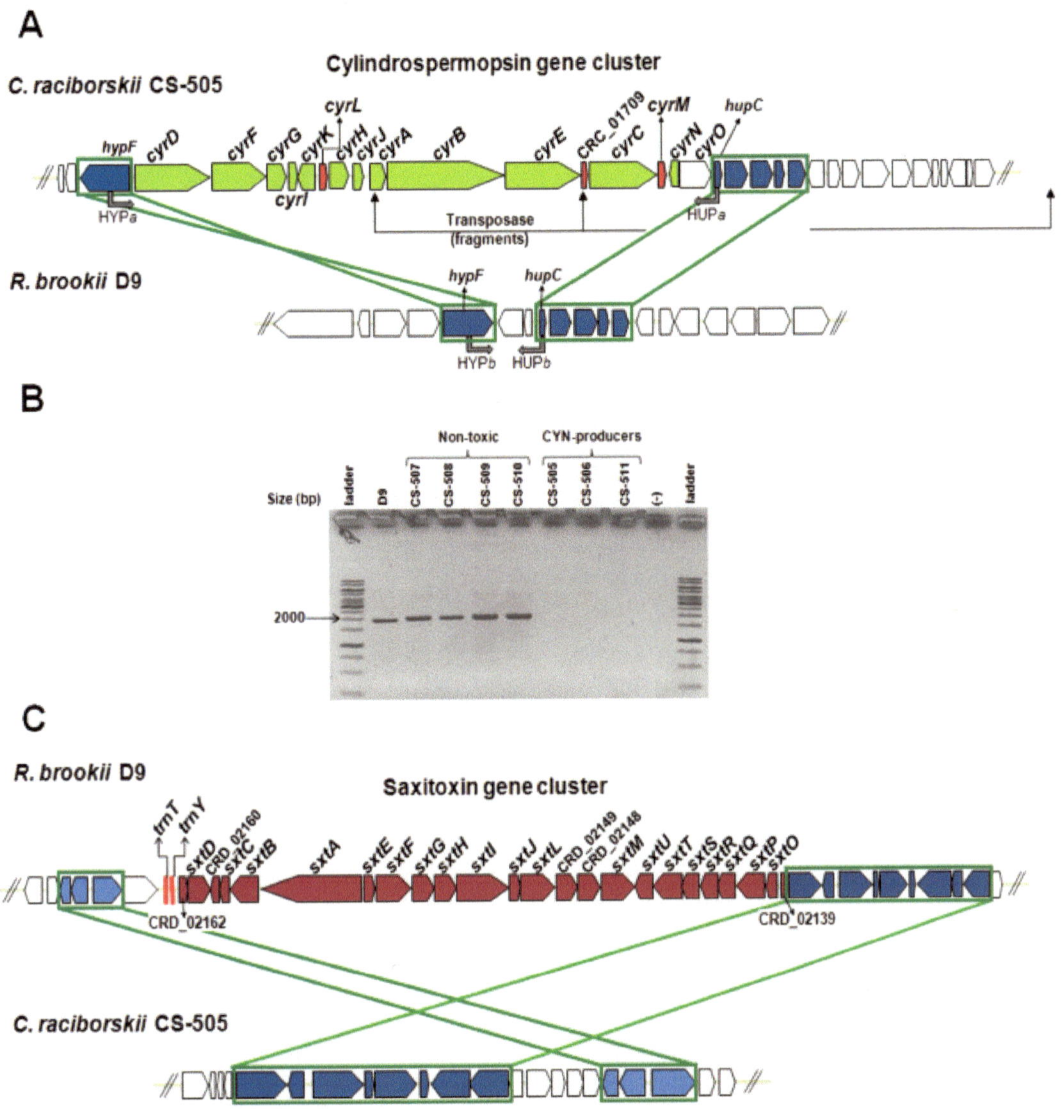
A
Cylindrospermopsin gene cluster
C. raciborskii CS-505
hypF
cyrD
cyrF
cyrG
cyrI
cyrK
cyrL
cyrH
cyrJ
cyrA
cyrB
cyrE
CRC_01709
cyrC
cyrM
cyrN
cyrO
hupC
HYPa
Transposase (fragments)
HUPa
R. brookii D9
hypF
hupC
HYPb
HUPb
B
Non-toxic
CYN-producers
Size (bp)
ladder
D9
CS-507
CS-508
CS-509
CS-510
CS-505
CS-506
CS-511
(-)
ladder
2000
C
R. brookii D9
Saxitoxin gene cluster
trnT
trnY
sxtD
CRD_02160
sxtC
sxtB
sxtA
sxtE
sxtF
sxtG
sxtH
sxtI
sxtJ
sxtL
CRD_02149
CRD_02148
sxtM
sxtU
sxtT
sxtS
sxtR
sxtQ
sxtP
sxtO
CRD_02162
CRD_02139
C. raciborskii CS-505

Figure 15

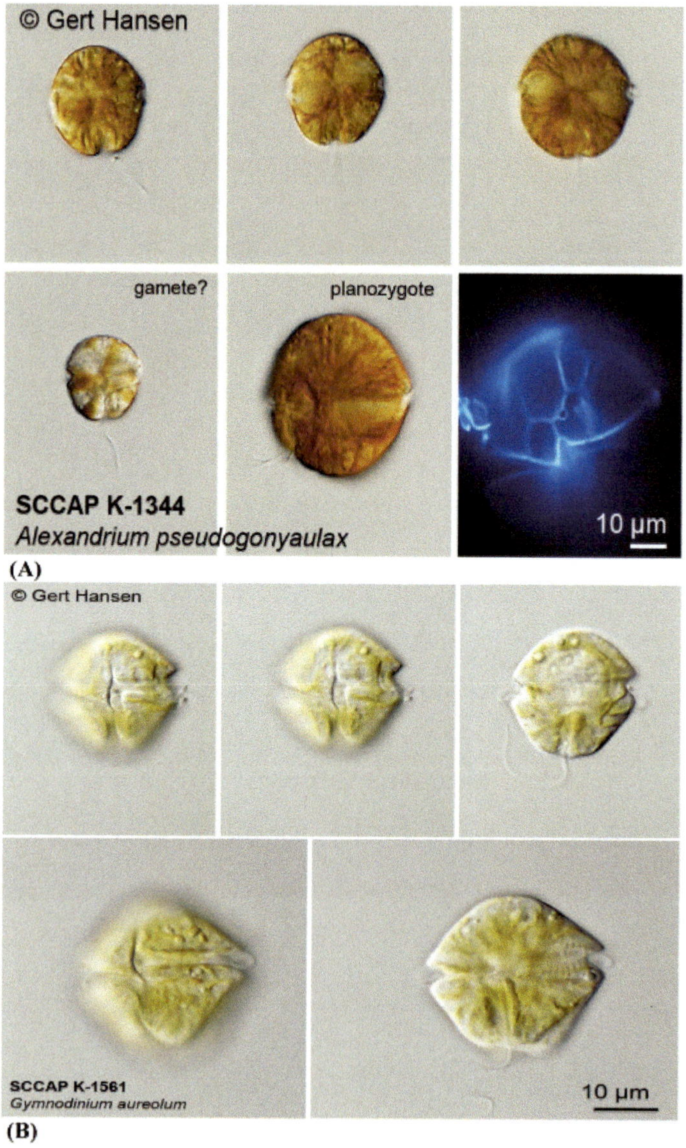
© Gert Hansen
gamete?
planozygote
SCCAP K-1344
Alexandrium pseudogonyaulax
10 μm
(A)
© Gert Hansen
SCCAP K-1561
Gymnodinium aureolum
10 μm
(B)

Figure 23

Figure 24

Figure 25

Figure 26

Chapter 10

Figure 1

(A)

(B)

(C)

Figure 10

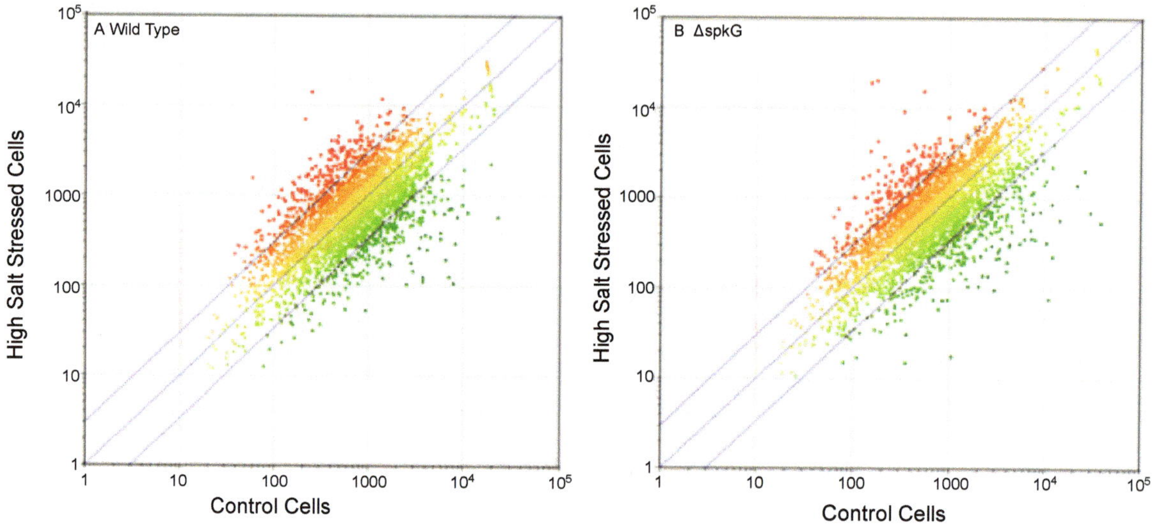

Figure 12

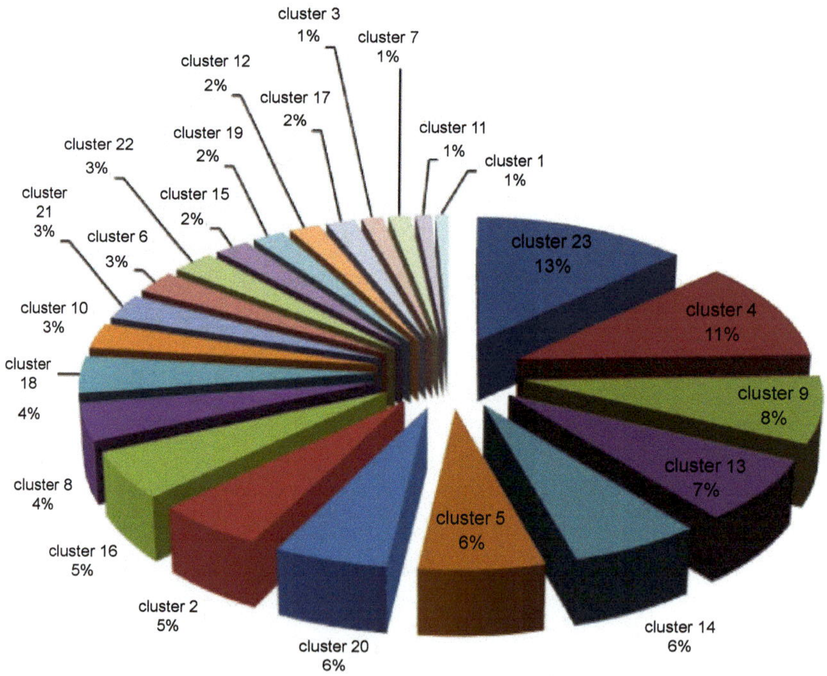

Figure 13

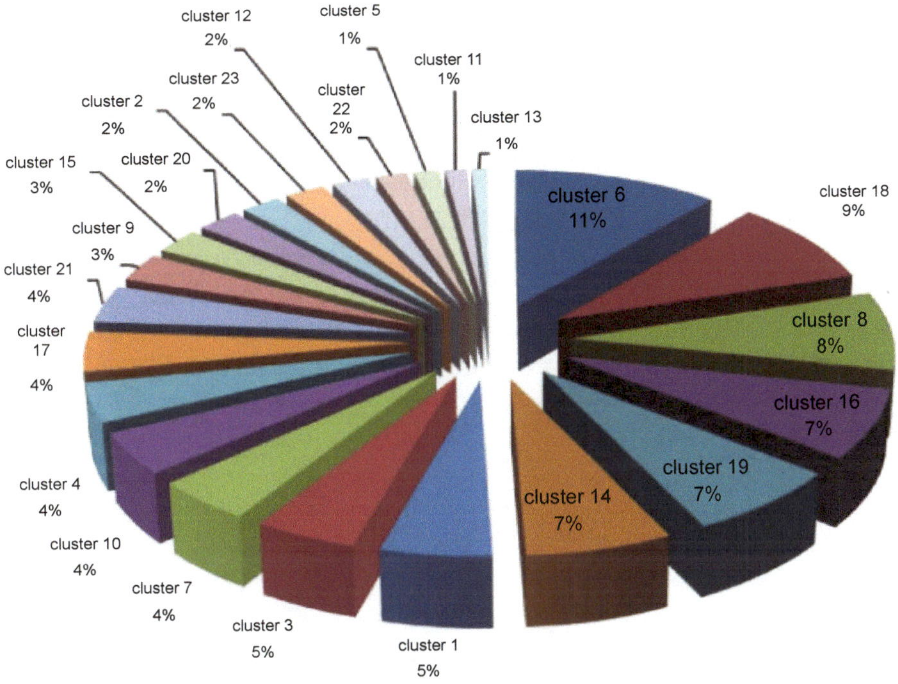

Figure 14

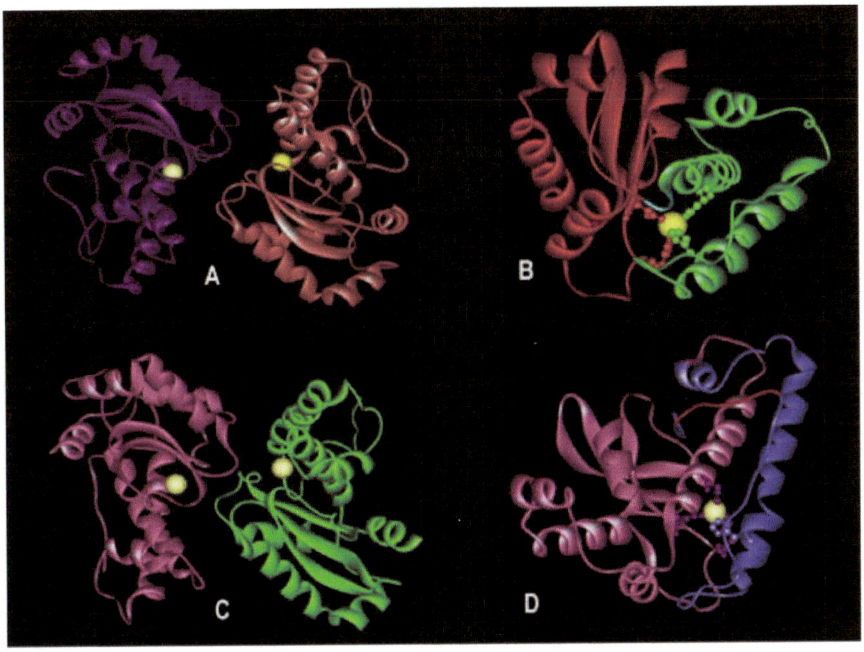

Figure 15

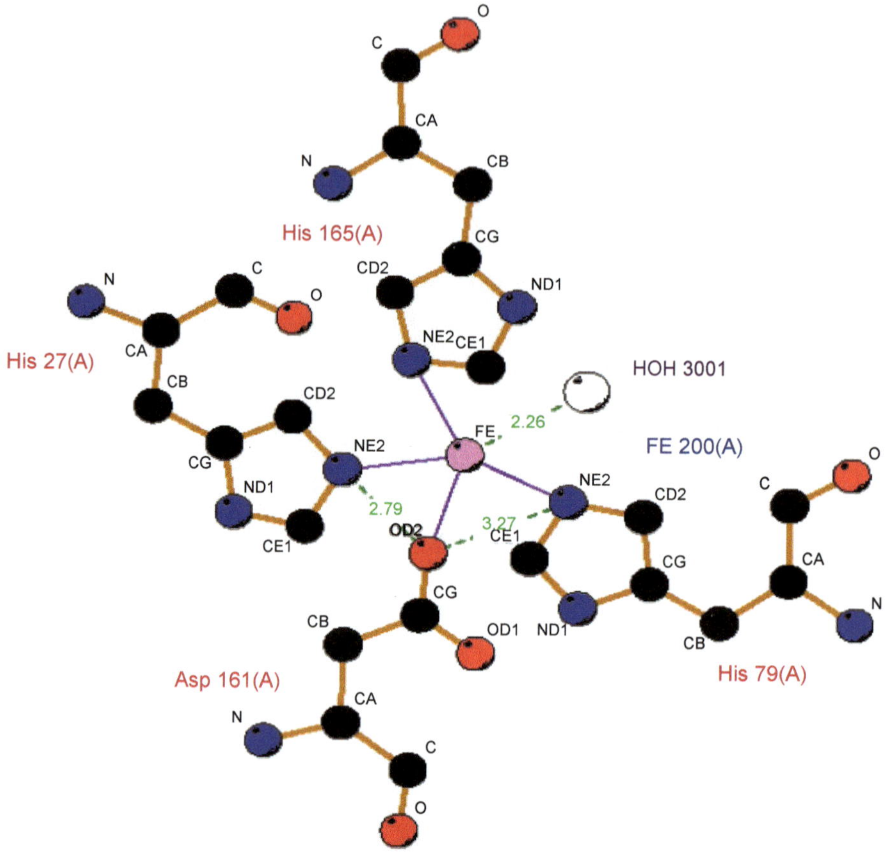
His 165(A)
His 27(A)
HOH 3001
FE 200(A)
Asp 161(A)
His 79(A)
2.26
2.79
3.27
FE
NE2
CE1
CD2
ND1
CG
CB
CA
N
C
O
OD1
OD2

Figure 16

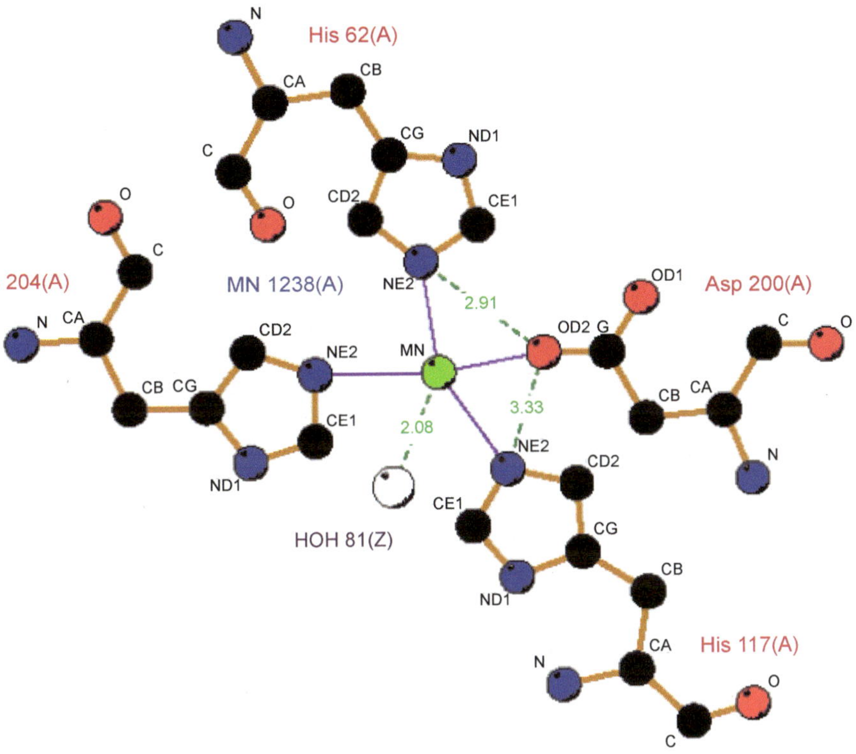

Figure 17

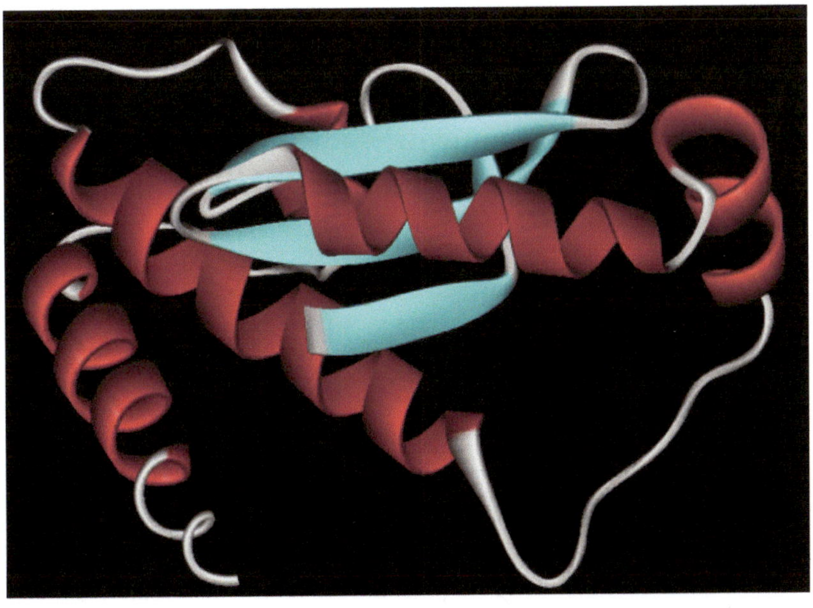

Figure 18

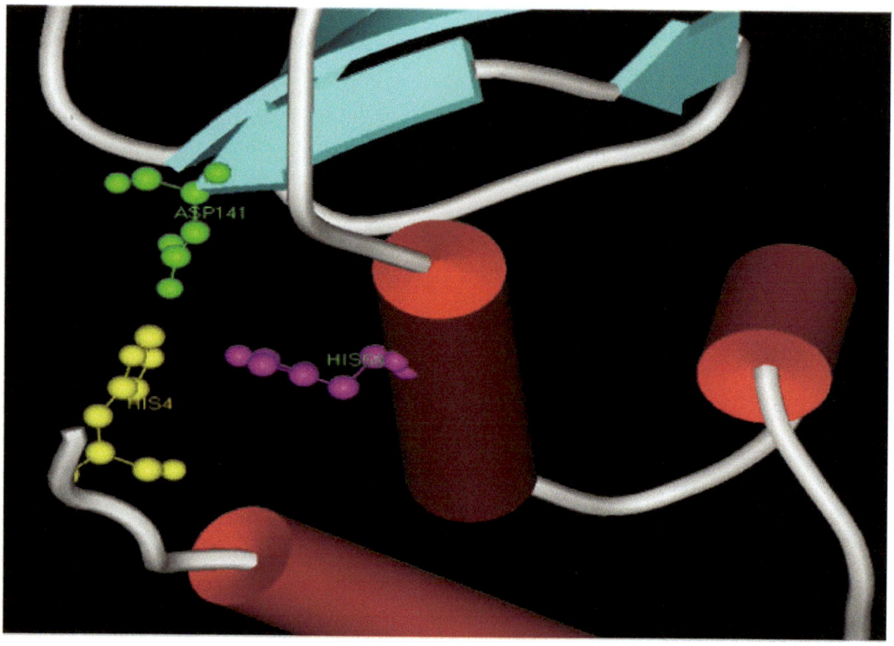

Figure 19

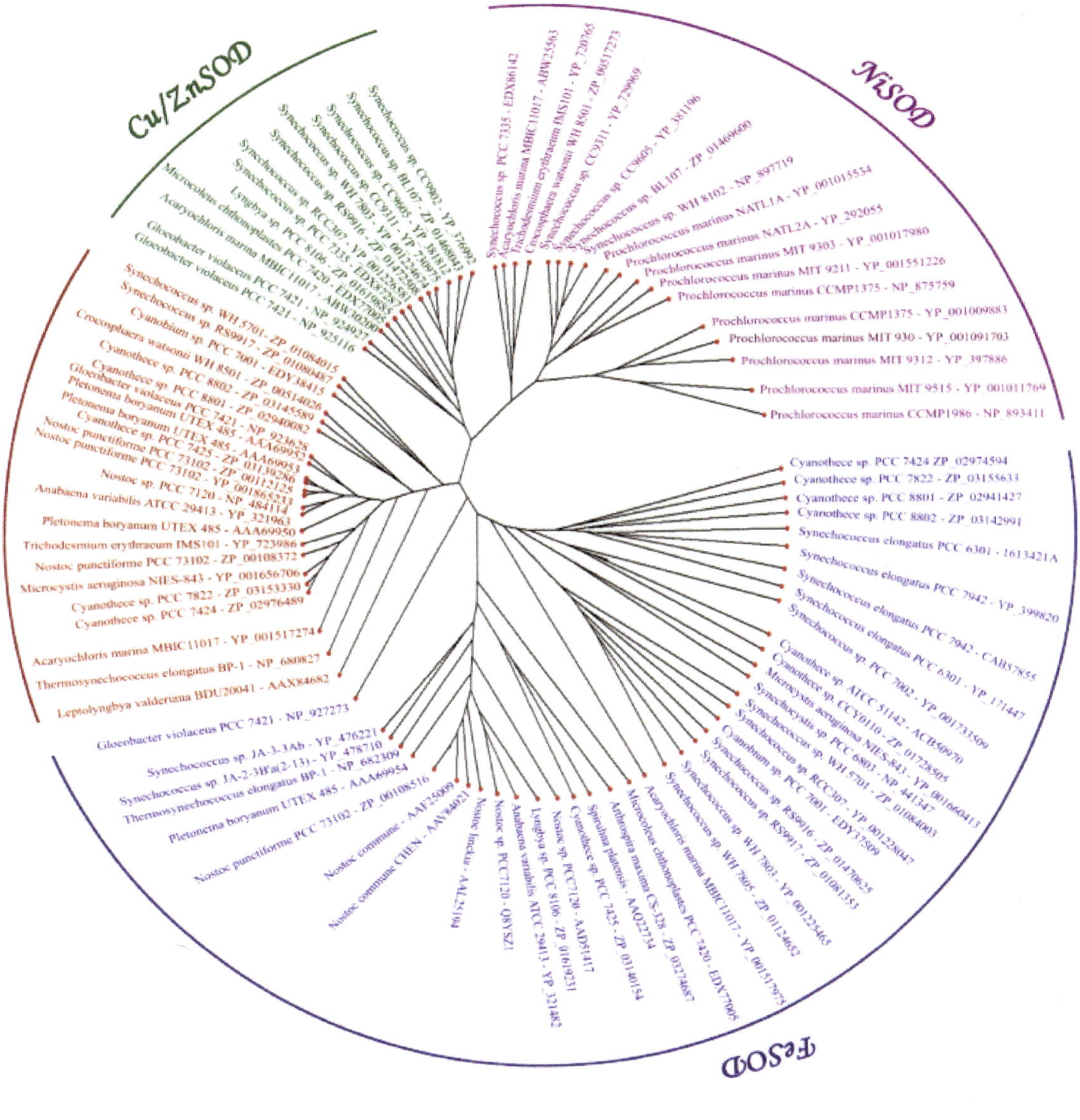
Cu/ZnSOD
NiSOD
MnSOD
FeSOD
Prochlorococcus marinus CCMP1375 - YP_001009883
Prochlorococcus marinus MIT 930 - YP_001091703
Prochlorococcus marinus MIT 9312 - YP_397886
Prochlorococcus marinus MIT 9515 - YP_001011769
Prochlorococcus marinus CCMP1986 - NP_893411
Cyanothece sp. PCC 7424 ZP_02974594
Cyanothece sp. PCC 7822 - ZP_03155633
Cyanothece sp. PCC 8801 - ZP_02941427
Cyanothece sp. PCC 8802 - ZP_03142991
Pletonema boryanum UTEX 485 - AAA69950
Trichodesmium erythraeum IMS101 - YP_723986
Nostoc punctiforme PCC 73102 - ZP_00108372
Microcystis aeruginosa NIES-843 - YP_001656706
Acaryochloris marina MBIC11017 - YP_001517274
Thermosynechococcus elongatus BP-1 - NP_680827
Leptolyngbya valderiana BDU20041 - AAX84682
Gloeobacter violaceus PCC 7421 - NP_927273

Figure 21